Fundamentals of Organic Chemistry

with Problems and Solutions

for IIT-JEE and other competitive examinations

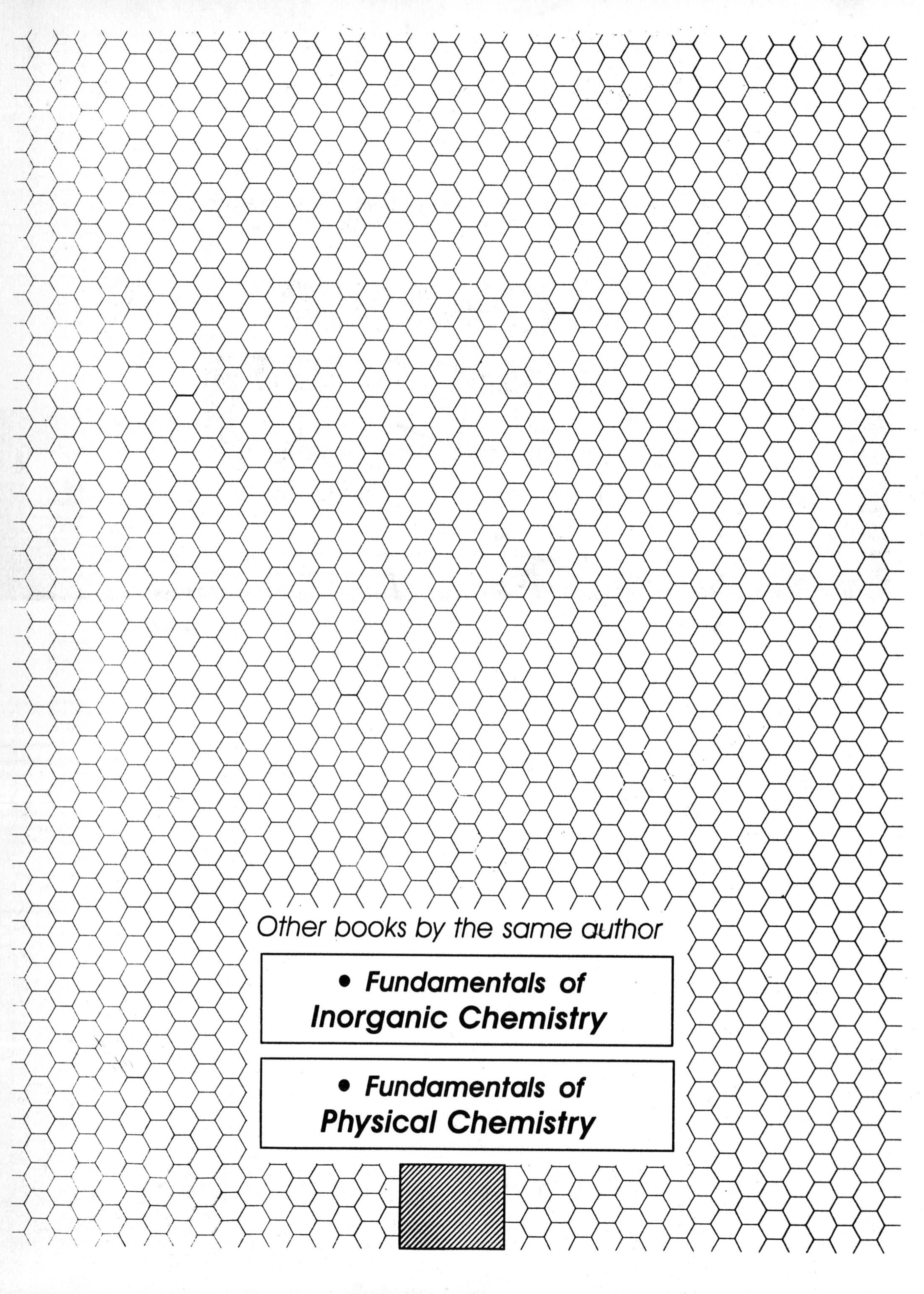
Other books by the same author
• Fundamentals of Inorganic Chemistry
• Fundamentals of Physical Chemistry

Fundamentals of Organic Chemistry

with Problems and Solutions

Ranjeet Shahi

Director
Ranjeet Shahi Chemistry Classes
SCO 48 Sec 20C, Chandigarh

CBS PUBLISHERS & DISTRIBUTORS

NEW DELHI • BANGALORE

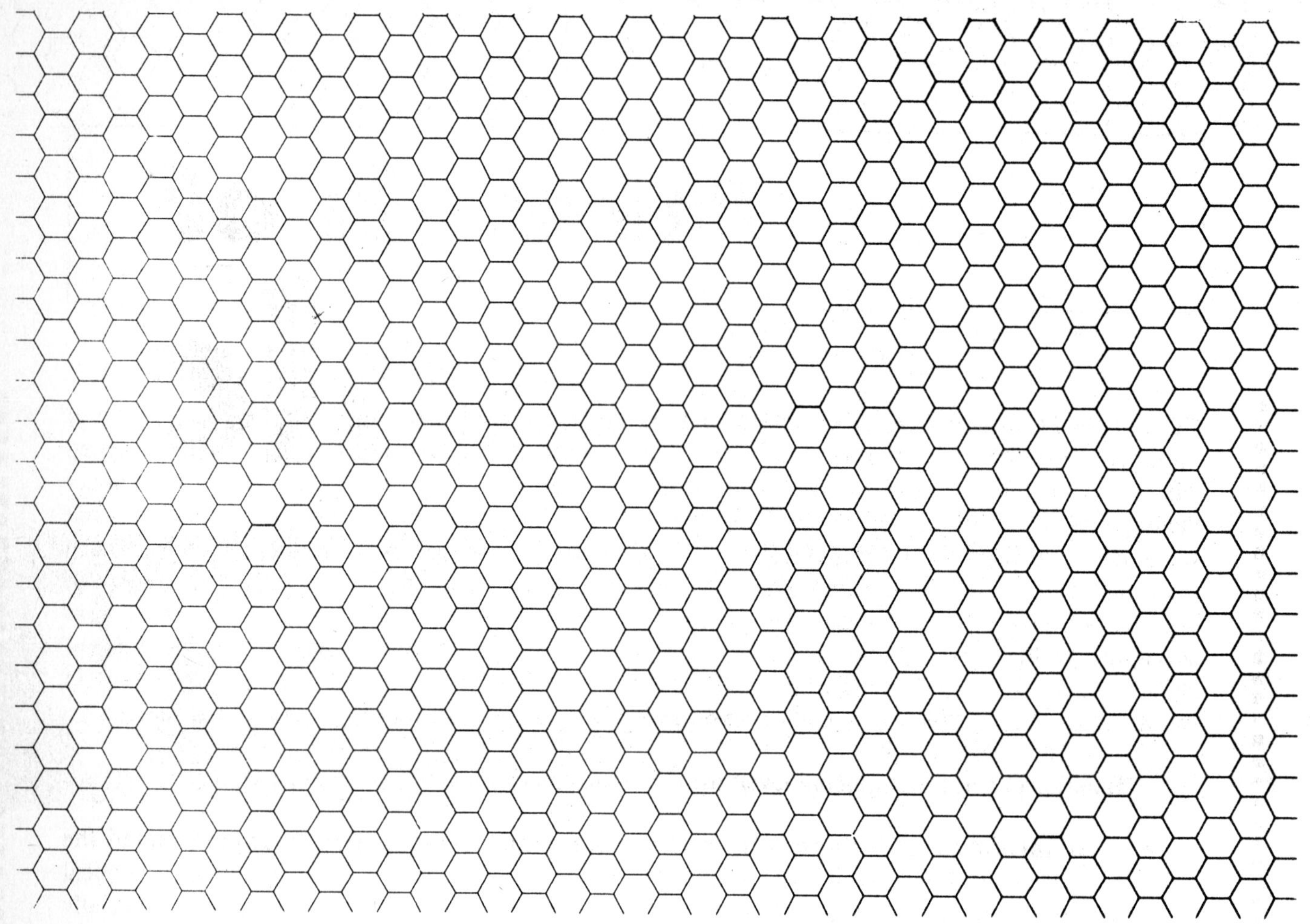

Fundamentals of
Organic Chemistry

ISBN: 81-239-1368-0

First Edition: 2006
Reprint : 2008

Published by :
Satish Kumar Jain for CBS Publishers & Distributors,
4596/1-A, 11 Darya Ganj, New Delhi - 110 002 (India)
E-mail : cbspubs@vsnl.com
Website : www.cbspd.com

Production Director : Vinod K. Jain

Branch Office :
Seema House, 2975, 17th Cross, K.R. Road,
Bansankari 2nd Stage, Bangalore - 560070
Fax : 080-26771680 • E-mail : cbsbng@vsnl.net

Printed at :
Asia Printograph, Shahdara, Delhi 110032 (India)

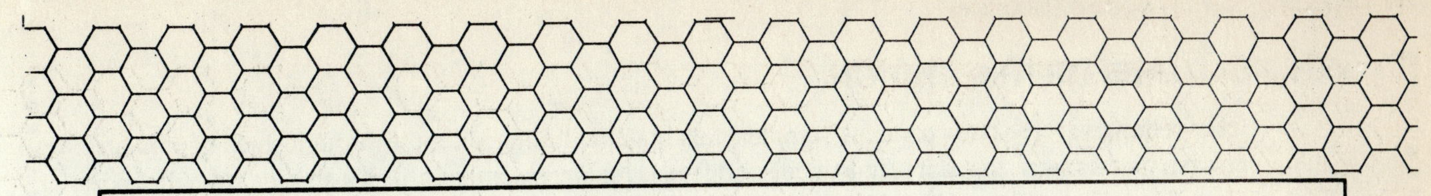

Preface

This book ***Fundamentals of Organic Chemistry*** is a comprehensive introduction to the subject of organic chemistry at senior-secondary and undergraduate level. The content of this book is designed with a viewpoint that the reader is about to begin his study of *Organic Chemistry*. He may have been under the impression that organic chemistry is a subject of memorizing the facts, but fortunately the fact is not so. Organic chemistry is a subject where every fact has very precise logic, and the student requires much more analytical aptitude to understand the logic than any other part of chemistry. The only limitation of organic chemistry is that it must be studied in a systematic order since every subsequent chapter has a very strong link to the previous chapter, i.e. at introductory level, a student must follow a specific order of chapters rather going to an intermediate chapter directly. Concepts of organic chemistry cumulate in ascending order and it must be done in that order to have the complete understanding of this subject.

Traditionally, at the beginning level, there are three principal areas of organic chemistry.

- **Isomerism**: This includes structural *isomerism* and *stereoisomerism*. To understand the stereochemical aspects of any organic compounds and its impact on the physical and chemical properties of that compound, a student must have a very sound background of isomerism, especially stereoisomerism. This book has a very detailed discussion of all the fundamental concepts of isomerism. All the fundamental terminologies of stereochemistry, which is used very frequently in rest of the organic chemistry, are defined very precisely and described exhaustively through illustrative examples.

- **Reaction Mechanism**: Mechanism of an organic reaction pertains to description of change in composition, constitution and sequence of such events that occur at molecular level, which finally decide the course of reaction leading to formation of a specific product. Therefore, mechanism of an organic reaction is equivalent to oxygen required for survival of living beings. *Reaction mechanism is the core of all organic reactions.* This book contains a complete chapter (Chapter 3) entitled *Reaction mechanism and reactive intermediates*. In this chapter, starting from bonding concepts, intermolecular force of attraction and its impact on physical properties, electronic displacement (*Inductive effect, Resonance effect, Electromeric effect* and *Hyperconjugation effect*) and its applications in describing relative acid base strength and relative stability of reactive intermediates, type of bond fission, nature of attacking reagents (*Electrophiles* and *Nucleophiles*), reactive intermediates (formations and their relative stability) are described systematically and elaborately. A through study of this chapter will enable a student to start the study of organic chemistry with complete understanding at every stage.

- **Classes of Organic Compounds**: This includes the individual family of organic compounds — their preparation, physical properties and chemical reactions. This book provides chapter-wise discussion of a specific class of organic compounds with emphasis on their method of preparation, physical properties and chemical reactions. Every reaction follows illustrative examples, based on that particular reaction, to enhance the analytical ability of a student for that reaction.

KEY FEATURES OF THIS EDITION

- **Comprehensive**: The book contains complete text material designed for the through study of organic chemistry at senior-secondary and undergraduate level. Instructors as well as students will find detailed materials that covers the core curriculum of typical organic chemistry course.
- **Versatile**: The content of this book is organized into chapters in ascending order of concepts, as per demand of this section of chemistry. Chapters of this book are the most common chapters prescribed in the syllabus of most of the organic chemistry courses at introductory level.
- **Extensively class tested**: The problems of this book have been experimented exhaustively on the top-brass students, preparing for the most competitive entrance examinations at the chemistry academy of the author, and fine-tuned for their efficacy and accuracy.

Highlights of Fundamentals of Organic Chemistry

- About **500 solved examples** within the chapters: A few sample examples are described below.

Chapter 2 : Isomerism

Example 18: Draw structures of all stereoisomers of 4-methyl-2-hexene.

Solution:

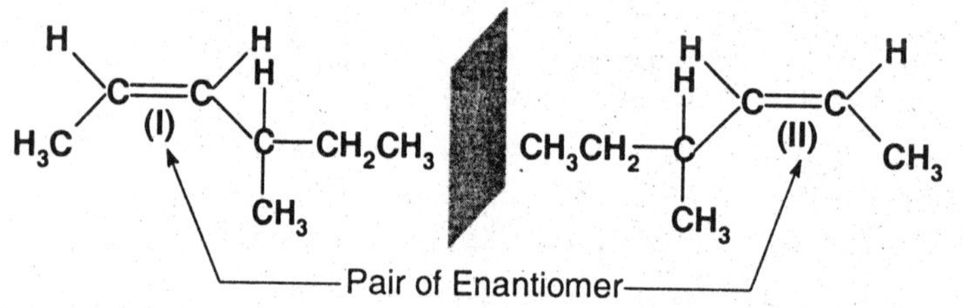

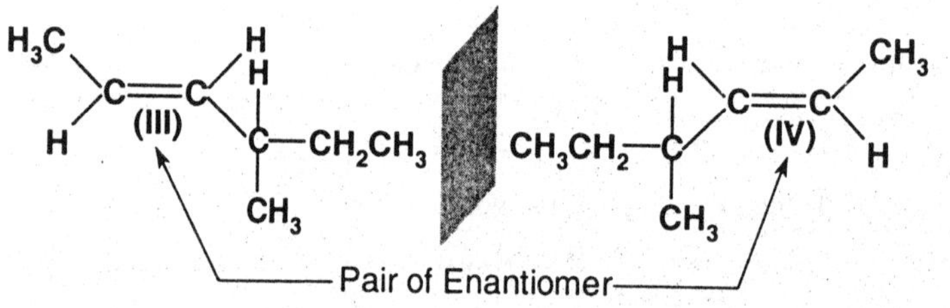

(I) (II) Pair of Enantiomer

(III) (IV) Pair of Enantiomer

Here I and II are pairs of enantiomers and III and IV are pairs of enantiomers. Pairs I and III, I and IV, II and III, and II and IV are pairs of distereomers.

Example 23: A sample of 2-bromobutane has an observed specific rotation of + 9.2°, what is the optical purity of (+) –2-bromobutane in this sample, if specific rotation of (+) –2-bromobutane is + 23.1°?

Solution: % Optical purity = $\frac{\text{Observed Optical Rotation}}{\text{Specific Rotation of Pure Enantiomer}} \times 100$

$\Rightarrow$ % Optical Purity = $\frac{9.20}{23.10} \times 100 = 40$

Chapter 3 : Reaction Mechanism and Reactive Intermediates

Example 64: α-H of an aldehyde is more acidic than α-H of a ketone, justify.

Solution: Conjugate base of both aldehyde and ketone, produced on deprotonation from α-position are stabilized by resonance as:

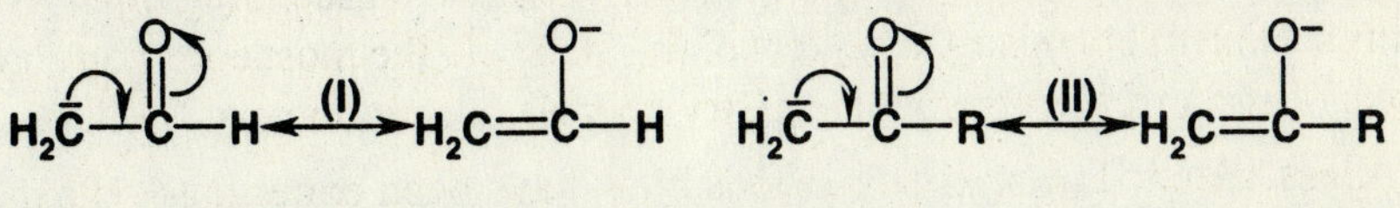

Extent of delocalization is more in (I) than in (II) since alkyl group (R) is also electron donating by +I effect, opposes delocalization of negative charge in ketone as:

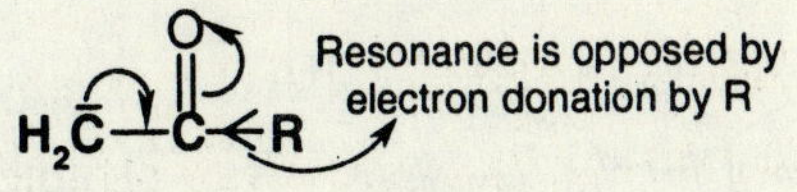

Resonance is opposed by electron donation by R

Therefore, conjugate base I is more stable and its conjugate acid (aldehyde) is stronger acid.

Chapter 5 : Alkyl Halides

Example 11: 3-bromocyclohexene is a secondary halide and benzyl bromide is primary halide. Both halides undergo S_N-1 substitution about as fast as most tertiary halides. Use resonance structure to explain this enhanced reactivity.

Solution: Both form resonance stabilized carbocation as:

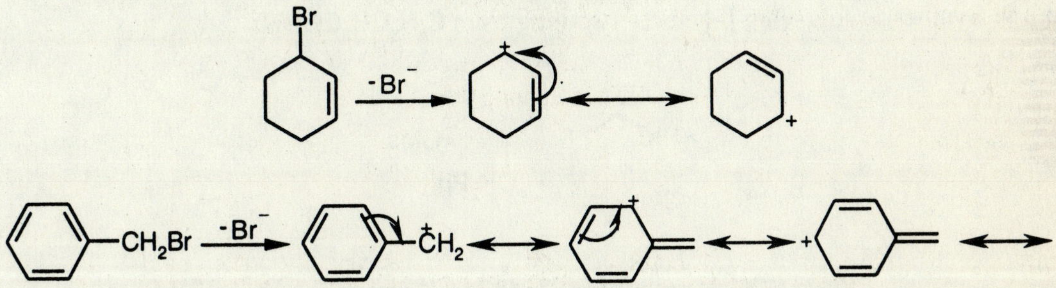

Chapter 6 : Alcohols, Ethers and Epoxides

Example 18: When the compound (I) is heated with 50% H_2SO_4, the product of the reaction is the compound (II) and not (III). Write a step-by-step mechanism showing how II is formed.

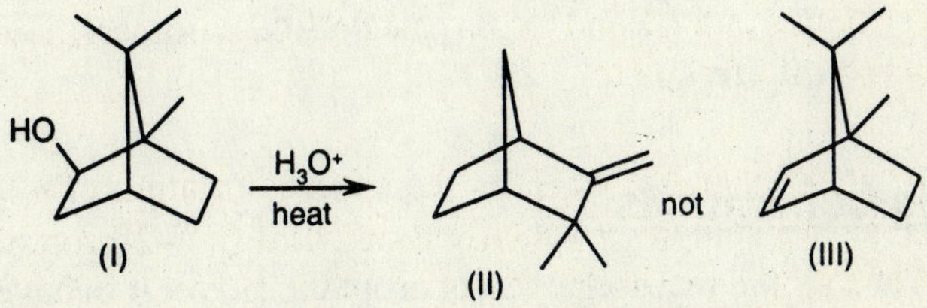

(I) $\xrightarrow[\text{heat}]{H_3O^+}$ (II) not (III)

Solution:

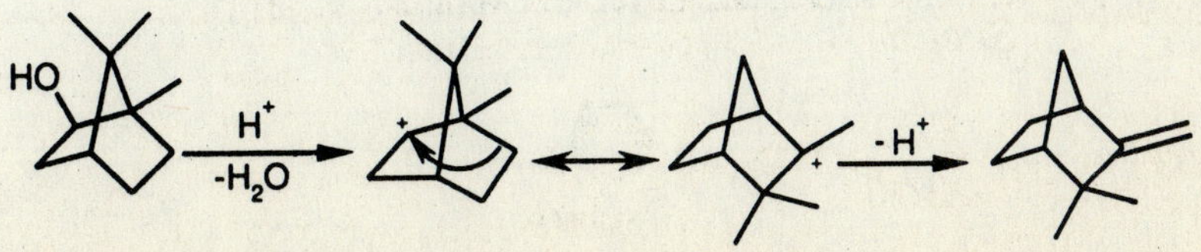

Chapter 7 : Aldehydes and Ketones

Example 12: Write all possible aldol products obtained by intramolecular cyclization of $CH_3-\overset{O}{\overset{\|}{C}}-(CH_2)_4-CHO$ assuming dehydration occurs in all instances.

Solution:

Chapter 8 : Carboxylic Acids and Derivatives

Example 9: Synthesize following β-ketoester starting from $C_6H_5CH_2COOH$:

Solution:

$PhCH_2COOH \xrightarrow{LiAlH_4} PhCH_2CH_2OH \xrightarrow{SOCl_2} PhCH_2CH_2Cl \xrightarrow{KCN(aq)} PhCH_2CH_2CN$

$PhCH_2CH_2CN \xrightarrow{H_3O^+} PhCH_2CH_2COOH \xrightarrow[H^+]{CH_3OH} PhCH_2CH_2COOCH_3$

$PhCH_2CH_2COOCH_3 \xrightarrow{CH_3ONa}$ (Claisen Condensation) $PhCH_2CH_2COCH(CH_2Ph)COOMe$

Chapter 10 : Aromatic Compounds I

Example 11: Thiophene (I) has all the requirements for aromaticity, yet it behaves much more like a non-aromatic 1,3-diene than an aromatic molecule. Offer an explanation.

Thiophene

Solution: Aromaticity can be considered to be a resonance effect, resulting from overlap of *p*-orbitals. Carbon and sulphur *p*-orbitals do not overlap well, because carbon is a 2nd row element and sulphur is a 3rd row element. This poor overlap reduces aromaticity in thiophene. Thiophene behaves more like a 1,3-diene with a sulphur atom at each end, and less like an aromatic molecule.

Example 18: Complete the following reactions:

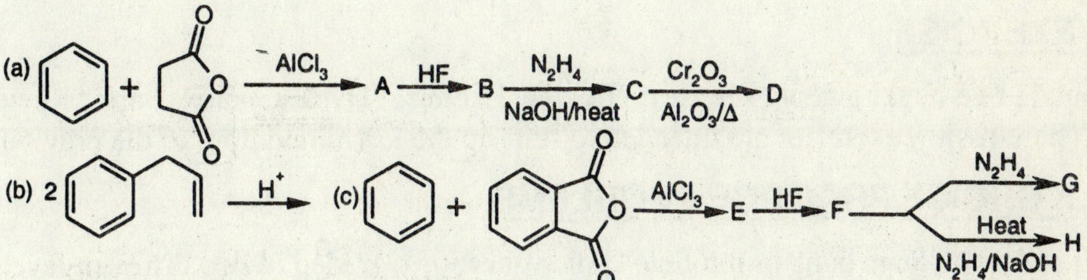

(a) benzene + succinic anhydride $\xrightarrow{AlCl_3}$ A $\xrightarrow{HF}$ B $\xrightarrow[NaOH/heat]{N_2H_4}$ C $\xrightarrow[Al_2O_3/\Delta]{Cr_2O_3}$ D

(b) 2 allylbenzene $\xrightarrow{H^+}$ (c) benzene + phthalic anhydride $\xrightarrow{AlCl_3}$ E $\xrightarrow{HF}$ F; F $\xrightarrow{N_2H_4}$ G; F $\xrightarrow[N_2H_4/NaOH]{Heat}$ H

Solution:

(a) A: $C_6H_5COCH_2CH_2COOH$; B; C; D

(b) CH₂Ph-substituted product

(c) E: HOOC; F; G; H

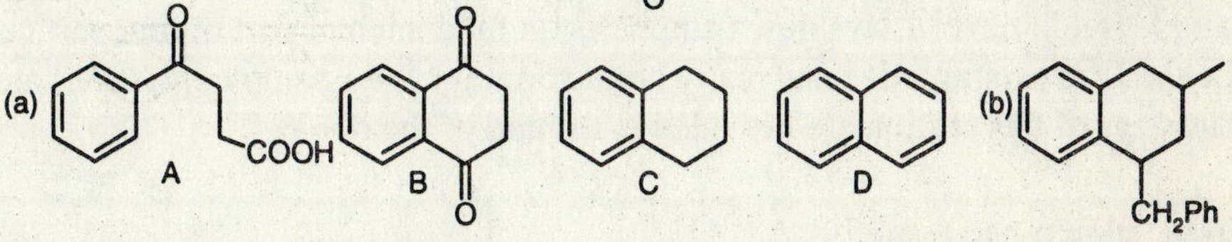

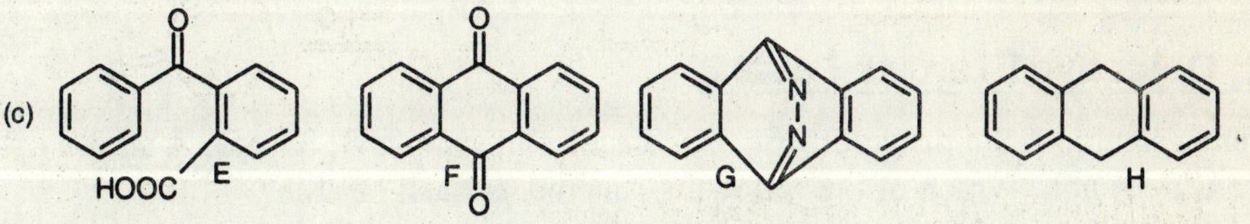

Chapter 11 : Aromatic Compounds II

Example 7: Bring about the following transformations in good yield: $C_6H_5CH_2NH_2$ → $C_6H_5NHCH_3$

Solution: $C_6H_5CH_2NH_2$ + HNO_2 → $C_6H_5CH_2OH$

$C_6H_5CH_2OH$ + $KMnO_4$ → C_6H_5COOH ; C_6H_5COOH + NH_3/heat → $C_6H_5CONH_2$

$C_6H_5CONH_2$ + KOH/Br_2 → $C_6H_5NH_2$; $C_6H_5NH_2$ + $CHCl_3/KOH$ → C_6H_5NC

C_6H_5NC + $LiAlH_4$ → $C_6H_5NHCH_3$

Example 10: In electrophilic aromatic substitution reaction, nitroso group (–NO) is ortho/para directing but deactivating. Provide mechanistic explanation.

Solution:

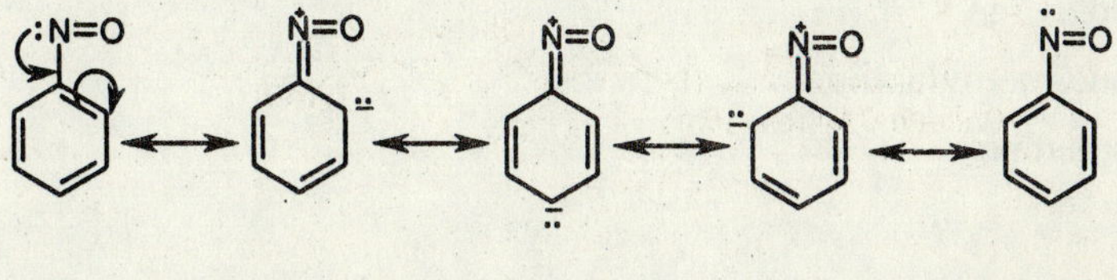

:N=O ⟷ N=O ⟷ N=O ⟷ N=O ⟷ N=O

The above electronic delocalization indicates that electrophile would preferably attach at ortho/para positions. At the same time, –NO is strong electron withdrawing group by both inductive as well as resonance effect which is responsible for deactivation of ring towards electrophilic substitution reaction as:

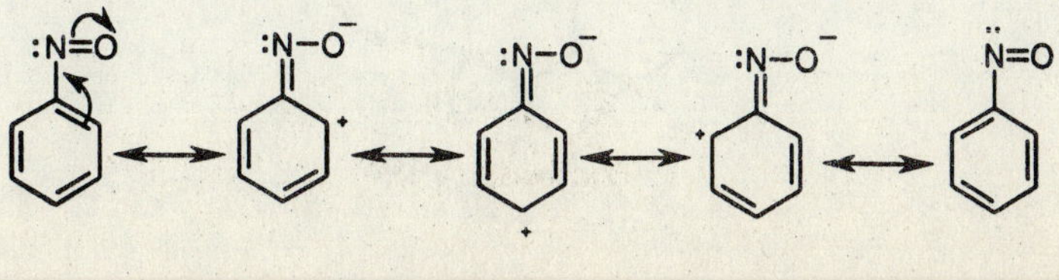

:N=O ⟷ :N–O⁻ ⟷ :N–O⁻ ⟷ :N–O⁻ ⟷ N=O

• END OF THE CHAPTER EXERCISE

A huge collection of problems for practice is provided at the end of every chapter. Exercise problems are provided into two different levels of difficulties. The first level exercise contains mainly the concept builder questions, through which one can consolidate the learning of that particular chapter. The second level exercise problems are really challenging and require an in-depth and high level of understanding to attempt them successfully. Chapterwise answers of the exercise problems are provided at the end of this book.

• REVIEW EXERCISE

The book contain two huge review exercises: "Review Exercise: HYDROCARBONS" and "Review Exercise: AROMATICS". These review exercises are intended to sum-up the learning output of the previous chapter.

• MULTIPLE CHOICE PRACTICE PROBLEMS

Every chapter ends with huge bank of multiple choice questions, graded in two different levels of difficulties: Level 1 and Level 2. Level 1 lays down emphasis on fundamental part of that particular chapter while questions of Level 2 are cummulative and really brainstorming, a few sample questions are provided below. Chapterwise answers of this section are provided at the end of the book.

• OBJECTIVE EXERCISES

Chapter 2: Objective Exercise Level 2

10. Molecules in sample **A** are distereomeric of molecules in sample **B**. Sample **A** shows optical activity, while sample **B** does not. Which of the following statements must be true?

(a) Sample **A** contains unequal amount of enantiomers while **B** is a racemic mixture.

(b) Molecules in **A** have stereocenter while molecules in **B** have no stereocentre.

(c) Molecules in **A** are chiral while molecules in **B** are meso compounds.

(d) Molecules in **A** and **B** are geometrical isomers.

Chapter 3: Objective Exercise Level 1

27. A basic substance **A** (C_7H_9N) reacts with sodium nitrite and hydrochloric acid at 0°C to give a salt, **B** ($C_7H_7ClN_2$). When a solution of **B** in dilute hydrochloric acid was heated with aqueous copper (I) cyanide and potassium cyanide, a product was obtained upon which complete hydrolysis gave an acid **C** ($C_8H_8O_2$). This acid could be oxidized to another acid which, on heating, formed an anhydride **D** ($C_8H_4O_3$). Compound **A** is:

(a) benzylamine

(b) *p*-toluidine

(c) mono-methylaniline

(d) *o*-toludine

Level 2

16. Which of the indicated **H** in the following is most acidic?

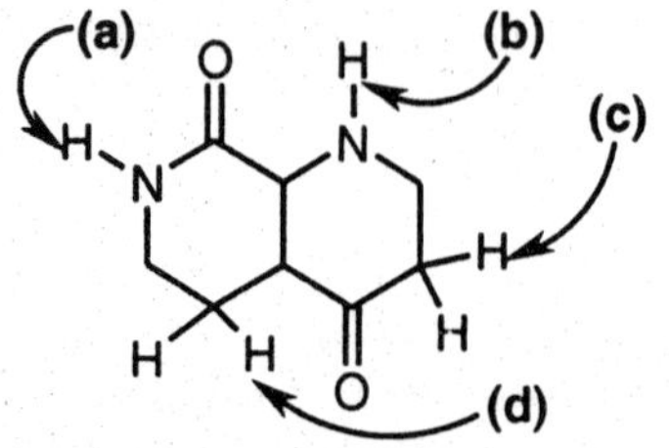

Chapter 4: Objective Exercise Level 2

19. Give the best reaction sequence to accomplish the following transformation:

(a) one eq HBr / ROOR → Li → (cyclopentenylmethyl bromide) →

(b) Bu-Li → (cyclopentenylmethyl bromide) → H_2/Lindlar's Cat →

(c) Bu-Li → (cyclopentenylmethyl bromide) → Na/liqNH_3 →

(d) H_2/Lindlar's Cat → Bu-Li → (cyclopentenylmethyl bromide) →

48. Choose the statement that is correct concerning catalytic hydrogenation:

(a) This reaction involves the direct reaction of H_2 with an alkene. The catalyst bonds only to alkene before the reaction occurs. Syn addition of H_2 must take place.

(b) This reaction involves the bonding of the catalyst first with alkene followed by bonding with H_2. Reaction takes place between alkene and hydrogen atoms on the surface of catalyst. Syn addition of H_2 must take place.

(c) This reaction proceeds first by the reaction of the catalyst with hydrogen. The alkene then bonds with the catalytic surface. Reaction between hydrogen and alkene, both bonded to surface take place. Syn addition of H_2 must take place.

(d) This reaction proceeds first by the reaction of the catalyst with hydrogen. The alkene then bonds with the catalytic surface. Reaction between hydrogen and alkene, both bonded to surface take place. Anti addition of H_2 must take place.

Chapter 5 : Objective Exercise

Level 1

8. Which statement best explains the partial racemization seen in some cases of SN-1 reactions?

(a) The C–X bond of the alkyl halide dissociates completely, and the dissociated and solvated carbonium ion does not become entirely planar but retains its configurational information in part.

(b) The alkyl halide reacts as an intimate ion pair or a solvent-separated ion pair, and the carbenium ion never really becomes fully planar; therefore, it retains its stereochemical information content.

(c) The alkyl halide reacts as an intimate ion pair or a solvent-separated ion pair, and one face of the planar carbonium ion is covered up by the leaving group.

(d) The partial racemization must be the result of contributions by the SN-2 reaction path.

Level 2

12. Which of the following is most nucleophilic?

(a) Et_2N^- (diethylamide)   (b) pyrrolidide anion   (c) pyrrolide anion   (d) EtOH

Chapter 6 : Objective Exercise

12. If the following *C labeled compound undergo the following reaction, which product would be obtained?

one mole of KCN
EtOH

(a) 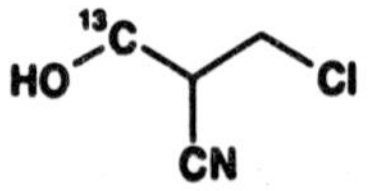 HO ¹³C CN Cl (b) 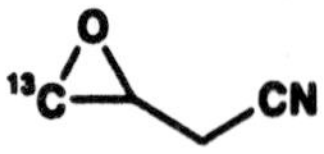 ¹³C CN (c) 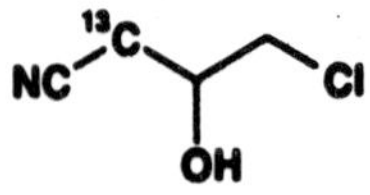 NC ¹³C OH Cl (d) 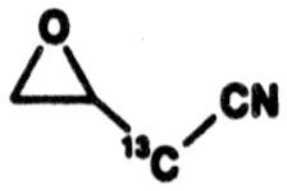 CN ¹³C

Chapter 7 : Objective Exercise Level 2

7. Give the product of the following reaction:

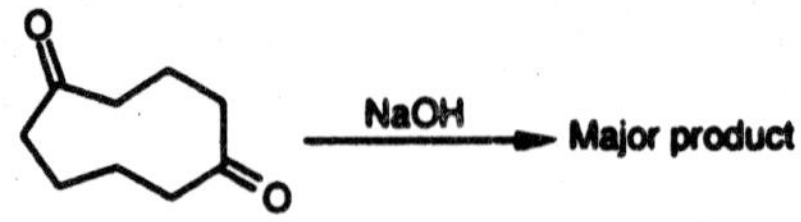

NaOH → Major product

(a) 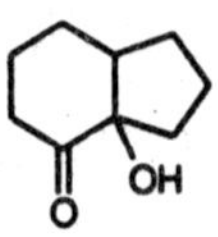 OH (b) 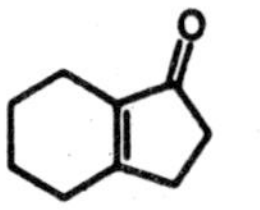 (c) 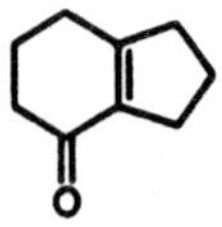 (d)

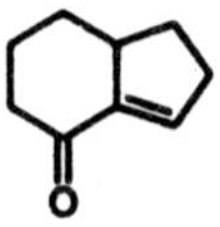

Chapter 8 : Objective Exercise

25. Treatment of two equivalents of ethylacetate with one equivalent of sodium ethoxide gives intermediate (1). Addition of an aldehyde to (1), followed by an acid quench, affords which product?

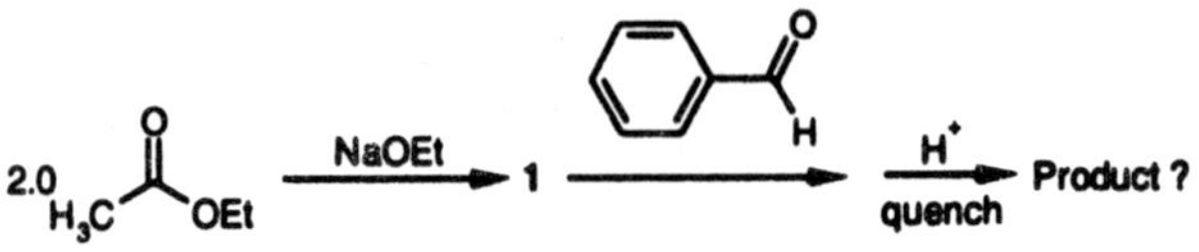

2.0 H_3C OEt —NaOEt→ 1 → H^+ quench → Product ?

(a) 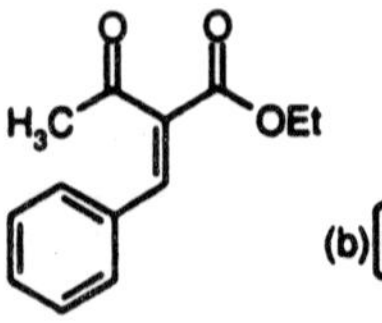 H_3C OEt (b) 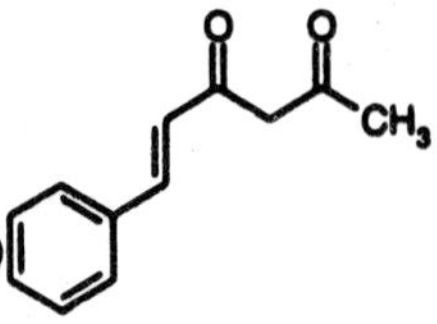 CH_3 (c) 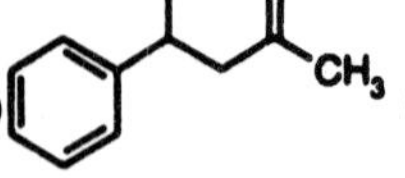 OH CH_3 (d) 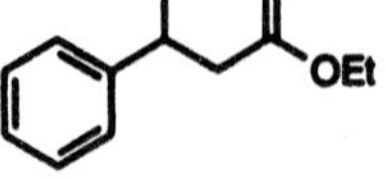 OH OEt

Chapter 11 : Objective Exercise

Level 1

23. Which one of the following compounds will be most readily hydrolyzed in aqueous alkali?

(a) 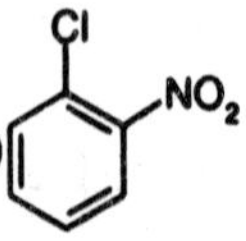 Cl NO_2 (b) 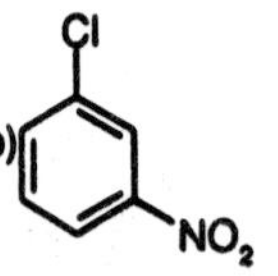 Cl NO_2 (c) 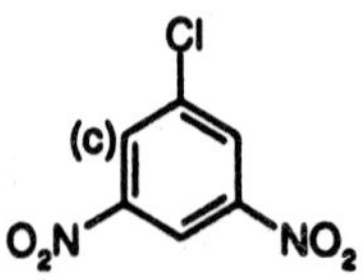 Cl O_2N NO_2 (d) 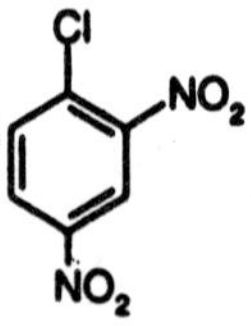 Cl NO_2 NO_2

Level 2

14. Choose the best reaction sequence to accomplish the transformation shown:

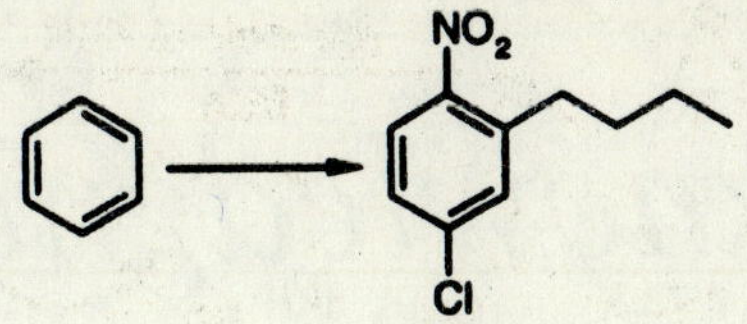

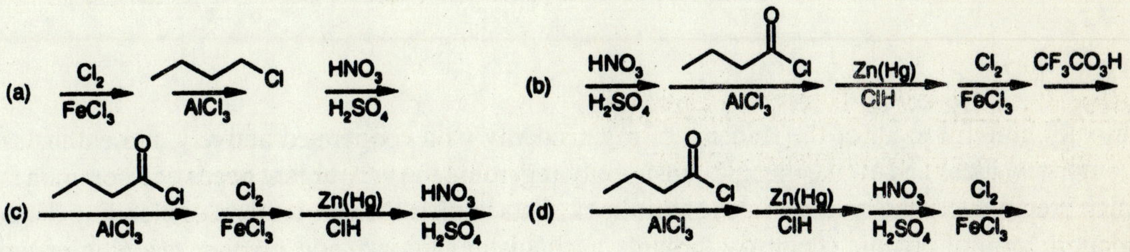

(a) $\xrightarrow[FeCl_3]{Cl_2}$ $\xrightarrow[AlCl_3]{\text{butyl chloride}}$ $\xrightarrow[H_2SO_4]{HNO_3}$

(b) $\xrightarrow[H_2SO_4]{HNO_3}$ $\xrightarrow[AlCl_3]{\text{butanoyl chloride}}$ $\xrightarrow[ClH]{Zn(Hg)}$ $\xrightarrow[FeCl_3]{Cl_2}$ $\xrightarrow{CF_3CO_3H}$

(c) $\xrightarrow[AlCl_3]{\text{butanoyl chloride}}$ $\xrightarrow[FeCl_3]{Cl_2}$ $\xrightarrow[ClH]{Zn(Hg)}$ $\xrightarrow[H_2SO_4]{HNO_3}$

(d) $\xrightarrow[AlCl_3]{\text{butanoyl chloride}}$ $\xrightarrow[ClH]{Zn(Hg)}$ $\xrightarrow[H_2SO_4]{HNO_3}$ $\xrightarrow[FeCl_3]{Cl_2}$

Ranjeet Shahi
Director
Ranjeet Shahi Chemistry Classes
SCO-48, Sector 20 C,
Chandigarh

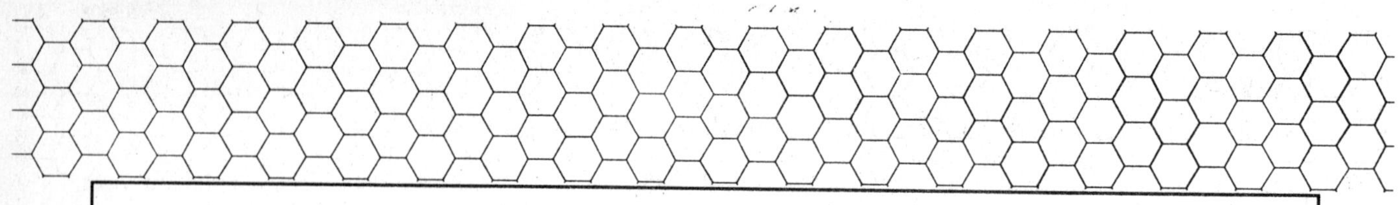

Acknowledgment

I am very grateful to all of the students of my academy who cooperated actively and enthusiastically through new ideas and awakening me consistently regarding the very urgent needs of a common student, which precipitated finally in the form of this book. This book is the end-product of blending the demand of student in term of organic chemistry through mechanistic approach and my vast teaching experience.

I am very grateful to all those students upon whom I experimented all the problems of this book, which helped me in bringing-up the contents of this book with a very high level of efficacy and accuracy.

I am greatly indebted to Dr. Shshank Deep, Assistant Professor, IIT–Delhi, for his painstaking contribution in reviewing this book thoroughly and providing significant suggestions in shaping the contents in the present format.

I am very thankful to the dedicated and resourceful team at ***CBS Publishers and Distributors***, New Delhi, India, who made this book come alive.

I pay a special thanks to Mr Y.N. Arjuna, Publishing Director, CBS Publishers and Distributors, for his vision, enthusiasm and regular suggestions regarding rendering of the content, that led to emerge this book in the best possible and most refined form.

I shall be thankful to all the potential readers who will give me suggestions or list the corrections of lacunae, if left inadvertently from my side. I will try my level best to incorporate all appropriate suggestions in the next edition of this book. Comments and suggestions may be sent to me at my e-mail address: *ranjeet_shahi@rediffmail.com.*

Ranjeet Shahi
Director
Ranjeet Shahi Chemistry Classes
SCO-48, Sector 20 C,
Chandigarh

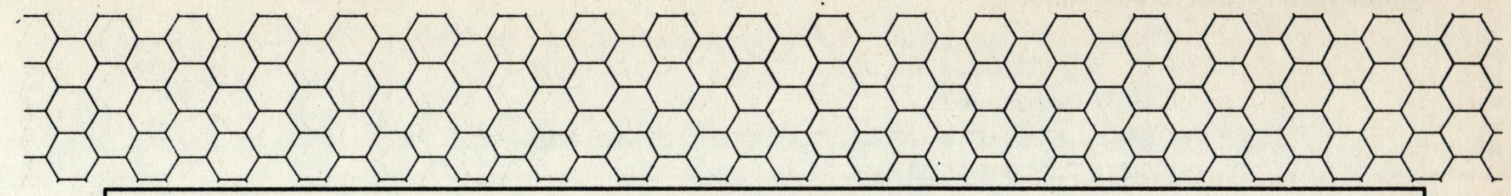

Contents

Overview of Chapters

Chapter Contents

Chapter 1

Nomenclature of Organic Compounds

1.1 INTRODUCTION

Because of infinitely large number of organic compounds existing already and several new compounds being either synthesized or discovered every day, each specific compound requires a unique name so that it can be recognized all over the world by that name.

Earlier, organic compounds were named on the basis of their source or after the name of the discoverer. Due to their large number, it was becoming difficult for the scientific world to keep record of all organic compounds having typical names originating from source, discoverer or place of discovery, etc.

In order to keep track of such a large number of organic compounds with a wide range in complexity of their skelton, the International Union of Pure and Applied Chemistry (IUPAC) has setup some standard conventions for systematic naming of organic compounds so that they can be easily recognized from their names and their structures can be drawn using IUPAC conventions. According to IUPAC conventions, organic compounds are classified into two broad categories for the purpose of naming: One being "hydrocarbons"—which contain only carbon and hydrogen and other being "derivatives of hydrocarbons" which, besides containing carbon and hydrogen, also contain some other atoms, viz. oxygen, nitrogen, halogen, sulphur, etc.

1.2 NAMING HYDROCARBONS ACCORDING TO IUPAC CONVENTIONS

1.2.1 Saturated Unbranched-chain Hydrocarbons and Univalent Radicals

I. *Saturated Unbranched-chain Hydrocarbons*

The first four saturated unbranched acyclic (non-cyclic) hydrocarbons are called methane, ethane, propane and butane. Names of higher members of this series consist of a numerical term followed by "ane". Examples of a few unbranched, saturated, acyclic hydrocarbons are shown in Table 1.1.

Table 1.1: (*n* = total number of carbon atoms)

n	*Name*	*n*	*Name*
5	Pentane	22	Docosane
6	Hexane	23	Tricosane
7	Heptane	24	Tetracosane
8	Octane	25	Pentacosane
9	Nonane	26	Hexacosane
10	Decane	27	Heptacosane
11	Undecane	28	Octacosane
12	Dodecane	29	Nonacosane
13	Tridecane	30	Triacontane
14	Tetradecane	40	Tetracontane
15	Pentadecane	50	Pentacontane
16	Hexadecane	60	Hexacontane
17	Heptadecane	70	Heptacontane
18	Octadecane	80	Octacontane
19	Nonadecane	90	Nonacontane
20	Icosane	100	Hectane
21	Henicosane		

II. *Univalent Radicals*

Univalent radicals derived from saturated, unbranched hydrocarbons by removal of one hydrogen atom from a terminal carbon atom are named by replacing –"ane" of the parent name by "yl". The carbon atom with the free valence is numbered as 1.

$$\overset{4}{C}H_3\overset{3}{C}H_2\overset{2}{C}H_2\overset{1}{C}H_2\text{—Butyl}$$

$$H_3\overset{10}{C}\text{–}(CH_2)_8\text{–}\overset{1}{C}H_2\text{ —Decyl, etc.}$$

1.2.2 Saturated Branched-chain Acyclic Hydrocarbon

In case of a saturated, branched-chain hydrocarbon, following rules are applicable in the same sequence:

Rule 1: Selection of the longest carbon chain: The longest carbon chain gives parent name to the compound.

Rule 2: Numbering the longest chain: Longest chain is numbered starting from a terminal carbon. Numbering is done in such a way that lowest possible number is given to the side chain as in (1).

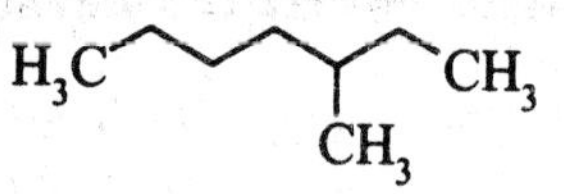

(1)

Longest chain has seven carbon atoms. If we start numbering from the left terminal, side chain methyl group comes at 5th carbon while if numbering starts from right terminal, side chain methyl group comes at 3rd carbon which is a lower number. Hence, the correct way of numbering will be to start from the right terminal.

When more than one locants (side groups) are present, than that number series is "lowest" which contains the lower number on the occasion of first difference as:

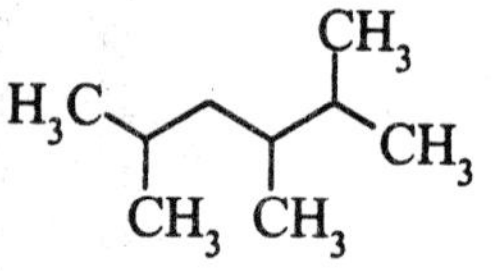

(2)

The two number series existing in compound (2) are: 2, 3, 5 and 2, 4, 5. The first one is correct and second is incorrect.

Rule 3: When writing names of branched, saturated hydrocarbon, locants are written first in alphabetical order (irrespective of their numbers) followed by name of parent chain. Locants in the name must be followed by its number separated by hyphen (-). If more than one locants of similar types are present at the same or different positions, di, tri, etc. are prefixed before the locants. Adjacent numbers must be separated by comma (,).

Applying the above mentioned rules, compounds (1) and (2) described just before can be named as:

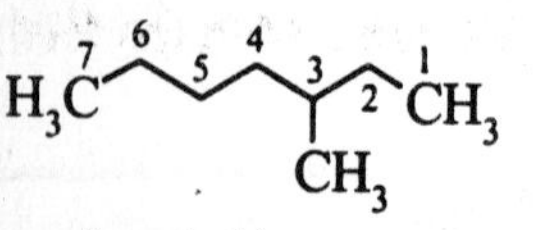

3-methyl heptane

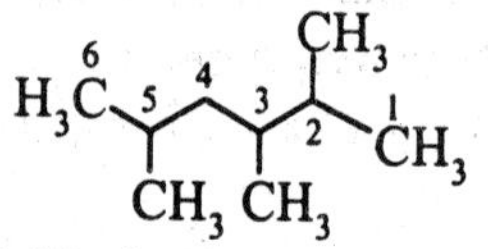

2,3,5-trimethyl hexane (not 2,4,5-trimethyl hexane)

Some other examples applying the above rules are:

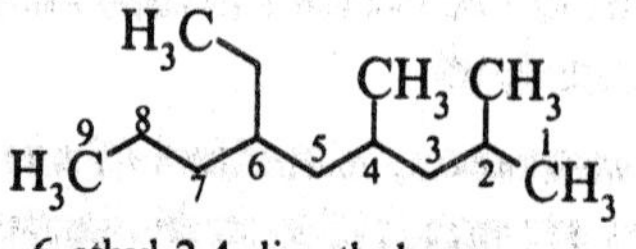

6-ethyl-2,4-dimethylnonane
not 2,4-dimethyl-6-ethylnonane

(In alphabetical order, initials of original name of the locant is considered, i.e. 'm' of methyl is considered here not 'd' of di)

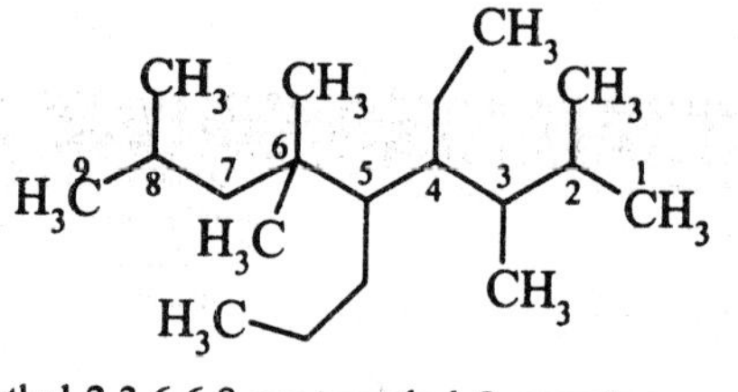

4-ethyl-2,3,6,6,8-pentamethyl-5-propylnonane
not 6-ethyl-2,4,4,7,8-pentamethyl-5-propylnonane

Rule 4: When two different alkyl locants are equidistant from the two ends of main chain, they are numbered in alphabetical order as:

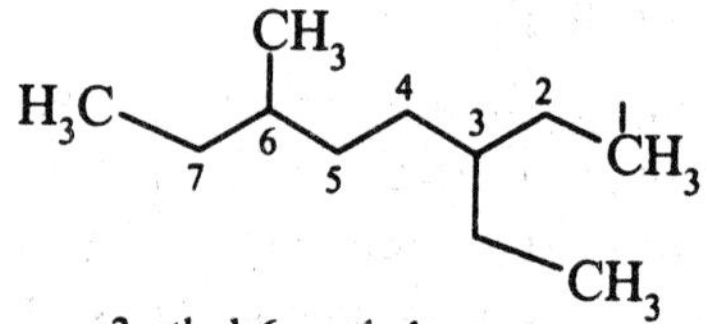

3-ethyl-6-methyloctane
not 3-methyl-6-ethyloctane

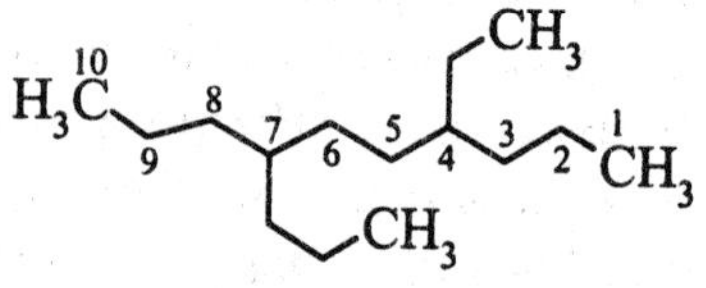

4-ethyl-7-propyldecane
not 7-ethyl-4-propyldecane

Rule 5: Following common names are retained in IUPAC system for unsubstituted hydrocarbons only.

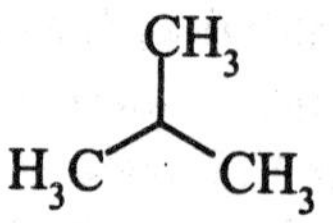

Isobutane

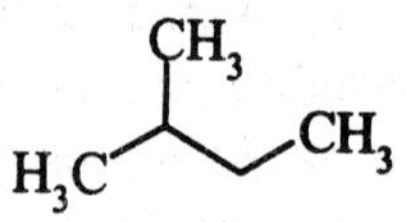

Isopentane

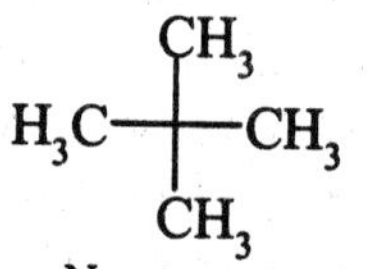

Neopentane

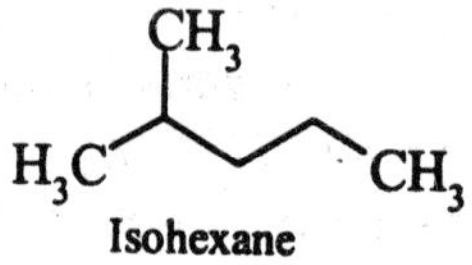

Isohexane

Rule 6: Branching in the branch: If there is a complex branch which is branched further, branch is numbered separately starting from the carbon directly bonded to parent chain and name of branch is written in small parentheses ( ) before the name of parent hydrocarbon as:

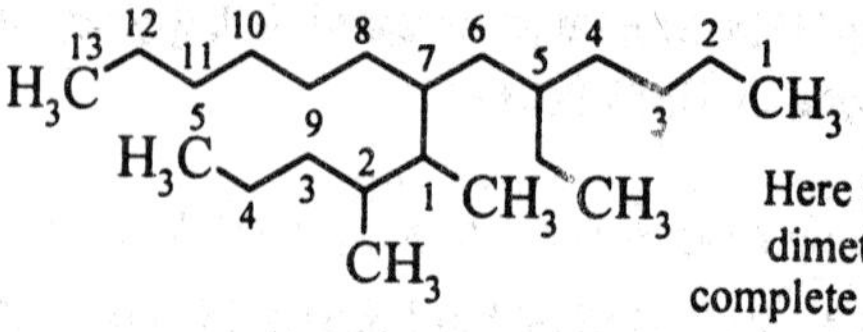

Here the branch is dimethylpentyl as complete single substituent

7-(1,2-dimethylpentyl)-5-ethyltridecane
not 5-ethyl-7-(1,2-dimethylpentyl) tridecane

In case where names of complex branch are composed of identical words, priority is given to that branch which has locants at the lowest number as:

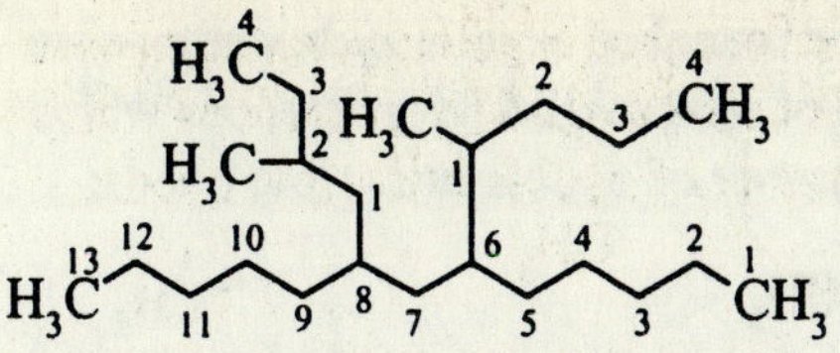

Both branches have the common word "butyl", therefore branch at position 6 will have priority since it has locant at C-1 while other has locant at C-2

6-(1-methylbutyl)-8-(2-methylbutyl)tridecane

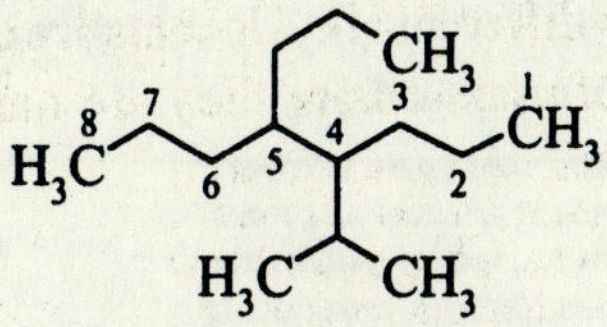

4-isopropyl-5-propyloctane

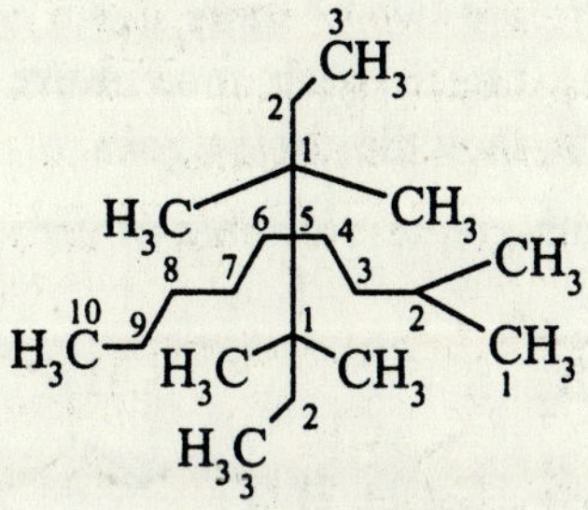

5,5-bis(1,1-dimethylpropyl)-2-methyldecane

Rule 7: The presence of identical complex branch, each substituted in the same way, may be indicated by the appropriate multiplying prefix bis, tris, tetrakis, pentakis, etc. The complete expression denoting such side chain may be enclosed in parentheses as:

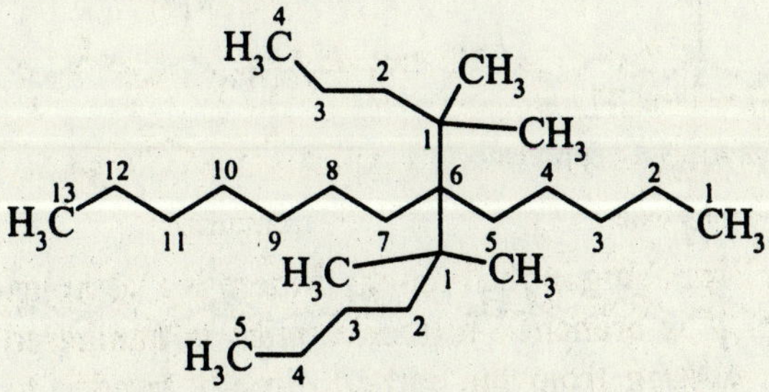

6-(1,1-dimethylbutyl)-6-(1,1-dimethylpentyl)tridecane

Rule 8: If chains of equal length are competing for selection as main chain in a saturated, branched acyclic hydrocarbons, the choice goes in series to:

(a) The chain, which has the greatest number of side chains:

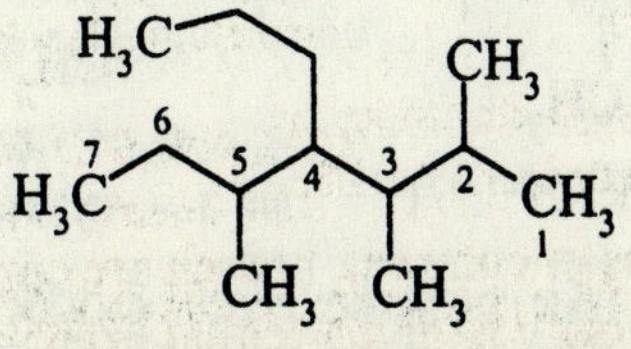

Correctly numbered (four locants)
2,3,5-trimethyl-4-propylheptane

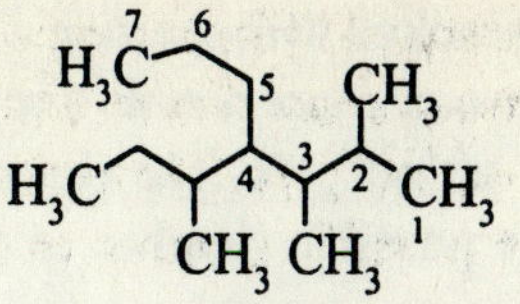

(three locants) 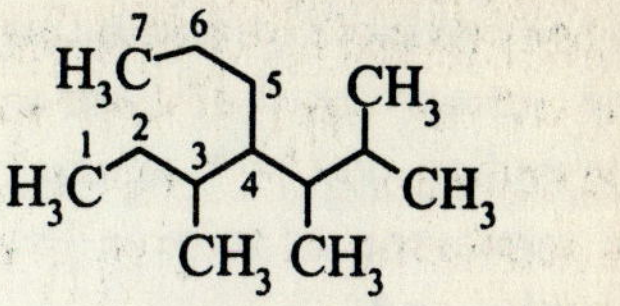 (two locants)
(wrongly numbered)

(b) The chain whose side chain has lowest numbered locants:

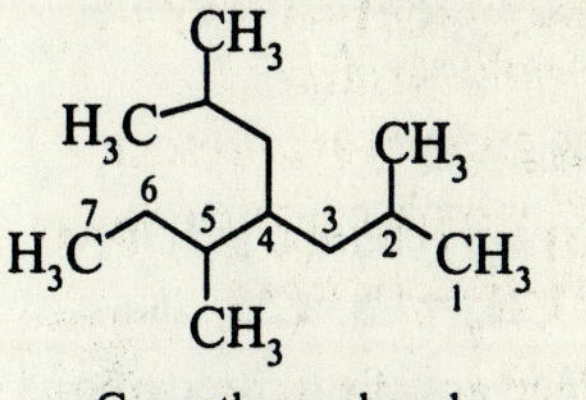

Correctly numbered
(locants are at 2,4,5 positions)
4-(2-methylpropyl)-2,5-dimethylheptane

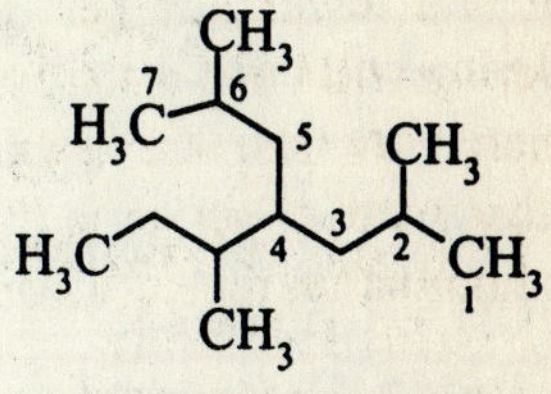

Wrongly numbered
(locants are at 2,4,6 positions)

(c) The chain having the greatest number of carbon atoms in the smaller side chain:

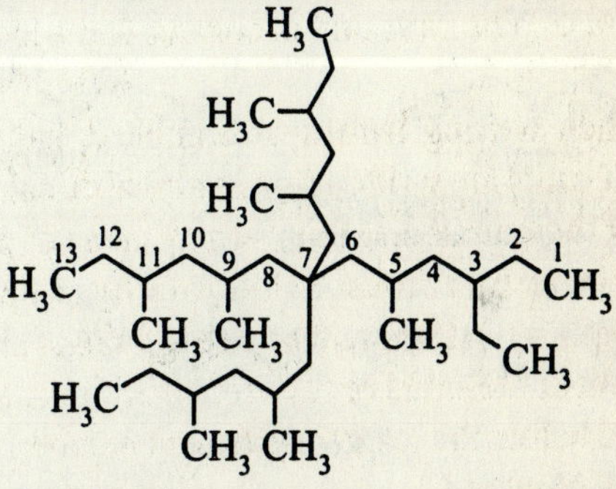

7,7-bis(2,4-dimethylhexyl)-3-ethyl-5,9,11-trimethyltridecane

(d) The chain having the least branched side chain:

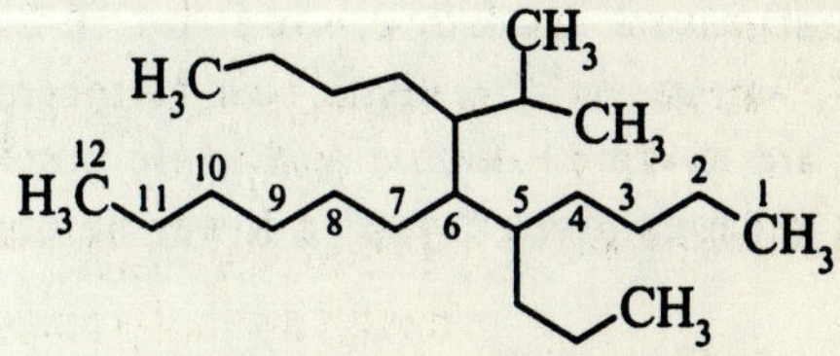

6-(1-isopropylpentyl)-5-propyldodecane

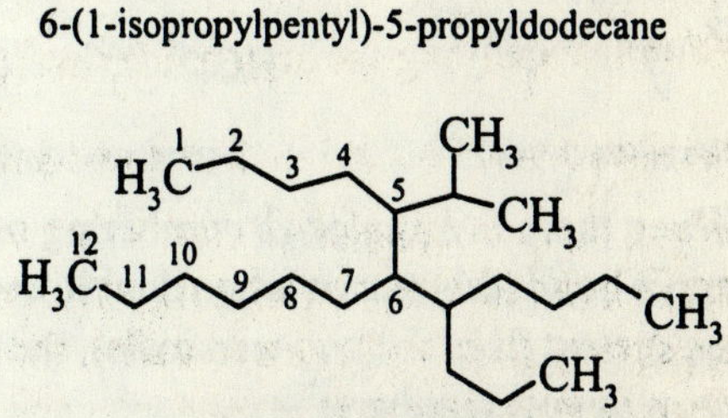

Wrongly numbered

1.2.3 Unsaturated Hydrocarbons

Rule 1: Unsaturated, unbranched acyclic hydrocarbons having one double bond are named by replacing "–ane"

of the name of corresponding saturated hydrocarbon with the ending "-ene". If there are two or more double bonds, the ending will be "-adiene", "-atrine", etc. The chain is so numbered as to give lowest possible number to the double bond.

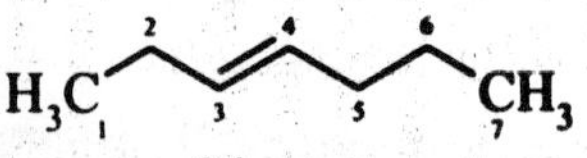

3-heptene

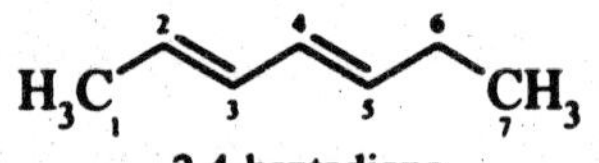

2,4-heptadiene

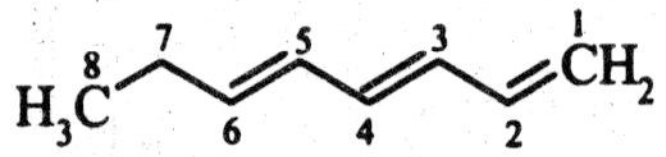

1,3,5-octatriene

Following unsystematic names are retained in IUPAC:

$CH_2{=}CH_2$ (ethene) $CH_2{=}C{=}CH_2$ (allene)

Rule 2: Unsaturated, unbranched acyclic hydrocarbons having one triple bond are named by replacing the ending "-ane" of corresponding saturated hydrocarbon with the ending "-yne". If there are two or more triple bonds, the ending will be "-adiyne", "-atriyne", etc. The chain is so numbered as to give lowest possible number to the triple bond:

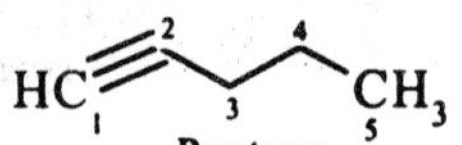

Pentyne
(Its triple bond is at terminal position, therefore it need not be numbered)

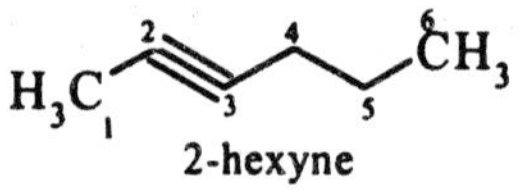

2-hexyne

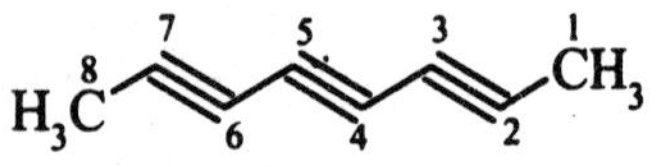

2,4,6-octatriyne

Rule 3: Unsaturated, unbranched acyclic hydrocarbons, having both double and triple bonds are named by replacing ending "-ane" from the name of the corresponding saturated hydrocarbons by "-enyne", "-adienyne", "-atrienyne" "-endiyne", etc. Numbers as low as possible are given to double and triple bonds even though this at times gives "-yne" a lower number than "-ene":

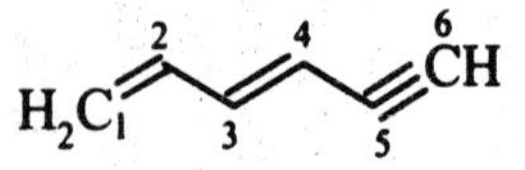

1,3-hexadien-5-yne

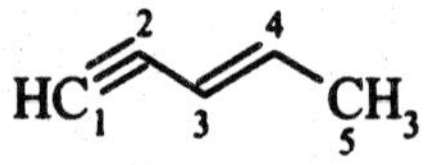

3-pentene-1-yne

Rule 4: When there is a choice in numbering between double and triple bond (this occurs when double and triple bonds are equidistant from the two terminals), the double bonds are given lowest numbers:

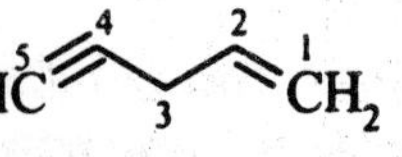

1-penten-4-yne

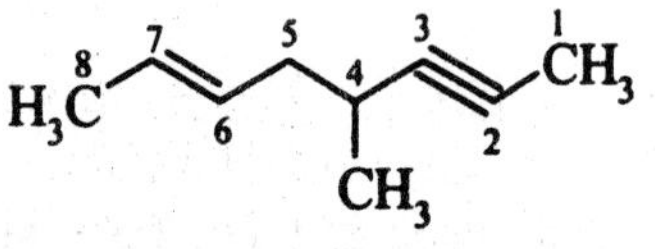

4-methyl-6-octen-2-yne

Rule 5: Unsaturated branched acyclic hydrocarbons are named as derivatives of unbranched hydrocarbons which contain maximum number of double and triple bonds:

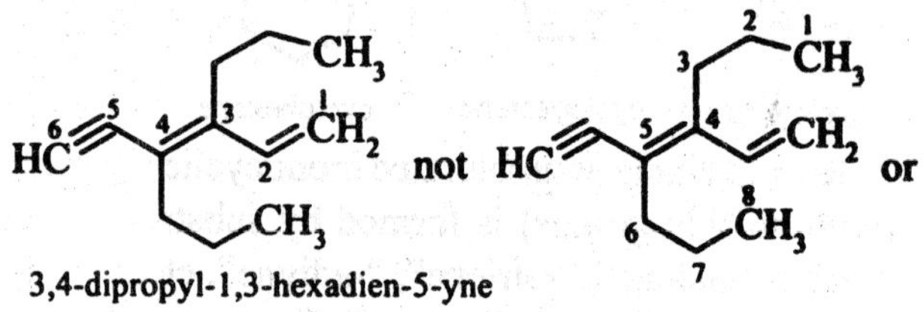

3,4-dipropyl-1,3-hexadien-5-yne (not ... or ...)

Carbon chain with the maximum number of double/triple bonds is selected as parent chain though it contains less no. of carbon atoms than other longer chain but with lower no. of double/triple bonds, existing in the molecule.

Rule 6: If there are two or more chain competing for selection as the chain with maximum number of unsaturated bonds, then the choice goes to:

(i) That one with greatest number of carbon atoms:

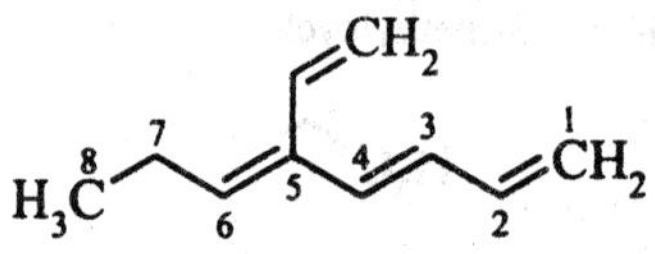

5-vinyl-1,3,5-octatriene

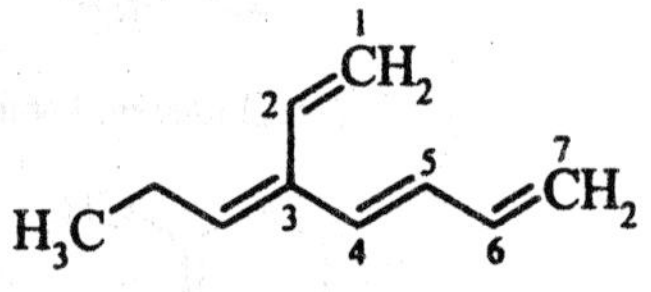

The above numbering system also has a main chain with three double bonds but one carbon less than other chain (mentioned on left).

(ii) If the number of carbon atoms being equal, the one containing maximum number of double bonds is selected as the parent chain.

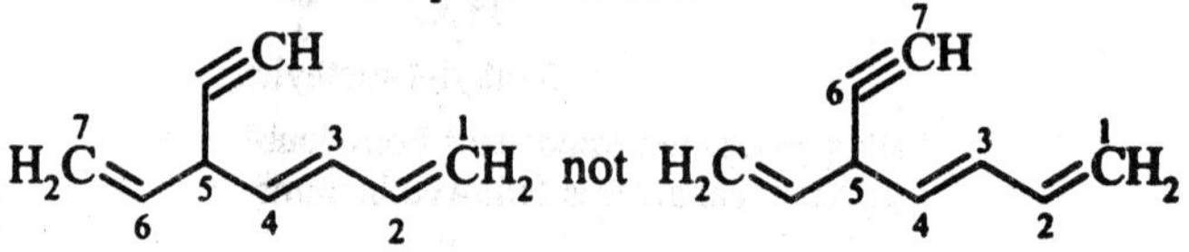

5-ethynyl-1,3,6-heptatriene (not ...)

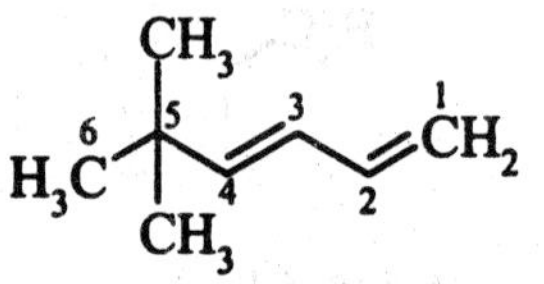

5, 5-dimethyl-1, 3-hexadiene

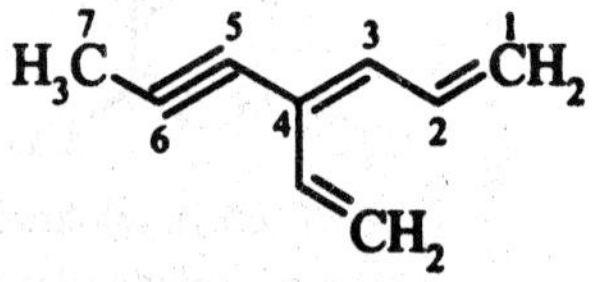

4-vinyl-1-hepten-5-yne

Following name is retained for unsubstituted compound only:

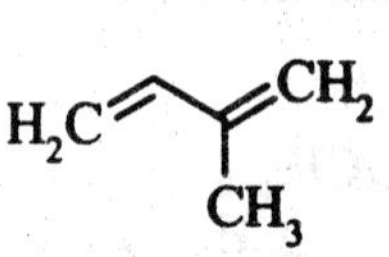

Isoprene

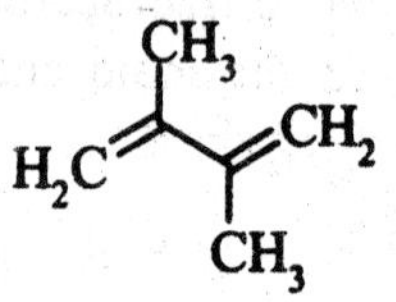

2,3-dimethyl-1,3-butadiene not 3-methylisoprene

1.2.4 Naming Monocyclic Hydrocarbons

Rule 1: Name of monocyclic, saturated hydrocarbons are formed by attaching the prefix "cyclo" to the names of acyclic, saturated hydrocarbon with same number of carbons:

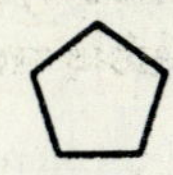
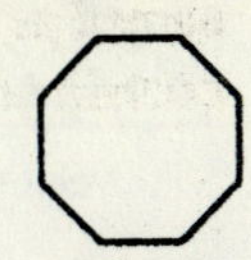

cyclopropane cyclopentane cyclohexane cycloctane

Rule 2: Name of unsaturated monocyclic hydrocarbons (with no side groups) is formed by substituting "-ene", "-yne", "-adiene", "-atriene", "-adiyne", etc. for "ane" in the name of corresponding cycloalkane, e.g.

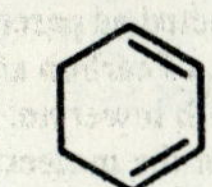
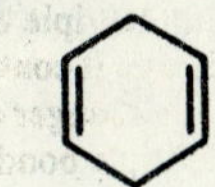

cyclohexene 1,3-cyclohexadiene 1,4-cyclohexadiene

1-cyclodecen-4-yne 3-methylcyclohexene

1,2,4-trimethylcyclohexene 4-ethyl-5-methylcyclohexene

not 5-ethyl-4-methylcyclohexene

If alkyl groups are equidistant from double bond, alphabetical order is followed in numbering

1,3,5-trimethylcyclohexene not 2,4,6-trimethylcyclohexene 5-methyl-1,3-cyclohexadiene

Rule 3: If a straight chain substituent attached to monocyclic hydrocarbon has greater number of carbon atoms in the main chain than the cyclic ring becomes substituent as:

1-cyclopropyl-2,3-diethyl-4-methylhexane

1-cyclopentyl-3-ethyl-1,4-pentadiene

1.2.5 Substituted Aromatic Hydrocarbons

Following names for the substituted monocyclic aromatic hydrocarbons are retained:

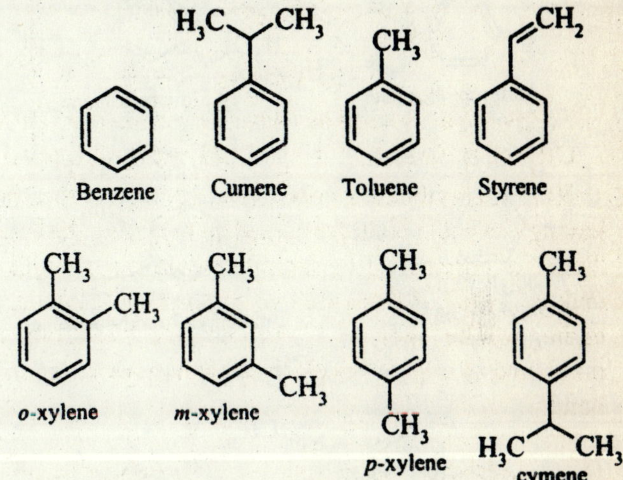

Benzene Cumene Toluene Styrene

o-xylene *m*-xylene *p*-xylene cymene

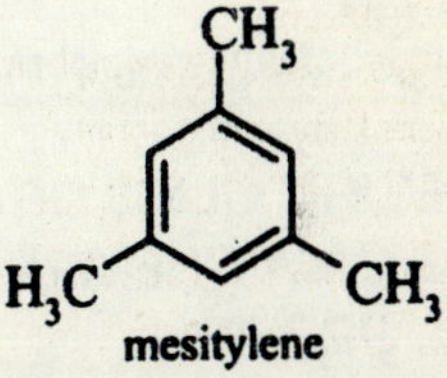

mesitylene

Rule 1: Other monocyclic substituted hydrocarbons are named as derivatives of benzene or one of the compounds listed above. However, if the substituent introduced into such a compound is identical with already present in that compound, then the substituted compound is named as derivatives of benzene as:

1,4-divinylbenzene 1-methyl-3-vinylbenzene

1,2,3-trimethylbenzene

Rule 2: The positions of substituents is indicated by numbers except *o*-(ortho), *m*-(meta), *p*-(para) may be used in place of 1,2- , 1,3- , and 1,4-, respectively, when only two substituents are present. The lowest possible number given to the substituent coming 1st in alphabetical order except that when names are based on those of the compounds mentioned in the beginning of section-E.

1-ethyl-4-pentylbenzene

1,4-dipropylbenzene

3,5-diethyltoluene

4-ethylstyrene
p-ethylstyrene

1,2-dimethyl-3-propylbenzene

1-butyl-3-ethyl-2-propylbenzene

1.3 NAMING SPECIFIC CLASS OF COMPOUNDS CONTAINING ONLY ONE TYPE OF PRINCIPAL FUNCTIONAL GROUPS

1.3.1 Halogen Compounds

Substitutive naming of halogen compounds are formed by adding the prefixes "fluoro-", "chloro-", "bromo-" or "iodo-" to the name of parent compound.

3-chlorohexane

3,4,7-trichlorononane

5-bromo-2-chloroctane

If there is a choice between the two halides for numbering, alphabetical order is followed, e.g.

2-bromo-3-chlorobutane

1-bromo-3-fluorocyclohexane

1-chloro-3-fluoro-5-iodocyclohexane

1.4 NAMING COMPOUNDS CONTAINING ONE OR MORE SIMILAR TYPE OF PRINCIPAL FUNCTIONAL GROUPS

Names of such compounds which contain principal functional group end with suffix of that group provided by IUPAC. Suffixes and prefixes for some of the common functional group are given in Table 1.2.

1.4.1 Aldehydes

Acyclic mono- and dialdehydes are named by adding suffix "-al" or "-dial" to the name of acyclic hydrocarbon with the same number of carbon atoms, eliding the final

Table 1.2: Suffixes and prefixes for common functional groups

Class	*Formula*	*Prefix*	*Suffix*
1. Acid halide	-CO-X	Halocarbonyl	Carbonyl halide
2. Alcohols	-OH	Hydroxy	-ol
3. Aldehydes	-CHO, -(C)HO	Formyl, Oxo	Carbaldehyde,-al
4. Amides	$-CONH_2$, $-(C)ONH_2$	Carbamoyl	Carboxamide,-amide
5. Amines	$-NH_2$	Amine	-amine
6. Carboxylic acid	-COOH-(C)OOH	Carboxy	-carboxylic acid,-oic acid
7. Ethers	-OR	Alkoxy	
8. Esters	-COOR-(C)OOR	R-oxycarbonyl	R-carboxylate, R-oate
9. Ketones	>(C)=O	Oxo	-one
10. Nitrile	-CN-(C)N	Cyano	-carbonitrile, nitrile
11. Sulphonic acid	$-SO_3H$	Sulpho	-Sulphonic acid

(Formulas containing carbon in parenthesis indicates that the carbon is part of main chain)

"e" of the hydrocarbon name before "a". Numbering starts from the carbon of the aldehyde group –CHO:

CH_3CHO Ethanal

Pentanal

4-ethyl-2,5-heptadienal

2-ethylhexandial

If the parent chain contain more than two –CHO groups, they are named as carbaldehyde and carbon of –CHO is not counted in main chain as:

Butan-1,2,4-tricarbaldehyde

Cyclohexancarbaldehyde

3-(formylmethyl)hexandial

4-cyclopentylbutanal

1.4.2 Ketones

Ketones are named substitutively by adding a suffix such as "-one", "-dione", etc. to the name of parent hydrocarbons:

2-pentanone

1-phenylpentan-2,3-dione

4,6-dimethylhept-3-en-2,5-dione

1.4.3 Carboxylic Acids

Simple (unsubstituted) acyclic mono- and dicarboxylic acids. Carboxylic acid group –COOH, that conceptually replace $–CH_3$ groups terminating an acyclic hydrocarbon chain are denoted by adding the suffix "-oic acid" or "-dioic acid" to the name of acyclic hydrocarbon with elision of its final "e" before "o".

CH_3-$[CH_2]_4$-COOH Hexanoic acid

HOOC-$[CH_2]_6$-COOH Octandioic acid

side group

2-(2-ethyl-1-methylpentyl) pentandioic acid

If an unbranched chain is directly linked to more than two carboxy groups, these carboxy groups are named from the parent hydride by using suffix carboxylic acid as:

Butan-1,2,4-tricarboxylic acid

4-ethyl-3-isopropyl-4-hexenoicacid

Cyclohexancarboxylic acid

When all carboxylic acid groups cannot be described in suffix, a carboxylic group is indicated in prefix as:

3-(carboxymethyl)heptandioic acid

Following names have been retained:

Benzoic acid

Ortho-toluic acid

Some trivial names of carboxylic acids are also very common as:

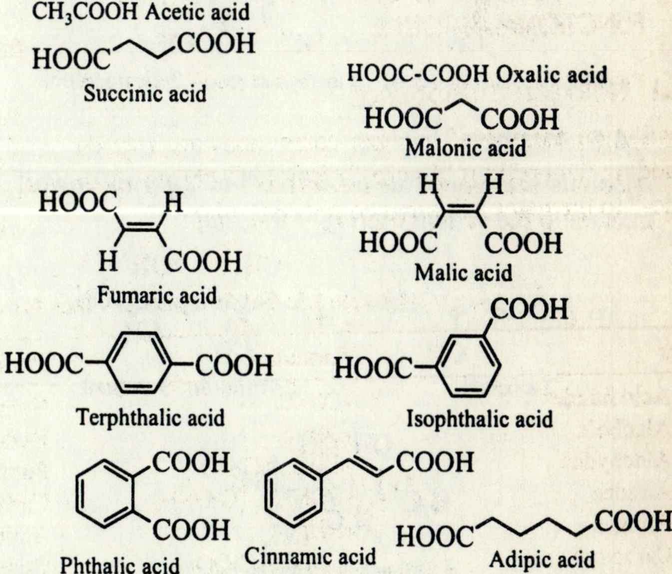

1.4.4 Nitriles

Acyclic, saturated compounds containing –CN functional group are named ending with "-nitrile", "-dinitrile" suffix as:

4-ethyl-3,5-dimethyl hexanitrile 3-(1-chloroethyl)-4-(1,2-dimethyl-propyl) hexandinitrile.

If there are more than two –CN groups directly linked to the parent chain, naming is done using prefix "-tricarbonitrile", etc. and carbon of the –CN group is not counted in the parent chain, e.g.

Butan-1,2,4-tricarbonitrile 1-ethyl-4-methyl-4-penten-1,2,3,5-tetracarbonitrile

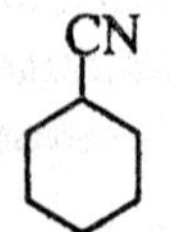

Cyclohexancarbonitrile 2-ethyl-6-methylcyclohexancarbonitrile

Monocyclic aromatic compound containing –CN group on ring are named as benzene carbonitrile or its derivatives as:

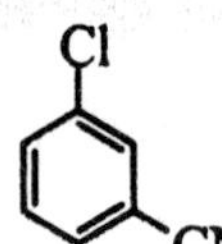

3-chlorobenzencarbonitrile 5-methylbenzene-1,3-dicarbonitrile

1.4.5 Alcohols

Alcohols are named using suffix "-ol", giving lowest number to the carbon bearing –OH group:

2-hexanol 3,6,7-trimethyl-3-octanol

4,5-dimethyl-2,4-hexandiol

4-(2-hydroxyethyl)-2-methyloctan-1, 8-diol 2-methyl-2,4-hexadien-1, 6-diol

Following names has been retained in IUPAC:

Phenol

1.4.6 Ethers

The generic IUPAC name of this family is "alkoxy-alkane", in which alkane part contains longer carbon chain in case of unsymmetrical ethers:

Ethoxy ethane Isopropoxy butane

1-chloroethoxy-2-hexene IUPAC: methoxy benzene common: Anisole

Cyclic ethers are named by giving typical name to the parent rings as:

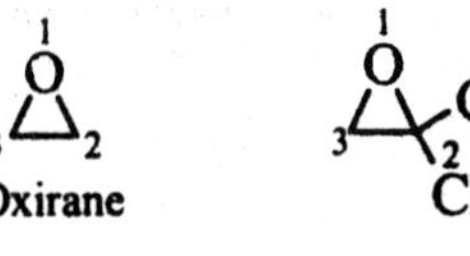

2,2-dimethyl oxirane Oxitane

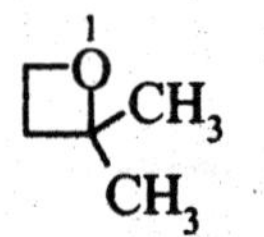

2,2-dimethyloxitane 2-chloro-3-bromo-4-methyloxitane

1.4.7 Amines

The generic IUPAC name of amines is "alkanamine".

4-methyl-2-butanamine 1,2,4-butantriamine

3-methyl-1,2-cyclohexandiamine

Aniline has been retained for $C_6H_5NH_2$:

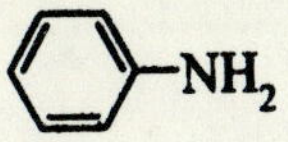

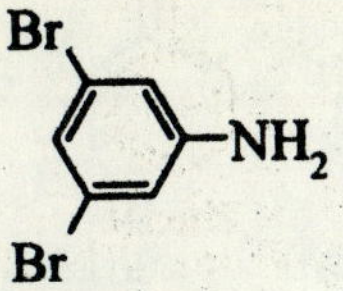

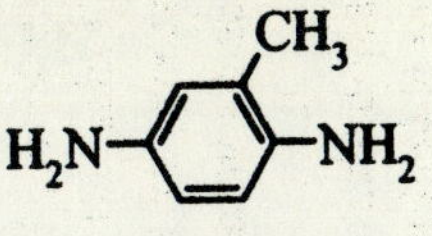

Aniline   *m*-dibromoaniline   2,5-diaminotoluene

Secondary and tertiary amines: Amines with substituent on nitrogen as –NHR and –NRR' are named by adding prefixe N-alkyl or N,N-dialkyl as:

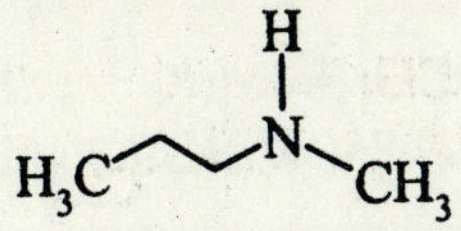

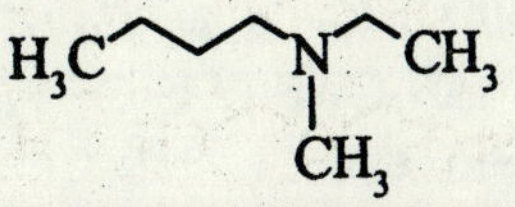

N-methylpropanamine   N-ethyl-N-methylbutanamine

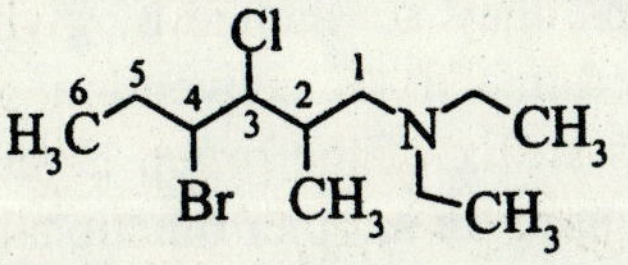

N,N-diethyl-4-bromo-3-chloro-2-methylhexanamine

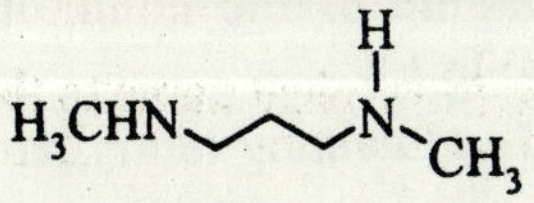

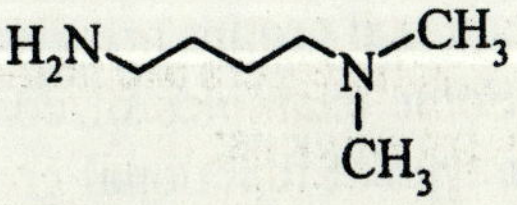

N,N'-dimethylpropan-1,3-diamine   N,N-dimethylbutan 1,4-diamine

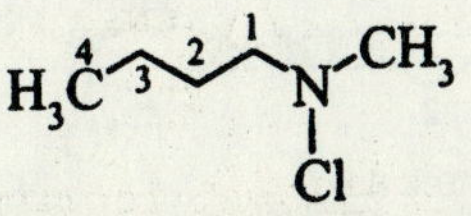

N-chloro-N-methylbutanamine

1.4.8 Esters

The generic name of ester is alkylalkanoate. Esters are named as:

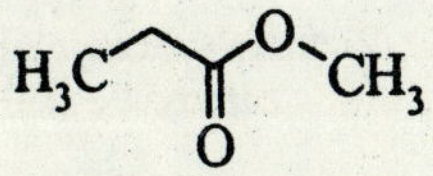

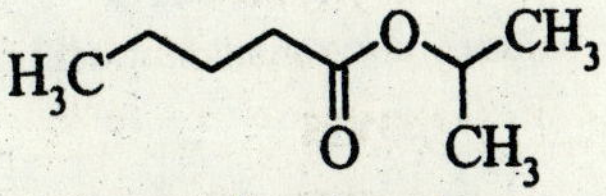

Methyl propanoate   Isopropyl pentanoate

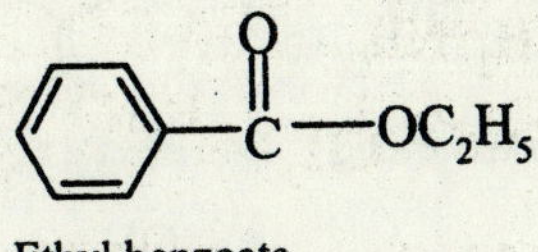

Ethyl benzoate

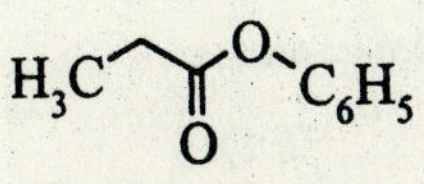

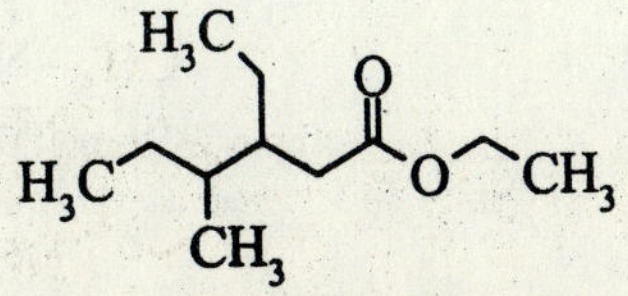

Phenyl propanoate   ethyl-3-ethyl-4-methylhexanoate

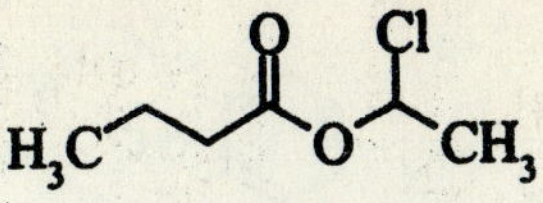

1-chloroethylbutanoate

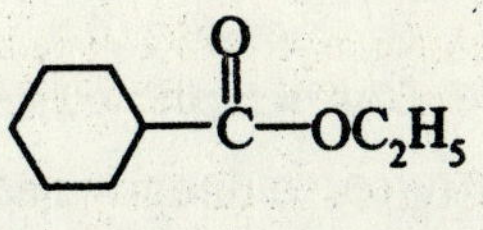

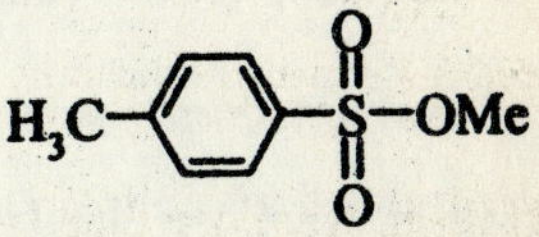

Ethyl cyclohexancarboxylate   methyl-4-methyl benzenesulphonate

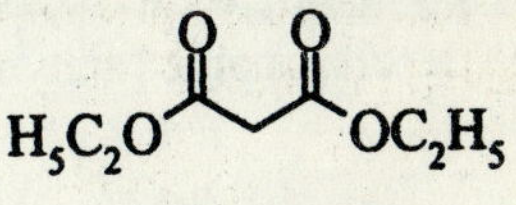

Diethyl malonate

1.4.9 Anhydrides

Anhydrides are compounds formally derived by loss of water molecules (by intermolecular dehydration of monocarboxylic acid or by intramolecular dehydration of suitable dicarboxylic acids) from two acid functional groups. Anhydrides are classified into two categories as symmetrical and unsymmetrical anhydrides:

(a) Symmetrical Anhydride

Symmetrical anhydrides of monobasic carboxylic acids, substituted or unsubstituted, are named by replacing the term "acid" of the acid name by the class name "anhydride" as:

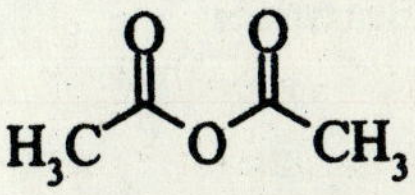

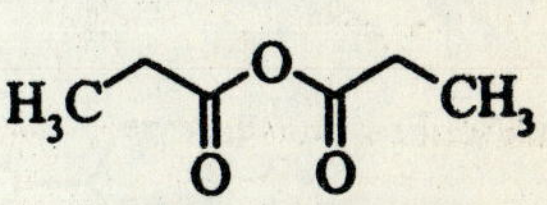

Ethanoic anhydride (acetic anhydride)   Propanoic anhydride

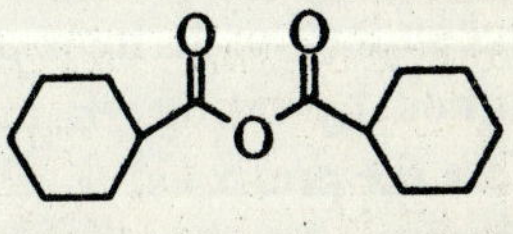

Cyclohexancarboxylic acid anhydride

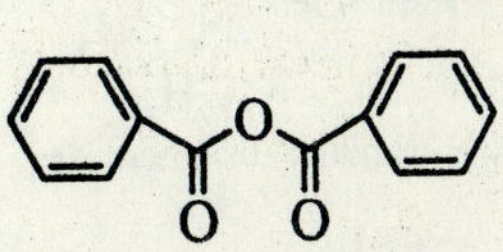

$(ClCH_2CH_2CO)_2O$

Benzoic acid anhydride   3-chloro propanoic acid anhydride

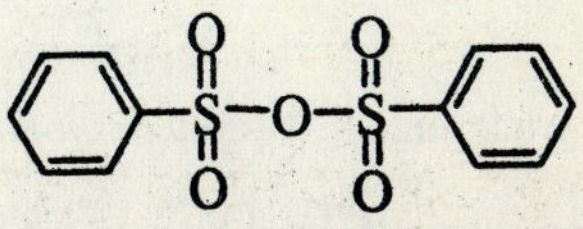

Benzenesulphonic acid anhydride

Cyclic anhydride derived by intramolecular dehydration of a dicarboxylic acid are named in the same way as:

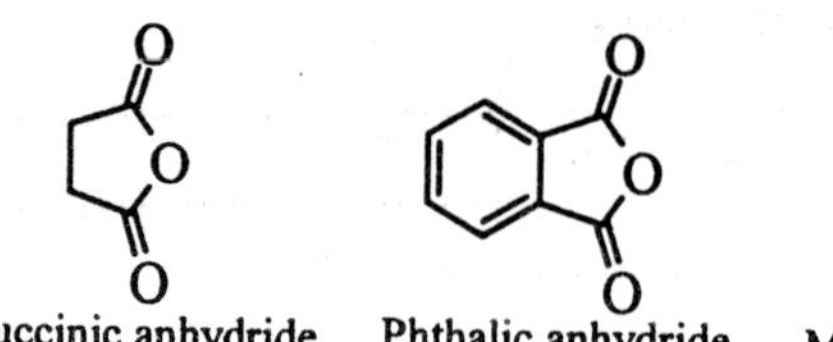

Succinic anhydride Phthalic anhydride Malic anhydride

(b) Unsymmetrical Anhydride (mixed anhydride)

Anhydrides derived from two different monobasic acids are named by citing the first parts of the name of the two acids (i.e. the parts preceding the term acid) in alphabetical order, followed by the class name "anhydride" as a separate word as:

Ethanoic propanoic anhydride Chloroethanoic ethanoic anhydride

Benzoic cyclohexancarboxylic anhydride

1.4.10 Amides

Primary amides are named by replacing the suffixes "oic acid" or "carboxylic acid" from the name of acid by "amide" or "carboxamide" as:

Hexanamide Benzamide

Cyclohexancarboxamide

Substituted, secondary or tertiary amides (-NHR or (-NRR')) are named by citing –R and –R' as prefix separately N-before the prefix as:

N-methyl propanamide N,N-dimethyl butanamide

N-ethyl-N-methyl pentanamide N-cyclohexyl-N-methyl benzamide

3-chloro-N-ethyl-4-methyl-N-propyl hexanamide

N-(1-chloropropyl)-N-methyl-3-chloropentanamide

N-methyl benzene sulphonamide

1.5 NAMING COMPOUNDS CONTAINING MORE THAN ONE TYPE OF PRINCIPAL FUNCTIONAL GROUPS

When a compound contains two or more types of principal functional groups, they are named by giving priority to one as prescribed in IUPAC convention. Name of such compound bears suffix of the group highest in priority and prefixes are used for another functional groups. Also lowest possible number is given to the group highest in priority. Relative priority of some of the common functional groups are given in Table 1.3.

Some examples of compounds containing more than one type of functional groups:

5-oxohexanoic acid

6-oxohexanoic acid here carbon of aldehyde group is part of parent chain

3-formyl-6-oxononanoic acid

2-cyano-4-oxocyclohexan carboxylic acid

4-hydroxy-5-oxohexanoic acid

7-methyl-4,6,8-trioxo octanoic acid

4-amino-5-ethoxy-3-hydroxy pentanal

4-formyl-2-oxo cyclohexan sulphonic acid

2-carbamoyl benzoic acid

Table 1.3: Relative priority of functional groups

Functional groups	*Formula*		
Carboxylic acid	-COOH	**Decreasing priority** ↓	Prefix and suffix for the groups are given in Table 1.2
Peroxy acid	$-CO_3H$		
Sulphonic acid	$-SO_3H$		
Acid anhydride	-COOCO-		
Esters	-COOR		
Acid halides	-COX		
Acid amides	$-CONH_2$		
Nitriles	-CN		
Aldehydes	-CHO		
Ketones	-CO-		
Alcohols, phenols	$-OH, C_6H_5OH$		
Amines	$-NH_2$		
Ethers	R-O-R		

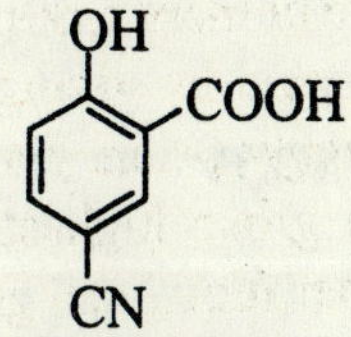

5-cyano-2-hydroxy benzoic acid

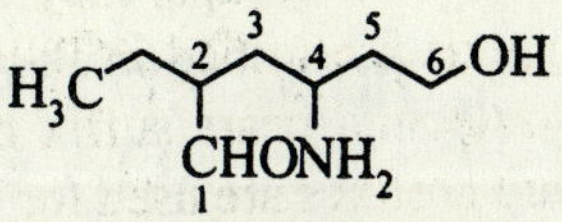

4-amino-2-ethyl-6-hydroxy hexanal

PRACTICE EXERCISE

1. Write names of the following compounds according to IUPAC convention.

(a) 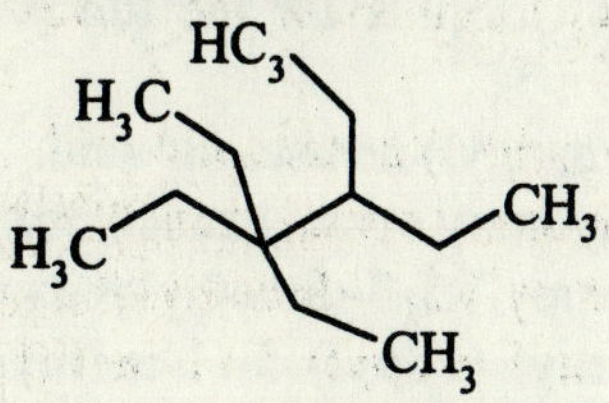 (b) 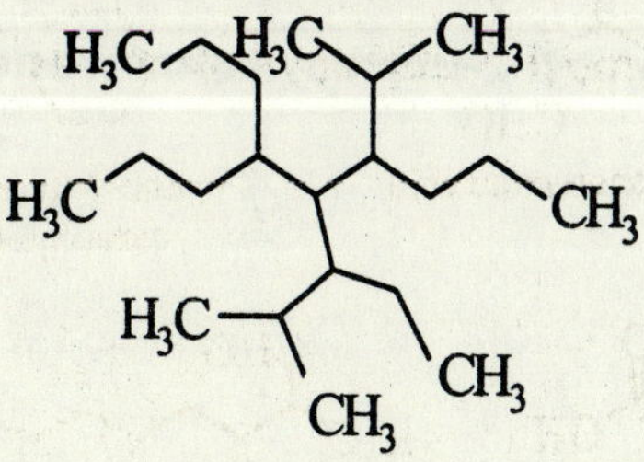 (c) 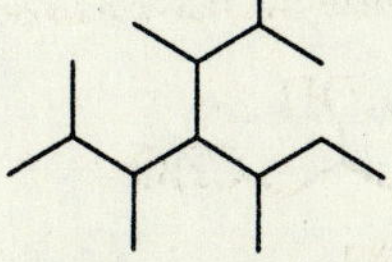 (d) 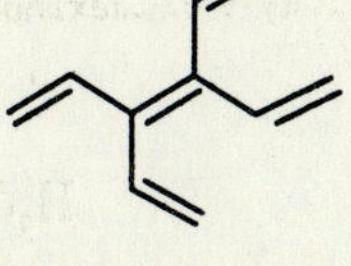 (e) 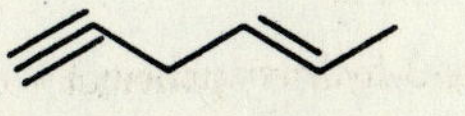 (f)  (g) 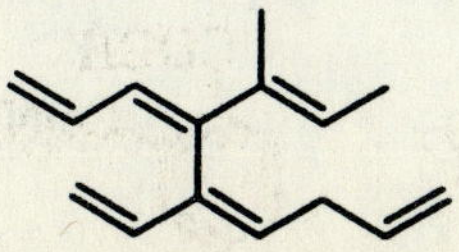 (h) 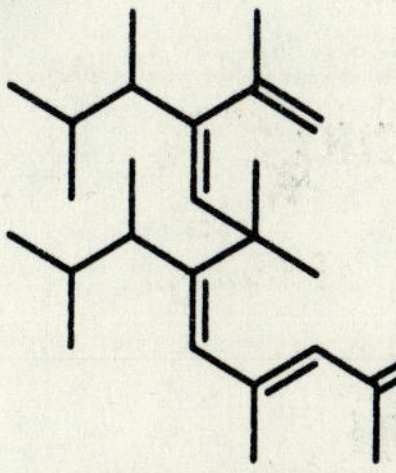 (i)

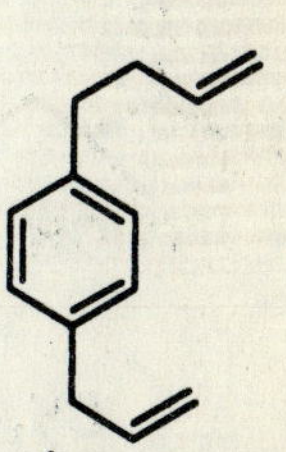

2. Write IUPAC names of the following compounds:

(a) 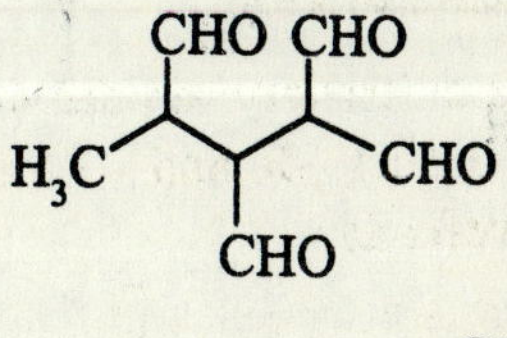 (b) 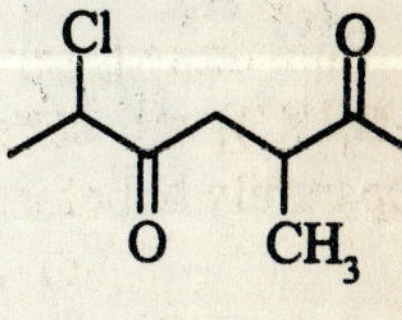 (c) 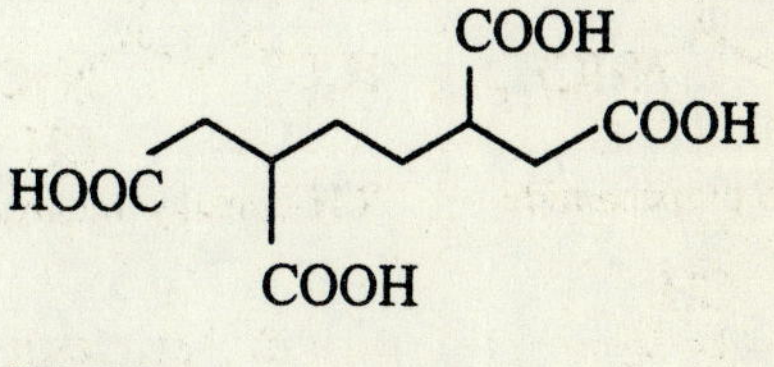 (d) 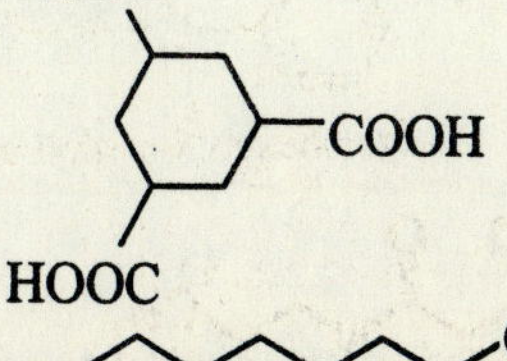 (e)

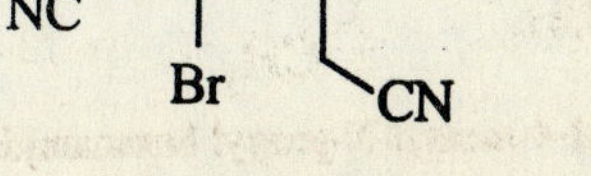

(f)  H_3OS — SO_3H

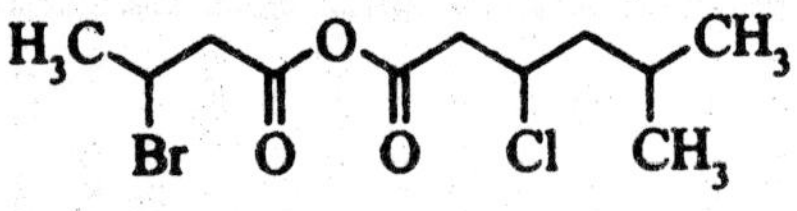

(g) H_3C O CH_3 Br O O Cl CH_3

(h) CH_3 COOH Cl Br

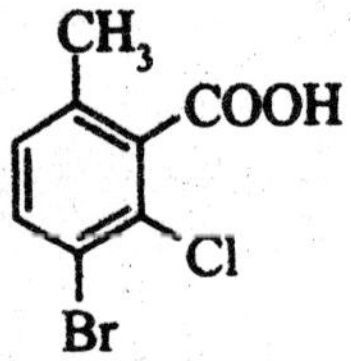

(i) 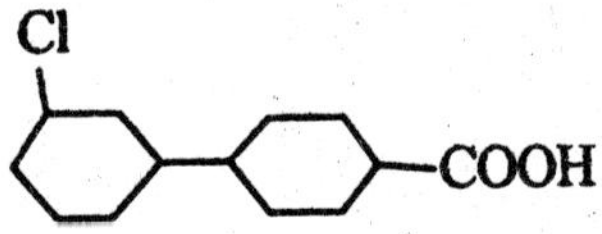 Cl COOH

(j) 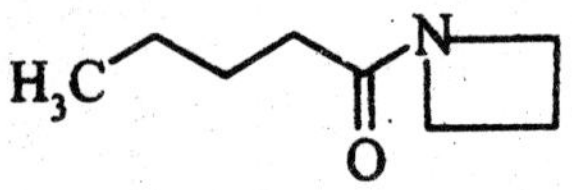 H_3C N O

(k) 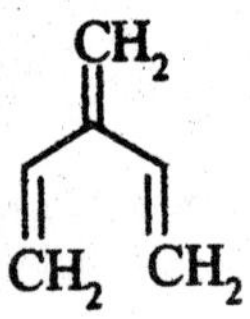 CH_2 CH_2 CH_2

(l) 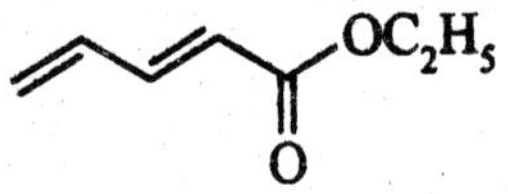 OC_2H_5 O

(e) COOH H_3C Cl COOH

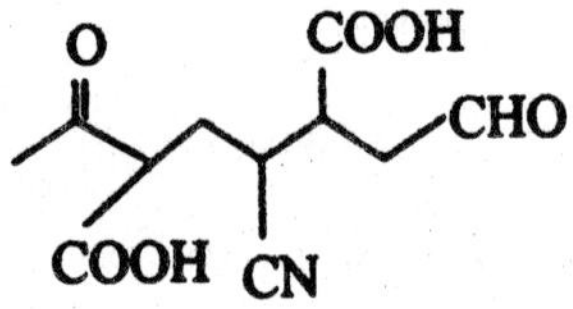

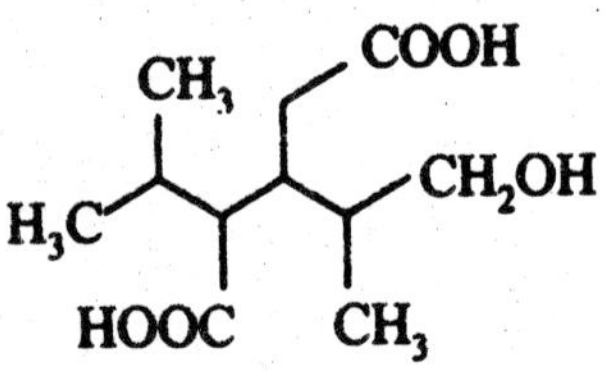

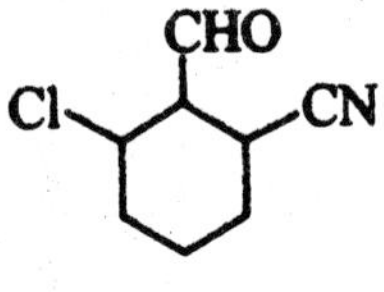

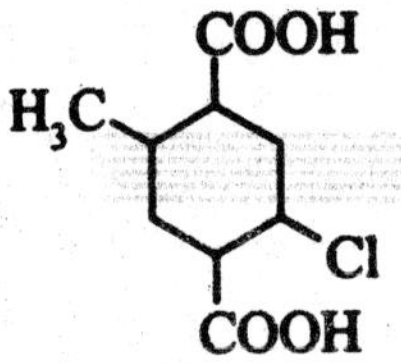

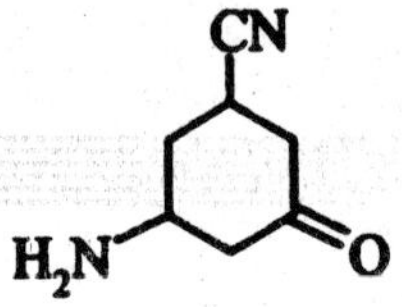

(f) CN H_2N O

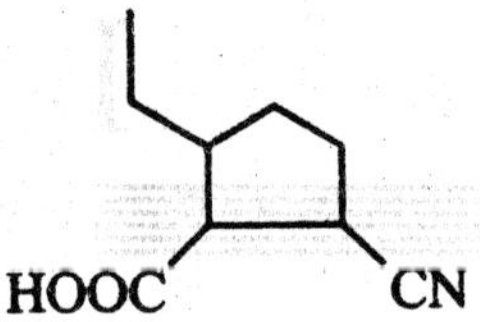

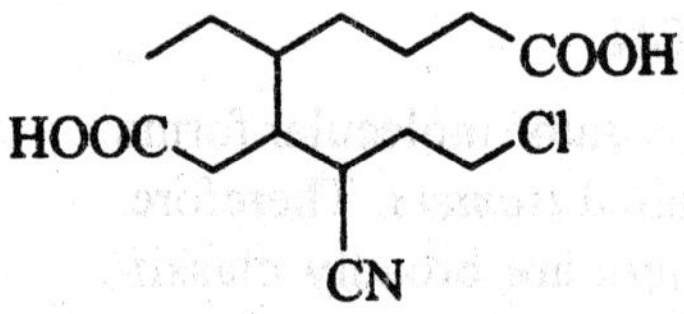

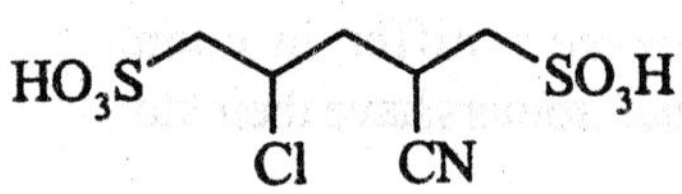

(g) HOOC CN

(h) HOOC COOH Cl CN

(i) HO_3S SO_3H Cl CN

(j) COOH NC

3. Write IUPAC name of the following compounds :

(a) 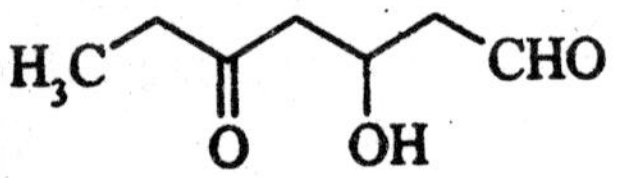 H_3C CHO O OH

(b) O COOH CHO COOH CN

(c) H_3C CH_3 COOH CH_2OH HOOC CH_3

(d) Cl CHO CN

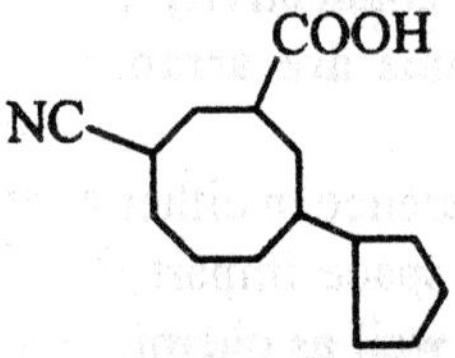

4. Provide structural formula for the following compounds:

(a) 3-(1-carboxymethyl) hexandioic acid.
(b) 3-formyl-5-hydroxy cyclol exancarboxylic acid.
(c) 2-(2-chloropropyl)-3, 3-dimethyloxirane.
(d) 5-(1,1-dimethyl propyl)-5-(1-methyl propyl) nonane.
(e) 4-cyclopropyl-2-isopropylbutanoic acid.

Chapter 2

Isomerism

2.1 INTRODUCTION

Compounds that have same molecular formula, but differ in properties, are called *isomers*. Therefore, isomers are not identical. Isomers are broadly classified into two categories, *constitutional* and *stereoisomers*.

Constitutional isomers differ in connectivity, i.e. different constitutional isomers have their atoms connected in different orders.

Stereoisomers have same connectivity of atoms, but differ in the way their atoms are arranged in three dimensional space.

The above mentioned difference in either connectivity or arrangement of atoms in space impart characteristic difference in the physical as well as chemical properties of isomers. These two main types of isomers are further subdivided into several categories which are summarized in Figure 2.1.

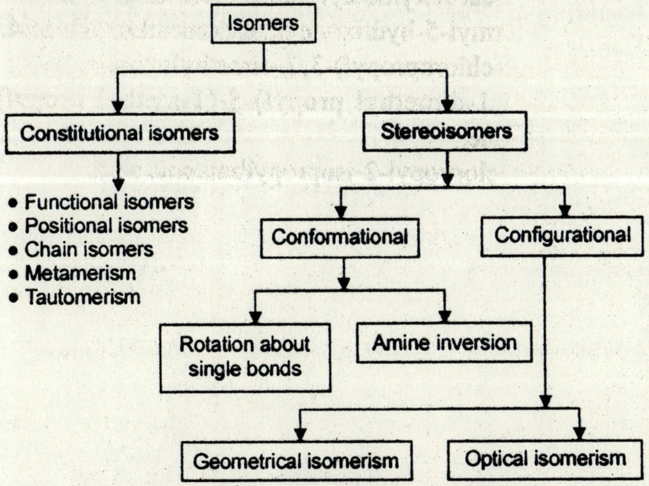

Figure 2.1: Classification of isomers

2.2 CONSTITUTIONAL ISOMERS

As described in the previous section, constitutional isomers have different bonding pattern (connectivity), i.e. isomers possesses different types of bonds as illustrated in the following examples:

$CH_3–CH_2–O–H$ and $CH_3–O–CH_3$

These two molecules have same molecular formula C_2H_6O, but different types of bond. Ethanol has only five C-H bond while dimethyl ether has six C-H bond, ethanol has only one C-O bond while dimethyl ether has two C-O bond, ethanol has one H-O bond, dimethyl ether has no H-O bond.

Example 1. Draw three constitutional isomers with their molecular formula C_3H_8O.

Solution

$CH_3CH_2CH_2OH$ (I) $CH_3OCH_2CH_3$ (II) $CH_3—CH(OH)—CH_3$ (III)

Example 2. Draw all constitutional isomers possible for $C_4H_{10}O$.

Solution: Isomers containing H-O bond:

$CH_3CH_2CH_2CH_2OH$ $CH_3CH_2CH(OH)CH_3$

$CH_3—CH(CH_3)CH_2OH$ $H_3C—C(CH_3)(OH)—CH_3$

Isomers without H—O bonds

$CH_3OCH_2CH_2CH_3$, $CH_3CH_2OCH_2CH_3$, $CH_3OCH(CH_3)—CH_3$

2.2.1 Functional Isomers

Functional isomer, are those isomers which have different functional group bonded to carbon atom as:

(i) Alcohols and ethers: Both alcohols and ethers have their general formula $C_nH_{2n+2}O$ but differs in connectivity and thus their properties are different.

$CH_3CH_2CH_2OH$ (Propanol) $CH_3OCH_2CH_3$ (Methoxy ethane)

$CH_3CH_2CH_2CH_2OH$ (Butanol) and $CH_3CH_2OCH_2CH_3$ (Ethoxy ethane);

$CH_3OCH_2CH_2CH_3$
Methoxy propane

(ii) Carboxylic acids and esters form functional isomers

$CH_3CH_2CH_2COOH$
Butanoic acid

$CH_3COOC_2H_5$ and $CH_3CH_2COOCH_3$
Ethylethanoate Methylpropanoate

$CH_3CH_2CH_2CH_2COOH$
Pentanoic acid

$CH_3CH_2CH_2COOCH_3$, $CH_3CH_2COOC_2H_5$,
Methylbutanoate Ethylpropanate

$CH_3COOCH_2CH_2CH_3$ and $HCOOC_4H_9$
Propylethanoate Butylmethanate

(iii) Aldehydes and ketones form functional isomers:

CH_3CH_2CHO $CH_3-\overset{O}{\overset{\|}{C}}-CH_3$ etc.
Propanal Propanone

2.2.2 Positional Isomers

Positional isomers have same functional group but different positions as:

(i) OH OH OH
1-Pentanol 2-Pentanol 3-Pentanol

(ii) O O
2-Hexanone 3-Hexanone

(iii) NO_2 NO_2
1-Nitropentane 3-Nitropentane

2.2.3 Chain Isomers

Chain isomers have different length of parent carbon chain, e.g.

CH_3
CH_3-C-CH_3
CH_3

n-pentane (chain of 5-C) isopentane (chain of 4-C) neopentane (chain of 3-C)

2.2.4 Metamerism

Metamers have different positions of functional groups in the main parent chain, e.g.

O O
I II

Here the functional group 'O' is part of main chain, therefore I and II are metamers. Metamerism is a special case of positional isomerism. All metamers are also positional isomers but reverse is not true, e.g.

OH OH OH
I II III

Here I, II, and III are all positional isomers but they are not metamers since -OH group is not part of main chain, it is a side group.

Example 3. Draw structures of all the isomers possible for C_6H_{14}.

Solution:

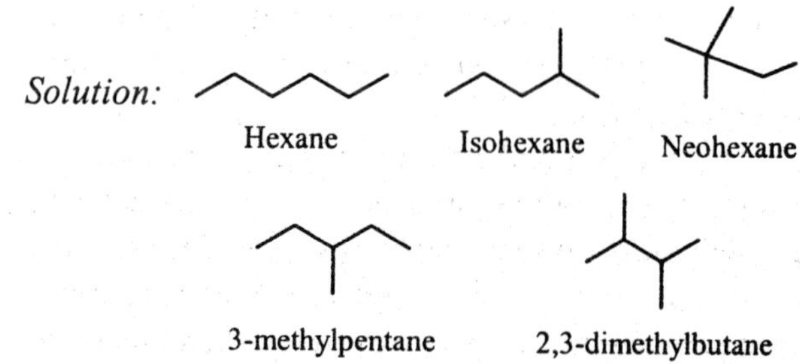

Hexane Isohexane Neohexane

3-methylpentane 2,3-dimethylbutane

Example 4. Draw structures of all isomers of C_7H_{16}.
Solution

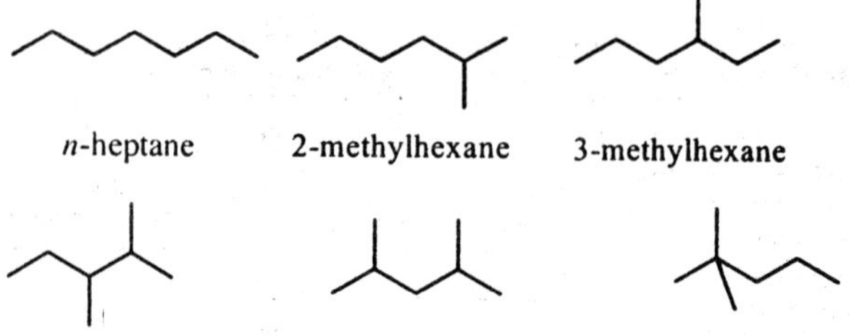

n-heptane 2-methylhexane 3-methylhexane

2, 3-dimethylpentane 2, 4-dimethylpentane 2, 2-dimethylpentane

CH_2CH_3

3, 3-dimethylpentane 3-ethylpentane 2, 2, 3-trimethylbutane

2.2.5 Tautomerism

Tautomerism is a phenomenon, in which more than one molecule having same molecular formula, remains in equilibrium. Therefore, tautomerism is a case of dynamic isomerism. An example of tautomerism is described below:

$$CH_3—\overset{O}{\overset{||}{C}}—CH_2COOC_2H_5 \rightleftharpoons CH_3—\overset{OH}{\overset{|}{C}}=CHCOOC_2H_5$$

I Acetoacetic ester; II

Experiments on solution of acetoacetic ester reveals that the above two compound exist together in equilibrium.

Evidence in favour of enol (II) form: (i) When acetoacetic ester is treated with sodium, hydrogen gas is evolved and sodium derivative is formed. This indicates the presence of a hydroxy functional group in the equilibrium mixture. (ii) When acetoacetic ester is treated with an ethanolic solution of bromine, brown colour of solution is discharged. This further indicates presence of unsaturated functional group in the equilibrium mixture. (iii) When acetoacetic ester is treated with ferric chloride, a reddish violet colour is produced. This is characteristic of compounds containing ($—\overset{|}{\underset{OH}{\underset{|}{C}}}=C—$) functional group.

Evidence in favour of keto (I) form: (i) Acetoacetic ester forms a bisulphate compounds with $NaHSO_3$ indicating presence of carbonyl (> C = O) functional group. (ii) Acetoacetic ester forms cyanohydrin with hydrogen cyanide further indicating presence of carbonyl functional group.

The above discussion suggest that tautomerism involves a dynamic equilibrium between two forms "tautomers' and in a given condition of temperature and pressure, one form predominate. That tautomer which predominate in a given condition is known as "stable" and the other form is known as "labile" tautomers. In practice, it is generally very difficult to say that which is labile, since very often a slight change in the conditions shift the equilibrium from one form to other form.

Tautomerism in solid state is rare, and hence, in the solid state one or other tautomer is normally stable.

Keto-Enol Tautomerism

Keto-enol tautomerism is one of the various type of tautomerism, which is exhibited by carbonyl compounds containing atleast one α-H.

$$CH_3—\overset{O}{\overset{||}{C}}—H \rightleftharpoons CH_2=\overset{OH}{\overset{|}{C}}—H$$

Keto form; Enol form

The two forms (keto and enol) remains in dynamic equilibrium and in most of the cases, keto forms, being thermodynamically more stable than enol form, dominate.

Although, it may not be possible to separate tautomers owing to the ease and rapidity of their interconversion or, as in many cases, one form being almost completely absent, the presence of more than one compound may be confirmed by specific properties of each isomer, e.g. α-H of aldehyde or ketone are exchanged by deuterium when they are dissolved in D_2O.

$$CH_3—\overset{O}{\overset{||}{C}}—H + D_2O \rightleftharpoons CD_3—\overset{O}{\overset{||}{C}}—H$$

No such H-exchange is observed with compounds lacking α-H:

$$(CH_3)_3—\overset{O}{\overset{||}{C}}—H + D_2O \rightarrow \text{No deuterium is incorporated.}$$

As stated above, keto form of tautomers are usually stable and enol form are labile, conjugation in enol form increases its stability and hence, equilibrium concentration as illustrated in the following example.

$$\underset{\text{Keto}}{CH_3—\overset{O}{\overset{||}{C}}—H} \rightleftharpoons \underset{\text{Enol}}{CH_2=\overset{OH}{\overset{|}{C}}—H} \quad \text{... (i)}$$

$$\underset{\text{Keto}}{CH_2=CH—CH_2—\overset{O}{\overset{||}{C}}—H} \rightleftharpoons \underset{\text{Enol}}{CH_2=CH—CH=\overset{OH}{\overset{|}{C}}—H} \quad \text{... (ii)}$$

Here enol formed in (ii) equilibrium is a conjugated system and it is more stable than non-conjugated enol formed in (i). Therefore, it can be concluded here that at equilibrium, enol content will be more in case of 3-butenal than in ethanal-ethenol equilibrium mixture.

Example 5. Draw enol tautomers of the following molecule and if more than one enols are possible, predict the major one.

(a) CH_3CH_2CHO (b) $CH_3—\overset{O}{\overset{||}{C}}—CH_3$

(c) $CH_3—\overset{O}{\overset{||}{C}}—CH_2—CH_3 \rightleftharpoons CH_3=\overset{OH}{\overset{|}{C}}—CH_2—CH_3 + CH_3—\overset{OH}{\overset{|}{C}}=CH—CH_3$

(d) 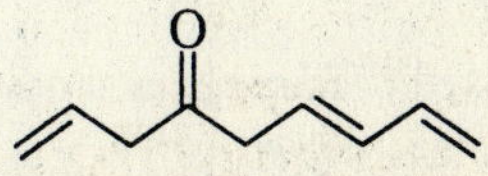 [skeletal structure with C=O]

Solution

(a) $CH_3-CH_2-\overset{O}{\overset{\|}{C}}-H \rightleftharpoons CH_3-CH=\overset{OH}{\overset{|}{C}}-H$

Keto — Enol

(b) $H_3C-\overset{O}{\overset{\|}{C}}-CH_3 \rightleftharpoons CH_2=\overset{OH}{\overset{|}{C}}-CH_3$

Keto — Enol

(c) $CH_3-\overset{O}{\overset{\|}{C}}-CH_2-CH_3 \rightleftharpoons CH_2=\overset{OH}{\overset{|}{C}}-CH_2-CH_3$

$+ CH_3-\overset{OH}{\overset{|}{C}}=CH-CH_3$

(more stable)

(d)

Example 6. Predict the relative amount of enol content in the following systems.

I — II — III ($CH_2=CH-\overset{O}{\overset{\|}{C}}-H$)

Solution

I

II

$CH_2=CH-\overset{O}{\overset{\|}{C}}-H \rightleftharpoons CH_2=C=\overset{OH}{\overset{|}{C}}-H$

III

In the above example, enol of (I) is most stable since it is phenol – a highly conjugated aromatic system. Enol of II is also stable due to conjugation, but less stable than phenol. Enol of III is least stable since it has consecutive double bond. Thus order of enol content is I > II > III.

Other than carbonyl compounds, there are several other classes of compounds which show tautomerism. They are:

(i) $CH-C=C \rightleftharpoons C=C-CH$, e.g.

$CH_3-CH=CH_2 \rightleftharpoons CH_2=CH-CH_3$

(ii) $CH-NO_2 \rightleftharpoons C=\overset{+}{N}(O^-)(OH)$

e.g. $CH_3NO_2 \rightleftharpoons CH_2=\overset{+}{N}(O^-)(OH)$

(iii) $HN-C=N \rightleftharpoons N=C-NH$

e.g. $CH(NH_2)=N-CH_3 \rightleftharpoons CH_2(NH)-N(NH)-CH_3$

(iv) $CH-N=O \rightleftharpoons C=N-OH$,

e.g. $CH_3-NO \rightleftharpoons CH_2=N-OH$

(v) $N=N-CH \rightleftharpoons HN-N=C$,

e.g. $CH_3-N=NH \rightleftharpoons CH_2=N-NH_2$

Example 7. Enol content in the aqueous solution of acetone is not more than 0.1%, but the enol content of 2, 4-pentandione in its aqueous solution is approximately 15%. Explain why this β-diketone has greater enol content?

Solution: As stated in the previous section that enols are thermodynamically less stable than its keto counterpart, but conjugation, aromaticity and factors like this, increases the equilibrium concentration of enol tautomers by lowering their potential energy. In the present case, enol of acetone is not stabilized by any such factor, but enol of β-diketone is stabilized by conjugation as well as by cyclization through intramolecular H-bonding as:

H-bond

H_3C CH_3 $\rightleftharpoons$ H_3C

2,4- pentandione (keto-form) — ≈15% (enolform)

2.3 CONFORMATIONAL ISOMERISM

Conformational isomerism in a molecule is invoked by free rotation of atoms or groups about a sigma bond, giving various temporary arrangements. These temporary arrangements arising due to unrestricted rotation of atoms or groups about sigma bond are called conformers.

Conformers are interconvertible and a continuous interchange from one form into other occurs in a molecule capable of showing conformation. Conformers differs in their energy by a very small amount and it is very difficult to isolate various conformers of a molecule under normal condition.

In organic compound, conformations arises due to rotation of atom or group around a carbon-carbon sigma bond as:

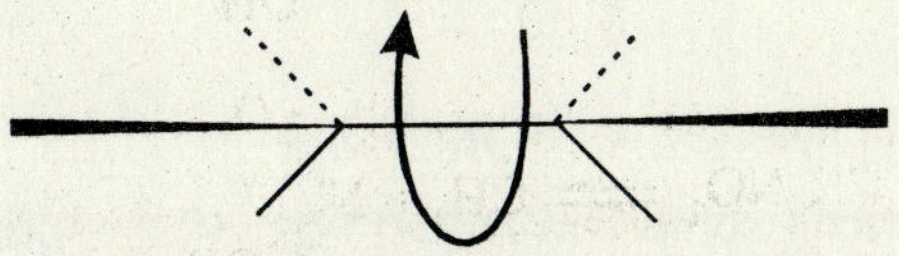

Rotation is free about the C – C single bond.

Sawhorse and Newman projections

Sawhorse and Newman projection are commonly used to show various conformers.

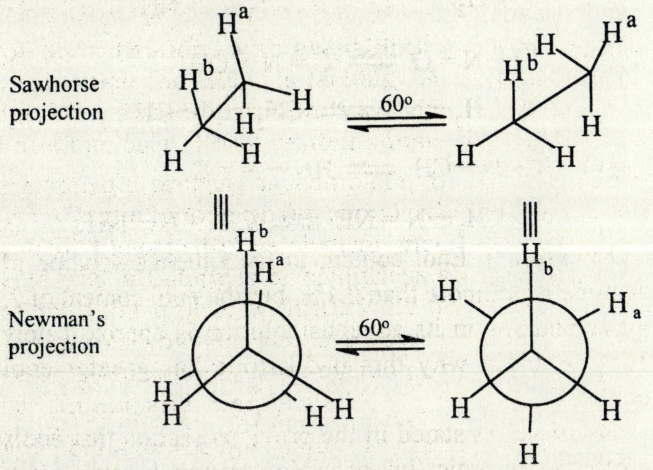

Energy concern of conformers

Stability of conformers depends on the extent of repulsion between shared pair of electrons. Smaller the repulsion, lower will be the potential energy and greater will be the stability. Extent of repulsion between shared pair of electrons increases as they pass closer to one-another. Hence, in the above example, conformer on left has maximum repulsion since each C—H electron pairs on one carbon are face-to-face to C—H electron pair on other carbon, i.e. they are closest to one another. Therefore, conformer on left in the above example has highest potential energy and least stable.

If we rotate the groups on rear carbon keeping groups on front carbon fix:

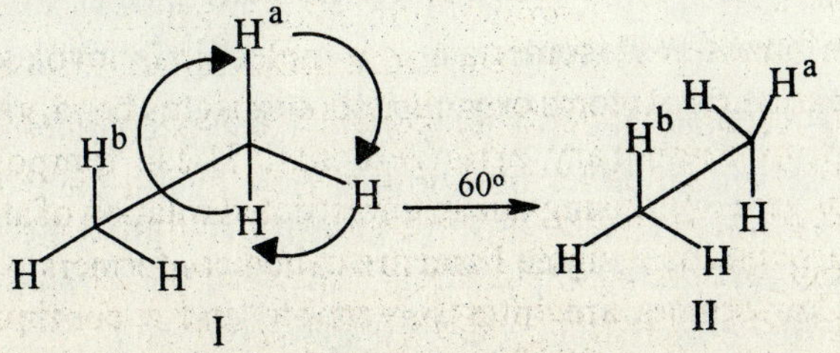

(Rotation of groups about C—C bond, on the rear carbon)

As we move from left to right, distance between C–H[a] and C—H[b] bonding electrons increases and repulsion decreases, increasing stability. The two bonding electron pairs suffer minimum repulsion when the groups on rear carbon is rotated by 60° angle. Any further rotation beyond 60°, increases the distance between C—H[a]—C—H[b] bonding pairs, but bring C—H[a] closer to other C–H bonding electrons on the second carbon, increasing repulsion. Thus, a 60° rotation brings ethane molecule from maximum potential energy state (I) to minimum potential energy state (II). Stability and energy of various other conformers will lie in between these two states.

Conformational analysis of ethane

In ethane molecule, rotation of hydrogen around C—C bond is free and gives rise to infinite temporary arrangement differing by a very small amount of energy. Two extreme case, the least stable and the most stable are described below:

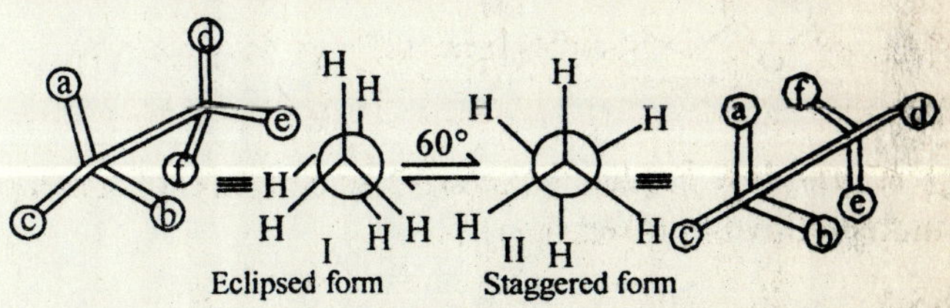

Energy change resulting from rotation about C—C bond is represented in diagram below:

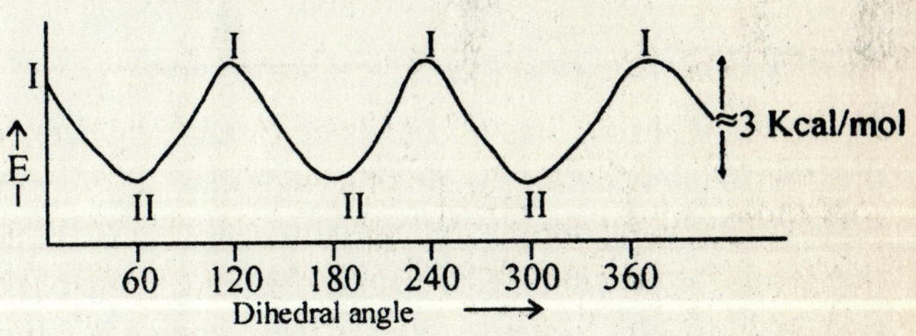

Conformational analysis of butane about C_2—C_3 bond

Important conformers of butane arising by 60° rotation about C_2—C_3 bonds are shown below:

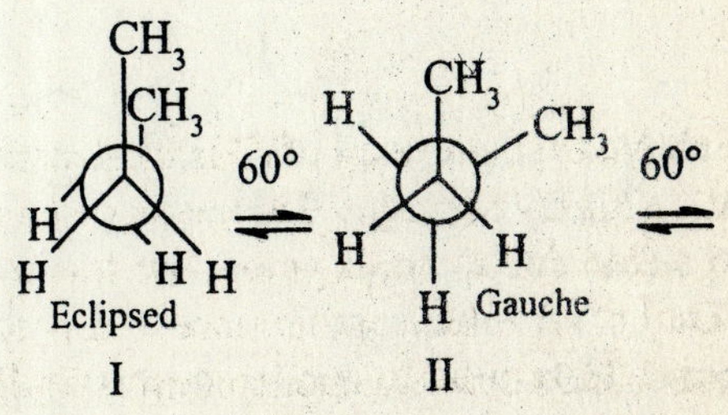

Eclipsed III $\underset{60°}{\rightleftharpoons}$ Anti IV $\underset{60°}{\rightleftharpoons}$

Eclipsed V $\underset{60°}{\rightleftharpoons}$ Gauche VI $\underset{60°}{\rightleftharpoons}$ Eclipsed VII

Relative potential energy is:

I = VII > III = V > II = VI > IV

Energy diagram of conformers of butane around C_2—C_3 bond is shown in diagram below:

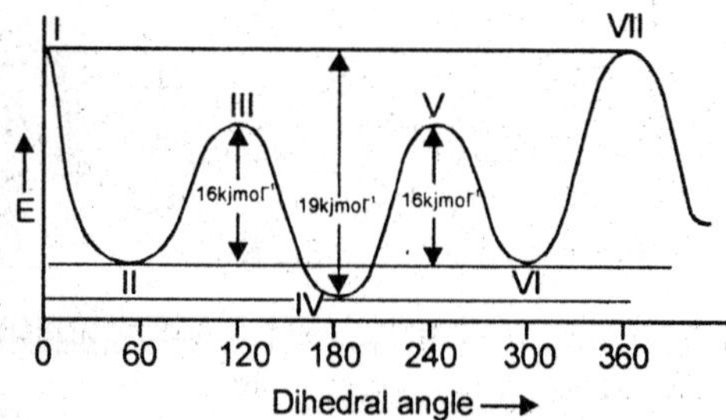

Conformational analysis of cyclohexane

Planar cyclohexane has all bonds eclipsed. However, in contrast to the situation for smaller rings, the C—C—C bond angles of planar cyclohexane are larger (120°) than the normal tetrahedral (109°) angle. Moving some of the carbon atoms out of plane reduces these angles and thus reduces angle strain.

Chair conformation: Moving one carbon atom up from the plane of cyclohexane ring, while the fourth carbon with respect to this carbon, down to the plane of ring yields a conformation with minimum strain, with ideal tetrahedral bond angle as:

Chair form

Boat conformation: If the above mentioned two carbons of cyclohexane are moved to the same side, a new conformer appearing like boat is formed:

Although boat form has relieved angle strain, the two C—H bonds, C-1 and C-4 are eclipsing one another. Therefore, boat conformer is slightly less stable than chair form.

2.4 GEOMETRICAL ISOMERISM

Geometrical isomerism is first type of stereoisomerism, in which isomers have identical connectivity. Here isomers have different arrangement of atoms or groups in three dimensional space.

Geometrical isomerism arises due to restricted rotation of groups about double bond. To illustrate the understanding of restricted rotation about double bond, let us consider an unsaturated molecule CHCl=CHCl. The two carbons are joined by a sigma and a π-bond.

XZ-plane; XY - plane; Rotation

I II

As shown in the above diagram, rotation of H—C—Cl (XY-plane) around C—C bond distort the π-plane (XZ-plane) as shown in Fig. II. Therefore, such rotation is not possible since it breaks the π-bond. Hence, it is concluded here that the positions held by H and Cl are fixed and non-interconvertible by rotation of groups about a bond. Due to this rigidity of relative positions, this type of molecule, if properly substituted, gives rise to geometrical isomerism. Thus, CHCl=CHCl can be represented by two different structures having same connectivity, but different arrangement in space and they are non-interconvertible by rotation:

non-interchangible by rotation

two H are on the same side of π-plane — two H are on the opposite side of π-plane

In organic compounds, geometrical isomerism can be observed in either unsaturated or a cyclic compounds. Unsaturation or cyclic structures are mere essential requirement but not sufficient. Geometrical isomerism is further classified into several sub-classes.

2.4.1 *Cis* and *trans* Isomerism

If identical atoms or groups are on the same side of π-plane, this isomer is known as *cis* isomer and if identical atoms or groups are on the opposite side of π-plane this is known as *trans* isomer.

(A)C=C(A)   (A)C=C(A)

Cis   *Trans*

Important: For any compound, *cis-trans* isomer must exist in pair, i.e. no molecule can have only *cis* or only *trans* form, both must have their independent existence. To understand *cis-trans* isomerism, unsaturated compounds can be classified into following categories:

(a) C_{AA} C_{AA}: Here all four groups are identical and both *cis* and *trans* form exist in the same structure, i.e. the two forms *cis-trans* cannot be isolated:

(A)(A)C=C(A)(A)

Identical groups 'A' are on the same as well as on opposite side of π-plane in the same structure, this type of molecule do not show geometrical isomerism. Examples of this category are C_2H_4, C_2Cl_4, 2, 3-dimethy, l-2-butene, etc.

(b) C_{AA} C_{AB}: Here three groups are identical and one is different.

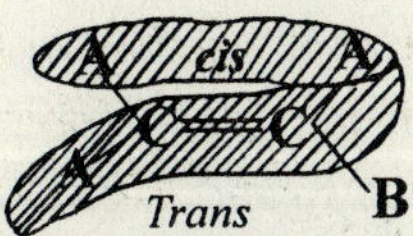

Both *cis* and *trans* appearing in the same structure. Hence, this type of molecule donot show *cis-trans* isomerism, examples are CH_2CHCl, C_2HCl_3, 2-methy-l-2-butene, etc.

(c) C_{AA} C_{BB}: Here identical groups are on same carbon:

(A)(A)C=C(B)(B)

We cannot have two identical atoms on the same side of π-plane. Therefore, this category of molecules also fails to show geometrical isomerism.

(d) C_{AB} C_{AB}: This is the 1st category of molecule for which two independent structures can be drawn as:

(A)(A)C=C(B)(B) *cis*   and   (A)(B)C=C(B)(A) *trans*

Therefore, this category of molecules do show geometrical isomerism. A few examples of this category are:

(Cl)(F)C=C(Cl)(F) *cis*   +   (Cl)(F)C=C(F)(Cl) *trans*

(H)(D)C=C(H)(D) *cis*   (H)(D)C=C(D)(H) *trans*

(C_6H_5)(H_3C)C=C(C_6H_5)(CH_3) *cis*   (C_6H_5)(H_3C)C=C(CH_3)(C_6H_5) *trans*

(e) $C_{AB}C_{AD}$: This category of molecule also show geometrical isomerism:

(A)(B)C=C(A)(D) *cis*   (A)(B)C=C(D)(A) *trans*

Identical group on both side is not necessary.

Examples are $CH_3CH{=}CHC_2H_5$:

(H)(H_3C)C=C(H)(C_2H_5) *cis*   (H)(H_3C)C=C(C_2H_5)(H) *trans*

(f) $C_{AB}C_{DE}$: Here all four groups are different, it can also have two different arrangement of A, B, D and E as:

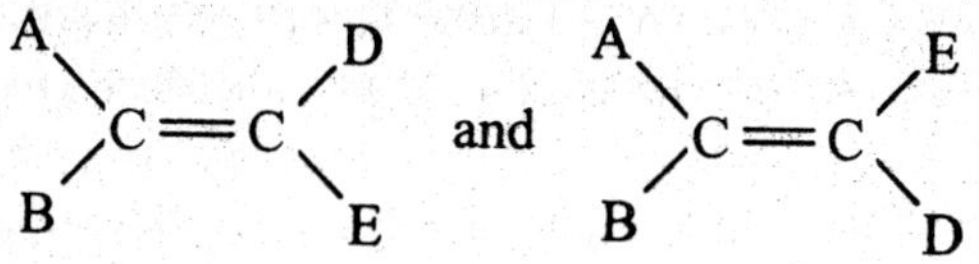

(A)(B)C=C(D)(E) and (A)(B)C=C(E)(D)

I II

I and II are geometrical isomers, but they are not *cis-trans*, they are E and Z isomers.

From the above discussion it can be concluded that if two identical atoms or groups are on the same carbon ($C_{AA}C$.....) it cannot show geometrical isomerism.

Example 8. Draw all geometrical isomers possible for 2, 4-hexadiene.

Solution

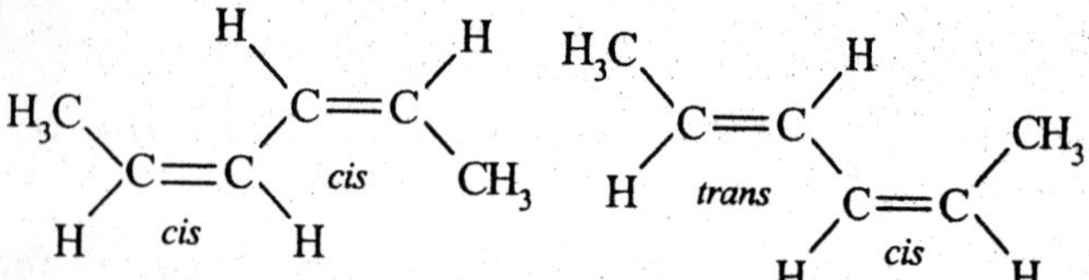

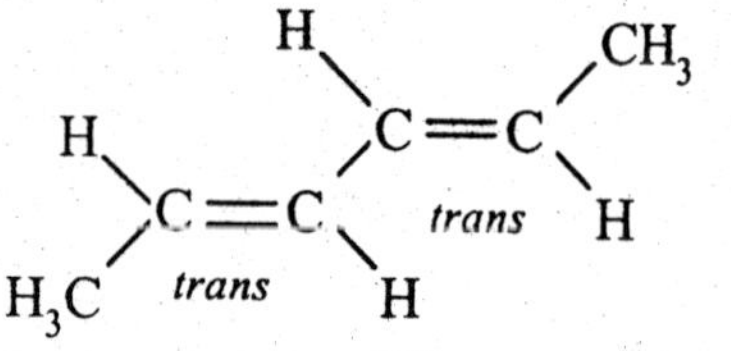

I (*cis*, *cis*) II (*trans*, *cis*)

III (*trans*, *trans*)

Above three are the possible geometrical isomers of 2, 4-hexadiene, we could expect one more *cis-trans* combination, but that infact will be same as II *trans-cis* since it is a symmetrical diene and can be numbered from either sides of the chain.

Example 9. Draw all geometrical isomers possible for 2, 4-heptadiene.

Solution

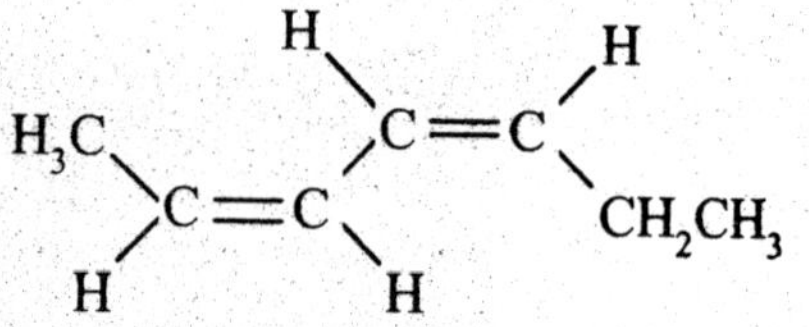

cis, cis-2, 4-heptadiene (i)

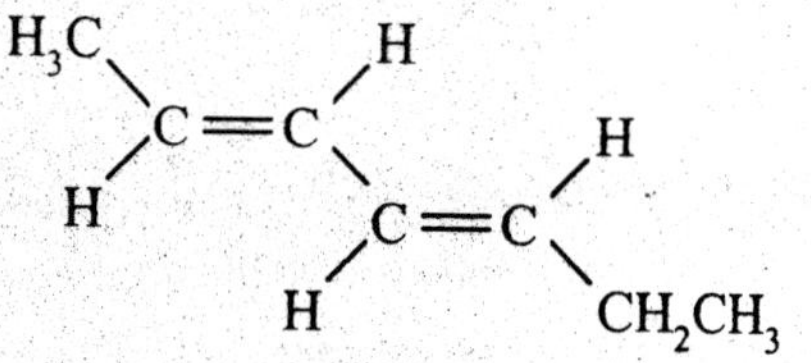

trans, trans-2, 4-heptadiene (ii)

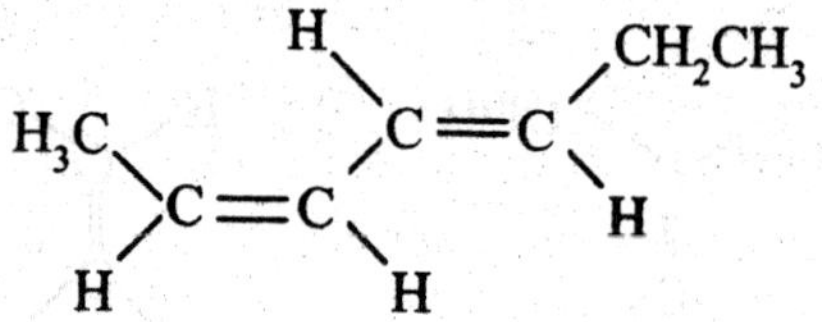

2-*cis*-4-*trans* heptadiene (iii)

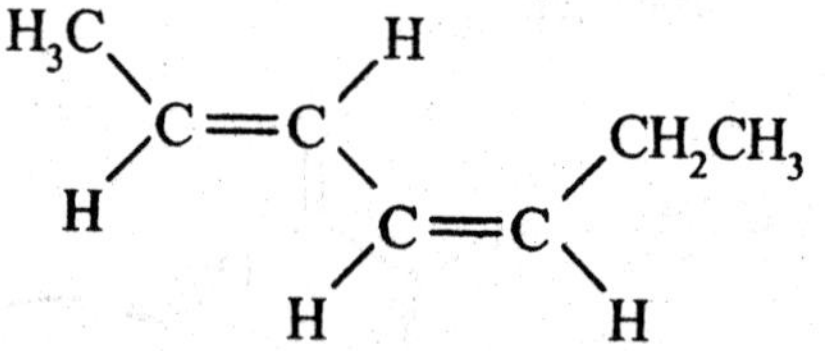

trans, cis-2, 4-heptadiene (ii)

Here III and IV are different stereoisomers.

Example 10. Draw all geometrical isomers possible for CFClBrI.

Solution

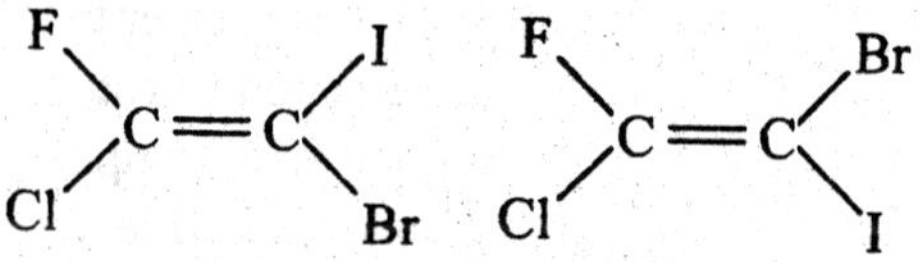

I II

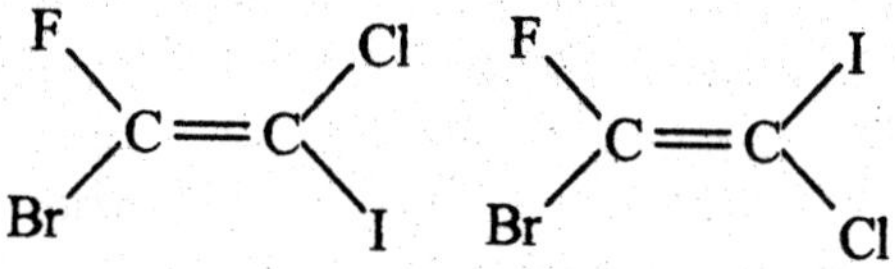

III IV

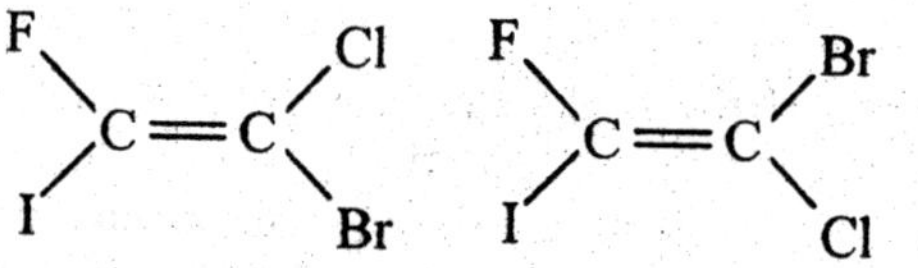

V VI

Here I and II are geometrical isomers since they have same connectivity.

III and IV are geometrical isomers and V and VI are geometrical isomers.

I and III are not stereoisomers since they have different connectivity, I has F and Cl bonded to one carbon and Br and I bonded to other carbon. While III has F and Br bonded to one carbon and Cl and I bonded to other carbon.

Example 11. HOOC—CH═CH—COOH exist in two stereoisomeric form A and B. A has greater Ka_1 than Ka_1 of B, but Ka_2 of B is greater than Ka_2 of A. Identify A and B and justify.

Solution

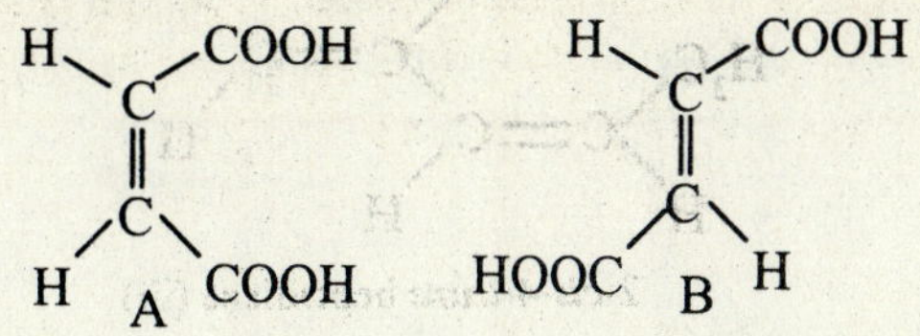

H, COOH / C=C / H, COOH (A) — Maleic acid

H, COOH / C=C / HOOC, H (B) — Fumaric acid

H- bond + H^+

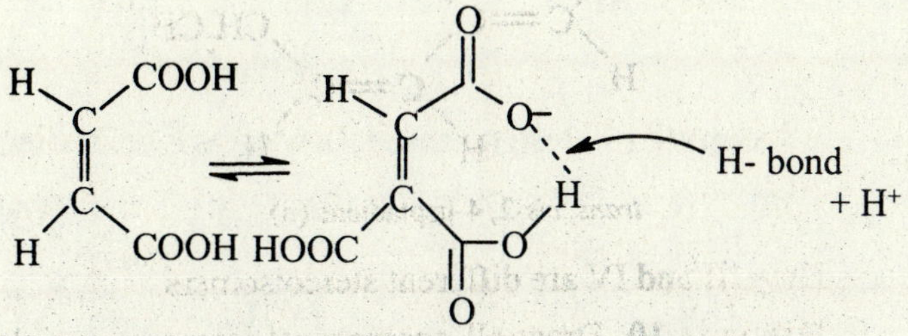

Conjugate base of maleic acid is stabilized by intramolecular H-bonding, but conjugate base of fumaric acid is not stabilized by H-bond since-COOH are *trans* to one another. Therefore Ka_1 of A is greater than Ka_1 of B. On the otherhand removal of second proton is difficult from conjugate base of A since H-is trapped by H-bond while H-in the conjugate base of B is free and can easily be ionized.

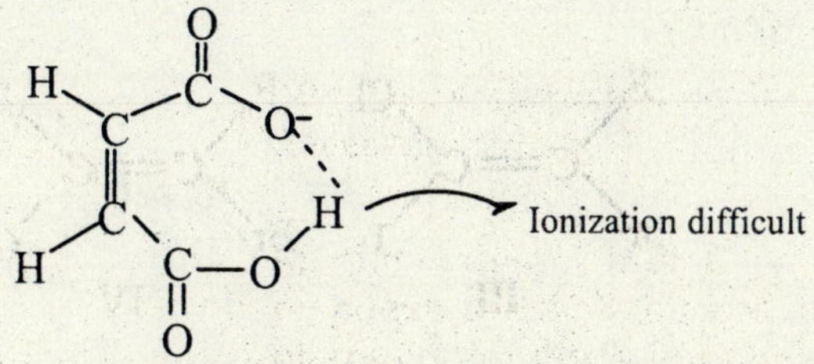

Ionization difficult

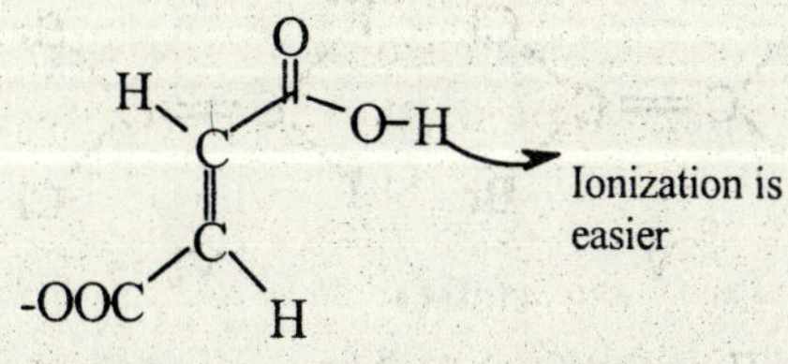

Ionization is easier

Therefore, Ka_2 of B is greater than Ka_2 of A.

2.4.2 Geometrical Isomerism in Cyclic Compound

Cycloalkanes with 3, 4 carbons in ring are known as rigid rings and rotation about C—C bond is restricted as shown below:

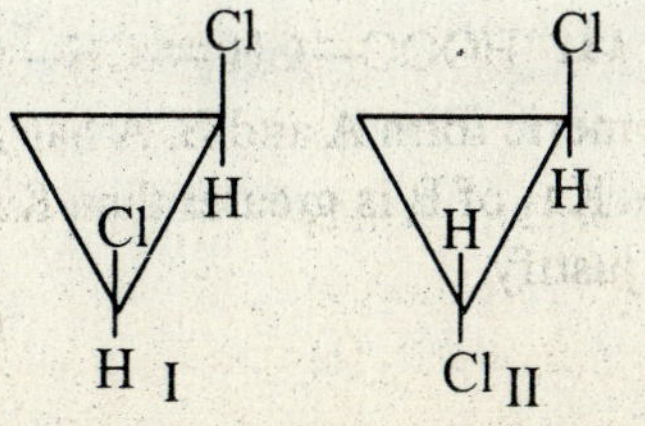

I II

I and II are two geometrical isomers of 1, 2-dichlorocyclopropane. I has identical chloro group on the same side of the plane of cyclopropyl ring. In II, two chlorine are on the opposite side of the ring plane. Hence, I is *cis*-1, 2-dichloro cyclopropane and II is *trans*-1, 2-dichlorocyclopropane.

Example 12. 1, 2-dichloro butane exist in two stereoisomeric form in which one is non-polar while other is polar. Draw the two structures and label them as polar and non-polar.

Solution

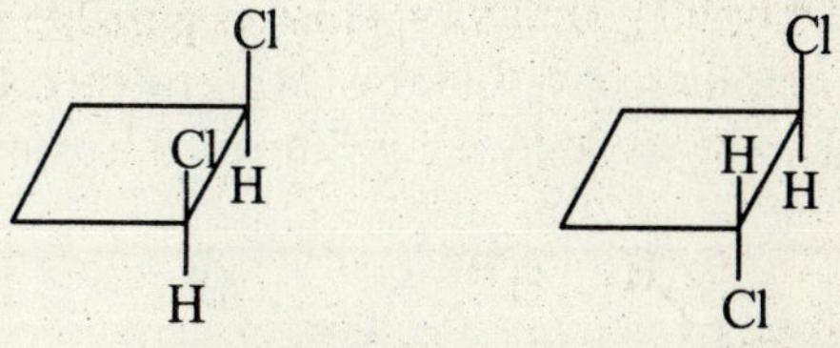

Cis-1,2-dichlorocyclobutane (Polar)

Trans-1,2-dichlorocyclobutane (Non-polar)

Example 13. Draw all different structures possible for dichlorocyclopentane and select stereo-isomer pair if any.

Solution

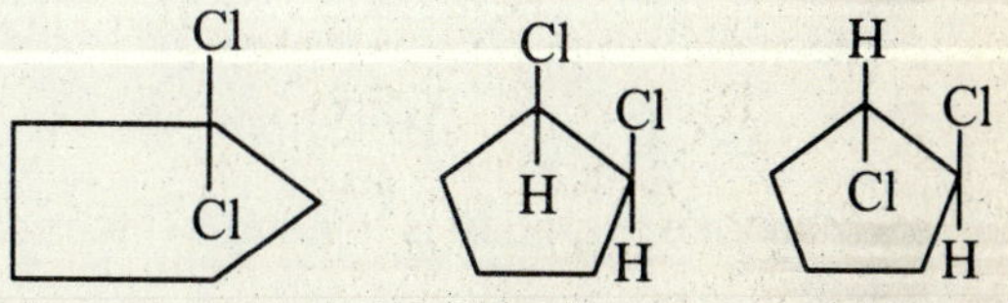

I II III

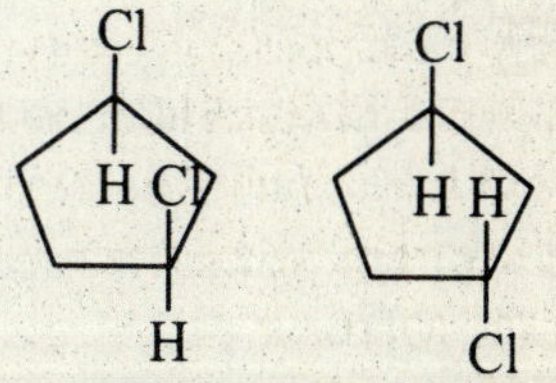

IV V

Here II and III are geometrical isomers, IV and V are geometrical isomers.

2.4.3 Geometrical Isomerism in Compounds other than Alkenes and Cycloalkane Derivatives

Molecule containing —C=N– group: Here again rotation about C=N is restricted due to same reason as in case of C=C bond.

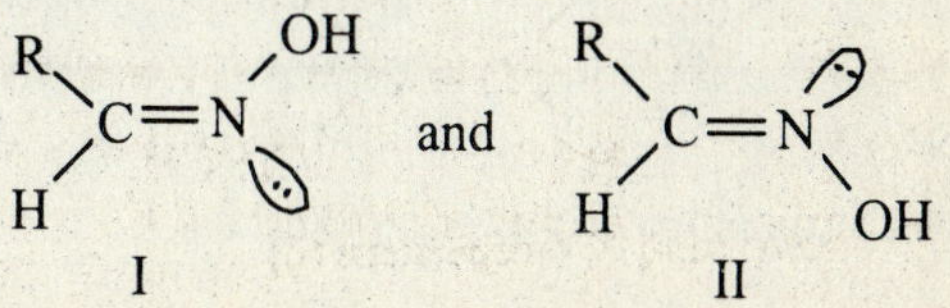

R, H / C=N / OH (I) and R, H / C=N / OH (II)

Here I and II are non interchangible stereoisomers of aldo-oxime. I has R and -OH on the same side of π-plane, while II has R and OH on the opposite side of π-plane.

Example 14. Discuss the geometrical isomerism in (a) N_2H_2, (b) $CH_3—CH{=}NH$, (c) $H_2C{=}NH$.

Solution

(a) 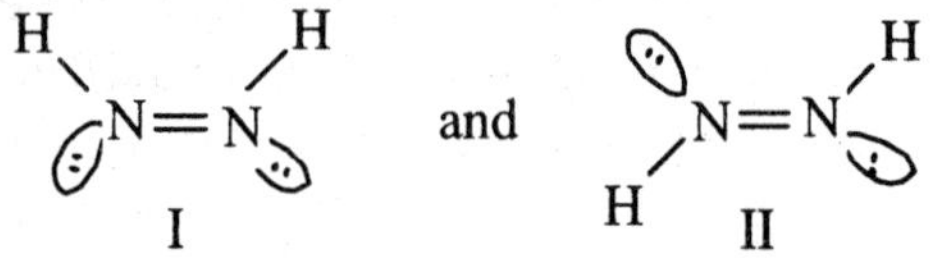 H–N=N–H (I, H and H on same side) and H–N=N–H (II, H and H on opposite sides)

I and II are non interconvertible by simple rotation about N=N bond and hence, they are geometrical isomers.

(b) 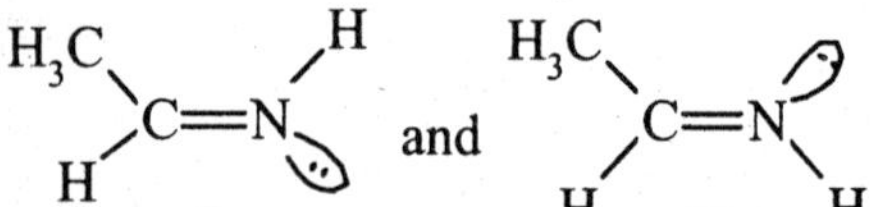 H_3C(H)C=N–H (I) and H_3C(H)C=N–H (II)

I and II are geometrical isomers since they have different arrangement of groups about C=N.

(c) 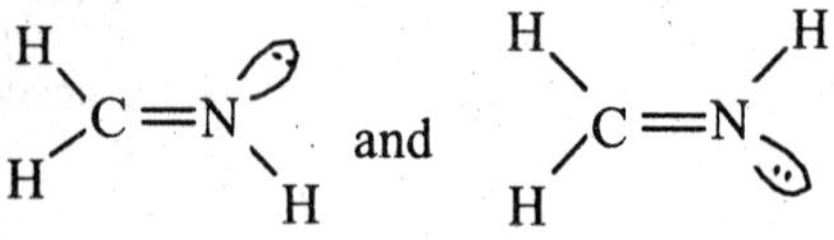 $H_2C{=}N{-}H$ and $H_2C{=}N{-}H$

are identical since carbon is bonded to identical atoms on both side.

2.4.4 E and Z Isomerism

This is a type of geometrical isomerism exhibited by molecules, in which doubly bonded carbon bears all four groups different and cannot be termed as *cis* or *trans* isomer, viz. $CH_3CH{=}CClF$. E and Z are the name given to two different geometrical isomers of this type of compounds.

To designate a geometrical isomer as E or Z, we assign priority to the groups separately at two doubly bonded carbon. Heavier atom directly bonded to doubly bonded carbon has higher priority than the lighter atom, e.g.

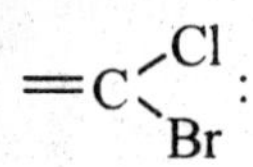

=C(Cl)(Br) : Here Br is heavier than Cl, therefore =C(Cl)(Br) — Cl (2), Br (1) — Br is given higher priority (1) and Cl is given lower priority 2.

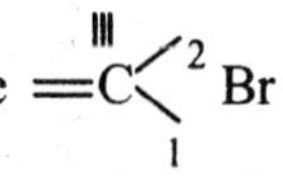

Similar procedure is repeated on other carbon.

If atoms or groups of identical priority are on the same side of π-plane, this isomers is known as Z-isomer, and if atoms or groups of identical priority are on opposite side of π-plane, this is known as E-isomer, e.g.

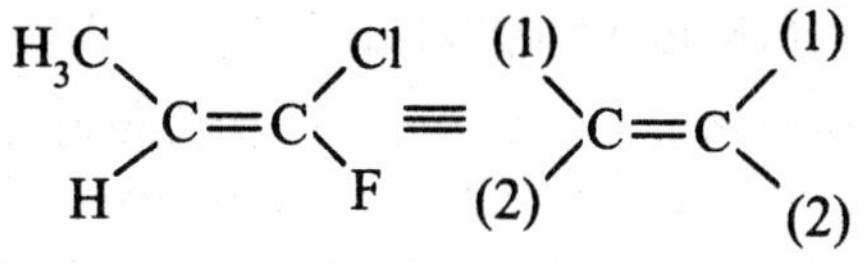

H_3C(H)C=C(Cl)(F) ≡ (1)(2)C=C(1)(2)

Z-isomer

H_3C(H)C=C(F)(Cl) ≡ (1)(2)C=C(2)(1)

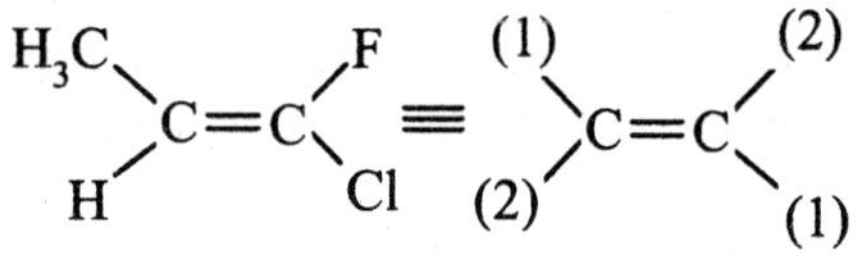

E-isomer

Example 15. Designate the following as E or Z isomers:

(a) 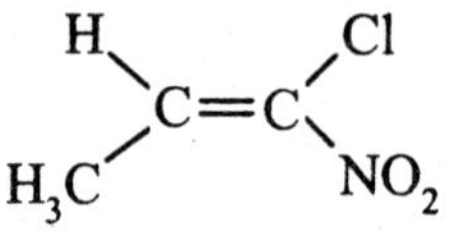 H(H_3C)C=C(Cl)(NO_2) (b) 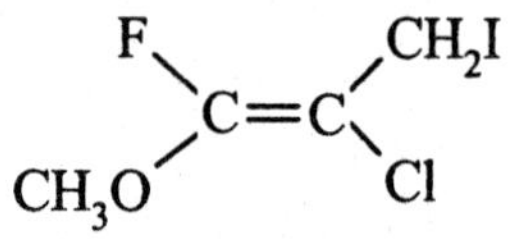 F(CH_3O)C=C(CH_2I)(Cl)

(c) D(H)C=C(CHO)(CH_3) (d) HOOC(Cl)C=C(CN)(NC)

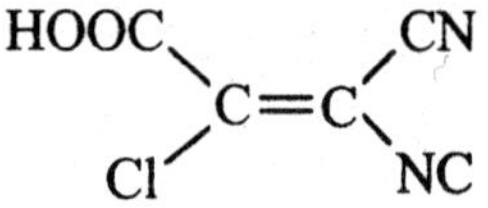

(e) 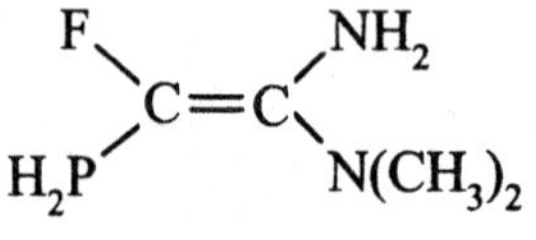 F(H_2P)C=C(NH_2)($N(CH_3)_2$)

Solution

(a) 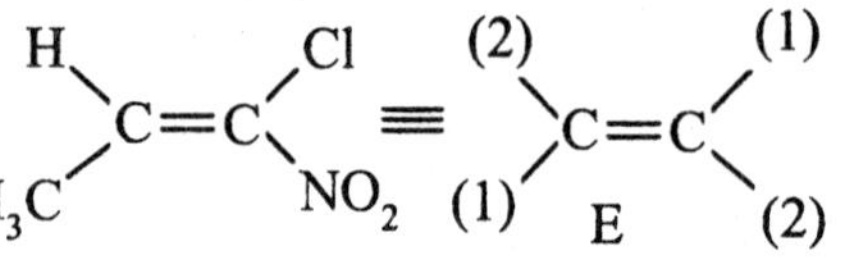 H(H_3C)C=C(Cl)(NO_2) ≡ (2)(1)C=C(1)(2) E

(b) 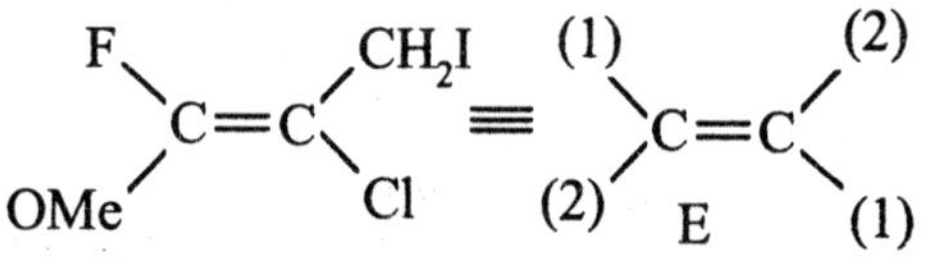 F(OMe)C=C(CH_2I)(Cl) ≡ (1)(2)C=C(2)(1) E

(c) D(H)C=C(CHO)(CH_3) ≡ (1)(2)C=C(1)(2) Z

(d) HOOC(Cl)C=C(CN)(NC) ≡ (2)(1)C=C(2)(1) Z

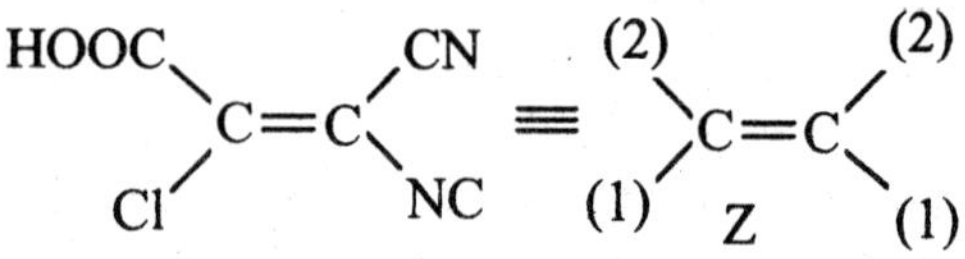

(e) 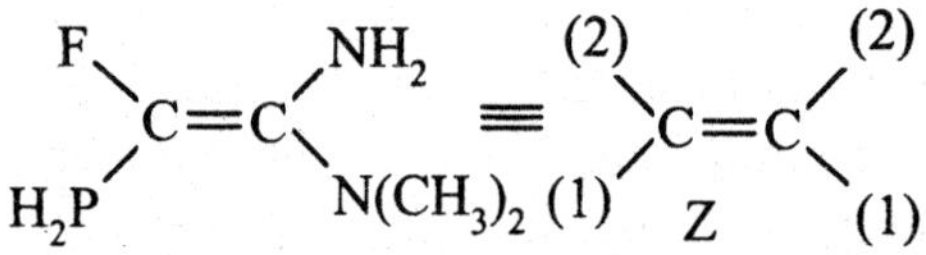 F(H_2P)C=C(NH_2)($N(CH_3)_2$) ≡ (2)(1)C=C(2)(1) Z

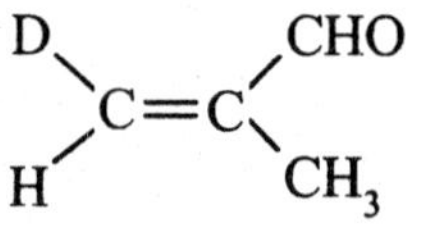

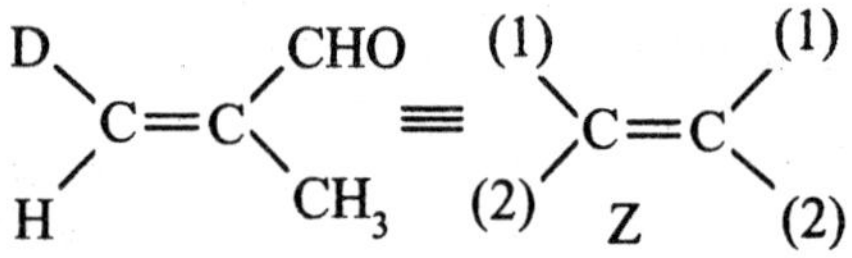

2.5 OPTICAL ISOMERISM

Optical isomerism is a type of stereoisomerism which is exhibited by dis-symmetric compounds. Optical isomers too differ in arrangement of atoms or groups in three dimensional space, but they have same connectivity.

Compounds whose solution rotate plane polarized light when passed through it, are said to be optically active and exhibit the phenomena of optical isomerism.

2.5.1 Plane Polarized Light

When all the light rays propagate in one single plane, such type of light rays are said to be plane polarized. Plane polarized light is produced by passing light from an incandescent source through a nicol-prism. Light rays coming out of nicol prism propagate in one single plane. This plane polarized light is then passed through a sample tube containing solution of compounds to be tested for optical activity. Three observations can be found as :

(i) *Plane polarized light passing undeflected:* If plane polarized light passes undeflected then the compound is called *optically inactive.*

(ii) *Plane polarized light deflected left to the normal*: If plane polarized light is deflected to left from the normal, compound is said to be *Levorotatory* and represented by symbol *l* or (–).

(iii) *Plane polarized light deflected to the right from normal*: If plane polarized light is deflected to right from the normal than the compound is said to be *Dextro rotatory* and represented by symbol *d* or (+). The experimental set-up used to detect and estimate optical activity is described in the following diagram.

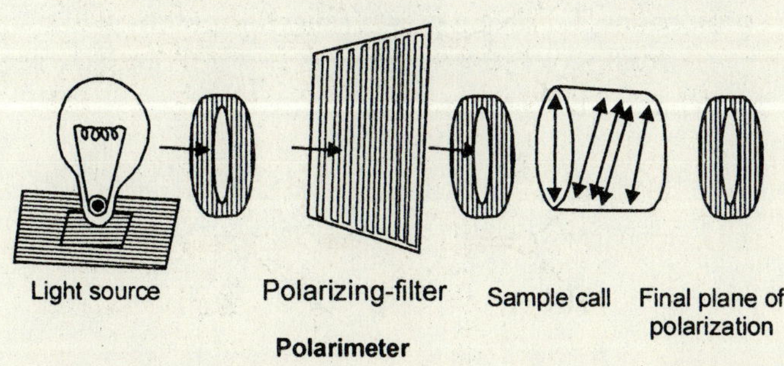

Polarimeter

The angle (α) by which the plane polarized light is rotated with respect to normal, is known as *observed rotation* and it is measured in degrees. The amount of rotation caused by the sample depends on the number of optically active molecules the light particle encounters in the sample tube. This inturn, depends on the concentration of the sample and length of sample tube. The observed rotation also depends on the experimental temperature and wavelengths of light source. Therefore

$$\alpha \propto lc$$

$$\Rightarrow \alpha = k\,lc$$

Where l = length of tube in dm.
c = density in gm^{-1}

$$\Rightarrow \boxed{k = \frac{\alpha}{lc} = [\alpha]_\lambda^T}$$

Here constant k, given the symbol $[\alpha]_\lambda^T$ is known as *specific-rotation* which is the number of degrees of rotation caused by a solution of density 1.0 g/ml placed in a sample-tube 1.0 dm long. Specific rotation is a characteristic constant of an optically active compound and it depends on temperature and wavelength of light.

Example 16. The observed rotation of 2.0 g of a compound dissolved in 10 ml of a solution in a polarimeter tube 10 cm long is 1.34°. What is specific rotation of this compound?

Solution: $$[\alpha]_\lambda^T = \frac{\alpha}{lc} = \frac{1.34 \times 10}{1 \times 2} = 6.7°$$

2.5.2 Chirality and Optical Isomerism

Chiral carbon: A carbon bonded to four different atoms or groups is known as *chiral carbon* as shown in the following molecules :

I	II	III
C* bonded to CH_3, Br, H, Cl	C* bonded to C_2H_5, H_3C, COOH, Cl	C* bonded to C_2H_5, H, CH_3, CN

All the above molecules has atleast one chiral carbon (C*).

Chiral molecule: A molecule is said to be chiral, if it is non-superimposable on its mirror image.

I	Mirror	II
C bonded to CH_3, Br, H, Cl		C bonded to CH_3, H, Br, Cl

Here I and II are mirror images of each other. Also I and II are non-superimposable since they have different configuration.

Configuration: Configuration pertains to a spatial arrangement of atoms or groups in three dimensional space. Here I represents a specific configuration of molecule, in which H → Cl → Br are arranged in clockwise order while order of these groups H → Cl → Br are anticlockwise in II and represent a configuration different from I. Rotation about a bond does not change the order in which groups are arranged and hence, rotation does not change the configuration.

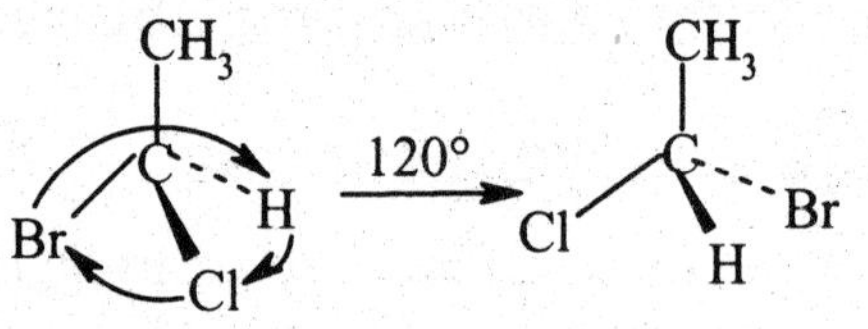

Here both structure represents the same configuration. Thus any rotation in I cannot give II, which has anti-clockwise arrangement of the same groups. Thus, configuration can be changed only by cleavage of bonds.

From the above discussion, it can be concluded that I and II have different configuration and they are non-superimposable mirror images of one another. Thus, non-superimposability on mirror image is the ultimate criteria for optical activtiy. Presence of chiral carbon does not guarantee the optical activity. For optical activity, the whole compound must be chiral. There are several example of compounds containing chiral carbon, but optically inactive and compounds whithout chiral carbon but optically active are also known.

"A compound containing only one chiral carbon is essentially chiral and show optical isomerism."

2.5.3 Enantiomers

A pair of stereoisomers which are non-superimposable mirror images of each other are known as *enantiomers*, e.g.

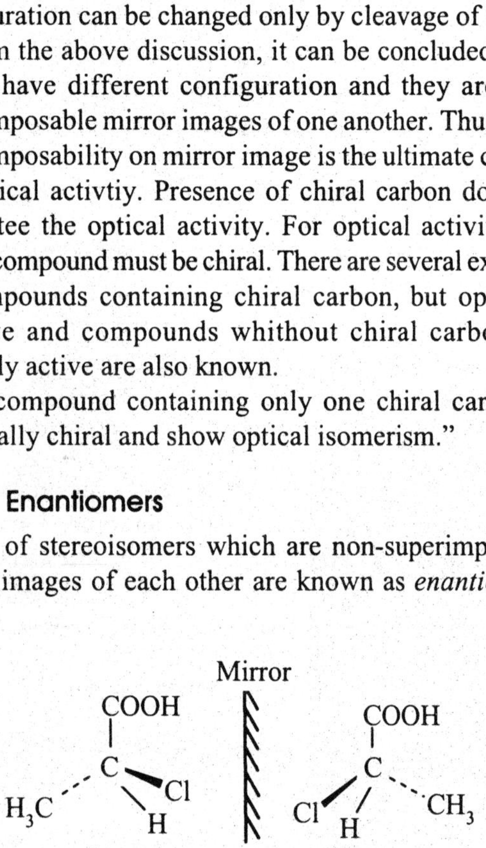

Here I and II are non-superimposable mirror images of each other hence they are enantiomers.

Properties of enantiomers

(i) Enantiomers have different configurations, i.e. arrangement of atoms or groups at chirality center are different for a pair of enantiomers. In the above shown example, I has Cl $\rightarrow$ H $\rightarrow$ CH_3 arranged in clockwise direction while the Cl $\rightarrow$ H $\rightarrow$ CH_3 have counter clockwise arrangement in II.

(ii) Enantiomers have same physical properties, i.e. they have same melting and boiling temperature, same refrective indices, same enthalpy and entropy of formations, same dipole moments, etc.

(iii) Enantiomers behave in similar manner with optically inactive substance, but behave differently with optically active substances.

(iv) Enantiomers have same magnitude of optical rotation, but opposite direction of rotation, i.e. if I is levorotatory, II must be destrorotatory but both will rotate plane polarized light by same angle, if their concentrations are same and analyzed in the same polarimeter.

2.5.4 Fischer Projection

It is a way of representing three dimensional molecule containing chirality center into two dimension. It helps in finding out the stereochemical relations between the pair of stereoisomers, whether they are identical, enantiomers, or distereomer. In Fischer projection, the chirality centers are represented by a point of intersection of two lines perpendicular to each other as (+). Groups on horizontal lines represents the groups out of plane in three dimensional structure and groups on vertical lines represents the groups present behind the plane in three dimensional structure.

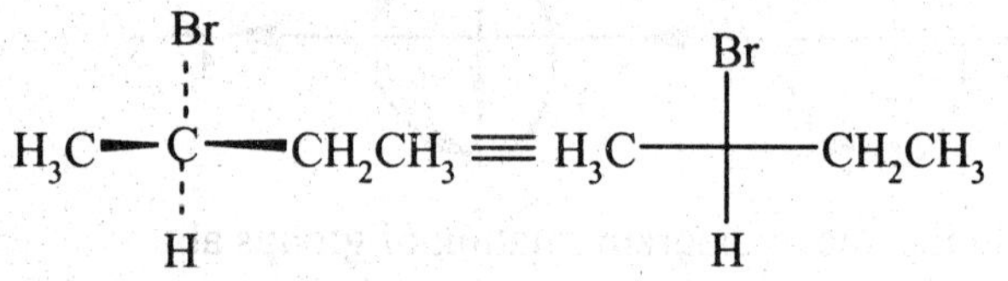

Wedge-dash representation I — Fischer projection II

Fischer projection represents a specific enantiomer of an optically active compound. It is always easier to compare the two stereoisomers by convertings them into Fischer projection. When comparing two Fischer projections, following points must be kept in mind.

(i) Rotating a Fischer projection by 180° in-plane does not change the configuration, e.g.

II is obtained by180° clockwise rotation of I in the plane of paper hence, I and II represents the same stereoisomers.

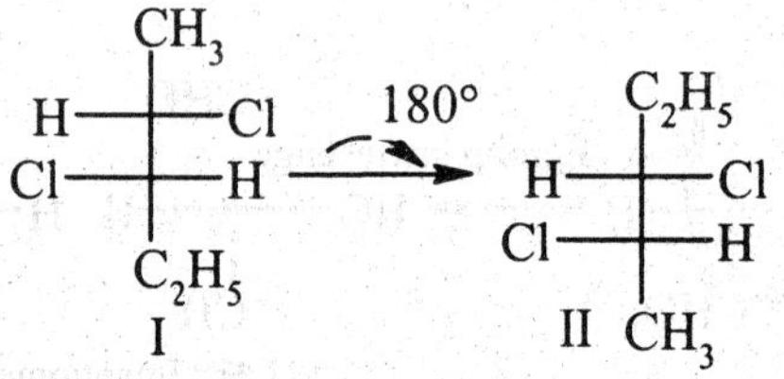

Here I and II are identical configuration since they are obtained by 180° rotation in plane of paper.

(ii) Rotating a Fischer projection by any other angle, e.g. rotation by 90° do not yield the same configuration.

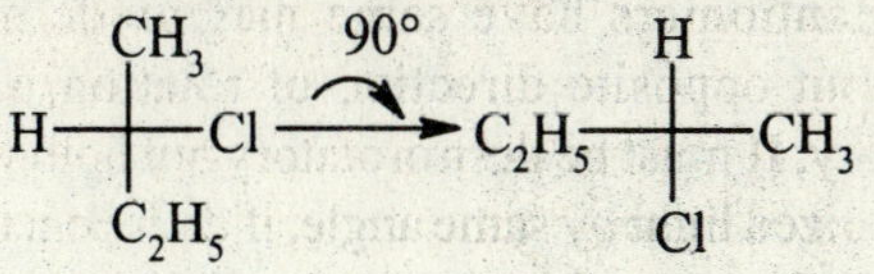

Here I and II are enantiomers.

(iii) Three group interchange in a cyclic manner, either clockwise or counter clockwise does not change the configuration as

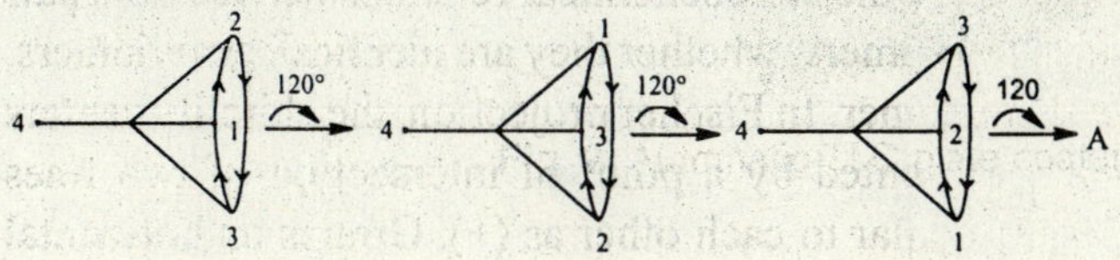

III(A) III(B) III(C)

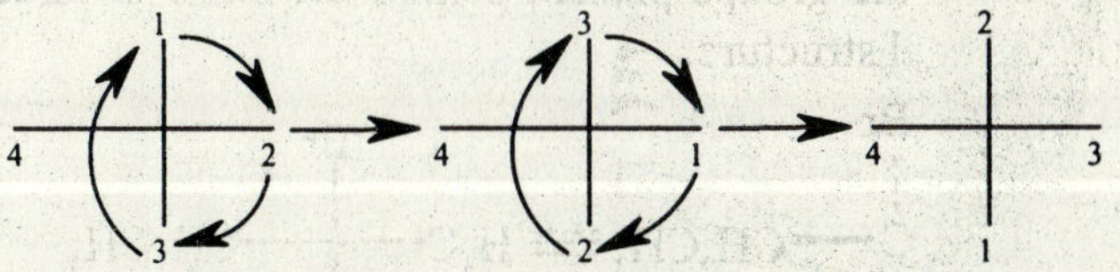

In the above diagram rotation of groups about a chiral carbon is shown. Since rotation of groups about chiral carbon donot change the configuration, A, B, and C represents the same configuration. Below to A, B, and C, are shown their Fisher projection equivalent showing rotation, if performed in two dimension. Let us illustrate this property by comparing stereochemical relationship between the following pairs of Fischer projections.

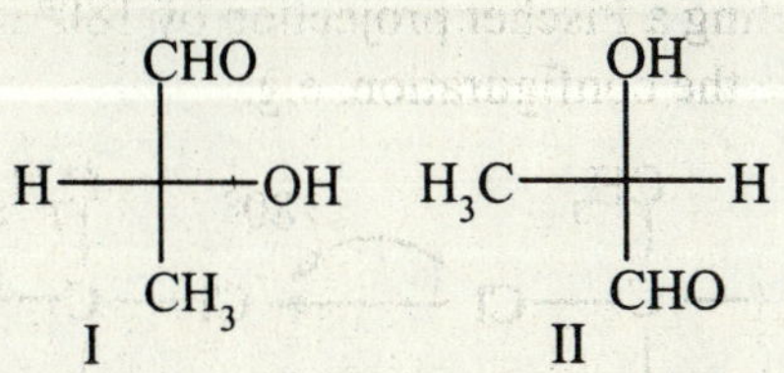

Here I and II are stereoisomers with a chiral carbon. To compare these two structures, keep I unchanged and perform rotation operation on II, so that identical groups are present on horizontal and vertical in I and II.

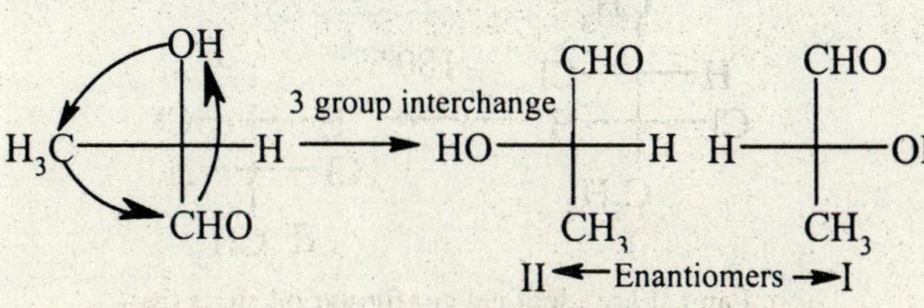

3 group interchange

II ← Enantiomers → I

I and II are non-superimposable mirror images.

2.5.5 Optical Isomerism in Compounds Containing more than One Chiral Centers

Let us consider $CH_3CHCH-CH_2CH_3$ (with Cl on each of the two CH carbons). This molecule has two chiral carbon and four Fischer projections (all different) drawn as:

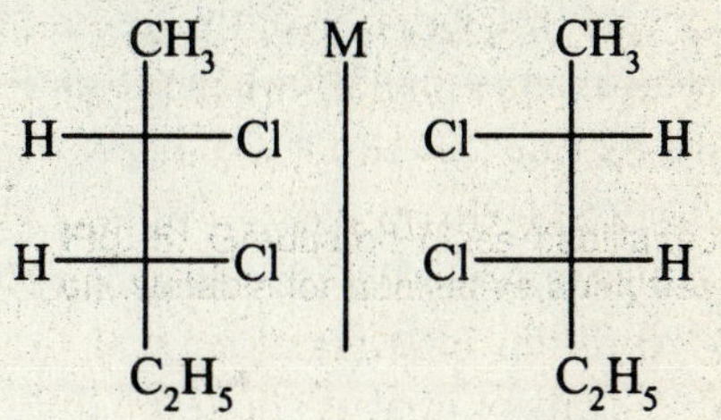

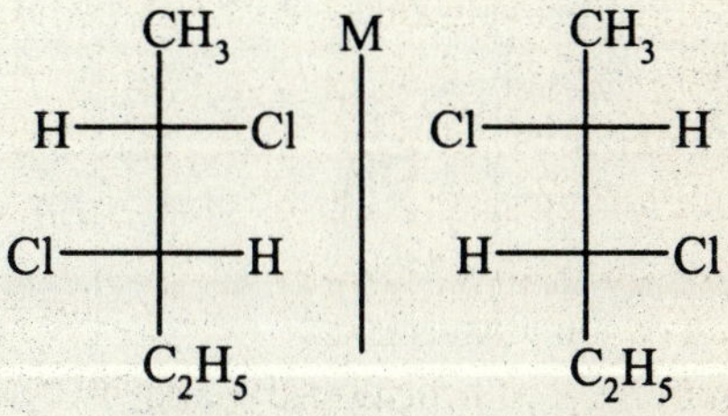

Here all four are stereoisomers. I and II are mirror image and they are non-superimposable, therefore I and II are enantiomers. Similarly III and IV are enantiomers. Also I and III, I and IV, II and III, II and IV are stereoisomers, but there is no mirror image relationship between them.

Stereoisomers which are not mirror image of one another are known as *Distereomers*. Therefore I and III, I and IV, II and III, II and IV are distereomers. Distereomers need not be essentially optically active.

- *Properties of distereomers*

(i) Distereomers have different physical properties.

(ii) Distereomers differs in their chemical properties with both optically active and optically inactive substances.

(iii) Optical rotations of distereomers have no relations either in magnitude or in directions.

2.5.6 Symmetry in a Compound Containing Chiral Carbon and its Effect on Optical Activity

For optical activity, the ultimate requirement is non-superimposibility on mirror image. Thus, if a molecule becomes superimposable on its mirror image, it will be optically inactive, no matter whether it has chiral carbon or not. To illustrate this, let us consider an example of tartaric acid.

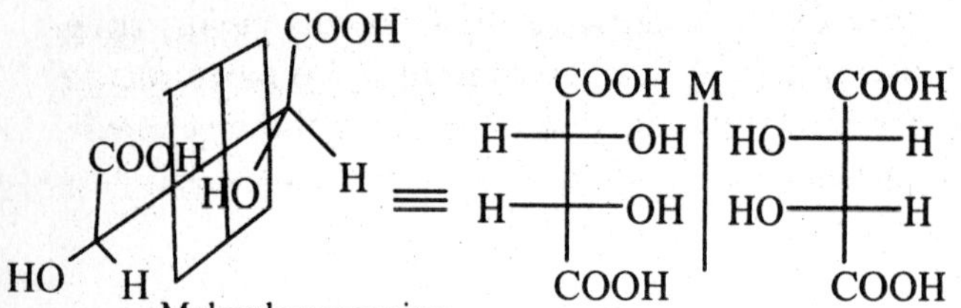

Molecule possessing a plane of symmetry

In the above case, the two mirror images are identical since they are superimposible. If molecule on left of the mirror is rotated by 180° in the plane of paper, it produces molecule on the right of the mirror. Hence, the above configuration of tartaric acid is optically inactive inspite of containing two chiral carbons.

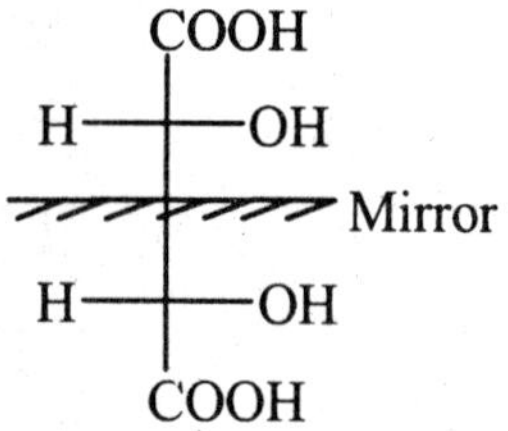

Mirror

(meso compound)

Here, the upper half is mirror image of lower half. Thus in meso compounds, enantiomerism is present within the single structure and hence, optical rotation caused by one part is exactly canceled by other part, since they rotate the plane polarized light by same angle but in opposite direction.

Racemic mixture: Racemic mixture are the mixture containing equal amount of enantiomers. Racemic mixture is optically inactive, but inactivity is temporary, i.e. optical inactivity is due to external compensation of optical rotation. Racemic mixture can be separated into active enantiomers by reacting with a enantiomerically pure reagent. Process of separating a racemic mixture into enantiomers is known as "Resolution of Racemic mixture".

Example 17. Predict the stereochemical relationship among the following pairs:

(a) 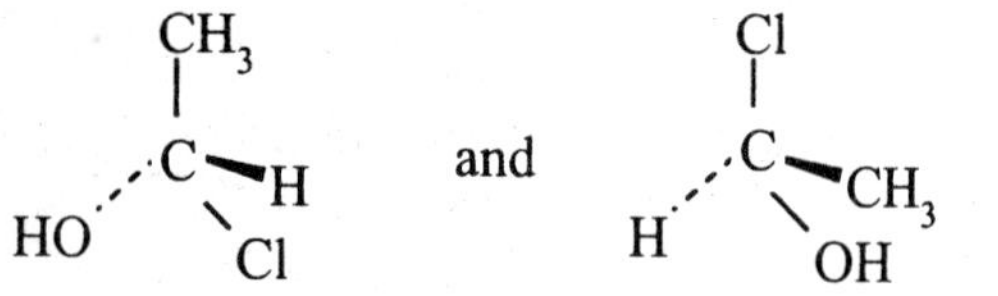 and

(b) 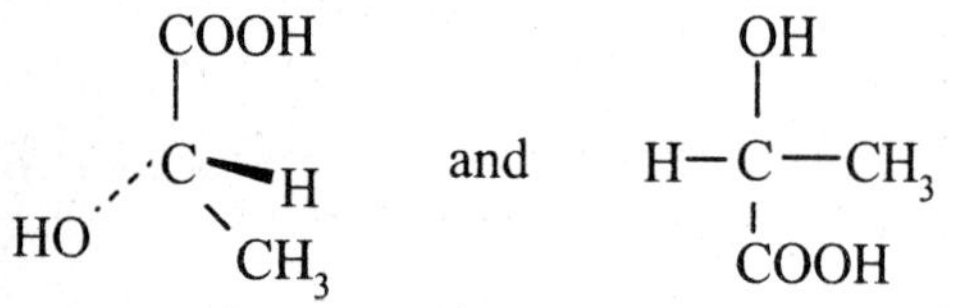 and

(c) 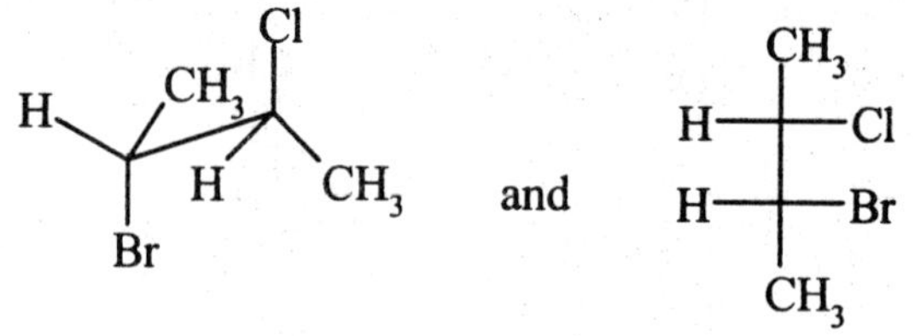 and

(d) 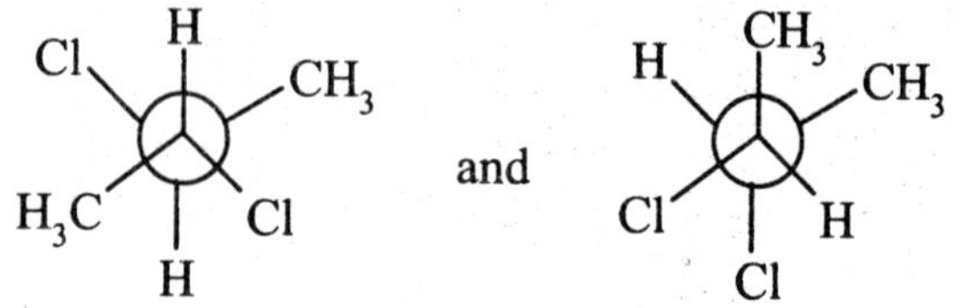 and

Solution

(a) 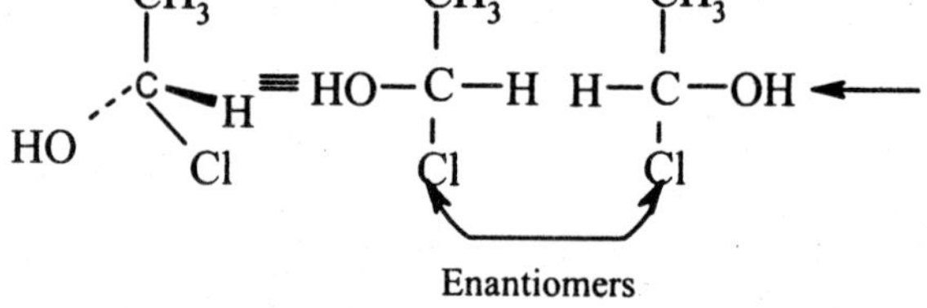 Enantiomers

(b) 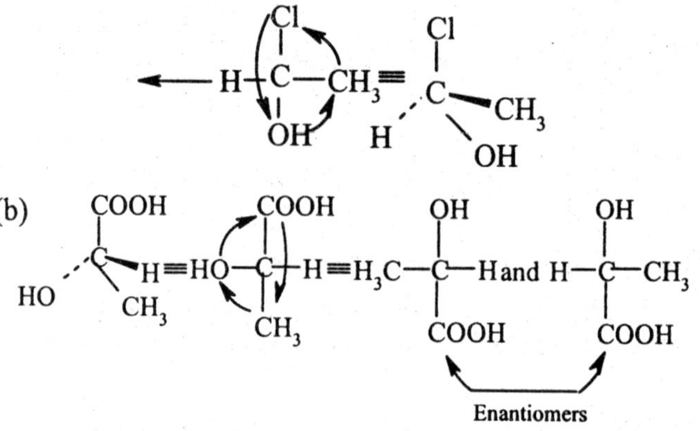 Enantiomers

(c) 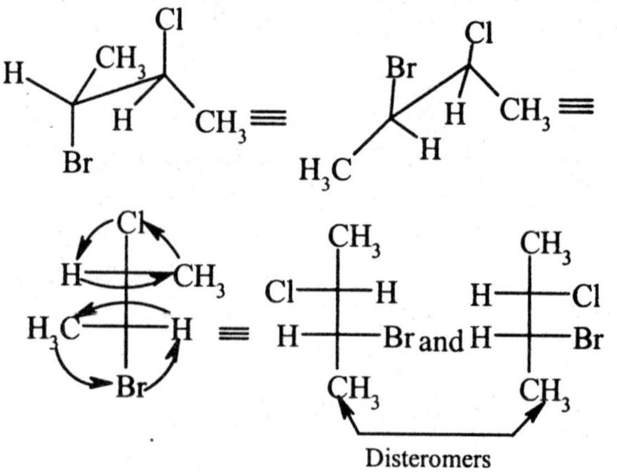 Disteromers

(d) 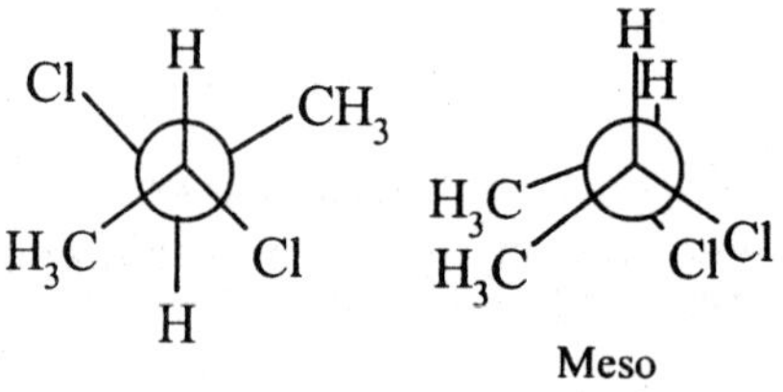 Meso

and

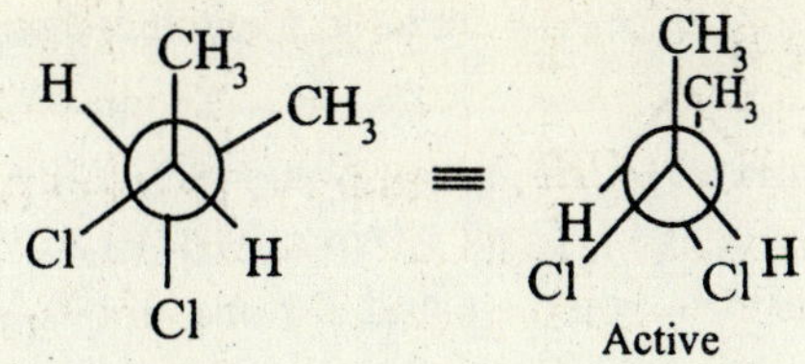

$$\text{Newman projections: (H, CH}_3\text{, CH}_3\text{; Cl, H, Cl)} \equiv \text{(CH}_3\text{, CH}_3\text{; H, Cl, Cl, H)}$$

Active

Hence, given pairs are distreomer.

Example 18. Draw structure of all stereoisomers of 4-methyl-2-hexene.

Solution

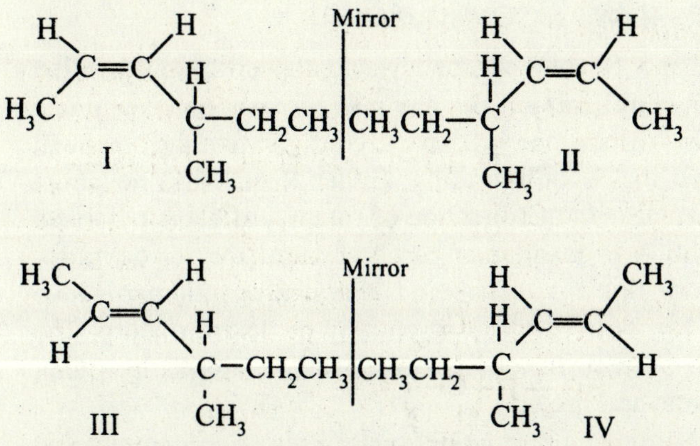

Mirror

I (H, H_3C C=C H; H–C(CH$_3$)–CH_2CH_3) | II (CH_3CH_2–C(H)(CH$_3$); H C=C H, CH_3)

Mirror

III (H_3C, H C=C H; H–C(CH$_3$)–CH_2CH_3) | IV (CH_3CH_2–C(H)(CH$_3$); H C=C CH_3, H)

I and II are enantiomers and III and IV are enantiomers. I and III, I and IV, II and III, II and IV are distereomers.

Example 19. A hydrocarbon (A) has molecular formula C_6H_8 and it rotate plane polarized light. A on treatment with H_2/Pt gives C_6H_{14} which is non-resolvable. Deduce structure of A.

Solution: A has three degree of unsaturation and on hydrogenation yields a completely saturated hydrocarbon indicate that all three degree of unsaturation is in the form of π-bonds. Three degree of unsaturation can be in the form of three double bonds or one triple and one double bond. Also A on hydrogenation yields non-resolvable hydrocarbon indicate that A has two different groups bonded to chiral carbon which becomes identical after hydrogenation. Based on these observation, structure of A can be predicted as:

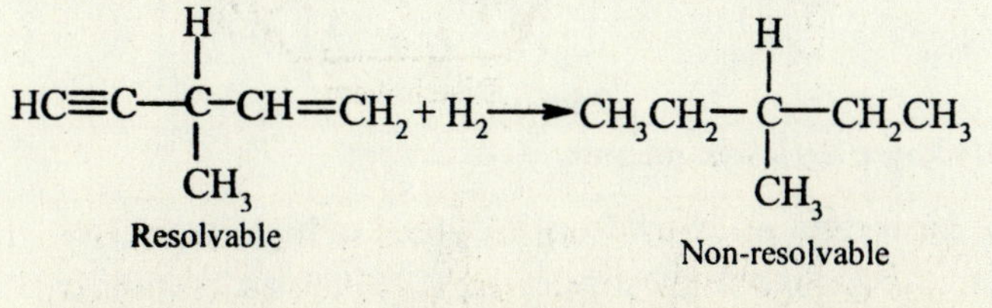

$$HC{\equiv}C{-}\underset{CH_3}{\overset{H}{C}}{-}CH{=}CH_2 + H_2 \longrightarrow CH_3CH_2{-}\underset{CH_3}{\overset{H}{C}}{-}CH_2CH_3$$

Resolvable — Non-resolvable

Example 20. An alkene exist in five isomeric form A—E (excluding stereoisomers) and all on addition of one mole of H_2 in presence of Pt metal yields same, smallest optically active hydrocarbon F. Deduce structure of A—F and label the alkene which can exhibit geometrical isomerism.

Solution

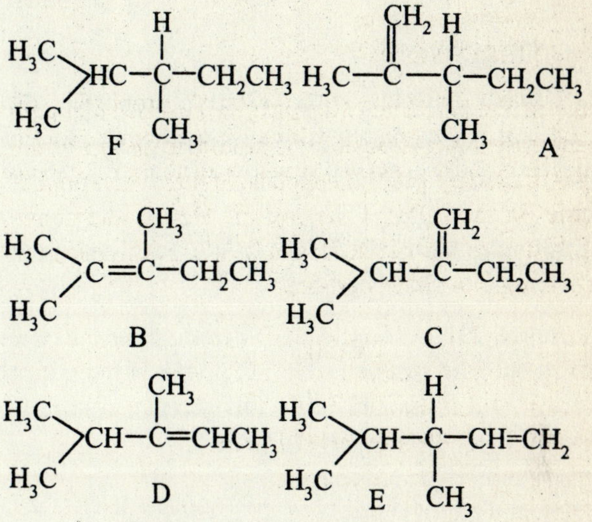

F: $(H_3C)_2CH{-}CH(CH_3){-}CH_2CH_3$

A: $H_3C{-}C({=}CH_2){-}CH(CH_3){-}CH_2CH_3$

B: $(H_3C)_2C{=}C(CH_3){-}CH_2CH_3$

C: $(H_3C)_2CH{-}C({=}CH_2){-}CH_2CH_3$

D: $(H_3C)_2CH{-}C(CH_3){=}CHCH_3$ (*cis-trans*)

E: $(H_3C)_2CH{-}CH(CH_3){-}CH{=}CH_2$

2.5.7 Optical Isomerism in Cyclic Compounds

Cyclo-alkane, if substituted properly, can show optical isomerism. In cyclic compound, a chiral carbon must have different atoms or groups at equal distance on the two side of ring as:

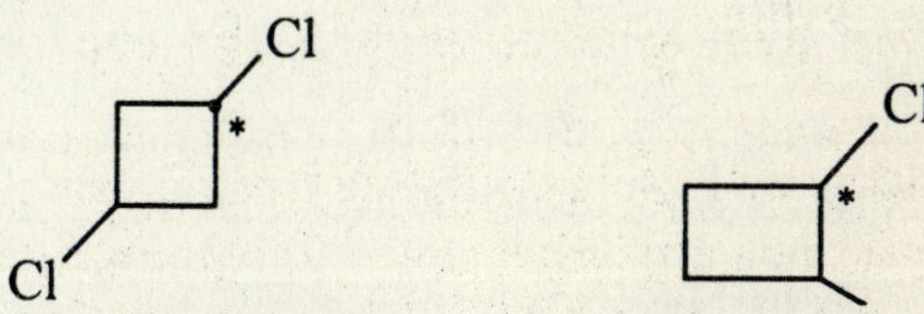

1,3-dichlorocyclobutane: It has identical groups at equal distance from the labeled carbon (Achiral)

Chlorocyclobutane with substituent: It has different groups on equal distance from the labeled carbon (*). (Chiral)

Example 21. In between 1, 4-dichlorocyclohexane and 1, 2-dichlorocyclohexane, only one can exhibit enantiomerism. Identify that and draw the structure of enantiomers.

Solution: 1, 4-dichlorocyclohexane has identical groups at equal distance on ring therefore it cannot show enantiomerism.

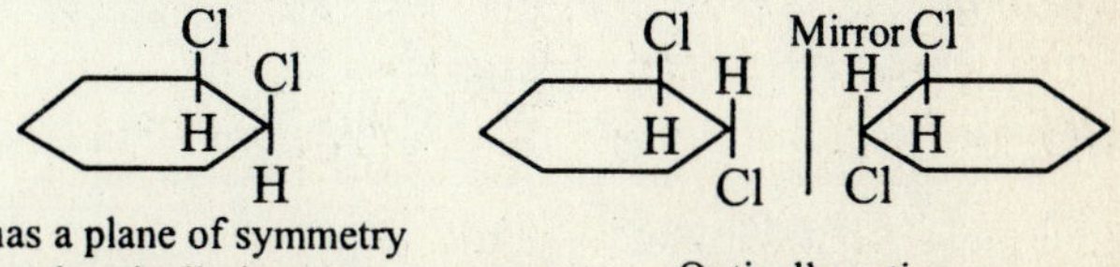

It has a plane of symmetry and optically inactive (Meso form)

Mirror

Optically active

Example 22. Draw structures of all stereoisomers of 2-methylchlorocyclopropane.

Solution

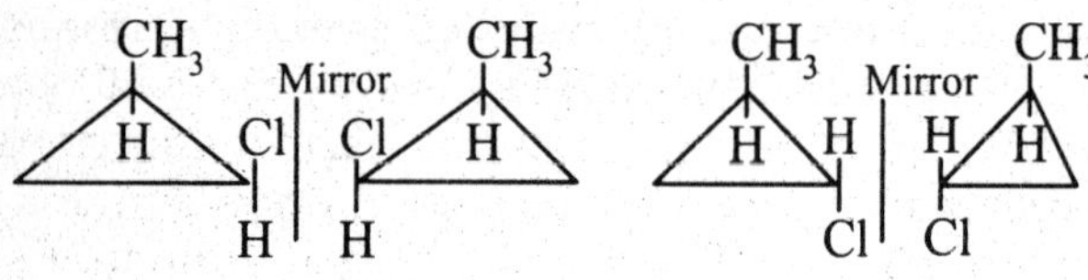

2.6 OPTICAL PURITY

Optical purity pertains to the relative proportion of a specific enantiomer in a mixture of enantiomers and it is determined by the observed specific rotation of mixture.

$$\% \text{ Optical purity} = \frac{\text{Observed specific rotation}}{\text{Specific rotation of pure enantiomer}} \times 100$$

Example 23. A sample of 2-bromobutane has an observed specific rotation of + 9.2°, what is the optical purity of (+) 2-bromobutane in this sample, if specific rotation of (+)2 bromobutane is + 23.1° ?

Solution: $\% \text{ optical purity} = \frac{9.2}{23.1} \times 100 = 40$

Example 24. (+) mandelic acid has a specific rotation of + 158°. What would be the observed specific rotation of each of the following mixtures?

(a) 25% (–) mandelic acid + 75% (+) mandelic acid.
(b) 50% (–) mandelic acid + 50% (+) mandelic acid.
(c) 75% (–) mandelic acid + 25 % (+) mandelic acid.

Solution

(a) 25% (–) mandelic acid + 75 % (+) mandelic acid indicate that 50% (+) mandelic acid is in excess.

$$\Rightarrow 50 = \frac{\text{Observed specific rotation}}{158} \times 100$$

$\Rightarrow$ specific rotation = +79°

(b) Observed specific rotation = 0.

(c) 50% (–) mandelic acid is in excess

$\Rightarrow$ observed specific rotation = –79°.

Example 25. The specific rotation of (–) glyceraldehyde is + 8.7°. If observed specific rotation of a mixture of enantiomers of glyceraldehyde is +1.4°, what percentage of glyceraldehyde is present as (+) glyceraldehyde ?

Solution: $\% \text{ optical purity} = \frac{1.4}{8.7} \times 100 = +\,16.09\%$

% of (+) glyceraldehyde in mixture

$$= 16.09 + \frac{(100 - 16.09)}{2} = 58.04\%$$

Example 26. A solution prepared by mixing 10 ml 0.1 M solution of (+) A and 30 ml of 0.1 M solution of (–) A was found to have an observed specific rotation of (–) 5°. What is the specific rotation of each enantiomer of A?

Solution: mmol of (+) A = 1, mmol of (–) A = 3.

$\Rightarrow$ 1.0 mmol of (+) A and 1.0 mmol of (–) A will form 2 mmol of racemic mixture and 2 mmol of (–) A will be in excess.

$\Rightarrow$ Percentage purity of solution = 50% in (–)A.

$$\Rightarrow 0.5 = -\frac{5}{x} \Rightarrow x = -10$$

Hence, specific rotation of (+) A = 10°, and of (–)A = –10°.

2.7 R AND S CONFIGURATION

'R' and 'S' are the name given to a specific configuration at a chiral carbon. For any pair of enantiomer with one chiral carbon, one will have R-configuration and other will have the S-configuration. Either R or S has no direct relation with the direction of rotation of plane polarized light. A R-enantiomer can be levorotatory or dextrorotatory, but it is also obvious that if R-enantiomer is levo its S-counterpart will essentially be dextrorotatory. The R/S system of nomenclature was developed by Cahn, Ingold and Prelog.

Naming an enantiomer under R and S system involve following steps:

1. Rank the groups that are bonded to chiral carbon in order of priority. The priority depends on the atomic weight of the atoms directly attached to the chiral carbon. Greater the atomic weight, higher the priority.

Br 1
C
CH_3CH_2 2
H 4
CH_3 3

Br– maximum atomic weight, highest priority between $-CH_3$, $-CH_2CH_3$, 1st atom bonded to chiral carbon are same-carbon. In such situation we see the next atom. Heavier atom coming at the first occasion of difference decide the priority.

$-CH_2-H$ < $-CH_2-CH_2-H$

Priority

Hydrogen has lowest priority.

2. Now draw an arrow from the group of highest priority (1) to the second highest priority, (2) if the arrow points in clockwise direction, the chiral center has R-configuration. If the arrow points in anticlockwise direction, the chirality center has the S-configuration.

S-2- brobutane R-2- brobutane

Priority of Some Specific Groups

- —CH_2OH and —CHO: —CHO is considered as —C(O)(O)—H, and thus has higher priority than —CH_2OH.
- —CHO and —COOH: —COOH is considered as —C(O)(O)—OH and has higher priority than —CHO.
- —CH_2CH_3 and —CH=CH_2: —CH=CH_2 is considered as —CH(C)(C) and has higher priority than —CH_2CH_3.

R and S Nomenclature in Fischer Projection Formula

For assigning R/S nomenclature in Fischer projection, the least priority group (4) must be on the vertical line. If fourth priority group is on horizontal line, bring it on vertical by three group interchange and then investigate for the direction of arrow starting from (1) to (3). If arrow head moves in clockwise direction, the chiral carbon has R-configuration and if arrow head moves in anticlockwise direction, the chiral carbon has S-configuration, e.g.

H-on horizontal anticlockwise S-configuration

Anticlockwise S-configuration

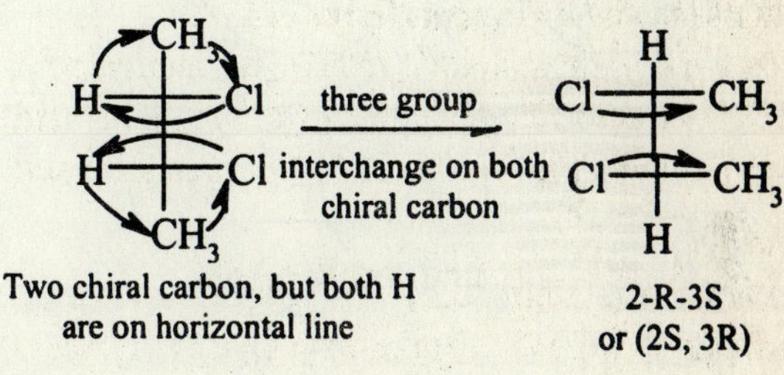

Two chiral carbon, but both H are on horizontal line

PRACTICE EXERCISE

1. Draw the enol tautomers for each of the following compounds. For those compounds that have more than one enol tautomer indicate one which is more stable.

(a) (b)

(c) (d) Ph—C(=O)—CH_3

(e) (f)

2. Draw structures of compound with molecular formula C_5H_{10}, which can exhibit stereoisomerism and draw the structure of their stereoisomers.
3. Draw structures of all stereoisomers possible for the following compounds :
   (a) 1-bromo-2-chlorocyclohexane.
   (b) 1, 2-dimethylcyclopropane.
   (c) 2-bromo-4-methylpentane.
   (d) 4-bromo-2-pentene.
   (e) 1, 2-dichlorocyclopentane.
   (f) 2-bromo-2-pentene.
   (g) 3-chloro-1-butene.
   (h) 1-bromo-2-chlorocyclobutane.
4. Draw the staggered and eclipsed conformers for rotation about C2–C3 bond of 1-chloropentane.

5. Indicate whether each of the following pair of compounds are identical, enantiomers, distereomers or constitutional isomers.

(a) 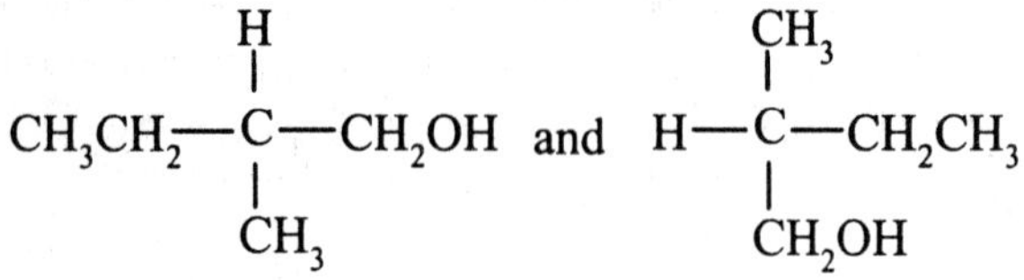 $CH_3CH_2-C(H)(CH_3)-CH_2OH$ and $H-C(CH_3)(CH_2OH)-CH_2CH_3$

(b) [Fischer projections: CH_3 top; HO–H; H–Cl; CH_2CH_3 bottom] and [C_2H_5 top; HO–H; H–Cl; CH_3 bottom]

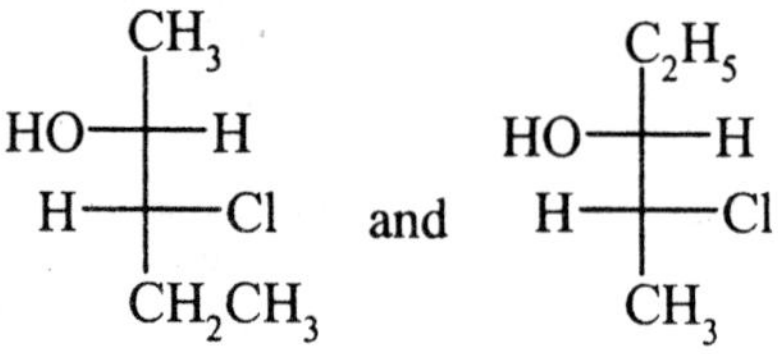

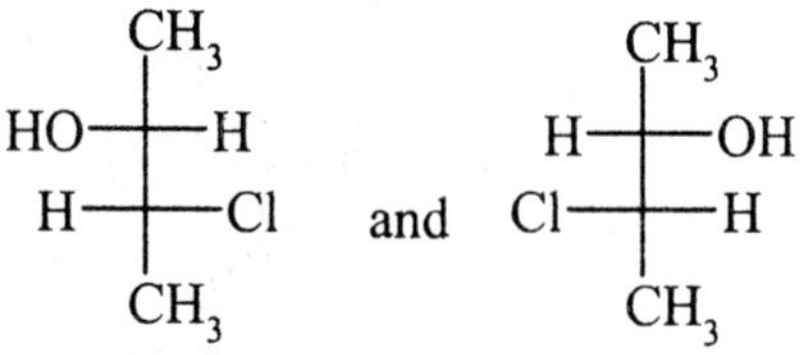

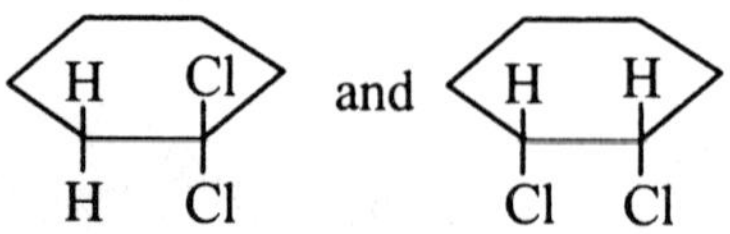

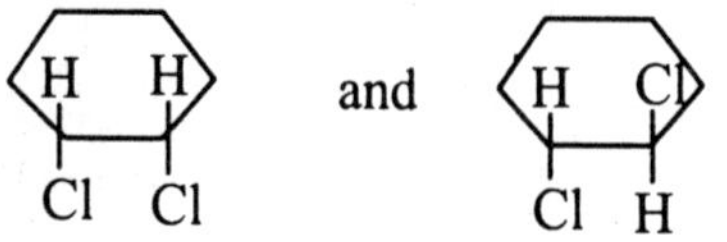

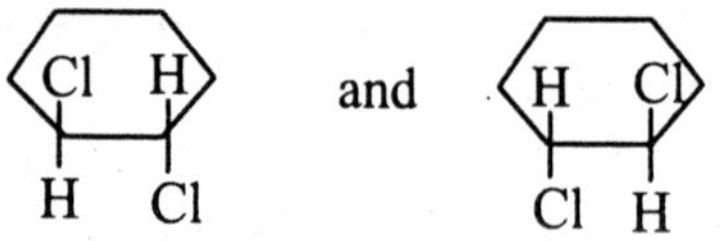

(c) [Fischer projections: CH_3 top; HO–H; H–Cl; CH_3 bottom] and [CH_3 top; H–OH; Cl–H; CH_3 bottom]

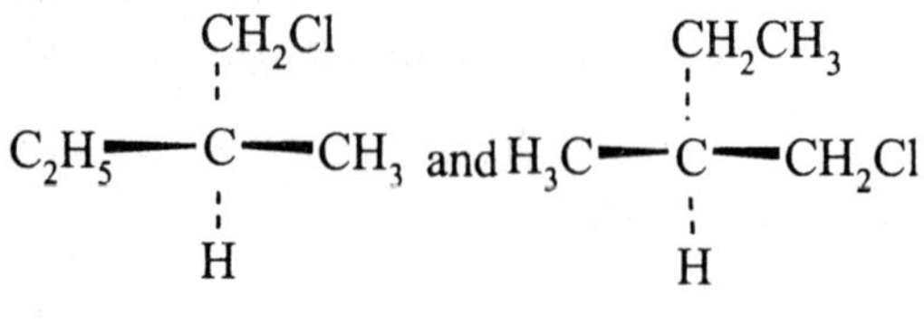

(d) [H Cl / H Cl] and [H H / Cl Cl]

(e) [H H / Cl Cl] and [H Cl / Cl H]

(f) [Cl H / H Cl] and [H Cl / Cl H]

(g) $C_2H_5-C(CH_2Cl)(H)-CH_3$ and $H_3C-C(CH_2CH_3)(H)-CH_2Cl$

(h) [cyclopropane: H, Br / Br, H] and [Br, H / H, Br]

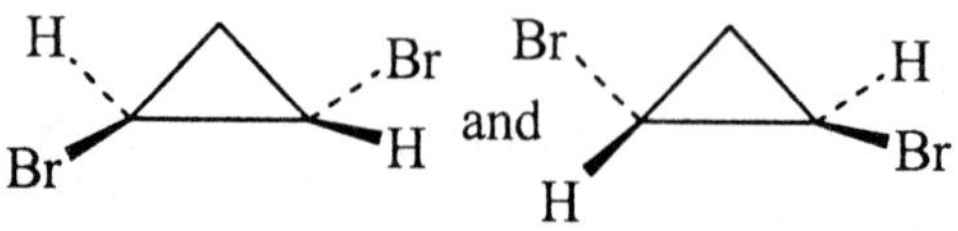

(i) 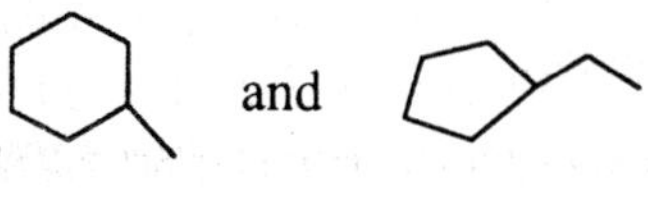 [methylcyclohexane] and [ethylcyclopentane]

(j) 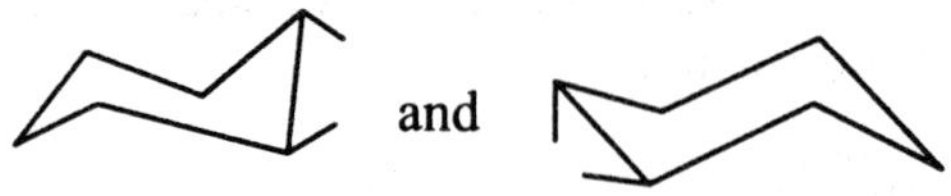 [chair cyclohexane] and [chair cyclohexane]

(k) 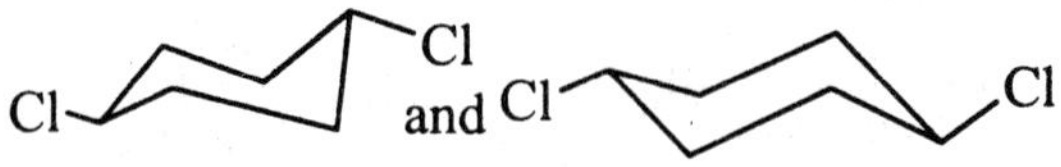 Cl–[chair]–Cl and Cl–[chair]–Cl

(l) 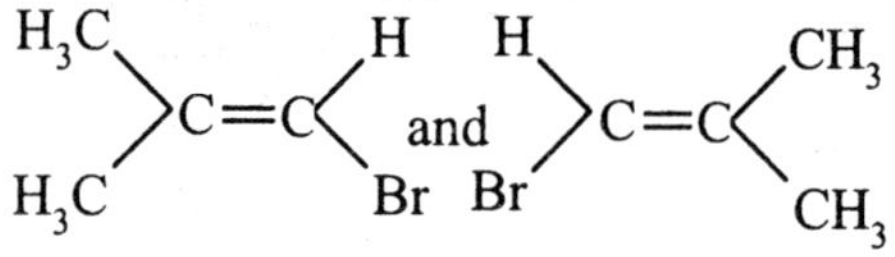 $(H_3C)_2C=C(H)(Br)$ and $(H)(Br)C=C(CH_3)_2$

(m) 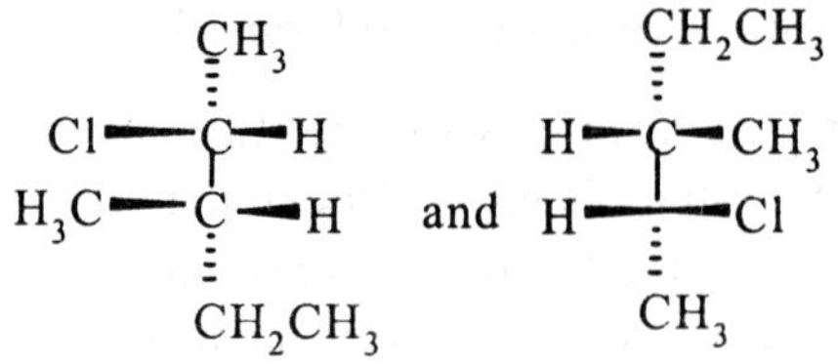 [CH_3; Cl–C–H; H_3C–C–H; CH_2CH_3] and [CH_2CH_3; H–C–CH_3; H–C–Cl; CH_3]

(n) 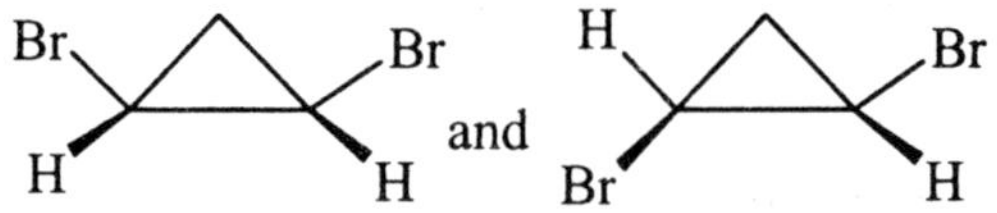 [cyclopropane: Br, H / Br, H] and [H, Br / Br, H]

6. Indicate whether following pairs are enantiomer or identical:

(a) 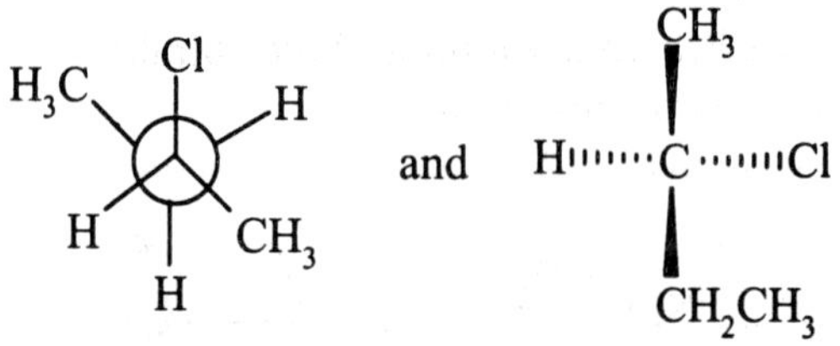 [Newman projection: Cl, H_3C, H / H, CH_3, H] and [CH_3; H···C···Cl; CH_2CH_3]

(b) 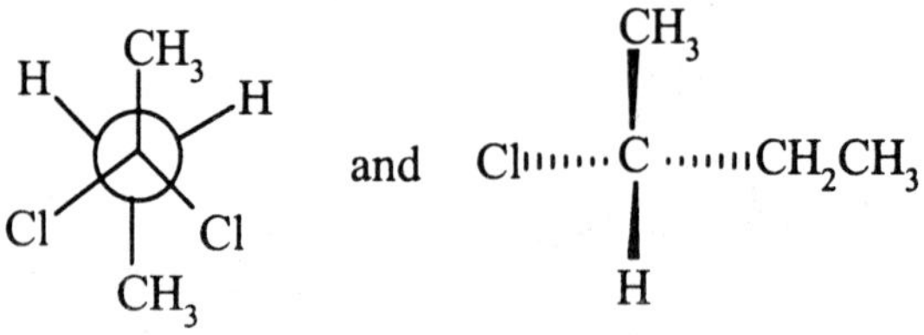 [Newman projection: CH_3, H, H / Cl, Cl, CH_3] and [CH_3; Cl···C···CH_2CH_3; H]

7. A solution of an unknown compound (3.0 g of X per 20 ml of solution), when placed in a polarimeter tube, 20 cm long, rotate the plane polarized light, 1.8° in anticlockwise direction. Determine specific rotation of X.

8. An optically active compound X has specific rotation of –39°. A solution of X (ρ = 0.187 g/ml) has an observed rotation of –6.25°, when placed in a polarimeter tube, 10 cm long. What is the percent optical purity of sample containing X.

9. A sample of d-lactic acid was found to have an optical purity of 72%. What percentage of ℓ-isomer is present in the sample?

10. Give major products indicating stereochemistry of the following reactions :

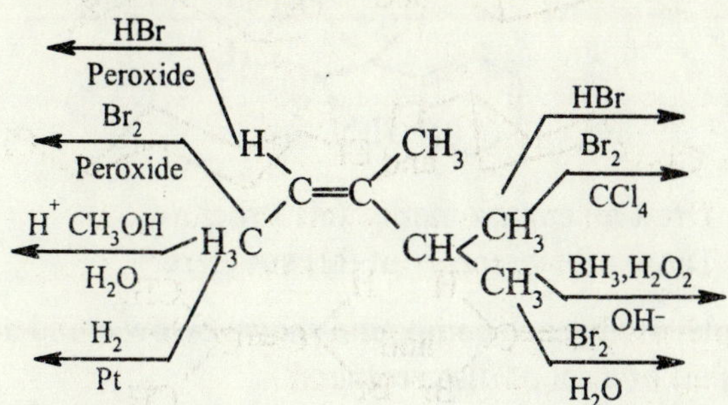

11. Cyclohexene exists only in *cis* form, while cyclodecene exist in both *cis* and *trans* forms. Explain.

12. Following compounds on treatment with HBr produces two stereoisomers. One of the stereoisomer is optically active while the other is optically inactive. Give structure of two products and label them as active and inactive.

CH_3; H—C—Br; H_3C—C—H; $CH{=}CH_2$ + HBr ⟶ Two stereoisomers.

13. Indicate R/S configuration of the chirality centers in the following molecules.

(a) O; CH_2CH_3; CH_3 (b) OH; H; Br; H

(c) H_3C; O

14. Does the following compound have any chirality. If yes, draw structures of both stereoisomer.

ClHC=C=CHCl

15. Discuss the stereochemical relationship between the following pairs of molecular structures as whether they are enantiomer, distereomers, constitutional isomers.

(a) H; D······C······CH_3; Cl and D; H······C······CH_3; Cl

(b) HO, Cl and Cl, OH

(c) H, H, HO, OH and HO, OH, H, H

(d) and

(e) Cl; F, C, Br; H and Cl; Br, C, H; F

(f) and

(g) HO, CH_3, H, D and H_3C, OH, D, H

(h) and

16. Draw Newman projections formula to represent each of the following conformers.

(a) Gauche conformers of hexane visualized down the C—3—C—4 bond.

(b) Anti conformer of pentane, visualized down the C2—C3 bond.

(c) Eclipsed conformer of butane, visualized down the C2—C3 bond.

(d) Chair conformation of methylcyclohexane with methyl group in an equatorial position.

17. Calculate the observed rotation expected in each of the following situation, assuming that a standard 10 cm cell is used.

(a) Ten grams of 75:25 mixture of enantiomer dissolved in 100 ml of solution. The major enantiomer has a specific rotation of +100°.

(b) 10 g of a 50:50 mixture of enantiomers dissolved in 100 ml solution.

(c) 10 grams of a 75:25 mixture of enantiomers dissolved in 100 ml solution. The major enantiomer has specific rotation of –50°.

(d) 10 grams of a 9:1 mixture of enantiomers dissolved in 100 ml solution. The major enantiomer has a specific rotation of +200°.

18. Fill in the blanks with appropriate term.

(a) Configurational isomers with two chirality centers are called .... if the configuration of both of the chirality centers in an isomer is the opposite of the configuration of the chirality centers in the other isomer.

(b) Configurational isomers with two chirality centers are called .... if the configuration of both of the chirality centers in one isomer is the same as the configuration of the chirality centers in the other isomer.

(c) Configurational isomers with two chirality centers are called .... if one of the chirality center has the same configuration in both isomer and the other chirality center has the opposite configuration in the two isomer.

19. Draw all the possible stereoisomers for each of the following compounds.

(a) 2-chloro-3-hexanol.
(b) 2, 3-dichloropentane.
(c) 2-bromo-4-chlorohexane.
(d) 1, 3-dibromopentane.

20. Draw structures of all stereoisomer possible for the following and discuss their optical activity.

(a) 1, 3-dichlorocyclohexane.
(b) 1, 4-dibromocyclohexane.
(c) 1, 2, 3-trichlorocyclopropane.

21. Deduce stereochemical relation between the following pairs:

(a) H_3C, Br, H / H, OH, CH_3 and CH_3 / HO—H / H—Br / CH_3

(b) CH_3 / H—OH / H—Br / CH_3 and H, HO, H_3C / CH_3, Br, H

(c) H_3C, H, Br / H, OH, CH_3 and CH_3 / HO—H / Br—H / CH_3

(d) H, HO, H_3C / CH_3, H, Br and CH_3 / H—OH / Br—H / CH_3

22. For the following structure:

H, OH, CH_3, H_2N, H

(a) Draw an enantiomer of this structure.
(b) Draw a distereomer of this structure.

23. Explain why each compound shown below could not be part of a set of distereomers?

(a) H_3C, H, Cl, Cl (b) H_3C, H, Br, Cl

(c) H, Cl

24. Why many drugs are sold as single enantiomers and not as a racemic mixture, despite the fact that it is usually less expensive to manufacture a racemic mixture than a single enantiomer.

25. An organic compound A has molecular formula $C_8H_{13}Br$, and it can be resolved into enantiomers. A on treatment with HBr produces B and C, both having molecular formula $C_8H_{14}Br_2$, in which B is resolvable while C is non resolvable. Also A on treatment with HBr in presence of a peroxide produces $C_8H_{14}Br_2$ for which four different structures, all chiral, can be drawn. A on treatment with $(CH_3)_3COK$ dissolved in $(CH_3)_3COH$ yields D, which on ozonolysis followed by work-up with Zn–H_2O yields 1, 3-cyclohexandione.

(a) Deduce structures of A to D.
(b) Draw structures of all four stereoisomers obtained on treatment of HBr/peroxide with A.

26. A hydrocarbon A has molecular formula "C_8H_{16}" and it can be resolved into enantiomers. A on catalytic hydrogenation yields B "C_8H_{18}" which cannot be resolved into enantiomers. B on treatment with chlorine gas (Cl_2) in presence of light produces four monochloroderivatives, in which only one can be resolved into enantiomers. Deduce structures of A, B and four monochloroderivatives indicating resolvable isomers.

27. A hydrocarbon A has molecular formula C_8H_{14} and it does not show stereoisomerism. A on treatment with HBr yields B(C_8H_{15}Br) which is still non-resolvable. On the otherhand A on treatment with HBr in presence of a perioxide yields a resolvable compound C, which is isomeric with B. A on treatment with O_3/$Zn–H_2O$ yields D($C_6H_{10}O$) and E (C_2H_4O). D on treatment with $ZnCl_2$/HCl yields F (C_6H_{12}), which on monochlorination yields $C_6H_{11}Cl$ as single isomers. Deduce structures of A to F.

28. An organic compound A has molecular formula $C_7H_{15}Cl$ and four different structures (all stereoisomers) can be drawn for this. A on treatment with ethanolic solution of KOH yield two alkenes B and C (B and C are structural isomers) with their molecular formula "C_7H_{14}". Both B and C can be resolved into enantiomers, but B does not show distereomerism while C does show. Oxidative cleavage of C with acidic solution of $KMnO_4$ yields D ($C_5H_{10}O_2$) as one product which still can be resolved into enantiomer. Heating D with sodalime yields C_4H_{10} which is non-resolvable. Identify A to D and draw structures of all stereoisomers of A to D.

OBJECTIVE EXERCISE

Level 1

1. How many isomers of $CH_3CH{=}CH{-}CH{=}NOH$ exist?

(a) Five (b) Two
(c) Three (d) Four

2. Which of the following compounds can exist in enantiomeric (i.e. D and L) forms?

(a) 3-methyl butanoic acid.
(b) *cis*-2-butene.
(c) isopropyl amine.
(d) 1-methylbutanamine.

3. Which of the following compounds can exist in two geometrically isomeric forms?

(a) 2-butyne (b) 2-butanol
(c) 2-pentene (d) 1-Pentene

4. The structures shown here are related as being:

H, Br, CH_3, Br, H, CH_3 H_3C, Br, H, H, H_3C, Br

(a) identical (b) enantiomorphous
(c) geometrical isomers (d) diastereomers

5. How many isomeric structures are possible for 2-methylcyclobutanone oxime?

(a) One only (no isomers) (b) Two
(c) Three (d) Four

6. One of the following completions (a) to (d) is incorrect. Choose the letter corresponding to the *incorrect* completion: Geometrical isomerism is exhibited by:

(a) 3-bromo-4-methylhex-3-ene
(b) 1-chloropropene
(c) 3-chloropropene
(d) 1,2-dibromoethene

7. Which of the following molecules are identical with that represented by:

CH_3, H_3C, OH, HO, H, H

(a) H, H, CH_3, H_3C, OH, OH   (b) OH, H_3C, H, H, CH_3, OH

(c) CH_3, H_3C, H, H, OH, OH   (d) OH, H_3C, OH, H, CH_3, H

8. Which of the following pairs are isomers?

(a) C_5H_{10} and $C_{10}H_{20}$
(b) $CH_3.(CH_2)_4.CH_3$ and $CH_3.(CH_2)_3.CH_3$
(c) $CH_3.CH(CH_3).(CH_2)_2.CH_3$ and $CH_3.(CH_2)_2.CH(CH_3)_2$
(d) $(CH_3)_3.CH$ and $CH_3.CH_2.CH_2.CH_3$

9. Which of the following pairs of compounds are identical?

(a) Cl, H, H, CH_3, CH_3, Cl + CH_3, Cl, Cl, H, H, CH_3

(b) [Sawhorse projections: Cl, Cl, I, H / I, H] + [Sawhorse projections: I, H, Cl / Cl, I, H]

(c) [Sawhorse projection: Br, Br, F, H, H / F] + [Newman projection: F, H, Br / F, Br, H]

(d) [Newman projection: H, H, H / Br, I, Cl] + [Newman projection: H, Br, Cl / H, I, H]

10. Which of the following *cannot* be written in an isomeric form?

(a) $CH_3—CH(OH)—CH_2—CH_3$
(b) $CH_3—CHO$
(c) $CH_2{=}CH—Cl$
(d) $CH_3—CH_2—CH{=}CH—CH_3$

11. Which of the following statements is not an essential feature of an optically active molecule? An optically active molecule will:

(a) Rotate the plane of polarized light.
(b) Have a non-superimposable mirror image.
(c) Have no element of symmetry.
(d) Have an asymmetric carbon atom.

12. Which of the following statements describes the product from the ionic addition of bromine to propene?

(a) Optically active.
(b) Optically inactive and capable of resolution.
(c) Optically inactive and incapable of resolution.
(d) A *meso*-compound.

13. A compound reacts *reversibly* with another to give two products which are isomeric. Isomer X is thermodynamically more stable than isomer Y, but Y is formed much faster. What would the reaction product consist of after reaction times, which were (1) short and (2) very long? Select from:

(a) (1) mainly X; (2) mainly X.
(b) (1) mainly Y; (2) mainly Y.
(c) (1) mainly X; (2) mainly Y.
(d) (1) mainly Y; (2) mainly X.

14. Which of the following is the correct structure for 4-methyl-3-chloro-pent-3-enal?

(a) $(CH_3)_2·C{=}C(Cl)·CH_2·CH_2·CHO$
(b) $CH_3·CH_2·CH{=}C(Cl)·CH_2·CHO$
(c) $(CH_3)_2·C{=}C(Cl)·CH_2·CHO$
(d) $(CH_3)_2·CH·C(Cl){=}CH·CHO$

15. How many pairs of diastereoisomers are possible in the following molecule?

Ph. CH(Cl). CH(Ph). CH(Cl). Ph

(a) None (b) Two
(c) Four (d) Eight

16. Which of the following is capable of exhibiting optical isomerism?

(a) 1, 2-dichloropropane
(b) 1, 3-dichloropropane
(c) 1, 1-dichloropropane
(d) 2, 2-dichloropropane

I	II	III
COOH	COOH	COOH
H—OH	H—OH	HO—H
H—OH	H—H	H—OH
H—OH	H—OH	H—OH
COOH	COOH	COOH

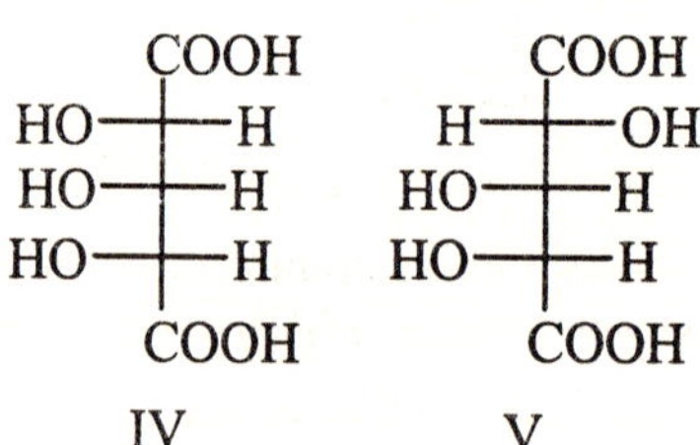

17. Which of the above formulae represent identical compounds?

(a) I and II (b) I and IV
(c) II and IV (d) III and V

18. Which of the compounds below are enontiomers?

I	II	III
COOH	COOH	COOH
H—OH	H—OH	HO—H
H—OH	H—H	H—OH
H—OH	H—OH	H—OH
COOH	COOH	COOH

IV	V
COOH	COOH
HO—H	H—OH
HO—H	HO—H
HO—H	HO—H
COOH	COOH

(a) II and III (b) III and IV
(c) III and V (d) I and V

19. Which of the following completions is *incorrect.* A pair of *enantiomers* are identical with each other with respect to their:
(a) chemical reactivity.
(b) melting point.
(c) direction of rotation of plane polarized light.
(d) solubility in a given solvent.

20. Which compounds are both *meso* forms?
(a) IV and II (b) I and III
(c) II and V (d) IV and V

```
 COOH          COOH          COOH
H—+—OH       H—+—OH      HO—+—H
H—+—OH       H—+—H        H—+—OH
H—+—OH       H—+—OH       H—+—OH
 COOH          COOH          COOH
   I             II            III

 COOH          COOH
HO—+—H        H—+—OH
HO—+—H       HO—+—H
HO—+—H       HO—+—H
 COOH          COOH
  IV             V
```

21. In the following sequence of standard stereochemical formulae, indicate that/those which is/are identical with **A.** *Note* that ONE or MORE of the formula **B, C** and **D** MAY be identical with **A.**

```
   NH2              NH2
H—+—CH3        H3C—+—H
  COOH             COOH
  (A)              (B)

     H              COOH
H2N—+—CH3      H3C—+—H
   COOH             NH2
   (C)              (D)
```

22. These compounds are:

```
   Ph           Ph
H—+—Br       H—+—Br
H—+—Br      Br—+—H
   Ph           Ph
```

(a) enantiomers (enantiomorphs)
(b) diastereoisomers
(c) conformers
(d) geometrical isomers

23. Which of the following compounds is optically active?
(a) $CH_3{\cdot}CH(OH){\cdot}CH_3$
(b) $CH_3{\cdot}CH(OCH_3){\cdot}CH_3$
(c) $CH_3{\cdot}CH(OH){\cdot}C_6H_5$
(d) $CH_3{\cdot}CO{\cdot}CH_3$

24. How many stereoisomers of the following molecule are possible? $HOOC{\cdot}CH{=}C{=}CH{\cdot}COOH$
(a) Two optical isomers.
(b) Two geometrical isomers.
(c) Two optical and two geometrical isomers.
(d) None.

25. In the following sequence of standard stereochemical formula, indicate that/those which is/are identical with **A.** *Note* that ONE or MORE of the formulae **B, C** and **D** MAY be identical with **A.**

```
          SH                        Ph
(a) Ph—+—CH3          (b) H3C◄—C—►SH
       CH2Cl                      CH2Cl

          CH3                       CH3
(c) ClH2C—+—SH        (d) ClH2C—+—SH
           Ph                        Ph
```

26. The following compounds differ in:

```
H       Cl          H       Cl
 \     /             \     /
  C = C               C = C
 /     \             /     \
H       Cl         Cl       H
```

(a) configuration (b) conformation
(c) structure (d) chirality

27. Which one of the following statements is true?
(a) Diastereoisomers are a pair of isomers relatated spatially as object and mirror image.
(b) Diastereoisomers can often be separated by fractional crystallization.
(c) Diastereoisomers have identical physical and chemical properties.
(d) Diastereoisomers rotate the plane of polarization of plane-polarized light to an equal but opposite extent.

28. Which of the following compounds is optically inactive?

(a) [skeletal structure: butan-2-ol, OH] (b) [skeletal structure: butadiene]

(c) H, Cl–C=C=C–Cl, H (allene)

(d) 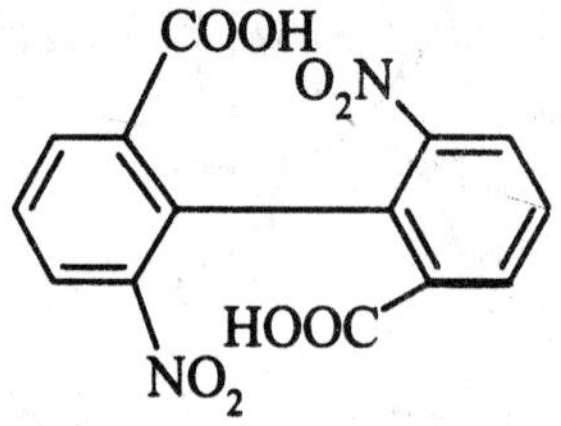 [structure: biphenyl with COOH, O_2N, HOOC, NO_2]

29. Which species exhibits a plane of symmetry?

(a) 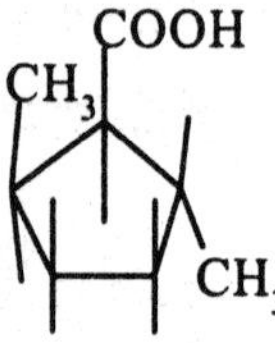 [structure: COOH, CH_3, CH_3] (b) 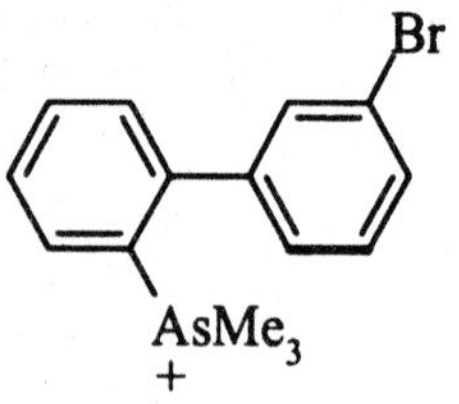 [structure: Br, $\overset{+}{As}Me_3$ biphenyl]

(c) 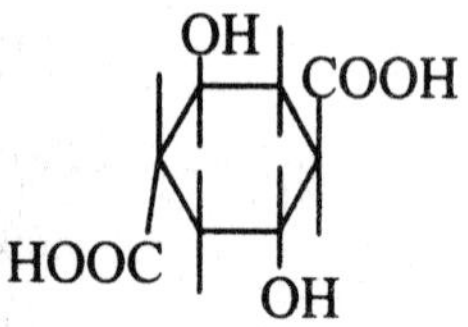 [structure: OH, COOH, HOOC, OH] (d) 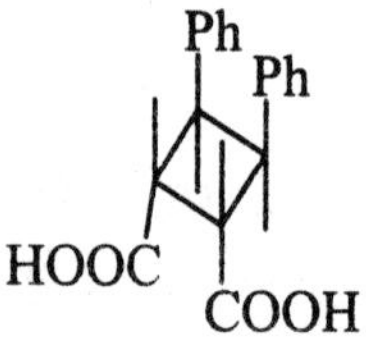 [structure: Ph, Ph, HOOC, COOH]

30. The specific rotation of alpha-D-glucose is $[alpha]_D$ = + 113°, while that for beta-D-glucose is $[alpha]_D$ = + 19°. When either anomer undergoes mutarotation the equilibrium solution exhibits $[alpha]_D$= + 52°. This solution contains:
(a) 65% of the anomer with the OH at C_1 Equatorial.
(b) 35% of the anomer with the OH at C_1 Equatorial.
(c) 80% of the anomer with the OH at C_1 Equatorial.
(d) 20% of the anomer with the OH at C_1 Equatorial.

31. Which one of the following compounds *cannot* exhibit optical activity?
(a) $Ph·CH_2·CH(OH)·COOH$
(b) $CH_3·CO·OCH(CH_3)·CH_2·CH_2·CH_3$
(c) $Ph·CH(Et)·CH_2·COOH$
(d) $CH_3·CO·OC(CH_3)(CH_2·CH_3)·CH_2·CH_3$

32. Which of the following compounds can exhibit optical activity?

(a) 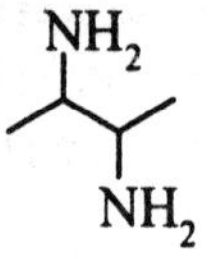 [structure: NH_2, NH_2] (b) 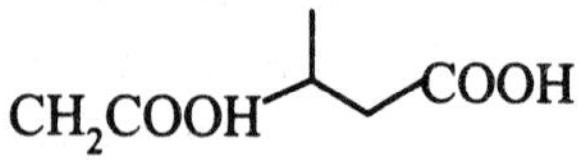 CH_2COOH [structure] COOH

(c) 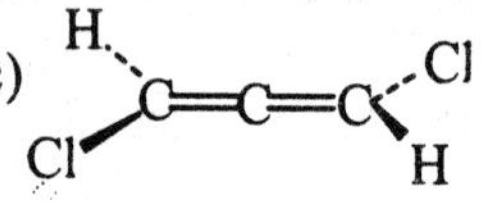 H, Cl, C=C=C, Cl, H (d) 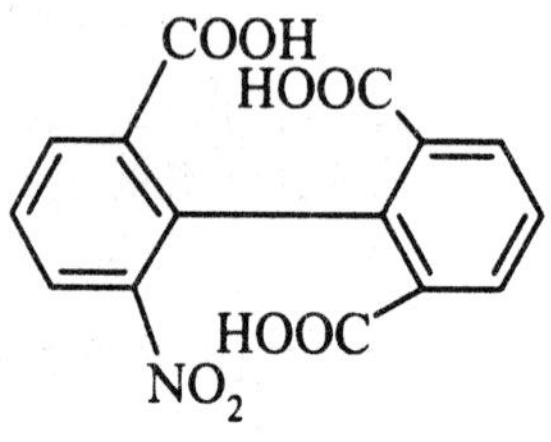 [structure: biphenyl with COOH, HOOC, HOOC, NO_2]

33. (i) Ethyl acetate will react with sodium. *BECAUSE.*
(ii) The enol molecules present form a sodium salt while the keto molecules remain unchanged.
(a) If both statements are true and (ii) is the explanation of (i).
(b) If both are true but (ii) is NOT the explanation of (i).
(c) If (i) is ture but (ii) is false.
(d) If (i) is false but (ii) is true.

34. The preferred conformation of *trans*-1,2-dibromo-cyclohexane is:
(a) diaxial (b) diequatorial
(c) axial/equatorial (d) neither A, B, or C

35. Which of the following completions is *incorrect.* Tautomerism is exhibited by :
(a) Ethyl 3-oxybutanoate
(b) Pentan-2,4-dione
(c) 1-Phenylprop-3-en-1-ol
(d) Propan-2-ol

36. The molecules below are:

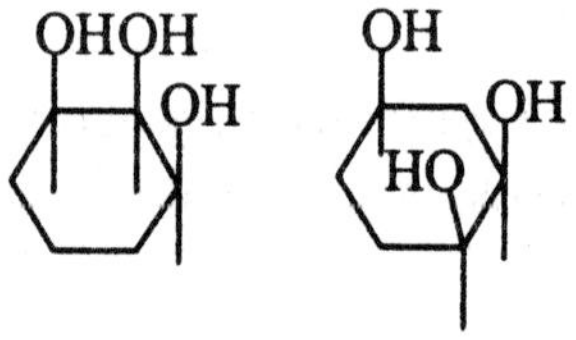

[structures: OH, OH, OH; OH, OH, HO]

(a) enantiomers (b) diastereoisomers
(c) geometrical isomerss (d) structural isomers

37. How many stereoisomers of the following compound are possible?
$CH_3·CH{=}CH·CHBr·CH_2·CH_3$
(a) Two (b) Three (c) Four (d) Six

38. The molecules of methane and ethylene have different shapes. Which of the following best highlights these differences?
(a) Methane is tetrahedral, ethylene is planar.
(b) Methane is formed by using sigma-bonds, while ethylene has no sigma-bonds.
(c) Molecular rotation is free in methane, but restricted in ethylene.
(d) The bond angles on ethylene are 120°, while in methane they are less than 120°.

39. *trans*-1,3-Di-*t*-butylcyclohexane exists primarily in the:
(a) racemic form (b) staggered form
(c) chair conformation (d) boat conformation

40. How many stereoisomers can exist for the acid:
HOOC—CH(OH)—CH(OH)—CH(OH)—COOH
(a) Two (b) Four
(c) Eight (d) Sixteen

41. An optically active carboxylic acid, X, has the molecular formula $C_6H_{10}O_2$ and can be shown by chemical and physical tests to contain the group $—CH{=}CH_2$ and at least one $—CH_3$ group. When X is reduced with hydrogen over platinum the product formed is optically inactive. Which of the following best represents the structure of X?
(a) $CH_2{=}CH—CH(CH_3)—CH_2—COOH$
(b) $CH_2{=}CH—CH(CH_2 \cdot CH_3)—COOH$
(c) $CH_2{=}CH—CH_2—CH(CH_3)—COOH$
(d) $CH_2{=}CH—C(CH_3)_2—COOH$

42. How many stereoisomers of 1,2-dichlorocyclohexane are optically active?
(a) None (b) One
(c) Two (d) Four

Level 2

1. Which of the following will produce chiral molecule after treatment with Lindlar's catalyst?

(a)  (b) 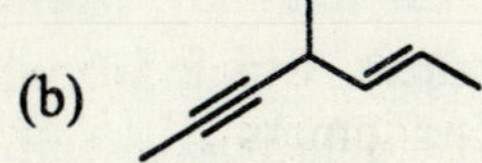

(c) 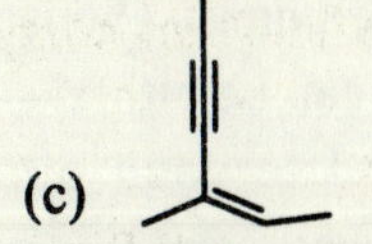 (d) 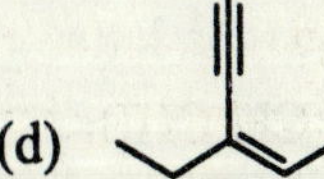

2. Molecules in sample A are distereomers of molecules in sample B. Neither sample shows optical isomerism in solution. Which of the following statements about them cannot be true?
(a) A and B contains *cis* and *trans* isomers of the same compound.
(b) A contains a meso compound and B contains a racemic mixture.
(c) Molecules in A have no stereogenic centers, and molecules in B have two chiral centers related by a symmetry plane.
(d) A and B contains equimolar mixture of enantiomers.

3. Which of the following Fischer projection is the enantiomer of the following molecule?

F, Br, F, HO, H, H_3C, Cl, CH_3

(a) F (top); H_3C—H; F—Br; Cl—CH_3; OH (bottom)

(b) F (top); H_3C—H; Br—F; H_3C—Cl; OH (bottom)

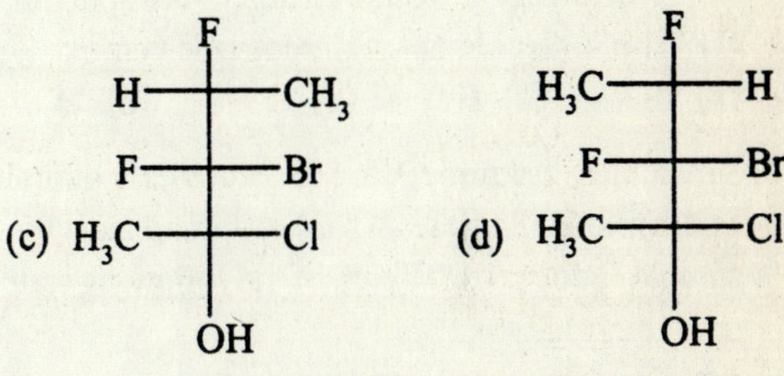

4. How many stereomers are possible for the following molecule?

Cl, Br

(a) 4 (b) 8 (c) 12 (d) 16

5. Which of the following molecule is (are) chiral?

I, Cl, (A); F, Br, F, Br, (B)

(a) A (b) B (c) both (d) none

6. Molecules of compound A dissolved in hexane showed [a] = –23°. What conclusion must be correct?
(a) Compound A contains at least one chiral center.
(b) Compound A cannot be a mixture on enantiomers.
(c) Compound A is a distereomer.
(d) A and its mirror image are non-superimposable.

7. How many stereomers are possible for the molecule 1, 3-dimethyl cyclohexane?
(a) 1 (b) 2 (c) 4 (d) 3

8. Which of the following molecules is (are) chiral?

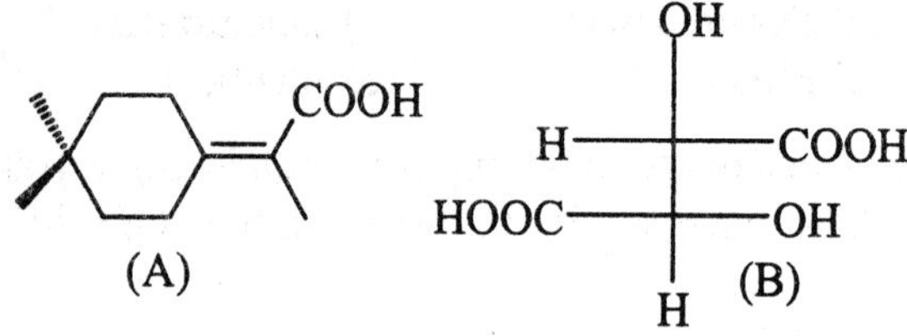

(a) A (b) B (c) Both (d) Neither

9. For the following : $H_2C{=}C{=}C(Br)CH_3$, which of the following statement is (are) correct?
(A) The molecule is chiral.
(B) The molecule has three degree of unsaturation.
(C) If H^+ were to add to the central carbon, its hybridization would change from sp to sp^2.
(D) The molecule has a stereogenic center.
(a) A (b) B (c) C (d) D

10. Molecules in sample A are distereomeric of molecules in sample B. Sample A shows optical activity, while sample B does not. Which of the following statements must be true?
(a) Sample A contains unequal amount of enantiomers while B is a racemic mixture.
(b) Molecules in A have stereocenter while molecules in B have no stereocentre.
(c) Molecules in A are chiral while molecules in B are meso compounds.
(d) Molecules in A and B are geometrical isomers.

11. How many stereomers are possible for the following molecules?

(a) 2 (b) 4 (c) 8 (d) 16

12. Consider the following energy profile where A reacts with reagent R and get converted into C, and B reacts with the same reagent R under identical conditions and gets converted into D. Compounds A and B are enantiomers of each other. Which statement about ***transition state*** for these transformation must be true?
(a) The two transition states are enantiomers.
(b) The two transition states are distereomers.
(c) The two transition states are geometrical isomers.
(d) The two transition states are superimposable mirror images of each other.

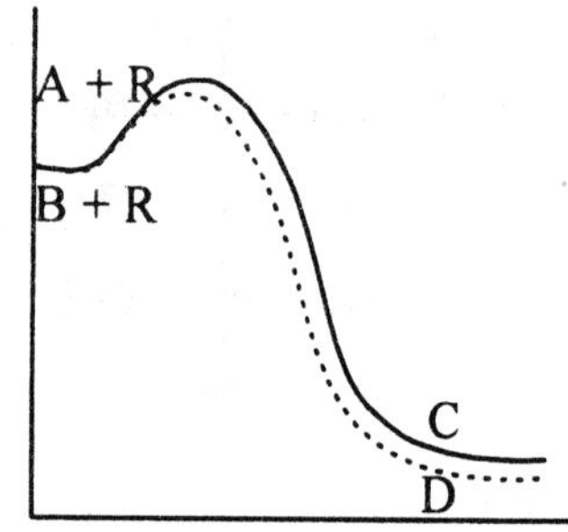

13. An aqueous solution containing A and B shows optical activity. A and B are stereomers. Which of the following possibilities cannot be correct?
(a) A has two chiral centers but B does not have any, because it has a symmetry plane.
(b) Both A and B contain at least one chiral carbon.
(c) A and B are distereomers.
(d) A and B are not presents in equal amounts.

14. Fill in the missing words in the passage below with the words from the provided list to preserve the best meaning of the paragraph. Not all words have to be used. Some words may be used more than once: Compounds that have the same atom connectivity, but differs from each other only in how their atoms are arranged in space are known as ———a———. These isomers may be classified into ———b——— isomers, which are interconvertible by rotation around single bonds, and ———c——— isomers, that have stereogenic centers and are not interconvertible by rotations around single bonds. The isomers that belong to the latter group can be either ——— d——— , which are related to each other as nonsuperimposable mirror images, or ——e——, which include all other ———f———. The last subgroup also include E/Z isomers. Except for optical activity, ——g—— have identical physical and chemical properties. In contrast ———h——— have different properties, and may not show optical activity, even if pure.

(a) enantiomers (b) constitutional
(c) distereomers (d) chiral
(e) racemic (f) conformational
(g) configurational (h) isomer
(i) stereomers

15. Which of the following is the enantiomer of the compound shown below?

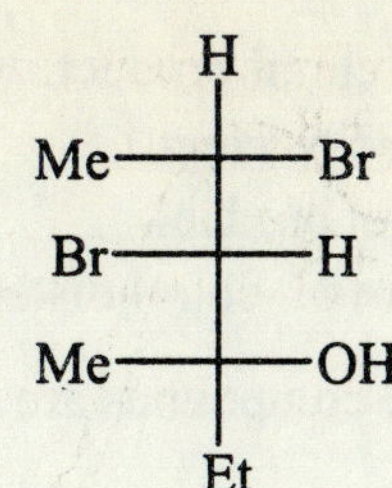

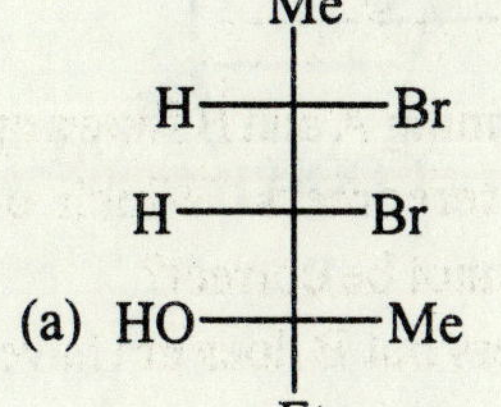

(b)

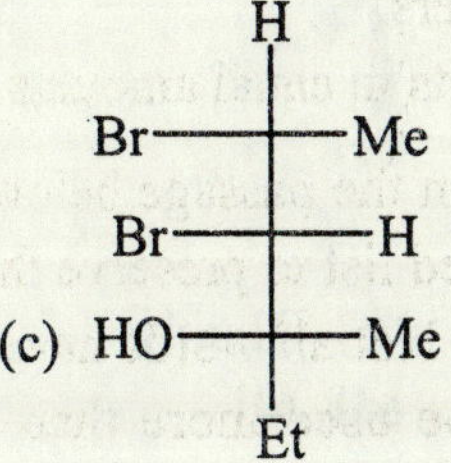

(d)

16. How many stereomers are possible for the following molecules?

COOH

(a) 2 (b) 4 (c) 6 (d) 8

17. Which of the following molecules is/are chiral?

(A) (B) (C)

(a) A, B (b) A, C (c) B, C (d) All

18. The following compounds are related as:

(a) Distereomers (b) Enantiomers
(c) Meso (d) Identical

19. Which of the following compounds might you use to resolve a racemic mixture of carboxylic acid?

(a) (b) (c) (d)

20. Three of the following Fischer's projections are identical but one. Select the structure which is different from others?

(a) (b) (c) (d)

21. Which of the following isomeric structures have lowest energy?

(a) (b) (c)

(d)

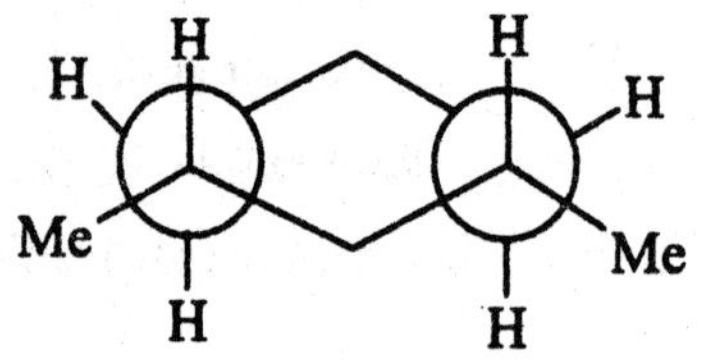

22. What is the lowest energy conformation of butane?

(a) 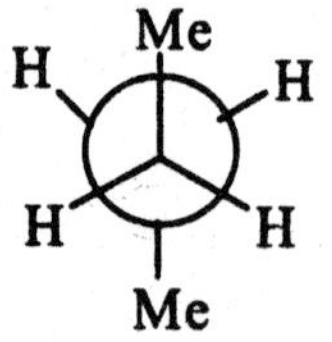 (b)

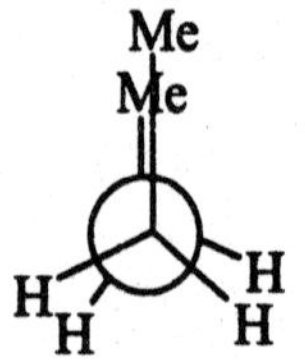

(c) 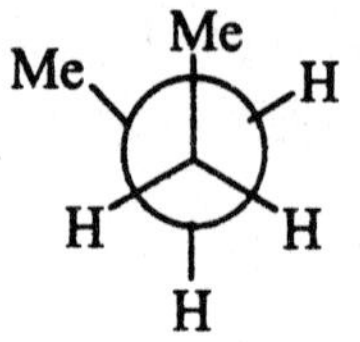 (d)

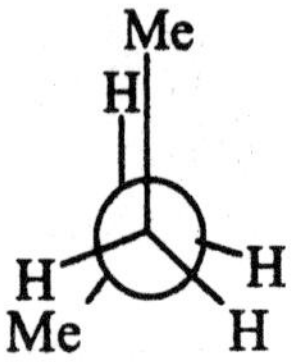

23. Alkylation of racemic (1) with allyl bromide provides:

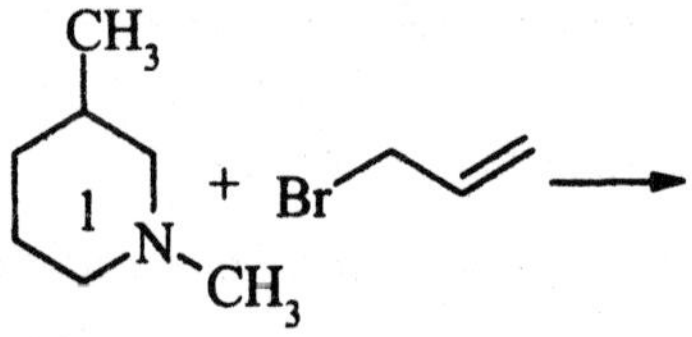

(a) Two achiral ammonium salts.
(b) Two enantiomeric ammonium salts.
(c) Two chiral distereomeric salts.
(d) Two racemic, distereomeric salts.

24. Which of the following amines is optically active?

(a) 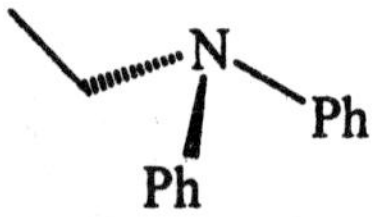 (b)

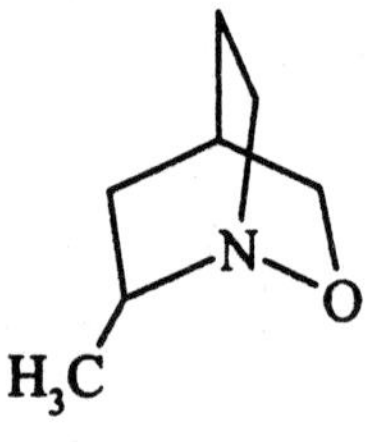

(c) 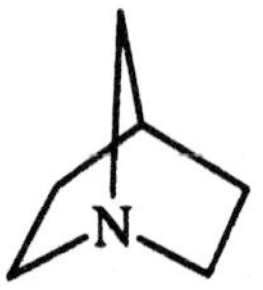 (d)

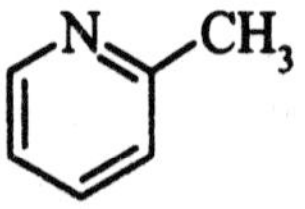

25. Reduction of a chiral carbohydrate (1) with $NaBH_4$ provides:

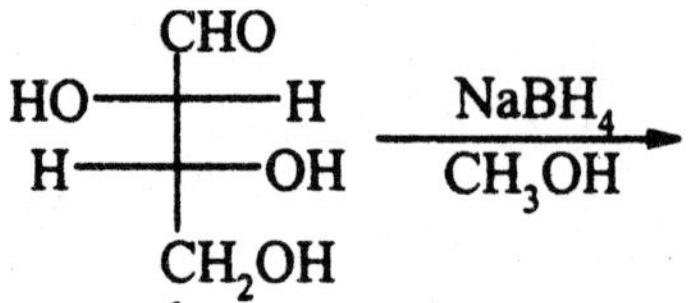

$\xrightarrow[CH_3OH]{NaBH_4}$

(a) A single chiral product.
(b) A racemic product.
(c) An achiral product.
(d) A mixture of chiral distereomeric product.

26. Which of the compounds are chiral?

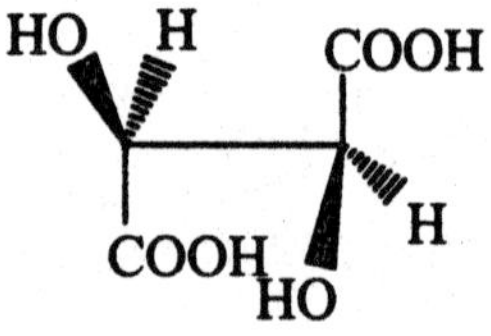

I 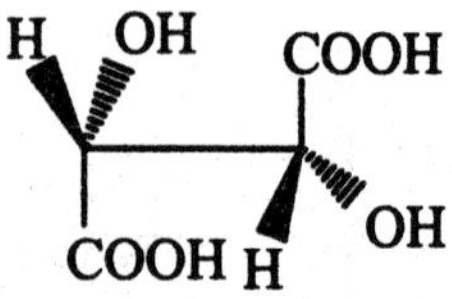 II

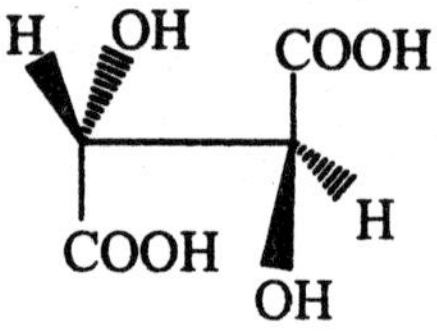

III

(a) I only (b) III only
(c) I and II only (d) All

27. On Pluto, where everything is frozen, astronauts discovered two forms of 1, 2-dibromoethane: gauche and anti. Assuming that there are no rotations around single bonds, which statement about the two forms is correct?

(a) They are enantiomers.
(b) They are distereomers.
(c) They are meso compounds.
(d) Both will show optical activity.

28. Which of the following will show optical activity?

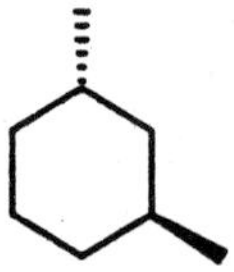

(A) 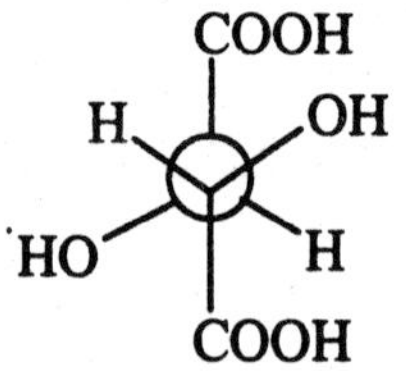 (B) 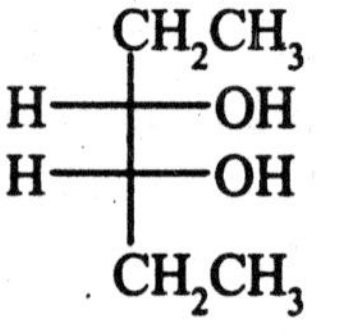 (C)

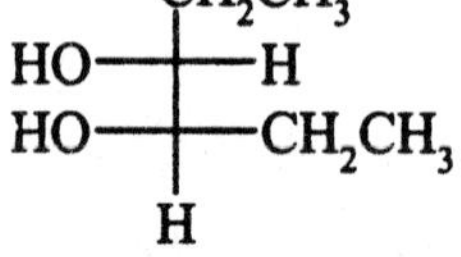

(D) 50/50 mix of C and D (E)

(a) A, D and E (b) A and E
(c) All (d) B, C and E

29. Which of the following is the most stable form of 1, 2, 4-trimethyl cyclohexane?

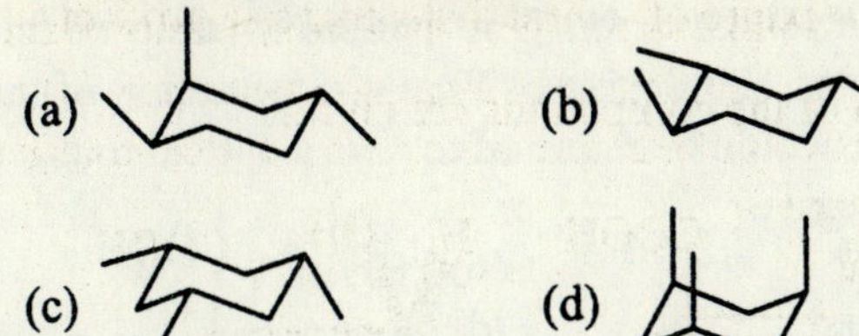

30. How many stereomers of the product will form in the reaction?

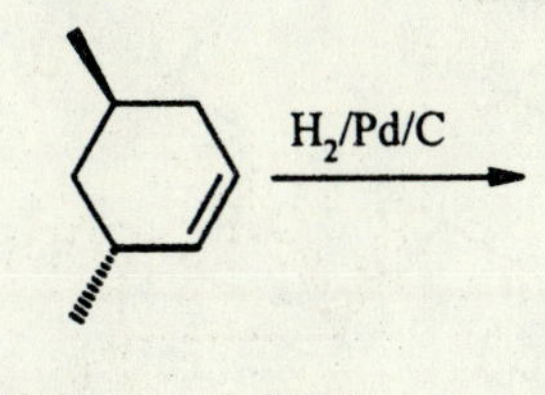

(a) 1 (b) 2 (c) 4 (d) 8

31. Which of the following molecules have dipole moment?

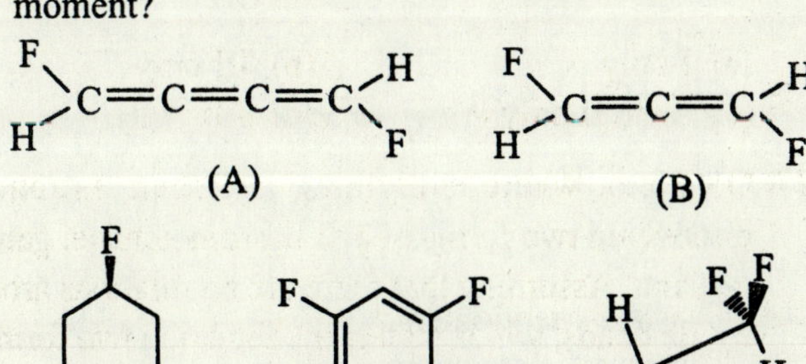

(a) A and D (b) B and C
(c) C and E (d) B and E

32. A solution of a compound A showed [a] = + 17°. Which of the following conclusion must be true?
(a) Compound A contains at least one chiral center.
(b) Compound A cannot be a mixture of enantiomer.
(c) Compound A is a distereomer.
(d) A and its mirror image are not superimposable.

33. Which of the following will show optical activity?

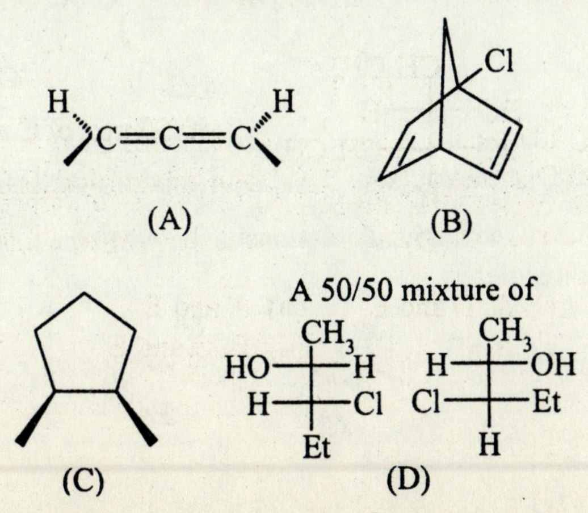

(a) All (b) A and D
(c) A and B (d) A and C

34. What is the relationship between the molecules in the following pairs?

CH_3 / HO—H / HO—Cl / COOH and HO, OH, H_3C—COOH, H, Cl

(A)

Cl, OH, Cl, CH_3, H_2N, CH_3 and Cl, Cl, OH, H_2N, CH_3, CH_3

(B)

(a) Enantiomer-Enantiomer.
(b) Distereomers-Enantiomer.
(c) Enantiomer-Distereomers.
(d) Distereomers-Distereomers.

35. Following pair of molecules are related as:

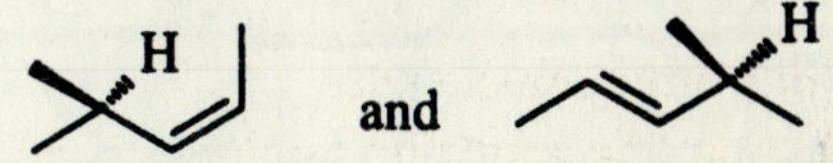

(a) Enantiomers (b) Distereomers
(c) Identical (d) Constitutional isomers

36. Which of the following posses a plane of symmetry?

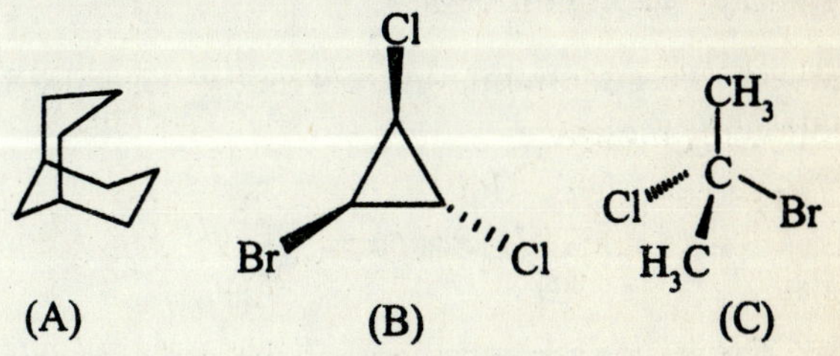

(a) A only (b) B only
(c) B and C (d) A and C

37. Which of the following molecule is expected to rotate the plane polarized light?

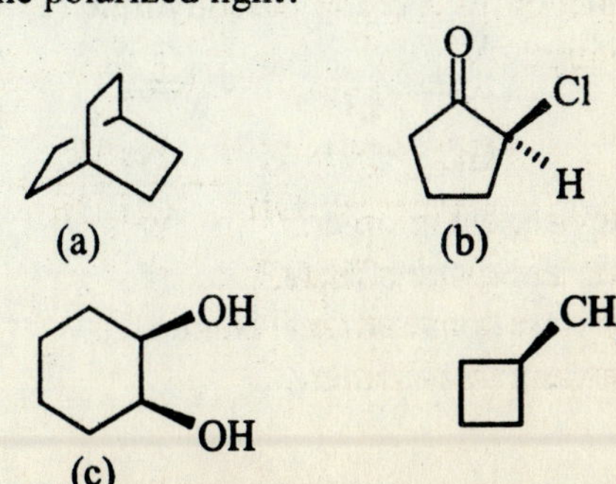

38. Which of the following statement is true?
    (a) Distereomers can usually be separated on the basis of their physical properties.
    (b) Enantiomers can usually be separated on the basis of their physical properties.
    (c) The dextrorotatory isomer of a pair of enantiomers has higher melting point.
    (d) The levorotatory isomer of a pair of enantiomers has higher melting point.

39. What is the relationship between the following two molecules?

    (a) They are non-isomeric.
    (b) They are identical.
    (c) They are constitutional isomers.
    (d) They are enantiomers.

40. The following molecules are:

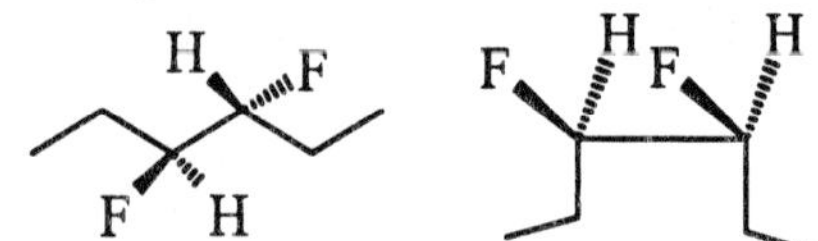

    (a) They are enantiomers.
    (b) They are distereomers.
    (c) They are constitutional isomers.
    (d) They are conformers.

41. Select the statement that correctly describe the following molecule:

    (a) There is no stereocenters and molecule is achiral.
    (b) There is one chiral center and molecule is achiral.
    (c) There are two chiral center and molecule is achiral.
    (d) There is no stereocenter and molecule is chiral.

42. The relationship between the following molecules?

    (a) They are enantiomers.
    (b) They are distereomers.
    (c) They are constitutional isomers.
    (d) They are non-isomers.

43. Suppose you have a solution of a chiral molecule at a concentration of 2 g/mL in a polarimeter sample cell with path length 1 dm. The optical rotation of this molecule is +30 degrees. What concentration of the molecule would be required to give a optical rotation of +60 degree?
    (a) 1 g/ml (b) 15 g/ml
    (c) 30 g/ml (d) None

44. Which of the following represents a meso compound?

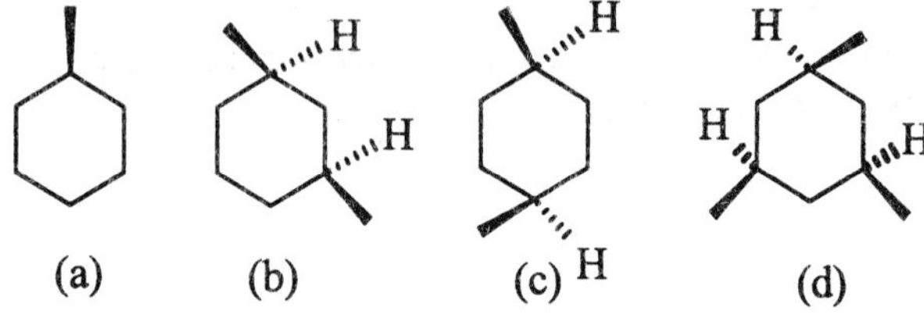

    (a) (b) (c) (d)

45. The following two molecules are related as:

    (a) They are enantiomers.
    (b) They are distereomers.
    (c) They are identical.
    (d) They are achiral.

46. The stereochemical relationship between the following molecules is:

    (a) Identical (b) Enantiomers
    (c) Distereomers (d) Conformers

47. The stereochemical relationship between the following molecules is:

    (a) Identical (b) Enantiomers
    (c) Distereomers (d) Constitutional isomer

48. The stereochemical relationship between the following molecules is:

(a) Identical (b) Enantiomers
(c) Distereomers (d) Constitutional isomer

49. The stereochemical relationship between the following molecules is:

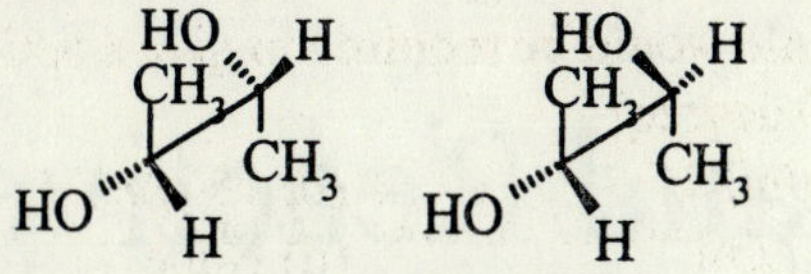

(a) Identical (b) Enantiomers
(c) Distereomers (d) Conformers

50. Which of the following structures below represents pair of enantiomers?

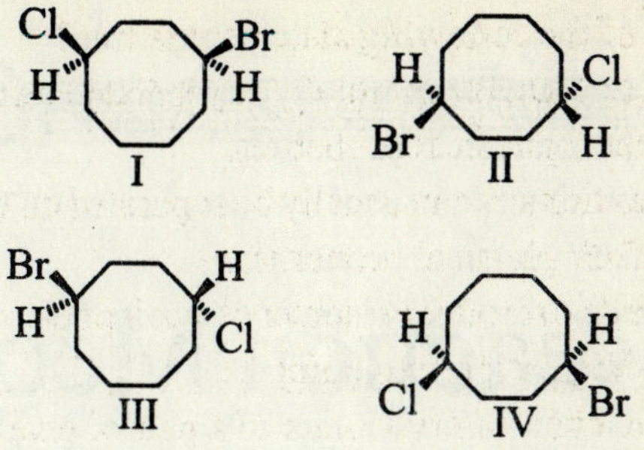

(a) I + II and III + IV (b) I + IV and II + III
(c) I + III and II + IV (d) Only I + IV

Chapter 3

Reaction Mechanism and Reactive Intermediates

3.1 BONDING AND HYBRIDIZATION IN ORGANIC COMPOUNDS

Most of the organic compound consists of covalent bonds, which is formed by sharing of electrons by the atoms forming chemical bonds. The simplest way to symbolize the bonding in a covalent molecule is to use Lewis-structure. In Lewis structure, each valence electron is denoted by a dot and a bonding pair of electrons is denoted by a dash (—) as shown below for methane and ethane.

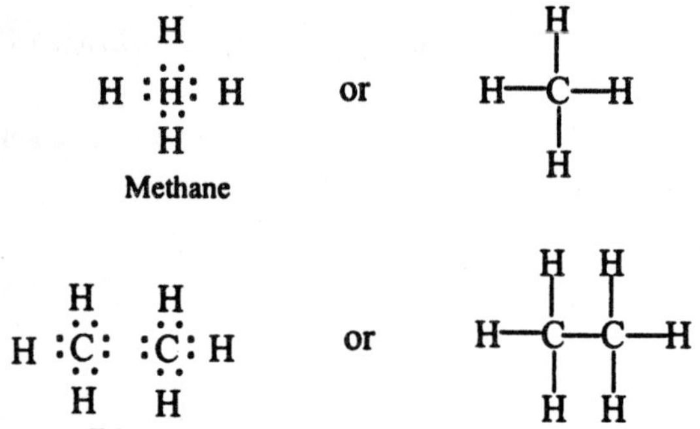

H :H: H (with H above and below) or H—C—H (with H above and below)

Methane

H :C: :C: H (with H above and below each C) or H—C—C—H (with H above and below each C)

Ethane

3.1.1 Multiple Bonds

In the above mentioned example, adjacent atoms are sharing one-one electron forming sigma (σ) covalent bonds. Many molecules contains adjacent atoms sharing two-two or three-three electrons forming double or triple bonds as:

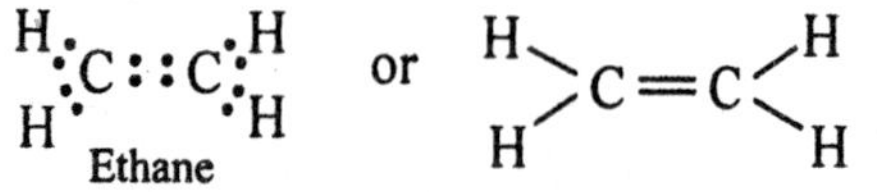

$H_2C::CH_2$ or $H_2C{=}CH_2$

Ethane

H:C::C:H or H—C≡C—H

Ethyne

It is also known to us that there cannot be more than one sigma bond between two adjacent atoms, it follow that in multiple bonds, there is one sigma and remaining are π-bonds.

3.1.2 Hybridization

As we know from the concepts of chemical bonding that only sigma bonds and unshared pair (Lone Pair) of electrons contribute to hybridization of an atom: hybridization in some common organic molecules.

Molecules	No. of (σ-bonds + LP)	Hybridization
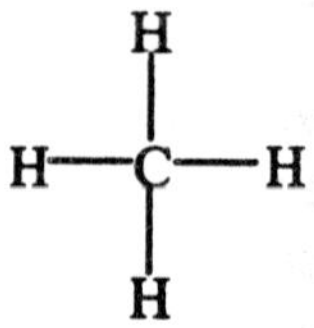 CH_4 (H—C—H with H above and below)	4	sp^3
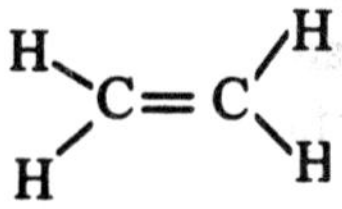 $H_2C{=}CH_2$	3	sp^2
H—C≡C—H	2	sp
$CH_3—\ddot{N}H_2$	3 + 1 = 4	sp^3
$R—\ddot{O}—H$	2 + 2 = 4	sp^3
 >C=O	3	sp^2

3.1.3 Electronegativity and Bond Polarity

A bond with the electrons shared equally between the two atoms is called non-polar bond, e.g. H_2, C—C in ethane, etc. In most bonds, between two different elements, the bonding electrons are attracted more strongly towards more electronegative atom. Such an unequally shared pair of bonding electrons is called a polar bond. The bond polarity is measured by its dipole moment (μ), defined to be the amount of charge separation multiplied by the bond length.

Greater the difference in electronegativities of elements constituting the bond, greater will be the magnitude of dipole moment. Direction of dipole vector is shown with arrow heading towards more electronegative atom, e.g.

$$\overrightarrow{CH_3-Cl}$$
$$\delta^+ \quad \delta^-$$

Example 1. Use electronegativity to predict the direction of dipole moments of the following bonds and arrange them in increasing order of their magnitude of dipole moment:

(a) C—Cl, C—O, C—N, C—S, C—B
(b) N—Cl, N—O, N—S, N—B, B—Cl

Solution

(a) C – B < C—S < C—N < C—C < C—O

(b) N—Cl < N—O < N—S < N—B < B—Cl

3.1.4 Effect of Dipole

Greater the magnitude of dipole moment, greater will be the force of attraction between the atoms forming this bond and therefore higher will be the bond energy and smaller will be the bond-length.

Electronegativity of an element in different hybridization state: A sp hybridized element is more electronegative than a sp^2 and a sp^2 element is more electronegative than a sp^3 atom. A very important consequence of this fact is C—H bond in acetylene is stronger than the same in ethylene and the same in ethylene is stronger than in ethane:

H—C≡ > H—C= > H—C< B·E

and bond length follow reverse order.

3.2 INTERMOLECULAR FORCES

When two melecules approaches, they attract or repel each other. In general, the forces are attractive until the molecules come so close that they infringe on each other's atomic radius. When this happens, the small attractive forces quickly becomes a large repulsive force and the molecules "bounces" off each other. Attractive forces are particularly important in solids and liquids. The melting points, boiling points and solubility of organic compounds show the effect of these forces. Three major kind of attractive forces prevelent cause their molecules to associate into solids and liquids.

3.2.1 Dipole—Dipole Forces

Each molecular dipole moment in a polar molecule has a positive end and a negative end. Molecules having a permanent dipole moment orient themselves in the more stable positive-to-negative end arrangement and an electrostatic force of attraction exist among such molecules as shown below.

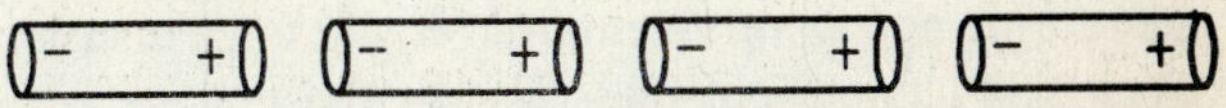

(– +) (– +) (– +) (– +)

This attraction must be overcome when the liquid vapourizes. Thus, a larger heat of vapourization is needed for strongly polar molecules and they have relatively high boiling points.

Example 2. CH_3F has higher boiling point than CH_3Cl. explain.

Solution: F is more electronegative than Cl, therefore C—F bond is more polar than C—Cl bond and attractive forces between the negative—positive ends of dipole is more in CH_3F than in CH_3Cl and former require higher heat of vapourization.

Example 3. Which is more polar NH_3 or NF_3?

Solution: NH_3 is more polar than NF_3, since fluorine is more electronegative than nitrogen as well as magnitude of dipole moment of N—F bond is smaller than magnitude of N—H dipole due to smaller electronegativity difference in the former case.

Example 4. Which has greater dipolemoment CH_2Cl_2 or $CHCl_3$?

Solution: CH_2Cl_2 has greater dipole moment as described in Figure 3.1.

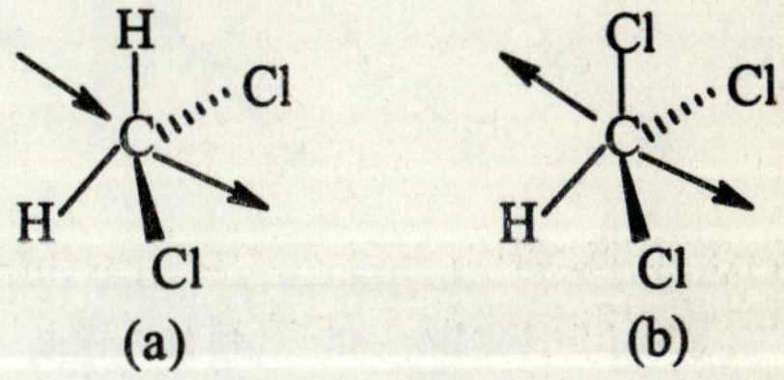

Figure. 3.1: Resultant dipole moment of (a) dichloro methane is larger than that of (b) trichloromethane

3.2.2 The London Dispersion Force or Vanderwaals Force of Attraction

In non-polar molecules, the principal attractive force is London force. It arises from temporary dipole moments that are induced in a molecule by other nearby molecules, e.g. CCl_4 has no permanent dipole moment, instead electrons are not always evenly distributed. A small temporary dipole moment is induced, when one molecule approaches other molecules, in which electrons are slightly displaced from a symmetrical arrangement. The electrons in the approaching molecules are slightly displaced so that an attractive dipole-dipole interaction result. This attractive forces become significant on close surface contact of two molecules. So it is roughly proportional to the molecular surface area. CCl_4 has a larger surface area than chloroform (a chlorine atom is much larger than a hydrogen atom), and therefore intermolecular force of attraction between CCl_4 molecules are stronger than they are between chloroform molecules. A very important effect of this London force can be seen in the boiling points of hydrocarbon. Increasing molecular weight increases their boiling points due to increased area of contact. Among isomeric alkanes, branching decreases the area of contact, thus branched alkanes have lower boiling point than linear chain isomeric, alkane, e.g. *n*-pentane has higher boiling point than isopentane and isopentane has higher boiling point than neopentane.

3.2.3 Hydrogen Bonding

A hydrogen bond is not a true bond but a particularly strong form of dipole-dipole attraction. A hydrogen atom can participate in hydrogen bonding if it is bonded to oxygen, nitrogen or fluorine. In hydrogen bond, hydrogen acts as a bridge between two strongly electronegative atom, viz. O, F or N as in case of H_2O, ROH, shown in Figure 3.2.

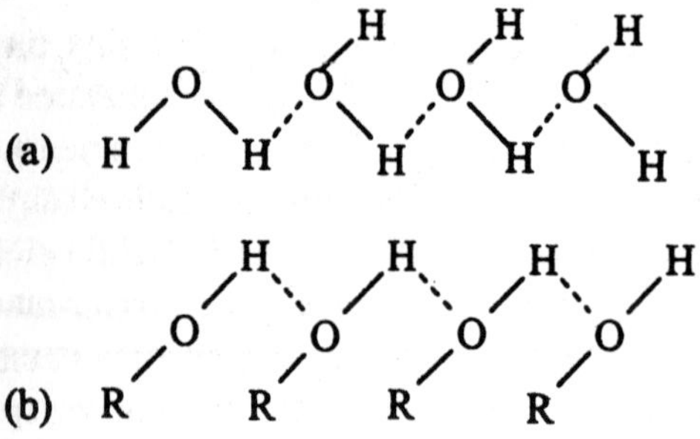

(a) H–O–H···O(H)–H···O(H)–H···O(H)–H

(b) R–O–H···O(R)–H···O(R)–H···O(R)–H

Figure. 3.2: H–bonding in (a) water, (b) alcohols. Here dotted lines showing H-bonds

H-bonding has large effect on the physical properties. Molecules exhibiting the phenomena of H-bonding have higher boiling point than the molecules of almost same molecular weight lacking H-bonds, e.g. ethanol has very high boiling point (78°C) whereas isomeric dimethyl ether has very low (–25°C) boiling point. The extent of H-bonding decreases with increase in molecular weight due to increased contribution of hydrocarbon part. H-bond has also very significant effect on solubility. Solubility of a solute depend on solute-solvent interaction. A molecule having H-bonded to either F, O or N may involve in H-bonding with solvent like H_2O, NH_3, etc. as

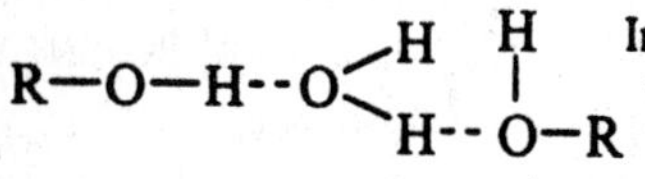

R—O—H··O(H)—H··O(H)—R

Interaction of solvent with solute through H-bonds

Interaction of solute alcohol with solvent water is responsible for fairly high solubility of alcohols in water. An increase in molecular weight of alcohols decreases its solubility.

Example 5. Arrange the following amines in increasing order of their boiling points and solubility in water.

(a) Trimethylamine
(b) N-methylethylamine
(c) Propylamine.

Solution: Trimethylamine has no H-bonded to N and therefore does not exhibit phenomena of H-bonding. N-methylethylamine has one H-bonded to N, and can involve in H-bonding with other amine molecule. Propylamine has two H-bonded to N and can involve in extensive H-bonding with other amine molecule. Thus, the order of their boiling points are:

(a) < (b) < (c) and same will be the order of their solubility in water due to the same reason.

Intermolecular and Intramolecular H-bonds

Hydrogen bonding is classified into two category as *intermolecular* and *intramolecular* hydrogen bonding. If H-bond is formed between hydrogen from one molecule and oxygen or other electronegative atoms from different molecule (similar or different) then the bond is known as intermolecular hydrogen bonds as described below in Figure 3.3.

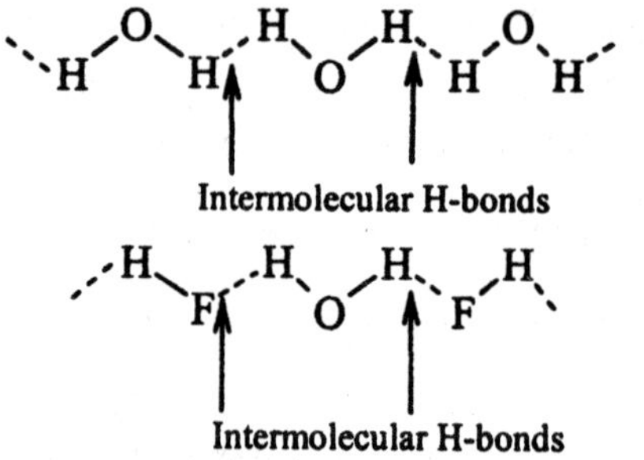

Intermolecular H-bonds

Intermolecular H-bonds

Figure. 3.3: Intermolecular H-bonds

On the otherhand if H-bond is formed bewteen H and electronegative atom from the same molecule, the bond is known as intramolecular H-bond as shown in Figure 3.4.

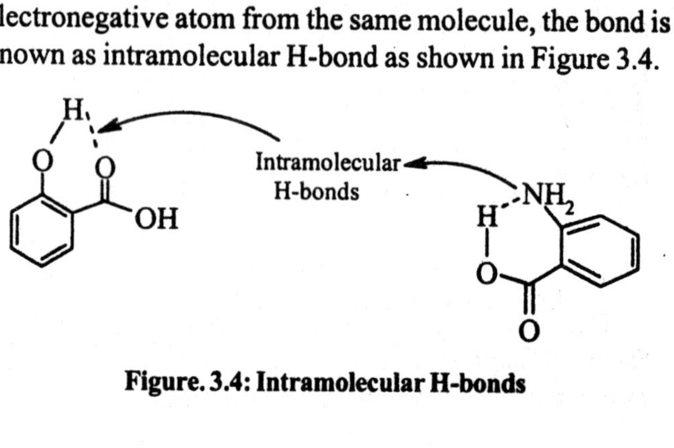

Intramolecular H-bonds

OH

NH_2

Figure. 3.4: Intramolecular H-bonds

Example 6. Show the two different types of intermolecular H-bonds possible between methanol and methylamine and explain, which will be the stronger H-bond.

Solution

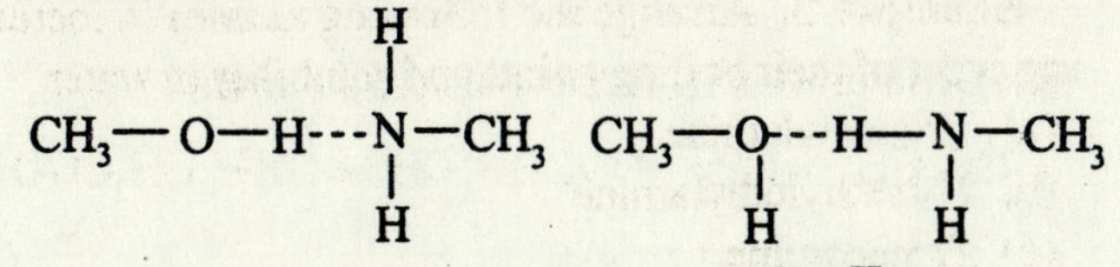

$$CH_3-O-H\cdots N(H)(H)-CH_3 \quad (I) \qquad CH_3-O(H)\cdots H-N(H)-CH_3 \quad (II)$$

II is stronger bond than I since H is bonded to more electronegative oxygen in II.

3.3 ELECTRONIC DISPLACEMENT IN ORGANIC COMPOUNDS

3.3.1 Induction or Inductive Effect

Inductive effect is the polarization of shared pair (covalently bonded pair) of electrons towards more electronegative atoms. Let us consider a molecule of ethane:

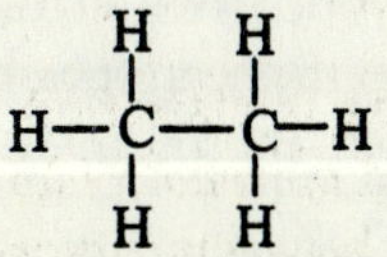

$$H_3C-CH_3$$

In this molecule, the two electrons involved in bonding between two carbons are centrally located, i.e. both the carbons have equal hold on this shared pair of electrons and no polarization of electron density is observed. This is due to the reason that both carbons have equal electronegativity, i.e. equal ability to attract the shared pair of electrons. Now consider another molecule, chloromethane:

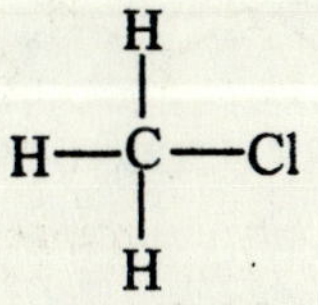

$$H_3C-Cl$$

In chloromethane, the shared pair of electrons between carbon and chlorine are attracted more towards chlorine since it has greater electronegativity. This results in polarization of shared pair of electrons and results in development of slight greater electron density around chlorine atom making the carbon slightly electron deficient. This observation can be represented as:

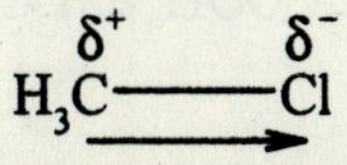

$$\overset{\delta^+}{H_3C} \xrightarrow{\quad} \overset{\delta^-}{Cl}$$

This effect of polarization of shared pair of electron in favour of more electronegative atom is known as *inductive effect*. The magnitude of induction depends on electronegativity difference between the atoms involved. Greater the electronegativity difference, stronger will be the induction.

The inductive effect is *permanent effect* since it depends on the permanent property of atoms "Electronegativity".

(+) and (–) Inductive Effect

If an atom or group donates (pushes) shared pair of electrons towards carbon then this group is said to be causing (+) Inductive effect (+I effect), and if an atom or groups withdraw electrons away from carbon then this group is said to be causing (–I) Inductive effect (–I effect). Now let us consider the following molecule:

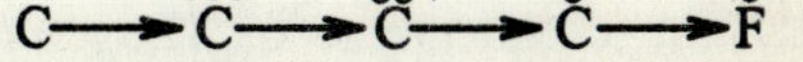

$$C \rightarrow \overset{\delta\delta\delta+}{C} \rightarrow \overset{\delta\delta+}{C} \rightarrow \overset{\delta+}{C} \rightarrow \overset{\delta-}{F}$$

In the above example, —F is withdrawing electrons away from the carbon, this is causing —I effect. Due to electron withdrawl by F, C–1 has become slightly electron deficient and would tend to acquire some electron density from bonds other than —C—F bond. As a result C—1 carbon withdraw electron density from C1—C2 bond. Here C—2 carbon is acting as electron donar (+I effect causing) group. Electron donation by C—2 to C—1 makes C—2 slightly electron deficient, but this time electron deficiency at C—2 is relatively less compared to the earlier electron deficiency at C—1. To compensate the electron-deficiency, C—2 withdraw some electron density from C2—C3 bond making C—3 slightly electron deficient and this process continue to the next terminal of the chain. The effect of polarization decreases drastically as we move away from the source atom or groups, responsible for induction.

Salient features of this effects

- Inductive effect operates through σ bonds.
- There occurs partial delocalization (movements) of shared pair of electrons.
- This is a very weak effect.
- This effect is additive in nature, i.e. two electronegative atoms exert greater induction effect than one atom of greater electronegativity from the same position:

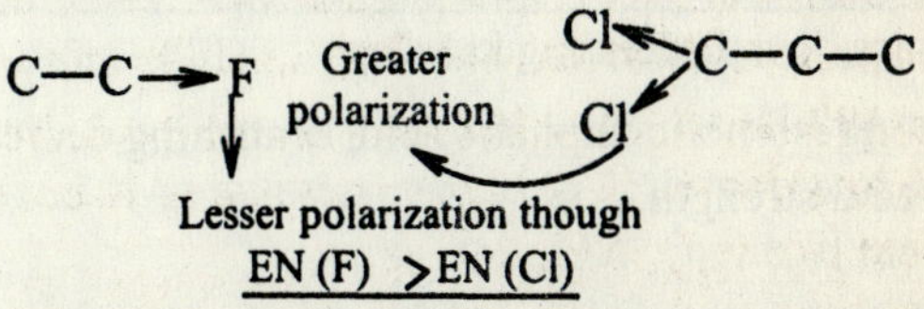

$$C-C\rightarrow F \qquad \text{Greater polarization} \qquad (Cl)_2C-C-C$$

Lesser polarization though EN (F) > EN (Cl)

- Some common +I and –I effect causing groups with their relative magnitude of induction are:

  + I effect causing group

$-O^- > -CO_2^- > R_3C- > R_2CH- > RCH_2- > -CH_3$

–I effect causing group

$-N^+R_3 > -SR_2^+ > -N^+H_3 > -NO_2 > -SO_2R > -CN >$
$-COOH > F > Cl > Br > I > OR > OH$

Applications of Inductive Effect

Among the various application of this inductive effect, the important are:

(i) Determining relative strength of acids and bases.

(ii) Determining relative stability of reactive intermediates formed during a chemical reactions.

Inductive Effect and Acid-Base Strength

Acid strength: A protic acid is defined as a compound which produces H^+ in a chemical reaction. The general mechanism of ionization of an organic acid can be represented as:

$$A-\overset{O}{\overset{||}{C}}-O-H \rightleftharpoons A-\overset{O}{\overset{||}{C}}-O^- + H^+$$

How does A influences the above ionization process-determines the relative acid strength. An atom or groups, which helps in withdrawing shared pair of electrons of O—H bond by oxygen and stabilize the product anion (conjugate base) will increase the acid strength and vice-versa. If A is –I effect causing group, it will help in ionization as well as stabilize the carboxylate ion by withdrawing electrons from negatively charged oxygen. Thus, any –I effect causing (A) will increase the acid strength by increasing ionizing power. In contrast, +I effect causing A will decrease the electron withdrawing ability of oxygen.

In general, electron withdrawing group bonded to –COOH increases the acid strength and electron donating group bonded to –COOH decreases the acid strength.

Acid Base Strength of Aliphatic and Substituted Aliphatic Acid

Aliphatic acid: R—COOH

(i) Increasing the size of R-group increases the +I effect, decreases the acid strength.

$$HCOOH > CH_3COOH > CH_3CH_2COOH > CH_3CH_2CH_2CHOOH\cdots$$

Decreasing ka values →

(ii) Among isomeric aliphatic acid, branching decreases the acid strength.

$$CH_3CH_2CH_2CH_2COOH > CH_3\underset{}{\overset{CH_3}{\overset{|}{C}H}}-CH_2-COOH > CH_3-CH_2-\overset{CH_3}{\overset{|}{C}H}-COOH \quad H_3C-\underset{CH_3}{\underset{|}{\overset{CH_3}{\overset{|}{C}}}}-COOH$$

Decreasing ka values →

(iii) Unsaturation increases the acid strength:

$$CH_3CH_2-CH_2-COOH < CH_2{=}CH-CH_2COOH$$
$$HC{\equiv}C-CH_2COOH$$

(vi) Substitution of H by electronegative atoms increases the acid strength:

$$CH_3COOH < ClCH_2COOH < Cl_2CHCOOH < Cl_3CCOOH$$

(v) Substitution by different electronegative atoms at the same position-greater the electronegativity stronger the acid:

$$\underset{I}{\underset{|}{C}H_2}-COOH < \underset{Br}{\underset{|}{C}H_2}COOH < \underset{Cl}{\underset{|}{C}H_2}COOH < \underset{F}{\underset{|}{C}H_2}COOH$$

(vi) Substitution by different atoms at different positions: Substitution at the nearest carbon has greatest effect on acid strength, irrespective of electronegativity:

$$CH_3\underset{Br}{\underset{|}{C}H}COOH > \underset{F}{\underset{|}{C}H_2}COOH > F_2CHCH_2COOH$$

Acid strength

$$H_3C-\underset{F}{\underset{|}{\overset{Cl}{\overset{|}{C}}}}HCOOH > \underset{F}{\underset{|}{C}H_2}\underset{F}{\underset{|}{C}H}COOH > CF_2HCH_2COOH$$

(vii) Stronger -I- effect causing group has greater effect on increasing acid strength from the same position.

$$CH_3\underset{NO_2}{\underset{|}{C}H}COOH > CH_3\underset{F}{\underset{|}{C}H}COOH > \underset{F}{\underset{|}{\overset{NO_2}{\overset{|}{C}}}}HCH_2COOH > \underset{F}{\underset{|}{\overset{F}{\overset{|}{C}}}}HCH_2COOH$$

Example 7. Arrange the following acids in increasing order of their acid strength.

(a) $CH_3CH_2CH_2COOH;\ CH_3-\overset{CH_3}{\overset{|}{C}H}COOH$
$CH_3CH_2CHOOH;\ CH_2{=}CHCH_2COOH$

(b) $CH_3CH_2CH_2COOH;\ CH_3\underset{F}{\underset{|}{C}H}\underset{F}{\underset{|}{C}H}COO;$

$CH_3\underset{F}{\underset{|}{C}H}-\underset{NO_2}{\underset{|}{C}H}-COOH;\ CH_3\underset{NO_2}{\underset{|}{C}H}-\underset{F}{\underset{|}{C}H}-COOH;$

(c) 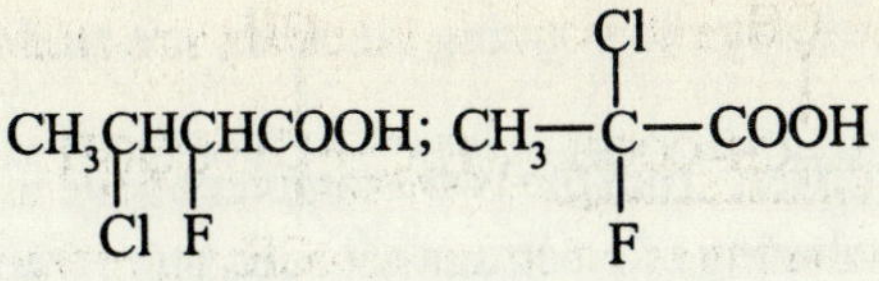 $CH_3CH(Cl)CH(F)COOH$; $CH_3-C(Cl)(F)-COOH$

Inductive Effect and Basic Strength

Basic strength of ammonia and its derivatives: Ammonia and its alkyl and aryl derivatives are nucleophile (a Lewis base) because its lone pair of non-bonded electrons can form a bond with an electrophile. An amine can also act as a Bronsted-Lowry base by accepting a proton from proton acid.

Reaction of an amine as a nucleophile:

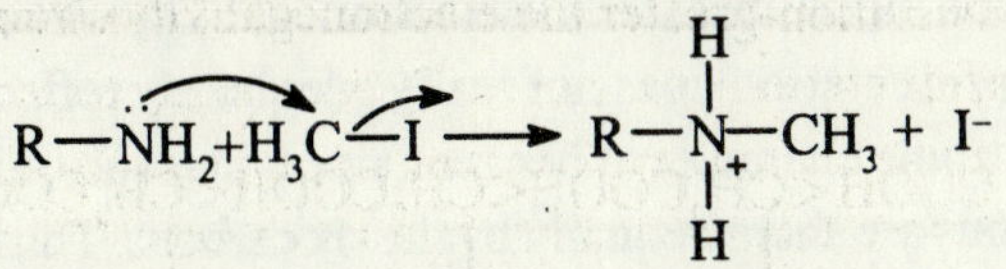

$$R-\ddot{N}H_2 + H_3C-I \longrightarrow R-\overset{+}{N}H_2-CH_3 + I^-$$

Reaction of an amine as a proton base:

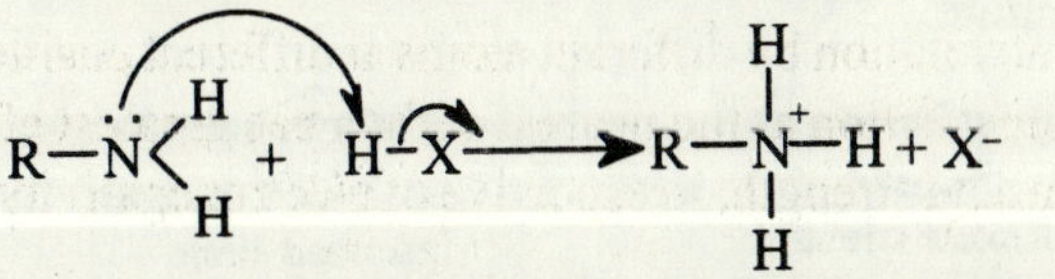

$$R-\ddot{N}H_2 + H-X \longrightarrow R-\overset{+}{N}H_3 + X^-$$

Because amines are fairly strong base, their aqueous solutions are basic. An amine can abstract a proton from water giving an ammonium ion and a hydroxide ion as:

$$R-NH_2 + H-O-H \rightleftharpoons R-\overset{+}{N}H_3 + H\bar{O}$$

Figure 3.5 describe the energy change during the reaction of amines with water:

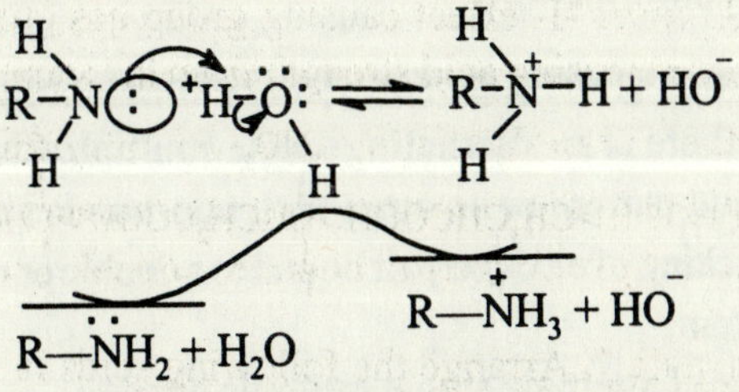

$$R-\ddot{N}H_2 + H-\ddot{O}-H \rightleftharpoons R-\overset{+}{N}H_3 + H\bar{O}$$

$R-\ddot{N}H_2 + H_2O$ → $R-\overset{+}{N}H_3 + H\bar{O}$

Figure. 3.5: Potential energy diagram of the base dissociation reaction of amine

Any structural feature that stabilizes the ammonium ion shift the reaction towards right, making the amine stronger base. Any feature that stabilizes the free amine shift the reaction towards left making amine-a weaker base. Stabilization of ammonium ion can be done by electron donating group either through inductive effect or by resenance effect.

Stabilization through induction: Alkyl groups are electron donating group through induction, they stabilize the charge on nitrogen in ammonium ion and therefore increases the basic strength. Thus, methyl amine ($pk_b = 3.36$) is a stronger base than ammonia ($pk_b = 4.47$).

On this basis, secondary amines is expected to be stronger base than a primary amine and a tertiary amine further stronger than both primary and secondary amine.

The actual situation is little more complex because of solvation effect. Ammonium ion is solvated by water molecule by means of H-bonds lowering potential energy of ammonium ion as shown in Figure 3.6.

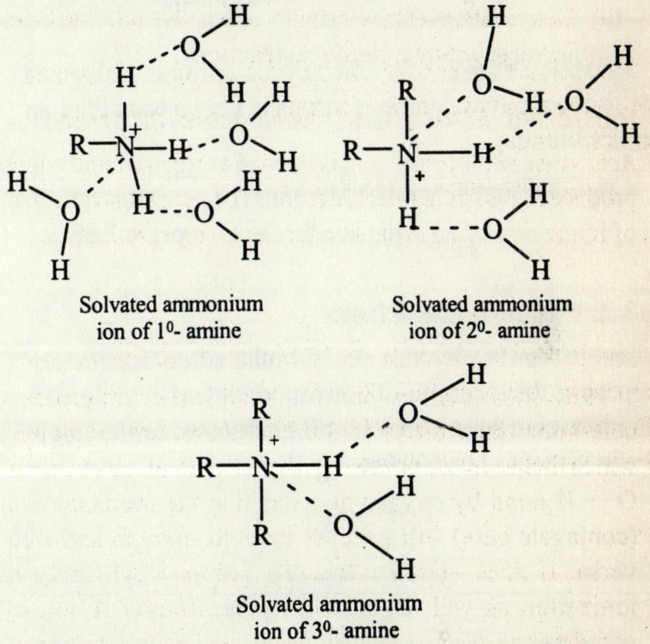

Solvated ammonium ion of 1°- amine

Solvated ammonium ion of 2°- amine

Solvated ammonium ion of 3°- amine

Figure. 3.6: Solvated ammonium ions of different degree

As can be seen above, substitution of H-by alkyl group increases the stabilization effect by induction, but decreases stabilization by solvation due to lesser number of water molecule attached with the ammonium ion of higher degree amine. Thus the two factors are opposing. Up to secondary amine inductive effect is dominating but in tertiary amine, solvation effect is more important and due to decreased solvation, its basic strength is decreased even from primary amine. Thus, in aqueous solution relative basic strength of amines are:

Secondary > Primary > Tertiary > NH_3

In gas phase, where there is no stabilization through solvation, the order of basic strength is solely determined by inductive effect and thus:

3° > 2° > 1° > NH_3 Basic strength in gas phase.

Hybridisation effect on basic strength: The relative electronegativity of carbon with different hybridization is

$$sp—c > sp^2—c > sp^3—c$$

Thus, basic strength of amine also depends on hybridization of carbon bonded to $—NH_2$ bonded to sp^3 carbon is more basic than $—NH_2$ bonded to sp^2 carbon and —NH_2 bonded to sp—c is less basic to both —NH_2 bonded to sp^3 and sp^2 carbon.

Basic Strength

(i) $R-CH_2-NH_2 > R-CH=CH-NH_2 > R-C\equiv C-NH_2$

(ii) Pyridine N: < piperidine N–H

sp^2—N sp^3—N

Relative strength of cyclic/acyclic amines: If degrees are same, a cyclic amine is stronger Lewis base than an acyclic amine.

$CH_3CH_2CH_2-NH-CH_2CH_2CH_3$ ($pk_b = 3$) < pyrrolidine N–H ($pk_b = 2.88$)

3.3.2 Resonance Effect

A single Lewis electron dot formula often adequately represents the locations of bonding electrons in molecule. For instance, ethyl alcohol is well represented by the single Lewis formula shown below:

$$H_3C-CH_2-\ddot{O}-H$$

(Lewis formula of ethanol)

There are, however, molecules whose electronic distribution is inadequately represented by a single Lewis formula. In these case two or more Lewis formulas may be necessary to satisfactorily represent the molecule.

Therefore, "Resonance is a phenomenon invoked when more than one Lewis formula is needed to describe the electronic distribution in molecule. Resonance is the use of contributing structures to represent the various Lewis formula of a single molecule needed for describing the complete electronic distribution in the molecule".

In other word, resonance may be defined as bonding or sharing of electrons between more than two atoms. Typical covalent or ionic bonds involve sharing (covalent) or transferring (ionic) electrons between two atoms as shown in case of ethane and NaCl below:

H_3C-CH_3 — Each electrons pair are shared between two nuclei

$\overset{+}{Na} \cdots \overset{-}{Cl}$ — Electron transfer process involves two atoms

In the above example bonding electrons are said to be localized.

Resonance differs from the two examples above in that, it involves the sharing of electrons between more than two atoms, via. *dilocalization.*

The classical example of resonance is provided by the pi-bonding system in benzene. Benzene is a six membered ring composed of six sp^2 hybridized carbon atoms in a plane and sharing six pi-electrons. It can be represented by "Kekule" structure shown below which suggests an "alternating" single bond-double bond pattern. On that basis, there should have been two different carbon-carbon bond distances. But x-ray analysis revealed that in benzene, all carbon-carbon bond distances are identical. Therefore, 'Kekule' representation does not really adequately reflect the true electronic character of benzene since, in reality, all six pi-electrons are shared equally by the six carbons. Thus, the "inscribed" circle representation shown below may be more accurate.

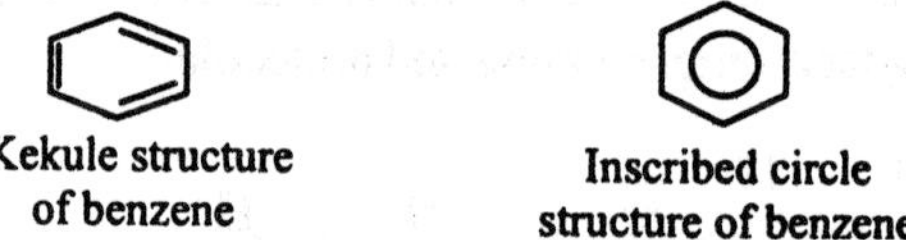

Kekule structure of benzene   Inscribed circle structure of benzene

Requirements for Resonance

One of the most basic requirement for resonance phenomena so exist a "conjugated" electronic system must be present. The second requirement is the atoms involved in resonance must be "coplanar" or capable of adopting coplanar conformation.

Understanding conjugation: An electronic configuration in which there is an alternating single-multiple (double or triple) bond pattern or an atom with non-bonded electron pairs or lacking of electron pair bound to a double or multiple bond system.

I. Case of alternate single multiple bonds.

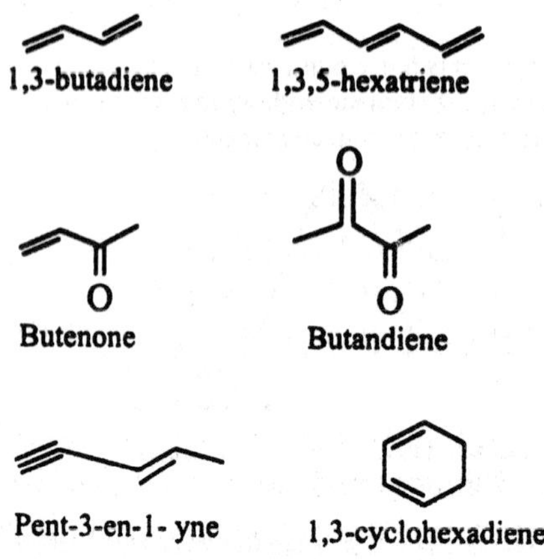

$CH_2{=}CH{-}NO_2$ (Nitroethene) $\quad CH_2{=}CH{-}C{\equiv}N$ (Vinyl cyanide)

All of the above molecules have alternate single multiple bonds and thus conjugated, capable of exhibiting phenomena of resonance.

II. Case with a non-bonded electron pair bound to a multiple bond:

$H_2C{=}CH{-}\ddot{Cl}:$ $\quad H_3C{-}\ddot{O}{-}CH{=}CH_2$

$H_2\dot{N}{-}CH{=}CH_2$

All of the above molecules have a lone-pair of electron on an atom bonded to double bond.

$CH_2{=}CH{-}BH_2$, here boron has only six shared electrons and a vacant *p*-orbital, and bonded to doubly bonded C-atom. This type of system is also conjugated, capable of exhibiting phenomena of resonance.

Representation of resonance-curved arrow mechanism: Let us consider a simple conjugated molecule-1, 3 butadiene.

Butadiene ⟷ Resonance structure

The above resonance structures allows delocalization (movement) of pi-eletrons.

Any such system which allows movement (dispersal or spreading-up) of pi-electrons or non-bonded pair of electrons, lowers the potential energy of the system providing extra stability. Thus, heat evolved during hydrogenation of 1, 3-butadiene is slightly less than the expected value, treating it as a system with two isolated double bonds. This observation is illustrated in the following potential energy Figure 3.7.

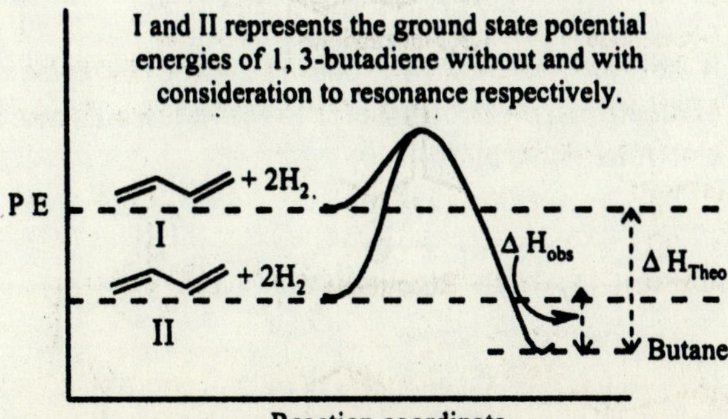

Figure. 3.7: Potential energy diagram for hydrogenation of 1, 3-butadiene

The difference in two enthalpies of hydrogenation (ΔH_{Theo} and ΔH_{obs}) is equivalent to the potential energy lowering of reactant due to delocalization and known as resonance energy (RE).

Some other systems exhibiting resonance phenomena: These are as follows.

(a) Propenone

$CH_2{=}CH{-}C(=O){-}H \longleftrightarrow \overset{+}{C}H_2{-}CH{=}C(O^-){-}H$

(b) Carboxylate ion: $R{-}C(=O){-}O^- \longleftrightarrow R{-}C(O^-){=}O$

(c) A carbanion: $R{-}\bar{C}H{-}C(=O){-}H \longleftrightarrow R{-}CH{=}C(O^-){-}H$

(d) Phenoxide ion:

(e) Vinyl chloride

$CH_2{=}CH{-}\ddot{Cl}: \longleftrightarrow \bar{C}H_2{-}CH{=}\overset{+}{Cl}$

(f) Vinyl cyanide:

$CH_2{=}CH{-}C{\equiv}N \longleftrightarrow \overset{+}{C}H_2{-}CH{=}C{=}\bar{N}$, etc.

Resonance Structures and Resonance Hybrid

Various Lewis formula arising by movements of π-electrons or non-bonded pair of electrons in a conjugated system are known as resonance structure or resonating structure or resonance contributor. For instance, all the Lewis formula drawn for phenoxide ion in the previous section are resonance structures of phenoxide ion. Resonance structures are separated by a double headed arrow (↔), e.g. following two are the resonance structure, of Kekule benzene:

I ⟷ II

Resonance hybrid

The actual molecule is NOT an equilibrium mixture of imaginary resonance structrues, but a single, unchanging composite structure called "Resonance Hybrid". Resonance hybrids are often represented by "Partial bonds". Resonance hybrid of some of the conjugated systems are shown below:

Butadiene

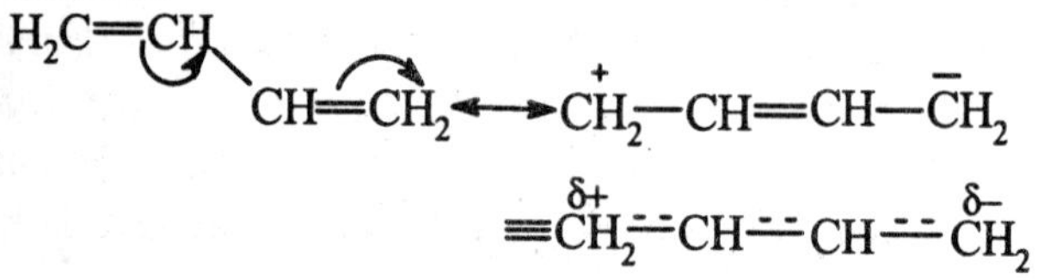

$$H_2C{=}CH{-}CH{=}CH_2 \longleftrightarrow \overset{+}{C}H_2{-}CH{=}CH{-}\overset{-}{C}H_2$$

$$\equiv \overset{\delta+}{C}H_2 \cdots CH \cdots CH \cdots \overset{\delta-}{C}H_2$$

Resonance hybrid

The above resonance hybrid structure indicates that all C—C bonds are equivalent. Since two π-bonds are dispersed over three sigma bonds, average π-bonds between any two C—C is 2/3 and the total bond order between any two carbon in butadiene is 5/3 (one sigma and 2/3 pi-bond). This reflect that in butadiene, all carbon-carbon bond lengths are identical and it is in between carbon-carbon bond length in ethane (bond order one) and ethene (bond order 2).

Phenoxide on

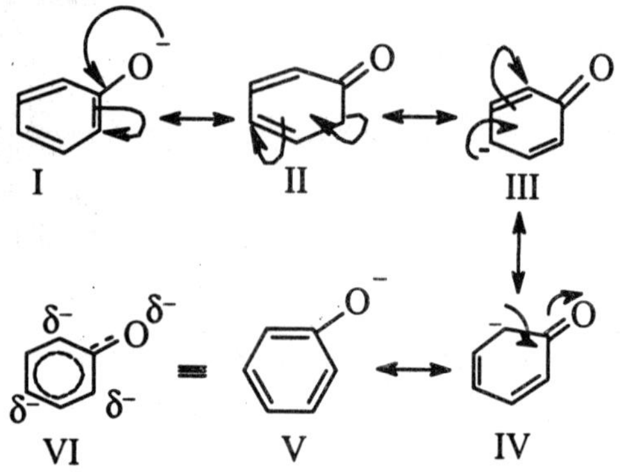

I ⟷ II ⟷ III ⟷ IV ⟷ V ≡ VI

Here, I to V are resonance structures of phenoxide ion and VI is the resonance hybrid.

Example 8. Carbon-chlorine bond in vinyl chloride is stronger than in ethylchloride explain.

Solution: Vinyl chloride is a conjugated system and can exhibit phenomena of resonance as:

$$CH_2{=}CH{-}\ddot{C}l{:} \longleftrightarrow \overset{-}{C}H_2{-}CH{=}\overset{+}{C}l \equiv \overset{\delta-}{C}H_2 \cdots CH \cdots \overset{\delta+}{C}l$$

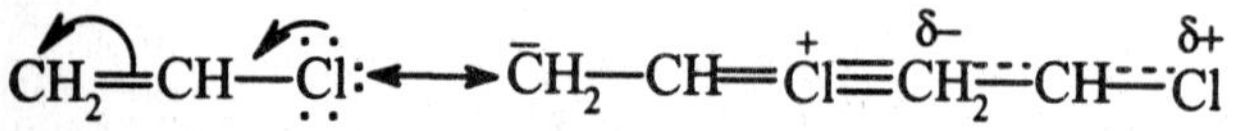

The above resonance structure indicate that C—Cl bond in vinyl chloride has acquired partial double bond character whereas in ethyl chloride, carbon-chlorine bond is pure σ-bond due to lack of resonance. Thus, C—Cl bond in vinyl chloride is stronger than pure single sigma covalent C—Cl bond in ethyl chloride.

Example 9. Carbon-oxygen bond length in phenol is shorter than in saturated aliphatic alcohol, explain.

Solution: Phenol is a conjugated system and can exhibit phenomena of resonance as follows:

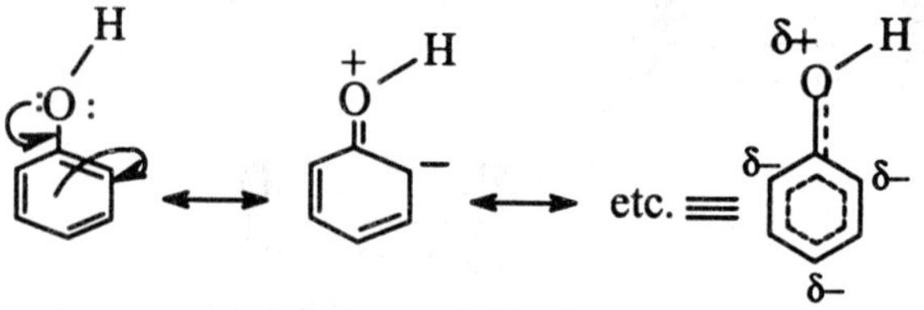

⟷ ⟷ etc. ≡ (δ+ on O, δ− on ring)

This gives partial double bond character to C—O bond in phenol. No such resonance is observed in aliphatic, saturated alcohol:

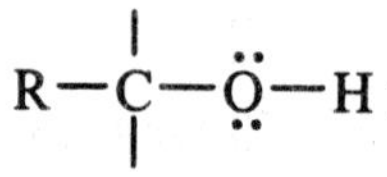

$$R{-}\overset{|}{\underset{|}{C}}{-}\ddot{\underset{\cdot\cdot}{O}}{-}H$$ No resonance

Therefore, in saturated aliphatic alcohol, C—O bond is purely single and little longer than C—O bond in phenol, where it has partial double bond character.

Rules for Writing Resonance Structure

1. Resonance involves movements of electrons, i.e. movement of nuclei is not allowed.
   "All resonating structure must have same positions of nuclei", e.g.

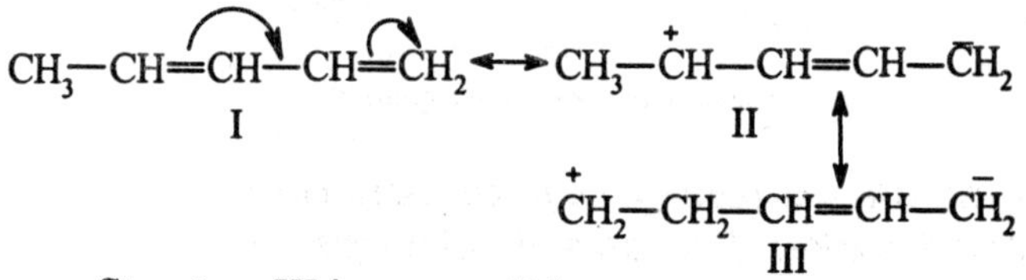

$$\underset{\text{I}}{CH_3{-}CH{=}CH{-}CH{=}CH_2} \longleftrightarrow \underset{\text{II}}{CH_3{-}\overset{+}{C}H{-}CH{=}CH{-}\overset{-}{C}H_2}$$

$$\updownarrow$$

$$\underset{\text{III}}{\overset{+}{C}H_2{-}CH_2{-}CH{=}CH{-}\overset{-}{C}H_2}$$

   Structure III is not a valid resonance structure of (I) since it is formed by movement of a nucleus (hydride ion has moved from CH_3– to adjacent carbon).

2. All resonance structure must have valid Lewis dot formula, i.e. octet should not be violated, e.g.

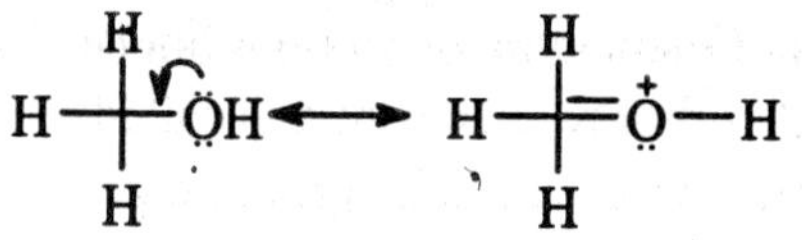

$$H{-}\underset{H}{\overset{H}{C}}{-}\ddot{O}H \longleftrightarrow H{-}\underset{H}{\overset{H}{C}}{=}\overset{+}{\ddot{O}}{-}H$$

   Not a valid Lewis structure since C has five covalent bond violating octet rule.

3. If there are more than one possible way of movement of electrons, electrons prefer to move towards more electronegative atom:

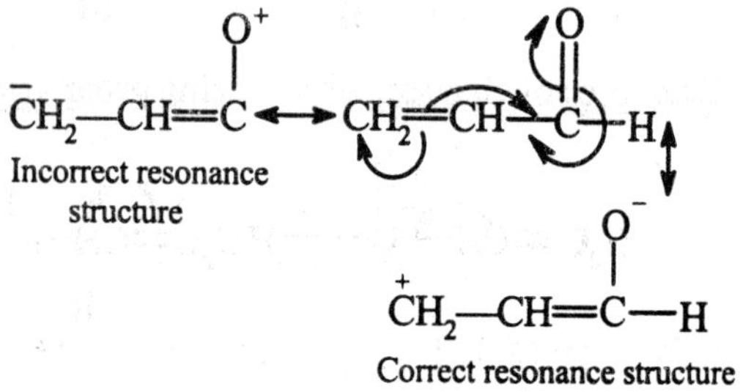

$$\underset{\text{Incorrect resonance structure}}{\overset{-}{C}H_2{-}CH{=}\overset{\overset{O^+}{|}}{C}} \longleftrightarrow CH_2{=}CH{-}\overset{\overset{O}{||}}{C}{-}H \longleftrightarrow \underset{\text{Correct resonance structure}}{\overset{+}{C}H_2{-}CH{=}\overset{\overset{O^-}{|}}{C}{-}H}$$

4. Atoms involved in resonance must lie in the same plane, i.e. in order for electrons to be delocalized over more than two atoms, the pi-orbitals of these atoms must lie in the same plane so that they can overlap:

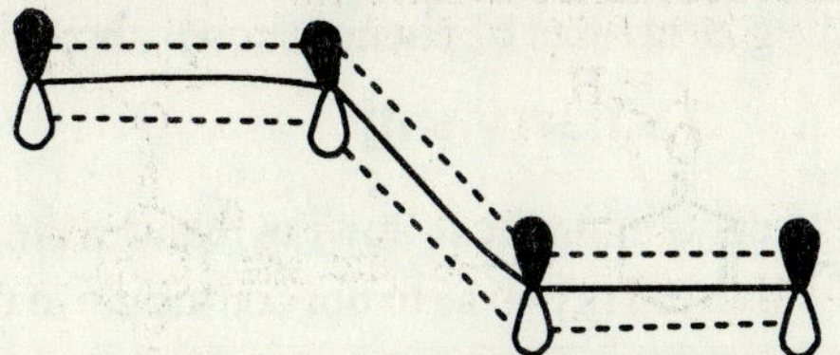

Co-planar pi-orbitals in 1, 3-butadiene

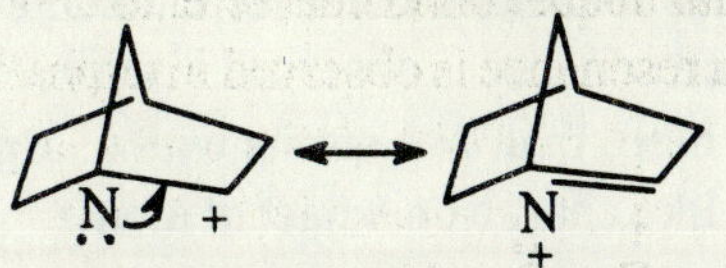

Here, in the above example nitrogen has a lone pair and adjacent carbon bearing a positive charge has vacant p-orbital and delocalization is expected to occur. In practice, it does not occur since N and said carbon are not in the same plane and their orbitals cannot overlap. Other example of non-planar arrangement is a bicyclic amide:

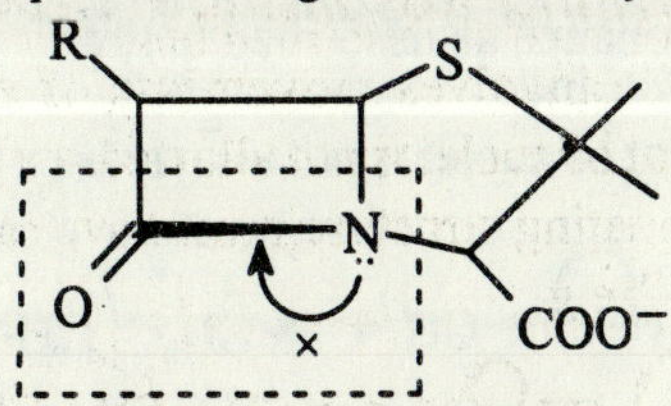

β-lactamamide- "non-coplanar"

Electron Donating (+R effect) and Electron Withdrawing (-R effect) Groups in Resonance

An atom or group with atleast one non-bonded electron pair on the atom involved in resonance (—OH, —OR, —X, —NH_2, etc.) are called electron donating group in resonance (+R effect causing group). On the other hand, if an electron deficient atom or group is bonded to the site of resonance, it withdraw electrons and called electron withdrawing (—R effect causing) group, e.g. —NO_2, COH, etc.

Examples of electron donating +R groups:

$H_2C{=}CH{-}\ddot{N}H_2$ (I), $H_2\ddot{N}$–C_6H_5 (II), C_6H_5–$\ddot{O}$–H (III), etc.

Examples of electron withdrawing groups (—R):

$H_2C{=}CH{-}C({=}O){-}H$ (I), $H_2C{=}CH{-}N^{+}({=}O){-}O^{-}$ (II)

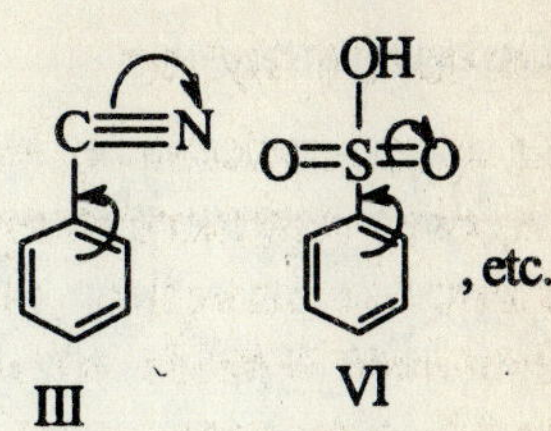

Rules for Assigning Relative Stability to Resonating Structure

Resonating structures of a molecule or ions may or may not contribute equally to resonance hybrid. Therefore, resonace structures may or may not have same potential energy (stability). Following rules summarizes the relative stability of resonating structures:

1. Contributors with more covalent bonds are more stable than those with fewer covalent bonds, e.g.

   Butadiene (Major) ⟷ $^{+}CH_2{-}CH{=}CH{-}CH_2^{-}$ (Minor)

   Here, the neutral butadiene molecule has one extra covalent bond than its charged dipolar ion, and former is more stable, contributing more to the resonance hybrid.

2. Contributors with complete octet of all the atoms have lower energy, more stable than those with an incomplete octet, e.g.

   I. $H_3C{-}\ddot{O}{-}\overset{+}{C}H_2 \longleftrightarrow H_3C{-}\overset{+}{\ddot{O}}{=}CH_2$
   Minor (Incomplete octet of CH_2) — Major (Complete octet of all atoms)

   II. $R{-}\overset{+}{C}{=}\ddot{O} \longleftrightarrow R{-}C{\equiv}\overset{+}{O}$
   Minor — Major

   III. $R{-}\overset{+}{C}H{-}\ddot{N}H_2 \longleftrightarrow R{-}CH{=}\overset{+}{N}H_2$
   Minor — Major

3. Equivalent resonance contributors have equal energies and therefore make equal contribution to the hybrid.

$$H_2C{=}CH{-}\overset{+}{C}H_2 \longleftrightarrow \overset{+}{C}H_2{-}CH{=}CH_2 \equiv \overset{\delta+}{C}H_2{-}CH{-}\overset{\delta+}{C}H_2$$

Equivalent resonating structures, making equal contribution to resonance hybrid.

Larger the number of equivalent resonance structures, more stable the molecule or ions. Some common examples where all resonating structures are equivalent.

I. Carbonate ion

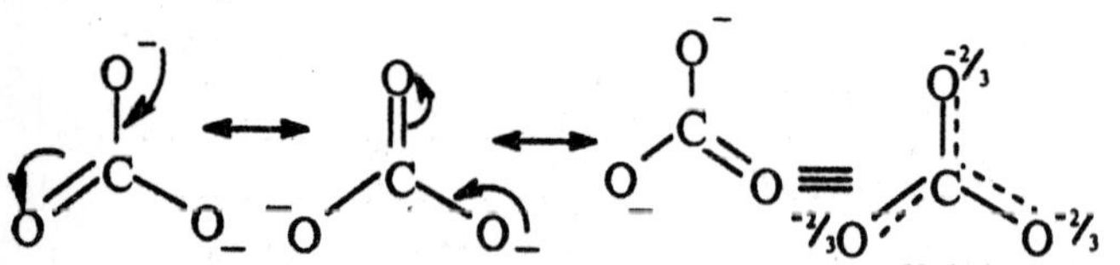

Hybrid

II. Sulphate ion

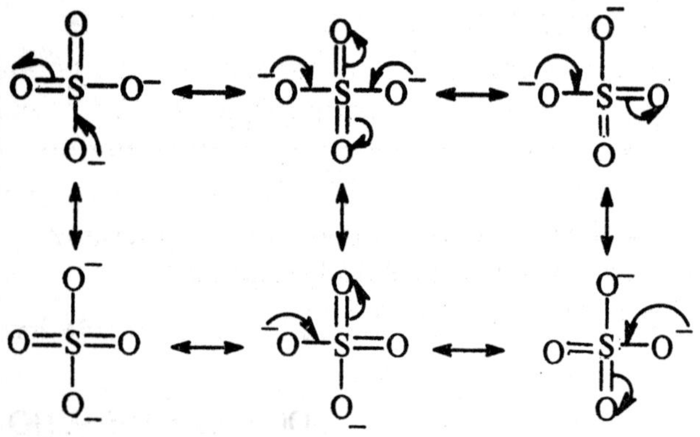

Due to such large numbers of equivalent resonance structure of sulphate ion, it is exclusively stable and its acid (H_2SO_4) is strongly acidic.

4. Resonating structure with negative charge on electronegative atoms or positive charge on electropositive atoms are more stable and contribute more to the resonance hybride, e.g.

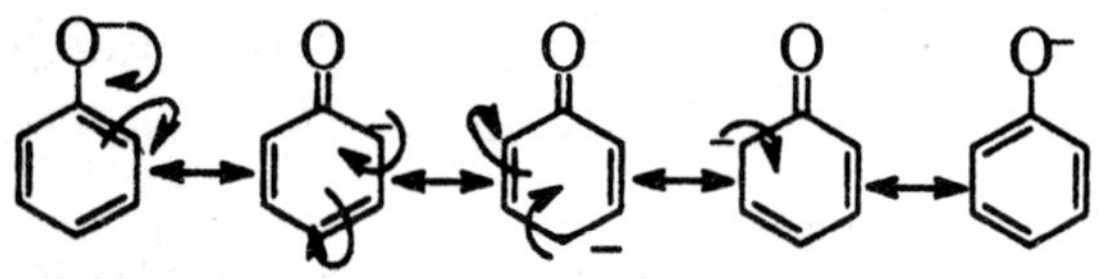

I II III IV V

Phenoxide ion

Among the various resonating structures of phenoxide ion, I and V have –ve charge on electronegative oxygen, whereas rest three (II, III, and IV) have –ve charge on electropositive carbon. Therefore, I and V are more stable than the remaining three and contributions of former to the resonance hybrid are more.

5. Charge separation decreases stability of resonating structure and therefore it's contribution to the resonance hybrid, e.g.

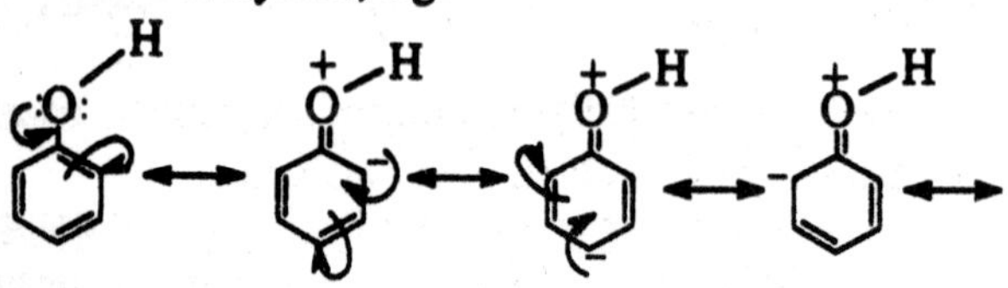

I II III IV

Among the four resonating structures of phenol molecule shown above, I-without any charge is most stable. Others are dipolar ion. Stability of the dipolar ions depends on the distance between the poles and larger the distance between opposite poles, smaller will be the force of attraction, higher will be the potential energy. Therefore, the order of stability of the resonating structures of phenol shown above is:

$$I > II = IV > III$$

6. Contributors containing the following features are of sufficiently high energy so as to not contribute to the hybrid.
   (a) Contain an expanded octet for elements of second period.
   (b) Contain more than two atoms with charge.
   (c) Contain like charge on adjacent atoms.

Example 10. Which of the following molecule has greater C=C bond energy:

$$CH_2{=}CH{-}\ddot{S}{-}CH_3 \quad \text{or} \quad CH_2{=}CH{-}\ddot{O}{-}CH_3$$

Solution: $CH_2{=}CH{-}SCH_3$ has greater C=C bond energy. Both molecule are capable of exhibiting phenomena of resonance as described below:

$$CH_2{=}CH{-}\ddot{S}{-}CH_3 \longleftrightarrow \bar{C}H_2{-}CH{=}\overset{+}{S}{-}CH_3$$

$$\text{and } CH_2{=}CH{-}\ddot{O}{-}CH_3 \longleftrightarrow \bar{C}H_2{-}CH{=}\overset{+}{O}{-}CH_3$$

In both cases, pi-bonds are pushed towards terminal carbon by resonance by the non-bonded electrons of 'S' and 'O', decreasing the C=C bond energy. Thus, the actual C=C bond energy will now depends on extent of delocalization.

It is also known that orbitals of same energy overlaps better than orbitals of different energy. In case of methyl vinyl ether, both the atoms "oxygen and carbon" involved in resonance are from the same row of periodic table, have same energy of *p*-orbitals (almost) involved in resonance, exert a greater +R -effect, decreasing C=C bond order more effectively than do the sulphur atom, since sulphur and carbons are using *p*-orbitals of different energies [(3p(S) and 2p(C)].

Example 11. Carbon-nitrogen bonds in aliphatic, saturated amines are longer than in aniline, explain.

Solution: In saturated aliphatic amine, like methyl amine, C—N bond is purely single bond and no delocalization of lone pair of electrons on nitrogen take place. On the other hand, lone pair of electron on nitrogen in aniline is in conjugation with the pi-bonds of benzene ring and are involved in delocalization giving partial double bond character to C—N bond as:

H–C(H)(H)–$\ddot{N}H_2$ $\ddot{N}H_2$ (benzene ring) ⟷ $\overset{+}{N}H_2$ (ring, –) ⟷, etc.

No-resonance   Aniline

Thus, C—N bond in saturated aliphatic amine is longer than in aniline.

Example 12. A carbon-nitrogen bond in aniline is shorter than in saturated aliphatic amine, but longer than C—N bond in amide (aliphatic) explain.

Solution: First part of this question is already answered in Example 11. IInd part: Both aniline and amide exhibit phenomena of resonance as:

$\ddot{N}H_2$ (ring) ⟷ $\overset{+}{N}H_2$ (ring, –) ⟷, etc. R—C(=O)—$\ddot{N}H_2$ ⟷ R—C(–O^-)=$\overset{+}{N}H_2$

Aniline   An amide

The +R effect of $—NH_2$ is more in amide since delocalization develops negative charge on electronegative oxygen. In case of aniline, delocalization develops negative charge on electropositive carbon (energetically less favoured). Therefore, C—N bond in amide acquire greater partial double bond character than in aniline, hence bond length is shorter in amide than in aniline.

Example 13. Hydrogenation of ethene is associated with evoluation of (approximately) 130 kJ of energy whereas in hydrogenation of 1, 3-butadiene to butane, energy less than 260 kJ is evolved. Justify.

Solution: Due to resonance phenomena in butadiene, it is slightly more stable than isolated diene and thus less energy is evolved in the hydrogenation reaction.

Example 14. Which has greater heat of hydrogenation-vinylamine or vinyl phosphine, explain with the help of resonance phenomena.

Solution: Vinyl phosphine has greater heat of hydrogenation. Both exhibit the phenomena of resonance as:

I. $CH_2=CH—\ddot{N}H_2 \longleftrightarrow \bar{C}H_2—CH=\overset{+}{N}H_2$

II. $CH_2=CH—\ddot{P}H_2 \longleftrightarrow \bar{C}H_2—CH=\overset{+}{P}H_2$

Resonance stabilization is more in case of I, since orbitals of same energies (both C and N uses 2p orbitals in overlap) are involved in delocalization. On the otherhand, stabilization by resonance is less in case of II, since in II *p*-orbitals differing in energies (2p of C and 3p of P) are involved in delocalization. Therefore, PE of vinyl amine is less than PE of vinyl phosphine. Hydrogenation of an alkene is an exothermic process, less heat will be evolved in hydrogenation of vinylamine.

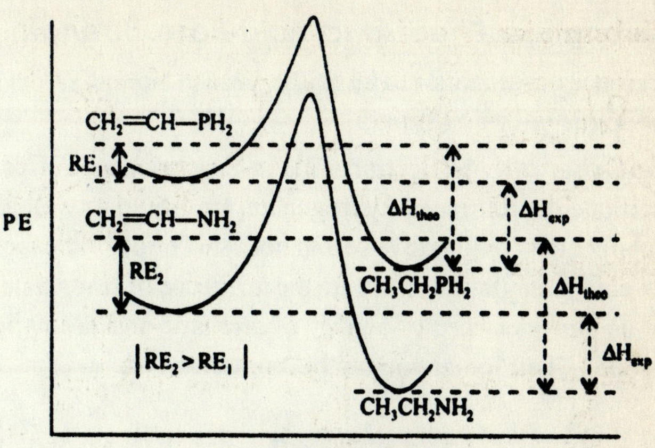

Figure. 3.8: Potential energy showing hydrogenation of vinylamine and vinyl phosphine

Example 15. Which of the following molecule has greater resonance stabilization:

$CH_3—O—CH=CH—CHO$ or $HOC—CH=CH—CHO$

I   II

Solution: Molecule I is more stabilized by resonance than II. The resonance structures of I and II are as follows:

I. $CH_3—O—CH=CH—C(=O)—H \longleftrightarrow$

$CH_3—\ddot{O}—\overset{+}{C}H—CH=C(O^-)—H$

↕

$CH_3—\overset{+}{O}=CH—CH=C(O^-)—H$

(Complete octet of all atom)

II. $H—C(=O)—CH=CH—C(=O)—H \longleftrightarrow H—C(=O)—\overset{+}{C}H—CH=C(O^-)—H$

Destabilized resonance structure due to electron withdrawing effect by CHO on adjacent positively charged carbon.

Example 16. Which of the following has greater C═C bond energy:

$CH_2=CH—CHO$ or $CH_2=CH—CONH_2$

I   II

Solution: **II has greater C═C bond energy due to lesser involvement of C═C π-electron in resonance:**

$CH_2=CH—C(=O)—\ddot{N}H_2 \longleftrightarrow \overset{+}{C}H_2—CH=C(O^-)—NH_2$

$\longrightarrow CH_2—CH—C(O^-)=\overset{+}{N}H_2$

Resonance Effect and Acid-Base Strength

Presence of resonance in an acid or base profoundly affect the acidic or basic strength.

Acid strength: An electron withdrawing resonance effect facilitate the release of hydrogen ion from acid as well as stabilize the conjugate base and both this effect increases the acid strength. To illustrate the influence of resonance, let us consider case of acidity of aliphatic and aromatic alcohol. Their ionization can be represented as:

$$R—O—H \rightleftharpoons R—\bar{O} + \overset{+}{H} \qquad \cdots \text{(i)}$$

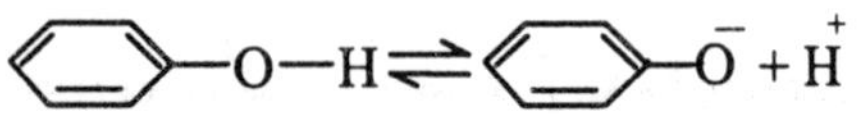

$$C_6H_5—O—H \rightleftharpoons C_6H_5—\bar{O} + \overset{+}{H} \qquad \cdots \text{(ii)}$$

Aliphatic alcohol do not possess conjugation and therefore either ionization process or stability of conjugate base (RO) is not influenced by resonance. Contrary to aliphatic alcohol, aromatic alcohol (phenol in the present case) is a conjugated system and exhibit following type of resonance:

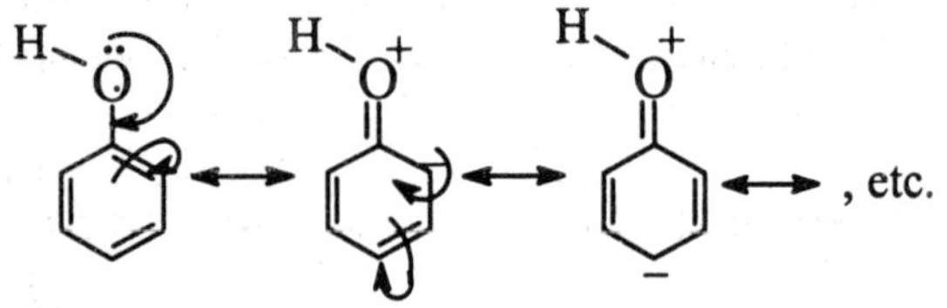

[Resonance structures of phenol] , etc.

The above resonance structures possess the positive charge on oxygen atom making it strongly electron withdrawing which makes release of H^+–bonded to oxygen, easier, favouring ionization process. On the other hand, conjugate base of phenol, i.e. phenoxide ion is also stabilized by resonance as:

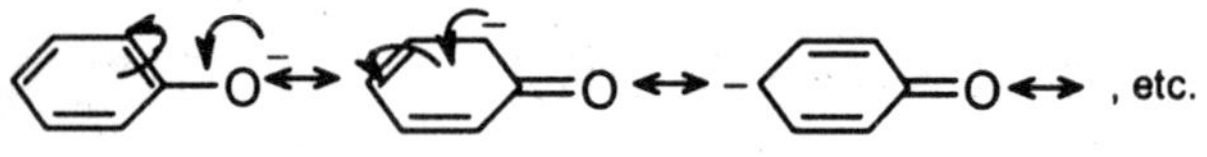

[Resonance structures] , etc.

(Phenoxide ion)

Both these factors favour ionization and drive the reaction (ii) in forward direction. Therefore an aliphatic alcohol is weaker acid than aromatic alcohol.

Comparing Acid Strength of Substituted Phenols

(i) *Nitrophenol:* Three isomeric nitrophenols are possible and their acidic strength varies significantly.

Para-nitrophenol and metanitrophenol

The relative acid strength can be explained either on the basis of electron withdrawing effect of substituents on the case of deprotonation or stabilization of conjugate base.

[Resonance structures of *p*-nitrophenol] etc.

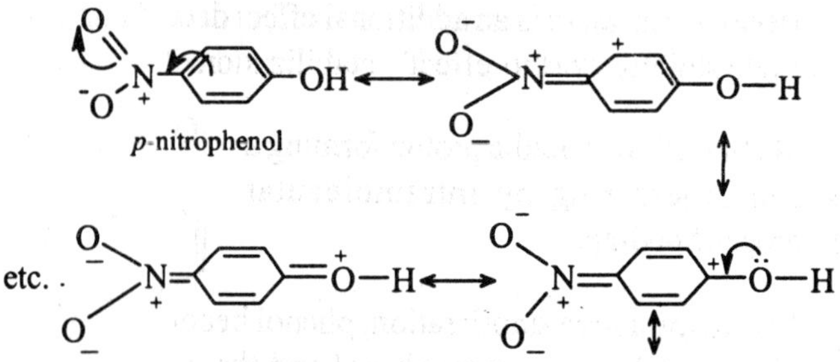

p-nitrophenol

The above resonating structures indicate that $-NO_2$ group is in direct resonance with hydroxyl group and one of the resonating structure (last one) has positive charge on oxygen favouring deprotonation and increase in acid strength. This increased acidic strength by nitrogroup from para position can also be explained on the basis of stabilization of conjugate base as:

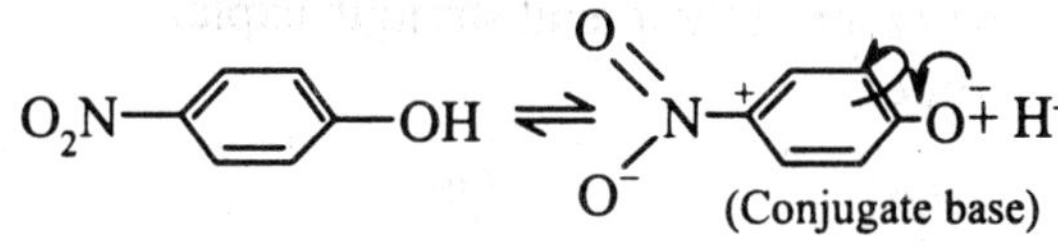

$$O_2N—C_6H_4—OH \rightleftharpoons {}^{-}O(O{=})\overset{+}{N}—C_6H_4—\bar{O} + H^+$$

(Conjugate base)

[Resonance structures of conjugate base] ,etc.

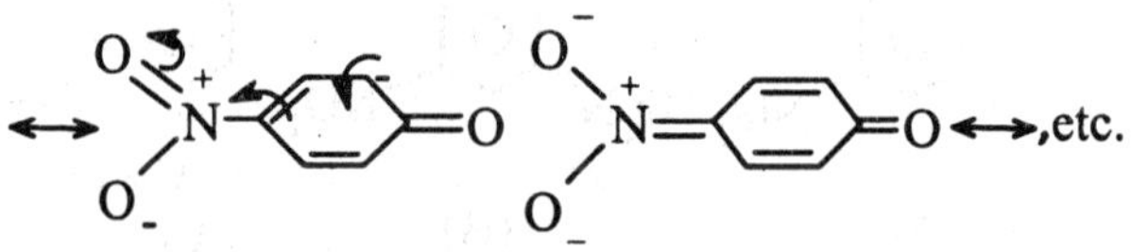

Here the last resonating structure has complete octet of all the atoms, highly stable, indicating greater acid strength of *p*-nitrophenol.

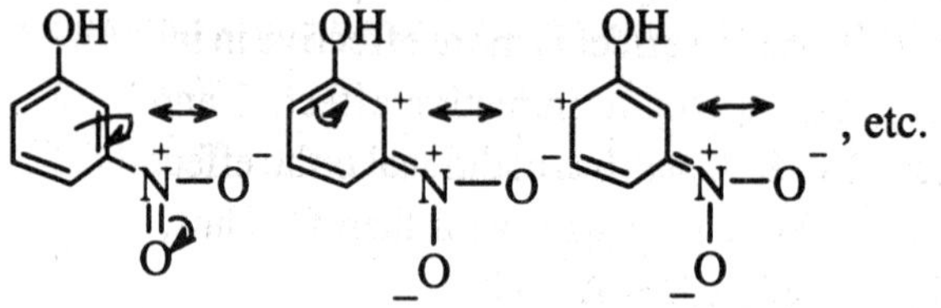

[Resonance structures of *m*-nitrophenol] , etc.

In meta nitrophenol, nitrogroup is not in direct resonance with—OH, i.e. it does not bring positive charge on hydroxy oxygen. Therefore, acidic strength of *m*-nitrophenol is not influenced by –R effect of nitrogroup. Thus, we conclude that:

Acid strength: p-nitrophenol > *m*-nitrophenol

At the same time, $-NO_2$ is also a very strong –I effect causing group, it increases acidic strength of *m*-nitrophenol by induction, and hence *m*-nitrophenol is stronger acid than phenol although weaker than *p*-nitrophenol.

In case of ortho nitrophenol, resonance effect on acidity is same as in case of *p*-nitrophenol as:

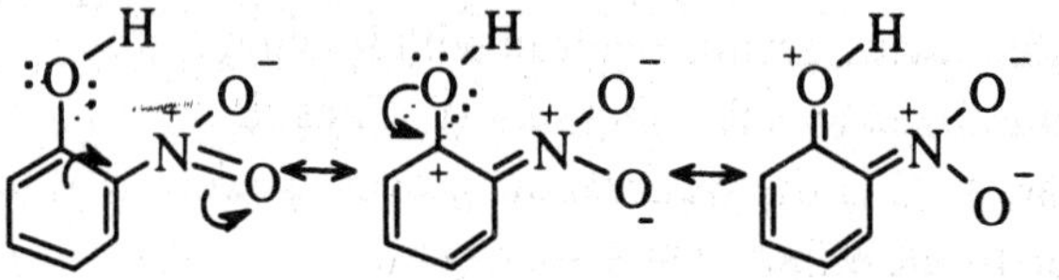

[Resonance structures of *o*-nitrophenol]

Besides this, there is an additional effect decreasing acid strength slightly, "ortho-effect" stabilization of acid as:

Stabilization of acidic proton forming a six memberd ring by intramolecular hydrogen bonding.

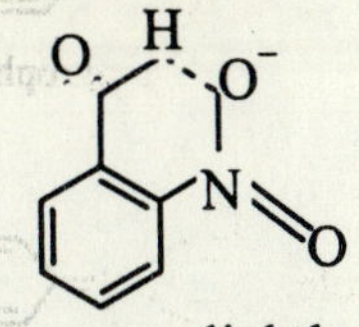

Due to the above stabilization, phenol becomes slightly weaker acid than para nitrophenol and the overall acidic strength order will be:

Phenol < *m*-nitrophenol < *o*-nitrophenol < *p*-nitrophenol

Increase in number of nitro substituent increases the acidic strength and a trinitrophenol (picric acid) is equivalent to common mineral acids in acidity.

Example 17. Arrange the following dinitrophenols in increasing order of their acid strength explaining reasons for your order.

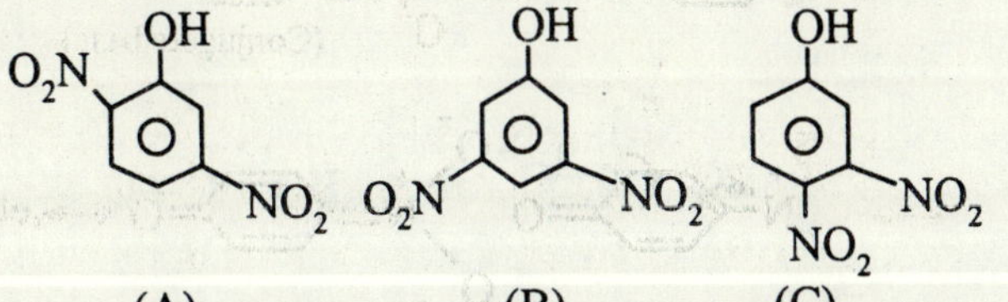

(A) (B) (C)

Solution: Nitro group from ortho and para position increases acid strength, via resonance whereas it also increases acid strength from meta position, via induction. Since resonance effect is more effective in increasing acid strength compared to inductive effect, C and A are more acidic than B. A also has additional ortho effect stabilization of acidic H, it is weaker acid than C. Thus, the order of acid strength is B < A < C.

Example 18. Rank the following substituted phenol in increasing order of their acid strength.

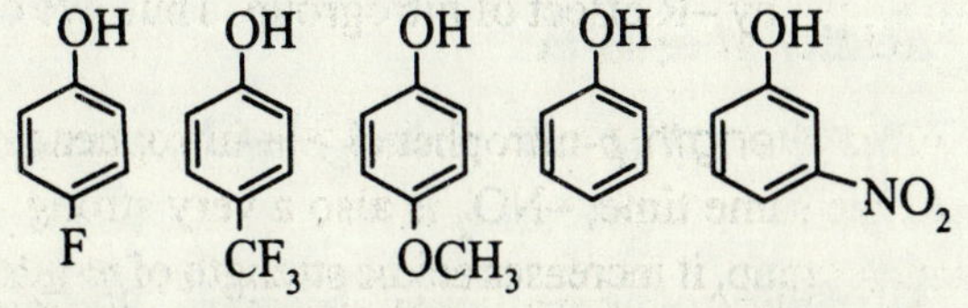

(A) (B) (C) (D) (E)

Solution: The strongest acid is E, since the strongest —I effect causing group (—NO_2) closest to —OH group. Both —F and —CF_3 in A and B respectively, are electron withdrawing group, electron withdrawing effect of —F is neutralized by +R effect of —F. Thus, B is stronger acid than A. C is the weakest acid, even weaker than phenol due to +R effect of —OCH_3 as:

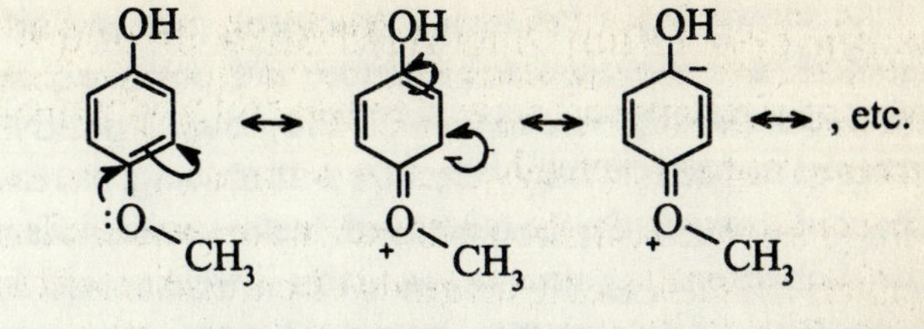

The last resonance structure above has a negative charge on carbon directly bonded to —OH group. A negatively charged carbon has very high electron donating tendency, it will resist the deprotonation (which require electron withdrawal from —O—H bond). Thus, the order of acid strength will be:

C < D < A < B < E

Acidity of Benzoic Acid and Substituted Benzoic Acid

In general, an unsubstituted benzoic acid is stronger acid than an unsubstituted aliphatic carboxylic acid. This is due to the following reason:

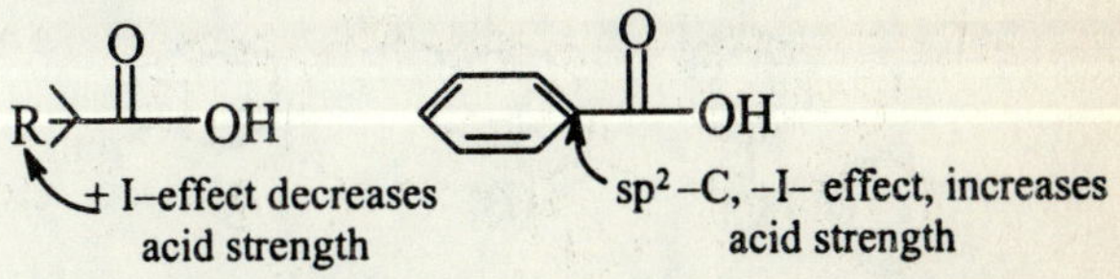

+I-effect decreases acid strength

sp^2–C, –I– effect, increases acid strength

In a substituted benzoic acid, substituent strength as well as its position decides the acidic strength. Electron withdrawing group increases the acid strength and electron donating group decreases the acid strength compared to unsubstituted benzoic acid. Let us consider the case of three isomeric mononitrobenzoic acid.

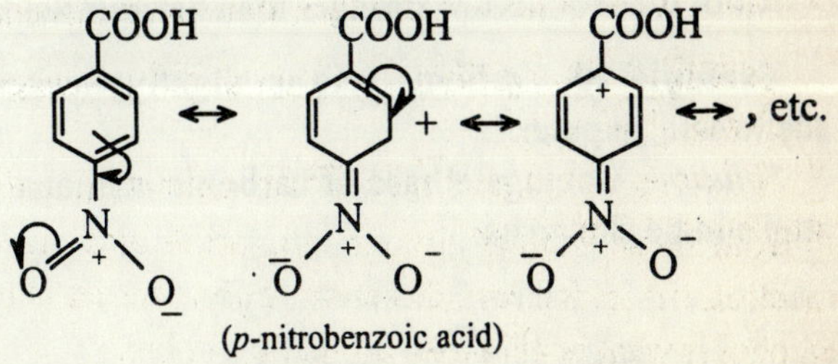

(*p*-nitrobenzoic acid)

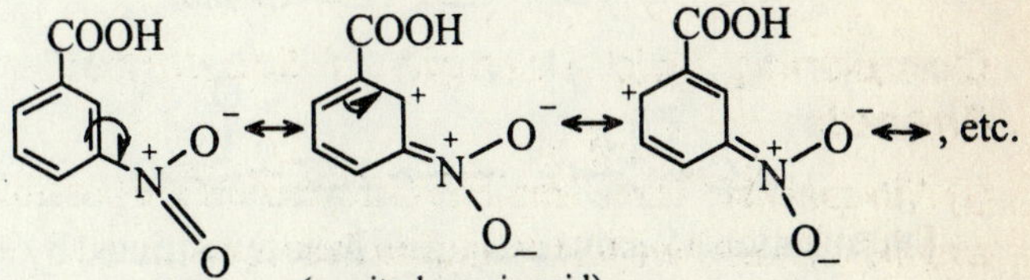

(*m*-nitrobenzoic acid)

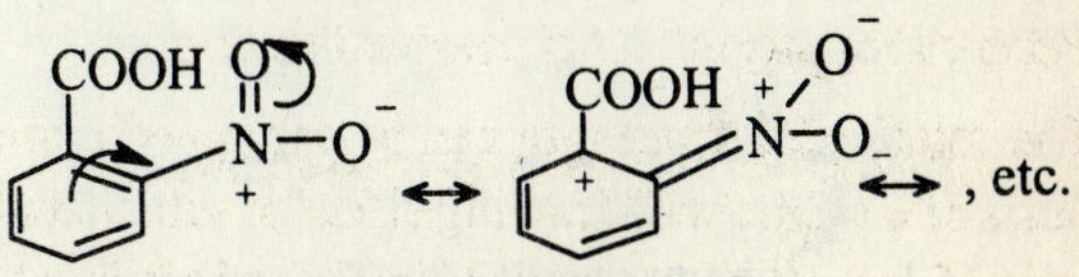

(*o*-nitrobezoic acid)

As shown in the resonance structures, para and ortho isomers have one resonance structure with positive charge on carbon directly bonded to —COOH group. A positively charged carbon is strongly electron withdrawing, increases the acid strength. On the othrehand, meta isomer, does not has such resonating structures and this is weaker acid than both ortho and para isomer. Conjugate base of ortho isomer is further stabilized by ortho effect, favouring ionization of its conjugate acid, hence stronger acid than para isomer:

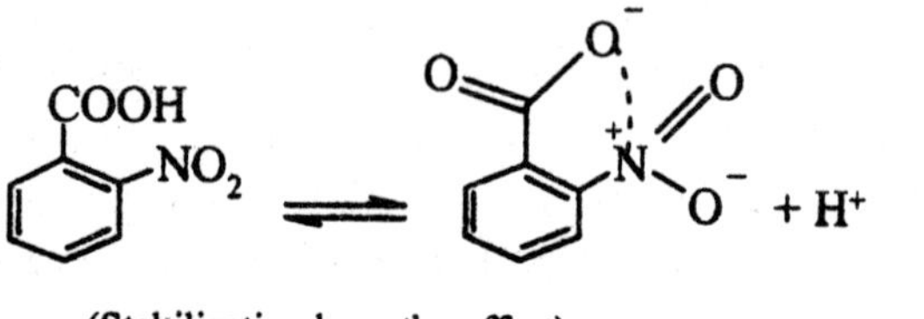

(Stabilization by ortho effect)

Thus, the order of acid strength is:

ortho > para > meta

Example 19. Rank the following acid in increasing order of acid strength:

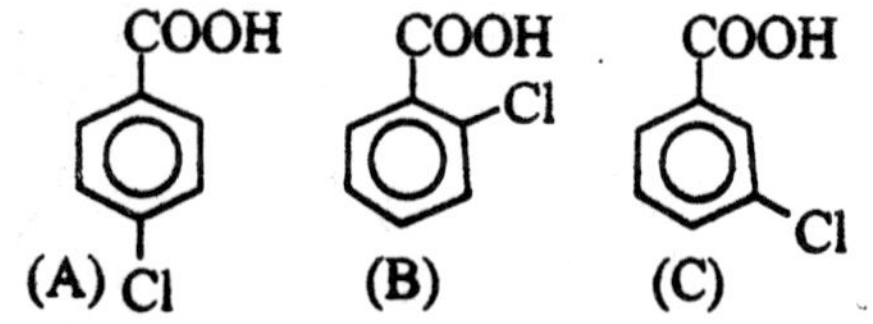

(A) (B) (C)

Solution: Chlorine has +R and –I effect, but the net effect is electron withdrawl. Since chlorine from ortho and para position are in direct resonance with —COOH but from meta position. Between ortho and para isomers, +R effect is same but —I effect is more from ortho position. Thus, the relative acid strength of these three acids are:

C > B > A

All of these acids are stronger than benzoic acid.

Example 20. Carbonic acid is slightly weaker than acetic acid, explain.

Solution: Conjugate base of carbonic acid and acetic acid can be shown as:

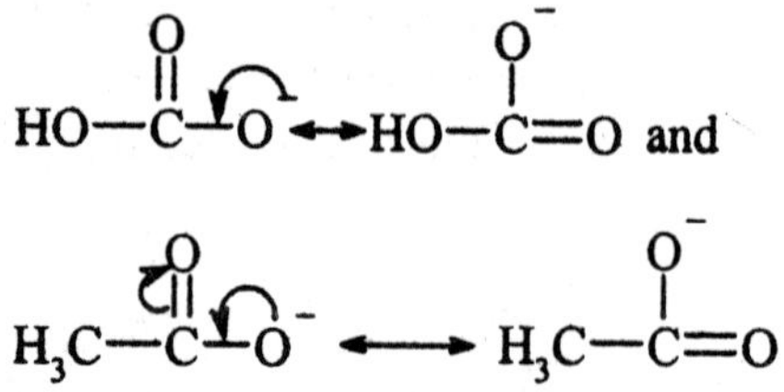

$HO-C(=O)-O^- \longleftrightarrow HO-C(-O^-)=O$ and

$H_3C-C(=O)-O^- \longleftrightarrow H_3C-C(-O^-)=O$

Stabilization of above conjugate base is restricted by +I effect of CH_3– group and +R effect of HO– group as:

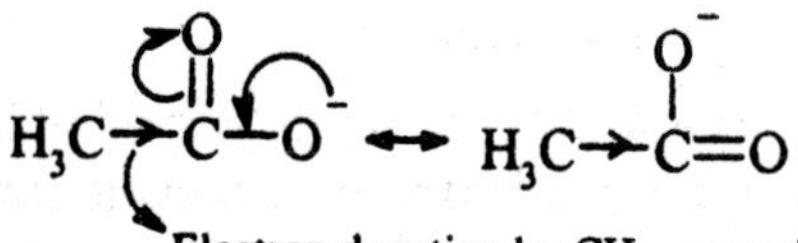

$H_3C\rightarrow C(=O)-O^- \longleftrightarrow H_3C\rightarrow C(-O^-)=O$

Electron donation by CH_3–opposing stabilization, lowers the stability of conjugate base.

$H-\ddot{O}-C(=O)-O^- \longleftrightarrow HO-C(-O^-)=O \longleftrightarrow H-\overset{+}{O}=C(-O^-)-O^-$

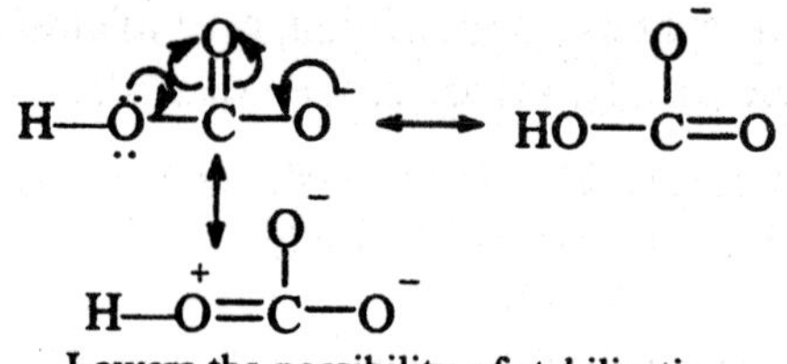

Lowers the possibility of stabilization decreasing stability of conjugate base. Since resonance effect is more significant, carbonic acid is weaker acid than acetic acid.

Example 21. Although conjugate base of carbonic acid has less number of resonating structures, than phenoxide ion, carbonic acid is stronger acid than phenol, justify.

Solution: Greater the stability of conjugate base, stronger will be the acid, but number of resonating structures is not the only criteria for assigning stability to the conjugate base. Conjugate base of carbonic acid have only two resonating structures as shown below:

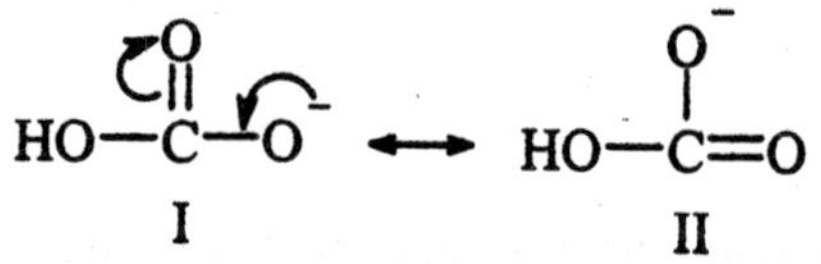

$HO-C(=O)-O^- \longleftrightarrow HO-C(-O^-)=O$

I II

Both these resonating structures are equivalent and have negative charge on electronegative atom, both favouring stability. On the otherhand, the resonating structures of phenoxide ion are:

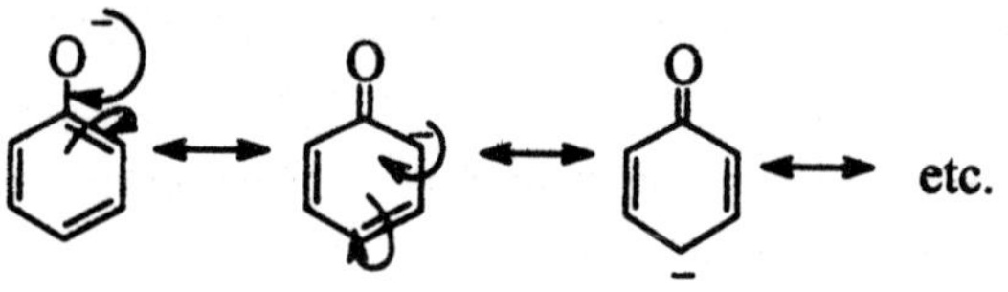

etc.

Though number of resonating structures are more than bicarbonate ion, but these are less-stable since they are non-equivalent and possesses negative charge on electropositive carbon, both against the stability requirement.

Acidity of α–H to

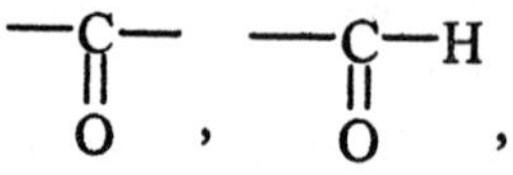

—C(=O)—, —C(=O)—H, —NO_2, —CN, ester group, etc.

All of the groups mentioned above are –R effect causing group and are capable of stabilizing conjugate base formed by deprotonation from their α–position, thus α–H of compounds containing these functional group is highly acidic, e.g.

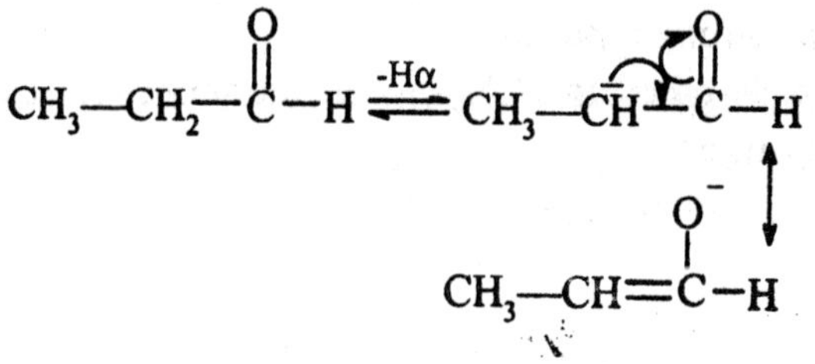

$CH_3-CH_2-C(=O)-H \xrightleftharpoons{-H\alpha} CH_3-\bar{C}H-C(=O)-H \longleftrightarrow CH_3-CH=C(-O^-)-H$

Due to above stabilization, α–H of aldehyde or ketones are most acidic proton of the molecule.

Example 22. Nitro methane has very high acidity compared to methane, explain.

Solution: $CH_4 \rightleftharpoons CH_3^- + H^+$, here (Conjugate base of CH_4) is not stabilized by resonance.

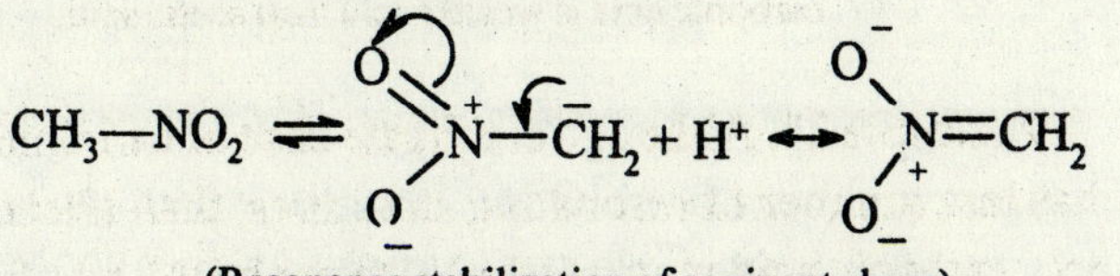

$$CH_3—NO_2 \rightleftharpoons \overset{+}{N}(O)(O^-)—\overset{-}{C}H_2 + H^+ \longleftrightarrow \overset{+}{N}(O^-)(O^-)=CH_2$$

(Resonance stabilization of conjugate base)

Example 23. Compare the acidity of labeled protons in the following compounds:

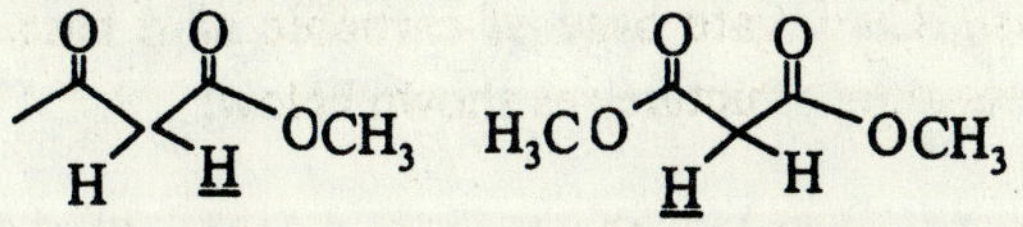

(A) $CH_3COCH_2COOCH_3$ (labeled H on CH_2) (B) $H_3CO–CO–CH_2–CO–OCH_3$ (labeled H on CH_2)

Solution: Both A and B have one $—\overset{O}{\overset{\|}{C}}—OCH_3$ in common so their acidity can be compared from the other part, which are different, both group will stabilize the negative charge on adjacent carbon, but its stabilizing power will be restricted by electron donating —OCH_3(+R effect) and CH_3 – (+I effect) groups. Since +R effect is more significant than +I effect, —OCH_3— is a stronger +R effect causing group than (CH_3) group and hence, labeled H in A is more acidic than in B.

Strength of Base and Resonance

Basic strength depends on how easily the base can react with an acid. Let us consider basic strength of an amine-amines are Lewis base and its basic strength depends on the electron donating ability of nitrogen. Resonance can influence the basic strength as follows:

Aliphatic and aromatic amine: In general, aromatic amines are less basic than aliphatic amines. In aliphatic amine, alkyl group increases electron density at nitrogen by +I effect making it stronger base as:

"R→NH_2, +I effect of R– increases the electron density at nitrogen making it a better electron donar and hence stronger base."

On the otherhand, NH_2 in aromatic amine is in resonance with ring as:

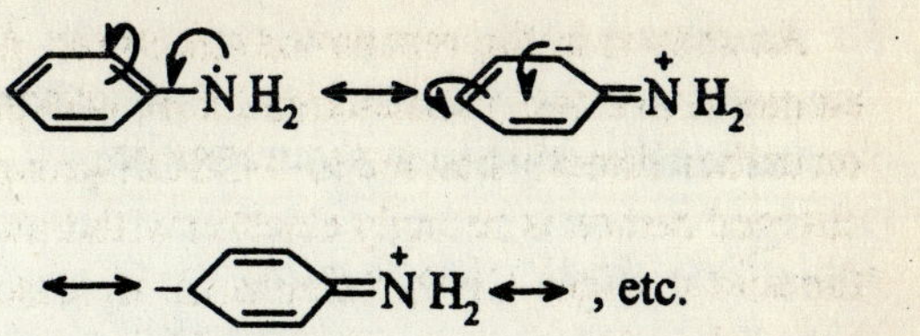

$$C_6H_5–\ddot{N}H_2 \longleftrightarrow \overset{-}{C_6H_5}=\overset{+}{N}H_2 \longleftrightarrow \ldots \longleftrightarrow {}^{-}C_6H_5=\overset{+}{N}H_2 \longleftrightarrow \text{, etc.}$$

Due to involvement of lone pair of nitrogen in resonance with the ring, its availability at nitrogen is deplated, reducing electron donating ability of nitrogen and hence a weaker Lewis base.

Relative basicity of substituted aniline: Any group which will increase electron donating power of —NH_2, will increase the basic strength and a group which decrease electron donating power of —NH_2, will decrease basic strength.

Let us consider three isomers of nitro aniline: Nitro group withdraw electron by both induction and by resonance.

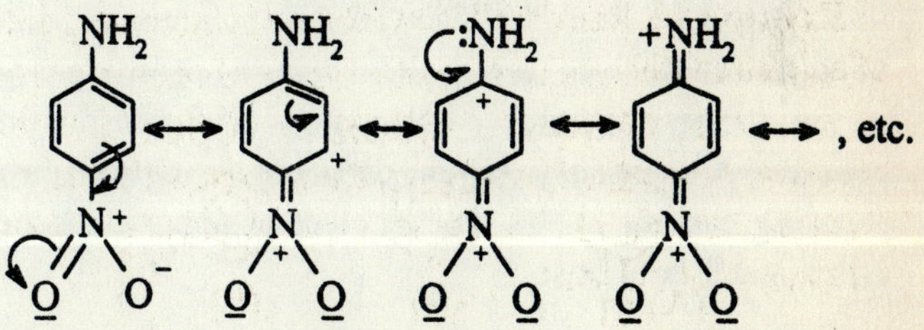

(*p*-nitroaniline)

Here —NO_2 is in direct resonance with —NH_2, a very weak base.

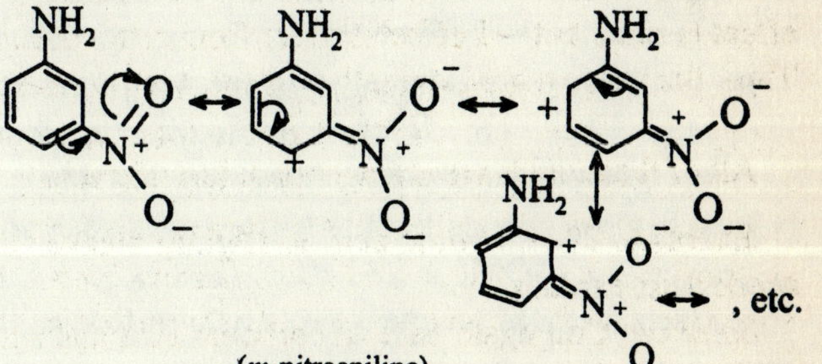

(*m*-nitroaniline)

Here —NO_2 is not withdrawing lone pair of —NH_2 by resonance and hence, stronger base than *p*-nitroaniline, but weaker than aniline due to electron withdrawn by induction.

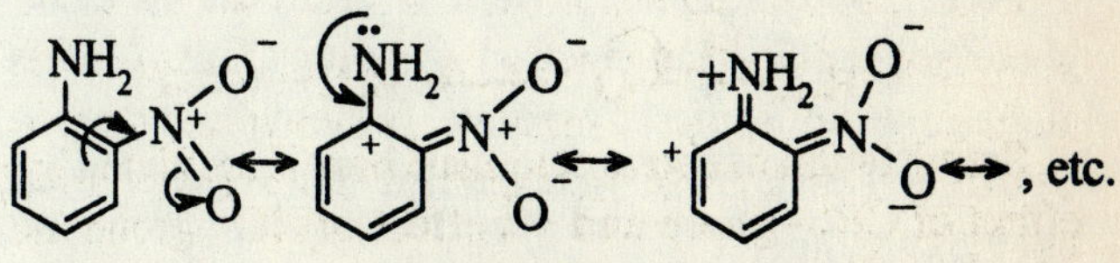

(*o*-nitroaniline)

Here —NO_2 is withdrawing lone pair of —NH_2 by resonance as in case of *p*-nitroaniline, but at the same time its electron withdrawing effect is more by induction compared to same in nitroaniline. Thus, orthonitroaniline is further weaker base than *p*-nitroaniline.

Relative Basic Strength

Aniline > *m*-nitroaniline > *p*-nitroaniline > orthonitroaniline. Similar effects on basic strength are observed when —R effect causing groups are present on the ring as —CN, —COH, COR, etc.

Aniline with a +R effect causing substituent, e.g. Methoxyaniline:

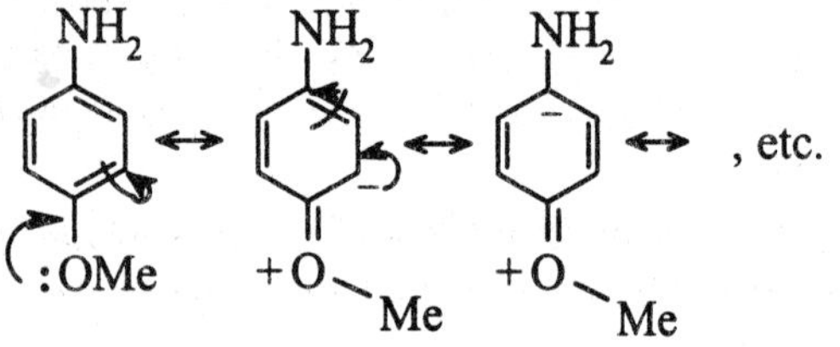

(*p*-methoxyaniline)

Here —OMe from para position bring a negative charge on the carbon bonded to —NH_2 in one of the resonating structure. A negatively charged carbon is strongly electron donating, making —NH_2 a better electron donar and hence a stronger Lewis base.

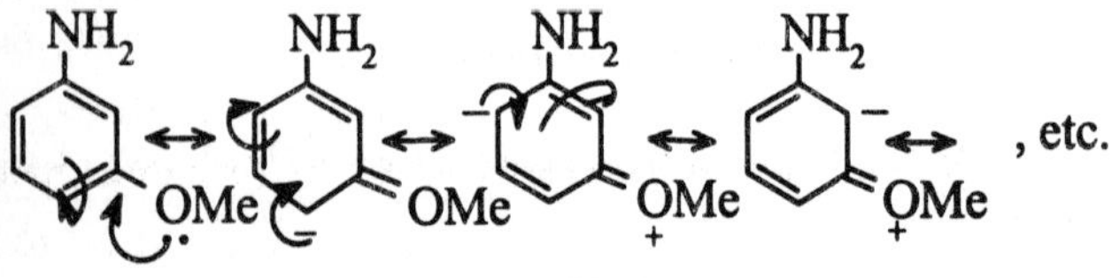

(*m*-methoxyaniline)

Here —OMe is not bringing negative charge on carbon bonded to —NH_2, i.e. —OMe is not in direct resonance with the carbon bonded to NH_2. Therefore, —OMe is not influencing basic property of NH_2 from meta position by resonance, but —OMe is a –I effect causing group, and hence meta methoxy aniline is a weaker base than aniline.

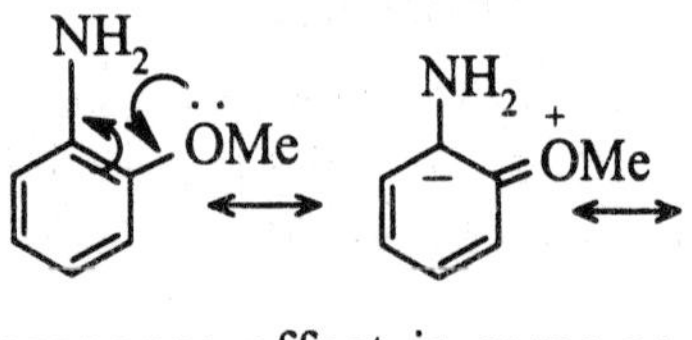

Here the resonance effect is same as in case of *p*-methoxyaniline but electron withdrawal by –I effect is more in ortho isomer. Therefore, orthomethoxyaniline is stronger Lewis base than aniline but weaker than *p*-methoxyaniline.

m-methoxyaniline < aniline < orthomethoxyaniline, < *p*-methoxyaniline.

Other similar group substituents, e.g. —OH, NH_3, etc. have similar effect on basic strength.

Case of chloroaniline: Chloro group is electron donating by resonance (+R), but electron withdrawing by induction. But the overall effect is electron withdrawl. Therefore chloroaniline is weaker base than aniline. Among isomeric aniline.

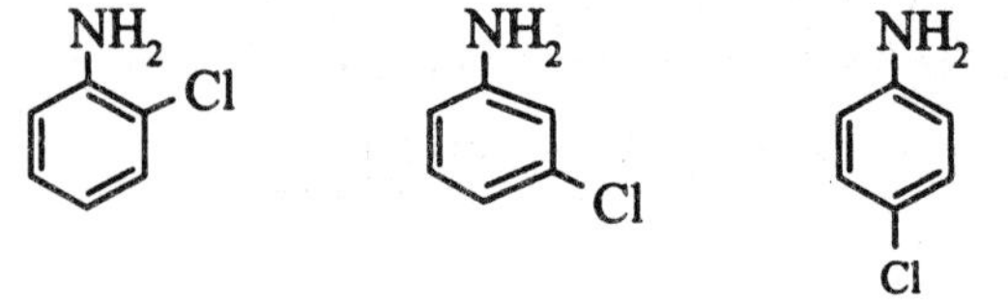

Ortho-chloroaniline Meta-chloroaniline Para-chloroaniline

Among the three chloroaniline, meta isomer is the weakest base since –Cl from meta position does not donate electron by resonance to the carbon bonded to —NH_2, and hence only its –I effect influences basic strength as:

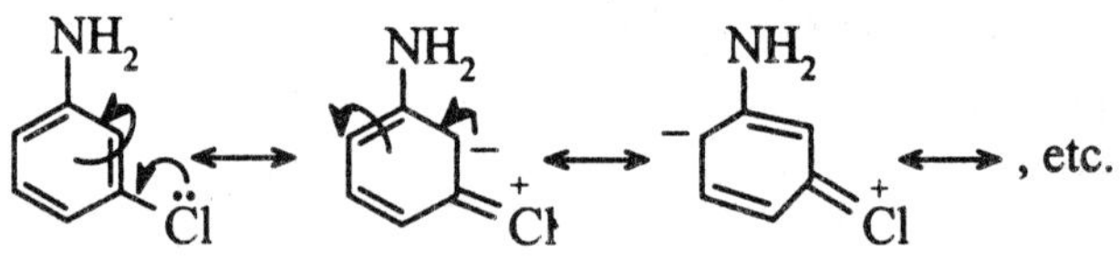

"Here —Cl is not in direct resonance with C-bonded to —NH_2."

Between ortho and para isomer, para-chloroaniline is stronger base than ortho-chloroaniline. It is due to the fact that Cl– brings negative charge on carbon bonded to —NH_2 from both ortho and para position, which favours basicity but withdraw electron from ring by induction (–I effect). Since inductive effect depends on distance, chloro group exert stronger –I effect from ortho position than from para position, decreasing basic strength more effectively from ortho position than from para position. Thus, overall order of basic strength is:

Aniline > *p*-chloroaniline > *o*-chloroaniline > *m*-chloroaniline.

Relative Basic Strength of Pyridine, Aniline, Pyrrole and Cyclohexylamine. Pyrrole is least basic amongst the above mentioned amines due to involvement of lone pair in aromaticity as:

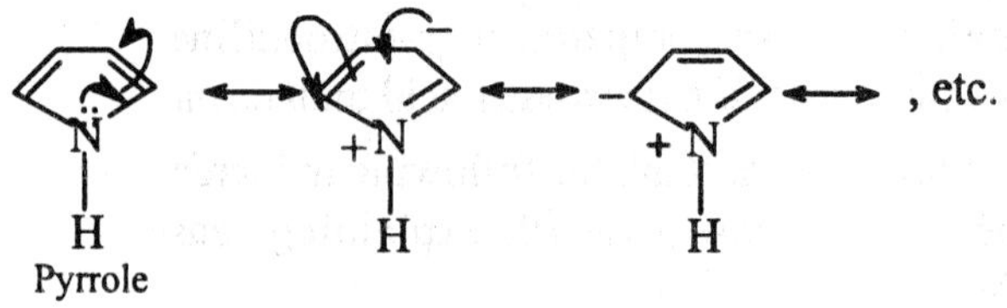

Pyrrole

Here the lone pair is part of $(4n + 2)$ π-electron aromatic system, its tendency to be donated to a Lewis acid is negligible. Between aniline and pyridine, Pyridine is a stronger base than aniline since lone pair of nitrogen in aniline is involved in delocalization into the ring, but the lone pair of nitrogen in pyridine is not delocalized, therefore more available for donation to a Lewis acid, and hence a stronger Lewis base.

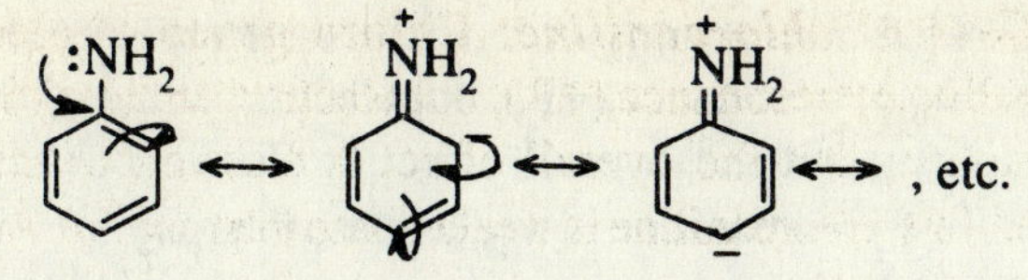

:NH_2 ↔ $\overset{+}{N}H_2$ ↔ $\overset{+}{N}H_2$ ↔ , etc.

[Lone pair is delocalized, but not part of aromaticity, therefore it is stronger base than pyrrole.]

Pyridine

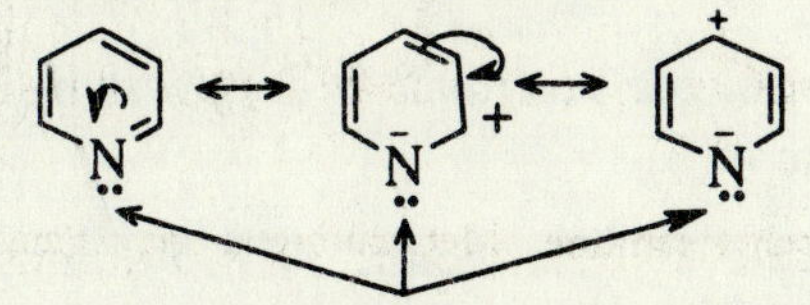

Not involved in delocalization, stronger base than aniline.

Example 24. *p*-nitroanilium chloride is a stronger acid than *m*-nitroanilium, chloride. Explain.

Solution: This can be better explained on the basis of relative stability of their conjugate base. A more stable conjugate base is formed from stronger acid.

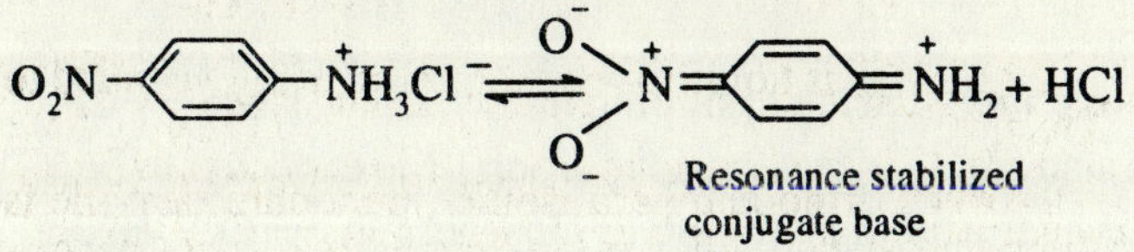

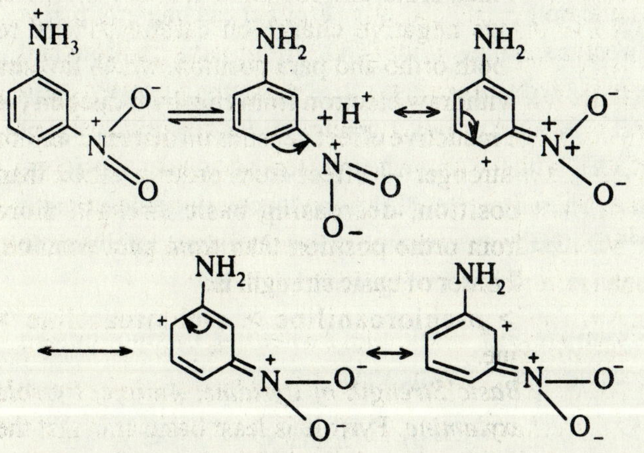

$$O_2N-C_6H_4-\overset{+}{N}H_3Cl^- \rightleftharpoons {}^-O(O)\overset{+}{N}=C_6H_4=\overset{+}{N}H_2 + HCl$$

Resonance stabilized conjugate base

Here, lone pair of electron, on nitrogen is not involved in resonance with $-NO_2$, and therefore less stable conjugate base compared to *p*-nitroaniline. Therefore, *p*-nitroanilium ion is stronger acid than its meta isomer.

Example 25. Rank the following in increasing order of their Lewis base strength explaining reason for your choice?

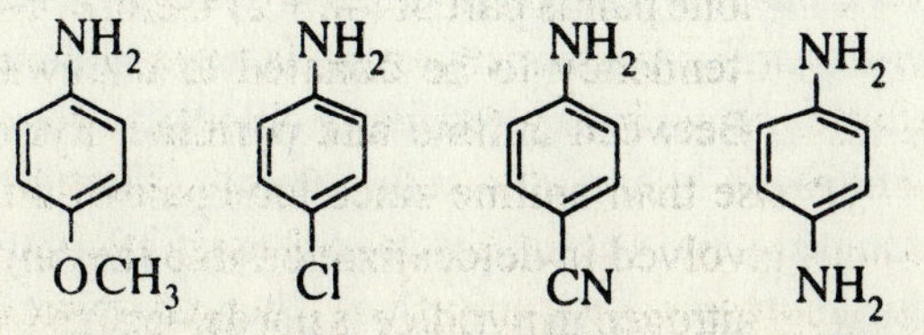

NH_2	NH_2	NH_2	NH_2
OCH_3	Cl	CN	NH_2
(A)	(B)	(C)	(D)

Solution: The increasing order of basic strength is C < B < A < D. Among the above mentioned bases, Cl and —CN have overall electron withdrawing effect and —OCH_3 and —NH_2 have electron donating (+R) effect. Both —OCH_3 and —NH_2 are para to —NH_2, and hence they are indirect resonance. Since —NH_2 is stronger +R effect giving group D is stronger base than A and both are stronger than B and C. Between B and C, —Cl is electron donating by resonance and electron withdrawing by induction (two effects are working in opposite direction cancelling part of effect) whereas —CN has electron withdrawing effect both by induction and resonance. Therefore, 'C' is weaker Lewis base than B and also both are weaker Lewis base than aniline.

Example 26. Which is a stronger Lewis base, aniline or phenyl phosphine explain.

Solution: Phenyl phosphine ($C_6H_5PH_2$) is a stronger Lewis base than aniline ($C_6H_5NH_2$). It is due to the reason that in aniline, lone pair is more delocalized into ring than in phenylphosphine. In aniline, both carbon and nitrogen, involved in delocalization are from the same row of periodic table, their *p*-orbitals have almost similar energy and better overlap is expected between orbitals of same energy. On the otherhand carbon and phosphorous are from different rows of periodic table, their *p*-orbitals involved in delocalization have different energy and an inferior overlap is expected. Therefore, lone pair of electrons are more available on phosphorous than on nitrogen indicating stronger Lewis base strength for donation to a Lewis acid.

Example 27. Which of the labeled nitrogen is protonated first when following compound is treated with one equivalent of H^+?

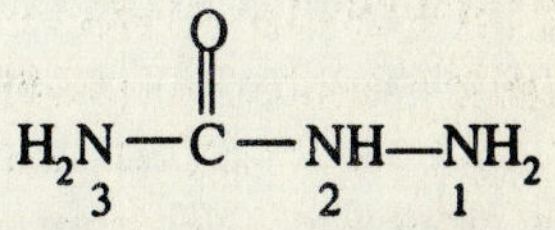

$$\underset{3}{H_2N}-\overset{O}{\overset{\|}{C}}-\underset{2}{NH}-\underset{1}{NH_2}$$

Solution: Nitrogen labeled 1 is protonated 1st since this is aminic nitrogen, rest two (2 and 3) are amidic nitrogen involved in resonance as:

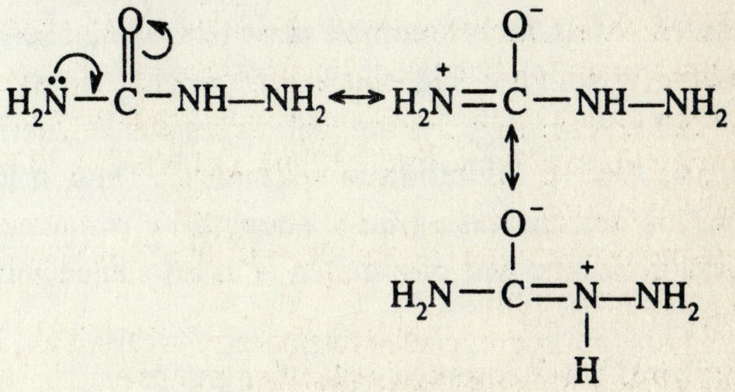

$$H_2\ddot{N}-\overset{O}{\overset{\|}{C}}-NH-NH_2 \leftrightarrow H_2\overset{+}{N}=\overset{O^-}{\overset{|}{C}}-NH-NH_2 \leftrightarrow H_2N-\overset{O^-}{\overset{|}{C}}=\overset{+}{N}(H)-NH_2$$

Example 28. Which of the two nitrogen will be protonated first when the following base is treated with two equivalent of H^+?

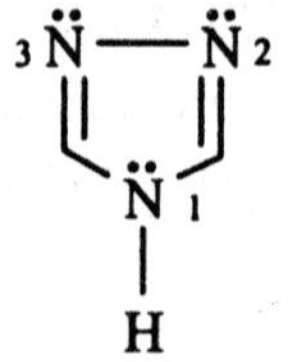

Solution: Nitrogen labeled 2 and 3 will be protonated first since their lone pair are not involved in delocalization whereas lone pair of N–1 is involved in resonance as well as part of stable (4n + 2) π-electron system required for aromaticity as:

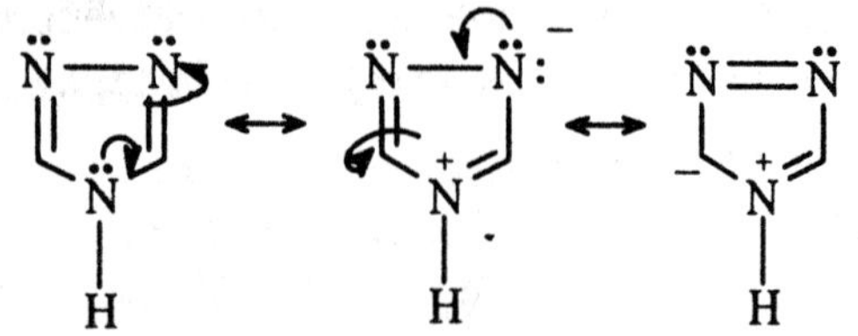

Therefore N–1 is least basic site.

Steric Inhibition to Resonance and Its Effect on Acid-Base Strength

Resonance ability of an atom is lost if it losses planarity with the other part of system, due to steric crowding by bulky groups from adjacent positions. To illustrate, let us consider following two compounds:

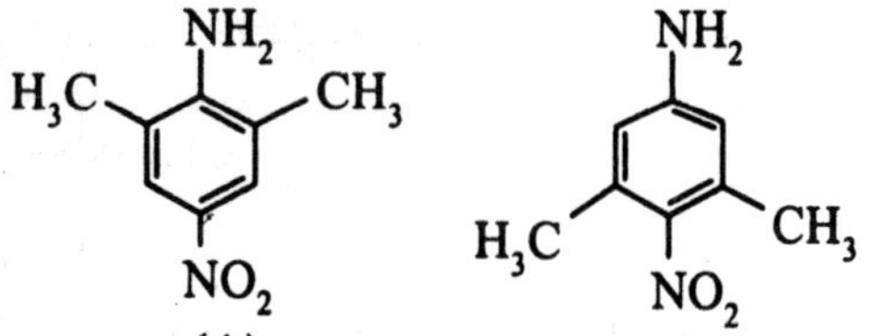

(A) (B)

The above two compounds A and B have everything identical except positions of methyl groups. It is expected that A should be stronger Lewis base than B due to closer +I effect causing groups to $-NH_2$, but it is not so. The significant factor influencing basic strength in these two compounds is –R effect of NO_2. In compound B, $—NO_2$ is sterically repelled by two adjacent bulky methyl groups and therefore inorder to minimize steric repulsion, it loses the planarity with the ring. Since now $—NO_2$ is not in plane of benzene ring it is not able to resonate. This is known as "steric, inhibition to resonance". Thus in B, $—NO_2$ is not decreasing basic strength by resonance whereas in A lone pair of nitrogen is in resonance with $—NO_2$.

Similar effect is observed with other groups as:

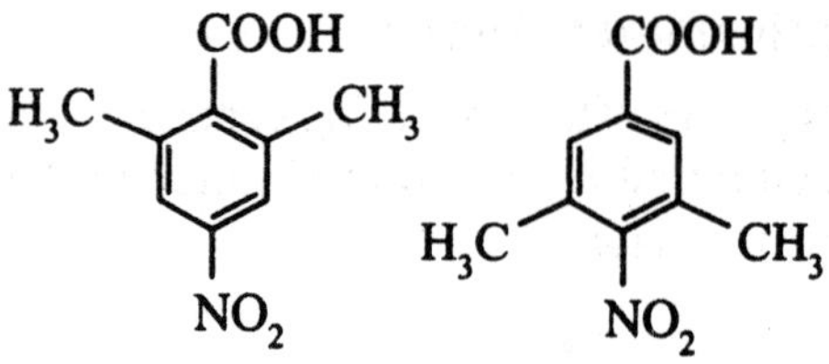

(A) (B)

A is stronger acid than B due to steric inhibition to resonance in B.

Example 29. Compound A is a stronger acid than compound B. Explain.

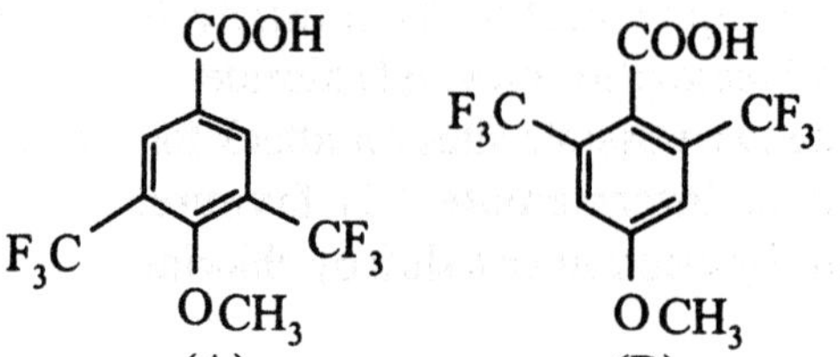

(A) (B)

Solution: On the basis of inductive effect, since $—CF_3$ is a strong –I effect causing group, B is expected to be a stronger acid than A, due to closeness to —COOH. At the same time $—OCH_3$ is a +R effect causing group, it is in direct resonance with COOH from para position, and thus expected to decrease the acid strength greatly. Also $—CF_3$ is a very bulky group, it destroy planarity of $—OCH_3$ with the ring when present on adjacent positions as in A. Therefore $—OCH_3$ is not decreasing acid strength of —COOH, via resonance in A due to steric inhibition to resonance, and it is stronger acid than B.

Example 30. Rank the following compounds in increasing order of basic strength:

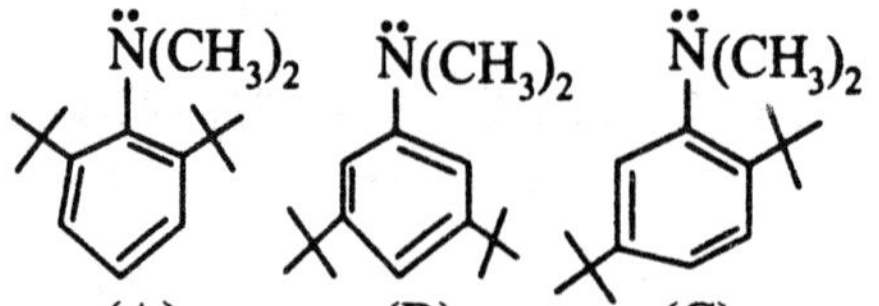

(A) (B) (C)

Solution: Lone pair on nitrogen is conjugated with the ring π-electron system and therefore may undergo delocalization, decreasing its availability on nitrogen and therefore basic strength. In A, two t-butyl groups are adjacent to amine functional group, destroying planarity of amine nitrogen with ring, and hence delocalization of lone pair of nitrogen in A is almost impossible. Therefore, A is the strongest Lewis base. Between B and C, C is stronger base due to greater inductive effect. Thus, the overall order of basic strength is:

$$B < C < A$$

3.3.3 Electromeric Effect

This is a temporary effect involving complete transfer of electrons to one or other atom joined by multiple bond, i.e. a double or triple bond, e.g.

$$A=B \rightleftharpoons \overset{+}{A}-\overset{-}{B}$$

The direction of electromeric effect is always that facilitate reaction. Therefore, electromeric effect comes in picture only during the course of reaction, i.e. during attack of electrophile, nucleophile, etc. This effect is neither seen in reactant nor in product, it appears only in the intermediate step of reaction. To illustrate, let us consider attack of an electrophile on vinyl chloride.

There are two possible sites for attack forming I and II intermediate. Intermediate I is favoured due to its stabilization by electromeric shift by chloride as:

$$\underset{\text{I}}{\overset{H}{CH_2}-\overset{+}{CH}-\ddot{Cl}:} \longleftrightarrow \overset{H}{CH_2}-CH=\overset{+}{Cl}$$

(Temporary arrangement)

$$\underset{\text{II}}{\overset{+}{CH_2}-CH_2-\ddot{Cl}:}$$

(Not stablized by electromeric shift)

In the above example, —Cl is stabilizing the intermediate (I) by electron donation. Such groups which stabilizes intermediate by electron donation are known as +E effect causing group. Examples of other such groups are —NH_2, —OCH_3, etc.

Now let us consider the case of vinyl cyanide.

$$CH_2=CH-C\equiv N + \overset{-}{Cl} \longrightarrow \underset{\text{I}}{\overset{Cl}{CH_2}-\overset{-}{CH}-C\equiv N}$$

(A-nucleophile)

$$\underset{\text{II}}{\overset{-}{CH_2}-\underset{Cl}{CH}-CN}$$

In the above example, intermediate I is favoured due to its stabilization from electron withdrawing electromeric effect of —CN as:

$$\underset{\text{I}}{\overset{Cl}{CH_2}-\overset{-}{CH}-C\equiv\ddot{N}} \longleftrightarrow \underset{\text{II}}{\overset{Cl}{CH_2}-CH=C=\ddot{N}^{-}}$$

Such group which stabilizes intermediates by electron withdrawing effect are known as –E effect causing groups. Examples of other such groups are —NO_2, —NO, —CHO, etc.

3.3.4 Hyper Conjugation Effect

Hyper conjugation involve movement of electrons of C—H bond, when it is attached to an unsaturated carbon, e.g.

$$C_6H_5-\underset{H}{\overset{H}{C}}-H \longleftrightarrow \overset{-}{C_6H_5}=\underset{H}{\overset{H^+}{C}}-H \longleftrightarrow \text{, etc.}$$

In the above example, all three hydrogen of —CH_3 group can involve in such delocalization. Due to this type of delocalization, C—H bond α–to unsaturated carbon becomes slightly more reactive than the other C—H bond, e.g.

$$CH_2=CH-\underset{\underset{a}{H}}{CH}-CH_2-\underset{\underset{b}{H}}{CH_2}$$

In this molecule C—H bond labeled a is weaker, can be attacked by a strong base more easily than the C—H bond labeled b.

$$\underset{\text{I}}{CH_2=CH-\underset{H}{\overset{H\;CH_3}{C}}-CH_2} \longleftrightarrow \underset{\text{II}}{\overset{-}{CH_2}-CH=\underset{H}{\overset{H+}{C}}-CH_2-CH_3}$$

$$\longleftrightarrow \underset{\text{III}}{\overset{-}{CH_2}-CH=\underset{H+}{\overset{H}{C}}-CH_2}$$

Hyperconjugation is a permanent effect. It is stronger than inductive effect, but weaker than resonance effect.

From the delocalization point of view, I, II, and III are hyper conjugation structure of pentene. Hyper conjugation may be regarded as "no-bond resonance". The hydrogen atoms α-to unsaturated carbon are not set free. The effect is first to increase the ionic character of C—H bond, the electron of which becomes partially delocalized through hyperconjugation.

Example 31. C—C single (σ) bond in propene is shorter than in ethane, explain.

Solution: In propene, following hyper conjugation effect is existing:

$$\underset{\text{I}}{H-\underset{H}{\overset{H}{C}}-HC=CH_2} \longleftrightarrow \underset{\text{II}}{H-\underset{H}{\overset{H+}{C}}=CH-\overset{-}{CH_2}}$$

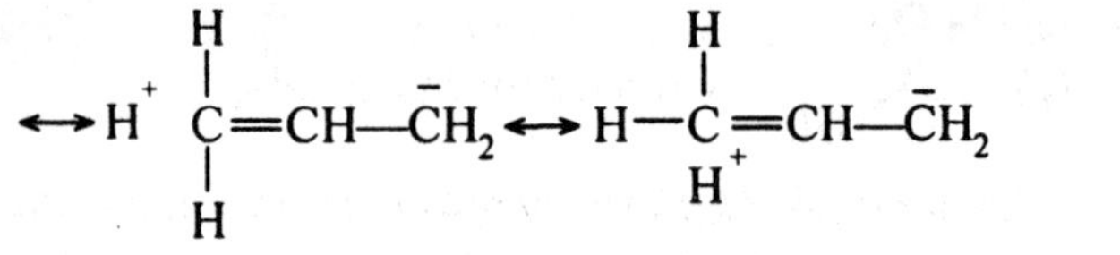

$$\leftrightarrow H^{+}\ \overset{H}{\underset{H}{C}}=CH-\bar{C}H_2 \leftrightarrow H-\overset{H}{\underset{H^{+}}{C}}=CH-\bar{C}H_2$$

III IV

Due to above hyper conjugation, C—C sigma bond has acquired some partial double bond character. In ethane ($CH_3—CH_3$), the C—C bond is purely single sigma covalent bond and bond order is one. Therefore, in propene C—C single bond is slightly stronger, and hence shorter than a pure single sigma C—C bond.

Example 32. Allyl bromide when treated with HBr, 1, 2-dibromopropane is formed as major product whereas –I effect of bromine indicate that 1, 3-dibromopropane should have been formed as major product, explain.

Solution: Electrophilic addition to allyl bromide can be examined by the following mechanism:

$$BrCH_2—CH=CH_2 + \overset{+}{H} \longrightarrow Br—CH_2—CH_2—\overset{+}{C}H_2 \quad (I)$$

$$+ Br—CH_2—\overset{+}{C}H—CH_3 \quad (II)$$

Inductive effect of Br (–I effect) suggest formation of I as stable intermediate but II is stabilized by hyper conjugation as:

$$Br—\overset{H}{C}H—\overset{+}{C}H—CH_3 \longleftrightarrow Br—\overset{H^{+}}{C}H=CH—CH_3 \text{, etc.}$$

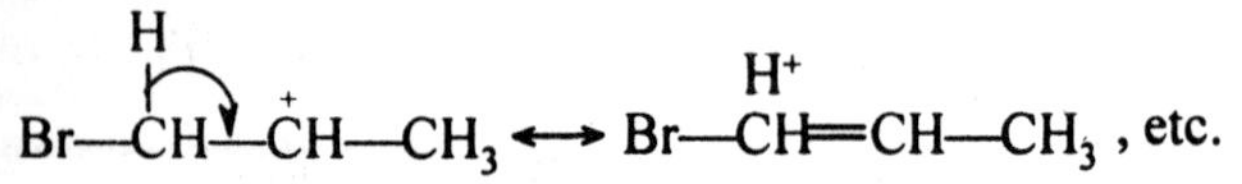

Since positively charged carbon in II has five α-H, which can be involved in hyper conjugation, II is favoured intermediate on the basis of hyperconjugation. Since hyperconjugation is stronger effect than inductive effect, II will be formed as major intermediate which will be attacked later by Br– ion giving 1, 2-dibromopropane as:

$$Br\text{-}CH_2—\overset{+}{C}H—CH_3 + \bar{Br} \longrightarrow Br\text{-}CH_2—\overset{Br}{C}H—CH_3$$

(1, 2-dibromopropane)

Example 33. Though tertiary butyl has greater +I effect than methyl group, compound I is more reactive than II towards SN-2 reaction with KCN, explain.

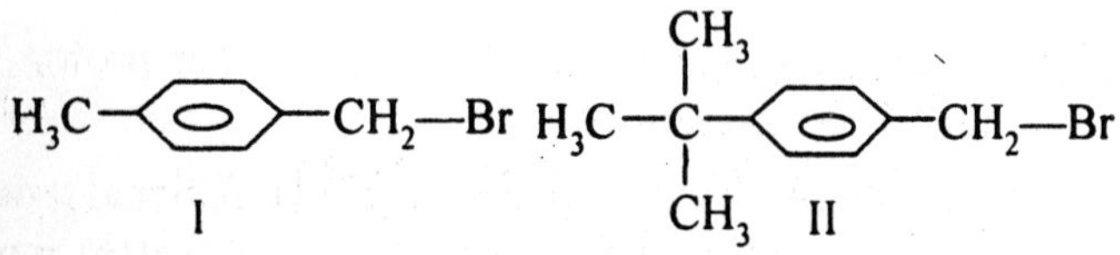

$$H_3C—C_6H_4—CH_2—Br \quad (I) \qquad H_3C—\underset{CH_3}{\overset{CH_3}{C}}—C_6H_4—CH_2—Br \quad (II)$$

Solution: The mechanism of substitution of Br^- by nucleophile (CN^-) is influenced by electron donating power of alkyl group as:

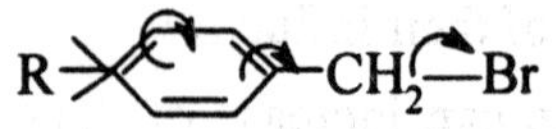

$$R—C_6H_4—CH_2—Br$$ The greater the +I effect of R— group, the faster should be substitution reaction. Besides +I effect, —CH_3 can involve in hyper conjugation facilitating substitution as shown below:

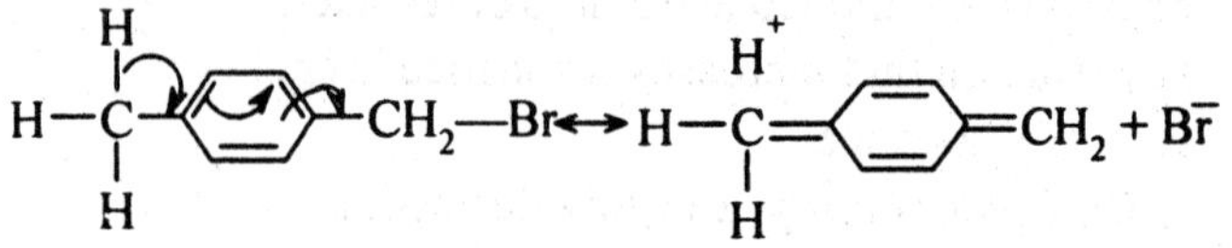

$$H—\underset{H}{\overset{H}{C}}—C_6H_4—CH_2—Br \longleftrightarrow H—\underset{H}{\overset{H^{+}}{C}}=C_6H_4=CH_2 + \bar{Br}$$

This hyper conjugation makes release of Br^- easier, therefore faster substitution is expected. No such hyper conjugation is possible with tertiary butyl group due to lack of α–H. Since hyper conjugation is stronger effect than inductive effect, substitution of Br^- in I will be faster than in II.

Example 34. Though tertiary butyl group has greater +I effect than methyl group, toluene is more reactive than tertiary butyl benzene towards aromatic electrophilic substitution reaction, explain.

Solution: Methyl group activate the benzene ring for aromatic electrophilic substitution reaction by hyper conjugation as:

$$C_6H_5—CH_2—H \longleftrightarrow \overset{CH_2\ H^{+}}{C_6H_5^{-}} \longleftrightarrow \text{, etc.}$$

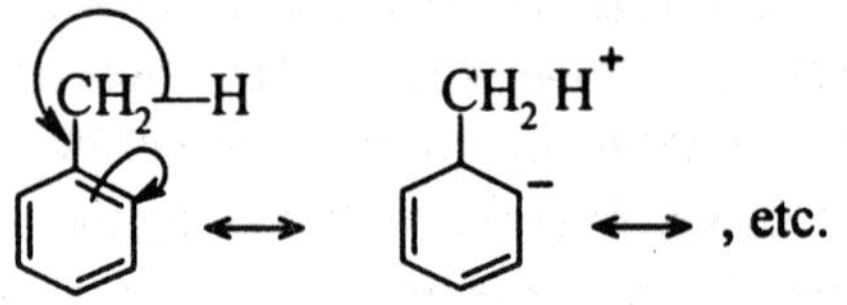

No such activation by tertiary butyl group can take place due to lack of α–H, and since hyper conjugation is a stronger effect then inductive effect, toluene is more reactive than tertiary butyl benzene towards aromatic electrophilic substitution reaction.

Some Revisionary Exercise on Electronic Displacements

Example 35. What is the single most important factor that influences the strength of any base?

Solution: A base is an electron donar, from this viewpoint, the single most important factor that influences strength of any base is the ability of that base to share electrons with an acid.

Example 36. List six structural features that influences basicity. Also rank these factors in order of most influential to least influential.

Solution: Each structural feature should be analyzed to see how it influences the ability or driving force of the base to donate electron density to an acid. Following factors are listed in decreasing order of their influence.

(a) *Resonance:* Resonance can increase or decrease electron density at the atom that shares electrons with acid. An alternate viewpoint may be useful since resonance is a stabilising feature, any reaction that causes a decrease in resonance is inhibited due to loss of resonance. Conversely, any reaction that increases resonance is enhanced by that gain of resonance.

(b) *Polarizability:* Loosely defined, polarizability is a measure of an atoms ability to distort its electron cloud. This definition works well for large atom that are easily distorted has low electronegativity, large size, but fails for small atoms that are not easily distorted. For reasons derived from quantum mechanics and electrostatics, atoms that has greater polarizability forms stronger bond with atoms with high polarizing power. Thus, the best bases are those in which a harder atom (high electronegativity and smaller size) shares electrons with protons.

(c) *Electronegativity:* Higher electronegativity of atom holding lone pair of electron mean the atom is less willing to share electron. Since the job of a base is to share its electron density with a base, higher electronegativity decrease basic strength.

(d) *Inductive effect*: An electron donating group bonded to atom holding lone pair, increases the basic strength and electron withdrawing group decreases the basic strength.

(e) *Charge:* Everything else being equal, a base with a negative charge has a greater driving force to share its electron density than a neutral base.

(f) *Hybridization:* The hybridization of an atom influences electronegativity, e.g. the relative electronegativity of carbon with different hybridization is:

$$sp > sp^2 > sp^3$$

therefore, hybridization of an atom affect the electron donating ability of an atom as:

$CH_2{=}NH$ is a weaker base than $CH_3{-}NH_2$ since nitrogen in former case has sp^2 hybridization, an inferior electron donor.

Example 37. For each pair of bases, briefly discuss the role of resonance in their relative basic strength.

(a)  cyclohexyl–O^- is stronger base than PhO^-

(b) PhO— is stronger base than o-nitrophenoxide (O^-, NO_2)

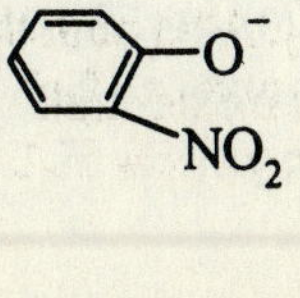

(c) 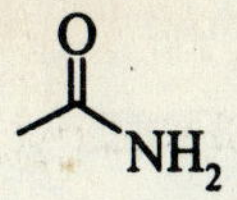 acetamide (CH_3CONH_2) is stronger base than acetone

Solution: (a) Negative charge on oxygen in phenoxide is delocalized into the ring and therefore less available for donation to a Lewis acid, hence a weaker Lewis base. On the otherhand cyclohexyl–O^- has localized negative charge on oxygen, greater electron donating power, a stronger Lewis base.

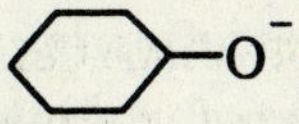

(b) Both phenoxide ion and nitrophenoxide ion has resonance delocalization as:

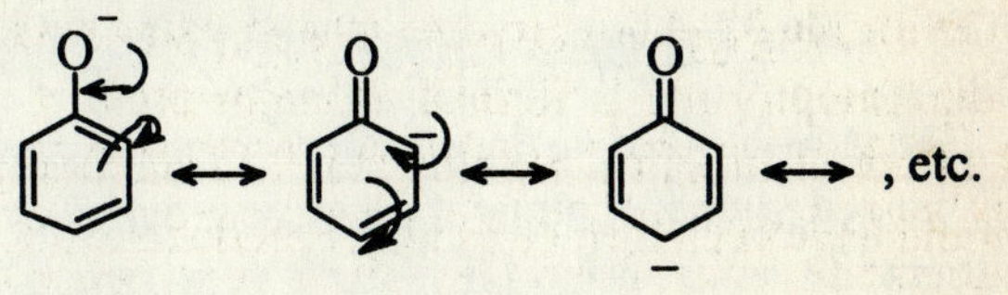

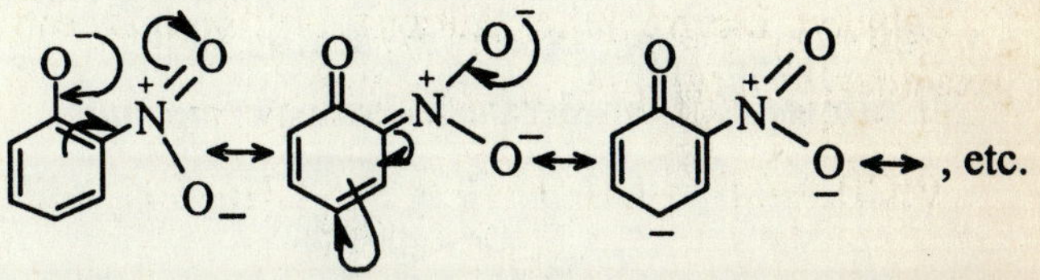

[resonance structures of phenoxide ion ↔ ↔ ↔ , etc.]

[resonance structures of o-nitrophenoxide ion ↔ ↔ ↔ , etc.]

Resonance delocalization is more in nitrophen oxide ion than in phenoxide ion due to (a) –R effect of $—NO_2$ intensify the delocalization of electron density of $—O^-$ group (b) both resonance structures have negative charge on electronegative atom, a more stable arrangement.

Therefore, nitrophenoxide is more stable, less basic. Stability of nitrophenoxide ion further intensified by ortho effect.

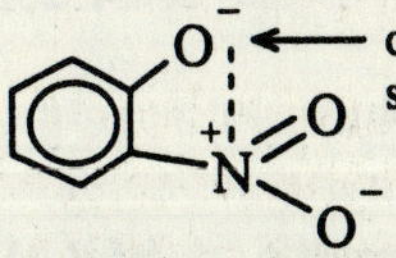

[structure: o-nitrophenoxide with ← ortho effect stabilization]

(c) Protonation of oxygen in ketone results in a cation without any other resonance contributor.

$$CH_3COCH_3 + H^+ \longrightarrow H_3C-\overset{+}{C}(OH)-CH_3$$

On the otherhand, protonation of oxygen atom in amide produces a resonance stabilized cation as:

$$H_3C-\overset{O}{\overset{\|}{C}}-NH_2 + H^+ \rightarrow H_3C-\overset{OH}{\overset{|}{\underset{+}{C}}}-\ddot{N}H_2 \leftrightarrow H_3C-\overset{OH}{\overset{|}{C}}=\overset{+}{N}H$$

Therefore, amide has greater affinity for proton, a stronger base than ketone.

Example 38. The two corboxylic acid functional group of following acid has different pk_a values. Select the most acidic —COOH functional group.

$$HOOC-\underset{}{\overset{NH_2}{\overset{|}{C}}}H_2-CH_2-COOH$$

Solution: —NH_2 is an electron withdrawing group by induction, therefore —COOH α – to —NH_2 is more acidic than β—COOH.

Example 39. Rank the following compounds in increasing order of acidic strength and briefly explain your reasoning.

$$CH_3CH_2SH,\ CH_3CH_2OH,\ CF_3CH_2SH$$

Solution: The increasing order of acid strength is

$$CH_3CH_2OH < CH_3CH_2SH < CF_3CH_2SH$$

Since H-bonded to larger atom is more acidic than H-bonded to smaller atom of the same group, if everything others are same. Also CF_3 being very strong electron withdrawing group, increases the acid strength.

Example 40. Consider the following compound:

[Structure: HO(a)–C(=O)–C(OH c)(SH b)–]

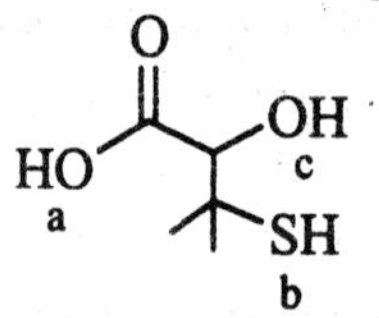

Rank the labeled proton in increasing order of acidity

Solution: Carboxy H is more acidic than hydroxy or thioxy hydroxy, between —OH and —SH, latter H is more acidic thus the order of acid strength is:

c < b < a

Example 41. Select the most acidic proton in the following molecule:

HO—CH_2—CH_2—COOH, HO—CH_2—CH_2—SH,

HO—CH_2—CH_2—NH_2, $HOCH_2CF_2CH_2CH_2OH$

Solution

$HOCH_2CH_2$—C(=O)—O—H; conjugate base will be resonance stabilized.

$HOCH_2CH_2S$—H; conjugate base with negative charge on S will be more stable.

H—$OCH_2CH_2NH_2$; H-bonded to oxygen is more acidic than H-bonded to nitrogen.

H—$OCH_2CF_2CH_2CH_2OH$; CF_2 is closer to labeled hydrogen.

3.4 ELECTROPHILES AND NUCLEOPHILES

Discussion: Consider the reaction of hydroxide ion with

$$HCl:\ \ H\ddot{O}:^- + \overset{\delta+}{H}-\overset{\delta-}{Cl} \longrightarrow H-O-H + :\ddot{Cl}:^-$$

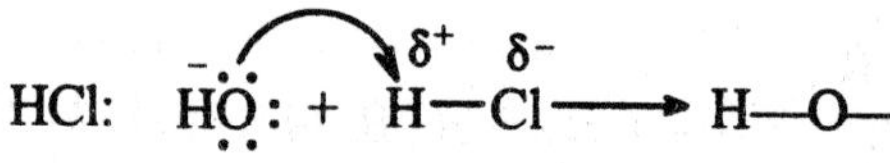

The oxygen of hydroxide ion bears a negative charge. The hydrogen of HCl bears a δ^+ charge, because chlorine is more electronegative than hydrogen and thus the H—Cl shared electron pair is unequally shared. We can envision the start of the acid-base reaction between hydroxide ion and HCl as electrostatic attraction between the opposite charges. As the reaction proceeds, the oxygen atom of the hydroxide ion shares a lone pair with the hydrogen atom of HCl as shown by curved arrow in the reaction above. This simple reaction shares one feature in common with the majority of reaction that we encounter in organic chemistry. One species in the reaction, shares an electron pair with another species having an electron deficient atom to make a new covalent bond.

Because frequent use of electron donor and acceptors used in organic chemistry, we assign a special and distinct term to these species.

A molecule or ion that accepts a pair of electron to make a new covalent bond is called *electrophile.* An electrophile is the same thing as a Lewis acid. Any molecule, ion or atom that is electron deficient in same way can behave as an electrophile. Electron deficiency may be in the form of a full positive charge, a partial positive charge or an open octet. "E" or "E^+" are the common abbreviations for generic electrophile. Some typical electrophiles are listed below:

(a) Electrophile with full positive charge.

CH_3^+, Br^+, H^+ R—C^+= O, etc.

(b) Electrophile with partial positive charge.

$\overset{\delta+}{H}-\overset{\delta-}{Cl}$, $\overset{\delta+}{H_3C}-\overset{\delta-}{Cl}$, etc.

(c) Eelectrophile with an open octet.

BH_3 (H—B(H)—H), BX_3 (X—B(X)—X), H—$\ddot{C}$—H, etc.

Carbone

A molecule or ion that donates a pair of electrons to form a new covalent bond is called a *nucleophile.* A nucleophile is the same thing as a Lewis base. Any molecule, ion or atom that has electrons that can be shared can be a nucleophile. The most common indications that electrons are available to be shared are a formal negative charge (δ^-) as on halide ions, hydroxide ion (HO^-), alkoxide ion (RO^-), etc. A partial negative charge (δ^-), usually in a polar bond, a π – bond or a neutral species with an unshared electron pair as H_2O, NH_3, etc.

$:\ddot{\underset{..}{I}}:^{-}$, $\overset{\delta -}{R}$—$\overset{\delta +}{Mg}$Br, $R_2C{=}CH_2$, $\ddot{N}H_3$, R—$\ddot{N}H_2$, $H_2\ddot{O}$, etc.

The study of reaction mechanism is central to the study of organic chemistry at any level. Therefore, identification of electrophiles and nucleophiles is a critical organic chemistry survival skill. Because electron flow from an electron source to a place of electron deficiency, a curved arrow points away from a nucleophile and to an electrophile.

Thus, direction of electron flow is always froms nucleophile to electrophile.

Nucleophile ⟶ Electrophile

Example 42. Using curved arrows, label each reactant as nucleophile or electrophile.

$$Cl^- + CH_3O{-}S(=O){-}Cl \rightarrow CH_3Cl + SO_2 + Cl^-$$

Solution

$$Cl^- + CH_3{-}O{-}S(=O){-}Cl \rightarrow CH_3{-}Cl + SO_2 + Cl^-$$

As shown in the above mechanism, chloride ion is donating a lone pair to form a new bond with carbon. The chloride ion is at the origin on the curved arrow that indicates this bond change. That the chloride ion bears a formal negative charge further suggest that it function as nucleophile. Because nucleophile must react with an electrophile that leaves other part of molecule (a chlorosulphite ester) to be the electrophile. The ester carbon accepting an electron pair from nucleophile to form new covalent bond C—Cl bond. The other electronic change within a chlorosulphite ester molecule are inconsequential when defining its role as nucleophile or electrophile. The chlorine atom that is expelled as chloride ion accept an electron pair from S—Cl bond, but it does not make any new covalent bond, so it is neither electrophile nor nucleophile, it is simply leaving group.

Example 43. Identify the nucleophiles and electrophiles in each mechanism shown below:

(a) $CH_3S^- + C(R)(CH_3)(I)(H) \rightarrow C(R)(CH_3)(H)(SCH_3) + I^-$

(b) $Ph{-}CH_2{-}Br \rightleftharpoons Ph{-}CH_2^+ + Br$

(c) Chlorocyclohexane $+ OH^- \rightarrow$ cyclohexene $+ H_2O + Cl^-$

(d) Benzene $+ \overset{+}{N}O_2 \rightarrow$ nitrobenzene ($C_6H_5NO_2$) $+ H^+$

Example 44. Decide if each of the molecule, or ion shown below will react as electrophile, nucleophile or both:

(a) $(CH_3)_3C^+$ (b) HO^- (c) $AlCl_3$ (d) $HC{\equiv}CH$

(e) H_2CO (f) CH_3Cl (g) $H_3C{-}C(=O){-}NH_2$

3.5 HOMOLYTIC AND HETEROLYTIC BOND FISSION IN A CHEMICAL REACTION

All chemical reactions involve some change in bondings as reactants are converted into product, e.g. some of the bonds in reactant are broken and some new bonds are formed in products.

A covalent bond may break into two fundamentally different ways—homolytic bond fission and heterolytic bond fission.

Homolytic bond fission: If a covalent bond is broken in such a way that each fragment takes away one of the shared electrons of the bond, this type of bond cleavage is known as homolytic bond fission.

$$A{:}B \longrightarrow A^{\circ} + B^{\circ} \text{ Homolytic bond fission.}$$

In homolytic bond fission, odd electron species are produced, commonly known as – "Radicals". Some of the examples of homolytic bond fission are described below:

$$H{-}H \longrightarrow 2H^{\circ}$$
$$Cl{-}Cl \longrightarrow 2Cl^{\circ}$$
$$HO{-}OH \longrightarrow 2HO^{\circ}$$
$$R{-}Br \longrightarrow R^{\circ} + Br^{\circ}$$

Heterolytic bond fission: If a covalent bond breaks in such a way that one fragment taken away both shared electrons of the bond leaving the other fragment with an empty orbital, the type of bond cleavage is known as heterolytic bond fission as:

$$A{-}B \equiv A{:}B \longrightarrow A^+ + B^-$$

In heterolytic bond fission, both shared electrons goes with more electronegative atom. Also, in heterolytic bond fission, ions are produced as the products:

$$H-Cl \longrightarrow H^+ + Cl^-$$

$$R-Cl \longrightarrow R^+ + Cl^-$$

$$I-Cl \longrightarrow I^+ + Cl^-$$

$$R-COOH \longrightarrow R-\overset{O}{\overset{\|}{C}}-\bar{O} + H^+$$

3.6 REACTIVE INTERMEDIATES PRODUCED IN A CHEMICAL REACTION

3.6.1 Introduction

A chemical reaction that complete in several stages include the formation of one or more reactive intermediates before the final, stable product is formed. The formation of an intermediate can be viewed on potential energy as shown in Figure 3.9.

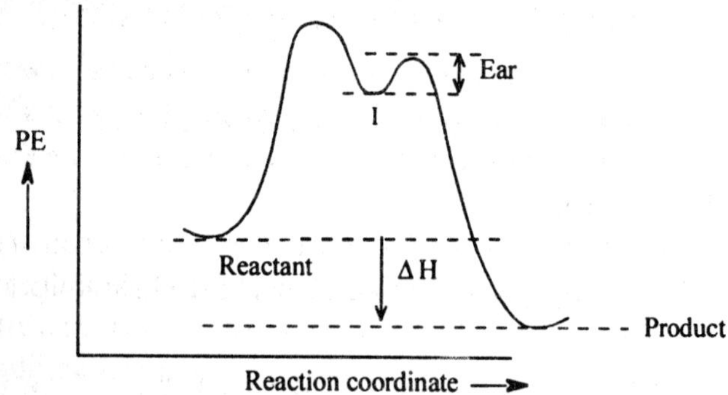

Figure 3.9: Potential energy showing formation of a reactive intermediate (I)

The species that exist at the bottom of the above potential energy well (I) is called reactive intermediate and thus have a real time existence, though very short lived. A reactive intermediate must overcome some energy barrier, however small (Ear), before it can follow the reaction pathway and proceed onward to product. It is this barrier that gives intermediate a measurable life-time.

Some of the salient features of a reactive intermediates are:

(a) It is produced in the intermediate step of a chemical reaction, i.e. it does not appear at the beginning nor at the end of a reaction.
(b) Reactive intermediates are real species, have very short, but measurable life time.
(c) Reactive intermediate cannot be preserved, i.e. once formed, it must be exhausted before the end of reaction.
(d) The most stable reactive intermediate is formed in largest fraction that leads to formation of major product.

Important type of ***Reactive Intermediates*** formed in organic reaction pathway.

3.6.2 Carbocation (or Carbonium Ion)

Introduction: Carbocation is an ionic species, in which a carbon atom bears a positive charge and three covalent bond, and positively charged carbon is short of two electron from the stable configuration, octet.

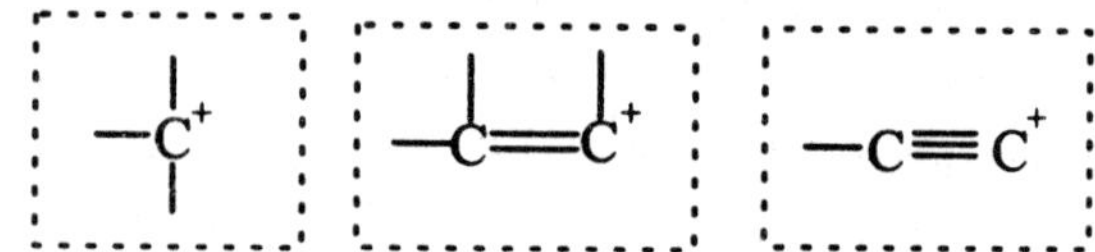

Alkyl carbocation Vinyl carbocation Acetylinic carbocation

One common thing in all the above mentioned carbocation is, they all have a positively charged carbon with open octet.

Since carbocations have one carbon with only six valence electron, they are highly electron-deficient species, and that is why they are highly reactive and termed as reactive intermediate. Also, carbocation comes in the family of electrophile, since they have very high affinity for electron. A carbocation can also be termed as Lewis acid, since it can accept a pair of electron from a Lewis base.

Hybridization of positively charged carbon in alkyl carbocation is sp^2, therefore it has triangular planar arrangement of bonds around the positively charged carbon as shown in Figure 3.10.

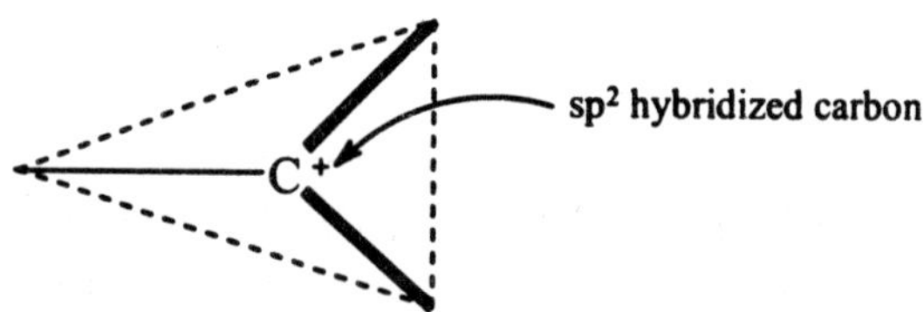

Figure 3.10: A carbocation has triangular planar arrangement of bonds around positively charged carbon

Classification: A carbocation is classified into following categories on the basis of bonding environment around the positively charged carbon.

(a) Methyl carbocation
(No C—C⁺ bonds)

$$H-\overset{H}{\underset{H}{C^+}} \qquad CH_3O-\overset{H}{\underset{H}{C^+}}, \text{ etc.}$$

(b) A primary carbocation
(One C—C⁺ bond, 1°)

$$CH_3CH_2-\overset{H}{\underset{H}{C^+}} \qquad C_6H_5-\overset{H}{\underset{H}{C^+}}, \text{ etc.}$$

(c) A secondary carbocation (2°)
(Two C—C⁺ bonds)

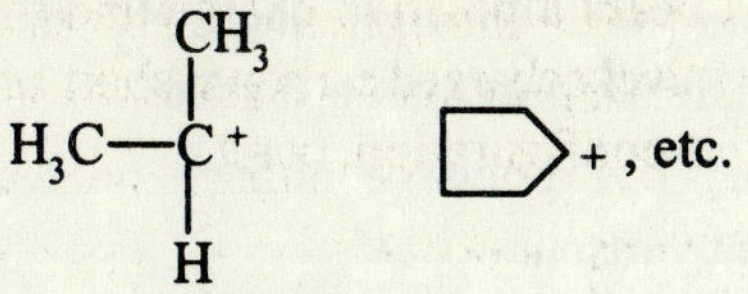

$$H_3C-\overset{CH_3}{\underset{H}{C^+}}$$, etc.

(d) A tertiary carbocation (3°)
(Three C—C⁺ bonds)

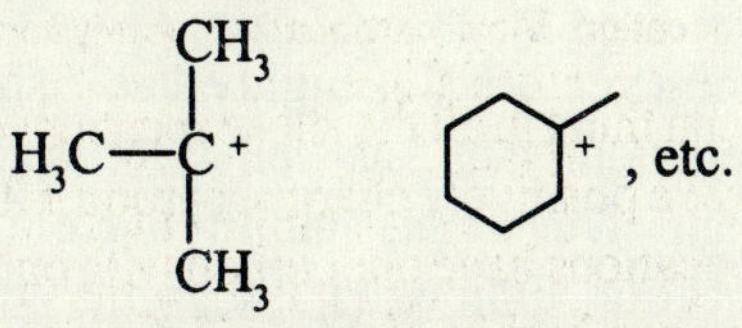

$$H_3C-\overset{CH_3}{\underset{CH_3}{C^+}}$$, etc.

(e) *Allyl and allylic carbocation:* If the positively charged carbon (C^+) is bonded to two 'H' and a vinyl group ($CH_2{=}CH{-}$), this is termed as allyl carbocation, and if positively charged carbon is bonded to one or two alkyl group and a vinyl group, these are termed as allylic carbocation as:

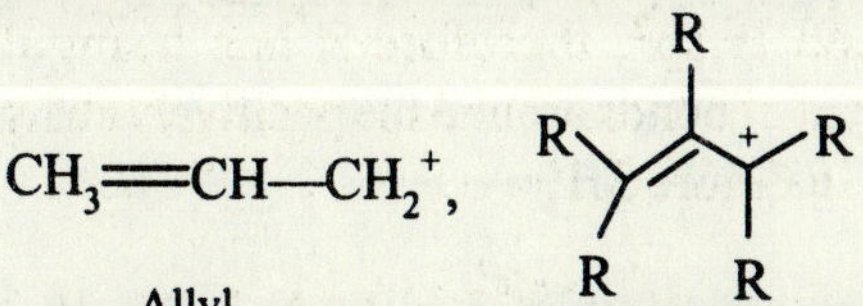

$$CH_3{=}CH{-}CH_2^+,$$

Allyl

Allylic

(f) *Benzyl and benzylic carbocation:* If a positively charged carbon is bonded to two 'H' and a phenyl ring, this is termed as benzyl carbocation, and if positively charged carbon is bonded to one or two alkyl group and a phenyl ring, these are termed as benzylic carbocation as:

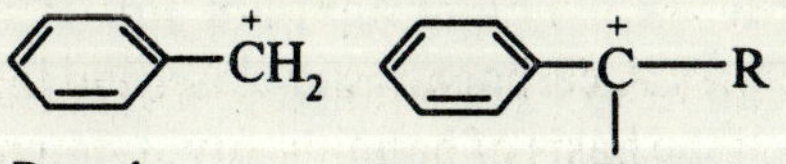

$$C_6H_5-\overset{+}{C}H_2 \qquad C_6H_5-\overset{+}{\underset{R}{C}}-R$$

Benzyl Benzylic

Relative Stability of Carbocations

Carbocations are electron deficient species, groups bonded to the positively charged carbon govern the relative stabilities as:

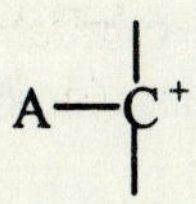

$$A-\overset{|}{\underset{|}{C^+}}$$

Group bonded to positively charged carbon (A) can be an electron donating group or it can be an electron withdrawing group.

A Electron Donating

An electron donating 'A' can share some of its electron-density with electron deficient carbon (C^+) compensating some of electron deficiency at C^+ and, therefore will increase the stability of carbocation. "Therefore electron donating atom or group bonded to C^+ increases the stability of carbocation intermediate". A can donate electron to C^+ by any of the following effect.

(i) Inductive effect.
(ii) Resonance effect.
(iii) Hyperconjugation effect, etc.

A Electron withdrawing

An electron withdrawing A, will take away some of the electron density from already electron deficient C^+, intensifying the electron-deficiency at C^+. Such group will rise-up the potential energy of carbocation, lowering stability.

Therefore, an electron withdrawing group bonded to C^+ lowers the stability. Electron withdrawal can occur by any one of the three effects mentioned above.

Relative Stability of Various Types of Carbocation

(a) Alkyl carbocation

Since an alkyl is electron donating group, donate electron by both inductive and hyper conjugation, when bonded to C^+, lowers the potential energy and increases stability of carbocation. Therefore, a methyl carbocation is lest stable among the alkyl carbocation since there is no alkyl group bonded to C^+.

Substitution of H– by alkyl group in CH_3^+ increases stability. The general order of stability of alkyl carbocation is:

$$CH_3^+ < \underset{1^\circ}{RCH_2^+} < \underset{2^\circ}{R_2CH^+} < \underset{3^\circ}{R_3C^+}$$

Example 45. Why a $O_2N{-}CH_2^+$ is less stable carbocation than CH_3^+?

Solution: Since $-NO_2$ is a very strong electron withdrawing group, it intensify the electron deficiency at C^+, decreasing stability compared to CH_3^+.

(b) Vinyl and acetylinic carbocation

In vinyl carbocation, the hybridization of C^+ is sp:

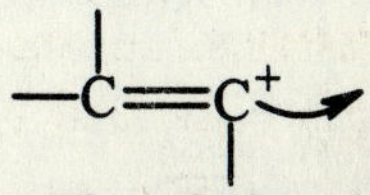

$$-\overset{|}{C}{=}\underset{|}{C^+} \rightarrow \text{sp hybridized carbon.}$$

The relative electronegativity of carbon in a compound depends on percentage of *s*-character. More the *s*-character, greater the electronegativity. This is due to the reason that nucleus of a carbon with greater *s*-character, will have greater electron attracting power.

A vinyl carbocation has 33.33% *s*-character at C^+ while an allkyl carbocation has 25% *s*-character at C^+. On the basis of relative electronegativity of carbon bearing positive charge, it can be concluded that greater the *s*-character, lower will be the stability due to greater electron withdrawing power of nucleus of carbon.

Alternatively, it can also be interprated on the basis of relative distance of positive charge from the nucleus of carbon bearing this positive charge. Since nucleus is also positively charged, greater the distance of this positive charge from the nucleus, smaller will be the repulsion, and hence greater will be the stability. Therefore, an alkyl carbocation is more stable than a vinyl carbocation.

On the same basis it can be established that an acetylinic carbocation ($RC{\equiv}C^+$) is further less stable than vinyl carbocation, since here the positive charge is closest to the nucleus of carbon and have maximum repulsion.

$$RCH_2^+ > RCH{=}CH^+ > RC{\equiv}C^+ \quad \text{Stability.}$$

(c) Stabilization of carbocation through resonance

If the positive charge of a carbocation is delocalized by resonance, the potential energy decreases and therefore stability increases. An allyl carbocation is more stable than a primary and secondary carbocation as:

$$CH_2{=}CH{-}\overset{+}{C}H_2 \longleftrightarrow \overset{+}{C}H_2{-}CH{=}CH_2$$

Allyl C+

An allylic carbocation is further more stable than allyl carbocation as:

$$R{-}CH{=}CH{-}\overset{+}{C}H{-}R \longleftrightarrow R{-}\overset{+}{C}H{-}CH{=}CH{-}R$$

There is stabilized by resonance as well as by +I effect of R— group. An allylic carbocation is even more stable than a tertiary alkyl C^+:

$$RCH_2^+ < R_2\overset{+}{C}H < RCH{=}CH{-}\overset{+}{C}H_2 < R_3\overset{+}{C} < R{-}CH{=}CH{-}\overset{+}{C}H{-}R$$

Increasing stability ⟶

A similar stabilization effect is observed with benzyl carbocation:

$$C_6H_5{-}\overset{+}{C}H_2 \longleftrightarrow (\text{ring}^+){=}CH_2 \longleftrightarrow (\text{ring}^+){=}CH_2 \longleftrightarrow (\text{ring}^+){=}CH_2$$

Benzyl C+

Therefore, benzyl carbocation is more stable than a primary, secondary and allyl carbocation.

Example 46. Though $ROCH_2^+$ is a primary carbocation, it is highly stable, explain.

Solution: The positive charge on carbon can be delocalized by resonance as:

$$\underset{\text{I}}{R{-}\ddot{O}{-}\overset{+}{C}H_2} \longleftrightarrow \underset{\text{II}}{R{-}\overset{+}{O}{=}CH_2}$$

Resonance structure II is highly stable since it has one extra covalent bond than I as well as it has complete octet of all the atoms. Due to this stability factor of II, it is very stable carbocation, although it is a primary carbocation.

Example 47. Which is a more stable carbocation, $ROCH_2^+$ or $RSCH_2^+$, explain.

Solution: $ROCH_2^+$ is more stable than $RSCH_2^+$. Both can be stabilized by resonance as:

$$R{-}\ddot{O}{-}\overset{+}{C}H_2 \longleftrightarrow R{-}\overset{+}{O}{=}CH_2 \quad \text{...(i)}$$

$$R{-}\ddot{O}{-}\overset{+}{C}H_2 \longleftrightarrow R{-}\overset{+}{S}{=}CH_2 \quad \text{...(ii)}$$

Resonance delocalization is more in I, because delocalization is taking place bewteen orbitals of same energy (2p of both carbon and oxygen). On the otherhand, extent of delocalization is less in II, because orbitals of different energies (2p of C and 3p of S) are invovled, stabilization in (ii) is less compared to (i).

Example 48. Though acylium ion ($R\overset{+}{C}{=}O$) has an electronegative oxygen bonded to C^+, it is a very stable carbocation, explain.

Solution: Acylium ion is a resonance stabilized species as:

$$\underset{\text{I}}{R{-}\overset{+}{C}{=}\ddot{O}} \longleftrightarrow \underset{\text{II}}{R{-}C{\equiv}\overset{+}{O}}$$

Structure II has an extra covalent bond as well as complete octet of all atoms, favouring stability.

Rearrangements of Carbocation

Rearrangement into a more stable carbocation is an important characteristic of this intermediate. A carbocation may undergo rearrangement by either hydride shift (H^- shift) or by alkyl shift (R^- shift) from an α-carbon to produce a more stable carbocation as:

(a) Hydride shift

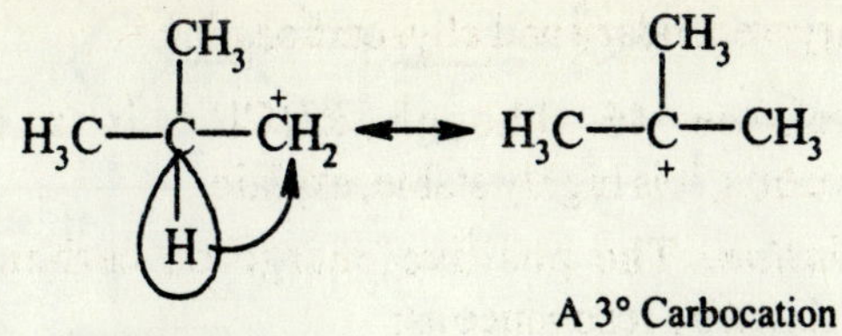

$$H_3C-\overset{CH_3}{\underset{H}{C}}-\overset{+}{C}H_2 \longleftrightarrow H_3C-\overset{CH_3}{\underset{+}{C}}-CH_3$$

A 1° Carbocation — A 3° Carbocation (more stable)

(b) Alkyl shift

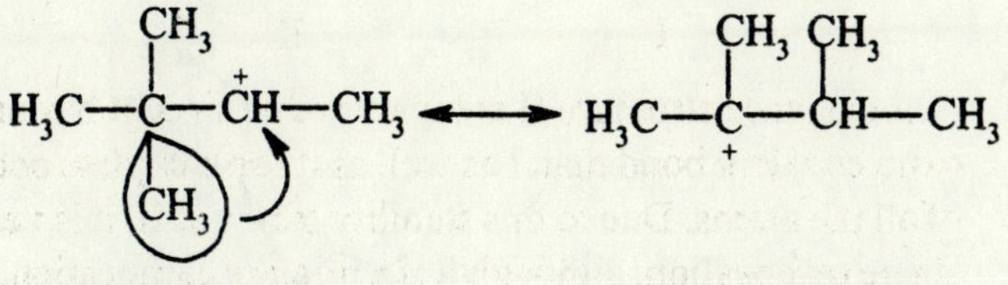

$$H_3C-\overset{CH_3}{\underset{CH_3}{C}}-\overset{+}{C}H-CH_3 \longleftrightarrow H_3C-\overset{CH_3}{\underset{+}{C}}-\overset{CH_3}{C}H-CH_3$$

A 2° Carbocation — A 3° Carbocation

The essential requirement for rearrangement is a more stable intermediate must be produced after rearrangement. Therefore, if a carbocation intermediate is formed in a chemical reaction and structure permits, the rearrangement into a more stable carbocation, as shown above in (a) and (b) occur the most stable carbocation formed after rearrangement dominate and gives rise to the major product by reacting in the subsequent step. An illustration of this rearrangement can be seen in the electrophilic addition of HCl to 3, 3-dimethyl-1-butene as:

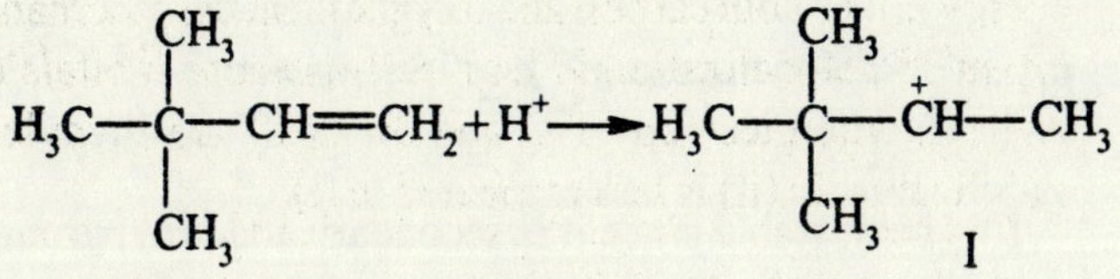

$$H_3C-\overset{CH_3}{\underset{CH_3}{C}}-CH{=}CH_2 + H^+ \longrightarrow H_3C-\overset{CH_3}{\underset{CH_3}{C}}-\overset{+}{C}H-CH_3 \quad \text{I}$$

A stable 2° C⁺ is formed in 1st step.

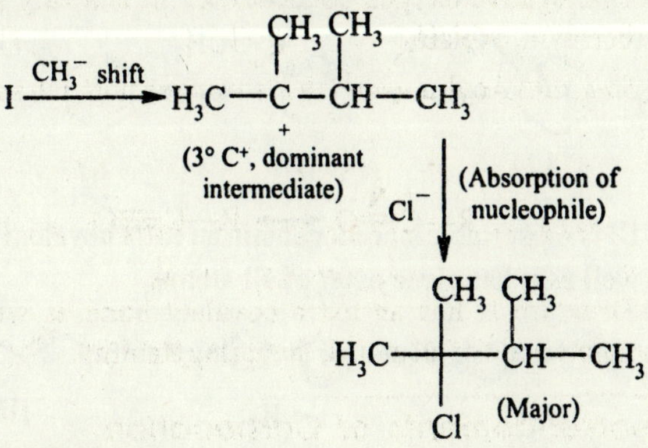

$$\text{I} \xrightarrow{CH_3^-\ \text{shift}} H_3C-\underset{+}{\overset{CH_3}{C}}-\overset{CH_3}{C}H-CH_3$$

(3° C⁺, dominant intermediate)

$\downarrow Cl^-$ (Absorption of nucleophile)

$$H_3C-\overset{CH_3}{\underset{Cl}{C}}-\overset{CH_3}{C}H-CH_3 \quad \text{(Major)}$$

Example 49

$$CH_3OCH_2CH{=}CH_2 + HCl \longrightarrow CH_3O\underset{Cl}{C}H-CH_2CH_3 \quad \text{(Major)}$$

Justify the formation of major addition product.

Solution: Mechanism of addition reaction:

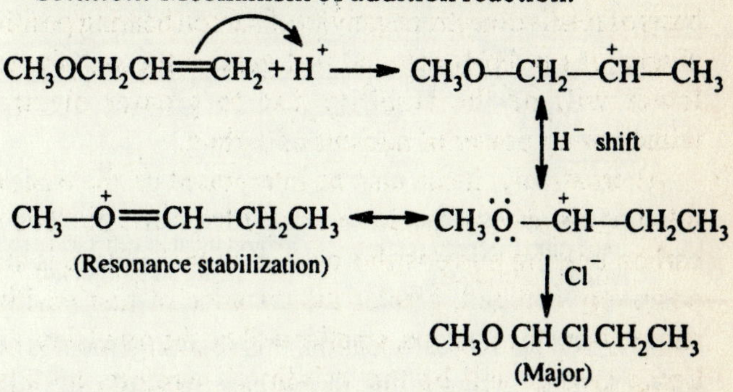

$$CH_3OCH_2CH{=}CH_2 + H^+ \longrightarrow CH_3O-CH_2-\overset{+}{C}H-CH_3$$

$\updownarrow$ H^- shift

$$CH_3-\overset{+}{O}{=}CH-CH_2CH_3 \longleftrightarrow CH_3\ddot{O}-\overset{+}{C}H-CH_2CH_3$$

(Resonance stabilization)

$\downarrow Cl^-$

$$CH_3O\,CH\,Cl\,CH_2CH_3 \quad \text{(Major)}$$

(c) Stabilization by ring expansion

Smaller rings like cyclopropyl, cyclobutyl rings are highly unstable due to angle strain as:

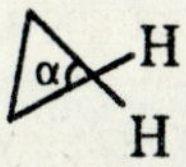

$\alpha = 60$, but C of ring are sp^3 hybridized, bond angle requirement is 109°. Therefore, these rings are highly strained due to forced compression of bond angle and may undergo stabilization by ring expansion releasing some of the strain and thereby increasing stability as:

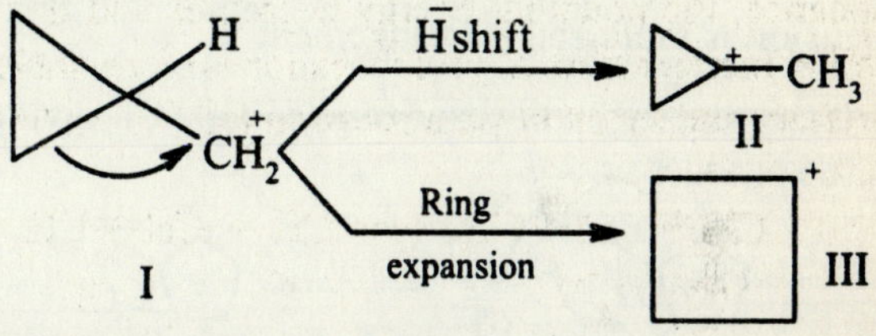

I (cyclopropyl(H)–$\overset{+}{C}H_2$) —$\bar{H}$ shift→ II ($\overset{+}{}$cyclopropyl–CH_3); I —Ring expansion→ III (cyclobutyl cation)

Here II is less stable than I, though II is a tertiary carbocation. This is due to increase in angle strain in II. Where as III is more stable than I due to release in same angle strain. To illustrate the effect of ring expansion, let us consider the following electrophilic addition reactions.

(i) (1-methyl-1-vinylcyclobutane) $\xrightarrow{HCl}$ (chloro-dimethylcyclopentane, Cl)

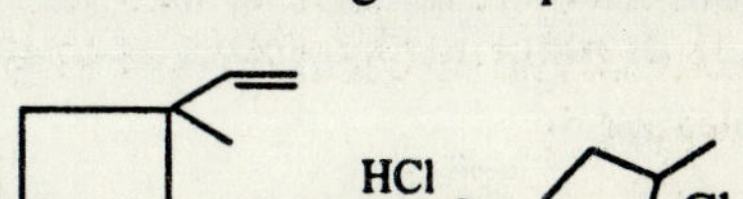

Formation of product in the above reaction can be explained by ring expansion mechanism of intermediate carbocation as:

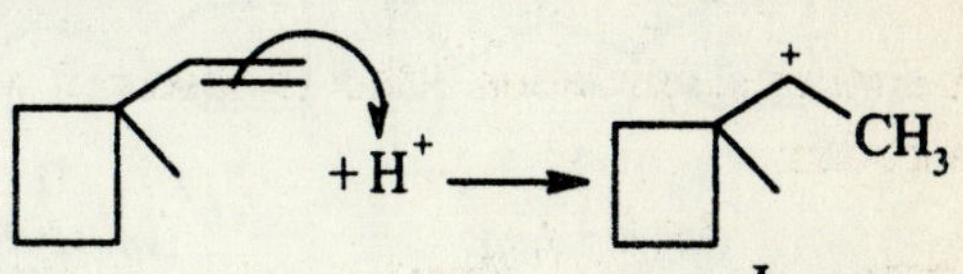

(1-methyl-1-vinylcyclobutane) $+ H^+ \longrightarrow$ (cyclobutyl–$\overset{+}{C}H$–CH_3) I

A 2° carbocation

Now I can undergo two type of rearrangement.

(a) Methyl shift to form a 3° carbocation as

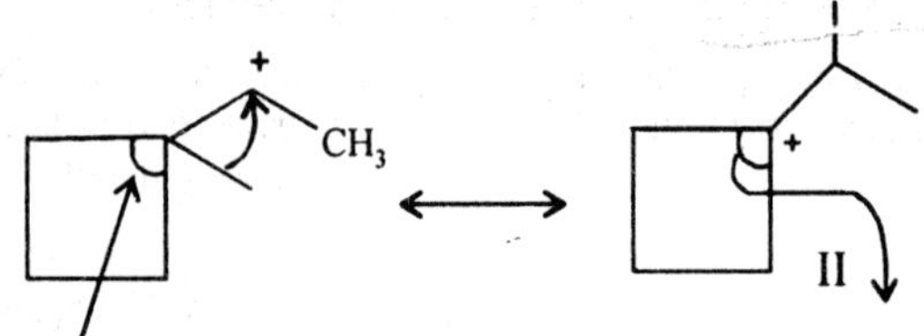

(Bond angle is 90° but the [A 2° carbocation (ii)] requirement is 109° due to sp^3C)

(Bond angle is 90° but the requirement is 120°due to sp^3C, hence strain is more)

Although (II) is a 3° carbocation, this rearrangement is not energetically favoured due to increase in angle strain as described above.

(b) Ring expansion

CH_3 III

(3° carbocation)

Ring expansion also producing a tertiary carbocation with less strain than the cyclobutane ring. Therefore ring expansion is preferred rearrangement.

At the end, (III) will absorb nucleophile Cl^- giving the desired product as:

(ii) + HCl ⟶ Cl

Formation of product in the above reaction can be explained as follows:

$+H^+$

I (More stable) II (Less stable)

I H + H^- shift Cl^- Cl

Example 50. Draw all significant resonance contributors for the following carbocations and select the most stable one.

(a) (b) OCH_3

(c) $(CH_2{=}CH)_2\overset{+}{C}H$

Solution

(a) I II III

II is most stable since it is secondary and vinyl groups are bonded on two sides.

(b) OCH_3 $\ddot{O}CH_3$ $\overset{+}{O}CH_3$ I II III

III is most stable since it contain an extra covalent bond as well as it has close octet of all atoms.

(c) I II III

I is most stable as explained in (a).

Example 51. For each of the following carbocation, draw the structure of best rearranged carbocation.

(a) 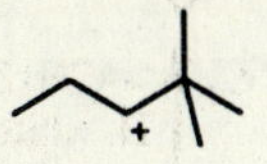 (b)

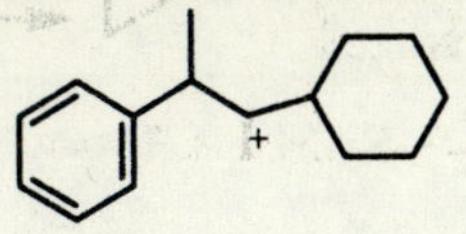

(c) 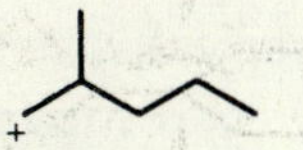 (d)

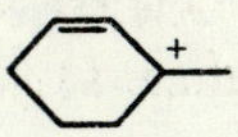

Solution

(a) 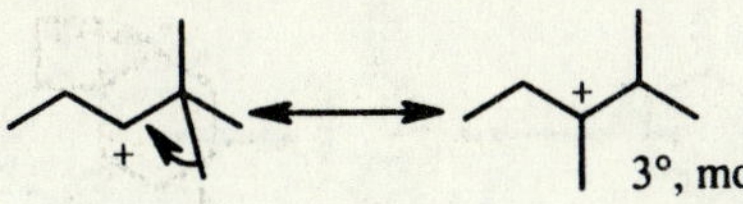 3°, most stable

(b) 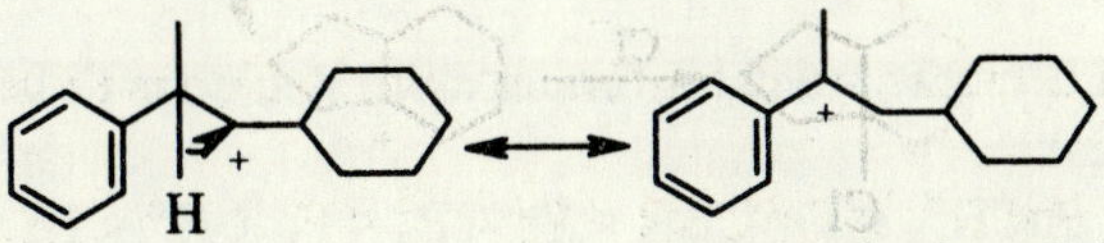 H (3° as well as benzylic)

(c) 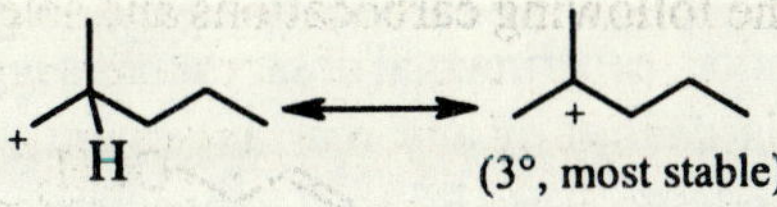 H (3°, most stable)

(d) Given carbocation is tertiary as well as allylic, cannot gain greater stability by any rearrangement.

Vinylic carbocations generally do not rearrange, even if rearrangement produces a more stable carbocation. For example, the rearrangement shown below does not occur, even though a secondary carbocation rearranging to more stable allylic carbocation.

$$R{-}\underset{}{\overset{H}{CH}}{-}\overset{+}{C}{=}CH_2 \longleftrightarrow R{-}\overset{+}{C}H{-}CH{=}CH_2$$

The reluctance to rearrangement is probably due to orbital alignment restrictions during the rearrangement transition state.

Example 52. Draw the rearranged carbocations based on the given curved arrows explaining, how the stability is increased by these rearrangement.

(a) 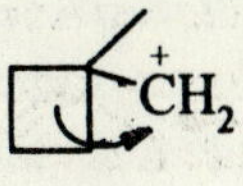 $\overset{+}{C}H_2$ (b)

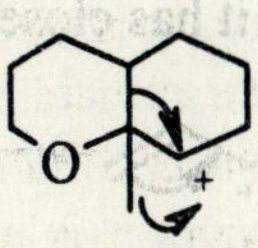

Solution

(a) Cyclobutane is an unstable ring due to angle strain. In the given rearrangement, rearranged carbocation has a five membered ring, which has less angle strain than cyclobutane, as well as the positively charged carbon is tertiary:

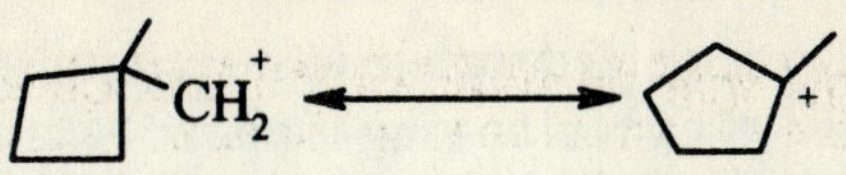

$\overset{+}{C}H_2$

1° carbocation with four membered ring (less stable)

3° carbocation with five membered ring (more stable)

(b) 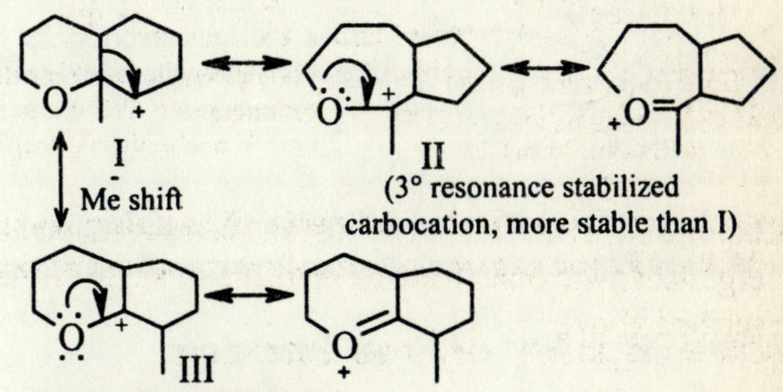 I — II (3° resonance stabilized carbocation, more stable than I)

Me shift

III (3° resonance stabilized carbocation, more stable than both I and II)

III is more stable than II, since II involve a ring contraction which lowers the stability.

3.6.3 Free Radical

Introduction

A free radical is species containing an unpaired electron on one of the atom. Free radicals are produced during the homolytic bond cleavage of covalent bonds as:

$$A{\bullet}{-}{\bullet}A \longrightarrow 2\ A^{\circ}$$

In organic reactions, most often are encountered a free readical with an unpaired electron on either carbon or on oxygen. Therefore, a free radical on carbon atom consist of a carbon with three covalent bonds and an unpaired electron as shown below:

$$-\overset{|}{\underset{|}{C^{\circ}}} \qquad -\overset{|}{C}{=}\overset{|}{C^{\circ}} \quad \text{or} \quad -C{\equiv}C^{\circ}$$

Likewise an oxygen free radical contain an oxygen atom with one sigma bond and an unpaired electron in addition to two lone pair of electrons as shown below:

$R{-}\ddot{\underset{..}{O}}{\cdot}$ An alkoxy radical

$R{-}\overset{O}{\overset{\|}{C}}{-}\ddot{\underset{..}{O}}{\cdot}$ A carboxy radical

Here we will confine our discussion mostly to free radicals at carbon atom. Therefore a free radical at carbon atom have three bonds (six shared electrons) and an unpaired electron making a total of seven electrons, one electron short of stable octet. Hence, like carbocation, free radicals are also electron deficient species, highly reactive and act as a strong electrophile in a chemical reaction. Hybridization of carbon bearing unpaired electron in an alkyl free radical is sp^2, therefore a carbon free radical of this category have triangular planar arrangement of bonds around the central carbon as shown in Figure 3.11.

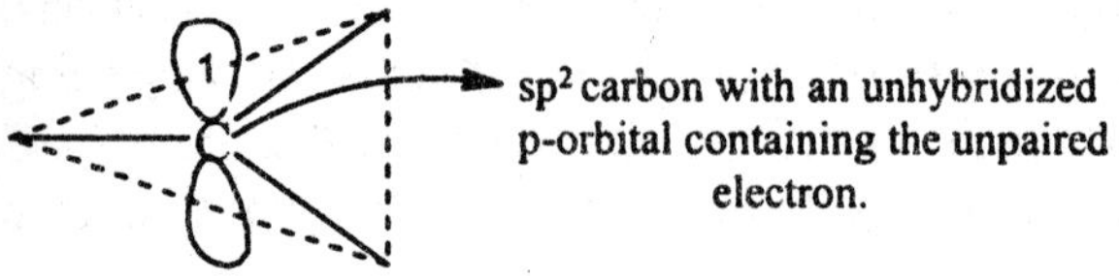

sp² carbon with an unhybridized p-orbital containing the unpaired electron.

Figure 3.11: A free radical with three bonds in a plane and a *p*-orbital holding an unpaired electron in perpendicular plane

Relative Stability of Free Radicals

Free radicals have an electron deficient atom (carbon with seven electrons). Therefore, an electron donating group bonded to electron deficient atom increases stability of free radical by compensating some of its electron deficiency. On the otherhand, an electron withdrawing group when bonded to electron deficient atom, decreases stability by intensifying the electron deficiency.

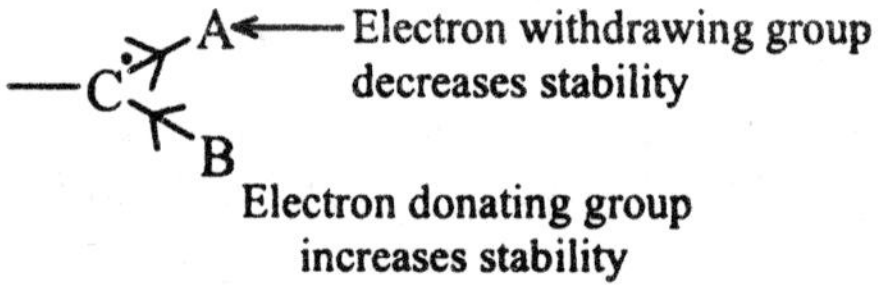

A ← Electron withdrawing group decreases stability

B Electron donating group increases stability

(a) Alkyl radicals

The order of stability of alkyl radicals are as follows:

$$CH_3^\circ < RCH_2^\circ < R_2CH^\circ < R_3C^\circ$$
$$1^\circ \quad 2^\circ \quad 3^\circ$$

The above order of stability is based on the fact that an alkyl group is electron donating group by +I effect.

Free radicals on carbon with greater *s*-character are less stable as:

$$CH_3CH_2^\circ > CH_2{=}CH^\circ > HC{\equiv}C^\circ$$

This is due to the some reason as with carbocation discussed in the previous section.

(b) Allyl and benzyl free radicals

Allyl radical: Allyl radicals are more stable than an alkyl radical due to resonance delocalization as:

$$CH_2{=}CH{-}\overset{\circ}{C}H_2 \longleftrightarrow \overset{\circ}{C}H_2{-}CH{=}CH_2$$

Allyl radical

$$CH_2{=}CH{-}\overset{\circ}{C}H{-}R \longleftrightarrow \overset{\circ}{C}H_2{-}CH{=}CH{-}R$$

Allylic radical

When compared with alkyl radicals, an allyl radical is more stable than a primary and secondary alkyl radicals, but less stable than tertiary alkyl radical. On the otherhand, an allylic free radical is even more stable a tertiary alkyl free radical.

$$RCH_2^\circ < R_2CH^\circ < CH_2{=}CHCH_2^\circ < R_3C^\circ < RCH{=}CHC^\circ HR$$
$$1^\circ \quad 2^\circ \quad \text{Allyl} \quad 3^\circ \quad \text{Allylic}$$

Benzyl free radical: A benzyl free radical is further more stable than allyl radical due to resonance-delocalization as:

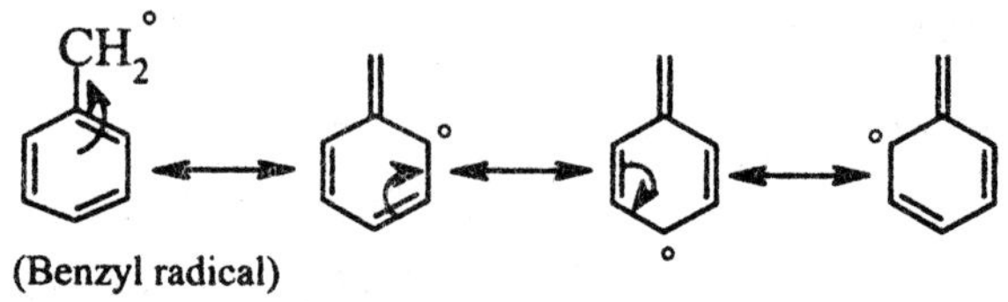

(Benzyl radical)

A benzylic radical is further more stable than a benzyl radical:

$$PhCH^\circ_2 < Ph^\circ CHR < Ph^\circ CR_2 < Ph_2CH^\circ < Ph_3C^\circ$$

Fate of a Free Radical

(i) Unlike carbocation, experimental observations suggest that a free radical is devoid of any rearrangements, i.e. free radical do not undergo H° or R° shift to produce more stable secondary or tertiary free radicals.

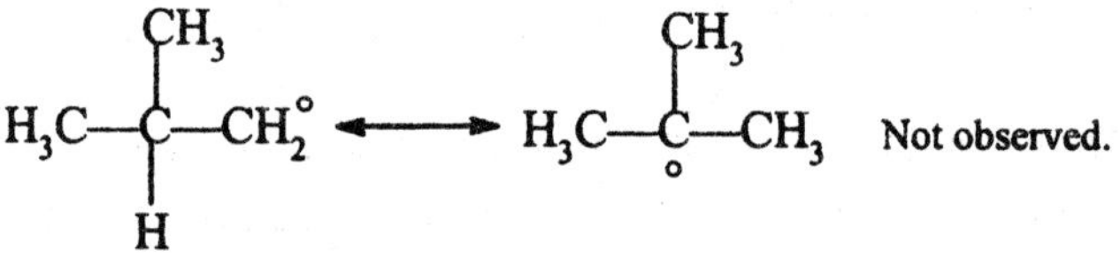

$$H_3C{-}C(CH_3)(H){-}CH_2^\circ \longleftrightarrow H_3C{-}\overset{\circ}{C}(CH_3){-}CH_3 \quad \text{Not observed.}$$

(1° free radical) (3° free radical)

Also, unlike carbocation, no ring expansion is observed with free radicals.

(ii) Combination of a free radical with another free radical producing stable neutral molecule with closed configuration. $R^\circ + {}^\circ R \rightarrow R{-}R$

(iii) Abstraction of a radical from stable molecule producing radical:

$$R^\circ + HBr \rightarrow R{-}H + Br^\circ$$
$$RO^\circ + HBr \rightarrow ROH + Br^\circ, \text{ etc.}$$

(iv) Disproportionation: Disproportionation is a characteristic property of a free radical. A multistep reaction involving free radical intermediate, produces some by products by disproportionation of intermediate free radical. Disproportionation is a process in which same species undergo simultaneous oxidation-reduction reactions to produce some unusual byproducts, e.g. during photochemical chlorination of cyclohexane, some cyclohexene is also produced as byproduct. Formation of cyclohexene can be explained by the disproportionation of cyclohexyl free radical produced during the reaction as:

Mechanism of Chlorination of Cyclohexane

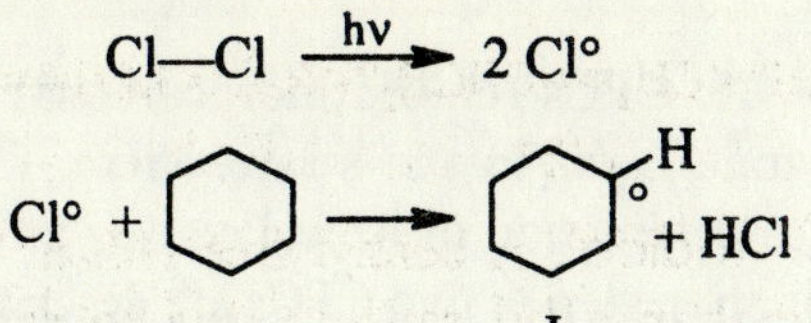

$$Cl—Cl \xrightarrow{hv} 2\,Cl^\circ$$

$$Cl^\circ + \text{cyclohexane} \longrightarrow \text{I (cyclohexyl radical)} + HCl$$

Disproportionation of I:

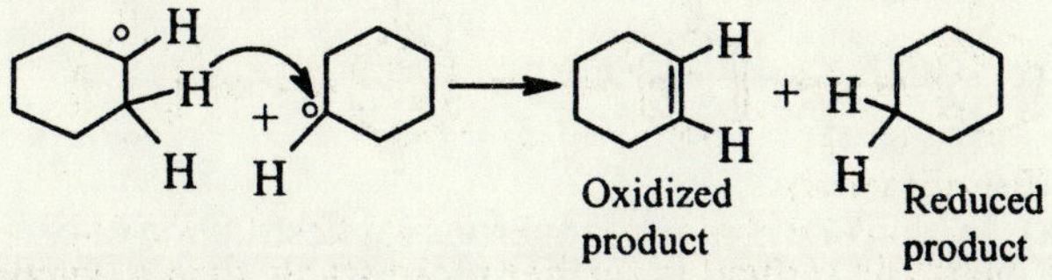

Oxidized product

Reduced product

Example 53. Briefly discuss the combustion of methane as a radical reaction.

Solution: The combustion of a hydrocarbon into CO_2, and water follow free radical mechanism as follows:

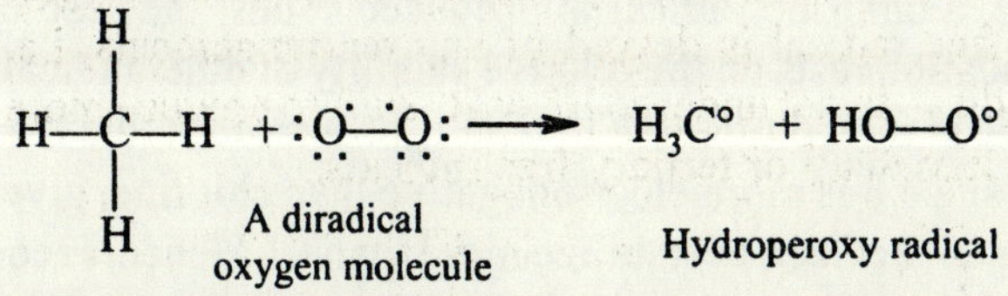

$$CH_4 + :\ddot{O}—\ddot{O}: \longrightarrow H_3C^\circ + HO—O^\circ$$

A diradical oxygen molecule

Hydroperoxy radical

The reminder of the mechanism is very complex, beyond the scope of this level.

Example 54. Discuss briefly the operations of biological antioxidants.

Solution: One of the most significant free radical present in cells of living beings is hydroxy radical (HO°) and superoxide (O_2°) radical. Biological system contain many molecules that are prone to damage by the attack of these free radicals. A free radical invoke a chain reaction, therefore one radical may result in damage to many biomolecueles. Antioxidants intercept radical chain reaction providing hydrogen atoms that are easily transferred to the propagating chain converting itself into a new, resonance stabilized, less reactive, free radicals. Therefore, an antioxidant makes the more reactive radical passive by H-transfer and convert themself into a stable radical. Two common biological antioxidants are vitamin E (α- tocopherol) and vitamin C (an ascorbate).

Example 55. Briefly discuss the role of Freon and chain reaction in depletion of ozone in atmosphere.

Solution: Freon is commercial name given to all compound containing only carbon, chlorine and fluorine (chlorofluorocarbons CFCs) which are used in a veriety of applications such as refrigeration, air-conditioning, etc. When CFCs escape to the atmosphere, they can undergo photolysis—a process of bond cleavage induced by light, from high energy UV-photon from the sun, which are normally absorbed by ozone layer. CFC photolysis produces a Cl° (Chlorine radical) which reacts with ozone in a chain reaction converting later into oxygen. This destruction of ozone may lead to fall of UV-light directly on living beings on earth surface causing a great damage. Therefore, CFCs represents a significant danger to the ozone layer of stratosphere.

Example 56. Discuss the high reactivity of nitricoxide NO.

Solution: The Lewis structure of NO is: $:\dot{N}═O$ with an odd electron on nitrogen atom. Therefore, NO is a free radical and all the properties of free radical (high reactivity, short life-time, etc.) are applicable to NO.

Example 57. Rank the following radical according to their relative stability explaining reasoning.

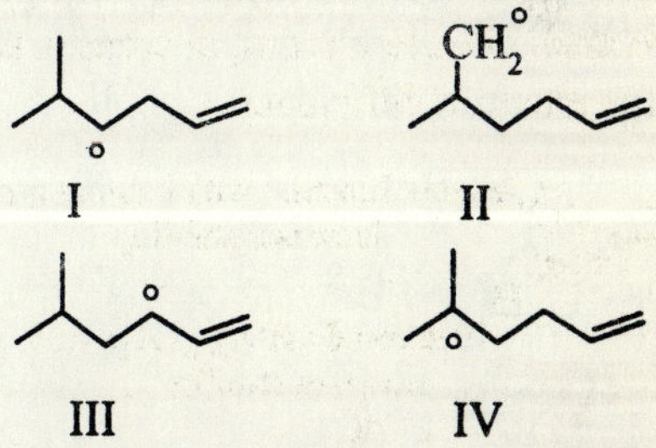

I

II

III

IV

Solution: As discussed earlier, radicals are electron deficient species. The carbon atom bearing open octet is one electron short from stability and hence, it can be stabilized by resonance or +I effect. Resonance stabilization outweigh stabilization by inductive effect even if the one stabilized by resonance is one degree inferior in substituent than the other being stabilized only by inductive effect. Hence, radical III (a 2° radical) is more stable than (IV) (a 3° radical) due to resonance effect. Thus, the overall stability order in the present case is:

$$II < I < IV < III$$

Example 58. Molecular oxygen is best described as a diradical. Using only oxygen, propene or products from the reaction of these two molecules, illustrate three fundamental radical mechanism step.

Solution: A radical may attack to π-bond as:

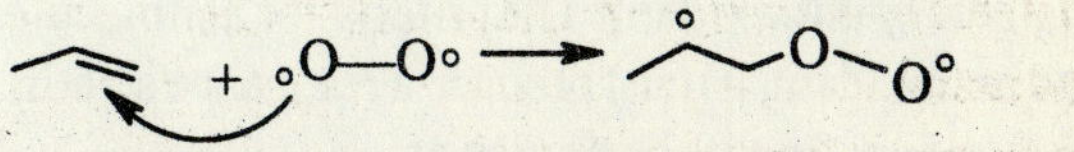

A radical may abstract a hydrogen as:

$$\text{propene} + \dot{O}—\dot{O} \longrightarrow \text{allyl}^\circ + H\dot{O}\dot{O}$$

The growing radical may combine in a termination step as:

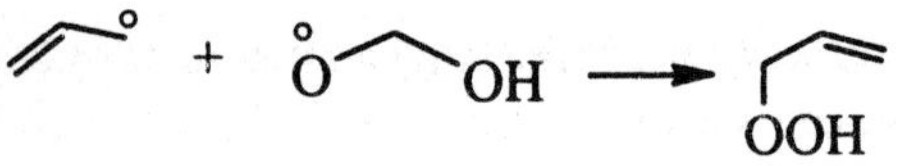

Example 59. Consider the following free radical bromination reaction.

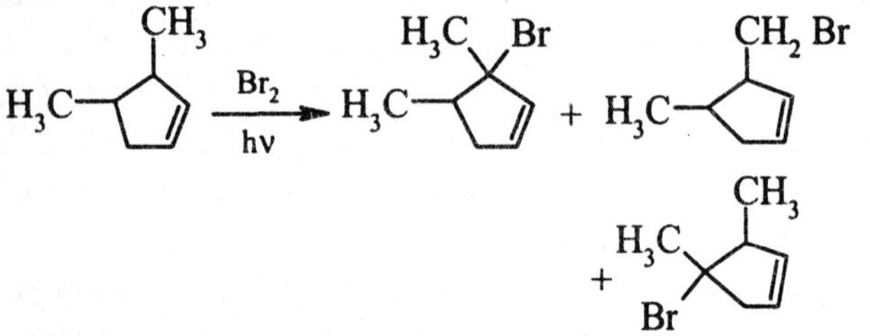

Discuss the mechanism of formation of all above mentioned products and select the major product explaining reason for your choice.

Solution

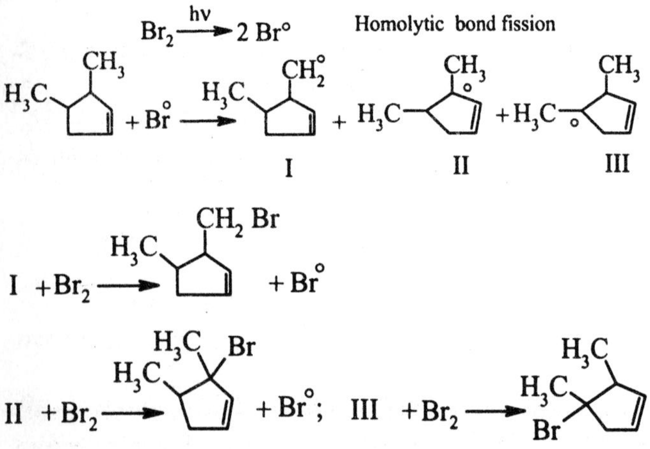

$Br_2 \xrightarrow{hv} 2\,Br^\circ$ Homolytic bond fission

I | II | III

I $+Br_2 \longrightarrow$ $+Br^\circ$

II $+Br_2 \longrightarrow$ $+Br^\circ$; III $+Br_2 \longrightarrow$

Here II is the most stable intermediate since it is tertiary as well as allylic, therefore bromination product from II will be the major product.

Example 60. Antioxidants are added to food products to prevent discolouration and spoilage due to reactions of radical present in the food and oxygen in air. Could the following molecule serve as such an antioxidant? Explain.

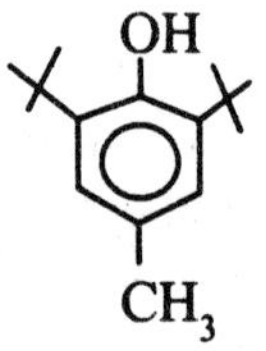

Solution: The above mentioned molecule cannot be used as a satisfactory antioxidant. The criteria for a molecule to be a good antioxidant, it must produce a resonance stabilized radical after donating its hydrogen as:

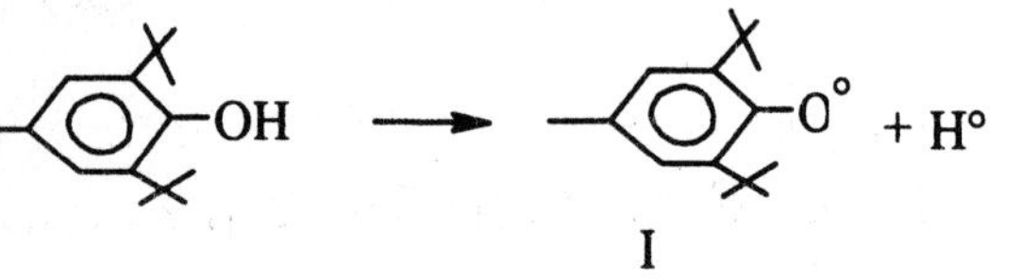

I

Here the radical on oxygen in I cannot be delocalized efficiently into the ring due to steric inhibition to resonance by the two bulky t-butyl groups present at adjacent positions.

Example 61. CFCs (chlorofluorocarbons) are major threat to ozone layer in the stratosphere. CFCs were implicated in destruction of ozon layer by a free radical process. Initiation step of two CFCs are described below:

I $CCl_2F_2 \xrightarrow{hv} {}^\circ CClF_2 + Cl^\circ$
(Freon 12)

II $CClF_3 \xrightarrow{hv} {}^\circ CF_3 + Cl^\circ$
(Freon 13)

(a) Which one is more dangerous for destruction of ozone layer?

(b) Design a CFC that is degraded more quickly then either of the above mentioned CFCs.

Solution: (a) A more easily degradable CFC can invoke faster depletion of ozone layer. In both cases mentioned above, a chlorine radical is produced. Thus, the ease of degradation rests on the relative stability of other radicals produced in addition to Cl° °$CClF_2$ is a more stable than °CF_3 since F is more electronegative than chlorine, gives stronger –I effect, destabilize more than Cl. Hence, Freon –12 can easily be degraded than Freon –13, and more dangerous for depleting ozone layer.

(b) A Freon $CFCl_3$ will further be easier to degrade than the previous two CFCs, since more stable carbon free radical (°CCl_2F) would be produced.

3.6.4 Carbanion

Carbanions are the species in which a carbon atom bears three covalent bonds and a negative charge as shown below:

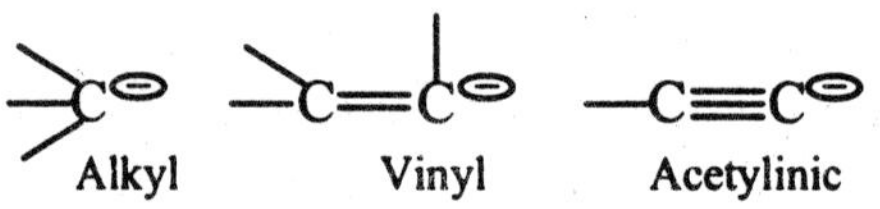

Alkyl | Vinyl | Acetylinic

In alkyl carbanion, carbon bearing negative charge is sp^3 hybridized and has pyramidal arrangement of bonds around the central carbon:

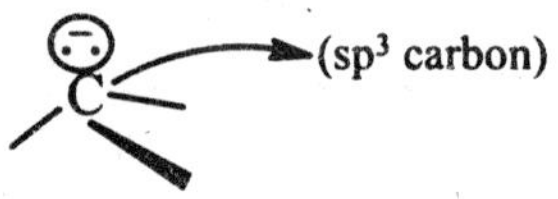

(sp^3 carbon)

Relative Stability

Carbanion possess a negative charge on one of its carbon, that makes carbanion highly reactive and very unstable. Dispersal of charge away from carbon will increase the stability of carbanion. On the other hand, if any factor intensifies negative charge on carbon, it will decrease the stability.

(i) Alkyl carbanion: Alkyl groups are electron donating by inductive effect, if bonded to carbon bearing negative charge, intensify the negative charge density decreasing stability. In that respect, CH_3^-–(methyl carbanion) is expected to be the most stable alkyl carbanion. The general order of stability of alkyl carbanion can be given as:

$$R_3C^- < R_2CH^- < RCH_2^- < CH_3^-$$

Increasing stability

(ii) Vinyl and acetylinic carbanion: Carbanion in which negatively charged carbon has multiple bonds (greater *s*-character), are more stable than the alkyl carbanion:

$$RCH_2^- < RCH{=}CH^- < R{-}C{\equiv}C^-$$

The above stability order can be explained on the basis of force of attraction between the negative charge and nucleus of carbon holding that negative charge. From alkyl to acetylinic carbanion, hybridization of carbon bearing negative charge, change from sp^3–sp^2–sp. Since order of size of these hybrid orbitals is:

$$sp^3 > sp^2 > sp$$

Therefore, in acetylinic carbanion, negative charge is closest to the nucleus of carbon, has maximum force of attraction, most stable situation. In alkyl carbanion, negative charge is in one of the sp^3 hybrid orbital, farest from the nucleus of carbon, minimum force of attraction, least stable situation.

(iii) Allylic and benzylic carbanion: Allylic and benzylic carbanions are more stable than normal alkyl carbanion due to resonance delocalization of negative charge as described below:

(a) 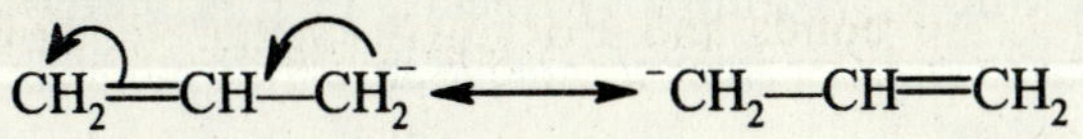 $CH_2{=}CH{-}\bar{C}H_2 \longleftrightarrow \bar{C}H_2{-}CH{=}CH_2$

(b) $CH_2{=}CH{-}CH{-}\bar{C}H{=}CH_2$ (with $\bar{C}H$)

$\updownarrow$

$CH_2{=}CH{-}CH{=}CH{-}\bar{C}H_2$

$\updownarrow$

$\bar{C}H_2{-}CH{=}CH{-}CH{=}CH_2$

(c) $C_6H_5{-}\bar{C}H_2 \longleftrightarrow$ (cyclohexadienyl anion)$=CH_2$

$\updownarrow$

$^-$(cyclohexadienyl)$=CH_2 \longleftrightarrow$, etc.

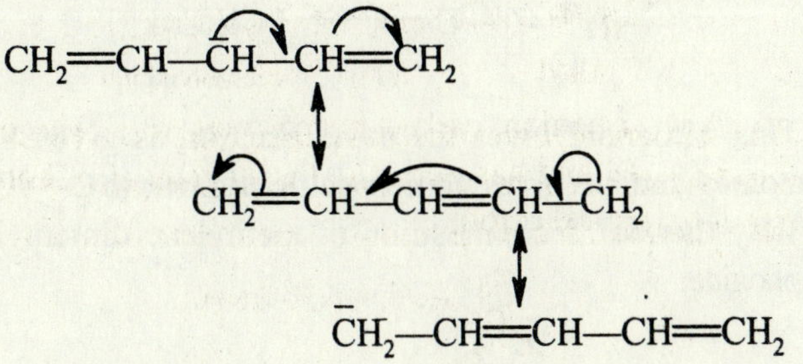

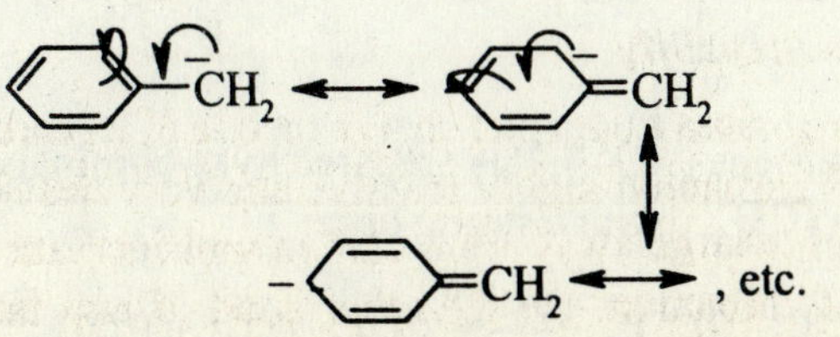

Example 62. Arrange the followings in increasing order of stability explaining reason:

$CH_2{=}CH{-}\bar{C}H_2$, (I) $\quad CH_2{=}CH{-}\bar{C}H{-}CH_3$, (II)

$CH_2{=}CH{-}\bar{C}(CH_3){-}CH_3$ (III)

Solution: The order of stability is III < II < I

Discussion: All three carbanions are resonance stabilized, but II is secondary and III is tertiary carbanion. Since alkyl group is electron donating by inductive effect, III is destabilized by +I effect of two methyl group, II is destabilized by +I effect of one methyl group, but no such destabilizing group is present in I. Hence, the order of stability is:

$$CH{=}CH{-}\bar{C}(CH_3)_2 < CH_2{=}CH{-}\bar{C}HCH_3 < CH_2{=}CH{-}\bar{C}H_2$$

Example 63. Arrange the following carbanions in incresing order of stability:

$O_2N{-}\bar{C}H_2$	$\bar{C}H_2F$	$\bar{C}HClF$	$\bar{C}HCl_2$
I	II	III	IV

Solution: The order of relative stability of the above mentioned carbanions is: II < III < IV < I

I is the only resonance stabilized carbanion where as all others are stabilized by electron withdrawing inductive effect as:

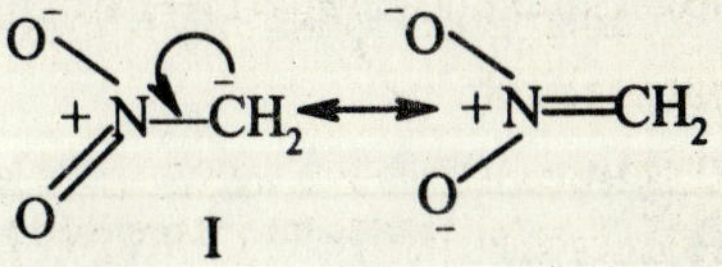

$(O^-)(O{=})N^+{-}\bar{C}H_2 \longleftrightarrow (^-O)(^-O)N^+{=}CH_2$

I

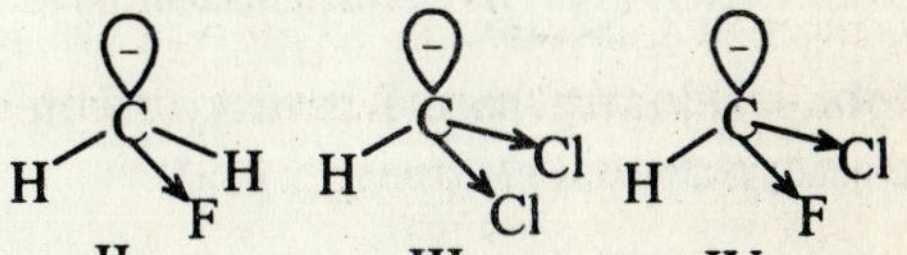

$H{-}\bar{C}H{-}F$ (II) $\quad H{-}\bar{C}{(Cl)}{-}Cl$ (III) $\quad H{-}\bar{C}(Cl){-}F$ (IV)

Electron withdrawal by one fluorine is less than electron withdrawal by two chlorine combined or one fluorine plus one chlorine. Also electron withdrawal by one chlorine and one fluorine combined is more than electron withdrawal by two chlorine.

(iv) Carbanion in which negatively charged carbons are α-to carbonyl, nitro, cyano groups are very stable due to resonance stabilization as:

(a) 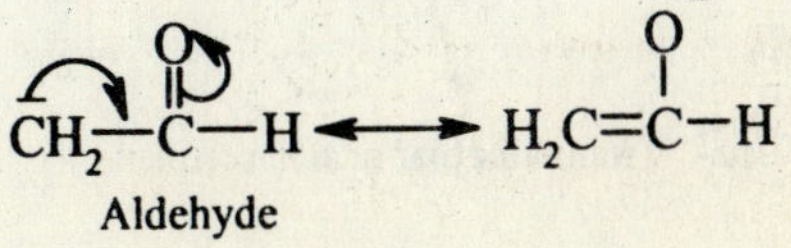 $\bar{C}H_2{-}C({=}O){-}H \longleftrightarrow H_2C{=}C(O^-){-}H$

Aldehyde

(b) $\overline{C}H_2-\overset{O}{\overset{\|}{C}}-CH_3 \longleftrightarrow CH=\overset{O^-}{\overset{|}{C}}-CH_3$

Ketone

(c) $\overline{C}H_2-\overset{O}{\overset{\|}{C}}-OCH_3 \longleftrightarrow H_2C=\overset{O^-}{\overset{|}{C}}-OCH_3$

Ester

(d) $\overline{C}H_2-C\equiv\ddot{N} \longleftrightarrow CH_2=C=\overline{N}$

Cyano

(e) $\overline{C}H_2-\overset{+}{N}(=O)-O^- \longleftrightarrow CH_2=\overset{+}{N}(O^-)(O^-)$

Nitro

Example 64. α-H of an aldehyde is more acidic than α-H of a ketone, justify.

Solution: Conjugate base of both aldehyde and ketones are stabilized by resonance as:

$\overline{C}H_2-\overset{O}{\overset{\|}{C}}-H \longleftrightarrow CH_2=\overset{O^-}{\overset{|}{C}}-H$

I

$\overline{C}H_2-\overset{O}{\overset{\|}{C}}-R \longleftrightarrow CH_2=\overset{O^-}{\overset{|}{C}}-R$

II

Extent of delocalization is more in I than in II since R– group is also electron donating by +I effect, opposes delocalization as:

$\overline{C}H_2-\overset{O}{\overset{\|}{C}}\leftarrow R$ Resonance is opposed by electron donation by R

Therefore, conjugate base I is more stable and its conjugate acid (aldehyde) is stronger acid.

3.6.5 Carbenes

Carbenes are the species in which one carbon bears two bonds and a non-bonded pair of electrons as:

$>C:$

Carbenes

Hybridization at carbon is sp^2, therefore, there is a planar arrangement of bonds around central carbon with one orbital holding lone pair.

Triangular planar arrangement

In carbene, the carbon bearing non-bonded pair of electrons also has a vacant *p*-orbital and there can be two different arrangement of electrons as shown below:

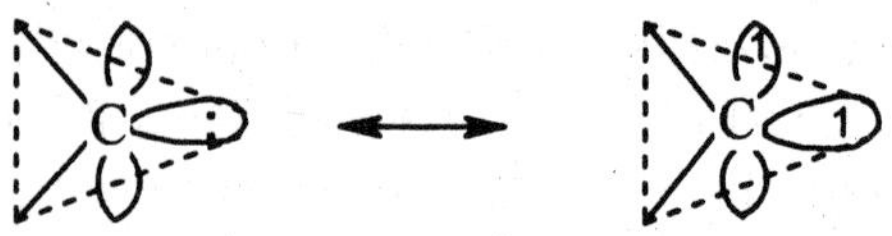

Singlet carbene: both electrons are in one of the sp^2 hybrid orbitals, and they are paired with opposite spin.

Triplet carbene: both *p*-orbital and sp^2 hybrid orbital have one electron each and the two electrons have parallel spin.

Singlet carbene has both electrons in sp^2 hybrid orbital, which impart nucleophilicity to carbene. On the other hand, triplet carbene has one electron each in vacant *p*-orbital and in one of the sp^2 hybrid orbital giving it radical like behaviour.

Thus, a carbene is ambiphilic – both electrophilic and nucleophilic in nature.

Formation of Carbenes

(i) By thermal decomposition of diazomethane:

$$CH_2N_2 \xrightarrow{\text{Heat}} :CH_2 + N_2$$

Diazomethane has the following resonance stabilized structure:

$$:\overline{C}H_2-\overset{+}{N}\equiv N: \longleftrightarrow CH_2=\overset{+}{N}=\overline{\ddot{N}}:$$

(ii) By thermal decomposition of ketene:

$$H_2C=C=O \xrightarrow{\text{Heat}} :CH_2 + CO$$

Ketene

(iii) The Simmons-Smith synthesis: In this method, diidomethane and a zinc/copper couple are stirred producing a carbenoid as:

$$CH_2I_2 + Zn(Cu) \longrightarrow I\,CH_2ZnI$$

A carbenoid

This carbanoid gives the same reaction as given by diazomethane, i.e. gives stereospecific addition to C=C.

(iv) By thermal decomposition of methylene dimethylsulphoxide:

$$CH_2=S(CH_3)_2=O \xrightarrow{\text{Heat}} :CH_2 + (CH_3)_2SO$$

(v) Dihalocarbene ($:CX_2$): It is prepared by α-elimination of haloform (CHX_3) with a strong base:

$$CHCl_3 + R\overset{-}{O}\overset{+}{K} \longrightarrow CCl_2 + KCl + ROH$$

3.7 CLASSIFICATION OF A CHEMICAL REACTION

When studying a new reaction we first investigate what is accomplished—whether the number of atoms in product differs from the number of atoms in the reactant, whether the positions of atom in product differs from their position in reactant, whether oxidation state of elements in product and reactant differs. Depending on the answer of these questions, we classify a given reaction as one of the seven major organic reaction types: addition, elimination, substitution, condensation, rearrangement, isomerization or a redox reaction.

(a) Addition reaction: An organic reaction is said to be addition reaction if two reactant molecule combines to form a product containing atoms of both reactant. Some of the common addition reactions are:

$$CH_2{=}CH_2 + H_2 \xrightarrow{Pt} C_2H_6 \quad \text{Hydrogenation}$$

$$CH_2{=}CH_2 + H_2O \xrightarrow{H_2SO_4} CH_3CH_2OH \quad \text{Hydration}$$

$$CH_2{=}CH_2 + HBr \longrightarrow CH_3CH_2Br \quad \text{Addition of H Br}$$

$$CH_2{=}CH_2 + Cl_2 \longrightarrow CH_2ClCH_2Cl \quad \text{Chlorination}$$

(b) Elimination reaction: It is opposite to addition reaction. When a single reactant molecule is converted into a product by loss of a smaller molecule as byproduct (usually H_2O, alcohols, HX, etc.), such reaction are known as *elimination reaction*. Some common elimination reactions are:

$$CH_3CH_2Cl \xrightarrow[-HCl]{EtONa} CH_2{=}CH_2$$

$$CH_3CH_2CH_2OH \xrightarrow[Heat]{H_2SO_4} CH_3CH{=}CH_2 + H_2O$$

(c) Substitution reaction: If an atom or a group of one reactant is replaced by another atom or group from a second reactant molecule, such type of reaction is known as *substitution reaction*.

$$CH_4 + Cl_2 \xrightarrow{hv} CH_3Cl + HCl$$

Cyclohexyl iodide + NaBr ⟶ Cyclohexyl bromide + NaI

(d) Condensation reaction: When two reactant molecules fuse together to form a bigger molecule with the loss of a smaller molecule as byproduct, such type of reaction is known as *condensation reaction*.

$$RCOOH + HOR \longrightarrow RCOOR + H_2O$$

$$RCOOH + H_2N{-}R \longrightarrow RCONHR + H_2O$$

(e) Rearrangement reaction: In a rearrangement reaction the molecular skeleton is changed, i.e. the connectivity is changed, e.g.

$$H_3C(H)C{=}N{-}OH \xrightarrow{H_3O^+} H_3C{-}\overset{O}{\overset{\|}{C}}{-}NH_2 \quad \text{Beckmann's rearrangement}$$

$$C_6H_5OCOCH_3 \xrightarrow{AlCl_3} p\text{-}HOC_6H_4OCH_3 \quad \text{Fries rearrangement}$$

$$R_2C(OH){-}C(OH)R_2 \xrightarrow{-H_3O^+} R{-}CR_2{-}\overset{O}{\overset{\|}{C}}{-}R \quad \text{Pinacol rearrangement}$$

(f) Isomerization reactions: Isomerization is a reaction, in which species with same molecular formula but different structures are interconverted as:

$$(CH_3)HC{=}C(H)CH_3 \longrightarrow (CH_3)HC{=}C(CH_3)H + CH_2{=}CHCH_2CH_3$$

$$CH_3CH_2CH_2OH \xrightarrow[H_2O]{H^+} CH_3\overset{OH}{CH}CH_3$$

(g) Redox reaction: Redox reactions are those reaction, in which some atoms of reactant undergo change in oxidation number, viz.

$$CH_3CH_2OH + CrO_4^{2-} \longrightarrow CH_3COOH + Cr^{+3} \quad \text{Oxidation of alcohol}$$

$$CH_3COOH + LiAlH_4 \longrightarrow CH_3CH_2OH \quad \text{Reduction of acid}$$

Example 65. Classify each of the following conversions according to reaction type:

(a) $(CH_3)_2C{=}C(CH_3)_2 + Cl_2 \longrightarrow Cl{-}C(CH_3)_2{-}C(CH_3)_2{-}Cl$

(b) $(CH_3)_2C{=}C(CH_3)_2 + Br_2/hv \longrightarrow (CH_3)_2C{=}C(CH_3)CH_2Br + HBr$

(c) $CH_3OH + HCl \longrightarrow CH_3Cl + H_2O$

(d) $CH_2{=}CHCH_2CH_3 + H_2 \xrightarrow{Pt} CH_3CH_2CH_2CH_3$

(e) 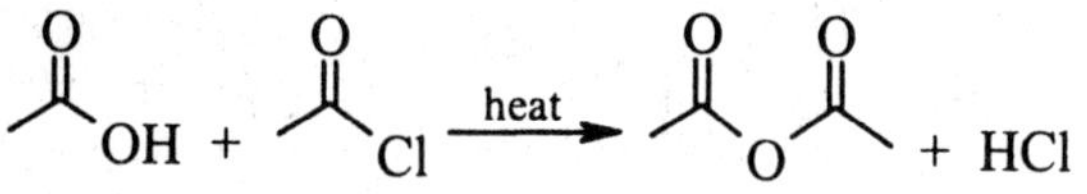 OH + Cl $\xrightarrow{\text{heat}}$ O + HCl

(f) 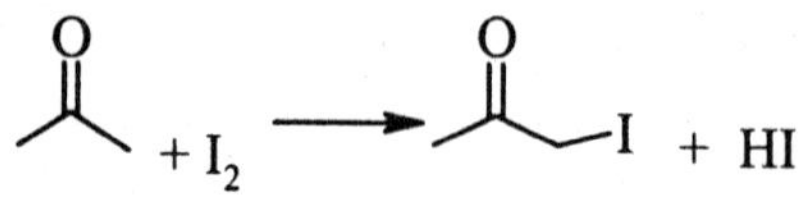 $+ I_2 \longrightarrow$ I + HI

(g) 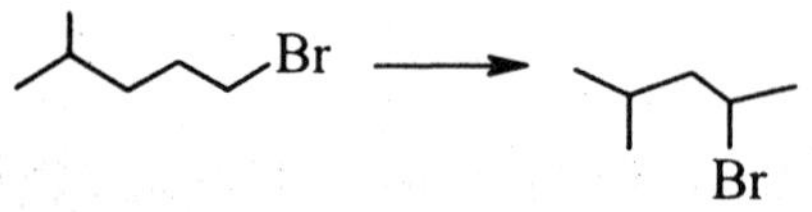 Br $\longrightarrow$ Br

(h) $CH_2{=}CH_2$ + $\longrightarrow$

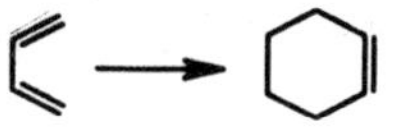

Solution

(a) Addition reaction.
(b) Substitution reaction: H is replaced by Br.
(c) Substitution: OH is replaced by Cl.
(d) Addition reaction.
(e) Condensation reaction.
(f) Substitution reaction.
(g) Isomerization.
(h) Addition.

PRACTICE EXERCISE

1. Identify the most acidic proton in the following molecule and then compare which member of each pair have lower pk_a.

(a) $H_3C{-}C({=}O){-}NH_2$ or $H_3C{-}C({=}O){-}N(CH_3)_2$

(b) $H_3C{-}C({=}O){-}CH_3$ or $H_3C{-}C({=}O){-}OH$

(c) CH_3NH_2 or $N(CH_3)_3$

(d) 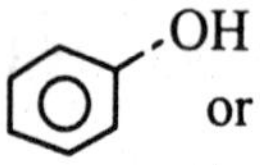 (phenol, OH) or (benzyl alcohol, OH)

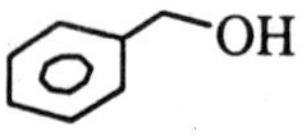

2. Predict which member of each of the following pairs of compound is more acidic and give reason for your choice:

(a) H_2S or PH_3 (b) CH_4 or SiH_4
(c) H_2O or H_2S (d) CH_3OH or CH_3NH_2
(e) HI or H_2Se

3. Rank each set of intermediates according to their relative stability explaining reason for your choice.

(a) $CH_3CH_2CH_2\overset{+}{C}H_2$; $CH_3\overset{+}{C}HCH_2CH_3$; $(CH_3)_2\overset{+}{C}CH_2CH_3$; $(CH_3)_3C^+$

(b) $CH_3CH_2CH_2O\overset{\circ}{C}H_2$; $CH_3\overset{\circ}{C}HCH_2CH_3$; $(CH_3)_2\overset{\circ}{C}CH_2CH_3$; $(CH_3)_3\overset{\circ}{C}$

(c) $:CH_2$, $CH_3CH_2:$; $C_6H_5CH:$; $(C_6H_5)_2C:$

(d) $H_2\overset{\circ}{C}{-}\bar{O}$; $(C_6H_5)_2\overset{\circ}{C}{-}\bar{O}$; $H_2C{=}CH{-}\overset{\circ}{C}{-}\bar{O}$

4. Upon treatment with strong base at low temperature, *cis*-1, 2,-diphenyl cyclopropane forms an anion. When quenched with D_2O, a mixture of 1-deutro-*cis*-1, 2-diphenylcyclopropane and 1-deutro-*trans*-1, 2-diphenyl cyclopropane are produced. Explain formation of products showing stereochemistry.

5. Reorder the following sets of compounds according to increasing pk_a.

(a) Cyclohexanol, phenol, cyclohexane carboxylic acid.
(b) 1-butyne, 1-butene, butane.
(c) Propanoic acid; 3-bromopropanoic acid, 2-nitropropanoic acid.
(d) Phenol, toluene, benzene.
(e) Dimethyl ether, ethanol, CH_3COOCH_3.
(f) Hexylamine, aniline, hexanamide.
(g) Benzoic acid, *p*-chlorobenzoic acid, 2, 4, 6-trichlorobenzoic acid.
(h) Ethanoic acid, oxalic acid, malonic acid.
(i) Protonated Pyrrole; pyridine, N-methyl pyrrole.

6. Crotonaldehyde $CH_3{-}CH{=}CH{-}CHO$ has a $pk_a{=}20$, despite the fact that it lacks enolizable hydrogen α-to-CHO. Determine which hydrogen is removed Ist by interaction with a base and why?

7. Octyl amine is insoluble in water but dissolves in dilute solution of sulphuric acid. Octanamide does not dissolve in either water or dilute sulphuric acid but dissolve in aqueous base. Propose an explanation for contrasting solubilities of the amines and amides.

8. For each of the following, determine the most acidic proton—that is, which one would be removed by treatment with one equivalent of base:

(a) 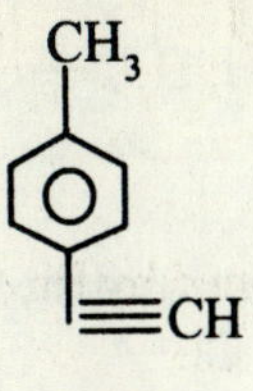 CH_3 ... $\equiv CH$ (b) 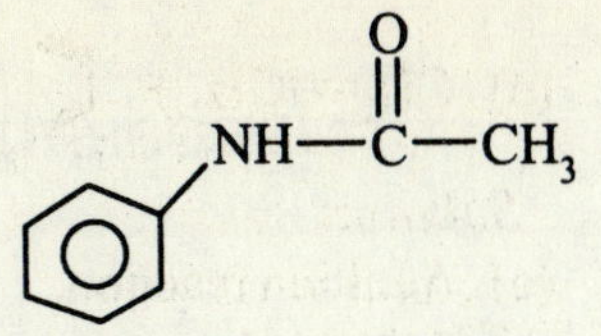 $NH-\overset{O}{\overset{\|}{C}}-CH_3$

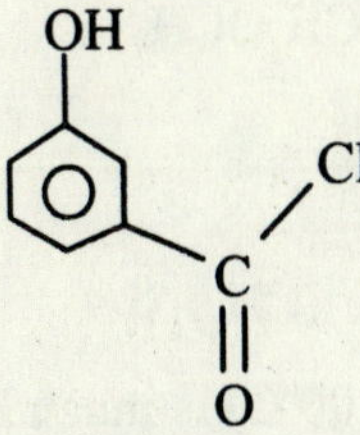

(c) OH ... CH_3, C=O (d) 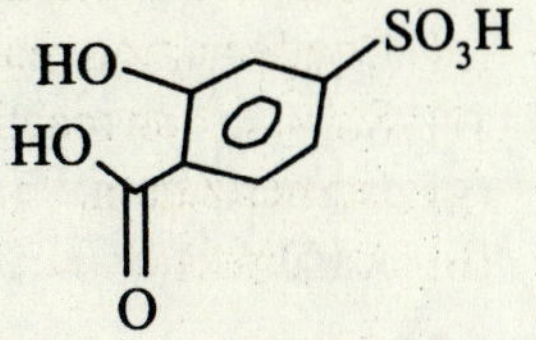 HO, HO, SO_3H, O

9. In the following molecules, determine which hydrogen is most acidic—that is which one would be removed by treatment with one equivalent of base:

(a) 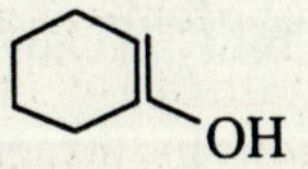 OH (b) 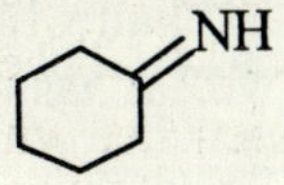 NH

(c) 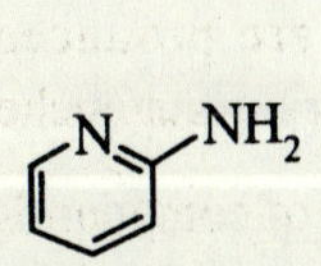 N, NH_2 (d) 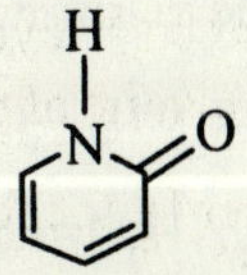 H, N, O

10. Draw all significant resonance cotributors for methylvinyl ketone and rank them in increasing order of stability.

11. Draw most stable protonated form of each of the anions:

(a) 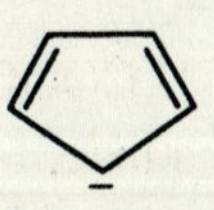 (b) E + O, O^-, O, OEt

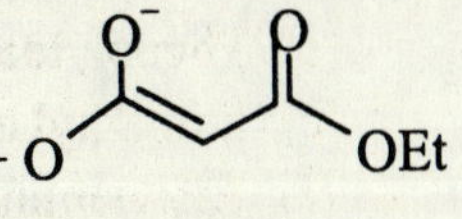

(c) 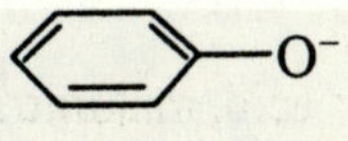 O^- (d) 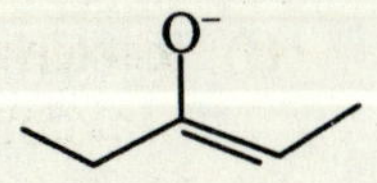 O^-

(e) 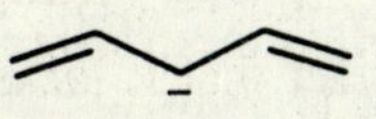 (f) 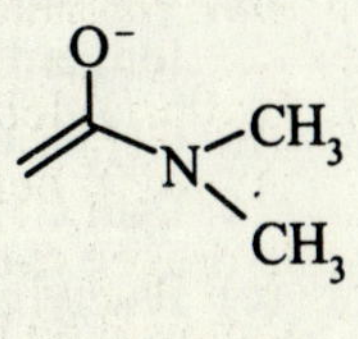 O^-, N, CH_3, CH_3

(g) 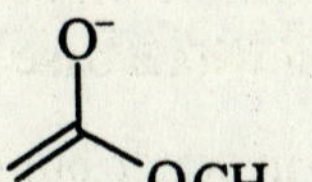 O^-, OCH_3

12. In each of the following pairs, choose the stronger base.

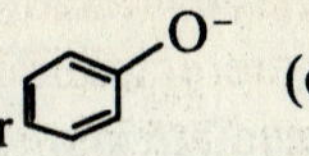

(a) Cl^- or F^- (b) Ph_3C^- or $C_6H_5^-$

(c) 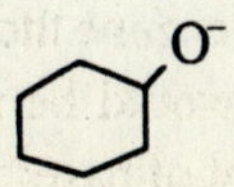 O^- or O^-

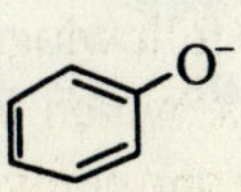

(d) 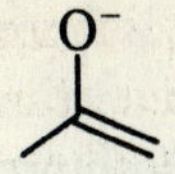 O^- or O^- (e) OH^- or O^-

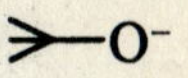

(f) $C_6H_5NH_2$ or Pyrrole

(g) NH_3 or $H_3C-\overset{O}{\overset{\|}{C}}-NH^-$

(h) $H_2\bar{C}-CN$ or $\bar{C}\equiv N$

13. Write conjugate base of each of the following sets of acid and arrange them in increasing order of their pk_a values.

(a) NH_3, H_2O, CH_3OH

(b) H_2O, $HC\equiv CH$, H_3O^+, RH

14. Starting with appropriate, unlabeled organic compounds, show synthesis of each of the following:

(a) $C_6H_5-C\equiv C-T$ (b) $CH_3-\underset{}{\overset{CH_3}{\overset{|}{C}}}H-OD$

(c) CH_3CH_2OD

15. Arrange the following compounds in increasing order of their pk_a values:

$CH_3CH_2NH_2$, CH_3CH_2OH, $CH_3CH_2CH_3$

16. Arrange the followings in decreasing order of acidity:

(a) $CH_3CH=CH_2$, $CH_3CH_2CH_3$, $CH_3C\equiv CH$

(b) CH_3CH_2OH, CH_3CH_2COOH, $CH_3\underset{\underset{Cl}{|}}{C}H-COOH$

(c) CH_3CH_2OH, $CH_3CH_2\overset{+}{O}H_2$, CH_3OCH_3

17. Arrange the followings in increasing order of basicity

(a) CH_3NH_2, $CH_3\overset{+}{N}H_3$, CH_3NH^-

(b) $CH_3CH=CH^-$, $CH_3CH_2CH_2^-$, $CH_3C\equiv C^-$

(c) CH_3O^-, CH_3NH^-, $CH_3CH_2^-$

18. Malonic acid is a diprotic acid $CH_2(COOH)_2$. The pk_a for the loss of Ist proton is 2.83, pk_a for the loss of second proton is 5.69. Explain why malonic acid is stronger than acetic acid ($pk_a = 4.75$). Also explain, why the conjugate base of malonic acid (HOOC CH_2COO^-) is so less acidic than malonic acid, though it still has one —COOH group.

19. At 25°C, the enthalpy of ionization of trichloroacetic acid is 6.3 kJ mol^{-1} and entropy change is 0.0084 kJ $mol^{-1}k^{-1}$. Determine pk_a of acetic acid in the given conditions.

20. The squaric acid (shown below), is diprotic with both proton being more acidic than acetic acid. In the dianion obtained after loss of both protons, all the C—C bonds have same length as well as well C—O bonds are of same length. Provide mechanistic explanation for this observations.

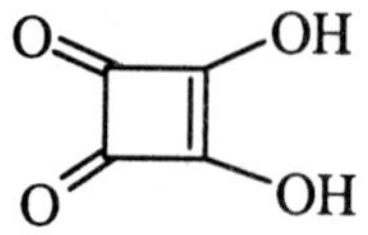

O= / \ OH, O= \ / OH Squaric acid

21. Label the protonation sites in order of increasing priority when the following compound is treated with dilute HCl solution:

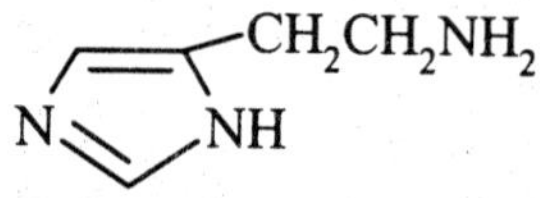

(imidazole ring with N, NH and side chain $CH_2CH_2NH_2$)

22. Complete the following equations and decide to which side the equilibrium will lie, and also indicate whether there is a decrease or increase in free energy for each reaction.

(a) 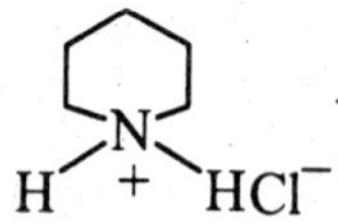 (piperidinium, $\overset{+}{N}$ with H, HCl^-) + NaOH ⟶

(b) $CH_3CH_2-\overset{+}{O}(H)-CH_2CH_3 + H_2O \longrightarrow$

(c) $CH_3COOH + CH_3CH_2SNa \longrightarrow$

(d) $Cl_3CCOOH + NaHCO_3 \longrightarrow$

(e) $CH_3CH_2COOH + NH_3 \longrightarrow$

(f) $CH_3NO_2 + CH_3CH_2ONa \longrightarrow$

(g) $CH_3CH_2SH + NaNH_2 \longrightarrow$

(h) $HC{\equiv}CH + NaHCO_3 \longrightarrow$

23. Arrange the following compounds in increasing order of acid strength.

(a) $\bar{C}lH_3\overset{+}{N}-CH_2COOH$, $ClCH_2COOH$, CH_3CH_2COOH

(b) $F_3C-COOH$, $Cl_3C-COOH$, $Cl-COOH$, CH_3COOH

(c) CH_3CH_2OH, FCH_2CH_2OH, $CH_3CHClOH$, $ClCH_2CH_2OH$

(d) CH_3CH_2SH, CH_3CH_2OH, $CH_3CH_2NH_2$, $CH_3CH_2PH_2$

(e) CH_3COOH, $H_3C-\underset{\overset{\|}{+OH}}{C}-OH$, $H_3C-\overset{\overset{O}{\|}}{C}-OCH_3$

24. Arrange each group of species in order of increasing basicity.

(a) Cl_3C^-; $(CH_3)_3N$; CH_3OCH_3

(b) $\ddot{N}F_3$, $\ddot{N}H_3$, NH_2OH

(c) NH_3, NH_2^-, CH_3NH_2

(d) CH_3SCH_3, $(CH_3)_3P$, $(CH_3)_4Si$

25. Boiling point of H_2O (100°C) is much higher than that of HF (–83°C) even though they both form H—bonds and have comparable molecular weight. Explain.

26. Explain why glycerol is a highly viscous liquid as well as has very high solubility in water.

27. Select the more stable electronic arrangement from the following pairs:

(a) $\overset{+}{C}H_2-\overset{-}{C}H_2 \longleftrightarrow CH_2{=}CH_2$

(b) $H_2C{=}O \longleftrightarrow H_2\overset{+}{C}-\overset{-}{O}$

(c) $R_2\overset{+}{C}-\underset{R}{C}{=}O \longleftrightarrow R_2C{=}\underset{R}{C}-\overset{+}{O}$

(d) $R-\overset{+}{C}{=}O \longleftrightarrow R-C{\equiv}\overset{+}{O}$

(e) $R-\underset{\underset{O}{\|}}{C}-O \longleftrightarrow R-\underset{\underset{\bar{O}}{|}}{C}{=}\overset{+}{O}H$

(f) $R-\overset{-}{C}H-\underset{\underset{O}{\|}}{C}-R \longleftrightarrow R-CH{=}\underset{\underset{\bar{O}}{|}}{C}-R$

28. (a) Carbon-carbon bond in $CH_2{=}CH-OCH_3$ is longer than in $CH_2{=}CH-SCH_3$. Explain.

(b) $CH_2{=}CH-SCH_3$ is a stronger Lewis base than $CH_2{=}CHOCH_3$ explain.

(c) Phenyl phosphine is a stronger Lewis base than aniline explain.

29. Account for the fact that isocyanic acid $HN{=}C{=}O$ and cyanic acid $N{\equiv}C-OH$ have same pk_a value.

30. Explain why each of the following structures is not a resonance form?

(a) $O{=}O$ and $\dot{O}-\dot{O}$

(b) $H_2\ddot{N}—O—H$ and $H_2N{=}\overset{+}{O}—H$

(c) $H_3C—C(OH){=}CH_2$ and $H_3C—C({=}O)—CH_3$

31. Consider the following mechanism:

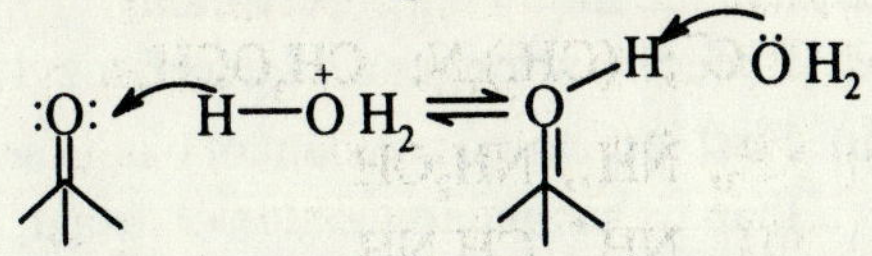

$$:O: \curvearrowleft H—\overset{+}{O}H_2 \rightleftharpoons O{-}H \curvearrowleft \ddot{O}H_2$$

List three significant things that are wrong in this mechanism.

32. Consider the following hypothetical enzyme.

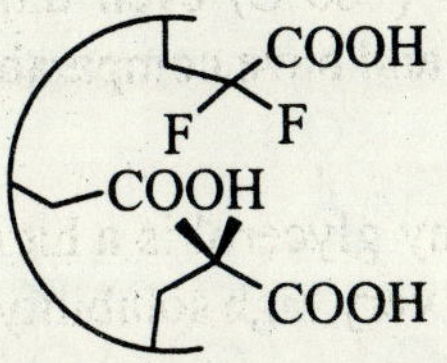

COOH, F, F, COOH, COOH

Write the stepwise ionization of acid.

33. Match the pk_a values:

(a) H_3O^+ (i) –9
(b) CH_3CO_2H (ii) –1
(c) NH_4^+ (iii) 4.76
(d) H_2SO_4 (iv) 9.4

34. Predict positions of equilibrium (left, right or at midpoint) for the following equilibrium reaction explaining reason for that:

$$CH_3\,NH^- + CH_3\,CH_3 \rightleftharpoons CH_3\,NH_2 + CH_3\,CH_2^-$$

35. Consider the following equilibrium reaction:

$$H_3C—C({=}S)—SH + H_3C—C({=}O)—O^- \rightleftharpoons H_3C—C({=}S)—S^- + CH_3\,COOH$$

Predict the position of equilibrium.

36. Consider the following equilibrium reaction:

$$CF_3CH_2OH + CH_3CH_2O^- \rightleftharpoons CF_3CH_2O^- + CH_3CH_2OH$$

Predict about the position of equilibrium.

37. Consider the following equilibrium reaction:

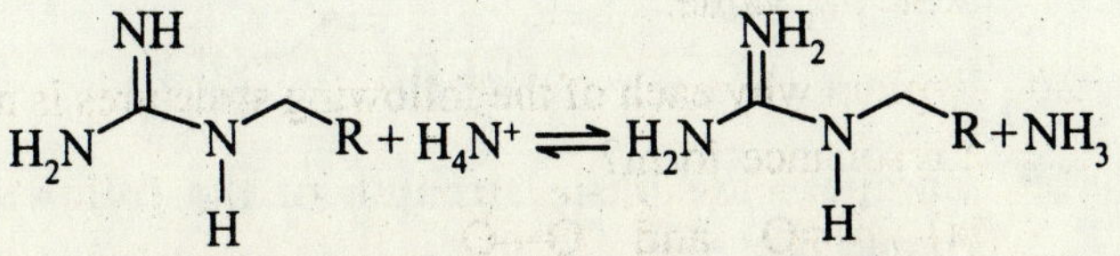

$$H_2N—C({=}NH)—N(H)—CH_2R + H_4N^+ \rightleftharpoons H_2N—C({=}\overset{+}{N}H_2)—N(H)—CH_2R + NH_3$$

Predict about the position of equilibrium.

38. Consider the following molecule:
Write stepwise protonation reaction justifying relative basic strength of the various basic sites.

39. Compare the relative acid strength of the following pairs of molecules:

(a) 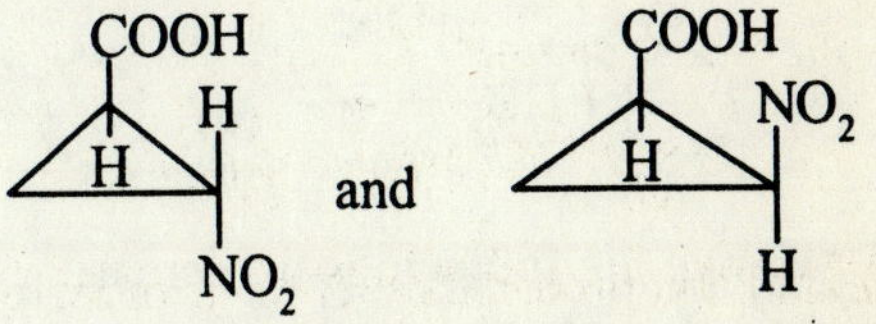 cyclopropane-COOH with H and NO_2 (cis) and cyclopropane-COOH with NO_2 and H (trans)

(b) 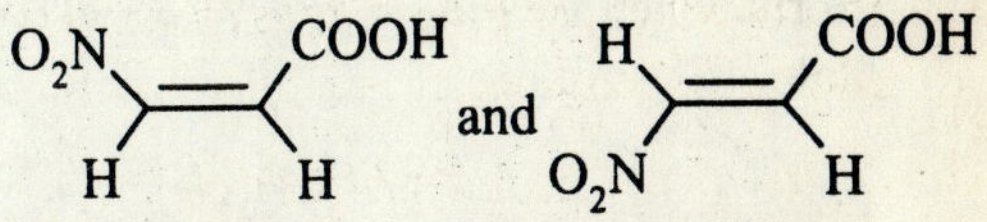 $O_2N(H)C{=}C(H)COOH$ and $H(O_2N)C{=}C(H)COOH$

(c) $H_2N—COOH$ and $F—COOH$

(d) $O_2N—COOH$ and $F—COOH$

40. Compare the acid strength of the following acids:

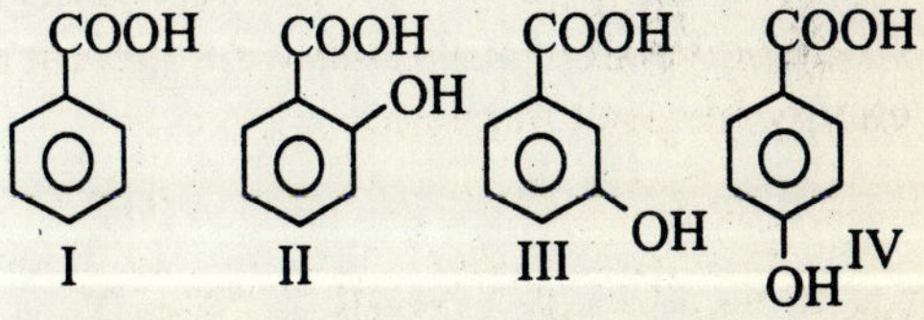

I: benzoic acid; II: o-hydroxybenzoic acid; III: m-hydroxybenzoic acid; IV: p-hydroxybenzoic acid

41. Which is more stable carbocation and why?

$$CH_3NHCH_2^+ \quad \text{or} \quad CH_3PHCH_2^+$$

42. Define the following terms:

(a) Homolytic and heterolytic bond fission.
(b) Free radical.
(c) Photolysis.
(d) Antioxidants.

43. Are radicals nucleophile or electrophile, justify?

44. Name and illustrate the three fundamental mechanistic way, by which a radical is consumed in the final step of a multistep reaction.

45. List and illustrate two structural features that influences the stability of a free radical.

46. HBr undergoes electrophilic addition to alkene and Markonikoff's product is formed. However, if the reactants are not carefully purified, the reaction can proceed to give an alternate isomer according to anti Markonikoff's rule as:

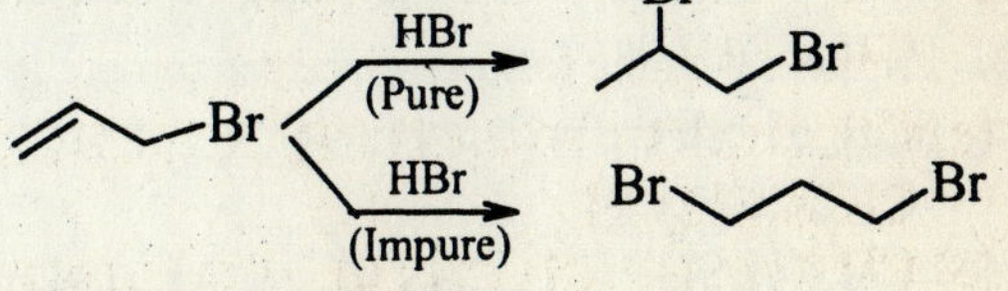

$CH_2{=}CHCH_2Br$ —HBr (Pure)→ $CH_3CHBrCH_2Br$
$CH_2{=}CHCH_2Br$ —HBr (Impure)→ $BrCH_2CH_2CH_2Br$

Explain the difference.

47. Provide the organic product(s) of the following reactions. If more than one products are produced, indicate the major one.

(a) [alkene] $\xrightarrow[H_2O_2]{HBr}$ (b) [alkene] $\xrightarrow[H_2O_2]{HBr}$

(c) [cyclohexene] + HBr ⟶ (d) [cyclohexene] $+ CBrCl_3 \xrightarrow{H_2O_2}$

48. Arrange the followings in increasing order of acid strength explaining reasons for your order:

(a) Phenol (OH) (b) HO–C$_6$H$_4$–OCH$_3$

(c) HO–C$_6$H$_4$–CH$_3$ (d) 3,5-dinitrophenol (OH, O_2N, NO_2)

(e) O_2N–C$_6$H$_4$–OH (f) o-nitrophenol (OH, NO_2)

49. Arrange the followings in increasing order of pk_a giving suitable explanation:

(i) $CH_2(COOEt)_2$, CH_3COOEt, CH_3CH_2OH

(ii) CH_3COCH_3, $CH(COOEt)_3$, CH_3COCH_2COOEt, CH_3COOH

50. Arrange the followings in increasing order of enol content:

CH_3COCH_3(I), CH_3COCH_2COOEt (II), $CH_3COCH_2COCH_3$ (III)

(IV) [cyclic structure with two C=O and ring O]

51. Arrange the followings in increasing order of their basic strength:

(a) *m*-nitrophenol (b) *p*-aminophenol

(c) Phenol (d) *p*-nitrophenol

52. Explain, why the following rearrangement is favoured though both are primary radicals

[hexenyl radical, $-CH_2^{\circ}$] ⟶ [cyclopentyl]$-CH_2^{\circ}$

53. Explain through mechanism, following bromination reaction.

[polyene] $+Br_2 \xrightarrow{Heat}$ Br–[bicyclic product]

54. Explain the following observations:

(a) Phenol is stronger acid than benzyl alcohol.

(b) First ionization constant of oxalic acid is greater than its second ionization constant.

(c) $CH_3OSO_3^-$ is stronger base than $CF_3SO_4^-$.

(d) C—C bond in $CH_2{=}CHOCH_3$ is longer than in ethene but shorter than in ethane.

55. Which of the following has greater dipole moment?

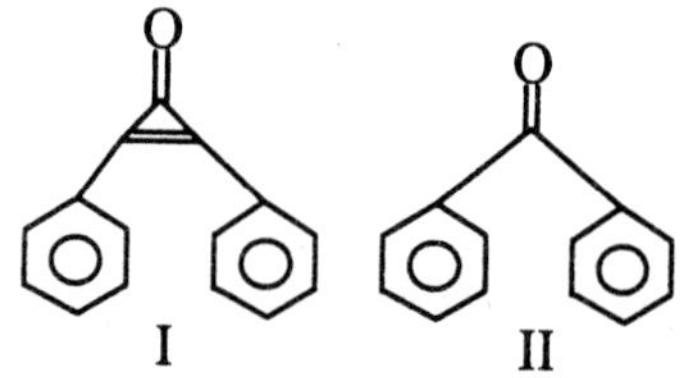

56. In 2, 4, 6-trinitro iodobenzene, two C—N bonds are of equal length and longer than the third C—N bond. Identify them and explain the difference.

57. Draw all resonance structures of the following species and rank them in order of increasing contribution to the resonance hybrid.

(a) $CH_3CH{=}CHOCH_3$ (b) $H-\overset{O}{\overset{\|}{C}}-NHCH_3$

(c) [cyclopentadienyl] (d) $CH_3CH_2\overset{O}{\overset{\|}{C}}OCH_2CH_3$

(e) $CH_2{=}CH\bar{C}H_2$

(f) $CH_3CH{=}CH—CH{=}CH—\overset{\circ}{C}H_2$

(g) $CH_3\bar{C}HCN$ (h) $CH_3\bar{C}H—NO_2$

(i) $CH_3\overset{O}{\overset{\|}{C}}—\bar{C}H\overset{O}{\overset{\|}{C}}—CH_3$ (j) $H\overset{O}{\overset{\|}{C}}—CH{=}CH\bar{C}H_2$

58. Rank the followings in increasing order of reactivity with HBr:

[alkene], [alkene]OCH_3, [alkene]$-OCH_3$

59. Compare the basic strength of the following compounds:

, $CH_3-CH(NH_2)-CH_3$, $CH_3-C(=NH)-NH_2$

60. Which of the following species have delocalized electron? Showing delocalization of electrons, compare their relative stability.

(a) $H_2C=CH-\overset{O}{\overset{\|}{C}}-CH_3$ (b)

(c) (d) $CH_3CH_2NHCH=CHCH_3$

(e) (f) $CH_3CH_2NHCH=CHCH_3$

(g) (h)

(i) $CH_2=CHCH_2CH=CH_2$

(j) $CH_3-\overset{+}{C}(CH_3)-CH_2-CH=CH_2$

(k) $CH_3CH=CHOCH_2CH_3$

61. Draw all the resonating structures of following species and predict the most stable one:

(a) (b)

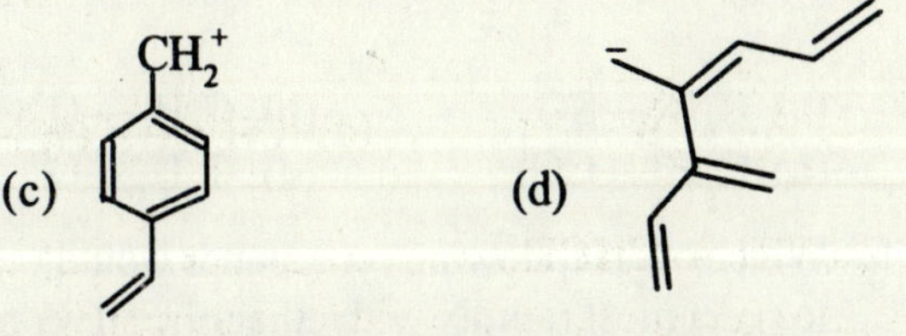

(c) CH_2^+ (d)

62. Which of the following compound has greater electron density on its nitrogen atom?

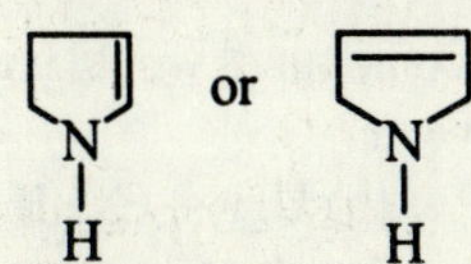

or

63. Which compound has greater electron density on oxygen atom?

—NH $COCH_3$ or —NH $COCH_3$

64. Rank the following compounds in increasing order of acidity:

(a) (b) (c)

65. Nitro group at para position has greater effect on the acidity of anilium ion than on bonzoic acid, justify.

66. K_{a1} of phthalic acid is greater than the K_{a1} of isophthalic acid while K_{a2} of later is greater than K_{a2} of phthalic acid, explain.

67. Are the following pairs of structures resonance contributors or different compounds?

(a) and

(b) $CH_3CH=CH-\overset{\circ}{C}H-CH=CH_2$ and $CH_3\overset{\circ}{C}HCH=CH-\overset{\circ}{C}H=CH_2$

(c) and OH

(d) + and +

68. Draw the resonance contributors for the following species, indicating major contributor to the resonance hybrid.

(a) (b) $H-\overset{O}{\overset{\|}{C}}-NHCH_3$

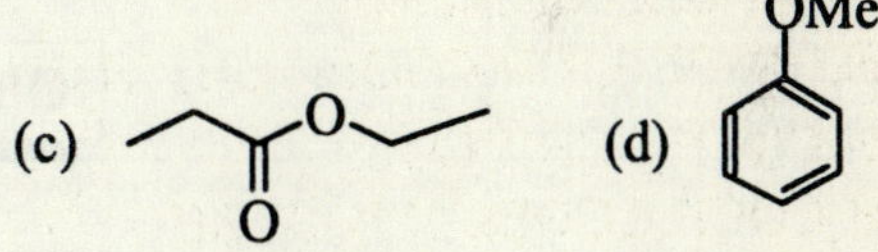

(c) (d) OMe

(e) $CH_3-\overset{\ominus}{C}H-CN$ (f) CH_3-NO_2

(g) $\overset{\ominus}{C}H_2-\overset{O}{\overset{\|}{C}}-OC_2H_5$ (h)

69. Which of the resonance contributor makes greater contribution to resonance hybrid?

(a) or

(b) or

(c) or

70. Rank the indicated C—H bonds in order of decreasing ease of homolytic cleavage.

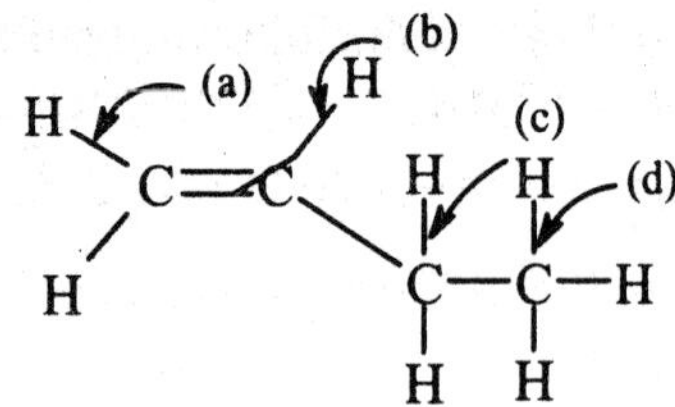

71. When this compound ionizes, is Cl^- or Cl^+ formed?

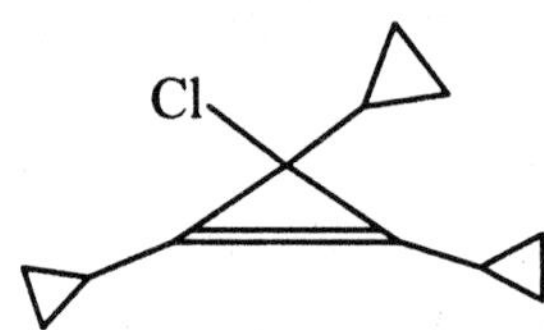

72. Draw resonance structures of the following species indicating the most and least stable.

(a) 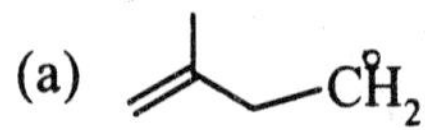

(b) $\overset{+}{O}H$
(c) +
(d) Br
(e) °
(f) +
(g) $\overset{+}{C}H_2$

73. From the following set of resonance structures, designate one that contribute most to the resonance hybrid and explain reason for your choice:

(a) $\overset{+}{C}H_2$ ↔ + (I) (II)

(b) $\overset{+}{C}H_2$ ↔ + (I) (II)

(c) $\overset{+}{C}H—N(CH_3)_2$ ↔ $CH=\overset{+}{N}(CH_3)_2$
I II

(d) 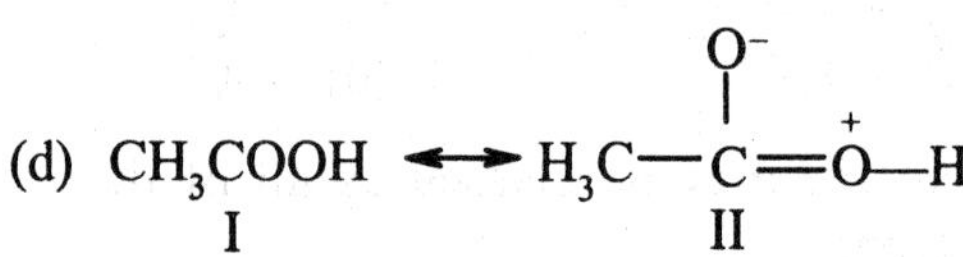

(e) $\overset{\circ}{C}H_2$ ↔ ↔ $\overset{\circ}{C}H_2$
I II III

(f) $H_2N—C\equiv\ddot{N}$ ↔ $H_2\overset{+}{N}=C=\overset{-}{\ddot{N}}$
I II

(g) $\overset{+}{C}H_2—\ddot{O}—CH_3$ ↔ $H_2C=\overset{+}{O}—CH_3$
I II

(h) $CH_3—\overset{+}{C}=\ddot{O}$ ↔ $CH_3—C\equiv\overset{+}{O}$

OBJECTIVE EXERCISE

Level 1

1. Which of the following is the strongest acid?
(a) $CH_3CH_2CO_2H$ (b) $ClCH_2CH_2CO_2H$
(c) $ClCH_2CO_2H$ (d) CH_3CO_2H

2. Which of the following molecules will not have a dipole moment?
(a) CH_3Cl (b) *trans*-CHCl:CHCl
(c) CH_2Cl_2 (d) $CCl_2=CH_2$

3. Which of the following may be classed as an electrophilic reagent?
(a) NH_3 (b) NH_4^+
(c) BCl_3 (d) $^-AlCl_4$

4. Which of the following reagent would you expect to react with 2-bromopropane to give a substitution product?
(a) $^-AlCl_4$ (c) C_2H_5OH
(c) $CH_3—COOH$ (d) NaCN

5. Which of the following anions will be the weakest base?
(a) $C_6H_5—SO_3^-$ (b) $C_6H_5—O^-$
(c) $C_6H_5—CO_2^-$ (d) $C_2H_5—O^-$

6. Which of the following compounds would you expect to react most rapidly with ethanolic silver nitrate?
(a) ethyl chloride (b) isopropyl chloride
(c) t-butyl chloride (d) vinyl chloride

7. Which of the following compounds would you expect to be most soluble in water?
(a) $C_2H_5—O—C_2H_5$ (b) $CH_3—CO_2H$
(c) $C_6H_5—NH_2$ (d) $C_6H_5—COOC_2H_5$

8. Which of the following olefins would you expect to react most readily with concentrated sulphuric acid?
(a) $CH_2=CH_2$ (b) $(CH_3)_2C=CH_2$
(c) $C_6H_5—CH=CH—NO_2$ (d) $CCl_2=CCl_2$

9. Which of the following carboxylic acids will have the largest k_a value?
(a) $CH_3—CO_2H$ (b) $ClCH_2—CO_2H$
(c) $C_6H_5—CO_2H$ (d) $CF_3—CO_2H$

10. Which of the following carbonium ions will be the most stable?

(a) Ph_3C^+ (b) $CH_3—CH_2^+$
(c) $(CH_3)_2CH^+$ (d) $CH_2{=}CH—CH_2^+$

11. Which of the following compounds may be classed as a protic solvent?

(a) diethyl ether (b) *n*-hexane
(c) acetone (d) ethanol

12. Which of the following groups has an electron-withdrawing mesomeric effect?

(a) $—CH_2—CH_3$ (b) —Cl
(c) $—OCH_3$ (d) —CN

13. Glycine may be classed as all of the following except.

(a) a chelating agent (b) an L-amino acid
(c) a base (d) a Zwitterion

14. Which of the following reactions is normally carried out using a heterogeneous catalyst?

(a) Hydration of a carbon-carbon triple bond.
(b) Hydrolysis of $R_3C—Cl$.
(c) Hydrogenation of $R_2C{=}CR_2$.
(d) Hydrochlorination of $R_2C{=}CR_2$.

15. An amino acid (1 mole) reacts with acetic anhydride (2 moles) and gives no N_2 on treatment with HNO_2. It is:

(a) N–H, COOH (b) HO, N–H, COOH
(c) COOH, NH_2 (d) NH_2, COOH, N–H

16. The decreasing order of electronegativity among hybridized carbon atoms should be:

(a) $sp > sp^2 > sp^3$ (b) $sp^3 > sp^2 > sp$
(c) $sp > sp^3 > sp^2$ (d) $sp^2 > sp^3 > sp$

17. One of the completions A to E is incorrect. Choose the letter corresponding to the *incorrect* completion. Glucose:

(a) contains six carbon atoms per molecule.
(b) is a reducing sugar.
(c) undergoes mutarotation.
(d) is a disaccharide.

18. The most basic of the following molecules is:

(a) $CH_3—CONH_2$ (b) $CH_3—CH_2—NH_2$
(c) $(CH_3—CH_2—)_2NH$ (d) $C_6H_5—NH_2$

19. Which of the following is *not* a nucleophile?

(a) H_2O (b) CH_3OH
(c) H_2 (d) NH_3

20. Which compound contains the amide functional group?

(a) CH_3OH
(b) $CH_3—COOCOCH_3$
(c) $CH_3—CH_2—CO—NH_2$
(d) $CH_3—NH_2$

21. Which compound is likely to have the highest boiling point:

(a) $CH_3—CH_2—CH_2—OH$
(b) $CH_3—CO—CH_2—CH_3$
(c) $CH_3—O—CH_2—CH_3$
(d) $CH_3—CH_2—CH_2—SH$

22. Which compound would *not* expected to react with sodium metal?

(a) $CH_3—O—CH_3$ (b) $CH_3—CH_2OH$
(c) $CH_3—CO—OH$ (d) C_6H_5OH

23. A compound $C_6H_{10}O$ shows a strong absorption in the ultraviolet and gives a precipitate with 2, 4-dinitrophenylhydrazine solution. What is the most likely structure of the compound?

(a) 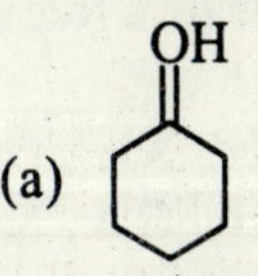

(b) 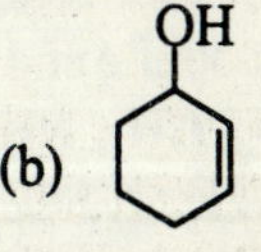

(c) 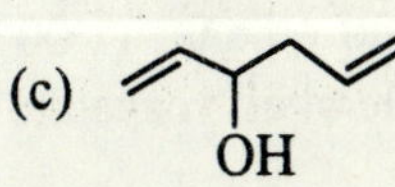

(d) O

24. Which of the following would not have a dipole moment?

(i) CCl_4 (ii) CH_2Cl_2
(iii) *trans*-1, 2-dichloroethene
(iv) 1, 2-dichloroacetylene

(a) (i) and (ii) (b) (i), (ii), and (iii)
(c) (i), (iii), and (iv) (d) (iii) and (iv)

25. The acid strength of substituted carboxylic acids is known to be dependent on the nature and position of the substituent. In the following examples an attempt has been made to arrange the acids in order of acid strength, the strongest first. One of the series is incorrect—which one?

(a) $CH_3—CH_2—CH(Cl)—COOH > CH_3—CH(Cl)—CH_2—COOH > CH_3—CH_2—CH_2—COOH$

(b) $NO_2—CH_2—COOH > HOCH_2—COOH > CH_3—COOH$

(c) $Cl_3C—COOH > BrCH_2—COOH > FCH_2—COOH$

(d) $CH_3—COOH > CH_3—CH_2—COOH > (CH_3)_3CCOOH$

26. An organic liquid, A, ($C_5H_{10}O$) gives a black precipitate on warming with Tollen's reagent. Upon treatment with dilute alkali, A is converted into a dimer which readily loses water to form an alpha,beta-unsaturated carbonyl compound, B ($C_{10}H_{18}O$). The liquid A is:

(a) pentanal (b) pentan-2-one
(c) 2-methyl-butanal (d) pentan-3-one

27. A basic substance, A (C_7H_9N), reacts with sodium nitrite and hydrochloric acid at 0°C to give a salt, B ($C_7H_7ClN_2$). When a solution of B in dilute hydrochloric acid was heated with aqueous copper (I) cyanide and potassium cyanide a product was obtained upon which complete hydrolysis gave an acid, C ($C_8H_8O_2$). This acid could be oxidized to another acid which, on heating, formed an anhydride, D ($C_8H_4O_3$). Compound A is:

(a) benzylamine (b) *p*-toluidine
(c) mono-methylaniline (d) *o*-toludine

28. Which of the following series contain a free radical, a nucleophile and an electrophile?

(a) Cl_2, CH_4, NH_3 (b) CH_3, NH_3, OH^-
(c) Cl, H_2O, SO_3 (d) CH_4, OH^-, NO_2^+

29. Acetic acid (0.50 mole) and ethanol (1.00 mole) reacted to produce 0.25 mole of ethyl acetate. What is the yield in this reaction?

(a) 12.5 (b) 25
(c) 37.5 (d) 50

30. A compound A (C_4H_6) on reduction with hydrogen and a deactivated palladium catalyst gives compound B(C_4H_8). Compound A gave a precipitate with ammoniacal silver nitrate solution.

(a) A is buta-1, 3-diene and B is but-1-ene.
(b) A is buta-1, 3-diene and B is but-2-ene.
(c) A is but-1-yne and B is but-1-ene.
(d) A is but-2-yne and B is but-2-ene.

31. Which of the following series of canonical forms is incorrect?

(a)  $(CH_3)_2C=O \longleftrightarrow (CH_3)_2\overset{+}{C}—O^-$

(b) benzene ⟷ benzene ⟷ (Dewar forms) ⟷ ⟷

(c) $H_3C—C(=O)—OH \longleftrightarrow H_3C—\overset{+}{C}(O^-)—OH \longleftrightarrow H_3C—C(O^-)=\overset{+}{O}H$

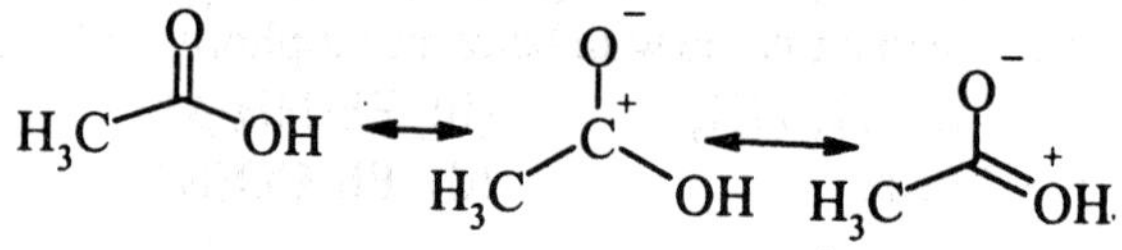

(d) $H_2C=CH_2 \longleftrightarrow \overset{-}{C}H_2—\overset{+}{C}H_2 \longleftrightarrow \overset{+}{C}H_2—\overset{-}{C}H_2 \longleftrightarrow \dot{C}H_2—\dot{C}H_2$

32. Which of the following series of bases shows a left to right increase in base strength?

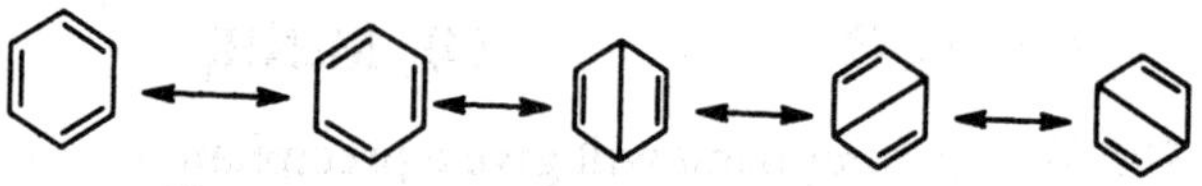

(a) Aniline, NH_3, CH_3NH_2
(b) NH_3, aniline, *p*-nitroaniline
(c) Aniline, Ph_2NH, Ph_3N
(d) *p*-nitroaniline, aniline, $(CF_3)_3N$

33. Which compound will not give a visible reaction product with cold aqueous potassium permanganate?

(a) methoxybenzene (b) propyne
(c) *trans*-1, 2-diphenylethene
(d) benzaldehyde

34. Which compound possesses the carbon-hydrogen bond with the lowest bond-dissociation energy?

(a) toluene (b) benzene
(c) ethane (d) 2, 2-dimethylpropane

35. Which of the folllowing compounds will give a positive iodoform test.

(a) $Ph.CH_2·CH_2·CO·CH_3$
(b) $Ph·CH_2·CH_2·CH_2·CHO$
(c) $Ph·CO·CH_2·CH_2·CH_3$
(d) $Ph·CH_2·CH_2·OCH_3$

36. A compound A ($C_8H_7ClO_2$) was hydrolysed with dilute aqueous sodium hydroxide to give, after acidification, a compound B ($C_8H_8O_3$). Oxidation of compound B with acidic sodium dichromate gave $C_8H_6O_4$ which, when heated alone, gave $C_8H_4O_3$. What is the most likely structure for A?

(a) 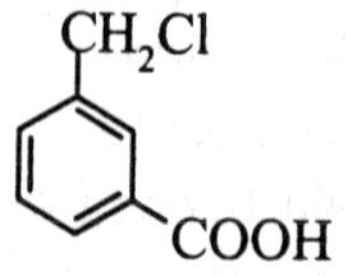 benzene ring with CH_2Cl and COOH (meta) (b) 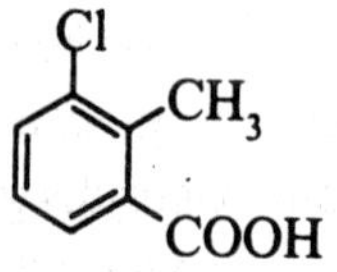 benzene ring with Cl, CH_3 and COOH

(c) 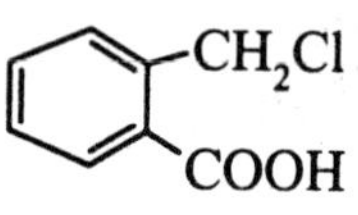 benzene ring with CH_2Cl and COOH (ortho) (d) 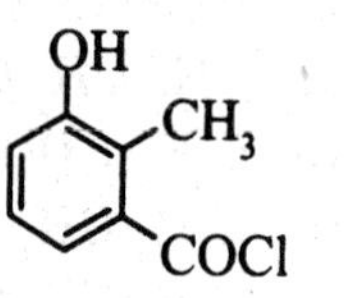 benzene ring with OH, CH_3 and COCl

37. Which is the acid with the lowest pk_a?
(a) *p*-nitroanilium ion (b) $PhNH_3^+$
(c) PhOH (d) $MeNH_3^+$

38. Which compound will give a precipitate soluble in alkali when treated with benzene sulphonyl chloride?
(a) $Ph \cdot CH_2 \cdot NH_2$ (b) $Ph \cdot NH \cdot Me$
(c) Me_3N (d) $Ph \cdot CO \cdot NH_2$

39. Which of the following compounds *CANNOT* exhibit keto-enol tautomerism?

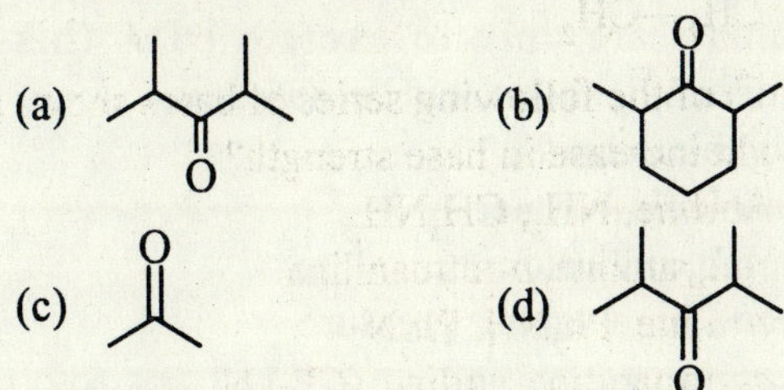

40. Which of the following solvents are inflammable?
(a) Diethyl ether (bp 37°)
(b) Benzene (bp 80°)
(c) Chloroform (bp 61°)
(d) Carbon tetrachloride (bp 78°)

41. The reaction of sodium ethoxide with iodoethane to form diethyl ether is termed:
(a) electrophilic substitution
(b) electrophilic addition
(c) nucleophilic substitution
(d) radical substitution

42. Which of the following compounds has the highest pk_a value?
(a) Phenol (b) *o*-nitrophenol
(c) 2, 4-dinitro phenol (d) Picric acid

43. The change in hybridization of carbon in the reaction is: Acetamide + $P_2O_5 \rightarrow CH_3CN$
(a) sp^3 to sp^2 (b) sp^2 to sp^3
(c) sp to sp^3 (d) sp^2 to sp

44. Compound Z reacts with hydrogen iodide; this product, on heating with sodium ethoxide in absolute ethanol, gives a compound isomeric with Z, and which on ozonolysis gives two products, one gives positive Fehling test while other gives negative Fehling test. Compound Z is:
(a) $CH_3 \cdot CH_2 \cdot CH:C(CH_3)_2$
(b) $C_6H_5 \cdot CH(OH) \cdot CH(CH_3)_2$
(c) $C_6H_5 \cdot CH_2 \cdot CH:CH_2$
(d) $C_6H_5 \cdot CH:C(CH_3)_2$

45. A low boiling point hydrocarbon reacts with bromine readily and gives a red precipitate with ammoniacal cuprous chloride solution. On treatment with mercuric sulphate/sulphuric acid solution a fragrant liquid, which reacts with sodium hydroxide and iodine to give iodoform an propanoic acid is obtained. The hydrocarbon is:
(a) pent-1-yne (b) but-1-yne
(c) but-2-yne (d) prop-1-yne

46. Which of the following set of reagents contains only nucleophile?
(a) H_2O, SO_3, H_3O^+ (b) NH_3, H_2O, ROH
(c) CN^-, SO_3, OH^- (d) NH_3, H_2O, $AlCl_3$

47. Which of the following will react with benzaldehyde *and* also with propylamine?
(a) 2-chloro-6-methylpyridine
(b) 4-chloro-6-methylisoquinoline
(c) 3-chloro-6-methylquioline
(d) 3-chloro-5-methylnitrobenzene
(e) 3-chloro-5-methylpyridine

48. An amino acid exhibits minimum solubility at pH = 6 and the pk_a for the carboxylic acid ionization was found to be 2.3. What is the second pk_a for this amino acid?
(a) 8.3 (b) 3.7 (c) 9.7 (d) 10.3

49. Which one of the following compounds gives a positive iodoform test with iodine and aqueous alkali and can be oxidized to a ketone containing the same number of carbon atoms?
(a) $CH_3 \cdot CH_2OH$
(b) $CH_3 \cdot CH_2 \cdot CO \cdot CH_2 \cdot CH_2OH$
(c) $CH_3 \cdot CH(OH) \cdot CH_2 \cdot CH_3$
(d) $Ph \cdot CH(OH) \cdot CH_2 \cdot CH_3$

50. Aniline has a basicity constant $k_b = 4.2 \times 10^{-10}$, thus the pk_a for its conjugate acid is:
(a) between 2 and 3 (b) between 3 and 4
(c) between 4 and 5 (d) between 5 and 6

51. An organic solid liberates carbon dioxide from an aqueous solution of sodium bicarbonate and on treatment with $Cl_2/AlCl_3$ gives only one monochloroderivative at very slow rate. The solid could be:
(a) phenol (b) acetic anhydride
(c) benzoic acid (d) propanoic acid

52. Which of the following is most readily decarboxylated by heat?
(a) $HOOC \cdot CH_2 \cdot COOH$
(b) $HOOC \cdot CH_2CH_2 \cdot COOH$

(c) $CH_3 \cdot COOH$ (d) PhCOOH

53. The haolgen compound most reactive with water is:
(a) *t*-butyl chloride (b) chlorobenzene
(c) acetyl chloride (d) *n*-butylchloride

54. An inductive effect is
(a) the displacement of electrons along a pi-bond.
(b) the transfer of electrons from a pi-bond to a sigma-bond.
(c) the displacement of electrons along a sigma-bond.
(d) the transfer of electrons across a benzene ring.

55. In terms of relative stability, which of the following is *not*, in geneal, correct?
(a) Tertiary radicals are more stable than secondary.
(b) Tertiary carbonium ions are more stable than primary.
(c) Primary radicals are less stable than secondary.
(d) Secondary carbonium ions are less stable than primary.

56. In order for a reagent to behave as a nucleophile it *must* possess.
(a) An overall positive charge.
(b) An overall negative charge.
(c) An unpaired electron.
(d) An unshared pair of electrons.

57. The inductive effects of the groups: $—CH_3$,$—CO_2$, —Br, $—NH_3^+$ are respectively.
(a) +I –I +I +I (b) +I +I –I –I
(c) –I –I +I +I (d) –I +I –I +I

58. The following list: Br_2, OH^-, HCl(gas), NH_3 contains:
(a) all nucleophiles.
(b) one nucleophile and three electrophiles.
(c) two nucleophiles and two electrophiles.
(d) three nucleophiles and one electrophile.

59. Ozonolysis of a compound X gives a single compound which produces iodoform on treatment with iodine and dilute sodium hydroxide solution. A possible structure is:
(a) $CH_3 \cdot CH_2 \cdot CH{=}CH \cdot CH_2 \cdot CH_3$
(b) $(CH_3)_2C{=}C(CH_3)_2$
(c) cyclopentene
(d) none of these

60. Which of the following compounds does *not* react with acetyl chloride?
(a) C_2H_5OH (b) $C_6H_6 + AlCl_3$
(c) $CH_3 \cdot COONa$ (d) C_2H_5Cl

61. The typical reaction of the olefinic bond is:
(a) electrophilic substitution.
(b) nucleophilic substitution.
(c) electrophilic addition.
(d) nucleophilic addition.

62. Free radical polymerized polyethylene has a lower softening point than catalytically polymerized polyethylene because it is:
(a) more crystalline (b) syndiotactic
(c) less chain branched (d) more chain branched

63. Mammals are able to excrete DDA (bis [*p*-chlorophenyl] acetic acid because it is:
(a) soluble in lipids.
(b) insoluble in water.
(c) soluble in water.
(d) has a high vapour pressure.

64. 3, 5-dimethyl-4-nitro phenol is less acidic than 2, 6-dimethyl-4-nitrophenol because?
(a) +I effect of two methyl group is less from 2, 6-positions.
(b) –I effect of NO_2 group is more effective in 2, 6-dimethyl-4-nitrophenol.
(c) Intra molecular hydrogen bonding.
(d) Steric inhibition to resonance of NO_2 group by two adjacent methyl group in 3, 5-dimethyl-4-nitro phenol.

65. The C—O—H bond angle in ethanol is approximately:
(a) 90° (b) 104° (c) 109° (d) 120°

66. Indicate which of the following assertions correctly applies to the electromeric effect.
(a) It can either facilitate or hinder attack on a molecule by a reagent.
(b) It results in the presence of an electronegative group within a molecule.
(c) It results from the polarisability of systems containing pi-orbitals.
(d) All molecules showing it have dipole moments.
(e) It makes paraffin hydrocarbons (alkanes) susceptible to attack by electrophiles.

67. The nitronium ion (NO_2^+) commonly functions in organic chemistry as:
(a) a nucleophile (b) an electrophile
(c) a free radical
(d) either (A) or (B), depending on the system

68. In which of the following ways does the hydride ion tend to function?

(a) An electrophile (b) A nucleophile
(c) A free radical (d) An oxidising agent

69. Which of the following statement regarding nucleophile and Lewis base is not true?
(a) All nucleophiles are Lewis base.
(b) Basic strength parallels nucleophilic strength.
(c) Nucleophile can be a neutral or negatively charged.
(d) Effectiveness of a nucleophile depends on solvent being used.

70. Which of the following series contains only nucleophiles:
(a) NH_3, H_2O, CN^-, I^-
(b) $AlCl_3$, NH_3, H_2O, I^-
(c) $AlCl_3$, BF_3, H_2O, NH_3
(d) $AlCl_3$, BF_3, NO_2^+, NH_3

71. An example of nucleophilic displacement is the reaction of:
(a) benzene with bromine.
(b) bromoethane with aqueous sodium hydroxide.
(c) chlorine with ethane.
(d) an aldehyde with potassium dichromate.

72. Which acid is most highly ionized in water?
(a) $(CH_3)_2CH{\cdot}COOH$ (b) $CH_3{\cdot}CH_2{\cdot}COOH$
(c) $Cl_2CH{\cdot}CH_2{\cdot}COOH$ (d) $CH_3{\cdot}CCl_2{\cdot}COOH$

73. A tertiary carbonium ion is more stable than either a secondary or primary carbonium ion because:
(a) it carries three positive charges
(b) it carries three negative charges
(c) it is trigonal planar
(d) it possesses three electron-donating groups

74. Which one of the following alcohols may be oxidized by chromium trioxide/sulphuric acid to a ketone having the same number of carbon atoms?
(a) Ethanol (b) Propan-1, 3-diol
(c) 2-methylbutan-2-ol (d) 1-phenylpropan-1-ol

75. Which one of the following compounds will *not* react (under normal conditions) with lithium aluminium hydride?
(a) Cyclohexene (b) Benzoic acid
(c) Acetonitrile (d) *n*-butanol

76. Which one of the following compounds contains the greatest number of sp^2 hybridized carbon atoms?
(a) Phenylethene (b) Phenylacetylene
(c) Cyclohexanone (d) Cyclohexanol

77. Which one of the following compounds would you expect to be the strongest carbon acid?

(a) [cyclohexane-1,3-dione structure] (b) $CH_2(COOEt)_2$

(c) CH_3COCH_2COOEt (d) [bicyclic diketone structure]

78. Which compound of the following would *not* give a positive reaction with iodine and sodium hydroxide?
(a) Pentan-3-one (b) Ethanal
(c) Pentan-2-one (d) Propan-2-ol

79. Which compound would be the strongest acid in aqueous solution?
(a) Benzoic acid (b) Formic acid
(c) Trichloroacetic acid (d) Phenol

80. A compound was found to be insoluble in dilute aqueous sodium hydroxide solution. A solution of the compound in dry benzene gave a colourless gas on the addition of sodium metal. On shaking with alcoholic silver nitrate solution a pale yellow precipitate was formed. Which of the following compound would satisfy these observations?

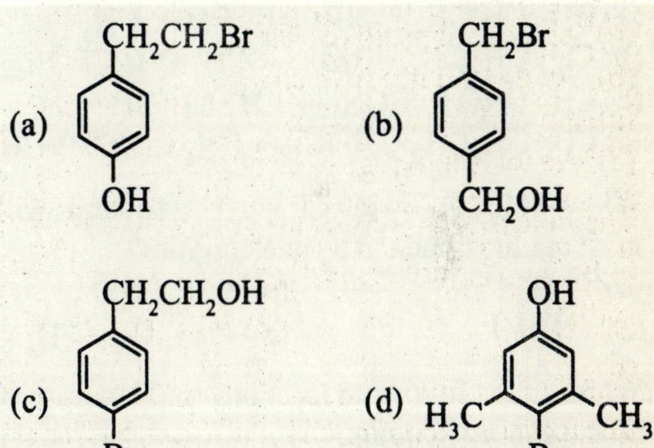

81. Which of the following is the *weakest* acid.
(a) Phenol (b) Benzoic acid
(c) Propan-2-ol (d) Myristic acid

82. Na^+ is not an electrophile because:
(a) It does not normally form a covalent bond with a nucleophile.
(b) It is an ion.
(c) It has a noble gas configuration.
(d) It has an empty orbital.

83. Which of the following tests would distinguish between an aldehyde and a ketone?
(a) Test with 2, 4-dinitrophenylhydrazine
(b) Iodoform test
(c) Add sodium
(d) Ammoniacal silver nitrate (Tollens Reagent)

84. In the reaction of phenylhydrazine with acetone:
(a) Acetone is reduced to propan-2-ol
(b) Phenylhydrazine acts as a nucleophile
(c) Acetone acts as a nucleophile
(d) Free radicals are involved

85. Which of the following would most readily migrate to the anode when submitted to electrophoresis in distilled water?
(a) $H_2NCH(CH_3)\cdot COOH$
(b) $H_2NCH(CH_2OH)\cdot COOH$
(c) $H_2NCH(COOH)(CH_2)_4NH_2$
(d) $CH_3\cdot CONH.CH_2\cdot COOH$

86. Formic acid reduces Fehling's solution and gives a positive "silver mirror test" because:
(a) It is a weak acid.
(b) It has a high proportion of oxygen in its structure.
(c) It is oxidised to $CO_2 + H_2O$.
(d) It contains a hydrogen atom joined to a carbonyl group.

87. Which of the following reactions involves free radical intermediates.
(a) $CH_3\cdot CH{:}CH_2$ + HBr $CH_3\cdot CHBr\cdot CH_3$
(b) $CH_3\cdot CH{:}CH_2 + HBr \rightarrow CH_3\cdot CH_2\cdot CH_2Br$
(c) $CH_3\cdot CHO + HCN \rightarrow CH_3\cdot CH(OH)\cdot CN$
(d) $CH_3\cdot CH\cdot CH_2 + HOCl \rightarrow CH_3\cdot CH(OH)\cdot CH_2Cl$

88. Which of the following carbonyl compounds will be most readily attacked by nucleophiles?
(a) $C_6H_5.CHO$ (b) $CH_3.CHO$
(c) HCHO (d) $C_6H_5.CO.C_6H_5$

89. Which of the following reactions could be described by the energy diagram:
(a) $C_6H_6 + NO_2^+ \rightarrow C_6H_5NO_2 + H^+$
(b) $C_2O_5Na^+ + CH_3—CH_2Br \rightarrow CH_3CH_2OCH_2CH_3 + NaBr$
(c) $(CH_3)_3CCl + H_2O \rightarrow (CH_3)_3COH + HCl$
(d) $CH_3—CH{=}CH_2 + HBr \rightarrow CH_3CHBrCH_3$

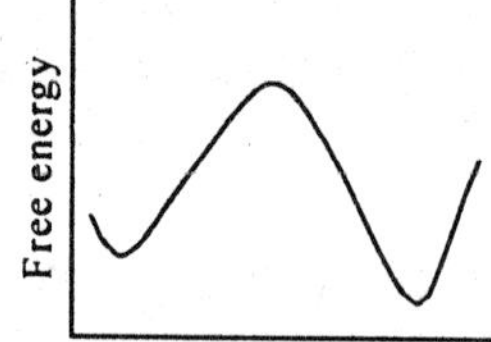

90. Which of the following reactions would you expect to occur mainly by an SN-2 mechanism?
(a) $(CH_3)_3CBr + C_2H_5ONa/C_2H_5OH \rightarrow (CH_3)_2C{=}CH_2 + C_2H_5OH + NaBr$
(b) $C_6H_5C(CH_3)_2I + CH_3OH$/heat $\rightarrow C_6H_5C(CH_3)_2OCH_3 + HI$
(c) CH_3NH_2 + Oxirane $\rightarrow CH_3NHCH_2CH_2OH$
(d) $C_6H_5C(CH_3)BrCH_2CH_3 + CH_3COOH$ /heat $\rightarrow C_6H_5C(CH_3)(OCOCH_3)CH_2CH_3 + HBr$

91. Fluorine reacts with methane by a free radical mechanism giving a variety of products. The formation of which of the following compounds points to a free radical machanism rather than a molecular process?
(a) CH_3F (b) CH_2F_2
(c) CHF_3 (d) C_2F_6

92. Which of the following reactions is classed as a nucleophilic aromatic substitution?
(a) Sulphonation of benzene.
(b) Reaction of a protein with dinitrofluorobenzene.
(c) Reaction of the diazonium cation with phenol.
(d) The conversion of pyridine to pyrrolidine.

93. In which of the following reactions may the *mode* of addition (*cis* or *trans*) be deduced from the stereochemistry of the product? Addition of:
(a) Br_2 to propylene
(b) H_2O to fumaric acid
(c) HBr to acetylene
(d) HBr to 1, 2-dimethylcyclohexene

94. When ICl adds to cyclohexene in the dark an unstable intermediate is formed. This is:
(a) $C_6H_{10}I^+$ (b) $C_6H_{10}I^-$
(c) $C_6H_{10}Cl^+$ (d) $C_6H_{10}Cl^-$

95. The *ionic* addition of HCl to one of the following compounds *does not* obey the MARKOWNIKOV rule. Which compound is it?
(a) $CH_3\cdot(CH_2)_3\cdot CH{=}CH_2$
(b) $CF_3\cdot(CH_2)_3\cdot CH{=}CH_2$
(c) $CH_3\cdot CH{=}CH_2$
(d) $CF_3\cdot CH{=}CH_2$
(e) $CH_3\cdot CH_2\cdot CH{=}CH_2\cdot CH_3$

96. One of the completions A to E is incorrect. Choose the letter corresponding to the *incorrect* completion. The Claisen condensation reaction is a base-catalysed reaction between:
(a) Two molecules of a ester possessing an alpha hydrogen atom.
(b) Molecules of two different aldehydes, one of which possesses an alpha hydrogen atom.
(c) A molecule of an aldehyde and a molecule of an ester possessing an alpha hydrogen atom.
(d) Molecules of two different esters, neither of which possesses an alpha hydrogen atom.

(e) Molecules of two different esters, both of which possess an alpha hydrogen atom.

97. One of the completions A to E is incorrect. Choose the letter corresponding to the *incorrect* completion. In an SN2 reaction between (d)-2-Bromooctane and hydroxyl ions:
(a) The rate of the reaction is independent of the concentration of hydroxyl ions.
(b) The product is (L)-2-octanol.
(c) If (D)-2-bromooctane is replaced by (2)-2-chlorooctane the reaction rate is decreased.
(d) The rate of reaction is increased, if the (D)-2-bromooctane is replaced by 1-bromooctane.
(e) A change of solvent from acetone to a mixture of acetone and water does not change the rate of reaction significantly.

98. An SN2 reaction involves:
(a) A two-step process.
(b) A radical intermediate.
(c) A carbonium ion intermediate.
(d) A one-step, synchronous process.

99. $CH_3\cdot CHO + NH_2\cdot NH_2 \rightarrow CH_3\cdot CH{=}N{-}NH_2$
Classify the reaction as:
(a) nucleophilic substitution
(b) elimination
(c) condensation
(d) electrophilic addition
(e) electrophilic substitution

100. Which one of the following statements is an incorrect description of heterolytic fission of a bond.
(a) One of the atoms leaves with both of the electrons from the bond.
(b) Polar solvents promote the fission.
(c) Ultraviolet light promotes the fission.
(d) Fission gives rise to the formation of ions.

101. Dehydration of 3, 3-dimethylbutan-2-ol with concentrated sulphuric acid gives, as principal product, 2, 3-dimethylbut-2-ene. Which of the following best describes the factor(s), which are involved in this reaction?
(a) Carbonium ion formation.
(b) Carbonium ion formation and a 1, 2-hydride shift.
(c) Carbonium ion formation and a 1, 2-alkyl shift.
(d) Carbonium ion formation, a 1, 2-alkyl shift and loss of a proton to give the most highly substituted olefin.

102. Which of the following alkyl halides undergoes a SN1 reaction the fastest?
(a) CH_3Cl (b) C_2H_5Cl
(c) $(CH_3)_2CHCl$ (d) $(CH_3)_3CCl$

103. In the rearrangement of a carbonium ion the migrating group leaves:
(a) with its electron pair (b) as a cation
(c) as a free radical (d) as a di-radical

104. Heterolysis of the C—Y bond in the compound CH_3Y gives:
(a) two free radicals. (b) a carbonium ion.
(c) a carbanion.
(d) either a carbonium ion or a carbanion.

105. The acid catalysed dehydration of alcohols to give olefins involves one of the following:
(a) A free radical intermediate.
(b) A carbanion intermediate.
(c) The expulsion of an OH^- ion.
(d) A carbonium ion intermediate.

106. Which of the following is the strongest acid?
(a) HOOC(H)C=C(H)NO$_2$ (b) HOOC(H)C=C(H)NO$_2$
(c) HOOC(H)C=C(H)Me (d) HOOC(H)C=C(H)CN

107. $HO^- +$ Et(Ph)(H$_3$C)C—Br →
(A) HO—C(Ph)(Et)(CH$_3$) + (B) Et(Ph)(H$_3$C)C—OH + (C) =C(Ph)(Et) + Br^-

If the above reaction was strictly SN2, the organic product(s) would be:
(a) A only (b) A and B
(c) C only (d) B only

108. $HO^- +$ Et(Ph)(H$_3$C)C—Br ↓

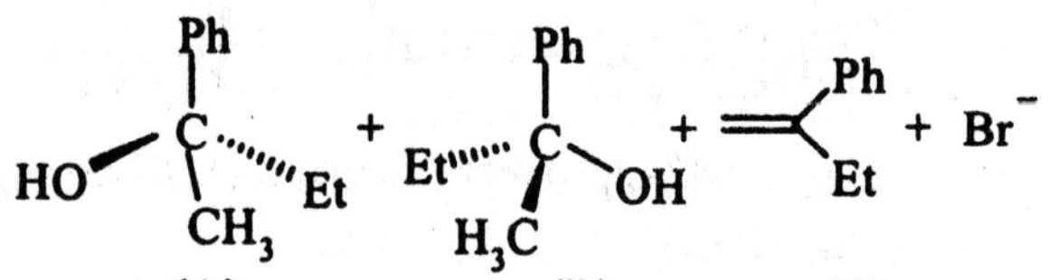

(A) + (B) + (C) + Br^-

If the above reaction was strictly SN1, the organic product(s) would be:

(a) A only (b) A and B
(c) A, B and C (d) B only

109. Which one of the following reactions does not occur by a free radical mechanism?
(a) chlorine and benzene in ultraviolet light.
(b) alkylation using diazomethane.
(c) Sandmeyer reaction.
(d) Chlorine and neopentane.

110. Electrophilic substitution in aromatic compounds may proceed, via sigma-complex formation. Indicate which one of the following statements is incorrect with regard to this mechanism.
(a) The electrophilic attacking species may be cation or a polarized molecule.
(b) When a cation attacks the aromatic ring it removes a pair of electrons from the sextet to form a carbanion.
(c) The sigma-complex is a highly reactive intermediate.
(d) In sigma-complex formation the C—H bond remains intact in the rate determining step.

111. Select the false statement concerning nucleophilic substitution at an aliphatic trigonal carbon atom.
(a) Substitution does not occur when the carbon atom is doubly bonded to a nitrogen atom.
(b) Some nucleophilic substitutions at a carbonyl carbon are catalyzed by nucleophiles.
(c) Simple SN2 mechanisms are rarely found in carbonyl substrates.
(d) Substitutions occurring by the tetrahedral mechanism are frequently speeded up by the addition of acid.

112. Which one of the following compounds is the most reactive towards SN2 displacement?
(a) $CH_3 \cdot CH_2 \cdot CH_2 \cdot CH_2 \cdot Cl$
(b) $(CH_3)_3CCl$
(c) $CH_3 \cdot CH(CH_2 \cdot CH_3)Cl$
(d) $(CH_3)_2CH \cdot CH_2Cl$

113. In SN1 reactions the order of reactivity of halides is:
(a) primary allyl > 3° > 2° > 1°
(b) 3° > primary allyl > 2° > 1°
(c) 1° > 2° > 3° > primary allyl
(d) primary allyl > 1° > 2° > 3°

114. (+)$PhCHClCH_3$ + OH^-/H_2O → $PhCH(OH)CH_3$, which statement is the most likely to be correct concerning the following SN1 reaction?
(a) The reaction proceeds with partial racemization.
(b) The stereochemistry of the halide is not inverted.
(c) The carbonium ion $Ph \cdot CH_2 \cdot CH_2^+$ is involved.
(d) The carbonium ion is attacked in each side to the same degree.

115. *t*-butyl bromide reacts with hydroxide ion by a first order process with the rate proportional to [*t*-Bu Br]. One rationale for this is that:
(a) The intermediate *t*-butyl radical is stabilized by solvation.
(b) The product *t*-butanol is thermodynamically stable.
(c) In this case a stereochemical inversion is favourable.
(d) Formation of a very stable intermediate *t*-butyl carbonium ion is facile.

116. Under certain conditions an important side reaction in the conversion of *t*-butyl bromide to *t*-butanol with OH^- ions is:
(a) Formation of isobutene by elimination.
(b) Formation of hexamethylethane by Wurtz coupling.
(c) Isomerization of *t*-butyl bromide.
(d) Stereochemical inversion of the *t*-butyl bromide.

117. Which of the following best represents the mechanism for removal of HBr from a bromo compound (dehydrobromination) by hydroxide ion?

(a) 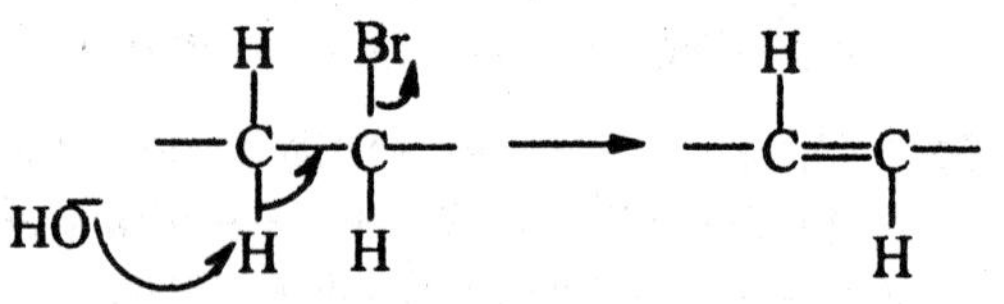 HO⁻ + H–C(H)(H)–C(Br)(H)– → –C(H)=C(H)–

(b) HO⁻ + H–C(H)(H)–C(Br)(H)– → –C(H)=C(H)–

(c) H–C(H)(H)–C(Br)(H)– → –C(H)=C(H)–

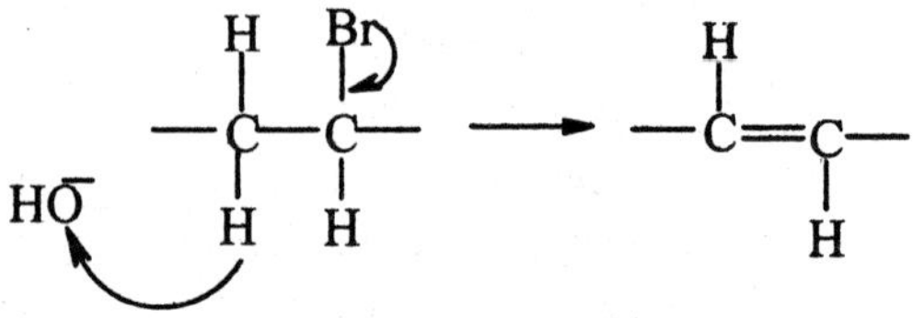

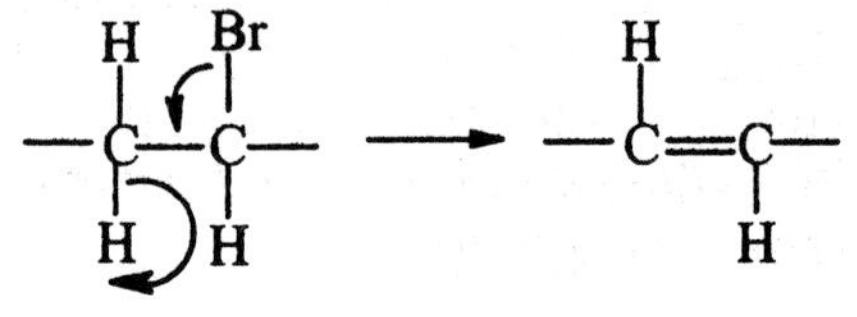

(d) H–C(H)–C(Br)(H) + HO^- ⟶ –C(H)=C(H)–

118. In the radical chlorination of methane at 250°, the reaction has three principal stages–initiation, propagation and termination. Which of the following represents a termination step?

(a) $Cl\cdot + CH_4 \rightarrow CH_3\cdot + HCl$

(b) $CH_3\cdot + HCl \rightarrow CH_4 + Cl\cdot$

(c) $CH_3\cdot + CH_3\cdot \rightarrow CH_3—CH_3$

(d) $CH_3\cdot + CH_4 \rightarrow CH_4 + CH_3\cdot$

119. Which of the following bases is most likely to promote SN1 reactions rather than SN2 reactions?

(a) OH^- (b) OEt^-

(c) H_2O (d) PhO^-

120. One of the following compounds readily undergoes SN1 reactions owing to the stability of its carbonium ion. It is:

(a) CH_3Cl (b) $CH_2{=}CHCl$

(c) $CH_3{\cdot}CH_2{\cdot}CH_2Cl$ (d) $CH_2{=}CH.CH_2Cl$

121. The complete inversion of configuration which occurs during the hydrolysis of a haloalkane suggests that the reaction is:

(a) SN1 (b) SN2 (c) E1 (d) E2

122. R–C(=O)–Z If Z in the this compound is electronegative, then the principal reactions of the compound will be:

(a) nucleophilic additions.

(b) electrophilic substitutions.

(c) nucleophilic substitutions.

(d) electrophilic additions.

123. In the electrophilic bromination of ethene it is believed that an intermediate can explain the stereochemistry of addition. This is:

(a) Br + (b) Br (c) Br + (d) + Br

124. Study the reactions below and decide, which compound (or intermediate) of those given would be the major product.

$(CH_3)_2C{=}CHCH_3 + H^+ \rightarrow$

(a) $(CH_3)_2C—{}^+CHCH_3$

(b) $(CH_3)_2C^+—CHCH_3$

(c) $(CH_3)_2C—{}^+CH_2—CH_3$

(d) $CH_3CH{=}C(CH_3)CH_2{}^+$

125. Study the reactions below and decide, which compound (or intermediate) of those given would be the major products. *cis*-2-butene + $Br_2 \rightarrow$

(a) CH_3 / H–C–Br / Br–C–H / CH_3

(b) CH_3 / H–C–Br / H–C–Br / CH_3

(c) CH_3 / H–C–Br / Br–C–H / CH_3 and CH_3 / Br–C–H / H–C–Br / CH_3

(d) CH_3 / H–C–Br / H–C–Br / CH_3 and CH_3 / Br–C–H / Br–C–H / CH_3

Level 2

1. Choose from the indicated protons, the one that is most acidic:

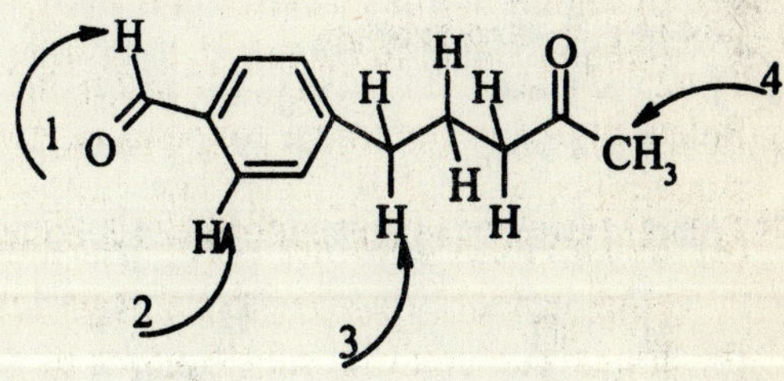

(a) 1 (b) 2 (c) 3 (d) 4

2. Give the statement that is true regarding diene stability:

(a) Conjugated dienes are more stable than non-conjugated dienes. Non-conjugated diene gives a greater exothermic reaction with H_2/Pt.

(b) Nonconjugated dienes are more stable than conjugated dienes. Conjugated diene gives a greater exothermic reaction with H_2/Pt.

(c) Conjugate dienes are more stable than non-conjugated dienes. Conjugated diene gives a greater exothermic reaction with H_2/Pt.

(d) Nonconjugated dienes are more stable than conjugated dienes. Non-conjugated diene gives a greater exothermic reaction with H_2/Pt.

3. Choose the statement that corresponds to structure of an allyl radical.
   (a) The carbon atoms of an allyl radical is sp^3 hybridized. P orbitals are involved in the delocalization of electrons.
   (b) The carbon atoms of an allyl radical is sp^2 hybridized. Hybrid orbitals are involved in the delocalization of *p*-electrons.
   (c) The carbon atom of an allyl radical is sp hybridized. P orbitals are involved in the delocalization of electrons.
   (d) The carbon atoms of an allyl radical is sp^2 hybridized. P orbitals are involved in the delocalization of electrons.

4. Which statement best describes 1, 3-butadiene?
   (a) 1, 3-butadiene is less stable than1, 4-pentadiene due to steric crowding.
   (b) 1, 3-butadiene is more stable than1, 4-pentadiene due to less steric crowding.
   (c) 1, 3-butadiene is more stable than1, 4-pentadiene due to resonance stabilization.
   (d) 1, 3-butadiene carbon atoms are sp^3 hybridized.

5. Rank the following compounds in order of decreasing stability.
   1. 3-methyl-1, 4-pentadiene.
   2. 2-methyl-1, 3-butadiene.
   3. 1, 3-hexadiene.
   4. 2-methyl-1, 4-pentadiene.

   (a) 1 > 2 > 3 > 4 (b) 4 > 3 > 2 > 1
   (c) 3 > 2 > 1 > 4 (d) 2 > 3 > 4 > 1

6. The major resonance contributor of allyl radical are:

1. 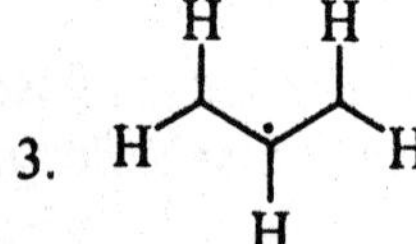

2. H H H H H

3. H H H H H

4. H H H H H H

5. H H H H H H

   (a) 1 and 2 (b) 4 and 5
   (c) 2 and 3 (d) 1 and 5

7. From the structure shown below, choose the indicated carbon-hydrogen bond that is weakest (the bond that undergoes easiest hemolytic cleavage):

(a) (b) (c) (d)

8. The most stable diene is:

(a) (b)

(c) (d)

9. Predict which of the following carbocation has the highest energy?

(a) $\overset{+}{C}H_2$ (b) 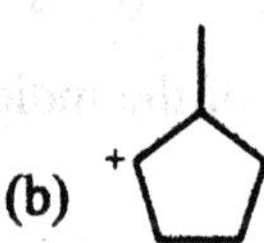

(c) (d)

10. Which of the following is a pair of resonance structure?

(a)

(b) 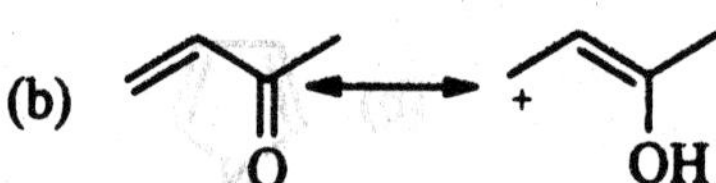

(c)

(d) 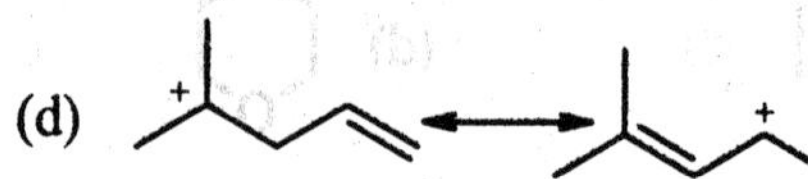

11. Which anion is the weakest base:
    (a) $C_2H_5O^-$ (b) NO_3^-
    (c) CH_3COO^- (d) CN^-

12. Which is the strongest acid in water?
    (a) Phenol (b) 3-nitro phenol
    (c) 4-nitro phenol (d) Hydroquinone

13. Which of the following carbocation would have the greatest stability?

(a) $CH_3—S—CH_2^+$ (b) $(CH_3)_2NCH_2^+$
(c) $CH_3OCH_2^+$ (d) FCH_2^+

14. Which of the following would react more rapidly with BF_3?

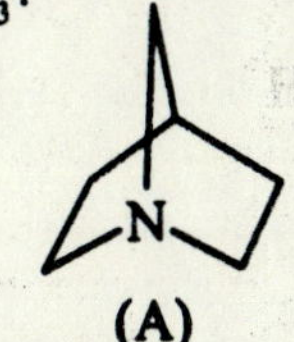

(A) (B)

(a) A (b) B
(c) Both have same basicity (d) None is basic

15. The equilibrium constant for the following reaction is:

$$(CH_3)_2NH + (CH_3)_2NH_2^+ \rightleftharpoons (CH_3)_2NH_2^+ + (CH_3)_2NH$$

is

(a) > 1 (b) 1 (c) < 1 (d) 0

16. Which of the indicated H in the following is most acidic?

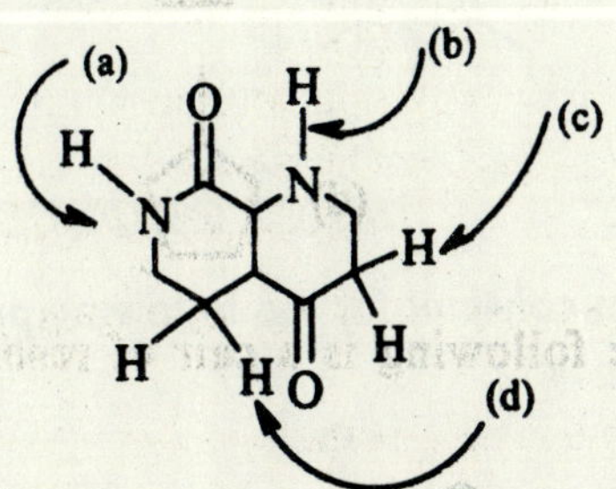

17. Which of the following is the strongest base?

(a) 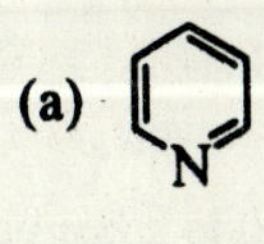 (b) 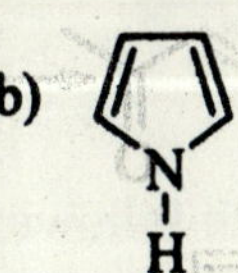

(c) 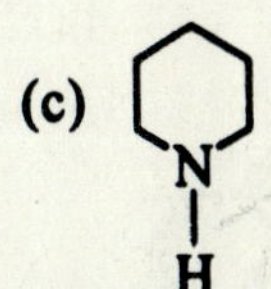 (d)

18. Rank the following compounds in increasing order of their acidity:

1. 3-fluorobutanoic acid.
2. 3-chlorobutanoic acid.
3. 2-fluorobutanoic acid.
4. Butanoic acid.
5. 4-chlorobutanoic acid.

(a) 3 < 1 < 2 < 4 < 5 (b) 4 < 3 < 1 < 2 < 5
(c) 5 < 2 < 1 < 3 < 4 (d) 4 < 5 < 2 < 1 < 3

19. Arrange the following in decreasing order of acidity:

1. 2. COOH
3. COOH 4.

(a) 1 > 3 > 4 > 2 (b) 3 > 2 > 1 > 4
(c) 2 > 4 > 3 > 1 (d) 2 > 3 > 4 > 1

20. Which of the following isomeric sugars is most stable?

(a) (b)

(c) (d)

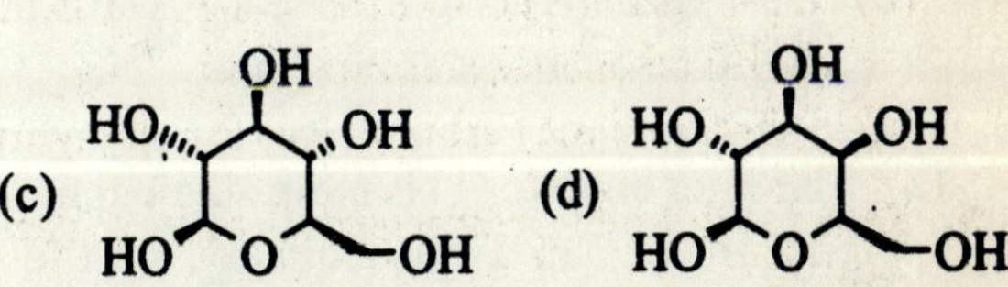

21. Which statement best describes the relative basicity of compounds A and B?

p-methoxy aniline (A) Aniline (B)

(a) Compound A is more basic than compound B, because the methoxy is electron donating, which makes the nitrogen lone pair more basic.
(b) Compound A is less basic than B, because there more resonance structures for A, which makes the nitrogen lone pair less available as base.
(c) Both compound are nearly equally basic.
(d) Compound B is more basic than A, because the methoxy group is electron withdrawing, which lowers the basic strength of nitrogen.

22. Which statement about amine basicity is false?

(a) All aromatic amines have similar basicity.
(b) All aliphatic amines are nearly equally basic.
(c) In general, aliphatic amines are more basic than aromatic amines.
(d) The more basic the amine, the less acidic the conjugate base.

23. Among the following molecules or ions, which has the longest C—N bond?

(a) H_2CNCH_3 (b) H_3CCN
(c) $(H_3C)_4N^+$ (d) $(H_2C)_2N^+$

24. Rank the following nitrogen containing compound from most basic to least basic:

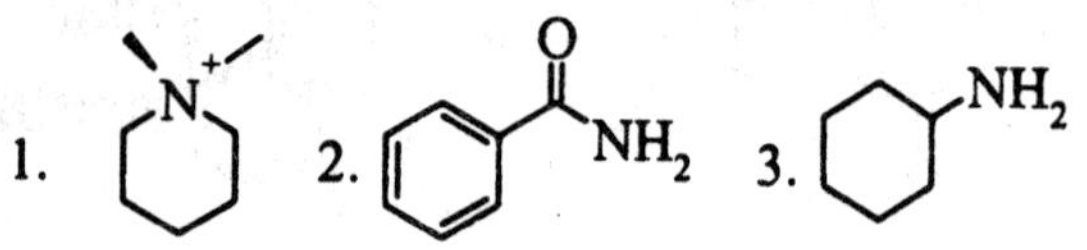

1. 2. NH_2 3. NH_2

(a) 1 > 2 > 3 (b) 1 > 3 > 2
(c) 3 > 1 > 2 (d) 3 > 2 > 1

25. The equilibrium constant for the following reaction:

$$(CH_3)_3N + C_6H_5\overset{+}{N}H_3 \underset{}{\overset{Keq}{\rightleftharpoons}} (CH_3)_3NH^+ + C_6H_5NH_2$$

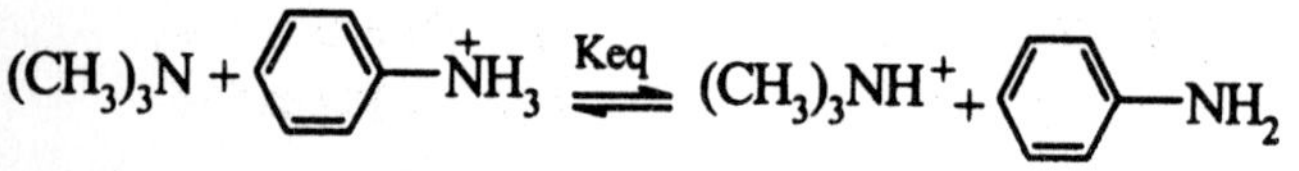

(a) > 1 (b) < 1 (c) 1 (d) 0

26. An ester is more acidic than amide. In view of the following conjugate base, the most appropriate explanation for this is:

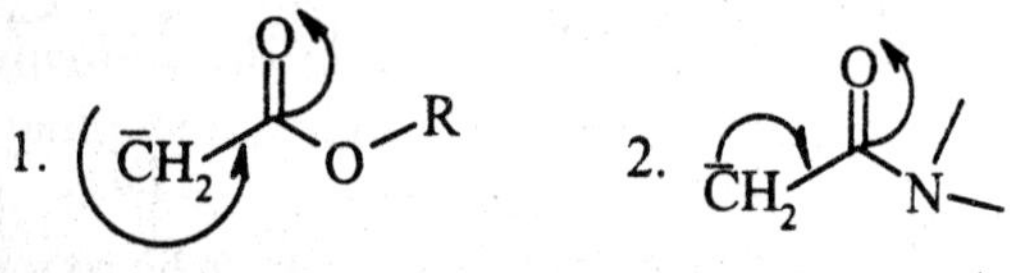

1. $\bar{C}H_2$ O R 2. $\bar{C}H_2$ O N

(a) The ester enolate (1) is more stable than amide enolate (2) as a consequence of the ester enolate's carbonyl being better acceptor of the enolate's electron density than is the amide enolate's carbonyl.
(b) The ester enolate (1) is more stable than amide enolate (2) as a consequence of the amide enolate's carbonyl being better acceptor of the enolate's electron density than is the ester enolate's carbonyl.
(c) The amide enolate (2) is more stable than ester enolate (1) as a consequence of the amide carbonyl being better acceptor of the enolate's electron density than is the ester enolate's carbonyl.
(d) The amide enolate (2) is more stable than ester enolate (2) as a consequence of the ester enolate's carbonyl being better acceptor of the enolate's electron density than is the amide enolate's carbonyl.

27. The equilibrium constant for the following reaction is:

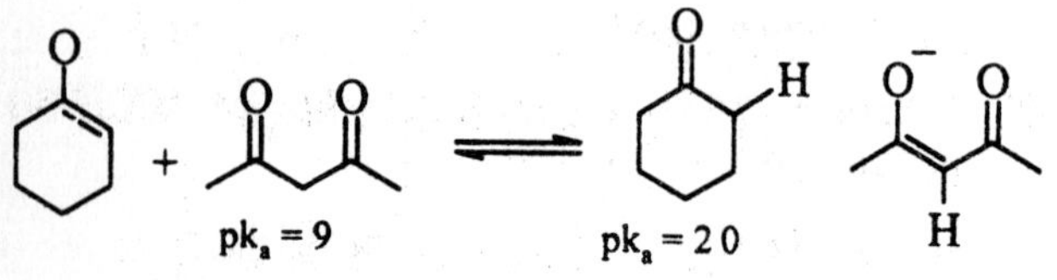

$pk_a = 9$ $pk_a = 20$ H

(a) 10^{11} (b) 10^{-11} (c) 10^{29} (d) 10^{-29}

28. Rank the esters 1–3 from most acidic to least acidic based on the acidity of the underlined H:

1. 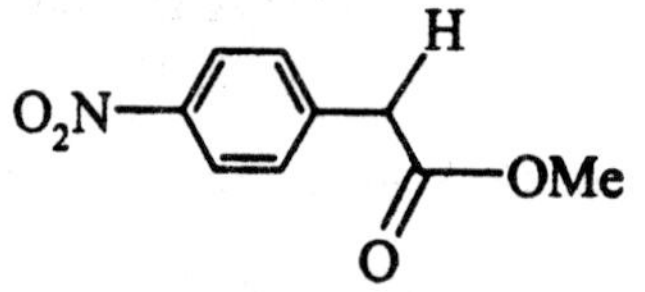 O_2N– H –OMe O
2. 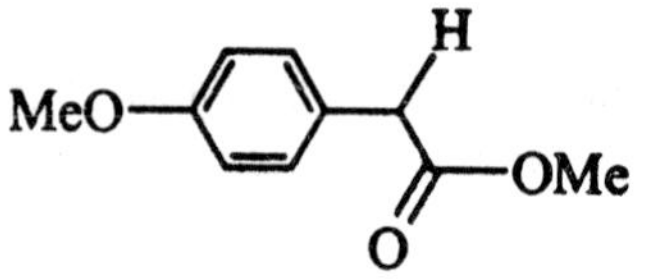 MeO– H –OMe O
.3. MeO– H –Me O

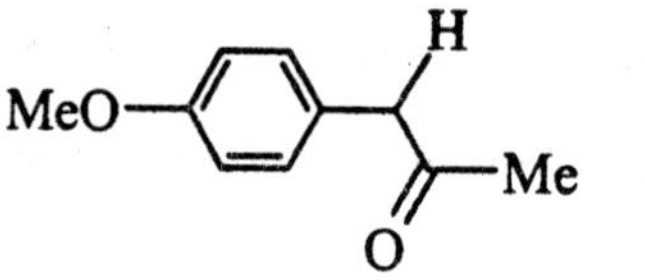

(a) 1 > 2 > 3 (b) 1 > 3 > 2
(c) 2 > 3 > 1 (d) 2 > 1 > 3

29. Which of the following set of carbon is electrophilic?

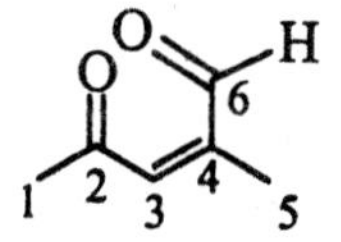

O O H 6 1 2 3 4 5

(a) C2, C4 (b) C3, C4
(c) C2, C6 (d) C2, C3, C4, C6

30. The equilibrium constant for the following proton transfer reaction is:

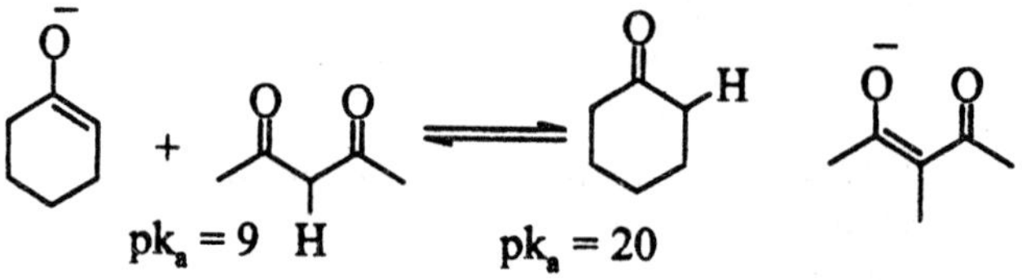

$pk_a = 9$ H $pk_a = 20$

(a) < 1 (b) > 1 (c) 1 (d) 0

31. All of the following are the resonance structure of one another except:

(a) 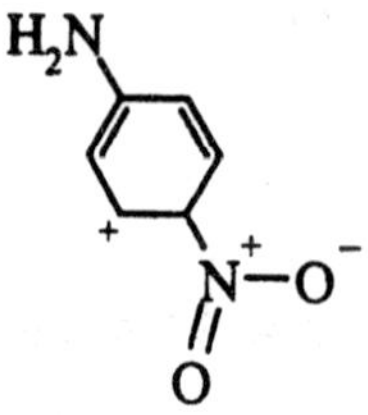 H_2N $\overset{+}{N}$–O^- O
(b) 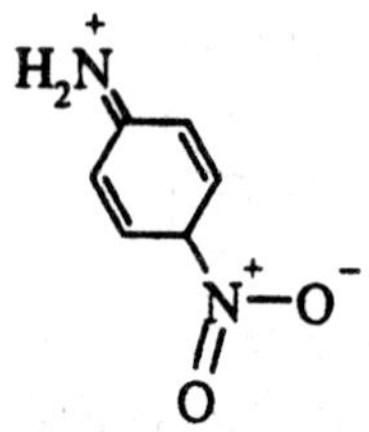 $H_2\overset{+}{N}$ $\overset{+}{N}$–O^- O
(c) 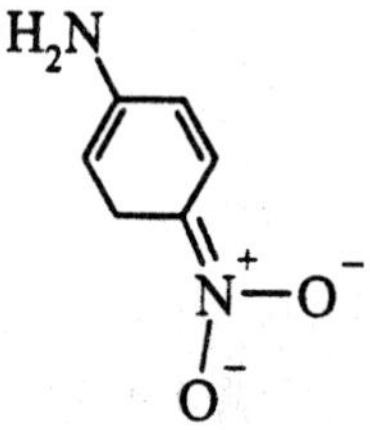 H_2N $\overset{+}{N}$–O^- O^-
(d) 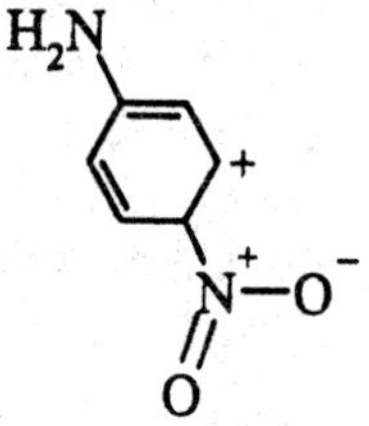 H_2N $\overset{+}{N}$–O^- O

32. Which of the following contains at least one sp^2 carbon?

(a) (b)

(c) (d) $=C=O$

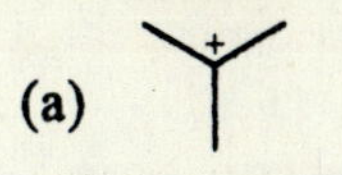

33. Which of the following carbocation would not likely rearrange to a more stable carbocation?

(a) (b)

(c) (d)

34. What is the true about the following equilibrium?

K^+

(a) It will be almost completely shifted to left.
(b) It will be almost completely shifted to right.
(c) The equilibrium constant is very close to one.
(d) The equilibrium constant is zero.

35. Consider an ion with the following structure: $[H_2C{=}C{=}NH_2]^+$. For this molecule, we can define two planes: one plane containing H—C—H group, the other plane containing H—N—H group. What is the relationship between these planes?
(a) They are at 120°.
(b) They are perpendicular to each other.
(c) They are in the same plane.
(d) More information is required.

36. Arrange the following compounds in order of increasing bond lengths of the shortest carbon-carbon bond in each molecule:
1. $H_2C{=}CH_2$
2. $H_2C{=}CH{-}CH{=}CH_2$
3. C_2H_2
4. $HC{\equiv}C{-}CH{=}CH_2$

(a) $3 < 4 < 1 < 2$ (b) $4 < 3 < 1 < 2$
(c) $3 < 1 < 4 < 2$ (d) $3 < 4 < 2 < 1$

37. Which of the following is not a valid resonance structure of others?

(a) OMe (b) OMe

(c) OMe (d) OMe

38. Which molecule has the largest dipole moment?

(a) Cl (b) Cl, Cl, Cl

(c) Cl, Cl, Cl, Cl, Cl (d) Cl, Cl, Cl, Cl, Cl, Cl

39. Some of the underlined atoms in the following molecule/ion have same hybridization. According to the hybridization type of those atoms, the molecule/ions can be divided into two categories. Which group assignment is correct?

(A) $^-\underline{C}H_3$ (B) $\underline{C}HCHCH_3$
(C) $CH_3\underline{O}H$ (D) $CH_3\underline{N}O_2$
(E) $^+\underline{N}H_4$ (F) $CH_3CH\underline{N}CH_3$

(a) A, B, C and D, E, F (b) A, C, E and B, D, F
(c) A, B, F and C, D, E (d) A, D, E and B, C, F

40. Which of the following is not a resonance structure of other?

(a) 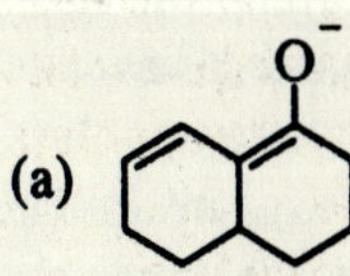

(b) 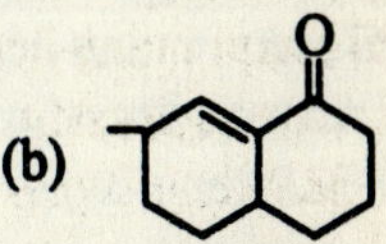

(c) 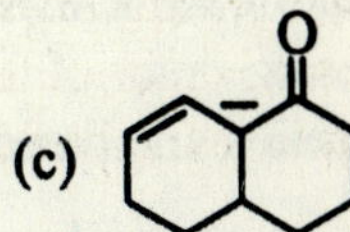

(d) 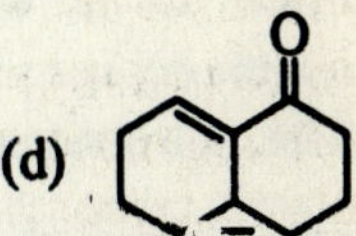

41. Which molecule has the lowest pk_a?

(a) O_2N, OH (b) OH, O_2N

(c) OH, EtO (d) NC, OH

42. Compound A could react to form compound B or C as shown in the following reaction diagram. Which of the following statement is true?

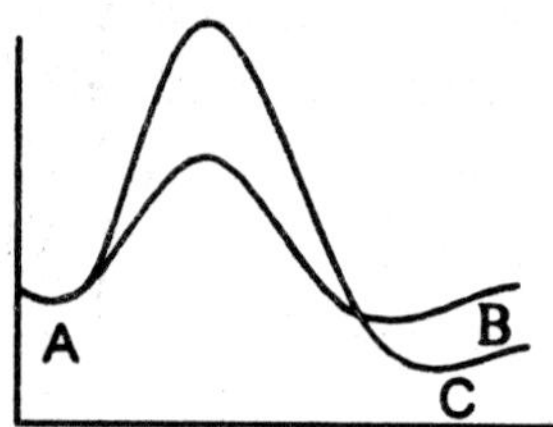

(a) Compound C will be the major product under kinetic condition.
(b) Compound B will be the major product under thermodynamic condition.
(c) Compound C will always be the major product, because it has lowest energy.
(d) Compound B will be the major product at low temperature.

43. Which of the following reaction is most exothermic?

(a) (tert-butyl bromide) ⟶ (tert-butyl cation) + Br^-

(b) (phenol)–OH + NaOH ⟶ (phenyl)–$\bar{O}\overset{+}{Na}$ + OH_2

(c) (propyne) + (cyclopentadiene, H) ⟶ (propynide)$^-$ + (cyclopentadiene, H H)

(d) $\dot{C}H_3$ + (benzene) ⟶ CH_4 + (phenyl radical)

44. Hyperconjugation is more important for stabilizing carbocation than neutral alkene because:
(a) The carbocation is flat.
(b) In alkene, hyperconjugation uses an antibonding π-orbital, which partly break the bond.
(c) There are no empty orbitals on an alkene, so hyperconjugation cannot occur.
(d) The carbocation is more stable than alkene.

45. Which carbonyl will be easily deprotonated by a bas EtONa?

(a) (cyclohexanone) (b) (methyl acetate, O, CH_3)

(c) (pentane-2,4-dione) (d) (3,3-dimethylpentane-2,4-dione)

46. Which of the following radical would be expected to be most selective?

(a) HC≡C·   (b) H_3C·
(c) Ph·   (d) $H_2C{=}CH{-}CH_2$·

47. Which of the following radical would be expected to be most reactive?

(a) HC≡C·   (b) H_3C·
(c) Ph·   (d) $H_2C{=}CH{-}CH_2$·

48. Which two of the following molecule has dipole moment?

(A) 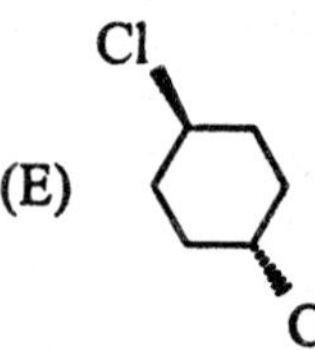

(B) (1,4-diiodobenzene, I, I)

(C) F(H)C=C=C(F)H

(D) CCl_4

(E) (1,4-dichlorocyclohexane, Cl, Cl)

(a) D and E   (b) C and E
(c) A and E   (d) A and C

49. Which statement about the following reaction is true?

(phenol, OH) + $CH_3CO\bar{O}$ ⇌ (phenoxide, O^-) + CH_3COOH

(a) The reaction will proceed towards left.
(b) The reaction will proceed towards right.
(c) The reaction can proceed towards any direction depending upon the initial composition.
(d) Direction of reaction cannot be predicted without K_a and K_b value.

50. Of the following bases, which can be used to create an acetylide ion?
(a) AcO^- (b) HO^-
(c) H_3CO^- (d) $(i\text{-}Pr)_2N^-$

51. Acetic acid reacts with sulphuric acid in an acid base reaction. What are the products of this reaction?

(a) $CH_3C(\overset{+}{O}H)OH$ + HSO_4^- (b) $CH_3\overset{+}{C}(OH)H$ + HSO_4^-

(c) $+ HSO_4^-$ (d) $+ HSO_4^-$

52. Which of the following is not a valid resonance structure for others?

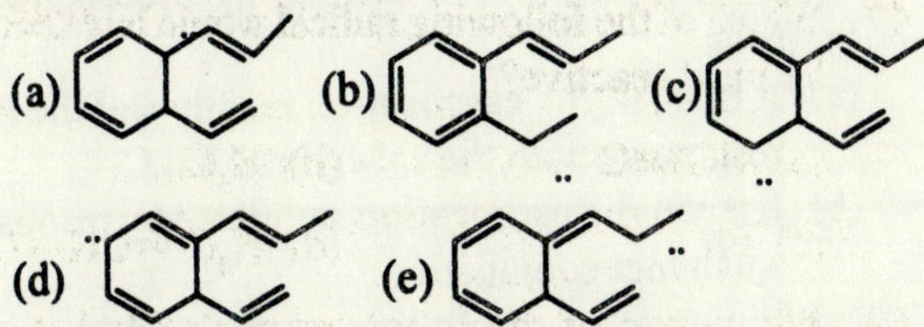

53. Arrange the following species in order of decreasing bond lengths of carbon-oxygen bonds.

I. H_2CO II. H_3COCH_3
III. $HCOO^-$ IV. CO_3^{2-}

(a) III > I > II > IV (b) II > IV > III > I
(c) I > IV > III > II (d) III > I > IV > II

54. Which of the following best define the term hyperconjugation:

(a) Transfer of H from a σ_{C-H} bond to the π-system.
(b) Participation of an H atom on terminal carbon in the electron sharing.
(c) Delocalization of electron density from a properly oriented σ-bond into π-system.
(d) Interaction between a properly aligned lone electron pair with the empty π-system.

55. The hydrogenation of some compounds to cyclohexane generates the following amounts of heat: (kcal/mol).

Cyclohexene	–28
1, 3-cyclohexadiene	–54
1, 4-cyclohexadiene	–56
Benzene	–49

Based on these data, the best estimate of the resonance energy of benzene:

(a) 7 (b) 18 (c) 26 (d) 35

56. Each of the structure below represents a planar compound that has a resonance form available with cyclic arrangements of π-electrons. Which compound will be most destabilized because of participation of that resonance structure in the description of compound:

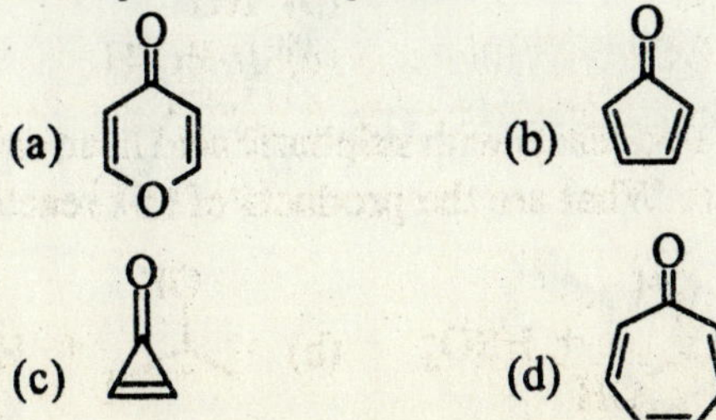

57. Radicals can be compared according to their selectivity in hydrogen abstraction based on the reactivity-selectivity principle. Which of the following radicals would be expected to have selectivity very similar to that of the H· radical?

(a) Cl. (b) F. (c) HO. (d) I.

58. Which of the following would produce the least stable radical upon abstraction of H atom?

(a) (b)

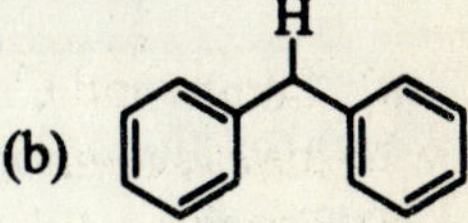

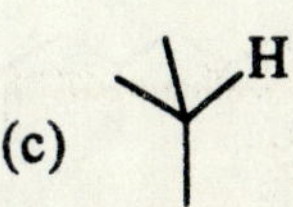

(c) (d)

59. Alkyne H are more stable than alkane or alkene H because?

(a) The alkyne C has greater *s*-character.
(b) The anion formed is more stable.
(c) The electron in the sp orbital are closer to nucleus.
(d) All of these.

60. Consider the following reaction profiles where A reacts with reagent R and gets converted into C and B reacts with same reagent R under identical condition and gets converted into D. Compounds A and B are stereomers of each other, and compounds C and D are stereomers of each other. Which statement about these compounds must be true?

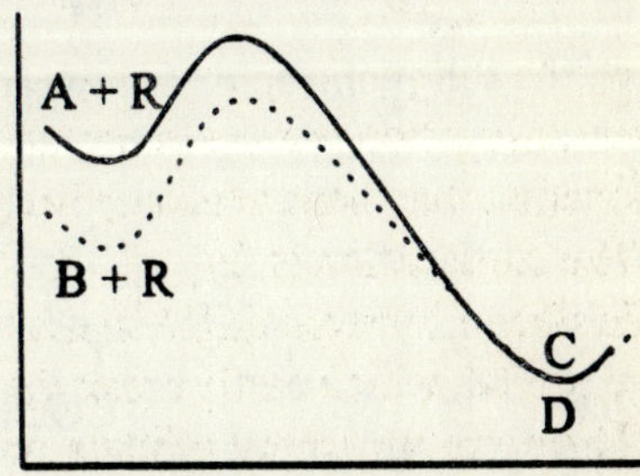

(a) A and B are geometrical isomers.
(b) A and B are distereomers.
(c) C and D are meso compounds.
(d) C and D are constitutional isomers.

61. Consider the following rearrangement reaction:

BrH, Br^-, Br

Which of the following reaction coordinates best represents the overall reaction?

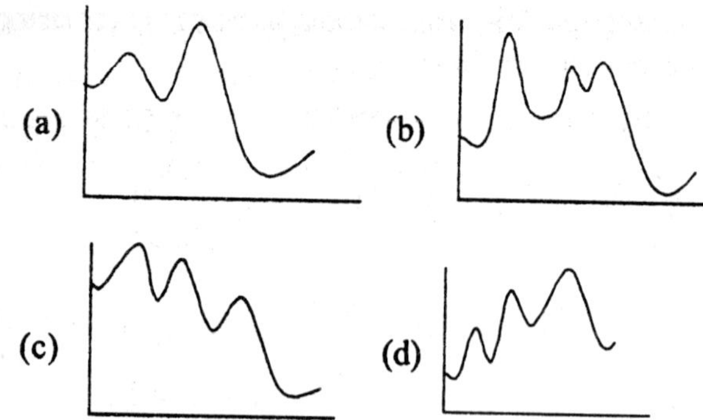

62. Which of the following is the most stabilized carbocation?

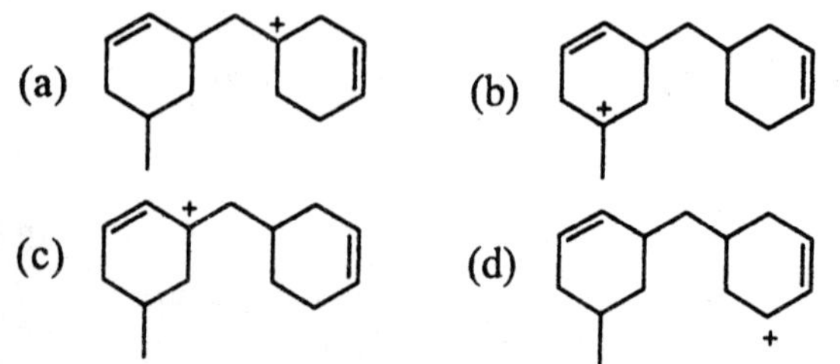

63. Which of the following structures represents the lowest energy form?

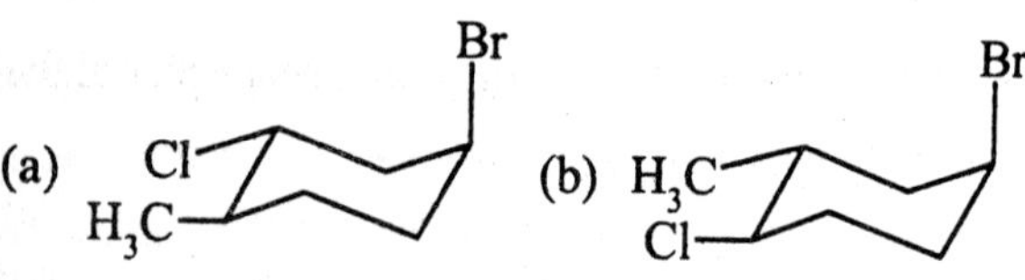

(c) Br Cl CH_3 (d) Cl CH_3 Br

64. Which of the following acid has lowest pk_a value?

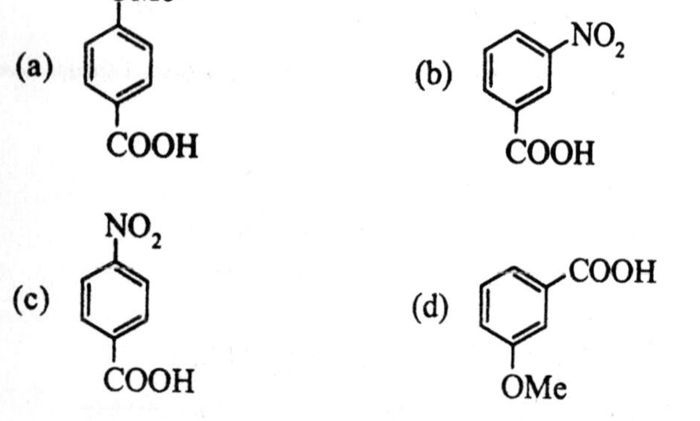

65. Which of the following is the most stabilized carbocation?

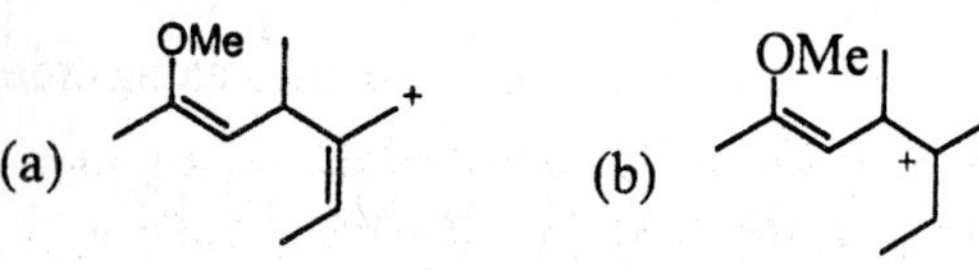

(c) OMe (d) OMe

66. Which of the following best describes the electronic structure of the cation formed by protonation of propyne:

(a) It is an allylic carbocation mainly stabilized by resonance with the sp^2-hybridized C.

(b) It is vinylic carbocation stabilized by resonance and hyper conjugation.

(c) It is a 2° carbocation stabilized by hyper conjugation and resonance with the sp^3 C.

(d) It is vinylic carbocation stabilized mainly by hyper conjugation with the sp-hybridized cationic carbon.

67. Which of the following is most likely to undergo a favourable hydride shift?

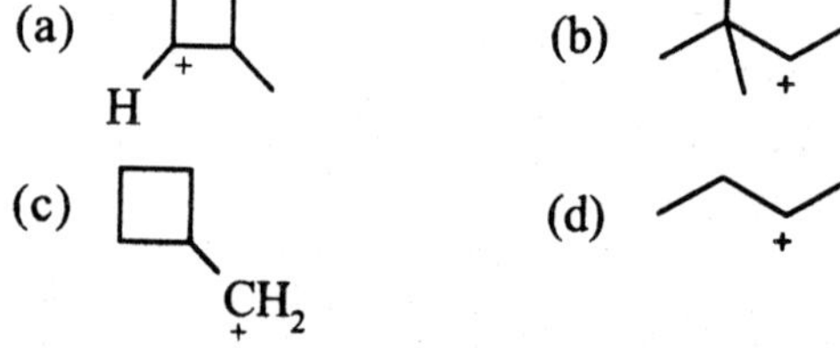

68. Which of the following is not a valid resonance structure of others?

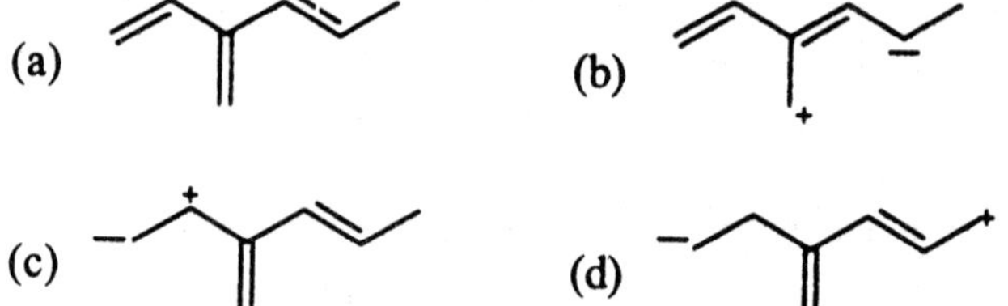

69. Which of the following molecules do not have dipole moment?

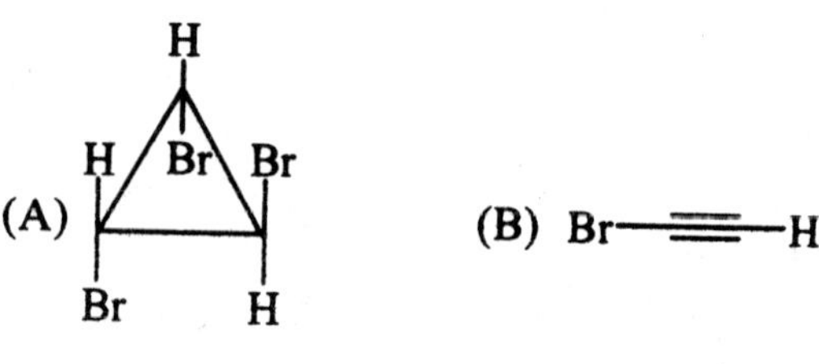

(C) CH_2Br_2 (D) Br H H Br

(E) Br Br

(a) A, D, E (b) D and E
(c) A, B and C (d) B,D and E

70. Which of the following statements about molecular strain is correct?
(a) *Cis* and *trans* stereomers of substituted cyclo-alkenes always have equal amount of strain.
(b) Gauche strain in butane is steric in origine.
(c) All staggered conformations of small alkenes have torsional strain.
(d) Strained alkane yield less heat when burned in oxygen than their strain free isomer.

71. Which statement about the following equilibrium is true?

C_6H_5COOH (I, $pk_a = +4$) + ClO_4^- (II) ⇌ $C_6H_5COO^-$ (III) + $HClO_4$ (IV, $pk_a = -10$)

(a) IV is acting as base.
(b) II is acting as base.
(c) The equilibrium favours reactants.
(d) I is 14 times less basic than IV.

72. Which of the following statement about resonance is false?
(a) Individual resonance structures are imaginary, not real.
(b) Resonance forms differ only in the placement of their π or nonbonding electrons.
(c) Different resonance structures of a substance donot have to be equivalent.
(d) In valid resonance structures, all atoms from the second row of the periodic table must have an octet of electrons.

73. Which of the following is not a valid resonance structure of the other three.

(a) MeO–C(=O)–$\overset{-}{C}H$–C(=O)–CH$_3$ (b) MeO–C(=O)–CH=C(O$^-$)–CH$_3$

(c) Me$\overset{+}{O}$=C(O$^-$)–CH=C(O$^-$)–CH$_3$ (d) MeO–C(=O)–CH$_2$–C(=O)–$\overset{-}{C}H_2$

74. Which of the following compounds is the strongest base?
(a) CH_3NH_2 (b) $H_2C{=}NH$
(c) $(H_2N)_2C{=}NH$ (d) $HN{=}C{=}NH$

75. Arrange the following carboxylic acids in decreasing order of their acid strength:

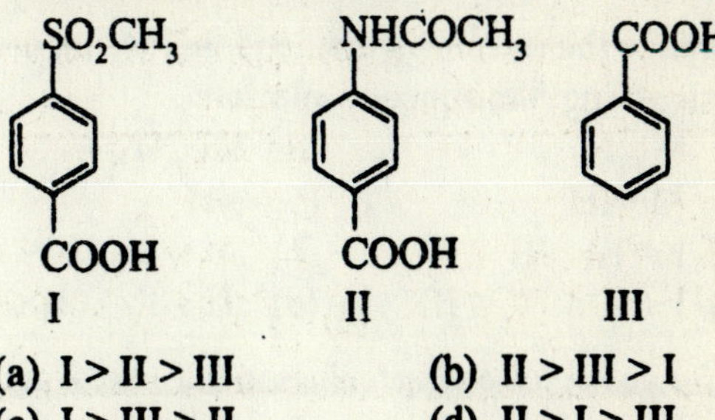

(a) I > II > III (b) II > III > I
(c) I > III > II (d) II > I > III

76. Arrange the following in the increasing order of their basic strength:

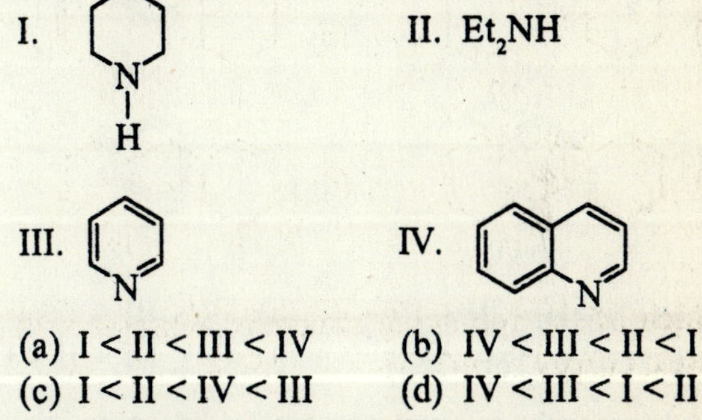

(a) I < II < III < IV (b) IV < III < II < I
(c) I < II < IV < III (d) IV < III < I < II

77. Arrange the following in the increasing order of their basic strength:

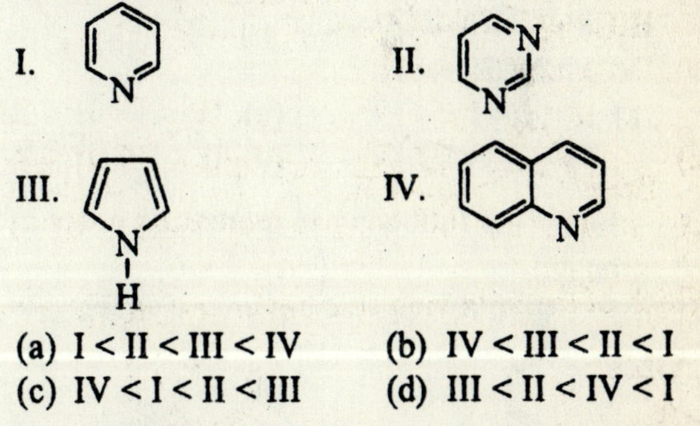

(a) I < II < III < IV (b) IV < III < II < I
(c) IV < I < II < III (d) III < II < IV < I

78. Arrange the following in the increasing order of their basic strength:

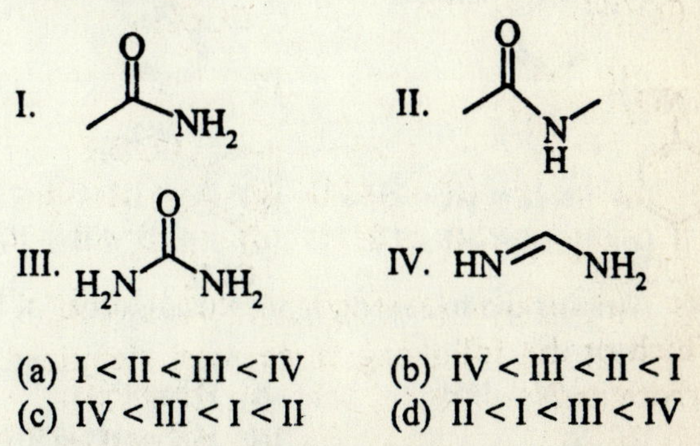

(a) I < II < III < IV (b) IV < III < II < I
(c) IV < III < I < II (d) II < I < III < IV

79. Arrange the following amines in the increasing order of their basic strength in gas phase:
I. Me_3N II. Me_2NH III. $MeNH_2$ IV. NH_3

(a) I < II < III < IV (b) IV < III < II < I
(c) I < II < IV < III (d) III < IV < II < I

80. Arrange the following amine in increasing order of basic strength in aqueous solution:

I. Me_3N II. Me_2NH
III. $MeNH_2$ IV. NH_3

(a) I < II < III < IV (b) IV < I < III < II
(c) I < II < IV < III (d) III < IV < II < I

81. Arrange the following amines in the increasing order of their basic strength:

I. 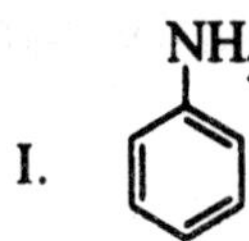

II. NH_2

III. 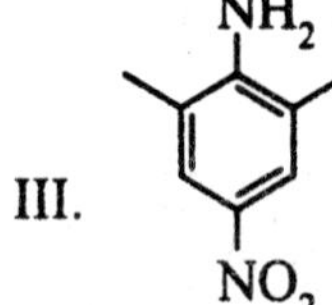

IV. NH_2, NO_2

(a) I < II < III < IV (b) IV < III < I < II
(c) III < IV < I < II (d) IV < III < II < I

82. Arrange the following amines in the increasing order of their basic:

I. Ph—NH_2 II. Ph_2—NH
III. Ph—NH—CH═CH_2
IV. Ph_2NCH═CH_2

(a) II < I < IV < III (b) IV < I < II < III
(c) IV < II < III < I (d) II < I < III < IV

83. Arrange the following in increasing order of acid strength:

I. COOH II. COOH

III. COOH, NO_2 IV. COOH, NO_2

(a) I < II < III < IV (b) IV < III < II < I
(c) II < I < III < IV (d) I < III < II < IV

84. Arrange the following in increasing order of basic strength:

I. H_2N—NH_2 II. HN═NH
III. N_2 IV. HN═N═NH

(a) III < II < IV < I (b) I < II < III < IV
(c) IV < III < II < I (d) I < III < IV < II

85. Arrange the following in increasing order of their acid strength:

I. OH II. OH, NO_2

III. OH, NO_2

IV. 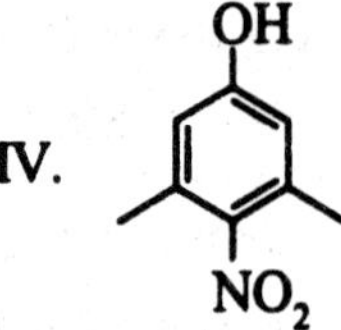

(a) I < II < III < IV (b) IV < III < II < I
(c) I < II < IV < III (d) I < III < II < IV

86. Consider an ion with the following structure: $[H_2C{=}C{=}NH_2]^+$. For this molecule, we can define two π-planes. What is the relationship between these planes?

(a) They are at 120°.
(b) They are perpendicular to each other.
(c) They are in the same plane.
(d) More information is required.

87. Arrange the following in the increasing order of their acid strength:

I. OH

II. 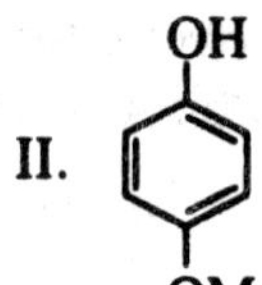

III. 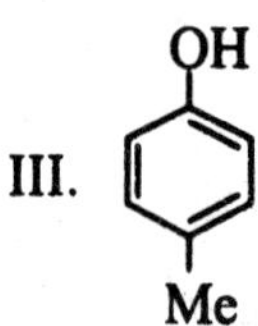

IV. 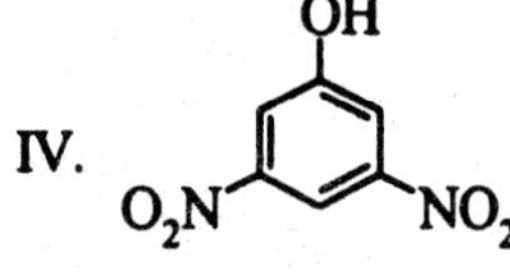

V. 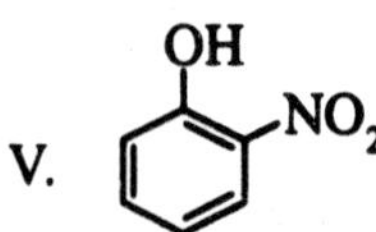

VI. 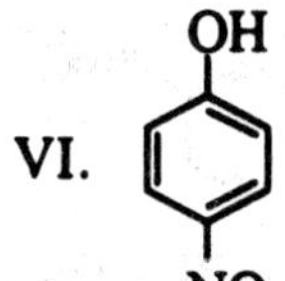

Ans: II < III < I < IV < VI < V

88. Arrange the following in the increasing order of pk_a giving suitable explanation:

(a) $CH_2(COOEt)_2$ (b) CH_3COOEt
(c) CH_3CH_2OH (d) CH_3COCH_3
(e) $CH(COOEt)_3$ (f) CH_3COCH_2COOEt
(g) CH_3COOH

Ans: g < c < e < f < a < d < b

89. Arrange the following in increasing order of acidic strength:

I. CF_3-SO_2-OH II. Cl_3CCOOH

III. Me–C(OH)(H)–Me IV. CH_3COOH

V. ∕∖∕∖OH

(a) I < II < III < IV < V (b) V < IV < III < II < I
(c) III < II < IV < I < V (d) III < V < IV < II < I

90. Arrange the following in increasing order of acidic strength of indicated proton:

I. II.

III. IV.

(a) I < II < III < IV (b) IV < III < II < I
(c) III < II < IV < I (d) II < III < IV < I

91. Arrange the following compounds in increasing order of their basic strength:

I. II.

III. IV.

V.

(a) I < II < III < IV < V (b) V < IV < III < II < I
(c) I < II < V < III < IV (d) IV < III < I < II < V

92. Arrange the following compounds in increasing order of their heat of combustion (–ΔH):

I. II.

III. IV.

V.

(a) IV <I < III < II < V (b) IV < I < III < V < II
(c) IV < III < I < V < II (d) I < II < III < IV < V

93. Arrange the following compounds in the increasing order of their rate of reaction with tertiary butanol: HBr(I), HCl(II), HI(III), HF(IV):

(a) I < II < III < IV (b) IV < III < II < I
(c) IV < II < I < III (d) I < II < IV < III

94. Arrange the following compound in the increasing order of their rate with HCl:

I. CH_3OH II.

III. IV.

V. $-CH_2OH$

(a) I < II < III < IV < V (b) V < IV < III < II < I
(c) I < V < III < IV < II (d) I < II < III < V < IV

95. Arrange the following radicals in the increasing order of their stability:

I. II.

III. IV.

(a) I < II < III < IV (b) IV < III < II < I
(c) II < III < IV < I (d) II < III < I < IV

96. Arrange the following in increasing order of their heat of hydrogenation (–ΔH):

I. 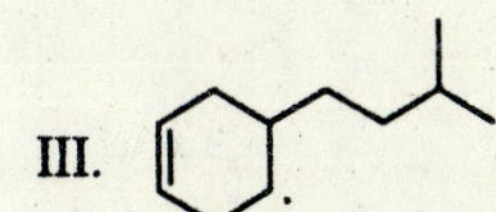 II. 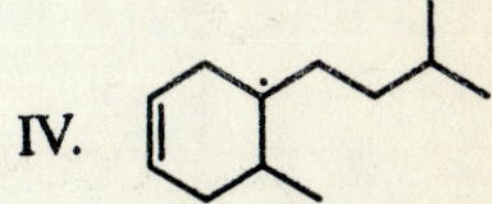

III. 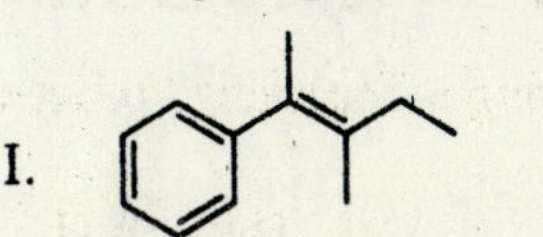 IV. 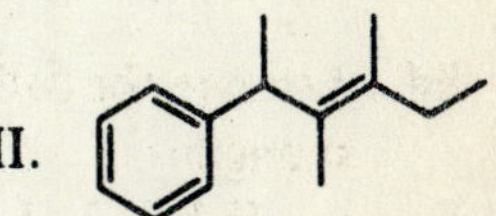

(a) I < II < III < IV (b) IV < III < II < I
(c) IV < III < I < II (d) I < II < IV < III

97. Arrange the following in increasing order of their heat of combustion:

I. 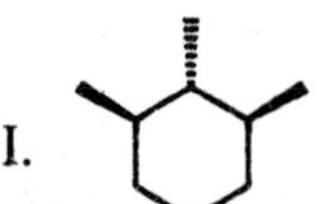 II.

III. IV. V.

(a) I < II < III < IV < V (b) V < IV < III < I < II
(c) IV < I < III < II < V (d) III < II < I < IV < V

98. Arrange the following in the increasing order of their basic strength:
Pyridine (I), Propyl amine (II), Butylnitrile (III), $CH_3CH_2CH_2NH^-$ (IV)

(a) I < II < III < IV (b) IV < III < II < I
(c) III < II < IV < I (d) III < I < II < IV

99. Arrange the following molecules in increasing order of electrophilicity at carbonyl carbon:

I. II. NH_2

III. Cl IV. OMe

(a) I < II < III < IV (b) IV < III < II < I
(c) II < IV < I < III (d) III < I < II < IV

100. Arrange the following in increasing order of pk_a:

I. CFH_2CH_2COOH II. CF_3CH_2COOH
III. CF_3CH_2OH IV. CH_3CF_2COOH

(a) IV < III < II < I (b) I < II < III < IV
(c) IV < III < I < II (d) IV < II < I < III

Chapter 4

Hydrocarbons

Alkanes

4.1 INTRODUCTION

Alkanes are saturated aliphatic hydrocarbon with their general formula C_nH_{2n+2}, general formula of cycloalkanes is C_nH_{2n}. All the carbons in alkane are sp^3 hybridized.

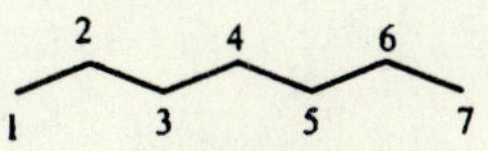

(Heptane)

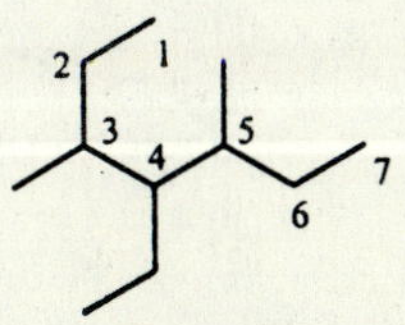

(4-ethyl-3, 5-dimethyl heptane)

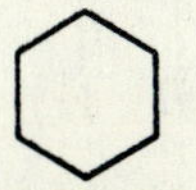

(Cyclohexane)

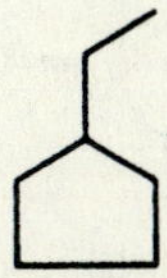

(Ethyl cyclopentane)

4.2 SYNTHESIS OF ALKANE

4.2.1 Hydrogenation of Alkenes and Alkynes

Alkenes and alkynes react with hydrogen in presence of a metal catalyst such as Ni, Pt, Pd, etc. to produce alkane. The mechanism of hydrogenation here involves adsorption of alkene or alkyne over the surface of metal containing adsorbed hydrogen. Metal surface then transfer hydrogen to alkene and alkyne forming alkane as shown below:

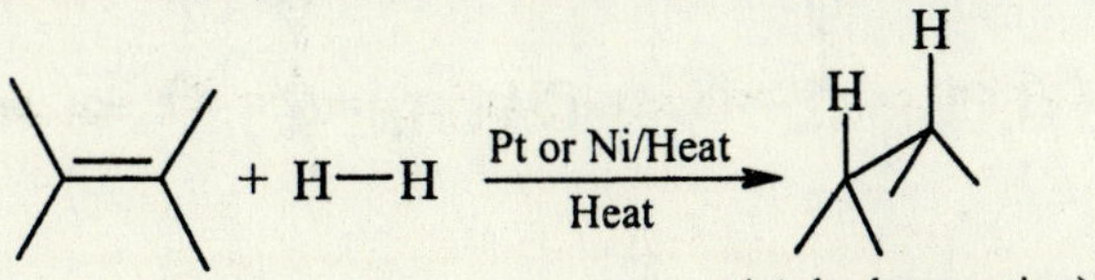

$$\text{C=C} + \text{H—H} \xrightarrow[\text{Heat}]{\text{Pt or Ni/Heat}}$$

(*cis* hydrogenation)

$$\text{C≡C} \xrightarrow{\text{Pt or Ni/Heat}}$$

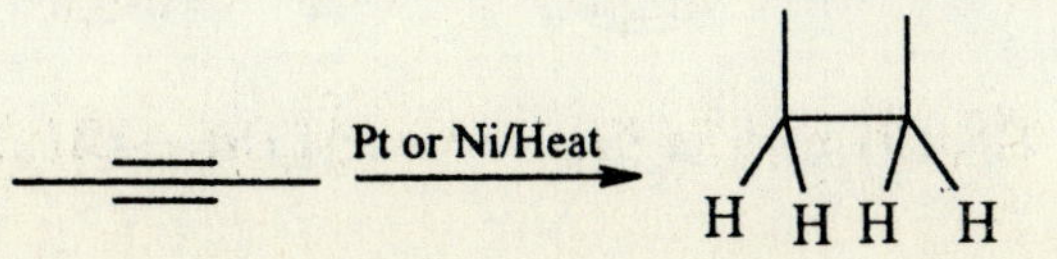

Detailed mechanism of hydrogenation is mentioned in the sub-chapter on Alkenes.

4.2.2 Reduction of Alkyl Halides

(a) *Reduction with Zn/H⁺:* An alkyl halide on treatment with strong electropositive metals like Zn, in acidic medium, undergo reduction to produce corresponding alkane as:

$$\text{R-X} + \text{Zn/H}^+ \rightarrow \text{RH} + \text{Zn}^{2+}$$

Here the solvent used should be either a dilute solution of acid or any protic solvent, e.g. ethanol, etc.

Mechanism

$$\text{Zn:} + \text{R-X} \rightarrow [\text{R}^-\text{Zn}^{2+}\text{X}^-] \xrightarrow{\text{H}^+} \text{RH} + \text{Zn}^{2+} + \text{X}^-$$

(b) *Reduction with complex hydride:* $LiAlH_4$ reduces primary and secondary alkyl halide to corresponding alkanes whereas tertiary alkyl halide is converted into alkene. Secondary and tertiary alkyl halides can also be conveniently reduced to corresponding alkane using $NaBH_4$, but primary alkyl halides are passive with this reagent. If triphenyl tinhydride (Ph_3SnH) is used as reducing agent, all types of alkyl halides are reduced to corresponding alkanes.

(c) *Reduction with HI/ red phosphorous:* This reagent is one of the most powerfull reducing agent, usually applied for reduction of a less reactive alkyl iodide. Besides this, the reagent also reduces acids, aldehydes, ketones, alcohols, etc. to alkane.

4.2.3 Hydrolysis of Grignard Reagent

Grignard reagent on treatment with a compound containing acidic proton, gives alkane as:

$$\text{RMgX} + H_2O \rightarrow \text{RH} + \text{Mg(OH)X}$$

$$\text{RMgX} + \text{HX} \rightarrow \text{RH} + \text{MgX}_2$$

$RMgX + R'OH \rightarrow RH + Mg(OR')X$. This is one of the most convenient method of deuteration as:

$$RX \xrightarrow[\text{Ether}]{Mg} RMgX \xrightarrow{D_2O} RD$$

4.2.4 Decarboxylation Reaction

A carboxylic acid on heating with soda lime (NaOH/CaO) loses CO_2 forming alkane with one carbon less than the parent acid as:

$$R\text{—}COOH + NaOH/CaO/\Delta \longrightarrow R\text{—}H + CO_2$$

Mechanism

$$RCOOH + NaOH \rightarrow RCOO^- + Na^+ + H_2O$$

$$R\text{—}\overset{O}{\overset{\|}{C}}\text{—}O^- \longrightarrow CO_2 + R^-$$

$$R^- + H_2O \longrightarrow RH + OH^-$$

Reactivity of acid towards decarboxylation depends on the stability of carbanion intermediate formed in the above mechanism.

Role of CaO: CaO absorbs CO_2 facilitating the formation of carbanion.

Example 1. Arrange the following carboxylic acids in the increasing order of their ease of decarboxylation:

I. CH_3CH_2COOH

II. $(CH_3)_2CHCOOH$

III. $(CH_3)_3CCOOH$

Solution: The order of stability of carbanion is 1° > 2° > 3°. Therefore in the given question, the relative stability of carbanion is $CH_3CH_2^- > (CH_3)_2CH^- > (CH_3)_3C^-$. Hence, the relative ease of decarboxylation is:

III < II < I

4.2.5 Wurtz's Reaction

This involves coupling of two molecules of alkyl halide in presence of sodium metal leading to formation of an alkane as:

$$R\text{—}X + 2Na + X\text{—}R \rightarrow R\text{—}R$$

Important: (a) This method is suitable only for the preparation of symmetrical alkane, i.e. when only one type of alkyl halide is used. If two different type of alkyl halides are used as R—X and R′—X, besides the cross products R—R′, other products R—R and R′—R′ are also formed by self-coupling of two R—X or two R′—X as:

$$R\text{—}X + 2Na + X\text{—}R' \rightarrow R\text{—}R'$$

$$R\text{—}X + 2Na + X\text{—}R \rightarrow R\text{—}R$$

$$R'\text{—}X + 2Na + X\text{—}R' \rightarrow R'\text{—}R'$$

Since alkanes have very close physical and chemical properties, they are difficult to separate.

(b) In Wurtz reaction, some unsaturated hydrocarbons are also formed. The two mechanism suggested for this reactions are:

(i) $C_2H_5\text{—}Br + 2Na \rightarrow C_2H_5^-Na^+ + NaBr$

$$C_2H_5^- + C_2H_5\text{—}Br \rightarrow C_2H_5\text{—}C_2H_5 + NaBr$$

Here a carbanion is formed which is confirmed by the failure of tertiary alkyl halide to undergo this reaction.

(ii) *Free radical mechanism:* Formation of unsaturated hydrocarbon can be explained by the formation of free radical intermediate and its subsequent disproportionation to give alkene as:

$$C_2H_5Br + Na^{\cdot} \rightarrow C_2H_5^{\cdot} + NaBr$$

$$H_3C\text{—}\dot{C}H_2 + {}^{\cdot}CH_2\text{—}CH_3 \longrightarrow H_2C{=}CH_2 + H_3C\text{—}CH_3$$

NB: Disproportionation is an important characteristics of free radical.

4.2.6 Kolbe's Electrolytic Method

This method of preparation of alkane involves electrolysis of a concentrated solution of potassium or sodium salt of carboxylic acid or mixture of carboxylic acid as:

$$2RCOOK + 2H_2O \rightarrow R\text{—}R + 2CO_2 + H_2 + 2KOH$$

The mechanism of this reaction involves free radical intermediate as:

$$CH_3CH_2\text{—}\overset{O}{\overset{\|}{C}}\text{—}O^- \longrightarrow CH_3CH_2\text{—}\overset{O}{\overset{\|}{C}}\text{—}\dot{O} + e \quad \text{At anode}$$

$$CH_3CH_2\text{—}\overset{O}{\overset{\|}{C}}\text{—}\dot{O} \longrightarrow CO_2 + CH_3CH_2^{\cdot}$$

$$CH_3CH_2^{\cdot} + {}^{\cdot}CH_2CH_3 \longrightarrow CH_3CH_2\text{—}CH_2CH_3$$

Since radical is formed as an intermediate, other products are also possible as:

$$CH_3CH_2-\overset{O}{\overset{\|}{C}}-\dot{O} + \dot{C}H_2CH_3 \longrightarrow CH_3CH_2COOCH_2CH_3 \text{ Ester}$$

$$CH_3CH_2-C-\dot{O} + \dot{O}-\overset{O}{\overset{\|}{C}}-CH_2CH_3$$

$$\downarrow$$

$$CH_3CH_2-\overset{O}{\overset{\|}{C}}-O-O-\overset{O}{\overset{\|}{C}}-CH_2CH_3$$

A peroxide

$$H_3C-\dot{C}H_2 + \dot{C}H_2-CH_3$$

$$\downarrow$$

$$H_2C=CH_2 + H_3C-CH_3$$

Just like Wurtz reaction, this method is also not suitable for preparation of unsymmetrical alkanes since mixture of alkane would be obtained as:

$$RCOOK + R'COOK \rightarrow R-R + R-R' + R'-R'$$

4.2.7 Corey House Synthesis

This is one of the most suitable method for preparation of unsymmetrical alkanes. The general mechanism of the reaction is:

$$R-X + Li \rightarrow R-Li \xrightarrow{Cux} R_2CuLi \xrightarrow{R'-X} R-R' + CuR + LiX$$

Condition here is, the R′—X should be preferably either primary alkyl halides or secondary alkyl halide or it should be less branched than R—X, because the last step involves bimolecular nucleophilic substitution (SN2 reaction, details of this reaction is described in alkyl halide chapter) of X^- by R^- as:

$$R-CuLi + R'-X \longrightarrow R-R' + CuX + RLi$$

4.3 PROPERTIES OF ALKANES

4.3.1 Physical Properties

(i) *Boiling points*: The boiling points of linear chain alkanes show a regular increase with increase in molecular weight. Branching of alkane chains, however, lowers the boiling point compared to its unbranched isomer. This decrease in boiling point due to branching is explained on the basis of vanderwaal's force of attraction. Since branching decreases the surface area, lowering the vanderwaal's force of attraction and hence boiling point.

(ii) *Melting points*: The trend in melting points of alkane is not very regular. In case of linear chain alkane, melting point increases with increase in molecular weight in general, but melting point of alkane with odd carbon is less than the adjacent homologue with even number of carbons, e.g. melting points of *n*-pentane is lower than *n*-butane as well as *n*-hexane. The reason behind this alternation in melting points from even to odd carbon alkanes is their ability of packing in crystalline state. X-ray diffraction studies revealed that alkane chain with even number of carbon pack more closely in crystalline state. As a result, attractive forces between individual chains are greater and melting points are higher. The effect of branching on melting points of alkanes is difficult to generalise. Generally, however, branching that produces highly symmetrical structure results in abnormally high melting points, e.g. 2, 2, 3, 3-tetramethylbutane melts at 100.7°C, which is very high for such a low molecular weight alkane. Cycloalkanes also have much higher melting points than their open chain counterparts.

(iii) *Density*: Alkanes are least dense organic compound. All alkanes have densities less than 1.0 g/cc, which is responsible for floating of petroleum on water.

(iv) *Solubility*: Alkanes and cycloalkanes are almost totally insoluble in water, because of their very low polarity and inability to form hydrogen bonds. Liquid alkanes and cycloalkanes are soluble in one another and they generally dissolves in solvent of low polarity.

4.3.2 Chemical Properties of Alkanes

(i) *Halogenation of alkanes:* Alkanes reacts with fluorine, chlorine and bromine either in presence of sun light or heat to give haloalkanes. This is an example of free radical substitution reaction. The general reaction to produce a mono-haloalkane can be represented as follows:

$$R-H + X_2 \rightarrow R-X + HX$$

Iodine is almost inert in this case. A complicating characteristics of free radical halogenation of alkane is multiple substitution. This happens because all hydrogen atoms attached to carbon are potentially capable of reacting with halogen. The extent of multiple substitution depends on two important factor.

(a) *Nature of halogen:* The general order of reactivity of halogen towards an alkane is $F_2 > Cl_2 > Br_2$. Therefore, the chances of multiple fluorination is maximum, next is chlorination and least is bromination.

(b) *Stoichiometry:* By mentaining appropriate stoichiometry, i.e. by taking excess of alkane, multiple halogenation can be controlled.

Mechanism

Chlorination of Ethane: Chlorination of alkane proceeds by a free radical intermediate.

$$C_2H_6 + Cl_2 \rightarrow C_2H_5Cl + HCl$$

Important: (a) Halogenation reaction is promoted by heat or light. At room temperature, halogenation of alkane does not occur at appreciable rate. However, halogenation occurs at room temperature, if the reactant is irradiated with UV light. The halogenations can also be done in dark but high temperature would be required.

(b) Light promoted reaction is more efficient in bringing about halogenation. Mechanism of halogenation involves three major steps as:

Step I: Homolytic cleavage of halogen:

$$Cl—Cl \xrightarrow{\text{UV light or heat}} 2Cl^{\cdot} \text{ (Radicals)}$$

Step II: $Cl^{\cdot} + C_2H_6 \rightarrow C_2H_5^{\cdot} + HCl$

Step III: $CH_3CH_2^{\cdot} + Cl_2 \rightarrow CH_3CH_2Cl + Cl^{\cdot}$

Steps II and III are known as chain propagating steps. From the above mechanism, it can be understood that once the reaction started, it can be continued for infinitely long time even in absence of light catalyst. However, it is seen that the continuous source of catalyst is required. This can be explained on the basis of a fact that the chain can be terminated by combination of radicals as:

$$\left.\begin{array}{l} CH_3CH_2^{\cdot} + Cl^{\cdot} \rightarrow CH_3CH_2Cl \\ CH_3CH_2^{\cdot} + {}^{\cdot}CH_2CH_3 \rightarrow CH_3CH_2—CH_2CH_3 \\ Cl^{\cdot} + Cl^{\cdot} \rightarrow Cl_2 \end{array}\right\} \text{Termination steps}$$

Since the above termination steps frequently exhaust the radicals, a continuous source of irradiation or heating is required for halogenations reaction.

Example 2. During light catalyzed chlorination of ethane, some 1-chloro and 2-chloro butane are also produced. Suggest mechanism of their formation.

Solution: During chlorination of ethane, the intermediate ethyl radical is formed. This ethyl radical may combine with other ethyl radical to form butane. This butane may also undergo free radical chlorination to produce 1-chlorobutane and 2-chlorobutane as byproduct.

Example 3. In synthesis of CCl_4 by halogenation of CH_4 a large excess of chlorinr is required, explain.

Solution: The basis of chlorination of alkane is collision of chlorine free radical (Cl·) with alkane in one of the propagation step forming an alkyl radical (R·), which abstract a chlorine atom from Cl_2 forming R—Cl. If a large excess of Cl_2 is taken, probability of collision of alkyl radical with Cl_2 and then formation of halogenated alkyl radical will increase as:

$$CH_3^{\cdot} + Cl_2 \rightarrow CH_3Cl + Cl^{\cdot}$$
$$CH_3Cl + Cl^{\cdot} \rightarrow {}^{\cdot}CH_2Cl + Cl^{\cdot}$$
$${}^{\cdot}CH_2Cl + Cl_2 \rightarrow CH_2Cl_2 + Cl^{\cdot} \rightarrow \text{, etc.}$$

Reactivity of Higher Alkanes Towards Chlorination

Reactivity of alkanes towards chlorination depends on the type of hydrogen being replaced, e.g.

$$\underset{\alpha}{CH_3}—\underset{\beta}{\overset{\overset{CH_3}{|}}{CH}}—\underset{\underset{\gamma}{\underset{H}{|}}}{CH}—CH_3$$

In this molecule we have primary (α) H, secondary (γ) H and tertiary (β) H. The rate determining step in chlorination (or in halogenation) is abstraction of H by Cl· (X·) radical. From the thermodynamic information and stability of radical formed after H-abstraction, it is evident that tertiary H has minimum activation energy for its abstraction. Therefore, the general order of reactivity of H to be replaced in a higher alkane is: 3° > 2° > 1° > CH_4. Here is β > γ > α. However, the activation energies for abstraction of 1°, 2°, 3° or CH_4 hydrogen by chlorine radical are not very different, chlorine does not discriminate among the different type of hydrogen atom. This fact is illustrated in the following activation energy Figure 4.1 for the above mentioned molecule.

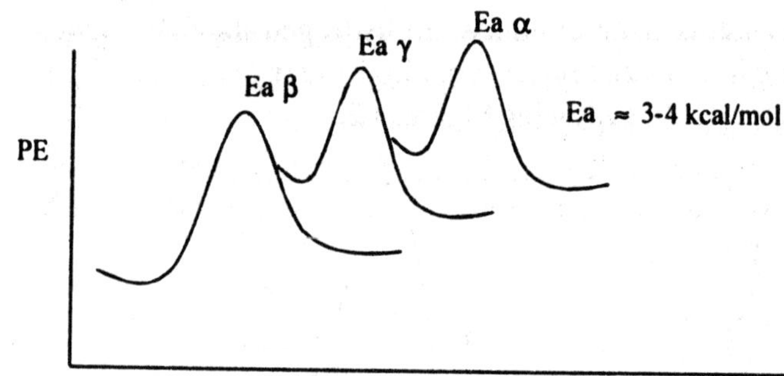

Figure 4.1: Potential energy showing activation energies for the abstraction of various type of H

With these finding it is concluded that chlorination of hydrocarbon is a fast, reactive process but *less selective*. Due to this reason, chlorination of higher alkanes have a very little synthetic application because, always mixture of alkyl halides are obtained. However, it is a very useful method for chlorination of hydrocarbon which contains only one type of hydrogen, viz. cyclopropane, cyclobutane and all unsubstituted cycloalkanes.

Example 4. A hydrocarbon has molecular formula C_8H_{18} and on monochlorination, yields single isomer with molecular formula $C_8H_{17}Cl$. Deduce the structure of hydrocarbon.

Solution: Molecular formula indicates that it is a saturated hydrocarbon. Single monochlorination product indicates that all H are equivalent. Therefore, it must be 2, 2, 3, 3-tetramethyl butane.

Example 5. An organic compound has molecular formula $C_6H_{10}O_2$(A). A on heating with sodalime yields a hydrocarbon B(C_5H_{10}). B on monochlorination yields a single isomeric alkyl chloride C(C_5H_9Cl). Deduce the structures of A, B and C.

Solution: It is obvious here that A is a carboxylic acid which on decarboxylation with sodalime producing hydrocarbon B. B has one degree of unsaturation, therefore it could be either alkene or cycloalkane. Since B on monochlorination yields single isomeric halide, all H in B must be equivalent. We know that all H in unsubstituted cycloalkane are equivalent. Therefore, B must be cyclopentane, and thus C is cyclopentyl chloride.

$$\underset{\text{(A)}}{C_5H_9\text{—COOH}} \xrightarrow{\text{NaOH/CaO}} \underset{\text{(B)}}{C_5H_{10}} \xrightarrow[h\nu]{Cl_2} \underset{\text{(C)}}{C_5H_9\text{—Cl}}$$

Bromination of Alkanes

In contrast to chlorination, the hydrogen atom abstraction step in bromination has very high energy of activation as shown in the following Figure. 4.2.

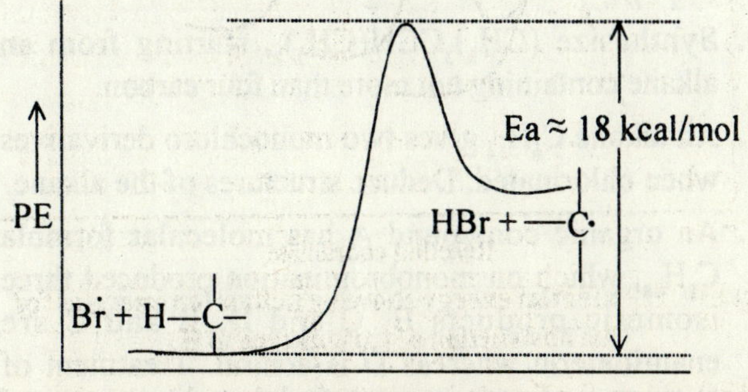

Figure 4.2: Potential energy diagram showing activation energy for the abstraction of H by Br · radical

Therefore, a very small fraction of all collision between bromine atom and hydrocarbon molecules are energetically effective in bringing about bromination even at a temperature of 300°C. As a result, bromine is much less reactive than chlorine towards alkane. However, bromine is much *more selective* than chlorine for hydrogen abstraction and leading to formation of a particular product in predominant amount. This is due to the reason that bromine is less reactive, whatever be the extent of bromination, it occurs at most reactive carbon, e.g.

$$H_3C\text{—}\underset{\text{H (tertiary H)}}{\overset{CH_3}{C}}\text{—}CH_3 + Br_2 \xrightarrow[130°C]{\text{Light}} \underset{>99\ \%}{H_3C\text{—}\underset{Br}{\overset{CH_3}{C}}\text{—}CH_3} + \underset{<10\%}{H_3C\text{—}\underset{H}{\overset{CH_3}{C}}\text{—}CH_2Br}$$

whereas chlorination of the same alkane produces significant amount of both isomers as:

$$H_3C\text{—}\underset{\text{H (tertiary H)}}{\overset{CH_3}{C}}\text{—}CH_3 + Cl_2 \xrightarrow[25°C]{h\nu} \underset{37\ \%}{H_3C\text{—}\underset{Cl}{\overset{CH_3}{C}}\text{—}CH_3} + \underset{63\ \%}{H_3C\text{—}\underset{H}{\overset{CH_3}{C}}\text{—}CH_2Cl}$$

Stereochemistry of Free Radical Halogenation

When an achiral molecules on halogenation produces a single tetrahedral stereocenter, the product will be obtained as a racemic mixture as:

$$CH_3CH_2CH_2CH_2CH_3 + Cl_2 \xrightarrow{h\nu} CH_3CH_2CH_2CH_2CH_2Cl \text{ (achiral)} + (\pm)\text{–2-chloropentane} + \text{3-chloropentane (achiral)}$$

The reason for obtaining racemic mixture lies in the fact that the intermediate alkyl radical has triangular planar shape and halogen may attack from either side with equal probability, producing equal amount of enantiomers, hence racemic mixture, e.g.

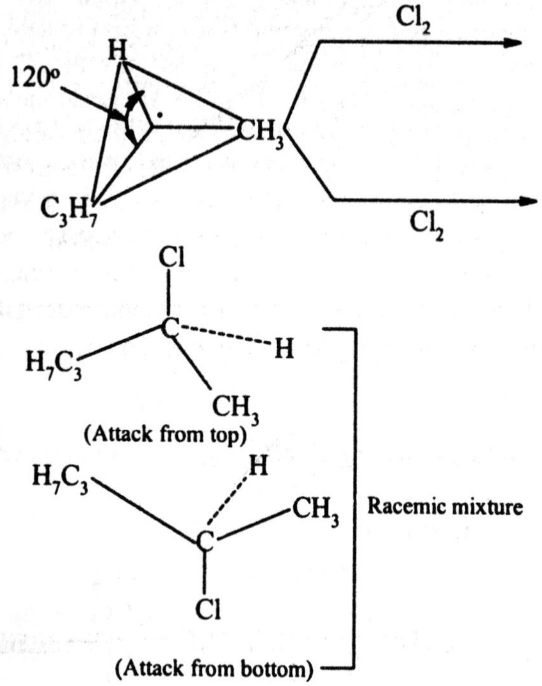

120° H C_3H_7 CH_3 Cl_2 Cl_2

Cl — C — H, H_7C_3, CH_3 (Attack from top)

H_7C_3, H, CH_3, C, Cl (Attack from bottom)

Racemic mixture

The two products obtained above are non-superimposable mirror images of one another, i.e. they are enantiomers. Also, these products are obtained in equimolar amount, the net result is racemization.

Iodination of Alkanes

Iodination of alkane is reversible, but it may be carried out in presence of an oxidising agent like HNO_3, HgO, etc. which destroy the HI as it is formed and consequently drives the reaction to the right, e.g.

$$CH_4 + I_2 \rightarrow CH_3I + HI$$

$$5HI + HIO_3 \rightarrow 3I_2 + 3H_2O$$

Nitration of Alkane

Under appropriate conditions, alkane reacts with nitric acid forming nitro alkanes. Nitration with nitric acid is carried out in vapour phase at about 500°C, where a complex mixture of mononitro alkanes are obtained. The mixture consists of all the possible mononitro derivatives and the nitro alkanes are also obtained by cleavage of C—C bond, e.g.

$$CH_3CH_2CH_3 \xrightarrow[500°C]{HNO_3} CH_3CH_2CH_2NO_2 + CH_3CH_2NO_2 + CH_3NO_2 + CH_3-\underset{NO_2}{\underset{|}{C}H}-CH_3$$

Sulphonation of Alkanes

Alkanes on treatment with oleum, converted into alkane sulphonic acid as:

$$RH + H_2SO_4/SO_3 \rightarrow RSO_3H$$

The reactivity of alkanes towards sulphonation is:

3° > 2° > 1°

PRACTICE EXERCISE I

1. Arrange the following alkanes in the increasing order of their boiling point explaining the reason: Pentane, hexane, 2, 3-dimethyl butane.
2. Select, from pairs of compounds, one with higher boiling point:
   (a) 2-mehtylhexane and 2, 2-dimethylpentane.
   (b) CF_4 and hexane.
3. Write structures of all alkene that can be hydrogenated to form 2-methylpentane.
4. Prepare 2-deutropropane from propane, in better yield.
5. Prepare butane starting from ethane.
6. Starting from 1-and 2-bromopropane, synthesize 2-methylpentane as the only product in good yield.
7. Compare the reactivity of halogens towards alkane.
8. Why iodination, which is otherwise very difficult, occurs in presence of an oxidising agent?
9. Write mechanism of chlorination of an alkane using sulphuryl chloride (SO_2Cl_2) and a free radical initiator (R_2O_2).
10. Synthesize $(CH_3)_3CCH(CH_3)_2$, starting from an alkane containing not more than four carbon.
11. An alkane C_6H_{14} gives two monochloro derivatives when chlorinated. Deduce structures of the alkane.
12. An organic compound A has molecular formula C_8H_{18}, which on monobromination produced three isomeric products B, C and D. B and C are enantiomeric whereas D is achiral. Treatment of either C or D with alcoholic KOH solution produced the same product E while B on similar treatment produced F(C_8H_{16}) an isomer of E as major product. Establish structures of A to F.

13. A hydrocarbon A(C_8H_{16}) does not decolourise Br_2-water. A on monochlorination produced four isomeric products B, C, D and E among which only C and D are enantiomeric. Compound B is resistant to dehydrohalogenation reaction. Treatment of C with alcoholic solution of KOH produced F(C_8H_{14}) as only product while similar treatment on E produced G, an isomer of F. D on treatment with alcoholic solution of KOH produces both F and G in comparable amount. Deduce structures of A to G.

14. A hydrocarbon A(C_8H_{16}) does not decolourise Br_2-water. A on monochlorination produced four isomeric products B, C, D and E among which only D is enantiomeric. Compound B is resistant to dehydrohalogenation reaction. Treatment of either C or D with alcoholic solution of KOH produced the same alkene F(C_8H_{14}) as major product, which does not rotate the plane polarized light while similar treatment on E produced G, an isomer of F which is enantiomeric. Deduce structures of A to G.

Alkenes

General

Alkenes are unsaturated hydrocarbons containing one double bond. They have the general formula C_nH_{2n}. C_nH_{2n} is also the general formula of cycloalkanes, general formula of cycloalkene is C_nH_{2n-2}:

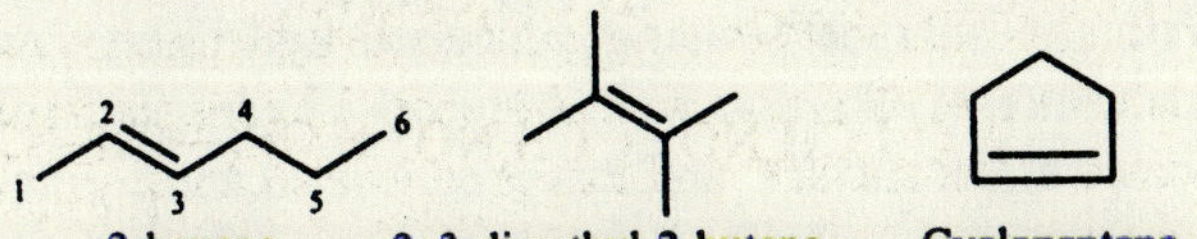

2-hexene   2, 3-dimethyl-2-butene   Cyclopentene

4.4 GENERAL METHODS OF PREPARATION OF ALKENES

4.4.1 By the Action of Concentrated Sulphuric Acid, at 180°C–200°C, on Primary Alcohols (Dehydration of Alcohols)

$$C_2H_5OH + \text{conc. } H_2SO_4 \xrightarrow{200°C} C_2H_4$$

Dehydration of secondary and tertiary alcohols is best carried out using dilute sulphuric acid since alkenes produced from these alcohols tend to polymerise under the influence of concentrated acid. Acid catalyzed dehydration gives mixture of alkenes because of rearrangement of carbocation intermediate formed (details of dehydration would be discussed in alcohol chapter), however rearrangement of carbocation intermediate can be avoided by dehydrating alcohols over heated alumina in the presence of a little pyridine.

4.4.2 Cyclic Elimination

(a) Pyrolysis of ester

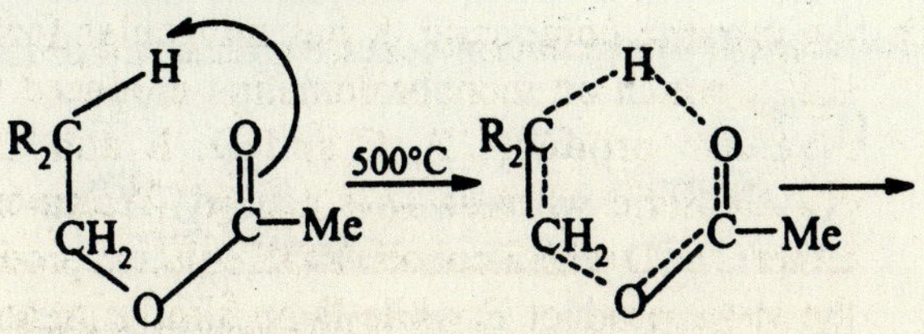

$$R_2C(H)-CH_2-O-C(=O)-Me \xrightarrow{500°C} [\text{cyclic transition state}] \longrightarrow$$

$$R_2C{=}CH_2 + CH_3COOH$$

(b) Cope reaction: By heating amine oxide-alkene is formed as:

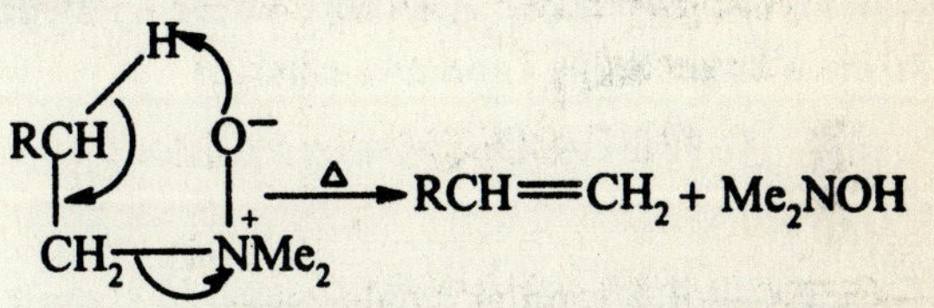

$$RCH(H)-CH_2-\overset{+}{N}Me_2-O^- \xrightarrow{\Delta} RCH{=}CH_2 + Me_2NOH$$

4.4.3 By the Action of Concentrated, Ethanolic Solution of KOH on Alkyl Halides

$$CH_3CH_2CH_2Br + KOH \xrightarrow{\text{Ethanol}} CH_3CH{=}CH_2$$

(Details of this reaction would be discussed in alkyl halide chapter.)

4.4.4 Dehalogenation of Vicinal or Geminal Dihalides

Vicinal or geminal dihalides on strong heating with Zn-dust undergoes dehalogenation to give alkenes as:

$$CH_3CH_2CHBr_2 \xrightarrow[\Delta]{Zn} CH_3CH{=}CH_2$$

$$CH_3CHBrCH_2Br \xrightarrow[\Delta]{Zn} CH_3CH{=}CH_2$$

4.4.5 By Heating Quaternary Ammonium Hydroxide

$$(C_2H_5)_4NOH \xrightarrow[\Delta]{Zn} C_2H_4 + (C_2H_5)_3N + H_2O$$

Mechanism: β-elimination

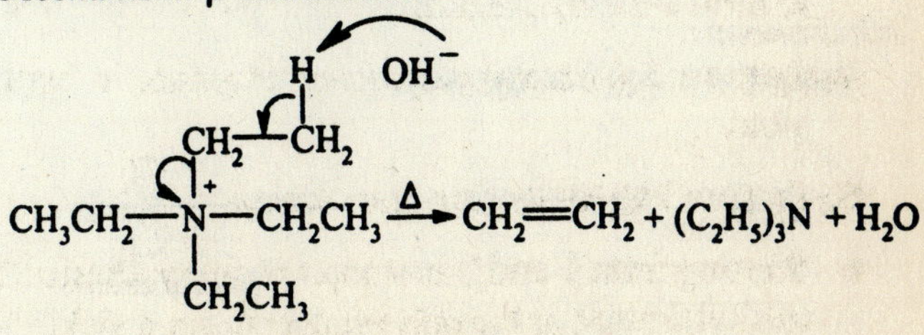

$$CH_3CH_2-\overset{+}{N}(CH_2CH_3)(CH_2CH_2H)-CH_2CH_3 \;\; OH^- \xrightarrow{\Delta} CH_2{=}CH_2 + (C_2H_5)_3N + H_2O$$

If more than one type of β-H are present, most acidic β-H is abstracted by OH^- as:

$$CH_3CH_2-\overset{+}{N}(CH_2CH_2CH_3)-CH_2CH_2CH_3 \xrightarrow{\Delta} CH_2{=}CH_2 + (C_3H_7)_2NC_2H_5 + H_2O$$

The most acidic H is one which on abstraction, produces most stable carbanion.

4.4.6 By Controlled Hydrogenation of Alkynes

Alkynes can be partially reduced to alkene, using a very specific catalyst as:

(a) *Cis hydrogenation:* Alkynes can be hydrogenated to *cis* alkenes using Lindlar's catalyst

$(Pd/CaCO_3/PbO)$ or $Pd/BaSO_4$

R—C≡C—R + Lindlar's catalyst → (R)(H)C=C(R)(H)

Mechanism

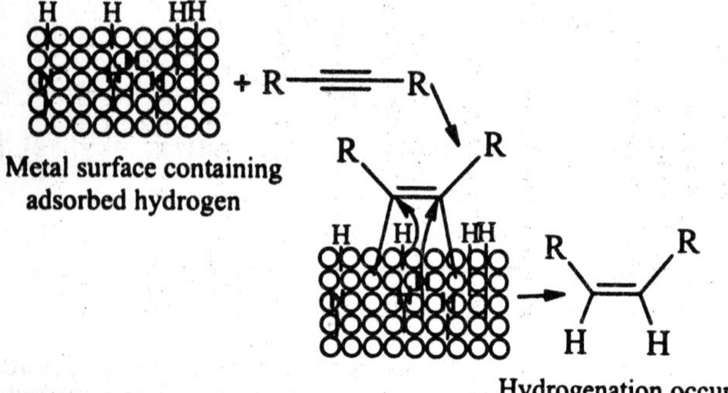

Figure 4.3: Mechanism of hydrogenation on the surface of metal containing adsorbed hydrogen gas. Hydrogenation from metal surface on the alkyne occurs from same side (syn hydrogenation)

(b) *Reduction with Na/liq NH_3 or Na/ethanol or Na/dil. HNO_3:* All these reagents reduces alkynes to *trans* alkenes as:

R—C≡C—R + Na/liq NH_3 → (R)(H)C=C(H)(R)

Mechanism

Step I $Na \rightarrow Na^+ + e^-$

Step II $R-C\equiv C-R + e^- \longrightarrow$ R–C=C–R (Radical anion)

$\xrightarrow{H^+}$ R–C=C(H)(R) $\xrightarrow{e^-}$ (R)C=C(H)(R) Stable anion

4.5 REACTIONS OF ALKENES

4.5.1 Hydrogenation

Alkenes are easily reduced to alkane under pressure in presence of a catalyst. Finely divided Pt, Pd are effective at room temperature, Ni require temperature between 200°C to 300°C, Raney nickel is effective at room temperature and atmospheric pressure:

$$CH_2{=}CH_2 \xrightarrow{\text{Catalyst}} C_2H_6$$

Mechanism of Hydrogenation and Relative Stability of Various Alkenes

Commonly used catalyst for hydrogenation of alkene (finely divided Pt, Ni, Pd or Rh) apparently serve to adsorbe hydrogen molecules on its surface. The collision of alkene with the surface bearing adsorbed hydrogen causes adsorption of alkene as well. A stepwise hydrogen transfer occurs at the surface, and this produces alkane before the organic molecule leaves the surface of catalyst. A schematic of hydrogenation reaction is described below in Figure 4.4.

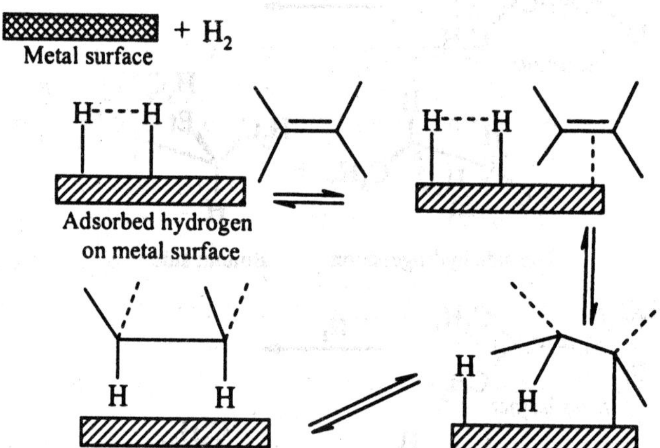

Figure 4.4: Hydrogenation at metal surface, occurs in syn orientation

As a consequence, both hydrogen atoms are added from the same side and therefore catalytic hydrogenation of alkene occurs with syn orientation.

Example 1. Catalytic hydrogenation of *cis*-3, 4-dimethyl-3-hexene yields a meso alkane, explain.

Solution: Hydrogenation occur from same side as:

$(H_3C)(Et)C{=}C(CH_3)(Et) \xrightarrow{H_2}$ (Syn hydrogenation) meso product

Example 2. Catalytic hydrogenation of *trans*-3, 4-dimethyl-3-hexene results in racemic mixture of alkanes, explain.

Solution: Hydrogenation occurs with equal probability from top as well as from bottom sides producing equal amount of enantiomers as:

$$(H_3C)(Et)C{=}C(Et)(CH_3) \xrightarrow{H_2}$$

Top side hydrogenation   Bottom side hydrogenation

Example 3. Catalytic hydrogenation of either *cis* or *trans* 3, 4-dimethyl-3-heptene results in racemic mixture of alkanes, explain.

Solution: Since 3, 4-dimethyl-3-heptene is an unsymmetrical alkene, hydrogenation from top side and bottom side will produce enantiomers. Due to equal probability of hydrogenation from two sides, net result will be formation of racemic mixture as:

$$(H_3C)(Et)C{=}C(CH_3)(C_3H_7) \xrightarrow{H_2}$$

cis isomer

Top side hydrogenation   Bottom side hydrogenation

$$(H_3C)(Et)C{=}C(C_3H_7)(CH_3) \xrightarrow{H_2}$$

trans isomer

Top side hydrogenation   Bottom side hydrogenation

Relative stability of alkenes towards hydrogenation can be seen from the steric hindrance point of view, at the metal surface. An alkene with bulky group substituent at doubly bonded carbon suffers steric hindrance when it tend to adsorbe on the metal surface. Therefore, more the substitution of bulky group at double bond, difficult will be the adsorption, and hence difficult will be their hydrogenation. Thus, an alkene with bulky groups at double bond are less reactive (more stable) towards hydrogenation. On the basis of this fact, the general order of stability of alkenes towards hydrogenation can be summarized as:

$$R_2C{=}CR_2 > R_2C{=}CHR > RHC{=}CHR\ (trans) > R_2C{=}CH_2 > RHC{=}CHR\ (cis) > RHC{=}CH_2 > H_2C{=}CH_2$$

Example 4. Arrange the above alkene in increasing order of their heat of hydrogenation.

Solution: Hydrogenation of an alkene is an exothermic process since molecule is transformed from unsaturated (unstable) to saturated (stable) form. The general energy profile for hydrogenation is described in the following Figure 4.5.

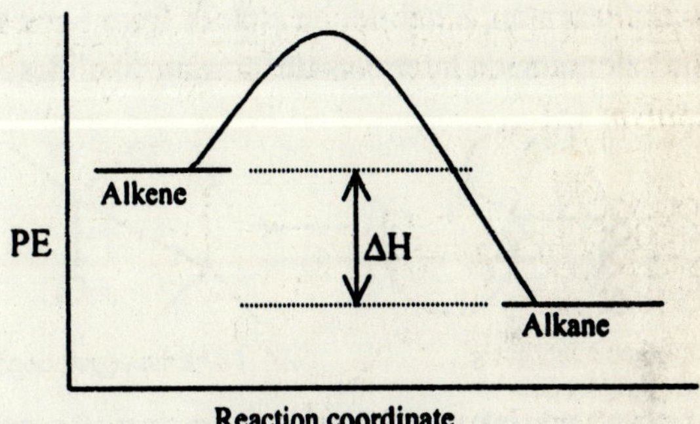

Figure 4.5: Potential energy diagram of hydrogenation of alkene

From the above energy Figure 4.5, it can be concluded that more unstable the alkene, higher will be its ground state potential energy and therefore higher will be the ΔH. Thus, heat of hydrogenation will be least for tetrasubstituted alkene and maximum for unsubstituted alkene (ethylene). Thus, the relative order of heat of hydrogenation of the above mentioned alkene will be reverse of their relative stability as:

$$R_2C{=}CR_2 < R_2C{=}CHR < RHC{=}CHR\ (trans) < R_2C{=}CH_2 < RHC{=}CHR\ (cis) < RHC{=}CH_2 < H_2C{=}CH_2$$

4.5.2 Halogenation

Alkenes forms addition product with chlorine, bromine, etc. as:

$$CH_2{=}CH_2 + Br_2 \rightarrow BrCH_2{-}CH_2Br$$

(Halogenation is either carried out in gas (dry) phase or in non aqueous solvent, e.g. CCl_4, CS_2, etc.)

Mechanism

Halogenation reaction is catalyzed by a Lewis acid catalyst $AlCl_3$ or $FeCl_3$, forming halonium ion (X^+) intermediate in the first step as:

$$Cl_2 + FeCl_3 \rightarrow Cl^+ + FeCl_4^-$$

In the next step, halonium ion adds to double bond forming an intermediate cyclic halonium ion as:

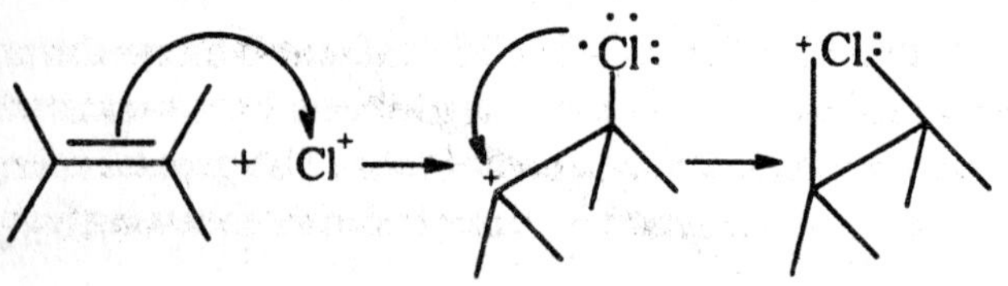

In the last step, a halide ion attacks from back side at cyclic halonium ion intermediate forming dihalides as:

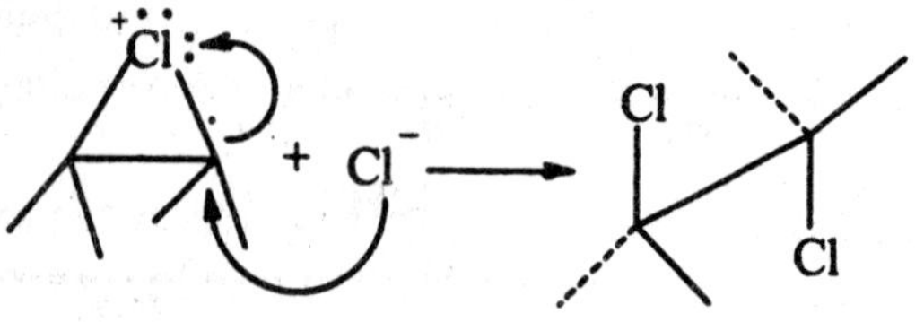

(Anti halogenation)

Stereochemistry of Halogenation Reaction

Halogenation of alkene is an example of stereospecific anti addition reaction, i.e. a specific stereomer after halogenation results a specific dihalides. To understand the stereochemistry of halogenation reaction, let us consider the bromination of *cis* and *trans*-2-butene:

Cis-2-butene:

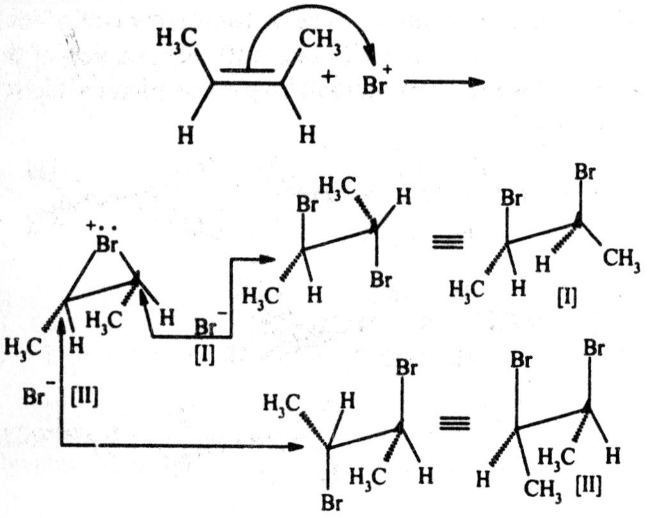

[I]

[II]

In the above reaction, products [I] and [II] are enantiomers. Also [I] and [II] are produced in equal amount, since there is equal probability of attack of bromide ion at the two carbon involved in cyclic bromonium ion, the net result will be a racemic mixture of [I] and [II].

Trans-2-butene

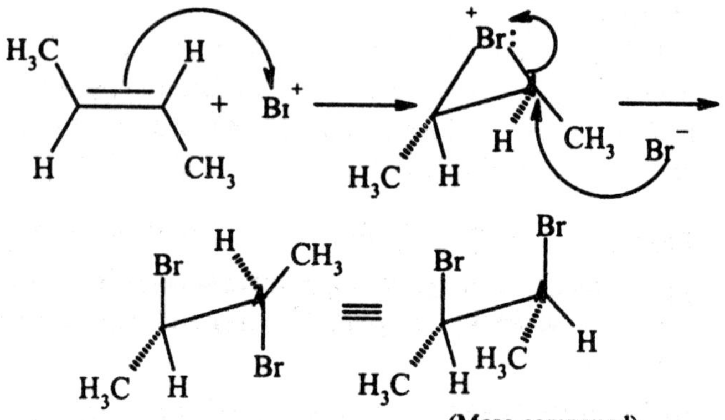

(Meso compound)

Halogenation of *trans*, symmetrical, butene results in meso dihalo compound.

Example 5. Discuss the stereochemistry of bromination of *cis* and *trans* 2-pentene.

Solution: Bromination at double bond of either *cis* or *trans* 2-pentene would result in formation of racemic mixture as:

(a) cis-2-pentene

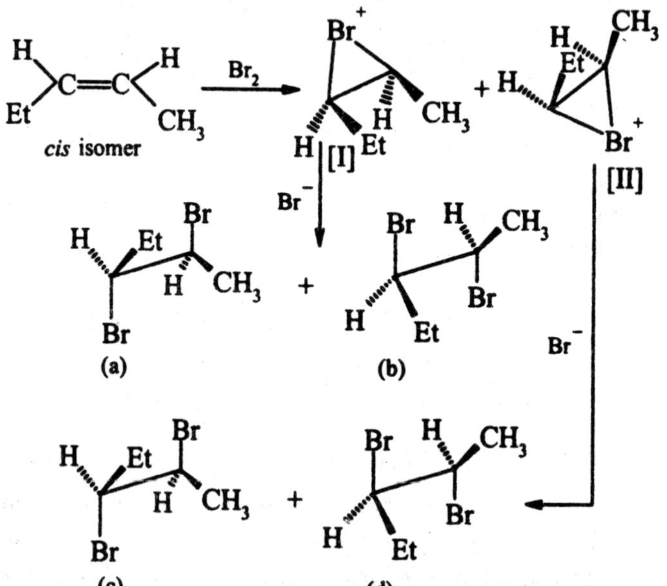

cis isomer

(a) (b) (c) (d)

In case of unsymmetrical alkene, enantiomeric cyclic halonium ions (I and II) are produced in equal amount. Out of four products, a and c are similar, b and d are similar. Also a and b, and, c and d are enantiomeric pairs. Due to difference in steric hindrance to the attack of Br^- ion at two bridging carbons of I and II, a and b and, c and d are not produced in equal amount but (a + c) = (b + d) giving rise to racemic products.

(B) Trans-2-pentene

Here again, enantiomeric cyclic brominium ions will be produced in equal amount giving racemic mixture of products in the subsequent attack by Br^- ion as in case of (A).

Evidence in support of this mechanism is obtained from the fact that when ethylene reacts with bromine in aqueous NaCl solution, ethylene dibromide and 1-bromo-2-chloro ethane are obtained, no ethylene dichloride is formed.

Fluorination of Alkene

Fluorination at double bond is done using PbF_4 as:

$$CH_2{=}CH_2 + PbF_4 \rightarrow FCH_2{-}CH_2F$$

4.5.3 Electrophilic Addition to Alkene

General discussion: The most common chemical transformation of a carbon-carbon double bond is the *addition reaction*. A large number of reagents, both inorganic and organic have been found to add to this functional group, and in this section we shall review many of these reactions. A majority of these reactions are exothermic, due to the fact that the C—C pi-bond is relatively weak (≈ 63 kcal/mole) relative to the sigma-bonds formed to the atoms or groups of the reagent. Remember, the bond energies of a molecule are the energies required to break (homolytically) all the covalent bonds in the molecule. Consequently, if the bond energies of the product molecules are greater than the bond energies of the reactants, the reaction will be exothermic. The following calculations for the addition of H-Br are typical. Note that by convention exothermic reactions have a negative heat of reaction.

$$>C{=}C< + H{-}Br \longrightarrow H\cdots\overset{|}{\underset{|}{C}}-\overset{|}{\underset{|}{C}}\cdots Br$$

Heat of reaction = (150.5) – (167.0) = –16.5 Kcal/mole

Bonds broken: = π C = C and H-Br 63 87.5	Sum = 150.5 Kcal/mole
Bonds formed: = HC and BrC	Sum = 167.0 Kcal/mole

Addition of Strong Bronsted Acids

As illustrated by the preceding general equation, strong bronsted acids such as HCl, HBr, HI and H_2SO_4, rapidly add to the C=C functional group of alkenes to give products, in which new covalent bonds are formed to hydrogen and to the conjugate base of the acid. Using the above equation as a guide, write the addition products expected on reacting each of these reagents with cyclohexene.

(Scheme: cyclohexene reacting with HCl, HBr, H_2SO_4, HI)

General mechanism of addition: Let us consider addition of a compound AB to an alkene:

Step I: Heterolytic cleavage of AB:

A-B → A^- (Nucleophile) + B^+ (Electrophile)

Step II: Addition of electrophile to double bond, formation of carbocation intermediate:

$$>C{=}C< + \overset{+}{B} \longrightarrow >\overset{+}{C}-\overset{B}{\overset{|}{C}}<$$

Example 6. Why electrophile adds first, not nucleophile, at double bond?

Solution: Double bond is a source of electrons due to presence of loosely held pi-electrons and thus double bond serves as Lewis base for attack of electrophile-a Lewis acid.

Step III: Rearrangement of carbocation to more stable carbocation (if possible): rearrangement is a characteristic feature of carbocation, and if possible, occurs before addition of nucleophile as described below:

(a) $CH_3-C(CH_3)(H)-CH{=}CH_2 + \overset{+}{B} \longrightarrow$

$CH_3-C(CH_3)(H)-\overset{+}{C}H-CH_2B \xleftrightarrow{\text{Hydride shift}} CH_3-\overset{+}{C}(CH_3)-CH(H)-CH_2B$

(Secondary carbocation) (Tertiary carbocation)

(b) $CH_3-C(CH_3)_2-CH{=}CH_2 + \overset{+}{B} \longrightarrow$

$CH_3-C(CH_3)(CH_3)-\overset{+}{C}H-CH_2B \xleftrightarrow{\text{Alkyl shift}} CH_3-\overset{+}{C}(CH_3)-CH(CH_3)-CH_2B$

(Secondary carbocation) (Tertiary carbocation)

Step IV: Addition of nucleophile:

$$\overset{-}{A} + -\overset{+}{C}-\overset{B}{\underset{|}{C}}- \longrightarrow -\overset{A}{\underset{|}{C}}-\overset{B}{\underset{|}{C}}-$$

(a) Addition of HX: $HX \rightarrow H^+ + X^-$

$$>C{=}C< + H^+ \longrightarrow >\overset{+}{C}-\overset{H}{C}< \xrightarrow{X^-} -\overset{X}{C}-\overset{H}{C}-$$

The order of reactivity for addition of halogen acid is:

$$HI > HBr > HCl > HF$$

Rules for addition (Markovnikov's Rule)

According to Markovnikov's rule, hydrogen of HX is added to doubly bonded carbon bearing more hydrogen. A more general rule for addition at C=C is, "positive part of addendum is added to carbon bearing more hydrogen". This rule is based on the fact that during addition of electrophile, more stable carbocation is formed, e.g.

$$CH_3{-}CH{=}CH_2 + HCl \longrightarrow CH_3{-}\overset{Cl}{\overset{|}{C}H}{-}CH_3$$

1-Methylcyclohexene + HCl ⟶ 1-chloro-1-methylcyclohexane

Empirical rules like the Markovnikov rule are useful aids for remembering and predicting experimental results. Indeed, empirical rules are often the first step toward practical mastery of a subject, but they seldom constitute true understanding. The Markovnikov rule, for example, suggests there are common and important principles at work in these addition reactions, but it does not tell us what they are. The next step in achieving an understanding of this reaction must be to construct a rational mechanistic model that can be tested by experiment.

All the reagents discussed here are strong Bronsted acids so, as a first step, it seems sensible to find a base with which the acid can react. Since we know that these acids do not react with alkanes, it must be the pi-electrons of the alkene double bond that serve as the base. As shown in Figure 4.6, the pi-orbital extends into the space immediately above and below the plane of the double bond, and the electrons occupying this orbital may be attracted to the proton of a Bronsted acid. The resulting acid-base equilibrium generates a carbocation intermediate (the conjugate acid of the alkene), which then combines rapidly with the anionic conjugate base of the bronsted acid.

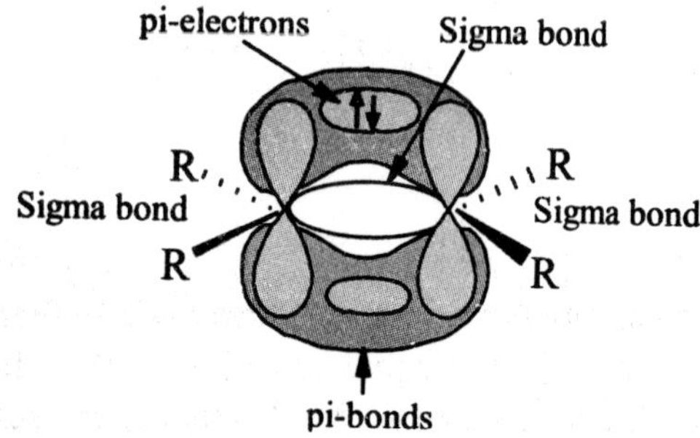

Figure 4.6: Pi-electron cloud of alkene

An energy diagram for this two-steps addition mechanism is shown in Figure 4.7. From this figure we see that the slow or rate-determining step (the first step) is also the product determining step (the anion will necessarily bond to the carbocation site), and it is here we must look for a rationalization of Markovnikov's rule.

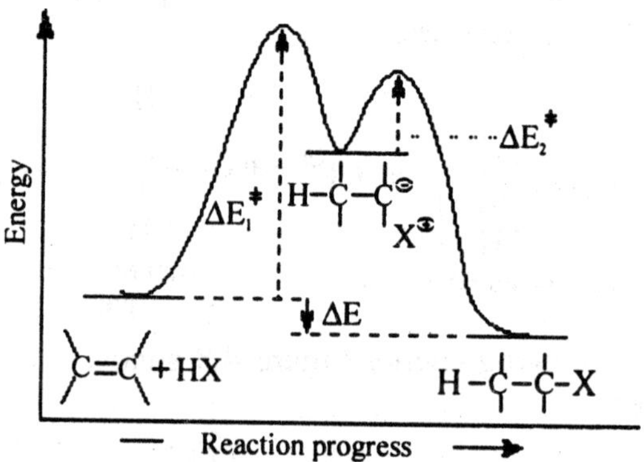

Figure 4.7: Potential energy diagram for addition of HX on alkene

The carbocation intermediate formed in the first step of the addition reaction now assumes a key role, in that it directly influences the activation energy for this step. Independent research shows that the stability of carbocations varies with the nature of substituents, in a manner similar to that seen for alkyl radicals. The exceptional stability of allyl and benzyl cations is the result of charge delocalization, and the stabilizing influence of alkyl substituents, although less pronounced, has been interpreted in a similar fashion.

From this information, we arrive at a plausible rationalization of Markovnikov's rule. *When an unsymmetrically substituted double bond is protonated, we expect the more stable carbocation intermediate to be formed faster than the less stable alternative*, because the activation energy of the path to the former is the lower of the two possibilities. This is illustrated by the following equation for the addition of hydrogen chloride to propene. Note that the initial acid-base equilibrium leads to a pi-complex, which immediately reorganizes to a sigma-bonded carbocation intermediate.

The more stable 2°-carbocation is formed preferentially, and the conjugate base of the bronsted acid (chloride anion in the example shown below) then rapidly bonds to this electrophilic intermediate to form the final product as shown below.

$$H_2C=CH-CH_3 + H-Cl \rightleftharpoons [\text{p-complex}]\ Cl^{\ominus} \longrightarrow$$

Propene p-complex

Favoured:

$$H_2\overset{\oplus}{C}... \quad CH_3-\overset{\oplus}{C}H-CH_3 + Cl^{\ominus} \longrightarrow H-\underset{H}{\overset{H}{C}}-\underset{CH_3}{\overset{H}{C}}-Cl \quad \text{Markonikov product}$$

Less-favoured:

$$\overset{\oplus}{C}H_2-CH_2-CH_3 + Cl^{\ominus} \longrightarrow Cl-\underset{H}{\overset{H}{C}}-\underset{CH_3}{\overset{H}{C}}-H \quad \text{Anti Markonikov's products}$$

The following energy Figure 4.8 summarizes these features.

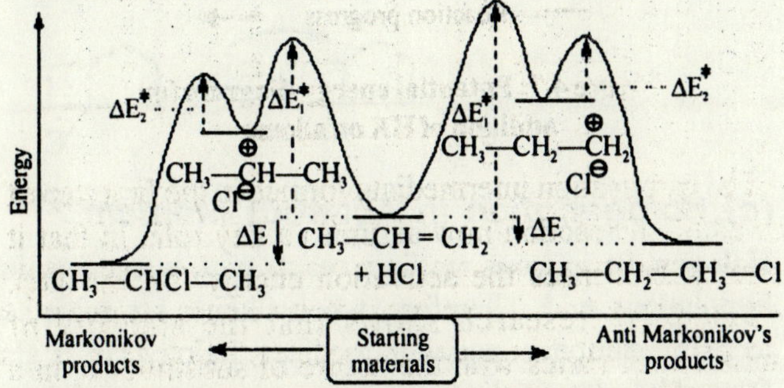

Figure 4.8: Potential energy diagram for the two possible orientation of addition to propene

Exception to Markovnikov's rule

In some alkenes, HCl is not directly added exactly to doubly bonded carbons due to rearrangement of carbocation to more stable one, e.g:

$$CH_3-\underset{}{\overset{CH_3}{C}}H-CH=CH_2 + HCl \longrightarrow CH_3-\underset{H}{\overset{CH_3}{C}}-\overset{+}{C}H-CH_3 \xleftrightarrow{H^- \text{ shift}}$$

(Secondary carbocation)

$$CH_3-\underset{H}{\overset{CH_3}{\underset{+}{C}}}-CH-CH_3 \longrightarrow CH_3-\underset{Cl}{\overset{CH_3}{C}}-\underset{H}{C}H-CH_3$$

(Tertiary carbocation) (Major product)

$$\text{3-methylcyclohexene} + HCl \longrightarrow \text{(Secondary carbocation)} \longrightarrow \text{(Tertiary carbocation)} \xrightarrow{Cl^-} \text{(Major product)}$$

(Secondary carbocation)

(Tertiary carbocation) (Major product)

$$CH_3-\underset{CH_3}{\overset{CH_3}{C}}-CH=CH_2 + HCl \longrightarrow CH_3-\underset{CH_3}{\overset{CH_3}{C}}-\overset{+}{C}H-CH_3 \xleftrightarrow{Me^- \text{ shift}}$$

(Secondary carbocation)

$$CH_3-\underset{+}{\overset{CH_3}{C}}-\underset{CH_3}{C}H-CH_3 \xrightarrow{Cl^-} CH_3-\underset{Cl}{\overset{CH_3}{C}}-\underset{H}{C}H-CH_3$$

(Tertiary carbocation) (Major product)

Therefore, the ultimate approach to electrophilic addition to an alkene is formation of carbocation then stabilization (if possible) by rearrangement followed by addition of nucleophile as:

Alkene + E⁺ (Electrophile) → Carbocation ⟷ Rearrangement → Addition of nucleophile

Stereochemistry of addition reaction

If addition of HX at double bond generate one chiral carbon, equal amount of enantiomers, i.e. racemic mixture would be obtained as:

$$CH_3-CH=CH-CH_3 + H^+ \longrightarrow$$

$$CH_3-\overset{+}{C}H-CH_2-CH_3 \xrightarrow{Cl^-} \text{(I)}$$

(sp²- planar arrangement) (Top side attack)

$$\xrightarrow{Cl^-} \text{(II)}$$

(Bottom side attack)

Here [I] and [II] are enantiomers and they are produced in equal amount because Cl⁻ can attack either from top or bottom side with equal probability. Therefore, if addition of HX generates one chiral carbon, net result will be racemization of product. On the other hand if, addition of HX makes both doubly bonded carbon chiral, all four stereomers would be produced as:

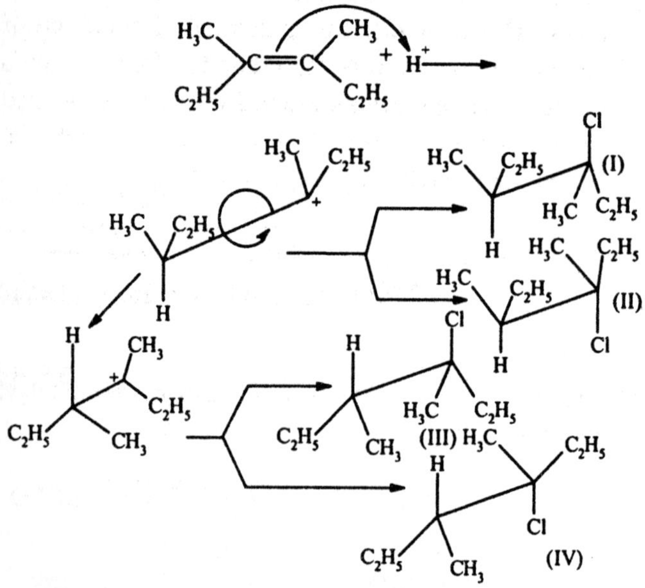

Here [I] and [III] are enantiomers as well as [II] and [IV] are enantiomers. Since all products are formed with equal probability, the net result is racemic mixture.

[b] Addition of ICl, IBr, BrCl, etc.

In addition of such compounds onto alkenes, the less electronegative species (atom) is considered to be electrophile and it is added in the first step of the reaction whereas more electronegative species (atom) is added in the second step, e.g.

$$CH_3—CH=CH_2 + ICl \rightarrow CH_3—CHCl—CH_2I$$

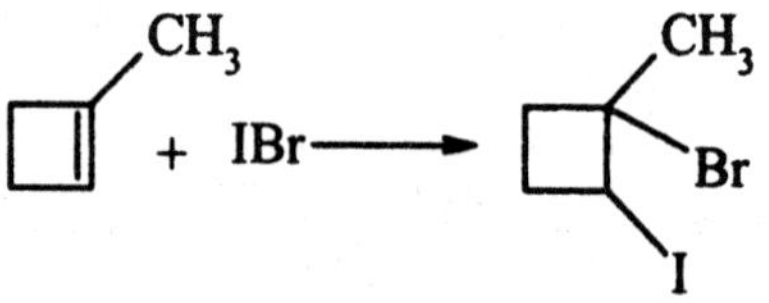

$$CH_3CH=C(CH_3)_2 + BrCl \rightarrow CH_3CHBr—C(CH_3)_2Cl$$

Example 7. Predict major addition products in the following reactions:

(a) $(CH_3)_3CCH=CH_2 + ICl \rightarrow$

(b) $C_6H_5—CH=CHCH_3 + BrCl \rightarrow$

(c) $C_6H_5—CH=CHCH_3 + I\ Br \rightarrow$

(d) $C_6H_5—C(CH_3)=C(C_6H_5)_2 + IBr \rightarrow$

Solution

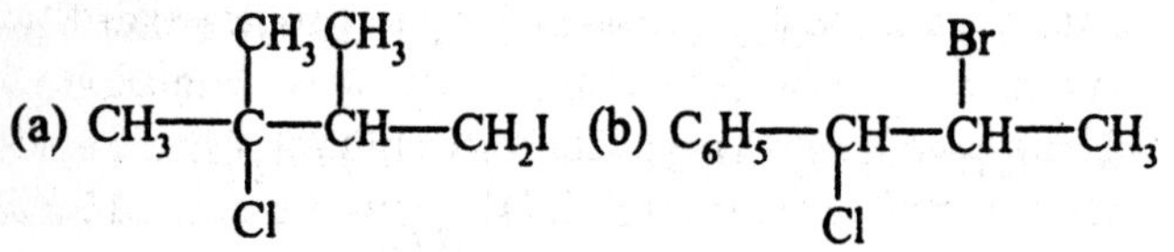

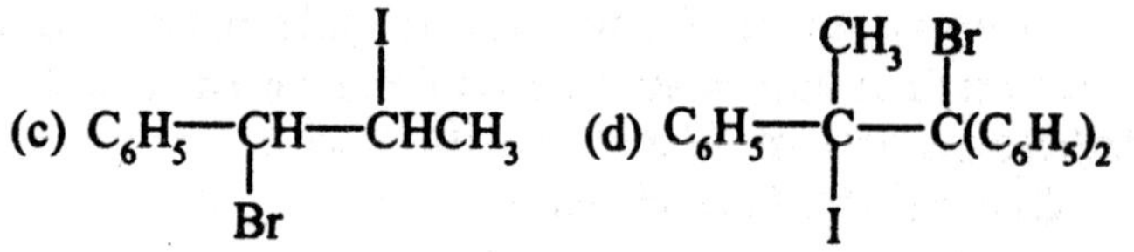

(a) $CH_3—C(CH_3)(Cl)—CH(CH_3)—CH_2I$ (b) $C_6H_5—CH(Cl)—CH(Br)—CH_3$

(c) $C_6H_5—CH(Br)—CH(I)CH_3$ (d) $C_6H_5—C(CH_3)(I)—C(Br)(C_6H_5)_2$

Example 8. Propose mechanism of formation of the indicated products in the following reactions:

(a) 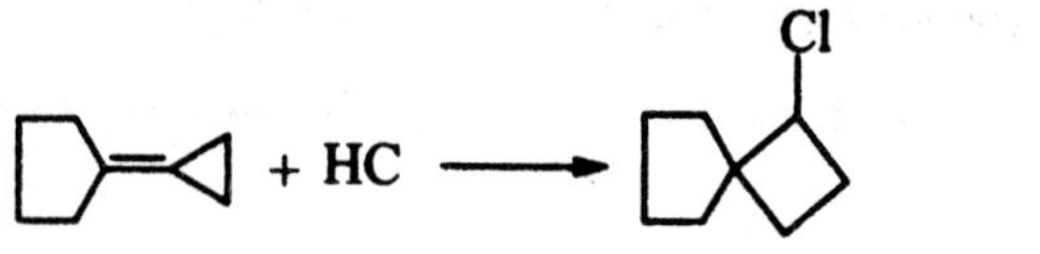 + HC →

(b) 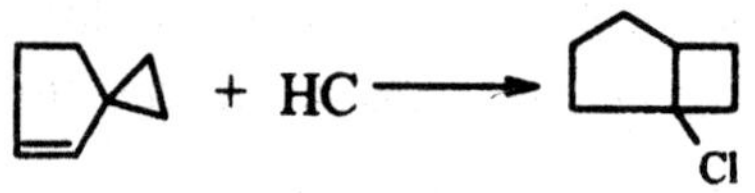 + HC →

Solution

(a) 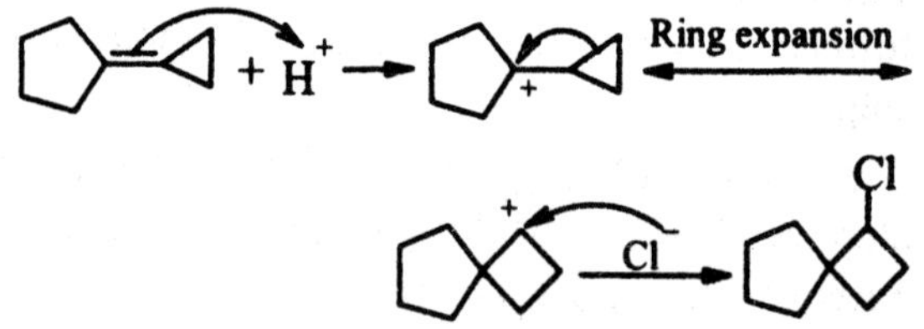 + H^+ → Ring expansion

(b) 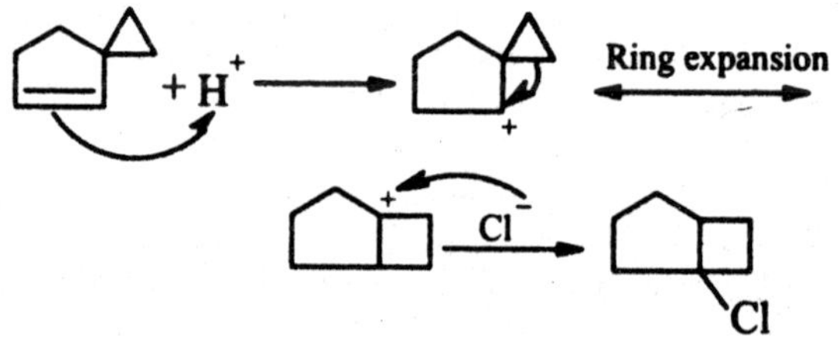 + H^+ → Ring expansion

(c) Halogenation in Aqueous Medium

Halogen, in aqueous medium reacts with water forming hypohalous acid. Therefore, when alkenes are subjected to halogenation in aqueous medium, halohydrin is also formed besides dihalide, viz.

$$CH_3—CH=CH_2 + Br_2\text{ -Water} \longrightarrow CH_3—CH(Br)—CH_2Br + CH_3—CH(OH)—CH_2Br$$

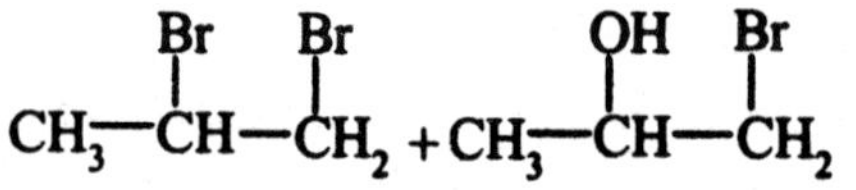

Halohydrin

Mechanism of formation of the halohydrin:

$$Br_2 + H_2O \rightarrow HOBr + HBr$$

$$HOBr \rightarrow HO^- + Br^+$$

$$CH_3—CH=CH_2 + Br^+ \longrightarrow CH_3—\overset{+}{C}H—CH_2Br \xrightarrow{OH^-} CH_3—CH(OH)—CH_2Br$$

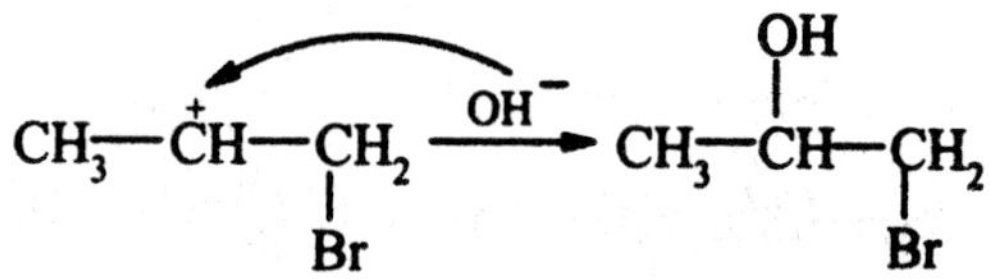

[d] Reaction of Alkenes with Cold, Concentrated H_2SO_4

Alkenes are absorbed by cold concentrated sulphuric acid to form alkyl hydrogen sulphates. Addition occurs according to Markovnikov's rule, e.g.

$$CH_3—CH{=}CH_2 + H_2SO_4 \rightarrow CH_3—\underset{\displaystyle OSO_3H}{\underset{|}{CH}}—CH_3$$

Alkanes are not absorbed by cold concentrated sulphuric acid and can be used to separate alkanes from alkenes.

Example 9. When excess of alkene is treated with cold concentrated sulphuric acid, some dialkyl sulphate is also formed. Discuss the mechanism of formation of dialkyl sulphate.

Solution: Alkyl hydrogen sulphate formed in the first step still has one acidic hydrogen and it may undergo heterolysis to produce electrophilic H^+ as:

$$ROSO_2—OH \rightarrow ROSO_2O^- + H^+$$

$$CH_3—CH{=}CH_4 + \overset{+}{H} \longrightarrow$$

$$CH_3—\overset{+}{C}H—CH_3 \xrightarrow{\ ^-O—O_2SOR\ } CH_3—\underset{}{\overset{\displaystyle OSO_2OR}{\overset{|}{CH}}}—CH_3$$

Here R-is isopropyl group.

[e] Hydration of Alkenes

(i) An alkene can be hydrated to alcohols by absorbing with cold, concentrated sulphuric acid followed by hydrolyzing the alkyl hydrogen sulphate as:

$$(CH_3)_2CHCH{=}CH_2 \xrightarrow{H_2SO_4}$$

$$CH_3—\overset{\displaystyle CH_3}{\overset{|}{\underset{\displaystyle OSO_3H}{\underset{|}{C}}}}—CH_2CH_3 \xrightarrow[OH_2]{\overset{+}{H}} CH_3—\overset{\displaystyle CH_3}{\overset{|}{\underset{\displaystyle OH}{\underset{|}{C}}}}—CH_2CH_3$$

Hydration, here proceeds through carbocation intermediate, therefore, if possible, rearranged alcohols may also be obtained as:

$$CH_3—\overset{\displaystyle CH_3}{\overset{|}{CH}}—CH{=}CH{=}CH_2 + H_2SO_4 \xrightarrow[OH_2]{\overset{+}{H}}$$

$$CH_3—\overset{\displaystyle CH_3}{\overset{|}{CH}}—\underset{\displaystyle OH}{\underset{|}{CH}}—CH_3 + CH_3—\overset{\displaystyle CH_3}{\overset{|}{\underset{\displaystyle OH}{\underset{|}{C}}}}—CH_2CH_3$$

(Rearranged product)

(ii) *Oxymercuration-demercuration reaction of alkenes:* This is an alternative method of hydration of alkene and reaction occurs according to Markovnikov's rule. The advantage of this method over acid catalyzed hydration is no rearranged product is obtained, i.e. hydration occurs at double bond only as:

$$(CH_3)_3CCH{=}CH_2 + (CH_3COO)_2Hg/THF \xrightarrow{NaBH_4} (CH_3)_3CCH(OH)CH_3 \text{ (only product)}$$

Mechanism

Mechanism of this reaction is believed to involve a cyclic intermediate as:

$$CH_3COO—Hg—OCOCH_3 \rightleftharpoons CH_3COO\overset{+}{H}g + CH_3COO^- \quad [\overset{+}{H}g(OAc)]$$

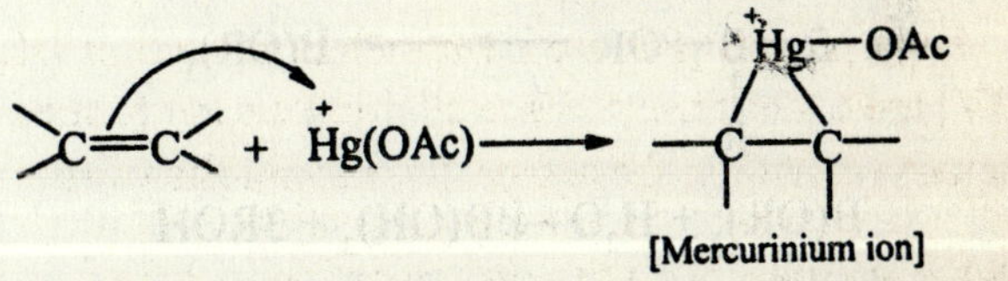

$$>C{=}C< + \overset{+}{H}g(OAc) \longrightarrow —\overset{|}{C}—\overset{|}{C}— \text{ (bridged by } \overset{+}{H}g—OAc)$$

[Mercurinium ion]

Formation of cyclic mercurinium ion prevents the rearrangement of carbocation. In the next step water attacks as nucleophile as:

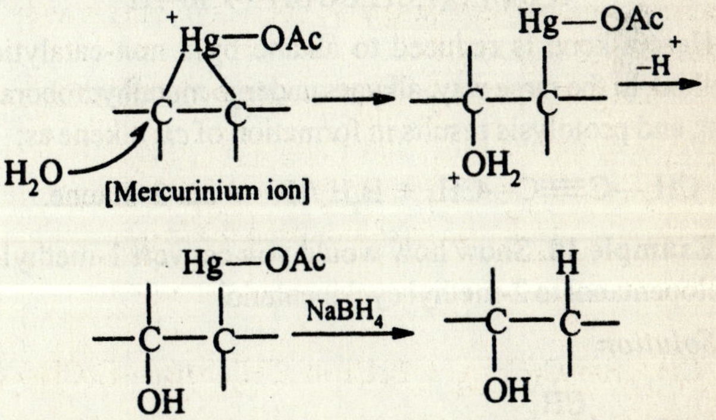

$$H_2O + \text{[Mercurinium ion]} \longrightarrow —\overset{\displaystyle Hg—OAc}{\overset{|}{C}}—\underset{\displaystyle \overset{+}{O}H_2}{\underset{|}{C}}— \xrightarrow{-\overset{+}{H}}$$

$$—\overset{\displaystyle Hg—OAc}{\overset{|}{C}}—\underset{\displaystyle OH}{\underset{|}{C}}— \xrightarrow{NaBH_4} —\overset{\displaystyle H}{\overset{|}{C}}—\underset{\displaystyle OH}{\underset{|}{C}}—$$

(iii) *Hydroboration of alkene:* This method is used for anti-Markovnikov's hydration of alkene. When an alkene is treated with diborane (B_2H_6), a trialkyl borane is formed in the first step of synthesis. This trialkyl borane on treatment with H_2O_2 in alkaline medium yields alcohol. The overall reaction is addition of water to alkene according to anti-Markovnikov's rule. No rearranged products are obtained, i.e. hydration occurs only at double bond.

$$3CH_3CH{=}CH_2 + B_2H_6 \rightarrow (CH_3CH_2CH_2)_3B \xrightarrow[OH^-]{H_2O_2} CH_3CH_2CH_2OH$$

Mechanism

Step I

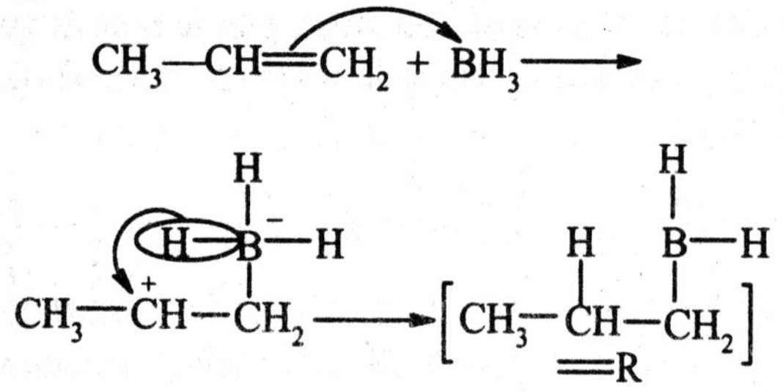

$$CH_3-CH=CH_2 + BH_3 \longrightarrow$$

$$CH_3-\overset{+}{C}H-CH_2 \cdot\cdot\cdot H-\overset{-}{B}H-H \longrightarrow [CH_3-CH(H)-CH_2-BH_2] = R$$

$$2CH_3CH=CH_2 + RBH_2 \rightarrow R_3B$$

Step II: $H-O-O-H + OH^- \rightarrow HOO^- + H_2O$

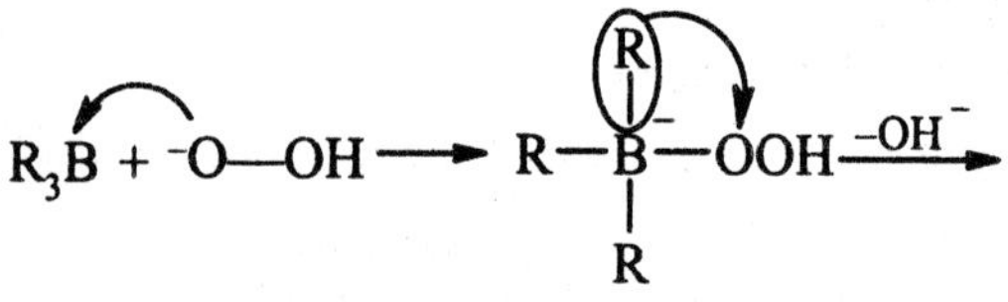

$$R_3B + {}^-O-OH \longrightarrow R-\overset{R}{\underset{R}{B^-}}-OOH \xrightarrow{-OH^-}$$

$$R-\underset{R}{B}-OR \longrightarrow B(OR)_3$$

$$B(OR)_3 + H_2O \rightarrow B(OH)_3 + 3ROH$$

Where ROH is $CH_3CH_2CH_2OH$

If trialkyl borane formed in the first step is treated with propionic acid, the corresponding alkane is produced as:

$$R_3B + CH_3CH_2COOH \rightarrow R-H$$

Here alkene is reduced to alkane by a non-catalytic method. In the same way, alkynes undergo monohydroboration, and protolysis results in formation of *cis* alkene as:

$$CH_3-C\equiv C-CH_3 + B_2H_6/H^+ \rightarrow cis\text{-2-butene.}$$

Example 10. Show how would you convert 1-methylcyclopentanol to 2-methyl cyclopentanol.

Solution

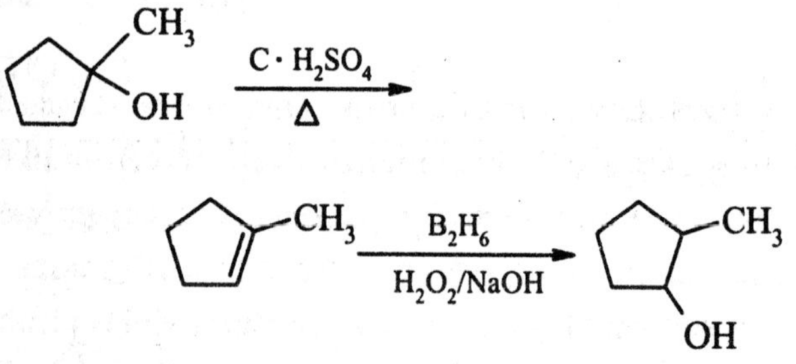

1-methylcyclopentanol $\xrightarrow[\Delta]{C\cdot H_2SO_4}$ 1-methylcyclopentene $\xrightarrow[H_2O_2/NaOH]{B_2H_6}$ 2-methylcyclopentanol

Example 11. Predict major products of the following reactions.

(a) Propene + BH_3/THF $\rightarrow$

(b) The product of part (a) + H_2O_2/NaOH $\rightarrow$

(c) 2-methyl-2-pentene + BH_3/THF $\xrightarrow{H_2O_2/NaOH}$

(d) 1-methylcyclohexene + BH_3/THF $\xrightarrow{H_2O_2/NaOH}$

Solution

(a) $(CH_3CH_2CH_2)_3B$

(b) $CH_3CH_2CH_2OH$

(c) 2-methyl-3-pentanol

(d) 2-methylcyclohexanol.

Example 12. Show how would you accomplish the following conversion:

(a) 1-butene $\rightarrow$ 1-butanol

(b) 1-butene $\rightarrow$ 2-butanol

(c) 2-bromo-2, 4-dimethylpentane $\rightarrow$ 2, 4-dimethyl-3-pentanol.

Solution

(a) 1-butene + BH_3/H_2O_2/NaOH $\rightarrow$ 1-butanol

(b) 1-butene + dil.H_2SO_4/heat $\rightarrow$ 2-butanol

(c) 2-bromo-2, 4-dimethylpentane + alcoholic KOH $\rightarrow$ 2, 4-dimethyl-2-pentene
2,4-dimethyl-2-pentene + BH_3/H_2O_2/NaOH $\rightarrow$ product.

4.5.4 Free Radical (*Anti-Markovnikov's*) Addition of HBr to Double Bond

When HBr is added to an alkene, in presence of a peroxide, addition occurs in a direction opposite to that predicted by Markovnikov's rule, i.e. H (hydrogen) is added to the carbon bearing less hydrogen, e.g.

$$CH_3CH=CH_2 + HBr \xrightarrow{\text{Peroxide}} CH_3CH_2CH_2Br$$

Mechanism

Step I: $R_2O_2 \rightarrow 2R\dot{O}$

Step II: $R\dot{O} + HBr \rightarrow ROH + Br^{\bullet}$

Step III

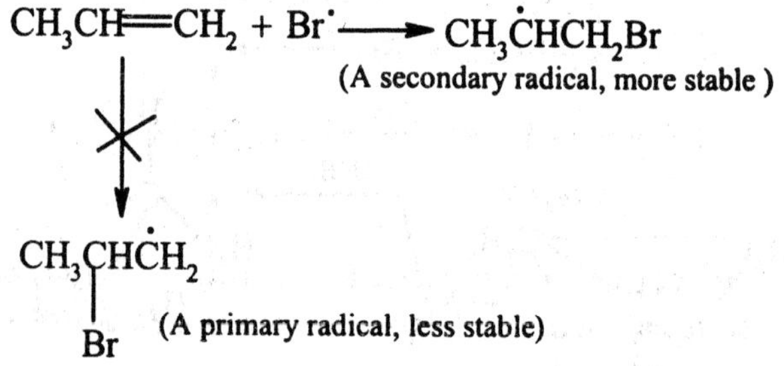

$$CH_3CH=CH_2 + Br^{\bullet} \longrightarrow CH_3\dot{C}HCH_2Br$$

(A secondary radical, more stable)

$$\not\rightarrow CH_3\underset{Br}{C}H\dot{C}H_2$$

(A primary radical, less stable)

Step IV: $CH_3\dot{C}HCH_2Br + HBr \rightarrow CH_3CH_2CH_2Br + Br$

Example 13. Why $R\dot{O}$ abstract $H^{\bullet}$ from HBr in Step II, not $\dot{Br}$?

Solution: ROH is more stable than ROBr, since –O—H bond is stronger than –O—Br bond.

This abnormal behaviour shown by HBr in presence of peroxide is known as peroxide effect and this effect is not shown by other hydrogen halide, viz. HF, HCl or HI. The reason for reluctance of these hydrogen halide to show peroxide effect can be explained with the help of thermodynamics of the reaction as: Followings are the enthalpies of reactions in the two step free radical addition reaction of HX:

Table 4.1: Enthalpies value for two step free radical addition of HX on alkene

HX	ΔH_1 $X^{\bullet} + CH_2{=}CH_2 \rightarrow XCH_2\dot{C}H_2$	ΔH_2 $XCH_2\dot{C}H_2 + HX \rightarrow XCH_2CH_2H$
HCl	–67 kJ/mol	+12.6kJ/mol
HBr	–25 kJ/mol	–50.2 kJ/mol
HI	+46 kJ/mol	–117 kJ/mol

It is evident from the above thermodynamic information, that both steps in the addition of HBr are exothermic, a favourable condition for reaction whereas in case of HCl and HI, one of the step is endothermic, which is thermodynamically not favourable for a chemical reaction. Thus, we conclude that only HBr can be efficiently added in presence of peroxide, not HCl or HI.

Tri and tetra halogenated methanes also adds on a double bond in presence of peroxide as:

$$RO^{\bullet} + CHCl_3 \rightarrow ROH + {}^{\bullet}CCl_3$$

$$RO^{\bullet} + CCl_4 \rightarrow ROCl + {}^{\bullet}CCl_3$$

$$RO^{\bullet} + CBrCl_3 \rightarrow ROBr + {}^{\bullet}CCl_3, \text{ etc.}$$

Example 14. Predict the product obtained during addition of HBr to 2-methyl-1-butene in presence of peroxide and discuss the stereochemistry of product.

Solution

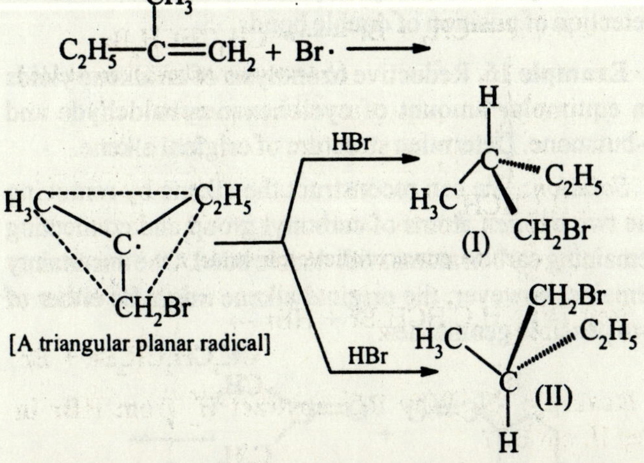

$$C_2H_5{-}\overset{CH_3}{\overset{|}{C}}{=}CH_2 + Br\cdot \longrightarrow$$

[A triangular planar radical] —HBr→ (I) ; —HBr→ (II)

Here I and II are obtained from attack of H radical from two sides of planar intermediate radical. Also I and II are obtained in equal amount since there is equal probability of abstraction of H-radical from top and bottom side, the net result is formation of racemic mixture.

4.5.5 Hydroxylation of Alkene

(a) When an alkene is treated with cold, dilute and alkaline solution of potassium permanganate (commonly known as Bayer's reagent), vicinal diol is formed.

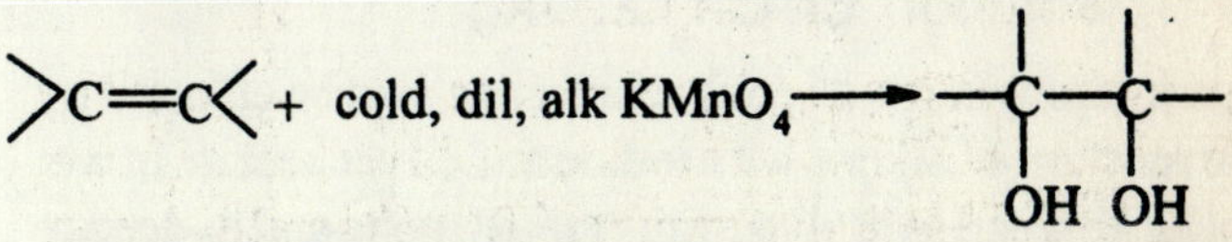

$$>C{=}C< + \text{cold, dil, alk } KMnO_4 \longrightarrow -\underset{OH}{\overset{|}{C}}-\underset{OH}{\overset{|}{C}}-$$

This reagent is very useful in detection of unsaturation in a compound, because the deep purple colour of reagent is decolourized after this reaction.

Mechanism

Syn hydroxylation occurs:

$$RHC{=}CHR + \text{cold, dil, alk } KMnO_4 \longrightarrow$$

(cyclic manganate ester intermediate) $\xrightarrow{OH^-}$ H–C(R)(OH)–C(R)(OH)–H

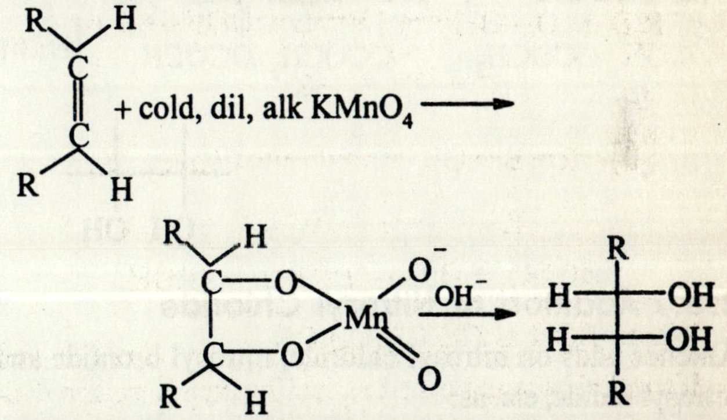

Similar reaction can also be carried out using cold, dilute alkaline solution of chromic acid or with OSO_4 followed by treatment of product with H_2O_2 or ethanol solution of $NaHSO_3$.

Example 15. *cis*-2-butene on treatment with Bayer's reagent gives meso diol whereas *trans*-2-butene on same treatment results in formation of racemic mixture. Accounts for the difference.

Solution: Hydroxylation occurs from same side as cyclic intermediate is formed:

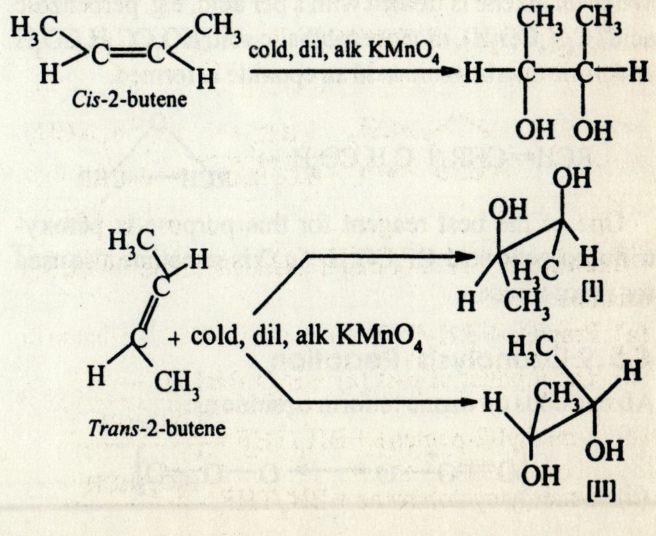

$$\underset{Cis\text{-2-butene}}{H_3C(H)C{=}C(H)CH_3} \xrightarrow{\text{cold, dil, alk } KMnO_4} H{-}\underset{OH}{\overset{CH_3}{C}}{-}\underset{OH}{\overset{CH_3}{C}}{-}H$$

Trans-2-butene + cold, dil, alk $KMnO_4$ → [I] + [II]

Here [I] and [II] are enantiomers, and produced in equal amounts due to equal probability of attack from either side, net result is racemization of products.

4.5.6 Reaction with Iodine in Acidic Solution of CH_3COOAg

Syn hydroxylation of double bond can also be effected by the reaction of alkene with iodine and silver acetate in wet acetic acid and hydrolysing the mixed mono and di-acetate with alkali:

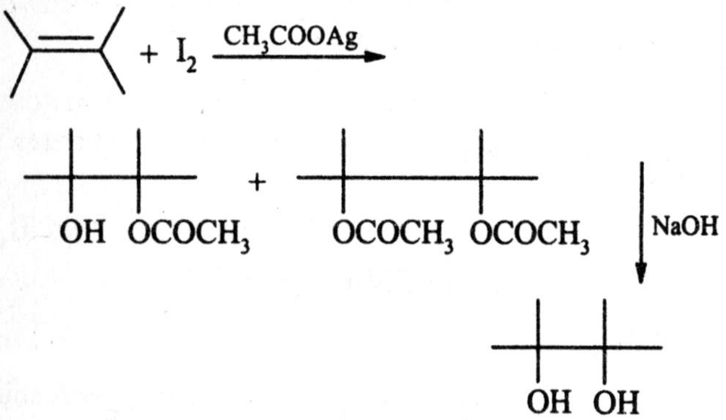

$$>C{=}C< + I_2 \xrightarrow{CH_3COOAg}$$

OH OCOCH$_3$ + OCOCH$_3$ OCOCH$_3$ $\xrightarrow{NaOH}$ OH OH

4.5.7 Addition of Nitrosyl Chloride

Alkenes adds on nitrosyl chloride, nitrosyl bromide and nitrosyl iodide, etc. as:

$$O{=}N{-}X \rightarrow X^- + [O{:=}N^+ \leftrightarrow \overset{+}{O}{\equiv}N{:}]$$

$$CH_3{-}CH{=}CH_2 + O{=}N^+ \rightarrow$$

$$CH_3{-}CH^+{-}CH_2NO \xrightarrow{X^-} CH_3{-}\underset{X}{CH}{-}CH_2NO$$

Addition occurs according to Markovnikov's rule.

4.5.8 Epoxidation of Alkene: (*Prileschaiev Reaction*)

When an alkene is treated with a per acid, e.g. perbenzoic acid ($C_6H_5CO_3H$), monoperphthalic acid $HO_2CC_6H_5CO_3H$ and *p*-nitroperbenzoic acid an epoxide is formed.

$$RCH{=}CHR + C_6H_5CO_3H \rightarrow RCH{-}CHR \text{ (bridged by O)}$$

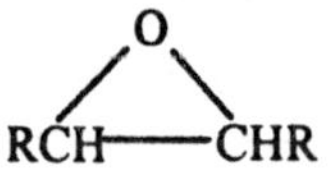

One of the best reagent for this purpose is peroxy-trifluoroacetic acid CF_3CO_3H. Ag_2O is sometime also used for this purpose.

4.5.9 Ozonolysis Reaction

Alkene adds on ozone to form ozonide as:

$$[O{=}\overset{+}{O}{-}\overset{-}{O} \longleftrightarrow \overset{-}{O}{-}\overset{+}{O}{=}O]$$

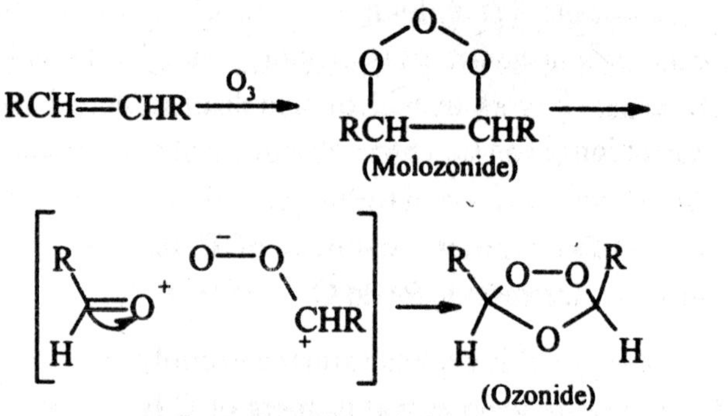

$$RCH{=}CHR \xrightarrow{O_3} \text{RCH—CHR (O–O–O ring)} \longrightarrow$$

(Molozonide)

$$\left[\text{R(H)C}{=}\overset{+}{O}\ \ \overset{-}{O}{-}O{-}\overset{+}{CHR} \right] \longrightarrow \text{R(H)C(O–O)(O)C(R)H}$$

(Ozonide)

Reduction of ozonide: Ozonides are reduced with Zn-H_2O to produce carbonyl compound as:

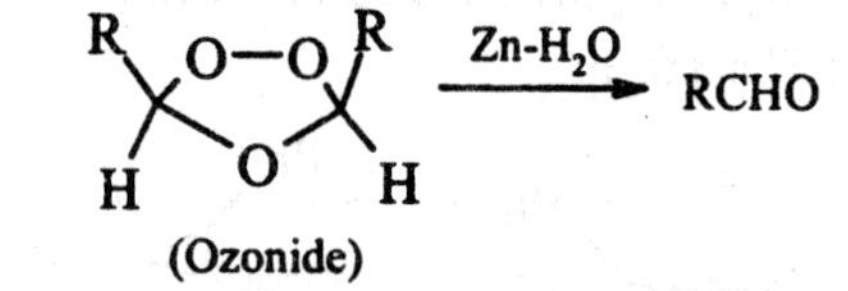

$$\text{(Ozonide)} \xrightarrow{Zn\text{-}H_2O} RCHO$$

Here, in place of Zn-H_2O, Zn-dimethylsulfide [$(CH_3)_2S$] can also be used to effect the reduction.

Oxidation of ozonide: Ozonide is oxidised by means of silver oxide, H_2O_2 or per acid, thereby producing acids and/or ketones as:

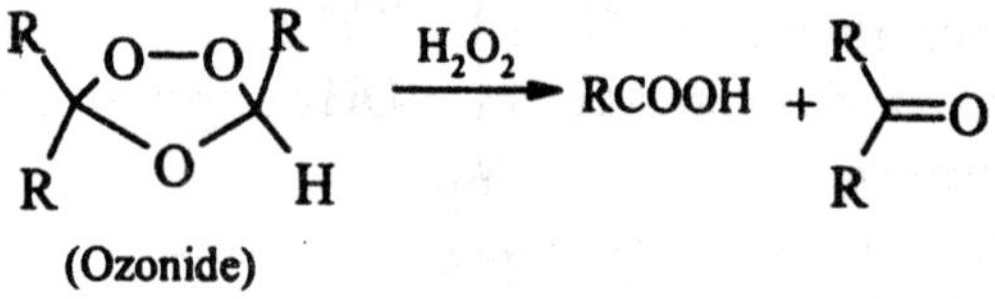

$$\text{(Ozonide)} \xrightarrow{H_2O_2} RCOOH + R_2C{=}O$$

Reduction of ozonide in presence of $LiAlH_4$ or $NaBH_4$: Alcohols are obtained.

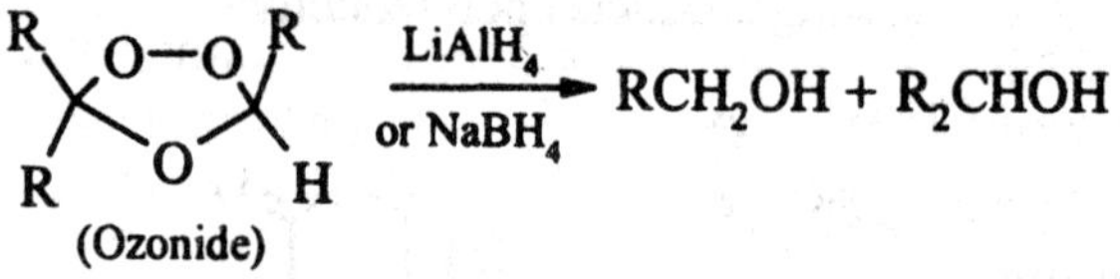

$$\text{(Ozonide)} \xrightarrow[\text{or } NaBH_4]{LiAlH_4} RCH_2OH + R_2CHOH$$

Ozonolysis is a very important reaction, used for detection of position of double bonds.

Example 16. Reductive ozonolysis of an alkene yields an equimolar amount of cyclohexancarbaldehyde and 2-butanone. Determine structure of original alkene.

Solution: We can reconstruct the alkene by removing the two oxygen atoms of carbonyl group and connecting remaining carbon atoms with double bond. One uncertainty remains, however, the original alkene might be either of two possible geometries.

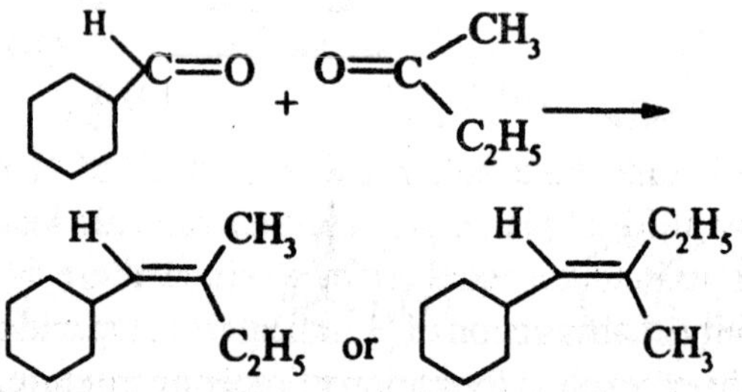

$$C_6H_{11}CH{=}O + O{=}C(CH_3)(C_2H_5) \longrightarrow$$

$$C_6H_{11}(H)C{=}C(CH_3)(C_2H_5) \text{ or } C_6H_{11}(H)C{=}C(C_2H_5)(CH_3)$$

Example 17: A hydrocarbon ($C_{14}H_{12}$) exist in two geometric forms A and B. Both A and B on ozonolysis followed by work-up with dimethylsulphide produced two molecules of benzaldehyde. Treatment of A with cold, dilute and alkaline solution of $KMnO_4$ results in a meso compound C whereas similar treatment of B results a racemic mixture. Identify A, B and C.

Solution: It is evident from the ozonolysis reaction that A and B are geometrical isomers of $C_6H_5CH{=}CHC_6H_5$. We know that cold, dilute and alkaline solution of $KMnO_4$ causes syn hydroxylation and *cis*-alkene gives meso compound. Therefore:

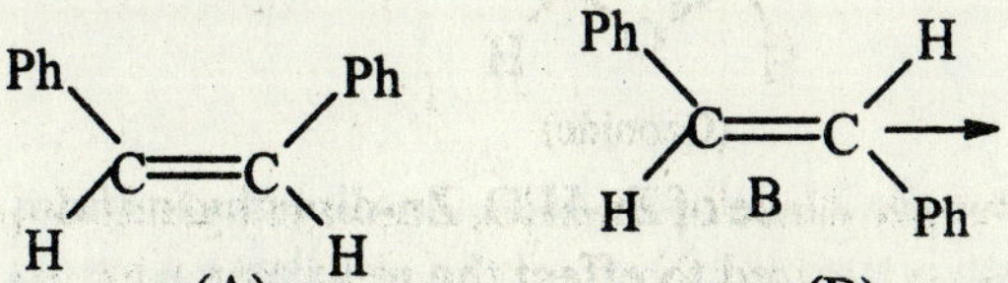

Ph, Ph, H, H: *cis* (A)   Ph, H, H, Ph: *trans-* (B)

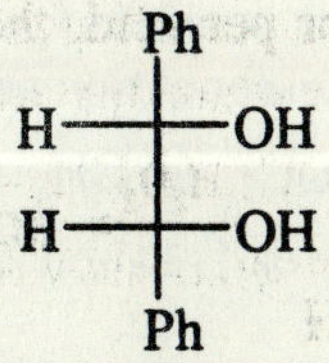

(C): Ph, H—OH, H—OH, Ph

Example 18: Give the structure of alkene that would give the following products upon reductive ozonolysis.

(a) 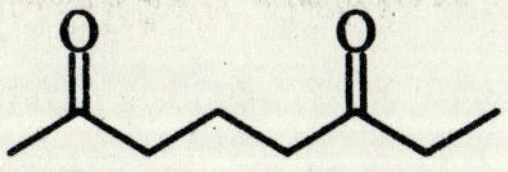 (b) 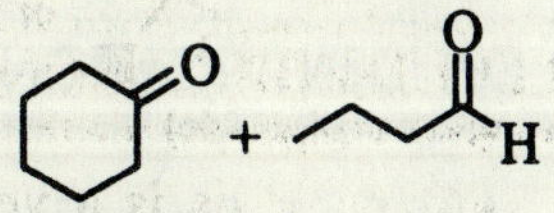 + H (c) 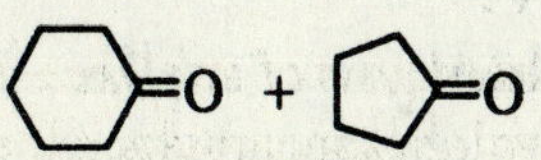 =O + =O (d) 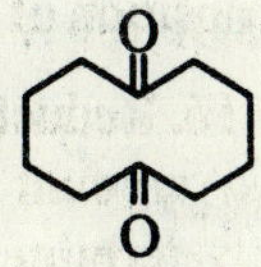 O O

(e) 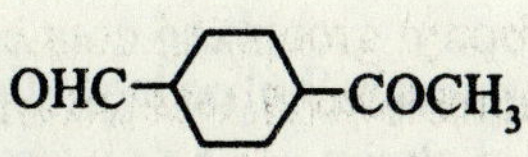 OHC—⬡—$COCH_3$ (f) 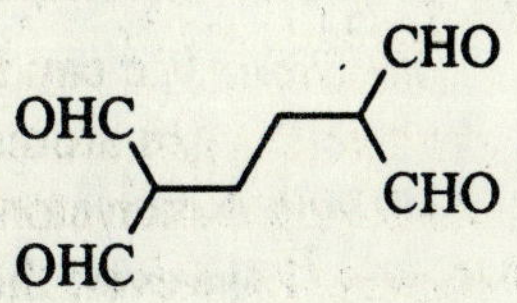 OHC, OHC, CHO, CHO

Solution

(a) 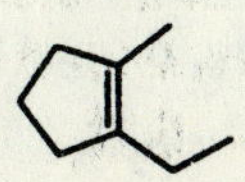 (b) 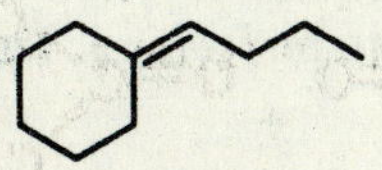 (c) 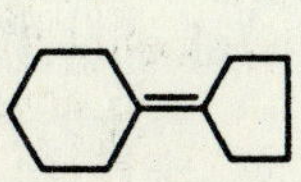 (d) 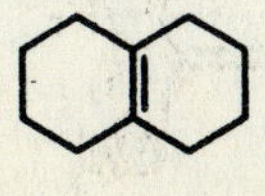 (e) 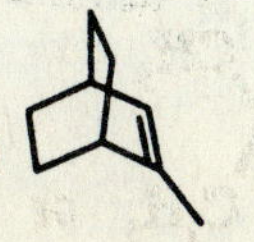 (f)

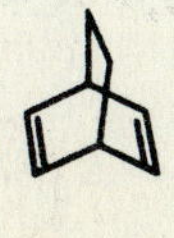

4.5.10 Oxidative Cleavage of Alkene with $KMnO_4$

If an alkene is treated with warm or acidic solution of $KMnO_4$, oxidative cleavage occurs at double bond forming ketones and/or acids as:

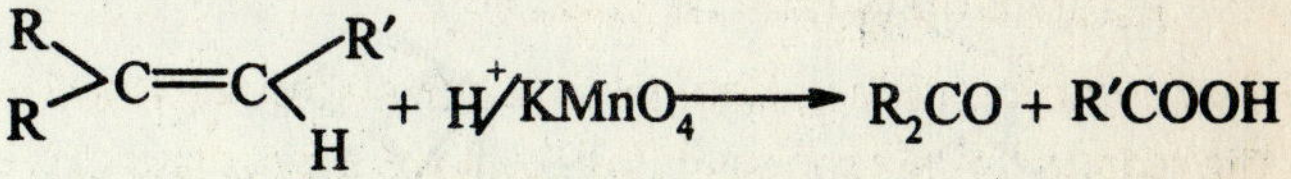

$$R_2C{=}CHR' + H^+/KMnO_4 \longrightarrow R_2CO + R'COOH$$

$CH_2{=}$ of terminal alkene is converted into CO_2 as:

$$CH_3{-}CH{=}CH_2 \rightarrow CH_3COOH + CO_2$$

4.5.11 Isomerization of Alkene

Alkenes, when heated to 500°C to 600°C or treated with a Lewis acid catalyst at lower temperature, isomerizes to yield another alkene as:

$$CH_3CH_2CH{=}CH_2 + AlCl_3 \rightarrow CH_3CH{=}CHCH_3$$

4.5.12 Addition of Carbenes on Double Bond

Carbene, being an electron deficient species (Lewis acid) adds to an alkene forming cyclic compound as:

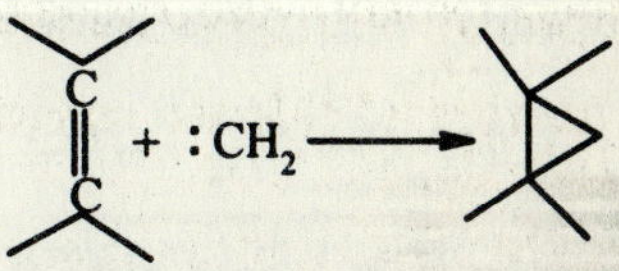

$$>C{=}C< + :CH_2 \longrightarrow \text{cyclopropane}$$

Carbenes are generated according to the following reaction:

(a) Simmon-Smith reaction

$$CH_2I_2 + Zn(Cu) \rightarrow ICH_2ZnI$$

when ICH_2ZnI is treated with an alkene, adds CH_2 leaving behind ZnI_2.

(b) By thermal decomposition of diazomethane

$$CH_2N_2 \xrightarrow{\Delta} :CH_2 + N_2$$

(c) By alpha elimination: When a haloform, CHX_3 is treated with a strong base, e.g. alcoholic solution of KOH, dihalocarbene is formed as:

$$CHBr_3 + KOH \rightarrow CBr_2 + KBr + H_2O$$

Example 19. Predict major products of the following reaction:

(a) 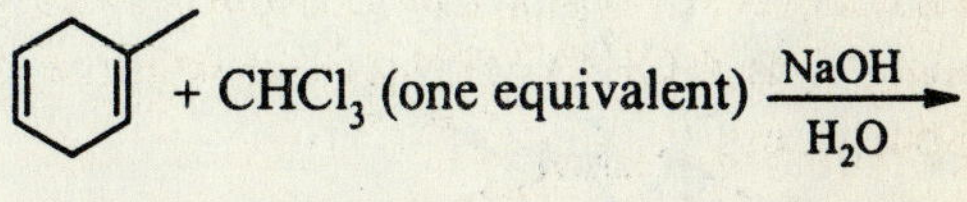 + $CHCl_3$ (one equivalent) $\xrightarrow[H_2O]{NaOH}$

(b) 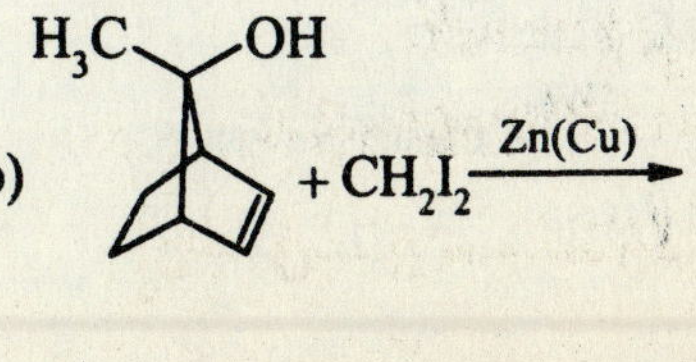 H_3C, OH + CH_2I_2 $\xrightarrow{Zn(Cu)}$ (c) 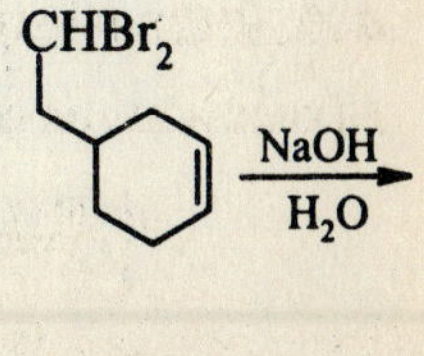 $CHBr_2$ $\xrightarrow[H_2O]{NaOH}$

Solution: (a) CCl_2 will add to more electrophilic double bond:

(a) [methylcyclohexadiene] + CCl_2 → [dichlorocyclopropane adduct with Cl, Cl]

(b) [bicyclic structure with H_3C, OH] (c) [structure with Br]

4.5.13 Reaction with Acetic Anhydride, Acid Chloride and α-haloether

$$CH_2{=}CH_2 + (CH_3CO)_2O \xrightarrow{ZnCl_2} CH_2{=}CHCOCH_3$$

$$CH_2{=}CH_2 + CH_3COCl \xrightarrow{AlCl_3} CH_3COCH_2CH_2Cl$$

$$CH_2{=}CH_2 + CH_3OCH_2Cl \xrightarrow{AlCl_3} CH_3OCH_2CH_2CH_2Cl$$

4.6.14 Alkylation and Dimerization of Alkenes

In presence of acid catalyst some alkenes either alkylate to alkene containing a tertiary hydrogen and some dimerizes as:

(a) Alkylation: $(CH_3)_2C{=}CH_2 + H^+ \rightarrow (CH_3)_3C^+$

$$\xrightarrow{(CH_3)_2C{=}CH_2} (CH_3)_3C{-}CH_2C^+(CH_3)_2 \xrightarrow{H^+} (CH_3)_3CCH_2CH(CH_3)_2$$

(b) Dimerization

$$Me_2C{=}CH_2 + H^+ \rightarrow Me_3C^+ \xrightarrow{Me_2C{=}CH_2}$$

$$Me_3CCH_2C^+Me_2 \xrightarrow{-H^+} Me_3CCH{=}CMe_2$$

Example 20. Predict products in the following reactions:

(a) Propene + $CH_3COCl/AlCl_3$ →

(b) $CH_3{-}CH{=}C(CH_3)_2 + CH_3COCH_2Cl/AlCl_3$ →

Solution: (a) $CH_3{-}CHClCH_2COCH_3$

(b) $CH_3{-}CH(CH_2OCH_3){-}C(CH_3)_2Cl$

Example 21. Why α-haloether in presence of a Lewis acid catalyst adds effectively to an alkene whereas β and γ haloether are very less reactive in this reaction?

Solution: α-haloether combines with a Lewis acid, e.g. $AlCl_3$ forming a stable carbocation as:

$$CH_3{-}O{-}CH_2Cl + AlCl_3 \rightarrow CH_3{-}O{-}CH_2^+ + AlCl_4^-$$

Here the intermediate carbocation is resonance stabilized as:

$$\left[CH_3{-}\ddot{O}{-}\overset{+}{C}H_2 \longleftrightarrow CH_3{-}\overset{+}{O}{=}CH_2\right]$$

No such resonance stabilization is possible with carbocation formed from β or γ haloether.

SOLVED PROBLEMS BASED ON IDENTIFICATION

Example 22. Alkenes (A) and (B) are isomeric and yields same alcohol (C) on hydration. On vigorous oxidation with $KMnO_4$, (A) gives a carbonyl compound (D) and an acid (E) each containing four carbon atoms. On the other hand, (B) gives an acid (F) and a carbonyl compound (G). Identify A to G considering no two identical groups are attached to carbon in G.

Solution: Form the oxidation of A, it is evident that arrangement at double bond is:

[H(C)C=C(C)(C), cleavage: Acid product (B), Carbonyl product (C)]

Also for A, four carbons are present on each side of the double bond, therefore A must be:

$$(CH_3)(C_2H_5)C{=}CH(CH_2CH_2CH_3) \longrightarrow H_3C{-}C(C_4H_9)(C_2H_5){-}OH$$

(A) (C)

Also both A and B produces same alcohol on hydration, therefore B also have the same skeleton of carbon chain, but different position of double bond. Also double bond cannot be moved away more than one carbon, because same alcohol (C) has to be formed. Therefore, B must be:

$$H_3C{-}CH{=}C(CH_3){-}C_4H_9 \longrightarrow CH_3CH_2{-}C(C_4H_9)(CH_3){-}OH$$

(B) (C)

⇒ D = 2-butanone, E = n-butanoic acid, F = Acetic acid, G = 2-hexanone.

Example 23. An organic compound $A(C_9H_{18})$ is enantiomeric, and decolourize bromine water. A on ozonolysis followed by work-up with H_2O_2 produces an acid $B(C_5H_{10}O_2)$ and a carbonyl compound $C(C_4H_8O)$. Compound B is still enantiomeric but C is optically inactive. C on reduction with $LiAlH_4$ produced $D(C_4H_{10}O)$, which is enantiomeric. Identify A to D.

Solution: Compound A is enantiomeric and decolourize bromine water, it confirms that A must contain one and only one double bond. Ozonolysis product indicates that the arrangement at C=C is:

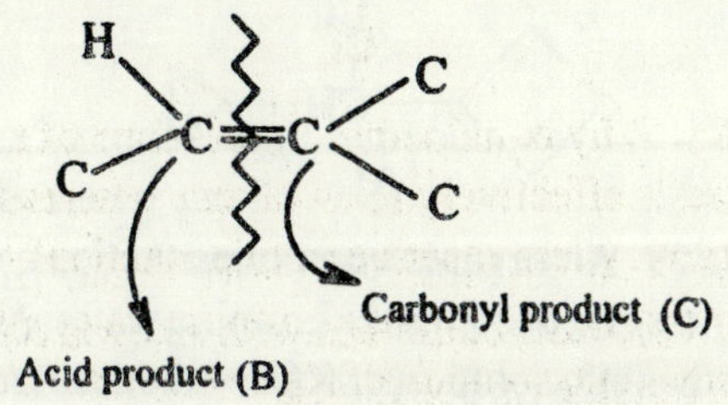

H, C on left carbon; C, C on right carbon of C=C

Carbonyl product (C)

Acid product (B)

Carbonyl compound must be unsymmetrical ketone, since it becomes optically active on reduction at >C=O. Also C contains four carbon, it must be:

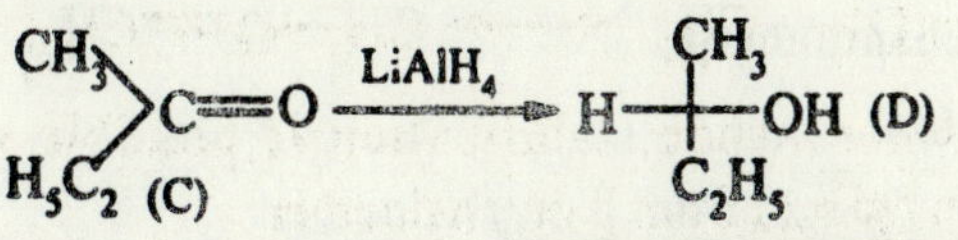

$$\underset{(C)}{CH_3(H_5C_2)C{=}O} \xrightarrow{LiAlH_4} H-C(CH_3)(C_2H_5)-OH \ (D)$$

Exist as enantiomers

Now acid contains four carbons and exist as enantiomers, it must contain a chiral carbon as: 2-methylbutanoic acid (B). Therefore, the starting compound A can be derived by combining the ozonolysis products B and C as:

2-methyl butanoic acid (B) + 2-butanone →

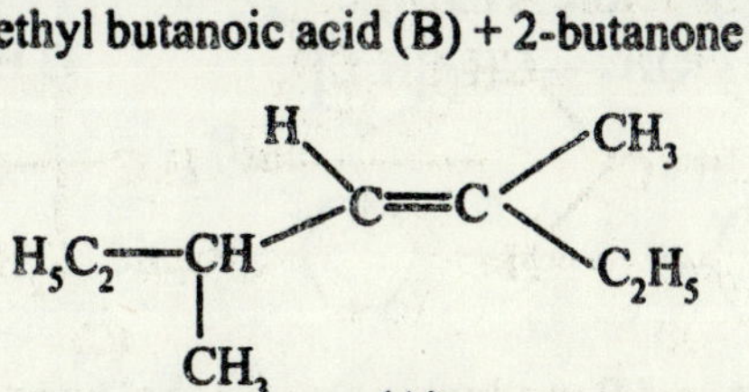

$$H_5C_2-CH(CH_3)-CH{=}C(CH_3)(C_2H_5) \quad (A)$$

Example 24. A hydrocarbon $A(C_6H_{10})$ on reduction first gives $B(C_6H_{12})$ and finally $C(C_6H_{14})$. A on ozonolysis followed by work-up with Zn-H_2O gives two molecules of aldehydes $C_2H_4O(D)$ and one molecule of aldehyde $E(C_2H_2O_2)$. Oxidation of B with acidified $KMnO_4$ gives an acid $F(C_4H_8O_2)$. Determine structures of A to F with proper reasoning.

Solution: Ozonolysis products shows that A is a symmetrical diene:

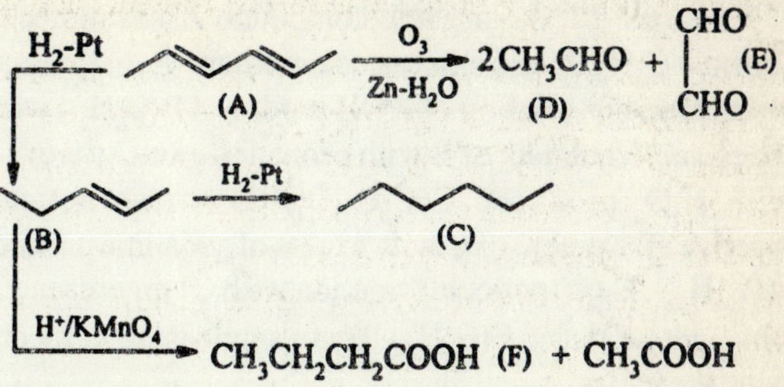

$$\text{(A)} \xrightarrow[Zn-H_2O]{O_3} 2CH_3CHO\ (D) + \underset{CHO}{\overset{CHO}{|}}\ (E)$$

(A) $\xrightarrow{H_2-Pt}$ (B) $\xrightarrow{H_2-Pt}$ (C)

$$\text{(B)} \xrightarrow{H^+/KMnO_4} CH_3CH_2CH_2COOH\ (F) + CH_3COOH$$

Example 25. A hydrocarbon exist in two stereomeric forms (A) and (B) with their molecular formula C_8H_{16}. A on treatment with cold, dilute and alkaline solution of $KMnO_4$ produces $C_8H_{18}O_2$(C), which is a meso form. B on the other hand, on similar treatment produces racemic mixture which is isomeric to C. Either A or B on treatment with O_3 followed by work with H_2O_2 produces $D(C_4H_8O)$ as the only product. Identify A to D.

Solution: Cold, dilute, alkaline $KMnO_4$ solution adds on C=C producing *cis*-diols. One of the stereomer A produces meso *cis*-diol, therefore it is confirmed that the starting compound is a symmetrical alkene.

Also oxidative ozonolysis of the starting compound gives no acid as product, it indicates that there is no hydrogen attached to doubly bonded carbon. From these finding it is now concluded that the starting compound has following structural formula-3, 4-dimethyl-3-hexene.

Also, A + Cold, dil-$KMnO_4$ → C(Meso) therefore A is *cis* isomer and B is thus *trans* one. Thus, the various compounds are: A: *cis*-3, 4-dimethyl-3-hexene. B: *trans*-3, 4-dimethyl-3-hexene. C: Meso-3, 4-dimethyl butane 3, 4-diol. D: 2-pentanone.

Example 26. A hydrocarbon has molecular formula $A(C_5H_{10})$ and known to decolourize the Bayer's reagent. A on treatment with B_2H_6 followed by work-up with H_2O_2/NaOH produces $B(C_5H_{12}O)$, which exist as enantiomers. B on treatment with HI in presence of red-P yields $C(C_5H_{12})$ which is optically inactive. Identify A to C.

Solution: A is an alkene with one double bond. B is an alcohol obtained by anti-Markovnikov's hydration of alkene A. Also, B is optically active, i.e. it contains a chiral carbon, which becomes achiral on reduction with HI/ red-P, it indicates that chiral carbon is attached with CH_3— and —CH_2OH, latter becomes —CH_3 after reduction, making C optically inactive. Therefore, B must be:

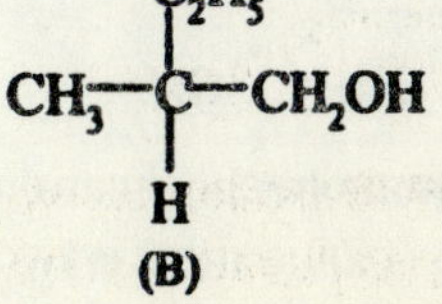

$$\underset{(B)}{CH_3-CH(C_2H_5)-CH_2OH} \quad \text{and A must be} \quad \underset{(A)}{(C_2H_5)(H_3C)C{=}CH_2}$$

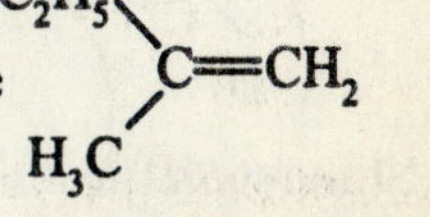

(A) and C is Methyl butane

Example 27. An organic compound A has molecular formula $C_8H_{13}Br$ and exist as enantiomers. A on treatment with ethanolic solution of KOH produced $B(C_8H_{12})$ which is achiral. Treatment of B with ozone followed by work-up with H_2O_2 produced $C(C_8H_{12}O_4)$. C is also optically inactive. Treatment of C with excess of sodalime produced D (C_6H_{12}). D on monochlorination with Cl_2 in presence of sun-light produced $E(C_6H_{11}Cl)$ as a single isomer. Identify A to E.

Solution: Working in backword direction, E must be a cyclohexylchloride, and therefore D is cyclohexane. D is obtained by decarboxylation of an acid C. Also C is optically inactive, indicates that C must be a symmetrical diacid as:

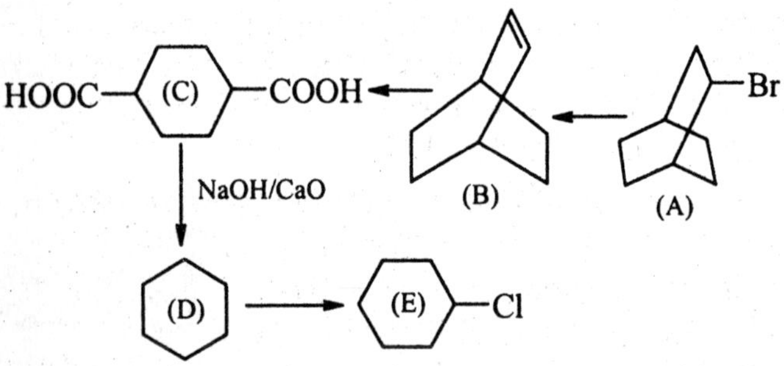

HOOC–(C)–COOH ← (B) ← (A) Br

NaOH/CaO

(D) ⟶ (E)–Cl

Example 28. An organic chloride $A(C_8H_{13}Cl)$ exist as enantiomers and does not decolourize Bayer's reagent. A on treatment with ethanolic solution of KOH produces $B(C_8H_{12})$ as major product, which does not show enantiomerism. B on ozonolysis followed by work-up with H_2O_2 gives two compounds $C(C_4H_6O)$ and $D(C_4H_6O_2)$, neither of them decolourize brown colour of bromine water. C on monochlorination in presence of sun light produces two isomeric products E and F with their molecular formula C_4H_5ClO of which only E is enantiomeric. On the other-hand, D on heating with NaOH/CaO, produced $G(C_3H_6)$. G does not decolourize bromine water and gives single monochlorination product. Identify A to G.

Solution: Working in reverse direction, G must be cyclopropane and therefore D must be cyclopropyl carboxylic acid. Ozonolysis followed by oxidative cleavage of B produced an acid D and a carbonyl compound C, later must be a ketone. Also C does not decolourise bromine water, it must be a cyclic ketone. Therefore, the possible structures of C are:

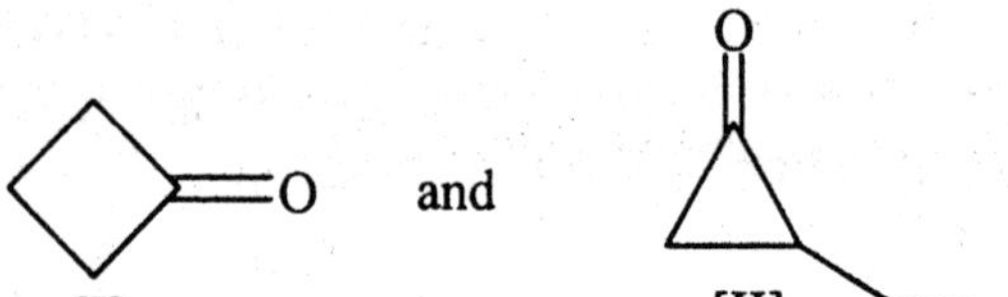

=O and O, CH_3

[I] [II]

Structure [II] is ruled out since it on monochlorination produces three isomeric halide whereas requirement is two which is possible with [I] as:

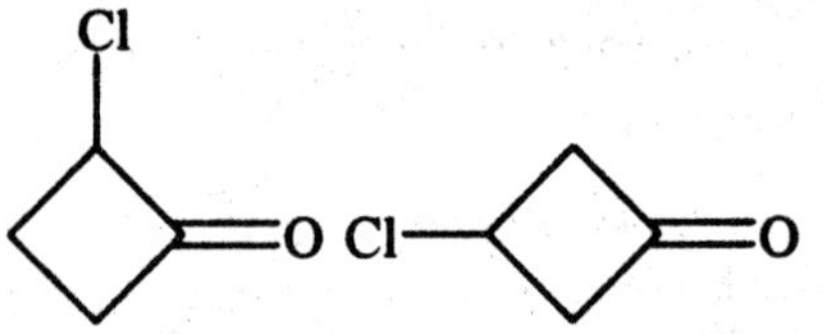

Cl, =O Cl–, =O

[III] [IV]

Structure [III] contains a chiral carbon and can exist as enantiomers whereas [IV] does not. Therefore, E is [III] and F is [IV] and C is [I]. Combining C and D gives-B as:

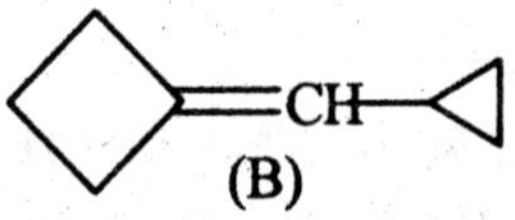

=CH–

(B)

B is obtained by dehydrohalogenation of enantiomeric A, it must be:

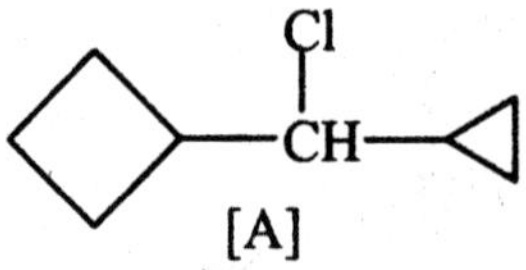

Cl, CH–

[A]

Example 29. An organic compound $A(C_{13}H_{23}Cl)$ exist as distereomers and decolourize bromine water. A on treatment with ethanolic solution of KOH produces isomeric B and C with their molecular formula $C_{13}H_{22}$. Treatment of either B or C with Rany Nickel produces 4-isopropyl-1-tertiarybutyl cyclohexane. A on oxidative ozonolysis gives acetone as one product. Identify A, B and C considering C to be enantiomeric.

Solution

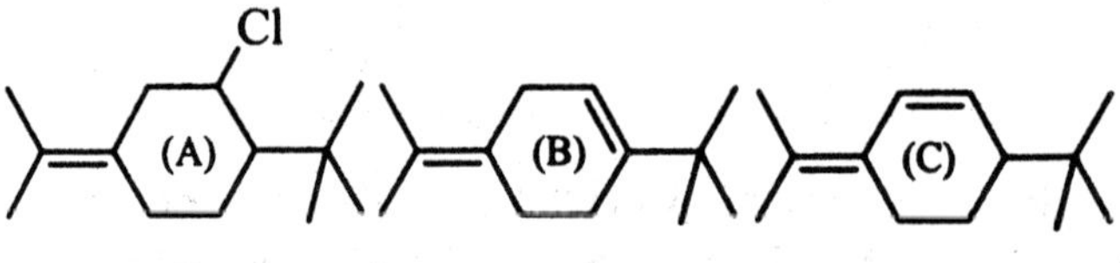

Cl

(A) (B) (C)

Example 30. Bring about the following conversions:

(a)  OH ⟶ OH

(b) 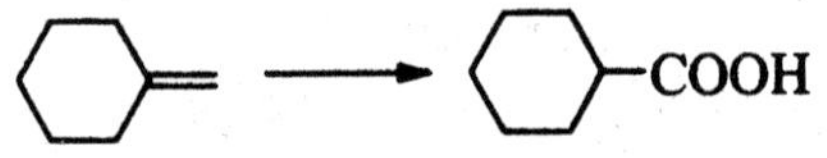 ⟶ COOH

(c)  ⟶

(d) 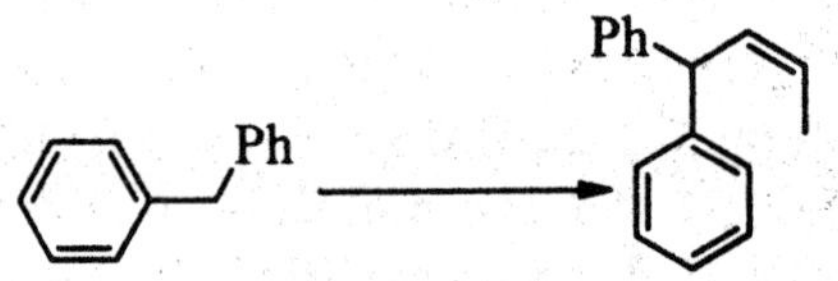 Ph ⟶ Ph

(e) 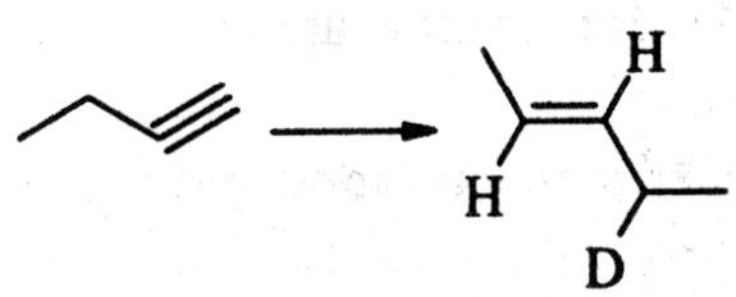 ⟶ H, H, D

(f) [phenylacetylene] → [3-phenyl-1-butene]

(g) [bromonorbornene] → [bromolactone]

(h) [diene] → $BrCH_2$–(tetrahydrofuran)–CH_2Br

Solution

(a) 1-methylcyclohexanol $\xrightarrow[\text{Heat}]{H_2SO_4}$ 1-methylcyclohexene $\xrightarrow[\text{NaOH}]{B_2H_6/H_2O_2}$ trans-2-methylcyclohexanol (OH)

(b) methylenecyclohexane $\xrightarrow[\text{NaOH}]{B_2H_6/H_2O_2}$ cyclohexylmethanol (OH) $\xrightarrow[H^+]{KMnO_4}$ Product

(c) methylcyclohexane $\xrightarrow[\text{Heat}]{Br_2}$ 1-bromo-1-methylcyclohexane (Br) $\xrightarrow[\text{Alcohol}]{KOH}$

1-methylcyclohexene $\xrightarrow{Br_2}$ (CH_3, Br, Br) $\xrightarrow[\text{Alcohol}]{KOH}$ diene

(d) (Ph) toluene derivative $\xrightarrow[h\nu]{Br_2}$ (Ph)(Br) $\xrightarrow{\equiv C^- Na^+}$

Ph–alkyne $\xrightarrow{Pd/H_2/BaSO_4}$ (Ph, H, H) cis-alkene

(e) 1-butyne $\xrightarrow[(2)\ CH_3Br]{(1)\ Na/heat}$ 2-pentyne $\xrightarrow{Na/liq\ NH_3}$

(H, H) trans-alkene $\xrightarrow{NBS}$ (H, Br, H) $\xrightarrow[D_2O]{Mg/ether}$ (H, D, H)

(f) phenylacetylene $\xrightarrow{Na/liq\ NH_3}$ $\xrightarrow{BrH}$

(1-bromoethyl)benzene (Br) $\xrightarrow{(CH_2{=}CH)_2Cu}$ 3-phenyl-1-butene

(g) (Br) bromonorbornene $\xrightarrow{KCN}$ $\xrightarrow[OH_2]{H^+}$ HOOC-norbornene $\xrightarrow{Br_2}$

HOOC–(bromonium ion, Br^+) $\xrightarrow{-H^+}$ bromolactone (Br, O, O)

(h) diene $\xrightarrow[OH_2]{Br_2}$ Product

PRACTICE EXERCISE II

1. (a) Briefly explain the origin of barrier of rotation about C2-C3 bond in 1,3-butadiene.
   (b) Which of the following has the highest and lowest barrier to rotation?

   [diene conformers] ⇌ [diene conformers] vs [enol, OH, O] ⇌ [enol, OH, O]

2. Rank the following molecule in term of their reactivity with HBr:

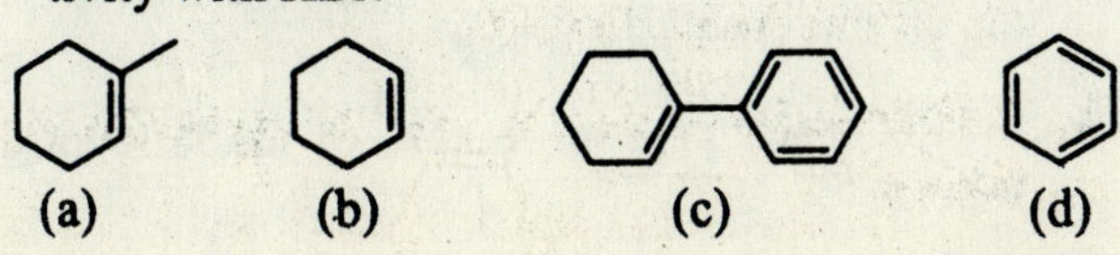

3. (a) Propose mechanism of formation of products in the following reaction:

   [2-(1,1-dimethylallyl)phenol, OH] $\xrightarrow{H_3O^+}$ [dihydrobenzofuran, O] + [HO, OH] + [OH]

   (b) Arrange the following alkenes in increasing order of their heat of hydrogenation explaining reason for your choice:

   (I) (II) (III) (IV) (V)

4. Bring about the following transformations:

(a)

(b)

(c)

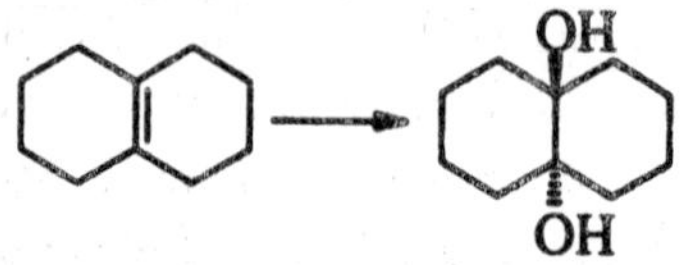

5. Predict major product in the following reaction describing mechanism of their formation:

(a) $C_6H_5-CH(CH_3)-CH=CH_2 + HCl \longrightarrow$

(b) $C_6H_5-C(CH_3)_2-CH=CH_2 + ICl \longrightarrow$

(c) $C_6H_5-C(C_6H_5)(CH_3)-CH=CH_2 + BrCl \longrightarrow$

(d) $CH_3CH=C(CH_3)_2 + CH_3Cl \longrightarrow$

(e) $CH_3CH=C(CH_3)_2 + CH_2ICl \longrightarrow$

6. Propose mechanism of formation of product in the following reactions:

(a) 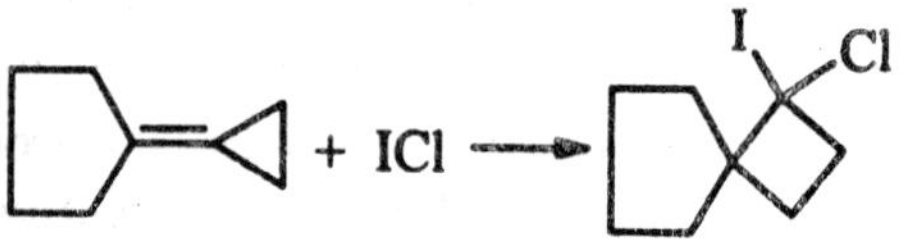 + ICl →

(b) 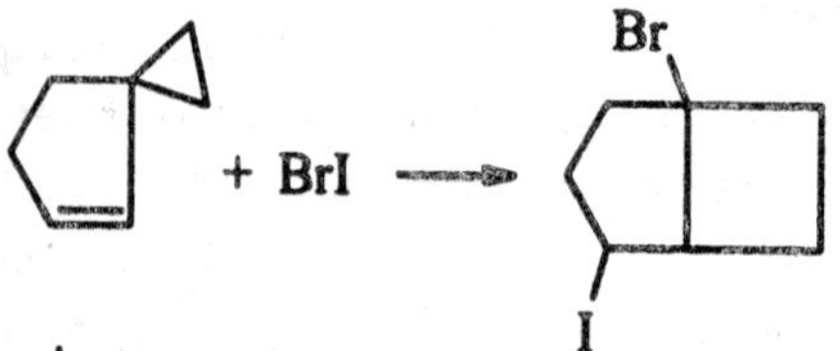 + BrI →

(c) 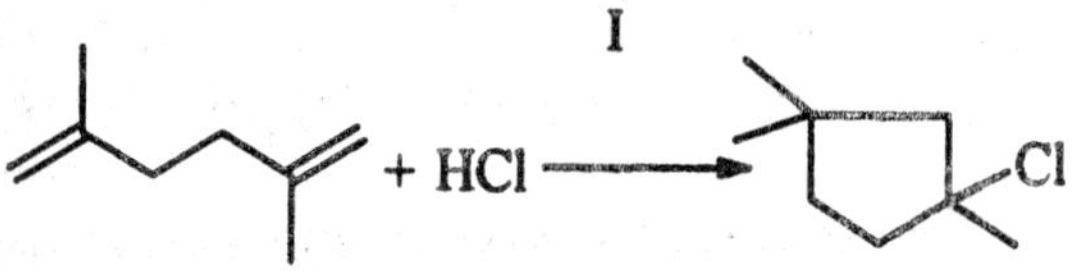 + HCl →

(d) 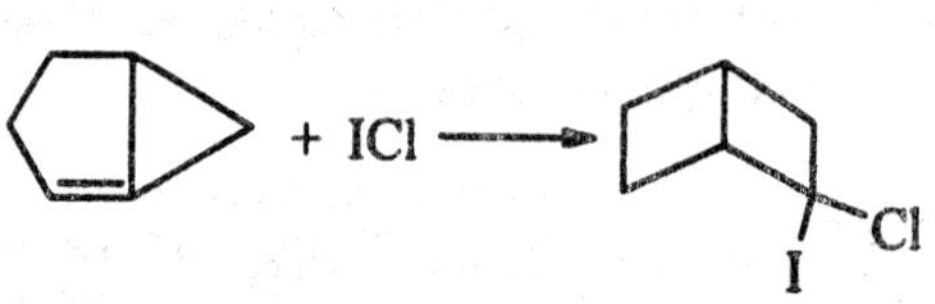 + ICl →

7. A hydrocarbon $A(C_5H_{10})$ on addition of HBr in presence of a peroxide forms $C_5H_{11}Br(B)$, which is found to be resolvable into enantiomers. A on addition of HBr in aqueous medium produces C, which is an isomer of B but cannot be resolved. A on reductive ozonolysis produced $C_4H_8O(D)$ as one product. D on catalytic reduction with $NaBH_4$ produced $E(C_4H_{10}O)$ which is resolvable. E on dehydration with concentrated sulfuric acid produced $F(C_4H_8)$, which on ozonolysis followed by work-up with $Zn-H_2O$ produced two molecules of ethanal. Identify A to F.

8. Predict major addition product when the following addition reaction carried out in presence of a peroxide:

(a) $CH_3CH=C(CH_3)_2 + HBr \longrightarrow$

(b)  + HBr →

(c) $CH_3-CH(CH_3)-CH=CH_2 + CHCl_3 \longrightarrow$

(d) $C_6H_5-CH=CH-CH_3 + CCl_4 \longrightarrow$

(e) $C_6H_5-C(CH_3)=C(C_6H_5)_2 + ClCl_3 \longrightarrow$

9. Discuss the stereochemistry of addition of DBr to cyclopentene.

10. Propose the mechanism of formation of the products in the following reaction explaining indicated relative proportions:

(a) 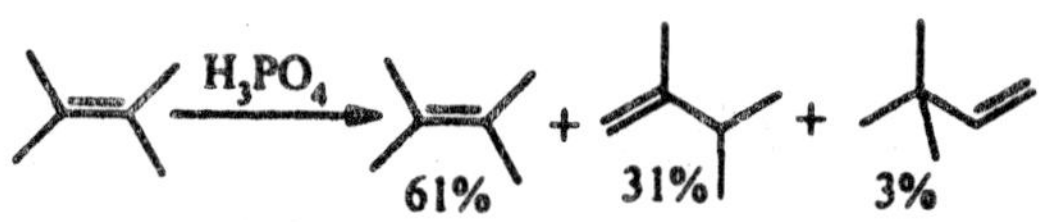 $\xrightarrow{H_3PO_4}$ 61% + 31% + 3%

(b)  $\xrightarrow{BrH}$

11. 2-methyl-1-phenyl-1-propene, when treated with N-bromo succinamide and water, gives a mixture of two halohydrines in roughly equal amounts. What are the structures of products? What factor is responsible for lack of regioselectivity in this case?

12. *Trans*-1-bromo-2-chlorocyclopentane is synthesized in laboratory by treating cyclopentene with bromine and dry HCl. Write the mechanism of reaction showing proper stereochemistry.

13. Predict major product in each of the following reactions:

(a) 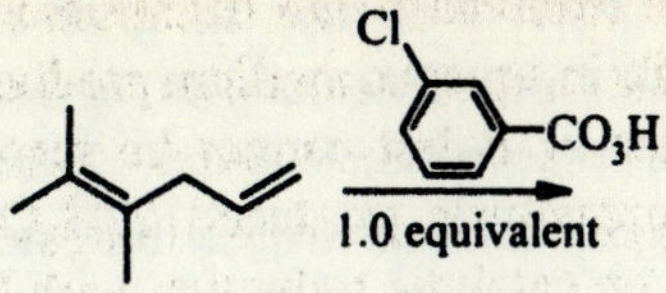 [alkene] $\xrightarrow[\text{1.0 equivalent}]{\text{m-}ClC_6H_4CO_3H}$

(b) 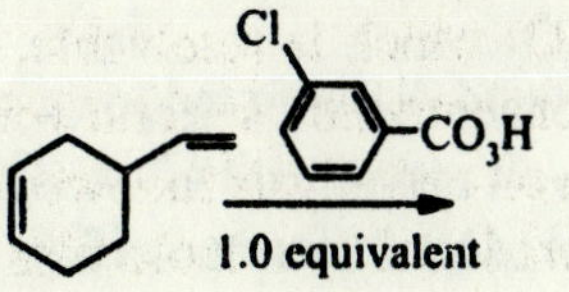 [3-vinylcyclohexene] $\xrightarrow[\text{1.0 equivalent}]{\text{m-}ClC_6H_4CO_3H}$

(c) 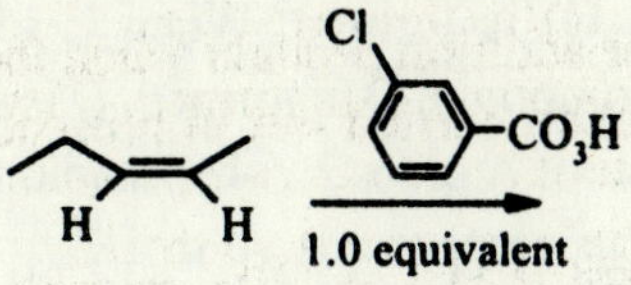 [cis alkene, H H] $\xrightarrow[\text{1.0 equivalent}]{\text{m-}ClC_6H_4CO_3H}$

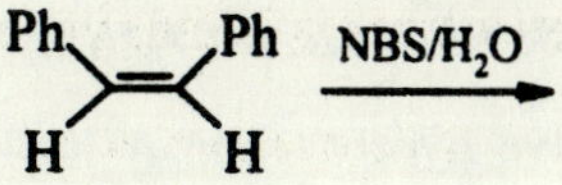

(d) 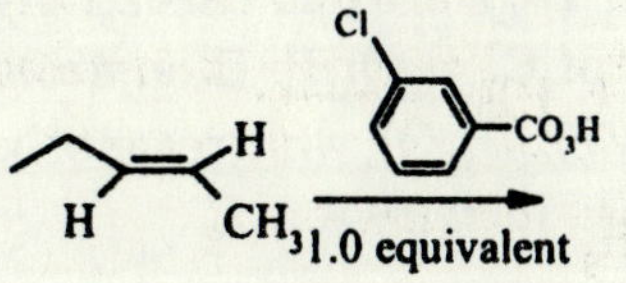 [alkene, H, H, CH_3] $\xrightarrow[\text{1.0 equivalent}]{\text{m-}ClC_6H_4CO_3H}$

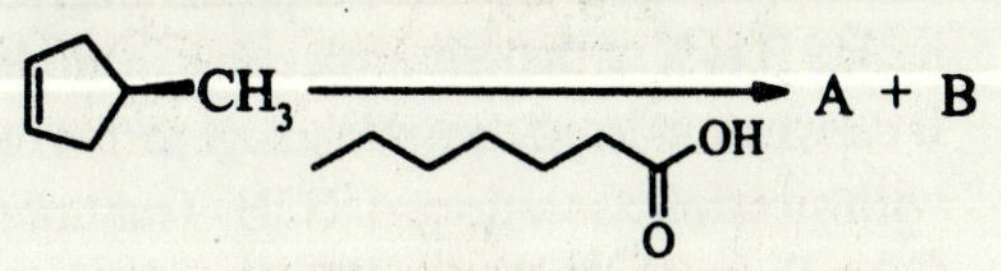

14. Identify the products in the following reactions (mention stereochemistry where applicable).

(a) 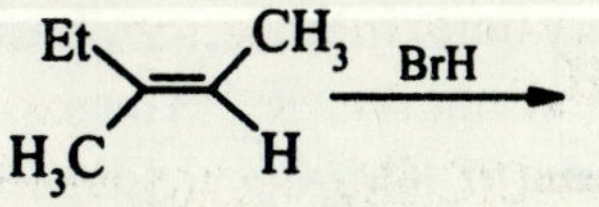 Et(H_3C)C=C(CH_3)(H) $\xrightarrow{BrH}$

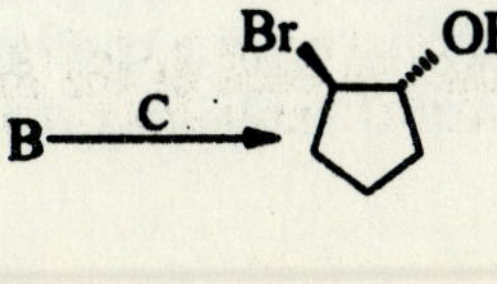

(b) 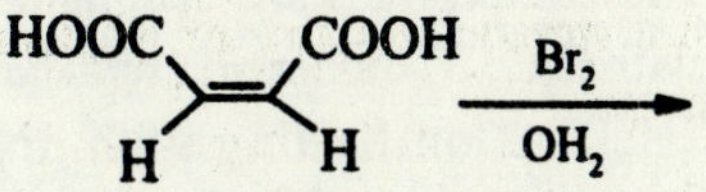 HOOC(H)C=C(COOH)(H) $\xrightarrow[OH_2]{Br_2}$

(c) Ph(H)C=C(Ph)(H) $\xrightarrow{NBS/H_2O}$

(d) [3-methylcyclopentene] $\xrightarrow{\text{peroxy acid (C(=O)OOH)}}$ A + B

(e) H(H_3C)C=C(H)CH$_2$CH$_2$OC(=O)CH$_3$ $\xrightarrow{BD_3}$ C + D $\xrightarrow[NaOH]{H_2O_2}$ E + F

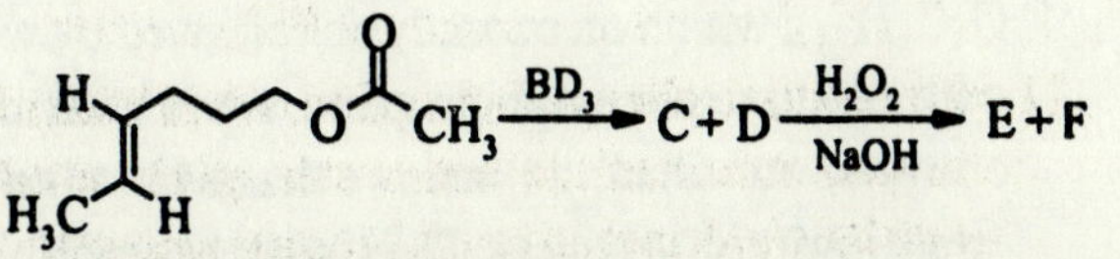

15. Fill in the structural formula for starting materials or reagents required for the following transformations:

(a) 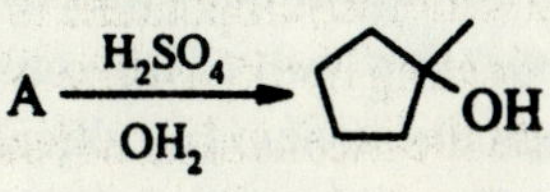 A $\xrightarrow[OH_2]{H_2SO_4}$ [1-methylcyclopentanol, OH]

(b) B $\xrightarrow{C}$ [trans-2-bromocyclopentanol, Br, OH]

(c) D $\xrightarrow{BD_3}$ $\xrightarrow[NaOH]{H_2O_2}$ [D, OH product]

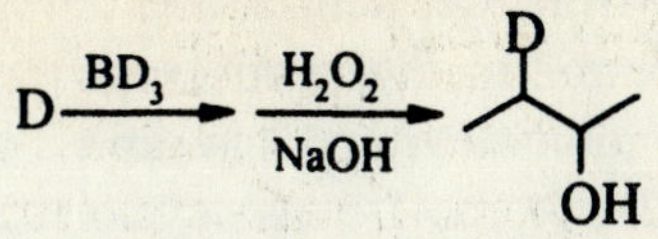

16. Arrange each of the following series of alkenes in order of increasing reactivity towards acid catalyzed hydration and explain the reason for your order:
(a) 1-hexene, 2-methyl-1-pentene, 2-hexene.
(b) 2-methyl propene, *cis*-2-butene, *trans*-2-butene.
(c) 1-phenyl-1-butene, 1-phenyl-2-butene, 2-phenyl-1-butene.

17. When 1-pentene is treated with an acid, *trans*-2-pentene is produced. Propose mechanism of this isomerization reaction and justify.

18. (a) Discuss the mechanism of addition of HBr_2/peroxide on 1,3-butadiene.
(b) Discuss the mechanism of reaction of aqueous solution of chlorine with *trans*-2-butene.
(c) Explain why 2-butyne is less reactive than butene towards electrophilic addition reaction, but reverse is true for hydrogenation reaction.

19. Provide chemical tests that can be used to distinguishes between the following pairs:
(a) 1-hexene and 1-hexyne.
(b) 2-hexene and 1-hexanol.
(c) 2-hexene and 2-bromo hexane.

20. A mixture of two alcohols 3-methyl-1-butanol and 2-methyl-1-butanol, gives chiefly 2-methyl-2-butene when dehydrated with acid. 2-methyl-2-butene is converted into 2-methyl-2-butanol with approximately 90% yield by treatment with 50% aqueous solution of sulphuric acid at 0°C. Write equations showing mechanism of conversions.

21. Compound A($C_{14}H_{12}$) could be hydrogenated to B($C_{14}H_{14}$) using one molecule of H_2 at 25°C over platinum catalyst. Treatment of A and B with hot $KMnO_4/H_2O$ led to benzoic acid as only product. Similarly ozonolysis of A, followed by reductive work-up gave benzaldehyde as only isolable organic product. Reaction of A with catalytic amount of HBr/Peroxide mixture led to C, an isomer of A, controlled hydrogenation of which gave B. Also C is more stable than A. Identify A to C.

22. Give all possible products of reaction of 3-methylmethylenecyclohexane with NBS.

23. Provide reaction for converting 1-methylcyclohexanol to:

(a) 1-bromo-1-methylcyclohexane.
(b) *cis*-1-bromo-2-methylcyclohexane.
(c) *trans*-2-bromomethylcyclohexane.
(d) *trans*-1, 2-dibromo-1-methylcyclobutane.

24. Devise a synthesis of (a) 3-bromocyclopentene and (b) 3, 5-dibromocyclopentene from cyclopentanol.

25. Three isomeric alkene A, B and C on hydrogenation yields same hydrocarbon 3-methyl pentane. Only A can exist as enantiomers, whereas B exist as *cis*-*trans* isomers. Deduce structure of A, B and C.

26. An alkene (C_6H_{12}) exist in two stereomeric forms A and B. A on treatment with cold dilute and alkaline $KMnO_4$ solution produces a meso diol, whereas B on same treatment produces racemic mixture. Deduce structure of A and B.

27. An organic compound has molecular formula A(C_7H_{10}) and known to decolourize Br_2-water solution. A on reduction with H_2/Pt produces B(C_7H_{14}). B on monochlorination with Cl_2/hv produces four isomeric monochloro derivatives $C_7H_{13}Cl$. A on ozonolysis followed by work-up with Zn-dimethyl sulphide affords two and only two products, one being methanal and other C($C_6H_8O_3$). C is oxidised with acidic solution of dichromate to yield D($C_6H_8O_5$), which is soluble in $NaHCO_3$. D is reduced with $NaBH_4$ to yield E($C_6H_{10}O_3$), which may be resolved into enantiomers. D on heating yields an optically active F(C_5HgO_3). Identify A to F.

28. Draw the structures of compounds A to C:

$(C_2H_5)_3N: + :CCl_2 \rightarrow$ A (an unstable adduct)

A $\rightarrow$ B + C_2H_4

$$\text{B} \xrightarrow{H_2O} \text{C}$$

29. Predict mechanism of formation of products in the following reactions:

(a) Ph + HCl $\longrightarrow$ (Ph, Cl)

(b) (O) + HCl $\longrightarrow$ (O, Cl)

(c) C_6H_5—CH=CH—CH_3 + HOCl $\longrightarrow$

C_6H_5—CH(OH)—CH(Cl)—CH_3

30. Cl_2 adds to *trans*-2-butene to give meso-2,3-dichlorobutane. Outline mechanism for this reaction. If the reaction is conducted in water, chlorohydrin is obtained. Suggest mechanism describing stereochemistry.

31. An organic compound A has found to have molecular formula C_7H_{12}. It reacts with dilute $KMnO_4$ and compound B($C_7H_{14}O_2$) is formed. If A is treated with perbenzoic acid in chloroform solution, compound C($C_7H_{12}O$) is formed. When C reacts with dilute H_2SO_4 compound D is formed. D is stereomer of B. Ozonolysis of A gives 2, 6-heptandione. Assign structures to the compounds A to D.

32. An alkene A($C_{10}H_{20}$) on oxidative ozonolysis yields two molecules of $C_5H_{10}O$(B). B on treatment with $LiAlH_4$ yields C($C_5H_{12}O$), which cannot be resolved into enantiomers. Identify A to C.

33. An optically active organic compound A has molecular formula $C_7H_{11}Br$, and it does not decolourize bromine water. A on treatment with potassium tertiary butoxide in tertiary butanol yields B with molecular formula C_7H_{10}. B on oxidation with acidified solution of $KMnO_4$ yields C($C_7H_{10}O_4$), which does not undergo intramolecular dehydration on heating with P_2O_5. C on heating with NaOH(CaO) gives D(C_5H_{10}). D on heating with dry bromine vapour gives E(C_5H_9Br) as only monobromo derivative. Deduce structures of A to E.

34. An optically active hydrocarbon A has molecular formula C_8H_{18}. A on monochlorination gives five alkyl halide B to F with their molecular formula $C_8H_{17}Cl$. B does not undergo dehydrohalogenation on treatment with alcoholic solution of KOH. Treatment of either C or D with alcoholic KOH yields same alkene G(C_8H_{16}), which on ozonolysis followed by work-up with Zn-dimethyl sulphide gives an optically inactive compound $C_6H_{12}O$ and ethanal. Also C is enantiomeric whereas D is distereomeric. E on dehydrohalogenation yields an alkene, which on reductive ozonolysis yields H($C_7H_{14}O$), which is optically inactive. H on treatment with $LiAlH_4$ yields I($C_7H_{16}O$) which can be resolved into enantiomers. F on dehydrohalogenation yields an alkene (C_8H_{16}), which on reductive ozonolysis yields J($C_7H_{14}O$), which is optically active and have same configuration as that of A. Identify A to J explaining the reactions involved.

35. A hydrocarbon (A) contains five carbon atoms, reacts with cold, dilute and alkaline solution of $KMnO_4$ to form B(resolvable). B on oxidation with hot concentrated $KMnO_4$ solution forms two compounds (C) and a neutral (D). C has molecular formula $C_2H_4O_2$ and turns blue litmus paper red. Compound D neither reacts with Fehling solution nor with Tollen's reagent and its formula is C_3H_6O. A on treatment with Cl_2 in CCl_4 yields another compound E($C_5H_{10}Cl_2$). E on treatment with alcoholic solution of KOH yields a stable compound F(C_5H_8). Identify A to F.

36. An alkene (A) on ozonolysis yields acetone and an aldehyde. The aldehyde is easily oxidised to an acid (B). When B is treated with Br_2/Red-P, it yields a compound (C), which on hydrolysis gives a hydroxy acid (D). This can also be obtained from acetone by reaction with HCN followed by hydrolysis. Identify A to D.

37. An organic compound A(C_8H_{14}) on treatment with H_2/Pt gives C_8H_{18}. Ozonolysis of A followed by work-up with Zn-dimethyl sulphide yields three products B, C and D. B reduces tollens reagent and gives a yellow precipitate on treatment with NaOH/I_2. C does not reduces Tollen's reagent but gives yellow precipitate with NaOH/I_2. D has molecular formula $C_2H_2O_2$ and on treatment with concentrated solution of NaOH followed by acidification of product yields E($C_2H_4O_3$). Deduce structures of A to E.

38. An organic compound (A) has molecular formula C_5H_{10}, and decolourize bromine-water solution. When A is treated with cold concentrated solution of sulfuric acid followed by refluxing of product, B($C_5H_{12}O$) was formed. B turns the orange colour of chromic acid solution to blue-green, converting itself into C($C_5H_{10}O$). B and C both reacts with alkaline solution of iodine producing yellow precipitate and salt of isobutyric acid. Identify A to C.

39. An organic compound (A) having molecular formula C_6H_{12} is treated with chlorine in CCl_4 to form B($C_6H_{12}Cl_2$). B on treatment with $NaNH_2$ yields C (C_6H_{10}). C on catalytic reduction with H_2/Pt yields 2-methylpentane. Compound C does not reacts with $NaNH_2$ and Tollen's reagent. Ozonolysis of A results in two aldehydes D and CH_3CHO. Deduce structures of A to D.

40. A hydrocarbon (A), molecular formula C_5H_{10}, yields 2-methylbutane on catalytic hydrogenation. A adds HBr to form B, which on reaction with $AgNO_3$ forms an alcohol C($C_5H_{12}O$). C on treatment with acidified solution of $KMnO_4$ yields D($C_5H_{10}O$). Deduce structural formula of A to D.

41. When isobutylene is dissolved in methanol containing a strong acid, a reaction takes place to produce ter-butyl methyl ether. Write mechanism that accounts the formation of this product.

42. Outline a mechanism that accounts the formation of following enantiomer, when cyclopentene is subjected to bromination in aqueous medium:

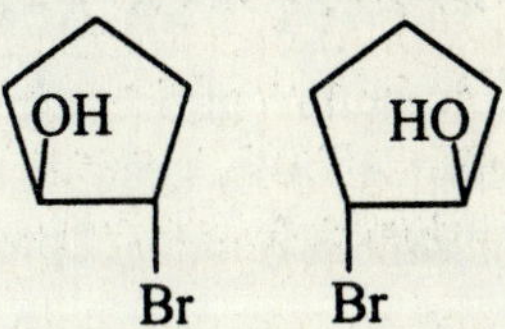

43. What products would you expect from each of the following reactions?

(a) *trans*-2-butene $\xrightarrow[CHCl_3]{KOC(CH_3)_3}$

(b) Cyclopentene $\xrightarrow[CHCl_3]{KOC(CH_3)_3}$

(c) *cis*-2-butene $\xrightarrow[\text{Diethyl ether}]{CH_2I_2/Zn(Cu)}$

44. Starting with cyclohexene and using any other needed reagents, outline a synthesis of the following compound:

Br Br

45. An unknown alkene with formula C_7H_{12} undergoes oxidation by hot alkaline $KMnO_4$ to yield, after acidification 6-ketoheptanoic acid. Identify the alkene.

46. By analogy with the mechanism of bromine addition to alkenes, draw the likely structures of A, B and C: Reaction of cyclopentene with bromine in water gives A. Reaction of A with aqueous KOH gives B(C_5H_8O). Heating B in methanol containing a catalytic amount of strong acid gives C($C_6H_{12}O_2$).

Alkynes

Alkynes are hydrocarbons that contains carbon-carbon triple bond as:

$$-C\equiv C-$$ (sp-carbon)

General formula of alkyne is C_nH_{2n-2} with two degree of unsaturation, i.e. two moles of hydrogen can be added per mole of alkyne. Remember C_nH_{2n-2} is also general formula of dienes and cycloalkene.

4.6 GENERAL METHODS OF PREPARATION

4.6.1 By Hydrolysis of CaC$_2$

$$CaC_2 + 2H_2O \rightarrow Ca(OH)_2 + HC\equiv CH$$

4.6.2 Higher Alkynes from Acetylene

Owing to the acidic nature of acetylinic hydrogen, it is converted into acetylide ion by treatment with strong base. Acetylide ion is subsequently used for preparation of higher alkynes as:

$$HC\equiv CH + Na \xrightarrow{\Delta} HC\equiv C^-Na^+$$

$$\xrightarrow[SN2]{R\text{-}X} R-C\equiv CH + NaX$$

Since alkylation of acetylide ion follow SN2 mechanism, R-X must be a primary halide otherwise, acetylide ion may abstract a proton, leading to formation of an alkene by E2 elimination reaction.

$$R-C\equiv C-Na^+ + CH_3CH_2CHBrCH_3 \rightarrow$$

$$R-C\equiv C-CH(CH_3)-CH_2CH_3 + CH_3CH=CHCH_3$$

Example 1. Show how to synthesize 3-octyne from acetylene and any necessary alkyl halide.

Solution: 3-octyne contains ethyl group on one side and butyl group on other side.

$$\Rightarrow HC\equiv CH \xrightarrow[\text{(ii) } C_4H_9Br]{\text{(i) } NaNH_2} CH_3-(CH_2)_2-CH_2-C\equiv CH \xrightarrow[\text{(ii) } C_2H_5Br]{\text{(i) } NaNH_2} \text{3-octyne}$$

4.6.3 Synthesis by Elimination Reactions

Under strongly basic conditions, vicinal or geminal dihalide gives alkyne:

$$R-CH_2-CX_2-R' \xrightarrow{Base} R-C\equiv C-R'$$

$$R-CHX-CHX-R' \xrightarrow{Base} R-C\equiv C-R'$$

Conditions for elimination

The double dehydrohalogenation occurs in extremely basic conditions, e.g. molten KOH or alcoholic KOH in a sealed tube, usually heated to 200°C, or using $NaNH_2$ as:

$$RCH_2CH_2-CHBr-CHBr-R' \xrightarrow[200°C]{KOH} RCH_2CH_2-C\equiv C-R'$$

$$CH_3CH_2CH_2CHCl_2 + NaNH_2 \rightarrow CH_3CH_2C\equiv CH$$

Rearrangement of alkynes in the presence of strong base: The double dehydrohalogenation is limited by severe conditions. Any functional groups that are sensitive to strong base cannot survive, also the alkyne products may rearrange under these extremely basic conditions as:

$$R-CH_2-C\equiv C-R' + B: \longleftrightarrow R-\overset{-}{C}H-C\equiv C-R' \longleftrightarrow R-CH=C=\overset{-}{C}-R' \xrightarrow{BH} R-CH=C=CH-R'$$

$$R-CH=C=CH-R' \xrightarrow{B:} R-\overset{-}{C}=C=CH-R' \longleftrightarrow R-C\equiv C-\overset{-}{C}H-R' \xrightarrow{BH} R-C\equiv C-CH_2-R'$$

Therefore, under extremely basic conditions, most stable isomer, i.e. most substituted alkyne predominate. However, in presence of $NaNH_2$, terminal alkyne predominate, because of the reason that terminal alkynes forms acetylide with $NaNH_2$, taking it out of the equilibrium.

Example 2

(a) Show with the help of mechanism, how 2-pentyne reacts with sodium amide to give 1-pentyne.

(b) Explain how this reaction converts a more stable alkyne to a less stable one.

(c) How would you accomplish the opposite reaction, converting 1-pentyne to 2-pentyne.

Solution: (a)

$$H_3C-CH_2-C\equiv C-CH_2(H) \xrightarrow{\bar{N}H_2}$$

$$H_3C-CH_2-\bar{C}=C=CH_2 \xrightarrow{NH_3} H_3C-CH_2-CH=C=CH(H) \xrightarrow{\bar{N}H_2}$$

$$H_3C-CH_2-\bar{C}H-C\equiv CH \xrightarrow{NH_3}$$

$$H_3C-CH_2-CH_2-C\equiv CH$$

(b) Since $NaNH_2$ is a stronger base than $RC\equiv CNa$, NH_2^- abstract proton from terminal alkynes forming $RC\equiv C^-$, which is insoluble in the reaction medium, comes out of equilibrium, favouring forward direction of reaction as:

$$R-C\equiv CH + \bar{N}H_2 \rightleftharpoons R-C\equiv C^- + NH_3$$

(c) 1-pentyne + KOH/fusion ⟶ 2-pentyne

Example 3. Show, how the following compounds could be synthesized in good yield by a doubledehydrohalogenation from a dihalide and strong base. Also show how your starting material might be synthsized from an alkene.

(a) 2-butyne (b) 1-octyne (c) Cyclodecyne.

Solution: (a) 2-butene + Cl_2 → 2, 3-dichlorobutane + KOH/fusion → 2-butyne. (b) 1-octene + Cl_2 → 1, 2-dichlorooctane + NaNH, followed by hydrolysis → 1-octyne. (c) Cyclooctene + Cl_2 → 1, 2-dichlorocyclooctane + KOH/fusion → Cyclooctyne.

4.7 PROPERTIES OF ALKYNES

4.7.1 Acidity

Terminal alkyne is acidic and its acidity is more than vinylic protons or protons attached to sp[3] carbon. Also acetylinic proton is more acidic than NH_3, but less acidic than H_2O and ROH. Therefore, the general order of acidic strength of these compounds are:

$$RH < CH_2=CH_2 < NH_3 < RC\equiv CH < ROH < H_2O$$

and the basic strength of their conjugate base is:

$$R^- > CH_2=CH^- > NH_2^- > RC\equiv C^- > RO^- > OH^-$$

Therefore, acetylinic protons cannot be abstracted by alkoxide or hydroxide ion.

Example 4. Explain the following:

(a) $CH_3CH_2C\equiv CH + CH_3Li \rightarrow CH_3CH_2C\equiv CLi + CH_4$ but its reverse does not occur.

(b) $C_6H_{11}-C\equiv CMgBr + C_2H_6 \longrightarrow C_6H_{11}-C\equiv CH + C_2H_5MgBr$ is not a feasible reaction.

Solution: (a) CH_3^- being stronger base than $RC\equiv C^-$. (b) $CH_3CH_2^-$ being stronger base than cyclohyhyl $-C\equiv C^-$

4.7.2 Catalytic Hydrogenation

The catalytic addition of hydrogen to 2-butyne not only serves as an example of such an addition reaction, but also provides heat of reaction data that reflect the relative thermodynamic stabilities of these hydrocarbons as shown in the Figure 4.9. From the heats of hydrogenation, in units of kcal/mole, it would appear that alkynes are thermodynamically less stable than alkenes to a greater degree than alkenes are less stable than alkanes. The standard bond energies for carbon-carbon bonds confirm this conclusion. Thus, a double bond is stronger than a single bond, but not twice as strong. The difference (36.7 kcal/mole) may be regarded as the strength of the pi-bond component.

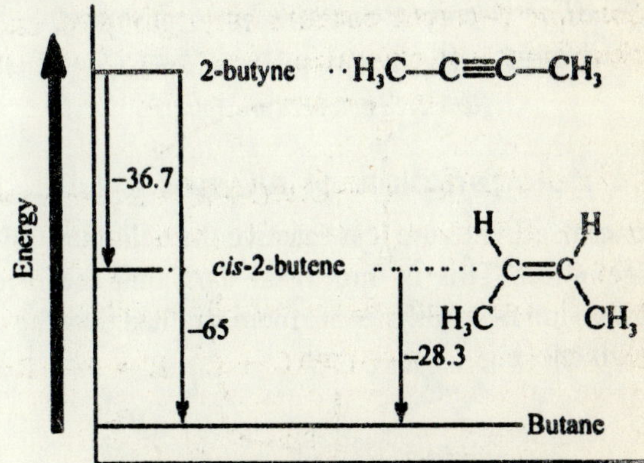

Figure 4.9: Stepwise hydrogenation of an alkyne

Similarly, a triple bond is stronger than a double bond, but not 50% stronger. Here the difference (28.3 kcal/mole) may be taken as the strength of the second pi-bond. The 8.4 kcal/mole weakening of this second pi-bond is reflected in the heat of hydrogenation (36.7 – 28.3 = 8.4).

Since alkynes are thermodynamically less stable than alkenes, we might expect addition reactions of the former to be more exothermic and relatively faster than equivalent reactions of the latter. This is true for the catalytic hydrogenation reaction; however, the usual Pt and Pd hydrogenation catalysts are so effective in promoting hydrogenation of both double and triple carbon—carbon bonds that the alkene intermediate formed by hydrogen addition to an alkyne cannot be isolated.

A less efficient catalyst, *Lindlar's catalyst*, prepared by deactivating (or poisoning) a conventional palladium catalyst by treating it with lead acetate and quinoline, permits alkynes to be converted to alkenes without further reduction to an alkane. The addition of hydrogen is stereoselectively syn (e.g. 2-butyne gives *cis*-2-butene). A complementary stereoselective reduction in the anti mode may be accomplished by a solution of sodium in liquid ammonia. Mechanism of this reaction has been discussed in previous section of preparation of alkene by controlled hydrogenation of alkyne.

$$R{-}C{\equiv}C{-}R + H_2 \text{ and Lindlar catalyst} \rightarrow \textit{cis}\ R{-}CH{=}CH{-}R$$

$$R{-}C{\equiv}C{-}R + 2Na \text{ in } NH_3 \text{ (liq)} \rightarrow \textit{trans}\ R{-}CH{=}CH{-}R + 2\,NaNH_2$$

4.7.3 Terminal Alkynes Reacts with Ag(I) and Cu(I) Salt to form Silver and Copper Acetylides

This allows us distinction of terminal alkynes from internal alkynes and other hydrocarbons:

$$R{-}C{\equiv}C{-}H + Ag(NH_3)_2^+ \rightarrow R{-}C{\equiv}CAg + H^+$$

(*Light coloured precipitate soluble in dilute acid*)

$$R{-}C{\equiv}C{-}H + Cu(I) \rightarrow R{-}C{\equiv}CCu+ + H^+$$

(*Brick red coloured precipitate*)

4.7.4 Halogenation of Alkynes

In general, alkynes are less reactive than alkenes towards halogenation. This is due to the fact that the bridged halonium ion from alkynes are more strained than the same from alkene, e.g.

$$-C{\equiv}C- + Br^+ \longrightarrow \text{C}{=}\text{C bridged by } Br^+$$

(*A less stable, cyclic halonium ion*)

Alkynes reacts with bromine water to produce dibromide:

$$R{-}C{\equiv}C{-}R + Br_2 \xrightarrow{H_2O} (R)(Br)C{=}C(Br)(R)$$

With liquid bromine and in absence of any solvent, tetrabromide is obtained.

$$R{-}C{\equiv}C{-}R + 2Br_2 \rightarrow R{-}CBr_2{-}CBr_2{-}R$$

4.7.5 Electrophilic Addition on Triple Bond

When the addition reactions of electrophilic reagents, such as strong bronsted acids and halogens, to alkynes are studied we find a curious paradox. The reactions are even more exothermic than the additions to alkenes, and yet the rate of addition to alkynes is slower by a factor of 100 to 1000 than addition to equivalently substituted alkenes. The reaction of one equivalent of bromine with 1-penten-4-yne, for example, gave 4,5-dibromo-1-pentyne as the chief product.

$$HC{\equiv}C{-}CH_2{-}CH{=}CH_2 + Br_2 \rightarrow HC{\equiv}C{-}CH_2{-}CHBrCH_2Br$$

Although these electrophilic additions to alkynes are sluggish, they do take place and generally display Markovnikov rule regioselectivity and anti-stereoselectivity. One problem, of course, is that the products of these additions are themselves substituted alkenes and can therefore undergo further addition.

Because of their high electronegativity, halogen substituents on a double bond act to reduce its nucleophilicity, and thereby decrease the rate of electrophilic addition reactions. Consequently, there is a delicate balance as to whether the product of an initial addition to an alkyne will suffer further addition to a saturated product. Although the initial alkene products can often be isolated and identified, they are commonly present in mixtures of products and may not be obtained in high yield.

The following reactions illustrate many of these features. In the last example, 1, 2-diodoethene does not suffer further addition inasmuch as vicinal-di-iodoalkanes are relatively unstable.

$$C_2H_5{-}C{\equiv}C{-}H \xrightarrow[\text{slow}]{HCl} (C_2H_5)(Cl)C{=}CH_2$$

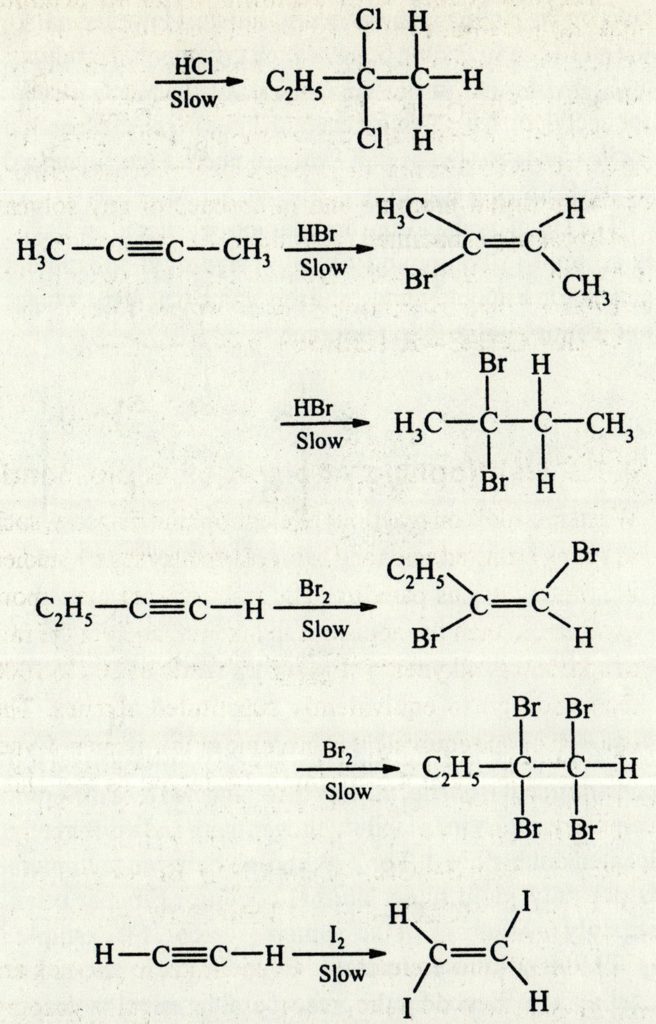

$$\xrightarrow[\text{Slow}]{HCl} C_2H_5-CCl_2-CH_2-H$$

$$H_3C-C\equiv C-CH_3 \xrightarrow[\text{Slow}]{HBr} (H_3C)(Br)C=C(H)(CH_3)$$

$$\xrightarrow[\text{Slow}]{HBr} H_3C-CBr_2-CH_2-CH_3$$

$$C_2H_5-C\equiv C-H \xrightarrow[\text{Slow}]{Br_2} (C_2H_5)(Br)C=C(Br)(H)$$

$$\xrightarrow[\text{Slow}]{Br_2} C_2H_5-CBr_2-CBr_2-H$$

$$H-C\equiv C-H \xrightarrow[\text{Slow}]{I_2} (H)(I)C=C(I)(H)$$

Why are the reactions of alkynes with electrophilic reagents more sluggish than the corresponding reactions of alkenes? After all, addition reactions to alkynes are generally more exothermic than additions to alkenes, and there would seem to be a higher pi-electron density about the triple bond (two pi-bonds versus one). Two factors are significant in explaining this apparent paradox. First, although there are more pi-electrons associated with the triple bond, the sp-hybridized carbons exert a strong attraction for the pi-electrons, which are consequently bound more tightly to the functional group than are the pi-electrons of a double bond. This is seen in the ionization potentials of ethylene and acetylene.

Acetylene $HC\equiv CH + \text{Energy} \rightarrow [HC\equiv CH]^{(+)} + e^{(-)}$
$\Delta H = +264$ kcal/mole

Ethylene $H_2C=CH_2 + \text{Energy} \rightarrow [H_2C=CH_2]^{(+)} + e^{(-)}$
$\Delta H = +244$ kcal/mole

Ethane $H_3C-CH_3 + \text{Energy} \rightarrow [H_3C-CH_3]^{(+)} + e^{(-)}$
$\Delta H = +296$ kcal/mole

As defined by the preceding equations, an **ionization potential** is the minimum energy required to remove an electron from a molecule or a compound. Since pi-electrons are less tightly held than sigma-electrons, we expect the ionization potentials of ethylene and acetylene to be lower than that of ethane, as is the case. Gas-phase proton affinities show the same order, with ethylene being more basic than acetylene, and ethane being less basic than either. Since the initial interaction between an electrophile and an alkene or alkyne is the formation of a pi-complex, in which the electrophile accepts electrons from alkyne and becomes weakly bonded to the multiple bond, the relatively slower reactions of alkynes becomes understandable.

The second factor is the stability of the carbocation intermediate generated by sigma-bonding of a proton or other electrophile to one of the triple bond carbon atoms. This intermediate has its positive charge localized on an unsaturated carbon, and such **vinyl cations** are less stable than their saturated analogs.

Therefore, alkynes are less reactive than alkene towards electrophilic addition reaction. The reason for this is the formation of a less stable intermediate, a vinyl carbocation as:

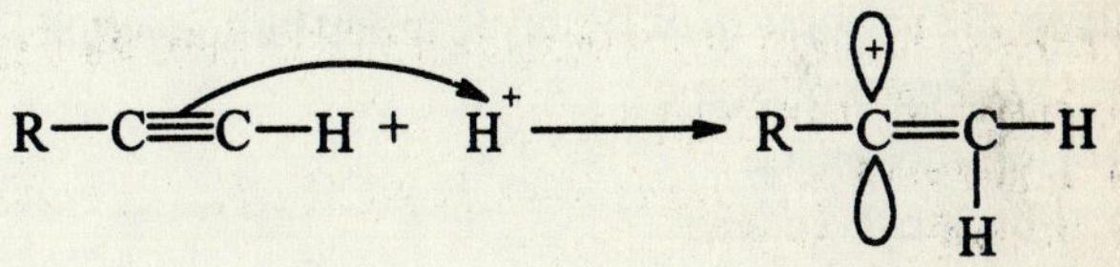

$$R-C\equiv C-H + H^+ \longrightarrow R-\overset{+}{C}=CH-H$$

Vinyl carbocation

Although it is less stable, addition follows Markovnikov's rule, because a secondary vinyl carbocation is more stable than a primary vinyl carbocation.

Example 5. Predict products in the following reaction sequence, indicating stereochemistry wherever possible:

(a) $H_3C-C\equiv C-CH_3 + Br_2$ (One equivalent) $\longrightarrow$

$$X \xrightarrow[CH_3CH_2COOH]{B_2H_6} Y$$

(b) $Ph-C\equiv C-CH_3 + ICl$ (One equivalent) $\longrightarrow$

$$X \xrightarrow{HCl} Y$$

(c) 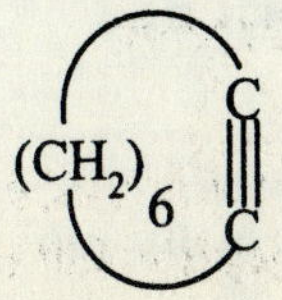 $(CH_2)_6$ [ring with $C\equiv C$] + Br_2 (One equivalent) $\longrightarrow X$

Solution

(a) 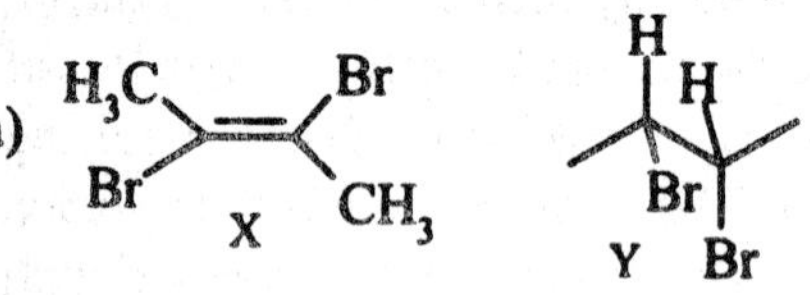 X, Y

(b) 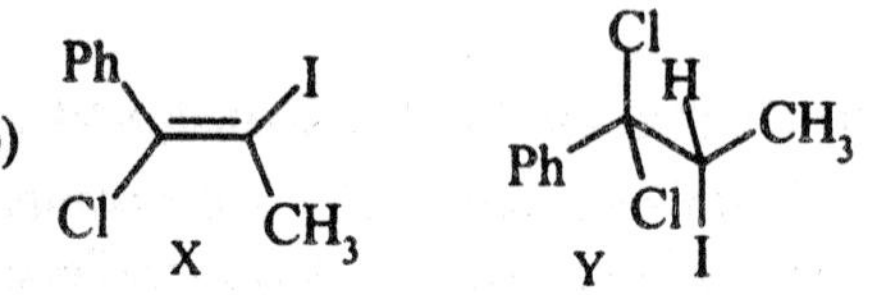 X, Y

(c)

4.7.6 Addition of HBr in Presence of Peroxide

The effect of peroxide on the addition of HBr to an alkyne is same as in the case of alkene:

$$RC{\equiv}CH + HBr \text{ (one equivalent)} \xrightarrow{H_2O_2 \text{ or } R_2O_2} R{-}CH{=}CHBr \; [\textit{cis-trans}]$$

Example 6: Show how 1-hexyne might be converted to:
(a) 1,2-dichlorohexene
(b) 1-bromohexane
(c) 2-bromo-1-hexene
(d) 1,2-dibromohexane

Solution

(a) 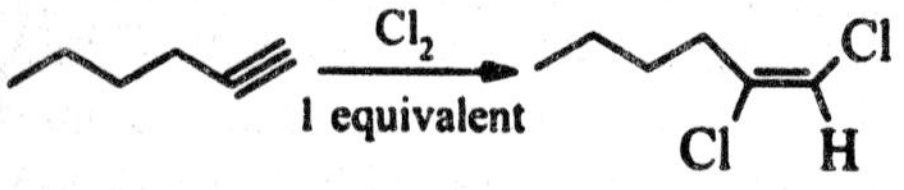 $\xrightarrow[\text{1 equivalent}]{Cl_2}$

(b) 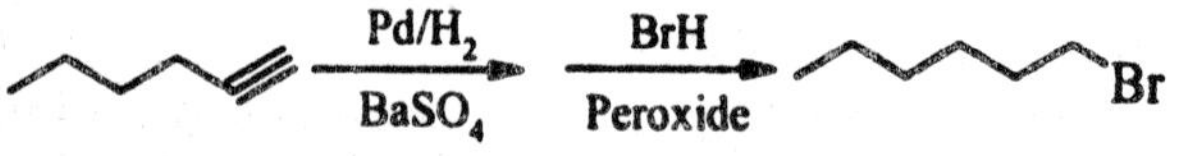 $\xrightarrow[BaSO_4]{Pd/H_2}$ $\xrightarrow[\text{Peroxide}]{BrH}$

(c) 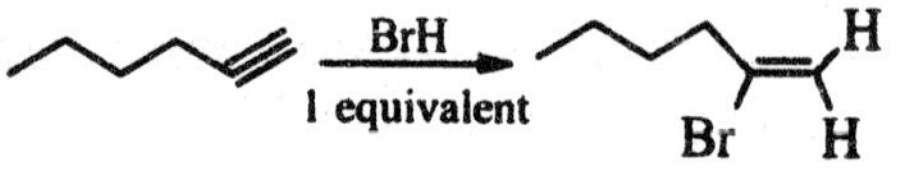 $\xrightarrow[\text{1 equivalent}]{BrH}$

(d) 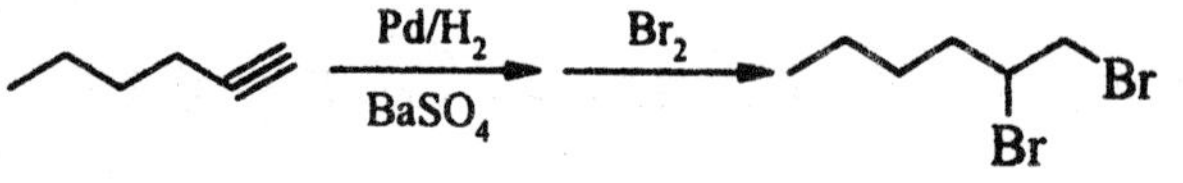 $\xrightarrow[BaSO_4]{Pd/H_2}$ $\xrightarrow{Br_2}$

4.7.7 Hydration of Alkynes to Ketones and Aldehydes

A general discussion: As with alkenes, the addition of water to alkynes requires a strong acid, usually sulfuric acid, and is facilitated by mercuric sulfate. However, unlike the additions to double bonds which give alcohol products, addition of water to alkynes gives ketone products (except for acetylene which yields acetaldehyde). The explanation for this deviation lies in enol-keto tautomerization, illustrated by the following equation.

The initial product from the addition of water to an alkyne is an enol (a compound having a hydroxyl substituent attached to a double-bond), and this immediately rearranges to the more stable keto tautomer.

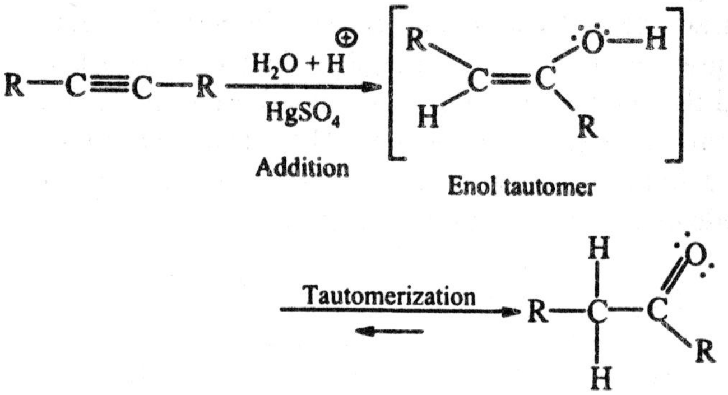

$$R{-}C{\equiv}C{-}R \xrightarrow[\text{HgSO}_4]{H_2O + H^{\oplus}} \left[(R)(H)C{=}C(\ddot{O}{-}H)(R) \right] \xrightleftharpoons{\text{Tautomerization}} R{-}CH_2{-}C({=}\ddot{O})R$$

Addition — Enol tautomer

Tautomers are defined as rapidly interconverted constitutional isomers, usually distinguished by a different bonding location for a labile hydrogen atom and a differently located double bond. The equilibrium between tautomers is not only rapid under normal conditions, but it often strongly favours one of the isomers (acetone, for example, is 99.999% keto tautomer). Even in such one-sided equilibria, evidence for the presence of the minor tautomer comes from the chemical behaviour of the compound. Tautomeric equilibria are catalyzed by traces of acids or bases that are generally present in most chemical samples. The three examples shown below illustrate these reactions for different substitutions of the triple-bond. For terminal alkynes the addition of water follows the Markovnikov rule, as in the second example below, and the final product is a methyl ketone (except for acetylene, shown in the first example). For internal alkynes (the triple-bond is within a longer chain) the addition of water is not regioselective. If the triple-bond is not symmetrically located (i.e. if R and R′ in the third equation are not the same) two isomeric ketones will be formed.

$$HC \equiv CH + H_2O + HgSO_4 \text{ and } H_2SO_4 \rightarrow [H_2C{=}CHOH] \rightarrow H_3C{-}CH{=}O$$

$$RC{\equiv}CH + H_2O + HgSO_4 \text{ and } H_2SO_4 \rightarrow [RC(OH){=}CH_2] \rightarrow RC({=}O)CH_3$$

$$RC{\equiv}CR' + H_2O + HgSO_4 \text{ and } H_2SO_4 \rightarrow [RHC{=}C(OH)R' + RC(OH){=}CHR'] \rightarrow RCH_2{-}C({=}O)R' + RC({=}O){-}CH_2R'$$

Two factors have an important influence on the enol-keto tautomerizations described here. The first is the potential energy difference between the tautomeric isomers. This factor determines the position of the equilibrium state.

The second factor is the activation energy for the interconversion of one tautomer to the other. This factor determines the rate of rearrangement. Since the potential energy or stability of a compound is in large part a function of its covalent bond energies, we can estimate the relative energy of keto and enol tautomers by considering the bonds that are changed in the rearrangement. From the following diagram, we see that only three significant changes occur, and the standard bond energies for those changes are given to the right of the equation. The keto tautomer has a 17.5 kcal/mole advantage in bond energy, so its predominance at equilibrium is expected.

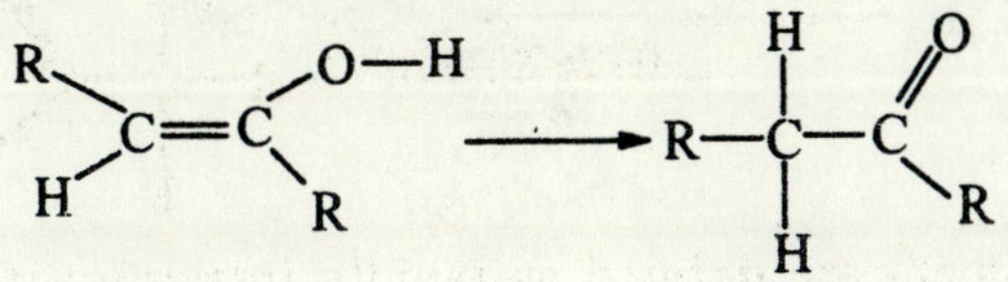

$$\underset{\text{Enol tautomer}}{R(H)C{=}C(R)\text{—}O\text{—}H} \longrightarrow \underset{\text{Keto tautomer}}{R\text{—}CH_2\text{—}C({=}O)\text{—}R}$$

Bond change		Change in bond energy
C=C (146) → C—C (83)		−63
C—O (85.5) → C=O (178)		+92.5
O—H (111) → C—H (99)		−12
		+17.5 k cal/mole

The rapidity with which enol-keto tautomerization occurs suggests that the activation energy for this process is low. We have noted that the rearrangement is acid and base catalyzed, and very careful experiments have shown that interconversion of tautomers is much slower if such catalysts are absent. A striking example of the influence of activation energy on such transformations may be seen in the following hypothetical rearrangement. Here we have substituted a methyl group for the proton of a conventional tautomerism, and the methyl shifts from oxygen to carbon just as the proton does in going from an enol to a ketone.

$$H_2C{=}CH\text{—}O\text{—}CH_3 \rightarrow CH_3\text{—}CH_2\text{—}CH{=}O$$

The potential energy change for this rearrangement is even more advantageous than for enol-keto tautomerism, being estimated at over 25 kcal/mole from bond energy changes.

Despite this thermodynamic driving force, the enol ether described above is completely stable to base treatment, and undergoes rapid acid-catalyzed hydrolysis with loss of methanol, rather than rearrangement. The controlling difference in this case must be a prohibitively high activation energy for the described rearrangement, combined with lower energy alternative reaction paths.

(a) Mercuration-Demercuration Reaction

Alkynes undergo acid catalyzed addition of water across triple bond in the presence of mercuric ion catalyst- addition occurs according to Markovnikov's rule but alcohols formed, being very unstable due to enolic structure, is not isolated and the final product is its stable keto form:

$$R\text{—}C{\equiv}C\text{—}H + HgSO_4/H_2SO_4 \rightarrow$$

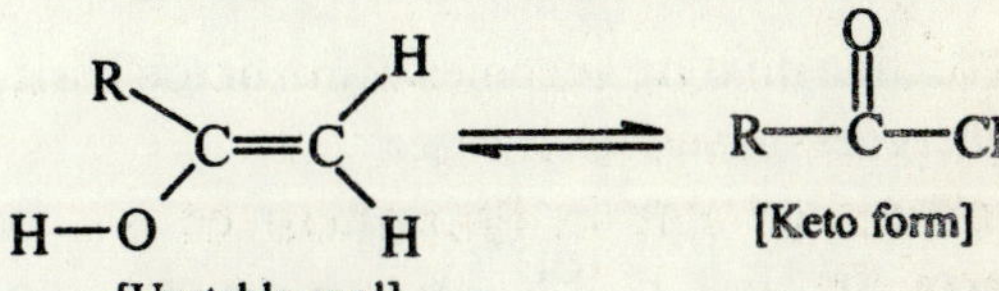

$$\underset{\text{[Unstable enol]}}{R(HO)C{=}CH_2} \rightleftharpoons \underset{\text{[Keto form]}}{R\text{—}C({=}O)\text{—}CH_3}$$

Mechanism of addition

$$R\text{—}C{\equiv}C\text{—}H + Hg^{2+} \rightarrow R\text{—}\overset{+}{C}{=}CH\text{—}Hg^{+} \xrightarrow{H_2O} R\text{—}C(\overset{+}{O}H_2){=}CH\text{—}Hg^{+} \xrightarrow{-H^{+}} R\text{—}C(OH){=}CH\text{—}Hg^{+} \xrightarrow{H_3O^{+}} \underset{\text{[Unstable enol]}}{R(HO)C{=}CH_2} \rightleftharpoons \underset{\text{[Keto form]}}{R\text{—}C({=}O)\text{—}CH_3}$$

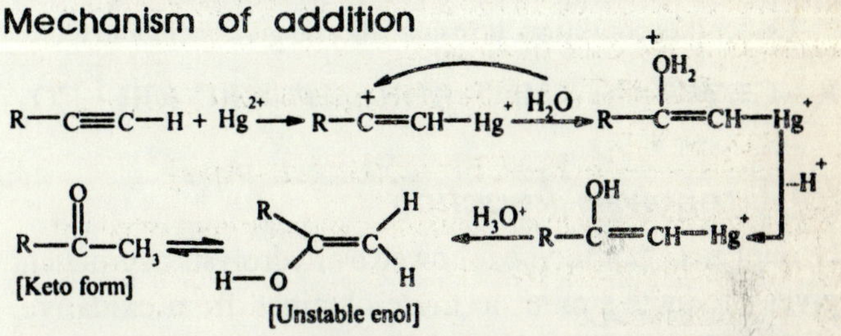

Example 7. An organic compound $A(C_{10}H_{18})$ reacts with H_2/Pt forming $C_{10}H_{22}(B)$. A on treatment with $HgSO_4$ in acidic medium produces $C(C_{10}H_{20}O)$ as single isomer. A on treatment with $Pd/BaSO_4$ produces stereomeric $D(C_{10}H_{20})$. D on oxidation with acidic solution with $KMnO_4$ produces $E(C_5H_{10}O_2)$ as only acid, which is almost inert towards decarboxylation reaction with NaOH/CaO. Identify A to E.

Solution: A is an alkyne since it on hydrogenation produces $C_{10}H_{22}(B)$, and reacts with $HgSO_4$ in acidic medium. Also A is a symmetrical alkyne since its partially hydrogenated product D gives E as only acid.

Therefore, A is:

$$C_4H_9\text{—}C{\equiv}C\text{—}C_4H_9\ (A) \xrightarrow{H_2/Pt} C_{10}H_{22}\ (B)$$

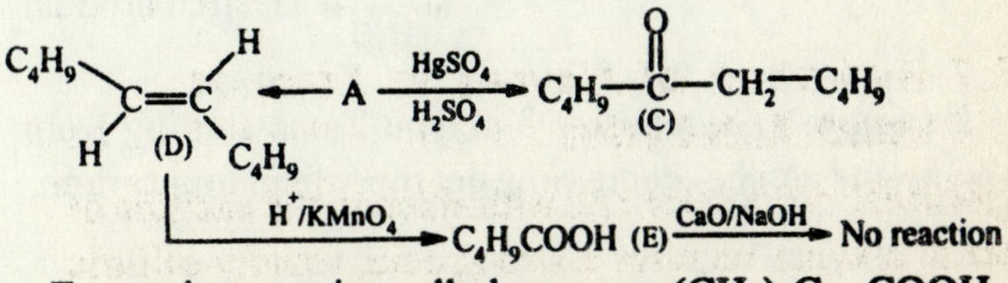

$$\underset{(D)}{C_4H_9(H)C{=}C(H)C_4H_9} \longleftarrow A \xrightarrow[H_2SO_4]{HgSO_4} \underset{(C)}{C_4H_9\text{—}C({=}O)\text{—}CH_2\text{—}C_4H_9}$$

$$(D) \xrightarrow{H^{+}/KMnO_4} C_4H_9COOH\ (E) \xrightarrow{CaO/NaOH} \text{No reaction}$$

E contains a tertiary alkyl group as $(CH_3)_3C\text{—}COOH$.

(b) Hydroboration-oxidation

Hydroboration-oxidation of alkynes leads to addition of water according to Anti-Markovnikov's rule as:

$$R—C{\equiv}C—H + B_2H_6 \xrightarrow[OH^-]{H_2O_2} [R—CH{=}CH—OH \longleftrightarrow R—CH_2CHO]$$

4.7.8 Oxidation of Alkynes

(a) Permanganate Oxidation

Under mild condition (aqueous $KMnO_4$ under neutral condition), alkynes reacts to give α-diketo compound:

$$R—C{\equiv}C—R' \xrightarrow[H_2O,\ neutral]{KMnO_4} R—\underset{\underset{O}{\|}}{C}—\underset{\underset{O}{\|}}{C}—R'$$

If the reaction mixture becomes warm or too basic, the diketo undergoes oxidative cleavage:

$$R—C{\equiv}C—R' + KMnO_4 \xrightarrow[Heat]{OH^-} R—COOK + R'—COOK$$

Under this condition, terminal alkyne gives acid and CO_2:

$$R—C{\equiv}C—H + KMnO_4/KOH \text{ or } H^+ \rightarrow RCOOH + CO_2$$

(b) Ozonolysis Reaction

Ozonolysis of an alkyne followed by hydrolysis of product, gives products similar to those obtained from oxidative cleavage by permanganate ion:

$$R—C{\equiv}C—R' + O_3/H_2O \rightarrow R—COOH + R'—COOH$$

$$R—C{\equiv}C—H + O_3/H_2O \rightarrow R—COOH + CO_2$$

Oxidative cleavage can help to determine position of triple bonds in alkyne.

Example 8. An unknown alkyne undergoes oxidative cleavage to give adipic acid and two equivalent of acetic acid. Propose a structure of alkyne.

Solution: Unknown alkyne $\xrightarrow[H_2O]{O_3}$ Acetic acid +

HOOC—$(CH_2)_4$—COOH + Acetic acid

⇒ Alkyne is $CH_3—C{\equiv}C—(CH_2)_4—C{\equiv}C—CH_3$

$\xrightarrow[H_2O]{O_3}$ Desired product

Example 9. Synthesize 3-octyne-2-one starting from alkyne and alcohol containing not more than four carbon.

Solution

$$n—C_4H_9Br + H—C{\equiv}CNa \rightarrow CH_3CH_2CH_2CH_2C{\equiv}CH$$

$CH_3CH_2CH_2CH_2OH \xrightarrow{PBr_3}$ n-C_4H_9Br

$CH_3CH_2CH_2CH_2C{\equiv}CH \xrightarrow{\text{(i) } NaNH_2\text{ (ii) } CH_3OH\text{ (iii) } H_3O^+}$ [structure: oct-3-yn-2-one]

Example 10. Develop synthesis for the following compounds, using acetylene and alcohols containing no more than four carbon as your starting materials.

(a) 3-methyl-4-nonyne-3-ol
(b) *cis*-ethyl-2-methylcyclopropane
(c) *cis*-2-methylhex-4-en-3-ol

Solution

(a) $C_2H_2 \xrightarrow{Na/heat} \xrightarrow{\text{[1-bromobutane]}}$ [1-hexyne]; [1-butanol] $\xrightarrow{PBr_3}$ [1-bromobutane]

$\xrightarrow{Na/heat} \xrightarrow{\text{[butan-2-one]}} \xrightarrow{H^+}$ [3-methylnon-4-yn-3-ol, with H_3C and OH]; [butan-2-ol] $\xrightarrow{[O]}$ [butan-2-one]

(b) $C_2H_2 \xrightarrow{Na/heat} \xrightarrow{CH_3Br} \xrightarrow[C_2H_5Br]{Na/heat} \xrightarrow[BaSO_4]{Pd/H_2}$

[cis-alkene: H_3C, C_2H_5; H, H] $\xrightarrow{CH_2N_2/heat}$ [cyclopropane: H, H; H_3C, C_2H_5]

(c) $C_2H_2 \xrightarrow{Na/heat} \xrightarrow{CH_3Br} \xrightarrow{Na/heat}$ [propynyl anion, C^-] $\xrightarrow{\text{[2-methylpropanal]}}$; [2-methylpropan-1-ol] $\xrightarrow{PCC}$ [2-methylpropanal, H, O]

[2-methylhex-4-yn-3-ol, OH] $\xrightarrow[BaSO_4]{Pd/H_2}$ [cis-alkene: H_3C, HO; H, H]

Example 11. A hydrocarbon $A(C_8H_{12})$ is optically active and on treatment with H_2/Pt gives (C_8H_{18}) which does not rotate plane polarized light. A does not give any precipitate on treatment with ammonical silver nitrate solution. A on treatment with Pd/$BaSO_4$ gives C(C_8H_{14}) which is also optically inactive but A on treatment with Na/liq NH_3 gives D, an isomer of C, which is able to show enantiomerism. Also A on treatment with ozone followed by hydrolysis gives $C_4H_6O_3$(E) which is optically active. Identify A to E representing their stereochemical structure.

Solution

$CH_3—C{\equiv}C—C(H)(CH_3)—C(H){=}C(H)—CH_3$ (A)

$\xrightarrow{H_2/Pt}$ 3-methylheptane (skeletal, CH_3 branch)

$\xrightarrow{Pd/BaSO_4}$ (Optically inactive C)

$\xrightarrow{Na/liq\ NH_3}$ (O A) D

$\xrightarrow{O_3}$ $HOOC—C(H)(CH_3)—CHO$ E(O A)

PRACTICE EXERCISE III

1. Which would you expect to be more stable, an internal or a terminal alkyne, explain?
2. An alkene is more reactive than an alkyne towards electrophilic addition reaction, a halo substituted alkene resulting from addition of HBr or HCl to alkyne, does not react further with HCl or HBr till all alkynes are exhausted, explain.
3. During addition of HBr in presence of a radical initiator peroxide, to an alkyne, the product, bromoalkene does not undergo addition reaction with HBr till all alkynes are exhausted, though alkene is more reactive than alkyne towards free radical addition reaction, explain.
4. Give major product of each of the following reactions:
   (a) $HC{\equiv}CCH_2CH_3 + HBr \rightarrow$
   (b) $HC{\equiv}CCH_2CH_3 + 2HBr \rightarrow$
   (c) $HC{\equiv}CCH_2CH_3$ + HBr/Peroxide $\rightarrow$
   (d) $HC{\equiv}CCH_2CH_3 + Br_2/CCl_4 \rightarrow$
   (e) $HC{\equiv}CCH_2CH_3 + Cl_2 \rightarrow$
5. Predict the configuration of product that would be obtained from the reaction of 2-butyne with:
   (a) One equivalent of Br_2 in CCl_4
   (b) One equivalent of HBr in presence of peroxide.
6. Starting from an alkyne, synthesize the following:
   (i) (a) Acetone, (b) 3-hexanone, (c) cyclohexyl–C(=O)–CH_3
   (ii) Draw all possible enol structure of the keto form in the above part of question.
7. Starting with 1-butyne, how could you make 3-hexanone?
8. Starting with acetylene, how could you make 1-bromopentane?
9. How could 2,6-dimethylheptane be prepared from an alkyne and a primary alkyl halide?
10. Bring about the following conversion starting from acetylene and any other organic/inorganic reagents:
    (a) $C_2H_2 \longrightarrow$ cyclohexyl–CH_2CH_2OH
    (b) $C_2H_2 \rightarrow$ *Trans*-2-pentene
    (c) $C_2H_2 \rightarrow$ 3, 3-dibromohexane
    (d) $C_2H_2 \rightarrow$ *cis*-2-butene
11. Starting with isopropyl acetylene, how you would prepare the following alcohols?
    (a) 2-methyl-2-pentanol (b) 4-methyl-2-pentanol
12. Which of the following pairs are keto-enol tautomers?
    (a) $CH_3CH_2CH{=}CHCH_2OH$ and $CH_3CH_2CH_2CH_2CHO$
    (b) 2-propanol and acetone
    (c) $CH_3CH_2CH{=}CHOH$ and $CH_3CH_2CH_2CHO$

(d) $CH_3CH_2CH_2C(OH){=}CH_2$ and $CH_3CH_2CH_2{-}C(=O){-}CH_3$

13. Provide reagents for the following transformation:

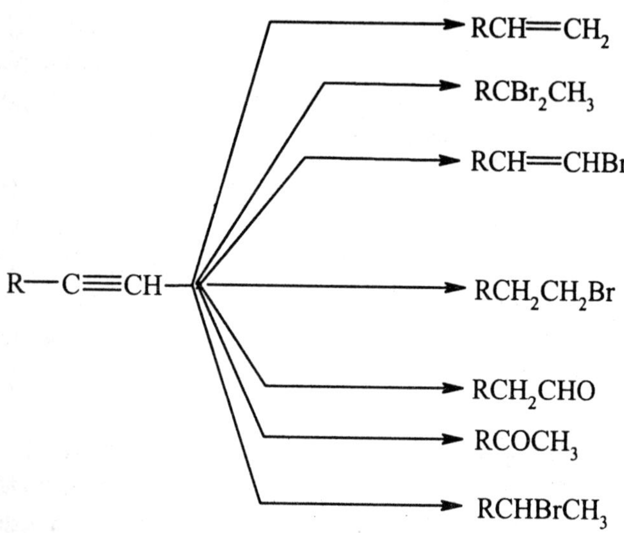

14. Provide products for the following reactions:

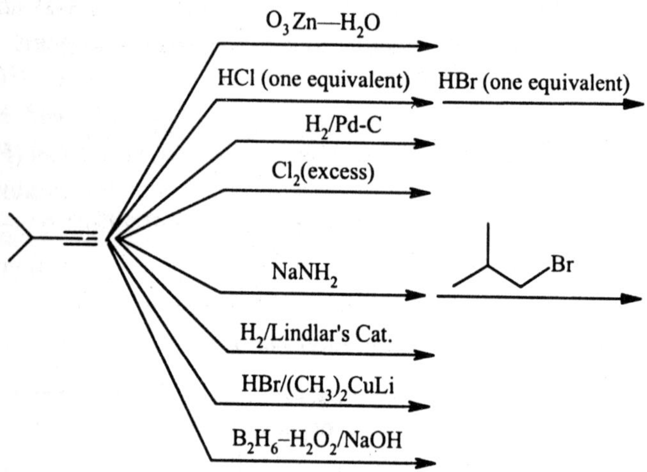

15. What will be the major product of the reaction of one mole of propyne with each of the following reagents?
(a) HBr (1 mole) (b) HBr (2 mole)
(c) Br_2 (1 mole)/CCl_4 (d) Br_2 (2 mole)/CCl_4
(e) aqueous $HgSO_4/H_2SO_4$
(f) $B_2H_6/H_2O_2/OH^-$ (g) HBr/H_2O_2
(h) H_2/Lindlar's catalyst (i) Na/liq NH_3

16. Answer question 15. Using 2-butyne as reagent in place of propyne.

17. Using acetylene as the starting material, synthesize the following compounds:
(a) 1,2-dibromobutane (b) *Trans*-2-pentene
(c) Ethanal (d) *Cis*-2-pentene

18. Discuss the stereochemistry of the reaction of 2-butyne with the following reagents:
(a) H_2/Lindlar's catalyst followed by Br_2/CCl_4
(b) Na/liq NH_3 followed by Br_2/CCl_4
(c) Cl_2/CCl_4 followed by Br_2/CCl_4

19. Explain why a single, pure product is obtained from hydroboration oxidation of 1-butyne, whereas two isomeric products are obtained from hydroboration oxidation of 2-pentyne.

20. Bring about the following conversion:
(a) 1-pentene → 1-pentyne
(b) 1-bromo propane → 2-hexyne
(c) Acetylene → 1-butanol
(d) Acetylene → 3-methoxy hexane
(e) *Trans*-1, 2-diphenyl ethylene → *cis*-1, 2-diphenyl ethylene
(f) 1-propanol → Acetone
(g) Propyne → $H_3C(H)C{=}C(D)H$

21. Using any necessary inorganic reagents, show how would convert acetylene and isobutyl bromide to:
(a) meso-2, 7-dimethyl-4, 5-octandiol
(b) d, l-2, 7-dimethyl-4, 5-octandiol

22. An organic compound A(C_8H_{10}) on hydrogenation with Pd/$BaSO_4$ yields B(C_8H_{12}). B on treatment with cold, dilute alkaline solution of $KMnO_4$ gives a meso compound C($C_8H_{14}O_2$). A on treatment with Na/liq NH_3 yields D, a stereomer of B. D on treatment with cold, dilute alkanline solution of $KMnO_4$ gives a racemic mixture of a compound with molecular formula $C_8H_{14}O_2$. Identify A to D.

23. An organic compound A(C_8H_{12}) gives a brick red precipitate on treatment with aqueous solution of Cu_2Cl_2. A reacts with $HgSO_4/H_2SO_4$ to give B($C_8H_{14}O$). B on treatment with NaOH/I_2 gives C($C_7H_{11}O_2Na$) and a yellow precipitate. C on heating with sodalime produces D(C_6H_{12}). D on monochlorination gives E($C_6H_{11}Cl$) as sole isomer. Identify A to E.

24. A hydrocarbon A(C_6H_{10}) on hydrogenation with H_2/Pt gives B(C_6H_{14}). Hydrogenation of A with Na/liq NH_3 produces C(C_6H_{12}). A on treatment with Pd/H_2 in presence of $BaSO_4$ produces D, a stereoisomer of C. D on chlorination in $CCl_4/FeCl_3$ produces a racemic mixture with molecular formula $C_6H_{12}Cl_2$, whereas chlorination of C produces nonresolvable compound. Deduce structure of A, B, C and D.

25. An organic compound A has molecular formula C_8H_{12} and is optically active. It reacts with H_2/Pt to give B(C_8H_{18}) which is optically inactive. Careful hydrogenation of A with Pd/$BaSO_4$ gives C(C_8H_{14}), which is still optically active. Compound A reacts with Na in liquid ammonia to give D, a stereomer of C, but it is optically inactive. Deduce structures of A to D.

26. An organic compound A has molecular formula $C_{12}H_{14}$ and it can be resolved into enantiomers. A on boiling with dilute alkaline $KMnO_4$ solution yield B($C_5H_{10}O_2$) and a white crystalline substance C. C on treatment with soda lime produced benzene. Compound B is still resolvable but decarboxylation of B with soda lime renders it non-resolvable. Deduce structures of A, B and C.

27. A hydrocarbon A has molecular formula $C_{10}H_{18}$. A on treatment with H_2/Pt gives B($C_{10}H_{22}$). B on monochlorination gives two isomer C and D with molecular formula $C_{10}H_{21}Cl$. Out of C and D, only C can undergo dehydrohalogenation with alcoholic solution of KOH as well as only C can be resolved into enantiomers. A on addition of one equivalent of Cl_2 in CCl_4 gives a stereomer E($C_{10}H_{18}Cl_2$). E on treatment with Bayer's reagent yields a racemic mixture of molecular formula ($C_{10}H_{20}Cl_2O_2$). Also A on partial hydrogenation with Pd/$BaSO_4$/quinone gives F($C_{10}H_{20}$), which on treatment with Bayer's reagent gives a meso compound G($C_{10}H_{22}O_2$). Deduce the structures of A to G.

28. A hydrocarbon A($C_{10}H_{12}$) has no chiral carbon. A gives a white precipitate with ammonical solution of silver nitrate. A on treatment with H_2/Pt gives B($C_{10}H_{20}$). A on ozonolysis gives C($C_8H_{12}O_4$) as one product, which on heating with soda lime gives D(C_6H_{12}). D on monochlorination with Cl_2/hv gives $C_6H_{11}Cl$ as sole isomer. Identify A to D.

29. An organic compound A(C_8H_6) gives brick red precipitate with ammonical solution of Cu_2Cl_2. A on treatment with acidic solution of $HgSO_4$ yields B(C_8H_8O). B gives positive iodoform test. B can also be prepared by the reaction of benzene with acetic anhydride in presence of $AlCl_3$. Identify A and B.

30. Potassium hydroxide is mixed with 2, 3-dibromohexane, the mixture is sealed in a fusion tube and heated to 200°C for one hour. The product mixture (A) is mixed with Cu(I)-ammonia complex and a precipitate forms. The precipitate (B) and liquid phase (C) are separated. The precipitate is acidified and the product (D) is distilled at 71°C. Product D is treated with $NaNH_2$, followed by acetone and then with dilute acid to give alcohol (F). The liquid phase (C) is distilled and the product alkyne is treated with $NaNH_2$ at 150°C for 1.0 hour and product mixture is distilled to give a pure alkyne (E) of boiling point 71°C. Identify A to F.

31. Propose structures for A to L:

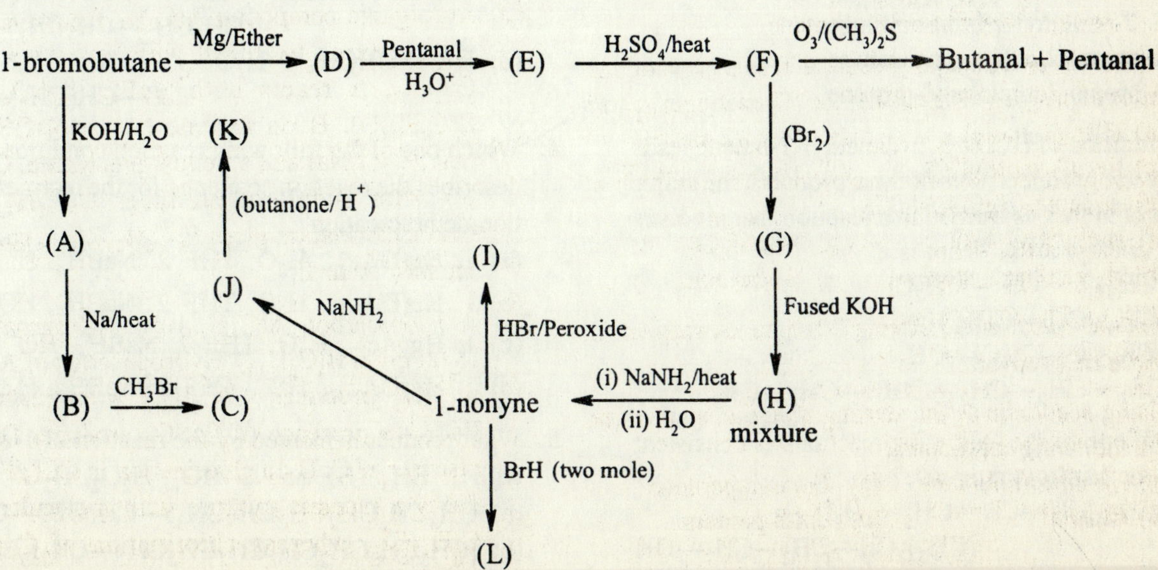

32. A hydrocarbon A containing 88.24% C (MW 68), when reacted with sodium followed by n-propyl chloride, produced B(C_8H_{14}). A gave a ketone C($C_5H_{10}O$) with $HgSO_4/H_2SO_4$. B on refluxing with $KMnO_4$ gives two isomeric acids D and E with their molecular formula $C_4H_8O_2$. Identify A to E.

33. A hydrocarbon was found to contain 85.7% C and 14.3% H. This compound consumes 1.0 molar equivalent of hydrogen to give a saturated hydrocarbon (B). One gram of A just decolourized 38.05 g 95% solution of Br_2 by weight. Compound A on oxidation with concentrated $KMnO_4$ gave a compound (C) with molecular formula C_4H_8O and acetic acid. Compound C can easily be prepared by the action of $HgSO_4/H_2SO_4$ on 2-butyne. Deduce structural formula of A, B and C.

34. Deduce the structures of compounds A—D. Draw the structures that show stereochemistry, where appropriate:

$C_6H_{10}O_4$(D) (Optically inactive) + [acetone] + [acetic acid, OH]

↑ (1) Hot $KMnO_4$, OH (2) H_3O^+

$C_{11}H_{20}$(B) (Optically active) ← H_2/Lindlar's catalyst — $C_{11}H_{18}$(A) — (a) Li/$EtNH_2$ (b) NH_4Cl → [diene structure with H, H]

↓ (a) O_3 (b) Zn, HOAc

$C_6H_{10}O_3$(C) (Optically active) + [acetone] + [acetic acid, OH]

OBJECTIVE EXERCISE I

1. What product is formed by HBr addition to 2-cyclobutylpropene?
(a) 2-bromo-2-cyclobutyl-propane
(b) 1, 2-dimethyl-1-bromocyclopentane
(c) 2-bromo-2-cyclopropyl-butane
(d) 1-bromo-2-cyclobutyl-propane

2. The addition of HCl to 3, 3-dimethyl-1-butene leads to several products. One of these products, the major product in fact, is formed in a reaction that involves a 1,2-methyl shift. What is it?
(a) $(CH_3)_3C—CH_2—CH_2Cl$
(b) $(CH_3)_2CCl—CH(CH_3)_2$
(c) $(CH_3)_3C—CHCl—CH_3$
(d) $CH_3—CH_2—CH_2—CHCl—CH_2—CH_3$

3. Which one of the following reactions is consistent with the Markovnikov rule?
(a) $(CH_3)_2CH—CH{=}CH_2 + H_2O \rightarrow (CH_3)_2CH—CH_2—CH_2—OH$
(b) $CH_3—CH_2—CH{=}CH_2 + H_2O \rightarrow CH_3—CH_2—CH_2—CH_2—OH$
(c) $CH_3CH{=}CHCH_3 + H_2O \rightarrow CH_3CH_2—CH(CH_3)—OH$
(d) $H_2C{=}CHCH_3 + H_2O \rightarrow H_3C—CH(CH_3)—OH$

4. Which one of the following reaction conditions best describes the reaction sequence for the oxymercuration-demercuration?
(a) 1. $Hg(OAc)_2$, H_2O, THF 2. $NaBH_4$, H^+
(b) 1. $Hg(OAc)_2$, H_2O, THF 2. $NaBH_4$, HO^-
(c) 1. $Hg(Ac)_2$, H_2O, THF 2. $NaBH_4$, HO^-
(d) 1. $Hg(OAc)_2$, HO^-, THF 2. $NaBH_4$, H_2O

5. What product is formed by the reaction of 1-octene with 1. BH_3/THF and 2. HO^-, H_2O_2, H_2O?
(a) $CH_3CH_2CH_2CH_2CH(CH_3)—CH(OH)—CH_3$
(b) $[CH_3CH_2CH_2CH_2CH_2CH_2CH_2—CH_2]_3B$

(c) $CH_3CH_2CH_2CH_2CH_2CH_2CH(OH)—CH_3$
(d) $CH_3CH_2CH_2CH_2CH_2CH_2CH_2CH_2—OH$

6. What is the product of the reaction of 2-butene and diazomethane?
(a) 1, 3-dimethylcyclopropane
(b) 1, 2-dimethylcyclopropane
(c) 1, 1-dimethylcyclopropane
(d) methylcyclopropane

7. Which reagent is needed to convert diiodomethane into the Simmons-Smith carbenoid?
(a) Sn/Zn couple (b) Sn/Cu couple
(c) Zn/Cu couple (d) Zn/Ca couple

8. The bromination involves so-called bromonium ions. Among the following statements, identify the one that does *not* apply.
(a) Bromonium ions are 3-membered rings.
(b) Bromonium ions are intermediates in the formation of halohydrins.
(c) Bromonium ions are intermediates in the formation of vicinal dibromides by addition of bromine to alkene.
(d) Bromonium ions of asymmetrical alkenes are attacked by nucleophiles at the least substituted alkene C-atom.

9. Which of the following compounds is peroxyformic acid?
(a) HCO_2H (b) H_3CCO_3H
(c) $PhCO_3H$ (d) HCO_3H

10. Consider the epoxidation of 1, 2-dimethylcyclohexene with peroxybenzoic acid. In what positions will the methyl groups be in the epoxide?
(a) Axial, equatorial (b) Equatorial, equatorial
(c) Axial, axial

11. What monomer undergoes hydroxide-catalyzed polymerization to form Super Glue?
(a) cyano *alpha*-methylacrylate
(b) methyl *beta*-methylacrylate
(c) methyl *alpha*-methylacrylate
(d) cyano *beta*-methylacrylate

12. Which statement correctly describes the reaction coordinate diagram of a typical electrophilic addition reaction?
(a) Two transition states and one intermediate
(b) No transition state and one intermediate
(c) Two transition states and two intermediates
(d) One transition state and one intermediate

13. What product is formed in the peroxide-catalyzed hydrobromination of 1-pentene?
(a) $(CH_3)_2CH—CH_2—CH_2—Br$
(b) $CH_3—CH_2—CH_2—CH_2Br—CH_3$
(c) $(CH_3)_2CH—CHBr—CH_3$
(d) $CH_3—CH_2—CH_2—CH_2—CH_2—Br$

14. What product is formed in the acid-catalyzed reaction between methanol, $CH_3—OH$, and propene, $CH_3—CH{=}CH_2$?
(a) $HO—C(CH_3)_3$
(b) $CH_3—CH_2—CH_2—O—CH_3$
(c) $CH_3—CH_2—CH_2—CH_2—OH$
(d) $(CH_3)_2CH(OCH_3)$

15. Identify the product of the reaction of water and 1,5-hexadiene.
(a) $CH_3—C(OH)H—CH_2—CH_2—CH_2—CH_2—OH$
(b) $(CH_3)_3C—C(OH)_2(CH_3)$
(c) $CH_3—C(OH)H—CH_2—CH_2—CH(OH)—CH_3$
(d) $HO—CH_2—CH_2—CH_2—CH_2—CH_2—CH_2—OH$

16. Select the correct alkene for the synthesis of 2-methyl-2-butanol via oxymercuration-demercuration.
(a) $H_3C—C(CH_3){=}CH—CH_3$
(b) $H_2C{=}C(CH_3)—CH_2—CH_3$
(c) $H_3C—CH(CH_3)—CH{=}CH_2$
(d) Both (a) and (b) are correct.

17. What product is formed by the reaction of 3-methyl-1-butene with 1. BH_3/THF and 2. HO^-, H_2O_2, H_2O?
(a) $(CH_3)_2CH—C(OH)H—CH_3$
(b) $CH_3—CH_2—CH_2—C(OH)H—CH_3$
(c) $CH_3—CH_2—CH_2—CH_2—CH_2—OH$
(d) $(CH_3)_2CH—CH_2—CH_2—OH$

18. What is the product of the reaction between 1, 2-dichloro-1-propene and diazomethane?
(a) 1, 2-dichloro-1-methylcyclopropane
(b) 2, 2-dichloro-1-methylcyclopropane
(c) 1, 2-dichloro-2-methylcyclopropane
(d) 1, 1-dichloro-2-methylcyclopropane

19. What is the product of the Simmons-Smith reaction with cyclopentene?
(a) bicyclo[3.1.0]pentane
(b) bicyclo[3.1.0]hexane
(c) bicyclo[5.3.0]hexane
(d) bicyclo[5.3.0]pentane

20. The bond dissociation energies for Cl—Cl, for the double bond of ethene, and for C—Cl bonds are 58, 64, and 82 kcal mol^{-1}. Which statement correctly describes the addition of Cl_2 to ethene?
(a) exothermic by 38 kcal mol^{-1}
(b) exothermic by 42 kcal mol^{-1}
(c) endothermic by 40 kcal mol^{-1}
(d) endothermic by 42 kcal mol^{-1}
(e) exothermic by 40 kcal mol^{-1}

21. What is the product of the acid-catalyzed opening of cyclopentene oxide?
(a) *trans*-1, 2-pentanediol, meso form
(b) *cis*-1, 2-pentanediol, meso form
(c) *cis*-1, 2-pentanediol, racemic
(d) *trans*-1, 2-pentanediol, racemic

22. Consider the epoxidation with peroxybenzoic acid. Which statement best describes the activation entropy of this reaction?
(a) Large and negative since the reaction is bimolecular and the transition state is highly ordered.
(b) Slightly negative and typical for bimolecular reactions.
(c) Activation entropy is negligible.
(d) Large and positive since the reaction is bimolecular and the transition state is highly ordered.

23. What reagent is used to convert the ozonide into the final products of ozonolysis?
(a) CH_3SH (b) $CH_3SO_2CH_3$
(c) DMSO (d) CH_3SCH_3

24. Identify the one alkene that will not rearrange in the course of an electrophilic addition.
(a) vinylcyclopropane
(b) 2-methyl-2-butene
(c) methylenecyclopropene
(d) 3, 3-dimethyl-1-butene

25. Which reaction obeys the Markovnikov rule?
(a) $(CH_3)_2CH—CH{=}CH—CH_3 + HBr \rightarrow (CH_3)_3C—CBrH—CH_3$
(b) $(CH_3)_3C—CH{=}CH_2 + HBr \rightarrow (CH_3)_3C—CBrH—CH_3$
(c) $(CH_3)_3C—CH{=}CH—C(CH_3)_3 + HBr \rightarrow (CH_3)_3C—CBrH—CH_2—C(CH_3)_3$
(d) $(CH_3)_3C—CH{=}CH_2 + HBr \rightarrow (CH_3)_3C—CH_2—CH_2—Br$

26. Which product is formed in the peroxide-catalyzed hydrobromination of 1-methyl cyclohexene?
(a) 5-bromo-1-methyl cyclohexane
(b) 1-bromo-1-methyl cyclohexane
(c) 2-bromo-1-methyl cyclohexane
(d) 4-bromo-1-methyl cyclohexane

27. The addition of HBr to 3-methyl-1-butene leads to several products. One of these products, the major product in fact, is formed via a rearranged carbocation. Identify that product.
(a) $(CH_3)_2CBr—CH_2—CH_3$
(b) $(CH_3)_2CH—CH_2—CH_2Br$
(c) $CH_3—CH_2—CH_2—CH_2—CH_2Br$
(d) $(CH_3)_2CH—CHBr—CH_3$

28. Which one of the following statements about the Markovnikov rule is *not* true?
(a) In Markovnikov additions, the alkene C-atom with the most H atoms will be protonated.
(b) When HX adds to an asymmetrical alkene, the halogen attaches to the least substituted alkene C-atom.
(c) When HX adds to an asymmetrical alkene, the halogen attaches to the most substituted alkene C-atom.
(d) When HX adds to an asymmetrical alkene, the protonation leads to the most stable carbocation intermediate.

29. At 25°C, the second-order rate constant for the acid-catalyzed hydration of 2-butene is 8.32×10^{-8} M^{-1} s^{-1}. What is the rate of reaction when [2-butene] = 0.1 M and $[H^+]$ = 0.01 M?
(a) 8.32×10^{-11} M s^{-1} (b) 8.32×10^{-11} M s
(c) 8.32×10^{-5} M s^{-1} (d) 8.32×10^{-11} M^{-1} s^{-1}

30. What product is formed in the acid-catalyzed reaction between ethanol $CH_3—CH_2—OH$ and *iso*-butene $(CH_3)_2C{=}CH_2$?
(a) $CH_3—O—(CH_2)_2—CH(CH_3)_2$
(b) $CH_3—CH_2—O—CH_2—CH(CH_3)_2$
(c) $(CH_3)_2CH—CH_2—CH_2—CH_2—OH$
(d) $CH_3—CH_2—O—C(CH_3)_3$

31. What hydrocarbon serves as the starting material for the synthesis of 1-methoxy-1-methylcyclohexane via an alkoxymercuration?
(a) 1-methoxy cyclohexene
(b) 1-methoxy cyclohexane
(c) 1-methyl cyclohexane
(d) 1-methyl cyclohexene

32. How does "borane" exist in THF solution?
(a) as the trimer B_3H_9
(b) as the complex between borane and THF
(c) as the dimer B_2H_6
(d) as the monomer BH_3

33. How many moles of BH_3/THF react with 1 mole of 1-pentene?
(a) 0.25 mol BH_3/THF (b) 0.33 mol BH_3/THF
(c) 0.75 mol BH_3/THF (d) 1 mol BH_3/THF

34. Consider the heterogeneous catalytic hydrogenation of (*Z*)-3, 4-dimethyl-3-hexene. What is the stereochemical outcome of the reaction?
(a) Distereomers (b) A single enantiomer
(c) Racemate (d) *meso* form

35. Identify a preparation for 1, 1-dichlorocarbene.
(a) Chlorine-catalyzed dehydrochlorination of chloroform.
(b) Hydroxide-catalyzed dehydrochlorination of chloroform.
(c) Sodium-catalyzed dehydrochlorination of chloroform.
(d) Acid-catalyzed dehydrochlorination of chloroform.

36. The photochemical reaction between diazomethane and an alkene gives several side products as the result of carbene insertion. Which of the following compounds cannot be formed by such a side reaction between diazomethane and propene?
(a) 1-butene (b) (*Z*)-2-butene
(c) 2-methylpropene (d) (*E*)-2-butene

37. What reagent is the best electrophile?
(a) Bromine cation, Br^+ (b) Tribromide, Br_3^-
(c) Bromide, Br^- (d) Bromine radical, Br

38. What product is formed by bromination of 2-methyl-1-pentene?
(a) 2-bromo-2-methyl-1-pentene
(b) 1, 1-dibromo-2-methylpentane
(c) 1, 2-dibromo-2-methylpentane
(d) 1-bromo-2-methyl-1-pentene

39. What product is formed in the treatment of cyclopentene with bromine water?
(a) *trans*-3-bromocyclopentanol
(b) *cis*-2-bromocyclopentanol
(c) *trans*-2-bromocyclopentanol
(d) *cis*-3-bromocyclopentanol

40. Consider the epoxidation of *cis*-butene with peroxybenzoic acid. Which statement describes the stereochemical outcome of the reaction?
(a) *cis*-1, 2-dimethyloxirane, mostly as (*S, S*) enantiomer
(b) *cis*-1, 2-dimethyloxirane, mostly as (*R, S*) enantiomer
(c) *cis*-1, 2-dimethyloxirane, a racemate
(d) *cis*-1, 2-dimethyloxirane, a meso form

41. What is the product of the acid-catalyzed opening of an epoxide?
(a) 1, 3-diol (b) 1, 1-diol
(c) enol (d) 1, 2-diol

42. What is the molecular formula of MCPBA?
(a) $C_7O_3H_5Cl$ (b) $C_6O_2H_5Cl$
(c) $C_7O_2H_5Cl$ (d) $C_7O_3H_5Cl$

43. What products are obtained in the oxidation of 2-hexene with warm concentrated permanganate solution?
(a) acetate and butanone
(b) acetate and butanal
(c) acetate and butanoate
(d) acetaldehyde and butanoate

44. Identify the alkene with the lowest molar heat of hydrogenation.
(a) ethene (b) 2, 3-dimethyl-1-butene
(c) propene (d) 2, 3-dimethyl-2-butene

45. What causes the rotational barrier in alkenes?
(a) C(sp^2)-C(p) overlap
(b) C(sp^3)-C(sp^3) overlap
(c) C(p)-C(p) overlap
(d) C(sp^2)-C(sp^2) overlap

46. Which one of the following cyclic hydrocarbons can actually exist as *cis* and *trans* geometrical isomers?
(a) cyclooctene (b) cyclopentene
(c) cyclohexene (d) cycloheptene

47. Which of the following statements concerning the ring strain in cyclopropene is correct?
(a) The C—C═C angle is reduced by 30 degrees compared to its normal value in propene.
(b) The C—C═C angle is reduced by 40 degrees compared to its normal value in propene.
(c) The C—C═C angle is reduced by 50 degrees compared to its normal value in propene.
(d) The C—C═C angle is reduced by 60 degrees compared to its normal value in propene.

48. How many of the following five 1-alkenes—propene, butene, pentene, hexene, and heptene—are liquid at room temperature?
(a) 1 (b) 2 (c) 3 (d) 4

49. The boiling point of *trans*-dichloroethene is 48 degrees celsius. Which number best approximates the boiling point of *cis*-dichloroethene?
(a) 36 degrees celsius (b) 42 degrees celsius
(c) 48 degrees celsius (d) 54 degrees celsius

50. Which one of the following compounds has the largest dipole moment?
(a) chloroethene (b) 1, 1-dichloroethene
(c) *cis*-1, 2-dichloroethene (d) trichloroethene

51. Consider the dehydrohalogenation of 2-bromo-2-methylbutane with ethoxide and t-butoxide. Which of the following statements is true?
(a) Saytzeff product is major product in both cases.
(b) Hofmann product is major product in both cases.
(c) Ethoxide gives more, t-butoxide give less Hofmann product.
(d) Ethoxide gives more, t-butoxide gives less Saytzeff product.

52. What conditions are best for the dehalogenation of vicinal dibromides?
(a) Zn in acetic acid (b) Zn in hydrochloric acid
(c) Zn in acetone (d) Zn in alcohol

53. Aside from conc. sulfuric acid, what other strong acid is employed in the dehydration of alcohols?
(a) conc. hydrochloric acid (b) conc. acetic acid
(b) conc. phosphoric acid (d) conc. acetone

54. In the dehydration of an alcohol with concentrated sulfuric acid, which reaction step is the rate-limiting step?
(a) Protonation of alcohol.
(b) Elimination of water from protonated alcohol.
(c) Elimination of proton.
(d) Unimolecular elimination of hydroxide from alcohol.

55. The following four molecules all are "isoelectronic" with ethene: $H_2C{=}O$, $H_2C{=}NH$, $HN{=}NH$, and $NH{=}O$. Isoelectronic means that the molecules have the same numbers of electrons in the same types of orbitals. Which one of the four molecules can form geometrical isomers?
(a) $H_2C{=}O$ (b) $H_2C{=}NH$
(c) $HN{=}NH$ (d) None

56. Identify the alkene with the highest molar heat of hydrogenation.
(a) propene (b) 2-methylpropene
(c) *cis*-2-pentene (d) 2-methyl-2-butene

57. Which of the following statements concerning the ring strain in cyclobutene is correct?
(a) The C—C=C angle is reduced by 20 degrees compared to its normal value in butene.
(b) The C—C=C angle is reduced by 30 degrees compared to its normal value in butene.
(c) The C—C=C angle is reduced by 40 degrees compared to its normal value in butene.
(d) The C—C=C angle is reduced by 50 degrees compared to its normal value in butene.

58. Which of the following statements about the physical properties of alkenes is *false*?
(a) Alkenes have a lower density than water.
(b) The density of alkenes increases with the number of C-atoms.
(c) *Cis/trans* isomerism affects density slightly.
(d) Boiling points and densities correlate.

59. Which one of the following compounds has the largest dipole moment?
(a) chlorobenzene
(b) 1, 4-dichlorobenzene
(c) 1, 2, 3-trichlorobenzene
(d) 1, 2, 4, 5-tetrachlorobenzene

60. Consider the dehydrohalogenation of 2-bromo-2-ethylhexane with methoxide and t-butoxide. Which of the following statements is true?
(a) Methoxide gives more, and t-butoxide gives less Hofmann product.
(b) Methoxide gives more, and t-butoxide gives less Saytzeff product.
(c) Saytzeff product is the major product in both cases.
(d) Hofmann product is the major product in both cases.

61. Which statement correctly describes the thermodynamics of the dehydration of the alcohols with concentrated sulfuric acid?
(a) It is highly endothermic since the cleavage of C—H and C—OH bonds require more energy than is gained by forming the C=C double bond and water.
(b) It is highly exothermic since the cleavage of C—H and C—OH bonds require less energy than is gained by forming the C=C double bond and water.
(c) The dehydration is exothermic, and the water's reaction with the concentrated acid also is exothermic.
(d) The dehydration is endothermic, while the water's reaction with the concentrated acid is exothermic, but less.

62. Which one of the following compounds is not formed in the "catalytic cracking" of octane?
(a) pentane (b) butene
(c) propene (d) hydrogen gas

63. Which one of the following compounds is *not* formed in the metal catalyzed thermal dehydrogenation of butane?
(a) hydrogen gas (b) *cis*-butene
(c) *trans*-butene (d) 1, 2-butadiene

OBJECTIVE EXERCISE II

1. Give the major product of the following sequence:

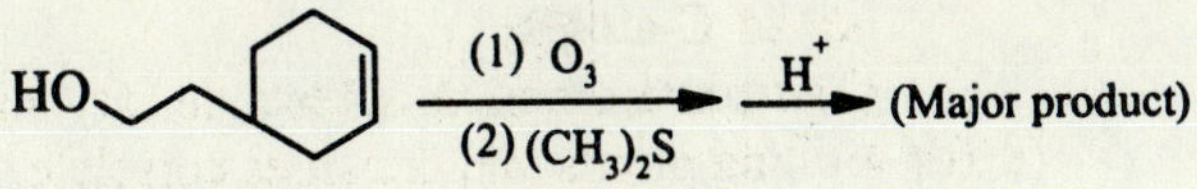

HO–CH_2CH_2–(cyclohexene) $\xrightarrow[(2)\ (CH_3)_2S]{(1)\ O_3}$ $\xrightarrow{H^+}$ (Major product)

(a) 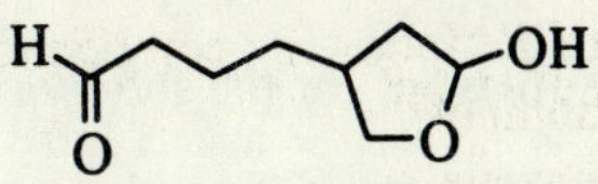 H, O, OH, O

(b) 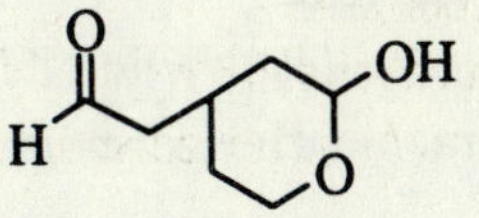 O, H, OH, O

(c) 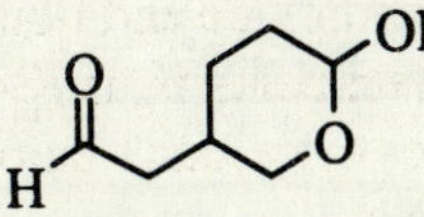 O, H, OH, O

(d) 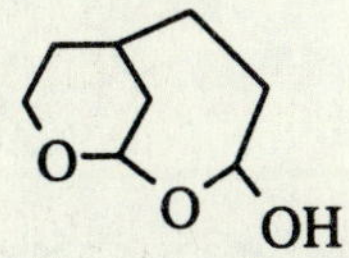 O, O, OH

2. Choose the best reagent for the following transformation:

1-Hexyne → Hexanal

(a) BH_3, H_2O_2, OH^-

(b) H^+/H_2O

(c) $HgSO_4/H_2SO_4$

(d) $(cyclohehyl)_2BH/H_2O_2/OH^-$

3. Give the major product from the following reaction:

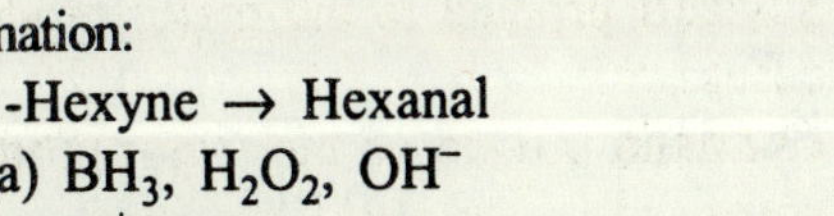

$\xrightarrow[(2)\ ROOR]{(1)\ NBS}$ (Major product)

(a) 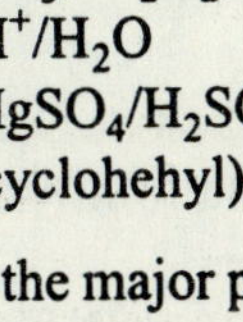 Br

(b) 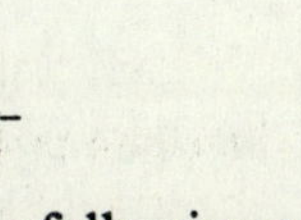 Br

(c) 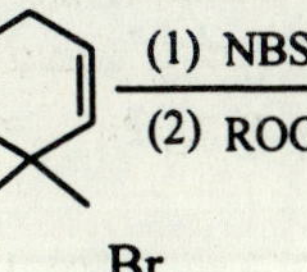 Br, Br

(d) 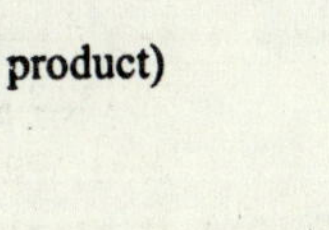 Br

4. How many different products (including stereomers) are obtained in the following reaction?

H_3C, CH_2CH_3, CH_3, H_3C, CH_3 $\xrightarrow{Light/NBS}$

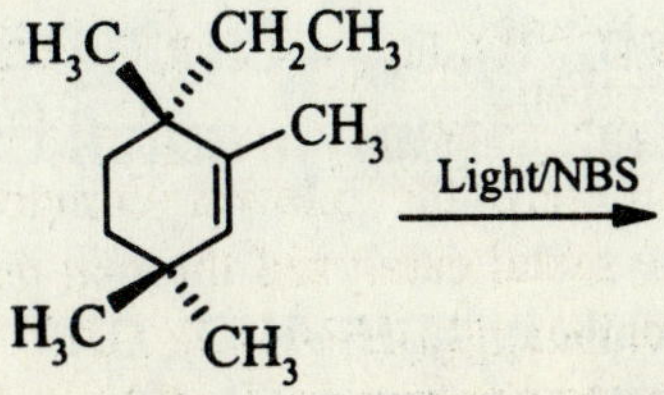

(a) 3 (b) 4 (c) 5 (d) more than 5

5. What is the major product of the reaction: methyl cyclopentene + $KMnO_4/H^+/Cold$→

(a) 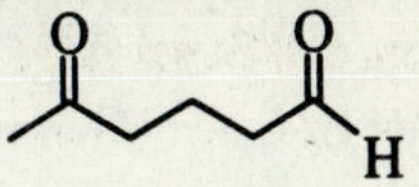 O, O, H

(b) 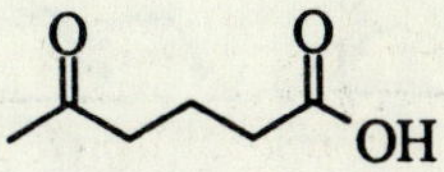 O, O, OH

(c) 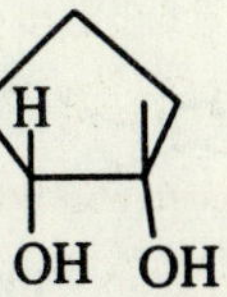 H, OH OH

(d) 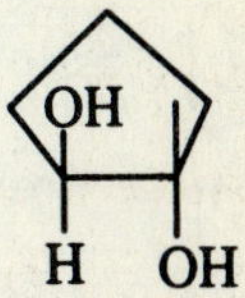 OH, H OH

6. Arrange the following reaction in decreasing order of reactivity with NBS/heat:

1. 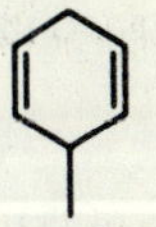 2.  3. 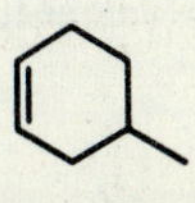 4.

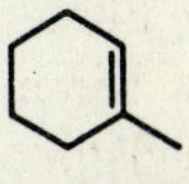

(a) 1 > 2 > 3 > 4 (b) 2 > 1 > 3 > 4

(c) 1 > 2 > 4 > 3 (d) 4 > 3 > 2 > 1

7. Give major product of the following reaction: $(CH_3)_3CCH{=}CHCH_3$ + NBS/ROOR → Major product(s)

1. 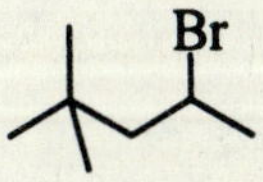 Br 2. 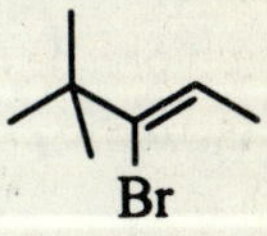 Br 3. 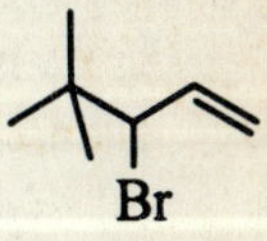 Br

4. 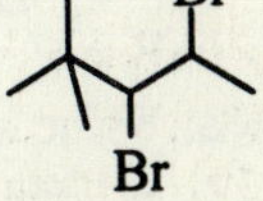 Br, Br 5. 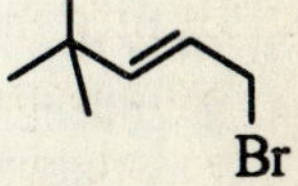 Br

(a) 1 only (b) 2 only

(c) 2 and 3 only (d) 3 and 5 only

8. Give the major products in the following reaction:

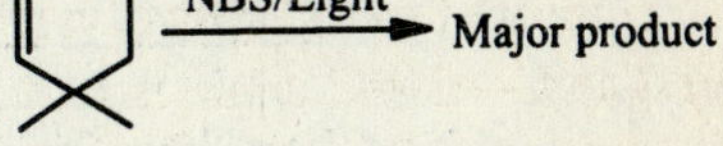

$\xrightarrow{NBS/Light}$ Major product

1. 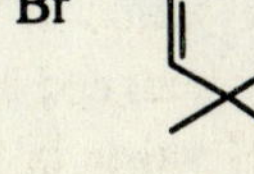 Br 2. 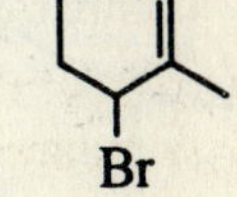 Br

3. 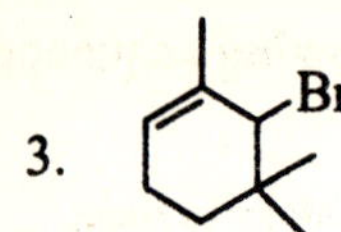

4. Br

(a) 1, 2 and 3 (b) 1, 2 and 4
(c) 1 and 3 (d) 2, 3 and 4

9. Give the major product of the following reaction:

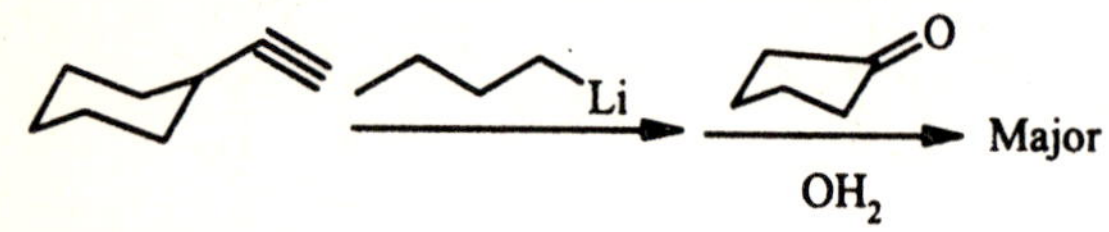

(a) 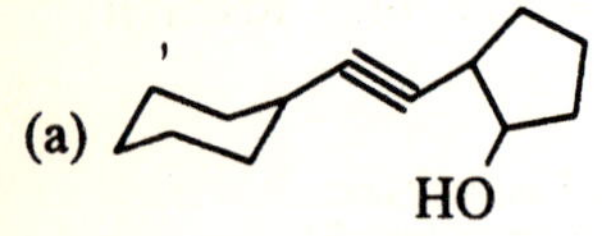

(b) 

(c) 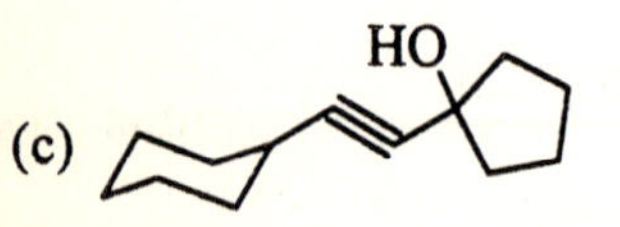

(d) 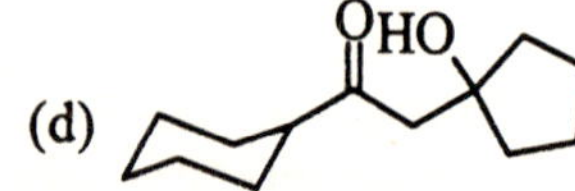

10. Give the intermediate(s) that is formed in the following reaction:

Na/ liq NH_3 → Intermediates

(a) (b) 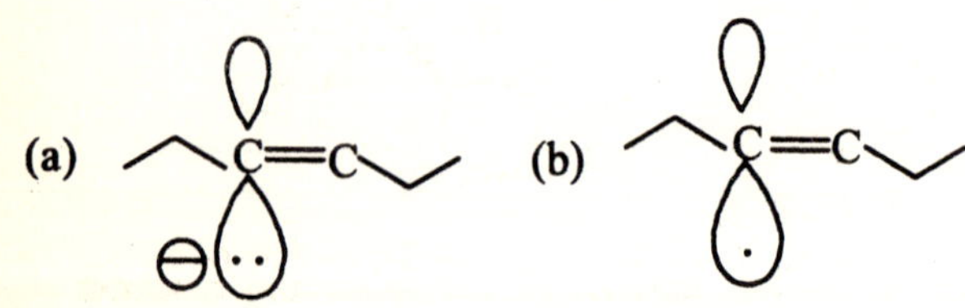

(c) 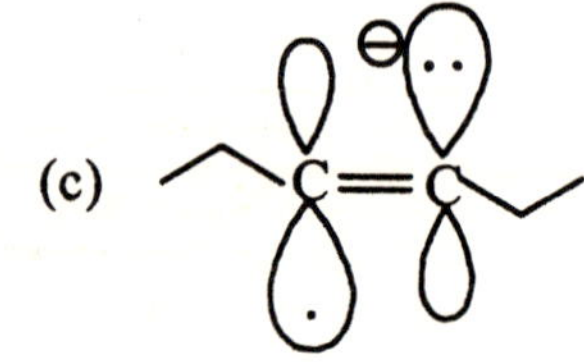

(a) only a (b) only c
(c) a and c (d) a, b and c

11. Give the major product of the following sequence of reaction:

one eq HBr → Li/CuI → (cyclohexyl-CH$_2$CH$_2$I) → Major product

(a)

(b)

(c) Br

(d) Br

12. Choose the best reagent for the following transformation:
2-heptyne → *cis*-2-heptene
(a) H_2/Pt (b) Na/liq NH_3
(c) H^+/H_2O (d) H_2/Lindlar's catalyst

13. Give the major product of the following reaction:

Bu-Li/EtBr → Major

(a) (b)

(c) (d)

14. Give the major product of the following reaction:

$[\text{cyclohexyl}]_2BH$; H_2O_2, OH^-

(a) C≡C (b)

(c) (d) CHO

15. Give the best reaction sequence that would accomplish the following conversion:
1-Pentyne → 4-methyloct-5-yn-4-ol
(a) $(cyclohexyl)_2BH \rightarrow H_2O_2$, $OH^- \rightarrow CH_3CH_2C{\equiv}CMgBr$, H^+/H_2O
(b) $HgSO_4$, $H_2SO_4 \rightarrow CH_3CH_2C{\equiv}CH$, H^+/H_2O
(c) $HgSO_4$, $H_2SO_4 \rightarrow CH_3CH_2C{\equiv}CMgBr \rightarrow H^+/H_2O$
(d) $(cyclohexyl)_2BH \rightarrow H_2O_2$, $OH^- \rightarrow CH_3CH_2C{\equiv}CH$, H^+/H_2O

16. Give the major product of the following reaction:

1-butyne + 1.0 eq of HBr/Peroxide $\xrightarrow{\text{Li}}$ $\xrightarrow[(CH_3)_2CHBr]{\text{CuI}}$

(a) 3-methylhex-3-ene (b) 2-methylhex-3-ene
(c) 2-methylhex-2-ene (d) 2-methyl-3-hexyne

17. Give the major product of the following reaction:
1-hexyne + dicyclohexylborane/H_2O_2/OH^- → Major

(a) 1-hexanal (b) Hex-1-en-2-ol
(c) Hex-1-yn-1-ol (d) 2-hexanone

18. Give the best reaction sequence to prepare the following compound:

(a) $\xrightarrow[\text{ROOR}]{\text{BrH}}$ $\xrightarrow{\text{Li}}$ $\xrightarrow{CuI/CH_3CH_2CH_2I}$

(b) $\xrightarrow[\text{ROOR}]{\text{ClH}}$ $\xrightarrow{\text{Li}}$ $\xrightarrow{CuI/CH_3CH_2CH_2I}$

(c) $\xrightarrow[CH_3CH_2CH_2Li]{\text{Dicyclohexylborane}}$ $\xrightarrow{H_3O+}$

(d) $\xrightarrow{HgSO_4,\ H_2SO_4}$ $\xrightarrow[H_3O+]{CH_3CH_2CH_2Li}$

19. Give the best reaction sequence to accomplish the following transformation:

H H

(a) $\xrightarrow[\text{ROOR}]{\text{One eq HBr}}$ $\xrightarrow{\text{Li}}$ $\xrightarrow{\text{Br}}$

(b) $\xrightarrow{\text{Bu-Li}}$ $\xrightarrow{\text{Br}}$ $\xrightarrow{H_2/\text{Lindlar's cat}}$

(c) $\xrightarrow{\text{Bu-Li}}$ $\xrightarrow{\text{Br}}$ $\xrightarrow{\text{Na/liq } NH_3}$

(d) $\xrightarrow{H_2/\text{Lindlar's cat}}$ $\xrightarrow{\text{Bu-Li}}$ $\xrightarrow{\text{Br}}$

20. Rank the following compounds with respect to increasing acidity:

1. 1-butene 2. Butane
3. 1-butyne 4. NH_3

(a) 4 < 3 < 1 < 2 (b) 2 < 1 < 4 < 3
(c) 2 < 1 < 3 < 4 (d) 3 < 1 < 2 < 4

21. Give the major product of the following sequence:

$\xrightarrow{Br_2}$ $\xrightarrow{NaNH_2,\ NH_3}$ Major product

(a) (b)

(c) NH_2 (d) NH_2 Br

22. Give the major product of the following reaction sequence:

$\xrightarrow[CH_3CH_2Br]{CH_3CH_2CH_2CH_2Li}$ $\xrightarrow{H_2/\text{Lindlar's cat}}$ $\xrightarrow{CH_3CO_3H}$ Major product

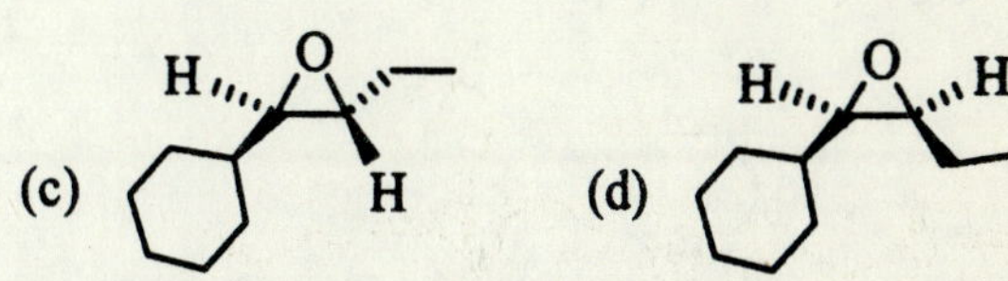

(c) H O H (d) H O H

23. Give the major product of the following reaction sequence:

$\xrightarrow{Br_2}$ $\xrightarrow[\text{liq } NH_3]{NaNH_2}$ $\xrightarrow[OH_2]{\text{Na, liq } NH_3}$ Major product

(a) Br (b)

(c) (d) Br

24. Give the major product of the following reaction sequence:

$\xrightarrow{H_2,\ \text{Lindlar's cat}}$ Major product

(a) (b)

(c) 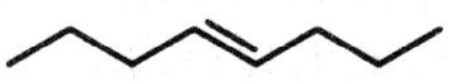 (d)

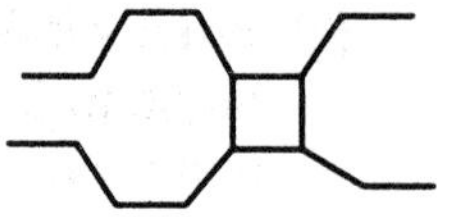

25. Predict major product of the following reaction sequence:

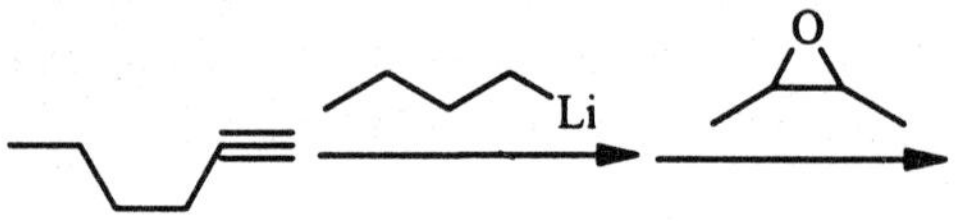

Li; Major

(a)  OH (b) 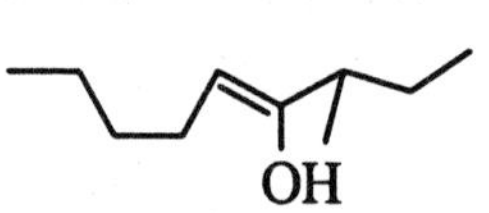 OH

(c) 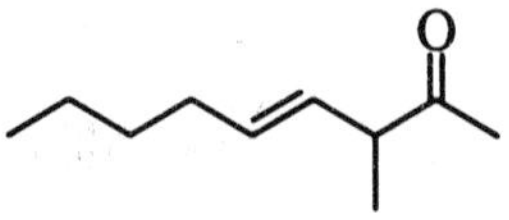 O (d)  OH

26. Give an intermediate that is formed in the following reaction:

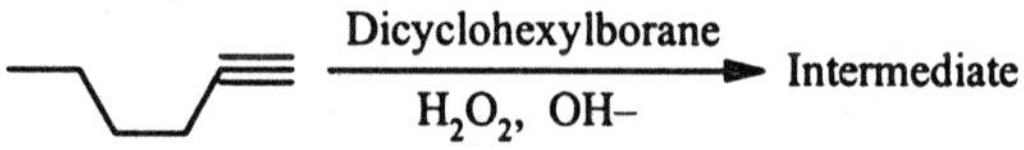

Dicyclohexylborane; H_2O_2, OH– Intermediate

(a) 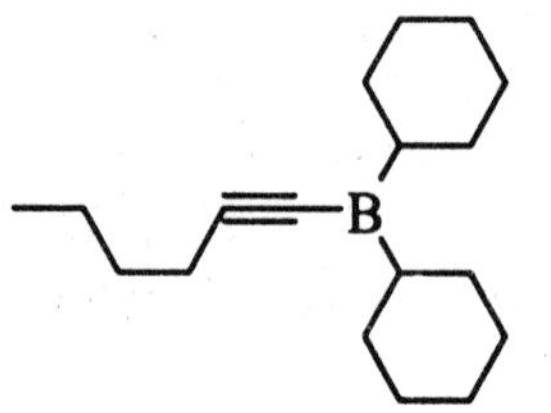 B (b) OH OH

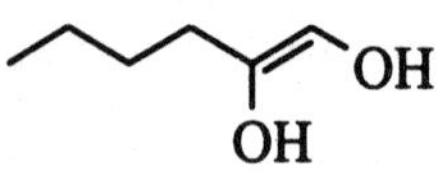

(c) 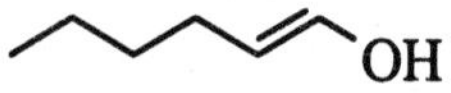 OH (d) 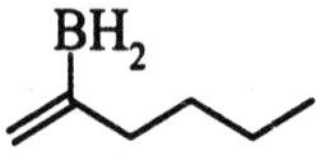 BH_2

27. Give the reagent that would best accomplish the following transformation:

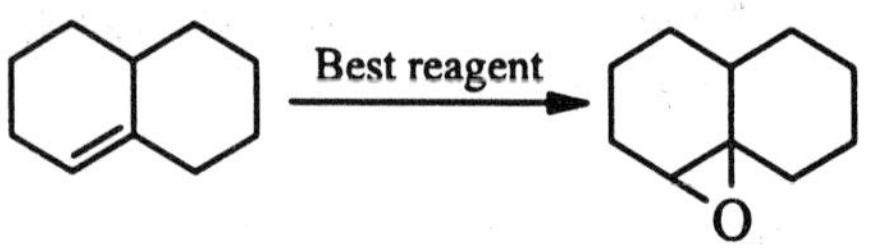

Best reagent; O

(a) MCPBA (b) CH_3COOH
(c) $(CH_3)_3COOC(CH_3)_3$ (d) Br_2/NaOH

28. Give the best reaction sequence for the preparation of indicated compound:

(a) 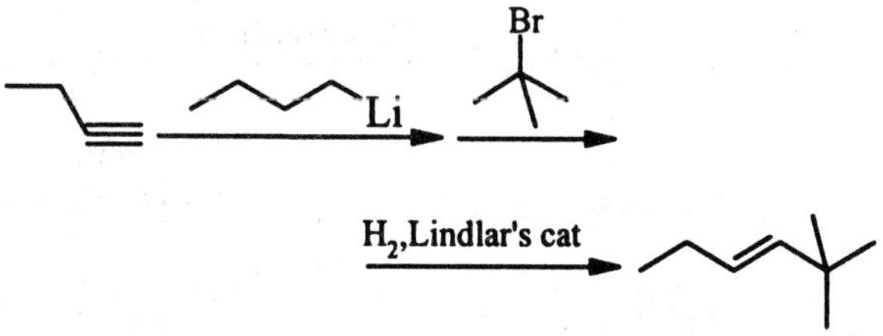 Li; Br; H_2,Lindlar's cat

(b) 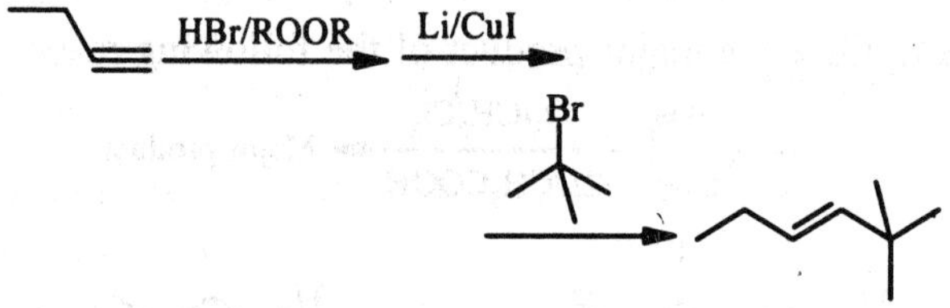 HBr/ROOR; Li/CuI; Br

(c) 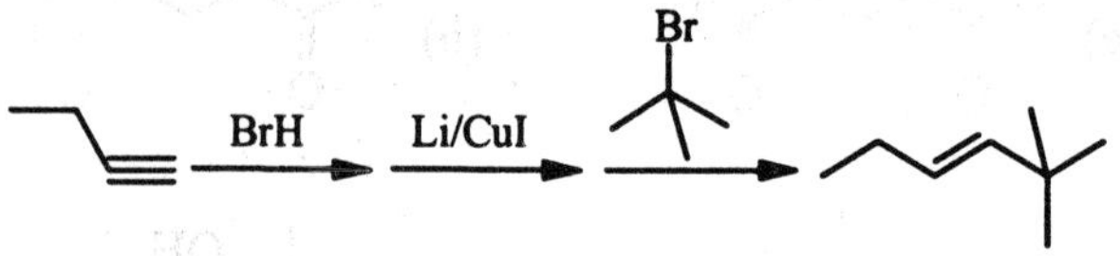 BrH; Li/CuI; Br

(d) 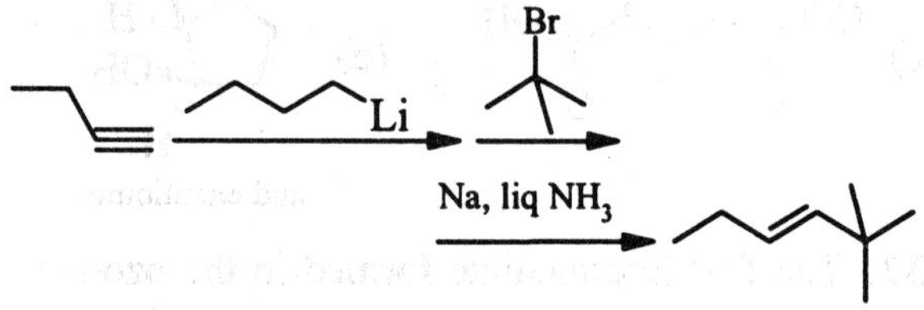 Li; Br; Na, liq NH_3

29. Give the major product of the following reaction:

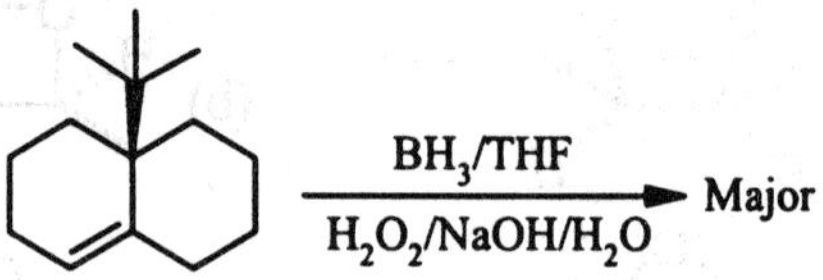

BH_3/THF; H_2O_2/NaOH/H_2O Major

(a) 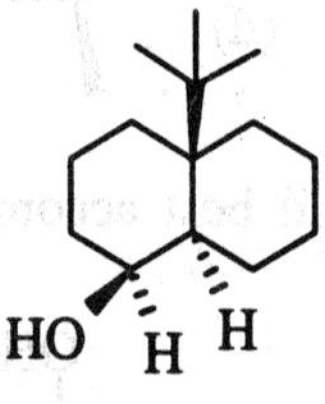 HO H H (b) 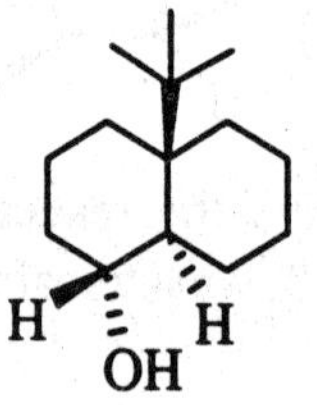 H OH H

(c) 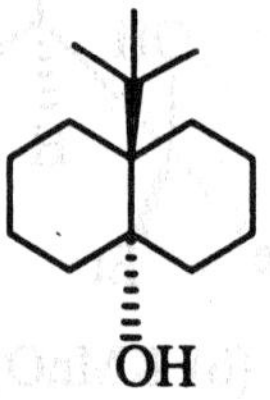 OH (d) 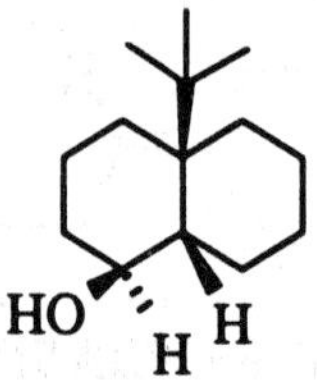 HO H H

30. Give the intermediate(s) formed in the mechanism of the following reaction:

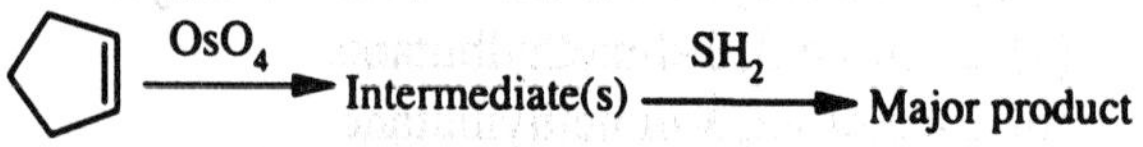

OsO_4 Intermediate(s) SH_2 Major product

(a) 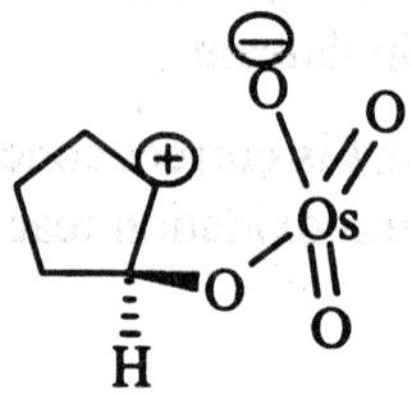 O Os O O O H (b) 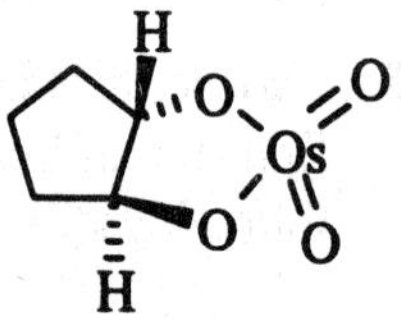 H O Os O O O H

(c) 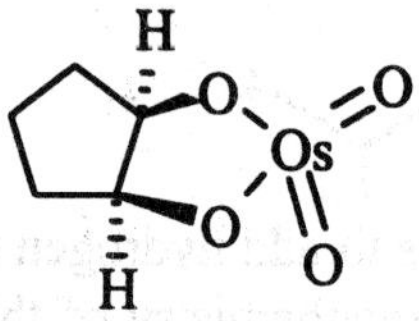 H O Os O O O H (d) Both (a) and (c)

31. Give the major product of the following reaction:

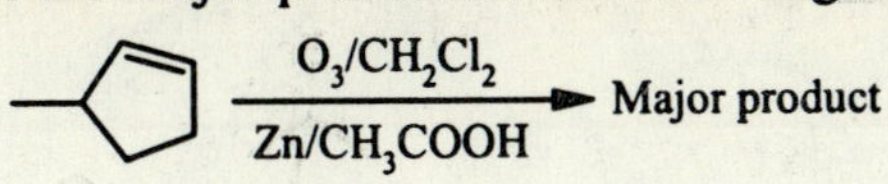

O_3/CH_2Cl_2; Zn/CH_3COOH → Major product

(a) 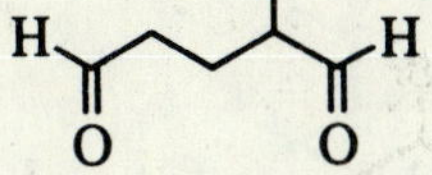 (b) 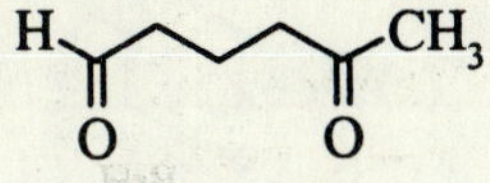 CH_3

(c) 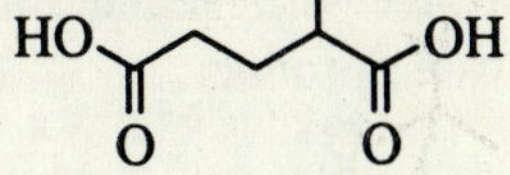 (d) 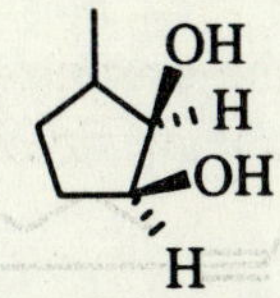 and enantiomer

32. The first intermediate formed in the ozonolysis of *trans*-2-butene is:

(a) 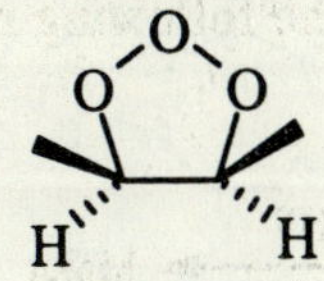 (b) 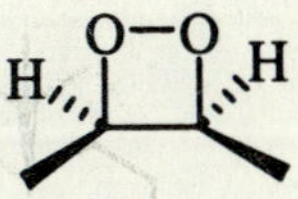 (c) 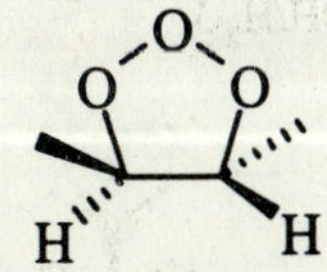 (d)

33. Give the reagent that would best accomplish the following transformation:

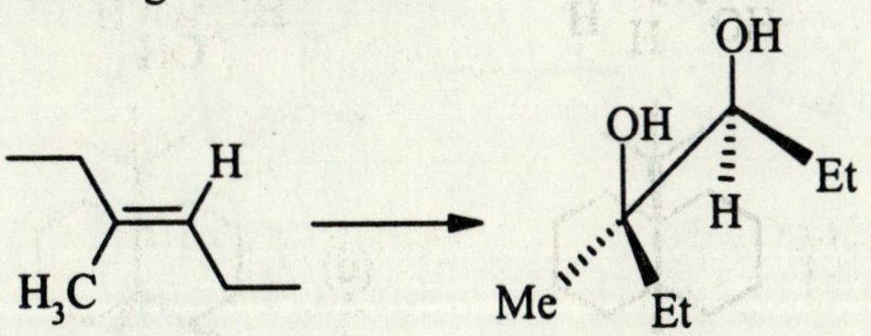

(a) $MCPBA/H^+/H_2O$ (b) $KMnO_4/H_2O$
(c) OsO_4/H_2S (d) All

34. Give the major product of the following reaction:
3, 3-dimethyl-1-butene + HCl → Major
(a) 2-chloro-3, 3-dimethylbutane.
(b) 2-chloro-2, 3-dimethylbutane.
(c) 1, 2-dichloro-3, 3-dimethylbutane.
(d) 1-chloro-3, 3-dimethylbutane.

35. Choose the statement that is correct concerning the first step of hydroboration oxidation reaction of the following alkene:

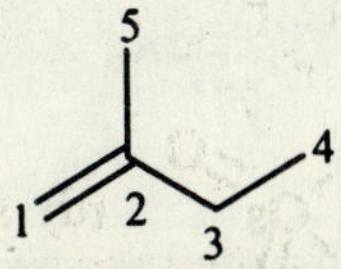

(a) Borohydride reacts to add hydrogen to C-1 and B to C-2. The stereochemistry of this reaction is controlled primarily by steric factor.
(b) Borohydride reacts to add hydrogen to C-2 and boron to C-1. The stereochemistry of this reaction is controlled primarily by steric factor.
(c) Borohydride reacts to add hydrogen to C-2 and boron to C-1. The stereochemistry of this reaction is controlled primarily by steric factor.
(d) Borohydride reacts to add hydrogen to C-1 and B to C-2. The stereochemistry of this reaction is controlled primarily by steric factor.

36. Give the reagent that best accomplishes the following transformation:

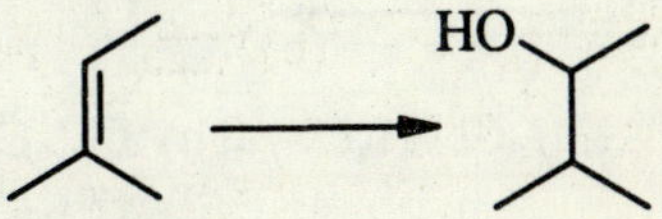

(a) H_2SO_4, Heat
(b) H_2SO_4, H_2O
(c) $(CH_3COO)_2Hg/NaBH_4/H_2O$
(d) BH_3,THF, H_2O_2, NaOH

37. Provide major product in the following reaction:

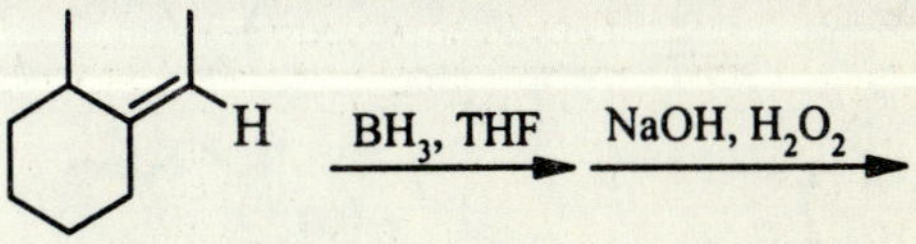

BH_3, THF; NaOH, H_2O_2 →

(a) 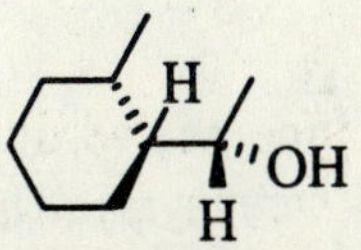 (and enantiomer) (b) 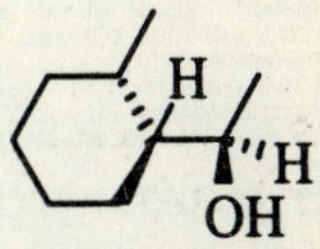 (and enantiomer)

(c) 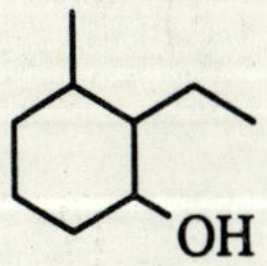 (d) 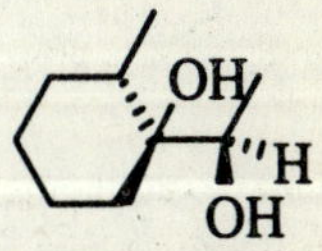 (and enantiomer)

38. Give the major product of the following reaction:

CH_3CO_3H → Major

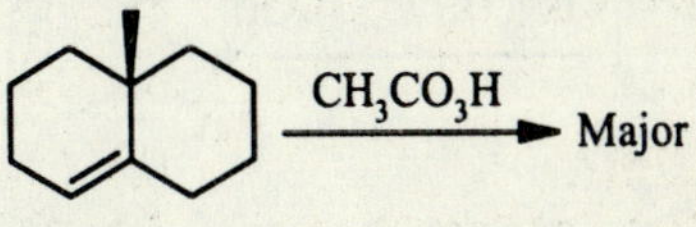

(a) 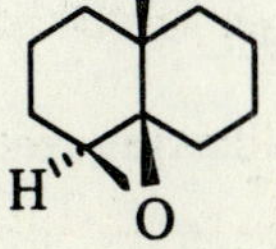 (b)

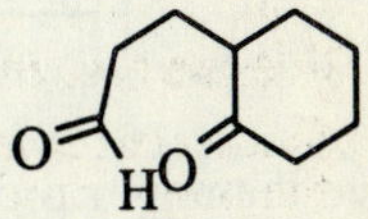

(c) 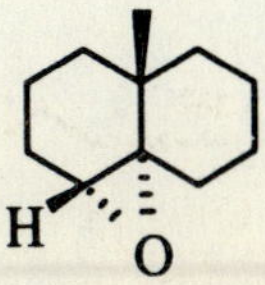 (d)

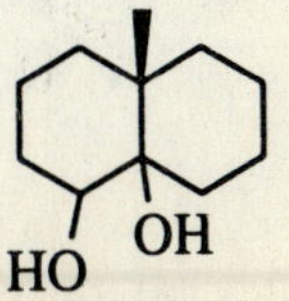

39. Which of the following compound is soluble in conc. sulphuric acid, decolourises bromine in CCl_4, and gives white ppt. with $AgNO_3$ in ammonia, but does not give any ppt. with $AgNO_3$ in ethanol?

(a) 1-pentene (b) 4-chloro-1-pentene
(c) 2-pentyne (d) 1-pentyne

40. Give the reagent that would best accomplish the following reaction:

HO OH H H

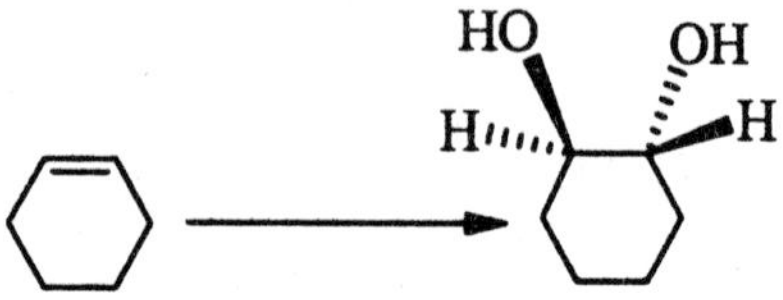

(a) MCPBA/CH_2Cl_2/H_3O^+ (b) CH_3CO_3H
(c) $KMnO_4$ (d) CH_3CO_2H

41. Give the major product of the following reaction:

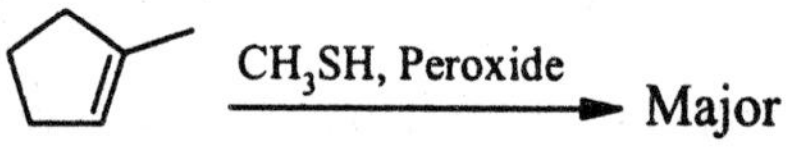

CH_3SH, Peroxide → Major

(a) 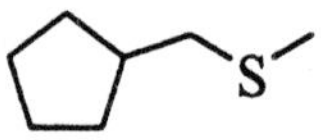 S (b)  S (c) 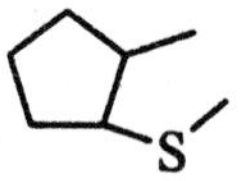 S (d) 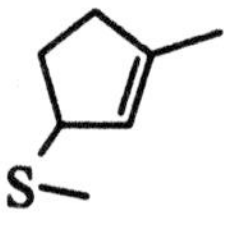 S

42. Choose the reaction sequence that would best accomplish the preparation of 2-methylcyclohexanol:

(a) 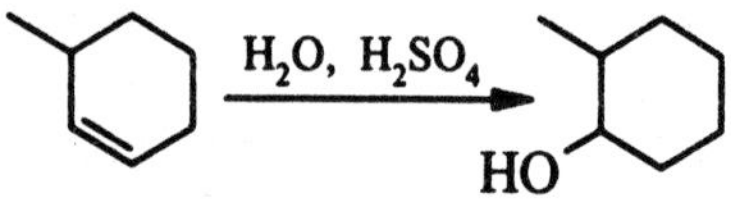 H_2O, H_2SO_4 → HO

(b) 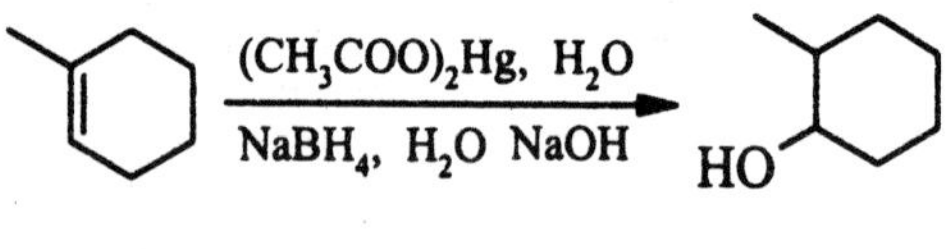 $(CH_3COO)_2Hg$, H_2O / $NaBH_4$, H_2O NaOH → HO

(c) 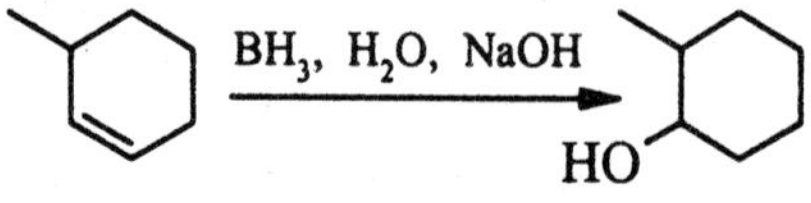 BH_3, H_2O, NaOH → HO

(d) 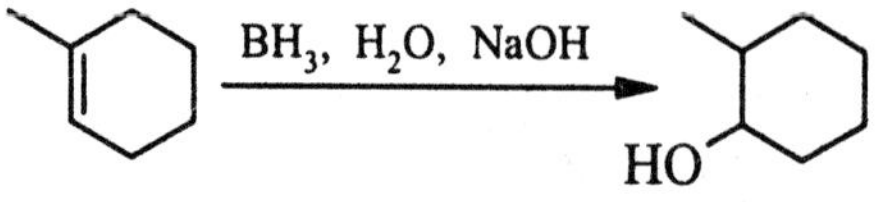 BH_3, H_2O, NaOH → HO

43. Give the major product of the following reaction:

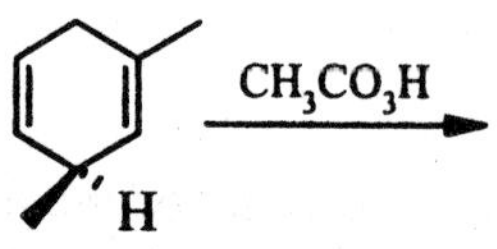

H CH_3CO_3H →

(a) 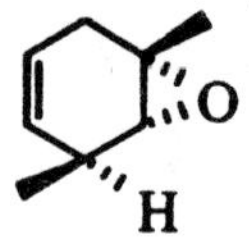 O H (b) 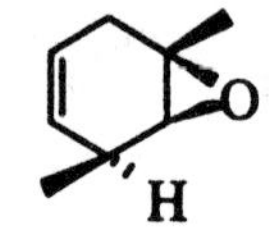 O H
(c) 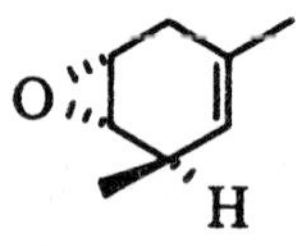 O H 1 (d) 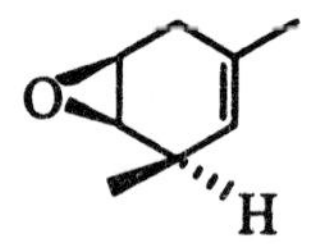 O H

44. Give the starting compound that when reacted as shown below, gives the indicated product:

Starting material? O_3/Zn, CH_3COOH → O + O

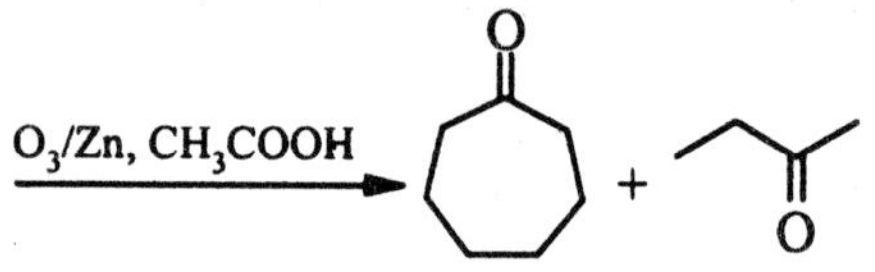

(a) 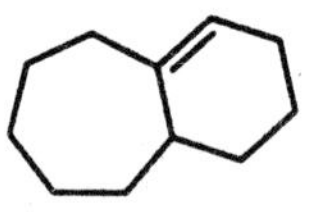 (b) 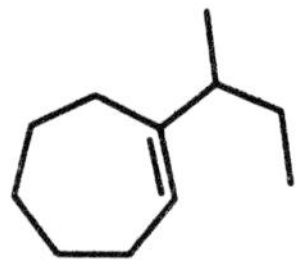 (c) 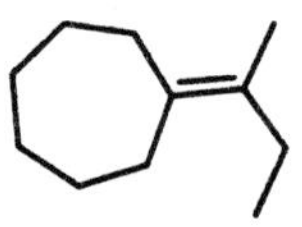 (d)

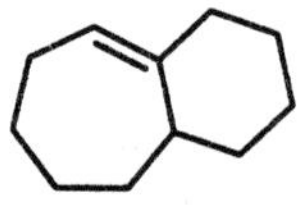

45. Give the best reagents that accomplish the following reaction:

→ H OH H HO

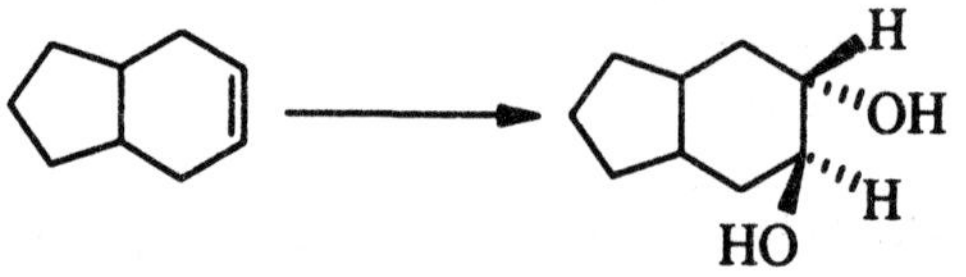

(a) OsO_4/H_2S (b) CF_3CO_3H/H^+
(c) H_2O_2/NaOH (d) MCPBA/H^+

46. Give the major product of the following reaction sequence:

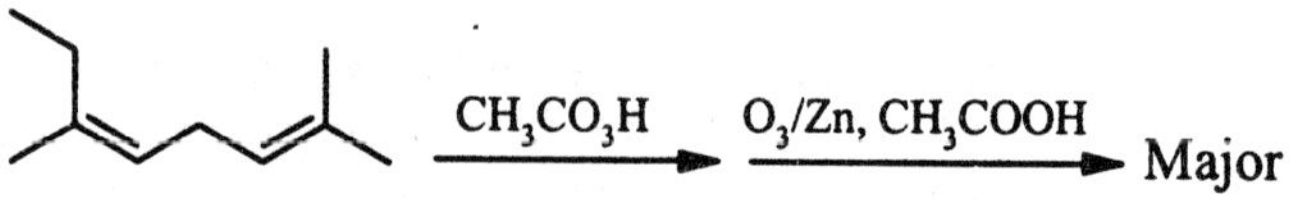

CH_3CO_3H → O_3/Zn, CH_3COOH → Major

(a) 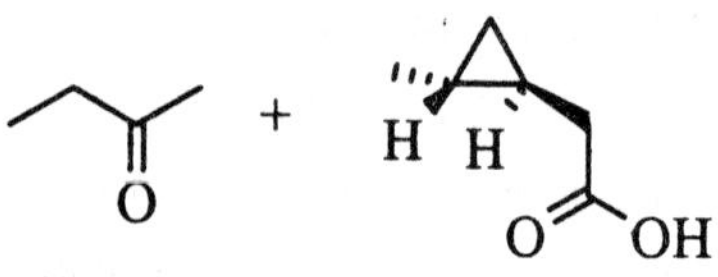 O + H H O OH
(and enantiomer)

(b) 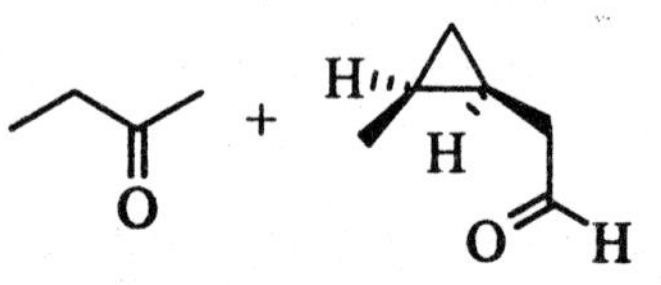 O + H H O H
(and enantiomer)

(c) + H, H, H, O, O

(and enantiomer)

(d) + H, H, COOH, O

(and enantiomer)

47. Give the major product of the following reaction:

ClH → Major

(a) Cl (b) Cl

(c) Cl (d) Cl

48. Choose the statement that is correct concerning catalytic hydrogenation.

(a) This reaction involves the direct reaction of H_2 with an alkene. The catalyst bonds only to alkene before the reaction occurs. Syn addition of H_2 must take place.

(b) This reaction involves the bonding of the catalyst first with alkene followed by bonding with H_2. Reaction takes place between alkene and hydrogen atoms on the surface of catalyst. Syn addition of H_2 must take place.

(c) This reaction proceeds first by the reaction of the catalyst with hydrogen. The alkene then bonds with the catalytic surface. Reaction between hydrogen and alkene, both bonded to surface take place. Syn addition of H_2 must take place.

(d) This reaction proceeds first by the reaction of the catalyst with hydrogen. The alkene then bonds with the catalytic surface. Reaction between hydrogen and alkene, both bonded to surface take place. Syn addition of H_2 must take place.

49. For the reaction in previous question, choose the reaction coordinate diagram that corresponds to this transformation:

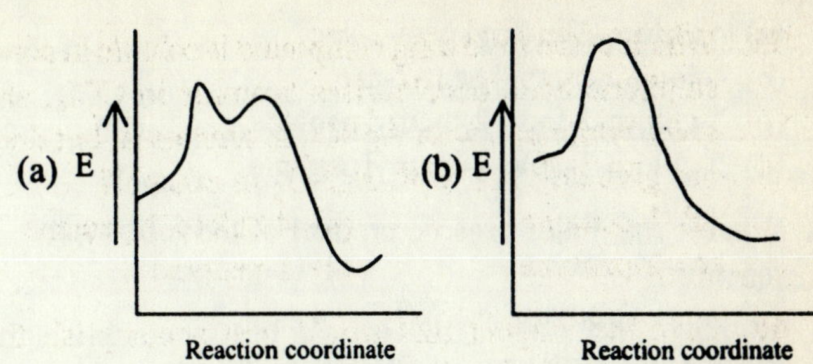

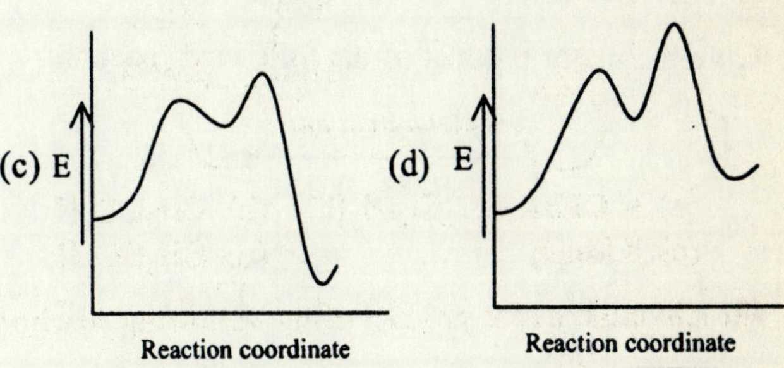

50. Give the major product of the following reaction:

Catalytic H^+ → Major product

(a) (b)

(c) (d)

51. Give the major product of the following reaction:

BrCN → Major

(a) Br, H, Br (b) Br, H, CN

(c) Br, CN, H (d) CN, H, Br

52. Give the major product of the following reaction:
2-methyl pent-2-ene + ICl → Major product

(a) 4-chloro-2-methyl pent-2-ene
(b) 4-iodo-2-methyl pent-2-ene.
(c) 3-chloro-2-iodo-2-methyl pentane
(d) 2-chloro-3-iodo-2-methyl pentane

53. What is the major organic product of the following reaction?

Propyne + NH_2^- →

(a) 1-amino propyne (b) $CH_3C{\equiv}C^-$
(c) $CH_3C{\equiv}CNH_3$ (d) $CH_3C{\equiv}C^+$

54. Give the major product of the following reaction:

$Hg(OAc)_2$ CH_3OH / $NaBH_4$, H_2O → Major

(a) H, OH (b) OH
(c) OMe (d) OMe

55. Give the major product of the following reaction:

CH_2CH_3, CH_3, CH_3, Br, H, H; EtONa →

(a) H (b)
(c) H (d) H, Br, H

56. Give the major product of the following reaction:

CH_3SCl →

(a) Cl, CH_3S, H (b) Cl, CH_3S, H
(c) Cl, H, CH_3S (d) Cl, H, SCH_3

57. Give the best reagent that would carry out the following conversion:

Br →

(a) NH_3 (b) EtONa/EtOH
(c) t-BuOK (d) $NaNH_2$

58. Give the product of the following reaction:

H, H; H_2/PtO_2 → Major

(a) H, H, H (b) H, H, H
(c) H, H, H (d) H, H, H, H

59. Give the intermediate that is formed in the following reaction:

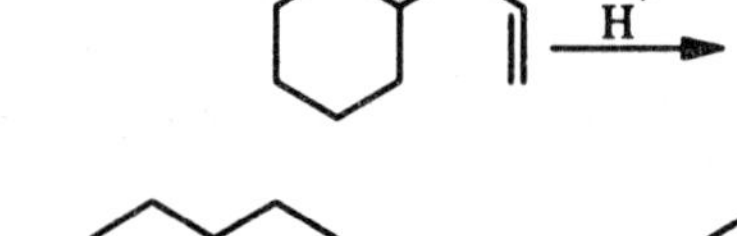

(a) (b)
(c) (d)

60. Give the major product of the following reaction:

OH; Br_2 → Major product

(a) Br, H, H, Br, OH (b)

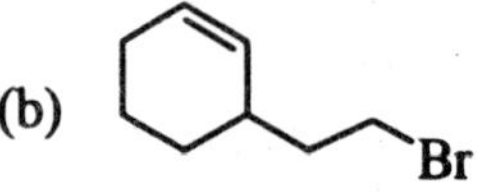

(c)

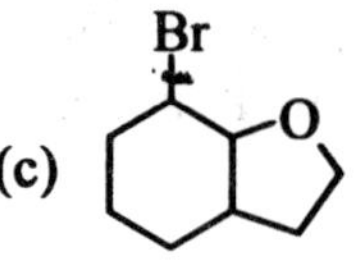

(d)

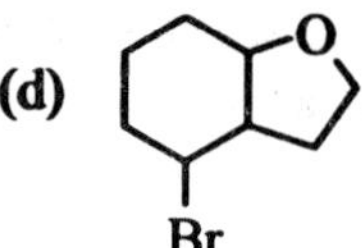

61. Give the statement that is true regarding the following reaction that generates two products.

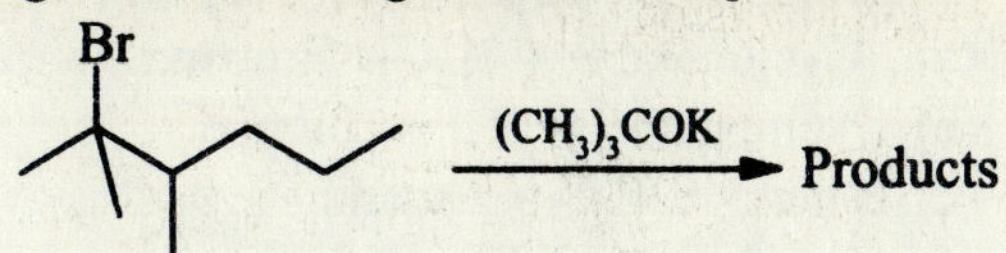

(a) The minor product that is formed faster is less stable than the major product.
(b) The major product that is formed faster is more stable than the minor product.
(c) The minor product is formed faster and is less stable than major product.
(d) The major product is formed faster and is less stable than the minor product.

62. Which of the following reacts with Tollen's reagent giving white precipitate?
(a) 2-pentene (b) 2-pentyne
(c) 1-butyne (d) 1-butanol

63. Which of the following compound has highest boiling point?
(a) n-hexane (b) 2, 3-dimethylbutane
(c) 2, 2-dimethylbutane (d) 2-methylbutane

64. The reagent that would distinguish between 1-hexyne and 1-hexene is:
(a) Ammonical silver nitrate (b) $KMnO_4$
(c) Br_2 in CCl_4 (d) H_2SO_4

65. The major monobromination product in the photochemical bromination of 2, 2-dimethylbutane is:

(a) Br (b) Br

(c) Br (d) Br

66. What is the major organic product of the following reaction?

Hg^{2+}, H_2SO_4

(a) O (b) H, OH, H

(c) H, O (d) H, H, HO

67. Which of the following reagent will bring about the following transformation?

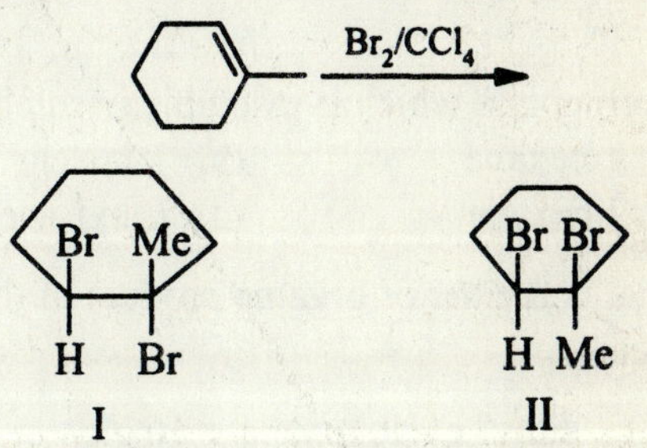

(a) Cl_2/H_2O (b) PBr_3/H^+
(c) $Hg(OAc)_2/H^+$ (d) Hydroboration oxidation

68. What are the major organic product of the following reaction?

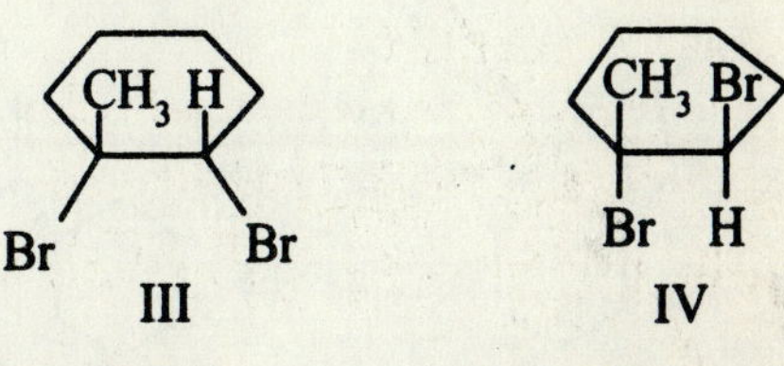

I: Br Me / H Br
II: Br Br / H Me
III: CH_3 H / Br Br
IV: CH_3 Br / Br H

(a) I and III (b) I and IV
(c) II and III (d) II and IV

69. Which of the following product is not formed in the following reaction?

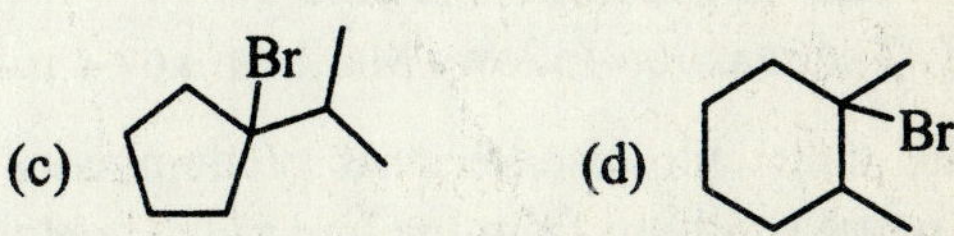

(a) Br (b) Br

(c) Br (d) Br

70. Which is the major addition product in the previous question?
(a) a (b) b (c) c (d) d

71. Which of the following species is an intermediate in the reaction shown below:

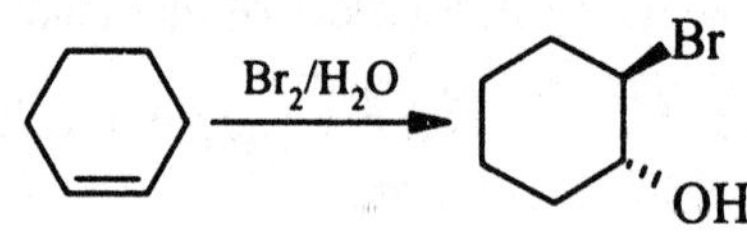

(a) 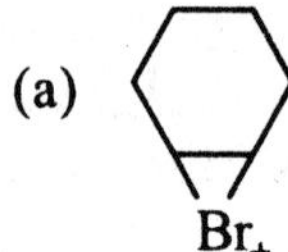

(b) 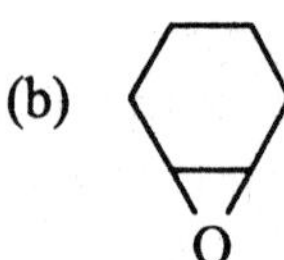

(c) 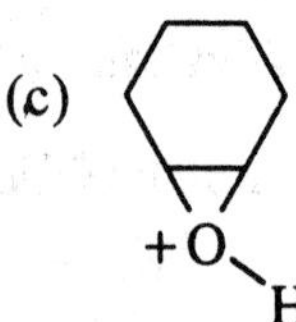

(d) 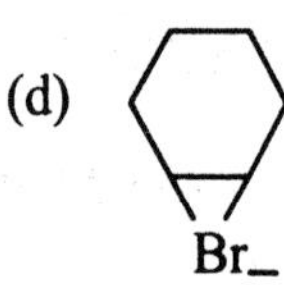

72. A compound which is insoluble in sulphuric acid is:
(a) *n*-hexane (b) 1-hexene
(c) 2-hexyne (d) Ethyl acetete

73. What is the major organic product of the following reaction:

Ethyl cyclohexane + $Br_2/h\nu \rightarrow$

(a) 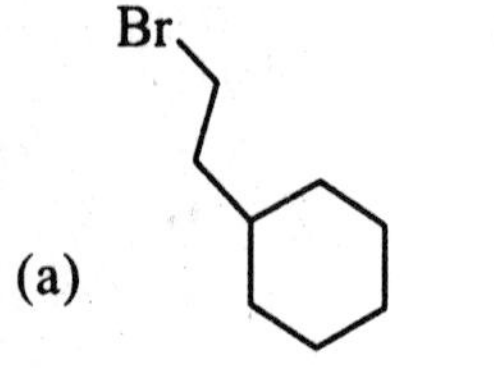

(b) 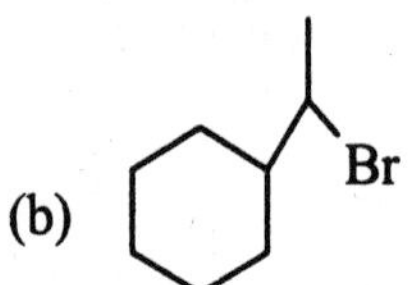

(c) 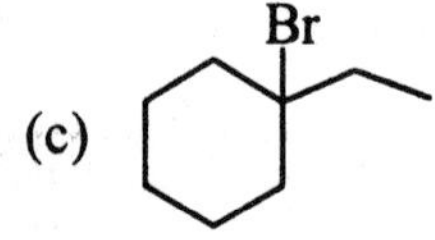

(d) Br

74. In methanol, Br_2 reacts with ethylene to yield $BrCH_2CH_2OCH_3$ in addition to 1, 2-dibromoethane because:
(a) MeOH solvates the bromine.
(b) The ion formed initially may react with Br^- or MeOH.
(c) This is a free radical reaction.
(d) The reaction follows Markovnikov's rule.

75. The monochloroisomer that predominates in the mixture resulting from the free radical chlorination of 1-pentene is:
(a) $ClCH_2CH_2CH_2CH{=}CH_2$
(b) $CH_3CHClCH_2CH{=}CH_2$
(c) $CH_3CH_2CHClCH{=}CH_2$
(d) $CH_3CH_2CH_2CH{=}CHCl$

76. Which metal catalyzed reaction would release the greatest amount of heat per CH_2 unit?
(a) Cyclohexene + $H_2 \rightarrow$ Propane
(b) Cyclobutane + $H_2 \rightarrow$ Butane
(c) Butene + $H_2 \rightarrow$ Butane
(d) Cyclohexane + $H_2 \rightarrow$ Hexane

77. Which of the following is not a characteristic of a free radical chain reaction?
(a) It produces a mole of product for a mole of free radical initiated.
(b) It gives the product derived from the most stable radical.
(c) It may be initiated by UV-light or high heat.
(d) It may be initiated by peroxide.

78. Which would yield 2-methyl-2-pentene when refluxed with Zn-dust in alcohol?
(a) 2, 3-dibromo-2-methyl pentane
(b) 3-bromo-2-methyl pentane
(c) 1, 2-dibromo-2-methyl pentane
(d) 3, 4-dibromo-2-methyl pentane

79. What is the major product obtained from the following reaction?

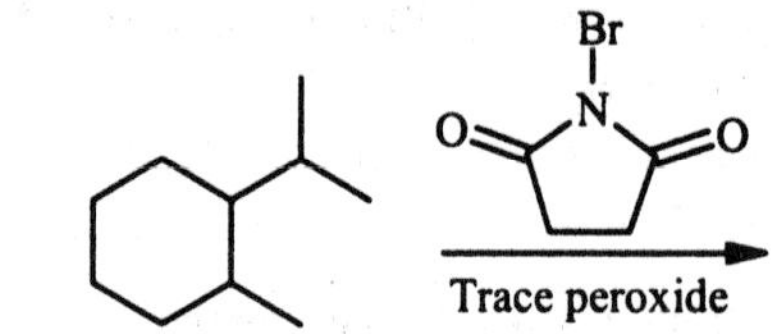

(a) 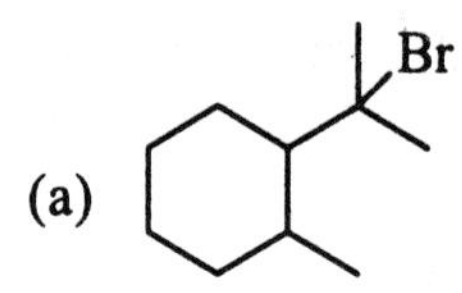

(b) 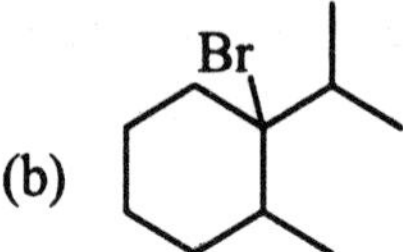

(c) 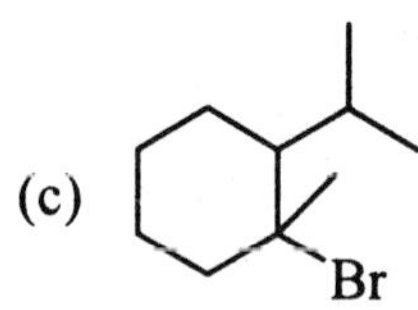

(d) 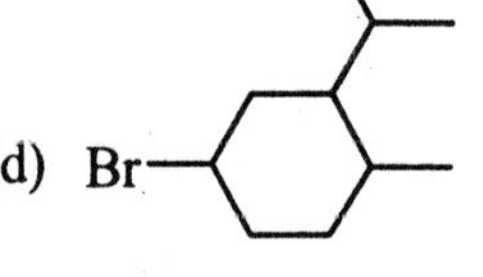

80. Which of the following compound would yield 5-keto-2-methylhexanal upon treatment with ozone:
(a) 1, 2-dimethyl cyclopentene
(b) 1, 3-dimethyl cyclopentene
(c) 1, 4-dimethyl cyclopentene
(d) 1, 6-dimethyl cyclopentene

81. Which of the following reaction would yield 2-pentyne in good yield:
(a) Oxitane + $CH_3MgBr/Et_2O \rightarrow H_3O^+ \rightarrow$
(b) $CH_3C{\equiv}CNa + CH_3CH_2Br \rightarrow$
(c) Propanal + $CH_3CH_2MgBr/H^+ \rightarrow$
(d) 1, 2-dibromo pentane + NaOH $\rightarrow$

82. Which of the following mechanism best describe the reaction between propene and BrCl?

(a) 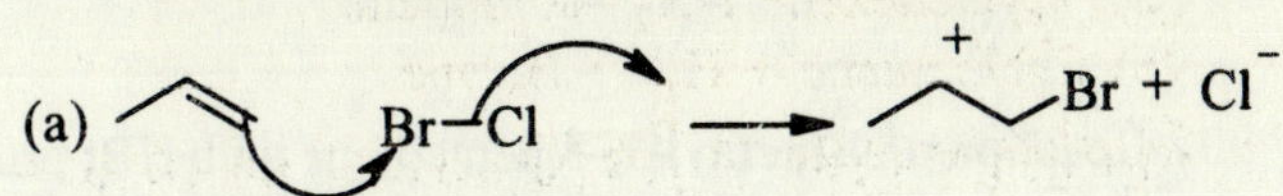

(b) 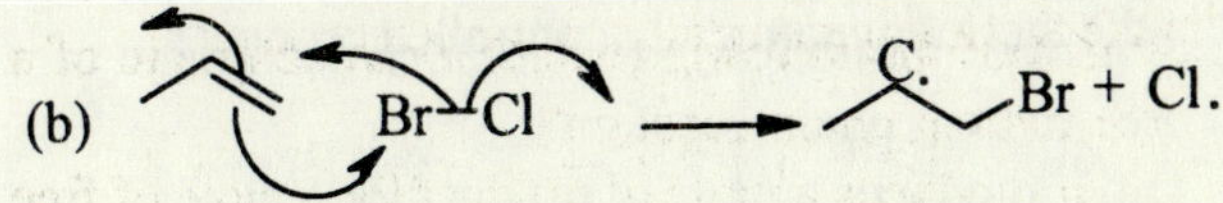

(c) Cl–Br ⟶ + Cl + Br^-

(d) Cl–Br + ⟶ Br + + Cl^-

83. From the list below, select the appropriate reagent for the following transformation:

Br ?→ ?→ H, O, O, H
↓ ?
O, O, H, H, Br, Br

(a) Br_2/CCl_4 (b) HBr
(c) KOH (d) 1 eq O_3; Zn, H^+
(e) Br_2/H_2O (f) 1 eq $KMnO_4$,H^+

84. Which molecule will give the following dicarboxylic acid upon treatment with acidic solution of $KMnO_4$?

HOOC, O, O, CH_2COOH

(a) (b)

(c) (d)

85. When 3-bromocyclohexene undergoes addition of HBr, the only intermediate formed is a bromonium ion. How many stereomers of that final product will be generated in that reaction:
(a) 2 (b) 4 (c) 6 (d) none.

86. When treated with acid, the strained ether shown below is protonated on oxygen, and ring opens to give a resonance-stabilized carbocation. Which diene listed below when treated with acid will give the same carbocation?

O

(a) HO (b) HO

(c) HO (d) HO

87. Choose the best reagent to carry out the following transformations:

Br I→ II→ III→ HO, OH

	I	II	III
(a)	Acetylene $NaNH_2/NH_3$	H_2/Pd	$OsO_4/NaHSO_3$
(b)	Acetylene $NaNH_2/NH_3$	H_2/Lindlar cat	$BH_3/NaOH$, H_2O_2
(c)	Acetylene $NaNH_2/NH_3$	Li/NH_3	$KMnO_4/NaOH$
(d)	Acetylene $NaNH_2/NH_3$	H_2/Lindlar cat	$OsO_4/NaHSO_3$

88. Which of the following reagent would best accomplish the following transformations?

?→

(a) Excess BH_3: $NaOH/H_2O_2$, followed by OsO_4: $NaHSO_3$.
(b) Excess $Hg(OAc)_2/H_2O$: $NaBH_4$ followed by H_2SO_4/heat.
(c) O_3: Zn, H^+, followed by $Hg(OAc)_2/H_2O$: $NaBH_4$.
(d) OsO_4: $NaHSO_3$ followed by NaOH.

89. What is the product of the following sequence of reaction?

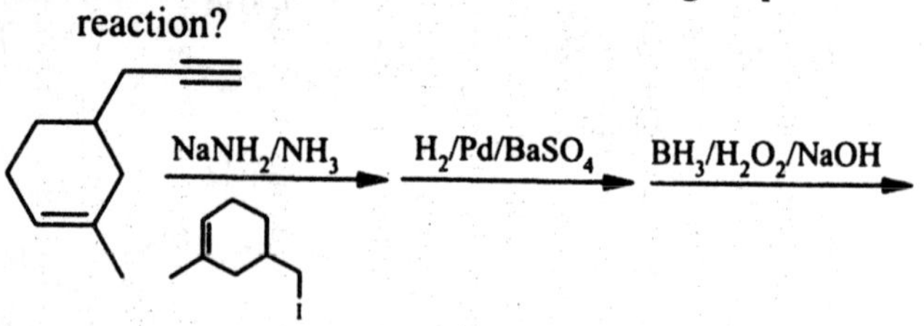

$NaNH_2/NH_3$ → $H_2/Pd/BaSO_4$ → $BH_3/H_2O_2/NaOH$ →

(a) 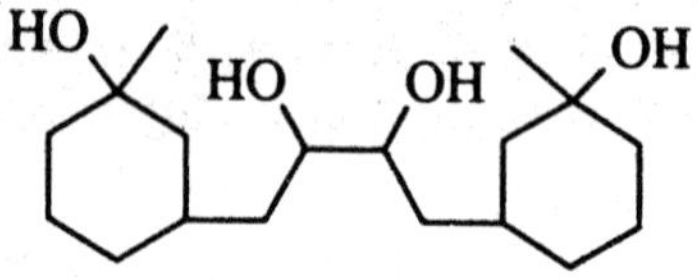 HO HO OH OH

(b) 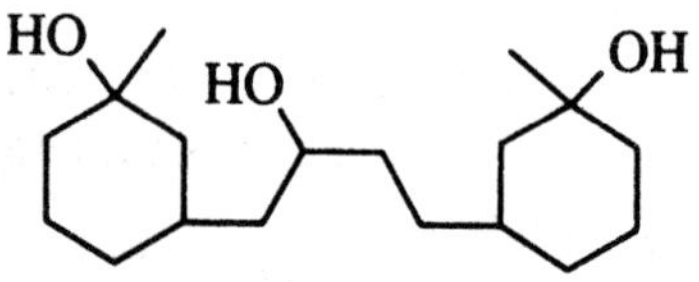 HO HO OH

(c) 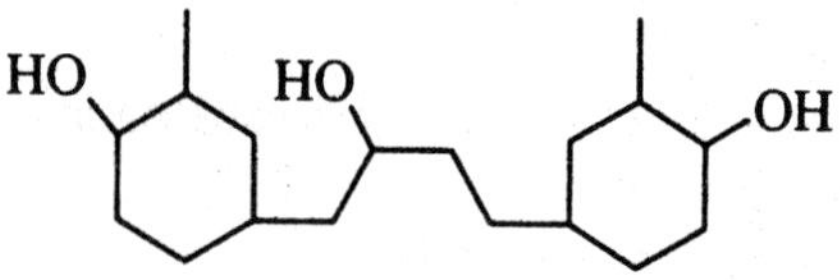 HO HO OH

(d) 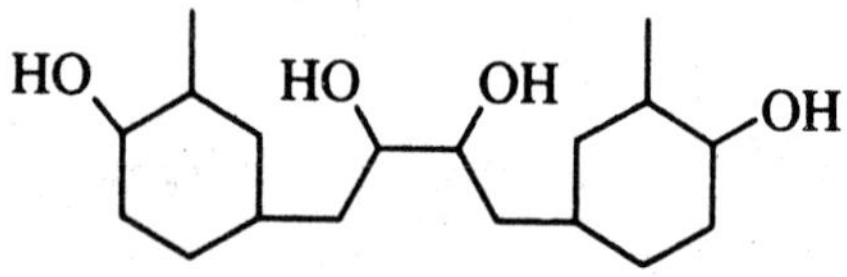 HO HO OH OH

90. For the following transformation choose the mechanistic step that best represents the reaction:

2-methyl-2-butene + BH_3/THF →

(a) 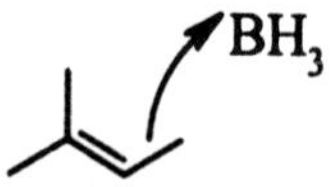 BH_3 (b) 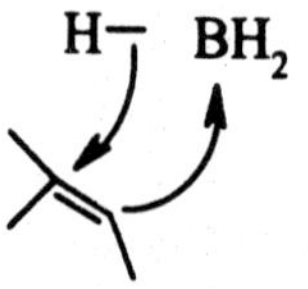 H— BH_2

(c) 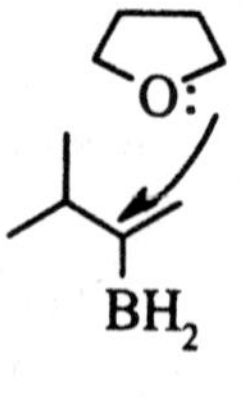 O: BH_2 (d) 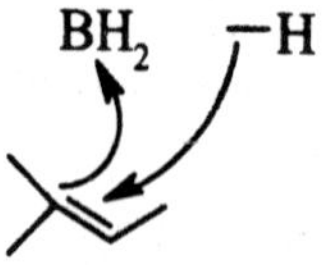 BH_2 —H

91. Of the following bases, which one may be used to create acetylide ion?

(a) EtO^- or OH^- (b) Et^-

(c) $CH_2{=}CH^-$ (d) both b and c

92. Which would produce chiral molecule after treatment with Lindlar's catalyst?

(a) 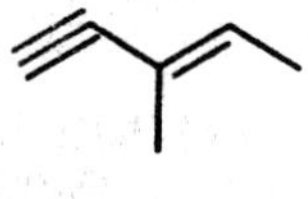 (b)

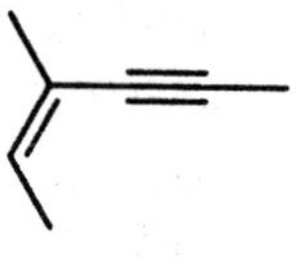

(c) 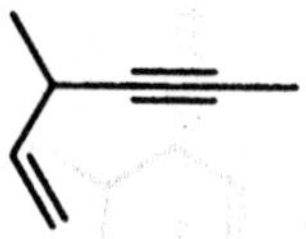 (d)

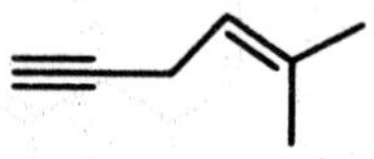

93. Reaction of 4-methyl-1, 3-pentadiene with HBr produces mostly the following products. Which one is the thermodynamically controlled product?

Br A Br B

(a) A (b) B (c) None (d) Both

94. Which of the above product in the previous question is kinetically controlled?

(a) A (b) B (c) None (d) Both

95. Which of the following is (are) the product(s) of the following reactions?

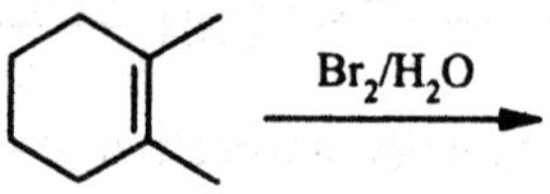

Br_2/H_2O →

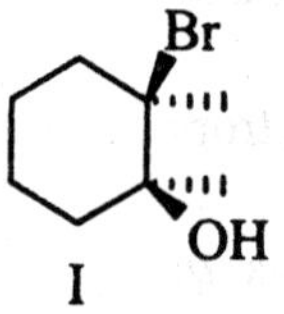

Br OH I Br OH II

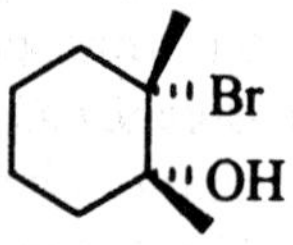

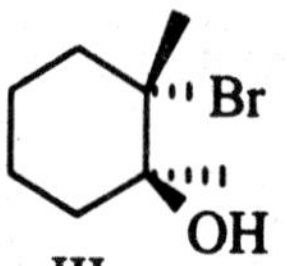

Br OH III Br OH IV

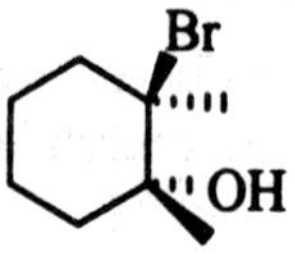

(a) I (b) II

(c) III (d) III and IV

96. How many stereomers of product will be formed in the following reactions?

$OsO_4/NaHSO_3$ →

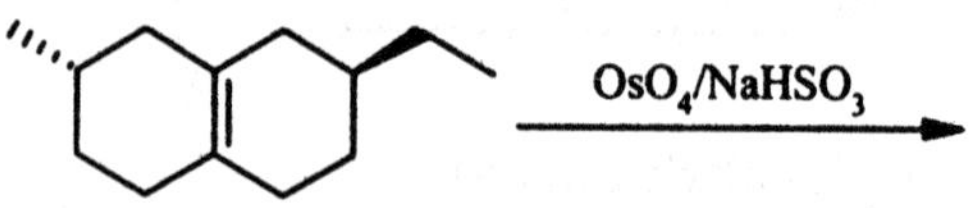

(a) 2 (b) 4 (c) 3 (d) None

97. Which of the following compounds was the starting material for the oxidation shown below?

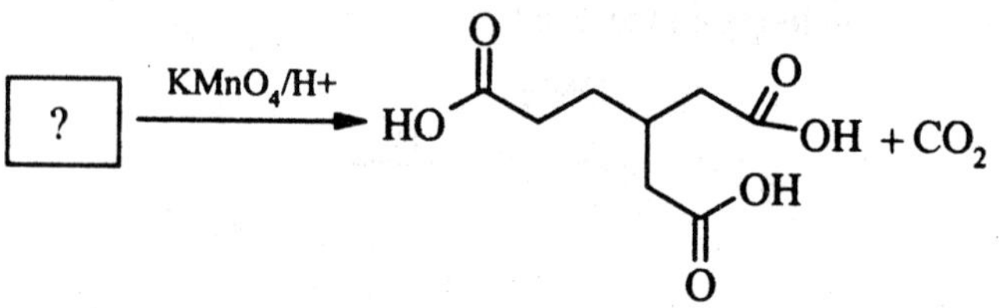

? $KMnO_4/H+$ → HO O O OH $+ CO_2$ OH O

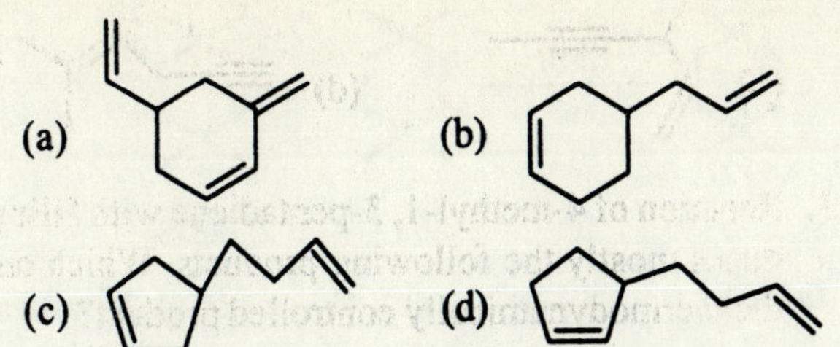

98. How is the following transformation best carried out?

(a) OsO_4: $NaHSO_3$ (b) H_2SO_4/H_2O
(c) $HgSO_4/H_2SO_4$ (d) HIO_4

99. Consider the following synthesis. What is wrong with this transformation?

BH_3-THF → BH_2, BH_2 → H_2O_2/NaOH

(a) BH_3 is actually unreactive towards alkyne.
(b) BH_3 should add to adjacent carbon rather than the same carbon.
(c) Borane can be used as an electrophile only with non-terminal alkyne.
(d) The borane addition follows regiochemistry opposite to that shown.

100. In which of the following solvent it is possible to create acetylide ion?
(a) AcOH (b) H_2O
(c) $(i\text{-}Pr)_2NH$ (d) Acetone

101. The following transformation is carried out in three steps. What is the appropriate reagent for the first step?

I → II → III

(a) H_2/Lindlar's catalyst
(b) $C_2H_2/NaNH_2/CH_3I$
(c) $NaNH_2/NH_3$: EtBr
(d) H_2/Pd/C

102. Where do you expect the best site of protonation for 2-methoxy-butadiene?

OMe, H^+

(a) (b) (c) (d)

103. Even though the rings are of different size, cyclopropane and cyclobutane have similar overall strain energy. This is because:
(a) They have similar angle compression from the optimal 109.5°.
(b) Strain energy is a kinetic phenomenon.
(c) Cyclobutane has more eclipsing interactions but less angle strain.
(d) Cyclopropane has a larger overall heat of combustion than cyclobutane.

104. Which of the following reactions involves a radical mechanism?

(a) HBr
(b) NBS, H_2O
(c) HBr, ROOR
(d) O_3, Zn, H_2O

105. The product(s) of the following reaction can best be described as:

HBr

(a) A racemic mixture
(b) A single enantiomer
(c) A pair of distereomers
(d) An achiral molecule

106. Which of the following reaction is most exothermic?
(a) Propane + Cl_2/hv → 2-chloro propane + HCl
(b) Propane + Br_2/hv → 2-bromopropane + HBr
(c) Isobutane + Cl_2/hv → *t*-butyl chloride + HCl
(d) Isobutane + Br_2/hv → *t*-butyl bromide + HBr

107. Which reaction is kinetically fastest (lowest DG#)?

(a) 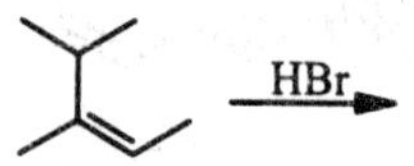 HBr (b) 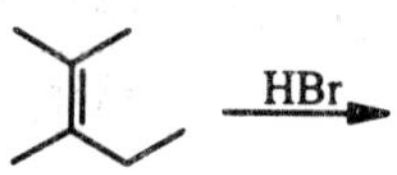 HBr

(c) 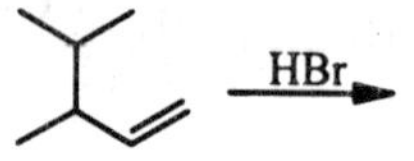 HBr (d) 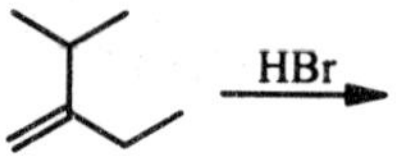 HBr

108. What is the product of the following reaction?

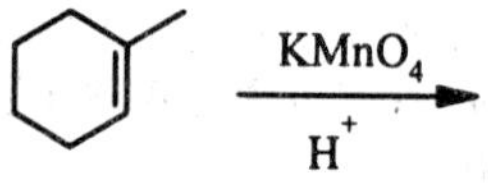

KMnO$_4$, H$^+$

(a) 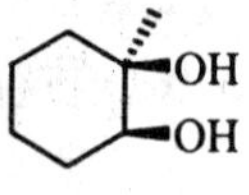 OH, OH (b) 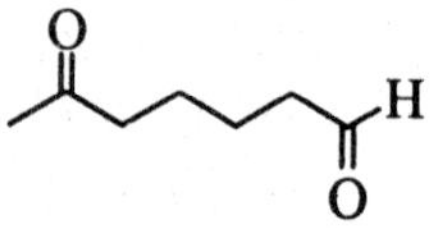 H, O

(c) 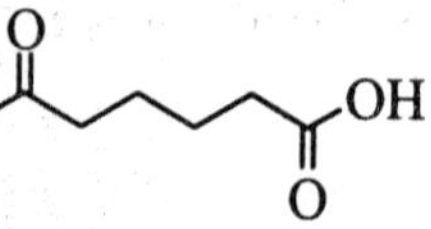 OH, O (d) 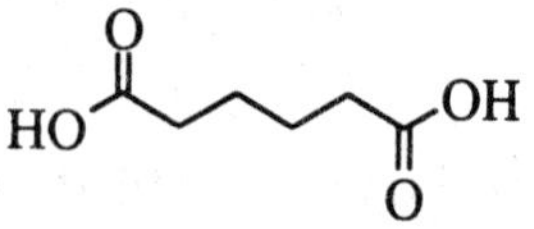 HO, OH, O

109. What is the best description of the first intermediate formed when 2-butyne reacts with HCl?
(a) Allylic cation (b) Vinylic cation
(c) Allylic anion (d) Chloronium ion

110. What is the final product of the following reactions?

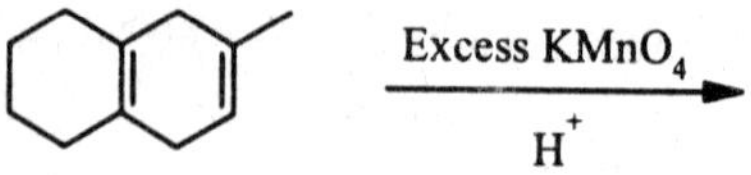

Excess KMnO$_4$, H$^+$

(a) 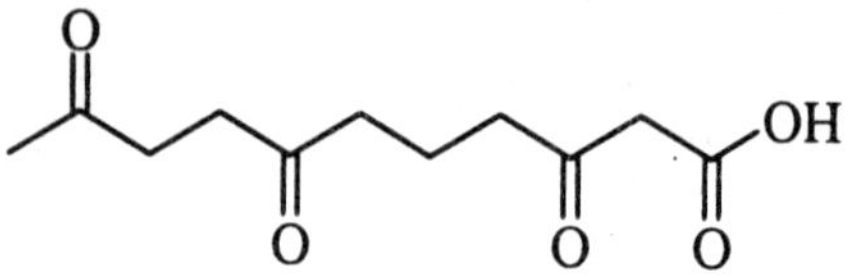 OH

(b) 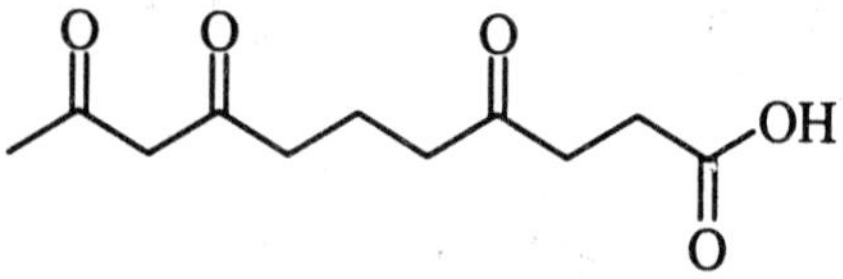 OH

(c) 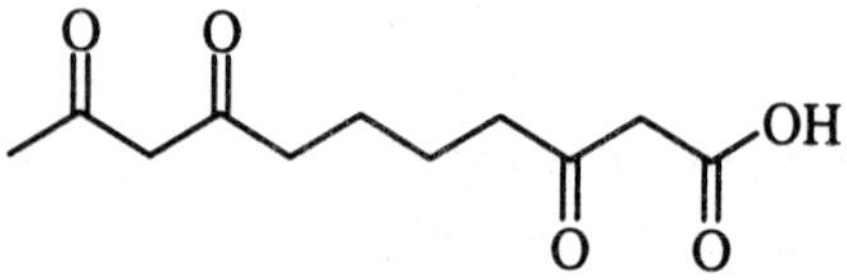 OH

(d) 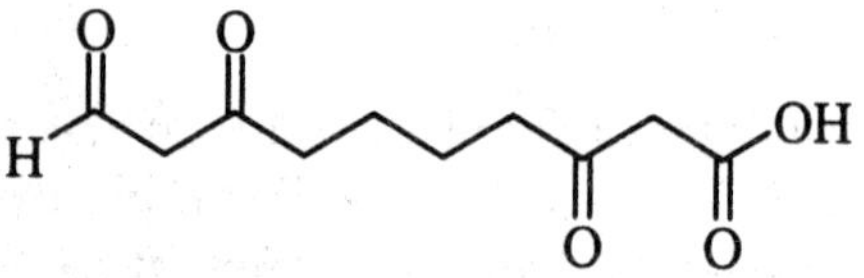 H, OH

111. Which radical is responsible for the observed selectivity in the following chain reaction?

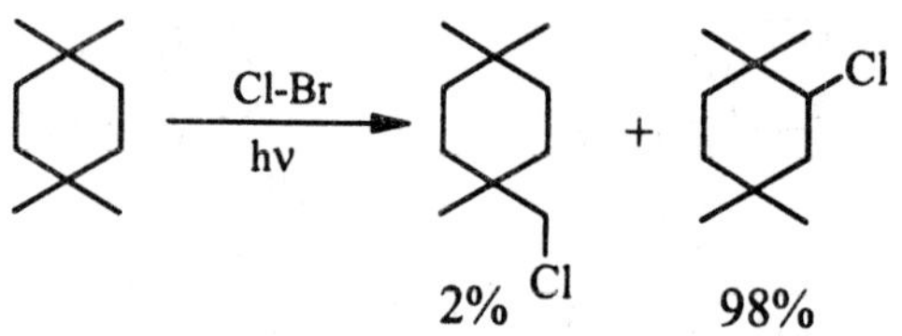

Cl-Br, hv → Cl (2%) + Cl (98%)

(a) Cl$^\cdot$ (b) Br$^\cdot$ (c) H$^\cdot$ (d) A 2° radical

112. For the following multistep reaction, which set of reagents would be more likely to give desired product in good yield?

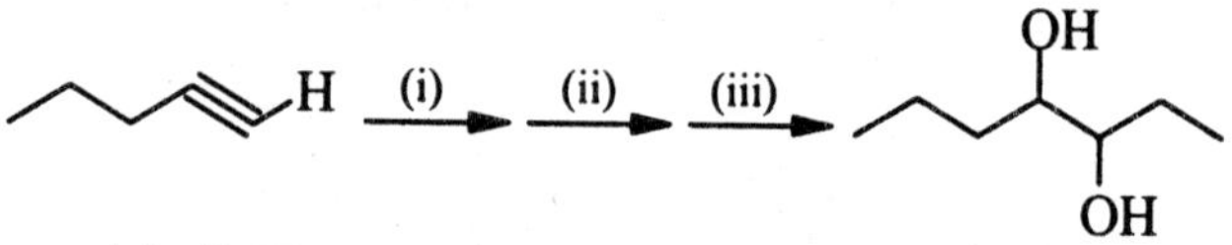

H (i) → (ii) → (iii) → OH, OH

(a) (i) HBr, (ii) O_3/Zn-H_2O, (iii) Li/NH_3.
(b) (i) $NaNH_2$/NH_3/EtI, (ii) Lindlar's catalyst/OsO_4/$NaHSO_3$.
(c) (i) H_2/Pd-C, (ii) $NaNH_2$/NH_3 followed by EtI, (iii) $KMnO_4$/OH^-.
(d) (i) $HgSO_4$/H_2SO_4, (ii) Lindlar's catalyst/H_2, (iii) OsO_4 followed by $NaHSO_3$.

113. What is the relationship between the two terminal $H_2C=$ groups in a cumulene containing three consecutive double bonds ($H_2C=C=C=CH_2$)?
(a) They are in the perpendicular plane.
(b) They are in the same plane.
(c) They are in two planes at 120°.
(d) They are in two planes at 109°.

114. What will be the product(s) of the following reactions?

OsO_4, $NaHSO_3$

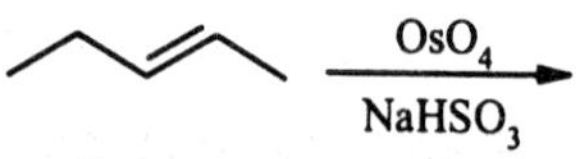

(a) A meso diol (b) Racemic mixture
(c) A pure enantiomer (d) Pair of distereomers

115. Which reaction condition would be best to perform the following transformations?

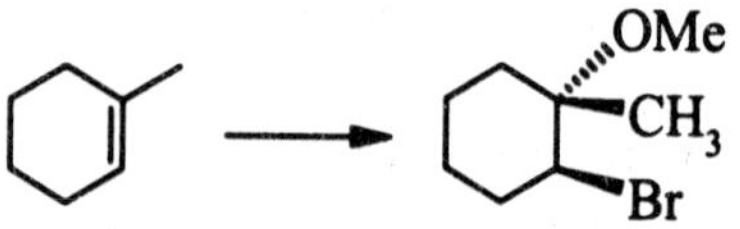

→ OMe, CH_3, Br

(a) NBS, MeOH
(b) OsO_4, HBr, NaOH, MeI
(c) Br_2, MeONa
(d) mCPBA, HBr, NaOH, MeI

116. What is the most likely product of the following reaction?

Cl_2, hv

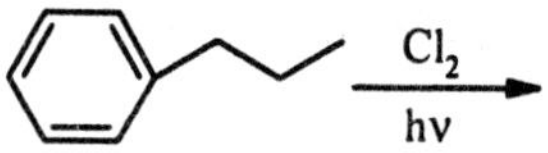

(a) (b)

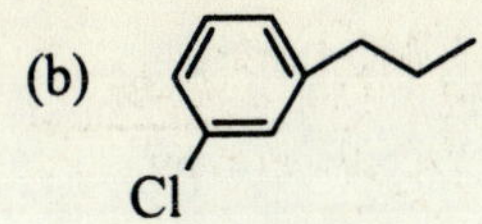

(c) (d)

117. Arrange the following in increasing order of reactivity towards HCl:

1. Ethyne 2. Ethene
3. 1, 3-butadiene 4. 2, 4-hexadiene

(a) 1 < 2 < 3 < 4 (b) 2 < 1 < 3 < 4
(c) 4 < 3 < 1 < 2 (d) 2 < 1 < 4 < 3

118. What is the major product of the reaction:
2, 3-pentadiene + HBr/ROOR →

(a) (b) (c) (d)

119. Which conditions would you use to perform the following transformation?

(a) (i) $Hg(OAc)_2/H_2O$, (ii) $NaBH_4$
(b) (i) OsO_4, (ii) $NaHSO_3$
(c) (i) BH_3, (ii) H_2O_2, (iii) NaOH
(d) (i) O_3, (ii) Zn, H_2O

120. The following transformation is carried in three steps (I → III). What are the appropriate reagents for these steps?

(a) (I) Br_2/H_2O, (II) $H^+/KMnO_4$, (III) $LiAlH_4$
(b) (I) Br_2/H_2O, (II) $LiAlH_4$, (III) $H^+/KMnO_4$
(c) (I) Br_2/CCl_4, (II) $H^+/KMnO_4$, (III) $LiAlH_4$
(d) (I) $Hg(OAc)_2/H_2O/NaBH_4$, (II) $H^+/KMnO_4$, (III) Br_2/hv

121. Which of the following is the major product of reaction?

NBS / DMSO, H_2O

(a) (b) (c) (d)

122. When reacted with an alkene, which of the following reagent would not introduce at least one hydroxyl (–OH) group into the final product?

(a) $Hg(OAc)_2$, H_2O/ $NaBH_4$
(b) $BH_3/H_2O_2/NaOH$
(c) HIO_4/H_2O
(d) $OsO_4/NaHSO_3$

123. Consider the following reaction of 1-methyl cyclohexene with NBS. Four likely products of the reaction are shown below. One of the likely products is not shown, however.

NBS / CCl_4

Which of the following is the missing product?

(a)

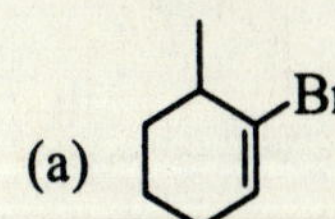

(b)

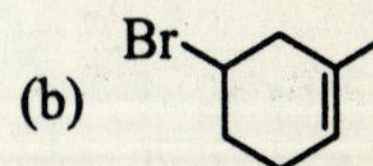

(c)

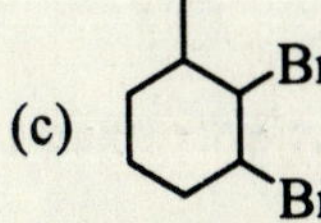

(d)

124. Which starting material should be used to produce the compound shown below?

? → O_3 / Zn-H_2O

(a)

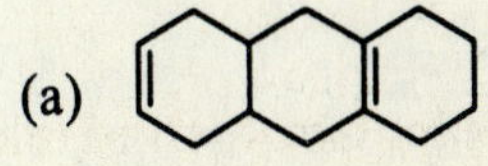

(b)

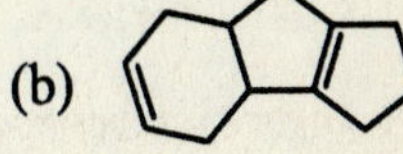

(c) (d)

125. What is the best way to carry out the following transformation?

1-pentyne → pentanal

(a) $HgSO_4/H_2SO_4$
(b) H_2/Lindlar's cat; O_3—Zn—H_2O
(c) HIO_4/H_2O (d) BH_3; H_2O_2/NaOH

126. Which is the least likely protonation site in the conjugated alkene shown below?

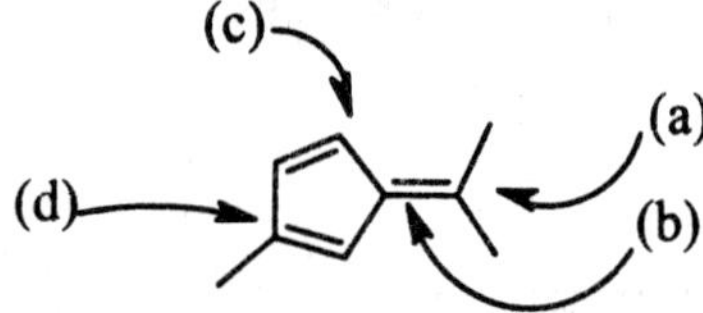

127. Compound X on catalytic hydrogenation gives 2, 6-dimethyloctane. On ozonolysis followed by treatment with Zn—H_2O, X yields formaldehyde, acetone and a dialdehyde shown below. The most likely structure of the starting compound X is:

H H O O O

(a) (b)

(c) (d)

128. Which statement best describes why 1-propyne can be deprotonated by ethyl anion?

(a) The acetylide ion is a stronger base than the ethyl anion.
(b) Ethane has a lower pk_a than acetylene.
(c) The lone pair orbitals in acetylide anion have more s-character than those in ethyl anion.
(d) The acetylide anion is a weaker base than ethyl anion.

129. What is the most likely product of the following reactions?

H_2SO_4 / OH_2

(a) (b) OH

(c) OH (d) OH

130. Which of the following product is least expected?

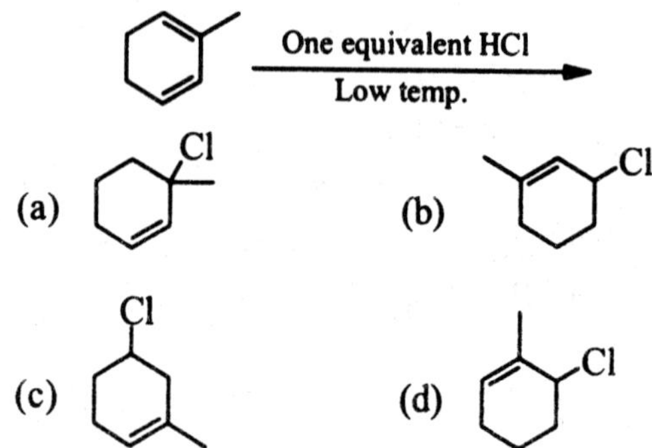

131. Which of the following is major product ?

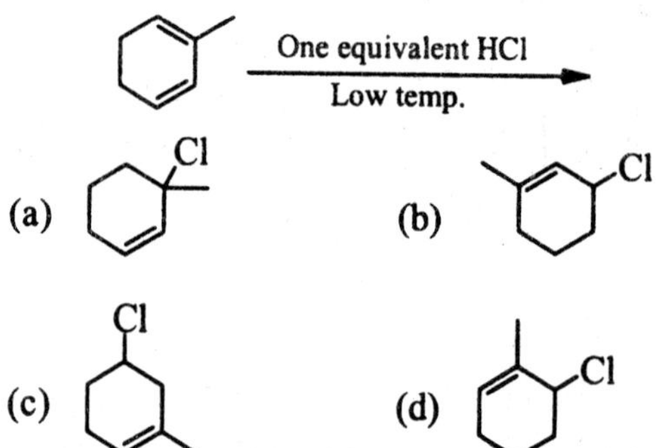

132. The alternative path for radical chain reactions of Br_2 with ethane may be represented as follows:

1. CH_3—CH_3 + Br_2 → CH_3—CH_2—Br + HBr
2. CH_3—CH_3 + Br_2 → $2CH_3Br$

(a) Reaction (1) is more likely, because it is more exothermic.
(b) Reaction (1) is more likely because it is more endothermic.
(c) Reaction (2) is more likely because it is more exothermic.
(d) Reaction (2) is more likely because it is more endothermic

133. Which of the following is the most likely mechanism for the transformation shown below?

Following three questions are based on addition of Iodine-azide (IN_3) to 1-methyl cyclohexene as:

134. What is the first intermediate formed in that reaction?

(a) a tertiary carbocation (b) iodonium ion
(c) azidonium ion (d) an allylic carbocation

135. What is the stereochemistry of addition?

(a) syn (b) anti
(c) syn/anti 50:50 (d) cannot be determined

136. What is the nucleophile in the second step?

(a) N_3^- (b) I^- (c) H^- (d) iodonium ion

137. What is the major product of the reaction:

2-hexyne + $HgSO_4/H_2SO_4/H_2O \rightarrow$

(a) 2-hexen-3-ol (b) 3-hexanone
(c) 2-hexanone (d) b and c

138. The main reason for the following rearrangement is:

(a) The initially formed carbocation is secondary and rearrange to more stable tertiary one.
(b) The reaction converts one tertiary carbocation into other to relieve torsional strain.
(c) The tertiary chloride is much more stable than a secondary chloride.
(d) This is a radical chain reaction, controlled by relative bond strength.

139. What is the major product of the following reactions?

EtO– (cyclohexene) $\xrightarrow{O_3/Zn\text{-}H_2O}$

(a) (b)

(c) H ... H (O, O) (d) H ... OEt (O, O)

140. Arrange the following in the increasing order of stability:

(a) A (b) B

(c) C (d) D

Following three questions are based on the following statement:

Hydrocarbon A(C_7H_{12}) was treated with BH_3, followed by H_2O_2/NaOH to produce B($C_7H_{14}O$) as the only product. Reaction of B with TsCl/pyridine followed by KOH gives C (isomeric with A) in addition to other olefinic products. Treatment of C with ozone, followed by Zn/AcOH produces only compound shown below:

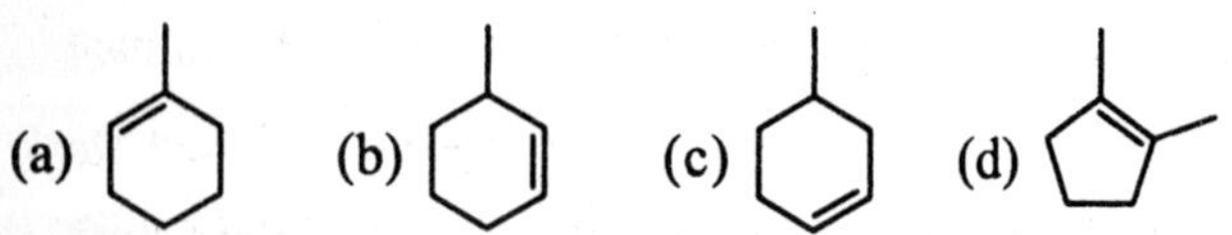

141. What is the correct structure of A?

(a) (b) (c) (d)

142. What is the correct structure of B?

(a) OH (b) OH (c) OH (d) OH

143. What is the correct structure of C?

(a) (b) (c) (d)

144. Addition of Br_2 to *cis*-2-hexene gives racemic mixture, even though attack of the bromide ion on unsymmetrical bromonium ion intermediate is not equally likely at both carbons. The reason for the racemic products are:

(a) The reaction is reversible, after a while both enantiomers are produced in equal amount.
(b) The bromonium ion formed has a plane of symmetry.
(c) Two enantiomeric bromonium ions are formed in equal amounts.
(d) Four different stereomers are formed, but all in equal amount.

145. Reaction of HBr with 3-bromocyclohexene (A) yields only *trans*-1, 2-dibromocyclohexane. Reaction of 3-methylcyclohexene (B) with HBr gives a mixture of *cis*- and *trans*-1-bromo-3-methylcyclohexane and *cis*- and *trans*-1-bromo-3-methylcyclohexane. Which statement explains the reactivity differences between A and B?

(a) There is no difference in reactivity. In the case of A the 1, 2 and 1, 3-dibromo products cannot be distinguished.
(b) The bromonium ion formed in the case of A can be attacked only in the anti manner.
(c) A has a chiral center and B has a plane of symmetry.
(d) In the case of B, there is a hydride shift following protonation.

146. For the following series of reaction, what can be predicted about stereochemistry of final product?

$$H_3C-C\equiv C-CH_3 \xrightarrow{H_2/\text{Lindlar's cat}} \xrightarrow{Br_2/CCl_4} \text{Product(s)}$$

(a) Meso compound (b) Achiral compound
(c) Racemic mixture (d) A pair of distereomers

147. The major product of the following sequence of reaction is:

$\xrightarrow[\text{DMSO}]{NBS/H_2O} \xrightarrow{NaOH}$

(a) (b) H, O, CH_3, H_3C, H

(c) H, O, H, H_3C, CH_3 (d) Br, OH, H, H_3C, H, CH_3

148. Consider the following reactions:

$\xrightarrow{HBr}$ Br (A) + Br (B)

The incorrect statement about this reaction is:

(a) In the intermediate, the positive charge on the secondary carbon is larger than on primary carbon, resulting in the more rapid formation of A.
(b) In B the double bond is more stabilized by hyperconjugation than in A.
(c) At higher temperature B is the major product.
(d) Under the kinetically controlled conditions, A and B can easily interconvert.

149. There are three common way to convert an alkene into alcohol: (A) acid catalyzed hydration, (B) hydroboration-oxidation, and (C) oxymercuration-demercuration. Which of the following statements about these reaction is incorrect?
(a) B results in the syn stereochemistry of –H and –OH groups, while A and C show no clear stereochemical preference.
(b) A and C gives Markovnikov's products, while B results in anti-Markovnikov's regioselectivity.
(c) Carbocation rearrangements are observed in A, but unrearranged products are obtained from B and C.
(d) In all three reactions, water serves as nucleophile.

150. Which sets of reagent would best bring about the following transformation?

Cyclohexane → 1, 3-cyclohexadiene

(a) (1) Cl_2/hv followed by KOH (2) Br_2/CCl_4 followed by KOH.
(b) (1) Br_2/heat followed by KOH, (2) NBS/DMSO/H_2O followed by KOH.
(c) (1) Cl_2/hv followed by NaOH (2) HBr followed by KOH.
(d) (1) Br_2/hv followed by NaOH (2) OsO_4/$NaHSO_3$ followed by HBr.

151. Which of the following synthetic scheme can be used to prepare the oct-2, 7-dione?
(a) React acetylene with $NaNH_2/NH_3$; add half of an equivalent of 1, 4-dibromobutane; then add $H_2O/H_2SO_4/HgSO_4$.
(b) React 1, 6-octadiene with O_3 followed by Zn-Acetic acid.
(c) React 1, 2-dimethylcyclohexene with $KMnO_4$/H^+.
(d) Both a and c may work.

152. Protonation of 1, 2-butadiene gives carbocation (A) and protonation of 2-butene gives carbocation (B). In both case the most stable carbocation possible is produced but no carbocation rearrangements take place. Which of the two carbocation is more stable and why?
(a) A is more stable because it is a terminal carbocation, according to Markovnikov's rule.
(b) A is more stable because it is stabilized by resonance and hyperconjugation.
(c) A is more stable because cationic carbon is sp-hybridized.
(d) B is more stable because it has two alkyl groups that can provide extra stabilization by hyperconjugation and inductive effect.

153. In the polar reaction of vinyl chloride with HBr, the only product formed is 1, 1-bromochloroethane:

$$ClCH{=}CH_2 + HBr \rightarrow ClBrCH{-}CH_3$$
(not: $ClCH_2{-}CH_2Br$)

The best explanation for this preference is:
(a) The electrophile prefers to add to chlorine substituted carbon.
(b) Chlorine is causing steric hindrance to the addition of bromine.
(c) The two possible carbocation have same stability but 1, 1-bromochloroethane is much more stable than 1-bromo-2-chloroethane.
(d) The carbocation formed by protonation on the unsubstituted carbon is stabilized by resonance.

154. Consider the following rearrangement. Select the structure of intermediate that has correct mechanistic arrows accounting properly for the transformation.

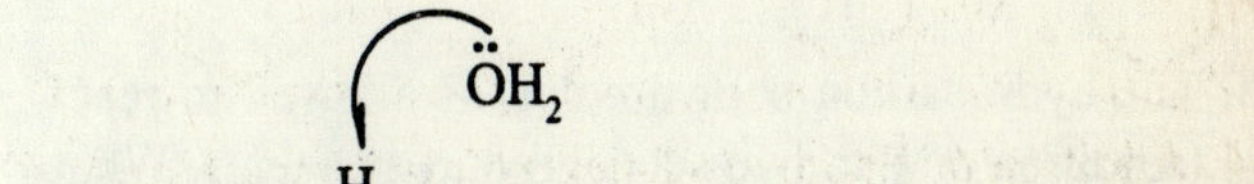

H^+, OH_2 → [ ] → $-H_3O^+$

(a) 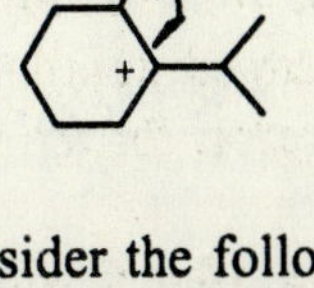 H (b) 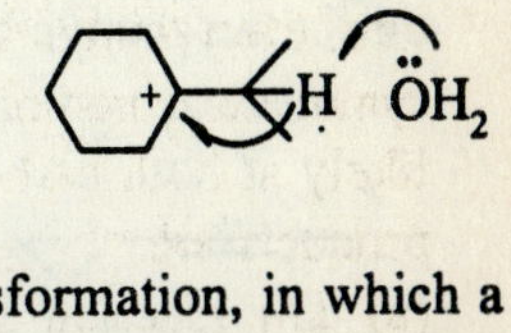 OH_2

$\ddot{O}H_2$ H

(c) 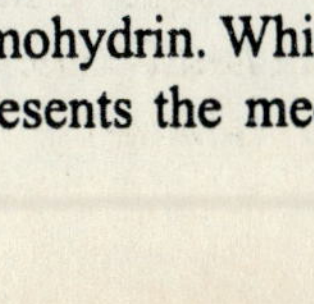 (d) H $\ddot{O}H_2$

155. Consider the following transformation, in which a cyclic bromoether is formed, rather than expected bromohydrin. Which of the electron pushing scheme represents the mechanism of this reaction?

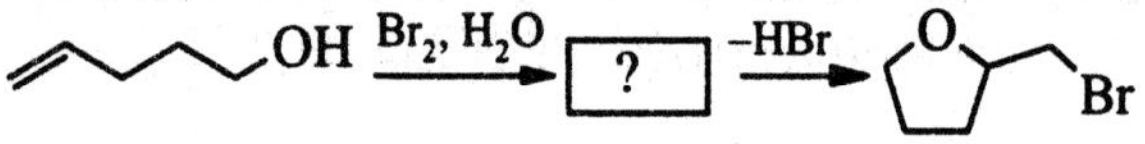

OH $\xrightarrow{Br_2, H_2O}$ [?] $\xrightarrow{-HBr}$ Br

(a) 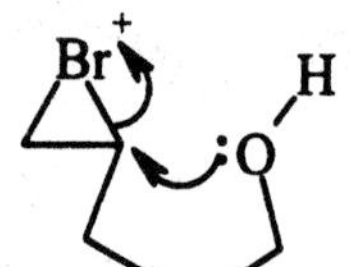 (b)

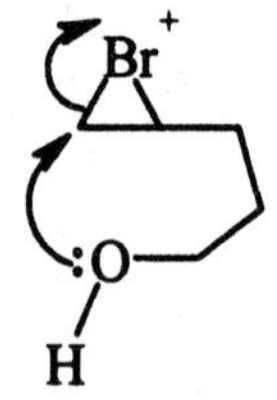

(c) 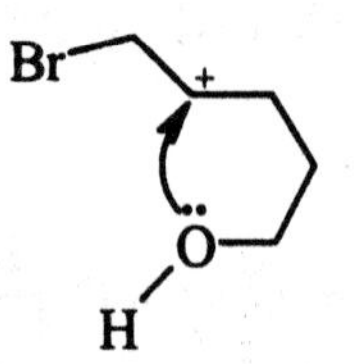 (d)

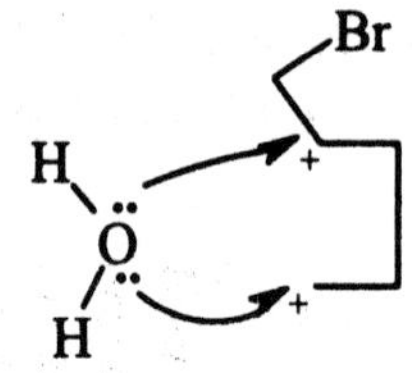

156. Consider only the following two major addition products of the addition reaction shown below, which statement is true about this reaction?

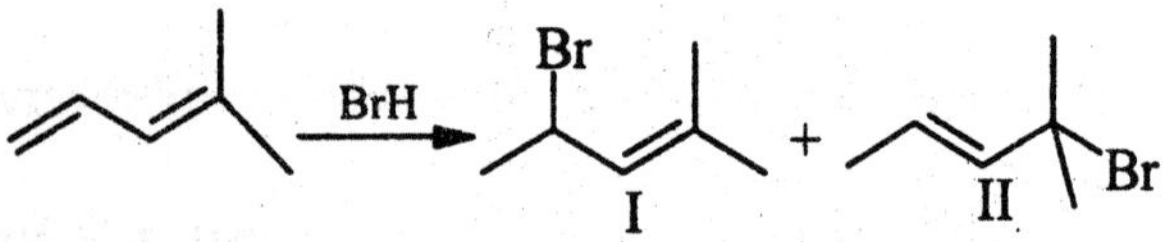

BrH → Br I + II Br

(a) II is the major product under thermodynamic control.
(b) I is the major product under kinetic control.
(c) I is the major product under thermodynamic control.
(d) There is no thermodynamic or kinetic control in this reaction.

157. On the reaction with acid 4-pyrone is protonated to give a very stable product. Which of the following structures shown below is the protonation product?

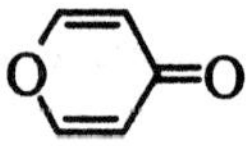

O =O 4-pyrone

(a) 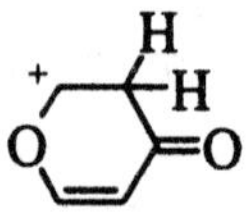 (b)

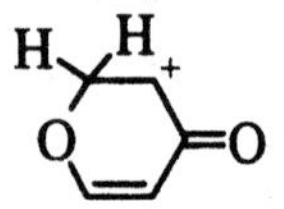

(c) 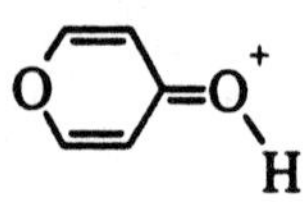 (d) 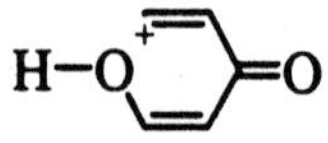 H–O

158. The hydrocarbon shown below is allowed to react (separately) with two sets of reducing agents. What are the respective products of these reactions?

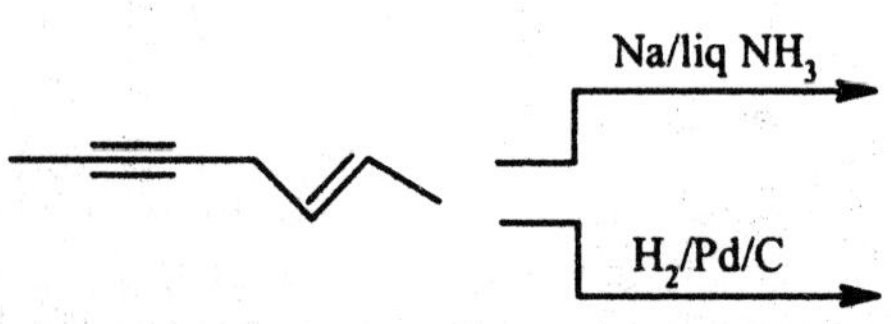

Na/liq NH_3

H_2/Pd/C

Na/liqNH_3	H_2/Pd/C
(a) Heptane	Heptane
(b) *Trans-trans* heptadiene	*Cis-trans* heptadiene
(c) *Trans-cis* heptadiene	*Cis*-heptene
(d) *Trans-trans* heptadiene	heptane

159. What would be the expected product of the reaction of propyne with Br_2/H_2O, if the mechanism of this reaction is analogous to that of propene?
(a) 2-bromopropenol (b) bromoacetone
(c) 2-bromo-2-propanol (d) bromopropenol

160. What are the products of the following reactions?

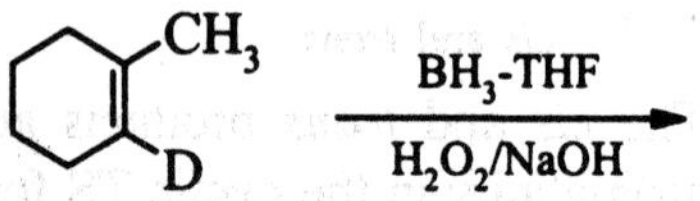

CH_3, D $\xrightarrow[H_2O_2/NaOH]{BH_3\text{-}THF}$

1. H, OH, D 2. H, D, OH
3. H, OH, D 4. 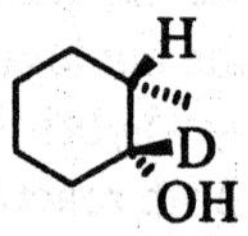 H, D, OH

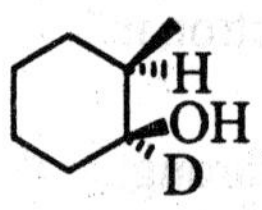

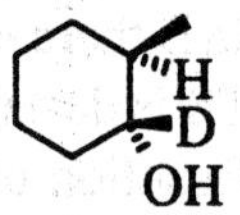

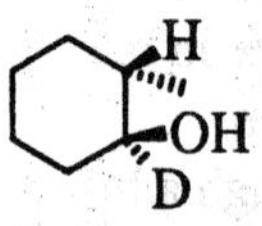

(a) 1 and 2 (b) 2 and 3
(c) 2 and 4 (d) 1, 2, 3 and 4

161. How many stereoisomers of the main product will form in the following reaction?

$\xrightarrow[NaOH]{BH_3/H_2O_2}$

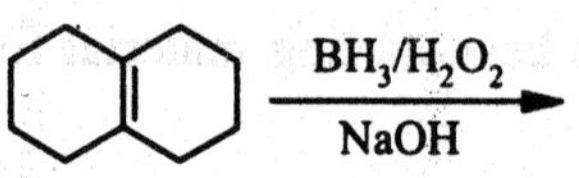

(a) 1 (b) 2 (c) 3 (d) 4

162. Consider the following reaction:
cis-2-pentene + HBr → Products:
The reaction occurs without rearrangement. What is (are) the organic product(s) of the reactions?
(a) 3-bromopentane
(b) Racemic mixture
(c) 3-bromopentane and an enantiomer of 2-bromopentane.
(d) Racemic mixture of 2-bromopentane, 3-bromopentane.

163. Select the reagents for the following transformations:

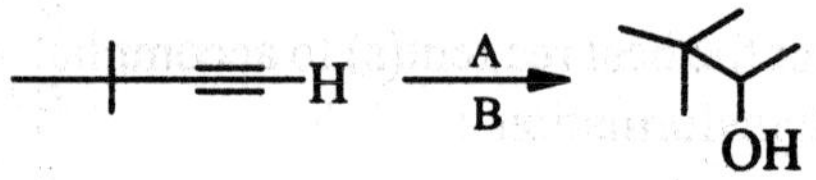

≡–H $\xrightarrow[B]{A}$ OH

(a) H_2/Lindlar's catalyst $Hg(OAc)_2/H_2O$; $NaBH_4$
(b) $H_2/Pd/CH_2O/H_2SO_4$
(c) $BH_3/H_2O_2/NaOHH_2/Pd/C$
(d) $HgSO_4/H_2SO_4/H_2OH_2$/Lindlar's catalyst

164. When cyclohexadiene (A) reacts with Br_2, a mixture of *cis* and *trans* 1, 2-addition products is formed (in addition to other products). However, when cyclohexene (B) reacts with Br_2 under identical conditions, only *trans* product is observed. What is the best explanation for the observed difference in stereochemistry of the addition?

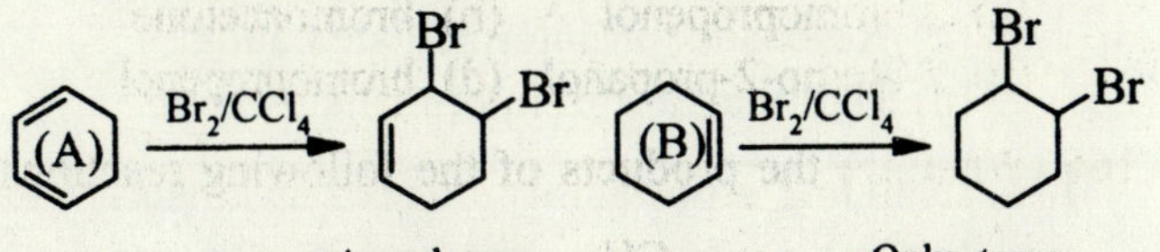

(A) $\xrightarrow{Br_2/CCl_4}$ 3,4-dibromocyclohexene (*cis* and *trans*) (B) $\xrightarrow{Br_2/CCl_4}$ 1,2-dibromocyclohexane (Only *trans*)

(a) The *cis* and *trans* products are the result of aromaticity in the cyclic TS for reaction of A. In B there is only four electron in TS, and cyclic TS is destabilized.
(b) Reaction of A proceeds through an intermediate that has an sp^3-hybridised carbocation, while the analogous intermediate in reaction of B has sp^2-hybridised carbocation.
(c) Both reaction occur through a bromonium ion, but because of planarity enforced by neighbouring double bond, *cis* addition is not sterically hindered in A.
(d) B reacts through a bromonium ion intermediate, while A does through an allyl cation.

165. Select the best starting material for the following reaction:

? $\xrightarrow{O_3/Zn\text{-}H_2O}$ [product: cyclohexane bearing CH(CH₃)CHO and CH₂CHO groups]

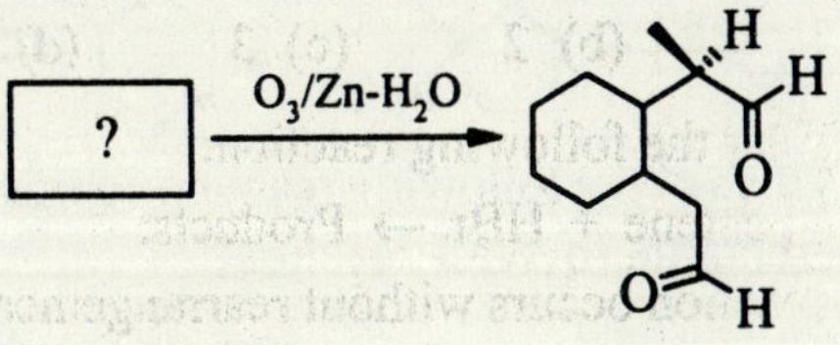

(a) 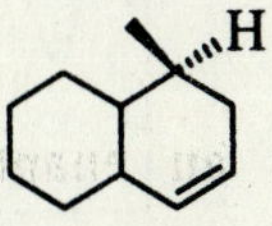 (b) 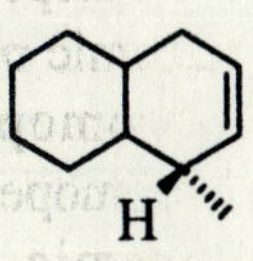 (c) 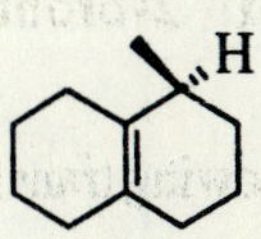 (d)

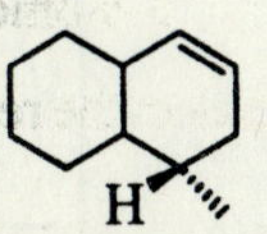

166. Select the best reagent(s) to accomplish the following transformations:

[enyne]–H ⟶ HO[chain]=O

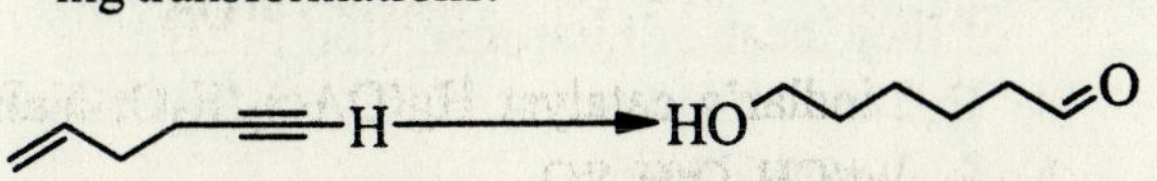

(a) O_3; Zn/AcOH (b) BH_3; $NaOH/H_2O_2$
(c) $Hg^{2+}/H_2SO_4/H_2O$ (d) $KMnO_4/H^+$

167. Which of the following will loose the optical activity on treatment with $Pd/BaSO_4/H_2$?

(a) 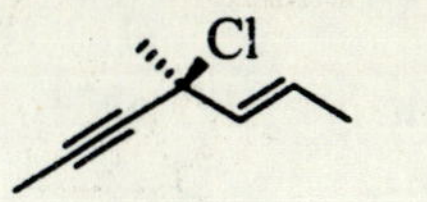 (b) 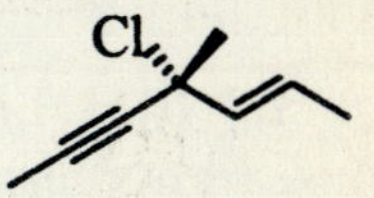 (c) 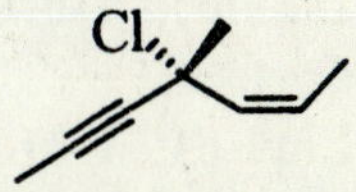 (d)

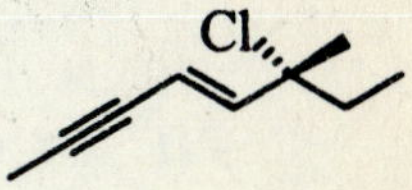

168. Consider the following reaction, which of the following statement is false?

$$CH_3\text{—}\underset{}{CH}(CH_3)\text{—}CH{=}CH_2 + Hg\,(CH_3COO)_2 \xrightarrow{H_2O}$$

(a) The mercurinium ion is electrophile in the reaction.
(b) Acetate ion is nucleophile in the addition reaction.
(c) Markovnikov regiochemistry will be observed in the reaction with water.
(d) To remove mercury from product, a follow up reaction with $NaBH_4$ is required.

169. What is true about stereochemistry of the product of the following reaction?

Methyl cyclohexene + $OsO_4/NaHSO_3$ →

(a) Since the addition is syn, the product will be meso.
(b) Since the product has two chiral centers, four stereoisomers of product will form.
(c) The product will be a racemic mixture of two enantiomers.
(d) Because of Markovnikov's rule only achiral alcohol will form.

170. Select the starting material for the following reaction:

? $\xrightarrow{Hg(OAc)_2/H_2O/NaBH_4}$ [1-methyl-2,2-dimethylcyclohexanol (OH)]

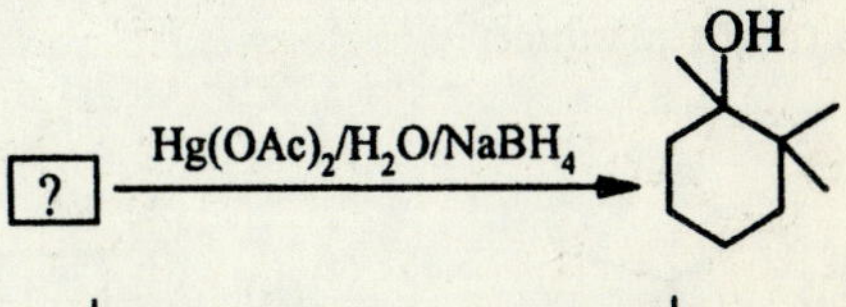

(a) 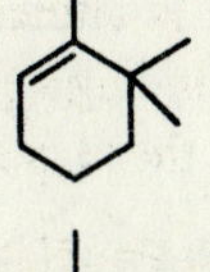 (b)

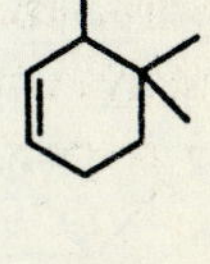

(c) 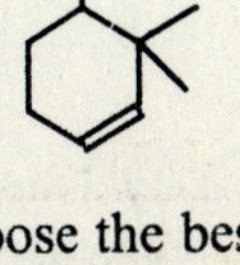 (d) both a and b

171. Choose the best reagent to carry out the following transformations:

$H_3C{-}\!\equiv\!{-}H \xrightarrow{I} \xrightarrow{II}$ (cis-alkene, H H)

(a) (I) H_2/Lindlar catalyst (II) $NaNH_2/NH_3$(Liquid); 1-bromopropane.
(b) (I) $NaNH_2/NH_3$(Liquid); 1-bromopropane (II) H_2/Lindlar catalyst.
(c) (I) $NaNH_2/NH_3$(Liquid); 1-bromopropane (II) Li/NH_3(liquid).
(d) (I) $NaNH_2/NH_3$(Liquid); 1-bromopropane (II) (I) H_2/Lindlar catalyst.

172. Examine the following reaction:

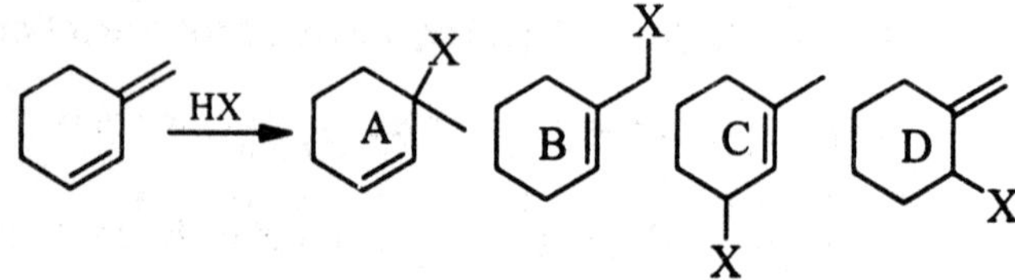

(a) A and B are formed under thermodynamic control.
(b) B and C are formed under thermodynamic control.
(c) A and C are formed under thermodynamic control.
(d) A and D are formed under thermodynamic control.

173. Rank the alkenes below in terms of their relative stability, starting with most stable:

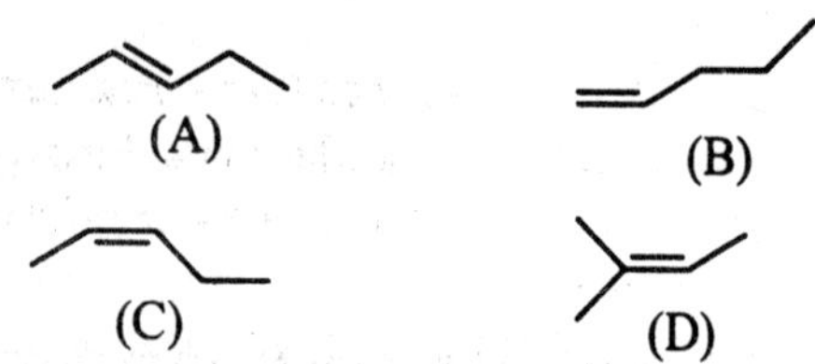

(a) D > A > C > B (b) B > C > A > D
(c) D > C > A > B (d) C > A > D > B

174. Vinyl cyclopropane on reaction with HBr yields — as major product:

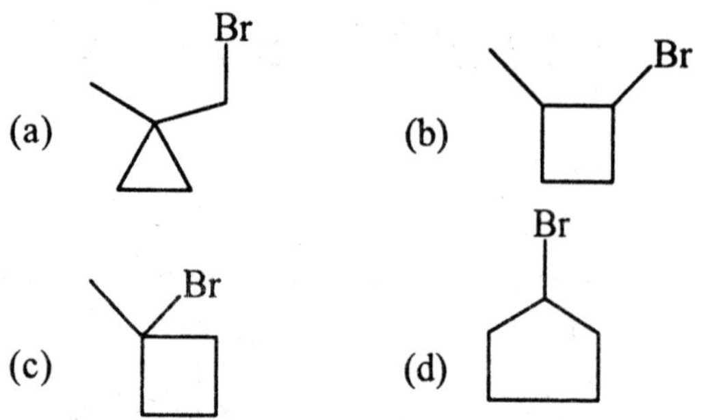

175. In the addition of DBr to 1, 3-cyclohexadiene, two products are obtained . What is the best explanation for the formation of B?

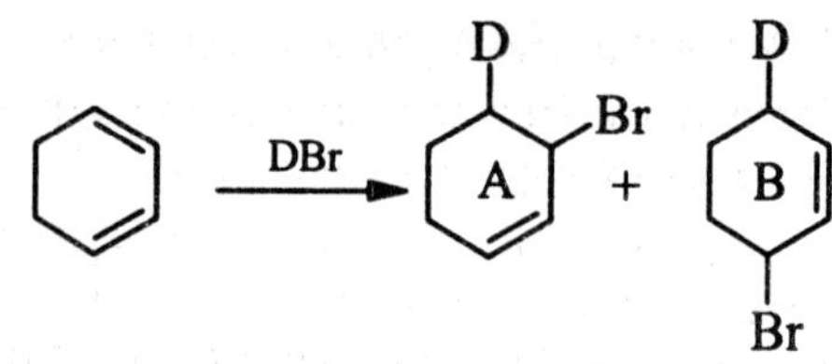

(a) Anti Markovnikov addition.
(b) Rearrangement in the intermediate cation.
(c) Resonance in intermediate cation.
(d) Hydride shift in intermediate cation.

176. Including a possibility of rearrangements, what is the major organic product of the reaction of 3-methylbutene with HBr?

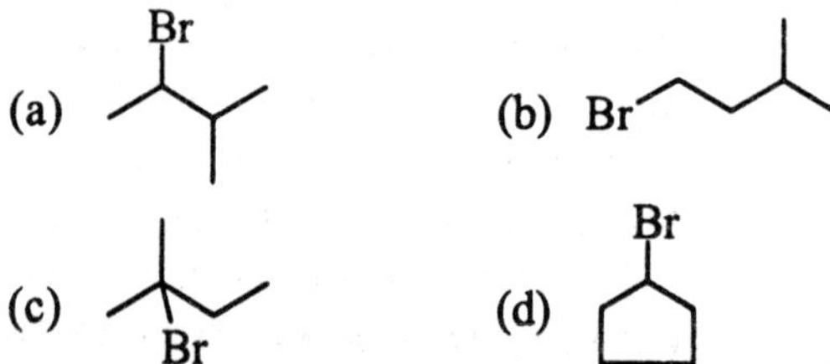

177. Which of the following will be most reactive in the addition reaction with HBr?

(a) (b)

(c) (d)

178. Consider the following rearrangement reaction:

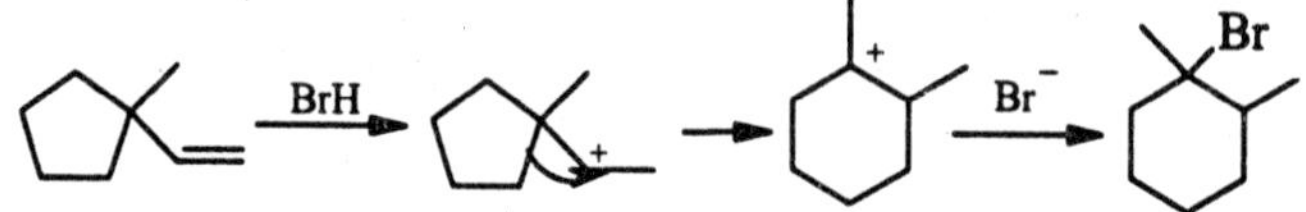

Which of the following reaction coordinate best represents the overall reaction?

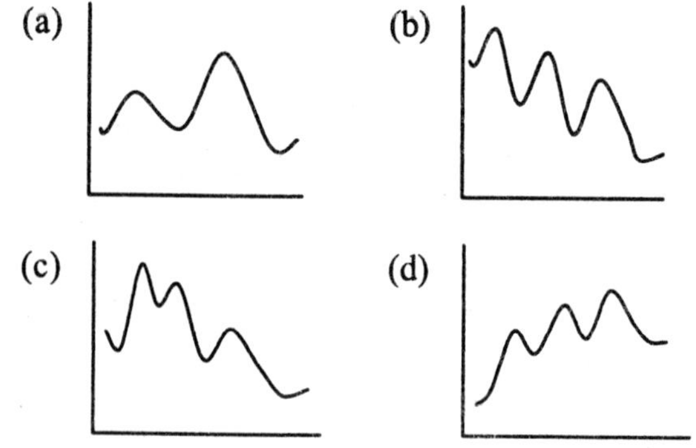

179. Arrange the following alkenes in the increasing order of their heat of hydrogenation ($-\Delta H$):
1. 2-methylbutene 2. *cis*-2-pentene
3. 3-methylbutene
(a) 1 < 2 < 3 (b) 2 < 1 < 3
(c) 3 < 1 < 2 (d) 3 < 2 < 1

180. How many isomers (including stereoisomers) can be obtained by free radical chlorination of 2, 2, 4-trimethylpentane?

(a) 3 (b) 4 (c) 5 (d) 6

181. What will be the major addition product of the following reaction?

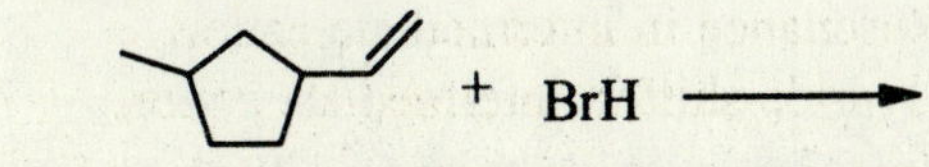

(a) 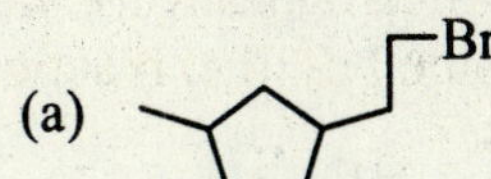

(b)

(c) 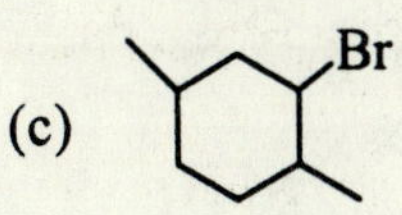

(d)

182. The tricyclic compound shown below reacts with HI to give a rearranged product. Based on the mechanistic arrows in the following sequence, what is the structure of product?

(a) 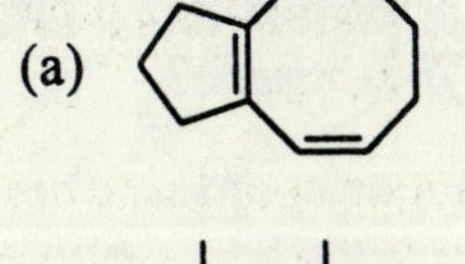

(b) 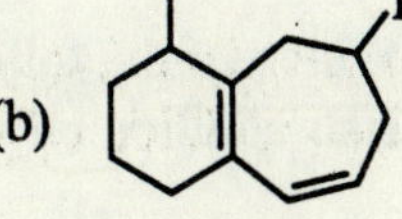

(c) 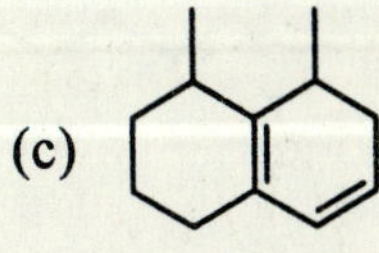

(d)

183. Under conditions where carbocation rearrangements are not observed, three products are obtained from the addition of HBr to 1, 3-pentadiene. Which of the following structures is not expected to be among the products?

(a) (b)

(c) (d)

184. Predict product from the following reactions:

(a) (b)

(c) (d)

185. Which of the following is not an oxidation reduction reaction?

(a) 2-butene + HCl → 2-chlorobutane

(b) 2-butene + Br_2 → 2, 3-dibromobutane

(c) 2-butene + H_2/Pt → butane

(d) 2-butene + NBS/H_2O → 3-bromo-2-butanol

186. Major product of the following reaction is:

Methylcyclohexene + OsO_4/$NaHSO_3$ →

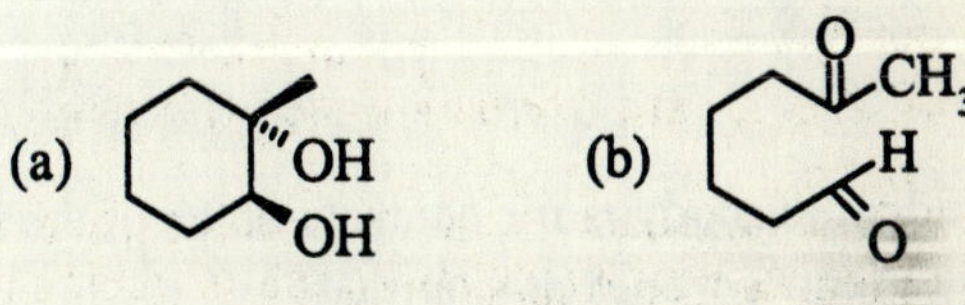

(c) 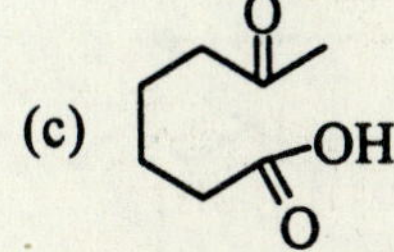

(d) 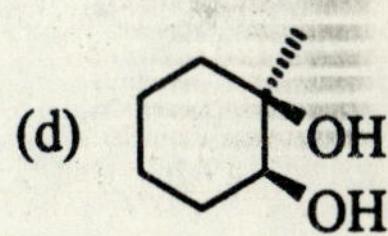

187. Compound A reacts with O_3/Zn—H_2O to produce compounds I and II. What is the structure of compound A?

$HOCCH_2CHO$(I) and CH_3COCH_2CHO(II)

(a) (b) 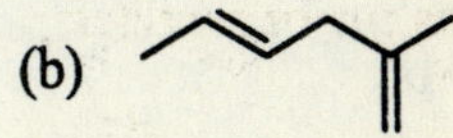

(c) 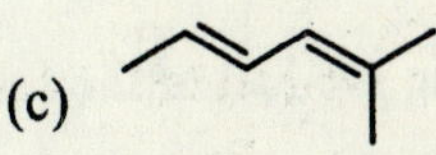

(d) 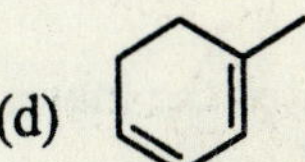

188. Which of the following reaction conditions would be best to perform the following reaction?

(a) HBr (one equivalent) then Br_2
(b) H_2, Pd/C then Br_2/hv
(c) (1) Br_2 (two equivalent) (2) H_2, Pd/C
(d) OsO_4/HIO_4

189. Which reagent would be best to perform the following transformations?

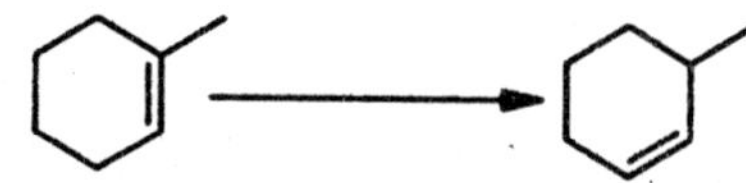

(a) (1) HBr (2) NaOH
(b) (1) HBr/ROOR/$(CH_3)_3$COK
(c) (1) BH_3 (2) NaOH/H_2O_2
(d) (1) Br_2 (2) NaOH

190. Which compound will most easily be eliminated to give an alkyne?

(a) 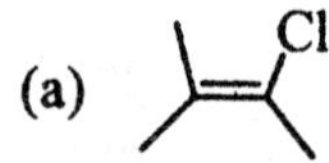

(b) 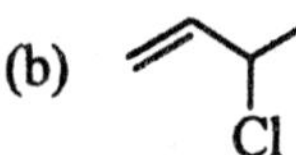

(c) –Cl (d) Cl

191. Under thermodynamic control, which of the following product would predominate?

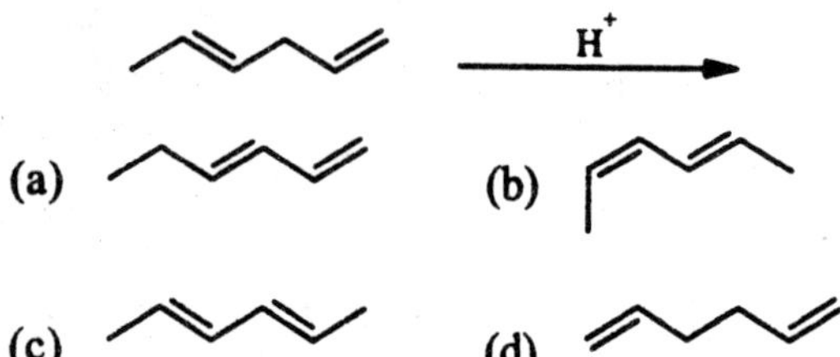

192. Which is the least likely product of the allylic bromination of 1, 4-hexadiene?

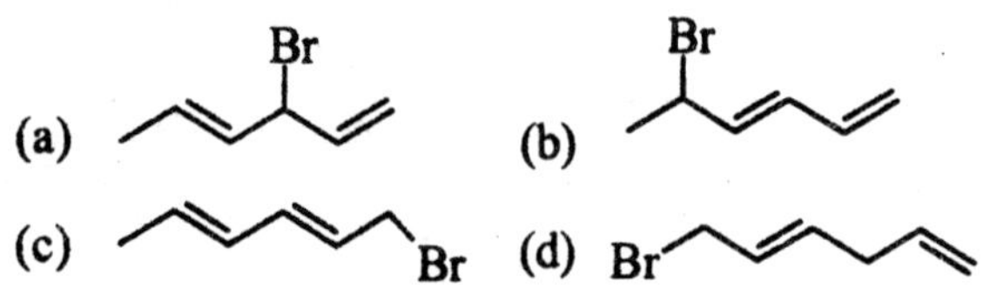

193. What is the major organic product of the following reactions?

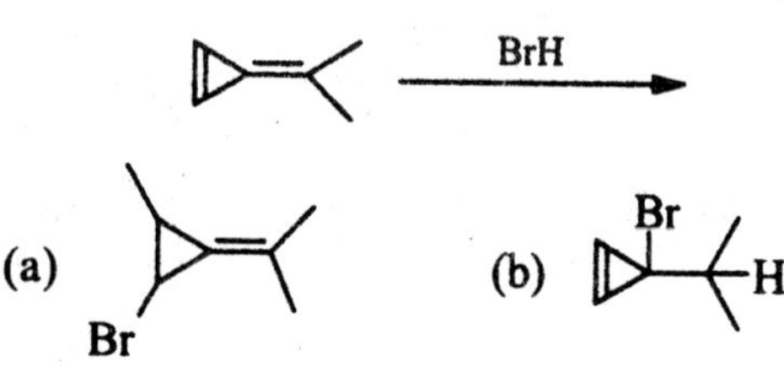

(c) H –Br (d) H –Br

194. Only one of the following reactions is likely to work as indicated. Which one?

(a) OMe + H_3C-NR_2 → H_3C-OMe + R_2N^-

(b) + Br_2 $\xrightarrow{hv}$ Br

(c) + BrH → H Br

(d) $\xrightarrow{NaNH_2/CH_3I}$ –CH_3

195. 1,3-butadiene reacts with one equivalent of HCl. How many different structures (including stereoisomers) will form?
(a) 6 (b) 2 (c) 3 (d) 4

196. Propene (I); methyl-vinyl ether (II); vinyl chloride (III) and *cis*-2-butene (IV). The increasing order reactivity towards addition reaction will be:
(a) I < II < III < IV (b) I < II < IV < III
(c) III < I < IV < II (d) IV < III < II < I

197. Which of the following is not a possible product of HBr addition on 4-methyl-2-hexene?

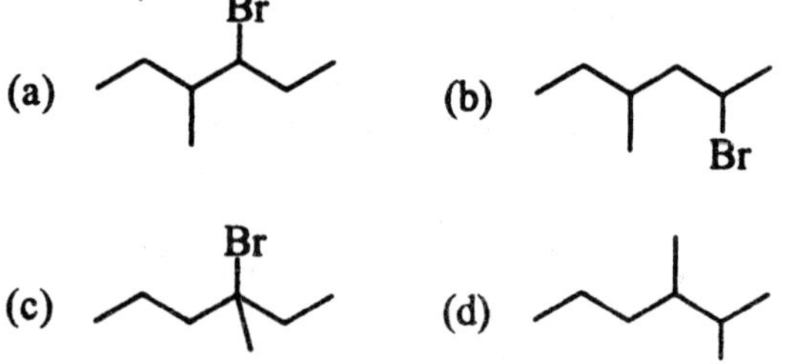

198. Which of the following product would result from this reaction?

• + CH_3Cl →

(a) H + •CH_2Cl (b) CH_2Cl + H•

(c) chlorobenzene + $\cdot CH_3$ (d) toluene + $\cdot Cl$

199. Which of the following reaction would occur at fastest rate?

(a) propene $\xrightarrow{BrH}$ 2-bromopropane (b) 2-methyl-1-butene $\xrightarrow{BrH}$ 1-bromo-2-methylbutane

(c) 1,4-cyclohexadiene $\xrightarrow{BrH}$ 4-bromocyclohexene (d) 2-methyl-1-butene $\xrightarrow{BrH}$ 2-bromo-2-methylbutane

200. What is the product of the following reaction?

3-methyl-2-butene + $Br_2/H_2O \rightarrow$

(a) HO, CH_3, Br, H (b) Br, CH_3, OH, H

(c) HO, CH_3, H, Br (d) Br, CH_3, H, OH

201. Arrange the following alkene in increasing order of their enthalpy of hydrogenation ($-\Delta H$):

I (vinylcyclohexane) II (methylenecyclohexane) III (ethylidenecyclohexane) IV (1-methylcyclohexene)

(a) III < IV < II < I (b) III < IV < I < II

(c) I < II < III < IV (d) I < II < IV < III

REVIEW EXERCISE (HYDROCARBONS)

1. Complete following:

(a) $CH_3I \xrightarrow{Li} X \xrightarrow{CuI} Y \xrightarrow{n\text{-}C_4H_9I} Z$

(b) iodocyclohexane + $Me_2CuLi \xrightarrow{Ether} X$

(c) $n\text{-}C_4H_9Br \xrightarrow{Li} X \xrightarrow{CuI} Y \xrightarrow{Iodobenzene} Z$

2. An alkane with molecular formula C_6H_{14} can be synthesised by treating (in separate reaction) five alkylchlorides $C_6H_{13}Cl$ with Zn in aq acid. Give the structure of alkane.

3. An alkane with molecular formula C_6H_{14} can be prepared by reduction of only two alkyl chloride $C_6H_{13}Cl$ and by hydrogenation of only two alkenes C_6H_{12}. Write the structure of alkane.

4. The C—C—C bond angles of isobutane are 111.5°, which is larger than normal tetrahedral bond angle where as H—C—C bond angle in the same compound is less than normal tetrahedral angle. Explain.

5. Convert CH≡CH into

(a) 1-butyne (b) 2-butyne (c) *cis*-2-butene

(d) *trans*-2-butene (e) $CH_3CH_2C{\equiv}CD$

(f) $H_3C(D)C{=}C(D)CH_3$

6. Write formula of products that form when1-pentene reacts with each of the following reagents:

(i) Br_2/CCl_4/RT (ii) $H_3O^+/H_2O/\Delta$

(iii) Cold. conc. H_2SO_4

7. Complete following:

1-octente + $CBrCl_3 \xrightarrow{ROOR}$?

1-octente + $CHCl_3 \xrightarrow{ROOR}$?

1-octente + $CCl_4 \xrightarrow{ROOR}$?

8. Complete following:

(a) $CH_3{-}C(CH_3)(OH){-}CH_2CH_2CH_3 \xrightarrow{H_3PO_4}$?

(b) (chloromethyl)cyclohexane $\xrightarrow[(CH_3)_3COH]{(CH_3)_3COK}$?

9. Propose mechanism:

(a) $CH_3{-}C(CH_3)_2{-}CH_2OH \xrightarrow{H^+}$ 2-methyl-2-butene + 2-methyl-1-butene

(b) 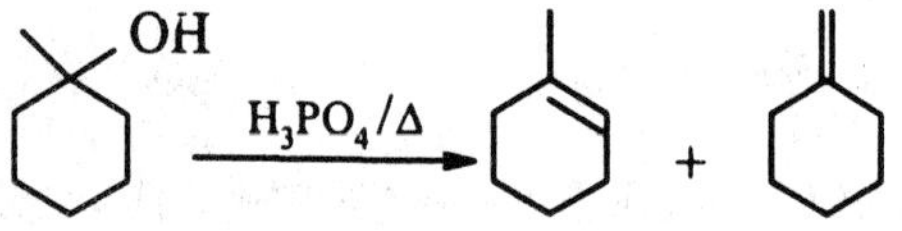 [1-methylcyclohexanol] $\xrightarrow{H_3PO_4/\Delta}$ [1-methylcyclohexene] + [methylenecyclohexane]

(c) 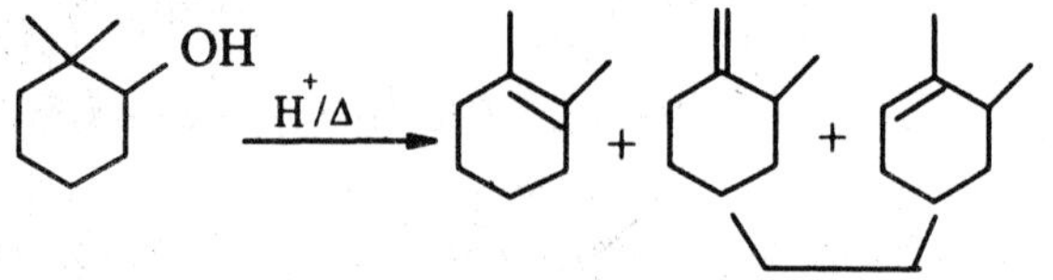 [2-methyl-1-(1-methylethyl)... cyclohexanol] $\xrightarrow{\overset{+}{H}/\Delta}$ [alkene] + [alkene] + [alkene]

Minors

10. Write mechanism of the reaction:

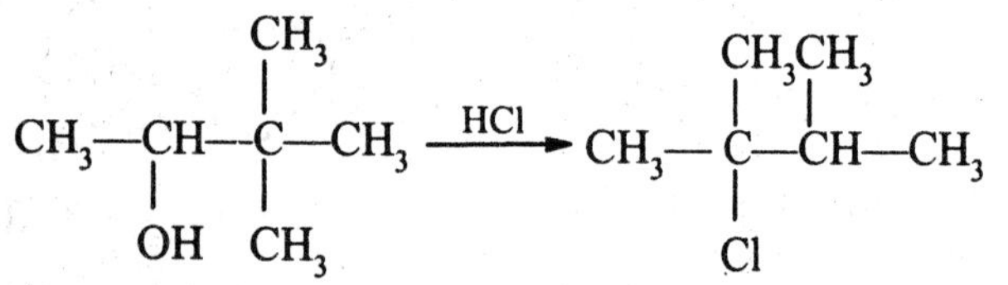

$$CH_3-\underset{OH}{CH}-\underset{CH_3}{\overset{CH_3}{C}}-CH_3 \xrightarrow{HCl} CH_3-\underset{Cl}{\overset{CH_3CH_3}{C}}-CH-CH_3$$

11. An optically active compound A has molecular formula $C_7H_{11}Br$. A reacts with HBr, in absence of peroxide to yield isomeric products, B and C with molecular formula $C_7H_{12}Br_2$. Compound B is optically active; C is not. Treating A with one mole of potassium-t-butoxide yields D(C_7H_{10}). Subjecting one mole of D to ozonolysis followed by treatment with Zn and water yield two moles of formaldehyde and one mole of 1,3-cyclopentandione. A on addition of HBr in presence of H_2O_2 produces another isomer of B which is E, and shows distereomerism. Deduce structure of A to E.

12. A hydrocarbon (A) has formula C_7H_{10}. On catalytic hydrogenation, A is converted to B (C_7H_{12}). On treatment with cold, dil, alkaline $KMnO_4$, A is converted to C($C_7H_{12}O_2$). When treated with hot $KMnO_4$ followed by acidification, either A or C produces meso form of 1, 3-cyclopentandicarboxylic acid. Give structural formula of A to C.

13. An alkane (A) with formula C_6H_{14} reacts with Cl_2 to yield three compounds with formula $C_6H_{13}Cl$, B, C and D, of these only C and D undergo dehydrahalogenation with C_2H_5ONa/EtOH to produce alkene (E). Also C can be resolved into enantiomers but D is non-resolvable. E on addition of HCl produces 'B', which on treatment with Zn-aceticacid produces A. Deduce structure of A to E.

14. Identify A to C.

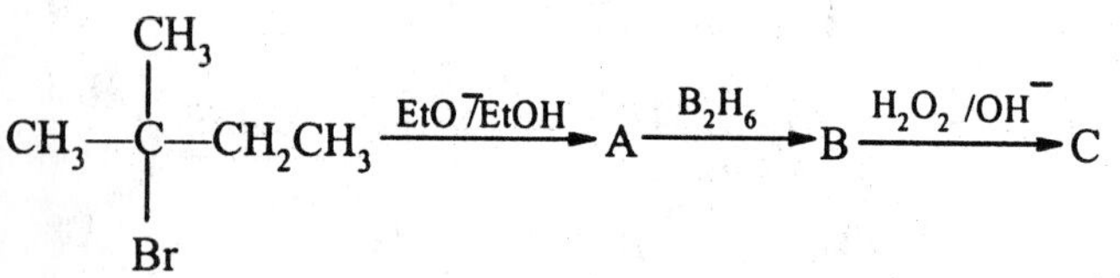

$$CH_3-\underset{Br}{\overset{CH_3}{C}}-CH_2CH_3 \xrightarrow{EtO^-/EtOH} A \xrightarrow{B_2H_6} B \xrightarrow{H_2O_2\ /OH^-} C$$

15. Complete the following (only major product):
(a) Cyclohexene + $CHCl_3$ (50% NaOH/H_2O) →

(b) 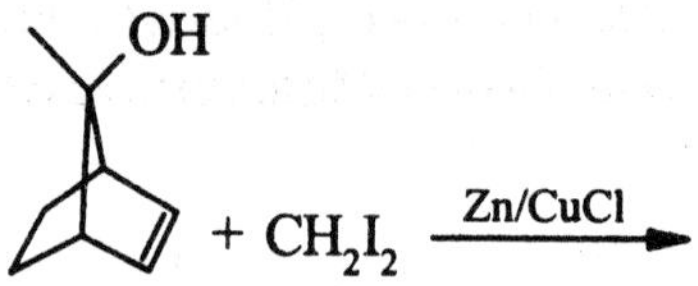 [bicyclic alkene with OH] + CH_2I_2 $\xrightarrow{Zn/CuCl}$

(c) 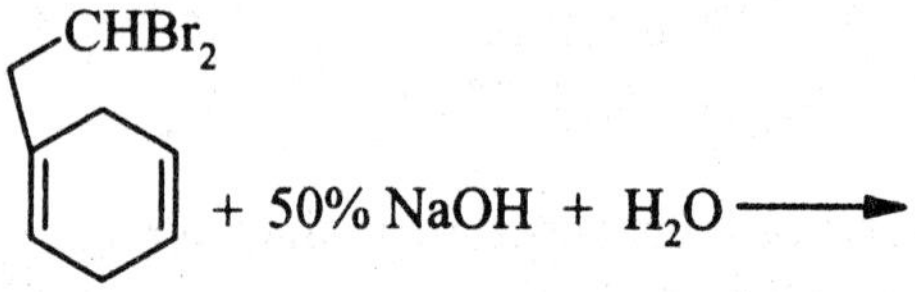 [$CHBr_2$-substituted cyclohexadiene] + 50% NaOH + H_2O ⟶

16. Convert:
(a) *Trans*-2-butene → *trans*-1,2-dimethylcyclopropane

(b) Cyclopentene ⟶ [bicyclo with H, Br]

(c) 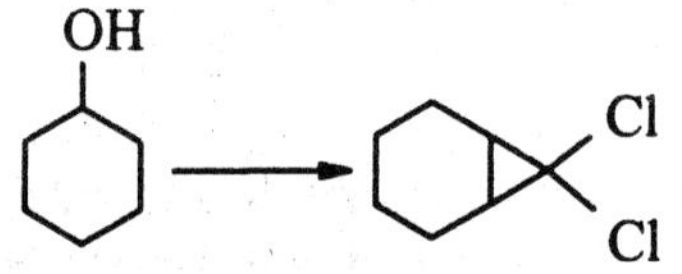 [cyclohexanol, OH] ⟶ [7,7-dichlorobicyclo, Cl, Cl]

17. When cyclohexene is treated with bromine in saturated aqueous solution of NaCl, mixture of 2-bromocyclohexanol and 1-bromo-2-chlorocyclohexane results. Give a mechanism to account for these products.

18. Predict major product and explain the mechanism of this formation

(a) 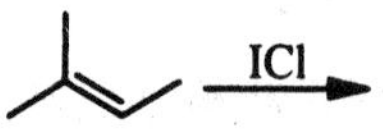 [2-methyl-2-butene] $\xrightarrow{ICl}$ (b) 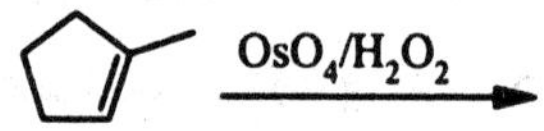 [1-methylcyclopentene] $\xrightarrow{OsO_4/H_2O_2}$

(c) 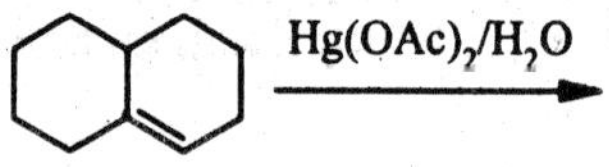 [octahydronaphthalene] $\xrightarrow{Hg(OAc)_2/H_2O}$

19. Propose mechanism of the formation of following products:

(a) 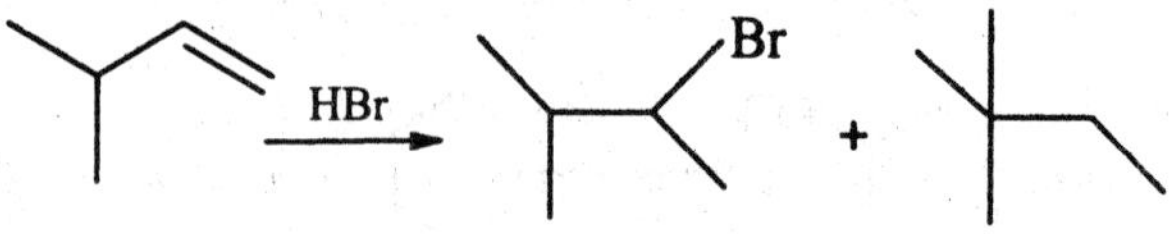 [3-methyl-1-butene] $\xrightarrow{HBr}$ [2-bromo-3-methylbutane, Br] + [2-bromo-2-methylbutane, Br]

(b) 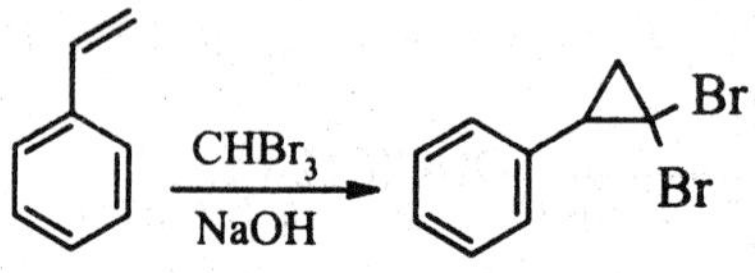 [styrene] $\xrightarrow[NaOH]{CHBr_3}$ [1,1-dibromo-2-phenylcyclopropane, Br, Br]

(c) 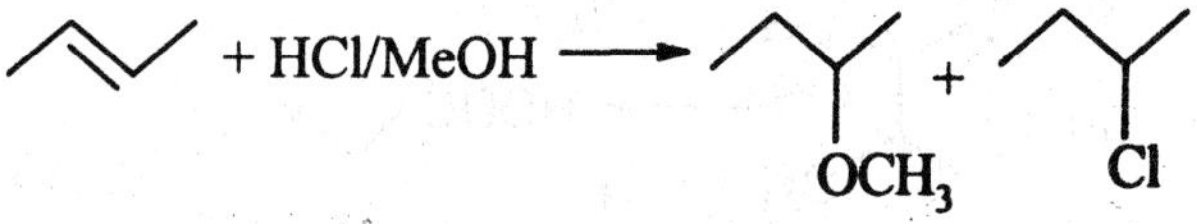 [propene] + HCl/MeOH ⟶ [OCH_3] + [Cl]

20. Convert:
(a) Fumaric acid ⟶ (d, l) tartaric acid.
(b) Maleic acid ⟶ (d, l) tartaric acid.
(c) Maleic acid ⟶ Meso tartaric acid.

21. Complete:

(a) 1-methylcyclohexene $\xrightarrow[\text{THF}]{BD_3/H_2O_2}$? (b) 1-methylcyclohexene $\xrightarrow{BH_3/D_2O_2}$?

22. Give principal organic product of reaction of butyne and 2-butyne with:
(a) Hot, aq, $KMnO_4$ (b) $B_2H_6/H_2O_2/OH^-$
(c) $HgSO_4/H_2SO_4$ (d) B_2H_6/CH_3COOD
(e) $CuCl/NH_4Cl$

23. Compound A has molecular formula C_7H_{12}. It reacts with dry HCl at –20°C to give B, $C_7H_{13}Cl$. Compound B reacts with K-t-butoxide in t-butanol to give small amount of A, and mainly C, which has formula C_7H_{12}. Ozonolysis of C gives cyclohexanone and an aldehyde, which does not respond to iodoform test. Identify A to C.

24. Compound A has formula C_8H_{12} and is optically active. It reacts with H_2 to give B, which has formula C_8H_{18} and is optically inactive. Careful hydrogenation of A with $Pd/CaCO_3/PbO$ gives C (C_8H_{14}), which is optically inactive. Compound A reacts with Na in liq. NH_3 to give D, which has formula C_8H_{14} and is optically active. Identify A to D.

25. Bring about the following conversions:

(a) $CH_3CH(Br)CH_3 \longrightarrow CH_3CH_2CH_2Br$

(b) $CH_3CH(Br)—CH_3 \longrightarrow CH_3CH_2CH_2OH$

(c) cyclohexane-1,2-diol (OH, OH) ⟶ cyclohexane

(d) cyclohexane-1,2-diol (OH, OH) ⟶ $OHC(CH_2)_4CHO$

(e) cyclohexane-1,2-diol (OH, OH) ⟶ HOOC–(CH₂)₄–COOH

(f) $CH_3CH_2CH(Br)—CH_2Br \longrightarrow CH_3CH_2CH(Br)—CH_2CBr_3$

(g) $CH_3CH_2—C(CH_3)=CH_2 \longrightarrow CH_3CH_2—CH(CH_3)—CH_2OCH_3$

(h) methylenecyclopentane ⟶ 1-deuterio-1-methylcyclopentane (D, CH_3)

(i) $CH_3CH_2CD(CH_3)CH_2OH \longrightarrow CH_3CH_2CH(CH_3)_2$

26. Write product:

(a) $CH_2{=}CHCH_2CH_3 \xrightarrow{NBS} (A) \xrightarrow[KOH]{C_2H_5OH} (B)$

(b) cyclohexene + NBS ⟶ (A) $\xrightarrow[KOH]{C_2H_5OH}$ (B) $\xrightarrow{ICl/–80°C}$ (C); (B) $\xrightarrow{ICl/+50°C}$ (D)

27. Light catalyzed reaction of cyclohexene with a low concentration of bromine gives exclusively 3-bromocyclohexene. Propose a mechanism to explain this preference for substitution at methylene group next to double bond.

28. Predict major product of free radical bromination of following:
(a) Decalin (b) Ethylbenzene

29. Three different alkenes yield 2-methylbutane when they are hydrogenated in the presence of a metal catalyst. Give the structures of alkenes.

30. Using propyne, how will you synthesize:
(a) $(CH_3)_2CHBr$ (b) CH_3COCH_3
(c) CH_3CH_2CHO (d) $CH_3(CH_2)_4CH_3$
(e) $H_3C(H)C{=}C(H)C_3H_7$ (trans) (f) $H_3C(Cl)C{=}CH_2$
(g) $CH_3CBr_2CBr_2D$ (h) $CH_3C{\equiv}C—C{\equiv}CCH_3$
(i) $H_3C(H)C{=}C(Br)H$ (cis) (j) $H_3C(H)C{=}C(H)Br$ (trans)

31. Write structural formula of missing reagents in each of the following synthesis:

(a) A + $CH_3CH_2CH_2Br \xrightarrow{?}$

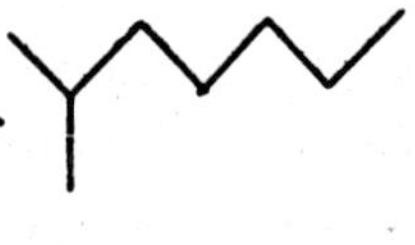

(b) B + $(CH_3)_2CuLi \longrightarrow$

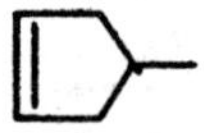

(c) C + $(CH_3CH_2)_2CuLi \longrightarrow$

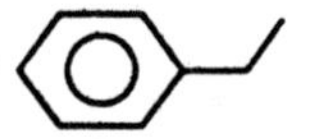

(d) $(CH_3CH_2)_2CuLi$ + D $\longrightarrow$

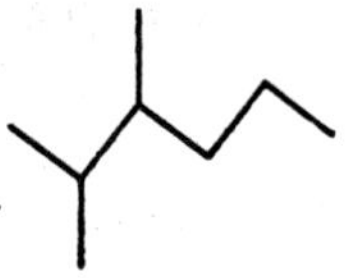

(e) E + Zn $\xrightarrow{H^+}$ Neopetane

(f) $\xrightarrow{F}$

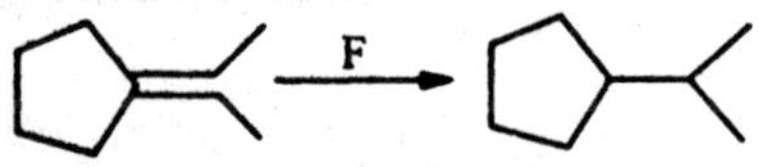

32. Addition of a small amount of iodine prevents the free radical chlorination of methane, explain.

33. Predict the elimination product of the following reaction:

(a) 1-bromo-1-methylcyclohexane + NaOH $\xrightarrow{\text{Acetone}}$

(b) 1-bromo-1-methylcyclohexane + Ethanol $\xrightarrow{\text{Heat}}$

(c) Chlorocyclohexane + NaOH $\xrightarrow{\text{Acetone}}$

(d) Chlorocyclohexane + $(C_2H_5)_3N \longrightarrow$

(e) 1-bromo-1-methylcyclohexane + $(C_2H_5)_3N \longrightarrow$

(f) Meso-1, 2-dibromo-1, 2-diphenylethane + NaOH $\xrightarrow{\text{Acetone}}$

(g) 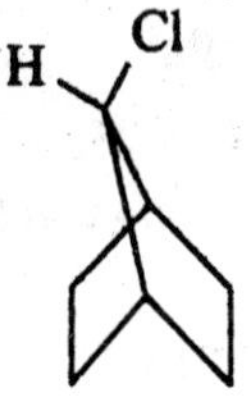 H, Cl + NaOH $\xrightarrow{\text{Acetone}}$

(h) Cl, H, D, H 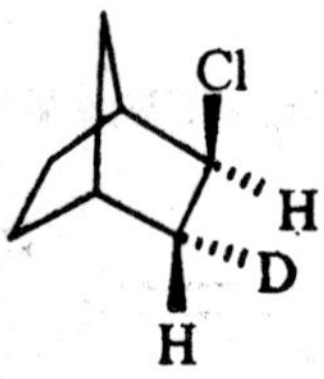 $\xrightarrow{(CH)_3 COK}$

34. Predict products of the following elimination reaction:

(a) 1,2-dibromodecane + $Zn/CH_3COOH \longrightarrow$

(b) *trans*-1, 2-dibromocyclohexane + NaI $\xrightarrow{\text{Acetone}}$

(c) 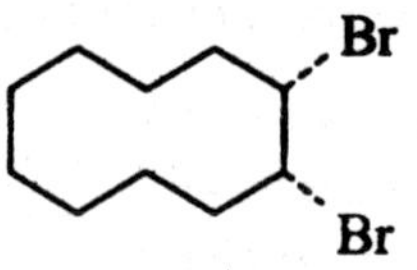 Br, Br + NaI $\xrightarrow{\text{Acetone}}$

35. Predict product of the following reactions, where more than one product is expected, predict the major one.

(a) 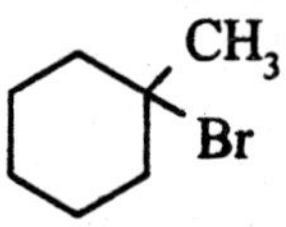 CH_3, Br $\xrightarrow{NaOCH_3}$

(b) 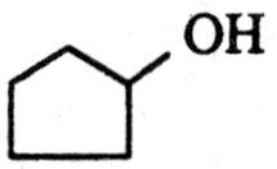 OH $\xrightarrow[\text{Heat}]{H_2SO_4}$ (c) 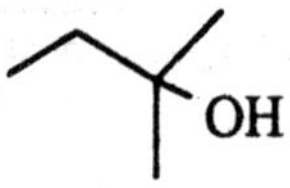 OH $\xrightarrow[\text{Heat}]{H_3PO_4}$

36. Propose mechanism:

(a) 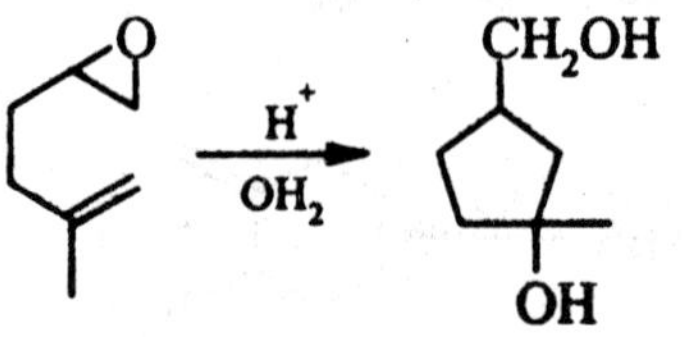 O $\xrightarrow[OH_2]{H^+}$ CH_2OH, OH

(b) 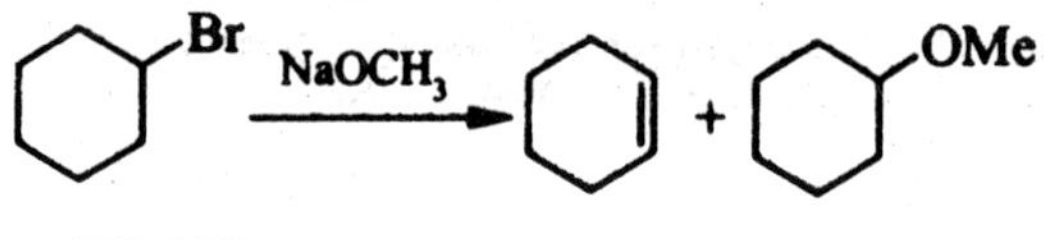 Br $\xrightarrow{NaOCH_3}$ + OMe

(c) 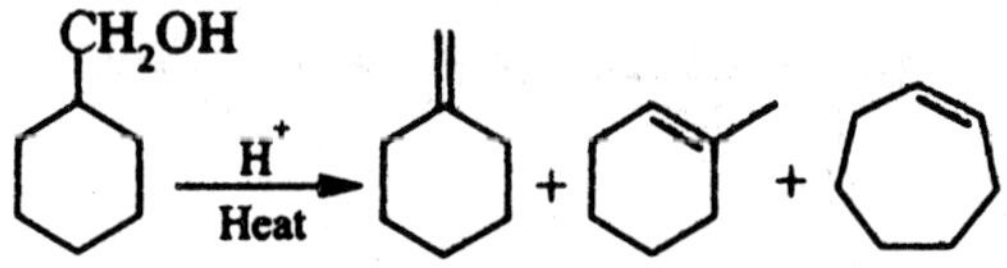 CH_2OH $\xrightarrow[\text{Heat}]{H^+}$ + +

37. Show, how would you prepare cyclopentene from:

(a) *trans*-1,2-dibromocyclopentane

(b) cyclopentanol

(c) cyclopentyl bromide

(d) cyclopentane

38. Predict product:

(a) 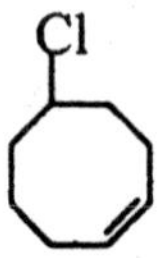 Cl + $Hg(OAc)_2/H_2O \xrightarrow{NaBH_4}$

(b) 1-methylcyclopentene + BH_3. THF then $H_2O_2/OH^- \longrightarrow$ (Stereochemistry)

39. Cyclohexene is dissolved in a solution of lithium chloride in chloroform. To this solution is added one equivalent of bromine. The material isolated from this reaction contains primarily a mixture of *trans*-1,2-dibromocyclohexane and *trans*-1-bromo-2-chlorocyclohexane. Use a detailed mechanism to show how these compounds are formed.

40. Unknown X(C_5H_9Br) does not react with Br_2 and does not rotate plane polarized light. Upon treatment with potassium-t-butoxide, X gives only one product Y(C_5H_8). Unlike X, Y decolourise Br_2-water. Ozonolysis followed by reduction with Zn-H_2O of Y gives Z, $C_5H_8O_2$, which rotate plane polarized light. Propose consistent structures for X,Y and Z.

41. Propose mechanism:

(a) 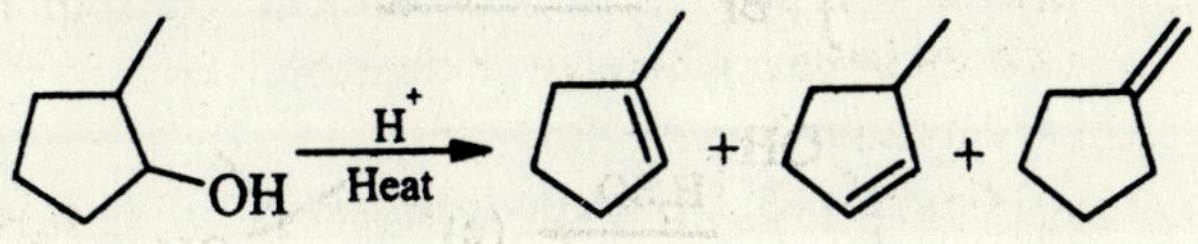 $\xrightarrow[\text{Heat}]{H^+}$ + +

(b) 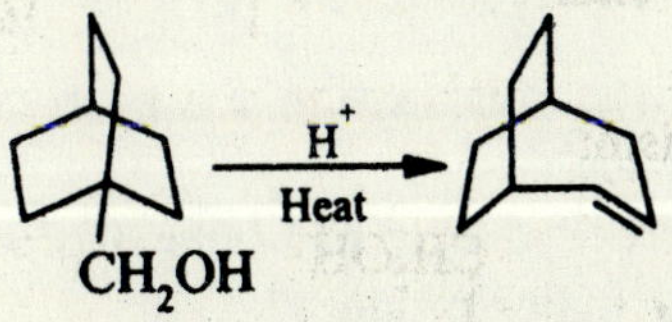 CH_2OH $\xrightarrow[\text{Heat}]{H^+}$

(c) 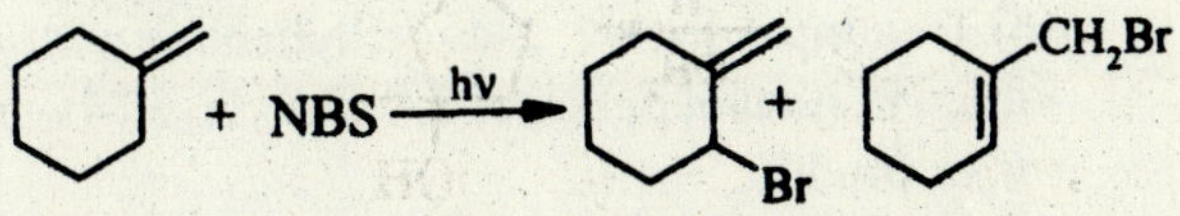 + NBS $\xrightarrow{h\nu}$ Br + CH_2Br

(d) 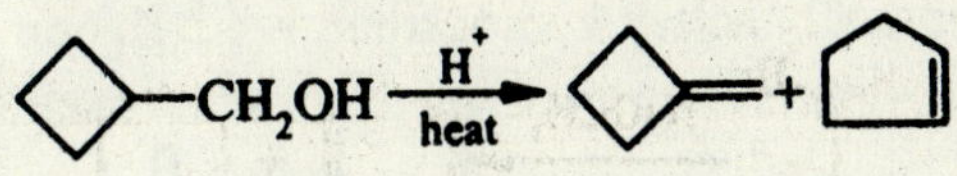 $-CH_2OH$ $\xrightarrow[\text{heat}]{H^+}$ +

(e) 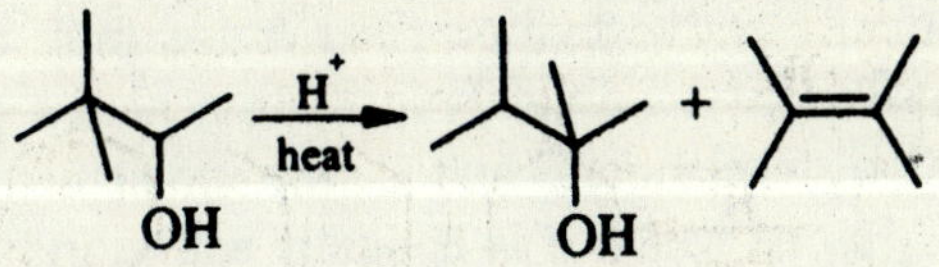 OH $\xrightarrow[\text{heat}]{H^+}$ OH +

42. Convert:

(a) 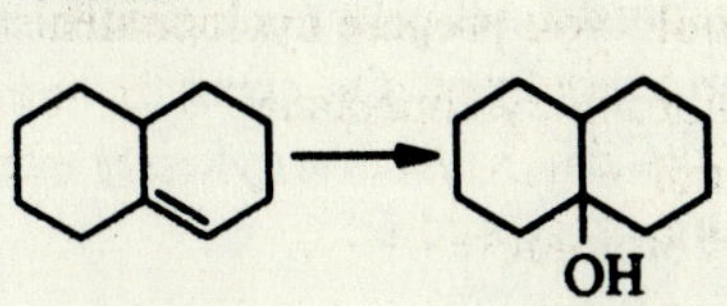 → OH

(b) 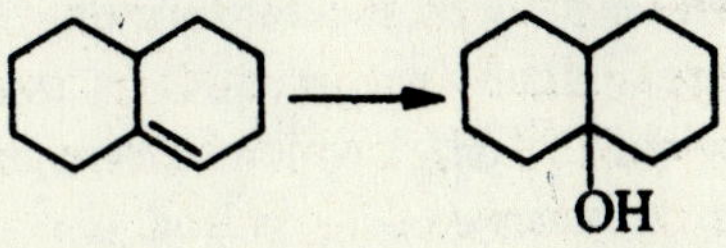 → OH

(c) 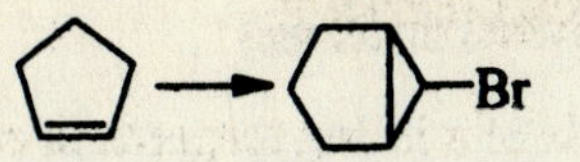 → Br (d) 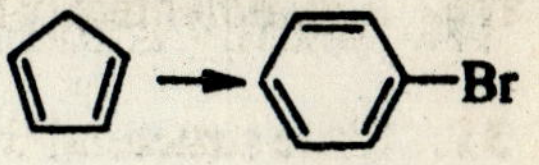 → Br

(e) *Trans*-2-butene → *Trans*–1, 2-dimethylcyclopropane

43. Predict major products of the following reactions (include stereochemistry) where ever possible.

(a) 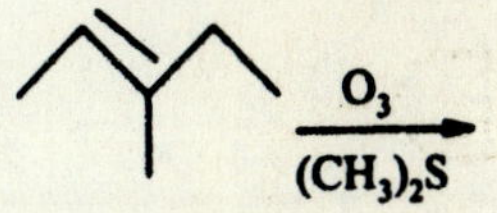 $\xrightarrow[(CH_3)_2S]{O_3}$ (b)  + HBr/R_2O_2

(c) 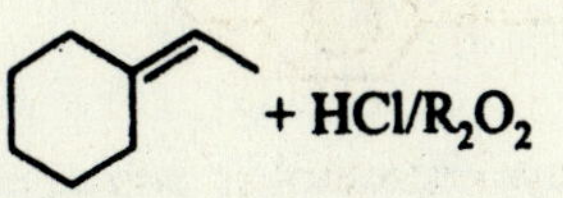 + HCl/R_2O_2 (d) 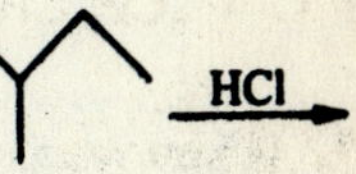 $\xrightarrow{HCl}$

(e) 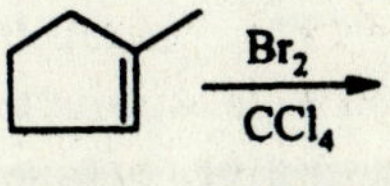 $\xrightarrow[CCl_4]{Br_2}$

(f) Methylcyclopentene + $BH_3/H_2O_2/OH^-$ →

44. Propose mechanism:

(a) 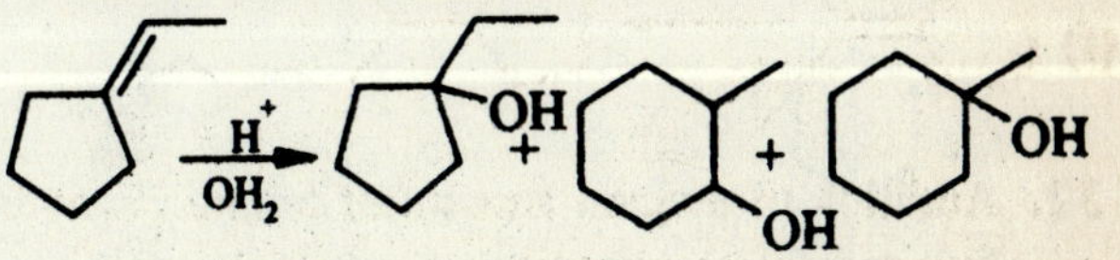 $\xrightarrow[OH_2]{H^+}$ OH + OH + OH

(b) 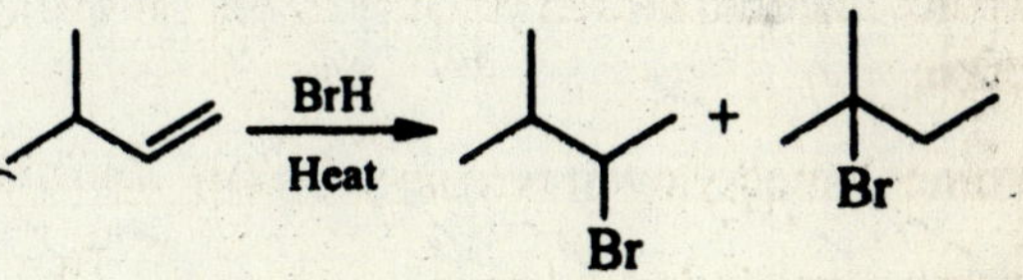 $\xrightarrow[\text{Heat}]{BrH}$ Br + Br

(c) 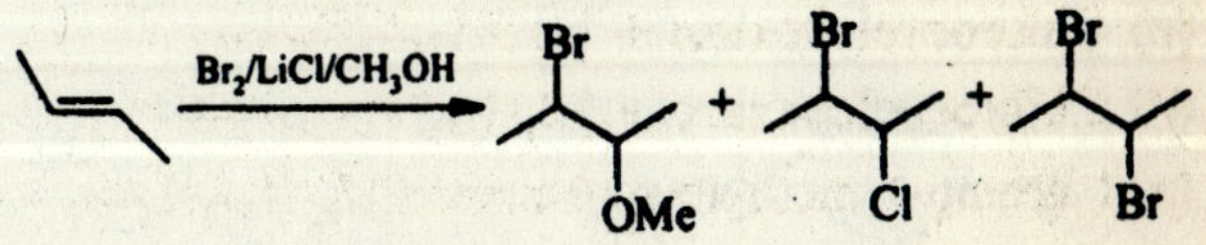 $\xrightarrow{Br_2/LiCl/CH_3OH}$ Br OMe + Br Cl + Br Br

(d) $C_6H_5CH{=}CH_2 \xrightarrow{H^+}$ Ph Ph

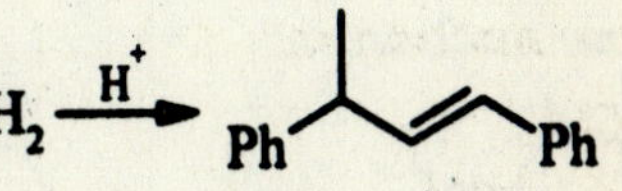

45. Starting with ethyne, outline synthesis of each of the following:

(a) 1-pentyne (b) $CH_3CH_2C{\equiv}CD$

(c) *trans*-2-butene (d) *cis*-2, 3-dideutero-2-butene.

46. Identify A to E (Provide structures)

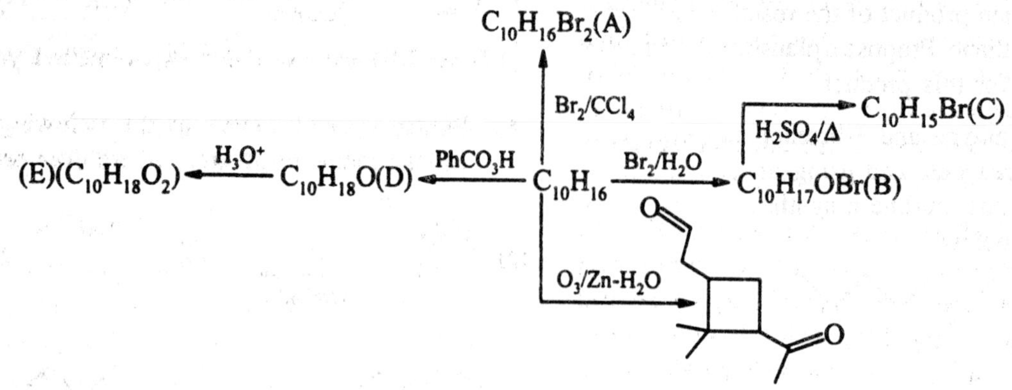

$C_{10}H_{16}Br_2$(A)

Br$_2$/CCl$_4$

H$_2$SO$_4$/Δ → $C_{10}H_{15}$Br(C)

(E)($C_{10}H_{18}O_2$) ←H$_3$O$^+$— $C_{10}H_{18}$O(D) ←PhCO$_3$H— $C_{10}H_{16}$ —Br$_2$/H$_2$O→ $C_{10}H_{17}$OBr(B)

O$_3$/Zn-H$_2$O

47. When *trans*-2-methylcyclohexanol is subjected to acid catalysed dehydration, the major product is 1-methylcyclohexene. However, when *trans*-1-bromo-2-methylcyclohexane is subjected to dehydrohalogenation, the major product is 3-methylcyclohexene. Explain this difference.

48. Show how would you synthesize each compound using methylcyclohexene as starting material.

(a) 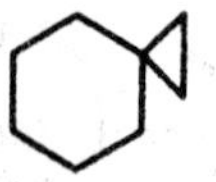 (b) 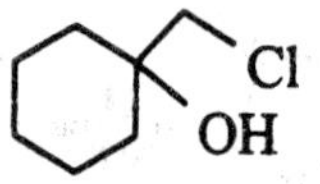 Cl OH

(c) 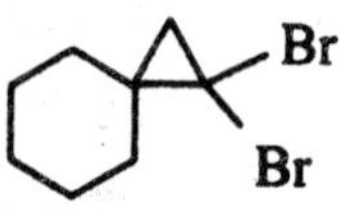 Br Br (d) 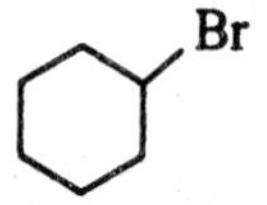 Br

(e) 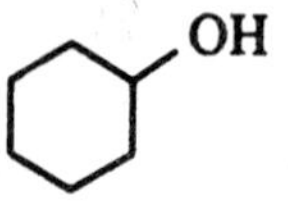 OH (f) 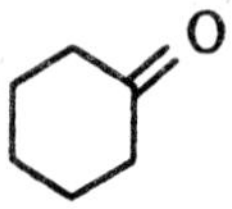 O (g) 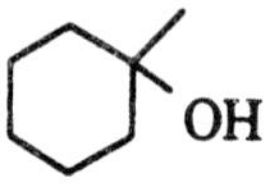 OH

(h) 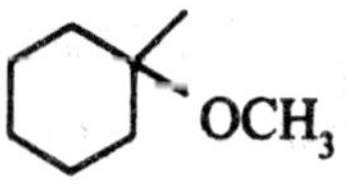 OCH$_3$ (i) 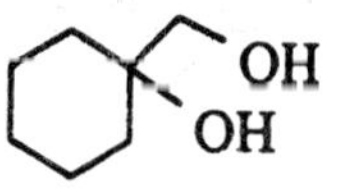 OH OH

49. Propose mechanism:

(a) 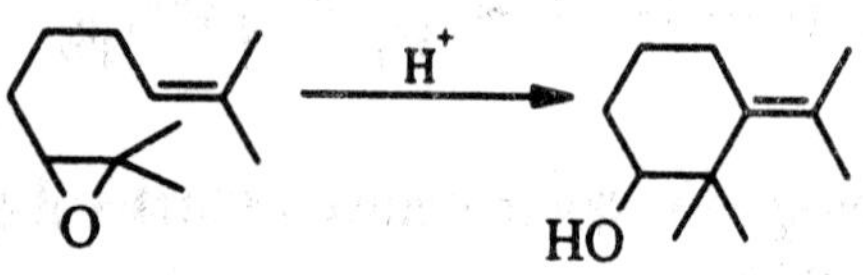 O —H$^+$→ HO

(b) 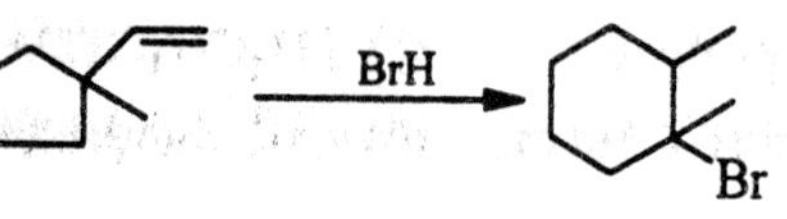 —BrH→ Br

(c) 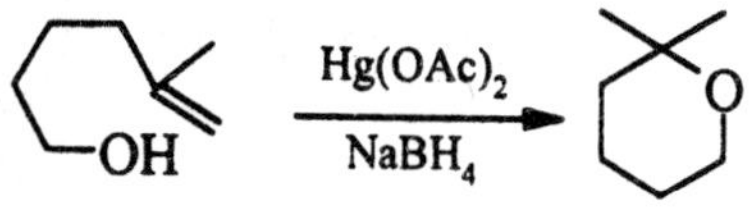 OH —Hg(OAc)$_2$ / NaBH$_4$→ O

50. Write structural formulas of all the products that would be obtained when the following alkylhalide is heated with potassium-t-butoxide in t-butanol. Indicate *cis-trans* where ever possible:
(a) 2-bromo-3-methylbutane
(b) 4-bromo-2, 2-dimethylpentane
(c) 1-bromo-1-methylcyclohexane.

51. Provide mechanistic explanation for each of the following occurrence:
(a) Acid-catalysed dehydration of 1-butanol produces *trans*-2-butene as major product.
(b) Treating 3-iodo-2, 2-dimethylbutane with silver nitrate in ethanol produces 2, 3-dimethyl-2-butene as major product.

52. An optically active hydrocarbon A (C_5H_8) on catalytic hydrogenation yield B(C_5H_{10}), which is optically inactive. Deduce structure of A and B.

53. An optically active hydrocarbon A has molecular formula C_5H_8. A on catalytic hydrogenation produces two stereomers B and C with molecular formula C_5H_{10} of which B is optically active and C is optically inactive. Deduce structure of A, B and C.

54. An organic compound A(C_6H_{10}) is optically active and contains no triple bond. On catalytic hydrogenation A yields B(C_6H_{14}), which is optically inactive. Deduce structure of A and B.

55. Compound A and B have molecular formula C_7H_{14}. Both A and B are optically active. Both on catalytic hydrogenation yield C(C_7H_{16}), which is also optically active. Deduce structures of A, B and C.

56. Starting with a compounds of two carbon, outline stereospecific synthesis of meso-3, 4-dibromohexane and racemic 3, 4-dibromohexane.

57. When 2, 2-diphenyl-1-ethanol is treated with aqueous HI, the main product of the reaction is 1-iodo-1, 1-diphenylethane. Propose a plausible mechanism that accounts for this product.

58. Starting with ethyne and 1-bromopentane as your only organic reagents and using any needed inorganic compounds, outline a synthesis of the compound shown below:

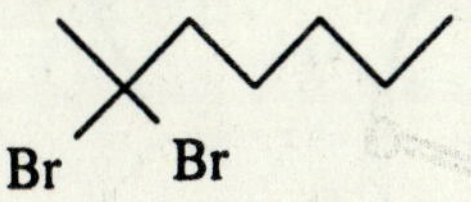

Br Br

59. When cyclohexene is allowed to react with bromine in an aqueous solution of NaCl, the products are *trans*-1, 2-dibromo cyclohexane, *trans*-2-bromocyclohexanol and *trans*-1-bromo-2-chlorocyclohexane. Write plausible mechanism that explain the formation of these products.

60. An organic compound A has molecular formula $C_{10}H_{16}$ and is known not to contain any triple bond. On catalytic hydrogenation of A, a new compound $B(C_{10}H_{22})$ is formed. Ozonolysis of A followed by treatment with zinc-H_2O yields two mole of CH_2O, one mole of acetone and a third compound $C(C_5H_6O_3)$. Deduce the structure of A, B and C considering C being optically active.

61. An organic compound $A(C_{10}H_{16})$ on catalytic hydrogenation produces 1-isopropyl-4-methylcyclohexane. Ozonolysis of A followed by workup with dimethylsulphide produces formaldehyde and Deduce structure of A.

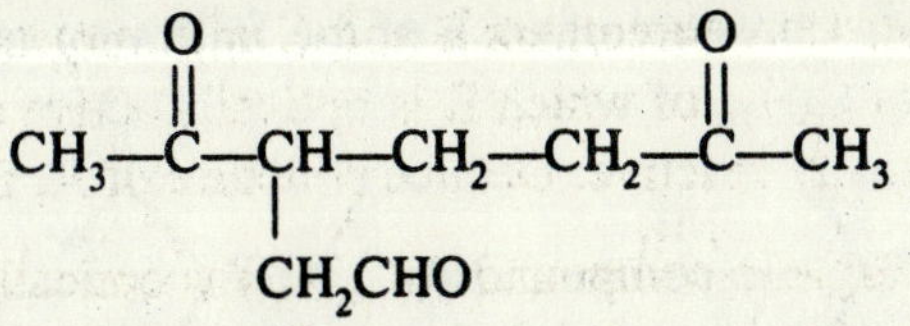

$$CH_3-\overset{O}{\overset{\|}{C}}-\underset{CH_2CHO}{\underset{|}{CH}}-CH_2-CH_2-\overset{O}{\overset{\|}{C}}-CH_3$$

62. Write stereochemical formulas for all of the products that you would expect from each of the following reactions.

(a) 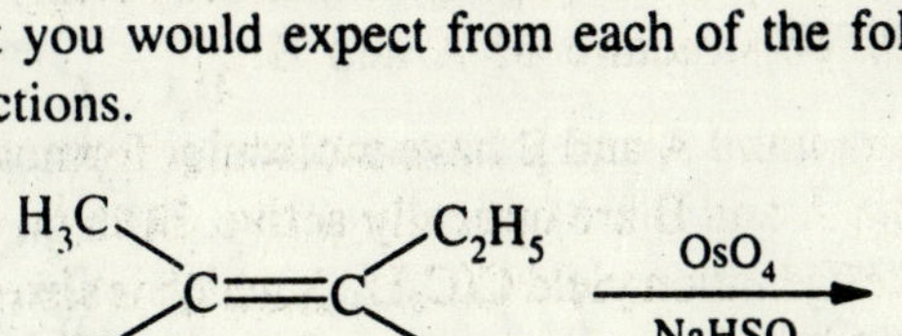 $H_3C(H)C{=}C(C_2H_5)(H)$ $\xrightarrow[NaHSO_3]{OsO_4}$

(b) $H_3C(H)C{=}C(H)(C_2H_5)$ $\xrightarrow{Br_2/CCl_4}$

(c) $H_3C(H)C{=}C(H)(C_2H_5)$ $\xrightarrow[NaHSO_3]{OsO_4}$

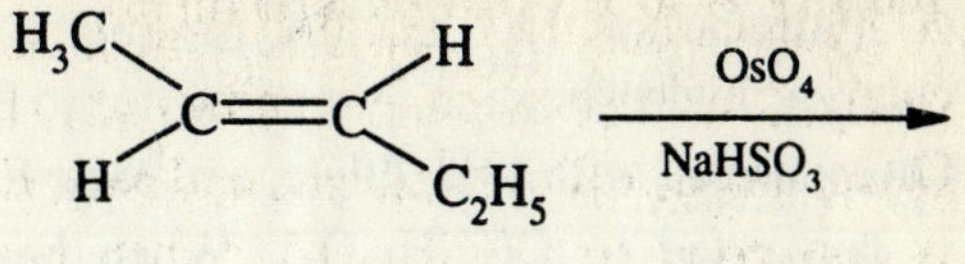

(d) 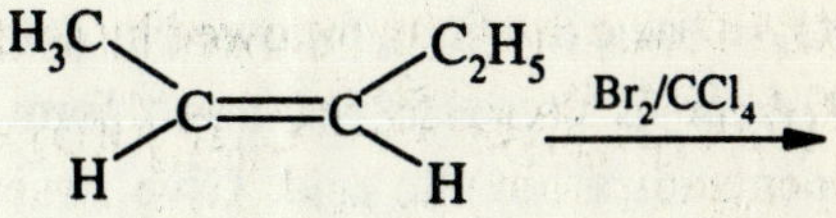 $H_3C(H)C{=}C(C_2H_5)(H)$ $\xrightarrow{Br_2/CCl_4}$

63. Three isomeric hydrocarbons A, B and C have molecular formula C_6H_{10}. All three compounds readily decolourise bromine in CCl_4. Compound A gives precipitate with $AgNO_3/NH_3$ solution but compound B and C do not. Compound A and B both yield hexane when treated with excess of H_2/Pt. Under these conditions, C absorb only one molar equivalent of H_2 and give product with formula C_6H_{12}. When A is oxidised with hot basic $KMnO_4$ and resulting solution is acidified, the only organic product that can be isolated is $CH_3(CH_2)_3COOH$. Similar oxidation of B gives only CH_3CH_2COOH and similar treatment of C gives only $HO_2C(CH_2)_4COOH$. Identify A, B and C.

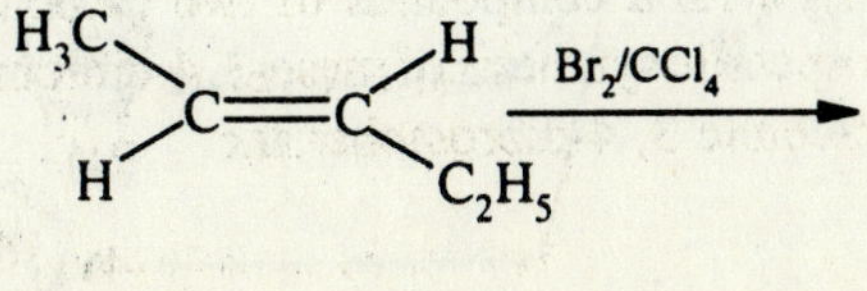

64. An optically active compound A has molecular formula C_6H_{10}. The compound gives a precipitate when treated with $Ag(NH_3)_2OH$. On catalytic hydrogenation, A yields $B(C_6H_{14})$, which is optically inactive. Deduce structure of A and B.

65. Propose mechanism:

(a) 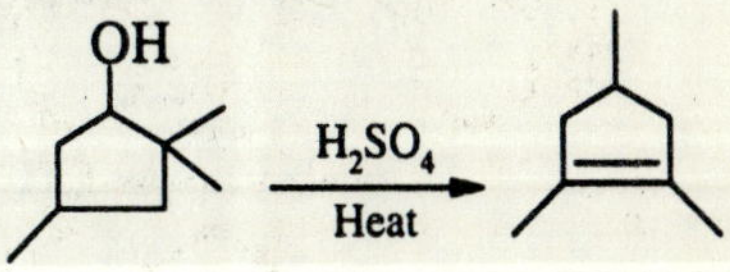 OH $\xrightarrow[Heat]{H_2SO_4}$

(b) 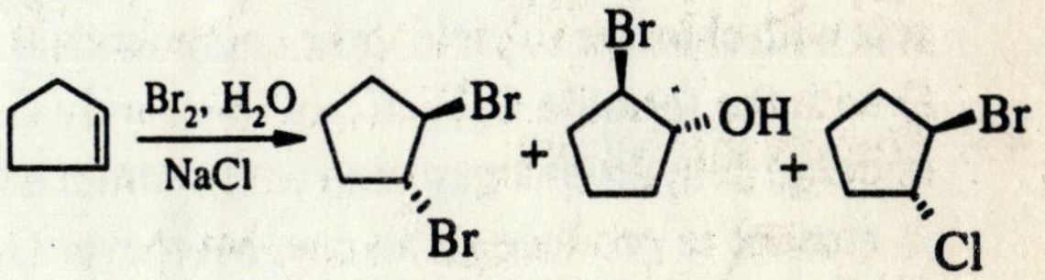 $\xrightarrow[NaCl]{Br_2, H_2O}$ Br, Br + Br, OH + Br, Cl

66. Identify A to C:

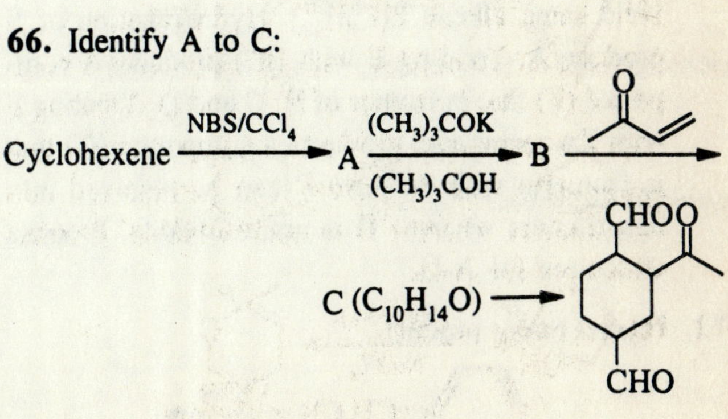

Cyclohexene $\xrightarrow{NBS/CCl_4}$ A $\xrightarrow[(CH_3)_3COH]{(CH_3)_3COK}$ B $\xrightarrow{}$

$C(C_{10}H_{14}O) \longrightarrow$ CHO, CHO

67. A hydrocarbon (A) has the formula C_7H_{10}. On catalytic hydrogenation, A is converted to B (C_7H_{12}). On treatment with cold, dilute, and basic $KMnO_4$, A is converted to C($C_7H_{12}O_2$). When heated with $KMnO_4$ in basic medium, followed by acidification, either A or C produces the meso form of 1, 3-cyclopentandicarboxylic acid. Give structural formula for A-C.

68. Compound A (C_4H_6) reacts with H_2/Pt to yield butane. A decolourise Br_2 in CCl_4 but does not react with $Ag(NH_3)_2^+$. On treatment with B_2H_6/CH_3COOH, A is converted to B(C_4H_8). When B is treated with OsO_4 and then with $NaHSO_3$, it is converted to C($C_4H_{10}O_2$). Compound C cannot be resolved. Provide structure to A, B and C.

69. Propose mechanism:

(a) [structure with C:] ⟶ [bicyclic structure]

(b) [structure]COOH + I_2 ⟶ [lactone structure with I]

70. Give structure for A-C.

[structure with Br] $\xrightarrow{EtO^-/EtOH}$ A $\xrightarrow[BH_3]{THF}$ $B(C_6H_{13})_2BH$ $\xrightarrow{H_2O_2/OH^-}$ C($C_6H_{14}O$)

71. An alkane (A) with molecular formula C_6H_{14} reacts with chlorine to yield three compounds B, C and D with the formula $C_6H_{13}Cl$, of these only C and D undergo dehydrohalogenation with sodium ethoxide in ethanol to produce an alkene. Moreover C and D yield same alkene E(C_6H_{12}). Hydrogenation of E produce A. Treating E with HCl produces a compound (F) that is isomer of B, C and D. Treating F with Zn-acetic acid produces a compound (G) that is isomeric with A. Also C can be resolved into enantiomers whereas D is nonresolvable. Propose structures for A-G.

72. Predict major product:

[benzene]$-CH_2CH_3$ $\xrightarrow{Cl_2/hv}$

73. Predict product:

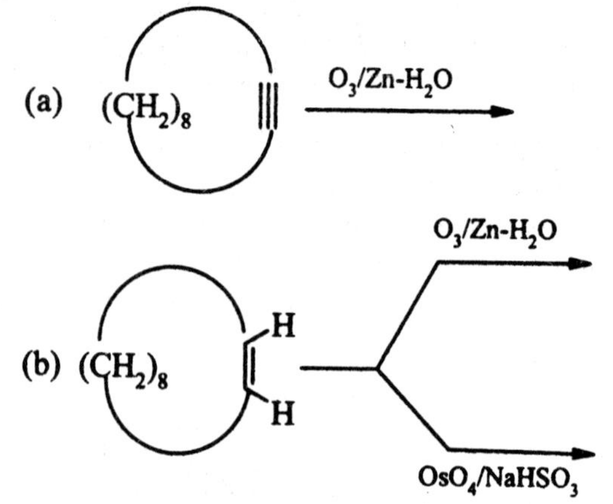

(c) Ph—CH=CH$_2$ + BrCl ⟶ ?

(d) Ph—CH=CH$_2$ + Br_2/H_2O ⟶

(e) Ph—C≡C—CH$_3$ + 1.0 Equivalent HCl?

(f) H_3C—C≡C—Ph + H_2O $\xrightarrow{Hg^{2+}/H^+}$

(g) Ph—C≡C—CH$_3$ + B_2H_6 $\xrightarrow{H_2O_2/H\bar{O}}$

74. Compound A ($C_8H_{15}Cl$) exist as racemic form. Compound A does not decolourize Br_2/CCl_4. When A is treated with Zn-acetic acid, and the mixture is separated by gas-liquid chromatography, two fractions B and C are obtained. The component of both fractions have formula C_8H_{16}. Fraction B consists of a racemic form whereas C cannot be resolved. Treating A with C_2H_5ONa/C_2H_5OH convert it into D(C_8H_{14}). Hydrogenation of D produces C. Ozonolysis of D followed by workup with $(CH_3)_2S$ yields 2, 7-octandione. Propose structures for A, B, C and D including stereochemistry where possible.

75. Predict product of each of the following reactions:

(a) [alkene with H] $\xrightarrow{NH_3}$ (b) [α,β-unsaturated ketone with H] $\xrightarrow{NH_3}$

76. Convert:

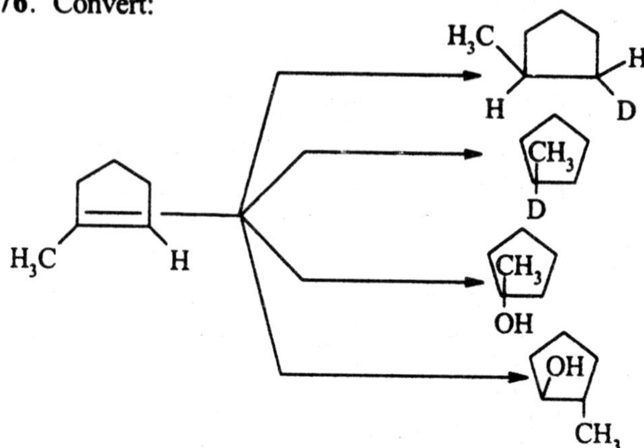

77. Predict major product:

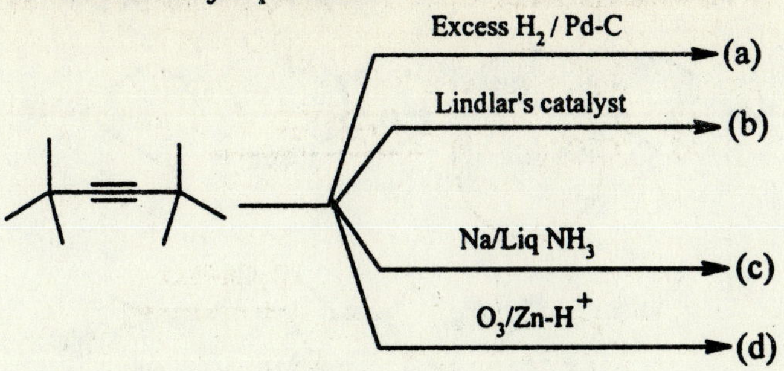

Reagents on the branches: Excess H_2 / Pd-C → (a); Lindlar's catalyst → (b); Na/Liq NH_3 → (c); O_3/Zn-H^+ → (d)

78. Predict major product:

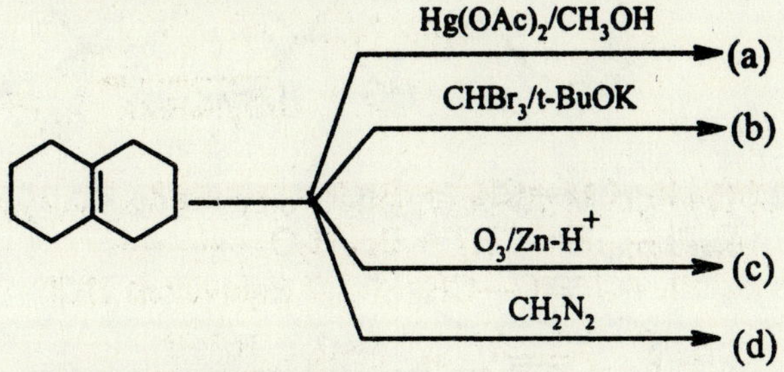

Reagents on the branches: $Hg(OAc)_2/CH_3OH$ → (a); $CHBr_3$/t-BuOK → (b); O_3/Zn-H^+ → (c); CH_2N_2 → (d)

79. Suggest mechanism:

(a) 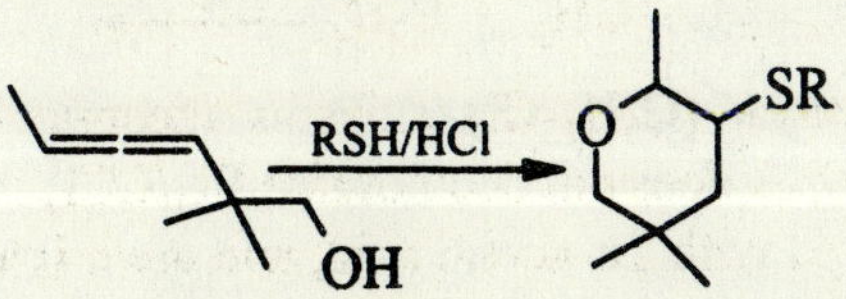 RSH/HCl; product labels: O, SR; starting material label: OH

(b) 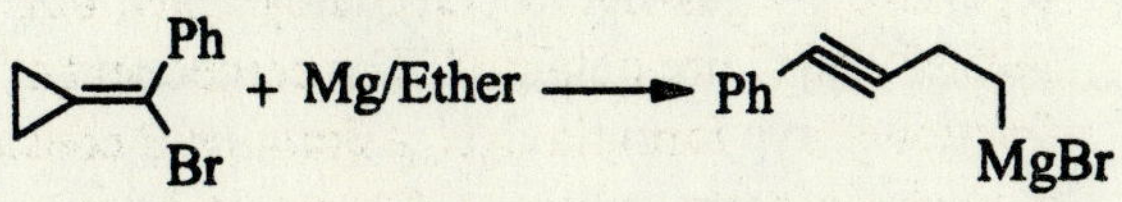 Ph, Br + Mg/Ether ⟶ Ph, MgBr

(c) 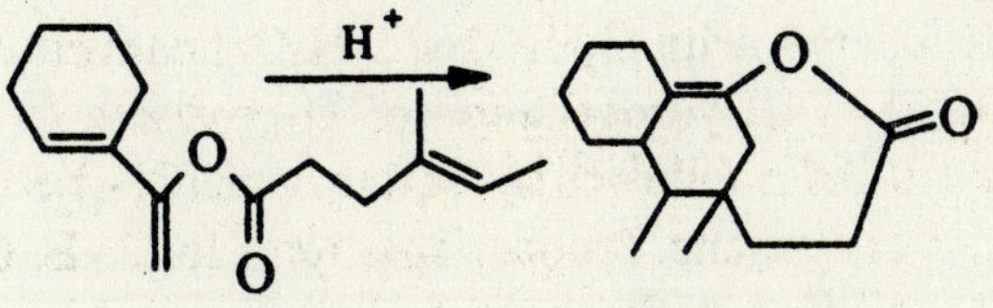 H^+; labels: O, O, O, O

80. An organic compound A has molecular formula C_8H_6 and it forms a white precipitate with ammonical solution of $AgNO_3$. A on treatment with H_2/Pt produces C_8H_{16}. Also A on treatment with liquid bromine produces a tetrabromide. Deduce structure of A.

81. Propose mechanism:

(a) 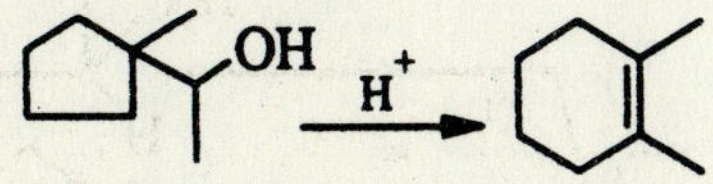 OH, H^+

(b) $(CH_3)_3CCH_2OH \xrightarrow{H^+}$

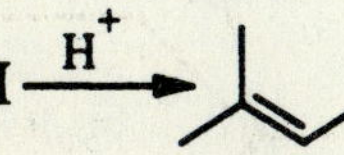

(c) 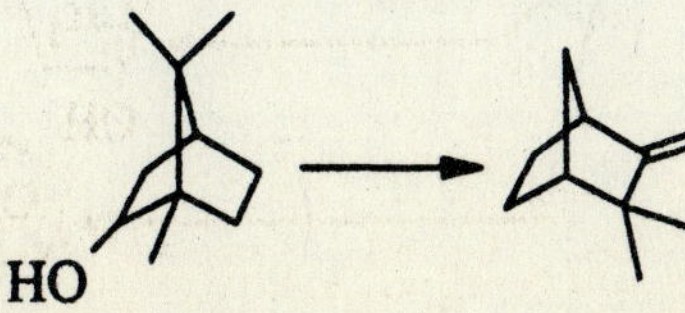 HO ⟶

82. Provide mechanistic explanation:

(a) The addition of HCl to 3-methyl-1-butene produces two products; 2-chloro-3-methylbutane and 2-chloro-2 methylbutane.

(b) The addition of HCl to 3, 3-dimethyl-1-butene produces two products; 3-chloro-2, 2-dimethylbutane and 2- chloro-2, 3-dimethylbutane.

83. When the following deuterium-labeled isomer (I) undergoes elimination, the reaction yields *trans*-2-butene (II) and *cis*-2-butene (III). The reaction does not yield *cis*-2-butene (IV) and *trans*-2-butene (V). Explain.

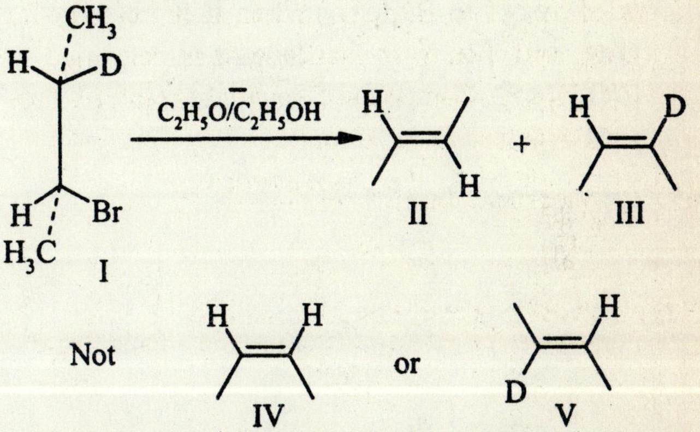

Labels: CH_3, H, D, H, Br, H_3C — I; $\xrightarrow{C_2H_5\bar{O}/C_2H_5OH}$ H, H — II + H, D — III

Not: H, H — IV or H, D — V

84. An organic compound A has molecular formula C_7H_{10}, which on treatment with H_2/Pt gives $B(C_7H_{14})$ which is nonresolvable. A on treatment with one equivalent of H_2O in acidic medium produces $C(C_7H_{12}O)$, which is resolvable while addition of second equivalent of water to C in acidic medium produces two isomeric compound D and E with their molecular formula $C_7H_{14}O_2$. D is resolvable whereas E is not. A on ozonolysis followed by work-up with H_2O_2 produced $F(C_4H_6O_3)$ and $C_3H_4O_4$. F on simple heating produced C_3H_6O, which does not change the colour of acidic solution of dichromate. Deduce structures of A to F.

85. Predict mechanism of formation of the products in the following reacitons:

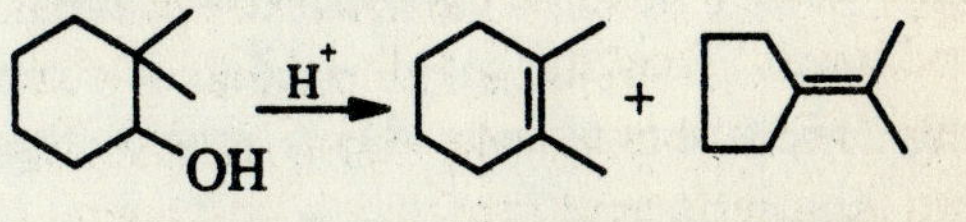

OH $\xrightarrow{H^+}$ +

86. (a) Give a mechanism to show how 2-pentyne reacts with sodium amide to give 1-pentyne.

(b) How would you accomplish the opposite reaction?

87. An optically active organic compound A has molecular formula $(C_{14}H_{18})$ and has only one chiral carbon. A on hydrogenation with H_2/Pt gives optically inactive compound $C_{14}H_{26}$. A on treatment with O_3/H_2O_2 gave B($C_{14}H_{18}O_3$), which still has one chiral carbon. B on heating with NaOH/CaO produced optically active C($C_{13}H_{18}O$). C on catalytic hydrogenation with $NaBH_4$ produced D($C_{13}H_{20}O$) which is distereomeric. D on heating with concentrated H_2SO_4 produced E($C_{13}H_{18}$), which on ozonolysis followed by work-up with Zn-H_2O gives acetophenone as one product. Deduce structures of A to E.

88. Give the dominant alkene for the each of the following reactions:

(a) 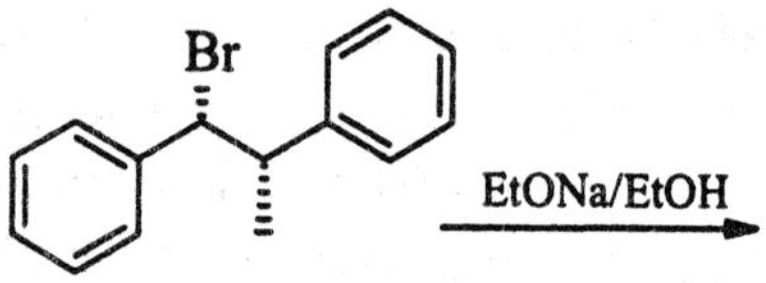 Br — EtONa/EtOH

(b) 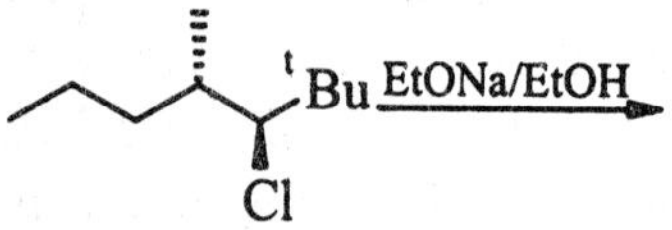 ^{t}Bu, Cl — EtONa/EtOH

(c) 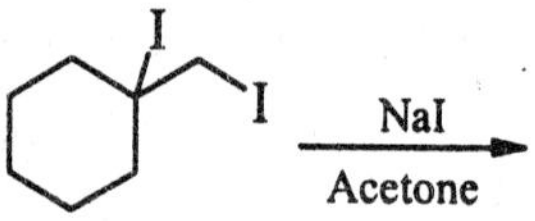 I, I — NaI / Acetone

(d) 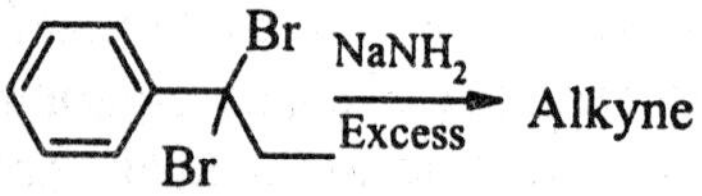 Br, Br — $NaNH_2$ / Excess → Alkyne

89. Predict major products:

(a) 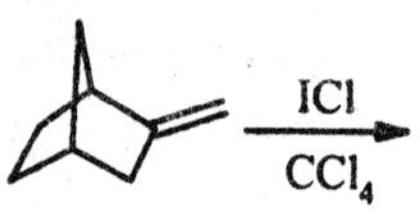 ICl / CCl_4 (b) 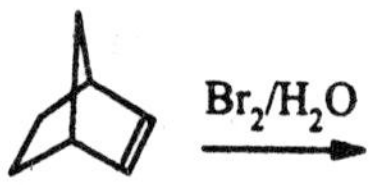 Br_2/H_2O

(c) 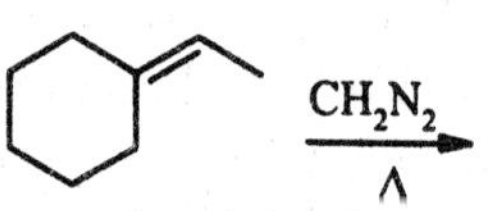 CH_2N_2 / Λ (d) 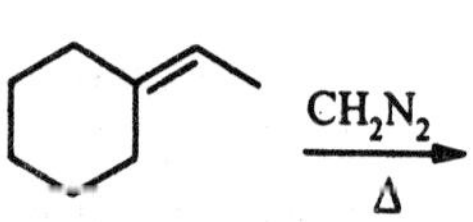 CH_2N_2 / Δ

(e) 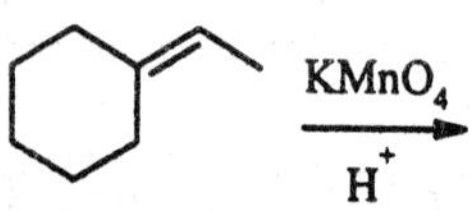 $KMnO_4$ / H^+

90. Consider the reaction:

(a) Write a detailed curved-arrow mechanism that shows how the major product is formed.

(b) Briefly explain why this isomer, instead of others, is formed?

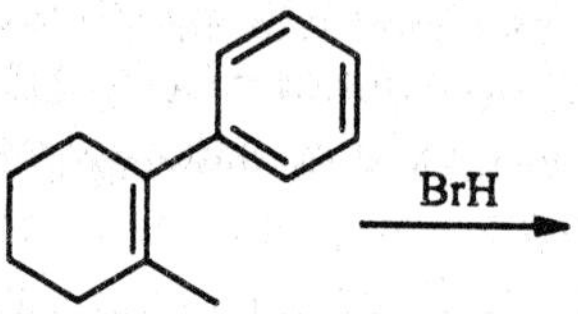

BrH

91. Propose mechanism indicating the major product:

(a) 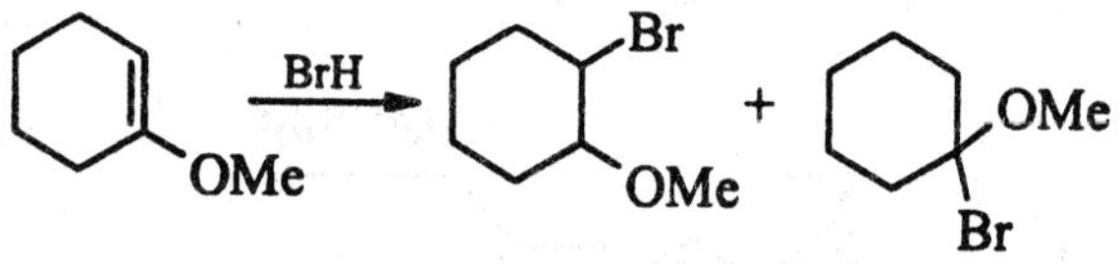 OMe — BrH → Br, OMe + OMe, Br

(b) 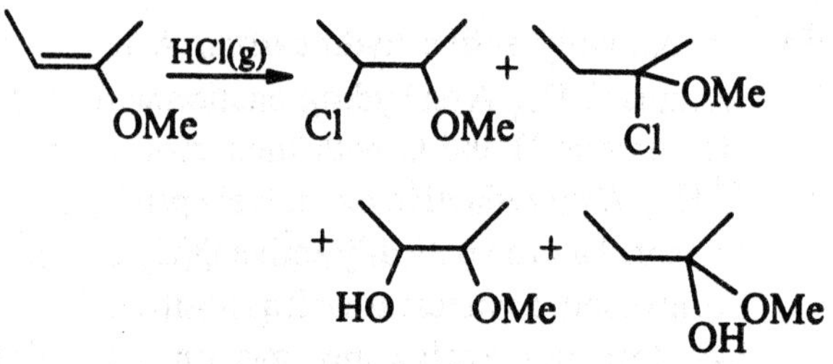 OMe — HCl(g) → Cl, OMe + OMe, Cl + HO, OMe + OMe, OH

92. Provide the major organic products of the following reaction:

(a) 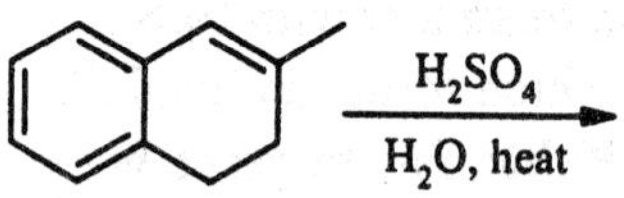 H_2SO_4 / H_2O, heat

(b) 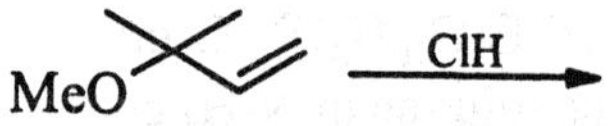 MeO — ClH

(c) 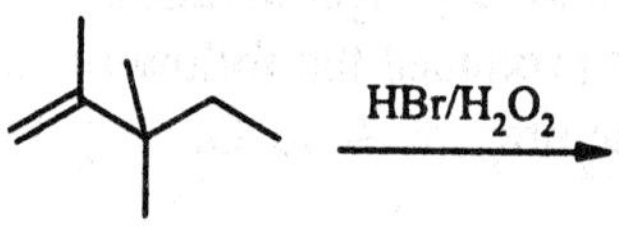 HBr/H_2O_2

(d) 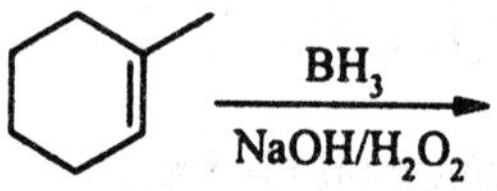 BH_3 / $NaOH/H_2O_2$

(e) 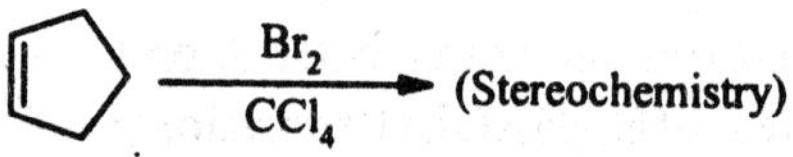 Br_2 / CCl_4 (Stereochemistry)

(f) 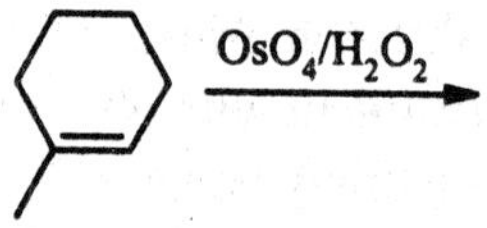 OsO_4/H_2O_2

(g) 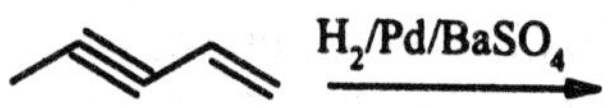 $H_2/Pd/BaSO_4$

(h) ClH

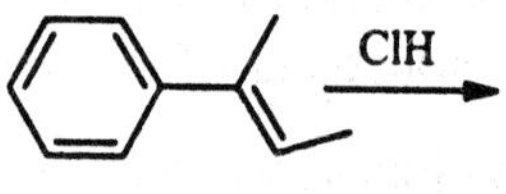

(i) 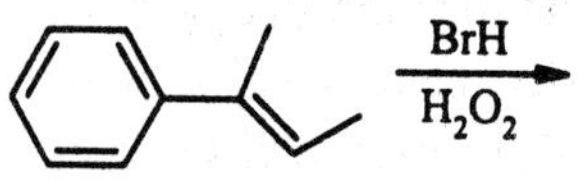 BrH / H_2O_2

(j) 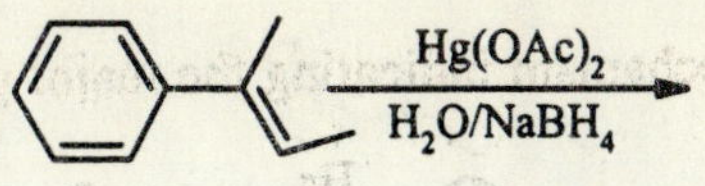 [structure] $\xrightarrow[H_2O/NaBH_4]{Hg(OAc)_2}$

(k) 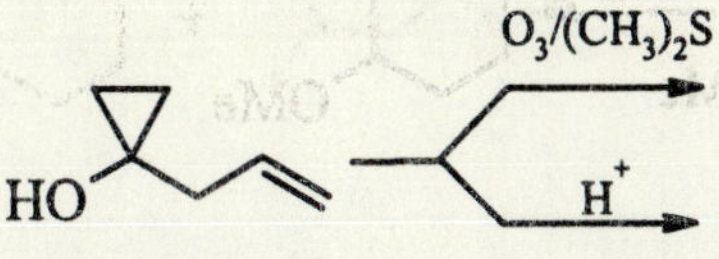 HO [structure] $\xrightarrow{O_3/(CH_3)_2S}$; $\xrightarrow{H^+}$

93. An optically active hydrocarbon A has molecular formula C_8H_{14}. A on hydrog`enation with H_2/Pt gives stereomeric B and C with their molecular formula C_8H_{18}. A on reductive ozonolysis produced ethanal, methanal and an optically active $D(C_5H_8O_2)$. D gives positive iodoform test as well as positive Tollen's test, but does not evolve any gas on treatment with $NaHCO_3$. Deduce structures of A to D.

94. An organic compound has molecular formula $C_{10}H_6(A)$. A on treatment with $HgSO_4/H_2SO_4$ gives $B(C_{10}H_{10}O_2)$, which forms an insoluble precipitate with 2, 4-DNP, but fail to react with ammonical solution of $AgNO_3$. B on refluxing with dilute NaOH solution produced $C(C_{10}H_8O)$. C on further treatment with alkaline solution of N_2H_4 gives $D(C_{10}H_{10})$, which decolourises Br_2-H_2O solution. D on treatment with H_2/Pt produced the following compound X. Deduce structures of A to D.

[structure X]

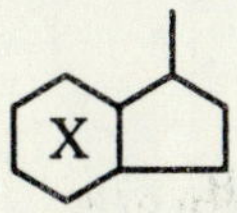

95. An optically active compound A has molecular formula $C_{10}H_{12}O$. A decolourises Bayer's reagent as well as it forms salt with NaOH. A on treatment with HBr forms $B(C_{10}H_{13}OBr)$ as major product which is still enantiomeric. B on treatment with alcoholic solution of KOH gives a stereomeric C, which is also an isomer of A. A on treatment with acidic solution of $KMnO_4$ gives ortho-hydroxy benzoic acid as one product. A on treatment with $B_2H_6/H_2O_2/NaOH$ produces $D(C_{10}H_{14}O_2)$, which on refluxing with dilute sulphuric acid produced $E(C_{10}H_{12}O)$. E neither decolourises Br_2-H_2O solution nor forms any salt with NaOH. Deduce structures of A to E.

96. An optically active hydrocarbon (A) was found to possess the molecular formula C_8H_{14}, hydrogenation of A gave an optically inactive substance $B(C_8H_{18})$. Treatment of A with $O_3/(CH_3)_2S$ gave ethanal and $C(C_4H_6O_2)$. C is non resolvable but gives positive Tollen's test and negative iodoform test. Deduce structures of A, B and C.

97. An optically active compound A has one chiral center and on subjecting to hydroboration oxidation provides mixture of two compounds B and C, both of which are optically active. Appropriate experiment showed that B is thermodynamically more stable than C. Hydrogenation of A provided mixture of two distereomeric dimethyl cyclohexane D and E, both of which are optically inactive. Compound D was found to be thermodynamically more stable than E. Deduce structures of A to E.

98. An organic compound A has molecular formula C_9H_{14} and on treatment with H_2/Pt gives C_9H_{20}. A exist in four stereomeric forms. A on treatment with O_3 followed by work-up with Zn-H_2O produced CO_2, $B(C_4H_6O_3)$ and $C(C_4H_8O)$. B rotates plane-polarized light but C does not. C changes the colour of acidic dichromate solution converting itself into an another optically inactive compound $D(C_4H_8O_2)$. D on treatment with I_2/Red-P produced an optically inactive compound $E(C_4H_7O_2I)$ as the only isomer. Deduce structures of A to E and draw the structures of all four stereomers of A.

99. (a) Reagent (two possible isomers) + $KMnO_4/H^+ \rightarrow$ 3, 4-diketo hexandioic acid. Provide structures of the reagent which will lead to the above product.

(b) Propose mechanism:

(i) 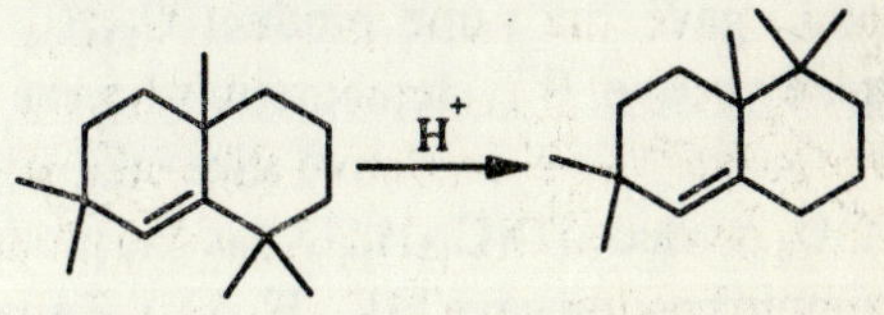 [structure] $\xrightarrow{H^+}$ [structure]

(ii) 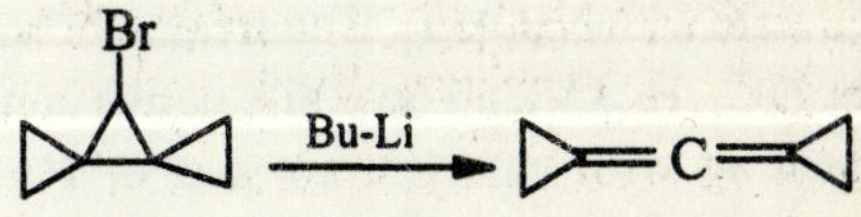 [structure with Br] $\xrightarrow{Bu\text{-}Li}$ [structure]=C=[structure]

(iii) 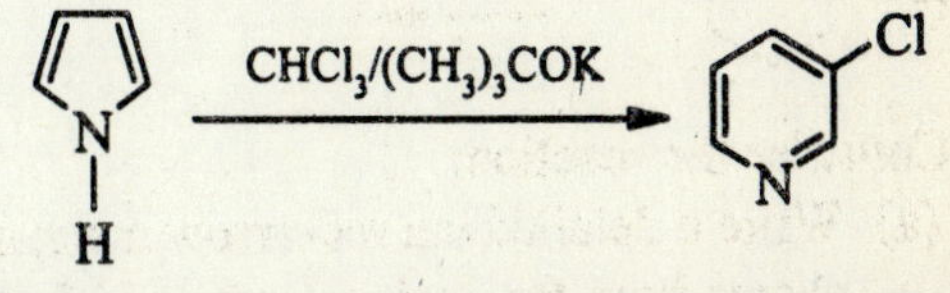 [pyrrole, N–H] $\xrightarrow{CHCl_3/(CH_3)_3COK}$ [3-chloropyridine]

100. An optically active compound A has molecular formula C_9H_{10}. A does not decolourize aqueous solution of Br_2. A on treatment with NBS produced another optically active compound $B(C_9H_9Br)$. B on hydrolyzing in aqueous KOH solution gave $C(C_9H_{10}O)$, which does not change the colour of acidic dichromate solution. B on treatment with C_2H_5ONa/C_2H_5OH produced $D(C_9H_8)$ which is optically inactive. D on oxidative cleavage produced E $(C_9H_8O_3)$

which is optically inactive. E on heating with sodalime gave C_8H_8O, which can also be obtained by Fridel Craft's acylation. Deduce the structures of A to E describing the reactions involved in each step.

101. (a) An organic compound A($C_{12}H_{20}$) on catalytic hydrogenation yields B($C_{12}H_{24}$). A on ozonolysis followed by work-up with H_2O_2 yields acetone and the following compound. Deduce the structures of A and B.

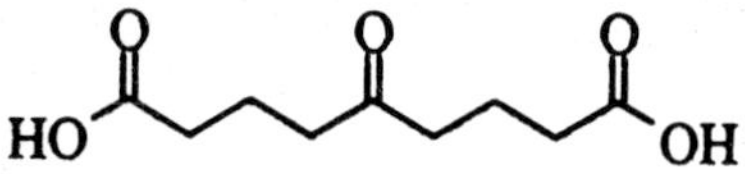

HO ... OH

(b) Predict product in the following sequence of reaction:

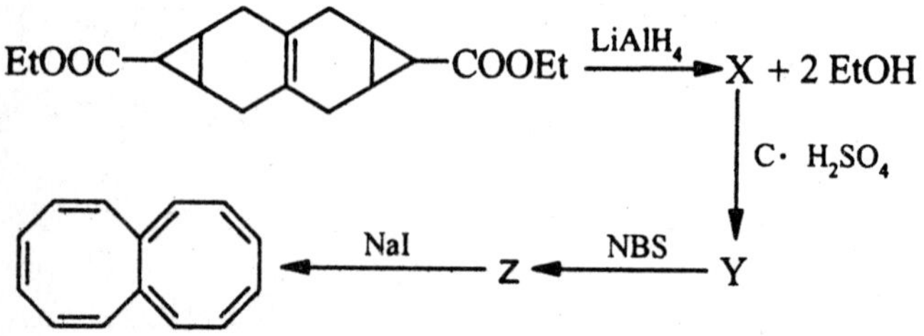

EtOOC— —COOEt $\xrightarrow{LiAlH_4}$ X + 2 EtOH $\xrightarrow{C \cdot H_2SO_4}$ Y $\xrightarrow{NBS}$ Z $\xrightarrow{NaI}$

102. An organic compound A ($C_{13}H_{21}Br$) is optically active and decolourise Bayer's reagent. A on treatment with EtONa/EtOH yields two and only two isomers B and C both having molecular formula $C_{13}H_{20}$ with B as major product. Catalytic hydrogenation of either B or C gave the same product $C_{13}H_{24}$ which is optically active. B is stereomeric whereas C is non-stereomeric. C on oxidative cleavage with acidic $KMnO_4$ produced D($C_{12}H_{18}O_5$) as one product which on simple heating gave E($C_{11}H_{18}O_3$). E gives yellow precipitate with NaOH/I_2. E on treatment with $NaBH_4$ gives F($C_{11}H_{20}O_3$), an optically active compound. F on heating with catalytic amount of H+ gave the following compound. Deduce structures of A to F.

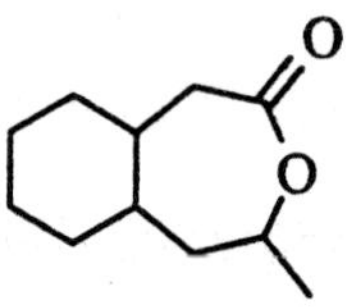

O
O

103. Convert:

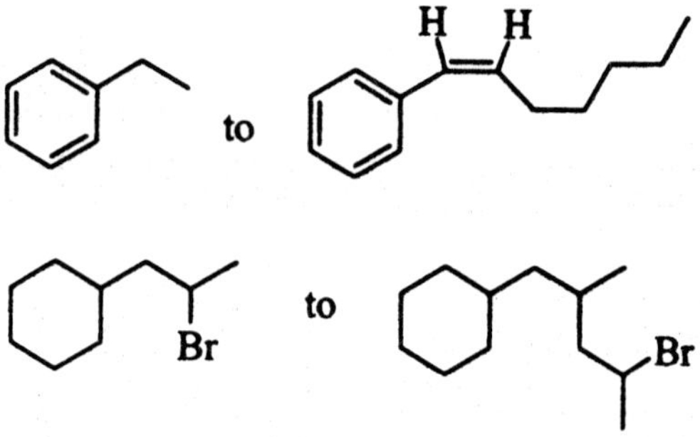

to

H H

Br to Br

104. Identify major product in the following reactions:

(a) 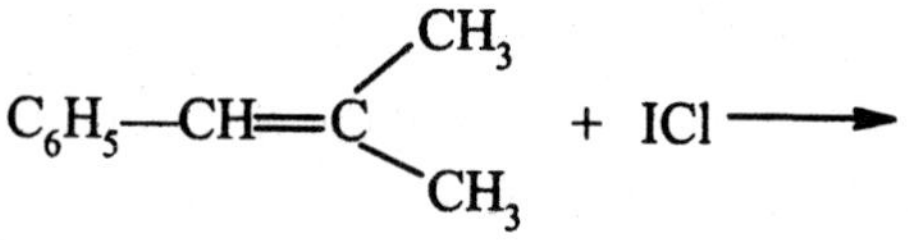 C_6H_5—CH=C(CH_3)(CH_3) + ICl ⟶

(b) 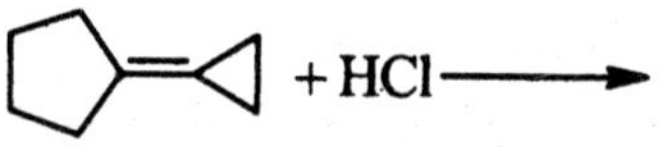 + HCl ⟶

(c) CH_2OH $\xrightarrow{H^+}$

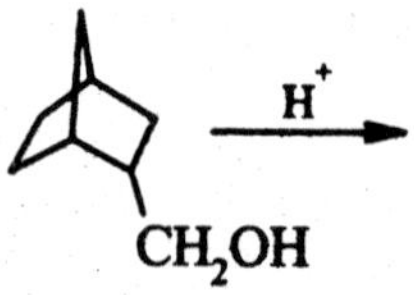

(d) 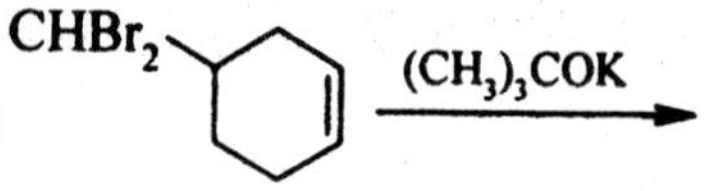 $CHBr_2$ $\xrightarrow{(CH_3)_3COK}$

(e) 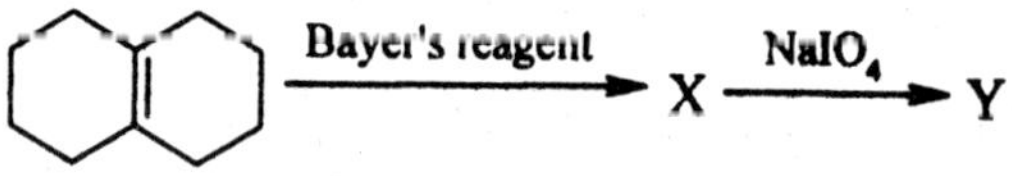 $\xrightarrow{\text{Bayer's reagent}}$ X $\xrightarrow{NaIO_4}$ Y

Chapter 5

Alkyl Halides

5.1 INTRODUCTION

The alkyl halides have general formula $C_nH_{2n+1}X$ or R-X where X-denotes chloride, bromide or iodide, fluoride behaves differently and is discussed separately.

5.2 GENERAL METHOD OF PREPARATION

I. *From alcohols:* It is the most general method and halogen acid is the most common reagent, but the condition depends on both structure of alcohol and the particular acid. In general:

$$R—OH + HX \xrightarrow{Heat} R—X + H_2O$$

Here HCl is used with $ZnCl_2$ (Grove's process) for primary and secondary alcohols, but tertiary reacts rapidly with concentrated acid even in absence of $ZnCl_2$.

Example 1. Why tertiary alkyl halides react rapidly even in absence of $ZnCl_2$?

Solution: The mechanism of reaction is:

$$R—\ddot{O}(H): + ZnCl_2 \longrightarrow R—\overset{+}{O}(H)—\bar{Z}nCl_2 \longrightarrow \overset{+}{R} + [HO\text{-}\bar{Z}nCl_2]$$

Therefore, the reaction proceeds via carbocation intermediate. In case of tertiary alkyl halide, it undergo rapid dissociation to produce 3° carbocation even in the absence of $ZnCl_2$ as:

$$R_3C—OH + HCl \longrightarrow$$

$$R_3C—\overset{+}{O}H_2 + Cl^- \underset{-H_2O}{\rightleftharpoons} R_3C^+ \text{ (a 3° carbocation)}$$

In this reaction, if structure permit, rearrangement of carbocation to more stable carbocation may also occur leading to formation of mixture of products as:

$$H_3C—CH(CH_3)—CH_2OH + HCl \longrightarrow H_3C—CH(CH_3)—CH_2Cl + H_3C—CCl(CH_3)—CH_3$$

Mechanism

$$H_3C—CH(CH_3)—CH_2OH + \overset{+}{H} \longrightarrow H_3C—CH(CH_3)—CH_2\overset{+}{O}H_2 \xrightarrow{-H_2O} H_3C—CH(CH_3)—\overset{+}{C}H_2 \text{ (Primary carbocation)}$$

$$\text{Primary carbocation} \xrightarrow{Cl^-} H_3C—CH(CH_3)—CH_2Cl$$

$$\text{Primary carbocation} \longleftrightarrow H_3C—\overset{+}{C}(CH_3)—CH_3 \text{ (More stable tertiary carbocation)} \xrightarrow{Cl^-} H_3C—CCl(CH_3)—CH_3$$

Phosphorous halides PCl_5, PCl_3, etc. reacts with alcohol resulting in formation of alkyl halide as:

$$ROH + PCl_5 \rightarrow R—Cl + HCl + POCl_3$$

$$3ROH + PCl_3 \rightarrow RCl + H_3PO_3$$

The important here is, straight-chain primary alcohols react with PCl_3 to give unrearranged alkyl halides, but many secondary alcohol undergo rearrangement. Thionyl chloride reacts with straight chain primary alcohols without rearrangement as:

$$ROH + SOCl_2 \rightarrow R—Cl + HCl + SO_2$$

Example 2. Why thionyl chloride is more suitable for halogenation of alcohol?

Solution: In halogenation by $SOCl_2$, the by-products obtained are gases, e.g. SO_2 and HCl and it need not be separated from product by putting any extra effort.

II. By halogenation of alkene with HX.

e.g. $>C=C< + ClH \longrightarrow -\underset{H}{C}-\underset{Cl}{C}-$

$CH_3-CH=CH_2 + HCl \xrightarrow{\text{Markovnikov's addition}} CH_3-\underset{Cl}{CH}-CH_3$

III. Halogenation of alkanes in presence of light, catalyst or heat.

$$CH_4 + Cl_2 / h\nu \rightarrow CH_3Cl + HCl$$
$$CH_4 + Cl_2 / \Delta \rightarrow CH_3Cl + HCl$$

IV. Hunsdiecker reaction:

$$RCOOAg + Br_2 \rightarrow RBr + CO_2 + AgBr$$

V. Allylic halogenation:

Allylic position (cyclohexene) $+ Br_2 \xrightarrow{h\nu}$ 3-bromocyclohexene

A large excess of Br_2 must be avoided, because it can add to the double bond. N-bromo succinammide is often used as source of Br_2, because it combines with HBr side product to regenerate a nearly constant low concentration of bromine.

Example 3. The light catalyzed reaction of 2, 3-dimethyl-2-butene with a small concentration of bromine gives two products. Give mechanism of formation of these two products.

Solution

$(H_3C)_2C=C(CH_3)_2 + Br\cdot \longrightarrow$ I $\longleftrightarrow$ II

Bromination on I and II yields the desired product

5.3 PHYSICAL PROPERTIES OF ALKYL HALIDES

I. *Boiling points:* The major factor influencing the boiling point of alkyl halides, are the vanderwaal's force of attraction and dipole-dipole attraction. Molecules with higher molecular weight are generally high boiling, because they are heavier and have greater surface area.

Example 4. Predict compound, which has higher boiling point.

(a) Isopropyl bromide or *n*-butyl bromide.
(b) Isopropyl chloride and *t*-butyl bromide.
(c) *n*-butyl bromide and *n*-butyl chloride.

Solution

(a) *n*-butyl bromide because it has higher surface area than isopropyl bromide.
(b) *t*-butyl bromide because it is heavier.
(c) *n*-butyl bromide because it is heavier.

II. *Densities:* Alkyl fluorides and alkyl chlorides are less dense than water. Alkyl halides with two or more halides are denser than water.

5.4 CHEMICAL REACTION

5.4.1 General Discussion on the Reaction of Alkyl Halides

The functional group of alkyl halides is a carbon-halogen bond, the common halogens being fluorine, chlorine, bromine and iodine. With the exception of iodine, these halogens have electronegativities significantly greater than carbon. Consequently, this functional group is polarized so that the carbon is electrophilic and the halogen is nucleophilic, as shown in the drawing below:

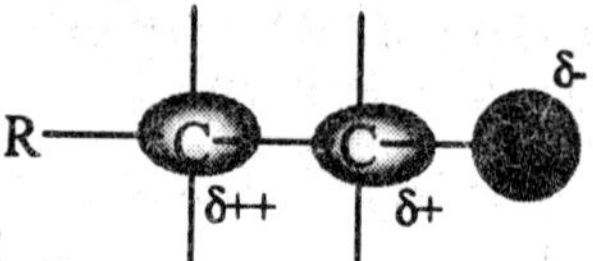

Two characteristics other than electronegativity also have an important influence on the chemical behaviour of these compounds. The first of these is covalent bond strength. The strongest of the carbon-halogen covalent bonds is that to fluorine. Remarkably, this is the strongest common single bond to carbon, being roughly 30 kcal/mole stronger than a carbon-carbon bond and about 15 kcal/mole stronger than a carbon-hydrogen bond. Because of this, alkyl fluorides and fluorocarbons in general are chemically and thermodynamically quite stable, and do not share any of the reactivity patterns shown by the other alkyl halides. The carbon-chlorine covalent bond is slightly weaker than a carbon-carbon bond, and the bonds to the other halogens are weaker still, the bond to iodine being about 33% weaker. The second factor to be considered is the relative stability of the corresponding halide anions, which is likely the form in which these electronegative atoms will be replaced. This stability may be estimated from the relative acidities of the H-X acids, assuming that the strongest acid releases the most stable conjugate base (halide anion). With the exception of HF ($pk_a = 3.2$), all the hydrohalic acids are very strong.

HCl < HBr < HI

Substitution and elimination

The two most common reactions of alkyl halides are "Nucleophilic substitution reaction" and "Elimination reaction".

The characteristics noted above lead us to anticipate certain types of reactions that are likely to occur with alkyl halides. In describing these, it is useful to designate the halogen-bearing carbon as alpha and the carbon atom(s) adjacent to it as beta, as noted in the first four equations shown below. Replacement or substitution of the halogen on the α-carbon by a nucleophilic reagent is a commonly observed reaction, as shown in equations 1, 2, 5, 6 and 7 below. Also, since the electrophilic character introduced by the halogen extends to the α-carbons, and since nucleophiles are also bases, the possibility of base induced H-X elimination must also be considered, as illustrated by equation 3. Finally, there are some combinations of alkyl halides and nucleophiles that fail to show any reaction over a 24 hour period, such as the example in equation 4. For consistency, alkyl bromides have been used in these examples. Similar reactions occur when alkyl chlorides or iodides are used, but the speed of the reactions and the exact distribution of products will change.

1. $CH_3CH_2CH_2CH_2Br + CN^- \xrightarrow[\text{(fast)}]{\text{Alcohol}} CH_3CH_2CH_2CH_2CN + Br^-$

2. 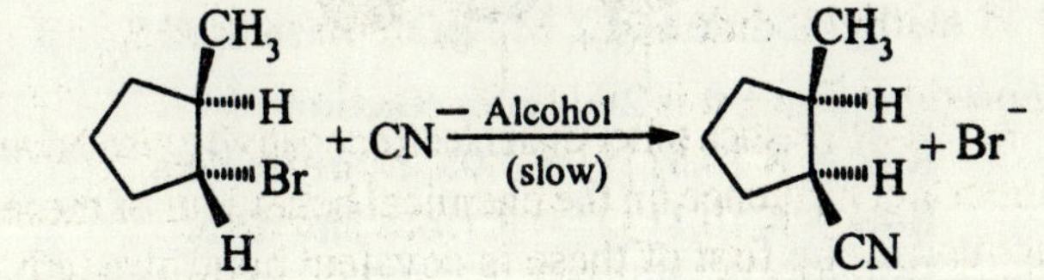 (cyclopentane with CH_3, H, Br) $+ CN^- \xrightarrow[\text{(slow)}]{\text{Alcohol}}$ (cyclopentane with CH_3, H, H, CN) $+ Br^-$

3. $H_3C-C(CH_3)_2-Br + CN^- \xrightarrow{\text{Alcohol}} CH_2{=}C(CH_3)_2 + HCN + Br^-$ (β carbons, α carbon)

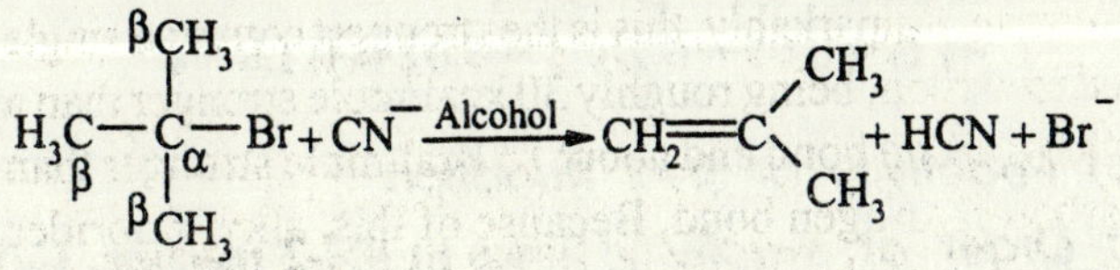

4. 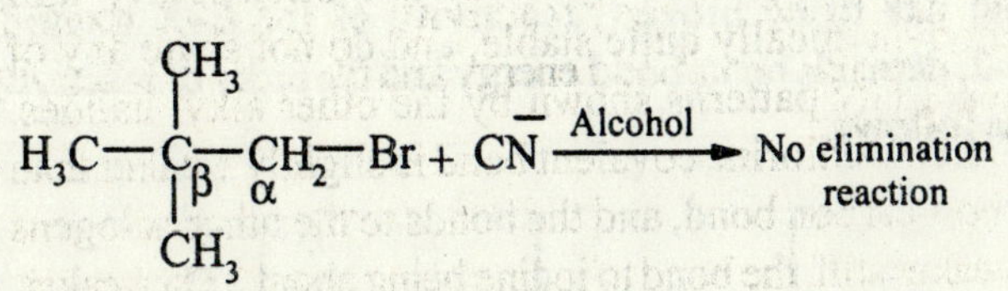 $H_3C-C(CH_3)_2-CH_2-Br + CN^- \xrightarrow{\text{Alcohol}}$ No elimination reaction (β, α)

5. 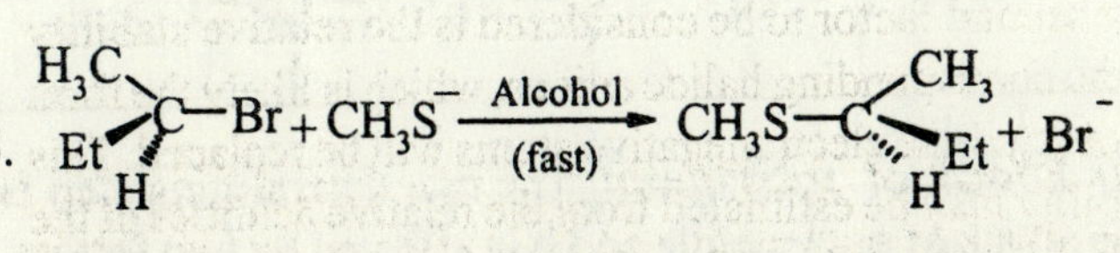 $(H_3C)(Et)(H)C-Br + CH_3S^- \xrightarrow[\text{(fast)}]{\text{Alcohol}} CH_3S-C(CH_3)(Et)(H) + Br^-$

6. 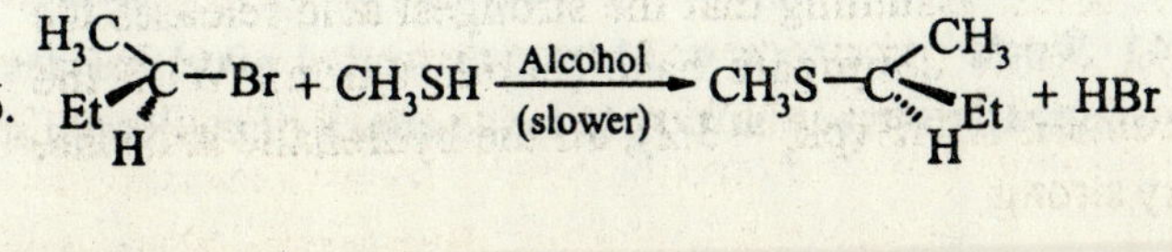 $(H_3C)(Et)(H)C-Br + CH_3SH \xrightarrow[\text{(slower)}]{\text{Alcohol}} CH_3S-C(CH_3)(Et)(H) + HBr$

7. 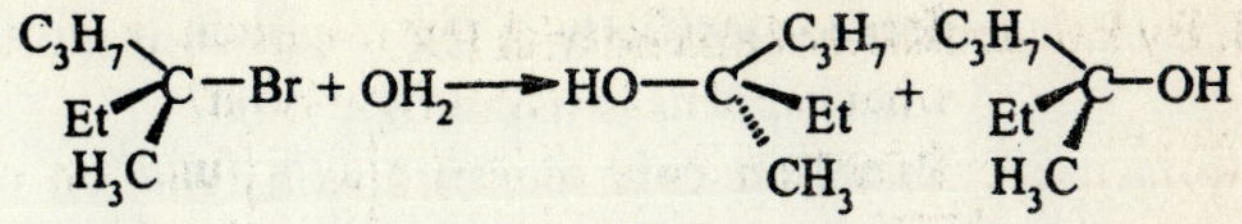 $(C_3H_7)(Et)(H_3C)C-Br + OH_2 \longrightarrow HO-C(C_3H_7)(Et)(CH_3) + (C_3H_7)(Et)(H_3C)C-OH$

To understand why some combinations of alkyl halides and nucleophiles give a substitution reaction, whereas other combinations give elimination, and still others give no observable reaction, we must investigate systematically the way in which changes in reaction variables perturb the course of the reaction. The following general equation summarizes the factors that will be important in such an investigation.

$$R-X + Nu: \xrightarrow{\text{Solvent}} \text{Product} \qquad \begin{array}{l} R = \text{Alkyl group} \\ X = Cl^-, Br^- \text{ or } I^- \\ Nu: = \text{Nucleophile} \end{array}$$

One conclusion, relating the structure of the R-group to possible products, should be immediately obvious. If R- has no beta-hydrogens an elimination reaction is not possible, unless a structural rearrangement occurs first. The first four halides shown on the left below do not give elimination reactions on treatment with base, because they have no β-hydrogens. The two halides on the right do not normally undergo such reactions, because the potential elimination products have highly strained double or triple bonds. It is also worth noting that sp^2 hybridized C-X compounds, such as the three on the right, do not normally undergo nucleophilic substitution reactions, unless other functional groups perturb the double bond(s).

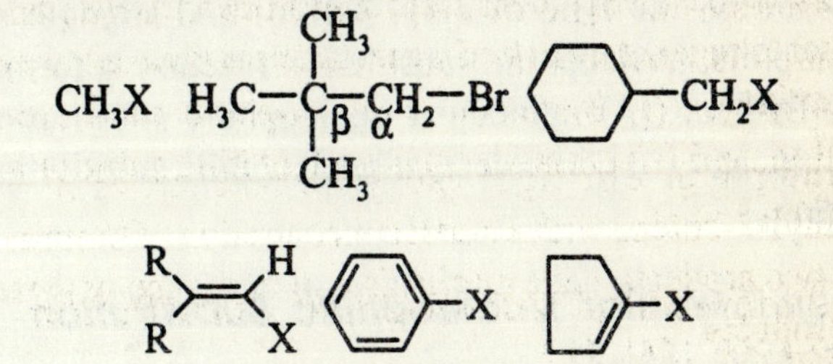

CH_3X $\quad H_3C-C(CH_3)_2-CH_2-Br$ (β, α) $\quad C_6H_{11}-CH_2X$

$R_2C{=}C(H)X$ $\quad C_6H_5-X$ $\quad$ (cyclohexenyl)$-X$

No β -hydrogen

Using the general reaction shown above as our reference, we can identify the following variables and observable.

Variables: **R** change α-carbon from 1° to 2° to 3°.
X change from Cl to Br to I (F is relatively unreactive).
Nu change from anion to neutral; change basicity; change polarizability.
Solvent polar vs. non-polar; protic vs. non-protic.

Observable: **Products substitution, elimination, no reaction.**

Stereospecificity if the α-carbon is chiral what happens to its configuration?

Reaction rate measure as a function of reactant concentration.

When several reaction variables may be changed, it is important to isolate the effects of each during the course of study. In other words, only one variable should be changed at a time, the others being held as constant as possible. For example, we can examine the effect of changing the halogen substituent from Cl to Br to I, using ethyl as a common R-group, cyanide anion as a common nucleophile, and ethanol as a common solvent. We would find a common substitution product.

C_2H_5—CN, in all cases, but the speed or rate of the reaction would increase in the order: Cl < Br < I. This reactivity order reflects both the strength of the C-X bond, and the stability of $X^{(-)}$ as a leaving group, and leads to the general conclusion that alkyl iodides are the most reactive members of this functional class.

5.4.2 Nucleophilic Substitution Reaction

In nucleophilic substitution reaction, a nucleophile (Nu) replaces a group (X^-) from carbon, using its lone pair of electron to form new bond to the carbon atom as:

$$-\overset{|}{\underset{|}{C}}-\overset{|}{\underset{|}{C}}-X + Nu \longrightarrow -\overset{|}{\underset{|}{C}}-\overset{|}{\underset{|}{C}}-Nu + X^-$$

Based on the type of alkyl halides and strength of nucleophile, nucleophilic substitution reaction is further classified as: (i) Bimolecular nucleophilic substitution reaction, and (ii) Unimolecular nucleophilic substitution reaction.

[A] Bimolecular Nucleophilic Substitution Reaction (SN-2)

(a) Mechanism

Back side attack of nucleophile occur to form a pentavalent transition state as:

$$HO^- + CH_3-I \longrightarrow \left[HO\overset{\delta^-}{\cdots}\underset{H\ \ H}{\overset{H}{C}}\cdots\overset{\delta^-}{I} \right] \longrightarrow HO-CH_3 + I^-$$

Transition state

Points to Remember

(i) HO^- and CH_3I are involved in a concerted reaction taking place, i.e. a single step with simultaneous making and breaking of bond.

(ii) The transition state is a point of maximum energy as:

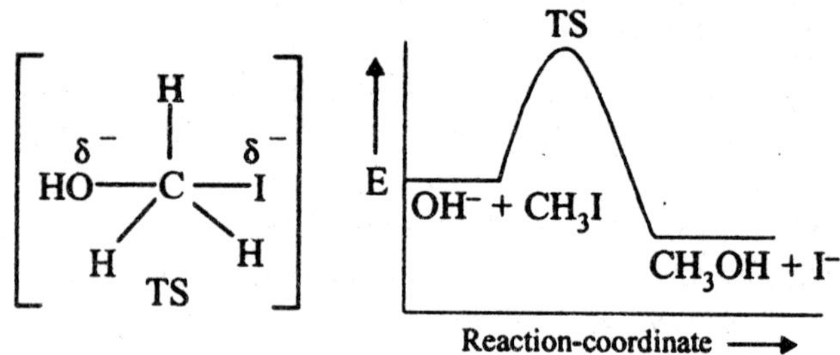

Example 5. In SN-2 reaction, nucleophile attacks from back, explain the reason.

Solution: Both nucleophile and leaving group have very high electronegativity and therefore they would tend to be at maximum distance in the TS. This is possible only when nucleophile attacks from backside to leaving group.

(b) Kinetics

The single step substitution mechanism is supported by kinetic information. Since both the nucleophile and substrate are involved in 1st step, which is also the rate determining step, its rate equation can be written as:

Rate = K [Nucleophile] [Substrate], i.e. order of reaction is two, one with respect to each nucleophile and substrate.

Example 6. Under certain condition, the reaction of 0.5 M methyl iodide with 1.0 M sodium hydroxide forms methanol at a rate of 0.05 M/s. What would be the rate, if 0.1 M methyl iodide and 2 M NaOH were used ?

Solution: Since it is 2nd order reaction:

Rate = K [0.5] [1] = 0.05, from 1st data.

Rate = K [0.1] [2] = R from 2nd data.

Dividing the two rate

R/0.05 = 0.1 × 2/0.5 × 1

⇒ R = 0.05 × 0.2/0.5 = 0.02 mol Lit^{-1} s^{-1}.

(c) Reactivity of SN-2 Reaction

(i) Order of reactivity: Since in SN-2 reaction, C-X bond has to be broken, reactivity of the C-X towards SN-2, depends on its bond energy and the order of reactivity is as follows:

$$-\overset{|}{\underset{|}{C}}-Cl < -\overset{|}{\underset{|}{C}}-Br < -\overset{|}{\underset{|}{C}}-I$$

(ii) Effect of substituent: The rate determining step is the attack of nucleophile and it is affected by two factors.

(A) Steric hindrance: More substituted alkyl halides pose greater steric hinderance to the attack of nucleophile

and slow down the rate. Therefore CH_3X, least substituted alkyl halide undergo fastest SN-2 reaction and the general reactivity order is:

$$CH_3X > 1° > 2° > 3°$$

(B) Electropositivity of carbon bearing halide: Rate of attack of nucleophile also depends on the affinity of carbon atom for negative charge. In case of alkyl substituted alkyl halide, e.g.

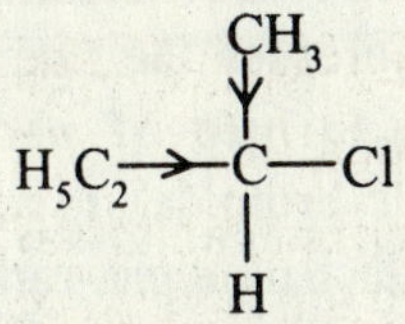

$$H_5C_2 \rightarrow \overset{\overset{\displaystyle CH_3}{\downarrow}}{\underset{\underset{\displaystyle H}{|}}{C}} - Cl$$

the two alkyl group decreases affinity of 'C' bearing halide, for negative nucleophile by inductive effect (+I). This further reduces the reactivity.

(iii) Strength of nucleophile: Nucleophile is an atom, ion or molecule that has an electron pair that may be donated informing a covalent bond to an electrophile (or Lewis acid).

If we use a common alkyl halide, such as methyl bromide, and a common solvent, ethanol, we can examine the rate at which various nucleophiles substitute the bromide at methyl carbon. Nucleophilicity is thereby related to the relative rate of substitution reactions at the halogen-bearing carbon atom of the reference alkyl halide. The more reactive nucleophiles are said to be more nucleophilic than less reactive members of the group. In general:

"Stronger the nucleophile, faster the rate of SN-2 reaction. Nucleophiles are all Lewis base and it can be either neutral or negatively charged". The relative strength of nucleophile can be correlated with two structural features:

- A negatively charged nucleophile is always stronger than its conjugate acid, i.e. HO^- is stronger than H_2O. RO^- is stronger than ROH, etc.
- In a group of nucleophiles, in which nucleophilic atom is same, nucleophilicity parallels basicity as:

$$RO^- > HO^- >> RCOO^- > ROH > H_2O$$

Overall: **Nucleophilicity**

$$CH_3CO_2^{(-)} < Cl^{(-)} < Br^{(-)} < N_3^{(-)} < CH_3O^{(-)} < CN^{(-)} < I^{(-)} < SCN^{(-)} < CH_3S^{(-)}$$

The reactivity range encompassed by these reagents is over 5,000 fold, thiolate being the most reactive. Note that by using methyl bromide as the reference substrate, the complication of competing elimination reactions is avoided.

The nucleophiles used in this study were all anions, but this is not a necessary requirement for these substitution reactions. Indeed reactions 6 and 7, presented at the beginning of this section, are examples of neutral nucleophiles participating in substitution reactions. The cumulative results of studies of this kind has led to useful empirical rules pertaining to nucleophilicity:

(i) For a given element, negatively charged species are more nucleophilic (and basic) than are equivalent neutral species.

(ii) For a given period of the periodic table, nucleophilicity (and basicity) decreases on moving from left to right.

(iii) For a given group of the periodic table, nucleophilicity increases from top to bottom (i.e. with increasing size), although there is a solvent dependence due to hydrogen bonding. Basicity varies in the opposite manner.

Example 7. For each pair of compounds, state which compound is better SN-2 substrate.

(a) 2-methyl-1-iodopropane or *t*-butyl iodide.
(b) Cyclohexyl bromide or 1-bromo-1-methyl cyclohexane.
(c) 2-bromobutane or isopropyl bromide.
(d) 2, 2 -dimethyl-1-chlorobutane or 2-chlorobutane.
(e) 1-iodo-2, 12-dimethyl propane or isopropyl iodide.

Solution

(a) 2-methyl-1-iodopropane since it is primary, less hindered than other.
(b) Cyclohexyl bromide since it is secondary and less hindered than other.
(c) Iso-propyl bromide. Both are secondary, but the size of substituent are smaller in isopropyl bromide.
(d) 2-chlorobutane. Though it is secondary, the size of substituent in other is very large instead it is primary.
(e) Isopropyl iodide. Same reason as in case of (d).

(d) Stereochemistry of SN-2 Reaction

We have just learned that in SN-2 reaction, backside attack occurs. Backside attack literally turns the tetrahedron of carbon inside out like an umbrella caught by the wind. As a result, nucleophile assumes a stereochemical position opposite to the leaving group originally occupied. This is called "Inversion of Configuration" or "Walden-Inversion". An example is I and II have opposite configuration.

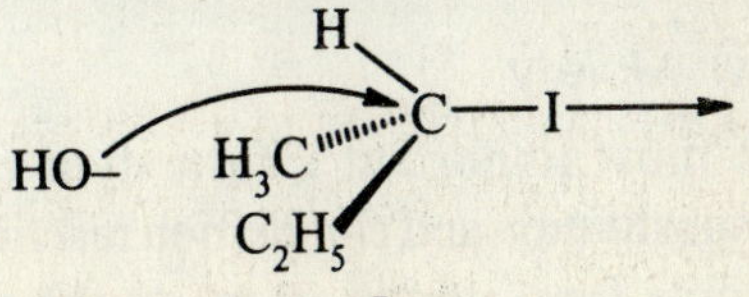

HO⁻ → H, H_3C, C_2H_5 C—I →

I

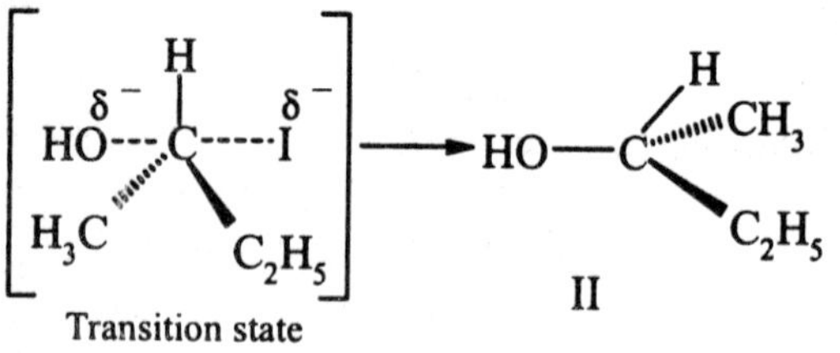

$$\left[\begin{array}{c} \overset{\delta^-}{HO}\cdots\overset{H}{C}\cdots\overset{\delta^-}{I} \\ H_3C \quad C_2H_5 \end{array} \right] \longrightarrow HO-C(H)(CH_3)(C_2H_5)$$

Transition state II

Therefore, SN-2 reaction is a good example of stereospecific reaction, in which a particular stereo-isomer react to give one specific stereoisomer of the product.

(e) Effect of Solvent

Solvation of nucleophilic anions markedly influences their reactivity. The nucleophilicities cited above were obtained from reactions in methanol solution. Polar, protic solvents such as water and alcohols solvate anions by hydrogen bonding interactions, as shown in Figure 5.1. These solvated species are more stable and less reactive than the unsolvated "naked" anions. Polar, aprotic solvents such as DMSO (dimethyl sulfoxide), DMF (dimethylformamide) and acetonitrile do not solvate anions nearly as well as methanol, but provide good solvation of the accompanying cations. Consequently, most of the nucleophiles discussed here react more rapidly in solutions prepared from these solvents. These solvent effects are more pronounced for small basic anions than for large weakly basic anions. Thus, for reaction in DMSO solution we observe the following reactivity order:

$$I^{(-)} < SCN^{(-)} < Br^{(-)} < Cl^{(-)} \sim N_3^{(-)} < CH_3CO_2^{(-)}$$
$$< CN^{(-)} \sim CH_3S^{(-)} < CH_3O^{(-)}$$

Note that this order is roughly the order of increasing basicity.

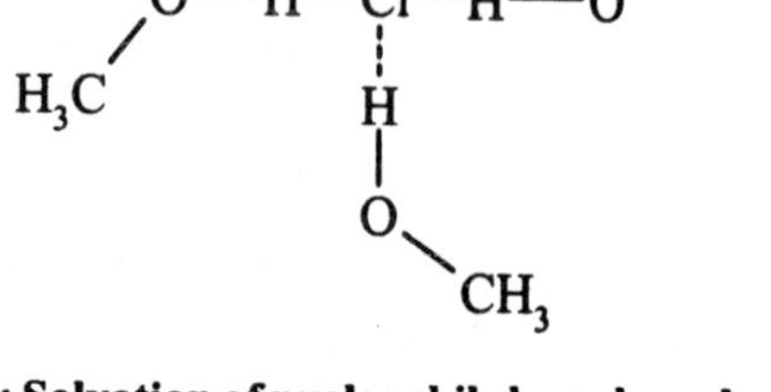

$$\begin{array}{c} H_3C-O \\ | \\ H \\ \vdots \\ H_3C-O-H\cdots \overset{-}{Cl}\cdots H-O-CH_3 \\ \vdots \\ H \\ | \\ O-CH_3 \end{array}$$

Figure 5.1: Solvation of nucleophile by polar solvents

(f) The Alkyl Moiety

Some of the most important information concerning nucleophilic substitution and elimination reactions of alkyl halides has come from studies, in which the structure of the alkyl group has been varied. If we examine a series of alkyl bromide substitution reactions with the strong nucleophile thiocyanide (SCN^-) in ethanol solvent, we find large decreases in the rates of reaction as alkyl substitution of the alpha-carbon increases. Methyl bromide reacts 20 to 30 times faster than simple 1°-alkyl bromides, which in turn react about 20 times faster than simple 2°-alkyl bromides, and 3°-alkyl bromides are essentially unreactive or undergo elimination reactions. Furthermore, beta-alkyl substitution also decreases the rate of substitution, as witnessed by the failure of neopentyl bromide, $(CH_3)_3CCH_2$-Br (a 1°-bromide), to react.

Alkyl halides in which the alpha-carbon is a chiral center provide additional information about these nucleophilic substitution reactions. Returning to the examples presented at the beginning of this section, we find that reactions 2, 5 and 6 demonstrate an inversion of configuration, when the cyanide nucleophile replaces the bromine. Other investigations have shown this to be generally true for reactions carried out in non-polar organic solvents, the reaction of (S)-2-iodobutane with sodium azide in ethanol being just one example. Inversion of configuration during nucleophilic substitution has also been confirmed for chiral 1°-halides of the type RCDH-X, where the chirality is due to isotopic substitution.

$$(S)\text{–}CH_3CHICH_2CH_3 + NaN_3 \rightarrow (R)\text{–}CH_3CHN_3CH_2CH_3 + NaI$$

We can now piece together a plausible picture of how nucleophilic substitution reactions of 1° and 2°-alkyl halides take place.

The nucleophile must approach the electrophilic alpha-carbon atom from the side opposite the halogen. As a covalent bond begins to form between the nucleophile and the carbon, the carbon halogen bond weakens and stretches, the halogen atom eventually leaving as an anion. Figure 5.2 shows this process for an anionic nucleophile. We call this description the SN-2 mechanism, where S stands for substitution, N stands for nucleophilic and 2 stands for bimolecular. In the SN-2 transition state the alpha-carbon is hybridized sp^2 with the partial bonds to the nucleophile and the halogen having largely *p*-character. Both the nucleophile and the halogen bear a partial negative charge, the full charge being transferred to the halogen in the products. The consequence of rear-side bonding by the nucleophile is an inversion of configuration about the alpha-carbon. Neutral nucleophiles react by a similar mechanism, but the charge distribution in the transition state is very different.

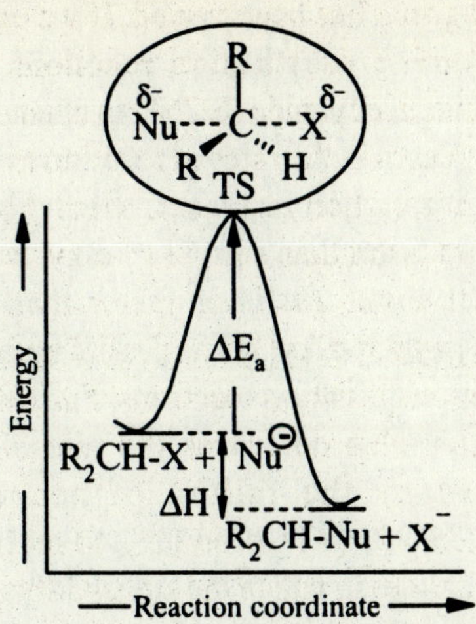

Figure 5.2: Potential energy diagram of an SN-2 reaction

This mechanistic model explains many aspects of the reaction. First, it accounts for the fact that different nucleophilic reagents react at very different rates, even with the same alkyl halide. Since the transition state has a partial bond from the alpha-carbon to the nucleophile, variations in these bond strengths will clearly affect the activation energy, ΔE_a, of the reaction and therefore its rate. Second, the rear-side approach of the nucleophile to the alpha-carbon will be subject to hindrance by neighbouring alkyl substituents, both on the alpha and the beta-carbons.

Example 8. Write product of the following SN-2 reaction showing change in stereochemistry.

(a) [cyclohexane ring with H, Br and H_3C substituents] $\xrightarrow{OH^-}$

(b) *Cis*-1-bromo-4-methyl cyclohexane + aq KOH ⟶ ?

(c) [Fischer projection: CH_3 top; Br—|—H; H—|—CH_3; C_2H_5 bottom] $\xrightarrow[\text{Acetone}]{\text{NaI}}$?

(d) [Br, H, F, CH_3 on C] + NaOH ⟶

(e) [Fischer projection: CH_3 top; H—|—Br; CH_2CH_3 bottom] + $NaOCH_3 \xrightarrow{CH_3OH}$

Solution: (a) [cyclohexane ring with H, OH, CH_3 and H substituents]

(b) [H_3C, H cyclohexane with H, Br] + HO⁻ ⟶ [transition state: H, H_3C, OH, H, Br]⁻ ⟶ Br^- + [H_3C, H cyclohexane with OH, H]

(c) [I: CH_3; Br—|—H; H—|—CH_3; C_2H_5] M [II: CH_3; H—|—Br; H—|—CH_3; C_2H_5] $\xrightarrow{I^-}$ [CH_3; H—|—I; H—|—CH_3; C_2H_5]

(d) [H, F, CH_3, Br on C] + $\bar{O}H$ ⟶ [H, F, Br, OH, CH_3 on C] TS $\xrightarrow{-Br^-}$ [F, H, CH_3, OH on C]

(e) [CH_3; H—|—Br; C_2H_5] M [CH_3; Br—|—H; C_2H_5] $\xrightarrow{C_2H_5O^-}$ [CH_3; C_2H_5O—|—H; C_2H_5]

NB: In this type of structure (Fischer projection), if SN-2 reaction occurs, it is advisable to take the mirror image of only that part of substrate, which has leaving group and then replace the leaving group by nucleophile without changing its position.

Unimolecular Nucleophilic Substitution Reaction

(SN-1), when tertiary butylbromide is placed in boiling methanol, methyl-*t*-butyl ether can be isolated from the reaction mixture:

$$(CH_3)_3CBr + CH_3OH \xrightarrow{\text{Boil}} (CH_3)_3COCH_3$$

Because this reaction takes place with the solvent acting as nucleophile, it is called solvolysis. Here CH_3OH is a weak nucleophile and *t*-butylbromide is a highly hindered tertiary halide–a poor SN-2 substrate. An interesting characteristic of this reaction is that, its rate does not depend on the concentration of nucleophile, it can be inferred here that the nucleophile is not present in the rate-determining step.

The SN-1 mechanism

Reaction 7, shown at the end of the previous section, is clearly different from the other cases we have examined. It not only shows first order kinetics, but the chiral 3°-alkyl bromide reactant undergoes substitution by the modest nucleophile water with extensive racemization. In all of these features this reaction fails to meet the characteristics of the SN-2 mechanism. A similar example is found in the hydrolysis of *t*-butyl chloride, shown below. Note that the initial substitution product in this reaction is actually a hydronium ion, which rapidly transfers a proton to the chloride anion. This second acid-base proton transfer is often omitted in writing the overall equation, as in the case of reaction 7 above.

$$(CH_3)_3C—Cl + H_2O \rightarrow (CH_3)_3C—OH_2^{(+)} + Cl^{(-)}$$
$$\rightarrow (CH_3)_3C—OH + HCl$$

The first order kinetics of these reactions suggests a two-step mechanism, in which the rate-determining step consists of the ionization of the alkyl halide, as shown in Figure 5.3. In this mechanism, a carbocation is formed as a high-energy intermediate, and this species bonds immediately to nearby nucleophiles. If the nucleophile is a neutral molecule, the initial product is an "onium" cation, as drawn above for *t*-butyl chloride, and presumed in the energy diagram. In evaluating this mechanism, we may infer several outcomes from its functioning.

First, the only reactant that is undergoing change in the first (rate-determining) step is the alkyl halide, so we expect such reactions would be unimolecular and follow a first-order rate equation. Hence, the name SN-1 is applied to this mechanism.

Second, since nucleophiles only participate in the fast second step, their relative molar concentrations rather than their nucleophilicities should be the primary product determining factor. If a nucleophilic solvent such as water is used, its high concentration will assure that alcohols are the major product. Any recombination of the halide anion with the carbocation intermediate simply reforms the starting compound. Note that SN-1 reactions, in which the nucleophile is also the solvent are commonly called solvolysis reactions. The hydrolysis of *t*-butyl chloride is an example.

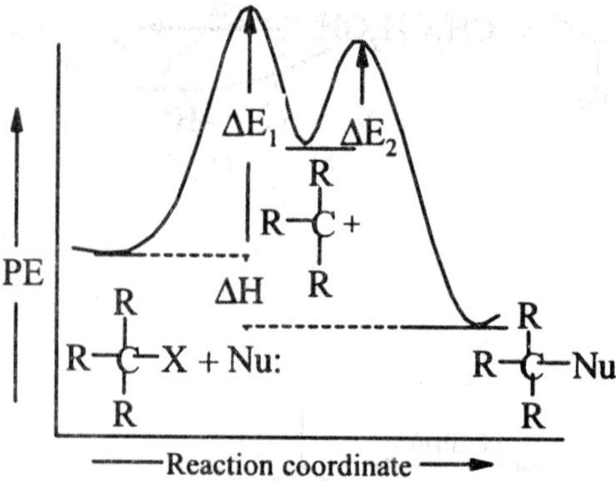

Figure 5.3: Potential energy diagram of a slow step SN-1 reaction

Mechanism

Step I: Formation of carbocation (rate determining step)

$$I(CH_3)_3CBr \rightarrow (CH_3)_3C^+ + Br^- \quad ....\text{Slow}$$

Step II: Nucleophilic attack:

$$(CH_3)_3C^+ + CH_3OH \xrightarrow{\text{Fast}} (CH_3)_3C\overset{+}{O}(H)CH_3 \xrightarrow{-H^+} (CH_3)_3COCH_3$$

Third, the Hammond postulate suggests that the activation energy of the rate-determining first step will be inversely proportional to the stability of the carbocation intermediate. The stability of carbocations was discussed earlier, and a qualitative relationship is given below.

Carbocation stability

$$CH_3^{(+)} < CH_3CH_2^{(+)} < (CH_3)_2CH^{(+)} \approx CH_2{=}CH\text{-}CH_2^{(+)}$$
$$< C_6H_5CH_2^{(+)} \approx (CH_3)_3C^{(+)}$$

Consequently, we expect that 3°-alkyl halides will be more reactive than their 2°- and 1°-counterparts in reactions that follow a SN-1 mechanism. This is opposite to the reactivity order observed for the SN-2 mechanism. Allylic and benzylic halides are exceptionally reactive by either mechanism.

Rearrangements in SN-1 reaction

Carbocation frequently undergo structural change called "Rearrangement" to form more stable ions. Since SN-1 reaction is a two step reaction and carbocation is formed in the 1st step, it may rearrange so that the nucleophile get attached to the 'C', which did not have leaving group. An illustration of rearrangement is:

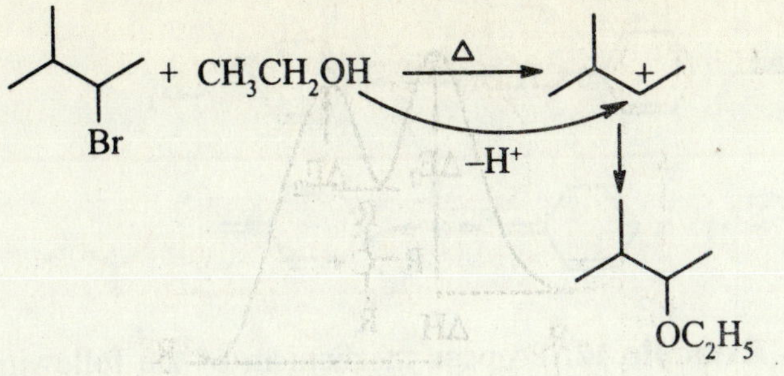

+ CH_3CH_2OH —Δ→ ... $-H^+$ ... OC_2H_5

or

H⁻ Shift (a more stable 3° carbocation) $C_2H_5\ddot{O}H$ ... OC_2H_5

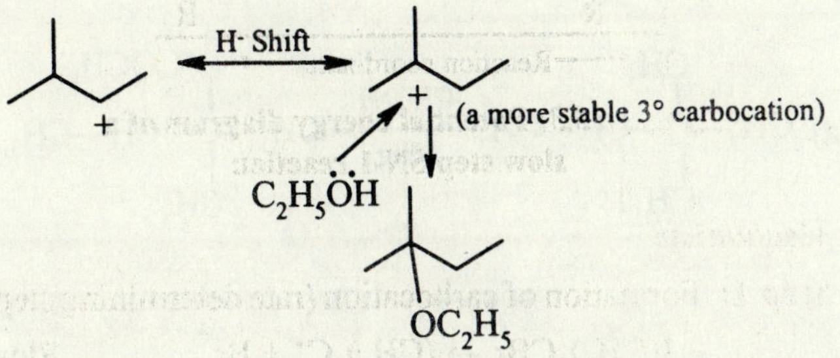

Fourth, in order to facilitate the charge separation of an ionization reaction, as required by the first step, a good ionizing solvent will be needed. Two solvent characteristics will be particularly important in this respect. The first is the ability of solvent molecules to orient themselves between ions so as to attenuate the electrostatic force one ion exerts on the other. This characteristic is related to the dielectric constant, ϵ, of the solvent. Solvents having high dielectric constants, such as water (ϵ = 81), formic acid (ϵ = 58), dimethyl sulfoxide (ϵ = 45) and acetonitrile (ϵ = 39) are generally considered better ionizing solvents than are some common organic solvents such as ethanol (ϵ = 25), acetone (ϵ = 21), methylene chloride (ϵ = 9) and ether (ϵ = 4). The second factor is solvation, which refers to the solvent's ability to stabilize ions by encasing them in a sheath of weakly bonded solvent molecules. Anions are solvated by hydrogen-bonding solvents, as noted earlier. Cations are often best solvated by nucleophilic sites on a solvent molecule (e.g. oxygen and nitrogen atoms), but in the case of carbocations these nucleophiles may form strong covalent bonds to carbon, thus converting the intermediate to a substitution product. This is what happens in the hydrolysis reactions described above.

Fifth, the stereospecificity of these reactions may vary. The positively charged carbon atom of a carbocation has a trigonal (flat) configuration (it prefers to be sp^2 hybridized), and can bond to a nucleophile equally well from either face. If the intermediate from a chiral alkyl halide survives long enough to encounter a random environment, the products are expected to be racemic (a 50:50 mixture of enantiomers). On the other hand, if the departing halide anion temporarily blocks the front side, or if a nucleophile is oriented selectively at one or the other face, then the substitution might occur with predominant inversion or even retention of configuration.

Stereochemistry of SN-1 reaction

In SN-1 mechanism, the intermediate carbocation is sp^2 hybridized and planar. A nucleophile can attack the carbocation from either face as:

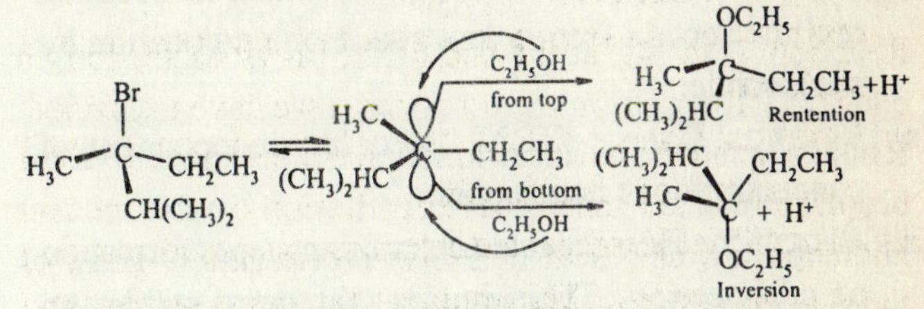

H_3C–C(Br)(CH_2CH_3)($CH(CH_3)_2$) ⇌ H_3C, $(CH_3)_2HC$, CH_2CH_3 carbocation; C_2H_5OH from top → H_3C–C(OC_2H_5)(CH_2CH_3)($(CH_3)_2HC$) + H^+ Rentention; C_2H_5OH from bottom → $(CH_3)_2HC$, CH_2CH_3, H_3C–C–OC_2H_5 + H^+ Inversion

The above reaction shows that if the nucleophile attack from the front, the product shows retention of configuration. Attack from backside gives inversion of configuration. A combination of two gives "Racemization".

Example 9. If you understand the factors and principles that influence the course of nucleophilic substitution reactions, try your hand at the following exercises. For each case the possibility of a substitution versus no reaction must be considered. If a substitution is predicted will it take place by a SN-1 or a SN-2 mechanism? If chiral centers are present will the configuration change?

(a) 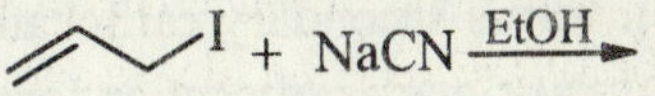 allyl iodide + NaCN —EtOH→

(b) 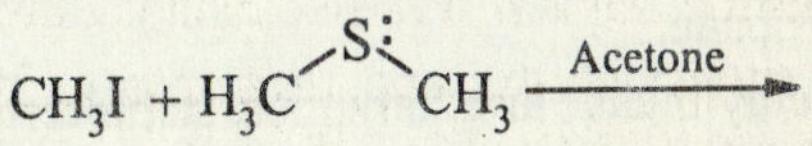 $CH_3I + H_3C-\ddot{S}-CH_3$ —Acetone→

(c) 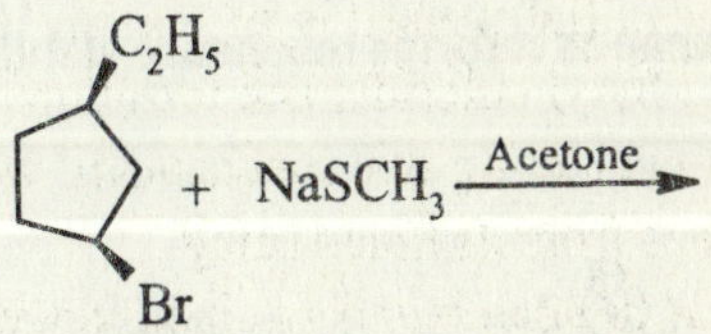 1-bromo-3-ethylcyclopentane (C_2H_5, Br) + $NaSCH_3$ —Acetone→

(d) 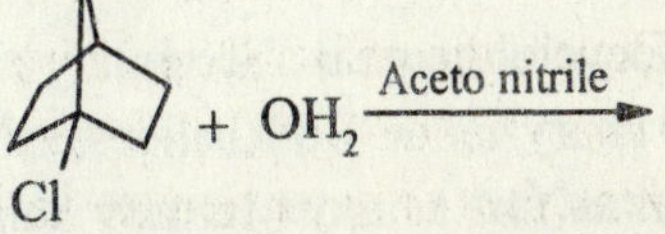 1-chloronorbornane (Cl) + OH_2 —Aceto nitrile→

(e) 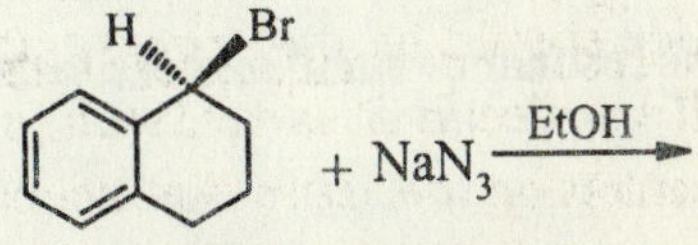 1-bromotetralin (H, Br) + NaN_3 —EtOH→

(f) 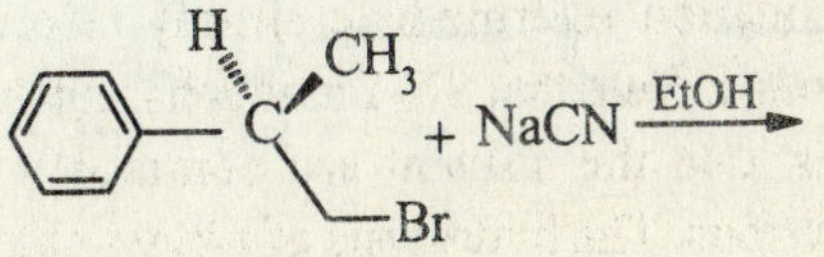 C_6H_5–C(H)(CH_3)–Br + NaCN —EtOH→

(g) [1-chloronorbornane] + NaI $\xrightarrow{\text{Acetone}}$

Points to Remember

(i) *Process:* The SN-1 mechanism is a two step process. The first step is a slow ionization to form carbocation and the second step is fast attack on carbocation by nucleophile.

(ii) *Kinetics:* Rate = K[RX] 1st order, independent of concentration of nucleophile.

(iii) *Reactivity:* Here the rate determining step is formation of carbocation. This implies that more stable the carbocation, faster will be the rate of substitution. Thus, reactivity towards SN-1 substitution follows the stability of carbocation.

SN-1 reactivity : $3° > 2° > 1° > CH_3X$

Example 10. Choose the member of each pair that will react faster by SN-1 mechanism.

(a) 1-bromopropane or 2-bromopropane.
(b) 2-bromo-2-methylbutane or 2-bromo-3-methylbutane.
(c) 3-bromocyclohexene or 4-bromocylohexene.

Solution

(a) 2-bromopropane as it forms secondary carbocation, which is more stable than primary carbocation formed by 1-bromopropane.
(b) 2-bromo-2-methylbutane as it form most stable tertiary carbocation.
(c) 3-bromocyclohexene as it forms resonance stabilized carbocation.

Br

Example 11. 3-bromocyclohexene is a secondary halide and benzyl bromide is primary halide. Both halides undergo SN-1 substitution about as fast as most tertiary halides. Use resonance structure to explain this enhanced reactivity.

Solution: Both forms resonance stabilized carbocation as:

Br

and $C_6H_5-CH_2Br \xrightarrow{Br^-} C_6H_5-\overset{+}{C}H_2$

Example 12. Propose mechanism of the following solvolysis reaction.

(a) $CH_3-C(CH_3)(CH_3)-CH(I)-CH_3 \xrightarrow[\text{Heat}]{CH_3OH} CH_3-C(CH_3)(CH_3)-CH(OCH_3)-CH_3$

$\downarrow$

$CH_3-C(OCH_3)(CH_3)-CH(CH_3)-CH_3$

(b) [1-chloro-2-methylcyclohexane] $\xrightarrow[\Delta]{CH_3CH_2OH}$ [2-methyl-1-ethoxycyclohexane, OCH_2CH_3, CH_3] + [1-methyl-1-ethoxycyclohexane, OCH_2CH_3, CH_3]

(c) [3-iodocyclohexene] + $CH_3-\overset{O}{\overset{\|}{C}}-OH$ $\xrightarrow{\text{Heat}}$ [cyclohex-2-enyl acetate, $O-\overset{O}{\overset{\|}{C}}-CH_3$] + [$O-\overset{O}{\overset{\|}{C}}-CH_3$ isomer]

(d) [cyclohexyl-CH_2I] $\xrightarrow[\text{Heat}]{CH_3CH_2OH}$ [1-methyl-1-ethoxycyclohexane, CH_3, OCH_2CH_3] + [cyclohexyl-$CH_2OCH_2CH_3$]

Solution

(a)

$\xrightarrow{CH_3OH}$ OMe + OMe

I

(b) [Reaction scheme: 1-chloro-2-methylcyclohexane $\xrightarrow{-Cl^-}$ secondary carbocation $\xrightarrow{C_2H_5OH}$ oxonium ion $\xrightarrow{-H^+}$ 1-ethoxy-2-methylcyclohexane; H-shift gives tertiary carbocation $\xrightarrow{C_2H_5OH, -H^+}$ 1-ethoxy-1-methylcyclohexane]

(c) [Reaction scheme: 3-iodocyclohexene derivative $\xrightarrow{-I^-}$ carbocation $\xrightarrow{CH_3COOH, -H^+}$ product with $OCOCH_3$; H-shift $\to$ allyl carbocation $\xrightarrow{CH_3COOH, -H^+}$ product with $OCOCH_3$]

(Allyl carbocation, more stable)

(d) [Reaction scheme: cyclohexyl-CH_2I $\xrightarrow{-I^-}$ primary carbocation $\xrightarrow{CH_3CH_2OH, -H^+}$ cyclohexyl-$CH_2OCH_2CH_3$; H-shift $\to$ tertiary carbocation (1-methylcyclohexyl) $\xrightarrow{C_2H_5OH, -H^+}$ 1-ethoxy-1-methylcyclohexane]

(A more stable tertiary carbocation)

Comparison of SN-1 and SN-2 reaction

(a) *Effect of nucleophile:* The nucleophile takes part in the slow step of SN-2 reaction but in the fast step of SN-1 reaction. Weak nucleophile fail to promote the SN-2 reaction; therefore, reaction with weak nucleophiles often go by SN-1 mechanism, if the substrate is secondary or tertiary.

(b) *Effect of substrate:* The structure of substrate is an important factor in determining, which of these substitution mechanism might operate. For SN-1; stable carbocation should be formed and for SN-2; carbon-bearing halide should be less hindered. This indicate that general reactivity in two mechanism are:

SN-1 : 3° > 2° > 1°; SN-2 : CH_3X > 1° > 2° > 3°

(c) *Effect of solvent:* SN-1 reaction is promoted by polar solvent due to stabilization of carbocation, the intermediate by polar solvent. In SN-2 reaction, polar solvent stabilize alkyl halide and nucleophile, thereby stabilizing substrate, and decreasing the rate.

(d) *Kinetics:* The rate of SN-1 reaction is a function of substrate (alkyl halide) only whereas rate of SN-2 reaction depends on concentration of both nucleophile and substrate.

5.4.3 Elimination Reaction

Elimination reaction involves the loss of two atoms or groups from the substrate as by product with formation of π-bond. On the basis of kinetics, elimination reaction is also classified into two categories as E1 and E2 reaction.

[A] E1(Elimination Unimolecular) Reaction

Unimolecular elimination reaction proceeds in two steps as:

Step I: Heterolysis of C-X bond and formation of carbocation:

$$CH_3-CH_2-CHBr-CH_3 \underset{}{\overset{Slow}{\rightleftharpoons}} CH_3-CH_2-\overset{+}{C}H-CH_3$$

Step II: Loss of proton to the solvent: In the second fast step, carbocation loses one of it's α-proton to the solvent forming alkene. The proton is lost from α-carbon bearing least number of protons to form most substituted (most stable) alkene:

$$CH_3-CH_2-\overset{+}{C}H-CH_3 \xrightarrow[Fast]{C_2H_5OH} CH_3-CH=CH-CH_3 + C_2H_5\overset{+}{O}H_2$$

NB: Carbocation has characteristic of rearrangement to more stable carbocation by hydride shift and alkyl shift if structure permits as:

$$CH_3-CH(CH_3)-CHBr-CH_2CH_3 \xrightarrow{-Br^-} CH_3-CH(CH_3)-\overset{+}{C}H-CH_2CH_3$$

(2° carbocation)

H-shift

$$CH_3-C(CH_3)=CH-CH_2CH_3 \xleftarrow{-H^+} CH_3-\overset{+}{C}(CH_3)-CH_2-CH_2CH_3$$

(A more stable 3° carbocation)

$$CH_3-\underset{CH_3}{\overset{CH_3}{C}}-\underset{Br}{\overset{H}{C}}-CH_2CH_3 \xrightarrow{-Br^-} CH_3-\underset{CH_3}{\overset{CH_3}{C}}-\overset{H}{\underset{+}{C}}-CH_2CH_3 \text{ (2° carbocation)}$$

$$\downarrow \text{H-shift}$$

$$CH_3-\overset{CH_3}{C}=\overset{CH_3}{C}-CH_2CH_3 \xleftarrow{-H^+} CH_3-\overset{CH_3}{\underset{+}{C}}-\underset{CH_3}{\overset{H}{C}}-CH_2CH_3$$

(A more stable 3° carbocation)

Summary of E1 reaction

- Base is not involved in the slow rate determining step, rate of elimination is independent of concentration of base.
- Since base is involved in the fast, second step, strength of base is unimportant.
- Rate determining step involves formation of carbocation intermediate, reactivity of E1 reaction depends on the stability of intermediate carbocation and general order of reactivity is:
  Tertiary halide > Secondary halide > Primary halide
- A polar solvent accelerate the rate of elimination by polarising the halide ion facilitating its release.

[B] E2 (Elimination Bimolecular) Reaction: or β-Elimination

In this reaction an alkyl halide is treated with an alcoholic solution of strong base leading to elimination of HX and formation of an alkene.

We have not yet considered the factors that influence elimination reactions, such as example 3 in the group presented at the beginning of this section (5.3.1).

(3) $(CH_3)_3C-Br + CN^{(-)} \rightarrow (CH_3)_2C{=}CH_2 + Br^{(-)} + HCN$

We know that t-butyl bromide is not expected to react by a SN-2 mechanism. Furthermore, the ethanol solvent is not sufficiently polar to facilitate a SN-1 reaction. The other reactant, cyanide anion, is a good nucleophile; and it is also a decent base, being about ten times weaker than bicarbonate. Consequently, a base-induced elimination seems to be the only plausible reaction remaining for this combination of reactants. To get a clearer picture of the interplay of these factors consider the reaction of a 2°-alkyl halide, isopropyl bromide, with two different nucleophiles.

$$(H_3C)_2CH-Br \xrightarrow[\text{Methanol}]{CH_3SNa \text{ (SN-2)}} (H_3C)_2CH-SCH_3$$

$$(H_3C)_2CH-Br \xrightarrow[\text{Methanol}]{CH_3ONa \text{ (E2)}} \text{Propene}$$

In the methanol solvent used here, methanethiolate has greater nucleophilicity than methoxide by a factor of 100. Methoxide, on the other hand is roughly 10^6 times more basic than methanethiolate.

As a result, we see a clearcut difference in the reaction products, which reflects nucleophilicity (bonding to an electrophilic carbon) versus basicity (bonding to a proton). Kinetic studies of these reactions show that they are both second order (first order in R-Br and first order in $Nu:^{(-)}$), suggesting a bimolecular mechanism for each. The substitution reaction is clearly SN-2. The corresponding designation for the elimination reaction is E2. An energy diagram for the single-step bimolecular E2 mechanism is shown in Figure 5.4. We should be aware that the E2 transition state is less well defined than is that of SN-2 reactions. More bonds are being broken and formed, with the possibility of a continuum of states, in which the extent of C-H and C-X bond-breaking and C=C bond-making varies.

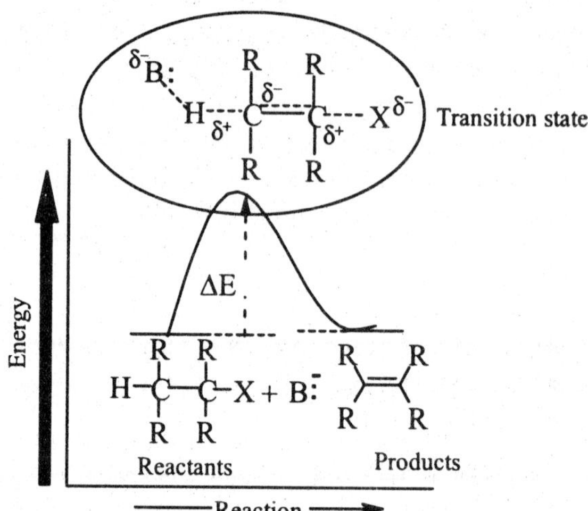

Figure 5.4: Potential energy diagram of E2 reaction

For example, if the R-groups on the beta-carbon enhance the acidity of that hydrogen, then substantial breaking of C-H may occur before the other bonds begin to be affected. Similarly, groups that favour ionization of the halogen may generate a transition state with substantial positive charge on the alpha-carbon and only a small degree of C-H breaking. For most simple alkyl halides, however, it is proper to envision a balanced transition state, in which there has been an equal and synchronous change in all the bonds. Such a model helps to explain an important regioselectivity displayed by these elimination reactions.

If two or more structurally distinct groups of beta-hydrogens are present in a given reactant, then several constitutionally isomeric alkenes may be formed by an E2 elimination. This situation is illustrated by the 2-bromobutane and 2-bromo-2, 3-dimethylbutane elimination examples given below.

$$H_3C{-}CH_2{-}CH(Br){-}CH_3 \rightarrow CH_3{-}CH_2CH{=}CH_2 + CH_3{-}CH{=}CH{-}CH_3$$

(β-H on both carbons adjacent to C–Br; 20% and 80%)

$$H_3C{-}CH(CH_3){-}C(Br)(CH_3){-}CH_3 \rightarrow (CH_3)_2CH{-}C(CH_3){=}CH_2 + (H_3C)_2C{=}C(CH_3)_2$$

(β-H; 21% and 79%)

By using the strongly basic hydroxide nucleophile, we direct these reactions toward elimination. In both cases there are two different sets of beta-hydrogens available to the elimination reaction. If the rate of each possible elimination was the same, we might expect the amounts of the isomeric elimination products to reflect the number of hydrogens that could participate in that reaction. For example, since there are three 1°-hydrogens and two 2°-hydrogens on beta-carbons in 2-bromobutane, statistics would suggest a 3 : 2 ratio of 1-butene and 2-butene in the products. This is not observed, and the latter predominates by 4 : 1. This departure from statistical expectation is even more pronounced in the second example, where there are six 1°-beta-hydrogens compared with one 3°-hydrogen. These results point to a strong regioselectivity favouring the more highly substituted product double bond, an empirical statement generally called the Sytzeff's rule.

Sytzeff's Rule of Elimination

If a strong base but relatively less bulky, e.g. CH_3O^-, $C_2H_5O^-$, etc. are used, proton is removed from β-carbon bearing least number of proton to result most substituted (most stable) alkene as:

(Skeletal structure with βH, Br, Hβ, CH_3, CH_3, CH_3^{β}, H) $\xrightarrow[ROH]{RO^-}$ (alkene with H_3C, CH_3) Major

(alkene with H, CH_3, H_3C, H_3C, H) Minor + (alkene) Minor

Exception to Sytzeff's Rule (Hofmann's rule)

If a very bulky base, e.g. $(CH_3)_3COK$ is used for elimination, least substituted alkene is obtained. This is due to the reason, a bulky base suffers steric hindrance in abstracting proton from more substituted β-carbon and prefer to abstract H from β-carbon, which is least substituted as:

More hindered: $(CH_3)_3C\bar{O}$ — Less hindered: $\bar{O}C(CH_3)_3$

$$H_3C{-}C(H)(CH_3){-}C(H)(Br){-}C(H)(H){-}H \rightarrow$$

$$H_3C{-}C(CH_3){=}C(H){-}C(H)(H){-}H + H_3C{-}C(H)(CH_3){-}C(H){=}C(H){-}H$$

(Minor) (Major)

The main factor contributing to Sytzeff's rule behaviour is the stability of the alkene. We noted earlier that carbon-carbon double bonds are stabilized (thermodynamically) by alkyl substituents, and that this stabilization could be evaluated by appropriate heat of hydrogenation measurements. Since the E2 transition state has significant carbon-carbon double bond character, alkene stability differences will be reflected in the transition states of elimination reactions, and therefore in the activation energy of the rate-determining steps. From this consideration we anticipate that if two or more alkenes may be generated by an E2 elimination, the more stable alkene will be formed more rapidly and will therefore be the predominant product. This is illustrated for 2-bromobutane by the energy Figure 5.5.

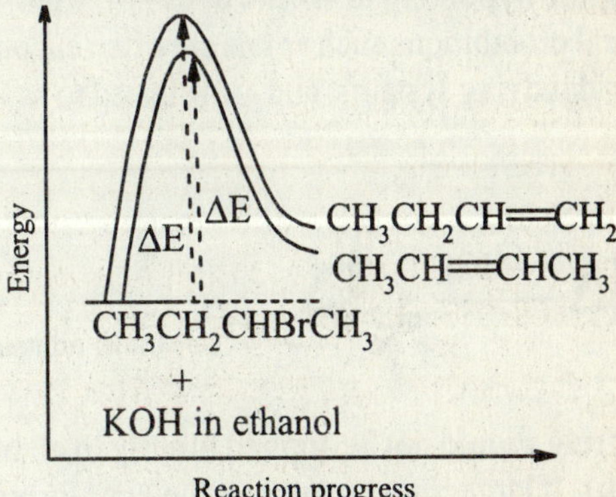

Figure 5.5: Potential energy diagram showing lower energy of activation for formation of more substituted alkene

The propensity of E2 eliminations to give the more stable alkene product also influences the distribution of product stereoisomers. In the elimination of 2-bromobutane, for example, we find that *trans*-2-butene is produced in a 6:1 ratio with its *cis*-isomer.

The Sytzeff's rule is a good predictor for simple elimination reactions of alkyl chlorides, bromides and iodides as long as relatively small strong bases are used. Thus hydroxide, methoxide and ethoxide bases give comparable results. Bulky bases such as tert-butoxide tend to give higher yields of the less substituted double bond isomers, a characteristic that has been attributed to steric hindrance. In the case of 2-bromo-2, 3-dimethylbutane, described above, tert-butoxide gave a 4:1 ratio of 2, 3-dimethyl-1-butene to 2, 3-dimethyl-2-butene (essentially the opposite result to that obtained with hydroxide or methoxide).

Bredt's Rule

The importance of maintaining a planar configuration of the trigonal double-bond carbon components must never be overlooked. For optimum pi-bonding to occur, the *p*-orbitals on these carbons must be parallel, and the resulting doubly-bonded planar configuration is more stable than a twisted alternative by over 60 kcal/mole. This structural constraint is responsible for the existence of alkene stereomerism when substitutuion patterns permit. It also prohibits certain elimination reactions of bicyclic alkyl halides, that might be favourable in simpler cases. For example, the bicyclooctyl 3°-chloride shown below appears to be similar to tert-butyl chloride, but it does not undergo elimination, even when treated with a strong base (e.g. KOH or KOC_4H_9). There are six equivalent beta-hydrogens that might be attacked by base, so an E2 reaction seems plausible. The problem with this elimination is that the resulting double bond would be constrained in a severely twisted (non-planar) configuration by the bridged structure of the carbon skeleton. Newman projection looking down the twisted bond is drawn below. Because a pi-bond cannot be formed, the hypothetical alkene does not exist.

Structural prohibitions such as this are often encountered in small bridged ring systems, and are referred to as Bredt's rule.

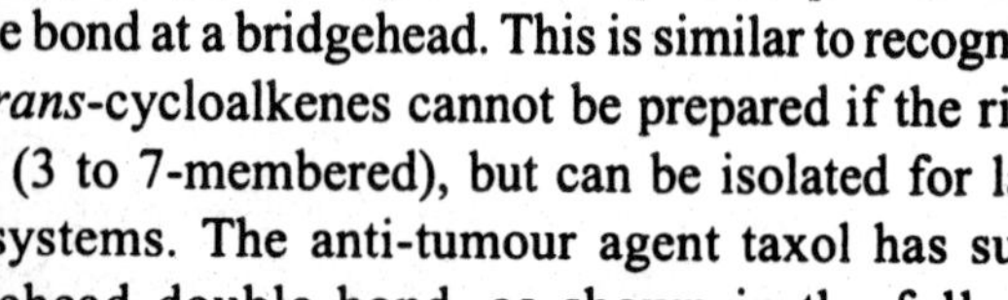

–HCl

Cl

Unparallel orbitals

Bredt's rule should not be applied blindly to all bridged ring systems. If large rings are present their conformational flexibility may permit good overlap of the *p*-orbitals of a double bond at a bridgehead. This is similar to recognizing that *trans*-cycloalkenes cannot be prepared if the ring is small (3 to 7-membered), but can be isolated for larger ring systems. The anti-tumour agent taxol has such a bridgehead double bond, as shown in the following illustration. The bicyclo [3.3.1] octane ring system is the smallest, in which bridgehead double bonds have been observed. The drawing next to the taxol shows this system. The bridgehead double bond has a *cis*-orientation in the six-membered ring, but a *trans*-orientation in the larger eight-membered ring.

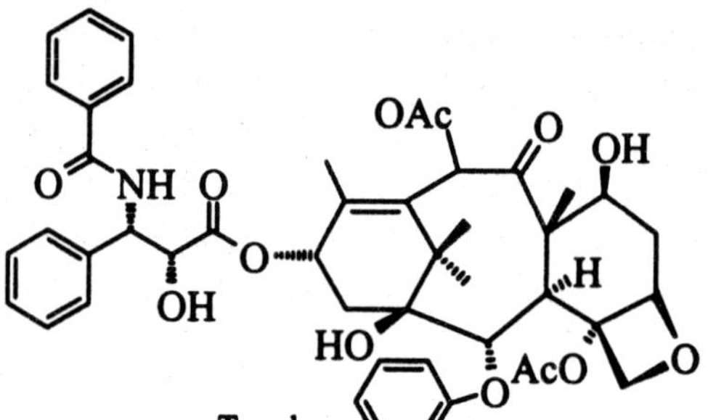

Taxol

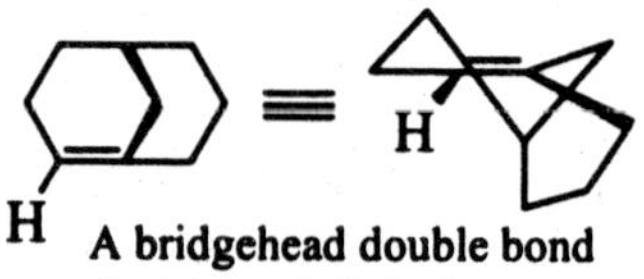

A bridgehead double bond in 1-bicyclo[3,3,1]octene

Mechanism of E2 reaction

Experimental results show that it is a single step concerted reaction. A strong base abstract a β-H from the position anti to halide and halide ion leaves the substrate in the simultaneous step as:

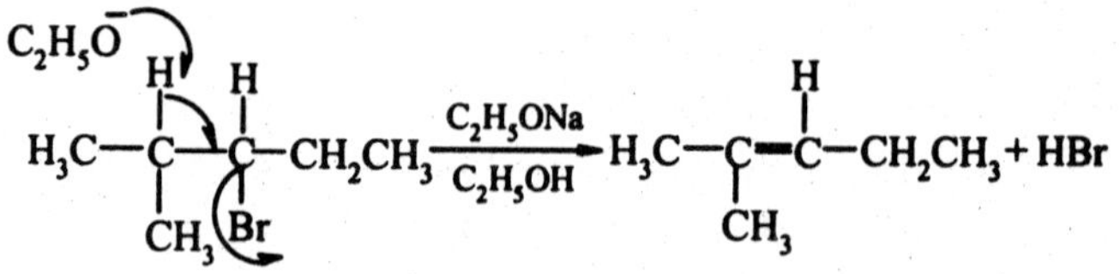

$$H_3C-CH(CH_3)-CHBr-CH_2CH_3 \xrightarrow[C_2H_5OH]{C_2H_5ONa} H_3C-C(CH_3)=CH-CH_2CH_3 + HBr$$

Here both base and substrate are involved in the single, slow rate-determining step, rate of elimination depends on concentration of both base and substrate.

Stereochemistry of E2 reaction

As we have seen in the mechanism of E2 reaction, base attacks from position anti to the halide, it is stereoselective, i.e. either of an enantiomer gives a specific stereomer after elimination as:

Let us consider E2 elimination in 1, 2-dibromo-1, 2-diphenyl ethane:

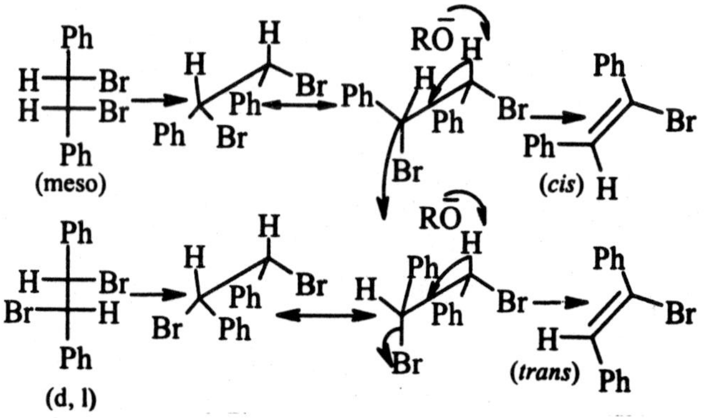

(meso) → (*cis*)

(d, l) → (*trans*)

Summary of E2 reaction

- E2 reaction is a single step reaction.
- Rate depends on concentration of base and substrate.
- Reactivity depends on both strength of base and nature of alkyl halide.
- Strong base is required to abstract proton. A more substituted alkyl halide gives more stable alkenes and therefore more reactive. General order of reactivity is:

Tertiary > Secondary > Primary

Comparison of E1 and E2 reaction

- E2 reaction requires a strong base, whereas strength of base is unimportant in E1 reaction.
- E2 reaction proceeds at room temperature, whereas for E1 reaction little higher temperature is required.
- Reactivity order of both E1 and E2 reaction is 3° > 2° > 1° alkyl halide.
- Polar protic solvent accelerate the rate of E1 reaction by solvating halide ion facilitating formation of carbonium ion. Polar protic solvent also solvates carbonium ion, which is a product, favouring forward shift in reaction. On the other hand rate of E2 reaction is retarded by polar protic solvent since it solvate the base.

Substitution vs Elimination

SN-2 vs E2: Reagents for both E2 and SN-2 are same but the reactivity order is opposite as:

SN-2 →
3° 2° 1°
← E2

Therefore, a primary alkyl halide undergoes predominantly SN-2 reaction and a tertiary alkyl halide undergoes predominantly E2 reaction in presence of strong base. A secondary alkyl halide gives mixture of substitution and elimination products.

SN-1 vs E1: Reagents and conditions for both SN-1 and E1 reactions are same as well as reactivity order is also same as:

← SN-1
3° 2° 1°
← E1

Therefore, E1 and SN-1 always completes, and it is difficult to predict major and minor product.

Example 13. Give the substitution and elimination products you would expect from the following reactions.

(a) 3-bromo-3-ethylpentane heated in methanol.
(b) 1-iodo-1-methylcyclopentane heated in ethanol.
(c) 3-bromo-2, 2-dimethylbutane heated in ethanol.
(d) Iodocyclohexane + silver nitrate in water.

Solution

(a) Br, C_2H_5 —(MeOH, Δ)→ C_2H_5 (Elimination) + MeO, C_2H_5 (Substitution)

(b) I, CH_3 —(Δ, $-I^-$)→ $^+$, CH_3 —(C_2H_5OH, $-H^+$)→ C_2H_5O, CH_3 (Substitution); —($-H^+$)→ CH_3 (Elimination)

(c) CH_3 Br, H_3C—CH_3, CH_3 H —(Δ)→ CH_3, H_3C—$^+$—CH_3, CH_3 H ←(Me-shift)→ CH_3, H_3C—$^+$—CH_3, CH_3 H

—(C_2H_5OH, SN-1)→ C_2H_5O CH_3, H_3C—CH_3, CH_3 H

—(E1)→ alkene

(d) I —($AgNO_3$)→ $^+$ + AgI; —(OH_2, SN-1)→ OH; —($-H^+$, E1)→ cyclohexene

5.5 SUMMARY OF FACTORS INFLUENCING ALKYL HALIDE REACTIONS

The following table summarizes the expected outcome of alkyl halide reactions with nucleophiles. It is assumed that the alkyl halides have one or more beta-hydrogens, making elimination possible, and that low dielectric solvents (e.g. acetone, ethanol, tetrahydrofuran and ethyl acetate) are used. Note that halogens bonded to sp^2 or sp hybridized carbon atoms do not usually undergo substitution or elimination reactions with nucleophilic reagents.

Nucleophile	Anionic nucleophiles (Weak bases) (RS^-, SCN^-, I^-, Br^-, N_3^-, CN^-, etc.)	Anionic nucleophiles (Strong Bases: HO^- RO^-)	Neutral nucleophiles (H_2O, ROH, RSH, R_3N)
Alkyl Group Primary RCH_2–	Rapid SN-2 substitution. The rate may be reduced by substitution of β-carbons, as in the case of neopentyl.	Rapid SN-2 substitution. E2 elimination may also occur. e.g. $ClCH_2CH_2Cl + KOH \rightarrow CH_2{=}CHCl$	SN-2 substitution ($N \cong S >> O$)
Secondary R_2CH–	SN-2 substitution plus E2 elimination (depends on basicity of the nucleophile). The rate of substitution may be reduced by branching at the β-carbons, and this will increase elimination.	E2 elimination will dominate.	SN-2 substitution. ($N \cong S >> O$). In high dielectric ionizing solvents, such as water, dimethyl sulfoxide and acetonitrile, SN-1 and E1 products may be formed slowly.
Tertiary R_3C–	E2 elimination will dominate with most nucleophiles (even if they are weak bases). No SN-2 substitution due to steric hindrance. In high dielectric ionizing solvents, such as water, dimethyl sulfoxide and acetonitrile, SN-1 and E1 products may be expected.	E2 elimination will dominate. No SN-2 substitution will occur. In high dielectric ionizing solvents SN-1 and E1 products may be formed.	E2 elimination with nitrogen nucleophiles (they are bases). No SN-2 substitution. In high dielectric ionizing solvents SN-1 and E1 products may be formed.
Allyl $H_2C{=}CHCH_2$–	Rapid SN-2 substitution for 1° and 2°- halides for 3°- halides a very slow SN-2 substitution or, if the nucleophile is moderately basic, E2 elimination. In high dielectric ionizing solvents, such as water, dimethyl sulfoxide and acetonitrile, SN-1 and E1 products may be observed.	Rapid SN-2 substitution for 1° halides (note there are no β hydrogens. E2 elimination will compete with substitution in 2°- halides. 3°- halides. In high dielectric ionizing solvents SN-1 and E1 products may be formed.	Nitrogen and sulfur nucleophiles will give SN-2 substitution in the case of 1° and 2°- halides. 3°- halides will probably give E2 elimination with nitrogen nucleophiles (they are bases). In high dielectric ionizing solvents SN-1 and E1 products may be formed. Water hydrolysis will be favourable for 2° and 3°- halides.
Benzyl $C_6H_5CH_2$	Rapid SN-2 substitution for 1° and 2°- halides. For 3°- halides a very slow SN-2 substitution or, if the nucleophile is moderately basic, E2 elimination. In high dielectric ionizing solvents, such as water, dimethyl sulfoxide and acetonitrile, SN-1 and E1 products may be observed.	Rapid SN-2 substitution for 1° halides (note there are no β hydrogens. E2 elimination will compete with substitution in 2°-halides, and dominate in the case of 3°-halides. In high dielectric ionizing solvents SN-1 and E1 products may be formed.	Nitrogen and sulfur nucleophiles will give SN-2 substitution in the case of 1° and 2°- halides. 3°- halides will probably give E2 elimination with nitrogen nucleophiles (they are bases). In high dielectric ionizing solvents SN-1 and E1 products may be formed. Water hydrolysis will be favourable for 2° and 3°- halides.

PRACTICE EXERCISE

1. Write products of the following substitution reactions:

   (a) *t*-butylbromide + $H_2O \xrightarrow{\Delta}$

   (b) *t*-butylchloride + MeOH $\xrightarrow{\Delta}$

   (c) Neopentylchloride + HCOOH $\xrightarrow{\Delta}$

2. Arrange the following alkyl halides in increasing order of their rate of SN-2 reaction providing explanation. CH_3Cl, C_2H_5Cl, $(CH_3)_2CHCl$, $(CH_3)_3CCH_2Cl$ and $(CH_3)_3CCl$.

3. Arrange the following in the increasing order of nucleophilic strength:
   (a) $RCOO^-$, ROH, H_2O, RO^-, OH^-
   (b) I^-, Br^-, F^-, Cl^- (in protic and aprotic solvents)

4. Which alkyl halide would you expect to react more rapidly by an SN-2 mechanism:
   (a) *n*-butyl bromide or 2-bromobutane.
   (b) *n*-chloropropane or *n*-bromopropane.
   (c) 3-methyl-1-bromobutane or 2-methyl-1-bromobutane.
   (d) ethylchloride or vinylchloride.

5. Which alkyl halide would you expect to react more rapidly by SN-2 mechanism in protic solvent?
   (a) *n*-propylbromide + CH_3OH or *n*-propylbromide + CH_3O^-
   (b) ethyliodide + OH^- or ethyliodide + SH^-
   (c) methylbromide + methanol or methylbromide + CH_3SH
   (d) CH_3CH_2I (1.0 M) + CH_3S^- (1.0 M) or CH_3CH_2I (1.0 M) + CH_3S^- (2 M)

6. Which of the following reaction would you expect to react more rapidly by SN-1 mechanism (provide explanation for your answer)?
   (a) $(CH_3)_3CI + CH_3OH \xrightarrow{\Delta}$
   or $(CH_3)_3CCl + H_2O \xrightarrow{\Delta}$
   (b) $(CH_3)_3CBr + H_2O \xrightarrow{\Delta}$
   or $(CH_3)_3CBr + CH_3OH \xrightarrow{\Delta}$
   (c) $(CH_3)_3CCl$ (1 M) + CH_3O^- (0.01M) $\xrightarrow{\Delta}$
   or $(CH_3)_3CCl$ (1 M) + CH_3O^- (0.001 M) $\longrightarrow$
   (d) $(CH_3)_3CCl + H_2O \longrightarrow$
   or $(CH_3)_2C{=}CHCl + H_2O \longrightarrow$

7. Following are the hypothetical SN-reaction, which fail to take place. Explain the reason for their failure:
   (a) $HO^- + CH_3CH_3 \rightarrow CH_3CH_2OH + H^-$
   (b) $HO^- + CH_2{=}CH_2 \rightarrow CH_2{=}CHOH + H^-$
   (c) Cyclopentane + $H_2O \rightarrow CH_3CH_2CH_2CH_2CH_2OH$
   (d) $CH_3CH{=}CHBr + CH_3S^- \rightarrow CH_3CH{=}CHSCH_3 + Br^-$
   (e) $NH_3 + CH_3CH_2OH \rightarrow CH_3CH_2NH_2 + H_2O$
   (f) $CH_3^- + CH_3CH_2OH \rightarrow C_3H_8 + OH^-$

8. Complete the following reactions:

   (a) *n*-butylbromide + $CH_3O^- \xrightarrow[50°C]{(CH_3)_3COH}$

   (b) *n*-butylbromide + $(CH_3)_3CO^- \xrightarrow[50°C]{(CH_3)_3COH}$

   (c) 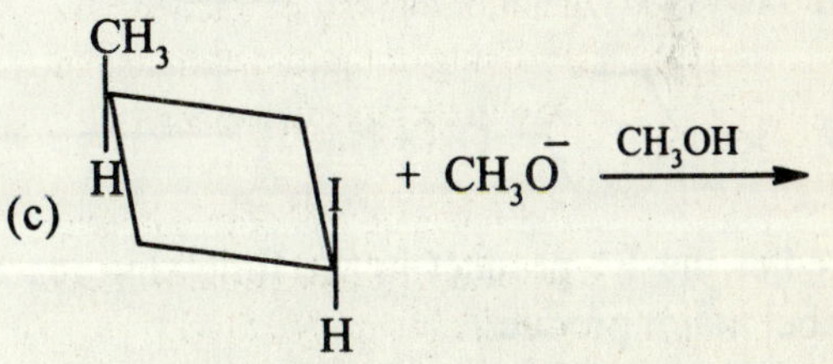

   (d) 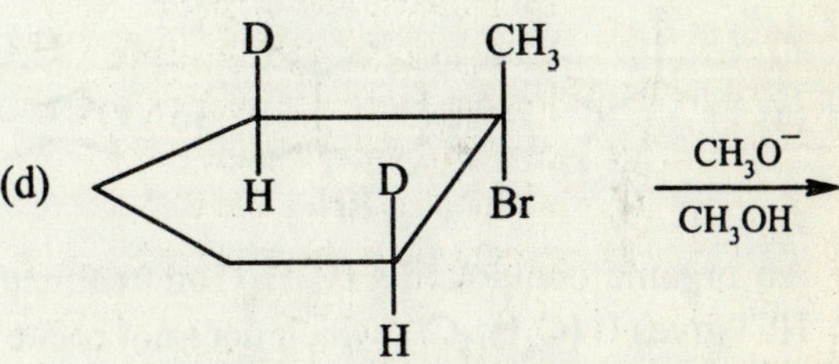

9. Write mechanism of the following reactions:

   (a) $BrCH_2CH_2OH \xrightarrow[OH_2]{HO^-}$ (epoxide)

   (b) $BrCH_2CH_2CH_2CH_2NH_2 \xrightarrow[OH_2]{HO^-}$ (pyrrolidine, N–H)

10. Give reason:
    (a) Hydrolysis of methylbromide takes place much faster in presence of NaI.
    (b) When tertiary butyl chloride undergo hydrolysis in aqueous NaOH, the rate of formation of tertiary butyl alcohol does not increase appreciably as OH^- concentration is increased, however a marked decrease in concentration of *t*-butyl chloride is observed.

(c) When methyl bromide reacts with NaCN, major product is CH_3CN, whereas when CH_3Br reacts with AgCN, major product is CH_3NC.

11. Show how each compound might be prepared by SN-2 displacement of an alkylhalide.

(a) 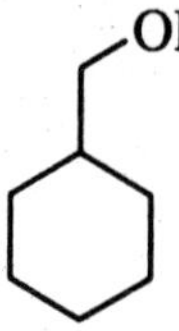 OH

(b) 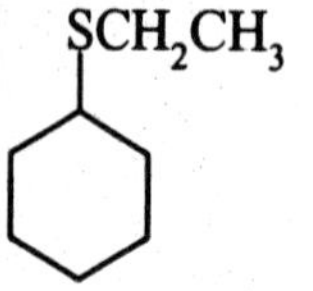 SCH_2CH_3

(c) $(CH_3)_3COCH_3$

(d) 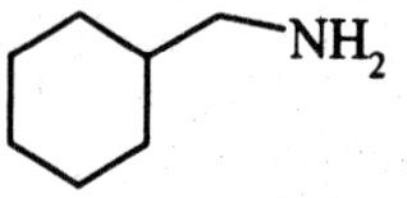 NH_2

12. Predict product of the following SN-2 reactions:

(a) $HO–(CH_2)_4–Cl + NaOH \longrightarrow$

(b) 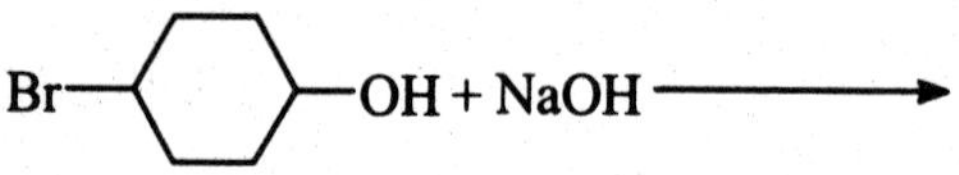 Br— —$OH + NaOH \longrightarrow$

13. Predict the E1 product in the following reactions: (label major product):

(a) 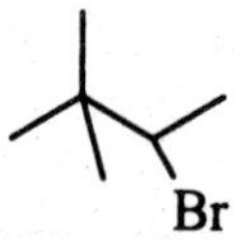 Br (b) 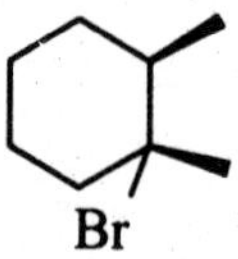 Br (c) 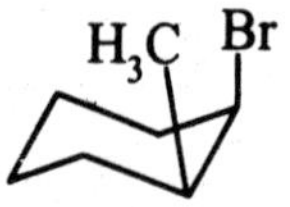 H_3C Br

14. An organic compound A (C_6H_{12}) on treatment with HCl gives B ($C_6H_{13}Cl$), which does not rotate plane polarized light. A on treatment with HBr/H_2O_2 gives C($C_6H_{13}Br$), which shows distereomerism. Oazonolysis of A followed by work-up with Zn-H_2O gives D (C_4H_8O) and E (C_2H_4O), both gives yellow precipitate on treatment with $NaOH/I_2$. D on catalytic reduction with $LiAlH_4$ gives F ($C_4H_{10}O$) which is optically active. Identify A to F.

15. Propose the mechanism for the formation of indicated products in the following reactions:

(a) 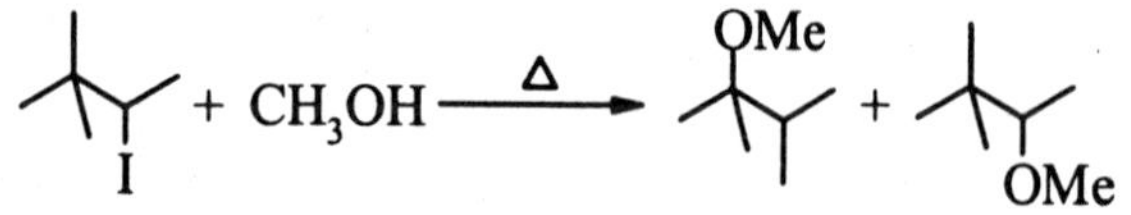 I $+ CH_3OH \xrightarrow{\Delta}$ OMe + OMe

(b) 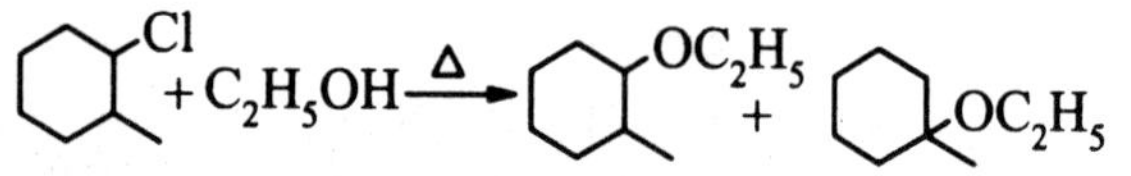 Cl $+ C_2H_5OH \xrightarrow{\Delta}$ OC_2H_5 + OC_2H_5

(c) I $+ CH_3COOH \longrightarrow$ $OCOCH_3$ + $OCOCH_3$

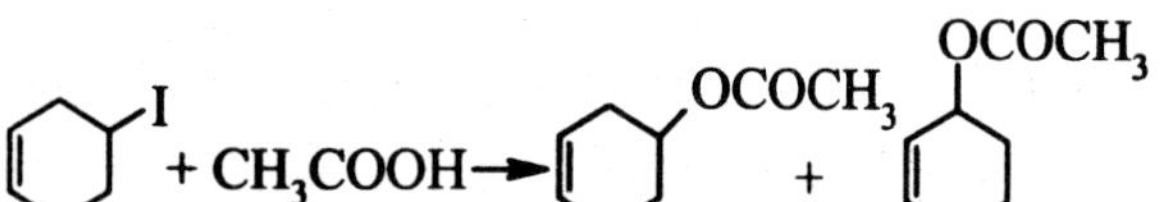

(d) 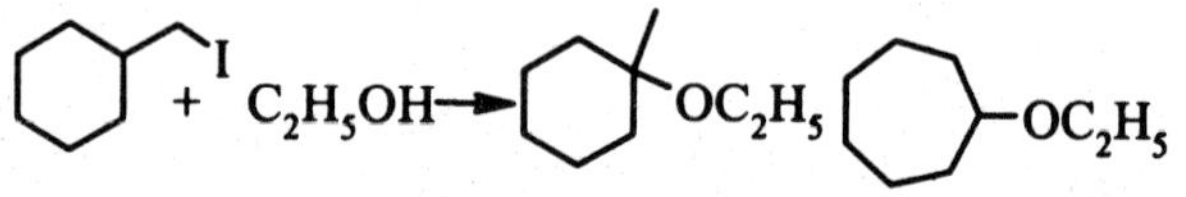 I $+ C_2H_5OH \longrightarrow$ OC_2H_5 OC_2H_5

(e) 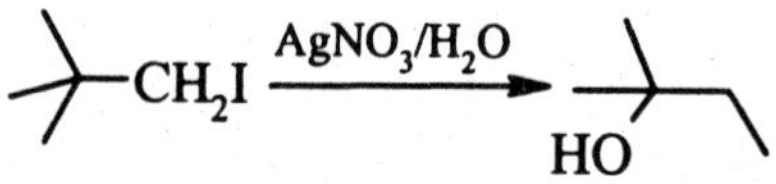 $CH_2I \xrightarrow{AgNO_3/H_2O}$ HO

(f) 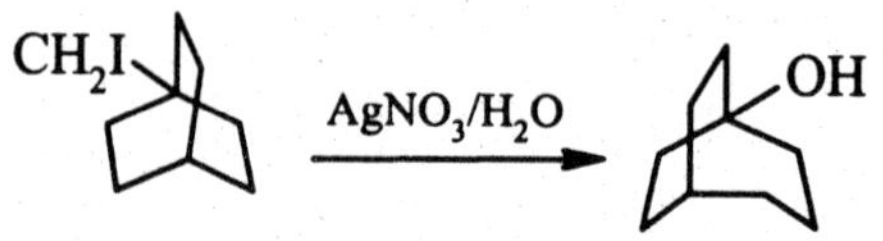 $CH_2I \xrightarrow{AgNO_3/H_2O}$ OH

16. Predict the compound in each pair that will undergo faster solvolysis reaction.

(a) $(CH_3CH_2)_2CHCl$ or $(CH_3)_3CCl$

(b) 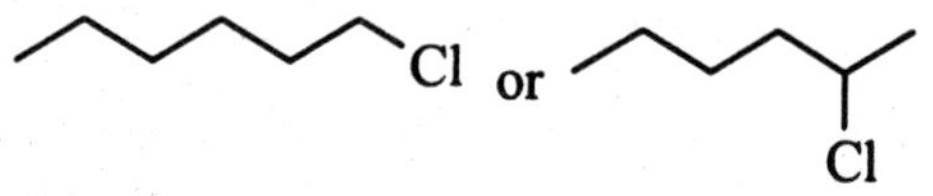 Cl or Cl

(c) 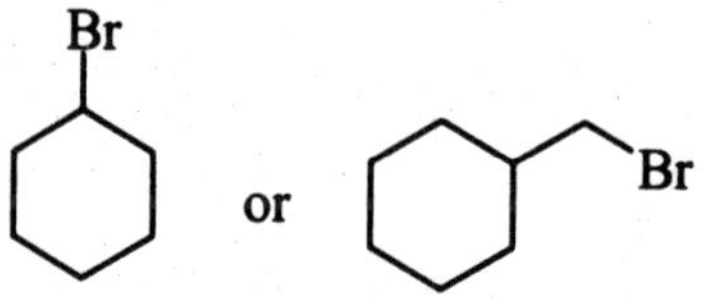 Br or Br

(d) 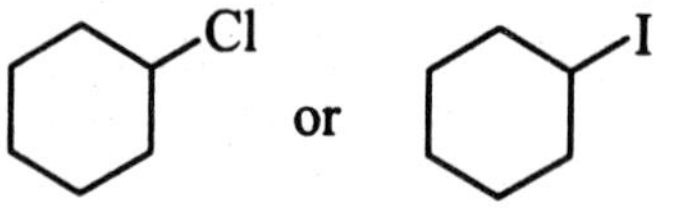 Cl or I

(e) 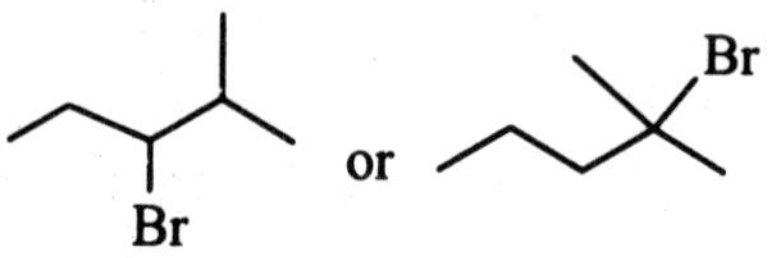 Br or Br

(f) 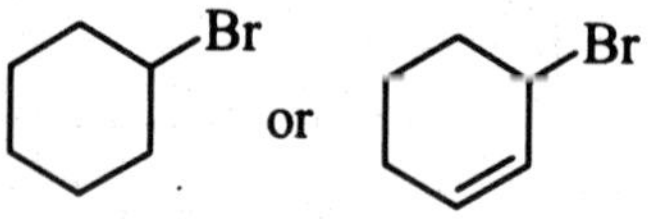 Br or Br

17. When tertiary butyl bromide undergoes solvolysis in a mixture of methanol and water, the rate of solvolysis increases when the percentage of water is increased. Explain.

18. Predict mechanism of formation of the products:

(a) 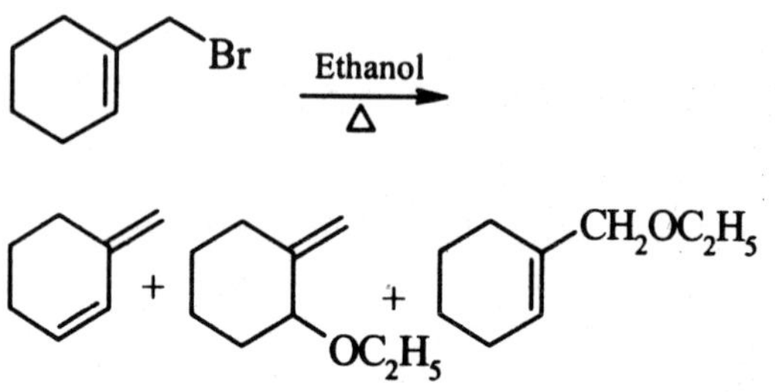 Br $\xrightarrow[\Delta]{\text{Ethanol}}$ + OC_2H_5 + $CH_2OC_2H_5$

(b) 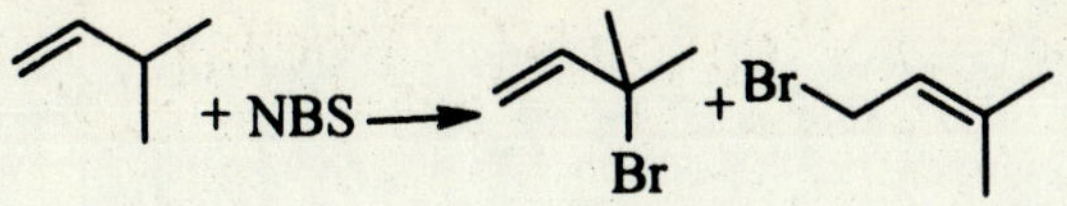 [structure] + NBS ⟶ [structure with Br] + [Br structure]

(c) 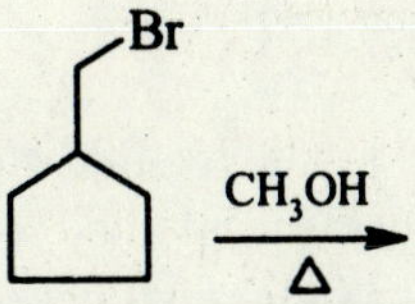 [bromomethylcyclopentane] $\xrightarrow[\Delta]{CH_3OH}$

[methylenecyclopentane] + [1-methylcyclopentene] + [MeO-methylcyclopentane] + [cyclohexene] + [OMe-cyclohexane]

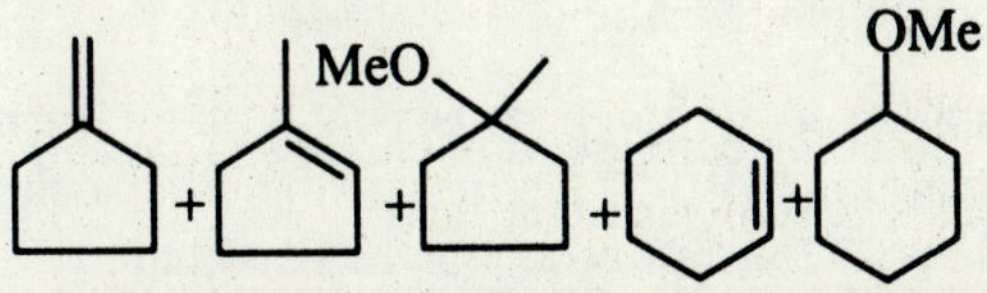

19. Write product(s), indicating major and minor, in the following reactions:

(a) $CH_3CH_2CH_2Br + CH_3O^- \xrightarrow[(CH_3)_3COH]{\Delta}$

(b) $CH_3CH_2CH_2Br + (CH_3)_3CO^- \xrightarrow[CH_3OH]{\Delta}$

(c)  [structure] Br + $HS^- \xrightarrow[CH_3OH]{\Delta}$

(d) $(C_2H_5)_3CBr + OH^- \longrightarrow$

(e) $(C_2H_5)_3CBr \xrightarrow[CH_3OH]{\Delta}$

20. (a) Which is convenient method of preparation of styrene:

$PhCH_2CH_2Br$ or $PhCHBrCH_3 \xrightarrow{C_2H_5OH/KOH}$

(b) Which is convenient method of preparation of isopropylmethylether:

$CH_3ONa + (CH_3)_2CHI \longrightarrow$ or

$CH_3I + (CH_3)_2CHONa \longrightarrow$

21. Predict product(s) in the following reaction mentioning mechanism (SN-1, SN-2, E1 or E2). Also mention whether the product be only product, a major product or a minor product.

(a) *n*-pentyl bromide + $CH_3CH_2O^- \xrightarrow[C_2H_5OH]{\Delta}$

(b) *n*-pentyl bromide + $(CH_3)_3CO^- \xrightarrow[CH_3OH]{\Delta}$

(c) $(CH_3)_3CBr + CH_3O^- \xrightarrow[(CH_3)_3COH]{\Delta}$

(d) 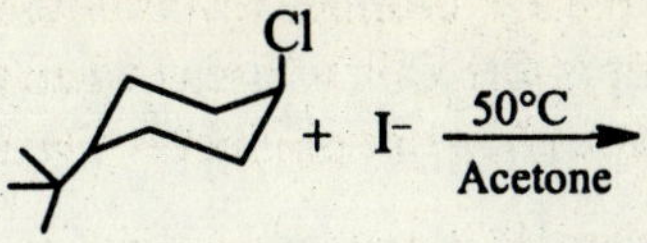 [Cl-substituted tert-butylcyclohexane] + $I^- \xrightarrow[Acetone]{50°C}$

(e) 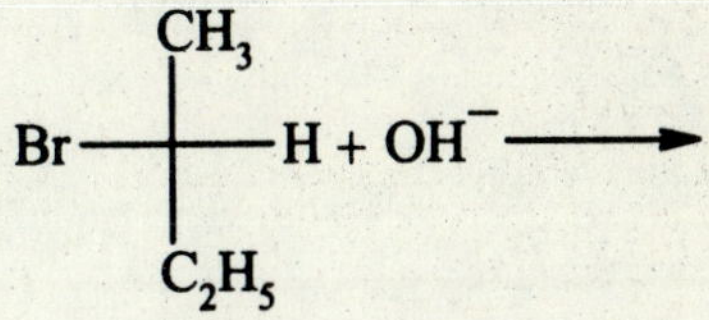 Br—C(CH_3 top, C_2H_5 bottom)—H + $OH^- \longrightarrow$

(f) 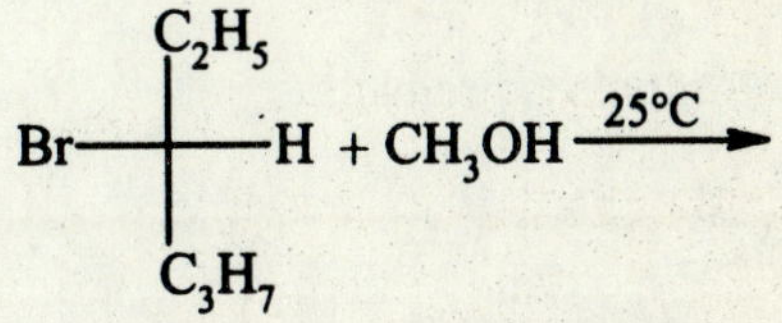 Br—C(C_2H_5 top, C_3H_7 bottom)—H + $CH_3OH \xrightarrow{25°C}$

(g) 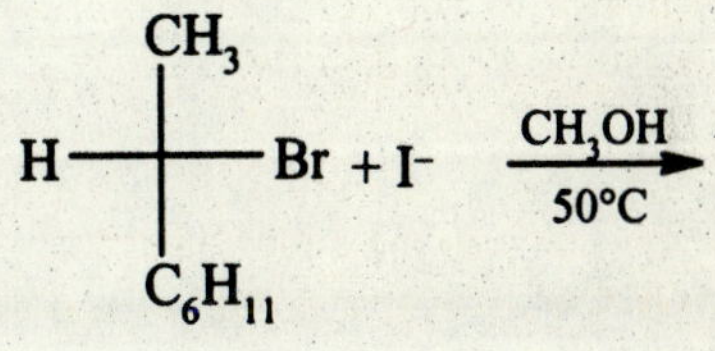 H—C(CH_3 top, C_6H_11 bottom)—Br + $I^- \xrightarrow[50°C]{CH_3OH}$

22. Although ethyl bromide and isobutyl bromide are both primary halides, C_2H_5Br undergoes SN-2 reactions more than 10 times faster than isobutyl bromide. When each compound is treated with a strong base, isobutyl bromide gives a greater yield of elimination product than substitution product, whereas with ethyl bromide, this behaviour is reversed. Explain these observations.

23. An alkyl halide X of formula $C_6H_{13}Cl$ on treatment with $(CH_3)_3COK$ gives two isomeric alkenes Y and Z of formula C_6H_{12}, in which Y on treatment with Bayer's reagent gives a meso diol. Both Y and Z on hydrogenation give 2, 3-dimethyl butane. Identify X, Y and Z.

24. Predict products:

(a) 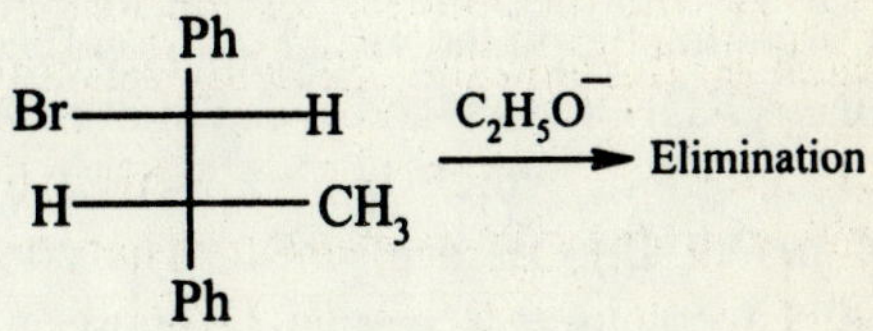 Fischer projection: Ph (top); Br—C—H; H—C—CH_3; Ph (bottom) $\xrightarrow{C_2H_5O^-}$ Elimination

(b) 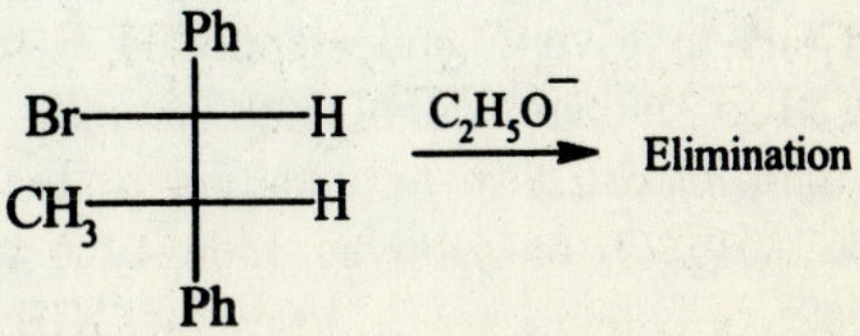 Fischer projection: Ph (top); Br—C—H; CH_3—C—H; Ph (bottom) $\xrightarrow{C_2H_5O^-}$ Elimination

(c) 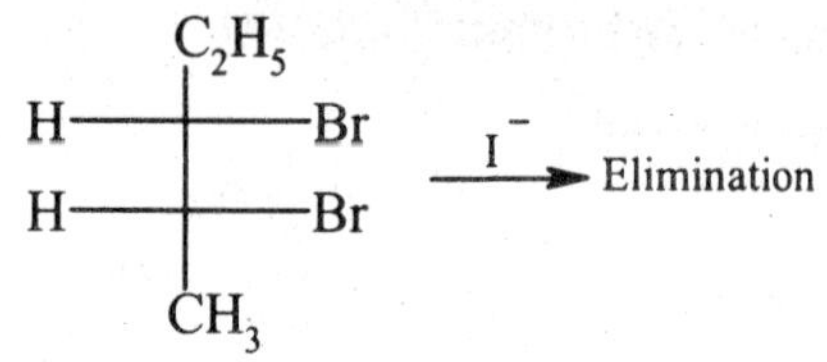 C_2H_5 / H—C—Br / H—C—Br / CH_3 $\xrightarrow{I^-}$ Elimination

(d) 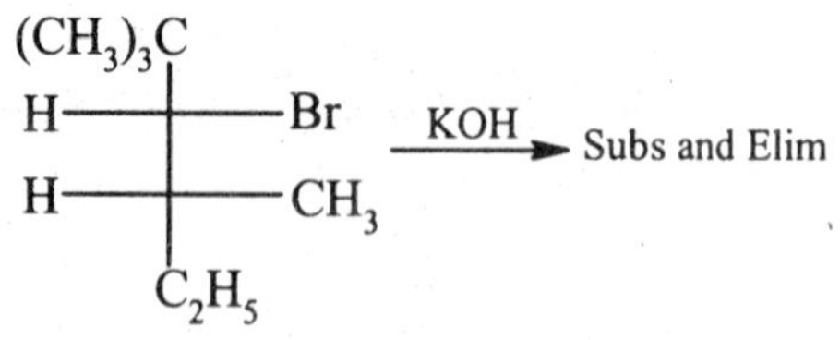 $(CH_3)_3C$ / H—C—Br / H—C—CH_3 / C_2H_5 $\xrightarrow{KOH}$ Subs and Elim

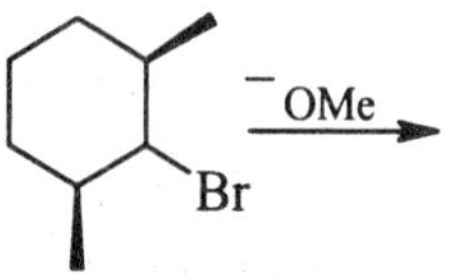

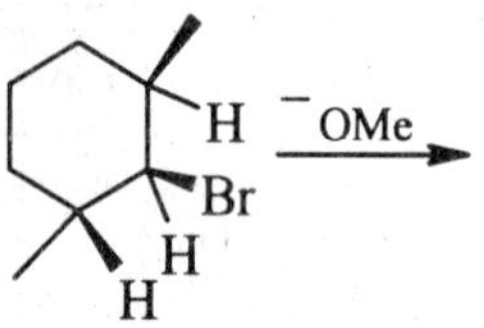

25. Predict elimination products:

(a) 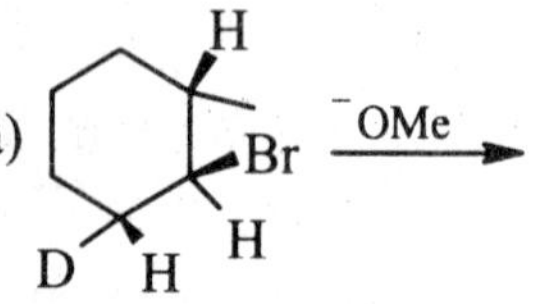 H, Br, H, D, H $\xrightarrow{^-OMe}$ (b) Br $\xrightarrow{^-OMe}$

(c) H, Br, H, H $\xrightarrow{^-OMe}$

(d) D, H, H, Br $\xrightarrow{^-OMe}$ Two products

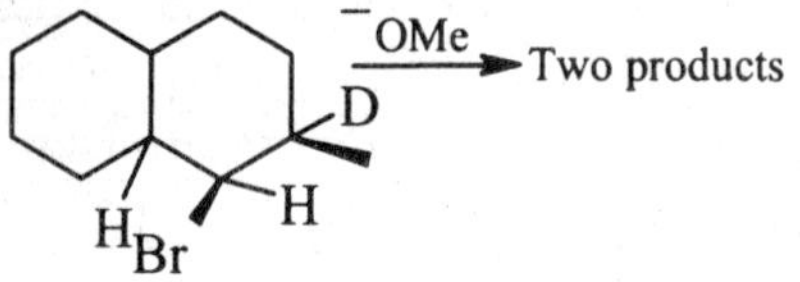

26. An organic compound A (C_5H_9Br) readily decolourize Bayer's reagent and gives B ($C_5H_{11}Br$) on treatment with Sn/HCl. B is found to be optically active. The reaction of A with $NaNH_2$ produces C, which neither reacts with Na nor forms any precipitate with $Ag(NH_3)_2^+$ but reacts with Lindlar's catalyst to give D and on reaction with Na/liq NH_3 produces E. Give structure to A to E.

27. An organic compound A(C_4H_9Cl) on reacting with aqueous KOH gives B and on reaction with alcoholic KOH gives C, which is also formed by passing vapours of B over heated Cu. The compound C readily decolourize Br_2-water. Ozonolysis of C gives two compounds D and E. Compound D reacts with NH_2OH to give F and compound E reacts with NaOH to give an alcohol G and sodium salt (H) of an acid. D can also be obtained by the action of $H_2SO_4/HgSO_4$ on propyne. Identify A to H.

28. Explain the mechanism of following reactions:

(a)  OH + HCl ⟶ Cl + Cl

(b) 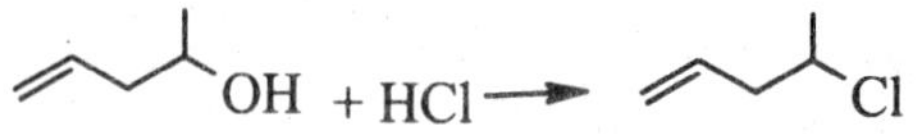 OH + HCl ⟶ Cl

29. Which alkyl halide can be made by free radical halogenation of alkanes? Why is this method seldom used for laboratory synthesis of alkyl halides? How may polyhalogenation be minimized?

30. Which hydrocarbon is consistent with following information? MW = 72, gives a single monochloride and two dichloride on photochlorination.

31. Account for the fact that reaction of RX with H_2O and MeOH yield the neutral final products ROH and ROMe, but reaction with NH_3 and $R'NH_2$ gives RNH_3^+ and $RR'NH_2^+$ as Cl^- salt.

32. Account for the following observations, when an acetone solution of optically active $CH_3CH_2CH_2CHDI$ is warmed with NaI, the enantiomer is racemized and with the excess of I^- (radioactive I), the rate of racemization is twice the rate of incorporation of I^- in the product.

33. Explain the following observations in SN-1 reactions:
(a) $AgNO_3$ increases the rate of solvolysis.
(b) The more acidic the nucleophilic solvent, faster is the rate of solvolysis.

34. Account for the following observations:
(a) *t*-BuF is solvolysed only in very acidic solution.
(b) *t*-BuCl is solvolysed more slowly than 2-chloro-2, 3, 3-trimethylbutane.
(c) *t*-BuCl is solvolysed much faster than 2-chloro-1, 1, 1-trifluoro-2-methylpropane.

35. Compare the rate of SN-1 and SN-2 reactions of cyclopropyl chloride and cyclopentylchloride.

36. Account for the rapid rate of ethanolysis of $ClCH_2OCH_2CH_3$, although the substrate is a 1° halide.

37. Provide structural formula of the alkenes from the reactions:
(a) Bromocyclopentane + CH_3CH_2OK
(b) 2-bromopentane + CH_3CH_2ONa
(Three products)

38. Write the stereochemical products formed in the following elimination reactions:

(a) 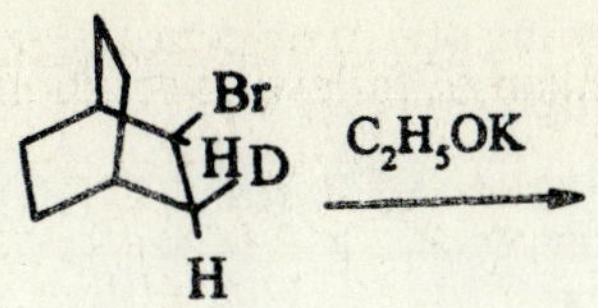 [bicyclic bromide with Br, H, D, H] $\xrightarrow{C_2H_5OK}$

(b) 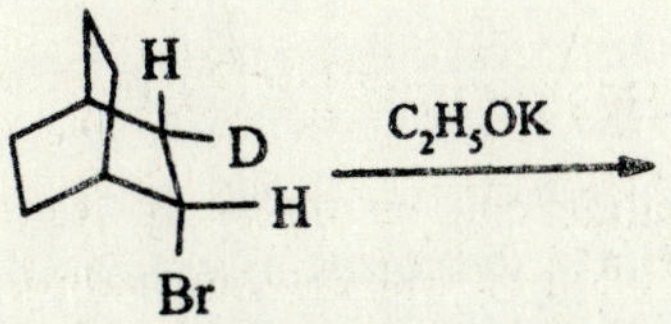 [bicyclic bromide with H, D, H, Br] $\xrightarrow{C_2H_5OK}$

39. Complete the following reaction sequence, showing the major product you expect for each stage. Also indicate stereochemistry where possible.

(a) $CH_3CH(OH)CH_3 \xrightarrow[Pyridine]{TsCl} A \xrightarrow{NaN_3} B$

(b) 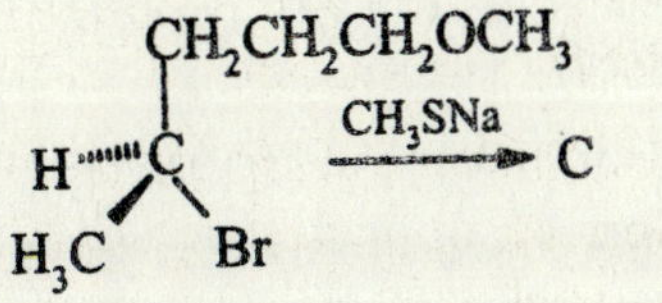 H_3C, H, Br, $CH_2CH_2CH_2OCH_3$ on C $\xrightarrow{CH_3SNa}$ C

40. Write equations for the reaction of *n*-butyl bromide as a typical primary alkyl halide, with the following reagents:

(a) KOH in ethanol/heat
(b) NaOH in ethanol
(c) NH_3
(d) NaN_3
(e) NaCN
(f) CH_3CH_2SNa
(g) CH_3COONa
(h) $NaCH(COOEt)_2$

41. Predict major elimination products if the following compounds are heated with strong base:

(a) 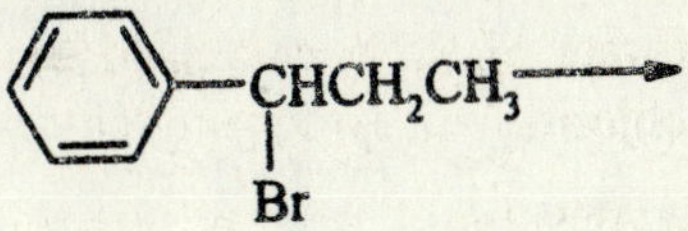 C_6H_5–$CH(Br)CH_2CH_3$ ⟶

(b) $CH_3CH_2CH_2CH(Br)COOH$ ⟶

(c) 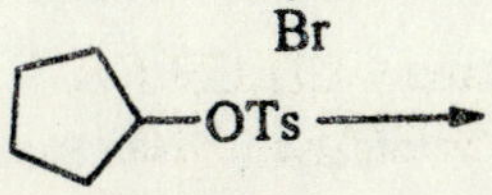 cyclopentyl–OTs ⟶

42. Bring about the following transformation:

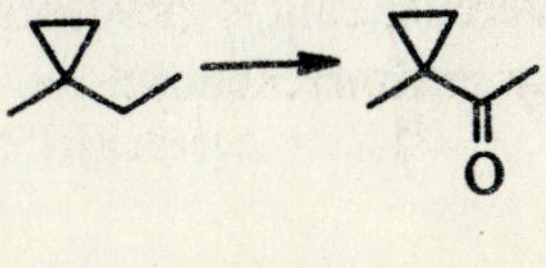

[cyclopropane derivative] ⟶ [cyclopropyl ketone]

43. Which is a better nucleophile, methyl-lithium or methylmagnesium bromide? Explain.

44. Provide mechanism:

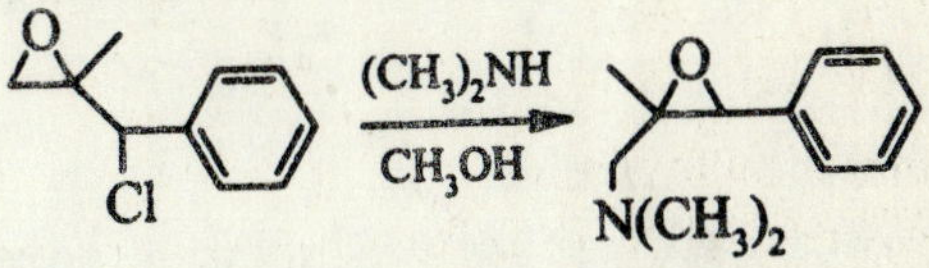

[epoxide with Cl and phenyl] $\xrightarrow[CH_3OH]{(CH_3)_2NH}$ [epoxide with $N(CH_3)_2$ and phenyl]

45. Explain the relative rate of E2 reaction of the following compounds:

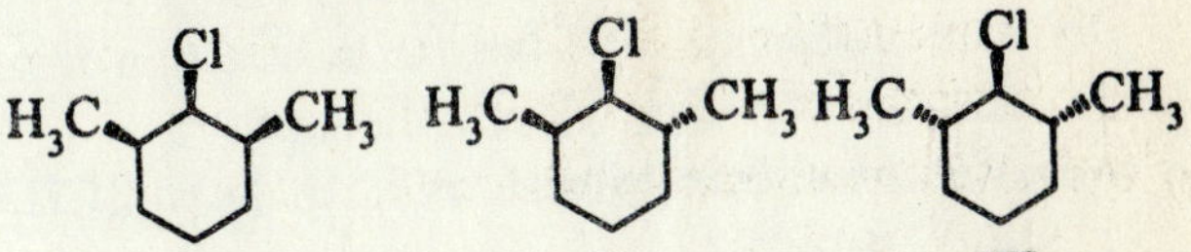

I, II, III: [Cl, H_3C, CH_3 substituted cyclohexanes]

46. Rank the following compounds in increasing order their reactivity towards reaction with CH_3ONa:

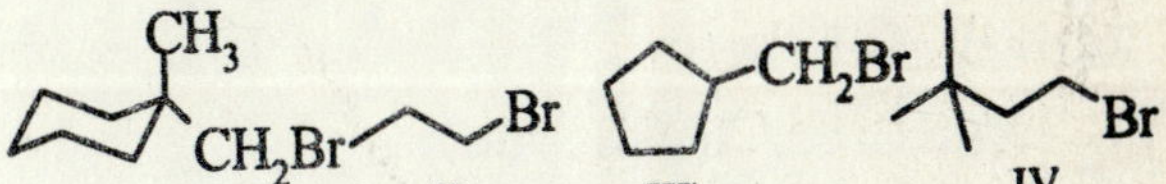

I: CH_3, CH_2Br cyclohexane; II: Br; III: cyclopentyl–CH_2Br; IV: Br

47. Predict product in each step of the following reaction:

(a) 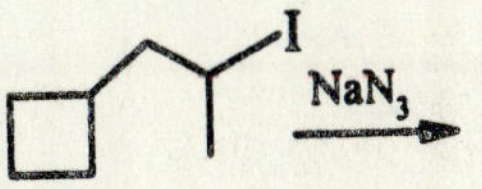 [cyclobutyl iodide] $\xrightarrow{NaN_3}$ (b) 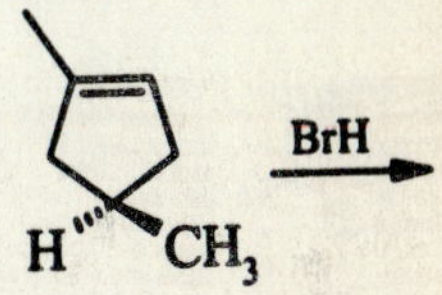 [methylcyclopentene with H, CH_3] $\xrightarrow{BrH}$

(c) 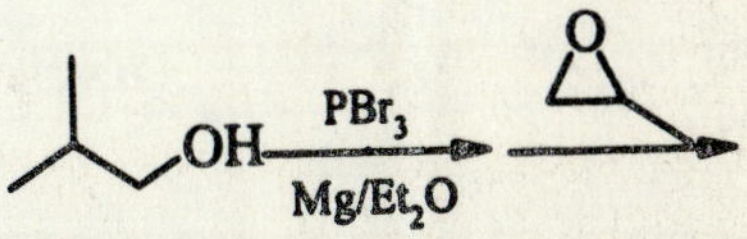 [2-methylpropanol] $\xrightarrow[Mg/Et_2O]{PBr_3}$ $\xrightarrow{\text{epoxide}}$

48. Predict substitution/elimination products that might be obtained when each of the following alkyl halides are treated with CH_3ONa in ethanol. Include stereochemistry where appropriate.

(a) 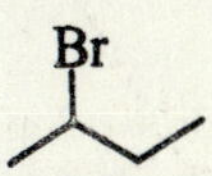 [2-bromobutane] (b) 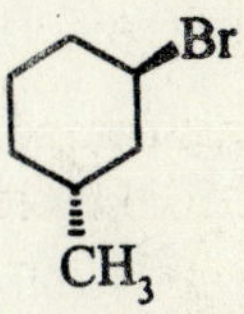 [3-methylcyclohexyl bromide, Br, CH_3]

(c) 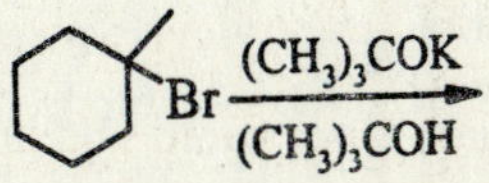 [1-methylcyclohexyl bromide] $\xrightarrow[(CH_3)_3COH]{(CH_3)_3COK}$ (d) 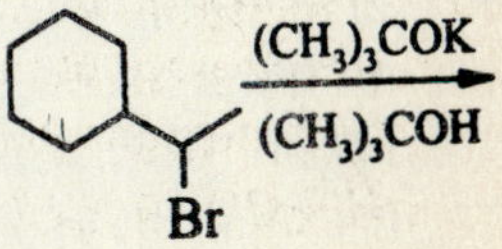 [cyclohexyl–CH(Br)CH_3] $\xrightarrow[(CH_3)_3COH]{(CH_3)_3COK}$

(e) 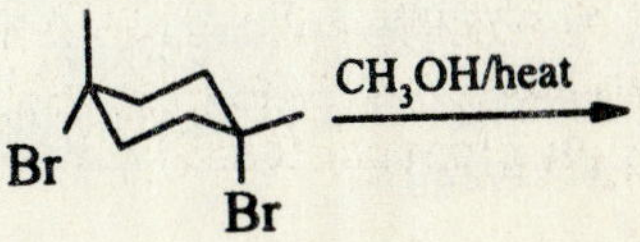 Br, Br [cyclohexane] $\xrightarrow{CH_3OH/heat}$

(f) 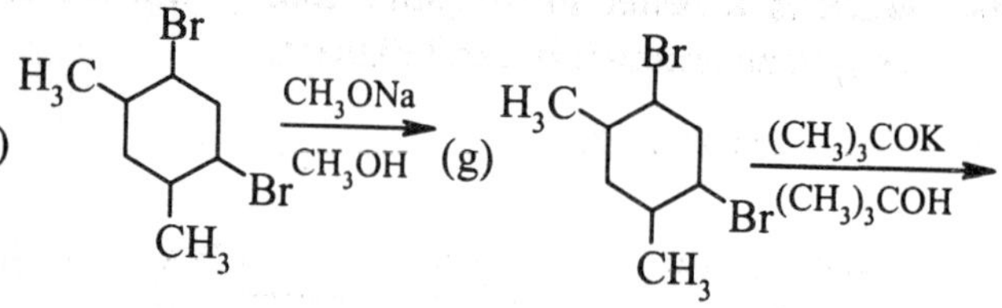 [structure] $\xrightarrow[CH_3OH]{CH_3ONa}$ (g) [structure] $\xrightarrow[(CH_3)_3COH]{(CH_3)_3COK}$

49. Explain the followings:
   (a) When CH_3Br is dissolved in ethanol, no reaction occurs at 25°C. When excess of C_2H_5ONa is added, a good yield of ethyl methyl ether is obtained.
   (b) Explain the role of solvent in the reaction of potassium ethanoate with ethyl iodide.
   (c) Which nucleophile reacts more rapidly with CH_3I: Et_3N or Et_3P?

50. For each of the species, put in the order of increasing basic strength:
   (a) Cl^-, I^-, F^-, Br^-
   (b) SH^-, OH^-, SeH^-
   (c) Me_2N^-, F^-, $CH_3CH_2^-$, EtO^-
   (d) $CH_3CH_2^-$, $CH_3CH{=}CH^-$, $HC{\equiv}C^-$
   (e) $CH_3SeCH_2^-$, $CF_3OCH_2^-$, CH_3COO^-

51. Predict product in the following reaction:

(a) 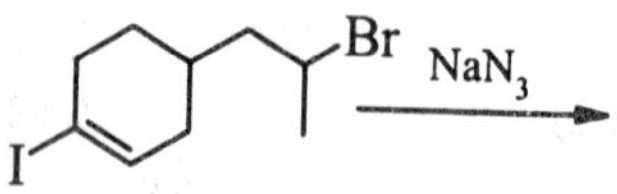 [structure] $\xrightarrow{NaN_3}$

(b) 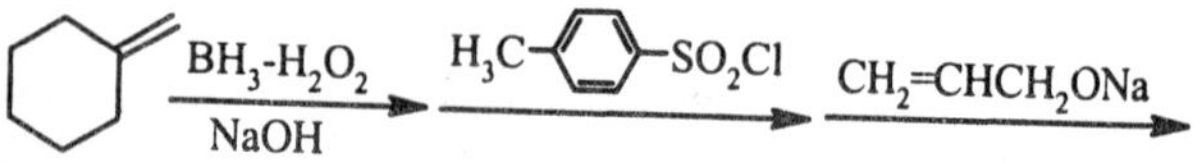 [structure] $\xrightarrow[NaOH]{BH_3\text{-}H_2O_2}$ $\xrightarrow{H_3C\text{-}C_6H_4\text{-}SO_2Cl}$ $\xrightarrow{CH_2{=}CHCH_2ONa}$

(c) [structure] $\xrightarrow[(CH_3)_3COH]{(CH_3)_3COK}$

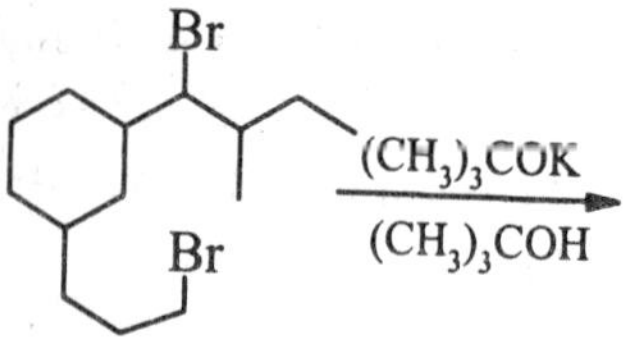

(d) 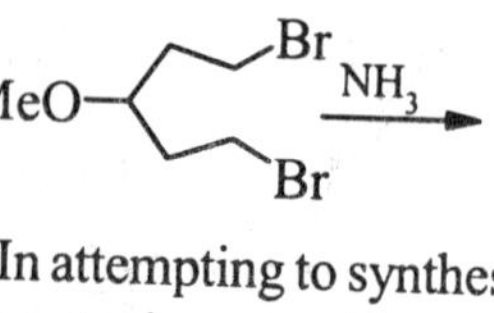 MeO–[structure] $\xrightarrow{NH_3}$ (e) 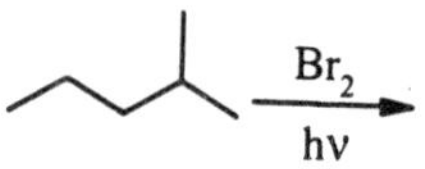 [structure] $\xrightarrow[hv]{Br_2}$

52. In attempting to synthesize 1-methylcyclohexene, you are trying reaction of *trans*-1-iodo-2-methyl cyclohexane with potassium tri-tertiary butoxide. The reaction fails to produce the desired product and the isolated product is 3-methylcyclohexene. Using a detailed mechanistic analysis, explain why this alternative product is formed?

53. Arrange the following sets of nucleophile in increasing order of nucleophilicity:
   (a) CH_3O^-, $(CH_3)_3CCH_2O^-$, $HCOO^-$
   (b) NH_3, PH_3, HF
   (c) H_2O, SH^-, OH^-

54. The two tosylate shown below react differently under identical conditions. One will undergo a facile SN-2 reaction while other will easily eliminate via E2 mechanism. Draw structures of major product for each reaction and provide mechanistic explanation for them:

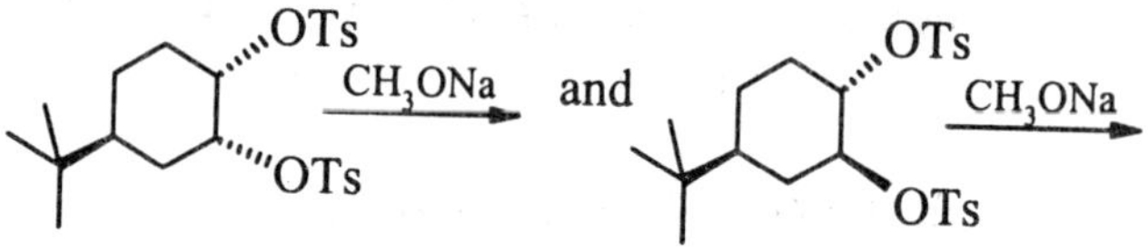

[structure] $\xrightarrow{CH_3ONa}$ and [structure] $\xrightarrow{CH_3ONa}$

55. In each of the following pair, predict which will react faster in the indicated reaction:
   (a) $MeOCH_2Cl$ or $MeOCH_2CH_2Cl \rightarrow$ SN-1 and SN-2
   (b) $MeOCH_2Cl$ or $MeSCH_2Cl$ with hot methanol$\rightarrow$ SN-1
   (c) $MeOCH_2CH_2Cl$ or $MeSCH_2CH_2Cl$ with KCN/EtOH $\rightarrow$ SN-2

56. Predict the major alkene formed in the following reaction giving mechanistic reasoning:

(a) 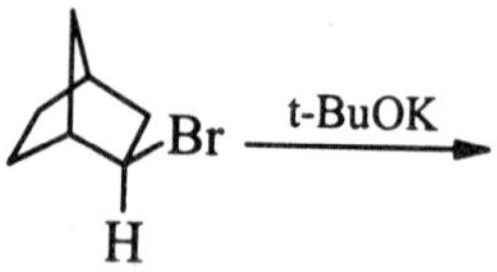 [structure] $\xrightarrow{t\text{-BuOK}}$ (b) 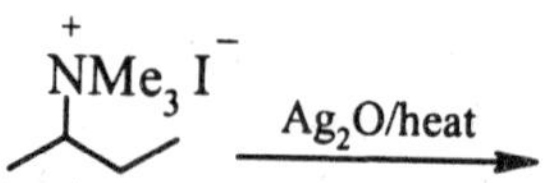 [structure] $\xrightarrow{Ag_2O/heat}$

(c) 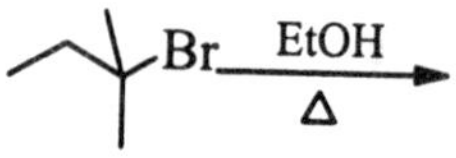 [structure] $\xrightarrow[\Delta]{EtOH}$ (d) 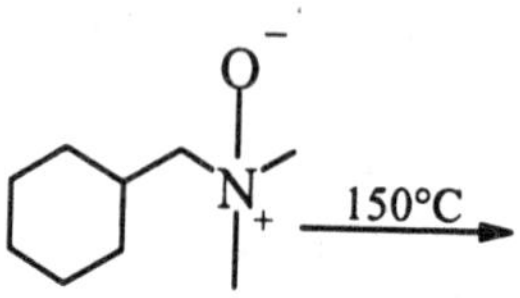 [structure] $\xrightarrow{150°C}$

57. Predict, giving reasons, which members of the following pairs will react at faster rate under the indicated reaction conditions?

(a)  [structure] or Me_3CBr solvolysis in aqueous HCOOH

(b) 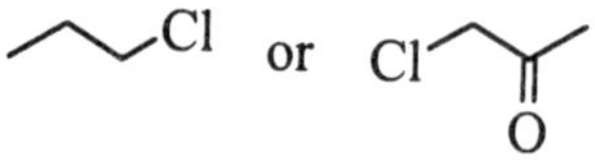 [structure] or [structure] with NaI/acetone

(c) 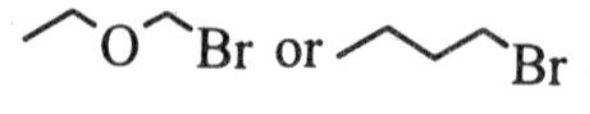 [structure] or [structure] solvolysis in aqueous HCOOH

(d) 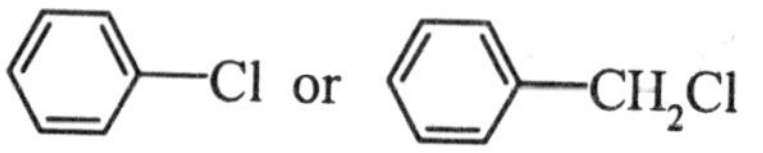 C_6H_5–Cl or C_6H_5–CH_2Cl with NaI/acetone

(e) Me_3CBr or $PhC(CH_3)_2CH_2Br$ solvolysis in aqueous HCOOH

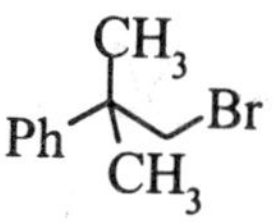

58. Give major substitution product of each of the following reaction:

(a) 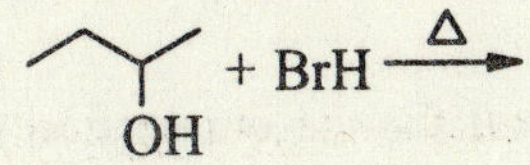 + BrH $\xrightarrow{\Delta}$ (b) 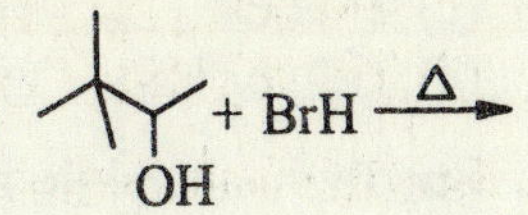 + BrH $\xrightarrow{\Delta}$

(c) 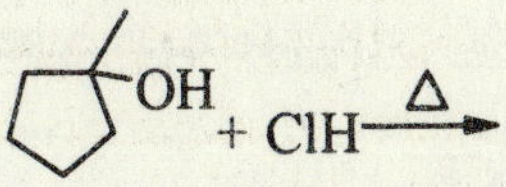 + ClH $\xrightarrow{\Delta}$ (d) 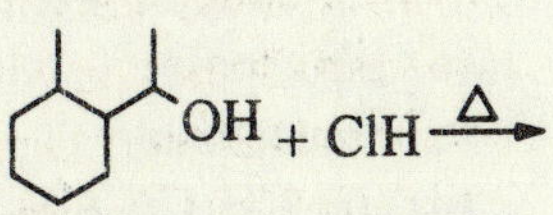 + ClH $\xrightarrow{\Delta}$

59. Give structures of product in the following reaction indicating their stereochemistry:

(a)  Me–C(D)(H)–OH $\xrightarrow{TsCl}$ $\xrightarrow{NaCN}$ X

(b) 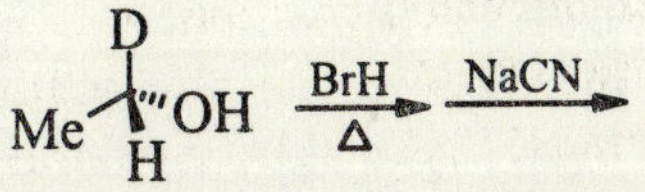 Me–C(D)(H)–OH $\xrightarrow[\Delta]{BrH}$ $\xrightarrow{NaCN}$

60. How 1-butanol can be converted efficiently into:

(a) 1-iodobutane (b) ethyl butanamine
(c) butyl propanoate (d) thio butanol
(e) methoxy butane

61. Why an acid catalysed dehydration of an alcohol a reversible reaction, whereas the base catalysed dehydrohalogenation of an alkyl halide is irreversible.

62. Heating an alcohol with sulphuric acid is a good method for preparing a symmetrical ether, but it is not suitable for preparing an unsymmetrical ether, explain. Also explain a suitable method for preparing unsymmetrical ether.

63. Synthesize the following from the given starting compounds:

(a) 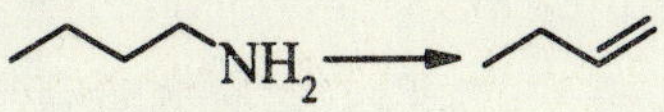 NH_2 ⟶

(b)  Br ⟶

64. Propose mechanism of the following reaction:

(a) 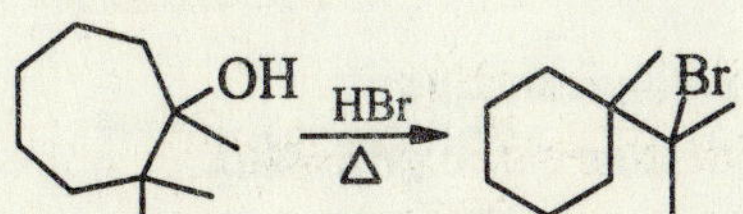 OH $\xrightarrow[\Delta]{HBr}$ Br

(b) 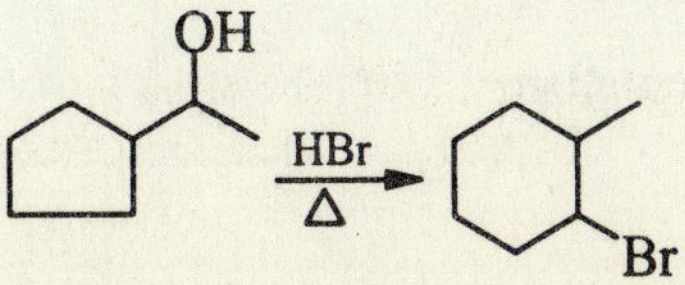 OH $\xrightarrow[\Delta]{HBr}$ Br

(c) 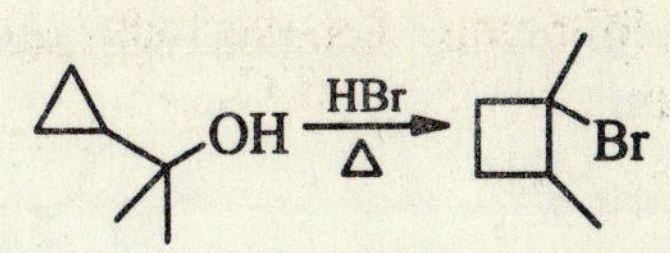 OH $\xrightarrow[\Delta]{HBr}$ Br

65. Give the major product of each of the following reactions:

(a) 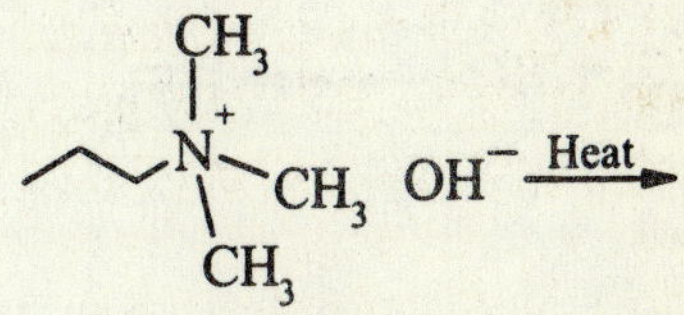 CH_3 $\overset{+}{N}$ CH_3 OH^- CH_3 $\xrightarrow{Heat}$

(b) 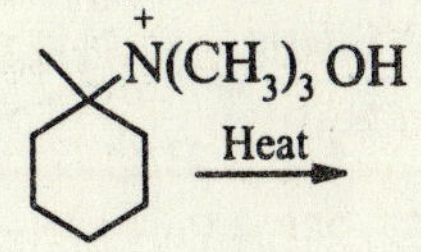 $\overset{+}{N}(CH_3)_3\ OH$ $\xrightarrow{Heat}$ (c) 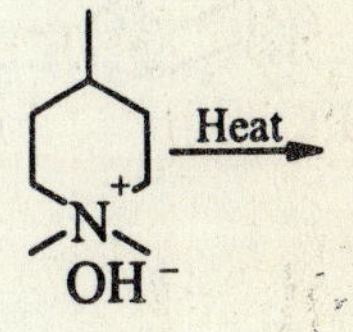 $\overset{+}{N}$ OH^- $\xrightarrow{Heat}$

(d) 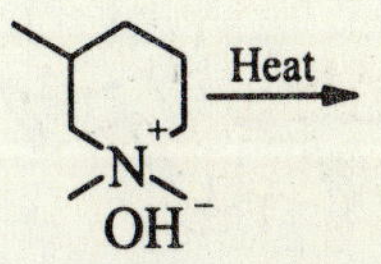 $\overset{+}{N}$ OH^- $\xrightarrow{Heat}$

66. Which of the following alkyl halide could be successfully used to synthesize a Grignard reagent?

(a) 4-bromobutanol
(b) N,N-dimethyl-3-bromopropanamine
(c) 4-bromobutanoic acid
(d) 3-bromopropanamine

67. Give product in the following reactions:

(a) 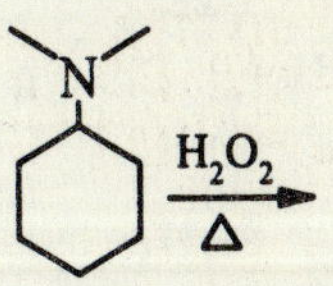 N $\xrightarrow[\Delta]{H_2O_2}$ (b) 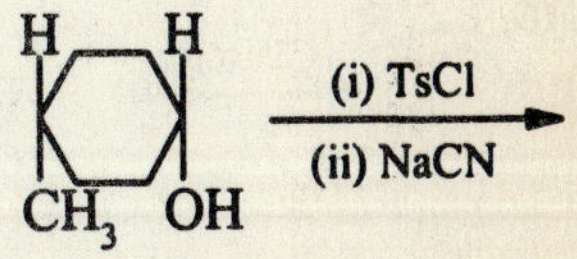 H H CH_3 OH $\xrightarrow[\text{(ii) NaCN}]{\text{(i) TsCl}}$

(c) 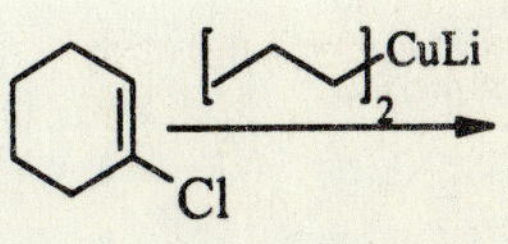 Cl $\xrightarrow{[\ \ ]_2CuLi}$

68. When piperidine undergoes the indicated series of reactions, 1, 4-pentadiene is obtained as product.

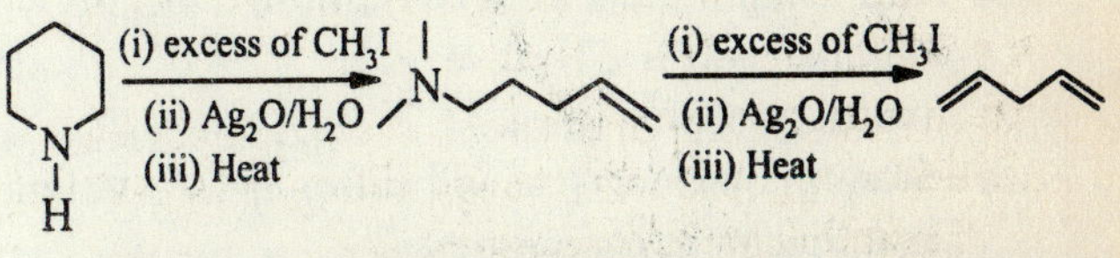

N–H $\xrightarrow[\text{(ii) } Ag_2O/H_2O \text{ (iii) Heat}]{\text{(i) excess of } CH_3I}$ N $\xrightarrow[\text{(ii) } Ag_2O/H_2O \text{ (iii) Heat}]{\text{(i) excess of } CH_3I}$

When four different methyl substituted piperidine is subjected to same series of reactions, each forms a different dienes-1, 5-hexadiene, 1, 4-pentadiene, 2-methyl-1, 4-pentadiene and 3-methyl-1, 4-pentadiene. Which methyl substituted piperidine yields which diene?

69. Which of the following reaction will occur more rapidly? Explain:

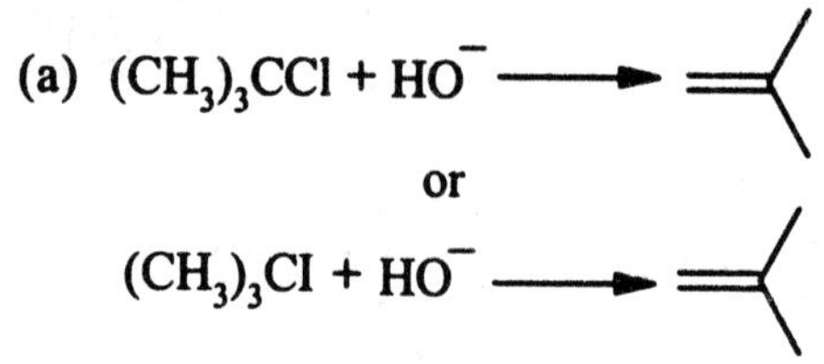

(a) $(CH_3)_3CCl + HO^-$ ⟶

or

$(CH_3)_3CI + HO^-$ ⟶

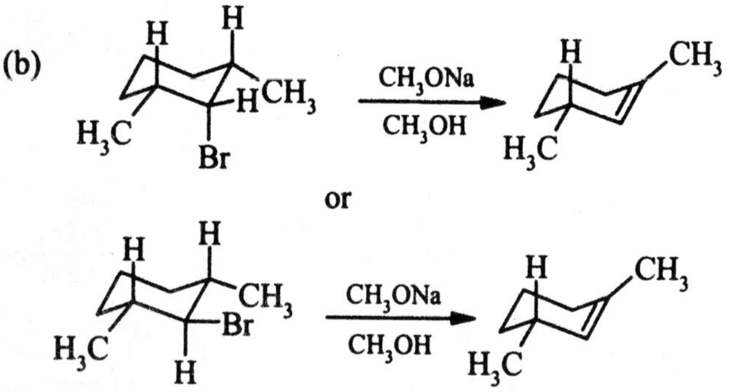

(b) [cyclohexane with H, H, CH_3, H_3C, Br] $\xrightarrow[CH_3OH]{CH_3ONa}$ [cyclohexene with H, CH_3, H_3C]

or

[cyclohexane with H, H, CH_3, Br, H_3C, H] $\xrightarrow[CH_3OH]{CH_3ONa}$ [cyclohexene with H, CH_3, H_3C]

70. Give substitution product mentioning their stereochemistry:

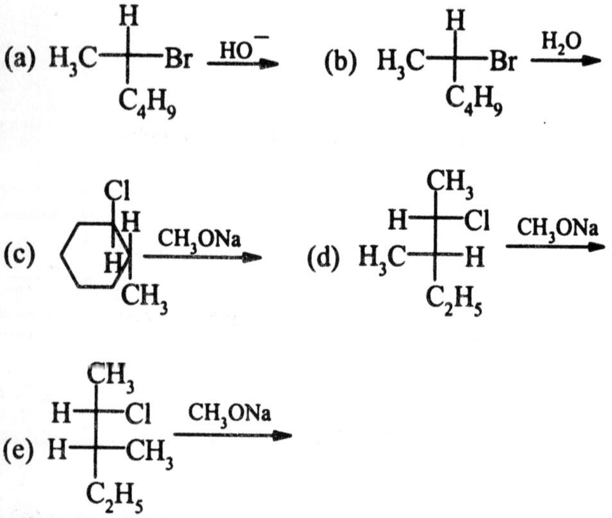

(a) H_3C–C(H)(Br)(C_4H_9) $\xrightarrow{HO^-}$ (b) H_3C–C(H)(Br)(C_4H_9) $\xrightarrow{H_2O}$

(c) [cyclohexane with Cl, H, H, CH_3] $\xrightarrow{CH_3ONa}$ (d) CH_3; H–C–Cl; H_3C–C–H; C_2H_5 $\xrightarrow{CH_3ONa}$

(e) CH_3; H–C–Cl; H–C–CH_3; C_2H_5 $\xrightarrow{CH_3ONa}$

71. Give structure of products obtained from the reaction of both enantiomers of *cis*-1-chloro-2-isopropyl cyclopropane with CH_3ONa.
   (a) Are all products chiral?
   (b) How the products would differ if the starting material were *trans* isomer?
   (c) Which pair of enantiomers forms substitution product more rapidly?
   (d) Which pairs of enantiomers forms elimination product more rapidly?

72. Which is a better nucleophile in ethanol?
   (a) H_2O or OH^- (b) H_2O or H_2S
   (c) I^- or Br^- (d) NH_3 or NH_2^-
   (e) HO^- or HS^-

73. Supply nucleophile required for the conversion of 1-bromobutane into the followings:
   (a) 1-butanol (b) Methoxy butane
   (c) Thiobutanol
   (d) $CH_3CH_2CH_2CH_2SCH_2CH_3$
   (e) $CH_3CH_2CH_2CH_2N(CH_3)_2$ (f) Pentanitrile
   (g) Butyl-ethanoate (h) 2-heptyne

74. Arrange the following in increasing order of nucleophilicity:
   (a) CH_3COO^-, $CH_3CH_2O^-$, $CH_3CH_2S^-$
   (b) H_2O, NH_3, PH_3 in methanol.
   (c) Br^-, Cl^-, I^- in methanol.
   (d) Br^-, Cl^-, I^- in acetone.
   (e) 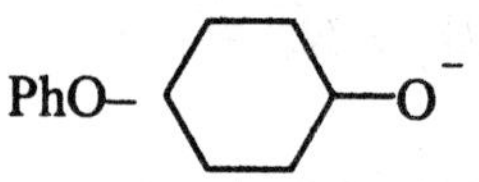 PhO– [cyclohexyl]–O^-

75. In the following pair of reactions, which will occur at faster rate-explain your choice:

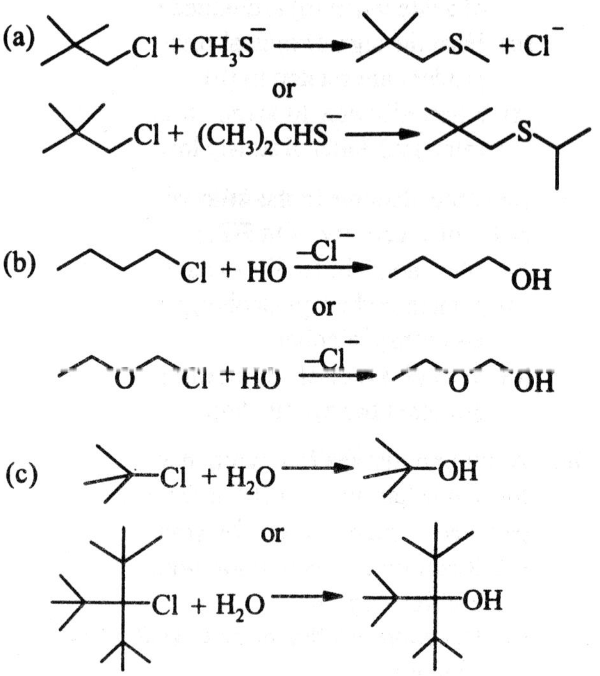

(a) [neopentyl]–Cl + CH_3S^- ⟶ [neopentyl]–S– + Cl^-

or

[neopentyl]–Cl + $(CH_3)_2CHS^-$ ⟶ [neopentyl]–S–[isopropyl]

(b) [butyl]–Cl + HO $\xrightarrow{-Cl^-}$ [butyl]–OH

or

[ethoxy-ethyl]–Cl + HO $\xrightarrow{-Cl^-}$ [ethoxy-ethyl]–OH

(c) [isopropyl]–Cl + H_2O ⟶ [isopropyl]–OH

or

[tri-tert-butyl-methyl]–Cl + H_2O ⟶ [tri-tert-butyl-methyl]–OH

76. Convert cyclohexane into:
   (a) Cyclohexane carboxylic acid.
   (b) 3-methoxy cyclohexene.
   (c) 3-cyclohexenol.

77. Propose mechanism:

(a) [cyclobutyl–C(CH₃)₂–Br] $+ H_2O \xrightarrow{\Delta}$ [2,2-dimethylcyclopentanol]

(b) [1-(1-bromoethyl)-1-methylcyclopropane] $+ H_2O \xrightarrow{\Delta}$ [cyclobutane product: OH, CH_3, H, CH_3] + [cyclobutane product: CH_3, H, OH, CH_3]

78. Explain the following observations:

(a) Azide ion (N_3^-) react with 2-bromopentane thousand times faster than with neopentyl bromide in a SN-2 reaction though former is a secondary halide while latter is primary.

(b) What will happen to the stereochemistry of product of the following reaction:

$$H-C(Br)(D)-CH_3 + N_3^- \xrightarrow{SN\text{-}2}$$

(c) What will happen to the rate if the concentration of alkyl bromide in (b) is doubled?

(d) What will happen to the rate if the concentration of azide ion in (b) is doubled?

(e) How the sign of optical rotation of reactant and product are related in (b)?

(f) When allowed to stand in dilute H_2SO_4, levo rotatory 2-butanol slowly loses optical activity.

79. Rank the alcohols in the following in the increasing order of reactivity with HBr:

(a) *t*-butanol, 2-butanol, 1-butanol.

(b) *p*-methoxybenzyl alcohol, *p*-nitrobenzyl alcohol and benzyl alcohol.

(c) benzyl alcohol, *p*-methoxy phenol, α, α-dimethyl benzyl alcohol.

80. Assuming that the following reaction takes place by SN-2 mechanism, select the faster reaction in each pair and explain reason for your choice:

(a) Reaction of cyanide ion with *n*-iodoheptane or *n*-chloroheptane.

(b) Reaction of ethanol or C_2H_5ONa with *n*-butyl bromide.

(c) Reaction of azide ion (N_3^-) with *n*-butyl tosylate or *s*-butyl tosylate.

(d) Reaction of isopropoxide withyl ethyl bromide or reaction of ethoxide with 2-bromopropane.

81. For each of the following reaction, predict the expected product. Also indicate change in stereochemistry if any:

(a) [phenol (OH)] + [2-bromobutane (Br)] $\xrightarrow{Na_2CO_3}$

(b) [cyclohexane] $\xrightarrow{\text{(i) } Br_2/\text{heat; (ii) KCN; (iii) } H_2O/H^+}$

(c) [phthalimide, N–H] $\xrightarrow{\text{(i) KOH; (ii) } CH_3CH_2CH_2CH_2Br\text{; (iii) } H_2NNH_2}$

(d) [2-butanol (OH)] $\xrightarrow{\text{(i) TsCl; (ii) NaCN}}$

(e) [H_3C–CH(NC)(H)–CH(I)(H)–CH_3] $\xrightarrow{\text{(ii) NaCN}}$

(f) [2-butanol (OH)] $\xrightarrow{\text{(i) } PCl_3\text{; (ii) } CH_3CH_2ONa}$

(g) [benzyl alcohol (OH)] $\xrightarrow{\text{(i) } PBr_3\text{; (ii) NaI}}$

(h) [2-phenylethanethiol (SH)] $\xrightarrow{\text{(i) Na; (ii) } CH_3Br}$

(i) [phenylacetylene] $\xrightarrow{\text{(i) Na; (ii) } H_2O}$

82. Synthesize the following compounds from indicated starting material:

(a) [1,4-dimethoxybenzene: OMe, OMe] from 1, 4-dihydroxy benzene

(b) [benzylamine: NH_2] from benzyl alcohol

(c) [phenyl propyl sulfide: S] from phenol

(d) [tetrahydrofuran: O] from [4-chloro-1-butene: Cl]

83. What reagents are required to accomplish the following transformations:

(a) [cyclohexyl–OH] ⟶ [cyclohexyl–NH_2]

(b) [benzyl alcohol (OH)] ⟶ [3-phenyl-1-propyne]

(c) 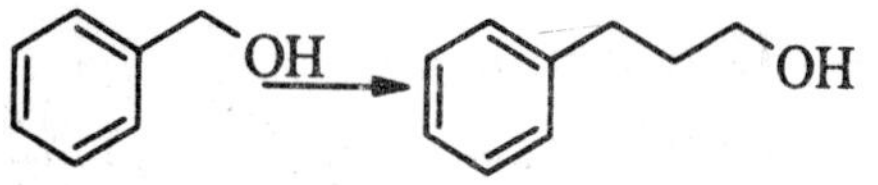 [structure] → [structure]

(d)  [structure] OH → [structure] $COOCH_3$

84. Write elimination product in the following reaction and compare their relative reactivity:

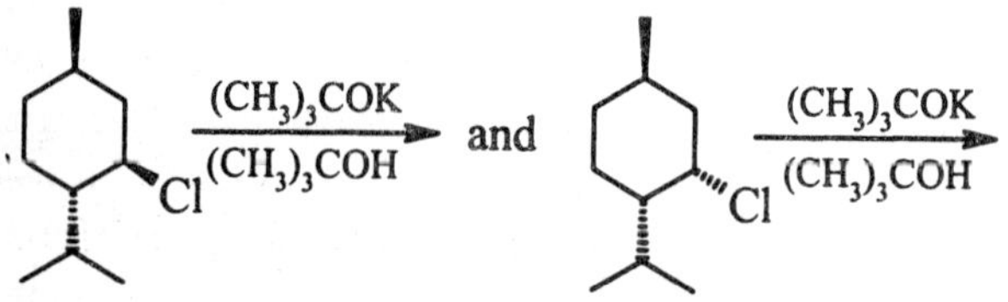

[structure] $\xrightarrow[(CH_3)_3COH]{(CH_3)_3COK}$ and [structure] $\xrightarrow[(CH_3)_3COH]{(CH_3)_3COK}$

85. Following reaction could occur by E2, SN-2, E1 or SN-1 mechanism. Predict major product in each case:

(a) $CH_3Br \xrightarrow[(ii)\ C_2H_5OH]{(i)\ KCN}$

(b) $CH_3CH_2OH \xrightarrow[Acetone]{KI}$

(c) 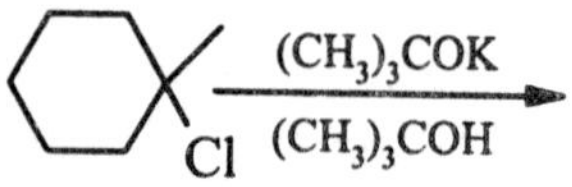 [structure] $\xrightarrow[(CH_3)_3COH]{(CH_3)_3COK}$

(d) 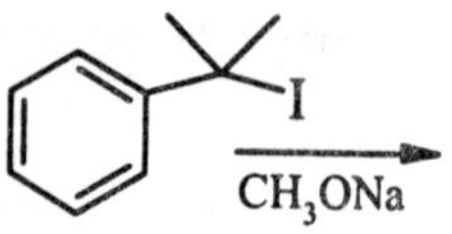 [structure] $\xrightarrow{CH_3ONa}$

(e) 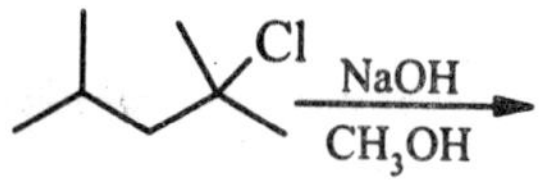 [structure] $\xrightarrow[CH_3OH]{NaOH}$

86. Compare the following compounds for their reactivity in SN-1 reaction:

(a) 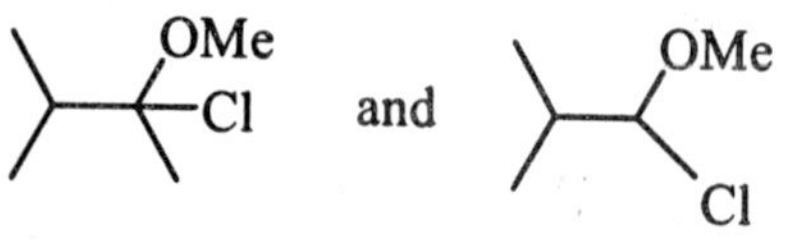 [structure] and [structure]

(b) 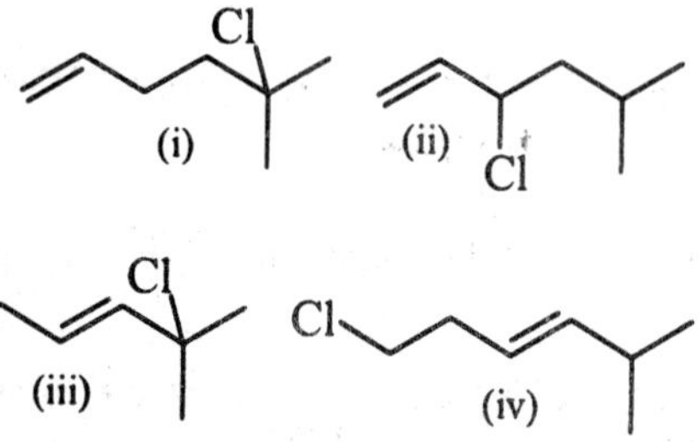 (i) (ii) (iii) (iv)

(c) 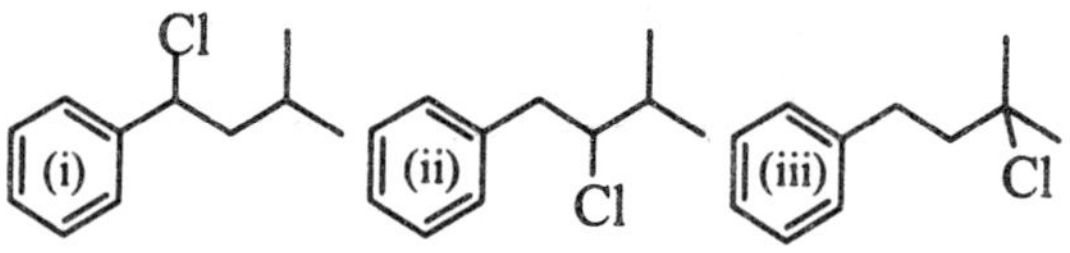 (i) (ii) (iii)

(d) 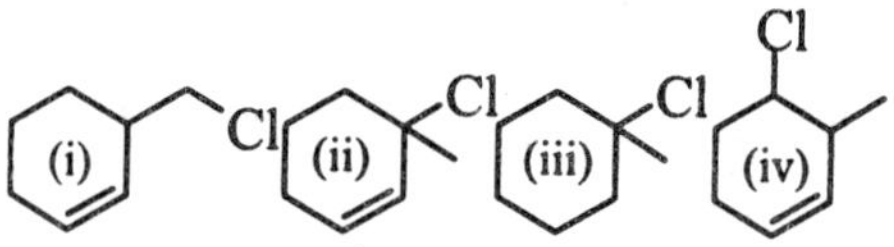 (i) (ii) (iii) (iv)

(e) 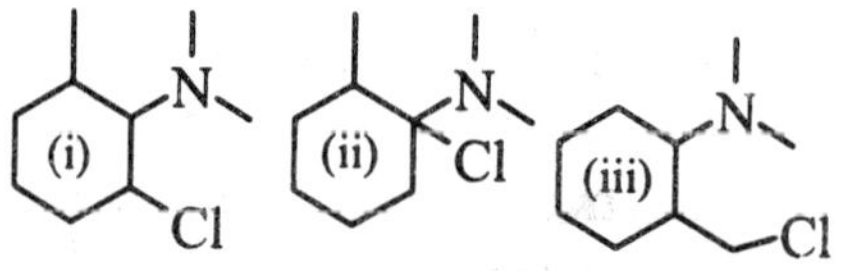 (i) (ii) (iii)

87. Propose mechanism:

[structure] $\xrightarrow[CH_3OH]{NaCl}$ [structure] + [structure] + [structure] + [structure]

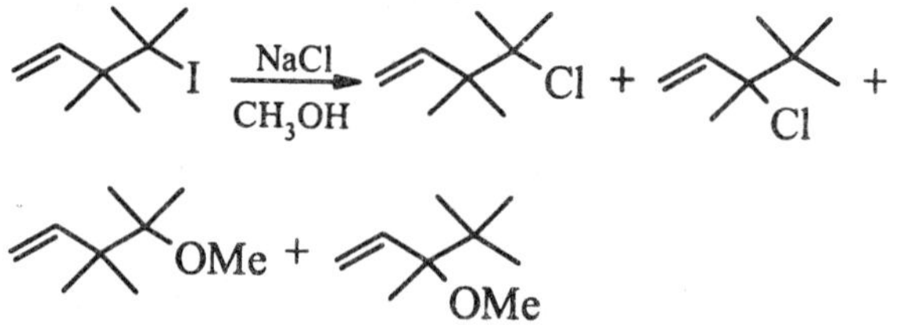

88. Explain why allyl bromide ($CH_2{=}CHCH_2Br$) undergoes rapid substitution under either SN-1 or SN-2 reaction condition.

89. For the following pair of reactions, predict the faster one and mechanism (SN-1 or SN-2), under which reaction will occur at faster rate:

(a) 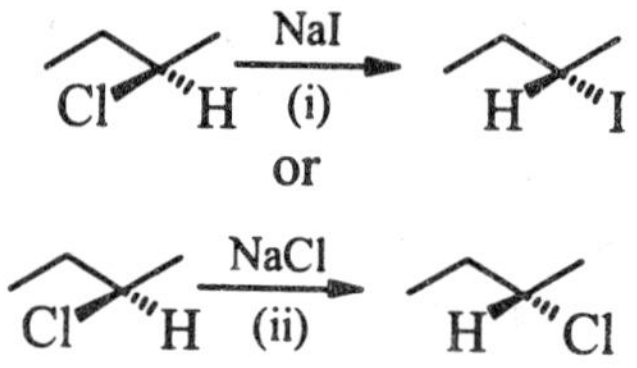 [structure] $\xrightarrow[(i)]{NaI}$ [structure]

or

[structure] $\xrightarrow[(ii)]{NaCl}$ [structure]

(b) 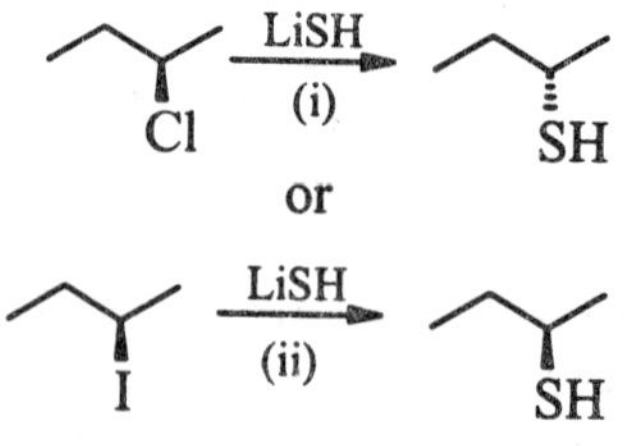 [structure] $\xrightarrow[(i)]{LiSH}$ [structure]

or

[structure] $\xrightarrow[(ii)]{LiSH}$ [structure]

(c) 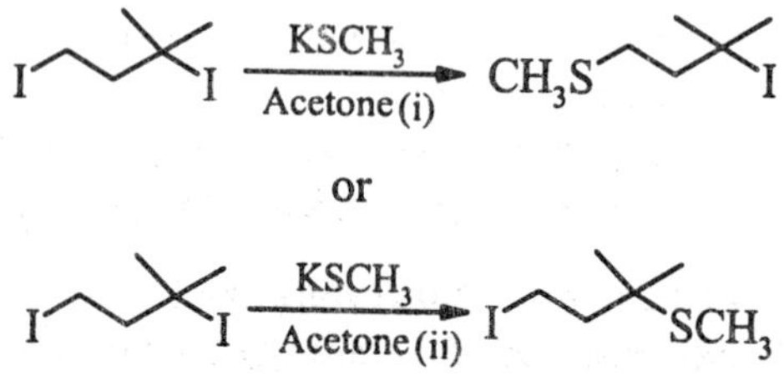 [structure] $\xrightarrow[Acetone\ (i)]{KSCH_3}$ [structure]

or

[structure] $\xrightarrow[Acetone\ (ii)]{KSCH_3}$ [structure]

(d) Cl $\xrightarrow[\text{Water (i)}]{\text{NaI}}$ I

or

Cl $\xrightarrow[\text{Acetone (ii)}]{\text{NaI}}$ I

90. Following reactions are not feasible as indicated. Briefly explain the reason for their failure:

(a) Br $\xrightarrow{\text{KCN}}$ Br, CN

(b) $^{+}$ ⟶ $\overset{+}{C}H_2$

(c) HO, $\overset{H}{\overset{|+}{O}}$–H, H_3C, CH_3 $\xrightarrow{-H^+}$ HO, OH, H_3C, CH_3

91. Provide mechanism:

(a) I, OH $\xrightarrow{OH^-}$ O

(b) HO Br $\xrightarrow{\text{KOH}}$ O

92. What is the most significant reason that transition state is the highest energy point between reactant and product in a SN-2 reaction?

93. Rank the followings in increasing order of their nucleophilicity in a polar aprotic solvent:

(a) $CH_3CH_2O^-$, $CH_3CH_2S^-$, $CH_3CS_2^-$

(b) I^-, Br^-, F^-

(c) HS^-, HSe^-, HO^-

(d) HS^-, HO^-, NH_2^-

(e) $CF_3CH_2S^-$, $CH_3CH_2S^-$, $CH_3CH_2O^-$, $CF_3CH_2O^-$

(f) $CF_3CO_2^-$, $CI_3CCO_2^-$, $CH_3CH_2CO_2^-$

(g) HO^-, CH_3COO^-, CH_3O^-, $CH_3C{\equiv}C^-$

94. Arrange the following in the increasing order of their ability as a leaving group:

(a) CH_3S^-, CH_3O^-, CF_3^- and F^-

(b) $CF_3SO_3^-$, $CH_3SO_3^-$ and CH_3COO^-

95. An organic compound A has molecular formula $C_6H_{11}Br$. A decolorizes brown colour of bromine water but it cannot be resolved into enantiomers. A on treatment with HBr produces $B(C_6H_{12}Br_2)$, which is still non-resolvable. Also A on treatment with HBr in presence of a peroxide (R_2O_2) yields C an isomer of B, but it is resolvable. B on heating with Zn-metal yields D (C_6H_{12}), which on treatment with ozone followed by hydrolysis of product in presence of Zn-powder yield acetone as the only organic compound. Also A on treatment with ethanolic KOH solution yields E (C_6H_{10}) as only possible product. E on treatment with excess of HBr/R_2O_2 in cold produces isomeric F and G with their molecular formula $C_6H_{12}Br_2$, in which F is a meso compound while G is resolvable. Also F on hydrolysis in aqueous NaOH yield a diol $C_6H_{14}O_2$ which on refluxing with dilute H_2SO_4 yields the following compound. Deduce structures of A to G.

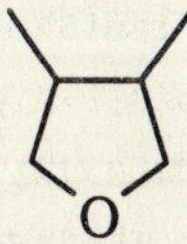

96. An organic compound is optically active and has molecular formula $C_{14}H_{23}Br$. A does not decolourise cold, dilute and alkaline solution of $KMnO_4$. A on treatment with potassium tertiary butoxide in tertiary butanol yields $B(C_{14}H_{22})$ as the major product. B on treatment with acidified permanganate solution yields $C(C_{14}H_{22}O_4)$. C on heating with soda lime yields $D(C_{12}H_{22})$, which can also be formed by dissolving cyclohexyl bromide in etheral solution of Mg. Identify A to D.

97. An optically active organic compound A has molecular formula $C_7H_{16}O$. A neither changes the colour of bromine water nor evolve any gas on reacting with MeMgBr. A on refluxing with dilute H_2SO_4 produces $B(C_3H_8O)$ and $C(C_4H_{10}O)$. B is non-resolvable and gives yellow precipitate with $NaOH/I_2$ while C is resolvable. C on heating with concentrated solution of H_2SO_4 yield stereoisomeric compounds D and E with their molecular formula C_4H_8. D on treatment with Br_2/CCl_4 yields a meso dibromide while E on similar treatment yields racemic mixture of products. Identify A to E.

98. An organic compound $A(C_5H_{11}Cl)$ is optically active and on treatment with ethanolic KOH solution yields $B(C_5H_{10})$ as major product, which does not show stereoisomerism. Also A on treatment with $(CH_3)_2Cu$ yields $C(C_6H_{14})$, which is optically inactive. Deduce structures of A to C.

99. An organic compound A has molecular formula $C_{10}H_{17}Br$ and it is non-resolvable. A does not decolourise brown colour of bromine water solution. A on treatment with $(CH_3)_3COK/(CH_3)_3COH$ yields $B(C_{10}H_{16})$ as major product. B on treatment with H_2/Pt yields $C_{10}H_{18}$, which on treatment with Cl_2/hv yields three monochloro derivative. Also B on boiling with acidic permanganate solution yields $C(C_{10}H_{16}O_3)$. C on heating with soda lime yields $D(C_9H_{16}O)$. D on reducing with $LiAlH_4$ followed by heating the product with concentrated H_2SO_4 yields $E(C_9H_{16})$ as major product. E on treatment with ozone followed by work-up with Zn-H_2O yields 6-keto nonanal. Deduce structures of A to E.

100. An organic compound A has molecular formula $C_5H_{10}O$ and it neither decolourise bromine water solution nor evolve any gas on heating with Na-metal. A on refluxing with aqueous HI yields $B(C_5H_{10}I_2)$, which is non-resolvable. B on treatment with NaCN in acetone yield $C(C_6H_{10}IN)$ in significant amount while B on treatment with aqueous Ag_2O yields $D(C_5H_{11}IO)$ in significant amount. D does not change the orange colour of acidic dichromate solution. Deduce structures of A to D.

101. An optically active organic compound A has molecular formula $C_{11}H_{16}O$ and it gives off a colourless gas on treatment with CH_3MgBr. A on heating with concentrated H_2SO_4 yields $B(C_{11}H_{14})$, which does not show stereoisomerism. B on treatment with HBr in absence of a peroxide yields $C(C_{11}H_{15}Br)$, which is resolvable while addition of HBr on B in presence of a peroxide yields D, an isomer of C but it is non-resolvable. Also B on treatment with ozone followed by work-up with Zn-H_2O yields C_8H_8O and C_3H_6O, in which former can also be obtained by Friedal Craft's acylation reaction of benzene. Both these ozonolysis products gives yellow precipitate on treatment with $NaOH/I_2$. Deduce structures of A to D.

102. Write product in the following SN-2 reaction:

(a) (cyclohexyl)–CH₂CH₂CH₂–OH $\xrightarrow{TsCl}$ $\xrightarrow{NaN_3}$

(b) C_6H_5–CH_2CH_2Cl $\xrightarrow{NaCN}$

(c) ~~~Br $\xrightarrow{C_2H_5SNa}$

(d) ~~~Cl $\xrightarrow{\text{~~~}\equiv \bar{C}\,\overset{+}{Na}}$

(e) (isobutyl)–Br $\xrightarrow{NH_3(\text{excess})}$

(f) ~~I $\xrightarrow{NaCH(COOCH_2CH_3)_2}$

(g) HO~~Cl $\xrightarrow{CH_3SNa}$

103. Recognize products in the following substitution reaction indicating their stereochemistry:

(a) $CH_2CH_2OCH_3$, H, H_3C, Br–C $\xrightarrow{CH_3SNa}$ (b) OTs, O=, OMe, ~~OH $\xrightarrow{NaN_3}$

(c) O, O, I $\xrightarrow{(CH_3)_3COK}$ (d) O, O, ~~OH $\xrightarrow[NaI]{TsCl}$

104. Bring about the following transformations:

(a) $C_6H_5CH_2OH \rightarrow C_6H_5CH_2SCH_2CH_3$

(b) *n*-pentanol → *n*-pentyl iodide

(c) $C_6H_5SH \rightarrow C_6H_5SCH_2CH_2C_6H_5$

(d) *n*-butyl bromide → 2-heptyne

(e) HO–C(CH_3)(H)(H_5C_2) → H–C(CH_3)(NH_2)(H_5C_2)

(f) (ester)~O~~Cl → (ester)~O~~N(H)–$CHCH_2OH$ / $CH{=}CH_2$

105. Explain the formation of products in the following reaction:

(a) $R–X + SCN^- \rightarrow RSCN + X^-$

(b) $R–X + OCN^- \rightarrow R–NCO + X^-$

OBJECTIVE EXERCISE

Level 1

1. Which one of the following statements describes the order of reaction of the bimolecular nucleophilic substitution?
   (a) The SN-2 reaction is second-order in alkyl halide.
   (b) The SN-2 reaction is second-order in nucleophile.
   (c) The SN-2 reaction is zero-order in nucleophile, but second-order in alkyl halide.
   (d) The SN-2 reaction is first-order in nucleophile and first-order in alkyl halide.

2. In an *aprotic* solvent, which relative ordering best describes the nucleophilicity of the halide ions?
   (a) iodide > bromide > chloride >> fluoride
   (b) iodide > bromide > chloride > fluoride
   (c) iodide < bromide < chloride < fluoride
   (d) iodide < bromide > chloride > fluoride

3. Which one of the following nucleophiles shows the highest nucleophilicity toward methyl iodide in methanol?
   (a) I^- (b) HS^-
   (c) CN^- (d) F^-

4. Consider the reaction coordinate diagram for an SN-2 reaction, that is, for a reaction, in which the charges are more distributed in the transition state than in the reactants. How does a change to a more polar solvent affect this reaction coordinate diagram?
   (a) Energy of reactants is raised, and transition state energy remains about the same.
   (b) Energies of reagents and transition state are lowered by about the same amount.
   (c) Energy of reactants are lowered slightly, and energy of transition state structure are lowered significantly more.
   (d) Energy of reactants is lowered, and energy of transition state structure is lowered significantly less.

5. The SN-2 reaction involves back-side attack and therefore results in a "Walden Inversion." For which one of the substrates shown would you be able to demonstrate that such back-side attack with "Walden Inversion" has in fact occurred?
   (a) methyl bromide (b) 1-bromopropane
   (c) 2-bromobutane (d) 2-bromopropane

6. **Which step is the rate-limiting slow step of the hydrolysis of *tert*-butyl bromide?**
   (a) Deprotonation of the addition product formed by addition of water to carbenium ion.
   (b) Unimolecular dissociation into carbenium ion and bromide.
   (c) Addition of water to carbenium ion.

7. When DMSO solvates a potassium cation, which atom(s) of DMSO is (are) coordinated to the cation?
   (a) The O-atom of the S═O bond.
   (b) One H-atom from each of the two methyl groups.
   (c) One H-atom of one of the methyl groups.

8. Which statement best explains the partial racemization seen in some cases of SN-1 reactions?
   (a) The C—X bond of the alkyl halide dissociates completely, and the dissociated and solvated carbenium ion does not become entirely planar, but retains its configurational information in part.
   (b) The alkyl halide reacts as an intimate ion pair or a solvent-separated ion pair, and the carbenium ion never really becomes fully planar; therefore, it retains its stereochemical information content.
   (c) The alkyl halide reacts as an intimate ion pair or a solvent-separated ion pair, and one face of the planar carbenium ion is covered up by the leaving group.
   (d) The partial racemization must be the result of contributions by the SN-2 reaction path.

9. Which one of the solvents listed is the solvent with the lowest dielectric constant?
   (a) hexane (b) methanol
   (c) acetone (d) diethyl ether

10. What is the best term to describe the rearrangement that causes the conversion of 3-bromo-2, 2-dimethylbutane to 2, 3-dimethyl-2-butanol?
    (a) 1, 2-methyl shift (b) 1, 3-hydride shift
    (c) 1, 3-methyl shift (d) 1, 2-hydride shift

11. Which statement properly describes the difference between the reaction coordinate diagrams of the SN-1 and SN-2 reactions?
    (a) The diagram for the SN-1 reaction contains two more transition states and one more intermediate compared to the SN-2 reaction.
    (b) The diagram for the SN-2 reaction contains two more transition states and one more intermediate compared to the SN-1 reaction.

(c) The diagram for the SN-2 reaction contains one more transition state and one more intermediate compared to the SN-1 reaction.
(d) The diagram for the SN-1 reaction contains one more transition state and one more intermediate compared to the SN-2 reaction.

12. For which one of the following substrates can you affect the stereochemical outcome by a simple change in the concentration of the nucleophile?
(a) Ethyl bromide reacts with methanol.
(b) 2-bromopropane reacts with methanol.
(c) 2-bromobutane reacts with methanol.
(d) 3-methyl-3-bromooctane reacts with methanol.

13. What does "secondary benzylic" mean when used to describe a carbenium ion?
(a) Ph—C(+)RR′ (b) Ph—HC(+)—R
(c) R—HC(+)R′ (d) Ph—C(+)H_2

14. What is the rate equation for the E1 reaction?
(a) Rate = k[alkyl halide][base]^2
(b) Rate = k[alkyl halide]^2[base]
(c) Rate = k[alkyl halide]
(d) Rate = k[alkyl halide][base]

15. How many products can be formed by the E1 reaction of 1-chloro-2, 2-diphenylpropane?
(a) 2 (b) 1 (c) 3 (d) 4

16. One classifies E2 reaction as, β-, 1, 2- ... eliminations depending on the relative positions of the two fragments that are being eliminated (e.g. proton H^+ and a leaving group Y^-). What is a γ-elimination?
(a) 1, 3-elimination (b) 1, 2-elimination
(c) 1, 1-elimination (d) 1, 4-elimination

17. Among the alkyl halides, which one will be *the least* reactive in the E2 reaction?
(a) H_3C—C(CH_3)Br—CH_2—CH_3
(b) Br—H_2C—CH_3
(c) Br—H_2C—CH_2—CH_2—CH_3
(d) H_3C—C(CH_3)Br—CH(CH_3)—CH_3

18. The E2 reaction of 1-bromo-1-methylcyclohexane with methoxide may result in methylenecyclohexane and 1-methylcyclohexene. Which one of the following statements correctly describes the regiochemistry?
(a) Only methylenecyclohexane is formed.
(b) Only methylcyclohexene is formed.
(c) Both are formed, and methylcyclohexene is the major product.
(d) Both are formed, and methylenecyclohexane is the major product.

19. Consider 2-bromo-3-deuterobicyclo[2.2.1]heptane. Assume that the bromine is on the same side as the "CH_2 bridge" of the 6-ring and D is on the other side. Will elimination remove HBr or DBr and why?
(a) HBr will be removed, because they are *anti*-coplanar.
(b) HBr will be removed, because of a kinetic isotope effect.
(c) DBr will be removed, because they are *syn*-coplanar.
(d) DBr will be removed, because they are *anti*-coplanar.

20. Which product is formed in the E1 reaction of 1-bromo-2, 3-dimethylcyclohexane?
(a) 3, 4-dimethylcyclohexene
(b) 2, 3-dimethylcyclohexene
(c) 4, 5-dimethylcyclohexene
(d) 1, 2-dimethylcyclohexene

21. Which alkene is formed as the major product in the E2 reaction of (1*S*,2*S*,6*R*)-1-bromo-2-methyl-6-deuterocyclohexane?
(a) (*S*)-3-methylcyclohex-(1)-ene
(b) 1-deutero-3-methylcyclohex-(1)-ene(c) 1-methyl-3-deuterocyclohex-(1)ene

22. What product(s) is (are) formed in the debromination of meso-2, 3-dibromobutane?
(a) *cis*- and *trans*-2-butene in equal amounts
(b) more *cis*- than *trans*-2-butene
(c) only *cis*-2-butene
(d) only *trans*-2-butene

23. Which statement best characterizes the polarity of a C—Cl bond?
(a) The bond is ionic and polar with chlorine carrying essentially a full positive charge.
(b) The bond is ionic and polar with chlorine carrying essentially a full negative charge.
(c) The bond is covalent and polar with chlorine carrying a partial positive charge.
(d) The bond is covalent and polar with chlorine carrying a partial negative charge.

24. Which one of the following items does not affect the mechanism of nucleophilic substitution in a significant fashion?
(a) Concentration of the nucleophile.
(b) Solvent polarity.
(c) Reactivity of the nucleophile.
(d) Structure of the alkyl halide.

25. Which rate law describes unimolecular nucleophilic substitution reactions?
(a) Rate = k[alkyl halide]
(b) Rate = k[alkyl halide]^2[nucleophile]
(c) Rate = k[alkyl halide]^2
(d) Rate = k[alkyl halide][nucleophile]

26. What nucleophile is needed to prepare a nitrile by way of an SN-2 reaction?
(a) RO^- (b) RCa═C^-
(c) CN^- (d) RS^-

27. Why do polar aprotic solvents increase the rates of SN-2 reactions?
(a) Aprotic solvents always have low polarity.
(b) Aprotic solvents raise the energy of the products and therefore accelerate the reaction.
(c) Aprotic solvents better solvate the leaving group and therefore accelerate the reaction.
(d) Aprotic solvents solvate the nucleophile less, and "naked" nucleophiles are more reactive.

28. Among the following solvents, which one is protic?
(a) acetic acid
(b) ethylacetate
(c) dimethyl sulfoxide, DMSO
(d) N, N-dimethylformamide, DMF

29. In a protic solvent, which relative ordering best describes the nucleophilicity of the halide ions?
(a) iodide < bromide < chloride < fluoride
(b) iodide < bromide > chloride > fluoride
(c) iodide > bromide > chloride > fluoride
(d) iodide < bromide < chloride > fluoride

30. Why is acetone the best solvent for the reaction between alkyl chloride and potassium iodide? [Think of the Le Châtelier principle].
(a) Alkyl iodide is insoluble in acetone.
(b) Alkyl chloride is less soluble in acetone.
(c) NaCl has a higher dipole moment than KI.
(d) NaCl formed is not soluble in acetone.

31. Which type of substrate shows the highest reactivity in SN-2 reactions?
(a) Methyl halide
(b) Seconday alkyl halide
(c) Quaternary alkyl halide
(d) Tertiary alkyl halide

32. Which ordering correctly describes the relative leaving group abilities of the halide ions?
(a) iodide < bromide < chloride < fluoride
(b) iodide < bromide < chloride > fluoride
(c) iodide > bromide > chloride > fluoride
(d) iodide < bromide > chloride > fluoride

33. Which statement best describes the mechanism of the SN-2 reaction?
(a) Back-side attack with inversion of configuration.
(b) Front-side attack with inversion of configuration.
(c) Back-side attack with retention of configuration.
(d) Front- and back-side attack with racemization.

34. Under SN-2 reaction conditions, what product will be formed in the nucleophilic substitution of bromine by methoxide in *cis*-1-bromo-4-methylcyclohexane?
(a) Equatorial, equatorial *trans*-4-methylcyclohexanol.
(b) Equatorial, equatorial *cis*-4-methylcyclohexanol.
(c) Axial, axial *trans*-4-methylcyclohexanol.
(d) Axial, equatorial *cis*-4-methylcyclohexanol.

35. Which step is the rate-limiting slow step of the hydrolysis of *tert*-butyl bromide?
(a) Addition of water to carbenium ion.
(b) Deprotonation of the addition product formed by addition of water to carbenium ion.
(c) Unimolecular dissociation into carbenium ion and bromide.

36. Which one of the bromides listed undergoes SN-1 reaction with the highest rate of reaction?
(a) 2-bromopropane
(b) 2-bromo-2-methylpropane
(c) 1-bromopropane
(d) 1-bromo-2-methylpropane

37. Which statement properly describes the reaction coordinate diagram of the hydrolysis of *tert*-butyl bromide?
(a) 1 transition state, 1 intermediate.
(b) 2 transition states, 2 intermediates.
(c) 1 transition state, no intermediate.
(d) 2 transition states, 1 intermediate.

38. Suppose the double-bonded methylene group of 1-bromo-2-propene is labeled with ^{13}C, $H_2{}^{13}C$═CH—CH_2—Br. What product will be formed by methanolysis of this substrate?
(a) $H_2{}^{13}C$═CH—CH_2—OCH_3 and H_2C═CH—$^{13}CH_2$—OCH_3
(b) H_2C═^{13}CH—CH_2—OCH_3
(c) $H_2{}^{13}C$═CH—CH_2—OCH_3
(d) H_2C═CH—$^{13}CH_2$—OCH_3

39. Consider the reaction coordinate diagram for an SN-1 reaction, that is, for a reaction in which the transition state is more highly charged than the reactants. How does a change to a more polar solvent affect this reaction coordinate diagram?

(a) Energy of reactants is lowered slightly, and energy of transition state structure is lowered significantly more.
(b) Energies of reagents and transition state are lowered by about the same amount.
(c) Energy of reactants is raised, and energy of transition state structure is lowered significantly.
(d) Energy of reactants is raised, and transition state energy remains about the same.

40. Why do some SN-1 reactions produce a bit more of the inverted product than of the product with retention?

(a) Reaction occurs in the intimate ion pair, and the leaving groups shields one side a bit while the other side is open for attack.
(b) The solvent-separated ion pair would give extra product with retention.
(c) Reaction occurs by way of a Walden inversion, and the product formed with retention comes from racemization of the product after the SN reaction.
(d) Reaction occurs for the free carbenium ion.

41. For substrates that can react by SN-1 and SN-2, there exists a simple way to obtain more SN-2 reaction product. How is this accomplished?

(a) Irradiate the reaction with UV light.
(b) Use higher nucleophile concentration.
(c) Use more polar solvent.
(d) Use sonication (ultrasound mixing).

Level 2

Following three questions are based on the diagram shown below:

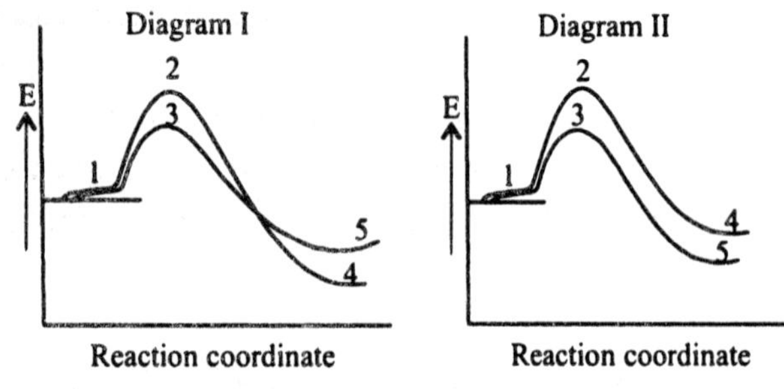

1. Choose the energy profile that corresponds to the following reaction:

Br + $^{-}$O ⟶ Minor product

(a) Diagram I, 1 → 5 (b) Diagram I, 1 → 4
(c) Diagram II, 1→ 5 (d) Diagram II, 1→ 4

2. Choose the energy profile that corresponds to the reaction.

Br + $^{-}$O ⟶ Major product

(a) Diagram I, 1 → 5 (b) Diagram I, 1 → 4
(c) Diagram II, 1 → 5 (d) Diagram II, 1 → 4

3. The structure of position 2 in diagram I is:

(a) (b) (c) (d)

4. Consider the following reaction:

—Cl + I^{-} $\xrightarrow{\text{SN-1}}$ —I + Cl^{-}

If the concentration of alkyl chloride is doubled, the rate of formation of alkyl iodide would be:

(a) Doubled
(b) Increases by a factor of 4
(c) Remain same
(d) Decreases to half of the initial rate

5. Give the major product of the following reaction:

Br, H, Et, Et, Me, Me + $^{-}$O ⟶ Major product

(a) Et, Et / Me, Me (b) Me, Et / Et, Me

(c) H_2C, H, Et, Me, Me (d) Me, HC, H, Et, Me, Me

6. Which of the following alkyl halide would you expect to undergo SN-1 most rapidly?
   (a) *n*-pentyl bromide
   (b) 2-bromohexane
   (c) 2-bromo-2-methylbutane
   (d) All react at the same rate

7. Which of the following will give only one alkene as product in E2 reaction?

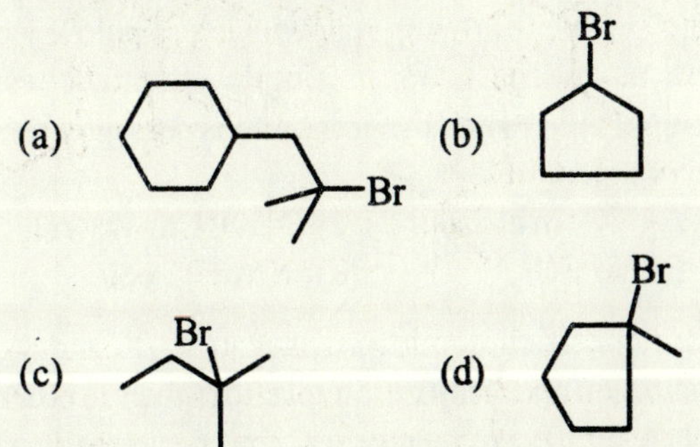

8. Which of the following statements is true about the reaction shown?

$$H-\underset{Me}{\overset{CH_2CH_2Cl}{C}}-OMe \xrightarrow{OH^-}$$

   (a) The product will have same configuration.
   (b) The product will have opposite configuration.
   (c) A meso compound is produced.
   (d) The reaction will happen with racemization.

9. Which of the following gives no reaction?
   (a) $CH_3^- + C_6H_5NH_2 \rightarrow$ (b) $CH_3CH_2^- + CH_3OH \rightarrow$
   (c) $OH^- + CH_4 \rightarrow$ (d) $OH^- + CH_3COOH \rightarrow$

10. Which of the following is the best synthesis of 2-ethoxybutane?
   (a) 2-bromobutane + sodium ethoxide.
   (b) 2-butanol + sodium ethoxide.
   (c) Sodium salt of 2-butanol + bromoethane.
   (d) Sodium salt of 2-butanol + ethanol.

11. Consider the rate equation of the following reaction. How would the rate be effected, if the concentration of KCN were doubled?

Chloropropane + KCN $\rightarrow$ Butylnitrile
   (a) No effect
   (b) The rate would be doubled
   (c) The rate would be halved
   (d) It depends on the concentration of starting material.

12. Which of the following is most nucleophilic?

(a) N: (triethylamine) (b) N: (pyrrolidine anion) (c) N: (pyrrole anion) (d) EtOH

13. Which reaction in the following pair will show the faster rate of disappearance of substrate?
   (a) *t*-butyl bromide + $EtOH/H_2O \rightarrow$
   (b) *t*-butyl bromide + $EtOH \rightarrow$

14. Which reaction in the following pair will show the faster rate of disappearance of substrate?

(a) Cl $\xrightarrow{CH_3S^-}$ (b) Cl $\xrightarrow{(CH_3)_2CHS^-}$

15. Which reaction in the following pair will show the faster rate of disappearance of substrate?
   (a) Bromopropane + $CN^- \rightarrow$
   (b) Iodopropane + $CN^- \rightarrow$

16. Which reaction in the following pair will show the faster rate of disappearance of substrate?
   (a) Benzyl bromide + $H_2O \rightarrow$
   (b) Bromobenzene + $H_2O \rightarrow$

17. When 2-bromo-2, 3-dimethylbutane reacts with any of the bases shown below under E2 conditions, two alkenes are produced: 2, 3-dimethyl-1-butene and 2, 3-dimethyl-2-butene. Which one of the base would produce the highest yield of 2, 3-dimethyl-1-butene?

(a) CH_3CH_2ONa (b) ONa

(c) ONa (d) ONa

18. When 2-bromo-2, 3-dimethylbutane reacts with any of the bases shown below under E2 conditions, two alkenes are produced: 2, 3-dimethyl-1-butene and 2, 3-dimethyl-2-butene. Which one of the above base would produce the highest yield of 2, 3-dimethyl-2-butene?

(a) CH_3CH_2ONa (b) ONa

(c) ONa (d) ONa

19. What is the major product of the following reaction?

Br —NaOH→

(a) 2-methyl-1-butene
(b) 2-methyl-2-butene
(c) 2-hydroxy-2-methylbutane
(d) 2-hydroxy-3-methylbutane

20. What is the product of the following SN-2 reaction?

OTs O, Ph, O, Me —NaBr, SN-2→

(a) O, OMe; Me—H; Br—H; Ph
(b) O, OMe; Me—H; H—Br; Ph
(c) O, OMe; H—Me; Br—H; Ph
(d) O, OMe; H—Me; H—Br; Ph

21. Which will be the major product of the following reaction?

H, OTs, Ph, H_3C, Cl, H —NaOH→

(a) Ph, H, Cl (b) TsO, Ph, H_3C, H

(c) Ph, H_3C, Cl (d) H_3C, Ph, Cl

22. The following reaction will likely follow which mechanism?

Br —NaCl→

(a) E1 (b) E2 (c) SN-1 (d) SN-2

23. Which of the following reagents listed below would serve as the basis for a simple chemical test to distinguish between the compounds *diethylether* and *2-bromobutane*?

(a) $AgNO_3$ in ethanol (b) NaOH in water
(c) Br_2 in CCl_4 (d) $KMnO_4$, cold

24. Under specified conditions, substrate X undergoes substitution and elimination reactions to give products A-D. A and B are stereomers, but not enantiomers. C and D are enantiomers. A is not an isomer of C. Which of the following could be starting material X?

$$X + H_2O/\Delta \rightarrow A + B + C + D$$

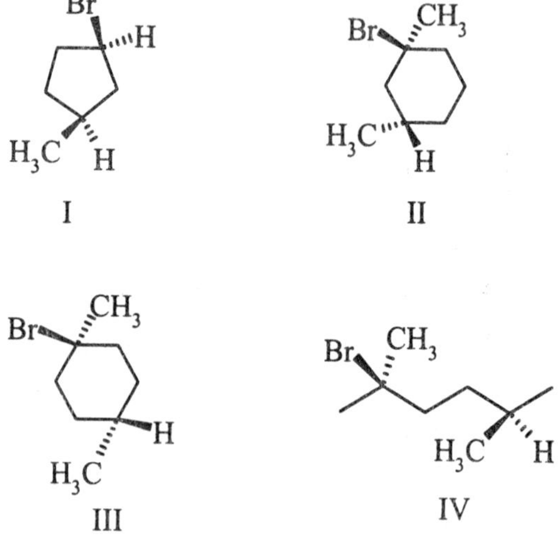

(a) I (b) II (c) III (d) IV

25. One can exchange ^{127}I with ^{128}I in some iodo compounds. When the reaction is run on an optically active 2-iodobutane, the rate of racemization is found to be exactly twice the initial rate of exchange (^{128}I for ^{127}I). Which reaction mechanism is most consistent with this observation?

Note: Optical rotation does not depends on the isotope present, i.e. optical rotation of 2-iodobutand will remain same whether it contain ^{127}I or ^{128}I.

(a) SN-1 (b) SN-2
(c) between SN-1 and SN-2 (d) E2

26. Addition of bromine to *cis*-2-hexene gives racemic products, even though attack of the bromide ion on the unsymmetrical bromonium ion intermediate is not equally likely at both carbons. The reason for the racemic products are:

(a) The reaction is reversible, after a while both enantiomers are produced in equal amounts.
(b) The product is a meso compound.
(c) The bromonium ion formed has a plane of symmetry.
(d) The two enantiomeric bromonium ions are formed in equal amount.

27. Reaction of HBr with 3-bromocyclohexene (A) yields only *trans*-1, 2-dibromocyclohexane. Reaction of 3-methyl cyclohexane (B) with HBr gives a mixture of *cis*- and *trans*-1-bromo-3-methyl cyclohexane and *cis*- and *trans*-1-bromo-2-methyl cyclohexene. Which statement explains the reactivity difference between A and B?

(a) There is no difference in the reactivity.
(b) The reaction of A and B have totally different mechanism, one is radical other is ionic.
(c) A bromonium ion formed in case of A can be attacked only in anti position.
(d) A has a chiral center and B has a plane of symmetry.

28. The best way to accomplish the following transformation is:

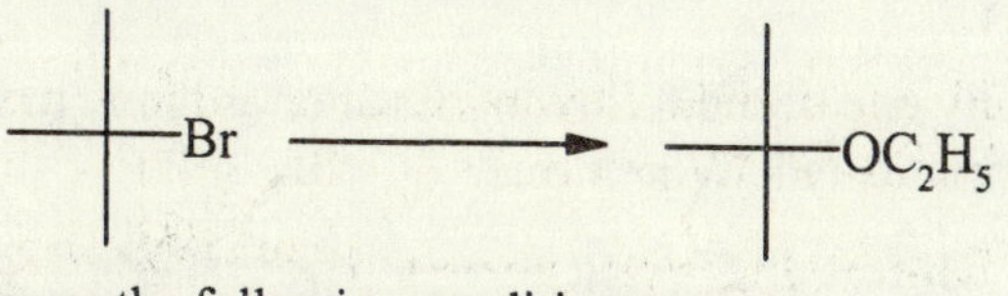

$$(CH_3)_3C\text{–}Br \longrightarrow (CH_3)_3C\text{–}OC_2H_5$$

to use the following conditions:

(a) C_2H_5ONa in ethanol (b) EtOH and heat
(c) $EtOH/H_2O$ (d) EtONa/DMSO

29. Questions 29-32 involve reactions of an alkyl bromide and two possible reagents. Which reaction in each pair will show the faster rate of disappearance of alkyl halide?

Bromopropane

(a) $(CH_3)_3B$ (b) $(CH_3)_3N$
(c) The reaction rate would be same

30. Methyl bromide:

(a) Sodium methoxide (b) CH_3S^-
(c) The reaction rate would be same

31. 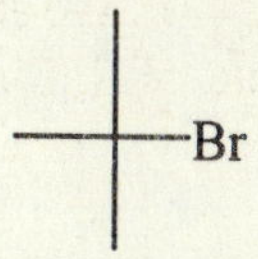 $(CH_3)_3C\text{–}Br$

(a) $EtOH/H_2O$ (b) EtOH
(c) The reaction rate would be same.

32. 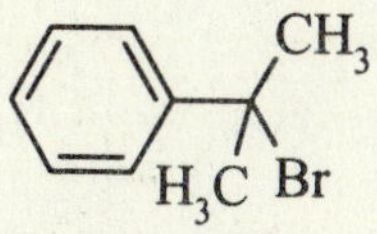 $C_6H_5C(CH_3)_2Br$ (CH₃, H₃C, Br on the benzylic carbon)

(a) CN^- (b) HS^-
(c) The reaction rate would be same

33. Tetrabutyl ammonium bromide (1) aids in the transformation of benzylbromide (2) into a nitrile (4) via phase transfer catalysis. Which of the following explanations best accounts for the role of (1) in this process?

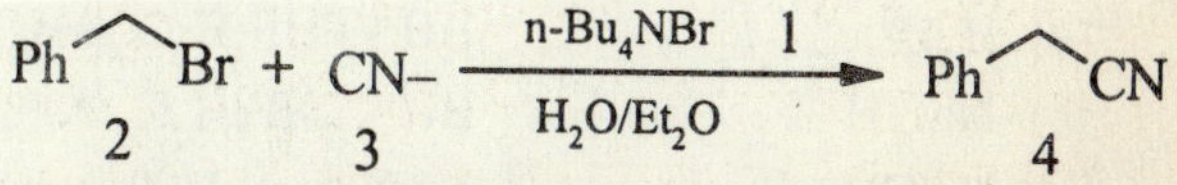

$$\underset{2}{PhCH_2Br} + \underset{3}{CN^-} \xrightarrow[H_2O/Et_2O]{n\text{-}Bu_4NBr\ \ 1} \underset{4}{PhCH_2CN}$$

(a) 1 transport the water insoluble benzyl bromide into aqueous layer, where it can react with cyanide.
(b) 1 makes water and Et_2O miscible.
(c) 1 transports the Et_2O insoluble cyanide into the organic layer, where it can react with benzyl bromide.
(d) 1 prevents the nucleophile water from competing with cyanide for the electrophilic benzyl bromide.

34. What is the mechanism of the following reaction?

(Reactant: $(H_3C)_2CH$–CH(H)–CH(Br)(H)(D) with $t\text{-}BuO^-K^+$ / $t\text{-}BuOH$ → alkene with Et, D, H_3C, H)

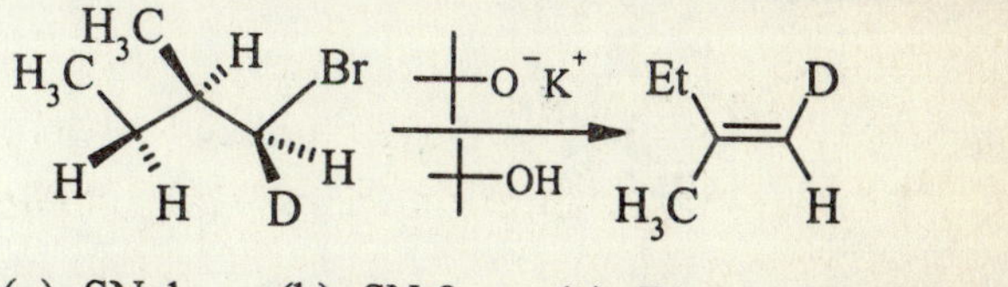

(a) SN-1 (b) SN-2 (c) E1 (d) E2

35. What happens to the reaction rate in the following reaction, if the concentration of *t*-butoxide is doubled?

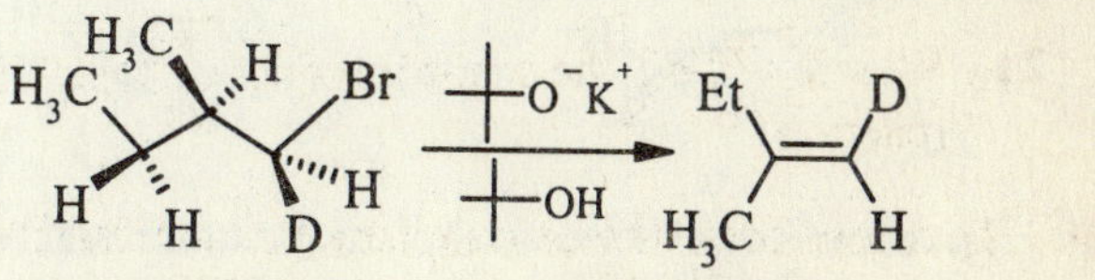

(Same reaction as in 34: $t\text{-}BuO^-K^+$ / $t\text{-}BuOH$ → alkene with Et, D, H_3C, H)

(a) No change (b) Doubled
(c) Halved (d) Tripled

36. What is the product of dehydrohalogenation of the following dibromide?

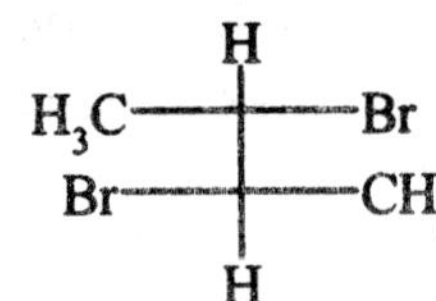

(a) H_3C, CH_3, Br, H (b) Br, CH_3, H_3C, H

(c) H_3C, Br, Br, CH_3 (d) H_3C, CH_3, Br, Br

37. Which of the reagent shown below would accomplish the following transformations?

Br ⟶ (A / B) OH

A	B
(a) H_3O^+	BH_3-THF;H_2O_2/NaOH
(b) NaOH	BH_3-THF;H_2O_2/NaOH
(c) $NaNH_2$	$Hg(OAc)_2/H_2O$;$NaBH_4$
(d) H_2/Pd/C	NBS/DMSO/H_2O

38. Treating 2, 2-dimethylpropanol with HCl produces:
(a) *n*-pentylchloride
(b) 2-methyl-1-chlorobutane
(c) 2-chloro-2-methylbutane
(d) 3-methyl-1-chlorobutane

39. The compound shown below undergoes a SN-2 reaction with hydroxide ion. What is the structure of product?

O, Br, H + $H\bar{O}$ ⟶

(a) HO, Br, H (b) O, Br, HO

(c) O, Br, H (d) O, H, HO

40. The competition between enolate alkylation and base-induced E2 elimination with ethyliodide electrophile is finely balanced. Amide enolate give a better ratio of alkylation-to-elimination compared with ketone enolate because:
(a) Ketone enolates are more nucleophilic, but both enolates are similarly basic.
(b) Amide enolates are more nucleophilic, but both enolates are similarly basic.
(c) Amide enolates are more basic, but both enolates are similarly nucleophilic.
(d) Ketone enolates are more basic, but both enolates are similarly nucleophilic.

41. Treatment of ammonia with excess of bromoethane in excess of a base will form:
(a) Ethanamine (b) Triethylamine
(c) $Et_4N^+Br^-$ (d) N^+ Br^-

42. Which of the following is most nucleophilic?
(a) Me_3N: (b) HO^-
(c) $Me_2P:^-$ (d) Me_2O

43. Which of the following SN-2 reaction would be fastest?

(a) H_3C—Br $\xrightarrow[H_3C\text{—}OH]{NaOH}$ H_3C—OH

(b) H_3C, H_3C —I $\xrightarrow[H_3C\text{—}C\equiv N]{NaOH}$ H_3C, H_3C —OH

(c) H_3C—I $\xrightarrow[H_3C\text{—}C\equiv N]{NaOH}$ H_3C—OH

(d) H_3C—I $\xrightarrow[H_3C\text{—}OH]{NaOH}$ H_3C—OH

44. Only one of the following reaction is likely to work as indicated. Which one?

(a) $^-OMe + H_3C\text{—}NR_2 \longrightarrow H_3C\text{—}OMe + {}^-NR_2$

(b) benzene $\xrightarrow{BrH}$ H, Br

(c) cyclohexene + Br_2 $\xrightarrow{hv}$ Br

(d) $\xrightarrow[CH_3I]{NaNH_2}$ $-CH_3$

45. Which of the following is not a nucleophile?
    (a) 2, 3-dimethyl-2-butene (b) CH_3OH
    (c) CH_3^- (d) Br_2

46. The synthesis of carboxylic acid (2) can best be accomplished from bromide (1) by:

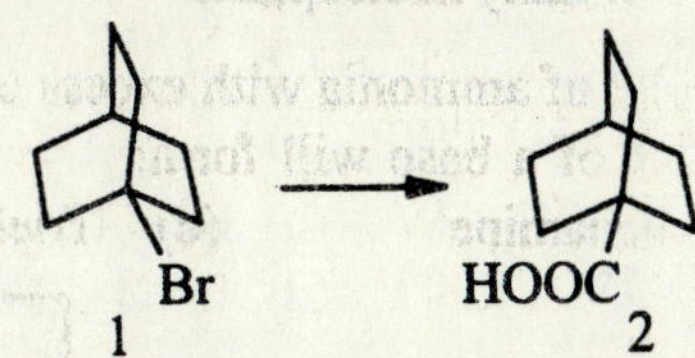

    (a) Grignard formation from (1) followed by CO_2 addition and acid quench.
    (b) Cyanide displacement of bromide in (1), followed by nitrile hydrolysis.
    (c) Grignard formation from (1) followed by cyanide addition and then nitrile hydrolysis.
    (d) Substitution of bromide in (1) with OH via SN-1, followed by CrO_3/H_3O^+ oxidation.

47. Which statement is correct for an SN-1 reaction at chiral carbon?
    (a) The product will be optically active, but have the opposite configuration.
    (b) The reaction will involve racemization.
    (c) The rate of reaction is a function of concentration of nucleophile.
    (d) The attacking group will be a strong electrophile.

48. Which of the following does not involve a carbocation?
    (a) $(CH_3)_3CCl + AgNO_3 \rightarrow$
    (b) $(CH_3)_3COH + HCl \rightarrow$
    (c) $(CH_3)_3CH + Cl_2 \rightarrow$
    (d) $CH_3CH{=}CHCH_2OH + H_3O^+ \rightarrow$

49. Sodium iodide in anhydrous acetone reacts most rapidly with:
    (a) Bromopropane (b) *Tertiary* butyl bromide
    (c) Vinyl bromide (d) 2-bromopropane

50. $CH_3CH{=}CHCH_2OH$ reacts with HBr to give which two bromobutenes?
    (a) $CH_3CH{=}CHCH_2Br + CH_3CH{=}CBrCH_3$
    (b) $CH_3CH{=}CHCH_2Br + CH_3CHBrCH{=}CH_2$
    (c) $BrCH_2CH_2CH{=}CH_2 + CH_2{=}CHBrCH{=}CH_2$
    (d) $BrCH_2CH{=}CHCH_3 + BrCH_2CH_2CH{=}CH_2$

51. A reaction at chiral carbon of an optically pure isomer which takes place exclusively by an SN-2 reaction, and in which priorities donot change, proceeds with:
    (a) Inversion and no racemization.
    (b) Inversion and some racemization.
    (c) Inversion and complete racemization.
    (d) Retention of configuration and some racemization.

52. Which compound would be the major product when $(CH_3)_2CHCHBrCH_2CH_3$ is treated with alcoholic KOH?
    (a) $(CH_3)_2CHCH_2CHOHCH_3$
    (b) $(CH_3)_2COCH_2CH_3$
    (c) $(CH_3)_2CHCH{=}CH_2$
    (d) $(CH_3)_2C{=}CHCH_2CH_3$

53. If tertiary butyl bromide and sodium amide ($NaNH_2$) react, the product would be?
    (a) *Tertiary* butyl amine
    (b) *Tertiary* butyl ammonium bromide
    (c) 2-methyl propene
    (d) A mixture of butyl amine

54. The rate of SN-2 reaction run in polar aprotic solvent relative to the same reaction in a polar protic solvent would be:
    (a) the same (b) slower
    (c) faster (d) unpredictable

55. Which compound will react most rapidly in a SN-1 reaction?
    (a) 2-chloropropene (b) 1-chloropropene
    (c) 2-bromopropene (d) 1-bromopropene

56. Which of the following would react most rapidly in an E2 reaction?
    (a) $CD_3CH_2CH_2Br$ (b) $CH_3CD_2CH_2Br$
    (c) CH_3CHDCH_2Br (d) $CH_3CD{=}CHBr$

57. The reason that vinyl chloride does not react with NaOH in a SN-2 reaction is that:
    (a) Hydroxide ion is a weak nucleophile.
    (b) The sp^2 C—Cl bond is stronger than a sp^3 C—Cl bond.
    (c) Steric hindrance at carbon bearing Cl.
    (d) Both b and c are correct.

58. The increasing order of reactivity of the following chlorides with aqueous KOH is:
    (i) Ethyl chloride (ii) Vinyl chloride
    (iii) Chlorobenzene (iv) $(CH_3)_2CHCH_2Cl$
    (a) i, ii, iii and iv (b) iii, ii, iv and i
    (c) iv, iii, ii and i (d) ii, iii, i and iv

59. The increasing order of solvolysis of the following halides is:

(i) iso propyl chloride (ii) Cyclopropyl chloride
(iii) Benzyl chloride (iv) Chlorobenzene
(a) i, ii, iii and iv (b) iv, iii, ii and i
(c) iv, ii, i and iii (d) iii, ii, i and iv

60. In the following sequence of reaction:

$$\text{(Cl-substituted alkane)} + (CH_3)_3COK/(CH_3)_3COH \longrightarrow X \xrightarrow[R_2O_2]{HBr} Y$$

Y is:
(a) 2-bromo-3-methylbutane
(b) 2-bromo-2-methylbutane
(c) 1-bromo-3-methylbutane
(d) 1-bromo-2-chloro-3-methylbutane

61. Which of the following are feasible reactions?
(i) $HC{\equiv}CH + CH_3Li \rightarrow HC{\equiv}CLi + CH_4$
(ii) $HC{\equiv}CH + NaOH \rightarrow HC{\equiv}CNa + H_2O$
(iii) $HC{\equiv}CNa + NH_3 \rightarrow HC{\equiv}CH + NaNH_2$
(iv) $H_2C{=}CH_2 + HC{\equiv}CNa \rightarrow H_2C{=}CHNa + HC{\equiv}CH$
(a) i, ii and iii (b) ii, iii and iv
(c) i, iii and iv (d) only i

62. Which of the following is inert towards SN-1 reaction?
(a) Ethyl chloride (b) Isopropyl chloride
(c) 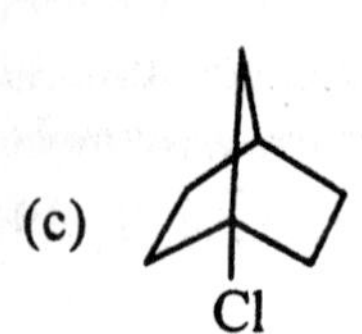

(d) 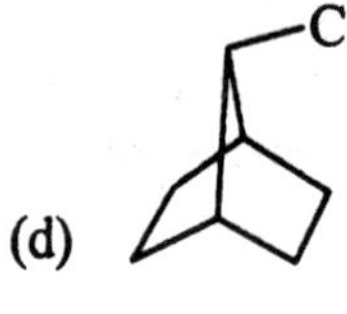

63. The reactivity of the following halides towards aqueous solution of KOH is:
CH_3CH_2Cl (I), $CH_2{=}CH{-}CH_2Cl$ (II), $CH_2{=}CHCl$ (III), C_6H_5Cl (IV)
(a) I < II < III < IV (b) I > II > III > IV
(c) II > I > III > IV (d) I < III < IV < II

64. Product of the following reaction:

(Cl, H, H, D substituted cyclohexane) $\xrightarrow[EtO^-]{EtOH}$ is

(a) (b) (c) (d) 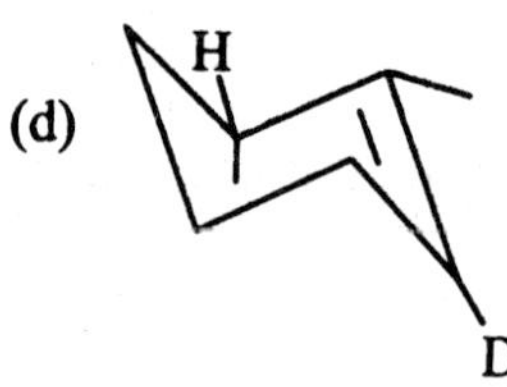

65. Arrange the following compounds in increasing order of their reactivity towards E2 elimination with $(CH_3)_3COK$ in *t*-butanol:
$PhCHClCH_3$ (I), $PhCH_2CH_2Cl$ (II), $CH_3COCH_2CH_2Cl$ (III) and $CH_3COCHClCH_3$ (IV)
(a) I < II < III < IV (b) III < II < IV < I
(c) III < IV < II < I (d) II < I < IV < III

66. The major product for the reaction:
$CH_3{-}CH_2{-}C(CH_3)_2Br + (CH_3)_3CONa \rightarrow$ would be
(a) $CH_3{-}CH_2{-}C(CH_3)_2{-}O{-}C(CH_3)_2{-}CH_3$
(b) $CH_3{-}CH{=}C(CH_3)_2$
(c) $CH_3{-}CH_2{-}C({=}CH_2){-}CH_3$
(d) $CH_2{=}CH{-}CH(CH_3)_2$

67. 5-bromo-2-pentanone $\xrightarrow[THF]{Mg} \xrightarrow{H_2O/H^+}$. The product of the above reaction is:

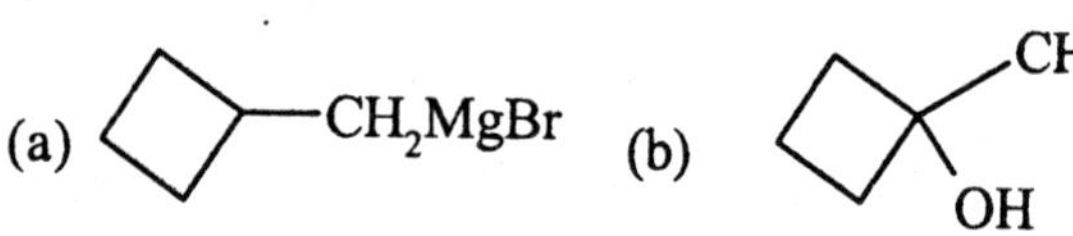

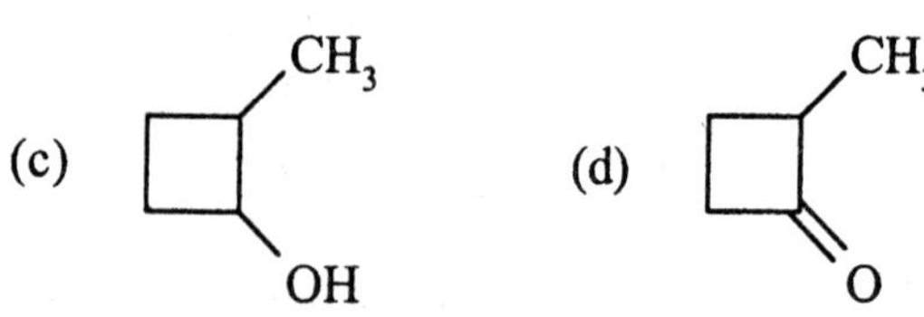

68. Arrange the following alkyl halides in the increasing order of their relative rates towards SN-2 reaction: 1-chloropropane (I), 2-chloropropane (II), Cyclopropylchloride (III) 2-methyl-1-chloropropane (IV)
(a) I < II < III < IV (b) I < III < II < IV
(c) II < III < IV < I (d) III < II < IV < I

Chapter 6

Alcohols, Ethers and Epoxides

Alcohols

6.1 INTRODUCTION

Alcohols are organic compound that contains one or more hydroxyl groups. They are classified according to the number of hydroxyl groups present as:

CH_3CH_2OH Ethanol, a monohydric alcohol.

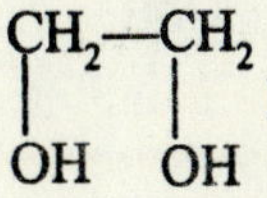

$CH_2(OH)—CH_2(OH)$ Ethan-1, 2-diol, a dihydric alcohol.

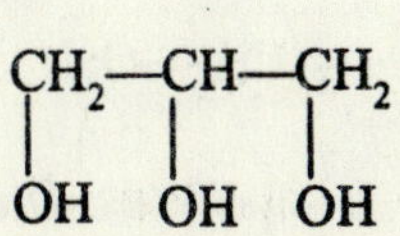

$CH_2(OH)—CH(OH)—CH_2(OH)$ Propan-1, 2, 3- triol (glycerol), a trihydric alcohol.

Alcohols are also classified as primary, secondary and tertiary alcohol as:

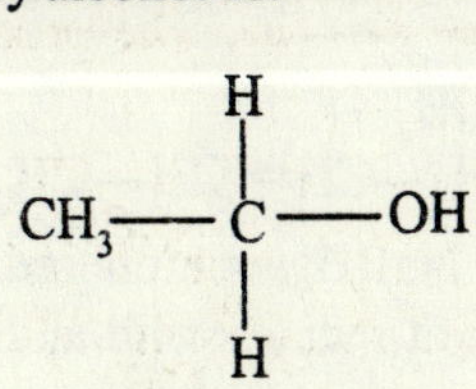

$CH_3—CH_2—OH$ (C bearing two H)

A primary (1°) alcohol

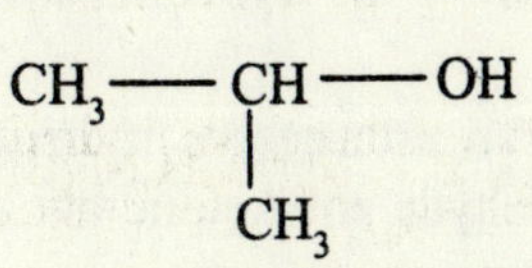

$CH_3—CH(CH_3)—OH$

A secondary (2°) alcohol

$CH_3—C(CH_3)_2—OH$

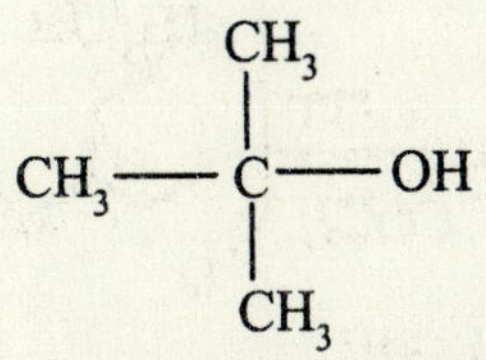

A tertiary (3°) alcohol

6.2 ALCOHOL NOMENCLATURE

In the IUPAC system of nomenclature, functional groups are normally designated in one of two ways. The presence of the functional group may be indicated by a characteristic suffix and a location number. This is common for the carbon-carbon double and triple bonds, which have the respective suffixes *ene* and *yne*. Halogens, on the other hand, do not have a suffix and are named as substituents, for example: $(CH_3)_2C{=}CHCHClCH_3$ is 4-chloro-2-methyl-2-pentene. Alcohols are usually named by the first procedure and are designated by an *ol* suffix, as in ethanol, CH_3CH_2OH (note that a locator number is not needed on a two-carbon chain). On longer chains the location of the hydroxyl group determines chain numbering. For example, $(CH_3)_2C{=}CHCH(OH)CH_3$ is 4-methyl-3-penten-2-ol. Other examples of IUPAC nomenclature are shown below, together with the common names often used for some of the simpler compounds. For the monofunctional alcohols, this common system consists of naming the **alkyl group** followed by the word **alcohol.**

	$H_3C—CH_2—CH_2—CH_2—OH$	$H_3C—CH_2—CH(OH)—CH_3$
IUPAC	1-butanol	2-butanol
common	(butyl alcohol)	(sec butyl alcohol)
class	a 1° alcohol	a 2° alcohol

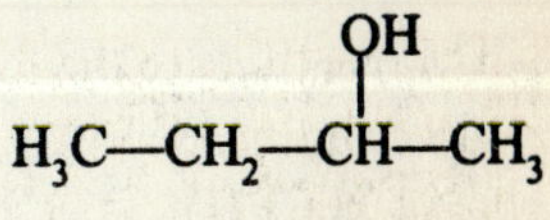

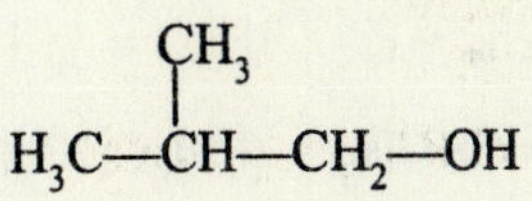

$H_3C—CH(CH_3)—CH_2—OH$	$H_3C—C(CH_3)_2—OH$
2-methyl-1-propanol	2-methyl-2-propanol
(iso-butyl alcohol)	(ter-butyl alcohol)
a 1° alcohol	a 3° alcohol

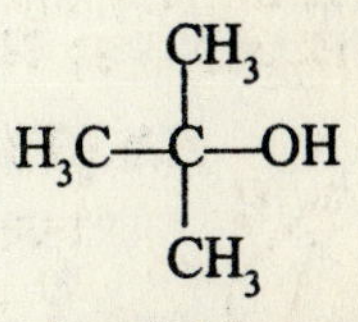

	$CH_2{=}CH—CH_2—OH$	cyclohexene ring with OH
IUPAC	2-propen-1-ol	3-cyclohexen-1-ol
Common	(allyl alcohol)	
class		

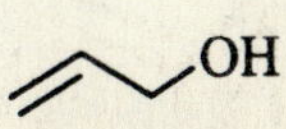

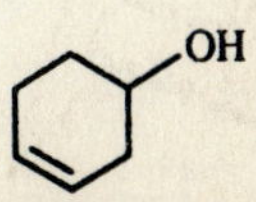

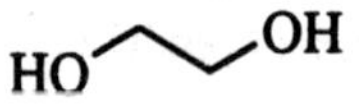

HO–CH₂–CH₂–OH

1, 2-ethandiol
(ethylene glycol)
(a dihydric alcohol)

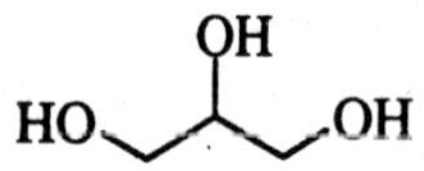

HO–CH₂–CH(OH)–CH₂–OH

1, 2, 3-propantriol
(glycerol)
(a trihydric alcohol)

Many functional groups have a characteristic suffix designator, and only one such suffix (other than *ene* and *yne*) may be used in a name. When the hydroxyl functional group is present together with a function of higher nomenclature priority, it must be cited and located by the prefix **hydroxy** and an appropriate number. For example, lactic acid has the IUPAC name 2-hydroxypropanoic acid.

Compounds incorporating a C—S—H functional group are named *thiols* or *mercaptans*. The IUPAC name of $(CH_3)_3C—SH$ is 2-methyl-2-propanethiol, commonly called tert-butyl mercaptan. The chemistry of thiols will not be described here, other than to note that they are stronger acids and more powerful nucleophiles than alcohols.

6.3 PREPARATION OF ALCOHOLS

(a) By hydrolysis of alkyl halides

By hydrolysis of alkyl halides with aqueous alkali or silver oxide suspended in water:

$$R—X + aq\ NaOH \rightarrow ROH + NaX$$

$$R—X + AgOH \rightarrow ROH + AgX$$

Silver oxide is a better reagent than aqueous alkali because former forms an insoluble precipitate AgX, which drives the reaction to right.

(b) By hydrolysis of esters with alkali

$$R—COOR' + aq\ OH^- \rightarrow RCOO^- + R'OH$$

This is an important method, specially used for preparation of tertiary alcohol from tertiary alkyl halide. Tertiary alkylhalide on treatment with aqueous alkali gives alkene as predominant product by E-2 elimination reaction.

$$(CH_3CH_2)_3CX + aq.\ NaOH \longrightarrow (CH_3CH_2)_2C{=}CHCH_3 + (CH_3CH_2)_3COH$$

(major product)

$$(CH_3CH_2)_3CX + CH_3COOAg \downarrow Agx + (CH_3CH_2)_3COOCCH_3$$

$$\xrightarrow{NaOH \mid H_2} (CH_3CH_2)_3COH + CH_3COONa$$

(major product)

(c) By acid hydrolysis of ether

$$(C_2H_5)_2O \xrightarrow[H^+]{OH_2} 2\,C_2H_5OH$$

(d) By reduction of carbonyl compounds

(i) With Na/EtOH (Bouveault-Blanc reduction)

$$RCHO \xrightarrow{Na/EtOH} RCH_2OH$$

$$R—\overset{O}{\overset{\|}{C}}—R \xrightarrow{Na/EtOH} R—\overset{OH}{\overset{|}{C}H}—R$$

Mechanism

(a) $Na \longrightarrow Na^+ + e$

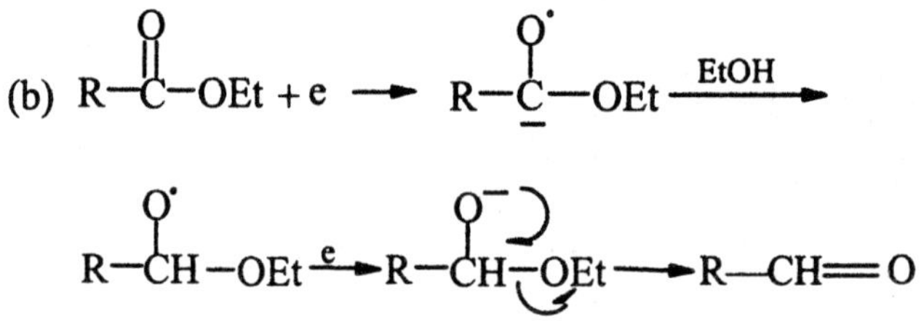

(b) $R—\overset{O}{\overset{\|}{C}}—OEt + e \rightarrow R—\overset{O^\bullet}{\overset{|}{C}^-}—OEt \xrightarrow{EtOH}$

$R—\overset{O^\bullet}{\overset{|}{C}H}—OEt \xrightarrow{e} R—\overset{O^-}{\overset{|}{C}H}—OEt \longrightarrow R—CH{=}O$

(c) $R—CH{=}O \xrightarrow{e} R—\bar{C}H—O^\bullet \xrightarrow{EtOH}$

De

$R—CH_2—O^\bullet \xrightarrow{(ii)\ EtOH} RCH_2—OH$

(ii) *Reduction with metalli c hydrides:* One of the most versatile complex hydride is $LiAlH_4$. This reduces most functional groups, viz. – aldehyde, ketone, acid, acid-derivatives, cyanide, nitrogroup, halide, etc. without affecting olefinic bond, e.g.

$$CH_3—CH{=}CH—CHO + LiAlH_4 \rightarrow CH_3—CH{=}CH—CH_2OH.$$

An alternative hydride is $NaBH_4$, which reduces aldehyde and ketone but does not reduces acid and acid derivatives, e.g.

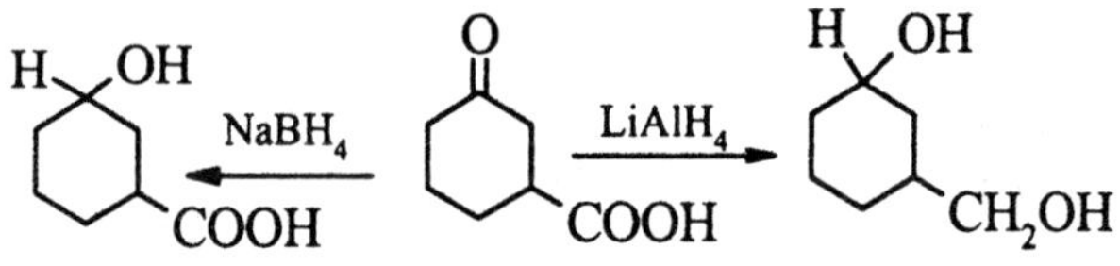

3-oxocyclohexanecarboxylic acid $\xrightarrow{NaBH_4}$ 3-hydroxycyclohexanecarboxylic acid (H, OH; COOH)

3-oxocyclohexanecarboxylic acid $\xrightarrow{LiAlH_4}$ 3-(hydroxymethyl)cyclohexanol (H, OH; CH_2OH)

Mechanism of reduction with $LiAlH_4$

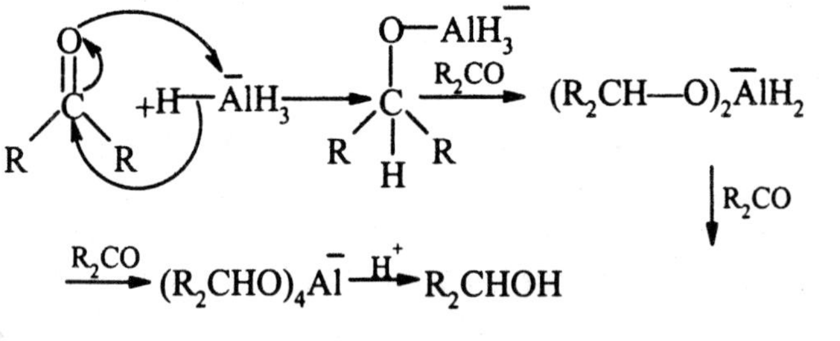

$$R_2C{=}O + H—\bar{A}lH_3 \longrightarrow R_2\overset{O—\bar{A}lH_3}{\overset{|}{C}}—H \xrightarrow{R_2CO} (R_2CH—O)_2\bar{A}lH_2$$

$$\xrightarrow{R_2CO} (R_2CHO)_4\bar{A}l \xrightarrow{H^+} R_2CHOH$$

Table 6.1 describes the role of some common reducing agents in selective reduction reactions.

Table 6.1: Common reducing agents and their role in organic chemistry

Reagent	*Preferred solvents*	*Functions reduced*	*Reaction work-up*
$NaBH_4$	Ethanol; aqueous ethanol 15% NaOH; diglyme, avoid strong acids	Aldehydes to 1°-alcohols ketones, to 2°-alcohols, reduction of enones is favoured by $CeCl_3$ inert to most other functions	1. Simple neutralization 2. Extraction of product
$LiAlH_4$	Ether; THF avoid alcohols and amines, avoid halogenated compounds, avoid strong acids	Aldehydes to 1°-alcohols, ketones to 2°-alcohols, carboxylic acids to 1°-alcohols, esters to alcohols, epoxides to alcohols, nitriles and amides to amines, halides and to tosylates to alkanes, most functions react	1. Addition of water 2. Dissolve Al salts 3. Extraction of product
Li-(*t*-butoxy)$_3$AlH $LiAlH(Ot\text{-}C_4H_9)_3$	Ether; THF avoid alcohols and amines avoid halogenated compounds, avoid strong acids	Fast: acid chlorides to aldehydes (at–78° C), 3°-amides to aldehydes (at–78° C), nitriles to aldehydes (at–78° C), slower: aldehydes to 1°-alcohols, ketones to 2°-alcohols	1. Addition of water 2. Dissolve Al salts 3. Extraction of product
$AlH[CH_2CH(CH_3)_2]_2$	THF; toluene avoid alcohols and amines avoid halogenated compounds, avoid strong acids	Fast: acid chlorides to aldehydes (at–78° C), 3°-amides to aldehydes (at–78° C), nitriles to aldehydes (at–78° C), slower: aldehydes to 1°-alcohols, ketones to 2°-alcohols	1. Addition of water 2. Dissolve Al salts 3. Extraction of product
$B_2H_6 = 2\,BH_3$	Ether; THF sulfide complex in CH_2Cl_2 complexes with amines, avoid alkenes and alkynes	Carboxylic acids to 1°-alcohols aldehydes to 1°-alcohols, ketones to 2°-alcohols, nitriles to amines, esters and epoxides slowly reduced	1. Dilute acid or H_2O_2 2. Extraction of product
H_2 and Pt, or Pd, or Ru, or Ni	Alcohols, ethers, hydrocarbons or carboxylic acids	Alkenes and alkynes to alkanes. Fast nitro groups to amines (fast) imines to amines (fast) aldehydes and ketones to alcohols (slow), nitriles to amines (slow) may remove benzylic groups	Filter to remove catalyst
Modified (poisoned) catalyst		Alkynes to alkenes, acyl chlorides to aldehydes	
Reactive metals Na, or Li, or K	Liq. ammonia and ether cosolvents or alcohols or amines	Ketones to 2°-alcohols alkynes to alkenes, conjugated π-systems (e.g. aromatic rings, dienes and enones), cleaves C-X and benzylic groups	1. Quench with aq. NH_4Cl 2. Extract product
Mg or Al or Zn or Fe	Water, alcohols, acetic acid or aqueous mineral acid	Cleaves activated substituents, nitro groups to amines, C=O (aldehyde/ketone) to CH_2	Extract product from salts.

Example 1. Complete the following reactions:

(a) $CH_3CHO + LiAlH_4 \xrightarrow{HCl}$

(b) $CH_3CHO + LiAlD_4 \xrightarrow{HCl}$

(c) $CH_3CHO + LiAlH_4 \xrightarrow{DCl}$

(d) $CH_3CHO + LiAlD_4 \xrightarrow{DCl}$

(e) $CH_3-\overset{O}{\overset{\|}{C}}-CH_2CH_2COOCH_3$ — (i) $LiAlH_4$, HCl; (ii) $NaBH_4$, HCl

Solution

(a) CH_3CH_2OH (b) CH_3CHDOH

(c) CH_3CH_2OD (d) CH_3CHDOD

(e) $CH_3CH(OH)CH_2CH_2CH_2OH + CH_3OH$ and $CH_3CH(OH)\ CH_2CH_2COOCH_3$

(e) Grignard reagent synthesis

This is a very usefull means for preparation of alcohol of desired skeleton. The organomagnesium halides of emperical formula R—Mg—X are called Grignard reagents (GR). Grignard reagents result from the reaction of an alkyl halide with Mg metal. The reaction is always carried out in ethereal medium, which is needed to solvate and stabilize the Grignard reagent as it is formed.

$$R—X + Mg \xrightarrow{C_2H_5OC_2H_5} RMgX$$

Grignard reagent may be made from primary, secondary and tertiary alkylhalides as well as from vinyl and aryl halides. Alkyl iodides are most reactive halide followed by bromides and chlorides. Alkyl fluorides generally do not react.

Addition of Grignard reagent to a carbonyl carbon: Grignard reagent contains R^- (one of the strongest nucleophile) as: $R^-(MgX)^+$. Therefore, Grignard reagent involves nucleophilic addition of R^- onto carbonyl carbon as:

$$\bar{R}\ (MgX)^+ + >C=O \longrightarrow R-\overset{|}{\underset{|}{C}}-OMgX \xrightarrow{H_3O^+} R-\overset{|}{\underset{|}{C}}-OH$$

(i) Addition of Grignard reagent to aldehyde:

$$\bar{R}\overset{+}{MgX} + R'CHO \longrightarrow R'-\underset{R}{\underset{|}{CH}}-OMgX \xrightarrow{H_3O^+} R'-\underset{R}{\underset{|}{CH}}-OH$$

(a 2° alcohol)

If formaldehyde is used, a primary alcohol is formed.

Example 2. Show, how would you synthesize the following alcohols by adding appropriate Grignard reagent to formaldehyde.

(a) cyclohexyl–CH_2OH (b) $(CH_3)_2CHCH_2CH_2OH$

(c) cyclopentenyl–CH_2OH

Solution

(a) cyclohexyl–Br $\xrightarrow{Mg/Et_2O} \xrightarrow{HCHO} \xrightarrow{H_2O}$ Product

(b) $(CH_3)_2CHCH_2$–Br $\xrightarrow{Mg/Et_2O} \xrightarrow{HCHO} \xrightarrow{H_2O}$ Product

(c) cyclopentenyl–Br $\xrightarrow{Mg/Et_2O} \xrightarrow{HCHO} \xrightarrow{H_2O}$ Product

Example 3. Show, how would you synthesize the following alcohols by adding an appropriate Grignard reagent to an appropriate carbonyl compound?

(a) $CH_3CH_2CH(CH_3)CH_2CH(OH)CH_3$ (b) $C_6H_5CH(OH)CH_2CH_3$

(c) $(C_6H_{11})_2C(H)OH$ (dicyclohexylmethanol)

Solution

(a) $(CH_3)_2CHCH_2$–Br $\xrightarrow{Mg/Et_2O} \xrightarrow{CH_3CHO} \xrightarrow{H_2O}$ Product

(b) $CH_3CH_2Br \xrightarrow{Mg/Et_2O} \xrightarrow{PhCHO} \xrightarrow{H_2O}$ Product

(c) cyclohexyl–Br $\xrightarrow{Mg/Et_2O} \xrightarrow{C_6H_{11}CHO} \xrightarrow{H_2O}$ Product

(ii) *Addition of Grignard reagent to ketone:* A tertiary alcohol is formed:

$$\bar{R}MgX^{+} + C_2C{=}O \longrightarrow R{-}CC_2{-}\bar{O}(MgX)^{+} \xrightarrow{H_3O^+} R{-}CC_2{-}OH + XMg(OH)$$

$$CH_3CH_2\bar{M}g\overset{+}{Br} + (CH_3CH_2)(CH_3)C{=}O \longrightarrow CH_3CH_2{-}C(CH_2CH_3)(CH_3){-}OMgBr \xrightarrow{H_3O^+} C_2H_5{-}C(CH_3)(C_2H_5){-}OH + BrMg(OH)$$

Example 4. How will you synthesize the following alcohol using compounds containing not more than four carbons?

(Cyclobutyl)–C(CH_3)(CH_2CH_3)–OH

Solution: To fulfill the requirement of starting compound, synthesis has to be designed as:

(Cyclobutyl)–C(CH_3)(CH_2CH_3)–OH ⟹ (Cyclobutyl)–MgX + (CH_3)(C_2H_5)C=O

Example 5. An organic compound $C_5H_{10}O$ can exist in two isomeric form A and B. Both A and B are optically inactive. Reduction of either A or B with $LiAlH_4$ makes them resolvable. Addition of CH_3MgBr followed by acidic hydrolysis makes them alcohols which cannot be resolved. Propose structures to A and B.

Solution

$$CH_3{-}\overset{O}{\overset{\|}{C}}{-}CH_2CH_2CH_3 \quad \text{and} \quad CH_3{-}\overset{O}{\overset{\|}{C}}{-}CH(CH_3)_2$$

(iii) *Addition of Grignard reagent to acid chloride and esters:* Acid chloride and esters reacts with two equivalents of Grignard reagents to give tertiary alcohol.

$$2R{-}MgX + R'{-}\overset{O}{\overset{\|}{C}}{-}Cl \xrightarrow{H_3O^+} R{-}C(R)(R'){-}OH$$

$$2R{-}MgX + R'{-}\overset{O}{\overset{\|}{C}}{-}OR'' \xrightarrow{H_3O^+} R{-}C(R)(R){-}OH$$

Mechanism

$$R{-}MgX + R'{-}\overset{O}{\overset{\|}{C}}{-}Cl \longrightarrow R{-}C(\bar{O})(R'){-}Cl \longrightarrow R'{-}\overset{O}{\overset{\|}{C}}{-}R$$

$$R{-}MgX + R'{-}\overset{O}{\overset{\|}{C}}{-}R \longrightarrow R'{-}C(R)(R){-}\bar{O} \xrightarrow{H_3O^+} R'{-}C(R)(R){-}OH$$

Example 6. Synthesize following alcohols starting from a Grignard reagent and an acid chloride:

(a) Ph_3COH

(b) 3-ethyl-2-methyl-3-pentanol

(c) Dicyclohexyl phenylmethanol.

Solution

(a) $PhCOCl \xrightarrow[\text{Excess}]{PhMgBr} \xrightarrow{H_2O} Ph_3COH$

(b) $(CH_3)_2CH{-}\overset{O}{\overset{\|}{C}}{-}Cl \xrightarrow[\text{Excess}]{C_2H_5MgBr} \xrightarrow{H_2O} (CH_3)_2CH{-}C(OH)(C_2H_5)_2$

(c) $PhCOCl \xrightarrow[\text{Excess}]{C_6H_{11}{-}MgBr} \xrightarrow{H_2O} (C_6H_{11})_2C(OH)Ph$

Example 7. Identify the compounds in following sequence of reaction and explain mechanism of their formation:

$$A + \text{Ethyl formate} \xrightarrow{H_3O^+} (CH_2{=}CH{-}CH_2)_2CH{-}OH \xrightarrow{H^+} \text{(cyclohexadiene)}$$

Solution: A: Allyl magnesium bromide.

$$\text{cyclic carbocation} \xrightarrow{H^+} \xrightarrow{-H^+} \xrightarrow[-H_2O]{H^+} \text{methylcyclohexene}$$

(iv) *Addition of GR to ethylene oxide:* Grignard reagents usually do not reacts with ether, but epoxide is unusually reactive ethers because of their ring strain. Ethylene oxide reacts with Grignard reagents to give, after protonation, primary alcohols with two additional carbon atoms.

$$RMgX + \text{(ethylene oxide)} \longrightarrow R{-}CH_2CH_2{-}OMgX \xrightarrow{H^+} R{-}CH_2CH_2{-}OH$$

In case of unsymmetrical epoxide, R^- attacks at less sterically hindered carbon as:

$$RMgX + \text{(2,2-dimethyloxirane)} \longrightarrow R{-}CH_2{-}C(CH_3)_2{-}OMgX \xrightarrow{H^+} R{-}CH_2{-}C(CH_3)_2{-}OH$$

Example 8. Show how would you synthesize the following alcohols by adding Grignard reagents to ethylene oxide:

(a) 2- phenylethanol (b) 4-methyl-1-pentanol

(c) 2-(2-methylcyclohexyl)ethanol (structure with OH)

Solution

(a) $PhMgBr + \text{(ethylene oxide)} \xrightarrow{H_3O^+}$ Product

(b) $(CH_3)_2CHCH_2MgBr + \text{(ethylene oxide)} \xrightarrow{H_3O^+}$ Product

(c) 2-methylcyclohexylmagnesium bromide $\xrightarrow{\text{ethylene oxide}} \xrightarrow{H_3O^+}$ Product

(f) *By hydration of alkenes*

Alkenes can be hydrated into alcohols by using several reagents as:

(i) *Hydration in presence of acid catalyst* (Markovnikov's addition of water):

$$CH_3CH{=}CH_2 + H_3O^+ \xrightarrow{\text{Boil}} CH_3{-}CH(OH){-}CH_3$$

The mechanism of hydration here involves formation of carbocation intermediate, therefore rearrangement is likely to occur, where possible:

$$H_3C{-}C(CH_3)_2{-}CH{=}CH_2 + H_3O^+ \longrightarrow H_3C{-}C(CH_3)_2{-}\overset{+}{C}H{-}CH_3$$

(secondary carbocation)

Methyl shift

$$H_3C{-}\overset{+}{C}(CH_3){-}CH(CH_3){-}CH_3$$

(tertiary carbocation)

$$\xrightarrow{OH_2} H_3C{-}C(CH_3)(\overset{+}{O}H_2){-}CH(CH_3){-}CH_3 \xrightarrow{-H^+} H_3C{-}C(CH_3)(OH){-}CH(CH_3){-}CH_3$$

(ii) *Oxymercuration-demercuration:* This is an alternative reagent of acidic hydration. The advantage of this reagent over the acidic hydration is its regioselectivity, i.e. addition occurs only at double bond and no rearranged product is obtained.

$$H_3C{-}CH(CH_3){-}CH{=}CH_2 + (CH_3COO)_2Hg \xrightarrow{OH_2} \xrightarrow{NaBH_4} H_3C{-}CH(CH_3){-}CH(OH){-}CH_3$$

(Markovnikov's product)

(iii) *Hydroboration reaction (anti Markovnikov's hydration):* This involves reaction of an alkene with B_2H_6 followed by treatment of alkyl borane with H_2O_2 in alkaline medium:

$$CH_3CH{=}CH_2 + B_2H_6 \rightarrow (CH_3CH_2CH_2)_3B \xrightarrow{H_2O_2/\overline{O}H} CH_3CH_2CH_2OH$$

(g) *Oxo process*

In this process, a mixture of alkene, CO and hydrogen, under pressure and elevated temperature, in the presence of a catalyst forms an aldehyde. A common catalyst is cobalt carbonyl hydride [$CoH(CO)_4$], and the product is a mixture of isomeric straight chain and branched aldehydes. These are reduced catalytically to the corresponding alcohols, e.g.

$$2MeCH{=}CH_2 + CO + H_2 \xrightarrow{[CoH(CO)_4]} MeCH_2CH_2CHO + Me_2CHCHO$$

$$\downarrow \text{Cu-Zn/H}_2$$

$$MeCH_2CH_2CH_2OH + Me_2CHCH_2OH$$

6.4 PHYSICAL PROPERTIES OF ALCOHOLS

Most of the common alcohols, up to 11 or 12 carbon atoms are liquids at room temperature. Methanol and ethanol are free flowing volatile liquids with characteristic fruity odours. Higher alcohols are somewhat viscous at room temperature.

(a) Boiling Points of Alcohols

Boiling point of any liquid depends on the strength of intermolecular force of attraction. In alcohols, there are two significant intermolecular forces of attraction responsible for high boiling points of alcohols compared to ether and alkane of comparable molecular weight, e.g.

C_2H_5OH(M W = 46) b. pt = 78 °C, CH_3OCH_3 (MW = 46) b. pt = –25 °C, $CH_3CH_2CH_3$(MW = 44) b. pt = –42°C

The first important force of attraction is H-bonding. The hydroxyl hydrogen of ethanol is strongly polarized by its bond to oxygen, and it can form a hydrogen bond with a pair of non-bonding electrons from the oxygen atom of another alcohol as:

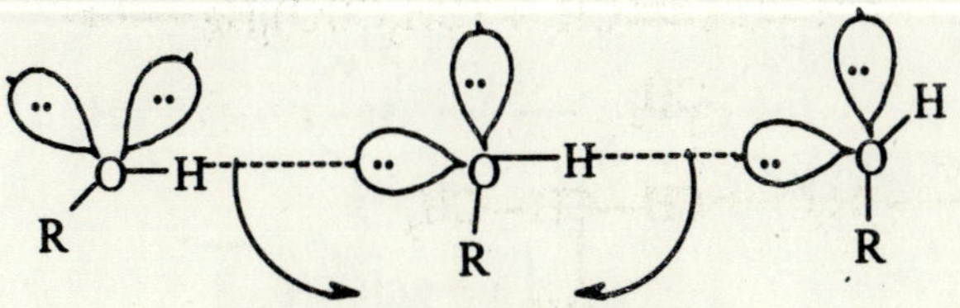

Intermolecular H-bonding

Dipole-dipole attraction also contribute to the relatively high boiling point of alcohols. Dipole moment in alcohols is much higher than any alkane, due to which alkanes have very low boiling point. When we compare alcohols with isomeric ether, dipole moment in the two category are comparable, but the later category lacks in its ability to form hydrogen bond. Due to absence of hydrogen bonding, ethers have very low boiling point compared to alcohols. Therefore, the prime factor here is hydrogen bonding. Among alcohols, increasing molecular weight increases boiling point. Branching decreases boiling point.

(b) Solubility Properties of Alcohols

Alcohols and water have almost similar properties, because one common thing in both compound is presence of hydroxyl functional group capable of forming hydrogen-bonds. Lower alcohols are completely miscible in water, but solubility decreases on increasing molecular weight. This is because of the reason.

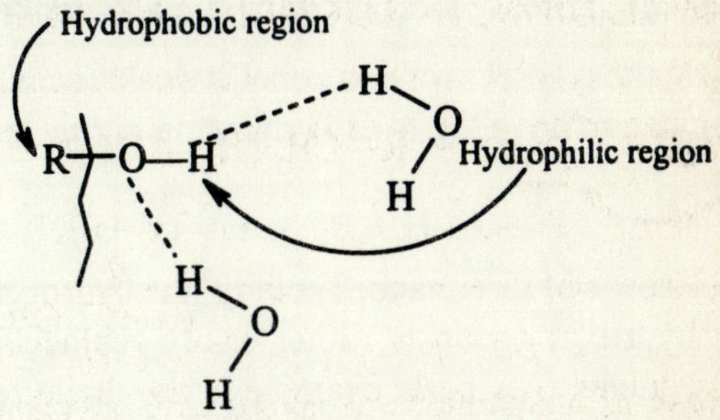

Hydrophobic region

Hydrophilic region

As shown above, the hydroxyl part of alcohol has affinity for water, known as hydrophilic part. Alkyl part of alcohol is called hydrophobic because it acts as alkane, disrupts the network of H-bonds and dipole-dipole attractions of a polar solvent such as water. But at the same time the alkyl part of alcohol has affinity for non-polar solvent, and therefore, alcohols have a wide range of polar as well as non-polar solvents. Therefore, solubility of alcohols in water decreases on increasing molecular weight, because it increases the size (contribution) of hydrophobic alkyl group. Also increasing the number of hydroxyl functional group in alcohol increases its solubility in water, e.g. ethylene glycol (HO—CH_2CH_2—OH—a dihydric alcohol) is highly soluble in water, glucose—a penta hydroxy compound is highly soluble in water because of its ability to form extensive H-bonds.

6.5 CHEMICAL PROPERTIES OF ALCOHOLS

The functional group of the alcohols is the hydroxyl group, —OH. Unlike the alkyl halides, this group has two reactive covalent bonds, the C—O bond and the O—H bond. The electronegativity of oxygen is substantially greater than that of carbon and hydrogen. Consequently, the covalent bonds of this functional group are polarized so that oxygen is electron rich and both carbon and hydrogen are electrophilic, as shown in the following diagram.

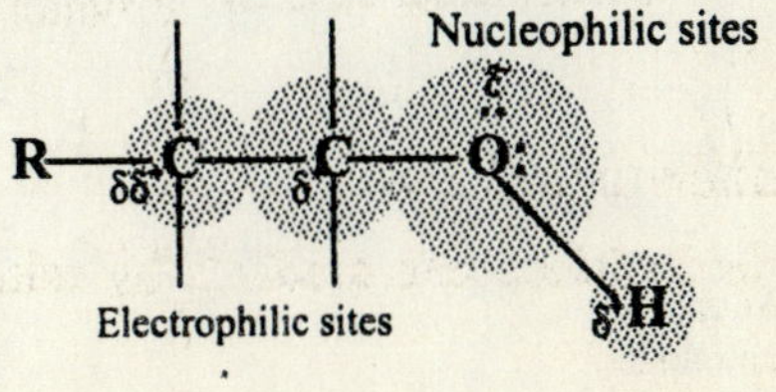

Nucleophilic sites

Electrophilic sites

Indeed, the dipolar nature of the O—H bond is such that alcohols are much stronger acids than alkanes (by roughly 10^{30} times), and nearly that much stronger than ethers (oxygen substituted alkanes that do not have an O—H group). The most reactive site in an alcohol molecule is the hydroxyl group, despite the fact that the O—H bond strength is significantly greater than that of the C—C, C—H and C—O bonds, demonstrating again the difference between thermodynamic and chemical stability.

6.6 ACIDITY OF ALCOHOLS AND PHENOLS

The hydroxyl proton of an alcohol is weakly acidic. A strong base can remove the hydroxyl proton giving an alkoxide ion:

$$R—O—H + B^- \rightarrow R—O^- + BH$$

Because of its enhanced acidity, the hydrogen atom on the hydroxyl group is rather easily replaced by other substituents. A simple example is the facile reaction of simple alcohols with sodium (and sodium hydride), as described in the first equation below. Another such substitution reaction is the isotopic exchange that occurs on mixing an alcohol with deuterium oxide (heavy water). This exchange, which is catalyzed by acid or base, is very fast under normal conditions, since it is difficult to avoid traces of such catalysts in most experimental systems.

$$2R—O—H + 2Na \rightarrow 2R—O^{(-)}Na^{(+)} + H_2$$

$$R—O—H + D_2O \rightleftharpoons R—O—D + D—O—H$$

The mechanism by which many substitution reactions of this kind take place is straightforward. The oxygen atom of an alcohol is nucleophilic and is therefore prone to attack by electrophiles. The resulting "onium" intermediate then loses a proton to a base, giving the substitution product. If a strong electrophile is not present, the nucleophilicity of the oxygen may be enhanced by conversion to its conjugate base (an alkoxide). This powerful nucleophile then attacks the weak electrophile. These two variations of the substitution mechanism are illustrated in the following diagram.

Substitution by a strong electrophile: R—Ö—H (Alcohol) + E^+ (Strong electrophile) ⇌ [R—$\overset{E}{O^+}$—H] (Onium intermediate) ⇌ R—Ö—E (Substitution product) + H^+ (Removed by a base)

Substitution by a weak electrophile: R—Ö—H (Alcohol) + base ⇌ R—Ö:$^-$ + $\overset{\delta+}{E}—\overset{\delta-}{X}$ (Weak electrophile) → R—Ö—E (Substitution product) + X^-

The preparation of tert-butyl hypochlorite from tert-butyl alcohol is an example of electrophilic halogenation of oxygen, but this reaction is restricted to 3°-alcohols, because 1° and 2°-hypochlorites lose HCl to give aldehydes and ketones. In the following equation the electrophile may be regarded as $Cl^{(+)}$.

$$(CH_3)_3C—O—H + Cl_2 + NaOH \rightarrow (CH_3)_3C—O—Cl + NaCl + H_2O$$

Alkyl substitution of the hydroxyl group leads to ethers. This reaction provides examples of both strong electrophilic substitution (first equation below), and weak electrophilic substitution (second equation). The latter SN-2 reaction is known as the Williamson ether synthesis, and is generally used only with 1°-alkyl halide reactants, because the strong alkoxide base leads to E2 elimination of 2° and 3°-alkyl halides.

Cyclohexyl—Ö—H + [$H_3C—\overset{+}{C}(CH_3)_2$] (from isobutylene + H^+) → [cyclohexyl—$\overset{+}{O}$(H)—C(H_3C)(CH_3)(CH_3)] → (−isobutylene) cyclohexyl—Ö—C(CH_3)(H_3C)(CH_3) + $H_3C—\overset{+}{C}(CH_3)_2$

$(C_2H_5)_2$CH—Ö—H →(NaH) $(C_2H_5)_2$CH—Ö:$^-$ + H_2

→(Br—CH_2—C(=CH_2)—CH_3...) $(C_2H_5)_2$CH—Ö—CH_2—(alkenyl) + NaBr

One of the most important substitution reactions at oxygen is **ester formation** resulting from the reaction of alcohols with electrophilic derivatives of carboxylic and sulfonic acids. The following illustration displays the general formulas of these reagents and their ester products, in which the R′—O— group represents the alcohol moiety.

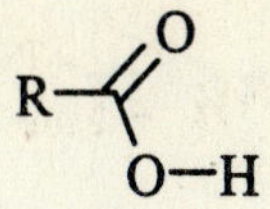

A carboxylic acid

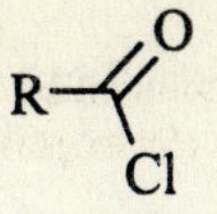

An acyl chloride

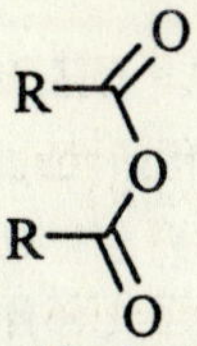

An acid anhydride

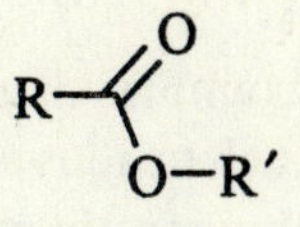

A carboxylate ester

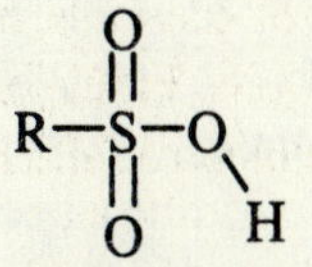

A sulphonic acid

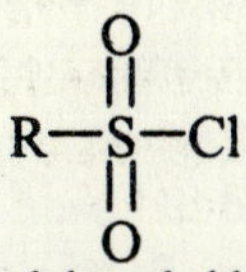

A sulphonyl chloride

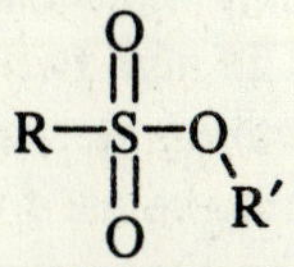

A sulphonate ester

With exception of methanol ($K_a = 3.2 \times 10^{-16}$), alcohols are slightly less acidic than water ($K_a = 1.8 \times 10^{-16}$). In homologous series, acidity decreases on increasing molecular weight. Substitution by electron withdrawing group increases the acid strength, e.g. $ClCH_2CH_2OH$ (2-chloroethanol) is stronger acid than water whereas ethanol is a weaker acid than water. Phenol is relatively stronger acid than water and aliphatic alcohol, because of electron withdrawing effect of benzene ring from hydroxyl group through resonance as:

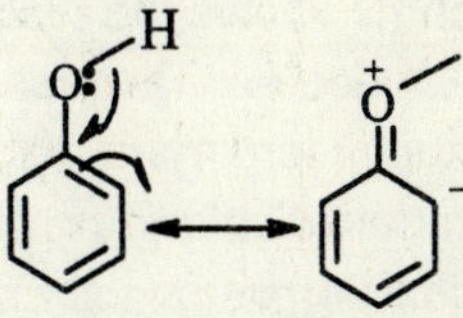

Positive charge on oxygen intensify the release of proton attached to it.

Using the chemical behaviour of alkyl halides as a reference, we are encouraged to look for analogous substitution and elimination reactions of alcohols. The chief difference, of course, is a change in the leaving anion from halide to hydroxide. Since oxygen is slightly more electronegative than chlorine (3.5 vs. 2.8 on the Pauling scale), we expect the C—O bond to be more polar than a C—Cl bond. Furthermore, an independent measure of the electrophilic character of carbon atoms from their nmr chemical shifts (both ^{13}C and alpha protons), indicates that oxygen and chlorine substituents exert a similar electron-withdrawing influence when bonded to sp^3 hybridized carbon atoms. Despite this promising background evidence, alcohols do not undergo the same SN-2 reactions commonly observed with alkyl halides. For example, the rapid SN-2 reaction of 1-bromobutane with sodium cyanide, shown below, has no parallel when 1-butanol is treated with sodium cyanide. In fact ethyl alcohol is often used as a solvent for alkyl halide substitution reactions such as this:

$$CH_3CH_2CH_2CH_2—Br + Na^{(+)}CN^{(-)} \rightarrow CH_3CH_2CH_2CH_2—CN + Na^{(+)}Br^{(-)}$$

$$CH_3CH_2CH_2CH_2—OH + Na^{(+)}CN^{(-)} \rightarrow \text{No reaction}$$

The key factor here is the stability of the leaving anion (bromide vs. hydroxide). We know that HBr is a much stronger acid than water (by more than 18 powers of ten), and this difference will be reflected in reactions that generate their conjugate bases. The weaker base, bromide anion, is more stable and its release in a substitution or elimination reaction will be much more favourable than that of hydroxide ion, a stronger and less stable base. Clearly, an obvious step toward improving the reactivity of alcohols in SN-2 reactions would be to modify the –OH functional group in a way that improves its stability as a leaving anion. One such modification is to conduct the substitution reaction in strong acid so that –OH is converted to $–OH_2^{(+)}$. Since the hydronium ion ($H_3O^{(+)}$) is a much stronger acid than water, its conjugate base (H_2O) is a better leaving group than hydroxide ion. The only problem with this strategy is that many nucleophiles, including cyanide, are deactivated by protonation in strong acid; effectively removing the nucleophilic co-reactant needed for the substitution. The strong acids HCl, HBr and HI are not subject to this difficulty, because their conjugate bases are good nucleophiles and are even weaker bases than alcohols. The following equations illustrate some substitution reactions of alcohols that may be effected by these acids. As was true for alkyl halides, nucleophilic substitution of 1°-alcohols proceeds by a SN-2 mechanism, whereas 3°-alcohols react by a SN-1 mechanism. Reactions of 2°-alcohols may occur by both mechanisms and often produce some rearranged products. The numbers in parentheses next to the mineral acid formulas represent the weight percentage of a concentrated aqueous solution, the form in which these acids are normally used.

$$CH_3CH_2CH_2CH_2—OH + HBr\ (48\%)$$

$$\rightleftharpoons$$

$$CH_3CH_2CH_2CH_2—OH_2^{(+)}\,Br^{(-)}$$

$$\rightleftharpoons$$

$$CH_3CH_2CH_2CH_2—Br + H_2O \quad \text{SN-2}$$

$$(CH_3)_3C—OH + HCl\ (37\%)$$

$$\rightleftharpoons$$

$$(CH_3)_3C—OH_2^{(+)}\,Cl^{(-)} \rightleftharpoons (CH_3)_3C^{(+)}\,Cl^{(-)} + H_2O$$

$$\rightleftharpoons$$

$$(CH_3)_3C—Cl + H_2O \quad \text{S N-1}$$

Although these reactions are sometimes referred to as "acid-catalyzed" this is not strictly correct. In the overall transformation a strong HX acid is converted to water, a very weak acid, so at least a stoichiometric quantity of HX is required for a complete conversion of alcohol to alkyl halide. The necessity of using equivalent quantities of very strong acids in this reaction limits its usefulness to simple alcohols of the kind shown above. Alcohols having acid sensitive groups would, of course, not tolerate such treatment. Nevertheless, the idea of modifying the —OH functional group to improve its stability as a leaving anion can be pursued in other directions. The following diagram shows some modifications that have proven effective. In each case the hydroxyl group is converted to an ester of a strong acid. The first two examples show the sulfonate esters described earlier. The third and fourth examples show the formation of a phosphite ester, and a chlorosulfite ester respectively. All of these leaving groups have conjugate acids that are much stronger than water (by 13 to 16 powers of ten), so the leaving anion is correspondingly more stable than hydroxide ion. The mesylate and tosylate compounds are particularly useful in that they may be used in substitution reactions with a wide variety of nucleophiles. The intermediates produced in reactions of alcohols with phosphorus tribromide and thionyl chloride (last two examples) are seldom isolated, and these reactions continue on to alkyl bromide and chloride products.

^{1}R, ^{2}R, ^{3}R–C–OH +

- CH_3SO_2Cl (mesyl chloride) → $^{1}R^{2}R^{3}R$C–OSO_2CH_3
- H_3C–C_6H_4–SO_2Cl (tosyl chloride) → $^{1}R^{2}R^{3}R$C–O–SO_2–C_6H_4–CH_3
- PBr_3 → $^{1}R^{2}R^{3}R$C–O–PBr_2
- $SOCl_2$ → $^{1}R^{2}R^{3}R$C–O–S(=O)Cl

The importance of sulfonate ester intermediates in general nucleophilic substitution reactions of alcohols may be illustrated by the following conversion of 1-butanol to pentanenitrile (butyl cyanide), a reaction that does not occur with the alcohol alone. The phosphorous and thionyl halides, on the other hand, only act to convert alcohols to the corresponding alkyl halides.

1-Butanol (OH) —CH_3SO_2Cl, Pyridine→ OSO_2CH_3 (a mesylate) —NaCN→ CN + CH_3SO_2ONa

Some examples of alcohol substitution reactions using this approach to activating the hydroxyl group are shown in the following diagram. The first two cases serve to reinforce the fact that sulfonate ester derivatives of alcohols may replace alkyl halides in a variety of SN-2 reactions. The next two cases demonstrate the use of phosphorous tribromide in converting alcohols to bromides. This reagent may be used without added base (e.g. pyridine), because the phosphorous acid product is a weaker acid than HBr. Phosphorous tribromide is best used with 1°-alcohols, since 2°-alcohols often give rearrangement by-products resulting from competing SN-1 reactions. Note that the ether oxygen in reaction 4 is not affected by this reagent; whereas, the alternative synthesis using concentrated HBr cleaves ethers. Phosphorous trichloride (PCl_3) converts alcohols to alkyl chlorides in a similar manner, but thionyl chloride is usually preferred for this transformation since the inorganic products are gases (SO_2 and HCl). Phosphorous triiodide is not stable, but may be generated *in situ* from a mixture of red phosphorous and iodine, and acts to convert alcohols to alkyl iodides. The last example shows the reaction of thionyl chloride with a chiral 2°-alcohol. The presence of an organic base such as pyridine is important, because it provides a substantial concentration of chloride ion needed for the final SN-2 reaction of the chlorosufite intermediate. In the absence of base, chlorosufites decompose on heating to give the expected alkyl chloride with ***retention of configuration***.

Tertiary alcohols are not commonly used for substitution reactions of the kind discussed here, because SN-1 and E1 reaction paths are dominant and are difficult to control.

(a) Cyclopentyl–OH —Pyridine, CH_3SO_2Cl→ Cyclopentyl–OSO_2CH_3 —CH_3CH_2SNa, SN-2→ Cyclopentyl–SCH_2CH_3 + CH_3SO_3Na

(b) Isobutyl–OH —Pyridine, $C_7H_7SO_2Cl$→ Isobutyl–$OSO_2C_7H_7$ (a tosylate) —NaI, SN-2→ Isobutyl–I + $C_7H_7SO_3Na$

(c) 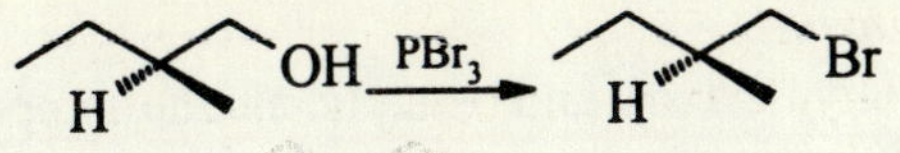 [H-substituted alcohol] –OH $\xrightarrow{PBr_3}$ –Br

(d) 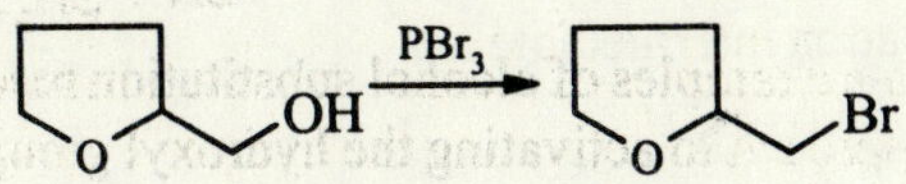 [tetrahydrofuran-2-ylmethanol] –OH $\xrightarrow{PBr_3}$ –Br

(e)  H, OH $\xrightarrow{SOCl_2}$ Cl, H

6.7 ELIMINATION OF THE HYDROXYL GROUP-(A GENERAL DISCUSSION)

In the discussion of alkyl halide reactions we noted that 2° and 3°-alkyl halides experienced rapid E2 elimination when treated with strong bases, such as hydroxide and alkoxides. Alcohols do not undergo such base-induced elimination reactions and are, in fact, often used as solvents for such reactions. This is yet another example of how leaving group stability often influences the rate of a reaction.

When an alcohol is treated with sodium hydroxide, the following acid-base equilibrium occurs. Most alcohols are slightly weaker acids than water so the left side is favoured.

$$R—O—H + Na^{(+)}OH^{(-)} \rightleftharpoons R—O^{(-)}\,Na^{(+)} + H—OH$$

The elimination of water from an alcohol is called dehydration. Recalling that water is a much better leaving group than hydroxide ion, it is sensible to use acid-catalysis rather than base-catalysis to achieve such reactions. Four examples of this useful technique are shown below. Note that hydrohalic acids (HX) are not normally used as catalysts, because their conjugate bases are good nucleophiles and may give substitution products. The conjugate bases of sulfuric and phosphoric acids are not good nucleophiles and do not give substitution under the usual conditions of their use.

(a) 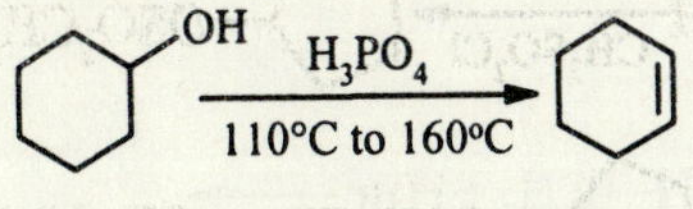 cyclohexanol $\xrightarrow[110°C\ to\ 160°C]{H_3PO_4}$ cyclohexene

(b) 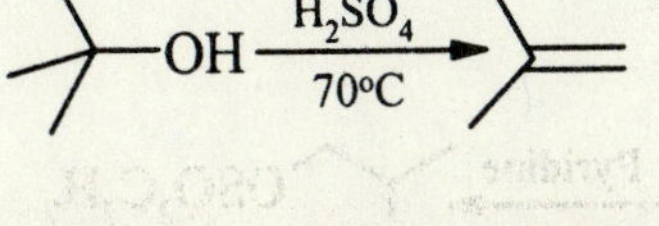 tert-butyl alcohol –OH $\xrightarrow[70°C]{H_2SO_4}$ isobutylene

(c) alcohol (OH) $\xrightarrow[Heat]{H_2SO_4}$ Major + Minor (*cis* < *trans*)

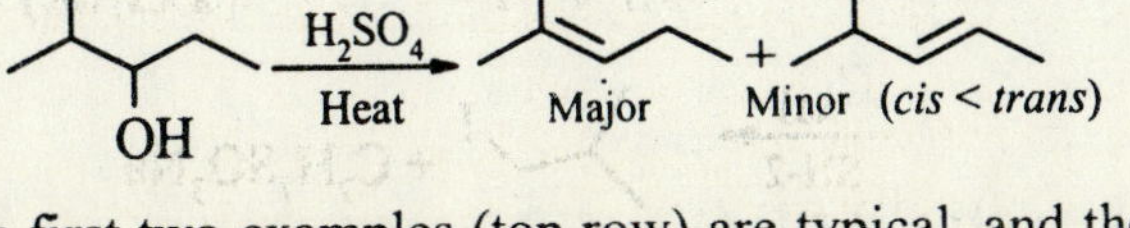

The first two examples (top row) are typical, and the more facile elimination of the 3°-alcohol suggests predominant E1 character for the reaction. This agrees with the tendency of branched 1°-and 2°-alcohols to give rearrangement products, as shown in the last example. The last two reactions also demonstrate that the Sytzebb's rule applies to alcohol dehydrations as well as alkyl halide eliminations. Thus, the more highly-substituted double bond isomer is favoured among the products.

It should be noted that the acid-catalyzed dehydrations discussed here are the reverse of the acid-catalyzed hydration reactions of alkenes. Indeed, for reversable reactions such as this the laws of thermodynamics require that the mechanism in both directions proceed by the same reaction path. This is known as the principle of microscopic reversibility.

To illustrate, the following diagram lists the three steps in each transformation. The dehydration reaction is shown by the dotted arrows; the hydration reaction by solid arrows. The intermediates in these reactions are common to both, and common transition states are involved.

Dehydration

Step I $R_2C(OH)—CH_2R \rightleftharpoons R_2C(\overset{+}{O}H_2)—CH_2R$ Step I

Step II $R_2C(\overset{+}{O}H_2)—CH_2R \underset{+H_2O}{\overset{-H_2O}{\rightleftharpoons}} R_2\overset{+}{C}—CH_2R$ Step II

Step III $R_2\overset{+}{C}—CH_2R \underset{H^+}{\overset{-H^+}{\rightleftharpoons}} R_2C{=}CHR$ Step III

Hydration

Base induced E2 eliminations of alcohols may be achieved if their sulfonate ester derivatives are used. This has the advantage of avoiding strong acids, which may cause molecular rearrangement and/or double bond migration in some cases. Since 3°-sulfonate derivatives are sometimes unstable, this procedure is best used with 1°-and 2°-mesylates or tosylates. Application of this reaction sequence is shown here for 2-butanol. The Sytzebb's rule favours formation of 2-butene (*cis* + *trans*) over 1-butene.

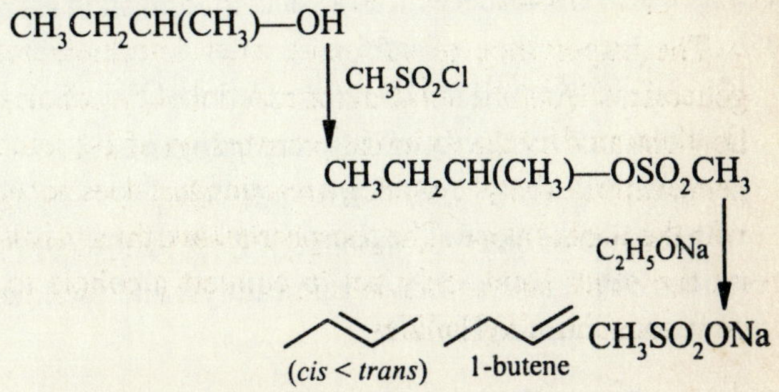

$CH_3CH_2CH(CH_3)—OH$

↓ CH_3SO_2Cl

$CH_3CH_2CH(CH_3)—OSO_2CH_3$

↓ C_2H_5ONa

2-butene (*cis* < *trans*) + 1-butene + CH_3SO_2ONa

The E2 elimination of 3°-alcohols under relatively non-acidic conditions may be accomplished by treatment with phosphorous oxychloride ($POCl_3$) in pyridine. This procedure is also effective with hindered 2°-alcohols, but for unhindered and 1°-alcohols an SN-2 chloride ion substitution of the chlorophosphate intermediate competes with elimination. Some examples of these and related reactions are given in the following diagram. The first equation shows the dehydration of a 3°-alcohol. The predominance of the non-Sytzebb's product (less substituted double bond) is presumed due to steric hindrance of the methylene group hydrogens, which interferes with the approach of base at that site. The second example shows two elimination procedures applied to the same 2°-alcohol. The first uses the single step $POCl_3$ method, which works well in this case because SN-2 substitution is retarded by steric hindrance. The second method is another example, in which an intermediate sulfonate ester confers halogen-like reactivity on an alcohol. In every case the anionic leaving group is the conjugate base of a strong acid.

HO–(alcohol) $\xrightarrow[\text{Pyridine}]{POCl_3}$ [Cl–P(=O)(Cl)–O–R] → alkene (Major product)

OH, CH_3, CH_3, H $\xrightarrow[\text{Pyridine}]{POCl_3}$ CH_3, CH_3, H; CH_3SO_2Cl → CH_3, CH_3, H

6.8 SPECIFIC REACTIONS OF ALCOHOLS

6.8.1 Reaction of Alcohols with an Organic Acid

Esterification: In presence of dilute acid, alcohol reacts with organic acid to form ester (detail of this reaction is dealt in acid derivative section).

$$R—COOH + R'OH \xrightarrow{H^+} R—COOR' + H_2O$$

6.8.2 Reaction of Alcohol with Halogen Acid

Halogen acid HX reacts with an alcohol to form alkyl halide as:

$$R—OH + HX \rightarrow R—X + H_2O$$

The order of reactivity of different halogen acid is:

$$HI > HBr > HCl$$

Presence of catalytic amount of $ZnCl_2$ activates the reaction. $ZnCl_2$ acts as Lewis acid catalyst and facilitate formation of carbocation as:

$$R—CH_2—\ddot{O}H + ZnCl_2 \rightarrow R—CH_2—\overset{+}{O}(H)—\overset{-}{Z}nCl_2 \rightarrow R—\overset{+}{C}H_2$$

Mechanism

Secondary, tertiary and benzylic alcohols appear to react by SN-1 mechanism which involves formation of carbocation intermediate as:

Step I

$$H_3C—C(CH_3)_2—\ddot{O}—H + H—\overset{+}{O}(H)—H \rightleftharpoons H_3C—C(CH_3)_2—\overset{+}{O}H_2$$

Step II

$$H_3C—C(CH_3)_2—\overset{+}{O}H_2 \rightleftharpoons H_3C—\overset{+}{C}(CH_3)_2$$

Step III

$$H_3C—\overset{+}{C}(CH_3)_2 + Cl^- \longrightarrow H_3C—C(CH_3)_2—Cl$$

Here, in step II, carbocation is formed, therefore, if possible rearrangement to a more stable carbocation may also occur in this step as:

$$H_3C—CH(CH_3)—CH_2\ddot{O}H + H—\overset{+}{O}(H)—H \rightleftharpoons H_3C—CH(CH_3)—CH_2\overset{+}{O}H_2$$

$\downarrow -H_2O$

$$H_3C—\overset{+}{C}(CH_3)—CH_3 \longleftarrow H_3C—C(H)(CH_3)—\overset{+}{C}H_2$$

(More stable carbocation)

Example 9. Predict major product in the following reactions:

(a) $H_3C—CH(CH_3)—CH(OH)—CH_3 + HBr \longrightarrow$

(b) $H_3C—C(CH_3)_2—CH_2OH + HCl \longrightarrow$

(c) $C_6H_5CH_2CH_2OH + HBr \longrightarrow$

(d) ▷—$CH_2OH + HBr \longrightarrow$

(e) 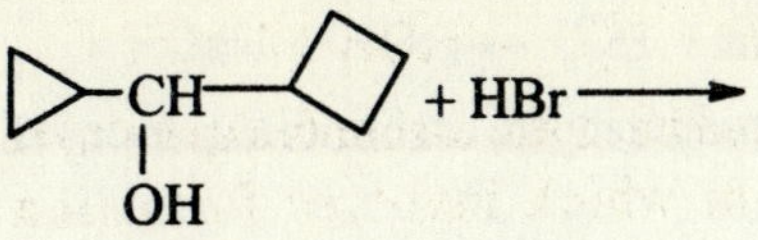 Cyclopropyl–CH(OH)–cyclobutyl + HBr ⟶

(f) 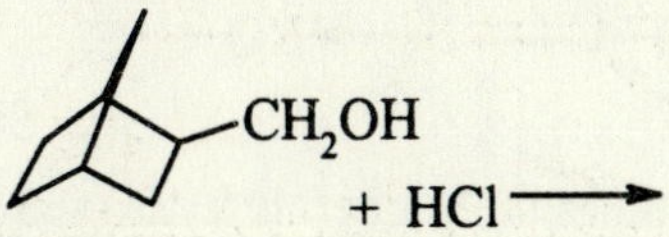 Bicyclic compound with CH_2OH + HCl ⟶

Solution

(a) 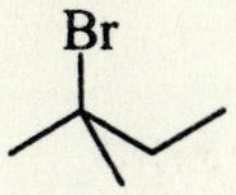 tertiary bromide (Br) (b) 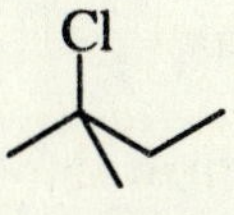 tertiary chloride (Cl)

(c) $PhCHBrCH_3$ (d) 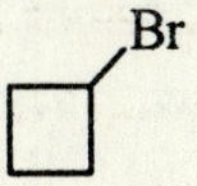 bromocyclobutane (Br)

(e) 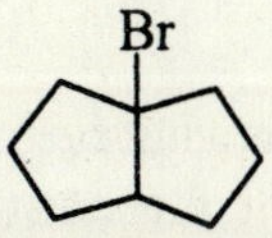 bicyclic bromide (Br) (f) 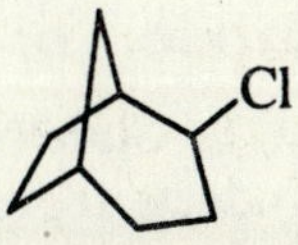 bicyclic chloride (Cl)

Primary alcohols and methanol apparently react through a SN-2 mechanism. In this reaction the function of acid is to produce a protonated alcohol. The halide ion then displaces a molecule of water from carbon as:

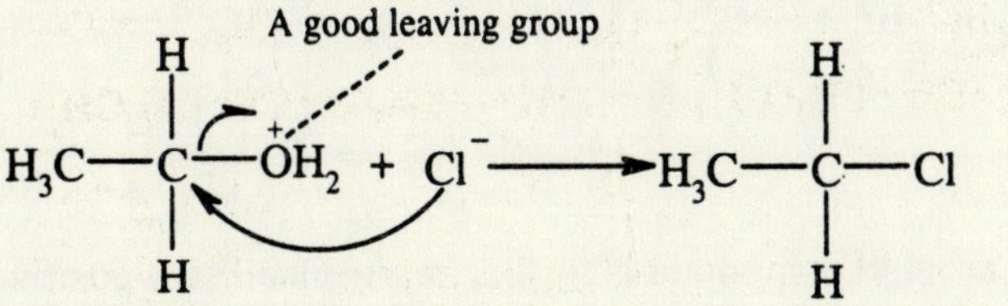

A good leaving group

$$H_3C-CH_2-\overset{+}{O}H_2 + Cl^- \longrightarrow H_3C-CH_2-Cl$$

Example 10. What factor explains the observation that tertiary alcohols react with HX faster than secondary alcohols? What factor explain the observation that methanol reacts with HX faster than a primary alcohols?

Solution: A 3°-alcohol forms stable intermediate (3°-carbocation) with HX.

Example 11. Outline a stepwise mechanism of the following transformations:

(a) 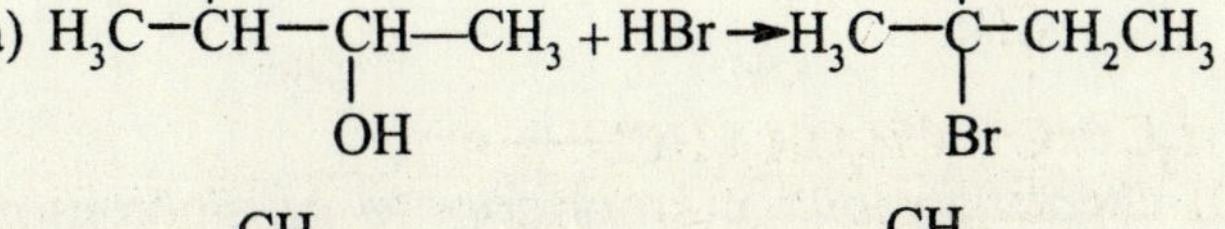 $H_3C-CH(CH_3)-CH(OH)-CH_3 + HBr \longrightarrow H_3C-CBr(CH_3)-CH_2CH_3$

(b) 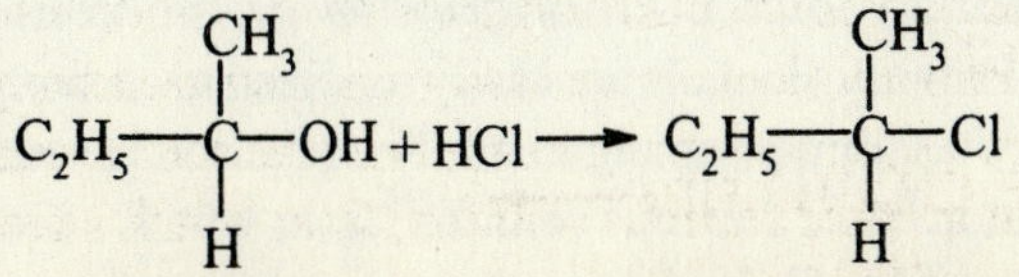 $C_2H_5-CH(CH_3)-OH + HCl \longrightarrow C_2H_5-CH(CH_3)-Cl$

(d-2-butanol) (d,l-2-chlorobutane)

Solution

(a) 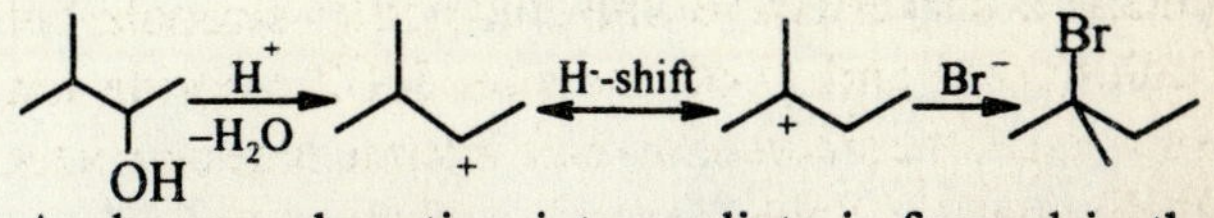 OH (H^+, $-H_2O$) ⟶ + (H⁻-shift) ⟶ + (Br^-) ⟶ Br

(b) A planar carbocation intermediate is formed in the intermediate step which gives racemic mixture of final product.

6.8.3 Reaction of Alcohols with Phosphorous Halides

Phosphorous pentachloride and trichloride reacts with alcohols to give alkyl halide. Unlike the reaction of alcohol with HBr, the reaction of alcohols with trihalide (usually PBr_3 is preferred) does not involve formation of a carbocation and usually occurs without rearrangement of carbon skeleton (especially if the reaction temperature is below 2 °C). Due to this reason, phosphorous tribromide has advantages over HX in formation of alkyl halide from alcohols.

Mechanism

Step I

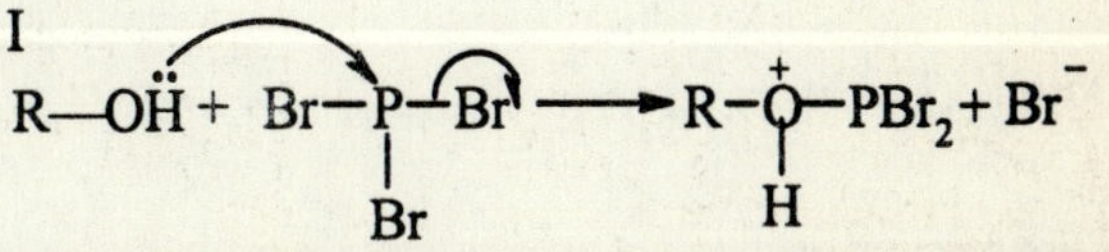

$$R-\ddot{O}H + Br-PBr-Br \longrightarrow R-\overset{+}{O}(H)-PBr_2 + Br^-$$

Step II

$$Br^- + R-\overset{+}{O}(H)-PBr_2 \longrightarrow R-Br + HOPBr_2$$

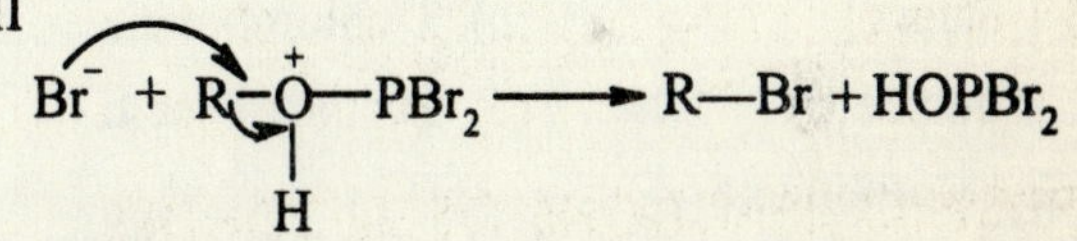

The $HOPBr_2$ can react with more alcohol, therefore the net result is the conversion of 3 moles of alcohol to alkyl bromide by one mole of PBr_3 as:

$$3R-OH + PBr_3 \rightarrow 3R-Br + H_3PO_3$$

6.8.4 Reaction of Thionyl Chloride with Alcohols

Thionyl chloride reacts with straight chain primary alcohols without rearrangement in the presence or absence of pyridine.

Mechanism

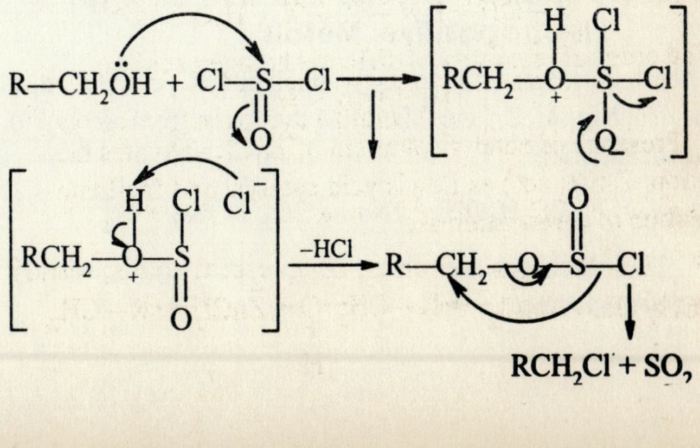

$$R-CH_2\ddot{O}H + Cl-S(=O)-Cl \longrightarrow [RCH_2-\overset{+}{O}(H)-S(Cl)(O^-)-Cl]$$

$$[RCH_2-\overset{+}{O}(H)-S(Cl)(=O)\ Cl^-] \xrightarrow{-HCl} R-CH_2-O-S(=O)-Cl \longrightarrow RCH_2Cl + SO_2$$

Example 12. Starting with a suitable alcohol, outline a synthesis of each of the following:

(a) Benzyl bromide
(b) Cyclohexyl chloride
(c) Butylbromide

Solution

(a) $C_6H_5CH_2OH \xrightarrow{TsCl} C_6H_5CH_2OTs \xrightarrow{NaBr} C_6H_5CH_2Br$

(b) Cyclohexanol $\xrightarrow{TsCl} \xrightarrow{NaCl}$ Cyclohexyl chloride

(c) $CH_3CH_2CH_2CH_2OH \xrightarrow{TsCl} \xrightarrow{NaBr} CH_3CH_2CH_2CH_2Br$

Example 13. Identify A to E in the following sequence of reaction:

$$A(C_4H_{10}O) \xrightarrow{SOCl_2} B(C_4H_9Cl)$$

$$A(C_4H_{10}O) \xrightarrow{H_2SO_4} C(C_4H_8)$$

$$D(C_3H_6O_2) \xleftarrow[H_2O_2]{O_3} C(C_4H_8) \xrightarrow{HCl} E(C_4H_9Cl)$$

Solution

(a) 1-butanol
(b) 1-chlorobutane
(c) 1-butene
(d) Propanoic acid
(e) 2-chlorobutane

6.8.5 Reaction of Alcohol with Isocyanate

Alcohols combine with phenyl or alkyl isocyanate to form urethane:

$$R{-}N{=}C{=}O + R'{-}OH \longrightarrow R{-}NH{-}\overset{O}{\overset{\|}{C}}{-}OR'$$

(Urethane)

$$Ph{-}N{=}C{=}O + R'{-}OH \longrightarrow Ph{-}NH{-}\overset{O}{\overset{\|}{C}}{-}OR'$$

6.8.6 Reaction of Alcohols with Strongly Electropositive Metals

Alcohols reacts with strongly electropositive metals like K, Na, Mg, Al, Zn, etc. liberating hydrogen from alcohol to form alkoxide.

e.g. $\quad 2C_2H_5OH + 2Na \rightarrow 2C_2H_5ONa + H_2$

The alkoxides are white, deliquescent solids, readily soluble in water.

$$RONa + H_2O \rightarrow ROH + NaOH$$

This alkoxide reacts with carbondisulphide to form xanthates:

$$R{-}ONa + S{=}C{=}S \longrightarrow RO{-}C({=}S){-}SNa$$

The fact here is that alcohols liberate hydrogen by the action of metals shows that alcohol can act as acids, but since they do not affect the pH of water, they are weaker acids than water.

6.8.7 Reaction of Alcohols with Acid Chloride

Alcohols reacts with acid chloride to form esters as:

$$R{-}COCl + R'OH \rightarrow R{-}COOR' + HCl$$

6.8.8 Oxidation of Alcohol

Alcohols may be oxidized by various oxidizing agents and oxidation products depends on type of alcohol and nature of oxidizing agents.

(a) Oxidation with halogens

$$Br{-}Br + H{-}\underset{R}{\overset{H}{C}}{-}\ddot{O}{-}H \xrightarrow[-HBr]{} Br^- + R{-}\overset{H}{C}{=}\overset{+}{O}{-}H \xrightarrow{H^+} R{-}CHO$$

Carbonyl compound is further halogenated at α-position, therefore, it is not a suitable method of oxidizing alcohols into aldehydes and ketoses.

(b) Oxidation of alcohols with $FeSO_4/H_2O_2$ to diol as

$$Fe^{2+} + H_2O_2 \longrightarrow Fe^{3+} + OH^- + {}^{\bullet}OH$$

$$CH_3{-}\underset{OH}{CH}{-}CH_3 + {}^{\bullet}OH \longrightarrow CH_3{-}\underset{OH}{CH}{-}\dot{C}H_2$$

$$\xrightarrow{\text{Dimerization}} CH_3{-}\underset{OH}{CH}{-}CH_2{-}CH_2{-}\underset{OH}{CH}{-}CH_3$$

(c) Oxidation with permanganate or dichromate solution: Primary alcohols are easily oxidized to carboxylic acids by potassium permanganate. The reaction is usually carried out in basic aqueous solution, from which a brown precipitate of MnO_2 is observed.

$$R{-}CH_2OH \xrightarrow[\bar{O}H/heat]{KMnO_4} R{-}COOK + MnO_2$$

$$\downarrow H_3O^+$$

$$R{-}COOH$$

Oxidation of secondary alcohols gives ketones. The reaction of an alcohols with either alkaline permanganate or acidic dichromate stops at the ketone stage, because further oxidation requires cleavage of carbon–carbon bond.

$$R{-}CH(R){-}OH \xrightarrow{[O]} R{-}C(R){=}O$$

Various other oxidizing agents based on Cr(VI) have been used to effect the oxidation reaction, e.g. chromic acid is reduced from Cr^{+6} (orange coloured solution) to Cr^{3+} (a blue green solution).

Mechanism of Chromic Acid Oxidation

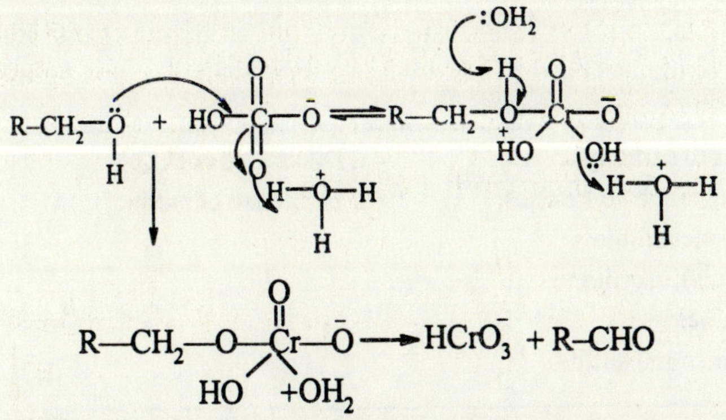

$$R{-}CH_2{-}\ddot{O}H + HO{-}CrO_2{-}\bar{O} \rightleftharpoons R{-}CH_2{-}\overset{+}{O}H{-}Cr(O)(HO)(OH){-}\bar{O}$$

$$R{-}CH_2{-}O{-}Cr(O)(HO)(+OH_2){-}\bar{O} \longrightarrow HCrO_3^- + R{-}CHO$$

(d) Oxidation with Jone's reagent: When CrO_3 is used in aqueous acetone solution for oxidation, double bonds, if any present, are not affected as:

$$R{-}CH{=}CH{-}CH_2OH + CrO_3/\text{aq Acetone} \rightarrow R{-}CH{=}CH{-}COOH$$

(e) Oxidation with PCC (Pyridinium chlorochromate): Primary alcohols can be oxidized to aldehydes and then carboxylic acid as:

$$R{-}CH_2OH + [O] \rightarrow R{-}CHO \xrightarrow{[O]} R{-}COOH$$

The oxidation of aldehydes to carboxylic acid, in aqueous medium usually takes place, with less powerful oxidizing agents than those required to oxidize primary alcohols to aldehydes. Therefore, it is difficult to stop the oxidation of primary alcohol to aldehyde stage. A special oxidizing agent used for oxidation of primary alcohol to aldehyde is PCC that is formed when CrO_3 is dissolved in HCl and then treated with pyridine.

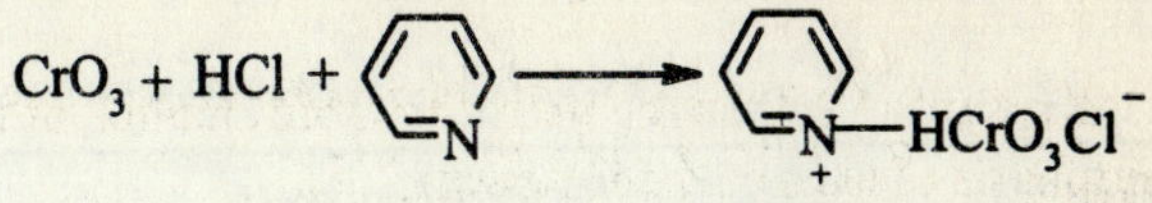

$$CrO_3 + HCl + C_5H_5N \longrightarrow C_5H_5\overset{+}{N}{-}HCrO_3Cl^- \text{ (PCC)}$$

The advantage of PCC over other common oxidizing agent is that, it also does not attacks the double bond.

(f) Dehydrogenation of alcohols: When a primary or secondary alcohol is treated with copper or copperoxide (CuO) catalyst at high temperature (≈ 300 °C), dehydrogenation occurs and corresponding aldehydes or ketones are formed as:

$$R{-}CH_2{-}OH + CuO/heat \rightarrow R{-}CHO$$

$$R_2CHOH + CuO/heat \rightarrow R_2CO$$

The primary limitation of dehydrogenation reaction is the inability of many organic compounds to survive a reaction at 300 °C. Dehydrogenation is not well suited for laboratory synthesis. Table 6.2 describes the role of common oxidizing agent used for selective oxidation.

Example 14. Give structure of the principal product(s) when each of the following alcohol reacts with:

(i) $Na_2Cr_2O_7/H_2SO_4$
(ii) PCC
(iii) $KMnO_4/OH^-$(cold, dilute)
(a) 1-octanol
(b) 3-octanol
(c) 2-cyclohexen-1-ol
(d) 1-methyl cyclohexanol

Solution

(a) (i) Octanoic acid, (ii) Octanal, (iii) Octanoic acid
(b) (i) 3-octanone, (ii) 3-octanone, (iii) 3-octanone
(c) (i) 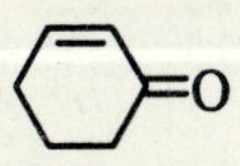 2-cyclohexen-1-one (ii) 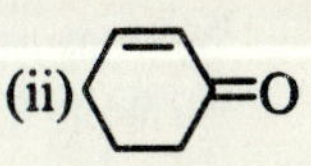 2-cyclohexen-1-one (iii) 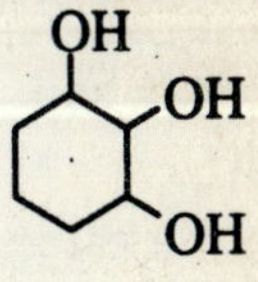 cyclohexane-1,2,3-triol (OH, OH, OH)
(d) No reaction

Example 15. Suggest the method that would work best for each of the following laboratory synthesis:

(a) 1-butanol → butanal.
(b) 1-butanol → botanoic acid.
(c) 2-butanol → 2-butanone.
(d) 2-buten-1-ol → 2-butenoic acid.
(e) 1-methyl cyclohexanol → 2-Methyl cyclohexanone.
(f) 2-buten-1-ol → 2-butenal.

Solution

(a) PCC
(b) Na_2CrO_4/H_2SO_4
(c) Na_2CrO_4/H_2SO_4

Table 6.2: Some common oxidizing agent and their role in organic chemistry

Reagent	*Preferred solvents*	*Functions oxidized*	*Reaction work-up*
Jones reagent H_2CrO_4	Aqueous sulfuric acid and acetone (avoid acid sensitive systems)	1°-alcohols to carboxylic acids saldehydes to carboxylic acids, 2°-alcohols to ketones avoid amines and sufides	1. Destroy excess reagent 2. Extract product
Collins reagent $CrO_3 \cdot 2\,C_5H_5N$	Methylene chloride (CH_2Cl_2)	1°-alcohols to aldehydes 2°-alcohols to ketones	1. Filter inorganic salts 2. Wash with aq. acid 3. Remove CH_2Cl_2 solvent
Pyridinium Chlorochromate $ClCrO_3 \cdot C_5H_5NH$	(CH_2Cl_2)	1°-alcohols to aldehydes 2°-alcohols to ketones	1. Filter inorganic salts 2. Wash with aqueous acid 3. Remove CH_2Cl_2 solvent
Dimethyl sulfoxide $(CH_3)_2S{=}O$ and DCC or Ac_2O or $(CF_3CO)_2O$or SO_3 or $(COCl)_2$	CH_2Cl_2 or ethers or DMSO	A mild procedure 1°-alcohols to aldehydes 2°-alcohols to ketones	1. Neutralize reactants 2. Remove solvents
Potassium permanganate $KMnO_4$	Water and aqueous solvent mixture, pyridine often used catalytically	Aldehydes to carboxylic acids, 2°-alcohols to ketones, alkenes to vicinal-diols (vic.-glycols), alkynes to carboxylic acids, avoid amines and sufides	1. Dilute acid or H_2O_2 2. Extraction of product
Osmium tetroxide OsO_4	Pyridine often used catalytically	Alkenes to vicinal-diols (vic.-glycols)	
Periodic acid HIO_4	Water or aqueous mixtures	vic.-glycols to carbonyl compounds	1. Destroy excess reagent 2. Extract product
Lead tetraacetate $Pb(OCOCH_3)_4$	Benzene or acetic acid	vic.-glycols to carbonyl compounds	1. Destroy excess reagent 2. Extract product
Peracids CH_3CO_3H $C_6H_5CO_3H$, etc.	CH_2Cl_2 or ethers	Alkenes to epoxides ketones to esters, avoid amines and sufides	1. Destroy excess reagent 2. Extract product
OzoneO_3	CH_2Cl_2 or $CHCl_3$ (sometimes alcohol)	Cleaves alkenes and alkynes avoid benzene derivatives amines and sufides	1. Destroy excess reagent and ozonides 2. Extract product

(d) PCC followed by treatment with Ag_2O/H_2O.

(e) [1-methylcyclohexanol $\xrightarrow{H_2SO_4}$ 1-methylcyclohexene $\xrightarrow[H_2O_2/NaOH]{B_2H_6}$ 2-methylcyclohexanol $\xrightarrow{PCC}$ 2-methylcyclohexanone]

(f) PCC

6.8.9 Reduction of Alcohols

(i) Reduction of Alcohol with HI/Red P

HI/Red P is one of the best reducing agent used for reduction of alcohol to alkane:

$$ROH + HI/red\ P \rightarrow RH$$

(ii) Reduction with $LiAlH_4$ in Presence of $TiCl_3$ or $AlCl_3$

Some alcohols are directly reduced to alkane with $LiAlH_4$ in presence of $TiCl_3$ or $AlCl_3$ catalyst.

[2-phenylcyclopentanol $\xrightarrow[TiCl_3]{LiAlH_4}$ phenylcyclopentane]

Unfortunately, this reduction does not work well with some alcohols. A more general method for reducing an alcohol involves converting the alcohol to the tosylate ester, then using hydride reducing agent to displace the tosylate leaving group.

$$ROH + Cl-SO_2-C_6H_4-CH_3 \xrightarrow{Pyridine} RO-SO_2-C_6H_4-CH_3 \xrightarrow{LiAlH_4} RH$$

Use of tosylate ester: Tosylate esters are easily made from alcohols in very high yields, often with tosylchloride as reagent and pyridine as solvent. The tosylate group is an excellent leaving group and alkyl tosylate undergo substitution and elimination much like alkyl halides.

[C–C(OH) $\xrightarrow[Pyridine]{TsCl}$ C–C(OTs) $\xrightarrow{Nu}$ C–C(Nu)]

[B⁻ abstracts H from C adjacent to C–OTs $\longrightarrow$ C=C + BH + $\overline{OTs}$ (elimination)]

The substitution of tosylate by a nucleophile follows SN-2 mechanism with inversion of configuration as:

[I^- + (Et)(H)(CH$_3$)C–OTs $\longrightarrow$ I–C(Et)(H)(CH$_3$) + OTs^-]

(inverted configuration)

Tosylate ion is a particularly stable anion with negative charge delocalized over three-oxygen atoms as:

[$H_3C-C_6H_4-S(=O)_2-O^-$ $\longleftrightarrow$ resonance structures with negative charge on each of the three oxygens]

Tosylate group can be conveniently substituted by OH^-, CN^-, X^-, RO^-, NH_2^-, H^-, (from $LiAlH_4$) through SN-2 mechanism. In each case, R- must be a primary or unhindered secondary alkyl group, if substitution is to predominate over elimination.

6.8.10 Dehydration of Alcohols (Formation of Alkenes)

An alcohol on treatment with a strong acid catalyst at around 200°C, undergo dehydration reaction (intramolecular) to produce alkene. Acidic catalyst protonate the hydroxyl group of alcohol and convert it into a good leaving group.

$$-\overset{H}{\underset{|}{C}}-\overset{OH}{\underset{|}{C}}- \overset{H^+}{\rightleftharpoons} >C=C< + OH_2$$

To drive the equilibrium to the right, we remove one or both of the products as they form: either by distilling the products out of the reaction mixture or by adding a dehydrating agent.

Mechanism

Dehydration of alcohols proceeds by E-I mechanism as:

[cyclopentanol $\overset{H^+}{\rightleftharpoons}$ protonated cyclopentanol ($\overset{+}{O}H_2$) $\rightleftharpoons$ cyclopentyl carbocation $\overset{-H^+}{\rightleftharpoons}$ cyclopentene]

(a carbocation)

Figure 6.1 shows the reaction energy profile for E1 dehydration of an alcohol.

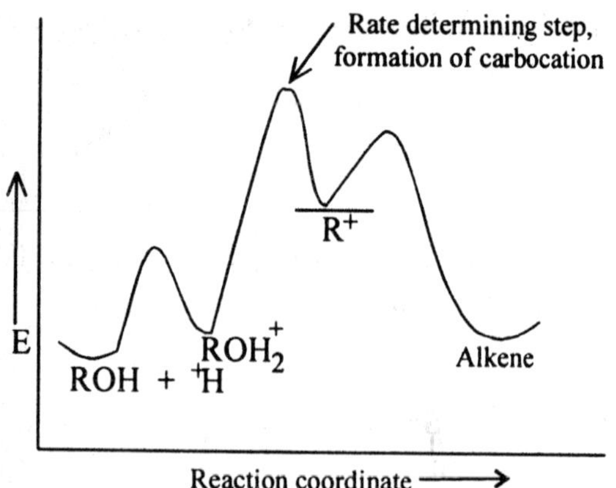

Figure 6.1: Potential energy diagram for dehydration of an alcohol

Since the rate-determining step is formation of carbocation, the ease of dehydration follows the same order as the ease of formation of carbocation: 3° > 2° > 1°. An important characteristic of carbocation is rearrangement to more stable carbocation. If structure permits, by hydride shift or alkyl shift, therefore during dehydration, carbon skeleton may be changed as:

$$H_3C-\underset{H}{\overset{CH_3}{C}}-\underset{OH}{CH}-CH_3 \xrightleftharpoons[-H_2O]{H^+}$$

$$\left[H_3C-\underset{H}{\overset{CH_3}{C}}-\overset{+}{C}H-CH_3 \text{ (sec. carbocation)} \xleftrightarrow{\text{Hydride shift}} H_3C-\underset{+}{\overset{CH_3}{C}}-\underset{H}{CH}-CH_3 \text{ (tertiary carbocation)} \right]$$

$$\Big\downarrow\!\!\Big\uparrow H^+$$

$$H_3C-\overset{CH_3}{C}=CH-CH_3 \text{ (major product)}$$

$$H_3C-\underset{CH_3}{\overset{CH_3}{C}}-\underset{OH}{CH}-CH_3 \xrightleftharpoons[-H_2O]{H^+}$$

$$\left[H_3C-\underset{CH_3}{\overset{CH_3}{C}}-\overset{+}{C}H-CH_3 \text{ (secondary carbocation)} \xleftrightarrow{\text{Methyl shift}} H_3C-\underset{+}{\overset{CH_3}{C}}-\underset{CH_3}{CH}-CH_3 \text{ (tertiary carbocation)} \right]$$

$$\Big\downarrow\!\!\Big\uparrow -H^+$$

$$(CH_3)_2C=C(CH_3)_2$$

If dehydration can produce a conjugated double bond, it will be the major product as:

$$C_6H_5-CH_2-\overset{OH}{CH}-\overset{CH_3}{CH}-CH_3 \rightleftharpoons$$

$$C_6H_5-CH_2-\overset{+}{C}H-\underset{CH_3}{CH}-CH_3 \quad \text{I}$$

$$C_6H_5-CH_2\ \ CH_2-\underset{+}{\overset{CH_3}{C}}-CH_3 \quad \text{II}$$

$$C_6H_5-\overset{+}{C}H-CH_2-\overset{CH_3}{CH}-CH_3 \quad \text{III}$$

II is a 3° carbocation but III is 2°-benzyl carbocation, relatively more stable carbocation, and formed in preference of II. Also III after deprotonation yields a conjugated alkene whereas II gives non-conjugated alkene after deprotonation. Therefore, III is preferred over II.

$$\text{(i)} \xrightleftharpoons{-H^+} C_6H_5-CH_2-CH=\overset{CH_3}{C}-CH_3$$

$$\text{(iii)} \xrightleftharpoons{-H^+} C_6H_5-CH=CH-\overset{CH_3}{CH}-CH_3$$

Ring Expansion During Dehydration of Alcohols

Certain strained rings like cyclopropyl, cyclobutyl, etc. prefer to expand during dehydration reaction to a less strained, larger ring as:

$$\triangleright-CH_2OH \xrightleftharpoons[-OH_2]{H^+} \triangleright-\overset{+}{C}H_2 \rightleftharpoons \square^{+} \xrightleftharpoons{-H^+} \text{cyclobutene}$$

Summary of Dehydration Reaction

1. Dehydration usually goes by E1 mechanism. Rearrangement may occur to form more stable carbocations.
2. Dehydration works best with tertiary alcohols and almost as well with secondary alcohols. Rearrangement and poor yields are common with primary alcohols.
3. Sytzebb's rule is followed in the deprotonation step, i.e. deprotonation occurs from α-carbon of carbocation, bearing least number of hydrogen.

Example 16. Predict major products of sulfuric acid catalyzed dehydration of following alcohols.

(a) 1-methyl cyclohexanol (b) Neopentyl alcohol
(c) 2-methyl-2-butanol (d) 1-pentanol
(e) 3-methyl cyclohexanol (f) 1-isopropyl cyclohexanol.

(g) 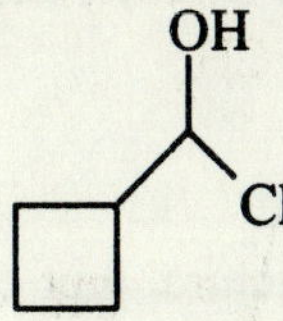 (h) 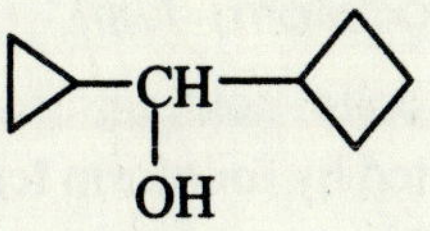 (i)

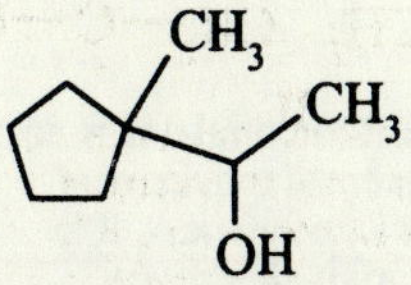

Solution

(a) Methyl cyclohexene
(b) 2-methyl-2-butene
(c) 2-methyl-2-butene
(d) 2-pentene
(e) 3-methyl cyclohexene and 4-methyl cyclohexene

(f) 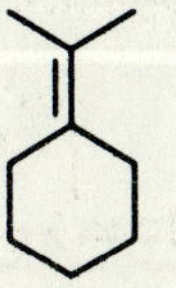 (g) 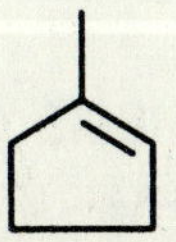 (h) 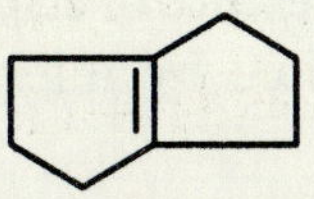 (i)

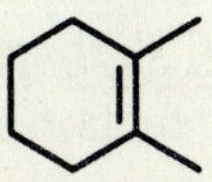

Example 17 Acid catalyzed dehydration of either 2-methyl-1-butanol or 3-methyl-1-butanol gives 2-methyl-2-butene as the major product. Write plausible mechanism that explains these results.

Solution

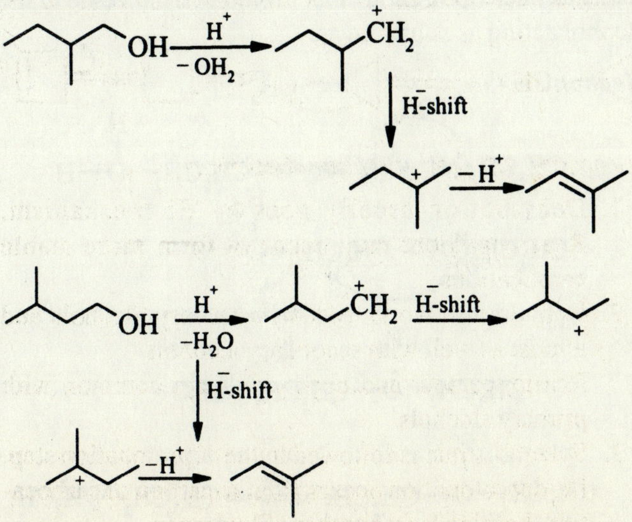

$\xrightarrow[-OH_2]{H^+}$ $\overset{+}{C}H_2$ $\xrightarrow{\bar{H}\text{-shift}}$ $\xrightarrow{-H^+}$

$\xrightarrow[-H_2O]{H^+}$ $\overset{+}{C}H_2$ $\xrightarrow{\bar{H}\text{-shift}}$ $\xrightarrow{\bar{H}\text{-shift}}$ $\xrightarrow{-H^+}$

Example 18. When the compound (I) is heated with 50% H_2SO_4, the product of the reaction is the compound (II) and not (III). Write a step-by-step mechanism showing how II is formed.

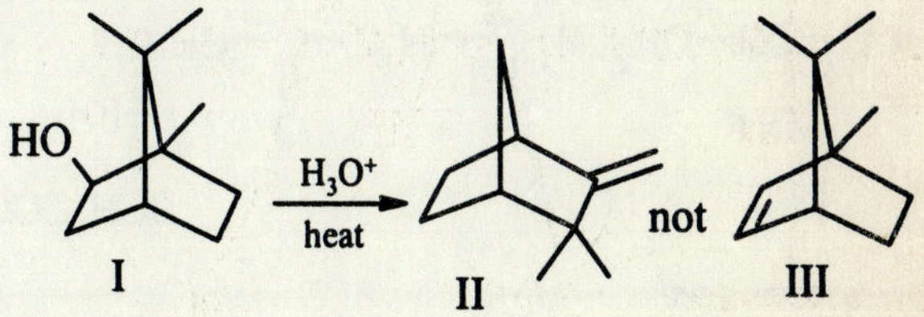

HO — I $\xrightarrow[heat]{H_3O^+}$ II not III

Solution

$\xrightarrow[-H_2O]{H^+}$ $\xrightarrow{-H^+}$

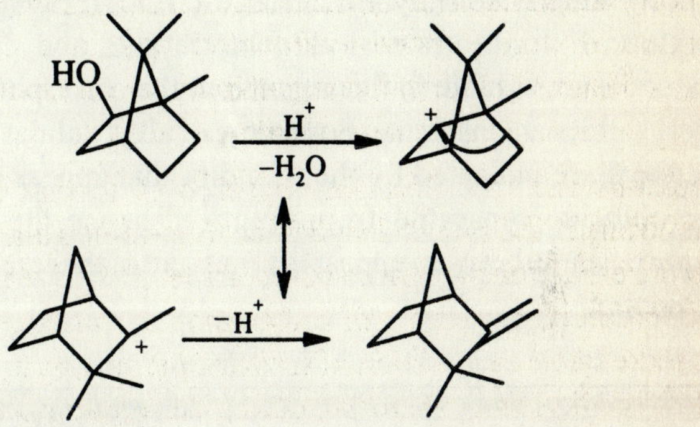

Some other important points regarding dehydration reaction: Dehydration of alcohols can also be done by passing alcohol over hot alumina Al_2O_3. Isomerization of product to a more substituted, more stable alkene is an important phenomena in dehydration of alcohols with acid catalyst. The isomerization can be suppressed by dehydration with alumina and a small amount of pyridine.

Primary alcohols are dehydrated by concentrated acid at 180°C, secondary and tertiary alcohol by boiling with dilute sulfuric acid, this is used to avoid polymerization of alkene.

6.9 METHODS OF DISTINGUISHING BETWEEN THE THREE CLASSES OF ALCOHOL

(a) By means of oxidation

The nature of oxidation products of alcohols depends on the type of alcohols.

(i) Primary alcohols on oxidation with acidic dichromate or permanganate gives an acid.

$$RCH_2OH + H^+/[O] \rightarrow RCOOH$$

(ii) A secondary alcohol on oxidation under mild condition gives ketone.

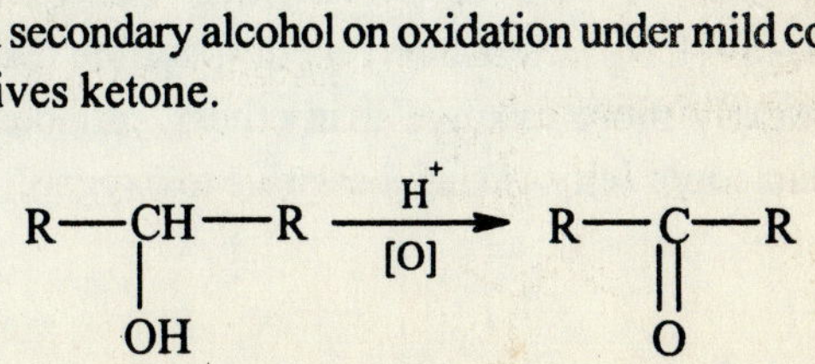

$$R-\underset{\displaystyle OH}{\underset{|}{CH}}-R \xrightarrow[{[O]}]{H^+} R-\underset{\displaystyle O}{\underset{||}{C}}-R$$

Ketones are fairly difficult to oxides, but prolonged action of oxidizing agents produces a mixture of acids. During oxidation of unsymmetrical ketone carbonyl group (C=O) goes with lower alkyl groups as:

$$H_3C-\underset{OH}{\underset{|}{CH}}-CH_2CH_3 \xrightarrow[{[O]}]{H^+} H_3C-\underset{O}{\underset{\|}{C}}-CH_2CH_3 \xrightarrow{[O]} 2CH_3COOH$$

(b) Lucas test

Distinction among primary, secondary and tertiary alcohol is conveniently done by the Lucas test. Lucas reagent is a mixture of concentrated hydrochloric acid and $ZnCl_2$. Alcohols are soluble in this regent but the corresponding alkyl halides are insoluble. Formation of alkyl halides from alcohols are indicated by the turbidity that appear when the chloride separated from solution, hence the time required for turbidity to appear is a measure of the reactivity of alcohol.

A tertiary alcohol reacts with Lucas reagent and an immediate turbidity is observed. Secondary alcohol reacts in a few minutes with Lucas reagent and delayed turbidity is observed. A primary alcohol does not react appreciably at room temperature. Ally alcohol reacts as rapidly as tertiary alcohol, but allyl chloride is water-soluble, therefore no turbidity is observed.

(c) Iodoform test

An alcohol containing a particular structural unit can be detected by iodoform test.

$$R-\underset{OH}{\underset{|}{CH}}-CH_3$$

This structural unit is an essential requirement for iodoform test, R is H or an alkyl group

e.g. $CH_3CH_2OH + NaOH/I_2 \rightarrow CHI_3$ (yellow precipitate) + HCOONa

$C_6H_5CH(OH)CH_3 + NaOH/I_2 \rightarrow CHI_3 + C_6H_5COONa$

Ethers and Epoxides

6.10 ETHER NOMENCLATURE

Ethers are compounds having two alkyl or aryl groups bonded to an oxygen atom, as in the formula R—O—R. The ether functional group does not have a characteristic IUPAC nomenclature suffix, so it is necessary to designate it as a substituent. To do so the common alkoxy substituents are given names derived from their alkyl component. Examples of ether nomenclature are provided below.

Alkyl group	*Name*	*Alkoxy group*	*Name*
CH_3–	Methyl	CH_3O–	Methoxy
CH_3CH_2–	Ethyl	CH_3CH_2O–	Ethoxy
$(CH_3)_2CH$–	Isopropyl	$(CH_3)_2CHO$–	Isopropoxy
$(CH_3)_3C$–	Tert-butyl	$(CH_3)_3CO$–	Tert-butoxy
C_6H_5–	Phenyl	C_6H_5O–	Phenoxy

Simple ethers are given common names, in which the alkyl groups bonded to the oxygen are named in alphabetical order followed by the word "ether". Sulfur analogs of ethers (R—S—R′) are called *sulfides*. For example, $(CH_3)_3C—S—CH_3$ is tert-butyl methyl sulfide. Sulfides are chemically more reactive than ethers, reflecting the greater nucleophilicity of sulfur relative to oxygen.

6.11 SYNTHESIS OF ETHERS

I. By Intermolecular Dehydration of Alcohols

When alcohols are distilled in presence of acid catalyst at about 140°C, ethers are formed by intermolecular dehydration reaction as

$$CH_3CH_2OH \xrightarrow[180°C]{H_2SO_4} CH_2=CH_2$$

$$CH_3CH_2OH \xrightarrow[140°C]{H_2SO_4} CH_3CH_2OCH_2CH_3$$

Ether is formed by SN-2 mechanism with one molecule acting as nucleophile and other protonated molecule of the alcohol acting as substrate as:

Mechanism

$$CH_3CH_2\ddot{O}H + H^+ \rightleftharpoons CH_3CH_2-\overset{+}{O}H_2$$

$$CH_3CH_2\ddot{O}H + CH_3CH_2-\overset{+}{O}H_2 \rightleftharpoons CH_3CH_2-\overset{+}{O}H-CH_2CH_3 + H_2O$$

$$CH_3CH_2-\overset{+}{O}H-CH_2CH_3 \xrightleftharpoons{-H^+} CH_3CH_2OCH_2CH_3$$

This method of synthesizing ethers has limited usefulness, because of the following reasons:

(i) Synthesis of ethers using secondary and tertiary alcohols leads to formation of alkenes as major products.

(ii) Unsymmetrical ether cannot be prepared in good yield by this method because, mixture of products are obtained as:

$$ROH + R'OH \overset{H^+}{\rightleftharpoons} ROR + R'OR' + ROR'$$

II. The Williamsons Synthesis

Ethers are usually prepared from alcohols or their conjugate bases. One important procedure, known as the *Williamson Ether Synthesis*, proceeds by an SN-2 reaction of an alkoxide nucleophile with an alkyl halide.

$$RO^- \overset{+}{Na} + R'-X \longrightarrow R-O-R' + NaX$$

Here the alkoxide ion (RO^-) acts as nucleophile and R′—X as substrate. The requirement is the substrate (R′—X) must be a primary alkyl halide otherwise elimination (E2) predominate leading to formation of alkene.

Reactions 1 and 2 below are two examples of this procedure. When applied to an unsymmetrical ether, as in this case, there are two different combinations of reactants are possible. Of these one is usually better than the other. Since alkoxide anions are strong bases, the possibility of a competing E2 elimination must always be considered. Bearing in mind the factors that favour substitution over elimination, a 1°-alkyl halide should be selected as a preferred reactant whenever possible. Thus, reaction 1 gives a better and cleaner yield of benzyl isopropyl ether than does reaction 2, which generates considerable elimination product.

1. $H-C(CH_3)_2-O^- + C_6H_5CH_2Br \xrightarrow{SN\text{-}2} C_6H_5-CH_2-O-C(CH_3)_2-H$

2. $H-C(CH_3)_2-Br + C_6H_5CH_2O^- \xrightarrow{SN\text{-}2} C_6H_5-CH_2-O-C(CH_3)_2-H + CH_3CH{=}CH_2$ (E2 product)

3. $C_2H_5-CH{=}CH_2 \xrightarrow[C_2H_5OH]{Hg(OCOCF_3)_2}$ (organomercury ether intermediate, Hg–O–CO–CF_3) $\xrightarrow{NaBH_4}$ (sec-butyl ethyl ether)

4. (cyclopentene) $\xrightarrow[(CH_3)_2CHCH_2OH]{Hg(OCOCF_3)_2}$ (trans-1-$OCH_2CH(CH_3)_2$-2-$HgOCOCF_3$-cyclopentane) $\xrightarrow{NaBH_4}$ (cyclopentyl–$OCH_2CH(CH_3)_2$)

Example 19. Outline two methods for preparing isopropyl ethyl ether by Williamson's method and explain, which is the better method.

Solution: Method I

$CH_3CH_2OH + Na \rightarrow C_2H_5ONa$ (1)

2-chloropropane $\xrightarrow{(1)} CH_3CH_2O-CH(CH_3)_2$

Method II

2-propanol + Na $\rightarrow (H_3C)_2CH-ONa \xrightarrow{C_2H_5Cl} CH_3CH_2O-CH(CH_3)_2$

Method II is better over method-I, because in the second step of method II, a primary alkyl halide is used, which undergo, predominantly substitution whereas in method I, secondary alkyl halide is used in the second step, which is less reactive towards SN-2 reaction and competes with E2 reaction.

Example 20. When 1-phenyl-2-propanol is treated with potassium followed by C_2H_5Br, 2-ethoxy-1-phenylpropane is obtained with retention of configuration. On the other hand, when the same alcohol is treated with TsCl followed by treatment with C_2H_5ONa gives the same product with inversion of configuration. Provide a mechanistic explanation of this observation.

Solution: Configuration at a chiral carbon can be changed only by cleavage of bond from the chiral carbon.

$$Ph-CH_2-\overset{*}{C}H(OH)-CH_3 \xrightarrow{K/heat} Ph-CH_2-\overset{*}{C}H(OK)-CH_3 \xrightarrow{EtBr} Ph-CH_2-\overset{*}{C}H(OEt)-CH_3$$

Here no bond from chiral carbon (starred) is broken, and therefore configuration at chiral carbon is retained.

$$Ph-CH_2-\overset{*}{C}H(OH)-CH_3 \xrightarrow{TsCl} Ph-CH_2-\overset{*}{C}H(OTs)-CH_3 \xrightarrow[SN\text{-}2]{EtONa} Ph-CH_2-\overset{*}{C}H(OEt)-CH_3$$

Here, last step is a SN-2, inversion of configuration is observed.

Example 21. Write a mechanism of formation of tetrahydro furan from the reaction of 4-chloro-1-butanol with aqueous KOH.

Solution

$$ClCH_2CH_2CH_2CH_2OH \xrightarrow[Heat]{KOH} {}^{-}O(CH_2)_4Cl \text{ (intramolecular displacement of } Cl) \longrightarrow \text{tetrahydrofuran}$$

III. By Alkylation of Alcohol

Tertiary ethers are conveniently prepared by dissolving an alcohol in a strong acid followed by addition of an alkene as:

$$CH_2=C(CH_3)_2 + H^+ \longrightarrow H_3C-\overset{+}{C}(CH_3)_2 \xrightarrow{RCH_2\ddot{O}H} H_3C-C(CH_3)_2-\overset{+}{O}(H)CH_2R \underset{}{\overset{-H^+}{\rightleftharpoons}} H_3C-C(CH_3)_2-OCH_2R$$

This method is also used to protect –OH group as:

$$HOCH_2CH_2CH_2Br \xrightarrow{HC\equiv CNa} NaOCH_2CH_2CH_2Br$$

$$HOCH_2CH_2CH_2Br \xrightarrow[H^+]{CH_2=C(CH_3)_2} H_3C-C(CH_3)_2-OCH_2CH_2CH_2Br \xrightarrow{HC\equiv CNa} H_3C-C(CH_3)_2-OCH_2CH_2CH_2-C\equiv CH \xrightarrow{H^+/H_2O} HOCH_2CH_2CH_2-C\equiv CH$$

6.12 REACTIONS OF ETHER

I. General discussion

Ethers are widely used as solvents for a variety of organic compounds and reactions, suggesting that they are relatively unreactive themselves. Indeed, with the exception of the alkanes, cycloalkanes and fluorocarbons, ethers are probably the least reactive, common class of organic compounds. The inert nature of the ethers relative to the alcohols is undoubtedly due to the absence of the reactive O—H bond. The most common reaction of ethers is cleavage of the C—O bond by strong acids. This may occur by SN-1 or E1 mechanisms for 3°-alkyl groups or by an SN-2 mechanism for 1°-alkyl groups. Some examples are shown in the following diagram. The conjugate acid of the ether is an intermediate in all these reactions, just as conjugate acids were intermediates in certain alcohol reactions.

(a) $C_6H_{11}-O-CH_3 \xrightarrow[SN\text{-}2]{HI} CH_3I + C_6H_{11}OH \xrightarrow[SN\text{-}2]{HI} C_6H_{11}I$

(b) $(CH_3)_2CHCH_2CH_2-O-CH_2CH_2CH(CH_3)_2 \xrightarrow[SN\text{-}2]{2\,HBr} (CH_3)_2CHCH_2CH_2Br$

(c) $C_6H_5-O-C(CH_3)_3 \xrightarrow[SN\text{-}2]{HBr} (CH_3)_3CBr + C_6H_5OH$

(d) 4,4-dimethylcyclohexyl tert-butyl ether $\xrightarrow{CF_3COOH}$ 4,4-dimethylcyclohexanol $+ (CH_3)_2C=CH_2$

The first two reactions proceed by a sequence of SN-2 steps, in which the iodide or bromide anion displaces an alcohol in the first step, and then converts the conjugate acid of that alcohol to an alkyl halide in the second step. Since SN-2 reactions are favoured at least hindered sites, the methyl group in example 'a' is cleaved first. The 3°-alkyl group in example 'c' is probably cleaved by an SN-2 mechanism, but the SN-1 alternative cannot be ruled out. The phenol formed in this reaction does not react further, since SN-2, SN-1 and E1 reactions do not take place on aromatic rings. The last example shows the cleavage of a 3°-alkyl group by a strong acid. Acids having poorly nucleophilic conjugate bases are often chosen for this purpose so that E1 products are favoured. The reaction shown here 'd' is the reverse of the tert-butyl ether preparation described earlier.

Ethers in which oxygen is bonded to 1°- and 2°-alkyl groups are subject to peroxide formation in the presence of air (gaseous oxygen). This reaction presents an additional hazard to the use of these flammable solvents, since peroxides decompose explosively when heated or struck. The mechanism of peroxide formation is believed to be free radical in nature.

$$R{-}O{-}CH(CH_3)_2 + O_2 \rightarrow R{-}O{-}C(CH_3)_2{-}O{-}O{-}H \quad \text{a peroxide}$$

From the above discussion it is concluded that ethers are very less reactive organic compound and used as inert solvent in many of the organic synthesis. Ether is absolutely inert in alkaline medium. Ether can react with proton donor to form oxonium ion as:

$$R{-}\ddot{O}{-}R + H^+ \rightleftharpoons R{-}\overset{H}{\overset{|}{O^+}}{-}R$$

Ethers on strong heating with concentrated solution of HI, HCl, HBr undergo reaction leading to formation of alkyl halide as:

$$C_2H_5OC_2H_5 + 2HBr \rightarrow 2C_2H_5Br + H_2O$$

Mechanism

Step I

$$CH_3CH_2\ddot{O}CH_2CH_3 + H^+ \rightleftharpoons$$

$$CH_3CH_2{-}\overset{H}{\overset{|}{O^+}}{-}CH_2CH_3 \xrightleftharpoons{Br^-} C_2H_5Br + C_2H_5OH$$

Step II

$$CH_3CH_2OH + H^+ \rightleftharpoons$$

$$CH_3CH_2\overset{+}{O}H_2 \xrightleftharpoons{Br^-} CH_3CH_2Br + H_2O$$

An exception is phenolic ether, phenol does not undergo substitution to halobenzene as

$$Ph{-}O{-}R + 2HBr \rightarrow PhOH + RBr$$

Example 22. Predict products of the following reaction:

(a) $CH_3OCH_2CH_2CH_2CH_3 + IH \longrightarrow$

(b) $(CH_3)_3COCH_2CH_3 + IH \xrightarrow{\text{Cold}}$ Provide mechanism

(c) [chroman structure] + IH ⟶

Solution

(a) $CH_3I + CH_3CH_2CH_2CH_2I$ (b) $(CH_3)_3CI + C_2H_5I$
(c) 2-(3-iodopropyl) phenol

II. Hydrolysis of ether

Ethers on distilling with dilute sulfuric acid solution, undergo hydrolysis to yields alcohols as:

$$R{-}O{-}R + H_3O^+ \rightarrow 2R{-}OH$$

6.13 EPOXIDES

Epoxides are cyclic ethers with three membered rings. In IUPAC system the three membered cyclic oxides are named as oxirane

(Oxirane) (2,2-dimethyl oxirane)

(2,2,3-trimethyl-oxirane)

6.14 Preparation

I. By the action of an alkene with an organic per acid

$$R{-}CH{=}CH_2 + R'CO_3H \rightarrow R{-}\text{(epoxide)}$$

Here peroxy acid transfers an oxygen atom to a double bond in a cyclic, single step mechanism. This results in syn addition of oxygen to alkene. This is an example of electrophilic addition of oxygen to an alkene. Therefore, if an alkene containing more than one double bond is treated with one equivalent of peroxy acid, epoxidation occurs preferentially at more electron rich double bond as:

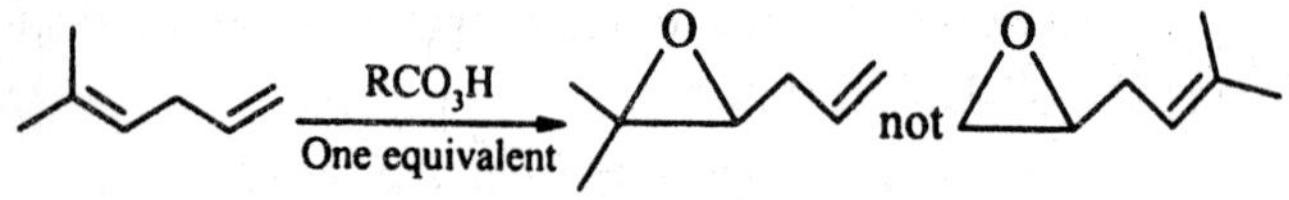

$\xrightarrow[\text{One equivalent}]{RCO_3H}$ not

The epoxidation reaction of an alkene is also stereo specific, i.e. a specific stereomer gives a specific stereomer after epoxidation as:

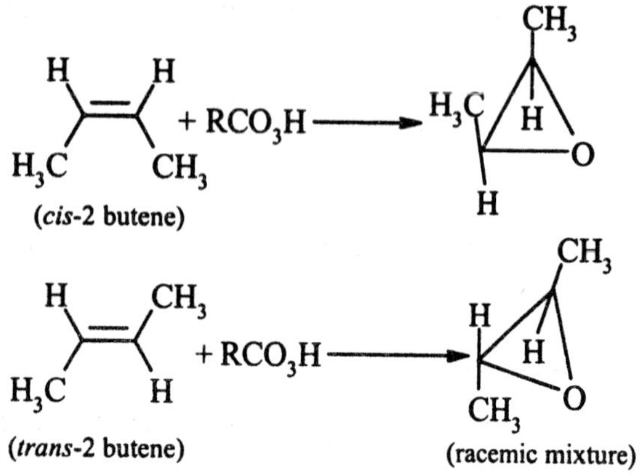

(*cis*-2 butene) + RCO_3H →

(*trans*-2 butene) + RCO_3H → (racemic mixture)

II. Preparation from halohydrin

A β-halohydrin, on treatment with a strong base gives an epoxide as:

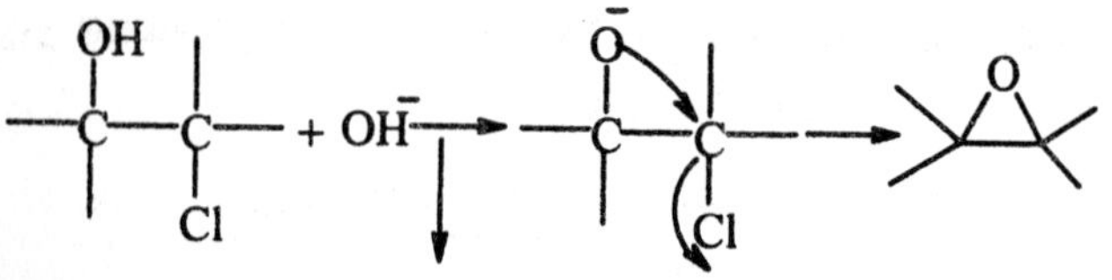

Example 23. Complete the following reaction:

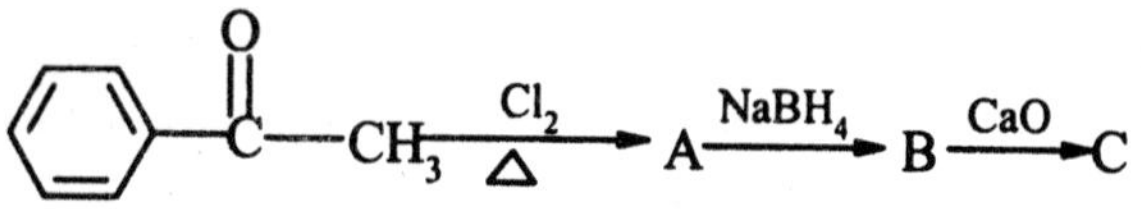

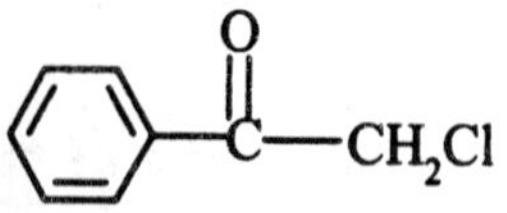

$C_6H_5COCH_3 \xrightarrow[\Delta]{Cl_2} A \xrightarrow{NaBH_4} B \xrightarrow{CaO} C$

Solution

$C_6H_5COCH_2Cl$ (A) $C_6H_5CH(OH)CH_2Cl$ (B)

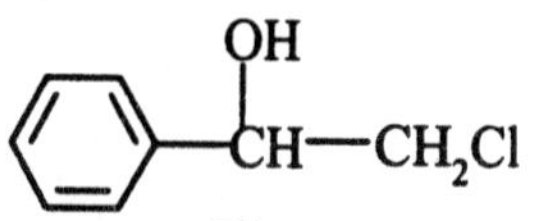

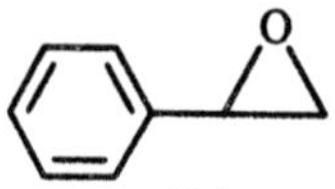

(C)

6.15 Reactions of Epoxides

Epoxides are cyclic ether. Instead of being ether, commonly known for its inertness, epoxides are highly reactive, because of very high angle strain.

I. *Acid catalyzed ring-opening reaction of an epoxide*

(i) *Hydrolysis*

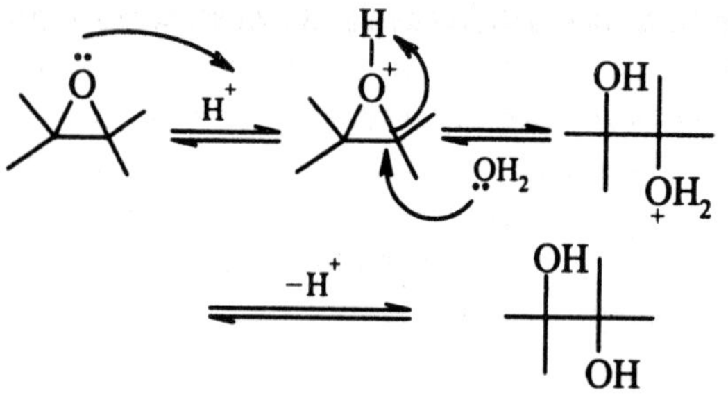

(Anti hydroxylation)

(ii) *Acid catalyzed alcoholysis*

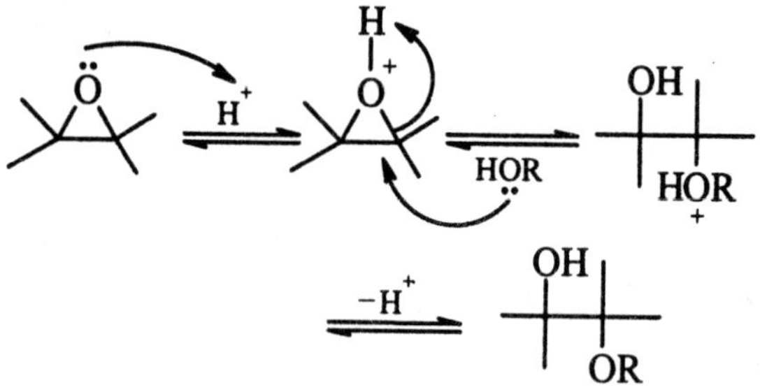

In acid catalyzed ring opening of an unsymmetrical epoxide, the nucleophile attacks primarily at the more substituted carbon as:

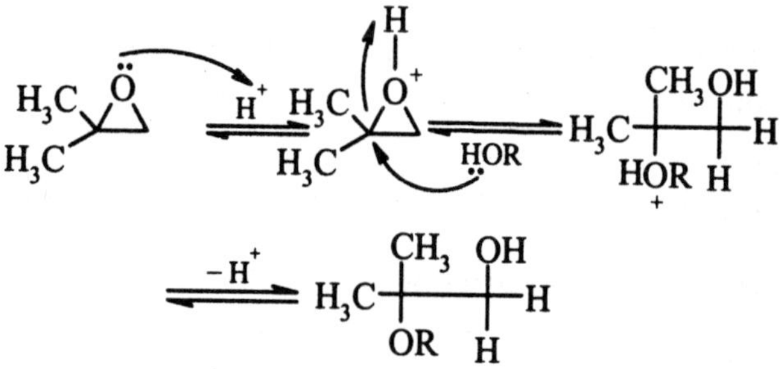

The reason is, in case of unsymmetrical, protonated epoxide, bond breaking step is faster than bond making step and the more substituted carbon acquires net positive charge, which is latter attacked by nucleophile as:

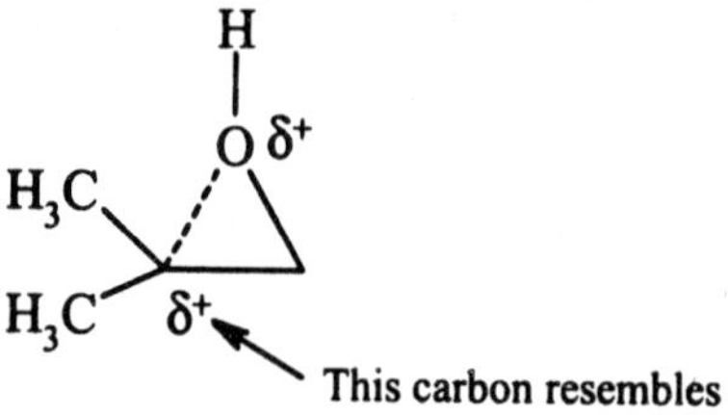

This carbon resembles a 3° carbocation

II. Base catalyzed reaction

In base catalyzed reaction, nucleophile attacks in the first step.

(i) *Hydrolysis*

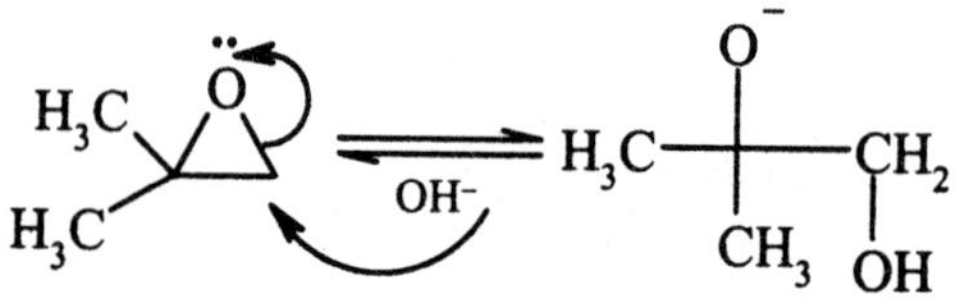

Therefore nucleophile attacks at less sterically hindered carbon.

(ii) *Base catalyzed alcoholysis*

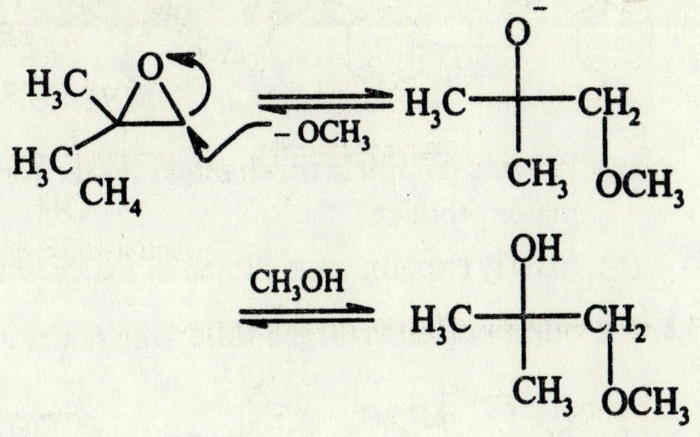

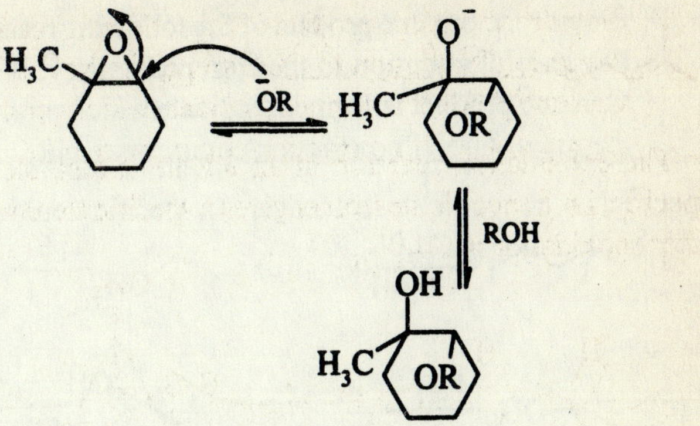

$$H_3C, H_3C \text{ epoxide} \xrightarrow{^-OCH_3} H_3C-C(O^-)(CH_3)-CH_2OCH_3 \xrightleftharpoons{CH_3OH} H_3C-C(OH)(CH_3)-CH_2OCH_3$$

$$\text{1-methylcyclohexene oxide} \xrightleftharpoons{^-OR} \text{(alkoxide, OR)} \xrightleftharpoons{ROH} \text{(OH, OR product)}$$

PRACTICE EXERCISE

Concept Builder Questions

Level 1

1. Why are the reactions of alcohols, ethers and thiols expected to have many similarities? What features make them different?

2. What is the single most important factor that controls the basicity of any alkoxide, or any base at all?

3. Give product of each reactions below, providing their mechanism:

(a) 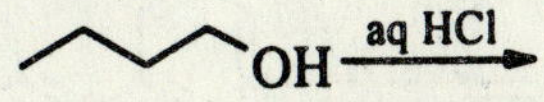 $CH_3CH_2CH_2CH_2OH \xrightarrow{\text{aq HCl}}$

(b) 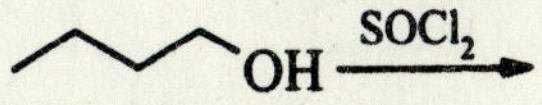 $CH_3CH_2CH_2CH_2OH \xrightarrow{SOCl_2}$

(c) 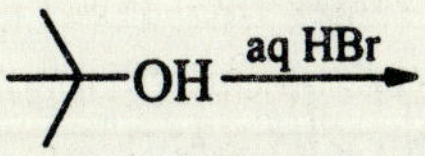 $(CH_3)_3C-OH \xrightarrow{\text{aq HBr}}$

(d) 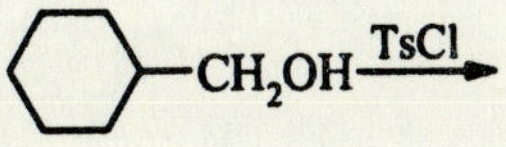 $C_6H_{11}-CH_2OH \xrightarrow{TsCl}$

(e) Product of reaction 'd' $\xrightarrow{NaCN}$

(f) 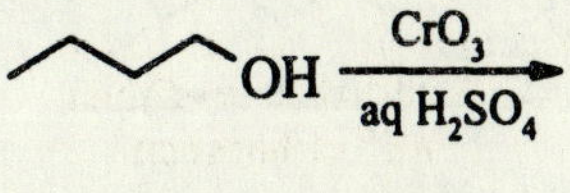 $CH_3CH_2CH_2CH_2OH \xrightarrow[\text{aq } H_2SO_4]{CrO_3}$

(g) 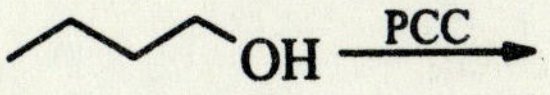 $CH_3CH_2CH_2CH_2OH \xrightarrow{PCC}$

(h) $CH_3CH_2CH(CH_3)CH_2OH \xrightarrow[\text{aq } H_2SO_4]{CrO_3}$

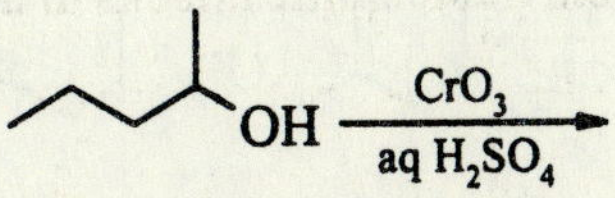

(i) 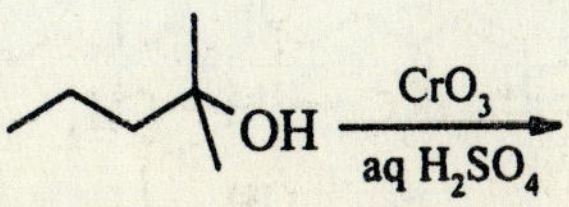 $CH_3CH_2CH_2C(CH_3)_2OH \xrightarrow[\text{aq } H_2SO_4]{CrO_3}$

(j) 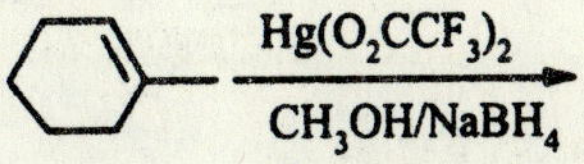 1-methylcyclohexene $\xrightarrow[CH_3OH/NaBH_4]{Hg(O_2CCF_3)_2}$

(k) 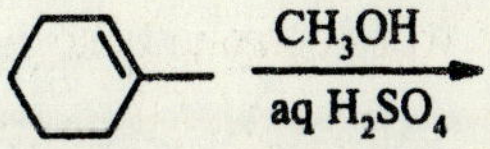 1-methylcyclohexene $\xrightarrow[\text{aq } H_2SO_4]{CH_3OH}$

4. Suggest reagents for the following transformations:

(a) $CH_3I \longrightarrow CH_3OH$

(b)  $CH_3CH_2CH=CH_2 \longrightarrow CH_3CH_2CH_2CH_2OH$

(c) 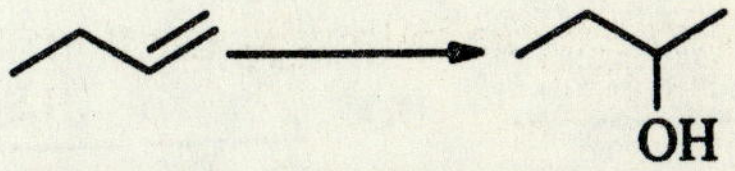 $CH_3CH_2CH=CH_2 \longrightarrow CH_3CH_2CH(OH)CH_3$

(d)  $CH_3CH_2CH_2CH_2OH \longrightarrow CH_3CH_2CH_2CH_2ONa$ (three ways)

5. Why does ethylene oxide react readily with nucleophiles such as ammonia, whereas THF is inert to nucleophilic attack by ammonia?

6. Give the products and mechanism of each of the reaction shown below:

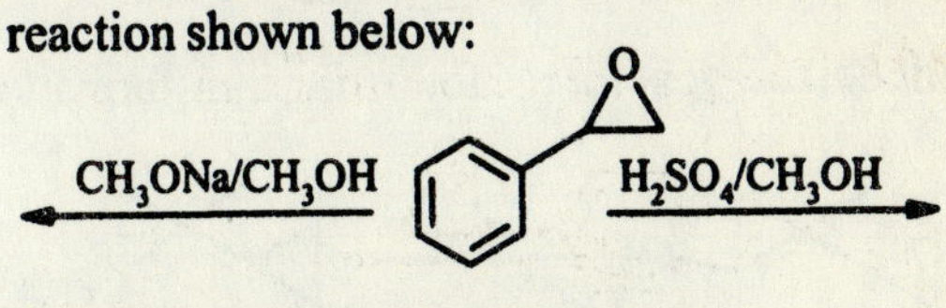

$$\xleftarrow{CH_3ONa/CH_3OH} \text{styrene oxide} \xrightarrow{H_2SO_4/CH_3OH}$$

7. Consider the following equilibrium:
Where does the equilibrium lie?

$$Cl_3CCH_2\bar{O} + CH_3CH_2OH \rightleftharpoons Cl_3CCH_2OH + CH_3CH_2\bar{O}$$

8. Consider the reaction:

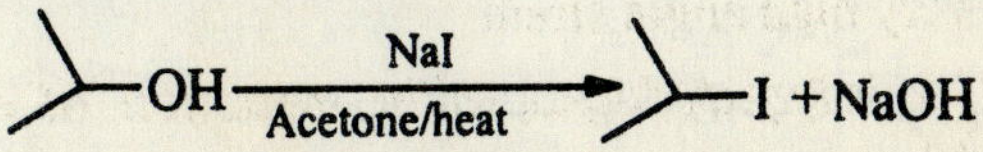

$$(CH_3)_2CH-OH \xrightarrow[\text{Acetone/heat}]{NaI} (CH_3)_2CH-I + NaOH$$

Explain briefly why reaction does not occur as written?

9. Provide the organic product of the following reactions. Pay careful attention to the stereochemistry. If more than one product is formed, indicate which product is the major one. If no reaction occur, write NR:

(a) 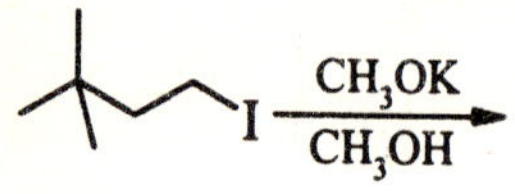 $\xrightarrow[CH_3OH]{CH_3OK}$ (b) 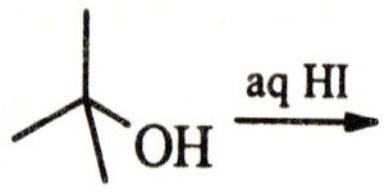 $\xrightarrow{\text{aq HI}}$

(c) 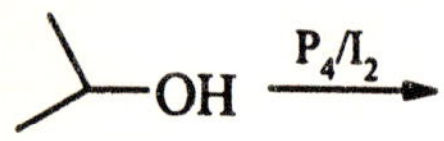 $\xrightarrow{P_4/I_2}$ (d) 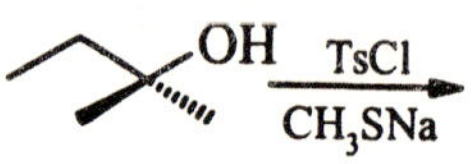 $\xrightarrow[CH_3SNa]{TsCl}$

(e) 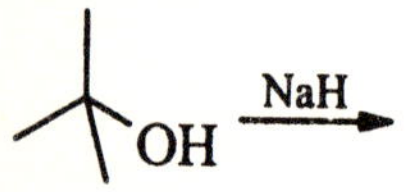 $\xrightarrow{NaH}$ (f) 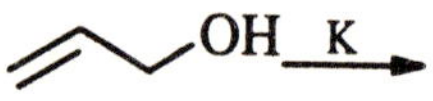 $\xrightarrow{K}$

(g) 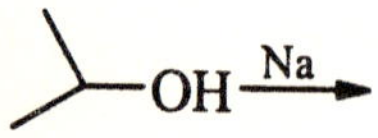 $\xrightarrow{Na}$ (h) 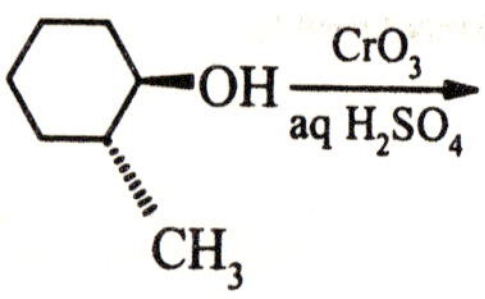 $\xrightarrow[\text{aq } H_2SO_4]{CrO_3}$

(i) 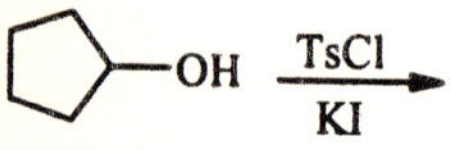 $\xrightarrow[KI]{TsCl}$ (j)  $\xrightarrow[NaOH]{BH_3/H_2O_2}$

(k) 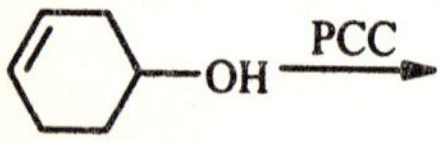 $\xrightarrow{PCC}$ (l) 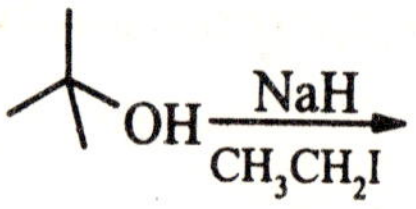 $\xrightarrow[CH_3CH_2I]{NaH}$

(m) 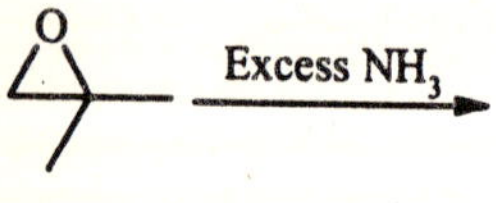 $\xrightarrow{\text{Excess } NH_3}$ (n) 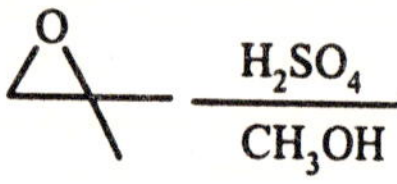 $\xrightarrow[CH_3OH]{H_2SO_4}$

(o) 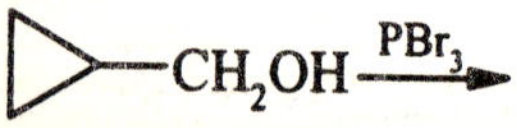 $CH_2OH \xrightarrow{PBr_3}$ (p) $\xrightarrow{SOCl_2}$

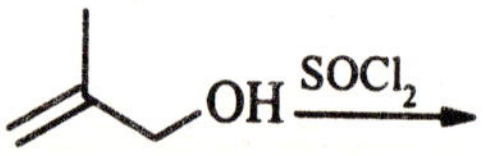

(q) 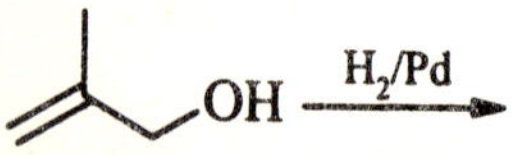 $\xrightarrow{H_2/Pd}$ (r) 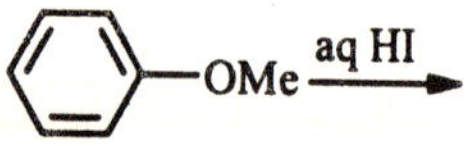 $\xrightarrow{\text{aq HI}}$

10. Provide best reagent for the following transformations:

(a)  OH ⟶ CHO

(b)  OH ⟶ Br

(c) 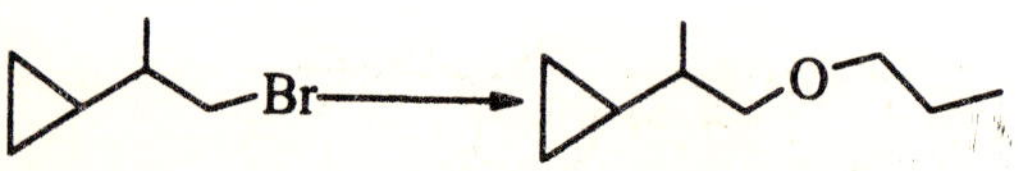 Br ⟶ O

11. Consider the following reaction:

$\xrightarrow[OH_2]{H^+}$ OH + OH + OH

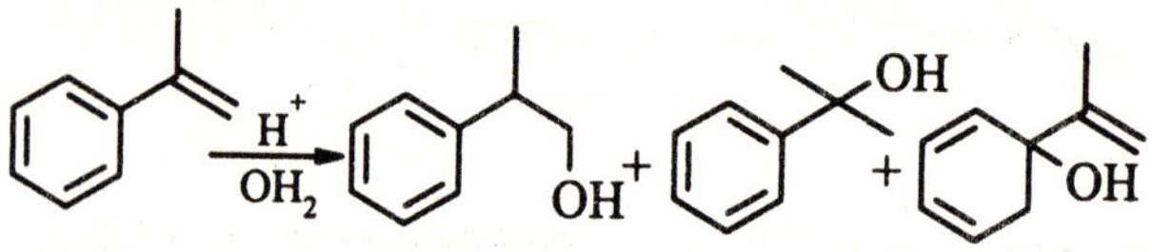

(a) Provide complete mechanism for the formation of major product.

(b) Briefly explain your choice of major product.

12. Provide mechanism of the following reaction:

$\xrightarrow{H_2SO_4}$ +

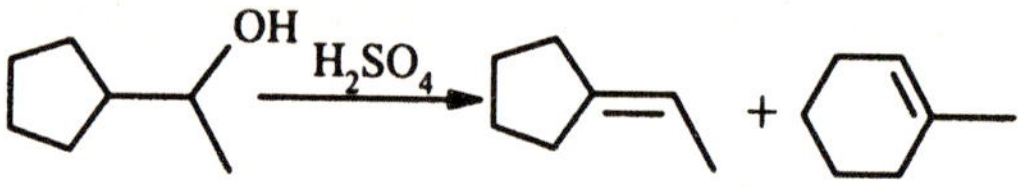

13. Which of the following two routes to allyl-isopropyl ether is more efficient?

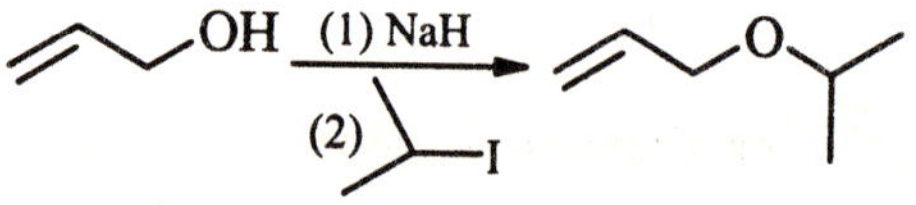

OH (1) NaH, (2) I ⟶ O or

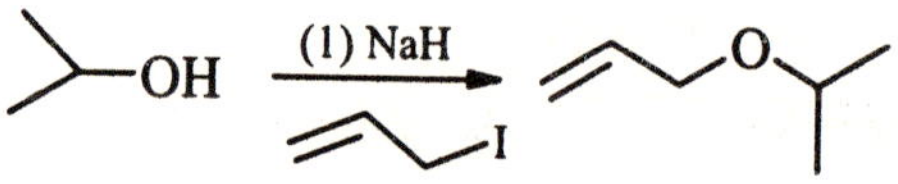

OH (1) NaH, I ⟶ O

14. Provide detailed mechanism for the following reactions:

(a) 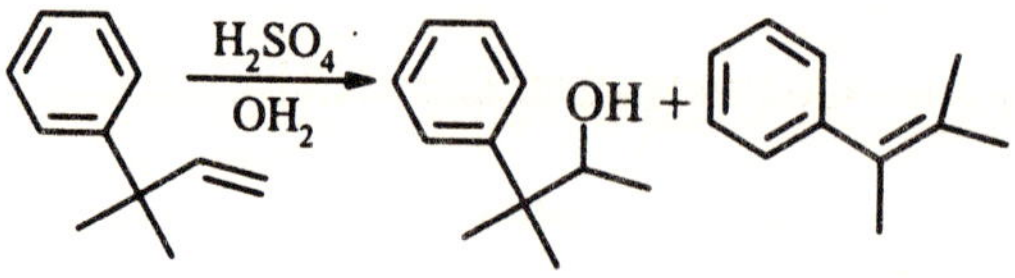 $\xrightarrow[OH_2]{H_2SO_4}$ OH +

(b) 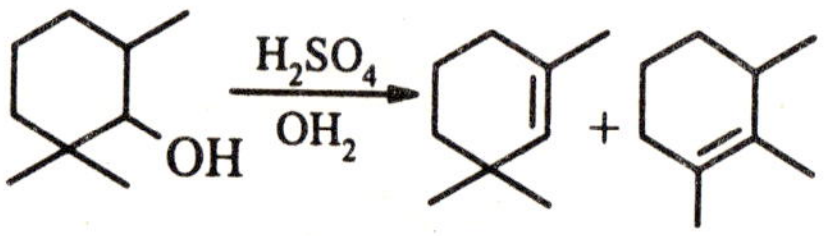 OH $\xrightarrow[OH_2]{H_2SO_4}$ +

(c) 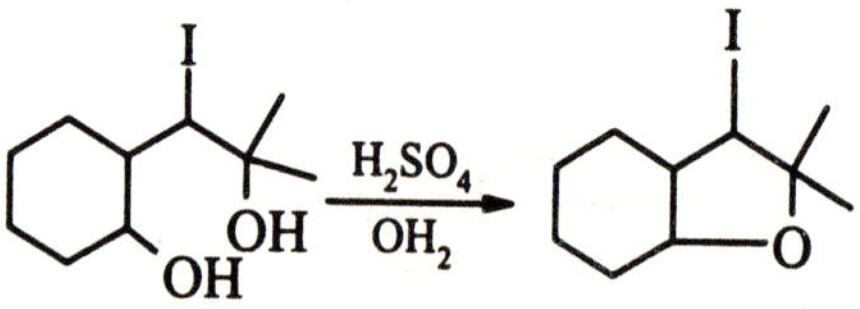 I, OH, OH $\xrightarrow[OH_2]{H_2SO_4}$ I, O

(d) 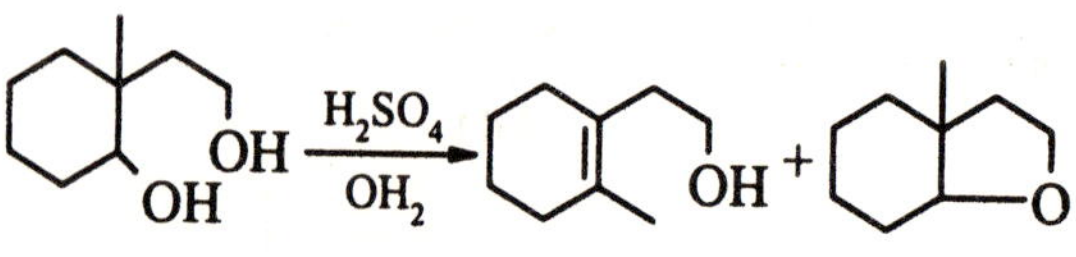 OH, OH $\xrightarrow[OH_2]{H_2SO_4}$ OH + O

(e) 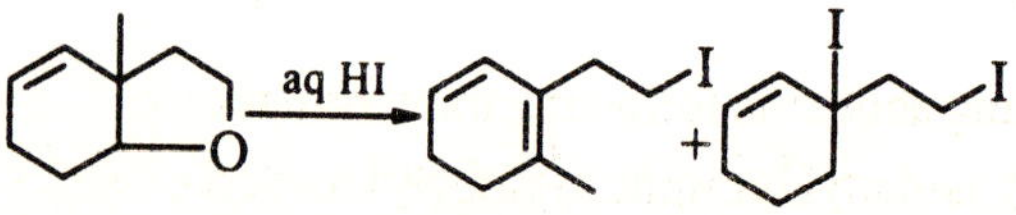 O $\xrightarrow{\text{aq HI}}$ I + I

15. Provide mechanism for the following reaction:

(2-methoxytetrahydropyran) + isopropanol $\xrightarrow{H^+}$ 2-$OCH(CH_3)_2$-tetrahydropyran

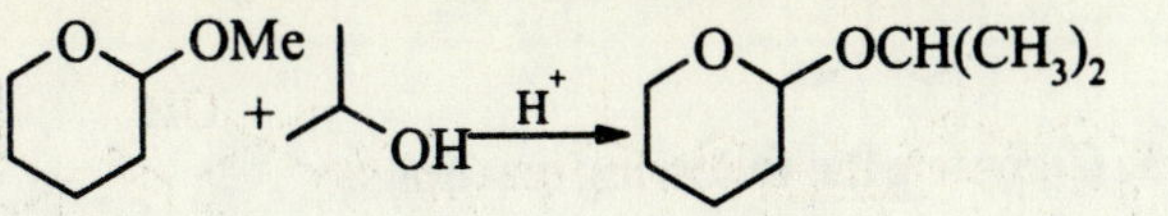

16. Supply oxidizing or reducing reagent for the following transformation:

(a) 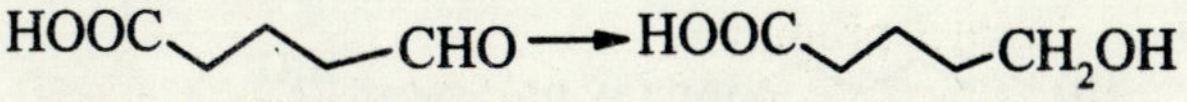 HOOC∼∼CHO ⟶ HOOC∼∼CH_2OH

(b) 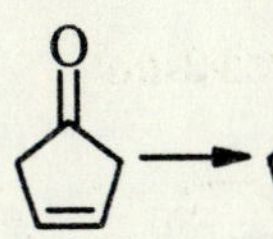 cyclopentenone ⟶ cyclopentenol (OH)

(c) 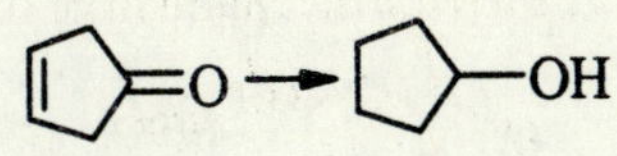 cyclopentenone (=O) ⟶ cyclopentanol (OH)

(d) 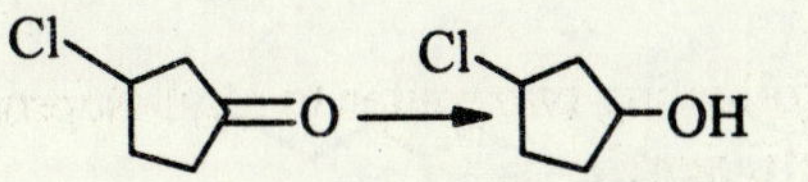 Cl-cyclopentanone (=O) ⟶ Cl-cyclopentanol (OH)

(e) 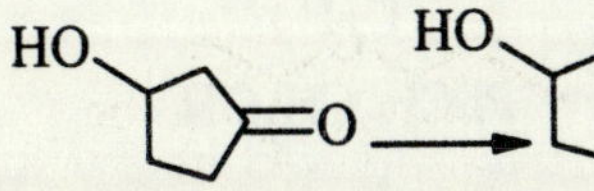 HO-cyclopentanone (=O) ⟶ HO-cyclopentane

(f) 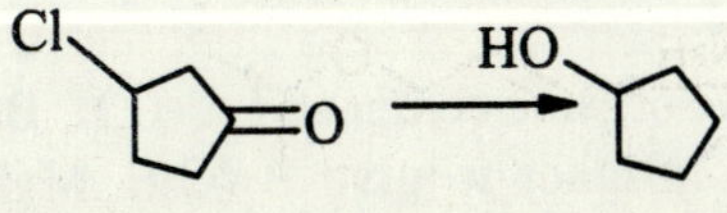 Cl-cyclopentanone (=O) ⟶ HO-cyclopentane

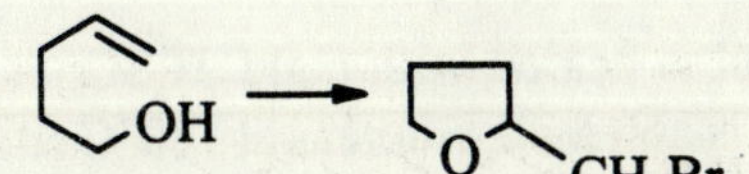

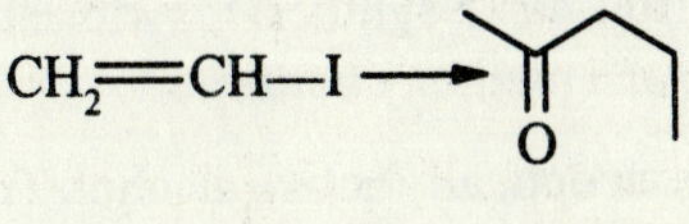

17. Bring about the following transformations:

(a) 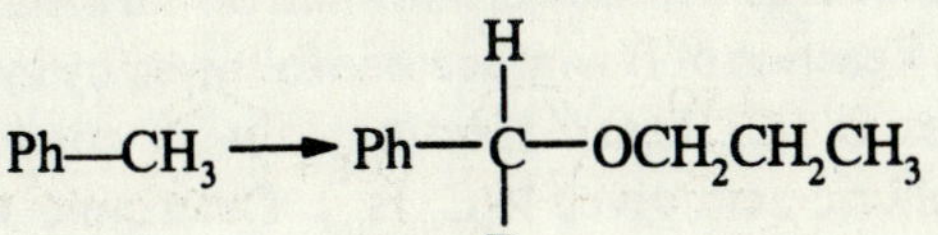 Ph—CH_3 ⟶ Ph—C(H)(D)—$OCH_2CH_2CH_3$

(b) (but-3-en-1-ol) OH ⟶ 2-(CH_2Br)tetrahydrofuran

(c) CH_2=CH—I ⟶ (2-butanone) O

(d) (pent-4-en-1-ol) OH ⟶ Br, OH

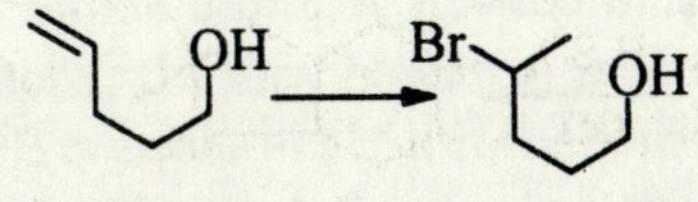

(e) 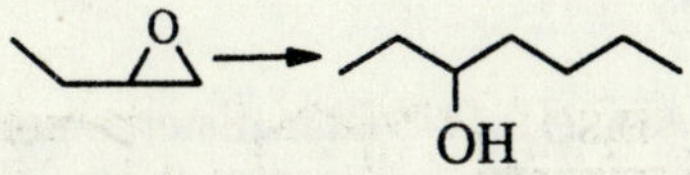 (ethyloxirane) ⟶ OH (heptan-3-ol)

(f) 2-methyl-1-butene ⟶ (ethoxy) O, COOH

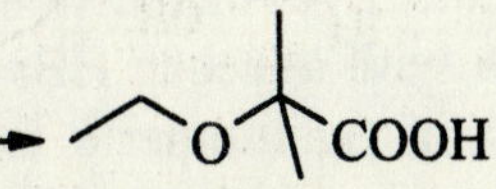

18. Bring about the following transformations:

(a) isobutyl bromide to isobutyl methyl ether.

(b) An alkyl halide containing not more than two carbon to 3-ethyl-2-pentanol.

(c) Ethene to CH_3OCH_2OH.

(d) 1-pentanol to 2-pentanone.

(e) PhCOOH to PhCH(OH)Ph.

19. Write product showing their stereochemistry:

(a) 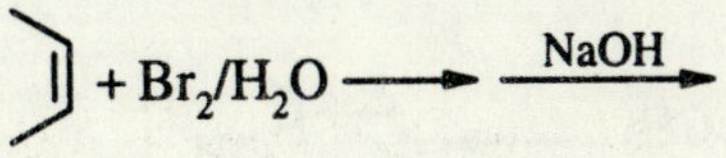 (cis-2-butene) + Br_2/H_2O ⟶ $\xrightarrow{NaOH}$

(b) 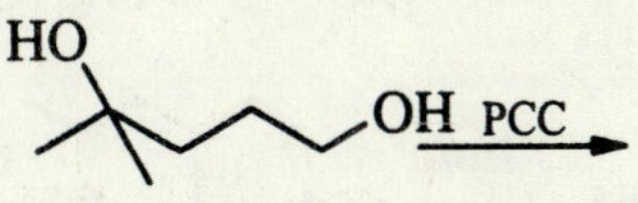 HO, OH $\xrightarrow{PCC}$

(c) 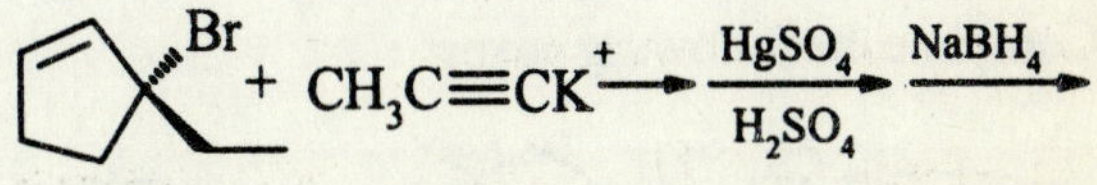 Br (cyclopentene) + $CH_3C\equiv CK^+$ ⟶ $\xrightarrow[H_2SO_4]{HgSO_4}$ $\xrightarrow{NaBH_4}$

(d) 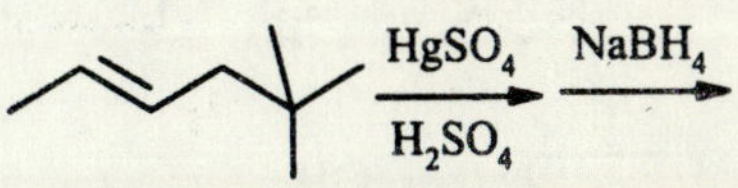 (alkene) $\xrightarrow[H_2SO_4]{HgSO_4}$ $\xrightarrow{NaBH_4}$

(e) 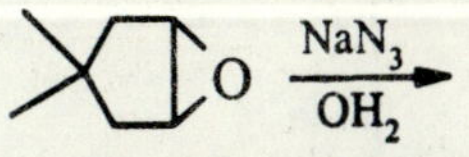 (epoxide) O $\xrightarrow[OH_2]{NaN_3}$

20. Propose mechanism of the following reaction:

(geraniol) OH ⟶ (p-menthadiene)

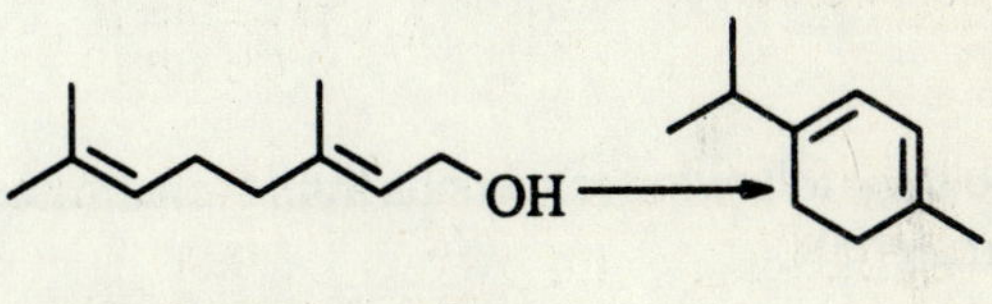

21. Write major product:

(a) 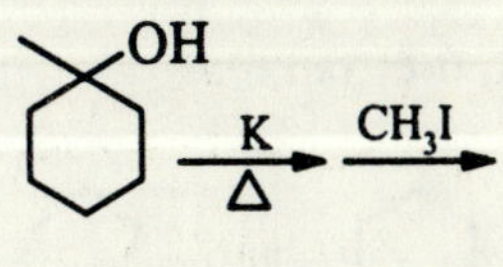 (1-methylcyclohexanol) OH $\xrightarrow[\Delta]{K}$ $\xrightarrow{CH_3I}$

(b) 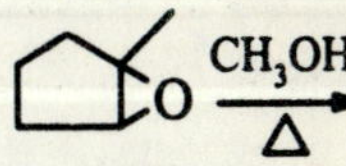 (epoxide) O $\xrightarrow[\Delta]{CH_3OH^+}$

(c) 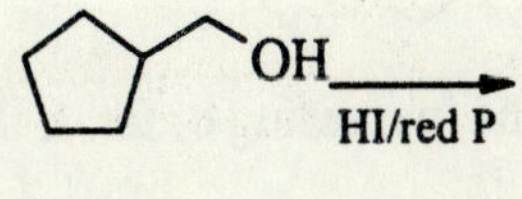 (cyclopentylmethanol) OH $\xrightarrow[HI/red\ P]{}$

(d) 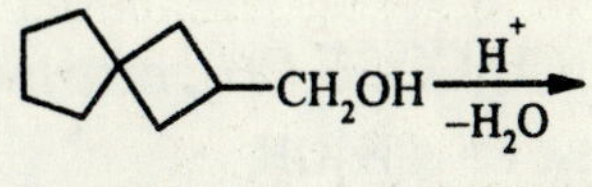 (spiro) $-CH_2OH$ $\xrightarrow[-H_2O]{H^+}$

22. Convert:

(a) 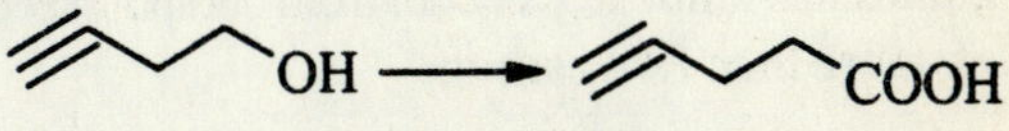 (but-3-yn-1-ol) OH ⟶ COOH

(b) 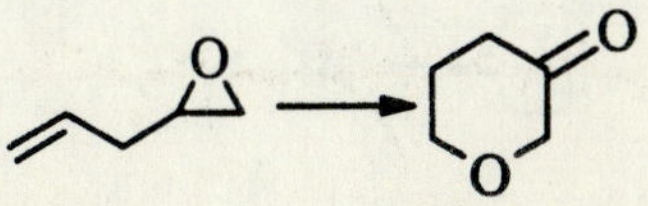 (allyl oxirane) O ⟶ O, O

(c) 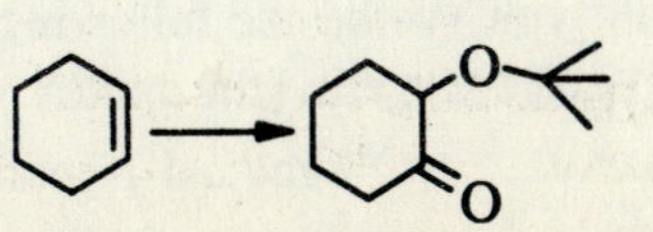 (cyclohexene) ⟶ O, O

23. Propose mechanism:

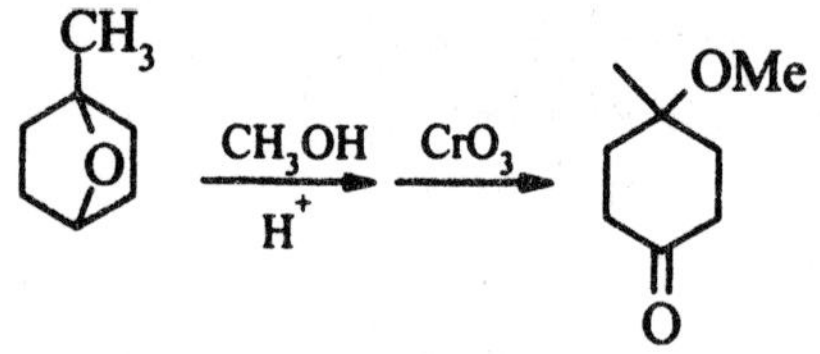

$\xrightarrow[H^+]{CH_3OH}$ $\xrightarrow{CrO_3}$

24. Design multiple synthesis of the followings:

(a) 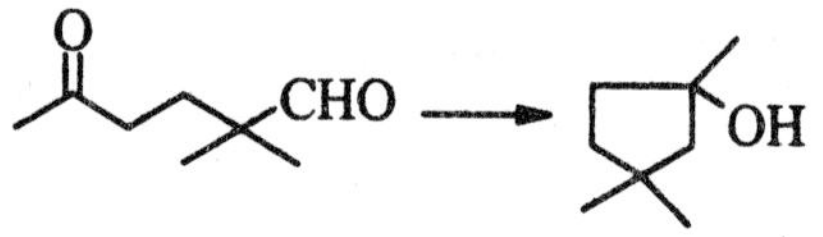 CHO → OH

(b) 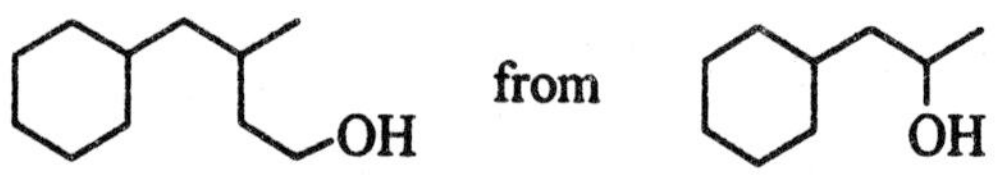 OH from OH

25. Complete the following boxes:

(a) 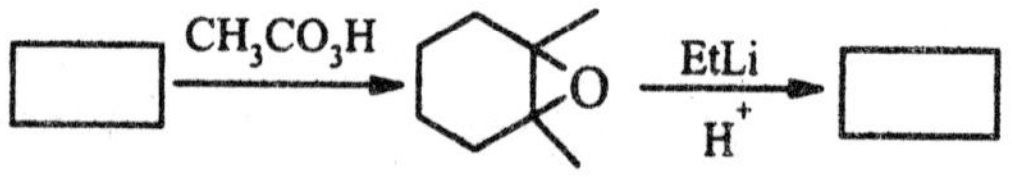 [ ] $\xrightarrow{CH_3CO_3H}$ O $\xrightarrow[H^+]{EtLi}$ [ ]

(b) 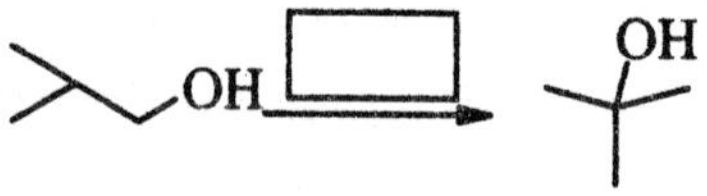 OH [ ] → OH

(c) 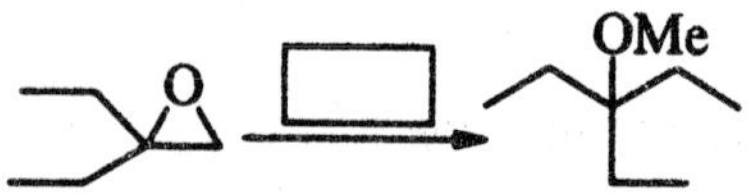 O [ ] → OMe

Level 2

1. Bring about the following conversion from the indicated starting materials:

(a) 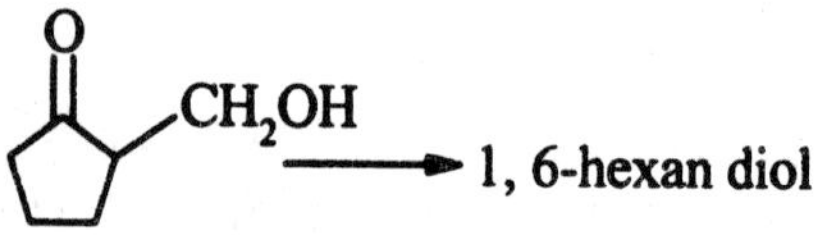 CH_2OH → 1, 6-hexan diol

(b) 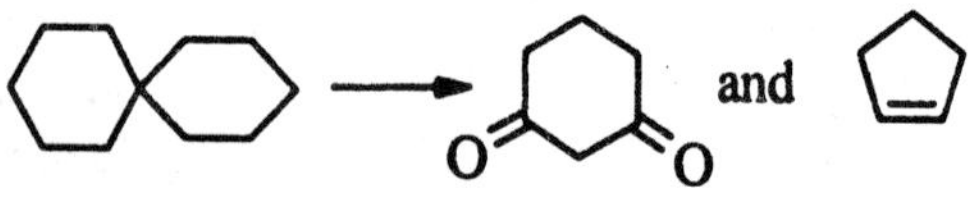 → and

2. Arrange the following in the increasing order of their acid strength:

CH_3CH_2OH, CF_3COOH, $CH_3COCH_2CH_3$, Propyne, CH_3COOCH_3, CF_3CH_2OH

3. Synthesize following alcohol from compound containing no more than five carbon:

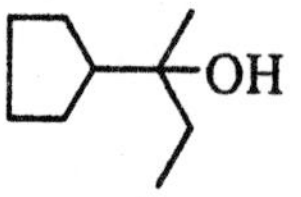

OH

4. Show how would you synthesize following alcohol adding Grignard reagent to ethylene oxide?

(a) 2-phenylethanol (b) 4-methyl-1-pentanol

(c) 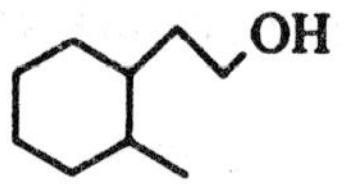 OH

5. Complete the following reactions:

(a) 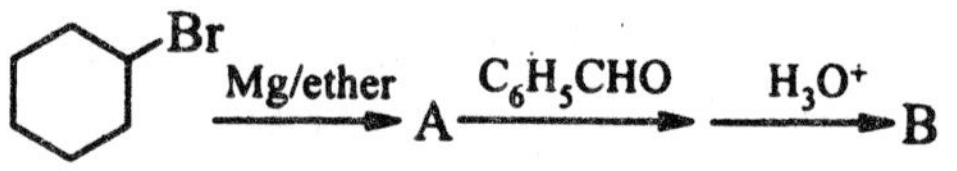 Br $\xrightarrow{Mg/ether}$ A $\xrightarrow{C_6H_5CHO}$ $\xrightarrow{H_3O^+}$ B

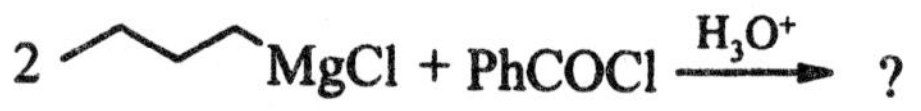

(b) 2 MgCl + PhCOCl $\xrightarrow{H_3O^+}$?

(c) OMe + $LiAlH_4$ → ?

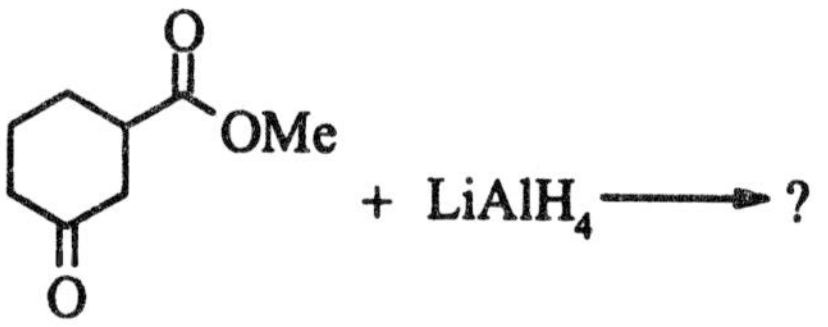

(d) 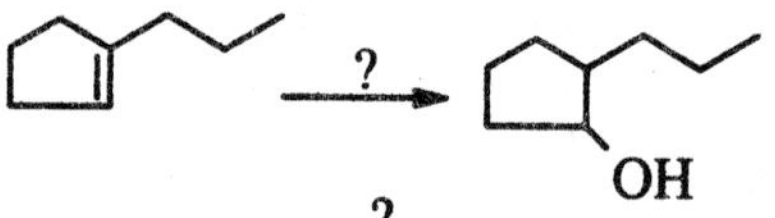 $\xrightarrow{?}$ OH

(e) $PhCH_2CH_2Cl \xrightarrow{?} PhCH_2CH_2OH$

(f) RMgX + O → ? $\xrightarrow{H_3O^+}$?

6. An optically active organic compound A ($C_7H_{11}Br$) is treated with Mg in ether to give B ($C_7H_{11}MgBr$), which react violently with D_2O to give methyl cyclohexene with deuterium atom on the methyl group (C). Reaction of B with acetone followed by hydrolysis gives D ($C_{10}H_{18}O$). Heating D with concentrated sulphuric acid gives E($C_{10}H_{16}$). Determine structure of A to E.

7. When ethanol is heated with conc. H_2SO_4 at 140°C, diethyl ether is obtained, whereas at 180°C ethylene is the major product. Suggest reason.

8. Using Williamson ether synthesis, synthesize butyl isopropyl ether and methoxy benzene.

9. Compound A is an optically active alcohol. Treatment with chromic acid convert A into a ketone B. In a separate reaction, A is treated with PBr_3, converting it into C. C on reaction with Mg is added to B to yield D, which after hydrolysis gives 3, 4-dimethyl-3-hexanol. Identify A to D.

10. A compound 'A' $C_{10}H_{14}O$ exist in distereomeric forms, liberates a gas on reaction with CH_3MgBr. Treatment of A with aqueous HBr gives B($C_{10}H_{13}Br$), which exist in enantiomeric forms. B on treatment with alcoholic solution of KOH gives a compound C. Ozonolysis of C followed by work-up with H_2O_2 gives a compound D and acetophenone. Identify A to D.

11. Synthesize the following from the indicated starting materials:

(a) 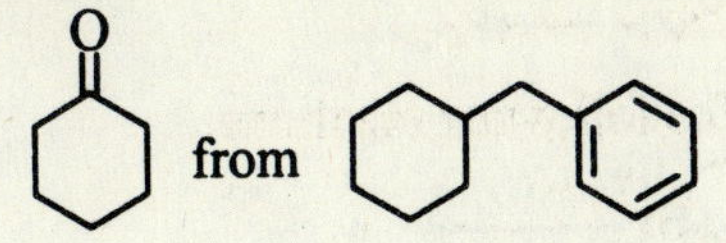 [cyclohexanone] from [benzylcyclohexane]

(b) 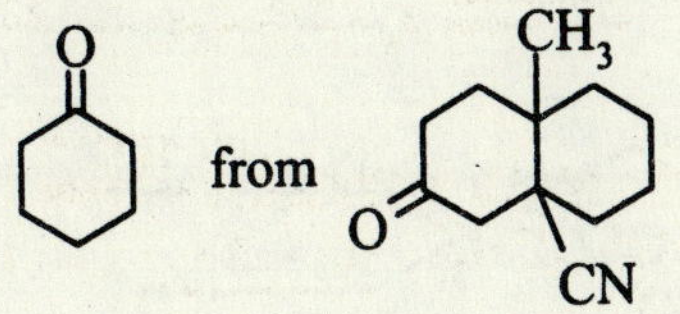 [cyclohexanone] from [bicyclic ketone with CH_3 and CN]

(c) 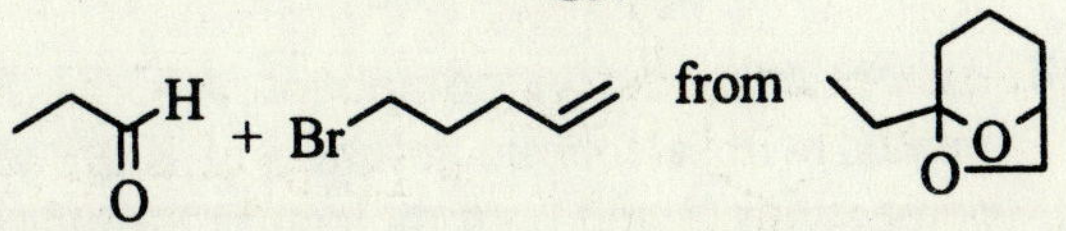 [propanal] + Br[pentene] from [cyclic ketal]

12. Synthesize:

(a) 3-methylpentan-2,3-diol from compound containing not more than three carbon.

(b) 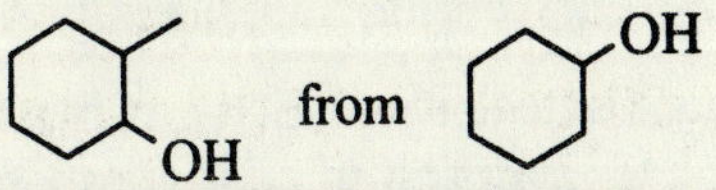 [2-methylcyclohexanol] from [cyclohexanol]

(c) 1-chloro-1-phenyl cyclohexane from cyclohexanol.

13. Complete the following reactions:

(a) 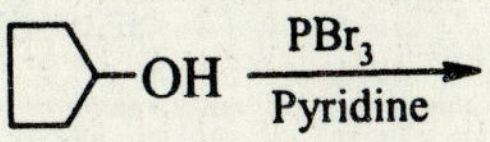 [cyclopentanol] $\xrightarrow[\text{Pyridine}]{PBr_3}$

(b) 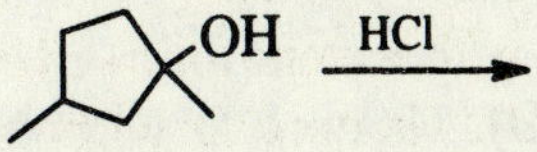 [1,3-dimethylcyclopentanol] $\xrightarrow{HCl}$

(c) $CH_3CH_2CH_2OH + HBr \xrightarrow{\Delta}$

14. Complete the following reactions:

[octahydronaphthalene] $\xrightarrow[H_2O_2]{OsO_4}$ $C_{10}H_{18}O_2$(A) $\xrightarrow{HIO_4}$ $C_{10}H_{16}O$(B)

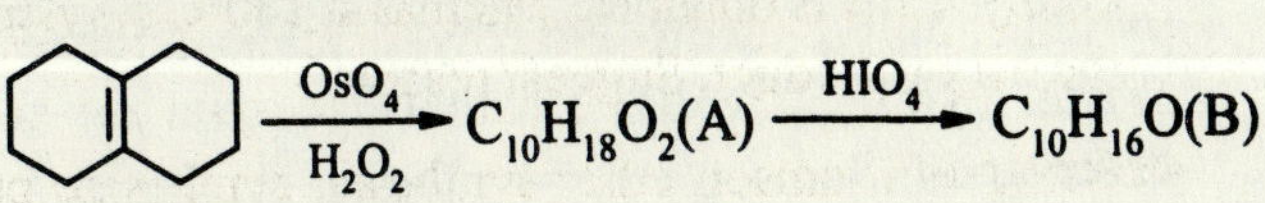

Identify A and B.

15. Determine the structure of compound A through H:

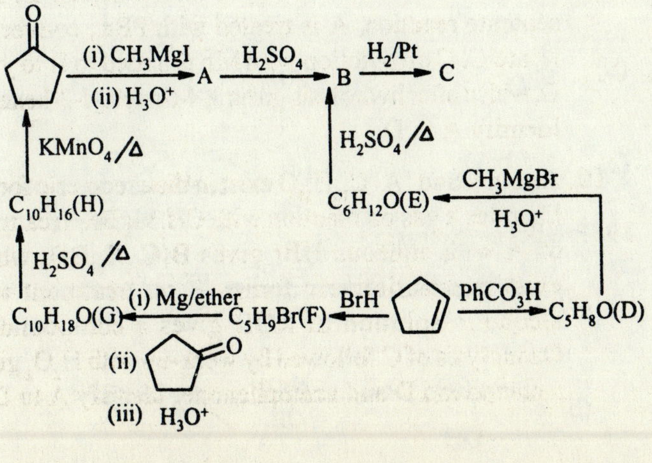

[cyclopentanone] $\xrightarrow[\text{(ii) } H_3O^+]{\text{(i) } CH_3MgI}$ A $\xrightarrow{H_2SO_4}$ B $\xrightarrow{H_2/Pt}$ C

$C_{10}H_{16}$(H) $\xrightarrow{KMnO_4/\Delta}$ [cyclopentanone]

$C_6H_{12}O$(E) $\xrightarrow{H_2SO_4/\Delta}$ B

$C_{10}H_{18}O$(G) $\xrightarrow{H_2SO_4/\Delta}$ $C_{10}H_{16}$(H)

[cyclopentene] $\xrightarrow{BrH}$ C_5H_9Br(F) $\xrightarrow{\text{(i) Mg/ether; (ii) [cyclopentanone]; (iii) } H_3O^+}$ $C_{10}H_{18}O$(G)

[cyclopentene] $\xrightarrow{PhCO_3H}$ C_5H_8O(D) $\xrightarrow{CH_3MgBr,\ H_3O^+}$ $C_6H_{12}O$(E)

16. Propose mechanism of the following reactions:

(a) 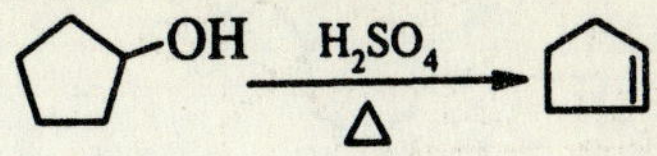 [cyclopentanol] $\xrightarrow[\Delta]{H_2SO_4}$ [cyclopentene]

(b) 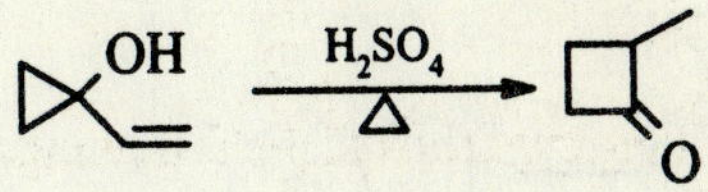 [1-vinylcyclopropanol] $\xrightarrow[\Delta]{H_2SO_4}$ [2-methylcyclobutanone]

(c) 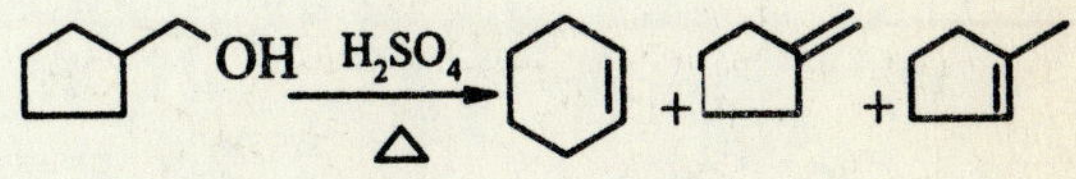 [cyclopentylmethanol] $\xrightarrow[\Delta]{H_2SO_4}$ [cyclohexene] + [methylenecyclopentane] + [1-methylcyclopentene]

(d) 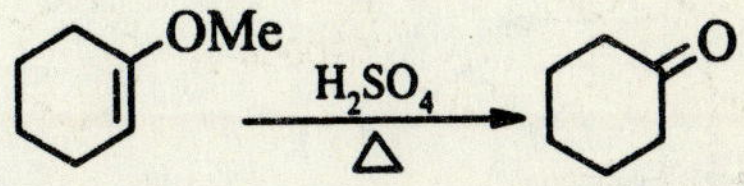 [1-methoxycyclohexene] $\xrightarrow[\Delta]{H_2SO_4}$ [cyclohexanone]

17. Convert:

(a) 2-methyl-2-butene to $(CH_3)_2C(OH)CHDCH_3$

(b) Synthesize 2, 5-dimethyl-3-hexanone using alcohol containing not more than four carbon.

(c) C_6H_5Br to $(C_6H_5)_2C(OH)CH_3$ using any additional compound containing not more than three carbon.

(d) 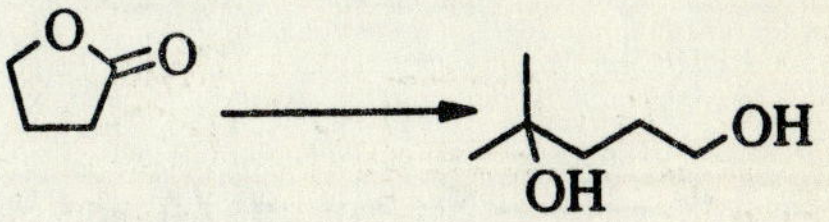 [γ-butyrolactone] ⟶ [4-methylpentane-1,4-diol]

18. (a) Convert:

[cis-2-methylcyclopentanol] ⟶ [2-methylcyclopentyl chloride]

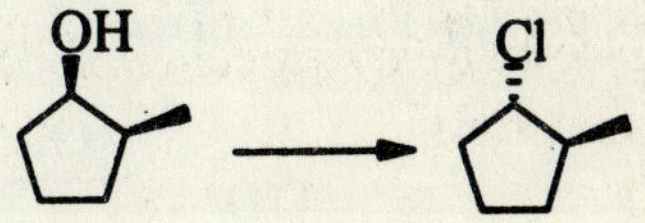

(b) Cyclohexyl bromide when treated with potassium ethoxide, the major product is a cyclohexene, whereas when ethyl bromide is treated with potassium salt of cyclohexanol, the major product is ethyl cyclohexyl ether. Explain the difference.

19. Bring about the following conversion:

(a) 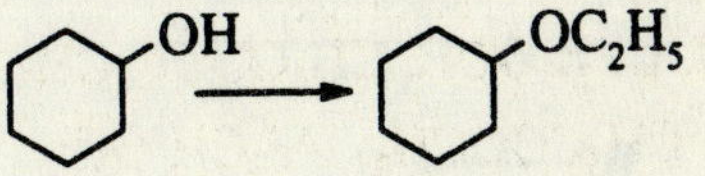 [cyclohexanol] ⟶ [cyclohexyl-OC_2H_5]

(b) 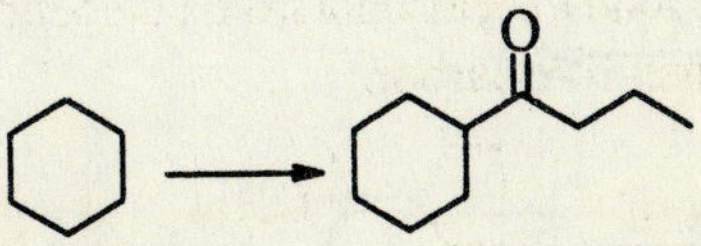 [cyclohexane] ⟶ [1-cyclohexylbutan-1-one]

(c) 2 [benzyl bromide] ⟶ [stilbene]

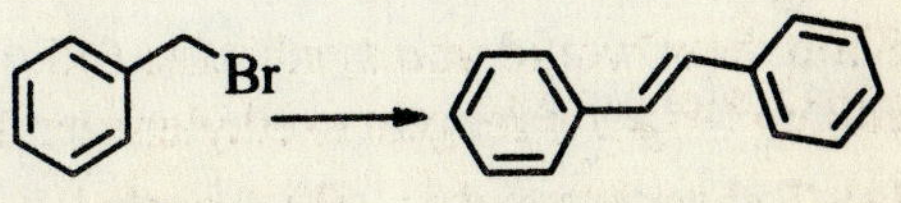

(d) 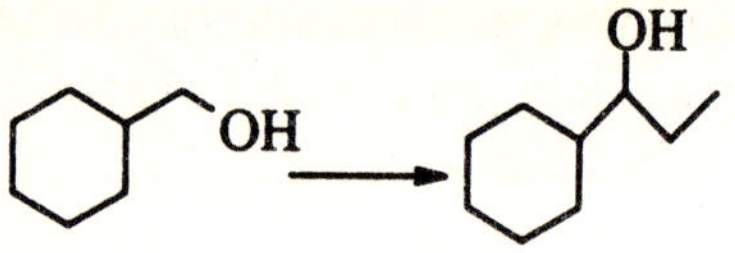 [cyclohexyl-CH₂CH₂OH → cyclohexyl-CH(OH)CH₂CH₃]

(e) 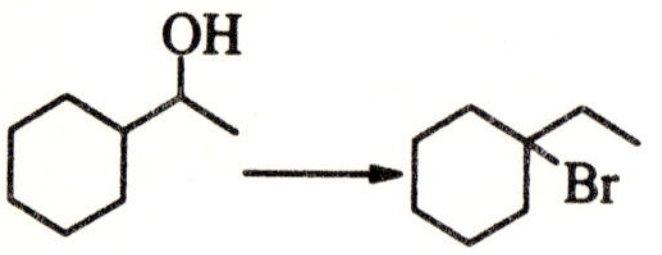 [cyclohexyl-CH(OH)CH₃ → 1-bromo-1-ethylcyclohexane]

(f) 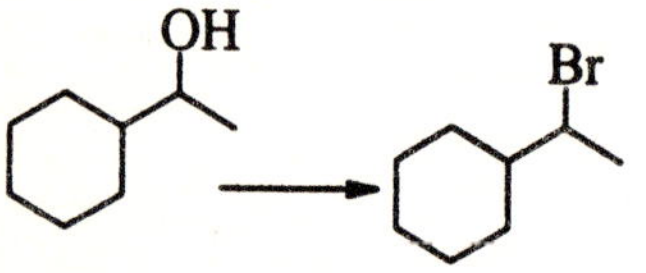 [cyclohexyl-CH(OH)CH₃ → cyclohexyl-CH(Br)CH₃]

20. Propose mechanism:

(a) 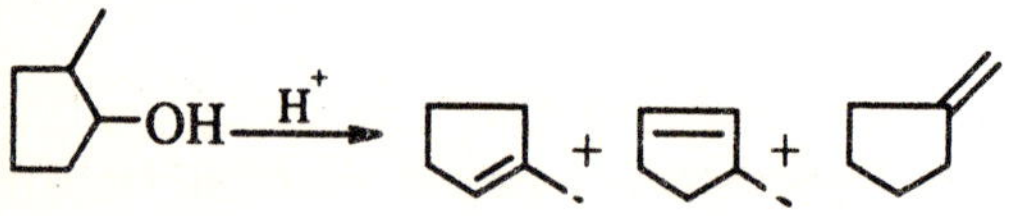 [cyclobutyl(methyl)-OH] $\xrightarrow{H^+}$ [alkenes] + [alkenes] + [methylenecyclopentane]

(b) 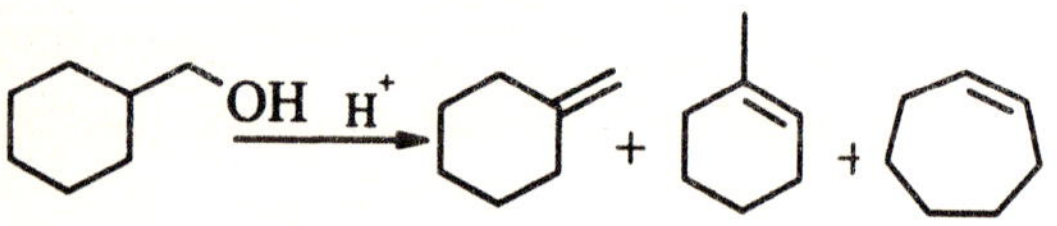 [cyclohexyl-CH₂-OH] $\xrightarrow{H^+}$ [methylenecyclohexane] + [1-methylcyclohexene] + [cycloheptene]

(c) 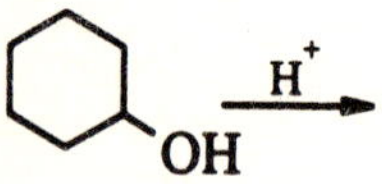 [cyclohexanol derivative, OH] $\xrightarrow{H^+}$ Stereomers (A and B)

21. Show major product for each of the following reactions:

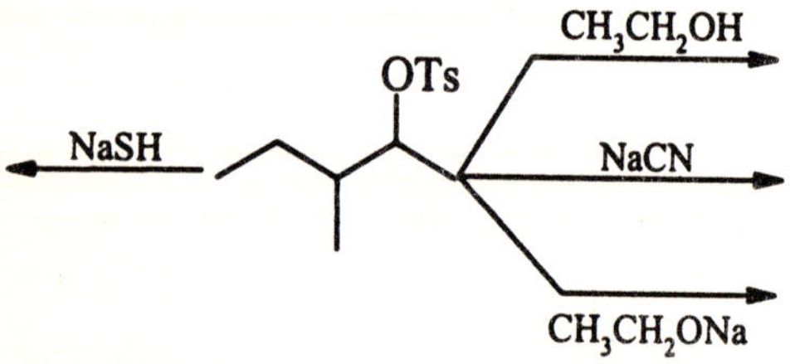

[tosylate substrate, OTs] reacts with NaSH; CH_3CH_2OH; NaCN; CH_3CH_2ONa

22. Provide structure of the compounds A to E in the following reaction:

(a)  [cyclohexene] $\xrightarrow{HOCl}$ A $\xrightarrow{NaOH\ (aq.)}$ B

(b) PhOMe + IH ⟶ C

(c)  [tetrahydrofuran] $\xrightarrow{ClH}$ D

(d) [2-methyltetrahydrofuran] $\xrightarrow{ClH}$ E

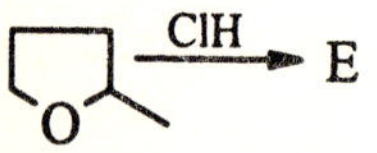

23. Predict major product:

(a) 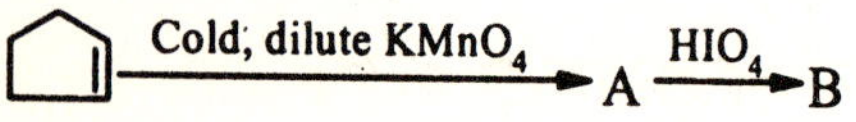 [cyclohexene] $\xrightarrow{\text{Cold; dilute } KMnO_4}$ A $\xrightarrow{HIO_4}$ B

(b) [cyclobutyl-CH(OH)-cyclopropyl] $\xrightarrow{H^+}$ A

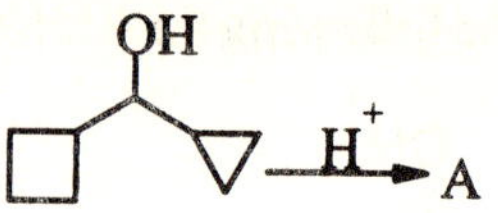

(c)  [cyclopentane with CH₂OH and OH] $\xrightarrow{HIO_4}$ A

(d) 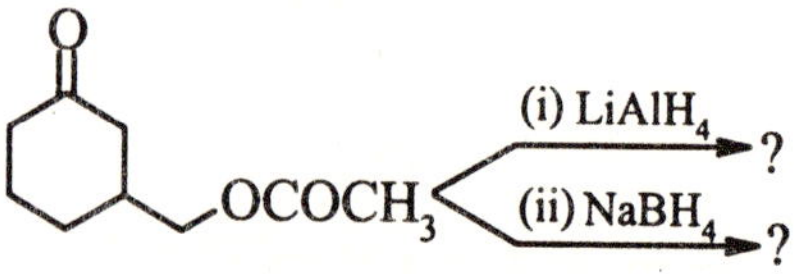 [3-substituted cyclohexanone, CH₂OCOCH₃] (i) $LiAlH_4$ → ? ; (ii) $NaBH_4$ → ?

24. An optically active compound $A(C_{10}H_{17}Br)$ when treated with alcoholic solution of KOH yields two compounds B and C of molecular formula $C_{10}H_{16}$ of which 'B' is optically active. Ozonolysis of A followed by treatment with $Zn\text{-}H_2O$ gives acetone as one of the products. Hydrogenation of either B or C yield 4-isopropyl-1-methyl cyclohexane, identify A, B and C.

25. When 2-butanol is treated with $SOCl_2$ in pyridine, there occurs complete inversion of configuration. Write a detailed mechanism that accounts for this fact.

26. When *cis*-2, 3-dimethyl oxirane is treated with water containing a trace of $HClO_4$, a racemic mixture of 2, 3-butandiol is formed, *trans*-2, 3-dimethyl oxirane give meso-2, 3-butandiol under identical conditions. Write mechanism for these reactions.

27. When 3-methyl-2-pentanol is treated with $ZnCl_2$ in concentrated solution of HCl, a mixture of chloroalkanes forms, including chiefly 2-chloro-3-methylpentane and 3-chloro-3-methylpentane. When 3-methyl-2-pentanol is treated with thionyl chloride in pyridine, only 2-chloro-3-methyl pentane is formed. Write detailed mechanism that accounts for the observation.

28. Synthesize the following by Williamson's method, choosing alkoxide anion and alkyl halide that will give the best yield:

(a) 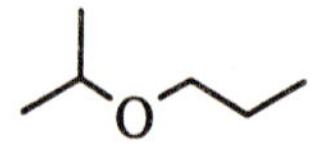 [isopropyl propyl ether] (b) 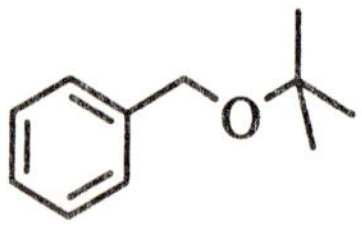 [benzyl tert-butyl ether]

(c) 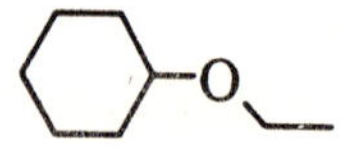 [cyclohexyl ethyl ether] (d)  [allyl ethyl ether]

29. Predict mechanism:

(a) 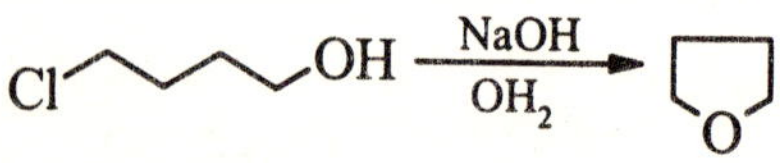 Cl~~~OH $\xrightarrow[OH_2]{NaOH}$ [tetrahydrofuran]

(b) 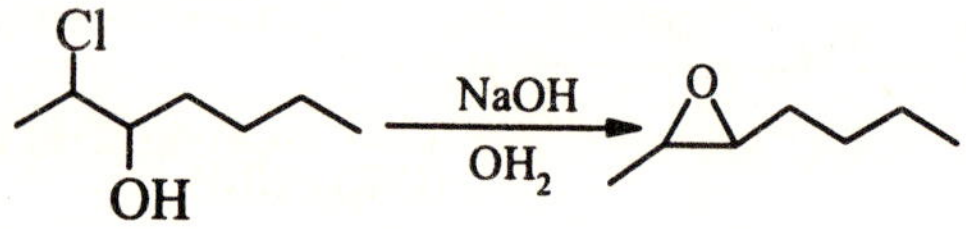 [Cl, OH chlorohydrin] $\xrightarrow[OH_2]{NaOH}$ [epoxide]

30. Suggest reagents for the following transformations:

(a) 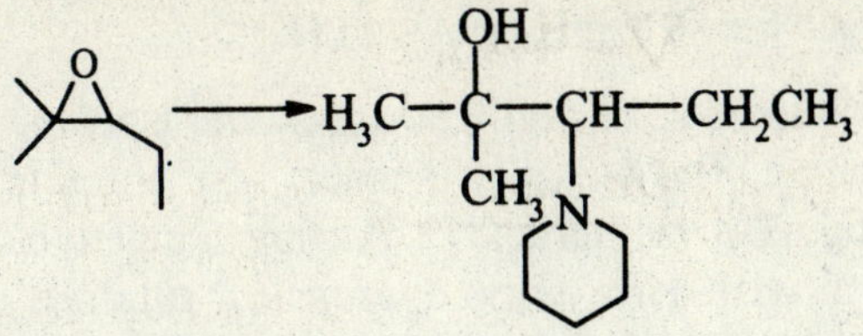 [epoxide] → $H_3C-C(OH)(CH_3)-CH(N\text{-piperidinyl})-CH_2CH_3$

(b) 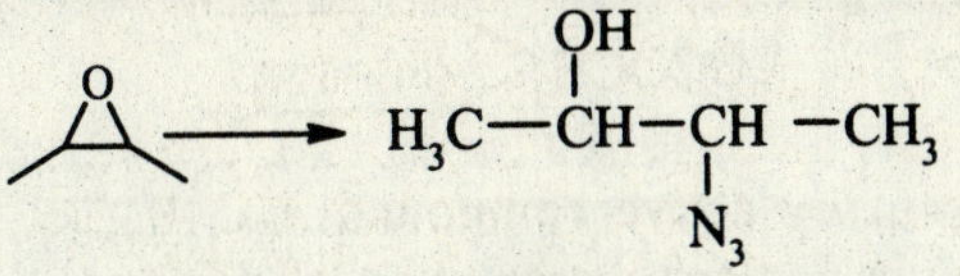 [epoxide] → $H_3C-CH(OH)-CH(N_3)-CH_3$

(c) [epoxide] → $H_3C-C(OH)(CH_3)-CH_2-{}^{18}OH$

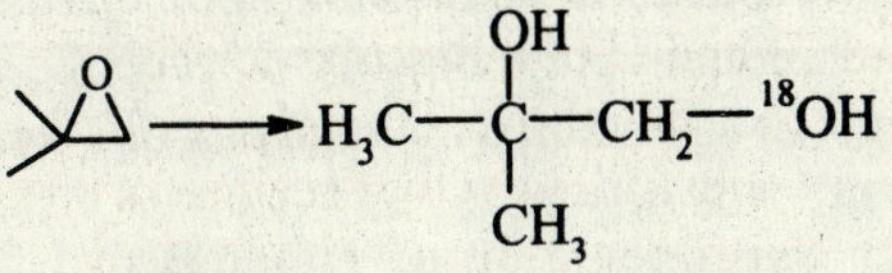

(d) 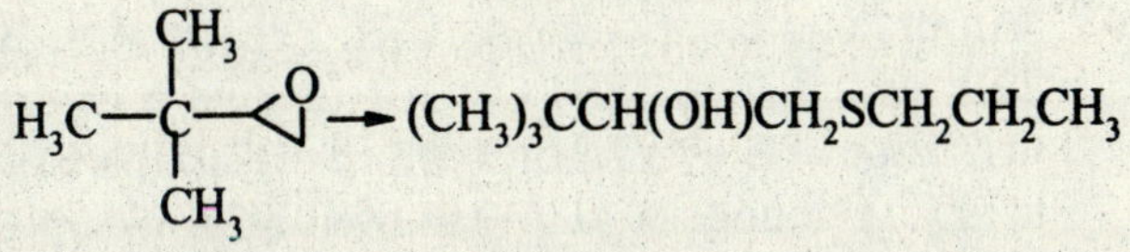 $H_3C-C(CH_3)_2-$[epoxide] → $(CH_3)_3CCH(OH)CH_2SCH_2CH_2CH_3$

(e) 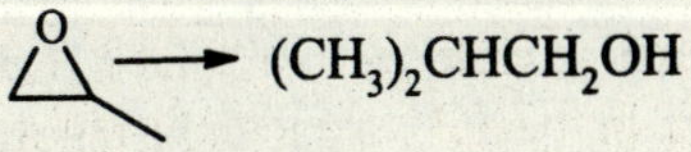 [epoxide] → $(CH_3)_2CHCH_2OH$

31. Predict reagents:

(a) 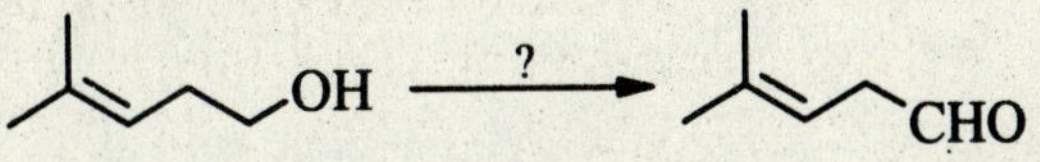 [allylic alcohol]–OH $\xrightarrow{?}$ [enal]–CHO

(b) 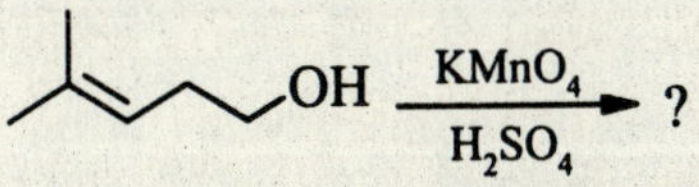 [allylic alcohol]–OH $\xrightarrow[H_2SO_4]{KMnO_4}$?

32. Treatment of 3-methyl-2-butanol with HCl, only a trace of 2-chloro-3-methyl butane is formed. An isomeric product was isolated in 97% yield. Suggest a reasonable structure of this product.

33. Suggest reasonable explanation for the following observations:

(a) The first order rate constant for the solvolysis of $(CH_3)_2C{=}CHCH_2Cl$ in ethanol is 600 times greater than that of allylchloride.

(b) After a solution of 3-buten-2-ol in aqueous H_2SO_4 be allowed to stand for a week, it was found to contain both 3-bute-2-ol and 2-buten-1-ol.

34. Give structural formula to the compounds indicated by the letters, being sure to show correct stereochemistry:

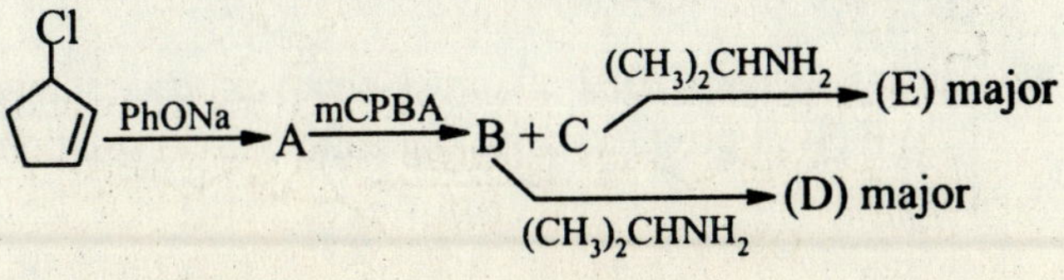

[3-chlorocyclopentene] $\xrightarrow{PhONa}$ A $\xrightarrow{mCPBA}$ B + C; B $\xrightarrow{(CH_3)_2CHNH_2}$ (E) major; C $\xrightarrow{(CH_3)_2CHNH_2}$ (D) major

Also D and E are stereomers. What is the stereochemical relationship between them? How many stereomers does D have all toge... ?

35. Convert:

(a) 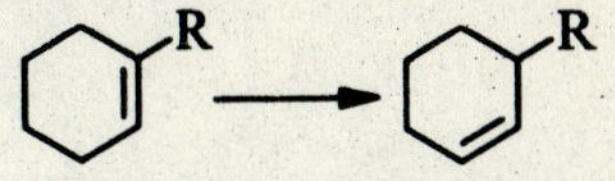 [1-R-cyclohexene] → [3-R-cyclohexene]

(b) Propose mechanism:

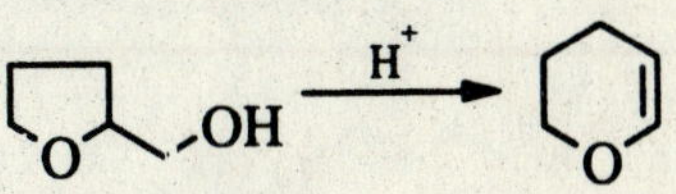

[2-(hydroxymethyl)tetrahydrofuran]–OH $\xrightarrow{H^+}$ [3,4-dihydro-2H-pyran]

(c) Write products:

(i) 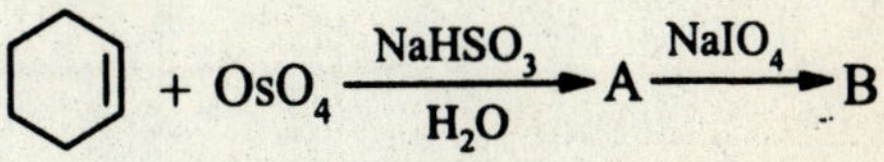 [cyclohexene] + OsO_4 $\xrightarrow[H_2O]{NaHSO_3}$ A $\xrightarrow{NaIO_4}$ B

(ii)  [epoxide] + PhSNa → C

(iii) 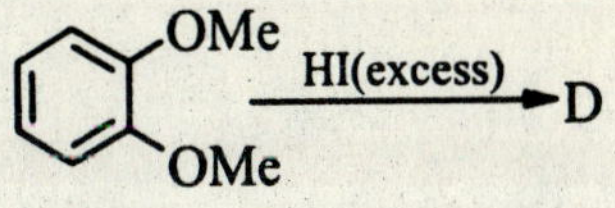 [1,2-dimethoxybenzene] $\xrightarrow{HI(excess)}$ D

(iv) 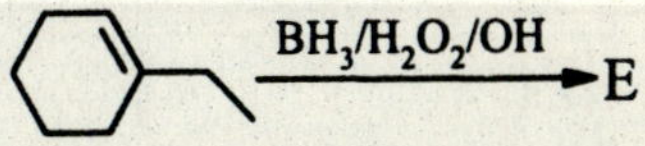 [1-ethylcyclohexene] $\xrightarrow{BH_3/H_2O_2/OH}$ E

36. Predict products in the following reactions:

(a) 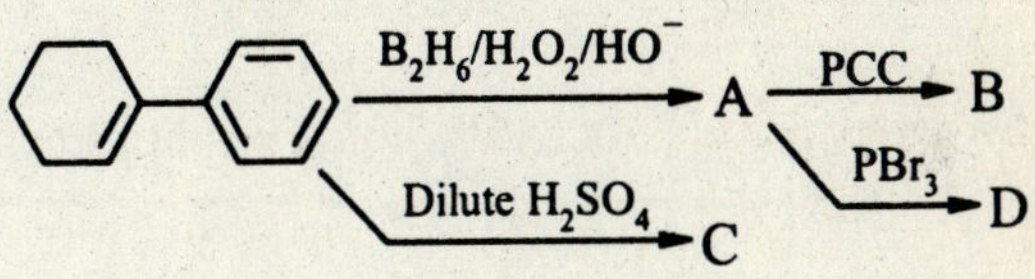 [4-(cyclohexenyl)biphenyl-type compound] $\xrightarrow{B_2H_6/H_2O_2/HO^-}$ A $\xrightarrow{PCC}$ B; A $\xrightarrow{PBr_3}$ D; [starting compound] $\xrightarrow{\text{Dilute } H_2SO_4}$ C

(b) [cyclopentyl-CH_2CH_2]–Br $\xrightarrow{HCCNa}$ E $\xrightarrow[H^+]{Hg^{2+}}$ F; [bromide] $\xrightarrow{\text{Alcoholic KOH}}$ G $\xrightarrow{Br_2/CCl_4}$ H; G $\xrightarrow{HBr}$ I $\xrightarrow[H_2O]{NaCN}$ J

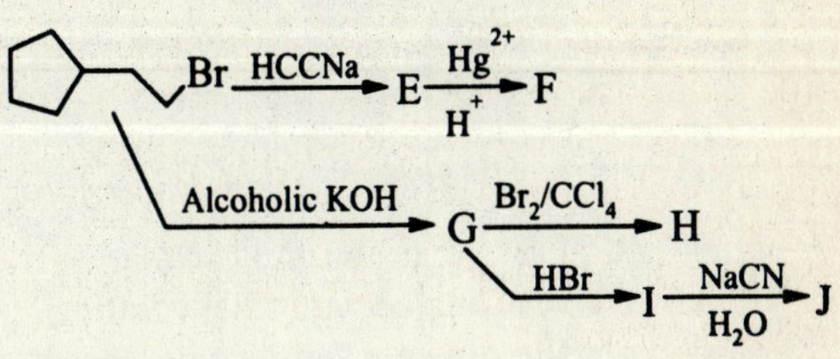

37. Provide structure of the compounds A to E in the following sequence of reaction:

$HC{\equiv}CNa$ + 1-bromo-5-methyl hexane $\xrightarrow{\text{Liq } NH_3}$

$A(C_9H_{16})$ $\xrightarrow[\text{Liq } NH_3]{NaNH_2}$ B : A + Na/heat $\xrightarrow{\text{1-bromobutane}}$

$C(C_{13}H_{24})$ $\xrightarrow{H_2/Pd/BaSO_4}$ $D(C_{13}H_{26})$ $\xrightarrow{CH_3CO_3H}$ $E(C_{13}H_{26}O)$

38. Carry out the following conversions:

(a) 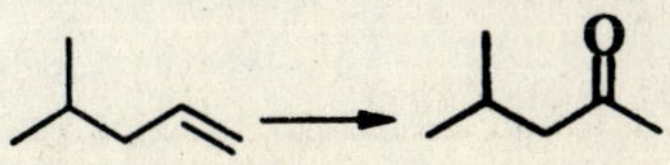 [4-methyl-1-pentene] → [4-methyl-2-pentanone]

(b)  → CHO

(c)  Ph → Ph OPh

(d)

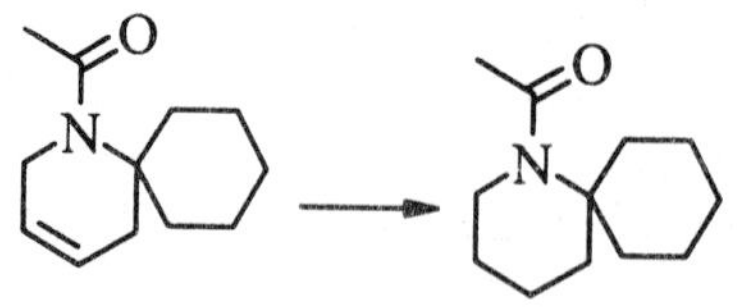

(e)

(f) 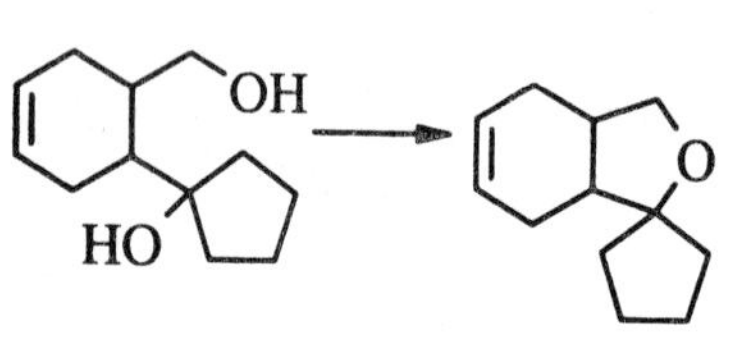 OH HO → O

39. Propose mechanism of the following reactions:

(a) 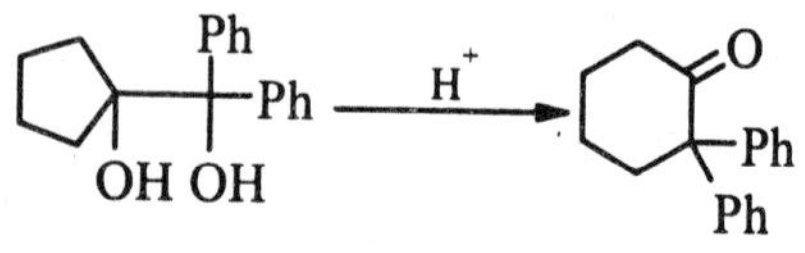 Ph, Ph, OH OH $\xrightarrow{H^+}$ O, Ph, Ph

(b) 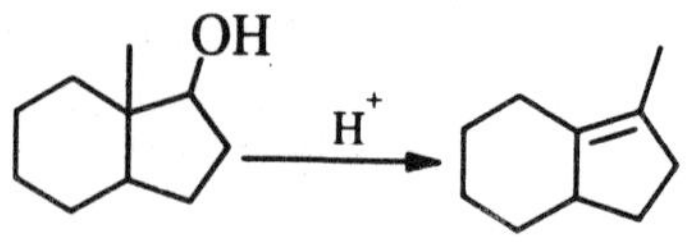 OH $\xrightarrow{H^+}$

(c) 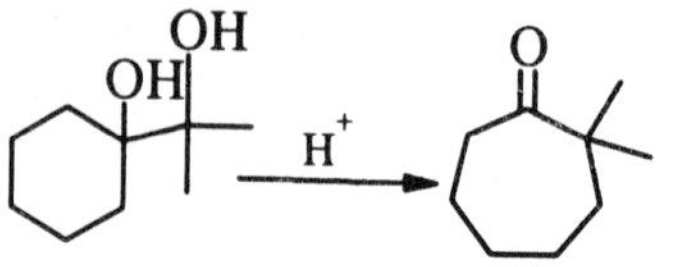 OH, OH $\xrightarrow{H^+}$ O

(d) 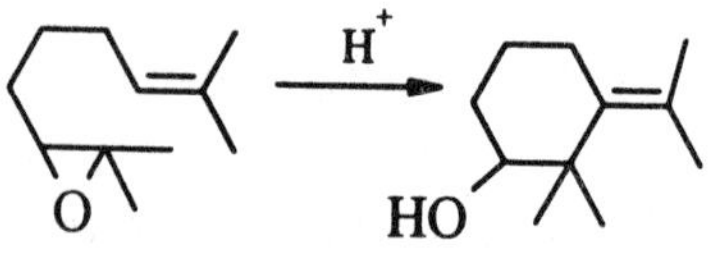 O $\xrightarrow{H^+}$ HO

40. A compound X ($C_{14}H_{14}O$) on mild oxidation yields $C_{14}H_{12}O$ (Y). If X is treated with a dehydrating agent, it loses a molecule of water and resulting product on vigorous oxidation yields two molecules of benzoic acid. Give structure of X and Y.

41. When 1-chloromethyl -4-methyl-1, 3-cyclopentadiene (A) is solvolysed in aqueous acetone, three isomeric alcohol are obtained. Write structure of alcohol predicting major product.

42. Two optically active compounds A and B have same molecular formula $C_9H_{12}O$ and do not decolourize bromine water solution. Both A and B are oxidised by $KMnO_4$ to give benzoic acid and react with Na to give a colourless, odourless gas. Compound A gives yellow precipitate with I_2/OH^- whereas B does not. Also A loses optical activity on treatment with PCC while B retains optical activity even on treatment with PCC. Deduce the structure of A and B.

43. Compounds A, B and C are isomeric alcohols with formula $C_5H_{12}O$. A and B reacts with chromic acid solution, B giving an acid D. The three isomeric alcohol reacts with HBr with decreasing relative rates C > A >> B, all giving same $C_5H_{11}Br$(E) in varying yields. A alone can be oxidised by I_2/OH^- to F. Write structure of A to F.

44. C_7H_{14} (A) decolourises Br_2 in CCl_4 and reacts with $Hg(OAc)_2$ in THF-H_2O followed by reduction with $NaBH_4$ to produce a resolvable compound B. A undergoes reductive ozonolysis to give the same compound (C) as obtained by oxidation of 3-hexanol with $KMnO_4$. Identify A, B and C.

45. An organic compound A (C_7H_{14}) reacts with BH_3-THF and then with H_2O_2/OH^- to give a chiral B. Oxidation of B with $KMnO_4$ affords a chiral carboxylic acid C. Ozonolysis of A followed by work-up with $(CH_3)_2S$ produces $C_6H_{12}O$(D) as one product. D on treatment with $LiAlH_4$ produces another compound E($C_6H_{14}O$) which is optically active. E on heating with conc H_2SO_4 produces F(C_6H_{12}) as major product. F on ozonolysis followed by work-up with DMS produces acetone as one product. Identify A to F.

46. Treating 3, 3-dimethyl-1-butene with dilute sulphuric acid is largely unsuccessful as a method of preparation of 3, 3-dimethyl-2-butanol, because an isomeric compound is major product. What is the isomeric product and how it is formed?

47. Explain the difference in the products from the following reactions of compound A.

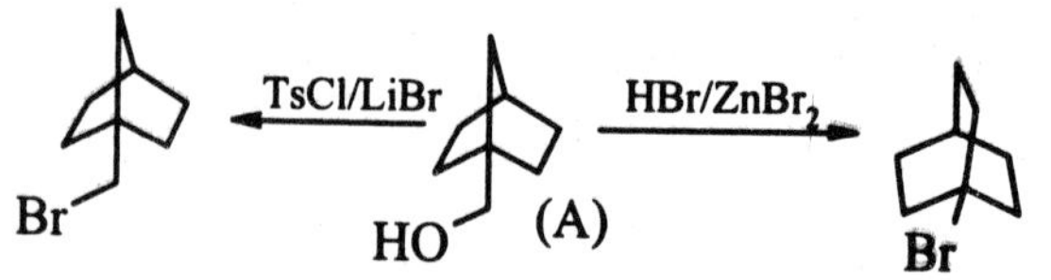

Br ← TsCl/LiBr — HO (A) — HBr/ZnBr$_2$ → Br

48. Starting with appropriate alkene and assuming you have deuterated acid (CH_3COOD), outline the synthesis of the following compound:

(a) $(CH_3)_2CHCH_2CH_2D$ (b) $(CH_3)_2CHCHDCH_3$

(c) 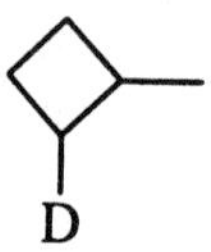 D (d) Convert → D, T

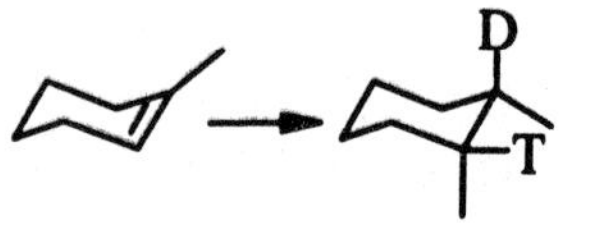

49. The two synthesis of 2-ethoxy-1-phenylpropane shown below gives products with opposite optical rotation. Explain.

Ph–CH₂–CH(OH)–CH₃ → (K) → Ph–CH₂–CH(OK)–CH₃ → (EtBr) → Ph–CH₂–CH(OEt)–CH₃, $[\alpha] = +23°$

$[\alpha] = +33°$ (OH) → (TsCl/Base) → Ph–CH₂–CH(OTs)–CH₃ → (C_2H_5OH/K_2CO_3) → Ph–CH₂–CH(OEt)–CH₃, $[\alpha] = -19°$

50. When C_2H_5ONa reacts with 1-(chloromethyl) oxirane, labeled with ^{14}C as shown by astric in I, the major product is an epoxide bearing the labeled carbon as shown in II. Provide mechanism to explain this fact.

Cl–CH₂–(oxirane)* (I) → EtONa → (oxirane)–CH₂*–OEt (II)

51. An unknown organic compound $A(C_4H_{10}O_2)$ reacts with sodium metal to liberate one mole of hydrogen gas per mole of A. Although A is inert towards periodic acid, it does reacts with CrO_3 to form $B(C_4H_6O_3)$. Identify A and B.

52. Convert:

(a) cyclopentyl–OH → cyclopentyl–D (b) $CH_2{=}CH_2$ → 1,4-dioxane (O O)

53. Starting with each of the following outline a practical synthesis of 1-butanol:

(a) 1-butene (b) 1-chlorobutane
(c) 2-chlorobutane (d) 1-butyne

54. Synthesize 2-bromobutane in good yield starting with each of the following compounds:

(a) 2-butanol (b) 1-butene
(c) 1-butanol (d) 1-butyne

55. Synthesize the following compounds starting with isobutane:

(a) *ter*-butyl bromide (b) 2-methyl propene
(c) *iso*-butyl bromide (d) *iso*-butyl methyl ether
(e) $(CH_3)_2CHCH_2OCOCH_3$
(f) $(CH_3)_2CHCH_2CN$

56. A neutral compound A has molecular formula $C_{10}H_{16}O_2$. A does not decolourize aqueous solution of Br_2 and evolve no gas on treatment with Na. A on acidic hydrolysis produces two compounds $B(C_4H_8O)$ and $C(C_6H_{10}O_2)$. B on treatment with $SOCl_2$ gives $D(C_4H_7Cl)$, which on further treatment with aqueous solution of KCN followed by hydrolysis of product yielded $E(C_5H_8O_2)$. E on reducing with $LiAlH_4$ gives $F(C_5H_{10}O)$, which on heating with concentrated sulfuric acid solution yields $G(C_5H_8)$. G on treatment with HCl yields $H(C_5H_9Cl)$. C on heating with soda lime gives $I(C_5H_{10})$. I on monochlorination with $Cl_2/h\nu$ give H as sole product. Identify A to I describing mechanism of their formation in each step.

57. Predict structures for each of the following products:

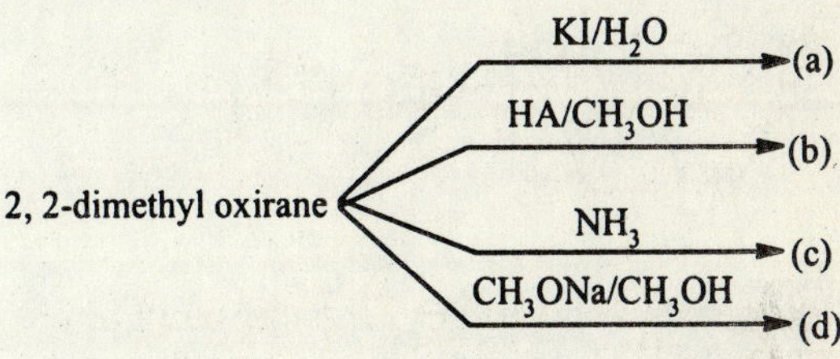

58. An organic compound A has molecular formula $C_6H_{10}O$ and known to decolourize aqueous solution of bromine. A is also resolvable and on catalytic hydrogenation with H_2/Pt yields $B(C_6H_{12}O)$ which is optically inactive. Also B does not change the orange colour of an acidic solution of dichromate ion. B on heating with concentrated sulfuric acid solution yields $C(C_6H_{10})$, which on reductive ozonolysis gives the following compound:

H–C(=O)–CH₂–CH₂–CH₂–C(=O)–CH₃

Deduce structures of A, B and C.

59. An organic compound $A(C_5H_8O_2)$ does not decolourize aqueous solution of bromine and evolve no gas on treatment with CH_3MgBr. A on reaction with concentrated solution of HI produced methylene iodide and an another compound $B(C_4H_6I_2)$. B on heating with sodium metal produced $C(C_4H_8)$, which on monochlorination gave a single isomeric product C_4H_7Cl. Deduce structures of A, B and C.

60. Starting with 2-methyl propene and using any other needed reagents, synthesize the following compounds:

(a) $(CH_3)_2CHCH_2OH$ (b) $(CH_3)_2CDCH_2T$
(c) $(CH_3)_2CHCH_2T$
(d) $(CH_3)_2CHCH_2OCH_2CH_3$

61. Propose mechanism:

$CH_2{=}CH–CH_2–CH_2–CH_2–OH \xrightarrow{H^+}$ (cyclopentene)

62. Give products with their stereochemical structures in the following reactions:

(a) 1-methyl cyclobutene $\xrightarrow[H_2O_2,\ OH^-]{THF/BH_3}$ A($C_5H_{10}O$)

$\xrightarrow[OH^-]{TsCl}$ B($C_{12}H_{16}SO_3$) $\xrightarrow{OH^-}$ C($C_5H_{10}O$)

(b) B + I^- → D(C_5H_9I)

(c) *Trans*-4-methyl cyclohexanol + CH_3SO_2Cl/OH^-

→ E($C_8H_{16}SO_3$) E $\xrightarrow{HC\equiv CNa}$ F(C_9H_{14})

63. Predict product in the following reaction indicating appropriate stereochemistry where applicable:

(a) 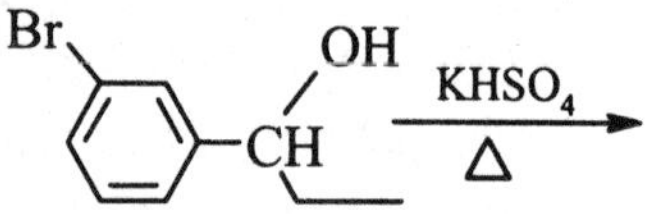 [structure: Br, OH, CH] $\xrightarrow[\Delta]{KHSO_4}$

(b) 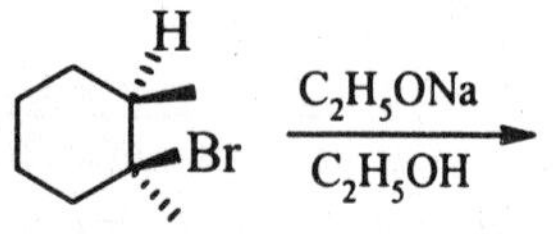 [structure: H, Br] $\xrightarrow[C_2H_5OH]{C_2H_5ONa}$

(c) [structure: HO, CN] $\xrightarrow[\Delta]{KHSO_4}$

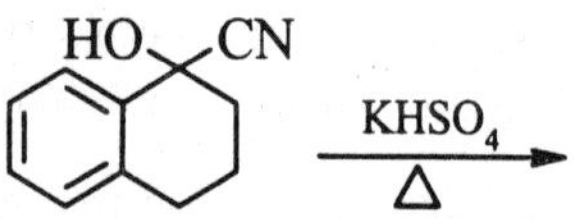

64. Propose mechanism of the following reactions:

(a) 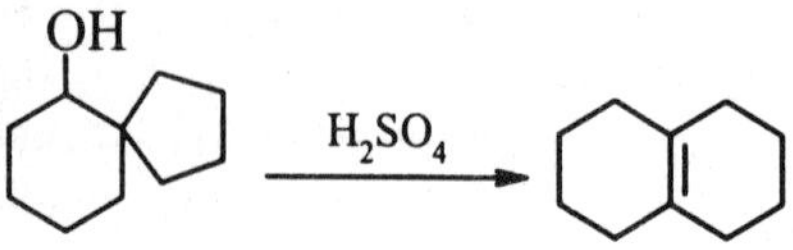 [structure: OH] $\xrightarrow{H_2SO_4}$

(b) 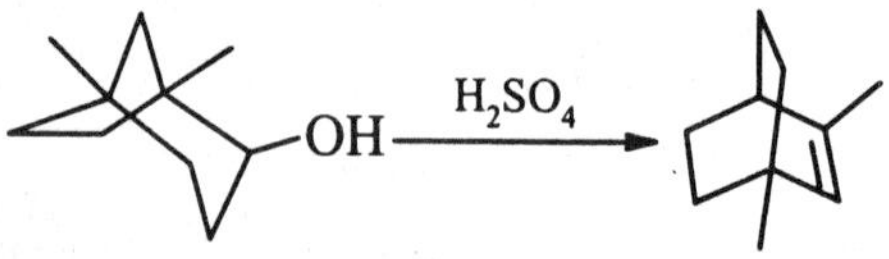 [structure: OH] $\xrightarrow{H_2SO_4}$

65. Predict major product indicating stereochemistry where applicable:

(a) 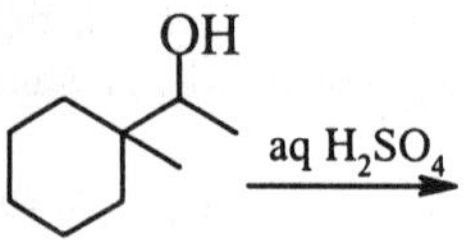 [structure: OH] $\xrightarrow{aq\ H_2SO_4}$ (b) 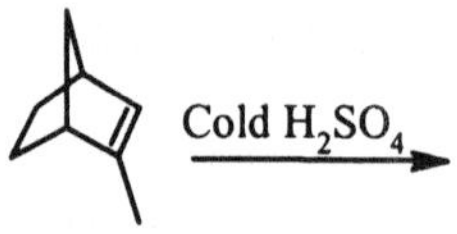 [structure] $\xrightarrow{Cold\ H_2SO_4}$

(c) 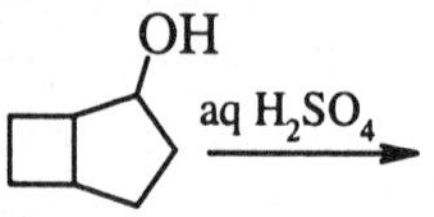 [structure: OH] $\xrightarrow{aq\ H_2SO_4}$

(d) 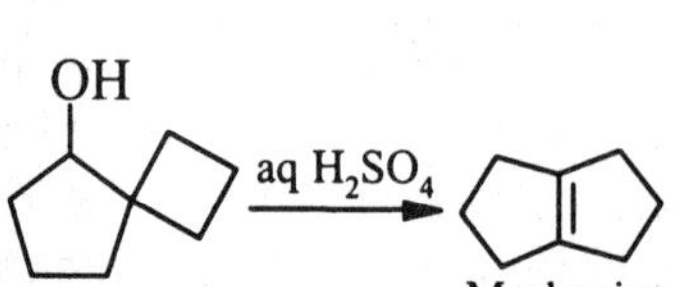 [structure: OH] $\xrightarrow{aq\ H_2SO_4}$ [structure]

Mechanism

66. Coniferyl alcohol (X) is not soluble in water or $NaHCO_3$, has moleculars formula $C_{10}H_{12}O_3$. A solution of Br_2 in CCl_4 is decolourized by X, forming $C_{10}H_{12}O_3Br_2$ (A). Upon reductive ozonolysis of X, 4-hydroxy-3-methoxy benzaldehyde and B($C_2H_4O_2$) is produced. X reacts with benzoylchloride C_6H_5COCl in presence of a base to form C($C_{24}H_{20}O_5$). This product rapidly decolourises aqueous solution of $KMnO_4$ and is insoluble in NaOH. X reacts with cold HBr to form D($C_{10}H_{13}O_2Br$). X reacts with HI to produce E($C_9H_9O_2I$) and CH_3I. In aqueous base, CH_3I and X forms F($C_{11}H_{14}O_3$), which is not soluble in strong base but decolourises aqueous solution of Br_2. Deduce structures of X and A to F.

67. An optically active organic compound A has molecular formula $C_7H_{12}O_2$ and does not decolourize cold, dilute and alkaline solution of potassium permanganate. A on treatment with concentrated HCl gives CH_2Cl_2 and B($C_6H_{10}Cl_2$). B on treatment with aqueous solution of KI gives C(C_6H_{10}) which decolourize bromine water. C on treatment with acidic solution of $KMnO_4$ gives 2-methyl pentan-1, 5-dioic acid. Deduce structures of A to C.

68. An optically active organic compound A has molecular formula $C_9H_{18}O$, which neither change colour of Bayer's reagent nor of acidic dichromate solution, but evolve a colourless gas on heating with Na-metal. A on dehydration with concentrated H_2SO_4 produces B(C_9H_{16}), which exist in two stereomeric forms. Ozonolysis of B followed by work-up with Zn-H_2O produced ethanal and C($C_7H_{12}O$). C on treatment with $LiAlH_4$ produced D($C_7H_{14}O$), which is enantiomeric and change the colour of acidic dichromate solution from orange to blue-green. D on dehydration produced E(C_7H_{12}), which on ozonolysis followed by hydrolysis of product in presence of Zn produced cyclopentanone. Deduce structures of A to E showing stereochemical structures of B.

69. (a) Propose mechanism of formation of products in the following reactions:

(i) 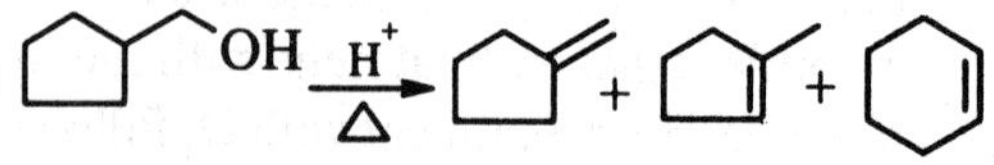 [structure: OH] $\xrightarrow[\Delta]{H^+}$ [structure] + [structure] + [structure]

(ii) 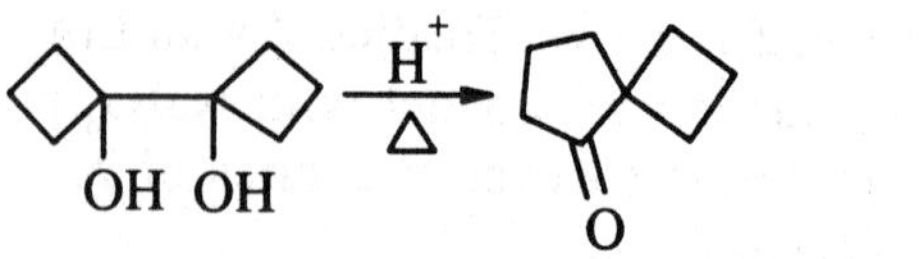 [structure: OH OH] $\xrightarrow[\Delta]{H^+}$ [structure: O]

(b) Complete the following reactions:

(i) 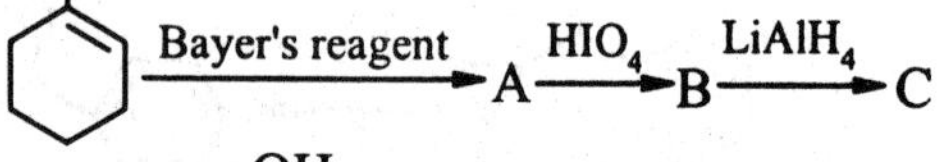 [structure] $\xrightarrow{\text{Bayer's reagent}}$ A $\xrightarrow{HIO_4}$ B $\xrightarrow{LiAlH_4}$ C

(ii) Convert [structure: OH] into [structure: OH]

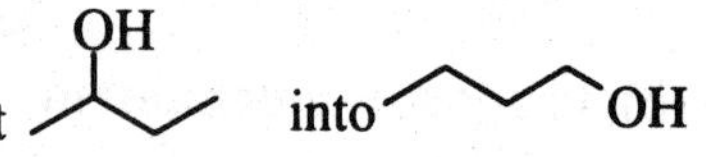

70. (a) Starting with bromobenzene and any other needed reagents, outline synthesis of the following compound in good yield: $C_6H_5CH_2—COCH_3$.

(b) Synthesize the following compound starting from 1-bromo-4-hydroxymethyl cyclohexane.

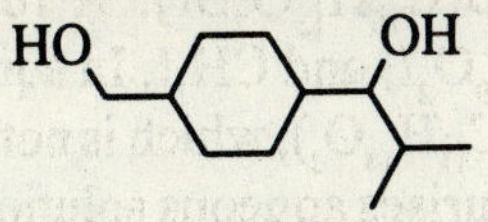

HO, OH

(c) Bring about the following conversion explaining mechanism: 1-pentanol to 2-pentanone.

71. A hydrocarbon C_7H_{14}(A) decolourizes bromine water and reacts with $(CH_3COO)_2Hg$ in THF followed by reduction with $NaBH_4$ to produce a resolvable compound (B). Compound A undergoes reductive ozonolysis to give the same compound obtained by oxidation of 3-hexanol with $KMnO_4$. C, an isomer of A reacts with $BH_3—H_2O_2/OH^-$ to give a chiral D. Oxidation of D with acidic permanganate gives a chiral carboxylic acid E, ozonolysis of C gives the same compound as obtained by the oxidation of 2-methyl-3-pentanol. Identify A to E.

72. An organic compound $A(C_6H_{10}O)$ does not decolourize Bayer's reagent and evolves no gas on treatment with sodium metal. A on treatment with HI solution yields two stereomers B and C with their molecular formula $C_6H_{10}I_2$, of which only B is optically active. Either B or C does not form alkene on heating with Zn metal. Treatment of either A or B with ethanolic KOH solution yields same D, which on treatment with NBS gives a single monobromo-derivative (E). E on further treatment with ethanolic KOH solution yields benzene. Deduce structures of A to E showing their stereochemical arrangements.

73. An organic compound X exist in four stereomeric forms and has molecular formula $C_5H_{10}O$. X decolourizes Bayer's reagent as well as evolves a colourless gas on treatment with methyl magnesium bromide. X on treatment with O_3 followed by work-up with Zn-H_2O produces Y and Z of which Y is inactive and Z is active. Treating Z with $LiAlH_4$ followed by oxidation of product with $NaIO_4$ produced Y and methanal. Deduce structures of X (all stereomers), Y and Z.

74. (a) Convert:

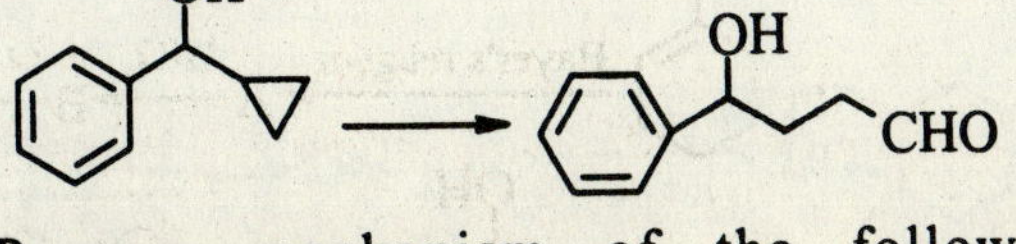

OH → OH, CHO

(b) Propose mechanism of the following transformation:

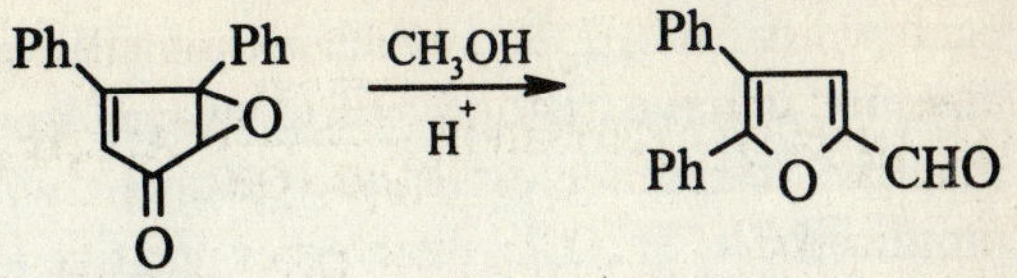

Ph, Ph, O, O $\xrightarrow[H^+]{CH_3OH}$ Ph, Ph, O, CHO

(c) Convert:

(i) O →

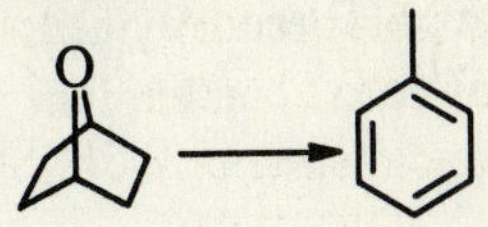

(ii) O → O

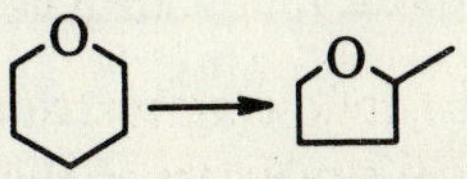

75. An organic compound $A(C_{11}H_{22})$ when ozonolyzed yields $B(C_6H_{12}O)$ and $C(C_5H_{10}O)$. Mild oxidation of C yields $D(C_5H_8O_2)$. Compound B reacts with NaOCl to yield chloroform and sodium salt of D. Reduction of B with $LiAlH_4$ yields $E(C_6H_{14}O)$. Dehydration of E gives $F(C_6H_{12})$. Ozonolysis of F yields $G(C_2H_4O)$ and $H(C_4H_8O)$. Compound H does not react with Tollen's reagent or Fehling solution, but does form an oxime and semicarbazide. Write structures of A to H describing the reaction involved.

76. An optically active organic compound $A(C_8H_{16}O)$ does not decolourize the brown colour of bromine-water, but turns orange colour of acidic dichromate solution to blue green converting itself into another optically active compound B. A on heating with conc. H_2SO_4 produced $C(C_8H_{14})$ as major product, which does not show stereomerism. C on reductive ozonolysis produced $D(C_5H_8O)$ as one of the product, which does not change the colour of acidic dichromate solution. D on reduction with $LiAlH_4$ produced $E(C_5H_{10}O)$, which on dehydration followed by reductive ozonolysis of product yields a non-resolvable compound $F(C_5H_8O_2)$. F gives positive Fehling's solution test but negative haloform test. Deduce structural formula of A to F.

77. An organic compound A $(C_{12}H_{20}O_2)$ neither decolourise bromine water nor evolves any gas on treatment with Na-metal. A on refluxing with dil H_2SO_4 produced B and C constitutional isomers, with their molecular formula $C_6H_{12}O_2$. B is optically active but C is inactive. B on treatment with acidic $NaIO_4$ produced $D(C_6H_{10}O_2)$. D on treatment with dil NaOH produced $E(C_6H_9O_2)$. Treating either C or E with HI/red-P yielded cyclohexane. Deduce structures of A to E.

78. An optically inactive organic compound A has molecular formula $C_7H_{12}O$ and does not produce any gas on treatment with Na. A also does not decolourise Bayer's solution. A on refluxing with dilute sulphuric acid yields B($C_7H_{14}O_2$), which is optically active and does not undergo periodic acid cleavage. An optically inactive stereomer of B, on treatment with HIO_4 produced C($C_7H_{12}O_2$), which gives positive iodoform test but negative Tollen's test. C is optically inactive and on treatment with HI/red P yields *n*-heptane. Deduce structures of A to C and explain mechanism of formation of B.

79. Propose mechanism of the following reactions:

(a) 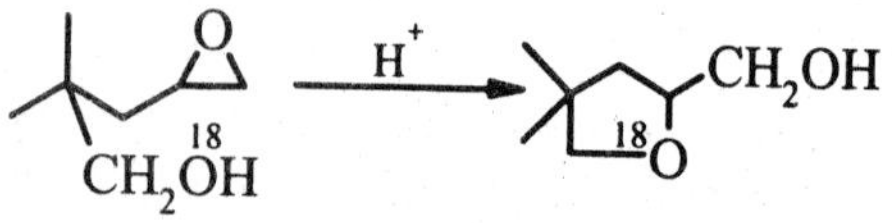 $\xrightarrow{H^+}$ CH_2OH, $CH_2{}^{18}OH$, ^{18}O

(b) 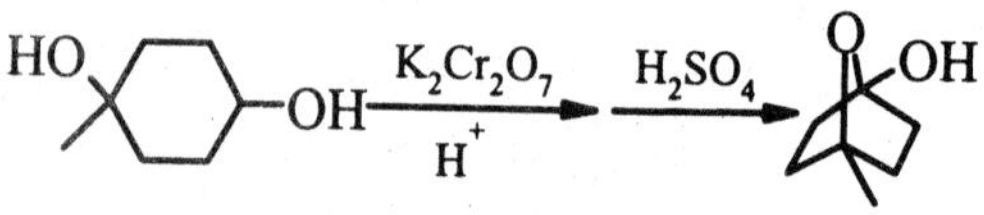 HO, OH $\xrightarrow[H^+]{K_2Cr_2O_7}$ $\xrightarrow{H_2SO_4}$ OH

80. (a) Bring about the following transformation:

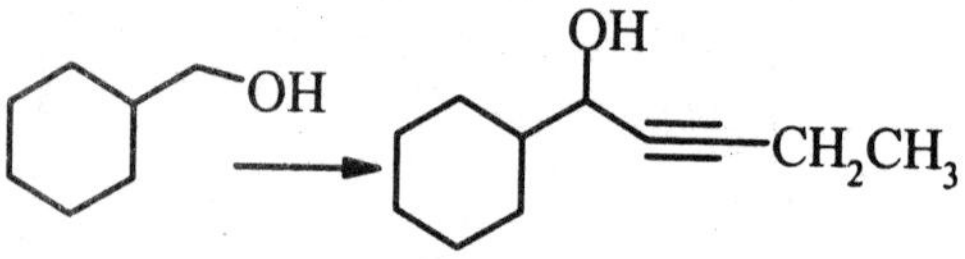

OH $\longrightarrow$ OH, CH_2CH_3

(b) Starting from benzo phenone and ethyl bromide, synthesize $Ph_2C{=}CHCH_3$.

81. In each pair of the reactions shown below, one reaction gives product while the other cannot proceed as written. Write the product of the reaction that proceed and briefly explain why the other fail to proceed.

(a)  OH $\xrightarrow{SOCl_2}$ and OH $\xrightarrow{SOCl_2}$

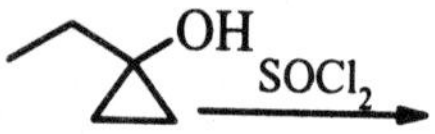

(b) 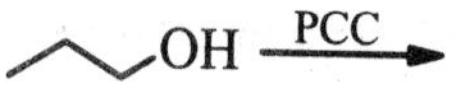 OH $\xrightarrow{PCC}$ and OH $\xrightarrow{PCC}$

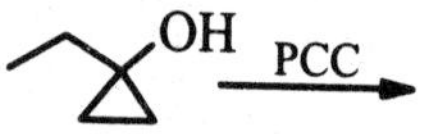

82. Provide missing reagents for each steps paying careful attention to stereochemistry:

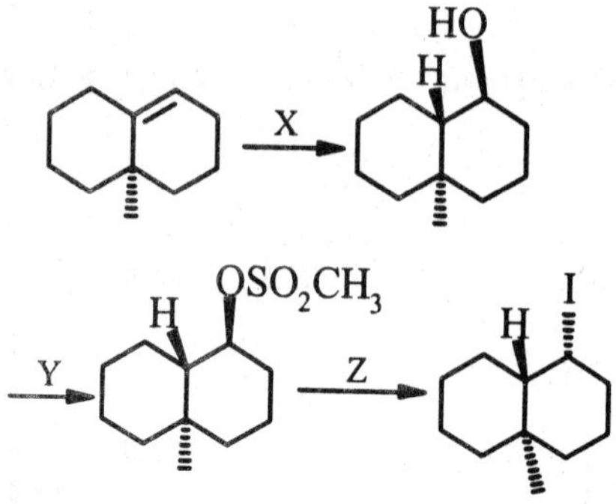

$\xrightarrow{X}$ HO, H

$\xrightarrow{Y}$ OSO_2CH_3, H $\xrightarrow{Z}$ I, H

83. (a) Provide mechanism of the following:

$\xrightarrow{CH_3OH/TsOH}$ OMe

(b) Acid catalyzed hydrolysis of tertiary butyl ether is easier than hydrolysis of diethyl ether, explain.

84. Bring about the following conversion:

(a) 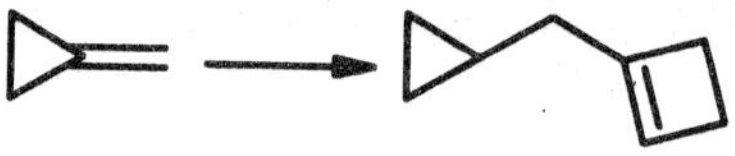 $\longrightarrow$

(b) 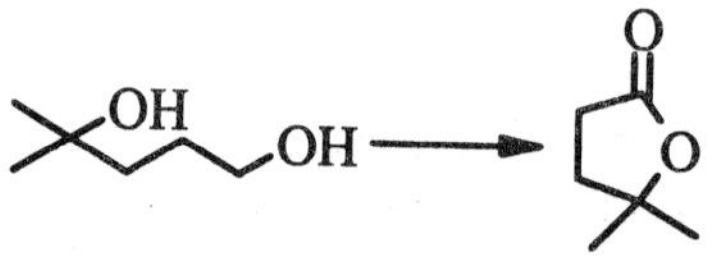 OH, OH $\longrightarrow$ O, O

(c) 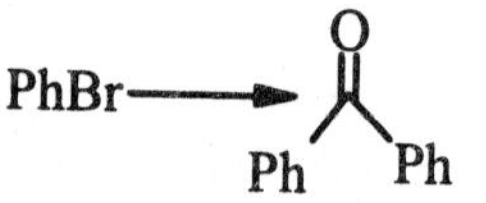 PhBr $\longrightarrow$ O, Ph, Ph

(d)  O, O, H $\longrightarrow$ COOH

(e)  Br $\longrightarrow$

(f) 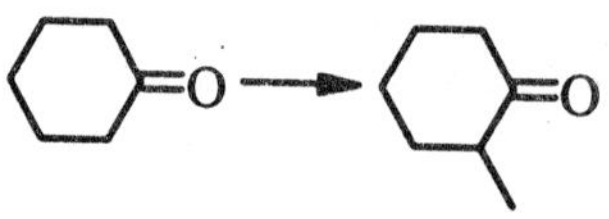 =O $\longrightarrow$ =O

85. (a) Propose mechanism OMe $\xrightarrow{IH}$

(b) Convert $\longrightarrow$ OMe

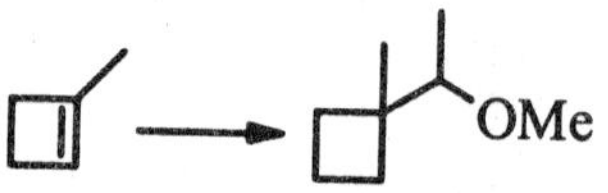

(c) Propose mechanism H, O, O, OMe $\xrightarrow{CH_3ONa}$ CHO, OH

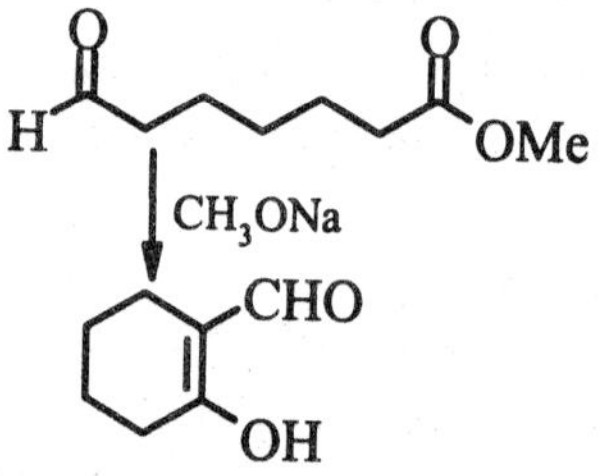

86. Predict product of the following reactions:

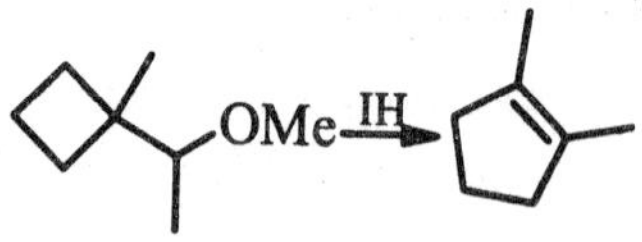

(a) 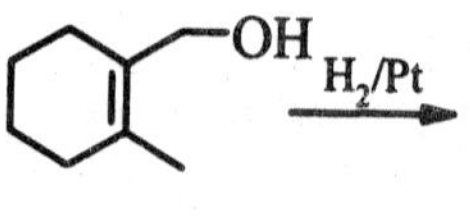 OH $\xrightarrow{H_2/Pt}$

(b) 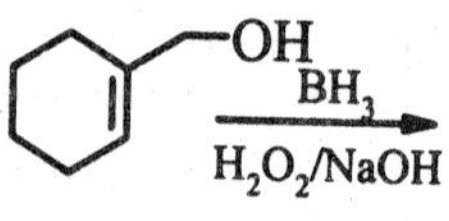 OH $\xrightarrow[H_2O_2/NaOH]{BH_3}$

(c) 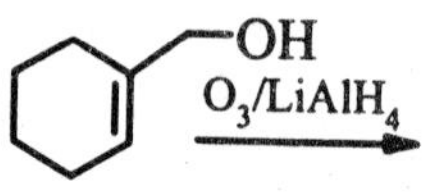 OH $\xrightarrow{O_3/LiAlH_4}$

(d) 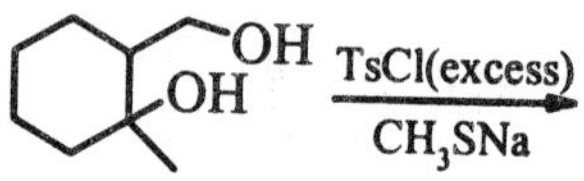 OH, OH $\xrightarrow[CH_3SNa]{TsCl(excess)}$

(e) 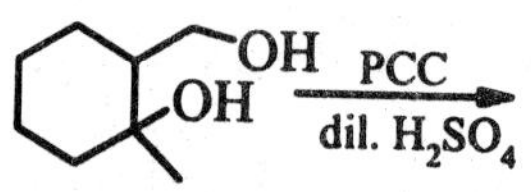 OH, OH $\xrightarrow[dil.\ H_2SO_4]{PCC}$

(f) 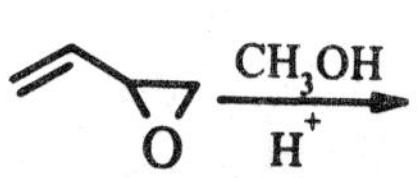 O $\xrightarrow[H^+]{CH_3OH}$

(g) 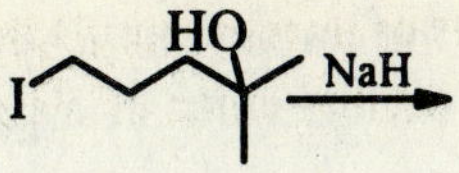 [structure] $\xrightarrow{NaH}$

(h) 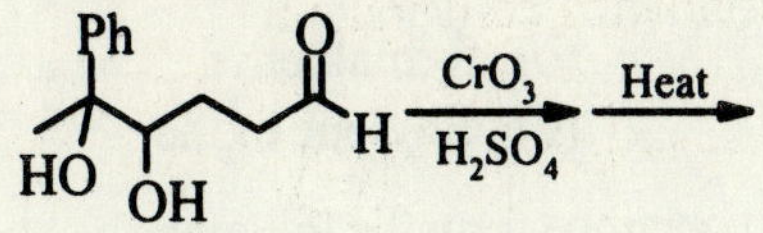 [structure] $\xrightarrow[H_2SO_4]{CrO_3}$ $\xrightarrow{Heat}$

(i) 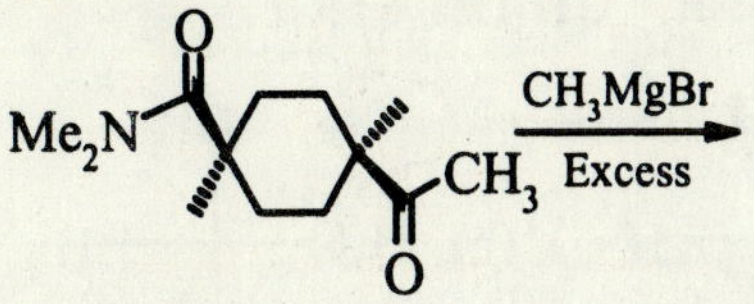 [structure] $\xrightarrow[Excess]{CH_3MgBr}$

(j) 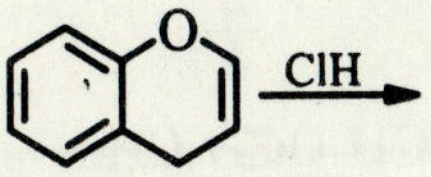 [structure] $\xrightarrow{ClH}$ (k) 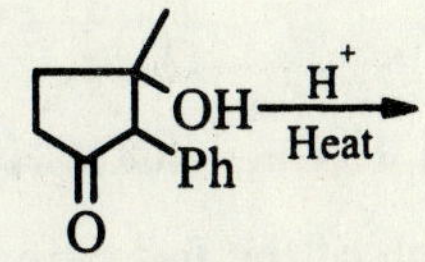 [structure] $\xrightarrow[Heat]{H^+}$

87. An optically active aldo pentose (A) has molecular formula $C_5H_{10}O_5$. A on treatment with Br_2-H_2O gives B($C_5H_{10}O_6$), which on oxidative degradation with $FeSO_4/H_2O_2$ gives a tetrose C($C_4H_8O_4$), which is still optically active. C on treatment with dilute HNO_3 gives meso tartaric acid. Deduce structures of A to C.

88. Bring about the following conversion:

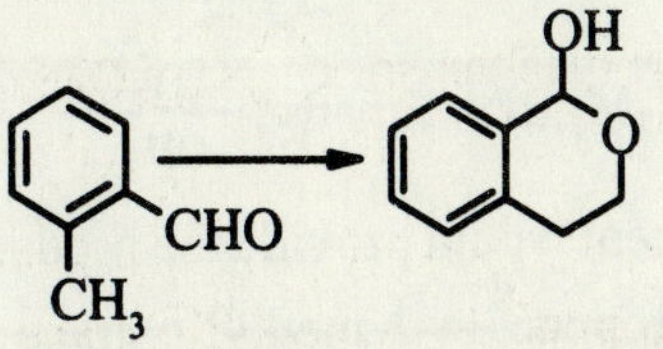

[structure: o-methylbenzaldehyde (CHO, CH_3) → cyclic hemiacetal (OH, O)]

89. An organic compound A($C_{11}H_{16}O$) is optically active but does not react with NaH. A on refluxing with dilute sulphuric acid produces optically active B($C_9H_{12}O$), which on treatment with cold, alkaline $K_2Cr_2O_7$ yields C($C_9H_{10}O$), which is neither resolvable nor yields any precipitate with $NaOH/I_2$. A on boiling with concentrated HI solution produced D($C_9H_{11}I$) and iodoethane. D on treatment with ethanolic solution of KOH gave E(C_9H_{10}) and E on reductive ozonolysis produced F(C_7H_6O). F reduces Tollen's reagent but not to Fehling's solution. Deduce structures of A to F.

90. Propose mechanism:

(i) 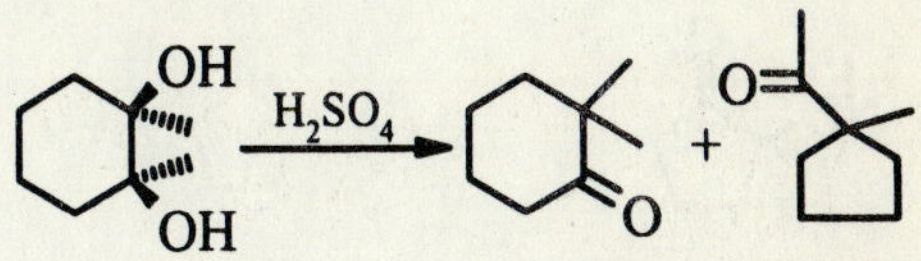 [structure] $\xrightarrow{H_2SO_4}$ [structure] + [structure]

(ii) 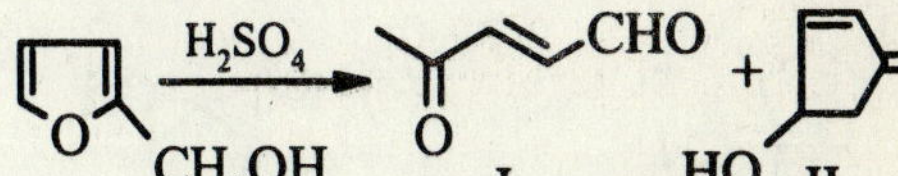 [structure, CH_2OH] $\xrightarrow{H_2SO_4}$ [structure, CHO] I + [structure, HO] II

91. (a) Explain the following observations:

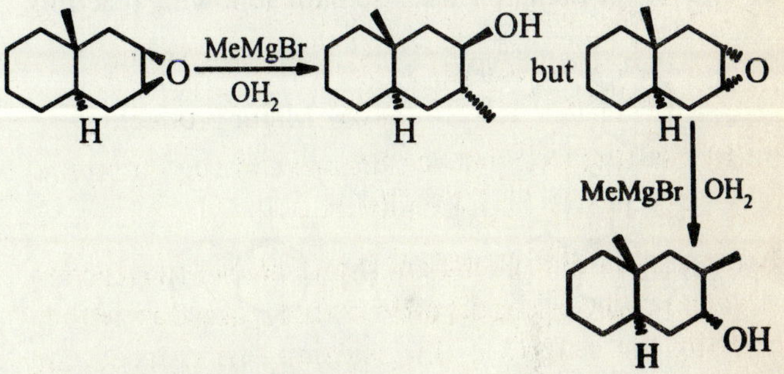

[structure] $\xrightarrow[OH_2]{MeMgBr}$ [structure, OH] but [structure] $\xrightarrow{MeMgBr / OH_2}$ [structure, OH]

(b) Propose mechanism:

[structure, OH, OH] $\xrightarrow[H_2SO_4\ (0°C)]{CH_3CN}$ [structure, O, N]

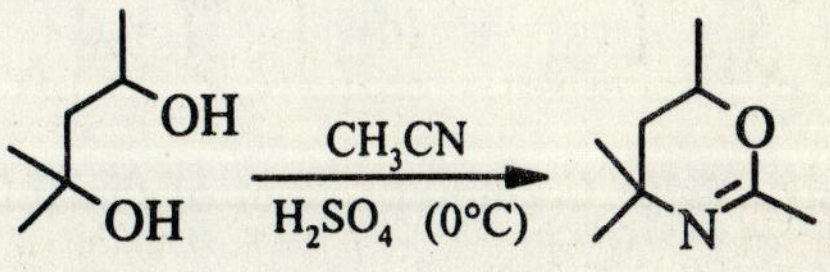

OBJECTIVE EXERCISE (ALCOHOLS)

1. Give the major product of the following reaction sequence:

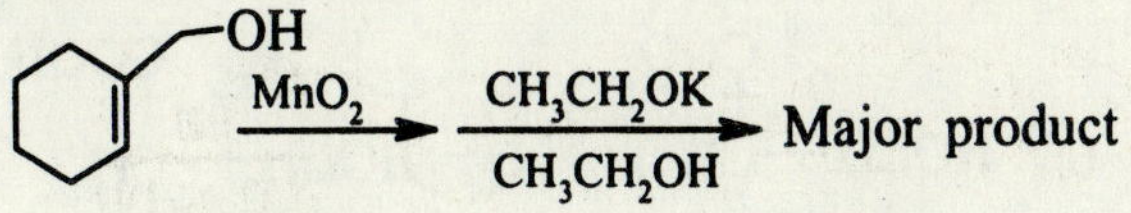

[structure, OH] $\xrightarrow{MnO_2}$ $\xrightarrow[CH_3CH_2OH]{CH_3CH_2OK}$ Major product

(a) 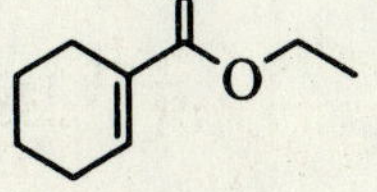 [structure] (b) 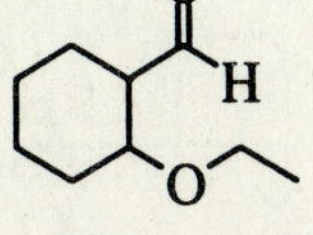 [structure, H]

(c) 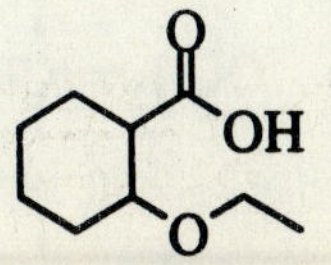 [structure, OH] (d) 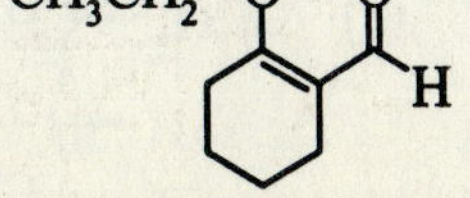 [structure, CH_3CH_2–O, H]

2. Give the major product of the following reaction:

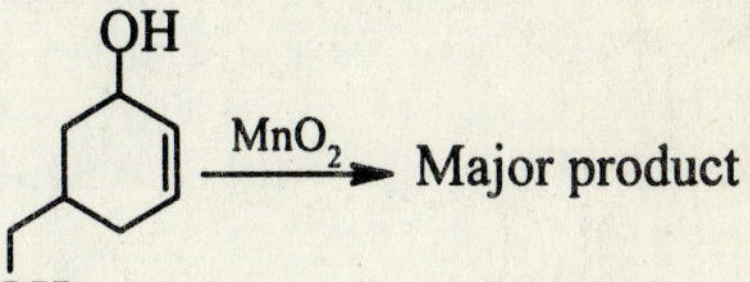

[structure, OH, OH] $\xrightarrow{MnO_2}$ Major product

(a) 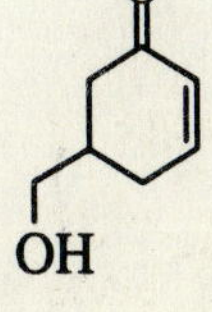 [structure, OH] (b) 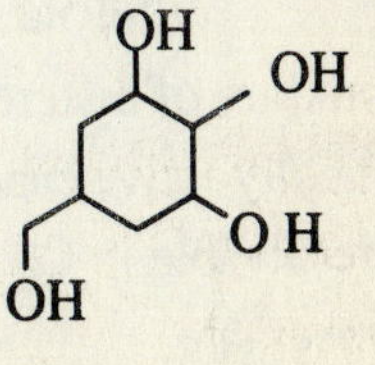 [structure, OH, OH, OH, OH]

(c) 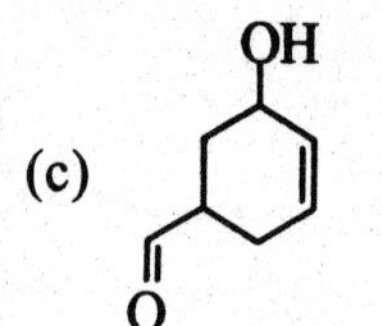

(d) 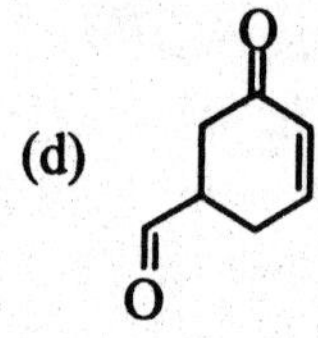

3. Give the major product of the following reaction:

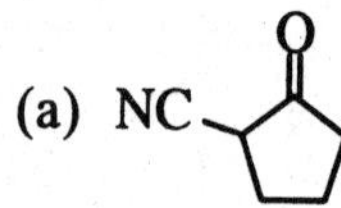 $\xrightarrow{CrO_3/H_2SO_4}$ $\xrightarrow{NaCN/HCI}$ Major product

(a) NC–(cyclopentanone)

(b) (cyclopentanone)–CI

(c) (cyclopentanone with CN and OH)

(d) NC–(cyclopentane)–OH

4. Choose the major product from the following reaction:

(cyclohexyl alcohol) $\xrightarrow[D]{H_2SO_4}$ Major product

(a) 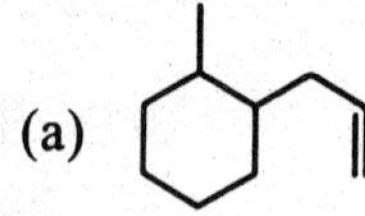

(b) 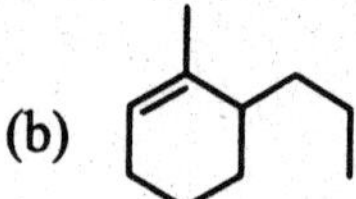

(c)

(d)

5. What is the major organic product of the following reaction?

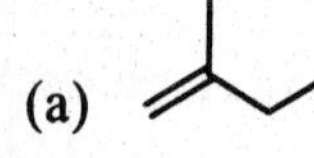 $\xrightarrow[\Delta]{H_2SO_4}$ Major product

(a)

(b)

(c)

(d)

6. In the reaction below, the product would be

2-butene + $H^+/H_2O \rightarrow$ 2-butanol

(a) A mixture of distereomers
(b) Optically active enantiomer
(c) Unresolvable
(d) A racemate

7. Which reagent will lengthen the carbon chain by two carbons in one step when reacted with an organolithium reagent?

(a) acetic acid (b) formaldehyde
(c) ethanol (d) ethylene oxide

8. The major product obtained form the following reaction:

Ethanal + CH_3OH + HCl →

(a) H_3C–C(OH)(OMe)–H (b) H_3C–C(OH)(Cl)–H

(c) H_3C—$COOCH_3$ (d) CH_3OCH_2CHO

9. Which of the following reactions would provide the best synthesis of 2-butanol?

(a) CH_3CH_2MgBr + (ethylene oxide) $\xrightarrow[OH_2]{H^+}$

(b) $CH_3CH_2CH_2MgBr \xrightarrow[H^+]{CO_2}$

(c) CH_3MgBr + Acetone $\xrightarrow[OH_2]{H^+}$

(d) CH_3CH_2MgBr + Ethanal $\xrightarrow[OH_2]{H^+}$

10. Which reagent would provide the product shown?

Propene → 2-methyl oxirane

(a) Peracetic acid (b) H_2SO_4
(c) CrO_3/pyridine, cold (d) $O_3/Zn/H_2O$

11. What is the major product from the following reaction?

$HOCH_2$—CH(COOH)—NH_3^+ Cl^- $\xrightarrow{\text{One mole NaOH}}$

(a) $HOCH_2$—CH(COO^-)—NH_3^+ Cl^- (b) $HOCH_2$—CH(COOH)—NH_2

(c) $^-OCH_2$—CH(COOH)—NH_3^+ Cl^- (d) $^-OCH_2$—CH(COO^-)—NH_3^+ Cl^-

12. If the following *C labeled compound undergo the following reaction, which product would be obtained?

One mole of KCN / EtOH

(a) (b) (c) (d)

13. What is the major product obtained from the following reaction?

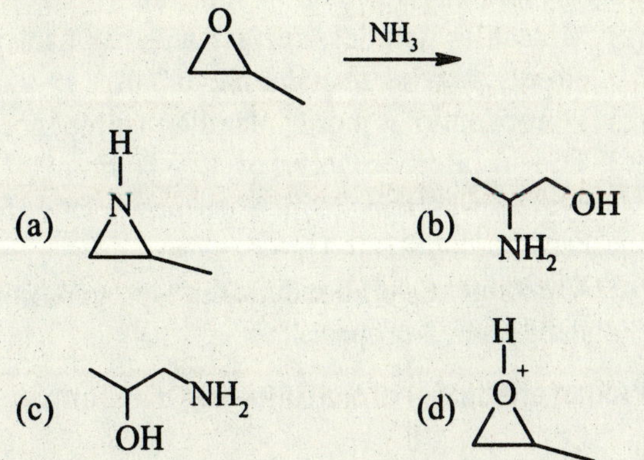

14. Treatment of a primary alcohol with CrO_3 will furnish an aldehyde. Inclusion of a key additive to this reaction mixture redirects the process, and a carboxylic acid is produced. What is the key additive?

(a) H_2O/H^+ (b) RMgX
(c) Pyridine (d) HCN

15. Select the best method for the preparation of the following compounds:

and

(a) React cyclohexanone with methyl lithium.
(b) React 1-methyl cyclohexene with $Hg(OAc)_2$/ $NABH_4$.
(c) Reacts cyclohexene with BH_3; $NaOH/H_2O_2$ followed by CH_3MgBr.
(d) Reacts cyclohexene with MCPBA, followed by CH_3MgBr.

16. When 1, 3, 9-tribromo-9-methyl-5-decanol is treated with a strong base, the major product will be?

(a) (b) (c) (d) None

17. Which reaction condition would be best to perform the following transformation?

Cyclohexanol → 1-methyl cyclohexene

(a) (i) PCC, (ii) MeMgBr, (iii) $POCl_3$
(b) TsCl/MeMgBr
(c) HBr/Mg/Mel
(d) NaOH/Mel

18. Which sets of reagents shown below would accomplish the following transformations?

(A) (B)

	A	B
(a)	BH_3, $H_2O_2/NaOH$	H_2SO_4
(b)	$H_2/Pd/C$	$BH_3/H_2O_2/NaOH$
(c)	H_3PO_4	$Hg(OAc)_2$, $NaBH_4$
(d)	EtOK	$BH_3/H_2O_2/NaOH$

19. When 4-penten-1-ol is treated with aqueous bromine, a cyclic bromo substituted ether is formed rather than the expected bromohydrin. Select the explanation that best accounts for this result.

Br_2/H_2O

(a) The ether is the result of hydride shift in the initially formed carbocation.
(b) The ether is formed in the intramolecular nucleophile attack on initially formed bromonium ion.
(c) The reaction is initiated by protonation of π-bond.
(d) The initially formed bromonium ion undergoes an alkyl shift to form new C—O bond.

20. Select the best starting material for the following reaction:

HIO_4

(a) (b)

(c) (d)

21. When treated with acid, the strained ether (I) is protonated on oxygen, and ring opens to give a resonance stabilized carbocation. Which diene listed below when treated with acid will give the same carbocation?

(I)

(a) (b)

(c) (d)

22. Choose the best reagent to carry out the following sequence of reaction:

Br —I→ —II→ —III→

	I	II	III
(a)	Acetylene/$NaNH_2/NH_3$	H_2/Pd	$OsO_4/NaHSO_3$
(b)	Acetylene/$NaNH_2/NH_3$	H_2/Lindlar's cat	$BH_3/NaOH/H_2O_2$
(c)	Pentyne/$NaNH_2/NH_3$	Li/NH_3	$KMnO_4$
(d)	Pentyne/$NaNH_2/NH_3$	H_2/Lindlar's cat	$OsO_4/NaHSO_3$

23. Which statement explains why thiols are stronger acid than alcohols?
   (a) The S—H bond is weaker than O—H bond.
   (b) The S—H bond is longer than O—H bond.
   (c) The conjugate base of thiols are weaker than those of alcohols.
   (d) All of the above.

24. Which set of reagents could be used to prepare sodium salt of cyclopentanol?
   (a) Cyclopentanol + Na
   (b) Cyclopentanol + $NaHSO_3$
   (c) Methoxy cyclopentane + Na
   (d) Methoxy cyclopentane + NaCl

25. Rank the following from lowest to highest boiling point: Dimethyl ether(A) Ethanol(B) Water(C) Dimethylsulphide(D)
   (a) DABC (b) ABCD (c) ABDC (d) DBAC

26. Which of the following statement is true?
   (a) Ethers are polar molecules that have low boiling points.
   (b) Ethers are nonpolar molecules that have high boiling point.
   (c) Ethers are nonpolar molecules that have low boiling point.
   (d) Ethers are nonpolar molecules that can exist only in liquid form.

27. What is the predominant reason why 1-propanol has a higher boiling point than 1-propanthiol?
   (a) Molecules with less surface area can pack more closely, making them harder to boil.
   (b) 1-propanthiol is more polar than 1-propanol.
   (c) The greater polarity of O—H bond allows hydrogen bonding to increase the attractive forces between 1-propanol molecules.
   (d) The longer S—H bond makes 1-propanthiol less stable than 1-propanol.

28. Predict product(s) of the following transformation:

H, OH —BrH, 25°C→

(a) H, Br + H, Br (b) H, Br

(c) H, Br + H, Br (d)

29. Predict the major product of the following reaction:

HO —H_3PO_4, D→

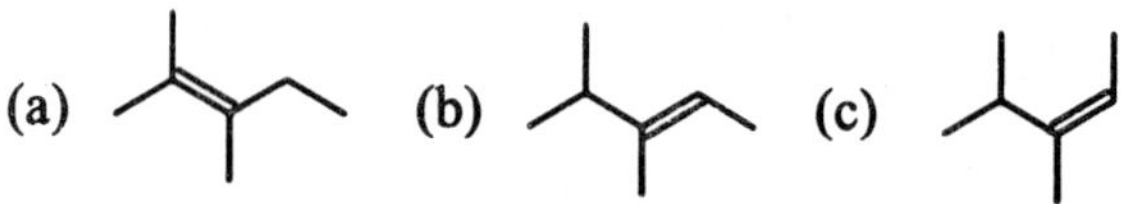

(a) (b) (c)

(d) All of the above formed in equal amounts.

30. Predict major product of the following reaction:

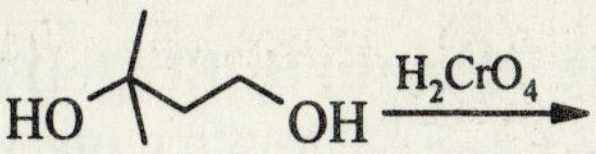

HO–C(CH$_3$)$_2$–CH$_2$CH$_2$–OH $\xrightarrow{H_2CrO_4}$

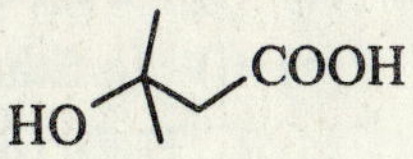

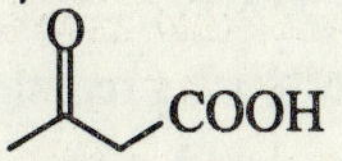

(a) [diene structure] (b) HO–C(CH$_3$)$_2$–CH$_2$COOH

(c) CH$_3$COCH$_2$COOH (d) 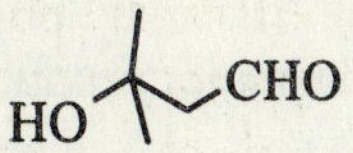 HO–C(CH$_3$)$_2$–CH$_2$CHO

31. What sequence of reagents will achieve the following transformation?

Cyclohexene → cis-cyclohexane-1,2-diol (OH, OH)

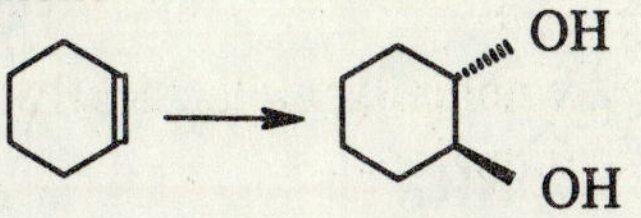

(a) OsO_4, $(CH_3)_3COH$ (b) NaOH, H_2CrO_4
(c) H_2SO_4, CH_3CO_3H (d) CH_3CO_3H, H^+/H_2O

32. Which of the following reactions represents an oxidation reaction?
(a) $CH_3CH_2OH + H_2CrO_4 \rightarrow CH_3COOH$
(b) $CH_3SH + O_2 \rightarrow CH_3S—SCH_3$
(c) Propene + $HCO_3H \rightarrow$ 2-methyl oxirane
(d) All of the above

33. Which statement explain why 2-propanol(A) is more soluble in water than dimethyl ether(B)?
(a) A has greater surface area than B.
(b) A is more symmetrical than B.
(c) A is more polar than B.
(d) A is more stable than B.

34. Which series of reagents can be used to achieve the following transformation?

(CH$_3$)$_2$CH–CH(CH$_3$)–OH ⟶ trimethyloxirane

(a) $SOCl_2$, HCO_3H (b) Na, HCO_3H
(c) H_3PO_4, HCO_3H (d) H_2CrO_4, HCO_3H

35. Which series of reagents can be used to achieve the following transformation?

Cyclohexene → Cyclohexanone

(a) HCl, $SOCl_2$ (b) H_2SO_4, PCC
(c) H_3PO_4, H_2CrO_4 (d) Both (b) and (c)

36. How many products are formed under the following conditions (include stereomers)?

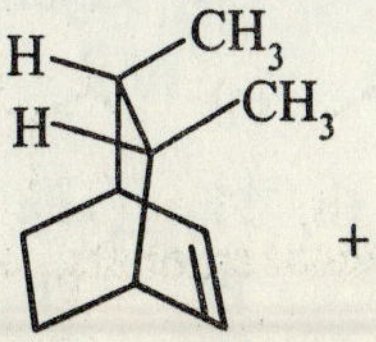

[bicyclic alkene with H, H, CH$_3$, CH$_3$] + CF_3CO_3H ⟶

(a) 1 (b) 2 (c) 3 (d) 4

37. A compound X(C_7H_8O) is insoluble in water, dilute HCl and aqueous $NaHCO_3$ but dissolve in dilute NaOH. When X is treated with bromine-water, it is converted into a compound of formula, $C_7H_5OBr_3$. Compound X is:

(a) 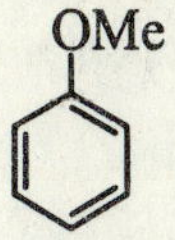 anisole (OMe on benzene) (b) 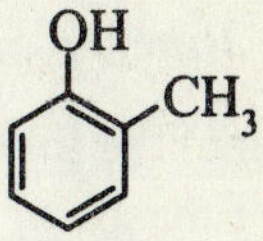 2-methylphenol (OH, CH$_3$ ortho)

(c) 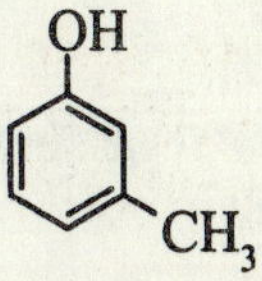 3-methylphenol (OH, CH$_3$ meta) (d) 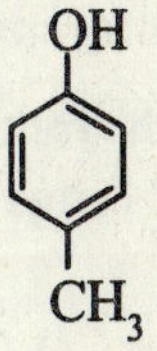 4-methylphenol (OH, CH$_3$ para)

38. 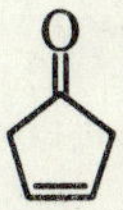 Cyclopent-3-en-1-one $\xrightarrow{CH_3CO_3H}$ Product of the reaction is:

(a)  [epoxy-cyclopentanone] (b) 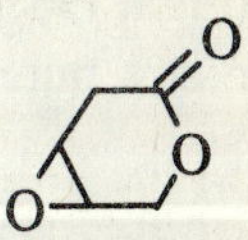 [lactone with O] (c) 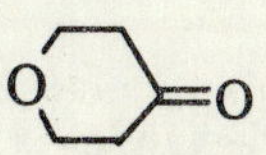 [cyclic lactone, O, =O] (d) 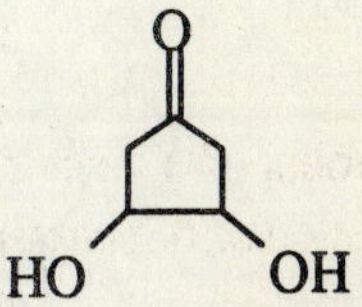 [cyclopentanone with HO, OH]

39. Which of the following is (are) not a feasible Grignard synthesis:

(i) $CH_3CH_2COOCH_3 + CH_3MgBr \xrightarrow{H_3O^+}$ [tertiary alcohol, HO]

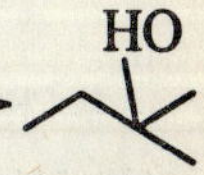

(ii) $HC{\equiv}C—CH_2CHO + CH_3MgBr \xrightarrow{H_3O^+}$ $HC{\equiv}C—CH_2CH(OH)CH_3$

(iii) $CH_3CH(Cl)CH_2CHO + CH_3MgBr \xrightarrow{H_3O^+}$ $CH_3CH(Cl)CH_2CH(OH)CH_3$

(iv)  [epoxide] + $CH_3MgBr \xrightarrow{H_3O^+}$ $CH_3CH_2CH_2OH$

(a) (i), (ii) (b) (i), (iii)
(c) (ii), (iii) and (iv) (d) (ii) and (iii)

40. Arrange the following alcohol in the increasing order of their ease of dehydration:

$C_6H_5CH_2CH_2OH$ (I) $CH_2{=}CH—CH_2OH$ (II)
$CH_2{=}CHCH_2CH_2OH$ (III) $C_6H_5CH(OH)CH_3$ (IV)

(a) I < II < III < IV (b) IV < III < II < I
(c) II < III < IV < I (d) II < III < I < IV

41. Which of the following compounds has highest boiling point?
(a) *n*-butanol (b) 2-butanol
(c) 2-methyl propanol (d) Neopentane

42. Which of the following is most acidic?
(a) Ethanol (b) 2-chloroethanol
(c) 2, 2, 2-tri fluoroethanol
(d) 2-methyl propanol

43. Which of the following statement regarding product Y of the following reaction is true?

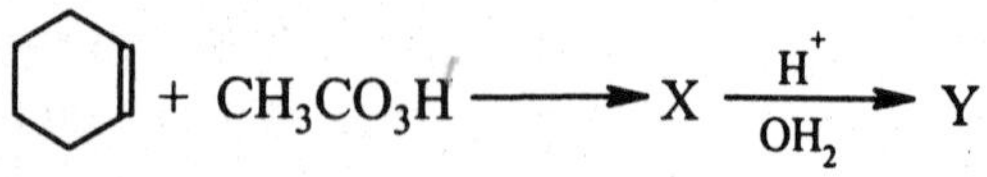

$$\text{Cyclohexene} + CH_3CO_3H \longrightarrow X \xrightarrow[OH_2]{H^+} Y$$

(a) Y is meso compound
(b) Y is a mixture of enantiomers
(c) Y is a racemic mixture
(d) Y is a pure enantiomer

44. Which of the following reagent will best bring about the transformation shown below?

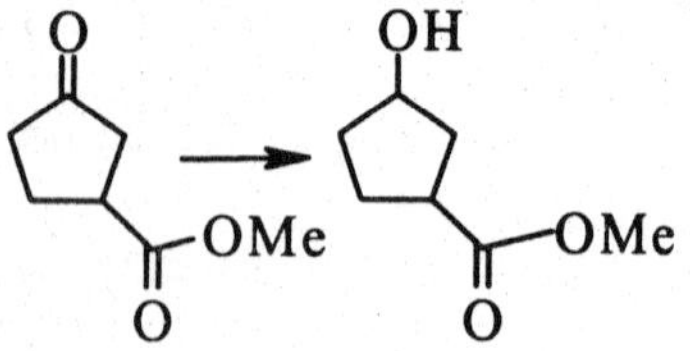

O ... OMe, O → OH ... OMe, O

(a) $NaBH_4$ (b) $LiAlH_4$
(c) Zn(Hg)/HCl (d) All

45. 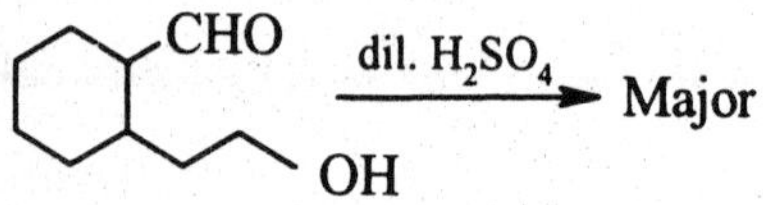 CHO, OH $\xrightarrow{\text{dil. } H_2SO_4}$ Major

(a) 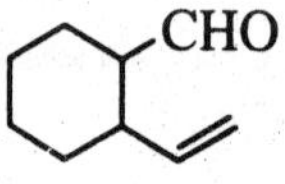 CHO (b) 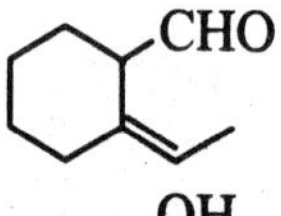 CHO
(c) 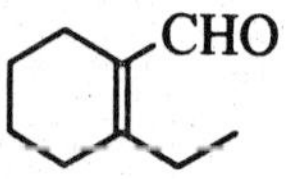 CHO (d) 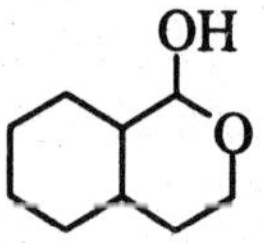 OH, O

46. Which reagent will convert cyclohexyl chloride to 2-cyclohexylethanol in best way?
(a) Mg/Ether → $CH_3CHO/H^+/H_2O$
(b) Mg/Ether → Oxirane/H^+/H_2O
(c) Mg/Ether → 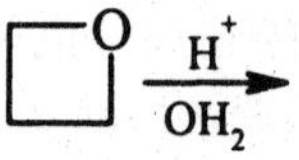 O $\xrightarrow[OH_2]{H^+}$
(d) All

47. Select the combination of reagents, which will bring about the following transformation in good yield:

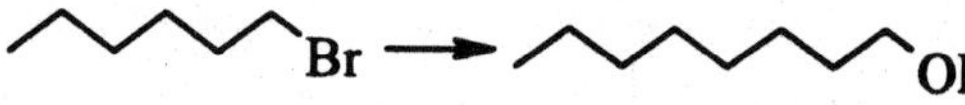

Br → OH

(a) (i) Mg/Ether, (ii) HCHO, (iii) H_3O^+
(b) (i) Na/Ether, (ii) HCHO, (iii) H_3O^+
(c) (i) Mg/Ether, (ii) Oxirane, (iii) H_3O^+
(d) (i) H_2O, (ii) Mg/Ether, (iii) H_3O^+

48. The major product of the following reaction is:

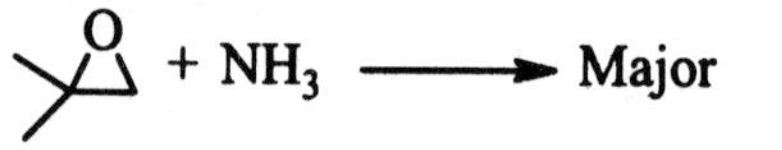

O + NH_3 ⟶ Major

(a)  O, NH_2 (b) 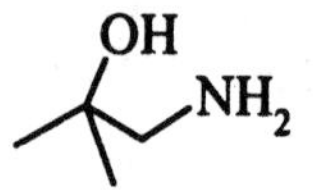 OH, NH_2
(c) 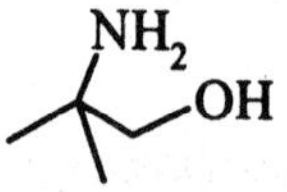 NH_2, OH (d)  H, N

49. The major product of the following reaction is:

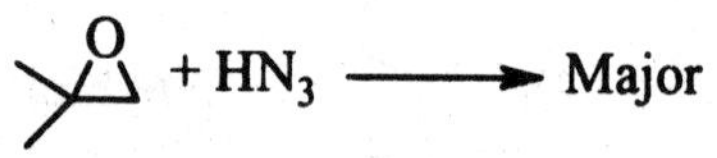

O + HN_3 ⟶ Major

(a) 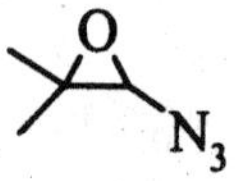 O, N_3 (b) 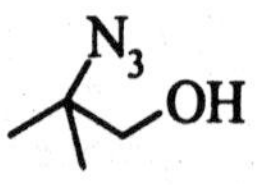 N_3, OH
(c) 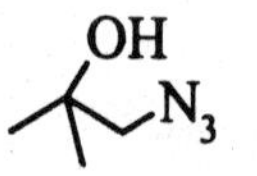 OH, N_3 (d)  N=N

50. Cyclopentanol on treatment with TsCl followed by reduction with $LiAlH_4$ produces mainly:
(a) Cyclopentene (b) Cyclopentyltosylate
(c) Cyclopentane (d) None

51. 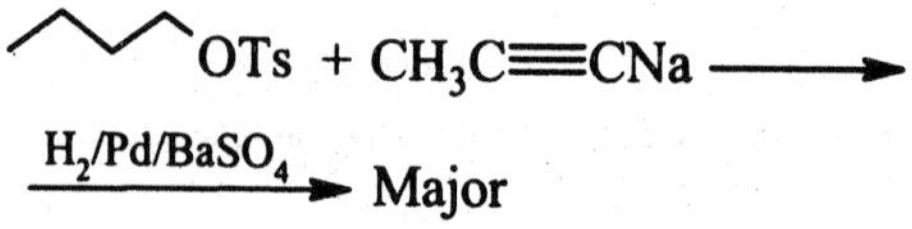 OTs + $CH_3C{\equiv}CNa$ ⟶ $\xrightarrow{H_2/Pd/BaSO_4}$ Major

(a)  (b) 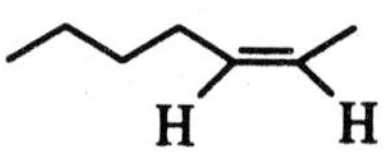 H H
(c) 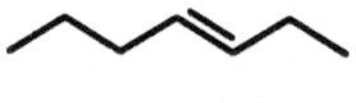 (d) 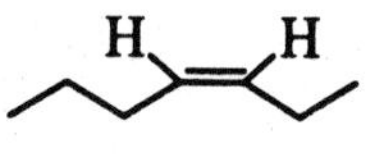 H H

52. The major product of the following reaction is:

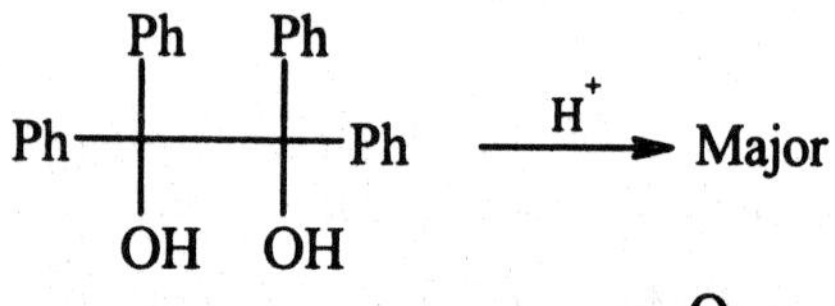

Ph Ph, Ph—C—C—Ph, OH OH $\xrightarrow{H^+}$ Major

(a) 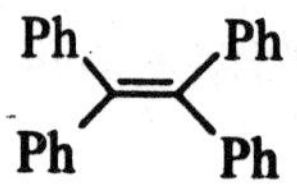 Ph Ph Ph Ph (b) 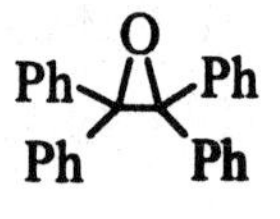 O, Ph Ph Ph Ph

(c) $Ph_3C{-}CO{-}Ph$ (d) None

53. The major dehydration product of the following reaction is:

$CH_3COCH(CH_3)CH(OH)CH(CH_3)_2 \xrightarrow[\Delta]{H^+}$ Major

(a) $CH_3COCH_2CH{=}C(CH_3)_2$ (b) $CH_3COCH_2CH_2C(CH_3){=}CH_2$

(c) $CH_3COCH{=}CHCH(CH_3)_2$ (d) $CH_3COC(CH_3){=}CHCH_2CH_3$

54. The major dehydration product of the following reaction is:

$PhCH_2CH(OH)CH(CH_3)_2 \xrightarrow[\Delta]{H^+}$ Major

(a) $PhCH_2CH{=}C(CH_3)_2$ (b) $PhCH_2C(CH_3){=}CHCH_3$

(c) $PhCH{=}CHCH(CH_3)_2$ (d) $PhCH{=}C(CH_3)CH_2CH_3$

55. Which combination of reagents will bring about the following conversion:

Cyclohexanone → 1-bromo-2-methylcyclohexane

(a) (i) MeMgl/H^+, (ii) H_2SO_4/Δ, (iii) HBr/R_2O_2,

(b) (i) MeMgl/H^+, (ii) H_2SO_4/Δ, (iii) HBr

(c) (i) MeMgl/H^+, (ii) HBr

(d) (i) HBr/R_2O_2, (ii) MeMgl/H^+

56. Cyclopentyl–CH_2OH is converted into

Cyclopentyl–$\underset{\displaystyle OH}{CH}CH_2CH_3$ by

(a) (i) Cu; 300°C, (ii) CH_3CH_2MgCl/H_3O^+

(b) (i) CrO_3, (ii) CH_3CH_2MgCl/H_3O^+

(c) (i) $KMnO_4$, (ii) CH_3CH_2MgCl/H_3O^+

(d) (i) $Na_2Cr_2O_7/H^+$, (ii) CH_3CH_2MgCl/H_3O^+

57. Which of the following alcohol will undergo easiest dehydration?

(a) $O_2N{-}C_6H_4{-}CH(OH)CH_3$ (b) $Cl{-}C_6H_4{-}CH(OH)CH_3$

(c) $C_6H_5{-}CH(OH)CH_3$ (d) $MeO{-}C_6H_4{-}CH(OH)CH_3$

58. An organic compound A ($C_6H_{12}O$) neither decolourise bromine water nor changes the colour of acidic dichromate solution. A on heating with H_2SO_4 produces an alkene which on oxidative ozonolysis gives B($C_6H_{10}O_3$), which gives an yellow precipitate with NaOH/I_2. The most probable structure of A is:

(a) Cyclopentyl–CH(OMe) (b) 1-methylcyclopentanol

(c) Cyclohexanol (d) 2,2-dimethylcyclobutanol type (cyclobutane with OH and methyl groups)

59. X(C_3H_8O) yields C_3H_6O on oxidation. X can be a:

(a) Aldehyde (b) Alkene
(c) 2°alcohol (d) 3°alcohol

60. Which of the following has greatest reactivity towards HBr?

(a) Propanol (b) 2-methylpropan-2-ol
(c) 2-propanol (d) 2-methylpropanol

61. The best reagent which will bring about the following transformation is:

$CH_3{-}CH{=}CH{-}CHO \rightarrow CH_3CH{=}CHCH_2OH$

(a) H_2/Pt
(b) H_2/Raney Ni
(c) $\{(CH_3)_2CHO\}_3Al/(CH_3)_2CHOH$
(d) $LiAlH_4$

62. Ethanol on treatment with excess of Cl_2(g) produces:

(a) C_2H_5Cl (b) Chloroform
(c) Chloral (d) Ethanol

63. An organic compound X produces Y in the first step of oxidation, which on further oxidation gave an acid. Y on treatment with ammonical silver nitrate solution gave a black precipitate. X is likely to be a:

(a) 1°-alcohol (b) 3°-alcohol
(c) Ethanal (d) 2-propanone

64. Which of the following would undergo most rapid hydrolysis with aqueous NaOH to furnish the corresponding hydroxy derivative?

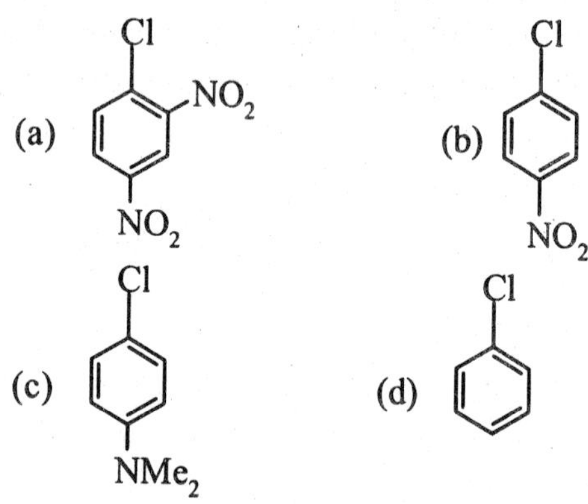

65. Bakelite is a polymer of:
(a) Phenol + HCHO (b) HCHO + Acetaldehyde
(c) Phenol + H_2SO_4 (d) HCHO + Acetone

66. Phenol gives tribromo derivative when brominated in aqueous medium, but monobromo derivative when brominated in CCl_4 medium because:
(a) In aqueous solution bromine is completely ionized.
(b) In aqueous solution, phenol exist in equilibrium with phenoxide ion, which is highly activated for electrophilic substitution.
(c) In CCl_4 the electrophilicity of Br_2 increases.
(d) In CCl_4, other positions of phenol is blocked by solvent molecules.

67. Which of the following statement about phenol is true?
(a) It is acidic in nature and gives effervescence with aqueous $NaHCO_3$.
(b) It is basic in nature due to presence of –OH functional group.
(c) It gives violet colour with $FeCl_3$ due to formation of a soluble complex, in which iron is in hexavalent state.
(d) Phenol as well as ethanol can be reduced to corresponding alkanes by heating with Zn dust.

68. The intermediate formed in the reaction of phenol with $CHCl_3$ in aqueous NaOH is:

(a) [cyclohexadienone with H and CCl_2] (b) [phenol with $CHCl_2$]

(c) [phenol with CH(Cl)(OH)] (d) [phenoxide with $CHCl_2$]

69. Choose the correct statement for the reaction:

$$C_6H_5OCOCH_3 \xrightarrow{AlCl_3\ (CS_2)} \text{o-HOC}_6H_4COCH_3\ (A) + H_3COC\text{-}C_6H_4\text{-}OH\ (B)$$

(a) B is formed more rapidly at higher temperature.
(b) B is more volatile than A.
(c) A is more volatile than B.
(d) A is formed in higher yield at lower temperature.

70. Proof spirit is:
(a) 57.1% aqueous alcohol.
(b) The minimum pure aqueous alcohol that can allow gunpowder to catch fire.
(c) A sample taken out for testing the alcohol percentage.
(d) 95% pure alcohol.

71. After the reaction of $C_6H_5SO_3Na$ with NaOH, it is hydrolysed by acid, the compound formed is:
(a) Phenol (b) Benzoic acid
(c) Benzene (d) C_6H_5CHO

72. The most steam volatile species is:

(a) [o-nitrophenol] (b) [p-nitrophenol]
(c) [3,5-dinitrophenol] (d) [2,4,6-trinitrophenol]

73. Aspirin is an acetylation product of:
(a) ortho-hydroxybenzoic acid.
(b) ortho-dihydroxybenzene.
(c) meta-hydroxybenzoic acid.
(d) para-dihydroxybenzene.

74. Reimer Tiemann reaction involves:
(a) Carbocation intermediate.
(b) Carbene intermediate.
(c) Carbanion intermediate.
(d) Free radical intermediate.

75. In the Liberman nitroso reaction, changes in the colour of phenol occurs as:

(a) Brown or red-green-red-deepblue.
(b) Red-deepblue-green.
(c) Red-brown-white.
(d) White-red-green.

76. The order of reactivity of the following alcohols towards concentrated HCl is:

I: F–CH₂–CH(OH)–CH₃   II: F–CH₂–CH₂–CH(OH)–CH₃

III: H_3C–CH(OH)–CH_3 IV: Ph–CH₂–OH

(a) I > II > III > IV (b) I > III > II > IV
(c) IV > III > II > I (d) IV > III > I > II

77. Cyclohexanone can be converted into methyl cyclohexene by:
(a) (i) CH_3Cl (ii) H_3O^+ (iii) Conc. H_2SO_4
(b) (i) CH_3OH (ii) H_3O^+ (iii) Conc. H_2SO_4
(c) (i) CH_3MgCl (ii) H_3O^+ (iii) Conc. H_2SO_4
(d) (i) CH_3MgCl (ii) Conc. H_2SO_4 (iii) H_3O^+

78. The reaction of HBr with the following compound would produce:

Ph–CH=CH–C₆H₄–OH

(a) Ph–CH(Br)–CH₂–C₆H₄–OH
(b) Ph–CH₂–CH(Br)–C₆H₄–OH
(c) Ph–CH(Br)–CH₂–C₆H₄–Br
(d) Ph–CH₂–CH(Br)–C₆H₄–Br

79. A crystalline solid C_7H_8O reacts with sodium to liberate hydrogen gas. It is insoluble in water but soluble in dilute solution of NaOH, upon treatment with Br_2, only one monobromo derivative C_7H_7OBr is produced. The most likely structure of the starting compound is:
(a) Benzyl alcohol (b) ortho-cresol
(c) *p*-cresol (d) *m*-cresol

80. In the following sequence of reaction:

$$C_6H_5CH{=}CHCH_3 + Hg(CH_3COO)_2 \xrightarrow[NaBH_4]{CH_3OH} X, \text{ X is}$$

(a) $C_6H_5CH_2CH(OH)CH_3$
(b) $C_6H_5CH(OH)CH_2CH_3$
(c) $C_6H_5CH(OCH_3)CH_2CH_3$
(d) $C_6H_5CH_2CH(OCH_3)CH_3$

81. In the following reaction:

$$(CH_3)_2C{=}CH{-}CH_2{-}CH{=}CH_2 + C_6H_5CO_3H(1\text{ equiv.}) \rightarrow X, \text{ X is}$$

(a) $(CH_3)_2C{=}CH{-}CH_2$–epoxide (terminal epoxide, trisubstituted alkene retained)
(b) bis-epoxide (both double bonds epoxidised)
(c) $>$CH–CH(–O–)CH–CH₃
(d) trisubstituted epoxide with terminal $CH{=}CH_2$ retained

82. A compound $X(C_7H_8O)$ is insoluble in water, dilute HCl, aqueous $NaCHO_3$ but dissolve in dilute NaOH. When X is treated with bromine-water, it is converted into a compound of formula C_7H_7OBr. Compound X is:

(a) anisole (C₆H₅–OMe) (b) o-cresol (OH, CH₃ ortho)
(c) m-cresol (OH, CH₃ meta) (d) p-cresol (OH, CH₃ para)

83. The alcohol which is most readily dehydrated is:
(a) 2-butanol (b) 1-phenyl-1-propanol
(c) $CH_3{-}CO{-}CH_2{-}CH_2OH$
(d) $CH_3C(=O){-}O{-}CH(OH)CH_3$

84. The major product for the reaction:
$CH_3—CH_2—C(CH_3)_2Br + (CH_3)_3CONa \rightarrow$
would be

(a) $CH_3—CH_2—C(CH_3)_2—O—C(CH_3)_2—CH_3$

(b) $CH_3—CH{=}C(CH_3)_2$

(c) $CH_3—CH_2—C({=}CH_2)—CH_3$

(d) $CH_2{=}CH—CH(CH_3)_2$

85. 5-bromo-2-pentanone $\xrightarrow[THF]{Mg}$. The product of the above reaction is:

(a) cyclobutyl—CH_2MgBr (b) 1-methylcyclobutanol (CH_3, OH)

(c) cyclobutane with CH_3 and OH on adjacent carbons (d) cyclobutanone with CH_3 (=O)

Chapter 7

Aldehydes and Ketones

7.1 GENERAL INTRODUCTION

With the exception of formaldehyde, all aldehydes have a common carbonyl group >C═O bonded on one side to a carbon and on the other side to a hydrogen.

$$R{-}\overset{\overset{\displaystyle O}{\|}}{C}{-}H$$

General structure of aldehyde

Formaldehyde, too has a carbonyl group, but on both side, it is attached to hydrogen.

$$H{-}\overset{\overset{\displaystyle O}{\|}}{C}{-}H$$

Formaldehyde

Ketones also contain a carbonyl group, but it is bonded to carbon on both ends as:

$$R{-}\overset{\overset{\displaystyle O}{\|}}{C}{-}R'$$

General structure of a ketone

7.2 NOMENCLATURE OF ALDEHYDES AND KETONES

Aldehydes and ketones are organic compounds, which incorporate a **carbonyl functional group,** C═O. The carbon atom of this group has two remaining bonds that may be occupied by hydrogen or alkyl or aryl substituents. If at least, one of these substituents is hydrogen, the compound is an **aldehyde**. If neither is hydrogen, the compound is a **ketone**.

$$R_1R_2C{=}O$$

Aldehyde R^1 and/or R^2 = H

Ketone R^1 and R^2 = alkyl or aryl

The IUPAC system of nomenclature assigns a characteristic suffix to these classes, **al** to aldehydes and **one** to ketones. For example, $H_2C{=}O$ is methanal, more commonly called formaldehyde. Since an aldehyde carbonyl group must always lie at the end of a carbon chain, it is by default position 1, and therefore defines the numbering direction. A ketone carbonyl function may be located anywhere within a chain or ring, and a locator number gives its position. Chain numbering normally starts from the end nearest to the carbonyl group. In cyclic ketones the carbonyl group is assigned position 1, and this number is not cited in the name, unless more than one carbonyl groups are present. In common names carbon atoms near the carbonyl group are often designated by Greek letters. The atom adjacent to the function is alpha, the next removed is beta and so on. Since ketones have two sets of neighbouring atoms, one set is labeled α, β, etc. and the other α′, β′, etc.

$$H{-}\overset{\overset{\displaystyle O}{\|}}{C}{-}H$$

IUPAC: Methanal
common: Formaldehyde

$$H_3C{-}\overset{\overset{\displaystyle O}{\|}}{C}{-}H$$

Ethanal
acetaldehyde

$$ClCH_2CH_2CH_2{-}\overset{\overset{\displaystyle O}{\|}}{C}{-}H$$

4-chlorobutanal

$$C_6H_5CH_2CHO$$

Phenyl ethanal
(Phenyl acetaldehyde)

Very simple ketones, such as propanone and phenylethanone (first two examples in the right column), do not require a locator number, since there is only one possible site for a ketone carbonyl function. Likewise, locator numbers are omitted for the simple dialdehyde at the bottom left, since aldehyde functions must occupy the ends of carbon chains. The hydroxy butanal and propenal examples (2nd and 3rd from the top, left column) and the oxopropanal example (bottom right) illustrate the nomenclature priority of IUPAC suffixes. In all cases the **aldehyde function has a higher status than either an alcohol, alkene or ketone and provides the nomenclature suffix.** The other functional groups are treated as

Aldehydes

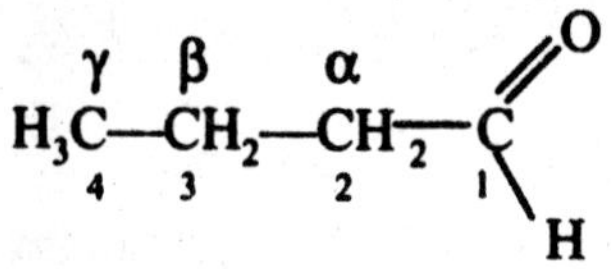

Butanal
butyraldehyde

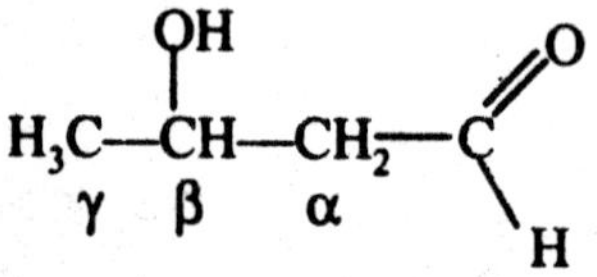

3-hydroxy butanal
β-hydroxybutyraldehyde
or aldol

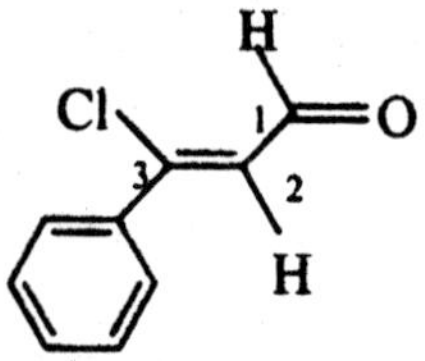

(Z)-3-chloro-3-phenyl-2-propenal

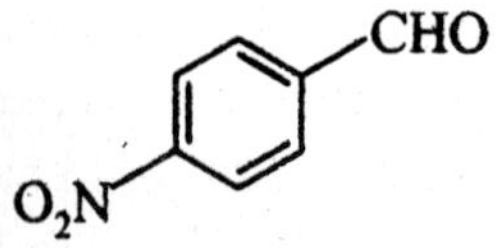

p-nitrobenzenecarbaldehyde
p-nitrobenzaldehyde

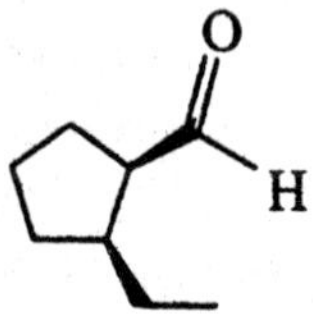

cis-2-ethyl cyclopentanecarbaldehyde

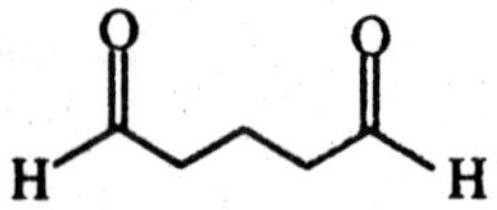

Pentanedial
Glutaraldehyde

Ketones

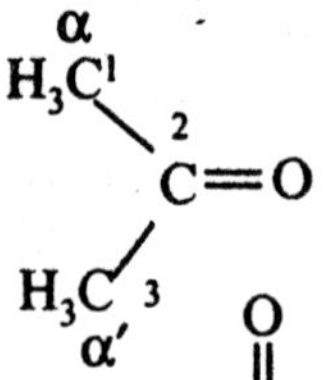

Propanone
acetone

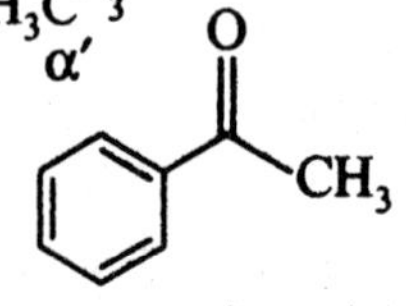

Phenylethanone
acetophenone
methyl phenyl ketone

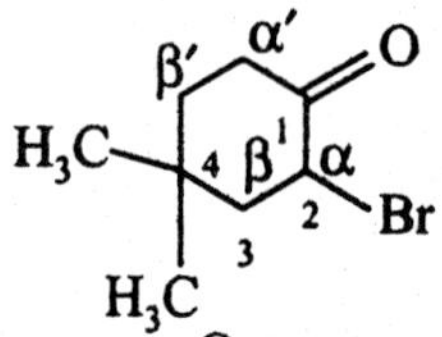

2-bromo-4, 4-dimethylcyclohexanone

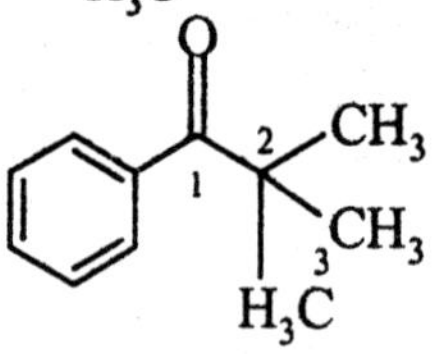

2, 2-dimethyl-1-phenyl propanone
t-butyl phenyl ketone

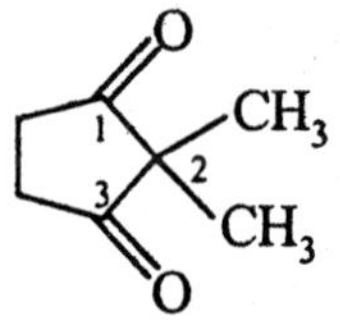

2, 2 -dimethyl -1, 3-cyclopentanedione

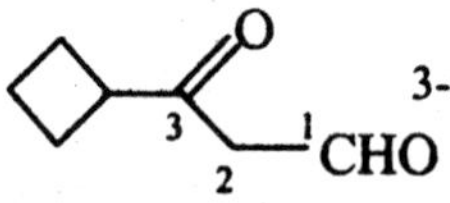

3-cyclobutyl-3-oxopropanal

substituents. Because ketones are just below aldehydes in nomenclature suffix priority, the "oxo" substituent terminology is seldom needed. Simple substituents incorporating a carbonyl group are often encountered. The generic name for such groups is **acyl**. Three examples of acyl groups having specific names are shown below:

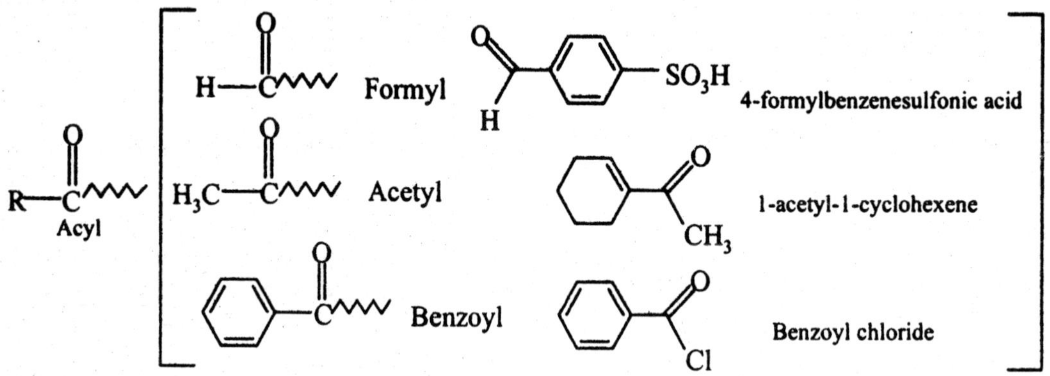

Acyl

Formyl — 4-formylbenzenesulfonic acid

Acetyl — 1-acetyl-1-cyclohexene

Benzoyl — Benzoyl chloride

7.3 STRUCTURES AND BONDING

Carbonyl carbon of aldehydes and ketones are sp^2-hybridized with a planar triangular shape as:

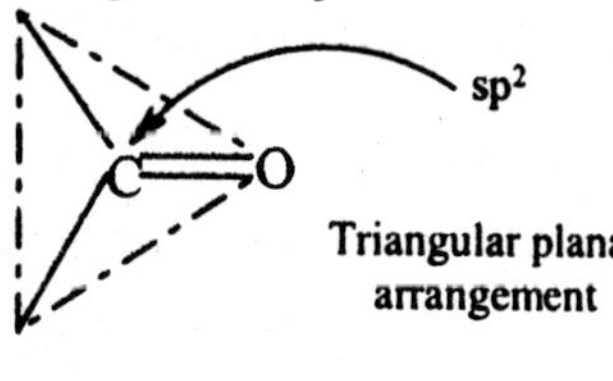

Triangular planar arrangement

Carbonyl group has polar character as: $>\overset{\delta^+}{C}=\overset{\delta^-}{O}$. Due to polar character aldehydes and ketones have higher boiling and melting points than hydrocarbons of comparable molar mass. However, since aldehydes and ketones cannot form strong intermolecular hydrogen bonds, they have lower boiling points than alcohol.

7.4 OCCURRENCE OF ALDEHYDES AND KETONES

Aldehydes and ketones are widespread in nature, often combined with other functional groups. Example are shown in the following diagram. The compounds in the top row are found chiefly in plants or microorganisms, those in the bottom row have animal origins. With the exception of the first three compounds (top row). These molecular structures are all chiral. When chiral compounds are found in nature they are usually enantiomerically pure, although different sources may yield different enantiomers. For example, carvone is found as its levorotatory (R)-enantiomer in spearmint oil, whereas, caraway seeds contain the dextrorotatory (S)-enantiomer. Note that the aldehyde function is often written as–**CHO** in condensed or complex formulas.

Cinnamaldehyde (cinnaman bark)

Vanillin (vanilla bean)

Citral (lemongrass)

Helminthosporal (a fungal toxin)

Carvone (spearmint and caraway)

Camphor (camphar tree)

Muscone (musk deer)

Testosterone

Progesterone

Cartisone

7.5 SYNTHETIC PREPARATIONS

7.5.1 Aldehydes

I. By partial oxidation of a primary alcohol

A primary alcohol on oxidation with PCC (Pyridinium chlorochromate) yields aldehyde as:

$$R{-}CH_2OH \xrightarrow{PCC} RCHO$$

$$\underset{\text{Butanol}}{CH_3CH_2CH_2CH_2OH} + \underset{\text{PCC}}{C_5H_5\overset{+}{N}HCr\bar{O}_3Cl} \xrightarrow{CH_2Cl_2} \underset{\text{Butanal}}{CH_3CH_2CH_2CHO}$$

II. By reduction of acid derivatives

(a) *By reduction of acid halide:* $LiAlH_4$ is strong reducing agents, it reduces aldehyde further to alcohols. Therefore, two common derivatives of $LiAlH_4$, which are less reactive than $LiAlH_4$, are tri-tertiary butoxy aluminium hydride and *di*-isobutyl aluminium hydride:

$Li\{[(CH_3)_3C\text{-}O]_3AlH\}$ — Lithium tri-ter-butoxy aluminium hydride.

$H{-}Al[CH_2{-}CH(CH_3)_2]_2$ — Di-isobutylaluminium hydride

The above-mentioned hydrides are much more sterically hindered and therefore, have difficulty in transferring hydride ion. An acyl chloride can be reduced to aldehyde by treating them with Lithium tri-tertiary butoxy aluminium hydride at 80° C as:

$$R{-}\overset{O}{\overset{\|}{C}}{-}Cl + Li\,[AlH(\text{t-butoxy})_3] \xrightarrow{-80°\,C} R{-}\overset{O}{\overset{\|}{C}}{-}H$$

An acyl chloride can also be reduced to corresponding aldehydes by treating them with $Pd/BaSO_4$/xylene (***Rosenmund reduction***):

$$R{-}\overset{O}{\overset{\|}{C}}{-}Cl + Pd/BaSO_4 \xrightarrow[\text{Xylene}]{\text{Quinoline}} R{-}CHO$$

(b) *Reduction of esters and nitrile:* Both esters and nitrile can be conveniently reduced into corresponding aldehydes using di-iso-butyl aluminium hydride as reducing agent:

$$R{-}\overset{O}{\overset{\|}{C}}{-}OR' + HAl(CH_2CH(CH_3)_2)_2 \longrightarrow RCHO$$

$$R-\overset{O}{\overset{\|}{C}}-OR' + HAl(CH_2CH(CH_3)_2)_2 \longrightarrow RCHO$$

III. By heating calcium salt of formic acid and any of its homologue

$$(RCOO)_2Ca + (HCOO)_2Ca \xrightarrow{\Delta} 2\,RCHO + 2\,CaCO_3$$

In the above reaction, $R-\overset{O}{\overset{\|}{C}}-R$ and $H-\overset{O}{\overset{\|}{C}}-H$ are also formed by self coupling of $(RCOO)_2Ca$ and $(HCOO)_2Ca$ respectively, as side products.

IV. By passing vapors of formic acid and any of its homologue over manganous oxide at 300°C

$$RCOOH + HCOOH \xrightarrow[300°\,C]{MnO} RCHO + CO_2 + H_2O$$

V. By ozonolysis of alkenes of typeR—CH═CHR′

$$R-CH{=}CHR' + O_3 \xrightarrow{Zn\text{-}H_2O} RCHO + R'CHO$$

VI. By the oxidation of vicinal diols with $(CH_3COO)_4Pb$ or $NaIO_4$ as

$$R-\underset{OH}{\underset{|}{CH}}-\underset{OH}{\underset{|}{CH}}-R' \xrightarrow[\text{or } NaIO_4]{Pb(CH_3COO)_4} RCHO + R'CHO$$

VII. By the mercuration-demercuration reaction of acetylene

$$HC{\equiv}CH + HgSO_4 \xrightarrow{H^+} CH_3CHO$$

VIII. By hydroboration of terminal alkynes

$$RC{\equiv}CH + B_2H_6/H_2O_2/OH^- \longrightarrow RCH_2CHO$$

IX. By Stephen's method

An alky cyanide on treatment with $SnCl_2$ in etheral medium followed by acidification of product yields an aldehyde as:

$$RCN + SnCl_2/HCl \xrightarrow{H_2O} RCHO$$

X. Oxo process

A mixture of alkene, CO and H_2, under pressure and elevated temperature, in the presence of a catalyst, forms aldehydes.

$$CH_3CH{=}CH_2 + CO + H_2 \xrightarrow{[CoH(CO)_4]} CH_3CH_2CH_2CHO + (CH_3)_2CHCHO$$

XI. Sommelet reaction

Aldehydes may be prepared by modified Sommelet reaction, in which hydrochloride of a primary amine is treated with hexamin in acetic acid as:

$$CH_3-(CH_2)_4-CH_2NH_2\cdot HCl + (CH_2)_6N_4 \xrightarrow[H_2O(v)]{CH_3OOH} CH_3(CH_2)_4-CHO$$

XII. By means of Grignard reagent

(a) *By the action of Grignard reagent on esters of formic acid:*

$$H-\overset{O}{\overset{\|}{C}}-OR' + RMgX \xrightarrow{H_3O^+} R-\overset{O}{\overset{\|}{C}}-H + Mg(OR')X$$

In this method, ester is taken in large excess so that the possibility of further addition of Grignard reagent on the product aldehyde is minimized.

(b) *By the action of Grignard reagents on HCN*

$$H-C{\equiv}N + RMgX \longrightarrow H-\overset{NMgX}{\overset{\|}{C}}-R \xrightarrow{H_3O^+} R-\overset{O}{\overset{\|}{C}}-H$$

7.5.2 Ketone

I. By oxidation of secondary alcohol

$$R-\underset{OH}{\underset{|}{CH}}-R \xrightarrow{[O]} R-\underset{O}{\underset{\|}{C}}-R$$

The oxidizing agent used for oxidation are acidic solution of chromate or dichromate, alkaline or cold, acidic permanganate or Cu/Δ. A very specific reagent is $[(CH_3)_3CO]_3Al$. In this reaction, a secondary alcohol is refluxed with aluminium *t*-butoxide and acetone is added:

$$R-\underset{OH}{\underset{|}{CH}}-R + [(CH_3)_3CO]_3Al \xrightarrow[Reflux]{Acetone} R-\underset{O}{\underset{\|}{C}}-R \quad \textbf{Oppenauer oxidation}$$

Advantage of this reagent over other oxidizing agents is, it does not affect the multiple bonds and can be used for oxidation of unsaturated alcohol, viz:

$$CH_3-CH=CH-CH(OH)-CH_3 + [(CH_3)_3CO]_3Al \xrightarrow{\text{Acetone}} CH_3-CH=CH-CH_2-\overset{O}{\overset{\|}{C}}-CH_3$$

MnO_2 also oxidizes unsaturated alcohol to corresponding carbonyl compounds without disturbing multiple bonds.

$$\text{cyclohex-2-en-1-ol} \xrightarrow{MnO_2} \text{cyclohex-2-en-1-one}$$

II. By heating calcium salts of any carboxylic acid other than formic acid.

$$(CH_3COO)_2Ca \xrightarrow{\Delta} CH_3-\overset{O}{\overset{\|}{C}}-CH_3 + CaCO_3$$

III. By passing vapours of any mono carboxylic acid other than formic acid over manganous oxide at 300°C.

$$2\ RCOOH \xrightarrow[300°\ C]{MnO} R-\overset{O}{\overset{\|}{C}}-R + CO_2 + H_2O$$

IV. By ozonolysis of alkenes of type $R_2C{=}CR'_2$

$$R_2C=CR'_2 + O_3 \xrightarrow{Zn-H_2O} R-\overset{O}{\overset{\|}{C}}-R + R'-\overset{O}{\overset{\|}{C}}-R'$$

V. By oxidative cleavage of alkene with Lemieux reagent

Lemieux reagent is an aqueous solution of $NaIO_4$ and a trace of $KMnO_4$. Alkene is first oxidized to *cis*-diol, which is then cleaved by periodate:

$$R-\overset{R}{C}=\overset{R'}{C}-R' + KMnO_4 \longrightarrow R-\overset{R}{\underset{OH}{C}}-\overset{R'}{\underset{OH}{C}}-R' \xrightarrow{IO_4^-} R-\overset{O}{\overset{\|}{C}}-R + R'-\overset{O}{\overset{\|}{C}}-R'$$

VI. By oxymercuration of alkynes

$$R-C\equiv CH + HgSO_4 \xrightarrow{H_2SO_4} R-\overset{O}{\overset{\|}{C}}-CH_3$$

Markovnikov's addition of water followed by tautomerization of unstable enol into stable keto form takes place.

VII. By means of Grignard reagent

(a) *By the action of Grignard reagent over acid chloride:*

$$R-\overset{O}{\overset{\|}{C}}-Cl + R'MgCl \longrightarrow R-\overset{O}{\overset{\|}{C}}-R' + MgCl_2$$

Excess of acid chloride is taken to avoid further reaction of Grignard reagent with product ketones.

(b) *By the action of Grignard reagent over an ester:*

$$R-\overset{O}{\overset{\|}{C}}-OR' + RMgX \longrightarrow R-\overset{O}{\overset{\|}{C}}-R + Mg(OR')X$$

Here, again excess of ester is taken to avoid further reaction of Grignard reagent with product ketones.

Example 1. A neutral organic compound has molecular formula $C_4C_8O_2$ and exists in two isomeric forms A and B. Treatment of A with one equivalent of ethyl magnesium bromide followed by hydrolysis of product yields C. C on reduction with $LiAlH_4$ yields a resolvable product $D(C_4H_{10}O)$. B on treatment with one equivalent of ethyl magesium bromide yields E, which on reduction with $LiAlH_4$ yields $F(C_5H_{12}O)$ which is non-resolvable. On treatment of B with one equivalent of methyl magnesium bromide yields C, identify A to F.

Solution

$$CH_3-\overset{O}{\overset{\|}{C}}-OC_2H_5(A) \qquad CH_3CH_2-\overset{O}{\overset{\|}{C}}-OCH_3(B)$$

$$CH_3-\overset{O}{\overset{\|}{C}}-C_2H_5(C) \qquad CH_3-\overset{OH}{\overset{|}{CH}}-C_2H_5(D)$$

$$CH_3CH_2-\overset{O}{\overset{\|}{C}}-CH_2CH_3(E) \qquad CH_3CH_2\overset{OH}{\overset{|}{C}H}CH_2CH_3(F)$$

VIII. By the action of Grignard reagent on alkylnitriles

$$R-C\equiv N + R'-MgX \longrightarrow R-\overset{\overset{\Large NMgX}{||}}{C}-R' \xrightarrow{H_3O^+} R-\overset{\overset{\Large O}{||}}{C}-R'$$

Example 2. Synthesize 3-pentanone using alkyl bromide containing not more than two carbons.

Solution

$$C_2H_5Br \xrightarrow{\text{aq NaCN}} C_2H_5-CN$$

$$C_2H_5Br \xrightarrow[\text{Ether}]{\text{Mg}} C_2H_5MgBr$$

$$\longrightarrow C_2H_5-\overset{\overset{\Large NMgBr}{||}}{C}-C_2H_5 \xrightarrow{H_3O^+} \text{3-pentanone}$$

Example 3. Convert:

(a) $Br(CH_2)_5Br \longrightarrow$ cyclohexanone (b) 1-bromo-2-methylbenzene $\longrightarrow$ benzocyclobutenone

Solution

(a) $Br(CH_2)_5Br \xrightarrow{\text{aq KCN}} NC(CH_2)_5CN \xrightarrow[HCl]{SnCl_2}$ OHC(CH$_2$)$_5$CHO $\xrightarrow{\text{Dilute NaOH}}$ 2-hydroxycyclohexanecarbaldehyde (CHO, OH) $\xrightarrow{Na_2Cr_2O_7}$ $\xrightarrow{\text{Heat}}$ Product

(b) 1-bromo-2-methylbenzene (CH_3, Br) $\xrightarrow{\text{Mg/ether}}$ $\xrightarrow{CH_2O}$ $\xrightarrow{H_3O^+}$ $\xrightarrow{PCC}$ 2-methylbenzaldehyde (CH_3, CHO) $\xrightarrow[\text{Heat}]{Br_2}$ $\xrightarrow{\text{Mg/ether}}$ $\xrightarrow{H_3O^+}$ $\xrightarrow{PCC}$ benzocyclobutenone

7.6 GENERAL PROPERTIES OF ALDEHYDES AND KETONES

A comparison of the properties and reactivity of aldehydes and ketones with those of the alkenes is warranted, since both have a double bond functional group. Because of the greater electronegativity of oxygen, the carbonyl group is polar, and aldehydes and ketones have larger molecular dipole moments (D) than do alkenes.

The resonance structures on the following diagram this polarity, and the relative dipole moments of formaldehyde, other aldehydes and ketones confirm the stabilizing influence that alkyl substituents have on carbocations (the larger the dipole moment the greater the polar character of the carbonyl group). We expect, therefore, that aldehydes and ketones will have higher boiling points than similar sized alkenes. Furthermore, the presence of oxygen with its non-bonding electron pairs makes aldehydes and ketones hydrogen-bond acceptors, and should increase their water solubility relative to hydrocarbons.

$$R_2C=\ddot{O}: \longleftrightarrow R_2\overset{+}{C}-\ddot{O}:^-$$

$$H_2C=O \quad (D = 2.29) \qquad HRC=O \quad (D = 2.5) \qquad R_2C=O \quad (D = 2.9)$$

Aldehydes and ketones have a net dipole, because of electronegativity difference between carbon and oxygen as:

$$\overset{\delta+}{C}=\overset{\delta-}{O}$$

Dipole moment measurements of aldehydes and ketones revealed that the magnitude of dipole moment is much larger than that expected only due to electronegativity difference. This can be accounted for by the resonance phenomena present in aldehydes and ketones as:

$$>C=O \longleftrightarrow -\overset{+}{C}-\overset{-}{O}$$

Thus, the carbon atom has a positive charge and can be attacked by nucleophile, a characteristic reaction of aldehydes and ketones. Oxygen atom of aldehydes and ketones possess two lone pairs of electrons and thus also exhibit basic properties, i.e. it is readily protonated by strong acid to form oxonium salts:

$$>C=\ddot{O}: + \overset{+}{H} \longrightarrow \underset{\text{(I)}}{>C=\overset{+}{O}-H} \longleftrightarrow \underset{\text{(II)}}{>\overset{+}{C}-OH}$$

Since oxygen is more electronegative than carbon, II is easily formed from I and can be easily attacked by a nucleophilic reagent. Thus, it is expected that acids catalyze nucleophilic addition. It should also be noted that, because of the positive charge on the carbonyl carbon, the >CO group has a strong –I effect.

The general scheme of nucleophilic addition to carbonyl carbon can be represented as:

$$\begin{matrix}R\\ R'\end{matrix}\!\!>C{=}O + H{-}Z \rightleftharpoons \begin{matrix}R & & OH\\ & C & \\ R' & & Z\end{matrix}$$

Here hybridization of carbon changes from sp^2 to sp^3. This results in increased crowding and so reaction can be expected to be increasingly sterically hindered with increasing size of R and R′. For aldehydes R′═H (for formal dehyde R═R′═H), and hence it can be anticipated that aldehydes are more reactive than ketones towards nucleophilic addition at carbonyl carbon. Moreover, in ketones, the two alkyl group, by its +I effect, partially neutralizes the positive charge on carbon atom decreasing its reactivity towards a nucleophile:

$$\begin{matrix}R\rightarrow (+I)\\ R\rightarrow (+I)\end{matrix}\!\!>C{=}O$$

Hence ketones, in general, are less reactive than aldehydes due to both steric and inductive factors.

7.7 CHEMICAL REACTIONS

7.7.1 Reactions Common to Aldehydes and Ketones

I. Catalytic hydrogenation

Catalytic hydrogenation with various reducing agents readily converts aldehydes and ketones into primary and secondary alcohol respectively as:

$$R{-}CHO + H_2/\text{Catalyst} \longrightarrow RCH_2OH$$

$$\begin{matrix}R\\ R\end{matrix}\!\!>C{=}O + H_2/\text{Catalyst} \longrightarrow RCH_2OH$$

(a) Hydrogenation with complex hydride

$$CH_3CH_2CHO \xrightarrow[\text{or } NaBH_4]{LiAlH_4} CH_3CH_2CH_2OH$$

The advantage of $LiAlH_4$ or $NaBH_4$ over H_2/Pt or H_2/Ni is that, they usually do not affect the olefinic bonds, e.g.

$$CH_3CH{=}CH{-}CH_2CHO \begin{cases}\xrightarrow{LiAlH_4} CH_3CH{=}CHCH_2CH_2OH\\ \xrightarrow{H_2/Ni} CH_3CH_2CH_2CH_2CH_2OH\end{cases}$$

$LiAlH_4$, besides aldehydes and ketones, also reduces acids and acid derivatives. Therefore, it cannot be used for specific reduction of aldehyde or ketonic functional group in a compound containing acid or acid derivative functional group. In contrast $NaBH_4$ does not reduces acid and acid derivatives, therefore, it can be used for selective reduction of aldehydic or ketonic functional group without affecting acid and acid derivative functional group as:

3-oxocyclopentane-COOH $\xrightarrow{LiAlH_4}$ 3-(CH$_2$OH)cyclopentanol (OH, CH$_2$OH)

3-oxocyclopentane-COOH $\xrightarrow{NaBH_4}$ 3-(COOH)cyclopentanol (OH, COOH)

$$H_2N{-}\overset{O}{\overset{\|}{C}}{-}CH_2{-}CH_2{-}CHO \begin{cases}\xrightarrow{LiAlH_4} H_2NCH_2CH_2CH_2CH_2OH\\ \xrightarrow{NaBH_4} H_2N{-}\overset{O}{\overset{\|}{C}}{-}CH_2CH_2CH_2OH\end{cases}$$

$$CH_3OCOCH_2CH_2{-}\overset{O}{\overset{\|}{C}}{-}CH_3 \begin{cases}\xrightarrow{LiAlH_4} HO{-}CH_2CH_2CH_2{-}\underset{OH}{\underset{|}{C}}H{-}CH_3\\ \xrightarrow{NaBH_4} CH_3OCOCH_2CH_2\overset{H}{\overset{|}{\underset{OH}{\underset{|}{C}}}}{-}CH_3\end{cases}$$

$LiAlH_4$ also reduces —Cl, —NO_2, —CN, etc. Therefore, compound containing these functional group cannot be reduced specifically at aldehydic or ketonic functional group with $LiAlH_4$.

(b) Meerwein Ponndorf-Verley reduction: A very specific reagent used for selective hydrogenation of aldehydic and ketonic functional group, without disturbing halide, nitro, cyano, olefinic group, is aluminium isopropoxide $[(CH_3)_2CHO]_3$ Al.

$$\text{cyclohexenone} + (CH_3-\underset{|}{C}HO)_3Al \longrightarrow \text{cyclohexenol} + CH_3-\overset{O}{\overset{\|}{C}}-CH_3$$

The reduction occurs by hydride ion transfer from aluminium isopropoxide to carbonyl compound via a cyclic transition state:

$$R_2C{=}O + (Me_2CH{-}O)AlX_2 \xrightarrow{\ X\ } \text{[cyclic transition state]} \rightleftharpoons R_2CH{-}O{-}AlX_2 + Me_2CO \xrightarrow{H_2O} R_2CHOH$$

X = isopropoxy

II. Clemmenson reduction

This alternative reduction involves heating a carbonyl compound with finely divided, amalgamated zinc in a hydroxylic solvent (often an aqueous mixture) containing a mineral acid such as HCl. The mercury alloyed with the zinc does not participate in the reaction, it serves only to provide a clean active metal surface. The first example below shows a common application of this reduction, the conversion of a Friedel-Crafts acylation product to an alkyl side-chain. The second example illustrates the lability of functional substituents alpha to the carbonyl group. Substituents such as hydroxyl, alkoxyl and halogens are reduced first, the resulting unsubstituted aldehyde or ketone is then reduced to the parent hydrocarbon.

$$>C{=}O + Zn(Hg) \xrightarrow{\text{Conc. HCl}} >CH_2$$

$$C_6H_6 + \text{succinic anhydride} \xrightarrow{AlCl_3} C_6H_5CO(CH_2)_2COOH \xrightarrow[\text{HCl in aq. alcohol}]{Zn(Hg)} C_6H_5(CH_2)_3COOH$$

$$\text{2-hydroxycycloheptanone} \xrightarrow[\text{HCl in aq. alcohol}]{Zn(Hg)} [\text{cycloheptanone}] \longrightarrow \text{cycloheptane}$$

This method can be used to convert aldehydic and ketonic functional group into CH_2 group, provided aldehydes or ketones do not contain any other functional group sensitive to concentrated HCl, viz. OH, olefinic, etc. groups, e.g.

$$\text{4-hydroxycyclohexanone} + Zn(Hg) \xrightarrow{\text{Con. HCl}} \text{cyclohexanol}$$ is not a suitable conversion, –OH will also be reduced.

Because concentrated HCl reacts with –OH functional group leading to substitution as well as elimination (dehydration) by products, Zn further reduces –OH to alkane.

III. Wolf-Kischner reduction

Reaction of an aldehyde or ketone with excess hydrazine generates a hydrazone derivative, which on heating with base gives the corresponding hydrocarbon. A high-boiling hydroxylic solvent, such as diethylene glycol, is commonly used to achieve the temperatures needed. The following diagram shows how this reduction may be used to convert cyclopentanone to cyclopentane. A second example, in which an aldehyde is similarly reduced to a methyl group, also illustrates again the use of an acetal protective group. The mechanism of this useful transformation involves tautomerization of the initially formed hydrazone to an azo isomer.

$$\text{Cyclopentanone} \xrightarrow[\text{Excess}]{N_2H_4} \left[\text{Cyclopentylidene}=N-NH_2\right] \xrightarrow[\text{DEG solution, heat}]{\text{KOH}} \text{Cyclopentane}$$

DEG = diethylene glycol (HO—CH_2CH_2—O—CH_2CH_2—OH)

$$H_3C-C(\text{ketal})-CH_2CH_2-CHO \xrightarrow[\text{Excess}]{N_2H_4} H_3C-C(\text{ketal})-CH_2CH_2-CH=N-NH_2 \xrightarrow[\text{DEG solution, heat}]{\text{KOH}} H_3C-C(\text{ketal})-CH_2CH_3$$

Here aldehydes or ketones is reacted with hydrazine in alkaline medium, N_2 is liberated as:

$$>C=O + H_2NNH_2 \longrightarrow >C=NNH_2 \xrightarrow[\Delta]{OH^-} >CH_2 + N_2$$

The advantages of this method is, it can be used for specific reduction of carbonyl functional group into $>CH_2$ without disturbing olefinic or hydroxyl functional group.

$$CH_2=CHCH_2CHO \xrightarrow[OH^-]{N_2H_4} CH_2=CHCH_2CH_3$$

$$HO-CH_2CH_2CHO \xrightarrow[OH^-]{N_2H_4} HO-CH_2CH_2CH_3$$

Limitations of this reagent, it cannot be used for specific reduction of carbonyl functional group in a compound that contains alkali sensitive functional group, viz. —COOH, —CN, —Cl, etc.

$$\text{4-Chlorocyclohexanone} \xrightarrow{N_2H_4/OH^-} \text{Cyclohexanol}$$

$$\text{4-Chlorocyclohexanone} \xrightarrow{\text{Zn (Hg)/HCl}} \text{Chlorocyclohexane}$$

$$HO-\overset{O}{\overset{\|}{C}}-CH_2CH_2-\overset{O}{\overset{\|}{C}}-H \xrightarrow{N_2H_4/OH^-} NaOC-\overset{O}{\overset{\|}{C}}-CH_2CH_2CH_3$$

$$HO-\overset{O}{\overset{\|}{C}}-CH_2CH_2-\overset{O}{\overset{\|}{C}}-H \xrightarrow{\text{Zn (Hg)/HCl}} HOOC-\overset{O}{\overset{\|}{C}}-CH_2CH_2CH_3$$

IV. Reaction of aldehydes and ketones with $NaHSO_3$

Aldehydes and ketones forms bisulphite with $NaHSO_3$:

$$>C=O + NaHSO_3 \longrightarrow >C<^{OH}_{SO_3Na}$$

Bisulphite

Bisulphite compounds are usually crystalline solids, insoluble in $NaHSO_3$ solution. Since they regenerate carbonyl compound when heated with dilute alkali or acid solution, it is a very convenient means of separating them from non-carbonyl compounds.

V. Nucleophilic addition of HCN

HCN adds on aldehydes and ketones forming cyano hydrin:

$$>C=O + HCN \rightleftharpoons >C<^{OH}_{CN}$$

(a cyanohydrin)

Cyanohydrins are very important compounds in organic synthesis, since it is readily hydrolyzed into α hydroxyl acid.

Example 4. A hydrocarbon $A(C_7H_{14})$ exist in two stereomeric forms. A on reductive ozonolysis yields $B(C_3H_6O)$ and $C(C_4H_8O)$. B on catalytic hydrogenation yields D(C_3H_8O) that cannot be resolved, whereas C on catalytic hydrogenation yields E($C_4H_{10}O$), which can be resolved into enantiomers. B on treatment with HCN followed by hydrolysis of product yields F($C_4H_8O_3$), which can be resolved into enantiomers. C on treatment with HCN followed by hydrolysis of product yields G($C_5H_{10}O_3$), which is also resolvable. Identity A to G and show the stereomeric structures of A.

Solution

(A) $C_2H_5(H)C{=}C(C_2H_5)(CH_3)$ and $H_3C(C_2H_5)C{=}C(C_2H_5)(H)$

(B) CH_3CH_2CHO C: 2-butanone

(D) Propanol E: 2-butanol

(F) $CH_3CH_2{-}C(OH)(H){-}COOH$ (G) $CH_3CH_2{-}C(OH)(CH_3){-}COOH$

VI. Reactions of aldehydes and ketones with hydrazine and its derivatives

(a) Reaction with hydrazine: Hydrazone is formed:

$$>C{=}O + N_2H_4 \longrightarrow >C{=}NNH_2$$

(b) Reaction with phenyl hydrazine: Phenylhydrazone is formed:

$$\underset{R'}{\overset{R}{>}}C{=}O + H_2NNHPh \longrightarrow \underset{R'}{\overset{R}{>}}C{=}NNHPh$$

A derivative of phenylhydrazine '2, 4-dinitro-phenylhydrazine' is used for separation of carbonyl compound from non-carbonyl compounds, because it forms an insoluble hydrazone:

$$\underset{R'}{\overset{R}{>}}C{=}O + H_2NHN{-}C_6H_3(NO_2)_2 \longrightarrow \underset{R'}{\overset{R}{>}}C{=}NNH{-}C_6H_3(NO_2)_2$$

'Hydrazone'

(c) Reaction with semicarbazide: A semicarbazone is formed:

$$R{-}CH{=}O + H_2N{-}NH{-}\overset{O}{\overset{\|}{C}}{-}NH_2 \longrightarrow R{-}CH{=}NNH{-}\overset{O}{\overset{\|}{C}}{-}NH_2$$

Semicarbazide

Carbonyl group here reacts with aminic $—NH_2$ group of semicarbazide, not with amide $—NH_2$ since former is more nucleophilic. The reason for greater nucleophilicity of aminic $—NH_2$ in semicarbazide is lone pair on nitrogen is in resonance with carbonyl group in later case, and not completely available for electrophilic attack whereas in aminic $—NH_2$, lone pair is completely available for electrophilic attack as:

This lone pair is not the part of resonance

$$H_2N{-}NH{-}\overset{O}{\overset{\|}{C}}{-}\ddot{N}H_2 \longleftrightarrow H_2\ddot{N}{-}NH{-}\overset{O^-}{\overset{|}{C}}{=}\overset{+}{N}H_2$$

VII. Reaction with ammonia and its derivatives

(a) Reaction with ammonia: Imine is formed:

$$R{-}\overset{O}{\overset{\|}{C}}{-}H + NH_3 \longrightarrow R{-}CH{=}NH$$

Imine can easily be reduced into amine, therefore, it is a satisfactory means of synthesizing amines from carbonyl compounds.

(b) Reaction with primary amines:

$$R{-}CH{=}O + H_2N{-}R' \longrightarrow R{-}CH{=}NR'$$

(c) Reaction with secondary amine: Secondary amine reacts with an aldehydes or ketones containing at least one α -H to form enamine as:

$$CH_3CH_2{-}\overset{O}{\overset{\|}{C}}{-}H + H{-}\overset{R}{\overset{|}{N}}{:}{-}R \rightleftharpoons$$

$$CH_3CH_2{-}\overset{\bar{O}}{\overset{|}{C}}(H){-}\overset{+}{N}H(R)(R) \rightleftharpoons CH_3CH_2{-}\overset{OH}{\overset{|}{C}}(H){-}NR_2$$

$$\downarrow -H_2O,\ H^+$$

$$CH_3CH_2{-}\overset{+}{C}(H){-}NR_2 \overset{-H^+}{\rightleftharpoons} CH_3CH_2{=}C(H){-}NR_2$$

Enamine

Example 5. Complete the following reactions:

(i) cyclohexanone + HN(pyrrolidine) $\xrightarrow{H^+}$

(ii) $CH_3NHCH_2CH_2CH_2CH_2CHO \xrightarrow{H^+}$

(iii) 3-hexanone + $H_2NHN{-}\overset{O}{\overset{\|}{C}}{-}NH_2 \longrightarrow$

Solution

(i) [1-cyclohexenyl pyrrolidine structure, N] (ii) [cyclic enamine structure, N–CH_3]

(iii) [semicarbazone structure: H, N, N, NH_2, O]

(d) Reaction with hydroxylamine: Oximes are formed:

$$\begin{matrix} H_3C \\ H \end{matrix}\!\!>C{=}O + H_2NOH \longrightarrow \begin{matrix} H_3C \\ H \end{matrix}\!\!>C{=}N{-}OH \;+\; \begin{matrix} H_3C \\ H \end{matrix}\!\!>C{=}N{-}OH$$

Antioxime Syn oxime

Symmetrical ketones give a single isomer of oxime as:

$$\text{cyclopentanone}{=}O + H_2NOH \longrightarrow \text{cyclopentane}{=}NOH \text{ Ketoxime}$$

VIII. Reaction with Grignard's reagent

Aldehydes and ketones adds a molecule of Grignard reagent, and the complex formed, when decomposed with acid, alcohols are produced.

$$R{-}CHO + R'MgX \xrightarrow{H^+} R{-}\underset{OH}{\underset{|}{CH}}{-}R'$$

$$R{-}\overset{O}{\overset{\|}{C}}{-}R + R'MgX \xrightarrow{H^+} R{-}\underset{R'}{\underset{|}{\overset{R}{\overset{|}{C}}}}{-}OH$$

IX. Reaction with PCl_5

Simple carbonyl compounds reacts with PCl_5 forming 1,1-dichloride as:

$$>C{=}O + PCl_5 \longrightarrow >CCl_2 + POCl_3$$

X. α-Halogenation

Chlorine and bromine replaces one or more α-H from aldehydes and ketones as:

$$CH_3{-}\overset{O}{\overset{\|}{C}}{-}CH_3 + Br_2 \longrightarrow CH_3{-}\overset{O}{\overset{\|}{C}}{-}CH_2Br$$

XI. Reaction with hydrazoic acid (Schmidt reaction)

Aldehydes and ketones combine with HN_3 in concentrated sulfuric acid medium. Aldehydes give a mixture of cyanide and formal derivatives of primary amine whereas ketone gives amide:

$$R{-}CHO + HN_3 \xrightarrow{H^+} RCN + RNHCHO + N_2$$

$$R{-}\overset{O}{\overset{\|}{C}}{-}R + HN_3 \xrightarrow{H^+} R{-}\overset{O}{\overset{\|}{C}}NHR + N_2$$

XII. Hydrate formation

Water adds to an aldehyde or to a ketone to form a hydrate:

$$R{-}\overset{O}{\overset{\|}{C}}{-}R(H) + OH_2 \rightleftharpoons R{-}\underset{OH}{\underset{|}{\overset{OH}{\overset{|}{C}}}}{-}R(H) \text{ a gem-diol}$$

An aldehyde or ketones (A hydrate)

Mechanism

Reaction involves nucleophilic attack by water on carbonyl carbon. Loss of a proton from oxonium ion and gain of a proton from oxyanion forms diol.

$$CH_3{-}\overset{O}{\overset{\|}{C}}{-}H + OH_2 \rightleftharpoons CH_3{-}\underset{+OH_2}{\underset{|}{\overset{O^-}{\overset{|}{C}}}}{-}H \rightleftharpoons CH_3{-}\underset{OH}{\underset{|}{\overset{OH}{\overset{|}{C}}}}{-}H$$

Acetaldehyde hydrate

Water is a poor nucleophile and therefore adds relatively slowly to a carbonyl compound, but an acid catalyst or an alkali catalyst can accelerate the rate of reaction:

$$R{-}\overset{O:}{\overset{\|}{C}}{-}H \overset{H^+}{\rightleftharpoons} R{-}\overset{\overset{+}{O}{-}H}{\overset{\|}{C}}{-}H \overset{OH_2}{\rightleftharpoons}$$

(more succeptible for nucleophilic attack)

$$R{-}\underset{+OH_2}{\underset{|}{\overset{OH}{\overset{|}{C}}}}{-}H \overset{-H^+}{\rightleftharpoons} R{-}\underset{OH}{\underset{|}{\overset{OH}{\overset{|}{C}}}}{-}H$$

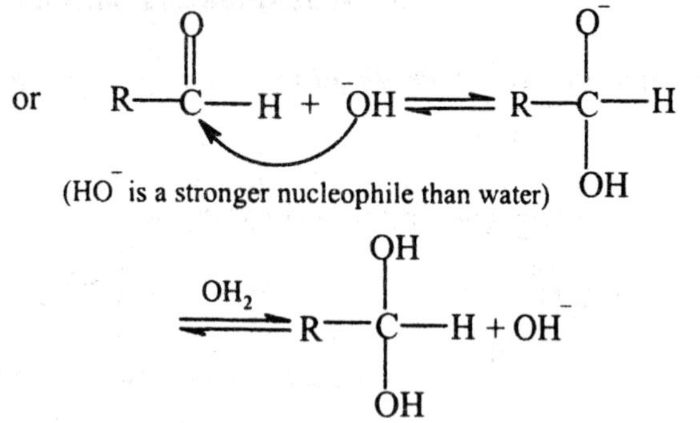

or $R-C(=O)-H + \bar{O}H \rightleftharpoons R-C(O^-)(OH)-H$

(HO$^-$ is a stronger nucleophile than water)

$\xrightleftharpoons{OH_2} R-C(OH)(OH)-H + \bar{O}H$

The extent to which an aldehydes or ketones are hydrated in aqueous solution depends on the aldehydes or ketones, e.g. only 0.2% of acetone is hydrated at equilibrium whereas 99.9% of formaldehyde is hydrated. The reason lies on two important factors:

(i) *Steric hindrance:* Alkyl groups in ketones poses greater steric hindrance to the attack of nucleophile (H_2O) on the carbonyl carbon, compared to aldehydes.

(ii) Alkyl group attached to carbonyl carbon in ketones, decreases affinity for nucleophile by partial neutralization of positive charge on carbonyl carbon by the +I effect:

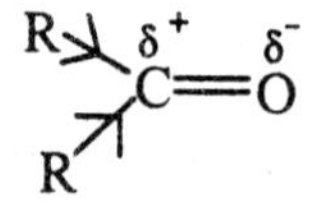

$R_2C^{\delta+}=O^{\delta-}$

+Inductive effect decreases the electropositivity of carbonyl carbon.

There are some vary stable hydrates of ketone derivatives, e.g. trichloroacetaldehyde is almost quantitatively converted into its hydrate form when dissolved in water as:

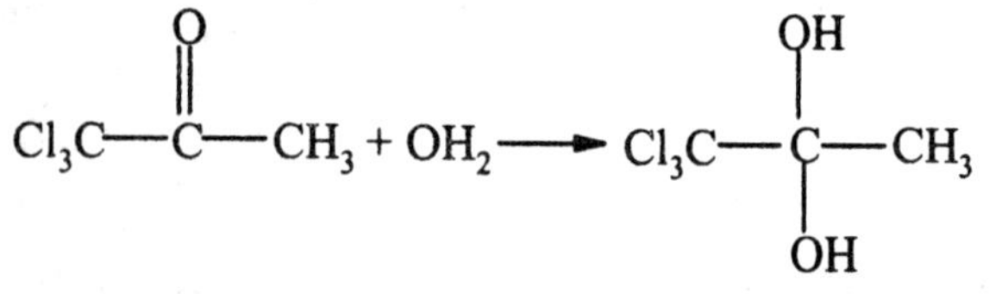

$$Cl_3C-C(=O)-CH_3 + OH_2 \longrightarrow Cl_3C-C(OH)(OH)-CH_3$$

The high stability of the above-mentioned hydrate is accounted for from the stablization of the hydrate by forming an intramolecular H-bonds with chlorine as:

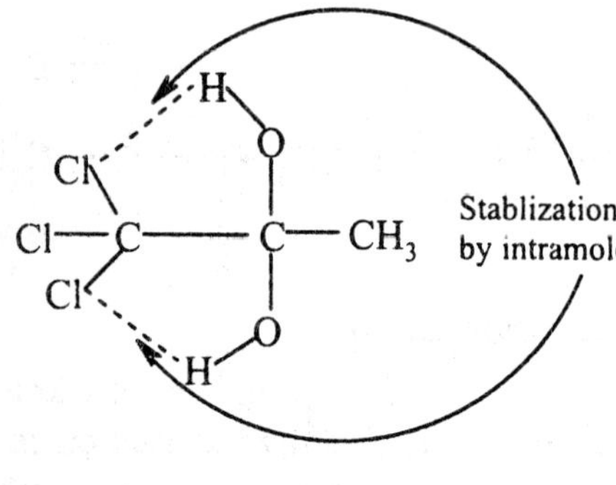

Stablization of chloral hydrate by intramolecular H-bonding

Example 6. Which of the following pairs has larger equilibrium constant for addition of water?

(a) 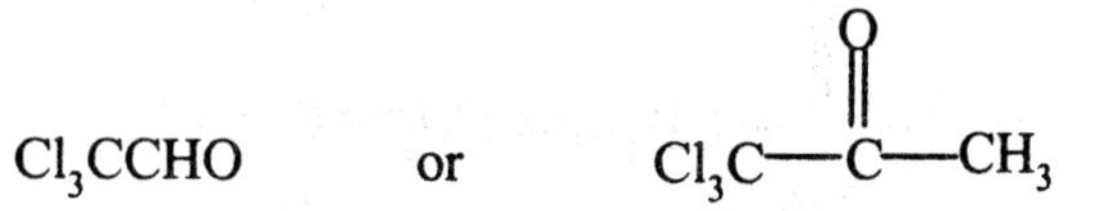 Cl_3CCHO or $Cl_3C-C(=O)-CH_3$

(b) $Cl_3CCH_2C(=O)-CH_3$ or $CH_3-CCl_2-C(=O)-CH_3$

(c) $C_6H_5-C(=O)-H$ or $MeO-C_6H_4-C(=O)-H$

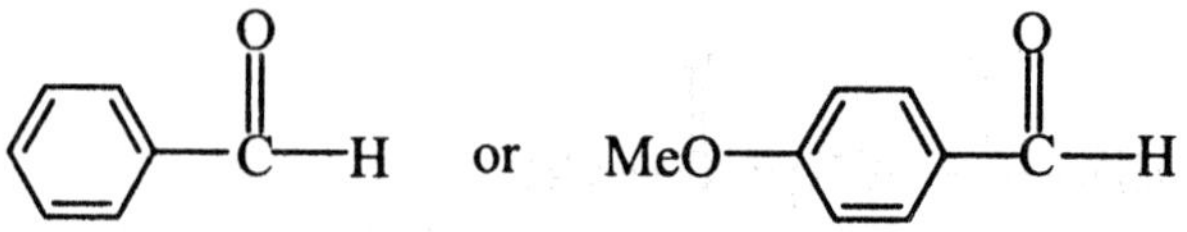

(d) 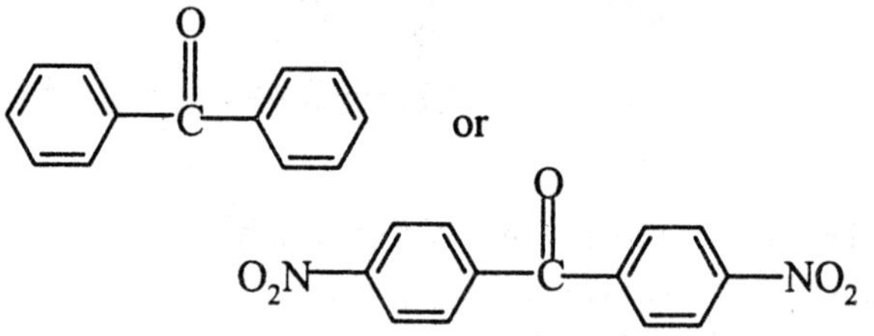 $C_6H_5-C(=O)-C_6H_5$ or $O_2N-C_6H_4-C(=O)-C_6H_4-NO_2$

(e) 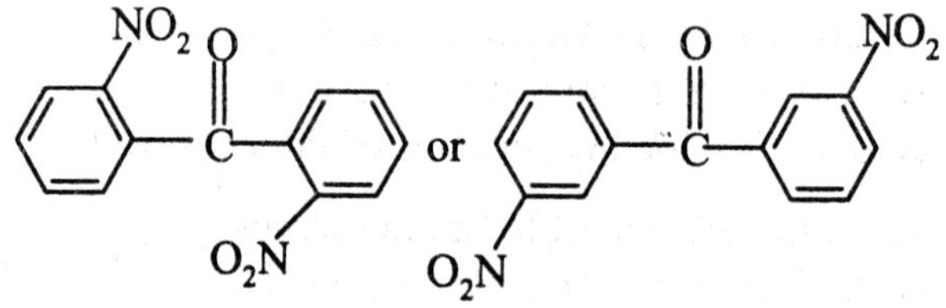 2,2'-dinitrobenzophenone or 3,3'-dinitrobenzophenone

Solution

(a) Cl_3CCHO (b) $CH_3CCl_2COCH_3$

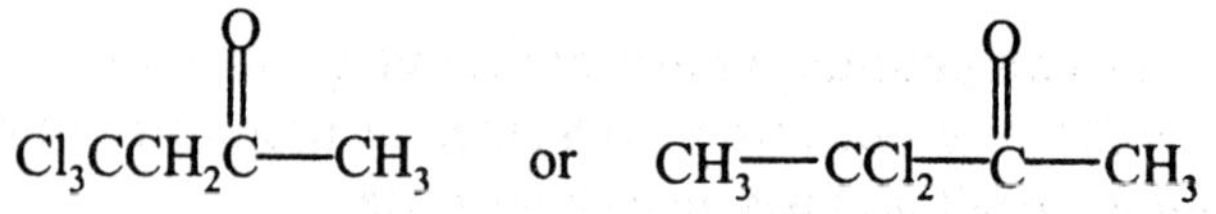

(c) PhCHO

(d) 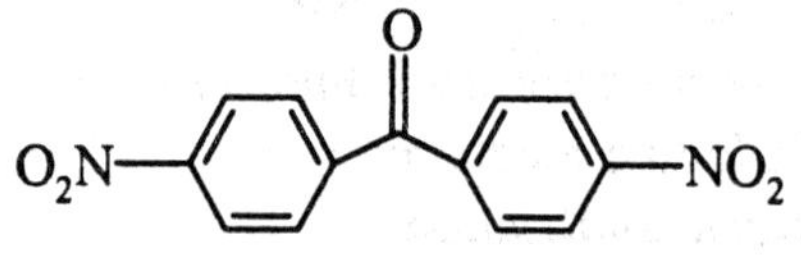 $O_2N-C_6H_4-C(=O)-C_6H_4-NO_2$

(e) 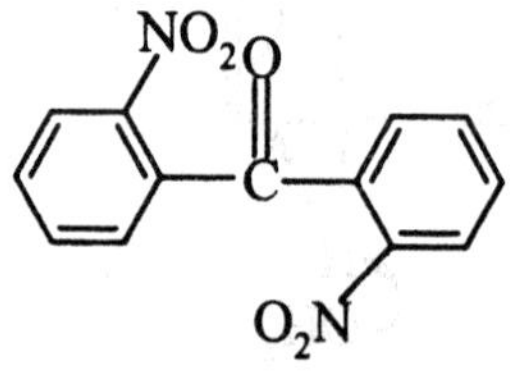 2,2'-dinitrobenzophenone

XIII. Keto-Enol tautomerism

Keto and enol forms of carbonyl compounds are constitutional isomers. There exist a dynamic equilibrium between keto and enol forms and in a particular circumstance, one form predominate.

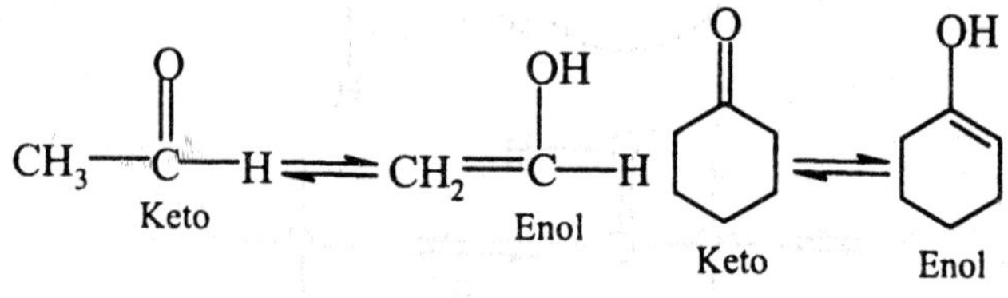

$$\underset{\text{Keto}}{CH_3-C(=O)-H} \rightleftharpoons \underset{\text{Enol}}{CH_2=C(OH)-H}$$

Cyclohexanone (Keto) $\rightleftharpoons$ Cyclohexenol (Enol)

$$CH_3-\overset{O}{\overset{\|}{C}}-CH_3 \rightleftharpoons CH_2=\overset{OH}{\overset{|}{C}}-CH_3$$

Keto Enol

In compounds whose carbonyl groups surrounds —CH_2— groups from both side, the proportion of enol form at equilibrium is much higher, viz:

$$\underset{24\%}{CH_3-\overset{O}{\overset{\|}{C}}-CH_2-\overset{O}{\overset{\|}{C}}-CH_3} \rightleftharpoons \underset{76\%}{CH_3-\overset{OH}{\overset{|}{C}}=CH-\overset{O}{\overset{\|}{C}}-CH_3}$$

The greater stability of enol forms in the above mentioned example can be attributed to the stability gained through resonance stabilization.

Example 7. The compound 2, 4-cyclohexadien-1-one exist totally in its enol form. What are the special factor that accounts for the stability of the enol forms?

Solution: The equilibrium lies almost to the left due to very high stability of enol form gained from aromaticity.

Phenol (OH) $\rightleftharpoons$ 2,4-cyclohexadien-1-one (O)

Mechanism of enol-formation: Enolization can occur either in acid or alkaline medium:

(a) *Acid catalyzed enolization:*

$$-\overset{|}{\underset{H}{\underset{|}{C}}}-\overset{O:}{\overset{\|}{C}}-H + H^+ \rightleftharpoons -\overset{|}{\underset{H}{\underset{|}{C}}}-\overset{\overset{+}{O}-H}{\overset{\|}{C}}-H \xrightleftharpoons[OH_2]{-H_3O^+} -C=\overset{OH}{\overset{|}{C}}-H$$

The reaction is reversible and an enol can be converted into its keto form as:

$$>C=\overset{OH}{\overset{|}{C}}- \xrightleftharpoons{H^+} -\underset{H}{\underset{|}{\overset{|}{C}}}-\overset{\overset{+}{O}{<}^H}{\overset{|}{C^+}}- \rightleftharpoons -\underset{H}{\underset{|}{\overset{|}{C}}}-\overset{^+O-H}{\overset{\|}{C}}- \xrightleftharpoons{-H^+} -\underset{H}{\underset{|}{\overset{|}{C}}}-\overset{O}{\overset{\|}{C}}-$$

(b) *Base catalyzed enolization*

$$-\underset{H}{\underset{|}{\overset{|}{C}}}-\overset{O}{\overset{\|}{C}}- \xrightleftharpoons{^-OH} -\overset{|}{C}=\overset{O^-}{\overset{|}{C}}- \xrightleftharpoons{OH_2} -C=\overset{OH}{\overset{|}{C}}- + OH^-$$

Example 8. When isopropyl phenyl ketone is treated with OD^- or D_3O^+ in D_2O medium, following compound is produced:

$CH_3-\overset{CH_3}{\overset{|}{CD}}-\overset{O}{\overset{\|}{C}}-C_6H_5$. Explain mechanism of its formation.

Solution: (a) *Base catalyzed*

$$CH_3-\underset{H}{\underset{|}{\overset{CH_3}{\overset{|}{C}}}}-\overset{O}{\overset{\|}{C}}-C_6H_5 \xrightleftharpoons{^-OD} CH_3-\overset{CH_3}{\overset{|}{C}}=\overset{O^-}{\overset{|}{C}}-C_6H_5 \longleftrightarrow CH_3-\overset{CH_3}{\overset{|}{\underset{-}{C}}}-\overset{O}{\overset{\|}{C}}-C_6H_5$$

$$CH_3-\underset{D}{\underset{|}{\overset{CH_3}{\overset{|}{C}}}}-\overset{O}{\overset{\|}{C}}-C_6H_5 \xleftrightarrow{D_2O} CH_3-\overset{CH_3}{\overset{|}{\underset{-}{C}}}-\overset{O}{\overset{\|}{C}}-C_6H_5$$

(b) *Acid catalyzed*

$$CH_3-\underset{H}{\underset{|}{\overset{CH_3}{\overset{|}{C}}}}-\overset{O:}{\overset{\|}{C}}-Ph \xrightleftharpoons{D_3O^+} CH_3-\underset{H}{\underset{|}{\overset{CH_3}{\overset{|}{C}}}}-\overset{\overset{+}{O}-D}{\overset{\|}{C}}-C_6H_5 \xrightleftharpoons{D_2\ddot{O}}$$

$$CH_3-\overset{CH_3}{\overset{|}{C}}=\overset{OD}{\overset{|}{C}}-C_6H_5 \xleftrightarrow{D_3O^+} CH_3-\underset{D}{\underset{|}{\overset{CH_3}{\overset{|}{C}}}}-\overset{\overset{D}{\diagdown}O:}{\overset{|}{C^+}}-C_6H_5$$

$$CH_3-\underset{D}{\underset{|}{\overset{CH_3}{\overset{|}{C}}}}-\overset{O}{\overset{\|}{C}}-C_6H_5 \xrightleftharpoons{-D+} CH_3-\underset{D}{\underset{|}{\overset{CH_3}{\overset{|}{C}}}}-\overset{\overset{+}{O}-D}{\overset{\|}{C}}-C_6H_5$$

XIV. The aldol condensation

A useful carbon-carbon bond-forming reaction known as the Aldol reaction or the Aldol condensation is an example of electrophilic substitution at the alpha carbon in enols or enolate anions. Three examples of the base-catalyzed aldol reaction are shown in the following diagram, and equivalent acid-catalyzed reactions also occur. The fundamental transformation in this reaction is a dimerization of an aldehyde (or ketone) to a beta-hydroxy aldehyde (or ketone) by alpha C-H addition of one reactant molecule to the carbonyl group of a second reactant molecule.

1. $2H_3C-C(=O)H \xrightleftharpoons{NaOH}$ (β-hydroxy aldehyde, skeletal structure)

2. 2 (butanal, skeletal structure) $\xrightleftharpoons{NaOH}$ (aldol product bearing C_2H_5, skeletal structure)

3. $2H_3C-CO-CH_3 \xrightleftharpoons{NaOH}$ (β-hydroxy ketone, skeletal structure)

Therefore, when an aldehyde or ketone is treated with dilute NaOH or K_2CO_3 or strong base or hydrochloric acid, condensation occurs to form Aldol, a β-hydroxy aldehyde or ketone, e.g.

$$CH_3CHO \xrightarrow{10\%\ NaOH,\ H_2O} CH_3CH(OH)-CH_2CHO$$

-hydroxy aldehyde

In the presence of acid or base catalysts the aldol reaction is reversible, and the beta-hydroxy carbonyl products may revert to the initial aldehyde or ketone reactants. In the absence of such catalysts these aldol products are perfectly stable and isolable compounds. Because of this reversibility, the yield of aldol products is related to their relative thermodynamic stability. In the case of aldehyde reactants (as in reactions 1 and 2 above), the aldol reaction is modestly exothermic and the yields are good. However, aldol reactions of ketones are less favourable (e.g. 3 above), and the equilibrium product concentration is small. A clever way of overcoming this disadvantage has been found. A comparatively insoluble base, $Ba(OH)_2$, is used to catalyze the aldol reaction of acetone, and the product is removed from contact with this base by filtration and recirculation of the acetone.

Mechanism of Base Catalyzed Aldol Condensation

Aldol condensation involves two step mechanism:

(i) Abstraction of α -H from aldehydes or ketone in the 1st step. Therefore presence of α -H is an essential requirement for aldol condensation.

$$H\bar{O} + H-CH_2-CHO \rightleftharpoons OH_2 + \left[\bar{C}H_2-C(=O)-H \longleftrightarrow CH_2=C(O^-)-H\right]$$

Enolate ion

[Resonance stabilization of enolate ion explains the acidity of α -H in aldehydes and ketones.]

(ii) Nucleophilic addition of enolate ion on the carbonyl carbon of another molecule of aldehydes or ketone:

$$CH_3-C(=O)-H + \bar{C}H_2-CHO \rightleftharpoons CH_3-CH(O^-)-CH_2CHO$$

An alkoxide ion

$$\downarrow\uparrow H_2O$$

$$H_3C-CH(OH)-CH_2CHO$$

Therefore, in aldol-condensation, the important points which we should always keep in mind are:

1. α -H is an essential requirement.
2. β -Hydroxy aldehyde or ketone should be the final product.
3. Aldehydes are more reactive than ketones in aldol condensation. This is because of the following two factors:
   (a) Carbonyl carbon of aldehydes are less sterically hindered than the ketone, because in aldehydes, carbonyl carbon always bears a smallest functional group (H), whereas carbonyl carbon of ketone bears bulky alkyl group on both side. Therefore, nucleophilic attack of enolate ion occurs at faster rate on the aldehyde molecule in step II than on ketone.

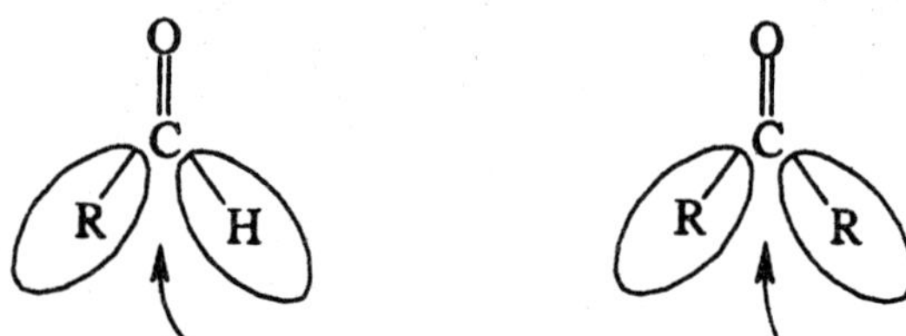

Less sterically hindered carbonyl carbon, poses less steric hinderance to the attack of enolate ion | More sterically hindered carbon, poses more steric hinderance to the attack of enolate ion

   (b) *Polarity factor:* In aldehydes, carbonyl carbon is attached with one alkyl group whereas in ketone, it is attached with two alkyl group. Since alkyl group has electron-donating tendency by + Inductive effect, carbonyl carbon of ketone is less electropositive than in aldehyed, therefore nucleophilic attack of enolate ion occurs at faster rate at the carbonyl carbon of aldehydes than on ketone:

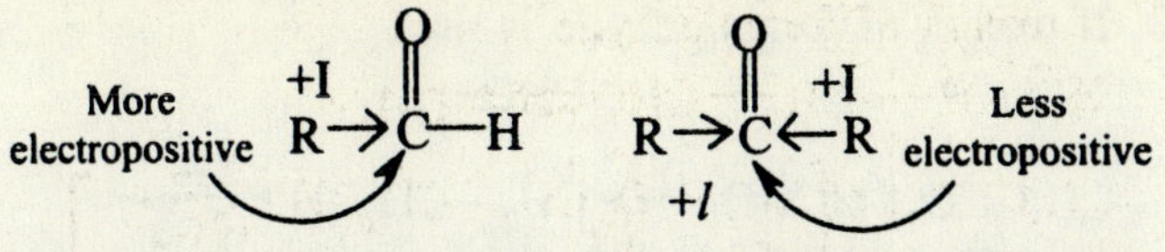

More electropositive $R \rightarrow \overset{O}{\overset{\|}{C}}-H$ (+I) $R \rightarrow \overset{O}{\overset{\|}{C}} \leftarrow R$ (+I, +I) Less electropositive

(iii) *Dehydration of aldol:* The products of aldol reactions often undergo a subsequent elimination of water, made up of an alpha-hydrogen and the beta-hydroxyl group. The product of this **beta-elimination** reaction is a conjugated, unsaturated aldehyde or ketone, as shown in Figure 7.1. Acid-catalyzed conditions are more commonly used to effect this elimination (examples 1, 2 and 5), but base-catalyzed elimination also occurs, especially on heating (examples 3, 4 and 5). The additional stability provided by the conjugated carbonyl system of the product makes some ketone aldol reactions thermodynamically favourable (4 and 5), and mixtures of stereoisomers (E and Z) are obtained from reaction 4. Reaction 5 is an interesting example of an intramolecular aldol reaction, such reactions create a new ring.

Reactions in which a larger molecule is formed from smaller components with the elimination of a very small by-product such as water are termed **condensations**. Hence, the following examples are properly referred to as **aldol condensations**. The dehydration step of an aldol condensation is also reversible in the presence of acid and base catalysts. Consequently, on heating with aqueous solutions of strong acids or bases, many α, β-unsaturated carbonyl compounds fragment into smaller aldehyde or ketones, a process known as the **retro-aldol reaction**.

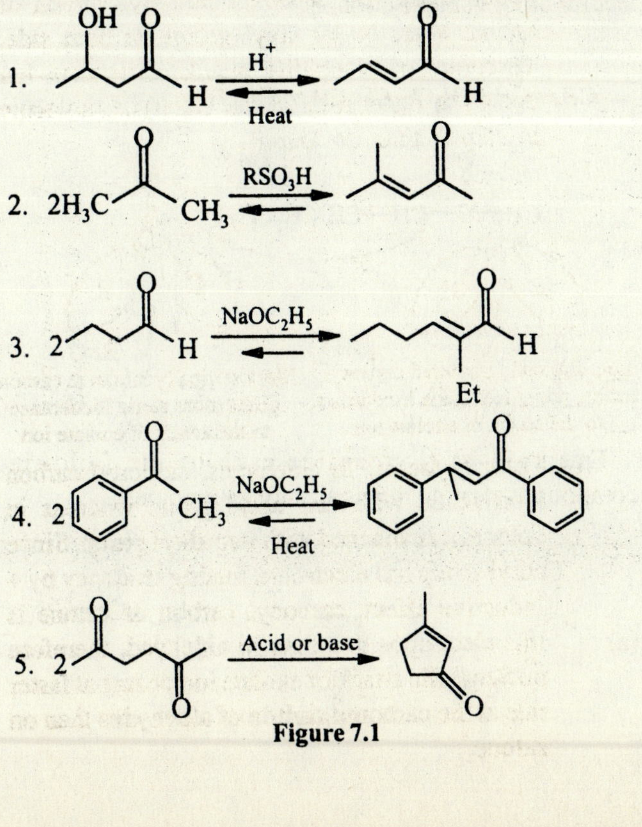

1. (3-hydroxybutanal) $\underset{\text{Heat}}{\overset{H^+}{\rightleftharpoons}}$ (but-2-enal)

2. $2H_3C-CO-CH_3$ $\overset{RSO_3H}{\rightleftharpoons}$ (4-methylpent-3-en-2-one)

3. 2 (butanal) $\overset{NaOC_2H_5}{\rightleftharpoons}$ (2-ethylhex-2-enal, Et)

4. $2\ C_6H_5COCH_3$ $\underset{\text{Heat}}{\overset{NaOC_2H_5}{\rightleftharpoons}}$ (1,3-diphenylbut-2-en-1-one)

5. 2 (hexane-2,5-dione) $\xrightarrow{\text{Acid or base}}$ (3-methylcyclopent-2-en-1-one)

Figure 7.1

Acid catalyzed aldol condensation: Acid catalyzed aldol condensation involves reaction between conjugate acid and enol form of the carbonyl compound as:

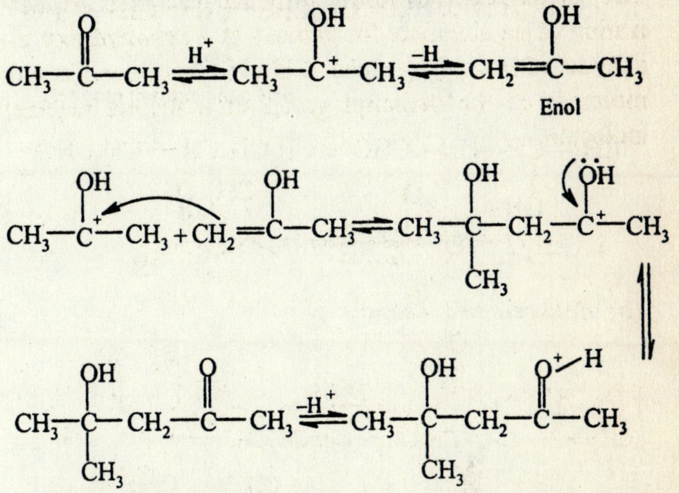

$$CH_3-\overset{O}{\overset{\|}{C}}-CH_3 \overset{H^+}{\rightleftharpoons} CH_3-\overset{OH}{\overset{|}{C^+}}-CH_3 \overset{-H}{\rightleftharpoons} \underset{\text{Enol}}{CH_2=\overset{OH}{\overset{|}{C}}-CH_3}$$

$$CH_3-\overset{OH}{\overset{|}{C^+}}-CH_3 + CH_2=\overset{OH}{\overset{|}{C}}-CH_3 \rightleftharpoons CH_3-\underset{CH_3}{\underset{|}{\overset{OH}{\overset{|}{C}}}}-CH_2-\overset{\ddot{O}H}{\overset{|}{C^+}}-CH_3$$

$$CH_3-\underset{CH_3}{\underset{|}{\overset{OH}{\overset{|}{C}}}}-CH_2-\overset{O}{\overset{\|}{C}}-CH_3 \overset{-H^+}{\rightleftharpoons} CH_3-\underset{CH_3}{\underset{|}{\overset{OH}{\overset{|}{C}}}}-CH_2-\overset{O^+-H}{\overset{\|}{C}}-CH_3$$

Example 9. Predict products in the following aldol-condensation reaction.

(a) $CH_3CH_2-CHO \xrightarrow{\bar{O}H}$

(b) $CH_3-\overset{O}{\overset{\|}{C}}-CH_2CH_3 \xrightarrow{\bar{O}H} A + B$

(c) 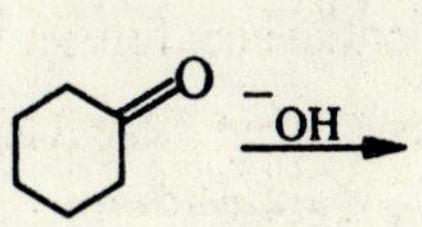 (cyclohexanone) $\xrightarrow{\bar{O}H}$

Solution: (a) 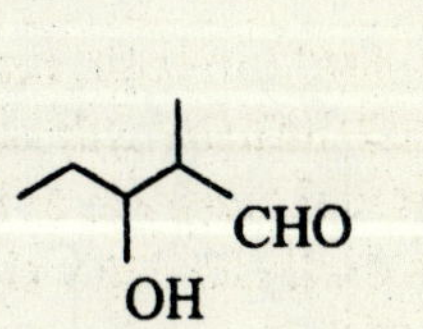 (3-hydroxy-2-methylpentanal: CHO, OH)

(b) 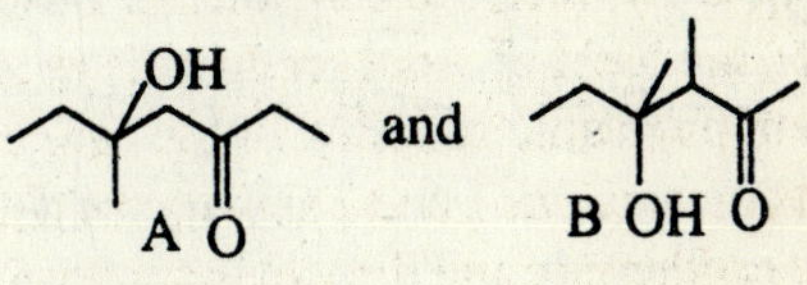 A (OH, O) and B (OH, O)

(c) 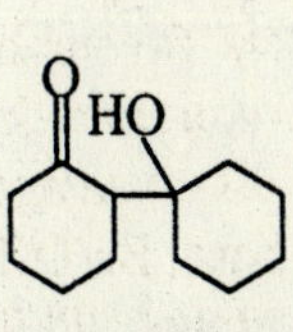 (2-(1-hydroxycyclohexyl)cyclohexanone: O, HO)

Cross aldol condensation: If two different aldehydes or ketones or a mixture of aldehyde and ketone, are subjected to aldol condensation, four products are obtained as:

(a) *Between two aldehydes:* **Acetaldehyde and** propanaldehyde:

$$CH_3CHO + CH_3CH_2CHO \xrightarrow{^-OH}$$

$$\underset{\displaystyle OH}{CH_3\underset{|}{C}HCH_2CHO} + CH_3CH_2-\underset{\displaystyle OH}{\underset{|}{C}H}-\overset{\displaystyle CH_3}{\overset{|}{C}H}-CHO$$

Self-condensation products

$$CH_3-\underset{\displaystyle OH}{\underset{|}{C}H}-\overset{\displaystyle CH_3}{\overset{|}{C}H}-CHO + CH_3CH_2\underset{\displaystyle OH}{\underset{|}{C}H}-CH_2CHO$$

Cross-condensation products

(b) *Between two ketones*

$$CH_3-\overset{\displaystyle O}{\overset{\|}{C}}-CH_3 + CH_3CH_2-\overset{\displaystyle O}{\overset{\|}{C}}-CH_2CH_3 \xrightarrow{^-OH}$$

$$CH_3-\underset{\displaystyle CH_3}{\overset{\displaystyle OH}{C}}-CH_2-\overset{\displaystyle O}{\overset{\|}{C}}-CH_3 + CH_3-\underset{\displaystyle CH_3}{\overset{\displaystyle OH}{C}}-\underset{\displaystyle CH_3}{CH}-\overset{\displaystyle O}{\overset{\|}{C}}-CH_2CH_3$$

$$CH_3CH_2-\underset{\displaystyle CH_2CH_3}{\overset{\displaystyle OH}{C}}-\overset{\displaystyle CH_3}{CH}-\overset{\displaystyle O}{\overset{\|}{C}}-CH_2CH_3 + CH_3CH_2-\underset{\displaystyle CH_2CH_3}{\overset{\displaystyle OH}{C}}-CH_2-\overset{\displaystyle O}{\overset{\|}{C}}-CH_3$$

If unsymmetrical ketones are condensed, further more aldol products are obtained.

(c) *Between an aldehyde and a ketone*

$$CH_3-\overset{\displaystyle O}{\overset{\|}{C}}-H + CH_3-\overset{\displaystyle O}{\overset{\|}{C}}-CH_3 \xrightarrow{^-OH}$$

$$CH_3-\overset{\displaystyle OH}{CH}-CH_2-\overset{\displaystyle O}{\overset{\|}{C}}-CH_3 + CH_3-\underset{\displaystyle CH_3}{\overset{\displaystyle OH}{C}}-CH_2CHO$$

$$\underset{\displaystyle OH}{CH_3\underset{|}{C}H}-CH_2CHO + CH_3-\underset{\displaystyle OH}{\overset{\displaystyle CH_3}{C}}-CH_2-\underset{\displaystyle O}{\overset{\displaystyle CH_3}{C}}-CH_3$$

(d) *Between an aldehyde with no α-H and a carbonyl compound with α-H*

$$CH_3CHO + HCHO \xrightarrow{^-OH} \underset{\displaystyle OH}{CH_3\underset{|}{C}HCH_2CHO} + H-\underset{\displaystyle OH}{\overset{\displaystyle H}{C}}-CH_2CHO$$

If excess of formaldehyde is used:

$$CH_3CHO + HCHO \xrightarrow{^-OH} \overset{\displaystyle CHO}{\overset{|}{C}H_2}-CH_2OH \xrightarrow[H_2CO]{^-OH}$$

$$H-\overset{\displaystyle O}{\overset{\|}{C}}-\underset{\displaystyle CH_2OH}{\overset{\displaystyle CH_2OH}{C}}-CH_2OH \xleftarrow[CH_2O]{^-OH} H-\underset{\displaystyle CH_2OH}{\overset{\displaystyle CHO}{C}}-CH_2OH$$

Example 10. Synthesize 2-methylenecyclohexanone (cyclohexanone with $=CH_2$ at C-2) via aldol condensation:

Solution

Cyclohexanone $+ \ ^-OH \rightleftharpoons$ enolate (carbanion $\longleftrightarrow$ O^- form); $\xrightarrow{CH_2O}$ 2-(CH_2-O^-)cyclohexanone $\xrightarrow{OH_2}$ 2-(CH_2OH)cyclohexanone $\xrightarrow{\Delta}$ 2-methylenecyclohexanone ($=CH_2$)

Intramolecular cyclization via aldol condensation: Aldol condensation offers a convenient means to synthesize five, six or sometimes larger ring via intramolecular aldol condensation as:

$$CH_3-\overset{\displaystyle O}{\overset{\|}{C}}-CH_2CH_2CH_2CH_2-\overset{\displaystyle O}{\overset{\|}{C}}-H$$

$$\downarrow ^-OH$$

$$CH_3-\overset{\displaystyle O}{\overset{\|}{C}}-\overset{-}{C}H-CH_2CH_2CH_2-\underset{H}{C}=O$$

$$\downarrow$$

cyclopentenyl$-\overset{\displaystyle O}{\overset{\|}{C}}-CH_3 \xleftarrow{\Delta}$ (2-hydroxycyclopentyl)$-\overset{\displaystyle O}{\overset{\|}{C}}-CH_3 \xleftarrow{H_2O}$ (2-oxidocyclopentyl, ^-O)$-\overset{\displaystyle O}{\overset{\|}{C}}-CH_3$

Example 11. Synthesize each of the following compounds starting from carbonyl compounds.

(a) 3-methylcyclopent-2-enone (ring with $C=O$, CH_3) (b) cyclohexenone (ring with $=O$) (c) 1-acetyl-2-methylcyclopentene (CH_3, $-\overset{\displaystyle O}{\overset{\|}{C}}-CH_3$)

Solution

(a) 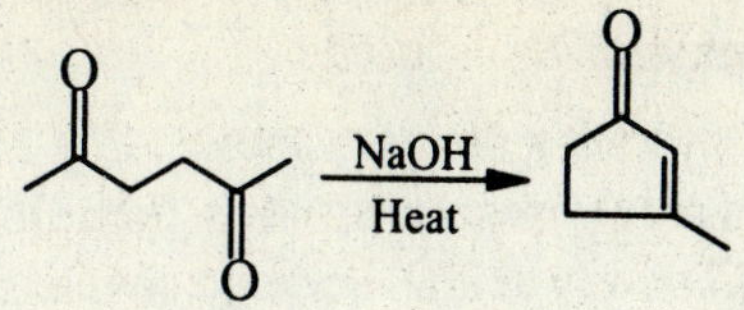 NaOH / Heat

(b) 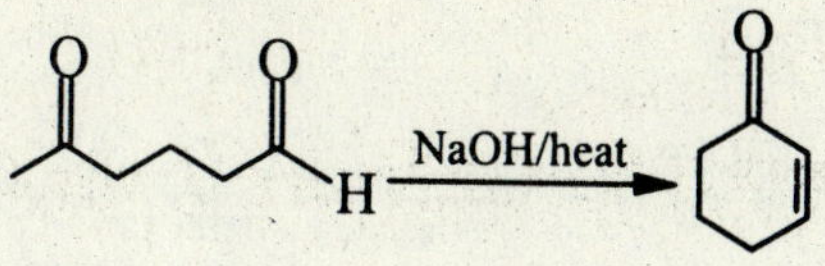 NaOH/heat

(c) 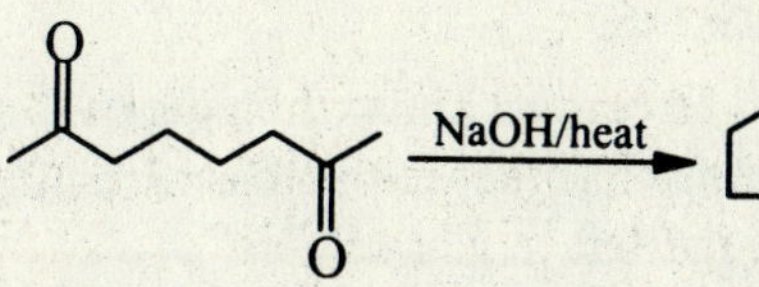 NaOH/heat

Example 12. Write all possible aldol products obtained by intramolecular cyclization of $CH_3-\overset{O}{\overset{\|}{C}}-(CH_2)_4-CHO$ assuming dehydration occurs in all instances.

Solution

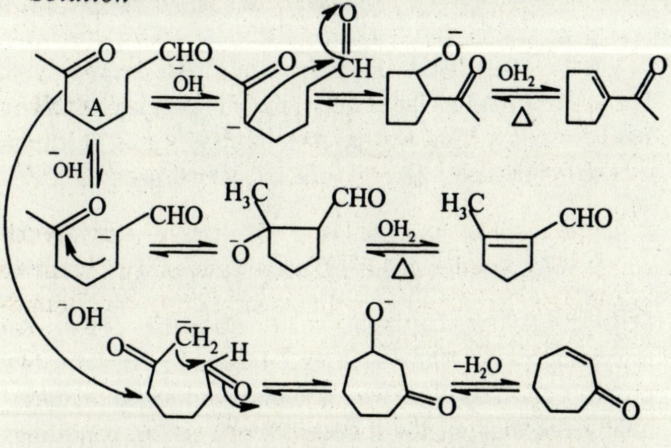

Example 13. Predict products in the following reactions:

(a) 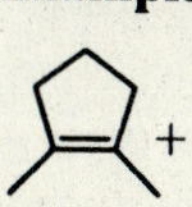 + cold, dil, alk $KMnO_4 \longrightarrow A \xrightarrow{HIO_4} B \xrightarrow{\bar{O}H} C$

(b) $A + OsO_4 \xrightarrow{H_2O_2} B \xrightarrow{HIO_4} C \xrightarrow{\bar{O}H}$

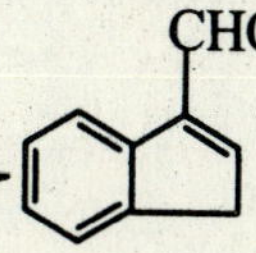

(c) 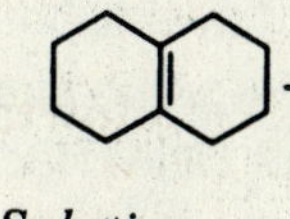 $\xrightarrow[Zn\text{-}H_2O]{O_3} A \xrightarrow[\Delta]{\bar{O}H} B$

Solution

(a) 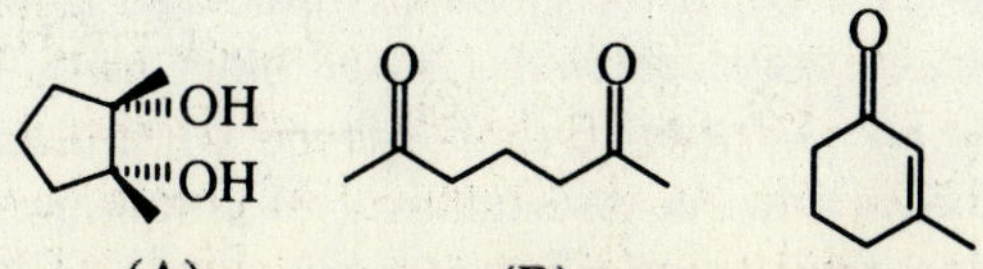 (A) (B) (C)

(b) 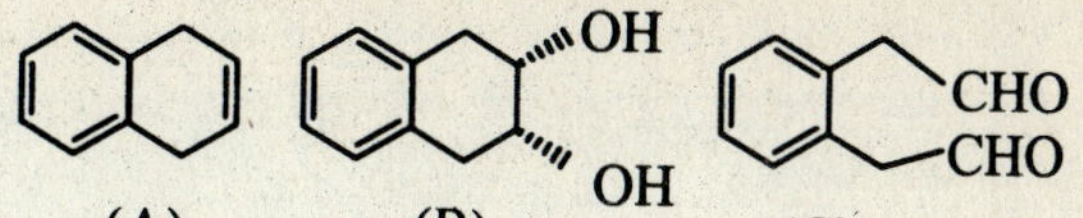 (A) (B) (C)

(c) 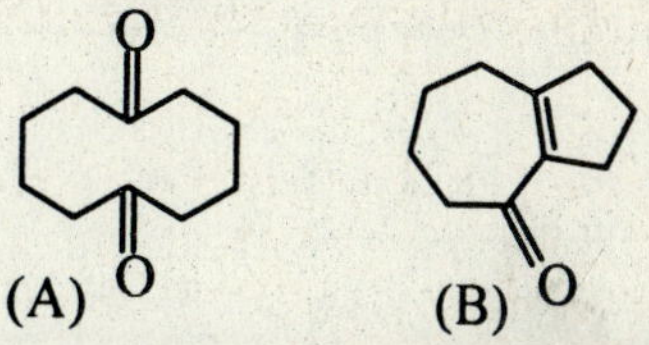 (A) (B)

Example 14. An organic compound $A(C_6H_{10})$ on catalytic hydrogenation with H_2/Pt yields C_6H_{12}. A on ozonolysis followed by work-up with $Zn\text{-}H_2O$ yields B $(C_6H_{10}O_2)$. B on refluxing with dilute NaOH solution gives $C(C_6H_8O)$. C on catalytic reduction yields $D(C_6H_{12}O)$. D on treatment with chromic acid yields E $(C_6H_{10}O_2)$, which on heating with soda lime yields $F(C_5H_{10})$. F on monochlorination yields a single isomer. Identify A to F.

Solution: A, $H-\overset{O}{\overset{\|}{C}}-(CH_2)_4-CHO$ (B)

C (CHO), D (CH_2OH), E (COOH), F

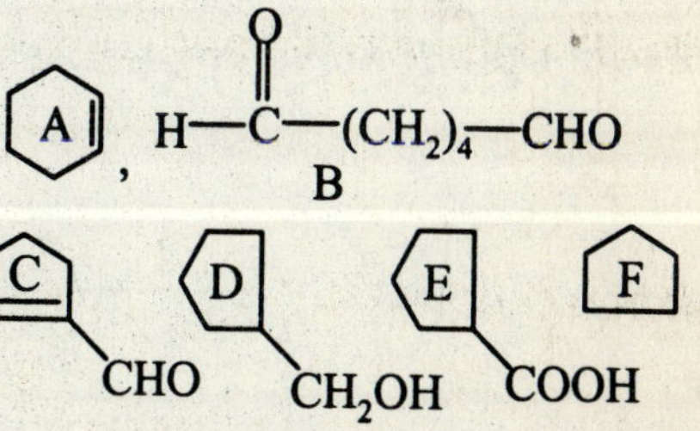

Example 15. Synthesize $CH_3CH_2CH=\underset{CH_3}{\underset{|}{C}}-CHO$ starting from ethyl bromide and any other inorganic reagents.

Solution

$$CH_3CH_2-Br \xrightarrow{aq\ KCN} CH_3CH_2CN \xrightarrow[Conc.\ HCl]{SnCl_2} CH_3CH_2CHO$$

$$\xrightarrow{OH^-} CH_3CH_2-\overset{OH}{\overset{|}{CH}}-\underset{CH_3}{\underset{|}{CH}}-CHO \xrightarrow{\Delta} CH_3CH_2-CH=\underset{CH_3}{\underset{|}{C}}-CHO$$

XV. Claisen-Schmidt reaction

When ketones are used as one component in cross-aldol condensation with a carbonyl compound lacking an α -H, it is called **Claisen Schmidt reaction**, e.g.

$$C_6H_5-\overset{O}{\overset{\|}{C}}-H + CH_3-\overset{O}{\overset{\|}{C}}-CH_3 \xrightarrow[100°\ C]{\bar{O}H} C_6H_5-CH=CH-\overset{O}{\overset{\|}{C}}-CH_3$$

(Benzalacetone)

These reactions are successful since ketones, because of its very low reactivity, do not undergo self-condensation. Other examples are:

$$C_6H_5-\overset{O}{\overset{\|}{C}}-H + CH_3-\overset{O}{\overset{\|}{C}}-C_6H_5 \xrightarrow{\bar{O}H} C_6H_5-CH=CH-\overset{O}{\overset{\|}{C}}-C_6H_5$$

(Benzalacetophenone)

In the Claisen Schmidt reaction, dehydration occurs readily because the double bond, that forms is in conjugation with benzene ring.

XVI. Condensation of aldehydes and ketones with nitroalkane containing an α-H

The α-H of nitroalkanes are appreciably acidic, much more acidic than the α-H of aldehydes and ketones. The acidity of α-H of a nitro alkane can be explained by the powerful electron withdrawing effect of nitro group as:

$$\underset{H}{CH_2NO_2} + \bar{O}H \rightarrow \bar{C}H_2-NO_2 \xrightarrow{R_2C=O} R-\underset{R}{\overset{O^-}{\overset{|}{C}}}-CH_2NO_2 \underset{H_2O/\Delta}{\rightleftharpoons} R-\underset{R}{\overset{|}{C}}=CH-NO_2$$

$$\bar{C}H_2-NO_2 \leftrightarrow H_2C=\overset{+}{N}\begin{smallmatrix}O^-\\O^-\end{smallmatrix}$$

Example 16. Bring about the following conversion:

(a) $C_6H_5CHO \longrightarrow C_6H_5CH_2CH_2NH_2$

(b) $HCHO \longrightarrow CH_3-\underset{CH_3}{\underset{|}{C}H}-NH_2$

Solution

(a) $$C_6H_5CHO + CH_3NO_2 \xrightarrow{\bar{O}H} C_6H_5CH=CH-NO_2 \xrightarrow{H_2/Ni} C_6H_5CH_2CH_2NH_2$$

(b) $$H-\overset{O}{\overset{\|}{C}}-H + CH_2-NO_2 \text{ (}CH_3\text{)} \xrightarrow{\bar{O}H} CH_2=\overset{CH_3}{\overset{|}{C}}-NO_2 \xrightarrow{H_2/Ni} CH_3-\overset{CH_3}{\overset{|}{C}H}-NH_2$$

XVII. Condensation of an aldehydes and ketones with cyanide

The α-hydrogen of cyanides are also appreciably acidic, but less than those of aldehydes and ketones. Some nitriles whose α-H is comparatively acidic, undergo condensation of aldol type, viz.

$$C_6H_5-\overset{O}{\overset{\|}{C}}-H + C_6H_5-CH_2-CN \xrightarrow[EtOH]{Et\bar{O}} C_6H_5CH=\underset{C_6H_5}{\underset{|}{C}}-CN$$

XVIII. Lithium enolate

The extent to which an enolate anion form depends on the strength of the base used. If the weaker base is employed, equilibrium lies to the left, e.g. with OH^-

$$CH_3-\overset{O}{\overset{\|}{C}}-CH_3 + \bar{O}H \rightleftharpoons \left[CH_3-\overset{O}{\overset{\|}{C}}-\bar{C}H_2 \longleftrightarrow CH_3-\overset{O^-}{\overset{|}{C}}=CH_2\right]$$

On the other hand, if a very strong base is employed, the equilibrium lies far to the right. One very useful strong base for converting ketone into enolate ion is "Lithium diisopropylamide." $(iC_3H_7)_2\,N^-Li^+$ abbreviated as LDA:

$$CH_3-\overset{O}{\overset{\|}{C}}-CH_3 \xrightarrow{LDA} CH_3-\overset{O}{\overset{\|}{C}}-\bar{C}H_2$$

Regioselective formation of enolate anion: An unsymmetrical ketone such as 2-butanone can form two possible enolates. Just which enolate is formed predominantly depends on the base used and on the conditions employed.

$$CH_3-CH_2-\overset{O}{\overset{\|}{C}}-CH_3 \xrightarrow{\bar{B}} CH_3CH_2-\overset{O}{\overset{\|}{C}}-\bar{C}H_2 \longleftrightarrow CH_3CH_2-\overset{O^-}{\overset{|}{C}}=CH_2 \quad \text{I}$$

$$\xrightarrow{\bar{B}} CH_3-\bar{C}H-\overset{O}{\overset{\|}{C}}-CH_3 \longleftrightarrow CH_3-CH=\overset{O^-}{\overset{|}{C}}-CH_3 \quad \text{II}$$

The enolate with more highly substituted double bond (II), is thermodynamically more stable enolate just as more substituted alkene is more stable. This enolate called "thermodynamic enolate" is formed predominantly under conditions that permit establishment of equilibrium. This will generally be the case, if the enolate is produced using a weaker base, in protic solvent. On the other hand, the enolate with less-substituted double bond (I) is usually formed at faster rate because removal of proton necessary to produce enolate ion is less sterically hindered. The enolate, called kinetic enolate, is formed predominantly when the reactions is kinetically controlled. The kinetically favoured enolate can be formed cleanly through the use of a very strong base, viz. LDA. This strong, sterically hindered base rapidly removes proton from less sterically hindered α-carbon, e.g.

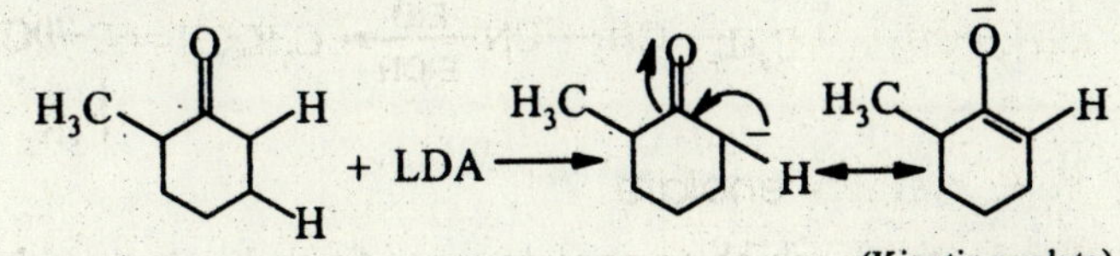

(Kinetic enolate)

Therefore, lithium enolate can be used in targeted aldol reaction as:

$$CH_3-\overset{O}{\overset{\|}{C}}-CH_2CH_3 + LDA \longrightarrow \bar{C}H_2-\overset{O}{\overset{\|}{C}}-CH_2CH_3$$

If very low temperature is used kinetic enolate predominate (Predominant)

$$\xrightarrow{CH_3CHO} CH_3-\underset{OH}{\underset{|}{CH}}-CH_2-\overset{O}{\overset{\|}{C}}-CH_2CH_3$$

Example 17. Starting with aldehydes and ketones of your choice, synthesize the following compounds using Lithium enolates.

(a) 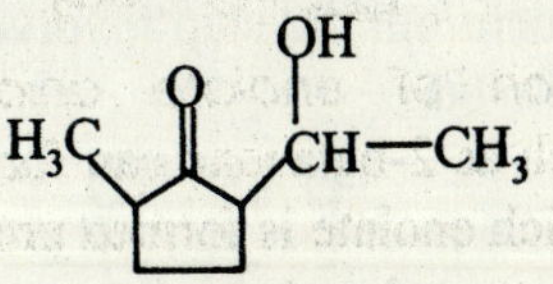 2-methyl-5-(1-hydroxyethyl)cyclopentanone (H_3C, $\overset{OH}{CH}-CH_3$)

(b) $CH_3-\underset{CH_3}{\underset{|}{CH}}-\overset{O}{\overset{\|}{C}}-CH_2-\overset{OH}{\overset{|}{CH}}-CH_2CH_3$

(c) $CH_3CH_2-\overset{O}{\overset{\|}{C}}-CH_2\overset{OH}{\overset{|}{CH}}-C_6H_5$

(d) $CH_3CH{=}CH-\overset{O}{\overset{\|}{C}}-CH_2-\overset{OH}{\overset{|}{CH}}-CH_3$

Solution

(a) 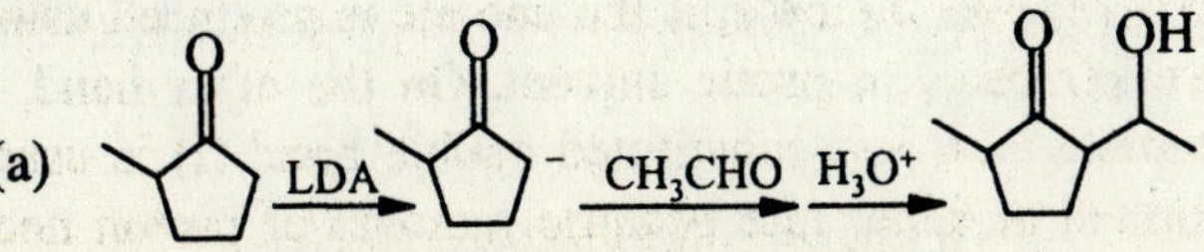 2-methylcyclopentanone $\xrightarrow{LDA}$ enolate $\xrightarrow{CH_3CHO} \xrightarrow{H_3O^+}$ aldol product

(b) 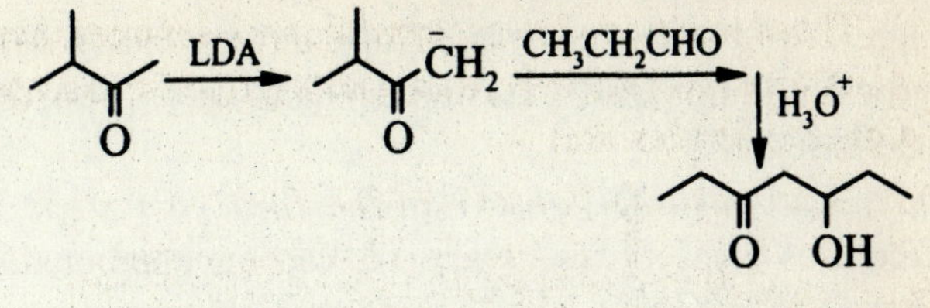 3-methyl-2-butanone $\xrightarrow{LDA}$ $\bar{C}H_2$ enolate $\xrightarrow{CH_3CH_2CHO}$ $\xrightarrow{H_3O^+}$ β-hydroxy ketone (O, OH)

(c) 2-butanone $\xrightarrow{LDA}$ $\bar{C}H_2$ enolate $\xrightarrow{PhCHO} \xrightarrow{H_3O^+}$ product (O, Ph, OH)

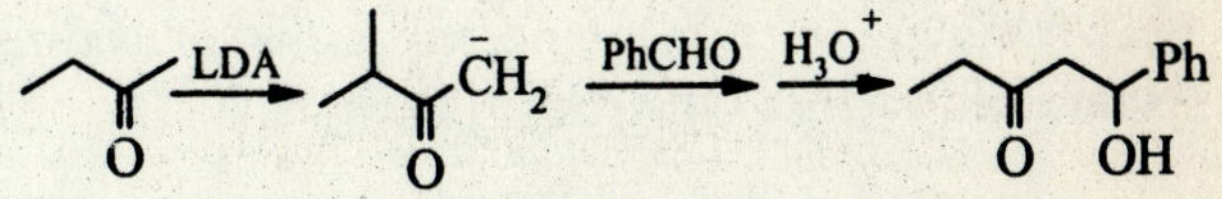

(d) 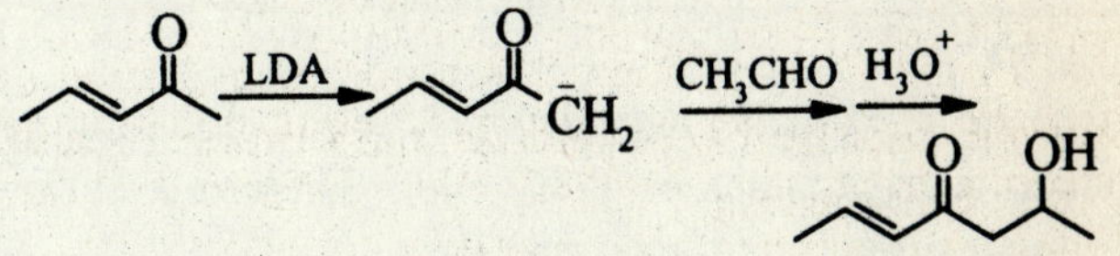 pent-3-en-2-one $\xrightarrow{LDA}$ $\bar{C}H_2$ enolate $\xrightarrow{CH_3CHO} \xrightarrow{H_3O^+}$ product (O, OH)

XIX. Cannizzaro reaction

Aldehydes that have no α-H atoms undergo Cannizzaro reactions, in which two molecules of aldehydes are involved, one molecule being converted into the corresponding alcohol and other into acid, therefore disproportionation takes place. The usual reagent for bringing about Cannizzaro reaction is 50% aqueous or ethanol solution of alkali, e.g.

$$HCHO + NaOH \rightarrow HCOONa + CH_3OH$$

Mechanism

The mechanism of Cannizzaro reaction involves hydride transfer:

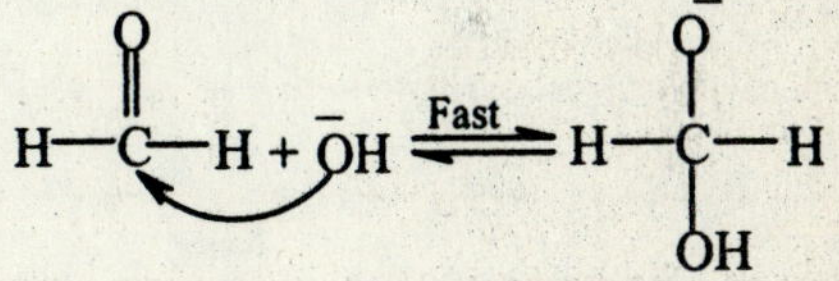

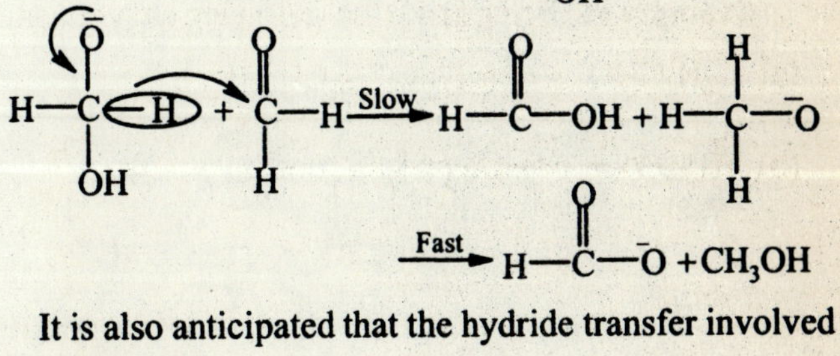

$$H-\overset{O}{\overset{\|}{C}}-H + \bar{O}H \xrightleftharpoons{Fast} H-\underset{OH}{\overset{\bar{O}}{C}}-H$$

$$H-\underset{OH}{\overset{\bar{O}}{C}}-H + \underset{H}{\overset{O}{C}}-H \xrightarrow{Slow} H-\overset{O}{\overset{\|}{C}}-OH + H-\underset{H}{\overset{H}{C}}-\bar{O}$$

$$\xrightarrow{Fast} H-\overset{O}{\overset{\|}{C}}-\bar{O} + CH_3OH$$

It is also anticipated that the hydride transfer involved dioxyanion as:

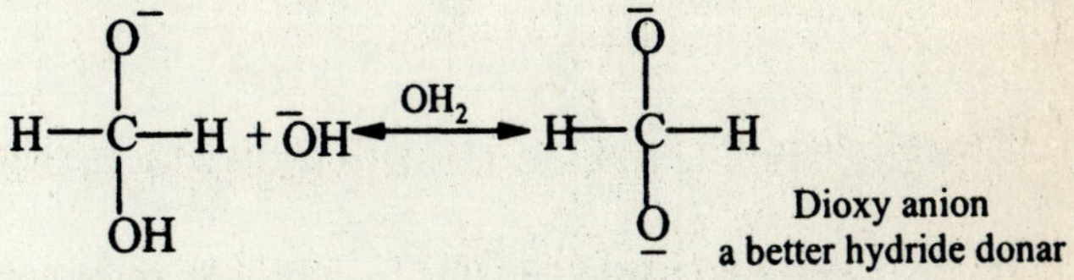

$$H-\underset{OH}{\overset{\bar{O}}{C}}-H + \bar{O}H \xrightleftharpoons{OH_2} H-\underset{\bar{O}}{\overset{\bar{O}}{C}}-H$$

Dioxy anion a better hydride donar

Other examples of Cannizzaro reactions are:

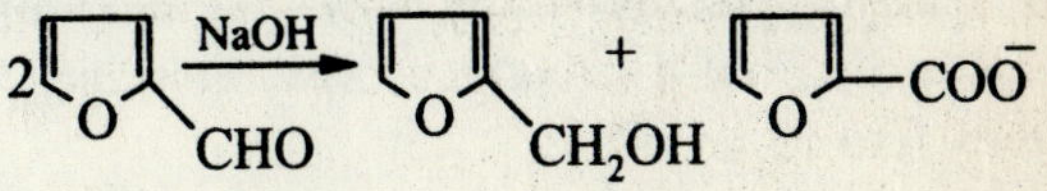

2 furfural (CHO) $\xrightarrow{NaOH}$ furfuryl alcohol (CH_2OH) + furoate ($CO\bar{O}$)

$$2C_6H_5CHO \xrightarrow{NaOH} C_6H_5CH_2OH + C_6H_5CO\bar{O}$$

Reactivity consideration: The reactivity of an aldehyde towards Cannizzaro reaction depends on two important factors.

(i) *Ease of attack of OH^- in the first, fast step:* This requires a less sterically hindered carbonyl carbon. Also, the aldehydes involved in the 1st step is oxidized, formaldehyde being least sterically hindered, is always oxidized to formic acid in cross-Cannizzaro reaction:

$$C_6H_5CHO + HCOH \xrightarrow{\bar{O}H} C_6H_5CH_2OH + HCO\bar{O}$$

(ii) *Ease of hydride transfer:* Electron donating group facilitate the attack of hydride ion. Therefore, in the following cross Cannizzaro reaction:

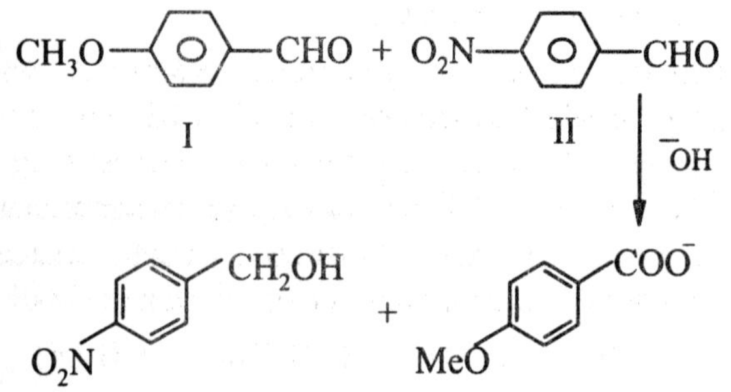

$CH_3O-C_6H_4-CHO$ (I) + $O_2N-C_6H_4-CHO$ (II) $\xrightarrow{\bar{O}H}$ $O_2N-C_6H_4-CH_2OH$ + $MeO-C_6H_4-CO\bar{O}$

Mechanism

$CH_3-O-C_6H_4-C(=O)-H + \bar{O}H \xrightarrow{Fast} CH_3-O-C_6H_4-C(O^-)(OH)-H$

$\xrightarrow[CH_3O]{+R} C_6H_4-C(O^-)(OH)-H + H-C(=O)-C_6H_4-NO_2 \xrightarrow{Slow} CH_3-O-C_6H_4-C(=O)-OH + \bar{O}CH_2-C_6H_4-NO_2$

$\xrightarrow{Fast} CH_3O-C_6H_4-C(=O)-\bar{O} + HOCH_2-C_6H_4-NO_2$

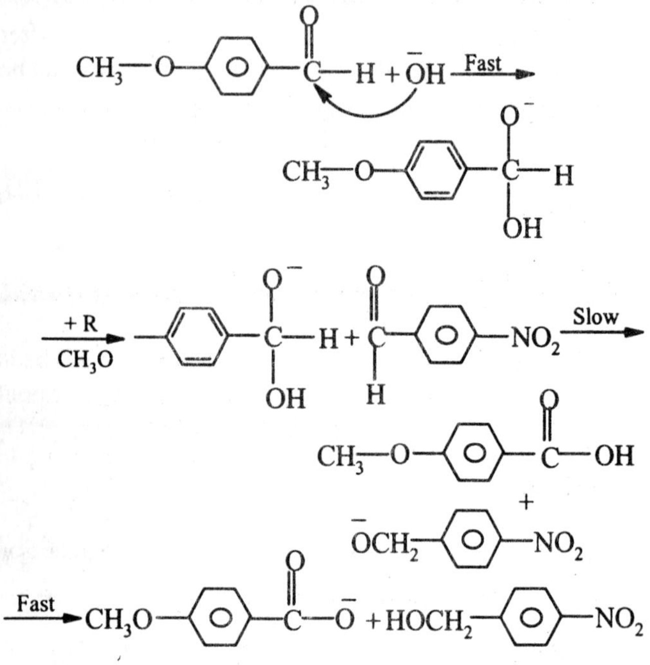

Here, electron donation by CH_3O—through resonance facilitate the release of H^- ion and electron withdrawal by —NO_2 though resonance increases the electropositive character of carbonyl carbon at para position, thus increases affinity for H^- ion.

Example 18. Predict products in the following Cannizzaro reactions:

(a) 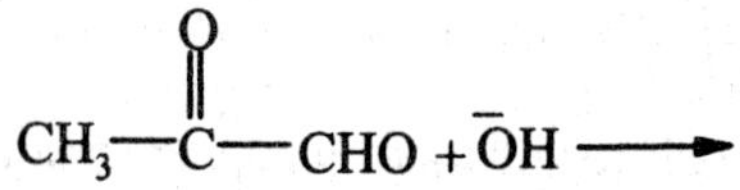 $CH_3-C(=O)-CHO + \bar{O}H \longrightarrow$

(b) 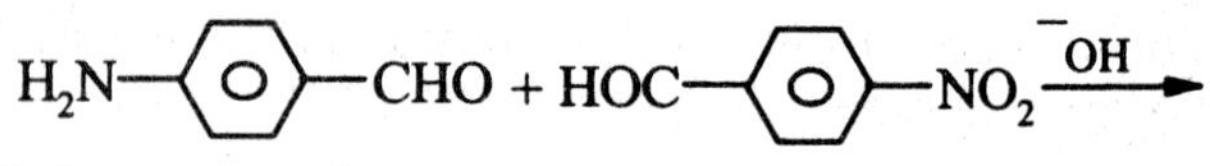 $H_2N-C_6H_4-CHO + HOC-C_6H_4-NO_2 \xrightarrow{\bar{O}H}$

Only cross product

(c) 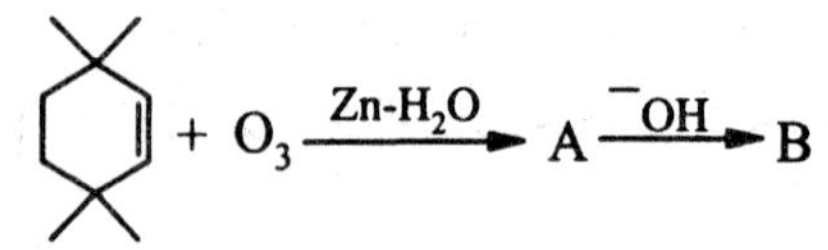 (1,1,4,4-tetramethylcyclohexene) $+ O_3 \xrightarrow{Zn\text{-}H_2O} A \xrightarrow{\bar{O}H} B$

Solution

(a) 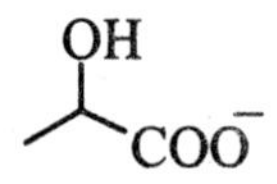 $CH_3-CH(OH)-CO\bar{O}$

(b) 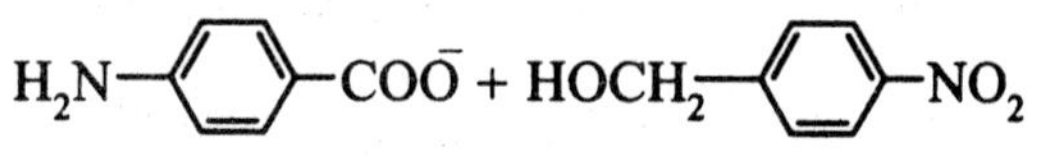 $H_2N-C_6H_4-CO\bar{O} + HOCH_2-C_6H_4-NO_2$

(c) A: $OHC-C(CH_3)_2-CH_2CH_2-C(CH_3)_2-CHO$; B: $\bar{O}OC-C(CH_3)_2-CH_2CH_2-C(CH_3)_2-CH_2OH$

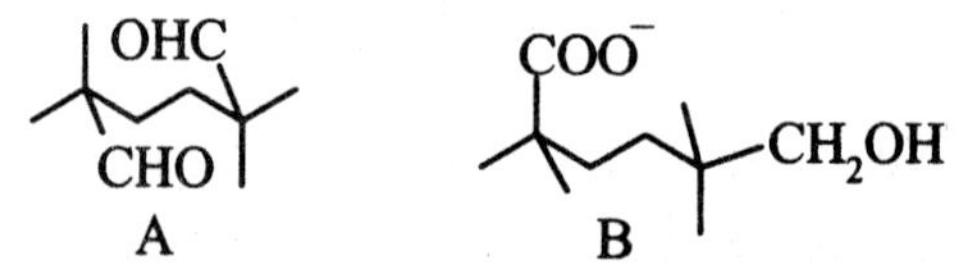

(iii) When reactions is run in D_2O, the recovered alcohol contained no α-deuterium, indicating that the hydrogen comes form the reagent and not from the medium.

(iv) *Kinetics of Cannizzaro reaction*: As mentioned in the mechanism slow, rate determining step involves H^- transfer from one molecule to other:

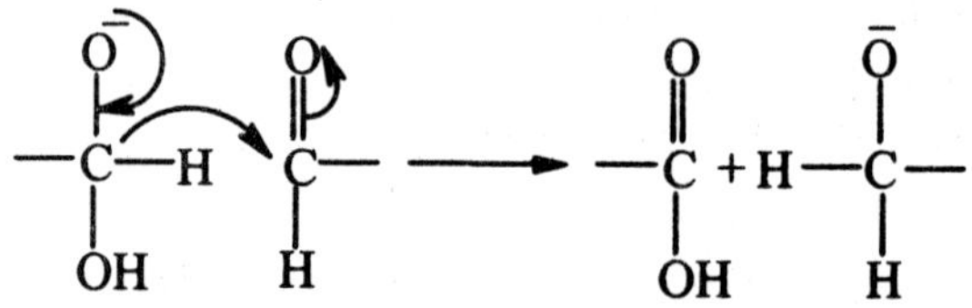

$-C(O^-)(OH)-H + C(=O)(H)- \longrightarrow -C(=O)(OH) + H-C(\bar{O})(H)-$

Therefore, reaction is seconds order in substrate. If the reaction is carried out in dilute alkaline medium, order with respect to base is one and it is two in concentrated medium as:

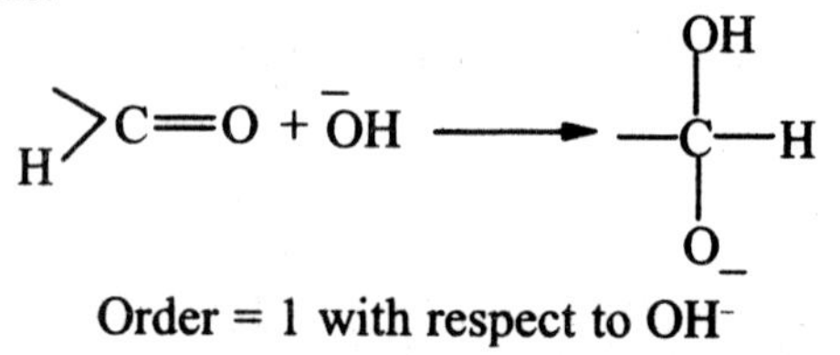

$H>C=O + \bar{O}H \longrightarrow -C(OH)(O^-)-H$

Order = 1 with respect to OH^-

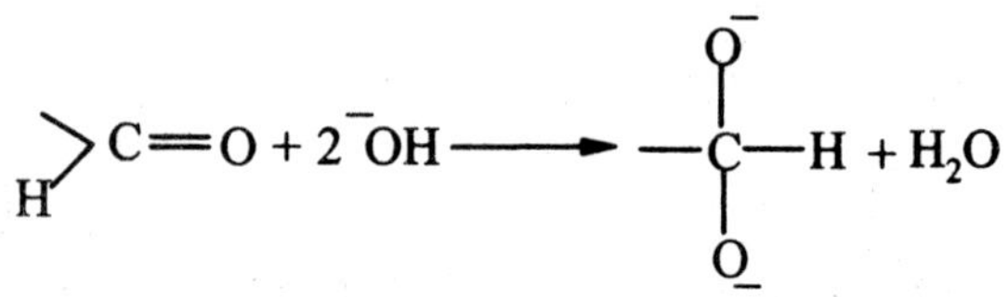

$H>C=O + 2\bar{O}H \longrightarrow -C(O^-)(O^-)-H + H_2O$

Order = 2 with respect to OH^-

XX. Tischenko reaction

All aldehydes can be made to undergo Cannizzaro reaction by treatment with aluminium ethoxide. Under these conditions, acid and alcohols are combined as ester, e.g. acetaldehyde gives ethyl acetate:

$$2CH_3CHO + Al(OEt)_3 \longrightarrow CH_3COOC_2H_5$$

XXI. Darzen glycidic ester condensation

When an α-halo ester is treated with an aldehydes or ketones in presence of a strong base, condensation occurs to produce an α, β-epoxy ester as:

$$ClCH_2COOEt + C_2H_5O^- \longrightarrow \bar{C}H(Cl)-COOEt \xrightarrow{>C=O} -C(O^-)-CH(Cl)-COOEt \longrightarrow -C(-O-)CH-COOEt$$

XXII. The Wittig's reaction

Formation of alkene from carbonyl compounds: Carbonyl compounds when allowed to react with phosphorous ylides, an alkene is produced as:

$$>C=O + Ph_3P=C< \longrightarrow >C=C< + POPh_3$$

Formation of Ylide: Ylides are prepared by the action of *tri*-phenyl phosphine with an alkyl halide containing at least one α-H, in presence of a strong base, e.g.

$$Ph_3P: + CH_2(Br)CH_3 \longrightarrow Ph_3\overset{+}{P}-CH(H)CH_3 \xrightarrow{PhLi} Ph_3P=CHCH_3 + PhH$$

Mechanism of Wittig's reaction

$$CH_3-CH(=PPh_3) + R-C(=O)-R \longrightarrow CH_3-CH(Ph_3P^+)-C(R)(O^-)-R \longrightarrow$$

$$CH_3-CH(Ph_3P)-C(R)(O)-R \longrightarrow POPh_3 + CH_3CH=C(R)-R$$

Example 19. Write products in the following reactions:

(a) $Ph-C(=O)-CH_2-CH_2-CH_2CH_2Cl \xrightarrow[BuLi]{PPh_3}$

(b) 3-(chloromethyl)cyclopentanecarbaldehyde (OHC, CH_2Cl) $\xrightarrow[BuLi]{PPh_3} A \xrightarrow[(ii)\ 50\%\ aq\ NaOH]{(i)\ O_3/Zn-H_2O} B$

Solution

(a) 1-phenylcyclopentene (Ph) (b) bicyclic alkene (A); $\bar{O}OC$–cyclopentane–CH_2OH (B)

XXIII. Reaction with alcohols

Aldehydes and ketones reacts with alcohols, in acidic medium forming acetals and ketals respectively.

$$R-CHO + 2R'OH \xrightarrow{H^+} R-CH<^{OR'}_{OR'} \quad \text{Acetal}$$

$$\underset{R}{\overset{R}{>}}C=O + 2R'OH \xrightarrow{H^+} \underset{R}{\overset{R}{>}}C<^{OR'}_{OR'} \quad \text{Ketal}$$

Both acetals and ketals have ether linkages, which are almost inert to most of the reagents, and thus, this reaction can be used for protection of aldehydic and ketonic functional groups in a polyfunctional compounds. After completion of synthesis, the carbonyl compounds can be recovered by refluxing in dilute acidic medium, viz:

$$H-C(=O)-CH_2CH_2-NO_2 \xrightarrow{R'OH/H^+} H-C(OR')(OR')-CH_2CH_2-NO_2$$

$$\xrightarrow{LiAlH_4} H-C(OR')(OR')-CH_2CH_2NH_2 \xrightarrow[Reflux]{H^+} OHC-CH_2CH_2NH_2$$

In the above reaction, had the carbonyl group not been protected before reduction, the strong reducing agent ($LiAlH_4$) would have reduced both –CHO and $-NO_2$ groups.

Mechanism

Mechanism of this reaction involves nucleophilic addition of alcohol on carbonyl carbon as:

$$R-C(=O)-H \overset{H^+}{\rightleftharpoons} R-C(=\overset{+}{O}-H)-H + H-\ddot{O}-R' \rightleftharpoons$$

$$R-C(OH)(H)-\overset{+}{O}(H)-R' \overset{-H^+}{\rightleftharpoons} R-C(OH)(H)-OR'$$

(Hemiacetal)

Hemiacetal $\xrightarrow{H^+}$ $R-\overset{\overset{+}{O}H_2}{\underset{H}{C}}-OR' \underset{}{\overset{-OH_2}{\rightleftharpoons}} R-\overset{+}{\underset{H}{C}}-\ddot{O}:-R' \rightleftharpoons R-\underset{H}{C}=\overset{+}{O}-R'$

$\xrightarrow{HOR'}$ $R-\overset{H}{C}(-OR')-\overset{+}{O}(H)-R' \underset{}{\overset{-H^+}{\rightleftharpoons}} R-\overset{H}{\underset{OR'}{C}}-OR'$ Acetal

Similarly if, ketone is used, hemiketal is formed in the first step and finally it is converted into ketal. Since the mechanism involves nucleophilic attack of alcohol at carbonyl carbon, bulky substituent at carbonyl carbon (as in case of ketones) poses steric hindrance to the attack and lower the reactivity. Therefore, if a compound containing both aldehydic and ketonic functional group is treated with acidic solution of alcohol, acetal formation takes places in preference of ketal as:

3-oxocyclohexanecarbaldehyde (CHO) $\xrightarrow[H^+]{ROH}$ (Acetal) 3-[CH(OR)$_2$]cyclohexanone not (Ketal) 3-CHO-1,1-(OR)$_2$-cyclohexane

A very well known protecting alcohol is ethylene glycol-

$$>C=O + \begin{matrix} HO-CH_2 \\ | \\ HO-CH_2 \end{matrix} \underset{}{\overset{H^+}{\rightleftharpoons}} >C\begin{matrix} O- \\ O- \end{matrix}\!]$$

Example 20. Predict product in the following reactions:

(a) $CH_3-\overset{O}{\overset{\|}{C}}-CH_2CH_2CHO + \underset{\text{Excess}}{CH_3OH} \xrightarrow{H^+} A \xrightarrow{LiAlH_4} B \xrightarrow[\text{Reflux}]{H^+} C$

(b) 2,6-bis(hydroxymethyl)cyclohexanone ($HOCH_2$, CH_2OH) $\xrightarrow{H^+}$

Solution

(a) A: $CH_3COCH_2CH_2CH(OMe)_2$; B: $CH_3CH(OH)CH_2CH_2CH(OMe)_2$; C: $CH_3CH(OH)CH_2CH_2CHO$

(b) (bicyclic internal ketal: cyclohexane with two O bridges)

XXIV. Beckmann's rearrangement

When an oxime is treated with reagents such as PCl_5, H_2SO_4, etc. rearrangement occurs to produce amide. Oxime can exist as two geometrical isomers as:

$$\begin{matrix}R \\ R'\end{matrix}\!>C=N\!\!<^{OH} \quad \text{and} \quad \begin{matrix}R \\ R'\end{matrix}\!>C=N\!\!\diagdown_{OH}$$

The rearrangement is very stereospecific, i.e. a specific stereomer gives a specific amide. The net result is anti-exchange of groups with hydroxide group as:

$$\begin{matrix}R \\ R'\end{matrix}\!>C=N^{\nearrow OH} \xrightarrow{H^+} \underset{\text{Enol}}{\begin{matrix}R \\ HO\end{matrix}\!>C=N^{\nearrow R'}} \longleftrightarrow R-\underset{O}{\underset{\|}{C}}-NHR'$$

$$\begin{matrix}R \\ R'\end{matrix}\!>C=N_{\searrow OH} \xrightarrow{H^+} \underset{\text{Enol}}{\begin{matrix}HO \\ R'\end{matrix}\!>C=N_{\searrow R}} \longleftrightarrow R'-\overset{O}{\overset{\|}{C}}-NHR$$

Mechanism

$$\begin{matrix}R \\ R'\end{matrix}\!>C=N^{\nearrow OH} \xrightarrow{H^+} \begin{matrix}R \\ R'\end{matrix}\!>C=N^{\nearrow \overset{+}{O}H_2}$$

$$\downarrow$$

$$\left[R-\overset{+}{C}=\ddot{N}:-R' \longleftrightarrow R-C\equiv\overset{+}{N}-R'\right]$$

$$\downarrow OH_2$$

$$R-\underset{O}{\underset{\|}{C}}-NHR' \longleftrightarrow \underset{\text{Enol}}{R-\underset{OH}{\underset{|}{C}}=NR'} \xleftarrow{-H^+} R-\underset{^+OH_2}{\underset{|}{C}}=N-R'$$

Example 21. Predict products in the following reactions:

(a) cyclohexanone $=O + H_2NOH \longrightarrow A \xrightarrow{H^+} B$

(b) $C_6H_5-\overset{O}{\overset{\|}{C}}-CH_3 \xrightarrow{NH_2OH} C + D \xrightarrow{H^+} G \xrightarrow{^-OH} H + I$

$C \xrightarrow{H^+} E \xrightarrow{OH^-} C_6H_5NH_2 + F$

(c) pyridine-2-carbaldehyde (CHO) $\xrightarrow{NH_2OH} J + K \xrightarrow{H^+} N \xrightarrow{OH^-} O + P$

$J \xrightarrow{H^+} L \xrightarrow{OH^-}$ 2-aminopyridine (NH_2) $+ M$

Solution

(a) 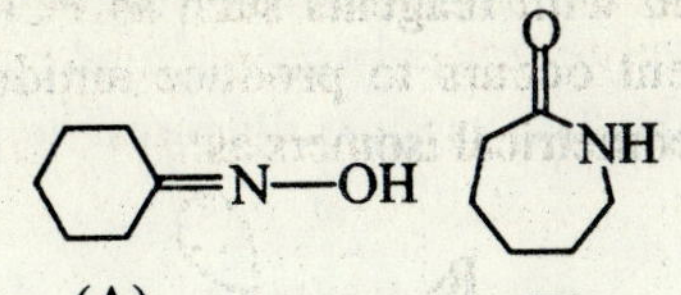 (A) cyclohexanone oxime: $C_6H_{10}{=}N{-}OH$; (B) caprolactam (cyclic amide, C=O, NH)

(b) 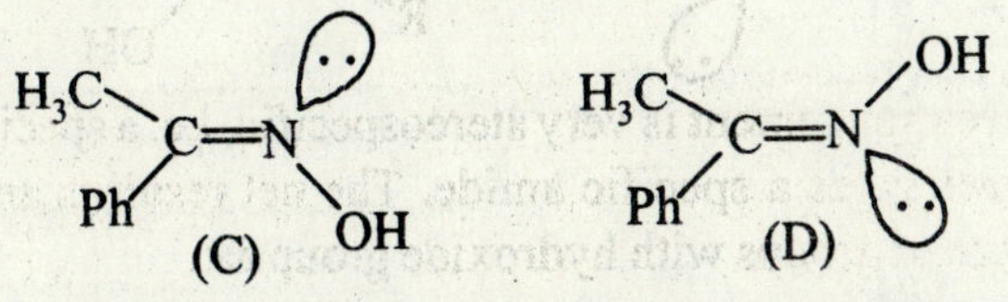 (C) $H_3C(Ph)C{=}N{-}OH$; (D) $H_3C(Ph)C{=}N{-}OH$ (stereoisomer)

$CH_3CONHPh$ (E) $CH_3CO\bar{O}$ (F) $PhCONHCH_3$ (G) $PhCO\bar{O}$ (H) CH_3NH_2 (I)

(c) 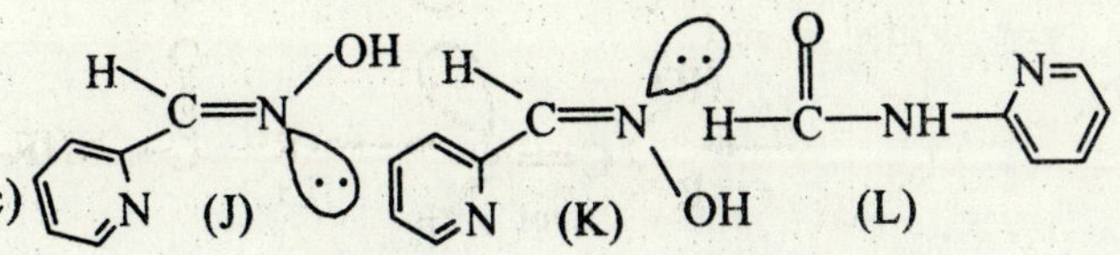 (J) $H(C_5H_4N)C{=}N{-}OH$; (K) $H(C_5H_4N)C{=}N{-}OH$; (L) $H{-}\overset{O}{\overset{\|}{C}}{-}NH{-}C_5H_4N$

$HCO\bar{O}$ (M) $C_5H_4N{-}CONH_2$ (N) $C_5H_4N{-}CO\bar{O}$ (O) 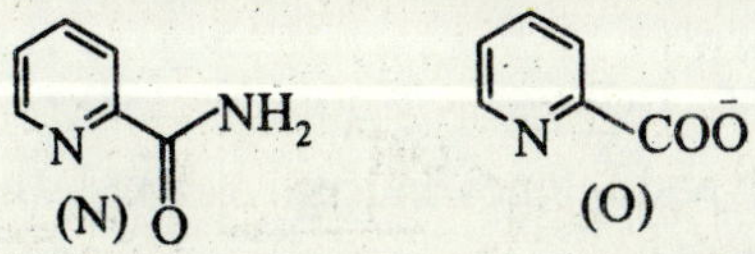 NH_3 (P)

Example 22. An organic compound A has molecular formula $C_{11}H_{14}O$. A on treatment with hydroxyl amine yields two stereomers B and C with their molecular formula $C_{11}H_{15}NO$. B on treatment with concentrated sulfuric acid solution yields D an isomer of B, where C on treatment with concentrated sulfuric acid solution yields E, another isomer of C. D on alkaline hydrolysis yields a white crystalline solid and an optically active amine $F(C_4H_{11}N)$. E on alkaline hydrolysis produces aniline as one product. Identify A to F.

Solution

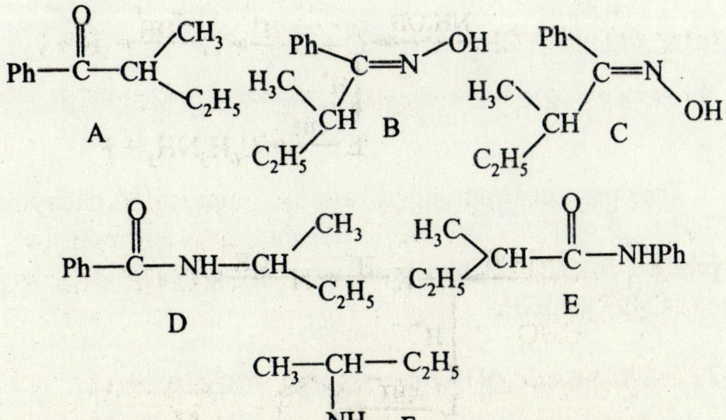

A: $Ph{-}\overset{O}{\overset{\|}{C}}{-}CH(CH_3)(C_2H_5)$

B: $Ph(CH(CH_3)(C_2H_5))C{=}N{-}OH$

C: $Ph(CH(CH_3)(C_2H_5))C{=}N{-}OH$

D: $Ph{-}\overset{O}{\overset{\|}{C}}{-}NH{-}CH(CH_3)(C_2H_5)$

E: $(H_3C)(C_2H_5)CH{-}\overset{O}{\overset{\|}{C}}{-}NHPh$

F: $CH_3{-}CH(NH_2){-}C_2H_5$

XXV. Favorsky rearrangement (Ring contraction)

When α-halo ketones are treated with alkali, ring contraction occurs as:

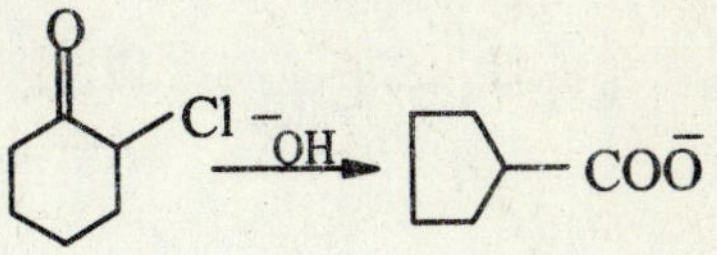

2-chlorocyclohexanone $\xrightarrow{\bar{O}H}$ cyclopentyl$-CO\bar{O}$

Mechanism

2-chlorocyclohexanone + $\bar{O}H$ ⟶ $H_2O + \bar{Cl}$ + bicyclic cyclopropanone

cyclopropanone + $\bar{O}H$ ⟶ ($\bar{O}$, OH adduct) ⟶ cyclopentyl carbanion $-C(=O)-OH$ ⟷ cyclopentyl $-C(=O)-\bar{O}$

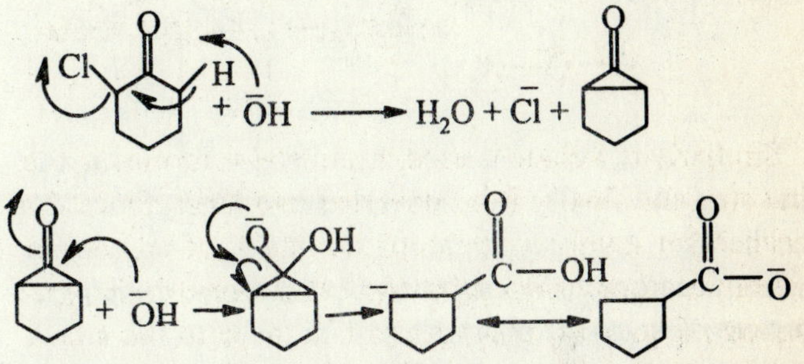

Example 23. When 2-chloro-6-methyl cyclohexanone is treated with NaOH, ring contraction occurs and compound I is formed not II, explain through mechanism.

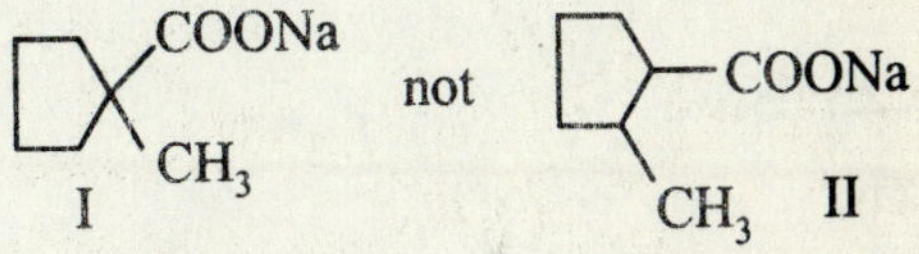

I: 1-methylcyclopentane-1-COONa ($COONa$, CH_3) not II: 2-methylcyclopentane-1-COONa ($COONa$, CH_3)

Solution

2-chloro-6-methylcyclohexanone $\xrightarrow{H\bar{O}}$ carbanion (Cl leaves) ⟶ bicyclic cyclopropanone $\xrightarrow{H\bar{O}}$ (HO, $\bar{O}$ adduct) ⟶

carbanion with $CO\bar{O}^-$ (more stable) + HOOC carbanion (less stable)

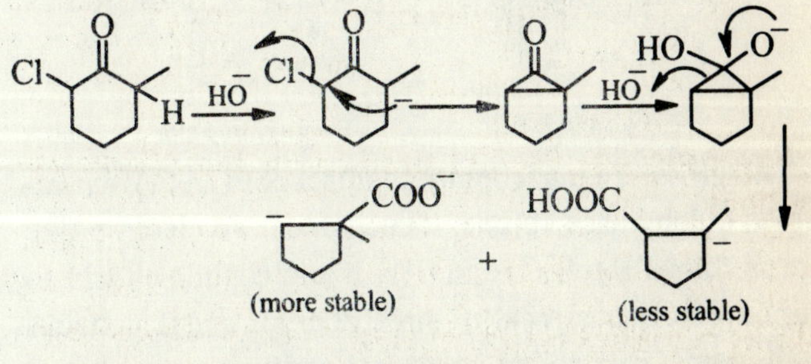

Treatment of acylic α-haloketones with CH_3CH_2ONa also results in Fovorsky rearrangement as:

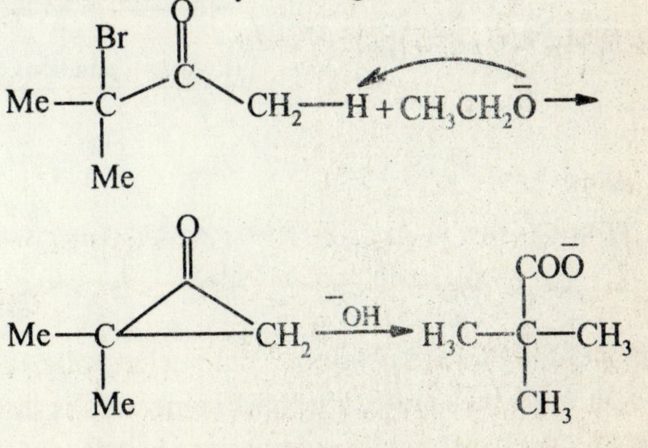

$Me{-}C(Br)(Me){-}\overset{O}{\overset{\|}{C}}{-}CH_2{-}H + CH_3CH_2\bar{O}$ ⟶

cyclopropanone (Me_2C, CH_2) $\xrightarrow{\bar{O}H}$ $H_3C{-}C(CO\bar{O})(CH_3){-}CH_3$

Example 24. Predict product in the following Favorsky rearrangement:

$$Ph-CH(Cl)-C(=O)-CH_2CH_3 \xrightarrow{\bar{O}H}$$

Solution

$$PhCH(Cl)C(=O)CH_2CH_3 \xrightarrow{HO^-} Ph\bar{C}(Cl)C(=O)CH_2CH_3 \xrightarrow{-Cl^-} \text{cyclopropanone (Ph, } CH_3) \xrightarrow{HO^-} PhCH_2CH(CH_3)COO^-$$

XXVI. Nucleophilic addition to α, β-unsaturated aldehydes and ketones

Direct verses conjugate addition: The contributing resonance structures for α, β-unsaturated carbonyl compound, show that the molecule has two electrophilic sites, the carbonyl carbon and the β-carbon:

$$R-CH=CH-C(=O)-R$$

$$\updownarrow$$

$$R-CH=CH\overset{+}{C}(O^-)R \longleftrightarrow R-\overset{+}{C}H-CH=C(O^-)R$$

I (Electrophilic site) II (Electrophilic site)

Therefore, if a nucleophile approaches to an α, β-unsaturated aldehyde or ketone, it encounter two electrophilic site as shown above (I) and (II). If nucleophile attacks to carbonyl carbon, it is called direct addition and if it attacks to β-carbon, it is called *conjugate addition*, e.g.

$$R-CH=CH-C(=O)-R + Y^- \begin{cases} \xrightarrow{\text{(Direct addition)}} I \\ \xrightarrow{\text{(Conjugate addition)}} II \end{cases}$$

$$R-CH=CH-C(O^-)(Y)-R \quad \text{I} \qquad R-CH(Y)-CH=C(O^-)-R \quad \text{II}$$

Whether the product obtained from nucleophilic addition to an α, β-unsaturated carbonyl compound is direct addition product or conjugate addition product depends on the nature of nucleophile. Nucleophiles that are very strong bases, such as organolithium reagents, hydride ion, Grignard reagent, tend to form direct addition products while nucleophiles that are relatively weaker bases such as cyanide ion, amines, thiols and halide ions, usually form conjugate addition product, e.g.

$$C_6H_5-C(=O)-CH=CH_2 \begin{cases} \xrightarrow{C_6H_5Li/H_3\overset{+}{O}} I \\ \xrightarrow[\text{Direct addition product}]{LiAlH_4} II \\ \xrightarrow{\text{aq KCN}} III \end{cases}$$

$$C_6H_5-C(OH)(C_6H_5)-CH=CH_2 \quad \text{I} \qquad C_6H_5-CH(OH)-CH=CH_2 \quad \text{II}$$

Direct addition product

$$C_6H_5-C(=O)-CH_2-CH_2CN$$

Conjugate addition product III

XXVII. The baeyer villiger oxidation

Aldehydes and ketones undergo oxidation on treatment with a peroxyacid as:

$$C_6H_5-C(=O)-CH_3 + R-CO_3H \longrightarrow C_6H_5-O-C(=O)-CH_3$$

Mechanism

$$H_3C-C(=O)-C_6H_5 + H-O-O-C(=O)-R \rightleftharpoons CH_3-C(OH)(C_6H_5)-O-O-C(=O)-R \overset{H^+}{\rightleftharpoons} CH_3-C(OH)(C_6H_5)-O-O-C(=\overset{+}{O}H)-R$$

$$\rightleftharpoons R-COOH + CH_3-\overset{+}{C}(OH)-O-C_6H_5 \overset{-H^+}{\rightleftharpoons} CH_3-C(=O)-OC_6H_5$$

The migratory aptitude of various groups is: H > phenyl > 3° > 2° > 1° > CH_3. Due to this migratory aptitude, the product formed in the above reaction in $CH_3COOC_6H_5$ not $C_6H_5COOCH_3$.

7.7.2 Chemical Analysis for Aldehydes

1. *Tollen's test (Silver Mirror Test):* Aldehydes are **very easily oxidized as compared to ketones. This**

gives a very convenient test that differentiate aldehydes from most of the ketones. A mixture of aqueous solution of silver nitrate in aqueous ammonia is known as Tollen's reagent. The reagent contains Ag $(NH_3)_2^+$ ion. This ion is a mild oxidizing agent, oxidizes aldehydes to carboxylate anion and silver ion itself is reduced to metallic silver. If the rate of reaction is slow and wall of the test tube, in which this test is been carried out is clean, silver deposits on the wall-giving rise to mirror on the wall of test tube as:

$$R-\overset{O}{\overset{\|}{C}}-H + Ag(NH_3)_2^+ \xrightarrow{OH_2} RCO\bar{O} + Ag$$

Tollen's reagent gives negative results with all ketones except α-hydroxy (secondary or primary) ketones as:

$$R-\overset{O}{\overset{\|}{C}}-\overset{OH}{\overset{|}{CH}}-R + Ag(NH_3)_2^+ \xrightarrow{OH_2} R-\overset{O}{\overset{\|}{C}}-CO\bar{O} + Ag$$

$$R-\overset{O}{\overset{\|}{C}}-\overset{OH}{\overset{|}{C}}H_2 + Ag(NH_3)_2^+ \xrightarrow{OH_2} R-\overset{O}{\overset{\|}{C}}-CO\bar{O}$$

2. Aldehyde restore the magenta colour of Schiff's reagent. To the rosaniline hydrochloride solution, SO_2 is passed till the magenta colour is discharged. Adding aldehyde restore the magenta colour.
3. Aldehydes reduces Fehling's solution into red coloured Cu_2O. Fehling's solution is an alkaline solution of cupric tartarate.
4. Aldehydes reduces Benedict's reagent (an alkaline solution containing a cupric citrate (complex) into brick red precipitate of Cu_2O.

PRACTICE EXERCISE

Level 1

1. Write structural formula for the products formed when propanal reacts with each of the following reagents:
   (a) $NaBH_4$ in aqueous NaOH
   (b) C_6H_5MgBr, then H_2O
   (c) Ag_2O, OH^- (d) $(C_6H_5)_3P{=}CH_2$
   (e) Semicarbazide (f) Cold, dilute $KMnO_4$
   (g) $HSCH_2CH_2SH$, H^+
   (h) Phenyl hydrazine (i) $HOCH_2CH_2OH/H^+$
   (j) H_2NOH
   (k) $BrCH_2COOC_2H_5$, Zn; H^+
2. Predict major product(s) of each of the following reactions:

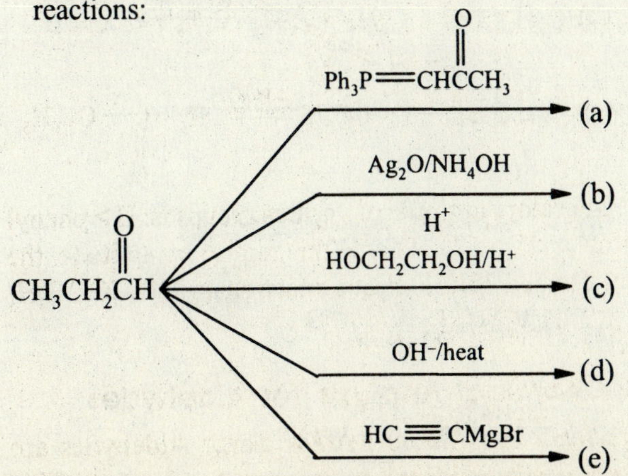

3. Give the structural formula for compounds A-E

$$\text{Cyclohexanol} \xrightarrow[\text{Acetone}]{H_2CrO_4} A(C_6H_{10}O) \xrightarrow[H^+]{CH_3MgI} B(C_7H_{14}O) \xrightarrow{\text{Heat},\ H^+} (C_7H_{12})\,C$$

$$(C_7H_{12})\,C \xrightarrow[O_3]{Zn\text{-}H_2O} (C_7H_{12}O)\,D \xrightarrow[OH^-]{Ag_2O} (C_7H_{12}O_3)\,E$$

4. Starting with benzyl chloride, how would you synthesize each of the following:
   (a) $C_6H_5CH_2CHOHCH_3$
   (b) $C_6H_5-CH{=}CH\text{-}C_6H_5$
   (c) $C_6H_5CH_2CH_2CHO$
   (d) $C_6H_5CH_2COCH_2CH_3$
5. Synthesize (a) 5-nonanone from 1-butanol as only starting organic compound.
   (b) 2-butanone from ethyl bromide.
6. Identify A to D in the following sequence of reaction:

$$\text{Ethanal} \xrightarrow[H^+]{BrCH_2COOEt/Zn} A(C_6H_{12}O_3) \xrightarrow[H^+]{\text{Heat}} B(C_6H_{10}O_2) \xrightarrow{H_2/Pt} (C_6H_{12}O_2)\,C$$

$$(C_6H_{12}O_2)\,C \xrightarrow{LiAlH_4} D \xrightarrow{PCC} \text{Butanal}$$

7. Complete the following reactions supplying missing reagent(s):

Cyclopentanone → A: cyclopentane; B: cyclopentanol; C: 1-phenylcyclopentanol; D: bicyclopentylidene; E: cyclopent-2-enone; F: δ-valerolactone; G: 2-methylenecyclopentanone

8. Outlined below is a synthesis of a compound used in perfumes. Provide all missing structures.

4-tert-butylbenzyl alcohol (CH_2OH) $\xrightarrow{PCC}$ A $(C_{11}H_{14}O)$ $\xrightarrow{\text{Propanal / NaOH}}$ B $(C_{14}H_{18}O)$ $\xrightarrow{H_2/Pd\text{-}C}$ C $(C_{14}H_{20}O)$

9. Compounds A and D do not give positive Tollen's tests, however compound C does. Give structures for A to D:

4-bromobutanal $\xrightarrow[H^+]{\text{Ethylene glycol}}$ A $(C_6H_{11}O_2Br)$ $\xrightarrow{\text{Mg / Ether}}$ B $\xrightarrow[H^+]{CH_3CHO}$ $C(C_6H_{12}O_2)$ $\xrightarrow[H^+]{MeOH}$ D

10. When excess of formaldehyde is treated with acetaldehyde in basic medium following reaction occur. Explain formation of products in the reaction.

$$CH_3CHO + CH_2O\,(\text{excess}) \xrightarrow{NaOH} C(CH_2OH)_4 + HCOONa$$

11. Provide missing reagents and intermediate in the following synthesis:

HO–C_6H_4–CH_2OH $\xrightarrow{(a)}$ MeO–C_6H_4–CH_2OH $\xrightarrow{(b)}$? $\xrightarrow{(c)}$ MeO–C_6H_4–CH(OH)–CH(CH_3)–COOEt $\xrightarrow{(d)}$ MeO–C_6H_4–CH(OH)–CH(CH_3)–CH_2OH

12. Deduce the structure of A to C and provide stereomeric structure of glyceraldehyde formed at the end:

Allyl alcohol (OH) $\xrightarrow{PCC}$ A (C_3H_4O) $\xrightarrow[H^+]{CH_3OH}$ B $(C_5H_{10}O_2)$ $\xrightarrow{KMnO_4 \text{ / Cold, dil, alkaline}}$ C $(C_5H_{12}O_4)$ $\xrightarrow[OH_2]{H^+}$ Glyceraldehyde

13. Write the structural formula of A to C:

$HC{\equiv}CH$ $\xrightarrow{\text{(i) } NaNH_2 \text{ (ii) } CH_3COCH_3 \text{ (iii) } NH_4Cl/H_2O}$ $A(C_5H_8O)$ $\xrightarrow{HgSO_4 \text{ / } H_2SO_4}$ $B(C_5H_{10}O_2)$ $\xrightarrow[HO^-]{C_6H_5CHO}$ $C(C_{12}H_{14}O_2)$

14. Show mechanism of the following reaction:

Br–$CH_2CH_2CH_2$–Br + ethyl acetoacetate $\xrightarrow{EtONa}$ ethyl 1-acetylcyclobutanecarboxylate

15. An organic compound $A(C_{12}H_{20})$ decolourize bromine water, and on reductive ozonolysis produces two molecules of B $(C_6H_{10}O)$. B on aldol condensation followed by dehydration of product gives $C(C_{12}H_{18}O)$, which on reduction with N_2H_4/OH^- regenerate A. B on reduction with Zn(Hg)/ C·HCl produces C_6H_{12}(D), which on monochlorination gives $C_6H_{11}Cl$(E) as a sole product. Identify A to E.

16. Complete the following reactions:

(a) C_6H_5CHO + acetone $\xrightarrow{\text{dil NaOH}}$ A + B

(b) $C_6H_5CHO + C_6H_5COCH_3 \xrightarrow{\text{dil NaOH}}$ A + B

(c) $C_6H_5CHO + CH_3NO_2 \xrightarrow{\text{dil NaOH}}$

(d) $CH_3CO(CH_2)_4$CHO $\xrightarrow{\text{dil NaOH}}$

(e) ? $\xrightarrow{\text{dil NaOH}}$ 3-methylcyclopent-2-enone

(f) ? $\xrightarrow{\text{dil NaOH}}$ cyclohex-2-enone

17. Carry out the following transformations:

(a) 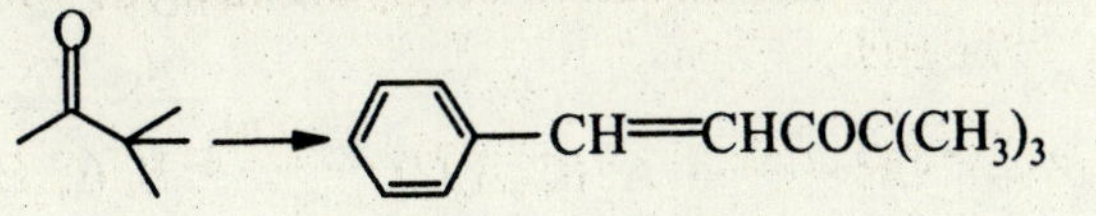 (tert-butyl methyl ketone) → C_6H_5–CH=CHCOC$(CH_3)_3$

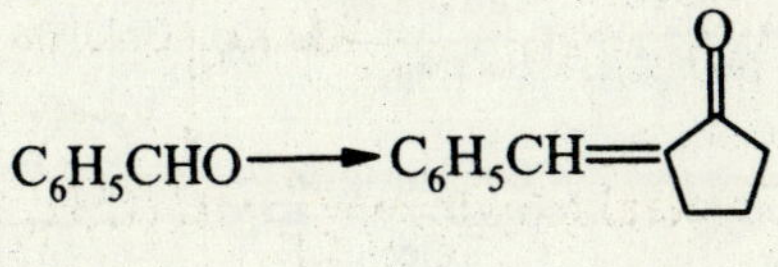

(b) C_6H_5CHO → C_6H_5CH=(cyclopentanone)

(c) C_6H_5CHO → $C_6H_5CH_2CH(NH_2)CH_3$

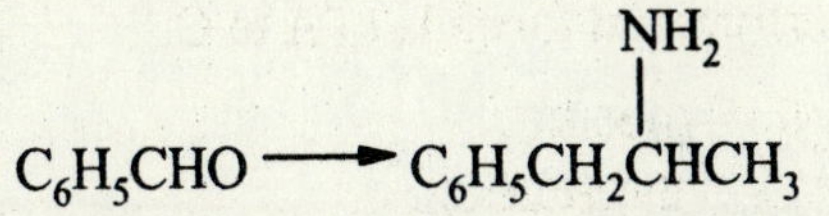

(d) 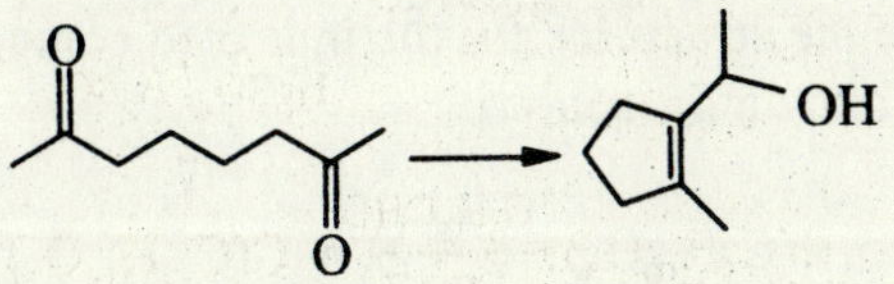 (2,6-diketone) → (cyclopentene alcohol) OH

(e) 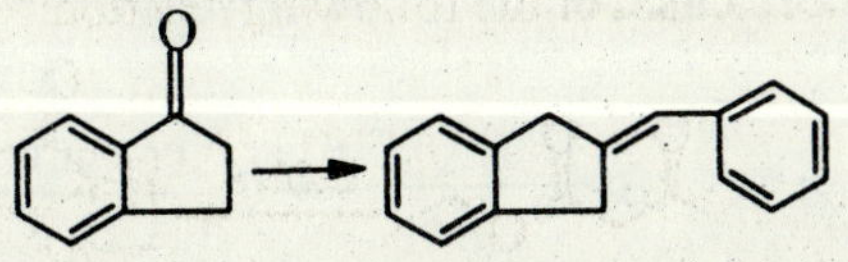 (indanone) → (benzylidene indane)

18. An alkene (A) on reductive ozonolysis yields acetone and an aldehyde. The aldehyde is easily oxidised to an acid B. B when treated with Br_2/P yields a compound C which on hydrolysis gives D. D can also be obtained from acetone by reaction with HCN/H_2O. Identify A to D.

19. A hydrocarbon A containing C = 88.24% (MW = 68) when reacted with Na followed by *n*-propylbromide, produces a compound B(C_8H_{14}). It gave a ketone $C_8H_{16}O$(C) with Hg^{2+}/H_2SO_4. B on refluxing with $H^+/KMnO_4$ gave two isomeric acid D and E. Identify A to E.

20. A compound having molecular formula C_6H_{12} is treated with Cl_2 in CCl_4 to form a compound B($C_6H_{12}Cl_2$). B is treated with alcoholic KOH followed by $NaNH_2$ resulting in the formation of C (C_6H_{10}). C when treated with Pt/H_2 gave 2-methyl pentane. Compound C does not react with $NaNH_2$ and with $NH_4OH/AgNO_3$. On ozonolysis, A gave two aldehydes, D and ethanal. Deduce the structure of A-D.

21. Compound A ($C_7H_{16}O$) can be resolved into enantiomers and react with $Na_2Cr_2O_7/H_2SO_4$ to give B($C_7H_{14}O$). When B is treated with NaOD at 25°C for several hours and then analyzed by mass spectroscopy, it is found to have MW = 116. Identify A and B.

22. An organic compound A(C_7H_6O) on treatment with hydroxyl amine produces two isomeric compounds B and C with molecular formula C_7H_7NO. B and C on acidification rearranges to D and E respectively. D on alkaline hydrolysis produces a white crystalline solid F($C_7H_6O_2$) and ammonia gas, whereas alkaline hydrolysis of E produces G and formic acid. Identify A to G describing mechanism of their formation.

23. The following road-map problem centers on the structure and properties of A, a key intermediate in these reaction. Give structures for the compounds A through J:

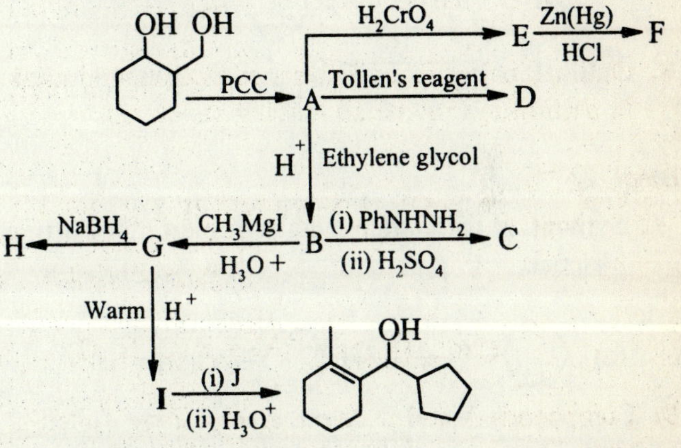

(cyclohexane diol, OH OH) $\xrightarrow{PCC}$ A; A $\xrightarrow{H_2CrO_4}$ E $\xrightarrow[HCl]{Zn(Hg)}$ F; A $\xrightarrow{\text{Tollen's reagent}}$ D; A $\xrightarrow{H^+,\ \text{Ethylene glycol}}$ B; B $\xrightarrow[(ii)\ H_2SO_4]{(i)\ PhNHNH_2}$ C; B $\xrightarrow[H_3O+]{CH_3MgI}$ G $\xrightarrow{NaBH_4}$ H; G $\xrightarrow{\text{Warm},\ H^+}$ I $\xrightarrow[(ii)\ H_3O^+]{(i)\ J}$ (product with OH)

24. Identify the products:

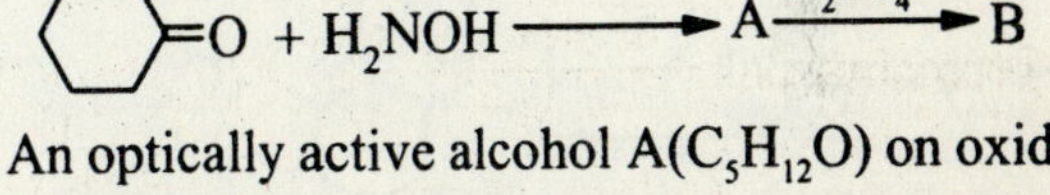

(cyclohexanone)=O + H_2NOH ⟶ A $\xrightarrow{H_2SO_4}$ B

25. An optically active alcohol A($C_5H_{12}O$) on oxidation with PCC gives an optically active aldehyde B($C_5H_{10}O$). B on treatment with dilute OH^- gives C($C_{10}H_{20}O_2$). C on heating with H_2SO_4 gives D($C_{10}H_{18}O$), which after reductive ozonolysis gives E($C_6H_{10}O_2$) and F(C_4H_8O). F on treatment with I_2/ OH^- gives an yellow precipitate. Identify A to F.

26. Provide reagents in the following transformations:

(a) 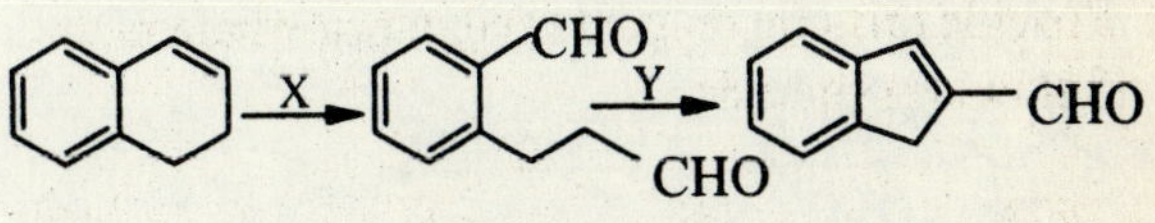 (dihydronaphthalene) $\xrightarrow{X}$ (CHO, CHO dialdehyde) $\xrightarrow{Y}$ (indene)–CHO

(b) 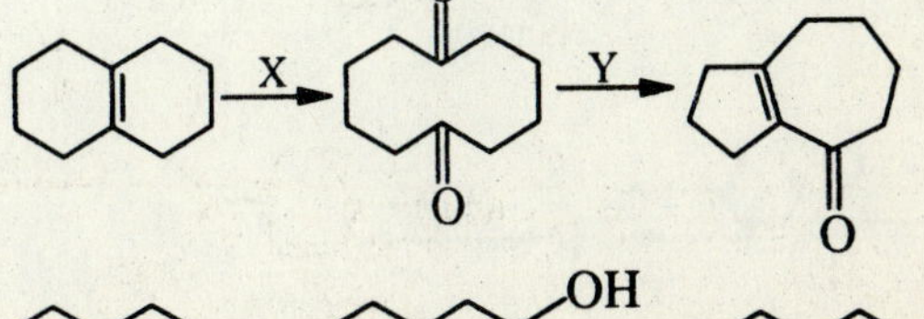 (octalin) $\xrightarrow{X}$ (cyclodecanedione) $\xrightarrow{Y}$ (bicyclic enone)

(c) 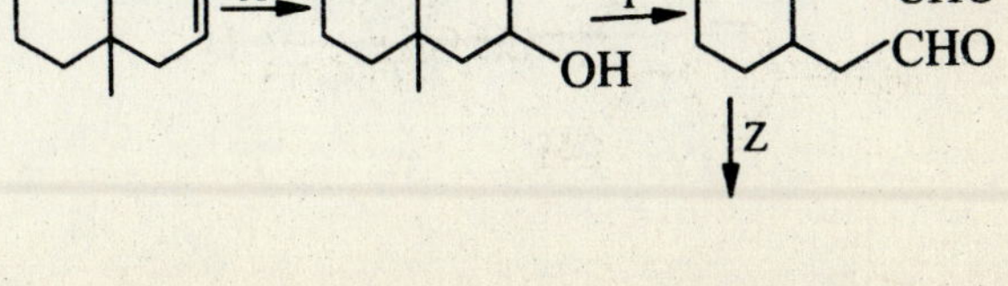 (methyl octalin) $\xrightarrow{X}$ (diol, OH OH) $\xrightarrow{Y}$ (dialdehyde, CHO CHO) $\xrightarrow{Z}$

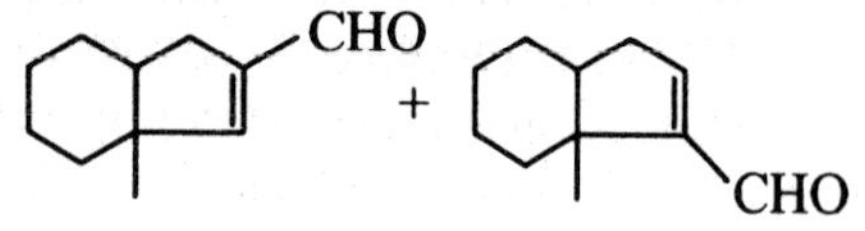

(d) 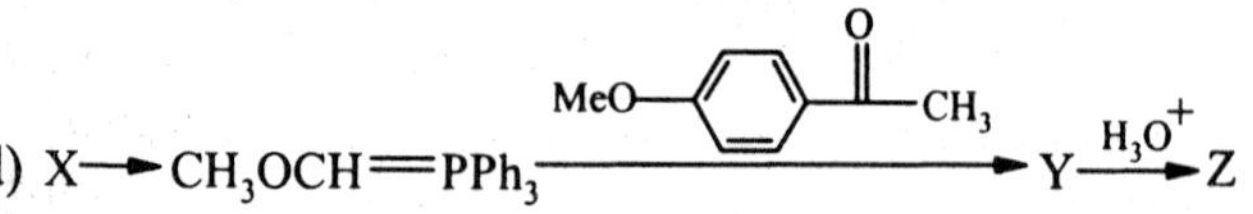 X → $CH_3OCH{=}PPh_3$ —(MeO–C₆H₄–COCH₃)→ Y —(H_3O^+)→ Z

27. Propose mechanism of the following reactions:

(a) 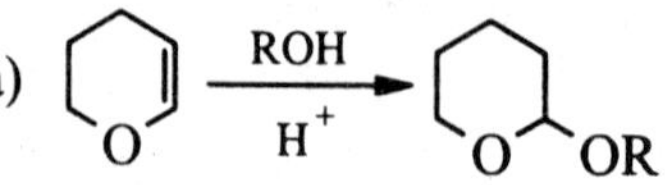 —(ROH, H^+)→ OR

(b) 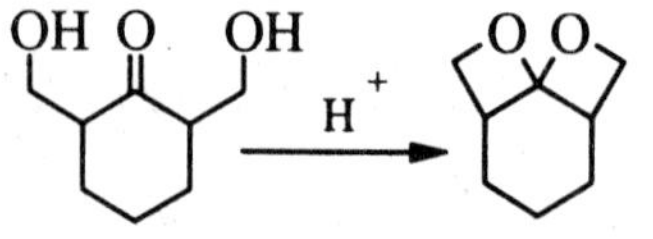 —(H^+)→

Level 2

1. Synthesize the compounds from indicated starting material:

(a) 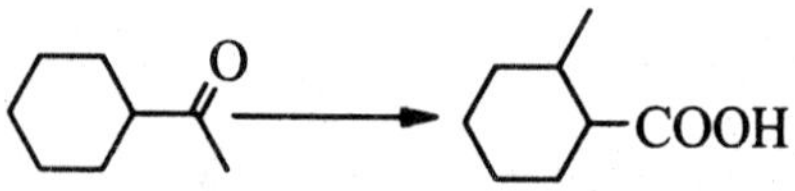 → COOH

(b) 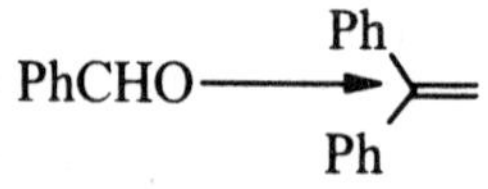 PhCHO → Ph, Ph

(c) 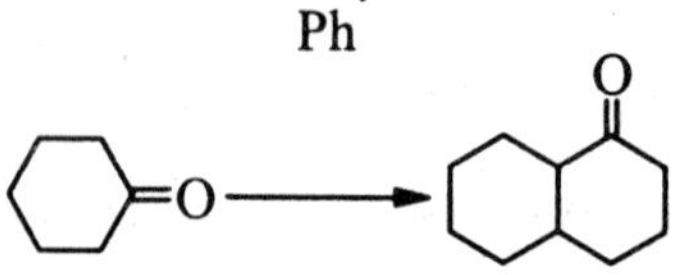 →

2. Predict mechanism of the following reaction:

$(CH_2)_{10}$ —(Dilute NaOH)→ $(CH_2)_9$ OH, O, O, OH $(CH_2)_9$

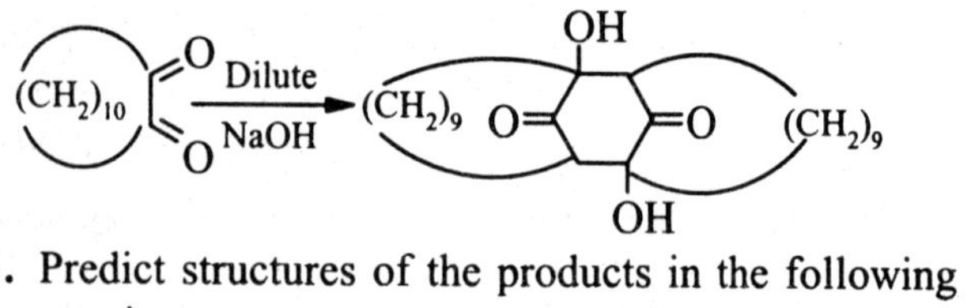

3. Predict structures of the products in the following reactions:

(a) 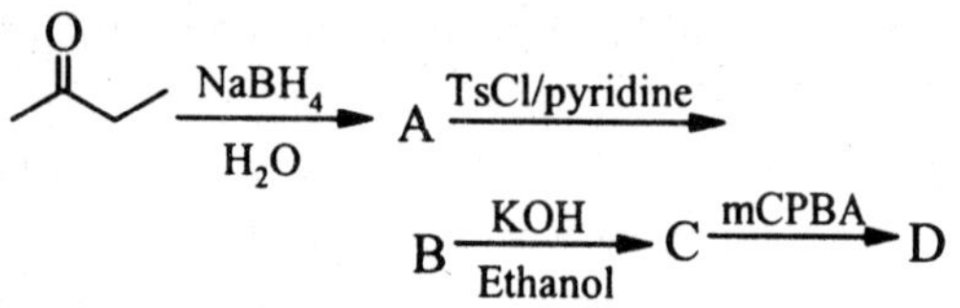 —($NaBH_4$, H_2O)→ A —(TsCl/pyridine)→ B —(KOH, Ethanol)→ C —(mCPBA)→ D

(b) —(C₆H₅–Li)→ E —(H_3O^+)→ F —(CrO_3, H_3O^+)→ G —($PhNHNH_2$, CH_3COOH)→ H

(c) —($B_2H_6/H_2O_2/NaOH$)→ I —($Na_2Cr_2O_7/H_2SO_4$, H_2O)→ J —(NaCN, H_3O^+)→ K

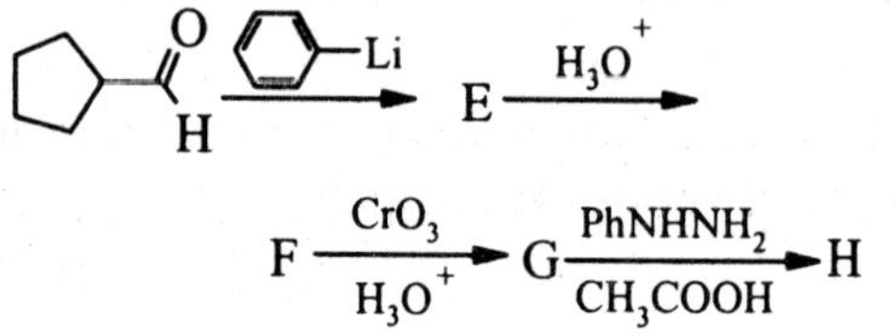

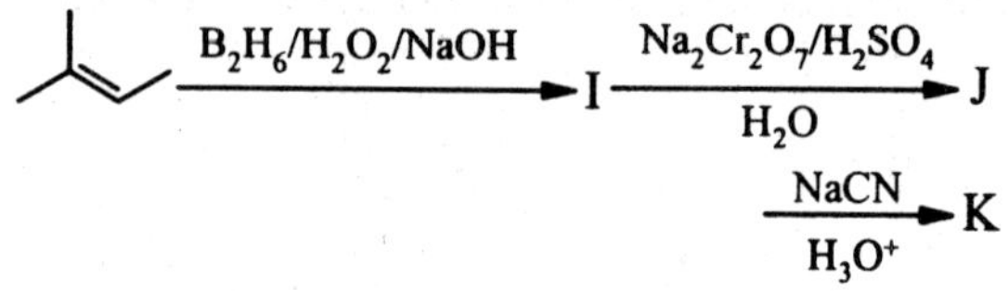

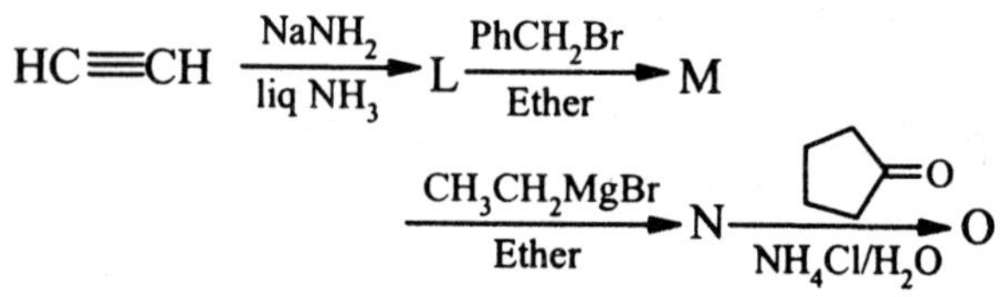

(d) HC≡CH —($NaNH_2$, liq NH_3)→ L —($PhCH_2Br$, Ether)→ M —(CH_3CH_2MgBr, Ether)→ N —(cyclopentanone, NH_4Cl/H_2O)→ O

(e) —(N_2H_4/KOH, Ethylene glycol)→ P

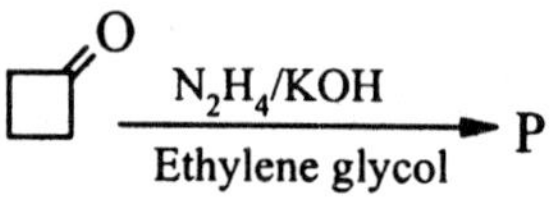

4. Give the structures for the indicated compounds in the following reactions:

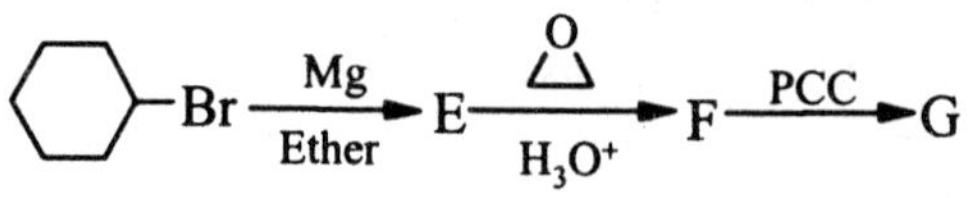

(a) 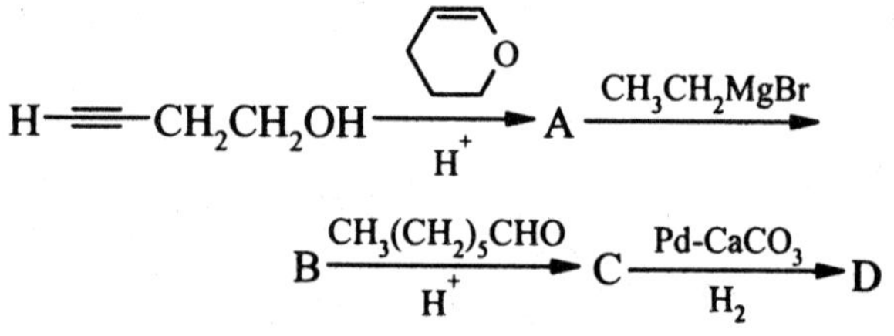 H–≡–CH_2CH_2OH —(dihydropyran, H^+)→ A —(CH_3CH_2MgBr)→ B —($CH_3(CH_2)_5CHO$, H^+)→ C —(Pd-$CaCO_3$, H_2)→ D

(b) –Br —(Mg, Ether)→ E —(ethylene oxide, H_3O^+)→ F —(PCC)→ G

(c) —($HOCH_2CH_2OH$, H^+)→ H —(BH_3/H_2O_2/KOH)→ I —(H^+)→ J

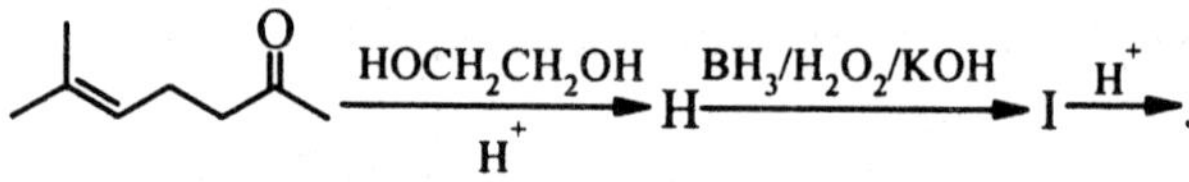

5. Plan a synthesis for each of the following compounds through shortest route. You have been supplied bromobenzene and any organic reagents containing three or fewer carbon atoms and any inorganic reagents you need.

(a)

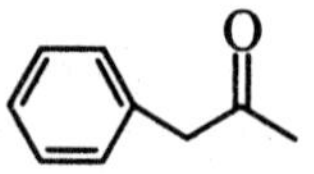

(b) 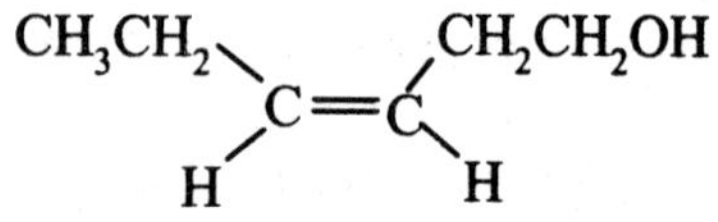 CH_3CH_2(H)C=C(H)CH_2CH_2OH

(c) Br

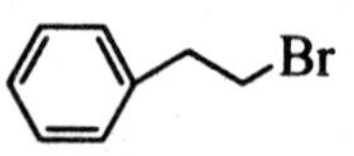

(d)

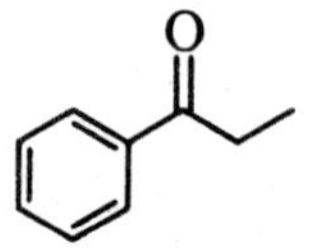

(e)

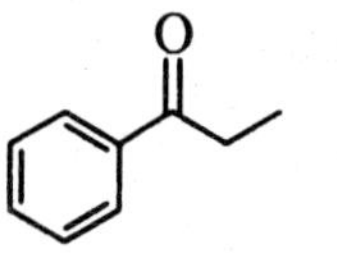

(f) 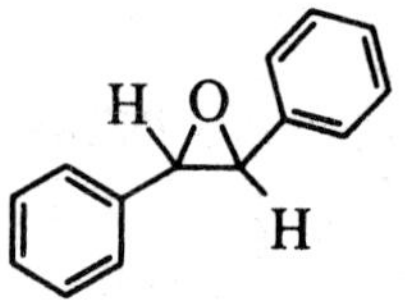 H, O, H

6. Predict mechanism of following exchange reaction:

$$CH_3C(=O^{16})CH_3 + {}^{18}OH_2 \rightleftharpoons CH_3C(=O^{18})CH_3 + {}^{16}OH_2$$

7. Assign structures to the compounds A to F in the following sequence of reaction:

$$Br(CH_2)_4OH \xrightarrow{PCC} A \xrightarrow[H^+]{HOCH_2CH_2OH} B \xrightarrow[Ether]{Mg}$$

$$C \xrightarrow{H_3C(H)C=C(H)CHO} D \xrightarrow{H_3O^+} E \xrightarrow{PCC} F\ (C_8H_{12}O_2)$$

8. Complete the following equations.

(a) Furfural (2-furaldehyde) + CH_3COCH_3 $\xrightarrow{NaOH}$ $\xrightarrow{H_3O^+}$

(b) C_6H_5CHO (2 equivalent) + CH_3COCH_3 (1 equivalent) $\xrightarrow{NaOH}$ $\xrightarrow{H_3O^+}$

(c) 2 $C_6H_5CH_2CHO$ $\xrightarrow{NaOH}$

9. Convert cyclohexanone to 2-methyl cyclohexanone.

10. Give products of each of the following reactions:

(a) $CH_3CH_2CH_2CHO + C_2H_5OH(\text{excess}) \xrightarrow{H^+}$

(b) $Ph\text{-}COCH_2CH_3 + H_2N\text{-}NH_2 \xrightarrow{H^+}$

(c) Product in (b) $\xrightarrow[\Delta]{\bar{O}H}$

(d) 3-pentanone + NaCN (Excess) $\xrightarrow{H^+}$

(e) 2-pentanone + Ethylene glycol $\xrightarrow{H^+}$

(f) 3-methylcyclohex-2-en-1-one + NaCN (excess) $\xrightarrow{HCl}$

(g) $C_6H_5\text{-}C(Et){=}NEt + OH_2 \xrightarrow{H^+}$

(h) Cyclohexanone + Ethanamine →

(i) Cyclohexanone + *di*-ethyl amine →

(j) Cyclopentanone + $Ph_3P{=}CHCH_3$ →

(k) 2-Pyrrolidinone (N–H) $\xrightarrow{LiAlH_4/H_2O}$

11. Arrange the following compounds in decreasing reactivity towards nucleophilic attack:

I: $CH_3CH_2CH(H)\text{—}CO\text{—}CH_2CH_3$

II: CH_3CH_2CHO

III: $CH_3CH_2CH(CH_3)\text{—}CO\text{—}C(CH_3)_2\text{—}CH_2CH_3$

IV: $CH_3\text{—}CH(CH_3)\text{—}CH_2\text{—}CO\text{—}CH_2CH_3$

V: $CH_3CH_2CH(CH_3)\text{—}CO\text{—}CH(CH_3)\text{—}CH_2CH_3$

VI: $CH_3CH_2\text{—}CO\text{—}CH_2CH_3$

12. Using cyclohexanone as starting material, describe how each of the following compounds could be synthesized:

(a) Cyclohexanol (OH)
(b) Cyclohexene
(c) Bromocyclohexane (Br)
(d) Vinylcyclohexane
(e) Cyclohexylamine (NH_2)
(f) Cyclohexane (two method)
(g) Cyclohexylmethylamine (CH_2NH_2)
(h) Ethylcyclohexane
(i) N,N-Dimethylcyclohexylamine ($N(CH_3)_2$)

13. List the following compounds in decreasing order of their K_{eq} value for hydrate formation: Ph—$COCH_3$, *p*-chloro acetophenone, *p*-nitro acetophenone, *p*-methoxy acetophenone.

14. Acetone is only 0.2% hydrated, acetaldehyde is 58%, while formaldehyde is 99% hydrated in aqueous medium, explain.

15. Chloral when dissolved in water, converted quantitatively into chloral hydrate, explain.

16. Fill in the boxes:

(a) $CH_3OH \xrightarrow{A} CH_3Br \xrightarrow{B} C \xrightarrow[E]{D} CH_3CH_2OH$

(b) $CH_4 \xrightarrow{A} CH_3Br \xrightarrow{B} C \xrightarrow[E]{D} CH_3CH_2CH_2OH$

(c) $CH_3CH_2CH_2OH \xrightarrow{A} CH_3CH_2CH_2Br \xrightarrow{B} C \xrightarrow[H^+]{Acetone} D$

17. Identify the compounds A and B:

$A \xrightarrow[(ii)\ H_3O^+]{(i)\ (CH_2=CH)_2CuLi} B \xrightarrow[H_3O^+]{CH_3Li} CH_2{=}CH{-}C(CH_3)_2{-}CH_2C(CH_3)(OH)CH_3$

18. Propose reasonable mechanism for each of the following reaction:

(a) $CH_3{-}CO{-}CH_2CH_2{-}CO{-}OCH_2CH_3 \xrightarrow{CH_3MgBr \mid H^+}$ (5,5-dimethyl lactone, H_3C, H_3C)

(b) 2-acetylbenzoic acid (CH_3, COOH) $\xrightarrow{H^+}$ (HO, CH_3 lactone)

(c) propiophenone $+ N_2H_4 \xrightarrow{OH^-}$ propylbenzene

(d) $CH_3CHO \xrightarrow{H^+}$ (trimer, 1,3,5-trioxane)

19. Propose mechanism:

(a) cyclohexanone $+ CH_2N_2 \xrightarrow{\Delta}$ cycloheptanone

(b) (bicyclic acetal) $+ H_3O^+ \rightarrow (HOCH_2CH_2CH_2)_2CHCHO$

20. Propose a mechanism for each of the following reactions:

(a) (bicyclic imine, N) $\xrightarrow{H_3\overset{+}{O}}$ (2-substituted cyclohexanone, NH_2, O)

(b) (1-methoxycyclohexene, OMe) $\xrightarrow{H^+}$ cyclohexanone

21. Write products (stereomers):

(a) $Ph{-}CO{-}CH_3 + Ph_3P{=}CHCH_3 \rightarrow$

(b) cyclohexenone $\xrightarrow[(ii)\ H_3O^+]{(i)\ (CH_3)_2CuLi}$ (Conjugate addition)

(c) 4-methylcyclohexanone $+ CH_3MgBr \xrightarrow{H^+}$

(d) propiophenone + pyrrolidine (N–H) $\longrightarrow$

22. Predict the structure of the products in the following reactions and show stereochemistry when it is known:

(a) $C_6H_5{-}CO{-}CH_2CH_2CH_2Cl \xrightarrow[(ii)\ TsOH/Benzene]{(i)\ HOCH_2CH_2OH} A \xrightarrow{[C_6H_5CH_2]_2NH} B \xrightarrow{H_3O^+} C$

(b) 2-butanone $\xrightarrow{NaBH_4/H_2O} A \xrightarrow[Pyridine]{TsCl} B \xrightarrow{KOH/Ethanol} C \xrightarrow{mCPBA} D$

(c) (2-methyl-2-butene) $\xrightarrow[\text{THF}]{BH_3}$ $\xrightarrow[\text{NaOH}]{H_2O_2}$ A $\xrightarrow{Na_2Cr_2O_7}$ B

(d) (cyclopentyl)–CHO $\xrightarrow[H^+]{\text{PhLi}}$ A $\xrightarrow{CrO_3\ H^+}$ B $\xrightarrow{\text{PhNHNH}_2}$ C

(e) HC≡CH $\xrightarrow{NaNH_2}$ D $\xrightarrow{C_6H_5CH_2Br}$ E $\xrightarrow{CH_3CH_2MgBr}$

(f) (cyclobutanone) $\xrightarrow{H_2NNH_2}$ X $\xrightarrow[\text{Ethylene glycol}]{KOH}$

23. Bring about the following conversions:

(a) PhCO–OMe ⟶ PhC(OH)(CH_3)–CH_3

(b) PhCO–OMe ⟶ Ph–CHO

(c) PhC(Br)(CH_3)–OMe ⟶ PhCO–CH_3

24. Bring about the following transformations:

(cyclohexanone)=O $\xrightarrow{A}$ (cyclopentyl)–OH

(cyclohexanone)=O $\xrightarrow{B}$ (cyclopentyl)–CH_2OH

(cyclohexanone)=O $\xrightarrow{C}$ (vinyl ether)–O–(chain)–COOMe

25. Complete each of the following equations, showing structures of the product that will be formed.

(a) (cyclohexanone)=O + Ph–NHNH$_2$ $\xrightarrow{\text{Acetic acid}}$

(b) (CH_3-cyclopentene) + $KMnO_4$/NaOH ⟶

(c) (cyclohexenyl)–CH_2OH $\xrightarrow{\text{(pyridinium)}\ \overset{+}{N}\text{–}HCrO_3\ \overset{-}{Cl}}$

(d) (cyclopentenyl)–CH_3 + BH_3 $\xrightarrow{H_2O_2,\ NaOH}$

(e) Ph–$COCH_3$ $\xrightarrow[\text{THF}]{CH_3CH_2CH_2CH_2Li}$ $\xrightarrow{NH_4Cl/H_2O}$

26. Complete the following reactions:

(a) (cyclopentanone) $\xrightarrow[H_2/Pt]{NH_3}$?

(b) (decalin triol: OH OH, OH) + (acetone) $\xrightarrow{H^+}$?

(c) ClOC–(cyclohexyl)–$COCH_3$ $\xrightarrow[\text{1 eq}]{MeNH_2}$?

(d) (dimethylcyclohexanone)=O $\xrightarrow[\text{(ii) } H_2/Pd\text{-}BaSO_4 \\ \text{(iii) H+/}H_2O]{\text{(i) KCN}}$?

27. (a) Bring about the following transformation:

(o-CHO, CH_2Br benzene) ⟶ (1-hydroxyisochroman: OH, O)

(b) Propose mechanism:

(bicyclic acetal, Et) $\xrightarrow{H^+}$ (keto diol: O, OH, OH)

28. Propose mechanism:

(a) (dimethylcyclohexane-1,2-diol: OH, OH) $\xrightarrow{H^+}$ (2,2-dimethylcyclohexanone: O)

(b) (dihydropyran, O) $\xrightarrow{H^+}$ (trimethylcyclohexenone: O)

(c) (furanyl)–CH_2OH $\xrightarrow{H^+}$ (dihydropyran: O)

29. Explain the formation of products in the following reaction:

(a) (epoxide, H) $\xrightarrow[OH_2]{MeMgBr}$ (ketone: O, H)

(b) 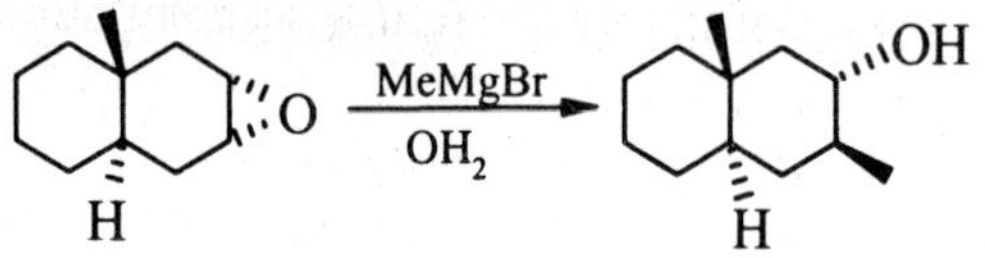 [structure] $\xrightarrow[OH_2]{MeMgBr}$ [structure]

30. Complete the followings:

(a) 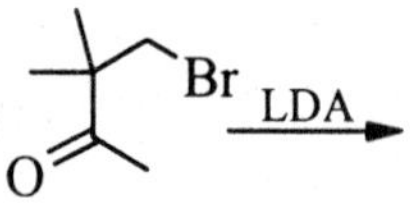 [structure] $\xrightarrow{LDA}$ (b) 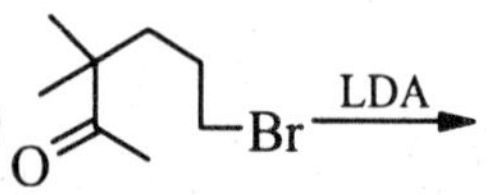 [structure] $\xrightarrow{LDA}$

31. Predict major product in the following reactions:

(a) 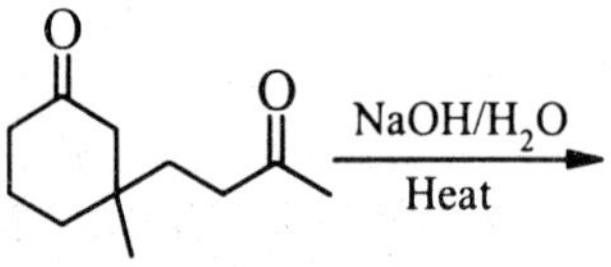 [structure] $\xrightarrow[Heat]{NaOH/H_2O}$ (b) 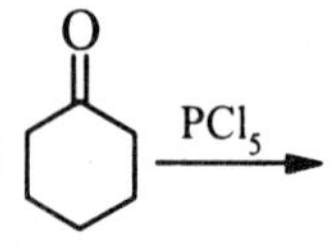 [structure] $\xrightarrow{PCl_5}$

32. Bring about the following conversions:

(a) 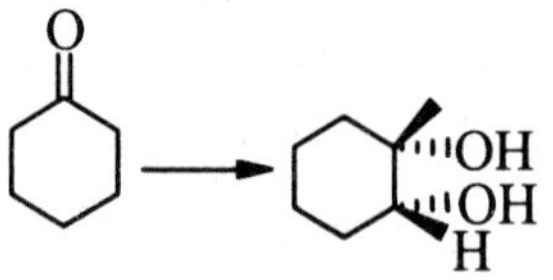 [structure] ⟶ [structure]

(b) 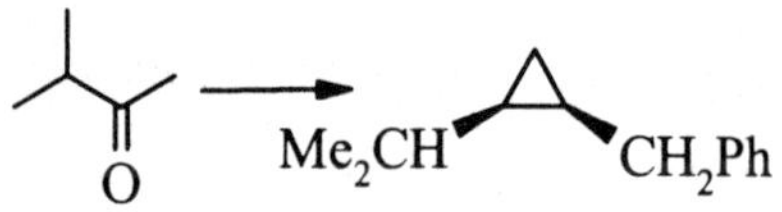 [structure] ⟶ Me_2CH–[cyclopropane]–CH_2Ph

33. Propose mechanism:

(a) 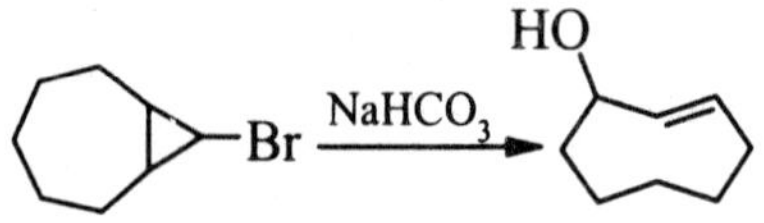 [structure] $\xrightarrow{NaHCO_3}$ [structure]

(b) 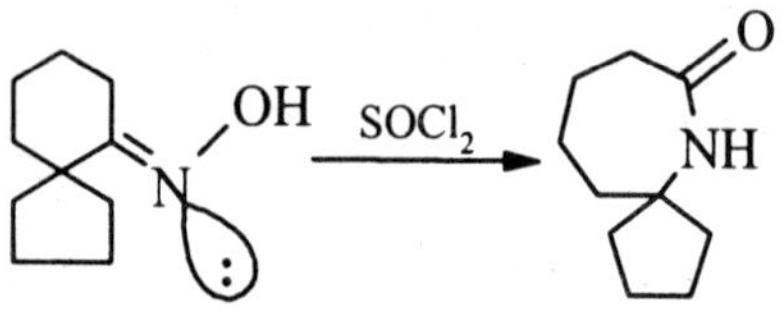 [structure] $\xrightarrow{SOCl_2}$ [structure]

34. Propose mechanism:

(a) 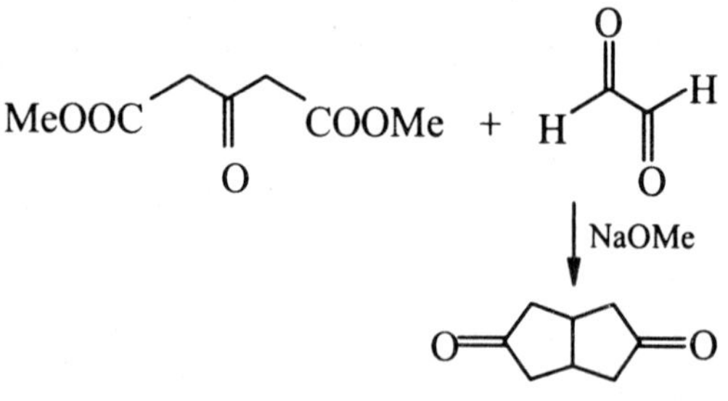 MeOOC–[structure]–COOMe + H–CO–CO–H $\xrightarrow{NaOMe}$ [structure]

(b) 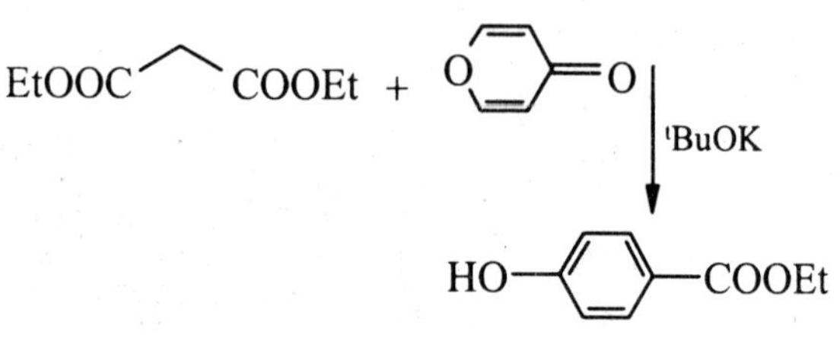 EtOOC–CH_2–COOEt + [structure] $\xrightarrow{^tBuOK}$ HO–C_6H_4–COOEt

(c) [structure] $\xrightarrow{MeNHOH}$ [structure]

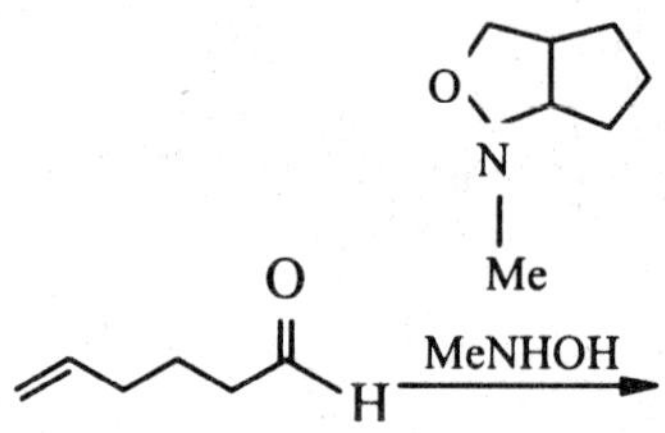

35. An organic compound A has molecular formula ($C_7H_{12}O$) and does not decolorize Bayer's reagent and evolves no gas with Na-metal. A also does not react with either ammonical silver nitrate solution or phenylhydrazine solution. A on refluxing with methanol in presence of trace amount of H_2SO_4 produces a resolvable B($C_8H_{16}O_2$). B on treatment with CrO_3/pyridine produced C($C_8H_{14}O_2$), which is non-resolvable. C on refluxing in HI solution produced another non-resolvable compound D($C_7H_{11}OI$). D on treatment with alcoholic KOH followed by oxidative ozonolysis of product produced E($C_7H_{10}O_4$), which does not show optical activity. E on simple heating produced $C_6H_{10}O_2$, which on refluxing with dilute NaOH gave:

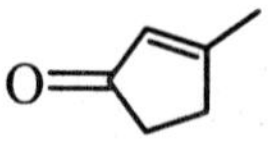

[structure] Deduce structures of A to E.

36. Starting from acetone, synthesize:

(a) 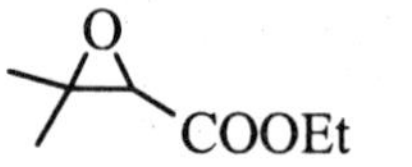 [structure] (b) 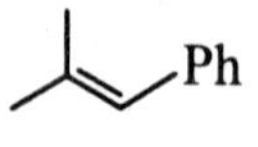 [structure]

(c) 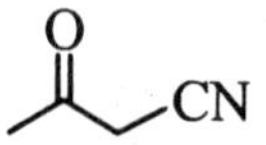 [structure] (d) 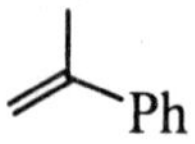 [structure]

(e) 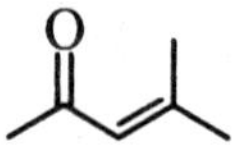 [structure]

37. A neutral, resolvable organic compound A has molecular formula ($C_8H_{16}O_2$). A on treatment with $LiAlH_4$ gives isomeric B and C($C_4H_{10}O$) of which only B is optically active. B on treatment with acidified dichromate solution gives C_4H_8O(D), which on refluxing with dilute NaOH followed by acidification of product gave E($C_8H_{14}O$). E on heating with N_2H_4 in alkaline medium affords F(C_8H_{16}). F on treatment with $B_2H_6/H_2O_2/OH^-$ produced a resolvable G($C_8H_{18}O$). G on treatment with acidified dichromate solution produced H($C_8H_{16}O$), which on treatment with "mcpba" produces A. Deduce structures of A to H.

38. Bring about the following conversions:

(a) 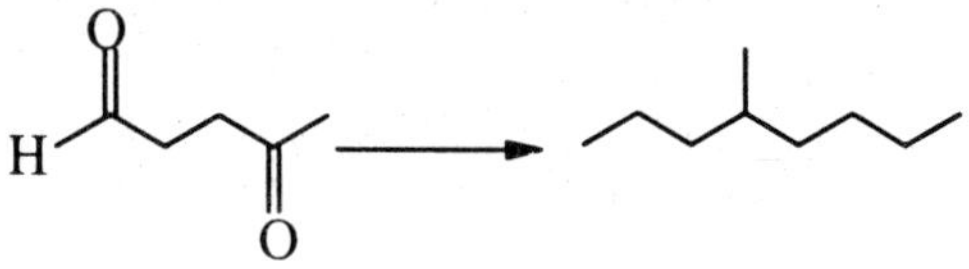 [structure] ⟶ [structure]

(b)

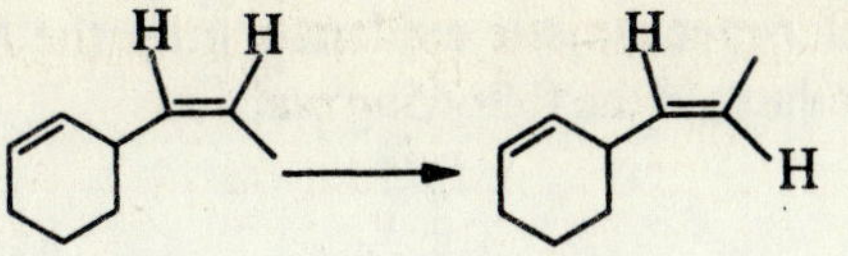

(c) 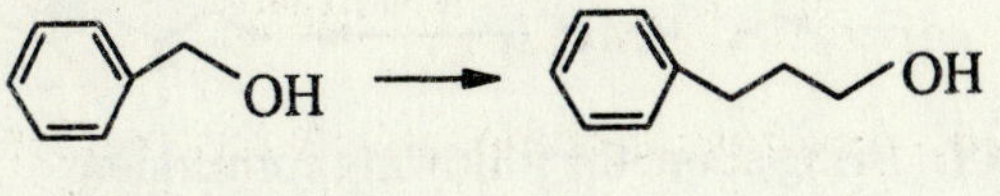 OH → OH

(d) 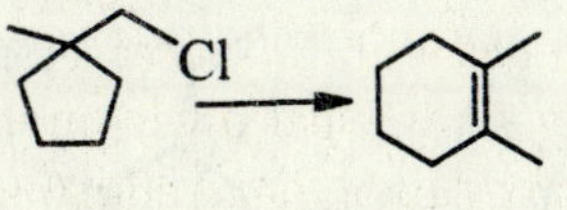 Cl →

(e) 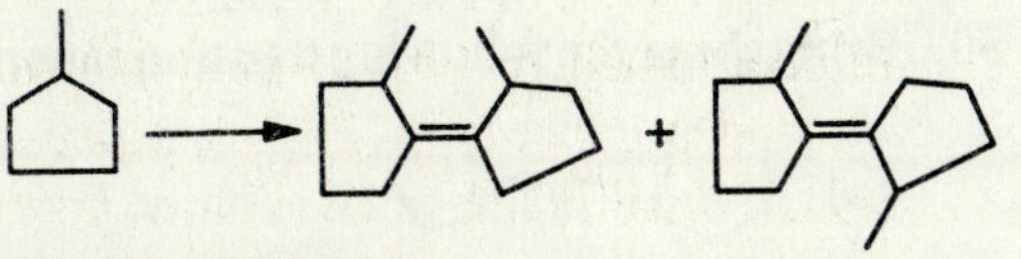 → +

39. Complete the following:

(a) 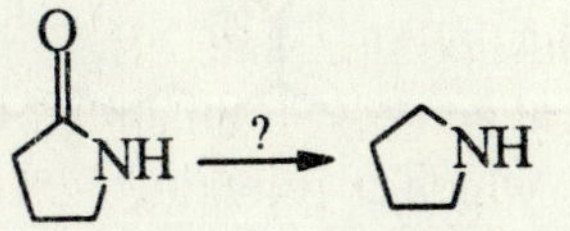 NH —?→ NH

(b) 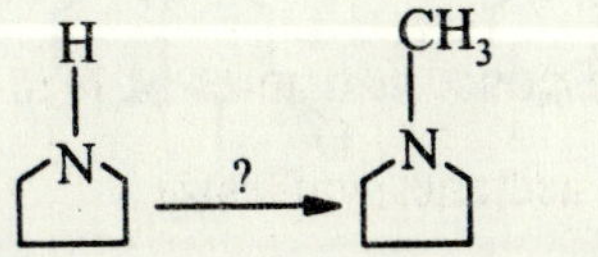 H, N —?→ CH_3, N

(c) 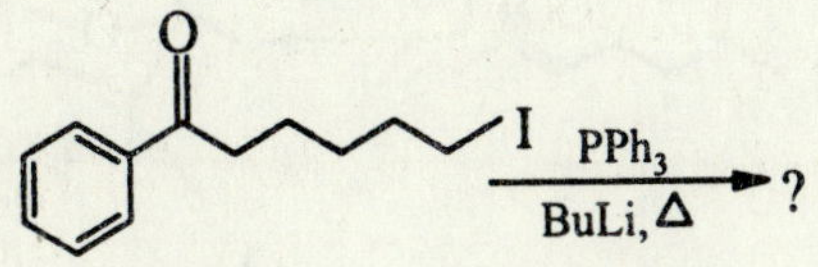 I —PPh_3, BuLi, Δ→ ?

(d) 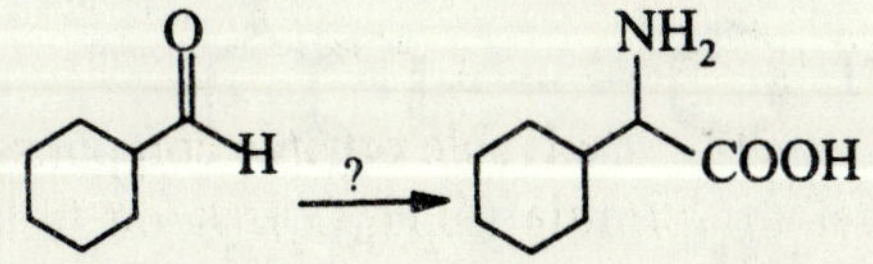 H —?→ NH_2, COOH

(e) 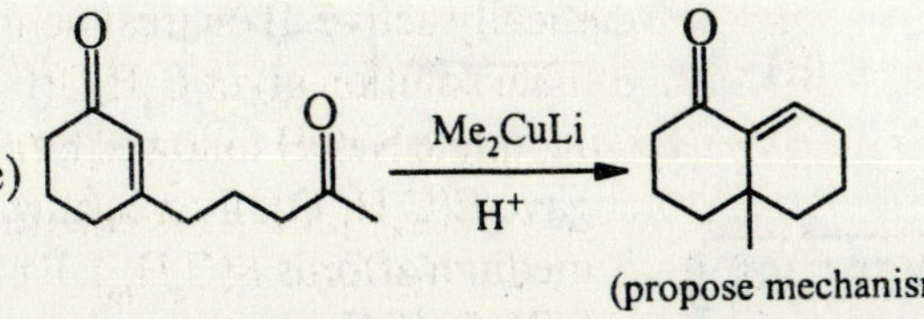 —Me_2CuLi, H^+→ (propose mechanism)

40. An organic compound has molecular formula $C_{10}H_6$(A). A on treatment with $HgSO_4/H_2SO_4$ gives B($C_{10}H_{10}O_2$), which forms an insoluble precipitate with 2, 4-DNP but fail to react with ammonical solution of $AgNO_3$. B on refluxing with dilute NaOH solution produced C($C_{10}H_8O$). C on further treatment with alkaline solution of N_2H_4 gives D($C_{10}H_{10}$), which decolourises Br_2-H_2O solution. D on treatment with H_2/Pt produced the following compound X. Deduce structures of A to D.

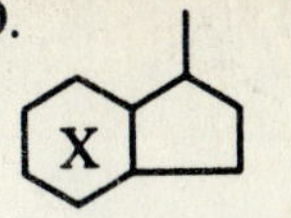

X

41. An optically active compound A has molecular formula $C_{10}H_{12}O$. A decolourises Bayer's reagent as well as it forms salt with NaOH. A on treatment with HBr forms B($C_{10}H_{13}OBr$) as major product, which is still enantiomeric. B on treatment with alcoholic solution of KOH gives a stereomeric C, which is also an isomer of A. A on treatment with acidic solution of $KMnO_4$ gives ortho hydroxy benzoic acid as one of the product. A on treatment with B_2H_6/H_2O_2/NaOH produces D($C_{10}H_{14}O_2$), which on refluxing with dilute sulphuric acid produced E($C_{10}H_{12}O$). E neither decolourises Br_2-H_2O solution nor forms any salt with NaOH. Deduce structures of A to E.

42. Provide products in the following reactions:

(a) 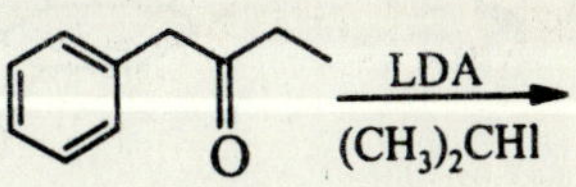 —LDA, $(CH_3)_2CHI$→ (b) 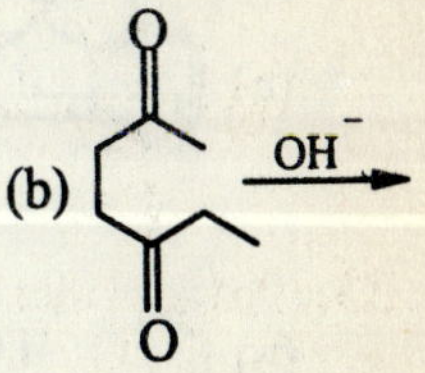 —OH^-→

43. Propose mechanism:

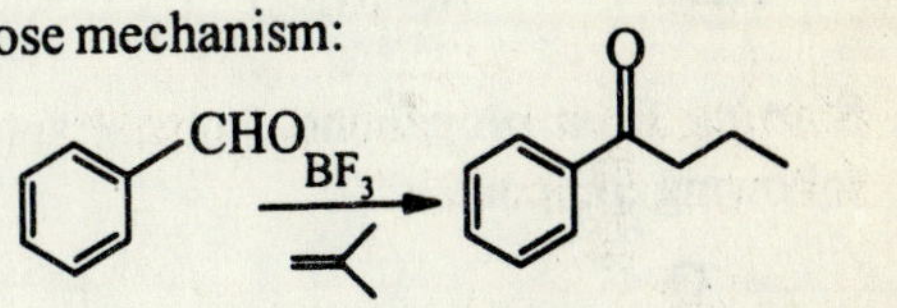

CHO —BF_3→

44. Convert: H_3C, H → Ph, OH, H, H, Ph

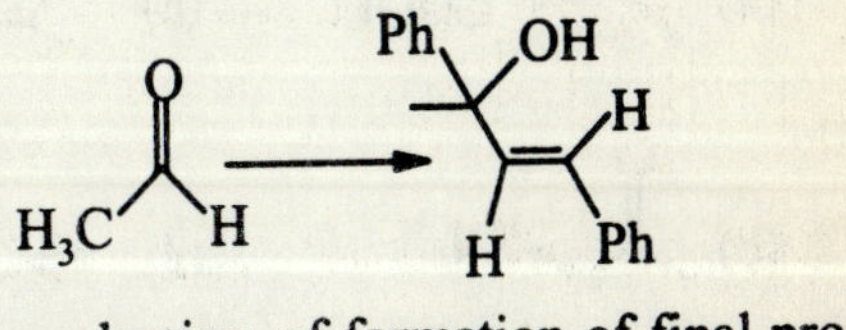

45. Propose mechanism of formation of final products in the following reactions and identify the labeled product:

(a) 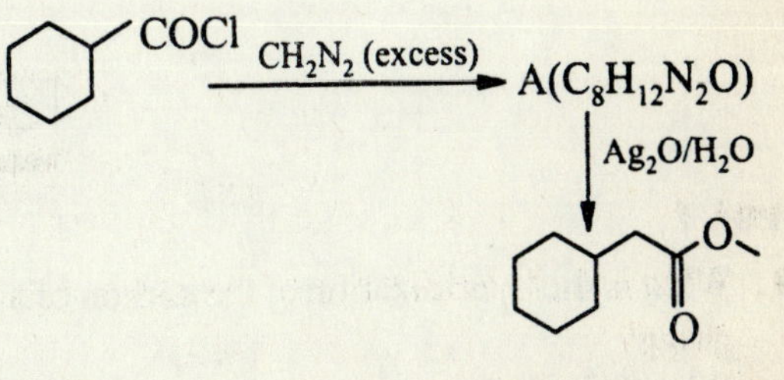 COCl —CH_2N_2 (excess)→ A($C_8H_{12}N_2O$) —Ag_2O/H_2O→

(b) 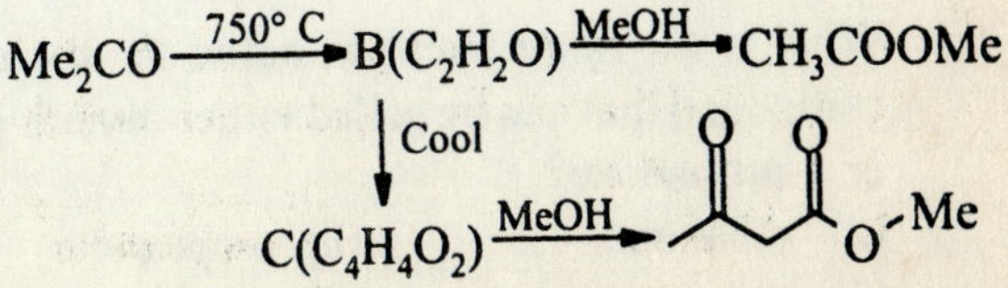 Me_2CO —750° C→ B(C_2H_2O) —MeOH→ CH_3COOMe

B —Cool→ C($C_4H_4O_2$) —MeOH→ Me

(c) 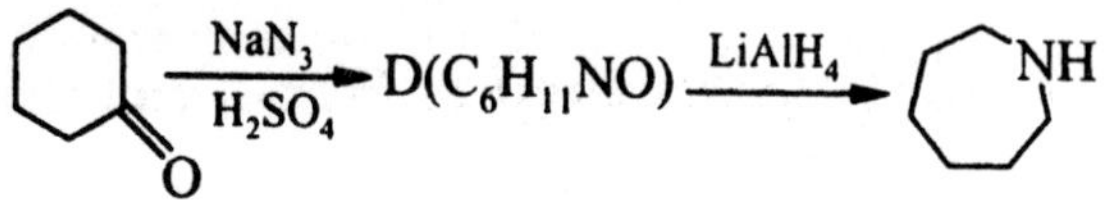 $\xrightarrow[H_2SO_4]{NaN_3}$ D($C_6H_{11}NO$) $\xrightarrow{LiAlH_4}$ NH

(d) $\xrightarrow{H_2SO_4}$ HO

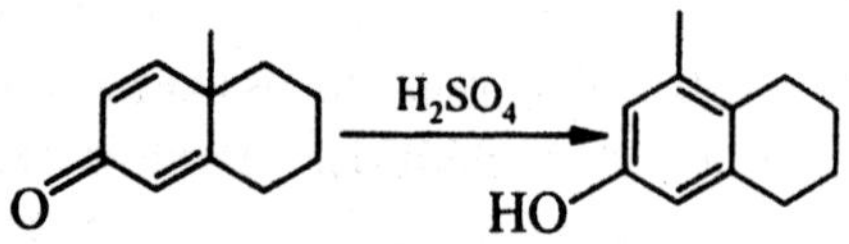

(e) 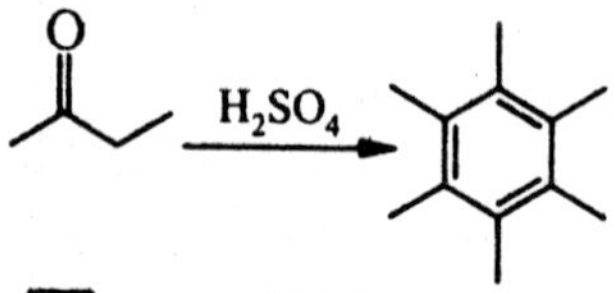 $\xrightarrow{H_2SO_4}$

(f) 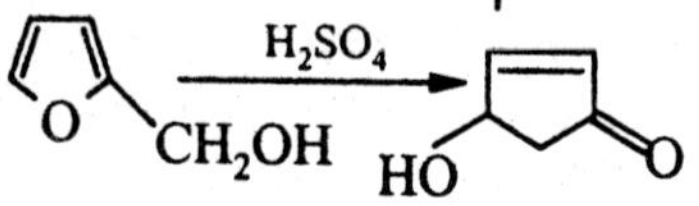 CH$_2$OH $\xrightarrow{H_2SO_4}$ HO

46. Propose mechanism:

(a) 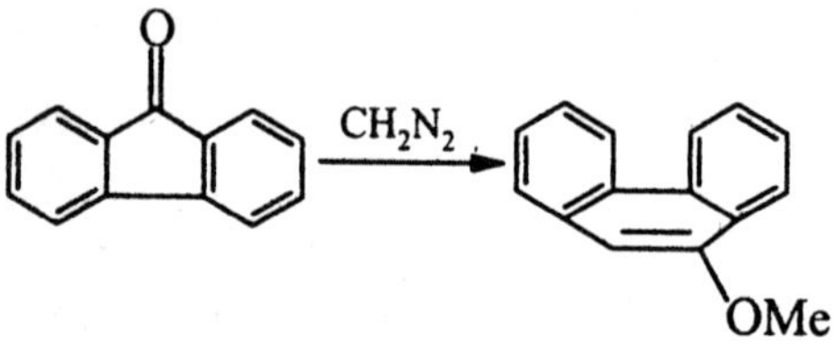 $\xrightarrow{CH_2N_2}$ OMe

(b) 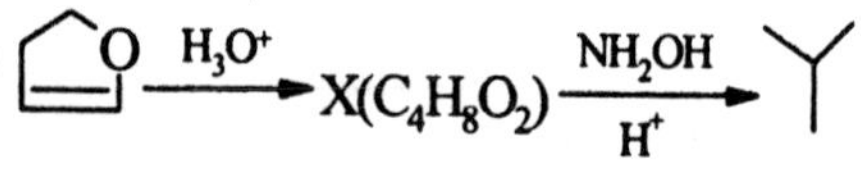 $\xrightarrow{H_3O^+}$ X($C_4H_8O_2$) $\xrightarrow[H^+]{NH_2OH}$

47. Starting from propanone, suggest synthesis of the following compounds:

(a) 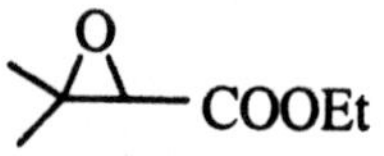 COOEt (b)  Ph

(c) 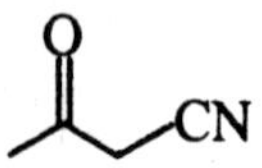 CN (d) 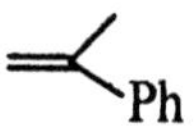 Ph

(e)

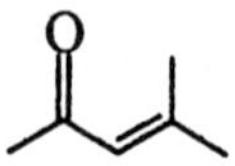

48. Offer mechanistic explanation for the formation of products in the following reaction:

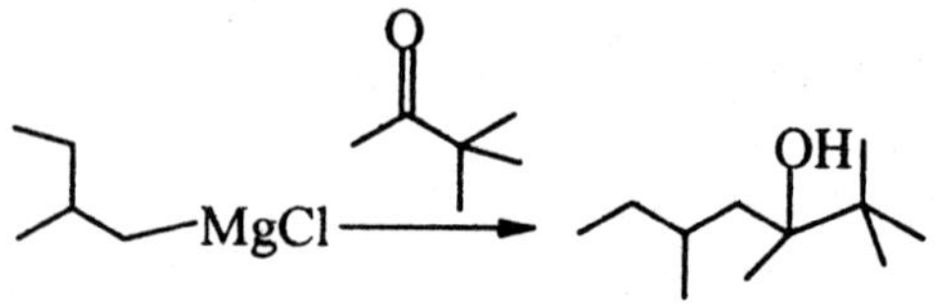

MgCl → OH

49. Bring about the following conversion:

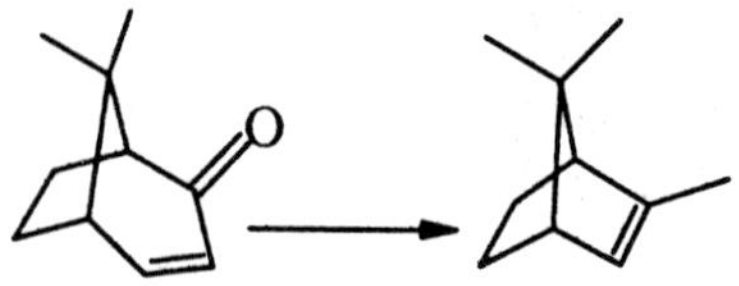

50. Bring about the following transformations:

(a)

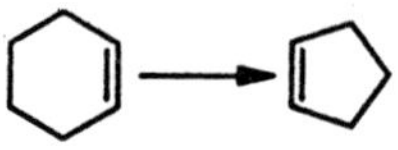

(b) 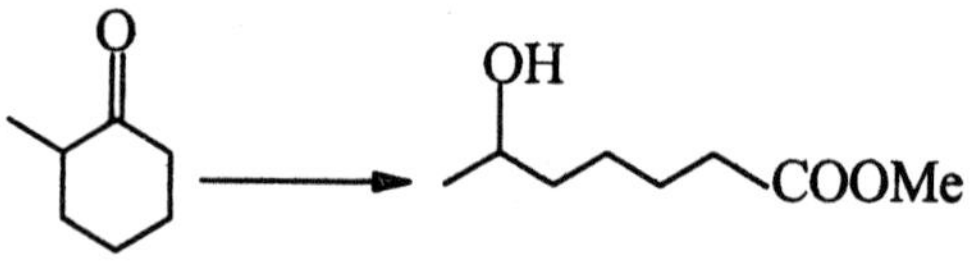 OH COOMe

(c) 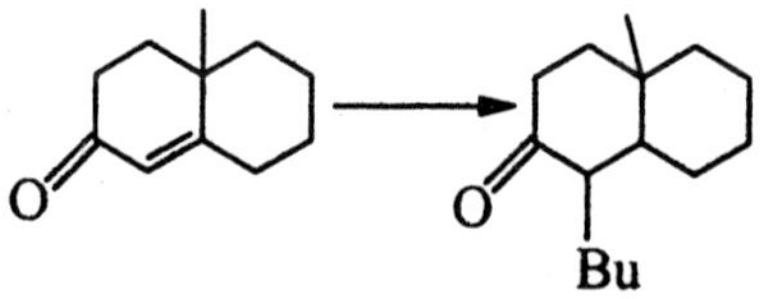 Bu

(d) 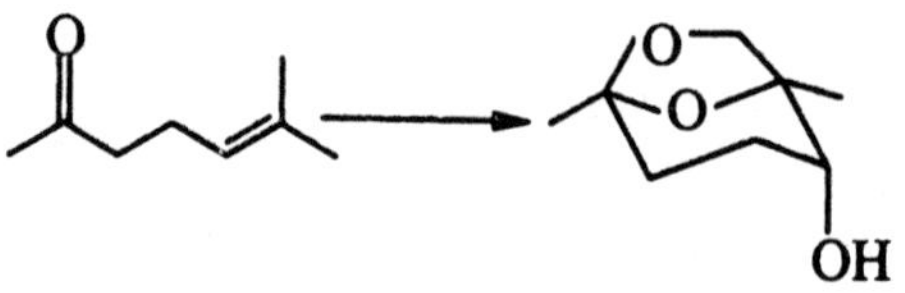 OH

(e)

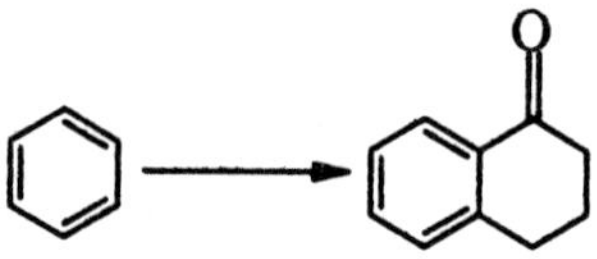

(f)

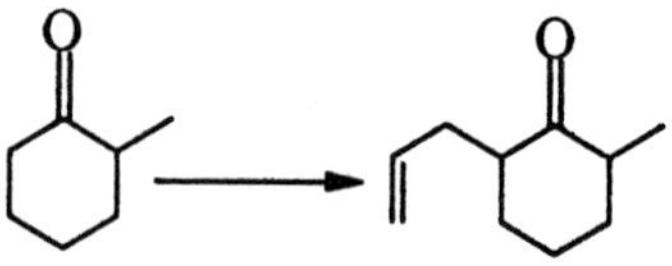

OBJECTIVE EXERCISE

Level 1

1. What is the hybridization of the carbon of a carbonyl group?
(a) s^3p^2 (b) sp^2 (c) sp (d) sp^3

2. What is the more common name for the carbonyl compound that can be called either dimethyl ketone or 2-propanone?
(a) ethanone (b) propionone
(c) acetanone (d) acetone

3. Why is the oxidation of a primary alcohol with a mixture of sodium dichromate and sulfuric acid *not* a good method for the preparation of the corresponding aldehyde?
(a) The product will be the corresponding ketone.
(b) Any aldehyde produced will be oxidized further.
(c) The product will be an alkyl sulfonate, R—SO_3H.

(d) A mixture of sodium dichromate and sulfuric acid will not oxidize a primary alcohol.

4. What reagent can be used to convert a carboxylic acid to a ketone?
   (a) Chromic acid
   (b) Acetyl chloride
   (c) Diborane, hydrogen peroxide
   (d) An organolithium

5. What represents the best method for converting a carboxylic acid to an aldehyde?
   (a) Convert the acid to an acid chloride, and react the acid chloride with water.
   (b) Reduce the acid with $LiAlH_4$.
   (c) Convert the acid to an acid chloride, and reduce the acid chloride with lithium aluminium tri (*t*-butoxy) hydride.
   (d) Convert the acid to an acid chloride, react the acid chloride with a Grignard reagent, and reduce the product with $LiAlH_4$.

6. What is the Wittig reaction most useful for?
   (a) synthesis of ketones
   (b) synthesis of alkenes
   (c) preparation of aldehydes
   (d) formation of aliphatic rings

7. Which best represents the product of the reaction of propanal with water?
   (a) $CH_3—CH_2—CHO$
   (b) $CH_3—CH_2—CH(OH)_2$
   (c) $CH_3—CH_2—CH_2—OH$
   (d) $CH_3—CH_2—C(OH)_3$

8. What carbonyl compound reacts with HCN to give the highest yield of cyanohydrin?
   (a) formaldehyde (b) propanal
   (c) 2-butanone (d) di-*t*-butyl ketone

9. What reacts with an aldehyde or ketone to produce an oxime?
   (a) hydroxylamine (b) hydrazine
   (c) semicarbazide (d) an amine

10. Which reagent is useful for protecting an aldehyde functional group in synthesis involving strong bases and nucleophiles?
   (a) NH_3 (b) B_2H_6
   (c) HCN (d) $HOCH_2CH_2OH$

11. What transformation does *Tollens' reagent* cause?
   (a) Reduction of acid chlorides.
   (b) Oxidation of aldehydes.
   (c) Oxidation of ketones.
   (d) Reduction of carboxylic acids.

12. What is the first step in the Wolf-Kischner reduction of a ketone?
   (a) Conversion of the ketone to a ketal.
   (b) Conversion of the ketone to a hydrazone.
   (c) Reduction of the ketone to a secondary alcohol.
   (d) Reduction of the carbonyl group to $—CH_2—$.

13. Which is the true statement?
   (a) The double bond of a typical carbonyl group is shorter, and stronger than the double bond of a typical alkene.
   (b) The double bond of a carbonyl group is longer, and stronger than the double bond of an alkene.
   (c) The double bond of a carbonyl group is shorter, but weaker than the double bond of an alkene.
   (d) The double bond of a carbonyl group is longer, but weaker than the double bond of an alkene.

14. Which one *cannot* be used to produce an aldehyde or ketone?
   (a) Hydration of 2-butyne with Hg^{2+}, H_2SO_4, H_2O.
   (b) Hydroboration-oxidation of 1-butyne.
   (c) Friedel-Crafts acylation of nitrobenzene.
   (d) Oxidation of 2-butanol with sodium dichromate in sulfuric acid.

15. What is the product of the reaction of propionic acid with ethyllithium?
   (a) dipropyl ketone (b) 3-pentanone
   (c) 3-ethyl-3-pentanol (d) ethyl propyl ketone

16. Which is most susceptible to attack by a nucleophile?
   (a) acetone (b) acetaldehyde
   (c) formaldehyde (d) methyl ethyl ketone

17. What product results from the reaction of 2-butanone with $Ph_3P^+—CH_2^-$?
   (a) 2-butene (b) 2-methyl-1-butene
   (c) 1-butene (d) 2-methyl-2-butene

18. Which one of the following carbonyl compounds forms the most stable hydrate?
   (a) $CH_3—CHO$ (b) CH_2O
   (c) $Cl_3C—CHO$ (d) $CH_3—CH_2—CHO$

19. In cyanohydrin formation, the hybridization of the carbonyl carbon:
   (a) Changes from sp^3 to sp.
   (b) Changes from sp^3 to sp^2.
   (c) Changes from sp^2 to sp.
   (d) Changes from sp^2 to sp^3.

20. The second step in the mechanism of imine formation is acid-catalyzed, yet the rate drops below pH 4.5. Why does the rate drop below this pH?

(a) The carbinolamine intermediate is stable at low pH.
(b) The imine product is hydrolyzed at low pH.
(c) Protonation of the amine decreases its nucleophilicity.
(d) The amine is hydrolyzed at low pH.

21. What pair of reagents should you use to form the following structure? CH_3CH_2O, OCH_2CH_3, H_3C, CH_3
(a) Acetone and ethyl alcohol.
(b) Isopropyl alcohol and ethyl alcohol.
(c) Acetone and propionaldehyde.
(d) Acetone and acetaldehyde.

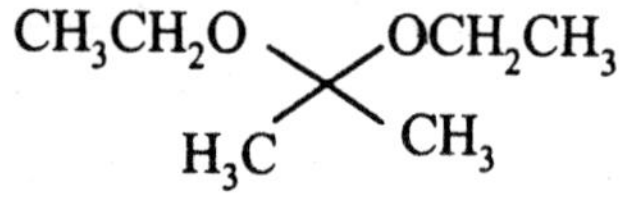

22. What is the product, C, of the following sequence of reactions?

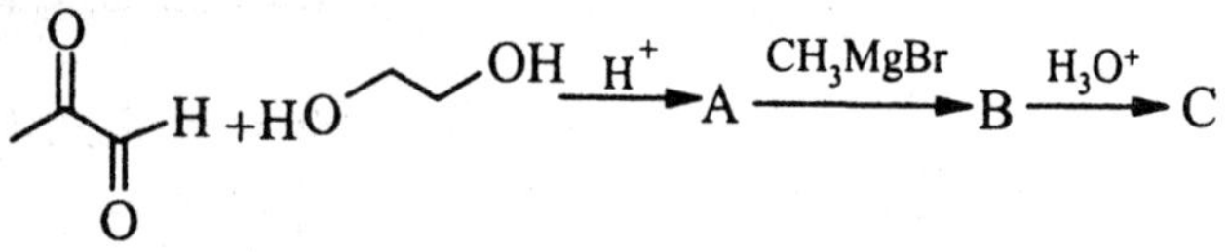

$$\text{(keto-aldehyde)} + HO\text{~~~}OH \xrightarrow{H^+} A \xrightarrow{CH_3MgBr} B \xrightarrow{H_3O^+} C$$

(a) 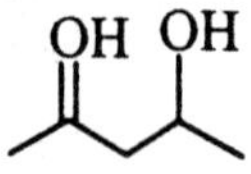 OH OH (b) 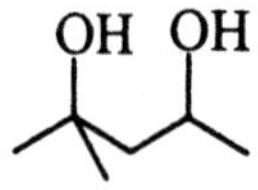 OH OH

(c) 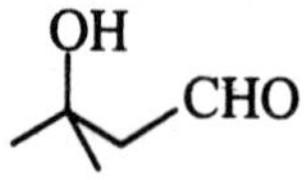 OH, CHO (d) 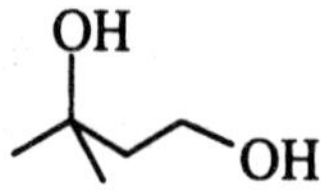 OH, OH

23. What is the product of the following reaction?

$$C_6H_5—CO—CH_3 \xrightarrow{Zn(Hg),\ HCl,\ H_2O}$$

(a) $C_6H_5—CH_3$
(b) $C_6H_5—CHOH—CH_3$
(c) $C_6H_5—CH_2—CH_3$
(d) $C_6H_5—CO_2H$

24. What conditions convert:
$CH_3—CH{=}CH—CH_2—CH{=}O$ to
$CH_3—CH{=}CH—CH_2—CH_2OH$?
(a) $NaBH_4$ (b) Zn(Hg), HCl, H_2O
(c) $Ag(NH_3)_2^+$
(d) $H_2N—NH_2$, then KOH and heat

25. Which one of the following has the highest boiling point?
(a) butanal (b) propionaldehyde
(c) methyl ethyl ketone (d) acctone

26. Which one is useful for converting a terminal alkyne to an aldehyde of the same carbon content?
(a) H_2O and a trace of acid.
(b) Hg^{2+}, H_2SO_4, H_2O.
(c) Hydroboration with disiamylborane, then hydrogen peroxide oxidation.
(d) Pyridinium chlorochromate (a complex of pyridine, HCl, and CrO_3).

27. What intermediate is formed in the reaction of a nitrile with a Grignard reagent to form a ketone?
(a) The magnesium salt of a carboxylic acid.
(b) An amine.
(c) The ammonium salt of a carboxylic acid.
(d) The magnesium salt of an imine.

28. What is the initial step in the reduction of acetone by sodium borohydride to produce 2-propanol?
(a) Attack of a hydride anion on the carbonyl carbon.
(b) Attack of a hydride anion on the carbonyl oxygen.
(c) Attack of the boron on the carbonyl oxygen.
(d) Attack of the sodium on the carbonyl oxygen.

29. In the reaction of acetaldehyde with HO^-, which atom of the acetaldehyde undergoes nucleophilic attack by the HO^-?
(a) The C of the CH_3 group.
(b) The carbonyl carbon.
(c) The H bonded to the carbonyl C.
(d) One of the H's bonded to the CH_3 group.

30. Which structural factor favours the formation of stable hydrates from aldehydes and ketones?
(a) The presence of electron-withdrawing groups on the alkyl carbons.
(b) The presence of basic substituents on the alkyl carbons.
(c) The presence of electron-donating groups on the alkyl carbons.
(d) The presence of acidic substituents on the alkyl carbons.

31. Which of the following carbonyl compounds reacts with HCN to give the lowest yield of cyanohydrin?
(a) formaldehyde (b) propanal
(c) 2-butanone (d) di-*t*-butyl ketone

32. What derivative of aldehydes and ketones is known as a *Schiff base*?
(a) hydrazone (b) cyanohydrin
(c) imine (d) oxime

33. What catalyst is used in acetal formation?
(a) base (b) acid
(c) pyridine (d) either acid or base

34. Which set of reagents would you use to form the following compound?

(a) HO—CH_2—CH_2—OH and CH_2=O
(b) HO—CH_2—CH=O and CH_2=O
(c) O=CH—CH=O and CH_2=O
(d) O=CH—CH=O and HO—CH_2—OH

35. Which reagent should you use to convert *p*—CH_3—(C_6H_4)—CH=O to *p*—CH_3—(C_6H_4)—CO_2H?
(a) Zn(Hg), HCl, H_2O
(b) $Ag(NH_3)_2^+$
(c) HO—CH_2—CH_2—OH, H^+
(d) $Na_2Cr_2O_7$ and H_2SO_4

36. What gas is evolved in a Wolf-Kischner reduction?
(a) N_2 (b) O_2 (c) SO_2 (d) CO_2

37. What term applies to an equilibrium between an enol and its corresponding carbonyl compound?
(a) Carbonyl condensation.
(b) Alpha rearrangement.
(c) Ketonic equilibrium.
(d) Keto-enol tautomerism.

38. What normally results from the reaction of a ketone and a halogen in the presence of a base?
(a) Reduction of the ketone to a 2° alcohol.
(b) Conversion of the ketone to a haloalkene.
(c) Conversion of the ketone to a haloalkane.
(d) Halogenation of the carbon α to the carbonyl group.

39. What represents the product of the Hell-Volhard-Zelinsky reaction of CH_3—CH_2—CO_2H, Br_2, PBr_3?
(a) CH_3—CH_2—Br
(b) CH_3CHBr—CO_2H
(c) CH_3—CHBr—C(=O)Br
(d) CH_3—CH_2—CO—Br

40. What reagents are used to generate an *enamine*?
(a) A 3° amine and an ester.
(b) Thionyl chloride and an amide.
(c) An acyl halide and a 3° amine.
(d) A 2° amine and an aldehyde or a ketone.

41. What is the product of the *aldol* condensation?
(a) A beta-hydroxy aldehyde or ketone.
(b) An aldehyde or ketone with two beta-alkyl groups.
(c) A beta-hydroxy amide.
(d) An alpha, beta-unsaturated ester.

42. What is normally produced on dehydration of the product of an aldol condensation?
(a) A lactone (b) An alkene
(c) A conjugated, unsaturated carbonyl compound.
(d) An unsaturated ester.

43. What is the effective nucleophile in a Dieckmann cyclization?
(a) The oxygen of a carbonyl group.
(b) The carbanion of the alpha-carbon of a diester.
(c) An alkoxide anion derived from an alcohol.
(d) Anion of an amide.

44. What is the most acidic of the hydrogens on propyl acetoacetate?

CH_3—CO—CH_2—CO_2—CH_2—CH_2—CH_3
(V) (W) (X) (Y)

(a) V (b) W (c) X (d) Y

45. Carboxylic acids that contain carbonyl groups in the beta position readily undergo, what reaction?
(a) Decarboxylation on heating.
(b) Amide formation on reaction with amines.
(c) Conversion to acyl bromides with solutions of Br_2 in water.
(d) Formation of acid anhydrides on reaction with another carboxylic acid A.

46. What is the final product of the *acetoacetic ester* synthesis?
(a) Substituted derivatives of malonic acid.
(b) Substituted derivatives of acetic acid.
(c) Substituted derivatives of acetone.
(d) Substituted derivatives of acetaldehyde.

47. In *conjugate addition,* a nucleophile attacks:
(a) A carbonyl carbon.
(b) A carbon alpha to the carbonyl carbon.
(c) An alpha-hydrogen.
(d) A carbon beta to the carbonyl carbon.

48. What has the lowest value for its pK_a?
(a) Benzaldehyde.
(b) A ketone bearing several alpha hydrogens.
(c) An alkane containing a 3° H.
(d) An amide containing a nitrogen with no alkyl groups bonded to it.

49. What are the products of the reaction of acetophenone with excess I_2 in the presence of base? (Ph– = C_6H_5)
(a) $PhCO_2^-$ and HCI_3 (b) *m*-iodoacetophenone
(c) PhC(O)—CH_2I (d) PhC(O)—CHI_2

50. The mechanism of the Hell-Volhard-Zelinsky reaction includes, which of the following?

(a) Electrophilic attack on the carbonyl carbon.
(b) Removal of an alpha-hydrogen by base.
(c) Decarboxylation of the carboxylic acid.
(d) Enolization of an acyl bromide.

51. What step occurs in an aldol condensation?
(a) A nucleophilic attack on a C=C double bond by a carbonyl carbon.
(b) Attack by a hydroxide anion on a carbonyl carbon.
(c) Nucleophilic attack by an enol or enolate carbon on a carbonyl carbon.
(d) Alkylation of an enolate with an alkyl bromide.

52. What is the total number of aldol products produced on all combinations of self-condensation and crossed-condensation of a mixture of propanal and butanal?
(a) 2 (b) 5 (c) 4 (d) 3

53. In the Claisen condensation:
(a) an ester serves as both the nucleophile and the acylating reagent.
(b) a keto ester serves as the nucleophile, and an ester serves as an acylating reagent.
(c) an aldehyde or ketone serves as the nucleophile, and an ester serves as the acylating reagent.
(d) an ester serves as the nucleophile, and an aldehyde or ketone serves as the acylating reagent.

54. Which set of carbonyl compounds will give $(C_6H_5)—CO—CH_2—CO—CH_3$ through a crossed Claisen condensation?
(a) Acetophenone and methyl acetate.
(b) Acetophenone and benzaldehyde.
(c) Methyl benzoate and acetaldehyde.
(d) Methyl benzoate and acetone.

55. Given: (X) diethyl malonate, (Y) acetone, and (Z) ethanol, what is the order in which their acidity increases?
(a) Least acidic Y < X < Z most acidic.
(b) Least acidic Y< Z < X most acidic.
(c) Least acidic X < Z < Y most acidic.
(d) Least acidic Z < Y < X most acidic.

56. What is the product of a malonic ester synthesis that employs $(C_6H_5)—CH_2—Br$ as the reactive halide?
(a) $(C_6H_5)—CO_2H$
(b) $(C_6H_5)—CH_2—CO_2H$
(c) $(C_6H_5)—CH_2—CH_2—CO_2H$
(d) $(C_6H_5)—CH_2—CH_2—CH_2—CO_2H$

57. What *cannot* be produced by either the malonic ester synthesis or the acetoacetic ester synthesis?
(a) 2-butanone
(b) $(C_6H_5)—CH_2—CH_2—CO_2H$
(c) 3-pentanone (d) none

58. In the resonance structures of an alpha, beta-unsaturated carbonyl compound such as $CH_2=CH—CO—CH_3$, where does the positive charge reside when there is a negative charge on the carbonyl oxygen?
(a) Only on the alpha carbon.
(b) On the alpha carbon and the beta carbon.
(c) On the carbonyl carbon and the beta carbon.
(d) On the carbonyl carbon and the alpha carbon.

59. Which is the *tautomer* of a ketone?
(a) alcohol (b) ketene (c) ketal (d) enol

60. What bases are preferred for the alkylation of enolate anions?
(a) $(i\text{-Pr})_2N^-\,{}^+Li$ (b) $Na^{+-}NH_2$
(c) NaOH (d) $Na^{+-}O—CH_2—CH_3$

61. An *enamine* is the nitrogen analog of which of the following?
(a) an alcohol (b) an ether
(c) a ketone (d) an enol

62. A crossed-aldol condensation will yield only one crossed-aldol product, if one of the carbonyl compounds:
(a) has no carbonyl group.
(b) has no beta hydrogens.
(c) has no alpha-hydrogens.
(d) is a methyl ketone.

63. What starting material gives 5- and 6-membered rings through aldol cyclizations?
(a) alpha, beta-unsaturated aldehydes.
(b) 1, 4- and 1, 5-diketones.
(c) diamides of dicarboxylic acids.
(d) esters of dicarboxylic acids.

64. What is the product of a Dieckmann condensation?
(a) A cyclic alpha-keto anhydride.
(b) A cyclic beta-keto ester.
(c) A cyclic alpha-keto ester.
(d) A cyclic diester.

65. A crossed Claisen condensation is normally carried out by adding the ester with alpha-hydrogens, slowly, to a mixture of the ester *without* alpha-hydrogens and the alkoxide base. Why is not the ester *without* alpha-hydrogens added to a mixture of the ester with alpha-hydrogens and the alkoxide base?

(a) The base would be consumed in a transesterification reaction.
(b) The ester with alpha-hydrogens would be converted to the unreactive enol.
(c) The base would be consumed in an aldol condensation.
(d) The base would induce a self-condensation of the ester with alpha-hydrogens.

66. What characteristic of acetoacetic ester and malonic ester makes them particularly useful in organic syntheses?
(a) They are stronger acids than other carbonyl compounds.
(b) They are more easily reduced than other carbonyl compounds.
(c) They are more readily attacked by weak nucleophiles than are other carbonyl compounds.
(d) They are more easily oxidized than other carbonyl compounds.

67. What is formed as an intermediate in the decarboxylation of a carboxylic acid containing a beta carbonyl group?
(a) a cation (b) a free radical
(c) an enol (d) an anion

68. What is always a product of the acetoacetic ester synthesis?
(a) acetone (b) CO
(c) H_2O (d) RCH_2COOH

69. Methyl ethyl ketone (2-butanone) can be produced easily, from which of the following sets of reactants?
(a) Methyl bromide and propanal.
(b) Methyl bromide and malonic ester.
(c) Ethyl bromide and acetaldehyde.
(d) Methyl bromide and acetoacetic ester.

70. What kind of reaction is represented by the following reaction? benzaldehyde + acetic anhydride → $(C_6H_5)—CH{=}CH—CO_2H$ + acetic acid
(a) Michael addition
(b) Dieckmann condensation
(c) Perkin condensation
(d) Claisen condensation

Level 2

1. Give the major product of the following reaction:

Et_2CuLi / H_2O → Major

(a) Et, =O (b) Et, OH (c) Et, =O, Et (d) Et, =O, Et

2. Choose the statement that is true for the following equilibrium:

(A) $\xrightarrow{OH^-}$ (B)

(a) The major intermediate, when A is treated with base, is an enol.
(b) This equilibrium lies to the left.
(c) This equilibrium lies to the right.
(d) The side that this equilibrium lies cannot be predicted.

3. The major product of the following reaction is?

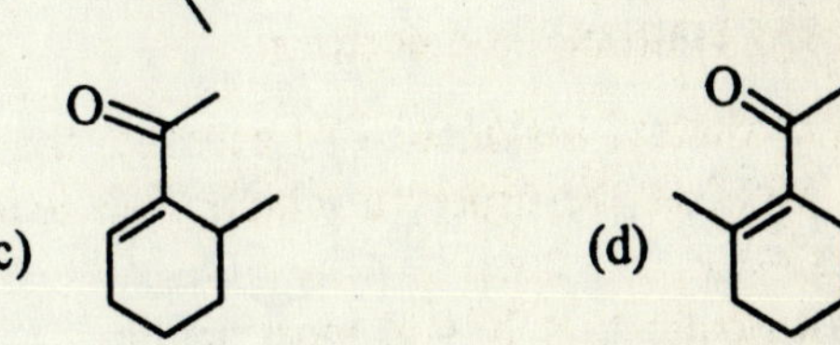

(a) (b) (c) (d)

4. Give the best reaction sequence to accomplish the transformation shown:
(a) $Me_2CuLi/H_3O^+/NaBH_4,MeOH$
(b) $MeMgBr/H_3O^+/LiAlH_4/H_3O^+$
(c) $Me_2CuLi/H_3O^+/MCPBA$
(d) $MeLi/H_3O^+/LDA,THF$

O → OH

5. Give the major product of the following reaction:

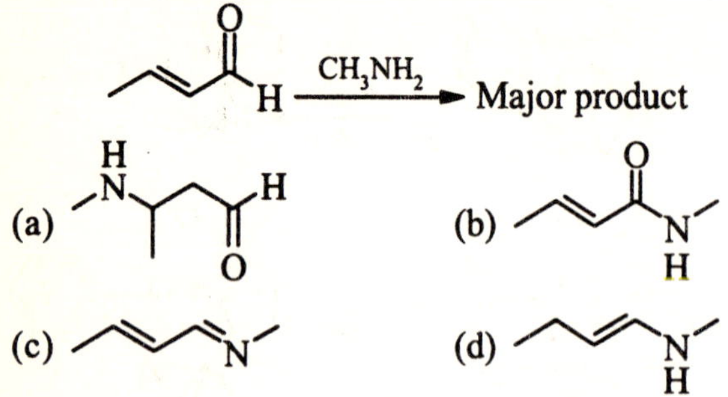

6. Give an intermediate that is formed in the following reaction: Cyclohexanone + $H_2NOH/HCl \rightarrow$

7. Give the product of the following reaction:

8. Give the major product of the following reaction:

9. Give the major product of the following reaction:

10. Give the major product of the following reaction:

11. Give the major organic product from the following reaction:

12. Give the best reagents for the following transformation:

(a) (1) cyclobutanamine; H^+ (2) MeI (3) H_3O^+

(b) (1) LDA (2) CH_3I

(c) (1) MeLi (2) H_3O^+

(d) (1) NaH (2) MeI (3) H_3O^+

13. Give the major product in the following reaction:

(a) $CH_3COOH + CH_3CH_2I_2$
(b) $CH_3CH_2COOH + CHI_3$
(c) $CH_3CH_2CHO + CHI_3$
(d) $CH_3CH_2COOH + CH_3I$

14. Give the major product of the following reaction:

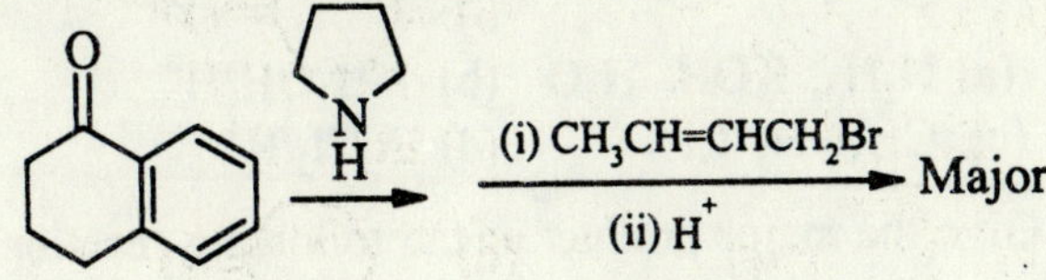

(a) (b)

(c) (d)

15. Give the major product of the following reaction:
Cyclohexanone + dil NaOH → Major

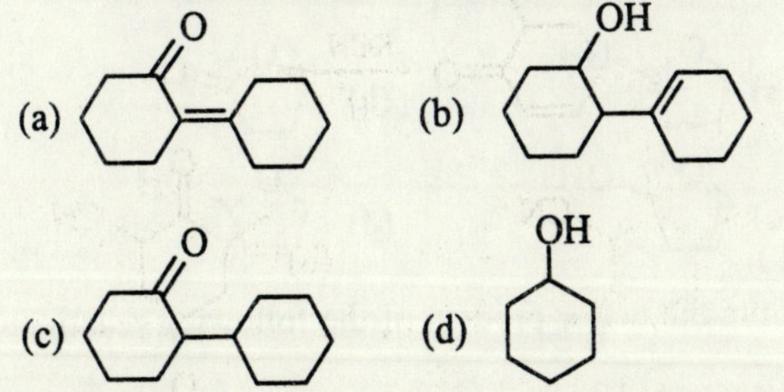

16. Give the major product of the following reaction sequence:

Excess EtBr → Major (H^+)

(a) (b)

(c) (d) All of these

17. Give the major product of the following reaction:

NaOH, Δ → Major

(a) (b)

(c) (d)

18. In the following reaction, what is the maximum number of different products formed that contain a hydroxyl functional group:
Benzaldehyde + Propanal + dil NaOH →
(a) Two (b) Four
(c) Three (d) Five or more

19. The intermediate formed in the following reaction:

CH_3COOH, Br_2

(a) (b)

(c) (d)

20. Give the major product of the following reaction:

NaOH → Major

(a) (b)

(c) (d)

21. Give an intermediate that is generated in the mechanism in the reaction of 2-butanone with excess methanol and anhydrous acid:

(a) 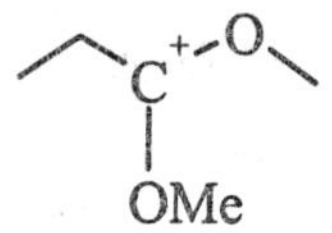 (b)

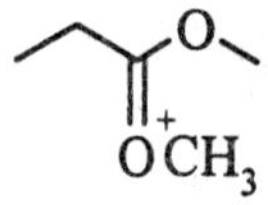

(c) 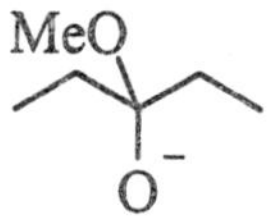 (d)

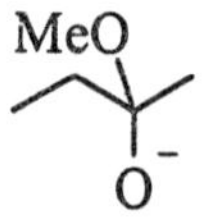

22. Give the product of the following reaction:

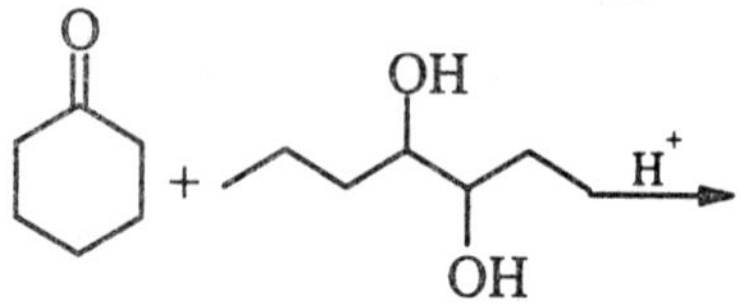

(a) 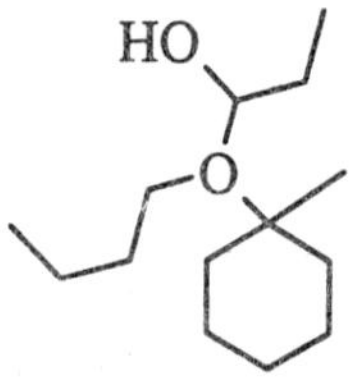 (b)

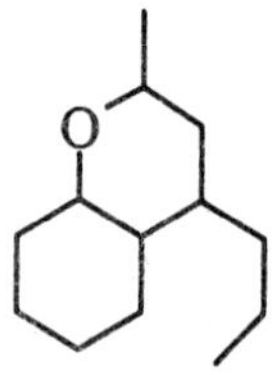

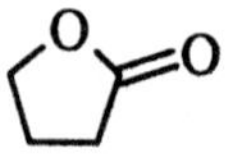

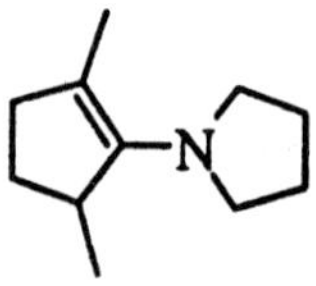

(c) 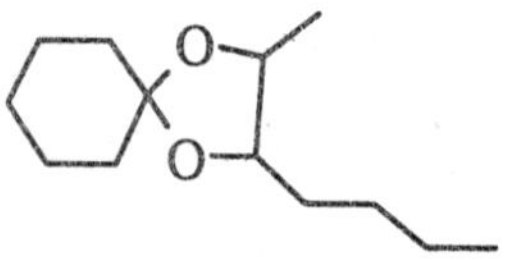 (d)

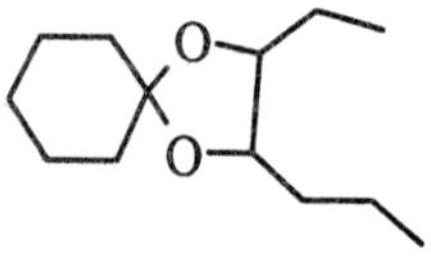

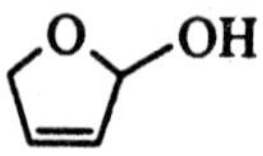

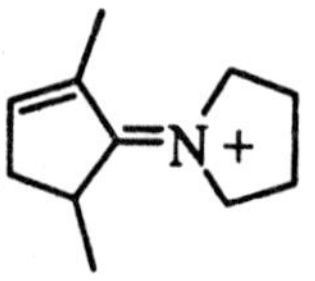

23. Give the product from the following reaction sequence:

Ethyleneglycol / OH H^+ → PCC → EtMgBr → H^+, H_2O

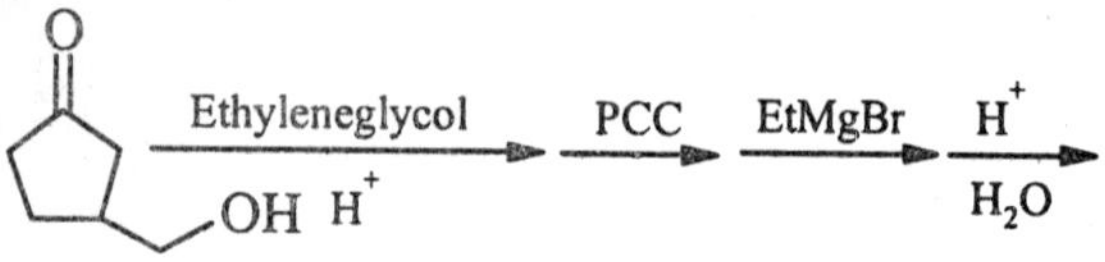

(a) 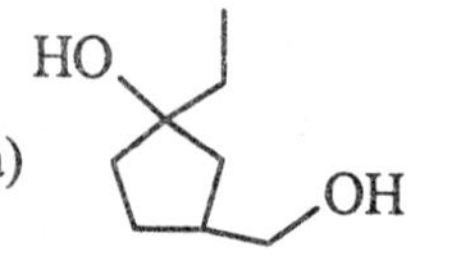 (b)

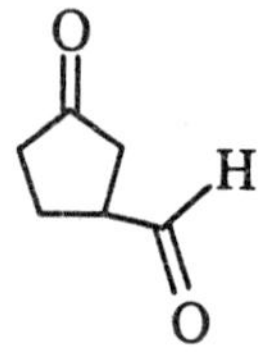

(c) 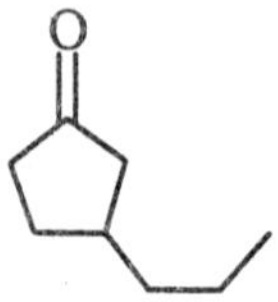 (d)

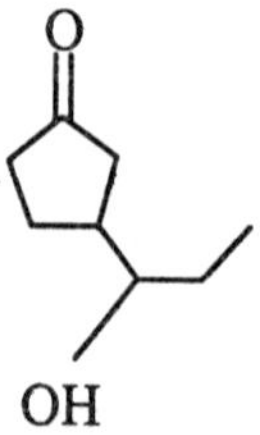

24. Give the major product of the following reaction sequence:

O_3 / Me_2S → $Ph_3P{=}CH_2$ → Major

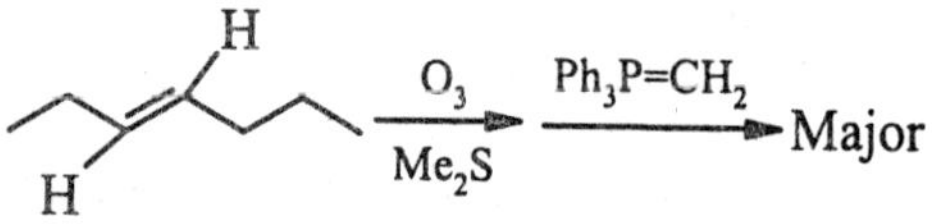

(a)  (b)

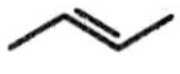

(c)  (d) Both (a) and (c)

25. Give the reagent that will best accomplish the following reaction:

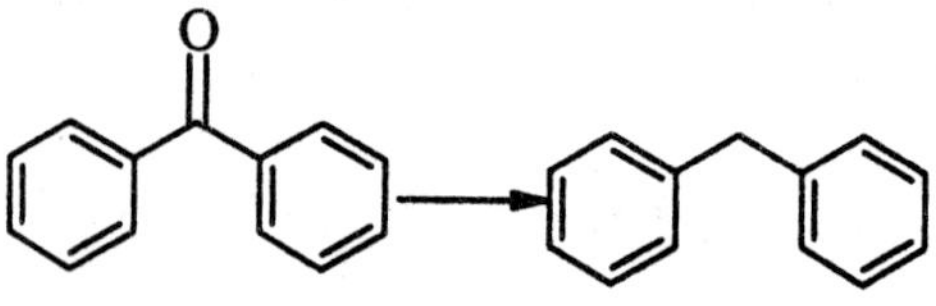

(a) N_2H_4, KOH, H_2O (b) NH_2OH/H^+
(c) $CH_3NH_3^+Cl^-$ (d) $AlCl_3/H^+$

26. Give the major product of the following reaction:

$C_6H_5CHO + H_2NHNCONH_2 \rightarrow$

(a) 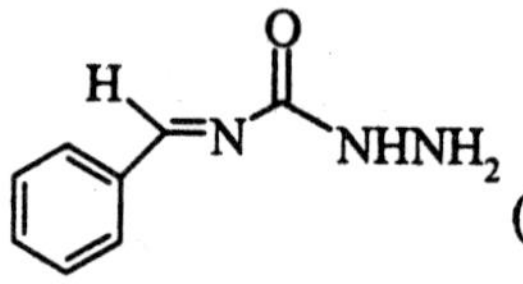 (b)

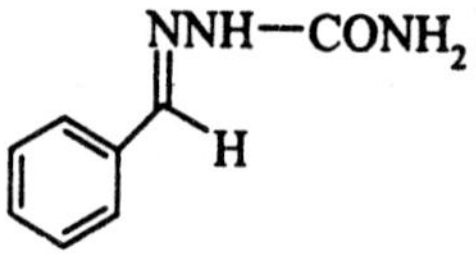

(c) 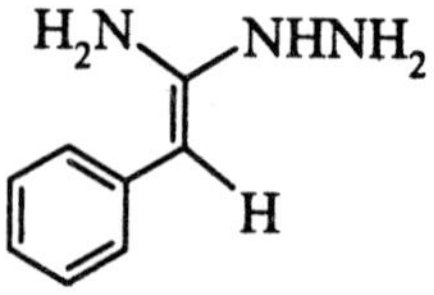 (d)

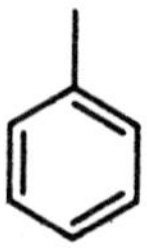

27. Give the major product of the following reaction:

H^+ → Major

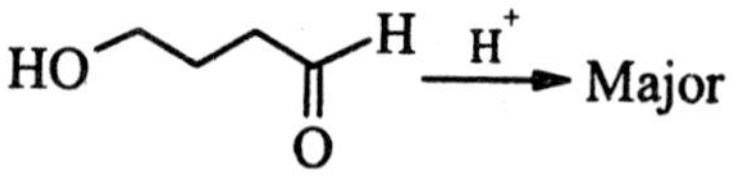

(a) 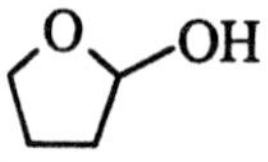 (b)

(c) 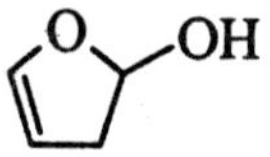 (d)

28. Give the major product of the following reaction:

H^+ → Major

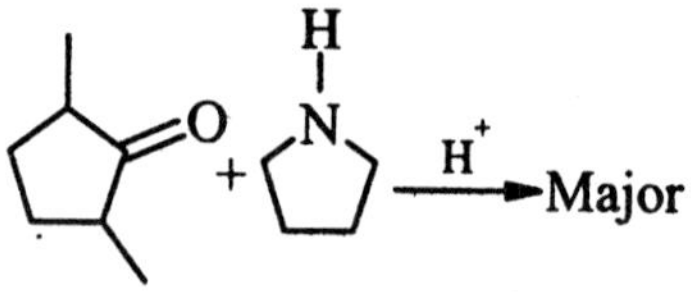

(a) 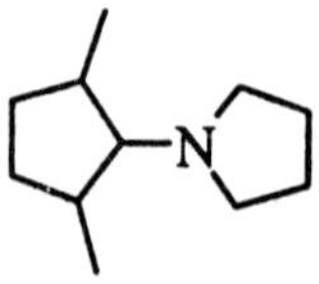 (b)

(c) 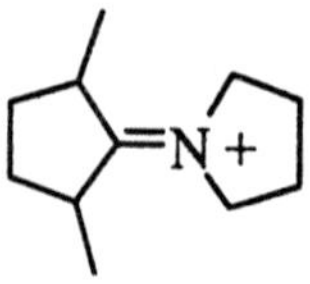 (d)

29. Give the major organic product of the following reaction:

CF_3CO_3H → Major

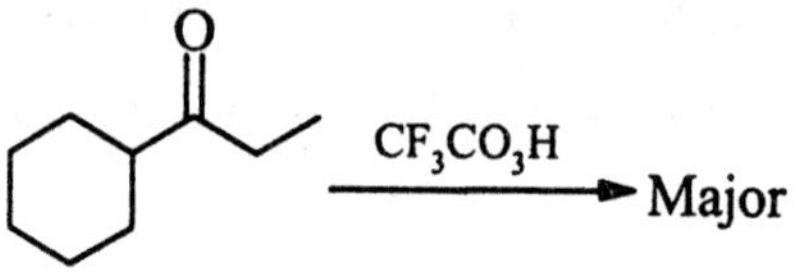

(a) (b) OH

(c) OH (d)

30. Give the major product of the following reaction:

Butanal + CH_3OH (excess) + H^+ → Major

(a) OMe (b) OMe OMe

(c) OMe OMe (d) OMe

31. Give the major organic product of the following reaction:

H_2NOHCl → Major

(a) NHOH (b) NHOH

(c) NH_2 (d) N–OH

32. Give the major organic product from the following sequence of reaction:

Br $\xrightarrow{Ph_3P}$ $\xrightarrow{BuLi}$ $\xrightarrow{CH_3CH_2CHO}$ Major

(a) OH (b)

(c) (d)

33. Give the major product of the following reaction:

Cyclohexanone + N_2H_4/OH^- → Major

(a) NNH_2 (b) $NHNH_2$

(c) (d) NH

34. Give the intermediate that is formed in the following reaction:

Acetone + C_2H_5OH (excess)/H^+ → Intermediate leading to major product

(a) O^-, OEt (b) OEt, C^+

(c) OH, OH (d) O^+–Et

35. Rank the following compounds with respect to increasing favourability of hydration with H_2O/H^+:

(1) (2) (3) Cl (4) F

(a) 4 < 3 <1 < 2 (b) 2 < 1 < 3 < 4
(c) 1 < 2 < 3 < 4 (d) 2 < 3 < 4 < 1

36. Rank the following compounds in order of decreasing reactivity towards addition of a nucleophile to the most electrophilic sp^2 hybridized carbon:

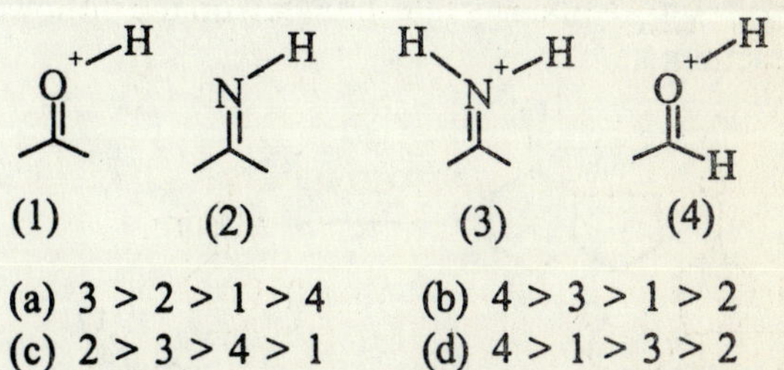

(a) 3 > 2 > 1 > 4 (b) 4 > 3 > 1 > 2
(c) 2 > 3 > 4 > 1 (d) 4 > 1 > 3 > 2

37. Give the reagent that would best accomplish the following transformations:

OH O —Reagent→ OH

(a) $NH_3/H_2O/NaOH$
(b) $N_2H_4/H_2O/NaOH$/heat

(c) $Pd/H_2O/NaOH$/heat
(d) Zn(Hg)/HCl

38. Give the major product of the following reaction:

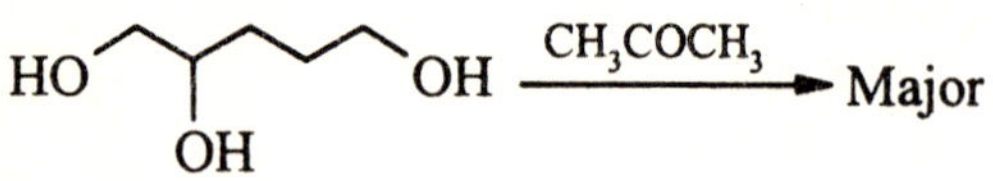

HO ... OH, OH —CH_3COCH_3→ Major

(a) 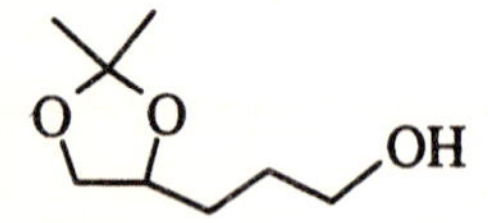 ... OH (b) 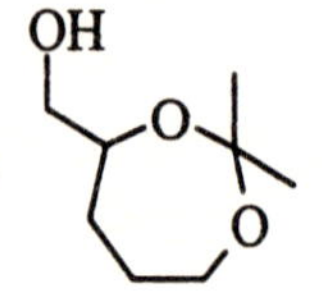 OH ... O ... O

(c) 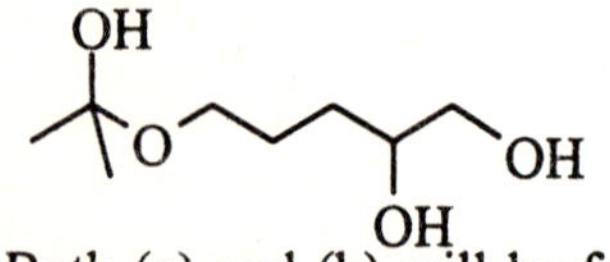 OH ... O ... OH, OH

(d) Both (a) and (b) will be formed in high yield

39. Give an intermediate that is required in the mechanism of the following acid catalyzed reaction:

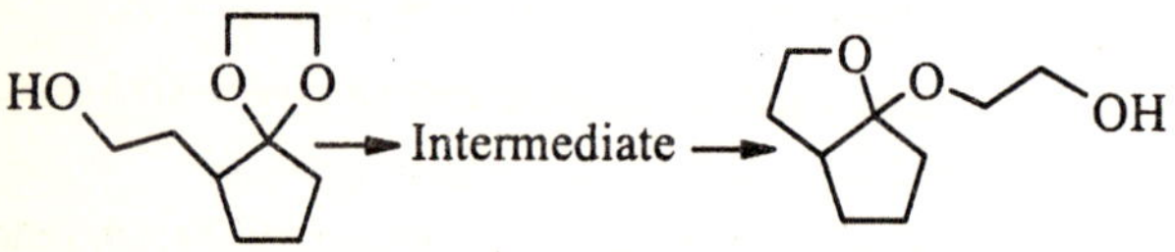

HO ... O O → Intermediate → O ... O ... OH

(a) 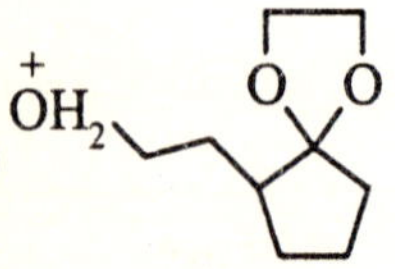 $\overset{+}{O}H_2$ ... O O (b) 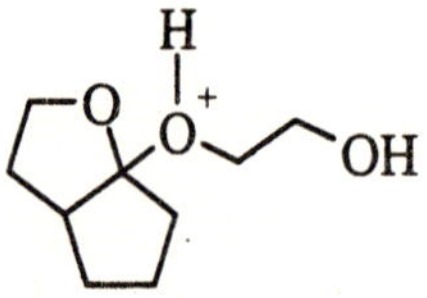 H, O ... $\overset{+}{O}$ ... OH

(c) 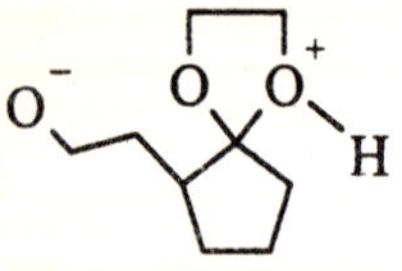 O^- ... O O^+ H (d) 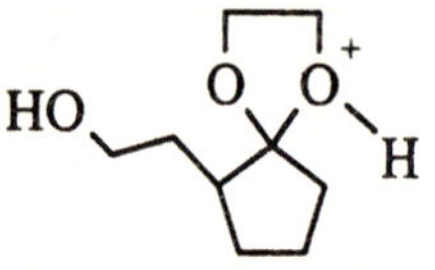 HO ... O O^+ H

40. Give the major organic product of the following reaction:

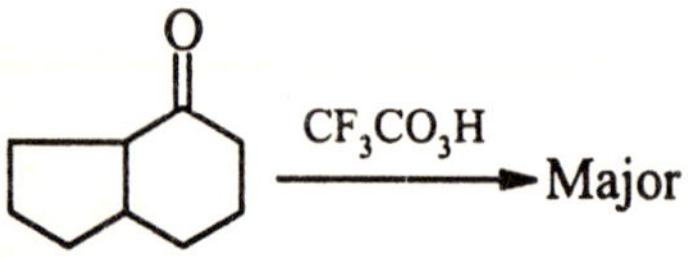

O —CF_3CO_3H→ Major

(a) 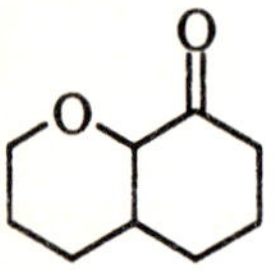 O, O (b) 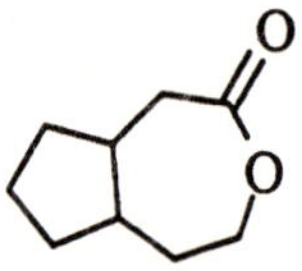 O, O

(c) 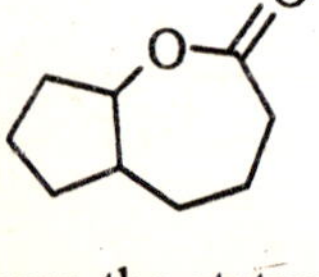 O, O (d) 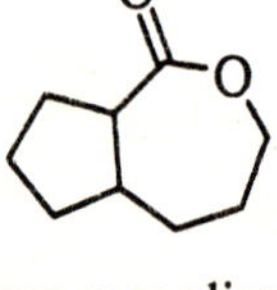 O, O

41. Choose the statement that is true regarding formation of hemiacetals:

(a) Hemiacetals can be formed in only acid catalyzed reactions. Six membered ring hemiacetals are usually isolable. However, acyclic hemiacetals prepared from ketones are usually not isolable.
(b) Hemiacetals can be formed only in base catalyzed reactions. Six membered ring hemiacetals are usually isolable, however acyclic hemiacetals prepared from ketones are usually not isolable.
(c) Hemiacetals can be formed in both acid catalyzed and base catalyzed reactions. Six membered ring hemiacetals are usually isolable, however acyclic hemiacetals prepared from ketones are usually not isolable.
(d) Hemiacetals can be formed in both acid and base catalyzed reactions. Seven membered ring hemiacetals are usually isolable, however acyclic hemiacetals prepared from aldehydes are usually not isolable.

42. Choose the best synthetic route for the synthesis of diene shown below:

(a) 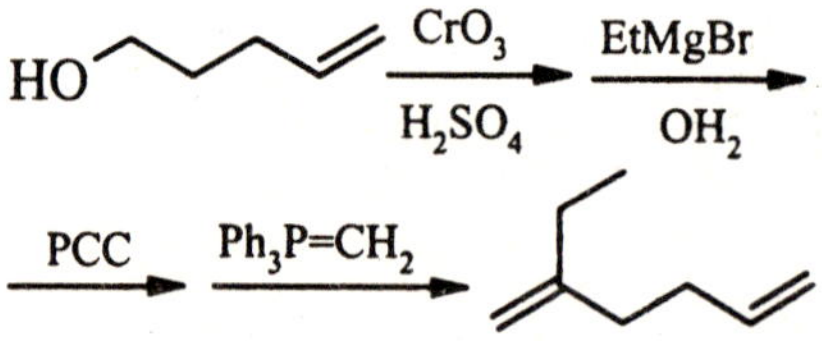 HO ... —CrO_3, H_2SO_4→ —EtMgBr, OH_2→ —PCC→ —$Ph_3P{=}CH_2$→

(b) 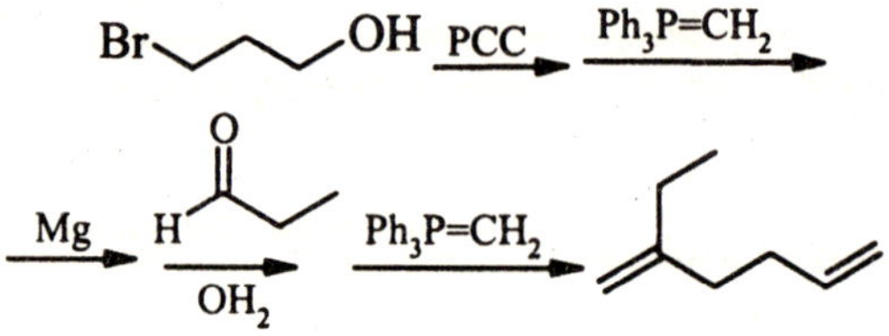 Br ... OH —PCC→ —$Ph_3P{=}CH_2$→ —Mg→ —H(C=O)..., OH_2→ —$Ph_3P{=}CH_2$→

(c) Br ... H(C=O) —$HOCH_2CH_2OH$, H^+→ —Mg→ —H(C=O)..., OH_2→ —CrO_3, H_2SO_4→ —H^+, OH_2→

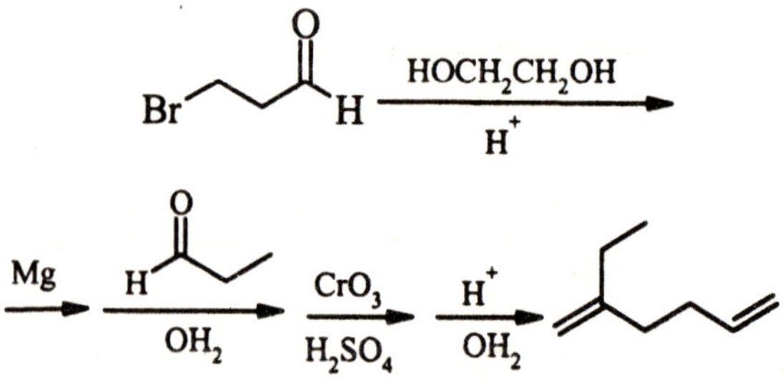

(d) 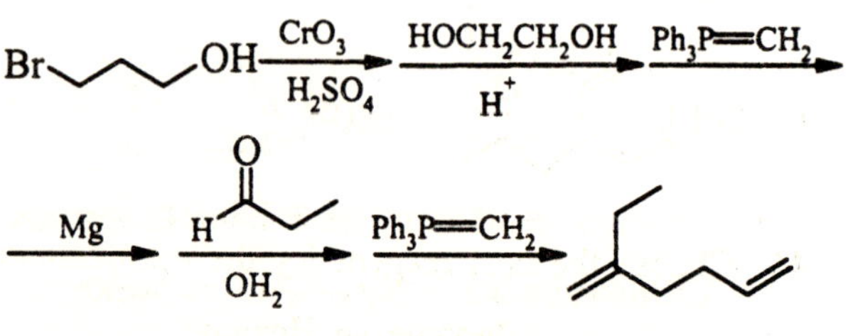 Br ... OH —CrO_3, H_2SO_4→ —$HOCH_2CH_2OH$, H^+→ —$Ph_3P{=}CH_2$→ —Mg→ —H(C=O)..., OH_2→ —$Ph_3P{=}CH_2$→

43. Choose the statement that is correct concerning the following reaction:

$\xrightarrow[DO^-]{D_2O \text{ (excess)}}$

(a) The product is racemic mixture, incorporation of four deuterium is observed for each stereoisomer.

(b) The product contain one stereoisomer, incorporation of one deuterium is observed.

(c) The product contain one stereomer and incorporation of three deuterium is observed.

(d) The product contain one stereoisomer and incorporation of four deuterium is observed.

44. Give the major product of the following reaction:

$\xrightarrow[CH_3COOH]{Br_2}$

(a) Br ... $+CHBr_3$ (b) Br

(c) Br (d) Br

45. Give the major product of the following reaction sequence:

OH $\xrightarrow{CrO_3, H_2SO_4}$ $\xrightarrow[H^+]{\text{(i) EtMgBr}}$

(a) O (b) HO

(c) O (d)

46. Choose the best reagent for the transformation:

1-hexyne → Hexanal

(a) H^+, H_2O (b) BH_3, H_2O_2, OH^-

(c) $HgSO_4$, H_2SO_4, H_2O (d) OH^-, H_2O

47. List all the electrophilic carbons in the molecule below:

(a) C2, C4

(b) C2, C6

(c) C3,C4

(d) C2, C3, C4, C6

48. Choose the statement that is correct concerning the following conversion:

(X)(Y)C=O ⟶ (X)(Y)C(OH)(OH)

(a) With H^+ and H_2O; this is an equilibrium reaction. If X = CH_3 and Y = H then the equilibrium constant K > 1.

(b) With H_2O and OH^-, this reaction is not an equilibrium reaction, and the reaction goes to completion with X = H and Y = H.

(c) Both acid and base can catalyze this reaction. If X = CF_3 and Y = H then K > 1.

(d) With H^+ and H_2O, this is an equilibrium reaction. If X = CH_3 and Y = CH_3, then K > 1.

49. Give the major product of the following reaction:

OH $\xrightarrow{H^+}$ Major

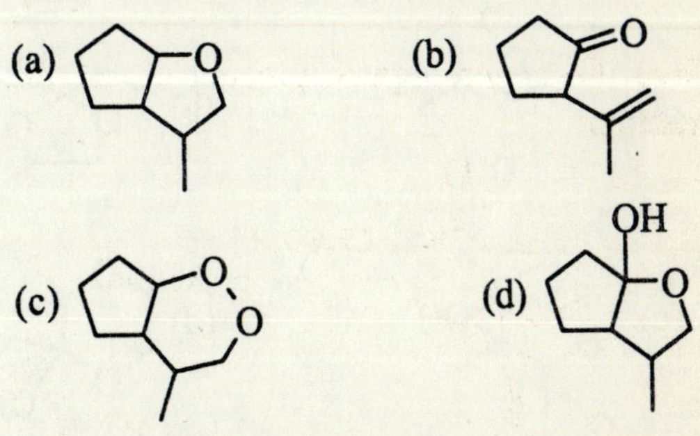

50. Give the major product of the following reaction sequence:

Br $\xrightarrow[H^+]{HOCH_2CH_2OH}$ $\xrightarrow{\text{i) Mg, (ii)}}$ $\xrightarrow[H_2O]{H^+}$ $\xrightarrow{HONH_3Cl}$ Major

(a) (b)

(c) (d)

51. Give the major isolable product of the following reaction:

$\xrightarrow[H^+]{^{18}OH_2\ (excess)}$ Major

(a) (b)

(c) (d)

52. Give the major product of the following reaction:

$\xrightarrow[H^+]{HO\text{–}CH_2CH_2\text{–}OH}$ Major

(a) (b)

(c) (d)

53. Which of the following compounds will have the largest equilibrium constant for the reversible reaction with HCN?

(a) Acetone (b) Ethanal
(c) CH_3COCF_3 (d) CF_3CHO

54. Give the major product of the following reaction:

$\xrightarrow[H^+]{H_2NNHCONH_2}$ Major

(a) (b)

(c) (d)

55. Give the intermediates that are formed in the following reaction:

Cyclopentanone + $N_2H_4/OH^-/heat \rightarrow$ Intermediates?

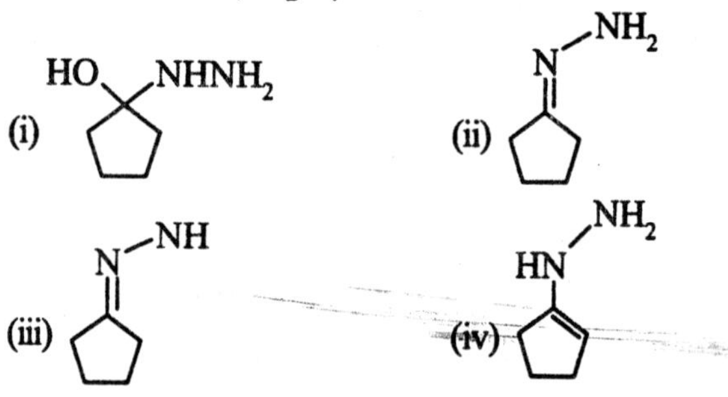

(a) 2, 3 and 4 (b) 1 and 2 only
(c) 2 and 3 only (d) 1, 2 and 3

56. Give the major product of the following reaction:

$\xrightarrow{CF_3CO_3H}$ Major

(a) (b)

(c) (d)

57. What is the major product obtained from the following reaction:

$\xrightarrow[Et_2O]{NaH}$ $\xrightarrow{EtI/Et_2O}$

(a) (b)

(c) (d)

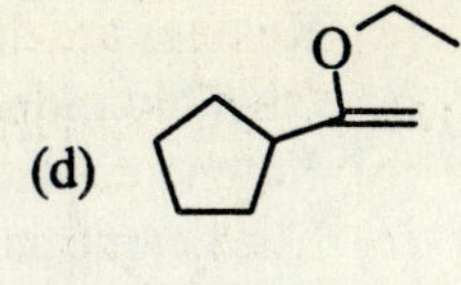

58. What is the major product of the reaction:

2-butanone + EtONa/heat → Major.

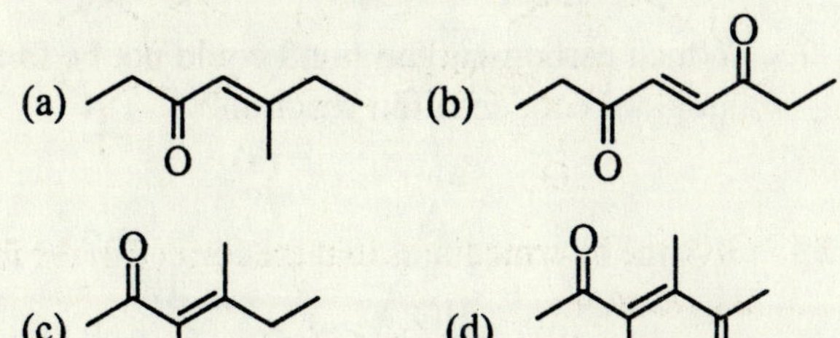

59. What is the major product of the following reaction?

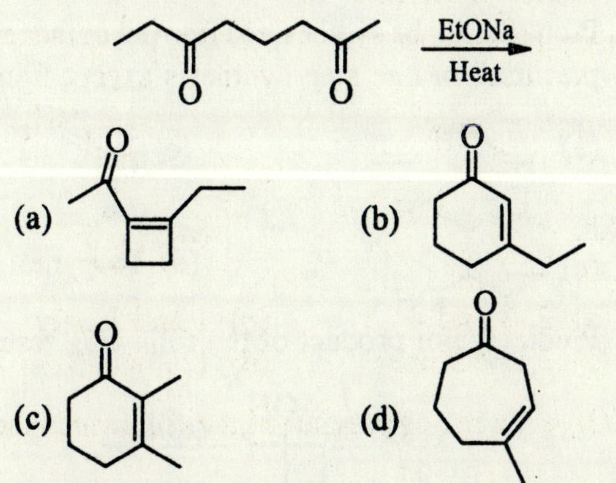

60. What is the major product obtained from the following reaction?

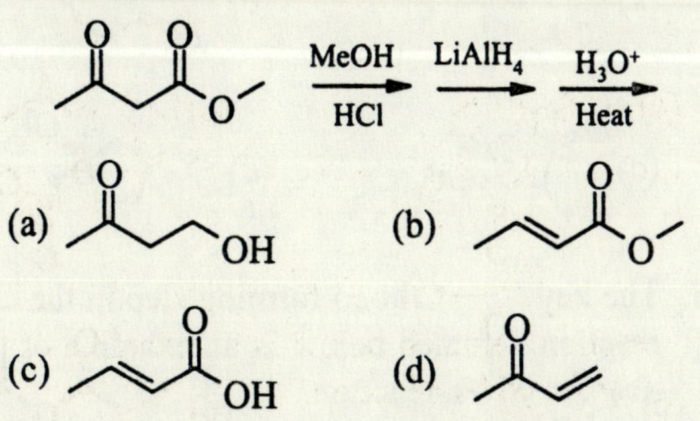

61. What is the major product of the following reaction?

Strong base / Heat

(a) (b)

(c) (d)

62. What is the major product obtained from the following reaction?

$\xrightarrow{Zn/Et_2O}$ $\xrightarrow{HCHO}$ $\xrightarrow[Heat]{H^+}$

(a) (b)

(c) (d)

63. Which sites on the following molecules are susceptible to attack of nucleophile?

(a) a and b
(b) b and c
(c) a and c
(d) all

64. What is the best reagent to use for:

Propanol → Propanal?

(a) Cold/H^+/$KMnO_4$
(b) Cold/OH^-/$KMnO_4$
(c) CrO_3/Pyridine/HCl
(d) O_3 followed by Zn/H_2O

65. Give the major product of the following reaction sequence:

$\xrightarrow{NaCN}$ $\xrightarrow[OH_2]{H^+}$

(a) Propanoic acid
(b) Butanoic acid
(c) 2-hydroxy butanoic acid
(d) 3-hydroxy butanoic acid

66. Which is the principal organic product when benzaldehyde is allowed to react with propionaldehyde in dilute aqueous solution of NaOH at room temperature?

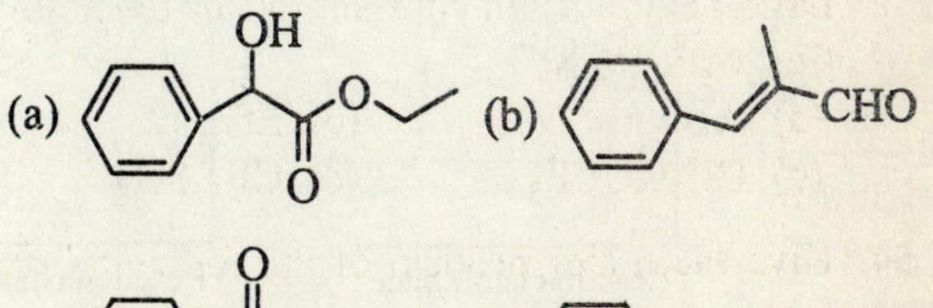

(c) (d)

67. Which statement about aldol condensation is correct?
   (a) A Lewis acid is generally used as catalyst.
   (b) The initial step is probably the formation of a carbanion.
   (c) The Lewis base is used to induce carbocation formation.
   (d) The carbon chain is lengthen through the elimination of one mole of water.

68. A compound gives positive test with I_2/NaOH and is extracted from benzene by saturated $NaHSO_3$. It may be:
   (a) Hexanal (b) 2-hexanone
   (c) 2-hexanol (d) Hexanol

69. In the reaction below, the product would be a:

CHO + HCN ⟶ OH, H CN

   (a) Racemate (b) Optically active
   (c) Meso compound (d) Unresolvable

70. An explanation of addition of sodium bisulphite to heptanal and 2-hexanone but not to 3-hexanone is:
   (a) Steric inhibition of resonance
   (b) Steric hindrance
   (c) Steric inhibition to solvation
   (d) Difference in resonance effect

71. Which reaction coordinate diagram best represents the energetics of the self-aldol condensation of acetone?

O, OH⁻, O⁻, O, O, O⁻, OH_2, O, OH, + OH-

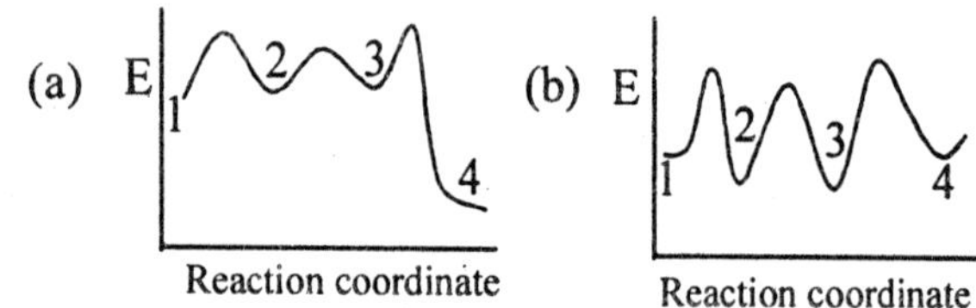

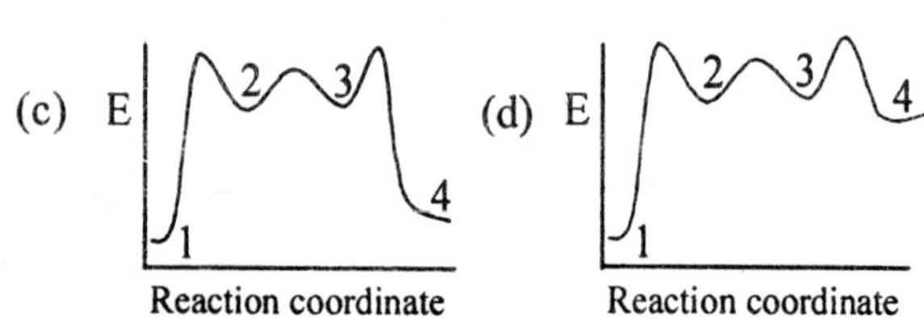

72. High yielding alkylation of ketone enolate with ethyl bromide likely to fail because:
   (a) Ketones are difficult to enolize with weak base ethyl bromide.
   (b) Ketone enolates are not good nucleophile.
   (c) The ketone enolate will attack ethyl bromide at the bromine atom, leading to formation of an α-bromoketone.
   (d) The ketone enolate will act as a base and promote E_2 elimination of HBr.

73. Which carbon-carbon bond could not be formed by an aldol condensation reaction?

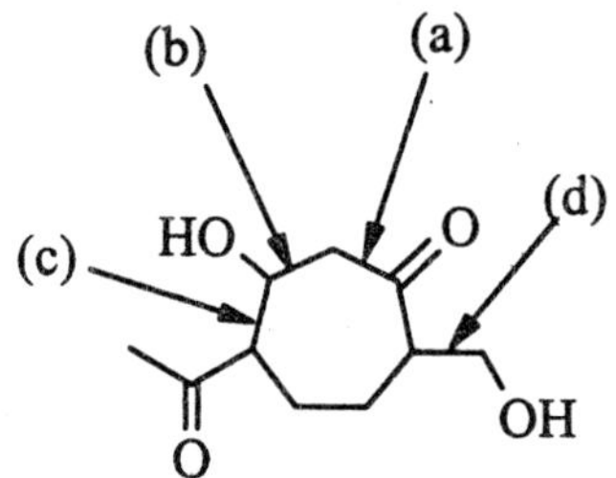

74. Propanal can be synthesized from all of the following precursor in one step synthesis except from:

(a) (b)

(c) OH, OH (d) Propyne

75. Predict major product of the following reaction:

O, OH, EtONa

(a)
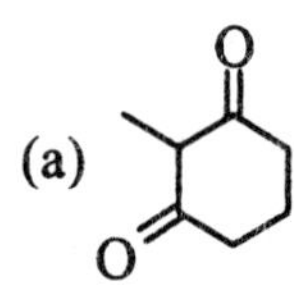

(b)
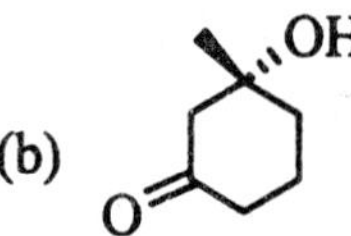

(c)
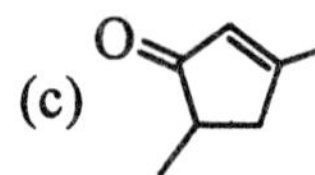

(d) O, OH

76. The key C—C bond forming step in the chemical reaction depicted below is an example of:
   (a) An SN-1 reaction
   (b) An SN-2 reaction
   (c) An enolization reaction
   (d) An aldol condensation

O ⟶ (^{t}BuO⁻K⁺, CH_3I) O

77. Treatment of triketone (1) with one equivalent of LDA and then excess of allyl bromide (2) will provide:

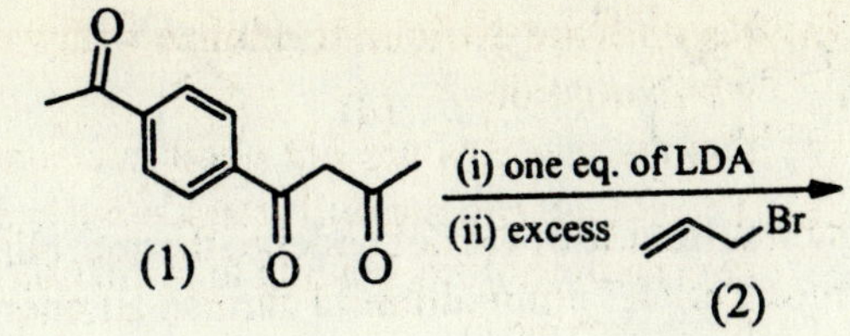

(1) (i) one eq. of LDA (ii) excess allyl bromide (2)

(a) 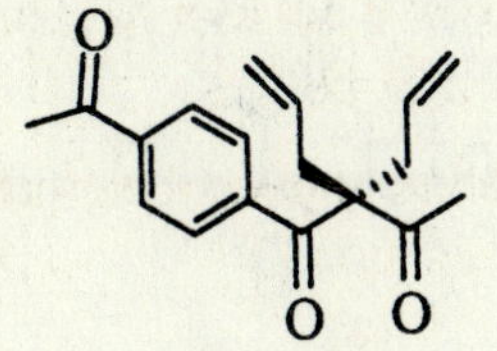 (b) 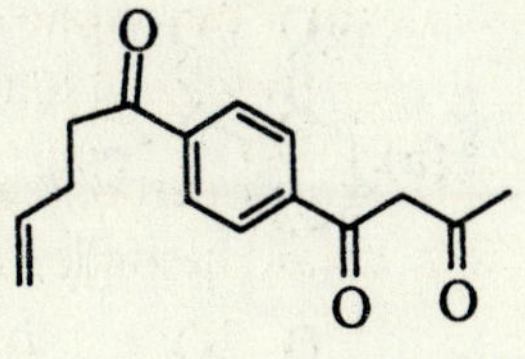 (c) 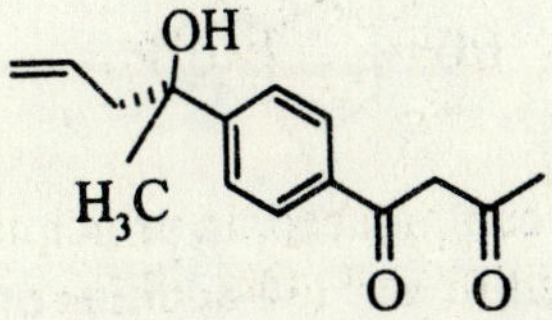 (d)

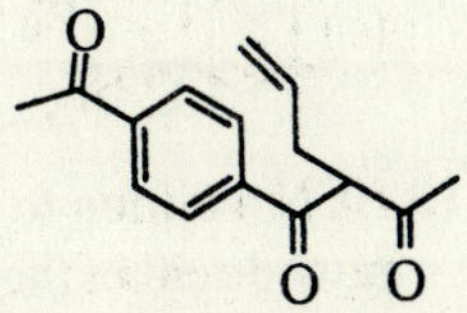

78. Which alkylating agent (2) is not to furnish good yields of alkylated ketone (3) from (1)?

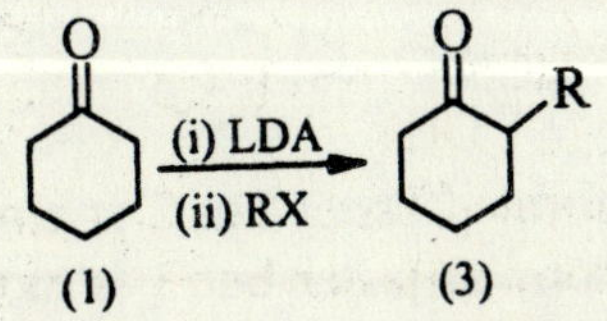

(1) (i) LDA (ii) RX → (3)

(a) 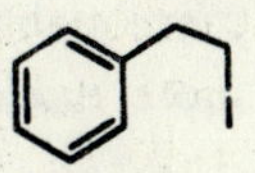 (b)

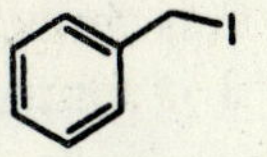

(c) CH_3I (d)

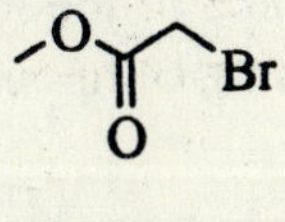

79. Addition of chiral alkyl lithium (1) to aldehyde (2) followed by aqueous acid quench, provides:

(1) 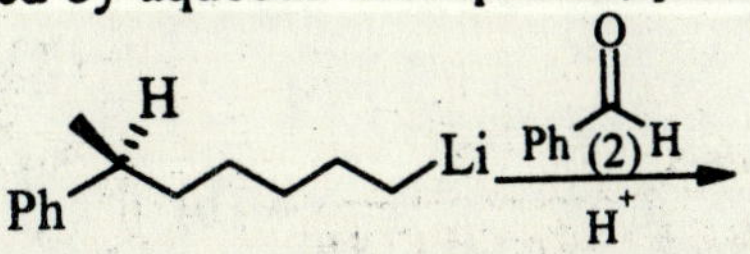 Ph(2)H, H^+

(a) A mixture of racemic distereomeric alcohols.
(b) A mixture of chiral distereomeric alcohols.
(c) A single racemic alcohol distereomer.
(d) A single chiral alcohol distereomer.

80. Which of the following will result in essentially quantitative deprotonation at α-position of cyclohexanone?

(a) LDA (b) CN^-

(c) EtO^- (d)

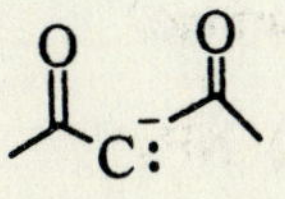

81. Which of the following compounds will have the smallest equilibrium constant for the reversible reaction with HCN?

(a) Acetone (b) Ethanal
(c) CH_3COCF_3 (d) CF_3CHO

82. Treatment of cyclohexanone with CH_3ONa in CH_3OD will provide:

(a) 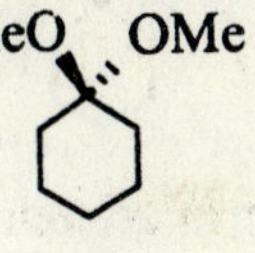 (b)

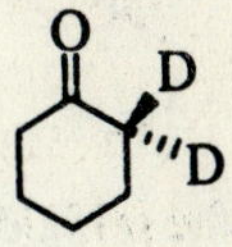

(c) 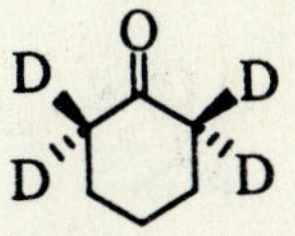 (d)

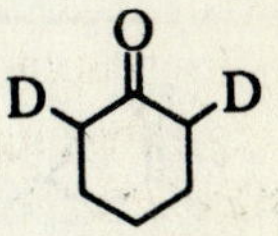

83. Which would not be a good method to prepare butanone?

(a) 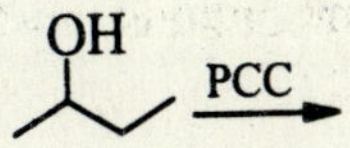 PCC (b) 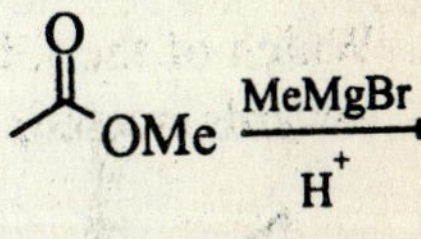 MeMgBr, H^+

(c) 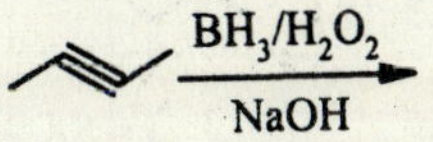 BH_3/H_2O_2, NaOH (d) 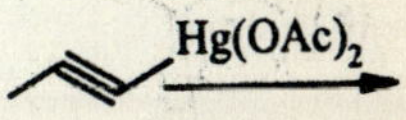 $Hg(OAc)_2$

84. Addition of methyl lithium to a chiral aldehyde (1), followed by acid quench, affords:

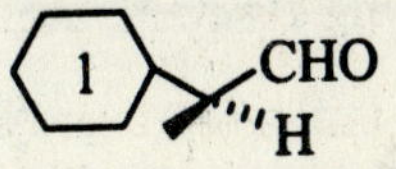

(a) 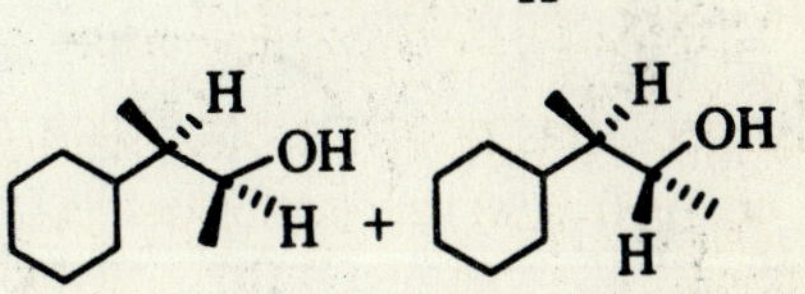 (b) 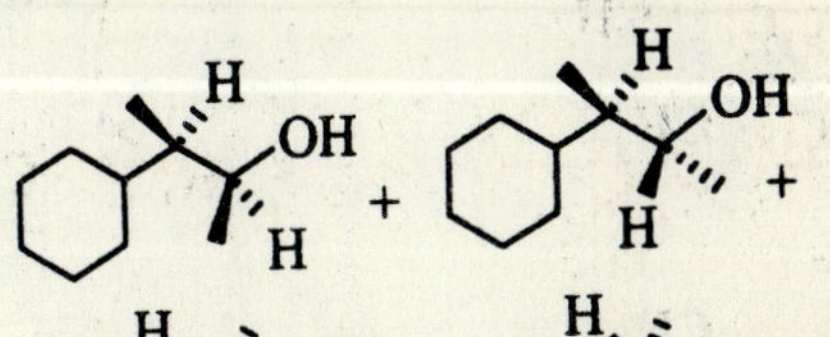 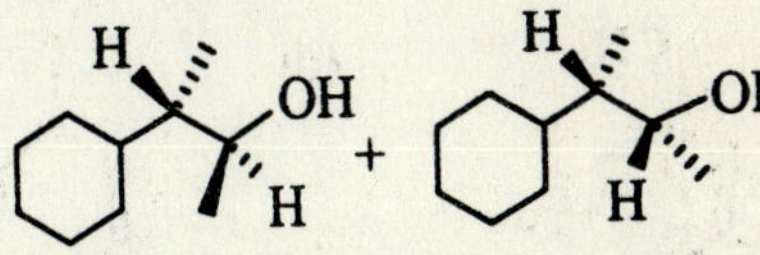 (c) 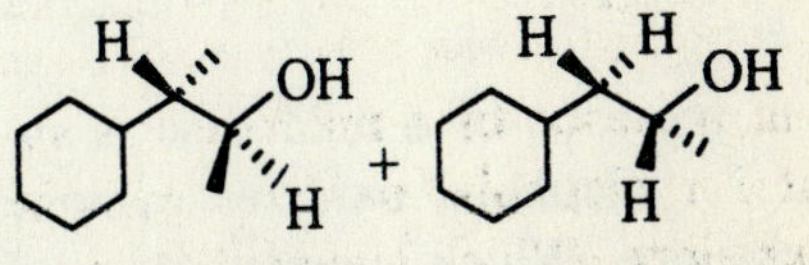 (d)

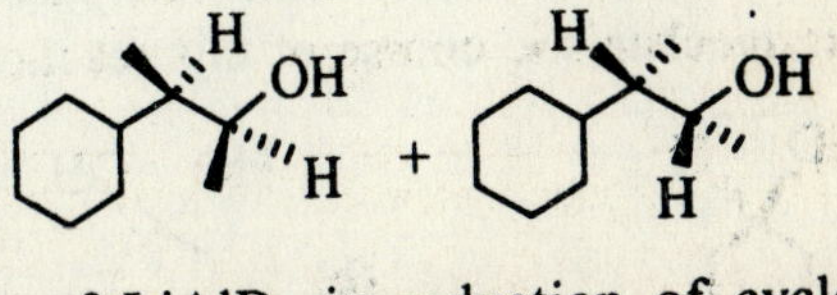

85. Use of $LiAlD_4$ in reduction of cyclohexanone followed by aqueous work-up furnishes:

(a) 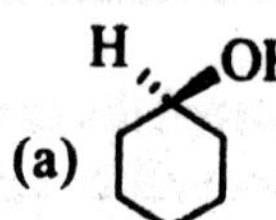
(b) 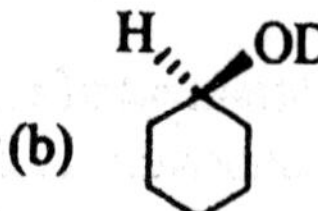

(c) 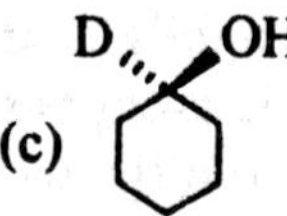
(d)

86. In the self aldol condensation of acetone, the nucleophilic component is:

(a) (b)

(c) NaOH (d)

87. Which of the following compound is most reactive towards nucleophilic attack?

(a) 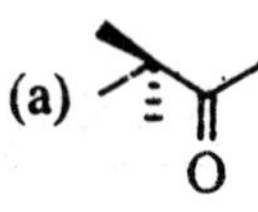
(b)

(c) (d)

88. In the following reaction, the product will be:

(a) 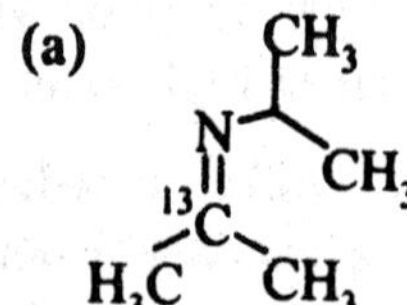
(b) 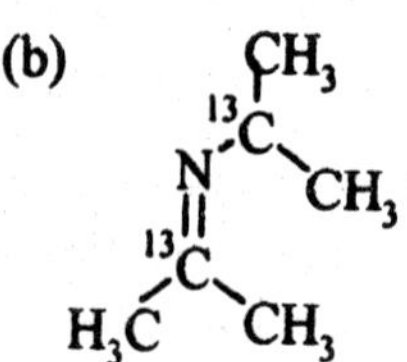

(c) (d)

89. HOCl in presence of a mild base is an effective oxidant for alcohols, oxidizes cyclohexanol to cyclohexanone. Which intermediate play a central role in mechanistic course of this reaction?

(a) (b)

(c) (d)

90. Which sulfoxide could not undergo thermal elimination of PhSOH elimination to furnish an enone?

(a) (b)

(c) (d)

91. HCN addition to enone can proceed, in principle to provide either the 1, 2-addition or 1, 4-addition product, which statement is true concerning the following reaction?

(a) 3 is formed faster but 2 is more stable.
(b) 2 is formed faster but 3 is more stable.
(c) 2 is formed faster and is more stable than 3.
(d) 3 is formed faster and is more stable than 2.

92. Which of the following is the best method of producing compound (1)?

(a) O_3, Zn/CH_3COOH
(b) O_3, Zn/CH_3COOH
(c) PCC
(d) $LiAlH_4$, H^+

93. Which of the following amines would not undergo reversible addition to acetone to form an amine or enamine?

(a) NH_3
(b) t-Bu–NH_2
(c) N–H
(d) H_3C–N(CH_3)–CH_3

94. Compound A was treated with a large excess of CH_3MgBr. The resulting product was exposed to $POCl_3$/Pyridine to give compound B, as one of many products. Which of the following compound can be A?

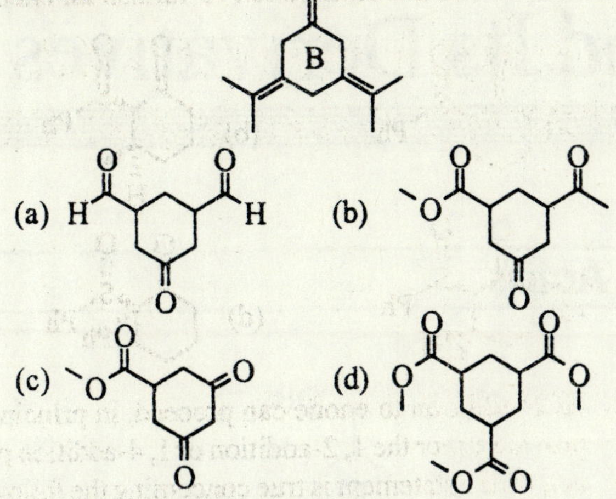

95. Which of the following is the product of the reaction between $AlCl_3$ and acetone?

(a) (b) (c) (d)

96. Which statement regarding the reaction below is true?

I + II → III + IV

(a) The reaction will proceed to the right.
(b) The reaction will proceed to the left.
(c) No reaction occurs: I and III are simply resonance structures.
(d) IV is a very acidic compound.

97. Successful mixed aldol condensation requires:
(a) Treatment of the electrophilic carbonyl components with stoichiometric strong base prior to addition of pre-nucleophilic carbonyl component.
(b) Treatment of a mixture of electrophilic carbonyl component and the pre-nucleophilic carbonyl component with stoichiometric strong base.
(c) Treatment of a mixture of electrophilic carbonyl component and the pre-nucleophilic carbonyl component with catalytic strong base.
(d) Treatment of a pre-nucleophilic carbonyl component with stoichiometric strong base prior to addition of the electrophilic carbonyl component.

98. Which of the following C—C bond is formed during an aldol condensation used to prepare the compound shown below?

(a) Ph (b) OH O (c) (d) CH_3

99. Consider the two alkylation reaction shown below:

1. (i) CH_3OLi (ii) CH_3I

2. LDA (ii) CH_3I

Which reaction is likely to proceed in high yield?
(a) Reaction (1) only.
(b) Reaction (2) only.
(c) Both reaction will proceed in poor yield.
(d) Both reaction will proceed in high yield.

100. In a reaction mechanistically similar to reductive amination, treatment of cyclohexanone with benzaldehyde and methanamine in presence of acid catalyst leads to formation the complex amine (A). A key intermediate en route to A is:

A: O, Ph, $NHCH_3$

(a) H, $\overset{+}{N}$, CH_3 (b) H, $\overset{+}{N}$, CH_3, Ph, H

(c) O, OH, Ph (d) O, $NHCH_3$

Chapter 8

Carboxylic Acids and Its Derivatives

Carboxylic Acids

8.1 INTRODUCTION

The combination of a carbonyl group and a hydroxyl group on the same carbon atom is called carboxyl group. Compound containing the carboxyl group are distinctly acidic and are called carboxylic acids.

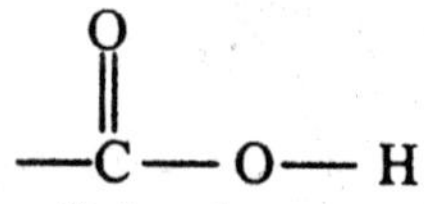

$-C(=O)-O-H$
Carboxyl group

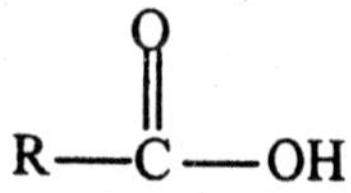

$R-C(=O)-OH$
Carboxyl acid

8.1.1 IUPAC Names

The IUPAC nomenclature for carboxylic acids uses the name of alkane that corresponds to the longest continuous chain of carbon atoms. The final 'e' in the alkane name is replaced by suffix-oic acid. The chain is numbered, starting with the carboxyl carbon atom. In polyfunctional carboxylic acids, the carboxyl group takes priority over any of the functional groups, e.g.

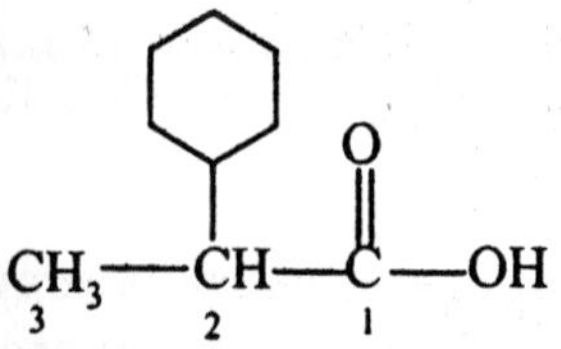

$\overset{3}{CH_3}-\overset{2}{CH}(C_6H_{11})-\overset{1}{C}(=O)-OH$
2-cyclohexyl propanoic acid

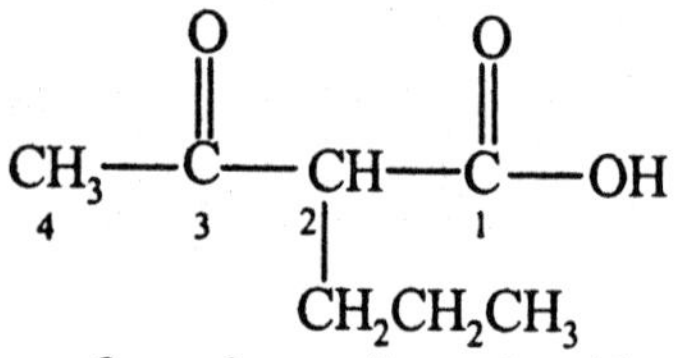

$\overset{4}{CH_3}-\overset{3}{C}(=O)-\overset{2}{CH}(CH_2CH_2CH_3)-\overset{1}{C}(=O)-OH$
3-oxo-2-propylbutanoic acid

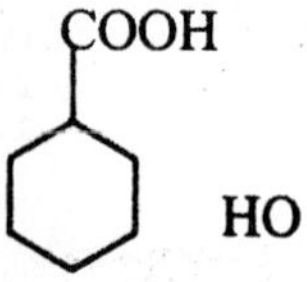

$C_6H_{11}-COOH$
Cyclohexane carboxylic acid

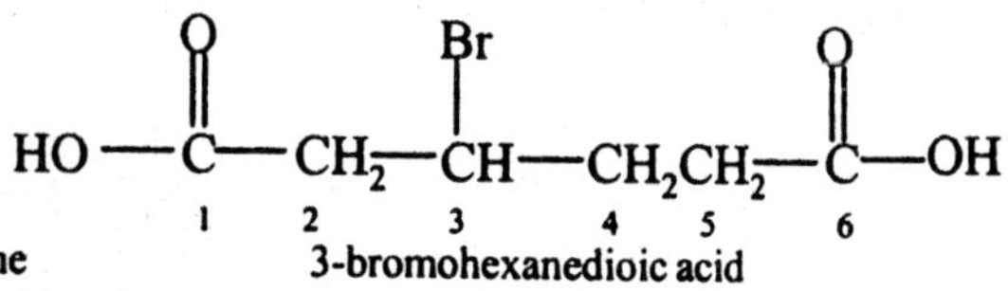

$HO-\overset{1}{C}(=O)-\overset{2}{CH_2}-\overset{3}{CH}(Br)-\overset{4}{CH_2}\overset{5}{CH_2}-\overset{6}{C}(=O)-OH$
3-bromohexanedioic acid

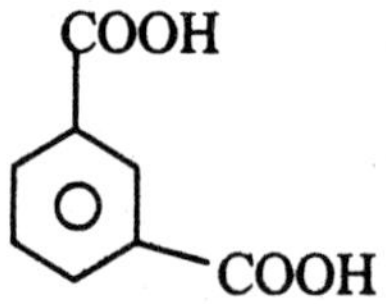

$C_6H_4(COOH)_2$
1,3-benzenedi carboxylic acid

8.2 PHYSICAL PROPERTIES

Boiling points: Carboxylic acids boil at considerably higher temperature than do alcohols, ketones or aldehydes of similar molecular weights. This high boiling points results from formation of a stable, hydrogen bonded dimer as:

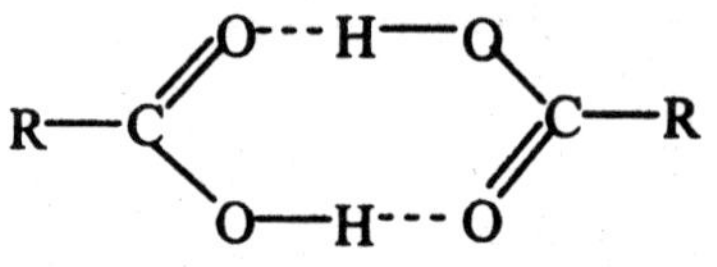

$R-C(=O\cdots H-O)(-O-H\cdots O=)C-R$
H-bonded acid dimer

Melting point: Acids containing more than eight carbons are generally solids, unless they contain double bonds. Increasing MW increases the melting points. Dicarboxylic acid have relatively high melting points.

Solubility: Carboxylic acid forms hydrogen bonds with water and the low MW carboxylic acid (up to 4-carbon) are miscible with water. As the length of hydrocarbon chain increases, solubility in water decreases. Diacids have higher solubility than monocarboxylic acids.

Carboxylic acids are very soluble in alcohols.

8.3 SYNTHESIS

Commercial source of carboxylic acids

I. The most important carboxylic acid from commercial point of view is acetic acid. It is prepared:

(a) By fermentation of sugars and starches:

$$\text{Sugar and starches:} \xrightarrow{\text{Fermentation}} CH_3CH_2OH \xrightarrow[O_2]{\text{Fermentation}} CH_3COOH$$

(b) From acetylene:

$$H-C\equiv C-H \xrightarrow{H_2O,\ HgSO_4,\ H_2SO_4} CH_3CHO$$

$$\xrightarrow[\text{Catalyst}]{O_2} CH_3COOH$$

II. Long-chain aliphatic acids are prepared from hydrolysis of fats and oils. The fatty acids found in fats and oils are generally straight-chain acids with even number of carbon atoms. The hydrolysis of animal fat gives mostly saturated fatty acids, while plant oils gives largely unsaturated fatty acids with one or more olefinic bonds, e.g.

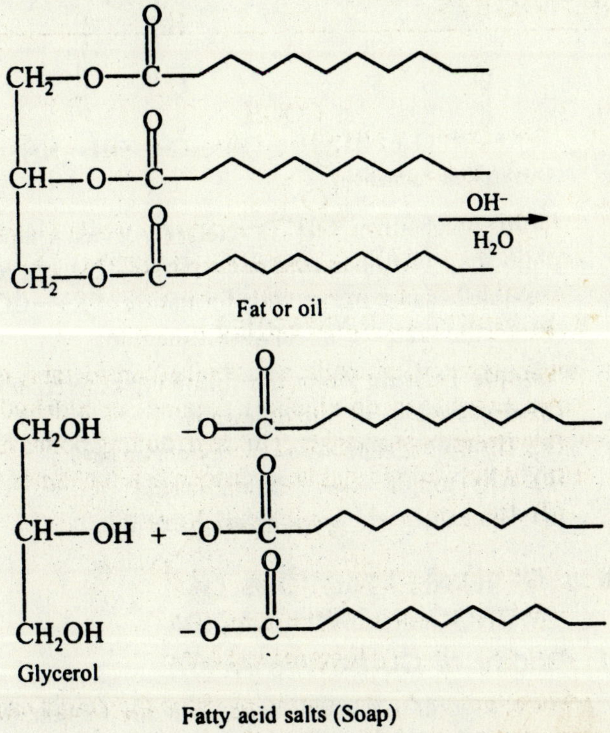

Fat or oil $\xrightarrow[H_2O]{OH^-}$ Glycerol + Fatty acid salts (Soap)

Fat or oil

Glycerol

Fatty acid salts (Soap)

III. Aromatic acids are prepared by oxidation of alkyl substituted benzene and its derivatives as:

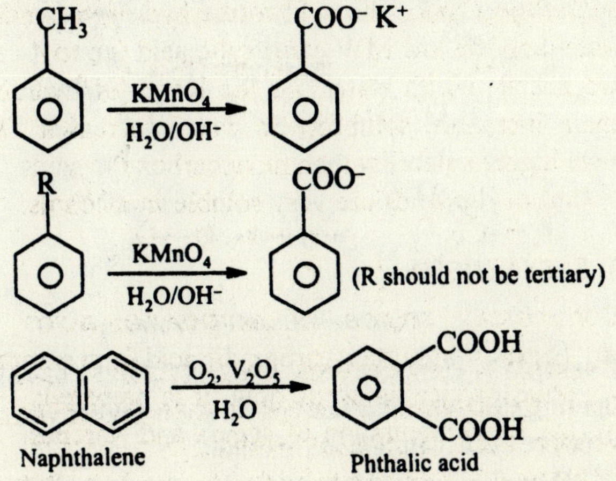

$$C_6H_5CH_3 \xrightarrow[H_2O/OH^-]{KMnO_4} C_6H_5COO^-K^+$$

$$C_6H_5R \xrightarrow[H_2O/OH^-]{KMnO_4} C_6H_5COO^- \quad \text{(R should not be tertiary)}$$

$$\text{Naphthalene} \xrightarrow[H_2O]{O_2,\ V_2O_5} \text{Phthalic acid}$$

Naphthalene

Phthalic acid

8.4 CHEMICAL METHOD OF PREPARATION OF CARBOXYLIC ACIDS

I. Oxidation of alcohols

Primary alcohols and aldehydes are oxidized to carboxylic acids by chromic acid (formed from Na_2CrO_4 and H_2SO_4):

$$R-CH_2OH + Na_2CrO_4 \xrightarrow{H_2SO_4} RCOOH$$

II. Oxidation of alkenes

Cold, dilute potassium permanganate gives glycols with alkene. Warm, concentrated solution of permanganate further oxidize glycols into carboxylic acid as:

$$R-CH=C\begin{matrix}R\\R\end{matrix} + \text{Conc. } KMnO_4 \xrightarrow{\Delta} RCOOH + O=C\begin{matrix}R\\R\end{matrix}$$

In the same way alkynes either on ozonolysis or oxidation with concentrated solution of potassium permanganate produces acids as:

$$R-C\equiv C-R' \xrightarrow[\text{or } O_3/H_2O_2]{\text{Conc. } KMnO_4} \left[R-\underset{O}{\underset{\|}{C}}-\underset{O}{\underset{\|}{C}}-R'\right] \longrightarrow RCOOH + R'COOH$$

Not isolated

III. Carboxylation of Grignard reagents

Grignard reagent attack on carbon dioxide forming magnesium salt of carboxylic acid, which on hydrolysis gives free carboxylic acids as:

$$R-X + Mg \xrightarrow{\text{Dry either}} RMgX \xrightarrow{O=C=O}$$

$$R-\overset{O}{\overset{\|}{C}}-OMgX \xrightarrow{H^+} RCOOH$$

IV. Hydrolysis of nitriles

A nitriles can be hydrolyzed into corresponding carboxylic acid in either acid or alkaline medium. Hydrolysis occurs in two steps—first nitriles are hydrolyzed into intermediate product—a primary amide which is finally hydrolyzed into acid in the subsequent step.

(a) *Hydrolysis to amides:* (i) ***In acid medium:***

Protonation:

$$R-C\equiv\ddot{N} + H^+ \longrightarrow R-C\equiv\overset{+}{N}H \xrightarrow{\ddot{O}H_2}$$

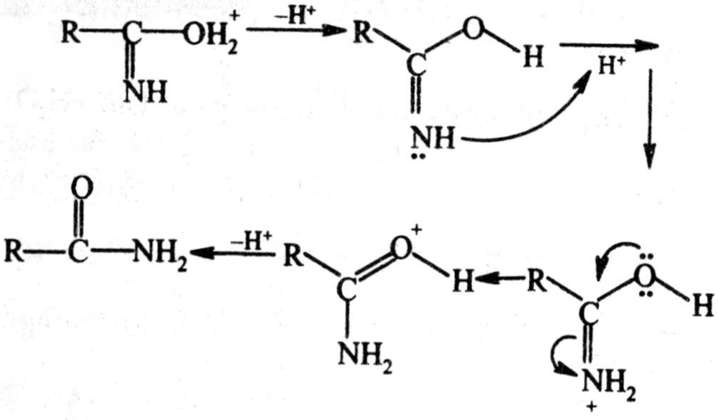

[In concentrated acidic medium, controlled hydrolysis occur and amide is isolated as the final, major product]

(ii) *In alkaline medium*

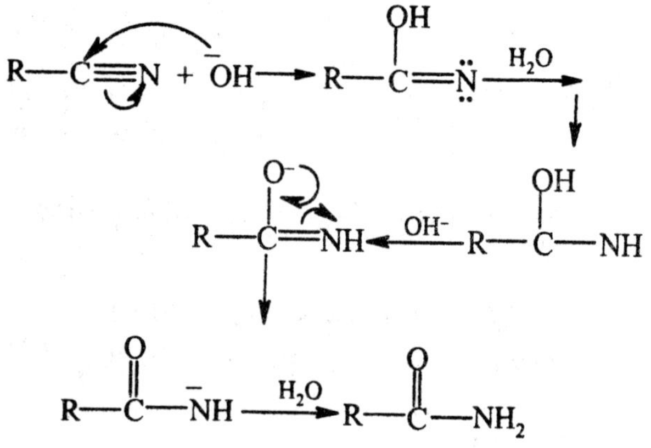

[In alkaline hydrolysis, controlling the reaction is almost impossible and final hydrolysis product (carboxylate ion) is isolated as major product]

(b) *Hydrolysis of amides:* (i) *In acidic medium:* Protonation occur at oxygen as:

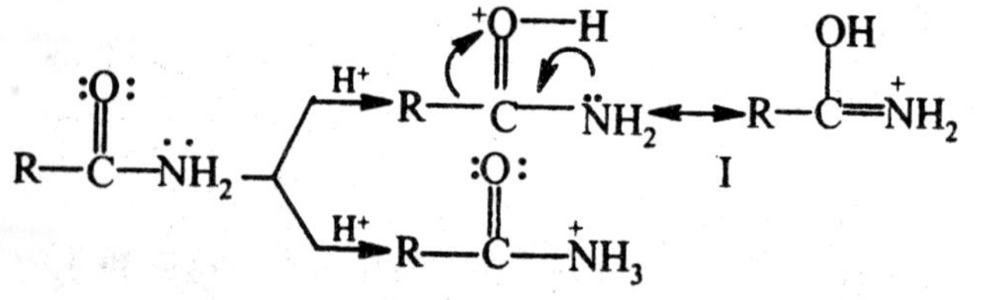

Not stabilized by resonance, therefore protonation does not take place at nitrogen.

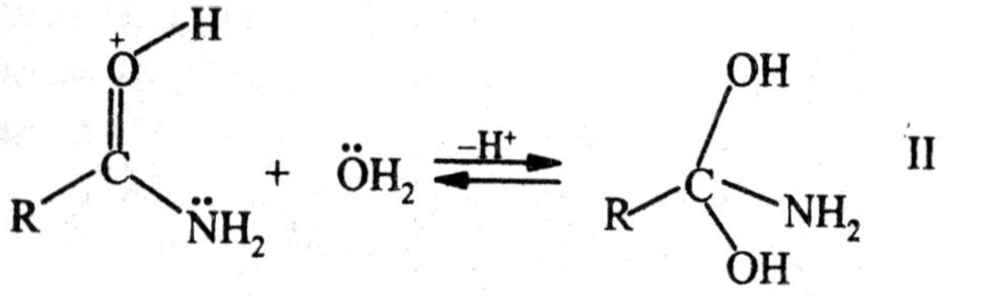

(a tetrahedral intermediate)

The amine nitrogen in tetrahedral intermediate is much more basic than the oxygen atoms, so now it gets protonated and $-^{+}NH_3$ becomes a very good leaving group as:

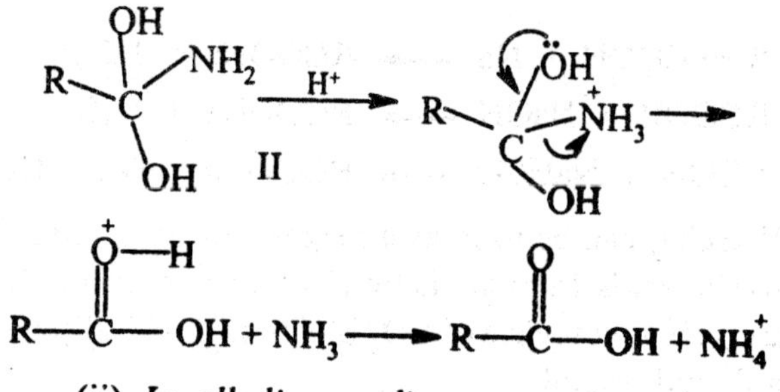

(ii) *In alkaline medium*

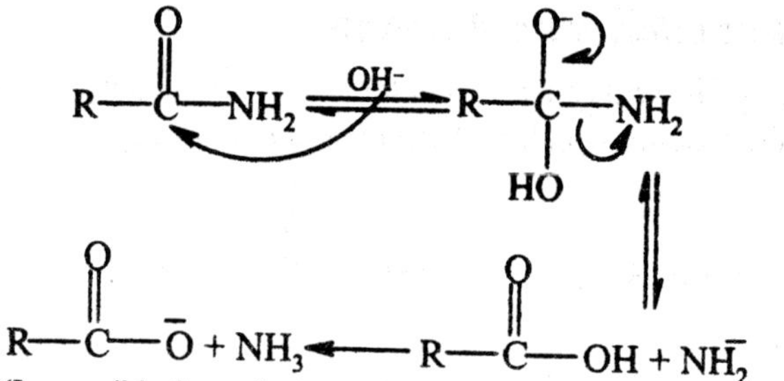

(Irreversible formation)

In this mechanism, OH^- (a relatively weaker nucleophile than NH_2^-) is replacing NH_2^-. This becomes possible because irreversible formation of R—COO^- drives the reaction in forward direction.

Example 1. Bring about the following conversions:
(a) 4-octyne ⟶ butanoic acid
(b) Bromobenzene ⟶ phenylacetic acid
(c) Allyl iodide ⟶ 3-butenoic acid
(d) Benzaldehyde ⟶ mandelic acid.

8.5 GENERAL, REACTIONS OF MONOCARBOXYLIC ACIDS

I. Acidity of Carboxylic Acid

Carboxylic acids are weak organic acid, weaker than mineral acids but stronger than alcohols, phenols and water. This is because of the formation of resonance stabilized carboxylate ion as conjugate base as:

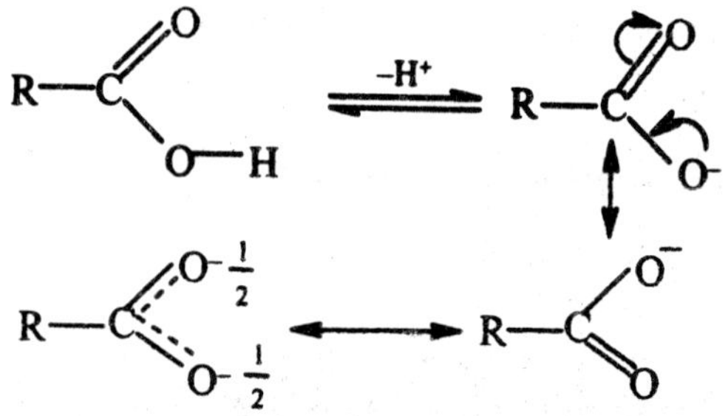

(Equivalent resonating structures)

Example 2. Though phenoxide ion ($C_6H_5O^-$) has more resonating structures than carboxylate ion, carboxylic acid is stronger acid than phenol, explain.

Carboxylic acid reacts with Na-metal, NaOH and $NaHCO_3$ producing its salts as:

$$R—COOH + Na \longrightarrow RCOONa + 1/2\ H_2$$
$$RCOOH + NaOH \longrightarrow RCOONa + H_2O$$
$$R—COOH + NaHCO_3 \longrightarrow RCOONa + H_2O + CO_2$$

$NaHCO_3$ can be used as a reagent for distinction of carboxylic acids from aliphatic alcohols and phenol because no efferevescence of CO_2 is obtained with aliphatic alcohols and phenol.

II. Reaction with Alcohols

Carboxylic acid reacts with alcohols in presence of dilute H_2SO_4 producing esters (Fischer esterification):

$$RCOOH + R'OH \xrightarrow{H^+} R—\overset{O}{\overset{\|}{C}}—OR' + H_2O$$

Mechanism

Fischer esterification mechanism is an acid catalyzed nucleophilic acyl substitution.

$$R—\overset{O}{\overset{\|}{C}}—OH + H^+ \rightleftharpoons R—\overset{\overset{+}{O}—H}{\overset{\|}{C}}—OH \xrightarrow{HOR'} R—\overset{OH}{\underset{H—\overset{+}{O}R'}{C}}—OH \xrightarrow{-H^+}$$

$$R—\overset{OH}{\underset{OR'}{C^+}} \xleftarrow{-H_2O} R—\overset{H—\overset{+}{O}—H}{\underset{OH}{C}}—OR' \xleftarrow{H^+} R—\overset{:OH}{\underset{OH}{C}}—OR'$$

$$\left[R—\overset{\overset{+}{O}—H}{\overset{\|}{C}}—OR' \longleftrightarrow R—\overset{OH}{C}=\overset{+}{O}—R' \right] \xrightarrow{-H^+} R—\overset{O}{\overset{\|}{C}}—OR'$$

Resonance stabilized

Methylesters can conveniently be prepared by the reaction of carboxylic acid with diazomethane as:

$$R—\overset{O}{\overset{\|}{C}}—OH + CH_2N_2 \longrightarrow R—\overset{O}{\overset{\|}{C}}—OCH_3$$

Mechanism

$$R—\overset{O}{\overset{\|}{C}}—O—H + \overset{-}{C}H_2—\overset{+}{N}\equiv N \longrightarrow$$

$$R—\overset{O}{\overset{\|}{C}}—O^- + CH_3—\overset{+}{N}\equiv N \longrightarrow R—\overset{O}{\overset{\|}{C}}—OCH_3 + N_2$$

From the above mechanism: It is evident that (i) alkoxy oxygen remain in the ester and –OH of acid is lost as:

$$R—\overset{O}{\overset{\|}{C}}—OH + R'—O^{18}—H \overset{H^+}{\rightleftharpoons} R—\overset{O}{\overset{\|}{C}}—\overset{18}{O}R' + H_2O$$

$$R—\overset{O^{18}}{\overset{\|}{C}}—OH + R'OH \overset{H^+}{\rightleftharpoons} R—\overset{O^{18}}{\overset{\|}{C}}—OR' + R—\overset{O}{\overset{\|}{C}}—OR'$$

Example 3: Propose mechanism of the following esterification reaction:

(a) $HO—CH_2CH_2CH_2CH_2COOH \overset{H^+}{\rightleftharpoons}$ (six-membered lactone)

(b) 3-(hydroxymethyl)cyclopentanone (CH_2OH) $\overset{H^+}{\rightleftharpoons}$ 4-hydroxycyclohexanone (OH)

III. Reaction with PCl_5, PCl_3, $SOCl_2$ and $(COCl)_2$

$$RCOOH + PCl_5 \longrightarrow RCOCl + POCl_3$$
$$3RCOOH + PCl_3 \longrightarrow 3RCOCl + H_3PO_3$$

$$R—\overset{O}{\overset{\|}{C}}—OH + SOCl_2 \longrightarrow RCOCl + SO_2 + HCl$$

$$RCOOH + Cl—\underset{O}{\underset{\|}{C}}—\underset{O}{\underset{\|}{C}}—Cl \longrightarrow RCOCl + HCl + CO + CO_2$$

Last two reagents are best suited because gaseous byproducts are produced in both cases.

Mechanism: (i) *With* PCl_5

$$R—\overset{O}{\overset{\|}{C}}—\underset{H}{\underset{|}{O}}: + PCl_4—Cl \longrightarrow R—\overset{O}{\overset{\|}{C}}—\underset{H}{\underset{|}{\overset{+}{O}}}—PCl_4$$

$$R—\overset{:OH}{\underset{Cl}{C}}—O—\overset{Cl}{\underset{Cl\ \ Cl}{P}}—Cl \overset{Cl^-}{\rightleftharpoons} R—\overset{\ddot{O}}{\overset{\|}{C}}—OPCl_4 \quad (H^+)$$

$$R—\overset{O}{\overset{\|}{C}}—Cl \xleftarrow{-H^+} R—\overset{\overset{+}{O}—H}{\overset{\|}{C}}—Cl + POCl_3 \longleftarrow$$

(ii) *With $SOCl_2$*

$$R-\overset{O}{\overset{\|}{C}}-OH + \underset{Cl\ \ Cl}{S=O} \rightarrow R-\overset{\overset{+}{O}H}{\overset{\|}{C}}-O-\overset{O}{\overset{\|}{S}}-Cl \quad Cl^-$$

$$\downarrow$$

$$R-\overset{\overset{+}{O}H}{\overset{\|}{C}}-Cl + SO_2 \leftarrow R-\underset{Cl}{\overset{:OH}{\overset{|}{C}}}-O-\overset{O}{\overset{\|}{S}}-Cl$$

$$\xrightarrow{-H^+} R-\overset{O}{\overset{\|}{C}}-Cl$$

IV. Condensation of Acid with Ammonia and Amines

Ammonia and amines condenses with carboxylic acid producing ammonium salts, which decomposes on heating, producing amides:

$$R-\overset{O}{\overset{\|}{C}}-OH + NH_3 \rightarrow R-\overset{O}{\overset{\|}{C}}-\bar{O}\overset{+}{N}H_4$$

$$\xrightarrow{\Delta} R-\overset{O}{\overset{\|}{C}}-NH_2 + H_2O$$

$$R-\overset{O}{\overset{\|}{C}}-OH + \underset{\text{(A2° amine)}}{HNR_2} \xrightarrow{\Delta} R-\overset{O}{\overset{\|}{C}}-NR_2 + H_2O$$

V. Reduction of Carboxylic Acids

(a) *With $LiAlH_4$:* $LiAlH_4$ reduces carboxylic acids to primary alcohols. The aldehydes is an intermediate reduction product, but it cannot be isolated because it is reduced more easily than the original acid.

$$R-\overset{O}{\overset{\|}{C}}-OH + LiAlH_4 \rightarrow \left[R-\overset{O}{\overset{\|}{C}}-H\right] \rightarrow RCH_2OH$$

Mechanism

$LiAlH_4$-being a strong base, first undergo acid-base reaction:

$$R-\overset{O}{\overset{\|}{C}}-OH + LiAlH_4 \rightarrow R-\overset{O}{\overset{\|}{C}}-OLi + H_2 + AlH_3$$

In the next step, H^--transfer occurs at carbonyl carbon:

$$R-\underset{OLi}{\overset{O}{\overset{\|}{C}}} + H-AlH_2 \rightarrow R-\underset{:OLi}{\overset{OAlH_2}{\overset{|}{C}}}-H$$

$$\downarrow$$

$$R-\overset{O}{\overset{\|}{C}}-H + \overset{+}{Li}\bar{O}-AlH_2 \rightarrow R-\underset{H}{\overset{OLi}{\overset{|}{C}}}-H$$

$$\xrightarrow{H_3O^+} R-CH_2OH$$

* $LiAlH_4$ cannot be used for reduction of —COOH functional group, if the compound contain other reducible group, viz. —CHO, —CO—, —CN, —NO_2, etc. Olefinic bonds are not reduced by $LiAlH_4$. In presence of any carbonyl functional group (—CHO or —CO—R) diborane can be used for selective reduction of —COOH functional group:

$$CH_3-\overset{O}{\overset{\|}{C}}-C_6H_{10}-COOH \xrightarrow{B_2H_6}$$

$$CH_3-\overset{O}{\overset{\|}{C}}-C_6H_{10}-CH_2OH$$

(b) *Reduction to aldehydes:* Reduction of carboxylic acid to aldehydes is difficult, because aldehydes are more reactive than carboxylic acid towards most reducing agent and alcohol is formed as final product. Instead, it is done via a derivative of carboxylic acid that is more reactive than an aldehydes, acid chloride is one such derivative. The reducing agent used for this purpose is also a derivative of $LiAlH_4$, which is lithium aluminium tri (tertiarybutoxy) hydride, a weaker reducing agent, reduces the more reactive acidchloride but does not reduces aldehydes:

$$R-\overset{O}{\overset{\|}{C}}-Cl + LiAl(OR)_3H \rightleftharpoons R-\underset{H}{\overset{OLi}{\overset{|}{C}}}-Cl + Al(OR)_3$$

$$\downarrow$$

$$R-\overset{O}{\overset{\|}{C}}-H + LiCl$$

e.g.

$$(CH_3)(H_3C)CH-COOH \xrightarrow{SOCl_2}$$

$$(CH_3)_2CHCOCl \xrightarrow{LiAl(OR)_3H} (CH_3)_2CHCHO$$

Example 4. Synthesize the following:

(a) 3-(hydroxymethyl)benzaldehyde (C_6H_4 with CHO and CH_2OH in meta positions) from 3-methylbenzaldehyde (C_6H_4 with CH_3 and CHO in meta positions)

(b) 3-(hydroxymethyl)cyclopentanone ($O=C_5H_7-CH_2OH$) from 3-oxocyclopentanecarboxylic acid ($O=C_5H_7-COOH$)

VI. Alkylation of Carboxylic Acid

A carboxylic acid on treatment with organolithium compound gives ketone as:

$$R-\overset{O}{\overset{\|}{C}}-OH \xrightarrow[H_2O]{2R'Li} R-\overset{O}{\overset{\|}{C}}-R' + R'H$$

e.g. $$C_6H_5COOH \xrightarrow[H_2O]{2CH_3CH_2Li} C_6H_5COCH_2CH_3 + CH_3CH_3$$

Mechanism

$$R-\overset{O}{\overset{\|}{C}}-OH \xrightarrow{2R'Li} R-\underset{R'}{\overset{OLi}{C}}-OLi \xrightarrow{H_2O}$$

$$R-\underset{R'}{\overset{OH}{C}}-OH \overset{-H_2O}{\rightleftharpoons} R-\overset{O}{\overset{\|}{C}}-R'$$

Example 5. Show how the following ketones might be synthesized from indicated carboxylic acid:

(a) Propiophenone from propionic acid and chlorobenzene.

(b) Methyl cyclohexylketone from cyclohexane carboxylic acid.

VII. Decarboxylation of Carboxylate Radicals

Hunsdiecker reaction: When a heavy metal salt of carboxylic acid is treated with halogen, alkyl halide is formed with the loss of one carbon from the starting acid, e.g.

$$R-COOAg + Br_2 \xrightarrow{Heat} R-Br + CO_2 + AgBr$$

$$C_6H_5-CH_2CH_2CH_2CH_2-\overset{O}{\overset{\|}{C}}-OAg + Br_2 \xrightarrow{\Delta} C_6H_5-CH_2CH_2CH_2CH_2-Br + AgBr + CO_2$$

$$C_5H_9-COOH \xrightarrow[Heat,\ I_2]{Pb(OAc)_4} C_5H_9-I$$

Mechanism

Carboxylate of heavy metal reacts with halogen forming metal halide and an acylhypohalite, which dissociate into two radicals as:

$$R-\overset{O}{\overset{\|}{C}}-OAg + Br_2 \longrightarrow R-\overset{O}{\overset{\|}{C}}-O-Br + AgBr$$

$$\downarrow$$

$$R-\overset{O}{\overset{\|}{C}}-\dot{O} + \dot{Br}$$

The carboxy radical then loses CO_2 forming alkyl radical R°, which reacts with second molecule of hypobromite as:

$$R-\overset{O}{\overset{\|}{C}}-\dot{O} \longrightarrow \dot{R} + CO_2$$

$$\dot{R} + Br-O-\overset{O}{\overset{\|}{C}}-\dot{R} \longrightarrow R-Br + R-\overset{O}{\overset{\|}{C}}-\dot{O} \xrightarrow{-CO_2} \dot{R}$$

VIII. α-Bromination (Halogenation) of Acids

(*Hell-Volhard-Zelinsky reaction*), when a carboxylic acid containing α-H is treated with halogen in presence of red phosphorous or PBr_3, α-halogenation occurs as:

$$R-\underset{H}{\underset{|}{CH}}-COOH + Br_2 \xrightarrow[H_2O]{PBr_3} R-\underset{Br}{\underset{|}{CH}}-COOH$$

Carboxylic Acid Derivatives

8.6 ACID DERIVATIVES

Carboxylic acid derivatives are compounds containing functional groups, which can be converted to carboxylic acid by simple acidic or alkaline hydrolysis. Common acid derivatives are esters, amides, nitriles, acid-halides and acid anhydrides.

$R-\overset{O}{\overset{\|}{C}}-X$ Acid-halide

$R-\overset{O}{\overset{\|}{C}}-O-\overset{O}{\overset{\|}{C}}-R$ Anhydride

$R-\overset{O}{\overset{\|}{C}}-OR$ Ester

$R-\overset{O}{\overset{\|}{C}}-NH_2$ Amide

$R-C\equiv N$ Nitrile

Preparations of these acid derivatives have already been discussed in the reactions of carboxylic acid.

8.7 PHYSICAL PROPERTIES

Boiling points of esters and acid-halides are near to the boiling points of straight chain hydrocarbon of similar molecular weights. Carboxylic acids are strongly hydrogen bonded in liquid phase resulting in elevated boiling points. The stable H-bonded dimer has greater effective molecular weight and boils at higher temperature. Nitrites also have higher boiling points than esters and acid chloride of similar molecular weight due to strong dipolar association between adjacent cyanide groups:

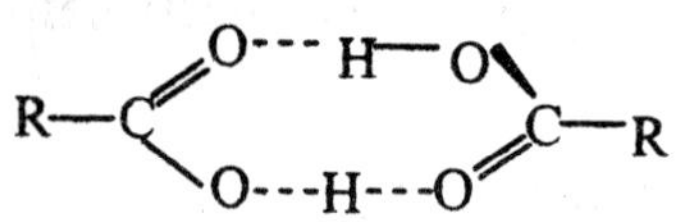

Dimer of carboxylic acid 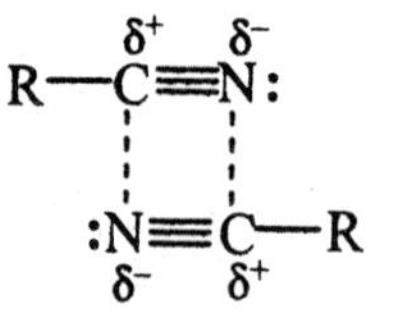   Dipolar association of nitriles

Amides have very high boiling and melting points. This is due to the resonance phenomena in amide giving a partial negative charge on oxygen and a partial positive charge on nitrogen. Due to this resonance phenomena, primary and secondary amides involves in a very strong intermolecular hydrogen bonding:

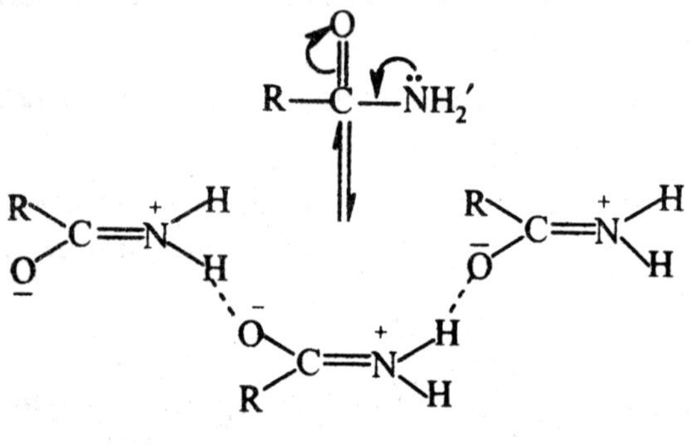

Intermolecular H-bonds in amide

8.8 CHEMICAL REACTIONS

8.8.1 Nucleophilic Acyl substitution of acid derivatives—a general discussion

Nucleophilic substitution at acyl carbon is the most important reaction of an acid derivatives. The general mechanism of addition-elimination reaction of acylcarbon can be visualized as:

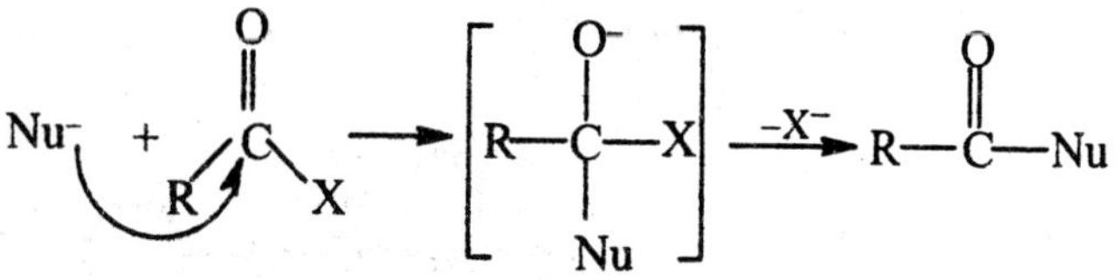

(A tetrahedral intermediate)

Depending on the nature of Nu^- and X^-, one can imagine converting any acid derivative into almost any other. Reaction that occur in actual practice generally convert a more reactive acid derivative into a stable acid derivative.

Relative reactivity: Any acid derivatives are more reactive than an alkyl halide towards nuclephilic substitution reaction due to the following reasons:

(a) Acylcarbon bears only three groups and poses less steric hindrance to the attack of nucleophile:

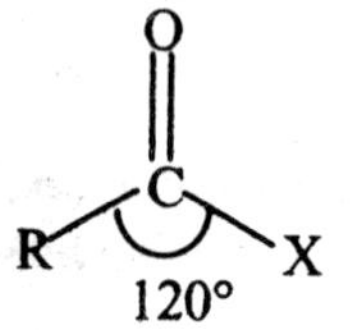

Less steric hindrance to the attack of Nu^- 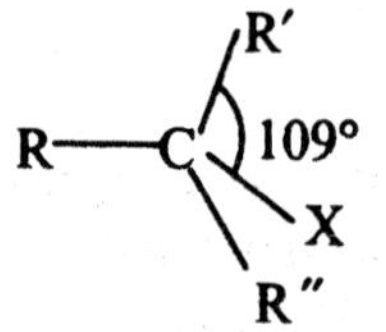   More steric hindrance at carbon bearing leaving group

(b) A tetrahedral intermediate is formed when Nu^- attacks at the acylcarbon of acid derivative while a pentavalent transition state is formed when Nu^- attack alkyl halide 'C' bearing leaving group.

The reactivity of acid derivatives towards nucleophilic attack depends on this structure and on the nature of attacking nucleophile. The general order of reactivity of acid derivatives towards nucleophilic substitution reaction is:

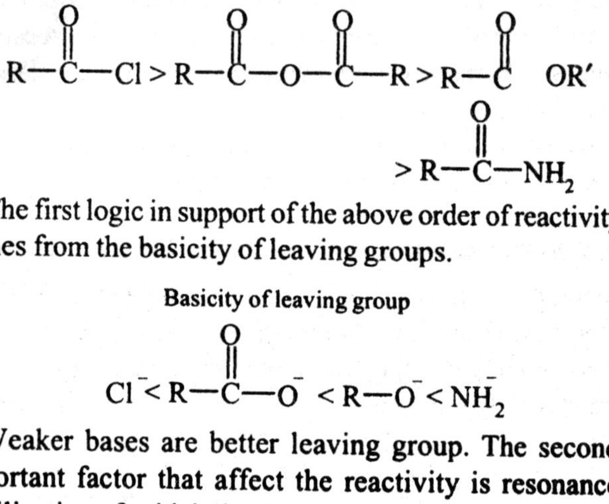

$$R-\overset{O}{\overset{\|}{C}}-Cl > R-\overset{O}{\overset{\|}{C}}-O-\overset{O}{\overset{\|}{C}}-R > R-\overset{O}{\overset{\|}{C}}\ OR' > R-\overset{O}{\overset{\|}{C}}-NH_2$$

The first logic in support of the above order of reactivity comes from the basicity of leaving groups.

Basicity of leaving group

$$Cl^- < R-\overset{O}{\overset{\|}{C}}-O^- < R-O^- < NH_2^-$$

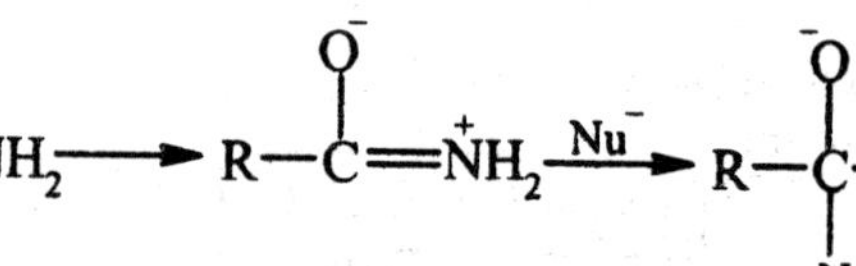

Weaker bases are better leaving group. The second important factor that affect the reactivity is resonance stabilization of acid derivatives. Amide has the maximum resonance stabilization and this stability is lost after attack of a nucleophile as:

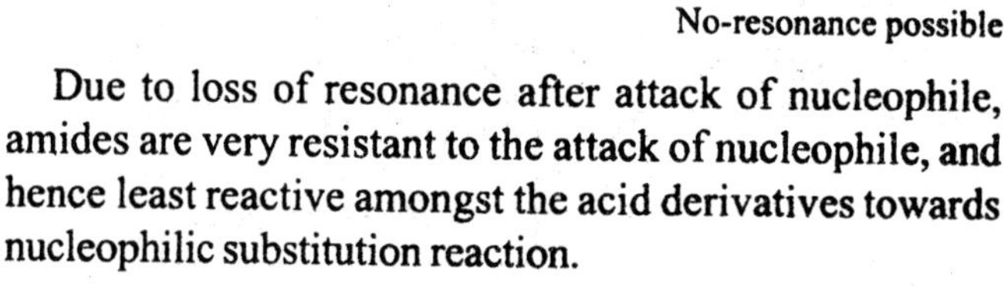

No-resonance possible

Due to loss of resonance after attack of nucleophile, amides are very resistant to the attack of nucleophile, and hence least reactive amongst the acid derivatives towards nucleophilic substitution reaction.

In case of an ester, resonance stabilization is less compared to amide. This is due to the reason that oxygen is more electronegative than nitrogen, shares less of its lone pair of electrons in resonance:

$$R-\overset{O}{\overset{\|}{C}}-\ddot{O}-R \longleftrightarrow R-\overset{O^-}{\overset{|}{C}}=\overset{+}{O}-R \xrightarrow{Nu^-} R-\overset{O^-}{\overset{|}{\underset{Nu}{\underset{|}{C}}}}-OR$$

Weak resonance stabilization

In case of an anhydride, resonance stabilization is like that in an ester, but the stabilization is shared between two carbonyl groups as:

$$R-\overset{O}{\overset{\|}{C}}-\ddot{O}-\overset{O}{\overset{\|}{C}}-R \longleftrightarrow R-\overset{O^-}{\overset{|}{C}}=\overset{+}{O}-\overset{O}{\overset{\|}{C}}-R \longleftrightarrow$$

$$R-\overset{O}{\overset{\|}{C}}-\overset{+}{O}=\overset{O^-}{\overset{|}{C}}-R \xrightarrow{Nu} R-\overset{O^-}{\overset{|}{\underset{Nu}{\underset{|}{C}}}}-O-\overset{O}{\overset{\|}{C}}-R$$

Shared, weak resonance stabilization

Due to shared resonance, each carbonyl group in anhydride receives less resonance stabilization than an ester carbonyl. Hence, an anhydride is more reactive (less stable) than esters and amides.

In acid chloride, resonance stabilization is very less and it is most reactive among acid derivatives.

8.8.2 Interconversions of Acid Derivatives

In general, interconversion of a more reactive acid derivatives into more stable acid derivative can easily be accomplished. Following diagram shows the feasibility of an interconversion reaction:

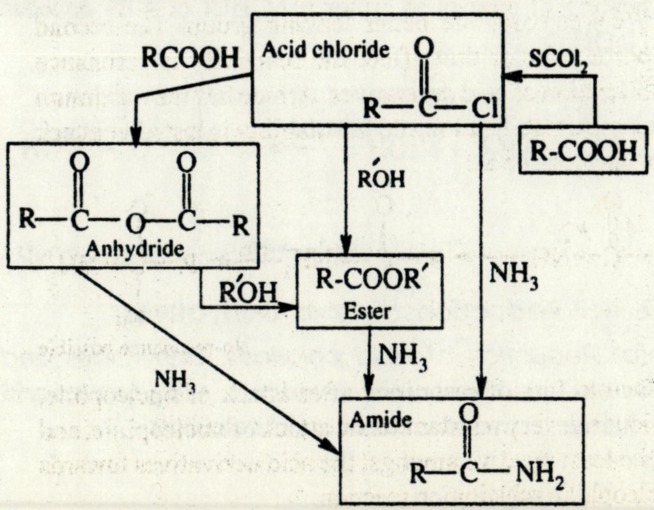

Example 6. Which of the following reaction will occur quickly under ordinary conditions?

(a) $CH_3CH_2-\overset{O}{\overset{\|}{C}}-NH_2 + NaCl \longrightarrow CH_3CH_2COCl + NaNH_2$

(b) $Ph-\overset{O}{\overset{\|}{C}}-Cl + CH_3NH_2 \longrightarrow Ph-\overset{O}{\overset{\|}{C}}-NHCH_3 + HCl$

(c) $C_6H_{11}-\overset{O}{\overset{\|}{C}}-OCH_3 + NH_3 \longrightarrow C_6H_{11}-\overset{O}{\overset{\|}{C}}-NH_2 + CH_3OH$

(d) $R-\overset{O}{\overset{\|}{C}}-O-\overset{O}{\overset{\|}{C}}-R + RNH_2 \longrightarrow R-\overset{O}{\overset{\|}{C}}-NHR + RCOOH$

Solution: (a) No—More stable amide cannot be converted into less stable acid chloride.

(b) Yes—More reactive acid chloride is being converted into more stable amide.

(c) Yes—More reactive ester is being converted into more stable amide.

(d) Yes—More reactive anhydride is being converted into more stable amide.

Example 7. Discuss the mechanism of synthesis of cyclohexylisobutyrate from isobutyryl chloride and cyclohexanol.

Solution: $CH_3-\overset{CH_3}{\overset{|}{C}H}-\overset{O}{\overset{\|}{C}}-Cl + H\ddot{O}-C_6H_{11} \longrightarrow$

$$CH_3-\overset{CH_3}{\overset{|}{C}H}-\overset{O^-}{\overset{|}{\underset{Cl}{\underset{|}{C}}}}-\overset{+}{\underset{H}{\underset{|}{O}}}-C_6H_{11} \longrightarrow$$

$$CH_3-\overset{CH_3}{\overset{|}{C}H}-\overset{O}{\overset{\|}{C}}-\overset{+}{\underset{H}{\underset{|}{O}}}-C_6H_{11} \xrightarrow{Cl^-}$$

$$CH_3-\overset{CH_3}{\overset{|}{C}H}-\overset{O}{\overset{\|}{C}}-O-C_6H_{11}$$

Example 8. Using appropriate acid chloride synthesize following compounds:

(a) Acetanilide (b) $C_6H_5-\overset{O}{\overset{\|}{C}}-N(CH_2)_5$ (piperidine)

(c) $CH_3-\overset{O}{\overset{\|}{C}}-N(CH_3)-CH_3$

Solution: (a) $CH_3-\overset{O}{\overset{\|}{C}}-Cl + H_2N-Ph \longrightarrow PhNH-\overset{O}{\overset{\|}{C}}-CH_3 + HCl$

(b) $C_6H_5-\overset{O}{\overset{\|}{C}}-Cl + H-N(CH_2)_5 \xrightarrow{-HCl} C_6H_5-\overset{O}{\overset{\|}{C}}-N(CH_2)_5$

(c) $CH_3-\overset{O}{\overset{\|}{C}}-Cl + HN(CH_3)_2 \xrightarrow{-HCl} CH_3-\overset{O}{\overset{\|}{C}}-N(CH_3)-CH_3$

8.8.3 Hydrolysis of Acid Derivatives

All acid derivatives hydrolyses to carboxylic acids. Hydrolysis of acid chloride occur by simply treating acid chloride with water, while hydrolysis of other acid derivatives require either acidic or basic catalyst.

Hydrolysis of acid chloride

$$R-\overset{O}{\overset{\|}{C}}-Cl + \ddot{O}H_2 \longrightarrow R-\overset{O^-}{\overset{|}{C}}(\overset{+}{O}H_2)-Cl \longrightarrow R-\overset{O}{\overset{\|}{C}}-\overset{+}{O}(H)-H \xrightarrow{Cl^-} R-\overset{O}{\overset{\|}{C}}-OH + HCl$$

Hydrolysis of esters

(a) *Acid catalyzed hydrolysis*

$$R-\overset{O}{\overset{\|}{C}}-OR' + H^+ \longrightarrow R-\overset{+O-H}{\overset{\|}{C}}-OR' \xrightarrow{\ddot{O}H_2}$$

$$R-\overset{\ddot{O}-H}{\overset{|}{C}}(\overset{+}{O}H_2)-\ddot{O}-R' \longrightarrow R-\overset{\ddot{O}-H}{\overset{|}{C}}(OH)-\overset{+}{O}(H)-R' \longrightarrow$$

$$R-\overset{+O-H}{\overset{\|}{C}}-OH + R'OH \xrightarrow{H_2O} R-\overset{O}{\overset{\|}{C}}-OH +$$

(b) *Base catalyzed hydrolysis*

$$R-\overset{O}{\overset{\|}{C}}-OR' + \bar{\ddot{O}}H \longrightarrow R-\overset{O^-}{\overset{|}{C}}(OH)-OR' \longrightarrow$$

$$R-\overset{O}{\overset{\|}{C}}-OH \xrightarrow{R'O^-} R-\overset{O}{\overset{\|}{C}}-O^- + R'OH$$

Hydrolysis of amides

Amides undergo hydrolysis in both acidic and basic medium. A typical hydrolysis condition involves prolonged heating of amides in either 6 M HCl or 40% equeous NaOH.

$$R-\overset{O}{\overset{\|}{C}}-NHR' + NaOH \xrightarrow{H_2O} R-\overset{O}{\overset{\|}{C}}-O^- + R'NH_2$$

$$+ H_3O^+ \longrightarrow R-\overset{O}{\overset{\|}{C}}-OH + R'\overset{+}{N}H_3$$

8.8.4 Reduction of Acid Derivatives

(a) Reduction of acid chlorides, anhydrides and esters: Lithium aluminium hydride ($LiAlH_4$) reduces acid chloride, anhydrides and esters into alcohols.

$$R-\overset{O}{\overset{||}{C}}-Cl + LiAlH_4 \xrightarrow{H_2O} R-CH_2OH$$

$$R-\overset{O}{\overset{||}{C}}-O-\overset{O}{\overset{||}{C}}-R \xrightarrow{LiAlH_4} 2RCH_2OH$$

$$R-\overset{O}{\overset{||}{C}}-OR' + LiAlH_4 \longrightarrow R-CH_2OH + R'OH$$

Acid derivatives can be reduced to aldehyde using a mild reducing agents like lithium tritertiary butoxy aluminium hydride or diisobutyl aluminium hydride:

$$R-\overset{O}{\overset{||}{C}}-Cl + Li(t\text{-}BuO)_3AlH \longrightarrow R-\overset{O}{\overset{||}{C}}-H$$

$$R-\overset{O}{\overset{||}{C}}-OR' + (CH_3-\overset{CH_3}{\overset{|}{C}H}-CH_2)AlH \longrightarrow R-\overset{O}{\overset{||}{C}}-H + R'OH$$

(b) *Reduction of amides:* $LiAlH_4$ reduces amides, azides and nitriles into amines:

$$R-\overset{O}{\overset{||}{C}}-NH_2 + LiAlH_4 \longrightarrow R-CH_2NH_2$$

$$R-\overset{O}{\overset{||}{C}}-\underset{R}{\underset{|}{N}}-R + LiAlH_4 \longrightarrow R-CH_2-N\begin{matrix}R\\R\end{matrix}$$

$$R-C\equiv N + LiAlH_4 \longrightarrow R-CH_2NH_2$$

$$R-N_3 + LiAlH_4 \longrightarrow R-NH_2$$

8.9 SOME SPECIFIC REACTIONS OF ACID DERIVATIVES

8.9.1 Reactions of Esters

I. Claisen Condensation

α-H of an ester is markedly acidic and can be abstracted to yield enolate ions. Esters are less acidic than aldehydes and ketones due to stabilization of carbonyl group by oxygen of alkoxy group:

$$R-\underset{H}{\underset{|}{CH}}-\overset{O}{\overset{||}{C}}-OR \rightleftharpoons R-\bar{C}H-\overset{O}{\overset{||}{C}}-O-R + H^+$$

$$R-CH_2-\overset{O^-}{\overset{|}{C}}=\overset{+}{O}R \qquad R-\underset{H}{\underset{|}{C}}=\overset{O^-}{\overset{|}{C}}-OR$$

Less stabilization than enolate of aldehydes and ketones.

When an ester containing α-H is treated with strong base, it undergo nucleophilic substitution of alkoxy group by an enolate ion producing a β-keto ester:

$$R-CH_2-\overset{O}{\overset{||}{C}}-OR \xrightarrow[C_2H_5OK]{C_2H_5OH} R-CH_2-\overset{O}{\overset{||}{C}}-\overset{R}{\overset{|}{C}H}-\overset{O}{\overset{||}{C}}-OR$$

β-ketoester

Mechanism

$$R-CH_2-\overset{O}{\overset{||}{C}}-OR + C_2H_5\bar{O} \longrightarrow R\bar{C}H-\overset{O}{\overset{||}{C}}-OR$$

$$R-\underset{OR}{\underset{|}{\overset{O}{\overset{||}{C}}}} + {}^-\overset{R}{\overset{|}{C}H}-\overset{O}{\overset{||}{C}}-OR \longrightarrow R-\underset{OR}{\underset{|}{\overset{O^-}{\overset{|}{C}}}}-\overset{R}{\overset{|}{C}H}-\overset{O}{\overset{||}{C}}-OR$$

$$\downarrow$$

$$R-\overset{O}{\overset{||}{C}}-\overset{R}{\overset{|}{C}H}-\overset{O}{\overset{||}{C}}-OR$$

The β-keto ester formed in claisen condensation are more acidic than simple aldehydes and ketones as well as normal esters because the negative charge of enolate is delocalized over both carbonyl groups.

$$2CH_3-\overset{O}{\overset{||}{C}}-OC_2H_5 \xrightarrow{CH_3CH_2ONa} CH_3-\overset{O}{\overset{||}{C}}-CH_2-\overset{O}{\overset{||}{C}}-OC_2H_5$$

Ethylacetoacetate (acetoacetic ester)

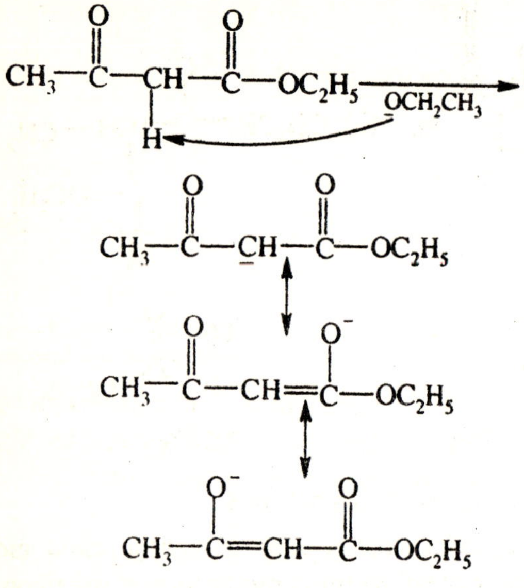

$$CH_3-\overset{O}{\overset{\|}{C}}-\underset{H}{\underset{|}{CH}}-\overset{O}{\overset{\|}{C}}-OC_2H_5 \xrightarrow{\bar{O}CH_2CH_3}$$

$$CH_3-\overset{O}{\overset{\|}{C}}-\bar{C}H-\overset{O}{\overset{\|}{C}}-OC_2H_5$$
$$\updownarrow$$
$$CH_3-\overset{O}{\overset{\|}{C}}-CH=\overset{O^-}{\overset{|}{C}}-OC_2H_5$$
$$\updownarrow$$
$$CH_3-\overset{O^-}{\overset{|}{C}}=CH-\overset{O}{\overset{\|}{C}}-OC_2H_5$$

Deprotonation of β-keto esters provide a driving force for Claisen condensation.

Example 9. Synthesize following β-keto ester starting from $C_6H_5CH_2COOH$.

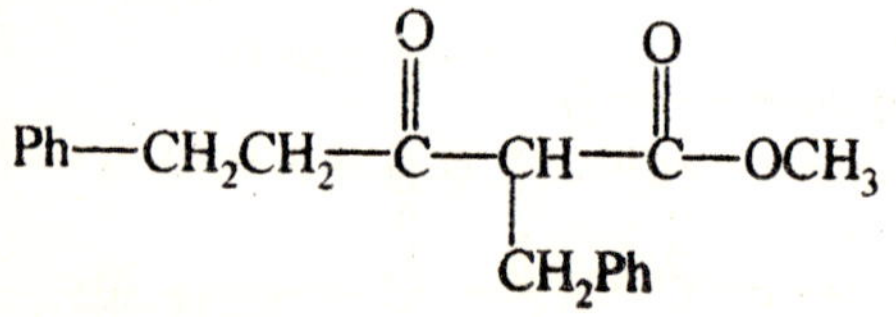

$$Ph-CH_2CH_2-\overset{O}{\overset{\|}{C}}-\underset{CH_2Ph}{\underset{|}{CH}}-\overset{O}{\overset{\|}{C}}-OCH_3$$

Solution

$$PhCH_2COOH \xrightarrow{LiAlH_4} PhCH_2CH_2OH \xrightarrow{SOCl_2}$$

$$PhCH_2CH_2Cl \xrightarrow{aq\ KCN} PhCH_2CH_2CN \xrightarrow[H_2O]{H_3O^+}$$

$$PhCH_2CH_2COOH \xrightarrow[H^+]{CH_3OH} Ph-CH_2CH_2COOCH_3$$

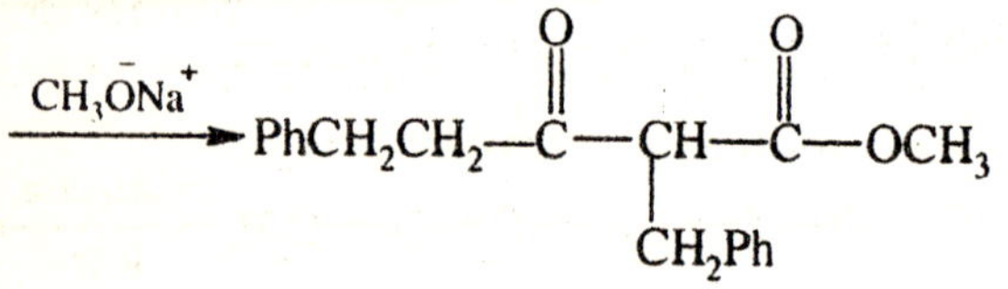

$$\xrightarrow{CH_3\bar{O}Na^+} PhCH_2CH_2-\overset{O}{\overset{\|}{C}}-\underset{CH_2Ph}{\underset{|}{CH}}-\overset{O}{\overset{\|}{C}}-OCH_3$$

Claisen cyclization (Dieckmann condensation)

An internal Claisen condensation of diester forming cyclic β-keto ester is known as *Dieckmann condensation*. The essential requirement for Dieckmann condensation is that a five or six membered ring must be formed, e.g.

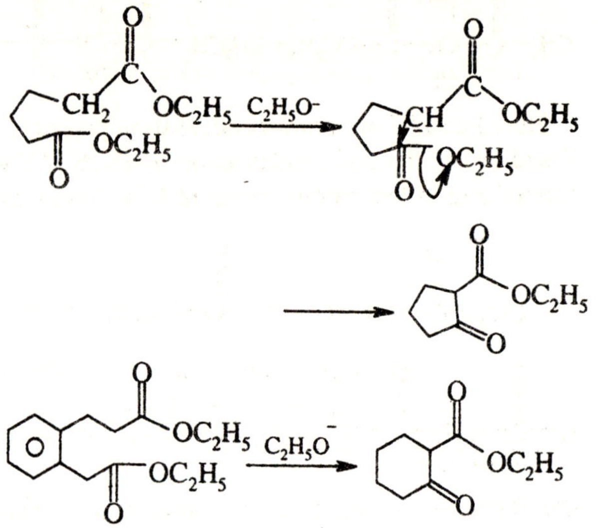

[Scheme: diethyl adipate (cyclic drawing, CH_2, OC_2H_5, OC_2H_5) $\xrightarrow{C_2H_5O^-}$ carbanion $\bar{C}H$ attacking ester C=O (loss of OC_2H_5) $\longrightarrow$ ethyl 2-oxocyclopentanecarboxylate]

[Scheme: benzene ring with two side-chain esters (OC_2H_5, OC_2H_5) $\xrightarrow{C_2H_5O^-}$ cyclic β-keto ester with OC_2H_5]

Example 10. Show how the following β-keto esters can be formed by Dieckmann's condensation reaction:

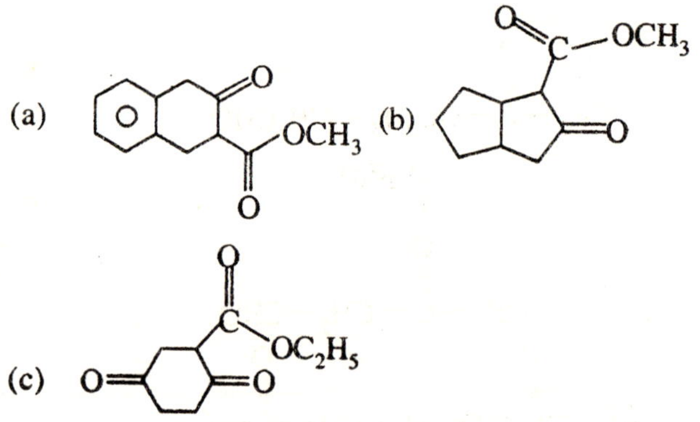

(a) [fused benzo six-membered ring β-keto ester with $C(=O)OCH_3$] (b) [bicyclic cyclopentane-fused cyclopentanone with $C(=O)OCH_3$]

(c) [cyclohexane-1,4-dione with $C(=O)OC_2H_5$ substituent]

Cross-Claisen condensation

(a) Between two esters: A cross-condensation may take place between two different types of esters. This type of condensation may be an efficient synthetic route if one of the ester is devoid of any α-H.

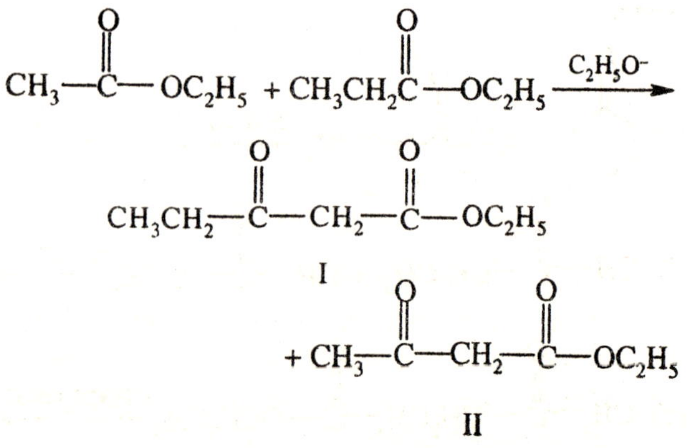

$$CH_3-\overset{O}{\overset{\|}{C}}-OC_2H_5 + CH_3CH_2\overset{O}{\overset{\|}{C}}-OC_2H_5 \xrightarrow{C_2H_5O^-}$$

$$\underset{\text{I}}{CH_3CH_2-\overset{O}{\overset{\|}{C}}-CH_2-\overset{O}{\overset{\|}{C}}-OC_2H_5}$$

$$+\ \underset{\text{II}}{CH_3-\overset{O}{\overset{\|}{C}}-CH_2-\overset{O}{\overset{\|}{C}}-OC_2H_5}$$

$$CH_3-\overset{O}{\overset{\|}{C}}-\underset{}{\overset{CH_3}{\overset{|}{CH}}}-\overset{O}{\overset{\|}{C}}-OC_2H_5 \;(\text{III}) + CH_3CH_2-\overset{O}{\overset{\|}{C}}-\overset{CH_3}{\overset{|}{CH}}-\overset{O}{\overset{\|}{C}}-OC_2H_5 \;(\text{IV})$$

Here I and III are **cross-condensation** products, while II and IV are the self-**condensation products**. All will be formed and in the product mixture II will **predominate**.

$$CH_3-\overset{O}{\overset{\|}{C}}-OC_2H_5 + H-\overset{O}{\overset{\|}{C}}-OC_2H_5 \longrightarrow$$

$$\underset{\text{Cross-product}}{H-\overset{O}{\overset{\|}{C}}-CH_2-\overset{O}{\overset{\|}{C}}-OC_2H_5} + CH_3-\overset{O}{\overset{\|}{C}}-CH_2-\overset{O}{\overset{\|}{C}}-OC_2H_5$$

(b) Between ester and an aldehyde or ketone: Since α-H of aldehyde and ketone are more acidic than ester, the major product will be formed by the attack of aldehyde or ketone enolate on ester carbonyl group:

$$CH_3-\overset{O}{\overset{\|}{C}}-OC_2H_5 + CH_3CHO \xrightarrow{C_2H_5O^-} CH_3-\overset{O}{\overset{\|}{C}}-CH_2-CHO$$

$$CH_3-\overset{O}{\overset{\|}{C}}-OC_2H_5 + CH_3-\overset{O}{\overset{\|}{C}}-CH_3 \xrightarrow{C_2H_5O^-} CH_3-\overset{O}{\overset{\|}{C}}-CH_2-\overset{O}{\overset{\|}{C}}-CH_3$$

Cross-products

Example 11. Predict the major product of the following reaction:

(a) Cyclohexanone $+ H-\overset{O}{\overset{\|}{C}}-OCH_3 \xrightarrow{CH_3ONa}$

(b) $CH_3-\overset{O}{\overset{\|}{C}}-CH_2CH_3 + CH_3-\overset{O}{\overset{\|}{C}}-OCH_3 \xrightarrow{NaOCH_3}$

(c) $CH_3-\overset{O}{\overset{\|}{C}}-CH_2CH_2-\overset{O}{\overset{\|}{C}}-OCH_2CH_3 \xrightarrow{CH_3CH_2ONa}$

Example 12. Show how Claisen condensation can be used to make the following compounds?

(a) 2-(benzoyl)cyclopentanone (cyclopentanone with C(=O)Ph at the α-carbon) (b) $CH_3CH_2-\overset{O}{\overset{\|}{C}}-\underset{\underset{\overset{\|}{O}}{C-OC_2H_5}}{\underset{|}{CH}}-CH_3$

(c) cyclohexane-1,3-dione (d) 2-($\overset{O}{\overset{\|}{C}}-OC_2H_5$)cyclohexane-1,3-dione

II. Malonic Ester Synthesis

α-Hydrogen of malonic acid is slightly acidic and this property is exploited for derivatization at α-carbon. The derivative is then hydrolyzed to β-dicarboxylic acid which can be easily decarboxylated to yield an acid as:

$$\underset{\text{(malonic ester)}}{C_2H_5-O-\overset{O}{\overset{\|}{C}}-\underset{\underset{H\,\leftarrow\,\text{Acidic hydrogen}}{|}}{CH}-\overset{O}{\overset{\|}{C}}-O-C_2H_5} \xrightarrow{C_2H_5O^-Na^+} C_2H_5O-\overset{O}{\overset{\|}{C}}-\underset{Na^+}{\bar{C}H}-\overset{O}{\overset{\|}{C}}-OC_2H_5 \xrightarrow{RBr} C_2H_5O-\overset{O}{\overset{\|}{C}}-\underset{R}{\underset{|}{CH}}-\overset{O}{\overset{\|}{C}}-OC_2H_5$$

$$\xrightarrow{\text{(i) } H_3O^+ \text{ (ii) Heat}} \underset{\text{R (a substituted acetic acid)}}{H-\underset{|}{CH}-\overset{O}{\overset{\|}{C}}-OH}$$

Second α-H can also be replaced by another alkyl group by repeating the same procedure:

$$C_2H_5O-\overset{O}{\overset{\|}{C}}-CH_2-\overset{O}{\overset{\|}{C}}-O-C_2H_5 \xrightarrow[RBr]{C_2H_5ONa} C_2H_5O-\overset{O}{\overset{\|}{C}}-\underset{R}{\underset{|}{CH}}-\overset{O}{\overset{\|}{C}}-OC_2H_5 \xrightarrow[R'Br]{C_2H_5ONa} C_2H_5O-\overset{O}{\overset{\|}{C}}-\underset{R}{\underset{|}{\overset{R'}{\overset{|}{C}}}}-\overset{O}{\overset{\|}{C}}-OC_2H_5$$

$$\xrightarrow{H_3O^+,\ \text{Heat}} R'-\overset{R}{\overset{|}{CH}}-COOH$$

Example 13. Show why the following acids cannot be prepared by malonic ester synthesis:

(a) $C_6H_5CH_2COOH$ (b) $CH_2{=}CHCH_2COOH$

Solution: Enolate of malonic ester acts as a nucleophile and its subsequent reaction with halides follow SN-2 mechanism. Both chlorobenzene and vinylchlorides are inert in SN-2 reactions due to partial C—Cl double bond character, above mentioned acids cannot be prepared by malonic ester synthesis.

Example 14. Show how the following compounds can be synthesized using malonic ester and alkyl bromides.

(a) 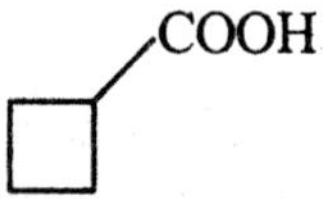 (cyclobutyl)–COOH (b) 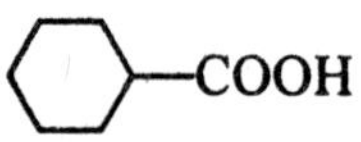 (cyclohexyl)–COOH

(c) 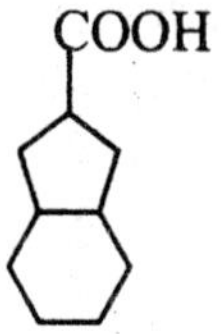 COOH on bicyclic (hydrindane) ring (d) 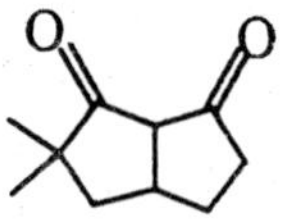 bicyclic diketone

Solution: (a) 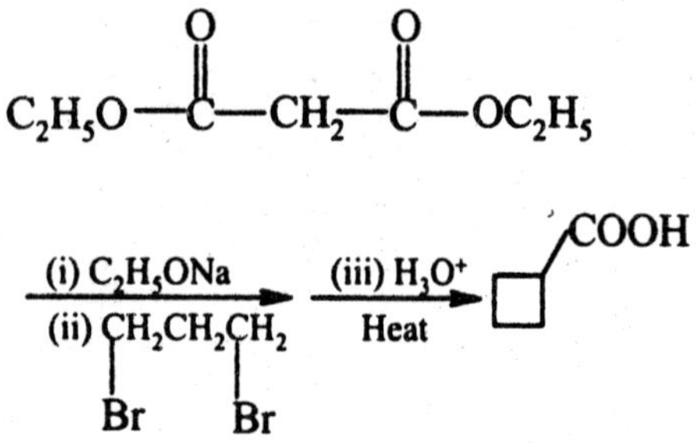 $C_2H_5O{-}\overset{O}{\overset{\|}{C}}{-}CH_2{-}\overset{O}{\overset{\|}{C}}{-}OC_2H_5$ $\xrightarrow[\text{(ii) }CH_2(Br)CH_2CH_2(Br)]{\text{(i) }C_2H_5ONa}$ $\xrightarrow[\text{Heat}]{\text{(iii) }H_3O^+}$ (cyclobutyl)–COOH

(b) $C_2H_5O{-}\overset{O}{\overset{\|}{C}}{-}CH_2{-}\overset{O}{\overset{\|}{C}}{-}OC_2H_5$ $\xrightarrow[Br{+}CH_2{\rightarrow_5}Br]{C_2H_5ONa}$ $\xrightarrow{H_3O^+}$ $\xrightarrow{\text{Heat}}$ (cyclohexyl)–COOH

(c) $C_2H_5O{-}\overset{O}{\overset{\|}{C}}{-}CH_2{-}\overset{O}{\overset{\|}{C}}{-}OC_2H_5$ $\xrightarrow[\text{(ii) dibromide}]{\text{(i) }C_2H_5ONa}$ $\xrightarrow{H_3O^+}$ $\xrightarrow{\text{Heat}}$ (bicyclic)–COOH

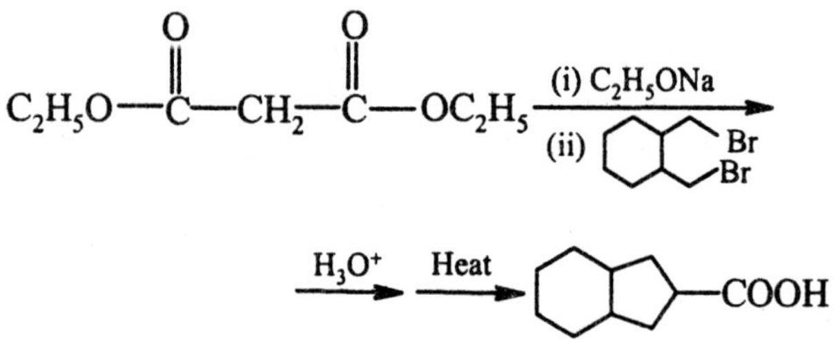

(d) 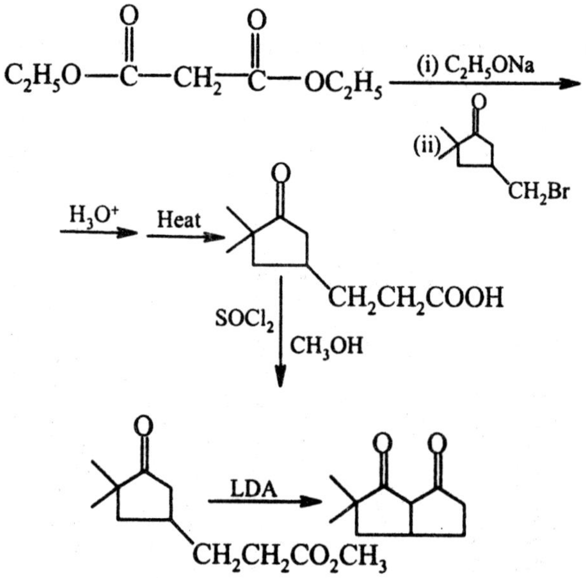 $C_2H_5O{-}\overset{O}{\overset{\|}{C}}{-}CH_2{-}\overset{O}{\overset{\|}{C}}{-}OC_2H_5$ $\xrightarrow[\text{(ii) bromide}]{\text{(i) }C_2H_5ONa}$ (ii) ketone bearing CH_2Br

$\xrightarrow{H_3O^+}$ $\xrightarrow{\text{Heat}}$ ketone–CH_2CH_2COOH $\xrightarrow{SOCl_2,\ CH_3OH}$ ketone–$CH_2CH_2CO_2CH_3$ $\xrightarrow{\text{LDA}}$ bicyclic diketone

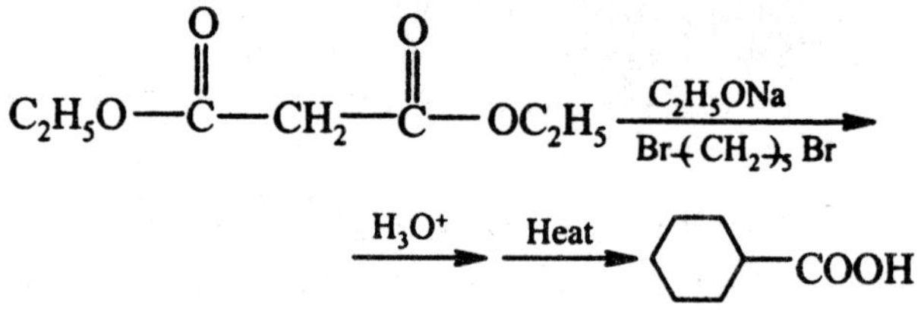

PRACTICE EXERCISES

Level 1

1. Predict the major product in each of the following reactions and show the mechanism of their formations:
   (a) $(CH_3)_3CCl + CH_3COOH \rightarrow$
   (b) $C_6H_5COOD + CH_2N_2 \rightarrow$

2. Complete the following reactions:

(a) 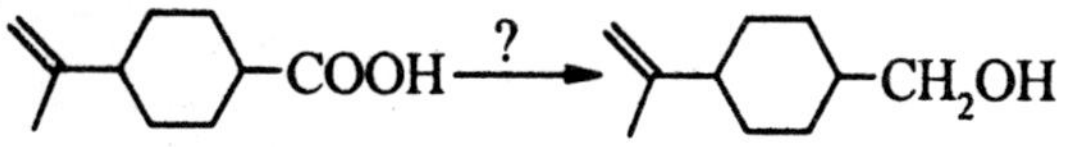 (isopropenyl)cyclohexyl–COOH $\xrightarrow{?}$ (isopropenyl)cyclohexyl–CH_2OH

(b) 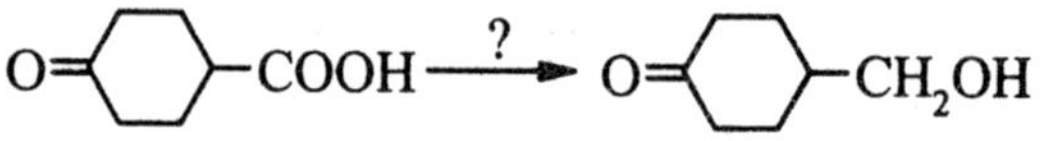 O=(cyclohexyl)–COOH $\xrightarrow{?}$ O=(cyclohexyl)–CH_2OH

(c) 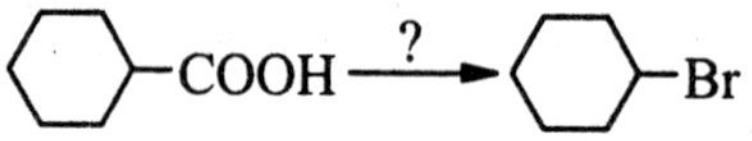 (cyclohexyl)–COOH $\xrightarrow{?}$ (cyclohexyl)–Br

(d)  2-oxocyclohexyl–COOH $\xrightarrow{\Delta}$?

3. Predict major product(s):

(a) $PhLi + CO_2/H^+ \rightarrow$

(b) 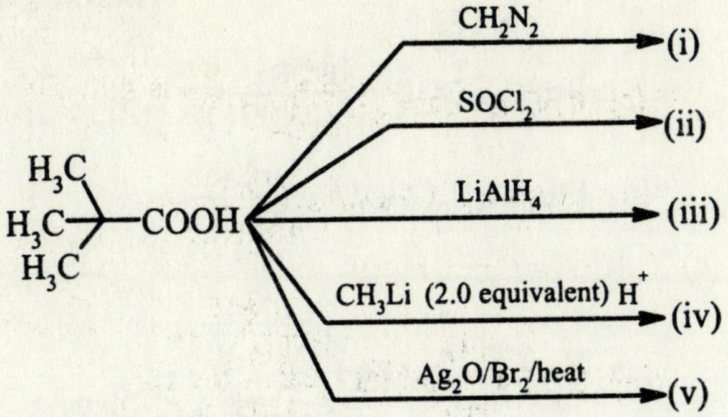 $(H_3C)_3C$–COOH: CH_2N_2 → (i); $SOCl_2$ → (ii); $LiAlH_4$ → (iii); CH_3Li (2.0 equivalent) H^+ → (iv); Ag_2O/Br_2/heat → (v)

(c) 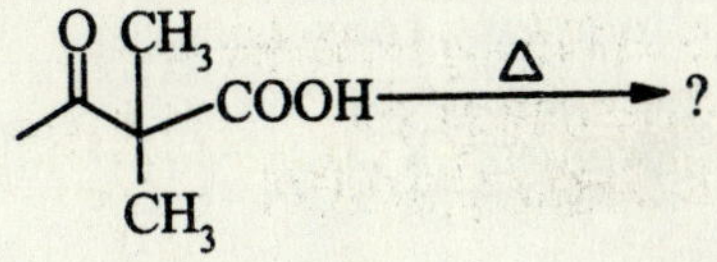 $CH_3COC(CH_3)_2COOH \xrightarrow{\Delta}$?

(d) 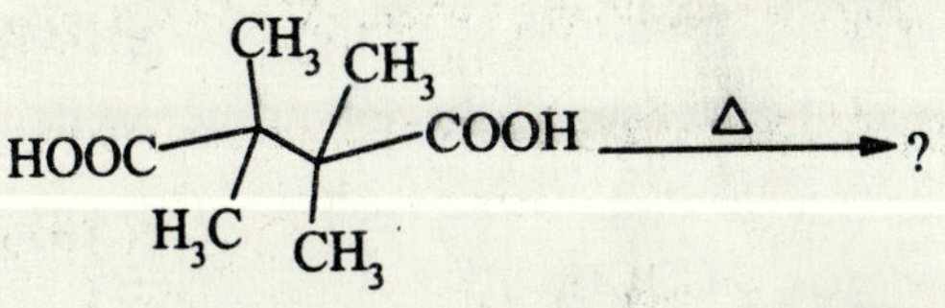 $HOOC–C(CH_3)_2–C(CH_3)_2–COOH \xrightarrow{\Delta}$?

4. Suggest a reason why β-keto acid shown below does not undergo decarboxylation reaction:

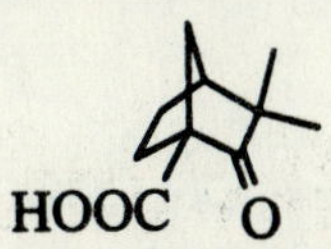

HOOC O

5. How could you use an acid base extraction technique to separate a mixture of carboxylic acid and phenol into its pure components?

6. A carboxylic acid (A) of unknown structure was found to contain only C, H and O. 150 mg of A required 11.9 ml of 0.22 N NaOH to reach the equivalence point. Gentle heating of A evolves CO_2 and a new carboxylic acid (B) was formed whose equivalent weight was found to be 74. Identify A and B.

7. Arrange the following esters in increasing order of their ease of hydrolysis giving suitable explanation.
(a) Phenyl acetate.
(b) *p*-nitrophenyl acetate
(c) *p*-methoxyphenyl acetate.
(d) Tolyl acetate.

8. Supply the missing reagents for each of the following transformations:

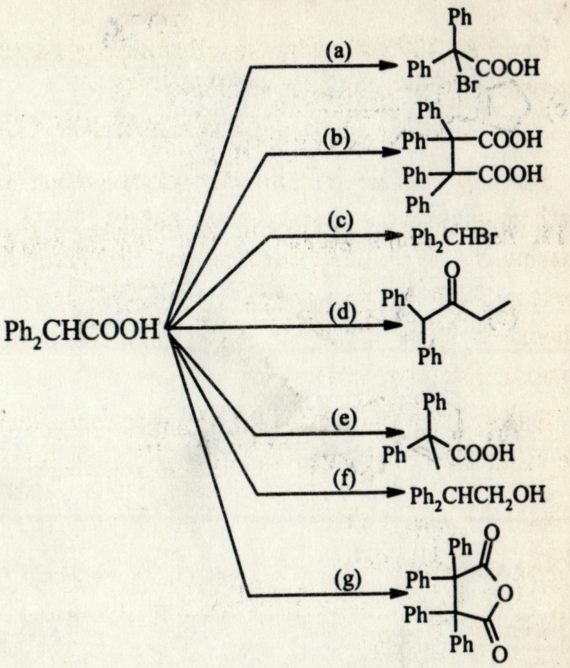

$Ph_2CHCOOH$ →
(a) $Ph_2C(Br)COOH$
(b) $Ph_2C(COOH)–C(COOH)Ph_2$
(c) Ph_2CHBr
(d) $Ph_2CH–CO–CH_2CH_3$
(e) $Ph_3C–COOH$
(f) Ph_2CHCH_2OH
(g) tetraphenylsuccinic anhydride (Ph, Ph / Ph, Ph, O, O)

9. For each of the following reactions, supply carboxylic acid starting material that would yield the indicated product under the conditions given:

(a) 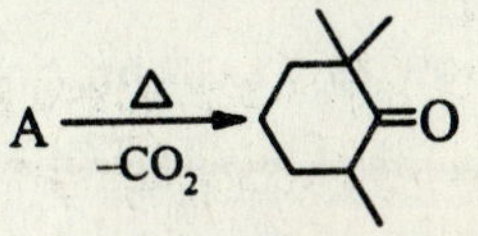 A $\xrightarrow[-CO_2]{\Delta}$ 2,2,6-trimethylcyclohexanone (=O)

(b) 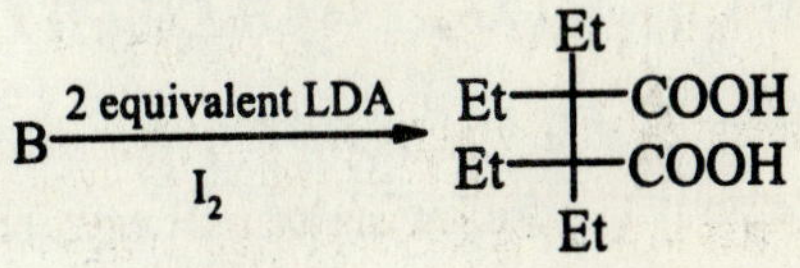 B $\xrightarrow[I_2]{\text{2 equivalent LDA}}$ $Et_2C(COOH)–C(COOH)Et_2$

(c) C $\xrightarrow[H^+]{B_2H_6}$ $CH_3CO–C_6H_4–CH_2OH$

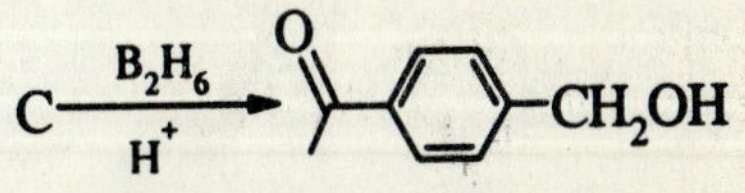

(d) 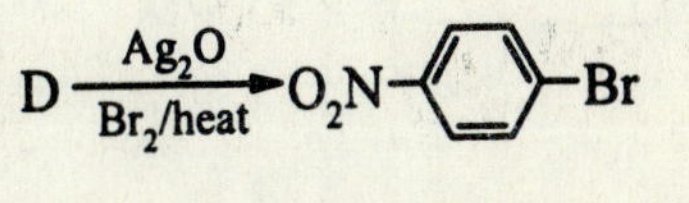 D $\xrightarrow[Br_2/heat]{Ag_2O}$ $O_2N–C_6H_4–Br$

10. Propose mechanism:

(a) 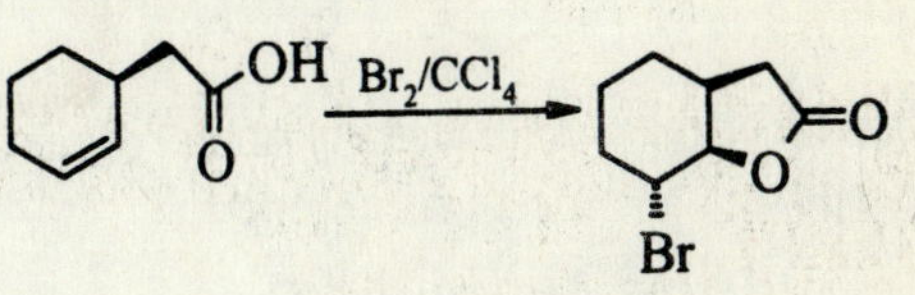 cyclohex-2-enylacetic acid (OH, O) $\xrightarrow{Br_2/CCl_4}$ bromo lactone (=O, O, Br)

(b) 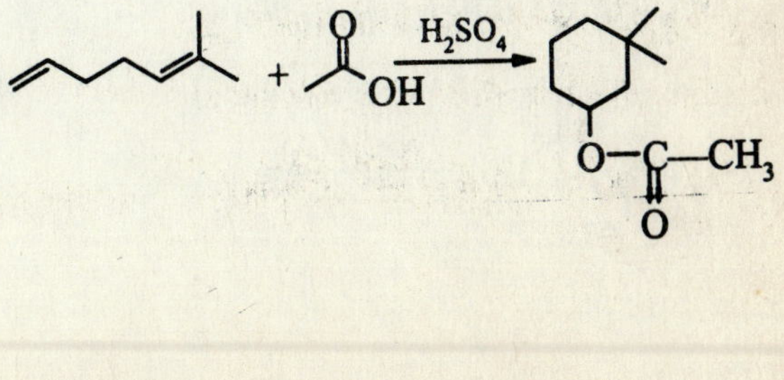 6-methylhepta-1,5-diene + CH_3COOH (O, OH) $\xrightarrow{H_2SO_4}$ 3,3-dimethylcyclohexyl acetate ($O–C(=O)–CH_3$)

(c) 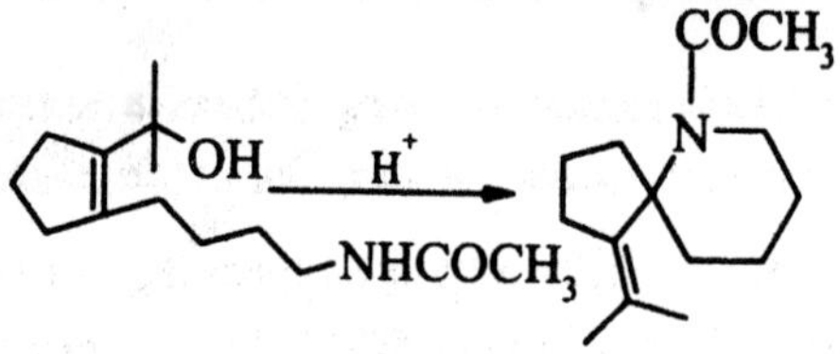 [structure] OH $\xrightarrow{H^+}$ [structure with $COCH_3$ on N] NHCOCH$_3$

11. Predict major product in the following reactions:

(a) 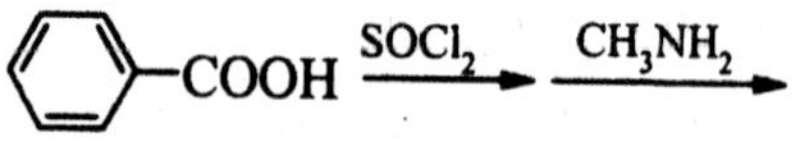 $C_6H_5COOH \xrightarrow{SOCl_2} \xrightarrow{CH_3NH_2}$

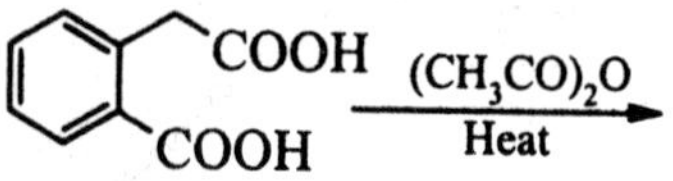

(b) [benzene with COOH and CH_2COOH] $\xrightarrow[\text{Heat}]{(CH_3CO)_2O}$

(c) [hydroquinone with OH, OH, OH] + $COCl_2 \rightarrow$

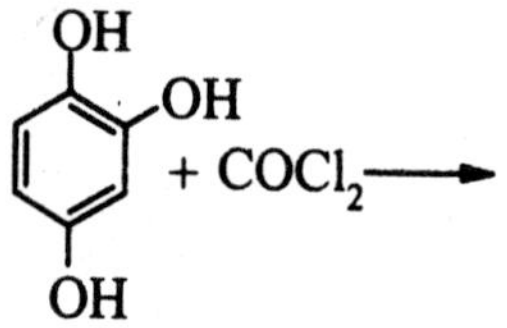

(d) 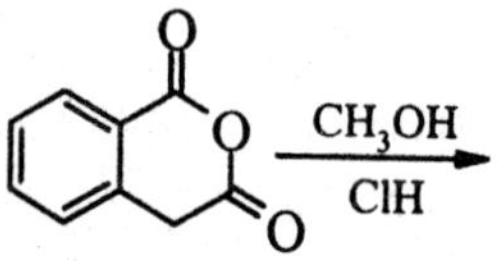 [cyclic anhydride] $\xrightarrow[\text{ClH}]{CH_3OH}$

12. Bring about the following transformations:

(a) $CH_3CH_2CONH_2 \Rightarrow CH_3CH_2COCl$

(b) $CH_3CH_2CH_2COOH \Rightarrow CH_3CH_2CH_2CH_2COOH$

(c) Toluene $\Rightarrow C_6H_5CH_2COOH$

13. Bring about the following transformations:

(a) 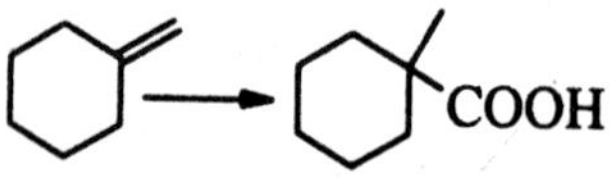 [methylenecyclohexane] $\rightarrow$ [1-methylcyclohexane] COOH

(b) 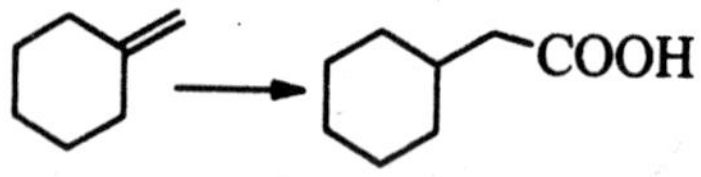 [methylenecyclohexane] $\rightarrow$ [cyclohexyl] COOH

(c) 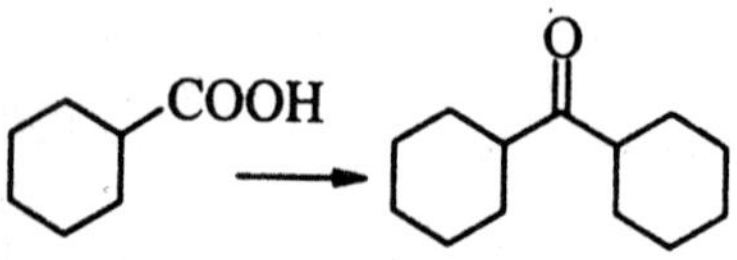 [cyclohexyl] COOH $\rightarrow$ [dicyclohexyl ketone]

14. Starting with ethyl acetate and any other reagent you need, propose a synthesis of ketone

$(CH_3)_3CCOCH_2CH_2Ph$:

15. Complete the following reactions:

(a) 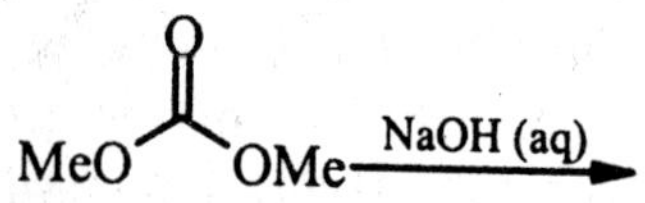 MeO–CO–OMe $\xrightarrow{\text{NaOH (aq)}}$

(b) 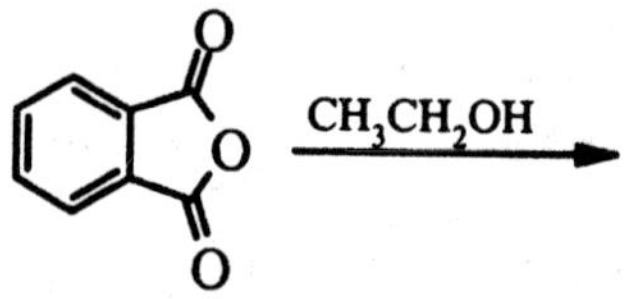 [phthalic anhydride] $\xrightarrow{CH_3CH_2OH}$

(c) $PhCH_2COOEt \xrightarrow[PhCOOEt]{EtONa}$

(d) $EtO_2CCH_2COOEt \xrightarrow[CH_3CH_2I]{EtONa}$

(e) 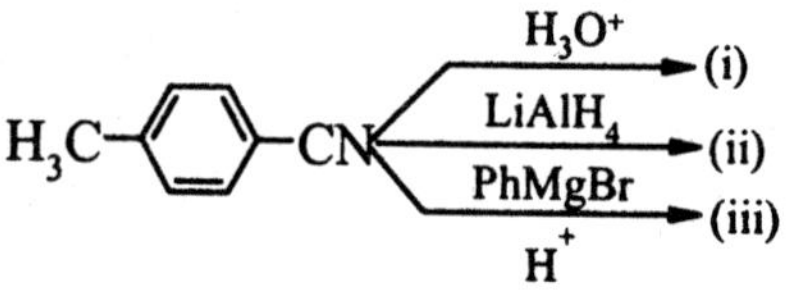 $H_3C-C_6H_4-CN$ $\xrightarrow{H_3O^+}$ (i); $\xrightarrow{LiAlH_4}$ (ii); $\xrightarrow[H^+]{PhMgBr}$ (iii)

16. Bring about the following conversions:

(a) $PhCH_2OH \rightarrow PhCH_2CN$

(b) $PhCOOH \rightarrow PhC(CH_3)_2OH$

(c) 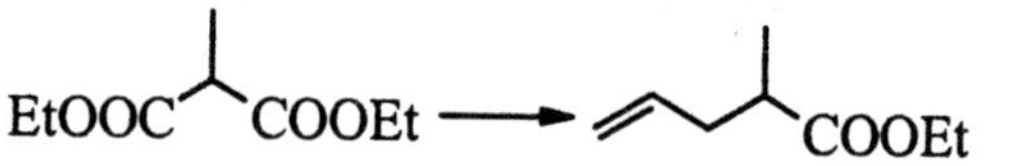 EtOOC–CH(CH$_3$)–COOEt $\rightarrow$ [alkene] COOEt

17. Suggest mechanism of the following reactions:

(a) 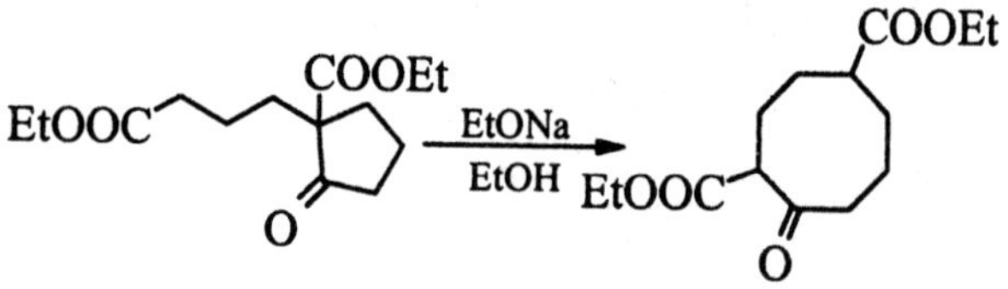 EtOOC [chain] [cyclopentanone with COOEt] $\xrightarrow[\text{EtOH}]{\text{EtONa}}$ EtOOC [cyclooctanone with COOEt]

(b) $CH_3CN \xrightarrow[H^+]{CH_3CH_2OH} CH_3COOC_2H_5$

18. Convert:

(a) 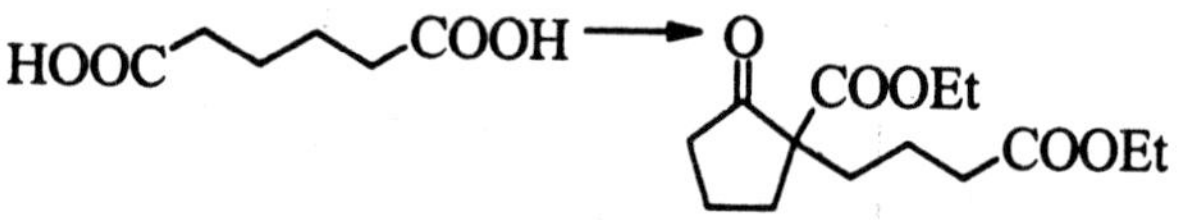 HOOC [chain] COOH $\rightarrow$ [cyclopentanone with COOEt and chain] COOEt

(b) 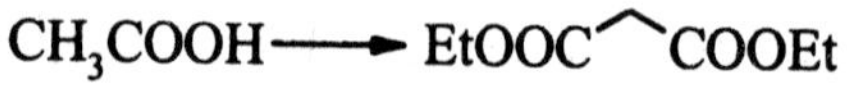 $CH_3COOH \rightarrow$ EtOOC–CH$_2$–COOEt

19. Show how would you accomplish the following synthesis in good yield?

(a) 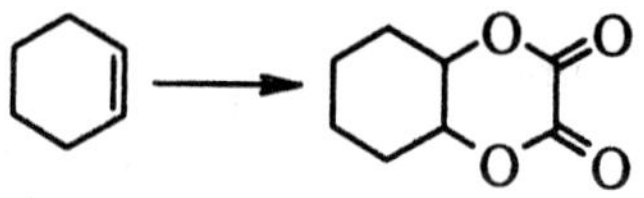 [cyclohexene] $\rightarrow$ [cyclic oxalate]

(b) 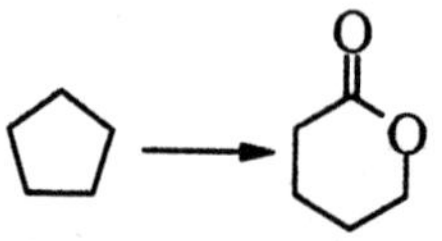 [cyclopentane] $\rightarrow$ [δ-valerolactone]

(c) 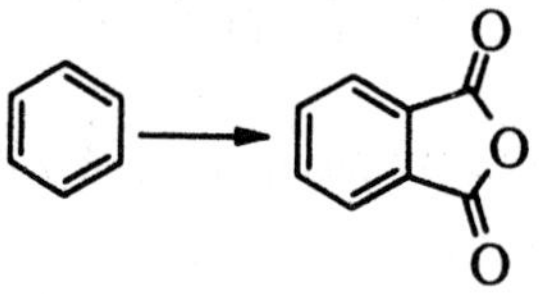 [benzene] $\rightarrow$ [phthalic anhydride]

(d) 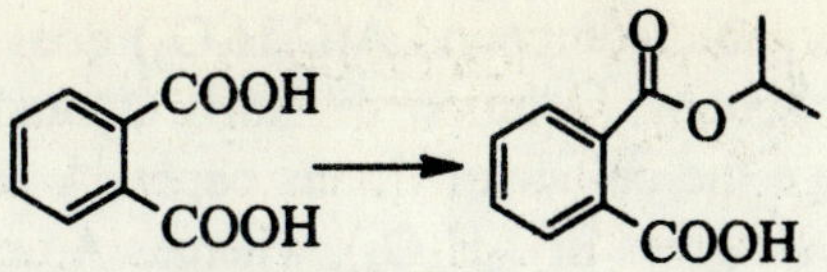 [Phthalic acid → monoisopropyl phthalate]

20. Propose mechanism:

(a) 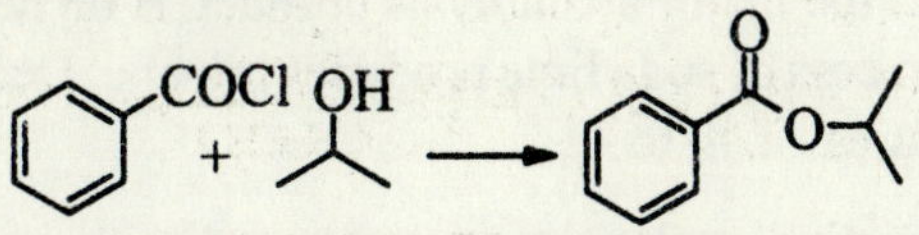 Benzoyl chloride (C_6H_5COCl) + isopropanol → isopropyl benzoate

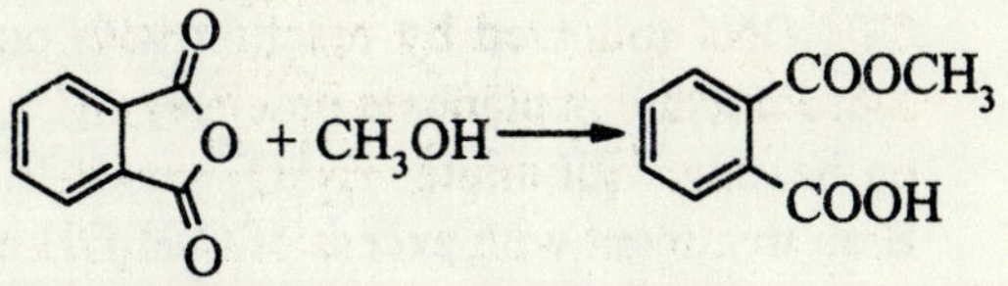

(b) Phthalic anhydride + CH_3OH → [benzene with $COOCH_3$ and $COOH$]

(c) Phthalic anhydride $\xrightarrow{RNH_2}$ N-R phthalimide

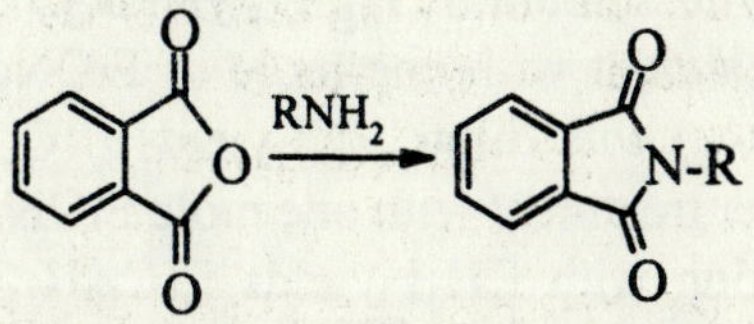

21. Show how would you accomplish the following multisteps synthesis?

(a) 6-hepten-1-ol → caprolactum

(b) 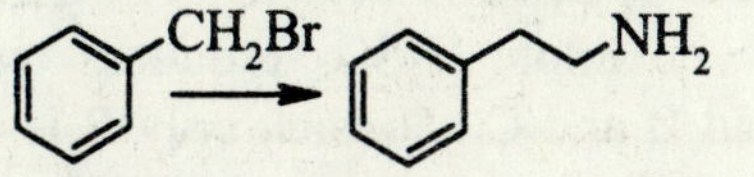 $C_6H_5CH_2Br$ → $C_6H_5CH_2CH_2NH_2$

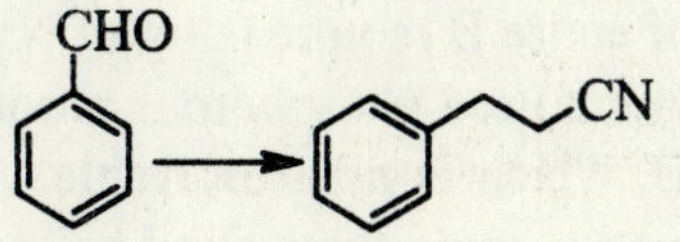

(c) C_6H_5CHO → $C_6H_5CH_2CH_2CN$

22. Synthesize:

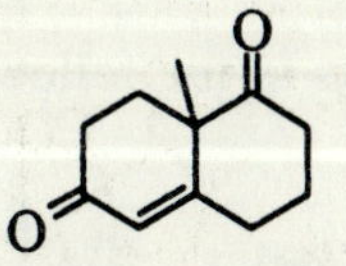

[bicyclic enedione] from [cyclohexane-1,3-dione]

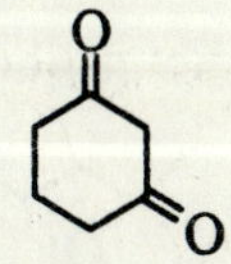

23. Synthesize the following compounds from indicated starting materials:

(a) 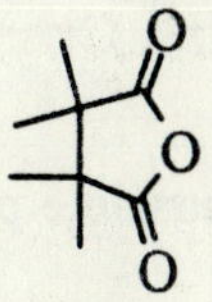 [tetramethylsuccinic anhydride] from [isobutyric acid, COOH]

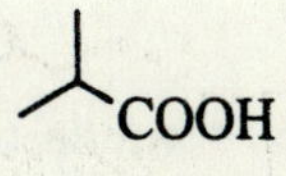

(b) 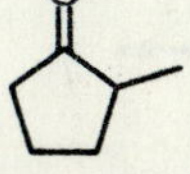 [2-methylcyclopentanone] from EtOOC~~~~COOEt

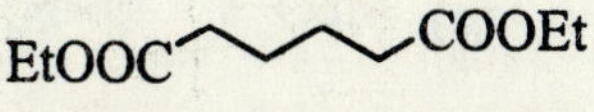

24. Show how would you accomplish the following synthesis:

(a) isobutyl amine → N-isobutyl formamide

(b) ethylacetate → 3-methyl-3-pentanol.

(c) Cyclohexyl amine → N-cyclohexylacetamide

(d) Bromocyclohexane → dicyclohexylmethanol

25. Bring about the following conversion:

(a) 1-hexanol → 1-heptanamine

(b) cyclohexancarboxamide → cyclohexyl ethylketone.

(c) 1-octanol → 2-decanone.

26. Provide reagents:

(a)  [propyl Br] → [propyl COOH]

(b) 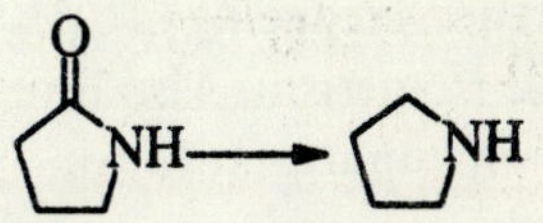 [2-pyrrolidinone, NH] → [pyrrolidine, NH]

27. Complete the following:

(a) 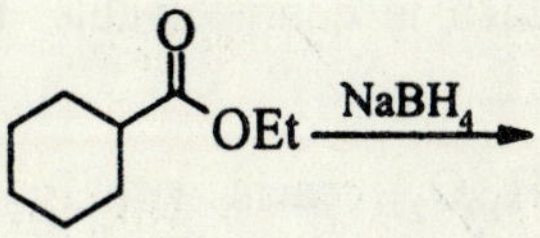 [cyclohexyl-COOEt] $\xrightarrow{NaBH_4}$

(b) 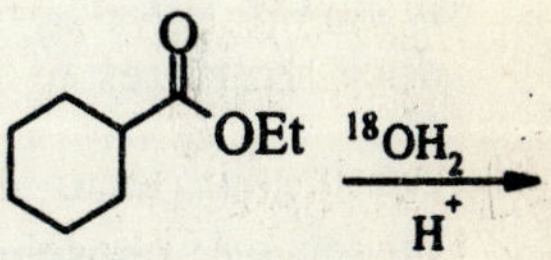 [cyclohexyl-COOEt] $\xrightarrow[H^+]{^{18}OH_2}$

(c) 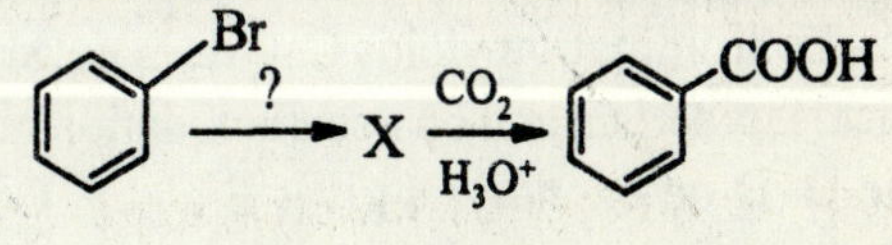 [bromobenzene] $\xrightarrow{?}$ X $\xrightarrow[H_3O^+]{CO_2}$ [benzoic acid, COOH]

(d) 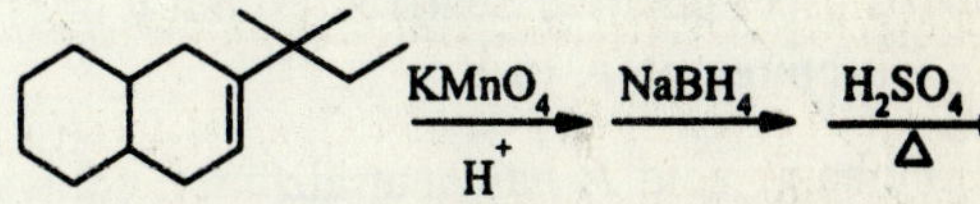 [substituted octahydronaphthalene] $\xrightarrow[H^+]{KMnO_4}$ $\xrightarrow{NaBH_4}$ $\xrightarrow[\Delta]{H_2SO_4}$

(e) 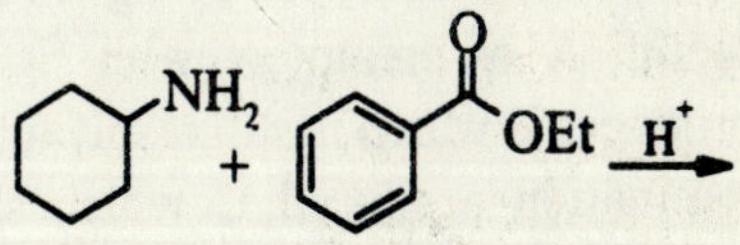 [cyclohexylamine, NH_2] + [ethyl benzoate, OEt] $\xrightarrow{H^+}$

28. Arrange the following compounds in increasing order of acidic strength:

(a)  (I) [pentane-2,4-dione] (II) [3,3-dimethylpentane-2,4-dione] (III) [3-($COOCH_3$)pentane-2,4-dione]

(b) 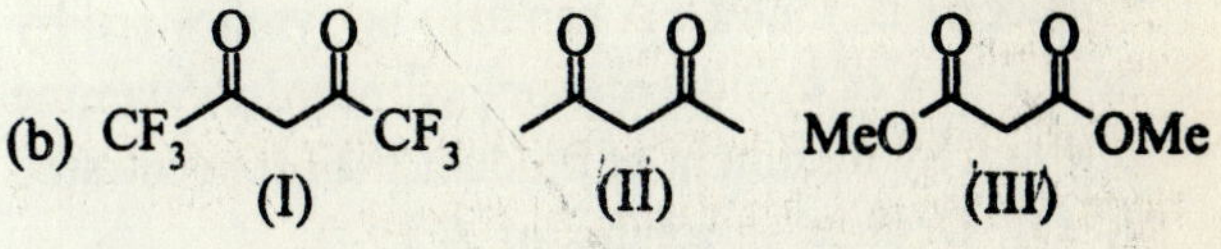 (I) $CF_3COCH_2COCF_3$ (II) [pentane-2,4-dione] (III) MeO–CO–CH_2–CO–OMe

Level 2

1. A neutral organic compound A has molecular formula $C_5H_8O_2$ and does not decolourise Bayer's reagent. A on acidic hydrolysis produced B($C_5H_{10}O_3$), which is resolvable. B on heating with concentrated H_2SO_4 undergo dehydration to produce C, an isomer of A. C decolourises Bayer's reagent and show stereo isomerism. Also B changes colour of acidic dichromate solution from orange to blue-green converting itself into $D(C_5H_8O_3)$, which is non-resolvable and forms an yellow precipitate on treatment with NaOI solution. Deduce structures of A to D.

2. A neutral organic compound $A(C_5H_8O_2)$ does not decolourise Bayer's reagent and on hydrolysis with dilute H_2SO_4 produces $B(C_5H_{10}O_3)$, which is distereomeric. B on heating with concentrated H_2SO_4 undergoes dehydration producing $C(C_5H_8O_2)$, which shows geometrical isomerism. Also B on treatment with acidic dichromate solution produced $D(C_5H_8O_3)$, which is enantiomeric and gives an yellow precipitate with NaOI. D on gentle heating produces $E(C_4H_8O)$, which is nonresolvable. Deduce structures of A to E.

3. An organic acid $A(C_5H_{10}O_2)$ reacts with Br_2 in presence of phosphorous to produce a resolvable B. B on dehydrobromination yields C. C does not show geometrical isomerism and on decarboxylation gives an alkene D. D on ozonolysis gives E and F. Compound E gives positive Schiff's test but F does not. Give structures of A to F.

4. A monobasic acid A has neutralization equivalent 116 and does not decolourise Bayer's reagent. Also A is enantiomeric. A on treatment with Br_2/red phosphorous produces B which is still resolvable. B on dehydrobromination produced C which show stereoisomerism. C on decarboxylation produced D, which does not show stereoisomerism. Deduce structures of A to D.

5. An organic lactum A on acid hydrolysis produced B, an amino acid. B on treatment with nitrous acid gives C. C on heating with concentrated H_2SO_4 produces a lactone D. A can also be synthesized by the reaction of cyclopentanone with hydroxylamine followed by treatment of product with concentrated H_2SO_4. Deduce structures of A to D.

6. A neutral organic compound A has formula $C_9H_{16}O_3$ and rotates plane polarized light. A on acid hydrolysis produces B and $C(C_3H_8O)$. C on partial oxidation with PCC gives $D(C_3H_6O)$, which does not gives iodoform test. D on treatment with dilute solution of NaOH gives $E(C_6H_{12}O_2)$, which is distereomeric. E on treatment with acidic solution of $KMnO_4$ produces B. Deduce structures of A to E.

7. An organic compound $A(C_9H_8O_3)$ does not reacts with aqueous solution of $NaHCO_3$ and does not change the colour of litmus paper. A on acid hydrolysis gives $B(C_9H_8O_4)$, whereas A on treatment with methanol in acid medium gives $C(C_{10}H_{14}O_4)$ as the major alcoholysis product. B on heating with excess of soda lime produces toluene. Deduce structures of A to C.

8. Diethyl malonate on treatment with one mole of C_2H_5ONa followed by reaction with one mole of 5-bromoethyl pentanoate produces $A(C_{14}H_{24}O_6)$. A on heating with dilute H_2SO_4 yields $B(C_7H_{12}O_4)$. B on treatment with excess of C_2H_5OH in presence of catalytic amount of H_2SO_4 yields $C(C_{11}H_{20}O_4)$. C on treatment with one mole of EtONa followed by work-up with dilute HCl yields $D(C_9H_{14}O_3)$. D on further treatment with one mole EtONa followed by reaction with CH_3I yields $E(C_{10}H_{16}O_3)$. E on heating with HCl yields 2-methyl cyclohexanone. Deduce structures of A to E.

9. An organic compound contain C, H and oxygen and its aqueous solution is neutral. 1.44 g of A on refluxing with dilute H_2SO_4 produces two organic compound B and C of which only B is resolvable and gives effervescence with $NaHCO_3$. Also neutralization of entire B require 0.4 g of NaOH. B on reduction with HI/red phosphorous produces a hydrocarbon D, which is non-resolvable. Also D on monochlorination yield three alkyl halide in which only one is resolvable. Deduce structures of A to D.

10. Predict major products in the following reactions:

(a) [dioxolane with COOH side chain] $\xrightarrow{CH_3Li\ (Excess)}$ A $\xrightarrow{H_3O^+}$ B

(b) HOOC–[substituted cyclopropane] $\xrightarrow{CH_3Li\ (Excess)}$ C $\xrightarrow{H_3O^+}$ D

11. Give structural formula of all possible products in the following reactions:

(a) MeO–C_6H_4–NH_2 $\xrightarrow{(CH_3CO)_2O}$]

(b) C_6H_5–COOEt $\xrightarrow{NH_2OH}$

(c) C_6H_4(CH_2COOH)(COOH) $\xrightarrow[Heat]{(CH_3CO)_2O}$

(d) Cyclohexenyl–CHO $\xrightarrow{Ag_2O/H_2O}$ X $\xrightarrow{H^+}$ Y

(e) 2-Furyl–Li + (methyloxirane) ⟶ X $\xrightarrow{(CH_3CO)_2O}$ Y

12. Complete the following reactions:

(a) (cyclopentanone bearing COOMe, gem-dimethyl) $\xrightarrow{NaBH_4}$ A $\xrightarrow{SOCl_2}$ B

(b) Ph–CO–CF_3 $\xrightarrow{NaCN}$ C $\xrightarrow{(CH_3)_2SO_4}$ D $\xrightarrow[Heat]{H^+}$ E

13. Complete the following sequence of reaction:

$HOCH_2$–C≡CH $\xrightarrow{TsOH}$ (dihydropyran) A $\xrightarrow{CH_3CH_2CH_2CH_2Li}$ B $\xrightarrow{BrCH_2CH_2Cl}$ C $\xrightarrow{NaCN}$ D $\xrightarrow[OH_2]{NaOH}$ $\xrightarrow{H^+}$ E

14. Complete the following sequence of reaction explaining the formations of products in the second step:

(isoprene) $\xrightarrow{mcpba}$ A $\xrightarrow{CH_3COCl/LiCl}$ H_3C–CO–O–CH_2–C(CH_3)=CH–CH_2Cl + H_3C–CO–O–CH_2–C(CH_3)(Cl)–CH=CH_2

15. Predict major organic product of the following reactions:

(a) $CH_3CH_2CH_2$–CO–N(CH_3)–OMe $\xrightarrow{CH_3MgBr}$ A $\xrightarrow{H_3O^+}$ B + C

(b) Cyclohexyl–CO–N(Me)–OMe + PhLi ⟶ D $\xrightarrow{H_3O^+}$ E + F

16. Predict the product formed on reaction of di-isobutyl aluminium hydride with the following reagents:

(a) CH_3CH_2–O–CH_2–CO–O–CH_2CH_3 (b) PhCOOH

(c) (1-cyano-1-(3,4-methylenedioxyphenyl)cyclopropane, CN)

17. Bring about the following conversions:

(a) Ph–CH(CH_3)–CH_2–COOH ⟶ Ph–CH(CH_3)–CH_2CH_2–CO–CH_3

(b) $(CH_3)_2C$=C($COOC_2H_5$)–CH_2–$COOC_2H_5$ ⟶ $(CH_3)_2CH$–CH(CH_2OCOCH_3)–CH_2CH_2–O–CO–CH_3

18. Complete the following reactions:

(a) (1,4-dioxane-2,5-dione) $\xrightarrow{NaOH/H_2O}$

(b) CH_3–CO–CH_2CH_2–COOH $\xrightarrow{NaBH_4}$ $\xrightarrow{H^+}$

(c) Cyclobutanone $\xrightarrow{mcpba}$ X $\xrightarrow[H_2O]{H^+}$ Y

19. Bring about the following transformations:

(bicyclic ketone, nopinone) ⟶ 3-(CH_2–CO–OMe)-cyclopentanone... (cyclohexanone bearing CH_2COOMe)

20. Complete the following reactions:

(a) (3,4-methylenedioxybenzaldehyde, CHO) $\xrightarrow{KMnO_4/H_2O}$ A $\xrightarrow{H_3O^+}$ B

(b) (2,4,6-trimethylbromobenzene; CH_3, Br, H_3C, CH_3) $\xrightarrow{Mg/Ether}$ C $\xrightarrow[CO_2]{H_3O^+}$ D

(c) Ph–CO–N(OMe)–CH_3 $\xrightarrow{PhC\equiv CLi}$ $\xrightarrow{H_3O^+}$ E

21. Bring about the following transformations:

(a) (H_3C)(H)C=C(H)(CH_2Cl) ⟶ (H_3C)(H)C=C(H)(CH_2CH_2Br)

(b) 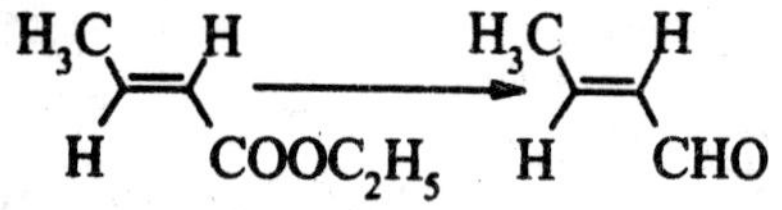 $H_3C(H)C{=}C(H)COOC_2H_5 \longrightarrow H_3C(H)C{=}C(H)CHO$

(c) → CH_2OH, HO

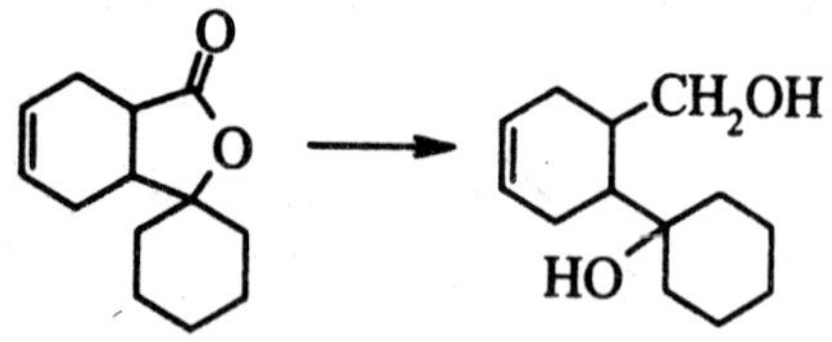

(d) 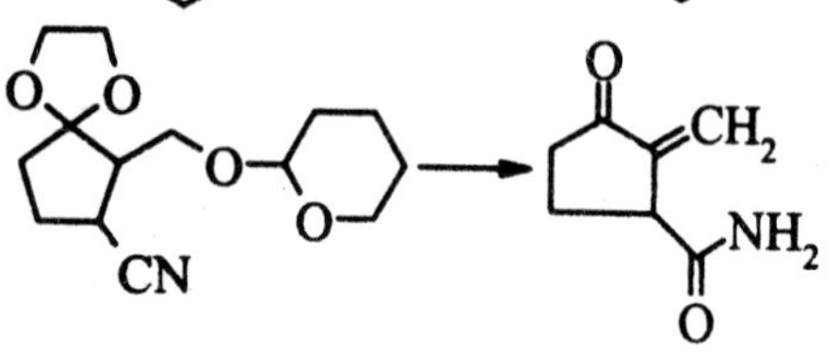 O, O, O, CN → O, CH_2, NH_2, O

(e) 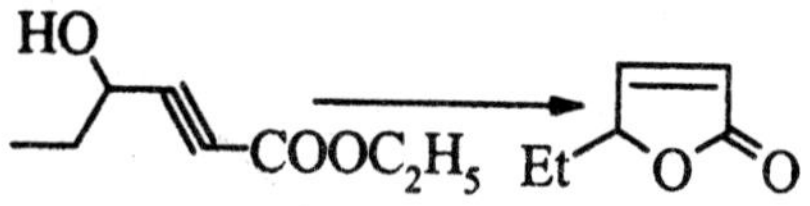 HO, $COOC_2H_5$ → Et, O, O

22. Bring about the following conversions:

(a) 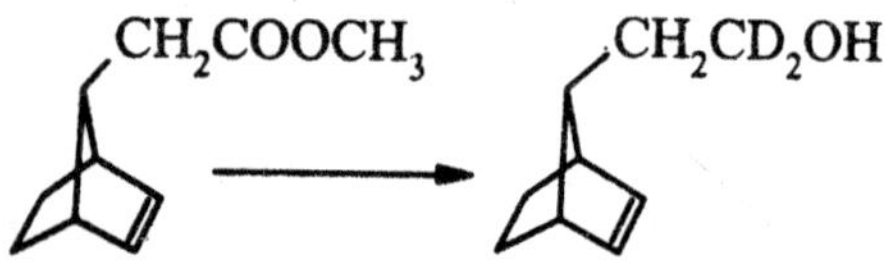 CH_2COOCH_3 → CH_2CD_2OH

(b) 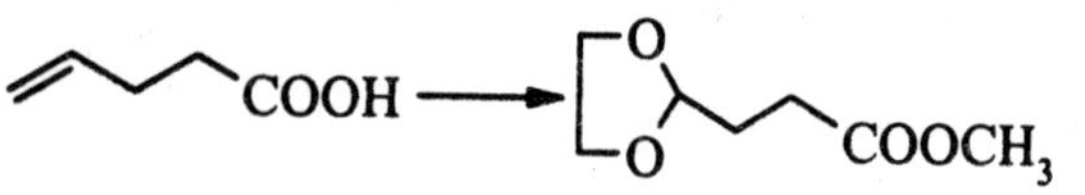 COOH → O, O, $COOCH_3$

23. Suggest mechanism for the following sequence of reactions:

(a) 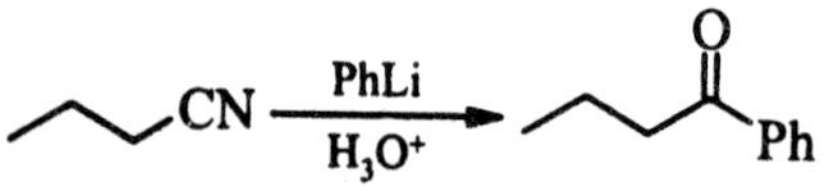 CN $\xrightarrow[H_3O^+]{PhLi}$ O, Ph

(b) 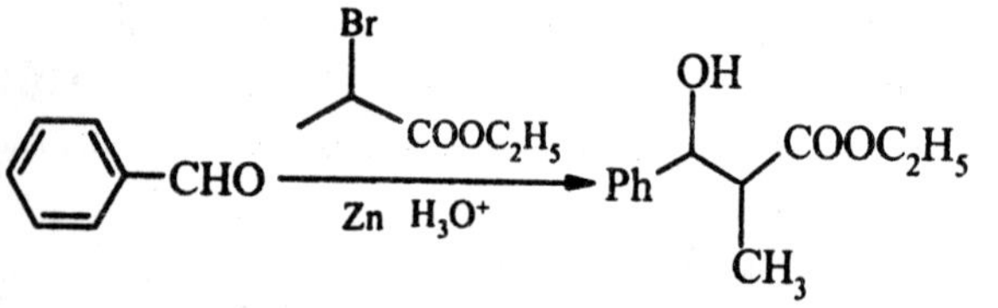 CHO $\xrightarrow[Zn\ H_3O^+]{Br,\ COOC_2H_5}$ Ph, OH, $COOC_2H_5$, CH_3

24. Bring about the following transformations:

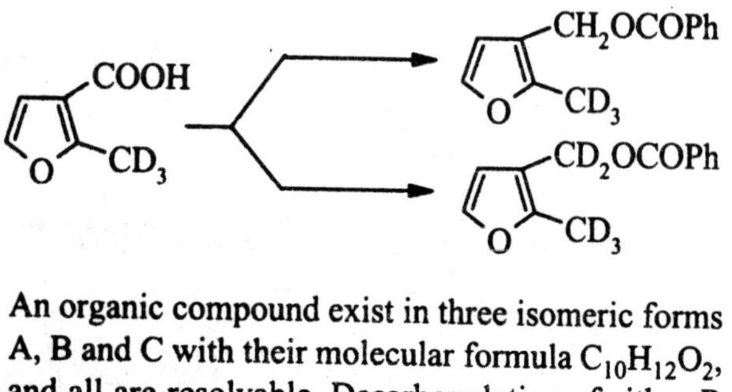

COOH, O, CD_3 → CH_2OCOPh, O, CD_3

→ CD_2OCOPh, O, CD_3

25. An organic compound exist in three isomeric forms A, B and C with their molecular formula $C_{10}H_{12}O_2$, and all are resolvable. Decarboxylation of either B or C gives the same $D(C_9H_{12})$, whereas decarboxylation of A gives E and isomer of D. Both D and E are non-resolvable. Treatment of either D or E with $KMnO_4$ in acidic medium produces benzoic acid as one product. Monobromination of E in presence of sunlight yields only two isomeric products, whereas D on monobromination yields three isomers. Also B is more reactive than C towards decarboxylation reaction. Deduce structures of A to E.

26. An organic compound $A(C_6H_{10})$ decolourises Bayer's reagent and it is non-resolvable. A on treatment with acidic permanganate solution gave $B(C_6H_{10}O_4)$ which is still non-resolvable. B on treatment with excess of ethanol in presence of acid catalyst gave $C(C_{10}H_{18}O_4)$ which is a neutral, sweet smelling liquid. C on refluxing with ethanolic solution of C_2H_5ONa gave $D(C_8H_{12}O_3)$ which is resolvable. D on treatment with acidic solution of ethylene glycol yields $E(C_{10}H_{16}O_3)$. E on treatment with one equivalent of Grignard reagent followed by reduction of product with Zn(Hg)—HCl and finally hydrolyzing product yields 2-ethyl cyclopentanoate. Deduce structures of A to E.

27. Identify A to D:

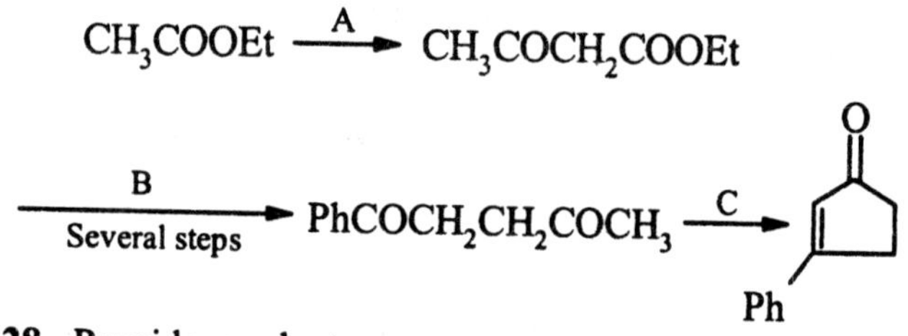

$CH_3COOEt \xrightarrow{A} CH_3COCH_2COOEt$

$\xrightarrow[\text{Several steps}]{B} PhCOCH_2CH_2COCH_3 \xrightarrow{C}$ O, Ph

28. Provide products:

(a) 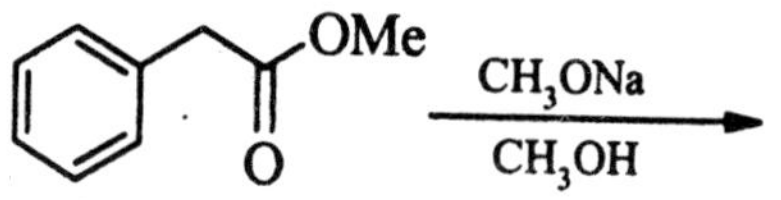 OMe, O $\xrightarrow[CH_3OH]{CH_3ONa}$

(b) 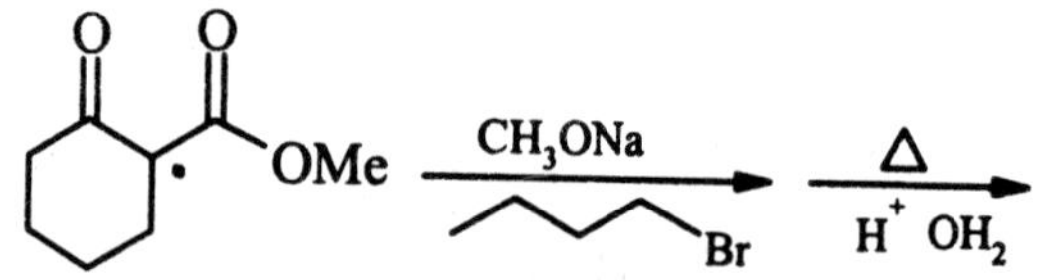 O, O, OMe $\xrightarrow[\text{Br}]{CH_3ONa} \xrightarrow[H^+\ OH_2]{\Delta}$

29. Predict products and write mechanism of their formations:

(a) 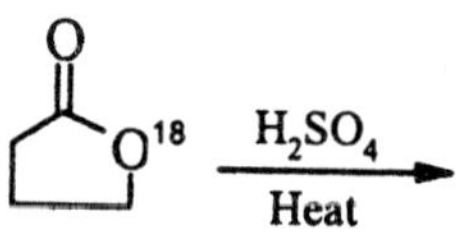 O, O^{18} $\xrightarrow[Heat]{H_2SO_4}$

(b) 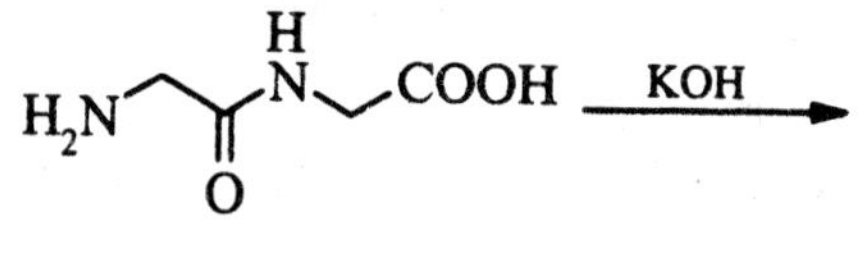 H_2N, O, H, N, COOH $\xrightarrow{KOH}$

(c) 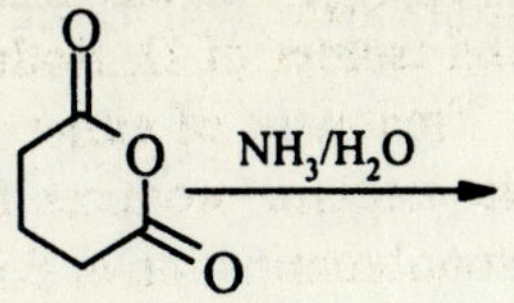 [glutaric anhydride] $\xrightarrow{NH_3/H_2O}$

30. Predict products:

(a) 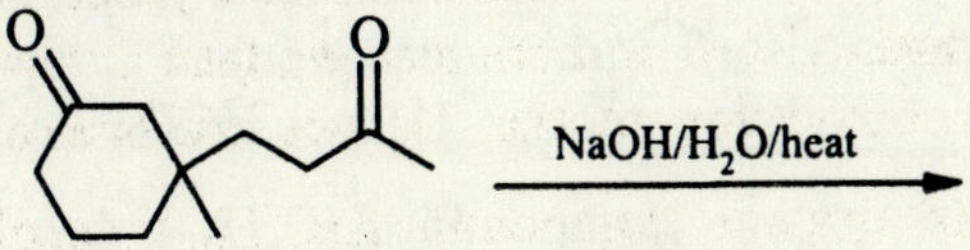 $\xrightarrow{NaOH/H_2O/heat}$

(b) 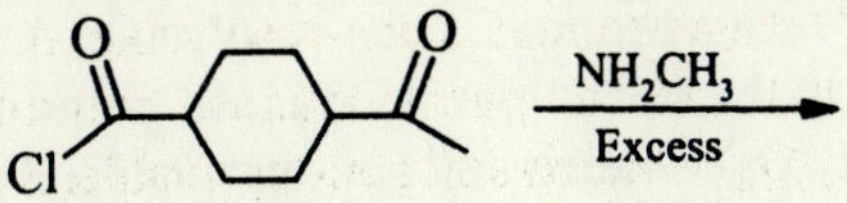 Cl $\xrightarrow[\text{Excess}]{NH_2CH_3}$

31. Propose mechanism:

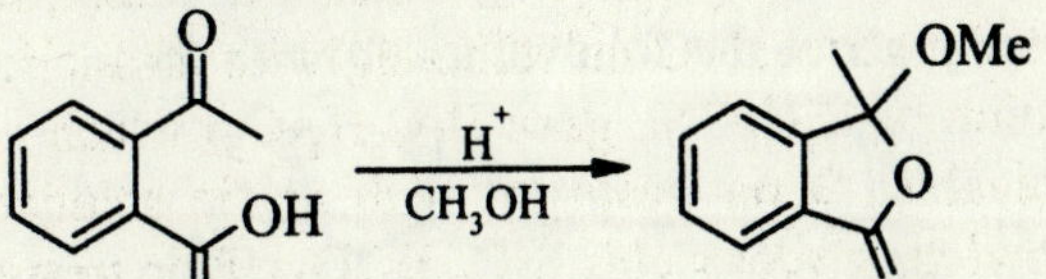

OH, O $\xrightarrow[CH_3OH]{H^+}$ OMe, O, O

32. Propose mechanism:

(a) 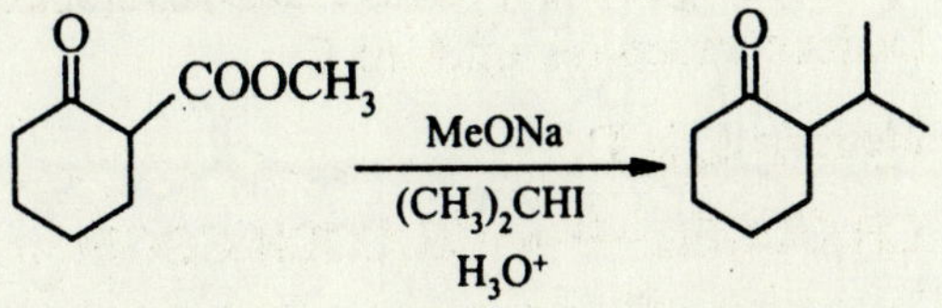 $COOCH_3$ $\xrightarrow[\substack{(CH_3)_2CHI \\ H_3O^+}]{MeONa}$

(b) 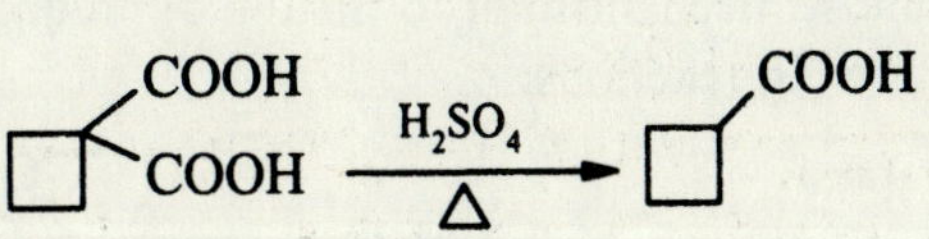 COOH, COOH $\xrightarrow[\Delta]{H_2SO_4}$ COOH

33. Provide organic products in the following reactions:

(a) 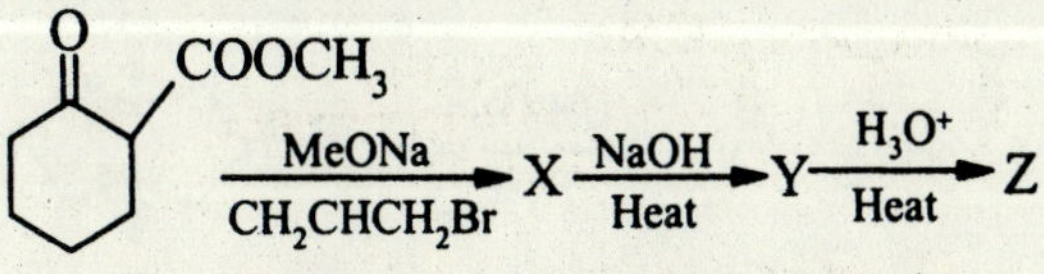 $COOCH_3$ $\xrightarrow[CH_2CHCH_2Br]{MeONa}$ X $\xrightarrow[\text{Heat}]{NaOH}$ Y $\xrightarrow[\text{Heat}]{H_3O^+}$ Z

(b) 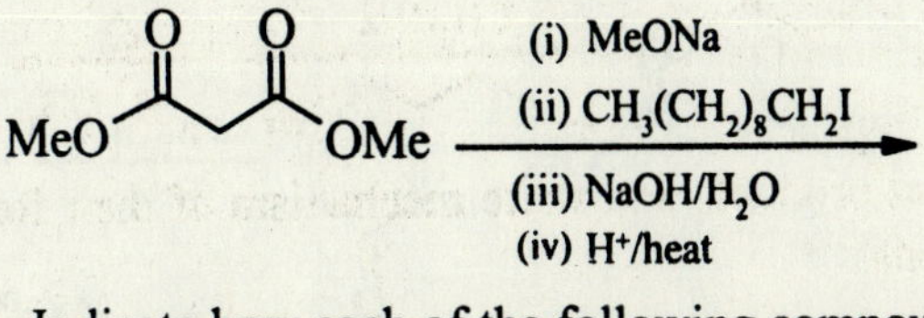 MeO, OMe $\xrightarrow{}$ (i) MeONa (ii) $CH_3(CH_2)_8CH_2I$ (iii) $NaOH/H_2O$ (iv) H^+/heat

34. Indicate how each of the following compounds can be synthesized from the given starting material?

(a) 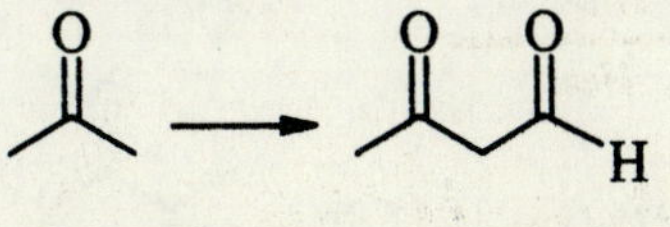 $\longrightarrow$ H

(b) 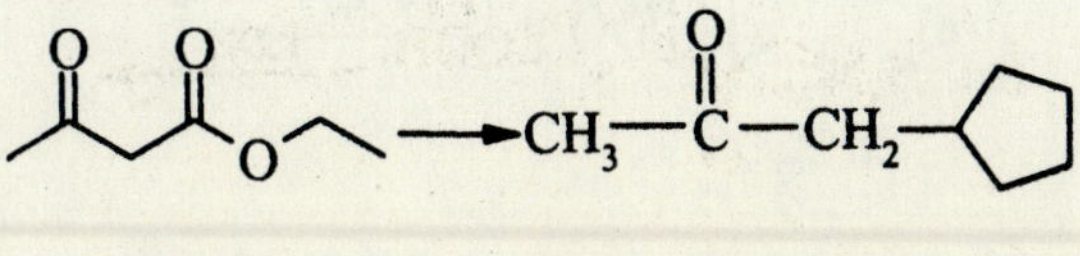 O $\longrightarrow CH_3-\overset{O}{\overset{\|}{C}}-CH_2-$[cyclopentyl]

(c) CH_2CH_3 $\longrightarrow$ $CH_2CH_2CH_2COOH$

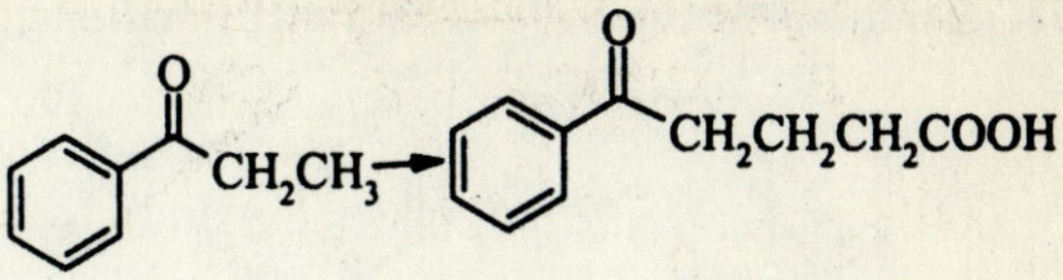

(d) 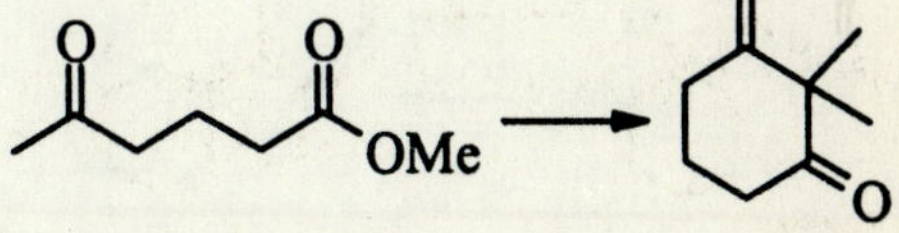 OMe $\longrightarrow$ O

(e) 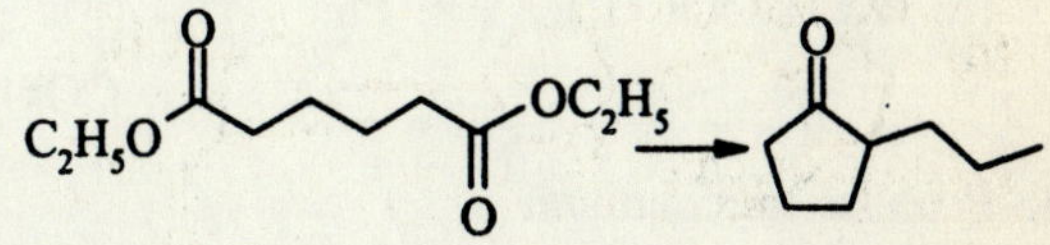 C_2H_5O, OC_2H_5 $\longrightarrow$

35. Propose mechanism for each of the following reactions:

(a) 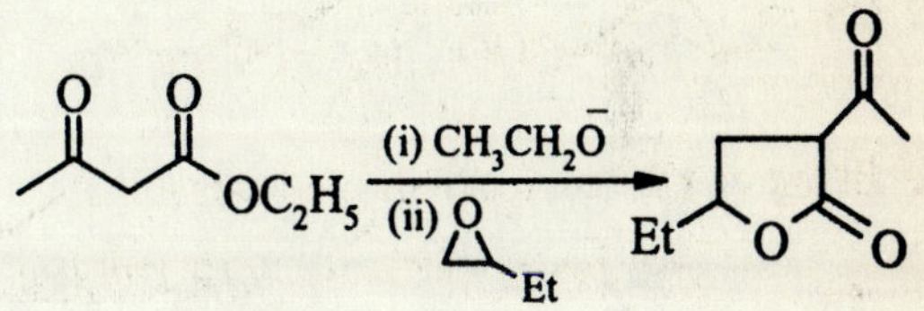 OC_2H_5 $\xrightarrow[\text{(ii) epoxide, Et}]{\text{(i) } CH_3CH_2O^-}$ Et, O, O

(b) 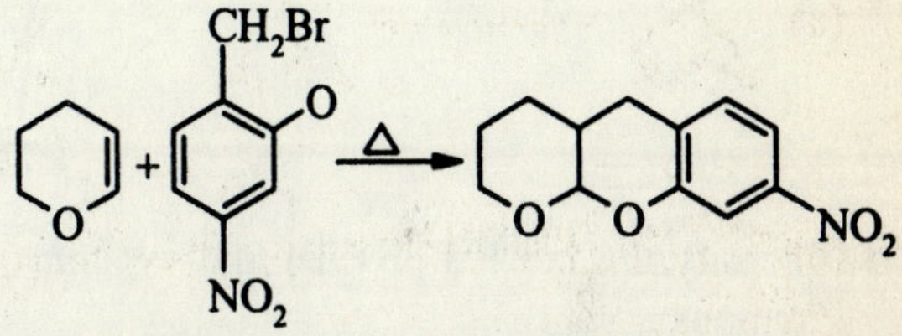 O + CH_2Br, O, NO_2 $\xrightarrow{\Delta}$ O, O, NO_2

36. Devise synthesis starting from diethyl malonate:

(a) 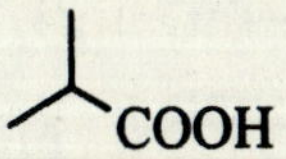 COOH (b)

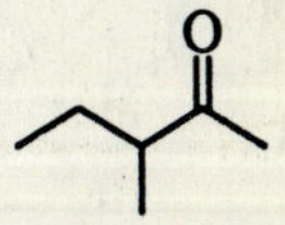

(c) 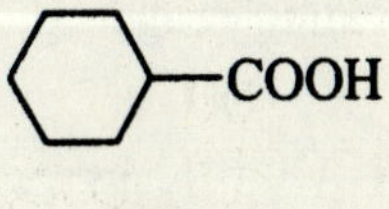 COOH (d) 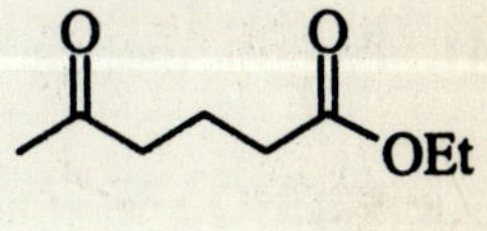 OEt

(e) 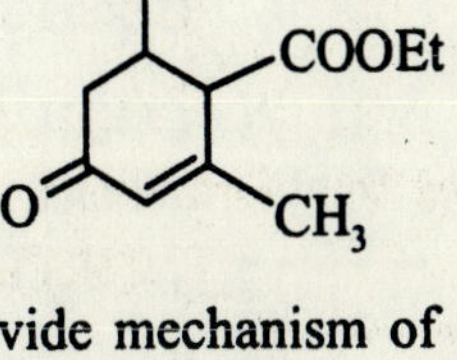 Ph, COOEt, CH_3, O

37. Provide mechanism of each step in the following reaction sequence:

(a) 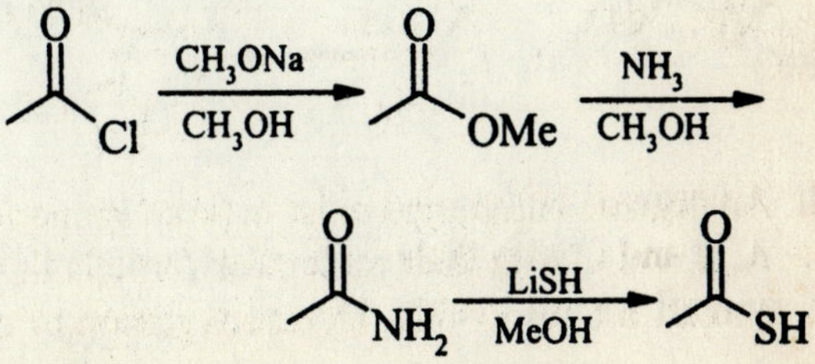 Cl $\xrightarrow[CH_3OH]{CH_3ONa}$ OMe $\xrightarrow[CH_3OH]{NH_3}$

NH_2 $\xrightarrow[MeOH]{LiSH}$ SH

38. Bring about the following transformations:

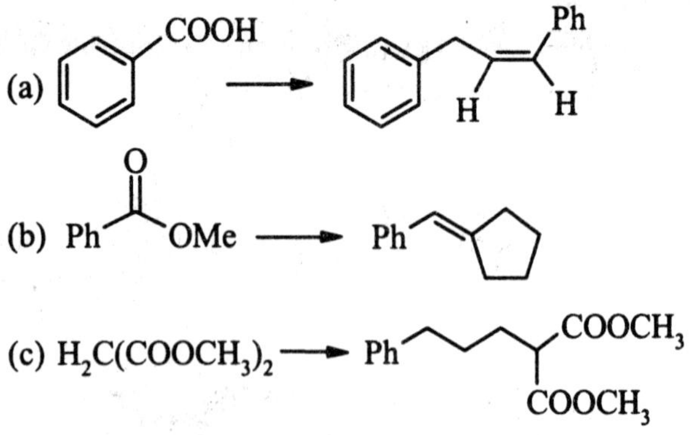

(a) [benzene ring with COOH] → [benzene ring–CH₂–CH=C(Ph)H, with H labels]

(b) Ph–C(=O)–OMe → Ph–CH=cyclopentane

(c) $H_2C(COOCH_3)_2$ → Ph–CH₂CH₂CH₂–CH(COOCH₃)(COOCH₃)

39. Propose mechanism:

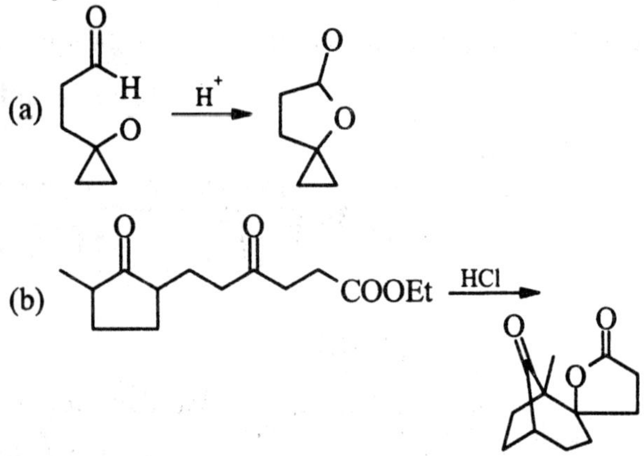

(a) [aldehyde with cyclopropane-bearing hydroxy group: O, H, O] $\xrightarrow{H^+}$ [cyclic product: O, O]

(b) [2-methylcyclopentanone chain with ketone and COOEt] $\xrightarrow{HCl}$ [bicyclic spiro lactone: O, O, O]

40. Identify the labeled product and explain their formations:

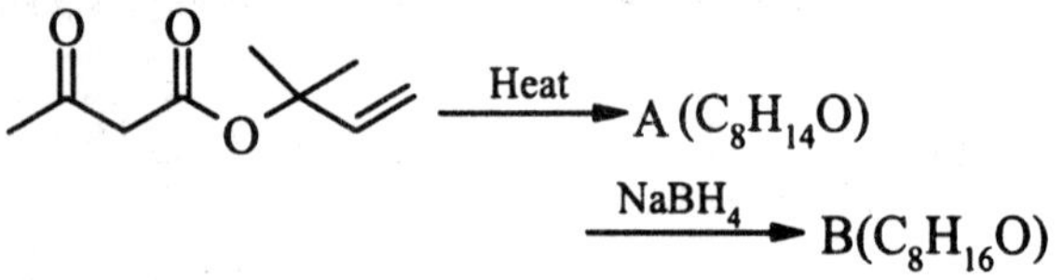

[allyl ester of acetoacetic acid] $\xrightarrow{Heat}$ A ($C_8H_{14}O$) $\xrightarrow{NaBH_4}$ B($C_8H_{16}O$)

41. Propose mechanism, providing reagents:

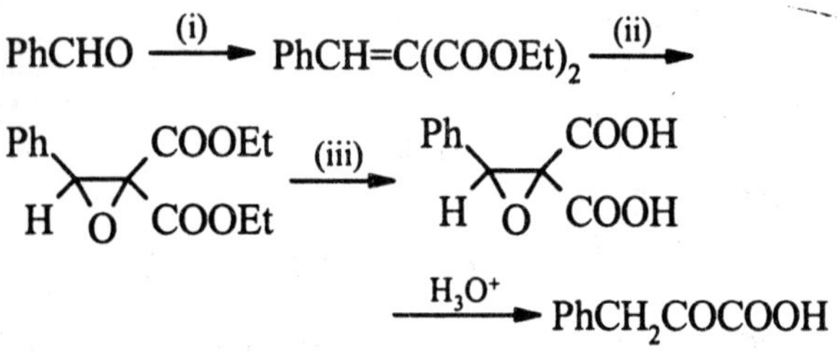

PhCHO $\xrightarrow{(i)}$ PhCH=C(COOEt)$_2$ $\xrightarrow{(ii)}$

[Ph, H, O epoxide with COOEt, COOEt] $\xrightarrow{(iii)}$ [Ph, H, O epoxide with COOH, COOH]

$\xrightarrow{H_3O^+}$ $PhCH_2COCOOH$

42. Provide mechanism:

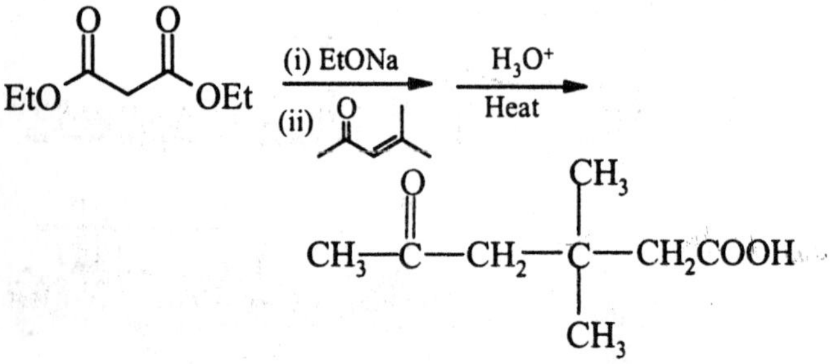

EtO–C(=O)–CH₂–C(=O)–OEt $\xrightarrow[(ii)\ (CH_3)_2C{=}CHCOCH_3]{(i)\ EtONa}$ $\xrightarrow[Heat]{H_3O^+}$

$CH_3-\overset{O}{\overset{\|}{C}}-CH_2-\underset{CH_3}{\overset{CH_3}{C}}-CH_2COOH$

43. Starting from ethyl propanoate synthesize:

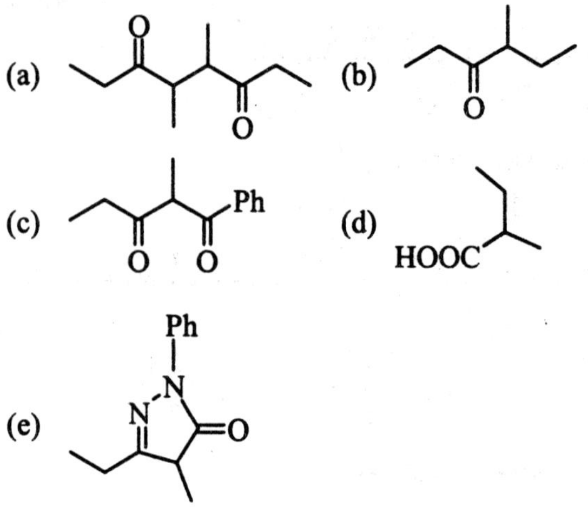

(a) [diketone: O, O] (b) [ketone: O]

(c) [diketone with Ph: O, O] (d) HOOC–[branched acid]

(e) [pyrazolone: Ph, N, N, =O]

44. Bring about the following conversions:

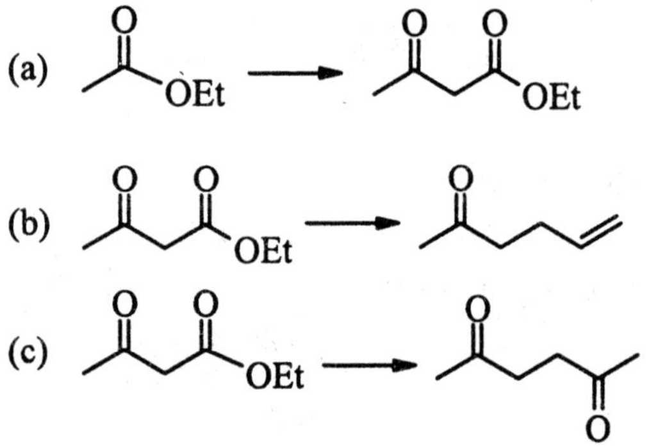

(a) CH₃C(=O)OEt → CH₃C(=O)CH₂C(=O)OEt

(b) CH₃C(=O)CH₂C(=O)OEt → CH₃C(=O)CH₂CH₂CH=CH₂

(c) CH₃C(=O)CH₂C(=O)OEt → CH₃C(=O)CH₂CH₂C(=O)CH₃

45. Propose mechanism of formation of products in the following reactions:

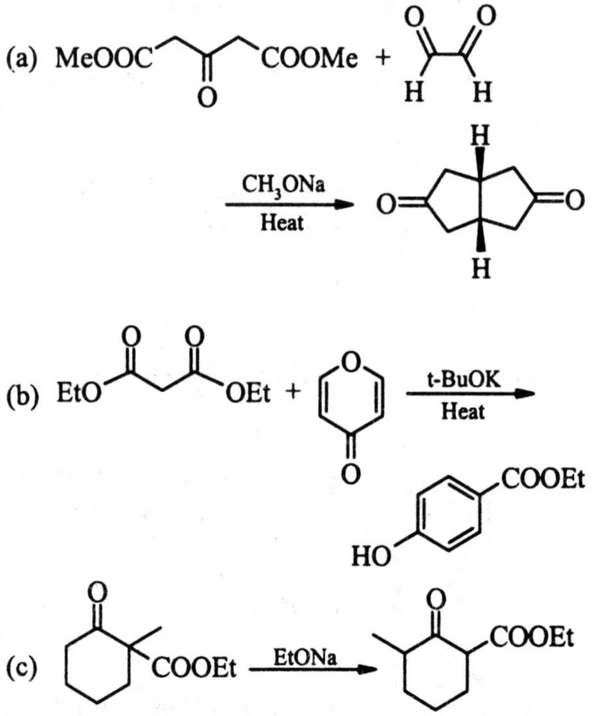

(a) MeOOC–CH₂–C(=O)–CH₂–COOMe + OHC–CHO (H, H) $\xrightarrow[Heat]{CH_3ONa}$ [bicyclic diketone O=, =O with H, H]

(b) EtO–C(=O)–CH₂–C(=O)–OEt + [4-pyrone: O, O] $\xrightarrow[Heat]{t\text{-}BuOK}$ HO–C₆H₄–COOEt

(c) [2-oxocyclohexane with COOEt and methyl] $\xrightarrow{EtONa}$ [2-methyl-6-oxocyclohexane with COOEt]

OBJECTIVE EXERCISE

1. What is true for the equilibrium reaction?

   Acetic acid + ethanol $\overset{H^+}{\rightleftharpoons}$ Ethyl acetate + H_2O

   (a) The use of equimolar quantities of ethanol and acetic acid will give greatest yield of ester at equilibrium.
   (b) Removal of water will increase the amount of ester at equilibrium.
   (c) Addition of ester at equilibrium will cause formation of equal amount of water.
   (d) Application of pressure increases the amount of ester at equilibrium.

2. The reaction between CO_2 and a Grignard reagent will yield:
   (a) An alkyl magnesium halide
   (b) An alcohol
   (c) Magnesium carbonate
   (d) A carboxylic acid

3. Select the final product from the following sequence of reaction:

   Malonic ester $\xrightarrow{NaOEt}$ $\xrightarrow{CH_3I}$ $\xrightarrow{NaOEt}$ $\xrightarrow{Br\text{—}CH_2CH_2CH_3}$

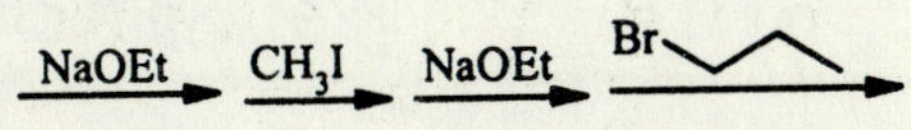

   (a) 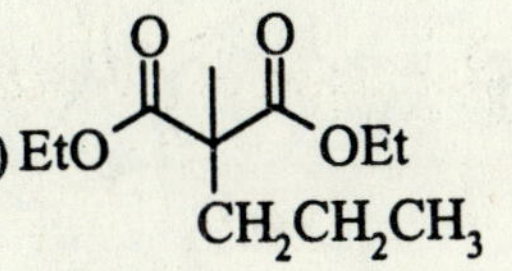 EtO(C=O)–C(CH_3)($CH_2CH_2CH_3$)–(C=O)OEt
   (b) H, O, $CH_2CH_2CH_3$
   (c) O
   (d) 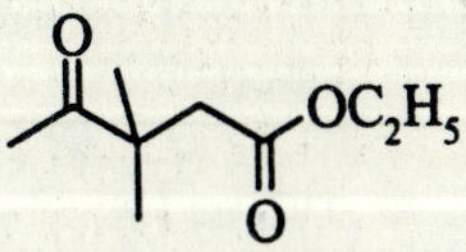 O, OC_2H_5, O

4. What is the major product of the following reaction:

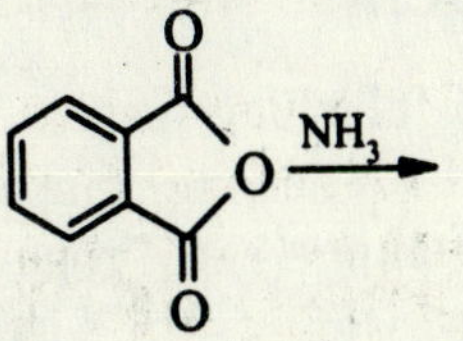

   Phthalic anhydride $\xrightarrow{NH_3}$

   (a) 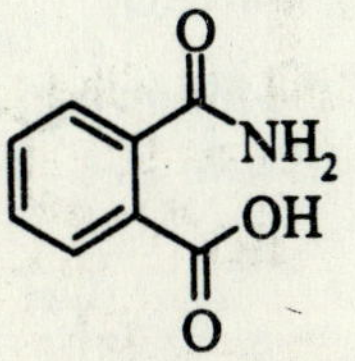 O, NH_2, OH, O
   (b) 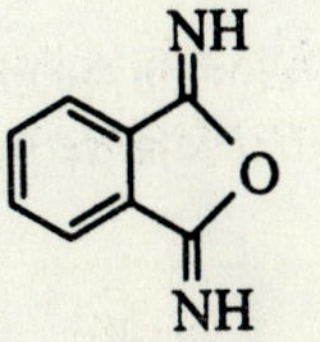 NH, O, NH
   (c) 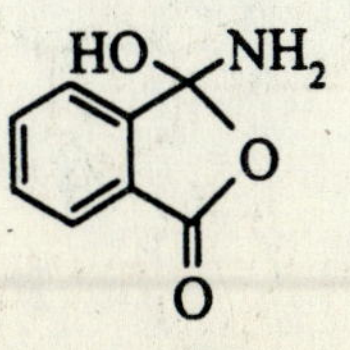 HO, NH_2, O, O
   (d) 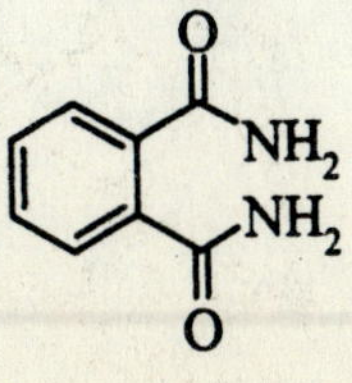 O, NH_2, NH_2, O

5. Which of the following is an intermediate in acid catalyzed hydrolysis of an ester?

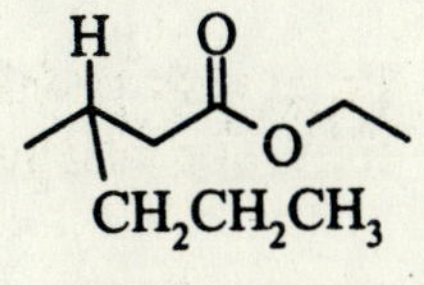

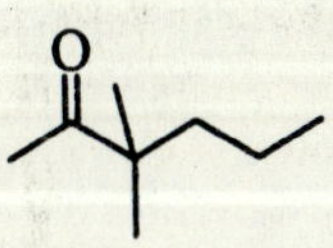

   (a) 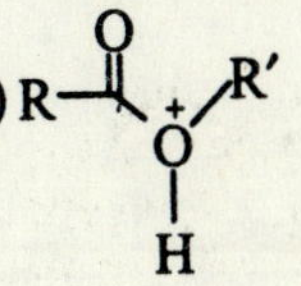 R—C(=O)—$\overset{+}{O}$(H)—R′
   (b) R—C(O^-)(OH)—O—R′
   (c) R—C(OH)($\overset{+}{O}H_2$)—O—R′
   (d) $R\overset{+}{C}=O$

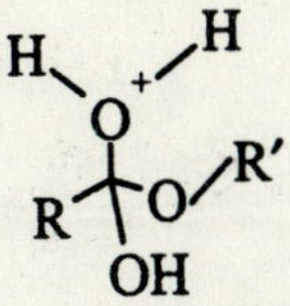

6. Which of the following would be the product of the initial protonation of an amide?

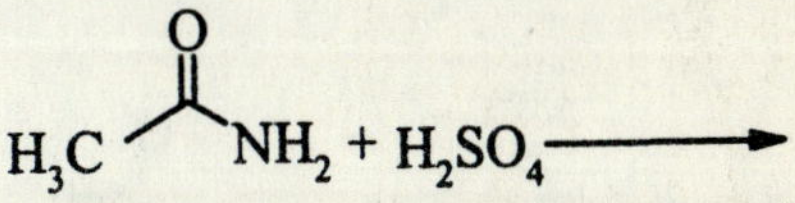

   H_3C–C(=O)–NH_2 + H_2SO_4 ⟶

   (a) H_3C–$\overset{+}{C}$(OH)–NH_2
   (b) H_3C–$\overset{+}{C}$=O
   (c) H_3C–C(=O)–$\overset{+}{N}H_3$
   (d) 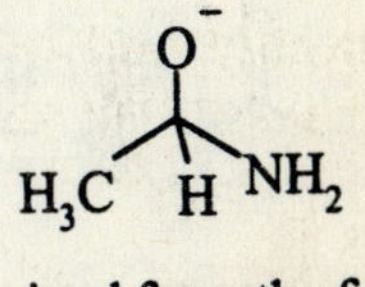 H_3C–C(O^-)(H)–NH_2

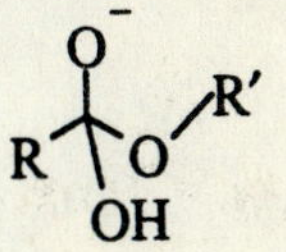

7. What is the major product obtained from the following reaction?

   HO–C(=O)–CH_2–OH $\xrightarrow{SOCl_2}$ $\xrightarrow{CH_3OH}$

   (a) HO–C(=O)–CH_2–OMe
   (b) MeO–C(=O)–CH_2–OMe
   (c) Cl–C(=O)–CH_2–OMe
   (d) 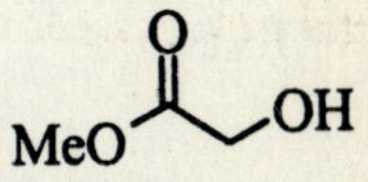 MeO–C(=O)–CH_2–OH

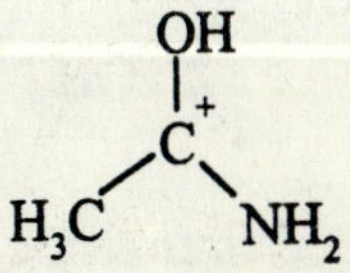

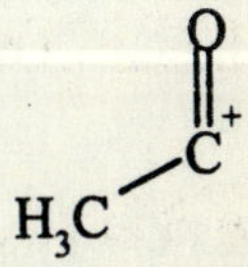

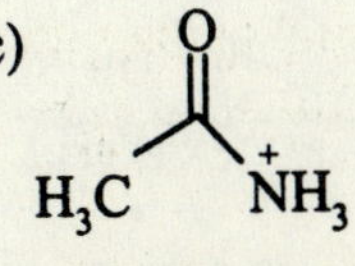

8. Which of the following would be the best synthesis of 2, 2-dimethylpropanoic acid?

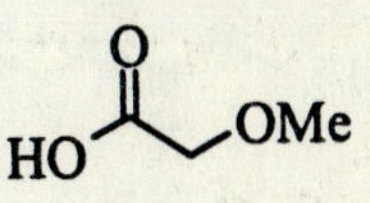

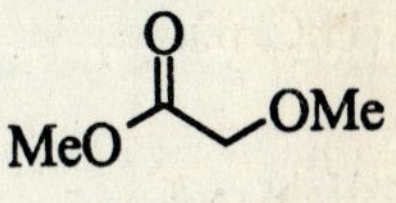

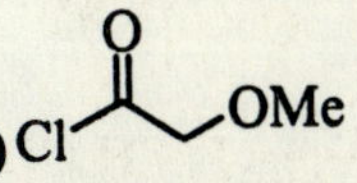

   (a) 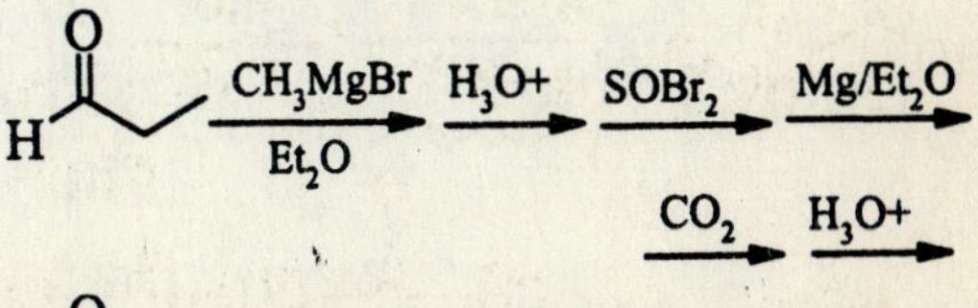 H–C(=O)–CH_2CH_3 $\xrightarrow[Et_2O]{CH_3MgBr}$ $\xrightarrow{H_3O+}$ $\xrightarrow{SOBr_2}$ $\xrightarrow{Mg/Et_2O}$ $\xrightarrow{CO_2}$ $\xrightarrow{H_3O+}$

   (b) 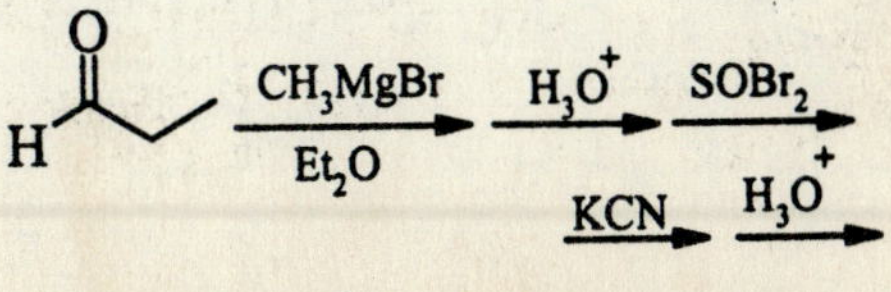 H–C(=O)–CH_2CH_3 $\xrightarrow[Et_2O]{CH_3MgBr}$ $\xrightarrow{H_3O^+}$ $\xrightarrow{SOBr_2}$ $\xrightarrow{KCN}$ $\xrightarrow{H_3O^+}$

(c) 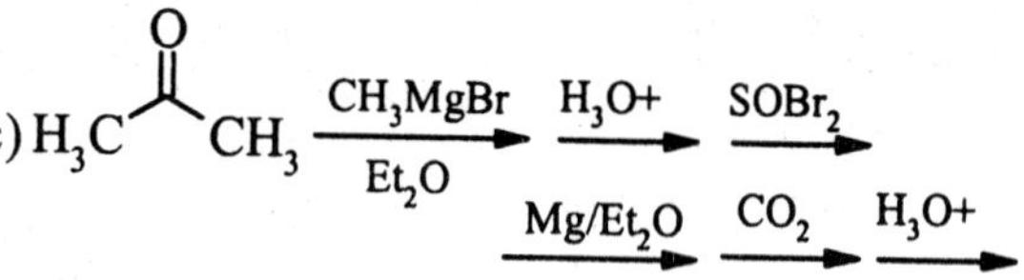 $H_3C-CO-CH_3$ $\xrightarrow[Et_2O]{CH_3MgBr}$ $\xrightarrow{H_3O+}$ $\xrightarrow{SOBr_2}$ $\xrightarrow{Mg/Et_2O}$ $\xrightarrow{CO_2}$ $\xrightarrow{H_3O+}$

(d) 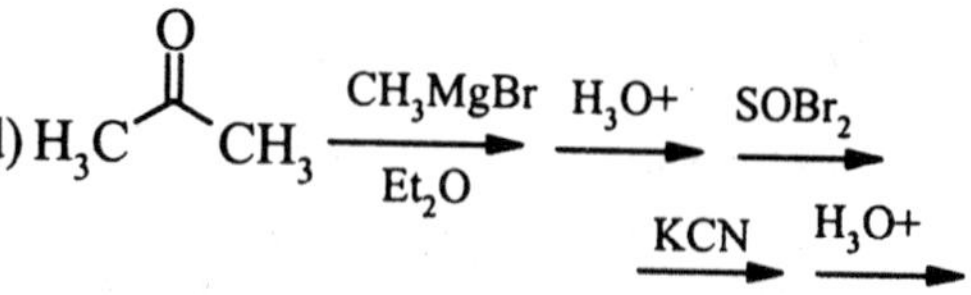 $H_3C-CO-CH_3$ $\xrightarrow[Et_2O]{CH_3MgBr}$ $\xrightarrow{H_3O+}$ $\xrightarrow{SOBr_2}$ $\xrightarrow{KCN}$ $\xrightarrow{H_3O+}$

9. What is the major product obtained from the following reaction?

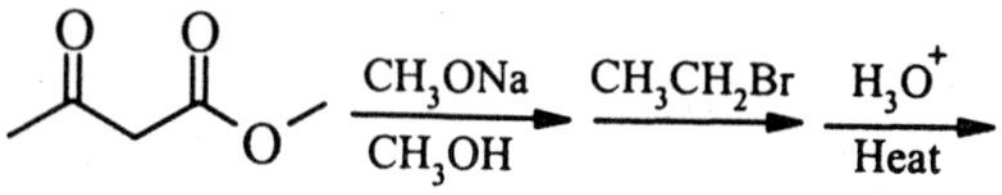

$\xrightarrow[CH_3OH]{CH_3ONa}$ $\xrightarrow{CH_3CH_2Br}$ $\xrightarrow[Heat]{H_3O^+}$

(a) 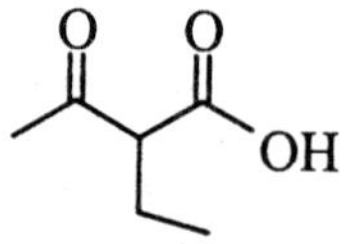 OH (b) 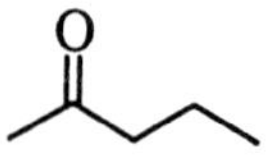 (c) 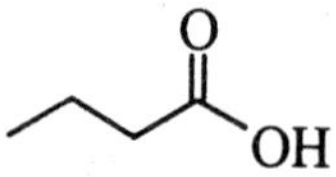 OH (d) 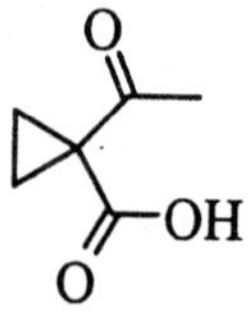 OH

10. What is the major product obtained from the following reaction?

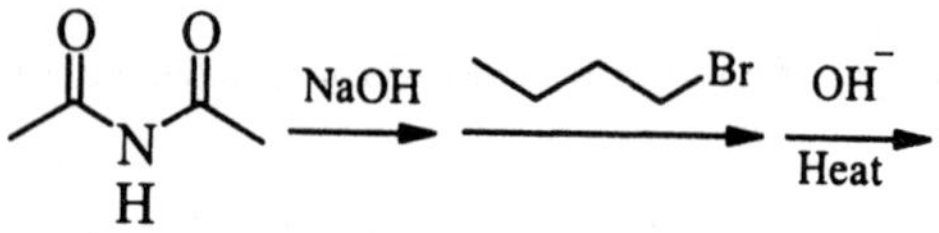

N H $\xrightarrow{NaOH}$ Br $\xrightarrow[Heat]{OH^-}$

(a) 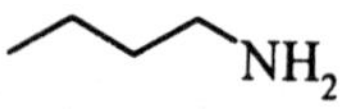 NH_2 (b) 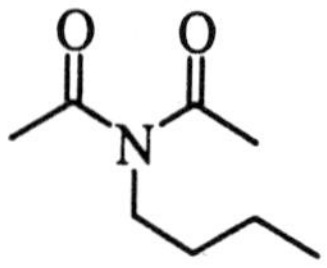 N (c) 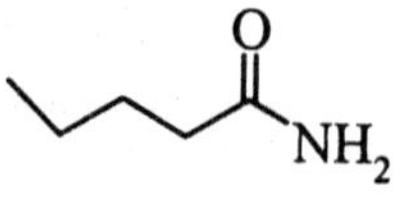 NH_2 (d) 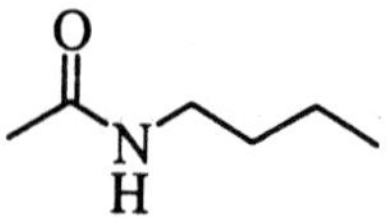 N H

11. What is the major product obtained from the following reaction?

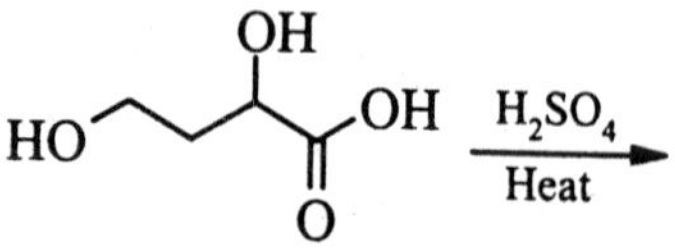

HO OH OH O $\xrightarrow[Heat]{H_2SO_4}$

(a) 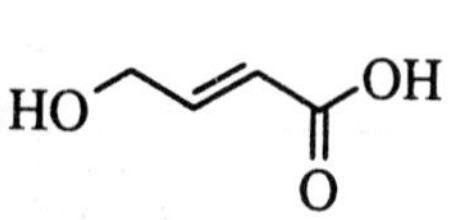 HO OH O (b) 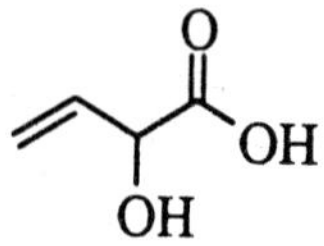 OH OH (c) 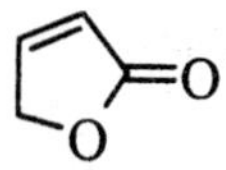 O O (d) 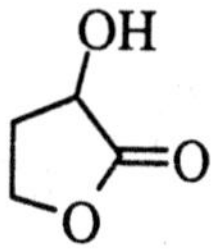 OH O O

12. What is the major product obtained from the following reaction?

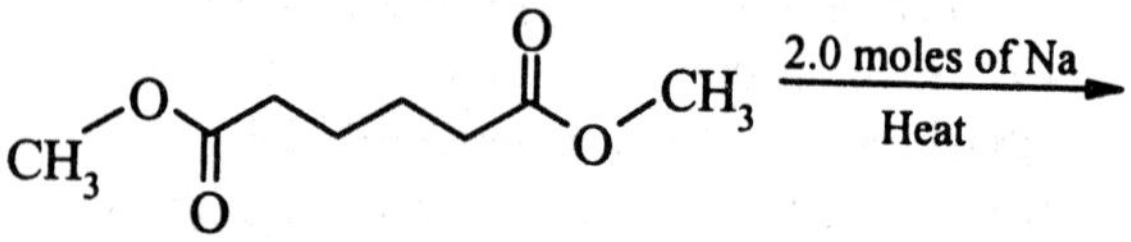

CH_3 O O O CH_3 $\xrightarrow[Heat]{2.0\ moles\ of\ Na}$

(a) 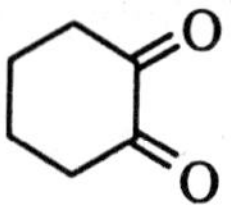 O O (b) 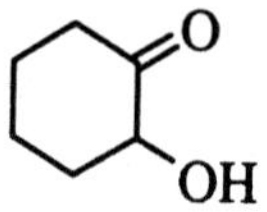 O OH (c) 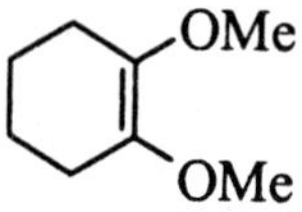 OMe OMe (d) 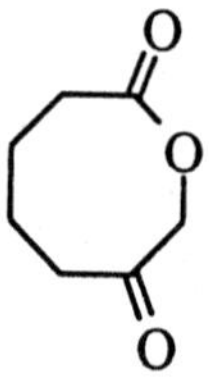 O O O

13. What is the major product of the following reaction?

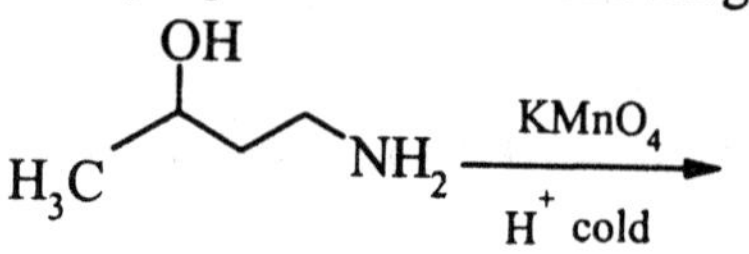

OH H_3C NH_2 $\xrightarrow[H^+\ cold]{KMnO_4}$

(a) 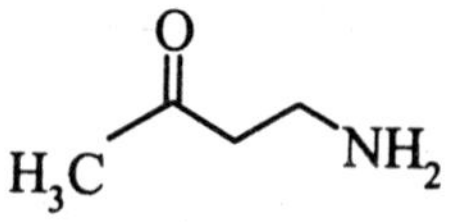 H_3C O NH_2 (b) 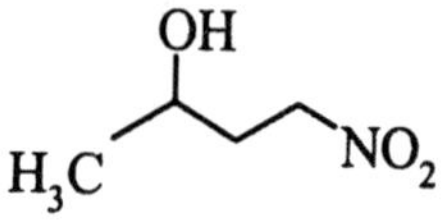 H_3C OH NO_2

(c) 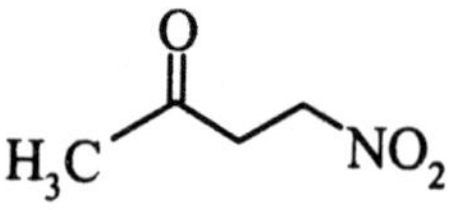 H_3C O NO_2 (d) 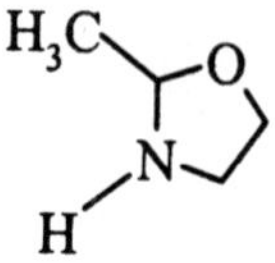 H_3C O N H

14. Which of the following reaction cannot produce acetamide?

(a) $(CH_3CO)_2O + NH_3 \rightarrow$
(b) $CH_3CHO + NH_3 \rightarrow$
(c) $CH_3COBr + NH_3 \rightarrow$
(d) $CH_3COOCH_3 + NH_3 \rightarrow$

15. Which of the following reagent could be used to differentiate the following compounds by a visible change? **2-butanone** and **Propanoic acid**:

(a) $Ag(NH_3)_2^+/OH^-$ (b) $LiAlH_4$
(c) $KMnO_4/H^+$/Cold (d) $NaHCO_3$

16. What is the major product of the following reaction?
Benzamide + H_2/Pd/high pressure →

(a) 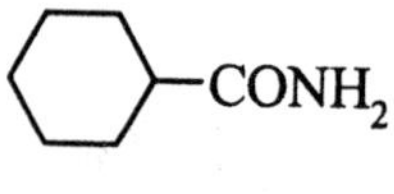 $-CONH_2$ (b) 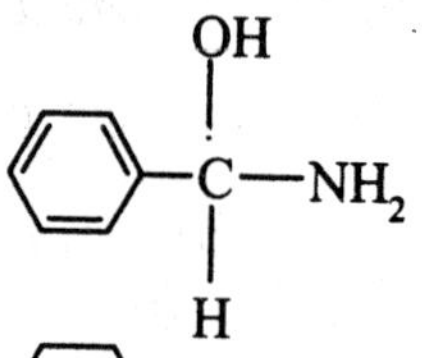 OH C NH_2 H (c) $-CH_2NH_2$ (d) 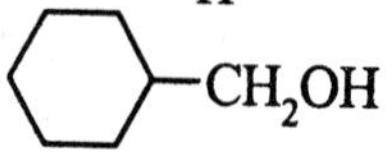 $-CH_2OH$

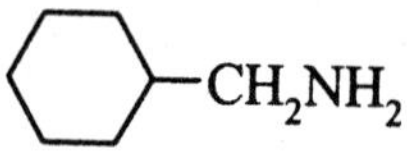

17. Which of the following reaction will produce isopropyl acetate in the best yield?
(a) Acetyl chloride + 2-propanol →
(b) Silver acetate + 2-chloropropane →
(c) Acetic acid + 2-propanol/H_2SO_4 →
(d) All of these reaction work equally well.

18. Give the major product of the following reaction:

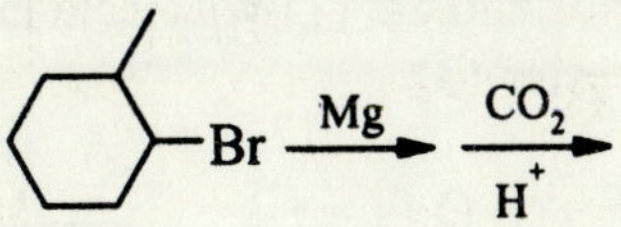

(Mg; CO_2, H^+)

(a) 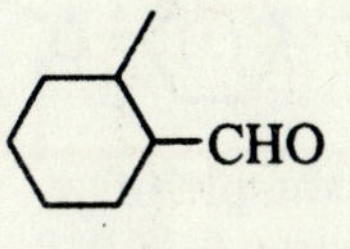 –CHO (b) 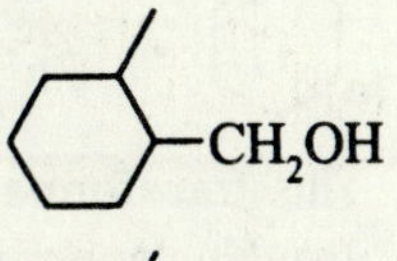 –CH_2OH
(c) 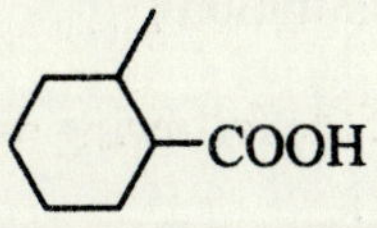 –COOH (d) 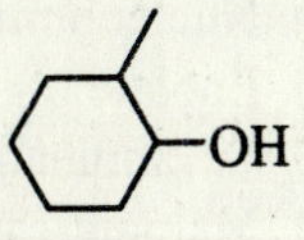 –OH

19. Give major product of the following reaction:

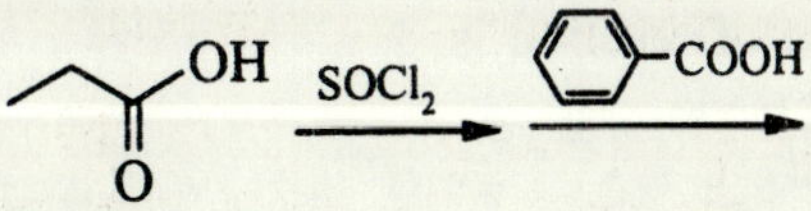

($SOCl_2$; –COOH)

(a) 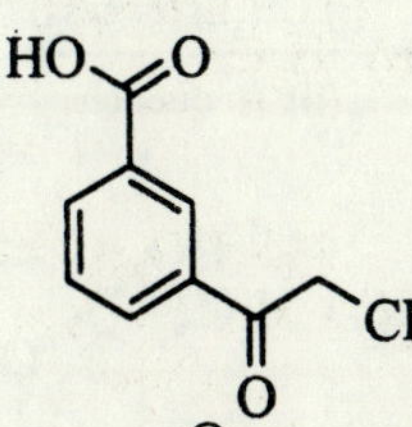 HO, O, CH_3, O (b) 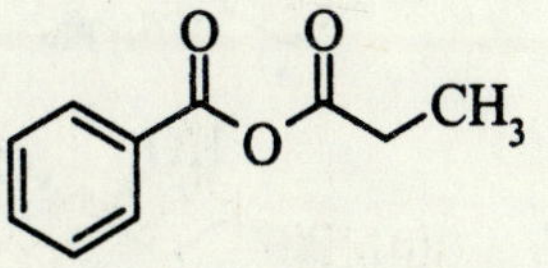 O, O, O, CH_3
(c) 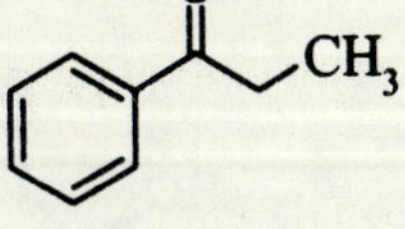 O, CH_3 (d) 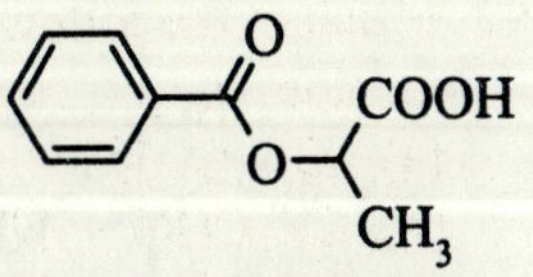 O, COOH, O, CH_3

20. Give the reagent that will best accomplish the following transformation:

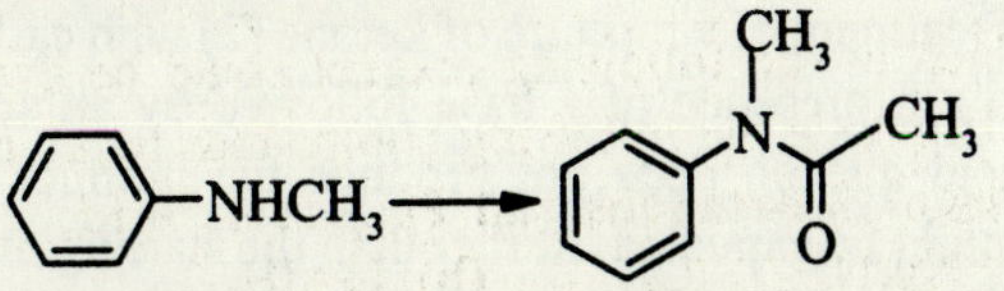

–$NHCH_3$ → CH_3, N, CH_3, O

(a) (1) NaOH, (2) CH_3CH_2OH, (3) CrO_3, H_2O, H_2SO_4
(b) CH_3COOH/heat
(c) Ethanal/heat
(d) (1) Ethanal/heat (2) CrO_3,H_2SO_4.

21. Give the major product of the following reaction:

O, O–CH_3 ($LiAlH_4$, H^+) (O, O; H^+)

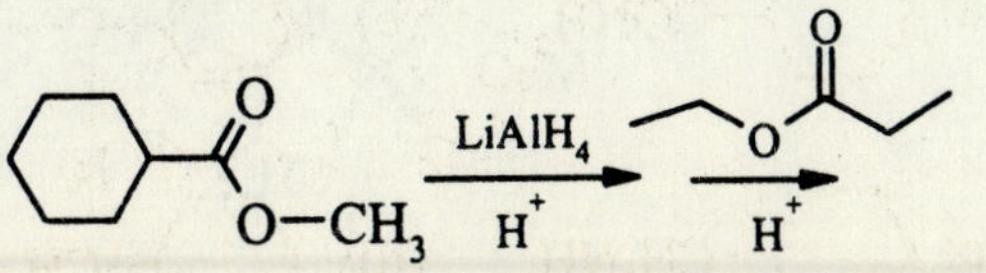

(a) 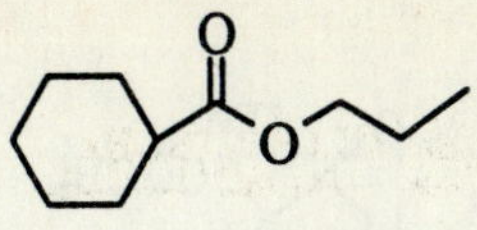 (b) 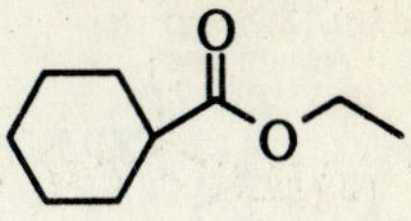 (c) 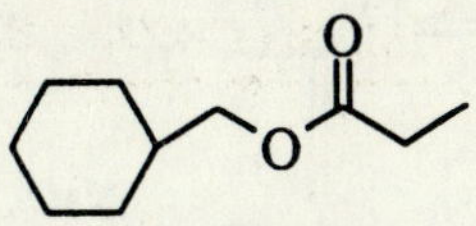 (d)

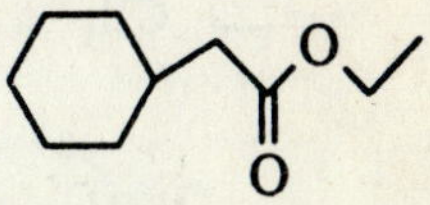

22. Predict product of the following reaction:

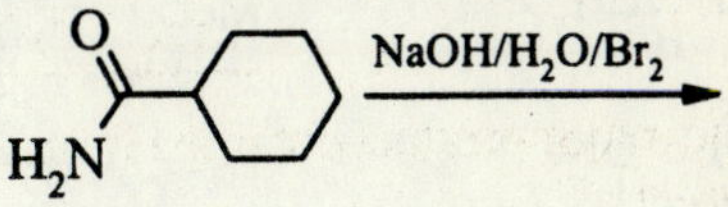

O, H_2N ($NaOH/H_2O/Br_2$)

(a) 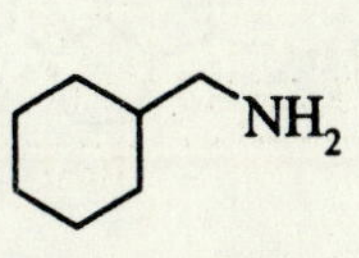 NH_2 (b) 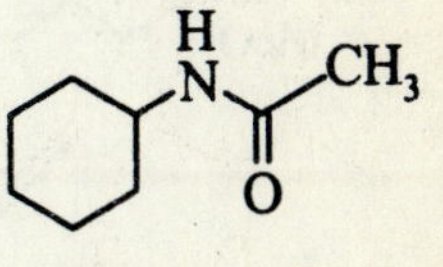 H, N, CH_3, O
(c) 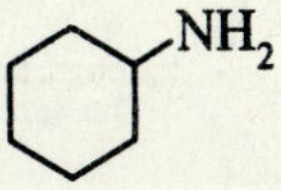 NH_2 (d) 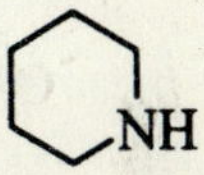 NH

23. Treatment of chiral ester (1) with one equivalent of LDA followed by allyl bromide (2) affords the alkylation product (3):

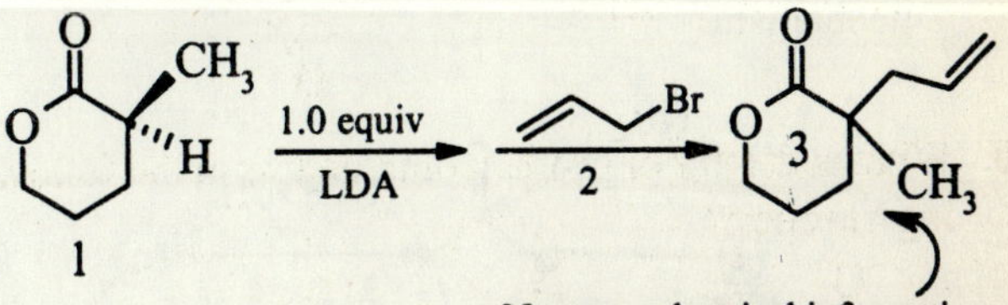

O, O, CH_3, H, 1 (1.0 equiv LDA) (Br, 2) O, O, 3, CH_3

No stereochemical information implied in this structural drawing

Which of the following statement is true?
(a) 3 is chiral (b) 3 is achiral
(c) 3 is a meso compound (d) 3 is racemic

24. Reaction of a β-ketoester (1) with $NaOCH_3$, followed by an aqueous acid quench, provides the rearranged product β-ketoester (2). Two possible mechanism for this rearrangement are shown below:

Mechanism A

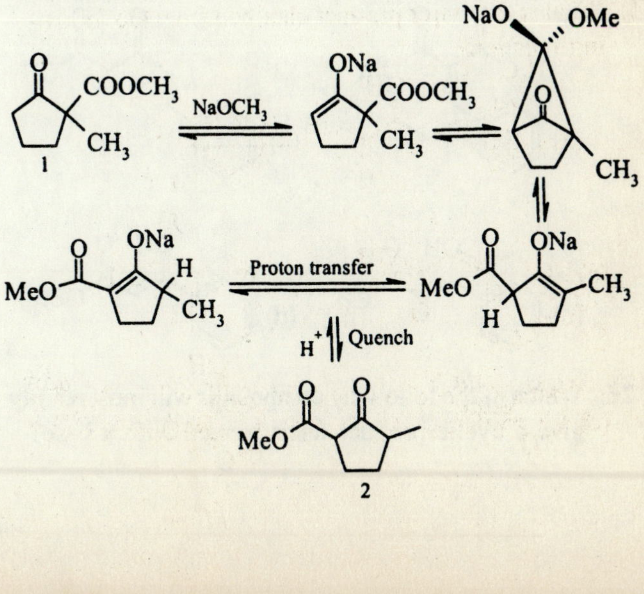

O, $COOCH_3$, CH_3, 1 ($NaOCH_3$) ONa, $COOCH_3$, CH_3 ⇌ NaO, OMe, O, CH_3

O, ONa, MeO, H, CH_3 (Proton transfer) O, ONa, MeO, H, CH_3

(H^+ Quench)

MeO, O, O, 2

Mechanism B

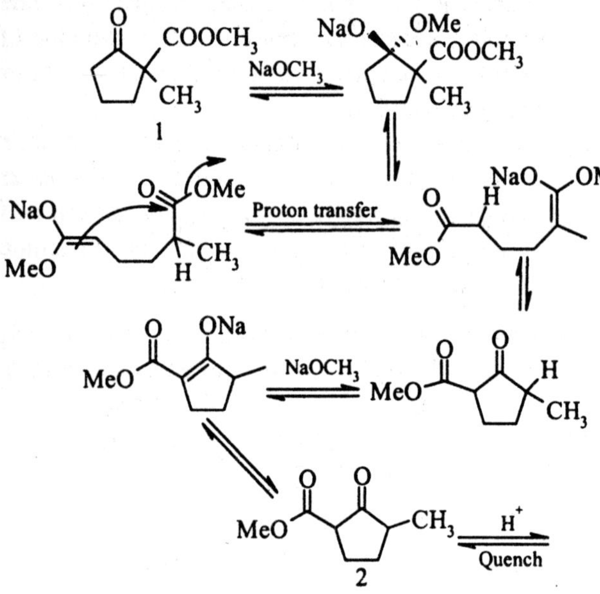

NaOCH$_3$ (1); Proton transfer; NaOCH$_3$; H^+ Quench (2)

A lebeling study can distinguish between these two mechanism. Which carbon in 1 would you lebel with ^{13}C to differentiate between these two possible mechanism?

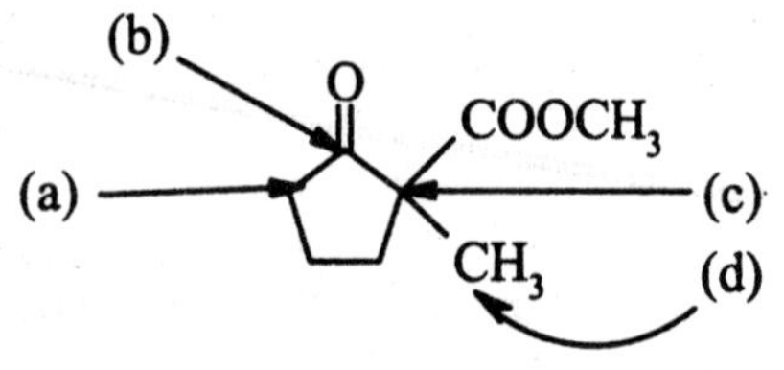

(a) (b) (c) (d)

25. Treatment of two equivalent of ethylacetate with one equivalent of sodium ethoxide gives intermediate (1). Addition of an aldehyde to (1), followed by an acid quench, affords which product?

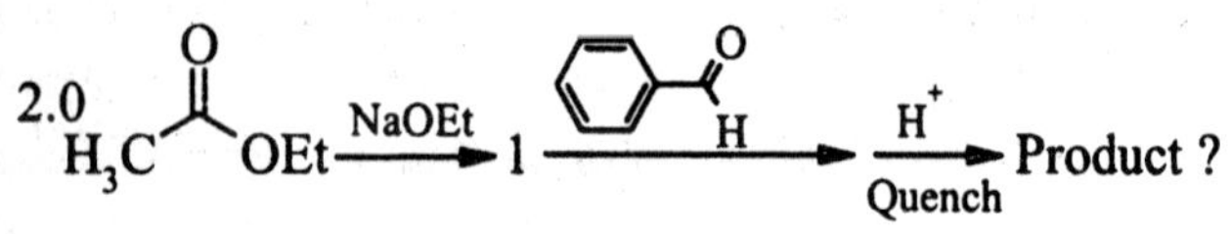

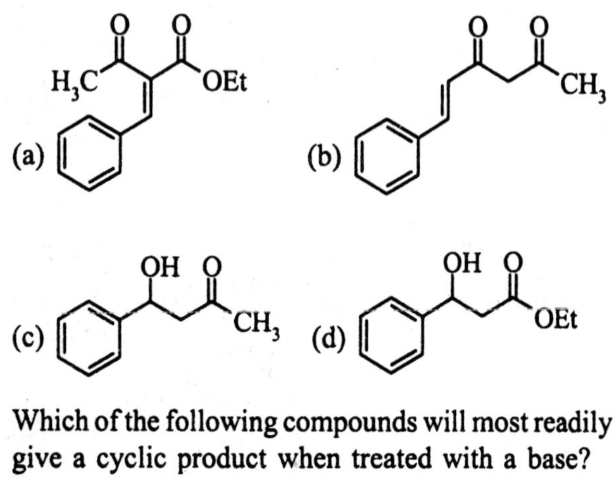

2.0 H_3C–CO–OEt $\xrightarrow{NaOEt}$ 1 $\xrightarrow{PhCHO}$ $\xrightarrow[Quench]{H^+}$ Product ?

(a) (b) (c) (d)

26. Which of the following compounds will most readily give a cyclic product when treated with a base?

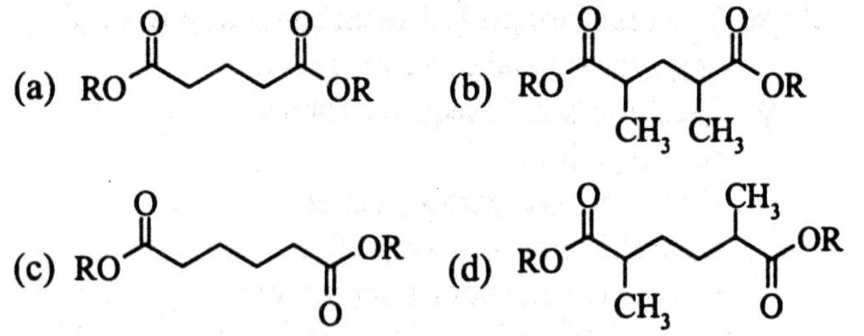

(a) (b) (c) (d)

27. Treatment of enolate (1) with acid chloride (2) delivers the ester (3).

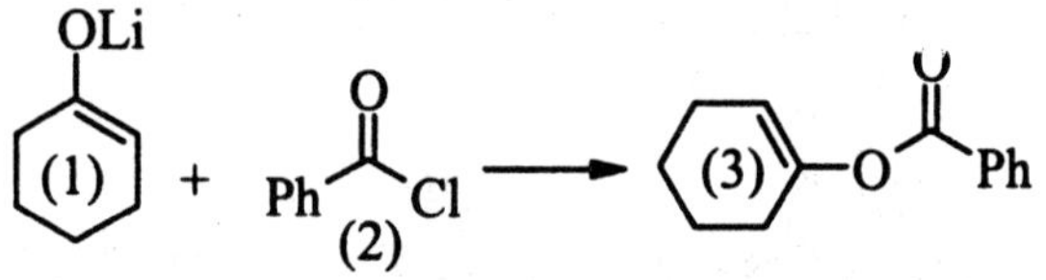

This transformation is an example of:
(a) Nucleophilic acyl substitution
(b) Enolate alkylation
(c) E2 elimination
(d) Malonic ester synthesis

28. The major organic product from the following sequence of reaction is:

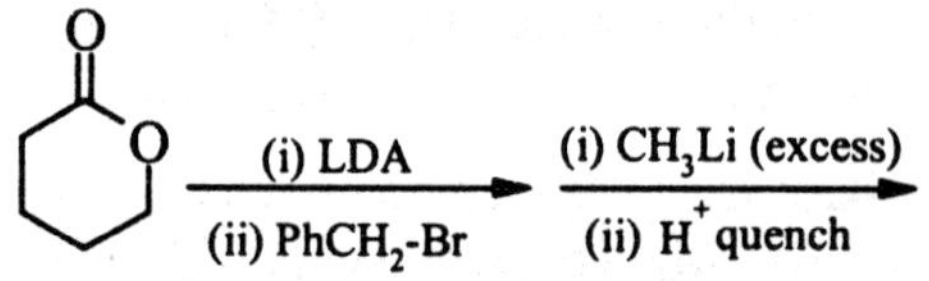

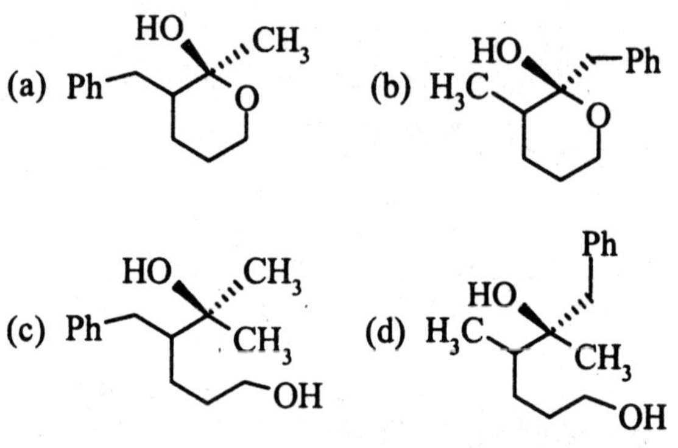

(i) LDA (ii) PhCH$_2$-Br; (i) CH$_3$Li (excess) (ii) H^+ quench

(a) (b) (c) (d)

29. Treatment of a mixture of ketone (2) with carbonate (1) in presence of a base followed by an aqueous acid quench does not produce the product (3) as might be expected. Rather than the starting material (1) and (2) are recovered. Why?

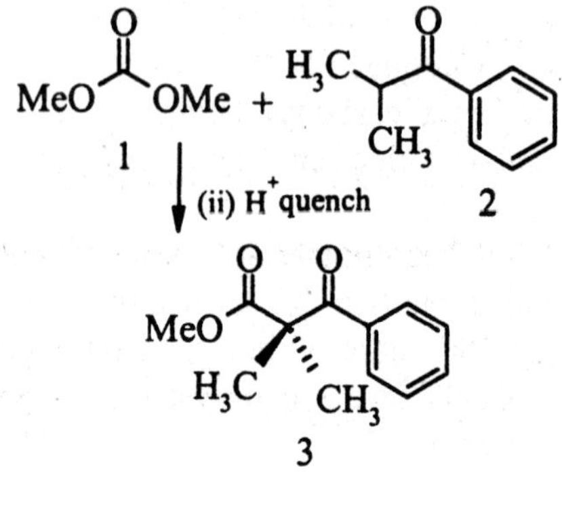

(ii) H^+ quench

(a) The first formed 1, 3-dicarbonyl product cannot deprotonate and drive the reaction forward to 3.
(b) Catalytic base is required for successful Claisen-condensation.
(c) Ester and not carbonates are required for successful Claisen-condensation.
(d) Ketone (2) can not be deprotonated by $NaOCH_3$ to form the required enolate.

30. The mechanism of β-ketoacid decarboxylation is shown below:

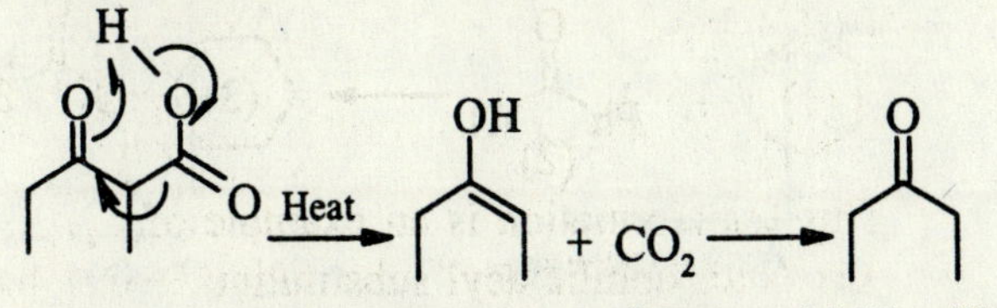

Heat, + CO_2

Heating a single enantiomer of a β-keto acid (1) will afford a decarboxylated product that is:

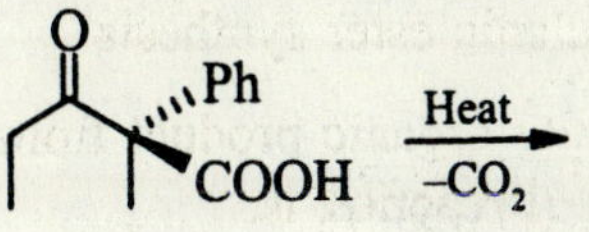

Ph, COOH, Heat, $-CO_2$

(a) A single enantiomer (b) Distereomer
(c) Achiral (d) Racemic

31. Which of the following compound can be prepared directly by Malonic-ester synthesis?

(a) 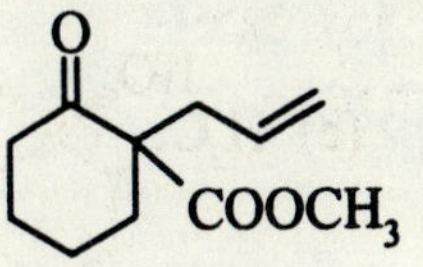 $COOCH_3$ (b) 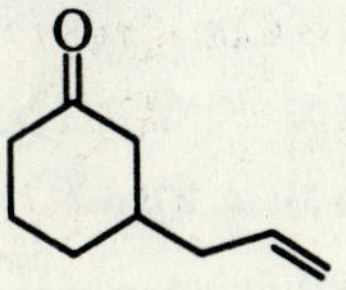
(c) 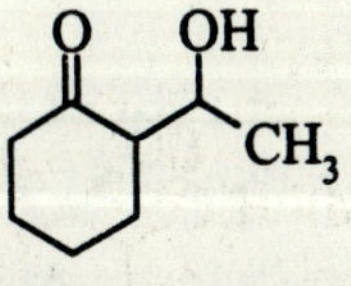 OH, CH_3 (d) 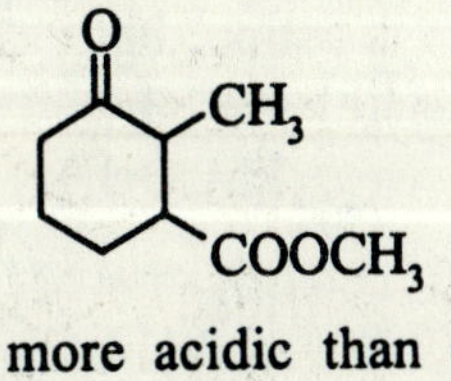 CH_3, $COOCH_3$

32. α-hydrogen of an ester is more acidic than α-hydrogen of an amide. The most succinet explanation for this observation is that:

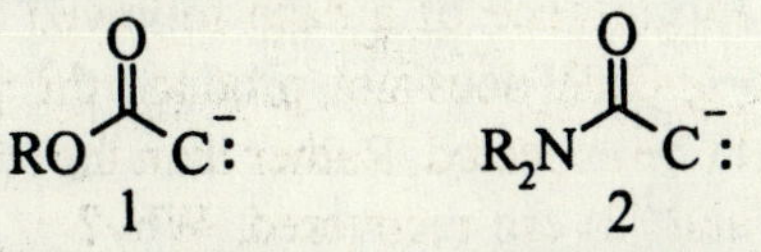

$RO-C(=O)-\bar{C}:$ (1) $R_2N-C(=O)-\bar{C}:$ (2)

(a) The ester enolate (1) is more stable than the amide enolate (2) as a consequence of the amide enolate's carbonyl being a better acceptor of enolate's electron density than the ester enolate's carbonyl.
(b) The ester enolate (1) is more stable than the amide enolate (2) as a consequence the ester enolate' carbonyl being a better acceptor of the enolate's electron density than is the amide enolate's carbonyl.
(c) The amide enolate (2) is more stable than the ester enolate (1) as a consequence of the amide enolate's carbonyl being a better acceptor of enolate's electron density than is the ester enolate's carbonyl.
(d) The amide enolate (2) is more stable than ester enolate (1) as a consequence of the ester enolate's carbonyl being a better acceptor of enolate's electron density than is the amide enolate's carbonyl.

33. Treatment of anhydride (1) with excess of $NaOCH_3$ in CH_3OH followed by an aqueous quench furnishes:

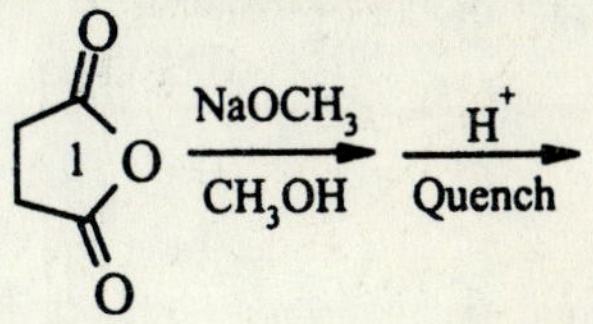

$NaOCH_3$ / CH_3OH → H^+ Quench →

(a) 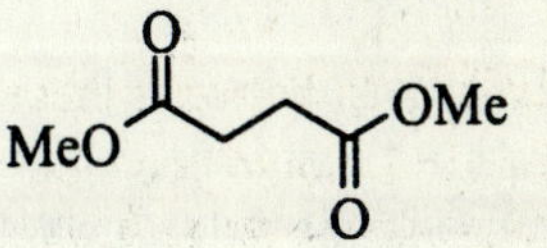 MeO ... OMe (b) 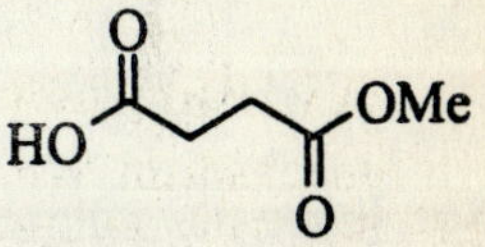 HO ... OMe
(c) 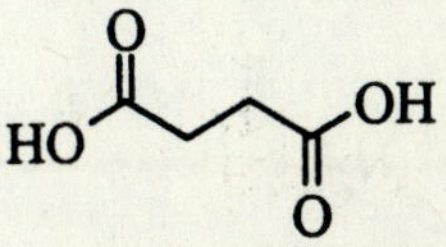 HO ... OH (d) 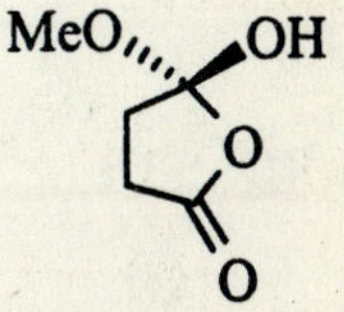 MeO, OH

34. Treatment of methyl ester (1) with one equivalent of NaOH in H_2O affords:

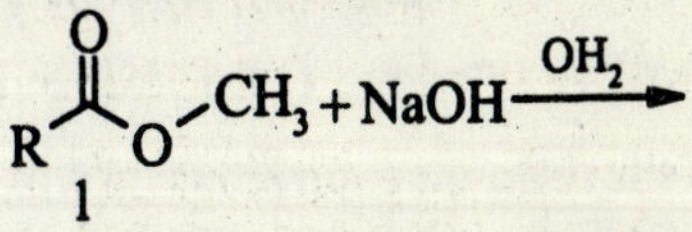

$RCOOCH_3 + NaOH \xrightarrow{OH_2}$ (1)

(a) 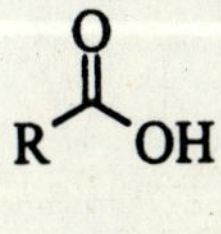 $RCOOH$ (b) 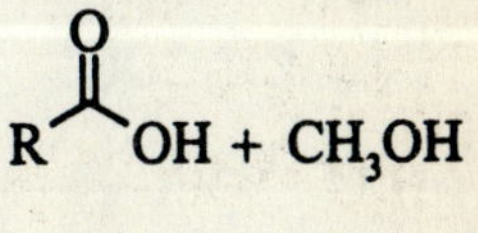 $RCOOH + CH_3OH$
(c) 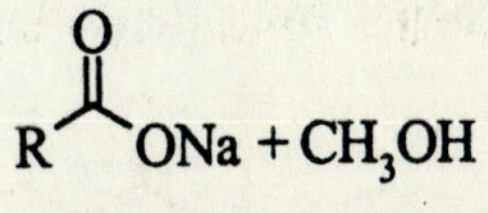 $RCOONa + CH_3OH$ (d) 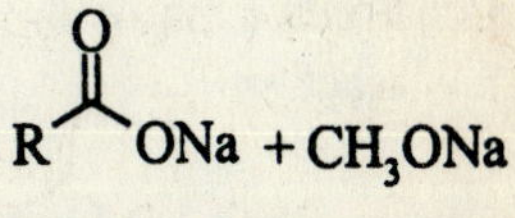 $RCOONa + CH_3ONa$

35. Which reaction would not proceed in good yield as indicated?

(a) 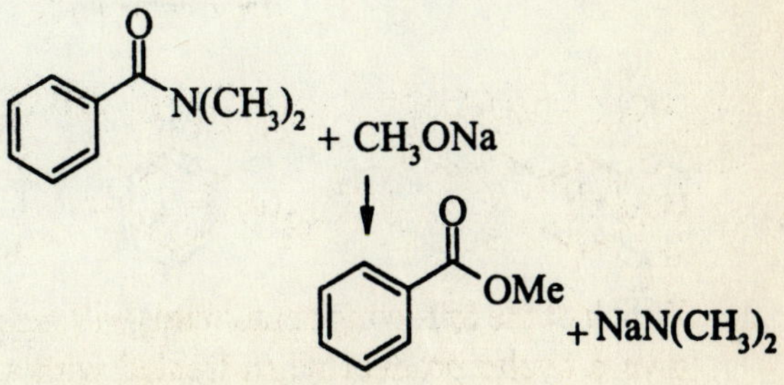 $PhC(=O)N(CH_3)_2 + CH_3ONa \rightarrow PhC(=O)OMe + NaN(CH_3)_2$

(b) PhCOCl + CH_3ONa → PhCOOMe + NaCl

(c) PhCOOMe + $NaN(CH_3)_2$ → $PhCON(CH_3)_2 + CH_3ONa$

(d) PhCOOCOPh + CH_3ONa → PhCOOMe + PhCOONa

36. Rank the following acid derivative in decreasing order of reactivity towards nucleophile:

I. ROCOCl   II. H_3CCOCl

III. ClCOCl   IV. ROCOOR

(a) IV > I > II > III   (b) II > III > I > IV
(c) III > II > I > IV   (d) I > II >III > IV

37. Which of the following nucleophile will add twice to ester RCOOEt:

(a) CH_3COO^- (b) OH^-
(c) Me_2N^- (d) PhMgBr

38. Predict major organic product from the following reaction:

H_3C–(cyclic keto lactone) $\xrightarrow{NaBH_4}$ A; $\xrightarrow{LiAlH_4}$ B

(a) A: OH, OH, O; B: OH, OH, OH   (b) A: O, O, OH; B: OH, OH
(c) A: OH, O, OH; B: OH, OH, OH   (d) A: OH, OH, O; B: O, O, OH

39. Which reaction coordinate diagram best describe the energetics of the following nucleophilic acyl substitution reaction?

RCOCl(1) + NH_3(2) → $RCONH_2$(3) + HCl(4)

(a)
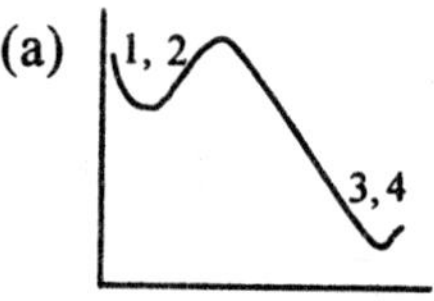

(b) 1, 2 … 3, 4 — Reaction coordinate

(c)
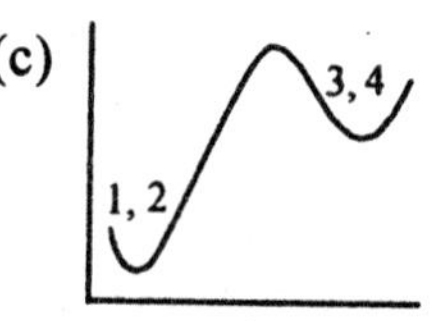

(d) 1, 2 … 3, 4 — Reaction coordinate

40. Treatment of acetic anhydride with 2.0 equivalents of Grignard reagent followed by an aqueous acid quench, furnishes:

(a) 2 $PhCOCH_3$
(b) $Ph_2C(OH)CH_3$ + CH_3COOH
(c) $PhCOCH_3$ + CH_3COOH
(d) 2 $Ph_2C(OH)CH_3$

41. What is the major product formed from the following reaction: PhCOCl + $LiAlH_4$ →

(a) PhCOH   (b) $PhCH_2OH$
(c) PhCOOH   (d) PhCHCl(OH)

42. Esterification of (–)–Lactic acid with methanol yields (+)–methyl lactate. Assuming that there are no side reactions, what is true about this reaction?

H_3C–CH(OH)–COOH (–) $\xrightarrow{CH_3OH/HCl}$ H_3C–CH(OH)–COOCH_3 (+)

(a) An SN-2 reaction has occurred, inverting the absolute configuration of the chiral center.
(b) An SN-1 reaction at the chiral center has inverted the optical rotation.
(c) A distereomer has been produced, distereomers have different physical properties, including optical rotation.

(d) Optical rotation is not directly related to absolute configuration, so the change in sign of rotation is merely a coincidence.

43. The keto acid (1) does not undergo facile expulsion of CO_2 upon heating. The reason for this reluctance is that:

HOOC 1 —Heat→ 2

(a) 1 is not a β-keto acid.
(b) The intermediate enol is too strained to be accessible.
(c) Loss of CO_2 is not thermodynamically favoured.
(d) The nucleophile and electrophile are too far apart to react.

44. Treatment of a β-keto ester 1 with a base and then with methyl iodide leads to product 3. The new carbon-carbon bond that is formed in this process is:

1 —CH_3ONa/CH_3I (2)→ 3 (a) (b) (c) (d)

45. Treatment of an acid chloride (1) with a non-nucleophilic base furnishes the unusual species Ketene (2), and not the expected enolate (3).

1 —Non-nucleophilic base→ 2 not 3

This reaction proceeds by a mechanism that mostly resembles:
(a) SN-2 reaction
(b) E2 reaction
(c) Nucleophilic acyl substitution
(d) Enolate alkylation

46. During synthesis of amine (2) from amide (1) by Hoffmann's rearrangement, the indicated carbon is lost as:

1 —NaOH, H_2O, Br_2→ 2

(a) CH_3OH (b) CO_2
(c) CH_4 (d) CH_3Br

47. Cr(VI) mediated oxidation of diol 1 furnishes the cyclic ester 2 rather than the expected dihaldehyde 3. The key intermediate enroute from 1 to 2 is:

HO~~~~~OH (1) —2 equivalents of CrO_3→ 2 and not 3

(a)

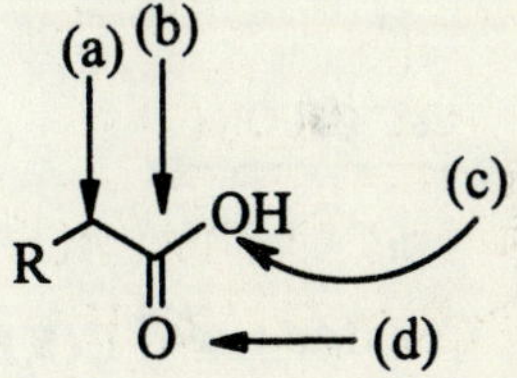

(c) (d)

48. Which atom is electrophilic in the generic carboxylic acid shown below?

(a) (b) (c) (d)

49. Cyclization of a diester 1 will yield a product with a ...... membered ring:

(a) 5 (b) 7 (c) 6 (d) 8

50. Which of the following hydrocarbon acids will be a constituent of the highest melting fat?

(a) (b) (c)

(d) 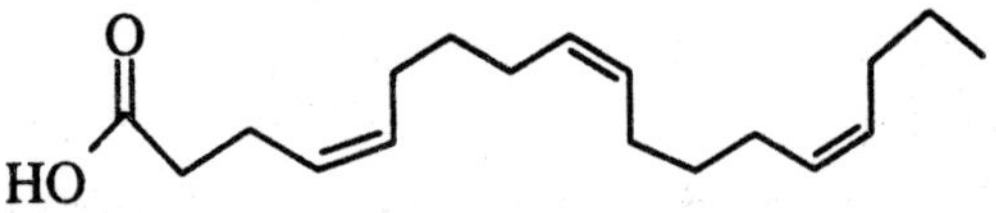 HO

51. Which of the following compound would be appropriate for the synthesis of amino acid (2)?

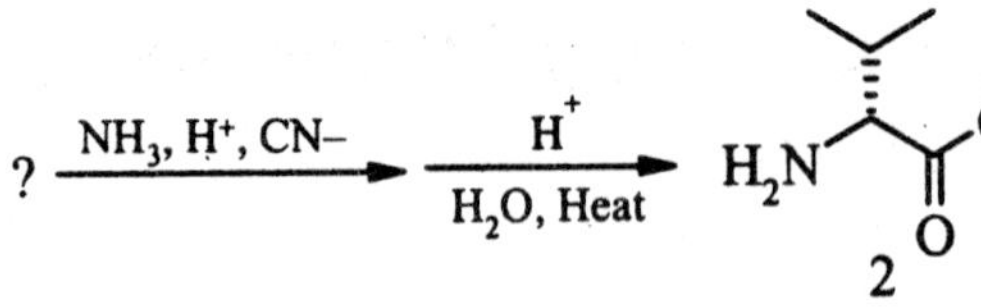

? $\xrightarrow{NH_3, H^+, CN^-}$ $\xrightarrow[H_2O, Heat]{H^+}$ H_2N ... OH (2)

(a) 2-methylpropanal
(b) 2-methylpropanoic acid
(c) 3-methylbutanal
(d) 3-methylbutanoic acid

52. Which is the best method to prepare compound 1?

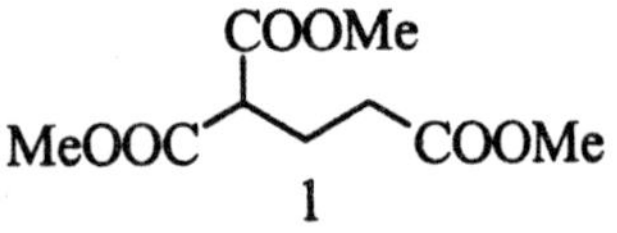

COOMe, MeOOC, COOMe (1)

(a) 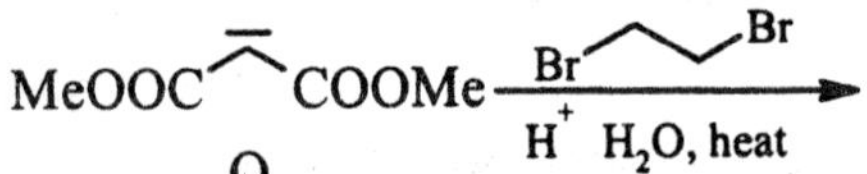 MeOOC COOMe $\xrightarrow[H^+\ H_2O,\ heat]{Br \quad Br}$

(b) 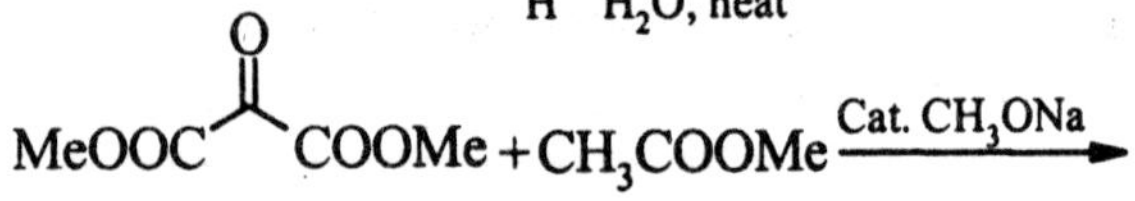 MeOOC COOMe + CH_3COOMe $\xrightarrow{Cat.\ CH_3ONa}$

(c) 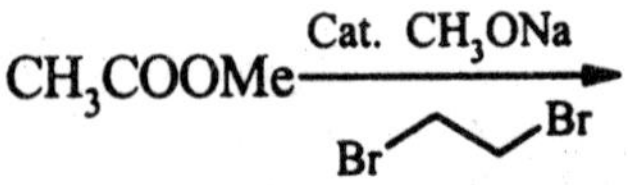 CH_3COOMe $\xrightarrow{Cat.\ CH_3ONa}$ Br Br

(d)  MeOOC COOMe + COOMe $\xrightarrow{Cat.\ CH_3ONa}$

53. Which of the following compound is NOT a possible product of the following reaction?

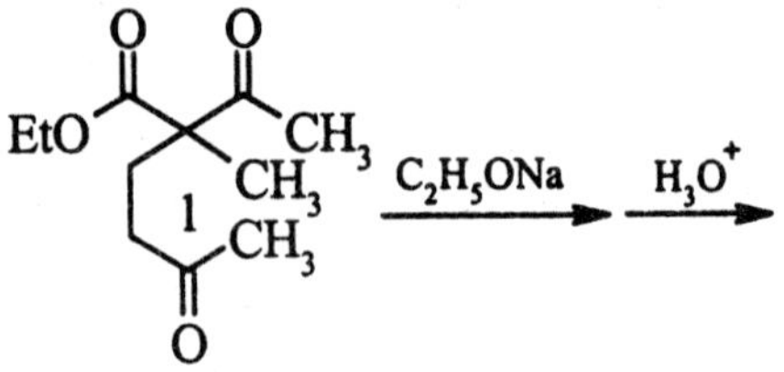

EtO, CH_3, CH_3, CH_3, 1 $\xrightarrow{C_2H_5ONa}$ $\xrightarrow{H_3O^+}$

(a) 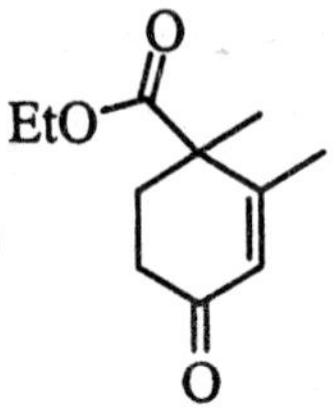 EtO (b) 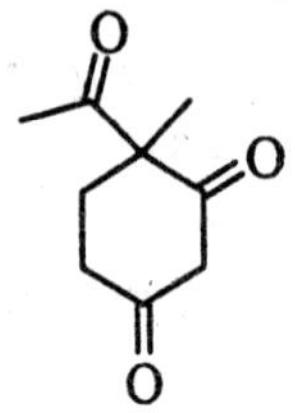 (c) 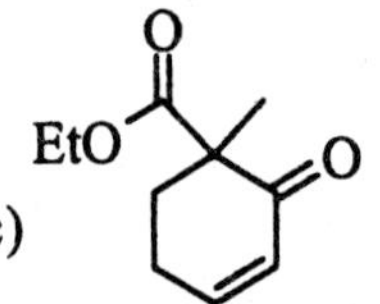 EtO (d) 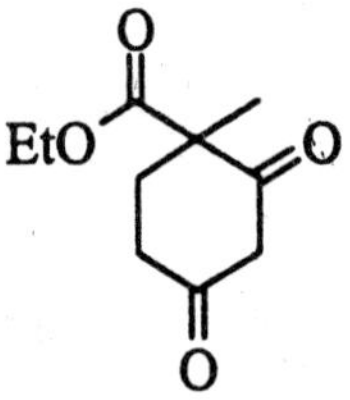 EtO

54. The following sequence of reaction affords:

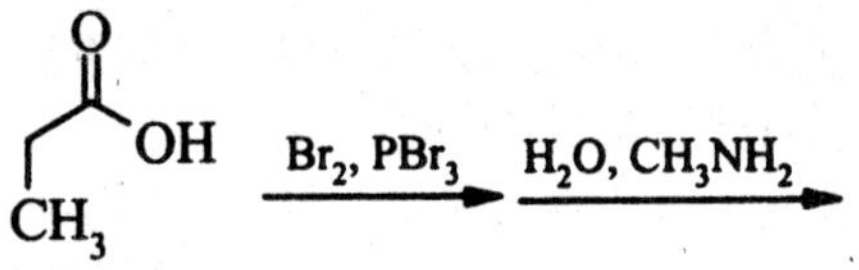

OH, CH_3 $\xrightarrow{Br_2, PBr_3}$ $\xrightarrow{H_2O, CH_3NH_2}$

(a) A mixture of chiral, distereomeric amino acids.
(b) A mixture of racemic, distereomeric amino acids.
(c) A chiral amino acid product.
(d) A racemic amino acid products.

55. Rank (fastest to slowest) the following carboxylic acid derivatives in terms of their reactivity towards nucleophile:

(I) Acetamide (II) Acetyl chloride
(III) Methyl acetate (IV) Acetic anhydride

(a) II > IV > III > I (b) I > III > IV > II
(c) II > III > IV > I (d) I > IV > III > II

56. The relative rate of nucleophilic acyl substitution for a given acid derivative (RCOX) depends on:

(a) Only the resonance effect of X.
(b) Both the resonance and inductive effect.
(c) Only the inductive effect of X.
(d) Neither the resonance effect nor the inductive effect.

57. In the acid catalyzed hydrolysis of ester(1), several possible intermediates can be envisioned, I → IV. Indicate in which these intermediates would appear during the conversion of 1 to 2:

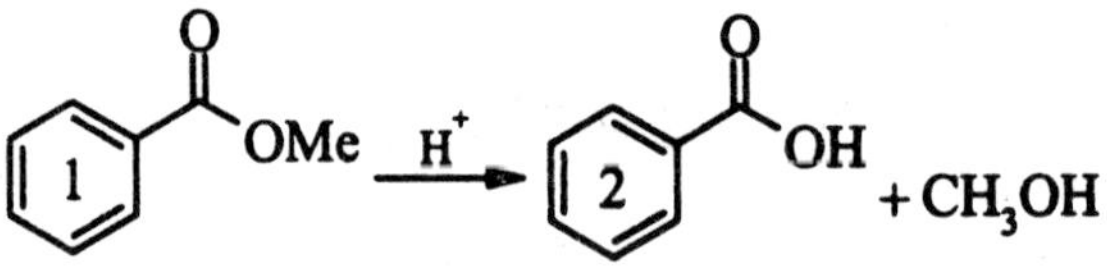

1 OMe $\xrightarrow{H^+}$ 2 OH + CH_3OH

I: HO, $\overset{+}{O}H_2$, OMe   II: $\overset{+}{O}$–H, OMe

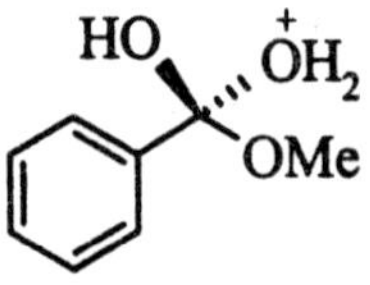

III: $\overset{+}{O}$–H, OH   IV: HO, OH, $\overset{+}{O}Me$–H

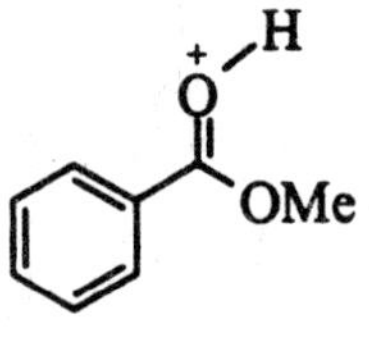

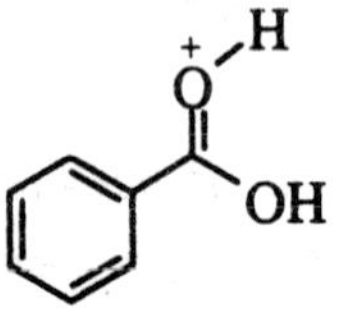

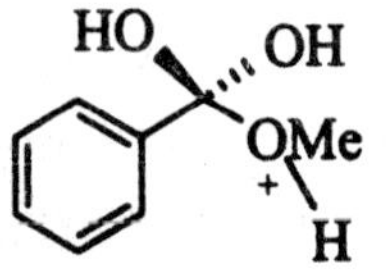

(a) 1 → II → I → III → IV → 2
(b) 1→ II → I → IV → III → 2
(c) 1 → II → IV → I → III → 2
(d) 1 → III → I → IV → II → 2

58. Which reaction coordinate diagram best expresses the difference in energy between adding OH^- to an acid chloride 1 vs to an ester 2?

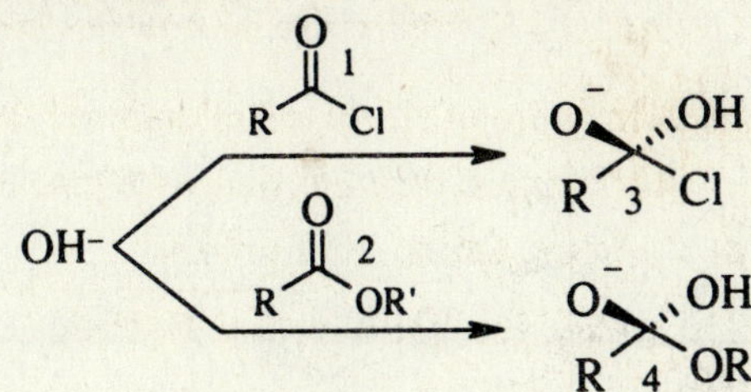

(a)

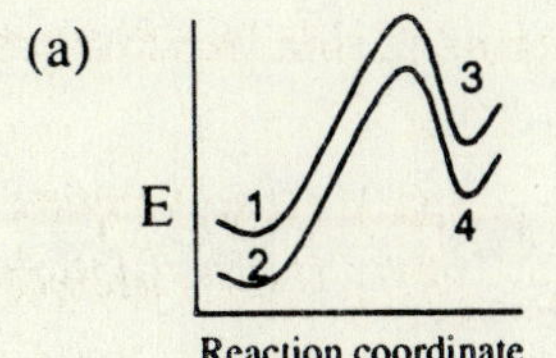

(b)

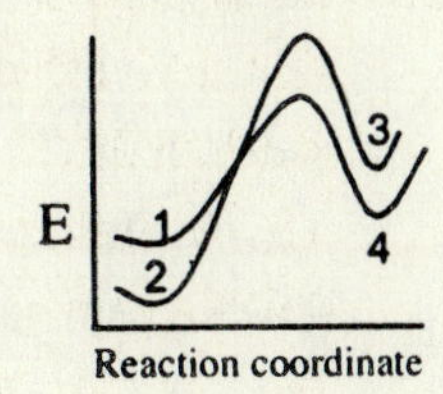

(c)

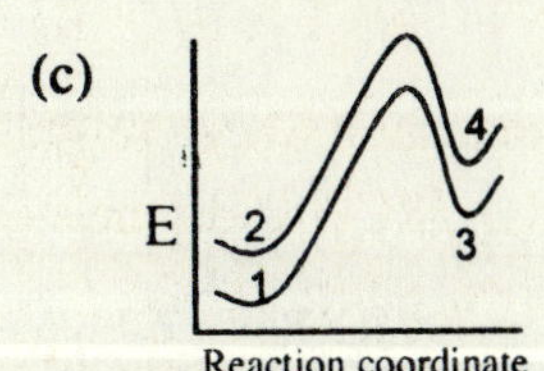

(d)

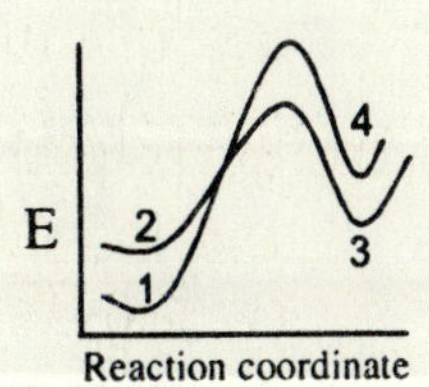

59. Predict the major organic product from the following reaction:

Methyl benzoate + 2.0 equiv. CH_3Li/H_3O^+, quench →

(a) $C_6H_5C(CH_3)_3$ (b) $C_6H_5C(CH_3)_2OMe$

(c) $C_6H_5COCH_3$ (d) $C_6H_5C(CH_3)_2OH$

60. Predict the major organic product from the following reaction:

$H_3C{-}CO{-}O{-}CO{-}CF_3 \xrightarrow{CH_3OH}$

I: CH_3COOMe II: $HOCOCF_3$ III: CH_3COOH IV: $MeOCOCF_3$

(a) I and II (b) I and IV

(c) III and IV (d) II and III

61. Product of the following reaction would be:

Diethylheptandioate + C_2H_5ONa followed by acid hydrolysis →

(a) Heptandioic acid (b) [cyclic anhydride]

(c) [cyclic anhydride] (d) [2-oxocyclohexanecarboxylic acid]

62. Product formed in the following sequence of reaction is:

Pentanoic acid + PBr_3/Br_2 followed by reduction of product with $Pd/BaSO_4$/Quinoline →

(a) Pentanol (b) Pentanal

(c) Pentanoic acid anhydride (d) Pentene

63. Major product formed in the following reaction is:

Acetone $+ CH_3COOC_2H_5 \xrightarrow{C_2H_5ONa} \xrightarrow[H_2O]{H^+}$

(a) [2,4-pentanedione] (b) [OH, $COOC_2H_5$]

(c) [COOH] (d) [OC_2H_5]

64. Predict product formed in the following reaction:

$C_2H_5O{-}CO{-}\ldots{-}CO{-}OC_2H_5 \xrightarrow{C_2H_5ONa} \xrightarrow[H_2O]{H^+}$

(a) [OH] (b) [OH]

(c) (d)

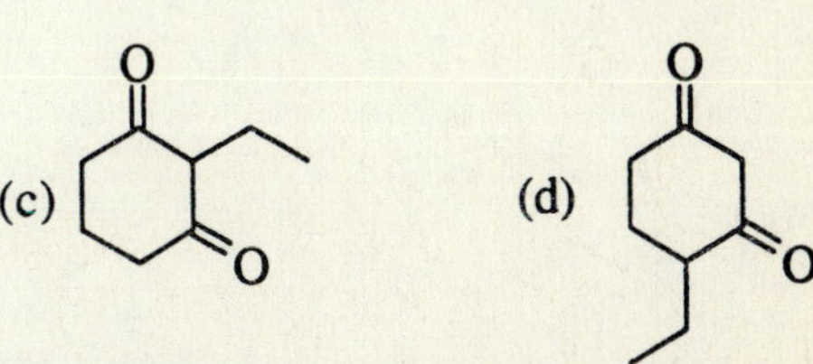

65. Major product formed in the following reaction sequence is?

Acetophenone $+ C_2H_5O{-}CO{-}OC_2H_5 \xrightarrow{C_2H_5ONa} \xrightarrow[H_2O]{H^+}$

(a) 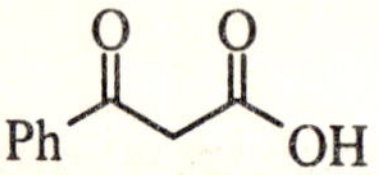 Ph ... OH (b) 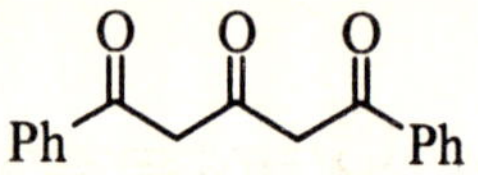 Ph ... Ph

(c) 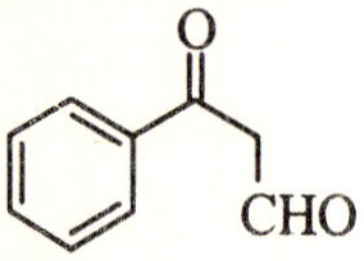 CHO (d)

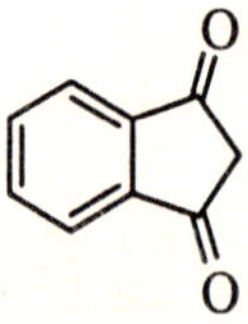

66. 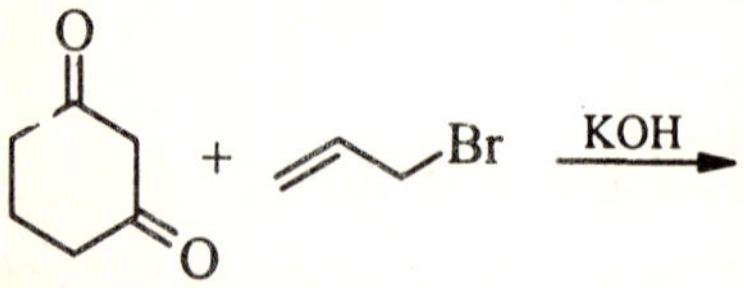 + Br $\xrightarrow{KOH}$ Major product:

(a) 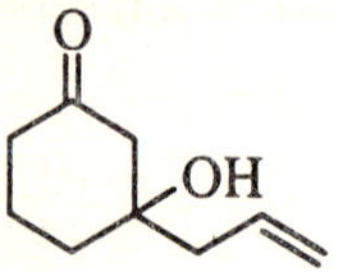 OH (b)

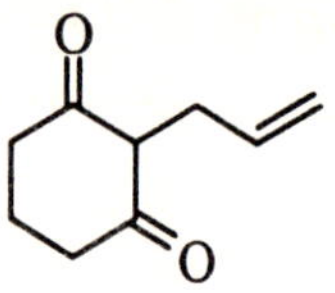

(c) 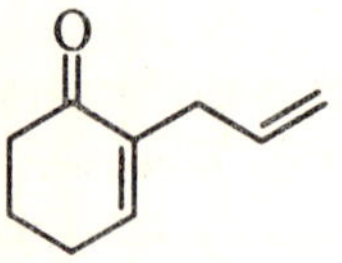 (d)

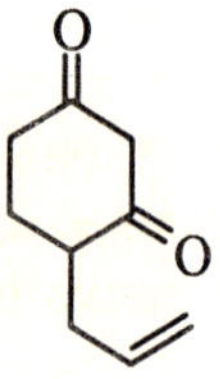

67. 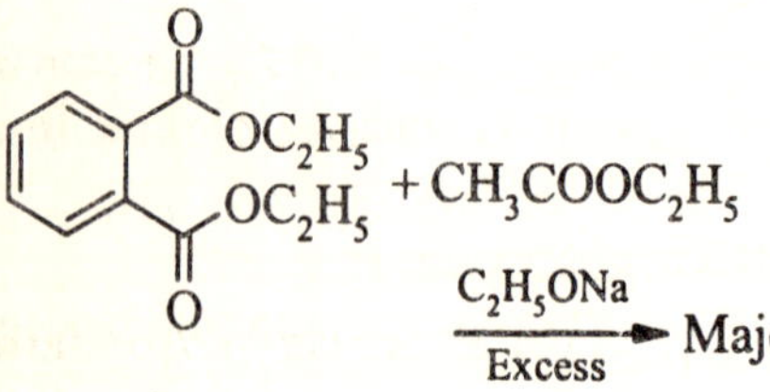 OC_2H_5, OC_2H_5 + $CH_3COOC_2H_5$ $\xrightarrow[Excess]{C_2H_5ONa}$ Major product

(a) 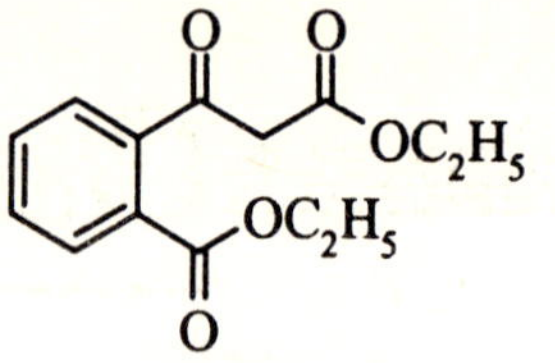 OC_2H_5, OC_2H_5 (b) 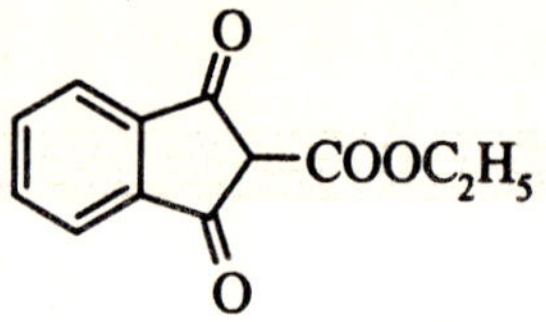 $-COOC_2H_5$

(c) 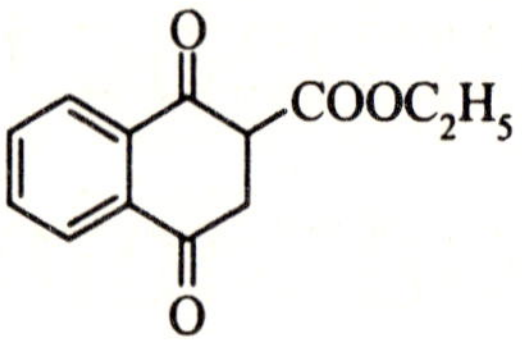 $COOC_2H_5$ (d) 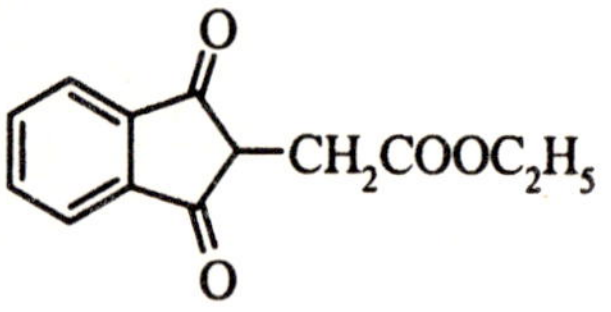 $-CH_2COOC_2H_5$

68. 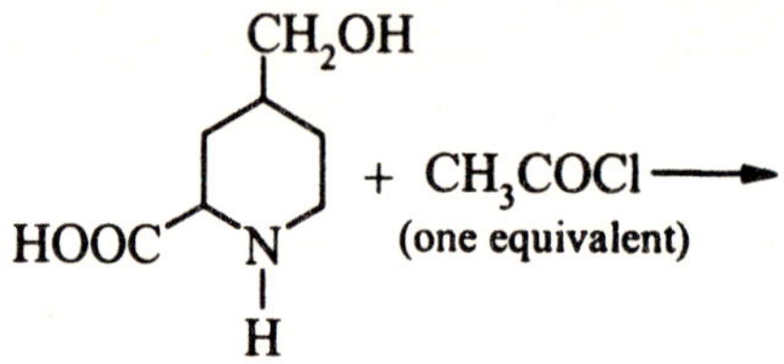 CH_2OH, HOOC, N–H + CH_3COCl (one equivalent) ⟶ Major product:

(a) 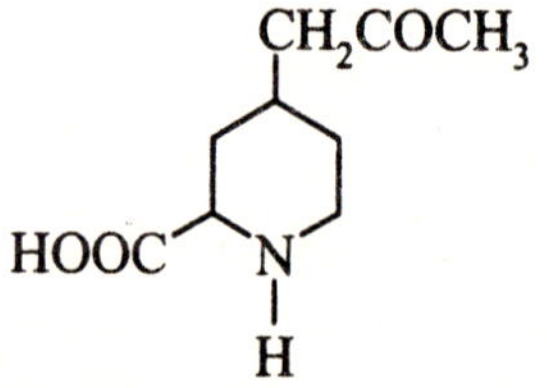 CH_2COCH_3, HOOC, N–H (b) 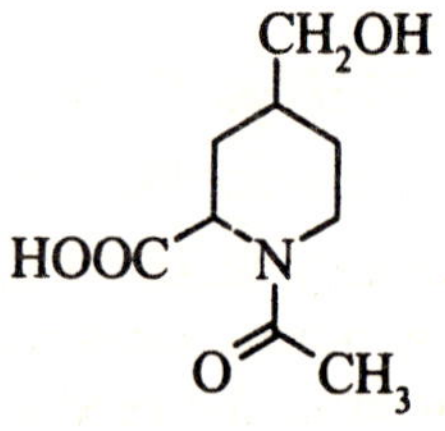 CH_2OH, HOOC, N, CH_3

(c) 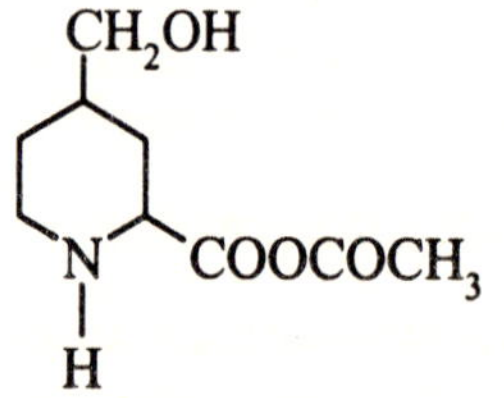 CH_2OH, N–H, $COOCOCH_3$ (d) None

Chapter 9

Aliphatic Compounds Containing Nitrogen

Nitro-alkanes

9.1 PREPARATIONS OF NITRO-ALKANES

I. By nucleophilic substitution of alkyl halides

Heating alkyl halides with silver nitrite in aqueous ethanolic solution, nitro-alkanes are produced. Alkyl nitrites are also produced in the reaction and proportions of the two products—nitro-alkanes and alkyl nitrite, depends on nature of substrate and reaction conditions:

$$RX + AgONO \rightarrow RNO_2 + RONO + AgX$$

This method is suitable for preparation of primary nitro-alkanes. With secondary halides, yield is about 15% and with tertiary halides 0–5%. Proportions of nitrite increases from primary to tertiary halides. If sodium nitrite is used in place of silver nitrite, yields of both nitro-alkanes and alkyl nitrites are improved.

II. Direct nitration

Nitro compounds are prepared industrially by direct nitration techniques. Direct nitration is done in two ways:

(i) *Liquid phase nitration:* In this method, the hydrocarbon is heated with concentrated nitric acid under pressure. Nitration under pressure occurs at very low rate and a large amounts of polynitro-alkanes are produced as by-products.

(ii) *Vapour phase nitration:* In this technique, hydrocarbon is heated with nitric acids at 150–475°C, each hydrocarbon has its optimum temperature:

$$CH_3CH_2CH_3 + HNO_3 \xrightarrow{400^\circ C} CH_3CH_2CH_2NO_2 + \underset{\displaystyle NO_2}{CH_3\underset{|}{C}HCH_3} + CH_3CH_2NO_2 + CH_3NO_2$$

Vapour phase nitration is more satisfactory than liquid phase nitration although a large amounts of by-products are obtained in both techniques and neither can be used for selective nitration at a specific carbon of hydrocarbon.

Under the above conditions (i.e. at 400° C), the ease of substitution by nitro group at various carbon in a hydrocarbon is:

t-H > s-H > prim-H

- Also, any alkyl group present in alkane can be substituted by nitro group, i.e. chain fission may also take place during nitration.

Example 1. How many nitro-alkanes can be produced in nitration of isopentane, draw structures of all of them.

Solution: Nine nitro-alkanes would be produced as:

H_3C–CH_2–$CH(CH_3)$–CH_2NO_2 H_3C–CH_2–$C(NO_2)(CH_3)$–CH_3 H_3C–$CH(NO_2)$–$CH(CH_3)$–CH_3

O_2N–CH_2–CH_2–$CH(CH_3)$–CH_3 H_3C–CH_2–$CH(CH_3)$–NO_2 CH_3NO_2

$CH_3CH_2NO_2$ $CH_3CH(NO_2)CH_3$ O_2N–CH_2–$CH(CH_3)$–CH_3

III. By oxidation of primary amines

Primary amines on oxidation with potassium permanganate yields nitro-alkanes as:

$$RNH_2 + KMnO_4 \rightarrow R\text{—}NO_2$$

- This method is suitable for preparation of tertiary nitro-alkane which is otherwise difficult to prepare by other method.

IV. By addition of oxides of nitrogen on alkenes

Addition of N_2O_4 on alkenes in presence of oxygen yields variety of nitration product and reaction is highly complicated:

$$R-CH=CH_2 + N_2O_4 \xrightarrow{O_2} R\text{-}\underset{X}{\underset{|}{C}}H\text{-}CH_2NO_2 + R\text{-}CO\text{-}CH_2NO_2$$

where X can be $-NO_2$, $-ONO$, $-ONO_2$, etc.

9.2 PROPERTIES OF NITRO-ALKANES

I. General properties

In pure form, nitroalkanes are colourless, pleasant smelling liquids and sparingly soluble in water. Nitroalkanes are mostly distilled at atmospheric pressure without decomposition.

II. Reduction reaction of nitro-alkanes

(i) Nitro-alkanes are reduced catalytically, by $LiAlH_4$ or by metal/acid to primary amines as:

$$R-NO_2 \xrightarrow[\text{or Zn-HCl}]{LiAlH_4} R-NH_2$$

(ii) Reduction of nitro-alkanes with metals in neutral medium yields N-alkyl hydroxy amine:

$$R-NO_2 \xrightarrow[NH_4Cl]{Zn} R-NHOH$$

(iii) Reduction of nitro-alkanes with stannous chloride in dilute acidic medium yields a mixture of N-alkyl hydroxyl amine and oximc:

$$RCH_2-NO_2 \xrightarrow[HCl]{SnCl_2} RCH_2-NHOH + RCH=NOH$$

III. Hydrolysis reaction of nitro-alkanes

Primary nitro-alkanes are hydrolysed by boiling hydrochloric acid or by 85% sulphuric acid into a carboxylic acid:

$$RCH_2-NO_2 \xrightarrow[Boil]{HCl} RCOOH + NH_2OH$$

Here hydroxyl amine is a by-product, and therefore this reaction is used for manufacturing hydroxyl amine. Secondary nitro-alkanes are hydrolysed into ketones under similar conditions of hydrolysis as:

$$R_2CH-NO_2 \xrightarrow[Boil]{HCl} R_2C=O + N_2O$$

Tertiary nitro compounds are generally inert in hydrolysis reactions.

IV. Tautomerism in nitro-alkanes

Nitro-alkanes containing atleast one α-H, exhibit tautomerism as:

$$-\overset{H}{\overset{|}{C}}-\overset{+}{N}(=O)-O^- \rightleftharpoons -C=\overset{+}{N}(-OH)-O^-$$

Tautomers

Tautomerism phenomena is exhibited by primary and secondary nitro-compounds due to the acidic behaviour of α-H. Acidity of α-H is explained by the strong inductive-effect of $-NO_2$ group and subsequent resonance stabilization of conjugate base as:

$$-\overset{H}{\overset{|}{C}}-\overset{+}{N}(=O)-O^- \rightleftharpoons H^+ + -\overset{..}{\overset{-}{C}}-\overset{+}{N}(=O)-O^- \longleftrightarrow -C=\overset{+}{N}(-O^-)-O^-$$

Due to the above mentioned acidic behaviour, nitro-alkanes containing α-H, are fairly soluble in aqueous alkali.

V. α-Halogenation of nitro-alkanes

Primary and secondary nitro-alkanes are readily halogenated in alkaline solution at α-position only:

$$R-\overset{H}{\underset{R}{C}}-\overset{+}{N}(=O)-O^- \xrightarrow{Br_2/NaOH} R-\overset{Br}{\underset{R}{C}}-\overset{+}{N}(=O)-O^-$$

$$H-\overset{H}{\underset{H}{C}}-\overset{+}{N}(=O)-O^- \xrightarrow{Cl_2/NaOH} Cl-\overset{Cl}{\underset{Cl}{C}}-\overset{+}{N}(=O)-O^-$$

Chloropicrin

VI. Reaction with nitrous acid

Different types of nitro compounds reacts differently with nitrous acid producing variety of products as:

Primary nitro compounds form nitrolic acid. Nitrolic acids are crystalline substance, which dissolve in sodium hydroxide to give red-coloured solution:

$$R-CH_2-\overset{+}{N}(=O)O^- + HNO_2 \rightarrow R-C(NO_2)=N-OH$$

Nitrolic acid

$$\xrightarrow{NaOH(aq)} R-C(NO_2)=N-O^-$$

Red coloured in water

Secondary nitro compounds forms pseudonitroles. Pseudonitroles are colourless, crystalline substances, which dissolve in aqueous alkali solution producing blue colouration:

$$R_2CH-\overset{+}{N}(=O)O^- + HNO_2 \rightarrow R_2C(NO_2)(N=O) \xrightarrow{NaOH\ (aq)} \text{Blue colouration}$$

Pseudonitrole

Tertiary nitro compounds are inert with nitrous acid due to lack of α-H. These reactions of nitro compounds with nitrous acid is the basis of distinction of primary, secondary and tertiary alcohols by "Victor Meyer Test".

VII. Condensation reaction of nitro-alkanes

Due to acidic nature of α-H, primary and up to some extent, secondary nitro-alkane, undergo condensation with aldehydes and ketones like ***aldol condensation*** reaction:

$$CH_3CHO + CH_3NO_2 \xrightarrow{NaOH\ (aq)} H_3C-CH(OH)-CH_2NO_2$$

Mechanism of condensation reaction

$$OH^- + R-CH_2-\overset{+}{N}(=O)O^- \xrightarrow{-H_2O} \left[R-\overset{..}{C}H^- -\overset{+}{N}(=O)O^- \longleftrightarrow R-CH=\overset{+}{N}(O^-)O^-\right]$$

$$R-CH=O + R-\overset{..}{C}H^- -\overset{+}{N}(=O)O^- \xrightarrow{H_2O} R-CH(OH)-CHR-\overset{+}{N}(=O)O^-$$

Example 2. Complete the following reactions:

(a) $H_3C-CO-CH_3 + CH_3NO_2 \xrightarrow{NaOH}$

(b) $C_6H_5CHO + CH_3NO_2 \xrightarrow{NaOH}$

(c) $C_6H_5-CO-C_6H_5 + CH_3CH_2NO_2 \xrightarrow{NaOH}$

Solution

(a) $CH_3NO_2 + OH^- \rightarrow \left[\overset{..}{C}H_2NO_2^- \longleftrightarrow H_2C=\overset{+}{N}(O^-)O^-\right]$

$$\xrightarrow{H_3C-CO-CH_3} H_3C-C(O^-)(CH_3)-CH_2NO_2 \xrightarrow{H_2O} H_3C-C(OH)(CH_3)-CH_2NO_2$$

(b) $CH_3NO_2 + OH^- \rightarrow \overset{..}{C}H_2NO_2^- \xrightarrow{C_6H_5CHO} C_6H_5CH(O^-)-CH_2NO_2 \xrightarrow{H_2O,\ -H_2O} C_6H_5CH=CHNO_2$

(c) $CH_3CH_2NO_2 + OH^- \rightarrow CH_3\overset{..}{C}HNO_2^- \xrightarrow[C_6H_5-CO-C_6H_5]{H_2O}$

$$C_6H_5-C(OH)(C_6H_5)-CH(CH_3)NO_2 \xrightarrow{-H_2O} C_6H_5-C(C_6H_5)=C(CH_3)NO_2$$

Example 3. Propose mechanism of formations of products in the following reactions:

(a) $HOCCH_2CH_2CHO + CH_3NO_2 \xrightarrow[Heat]{NaOH}$ (cyclopentadiene ring)$-NO_2$

(b) $(CH_3)_2C=CH-CO-CH_3 + CH_3NO_2 \xrightarrow{NaOH,\ Heat} (CH_3)_2C(CH_2NO_2)-CH_2COCH_3$

Solution

(a) $CH_3NO_2 + OH^- \longrightarrow H_2O + {}^-CH_2NO_2$

$$\text{OHC-CH}_2\text{CH}_2\text{-CHO} + {}^-CH_2NO_2 \rightarrow \text{intermediate alkoxide} \xrightarrow{H_2O} \text{hydroxy nitro aldehyde} \xrightarrow{OH^-} \text{carbanion} \rightarrow \text{cyclopentanolate-NO}_2 \xrightarrow{H_2O} \text{2-nitrocyclopentane-1,3-diol} \xrightarrow[\text{Heat}]{-2H_2O} \text{nitrocyclopentadiene}$$

(b) $CH_3NO_2 + OH^- \longrightarrow H_2O + {}^-CH_2NO_2$

$$(CH_3)_2C{=}CH{-}\overset{O}{\overset{\|}{C}}{-}CH_3 + {}^-CH_2NO_2$$

$$(CH_3)_2\underset{CH_2NO_2}{\underset{|}{C}}{-}\bar{C}H{-}\overset{O}{\overset{\|}{C}}{-}CH_3 \xrightarrow{H_2O} (CH_3)_2\underset{CH_2NO_2}{\underset{|}{C}}{-}CH_2COCH_3$$

Aliphatic Amines

9.3 INTRODUCTION

Amines are organic derivatives of ammonia in which one, two or three hydrogens have been replaced by alkyl groups (R). Amines are classified as primary, secondary and tertiary amines according to the number of alkyl group bonded to nitrogen atom as:

$R{-}NH_2$	Primary amine	"one alkyl group is bonded to N"
$R{-}\underset{H}{\underset{\|}{N}}{-}R$	Secondary amine	"two alkyl groups are bonded to N"
$R{-}\underset{R}{\underset{\|}{N}}{-}R$	Tertiary amine	"three alkyl groups are bonded to N"

9.4 GENERAL METHODS OF PREPARATION OF AMINES

9.4.1 Methods of Preparation of all Three Types of Amines

I. Hofmann's method

Heating an alkyl halide with ethanolic solution of ammonia in a sealed tube at around 100°C gives a mixture of all three types of amines as:

$$RX + NH_3 \rightarrow RNH_2 + HX$$

$$RNH_2 + RX \rightarrow R_2NH + HX$$

$$R_2NH + RX \rightarrow R_3N + HX$$

$$R_3N + RX \rightarrow [R_4N^+]X^- \text{ "quaternary ammonium halide"}$$

Mechanism of ammonolysis of alkyl halides proceeds by bimolecular substitution reaction (SN-2) of halide (X^-) by $-NH_2$ as:

$$R{-}\underset{H_2}{C}{-}X + \ddot{N}H_3 \rightarrow R\text{-}CH_2\text{-}\overset{+}{N}H_2\text{-}H \xrightleftharpoons{X^-} RCH_2NH_2$$

Therefore, ammonolysis reaction works well with primary alkyl halides, fairly well with secondary alkyl halides, but tertiary alkyl halides gives mainly alkene by E2 reaction as:

$$Cl{-}\underset{R}{\overset{R}{C}}{-}\underset{R}{CH}{-}H + \ddot{N}H_3 \rightarrow NH_4Cl + RCH{=}C\langle^R_R$$

With different halides but the same alkyl group, the order of reactivity in ammonolysis reaction is:

$$RI > RBr > RCl$$

II. A mixture of three types of amines are prepared by amonolysis of alcohols. In this method, alcohols and ammonia are heated under pressure in presence of a catalyst, e.g. copper chromite or alumina:

$$ROH + NH_3 \xrightarrow[\text{Heat}]{Al_2O_3} OH_2 + RNH_2 \xrightarrow[Al_2O_3,\ \text{Heat}]{ROH} R_2NH \xrightarrow[Al_2O_3,\ \text{Heat}]{ROH} R_3N$$

9.4.2 Preparation of Primary Amines

I. By reduction of nitro-alkanes

Nitro-alkanes, on reduction with H_2/Ni or with $LiAlH_4$ or with electropositive metals, e.g. Zn, Fe, etc. in dilute acidic medium, yields primary amines:

$$R{-}NO_2 \xrightarrow[\text{or LiAlH}_4 \text{ or Zn/HCl}]{H_2/Ni} R{-}NH_2$$

H_2/Ni being a very strong reducing agent, reduces all reducible groups present in the compound while $LiAlH_4$ does not reduces olefins, it can be used for selective reduction of $-NO_2$ group leaving double or triple bonds intact.

$$CH_2{=}CH{-}CH_2{-}NO_2 \xrightarrow{H_2/Ni} CH_3CH_2CH_2NH_2$$

$$CH_2{=}CH{-}CH_2{-}NO_2 \xrightarrow{LiAlH_4} CH_2{=}CHCH_2NH_2$$

II. By reduction of alkyl cyanides

Alkyl cyanides (RCN) undergo reduction with either $LiAlH_4$ or Na/ethanol or H_2/Ni to produce primary amines:

$$R{-}CN \xrightarrow{H_2/Ni \text{ or } Na/EtOH \text{ or } LiAlH_4} RCH_2NH_2$$

Example 4. During reduction of alkyl cyanide, some secondary amine is produced as by-product. Propose mechanism of formations of secondary amine.

Solution: Alkyl cyanide is reduced into amine in a two steps reduction process via imine intermediate as:

$$R{-}CH_2CN \xrightarrow{H_2/Ni} \underset{\text{Imine}}{R{-}CH_2CH{=}NH} \xrightarrow{H_2/Ni} RCH_2CH_2NH_2$$

During reduction process, some intermediate imine undergo nucleophilic attack by the product amine which results in formation of secondary amine as:

$$R{-}CH_2CH{=}\ddot{N}H + RCH_2CH_2\ddot{N}H_2 \longrightarrow$$

$$RCH_2{-}\underset{\underset{\text{NH}}{|}}{CH}{-}\overset{+}{N}H_2{-}CH_2CH_2R \rightleftharpoons RCH_2{-}\underset{\underset{NH_2}{|}}{CH}{-}NH{-}CH_2CH_2R$$

$$\xrightarrow{-NH_3} RCH_2{-}CH{=}N{-}CH_2CH_2R \xrightarrow{H_2/Ni} (RCH_2CH_2)_2NH$$

III. By reduction of an oxime or a primary amide

Reduction of either oxime or a primary amide ($RCONH_2$) with either Na/EtOH or $LiAlH_4$ yields primary amine:

$$RCH{=}NOH \xrightarrow[\text{or LiAlH}_4]{Na/C_2H_5OH} RCH_2NH_2 \xleftarrow[\text{or LiAlH}_4]{Na/C_2H_5OH} R{-}\overset{O}{\overset{\|}{C}}{-}NH_2$$

IV. From aldehydes and ketones (reductive amination of aldehydes and ketones)

Passing mixture of aldehydes or ketones with a large excess of ammonia under pressure yields imine:

$$RCHO + NH_3(\text{excess}) \longrightarrow RCH{=}NH \xrightarrow{H_2/Ni} RCH_2NH_2$$

$$R_2C{=}O + NH_3(\text{excess}) \longrightarrow R_2C{=}NH \xrightarrow{H_2/Ni} R_2CH{-}NH_2$$

Reduction of imines with H_2/Ni yields amines.

V. Leucart reaction

Aldehydes and ketones react with ammonium formate or with formamide to give formyl derivatives of amine, which when hydrolysed, yields a primary amine:

$$R_2C{=}O + HCOONH_4 \longrightarrow R_2CHNHCHO \xrightarrow[H_2O]{H^+} R_2CHNH_2$$

$$R_2C{=}O + HCONH_2 \longrightarrow R_2CHNHCHO \xrightarrow[H_2O]{H^+} R_2CHNH_2$$

VI. Gabriel's phthalimide synthesis

This is very useful method since it gives a pure, primary amine. In this method, phthalimide is first treated with NaOH converting it into sodium salt of phthalimide, which is then heated with an alkyl halide in subsequent step, producing N-alkyl phthalimide. In the last, final step, N-alkyl phthalimide is hydrolysed or treated with hydrazine, producing a primary amine:

$$\text{Phthalimide (N—H)} \xrightarrow{NaOH} \text{Phthalimide } N^- Na^+ \xrightarrow{RX} \text{N—R phthalimide}$$

$$\text{N—R phthalimide} \xrightarrow[H_2O]{H^+} RNH_2 + \text{phthalic acid (}C_6H_4(COOH)_2\text{)}$$

$$\text{N—R phthalimide} \xrightarrow{N_2H_4} RNH_2 + \text{phthalhydrazide (cyclic C(=O)NH–NHC(=O))}$$

VII. By means of Grignard's synthesis

Reaction of Grignard's reagent with chloramines yields a primary amine:

$$RMgCl + ClNH_2 \rightarrow RNH_2 + MgCl_2$$

VIII. Ritter reaction

When an alcohol or an alkene is treated with NaCN in presence of sulphuric acid, a primary amine is produced as:

$$R_3COH \xrightarrow{H^+} R_3\overset{+}{C} \xrightarrow{NaCN} R_3C{-}\overset{+}{N}{\equiv}C$$

$$\downarrow OH_2$$

$$R_3CNHCHO \xrightarrow{NaOH\,(aq)} R_3CNH_2 + HCOONa$$

This reaction is very useful for preparing primary amine with tertiary alkyl group bonded to nitrogen. This reaction depends on the formation of a carbocation intermediate, hence, mainly used with a tertiary alcohols or an alkene which produces a tertiary carbocation on protonation.

IX. By Hofmann's method

When a primary amide is treated with bromine or chlorine in alkaline medium, a primary amine with one carbon less than starting amide is produced:

$$R{-}\overset{O}{\overset{\|}{C}}{-}NH_2 \xrightarrow{Br_2/NaOH} R{-}NH_2 + 2NaBr + Na_2CO_3$$

X. Schmidt reaction

A carboxylic acid when treated with hydrazoic acid, primary amine is formed:

$$RCOOH + HN_3 \rightarrow RNH_2 + N_2 + CO_2$$

9.4.3 Preparation of Secondary Amines

I. Hofmann's amonolysis of a primary amine

Heating a primary amine with calculated amount of alkyl halide, secondary amine is produced as:

$$RNH_2 + R'{-}Cl \rightarrow RR'NH + HCl$$

II. From aniline

Aniline when heated with excess of an alkyl halide, a dialkyl aniline is produced in the first step. This N, N'-dialkyl aniline is then treated with nitrous acid to give *p*-nitroso dialkyl aniline. *p*-nitroso dialkylaniline on boiling with dilute solution of NaOH undergo hydrolysis producing *p*-nitrosophenol and secondary amine as:

$$C_6H_5NH_2 \xrightarrow{2RI} C_6H_5NR_2 \xrightarrow{HNO_2} p\text{-}ON{-}C_6H_4{-}NR_2 \xrightarrow{NaOH\quad H^+} R_2NH + p\text{-}ON{-}C_6H_4{-}OH$$

III. From calcium cyanamide

$$CaNCN + 2NaOH \rightarrow Ca(OH)_2 + Na_2NCN$$

$$Na_2NCN \xrightarrow{2RI} R_2NCN \xrightarrow{H^+} R_2NH + CO_2 + NH_3$$

IV. Reductive amination of a carbonyl compounds with a primary amine

$$R_2C{=}O + RNH_2 \longrightarrow R_2C{=}NR \xrightarrow{H_2/Ni} R_2CH{-}NHR$$

9.4.4 Preparation of Tertiary Amines

I. Heating ethanolic solution of ammonia with excess of an alkyl halide, tertiary amine is produced.

$$3R{-}X + NH_3 \rightarrow R_3N + 3HX$$

II. By reductive amination of a carbonyl compound with secondary amine:

$$R_2NH + R{-}CH_2{-}CHO \xrightarrow{Heat}$$

$$R_2N{-}CH{=}CH{-}R \xrightarrow{H_2/Ni} R_2NCH_2CH_2R$$

9.5 DISTINCTION AND SEPARATION OF THREE CLASSES OF AMINES

I. Distinction of primary, secondary and tertiary amines using nitrous acid

(a) *Primary amine:* Primary amine reacts with nitrous acid librating nitrogen gas.

$$RNH_2 + HNO_2 \rightarrow ROH + N_2(g) + H_2O$$

This reaction is highly complicated and the products may contain certain isomeric alcohols, alkenes and other compounds. At the same time, whatever are the products formed by the action of nitrous acid on a primary amine, nitrogen gas is always evolved. None other classes of amines evolve nitrogen gas with nitrous acid.

(b) *Secondary amine:* When treated with nitrous acid, secondary amine forms a *yellow* coloured, neutral, oily nitrosamine:

$$R{-}NHR + HNO_2 \longrightarrow \underset{\text{Nitrosamine}}{R_2N{-}NO}$$

The above solution containing nitrosamine when warmed with crystals of phenol and a few drops of sulphuric acid, a green solution is formed, which when made alkaline with aqueous NaOH, turns deep blue. This test is a common ***Libermann's nitrosamine*** test of secondary amine. This test is not given by either primary or a tertiary amines. These reactions can be used for distinguishing three classes of amines as:

Amine $\xrightarrow{HNO_2}$ Effervescence of N_2 (1° amine confirmed)

Amine $\xrightarrow{HNO_2}$ No gas "either 2° or 3° amines but not 1°"

$\downarrow C_6H_5OH/H_2SO_4$

Green solution → No green solution obtained "3° amine confirmed"

Green solution $\xrightarrow{NaOH}$ Blue colouration "2° amine confirmed"

II. Hinsberg's method

This method of separation of three classes of amine is based on the reaction of sulphonyl chloride with amines forming sulphonamides. The sulphonyl chloride used here is *p*-toluene sulphonyl chloride. Solution containing primary, secondary and tertiary amines are treated with *p*-toluene sulphonyl chloride, primary and secondary amines forms N-alkyl and N, N-dialkyl sulphonamide respectively which are insoluble, precipitate out leaving behind tertiary amine in solution. The solution is filtered-off and filtrate is boiled, where tertiary amine distils off. The solid residue is then treated with aqueous KOH solution where a potassium salt of N-alkyl sulphonamide is formed and comes in solution while N, N-dialkyl disulphonamide does not reacts with KOH, remains insoluble. The amine is regenerated from the sulphonamides by refluxing with 70% sulphuric acid solution.

RNH_2, $RNHR$, R_3N + CH_3–C_6H_4–SO_2Cl →

- $RNHSO_2$–C_6H_4–CH_3 N-alkyl sulphonamide
- $RN(R)SO_2$–C_6H_4–CH_3 N, N-dialkyl sulphonamide

Insoluble

$R–N(R)–R$ Remains in solution

$\downarrow$ KOH

$R_2N–SO_2$–C_6H_4–CH_3 Remains insoluble

$R–\bar{N}(K^+)–SO_2$–C_6H_4–CH_3 Comes in solution

9.6 CHEMICAL REACTIONS OF AMINES

9.6.1 Reactions Given by all Three Types of Amines

I. Basic properties

Amines are Lewis base due to presence of a lone pair of electron on nitrogen atom. All amines are stronger base than ammonia.

In aqueous solution, basic strength of amine depends on two prime factors availability of lone pair of electron on nitrogen atom and stability of conjugate acid formed after protonation of base. Since alkyl groups are electron donor (+I effect), this increases the availability of lone pair and therefore basic strength:

$R \rightarrow NH_2$ (+I) $\quad R \rightarrow NH(\leftarrow R)$ (+I, +I) $\quad R \rightarrow N(\leftarrow R)(\leftarrow R)$ (+I, +I, +I)

Increasing order of electron density on nitrogen atom →

On the other hand conjugate acids of the above bases are stabilized by hydration (by forming intermolecular H-bonds with water molecules) as:

$R–\overset{+}{N}(H\cdots OH_2)_3$ 3-H bonds $\quad R_2\overset{+}{N}(H\cdots OH_2)_2$ 2-H bonds $\quad R_3\overset{+}{N}–H\cdots OH_2$ 1-H bond

← Increasing order of stability of conjugate acids in aqueous solution

Experimental observations revealed that up to secondary amines, electron donating inductive effect controls the basic strength but with tertiary amines, very less stability of conjugate acid due to fewer number of H-bond, makes this amine very less basic, even less basic than primary amine. Hence, between primary amine and tertiary amine, hydrogen bonding is the dominant factor. Therefore, the over all order of basic strength is:

$$NH_3 < R_3N < RNH_2 < R_2NH$$

II. Salt of amines and its separation from other compound

Treatment of amines with a very strong acid, forms amine salt as:

$$RNH_2 + HCl \rightarrow RNH_3Cl$$
$$R_2NH + H_2SO_4 \rightarrow R_2NH_2HSO_4$$

$$C_5H_5N + CH_3COOH \longrightarrow C_5H_5N^{+}\text{-}H + {}^{-}O\text{-}\overset{O}{\overset{\|}{C}}\text{-}CH_3$$

Amine salts are ionic, have very high melting points and have very high solubility in water compared to its parent amine, but very less soluble in non-polar solvents like carbon tetrachloride, ethers, etc.

Formations of amine salts may be used to isolate amines from other organic compounds and characterize them. Most of the amines containing more than six carbons atoms are relatively insoluble in water, but their ammonium salts formed by the action of acids are fairly soluble in water.

$$\underset{\text{Free amine}}{RNH_2} + HCl \rightarrow \underset{\substack{\text{Amine salt} \\ \text{(water soluble)}}}{RNH_3Cl}$$

Isolation and purification of amines via amine salt: Following are the steps followed in laboratory to isolate amines from mixture containing amine, organic and inorganic impurity.

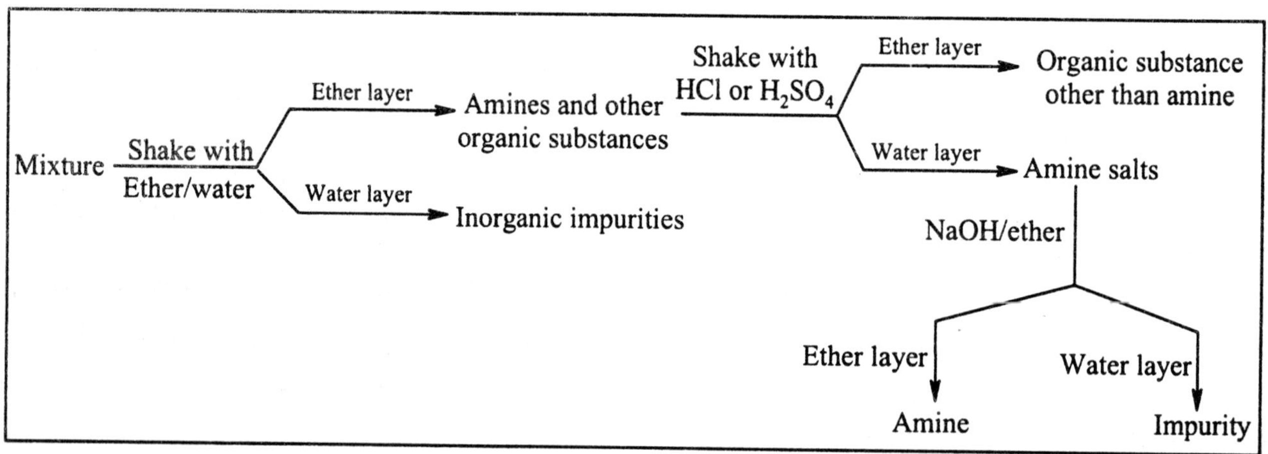

Separation of amines from other organic and inorganic compounds are shown diagrammatically in Figure 9.1.

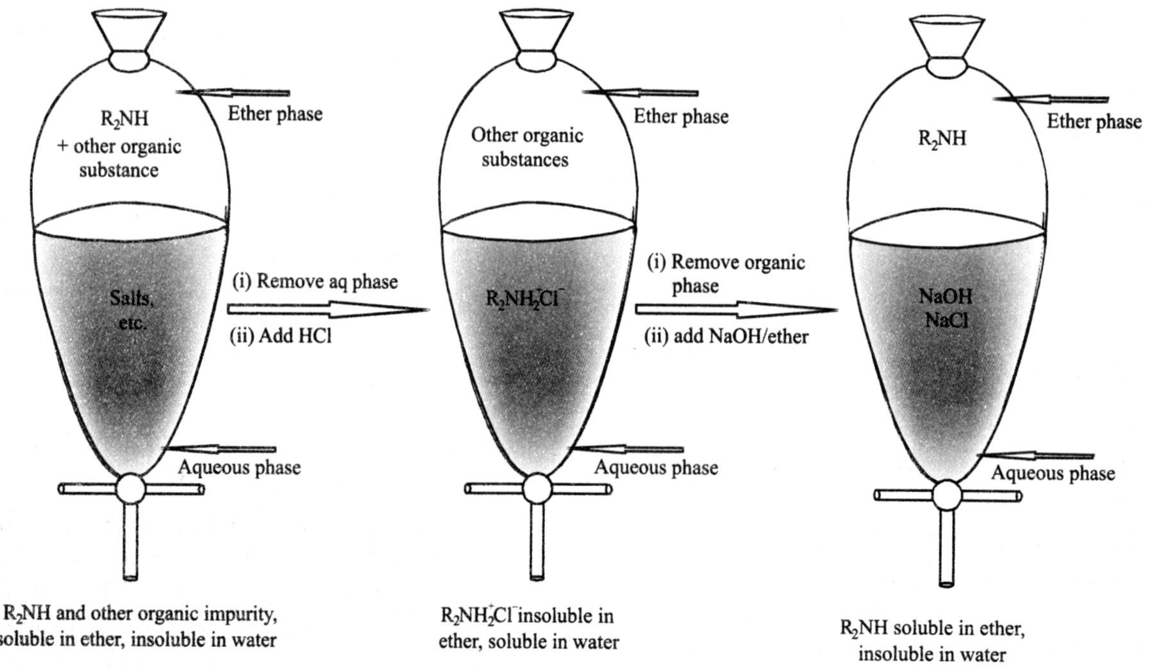

Figure 9.1: Systematic method of separation of three types of amines by extraction technique

9.6.2 Reactions given by Both Primary and Secondary Amines

I. Reaction with acid chlorides, acid anhydrides and esters-amides are produced

$$RNH_2 \xrightarrow{RCOCl} RNHCOR + HCl$$
$$RNH_2 \xrightarrow{(RCO)_2O} RNHCOR + RCOOH$$
$$RNH_2 \xrightarrow{RCOOR} RNHCOR + ROH$$

$$R_2NH \xrightarrow{RCOCl} R_2NCOR + HCl$$
$$R_2NH \xrightarrow{(RCO)_2O} R_2NCOR + RCOOH$$
$$R_2NH \xrightarrow{RCOOR} R_2NCOR + ROH$$

II. Reaction with isocyanates

Phenyl isocyanate reacts with primary and secondary amines forming disubstituted urea:

$$Ph{-}NCO + RNH_2 \longrightarrow Ph{-}NH{-}\overset{O}{\overset{\|}{C}}{-}NHR$$

$$Ph{-}NCO + R_2NH \longrightarrow Ph{-}NH{-}\overset{O}{\overset{\|}{C}}{-}NR_2$$

III. Reaction with formaldehyde

Primary and secondary amine are methylated by heating with formaldehyde and an excess of formic acid at 100° C.

$$RNH_2 + CH_2O + HCOOH \xrightarrow{Heat} RNHCH_3 + CO_2 + H_2O$$

IV. Reaction with halogens

Primary and secondary amines are halogenated at nitrogen giving, N-halogen compounds, when treated with halogen in weakly alkaline medium.

$$RNH_2 + Cl_2 \xrightarrow{NaOH} RNHCl$$
$$R_2NH + Cl_2 \xrightarrow{NaOH} R_2NCl$$

9.6.3 Reactions given by Primary Amines

I. Carbyl amine reaction

Primary amine forms isocyanide when treated with alkaline solution of chloroform:

$$RNH_2 + CHCl_3 + 3KOH \rightarrow RNC + 3KCl + 3H_2O$$

Isocyanides have characteristic fouling smell and therefore this reaction can be used to distinguish primary amines from secondary and tertiary amines.

II. Reaction with carbon disulphide (Hofmann's mustard oil test)

Primary amine on warming with CS_2, forms a dithiocarbamic acid, which when decomposed with mercuric chloride, yields alkyl isothiocyanide, which smells like mustard oil. This reaction is not given by secondary and tertiary amines.

$$RNH_2 + CS_2 \longrightarrow S{=}C\langle{}^{NHR}_{SH} \xrightarrow{HgCl_2} RNCS + HgS + 2HCl$$

III. Primary amine combines with aldehydes forming imine

$$C_6H_5CHO + H_2N{-}R \rightarrow C_6H_5CH{=}NR + H_2O$$

IV. Reaction with nitrous acid

Alcohols are produced with evolution of nitrogen gas:

$$RNH_2 + HNO_2 \rightarrow ROH + N_2 + H_2O$$

V. Oxidation of primary amine

Oxidation products of primary amine depends on the nature of oxidizing agents as:

$$RCH_2NH_2 + KMnO_4 \longrightarrow RCH{=}NH \xrightarrow{H^+} RCHO$$
$$R_2CHNH_2 + KMnO_4 \longrightarrow R_2C{=}NH \xrightarrow{H^+} R_2CO$$
$$R_3C{-}NH_2 + KMnO_4 \longrightarrow R_3C{-}NO_2$$

9.6.4 Reactions given by Secondary Amine

I. With nitrous acid

Secondary amine reacts with nitrous acid forming a yellow coloured, oily liquid "nitrosamine".

$$R_2NH + HNO_2 \rightarrow R_2N{-}NO + H_2O$$

II. With carbon disulphide

Secondary amine, on warming with CS_2 forms a dithiocarbamic acid, but it is not further decomposed with $HgCl_2$ and no alkyl isothiocyanate is formed.

$$CS_2 + R_2NH \longrightarrow S{=}C\langle{}^{NR_2}_{SH} \xrightarrow{HgCl_2} \text{no reaction}$$

9.6.5 Reactions given by Tertiary Amines

I. With nitrous acid

Tertiary amine dissolves in cold nitrous acid solution forming nitrite salt ($R_3N^+HNO_2^-$). If the amine is of the type $(RCH_2)_3N$ or $(R_2CH)_3N$, then, when the salt solution is heated, former produces an aldehyde and latter a ketone, and in both cases a nitrosamine and nitrous oxide.

$$(RCH_2)_3N + HNO_2 \longrightarrow (RCH_2)_3\overset{+}{N}HN\bar{O}_2 \xrightarrow{\text{Heat}} RCHO + RCH_2-N(CH_2R)-NO$$

$$(R_2CH)_3N + HNO_2 \longrightarrow (R_2CH)_3\overset{+}{N}HN\bar{O}_2 \xrightarrow{\text{Heat}} R-C(=O)-R + R_2CH-N(CHR_2)-NO$$

II. Oxidation

Tertiary amine are not oxidized by $KMnO_4$, but are oxidized to amine oxide by Caro's acid or H_2O_2 or Ozone as:

$$R_3N + [O] \longrightarrow R_3\overset{+}{N}-O^-$$

9.6.6 QUATERNARY AMMONIUM COMPOUNDS

Quaternary ammonium salts are white crystalline solids, soluble in water and completely ionized in solution. When a quaternary ammonium halide is heated, it gives tertiary amine:

$$R_4N^+Cl^- + \text{heat} \rightarrow R_3N + RCl$$

Elimination reaction of quaternary ammonium salts

Quaternary ammonium hydroxide undergo thermal decomposition to form alkenes and reaction follow E2 mechanism. Exception occurs in case of tetramethyl ammonium hydroxide, which produces methanol and trimethyl amine:

$$(CH_3)_4\overset{+}{N}\overset{-}{Cl} \xrightarrow[H_2O]{Ag_2O} AgI + (CH_3)_4\overset{+}{N}\overset{-}{OH} \xrightarrow{\text{Heat}} CH_3OH + (CH_3)_3N$$

If alkyl group other than methyl are bonded to nitrogen, β-elimination takes place forming alkene:

$$(C_2H_5)_4\overset{+}{N}\overset{-}{Cl} \xrightarrow[H_2O]{Ag_2O} AgI + (C_2H_5)_4\overset{+}{N}\overset{-}{OH} \xrightarrow{\text{Heat}} CH_2{=}CH_2 + (C_2H_5)_3N$$

If there is a choice of β-proton to be removed, most acidic β-proton is removed, producing least substituted alkene as major product as:

$$CH_3CH(H)CH_2-\overset{+}{N}(CH_3)_2-CH_2CH_2-H \ \overset{-}{OH} \xrightarrow{\text{Heat}} CH_3CH_2CH_2-N(CH_3)-CH_3 + CH_2{=}CH_2$$

(H on CH_3CHCH_2: less acidic β-proton; H on CH_2CH_2: more acidic β-proton)

$$CH_3CH_2(H)-\overset{+}{N}(CH_3)_2-CH_2CH(C_6H_5)-H \ \overset{-}{OH} \xrightarrow{\text{Heat}} CH_3CH_2-N(CH_3)-CH_3 + C_6H_5CH{=}CH_2$$

(H on CH_3CH_2: less acidic β-proton; H on $CH_2CH(C_6H_5)$: more acidic β-proton)

Example 5. What products you expect to obtain by the Hofmann's exhaustive methylation of:

(a) $Me_2CHCH_2CH_2NH_2$ (b) Me_2CHCH_2NHMe

(c) $Me_2CHCH_2NHCH_2CH_2Me$ (d) $CH_3CH_2NHCH_2Cl$

(e) $(H_3C)_2N$–(piperidin-4-yl)–N–H

Solution

(a) $Me_2CHCH_2CH_2-\overset{+}{N}(CH_3)_3$

(b) $Me_2CHCH_2-\overset{+}{N}(CH_3)_3$

(c) $Me_2CHCH_2-\overset{+}{N}(CH_3)_2-CH_2CH_2CH_3$

(d) $CH_3CH_2-\overset{+}{N}(CH_3)_2-CH_2CH_2Cl$

(e) 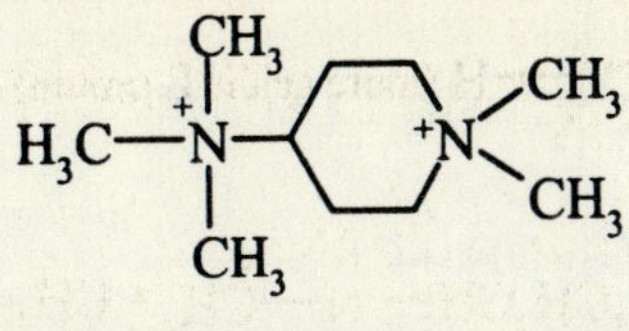 $(CH_3)_3\overset{+}{N}$–(piperidine ring)–$\overset{+}{N}(CH_3)_2$

Example 6. t-$BuNH_2$ and $(CH_3)_3CCH_2NH_2$ cannot be prepared by the action of NH_3 on corresponding alkyl bromides, why? How are they prepared?

Solution: Tertiary alkyl bromide undergo elimination reaction (E2) more rapidly than substitution reaction (SN-2):

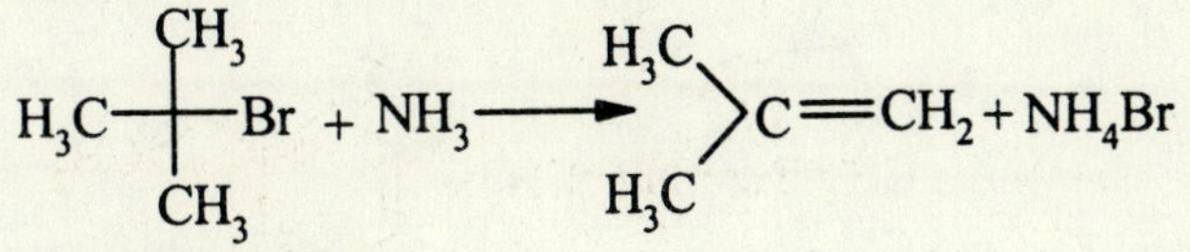

$$(CH_3)_3C{-}Br + NH_3 \longrightarrow (H_3C)_2C{=}CH_2 + NH_4Br$$

Has greater reactivity towards E2 reaction

Neopentyl bromide undergo SN-1 reaction much faster than SN-2 reaction as:

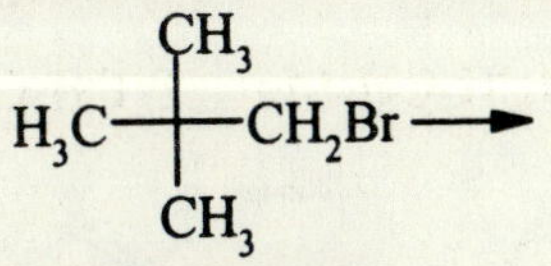

$$(CH_3)_3C{-}CH_2Br \longrightarrow$$

Has greater reactivity towards SN-1 reaction

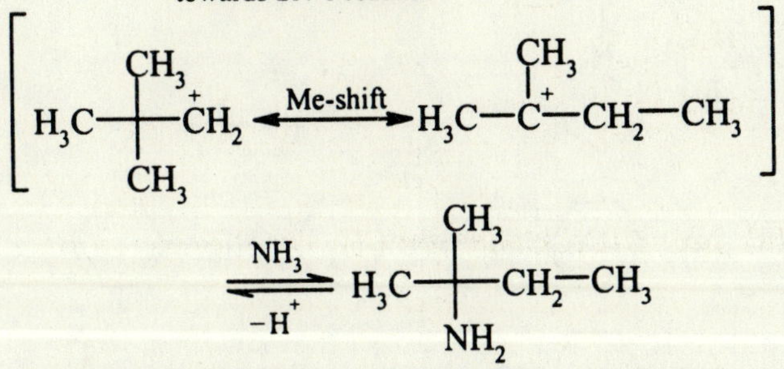

$$\left[(CH_3)_3C{-}\overset{+}{C}H_2 \xrightleftharpoons{\text{Me-shift}} H_3C{-}\overset{+}{C}(CH_3){-}CH_2{-}CH_3\right]$$

$$\underset{-H^+}{\overset{NH_3}{\rightleftharpoons}} H_3C{-}C(CH_3)(NH_2){-}CH_2{-}CH_3$$

These amines are conveniently prepared from corresponding carboxylic acids as:

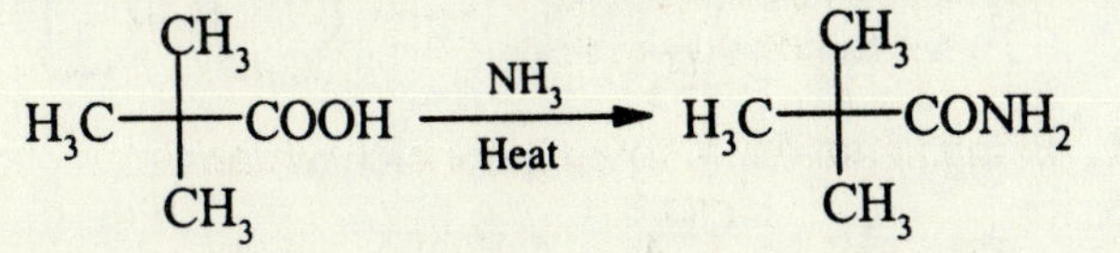

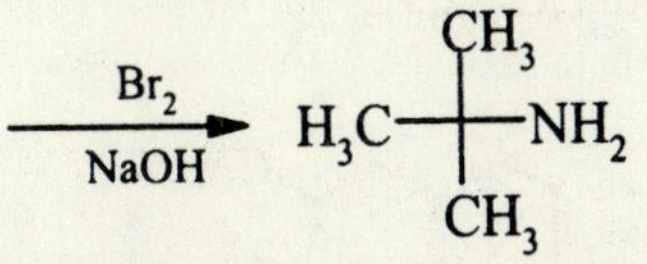

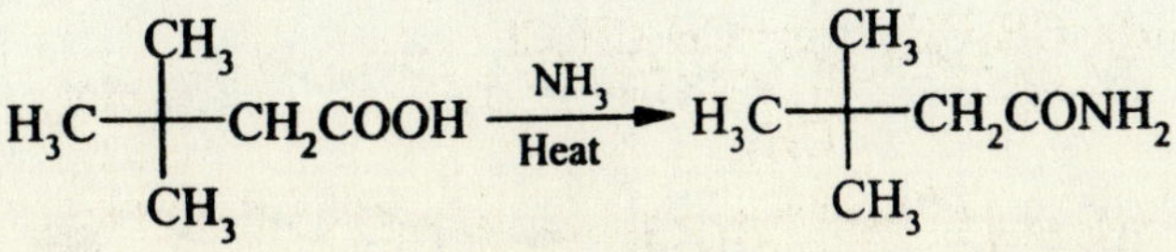

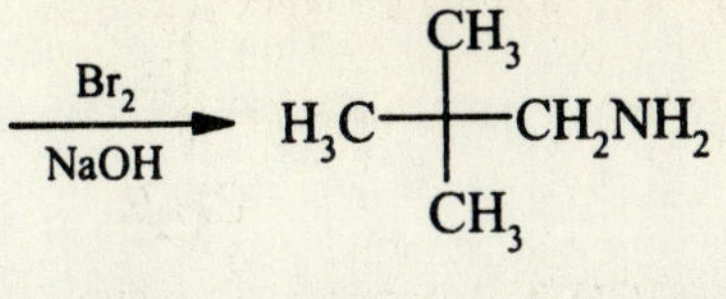

$$(CH_3)_3C{-}COOH \xrightarrow[\text{Heat}]{NH_3} (CH_3)_3C{-}CONH_2$$

$$\xrightarrow[NaOH]{Br_2} (CH_3)_3C{-}NH_2$$

$$(CH_3)_3C{-}CH_2COOH \xrightarrow[\text{Heat}]{NH_3} (CH_3)_3C{-}CH_2CONH_2$$

$$\xrightarrow[NaOH]{Br_2} (CH_3)_3C{-}CH_2NH_2$$

Example 7. In the reaction below:

$$CH_3COCH_2Cl + CH_2N_2 \rightarrow C_4H_7ClO(X) + C_4H_7ClO(Y)$$

X is a ketone while Y is an epoxide. Deduce structures of X and Y and explain mechanism of their formations.

Solution

$$CH_2N_2 \xrightarrow{\text{Heat}} :CH_2 + N_2$$

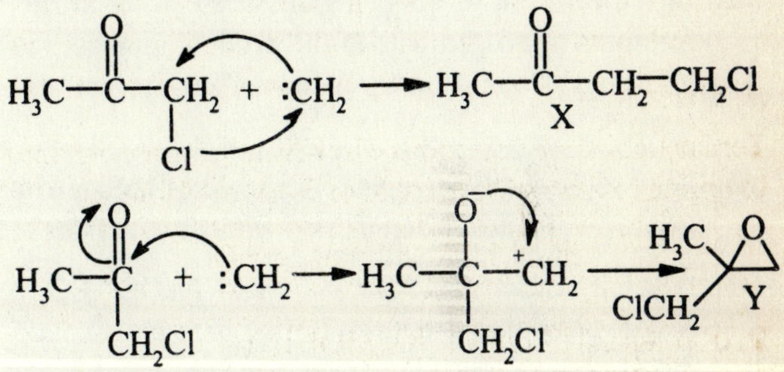

$$H_3C{-}\overset{O}{\overset{\|}{C}}{-}CH_2Cl + :CH_2 \longrightarrow \underset{X}{H_3C{-}\overset{O}{\overset{\|}{C}}{-}CH_2{-}CH_2Cl}$$

$$H_3C{-}\overset{O}{\overset{\|}{C}}{-}CH_2Cl + :CH_2 \longrightarrow H_3C{-}C(\overset{-}{O})(CH_2Cl){-}\overset{+}{C}H_2 \longrightarrow \underset{Y}{(H_3C)(ClCH_2)C{<}\overset{O}{|}{>}CH_2}$$

Example 8. Complete the following reaction: ($^*C = {}^{13}C$)

$$RMgBr + {}^*CO_2 \longrightarrow A \xrightarrow{H^+} B \xrightarrow[\text{(ii) } CH_2N_2]{\text{(i) } SOCl_2} C \xrightarrow{Ag_2O} D \xrightarrow{H_2O} E$$

Solution

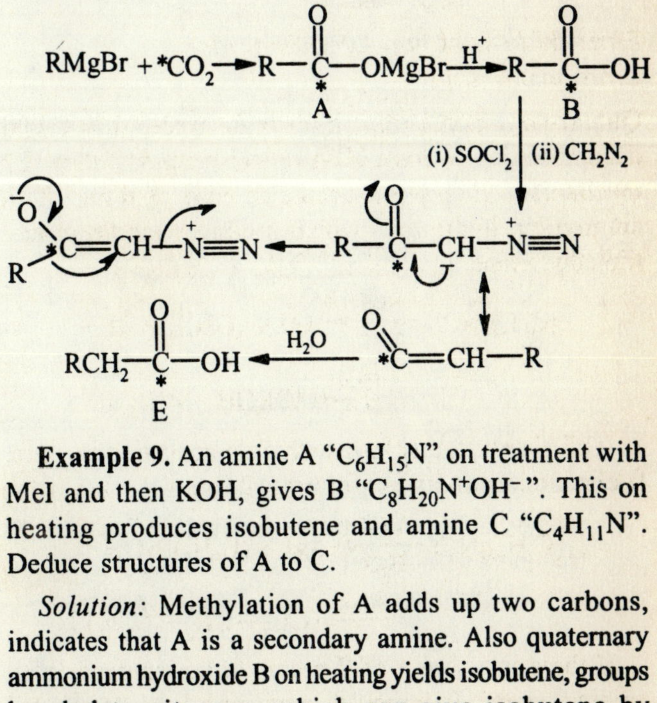

$$RMgBr + {}^*CO_2 \longrightarrow \underset{A}{R{-}\overset{O}{\overset{\|}{{}^*C}}{-}OMgBr} \xrightarrow{H^+} \underset{B}{R{-}\overset{O}{\overset{\|}{{}^*C}}{-}OH}$$

$$\xrightarrow{\text{(i) } SOCl_2 \text{ (ii) } CH_2N_2} R{-}\overset{O}{\overset{\|}{{}^*C}}{-}CH{-}\overset{+}{N}{\equiv}N \longrightarrow (R)(\overset{-}{O}){}^*C{=}CH{-}\overset{+}{N}{\equiv}N$$

$$\longleftrightarrow O{=}{}^*C{=}CH{-}R \xrightarrow{H_2O} \underset{E}{RCH_2{-}\overset{O}{\overset{\|}{{}^*C}}{-}OH}$$

Example 9. An amine A "$C_6H_{15}N$" on treatment with MeI and then KOH, gives B "$C_8H_{20}N^+OH^-$". This on heating produces isobutene and amine C "$C_4H_{11}N$". Deduce structures of A to C.

Solution: Methylation of A adds up two carbons, indicates that A is a secondary amine. Also quaternary ammonium hydroxide B on heating yields isobutene, groups bonded to nitrogen, which can give isobutene by β-elimination are:

$CH_3-CH(CH_3)-CH_2-\overset{+}{N}$ or $H_3C-C(CH_3)_2-\overset{+}{N}$

Hence, two structures of A satisfy the above conditions. A can be either of these:

$$(CH_3)_2CHCH_2-NH-CH_2CH_3$$

or

$$(CH_3)_3C-NH-CH_2CH_3$$

PRACTICE EXERCISE

1. Why secondary and tertiary alkyl halides give poor yields of nitro-alkanes, when treated with silver nitrite?

2. In the reaction of alkyl halides with silver nitrite, fractions of alkyl nitrite in the product increases as we move from primary to tertiary alkyl halides, explain.

3. Why sodium nitrite gives better yield of both nitrite and nitro alkanes, when treated with alkyl halides?

4. When 2-nitropropane is treated with $NaOD/D_2O$, it is converted into 2-deutro-2-nitro propane, explain mechanism of this conversion reaction.

5. How a primary nitro compound can be separated out from a mixture of primary and tertiary nitro alkanes?

6. How primary, secondary and tertiary nitro-alkanes are distinguished? Provide chemical reactions involved in distinction.

7. Complete the following reactions:

(a) Cyclohexyl$-NH_2$ $\xrightarrow[\text{(ii) } Ag_2O]{\text{(i) } CH_3I \text{ (excess)}}$ A $\xrightarrow{\text{Heat}}$ B + $(CH_3)_3N$

(b) 2-Methylpyrrolidine (N–H, CH_3) $\xrightarrow[\text{(ii) } Ag_2O]{\text{(i) } CH_3I \text{ (excess)}}$ A $\xrightarrow{\text{Heat}}$ B

$\xrightarrow[\text{(ii) } Ag_2O]{\text{(i) } CH_3I \text{ (excess)}}$ C $\xrightarrow{\text{Heat}}$ D + $(CH_3)_3N$

8. Predict major products obtained when the following amines undergo exhaustive methylation followed by treatment with Ag_2O and finally heating:

(a) 2-hexanamine (b) 2-methyl piperidine

(c) [decahydroquinoline, N–H] (d) [bicyclic amine, N:] (e) [bicyclic amine, N–H]

9. Predict the products expected when the following compound is treated with H_2O_2 and finally heated:

Cyclohexyl$-N(CH_3)_2$ with CH_3 $\xrightarrow[\text{Heat}]{H_2O_2}$

10. Bring about the following transformations:

(a) Benzene$-CHO$ ⟶ Benzene$-CH_2NHCH_3$

(b) $C_6H_5CH_2C(=O)CH_3$ ⟶ $C_6H_5CH_2CH(NH_2)CH_3$

(c) Piperidine (N–H) ⟶ N-Phenylpiperidine

(d) Cyclohexanone (=O) ⟶ Cyclohexyl$-NH-$Phenyl

(e) Cyclohexanone (=O) ⟶ Cyclohexyl$-NH_2$

(f) Compounds containing not more than five carbon ⟶ N-Cyclopentylpiperidine

11. Bring about the following transformations:

(a) Benzene$-Br$ ⟶ Benzene$-CH_2NH_2$

(b) Benzene$-CH_2Br$ ⟶ Benzene$-CH_2CH_2CH_2NH_2$

(c) $H_3C\text{~~~}COOH$ ⟶ $H_3C\text{~~~}CH_2NH_2$

(d) $H_3C\text{~~~}COOH$ ⟶ $H_3C\text{~~~~}NH_2$

(e) H_3C–$CH_2CH_2CH_2$–CO–CH_3 ⟶ H_3C–$CH_2CH_2CH_2$–C(CH_3)(OH)–NH_2

(f) Br–C(C_2H_5)(H)(CH_3) ⟶ H_2N–C(C_2H_5)(H)(CH_3)

12. Arrange the amines in each set in increasing order of their basic strength:

(a) $C_6H_5NH_2$, cyclohexylamine, diphenylamine

(b) imidazole, pyrrole, pyrrolidine, pyridine

(c) piperidine, pyridine, pyrrole

13. Select the compounds being capable of resolutions into enantiomers:

(a) N-ethyl-N-methyl aniline.

(b) 2-methylpepridine (c) 1-methylpepridine

(d) N-ethyl-N-methylpyrrolidinium Cl^- (e) N-ethyl-N-methyl-3-pyrrolinium Cl^-

14. Predict products in the following reactions:

(a) $C_6H_5CH_2CH_2CH_2Br$ $\xrightarrow{NH_3\text{ (excess)}}$

(b) C_6H_5COCl $+ CH_3CH_2NH_2$ ⟶

(c) N-methylpiperidine + 3-chloroperbenzoic acid ⟶

(d) Product of "C" $\xrightarrow{Heat}$

(e) octahydroquinoline (decahydroquinoline) $\xrightarrow{\text{(i) } CH_3I \text{ (excess)}}_{\text{(ii) } Ag_2O}$ $\xrightarrow{Heat}$

(f) Product of "e" $\xrightarrow{\text{(i) } CH_3I \text{ (ecxess)}}_{\text{(ii) } Ag_2O}$ $\xrightarrow{Heat}$

(g) piperidine $+ NaNO_2 + HCl$ ⟶

(h) CH_3CH_2–C(=NCH_3)–CH_3 $\xrightarrow{LiAlH_4}$

(i) C_6H_5–CH(CN)–CH_2CH_3 $\xrightarrow{LiAlH_4}$

(j) H_3C–CO–$CH_2CH_3 + (C_2H_5)_2NH$ $\xrightarrow{NaBH_3CN}$

(k) pyrrole (N–H) $\xrightarrow{HNO_3}$

(l) H_3C–$CH_2CH_2CH_2CH_2$–Br $\xrightarrow{\text{(i) } NaN_3}_{\text{(ii) } LiAlH_4}$

15. A student was trying to prepare 1-octanamine by the following process:

$CH_3(CH_2)_6COOH + SOCl_2$ ⟶ $\xrightarrow{NH_3}$ $H_3C(CH_2)_6CONH_2$ $\xrightarrow{LiAlH_4}$ $CH_3(CH_2)_6CH_2NH_2$

The $LiAlH_4$ was of poor quality and the final product was contaminated by 20% octanamide. Give explicit method, which can effect the separation of products.

16. Bring about the following transformations:

(a) Benzoic acid to benzyl amine.

(b) Benzaldehyde to benzyl amine.

(c) Pyrrolidine to N-ethylpyrrolidine.

(d) Pentandioic acid to 1, 5-diamino pentane.

(e) Cyclohexanone to N-cyclohexyl pyrrolidine.

17. Complete the following reactions:

(a) 4-(nitromethyl)cyclohexanecarbaldehyde (CHO, CH_2NO_2) $\xrightarrow[Heat]{NaOH}$ X (b) 2-(2-nitroethyl)benzaldehyde (CHO, NO_2) $\xrightarrow[Heat]{NaOH}$ X

18. A neutral organic compound A has molecular formula "$C_9H_{18}O_2$" and it can be resolved into enantiomers.

Refluxing A with dilute sulphuric acid yields B "$C_5H_{10}O_2$" and C "$C_4H_{10}O$", both resolvable. C gives yellow precipitate with alkaline solution of iodine. B on heating with ammonia yields another neutral compound D. D on treatment with alkaline solution of bromine yields a new compound which, when heated with nitrous acid yields C. Identify A to D and explain formation of C from D.

19. Provide appropriate reagents for the following transformations:

(a) ROH ⟶ RCH_2NH_2

(b) Cyclohexyl–OH ⟶ Cyclohexyl–NH_2

(c) C_6H_5–NH–CH_3 ⟶ C_6H_5–N(CH_3)–CH_2CH_3

(d) C_6H_5–NH_2 ⟶ C_6H_5–CH_2–NH–CH_3

20. An organic compound A has molecular formula "$C_9H_{13}NO$" and can be resolved into enantiomers. A does not decolourises Br_2-water solution. A on refluxing with dilute H_2SO_4 solution yields another resolvable compound B "$C_9H_{14}O_3$", which gives effervescence with $NaHCO_3$. B on treatment with $NaBH_4$ yields C "$C_9H_{16}O_3$". C on heating with concentrated H_2SO_4 solution yields a sweet smelling, neutral compound D "$C_9H_{14}O_3$". Also A on reduction with $LiAlH_4$ yields E " $C_9H_{19}O$", which on treatment with H_2SO_4 yields the following compound:

(decahydroquinoline, N–H)

Identify A to E.

21. An organic compound A has molecular formula "C_7H_9N" and it forms a clear solution when dissolved in aqueous KOH solution of benzene sulphonyl chloride. A on treatment with $NaNO_2$ and HCl at 0°C, and then with 2-naphthol forms an intensely coloured compound. Also A on treatment with CH_3COCl followed by electrophilic substitution yields a single product. Identify A.

22. An organic compound A has molecular formula "$C_5H_{13}N$" and it reacts with benzene sulphonyl chloride in aqueous potassium hydroxide solution to give a clear solution, which on acidification gives a precipitate. Also A can be resolved into enantiomers. A on refluxing with nitrous acid gives B "$C_5H_{12}O$", which is still resolvable. B on treatment with CrO_3 in pyridine solution of HCl yields another resolvable compound C "$C_5H_{10}O$", which on further treatment with Zn(Hg) in concentrated HCl solution yields a non-resolvable compound D "C_5H_{12}". Identify A to D.

23. Provide structures of the product in the following sequence of reaction:

N-Methylpiperidine $\xrightarrow{CH_3I}$ A($C_7H_{16}NI$) $\xrightarrow[H_2O]{Ag_2O}$

B($C_7H_{17}NO$) $\xrightarrow{Heat}$ C($C_7H_{15}N$) C $\xrightarrow{CH_3I}$ D($C_8H_{18}NI$)

$\xrightarrow[H_2O]{Ag_2O}$ E($C_8H_{19}NO$) $\xrightarrow{Heat}$ F(C_5H_8) + H_2O + $N(CH_3)_3$

24. Provide structures of the product in the following sequence of reaction:

(a) 3,6-Dibromocyclohexene $\xrightarrow[\text{2 equivalents}]{CH_3NHCH_3}$ A $\xrightarrow{BaO}$ B $\xrightarrow[\text{2 equivalents}]{CH_3I}$

C $\xrightarrow{Ag_2O/H_2O}$ D $\xrightarrow{Heat}$ E + $2(CH_3)_3N$

(b) 2,2-Diphenylcyclohexanone $\xrightarrow{NH_2OH}$ A $\xrightarrow{LiAlH_4}$ B $\xrightarrow[\text{2 equivalents}]{CH_3I}$

C $\xrightarrow{Ag_2O/H_2O}$ D $\xrightarrow{Heat}$ E + $(CH_3)_3N$

25. In the following sequence of reaction, deduce structures of A and B and explain formation of C:

3,6-Dibromocyclohexene $\xrightarrow[\text{2 equivalents}]{CH_3NHCH_3}$ A $\xrightarrow{BaO}$

B $\xrightarrow[\text{2 equivalents}]{CH_3I}$ $\xrightarrow[Heat]{KOH}$ Benzene (C) + $2(CH_3)_3N$

26. Propose mechanism of formation of products in the following reaction:

(cyclohexenone with side chain bearing N_3 and CH_3) $\xrightarrow[H_2]{Pd/CaCO_3}$ (spiro product with N–H and CH_3)

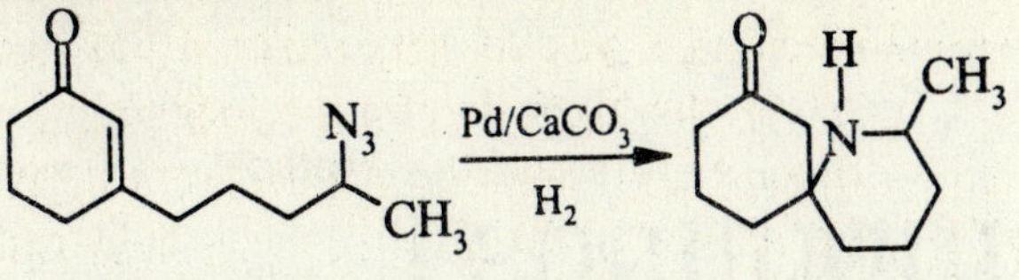

27. How butyl amine and dibutyl amine can be separated?

28. An organic compound A has molecular formula "$C_5H_{10}N_2$" and it can be resolved into enantiomers. A on hydrogenation yields optically inactive B ($C_5H_{12}N_2$). B on treatment with nitrous acid forms an oily liquid compound C "$C_5H_{10}N_2O_2$" which is yellow coloured. C, when warmed with crystals of phenol and a few drops of concentrated sulphuric acid, turns green which when made alkaline, turns deep blue. Also A on treatment with excess of iodomethane forms D "$C_{10}H_{22}N_2I_2$". D on treatment with aqueous Ag_2O followed by heating yields pyridine. Identify A to D.

29. Complete the following reactions:

(a) 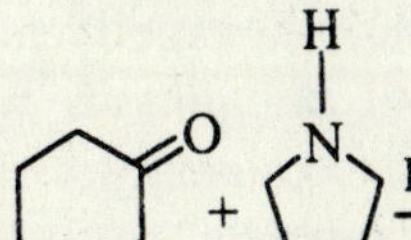 cyclohexanone + pyrrolidine $\xrightarrow{Heat}$ A $\xrightarrow{CH_2=CHCOOEt}$ B $\xrightarrow{H_3O^+}$ C

(b) $HN_3 \xrightarrow[CH_2=CH_2]{h\nu}$ D

(c) $BrCH_2CH_2Br \xrightarrow{KCN} \xrightarrow{Na/EtOH}$ E

30. Bring about the following transformations:

(a) R—CN → R—NC (b) R—NC → R—CN

31. Complete the following reactions:

(a) $CH_2=CH-CN$ + (benzyl alcohol, $C_6H_5CH_2OH$) $\xrightarrow{H_2SO_4}$

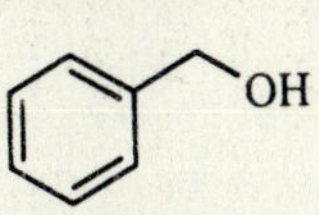

(b) 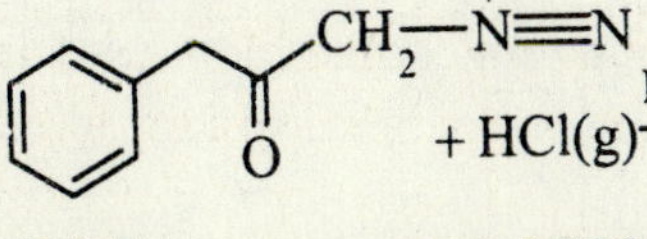 $C_6H_5CH_2COCH_2-\overset{+}{N}\equiv N$ + HCl(g) $\xrightarrow{Et_2O}$

(c) $MeNH_2$ + ? ⟶ MeNHCOOEt

(d) $HOCH_2CH_2CN \xrightarrow[Heat]{HBr}$

32. Complete the following reactions:

(a) $CH_3CH_2CH=CH_2 \xrightarrow{B_2H_6}$ A $\xrightarrow{ClNH_2}$ B

(b) 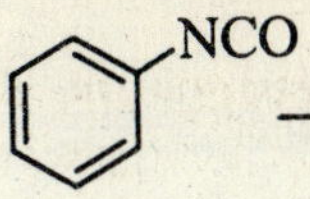 (phenyl isocyanate, C_6H_5NCO) $\xrightarrow{H_2O}$ C $\xrightarrow{-CO_2}$ D $\xrightarrow{PhNCO}$ E

(c) $C_2H_5NH_2 + KCN + Br_2 \xrightarrow{KOH}$ F $\xrightarrow{LiAlH_4}$ G

33. Complete the following reactions:

(a) $CH_3CH_2NO_2 + 2CH_2O \xrightarrow{KHCO_3}$ A

(b) $RCH_2NH_2 + 2CH_2O \xrightarrow{HCOOH}$ B

(c) $H_2C=C=O + HCN \longrightarrow C \xrightarrow{H_2C=C=O}$ D

34. An amine A "$C_6H_{15}N$", on treatment with CH_3I and then KOH gives B "$C_8H_{20}N^+OH^-$". B on heating produces C "C_4H_8" and D "$C_4H_{11}N$". C on treatment with aqueous $(CH_3COO)_2Hg$ followed by reduction with $NaBH_4$ yields E ($C_4H_{10}O$), which does not change the colour of chromic acid solution. Deduce structures of A to E.

35. Synthesize the following compounds starting from cyclohexanone and any amines needed:

(a) 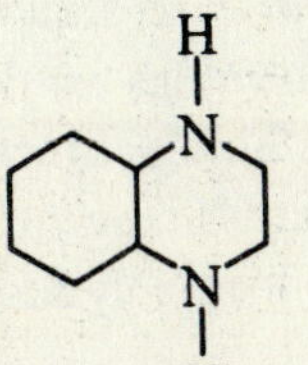 (decahydroquinoxaline, N–H at both nitrogens) (b) 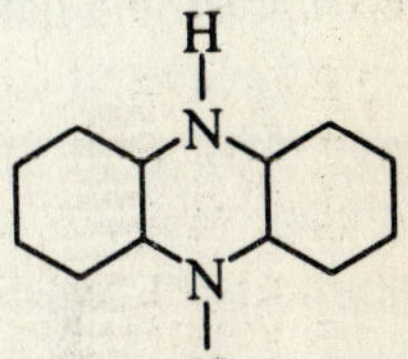 (perhydrophenazine, N–H at both nitrogens)

36. Write chemical equations that could be used in laboratory for distinction between the following pairs of compounds:

(a) Benzyl amine and benzamide.
(b) Allyl amine and propyl amine.
(c) Tetrapropyl ammonium chloride and tetrapropyl ammonium hydroxide.
(d) Tripropylammonium chloride and tetrapropyl ammonium chloride.
(e) Triethyl amine and diethyl amine.
(f) *p*-toludine and N-methyl aniline.
(g) Cyclohexyl amine and piperidine.
(h) Pyridine and benzene.
(i) Cyclohexyl amine and aniline.

37. Propose mechanism of the following reaction:

2-(2-cyanoethyl)cyclohexanone $\xrightarrow{H_2/Pd}$ (decahydroquinoline, N–H)

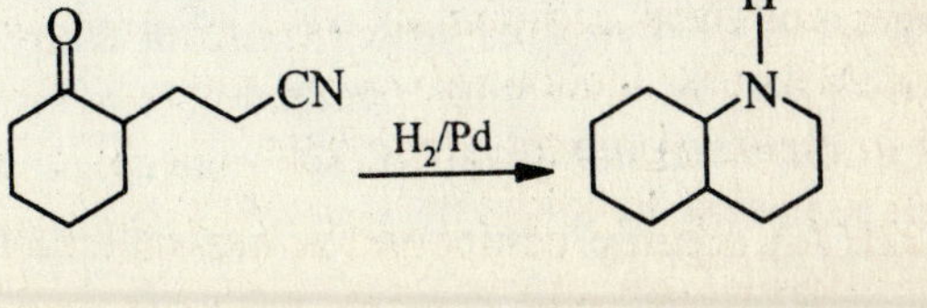

Chapter 10

Aromatic Compounds I

10.1 AROMATICITY

10.1.1 Benzene and Other Aromatic Compounds

The adjective "aromatic" is used by organic chemists in a rather different way than it is normally applied. It has its origin in the observation that certain natural substances, such as cinnamon bark, wintergreen leaves, vanilla beans and anise seeds, contained fragrant compounds having common but unexpected properties. Cinnamon bark, for example, yielded a pleasant smelling compound, formula C_9H_8O, named cinnamaldehyde. Because of the low hydrogen to carbon ratio in this and other aromatic compounds (note that the H : C ratio in an alkane is > 2), chemists expected their structural formulas would contain a large number of double or triple bonds. Since, double bonds are easily cleaved by oxidative reagents such as potassium permanganate or ozone, and rapidly add bromine and chlorine, these reactions were applied to these aromatic compounds. Surprisingly, products that appeared to retain many of the double bonds were obtained, and these compounds exhibited a high degree of chemical stability compared with known alkenes and cycloalkenes. On treatment with hot permanganate solution, cinnamaldehyde gave a stable, crystalline $C_7H_6O_2$ compound, now called benzoic acid. The H : C ratio in benzoic acid is < 1, again suggesting the presence of several double bonds. Benzoic acid was eventually converted to the stable hydrocarbon benzene, C_6H_6, which also proved unreactive to common double bond transformations, as shown below. For comparison, reactions of cyclohexene, a typical alkene, with these reagents are shown in the box. As experimental evidence for a wide assortment of compounds was acquired, those incorporating this exceptionally stable six-carbon core came to be called "aromatic".

COOH
CaO / Heat
Benzene C_6H_6
H_2SO_4 → No reaction
Br_2 → No reaction
$KMnO_4$ → No reaction

Benzene C_6H_6
H_2SO_4 / Heat → SO_3H
Br_2 / $FeBr_3$ → Br
H_2SO_4/HNO_3 → NO_2

H_2SO_4 → OSO_3H
$KMnO_4$ → COOH COOH
Br_2 → Br Br

If benzene is forced to react by increasing the temperature and/or by addition of a catalyst, it undergoes **substitution reactions** rather than the addition reactions that are typical of alkenes. This further confirms the previous indication that the six-carbon benzene core is unusually stable to chemical modification. The conceptual contradiction presented by a high degree of unsaturation (low H : C ratio) and high chemical stability for benzene and related compounds remained an unsolved puzzle for many years. Eventually, the presently accepted structure of a regular-hexagonal, planar ring of carbons was adopted, and the exceptional thermodynamic and chemical stability of this system was attributed to resonance stabilization of a conjugated cyclic triene.

Benzene Six delocalized pi-electrons

Here, two structurally and energetically equivalent electronic structures for a stable compound are written, but no single structure provides an accurate or even an adequate representation of the true molecule. The six-membered ring in benzene is a perfect hexagon (all carbon-carbon bonds have an identical length of 1.40 Å). The cyclohexatriene contributors would be expected to show alternating bond lengths, the double bonds being shorter (1.34 Å) than the single bond (1.54 Å). An alternative representation for benzene (circle within a hexagon) emphasizes the pi-electron delocalization in this molecule, and has the advantage of being a single diagram. In cases such as these, the electron delocalization described by resonance enhances the stability of the molecules, and compounds composed of such molecules often show exceptional stability and related properties.

Evidence for the enhanced thermodynamic stability of benzene was obtained from measurements of the heat released, when double bonds in a six-carbon ring are hydrogenated (hydrogen is added catalytically) to give cyclohexane as a common product. In the following diagram cyclohexane represents a low-energy reference point. Addition of hydrogen to cyclohexene produces cyclohexane and releases heat amounting to 28.6 kcal per mole.

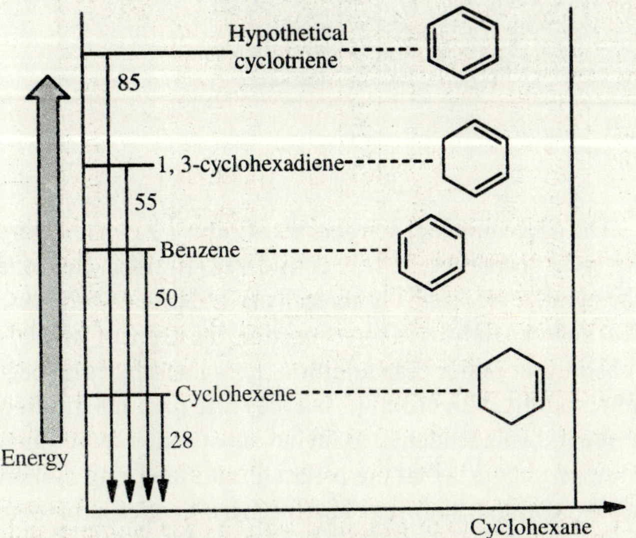

Energy released in hydrogenation reaction

If we take this value to represent the energy cost of introducing one double bond into a six-carbon ring, we would expect a cyclohexadiene to release 57.2 kcal per mole on complete hydrogenation, and 1, 3, 5-cyclohexatriene to release 85.8 kcal per mole. These **heats of hydrogenation** would reflect the relative thermodynamic stability of the compounds. In practice, 1, 3-cyclohexadiene is slightly more stable than expected by about 2 kcal, presumably due to conjugation of the double bonds. **Benzene, however, is an extraordinary 36 kcal/mole more stable than expected.** This sort of stability enhancement is now accepted as a characteristic of all aromatic compounds.

10.1.2 Fused Benzene Ring Compounds

Benzene rings may be joined together (fused) to give larger polycyclic aromatic compounds. A few examples are drawn below, together with the approved numbering scheme for substituted derivatives. The peripheral carbon atoms (numbered in all but the last three examples) are all bonded to hydrogen atoms. The six benzene rings in coronene are fused in a planar ring.

Naphthalene $C_{10}H_8$ Anthracene $C_{14}H_{10}$ Phenanthrene $C_{14}H_{10}$

Chrysene $C_{18}H_{12}$

Pyrene $C_{16}H_{10}$ Corannulene $C_{20}H_{10}$ Coronene $C_{24}H_{12}$

As these extended aromatic compounds become larger, the ratio of hydrogen to carbon decreases. For example, the symmetrical hexacyclic compound coronene has a H/C ratio = 1/2, compared with 1 for benzene. If we were to imagine fused ring systems of this kind to be further extended in space, the H/C ratio would approach zero, and the resulting compound would be a form of carbon. Such a carbon allotrope exists and is called **graphite**. Another well-characterised carbon allotrope is **diamond**. The structures for these two forms of carbon are very different. Diamond is an extended array of sp^3 hybridized carbon atoms; whereas, graphite consists of overlapping sheets of sp^2 hybridized carbon atoms arranged in a hexagonal pattern.

If we extend the structure of corannulene by adding similar cycles of five benzene rings, the curvature of the resulting molecule should increase, and eventually close into a sphere of carbon atoms. The archtypical compound of this kind (C_{60}) has been named **buckminsterfullerene**, because of its resemblance to the geodesic structures created by Buckminster Fuller. It is a member of a family of similar carbon structures are called **fullerenes**. These materials represent a third class of carbon allotropes.

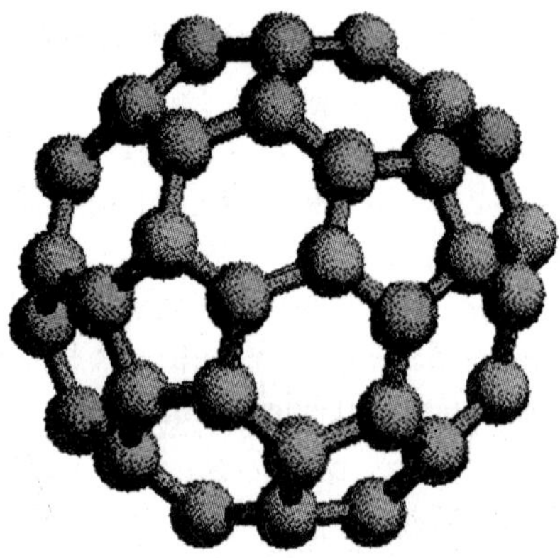

Buckminsterfullerene

Although C_{60} is composed of fused benzene rings its chemical reactivity resembles that of the cycloalkenes more than benzene. Indeed, exposure to light and oxygen slowly degrade fullerenes to cage opened products. Most of the reactions thus far reported for C_{60} involve addition to, rather than substitution of the core structure. These reactions include hydrogenation, bromination and hydroxylation. Strain introduced by the curvature of the surface may be responsible for the enhanced reactivity of C_{60}.

Larger fullerenes, such as C_{70}, C_{76}, C_{82} and C_{84} have elipsoidal or distorted spherical structures, and fullerene-like assemblies up to C_{240} have been detected. A fascinating aspect of these structures is that the space within the carbon cage may hold atoms, ions or small molecules. Such species are called **endohedral fullerenes**.

The cavity of C_{60} is relatively small, but encapsulated helium, lithium and atomic nitrogen compounds have been observed. Larger fullerenes are found to encapsulate lanthanide metal atoms.

Interest in the fullerenes has led to the discovery of a related group of carbon structures referred to as nanotubes. As shown in the following illustration, nanotubes may be viewed as rolled up segments of graphite. The chief structural components are six-membered rings, but changes in tube diameter, branching into side tubes and the capping of tube ends is accomplished by fusion with five and seven-membered rings. Many interesting applications of these unusual structures have been proposed.

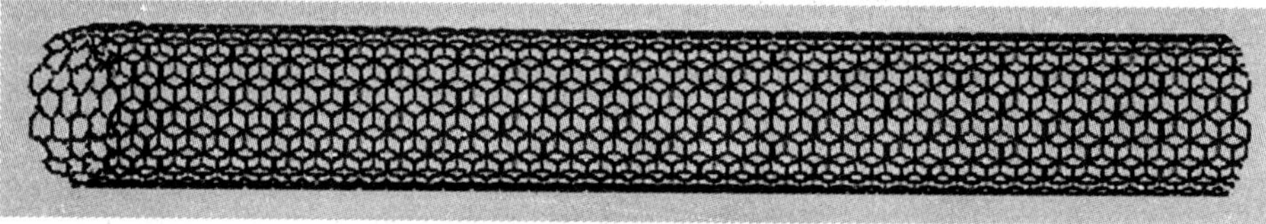

(A Nanotube)

10.1.3 Other Aromatic Systems

Many unsaturated cyclic compounds have exceptional properties that we now consider characteristic of "aromatic" systems. The following cases are illustrative:

The first three compounds (cyclic polyenes) of table 10.1 have properties associated with alkenes in general. Each reacts readily with bromine to give addition products, as do most alkenes. The thermodynamic change on introducing double bonds into the carbon atom ring is also typical of alkenes (a destabilization of 26 kcal/mol for each double bond). Conjugation offsets this increase in energy by a small amount (4-6 kcal/mol).

The remaining four compounds of Table 10.1 exhibit very different properties, and are considered aromatic. Benzene and pyridine are relatively unreactive with bromine, requiring heat and/or catalysts to force reaction, the result of which is substitution rather than addition. Furan and pyrrole react more rapidly with bromine, but they also give substitution products. This tendency to favour substitution rather than addition suggests that the parent unsaturated ring system has exceptional stability. Thermodynamic measurements support this conclusion. The enhanced stability, often referred to as **aromatic stabilization**, ranges (in the above cases) from a low of 16 kcal/mol for furan to 36 kcal/mol for benzene (Summarized in table 10.1).

Table 10.1: Summary of thermodynamic stability of some conjugated system

Compounds	*Structural formula*	*Reaction with Br_2*	*Thermodynamic stabilization*
1, 3-cyclo-pentadiene		Addition (0°C)	Slight
1, 3, 5-cyclo-heptatriene		Addition (0°C)	Slight
1, 3, 5, 7-cyclo-octatetraene		Addition (0°C)	Slight
Benzene		Substitution	Large
Pyridine	N	Substitution	Large
Furan	O	Substitution (0°C)	Moderate
Pyrrole	N–H	Substitution	Moderate

10.1.4 Factors Required for Aromaticity "The Hückel Rule"

1. A planar (or near planar) cycle of sp² hybridized atoms, the *p*-orbitals of which are oriented parallel to each other. These overlapping *p*-orbitals generate an array of π-molecular orbitals.
2. These π-orbitals are occupied by $4n + 2$ electrons (where n is an integer or zero). This requirement is known as **The Hückel rule**. All the aromatic compounds discussed above have 6 π-electrons ($n = 1$).

1, 3-cyclo-pentadiene and 1, 3, 5-cyclo-heptatriene both fail to meet the first requirement, since one carbon atom of each ring is sp³ hybridized and has no *p*-orbital. Cyclo-octatetraene fails both requirements, although it has a ring of sp² hybridized atoms. This molecule is not planar (a geometry that would have 135° bond angles). Angle strain is relieved by adopting a tub-shaped conformation; consequently, the *p*-orbitals can only overlap as isolated pairs, not over the entire ring. Furthermore, cyclo-octatetraene has 8 π-electrons, a number not consistent with the Hückel rule.

Benzene is the archetypical aromatic compound. It is planar, bond angles = 120°, all carbon atoms in the ring are sp² hybridized, and the pi-orbitals are occupied by 6 electrons. Pyridine is similar to benzene, but it is essential to recognize that the non-bonding electron pair on nitrogen occupies a sp² orbital lying in the plane of the ring, and is not part of the π-electron system. Thus, pyridine has 6 π-electrons and 2 non-bonding sp² electrons. Furan and pyrrole have five-membered rings, in which one atom is not carbon (a heteroatom) and has at least one pair of non-bonding valence shell electrons. By hybridizing this heteroatom to a sp² state, a *p*-orbital occupied by a pair of electrons and oriented parallel to the carbon *p*-orbitals is created. The resulting planar ring meets the first requirement for aromaticity, and the π-system is occupied by 6 electrons, 4 from the two double bonds and 2 from the heteroatom, thus satisfying the Hückel rule.

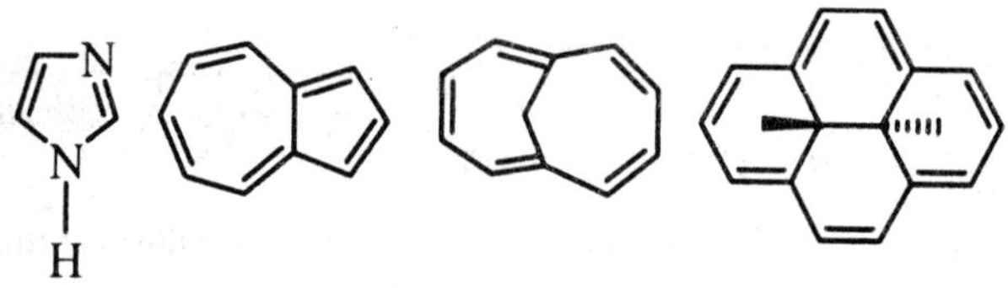

Four other examples of aromatic compounds are shown above. The first example is imidazole, a heterocycle having two nitrogen atoms. Note that only one of the nitrogen non-bonding electron pairs is used for the aromatic π-electron sextet. The other electron pair behaves similarly to the electron pair in pyridine. The second and third compounds have ten-membered rings, which are forced to be planar, or nearly planar, by the bridging bond or methylene group. These are 10 π-electron systems that fit the aromaticity requirements. Finally, the last example has a fourteen-membered ring with 7 double bonds. The bridging atoms and bonds force it to be planar. The number of π-electrons in this aromatic system is 14. Monocyclic compounds made up of alternating conjugated double bonds are called **annulenes**. Benzene and 1, 3, 5, 7-cyclooctatetraene are examples of annulenes; they are named [6] annulene and [8] annulene respectively, according to a general nomenclature system, in which the number of pi-electrons in an annulene is designated by a number in brackets. Some annulenes are aromatic (e.g. benzene), but many are not due to non-planarity or a failure to satisfy the Hückel rule. Non-planarity of a simple ring may be corrected by bridging bonds or atoms, as noted above.

Aromatic Ions

Carbanions and carbocations may also show aromatic stabilization. Some examples are:

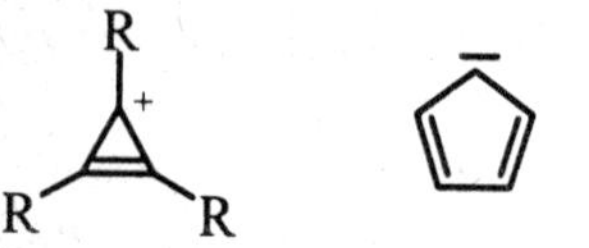

The three-membered ring cation has 2 π-electrons and is surprisingly stable, considering its ring strain. Cyclopentadiene is as acidic as ethanol, reflecting the stability of its 6 π-electrons conjugate base. Salts of cycloheptatrienyl cation (tropylium ion) are stable in water solution, again reflecting the stability of this 6 π-electrons cation.

Concept Builder Questions on Aromaticity

Example 1. A hydrocarbon has molecular formula C_6H_6. Draw structures of five isomers of this compound.

Solution

I II III IV V

Example 2. Treatment of compound (I) of the previous question with sulphuric acid affords benzene. Propose mechanism and explain the driving force for this reaction.

Solution

Driving forces for this reaction are:

(a) Release of significant amount of ring strain.

(b) Formation of stable ring with the property of aromatic compound.

Example 3. Suggest how each of the following empirical observations, suggest that benzene is not an alkene.

(a) Benzene does not react with Br_2.

(b) The x-ray crystal structure of benzene shows all C—C bonds to be equal in length.

(c) The heat of hydrogenation of benzene is 36 kcal mol^{-1} less than three times the heat of hydrogenation of cyclohexene.

Solution: (a) A normal alkene (such as ethylene) reacts readily with Br_2 to form addition product 1, 2-dibromo ethane. Because of resonance delocalization of π-electrons of benzene, it is inert to addition reaction. Addition of bromine to benzene will destroy the aromaticity which is not energetically favourable. Therefore, benzene does not behave like normal alkene in addition reaction.

(b) If the actual structure of benzene contained alternating single and double bonds, these bond would also alternate in length. The x-ray crystal structure does not agree with this prediction, so benzene cannot be a simple alkene.

(c) If benzene was a normal alkene, with three normal C=C bonds, then the heat of hydrogenation for benzene would be three times the heat of hydrogenation of cyclohexene. The fact that the heat of hydrogenation for benzene is 36 kcal mol^{-1} lower than three times the heat of hydrogenation of cyclohexene indicates that benzene is more stable than the expected stability. This stability arises due to involvement of three alternating double bonds in resonance delocalization. So, benzene cannot be a normal alkene.

Example 4. Discuss the resonance model prediction for the structure of benzene.

Solution: The resonance hybrid that best describe the structure of benzene is:

Evaluation of this resonance hybrid suggests the following structural features for benzene:

- All C—C bonds are equal, and half-way between a C—C and a C=C in length.
- All C—C—C bond angles are equal, and all H—C—C bond angles are equal at 120°.
- All C—H bond lengths are equal.
- Each carbon is sp^2 hybridised, forming a continuous loop of π-electron density. That the molecule is planar (required for resonance interaction of all six sets of p_z orbitals.

Example 5. 1, 3-cyclobutadiene has the same 1 : 1 carbon to hydrogen ratio as benzene, yet unlike benzene it is very reactive. Explain.

Solution: 1, 3-cyclo-butadiene is much more reactive than benzene, because it does not have the aromatic stability of benzene. 1, 3-cyclo-butadiene does not satisfy the Hückel $4n + 2$ π-electron rule required for aromaticity. Examination of the molecular orbital description of this molecule suggests that it should behave like a diradical, a species with two unpaired electrons that is generally very reactive.

Example 6. 1 , 3, 5, 7-cyclo-octatetraene has the same 1 : 1 carbon to hydrogen ratio as benzene, yet unlike benzene, it is very reactive. Explain.

Solution: 1, 3, 5, 7-cyclo-octatetraene is not a planar molecule. There is no overlap between p_z- orbitals on the terminal carbons of adjacent carbon-carbon double bonds. Because, these π-bonds cannot share electrons via resonance, these C=C groups behave as normal, isolated alkenes.

Example 7. List four structural features necessary for a molecule to be considered aromatic.

Solution: The four predictors for aromaticity are:

(i) The π-bonds of aromatic system must all lie within the cyclic structure.

(ii) Each atom in the cycle must have a *p*-orbital. Each atom must use atomic orbitals or be sp or sp^2 hybridised, forming a loop of p_z orbitals.

(iii) All p_z-orbitals in the loop must overlap. This usually means the p_z loop must be planar.

(iv) This orbital arrangement must result in a lowering of energy. This occurs when there are $4n + 2$ π-electrons in the loop, i.e. they must satisfy Hückel rule.

Example 8. Classify each of the structures shown following as aromatic or not aromatic. Explain.

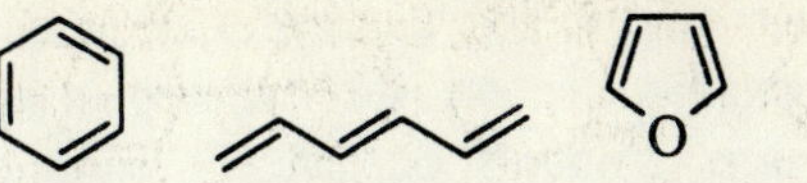

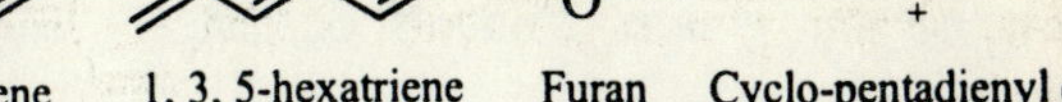

Benzene 1, 3, 5-hexatriene Furan Cyclo-pentadienyl cation

Solution: ***Benzene:*** All three π-bonds lie within a six membered ring. Each atom of this ring is sp^2 hybridised, forming a continuous overlapping loop of p_z orbitals. There is no reason for the ring not to be planar. It has 6 π-electrons and satisfy Hückel rule. Therefore, benzene is aromatic.

1, 3, 5-hexatriene: This molecule does not have a closed loop of p_z-orbitals, so it is not aromatic.

Furan: The oxygen must be sp^2 hybridised to contribute a p_z-orbitals, and so that the ring may be flat. Then we see that furan meets all the requirements for aromaticity, including six π-electrons. The second lone pair on oxygen is in sp^2 orbital, perpendicular to the π-electron cloud. This lone pair cannot be involved in the ring π-electron cloud.

Cyclo-pentadienyl cation: This carbocation has only four π-electrons, and cannot be aromatic.

Example 9. Cyclohexene and benzene are both six membered rings containing π-bonds. Catalytic hydrogenation of cyclohexene occurs readily at room temperature with a low pressure of hydrogen. Similar reaction of benzene requires heat, high pressure and a more reactive catalyst:

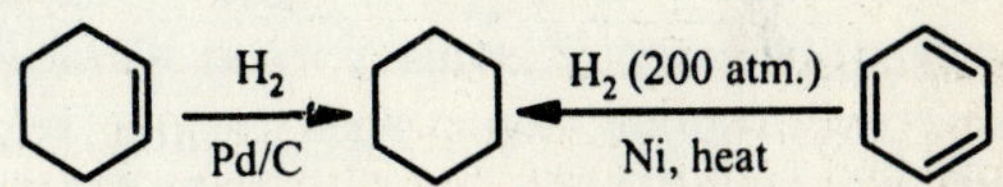

(a) Offer an explanation for this difference in reactivity.

(b) In one experiment, the hydrogenation of benzene was interrupted before all the benzene had reacted. Analysis of the reaction mixture gave the following composition: benzene 47%, 1, 3-cyclo-hexadiene 2%, cyclohexene 1%, and 50% cyclohexane. Account for the relative ratio of the three reduction products.

Solution: (a) Addition of hydrogen to benzene is energetically expensive, because it disrupts aromaticity. Thus, hydrogenation of benzene is slower than hydrogenation of cyclohexene and requires a more drastic condition for reduction.

(b) The product ratio can be taken as an indication of how quickly steps within the overall process occur. The low concentration of cyclo-hexadiene and cyclohexene mean that these are more reactive towards hydrogenation than benzene and are reduced quickly. The higher concentration of cyclo-hexadiene compared to cyclohexene is expected, as reduction of cyclo-hexadiene mean a loss of conjugation. There is no such loss for reduction of cyclohexene, so it reacts more rapidly than cyclo-hexadiene. Benzene to cyclo-hexadiene is the slowest step in the process, so the greater amounts of benzene remains unreacted.

Example 10. Classify the following as aromatic or not aromatic. Briefly explain your reasoning.

(a) 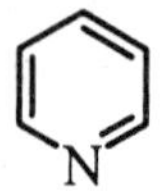 Pyridine (b) 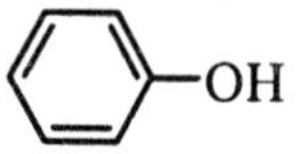 Phenol

(c) 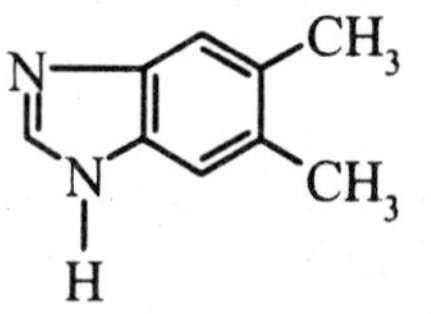 Dimethylbenzimidazole (d)  Cyclopentadienyl anion

(e)  Imidazole (f) 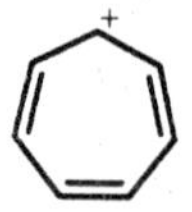 Tropylium cation

Solution: (a) *Pyridine* has $4n + 2\pi$-electrons, from three double bonds. This molecule is planar, and has a closed loop of p_z-orbitals, so we expect pyridine to be aromatic.

(b) *Phenol:* The —OH group of phenol is acyclic, so the lone pairs of the oxygen atom is not part of aromaticity. This leaves only a normal atomatic benzene ring, so phenol is aromatic.

(c) *Dimethylbenzimidazole:* The nitrogen bearing the hydrogen has four attachments, and thus would normally be sp^3 hybridised so as to have the least electronic repulsion. An sp^3 nitrogen atom cannot participate in aromaticity because it has no *p*-orbital. If this nitrogen atom was sp^2 hybridised there would be more electronic repulsion but the molecule could be aromatic. Aromaticity usually wins out, so this nitrogen is sp^2 hybridised and the closed loop of *p*-orbitals requirement is met. The closed loop contain four double bonds, each of which contribute 2π-electrons, for a total of 8. This is not a $4n + 2$ number. If the sp^2 nitrogen lone pair is in a *p*-orbital, it can participate in the π-electron cloud as well. Four double bonds plus one lone pair gives a total of 10 delocalised electrons satisfying $4n + 2\pi$-electron rule. Therefore, this molecule appears to be aromatic.

(d) *Cyclcpentadienyl anion:* The cyclopentadienyl anion has $4n + 2\pi$-electron and satisfy all the requirements for aromaticity. Therefore this anion is aromatic.

(e) *Imidazole:* This molecule has a continuous loop of *p*-orbitals because both nitrogen would be sp^2 hybridised. This molecule has 6π-electrons and satisfy all other requirements needed for aromaticity. Therefore imidazole is aromatic.

(f) *Tropylium cation:* Every atom in the ring is sp^2 hybridised, so the structure has a closed loop of *p*-orbitals. The ion has 6π-electrons and it is aromatic.

Example 11. Thiophene has all the requirements for aromaticity, yet it behaves much more like a non-aromatic 1, 3-diene than an aromatic molecule. Offer an explanation.

Thiophene

Solution: Aromaticity can be considered to be a resonance effect, resulting from overlap of *p*-orbitals. Carbon and sulphur *p*-orbitals do not overlap well, because carbon is a 2nd row element and sulphur is a 3rd row element. This poor overlap reduces aromaticity in thiophene. Thiophene behaves more like a 1, 3-diene with a sulphur atom at each end, and less like an aromatic molecule.

Example 12. Which has more acidic proton and why, 1, 3-cyclo-butadiene or 1, 3-cyclo-pentadiene?

Solution: 1, 3-cyclo-pentadiene, since its conjugate base is aromatic, more stable.

Example 13. Discuss the aromaticity of the following species:

I 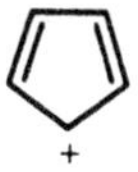 II  III

Solution: (I) and (III) are aromatic while (II) is not.

10.2 BENZENE: PREPARATION/ISOLATION AND ITS REACTIONS

10.2.1 Preparation

I. Benzene was first isolated by Faraday from cylinders of compressed illuminating gas obtained from pyrolysis of whale oil. Hofman separated benzene from coal-tar and this is still a source of benzene and its derivatives.

Light oil fraction collected from destructive distillation of coal is first washed with NaOH (to remove acids) and then with water. The washed oil is now distilled. The fraction collected up to 110°C is known as 90% benzol (70% benzeze, 24% toluene and some xylene). Careful fractionation of benzol yields pure benzene.

II. From acetylene: Benzene was first synthesized by passing acetylene gas through a red hot copper tube as:

$3C_2H_2 \xrightarrow{\text{Red hot Cu}}$ (benzene) and other products.

III. By decarboxylation of benzoic acid, phthalic acid, etc. When aromatic acids are heated with soda lime, decarboxylation occur as:

Benzoic acid (COOH) $\xrightarrow[\text{Heat}]{\text{CaO/NaOH}}$ Benzene $\xleftarrow[\text{Heat}]{\text{CaO/NaOH}}$ Phthalic acid (COOH, COOH)

IV. From petroleum

(a) *Hydroforming or catalytic reforming:* This method is based on dehydrogenation, cyclization and isomerization reactions and aromatic compounds obtained contain the equal numbers of carbons as the starting material. Hydroforming is carried out under pressure, at high temperature in presence of Cr_2O_3/Al_2O_3 catalyst as:

n-Hexane $\xrightarrow[\text{P/ 500°C}]{Cr_2O_3/Al_2O_3}$ Benzene $+ 4H_2$

n-Heptane $\xrightarrow[\text{P/ 500°C}]{Cr_2O_3/Al_2O_3}$ Toluene $+ 4H_2$

n-Octane $\xrightarrow[\text{P/ 500°C}]{Cr_2O_3/Al_2O_3}$ Three isomers of xylene + Ethyl benzene

This reaction is commonly known as aromatization reaction.

(b) *High temperature cracking in presence of a catalyst:* The charging stocks are cracked at about 650–680°C in tube packed with metallic dehydrogenation catalyst (same as those used in hydroforming). By this technique, benzene, toluene, xylene, naphthalene, anthracene and many other polynuclear hydrocarbons have been isolated.

10.2.2 Properties and Reactions of Benzene

I. General

Benzene is a colourless liquid with a peculiar smell. It is highly inflammable burning with a smoky flame.

II. Reduction reaction of benzene

Benzene can be reduced completely into cyclohexane or partially into cyclo-hexadiene choosing suitable catalyst.

(a) *Complete reduction:* Benzene on treatment with H_2 in presence of Pt, Ni catalyst, reduced into cyclohexane as:

Benzene $\xrightarrow{H_2/Pt}$ Cyclohexane   Styrene $\xrightarrow{H_2/Pt}$ Ethylcyclohexane

If there is any unsaturated side chain, it is also reduced with H_2/Pt as shown above. Unsaturated side chain is more reactive towards reduction reaction and when a slightly milder reducing agent is used like copper chromite, copper-bariumchromite, only side chain is reduced.

Styrene $\xrightarrow[H_2]{\text{Cu-chromite}}$ Ethylbenzene

(b) *Birch reduction (Partial reduction):* Benzene when treated with sodium metal in liquid ammonia in presence of ethanol, partial hydrogenation occur producing 1,4-cyclohexadiene:

Benzene $\xrightarrow[\text{EtOH}]{\text{Na/liq } NH_3}$ 1,4-Cyclohexadiene

Mechanism of birch reduction

$Na \longrightarrow Na^+ + e$

Benzene $\xrightarrow{e}$ radical anion (H, H) $\xrightarrow{\text{EtOH}}$ (H, H; H) $\xrightarrow{e}$ carbanion (H, H; H) $\xrightarrow{\text{EtOH}}$ 1,4-cyclohexadiene (H, H; H, H)

If the benzene ring possesses a substituent, electronic nature of the substitutent decides the site of hydrogenation. If the substituent is electron withdrawing group, carbon of the ring bearing that substituent is reduced:

Nitrobenzene (NO_2) $\xrightarrow{e}$ radical anion (NO_2; H) $\xrightarrow{\text{EtOH}}$ (O_2N, H; H) $\xrightarrow{e}$

Negative charge is stabilized by electron withdrawing group

(O_2N, H; H) $\xrightarrow{\text{EtOH}}$ (O_2N, H; H, H)

If the ring possesses an electron donating group, carbon, next to the carbon bearing that group, is reduced in birch reduction:

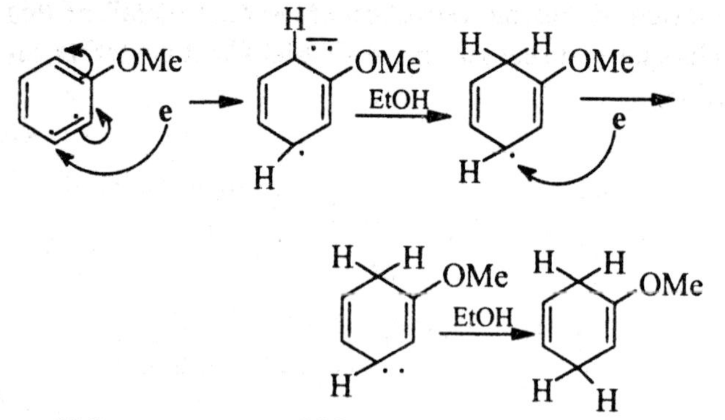

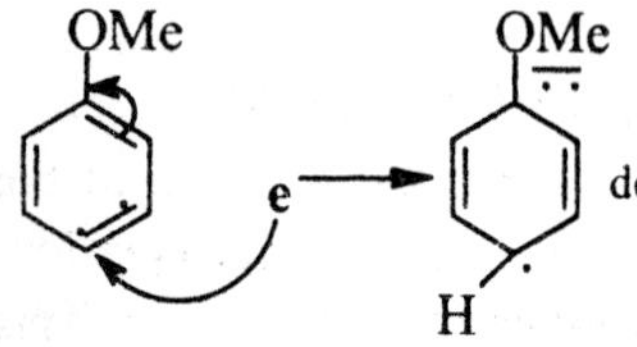

Negative charge is destabilized by +R effect of -OMe (Not formed)

III. Friedal Craft's Reaction

This reaction involves incorporation of an alkyl or acyl group in the benzene ring in presence of a catalyst.

(a) *Friedal Craft's alkylation*: Benzene or derivative of benzene on treatment with an alkyl halide (alcohols, alkenes are also used) in presence of a Lewis acid catalyst undergoes alkylation on the benzene ring.

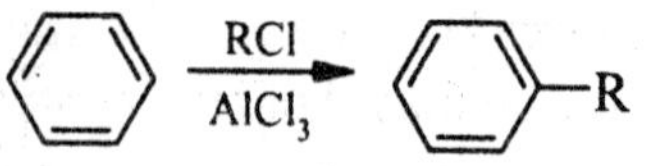

Friedal Craft alkylation

Mechanism

Step 1: R—Cl $\xrightarrow{AlCl_3}$ $\overset{-}{AlCl_4}$ + $\overset{+}{R}$

A carbocation intermediate is formed in the first step, which may undergo rearrangement to more stable carbocation.

Step 2:

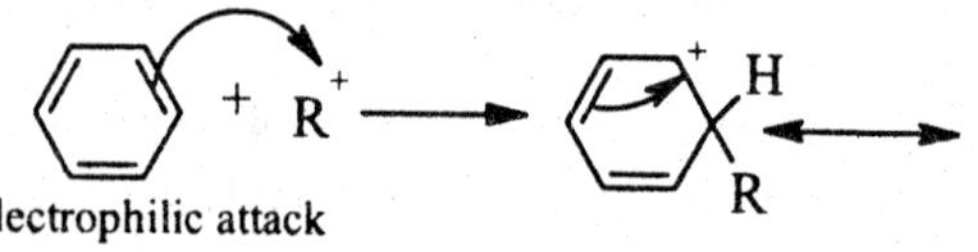

Electrophilic attack of carbocation

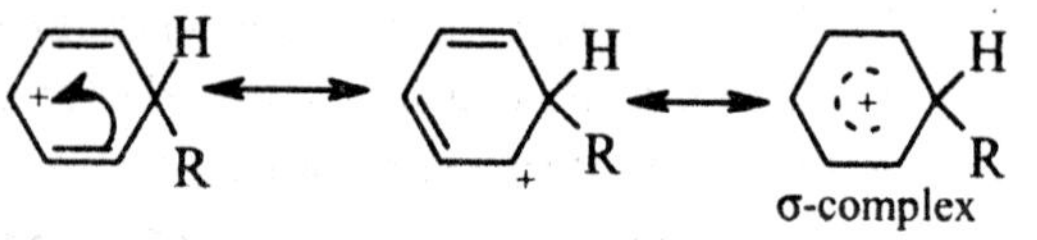

σ-complex

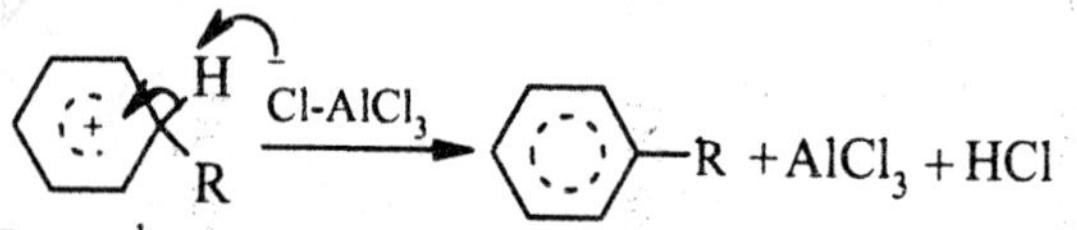

σ-complex

Reactivity of alkylation reaction depends on several factors as described below:

(i) *Strength of a catalyst:* A stronger Lewis acid catalyst is more efficient in bringing about the alkylation reaction and general order of reactivity of common Lewis acid is:

$$AlCl_3 > BF_3 > SbCl_5 > FeCl_3 > SnCl_4 > ZnCl_2$$

(ii) *Nature of alkyl-halides:* For a given halogen atom, the order of reactivity follow the order of stability of carbocations formed from the alkyl halides as:

$$t\text{-R-X} > 2° \text{ R-X} > 1° \text{ R-X}$$

For a given alkyl group, the order of reactivity is:

$$\text{R-F} > \text{R-Cl} > \text{R-Br} > \text{R-I}$$

The above order of reactivity is based on the relative basic strength of X^- and their affinity for a Lewis acid. A stronger base (like F^-) has greater affinity for Lewis acid and carbocation is formed more easily from this alkyl halide.

(iii) *Nature of substituent:* As it will be discussed in detail in the subsequent topic "Aromatic electrophilic substitution reaction" an electron donating substituent on the benzene nucleus increases the reactivity and electron withdrawing substituent on the benzene nucleus decreases the reactivity towards Friedal Craft reaction, since it is also an example of aromatic electrophilic substitution reaction. Due to this reason, nitrobenzene, benzoic acid, $C_6H_5NR_3^+$, etc. are almost inert in Friedal Craft reaction.

Presence of some strongly electron donating group on the ring also debar it from Friedal Craft reaction, e.g. —OH, —NH_2, etc. This is due to the reason that these groups are themselves basic in nature, combines with Lewis acid catalyst employed in the reaction.

(iv) Some heterocyclic aromatic compounds are also reluctant to Friedal Craft reaction, e.g. pyridine, imidazole, etc. fail in Friedal Craft reaction. This is due to the basic nature of nitrogen in the ring, which combines with Lewis acid catalyst.

Example 14. Both pyridine and pyrrole are nitrogen containing heterocyclic aromatic compounds, pyrrole undergoes Friedal Craft reaction comfortably while pyridine does not. Explain.

Solution: Both pyridine and pyrrole has a nitrogen atom in the aromatic ring, but the lone pair of electron on nitrogen in pyrrole is part of aromatic sextet, not available for donation to a Lewis acid catalyst while lone pair on nitrogen in pyridine is not involved in resonance, available for donation to Lewis acid catalyst. Hence, lone pair of electron on nitrogen in pyrrole does not interfere in its Friedal Craft reaction while the same reaction is interfered by lone pair of electron on nitrogen in pyridine.

Example 15. Which is more reactive in Friedal Craft reaction, Pyrrole of Furan?

Solution: Furan has less resonance energy compared to pyrrole, lone pair on oxygen in furan is more available for reaction with a Lewis acid catalyst, hence less reactive in FC reaction.

Example 16. Which is more reactive in Friedal Craft reaction: Furan or Thiophene.

Solution: Thiophene has less resonance energy than furan, lone pair on sulphur in thiophene is more available for reaction with Lewis acid catalyst, less reactive in FC reaction.

Rearrangement of Carbocations in Friedal Craft Alkylation Reaction

The structure of alkyl halide plays an important role in determining the fate of end product in Friedal Craft alkylation reaction. It is easier to introduce a methyl, ethyl group isopropyl group, tertiary butyl group but usually difficult to introduce *n*-propyl or *n*-butyl group. The reason for this is formation of carbocation intermediate, which has characteristic property of rearrangement into more stable carbocation either by hydride (H^-) shift or by alkyl (R^-) shift. Following examples show these drawbacks:

(i) Benzene + $CH_3CH_2CH_2Cl/AlCl_3 \rightarrow C_6H_5CH(CH_3)_2$ isopropyl benzene

This is due to formation of $CH_3CH_2CH_2^+$ (1° carbocation) followed by rearrangement into a more stable $(CH_3)_2CH^+$ secondary carbocation.

(ii) $C_6H_6 + (CH_3)_2CHCH_2Cl/AlCl_3 \rightarrow C_6H_5C(CH_3)_3$ due to the following rearrangement reaction:

$$H_3C-CH(CH_3)-CH_2Cl \xrightarrow{AlCl_3} H_3C-CH(CH_3)-\overset{+}{C}H_2$$

(a primary carbocation)

$$H_3C-\overset{+}{C}(CH_3)-CH_3$$

More stable (a tertiary carbocation)

• Carrying the reaction in cold conditions can minimize the rearrangement of carbocation:

$$C_6H_6 + CH_3CH_2CH_2Cl \xrightarrow[\text{Cold}]{AlCl_3} C_6H_5CH_2CH_2CH_3$$

Even though low temperature offer a better yield of unrearranged product, rearrangement product is always produced up to some extent. Rearrangement can be avoided completely by an indirect two step synthesis-acylation followed by ketonic reduction of product (detail of this technique is discussed in the Friedal Craft acylation reaction).

$$C_6H_6 + CH_3CH_2COCl \xrightarrow{AlCl_3} C_6H_5COCH_2CH_3 \xrightarrow[\text{Wolf-Kischner reduction}]{N_2H_4,\ \overline{O}H\ \text{Heat}} C_6H_5CH_2CH_2CH_3$$

Poly Alkylation in Friedal Craft's Reaction

It is not always possible to control the reaction at the required stage and over alkylation, i.e. dialkylation, trialkylation always accompany the monoalkylation. This occurs due to the reason that alkyl benzene is always more reactive than benzene towards electrophilic aromatic substitution reaction:

$$C_6H_6 \xrightarrow[AlCl_3]{RCl} C_6H_5R$$

Product is more reactive than benzene, involves in further alkylation with $RCl/AlCl_3$, giving dialkyl substitution product

• Over alkylation can be minimized by taking excess of benzene and removing alkyl benzene continuously.

Example 17. In Friedal Craft alkylation reaction, it is always recommended to use excess of benzene and the product alkyl benzene should be continuously removed. Why?

Solution: Excess of benzene increases the probability of collision of carbocation intermediate with benzene rather with alkyl benzene (product) and continuous removal of alkyl benzene further decreases the chances of over alkylation.

Disproportionation of Products in Friedal Craft Reaction

In Friedal Craft alkylation reaction, an alkyl group may be removed from one molecule and transferred to another or may migrate from one position to other position as:

$$C_6H_5Et + C_6H_5Et \xrightarrow{AlCl_3} C_6H_6 + C_6H_4Et_2$$

and

$$o\text{-}C_6H_4R_2 \xrightarrow{AlCl_3} m\text{-}C_6H_4R_2$$

(a) Friedal Craft Reactions with other reagents

(i) *With alcohols:* Benzene when treated with alcohols in presence of a Lewis acid or concentrated sulphuric acid catalyst, alkyl benzene is produced as:

$$C_6H_6 + ROH + BF_3 \rightarrow C_6H_5—R$$

$$H_3C—C(CH_3)(H)—CH_2OH \xrightarrow{BF_3} H_3C—C(CH_3)(H)—\overset{+}{C}H_2$$

(a primary carbocation)

$$\longleftrightarrow H_3C—\overset{+}{C}(CH_3)—CH_3 \xrightarrow{C_6H_6} C_6H_5C(CH_3)_3$$

More stable
(a tertiary carbocation)

(ii) *With alkenes:* Alkene in presence of an acid catalyst attack benzene forming alkyl benzene as:

$$(CH_3)_2C{=}CH_2 + H^+ \rightarrow (CH_3)_3C^+$$

This carbocation attacks on benzene in the same manner as in case of other Friedal Craft alkylation reaction.

(b) Friedal Craft Acylation Reaction

Benzene or its derivative when treated with an acid chloride or acid anhydride in presence of a Lewis acid catalyst, acyl group (R—C=O) is introduced into the benzene nucleus as:

$$C_6H_6 \xrightarrow[AlCl_3]{RCOCl \text{ or } (RCO)_2O,\ AlCl_3} C_6H_5COR$$

Mechanism

With acid chloride

$$R—\overset{O}{\overset{\|}{C}}—Cl \xrightarrow[-AlCl_4^-]{AlCl_3} R—\overset{+}{C}{=}\ddot{O} \longleftrightarrow R—C{\equiv}\overset{+}{O}$$

(a resonance stabilized acylium ion)

$$C_6H_6 + R—\overset{+}{C}{=}\ddot{O} \longrightarrow [C_6H_6(H)(COR)]^+ \xrightarrow{AlCl_4^-} C_6H_5COR + AlCl_3 + HCl$$

With acid anhydride

$$R—\overset{O}{\overset{\|}{C}}—O—\overset{O}{\overset{\|}{C}}—R \xrightarrow{AlCl_3} R—\overset{+}{C}{=}\ddot{O} \longleftrightarrow Cl_3\bar{A}l(RCOO) + R—C{\equiv}\overset{+}{O}$$

(a resonance stabilized acylium ion)

$$C_6H_6 + R—\overset{+}{C}{=}\ddot{O} \longrightarrow [C_6H_6(H)(COR)]^+ \xrightarrow{Cl_3\bar{A}l(RCOO)} C_6H_5COR + AlCl_3 + RCOOH$$

Important points regarding acylation reaction

(i) No rearrangement is observed due to formation of a resonance stabilized acylium ion.

(ii) Over acylation does not occur due to deactivation of ring by acyl group for electrophilic substitution reaction.

(iii) This is a convenient means of introducing long straight chain alkyl group in benzene ring, which otherwise undergo rearrangement in direct alkylation reaction, e.g. *n*-pentyl benzene can conveniently be synthesized via acylation reaction as:

$$C_6H_6 + CH_3CH_2CH_2CH_2COCl \xrightarrow{AlCl_3} C_6H_5CO(CH_2)_3CH_3 \xrightarrow[\text{Heat, } N_2H_4/NaOH]{Zn(Hg)/HCl} C_6H_5(CH_2)_4CH_3$$

Example 18. Complete the following reactions:

(a) C_6H_6 + succinic anhydride $\xrightarrow{AlCl_3} A \xrightarrow{HF} B \xrightarrow[NaOH/heat]{N_2H_4} C \xrightarrow[Al_2O_3/\Delta]{Cr_2O_3} D$

(b) 2 $C_6H_5CH_2CH{=}CH_2 \xrightarrow{H^+}$

(c) 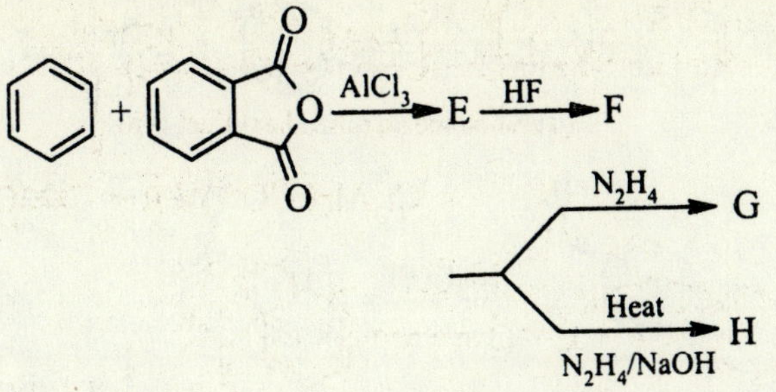 Benzene + phthalic anhydride $\xrightarrow{AlCl_3}$ E $\xrightarrow{HF}$ F

$\xrightarrow{N_2H_4}$ G

$\xrightarrow[N_2H_4/NaOH]{Heat}$ H

Solution

(a) 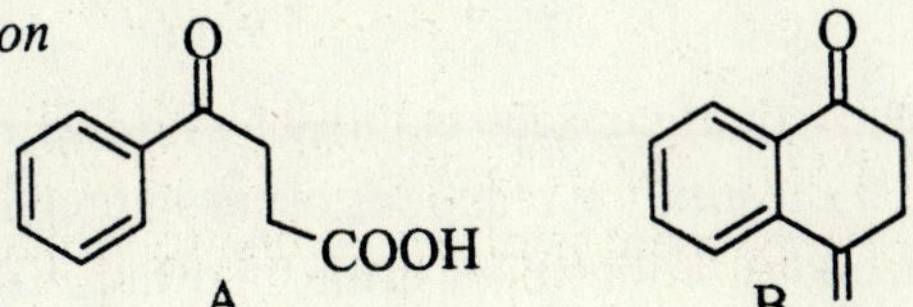 A: COOH; B: O, O

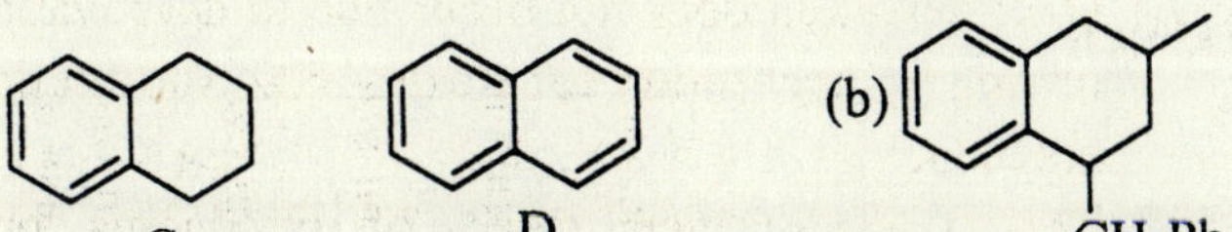

C D (b) CH$_2$Ph

(c) 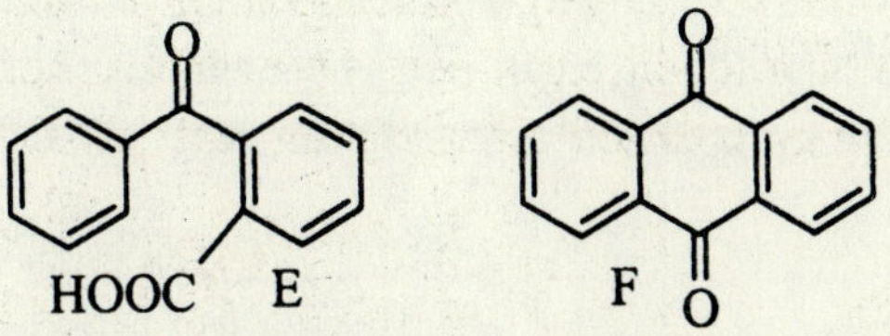 HOOC E F O

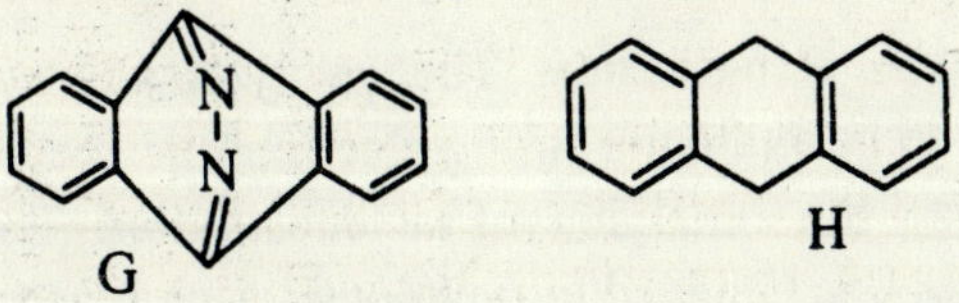

G H

IV. Wurtz-Fittig reaction

Homologues of benzene can be prepared by warming an etheral solution of aryl halide and alkyl halide with sodium metal:

$$ArBr + RBr + Na/heat \rightarrow Ar—R + 2NaBr$$

Biphenyl and some R—R are also produced as by-products.

Example 19. Write products formed in the following Friedal Craft reaction:

(a) 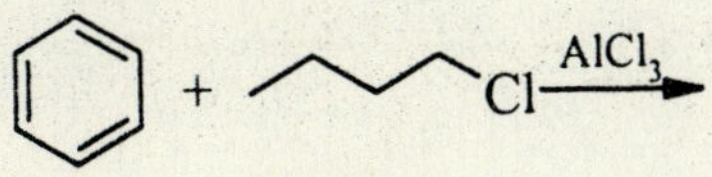 Benzene + Cl $\xrightarrow{AlCl_3}$

(b) 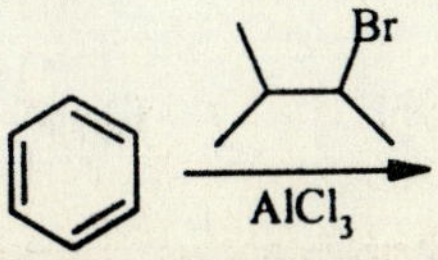 Benzene $\xrightarrow[AlCl_3]{Br}$ (c) 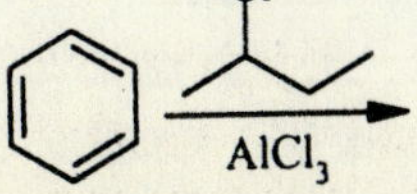 Benzene $\xrightarrow[AlCl_3]{Cl}$

(d) 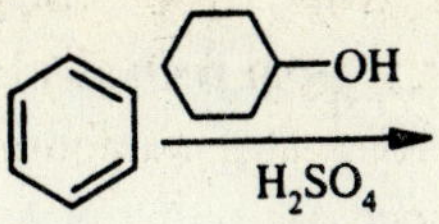 Benzene $\xrightarrow[H_2SO_4]{\text{cyclohexyl}-OH}$ (e) 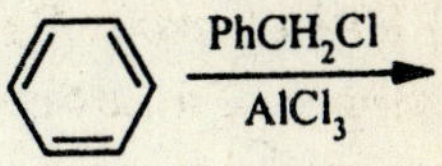 Benzene $\xrightarrow[AlCl_3]{PhCH_2Cl}$

(f) 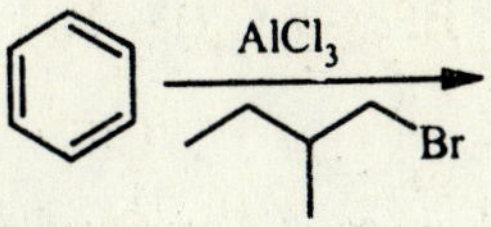 Benzene $\xrightarrow[Br]{AlCl_3}$

Solution

(a) 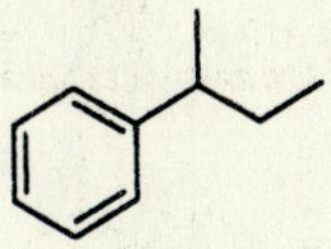 (b)

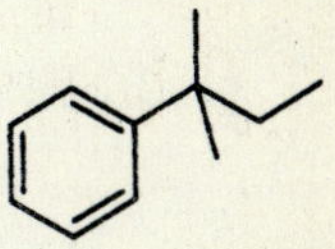

(c) 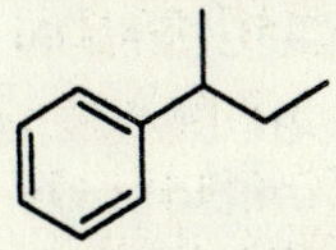 (d)

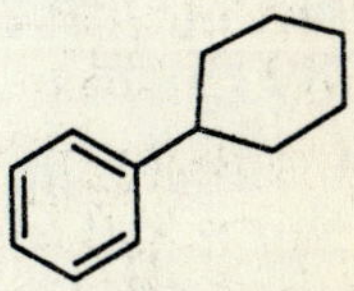

(e) 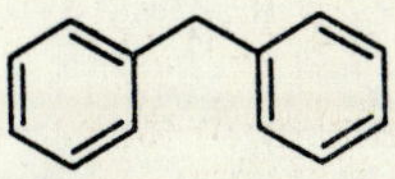 (f)

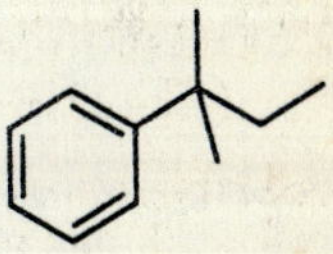

Example 20. Device synthesis of the following compounds starting from benzene and other reagents needed:

(a) 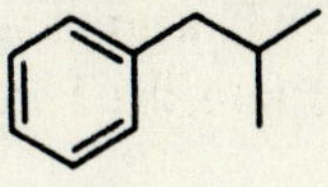 (b)

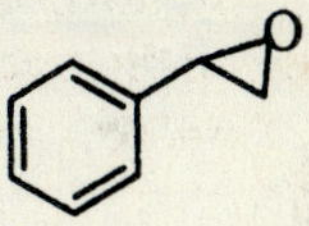

(c) 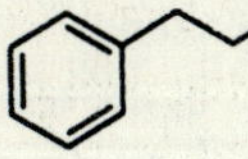 (d) 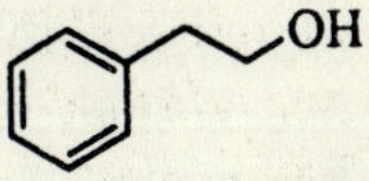 OH (e) 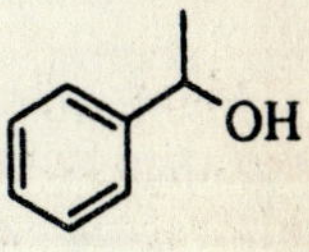 OH

Solution

(a) 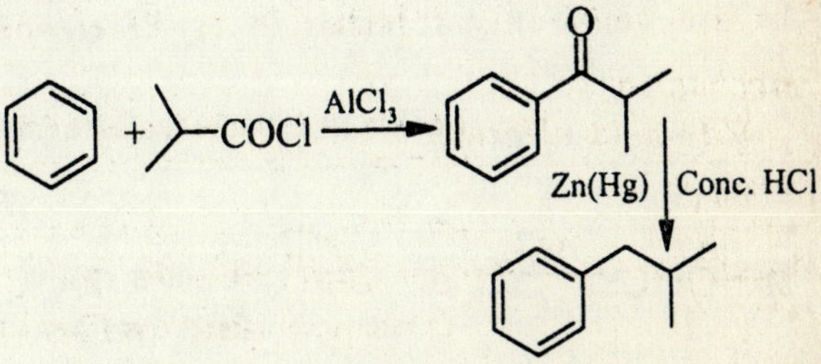 Benzene + COCl $\xrightarrow{AlCl_3}$ O $\xrightarrow{Zn(Hg) \mid Conc. HCl}$

(b) 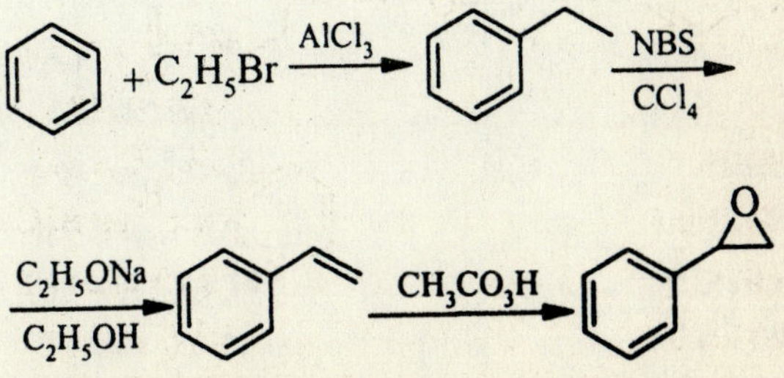 Benzene + C_2H_5Br $\xrightarrow{AlCl_3}$ $\xrightarrow[CCl_4]{NBS}$

$\xrightarrow[C_2H_5OH]{C_2H_5ONa}$ $\xrightarrow{CH_3CO_3H}$ O

(c) + $AlCl_3$ → Cl; Zn(Hg) Conc.HCl

(e) + C_2H_5Br $\xrightarrow{AlCl_3}$ $\xrightarrow[CCl_4]{NBS}$ Br $\xrightarrow{aq.\ KOH}$ OH

(d) + Br_2/Fe → Br $\xrightarrow{Mg/Et_2O}$ O $\xrightarrow{}$ $\xrightarrow{H_3O^+}$

10.3 AROMATIC ELECTROPHILIC SUBSTITUTION REACTIONS: A GENERAL DISCUSSION

10.3.1 Substitution Reactions of Benzene and Other Aromatic Compounds

The remarkable stability of the unsaturated hydrocarbon benzene has been discussed in an earlier section. The chemical reactivity of benzene contrasts with that of the alkenes in that substitution reactions occur in preference to addition reactions, as illustrated in the following diagram (some comparable reactions of cyclohexene are shown in the box).

Br; $Br_2 + FeBr_3$ Heat; HO_3S; H_2SO_4 Heat; Br_2 → No reaction; H_2SO_4 → No reaction; $KMnO_4$ → No reaction

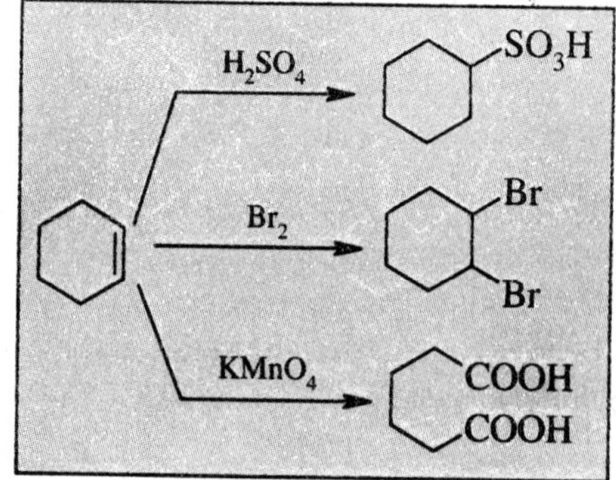

Many other substitution reactions of benzene have been observed, the five most useful are listed in the table (chlorination and bromination are the most common halogenation reactions). Since the reagents and conditions employed in these reactions are electrophilic, these reactions are commonly referred to as ***Electrophilic Aromatic Substitution***. The catalysts and co-reagents serve to generate the strong electrophilic species needed to effect the initial step of the substitution. The specific electrophile believed to function in each type of reaction is listed table 10.2.

Table 10.2: Some typical aromatic electrophilic reactions and electrophiles

Reaction type	*Typical equation*						*Electrophile $E^{(+)}$*
Halogenation:	C_6H_6	+	Cl_2 and heat $FeCl_3$ catalyst	→	C_6H_5Cl Chlorobenzene	+ HCl	$Cl^{(+)}$ or $Br^{(+)}$
Nitration:	C_6H_6	+	HNO_3 and heat H_2SO_4 catalyst	→	$C_6H_5NO_2$ Nitrobenzene	+ H_2O	$NO_2^{(+)}$
Sulfonation:	C_6H_6	+	$H_2SO_4 + SO_3$ and heat	→	$C_6H_5SO_3H$ Benzenesulfonic acid	+ H_2O	$SO_3H^{(+)}$
Alkylation: Friedel-Crafts	C_6H_6	+	R—Cl and heat $AlCl_3$ catalyst	→	C_6H_5—R An Arene	+ HCl	$R^{(+)}$
Acylation: Friedel-Crafts	C_6H_6	+	RCOCl and heat $AlCl_3$ catalyst	→	C_6H_5COR An Aryl Ketone	+ HCl	$RCO^{(+)}$

I. A Mechanism for Electrophilic Substitution Reactions of Benzene

A two-step mechanism has been proposed for these electrophilic substitution reactions. In the first, slow or rate-determining step, the electrophile forms a sigma-bond to the benzene ring, generating a positively charged **benzenonium intermediate**. In the second, fast step, a proton is removed from this intermediate, yielding a substituted benzene ring. The following four-part illustration shows this mechanism for the bromination reaction.

Bromination of Benzene—An Example of Electrophilic Aromatic Substitution

+ Br—Br ⟶ No reaction

Weak nucleophile

Moderate electrophile

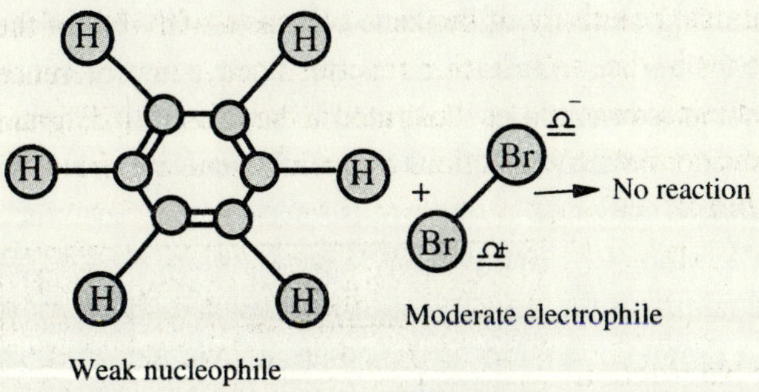

Bromine molecule is polarized so that one end is positively charged and other end is negatively charged. Although the positive end reacts easily with alkene and diene, but fail to react with more stable and weaker nucleophilic π-electron system of benzene.

Bromination in Presence of a Catalyst

Bromination of benzene ring can be done in presence of a Lewis acid catalyst and mechanism can be explained as follows:

$$Br_2 + FeBr_3 \longrightarrow Br^+ + FeBr_4^-$$

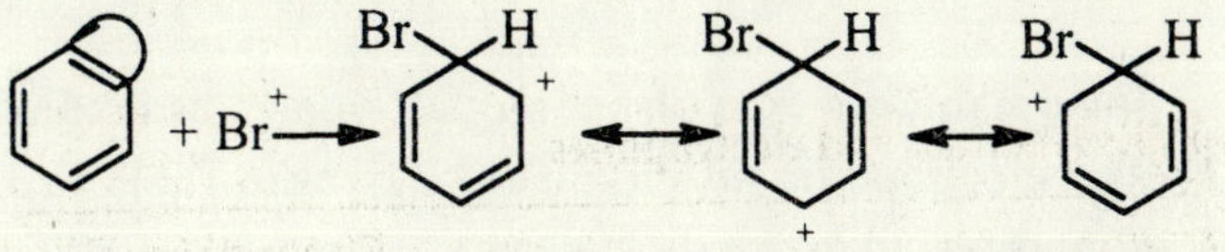

Resonance stabilized benzenonium ion

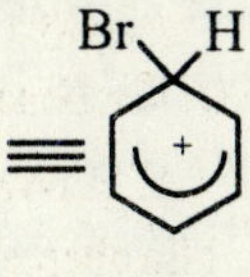

σ-complex

In the fast, final step, this σ-complex deprotonate to revert back to aromaticity as:

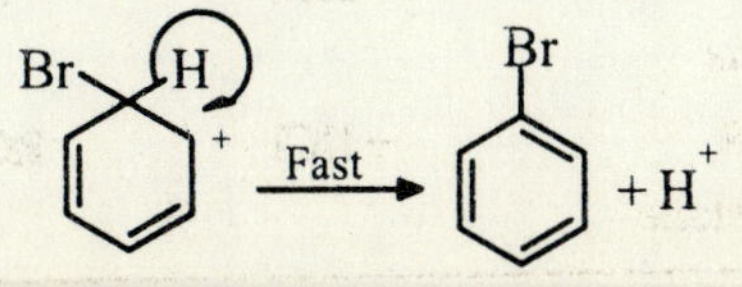

Fast ⟶ $+ H^+$

The proton is not set free as such but is removed by some base present, in the present case, it is bromide ion. The carbocation intermediate in electrophilic aromatic substitution (the benzenonium ion) is stabilized by charge delocalization (resonance), so it is not subject to rearrangement. The energetic advantage of reforming an aromatic ring leads to exclusive reaction by proton loss.

II. Substitution Reactions of Benzene Derivatives "Orientation and Reactivity"

When substituted benzene compounds undergo electrophilic substitution reactions of the kind discussed above, two related features must be considered:

(a) Relative reactivity

The first is the relative reactivity of the compound compared with benzene itself. Reactivity relates to rate of electrophilic substitution reaction of substituted benzene compared to unsubstituted benzene. If the rate of substitution in a substituted benzene is more than benzene it self, then the substitutent already present in the ring is known as activating group or **activator**. Examples of some common activating groups are $—NH_2$, —OH, —OR, —R, etc.

On the other hand if the rate of substitution in a substituted benzene is less than the benzene itself, the group already present in the ring is known as deactivating group or **deactivator**, e.g. $—NO_2$, —COOH, $—SO_3H$, —COR, etc.

Experiments have shown that substituents on a benzene ring can influence reactivity in a profound manner. For example, a hydroxy or methoxy substituent increases the rate of electrophilic substitution about ten thousand fold, in contrast, a nitro substituent decreases the ring's reactivity by roughly a million. This **activation** or **deactivation** of the benzene ring toward electrophilic substitution may be correlated with the electron donating or electron withdrawing influence of the substituents, as measured by molecular dipole moments. In the following diagram, we see that electron donating substituents activate the benzene ring toward electrophilic attack, and electron withdrawing substituents deactivate the ring (make it less reactive to electrophilic attack).

Activating substituents

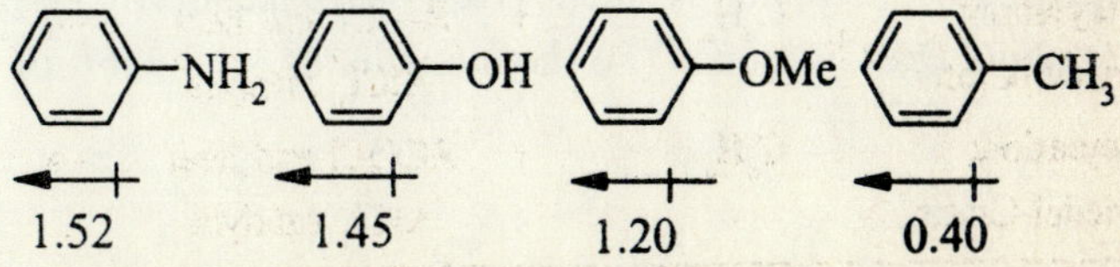

$C_6H_5—NH_2$	$C_6H_5—OH$	$C_6H_5—OMe$	$C_6H_5—CH_3$
1.52	1.45	1.20	0.40

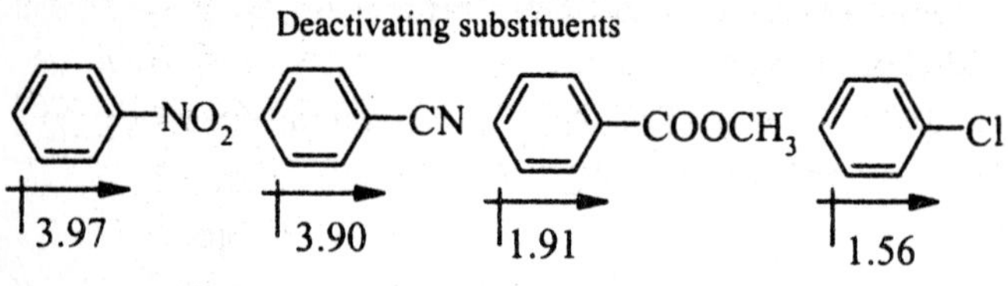

Deactivating substituents

C_6H_5—NO_2 (3.97), C_6H_5—CN (3.90), C_6H_5—$COOCH_3$ (1.91), C_6H_5—Cl (1.56)

The influence a substituent that it exerts on the reactivity of a benzene ring may be explained by the interaction of two effects:

The first is the **inductive effect** of the substituent. Most elements other than metals and carbon have a significantly greater electro negativity than hydrogen. Consequently, substituents in which nitrogen, oxygen and halogen atoms form sigma-bonds to the aromatic ring exert an inductive electron withdrawal, which deactivates the ring (left-hand diagram below). The second effect involves electron donation or withdrawal to or from the benzene ring as a result of **conjugation**. If the atom bonded to the ring has one or more non-bonding valence shell electron pairs, as do nitrogen, oxygen and the halogens, electrons may flow into the aromatic ring by *p*-pie conjugation (resonance), as in the right-hand diagram. Finally, polar double and triple bonds conjugated with the benzene ring may withdraw electrons, as in the bottom of diagram. Note, that in the resonance examples all the contributors are not shown. In both cases the charge distribution in the ring is greatest at sites ortho and para to the substituent.

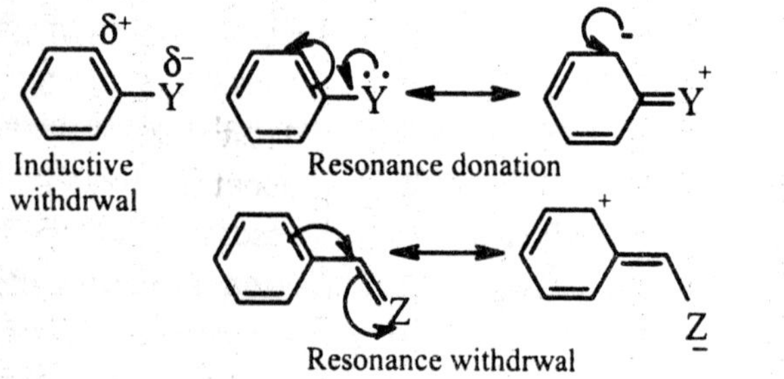

δ^+ δ^- Y

Inductive withdrwal

Resonance donation

Resonance withdrwal

In the case of the nitrogen and oxygen activating groups displayed in the top row of the previous diagram, electron donation by resonance dominates the inductive effect and these compounds show exceptional reactivity in electrophilic substitution reactions. Although halogen atoms have non-bonding valence electron pairs that participate in *p*-pie conjugation, their strong inductive effect predominates, and compounds such as chlorobenzene are less reactive than benzene. The three examples on the left of the bottom row (in the same diagram) are examples of electron withdrawal by conjugation to polar double or triple bonds, and in these cases the inductive effect further enhances the deactivation of the benzene ring. Alkyl substituents such as methyl increase the nucleophilicity of aromatic rings in the same fashion as they act on double bonds.

(b) Orientation

The second factor that becomes important in reactions of substituted benzenes concerns the site, at which electrophilic substitution occurs. Since a mono-substituted benzene ring has two equivalent ortho-sites, two equivalent meta-sites and a unique para-site, three possible constitutional isomers may be formed in such a substitution. If reaction occurs equally well at all available sites, the expected statistical mixture of isomeric products would be 40% ortho, 40% meta and 20% para. Again, we find that the nature of the substituent influences this product ratio in a dramatic fashion. Bromination of methoxy-benzene (anisole) is very fast and gives mainly the para-bromo isomer, accompanied by 10% of the ortho-isomer and only a trace of the meta-isomer. Bromination of nitrobenzene requires strong heating and produces the meta-bromo isomer as the chief product.

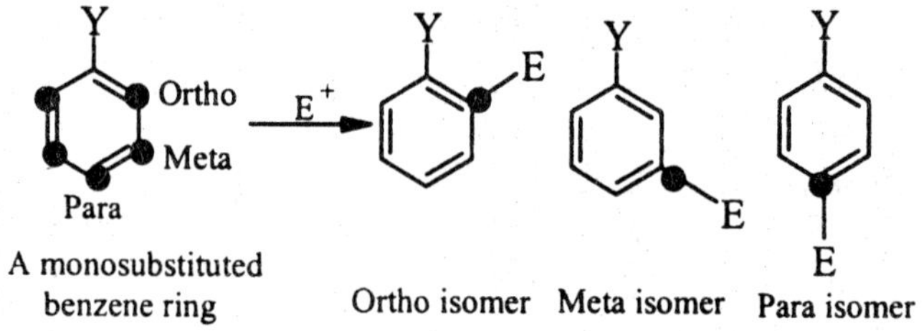

Y, Ortho, Meta, Para, E^+

A monosubstituted benzene ring — Ortho isomer — Meta isomer — Para isomer

Some additional examples of product isomer distribution in other electrophilic substitutions are given in Table 10.3. It is important to note here that the reaction conditions for these substitution reactions are not the same, and must be adjusted to fit the reactivity of the reactant C_6H_5—Y. The high reactivity of anisole, for example, requires that the first two reactions be conducted under very mild conditions (low temperature and little or no catalyst). The nitrobenzene reactant in the third example is very unreactive, so rather harsh reaction conditions must be used to accomplish that reaction.

Table 10.3: Relative proportions of various isomers produced in aromatic electrophilic substitution reaction of substituted benzene

Y in C_6H_5—Y	*Reaction*	*% Ortho-product*	*% Meta-product*	*% Para-product*
—O—CH_3	Nitration	30–40	0–2	60–70
—O—CH_3	F-C Acylation	5–10	0–5	90–95
—NO_2	Nitration	5–8	90–95	0–5
—CH_3	Nitration	55–65	1–5	35–45
—CH_3	Sulfonation	30–35	5–10	60–65
—CH_3	F-C Acylation	10–15	2–8	80–85
—Br	Nitration	35–40	0–4	55–65
—Br	Chlorination	40–45	5–10	50–60

The following mechanism summarizes orientation of electrophilic substitution in some substituted benzene:

(i) *Aromatic ketones*

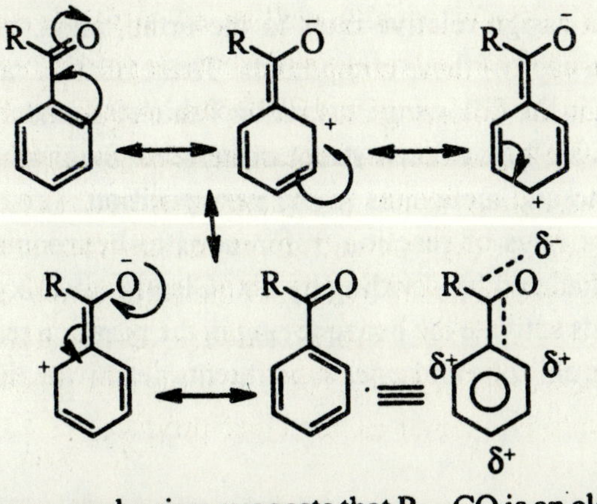

The above mechanism suggests that R—CO is an electron withdrawing group and electron withdrwal is more from ortho and para position giving slight positive charges at these positions as shown in the resonance hybrid. Therefore, we conclude that in aromatic electrophilic substitution of aromatic ketones, meta positions are relatively more electron dense than ortho and para positions and electrophiles prefer to attack on meta positions.

(ii) *Nitro benzene*

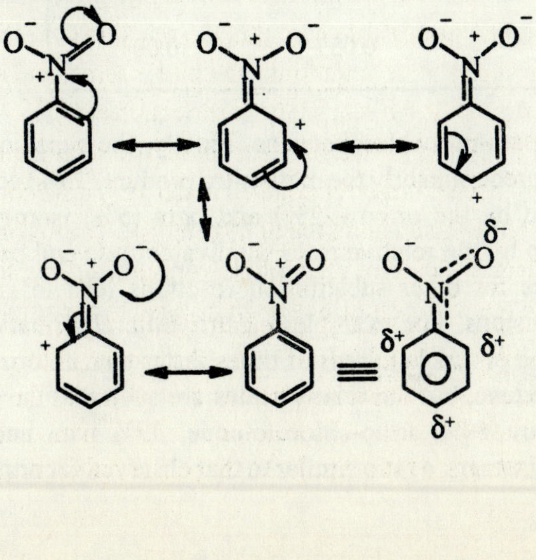

Similar observations as in case of aromatic ketone is observed here, and therefore —NO_2 is also deactivating, meta directing group.

(iii) *Anisole*

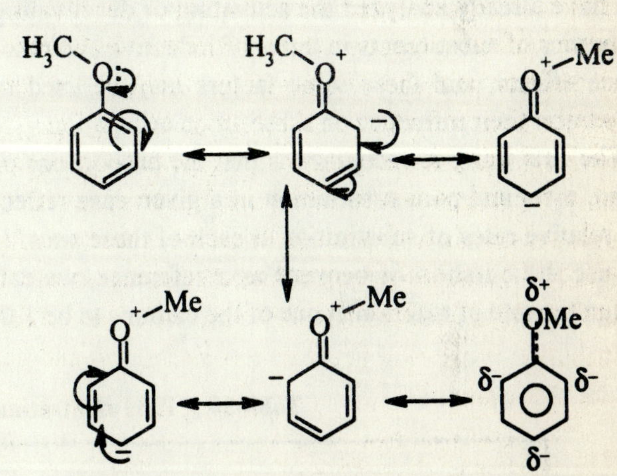

Here the methoxy group is electron donating and various resonating structures drawn above suggests that in anisole, ortho and para positions are more electron rich than meta positions and during attack of electrophile, ortho and para positions are preferred sites. Therefore, —OMe (or OR) is ortho/para directing, activating group.

These observations, and many others like them, have led chemists to formulate an empirical classification of the various substituent groups commonly encountered in aromatic substitution reactions. Thus, substituents that activate the benzene ring toward electrophilic attack generally direct substitution to the ortho and para locations. With a some exceptions, such as the halogens, deactivating substituents direct substitution to the meta location. The following Table 10.4 summarizes this classification.

Table 10.4: Orientation and reactivity effects of ring substituents

Activating substituents ortho and para-orientation		*Deactivating substituents meta-orientation*		*Deactivating substituents ortho and para-orientation*
$—O^-$	$—NH_2$	$—NO_2$	$—CO_2H$	—F
—OH	$—NR_2$	$—NR_3^{(+)}$	$—CO_2R$	—Cl
—OR	$—NHCOCH_3$	$—PR_3^{(+)}$	$CONH_2$	—Br
$—OC_6H_5$	—R	$—SR_2^{(+)}$	—CHO	$—CH_2Cl$
$—OCOCH_3$	$—C_6H_5$	$—SO_3H$	—COR	$—CH{=}CHNO_2$
		$—SO_2R$	—CN	

The information summarized in the Table 10.4 is very useful for rationalizing and predicting the course of aromatic substitution reactions, but in practice most chemists find it desirable to understand the underlying physical principles that contribute to this empirical classification. We have already analyzed the activating or deactivating properties of substituents in terms of inductive and resonance effects, and these same factors may be used to rationalize their influence on substitution orientation.

The first thing to recognize is that the proportions of ortho, meta and para substitution in a given case reflect the relative rates of substitution at each of these sites. If we use the nitration of benzene as a reference, we can assign the rate of reaction at one of the carbons to be 1.0. Since, there are six equivalent carbons in benzene, the total rate would be 6.0.

If we examine the nitration of toluene, tert-butylbenzene, chlorobenzene and methyl benzoate in the same manner, we can assign relative rates to the ortho, meta and para sites in each of these compounds. These relative rates are shown in the following illustration Table 10.5, and the total rate given below each structure reflects the 2 to 1 ratio of ortho and meta sites to the para position. The overall relative rates of reaction, referenced to benzene as 1.0, are calculated by dividing by six. Clearly, the alkyl substituents activate the benzene ring in the nitration reaction, and the chlorine and ester substituents deactivate the ring.

Table 10.5: Rate of nitration at different sites on benzene ring

	(benzene)	CH_3	$C(CH_3)_3$	Cl	$COOCH_3$
	1.0, 1.0, 1.0, 1.0, 1.0, 1.0	43, 3, 55	8, 4, 75	0.03, 0.0, 0.14	0.0025, 0.008, 0.001
Total rate	6.0	147	99	0.20	0.022
Relative rate	1.0	24.5	16.5	0.033	0.004

From rate data of this kind, it is a simple matter to calculate the proportions of the three substitution isomers. Toluene gives 58.5% ortho-nitrotoluene, 37% para-nitrotoluene and only 4.5% of the meta isomer. The increased bulk of the tert-butyl group hinders attack at the ortho-sites, the overall product mixture being 16% ortho, 8% meta and 75% para-nitro product. Although chlorobenzene is much less reactive than benzene, the rate of ortho and para-substitution greatly exceeds that of meta-substitution, giving a product mixture of 30% ortho and 70% para-nitrochlorobenzene. Finally, the benzoic ester gave predominantly the meta-nitro product (73%) accompanied by the ortho (22%) and para (5%) isomers, as shown by the relative rates. Equivalent rate and product studies for other substitution reactions lead to similar conclusions. For example, electrophilic chlorination of toluene occurs hundreds of times faster than chlorination of benzene, but the relative rates are such that the products are 60% ortho-chlorotoluene, 39% para and 1% meta-isomers, a ratio similar to that observed for nitration.

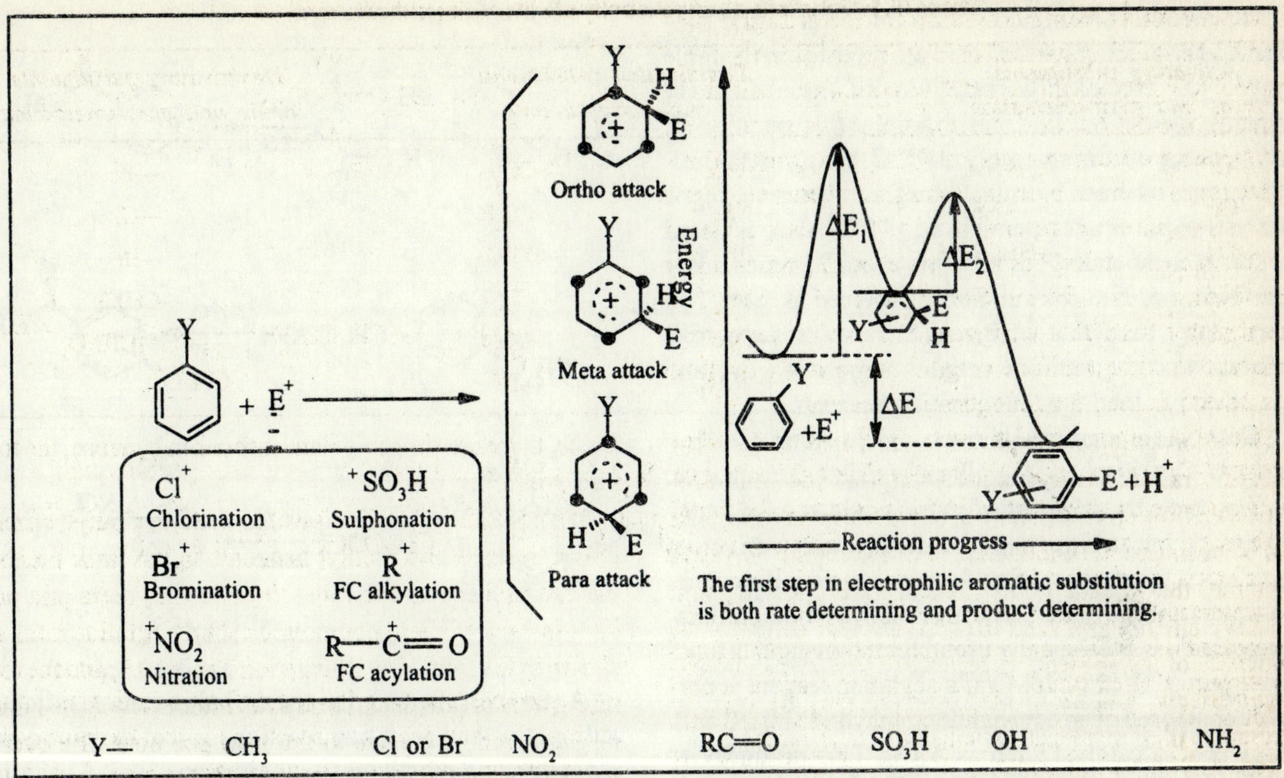

Figure 10.1: Potential energy diagram for aromatic electrophilic substitution reaction

In the case of alkyl substituents, charge stabilization is greatest, when the alkyl group is bonded to one of the positively charged carbons of the benzenonium intermediate. This happens only for ortho and para electrophilic attack, so such substituents favour formation of those products.

Interestingly, primary alkyl substituents, especially methyl, provide greater stabilization of an adjacent charge than do more substituted groups (note the greater reactivity of toluene compared with tert-butylbenzene).

Nitro (NO_2), sulfonic acid (SO_3H) and carbonyl (C=O) substituents have a full or partial positive charge on the atom bonded to the aromatic ring. Structures in which like-charges are close to each other are destabilized by charge repulsion, so these substituents inhibit ortho and para substitution more than meta substitution. Consequently, meta-products preominate when electrophilic substitution is forced to occur.

Halogen (X), OR and NR_2 substituents all exert a destabilizing inductive effect on an adjacent positive charge, due to the high electronegativity of the substituent atoms. By itself, this would favour meta-substitution; however, these substituent atoms all have non-bonding valence electron pairs, which serve to stabilize an adjacent positive charge by pi-bonding, with resulting delocalization of charge. Consequently, all these substituents direct substitution to ortho and para sites. The balance between inductive electron withdrawal and *p*-pi conjugation is such that the nitrogen and oxygen substituents have an overall stabilizing influence on the benzenonium intermediate and increase the rate of substitution markedly; whereas halogen substituents have an overall destabilizing influence.

III. Characteristics of Specific Substitution Reactions

Halogenation	C_6H_6	+	Cl_2 and heat $FeCl_3$ catalyst	→	C_6H_5Cl Chlorobenzene	+ HCl
Nitration	C_6H_6	+	HNO_3 and heat H_2SO_4 catalyst	→	$C_6H_5NO_2$ Nitrobenzene	+ H_2O
Sulfonation	C_6H_6	+	H_2SO_4 + SO_3 and heat	→	$C_6H_5SO_3H$ Benzenesulfonic acid	+ H_2O
Alkylation Friedel-Crafts	C_6H_6	+	R—Cl and heat $AlCl_3$ catalyst	→	C_6H_5—R An Arene	+ HCl
Acylation Friedel-crafts	C_6H_6	+	RCOCl and heat $AlCl_3$ catalyst	→	C_6H_5COR an aryl ketone	+ HCl

The conditions commonly used for the aromatic substitution reactions discussed here are repeated in the table above. The electrophilic reactivity of these different reagents varies. We find, for example, that nitration of nitrobenzene occurs smoothly at 95°C, giving meta-dinitrobenzene, whereas bromination of nitrobenzene (ferric catalyst) requires a temperature of 140°C. Also, as noted earlier, toluene undergoes nitration about 25 times faster than benzene, but chlorination of toluene is over 500 times faster than that of benzene. From this we may conclude that the nitration reagent is more reactive and less selective than the halogenation reagents.

Both sulfonation and nitration yield water as a by-product. This does not significantly affect the nitration reaction (note the presence of sulfuric acid as a dehydrating agent), but sulfonation is reversible and is driven to completion by addition of sulfur trioxide, which converts the water to sulfuric acid. The reversibility of the sulfonation reaction is occasionally useful for removing this functional group. The Friedel-Crafts acylation reagent is normally composed of an acyl halide or anhydride mixed with a Lewis acid catalyst such as $AlCl_3$. This produces an acylium cation, $R—C{\equiv}O^{(+)}$, or a related species. Such electrophiles are not exceptionally reactive, so the acylation reaction is generally restricted to aromatic systems that are at least as reactive as chlorobenzene. Carbon disulfide is often used as a solvent, since it is unreactive and is easily removed from the product. If the substrate is a very reactive benzene derivative, such as anisole, carboxylic esters or acids may be the source of the acylating electrophile. Some examples of Friedel-Crafts acylation reactions are shown in the following diagram. The first demonstrates that unusual acylating agents may be used as reactants. The second makes use of an anhydride acylating reagent, and the third illustrates the ease with which anisole reacts, as noted earlier. The $H_4P_2O_7$ reagent used here is an anhydride of phosphoric acid called *pyrophosphoric acid.*

Finally, the fourth example illustrates several important points. Since, the nitro group is a powerful deactivating substituent, Friedel-Crafts acylation of nitrobenzene does not take place under any conditions. However, the presence of a second strongly-activating substituent group permits acylation, the site of reaction is that favoured by both substituents.

1. (pyridine-3-COCl) + (benzene) —$AlCl_3$, Reflux→ (3-benzoylpyridine)

2. (p-bromobenzene) + $(H_3C-CO)_2O$ —$AlCl_3$, CS_2→ Br-C_6H_4-$COCH_3$

3. CH_3O-C_6H_5 + CH_3COOH —H_3PO_4, Heat→ CH_3O-C_6H_4-$COCH_3$

4. (2-nitroanisole: OMe, NO_2) + CH_3COCl —$AlCl_3$, Reflux→ (OMe, NO_2, $COCH_3$ para to OMe)

A common characteristic of the halogenation, nitration, sulfonation and acylation reactions is that they introduce a deactivating substituent on the benzene ring. As a result, we do not normally have to worry about disubstitution products being formed. Friedel-Crafts alkylation, on the other hand, introduces an activating substituent (an alkyl group), so more than one substitution may take place. If benzene is to be alkylated, as in the following synthesis of tert-butylbenzene, the mono-alkylated product is favoured by using a large excess of this reactant. When the molar ratio of benzene to alkyl halide falls below 1:1, para-ditertiary-butylbenzene becomes the major product.

$$C_6H_6 \text{ (large excess)} + (CH_3)_3C—Cl + AlCl_3 \rightarrow C_6H_5—C(CH_3)_3 + HCl$$

The carbocation electrophiles required for alkylation may be generated from alkyl halides (as above), alkenes + strong acid or alcohols + strong acid. Since, 1°-carbocations are prone to rearrangement, it is usually not possible to introduce 1°-alkyl substituents larger than ethyl by Friedel-Crafts alkylation. For example, reaction of excess benzene with 1-chloropropane and aluminum chloride gives a good yield of isopropylbenzene (cumene).

$$C_6H_6 \text{ (large excess)} + CH_3CH_2CH_2—Cl + AlCl_3 \rightarrow C_6H_5—CH(CH_3)_2 + HCl$$

Additional examples of Friedel-Crafts alkylation reactions are shown in the following examples:

1. (benzene) + (cyclohexene) —H_2SO_4, 0° C→ (cyclohexylbenzene)

2. $2\,C_6H_6 + CCl_4 \xrightarrow{AlCl_3} (C_6H_5)_2CCl_2$

3. $C_6H_5OH + (C_6H_5)_3COH \xrightarrow{H_3PO_4}$ p-$HOC_6H_4C(C_6H_5)_3$

4. m-xylene (H_3C, CH_3) $+ (CH_3)_3CCl \xrightarrow[\text{Reflux}]{AlCl_3}$ 1-tert-butyl-3,5-dimethylbenzene (H_3C, CH_3; H_3C, CH_3, CH_3)

The first and third examples show how alkenes and alcohols may be the source of the electrophilic carbocation reactant. The triphenylmethyl cation generated in the third case is relatively unreactive, due to extensive resonance charge delocalization, and only substitutes highly activated aromatic rings. The second example shows an interesting case, in which a polychlororeactant is used as the alkylating agent. A four-fold excess of carbon tetrachloride is used to avoid tri-alkylation of this reagent, a process that is retarded by steric hindrance. The fourth example illustrates the poor orientational selectivity often found in alkylation reactions of activated benzene rings. The bulky tert-butyl group attaches to the reactive meta-xylene ring at the least hindered site.

IV. Electrophilic Substitution of Disubstituted Benzene Rings

When a benzene ring has two substituent groups, each exerts an influence on subsequent substitution reactions. The activation or deactivation of the ring can be predicted more or less by the sum of the individual effects of these substituents. The site at which a new substituent is introduced depends on the orientation of the existing groups and their individual directing effects. We can identify two general behaviour categories, as shown in the following table. Thus, the groups may be oriented in such a manner that their directing influences act in concert, reinforcing the outcome, or are opposed (antagonistic) to each other. Note that the orientations in each category change depending on whether the groups have similar or opposite individual directing effects.

Orientational Interaction of Substituents

Antagonistic or non-cooperative: o-D,D ; p-D,D ; m-D,D ; o-W,W ; p-W,W

Reinforcing or cooperative: m-D,D ; m-W,W ; o-D,W ; p-D,W

D = Electron donating group (ortho/para-directing) W = Electron withdrawing group (meta-directing)

The products from substitution reactions of compounds having a reinforcing orientation of substituents are easier to predict than those having antagonistic substituents. For example, the six equations shown below are all examples of reinforcing or cooperative directing effects operating in the expected manner. Symmetry, as in the first two cases, makes it easy to predict the site at which substitution is likely to occur. Note that if two different sites are favoured, substitution will usually occur at the one that is least hindered by ortho groups.

1. m-xylene $\xrightarrow[AlCl_3]{(CH_3CO)_2O}$ 2,4-dimethylacetophenone (O)

2. 1,3-dichlorobenzene (Cl, Cl) $\xrightarrow[H_2SO_4]{HNO_3}$ 2,4-dichloronitrobenzene (Cl, Cl, NO_2)

3. O_2N–C_6H_4–COOH (meta) $\xrightarrow[H_2SO_4]{HNO_3}$ 3,5-dinitrobenzoic acid (O_2N, COOH, NO_2)

4. o-bromonitrobenzene (Br, NO_2) $\xrightarrow[SO_3]{H_2SO_4}$ 4-bromo-3-nitrobenzenesulfonic acid (Br, NO_2, SO_3H)

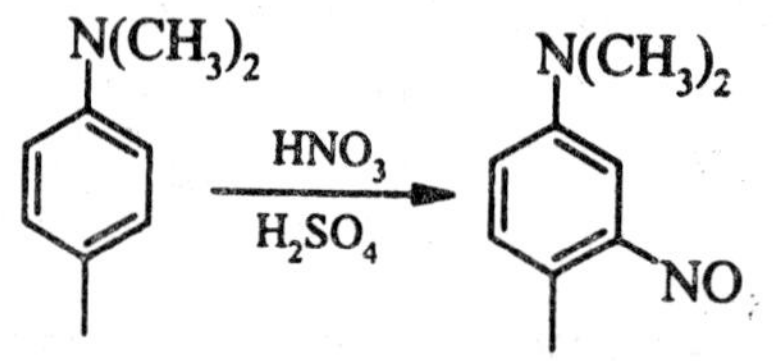

The first three examples have two similar directing groups in a meta-relationship to each other. In examples 4 through 6, oppositely directing groups have an ortho or para-relationship. The major products of electrophilic substitution, as shown, are the sum of the individual group effects. The strongly activating hydroxyl (OH) and amino (NH_2) substituents favour dihalogenation in examples 5 and 6.

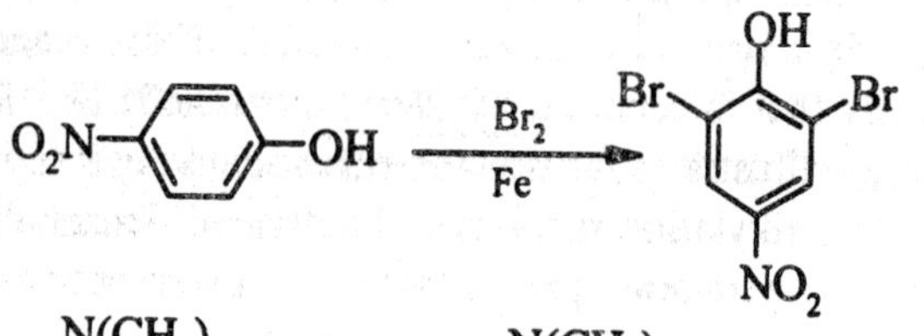

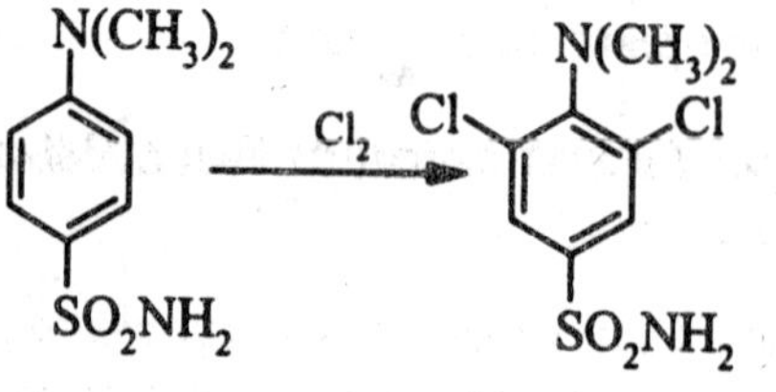

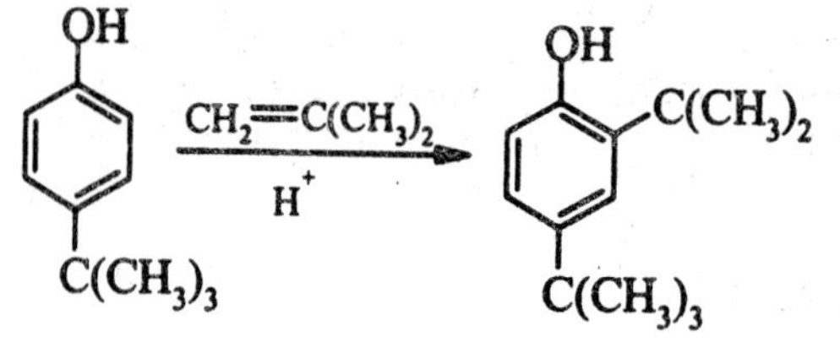

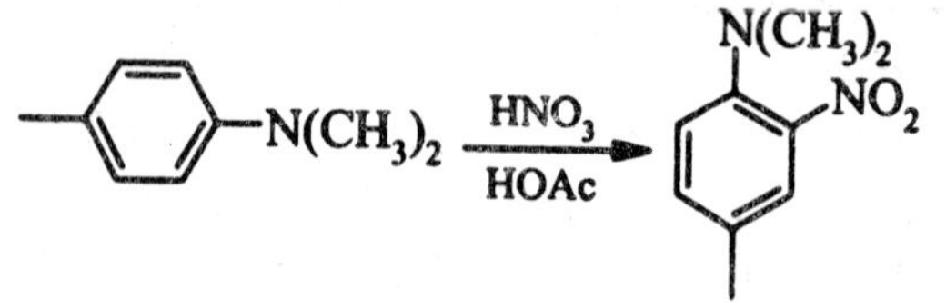

Substitution reactions of compounds having an antagonistic orientation of substituents require a more careful analysis. If the substituents are identical, as in example 1 below, the symmetry of the molecule will again simplify the decision. When one substituent has a pair of non-bonding electrons available for adjacent charge stabilization, it will normally exert the product determining influence, examples 2, 4 and 5, even though it may be overall deactivating (case 2). Case 3 reflects a combination of steric hindrance and the superior innate stabilizing ability of methyl groups relative to other alkyl substituents. Example 6 is interesting in that it demonstrates the conversion of an activating ortho/para-directing group into a deactivating meta-directing "onium" cation [—$NH(CH_3)_2^{(+)}$] in a strong acid environment.

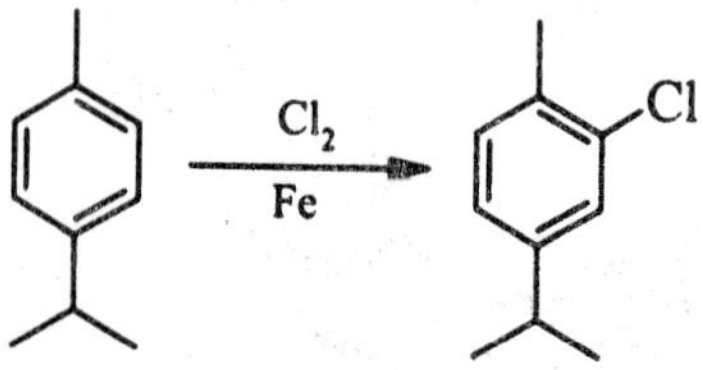

10.4 REACTIONS OF SUBSTITUENT GROUPS

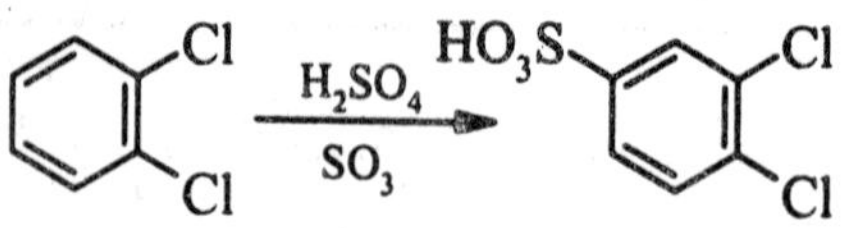

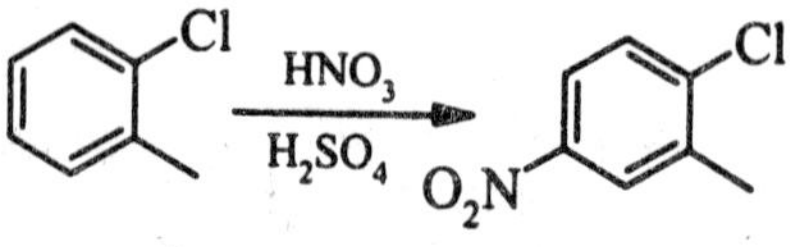

I. Oxidation of Alkyl Side-Chains

The benzylic hydrogens of alkyl substituents on a benzene ring are activated toward free radical attack, as noted earlier. Furthermore, SN-1, SN-2 and E1 reactions of benzylic halides, show enhanced reactivity, due to the adjacent aromatic ring. The possibility that these observations reflect a general **benzylic activation** is supported by the susceptibility of alkyl side-chains to oxidative degradation, as shown in the following examples. Such oxidations are normally effected by hot acidic pemanganate solutions, but for large scale industrial operations catalysed air-oxidations are preferred. Interstingly, if the benzylic position is completely substituted this oxidative degradation does not occur.

$$C_6H_5—CH_2CH_2CH_2CH_3 + KMnO_4 + H_3O^{(+)} \text{ and heat} \rightarrow C_6H_5—CO_2H + CO_2$$

$$\text{p—}(CH_3)_3C—C_6H_4—CH_3 + KMnO_4 + H_3O^{(+)} \rightarrow \text{p—}(CH_3)_3C—C_6H_4—CO_2H$$

These equations are not balanced. The permanganate oxidant is reduced, usually to Mn(IV) or Mn(II). Two other examples of this reaction are given below, and illustrate its usefulness in preparing substituted benzoic acids.

KMnO$_4$ / H$^+$; KMnO$_4$ / H$^+$

COOH, NO$_2$, NO$_2$, COOH, COOH

II. Reduction of Nitro Groups and Aryl Ketones

Electrophilic nitration and Friedel-Crafts acylation reactions introduce deactivating, meta-directing substituents on an aromatic ring. The attached atoms are in a high oxidation state, and their reduction converts these electron withdrawing functions into electron donating amino and alkyl groups. Reduction is easily achieved either by catalytic hydrogenation (H_2 + catalyst), or with reducing metals in acid. Examples of these reductions are shown here, equation 6 demonstrating the simultaneous reduction of both functions. Note that the butylbenzene product in equation 4 cannot be generated by direct Friedel-Crafts alkylation due to carbocation rearrangement. The zinc used in ketone reductions, such as 5, is usually activated by alloying with mercury (a process known as amalgamation).

Nitro group reduction

1. COOH, NO$_2$ —Fe/HCl→ COOH, NH$_2$

Strong reducing agent may also affect —COOH

2. NO$_2$, NO$_2$ —Fe/HCl→ NH$_2$, NH$_2$

3. CHO, NO$_2$ —Sn/HCl→ CHO, NH$_2$

Carbonyl group reduction

4. O —Zn(Hg) / Conc. HCl→

5. O, HO, OH —Zn(Hg) / Conc. HCl→ HO, OH

6. O, OH, CH$_3$, O_2N —Zn(Hg) / Conc. HCl→ OH, CH$_3$, O_2N

Several alternative methods for reducing nitro groups to amines are known. These include zinc or tin in dilute mineral acid, and sodium sulfide in ammonium hydroxide solution. The procedures described above are sufficient for most cases.

III. Conversion of Halogens to Organometallic Reagents

The reaction of alkyl and aryl halides with reactive metals (usually Li and Mg) to give nucleophilic reagents has been noted. This provides a powerful tool for the conversion of chloro, bromo or iodo substituents into a variety of other groups. Many reactions of these aryl lithium and Grignard reagents will be discussed in later sections, and the following equations provide typical examples of carboxylation, protonation and Gilman coupling.

1. Br —Li→ Li —CO_2→ COOLi —H_3O^+→ COOH

2. Br, Br —Mg / Ether→ BrMg, BrMg —D_2O→ D, D

3. $CH_2N(CH_3)_2$, Br —Li→ $CH_2N(CH_3)_2$, Li —CuI→ ; Br ↓ $CH_2N(CH_3)_2$

IV. Hydrolysis of Sulfonic Acids

The following equation illustrates how this characteristic of the sulfonic acids may be used to prepare the 3-bromo derivative of ortho-xylene. Direct bromination would give the 4-bromo derivative.

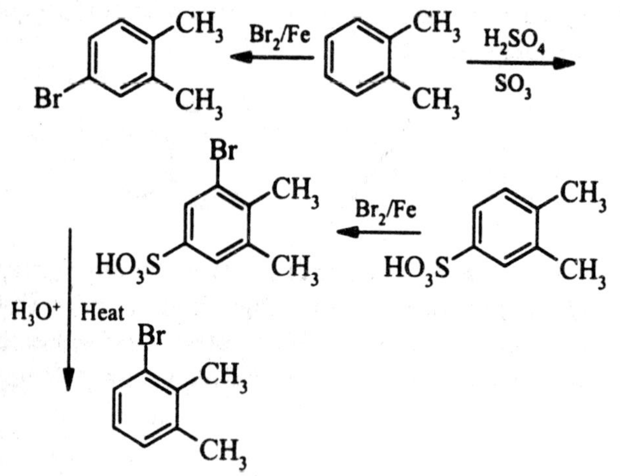

Br_2/Fe H_2SO_4 SO_3 CH_3 CH_3 Br CH_3 CH_3

Br CH_3 Br_2/Fe CH_3 HO_3S CH_3 HO_3S CH_3

H_3O^+ Heat

Br CH_3 CH_3

V. Modifying the Influence of Strong Activating Groups

The strongest activating and ortho/para-directing substituents are the amino (—NH_2) and hydroxyl (—OH) groups. Direct nitration of phenol (hydroxybenzene) by dilute nitric acid gives modest yields of nitrated phenols and considerable oxidative decomposition to tarry materials, aniline (aminobenzene) is largely destroyed. Bromination of both phenol and aniline is difficult to control, with di- and tri-bromo products forming readily. Because of their high nucleophilic reactivity, aniline and phenol undergo substitution reactions with iodine, a halogen that is normally unreactive with benzene derivatives. The mixed halogen iodine chloride (ICl) provides a more electrophilic iodine moiety, and is effective in iodinating aromatic rings having less powerful activating substituents.

$$C_6H_5—NH_2 + I_2 + NaHCO_3 \rightarrow$$
$$p—I—C_6H_4—NH_2 + NaI + CO_2 + H_2O$$

By acetylating the heteroatom substituent on phenol and aniline, its activating influence can be substantially attenuated. For example, acetylation of aniline gives acetanilide (first step in the following equation), which undergoes nitration at low temperature, yielding the para-nitro product in high yield. The modifying acetyl group can then be removed by acid-catalyzed hydrolysis (last step), to yield para-nitroaniline. Although the activating influence of the amino group has been reduced by this procedure, the acetyl derivative remains an ortho/para-directing and activating substituent.

$$C_6H_5NH_2 + (CH_3CO)_2 \longrightarrow C_6H_5NHCOCH_3$$

HNO_3 | 5-10° C

$NHCOCH_3$ NH_2

H_3O^+

NO_2 NO_2

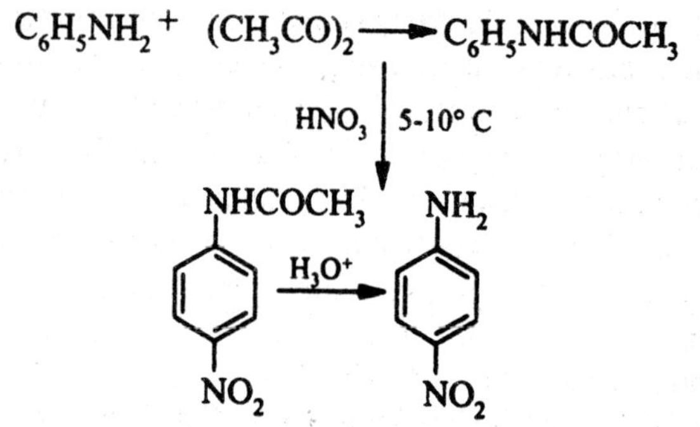

The following diagram illustrates how the acetyl group acts to attenuate the overall electron donating character of oxygen and nitrogen. The non-bonding valence electron pairs that are responsible for the high reactivity of these compounds are diverted to the adjacent carbonyl group. However, the overall influence of the modified substituent is still activating and ortho/para-directing.

OH $OCOCH_3$

$(CH_3CO)_2O$ Pyridine

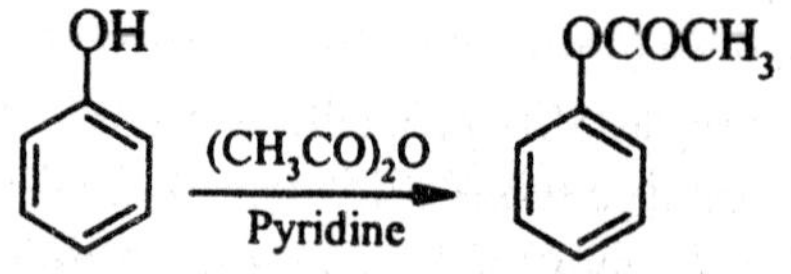

Phenol
Strongly activated
aromatic ring

Phenyl acetate
A less activated
aromatic ring

NH_2 $NHCOCH_3$

$(CH_3CO)_2O$ Pyridine

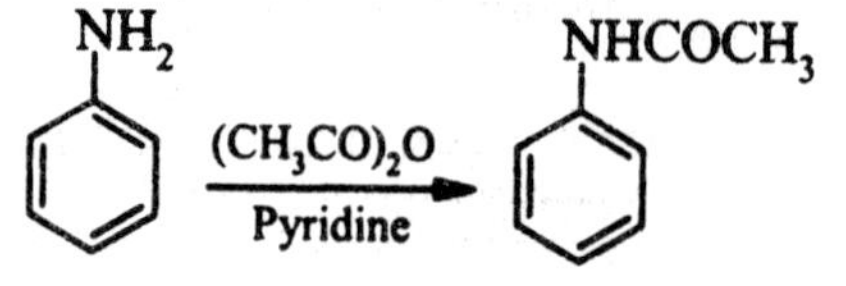

Aniline
Strongly activated
aromatic ring

Acetanilide
A less activated
aromatic ring

10.5 REACTIONS OF FUSED BENZENE RINGS

Compounds in which two or more benzene rings are fused together were described in an earlier section, and they present interesting insights into aromaticity and reactivity. The smallest such hydrocarbon is naphthalene. Naphthalene is stabilized by resonance. Three canonical resonance contributors may be drawn, and are displayed in the following diagram.

Naphthalene

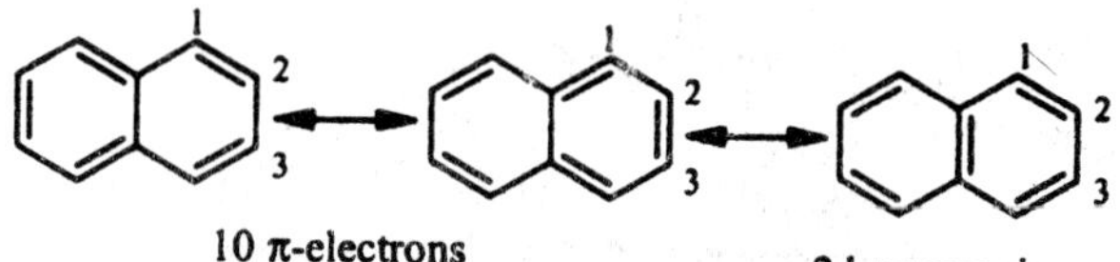

1 2 3 1 2 3 1 2 3

10 π-electrons

2 benzene rings

Resonance energy = 61 kcal/mole

The two structures on the left have one discrete benzene ring each, but may also be viewed as 10 pi-electron annulenes having a bridging single bond. The structure on the right has two benzene rings which share a common double bond. From heats of hydrogenation or combustion, the resonance energy of naphthalene is calculated to be 61 kcal/mole, 11 kcal/mole less than that of two benzene rings (2 × 36). As expected from an average of the three resonance contributors, the carbon-carbon bonds in naphthalene show variation in length, suggesting some localization of the double bonds. The C1:C2 bond is 1.36 Å long, whereas the C2:C3 bond length is 1.42 Å. This contrasts with the structure of benzene, in which all the C—C bonds have a common length, 1.39 Å. Naphthalene is more reactive than benzene, both in substitution and addition reactions, and these reactions tend to proceed in a manner that maintains one intact benzene ring.

The following diagram shows three oxidation and reduction reactions that illustrate this feature. In the last example, catalytic hydrogenation of one ring takes place under milder conditions than those required for complete saturation (the decalin product exists as *cis/trans* isomers). Electrophilic substitution reactions take place more rapidly at C1, although the C2 product is more stable and predominates at equilibrium.

Oxidation and reduction reaction

CrO_3, HOAc 25°C

O_2, V_2O_5, 450–500°C

$H_2/RhCl_3$, 1.0 atm. 30°C

Example 21. Consider the following substitution reaction:

CH_3 —(Br_2, $AlBr_3$)→ Ortho bromotoluene + Meta bromotoluene + Para bromotoluene

(a) Provide mechanism of formation of major product of this reaction.

(b) Briefly explain reason for your choice.

(c) What is the purpose of $AlBr_3$?

(d) What is rate determining step in this reaction?

(e) Draw a potential energy diagram for formation of ortho product.

(f) Explain why reaction gives substitution product not addition product?

Solution

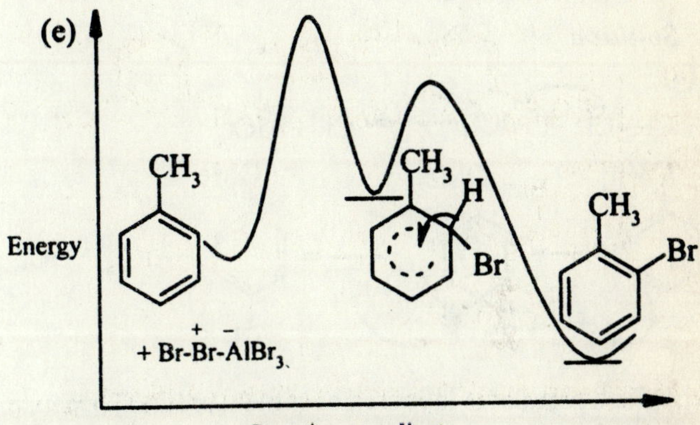

Example 22. Explain why halogens are ortho/para directing but deactivating group?

Solution: The halogens are ortho/para directing due to small resonance contribution. This small resonance contribution is a combination of high electronegativity and poor overlap with carbon *p*-orbitals by elements not in the same period as carbon. When the electrophile attacks either at ortho or para to halogen, the carbocation can be stabilized to small extent by resonance contributor, which places the positive charge on the halogen giving close octet of all the atoms in this structure. The electronegativity of the halogen deactivate the ring due to inductive effect, since they are more electronegative than carbon. The two effect counter each other, but the inductive effect is more significant in this case. This is an uncommon example of any effect overwhelming resonance effect. Therefore, halogens are ortho/para directing due to resonance effect but deactivating due to inductive effect.

Complete octet of all atoms

Example 23. Consider the following substitution reactions:

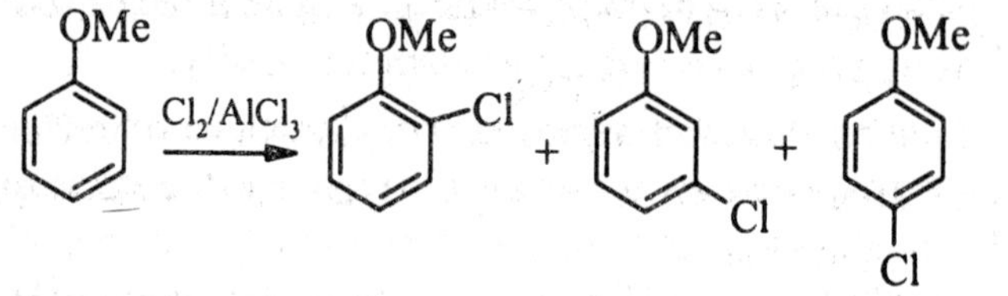

OMe, $Cl_2/AlCl_3$ → OMe (o-Cl) + OMe (m-Cl) + OMe (p-Cl)

(a) Draw a curved arrow mechanism how major product is formed?

(b) Discuss in brief why the other products are formed in lesser amount.

Solution

(a)

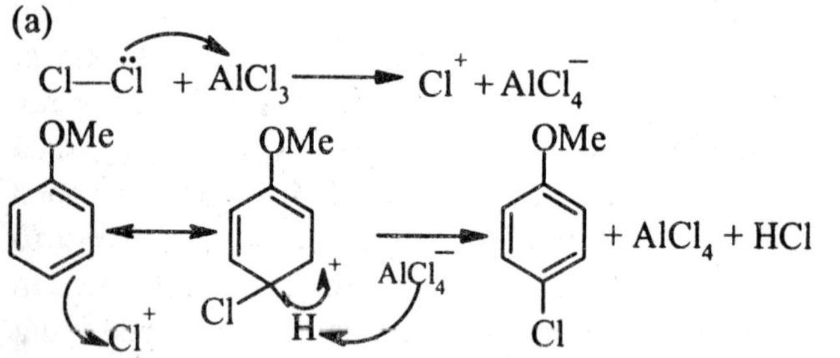

$$Cl—\ddot{Cl} + AlCl_3 \longrightarrow Cl^+ + AlCl_4^-$$

OMe ⟷ OMe (Cl, H, +) $\xrightarrow{AlCl_4^-}$ OMe (Cl) + $AlCl_4$ + HCl

(b) Para product is favoured over meta product, because the methoxy group can provide resonance stabilization to carbocation intermediates resulting from ortho or para attack of the electrophile, but not from meta attack. Para product is favoured over ortho product due to steric reason.

Example 24. Provide a detailed curved arrow mechanism for this reaction:

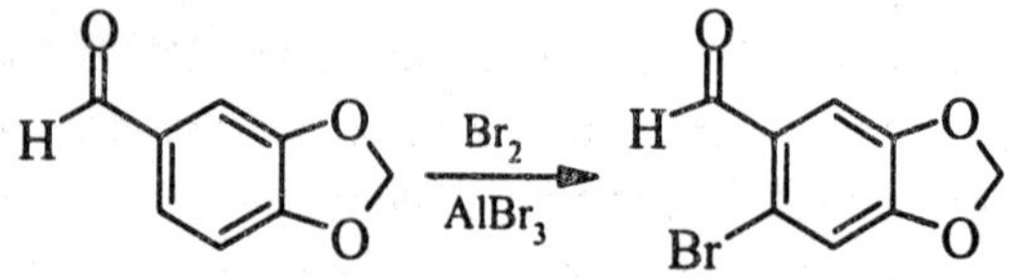

$\xrightarrow[AlBr_3]{Br_2}$ Br

Solution

$$Br—\ddot{Br} + AlBr_3 \longrightarrow Br^+ + AlBr_4^-$$

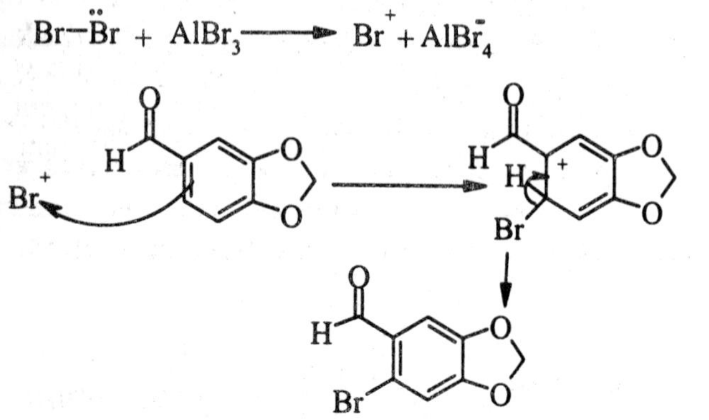

One may think that the above product is a wrong one. Ether is an ortho/para directing and aldehyde is a meta directing group, one might predict the major product to have new C—Br bond meta to aldehyde group. However, the given product is correct and its explanation comes from resonance. Consider the following resonance contributors:

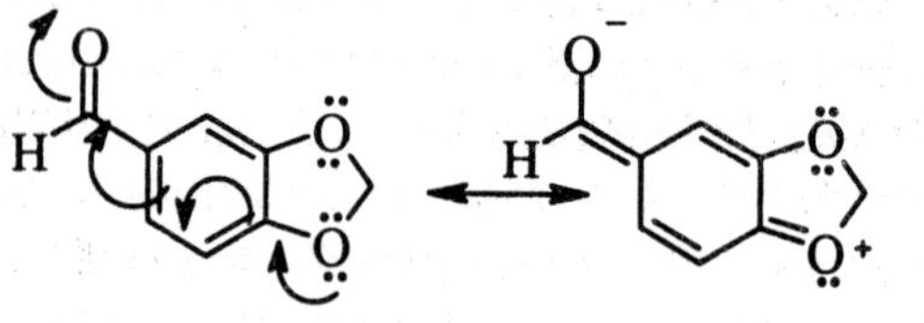

This resonance ties up some of the electron lone pair density on the ether oxygen reducing its ability to stabilize an adjacent carbocation, and thus weakening its influence as an ortho/para director. This same resonance adds electron density to the aldehyde group, decrease its electron withdrawing influence responsible for its meta directing ability. This leaves the other ether oxygen as the dominant directing group. The net effect is electrophilic aromatic substitution para to this strongest group (ortho to aldehyde as shown).

Example 25. Consider the Friedal Craft methylation of toluene.

(a) Predict the major product showing mechanism of their formation.

(b) Give two reasons why this is the major product.

(c) Would the reaction of *p*-methyl anisole with $CH_3Cl/AlCl_3$ be faster or slower than the reaction with toluene? Explain.

Solution

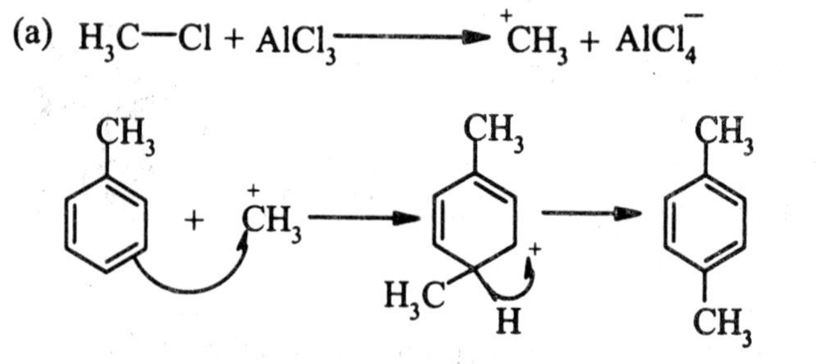

(a) $H_3C—Cl + AlCl_3 \longrightarrow \overset{+}{C}H_3 + AlCl_4^-$

CH_3 (toluene) + $\overset{+}{C}H_3$ ⟶ (CH_3, H_3C, H, +) ⟶ CH_3 (CH_3)

(b) (i) The methyl group disfavour ortho attack due to steric repulsion of the incoming electrophile. This repulsion is not a restricting factor in meta and para attack.

(ii) Ortho and para attack are favoured because both have an intermediate carbocation with three resonance contributors. Two of these resonance contributors are secondary carbocation and other is a tertiary one. Meta attack yields an intermediate that also has three resonance structure, each of which is a secondary carbocation. The resonance hybrid resulting from ortho or para attack are more stable than the resonance hybrid from meta attack, hence ortho/para attack is favoured over meta attack.

(c) Kinetics of a reaction depends on rate determining step of the reaction mechanism. In aromatic electrophilic substitution reaction, the rate determining step is the initial attack of electrophile on the ring. Factors which make the ring a better nucleophile will increase the rate of reaction. The —OCH_3 group is a strong electron donating group due to its +R effect. This increases the electron density of the benzene ring making it a better nucleophile. A methyl group is a weak electron donating group, so *p*-methyl anisole is a better nucleophile than toluene. Therefore, the reaction of *p*-methyl anisole will be faster than the reaction of toluene.

Example 26. What is the major product in the following reaction and why?

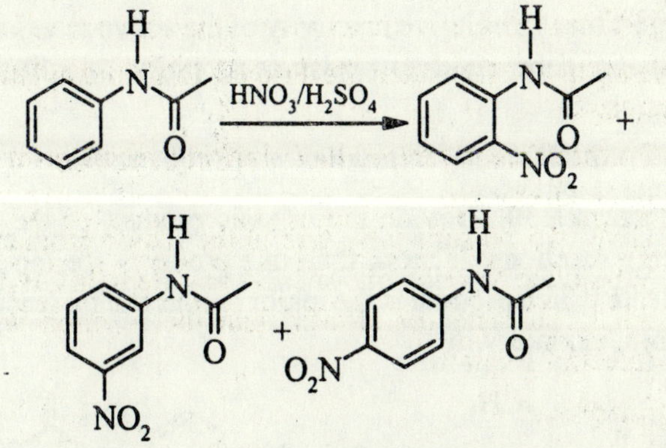

HNO_3/H_2SO_4

NO_2

O_2N

NO_2

Solution: The major product of this reaction is para isomer. There are two major factors favouring this product.

(a) *Stability of carbocation intermediate:* Initial electrophilic attack on benzene ring occurs so as to afford the most stable carbocation. Attack at the ortho or para positions affords a carbocation that has four resonance contributors. In one of these contributors all atoms have complete octet. This is due to the lone pair on the nitrogen atom, adjacent to the benzene ring. Attack at meta position affords a carbocation that is less stable, as it only has three resonance contributors, and none of these have a complete octet on all atoms. Thus, attack at the ortho and para positions is favoured over attack at the meta position.

(b) *Steric factors:* The amide hinders the incoming electrophile when attack occurs at the ortho position. There is significantly less hindrance to electrophilic attack at the para position. Thus, steric effects favour para attack over ortho attack.

Example 27. The nitro group is a meta director, but nitroso group is an ortho/para director, justify.

Solution: Whether a substituent is ortho/para director or meta director depends on how it influences the stability of arenium ion intermediate formed in the initial step of electrophilic attack. If a group stabilize the carbocation by resonance or by inductive electron release, then it enhances the ortho/para attack, because this forms an arenium ion whose positive charge may be stabilized. If the group destabilizes adjacent positive charge by inductive electron withdrawing effect then it inhibit ortho/para attack but not meta attack, so meta attack is preferred (note that a meta director does not enhance meta attack, but instead exerts it directing influence by inhibiting ortho/para attack). A nitro group bears a positively charged nitrogen atom and therefore destabilizes an adjacent open octet on carbon. The nitroso group bears a nitrogen lone pair, so it can stabilize an adjacent open octet through resonance. The nitroso group is thus an ortho/para director.

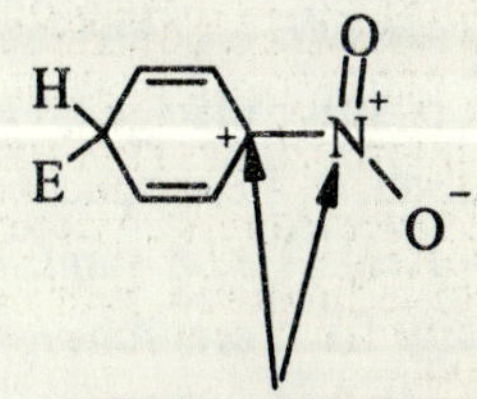

Adjacent positive charge (destabilization)

Nitroso group stabilize adjacent positive charge by resonance

Example 28: Carbocation can be stabilized by resonance involving adjacent π-bonds, so one may expect that all functional groups that contain a π-bond directly attached to benzene ring are ortho/para directors. This is true if the substituent is a benzene ring or an olefin, but not true if the substituent is a ketone or nitrile. Explain.

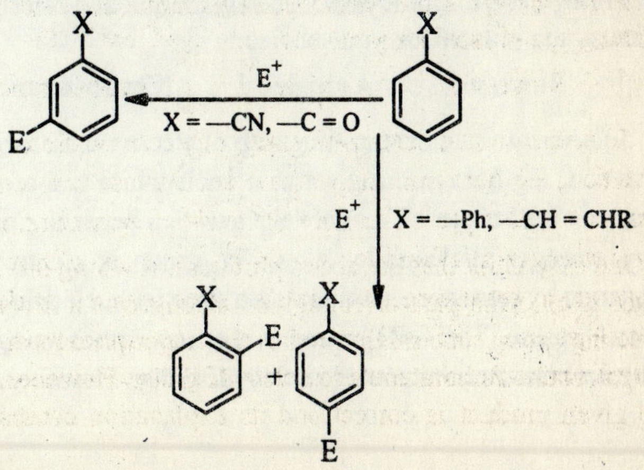

E^+

X = —CN, —C = O

E^+ X = —Ph, —CH = CHR

Solution: In electrophilic aromatic substitution reaction, directing influence of a substituent is controlled by its effect on the stability of arenium ion intermediate. If the substituent stabilizes the arenium ion, then it is an ortho/para director. If it destabilizes the arenium ion then it is a meta director. A benzene ring provides three additional resonance contributors, thereby increasing the stability of the arenium ion intermediate. Therefore a benzene ring is ortho/para director.

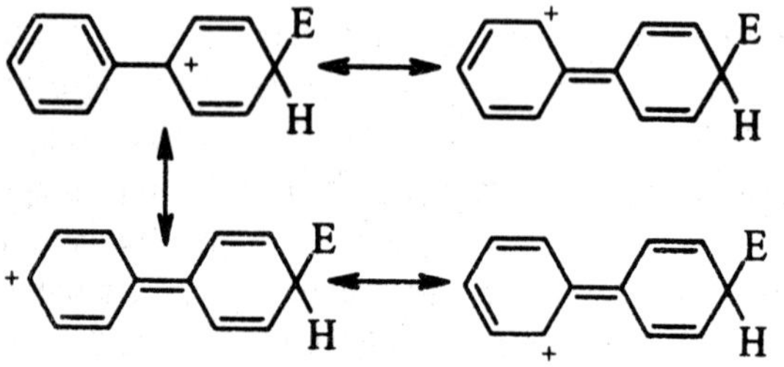

Functional groups that includes a carbon-heteroatom π-bond conjugated with the benzene ring allows for at least one more arenium ion resonance contributor. However, this contributor bears a positive charge on an atom more electronegative than carbon, so these contributors are very less stable and unimportant. These same hetero atoms withdraw electron density from arenium ion through the inductive effect, resulting in destabilization. The unimportant resonance contributor is outweighed by the inductive effect, so carbonyl and nitrile functional groups destablilze arenium ions and are thus meta director.

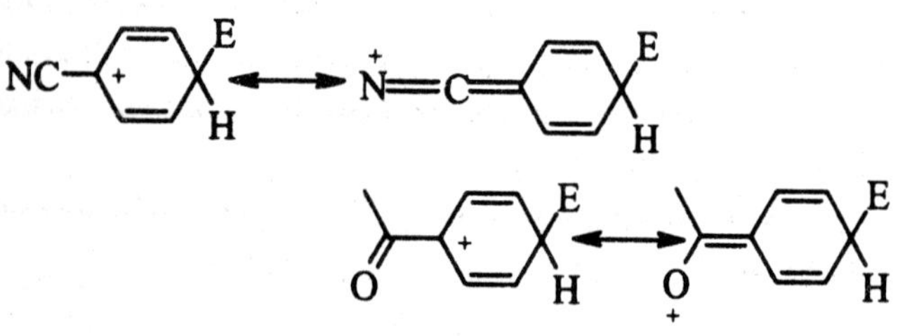

Example 29. Rank the following compounds of relative rate of electrophilic aromatic substitution and briefly discuss the reason for your choice:

Benzene **Anisole** **Nitrobenzene**

Solution: In rate determining step of electrophilic substitution, the benzene ring act as a nucleophile and aromaticity is disrupted. The more electron rich benzene ring is a stronger nucleophile. A $-OCH_3$ group is electron donating by resonance, so anisole is a stronger nucleophile than benzene. The $-NO_2$ group is electron withdrawing, so nitrobenzene is a poorer nucleophile than benzene.

Also, formation of a more stable carbocation in the rate determining step results in faster substitution reaction. Electrophilic attack on anisole affords an arenium ion with four resonance contributors, one of which has a close octet (complete) of all the atom indicating a more stable carbocation. Similar electrophilic attack on benzene affords an arenium ion with three resonance contributors, none of which has a close octet on every atom. Hence, anisole is more reactive than benzene. In the nitrobenzene, electrophilic attack affords an arenium ion with three resonance contributors, none of which have a close octet on each atom. The electron withdrawing effect of nitro group further destabilize the arenium ion. Because, the arenium ion derived from nitro benzene is less stable than the arenium ion derived from benzene, the rate of electrophilic attack on nitrobenzene is less than the rate of attack on benzene. The combination of these effects suggest that the relative reactivity of the above mentioned aromatic compound towards electrophilic substitution reaction is:

Anisole > Benzene > Nitrobenzene

Example 30. Provide the organic product of the following reactions. If more than one products are formed, indicate which product is the major product. If no reaction occurs, explain why?

(a) 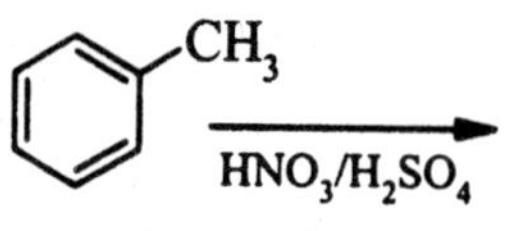 $\xrightarrow{HNO_3/H_2SO_4}$ (b) 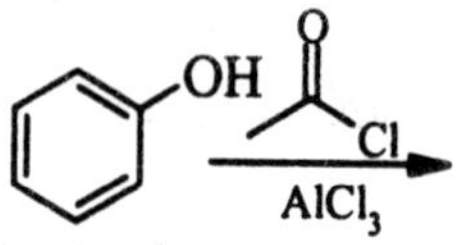 $\xrightarrow{AlCl_3}$

(c) 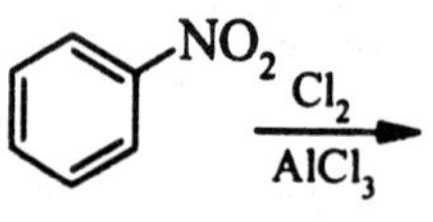 $\xrightarrow[AlCl_3]{Cl_2}$ (d) 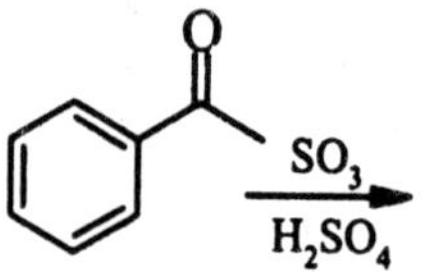 $\xrightarrow[H_2SO_4]{SO_3}$

(e) 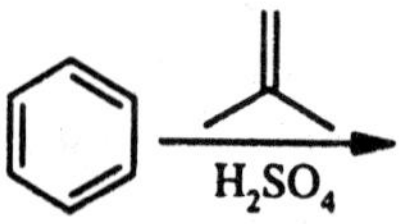 $\xrightarrow{H_2SO_4}$

Solution

(a) 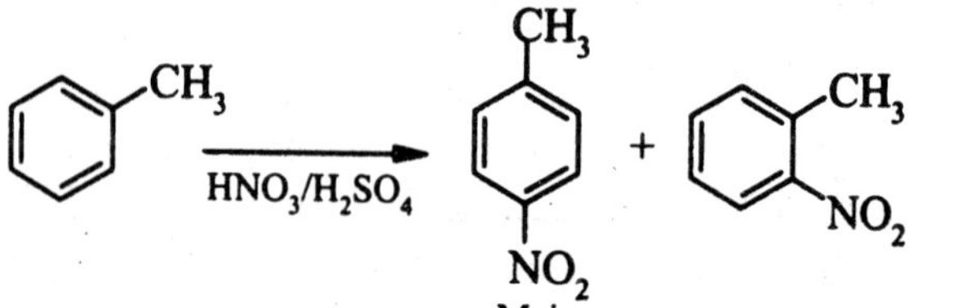 $\xrightarrow{HNO_3/H_2SO_4}$ + Major

(b) 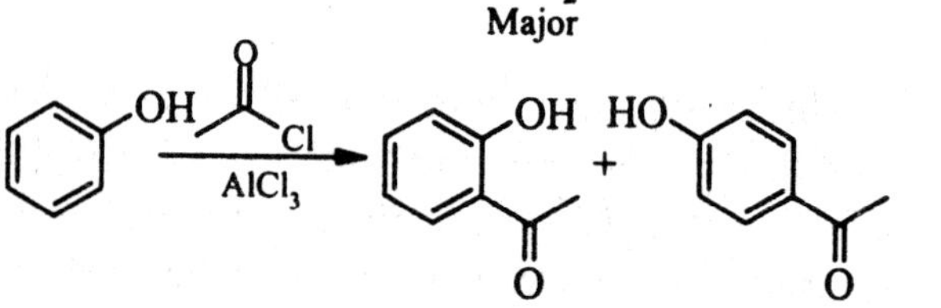 $\xrightarrow{AlCl_3}$ + Major

(c) Nitrobenzene $\xrightarrow[AlCl_3]{Cl_2}$ m-chloronitrobenzene

(d) Acetophenone $\xrightarrow[H_2SO_4]{SO_3}$ m-acetylbenzenesulfonic acid (SO_3H)

(e) Benzene + isobutylene $\xrightarrow{H_2SO_4}$ tert-butylbenzene

Example 31. Select the major product of the following reaction explaining reason for your choice:

Furan $\xrightarrow{NO_2BF_4}$ 2-nitrofuran ($\overset{+}{N}O_2$) or 3-nitrofuran (NO_2)

Solution: Furan is aromatic and one of its H is being replaced by $-NO_2$. Therefore, this seems to be an electrophilic aromatic substitution reaction. The major product of an electrophilic aromatic substitution reaction is derived from more stable carbocation intermediate.

Attack at 2-C:

Furan + $\overset{+}{N}O_2$ → (H, NO_2 at C-2, + on ring) ↔ (H, NO_2, + at C, $\ddot{O}$) ↔ (H, NO_2, $\overset{+}{O}$)

This attack leads to formation of a carbocation that has three significant resonance contributors, one of which features a full octet on every atom.

Attack at 3-C:

Furan + $\overset{+}{N}O_2$ → (NO_2, H at C-3, + on ring, $\ddot{O}$) ↔ (NO_2, H, $\overset{+}{O}$)

This attack leads to formation of a carbocation with two significant resonance contributors, one of which features a full octet on every atom. Since a carbocation with more resonance contributor is generally more stable, attack at C-2 yields a more stable carbocation and hence 2-nitrofuran will the major substitution product.

Example 32. Discuss the relative reactivity of three isomers of xylene towards electrophilic aromatic substitution reaction.

Solution: Meta xylene is the most reactive towards electrophilic substitution reaction due to the reinforced activation by two methyl groups as:

[Resonance structures of m-xylene: H_2C–H hyperconjugation giving $\overset{+}{CH_2H}$ structures with CH_3, and vice versa from the other methyl group]

Hyperconjugation by both $-CH_3$ from the meta position increases electron density at the same carbon indicating high reactivity of meta xylene towards aromatic electrophilic substitution reaction.

Ortho xylene is less reactive than meta xylene due the following reason:

[o-Xylene (H_2C–H, CH_3) ↔ structure with $-$ adjacent to CH_3: +I effect]

Here, + I effect destabilizes the carbanion decreasing its reactivity. Same is the reason for lower reactivity of para isomer.

10.6 NUCLEOPHILIC SUBSTITUTION, ELIMINATION AND ADDITION REACTIONS

Nucleophilic substitution, elimination and addition reactions of benzene derivatives.

Substitution

An early method of preparing phenol (the Dow process) involved the reaction of chlorobenzene with a concentrated sodium hydroxide solution at temperatures above 350°C. The chief products are phenol and diphenyl ether (see below). This apparent nucleophilic substitution reaction is surprising, since aryl halides are generally incapable of reacting by either an SN-1 or SN-2 pathway.

$$C_6H_5\text{—}Cl + NaOH \text{ solution} \xrightarrow{350°\,C} C_6H_5\text{—}OH + C_6H_5O\text{—}C_6H_5 + NaCl$$

The presence of electron-withdrawing groups (such as nitro) ortho and para to the chlorine substantially enhance the rate of substitution, as shown in the set of equations presented on right hand. To explain this, a third mechanism for nucleophilic substitution has been proposed. This two-step mechanism is characterized by initial addition of the nucleophile (hydroxide ion or water) to the aromatic ring, followed by loss of a halide anion from the negatively charged intermediate. The ability of nitro, and other electron withdrawing, groups to stabilize adjacent negative charge accounts for their rate enhancing influence at the ortho and para locations. Such addition-elimination processes generally occur at sp^2 or sp hybridized carbon atoms, in contrast to SN-1 and SN-2 reactions. When applied to aromatic halides, as in the present discussion, this mechanism is called SNAr.

Cl, NO_2 (para) —1.0 M NaOH, 150°C→ OH, NO_2

Cl, NO_2, NO_2 —10^{-3} M NaOH, 100° C→ OH, NO_2, NO_2

Cl, O_2N, NO_2, NO_2 —H_2O, 40°C→ OH, O_2N, NO_2, NO_2

Some distinguishing features of the three common nucleophilic substitution mechanisms are summarized in Table 10.6.

Table 10.6: Features of nucleophilic substitution reactions

Mechanism	*Number of steps*	*Bond formation timing*	*Carbon hybridization*
SN-1	Two	After bond breaking	Usually sp^3
SN-2	One	Simultaneous with bond breaking	Usually sp^3
SNAr	Two	Prior to bond breaking	Usually sp^2

PRACTICE EXERCISE

1. Predict major product expected from aromatic electrophilic substitution of the following compounds when treated with $Cl_2/AlCl_3$:

(a) CH_3 (b) OH, Cl (c) NO_2 (d) O (e) H, N, O (f) NH_2, O (g) CH_3, Cl (h) OMe

2. Write major product in the following reactions:

(a) CH_3 —SO_3, H_2SO_4→ —Br_2, $AlBr_3$→

(b) Cl —Br_2, $AlBr_3$→ —SO_3, H_2SO_4→

(c) [Benzaldehyde; $Cl-CO-CH_2CH_3$ / $AlCl_3$; Sn/HCl; Br_2 / $AlBr_3$]

(d) [Benzoic acid; $Cl-CH_2CH_2CH_3$ / $AlCl_3$; NaOH; $Cl-CH_2CH_2CH_3$]

(e) [2-Phenylpyridine; Br_2 / $AlBr_3$; Mg/Et_2O; ethylene oxide / H_3O^+]

(f) [1-Phenylpropene; Cl_2; $NaNH_2$]

(g) [Phenylpentene; H^+]

3. Write mechanism of bromination of phenyl benzoate and compare reactivity with benzene towards aromatic electrophilic substitution reaction.

4. Bring about the following transformation starting from toluene:
   (a) *p*-toluene sulphonic acid
   (b) *p*-nitro benzyl alcohol
   (c) *o*-nitro benzoic acid
   (d) *p*-nitro benzoic acid

5. Provide mechanism of the following reaction:

(a) Ethylbenzene + isobutyl chloride, $AlCl_3$ → *p*-tert-butyl ethylbenzene (Et, $C(CH_3)_3$)

(b) Anisole (OMe), HNO_3 / H_2SO_4 → *p*-nitroanisole (OMe, NO_2)

(c) Nitrobenzene (NO_2), Br_2 / $AlBr_3$ → *m*-bromonitrobenzene (NO_2, Br)

6. Propose mechanism of bromination of $F_3C—Ph$ and compare its reactivity with benzene towards bromination reaction.

7. Predict major product obtained from nitration of each of the following compounds:

(a) Acetanilide (H, N, CH_3, O) (b) Fluorobenzene (F)

(c) 1-Methylnaphthalene (CH_3) (d) Methyl benzoate (O, OMe)

8. Synthesize the following compounds starting from benzene or toluene:

(a) *N,N*-dimethyl-*p*-toluidine (CH_3, N) (b) *p*-bromobenzenesulphonic acid (Br, SO_3H)

(c) *p*-bromobenzyl chloride (Br, Cl) (d) *o*-nitrostyrene (NO_2)

(e) *m*-nitroacetophenone (NO_2, CH_3, O) (f) *p*-chloronitrobenzene (Cl, NO_2)

(g) 3-(*p*-tolyl)propanol (OH) (h) Terephthalic acid (COOH, COOH)

(i) Butylbenzene

9. Chlorination of alkyl benzene produces mainly mixture of ortho and para isomer, which on further chlorination yields mainly the 2, 4-dichloro derivatives. This compound is converted into 3, 5-dichloro derivatives on heating with $AlCl_3$ for prolonged period. Give mechanistic explanation for this isomerization reaction.

10. Reaction of *p*-xylene with CH_3Cl in presence of $AlCl_3$ catalyst at 25°C yields 1, 2, 4-trimethyl benzene. However, when this product is treated further with $AlCl_3$ at 100°C, majority of 1, 2, 4-trimethyl

benzene is isomerized to 1, 3, 5-trimethyl benzene. Discuss in detail, the mechanism of this isomerization reaction.

11. *m*-xylene on treatment with 80% sulphuric acid at 25°C yields mainly 2, 4-dimethyl benzene sulphonic acid as major product. On the other hand ortho/para xylene are inert in this conditions and sulphonation of these isomers require a drastic condition. Justify.

12. Draw structures of all isomers of tetra chloro benzene and predict, which will be formed in maximum yield when benzene is treated with excess of chlorine in presence of $AlCl_3$. Reaction of tetrachloro benzene with NaOH at elevated temperature produces 2,4,5-trichloro phenol. Propose a mechanism for this conversion and explain, why does this reaction not proceed to replace all of the chlorine substituents with hydroxy groups?

13. Device a synthesis of 1,3,5-benzene tricarboxylic acid starting from benzene.

14. Arrange the following compounds in increasing order of electrophilic aromatic substitution reaction:

(a) 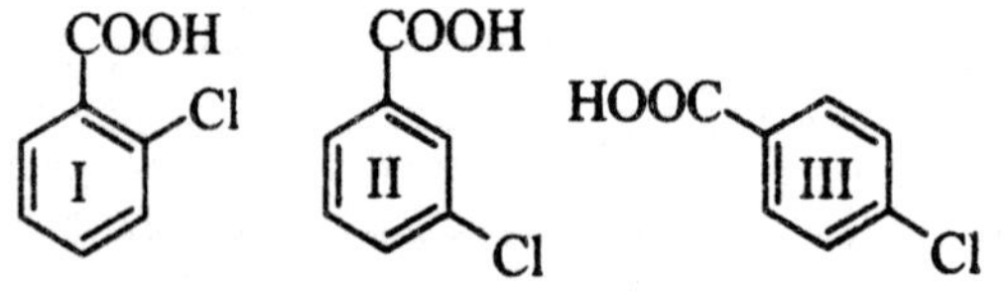 I: COOH, Cl; II: COOH, Cl; III: HOOC, Cl

(b) 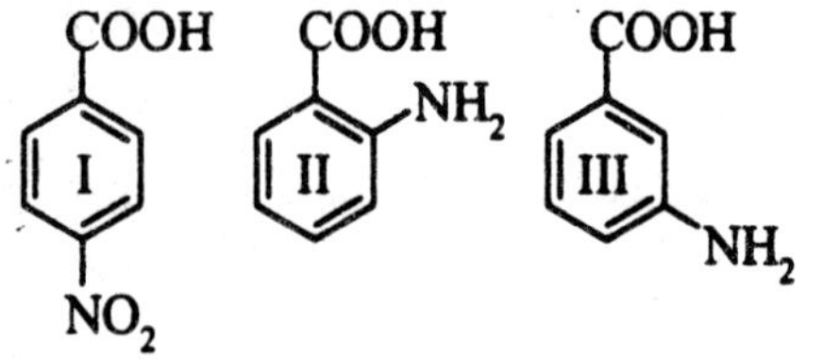 I: COOH, NO_2; II: COOH, NH_2; III: COOH, NH_2

(c) 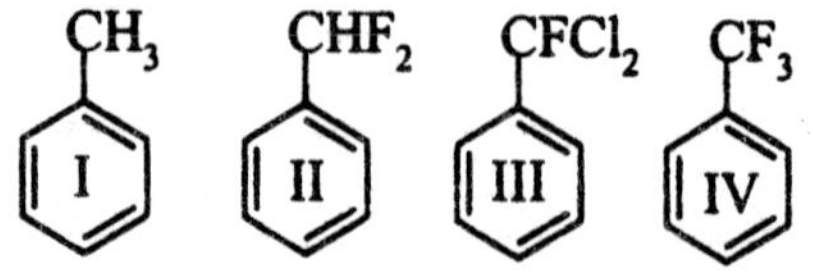 I: CH_3; II: CHF_2; III: $CFCl_2$; IV: CF_3

(d) 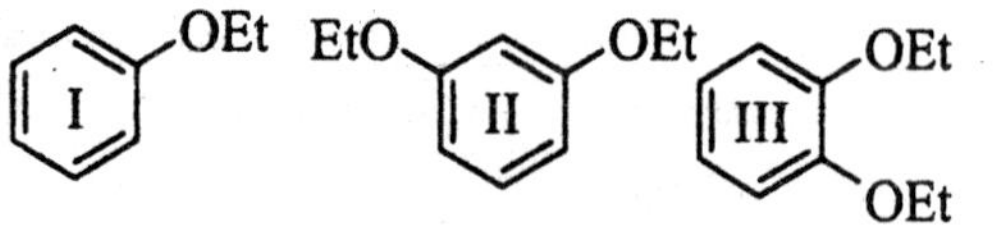 I: OEt; II: EtO, OEt; III: OEt, OEt

15. For each of the substituent in column I, choose the most suitable statement from column II.

I –OH (a) Donate electrons inductively, but does not donate/withdraw electrons by resonance.

II –Br (b) Withdraw electrons by both inductive and resonance effect.

III $-NO_2$ (c) Deactivate the ring and direct ortho/para in electrophilic substitution.

IV $-NH_3^+$ (d) Withdraw electrons inductively but donate by resonance effect.

V C_2H_5- (e) Withdraw electrons inductively, but does not donate/withdraw electron by resonance.

16. For each of the following compounds, indicate the ring carbon that would be brominated, if the compounds were treated with $Br_2/FeBr_3$.

(a) 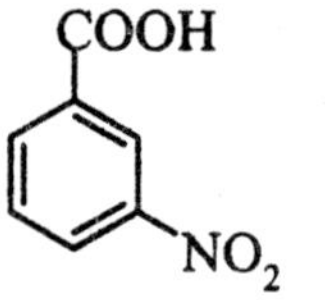 COOH, NO_2 (b) 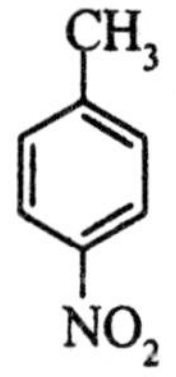 CH_3, NO_2

(c) 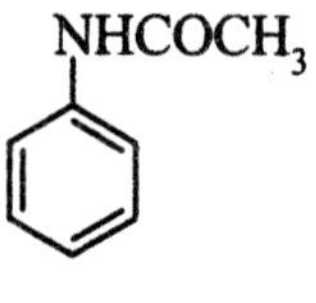 $NHCOCH_3$ (d) 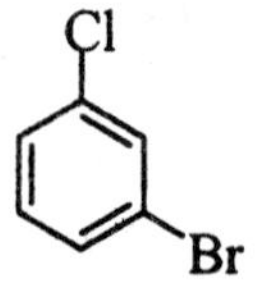 Cl, Br

(e) 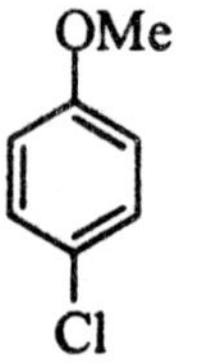 OMe, Cl (f) 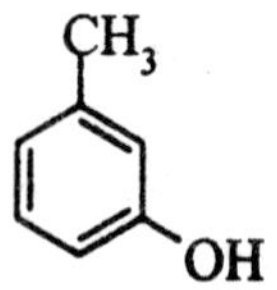 CH_3, OH

(g) 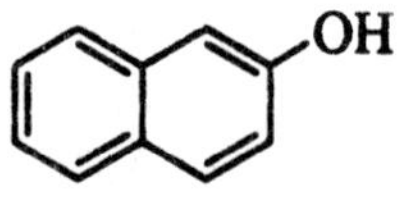 OH (h) 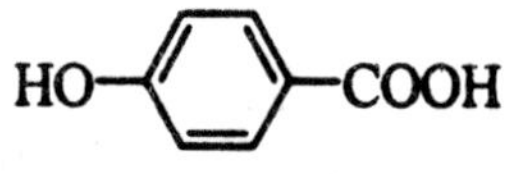 HO–⟨⟩–COOH

(i) 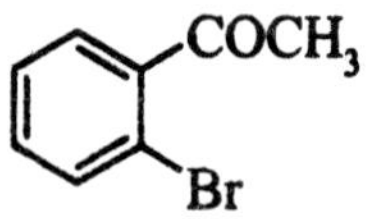 $COCH_3$, Br (j) 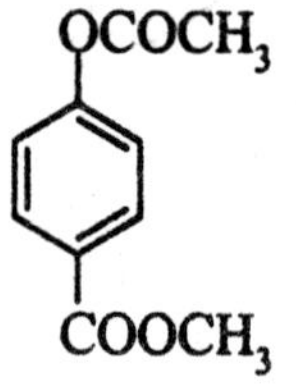 $OCOCH_3$, $COOCH_3$

17. Predict major product in the following reactions:

(a) 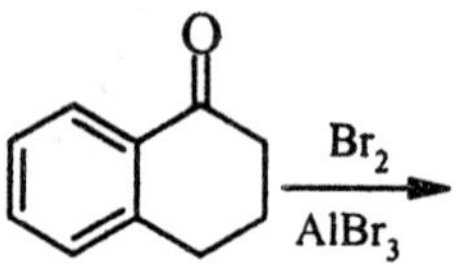 O $\xrightarrow[AlBr_3]{Br_2}$ (b) 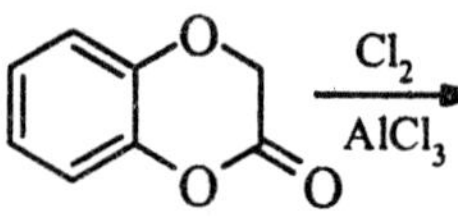 O, O, O $\xrightarrow[AlCl_3]{Cl_2}$

(c) 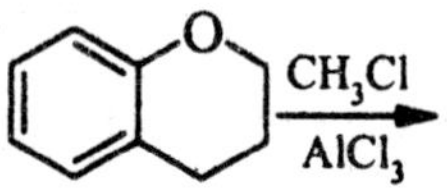 O $\xrightarrow[AlCl_3]{CH_3Cl}$

(d) 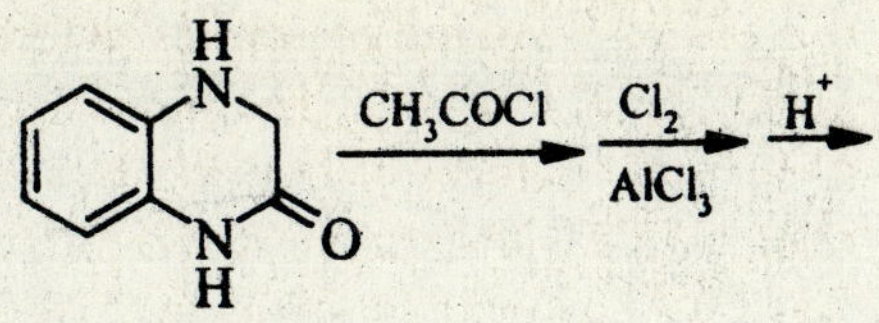 [quinoxalinone] $\xrightarrow{CH_3COCl}$ $\xrightarrow[AlCl_3]{Cl_2}$ $\xrightarrow{H^+}$

18. Propose mechanism of formation of product in the following reactions:

(a) 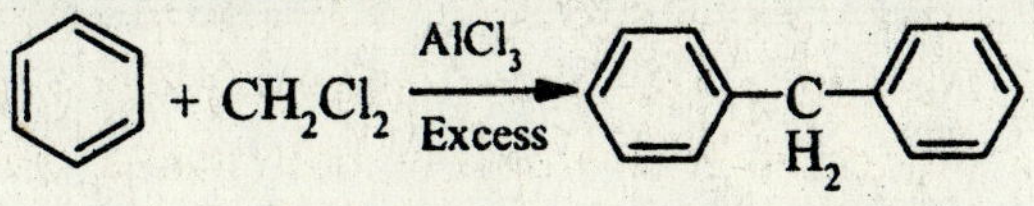 Benzene + CH_2Cl_2 $\xrightarrow[\text{Excess}]{AlCl_3}$ $C_6H_5-CH_2-C_6H_5$

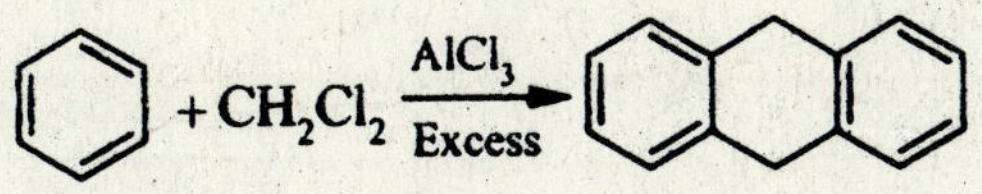

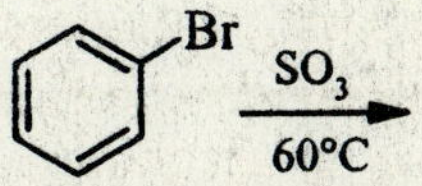

(b) Benzene + CH_2Cl_2 $\xrightarrow[\text{Excess}]{AlCl_3}$ 9,10-dihydroanthracene

(c) [α-methylstyrene] $\xrightarrow{H^+}$ [indane with Ph, CH_3]

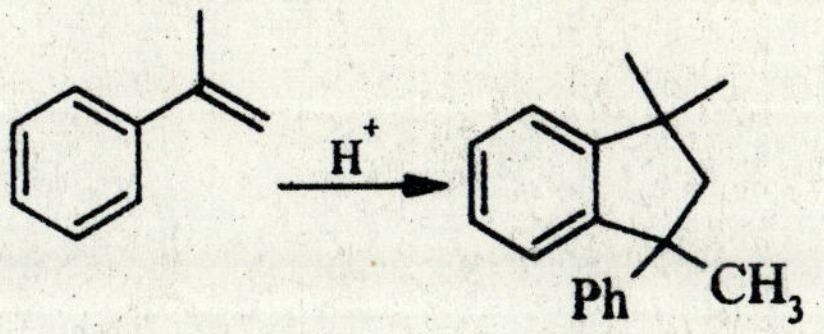

19. Propose mechanism:

[tetrahydropyrene] $\xrightarrow{HNO_3}$ [nitro product] $-NO_2$

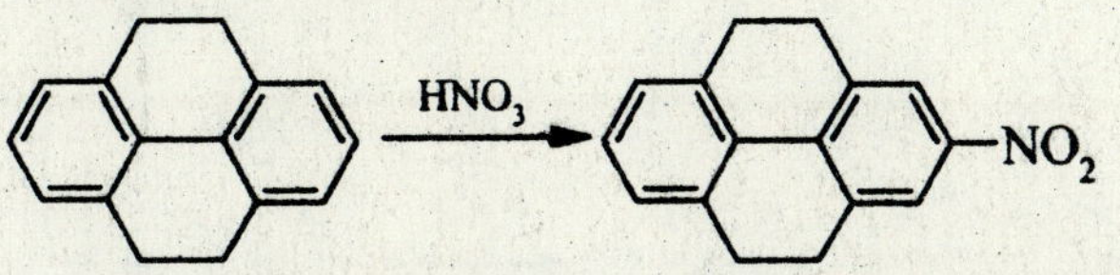

20. Discuss the relative basic strength of the following bases:

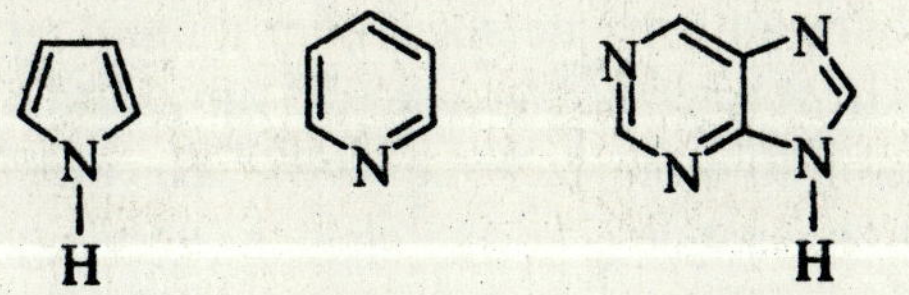

I (pyrrole, N–H) II (pyridine) III (purine, N–H)

21. Discuss the relative reactivity of the following compounds towards electrophilic aromatic substitution reaction:

Toluene Ethyl benzene, *iso*-propyl benzene, *tertiary*- butyl benzene.

22. Complete the following reactions:

(a) 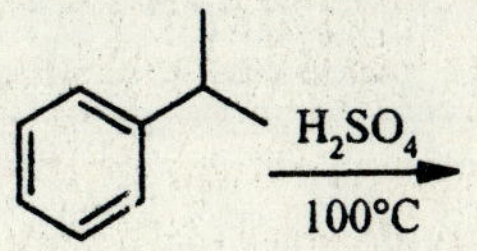 [isopropylbenzene] $\xrightarrow[100°C]{H_2SO_4}$

(b) [bromobenzene] $\xrightarrow[60°C]{SO_3}$ (c) [o-xylene, CH_3, CH_3] $\xrightarrow[\text{Heat}]{H_2SO_4}$

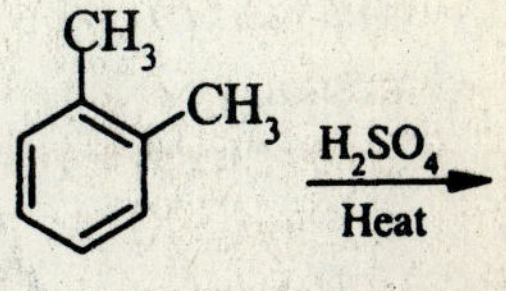

23. When toluene is sulphonated with concentrated sulphuric acid at 0°C, the product mixture consists of 53% *p*-toluene sulphonic acid, 43% of ortho isomer and rest 4% meta isomer. If the sulphonation is carried out at 100° C, *p*-toluene sulphonic acid constitute 79% of the product mixture. How do you explain the correlation of change in composition of the product mixture with temperature of reaction? (*Hint:* Thermodynamic vs Kinetic control of the reaction mechanism.)

24. 1, 3, 5-trimethyl benzene undergoes electrophilic substitution with ICl. Write detail mechanism for this reaction justifying the product formed.

25. Allyl alcohol on treatment with HF/Benzene yield two products: 3-phenyl-1-propene and 1, 2-diphenyl propane. Write mechanism of formation of products in this reaction.

Chapter 11

Aromatic Compounds II

11.1 NITRO DERIVATIVES OF BENZENE

11.1.1 Preparation

I. Direct Nitration of Benzene

Benzene or its derivatives, on treatment with a mixed solution of concentrated nitric acid and concentrated sulphuric acid, undergo nitration via electrophilic aromatic substitution reaction as:

$$C_6H_6 + HNO_3 \xrightarrow{H_2SO_4} C_6H_5NO_2$$

(Nitrobenzene)

Mechanism of nitration: H_2SO_4 and HNO_3 react to produce nitronium ion, which act as an electrophile. Nitronium ion involve in the subsequent step resulting in formation of nitrobenzene or its derivative, depending on starting aromatic compound as:

Step I: $HNO_3 + 2H_2SO_4 \rightarrow 2HSO_4^- + H_3O^+ + NO_2^+$

Nitronium ion: It is a resonance stabilized intermediate as:

$$O{=}\overset{+}{N}{\rightarrow}O \leftrightarrow [O{=}N{=}O]^+$$

Step II: $C_6H_6 + O{=}\overset{+}{N}{=}O \longrightarrow$ [arenium ion, H and NO_2 on same carbon] $\xrightarrow{-H^+} C_6H_5NO_2$

Other reagents, viz. fuming nitric acid, nitric acid in organic solvents like acetic acid, nitromethane, etc. are also used for nitration of benzene and its homologue.

Dinitration: Benzene or mononitrobenzene, when treated with fuming nitric acid, fuming sulphuric acid mixture, *m*-dinitrobenzene is produced as the major product:

$$C_6H_5NO_2 \xrightarrow[H_2SO_4\ (Fuming)]{HNO_3} m\text{-}C_6H_4(NO_2)_2$$

Other isomers of dinitrobenzene are produced by indirect method as:

p-dinitrobenzene

$$C_6H_5NHCOCH_3 \xrightarrow[H_2SO_4]{HNO_3} p\text{-}O_2NC_6H_4NHCOCH_3 \xrightarrow{H^+} p\text{-}O_2NC_6H_4NH_2 \xrightarrow[HCl/0°C]{NaNO_2} \xrightarrow[Cu]{NaNO_2} p\text{-}C_6H_4(NO_2)_2$$

ortho-dinitrobenzene

$$C_6H_5NHCOCH_3 \xrightarrow{C\cdot H_2SO_4} p\text{-}HO_3SC_6H_4NHCOCH_3 \xrightarrow[C\cdot HNO_3]{C\cdot H_2SO_4} \text{2-nitro-4-sulpho-acetanilide} \xrightarrow[H_2O]{H^+} o\text{-}O_2NC_6H_4NH_2 \xrightarrow[HCl/0°C]{NaNO_2} \xrightarrow[Cu]{NaNO_2} o\text{-}C_6H_4(NO_2)_2$$

Trinitration: Trinitro benzene cannot be prepared by direct nitration of benzene since after dinitration, the ring becomes inert towards electrophilic substitution reaction. Inertness of dinitrobenzene towards further electrophilic substitution reaction is attributed to the very high electron withdrawing effect given by two nitro groups. However, trinitrobenzene is synthesized but by indirect method. An activated benzene ring, viz. toluene is trinitrated first and then the substituent (here methyl group) is removed as shown in the following synthesis:

$$C_6H_5CH_3 + HNO_3 \xrightarrow[Heat]{H_2SO_4} \text{2,4,6-}(O_2N)_3C_6H_2CH_3$$

(Trinitrotoluene)
(an explosive)

$$\xrightarrow[NaOH,\ Heat]{KMnO_4} \text{2,4,6-}(O_2N)_3C_6H_2COONa \xrightarrow[CaO]{NaOH} \text{1,3,5-}C_6H_3(NO_2)_3$$

II. Nitration of Substituted Benzene

Reactivity and orientation of nitration of a substituted benzene depends on the nature of substituent already present in the ring, e.g.

Phenol $\xrightarrow[\text{Dilute}]{HNO_3}$ o-nitrophenol + p-nitrophenol

Phenol is nitrated with dilute nitric acid. Nitration of phenol takes place in presence of dilute nitric acid due to very high activating power of the —OH functional group. If phenol is nitrated with mixed acid (mixed solution of concentrated nitric acid and sulphuric acid), trinitrophenol is produced as:

Phenol $\xrightarrow[H_2SO_4]{HNO_3}$ (Both concentrated) 2,4,6-trinitrophenol

(Trinitrophenol)
Picric acid

11.1.2 General Properties

Most of the aromatic nitro compounds are yellow coloured, crystalline solid and a few including nitrobenzene, are yellow liquids at room temperature. Most of the aromatic nitro compounds are steam volatile and cannot be distilled under atmospheric pressure because, when strongly heated, they decompose with violence. Most of the aromatic nitro compounds are denser than water and insoluble in water.

Since, nitro group is attacked by Grignard reagent, nitro substituted aryl halides cannot be used in Grignard synthesis.

Example 1. Ortho nitrophenol being more volatile than its para isomer, can be separated from the mixture by steam distillation, justify.

Solution: Ortho nitrophenol undergo intramolecular H-bonding (chelation), which makes them to behave like a monomer, and distills at lower temperature:

intramolecular H-bonding
(Chelation)

On the other hand, *p*-nitrophenol undergo intermolecular H-bonding, which makes them to behave like a polymer and distills at higher temperature:

(Intermolecular H-bonding)

Example 2. Ortho nitrophenol has lower solubility in water than its para isomer, explain.

Solution: Solubility of a compound in hydroxylic solvents like water depends on, among other factors, the power to form H-bonds with the solvent. In ortho nitrophenol, H-bonding with the solvent is hindered due to chelation and hence, its solubility is lowered.

Example 3. Synthesize *m*-nitrotoluene from benzene in good yield, keeping in mind that nitrobenzene fails in Friedel Craft reaction.

Solution

Benzene $\xrightarrow[AlCl_3]{CH_3Cl}$ toluene $\xrightarrow[H_2SO_4]{HNO_3}$ p-nitrotoluene $\xrightarrow{Zn/HCl}$ p-toluidine $\xrightarrow{Ac_2O}$ $\xrightarrow[H_2SO_4]{HNO_3}$ 4-acetamido-3-nitrotoluene ($NHCOCH_3$, NO_2) $\xrightarrow{H^+}$ $\xrightarrow[HCl\text{-}0°C]{NaNO_2}$ $\xrightarrow{H_3PO_2}$ m-nitrotoluene

11.1.3 Chemical Reactions

I. Reduction

Nitrobenzene can be reduced to aniline using following reducing agents:

Nitrobenzene $\xrightarrow{LiAlH_4 \;/\; Zn/HCl \;/\; Sn/HCl}$ Aniline (NH_2)

If more than one nitro groups are present on the ring, they can be reduced all together or stepwise using selective reducing agent as:

m-Dinitrobenzene $\xrightarrow{Na_2S}$ m-nitroaniline

m-Dinitrobenzene $\xrightarrow{Zn/HCl}$ m-phenylenediamine

Besides Na_2S, NH_4SH and NaHS also reduces one of the several nitro groups present on the benzene ring, leaving other intact. If reduction is done with Zn dust in ammonium chloride solution, phenyl hydroxyl amine is produced as:

$$C_6H_5NO_2 \xrightarrow[NH_4Cl\ (aq)]{Zn} C_6H_5NHOH$$

In alkaline medium, reduction product depends on the nature of reducing agents used as:

$$C_6H_5NO_2 \xrightarrow[NaOH]{Zn/MeOH} C_6H_5-N{=}N-C_6H_5$$

$$C_6H_5NO_2 \xrightarrow{Zn/NaOH} C_6H_5-NH-NH-C_6H_5$$

In case of a 2, 4-dinitroalkyl benzene, if NH_4HS or Na_2S is used as reducing agent, nitro group at para position is reduced while, if $SnCl_2/HCl$ is used as reducing agent, nitro group ar ortho position is reduced as:

(R, 2-NO_2, 4-NO_2 benzene) $\xrightarrow[\text{or } Na_2S]{NH_4HS}$ (R, 2-NO_2, 4-NH_2 benzene)

(R, 2-NO_2, 4-NO_2 benzene) $\xrightarrow{SnCl_2—HCl}$ (R, 2-NH_2, 4-NO_2 benzene)

Reduction of nitro group in presence of other reducible functional group on the ring: If nitro compound contain an unsaturated side chain or an aldehyde/ketone group, alkaline solution of ferrous sulphate reduces only nitro group leaving other group intact. If only aldehyde group is present, $SnCl_2$ may be used:

(m-NO_2 benzaldehyde, CHO) $\xrightarrow{SnCl_2—HCl}$ (m-NH_2 benzaldehyde, CHO)

II. Electrophilic Aromatic Substitution Reaction of Nitrobenzene

Nitro group is a very strong electron withdrawing group both by inductive effect and resonance effect as:

(Resonance structures of nitrobenzene showing + charge at ortho and para positions; resonance hybrid with δ− on the oxygens and δ+ at ortho and para positions)

Because, the electron withdrawing effect of $-NO_2$, benzene ring is deactivated for the attack by electrophile and hence, nitrobenzene is very less reactive towards electrophilic aromatic substitution reaction.

Also, it appears from electronic displacement that electron withdrawing effect of nitro group is more intense at ortho and para position, therefore electrophilic substitution, if occur at all, takes place at meta positions.

$$C_6H_5NO_2 \xrightarrow{Cl_2/AlCl_3} m\text{-}ClC_6H_4NO_2$$

Nitrobenzene is inert towards Friedal-Craft reaction.

III. Nucleophilic Substitution Reaction in Nitro Substituted Ring

In general, nitro group activate the benzene ring towards nucleophilic substitution reaction, but position of nitro group with respect to the leaving group has profound effect on the reactivity as described below:

Ortho nitro chlorobenzene

(o-chloronitrobenzene + ^-OH → HO, Cl σ-complex with NO_2 ↔ resonance structures, I → o-nitrophenol (OH, NO_2))

Para nitro chlorobenzene

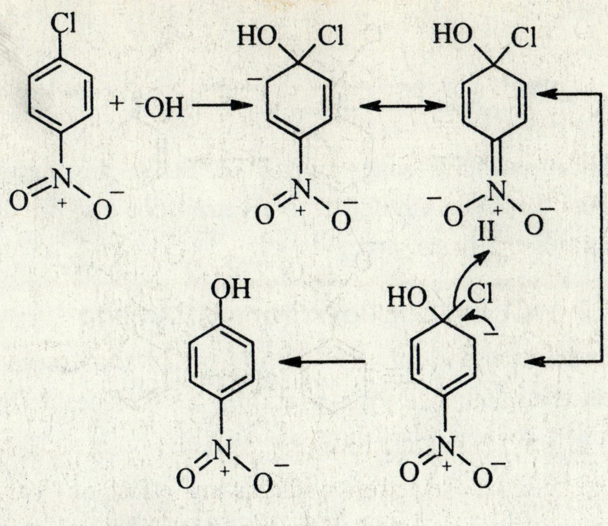

Meta nitro chlorobenzene

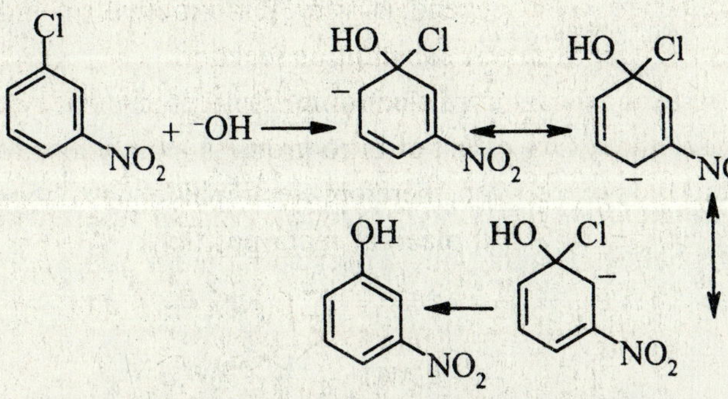

Inspections of these canonical structures shows that in both ortho and para isomers, carbanion intermediates have contributions from structures involving extended conjugation, resulting in spreading of charge which resides on most electronegative atom (structures I and II). These are the most stable contributing structures in ortho and para isomers, and no such structure can be drawn for the *m*-isomer, the latter is less stable than the formers.

Example 4. Write down the major substitution product in the following reactions:

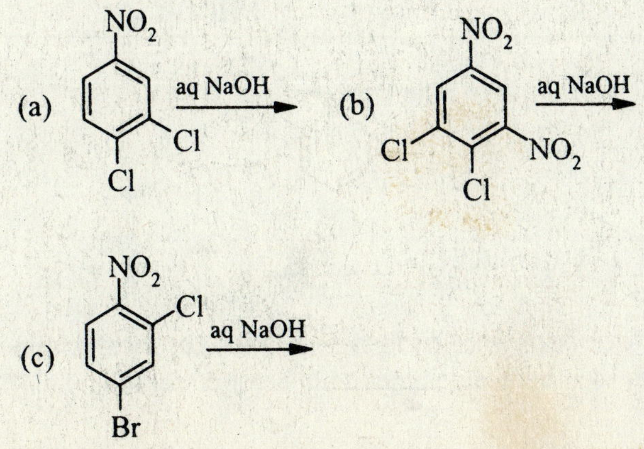

(a) aq NaOH (b) aq NaOH (c) aq NaOH

Solution

(a) (b) (c)

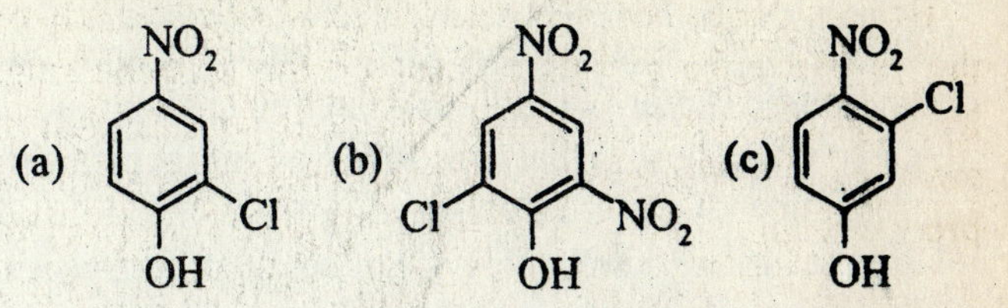

Example 5. Hydrogen ortho to a nitro group is highly activated towards nucleophilic substitution reaction. On the basis of above fact, predict the major product of the following reaction and justify.

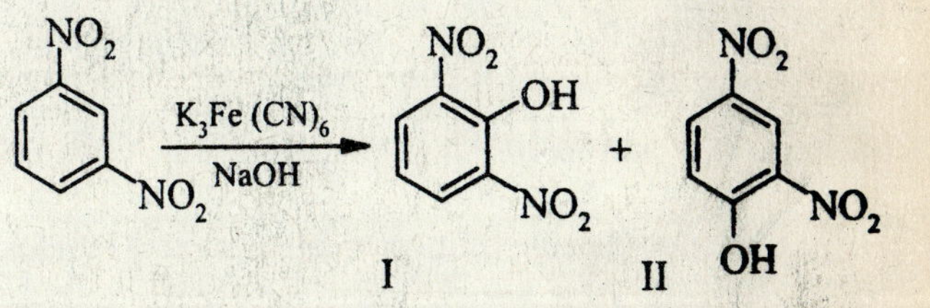

$K_3Fe(CN)_6$, NaOH → I + II

Solution: II will be the major product, although substitution site in I is more activated but at the same time, highly sterically hindered.

Example 6. An aromatic compound A has molecular formula C_8H_9Br and on treatment with diethylmalonate/ sodium ethoxide, followed by refluxing with H_2SO_4 gave B($C_{10}H_{12}O_2$). B on vigorous oxidation, gave C($C_8H_6O_4$), which on mono nitration gave only one mono-nitro derivative D. Identify A-D.

Solution: C must be a benzene dicarboxylic acid. Also, since C gave single mono-nitration product, it must be 1, 4-dicarboxylic acid.

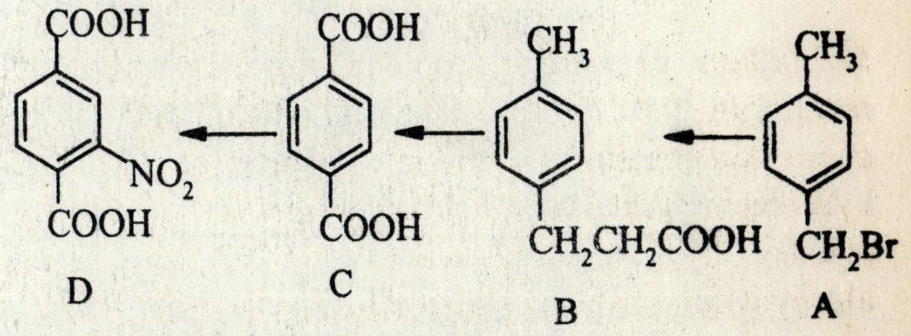

D ← C ← B ← A

11.2 ANILINE AND ITS DERIVATIVES

11.2.1 Preparation

I. Primary aromatic amines (aniline and substituted aniline) are usually prepared by reduction of nitro compounds. Various reducing agents are used for this purpose, but the most common one is Zn (Other electropositive metals, viz. Sn, Fe are also used) in dilute hydrochloric acid or acetic acid medium:

$$C_6H_5NO_2 + Zn/HCl \rightarrow C_6H_5NH_2$$

Zn/HCl is the most common reducing agent here, but Zn/acetic acid is more usefull for reduction of compounds which contain a group that is attacked by inorganic acid, e.g. reduction of *p*-nitroacetanilide in hydrochloric acid solution leads to some hydrolysis of acetyl group, acetic acid prevent such hydrolysis reaction:

p-Nitroacetanilide (NHCOCH₃ / NO₂ on benzene) —Zn/HCl→ $H_2N-C_6H_4-NHCOCH_3$ → $H_2N-C_6H_4-NH_2$ (hydrolysis product)

p-Nitroacetanilide —Zn/AcOH→ $H_2N-C_6H_4-NHCOCH_3$

II. By nucleophilic substitution reaction of aryl halides:

$$C_6H_5Cl + NH_3 \xrightarrow[\text{High pressure}]{\text{High temperature}} C_6H_5NH_2$$

III. By hydrolysis of benzene isocyanide:

$$C_6H_5NC + OH_2 \xrightarrow{H^+} C_6H_5NH_2 + HCOOH$$

IV. By Beckmann's rearrangement of suitable oxime:

$$\underset{Ph}{\overset{R}{}}\!\!>C{=}N{-}OH \xrightarrow{H^+} R-\underset{\|}{\overset{}{C}}(=O)-NHPh \xrightarrow{H_3O^+ \mid H_2O} PhNH_2 + RCOOH$$

11.2.2 Properties of Aniline

General Properties

Aniline is, when freshly prepared, a colourless liquid with unpleasant odour, and poisonous in nature. On exposure to atmosphere, it darkens since it is highly sensitive to oxidation.

Aniline, being basic in nature, forms crystalline salt with strong inorganic acid, and can be separated from its etheral solution by extraction with an aqueous solution of strong inorganic acid.

Aniline is less basic than an aliphatic amine due to delocalization of lone pair of electron on nitrogen into the benzene ring as:

(Resonance structures of aniline: $\ddot{N}H_2$ ↔ $\overset{+}{N}H_2$ (ortho −) ↔ $\overset{+}{N}H_2$ (para −) ↔ $\overset{+}{N}H_2$ (ortho −) ↔ NH_2)

Electron withdrawing group in the benzene ring further lowers the basic strength and electron donating group increases the basic strength.

11.2.3 Chemical Reactions of Aniline

I. Replacement of hydrogen of amino group: Hydrogen atom of NH_2– in aniline are replaceable by halogen, sodium, potassium, etc. as:

$$C_6H_5NH_2 \xrightarrow[\text{Heat}]{Cl_2} C_6H_5NHCl$$

$$C_6H_5NH_2 \xrightarrow[\text{Heat}]{\text{Na or K}} C_6H_5\overset{-}{N}H\overset{+}{Na}$$

II. Reaction with alkyl halides: Aniline reacts with alkyl halides forming quaternary ammonium salts as:

$$C_6H_5NH_2 + \underset{\text{Excess}}{CH_3I} \longrightarrow C_6H_5\overset{+}{N}(CH_3)_3\,I^-$$

III. Carbyl amine reaction: Like primary aliphatic amine, aniline also forms isocyanide when treated with alkaline solution of chloroform as:

$$C_6H_5NH_2 + CHCl_3 + 3KOH \rightarrow C_6H_5NC + 3KCl + 3H_2O$$

IV. Reaction with acid chloride, anhydride: Aniline condenses with acid chloride or anhydride forming anilide as:

$$C_6H_5NH_2 + RCOCl \longrightarrow C_6H_5NHCOR + HCl$$

$$C_6H_5NH_2 + (RCO)_2O \longrightarrow C_6H_5NHCOR + RCOOH$$

V. Reaction with aldehydes and ketones: Aniline condenses with aldehydes and ketones forming Schiff base:

$$C_6H_5NH_2 + C_6H_5CHO \rightarrow C_6H_5N{=}CHC_6H_5 + H_2O$$

This reaction can be used to protect —NH_2 group, especially during nitration, otherwise nitric acid tend to oxidize —NH_2 group.

VI. Reaction with nitrous acid: Aniline forms diazonium salt with nitrous acid:

$$C_6H_5NH_2 + HNO_2 + HCl/0°C \rightarrow C_6H_5N_2Cl + 2H_2O$$

VII. Electrophilic substitution reaction in aniline: —NH_2 is a highly activating, ortho/para directing group in electrophilic substitution reaction. The activating power of —NH_2 is so intense that if, aniline is treated with an aqueous solution of chlorine or bromine, an immediate precipitate of trichloride or tribromide is produced, otherwise chlorination or bromination require a Lewis acid catalyst:

$$C_6H_5NH_2 + Br_2/H_2O \longrightarrow \text{2,4,6-tribromoaniline}$$

The yield is quantitative, can be used for estimation of aniline. For monohalogenation, the activating power of —NH_2 group is reduced by acetylation:

$$C_6H_5NH_2 + RCOCl \longrightarrow C_6H_5NHCOR \xrightarrow{Br_2/Fe} \text{o-BrC}_6H_4NHCOR + \underset{\text{Major}}{\text{p-BrC}_6H_4NHCOR}$$

Above products on hydrolysis yields corresponding monobromo aniline. Ortho bromo aniline may conveniently be prepared by first blocking the para position by sulphonation as:

$$C_6H_5NHCOR + H_2SO_4 \longrightarrow \text{p-}HO_3SC_6H_4NHCOR \xrightarrow{Br_2/Fe} \text{2-Br-4-}SO_3H\text{-}C_6H_3NHCOR \xrightarrow{\text{Steam distill}} \text{o-BrC}_6H_4NH_2$$

Para iodo aniline is usually prepared by the action of iodine on aniline in presence of aqueous $NaHCO_3$ solution. This is the very high activating power of —NH_2 that molucular iodine is capable of direct iodination:

$$C_6H_5NH_2 + I_2 \xrightarrow[\text{Aqueous}]{NaHCO_3} \text{p-}IC_6H_4NH_2$$

To some extent, meta substitution in aniline can be done in presence of concentrated solution of inorganic acid. The meta orienting effect is being due to the presence of —NH_3^+ group.

$$C_6H_5NH_2 \xrightarrow{\text{Conc. } H_2SO_4} C_6H_5\overset{+}{N}H_3 \xrightarrow{Br_2/FeBr_3} \text{m-BrC}_6H_4\overset{+}{N}H_3$$

Strong -I effect of —NH_3^+ controls the orientation and it is *m*-directing

VIII. Reaction with phenol: Aniline, on heating with phenol in presence of $ZnCl_2$ forms diphenylamine as:

$$C_6H_5NH_2 + C_6H_5OH \xrightarrow[\text{Heat}]{ZnCl_2} (C_6H_5)_2NH$$

IX. Oxidation of —NH_2 group in aniline: —NH_2 group of aniline can be oxidized stepwise, first to nitroso (—NO) and finally to (—NO_2) group or directly to nitro group using selective oxidizing agents:

$$C_6H_5NH_2 + \xrightarrow[\text{Caro acid}]{H_2SO_5} C_6H_5NO \xrightarrow[H_2O_2]{HNO_3} C_6H_5NO_2$$

$$\xrightarrow{CH_3CO_3H} C_6H_5NO_2$$

Example 7. Bring about the following transformations in good yield:

$$C_6H_5CH_2NH_2 \rightarrow C_6H_5NHCH_3$$

Solution: $C_6H_5CH_2NH_2 + HNO_2 \rightarrow C_6H_5CH_2OH$

$C_6H_5CH_2OH + KMnO_4 \rightarrow C_6H_5COOH;$
$C_6H_5COOH + NH_3/\text{heat} \rightarrow C_6H_5CONH_2$

$C_6H_5CONH_2 + KOH/Br_2 \rightarrow C_6H_5NH_2;$
$C_6H_5NH_2 + CHCl_3/KOH \rightarrow C_6H_5NC$

$C_6H_5NC + LiAlH_4 \rightarrow C_6H_5NHCH_3$

Example 8. Synthesize the following compounds starting from aniline:

(a) o-phenylenediamine ($C_6H_4(NH_2)_2$, NH_2 groups ortho) (b) $(C_6H_5)_3N$ (c) $Ph_2N—NPh_2$

Solution

(a) NH_2 (aniline) $\xrightarrow{Ac_2O}$ NHAc $\xrightarrow{H_2SO_4}$ NHAc / SO_3H $\xrightarrow[H_2SO_4]{HNO_3}$ NHAc / NO_2 / SO_3H $\xrightarrow[Heat]{OH_2}$ NH_2 / NO_2 $\xrightarrow{Zn/HCl}$ NH_2 / NH_2

(b) $C_6H_5NH_2 + C_6H_5OH \xrightarrow[Heat]{Zn/HCl} (C_6H_5)_2NH \xrightarrow[K_2CO_3/Cu]{C_6H_5I} (C_6H_5)_3N$

(Ullamann reaction)

(c) $(C_6H_5)_2NH$ (from b) $\xrightarrow{Na/heat} (C_6H_5)_2\bar{N}\,\overset{+}{Na} \xrightarrow{I_2/heat} (C_6H_5)_2N—N(C_6H_5)_2$

Example 9. Complete the following reactions:

(a) NH_2 (aniline) $\xrightarrow{Ac_2O}$ A $\xrightarrow{H_2SO_4}$ B $\xrightarrow{HNO_3}$ C $\xrightarrow[Heat]{OH_2}$ D

(b) $Ph_2NH \xrightarrow{Li} E \xrightarrow[CuI]{PhI} F$

(c) $G \xleftarrow{LiAlH_4} PhNO_2 \xrightarrow[Pd\text{-}C]{NaBH_4} H$

Solution

(a) A: $NHCOCH_3$ (acetanilide); B: $NHCOCH_3$ / SO_3H; C: $NHCOCH_3$ / NO_2 / SO_3H; D: NH_2 / NO_2

(b) E: $Ph_2\bar{N}\,\overset{+}{Li}$, F: Ph_3N (c) G: PhN═NPh, H: $PhNH_2$

Example 10. In electrophilic aromatic substitution reaction, nitroso group (—NO) is ortho/para directing but deactivating. Provide mechanistic explanation.

Solution

:N═O ⟷ $\overset{+}{N}$═O ⟷ $\overset{+}{N}$═O ⟷ $\overset{+}{N}$═O ⟷ $\ddot{N}$═O

The above electronic delocalization indicates that electrophile would preferably attach at ortho/para positions. At the same time, —NO is strong electron withdrawing group by both inductive as well as resonance effect, which is responsible for deactivation of ring towards electrophilic substitution reaction as:

:N═O ⟷ :N—$\bar{O}$ ⟷ :N—$\bar{O}$ ⟷ :N—$\bar{O}$ ⟷ $\ddot{N}$═O

Example 11. An organic compound A($C_9H_{11}ON$) evolves H_2 gas on heating with sodium metal. A on refluxing with alkaline solution of Br_2 yields B($C_8H_{11}N$). B on oxidation yields C($C_7H_7O_2N$). C on subjecting to aromatic electrophilic bromination with $Br_2/FeBr_3$, yields D($C_7H_6O_2NBr$) as the single mono bromo isomer. Deduce structures of A to D.

Solution

A ($CONH_2$ / CH_2CH_3) $\xrightarrow{KOH/Br_2}$ B (NH_2 / CH_2CH_3) $\xrightarrow{[O]}$ C (NH_2 / COOH) $\xrightarrow{Br_2/FeBr_3}$ D (NH_2 / Br / COOH)

A $\xrightarrow[Heat]{Na}$ $CO\bar{N}H\,\overset{+}{Na}$ / CH_2CH_3

11.3 BENZENE DIAZONIUM SALT

11.3.1 Preparation of Diazonium Salts

When a primary aromatic amine is treated with nitrous acid in ice cooled conditions, the product formed is an unstable compound, known as diazonium salt:

$$C_6H_5NH_2 + HNO_2 + HCl/0°C \rightarrow C_6H_5N_2^+Cl^- + H_2O$$

Commonly used reagent for diazotization is sodium nitrite or potassium nitrite, an inorganic acid (HCl or H_2SO_4 or HBF_4) and a primary aromatic amine, in ice cooled solution:

NH_2 (benzene) $\xrightarrow[HCl\ 0°C]{NaNO_2}$ $\overset{+}{N_2}\overset{-}{Cl}$ (benzene)

Benzene diazonium chloride

Diazonium salt is electrolytic in aqueous solution, colourless solid at room temperature and highly soluble in water.

The solubility of diazonium salt is attributed to resonance as:

11.3.2 Reactions of Diazonium Salts

Reactions of diazonium salts can be divided into two groups, one that involve liberation of nitrogen gas and replacement of diazo group by other univalent group and other, in which diazo group is retained.

[A] Replacement Reaction

I. Replacement by hydrogen: Treatment of diazonium salt with hypophosphorus acid (H_3PO_2) leads to replacement of diazo group by hydrogen as:

$$ArN_2^+Cl^- + H_3PO_2 \rightarrow C_6H_6$$

Example 12. Synthesize *m*-bromotoluene, starting from bromo benzene.

Solution: Since bromo group is ortho/para directing, direct methylation will give ortho/para bromo toluene, hence direct methylation will not work.

Br (benzene) $\xrightarrow[H_2SO_4\ Concentrated]{HNO_3}$ Br, NO_2 $\xrightarrow{Sn/HCl}$ Br, NH_2 $\xrightarrow{Ac_2O}$ $\xrightarrow{CH_3Cl\ AlCl_3}$ Br, CH_3, $NHCOCH_3$ $\xrightarrow[Heat]{H_2O}$ $\xrightarrow[HCl\ 0°C]{NaNO_2}$ $\xrightarrow{H_3PO_2}$ Br, CH_3

Example 13. Synthesize 4-fluoro-3-nitro toluene, starting from toluene.

Solution

CH_3 (benzene) $\xrightarrow[H_2SO_4\ Concentrated]{HNO_3}$ CH_3, NO_2 $\xrightarrow{Sn/HCl}$ CH_3, NH_2 $\xrightarrow{HNO_3}$ CH_3, NO_2, NH_2 $\xrightarrow{NaNO_2\ HCl\ 0°C}$ $\xrightarrow[Heat]{NaBF_4}$ CH_3, NO_2, F

II. Replacement by hydroxyl group: Benzene diazonium salt on boiling with water gives diazo group replaced by hydroxy group as:

$$ArN_2^+HSO_4^- + H_2O/hcat \rightarrow ArOH + N_2 + H_2SO_4$$

Example 14. Synthesize *m*-Cresol, starting from toluene.

Solution

CH_3 (benzene) $\xrightarrow[H_2SO_4\ Concentrated]{HNO_3}$ CH_3, NO_2 $\xrightarrow{Sn/HCl}$ CH_3, NH_2 $\xrightarrow{HNO_3}$ CH_3, NO_2, NH_2 $\xrightarrow[HCl\ 0°C]{NaNO_2}$ $\xrightarrow{H_3PO_2}$ CH_3, NO_2 $\xrightarrow{Zn/HCl}$ CH_3, NH_2 $\xrightarrow[HCl\ 0°C]{NaNO_2}$ $\xrightarrow{H_2O/boil}$ CH_3, OH

Example 15. Synthesize resorcinol starting from benzene:

Solution

Benzene → (Fuming HNO_3 / Fuming H_2SO_4) → m-dinitrobenzene (NO_2, NO_2) → (Zn/HCl) → m-phenylenediamine (NH_2, NH_2) → ($NaNO_2$/HCl, 0°C) → bis-diazonium salt ($\overset{+}{N}_2\overset{-}{Cl}$, $\overset{+}{N}_2\overset{-}{Cl}$) → ($H_2O$, Boil) → resorcinol (OH, OH)

Example 16. Synthesize *m*-aminophenol starting from benzene.

Solution

Benzene → (Fuming HNO_3 / Fuming H_2SO_4) → m-dinitrobenzene → (Sn/HCl) → m-nitroaniline (NO_2, NH_2) → ($NaNO_2$/HCl, 0°C) → (NO_2, $\overset{+}{N}_2\overset{-}{Cl}$) → ($H_2O$, Boil) → m-nitrophenol (NO_2, OH) → (Sn/HCl) → m-aminophenol (NH_2, OH)

III. Replacement by halogens (Sandmeyer reaction): When a diazonium salt is run into a solution of cuprous halide, dissolved in corresponding halogen acid, the diazo group is substituted by halogen, e.g.

m-toluidine (CH_3, NH_2) → ($NaNO_2$/HCl, 0°C) → (CH_3, $\overset{+}{N}_2\overset{-}{Cl}$) → (CuCl, ClH) → m-chlorotoluene (CH_3, Cl); → (CuBr, BrH) → m-bromotoluene (CH_3, Br)

Substitution by iodide is achieved by boiling diazonium salt with aqueous potassium iodide solution:

(CH_3, NH_2) → ($NaNO_2$/HCl, 0°C) → (CH_3, $\overset{+}{N}_2\overset{-}{Cl}$) → (KI (boil)) → ($CH_3$, I)

Substitution by fluoride is conveniently done by heating diazonium salt with $NaBF_4$ as:

$$ArN_2^+Cl^- + NaBF_4 \rightarrow ArF + N_2 + NaCl + BF_3$$

Example 17: Synthesize 4-bromophenol from chloro -benzene.

Solution

Chlorobenzene → (HNO_3 / H_2SO_4) → p-chloronitrobenzene (Cl, NO_2) → (aq KOH/heat / SNAr) → p-nitrophenol (OH, NO_2) → ($SnCl_2$ / ClH) → p-aminophenol (OH, NH_2) → ($NaNO_2$ / HCl-0°C) → (CuBr / BrH) → 4-bromophenol (OH, Br)

Example 18: Synthesize 3-fluorophenol, starting from benzene.

Solution

Benzene → (Fuming HNO_3 / Fuming H_2SO_4) → m-dinitrobenzene (NO_2, NO_2) → (NH_4HS) → m-nitroaniline (NO_2, NH_2) → ($NaNO_2$ / HCl-0°C) → ($NaBF_4$ / Heat) → (NO_2, F) → ($SnCl_2$ / ClH) → ($NaNO_2$ / HCl-0°C) → (H_2O/heat) → 3-fluorophenol (OH, F)

IV. Replacement by a cyano group: This is a modification of Sandmeyer reaction. When diazonium salt is treated with a solution of cuprous cyanide dissolved in potassium cyanide, diazo group is replaced by a cyano group:

$$C_6H_5N_2^+Cl^- + CuCN + KCN \rightarrow C_6H_5CN$$

Example 19. Synthesize benzyl amine starting from benzene.

Solution

Benzene → (HNO_3 / H_2SO_4) → nitrobenzene (NO_2) → ($SnCl_2$ / ClH) → aniline (NH_2) → ($NaNO_2$ / HCl-0°C) → (CuCN / KCN) → benzonitrile (CN) → ($LiAlH_4$) → benzylamine (CH_2NH_2)

Example 20. Synthesize meta cyano toluene starting from toluene.

Solution

Toluene (CH_3) → (HNO_3 / H_2SO_4) → p-nitrotoluene (CH_3, NO_2) → (Zn/HCl) → p-toluidine (CH_3, NH_2) → (HNO_3)

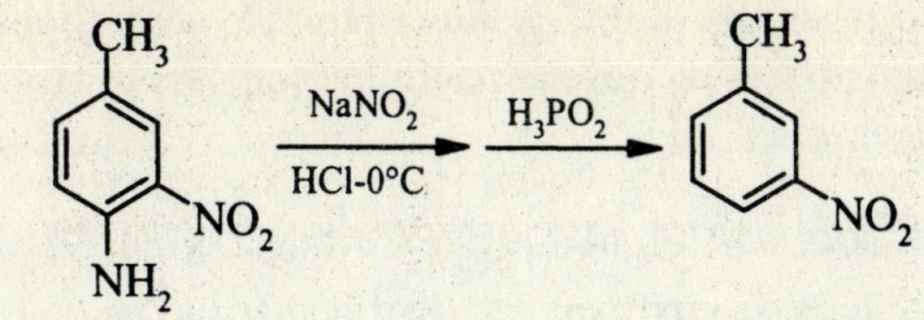

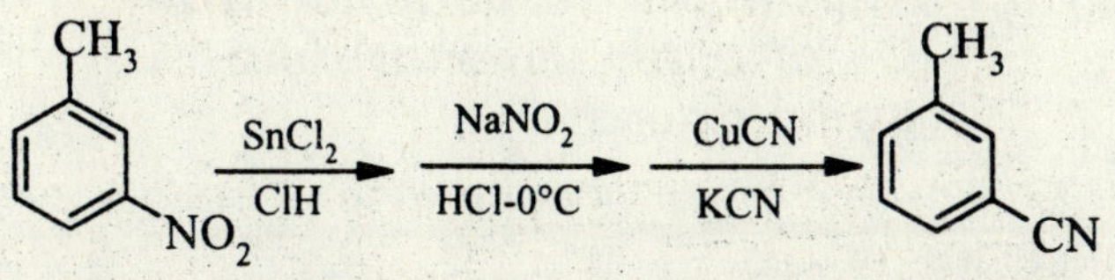

V. Replacement by nitro group: Diazonium salt on treatment with sodium nitrite in presence of copper powder, at little elevated temperature, undergo replacement of diazo group as:

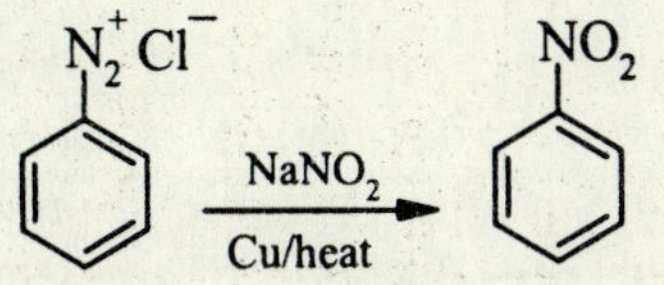

VI. Replacement by a phenyl group: Diazo group is replaced by an aryl group by two common method:

(a) By treating diazonium salt with ethanol and copper powder:

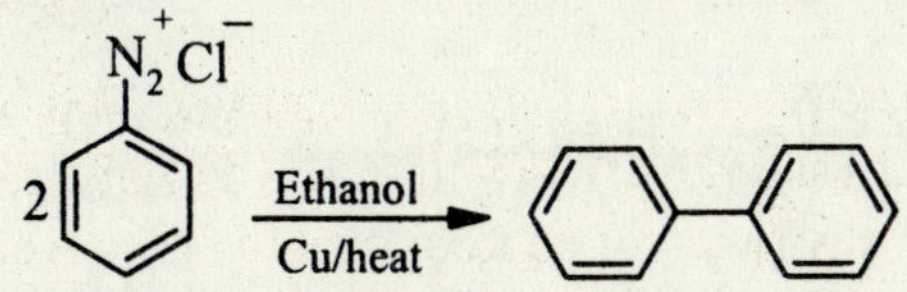

(b) By adding an aromatic compound to an alkaline solution of diazonium salt:

$$C_6H_5N_2^+Cl^- + C_6H_6 + NaOH \rightarrow Ar\text{-}Ar + N_2 + NaCl + H_2O$$

In case of substituted aromatic compound, *p*-substitution occurs predominantly:

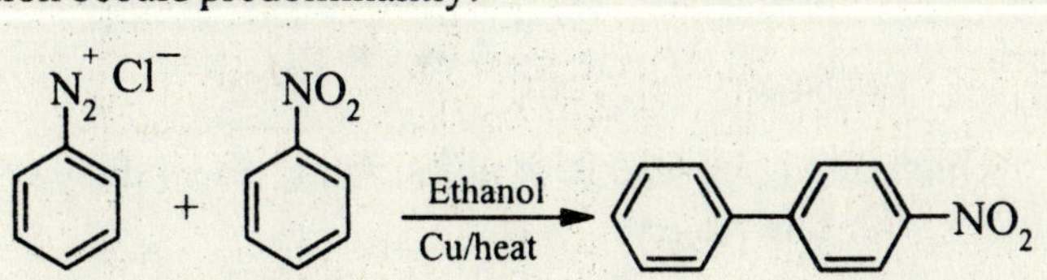

Example 21. Bring about the following transformation by two methods:

Nitro benzene → ortho nitro toluene

Solution

Method I

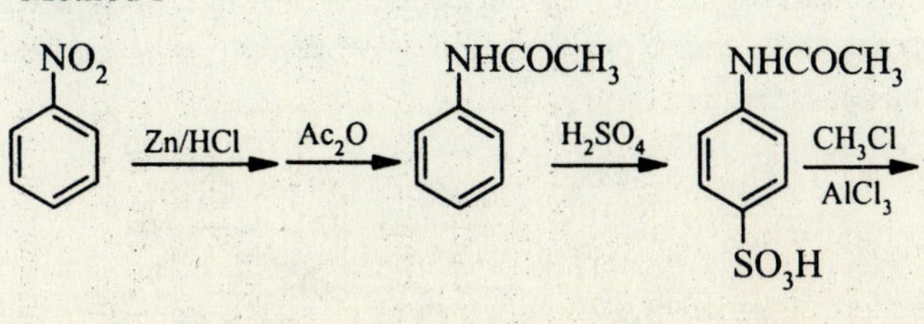

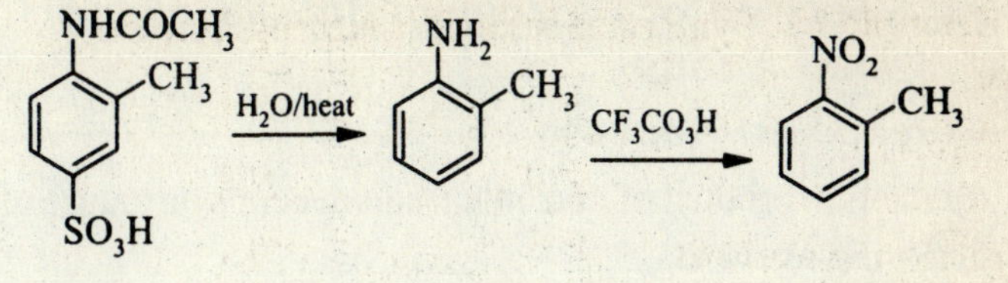

Method II

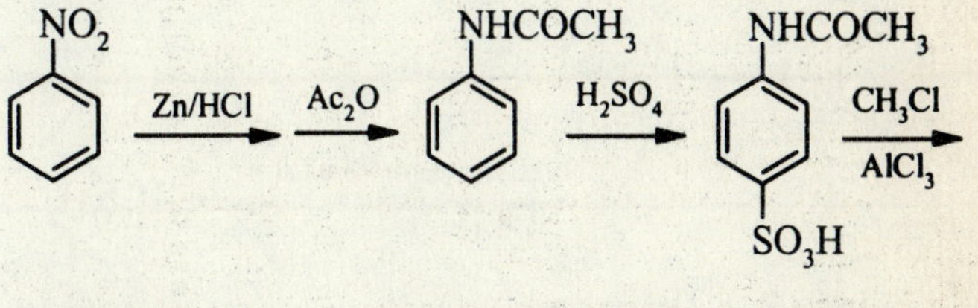

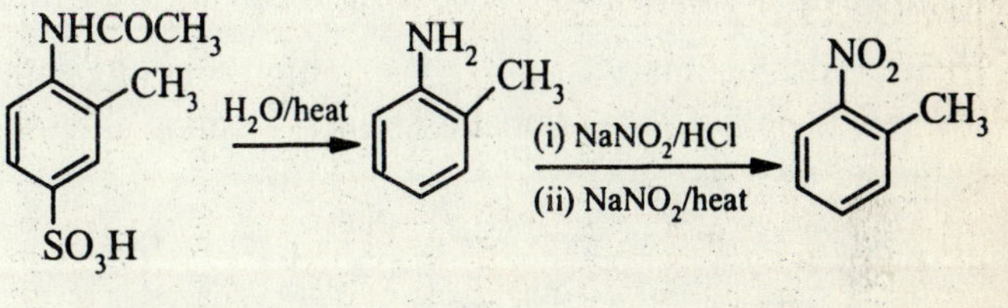

Example 22. Bring about the following transformations:

Benzene → meta propoxy aniline

Solution

Method I

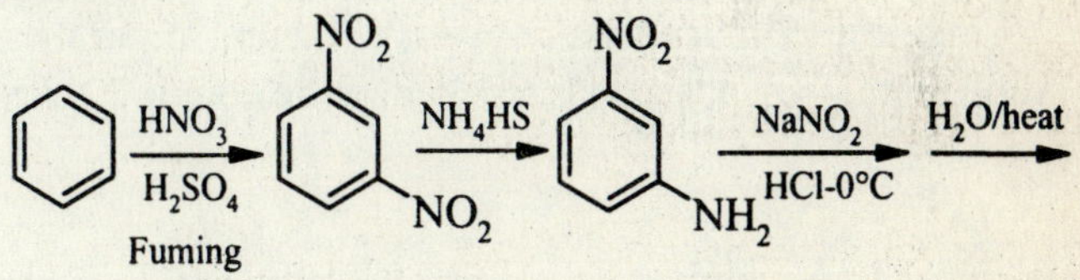

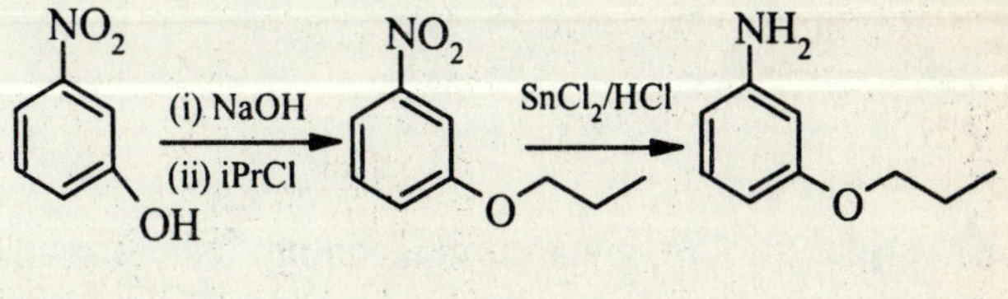

Method II

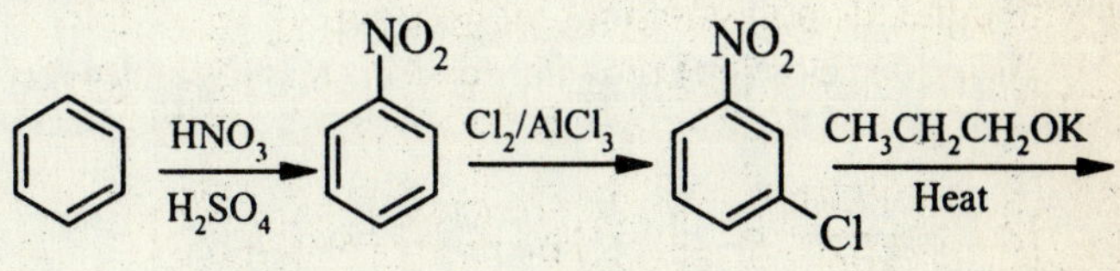

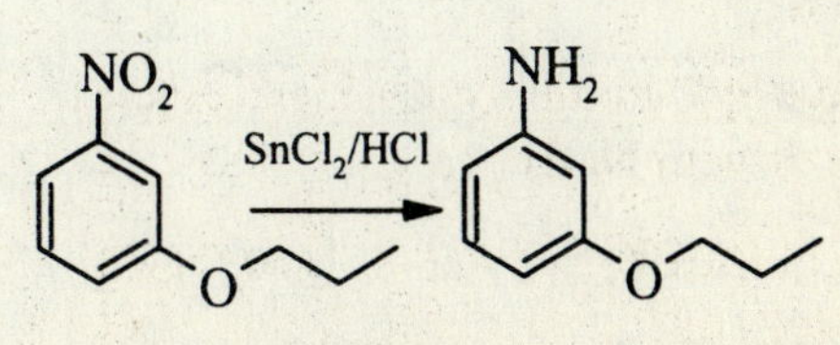

[B] Reaction of Diazonium Salt Retaining Diazo Group

I. Reduction Reaction

Reduction product of diazonium salt depends on nature of reducing agent as:

$$C_6H_5N_2^+Cl^- \xrightarrow[\text{or } NaHSO_3 \text{ or } Na_2SO_3]{SnCl_2/HCl} C_6H_5NHNH_2$$

$$C_6H_5N_2^+Cl^- \xrightarrow{Zn/HCl \text{ or metal/acid}} C_6H_5NH_2$$

II. Coupling Reaction

Benzene diazonium salts undergo very rapid coupling with phenol, naphthol and aromatic amines forming intense coloured dyes as:

$$HO\text{-}C_6H_4\text{-}N{=}N\text{-}C_6H_5 \xleftarrow{PhOH} C_6H_5N_2^+Cl^- \ (I)$$

$$I \xrightarrow{PhNMe_2} C_6H_5\text{-}N{=}N\text{-}C_6H_4\text{-}N(CH_3)_2$$

Coupling with benzene derivatives occurs at para position, but if para position is blocked, ortho coupling takes place:

$$C_6H_5N_2^+Cl^- + (4\text{-}R)C_6H_4OH \longrightarrow C_6H_5\text{-}N{=}N\text{-}C_6H_3(OH)(R)$$

Example 23. Why benzene diazonium salt is stable while aliphatic diazonium salt is unstable, cannot be isolated?

Solution: Benzene diazonium ion is stabilized by resonance as:

(Resonance structures of benzene diazonium ion: $N^+{\equiv}N$ ⟷ $N^+{=}\ddot{N}^-$ with ring + charges ⟷ … ⟷ $N^+{\equiv}N$)

No resonance stabilization is possible with aliphatic diazonium ion, they are unstable, decomposes to give off nitrogen gas.

11.4 AROMATIC ALDEHYDES AND KETONES

11.4.1 Preparation of Benzaldehyde (Oil of bitter almonds) and its Homologues

I. By hydrolysis of benzylidene chloride with aqueous acid:

$$C_6H_5CHCl_2 + H_2O/H^+ \rightarrow C_6H_5CHO + 2HCl$$

II. Boiling benzyl chloride with aqueous solution of copper or lead nitrate in current of dry CO_2 yields benzaldehyde:

$$C_6H_5CH_2Cl + Pb(NO_3)_2 \rightarrow C_6H_5CHO$$

III. By controlled oxidation of toluene: Toluene, on oxidation with chromium trioxide in presence of acetic anhydride yields benzaldehyde. Benzaldehyde so formed, is converted into benzylidene acetate preventing further oxidation. Hydrolysis of this acetate with dilute sulphuric acid yields benzaldehyde:

$$C_6H_5CH_3 \xrightarrow[Ac_2O]{CrO_3} C_6H_5CH(OAc)_2$$

$$\xrightarrow[H_2O]{H^+} C_6H_5CHO + 2\,CH_3COOH$$

Etard's reaction: A better means of oxidizing toluene to benzaldehyde is treating toluene with chromyl chloride in CCl_4 solution.

$$C_6H_5CH_3 \xrightarrow[CCl_4]{CrO_2Cl_2} C_6H_5CHO$$

Other alkyl benzene is oxidized to corresponding ketones with this reagent:

$$C_6H_5CH_2CH_2CH_3 \xrightarrow[CCl_4]{CrO_2Cl_2} C_6H_5COCH_2CH_3$$

This reagent has an additional advantage, that when more than one methyl groups are present, only one can be oxidized as:

$$m\text{-}H_3C\text{-}C_6H_4\text{-}CH_3 \xrightarrow[CCl_4]{CrO_2Cl_2} m\text{-}H_3C\text{-}C_6H_4\text{-}CHO$$

IV. Gattermann's Koch synthesis: Passing a gaseous mixture of CO and HCl through solution of benzene in presence of a Lewis acid catalyst, benzaldehyde is formed:

$$C_6H_6 \xrightarrow[AlCl_3]{CO\ +\ HCl} C_6H_5CHO$$

Mechanism

$$\overset{+}{H} + CO \longrightarrow \left[H-\overset{+}{C}=O \longleftrightarrow H-C\equiv\overset{+}{O}\right] \xrightarrow{C_6H_6}$$

$$[C_6H_6(H)(CHO)]^{+} \xrightarrow{-\overset{+}{H}} C_6H_5CHO$$

Gattermann's Koch synthesis can also be done with substituted benzene, e.g. toluene is formylated mainly at para position. The Gattermann's Koch synthesis is not applicable with phenol or benzene ring containing highly deactivating group, e.g. nitrobenzene, benzene sulphonic acid, etc. fails in Gattermann Koch synthesis.

V. Gattermann's aldehyde synthesis: Benzene on treatment with a mixture of HCN and HCl, in presence of a Lewis acid catalyst, gives benzaldehyde as:

$$C_6H_6 + HCN + HCl + AlCl_3 \rightarrow C_6H_5CHO$$

This synthesis is applicable to phenol and phenolic ethers but fails with nitrobenzene.

VI. Rosenmund reduction: Catalytic reduction of benzoyl chloride with $Pd/BaSO_4$ in the presence of quinoline yields benzaldehyde:

$$C_6H_5COCl \xrightarrow[Quinoline]{Pd/BaSO_4} C_6H_5CHO$$

VII. Stephen's reaction: When phenyl cyanide is reduced with $SnCl_2$ in presence of concentrated HCl solution in etheral medium, benzaldehyde is produced:

$$C_6H_5CN \xrightarrow[Ether]{SnCl_2/HCl} C_6H_5CHO$$

VIII. Grignard's synthesis

$$C_6H_5MgBr + HCOOR \rightarrow C_6H_5CHO + Mg(OR)Br$$

11.4.2 Preparation of Aromatic Ketones

I. By Friedel Craft's reaction: Treating benzene or its homologue with acid chloride or acid anhydride in presence of a Lewis acid catalyst, aromatic ketone is produced:

$$C_6H_6 \xrightarrow[AlCl_3]{RCOCl} C_6H_5COR$$

$$C_6H_6 \xrightarrow[AlCl_3]{(RCO)_2O} C_6H_5COR$$

II. By oxymercuration-demercuration of aromatic alkyne

$$C_6H_5C\equiv C-R \xrightarrow[H_2SO_4]{HgSO_4} C_6H_5COCH_2R$$

III. By Grignard's Synthesis

$$C_6H_5MgBr + RCOOR \longrightarrow C_6H_5COR$$

11.4.3 Properties of Aromatic Aldehydes and Ketones

I. General properties: Benzaldehyde is a colourless liquid with a smell of almond. Benzaldehyde resembles aliphatic aldehyde in the following reactions:

(i) It gives Schiff's reaction.

(ii) It is easily oxidized to benzoic acid.

(iii) Benzaldehyde forms bisulphate with $NaHSO_3$

$$C_6H_5CHO + NaHSO_3 \rightarrow C_6H_5CH(OH)SO_3Na$$

(iv) Benzaldehyde forms cyanohydrin with HCN and oxime with NH_2OH:

$$C_6H_5CHO + HCN \rightarrow C_6H_5CH(OH)CN$$

$$C_6H_5CHO + NH_2OH \rightarrow C_6H_5CH{=}NOH$$

These reactions are also given by aromatic ketones.

Benzaldehyde and other aromatic aldehydes differs from aliphatic aldehydes in the following ways:

(i) It does not reduce Fehling solution.

(ii) It does not polymerize easily.

II. Reaction with ammonia: Benzaldehyde forms a complex product "hydrobenzamide" with ammonia:

$$C_6H_5CHO + NH_3 \rightarrow [C_6H_5CHOHNH_2] \xrightarrow{-H_2O}$$

$$C_6H_5CH{=}NH \xrightarrow{3.0\ moles} C_6H_5CH{=}NCH(C_6H_5)N{=}CHC_6H_5$$

III. Reduction of benzaldehyde: Benzaldehyde on reduction with Zn/HCl or Na(Hg) gives hydrobenzoin as well as benzyl alcohol:

$C_6H_5CHO + Na(Hg) \rightarrow C_6H_5CH(OH)CH(OH)C_6H_5 + C_6H_5CH_2OH$

IV. Claisen reaction: Benzaldehyde, in presence of dilute alkali, condenses with aliphatic aldehyde or ketones containing an α -H, to produce α, β-unsaturated aldehyde or ketone:

$C_6H_5CHO + CH_3CHO + NaOH \text{ (dilute)} \rightarrow C_6H_5CH{=}CHCHO$ 'Cinnamaldehyde'

Mechanism

$$CH_3CHO + {}^-OH \rightarrow {}^-CH_2CHO + OH_2$$

$$C_6H_5CHO + {}^-CH_2CHO \rightarrow$$

$$C_6H_5{-}CH(OH){-}CH_2CHO \xrightarrow{-H_2O} C_6H_5{-}CH{=}CHCHO$$

Example 24. Predict product of the reaction of benzaldehyde with acetone in dilute alkaline medium. What will happen, if the above product is further condensed with excess of benzaldehyde in ethanolic solution of NaOH?

Solution

$$H_3C{-}C(=O){-}CH_3 + {}^-OH \rightarrow H_2\bar{C}{-}C(=O){-}CH_3 + H_2O$$

$$H_2\bar{C}{-}C(=O){-}CH_3 + C_6H_5CHO \rightarrow \underset{(A)}{C_6H_5CH{=}CHCOCH_3}$$

$$(A) + {}^-OH \rightarrow C_6H_5CH{=}CH{-}C(=O){-}\bar{C}H_2 \xrightarrow[H_2O]{C_6H_5CHO} C_6H_5CH{=}CH{-}C(=O){-}CH{=}CHC_6CH_5$$

Example 25. Predict products in the following condensation reaction:

(a) C_6H_5CHO + Acetophenone + NaOH →

(b) $C_6H_5CHO + CH_3CH_2NO_2 + NaOH \rightarrow$

Solution

(a) $C_6H_5CH{=}CHCOC_6H_5$

(b) $C_6H_5CH{=}C(NO_2)CH_3$

V. Perkin reaction: When an aromatic aldehyde is heated (200°C) with the anhydride of an aliphatic acid (containing two α-H), in presence of its sodium salt, condensation takes place and a β-aryl acrylic acid is produced, e.g.

$C_6H_5CHO + (CH_3CO)_2O + CH_3COONa \rightarrow C_6H_5CH{=}CHCOOH + CH_3COONa$

Mechanism

$$H_3C{-}CO{-}O{-}CO{-}CH_3 + CH_3COONa \rightarrow H_3C{-}CO{-}O{-}CO{-}\bar{C}H_2$$

$$C_6H_5{-}C(=O){-}H + H_2\bar{C}{-}CO{-}O{-}CO{-}CH_3 \rightarrow$$

$$C_6H_5{-}CH(O^-){-}H_2C{-}CO{-}O{-}CO{-}CH_3$$

$$\xrightarrow{H^+}\xrightarrow{-H_2O} C_6H_5{-}CH{=}CH{-}CO{-}O{-}CO{-}CH_3$$

$$\xrightarrow{H_3O^+} C_6H_5{-}CH{=}CH{-}COOH + CH_3COOH$$

Example 26. Predict products in the following reactions:

(a) $C_6H_5CHO + (CH_3CH_2CO)_2O \xrightarrow[Heat]{CH_3CH_2COONa}$

(b) o-$C_6H_4(CHO)_2 + (CH_3CO)_2O \xrightarrow[Heat]{CH_3COONa}$

Solution

(a) $C_6H_5CH{=}C(CH_3){-}COOH$

(b) o-$C_6H_4(CH{=}CHCOOH)_2$

Example 27. Predict product in the following reactions:

(a) 4-nitrobenzene-1,3-dicarbaldehyde (NO_2, CHO, CHO on benzene ring) $+ (CH_3CO)_2O \xrightarrow[Heat]{CH_3COONa}$

(b) C_6H_5CHO + succinic anhydride $\xrightarrow[Heat]{CH_3COONa}$

Solution

(a) Since Perkin reaction involves nucleophilic addition of anhydride anion at carbonyl carbon, –CHO para to $-NO_2$ is more activated for nucleophilic attack:

$$\text{(2-formyl-4-nitrobenzaldehyde)} + (CH_3CO)_2O \xrightarrow[\text{Heat}]{CH_3COONa} \text{(}NO_2\text{, CHO, }CH{=}CHCOOH\text{ substituted benzene)}$$

(b) $C_6H_5CH{=}CHCH_2COOH$ + (5-phenyl-γ-butyrolactone)

VI. Benzoin condensation: Benzaldehyde when refluxed with aqueous, ethanolic solution of potassium cyanide, benzoin is formed:

$$2C_6H_5CHO + KCN \rightarrow C_6H_5CHOHCOC_6H_5$$

Mechanism

$$C_6H_5{-}C(=O){-}H + CN^- \rightarrow \left[C_6H_5{-}CH(O^-){-}CN \leftrightarrow C_6H_5{-}C^-(OH){-}CN\right]$$

$$\xrightarrow{C_6H_5CHO} C_6H_5{-}C(OH)(CN){-}C(O^-)(H){-}C_6H_5 \rightleftharpoons$$

$$C_6H_5{-}C(O^-)(CN){-}C(OH)(H){-}C_6H_5 \xrightarrow{-CN^-} C_6H_5{-}C(=O){-}C(OH)(H){-}C_6H_5 \text{ (Benzoin)}$$

VII. Reaction of benzaldehyde with phenols and tertiary aromatic amines: Benzaldehyde condenses with phenols and tertiary aromatic amines in presence of sulphuric acid or $ZnCl_2$ to form triphenyl methane derivatives as:

$$C_6H_5{-}CHO + 2\,C_6H_5{-}N(CH_3)_2 \xrightarrow{H^+} C_6H_5{-}CH(C_6H_4{-}NMe_2)_2$$

11.5 AROMATIC ACIDS

11.5.1 Preparation

I. By oxidation of corresponding alcohols or aldehydes, e.g.

$$C_6H_5CH_2OH + KMnO_4 \rightarrow C_6H_5COOH$$
$$C_6H_5CHO + KMnO_4 \rightarrow C_6H_5COOH$$

II. By hydrolysis of corresponding cyanides

$$R{-}C_6H_4{-}CN \xrightarrow[H_2O]{H_2SO_4} R{-}C_6H_4{-}COOH$$

III. By means of Grignard's reagent

$$C_6H_5Br + Mg \xrightarrow{\text{Dry ether}} C_6H_5MgBr \xrightarrow{CO_2}$$

$$C_6H_5COOMgBr \xrightarrow{H^+} C_6H_5COOH$$

IV. By means of Friedal Craft's Acylation reaction

$$C_6H_6 \xrightarrow[AlCl_3]{COCl_2} C_6H_5COCl \xrightarrow{H_2O} C_6H_5COOH$$

Excess of phosgene ($COCl_2$) gas is used to avoid formation of benzophenone by further reaction of benzoyl chloride with benzene.

V. By oxidation of benzene homologues: Alkyl benzene can be oxidized to benzoic acid using dilute nitric acid or acidic solution of dichromate or slightly alkaline solution of potassium permanganate or CrO_3 in glacial acetic acid as an oxidizing agent:

$$C_6H_5R \xrightarrow[NaOH]{KMnO_4} C_6H_5COOH$$

Alkyl group must have at least one α-H otherwise oxidation does not take place, e.g. tertiary butyl benzene is not oxidized to benzoic acid.

11.5.2 Reactions of Benzoic Acid

I. Reduction: Benzoic acid can be reduced to benzyl alcohol or toluene using suitable catalyst as:

$$C_6H_5COOH \xrightarrow{LiAlH_4} C_6H_5CH_2OH$$

$$C_6H_5COOH \xrightarrow{HI/\text{red P}} C_6H_5CH_3$$

II. Electrophilic substitution reaction: Carboxy group is a strong deactivating, meta directing functional group and ring is attacked by usual electrophile, e.g.

Benzoic acid (COOH) → $Br_2/FeBr_3$ → m-bromobenzoic acid (COOH, Br)

Benzoic acid (COOH) → HNO_3 / H_2SO_4 → m-nitrobenzoic acid (COOH, NO_2)

Example 28. Synthesize the following compounds as exclusive product, starting from benzene:

(i) o-chlorobenzoic acid (COOH, Cl) (ii) p-aminobenzoic acid (COOH, NH_2)

(iii) isophthalic acid (COOH, COOH) (iv) phthalic acid (COOH, COOH)

Solution

(i) Benzene → CH_3Cl / $AlCl_3$ → toluene (CH_3) → H_2SO_4 / Heat → p-toluenesulphonic acid (CH_3, SO_3H) → Cl_2 / $AlCl_3$ → (CH_3, Cl, SO_3H) → H_2O / Heat → (CH_3, Cl) → $K_2Cr_2O_7$ / H^+ → (COOH, Cl)

(ii) Benzene → HNO_3 / H_2SO_4 → (NO_2) → Zn/HCl → CH_3COCl → ($NHCOCH_3$) → CH_3Cl / $AlCl_3$ → ($NHCOCH_3$, CH_3) → $K_2Cr_2O_7$ / H^+ → ($NHCOCH_3$, COOH) → H_2O / Heat → (NH_2, COOH)

(iii) Benzene + HCN + HCl → $AlCl_3$ → (CHO) → CH_3Cl / $AlCl_3$ → (CHO, CH_3) → $KMnO_4$ / NaOH → (COOH, COOH)

(iv) Benzene → CH_3Cl / $AlCl_3$ → (CH_3) → H_2SO_4 / Heat → (CH_3, SO_3H) → CH_3Cl / $AlCl_3$ → (CH_3, CH_3, SO_3H) → H_2O / Heat → (CH_3, CH_3) → $KMnO_4$ / NaOH → (COOH, COOH)

III. Reaction with PCl_5, PCl_3 or $SOCl_2$: Benzoyl chloride is formed.

Benzoic acid (COOH) → PCl_5 / PCl_3 / $SOCl_2$ → Benzoyl chloride (COCl)

IV. Intermolecular and intramolecular dehydration: Heating benzoic acid or its derivatives with P_2O_5 or V_2O_5 yields anhydride as:

2 Benzoic acid (COOH) → P_2O_5 / Heat → Benzoic acid anhydride

Benzoic acid anhydride
(Intermolecular dehydration)

An α, β-diacid, under similar reaction conditions, yields anhydride by intramolecular dehydration as:

Phthalic acid (COOH, COOH) → P_2O_5 / Heat → Phthalic acid anhydride

Phthalic acid anhydride
(Intramolecular dehydration)

Other isomers of diacids undergo intermolecular dehydration, rather intramolecular dehydration due to lack of ability to form cyclic anhydride ring.

V. Esterification: Benzoic acid or its derivatives, on refluxing with an alcohol in acid medium forms ester:

$$C_6H_5COOH \xrightarrow[H^+\ heat]{ROH} C_6H_5COOR$$

VI. Reactions with ammonia or amines: Ammonium salts are formed, which on heating undergo dehydration and amides are formed:

$$C_6H_5COOH \xrightarrow{NH_3} [C_6H_5COONH_4] \xrightarrow{Heat} C_6H_5CONH_2$$

$$C_6H_5COOH \xrightarrow{RNH_2} [C_6H_5COONH_3R] \xrightarrow{Heat} C_6H_5CONHR$$

11.6 ARYL HALIDES

11.6.1 Preparation

I. Direct halogenation: Chlorination or bromination of benzene can be carried out conveniently at room temperature, in presence of a Lewis acid catalyst:

$$C_6H_6 \xrightarrow{Cl_2/AlCl_3} C_6H_5Cl$$

$$C_6H_6 \xrightarrow{Br_2/FeBr_3} C_6H_5Br$$

II. From diazonium salt

$$C_6H_5\overset{+}{N_2}Cl^- \xrightarrow{CuCl} C_6H_5Cl$$

$$C_6H_5\overset{+}{N_2}Cl^- \xrightarrow{CuBr} C_6H_5Br$$

$$C_6H_5\overset{+}{N_2}Cl^- \xrightarrow{KBF_4/heat} C_6H_5F$$

$$C_6H_5\overset{+}{N_2}Cl^- \xrightarrow[Cu/heat]{KI} C_6H_5I$$

III. From silver salt of benzoic acid: Heating silver salt of benzoic acid with bromine yields bromobenzene:

$$C_6H_5COOAg \xrightarrow[Heat]{Br_2} C_6H_5Br + CO_2 + AgBr$$

IV. Iodination: Iodination of benzene or its derivatives can be carried out in presence of an oxidizing agent, viz. dilute nitric acid or HgO:

$$C_6H_6 + I_2 + [O] \rightarrow C_6H_5I$$

11.6.2 Reactions of Aryl Halides

I. Grignard's synthesis: Alkyl halides forms Grignard reagent with Mg in dry ether medium:

$$C_6H_5X + Mg/dry\ ether \rightarrow C_6H_5MgX$$

Example 29. Synthesize the following compounds starting from bromobenzene:

(a) $C_6H_5CH_2CH_2OH$ (b) $C_6H_5CH(OH)CH_3$

(c) $C_6H_5C(CH_3)_2OH$

Solution

$$C_6H_5Br + Mg \xrightarrow{Dry\ ether} C_6H_5MgBr$$

$$C_6H_5MgBr \xrightarrow{\text{ethylene oxide}} \xrightarrow[H_2O]{H^+} C_6H_5CH_2CH_2OH\ (a)$$

$$C_6H_5MgBr \xrightarrow{CH_3CHO} \xrightarrow[H_2O]{H^+} C_6H_5CH(OH)CH_3\ (b)$$

$$C_6H_5MgBr \xrightarrow{CH_3COCH_3} \xrightarrow[H_2O]{H^+} C_6H_5-C(CH_3)_2-OH\ (c)$$

II. DDT synthesis: Chlorobenzene on treatment with chloral, in presence of concentrated sulphuric acid yields DDT:

$$2\,C_6H_5Cl + \underset{\text{Chloral}}{Cl_3CCHO} \xrightarrow{H^+} Cl_3C-CH(C_6H_4Cl)_2$$

DDT

(p, *p*-dichlorodiphenyltrichloro ethane)

III. Electrophilic substitution: Halides are weakly deactivating but ortho/para directing group in electrophilic aromatic substitution reaction. It undergo normal electrophilic substitution reaction as:

$$C_6H_5Cl \xrightarrow{Cl_2/AlCl_3} p\text{-}Cl-C_6H_4-Cl + o\text{-}Cl-C_6H_4-Cl$$

$$C_6H_5Cl \xrightarrow[H_2SO_4]{HNO_3} o\text{-}Cl-C_6H_4-NO_2 + p\text{-}Cl-C_6H_4-NO_2$$

$$C_6H_5Cl \xrightarrow[AlCl_3]{RCl} o\text{-}Cl-C_6H_4-R + p\text{-}Cl-C_6H_4-R$$

IV. Coupling reactions, Ullamann reaction: Bromo benzene and iodo benzene undergo coupling reaction when heated in presence of Cu-powder:

$$C_6H_5I + Cu/heat \rightarrow C_6H_5—C_6H_5$$

V. Reaction with phenol: Bromo and iodo benzene forms diphenyl ether when heated with phenol in strong alkaline medium:

$$C_6H_5Br + C_6H_5OH + KOH \rightarrow C_6H_5—O—C_6H_5 + KBr + H_2O$$

VI. Nucleophilic aromatic substitution reaction: Aryl halides, in general, do not undergo nucleophilic substitution reactions under normal conditions. This is because of involvement of lone pair of electrons on halide into resonance delocalization with the benzene nucleus. This gives partial double bond character to the C—X bond and makes removal of halide very difficult:

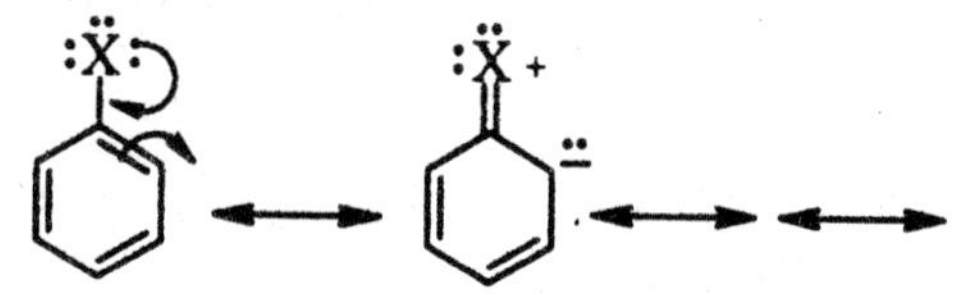

However, if the ring has strong electron withdrawing groups like nitro, carboxy, sulphonic, cyano, etc. nucleophilic substitution becomes possible otherwise drastic reaction conditions are required as described below:

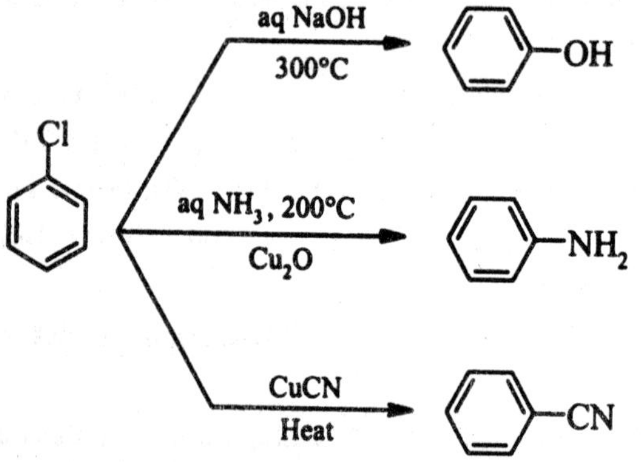

aq NaOH, 300°C → OH

aq NH_3, 200°C, Cu_2O → NH_2

CuCN, Heat → CN

With the electron withdrawing substituents in the ring, substitution occur in milder conditions:

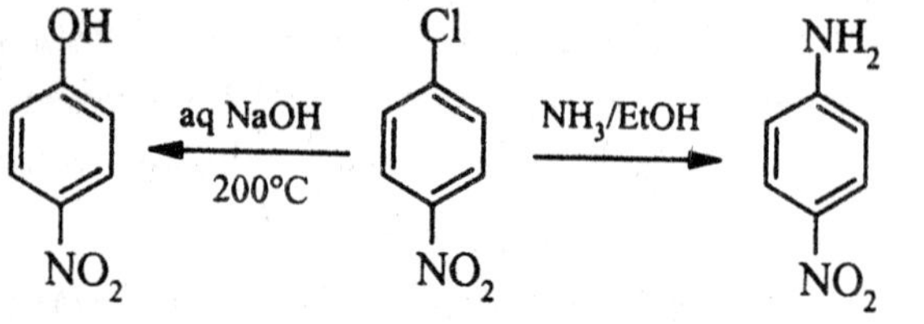

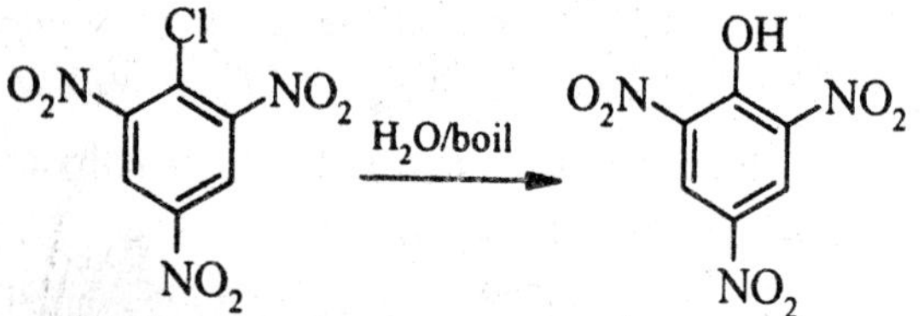

aq NaOH, 200°C; NH_3/EtOH

OH, NO_2 ← Cl, NO_2 → NH_2, NO_2

Cl, O_2N, NO_2, NO_2 — H_2O/boil → OH, O_2N, NO_2, NO_2

Highly activated (Picric acid)

Mechanism of Nucleophilic Substitution

Elimination addition mechanism: In a number of cases of nucleophilic substitution reactions, it has been observed that the entering nucleophile does not occupy the position vacated by the leaving halide. Reactions of this type have been called **Cine Substitution**, which involves formation of an intermediate "Benzyne":

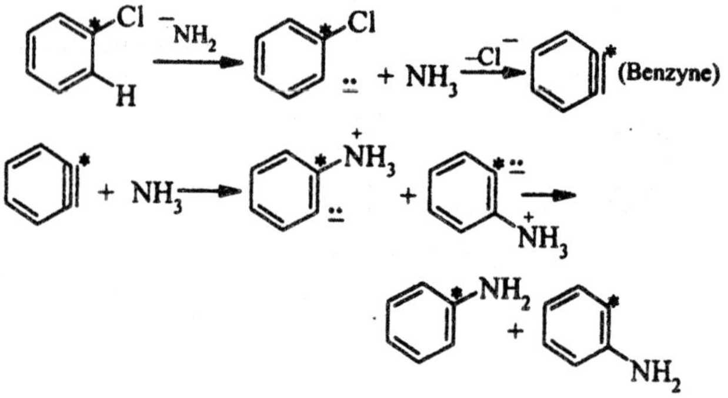

Cl, H — $\bar{N}H_2$ → Cl + NH_3 — $-Cl^-$ → (Benzyne)

+ NH_3 → $\overset{+}{N}H_3$ + $\overset{+}{N}H_3$ → NH_2 + NH_2

Orientation of benzyne mechanism: The possible benzynes are:

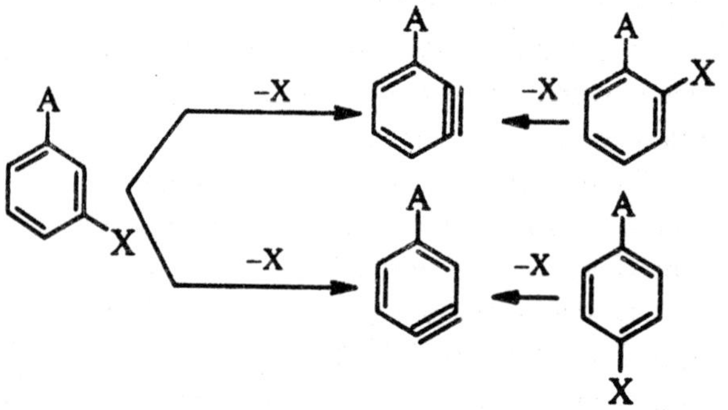

A, X — −X; −X

The ortho and para substituted aryl halides can give only one type of benzyne while meta can give both, and which will predominate depends on stability of carbanion which precedes the benzyne:

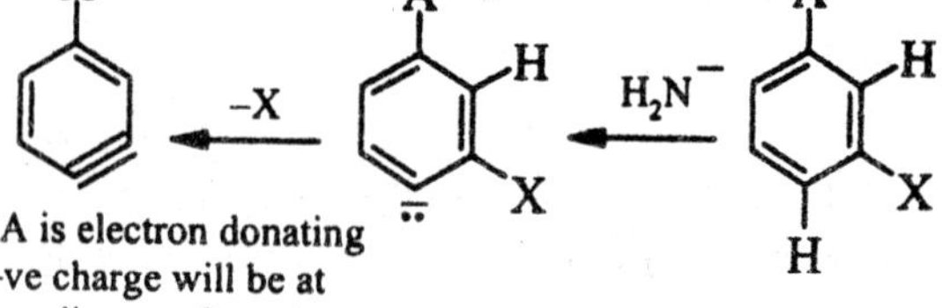

A — −X ← H, X ← H_2N^- H, X, H

Formed if A is electron donating so that –ve charge will be at maximum distance from A

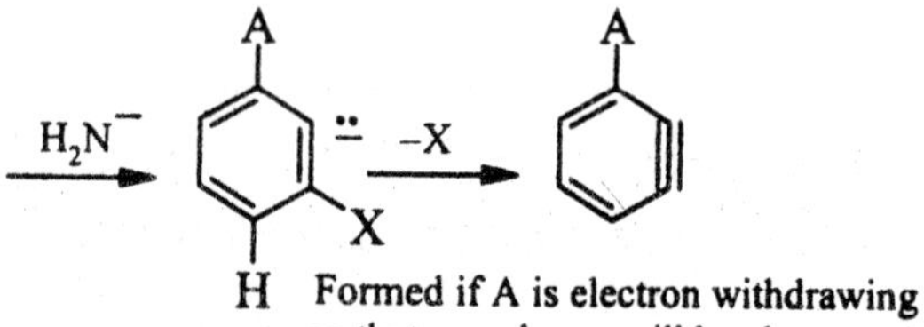

H_2N^- → A, X, H — −X → A

H Formed if A is electron withdrawing so that –ve charge will be closer to A

Orientation of addition: Orientation of addition also depends on the stability of carbanion:

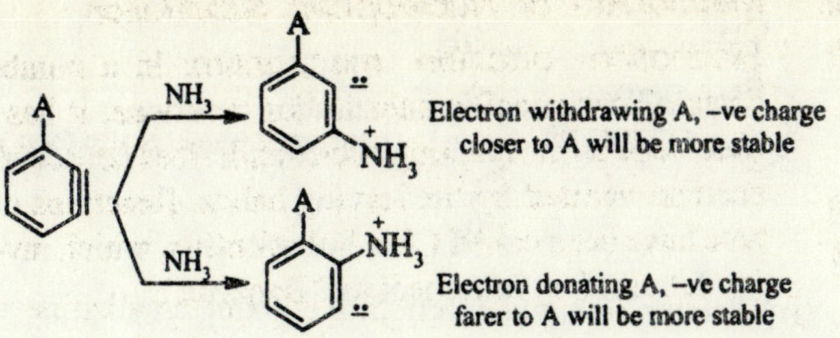

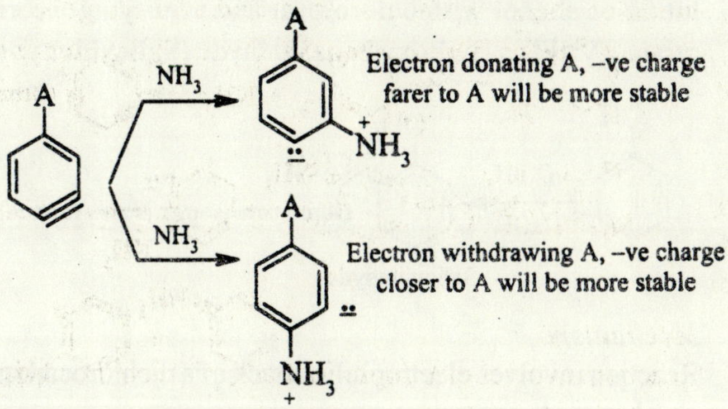

Electron withdrawing A, –ve charge closer to A will be more stable

Electron donating A, –ve charge farer to A will be more stable

Electron donating A, –ve charge farer to A will be more stable

Electron withdrawing A, –ve charge closer to A will be more stable

11.7 PHENOL AND PHENOLIC ETHER

11.7.1 Preparations

I. From natural sources: Various types of monohydric phenols and its derivatives occurs in coal-tar and separated commercially by extraction:

(a) *From middle oil:* Middle oil contain mainly naphthalene and phenol/derivatives of phenol. Middle oil is cooled where naphthalene crystallizes out. The oil is then treated with an aqueous solution of NaOH, which dissolves phenol as sodium phenoxide. To the aqueous solution of sodium phenoxide, CO_2 is bubbled, where sodium phenoxide decomposes into free phenol and sodium carbonate formed, goes into aqueous layer. The aqueous layer is drawn off and crude phenol is separated by fractionation.

(b) *Heavy oil:* Heavy oil is treated in the same way as middle oil. The residual oil contain phenol, cresol, naphthol and higher phenols.

II. From aryl sulphonates: Sodium salt of benzene sulphonic acid on heating with NaOH yields sodium phenoxide, which on acidification yields phenol:

$$C_6H_5SO_3Na + NaOH/heat \rightarrow C_6H_5ONa,$$

$$C_6H_5ONa + H^+/H_2O \rightarrow C_6H_5OH$$

III. By steam distillation of diazonium salt

$$C_6H_5N_2^+HSO_4^- + H_2O/heat \rightarrow C_6H_5OH + N_2 + H_2SO_4$$

IV. By nucleophilic substitution of aryl halide

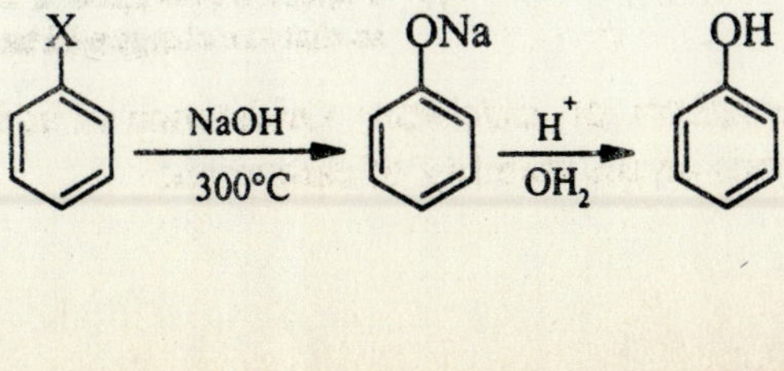

X (aryl halide) —NaOH, 300°C→ ONa —H⁺, OH₂→ OH

V. By decarboxylation of phenolic acids

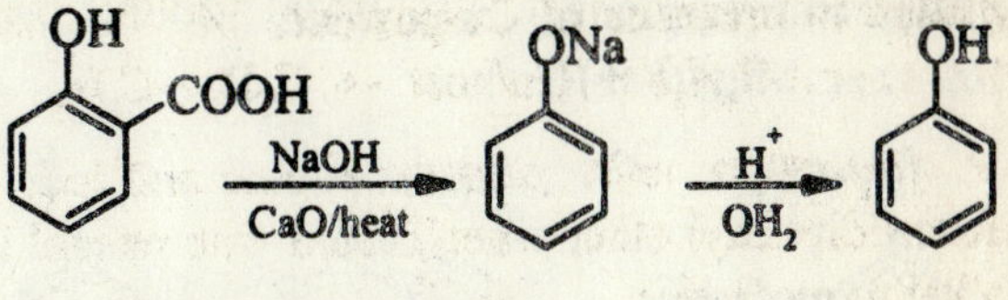

OH, COOH —NaOH, CaO/heat→ ONa —H⁺, OH₂→ OH

VI. By Grignard synthesis

$$ArMgBr + O_2 \rightarrow ArO_2MgBr$$

$$ArO_2MgBr + ArMgBr \rightarrow 2ArOMgBr;$$

$$ArOMgBr + H_2O \rightarrow ArOH$$

VII. Cumene process: Phenol is prepared industrially by oxidation of cumene as:

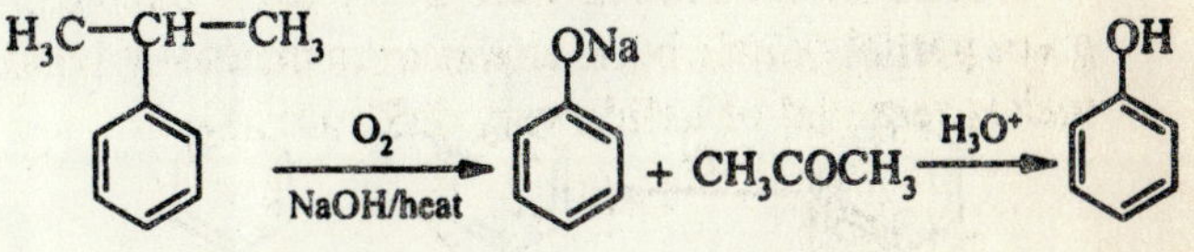

$H_3C—CH—CH_3$ (cumene) —O_2, NaOH/heat→ ONa + CH_3COCH_3 —H_3O^+→ OH

11.7.2 Properties of Phenol and its Derivatives

(A) General properties: Phenol is a colourless, crystalline solid, which turns to pink on exposure to air. Phenol is moderately soluble in water but fairly soluble in ethanol and ether. Phenol is distinguished from other organic compounds by:

(a) *Liebermann's test:* When phenol is dissolved in concentrated sulphuric acid and a few drops of $NaNO_2(aq)$ is added, red colouration is observed, which turns blue on making solution alkaline with NaOH.

(b) *Ferric chloride test:* Phenol gives violet colouration with $FeCl_3$ solution.

Phenol is used as antiseptic and disinfectant and in preparation of dyes, drugs, bakelite, etc.

(B) Acidity: Phenol is a weak acid, weaker than a carboxylic acid but stronger than water and aliphatic alcohols, hence phenol forms salt with NaOH, but it does not forms salt with $NaHCO_3$.

(C) Chemical reactions

(i) In strongly alkaline medium, it forms phenolic ether when an alkyl halide is added dropwise:

$$PhONa + R—X \rightarrow PhOR + NaX$$

(ii) Reaction with acid chloride or anhydride-Phenolic esters are formed:

$$C_6H_5OH + RCOCl \rightarrow C_6H_5COOR + HCl$$

$$C_6H_5OH + (RCO)_2O \rightarrow C_6H_5COOR + RCOOH$$

Phenolic ester when treated with a Lewis acid, undergo rearrangement to produce ortho/para phenolic ketones **(Fries rearrangement)**.

OCOR —($AlCl_3$)→ o-hydroxyphenyl ketone (COR) + p-hydroxyphenyl ketone (COR)

(iii) *Electrophilic substitution reaction:* Phenol is highly activated for electrophilic substitution reaction and substitution occurs at ortho and para positions:

OH —(E^+)→ o-E-phenol + p-E-phenol

Very high activating power of the hydroxy group in aqueous medium is due to its deprotonation, forming phenoxide ion (PhO^-)which takes part in the substitution reaction rather unionized phenol:

Phenol —(H_2O)→ phenoxide (O^-) + H_3O^+

Therefore, in aqueous medium, phenol is brominated at all three positions simultaneously even in absence of a Lewis acid catalyst, nitrated in dilute nitric acid medium which otherwise require a mixed solution of concentrated nitric and sulphuric acids, e.g.

Phenoxide —(Br_2, H_2O)→ 2,4,6-tribromophenoxide (Br, Br, Br)

Phenoxide —(HNO_3, Dilute)→ o-nitrophenoxide (NO_2) + p-nitrophenoxide (NO_2)

For monobromination of phenol, non-aqueous solvents are preferred:

Phenol —($Br_2/FeBr_3$, CS_2)→ p-bromophenol (Br) (some ortho isomer is also formed)

(iv) *Reimer Tiemann reaction:* Refluxing an alkaline solution of phenol with chloroform and acidifying the end product yields *o*-hydroxy benzaldehyde (**Salicyldehyde**):

Phenol —($CHCl_3$, NaOH)→ o-hydroxybenzaldehyde (CHO) (some para isomer is also formed)

Salicyldehyde

Mechanism

Reaction involves electrophilic attack of a dichlorocarbene on phenoxide ion as:

$$CHCl_3 + {}^-OH \longrightarrow :CCl_2 + H_2O + Cl^-$$

(Dichlorocarbene)

Phenoxide (O^-) + $:CCl_2$ → cyclohexadienone (H, $\bar{C}Cl_2$) —(H_2O)→ cyclohexadienone (H, $CHCl_2$)

—($-H^+$)→ o-($CHCl_2$)phenoxide —(${}^-OH$, H_2O)→ o-(CHO)phenoxide

In Reimer Tiemann reaction, ortho product predominate, but if ortho position is blocked, para substitution takes place:

2-R-phenol —($CHCl_3$, NaOH)→ 2-R-4-CHO-phenoxide (O^-)

In Reimer Tiemann reaction, if CCl_4 is used in place of chloroform, phenolic acid is formed:

Phenol —(CCl_4, NaOH)→ o-(COO^-)phenoxide —(H_3O^+)→ o-hydroxybenzoic acid (COOH)

Salicylic acid

Salicylic acid is used in synthesis of pain killer Aspirin as:

Salicylic acid $\xrightarrow[\text{Heat}]{Ac_2O}$ Acetyl salicylate (Aspirin)

(v) *Reaction with formaldehyde:* Phenol condenses with formaldehyde to form hydroxy benzyl alcohol in the beginning, which finally polymerizes into Bakelite.

Phenol + HCHO $\xrightarrow{NaOH}$ $HOCH_2$–C$_6$H$_3$(OH)–CH_2OH + $HOCH_2$–C$_6$H$_3$(OH)–CH_2OH $\xrightarrow{\text{Polymerization}}$ Bakelite

(vi) *Reduction:* Phenol on heating with Ni/H_2 at 160°C, reduced to cyclohexanol:

Phenol $\xrightarrow[H_2\text{-}160°C]{Ni}$ Cyclohexanol

Example 30. Provide organic products of the following reactions:

(a) Nitrobenzene $\xrightarrow[AlCl_3]{Cl_2}$

(b) 4-Methylanisole (CH_3, OMe) $\xrightarrow[FeBr_3]{Br_2}$

(c) 4-Methylacetophenone ($COCH_3$, CH_3) $\xrightarrow[HgO]{I_2}$

(d) p-Toluidine (NH_2, CH_3) $\xrightarrow[\text{(ii) aq } HBF_4]{\text{(i) } NaNO_2\text{, HCl-0°C}}$

(e) 3-Aminoacetophenone (NH_2, $COCH_3$) $\xrightarrow[\text{HCl-0°C (ii) KI}]{\text{(i) } NaNO_2}$

(f) 5-Amino-tetralin (NH_2) $\xrightarrow[\text{(ii) CuBr}]{\text{(i) } NaNO_2\text{, HCl-0°C}}$

(g) 4-Bromoaniline (Br, NH_2) $\xrightarrow[\text{(ii) CuCl}]{\text{(i) } NaNO_2\text{, HCl-0°C}}$

(h) 2-Bromotoluene (CH_3, Br) $\xrightarrow[Ether]{Mg}$ $\xrightarrow[H_3O^+]{H_2C=O}$

(i) 3-Iodotoluene (CH_3, I) $\xrightarrow[NH_3\text{ (aq)}]{KNH_2}$

(j) Chlorobenzene $\xrightarrow[350°C]{\text{aq NaOH}}$ $\xrightarrow{H_3O^+}$

(k) Iodobenzene $\xrightarrow[CH_3OH/heat]{KSCH_3}$

(l) 2,6-Dinitro-3-fluorotoluene (CH_3, O_2N, NO_2, F) $\xrightarrow[Heat]{\text{aq NaOH}}$

(m) Phenol $\xrightarrow[CH_3OH]{K_2CO_3}$

(n) 2-Chlorophenol (OH, Cl) $\xrightarrow[BrCH_2CH=CH_2]{NaH}$

(o) 3,5-Dimethylphenol (OH, H_3C, CH_3) $\xrightarrow[\text{aq } H_2SO_4]{Na_2Cr_2O_7}$

Solution

(a) 3-Chloronitrobenzene (NO_2, Cl)

(b) 2-Bromo-4-methylanisole (CH_3, Br, OMe)

(c) 3-Iodo-4-methylacetophenone ($COCH_3$, I, CH_3)

(d) 4-Fluorotoluene (F, CH_3)

(e) 3-Iodoacetophenone (I, $COCH_3$)

(f) 5-Bromotetralin (Br)

(g) 4-Bromochlorobenzene (Br, Cl)

(h) 2-Methylbenzyl alcohol (CH_3, CH_2OH)

(i) 3-Methylaniline (CH_3, NH_2)

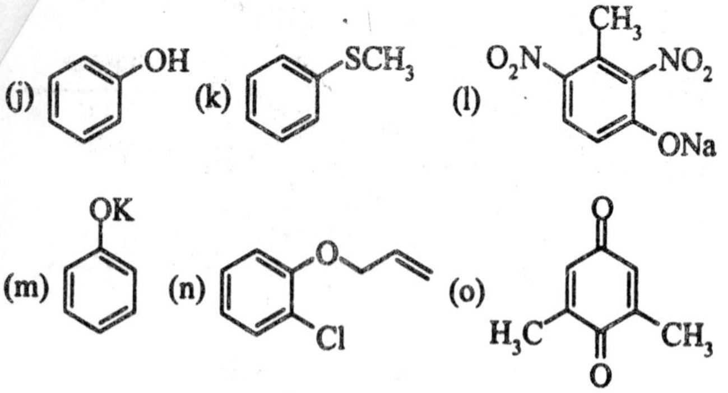

(j) (k) (l) (m) (n) (o)

Example 31. Consider the following transformation:

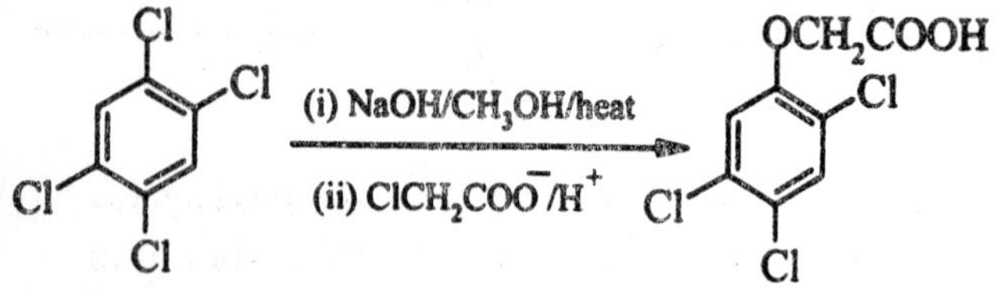

(i) NaOH/CH_3OH/heat

(ii) $ClCH_2COO^-/H^+$

(a) Propose mechanism and explain why this reaction succeed at about 150°C, whereas similar reaction of chloro benzene require 350°C.

(b) A side product of this reaction is:

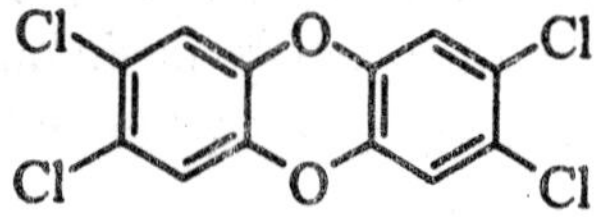

Propose mechanism of its formation.

Solution:

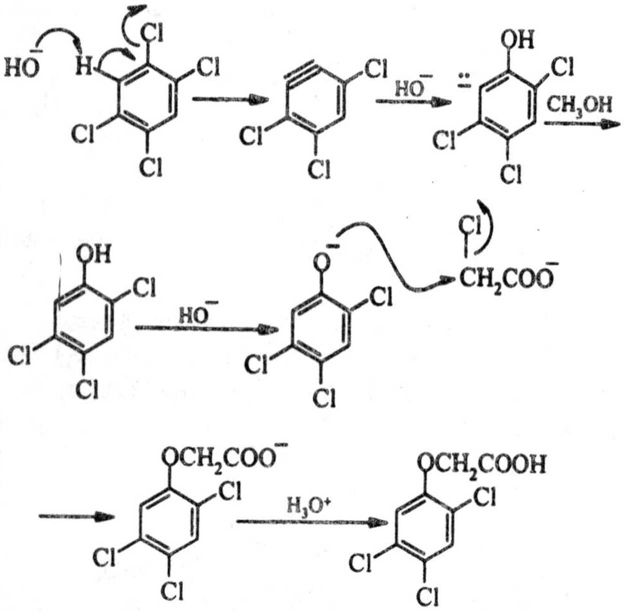

The inductive effect (–I effect) of chlorine gives each phenyl hydrogen atom a δ^+ charge making them slightly acidic. Thus, tetrachlorobenzene undergo nucleophilic substitution under milder condition while chlorobenzene undergo the similar reaction under drastic conditions.

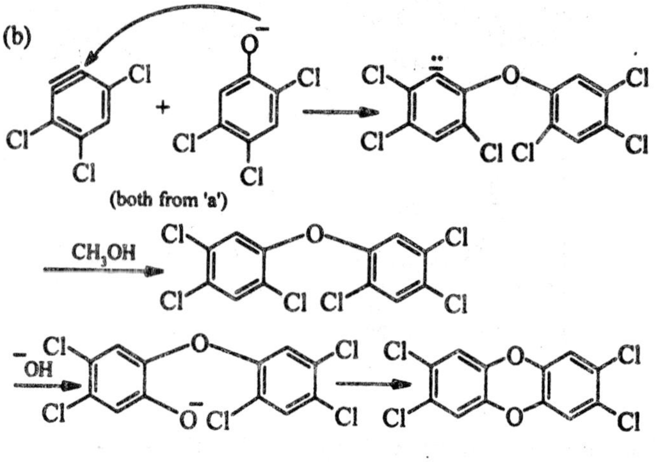

(b)

(both from 'a')

Example 32. Consider the following substitution reaction:

Ortho nitro fluorobenzene +
CH_3CH_2OK/CH_3CH_2OH/heat →

(a) Write a mechanism that would lead to a single product.

(b) Write a mechanism that would lead to two products for this reaction and compare the two mechanism.

Solution: (a) Addition-elimination:

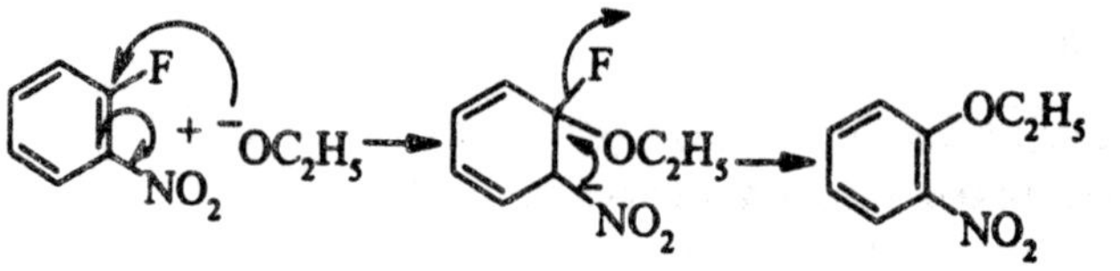

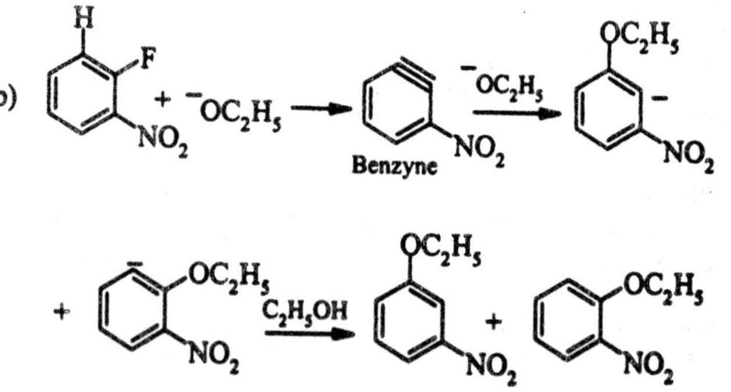

(b)

Benzyne

Fluorine is a poor leaving group, hence benzyne mechanism is not favoured for the above substitution reaction.

Example 33. Fluorine is normally considered to be a poor leaving group, yet it leaves rapidly in addition elimination nucleophilic substitution reaction if the ring contain electron withdrawing group at ortho or para position. Explain.

Solution: In order to understand why fluoride is substituted easily compared to other halide viz. iodide, by nucleophile, let us consider the following aromatic nucleophilic substitution reaction:

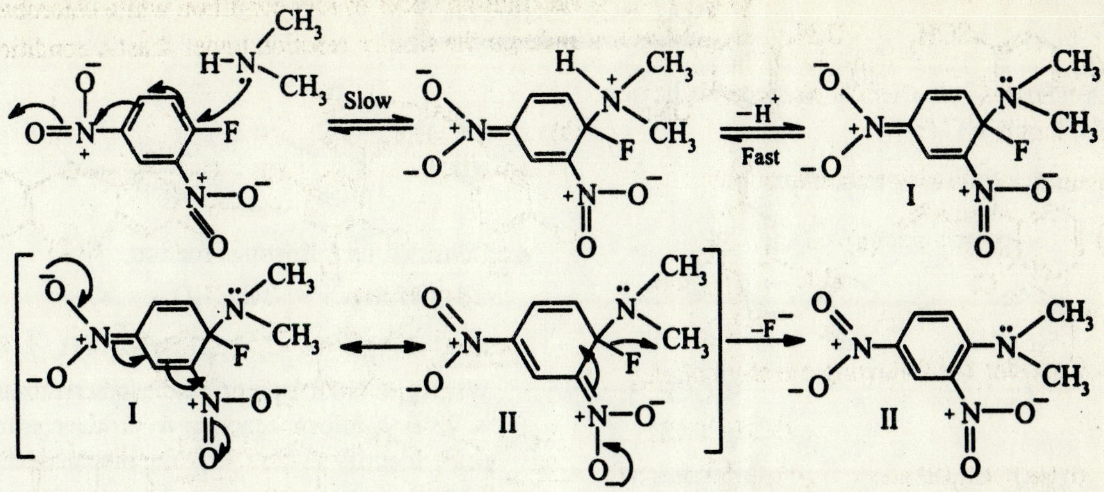

The rate determining step of the above reaction is formation of tetrahedral intermediate I by nucleophilic attack of amine at the carbon bearing fluorine atom. A fluoro compound reacts much faster than an iodo compound for two reasons. First, fluorine being the most electronegative, carbon bonded to fluorine has much higher positive charge, activate the most for nucleophilic attack. Second, fluorine being the smallest halogen, thus offer very less steric hindrance to an approaching nucleophile that will bond to carbon atom to give a tetrahedral intermediate.

Example 34. Provide mechanism for the formation of products in the following reaction:

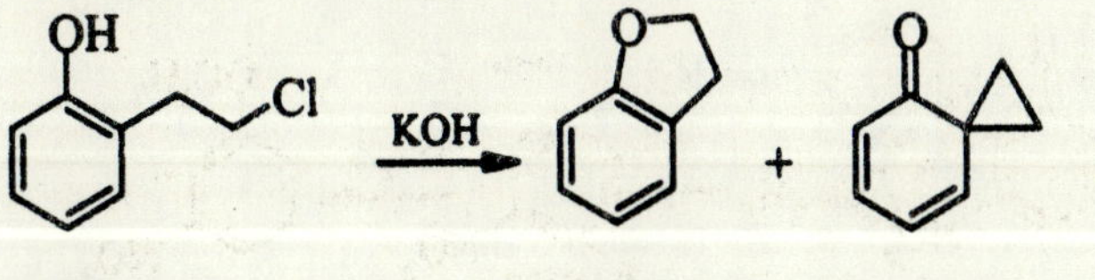

Solution

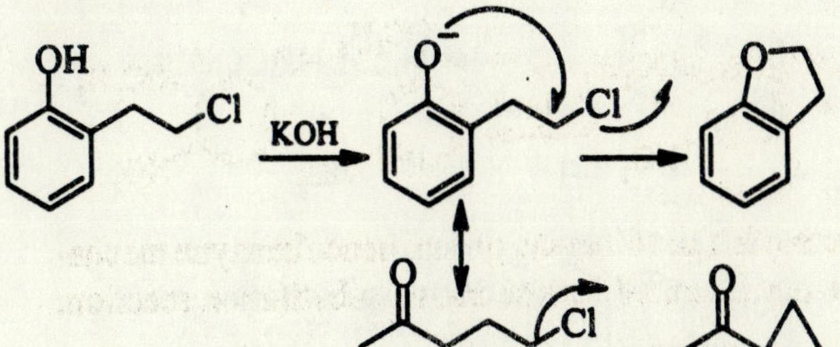

Example 35. Does this equilibrium lies to left or right?

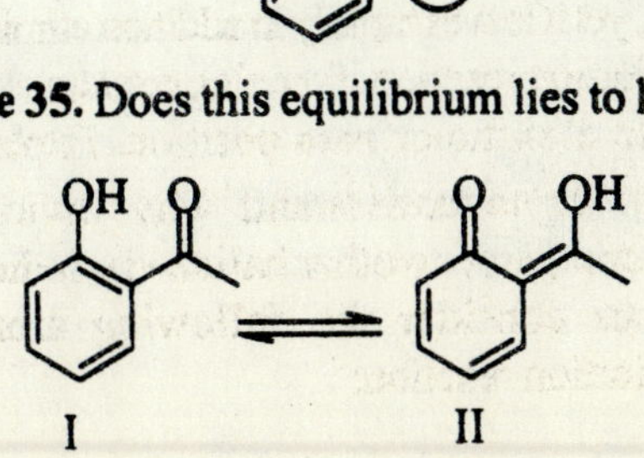

Solution: Equilibrium favours to more thermodynamically stable product side. Since, I is aromatic while II is not, hence equilibrium lies predominantly to the left.

Example 36. Ortho nitrophenol is less acidic than para nitrophenol while reverse is true for nitrobenzoic acid. Rationalize.

Solution: Acidic proton of ortho nitrophenol is involved in intramolecular hydrogen bonding, deprotonated with difficulty as compared to the deprotonation of free acidic proton of *p*-nitro phenol as:

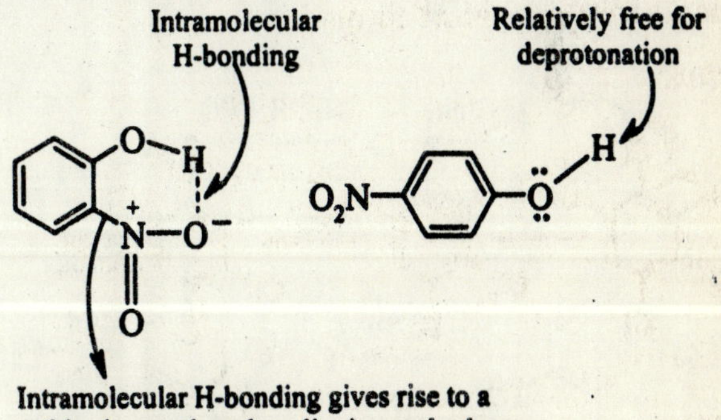

Intramolecular H-bonding

Relatively free for deprotonation

Intramolecular H-bonding gives rise to a stable six membered cyclic ring and release of H^+ becomes difficult

On the other hand, no such intramolecular H-bond prevent deprotonation from ortho nitrobenzoic acid and conjugate base of ortho nitrobenzoic acid is stabilized by forming a stable ring as:

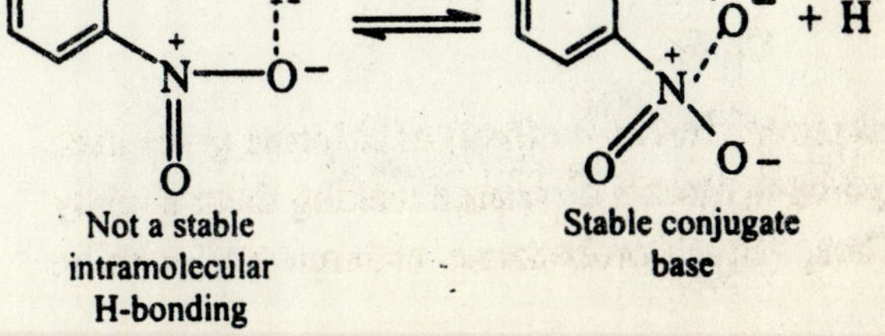

Not a stable intramolecular H-bonding

Stable conjugate base

PRACTICE EXERCISE

1. Write structures of all aromatic compounds having molecular formula C_7H_7Cl.

2. Bring about the following transformations:

(a) 1,4-dichlorobenzene → 4-chlorobenzoic acid (Cl, Cl → COOH, Cl)

(b) Benzene → p-Br–C_6H_4–$COCH_2Br$

(c) Benzene → 2-bromotoluene (CH_3, Br)

(d) Benzene → p-Br–C_6H_4–$CH{=}CH_2$

3. Complete the following reactions:

(a) 2-chlorotoluene (CH_3, Cl) $\xrightarrow[Cl_2]{h\nu}$ A $\xrightarrow{B}$ 2-chlorobenzaldehyde (CHO, Cl)

(b) $2\,Ph_2CCl_2 \xrightarrow[C_6H_6]{Cu}$ C

(c) 1-bromo-3-chlorobenzene (Cl, Br) $\xrightarrow{Mg/ether}$ D $\xrightarrow[(ii)\ aq\ NH_4Cl]{(i)\ CH_3CHO}$ E

(d) o-xylene (CH_3, CH_3) $\xrightarrow{Br_2/Fe}$ F

(e) 4-chlorotoluene (Cl, Me) $\xrightarrow[Heat]{NaSH}$ G

(f) $PhCH_3 + CBrCl_3 \xrightarrow{h\nu}$ H

(g) Bromobenzene (Br) $\xrightarrow[liq\ NH_3]{NaNH_2}$ I $\xrightarrow{[EtCOCHMe]^- Na^+}$ J

(h) 2,4-dinitroanisole (OMe, NO_2, NO_2) $\xrightarrow[C_2H_5OH]{KOH}$ K

(i) 2-bromo-1-methyl-3-nitrobenzene (H_3C, Br, NO_2) $\xrightarrow[liq\ NH_3]{NaNH_2}$ L

4. Consider the following reaction:

$C_6H_5CH_2MgCl + CH_3CHO \rightarrow X,$

$X + H^+/H_2O \rightarrow C_6H_5CH_2CHOHCH_3 + A$

Where, A is an isomer of the other product. Also A gives haloform reaction to produce ortho toluic acid. Identify A and explain mechanism of their formation.

5. Complete the following reactions:

(a) 4-bromoaniline (Br, NH_2) $\xrightarrow{NaNCO}$ $\xrightarrow{CH_3COOH}$ A

(b) Aniline (NH_2) + B $\xrightarrow{Et_2O}$ $PhNHCOCH{=}CHCOOH$

(c) 4-nitroaniline (NO_2, NH_2) + 2ICl ⟶ C

(d) Benzylamine (CH_2NH_2) + D ⟶ $PhCH_2NHCOOEt$

(e) Diphenylamine (H–N) $\xrightarrow{Li}$ E $\xrightarrow[CuI]{C_2H_5I}$ F

(f) 4-ethoxyaniline (NH_2, OEt) $\xrightarrow[Urea]{HCl}$ $\xrightarrow{Heat}$ G

(g) 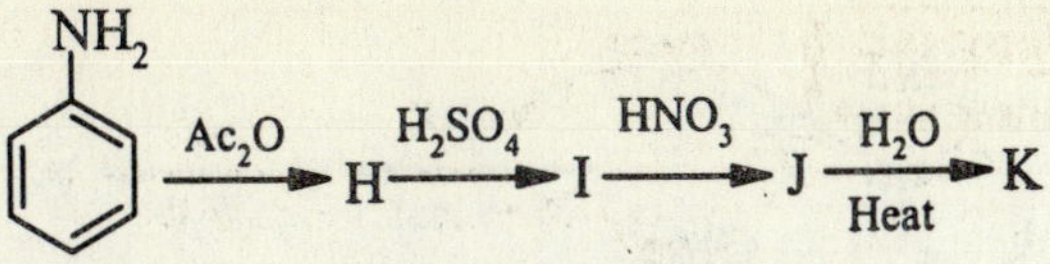 Aniline (NH_2) $\xrightarrow{Ac_2O}$ H $\xrightarrow{H_2SO_4}$ I $\xrightarrow{HNO_3}$ J $\xrightarrow[\text{Heat}]{H_2O}$ K

6. Bring about the following transformations:
   (a) Benzene → ortho ethyl aniline
   (b) Toluene → 4-amino-3-bromo toluene
   (c) Benzene → *m*-chloro aniline
   (d) Benzene → *p*-iodo aniline
   (e) Toluene → *p*-amino benzoic acid
   (f) Benzene → 2, 5-dichloro aniline.

7. Complete the following reactions:

(a) 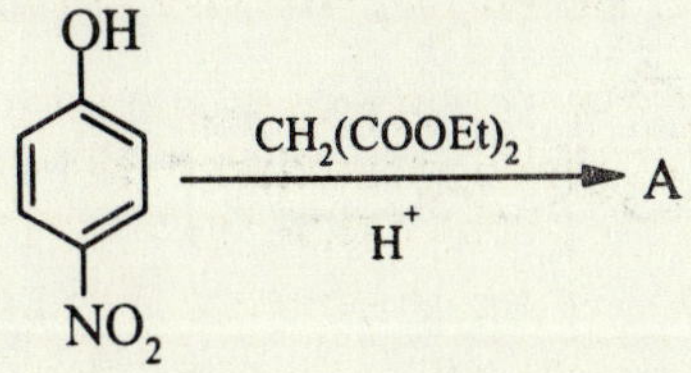 p-Nitrophenol (OH, NO_2) $\xrightarrow[H^+]{CH_2(COOEt)_2}$ A

(b) 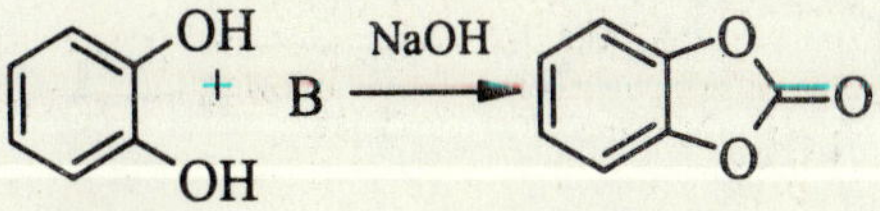 Catechol (OH, OH) + B $\xrightarrow{NaOH}$ 1,3-benzodioxol-2-one (cyclic carbonate, O–C(=O)–O)

(c) Phenol (OH) + $CH_2{=}CHCH_2Br$ ⟶ C

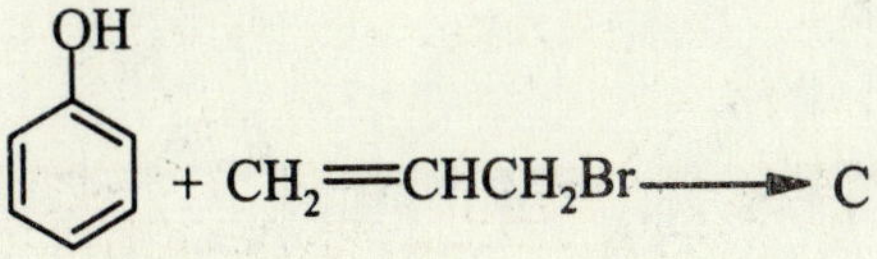

(d) 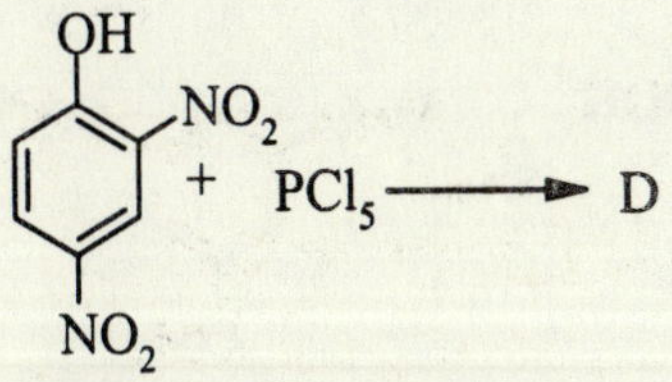 2,4-Dinitrophenol (OH, NO_2, NO_2) + PCl_5 ⟶ D

(e) 2,4-Dinitroanisole (OMe, NO_2, NO_2) $\xrightarrow[EtOH]{KOH}$ E

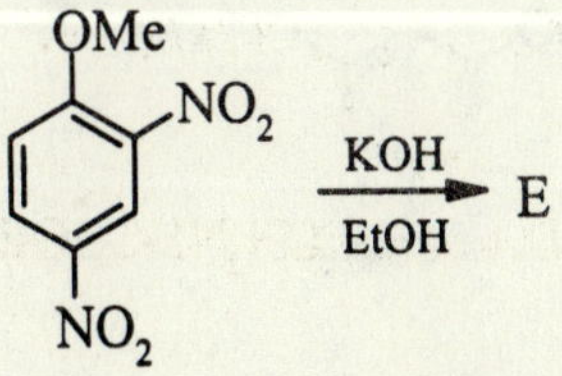

8. Bring about the following transformations:

(a) 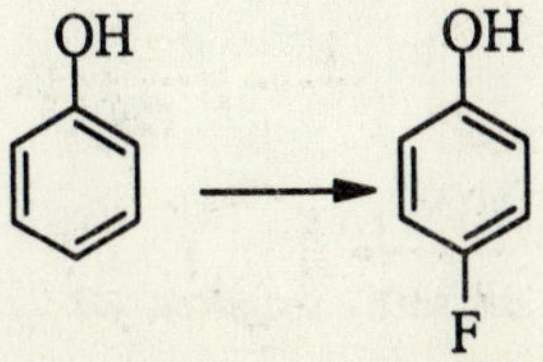 Phenol (OH) ⟶ p-Fluorophenol (OH, F)

(b) 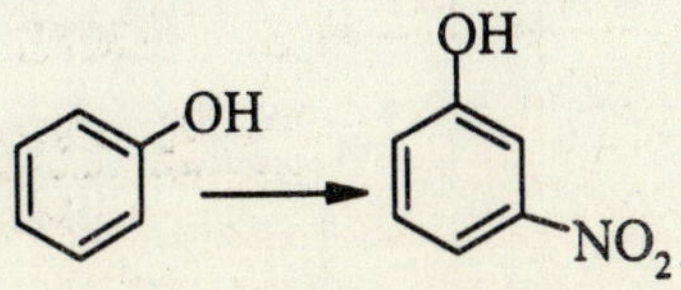 Phenol (OH) ⟶ m-Nitrophenol (OH, NO_2)

(c) 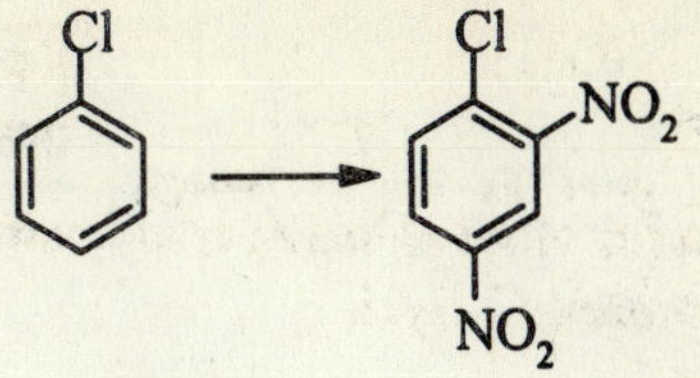 Chlorobenzene (Cl) ⟶ 1-Chloro-2,4-dinitrobenzene (Cl, NO_2, NO_2)

(d) 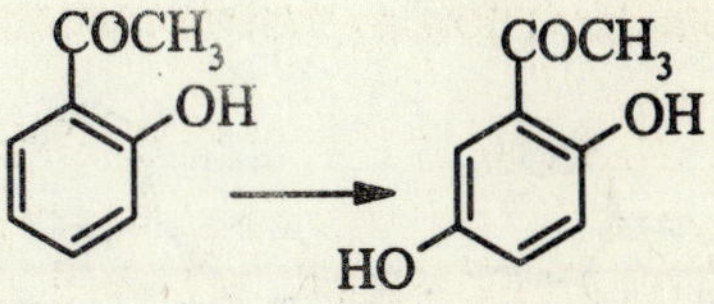 o-Hydroxyacetophenone ($COCH_3$, OH) ⟶ 2,4-Dihydroxyacetophenone ($COCH_3$, OH, HO)

9. Complete the following reactions:

(a) 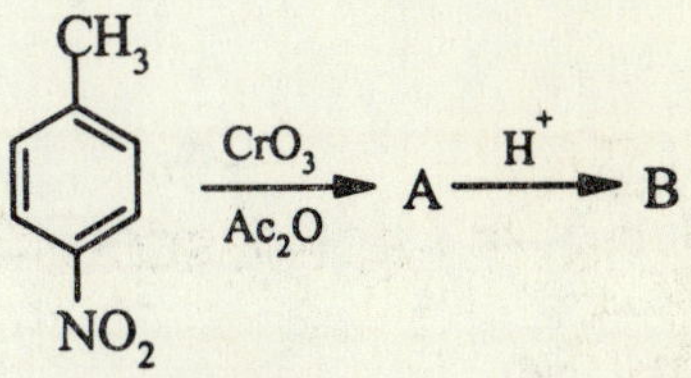 p-Nitrotoluene (CH_3, NO_2) $\xrightarrow[Ac_2O]{CrO_3}$ A $\xrightarrow{H^+}$ B

(b) Ph_2CO $\xrightarrow[NaOH]{Zn}$ C

(c) $PhCH_2OH$ + D ⟶ $PhCH_2OCOCl$

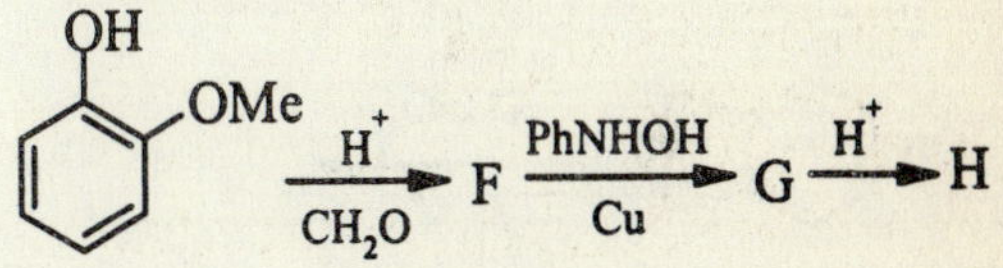

(d) 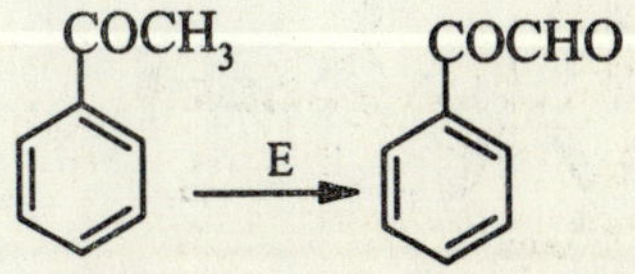 Acetophenone ($COCH_3$) $\xrightarrow{E}$ Phenylglyoxal ($COCHO$)

(e) Guaiacol (OH, OMe) $\xrightarrow[CH_2O]{H^+}$ F $\xrightarrow[Cu]{PhNHOH}$ G $\xrightarrow{H^+}$ H

(f) Phenyl butanoate ($OCOCH_2CH_2CH_3$) $\xrightarrow{AlCl_3}$ I

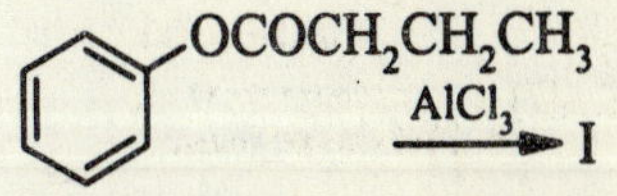

(g) PhCOOH $\xrightarrow{SOCl_2}$ J $\xrightarrow{Me_2NH}$ K $\xrightarrow{L}$ PhCHO

10. Convert:
    (a) Benzene → *m*-nitro acetophenone.
    (b) Nitrobenzene → *m*-bromobenzaldehyde.
    (c) Benzene → Benzene-1, 4-dicarbaldehyde.

11. Complete the following reactions:

(a) 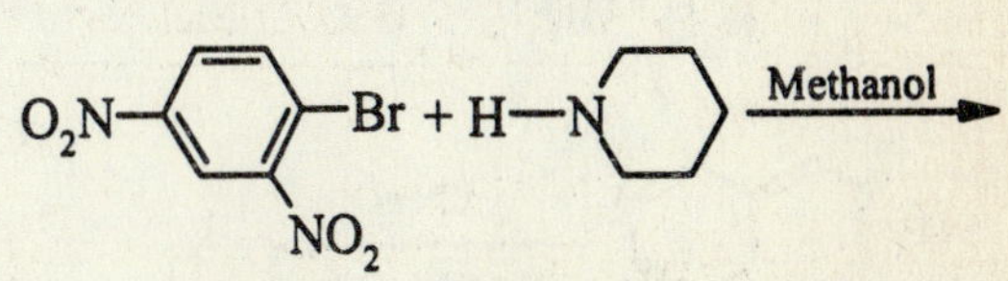 O_2N–C$_6$H$_3$(NO_2)–Br + H–N(piperidine) $\xrightarrow{\text{Methanol}}$

(b) 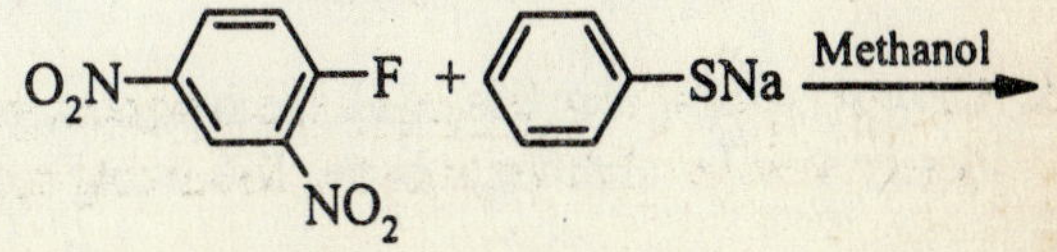 O_2N–C$_6$H$_3$(NO_2)–F + C$_6$H$_5$–SNa $\xrightarrow{\text{Methanol}}$

(c) O_2N–(2,6-dinitrophenyl)–OTs + $C_6H_5NH_2$ $\xrightarrow[\text{Heat}]{\text{Benzene}}$

(d) O_2N–(2,6-dinitrophenyl)–Cl + H_2N–$N(C_6H_5)_2$ $\xrightarrow{CHCl_3}$

(e) O_2N–(2-nitrophenyl)–F + $H_3\overset{+}{N}$–$CH(CH_3)COO^-$ $\xrightarrow{H_2O}$

12. Complete the following reactions:

(a) $C_6H_5CH(OH)C_6H_5$ $\xrightarrow{C_5H_5\overset{+}{N}H\ CrO_3Cl^-}$

(b) Indene $\xrightarrow[H_2O,\ 0°C]{KMnO_4}$

(c) $C_6H_5CH_2CH_3$ $\xrightarrow[\text{NaOH (aq)/heat}]{KMnO_4}$

(d) m-Br$C_6H_4CH_3$ $\xrightarrow[\text{NaOH (aq)/heat}]{KMnO_4}$

(e) 1,1,3,3-tetramethylindane (H_3C, CH_3, H_3C, CH_3) $\xrightarrow[\text{NaOH (aq)/heat}]{KMnO_4}$

(f) Naphthalene $\xrightarrow[\text{NaOH (aq)/heat}]{KMnO_4}$

13. Give structural formula of all the reagents, products designated by alphabets in the following reactions:

(a) o-(CHO)C_6H_4–$CH(OMe)_2$ $\xrightarrow[\text{Methanol}]{NaBH_4}$ A $\xrightarrow[H_2O]{H^+}$ B

(b) 2-(OH)-4-isopropyl-C_6H_3–$CH(CH_3)CH=CH_2$ $\xrightarrow[\text{THF}]{NaH}$ $\xrightarrow{CH_3I}$ C $\xrightarrow{BH_3}$ $\xrightarrow[NaOH/H_2O]{H_2O_2}$ D

(c) D $\xrightarrow[H_2O/\text{heat}]{CrO_3/H_2SO_4}$ E $\xrightarrow[\text{Heat}]{H_3PO_4}$ F $\xrightarrow{LiAlH_4}$ $\xrightarrow{H_2O}$ G

(d) HOOC~~(m-C_6H_4)~~COOH $\xrightarrow{H_3PO_4\ \text{heat}}$ H + I

(e) 2-(CHO)-6-(OH)-(OMe)-C_6H_3 $\xrightarrow[\text{Methanol}]{NaBH_4}$ J $\xrightarrow[\text{THF}]{NaH}$ K $\xrightarrow{CH_3I}$ L

(f) $H_2C(COOEt)_2$ $\xrightarrow[Et_2O]{Na}$ M —Br–(2,6-dinitrophenyl)–Cl ⟶ N

(g) 3,5-(HO)(OH)C_6H_3COOH $\xrightarrow{O}$ 3,5-(MeO)(OMe)C_6H_3COOMe $\xrightarrow{P}$ 3,5-(MeO)(OMe)$C_6H_3CH_2OH$

14. Identify the products, intermediates and reagents in the following reactions:

(a) A $\xrightarrow{HNO_3}$ B $\xrightarrow{Sn,\ HCl}$ $\xrightarrow[HCl\text{-}0°C]{NaNO_2}$ 4-methoxy-1-naphthyl-$\overset{+}{N_2}Cl^-$ (OCH_3) $\xrightarrow{Na_2CO_3}$ $\xrightarrow[\text{Heat}]{CuCN}$ C

(b) O_2N–C₆H₃(NO_2)–O–C₆H₄–NO_2 $\xrightarrow{Cl_2,\ AlCl_3}$ D ; $\xrightarrow{\text{H–N(piperidine)}}$ E

(c) O_2N–C₆H₄–NH_2 + ICl $\xrightarrow{CH_3COOH}$ F $\xrightarrow[H_2SO_4\text{-}0°C]{NaNO_2}$ $\xrightarrow{KI}$ G

(d) CH_3CH_2O–C₆H₄–NH_2 $\xrightarrow[\text{Acetic acid}]{Br_2}$ H $\xrightarrow[HCl\text{-}0°C]{NaNO_2}$ I $\xrightarrow{H_3PO_2}$ J

(e) 1-nitronaphthalene (NO_2) $\xrightarrow{Sn\text{-}HCl}$ K $\xrightarrow[HCl\text{-}0°C]{NaNO_2}$

L $\xrightarrow{HBF_4}$ M $\xrightarrow{Heat}$ N

(f) O_2N–C₆H₃(CH_3)–CH_3 $\xrightarrow{H_2/Ni}$ O $\xrightarrow[H_2SO_4\text{-}0°C]{NaNO_2}$

P $\xrightarrow[Heat]{CuCN}$ Q $\xrightarrow[H_2O]{H^+}$ R

15. Predict structural formula for all products and intermediates designated by alphabets in the following reactions:

(a) C₆H₅CHO $\xrightarrow[H_2SO_4]{HNO_3}$ A $\xrightarrow{Sn\text{-}HCl}$ B $\xrightarrow[HCl\text{-}0°C]{NaNO_2}$ C $\xrightarrow[HCl/heat]{CuCl}$ D

(b) 3,5-dinitroaniline (NH_2, O_2N, NO_2) $\xrightarrow[H_2SO_4\text{-}0°C]{NaNO_2}$ E $\xrightarrow{H_2SO_4/H_2O}$ F

(c) Br–C₆H₄–OMe $\xrightarrow[NH_3(aq)]{NaNH_2}$ G (major) + H (minor)

(d) C₆H₄(COOH)(NH_2) $\xrightarrow[HCl\text{-}0°C]{NaNO_2}$ I $\xrightarrow{\text{C}_6\text{H}_5\text{–NMe}_2}$ J

(e) $PhNHCH_3$ $\xrightarrow[HCl\text{-}0°C]{NaNO_2}$ K

(f) Pyrrolidine (N–H) $\xrightarrow[Excess]{CH_3I}$ L $\xrightarrow[Heat]{KOH}$ M $\xrightarrow[Excess]{CH_3I}$ N $\xrightarrow[Heat]{KOH}$ O + P

16. Propose mechanism of the following reaction:

O_2N–C₆H₃(NO_2)–Cl + NC–CH_2–C(=O)–O–CH_2CH_3 $\xrightarrow[Ethanol]{CH_3CH_2ONa}$ O_2N–C₆H₃(NO_2)–CH(CN)–$COOC_2H_5$

17. Arrange each sets of compounds in increasing order of acid strength:

(a) (i) p-CH_3–C₆H₄–COOH (ii) p-CH_3–C₆H₄–NH_2 (iii) p-CH_3–C₆H₄–OH (iv) C₆H₅–CH_2OH

(b) (i) C₆H₅–OH (ii) p-CH_3–C₆H₄–OH (iii) p-F–C₆H₄–OH (iv) p-NO_2–C₆H₄–OH

(c) (i) p-OMe–C₆H₄–COOH (ii) p-NO_2–C₆H₄–COOH (iii) C₆H₅–COOH (iv) p-Cl–C₆H₄–COOH

18. Following are the relative rates of nucleophilic substitution of fluoride from *p*-nitrofluorobenzene, with various nucleophiles:

$O_2N-C_6H_4-F$ CH_3O^- $C_6H_5S^-$

Relative rates: 11600 10300

$C_6H_5O^-$ $C_6H_5NH_2$

65 1

Explain the above experimental observations:

19. The nucleophilic substitution reactions of 2, 4-dinitro chlorobenzene with a series of para substituted aniline (shown below) were studied. Write equations for substitution reaction and predict their order of reactivity:

Aniline (I) *p*-toludine (II)
p-nitroaniline (III) *p*-chloroaniline (IV)
p-methoxyaniline (V)

20. Bring about the following transformations:
(a) Toluene → *p*-fluoro ethyl benzoate
(b) Toluene → 3-bromo-4-hydroxy toluene
(c) Toluene → $C_6H_5CH_2CH(C_2H_5)CH_2NH_2$
(d) Toluene → *p*-(N, N-dimethyl amino) benzoic acid
(e) Toluene → *p*-iodo toluene
(f) Chlorobenzene → *p*-chloroaniline

OBJECTIVE EXERCISE

Level 1

1. Which one of the following groups X in the compound (C_6H_5X) is *m*-directing in electrophilic substitution reactions?
(a) $—OCH_3$ (b) $—NH·CO·CH_3$
(c) $—COCH_3$ (d) $—N(CH_3)_2$

2. Where does the compound $C_6H_5COC_2H_5$ substitute on electrophilic attack?
(a) meta (b) ortho
(c) ortho and para (d) para

3. Which of the following compounds undergoes nucleophilic substitution most readily?

(a) Cl, $COCH_3$ (b) Cl
(c) Cl, H_3C (d) Cl, O_2N

4. Which of the following compounds yields *n*-propylbenzene when it is treated with zinc amalgum in concentrated hydrochloric acid?
(a) $PhCH_2CH_2CO_2H$
(b) $PhCH_2CH{=}CH_2$
(c) $PhCH_2CH(OH)CH_3$
(d) $PhCOCH_2CH_3$

5. The reaction of toluene with chlorine in the presence of light gives:

(a) Cl (b) Cl
(c) Cl, Cl, H_3C, Cl (d) ClH_2C

6. One of the completions A to D is incorrect. Choose the letter corresponding to the *incorrect* completion. In the nitration of benzene by concentrated sulphuric and concentrated nitric acid.
(a) The reaction is irreversible.
(b) The rate determining step is the removal of H^+ from the intermediate.
(c) The attacking electrophile is NO_2^+.
(d) The second nitro group enters the benzene ring meta to the first nitro group.

7. In which of the following compounds is there the least delocalization of the lone pair of electrons on the nitrogen atom ?

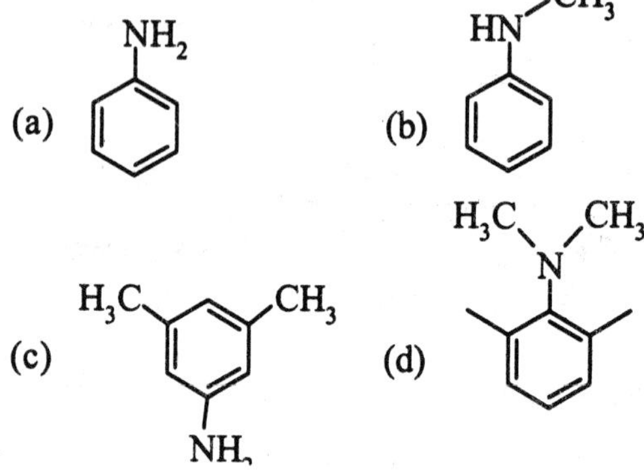

8. One of the completions A to E is incorrect. Choose the letter corresponding to the *incorrect* completion. Pyrrole will undergo the following reactions to give the stated products.
   (a) Sulphonation with $C_5H_5N \cdot SO_3$ yields mainly pyrrole-2-sulphonic acid.
   (b) Benzene diazonium chloride reacts to give benzeneazo-2-pyrrole.
   (c) Acetic anhydride yields methyl-2-pyrryl ketone.
   (d) Chloroform and potassium hydroxide react to give mainly 3-chloropyridine.

9. The major organic product from the reaction $C_6H_6 + CH_3COCl/AlCl_3$ is:
   (a) C_6H_5Cl (b) C_6H_5COCl
   (c) $C_6H_5COCH_3$ (d) [benzene ring with CH_3 and ortho C(=O)Cl]

10. The major organic products from the reaction: $C_6H_5 \cdot CO \cdot CH_3 + I_2/NaOH$ followed by acidification is:
   (a) $C_6H_5CI_3 + CH_3OH$ (b) $C_6H_5 \cdot CHO + CH_3I$
   (c) $C_6H_5COOH + CHI_3$ (d) [benzene ring with CH_3 and ortho I]

11. In the compound C_6H_5X which of the following is predominantly *ortho/para* directing?
   (a) X = —NO_2, X = —Cl, X = —OH
   (b) X = —OMe, X = —CN, X = —NH_2
   (c) X = —$NHCOCH_3$, X = —Cl, X = —CO_2H
   (d) X = —$NHCOCH_3$, X = —CH_3, X = —Br

12. The relative rates of mononitration of R—C_6H_5, where R = —CH_3, —NO_2, —OH, —Cl are:
   (a) $CH_3 > OH > NO_2 > Cl$
   (b) $OH > Cl > CH_3 > NO_2$
   (c) $OH > CH_3 > NO_2 > Cl$
   (d) $OH > CH_3 > Cl > NO_2$

13. Which statement is most relevant to the fact that pyridine reacts with sodamide, $NaNH_2$, in liquid ammonia to yield 2-amino pyridine.
   (a) Pyridine readily undergoes electrophilic substitution.
   (b) Pyridine resists attack by electrophiles.
   (c) The amide ion is a strong base in liquid ammonia.
   (d) Pyridine is susceptible to nucleophilic attack.

Following questions (14 to 17) are based on the following reaction road-map:

Benzene —I→ $C_6H_5COCH_3$ —II→ C_6H_5COOH —III→ C_6H_5COCl —IV→ $C_6H_5CONH_2$ —V→ NC–C_6H_5 (A) —VI→ H_2NCH_2–C_6H_5 (B)

14. Which combination of reagents could be used to bring about Step I (the use of water, dilute acid or alkali can be assumed at any stage)?
   (a) CH_3COOH/Zn
   (b) $CH_3COCl/AlCl_3$
   (c) $CH_3CH_2Br/AlCl_3$ then $KMnO_4$
   (d) CH_3CHO/OH^-

15. Which reaction type would be involved in Step II?
   (a) Mannich reaction
   (b) Clemmensen reaction
   (c) Iodoform reaction
   (d) Friedel-Crafts reaction

16. Which reagent could be used in Step VI ?
   (a) $NaBH_4$
   (b) Zn + conc. HCl at reflux
   (c) Na/EtOH (d) $LiAlH_4$

17. On treatment with an excess of nitrous acid at 0°C, followed by heating of the resulting solution an organic product is obtained. The product is soluble in dilute aqueous caustic soda solution, but not in aqueous sodium bicarbonate solution would this behaviour be shown by:
   (a) A and B? (b) A only?
   (c) B only? (d) Neither A nor B?

18. Which of the following structures is *least likely* to be a resonance form of the others in the group?
   (a) [cyclohexadienone, C=O] (b) [ring with $\overset{+}{O}$–H and negative charge on ring]
   (c) [phenol, OH] (d) [ring with $=\overset{+}{O}H$ and negative charge at para position] (e) [ring with $=\overset{+}{O}H$ and negative charge at ortho position]

19. When benzaldehyde is refluxed with potassium cyanide in ethanolic solution the product isolable on cooling will be:
    (a) PhCH(OH)CN
    (b) PhCH(OH)CH(OH)Ph
    (c) PhCOCH(OH)Ph
    (d) $PhCONH_2$

20. When phenol is refluxed with allyl bromide in acetone solution in the presence of anhydrous potassium carbonate a product may be isolated which, on heating to 200° is converted mainly to:

    (a) O (b) OH

    (c) OH (d) OH

21. A mixture of nitric and sulphuric acids is in common use for the nitration of aromatic compounds. Which one of the following statements concerning this reagent is incorrect?
    (a) Sigma-complexes are involved in nitration.
    (b) In the absence of sulphuric acid, the nitric acid is about 50 ionized therefore niration is slower.
    (c) A solution of nitric acid in sulphuric acid shows a freezing point depression by a factor of 4.
    (d) The rate-determining step is the initial attack by NO_2^+.

22. Which one of the following compounds undergoes ready Friedel-Crafts acylation reaction at carbon atom?
    (a) nitrobenzene (b) pyridine
    (c) pyrimidine (d) acetophenone

23. Which one of the following compounds will be most readily hydrolysed in aqueous alkali?

    (a) Cl, NO_2 (b) Cl, NO_2

    (c) Cl, O_2N, NO_2 (d) Cl, NO_2, NO_2

24. Where does the above compound substitute on electrophilic attack?

    $CONHCH_3$

    (a) meta (b) ortho
    (c) ortho and para (d) para

25. The nitration of 1-nitronaphthalene gives the following product(s):
    (a) 1, 4-dinitronaphthalene
    (b) mainly 1, 3- and some 1, 5-dinitronaphthalene
    (c) 1, 3-dinitronaphthalene
    (d) mainly 1, 5- and some 1, 8-dinitronaphthalene

26. Which of the following sets of substituents, when separately attached to the aromatic ring, is most likely to 'activate' nucleophilic aromatic substitution?
    (a) NO_2, CH_3, CN (b) NO, $COCH_3$, $^+N(CH_3)_3$
    (c) COOH, Br, NH_2 (d) CF_3, CHO, OCH_3

27. The conversion of aqueous benzene diazonium-sulphate into bromobenzene is best effected by heating with:
    (a) HBr (b) Br_2
    (c) N-bromosuccinimide (d) HBr/cuprous bromide

28. Which of the following reagents is necessary to convert 2-methylpyridine into the compound:

    CH_2CH_3, H_3C, N

    (a) $CH_3CH_2Cl/AlCl_3$ (b) $CH_2{=}CH_2/H^+$
    (c) CH_3CHO/OH^- (d) CH_3CH_2MgBr

29. Is the main product from the reaction between pyridine and sodamide at 100°C, followed by hydrolysis:
    (a) 1-aminopyridine (b) 2-aminopyridine
    (c) 3-aminopyridine (d) 4-aminopyridine

30. Which one of the following cyclic olefins would you predict to be aromatic?

    (a) (b)

    (c) 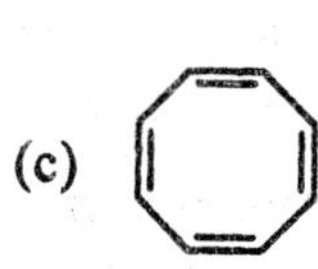 (d) 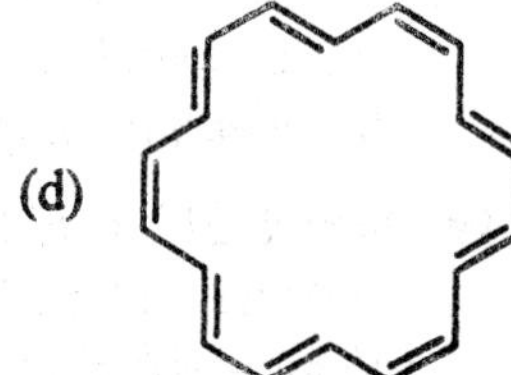

31. Select the correct order of basicities of thiophene, pyridine, and piperidien from the following:
(a) thiophene > pyridine > piperidine
(b) piperidine > pyridine > thiophene
(c) piperidine > thiophene > pyridine
(d) pyridine > piperidine > thiophene

32. Benzene reacts with ethyl chloride and aluminium chloride in nitrobenzene solvent to give mainly:
(a) isopropylbenzene (b) naphthalene
(c) ethylbenzene (d) *m*-diethylbenzene

33. The orientation of the products (*o*-, *m*-, *p*-) from aromatic substitution reactions is governed by:
(a) the nature of the reaction medium.
(b) the relative stabilities of the products.
(c) the type of substituent already present in the aromatic ring.
(d) the delocalization of the pi-electron system.

34. Which species is *not* aromatic?

(a) [pyridine] (b) [pyrrole, N–H] (c) [cyclooctatetraene] (d) [cyclopentadienyl anion]

35. Which of the following would be expected to react most readily with sodium methoxide in methanol?

(a) [fluorobenzene] (b) [4-fluorotoluene, F / CH_3] (c) [4-fluoronitrobenzene, F / NO_2] (d) [3-fluoronitrobenzene, F / NO_2]

36. The relative reactivities of benzene, aniline, nitrobenzene, and toluene towards aromatic electrophilic substitution are:
(a) benzene > toluene > aniline > nitrobenzene
(b) nitrobenzene > aniline > toluene > benzene
(c) toluene > aniline > nitrobenzene > benzene
(d) aniline > toluene > benzene > nitrobenzene

37. In the nitration of aromatic compounds, the nitrating species is:
(a) NO_2^- (b) NO_2^+ (c) NO_3^+ (d) NO_3^-

38. The principal reactions of aromatic compounds are:
(a) nucleophilic substitution
(b) nucleophilic addition - with -elimination
(c) electrophilic addition - with -elimination
(d) electrophilic addition

39. Which of the following sequences correctly represents the order of activation, by ring substituents of an aromatic ring to electrophilic attack.
(a) $CH_3 > N(CH_3)_2 > OCH_3 > NO_2 > Cl$
(b) $NO_2 > Cl > CH_3 > OCH_3 > N(CH_3)_2$
(c) $OCH_3 > NO_2 > N(CH_3)_2 > Cl > CH_3$
(d) $N(CH_3)_2 > OCH_3 > CH_3 > Cl > NO_2$

40. Sulphonation of naphthalene with concentrated sulphuric acid at 160°C gives as the major product:

(a) [1-naphthalenesulphonic acid, SO_3H] (b) [1,8-naphthalenedisulphonic acid, SO_3H SO_3H]
(c) [1,2-naphthalenedisulphonic acid, SO_3H, SO_3H] (d) [2-naphthalenesulphonic acid, SO_3H]

41. Which diagram does *NOT* show the initial site of electrophilic substitution in the ring?

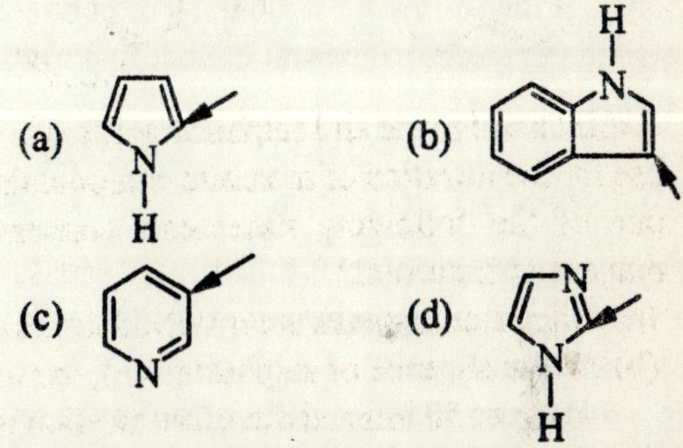

42. Which one of the following reactions is the first step in the sodium hydroxide catalysed conversion of benzaldehyde to benzyl alcohol and the benzoate ion?

(a) [benzene]–CHO + $\overline{OH}$ ⟶ [ring anion with OH and CHO]
(b) [benzene]–CHO + $\overline{OH}$ ⟶ [benzene]–$\overline{CH}$–O–OH
(c) [benzene]–CHO + $\overline{OH}$ ⟶ [benzene]–CH(OH)–$\overline{O}$
(d) [benzene]–CHO + $\overline{OH}$ ⟶ [benzene]–C^+(OH)(OH)

43. The following list shows possible yields of products from the mononitration of (trichloromethyl)-benzene. From theoretical considerations predict the most probable proportions. Percentage yield of nitrated product:

	ortho	meta	para
(a)	25	7	68
(b)	20	66	14
(c)	46	8	46
(d)	7	64	29

44. Irradiation (UV) of a mixture of toluene and chlorine (2 moles) with subsequent mild alkaline hydrolysis of the product gives:
(a) product which gives effervescence with $NaHCO_3$.
(b) a product which gives a precipitate of silver chloride on treatment with aqueous silver nitrate.
(c) a product which has $pk_a = 4.0$.
(d) product will give deep violet colouration on adding aqueous $FeCl_3$.

45. Which of the following is the most effective catalyst for the reaction of benzene and Br_2 to produce bromobenzene?
(a) KBr (b) HNO_3 (c) HBr
(d) $Fe(OH)_3$ (e) $FeBr_3$

46. What ring position(s) of toluene are most reactive toward electrophilic aromatic substitution?
(a) meta and para (b) meta
(c) ortho and para (d) ortho and meta

47. What characteristic of the methoxy group causes it to direct electrophiles to the ortho and para positions of anisole?
(a) Inductive stabilization.
(b) Removal of ring electrons through resonance.
(c) Interaction of the hydrogens of the methyl group.
(d) Donation of the oxygen's non-bonding electrons to the ring through resonance.

48. Which one of the following is least reactive toward electrophilic aromatic substitution?
(a) the phenoxide anion (b) phenol
(c) ethylbenzene (d) nitrobenzene

49. What unusual characteristic does a halogen substituent exhibit in electrophilic aromatic substitutions?
(a) It is activating but meta directing.
(b) It is activating and directs to all positions equally.
(c) It is deactivating but ortho/para directing.
(d) It is activating but exclusively ortho directing.

50. How many isomeric, trisubstituted benzenes, each bearing one $—SO_3H$, one $—OCH_3$, and one $—NO_2$, would be found among the major products of the reaction of *m*-nitroanisole with fuming sulfuric acid?
(a) 1 (b) 3 (c) 2 (d) 4

51. What is the major product of Friedel-Crafts acylation of toluene with acetyl chloride?
(a) $C_6H_5—CH_2—CO—CH_3$
(b) para $CH_3—C_6H_4—CO—CH_3$
(c) $C_6H_5—CO—CH_3$
(d) ortho $CH_3—C_6H_4—CO—CH_3$

52. Which one of the following occurs by an addition-elimination mechanism?
(a) toluene, Br_2, $FeBr_3$ → *p*-bromotoluene
(b) benzene, HNO_3, H_2SO_4 → nitrobenzene
(c) benzene, CO, HCl, $AlCl_3$, CuCl → benzaldehyde
(d) 2, 4-dinitrochlorobenzene, NaOH, heat, then H^+ → 2, 4-dinitrophenol

53. What is produced by the reaction of toluene with molecular hydrogen at high pressures and in the presence of a catalyst such as Pt, Pd, or Ni?
(a) benzene (b) 1-methyl-1, 3-cyclohexadiene
(c) cyclohexane (d) methyl cyclohexane

54. In SN-1 reactions, benzyl chloride, $C_6H_5—CH_2—Cl$, is about as reactive as a *t*-alkyl chloride. What is the reason for this reactivity?
(a) The $—CH_2—Cl$ group is an activating group.
(b) The ring is strongly electron withdrawing.
(c) The planar ring offers little resistance to the approach of the nucleophile.
(d) The SN-1 reaction takes place via carbocation intermediate.

55. What is the chemical name of common aspirin?
(a) anethole (b) hydroquinone
(c) acetylsalicylic acid
(d) 2-methyl-1, 4-benzoquinone

56. When a benzene ring undergoes electrophilic aromatic substitution, a sigma complex forms as an intermediate. How many pi-electrons does this sigma complex contain?
(a) 5 (b) 6 (c) 3 (d) 4

57. What function does HNO_3 serve in the reaction of benzene with I_2 to produce iodobenzene?
(a) The HNO_3 converts the I^- to I_2.
(b) The HNO_3 serves as a catalyst.
(c) The HNO_3 converts I_2 to HI.
(d) The HNO_3 oxidizes I_2 to I^+.

58. Benzene reacts with a mixture of nitric acid and sulfuric acid to form nitrobenzene. Which one of the

following best represents the final products of the reaction of the nitric acid with the sulfuric acid?
(a) $H_3SO_4^+ + NO_3^-$ (b) $HSO_4^- + H_2O + NO_2^+$
(c) $H_2NO_3^+ + HSO_4^-$ (d) $H_2O + HSO_4^- + NO_2$

59. In the reaction of benzene with fuming sulfuric acid to produce benzenesulfonic acid, what is the effective electrophile?
(a) HSO_4^- (b) SO_2^+ (c) HSO_3^+ (d) SO_3

60. Which one of the following factors is most important in determining the position of electrophilic attack on the ring of toluene?
(a) The presence of a catalyst.
(b) Temperature.
(c) The concentration of the toluene.
(d) The concentration of the electrophile.

61. What effect does a —NH—CO—CH_3 substituent have on electrophilic aromatic substitution?
(a) It is meta directing but neither activating nor deactivating.
(b) It is activating and ortho/para directing.
(c) It is deactivating and ortho/para directing.
(d) It is deactivating and meta directing.

62. The —NO_2 group deactivates the benzene ring towards electrophilic aromatic substitution. At which ring position(s) is this deactivation least effective?
(a) meta (b) ortho and para
(c) ortho and meta (d) ortho

63. Which one of the following combinations would you expect in the nitration of bromobenzene?
(a) Ortho/para product and a faster reaction than benzene.
(b) Para product only and a faster reaction than benzene.
(c) Meta product and a slower reaction than benzene.
(d) Ortho/para product and a slower reaction than benzene.

64. Which one of the following compounds will produce one and only one monobromination product?
(a) *p*-methyoxyphenol (b) *m*-nitroanisole
(c) *m*-dinitrobenzene (d) *o*-chlorotoluene

65. What is the product of the Friedel-Crafts reaction of benzene with 1-chloropropane?
(a) *n*-propylbenzene (b) ethylbenzene
(c) toluene (d) isopropylbenzene

66. What is the best method for converting benzene to *n*-propylbenzene?
(a) toluene, ethyl chloride, $AlCl_3$
(b) benzene, CH_3—CH_2—CO—Cl, $AlCl_3$, then H_2O, then Zn(Hg) and aq HCl
(c) toluene, Cl—CO—CH_3
(d) benzene, Cl—CH_2—CH_2—CH_3, $AlCl_3$

67. What term describes the reactive intermediate formed in the following reaction? *p*-bromotoluene, $NaNH_2$, NH_3, —33°C → CH_3—(C_6H_4)—NH_2 in a ratio 50% para isomer, 50% meta isomer.
(a) benzyne (b) aryl radical
(c) sigma complex (d) benzocation

68. What kind of compound results from the Birch reduction of an aromatic ring?
(a) an open-chain hexane (b) a cyclic diene
(c) an open-chain hexane (d) a cyclohexene

69. What represents the best method for producing salicylic acid?
(a) Benzoic acid, strong base, high temperature
(b) Toluene, strong base, water, high temperature
(c) Phenol, base, carbon dioxide, then acid
(d) Bromobenzene, carbonic acid, high temperature

70. What are the ring-carbon hybridizations in the sigma complex that form as an intermediate in electrophilic aromatic substitution?
(a) sp^2 and sp only (b) sp^2 only
(c) sp^2 and sp^3 only (d) sp and sp^3 only

71. What is the rate-determining step in the reaction of Br_2 with benzene, in the presence of $FeBr_3$, to produce bromobenzene?
(a) The reaction of benzene with $FeBr_3$—Br_2 to produce the sigma complex, $C_6H_6^+$—Br.
(b) The reaction of Br_2 with $FeBr_3$ to produce Br^+.
(c) The reaction of Br_2 with $FeBr_3$ to produce $FeBr_3$—Br_2.
(d) The loss of a proton from the sigma complex, $C_6H_6^+$—Br.

72. Benzene is converted to nitrobenzene by a mixture of sulfuric acid and nitric acid. What is the catalyst for this reaction?
(a) the nitronium ion (b) water
(c) sulfuric acid (d) nitric acid

73. What is produced by the reaction of benzene with a solution of SO_3 in D_2O?
(a) C_6H_5-OH (b) $C_6(SO_3H)_6$
(c) benzenesulfonic acid (d) C_6H_5-OD

74. If all the ring-carbons of toluene were to undergo electrophilic attack at the same rate, what would be

the major product(s) of the electrophilic nitration of toluene?
(a) *o*-nitrotoluene
(b) *o*-nitrotoluene and *m*-nitrotoluene
(c) *o*-nitrotoluene and *p*-nitrotoluene
(d) *p*-nitrotoluene

75. Which one of the following structures reacts fastest in electrophilic aromatic substitution?
(a) $C_6H_5—N(CH_3)_2$ (b) $C_6H_5—CH_3$
(c) $C_6H_5—O^-$ (d) $C_6H_5—OH$

76. Why does an electron withdrawing substituent on a benzene ring tend to be meta directing in electrophilic substitution?
(a) Its electron withdrawing ability is least effective in the meta position.
(b) It withdraws electrons most effectively from the meta position.
(c) It withdraws electrons from the meta position, but releases them to the ortho and para positions.
(d) It releases electrons to the meta position.

77. Halogens are deactivating yet ortho/para directing in electrophilic aromatic substitution. How do we explain this?
(a) A combination of inductive electron withdrawal and resonance electron release.
(b) Inductive electron withdrawal with no resonance effect at all.
(c) A combination of inductive electron release and resonance electron withdrawal.
(d) A combination of inductive electron release and resonance electron release.

78. Which one of the following will *not* undergo a Friedel-Crafts alkylation?
(a) toluene (b) *p*-chlorotoluene
(c) bromobenzene (d) benzoic acid

79. Which pair is usually required for a nucleophilic aromatic substitution reaction?
(a) An $—NO_2$ substituent and a strong electrophile.
(b) A ring bearing a strong activating group and a strong acid.
(c) An aryl halide with an $—NO_2$ and a strong nucleophile.
(d) An unsubstituted benzene ring and a strong electrophile.

80. What starting material and conditions would produce benzene hexachloride?
(a) Cyclohexane, excess HCl, heat, pressure.
(b) Benzene, excess Cl_2, $FeCl_3$.
(c) Benzene, excess Cl_2, $AlCl_3$.
(d) Benzene, excess Cl_2, and either heat and pressure or ultraviolet light.

81. What refers to an sp^3 hybridized carbon bonded directly to a carbon of a benzene ring?
(a) benzylic (b) phenylic
(c) arylic (d) allylic

82. What reaction does hydroquinone undergo most readily?
(a) SN-2 substitution (b) hydrolysis
(c) oxidation (d) reduction

83. What is product B of the following 2 step reaction?
Step I: Ethylbenzene + $CH_3—CO—Cl$, $AlCl_3$, then $H_2O \rightarrow$ Product A
Step II: Product A, Zn(Hg), and aq HCl $\rightarrow$ Product B
(a) *o*-diethylbenzene (b) *m*-diethylbenzene
(c) *p*-diethylbenzene (d) *m*-ethylacetophenone

Level 2

1. Which of the following carbanion is most stable?
(a) (b) $Ph—C≡C^-$
(c) (d) $H_2\bar{C}—CH=CH_2$

2. Which of the following is anti aromatic?
(a) (b)
(c) 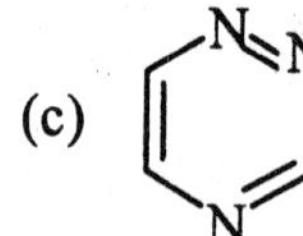

(d)

3. Predict the predominant site(s) of mono-nitration in the following compound:

OMe, Br, NO_2, 1, 2, 3

(a) 1 only (b) 2 only
(c) 3 only (d) 2 and 3

4. Choose the statement that is correct concerning the following reaction:

Phenol + $CH_3I/NaOH$ → Major product

(a) This reaction results in methylation of oxygen in phenol. It is utilized frequently, because the hydroxide group is often highly activated. This reaction can be reversed with HI.

(b) This reaction results in methylation of oxygen in phenol. The resulting product is activated approximately the same amount as in phenol. The reaction can be reversed with HI.

(c) This reaction results in methylation of oxygen in phenol. The resulting product is more highly activated than phenol. This reaction can be reversed with HI.

(d) This reaction results in an SN-2 reaction between methyl iodide and NaOH. The resulting iodide ion substitutes at ortho and para positions of phenol. Treatment of this product with Mg will give Grignard reagent.

5. Rank the following compounds in decreasing order of reactivity in electrophilic aromatic substitution reaction:

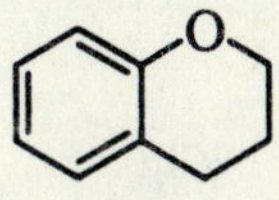

I 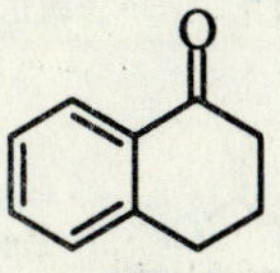   II 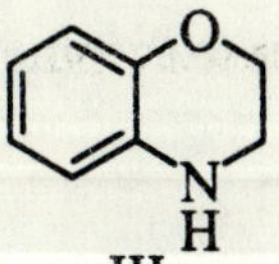   III 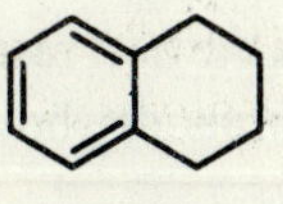   IV

(a) IV > I > III > II   (b) III > I > IV > II
(c) I > III > IV > II   (d) III > II > I > IV

6. Select the best reaction sequence to accomplish the following transformation:

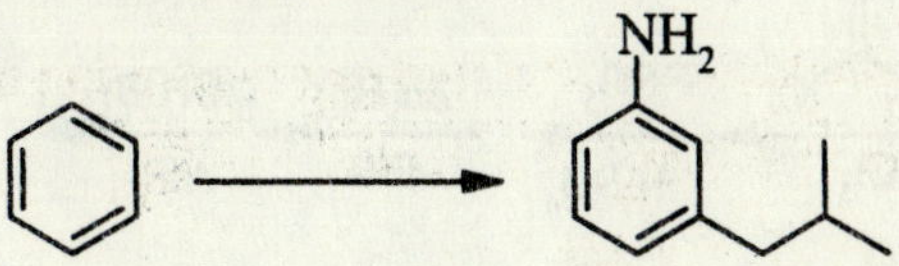

(a) 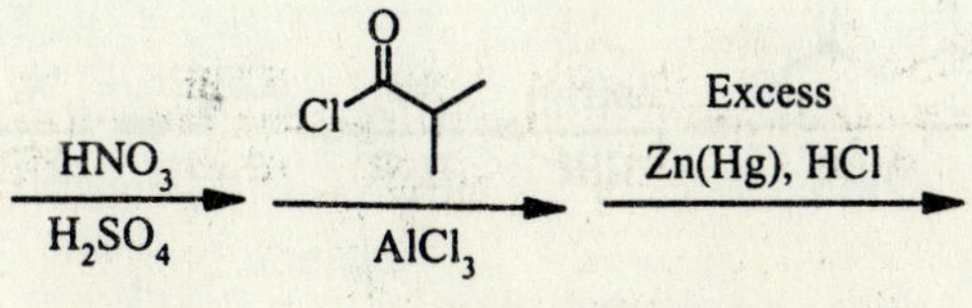 $\xrightarrow[H_2SO_4]{HNO_3}$ $\xrightarrow[AlCl_3]{Cl\text{-CO-CH(CH}_3)_2}$ $\xrightarrow{\text{Excess Zn(Hg), HCl}}$

(b) 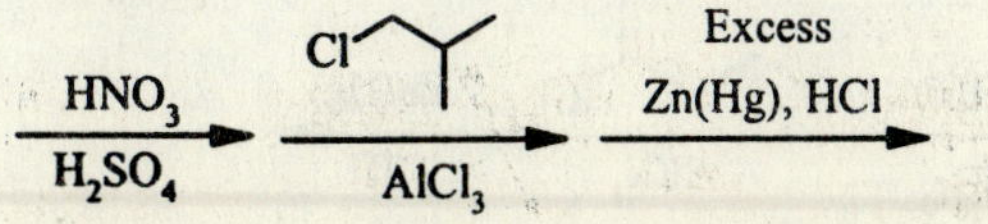 $\xrightarrow[H_2SO_4]{HNO_3}$ $\xrightarrow[AlCl_3]{Cl\text{-CH}_2\text{CH(CH}_3)_2}$ $\xrightarrow{\text{Excess Zn(Hg), HCl}}$

(c) 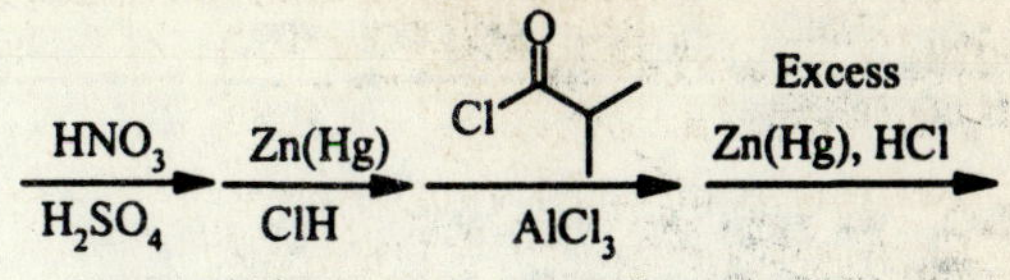 $\xrightarrow[H_2SO_4]{HNO_3}$ $\xrightarrow[ClH]{Zn(Hg)}$ $\xrightarrow[AlCl_3]{Cl\text{-CO-CH(CH}_3)_2}$ $\xrightarrow{\text{Excess Zn(Hg), HCl}}$

(d) 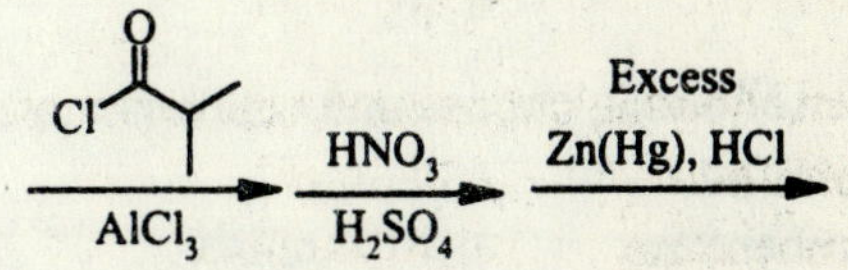 $\xrightarrow[AlCl_3]{Cl\text{-CO-CH(CH}_3)_2}$ $\xrightarrow[H_2SO_4]{HNO_3}$ $\xrightarrow{\text{Excess Zn(Hg), HCl}}$

7. Give the major product of the following reaction:

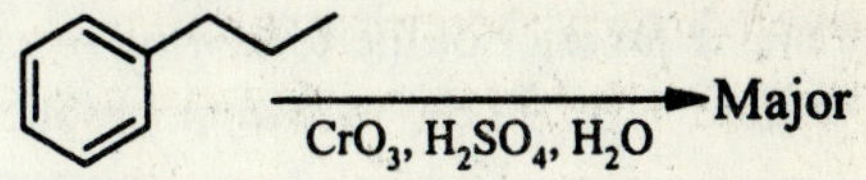

$\xrightarrow{CrO_3,\ H_2SO_4,\ H_2O}$ Major

(a) 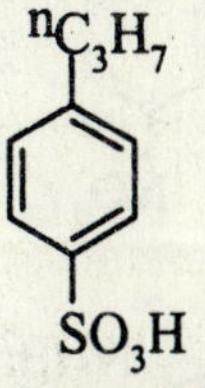 $^nC_3H_7$, SO_3H (b) 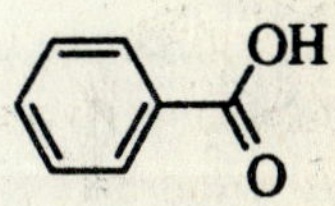 OH, O

(c) 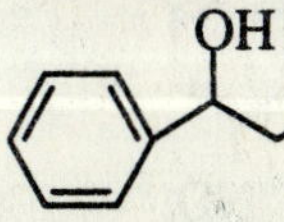 OH   (d) 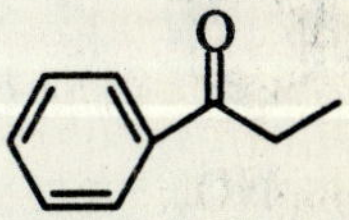 O

8. Give the major product of the following reaction:

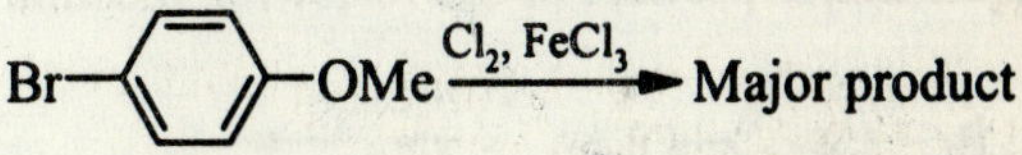

Br—⟨⟩—OMe $\xrightarrow{Cl_2,\ FeCl_3}$ Major product

(a) 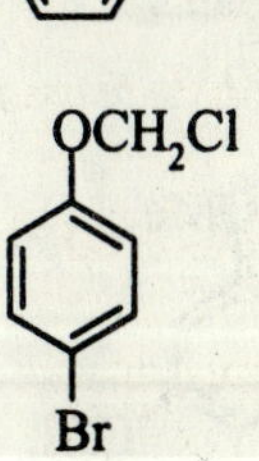 OCH_2Cl, Br   (b) 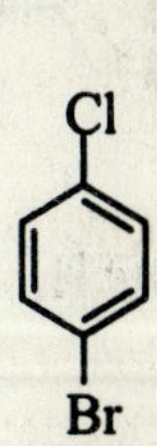 Cl, Br

(c) 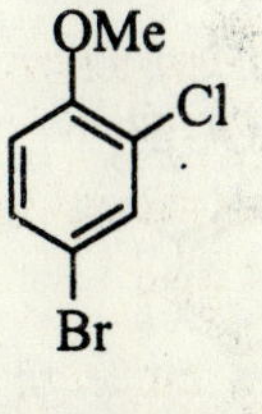 OMe, Cl, Br   (d) 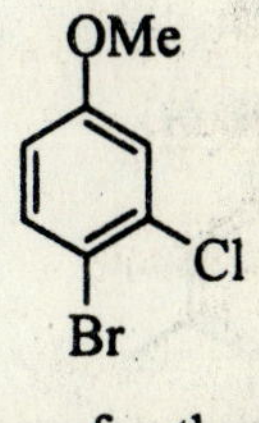 OMe, Cl, Br

9. Give the best reaction sequence for the following transformation:

NO_2, NO_2

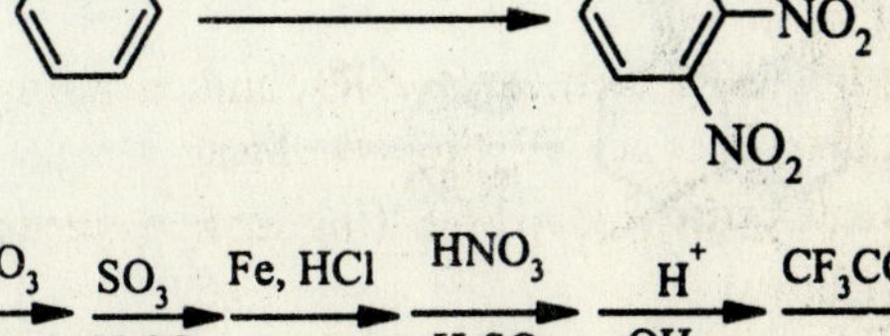

(a) $\xrightarrow[H_2SO_4]{HNO_3}$ $\xrightarrow[H_2SO_4]{SO_3}$ $\xrightarrow{Fe,\ HCl}$ $\xrightarrow[H_2SO_4]{HNO_3}$ $\xrightarrow[OH_2]{H^+}$ $\xrightarrow{CF_3CO_3H}$

(b) $\xrightarrow[H_2SO_4]{HNO_3}$ $\xrightarrow{Fe,\ HCl}$ $\xrightarrow{CH_3COCl}$ $\xrightarrow[H_2SO_4]{SO_3}$ $\xrightarrow[H_2SO_4]{HNO_3}$ $\xrightarrow[OH_2]{H^+}$ $\xrightarrow{CF_3CO_3H}$

(c) $\xrightarrow[H_2SO_4]{SO_3}$ $\xrightarrow[H_2SO_4]{HNO_3}$ $\xrightarrow{Fe, HCl}$ $\xrightarrow{CH_3COCl}$ $\xrightarrow[H_2SO_4]{HNO_3}$ $\xrightarrow[OH_2]{H^+}$ $\xrightarrow{CF_3CO_3H}$

(d) $\xrightarrow[H_2SO_4]{HNO_3}$ $\xrightarrow{Fe, HCl}$ $\xrightarrow{CH_3COCl}$ $\xrightarrow[H_2SO_4]{HNO_3}$ $\xrightarrow{CF_3CO_3H}$

10. From the following choose one which is most reactive to nitration:

(a) Nitrobenzene (b) Aniline
(c) *p*-nitroaniline (d) *m*-diaminobenzene

11. Give the major product of the following reaction:
m-nitroaniline + $Br_2/FeBr_3$ → Major product

(a) 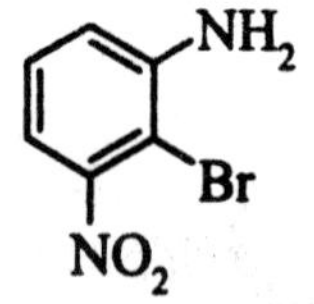 NH_2, Br, NO_2 (b) 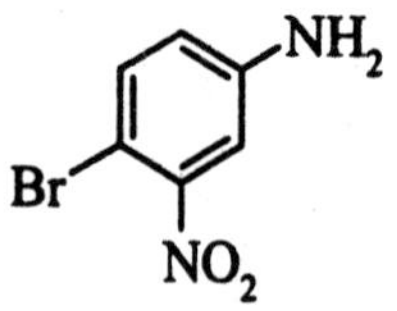 NH_2, Br, NO_2

(c) 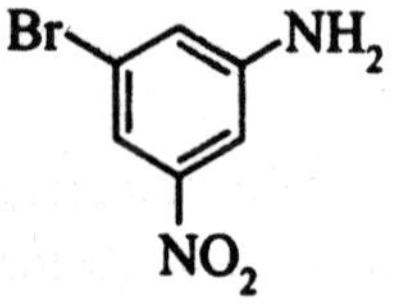 Br, NH_2, NO_2 (d) 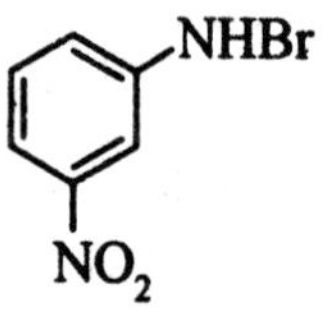 NHBr, NO_2

12. Give the major product of the following reaction:

Br–C₆H₄–OMe $\xrightarrow[AlCl_3]{CH_3CH_2Cl}$

(a) 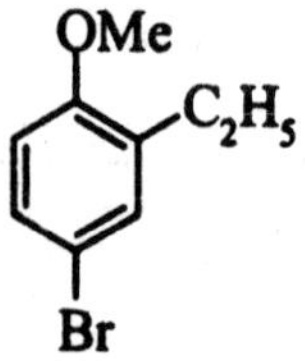 OMe, C_2H_5, Br (b) OMe, C_2H_5, Br

(c) OC_2H_5, Br (d) 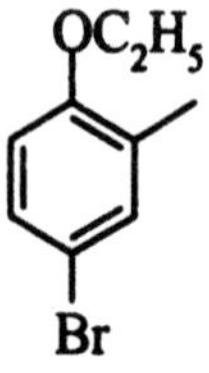 OC_2H_5, Br

13. Give the major product of the following reaction:

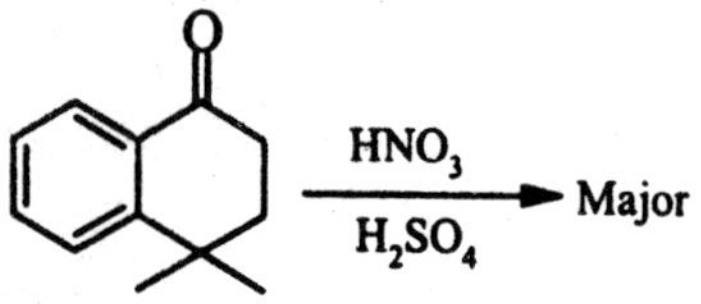

$\xrightarrow[H_2SO_4]{HNO_3}$ Major

(a) 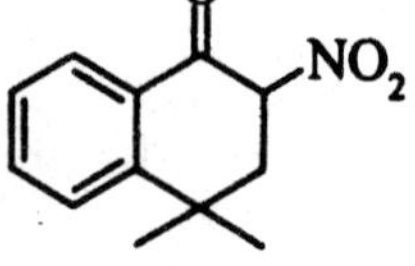 NO_2 (b) 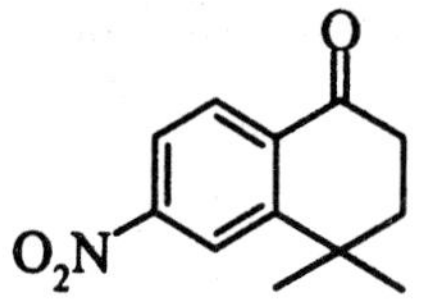 O_2N

(c) 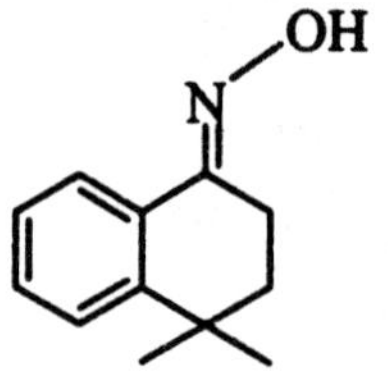 N, OH (d) 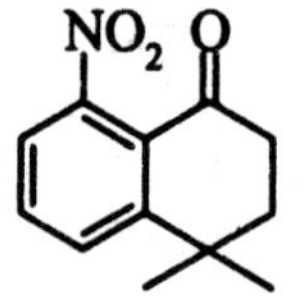 NO_2, O

14. Choose the best reaction sequence to accomplish the transformation shown:

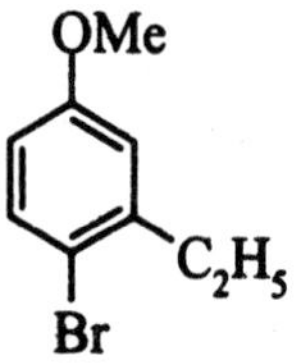

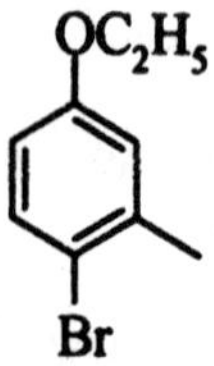

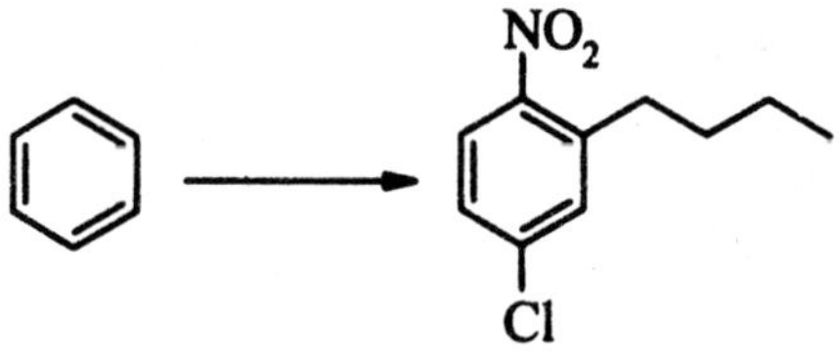

→ NO_2, Cl

(a) 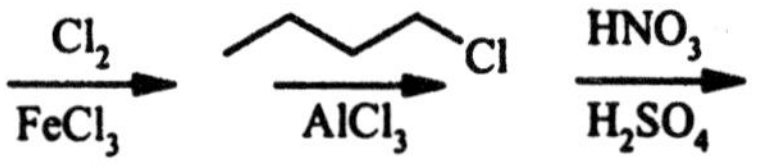 $\xrightarrow[FeCl_3]{Cl_2}$ $\xrightarrow[AlCl_3]{Cl}$ $\xrightarrow[H_2SO_4]{HNO_3}$

(b) 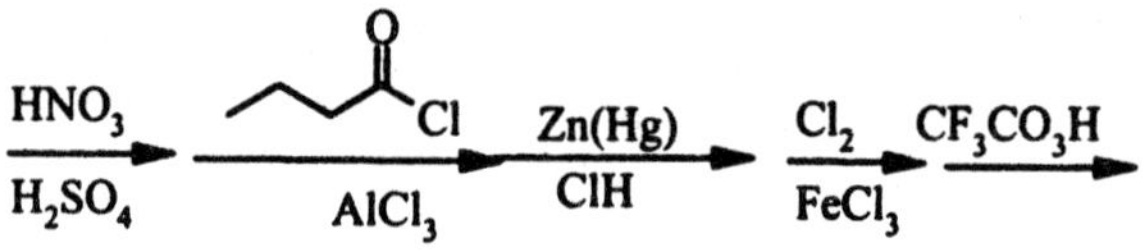 $\xrightarrow[H_2SO_4]{HNO_3}$ $\xrightarrow[AlCl_3]{Cl}$ $\xrightarrow[ClH]{Zn(Hg)}$ $\xrightarrow[FeCl_3]{Cl_2}$ $\xrightarrow{CF_3CO_3H}$

(c) 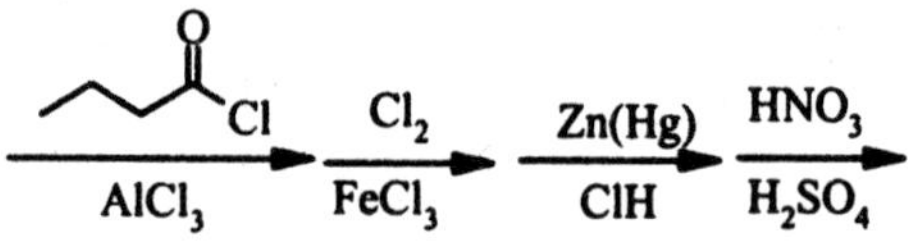 $\xrightarrow[AlCl_3]{Cl}$ $\xrightarrow[FeCl_3]{Cl_2}$ $\xrightarrow[ClH]{Zn(Hg)}$ $\xrightarrow[H_2SO_4]{HNO_3}$

(d) 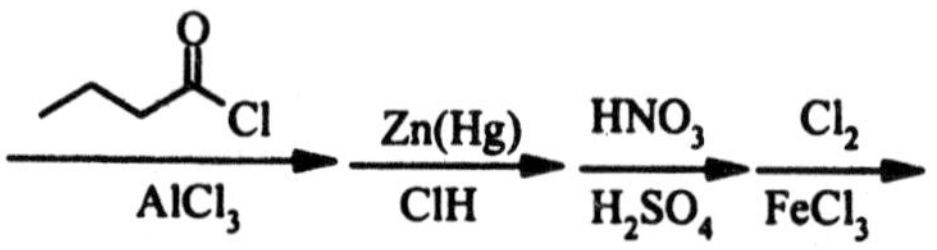 $\xrightarrow[AlCl_3]{Cl}$ $\xrightarrow[ClH]{Zn(Hg)}$ $\xrightarrow[H_2SO_4]{HNO_3}$ $\xrightarrow[FeCl_3]{Cl_2}$

15. Choose the best reaction sequence to accomplish the transformation shown:

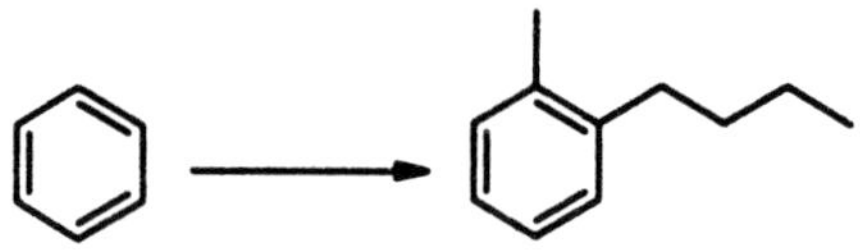

→

(a) 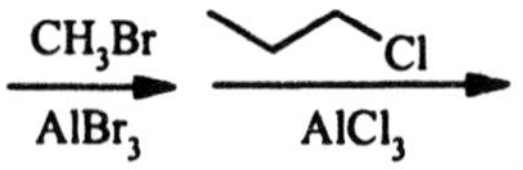 $\xrightarrow[AlBr_3]{CH_3Br}$ $\xrightarrow[AlCl_3]{Cl}$

(b) 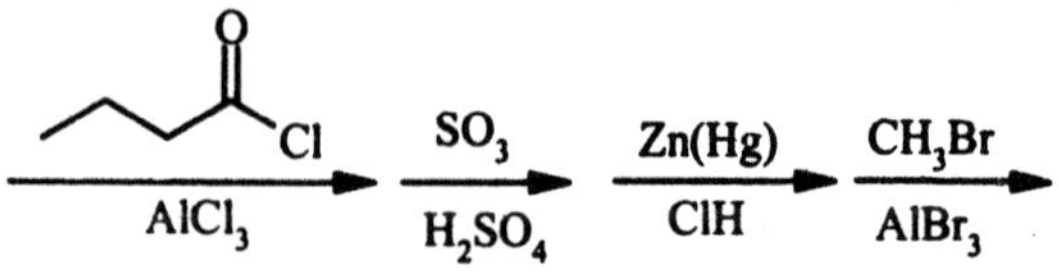 $\xrightarrow[AlCl_3]{Cl}$ $\xrightarrow[H_2SO_4]{SO_3}$ $\xrightarrow[ClH]{Zn(Hg)}$ $\xrightarrow[AlBr_3]{CH_3Br}$

(c) 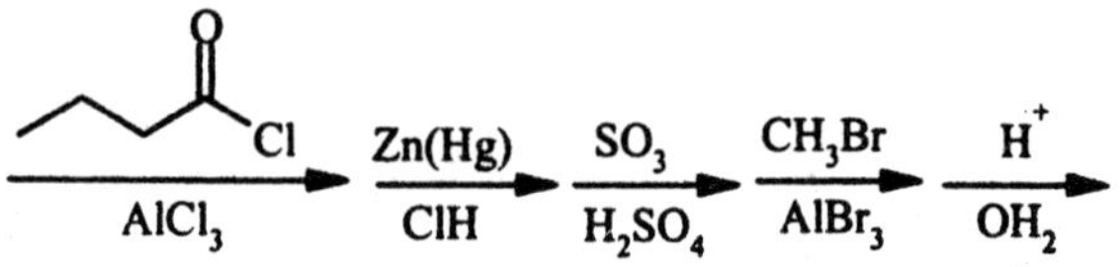 $\xrightarrow[AlCl_3]{Cl}$ $\xrightarrow[ClH]{Zn(Hg)}$ $\xrightarrow[H_2SO_4]{SO_3}$ $\xrightarrow[AlBr_3]{CH_3Br}$ $\xrightarrow[OH_2]{H^+}$

(d) 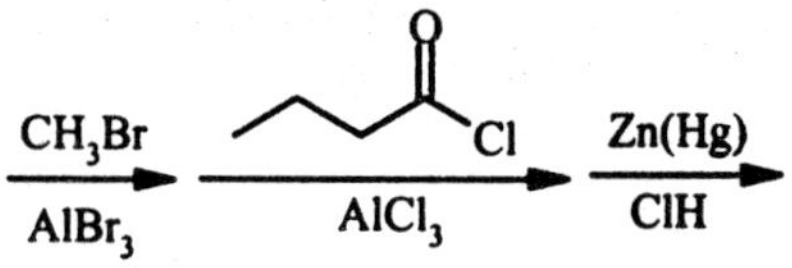 $\xrightarrow[AlBr_3]{CH_3Br}$ $\xrightarrow[AlCl_3]{Cl}$ $\xrightarrow[ClH]{Zn(Hg)}$

16. Give the major product of the reaction:
phenol + $CH_3I/NaOH \rightarrow$ Major product

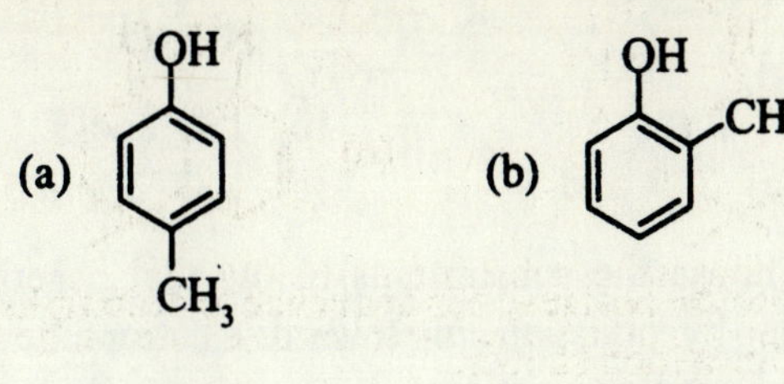

(a) OH, CH_3 (b) OH, CH_3

(c) OH, CH_3 (d) OMe

17. Give the major product from the following reaction sequence:

OH —(CH_3I / NaOH)→ —(HNO_3 / H_2SO_4)→ —(Zn(Hg), HCl)→ —(CH_3COCl)→ Major product

(a) OH, $\overset{H}{N}$, CH_3, O, CH_3 (b) OMe, HN, O, CH_3

(c) OH, NO_2, CH_3 (d) $OCOCH_3$, H_3C, NO_2

18. Give the major product from the following reaction sequence:

—(Br_2 / $FeBr_3$)→ —(SO_3 / H_2SO_4)→ —(Br / $FeBr_3$)→ —(H^+ / OH_2)→ Major product

(a) Br (b) Br, SO_3H

(c) Br (d) Br, SO_3H

19. Choose the best reaction sequence for the following conversion:

→ H_2N, Cl

(a) —(HNO_3,H_2SO_4)→ —(Zn(Hg), HCl)→ —($CH_3CH_2CH_2Cl$ / $AlCl_3$)→ —(Cl_2, $FeCl_3$)→

(b) —(HNO_3,H_2SO_4)→ —(Zn(Hg), HCl)→ —(CH_3CH_2COCl / $AlCl_3$)→ —(Zn(Hg), HCl)→ —(Cl_2, $FeCl_3$)→

(c) —(CH_3CH_2COCl / $AlCl_3$)→ —(Zn(Hg), HCl)→ —(HNO_3,H_2SO_4)→ —(Cl_2, $FeCl_3$)→ —(Zn(Hg), HCl)→

(d) —(HNO_3,H_2SO_4)→ —(Zn(Hg), HCl)→ —(CH_3CH_2COCl / Pyridine)→ —(Zn(Hg), HCl)→ —(Cl_2, $FeCl_3$)→

20. From the following choose the intermediate electrophile that is generated in when benzene is brominated:

(a) $\overset{+}{Br}{=}\overset{-}{Fe}Br_3$ (b) Br, Br $>FeBr_3$

(c) $Br{-}\overset{+}{Br}{-}\overset{-}{Fe}Br_3$ (d) $Br{-}\overset{+}{Br}{=}\overset{-}{Fe}Br_3$

21. Give major product of the reaction:
Benzene + $SO_3/H_2SO_4 \rightarrow$ Major product

(a) SO_2 (b) SO_3 (c) SO_3H (d) SOHO

22. One of the intermediate formed in the nitration reaction is:

(a) H, NO_3H, + (b) $H_2\overset{+}{O}{-}NO_2$ (c) H, $\overset{+}{N}O_3$, + (d) NO_2

23. Give the reagent that would best accomplish the following transformation:

—(Best reagent)→

(a) Cl + ClH (b) + $AlCl_3$

(c) Cl + $AlCl_3$ (d) Cl + H_3PO_4

24. Give the major product of the following reaction:

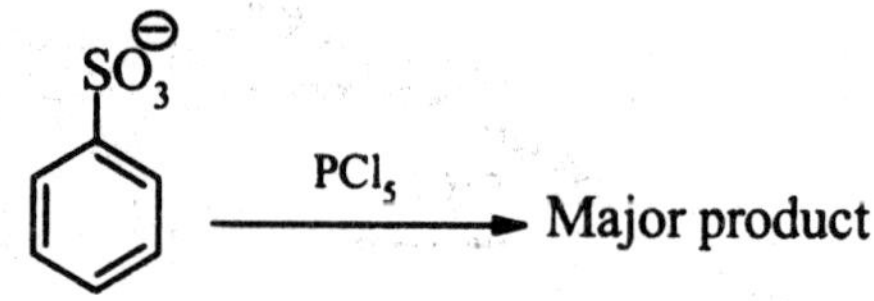

(a) SO_2Cl (b) SO_3H, Cl (c) SO_2, Cl (d) SO_3H, Cl

25. Consider the following reaction coordinate diagram for an electrophilic aromatic substitution reaction and choose the statement that is true:

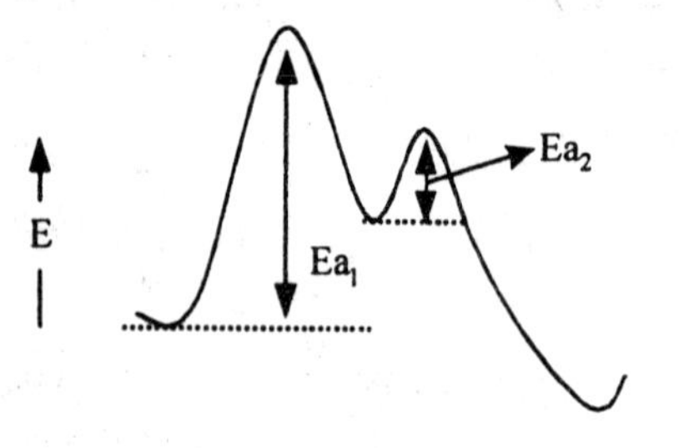

(a) The nitration of benzene has a higher Ea_1 than that of nitrobenzene.
(b) The nitration reaction of chlorobenzene has a higher Ea_1 than that of benzene.
(c) The rate of reaction of the nitration of benzene is dependent on Ea_2.
(d) The nitration of phenol has a higher Ea_1 than that of benzene.

26. Rank the following compounds of decreasing reactivity towards electrophilic aromatic substitution reaction:

(a) $\overset{+}{N}Me_3$ (b) CH_3 (c) Cl (d) OMe

(a) IV > II > I > III (b) II > IV > I > III
(c) IV > I > III > II (d) IV > II > III > I

27. Choose the substituents below that is ortho-para directors due to resonance:

1. —CH_3 2. —Cl 3. —OH
4. —NO_2 5. —NH_2

(a) 3 and 5 (b) 1, 3 and 5
(c) 2, 3 and 5 (d) 1, 2, 3 and 5

28. Choose the substituents below that deactivate the ring by induction only towards electrophilic aromatic substitution reaction:

1. —CH_3 2. —Cl 3. —CF_3
4. —NO_2 5. —N^+Me_3

(a) 2, 4, 5 (b) 1, 3, 5
(c) 2, 3, 4, 5 (d) 2, 3, 5

29. Give the major product of the following reaction:

CF_3 $\xrightarrow{HNO_3, H_2SO_4}$ Major

(a) CF_3, NO_2 (b) CF_3, NO_2 (c) CF_3, O_2N

(d) All are the major products.

30. Give the major product of the following reaction:

H N O $\xrightarrow[AlCl_3]{CH_3COCl}$ Major

(a) H N O $COCH_3$

(b) H N O $COCH_3$

(c) H_3COC H N O

(d) H N O COCl

31. From the following, choose the intermediate electrophile that is generated when benzene is sulphonated:

(a) HSO_3^- (b) HSO_3 (c) SO_3 (d) SO_3^+

32. Give the major monosubstitution product of the following reaction:

HF → Major

(a) (b)

(c) F (d)

33. Give the reactants that complete the following reaction:

+ Reactant ? →

(a) C_6H_5COOH/H^+ (b) $C_6H_5CHO/AlCl_3$
(c) $(PhCO)_2O/OH^-$ (d) $C_6H_5COCl/AlCl_3$

34. Give the major product of the following reaction:

+ OH, OH; BF_3 → Major product

(a) OH (b)

(c) (d)

35. Rank the following of decreasing electrophilic aromatic substitution reactivity.

OH (1), OMe (2), $COCH_3$ (3), $OCOCH_3$ (4)

(a) 1 > 2 > 3 > 4 (b) 4 > 3 > 2 > 1
(c) 1 > 2 > 4 > 3 (d) 2 > 1 > 4 > 3

36. Give the major product of the following reaction:

COOPh; Br_2, $FeBr_3$ → Major product

(a) COOPh, Br (b) COOPh, Br

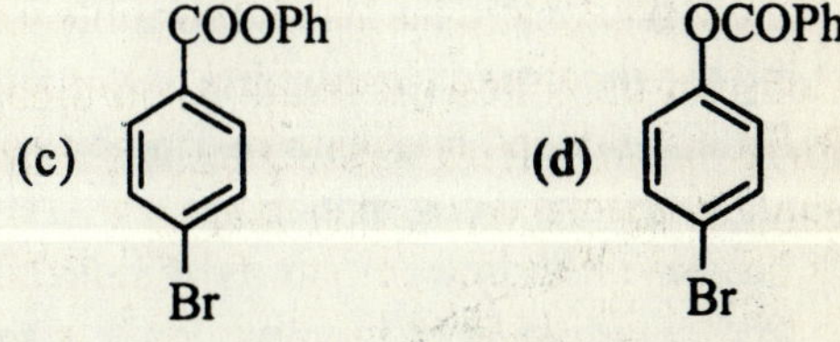

37. Give the major product of the following reaction:

SO_3H; H^+, OH_2 → Major product

(a) SO_3H, OH (b) SO_3H, OH

(c) SO_3H, OH (d)

38. Choose the most important resonance structure of the intermediate that is formed, in the mechanism of the reaction, when an electrophile (E^+) attacks dimethylaniline at the ortho position.

$N(CH_3)_2$; E^+, OH_2 → Major product

(a) 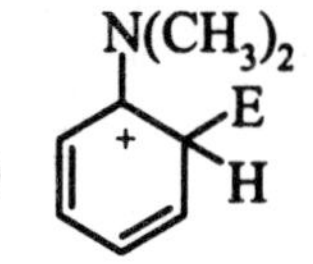 (b)

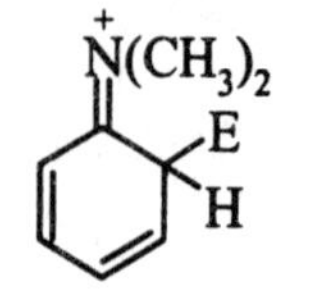

(c) 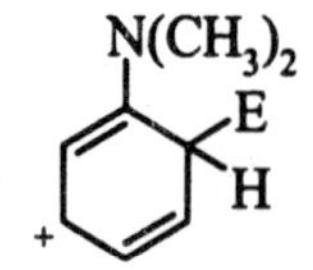 (d)

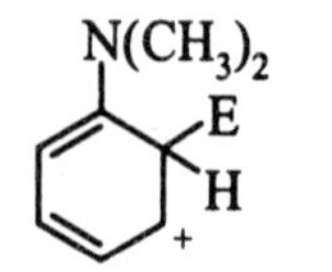

39. Choose the statement that is true concerning bromobenzene and electrophilic aromatic substitution:

(a) Electron donation by resonance from bromine to benzene ring is responsible for the deactivation of benzene ring.

(b) Bromine deactivate the benzene ring by electron donation from bromine to benzene ring. Electron donation is also responsible for directing the incoming electrophile.

(c) Bromine deactivate benzene ring by withdrawing electrons using induction.

(d) Bromine withdraws electrons by induction. Electron donation from bromine to the ring directs the incoming electrophile.

40. Choose the intermediate that is generated in the mechanism of reaction of $FeBr_3/Br_2$ with benzene:

(a) 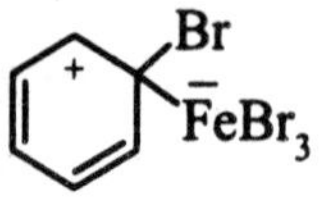 (b)

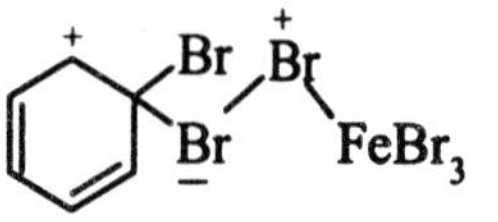

(c) 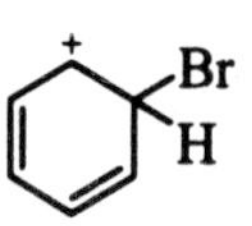 (d)

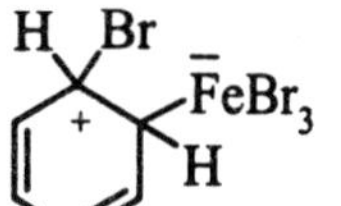

41. From the following choose the intermediate electrophile that is generated when benzene is treated with CH_3COCl and $AlCl_3$:

(a) $CH_3\overset{|}{C}{=}\overset{+}{O}$ with H   (b) $CH_3C{\equiv}\overset{+}{O}$

(c) $CH_3C{=}O$ with $^{+}Cl{-}\bar{A}lCl_3$ (d) $\overset{+}{C}H_2CH{=}O$

42. Give the major product of the following reaction:

$\xrightarrow{AlCl_3}$ Major

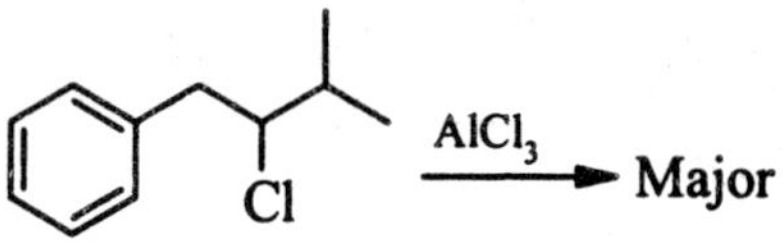

(a) 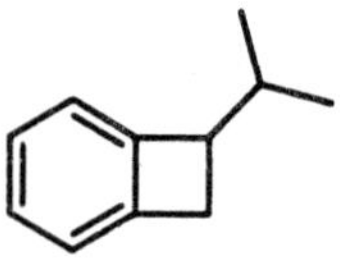 (b)

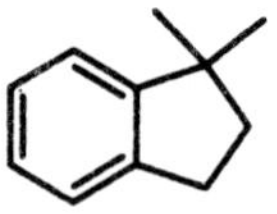

(c) 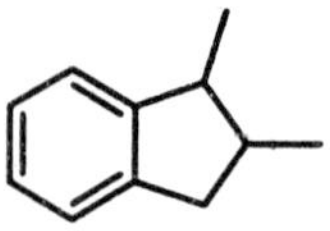 (d)

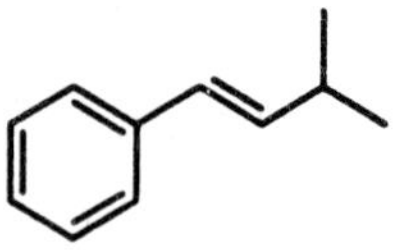

43. Rank the following compounds of increasing reactivity with respect to electrophilic aromatic substitution:

I. 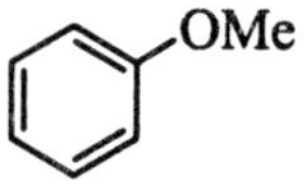 (OMe)   II. 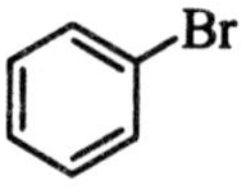 (Br)

III. 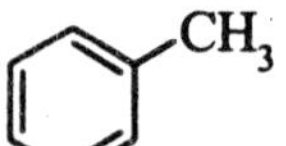 (CH_3)   IV. 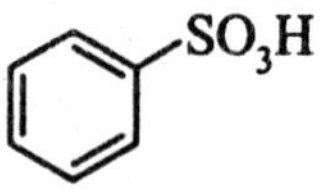 (SO_3H)

(a) I < II < III < IV   (b) IV < III < II < I

(c) IV < II < III < I   (d) IV < III < I < II

44. Give the major product of the following reaction:

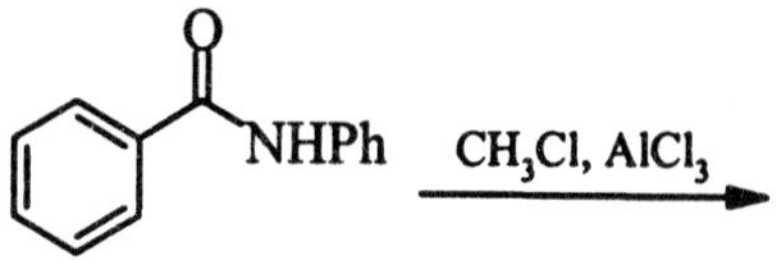

NHPh $\xrightarrow{CH_3Cl,\ AlCl_3}$

(a) 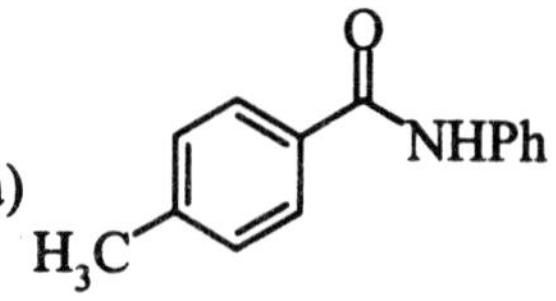 NHPh, H_3C (b) 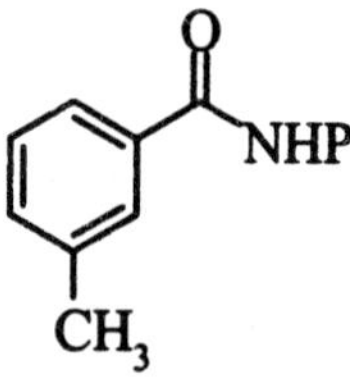 NHPh, CH_3

(c) 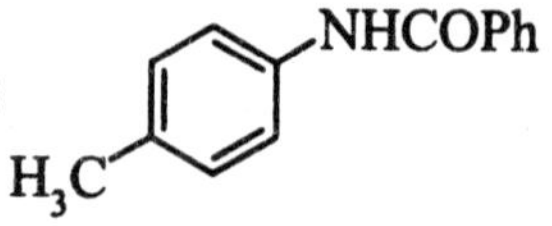 NHCOPh, H_3C (d) 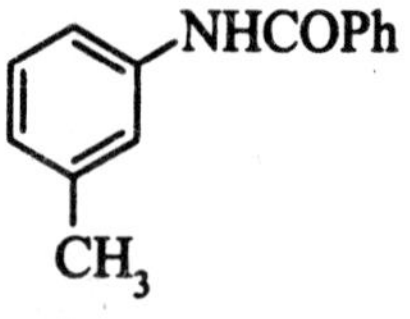 NHCOPh, CH_3

45. Choose from the below, the mechanism of attack of an electrophile (E) on benzene ring during aromatic electrophilic substitution:

(a) 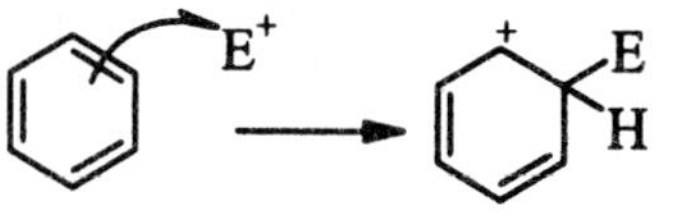 E^+ ⟶ E, H

(b)

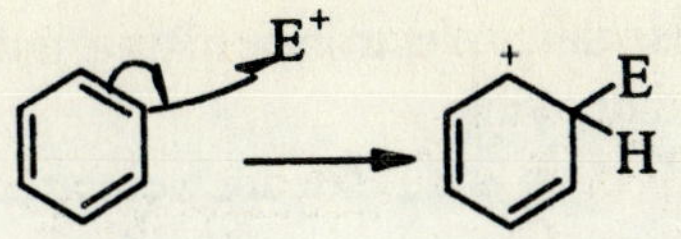

(c)

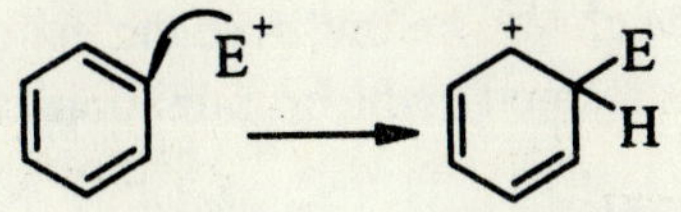

(d)

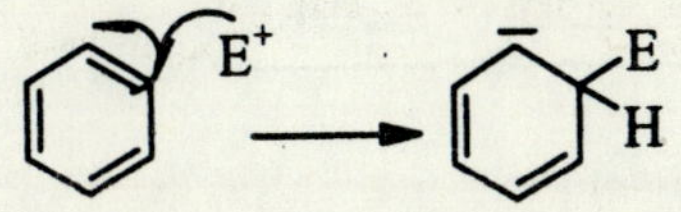

46. Predict the major product of the following reaction:

$\xrightarrow{HNO_3,\ H_2SO_4}$

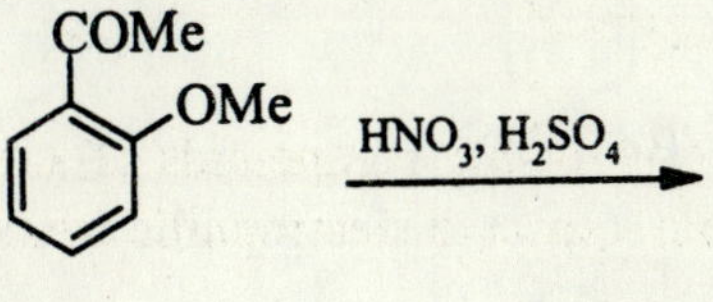

(a) 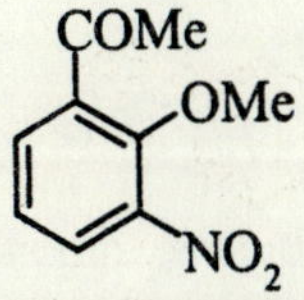 (b)

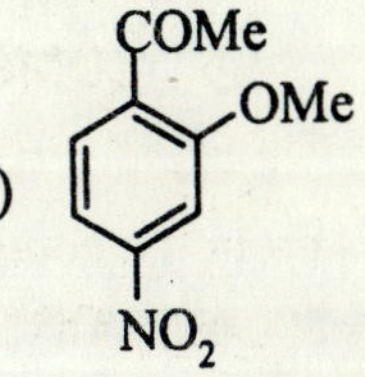

(c) 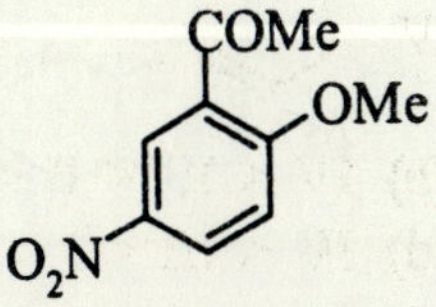 (d)

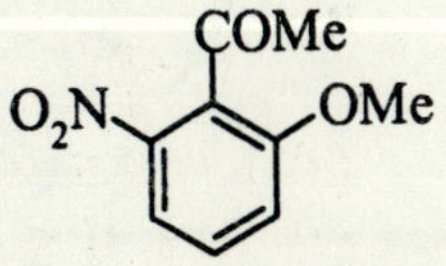

47. Rank the following compounds in the decreasing order of reactivity towards aromatic electrophilic substitution reaction:

I. 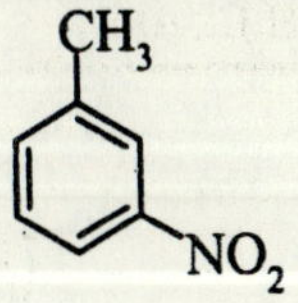 II.

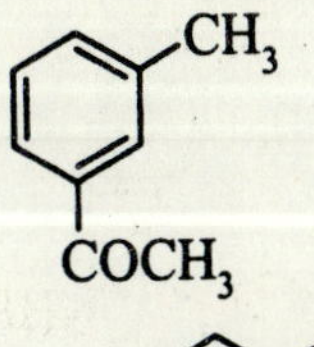

III. 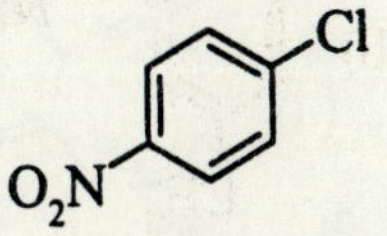 IV.

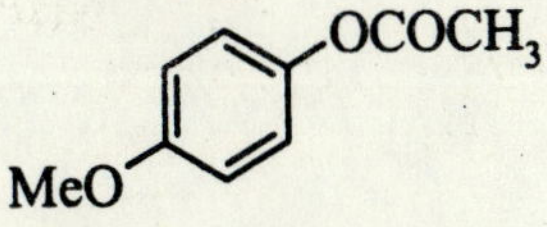

(a) I > II > III > IV (b) IV > III > II > I
(c) IV > I > III > II (d) IV > II > I > III

48. Give the intermediate formed in the following reaction:

$\xrightarrow[AlCl_3]{}$ Intermediate

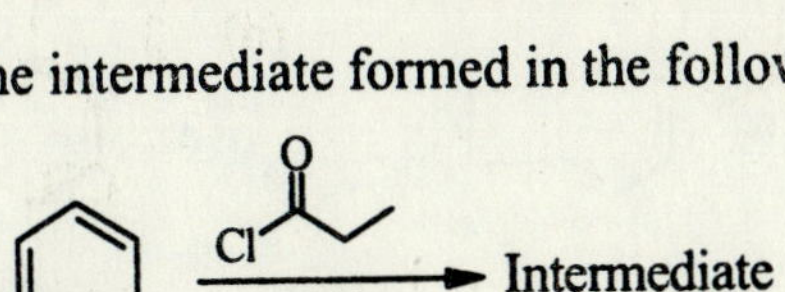

(a) 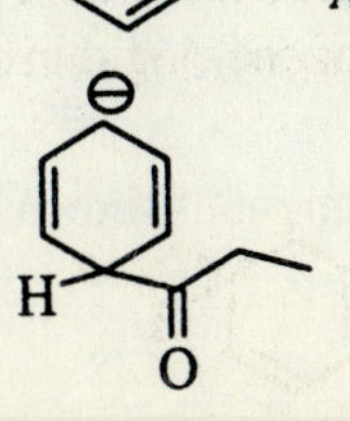 (b)

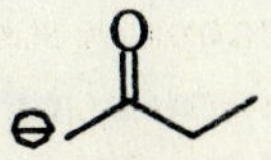

(c) 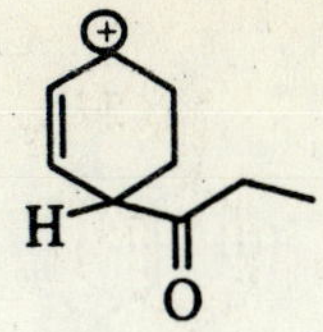 (d)

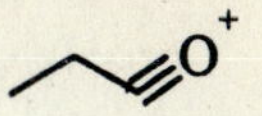

49. Give the major product of the following reaction:

$\xrightarrow{H_3PO_4}$

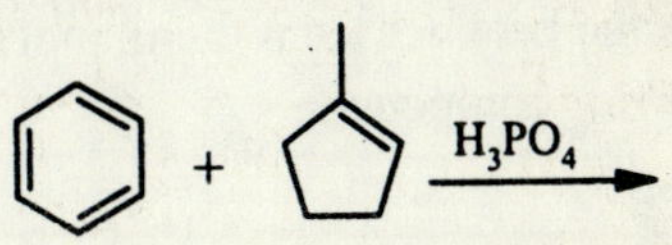

(a) 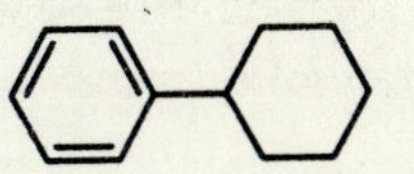 (b)

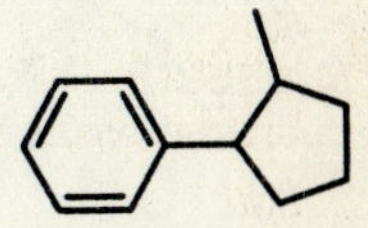

(c) 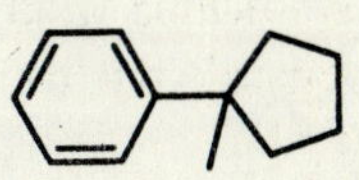 (d)

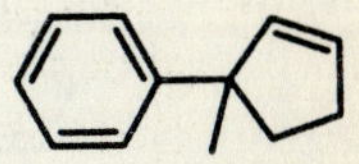

50. Rank the following compounds in the decreasing order of reactivity towards aromatic electrophilic substitution:

1. Benzenesulphonic acid 2. Acetanilide
3. Aniline 4. Bromobenzene

(a) 4 > 3 > 1 > 2 (b) 3 > 2 > 4 > 1
(c) 2 > 3 > 4 > 1 (d) 1 > 4 > 3 > 2

51. Which statement best characterize the properties of —NO_2 as a substituent to benzene ring:

(a) Inductive electron donating, resonance electron withdrawing.
(b) Inductive electron donating and no resonance effect.
(c) No inductive effect, resonance electron withdrawing.
(d) Inductive and resonance electron withdrawing.

52. Give the most important intermediate resonance structure that represents the directing effect of the methoxy group in monochlorination of methoxy benzene:

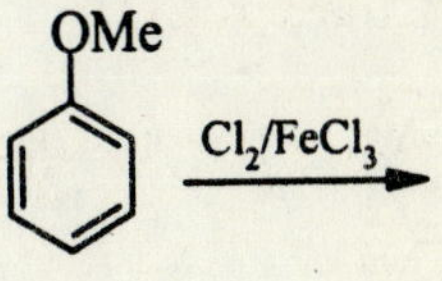

$\xrightarrow{Cl_2/FeCl_3}$

(a) 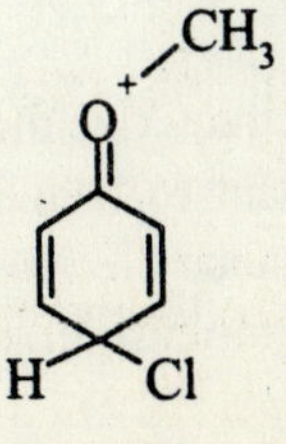 (b)

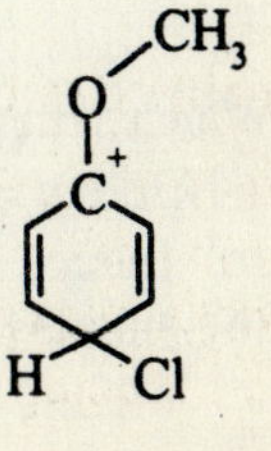

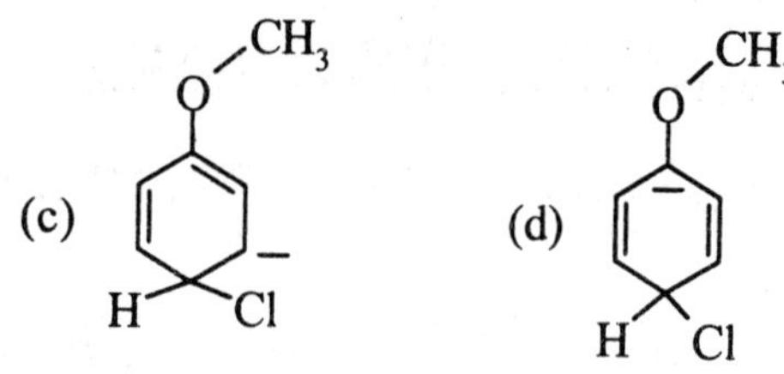

53. Give the product from the monochlorination of the following compound:

$\xrightarrow{Cl_2/FeCl_3}$ Major substitution product

(a)

(b)

(c)

(d)

54. Give the major product from the following reaction sequence:

$\xrightarrow{HNO_3,\ H_2SO_4}$ $\xrightarrow{Br_2/FeBr_3}$ Major substitution product

(a) Br, O_2N

(b) Br, O_2N

(c) Br, NO_2

(d) Br, SO_3H

55. Which st1atement best characterize the properties of —$N(CH_3)_2$ as a substituent to benzene ring?

(a) Inductive and resonance electron donating.

(b) Inductive electron donating, no resonance effect.

(c) Resonance electron donating, inductive electron withdrawing.

(d) Inductive and resonance electron withdrawing.

56. From the choice below, give the major intermediate which is formed in the bromination of nitrobenzene:

$\xrightarrow{HNO_3,\ H_2SO_4}$ $\xrightarrow{Br_2/FeBr_3}$ Major substitution product

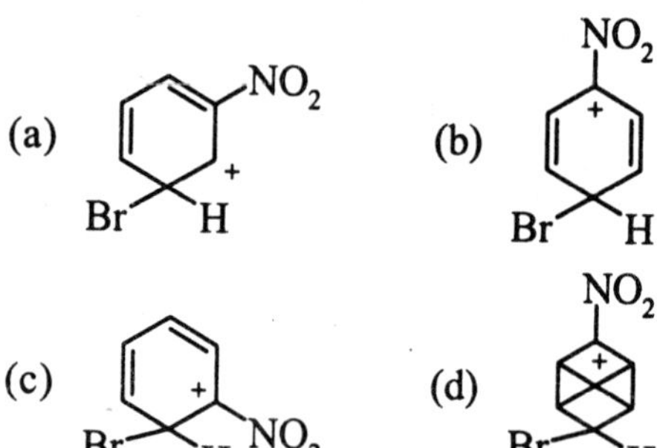

57. From the compound shown below, choose which is not aromatic:

(a) 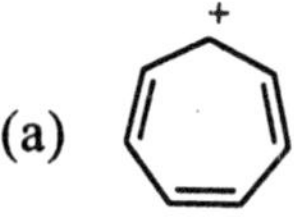

(b) 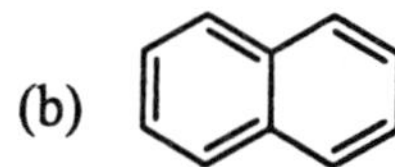

(c)

(d) all are aromatic

58. Choose the compound below that is aromatic:

(a) 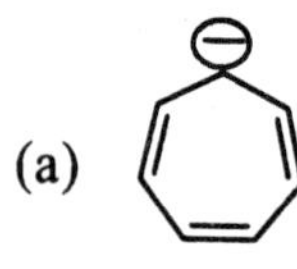

(b) 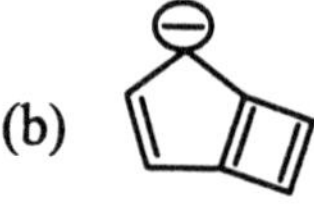

(c) 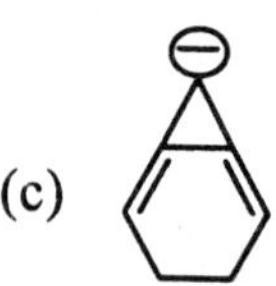

(d) 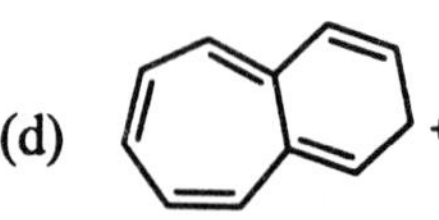

59. From the compounds shown below, choose which is(are) not aromatic:

(a)

(b) 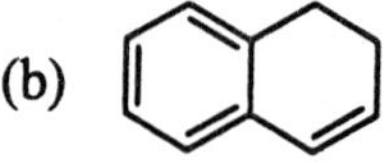

(c)

(d) 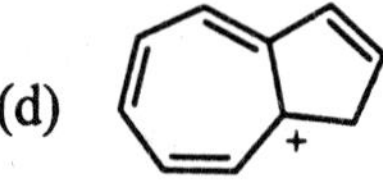

60. From the following compounds shown below, choose that is(are) aromatic:

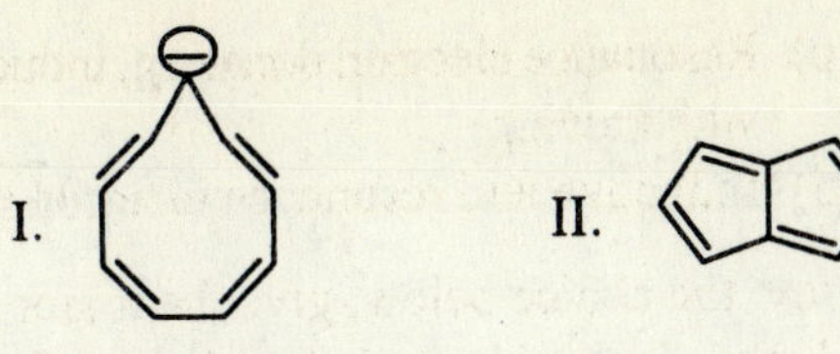

I. II.

III. IV.

(a) II and IV (b) I, II and IV
(c) I and IV (d) I and III

61. From the compounds shown below, choose the one that is aromatic:

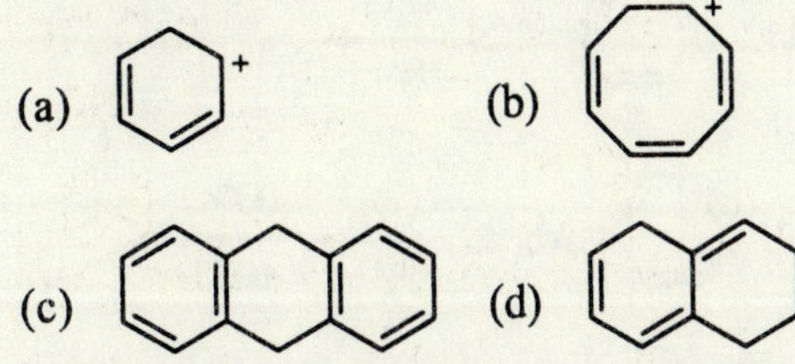

(a) (b)

(c) (d)

62. Which could not be directly prepared from 4-bromobenzenediazonium ion:
(a) Bromobenzene (b) 1, 4-dibromobenzene
(c) 4-bromofluorobenzene
(d) 4-bromoacetophenone

63. Which represents an intermediate formed in the reaction of toluene and chlorine at elevated temperature in sunlight:

(a) (b)

(c) $C_6H_5\dot{C}H_2$ (d) $C_6H_5\overset{+}{C}H_2$

64. The reaction of benzyl chloride with sodium isopropoxide yields:
(a) Benzaldehyde and 2-chloropropane
(b) 1, 2-diphenyl ethane and acetone
(c) Benzyl isopropyl ether
(d) Acetone and toluene

65. An aromatic hydrocarbon yielding single mononitro derivative is:
(a) Ethyl benzene (b) *o*-xylene
(c) *p*-xylene (d) naphthalene

66. Which is the best way to prepare meta-bromoaniline?

(a) Aniline $\xrightarrow{H_2SO_4}$ $\xrightarrow{Br_2}$ $\xrightarrow[H_2O]{H^+}$

(b) Benzene $\xrightarrow{Br_2}$ $\xrightarrow{HNO_3}$ $\xrightarrow{Sn/HCl}$

(c) Aniline $\xrightarrow[HNO_3]{H_2SO_4}$ $\xrightarrow{(NH_4)_2S}$ $\xrightarrow[BrH]{NaNO_2}$ $\xrightarrow{CuBr}$ $\xrightarrow{Sn/HCl}$

(d) Nitrobenzene $\xrightarrow{H_2SO_4}$ $\xrightarrow{NaOH}$ $\xrightarrow[Fuse]{NaBr}$ $\xrightarrow{Sn/HCl}$

67. Which is the principal product when benzaldehyde is allowed to react with propionaldehyde in the presence of dilute aqueous sodium hydroxide at room temperature?
(a) $C_6H_5CH(OH)COOC_2H_5$
(b) $C_6H_5CH{=}C(CH_3)CHO$
(c) $C_6H_5COCOC_2H_5$
(d) $C_6H_5CH{=}CH{-}CH_2CHO$

68. Select the final product of the following sequence of reaction:

$C_6H_5CCl_3$ $\xrightarrow{NaOH/H_2O}$ $\xrightarrow[H_2O]{H^+}$ $\xrightarrow[H_2SO_4]{HNO_3}$ $\xrightarrow{Sn/HCl}$ $\xrightarrow{HNO_2}$ $\xrightarrow{C_6H_5OH}$

(a) HO, —N=N—

(b) HOOC, —N=N—, OH

(c) —COOPh, HO

(d) COOH, —N=N—, OH

69. In the reaction below, the product would be:
$C_6H_5CHO + HCN \rightarrow C_6H_5CH(OH)CN$
(a) a racemate (b) optically active
(c) a meso compound (d) unresolvable

70. Which is a principal method for preparing iodobenzene:
(a) Iodine + Benzene + Iron
(b) Iodine + Benzene + UV-light
(c) KI + C_6H_5Cl (d) KI + $C_6H_5N_2^+Cl^-$

71. Which reaction sequence would be best to prepare 3-chloro aniline from benzene:
(a) Chlorination → nitration → reduction
(b) Nitration → chlorination → reduction
(c) Nitration → reduction → acetylation → chlorination → hydrolysis
(d) Nitration → chlorination

72. Which aromatic compound behaves most like benzoic acid when it undergoes reaction with concentrated nitric acid and sulphuric acid:
(a) Phenol (b) Chlorobenzene
(c) Phenyl methyl ketone (d) Toluene

73. Which reaction sequence would be best to prepare 1, 4-diaminobenzene from benzene:
(a) Nitration → reduction → nitration → reduction
(b) Nitration → reduction → acetylation → nitration → reduction
(c) Nitration → reduction → acetylation → nitration → hydrolysis → reduction
(d) Nitration → reduction

74. Which of the following would react most rapidly with sodium ethoxide to produce ether?
(a) Chlorobenzene (b) *p*-nitrotoluene
(c) *p*-nitrochlorotoluene
(d) *m*-(chloromethyl)-toluene

75. A mixture of toluene, benzoic acid, aniline and phenol was dissolved in diethyl ether and extracted with 5% $NaHCO_3$ solution. This extract was acidified until a white crystalline compound appeared. What is the identity of this compound:
(a) Toluene (b) Benzoic acid
(c) Aniline (d) Phenol

76. Which of the following compound would exhibit aromatic properties:
(a) B—H (b) B—H
(c) N—H (d) H—N N—H

77. What is the major product obtained from the following reaction:

Cl, O, $AlCl_3$

(a) Cl, O, Cl (b) O
(c) O (d) O

78. Which of the following reactions will not yield benzoic acid as the major product?
(a) —CN $\xrightarrow[\text{Heat}]{H_3O^+}$
(b) —MgBr $\xrightarrow[Et_2O]{CO_2}$ $\xrightarrow[\text{Heat}]{H_3\overset{+}{O}}$
(c) —$COOCH_3$ $\xrightarrow[Et_2O]{CH_3MgBr}$ $\xrightarrow[\text{Heat}]{H_3O^+}$
(d) —$COOCH_3$ $\xrightarrow[\text{Heat}]{H_3\overset{+}{O}}$

79. Give the reagents that will best accomplish the following transformation:

H, N, CH_3 ⟶ CH_3, N, CH_3, O

(a) $NaOH/CH_3CH_2OH/CrO_3$, H_2O/H_2SO_4
(b) Acetone/heat
(c) CH_3CHO/heat
(d) Ethanol/heat/CrO_3,H_2O/H_2SO_4

80. Give the major product of the following reaction sequence:

OH, O $\xrightarrow[\text{Trace of P}]{Br_2}$

(a) $C_6H_5CHBrCOOH$ (b) $C_6H_5CH_2COBr$
(c) *p*-Br-$C_6H_4CH_2COOH$ (d) *m*-Br-$C_6H_4CH_2COOH$

81. Which of the following is not aromatic?
(a) N (b)
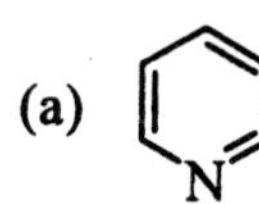

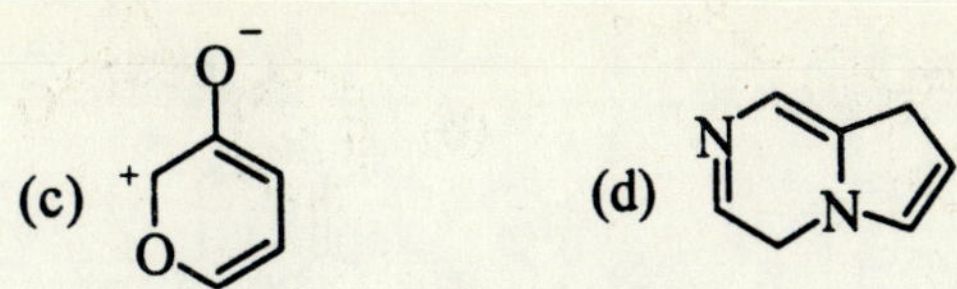

82. Which reaction condition would be best to perform the following transformation?

(benzene → 1-bromo-3-ethylbenzene)

(a) (1) Br_2, $FeBr_3$ (2) EtCl, $AlCl_3$
(b) (1) EtCl, $AlCl_3$ (2) Br_2, $FeBr_3$
(c) (1) CH_3COCl, $AlCl_3$ (2) Br_2, $FeBr_3$ (3) H_2/Pd
(d) (1) Br_2, $FeBr_3$ (2) CH_3COCl, $AlCl_3$, H_2/Pd

83. Which of the following heterocyclecs is not aromatic?

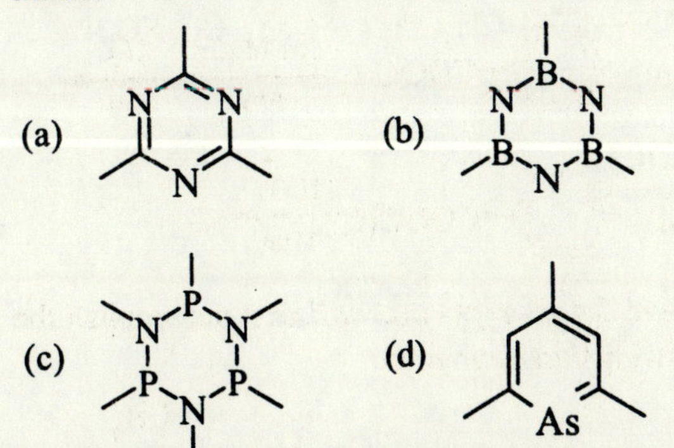

84. Which of the following is the best reaction sequence for preparation of 2, 4-dinitrobenzoic acid from benzene:

(a) HNO_3/H_2SO_4 → $CH_3Br/AlCl_3$ → HNO_3/H_2SO_4 → $KMNO_4/H^+$
(b) $CH_3Br/AlCl_3$ → HNO_3/H_2SO_4 → $KMNO_4/H^+$ → HNO_3/H_2SO_4
(c) $CH_3Br/AlCl_3$ → $KMNO_4/H^+$ → HNO_3/H_2SO_4(excess)
(d) $CH_3Br/AlCl_3$ → HNO_3/H_2SO_4(excess) → $KMNO_4/H^+$

85. Compound A shows a large dipole moment. Which structure should be used to adequately, explain this observation?

(a) (b) (c) (d)

86. A ($C_{10}H_{14}$) + Cl_2/Δ → $C_{10}H_{13}Cl$ (Single isomer)
A + $Cl_2/FeCl_3$ → $C_{10}H_{13}Cl$ (Two isomers).
Possible structure of A is:

(a) $C(CH_3)_3$ (b) CH_2CH_3, H_3C, CH_3
(c) CH_3, H_3C, CH_3, CH_3 (d) C_2H_5, C_2H_5

87. A compound $X(C_7H_8O)$ is insoluble in water, dilute HCl and aqueous $NaHCO_3$ but dissolve in dilute NaOH. When X is treated with bromine-water, it is converted into a compound of formula $C_7H_5OBr_3$. Compound X is:

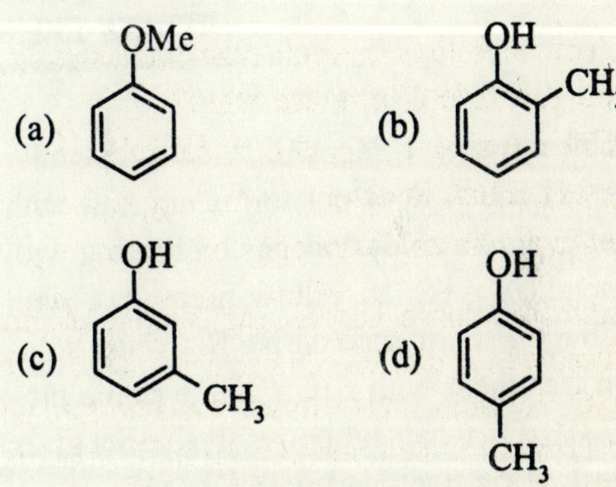

88. Arrange the following compounds in increasing order of their reactivity towards aromatic electrophilic substitution reaction:
$C_6H_5NO_2$ (I), $C_6H_5CH_2NO_2$ (II), $C_6H_5NH_2$ (III), $C_6H_5CH_2NH_2$ (IV)
(a) I < II < III < IV (b) I < II < IV < III
(c) I < IV < II < III (d) I < IV < III < II

89. Which of the following organic compound has activated benzene ring for aromatic electrophilic substitution reaction:

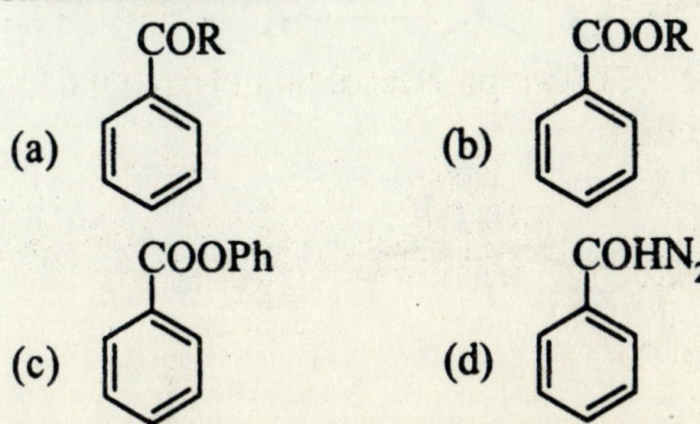

90. A (C_7H_8O) + Na/Δ → No gaseous product;
A + $Br_2/FeBr_3$ → C_7H_7OBr (Two isomers), A can be

(a) OH, CH_3 (b) OH, CH_3

(c) OH, CH_3 (d) OMe

91. A ($C_{10}H_{14}$) + Cl_2/Δ → $C_{10}H_{13}Cl$ (Single isomer);
A + $Cl_2/FeCl_3$ → $C_{10}H_{13}Cl$ (Two isomers).
Possible structure of A is:

(a) $C(CH_3)_3$ (b) CH_2CH_3, H_3C, CH_3

(c) CH_3, H_3C, CH_3, CH_3 (d) C_2H_5, C_2H_5

92. Which of the following has maximum enol proportion:

(a) $CH_3COCH_2COOC_2H_5$ (b) O

(c) NC—CO—CH_2CH_3 (d) CH_3NO_2

REVIEW EXERCISE (AROMATIC)

1. A resolvable organic compound A has molecular formula $C_{10}H_{14}O$. A does not forms any salt with NaOH, but evolve a colourless gas on heating with sodium metal. A gives an yellow precipitate with $NaOH/I_2$, but does not decolourise Br_2-water solution. A on treatment with CrO_3/HCl/pyridine produces B($C_{10}H_{12}O$), which is non-resolvable but gives idoform test. B on treatment with $Cl_2/AlCl_3$ gives C($C_{10}H_{11}OCl$) as the only possible isomer. Also A on heating with alkaline $KMnO_4$ solution yields $C_8H_6O_4$(D) as one of the product which evolves a gas with $NaHCO_3$. D on heating with ethan-1, 2-diol in presence of an acid catalyst forms a polyester E. Deduce structures of A to D and write the formula of repeat unit of E.

2. Predict products/reagents:

(a) O, CHO + O $\xrightarrow[\text{Heat}]{\text{NaOH}}$

(b) COOH, Cl, COOH $\xrightarrow[H_2SO_4]{HNO_3}$

(c) OH $\xrightarrow{Br_2/CHCl_3}$

(d) COOH, S $\xrightarrow{Br_2/AlBr_3}$

(e) CH_3, O, CH_3, CH_3 $\xrightarrow[H_2SO_4]{HNO_3}$

(f) OMe, CH_3 + $\xrightarrow{H_2SO_4}$

(g) O, C, CF_3

X → O_2N, CH_2, CF_3

Y → O_2N—, CH_2—, CF_3

3. Propose mechanism:

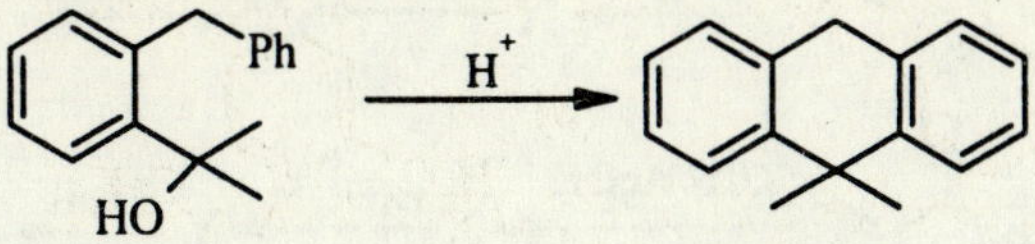

Ph ——H^+——→

HO

4. Bring about the following transformation:

——→ Ph

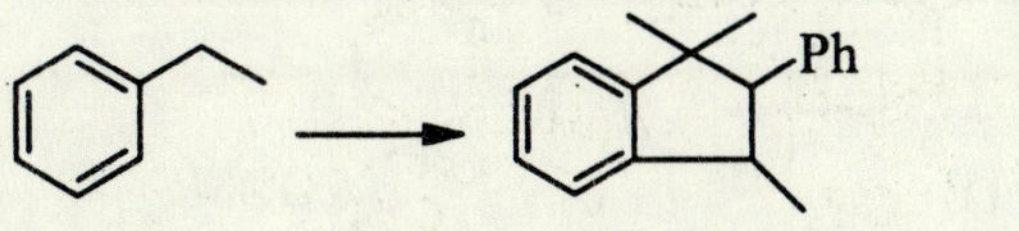

5. An organic compound A($C_{11}H_{14}O$) exist in two stereoisomeric forms, none of which is resolvable. A does not evolve any gas with sodium metal but decolourise bromine water solution. A on treatment with ozone followed by work-up with $(CH_3)_2S$ yields B($C_9H_{10}O_2$) as one of the product. B neither forms any salt with NaOH nor does it evolve any gas with sodium metal. Also B does not react with Tollen's reagent. B on treatment with concentrated solution of HI forms C($C_8H_8O_2$) as one of the product, which forms salt with NaOH but does not evolve any gas with $NaHCO_3$. B on treatment with $AlCl_3/Cl_2$ gives D($C_9H_9O_2Cl$) as the only possible product. C can also be obtained by treatment of an ester E ($C_8H_8O_2$) with $AlCl_3$. Deduce structures of A to E and explain the formation of C from E.

6. Bring about the following transformations:

(a) 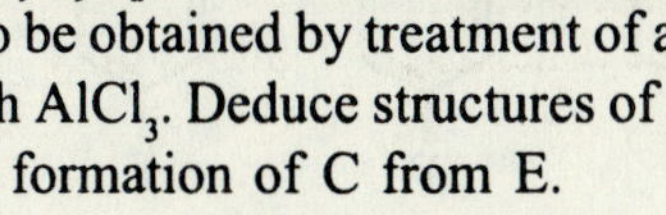 ——→ $PhCH_2CH_2$, Cl

(b) 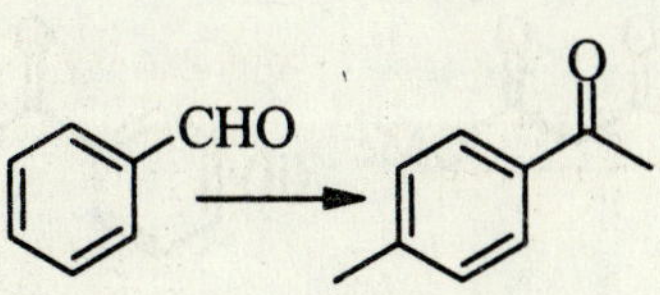 CHO ——→ O

(c) 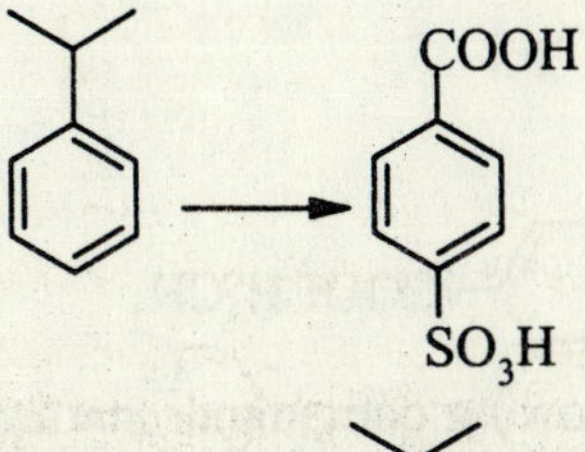 ——→ COOH, SO_3H

(d) 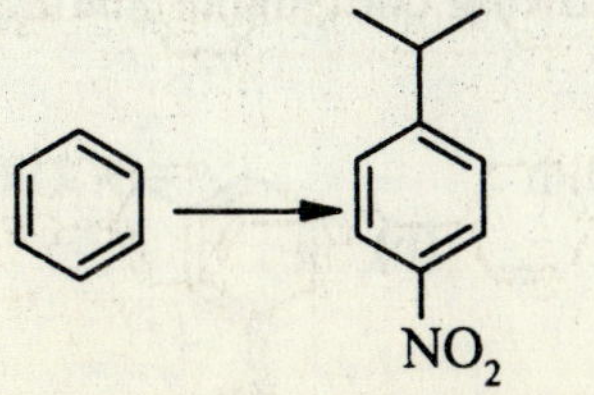 ——→ NO_2

7. Provide missing reagents/products for the following reactions:

(a) 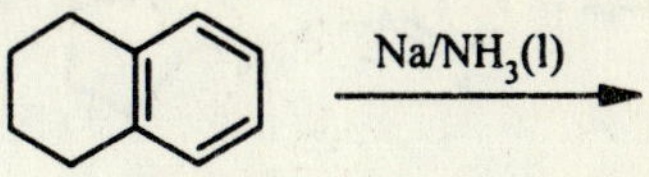 ——$Na/NH_3(l)$——→

(b) 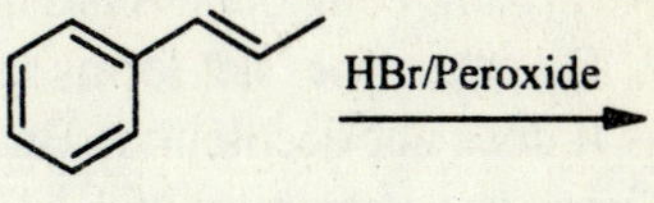 ——HBr/Peroxide——→

(c) 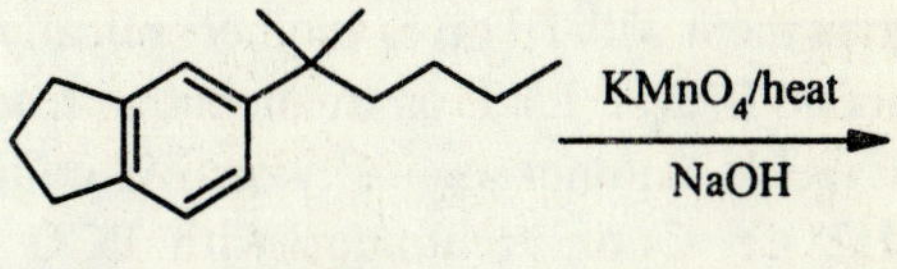 ——$KMnO_4$/heat, NaOH——→

(d) 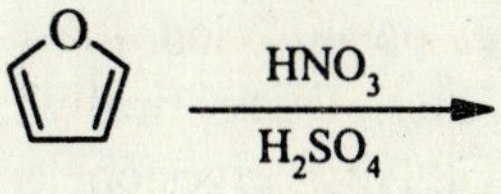 O ——HNO_3, H_2SO_4——→

(e) 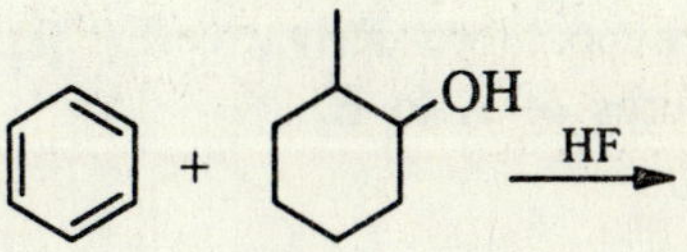 + OH ——HF——→

(f) 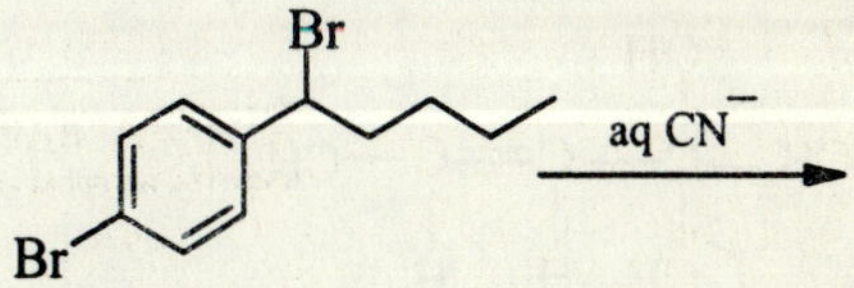 Br, Br ——aq CN^-——→

(g) 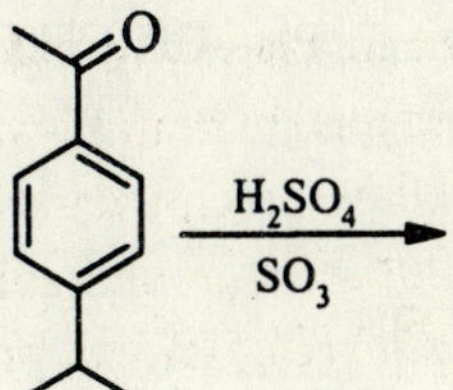 O ——H_2SO_4, SO_3——→

(h) Cl ——$AlCl_3$——→

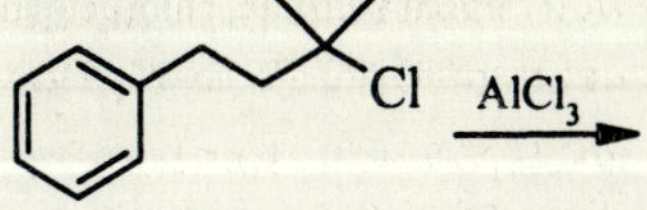

8. Bring about the following transformations:

(a) 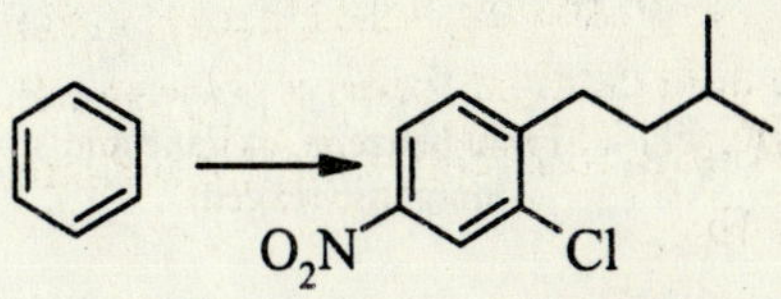 ——→ O_2N, Cl

(b) 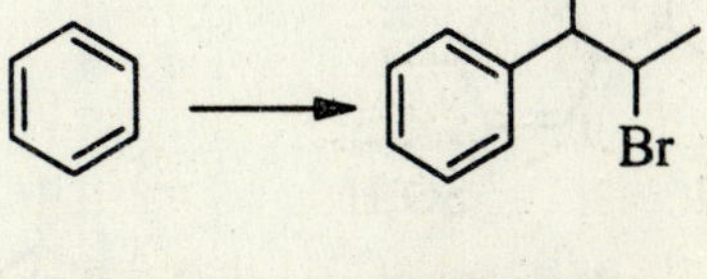 ——→ OH, Br

(c) 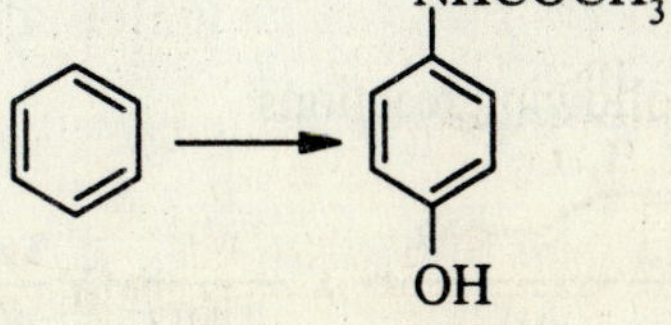 ——→ $NHCOCH_3$, OH

(d) [Bromobenzene → N-benzylidene cyclopentylamine structure]

9. An optically active organic compound A has molecular formula $C_{11}H_{16}O_2$, and it does not forms salt with NaOH as well as it does not decolourise Br_2—H_2O solution, but evolve a gas on treatment with MeMgBr. A on treatment with HI gives another optically active compound B($C_9H_{10}I_2$). B on treatment with aqueous KOH yields another optically active compound C($C_9H_{12}O_2$). C on treatment with PCC yields D($C_9H_8O_2$), which gives positive iodoform test as well as positive Tollen's test. D is optically inactive and on treatment with $Cl_2/AlCl_3$ gives one and only one possible monochloro derivative E($C_9H_7O_2Cl$). Deduce structures of A to E.

10. Synthesize:

(a) Ph—CH_2—C(OH)(CH_3)—CH=CH—CH_3

From Ph—$CH_2CH_2CH_2Br$

(b) O_2N—C_6H_4—CH(Br)—CH_2OH

From benzene, chloroethane and any inorganic reagent

(c) 4-Cl-C_6H_4—CO—cyclopentyl From benzene

(d) $C_6H_5CH_2COC_6H_5$ From benzene, oxirane and inorganic reagent

(e) $(CH_3)_3CCH_2$—CO—C_6H_3(I)(SO_3H)

From iodo benzene and $(CH_3)_3C$—CH=CH_2

11. Complete the following reactions:

(a) Ph—CO—CH=CH_2 $\xrightarrow[OH^-]{CH_3CH_2CH_2NO_2}$ X $\xrightarrow[DMG]{TiCl_4}$ Y $\xrightarrow[\Delta]{TsOH}$ Z

(b) Ph—CO—CH_3 $\xrightarrow{LDA}$ X $\xrightarrow{CuCl_2}$ Y $\xrightarrow{H_3PO_4}$ Z

(c) Ph—CO—CH=CH_2 $\xrightarrow{KCN/AcOH}$ X $\xrightarrow{DIBL-H}$ Y $\xrightarrow[\Delta]{TsOH}$ Z

12. Bring about the following transformations:

(a) Furan → 5-bromo-2-furyl—CH=CH—COOMe

(b) Ph—CO—CH=CH_2 → 2-phenyl-5-methylthiophene

(c) Thiophene → O_2N-substituted thiophene bearing —CH(OH)CH_3

13. (a) Synthesize the following compounds starting from benzene:

Fluorobenzene; 4-F-C_6H_4-CH_3; 4-F-C_6H_4-COOH; 4-F-C_6H_4-NO_2

(b) Write mechanism of the reaction of *p*-nitro fluoro benzene with $(C_2H_5)_2NH$. Explain, why *m*-nitro fluoro benzene does not react with diethyl amine under identical conditions.

(c) Synthesize the following compounds starting from benzene and any other reagents needed:

(i) Ph—CO—CH_2—CO—OMe (ii) Ph—CO—CH(R)R'

(iii) Cyclohexyl—CO—Ph

(iv) H_2N—C_6H_4—$COOCH_2CH_3$

14. Synthesize the following compounds starting from benzene:

(a) O_2N—C_6H_4—OCOPh (b) 3-Br—C_6H_4—OCOPh

(c) 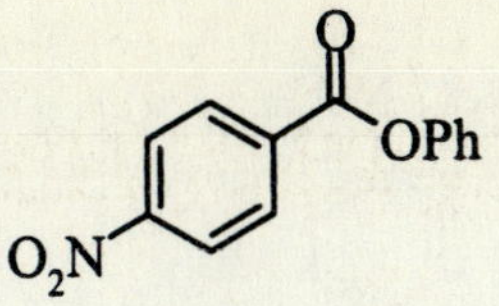 (d) 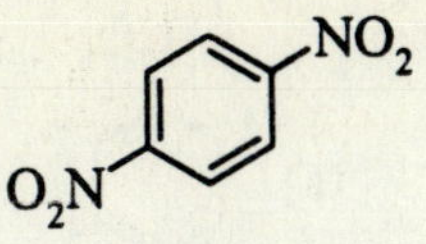 (e) 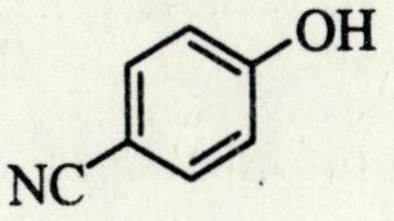 (f)

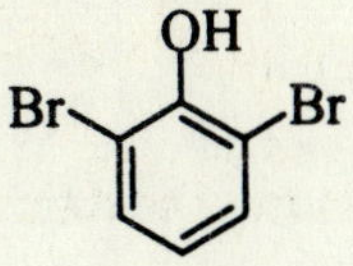

15. Synthesize the following compounds starting from benzene:

(a) 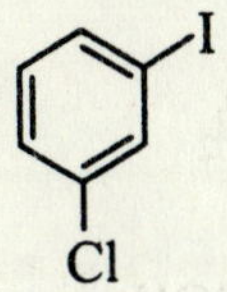 (b) 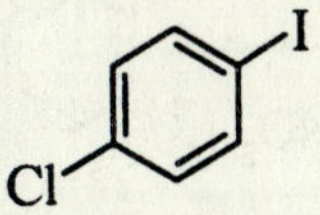 (c)

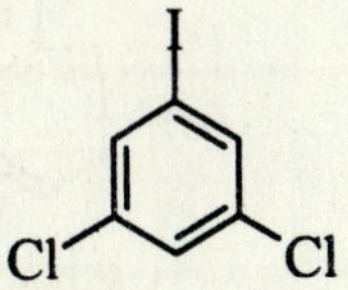

16. Bring about the following transformations:

(a) 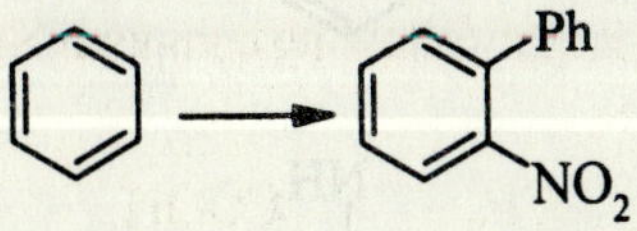 (b) 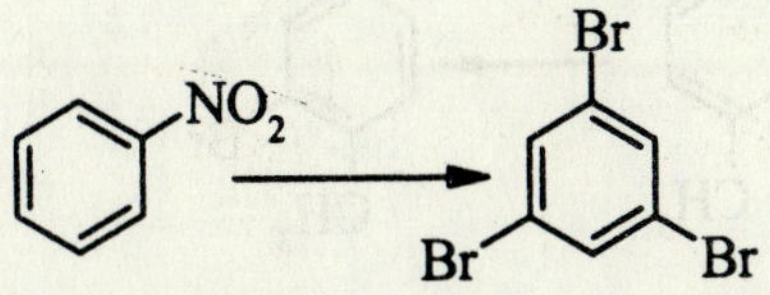 (c)

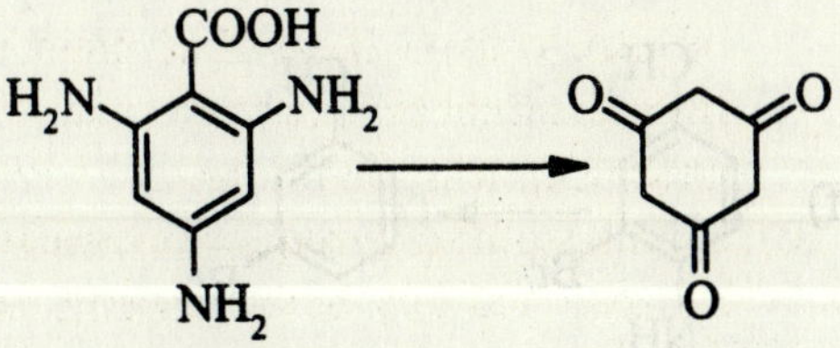

17. Propose mechanism:

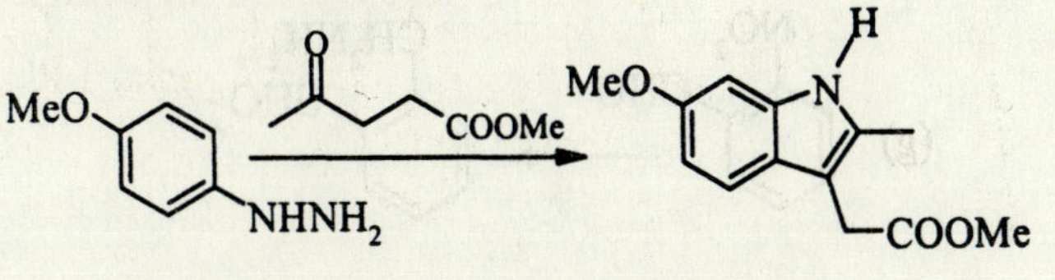

MeO, NHNH₂, COOMe → MeO, N, H, COOMe

18. Bring about the following transformations:

(a) 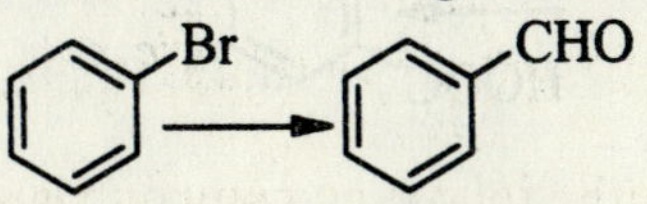 (b) 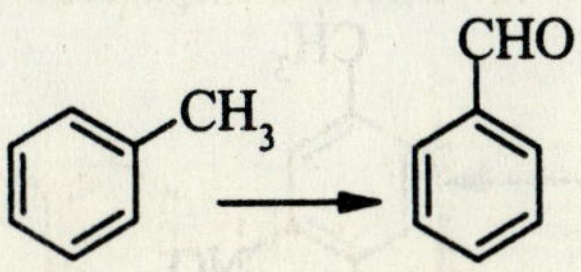 (c) 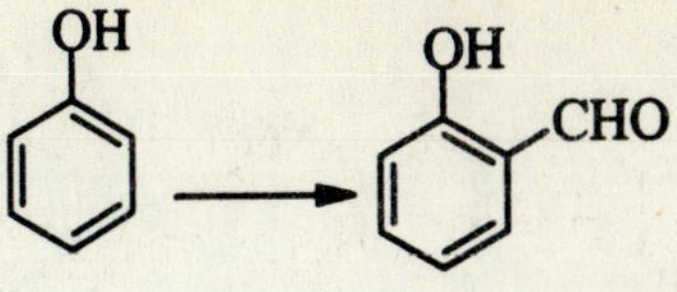 (d) 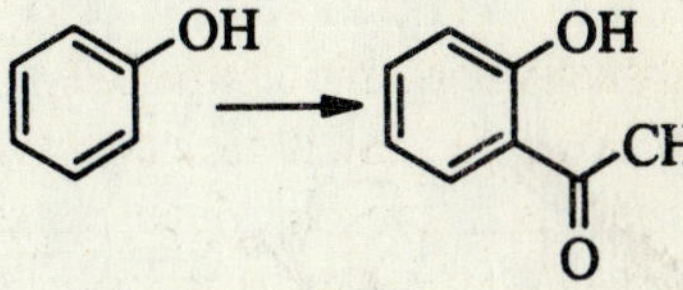 (e) 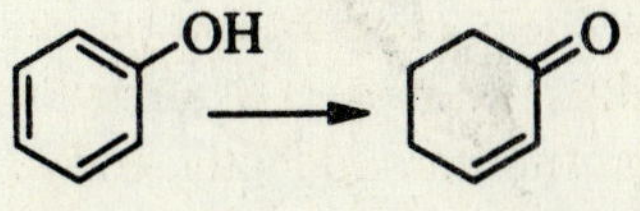 (f) 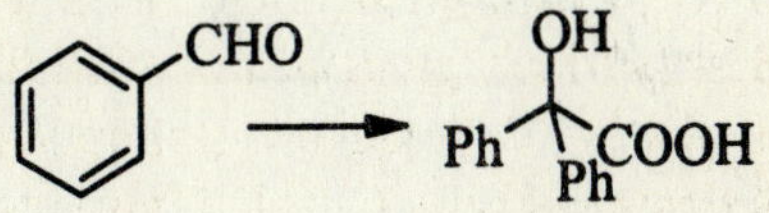 (g) 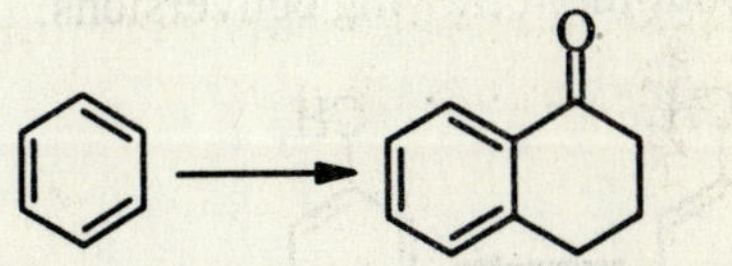 (h)

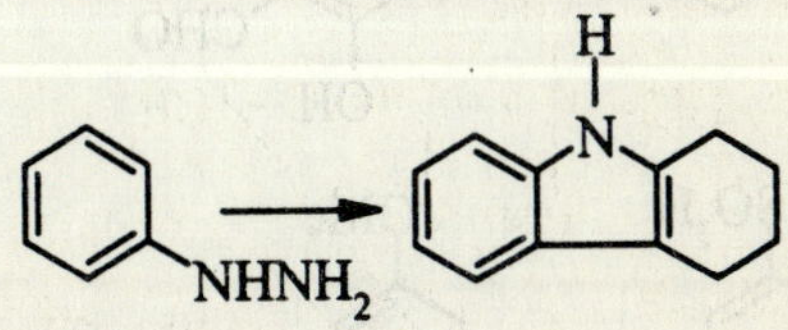

19. Prepare the following compounds starting with either benzene or toluene:

(a) 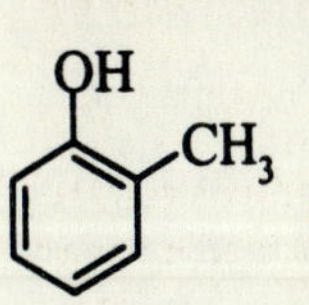 (b) 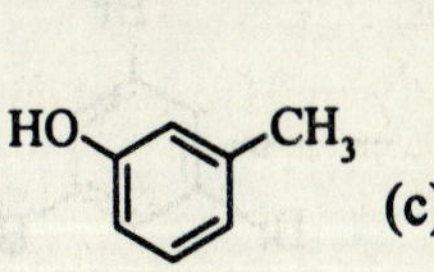 (c) 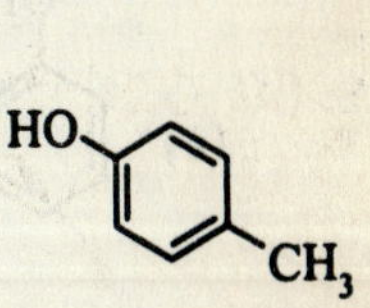 (d) 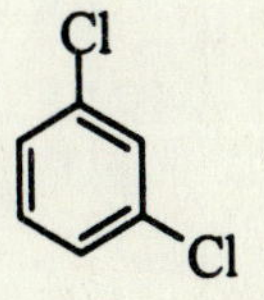 (e) 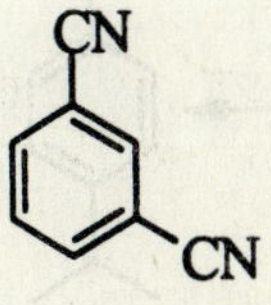 (f) 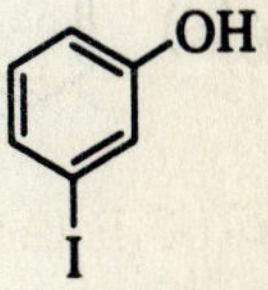 (g) 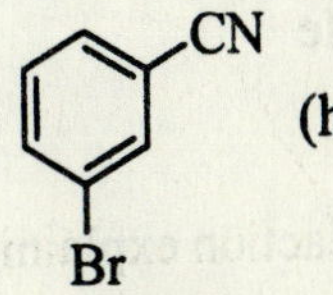 (h) 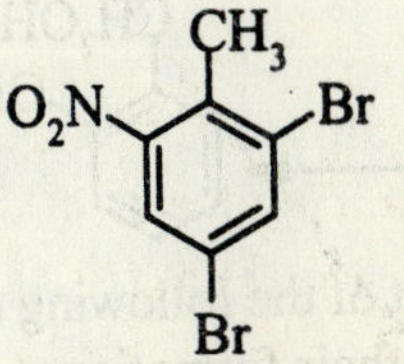 (i) 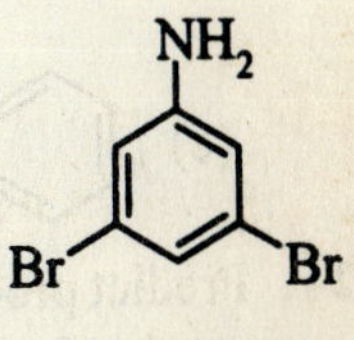 (j) 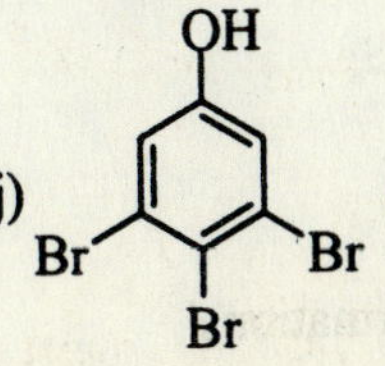 (k) 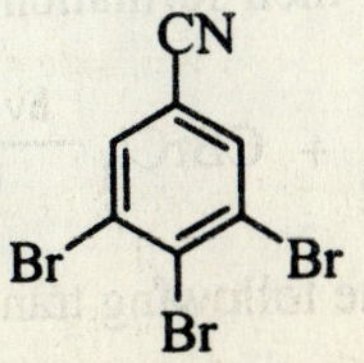 (l)

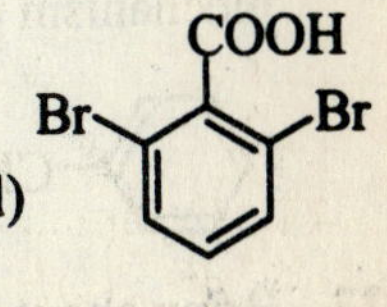

(m) 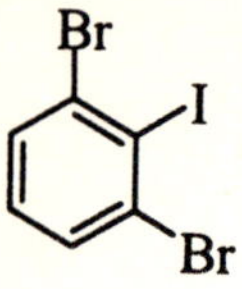 Br, I, Br

(n) 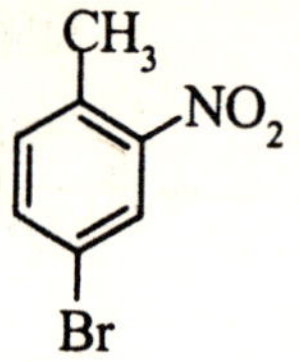 CH_3, NO_2, Br

(o) 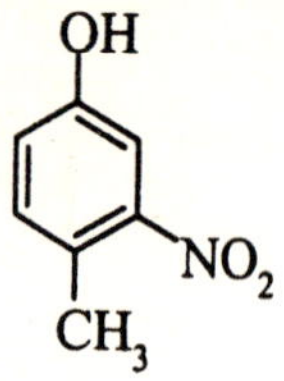 OH, NO_2, CH_3

(p) 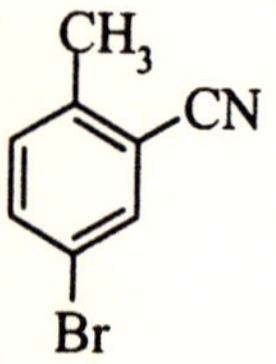 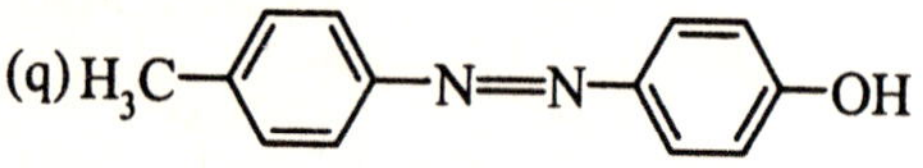 CH_3, CN, Br

(q) H_3C–C_6H_4–N=N–C_6H_4–OH

(r) H_3C–C_6H_4–N=N–, CH_3, HO

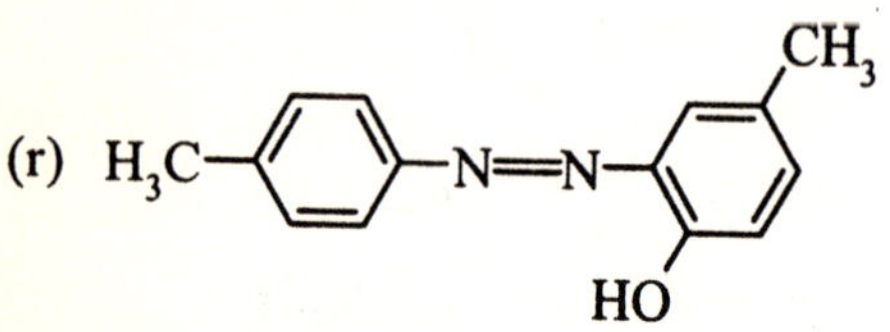

20. Bring about the following conversions:

(a) 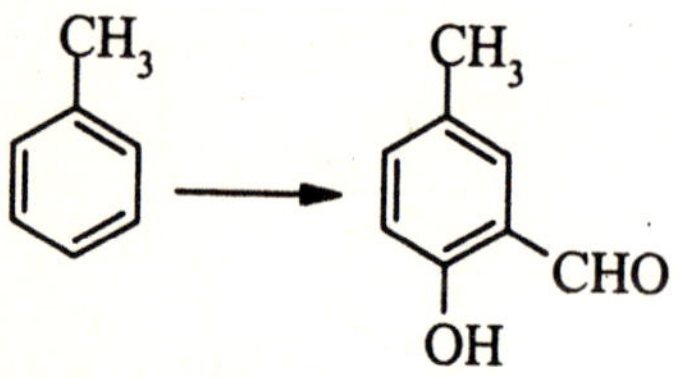 CH_3 → CH_3, CHO, OH

(b) 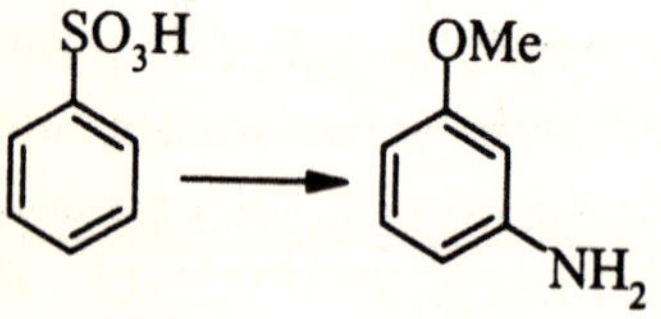 SO_3H → OMe, NH_2

(c) 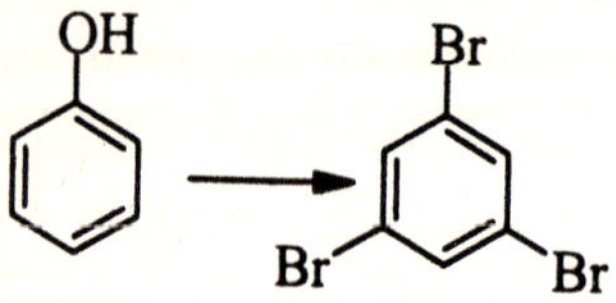 OH → Br, Br, Br

(d) 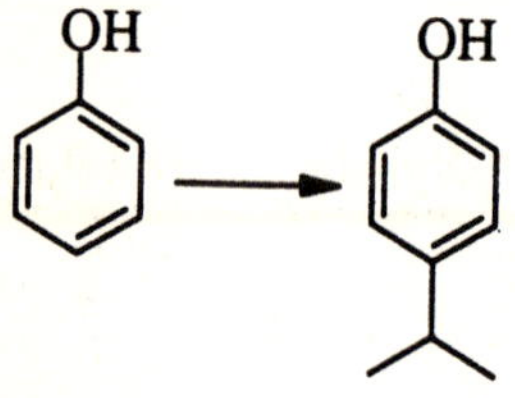 OH → OH

(e) 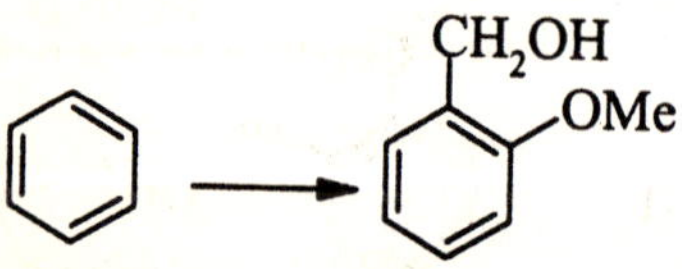 → CH_2OH, OMe

21. Predict product of the following reaction explaining mechanism of their formation:

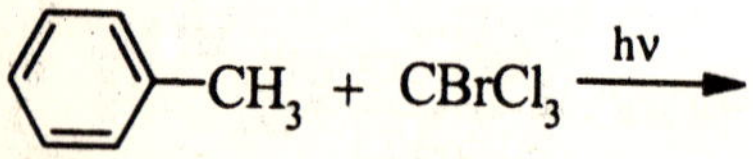

$$C_6H_5\text{–}CH_3 + CBrCl_3 \xrightarrow{h\nu}$$

22. Bring about the following transformations:

(a) 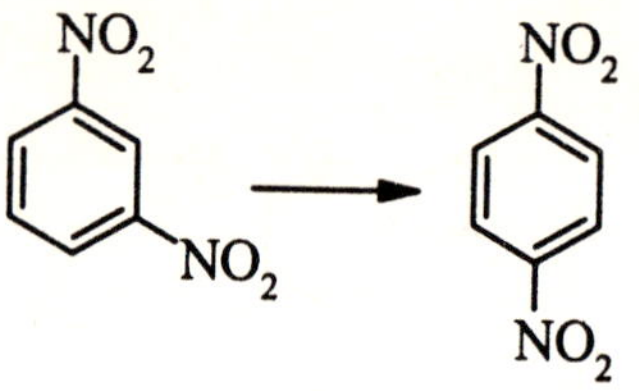 NO_2, NO_2 → NO_2, NO_2

(b) 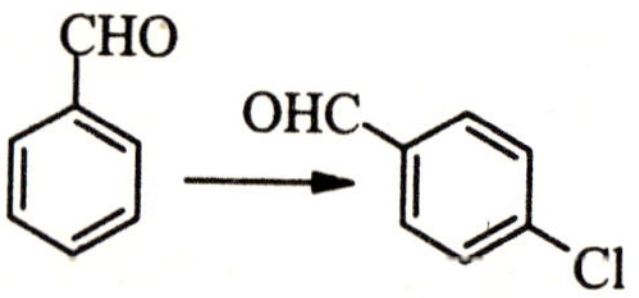 CHO → OHC, Cl

(c) 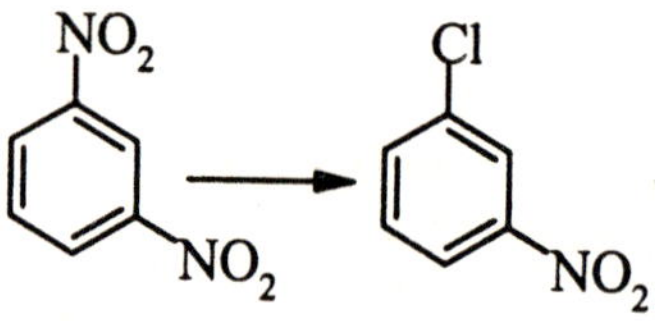 NO_2, NO_2 → Cl, NO_2

(d) 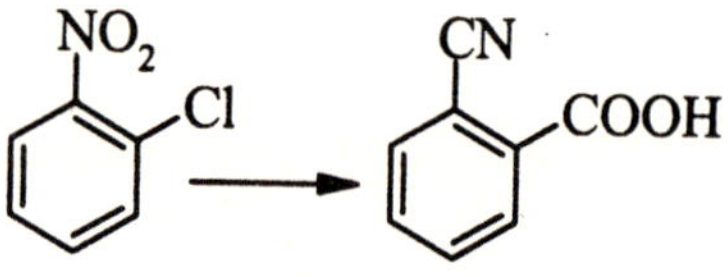 NO_2, Cl → CN, COOH

(e) 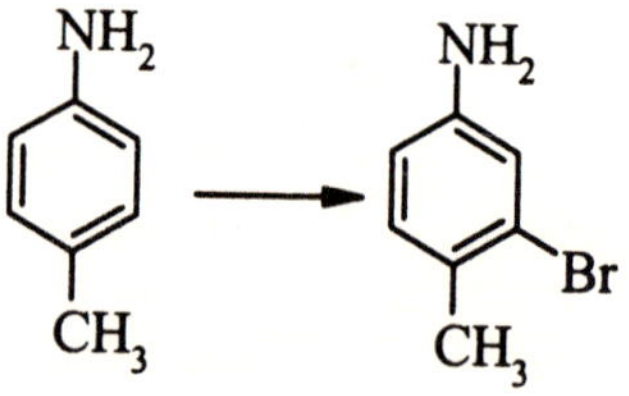 NH_2, CH_3 → NH_2, Br, CH_3

(f) 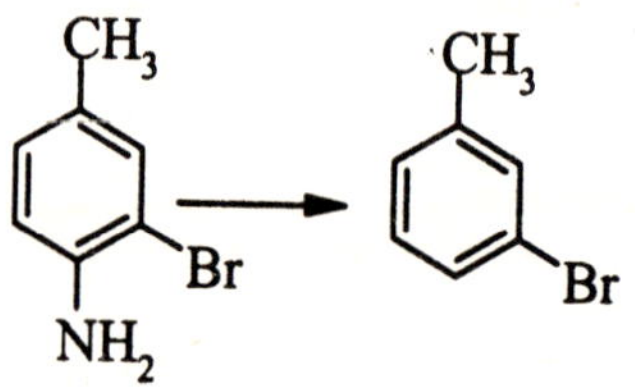 CH_3, Br, NH_2 → CH_3, Br

(g) 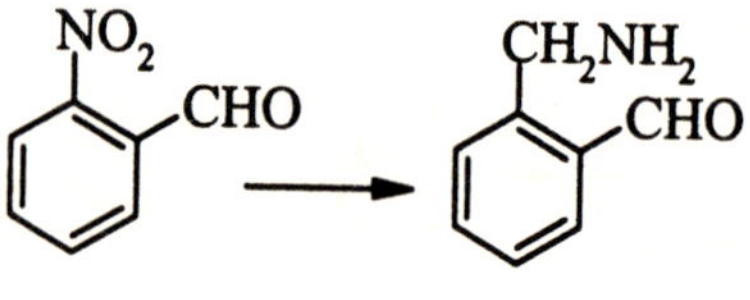 NO_2, CHO → CH_2NH_2, CHO

(h) 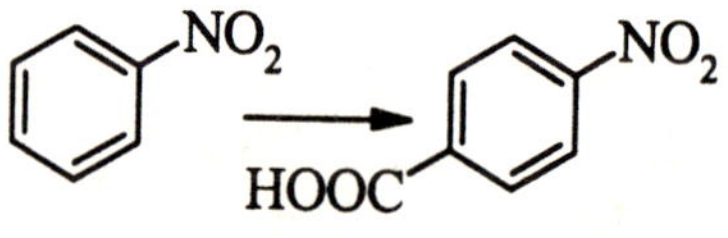 NO_2 → NO_2, HOOC

(i) 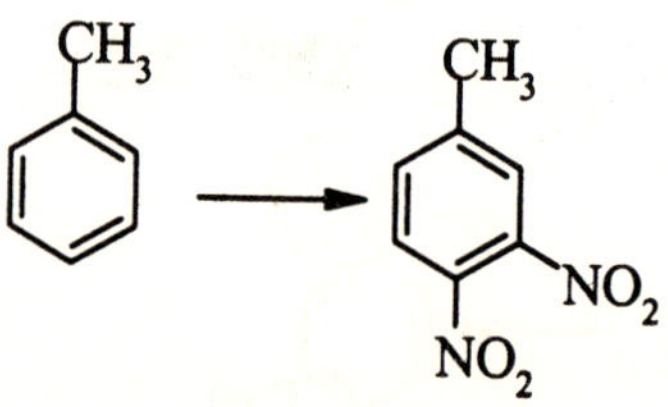 CH_3 → CH_3, NO_2, NO_2

(j) 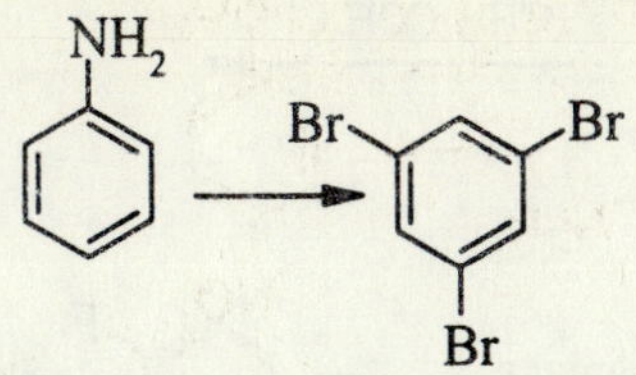 [Aniline → 1,3,5-tribromobenzene]

NH$_2$ → Br, Br, Br

23. Br_2 adds to styrene to give ring substitution rather than addition to C=C, in presence of a Lewis acid catalyst. Explain.

24. Bring about the following transformations:

(a) 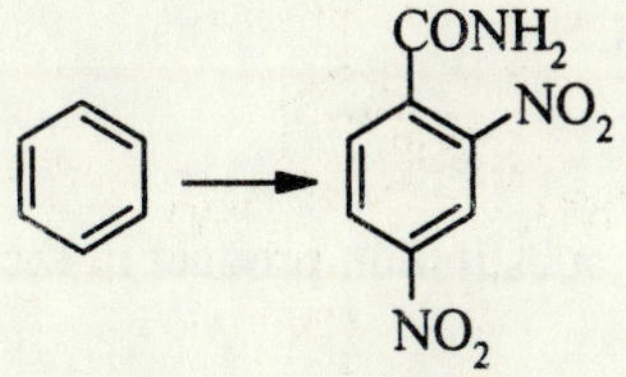 Benzene → CONH$_2$, NO$_2$, NO$_2$

(b) 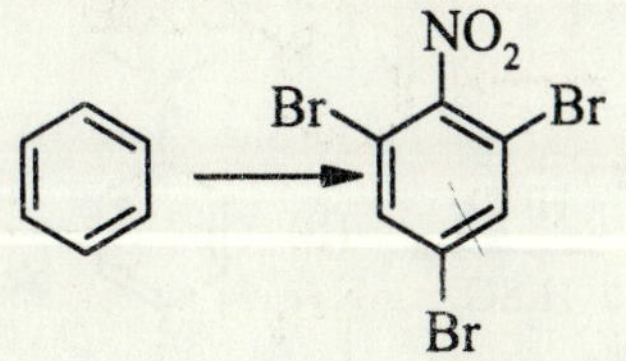 Benzene → NO$_2$, Br, Br, Br

(c) 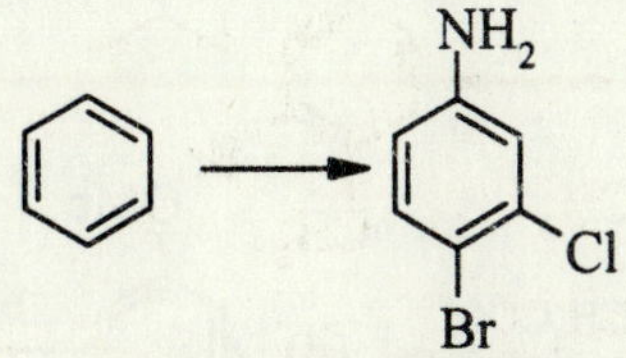 Benzene → NH$_2$, Cl, Br

(d) 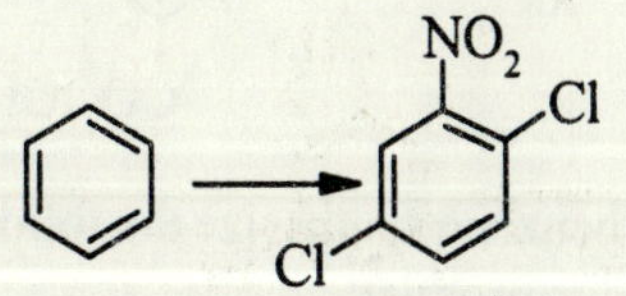 Benzene → NO$_2$, Cl, Cl

25. Bring about the following transformations:

(a) 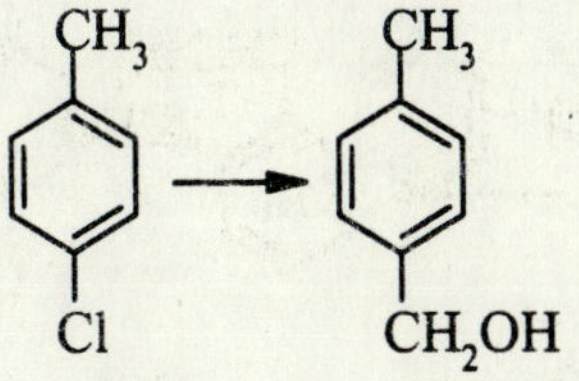 CH$_3$, Cl → CH$_3$, CH$_2$OH

(b) 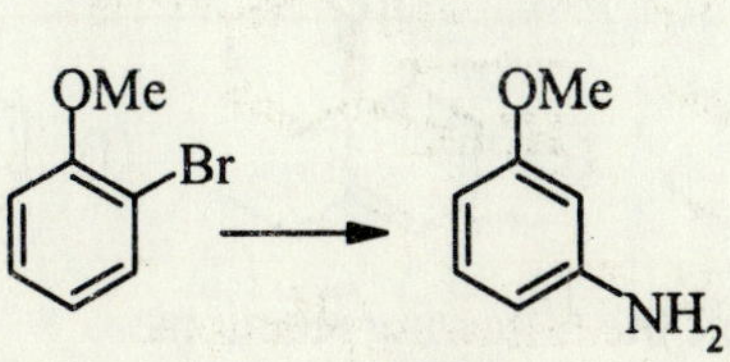 OMe, Br → OMe, NH$_2$

(c) 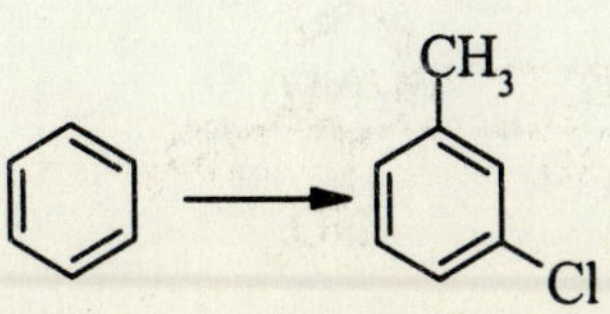 Benzene → CH$_3$, Cl

(d) 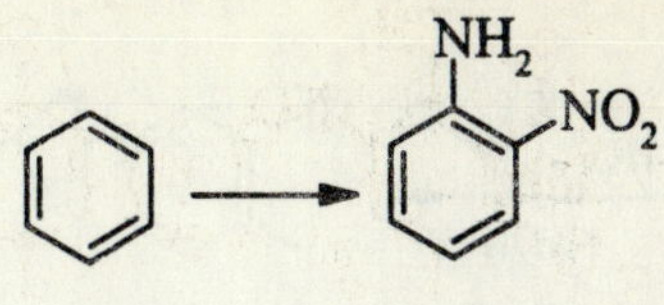 Benzene → NH$_2$, NO$_2$

(e) 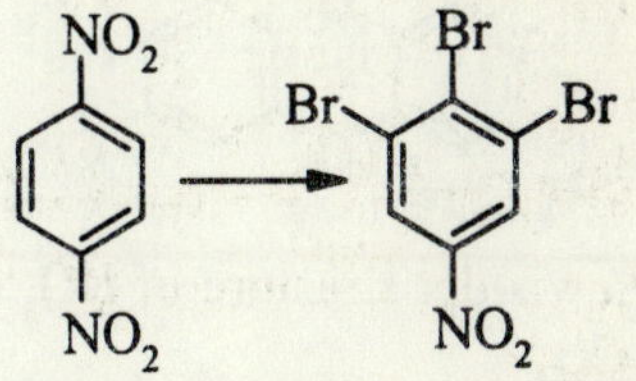 NO$_2$, NO$_2$ → Br, Br, Br, NO$_2$

(f) 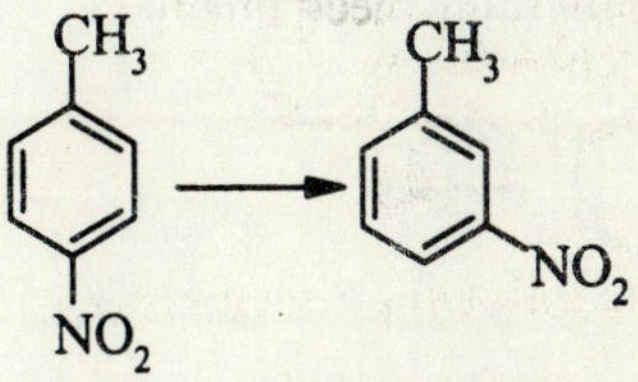 CH$_3$, NO$_2$ → CH$_3$, NO$_2$

(g) 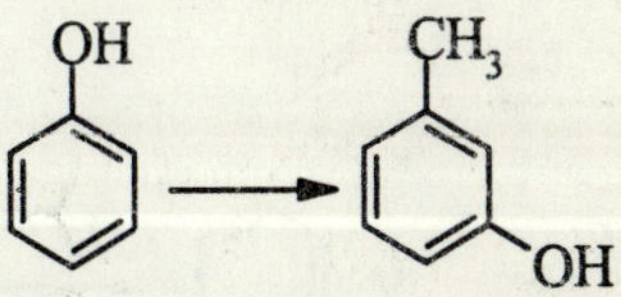 OH → CH$_3$, OH

26. An organic compound A(C_8H_{10}) on treatment with fuming sulphuric acid yields two isomeric products B and C. B on fusion with KCN gives D(C_9H_9N), which on hydrolysis yields E($C_9H_{10}O_2$). E on treatment with hot alkaline $KMnO_4$ yields F($C_8H_6O_4$). F on heating with P_2O_5 undergo intramolecular dehydration to yield G($C_8H_4O_3$). Identify A to G.

27. Identify the aromatic product formed in the following reactions:

(a) 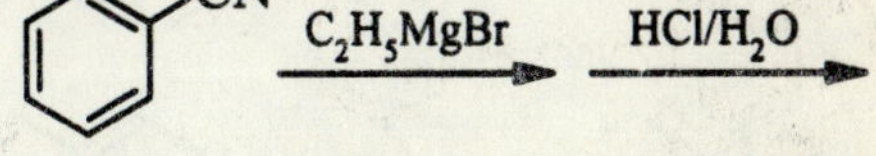 Benzonitrile (CN) $\xrightarrow{C_2H_5MgBr}$ $\xrightarrow{HCl/H_2O}$

(b) 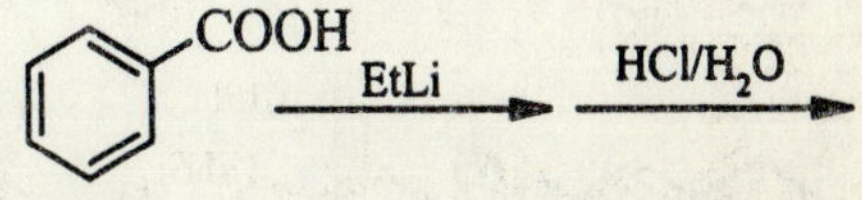 Benzoic acid (COOH) $\xrightarrow{EtLi}$ $\xrightarrow{HCl/H_2O}$

(c) Benzoic acid (COOH) $\xrightarrow{SOCl_2}$ $\xrightarrow{(CH_3)_2Cd}$

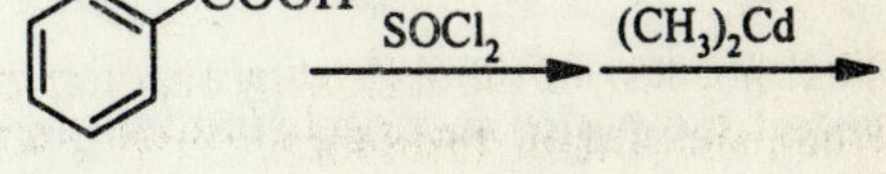

(d) 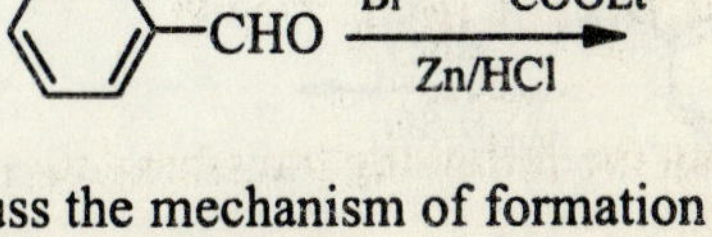 Benzaldehyde (CHO) $\xrightarrow[Zn/HCl]{Br\text{–}CH_2\text{–}COOEt}$

28. Discuss the mechanism of formation of products in the following reaction:

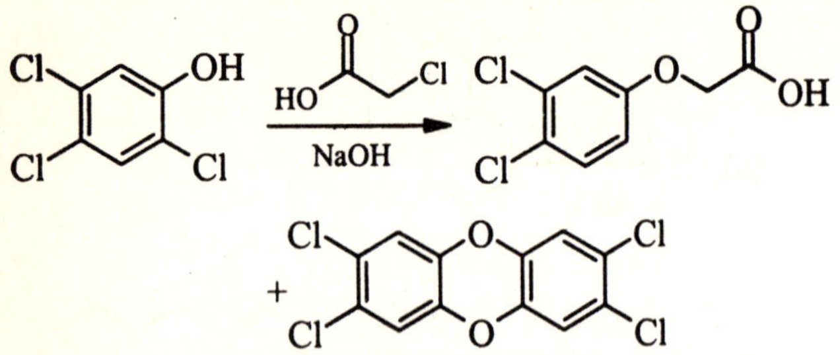

29. Discuss the Lewis dot structure of NO_2^+ and compare it with CO_2.

30. Write product providing mechanism:

(a) 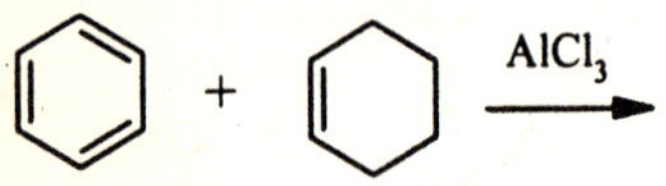 + $AlCl_3$

(b) 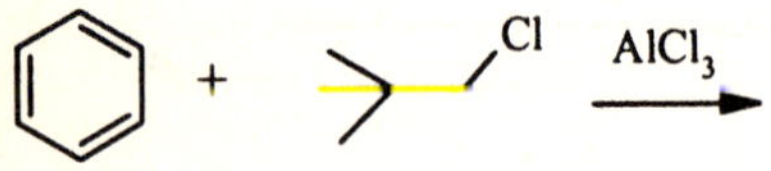 + $AlCl_3$

(c) 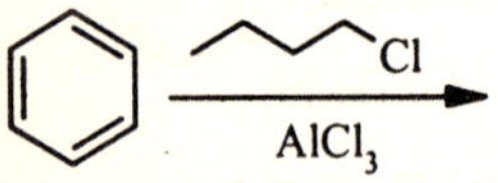 $AlCl_3$ (d) 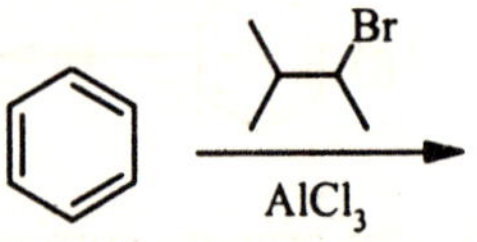 $AlCl_3$

(e) 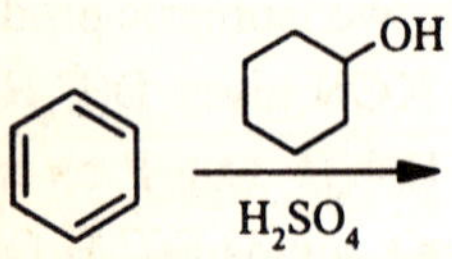 H_2SO_4 (f) 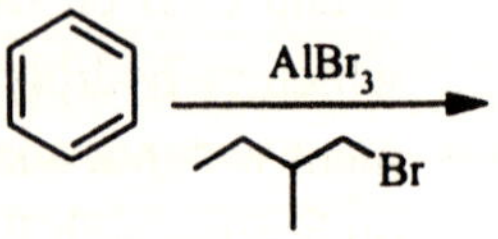 $AlBr_3$

31. Synthesize the following compounds starting from benzene:

(a) 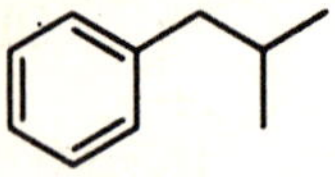 (b)

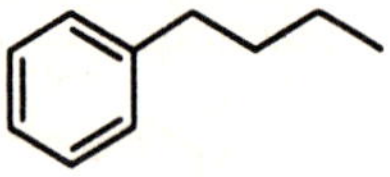

(c) 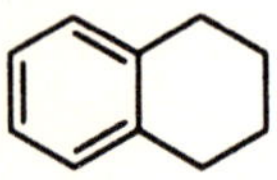 (d)

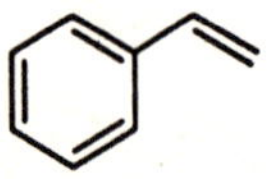

(e) 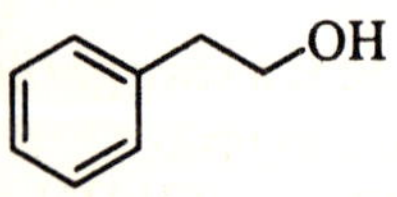 (f)

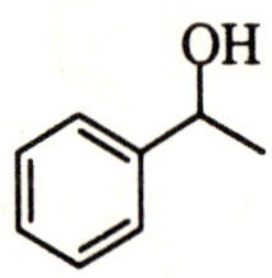

32. Predict the major monosubstitution product in the following reactions:

(a) 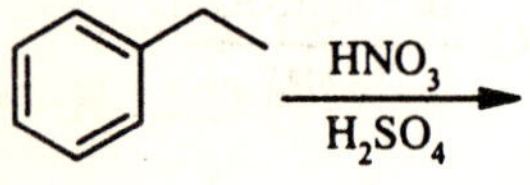 HNO_3 / H_2SO_4

(b) 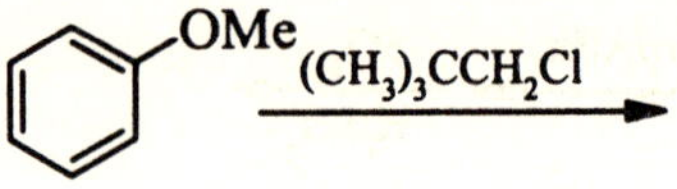 $(CH_3)_3CCH_2Cl$

(c) 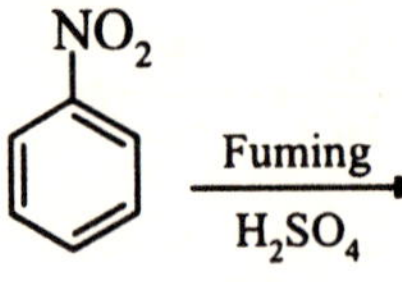 Fuming H_2SO_4 (d) 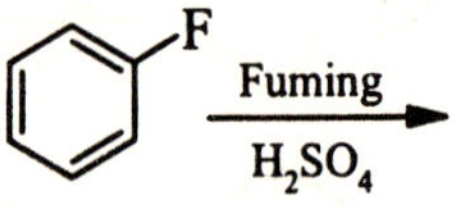 Fuming H_2SO_4

(e) 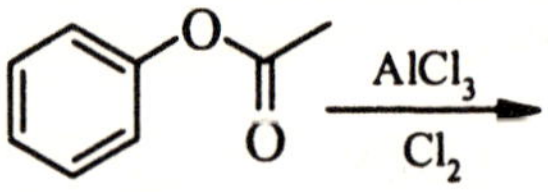 $AlCl_3$ / Cl_2

(f) 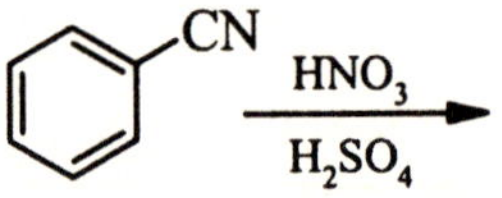 HNO_3 / H_2SO_4

33. Predict major substitution product in the following reactions:

(a) 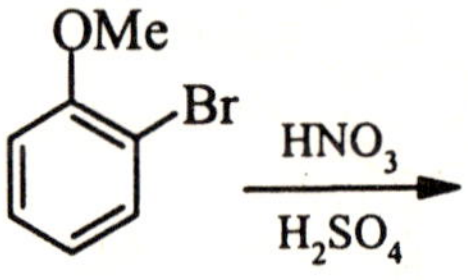 HNO_3 / H_2SO_4 (b) 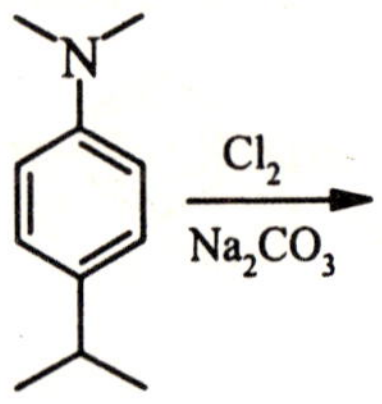 Cl_2 / Na_2CO_3

(c) 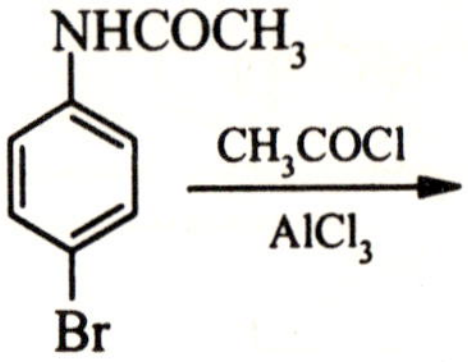 CH_3COCl / $AlCl_3$ (d) 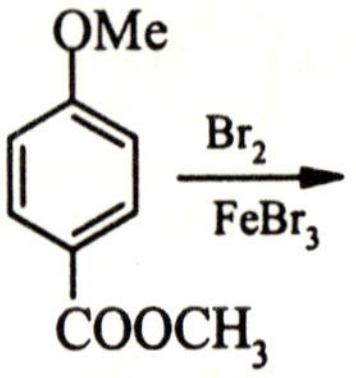 Br_2 / $FeBr_3$

34. Starting from benzene synthesize the following compounds as major product:

(a) 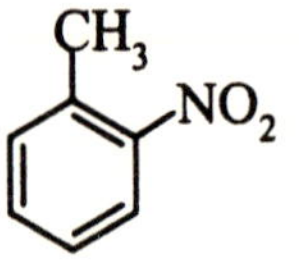 (b) 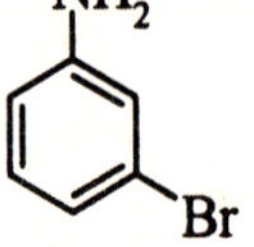 (c)

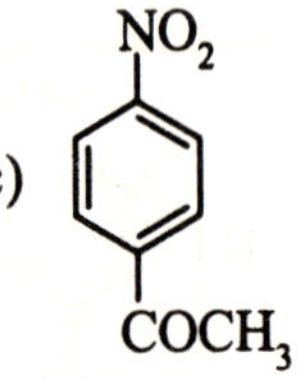

(d) 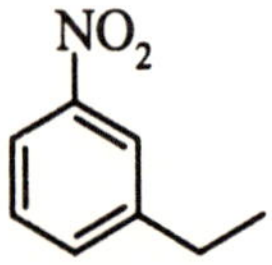 (e) 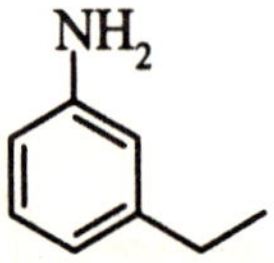 (f)

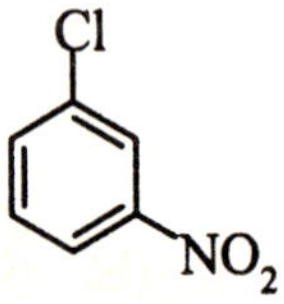

35. Complete the following reactions:

(a) 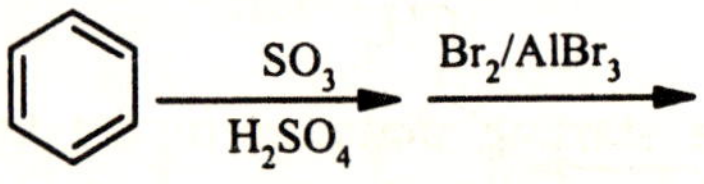 SO_3 / H_2SO_4 → $Br_2/AlBr_3$ →

(b) 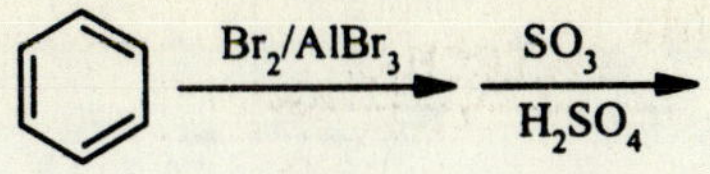 Benzene $\xrightarrow{Br_2/AlBr_3}$ $\xrightarrow[H_2SO_4]{SO_3}$

(c) 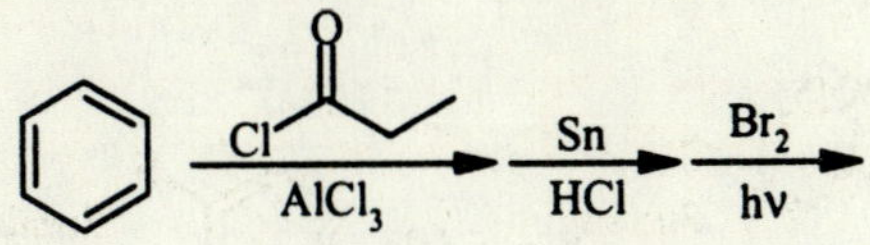 Benzene $\xrightarrow[AlCl_3]{CH_3CH_2COCl}$ $\xrightarrow[HCl]{Sn}$ $\xrightarrow[h\nu]{Br_2}$

(d) 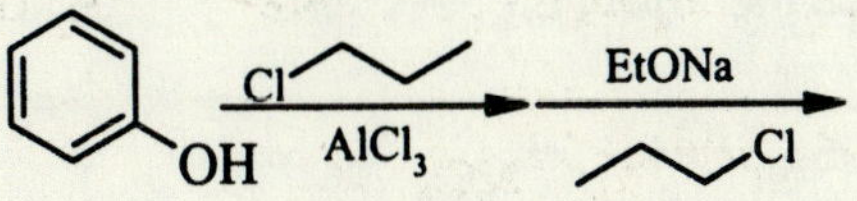 Phenol $\xrightarrow[AlCl_3]{CH_3CH_2CH_2Cl}$ $\xrightarrow[CH_3CH_2CH_2Cl]{EtONa}$

(e) 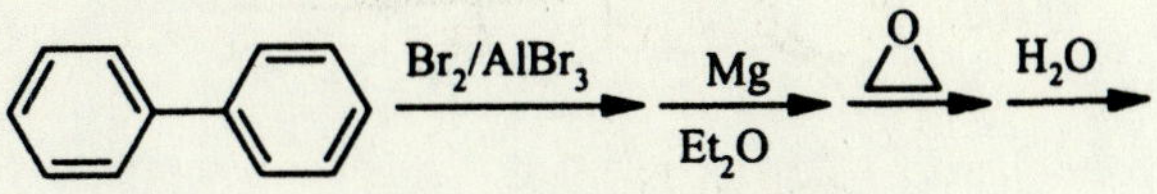 Biphenyl $\xrightarrow{Br_2/AlBr_3}$ $\xrightarrow[Et_2O]{Mg}$ $\xrightarrow{\text{ethylene oxide}}$ $\xrightarrow{H_2O}$

(f) 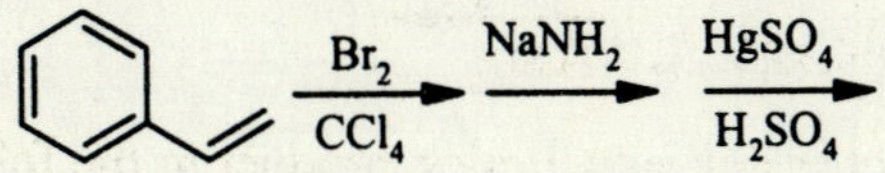 Styrene $\xrightarrow[CCl_4]{Br_2}$ $\xrightarrow{NaNH_2}$ $\xrightarrow[H_2SO_4]{HgSO_4}$

36. Propose synthesis of the following compounds starting from benzene:

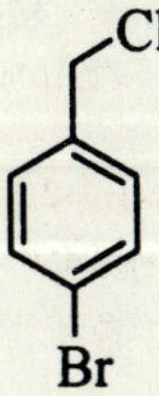

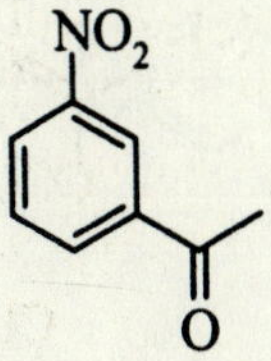

(a) 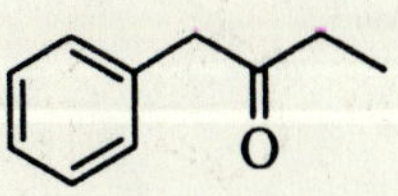 $PhCH_2COCH_2CH_3$ (b) 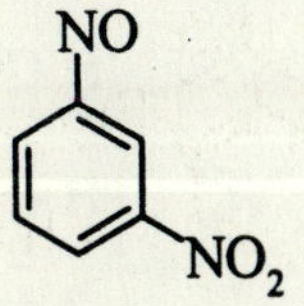 1-NO-3-NO_2-benzene

(c) 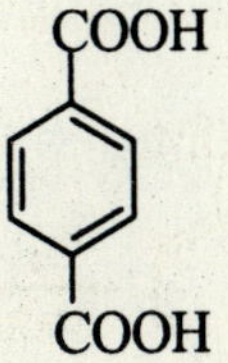 Terephthalic acid ($HOOC-C_6H_4-COOH$, para) (d) 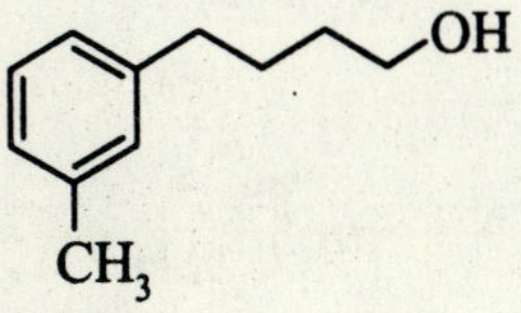 5-(3-methylphenyl)pentan-1-ol (CH_3, OH)

(e) 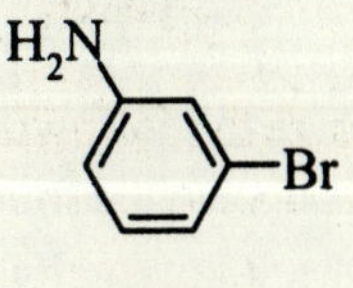 H_2N–, –Br (m-bromoaniline) (f) p-Br–C_6H_4–CH_2Cl

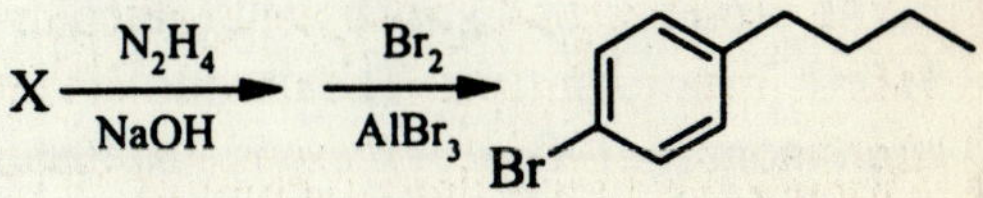

(g) m-NO_2-acetophenone (h) o-nitrostyrene (NO_2)

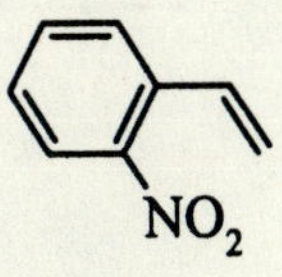

(i) 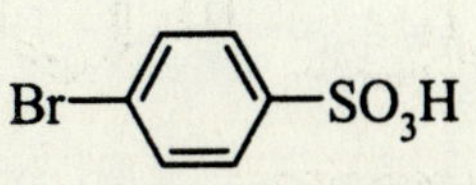 Br–C_6H_4–SO_3H (para) (j) 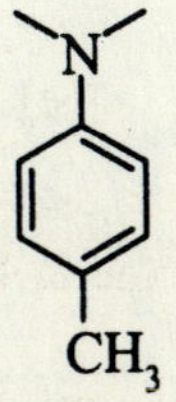 p-CH_3–C_6H_4–$N(CH_3)_2$

37. Identify the starting material in the following sequence of reactions:

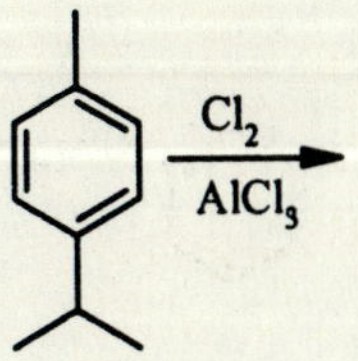

(a) 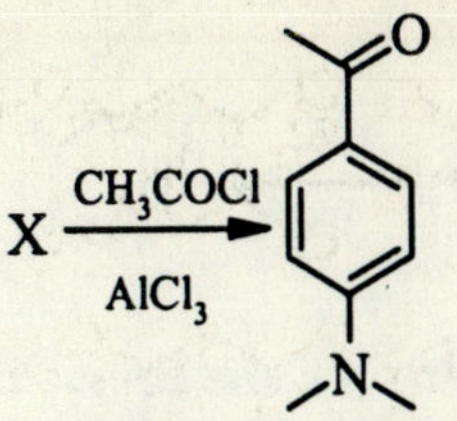 X $\xrightarrow[AlCl_3]{CH_3COCl}$ p-$(CH_3)_2N$–C_6H_4–$COCH_3$

(b) X $\xrightarrow[NaOH]{N_2H_4}$ $\xrightarrow[AlBr_3]{Br_2}$ p-Br–C_6H_4–butyl (Br)

(c) X $\xrightarrow[AlBr_3]{Br_2}$ $\xrightarrow{Sn/HCl}$ m-Br–C_6H_4–propyl (Br)

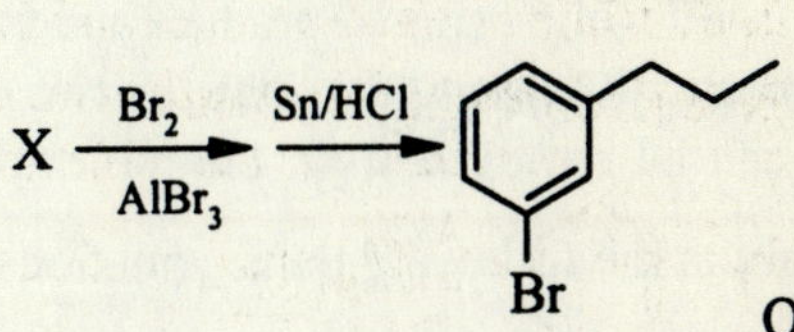

(d) 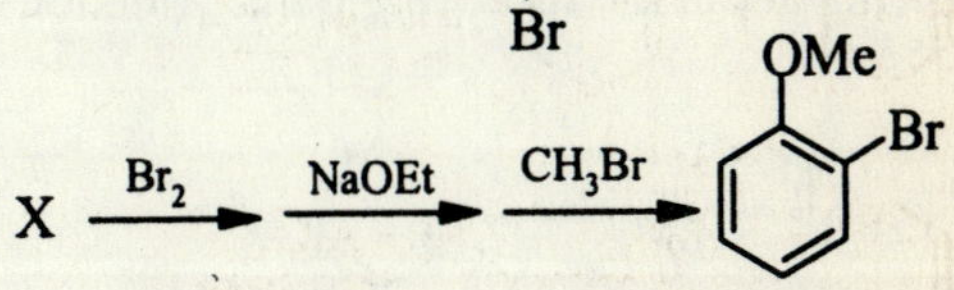 X $\xrightarrow{Br_2}$ $\xrightarrow{NaOEt}$ $\xrightarrow{CH_3Br}$ o-Br–C_6H_4–OMe (OMe, Br)

(e) 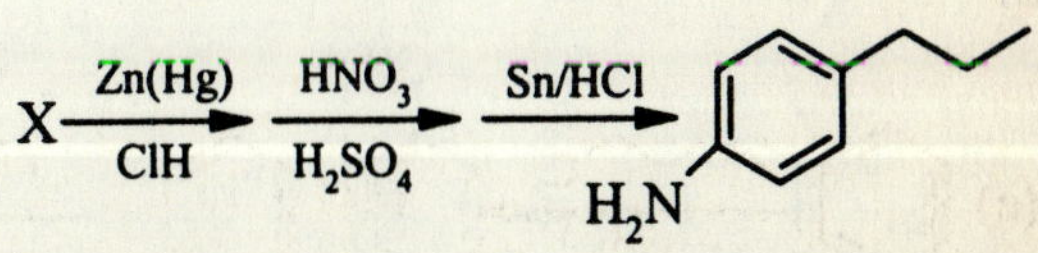 X $\xrightarrow[ClH]{Zn(Hg)}$ $\xrightarrow[H_2SO_4]{HNO_3}$ $\xrightarrow{Sn/HCl}$ p-H_2N–C_6H_4–propyl (H_2N)

38. Predict the major product in the following reactions:

(a) 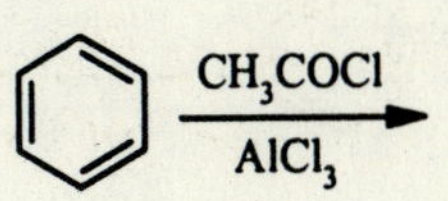 Benzene $\xrightarrow[AlCl_3]{CH_3COCl}$ (b) 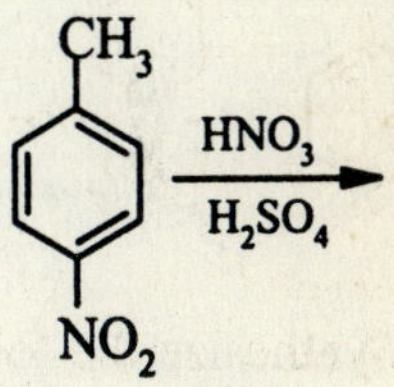 p-CH_3–C_6H_4–NO_2 $\xrightarrow[H_2SO_4]{HNO_3}$

(c) p-Isopropyltoluene $\xrightarrow[AlCl_3]{Cl_2}$ (d) 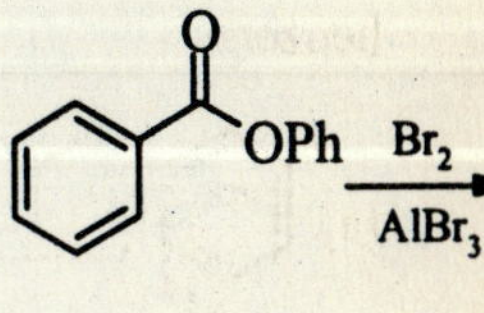 $PhCOOPh$ (OPh) $\xrightarrow[AlBr_3]{Br_2}$

(e) 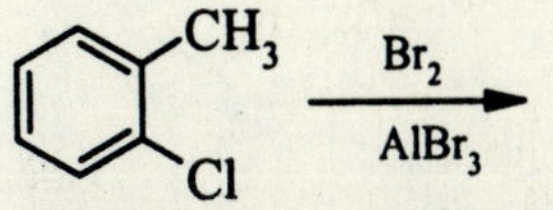 o-Chlorotoluene (CH_3, Cl) $\xrightarrow[AlBr_3]{Br_2}$

39. Trimethyl ammonium ion is a strongly deactivating and meta directing group, explain.

40. Predict whether the following groups are activating/deactivating and ortho, para or meta directing:

(a) CF_3– (b) $(CH_3)_3P^+$–

(c) O_2NCH_2– (d) $(CH_3)_2S^+$–

(e) C_6H_5– (f) $CH_3CH{=}CH$–

41. How would you rationalize the following experimental result:

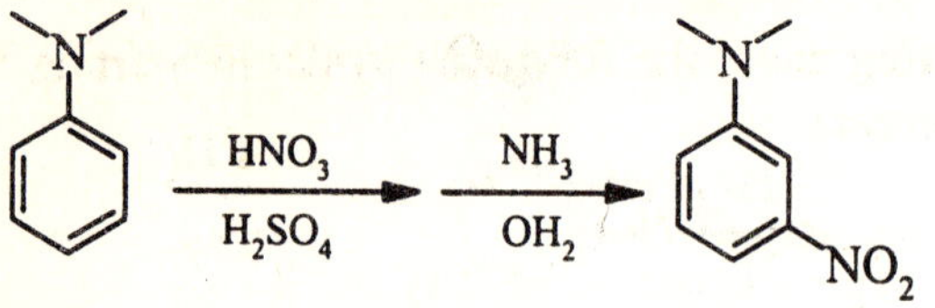

$\xrightarrow[H_2SO_4]{HNO_3}$ $\xrightarrow[OH_2]{NH_3}$ NO_2

42. *p*-nitrophenol is shaken with one equivalent of D_2O in the presence of a strong acid (perchloric acid) at 100°C for a long time. The product has two deuterium atom in it. One of these is lost instantly when the product is treated with ordinary water. Propose a structure for the original product and explain the difference in the ease with which the two deuterium atoms can be removed from the molecule.

43. Bring about the following transformations:

(a) 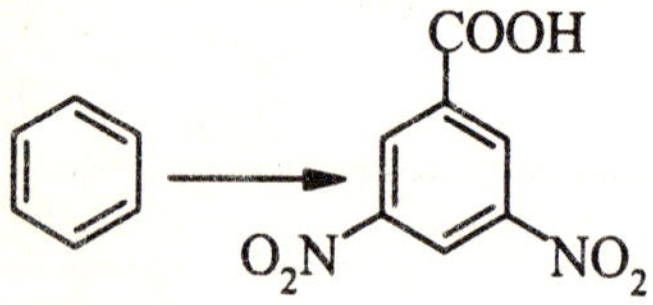 COOH, O_2N, NO_2

(b) 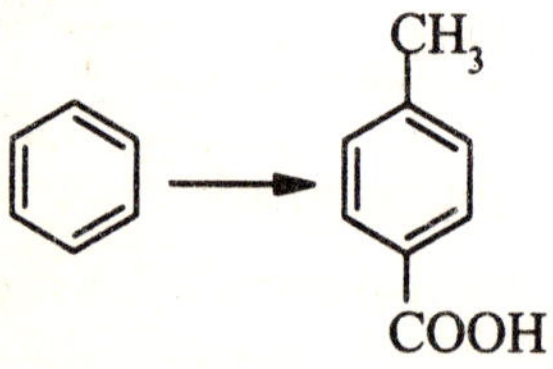 CH_3, COOH

(c) 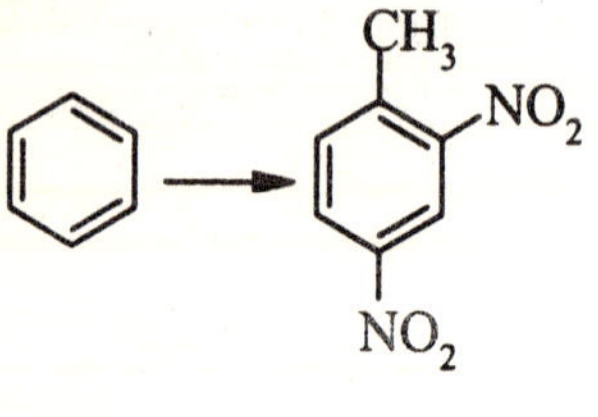 CH_3, NO_2, NO_2

(d) 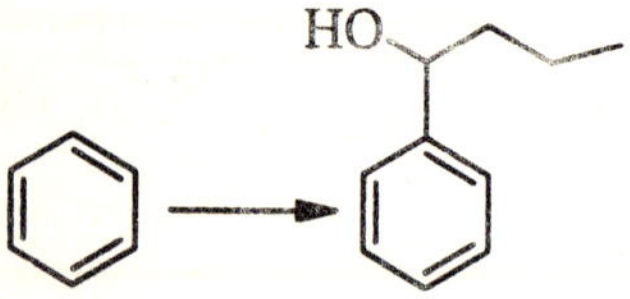 HO

(e) 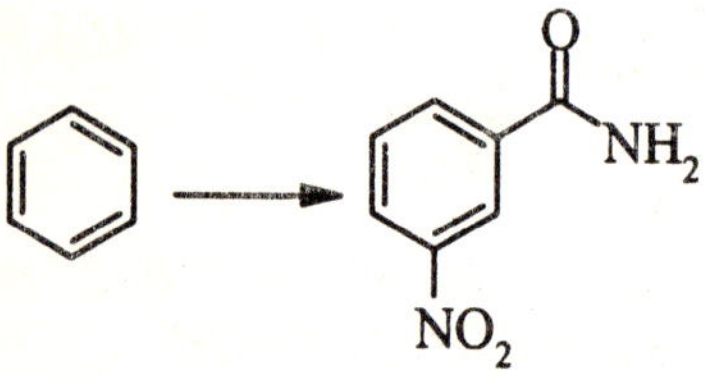 O, NH_2, NO_2

(f)

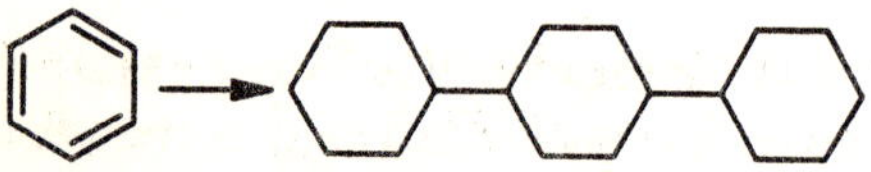

44. Complete the following equations:

(a) 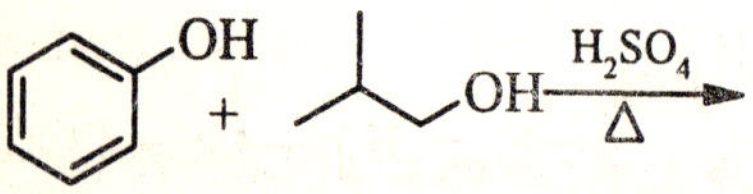 OH + OH $\xrightarrow[\Delta]{H_2SO_4}$

(b) 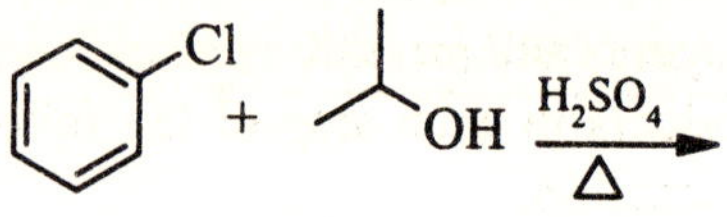 Cl + OH $\xrightarrow[\Delta]{H_2SO_4}$

(c) 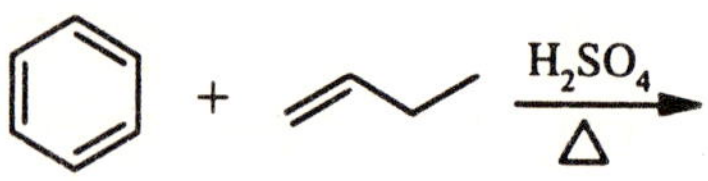 + $\xrightarrow[\Delta]{H_2SO_4}$

45. Complete the following reactions:

(a) 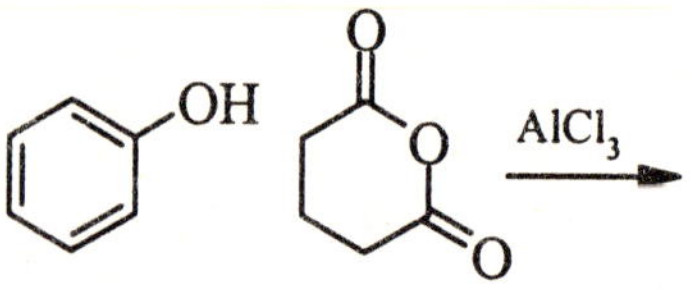 OH, O, O, O $\xrightarrow{AlCl_3}$

(b) 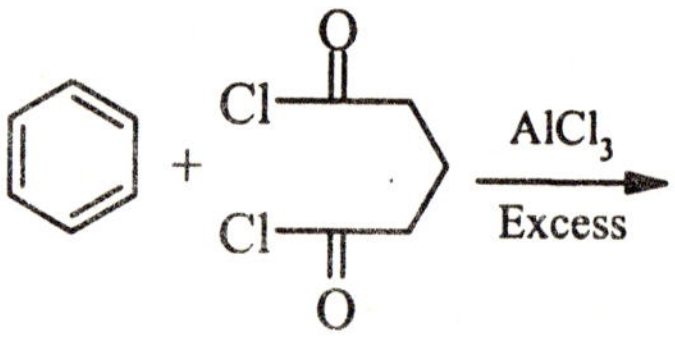 + Cl, O, Cl, O $\xrightarrow[Excess]{AlCl_3}$

(c) 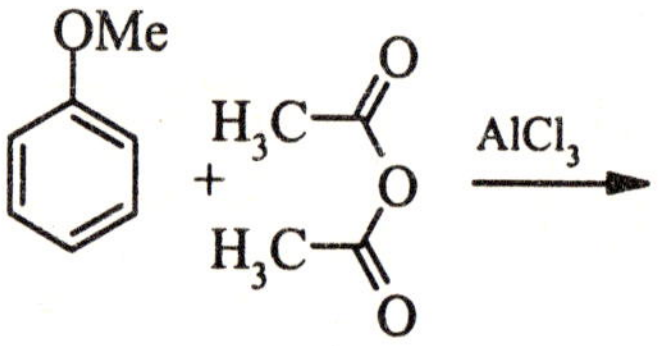 OMe + H_3C, O, O, H_3C, O $\xrightarrow{AlCl_3}$

46. Synthesize the following compounds starting from benzene:

(a) 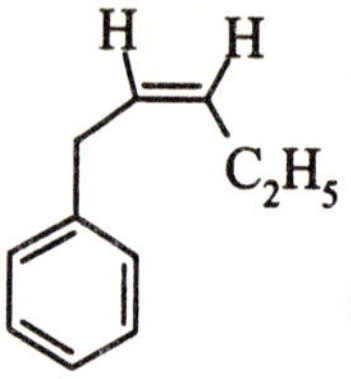 H, H, C_2H_5 (b) 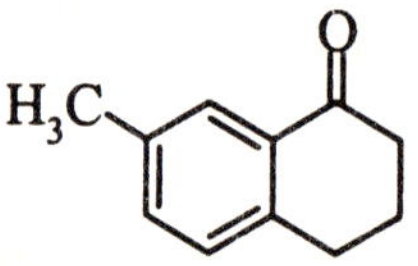 H_3C, O (c) Ph, OH

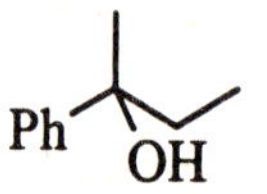

47. Predict the major substitution product in the following sequence of reaction:

(a) 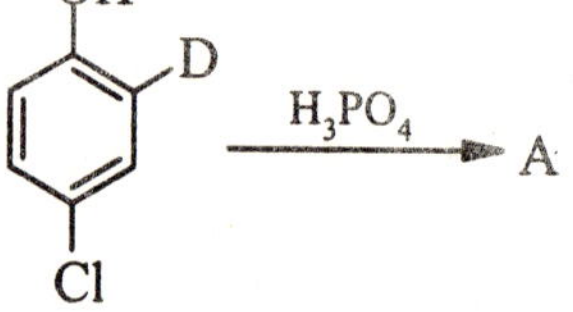 OH, D, Cl $\xrightarrow{H_3PO_4}$ A (b) 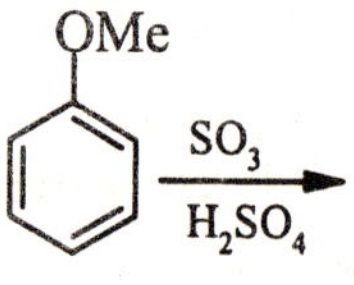 OMe $\xrightarrow[H_2SO_4]{SO_3}$

(c) 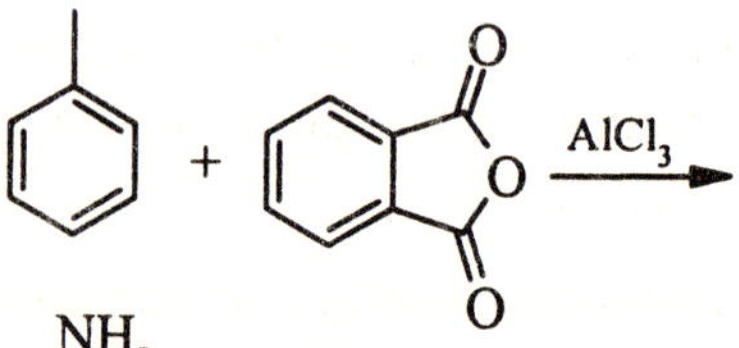 + O, O, O $\xrightarrow{AlCl_3}$

(d) NH_2, CH_3 $\xrightarrow{(CH_3CO)_2O}$ $\xrightarrow[H_2SO_4]{HNO_3}$ $\xrightarrow{H_3O^+}$

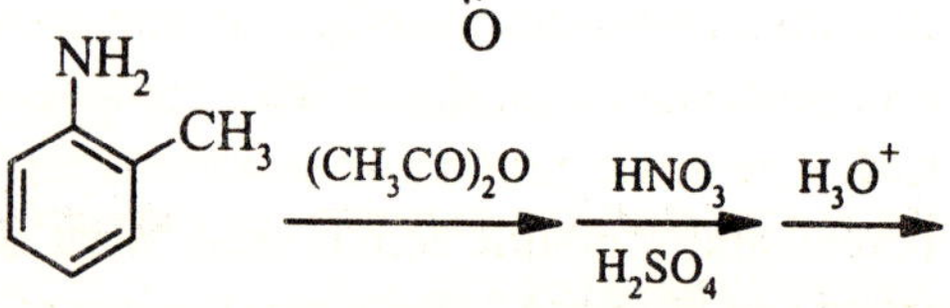

48. Identify reagents and propose mechanism of formation of products in each step of the following sequence of reaction:

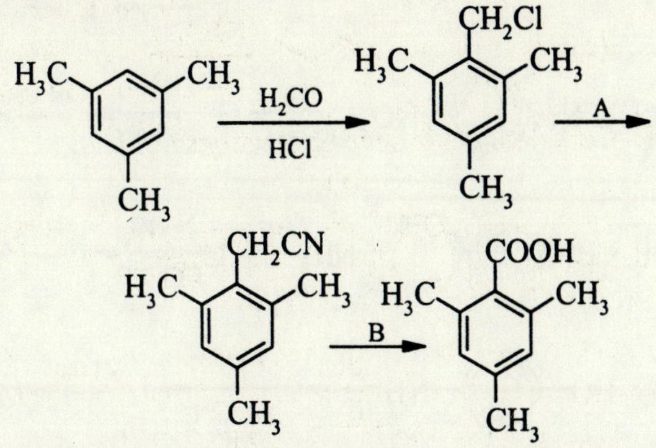

H_3C, CH_3, CH_3 → (H_2CO, HCl) → CH_2Cl, H_3C, CH_3, CH_3 → (A) → CH_2CN, H_3C, CH_3, CH_3 → (B) → COOH, H_3C, CH_3, CH_3

49. Predict synthesis of the following products starting from benzene:

(a) 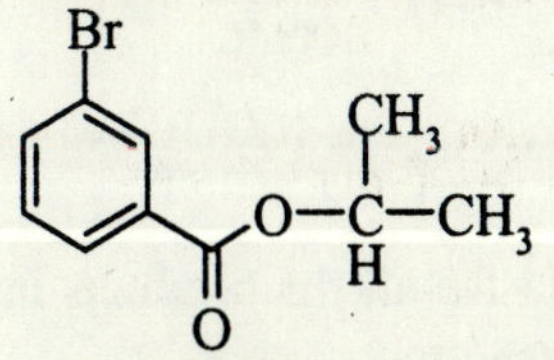 Br, CH_3, O–C–CH_3, H, O

(b)

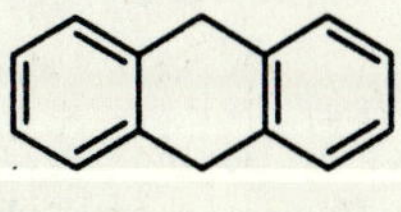

(c) 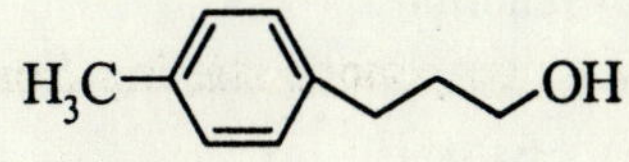 H_3C–, OH

(d) 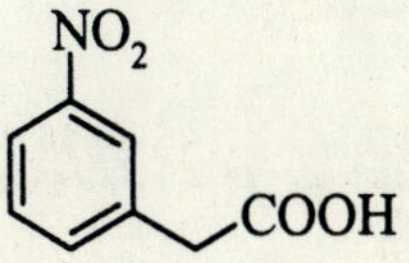 NO_2, COOH

(e)

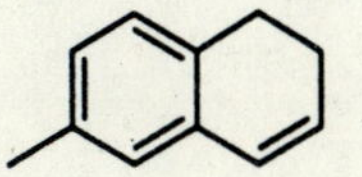

(f) 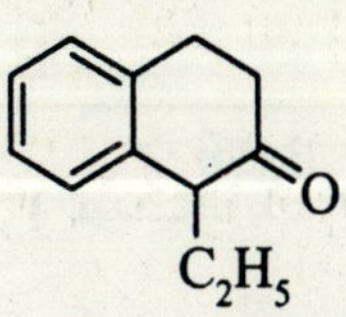 O, C_2H_5

(g) 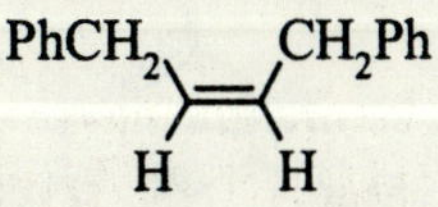 $PhCH_2$, CH_2Ph, H, H

50. 1, 3, 5-trimethyl benzene undergoes electrophilic aromatic substitution with iodine monochloride (ICl). Write mechanism for this reaction showing major product.

51. When allyl alcohol is treated with HF in presence of benzene, two products are formed: 3-phenyl-1-propene and 1, 2-diphenyl propane. Write equations showing the mechanism for the formations for these products.

52. When 2-hydroxy benzoic acid is heated with isobutyl alcohol in presence of sulphuric acid, compound A is formed. The same product is formed if tertiary butyl alcohol and sulphuric acid is used. What is the structure of compound A? Write mechanism of formation of this product.

53. Bring about the following synthesis starting from benzene:

(a) 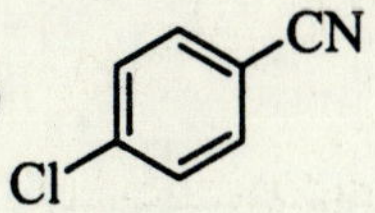 CN, Cl

(b)* 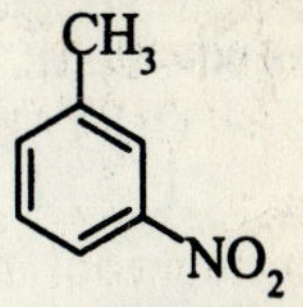 CH_3, NO_2

(* direct FC alkylation of nitro benzene does not succeed)

(c) 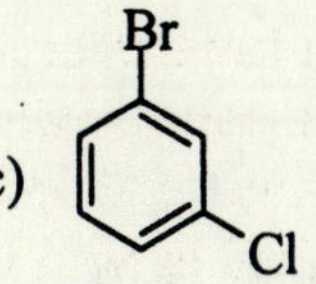 Br, Cl

(d) 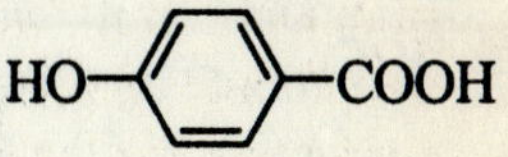 HO–, –COOH

(e) 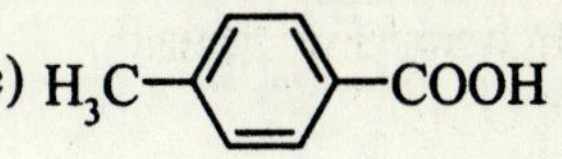 H_3C–, –COOH

(f) F, NO_2, NO_2

(g) OMe, NH_2

(h) HO–, –I

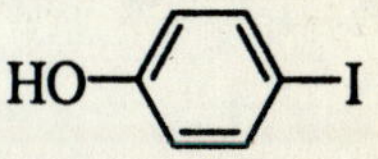

54. Complete the following reactions:

(a) 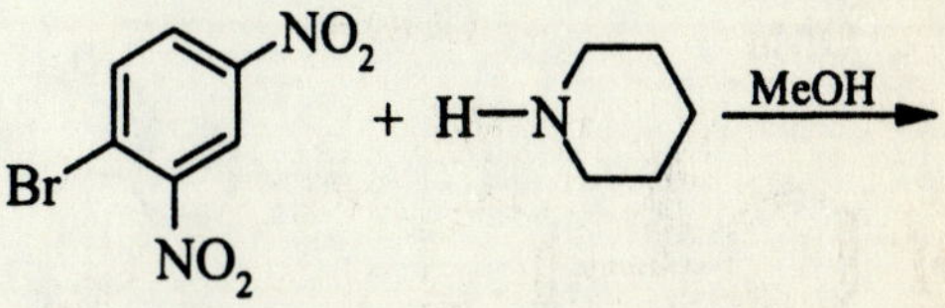 NO_2, Br, NO_2 + H–N (piperidine) → (MeOH)

(b) 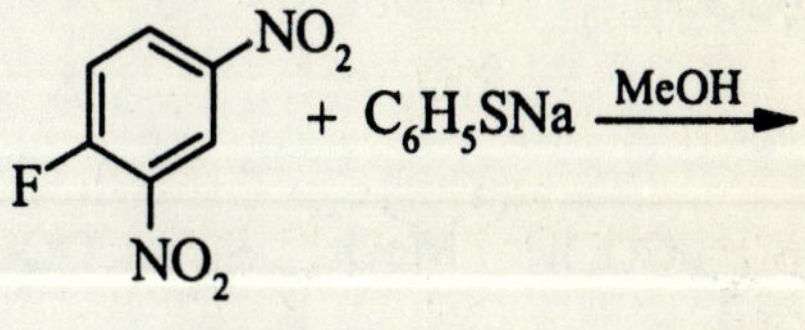 NO_2, F, NO_2 + C_6H_5SNa → (MeOH)

(c) NO_2, OTs, O_2N, NO_2 + –NH_2 → (Heat)

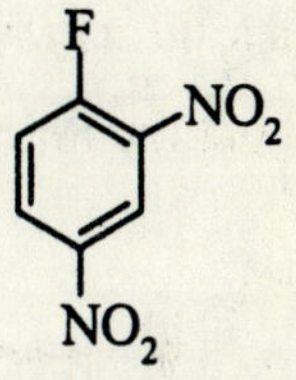

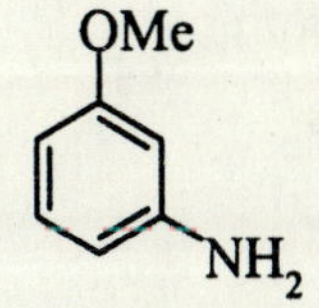

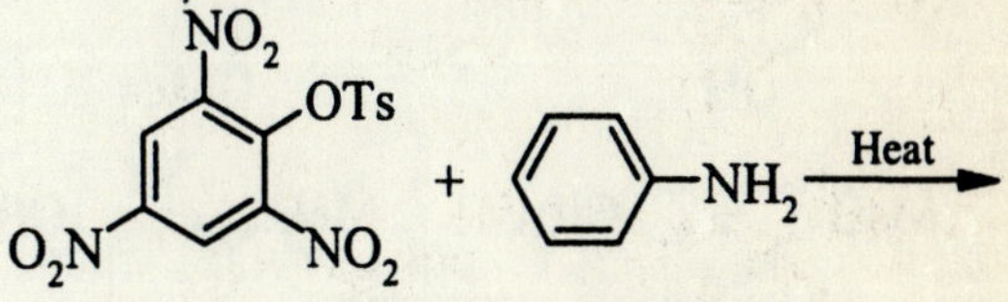

(d) 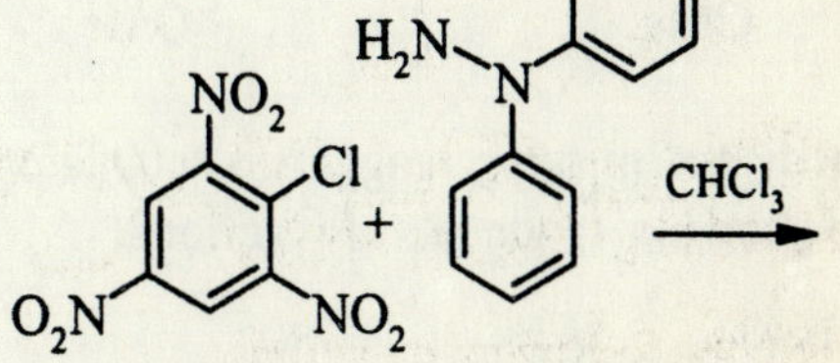 NO_2, Cl, O_2N, NO_2 + H_2N–N(Ph)(Ph) → ($CHCl_3$)

(e) 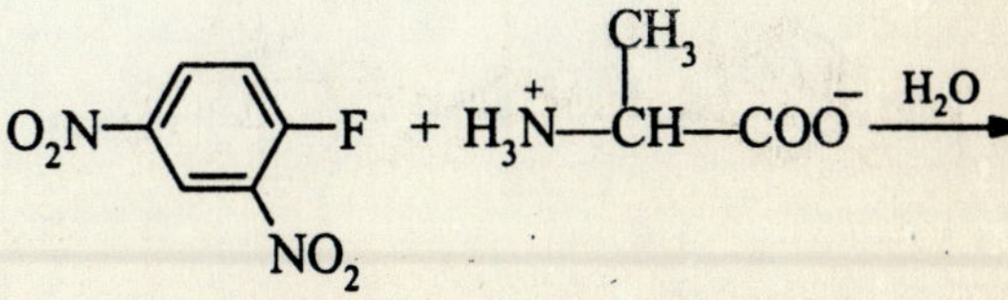 O_2N–, –F, NO_2 + $H_3\overset{+}{N}$–CH(CH_3)–COO^- → (H_2O)

55. Give structural formula for all reagents, intermediates and products designated by letters in the following equations:

(a) [1-CHO-2-CH(OMe)OMe naphthalene] $\xrightarrow{NaBH_4}$ A $\xrightarrow{H_3O^+}$ B

(b) HOOC–(CH₂)₂–C₆H₄–(CH₂)₂–COOH $\xrightarrow{H_3PO_4}$ C + D

(c) [phenol with CH(CH₃)CH=CH₂ and isopropyl substituents] $\xrightarrow{NaH}$ $\xrightarrow{CH_3I}$ E $\xrightarrow[H_2O_2/NaOH]{B_2H_6}$ F $\xrightarrow[H_2SO_4]{CrO_3}$ G $\xrightarrow{H_3PO_4}$ H $\xrightarrow{LiAlH_4/H_2O}$ I

(d) [benzene with CHO, OH, OMe] $\xrightarrow{NaBH_4/H_2O}$ J $\xrightarrow{NaH}$ K $\xrightarrow{CH_3I}$ L

(e) H—CH(COOEt)$_2$ $\xrightarrow{Na}$ M $\xrightarrow{\text{Br-C}_6\text{H}_2(\text{NO}_2)_2\text{-Cl}}$ O

(f) [3,5-dihydroxybenzoic acid (HO, COOH, OH)] $\xrightarrow{P}$ [MeO, COOCH₃, OMe benzene] $\xrightarrow{Q}$ [MeO, CH₂OH, OMe benzene] $\xrightarrow{R}$ [MeO, Br, CH₂OH, OMe benzene]

56. Identify the missing reagents/reactants/products in the following sequence of reactions:

(a) A $\xrightarrow{HNO_3}$ B $\xrightarrow{Sn/HCl}$ C $\xrightarrow[HCl\backslash 0°C]{NaNO_2}$ D $\xrightarrow{Na_2CO_3}$ $\xrightarrow{CuCN}$ NC–C₆H₄–OMe

(b) O₂N–C₆H₃(NO₂)–O–C₆H₄–NO₂ + H—N(piperidine) $\xrightarrow{OH_2}$ E + F

(c) H₂N–C₆H₄–NO₂ $\xrightarrow[Excess]{ICl}$ G $\xrightarrow[HCl\backslash 0°C]{NaNO_2}$ $\xrightarrow{KI\ (aq)}$ H

(d) CH₃CH₂O–C₆H₄–NH₂ $\xrightarrow{Br_2}$ I $\xrightarrow[HCl\backslash 0°C]{NaNO_2}$ J $\xrightarrow{H_3PO_2}$ K

(e) HO–C₆H₄–NH₂ $\xrightarrow{Br_2}$ $\xrightarrow[HCl\backslash 0°C]{NaNO_2}$ L $\xrightarrow{KI\ (aq)}$ M

(f) O₂N–C₆H₃(CH₃)–CH₃ $\xrightarrow{H_2/Ni}$ N $\xrightarrow[HCl\backslash 0°C]{NaNO_2}$ O $\xrightarrow{CuCN/heat}$ P $\xrightarrow{H_3O^+}$

57. Write structural formula for all the products in the following sequence of reaction:

(a) C₆H₅CHO $\xrightarrow[H_2SO_4]{HNO_3}$ A $\xrightarrow{Sn\text{-}HCl}$ B $\xrightarrow[HCl\backslash 0°C]{NaNO_2}$ C $\xrightarrow{CuCl/HCl}$ D

(b) [3,5-dimethoxyaniline (NH₂, MeO, OMe)] $\xrightarrow[H_2SO_4\ 0°C]{NaNO_2}$ E $\xrightarrow{H_2O/H_2SO_4}$ F

(c) [2,6-dinitrotoluene (O₂N, CH₃, NO₂)] $\xrightarrow{H_2/Ni}$ G $\xrightarrow[HCl\backslash 0°C]{NaNO_2}$ H $\xrightarrow{H_2O/H_2SO_4}$ I

(d) [anthranilic acid (COOH, NH₂)] $\xrightarrow[HCl\backslash 0°C]{NaNO_2}$ J $\xrightarrow{C_6H_5N(CH_3)_2}$ K

(e) C_6H_5—N(CH₃)—C_6H_5 $\xrightarrow[HCl\backslash 0°C]{NaNO_2}$ L

(f) Pyrrolidine (N–H) $\xrightarrow{CH_3I}$ M $\xrightarrow{KOH}$ N $\xrightarrow{CH_3I}$ O $\xrightarrow[\Delta]{KOH}$ P + Q

58. Complete the following series of reactions:

C_6H_5–CH_2–C(=O)–OH $\xrightarrow{2CH_3Li}$ $\xrightarrow{OH_2}$ A $\xrightarrow{LiAlH_4}$ $\xrightarrow{H_3O^+}$ B

B $\xrightarrow{TsCl}$ C $\xrightarrow{HCOOH}$ D $\xrightarrow{LiAlH_4}$ $\xrightarrow{H_3O^+}$ B

B $\xrightarrow{HBr}$ E, B $\xrightarrow{SOCl_2}$ F, B $\xrightarrow[NaOH/heat]{KMnO_4}$ G + H

59. Complete the following reaction showing their mechanism:

o-CH_3-C_6H_4-SO_2Cl $\xrightarrow{(NH_4)_2CO_3}$ A $\xrightarrow[NaOH/heat]{KMnO_4}$ B $\xrightarrow{ClH}$ (benzene ring fused with C(=O)–N–H–SO_2)

60. Arrange the following group of compounds in increasing order of their acid strength:

(a) H_3C–C_6H_4–COOH (I) H_3C–C_6H_4–NH_2 (II)

H_3C–C_6H_4–OH (III) C_6H_5–CH_2OH (IV)

(b) C_6H_5–OH (I) H_3C–C_6H_4–OH (II)

F–C_6H_4–OH (III) O_2N–C_6H_4–OH (IV)

(c) MeO–C_6H_4–COOH (I) O_2N–C_6H_4–COOH (II)

C_6H_5–COOH (III) Cl–C_6H_4–COOH (IV)

61. Bring about the following transformations:

(a) H_3C–C_6H_5 ⟶ F–C_6H_4–COOEt

(b) H_3C–C_6H_5 ⟶ (HO)$_2$$C_6H_3$–$CH_3$ (HO, HO, CH_3)

(c) C_6H_5–CH_2Br ⟶ C_6H_5–$CH_2CH(CH_2CH_3)CH_2NH_2$

(d) C_6H_5–Br ⟶ I–C_6H_4–Br

(e) C_6H_5–Cl ⟶ H_2N–C_6H_4–Cl

(f) C_6H_5–CH_3 ⟶ H_3C–N(CH_3)–C_6H_4–COOH

(g) C_6H_6 ⟶ (isobutyl)–C_6H_4–CH_2–C(=O)–OMe

62. For each of the following arrange them in increasing order of reactivity towards electrophilic substitution reaction:

(a) $C_6H_5CH_3$ (I), $C_6H_5CH_2F$ (II), $C_6H_5CF_3$ (III)

(b) $C_6H_5N^+(CH_3)_3$ (I), $C_6H_5CH_2N^+(CH_3)_3$ (II), $C_6H_5CH_2CH_2N^+(CH_3)_3$ (III)

(c) $C_6H_5OCH_2CH_3$ (I), $C_6H_5CH_2OCH_3$ (II), $C_6H_5COOCH_3$ (III)

(d) C_6H_6 (I), C_6H_5COOH (II), C_6H_5OH (III), $C_6H_5CH_2CH_2CH_3$ (IV)

(e) C_6H_6 (I), $C_6H_5C_2H_5$ (II), C_6H_5Cl (III), $C_6H_5NO_2$ (IV), $C_6H_5OCH_3$ (V)

(f) I: 1-Cl-2,4-(NO_2)$_2$$C_6H_3$; II: 1-OH-2,4-($NO_2$)$_2$$C_6H_3$; III: 1-$CH_3$-2,4-($NO_2$)$_2$$C_6H_3$

(g) I: $C_6H_5CH_3$; II: p-CH_3-C_6H_4-OH; III: p-CH_3-C_6H_4-CH_3; IV: C_6H_6

(h) I: p-NO_2-C_6H_4-CH_3; II: 1-CH_3-2-Cl-4-NO_2-C_6H_3; III: 1-CH_3-2,4-(NO_2)$_2$$C_6H_3$; IV: p-$CH_3$-$C_6H_4$-Cl

(i) 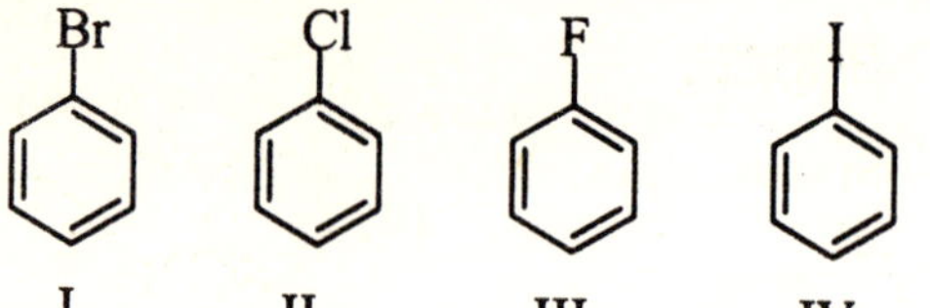 Br, Cl, F, I substituted benzenes: I, II, III, IV

63. Give product obtained from the reaction of each of the following compounds with $Br_2/FeCl_3$:

(a) PhO-COPh (b) $PhCH_2OPh$

(c) 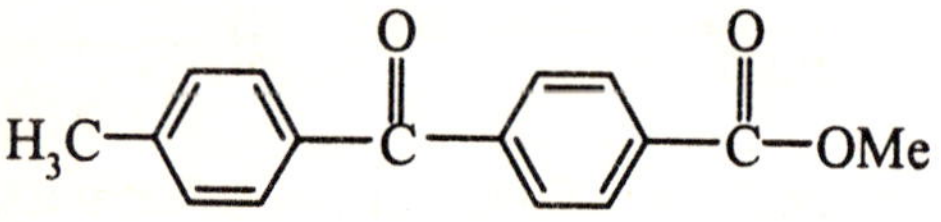 $H_3C-C_6H_4-CO-C_6H_4-COOMe$

(d) 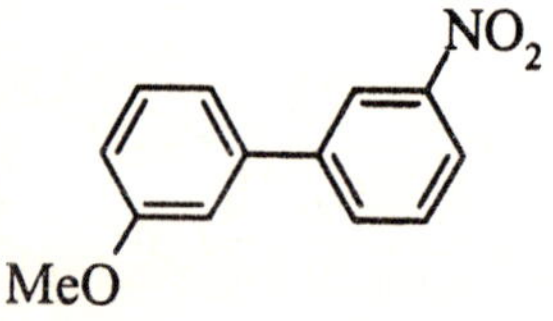 3-methoxy-3'-nitrobiphenyl (NO_2, MeO)

64. Show how the following compounds can be synthesized starting from benzene?

(a) *m*-chlorobenzene sulphonic acid
(b) *m*-chloro ethyl benzene
(c) Benzyl alcohol
(d) *m*-bromo benzonitrile
(e) *p*-nitrophenol
(f) Benzyl methyl ether
(g) *p*-benzyl chloro benzene
(h) 3-phenyl pentane
(i) *m*-hydroxy benzoic acid
(j) *o*-chloro anisole
(k) *p*-nitroaniline
(l) *m*-bromo iodo benzene
(m) *p*-dideutro benzene

65. Provide necessary reagents in the following road-map:

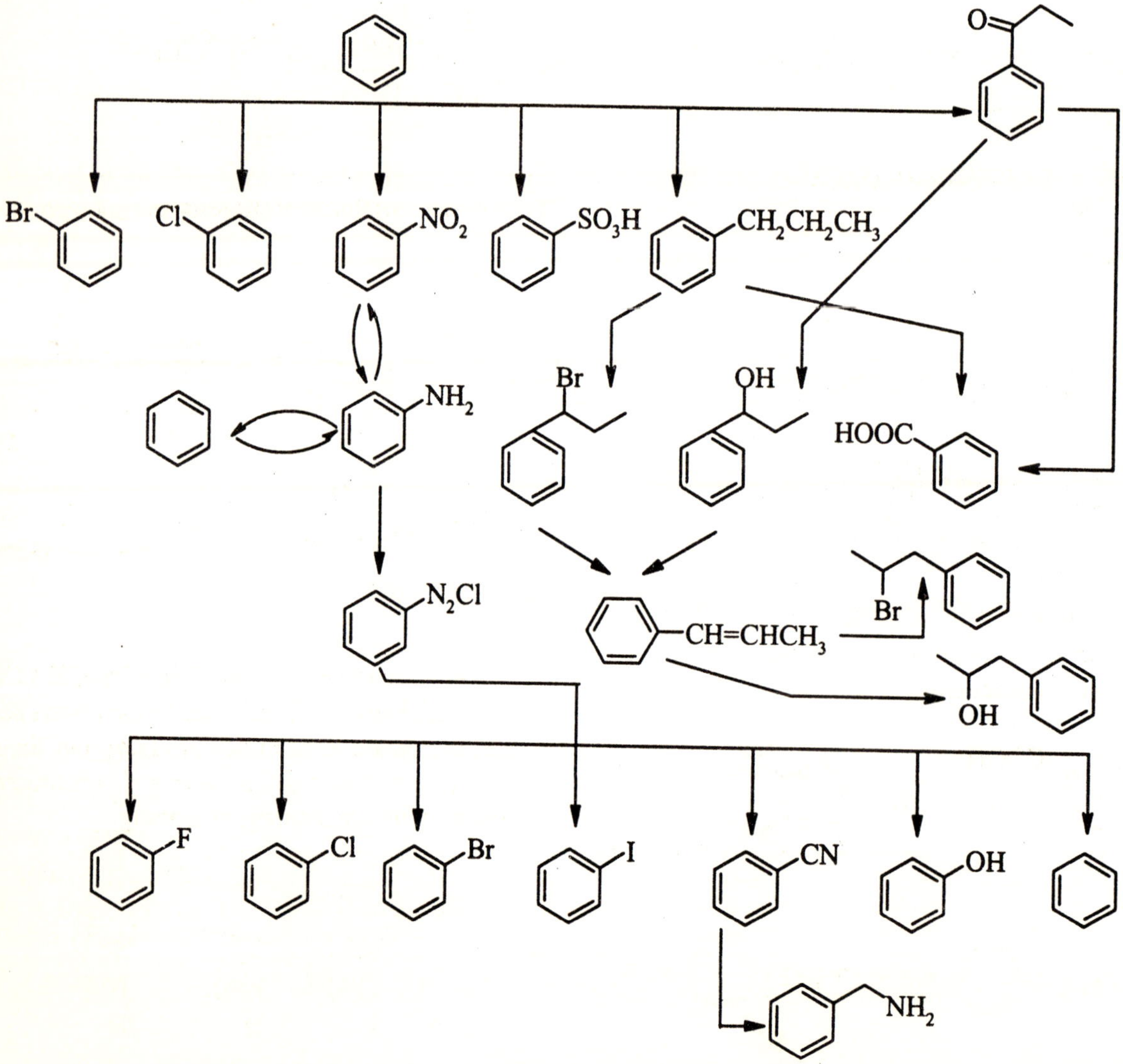

Br, Cl, NO_2, SO_3H, $CH_2CH_2CH_3$, NH_2, Br, OH, HOOC, N_2Cl, $CH=CHCH_3$, Br, OH, F, Cl, Br, I, CN, OH, NH_2

66. Propose mechanism of the following reactions:

(a) 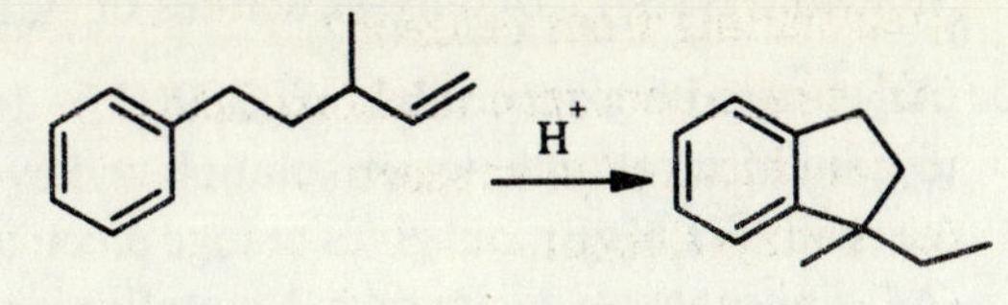 $\xrightarrow{H^+}$

(b) 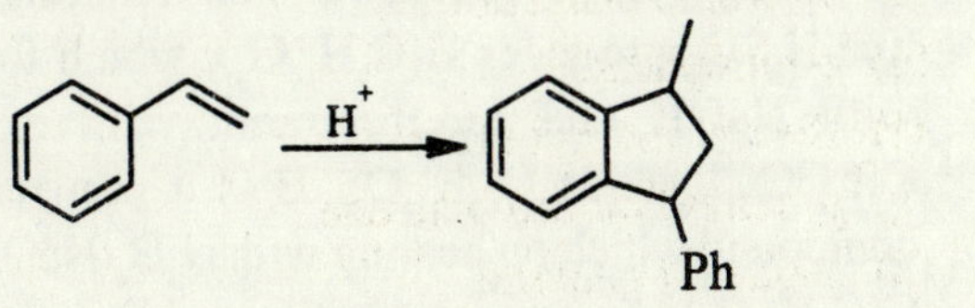 $\xrightarrow{H^+}$ Ph

NHCOCH$_3$ CH$_3$ (c)

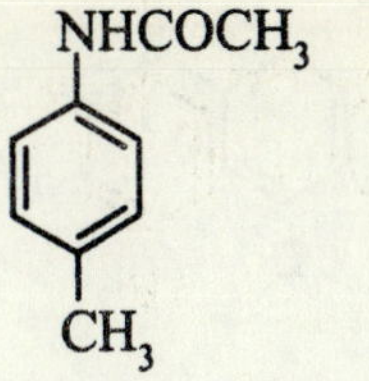

(d) 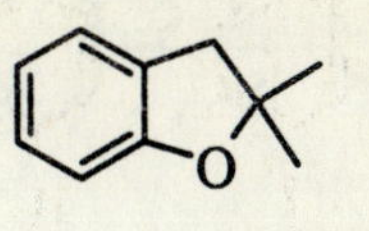 O

OH (e) 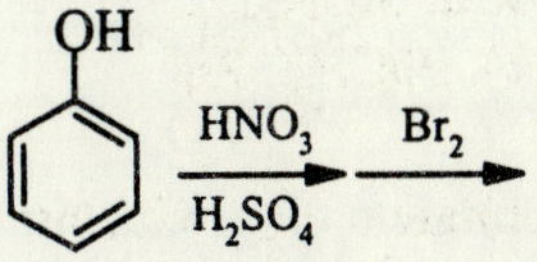 $\xrightarrow[H_2SO_4]{HNO_3}$ $\xrightarrow{Br_2}$

Br (f) 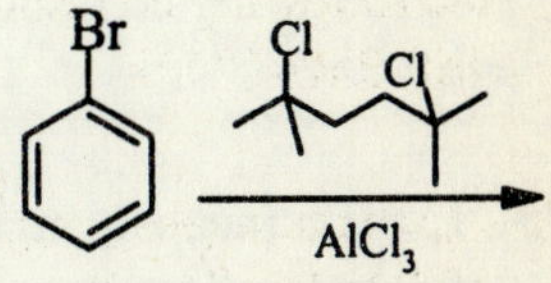 Cl Cl $\xrightarrow{AlCl_3}$

67. Show how the following compounds can be synthesized starting from benzene:

(a)

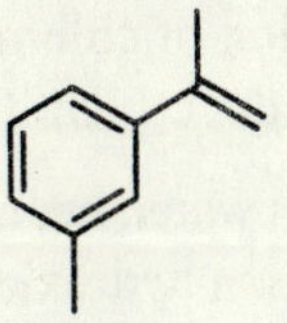

(b) 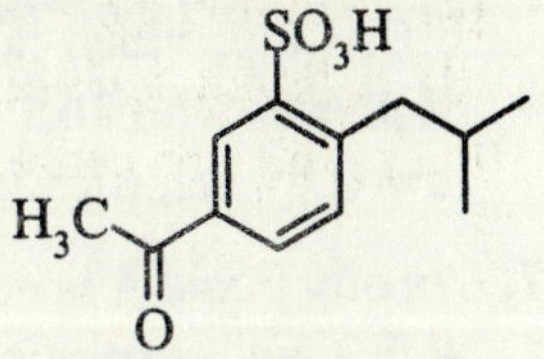 SO$_3$H H$_3$C O

(c) OMe Br NO$_2$

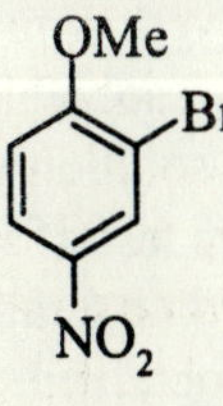

(d) 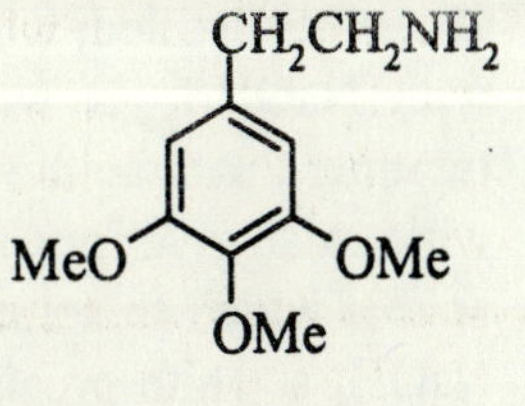 CH$_2$CH$_2$NH$_2$ MeO OMe OMe

68. Bring about the following transformations:

(a) 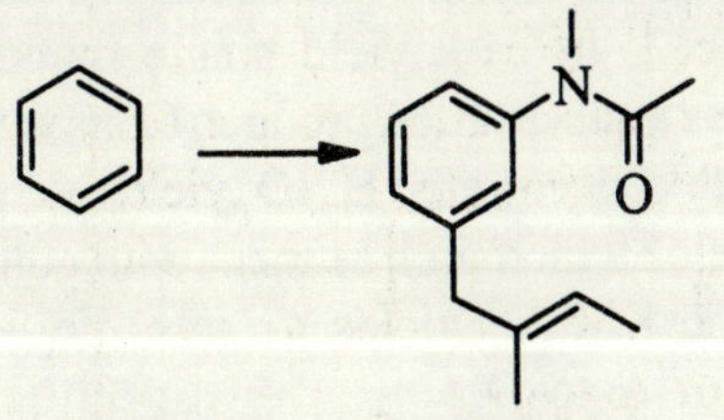 → N O

(b) 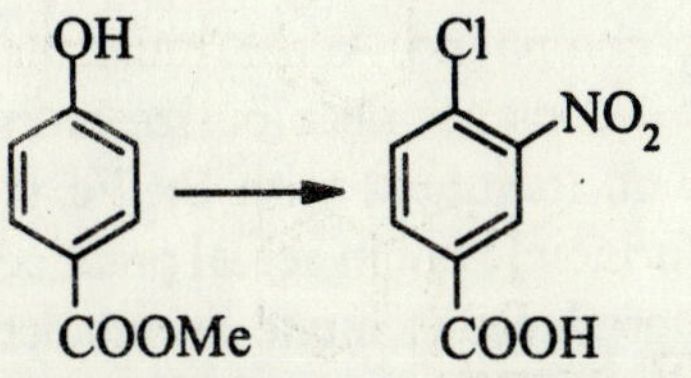 OH COOMe → Cl NO$_2$ COOH

69. Predict the major mono-nitration product in the following sets of reaction:

(a) 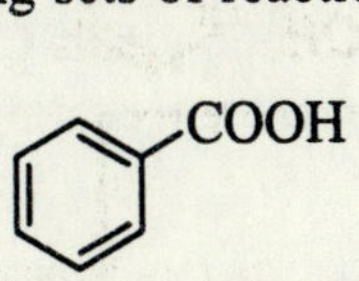 COOH

(b)

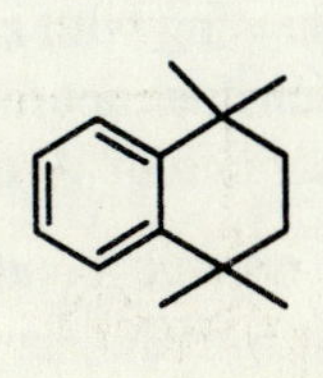

70. Write mechanism for the following reaction justifying position of substituents:

(a) 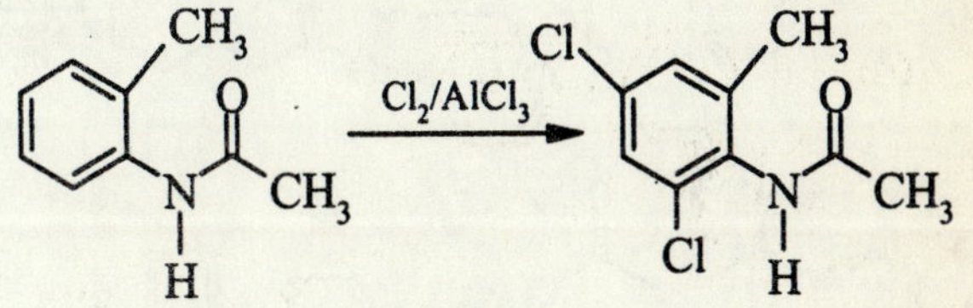 CH$_3$ O N H CH$_3$ $\xrightarrow{Cl_2/AlCl_3}$ Cl CH$_3$ O N H CH$_3$ Cl

(b) 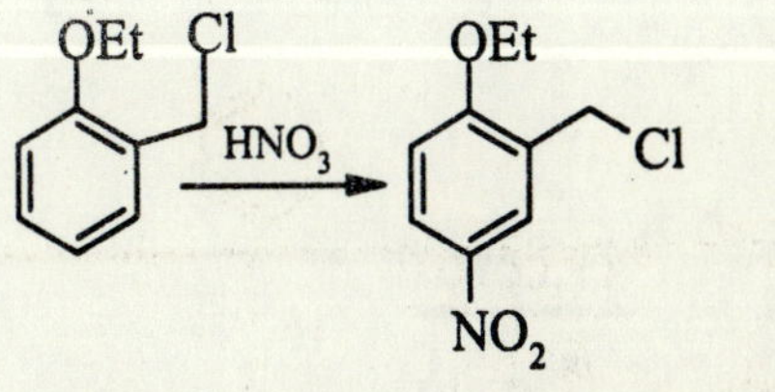 OEt Cl $\xrightarrow{HNO_3}$ OEt Cl NO$_2$

71. Provide structures of missing reagents/products:

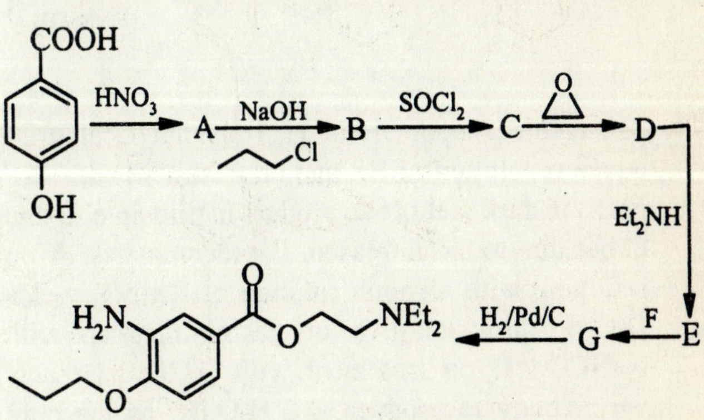

COOH OH $\xrightarrow{HNO_3}$ A $\xrightarrow[\text{Cl}]{NaOH}$ B $\xrightarrow{SOCl_2}$ C $\xrightarrow{\text{O (epoxide)}}$ D $\xrightarrow{Et_2NH}$ E $\xrightarrow{F}$ G $\xrightarrow{H_2/Pd/C}$ H$_2$N O O NEt$_2$ O

72. In Friedal craft acylation of C_6H_6 with $(CH_3)_3CCOCl$, expected ketone (A) and some tertiary butyl benzene (B) are obtained as minor products, but the major product of this reaction was a disubstituted (C). Justify with mechanistic reasoning.

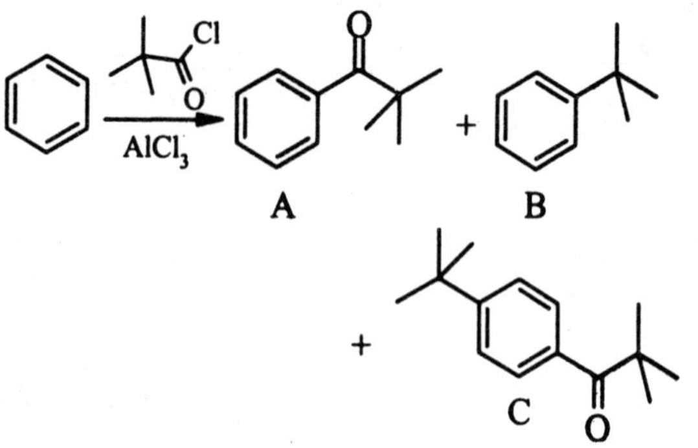

$AlCl_3$ A + B + C

73. Explain mechanism of formation of product in the following reactions:

(a) 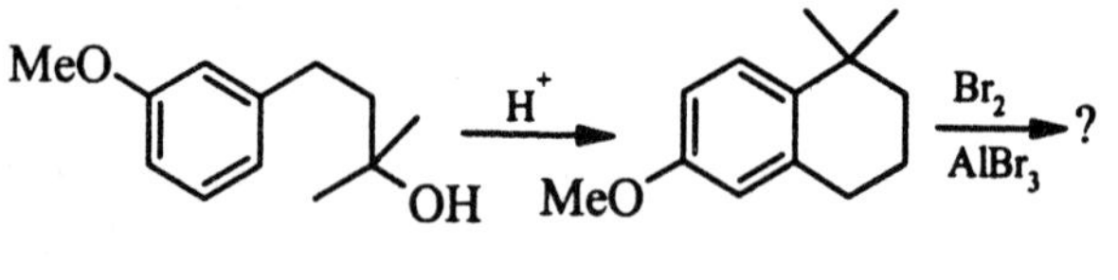 MeO ... OH $\xrightarrow{H^+}$ MeO ... $\xrightarrow[AlBr_3]{Br_2}$?

(b) 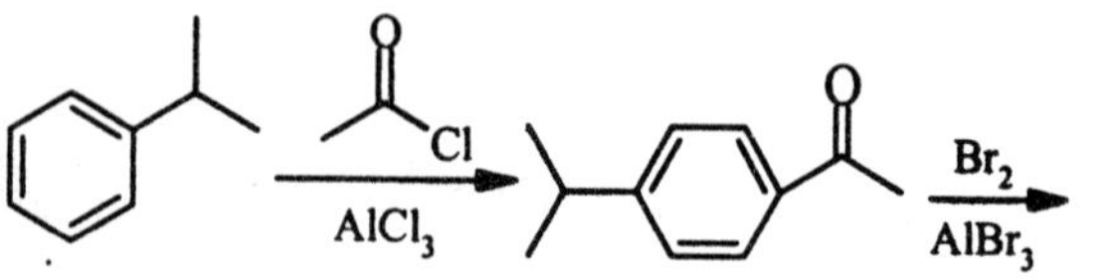 $\xrightarrow{AlCl_3}$ ... $\xrightarrow[AlBr_3]{Br_2}$

(c) 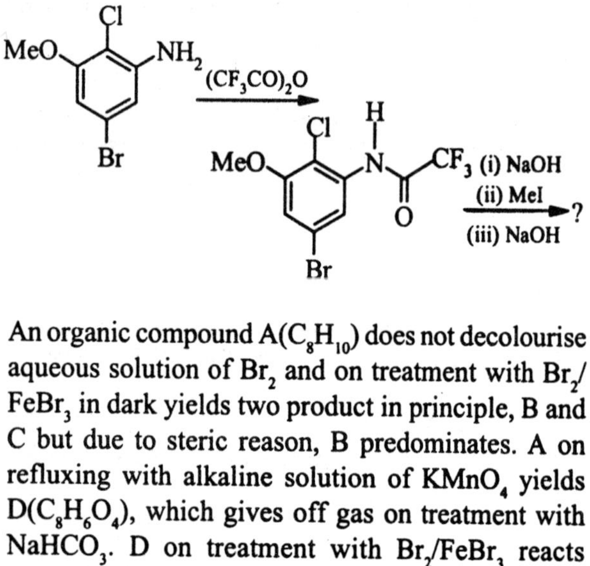 MeO, Cl, NH_2, Br $\xrightarrow{(CF_3CO)_2O}$ MeO, Cl, N–H, CF_3, O, Br $\xrightarrow{\text{(i) NaOH (ii) MeI (iii) NaOH}}$?

74. An organic compound $A(C_8H_{10})$ does not decolourise aqueous solution of Br_2 and on treatment with $Br_2/FeBr_3$ in dark yields two product in principle, B and C but due to steric reason, B predominates. A on refluxing with alkaline solution of $KMnO_4$ yields $D(C_8H_6O_4)$, which gives off gas on treatment with $NaHCO_3$. D on treatment with $Br_2/FeBr_3$ reacts very slowly to produce $E(C_8H_5O_4Br)$ as the only isomer. Also D on treatment with excess of $SOCl_2$ followed by work-up with benzene solution of $AlCl_3$ produces $F(C_{14}H_8O_2)$, which does not yields any gas with $NaHCO_3$. F on heating with N_2H_4 produces $G(C_{14}H_8N_2)$. Deduce structures of A to G.

75. An organic compound A $(C_9H_{10}O_2)$ does not change colour of Br_2–H_2O and produces no gas with NaH. A on treatment with Br_2 in presence of $FeBr_3$ yields $B(C_9H_9O_2Br)$ as only isomer. Also A gives orange precipitate with 2, 4-dinitrophenyl hydrazine and produces a resolvable compound $C(C_9H_{12}O_2)$ on reduction with $NaBH_4$. Deduce structures of A, B and C.

76. An organic compound $A(C_9H_8O_2)$ does not decolourise bromine water solution and evolves no gas with CH_3MgBr, but gives orange precipitate with 2,4-dinitrophenyl hydrazine. A on refluxing with dilute H_2SO_4 produces $B(C_9H_{10}O_3)$, which forms salt with NaOH and on treatment with excess of CH_3COCl yields $C_{13}H_{14}O_5$. B is a non resolvable compound which on heating with N_2H_4/NaOH yields $C(C_9H_{12}O_2)$. C on dehydrating with concentrated H_3PO_4 yields $D(C_9H_{10}O)$ as major product. D on ozonolysis followed by work-up with $(CH_3)_2S$ yields $E(C_7H_6O_2)$, which can also be obtained by the action of phenol with alkaline solution of chloroform followed by acidification of product. Identify A to E.

77. An oily liquid A is insoluble in water, but on heating with aqueous solution of sodium hydroxide for one hour, it dissolves. From the reaction mixture, a liquid B can be distilled, which gives a yellow precipitate with NaOH/I_2 as well as it is resolvable. B on treatment with acidic dichromate solution yields C, which also gives positive iodoform test. If sulphuric acid is added to solution obtained on heating with NaOH, a white precipitate D is obtained. D gives effervescence with $NaHCO_3$ and heating D with soda lime converts it into toluene. Also A on treatment with $Br_2/FeCl_3$ in dark yield single mono bromo derivative as a substitution product. Compound B on heating with concentrated H_2SO_4 yields stereomeric alkene, one of which on treatment with cold, dilute and alkaline $KMnO_4$ yields a meso diol. Deduce structures of A to D.

78. An organic compound A has molecular formula $C_{10}H_{14}$, and it does not decolourize bromine water solution. A on treatment with Br_2/Fe yields three products in principle but in actual practice, only two of these are produced as mono-bromo derivative. A on heating with Br_2 yields B $(C_{10}H_{13}Br)$ as the monobromo derivative which is optically inactive. A on heating with alkaline $KMnO_4$ yields $C(C_8H_6O_4)$, which does not forms anhydride on heating. Deduce structures of A to C.

79. An organic compound $A(C_8H_8O_3)$ evolves a gas with $NaHCO_3$ as well as forms salt with NaOH. A on treatment with HI yields $B(C_7H_6O_3)$, which when treated with excess of acetyl chloride (CH_3COCl) yields $C(C_{11}H_{10}O_5)$. B on treatment with $Br_2/FeBr_3$

in CCl_4 should yield three products D, E and F in principle, but only two products E and F are produced in reality and E is the major product. Deduce structures of A to E.

80. An oily liquid is insoluble in water, but on heating with aqueous solution of sodium hydroxide for one hour, it dissolves. From the reaction mixture, a liquid B can be distilled, which gives a yellow precipitate with iodine and NaOH solution. On careful oxidation, B gives an aldehyde C, which also gives positive iodoform test. If sulphuric acid solution is added to the solution obtained from heating A with NaOH, a white crystal D is obtained. D liberates a gas with $NaHCO_3$. Heating D with NaOH/CaO, converts it into C_6H_6. Identify A to D.

81. Complete the following reactions:

(a) catechol (o-dihydroxybenzene) $\xrightarrow{NaOH}$ A $\xrightarrow{(ClCH_2CH_2)_2O}$ B

(b) p-nitrophenol $\xrightarrow[EtBr]{NaOH}$ A $\xrightarrow{Sn/HCl}$ B $\xrightarrow[HCl/0°C]{NaNO_2}$ C $\xrightarrow{PhOH}$

D $\xrightarrow[NaOH]{(C_2H_5)_2SO_4}$ E $\xrightarrow{Sn/HCl}$ F $(C_8H_{11}ON)$ $\xrightarrow{CH_3COCl}$ $(C_{10}H_{13}O_2N)$ G

(c) 2-(2-hydroxyethyl)phenol $\xrightarrow{BrH}$ A $\xrightarrow{KOH}$ B (C_8H_8O)

(d) phenol + $ClCH_2COOH$ $\xrightarrow{NaOH \quad ClH}$ A$(C_8H_8O_3)$

$\xrightarrow{SOCl_2}$ B $(C_8H_7O_2Cl)$ $\xrightarrow{AlCl_3}$ $C_8H_6O_2$(C)
Insoluble in NaOH

82. Complete the following reactions:

(a) toluene (C_6H_5–CH_3) + Conc. H_2SO_4 ⟶ A + B

(b) MeO–C_6H_4–CH=CH–CH_3 (p-methoxy) $\xrightarrow{BrH}$ A $(C_{10}H_{13}OBr)$ $\xrightarrow{Mg}$ B

(c) catechol (o-dihydroxybenzene) $\xrightarrow[KOH]{CHCl_3}$ C$(C_7H_6O_3)$ $\xrightarrow{NaOH \quad CH_2I_2}$

D$(C_8H_6O_3)$ $\xrightarrow[NaOH]{CH_3CHO}$ E$(C_{10}H_8O_3)$ $\xrightarrow{(CH_3CO)_2O}$ F $(C_{12}H_{10}O_4)$

83. Complete the following reaction sequence identifying A to G:

p-toluic acid (CH_3–C_6H_4–COOH) + Conc. H_2SO_4 ⟶ A $\xrightarrow[KOH]{Fusion}$ B $(C_8H_8O_3)$

$\xrightarrow[EtOH]{Na}$ C$(C_8H_{14}O_3)$ $\xrightarrow{HBr}$ D$(C_8H_{13}O_2Br)$ $\xrightarrow{NaOH \ \Delta}$ E$(C_8H_{12}O_2)$ ⟶ F$(C_{10}H_{16}O_2)$ $\xrightarrow{CH_3MgCl}$ $\xrightarrow{H_2O}$ G$(C_{10}H_{18}O)$

84. Identify products in the following reactions:

(a) hydroquinone (HO–C_6H_4–OH) + $(CH_3)_2SO_4$ $\xrightarrow{KOH}$ A $\xrightarrow[AlCl_3]{(CH_3)_3C\text{–}Cl}$ B $\xrightarrow[H_2SO_4]{HNO_3}$ C

(b) p-toluidine (NH_2–C_6H_4–CH_3) + Conc. H_2SO_4 $\xrightarrow{Heat}$ A $\xrightarrow[0°C]{HNO_2}$ B

$\xrightarrow{H_3PO_4}$ C $\xrightarrow{KOH}$ $\xrightarrow[Heat]{H^+}$ D

(c) $C_6H_5CH{=}CHCOOH \xrightarrow{H_2/Ni} A \xrightarrow{SOCl_2} B \xrightarrow{FH} C$

85. Identify products in the following sequence of reaction:

Benzene + succinic anhydride $\xrightarrow{AlCl_3} A \xrightarrow[HCl]{Zn\text{-}Hg} B \xrightarrow{SOCl_2} C \xrightarrow{AlCl_3} D(C_{10}H_{10}O) \xrightarrow[HCl]{Zn\text{-}Hg} E(C_{10}H_{12}O) \xrightarrow[Heat]{H_2SO_4} (C_{10}H_{10})F$

86. Complete the following reactions:

(a) Phthalic anhydride $\xrightarrow[Toluene]{AlCl_3} A \xrightarrow{H_2SO_4} B$

(b) α-Tetralone $\xrightarrow[OH_2]{PhMgBr} A \xrightarrow[Heat]{H^+} B \xrightarrow{H_2/Ni} C$

(c) $C_6H_5COCH_2CH_3 \xrightarrow[Heat]{Br_2} A \xrightarrow{NaBH_4} B \xrightarrow{CaO} C$

(d) $C_6H_5C{\equiv}C{-}CH_3 \xrightarrow[H_2SO_4]{HgSO_4} A \xrightarrow{NH_2OH} B \xrightarrow{H^+} C + D$

87. Identify products in the following sequence of reaction:

(a) $I{-}C_6H_4{-}CH_3 \xrightarrow[Heat]{Cu} A$

(b) Anisole (C₆H₅OMe) + succinic anhydride $\xrightarrow{AlCl_3} A \xrightarrow[HCl]{Zn\text{-}Hg} B \xrightarrow{H_3PO_4} C \xrightarrow{Zn\text{-}Hg \mid HCl} D$

88. Bring about the following transformation in good yield:

(a) $Cl{-}C_6H_4{-}Cl \longrightarrow Cl{-}C_6H_4{-}COOH$

(b) Benzene $\longrightarrow$ $Br{-}C_6H_4{-}COCH_2Br$

(c) Benzene $\longrightarrow$ $Br{-}C_6H_4{-}NO_2$

(d) $C_6H_5CH_3 \longrightarrow$ 2,4-dinitrobenzoyl chloride (COCl, NO_2, NO_2)

89. Identify products in the following reactions:

(a) Toluene + isobutylene $\longrightarrow A \xrightarrow{CH_3Cl/AlCl_3} B$

(b) Nitrobenzene $\xrightarrow{Cl_2/FeCl_3} A \xrightarrow{Sn/HCl} B \xrightarrow[(ii)\ NaOH]{(i)\ HCl/heat} C$

(c) Toluene $\xrightarrow{A}$ m-$CH_3C_6H_4SO_2Cl \xrightarrow{NH_3} B \xrightarrow{KMnO_4} C$

90. Bring about the following transformations:

(a) Benzene $\longrightarrow$ 1,3,5-tribromobenzene (Br, Br, Br)

(b) Benzene $\longrightarrow$ 1,3-difluorobenzene (F, F)

(c) Benzene $\longrightarrow$ 2,4-dinitrotoluene (CH_3, O_2N, NO_2)

(d) Benzene $\longrightarrow$ 1-chloro-3-fluorobenzene (Cl, F)

(e) Benzene $\longrightarrow$ 3,5-dihydroxytoluene (CH_3, HO, OH)

91. Convert:

(a) 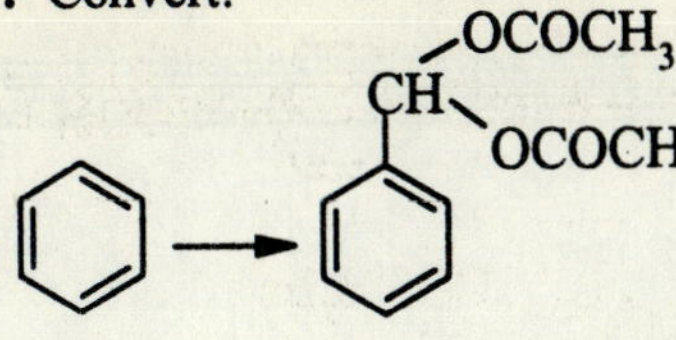 Benzene → $C_6H_5CH(OCOCH_3)_2$

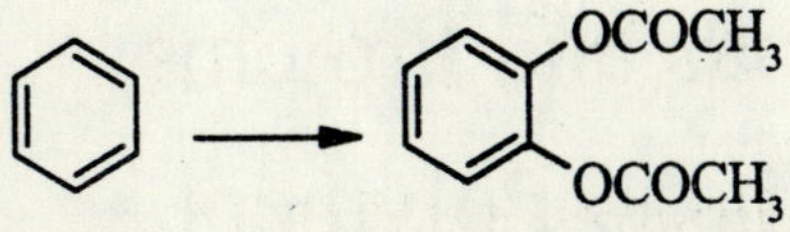

(b) Benzene → 1,2-$C_6H_4(OCOCH_3)_2$

(c) Benzene → 2-($COOCH_3$)C_6H_4COOH

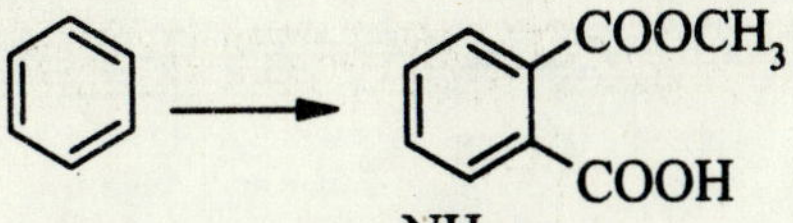

(d) 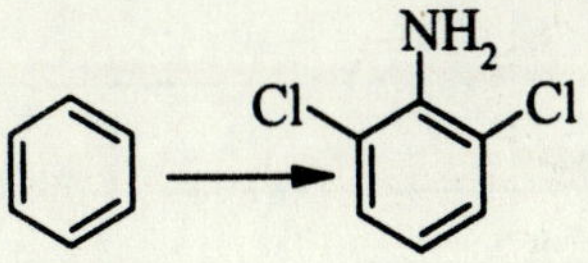 Benzene → 2,6-dichloroaniline (NH_2, Cl, Cl)

(e) 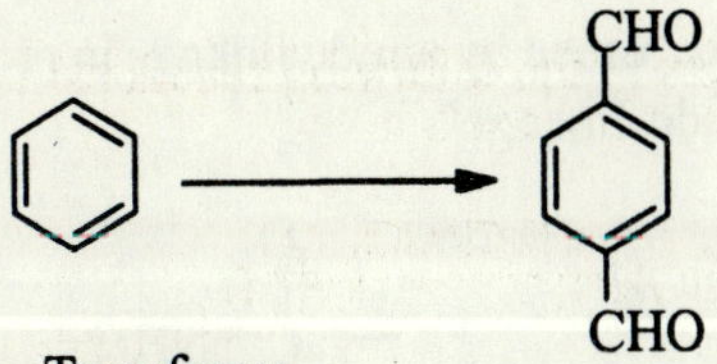 Benzene → 1,4-$C_6H_4(CHO)_2$

92. Transform:

(a) 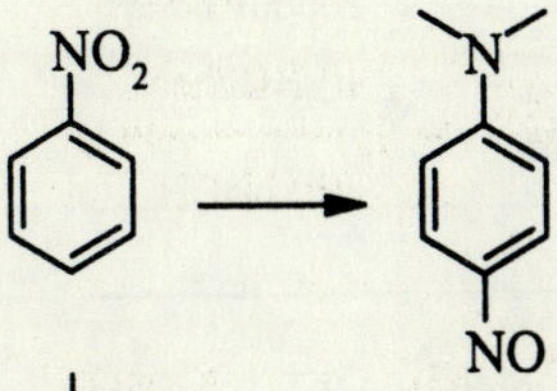 Nitrobenzene (NO_2) → 4-nitroso-N,N-dimethylaniline (NO)

(b) 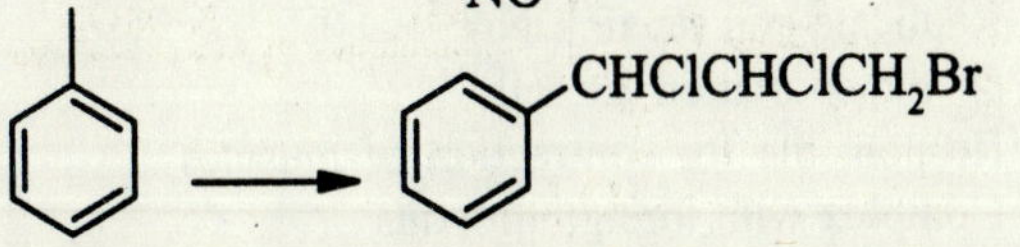 Toluene → $C_6H_5CHClCHClCH_2Br$

(c) Benzene → naphthalene

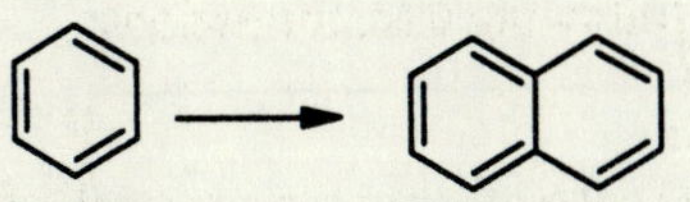

(d) 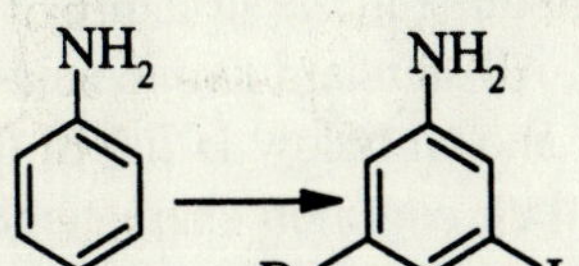 Aniline (NH_2) → 3-bromo-5-iodoaniline (NH_2, Br, I)

(e) 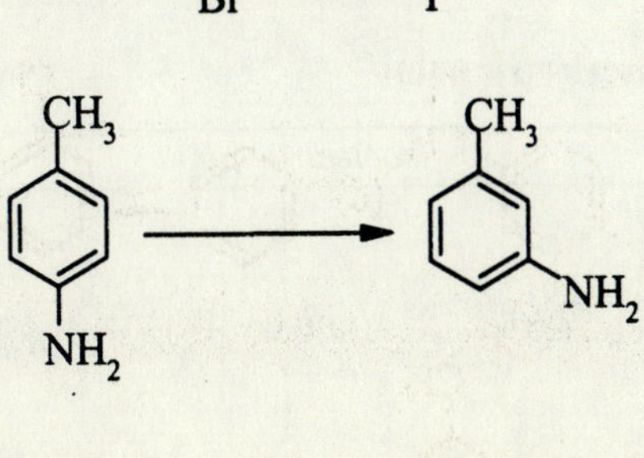 p-Toluidine (CH_3, NH_2) → m-toluidine (CH_3, NH_2)

93. Convert:

(a) 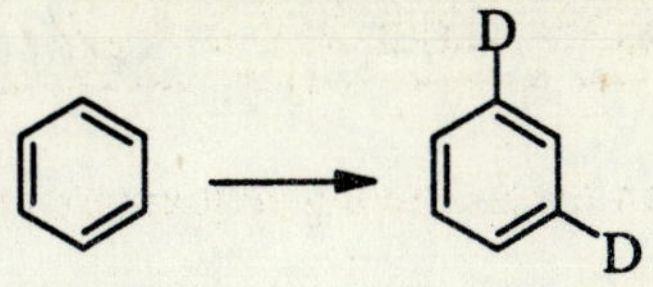 Benzene → 1,3-dideuteriobenzene (D, D)

(b) Phenol (OH) → $C_6H_5NHCOCH_3$

(c) Toluene → 4-(4-tert-butyl-2,6-dimethylphenylazo)-N,N-dimethylaniline (N=N)

(d) Aniline (NH_2) → 3-bromo-5-iodoaniline (NH_2, Br, I)

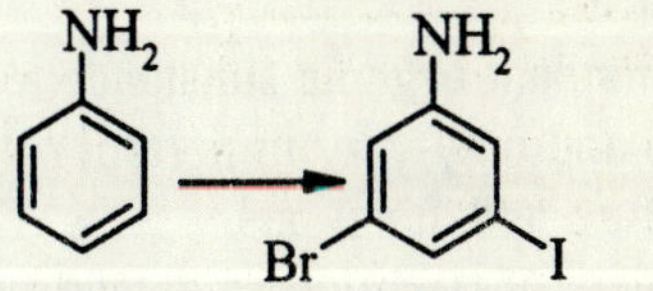

(e) 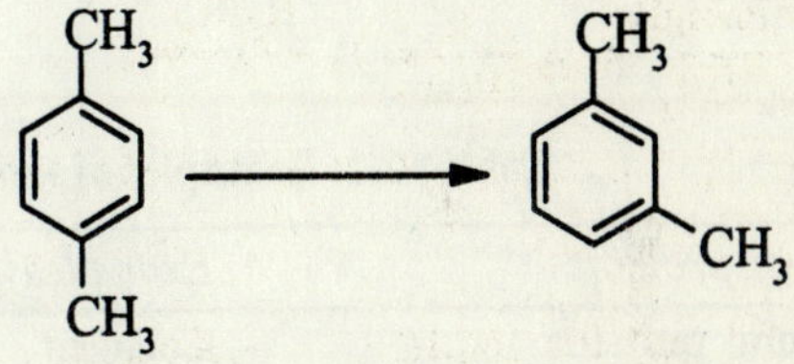 p-Xylene (CH_3, CH_3) → m-xylene (CH_3, CH_3)

(f) 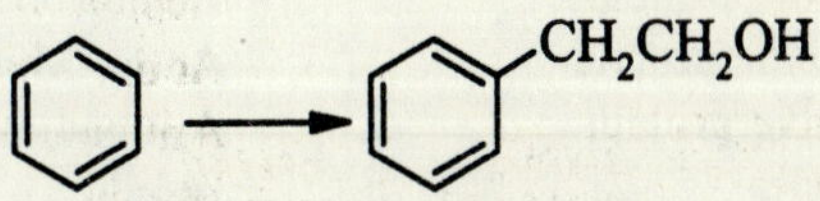 Benzene → $C_6H_5CH_2CH_2OH$

(g) 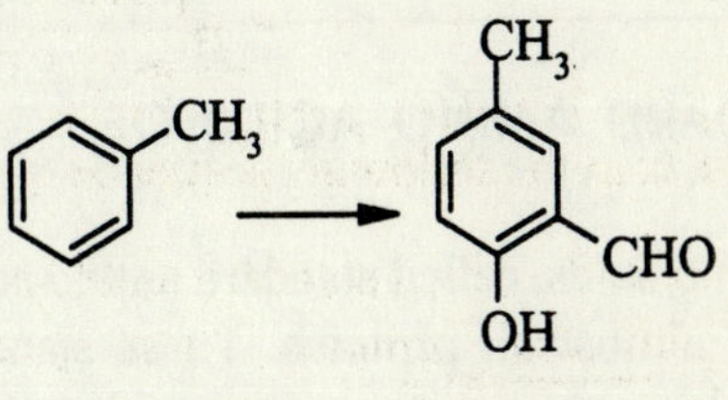 Toluene (CH_3) → 2-hydroxy-5-methylbenzaldehyde (CH_3, CHO, OH)

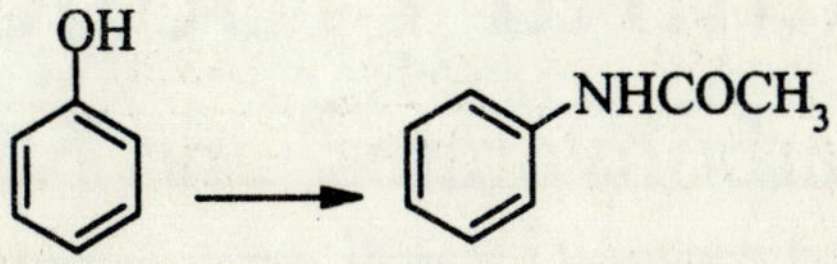

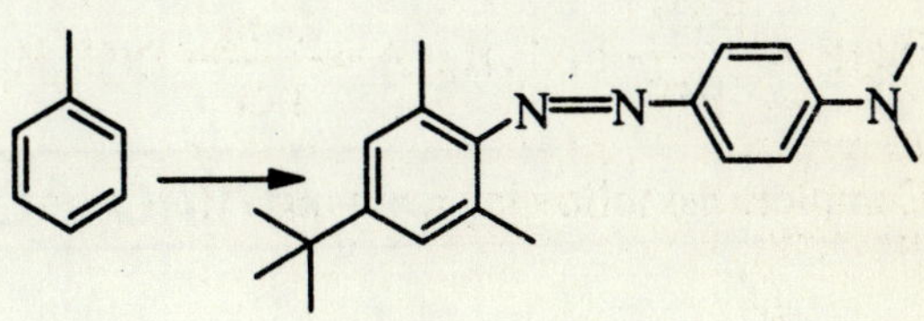

(h) 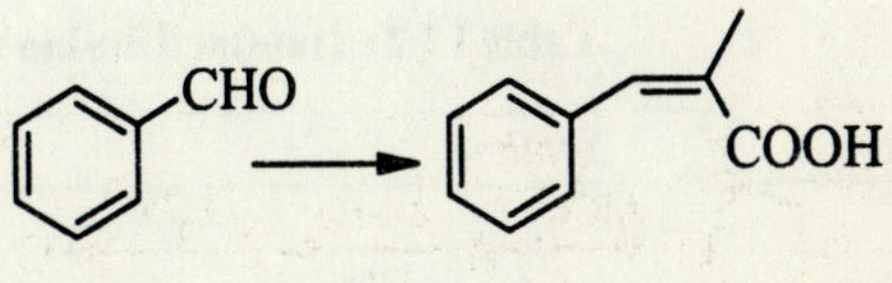 Benzaldehyde (CHO) → $C_6H_5CH=C(CH_3)COOH$ (COOH)

Chapter 12

Natural Products Amino Acids and Proteins, Carbohydrates, and Natural and Synthetic Polymers

Amino Acids and Proteins

12.1 INTRODUCTION

Proteins are the most abundant organic substance available in the living beings and play very important role in all aspects of cell structure and functions.

Proteins are biopolymers of α-amino acids and properties of a protein largely depend on constituent amino acids. Amino acids are joined by amide linkage in protein and known as "Peptide linkage".

$$H_2N-\underset{\displaystyle R}{\underset{|}{CH}}-COOH$$

An α-amino acid

$$-\!\!\!\!+\overset{\displaystyle O}{\overset{\|}{C}}-NH\!+\!\!\!\!-$$

Amide linkage
"Peptide bond"

Table 12.1: Examples of some proteins and their biological functions

Class of proteins	*Examples*	*Functions*
Structural proteins	Collagen	Connective tissue
Enzymes	DNA polymerase	Replicate and repair DNA
Transport proteins	Hemoglobin	O_2-transportation to cells
Contractile protein	Actin, Myosin	Responsible for contraction of muscles
Protective protein	Antibodies	Complex with foreign proteins
Hormones	Insulin	Regulate glucose metabolism
Toxins	Snake venom	Incapcitate prey

12.2 THE STANDARD AMINO ACIDS OF PROTEINS

There are 20 α-amino acids, called standard amino acids, that are present in almost all proteins. These standard amino acids differ from each other in the structure of side chains bonded to α-carbon. All standard amino acids are L-amino acids. Table 12.2 shown below is list of these amino acids with some of their common characteristics:

Table 12.2: Standard amino acids, their structures and isoelectric point

Name	*Symbol*	*Structure*	*Isoelectric point*
Glycine	G	$H_2N-\underset{\boxed{H}}{\underset{|}{CH}}-COOH$	6.0
Alanine	A	$H_2N-\underset{\boxed{CH_3}}{\underset{|}{CH}}-COOH$	6.0

(Contd.)

Name	*Symbol*	*Structure*	*Isoelectric point*
Valine	**V**	$H_2N-CH-COOH$; side chain: $CH(CH_3)_2$ [H_3C–CH–CH_3]	6.0
Leucine	**L**	$H_2N-CH-COOH$; side chain: $CH_2-CH(CH_3)-CH_3$	6.0
Isoleucine	**I**	$H_2N-CH-COOH$; side chain: $H_3C-CH-CH_2-CH_3$	6.0
Phenyl alanine	**F**	$H_2N-CH-COOH$; side chain: $CH_2-C_6H_5$	5.5
Proline	**P**	$HN-CH-COOH$; ring: H_2C, CH_2, CH_2	6.3
Serine	**S**	$H_2N-CH-COOH$; side chain: CH_2OH	5.7
Threonine	**T**	$H_2N-CH-COOH$; side chain: $H_3C-CH-OH$	5.6
Tyrosine	**Y**	$H_2N-CH-COOH$; side chain: $CH_2-C_6H_4-OH$	5.7
Asparagines	**N**	$H_2N-CH-COOH$; side chain: $CH_2-C(=O)-NH_2$	5.4
Glutamine	**Q**	$H_2N-CH-COOH$; side chain: $CH_2-CH_2-C(=O)-NH_2$	5.7

(*Contd.*)

Name	Symbol	Structure	Isoelectric point
Tryptophan	W	$H_2N-CH(COOH)-CH_2$–(indol-3-yl)	5.9
Cysteine	C	$H_2N-CH(COOH)-CH_2SH$	5.0
Methionine	M	$H_2N-CH(COOH)-CH_2CH_2-SCH_3$	5.7
Aspartic acid	D	$H_2N-CH(COOH)-CH_2COOH$	2.8
Glutamic acid	E	$H_2N-CH(COOH)-CH_2CH_2COOH$	3.2
Lysine	K	$H_2N-CH(COOH)-CH_2CH_2CH_2CH_2NH_2$	9.7
Arginine	R	$H_2N-CH(COOH)-CH_2CH_2CH_2NH-C(=NH)-NH_2$	10.8
Histidine	H	$H_2N-CH(COOH)-CH_2$–(imidazol-4-yl)	7.6

12.3 ESSENTIAL AMINO ACIDS

Human beings can synthesize about ten standard amino acids required for the synthesis of proteins. Other ten amino acids called *essential amino acids* are supplied in the diet. The essential amino acids are:

(i) Arginine
(ii) Valine
(iii) Methionine
(iv) Leucine
(v) Threonine
(vi) Phenylalanine
(vii) Histidine
(viii) Isoleucine
(ix) Lysine
(x) Tryptophan

12.4 COMPLETE AND INCOMPLETE PROTEINS

Proteins that provides all the essential amino acids in about right proportion as required for the human nutrition, are called complete protein. Some of the very common complete proteins are meat, fish, milk, eggs, etc.

Proteins, that are severely deficient in one or more of the essential amino acids are called incomplete proteins.

12.5 AMINO ACIDS AS ACID AND BASE

Writing amino acid with an intact carboxyl group (—COOH) and an amino group (—NH_2) is not always appropriate. Actual structure of an amino acid is ionic, which has been revealed by high electrical conductivity of its aqueous solution. In fact, the acidic carboxyl group loses it proton, giving a carboxylate ion and the amino group is protonated to an ammonium ion. Since both, negatively charged carboxylate ion (—CO_2^-), and positively charged ammonium ion (—NH_3^+) are present in the same formula unit, this structure is called a "dipolar ion" or "Zwitterion".

$$\underset{\text{Neutral amino acid}}{H_2N-\underset{\displaystyle R}{\underset{|}{CH}}-COOH} \rightleftharpoons \underset{\text{Dipolar ion "Zwitterion"}}{H_3\overset{+}{N}-\underset{\displaystyle R}{\underset{|}{CH}}-CO\bar{O}}$$

Therefore in solid state, amino acids predominantly remains in the form of Zwitterions, which gives them some unusual physical and chemical behaviour as:

I. *Melting points:* Amino acids behaves like ionic compounds (salts) in solid state and a large interionic attraction (electrostatic attraction) gives them high melting points compared to normal covalent compounds, viz. glycine melts at 262°C.
Besides the interionic attractions, intermolecular hydrogen bonding between the amino acid molecules further add-up to the melting point.

II. *Dipole moments:* Amino acids have much higher dipole moment compared to simple amine or acids. This is again due to its existence in the form of Zwitterion.

III. *Solubility:* Amino acids have very high solubility in water but less soluble in ether, carbon tetrachloride and other common organic solvents.

IV. Amino acids are less acidic than most of the carboxylic acids and less basic than most of the amines. Due to their existence as Zwitterions, the acidic moiety of amino acid is —NH_3^+ and not —COOH, and basic part is —COO^-, not —NH_2.

Due to presence of both acidic and basic groups, amino acids are "amphoteric", reacts with strong acids as well as strong base. The predominant form of an amino acid in its aqueous solution is determined by pH of the solution. In strongly basic medium, —NH_3^+ group is deprotonated into free —NH_2 group and the molecule possess an overall negative charge. In strongly acidic solution, the —COO^- group is protonated and the molecule acquire an overall positive charge:

$$H_2N-\underset{\displaystyle R}{\underset{|}{CH}}-CO\bar{O} \underset{\text{High pH}}{\overset{^-OH}{\rightleftharpoons}} \underset{\text{Dipolar ion Zwitterion}}{H_3\overset{+}{N}-\underset{\displaystyle R}{\underset{|}{CH}}-CO\bar{O}}$$

$$\xrightarrow[\text{Low pH}]{H^+} H_3\overset{+}{N}-\underset{\displaystyle R}{\underset{|}{CH}}-COOH$$

12.6 ISOELECTRIC POINTS AND ELECTROPHORESIS

Let us consider an amino acid dissolved in highly acidic solution, where all amino acid are in cationic form. As we add a base to the above solution following neutralization start to take place:

$$H_3\overset{+}{N}-\underset{\displaystyle R}{\underset{|}{CH}}-COOH \overset{^-OH}{\rightleftharpoons} H_3\overset{+}{N}-\underset{\displaystyle R}{\underset{|}{CH}}-CO\bar{O} + H_2O$$

On adding further base, the following neutralization start to take place:

$$H_3\overset{+}{N}-\underset{\displaystyle R}{\underset{|}{CH}}-COO^- \overset{^-OH}{\rightleftharpoons} H_2N-\underset{\displaystyle R}{\underset{|}{CH}}-CO\bar{O} + H_2O$$

Thus, we observe here that, as we increase the pH of this solution, cationic form is gradually converted into Zwitterionic form and finally into anionic form if excess of base is added.

Isoelectric point is some intermediate pH, where the two forms, cationic and anionic forms, of amino acid is evenly balanced as dipolar ion with a net charge of zero. Therefore, at isoelectric point, amino acid completely exist as Zwitterions.

Isoelectric point of various amino acids depends on the nature of other functional groups present. Amino acids having additional acidic functional group have lower isoelectric point.

Aspartic acid: isoelectric pH = 2.8
Glutamic acid: isoelectric pH = 3.2

Amino acids having basic functional group have higher isoelectric pH, e.g.

Lysine: isoelectric pH = 9.7
Arginine: isoelectric pH = 10.8
Histadine: isoelectric pH = 7.6

Electrophoresis: Electrophoresis is an experimental method of separating amino acids from their mixture. Difference in isoelectric pH of amino acids are exploited in this method. A streak of the amino acid mixture is placed in the center of a layer of acrylamide gel. The two electrodes are placed in contact with the edges of gel and a potential difference of around 1000 V is applied across the electrodes. Positively charged amino acids are attracted towards negative electrode (cathode) and negatively charged amino acids are attracted towards anode. An amino acid at its isoelectric point, has no net charge and it does not move.

12.7 SYNTHESIS OF α-AMINO ACIDS

I. By Reductive Amination

When a α-keto acid is treated with ammonia, imine is formed, which on selective reduction with H_2/Pd-C yields α-amino acid:

$$R-\underset{\|}{\overset{O}{C}}-COOH + NH_3(\text{excess}) \longrightarrow R-\overset{NH}{\overset{\|}{C}}-COONH_4 \xrightarrow{H_2/Pd/C} R-\overset{NH_2}{\overset{|}{C}}-COO^-$$

II. By HVZ (Hell Volhard Zellinsky) Reaction

Carboxylic acid containing α-H, when treated with Br_2 in presence of red phosphorous, gives a α-halo acid. This α-halo acids on treatment with large excess of ammonia yields a α-amino acid:

$$R-CH_2-COOH \xrightarrow{Br_2/\text{red P}} R-\underset{Br}{\underset{|}{CH}}-COOH \xrightarrow{NH_3(\text{excess})} \xrightarrow{H_3O^+} R-\underset{NH_2}{\underset{|}{CH}}-COOH$$

III. The Strecker Synthesis

Treating an aldehyde with ammonia and HCN yields a α-amino nitrile, which on hydrolysis gives α-amino acid:

$$R-CHO \xrightarrow[HCN]{NH_3} R-\underset{NH_2}{\underset{|}{CH}}-CN \xrightarrow{H_3O^+} R-\underset{NH_2}{\underset{|}{CH}}-COOH$$

12.8 REACTIONS OF AMINO ACIDS

I. Esterification of Carboxyl Group

Like normal, monobasic carboxylic acid, amino acids are esterified by treatment with a large excess of an alcohol and an acid catalyst:

$$R-\underset{\overset{+}{N}H_3}{\underset{|}{CH}}-COO^- + R'OH \xrightarrow{HCl} R-\underset{\overset{+}{N}H_3}{\underset{|}{CH}}-COOR'$$

Esterification of amino acid is oftenly used as a means of protection to prevent undesirable reaction of —COOH group.

II. Acylation Reaction

Amino acids on treatment with an acid chloride or acid anhydride, undergo acylation reaction at amine group, forming amides as:

$$H_2N-\underset{CH_2-\text{(imidazolyl)}}{\underset{|}{CH}}-COOH + Ac_2O \xrightarrow{HCl} H_3C-\overset{O}{\overset{\|}{C}}-HN-\underset{CH_2-\text{(imidazolyl)}}{\underset{|}{CH}}-COOH$$

III. Ninhydrin Test

Ninhydrin forms a deep violet complex called "Ruhemann's Purple", when treated with an amino acid, irrespective of the structure of original amino acid:

$$H_2N-\underset{R}{\underset{|}{CH}}-COOH + \text{Ninhydrin} \longrightarrow \text{Ruhemann's purple}$$

Ninhydrin

Ruhemann's purple

12.9 PEPTIDE LINKAGE AND PROTEINS

Amines and acids condenses to form amides as:

$$RNH_2 + HOOC—R' \xrightarrow{Heat} R'—CONHR + OH_2$$

An amino acid has both amine and acid groups within the same molecule, under suitable reaction conditions, amine group of one molecule may condenses with acid group of other molecule. The product formed is an amide, called a "dipeptide" due to the reason that it contain two amino acids. The amide linkage between the amino acids is called "Peptide Bond".

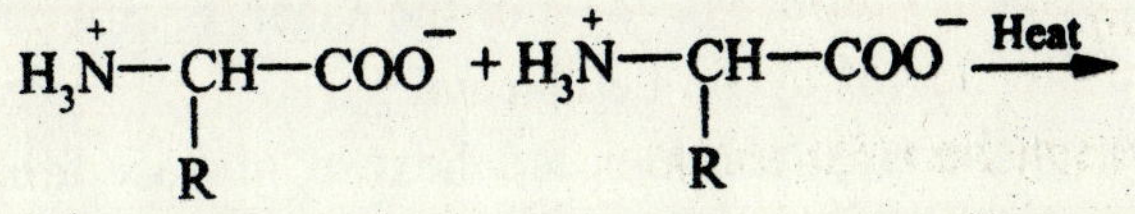

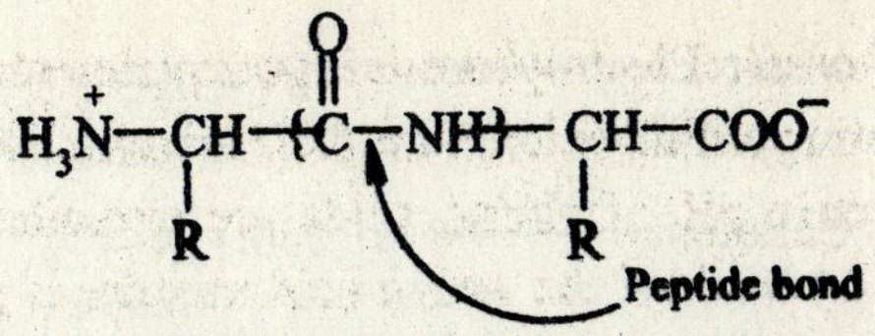

$$H_3\overset{+}{N}—\underset{R}{\underset{|}{CH}}—COO^- + H_3\overset{+}{N}—\underset{R}{\underset{|}{CH}}—COO^- \xrightarrow{Heat}$$

$$H_3\overset{+}{N}—\underset{R}{\underset{|}{CH}}—(\overset{O}{\overset{\|}{C}}—NH)—\underset{R}{\underset{|}{CH}}—COO^-$$

Peptide bond

Therefore, under suitable reaction conditions, any number of amino acids can be joined together forming a long chain polymer. A polymer of amino acid having peptide linkage and molecular weight in the range of 5000 are known as peptide. Proteins contain more amino acid unit with their molecular weight in the range of 6000 to 40,000,000.

The end of the peptide with the free amine group ($—NH_3^+$) is called "N-terminal" and the end with free carboxy group is called "C-terminal".

Carbohydrates

12.10 INTRODUCTION

Carbohydrates are the most abundant organic substance available in nature. Carbohydrates are synthesized and consumed by most plants and animals of nature, which use them to store energy and deliver on requirement. In metabolic process in the living beings, glucose is usually oxidized into carbon dioxide and water to provide energy to the cell for their functioning. Plants store energy in the form of starch while animals store energy in the form of glycogen. Cellulose, an important carbohydrate, makes up the cell walls of plants and forms their structural network.

Every moments of human life involves carbohydrates in one or other form, e.g. provides energy through metabolism, provides fabrics for clothing, provides woods for furniture, etc.

12.11 CLASSIFICATION OF CARBOHYDRATES

According to the modern definition, carbohydrates are the compounds which include polyhydroxy aldehydes, polyhydroxy ketones and compounds that are hydrolyzed to them. On the basis of hydrolysis properties of carbohydrates, they are classified into three categories:

I. Monosaccharides (Simple Sugar)

Monosaccharides are carbohydrates that are not hydrolysed to further simpler carbohydrates. Glucose, fructose, mannose, glactose, etc. are the examples of monosaccharides:

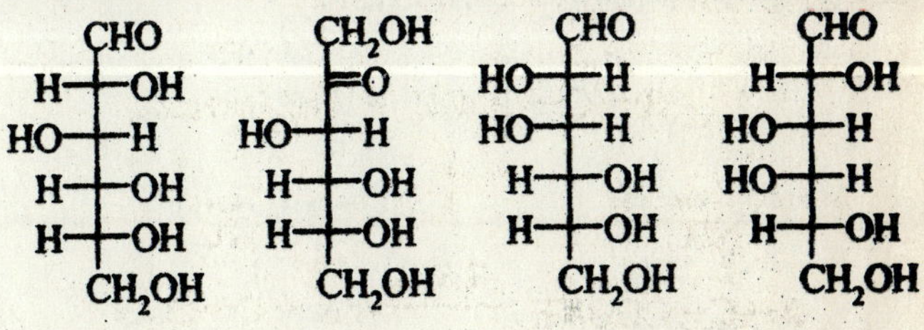

D-glucose	D-fructose	D-mannose	D-glactose
CHO	CH_2OH	CHO	CHO
H—C—OH	C=O	HO—C—H	H—C—OH
HO—C—H	HO—C—H	HO—C—H	HO—C—H
H—C—OH	H—C—OH	H—C—OH	HO—C—H
H—C—OH	H—C—OH	H—C—OH	H—C—OH
CH_2OH	CH_2OH	CH_2OH	CH_2OH

II. Disaccharides

A disaccharide is a sugar that can hydrolyzes to two monosaccharides, e.g. sucrose (table sugar) is a disaccharide that is hydrolyzed into glucose and fructose:

$$\underset{\text{Sucrose}}{C_{12}H_{22}O_{11}} + OH_2 \longrightarrow \underset{\text{Glucose}}{C_6H_{12}O_6} + \underset{\text{Fructose}}{C_6H_{12}O_6}$$

Other common examples of disaccharides are cellobiose, maltose, lactose, etc.

III. Polysaccharides

Polysaccharides are carbohydrates that on hydrolysis produces many monosaccharide units, e.g. starch, cellulose-both biopolymers of glucose.

Starch: Starch is a polysaccharide whose carbohydrate units are easily added to store energy or removed to provide energy to cell.

Cellulose: The polysaccharide cellulose is a major structural component of plants. Hydrolysis of either starch or cellulose gives many molecule of glucose:

$$\text{Starch} \xrightarrow{H_3O^+} < 1000\ \underset{\text{Glucose}}{C_6H_{12}O_6}$$

$$\text{Cellulose} \xrightarrow{H_3O^+} < 1000\ \underset{\text{Glucose}}{C_6H_{12}O_6}$$

12.12 THE D AND L CONFIGURATION OF SUGAR

D and L are the relative configurations given to carbohydrates to distinguish between the two different stereoisomers. The study started with degradation of sugar to lower molecules. It was observed that any natural sugar degrade to dextro rotatory (+) glyceraldehydes. Also degradation study of some synthetic carbohydrates revealed that they degrade to levororatory (–) glyceraldehydes. Thus, carbohydrates that degrade to (+) glyceraldehydes are now designated as D-sugar and those degrading to (–) glyceraldehydes are designated as L-sugar. These D and L do not reflect the absolute configuration of sugars. We know the absolute configuration of glyceraldehydes as shown in Figure 12.1.

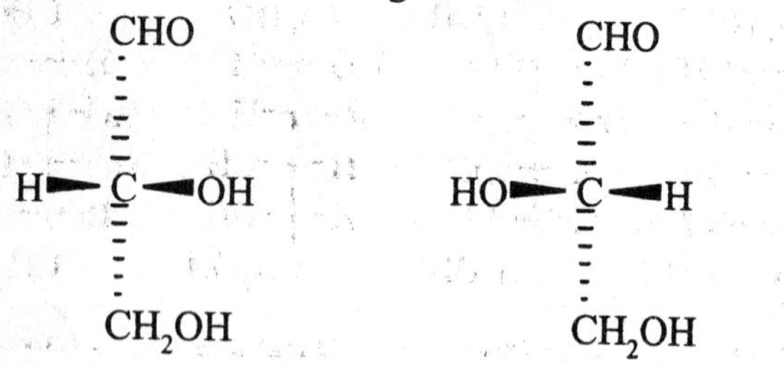

CHO, H–C–OH, CH_2OH — (+) Glyceraldehyde, D-Series of sugar, I

CHO, HO–C–H, CH_2OH — (–) Glyceraldehyde, L-Series of sugar, II

Figure 12.1: D and L representation of sugar

If the positions of H and OH on the second carbon from bottom (next to —CH_2OH) in Fischer projection are identical to that in I, the sugar are called D-sugar otherwise L-sugar as:

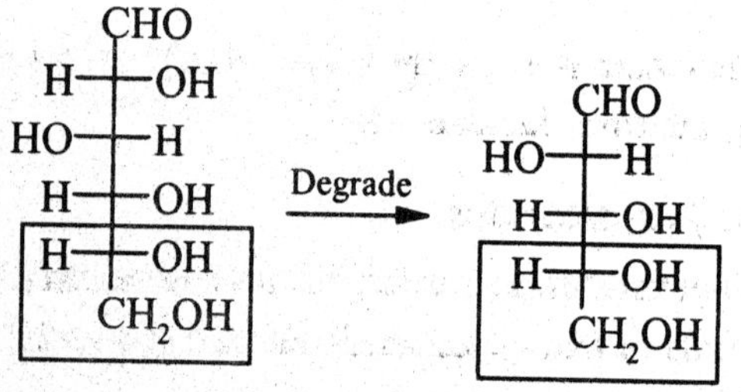

(+) D-glucose (CHO; H–OH; HO–H; H–OH; H–OH; CH_2OH) —Degrade→ D (–) Arabinose (CHO; HO–H; H–OH; H–OH; CH_2OH) —Degrade→ D (–) Erythrose (CHO; H–OH; H–OH; CH_2OH) —Degrade→ D (+) Glyceraldehyde (CHO; H–OH; CH_2OH)

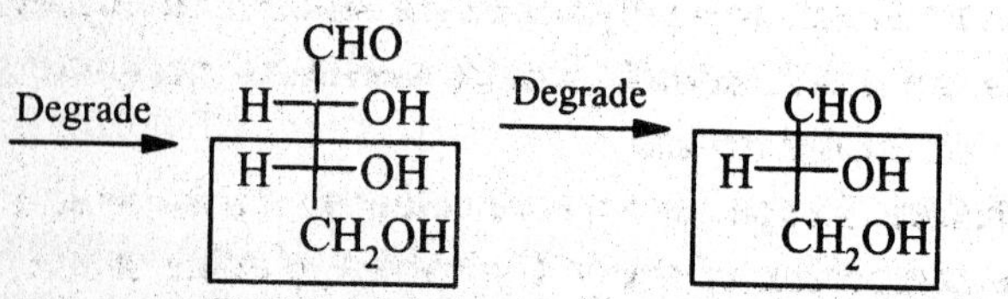

12.13 CYCLIC STRUCTURE OF MONOSACCHARIDES

As described in the previous chapters, a carbonyl group reacts with alcohols forming a hemiacetal and finally acetals. Since, hemiacetals are not as stable as acetals, hemiacetals usually decomposes spontaneously into aldehydes/ketones and alcohol. Since carbohydrates (monosaccharides) are either aldose or ketose, containing both aledhyde/ketone and alcohol functional groups in the same molecule. The solid, crystalline form of an aldose/ketose are normally a cyclic hemiacetals. In solution, they exist as an equilibrium mixture of the cyclic hemiacetal and open chain (acyclic) form. For most of the sugar, equilibrium lies to the side of cyclic hemiacetal.

One of the most common aldohexose, glucose forms a six membered ring with a hemiacetal linkage between carbonyl carbon and the hydroxy group at C5 as shown in Figure 12.2.

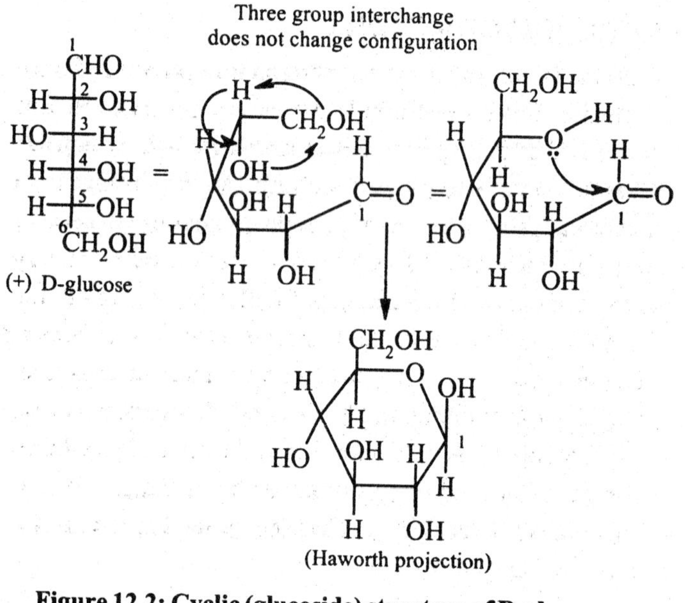

Three group interchange does not change configuration

(+) D-glucose

(Haworth projection)

Figure 12.2: Cyclic (glucoside) structure of D-glucose

It can be noticed from the above hemiacetal structure that a new chiral carbon at C-1 has been generated. In the above shown hemiacetal, —OH group has been shown up, but another possible stereoisomer would have this —OH group down.

A cyclic structure of monosaccharides are named according to their five or six membered ring. A six membered cyclic hemiacetal is called a "Pyranose" and a five membered cyclic hemiacetal is called "Furanose". Furanose are formed by ketose as shown in the following Figure 12.3.

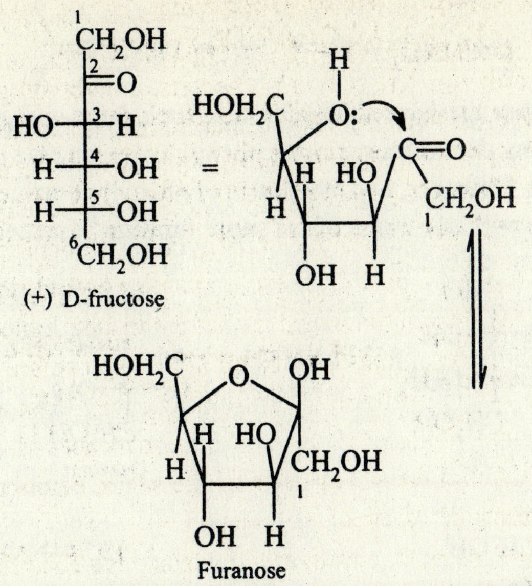

```
       1
      CH2OH
      |2
      C=O                        H
      |3                HOH2C    |
HO ---+--- H                  \  O:
      |4              =   H HO   \ C=O
H  ---+--- OH           H         CH2OH
      |5                          1
H  ---+--- OH             OH  H
      6
      CH2OH

(+) D-fructose

HOH2C     O     OH

   H  HO
H          CH2OH
           1
   OH  H
   Furanose
```

Figure 12.3: Cyclic structure (furanose) of D-fructose

12.14 ANOMERS OF MONOSACCHARIDES: "MUTAROTATION"

As we have seen in the previous section, when open chain structure cyclizes into a pyranose or furanose, a new chiral carbon is generated. Depending on which face of carbonyl carbon is attacked, the hemiacetal —OH group can be oriented either up or down giving distereomeric products called "anomers" as shown in Figure 12.4.

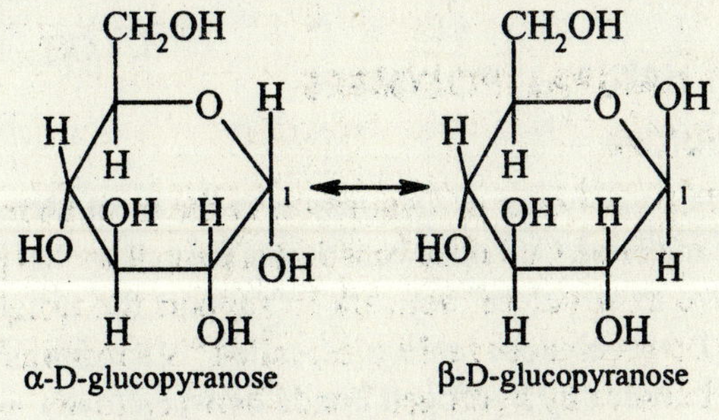

```
      CH2OH                      CH2OH
          O   H                      O   OH
H    H              <-->   H    H
   OH   H      1              OH   H      1
HO          OH             HO           H
    H   OH                      H   OH
α-D-glucopyranose          β-D-glucopyranose
```

Figure 12.4: Anomers of D-glucose

Mutarotation "a property of anomers"

Since anomers are distereomeric, they have different physical characteristics, viz. α-D-glucopyranose melts at 146°C and have specific rotation of +112.2° while β-D-glucopyranose melts at 150°C and have specific rotation of +18.7°.

If glucose is crystallized from water at room temperature, pure crystalline α-D-glucopyranose results. If glucose is crystallized by evaporation of water above 98°C, pure crystalline β-D-glucopyranose results.

Thus, in aqueous solution, the two anomers are in equilibrium with a little amount of open chain form and this equilibrium continues to supply more of the anomer crystallizing out of solution.

When a pure anomer is dissolved in water, equilibrium is established between anomers by interconversion through open-chain form. As an example, when α-anomer is dissolved, specific rotation start to decrease from its initial value of +112.2° and acquire a constant value of +52.6° at equilibrium. This decrease in specific rotation is due to conversion of α-anomer into its β-anomer which has a very less specific rotation. This change in specific rotation to a constant value, when a pure crystalline anomer is dissolved in solvent is known as **Mutarotation.**

12.15 REACTIONS OF MONOSACCHARIDES

I. Reduction Reaction

Sugars are reduced with either $NaBH_4$ of H_2/Ni into sugar alcohols or alditol as:

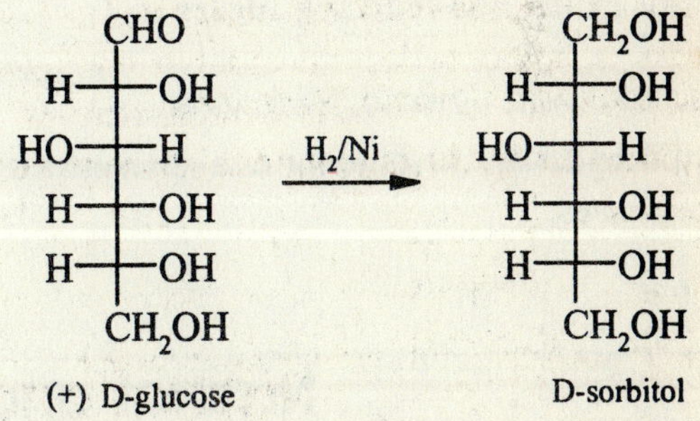

```
     CHO                       CH2OH
H ---+--- OH              H ---+--- OH
HO---+--- H     H2/Ni     HO---+--- H
H ---+--- OH   ------>    H ---+--- OH
H ---+--- OH              H ---+--- OH
     CH2OH                     CH2OH
(+) D-glucose                D-sorbitol
```

II. Oxidation Reaction

Reducing and non-reducing sugars

(i) Oxidation with bromine water: Bromine water is a mild oxidizing agent, oxidizes aldehyde group into carboxylic group leaving alcohol groups intact:

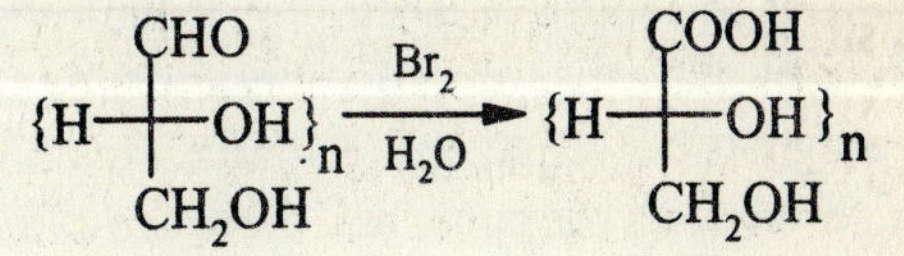

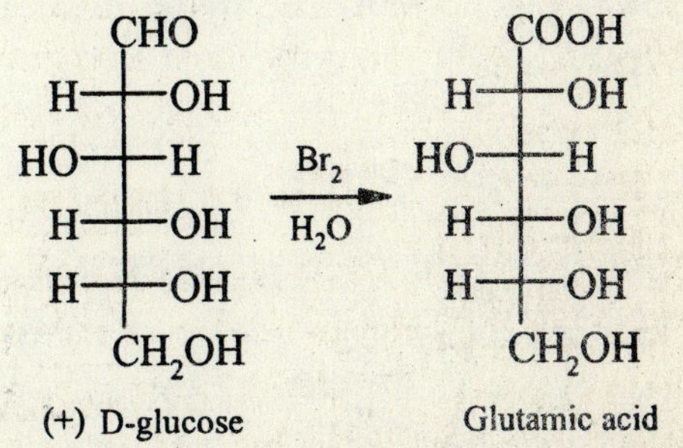

```
     CHO          Br2           COOH
{H---+---OH}n   ------>  {H---+---OH}n
     CH2OH        H2O          CH2OH

     CHO                       COOH
H ---+--- OH              H ---+--- OH
HO---+--- H      Br2      HO---+--- H
H ---+--- OH   ------>    H ---+--- OH
H ---+--- OH     H2O      H ---+--- OH
     CH2OH                     CH2OH
(+) D-glucose               Glutamic acid
```

Also Br_2-H_2O solution is acidic, it does not causes epimerization (movement of carbonyl group), it oxidizes an aldose but not ketose.

(ii) Nitric acid oxidation: Nitric acid is a stronger oxidizing agent, oxidizes aldehyde functional group as well as primary alcohol to carboxyl group. Sugars are oxidized at both ends producing a dicarboxylic acid as:

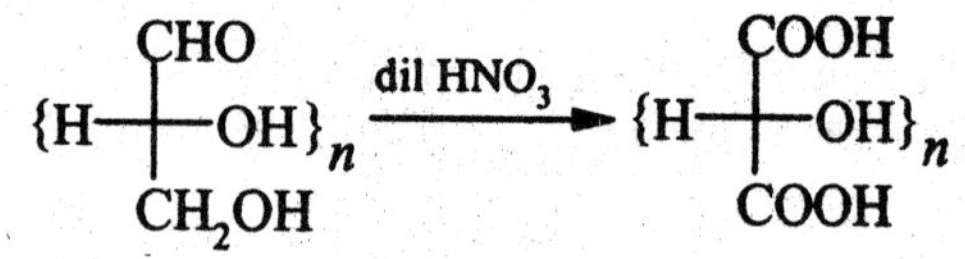

$$\{H-\underset{CH_2OH}{\overset{CHO}{C}}-OH\}_n \xrightarrow{\text{dil } HNO_3} \{H-\underset{COOH}{\overset{COOH}{C}}-OH\}_n$$

Aldose Aldaric acid

(iii) Tollen's test: Tollen's reagent oxidizes —CHO to —COOH. In its open chain form, an aldose has a —CHO group, and oxidized by Tollen's reagent giving aldonic acid. Also, since Tollen's reagent is basic in nature, it promotes epimerization (a ketose is epimerized to aldose in basic medium). Hence, Tollen's test is given by both aldose and a ketose.

Reducing sugar: Sugars that reduces Tollen's reagent are known as reducing sugar. Therefore glucose, fructose are all reducing sugar.

Non-reducing sugar: Sugars that does not reduces Tollen's reagent are called non-reducing sugars.

III. Reaction with Phenyl Hydrazine

Normal aldehydes and ketones forms hydrazone with phenyl hydrazine as:

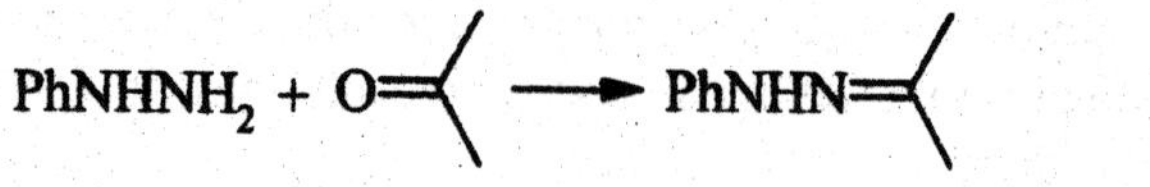

$$PhNHNH_2 + O{=}C\langle \longrightarrow PhNHN{=}C\langle$$

Sugars also have aldehyde and ketonic functional group, but, they do not form simple phenyl hydrazone we might expect. However, two molecules of phenylhydrazine condenses with one molecule of sugar forming "osazone" as:

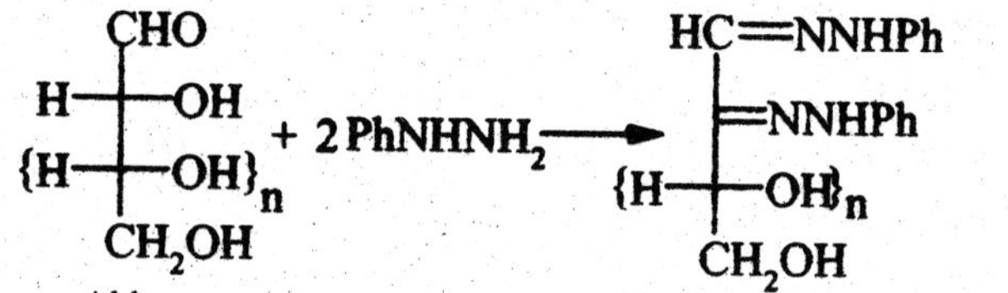

$$\underset{\text{Aldose}}{\begin{array}{c} CHO \\ H-C-OH \\ \{H-C-OH\}_n \\ CH_2OH \end{array}} + 2\,PhNHNH_2 \longrightarrow \underset{\text{Osazone}}{\begin{array}{c} HC{=}NNHPh \\ C{=}NNHPh \\ \{H-C-OH\}_n \\ CH_2OH \end{array}}$$

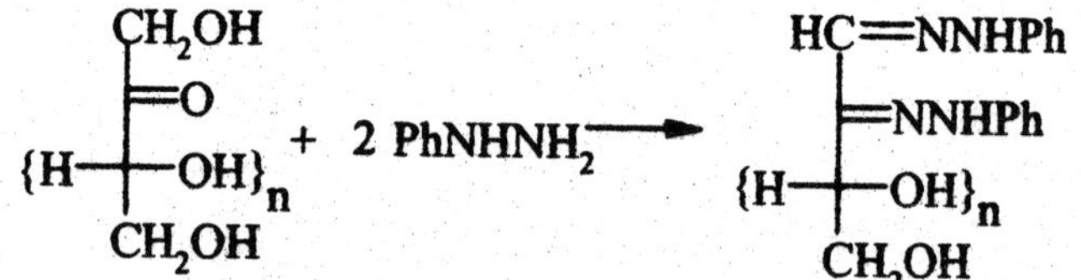

$$\underset{\text{Ketose}}{\begin{array}{c} CH_2OH \\ C{=}O \\ \{H-C-OH\}_n \\ CH_2OH \end{array}} + 2\,PhNHNH_2 \longrightarrow \underset{\text{Osazone}}{\begin{array}{c} HC{=}NNHPh \\ C{=}NNHPh \\ \{H-C-OH\}_n \\ CH_2OH \end{array}}$$

Natural and Synthetic Polymers

12.16 INTRODUCTION

Polymer is a large molecule composed of many smaller repeating units called monomers. A large number of polymers exist in nature as such, e.g. cellulose, natural rubber, proteins and other biopolymers and they are called "Natural polymer". Human being synthesizes a verity of polymers, for various purposes and these are called "Synthetic Polymer", e.g. polyethylene, nylon, decaron, PVC, Teflon, etc.

12.17 NATURAL POLYMERS

I. Cellulose

Cellulose, a polymer of D-glucose, is the most abundant organic material. Cellulose constitutes the cell wall of plant and serve as structural network to support the weight of a plant. Long cellulose molecules, called "Microfibrils" are held in bundles by hydrogen bonds between many —OH groups of glucose ring. In Figure 12.5 represents a partial structure of cellulose:

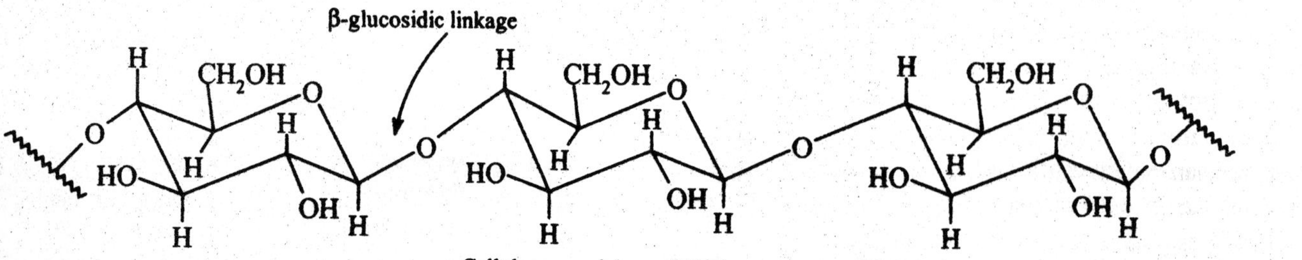

Cellulose, a polymer of D-glucose

Figure 12.5: Structure of cellulose showing a β-glucosidic linkage

Human beings lacks the β-glucoside enzyme needed to hydrolyse the cellulose and they cannot use it directly in their diet.

One of the most important application of the cellulose is in cloth industry. "Rayon" is a fiber made from cellulose, that has been converted to a soluble derivative and then regenerated by "Viscose process".

Viscose Process

In viscose process used for manufacturing rayon, wood pulp is treated with CS_2 and NaOH (aq) to convert free hydroxy groups to xanthate, which are water soluble:

$$\underset{\text{Cellulose}}{R{-}OH} + CS_2 + NaOH \longrightarrow \underset{\text{Xanthate derivatives viscose}}{[RO{-}\overset{\overset{S}{\|}}{C}{-}SNa]}$$

The viscose xanthate solution is then forced through a spinneret into an aqueous sodium bisulphate solution, where fiber of insoluble cellulose is regenerated:

$$\underset{\text{Xanthate derivatives}}{[RO{-}\overset{\overset{S}{\|}}{C}{-}SNa]} \xrightarrow{NaHSO_4} ROH + CS_2 + Na_2SO_4$$

Alternatively, the viscose solution is extracted into sheets to give "Cellophane" film.

II. Natural Rubber

Natural rubber is a polymer of isoprene, and obtained from natural source-latex tree. In natural rubber, isoprene units are joined together in head-to-tail fashion and all double bonds in the polymer chain have *cis* configuration as shown in the Figure 12.6.

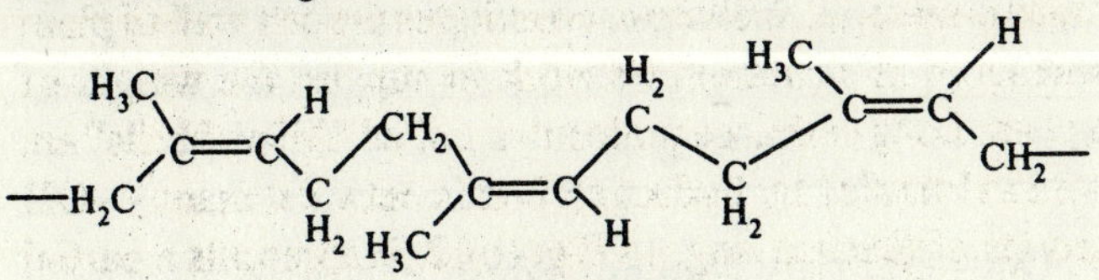

cis-1, 4-polyisoprene
(Natural rubber)

Figure 12.6: *cis* configuration of natural rubber, —CH_3 and —H are *cis* at every double bond

cis polyisoprene is also synthesized by addition polymerization of isoprene and stereochemistry of polymerization is controlled by using "Zigler Natta" catalyst.

Pure natural rubber is soft and tacky, they have very low mechanical strength and are of no use for any commercial purpose. To enhance the mechanical properties and stiffness, natural rubber is vulcanized. Vulcanization is a process in which natural rubber is heated with sulphur, where it undergo cross-linking and connect the isolated rubber chain through a network making them strong and stiff.

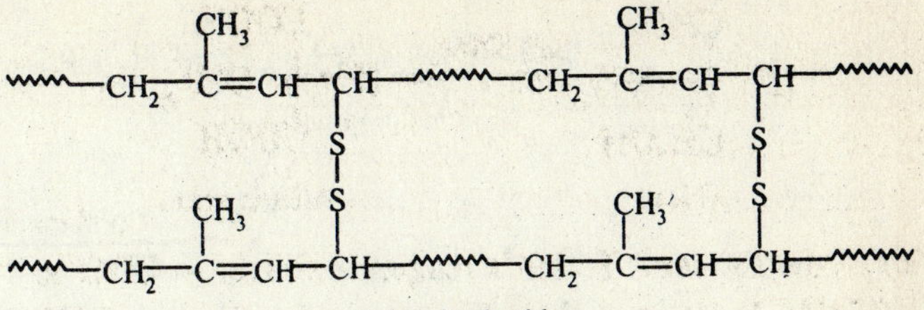

Vulcanized rubber

12.18 SYNTHETIC POLYMERS

Polymers that are synthesized in laboratory or industry by polymerizing monomers, under suitable reaction conditions are called synthetic polymers or man made polymers.

Based on the techniques of polymerization, synthetic polymers are grouped together in two class—addition polymers and condensation polymers.

I. Addition Polymers (Chain growth polymers)

Addition polymers results from free radical polymerization (also by cationic and anionic polymerization) of monomer unit.

Free Radical Polymerization of some Specific Monomers

(a) Vinyl Chloride

Free radical polymerization is initiated by a free radical initiator, usually an organic peroxide as:

$$\underset{\substack{\text{Benzoyl peroxide}\\ \text{(a very common}\\ \text{free radical initiator)}}}{C_6H_5{-}\overset{\overset{O}{\|}}{C}{-}O{-}O{-}\overset{\overset{O}{\|}}{C}{-}C_6H_5} \xrightarrow{\text{Heat}} 2\,C_6H_5{-}\overset{\overset{O}{\|}}{C}{-}\dot{O} \equiv R\dot{O}$$

In the next subsequent step, free radical attacks at the π-bonds of an olefin, start the polymerization process as:

$$R\dot{O} + \underset{\text{Vinyl chloride}}{\overset{Cl}{\overset{|}{C}H}{=}CH_2} \longrightarrow RO{-}\overset{Cl}{\overset{|}{C}H}{-}\dot{C}H_2 \quad \text{Initiation}$$

$$RO{-}\overset{Cl}{\overset{|}{C}H}{-}\dot{C}H_2 + \overset{Cl}{\overset{|}{C}H}{=}CH_2 \longrightarrow RO{-}\overset{Cl}{\overset{|}{C}H}{-}CH_2{-}\overset{Cl}{\overset{|}{C}H}{-}\dot{C}H_2 \xrightarrow{\text{Propagation}} \longrightarrow$$

The growing polymer chain finally terminate to stable polymer molecule and termination may occur in a number of ways as:

$$\mathrm{RO{-}\!\!\left[CH(Cl){-}CH_2\right]_n\!{-}CH(Cl){-}\dot{C}H_2 + RO^{\bullet} \longrightarrow RO{-}\!\!\left[CH(Cl){-}CH_2\right]_n\!{-}CH(Cl){-}CH_2{-}OR}$$

Growing polymer chain

$$\xrightarrow{\text{Combination of growing chain}} \mathrm{RO{-}\!\!\left[CH(Cl){-}CH_2\right]_{2n+2}\!{-}OR}$$

$$\xrightarrow{\text{Disproportionation}} \mathrm{RO{-}\!\!\left[CH(Cl){-}CH_2\right]_n\!{-}C(Cl){=}CH_2}$$

Poly vinyl chloride (PVC) is a low melting, flexible polymer, used in a large number of applications-ranging from day-to-day domestic applications to high-tech appliances. One of the most common application of PVC is in electrical insulation, since it is non-conductor of electric current. Besides this, PVC is largely used for packaging, making house-hold furnitures as well as high strength shock absorber in automobile industry.

(b) Teflon

"Poly tetrafluoro ethylene" Teflon is another very important synthetic polymers, synthesized by free radical polymerization of C_2F_4 as:

$$n\,\mathrm{CF_2{=}CF_2} \xrightarrow{\text{Free radical initiator}} \mathrm{{-}\!\left(CF_2{-}CF_2\right)_n\!{-}}$$

Polytetra fluoro ethylene (Teflon)

Teflon has very high mechanical strength and its most important feature is its high thermal stability. Teflon melts at very high temperature, therefore, it is widely used for making surgical tools where it can be steam sterilized without any deformations. These days, Teflon is extensively used in coating for the cooking utensils (non-stick pan), since it gives a very smooth surface and withstand high temperature as well as does not undergo abrasion. Some of the other vinylic polymers are tabulized below in Table 12.3.

Table 12.3: Some common synthetic polymers, monomers and their uses

Polymers	*Monomers*	*Uses*
Polypropylene	$CH_3CH{=}CH_2$	Bottles, bags, packaging, etc.
Polyethylene	$CH_2{=}CH_2$	Same as polypropylene.
Polystyrene	$C_6H_5CH{=}CH_2$	Foam, rubber, insulation, adhesive, etc.
Polyisobutylene	$(CH_3)_2C{=}CH_2$	Specific rubber.
Polyacrylonitrile	$CH_2{=}CHCN$	Orlon, etc.
Poly(methylmethacrylate)	$CH_2{=}C(CH_3)COOCH_3$	Acrylic fibers, plexiglass, lucite paints, etc.
Poly(methyl-α-cyanoacrylate)	$CH_2{=}C(CN)COCH_3$	Super glue

II. Condensation Polymers (Step growth polymers)

Condensation polymers result from condensation of two functional groups of the monomer unit, e.g.

$$\mathrm{MeO{-}C({=}O){-}C_6H_4{-}C({=}O){-}OMe + HO{-}CH_2CH_2{-}OH}$$

Dimethyl terphthalate — Ethylene glycol

$$\longrightarrow \mathrm{\left[{-}C({=}O){-}C_6H_4{-}C({=}O){-}O{-}CH_2CH_2O{-}\right]_n + CH_3OH}$$

Dacron
a polyester

One of the most common family of condensation polymers is polyamides (Nylons), formed by the condensation of an acid functional group and an amine functional group as:

$$n\,\mathrm{HOOC{-}R{-}COOH} + n\,\mathrm{H_2N{-}R'{-}NH_2}$$

$$\longrightarrow \mathrm{\left[{-}C({=}O){-}R{-}C({=}O){-}NH{-}R'{-}NH{-}\right]_n} + 2n\,\mathrm{H_2O}$$

Nylons are further classified according to the number of carbons present in the starting diacid and diamine as:

Nylon-6, 6

Synthesized by the condensation of hexane-1, 6-diamine and hexandioic acid (Adipic acid) as:

$$HOOC-(CH_2)_4-COOH + H_2N-(CH_2)_6-NH_2$$

Adipic acid

$$\downarrow$$

$$\left[-\overset{O}{\overset{\|}{C}}-(CH_2)_4-\overset{O}{\overset{\|}{C}}-NH-(CH_2)_6-NH-\right]_n + 2n\ H_2O$$

Nylon-6, 6

Here, the first number represents the number of carbon in diamine and second in diacid.

Nylon-6, 10

$$HOOC-(CH_2)_8-COOH + H_2N-(CH_2)_6-NH_2$$

Sebacic acid

$$\downarrow$$

$$\left[-\overset{O}{\overset{\|}{C}}-(CH_2)_8-\overset{O}{\overset{\|}{C}}-NH-(CH_2)_6-NH-\right]_n + 2n\ H_2O$$

Nylon-6, 10

Nylon-6: Obtained by the condensation of ε-amino caproic acid or by the ring opening polymerization of caprolactum:

$$H_2N-(CH_2)_5-COOH \xrightarrow{\text{Heat}}$$

$$\left[-\overset{O}{\overset{\|}{C}}-(CH_2)_5-NH-\right]_n \xleftarrow[\text{Heat}]{H_2O/H^+} \text{Caprolactum}$$

Nylon-6 Caprolactum

Properties of nylons: Nylons are the most common polymers of commercial applications. Nylons are used in making fibers for clothing, as an engineering plastics, in packaging bags, etc. Nylons are usually high melting polymer due to very strong intermolecular hydrogen bonding. Nylon has affinity for water due to its ability to form H-bonds.

12.19 STEREOCHEMISTRY OF POLYMERS

Free radical polymerization of olefin, usually gives a head to tail arrangement with substituents appearing on alternate carbon of polymer chain. The bonding arrangement for a generic polyalkene can be shown as:

$$CH_2{=}CH{-}R \xrightarrow{\text{Polymerize}} \ast\!\left[\ \ \right]_n\!\ast$$

The stereochemistry of side group (R) in polymer has very great effect on the physical properties of polymers.

Polymers are classified into three major categories according to the predominant stereochemistry of group.

I. Isotactic

If the side groups are all lying on the same side of polymer chain then the polymer is known as *isotactic*.

II. Syndiotactic

If the side group alternate sides, i.e. if one group is out of plane and the next one is behind the plane and so on, the polymer is known as *syndiotactic*.

III. Atactic

If the side groups occur randomly on either side of the polymer chain, they are called *atactic polymers*. These three type of polymers are shown in Figure 12.7.

An isotactic polymer (side groups on the same side of the backbone)

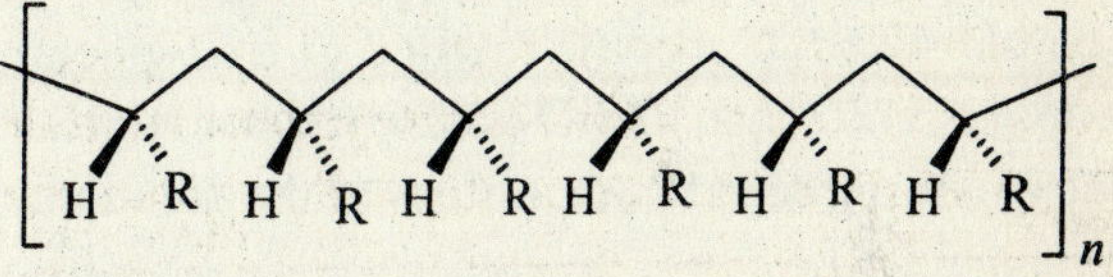

A syndiotactic polymer (side groups on alternating sides of the backbone)

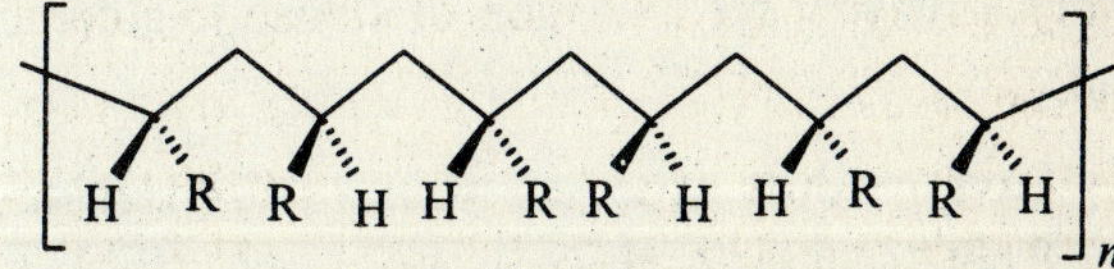

An atactic polymer (side groups on random sides of the backbone)

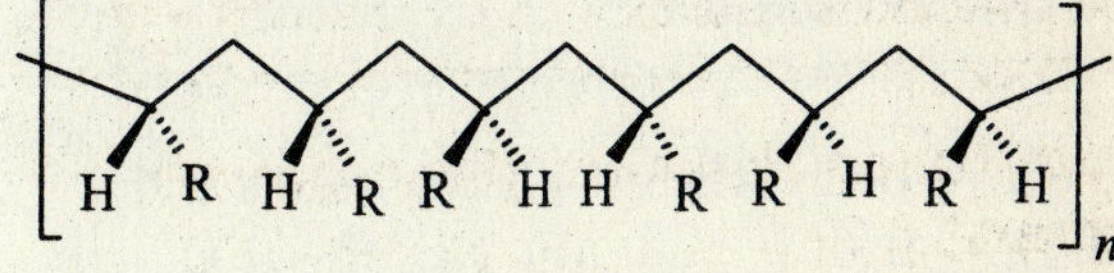

Figure 12.7: Arrangements of group at chiral carbon

PRACTICE EXERCISE

1. Draw the structure of predominant form of:
   (a) Valine at pH = 11 (b) Proline at pH = 2
   (c) Arginine at pH = 7 (d) Glutamic acid at pH = 7
   (e) A mixture of alanine, lysine and aspartic acid at
   (i) pH = 6 (ii) pH = 2
2. Synthesize following amino acids by reductive amination techniques of an α-keto acids:
   (a) Alanine (b) Leucine
   (c) Serine (d) Glutamine

3. Synthesize following amino acids, starting from a carboxylic acid:
   (a) Glycine (b) Leucine
   (c) Valine (d) Glutamic acid

4. Predict the products of the following reactions:
   (a) $PhCH_2OCONHCH(CH_3)COOH \xrightarrow{H_2/Pd}$
   (b) Lysine + $(CH_3CO)_2O \longrightarrow$
   (c) Proline $\xrightarrow[\text{Excess}]{(CH_3CO)_2O}$
   (d) $CH_3CH_2CH(CH_3)CHO + NH_3 + HCN \xrightarrow{OH_2} \xrightarrow{H_3O^+}$
   (e) Isovaleric acid + $Br_2 \xrightarrow{PBr_3} \xrightarrow[\text{Excess}]{NH_3}$

5. Convert alanine into:
   (a) alanine isopropyl ester
   (b) N-benzoylalanine
   (c) t-butyloxy carbonylalanine

6. An aqueous solution of glucose has a specific rotation of +52.6°. If specific rotation of α-D-glucopyranose is 112.2° and of β-D-glucopyranose is +18.7°, determine anomeric composition of the solution.

7. Explain why bromine water is better reagent than Tollen's reagent for oxidation of aldose to aldonic acid?

8. (a) Give product expected when (+) glyceraldehyde react with HCN.
   (b) What is the relationship between the products and how might they be separated?
   (c) Discuss the stereochemistry of products.

9. Predict products when D-glucose reacts with:
   (a) Br_2—H_2O (b) NaOH—H_2O
   (c) CH_3OH/H^+ (d) $Ag(NH_3)_2OH$
   (e) H_2/Ni (f) Ac_2O
   (g) Excess of CH_3I/Ag_2O
   (h) Excess of HIO_4.

10. Define and discuss Mutarotation.

11. Why rayon is called regenerated cellulose and how it is extracted from plant cells?

12. Free radical polymerization of vinyl chloride can proceed in the following three ways:

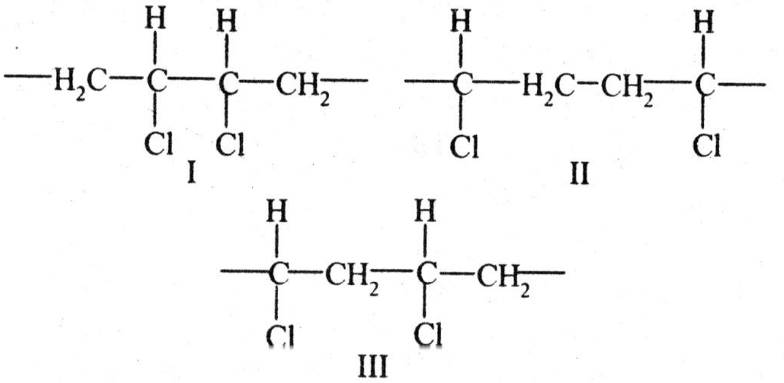

    I: $—H_2C—CH(Cl)—CH(Cl)—CH_2—$

    II: $—CH(Cl)—H_2C—CH_2—CH(Cl)—$

    III: $—CH(Cl)—CH_2—CH(Cl)—CH_2—$

    Out of the above three ways, only one type of polymer is formed. Which is the most preferred way of polymerization and why? Provide mechanistic explanation.

13. What is the role of Zigler Natta catalyst in free radical polymerization. Provide detailed mechanism of polymerization of propylene using Zigler Natta catalyst.

14. What do you understand by tacticity? Taking a suitable example of polymers, explain isotactic, syndiotactic and atactic polymers.

15. During free radical polymerization with benzoyl peroxide initiator, some CO_2 gas is also evolved. Discuss the source and mechanism of formations of CO_2.

16. Free radical polymerization is usually exothermic in nature. Explain.

ANSWERS

CHAPTER 1

Practice Exercise

1. (a) 3, 3, 4-triethyl hexane.
   (b) 5-(1-ethyl-2-methyl propyl)-4-isopropyl-6-propylnonane.
   (c) 2, 3 ,5, 6-tetramethyl-4-(1-methyl propyl) heptane.
   (d) 3, 4-divinyl-1, 3 ,5-hexatriene.
   (e) hex-4-en-1-yne or 4-hexen-1-yne.
   (f) 1-hexen-3, 5-diyne.
   (g) 4-(1-methyl propenyl)-5-vinyl-1, 3, 5, 8-non-tetraene.
   (h) 2,4,7,7,10-pentamethyl-6,9-bis(1,2-dimethyl-propyl)-1, 3, 5, 8, 10-undectetraene.
   (i) 1-(3-butenyl)-4-(2-propenyl) benzene.
2. (a) Butan-1, 1, 2, 3-tetracarbaldehyde.
   (b) 6-chloro-3-methylheptan-2, 5 dione.
   (c) Hexan-1, 2, 5, 6-tetracarboxylic acid.
   (d) Cyclohexan-1, 3, 5-tricarboxylic acid.
   (e) 3-bromo-5-cyanomethyl octandinitrile.
   (f) Cyclohexan-1,4-disulphonic acid.
   (g) 3-bromobutanoic-3-chloro-5-methylhexanoic anhydride.
   (h) 3- bromo-2-chloro-5-methyl benzoic acid.
   (i) 4-(3-chlorocyclohexyl) cyclohexan carboxylic acid.
   (k) 3-methylene-1, 4-pentadiene.
   (l) ethylpent-2, 4-dienoate.
3. (a) 3-hydroxy-5-oxoheptanal.
   (b) 2-formyl methyl-3-cyano-5-oxoethylhexandioic acid.
   (c) 2-isopropyl-3-(1-hydroxymethylethyl) pentandioic acid.
   (d) 3-chloro-2-formyl cyclohexan carbonitrile.
   (e) 2-chloro-5-methylcyclohexan-1,4-dicarboxylic acid.
   (f) 3-amino-5-oxocyclohexan carbonitrile.
   (g) 2-cyano-5-ethylcyclopentancarboxylic acid.
   (h) 3-(3-chloro-1-cyanopropyl)-4-ethyloctandioic acid.
   (i) 2-chloro-4-cyanopentan-1, 5-disulphonic acid.
   (j) 3-cyano-7-cyclopentyl cycloctancarboxylic acid.
4. (a) HOOC COOH COOH

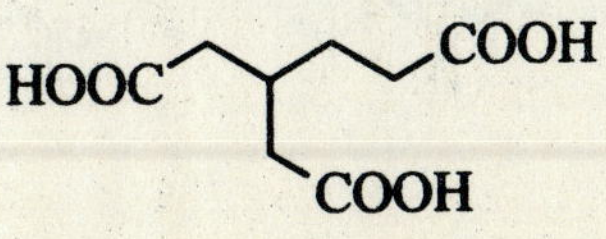

   (b) COOH OHC OH

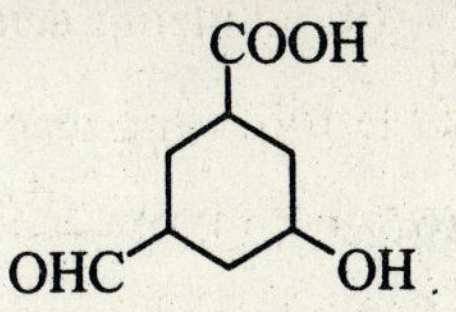

   (c) O Cl

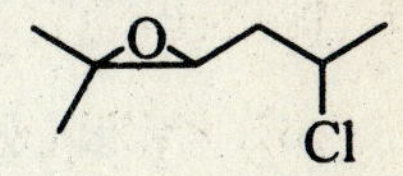

   (d)

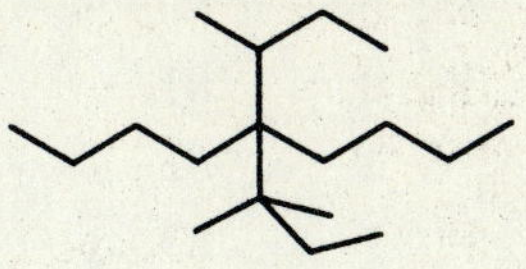

   (e) COOH

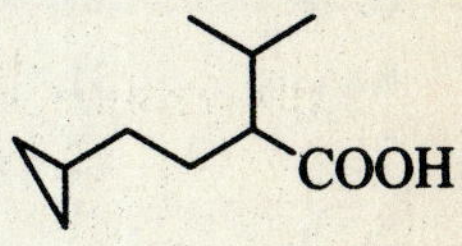

CHAPTER 2

Objective Exercise

Level 1

1. d	**11.** d	**21.** d	**31.** d	**41.** b
2. d	**12.** b	**22.** b	**32.** ac	**42.** c
3. c	**13.** d	**23.** c	**33.** a	
4. d	**14.** c	**24.** a	**34.** a	
5. d	**15.** a	**25.** c	**35.** d	
6. c	**16.** a	**26.** c	**36.** d	
7. a, d	**17.** b	**27.** b	**37.** c	
8. d	**18.** c	**28.** b	**38.** a	
9. a	**19.** c	**29.** d	**39.** c	
10. c	**20.** a	**30.** b	**40.** b	

Level 2

1. b	**15.** a	**28.** a	**41.** d
2. c	**16.** a	**29.** b	**42.** b
3. a	**17.** a	**30.** a	**43.** d
4. d	**18.** a	**31.** d	**44.** b
5. b	**19.** c	**32.** d	**45.** b
6. d	**20.** b	**33.** b	**46.** d
7. d	**21.** b	**34.** b	**47.** a
8. d	**22.** a	**35.** b	**48.** b
9. c	**23.** d	**36.** d	**49.** c
10. c, a	**24.** b	**37.** b	**50.** b
11. c	**25.** a	**38.** a	
12. b	**26.** c	**39.** b	
13. a	**27.** b	**40.** d	

14. a (I), b (F), c (G), d (A), e (C), f (I), g (A), h (C).

CHAPTER 3

Objective Exercise

Level 1

1. c	26. a	51. c	76. a	101. d
2. b	27. d	52. a	77. a	102. d
3. c	28. c	53. c	78. a	103. a
4. d	29. d	54. c	79. c	104. d
5. a	30. c	55. d	80. b	105. d
6. c	31. d	56. d	81. c	106. a
7. b	32. a	57. b	82. a	107. a
8. b	33. a	58. c	83. d	108. b
9. d	34. a	59. b	84. b	109. c
10. a	35. a	60. d	85. c	110. b
11. d	36. c	61. c	86. c	111. c
12. d	37. a	62. d	87. b	112. a
13. b	38. a	63. a	88. c	113. b
14. c	39. d	64. d	89. a	114. d
15. b	40. a	65. b	90. c	115. d
16. a	41. c	66. a	91. d	116. a
17. d	42. a	67. b	92. b	117. a
18. c	43. d	68. b	93. -	118. c
19. c	44. c	69. b	94. a	119. c
20. c	45. b	70. a	95. d	120. d
21. a	46. b	71. b	96. d	121. b
22. a	47. c	72. d	97. a	122. c
23. d	48. c	73. d	98. d	123. c
24. c	49. a	74. c	99. c	124. c
25. b, c	50. c	75. a	100. c	125. c

Level 2

1. c	16. a	31. c	46. d	61. c
2. a	17. c	32. a, d	47. c	62. c
3. d	18. d	33. b	48. d	63. d
4. c	19. d	34. a	49. a	64. c
5. d	20. c	35. b	50. d	65. d
6. a	21. a	36. a	51. a	66. d
7. b	22. a	37. b	52. b	67. a
8. a	23. c	38. b	53. b	68. d
9. a	24. d	39. b	54. c	69. b
10. a	25. a	40. d	55. d	70. b
11. b	26. a	41. b	56. b	71. b
12. c	27. a	42. c	57. d	72. d
13. b	28. b	43. b	58. d	73. d
14. b	29. d	44. a	59. d	74. c
15. b	30. a	45. c	60. b	75. c

76. b	81. c	86. b	91. c	96. a
77. d	82. c	87. -	92. b	97. b
78. a	83. c	88. -	93. c	98. d
79. b	84. a	89. d	94. c	99. c
80. b	85. d	90. c	95. c	100. d

CHAPTER 4

Practice Exercise I (Alkanes)

1. Pentane < 2, 3-dimethyl butane < hexane
2. (a) 2-methyl hexane (b) hexane
3.
4. $\xrightarrow{Br_2/heat}$ Br $\xrightarrow{Mg/ether}$ $\xrightarrow{D_2O}$ D
5. Ethane + $Cl_2/h\nu \rightarrow$ chloroethane,
   Chloroethane + $Na/\Delta \rightarrow$ Butane
6. Corey House synthesis:

   Br $\xrightarrow{Li}$ Li $\xrightarrow{CuBr}$ $(\;)_2CuLi$ $\xrightarrow{\text{Br}}$

7. Reactivity of different halogenations with alkane is discussed in the text.
8. Oxidizing agent consumes HI formed in the iodination reaction which otherwise reduces the alkyl iodide back into alkane.
9. $$SO_2Cl_2 \xrightarrow{RO^{\bullet}} ROSO_2Cl + Cl^{\bullet}$$
   $$RH + Cl^{\bullet} \longrightarrow R^{\bullet} + ClH$$
   $$R^{\bullet} + SO_2Cl_2 \longrightarrow R{-}Cl + SO_2Cl^{\bullet}$$
10. $$(CH_3)_3CH \xrightarrow[Heat]{Br_2} (CH_3)_3CBr \xrightarrow{Li} \xrightarrow{CuBr} \times$$
    $$(CH_3)_2CH_2 \xrightarrow[Heat]{Br_2} (CH_3)_2CHBr \xrightarrow{\times} \longrightarrow \text{Product}$$
11. 2, 3-dimethyl butane
12. (A) (B) Br (C) Br

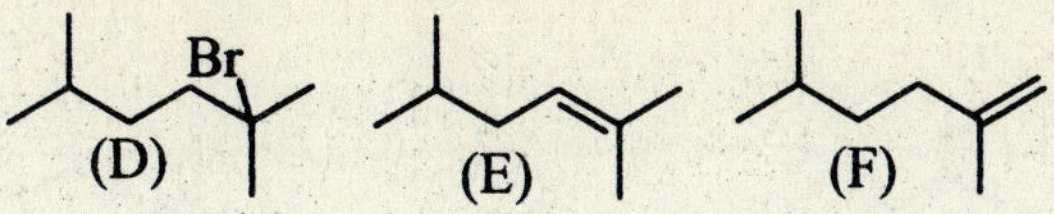

(D) (E) (F)

13. (A) (B) (C) (D) (E) (F) (G)

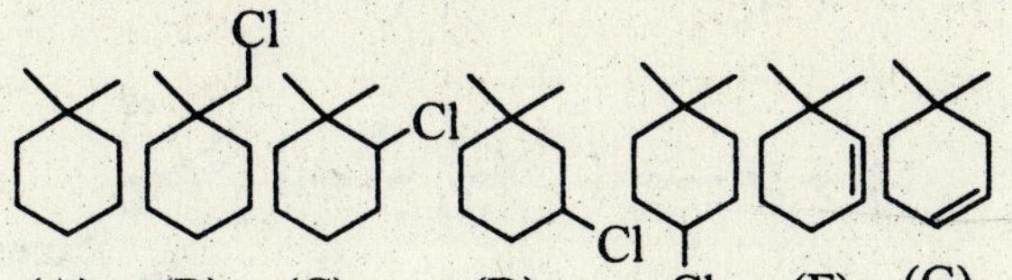

14. (A) (B) (C) (D) 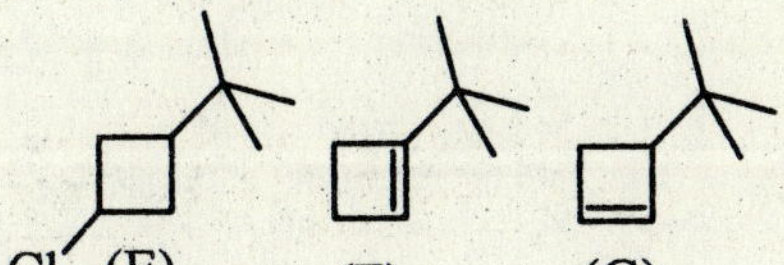 (E) (F) (G)

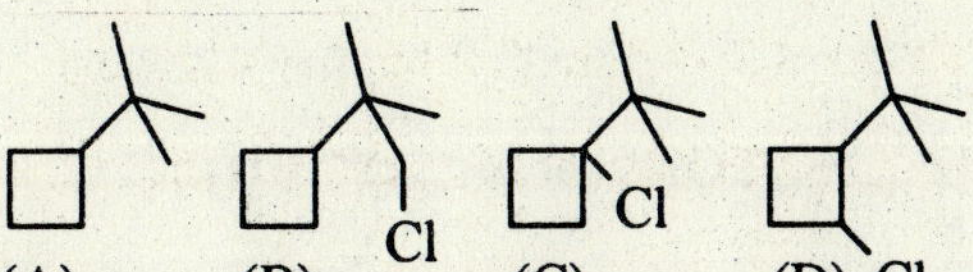

Practice Exercise II (Alkenes)

1. (a) In 1, 3-butadiene, rotation about C2 - C3 bond is not free as in case of butane. The restriction to rotation arises due to delocalization of pi-electrons.

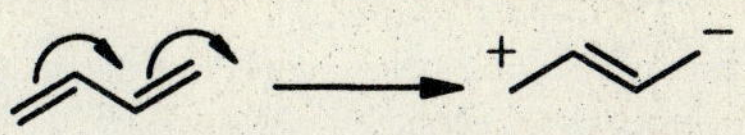

Due to the above delocalization, C2-C3 bond acquire some partial double bond character and free rotation is prevented.

(b) I ⇌ II

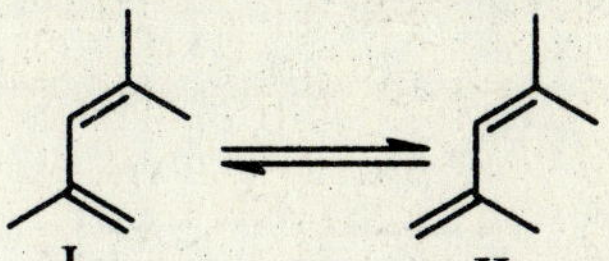

I has greater barrier to rotation due to interaction between the pi-bonds since they are on the same side.

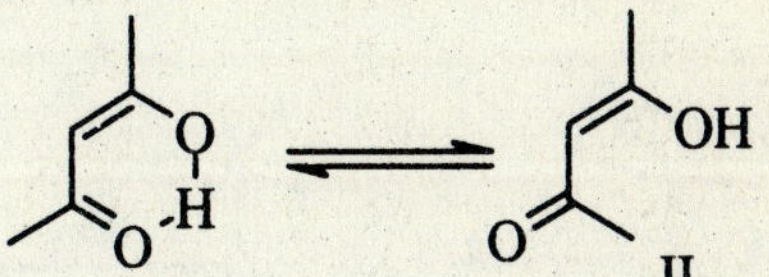

I ⇌ II

I has greater barrier to rotation due to formation of intramolecular H-bond as shown.

2. d < b < a < c

3. (a) $\xrightarrow{H^+}$ $\xrightarrow{-H^+}$ Me-shift $\xrightarrow{-H^+}$ $\xrightarrow[-H^+]{H_2O}$

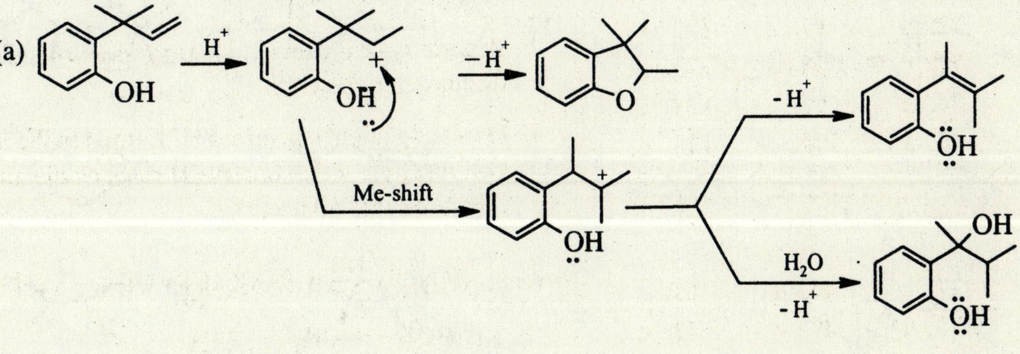

(b) IV < II < V < III < I "It is the order from least negative to most negative".

4. (a) $\xrightarrow[\text{Ether}]{\text{Mg}}$ $\xrightarrow[OH_2]{CH_2O}$ $\xrightarrow[\text{Heat}]{H^+}$ Product

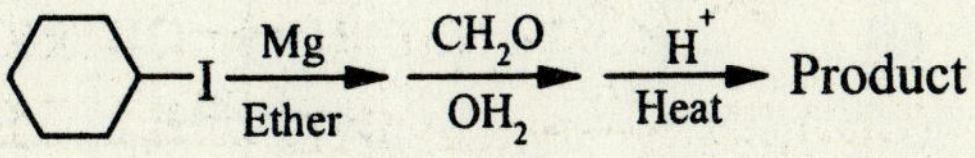

(b) $\xrightarrow{H^+}$ $\xrightarrow{H^- \text{ shift}}$ $\xleftarrow{-H^+}$

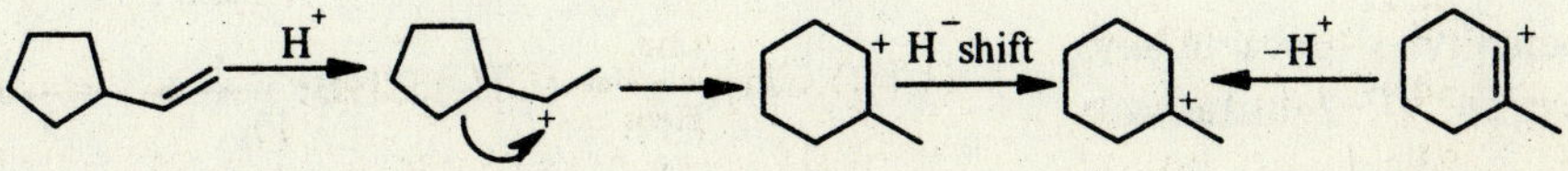

(c) $\xrightarrow{RCO_3H}$ $\xrightarrow{H_3O^+}$

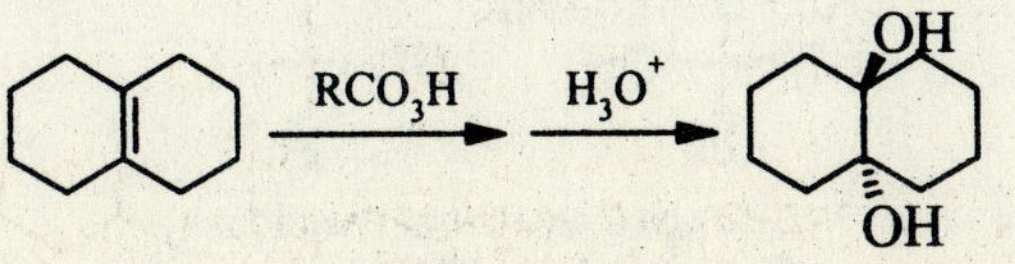

5. (a) 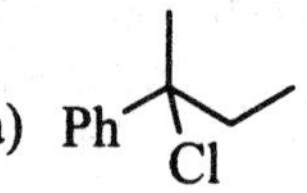 Ph, Cl (b) 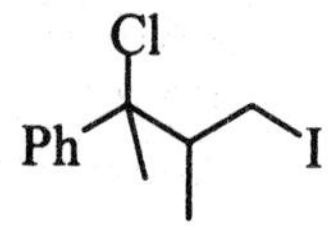 Ph, Cl, I (c) 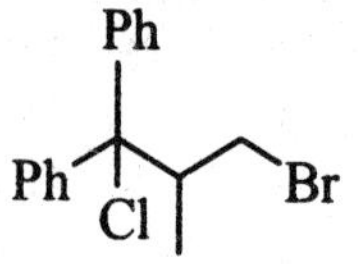 Cl, $CHCl_2$ (d) 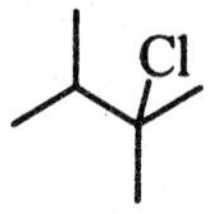 Ph, CCl_3, Cl (e) 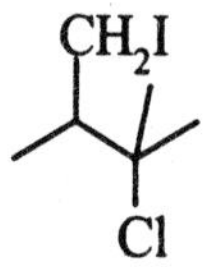 Ph, CCl_3, Ph, Ph, I

(c) Ph, Ph, Cl, Br (d) Cl (e) CH_2I, Cl

6. (a) 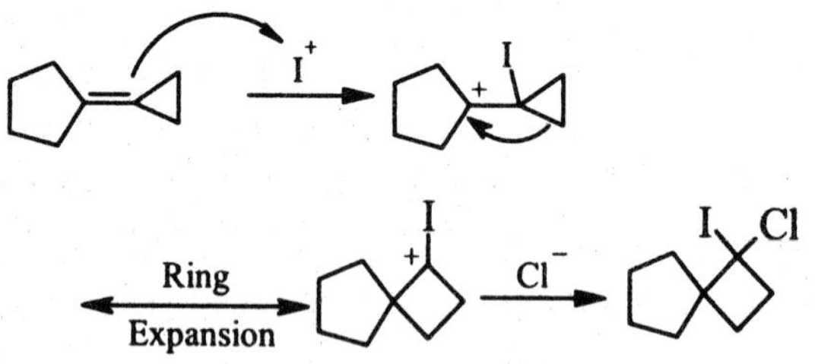 I^+ → Ring Expansion → Cl^- → I, Cl

(b) 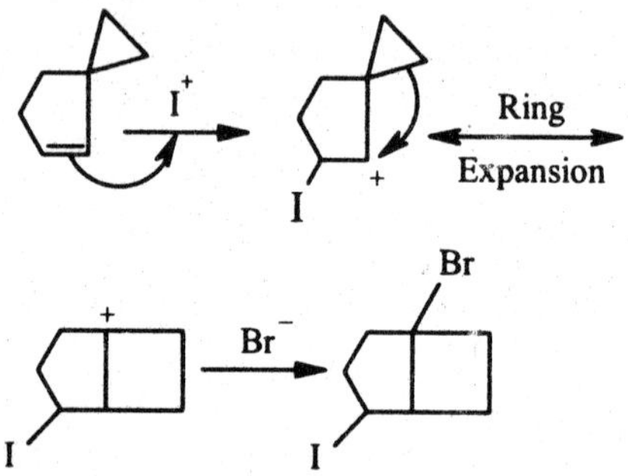 I^+ → Ring Expansion → Br^- → Br, I

(c) 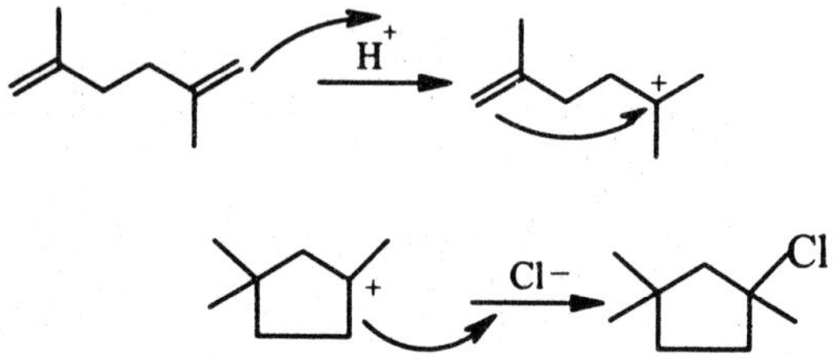 H^+ → Cl^- → Cl

(d) 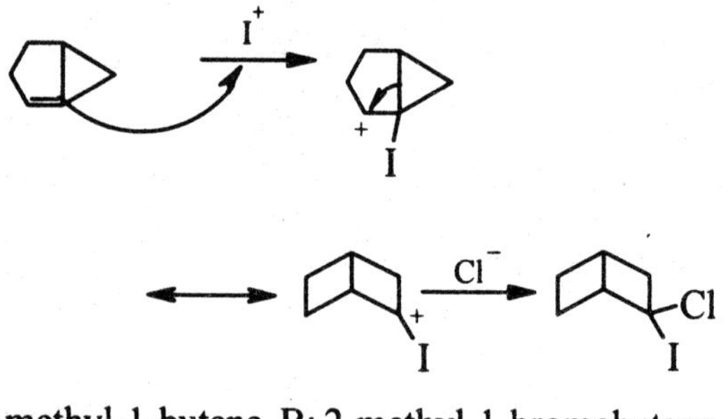 I^+ → Cl^- → Cl, I

7. A: 2-methyl-1-butene, B: 2-methyl-1-bromobutane, C: 2-bromo-2-methylbutane, D: 2-butanone, E: 2-butanol, F: 2-butene.

8. (a) 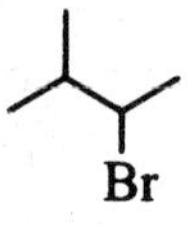 Br (b) 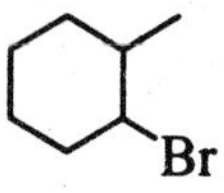 Br

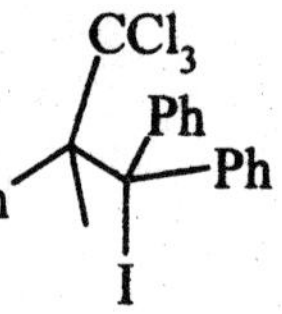

9. 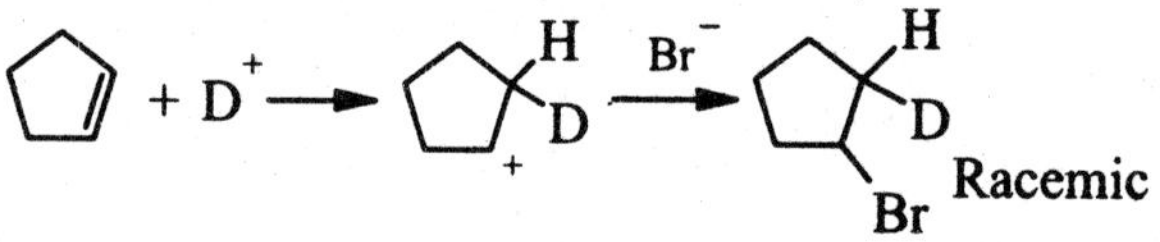 + D^+ → H, D, + → Br^- → H, D, Br Racemic

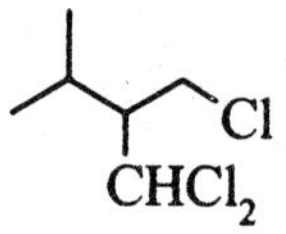

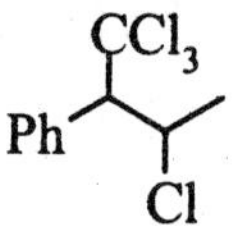

10. (a) Proportion of products reflect their relative stability.

(b) 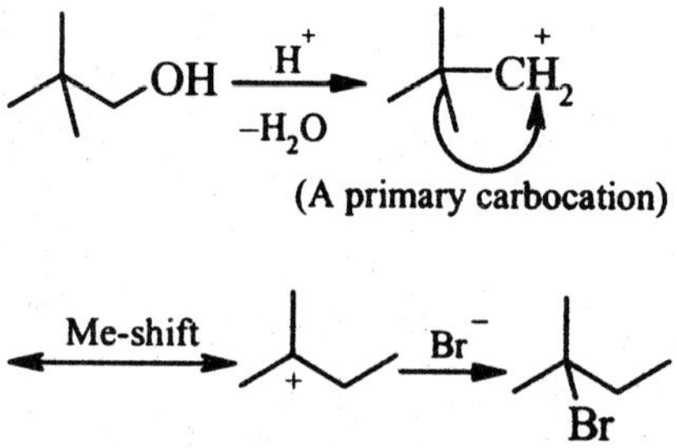 OH $\xrightarrow[-H_2O]{H^+}$ $\overset{+}{C}H_2$ (A primary carbocation)

Me-shift → + → Br^- → Br

11. 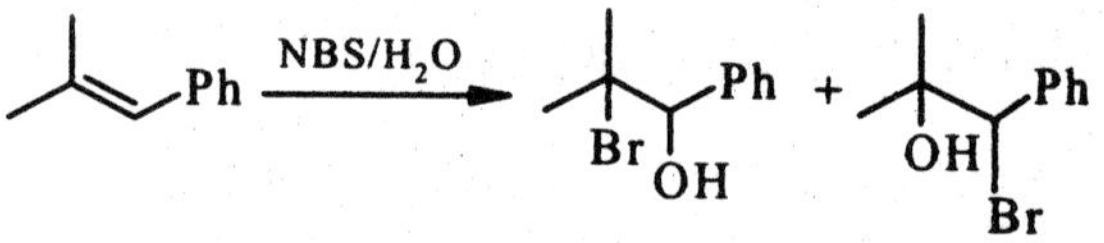 Ph $\xrightarrow{NBS/H_2O}$ Ph, Br, OH + Ph, OH, Br

12. 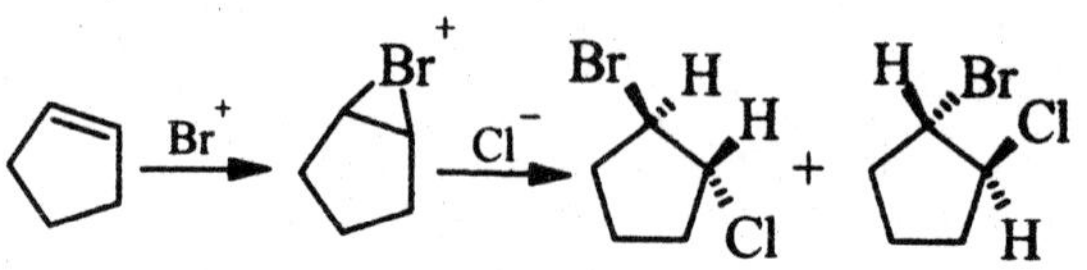 Br^+ → Br^+ → Cl^- → Br, H, H, Cl + H, Br, Cl, H

13. (a) 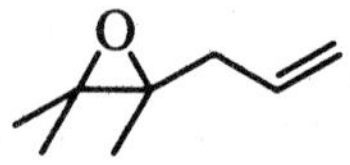 (b) 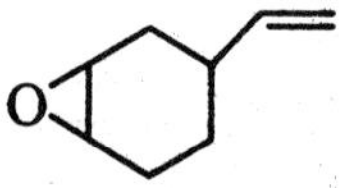 O (c) 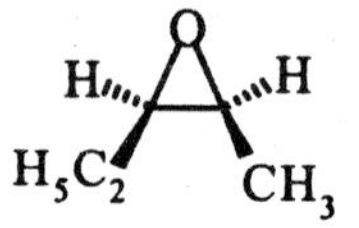 H, H, H_5C_2, CH_3 (d) 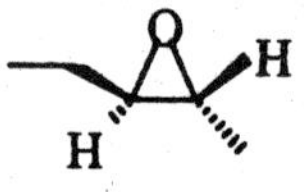 H, H

14. (a) 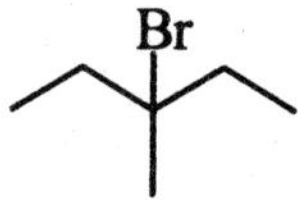 Br

(b) 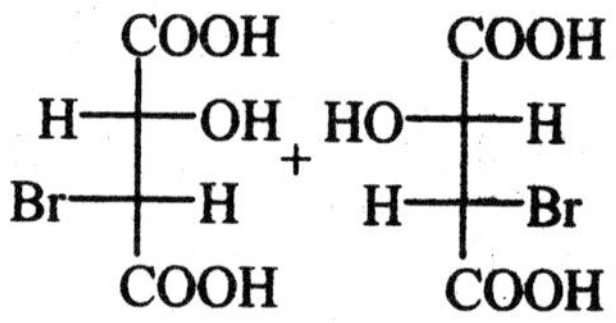 COOH, H—OH, Br—H, COOH + COOH, HO—H, H—Br, COOH

(c) 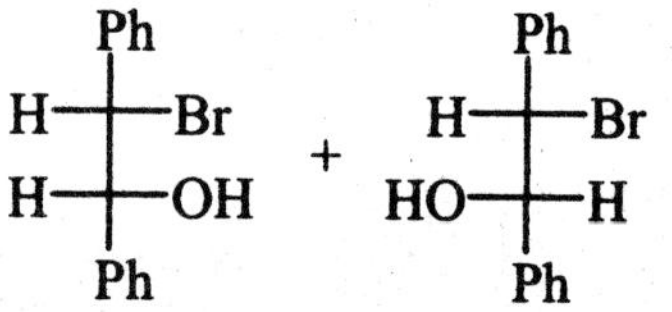 Ph, H—Br, H—OH, Ph + Ph, H—Br, HO—H, Ph

(Also equal amount of enantiomer of both will be produced)

(d) 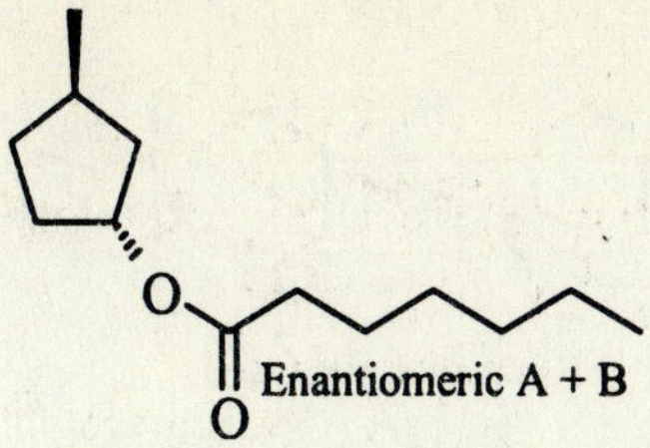 Enantiomeric A + B

(e) 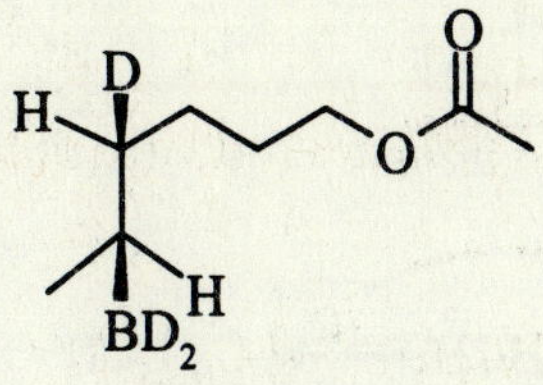 and its enantiomer C + D

(f) 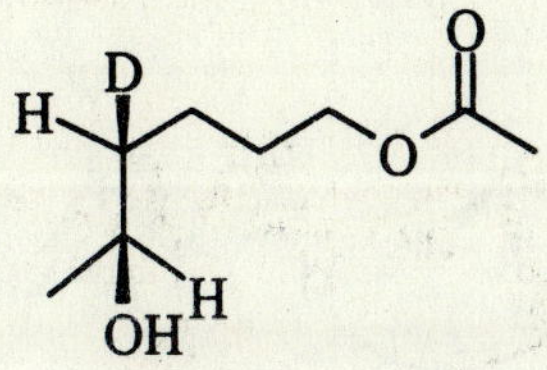 and its enantiomer E + F

15. (a) 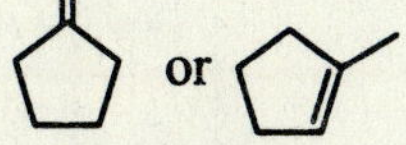 or (A) (b)  (B)

Br_2/H_2O (C) (c)  (D)

16. (a) 1-hexene < 2-hexene < 2-methyl-1-pentene.

(b) *trans*-2-butene < *cis*-2-butene < 2-methyl propene.

(c) 1-phenyl-2-butene < 1-phenyl-1-butene < 2-phenyl-1-butene.

17. 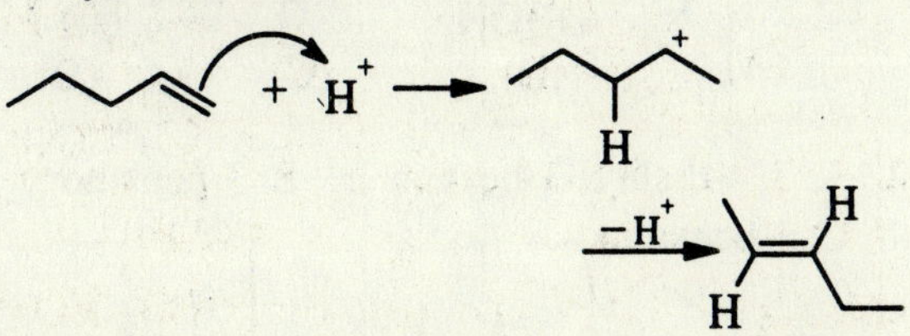 $+ H^+ \rightarrow$; $-H^+$

18. (a) 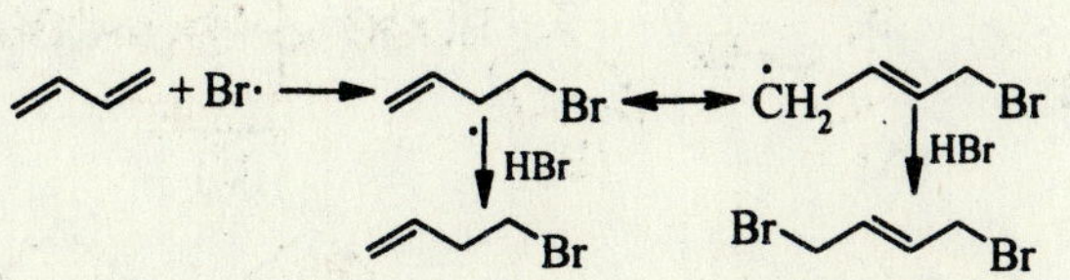 $+ Br\cdot \rightarrow$; HBr ; HBr

(b) 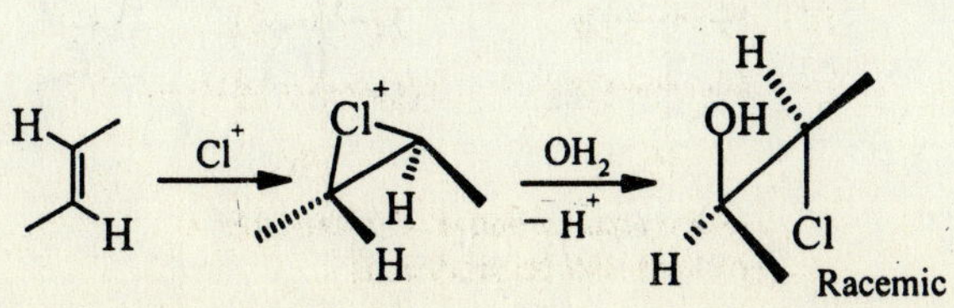 Cl^+ ; OH_2, $-H^+$; Racemic

(c) During electrophilic addition to 2-butyne, a less stable vinylic carbocation is produced, therefore it is less reactive, on other hand electrophilic addition to alkene produces a relatively stable alkyl carbocation, therefore alkene is more reactive. The mechanism of hydrogenation is different from electrophilic addition reaction. Hydrogenation reaction is retarded by steric hindrance from the bulky substituent at unsaturated carbon. In that respect an alkyne contain only two substituent at unsaturated carbons, whereas alkene contain four substituents at unsaturated carbons. Hence, alkyne is more reactive than alkene towards hydrogenation reaction.

19. (a) Tollen's test distinguishes terminal alkynes from other compounds.

(b) 2-hexene will decolourise the brown colour of bromine water.

(c) Same as (b).

20. 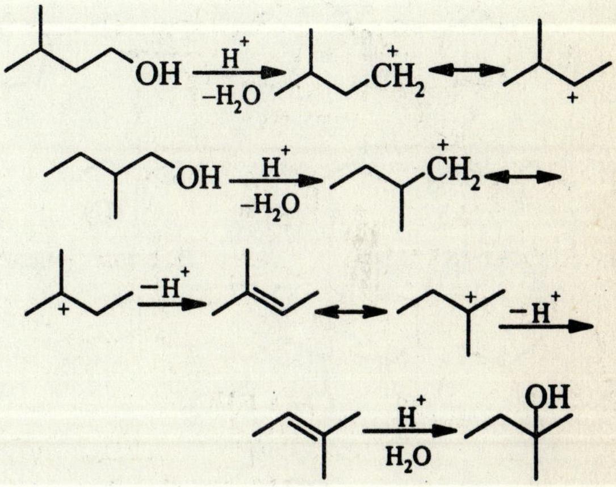 $\xrightarrow[-H_2O]{H^+}$; $-H^+$; $\xrightarrow[H_2O]{H^+}$

21. A: *cis*-1, 2-diphenylethene B: $PhCH_2CH_2Ph$
C: *trans*-1, 2-diphenylethene

22.

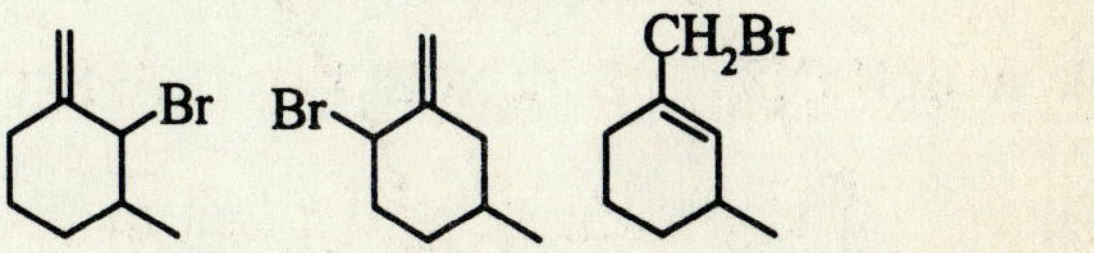

23. (a) 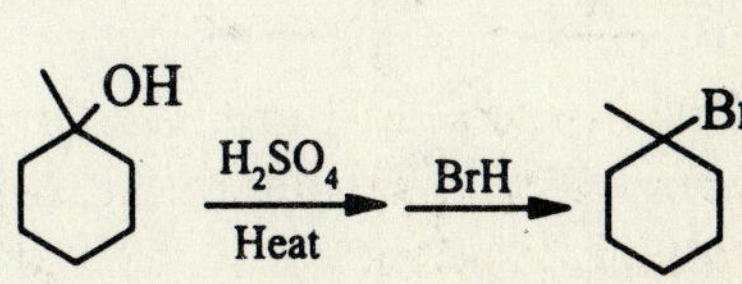 $\xrightarrow[Heat]{H_2SO_4}$ $\xrightarrow{BrH}$

(b) 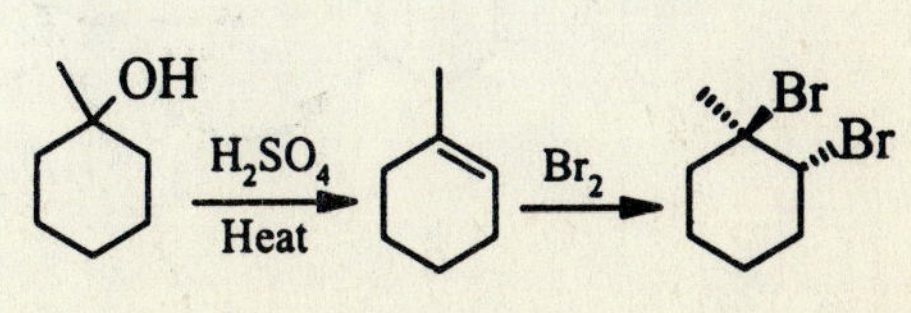 $\xrightarrow[Heat]{H_2SO_4}$ $\xrightarrow{Br_2}$

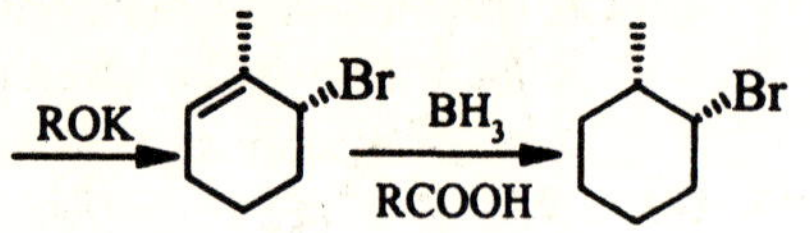

ROK, BH_3, RCOOH

(c) 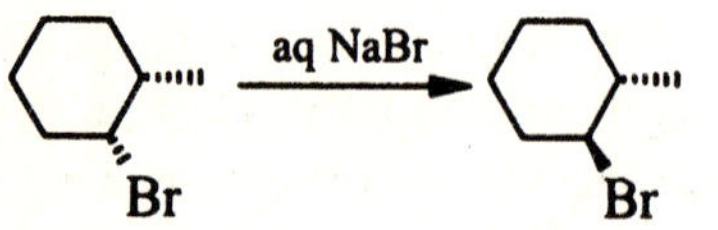 aq NaBr

Br Br

(d) 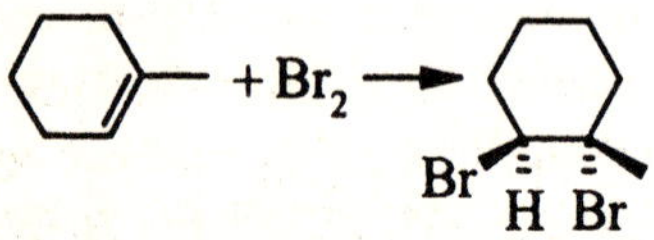 $+ Br_2 \rightarrow$

Br H Br

from (b) (isomerizes into more stable *trans* isomer.)

24. (a) 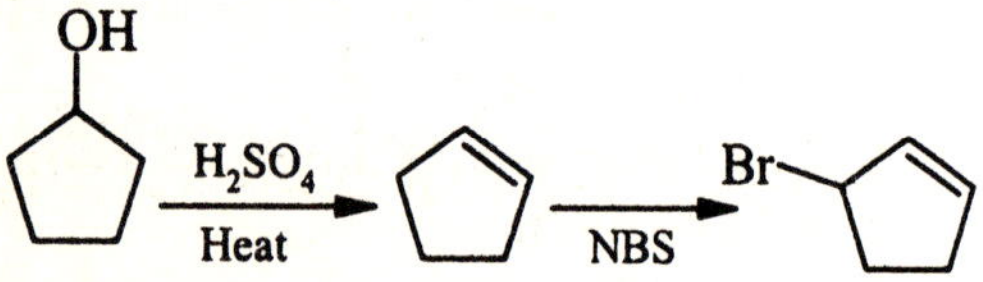 OH $\xrightarrow[Heat]{H_2SO_4}$ $\xrightarrow{NBS}$ Br

(b) 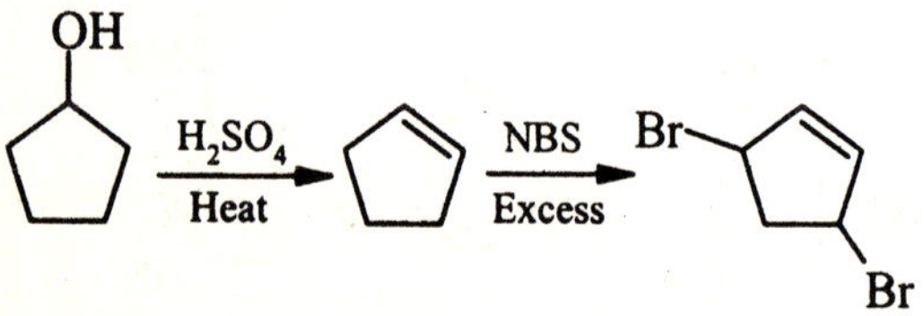 OH $\xrightarrow[Heat]{H_2SO_4}$ $\xrightarrow[Excess]{NBS}$ Br, Br

25. 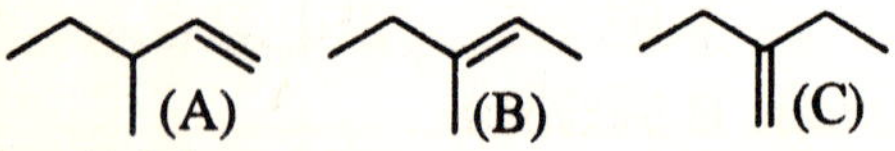 (A) (B) (C)

26. A: *cis*-3-hexene B: *trans*-3-hexene

27. 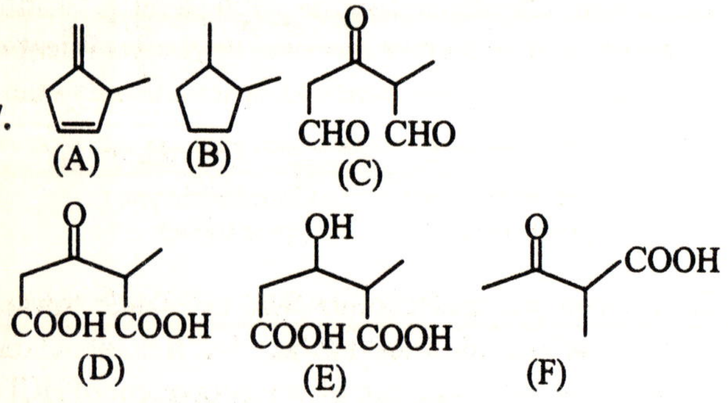 (A) (B) CHO CHO (C)

COOH COOH (D) OH COOH COOH (E) COOH (F)

28. A: $Et_3N^+CCl_2^-$ B: Et_2NCHCl_2 C: $Et_2N–CHO$

29. (a) 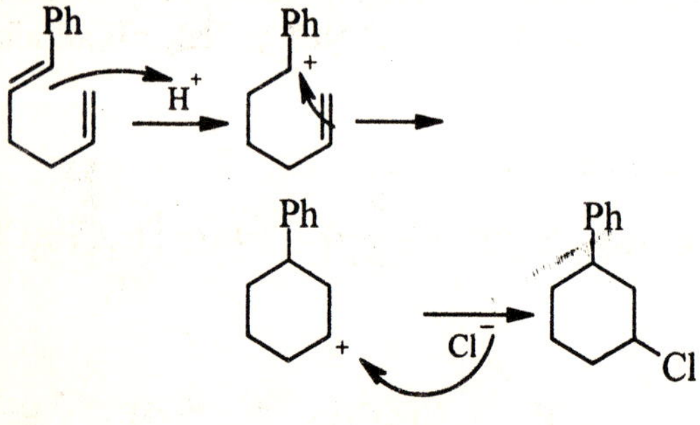 Ph $\xrightarrow{H^+}$ Ph + $\rightarrow$

Ph + $\xrightarrow{Cl^-}$ Ph Cl

(b) 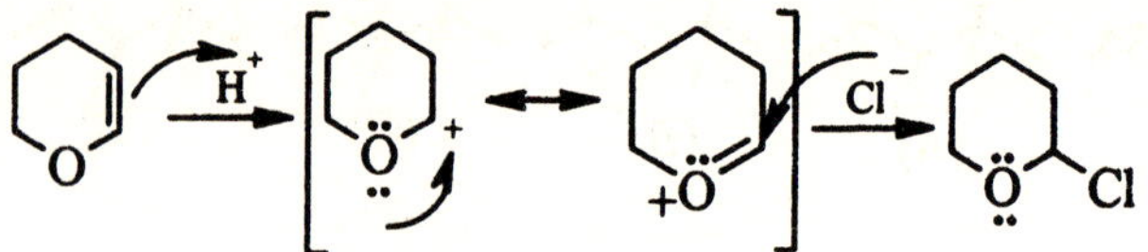 $\xrightarrow{H^+}$ [Ö + ↔ +Ö] $\xrightarrow{Cl^-}$ Ö Cl

Resonance stabilized

(c) $HOCl \rightarrow HO^- + Cl^+$

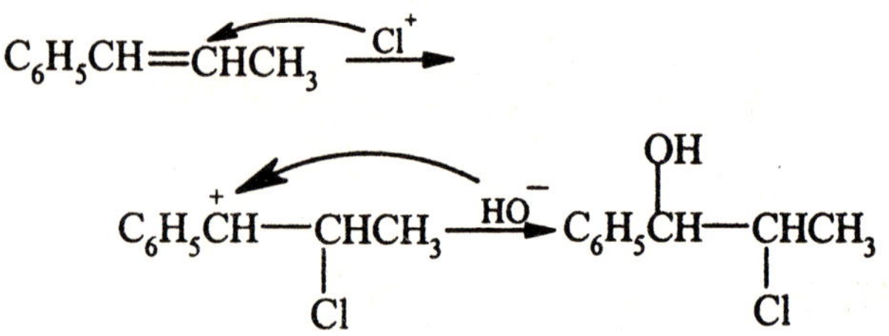

$C_6H_5CH{=}CHCH_3 \xrightarrow{Cl^+}$

$C_6H_5\overset{+}{C}H–\underset{Cl}{C}HCH_3 \xrightarrow{HO^-} C_6H_5\overset{OH}{C}H–\underset{Cl}{C}HCH_3$

Stabler carbocation

30. 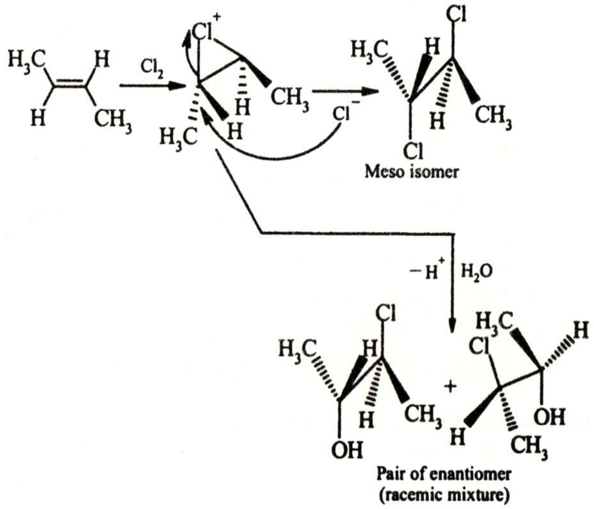 H_3C H H CH_3 $\xrightarrow{Cl_2}$ Cl^+ H CH_3 H_3C H $\xrightarrow{Cl^-}$ H_3C H Cl H CH_3 Cl

Meso isomer

$-H^+$ H_2O

H_3C H Cl H CH_3 OH + H_3C H Cl H OH CH_3

Pair of enantiomer (racemic mixture)

31. 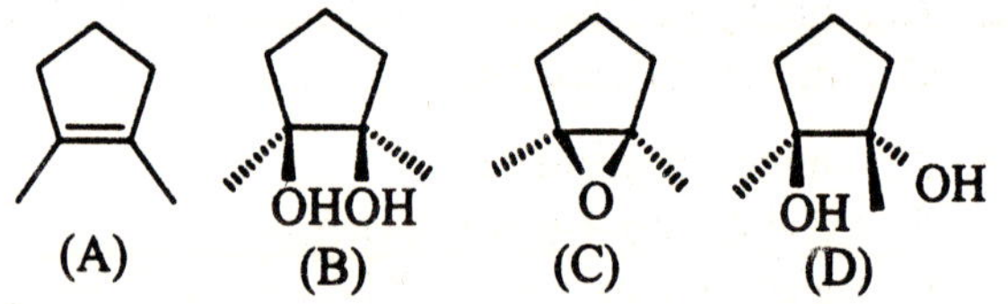 (A) OHOH (B) O (C) OH OH (D)

32. A: 3, 4-diethyl-3-hexene B: 3-pentanone
C: 3-pentanol

33. 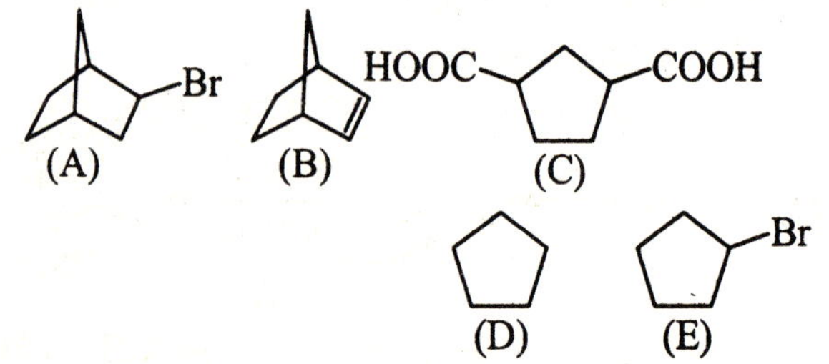 Br (A) (B) HOOC COOH (C)

(D) Br (E)

34. (A) (B) (C) (D) (E) (F) (G) (H) (I) (J)

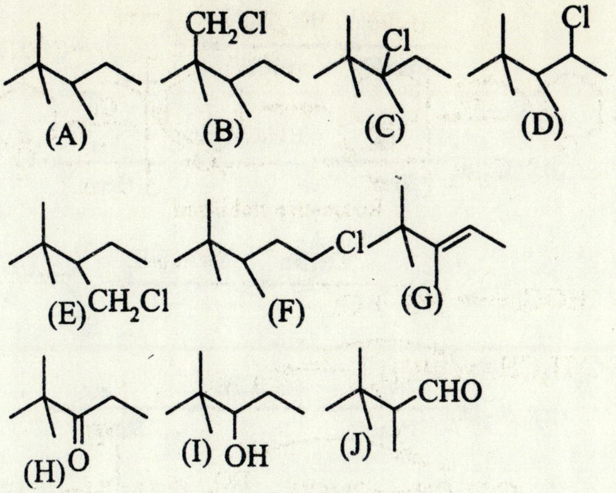

35. (A) (B) CH_3COOH (C) (D) (E) (F)

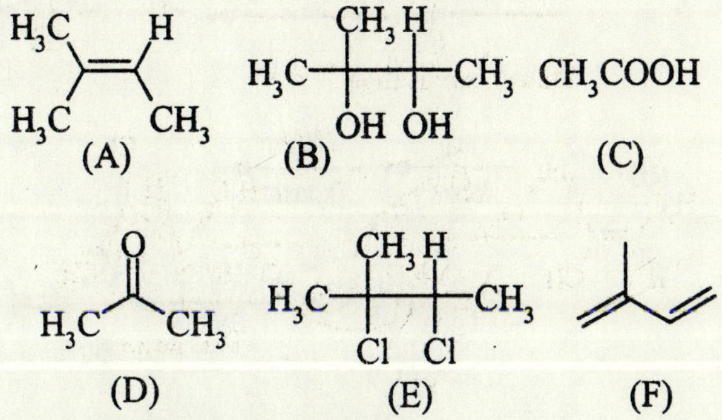

36. A: 2, 4-dimethyl-2-pentene
B: 2-methylpropanoic acid
C: 2-bromo-2-methyl propanoic acid
D: 2-hydroxy-2-methyl propanoic acid

37. A: 5-methyl-2, 4-heptadiene
B: CH_3CHO C: 2-butanone
D: Glyoxal E: 2-hydroxyethanoic acid

38. A: 3-methyl-1-butene B: 3-methyl-2-butanol
C: 3-methyl-2-butanone

39. A: 4-methyl-2-pentene
B: 2, 3-dichloro-4-methylpentane
C: 4-methyl-2-pentyne D: 2-methylpropanal

40. A: 3-methyl-1-butene
B: 2-bromo-3-methylbutane
C: 3-methyl-2-butanol
D: 3-methyl-2-butanone

41. $\xrightarrow{H^+}$ $\xrightarrow{-H^+}$:OMe-H; OMe

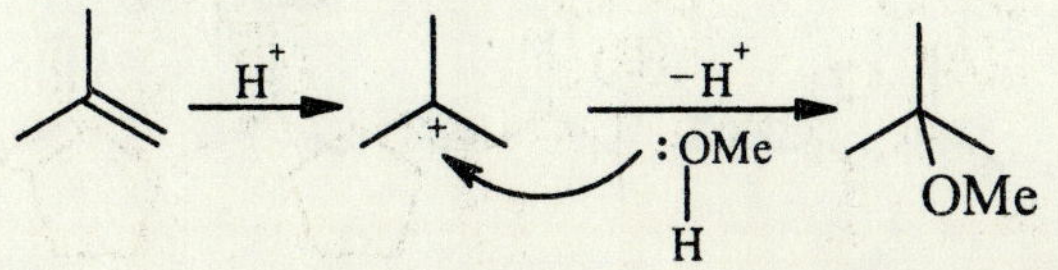

42. $\xrightarrow{Br_2}$ Cyclic bromonium ion $\xrightarrow{H_2O}$ (I, II) Br OH + OH Br

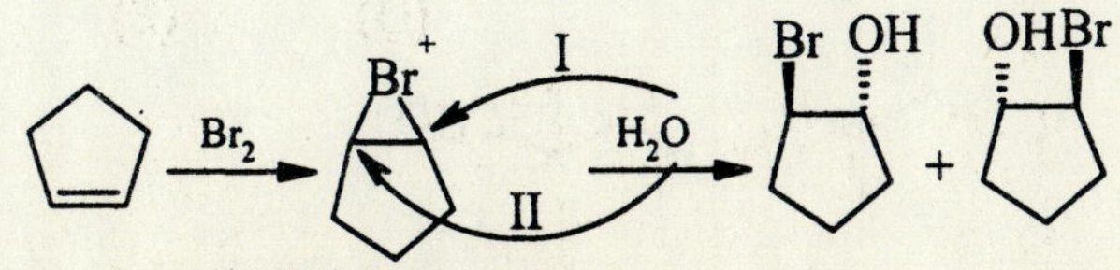

43. (a) (b) (c)

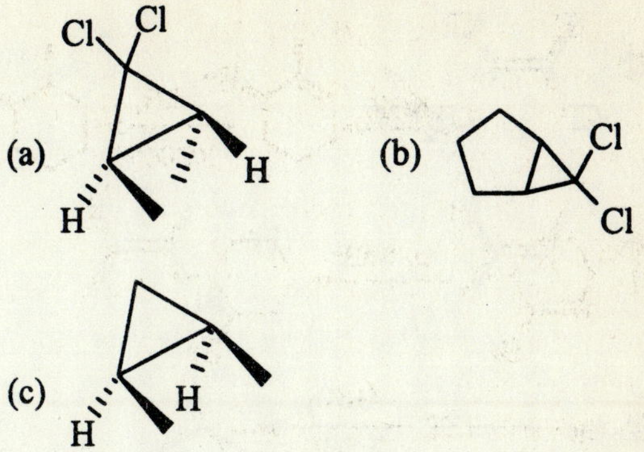

44. $\xrightarrow{Cl_2}$ $\xrightarrow[\text{Excess}]{\text{alc KOH}}$ $\xrightarrow[(CH_3)_3COK]{CHBr_3}$ $\xrightarrow{H_2/Pd\text{-}C}$

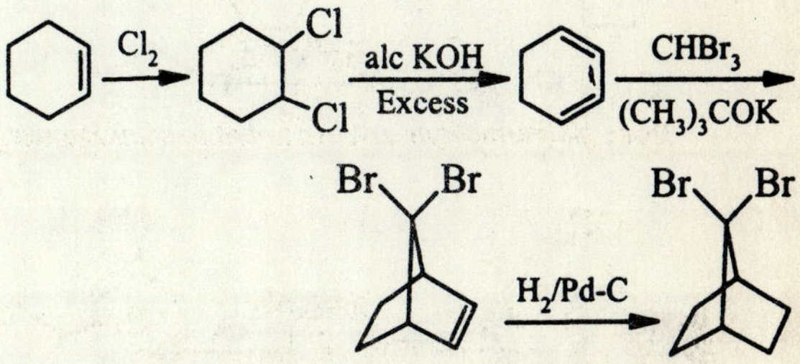

45. Methyl cyclohexene.

46. (A) (B) (C)

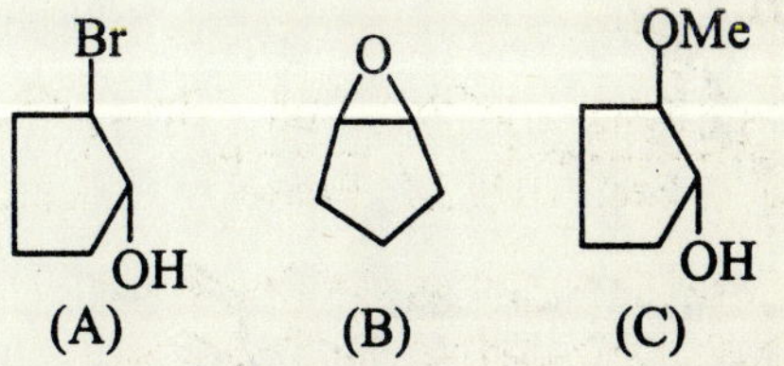

Practice Exercise III (Alkynes)

1. Internal alkynes are more stable than a terminal alkynes since former has two alkyl substituent on unsaturated carbons whereas latter has only one:

$R-C\equiv C-R$ An internal alkyne (more stable) $R-C\equiv C-H$ A terminal alkyne (less stable)

2. Alkynes are less reactive than an alkene towards electrophilic addition reaction, but the halo alkene produced after addition of one equivalent of HCl or HBr to alkyne are less reactive than both alkyne and alkene due to –I effect of halo group. Therefore, halo-alkene do not add HCl or HBr till all alkynes are exhausted.

3. Same as discussed in Ans. 2.

4. (a) $CH_2=C(Br)CH_2CH_3$ (b) $CH_3CBr_2CH_2CH_3$

(c) (d)

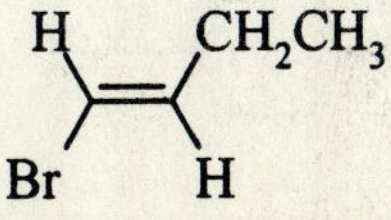

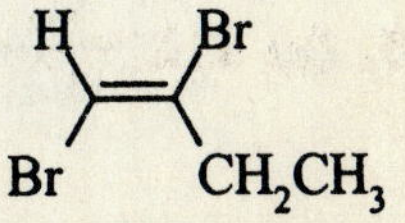

(e) 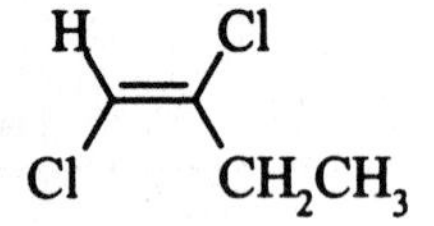 H, Cl, Cl, CH_2CH_3

5. (a) 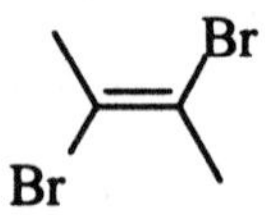 Br, Br (b) 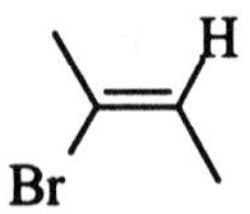 H, Br

6. (i) (a) 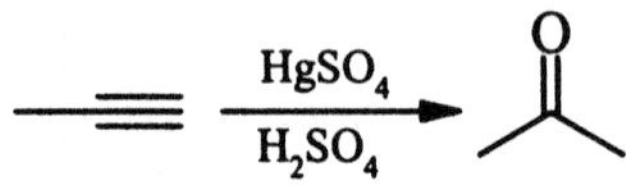 $\xrightarrow[H_2SO_4]{HgSO_4}$

(b) 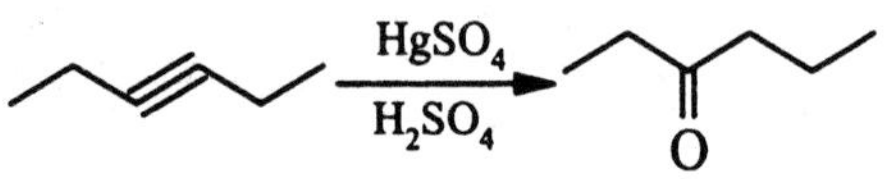 $\xrightarrow[H_2SO_4]{HgSO_4}$ O

(c) 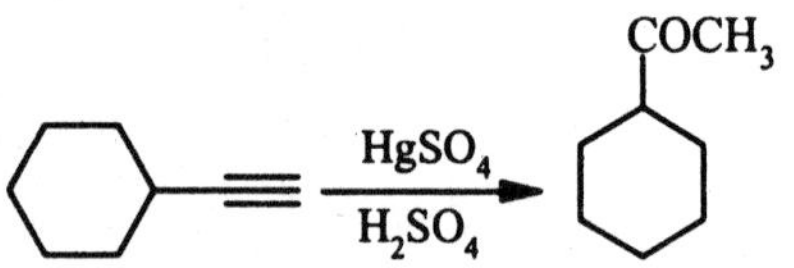 $\xrightarrow[H_2SO_4]{HgSO_4}$ $COCH_3$

(ii) (a) 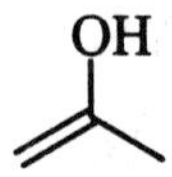 OH

(b) 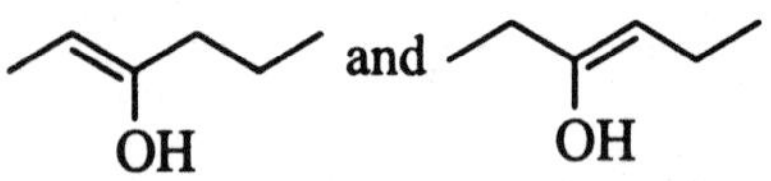 OH and OH

(c) 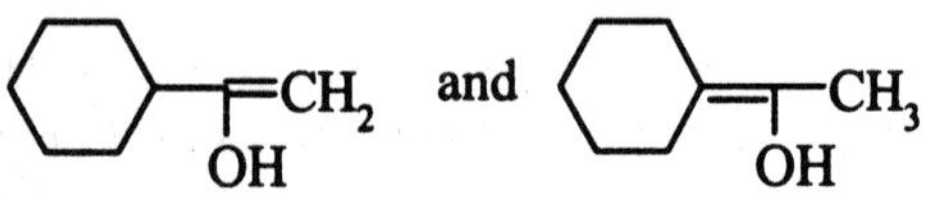 $=CH_2$ OH and $-CH_3$ OH

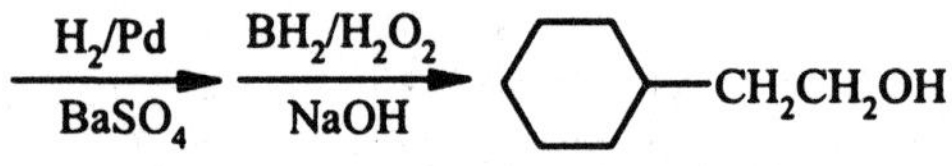

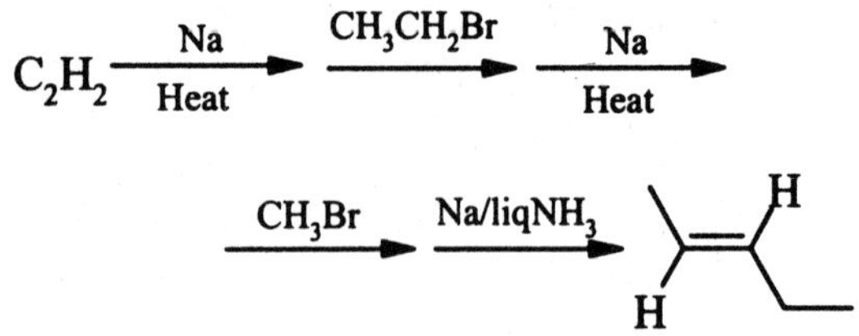

7. 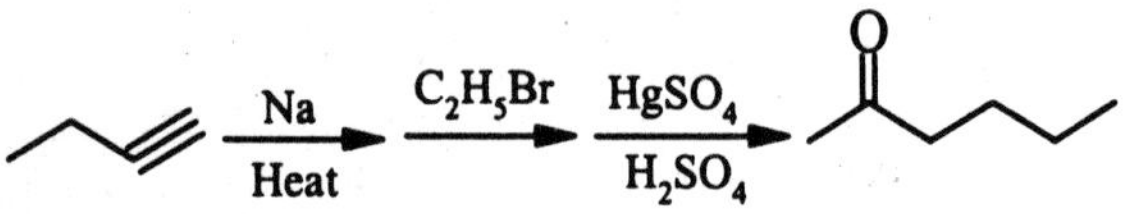 $\xrightarrow[Heat]{Na}$ $\xrightarrow{C_2H_5Br}$ $\xrightarrow[H_2SO_4]{HgSO_4}$ O

8. 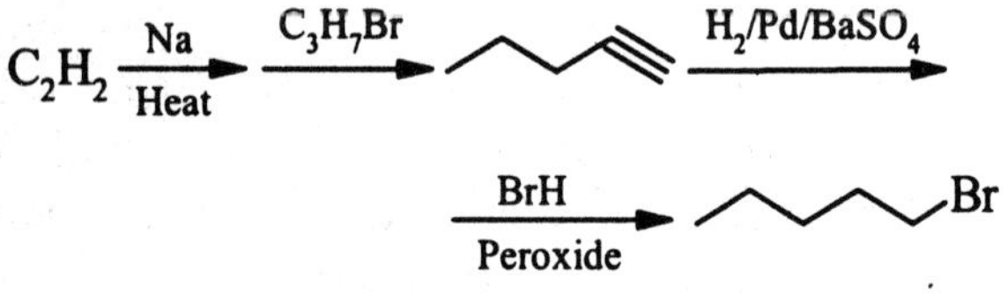 $C_2H_2 \xrightarrow[Heat]{Na}$ $\xrightarrow{C_3H_7Br}$ $\xrightarrow{H_2/Pd/BaSO_4}$

$\xrightarrow[Peroxide]{BrH}$ Br

9. 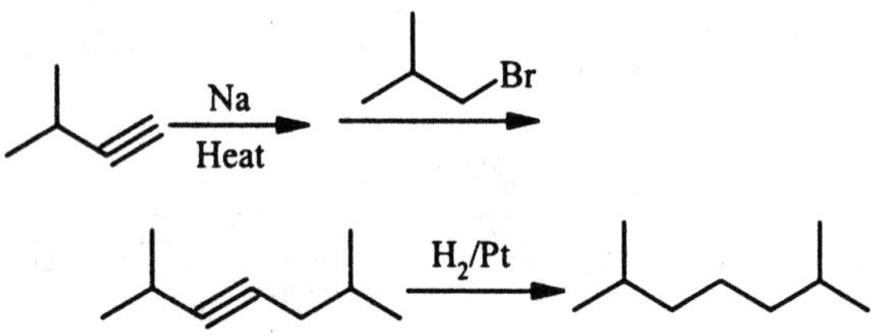 $\xrightarrow[Heat]{Na}$ $\xrightarrow{\text{Br}}$

$\xrightarrow{H_2/Pt}$

10. (a) 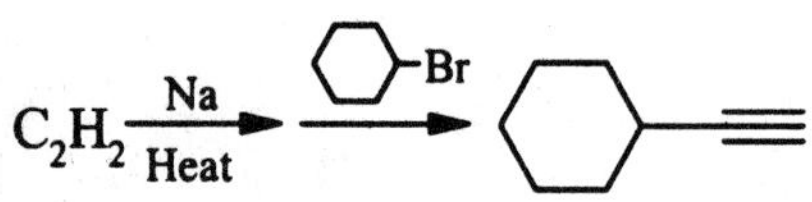 $C_2H_2 \xrightarrow[Heat]{Na}$ $\xrightarrow{\text{Br}}$

$\xrightarrow[BaSO_4]{H_2/Pd}$ $\xrightarrow[NaOH]{BH_3/H_2O_2}$ $-CH_2CH_2OH$

(b) $C_2H_2 \xrightarrow[Heat]{Na}$ $\xrightarrow{CH_3CH_2Br}$ $\xrightarrow[Heat]{Na}$

$\xrightarrow{CH_3Br}$ $\xrightarrow{Na/liqNH_3}$ H, H

(c) $C_2H_2 \xrightarrow[Heat]{Na}$ $\xrightarrow{C_3H_7Br}$ $\xrightarrow[Heat]{Na}$

$\xrightarrow{CH_3Br}$ $\xrightarrow{HBr\ (excess)}$ Br, Br

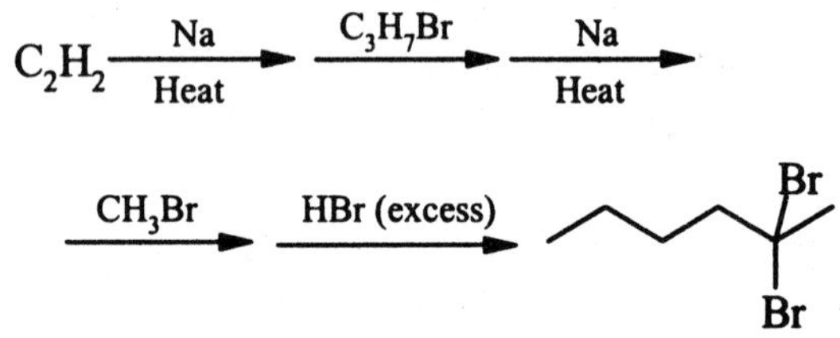

(d) 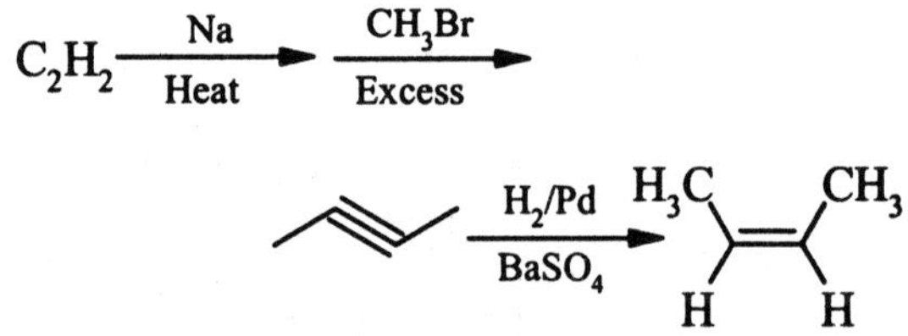 $C_2H_2 \xrightarrow[Heat]{Na}$ $\xrightarrow[Excess]{CH_3Br}$

$\xrightarrow[BaSO_4]{H_2/Pd}$ H_3C, CH_3, H, H

11. (a) 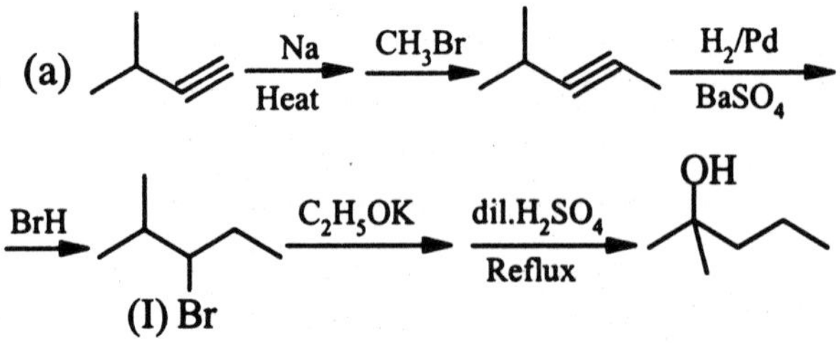 $\xrightarrow[Heat]{Na}$ $\xrightarrow{CH_3Br}$ $\xrightarrow[BaSO_4]{H_2/Pd}$

$\xrightarrow{BrH}$ (I) Br $\xrightarrow{C_2H_5OK}$ $\xrightarrow[Reflux]{dil.H_2SO_4}$ OH

(b) (I) 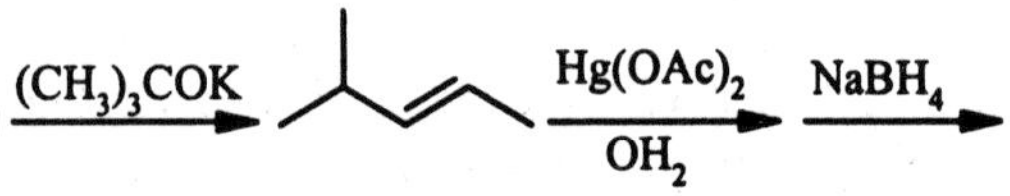 $\xrightarrow{(CH_3)_3COK}$ $\xrightarrow[OH_2]{Hg(OAc)_2}$ $\xrightarrow{NaBH_4}$

12. c and d only.

13. $H_2/Pd/BaSO_4$, HBr (excess), HBr/Peroxide, $H_2/Pd/BaSO_4$ then HBr/Peroxide, $BH_3/H_2O_2/NaOH$, $HgSO_4/H_2SO_4$, $H_2/Pd/BaSO_4$ then HBr.

14. 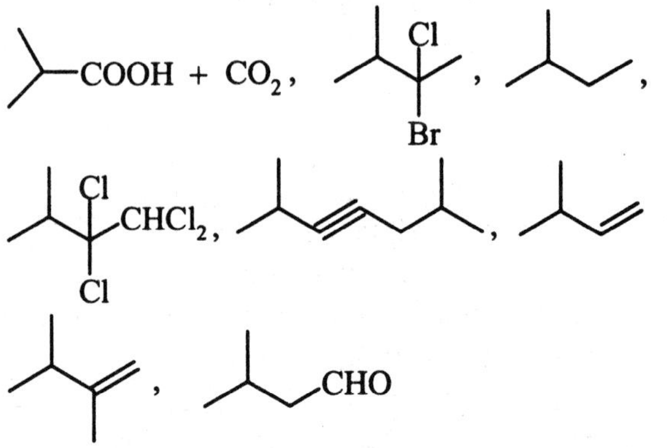 $-COOH + CO_2$, Cl, Br, , ,

Cl, $CHCl_2$, Cl, , ,

, CHO

15. (a) $CH_3CBr{=}CH_2$ (b) $CH_3CBr_2CH_3$
(c) $CH_3CBr{=}CHBr$ (d) $CH_3CBr{=}CHBr$
(e) CH_3COCH_3 (f) CH_3CH_2CHO
(g) $CH_3CH{=}CHBr$ (h) $CH_3CH{=}CH_2$
(i) $CH_3CH{=}CH_2$

16. (a) $CH_3CBr{=}CHCH_3$ (b) $CH_3CBr_2CH_2CH_3$
(c) $CH_3CBr{=}CBrCH_3$ (d) $CH_3CBr{=}CBrCH_3$
(e) $CH_3COCH_2CH_3$ (f) $CH_3COCH_2CH_3$
(g) $CH_3CH{=}CBrCH_3$ (h) *cis*-2-butene
(i) *trans*-2-butene

17. (a) 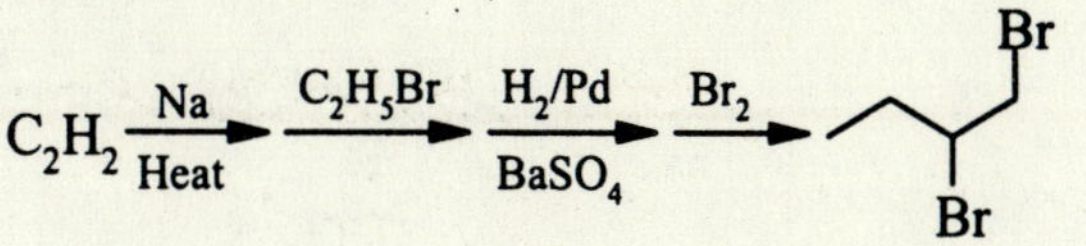 $C_2H_2 \xrightarrow[\text{Heat}]{Na} \xrightarrow{C_2H_5Br} \xrightarrow[BaSO_4]{H_2/Pd} \xrightarrow{Br_2}$ Br, Br

(b) 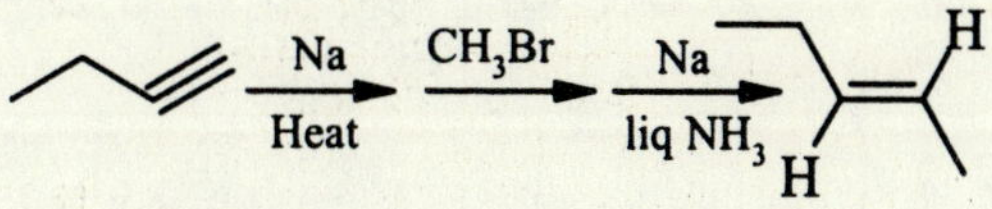 $\xrightarrow[\text{Heat}]{Na} \xrightarrow{CH_3Br} \xrightarrow[\text{liq } NH_3]{Na}$ H, H

(c) $C_2H_2 \xrightarrow[H_2SO_4]{HgSO_4} CH_3CHO$

(d) 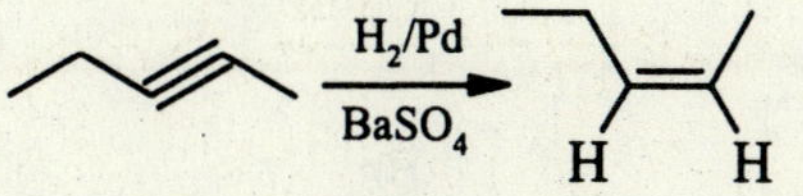 $\xrightarrow[BaSO_4]{H_2/Pd}$ H H

18. (a) 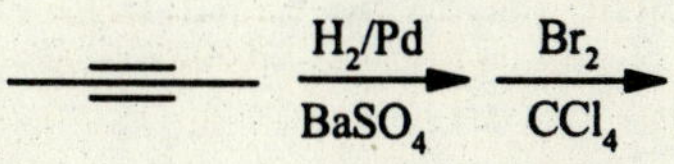 $\xrightarrow[BaSO_4]{H_2/Pd} \xrightarrow[CCl_4]{Br_2}$

Racemic 2, 3-dibromobutane

(b) 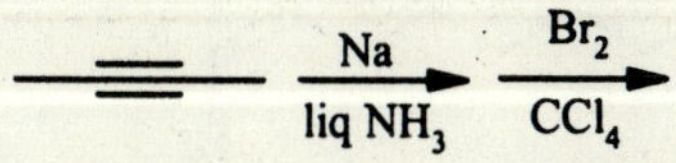 $\xrightarrow[\text{liq } NH_3]{Na} \xrightarrow[CCl_4]{Br_2}$

Meso 2, 3-dibromobutane

(c) 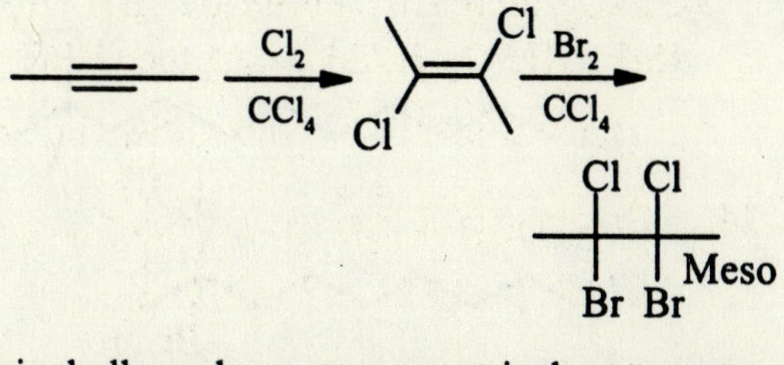 $\xrightarrow[CCl_4]{Cl_2}$ Cl, Cl $\xrightarrow[CCl_4]{Br_2}$

Cl Cl
Br Br Meso

19. Terminal alkyne have unsymmetrical arrangement of groups at triple bond, gives regioselective product, whereas internal alkynes have carbon bonded on both sides of triple bond, no regioselectivity is observed during the reaction.

20. (a) Pentene $\xrightarrow[CCl_4]{Br_2} \xrightarrow{NaNH_2} \xrightarrow[OH_2]{H^+}$ 1-Pentyne

(b) 1-bromopropane $\xrightarrow{C_2H_5OK} \xrightarrow[CCl_4]{Br_2} \xrightarrow{NaNH_2}$

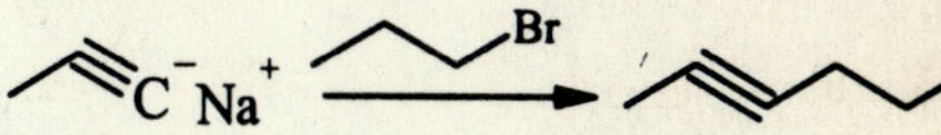

C^-Na^+ Br $\longrightarrow$

(c) $C_2H_2 \xrightarrow{\text{Na/heat}} \xrightarrow{C_2H_5Br} \xrightarrow[NaOH]{B_2H_6/H_2O_2}$

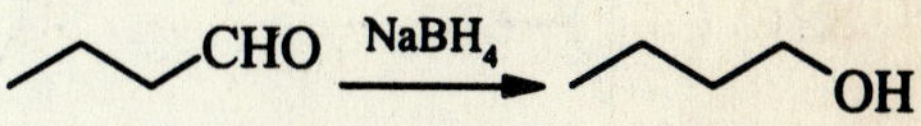

CHO $\xrightarrow{NaBH_4}$ OH

(d) 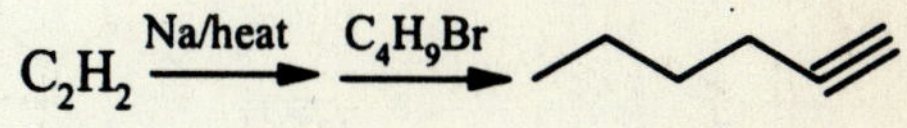 $C_2H_2 \xrightarrow{\text{Na/heat}} \xrightarrow{C_4H_9Br}$

$\xrightarrow[BaSO_4]{H_2/Pd} \xrightarrow{NBS}$ Br $\xrightarrow[RCOOH]{BH_3}$

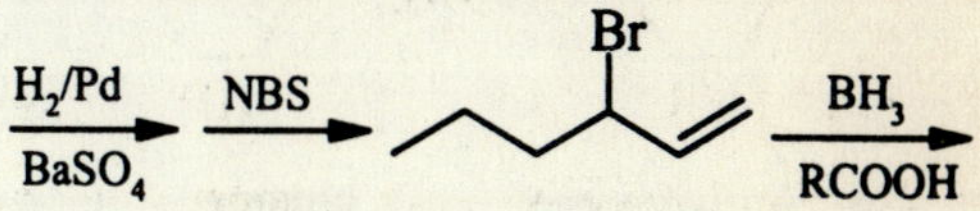

(e) 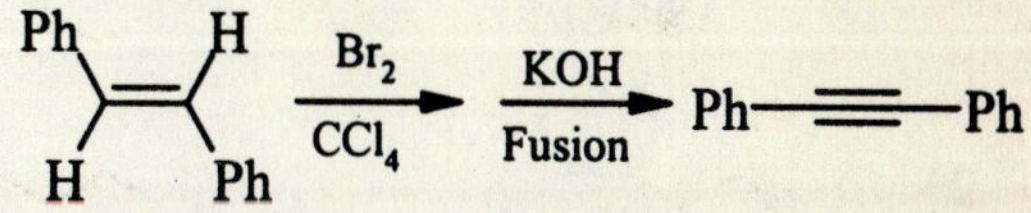 Ph H / H Ph $\xrightarrow[CCl_4]{Br_2} \xrightarrow[\text{Fusion}]{KOH}$ Ph—≡—Ph

$\xrightarrow[BaSO_4]{H_2/Pd}$ Ph Ph / H H $\xrightarrow{CH_3ONa}$ OMe

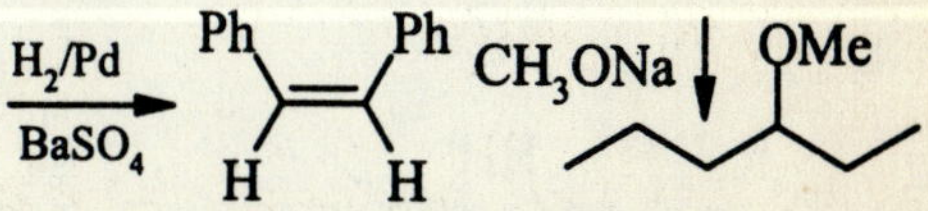

(f) 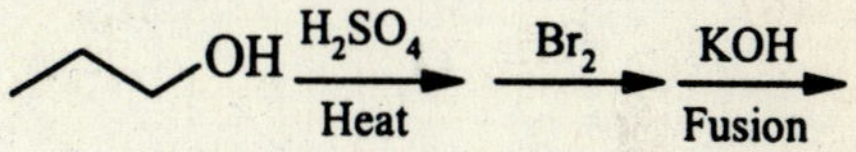 OH $\xrightarrow[\text{Heat}]{H_2SO_4} \xrightarrow{Br_2} \xrightarrow[\text{Fusion}]{KOH}$

Propyne $\xrightarrow[HgSO_4]{H_2SO_4}$ Acetone

(g) Propyne $\xrightarrow{\text{Na/heat}} \xrightarrow{D_2O}$

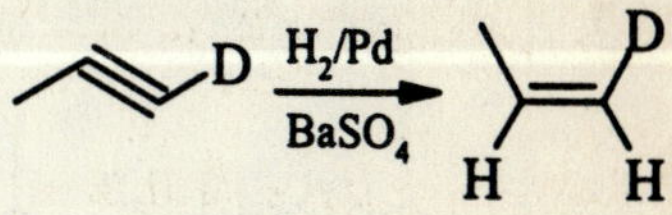

D $\xrightarrow[BaSO_4]{H_2/Pd}$ D / H H

21. (a) 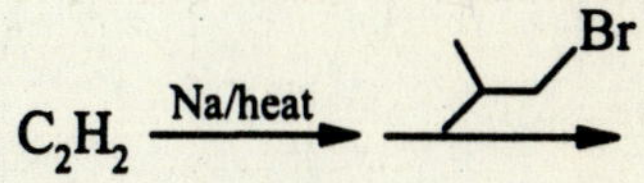 $C_2H_2 \xrightarrow{\text{Na/heat}}$ Br $\longrightarrow$

$\xrightarrow[BaSO_4]{H_2/Pd} \xrightarrow[NaOH]{\text{Cold, dil. } KMnO_4}$ Product

Na | liq NH_3 $\xrightarrow[NaOH]{\text{Cold, dil. } KMnO_4}$ Product (b)

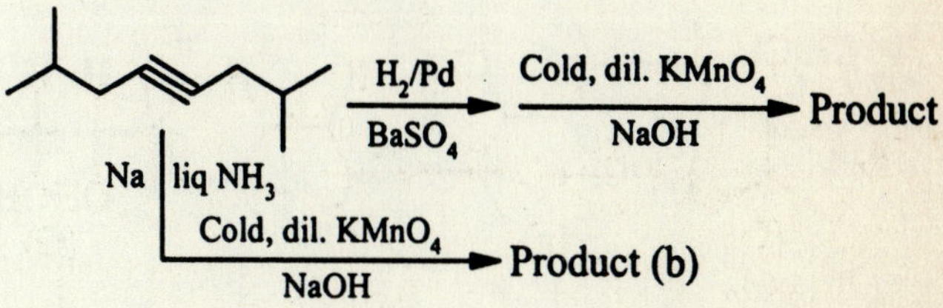

22. 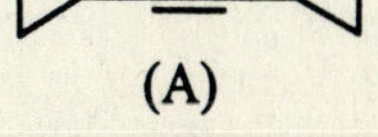 (A) 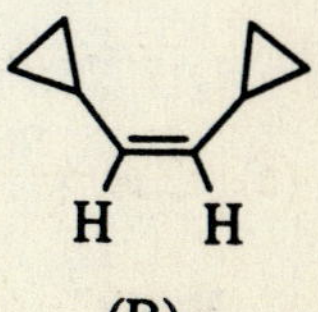 (B) H H

(C) (D)

29. (A) (B)

23. (A) (B)

30. Br Br A Cu(I) C

(C) COONa (D) (E) Cl

B $C^- Cu^+$ H^+ D═A

HO F E═A

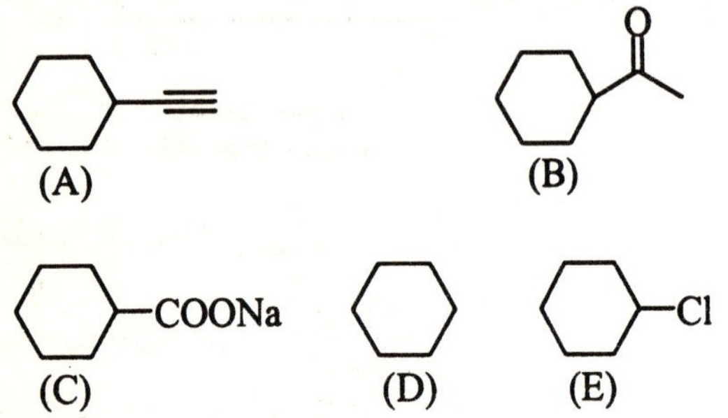

24. A: 3-hexyne, B: hexane, C: *trans*-3-hexene
D: *cis*-3-hexene

25. (A) (B)

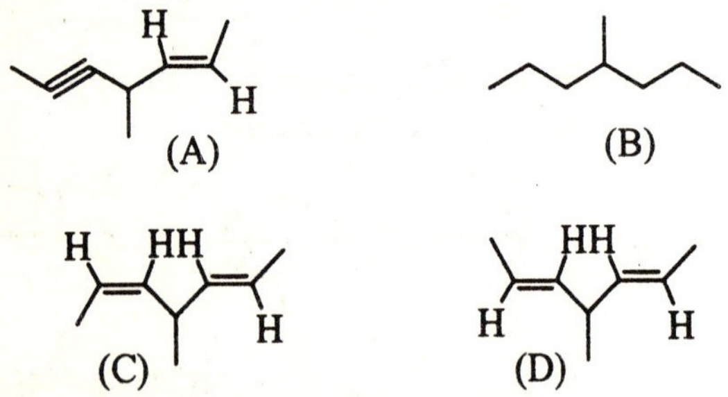

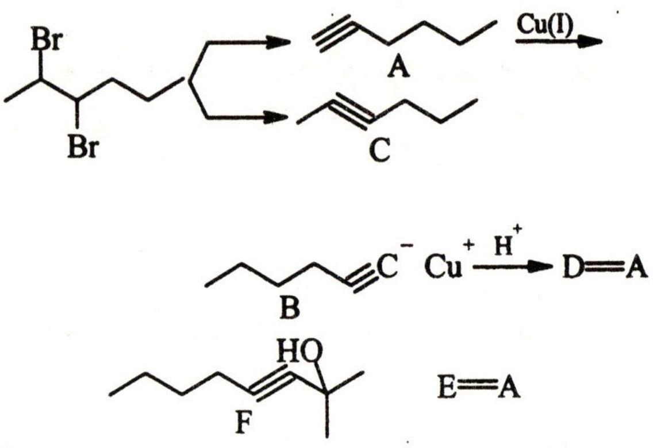

31. (A) OH (B) ONa
(C) OMe (D) MgBr
(E) OH (F)

(C) (D)

(G) Br Br

26. $Ph—\equiv—CH(CH_3)CH_2CH_3$ COOH COOH
(A) (B) (C)

(H)
(I) Br

27. (A) (B) Cl (C)

(J) $\bar{C}Na^+$

(K) OH

(D) CH_2Cl (E) Cl Cl (F) (G) OH OH

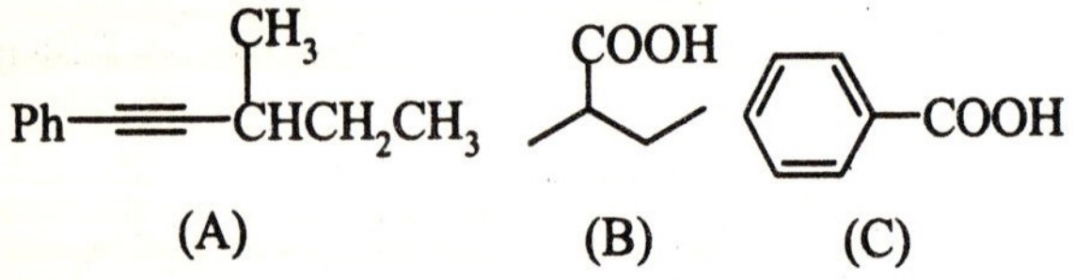

(L) Br Br

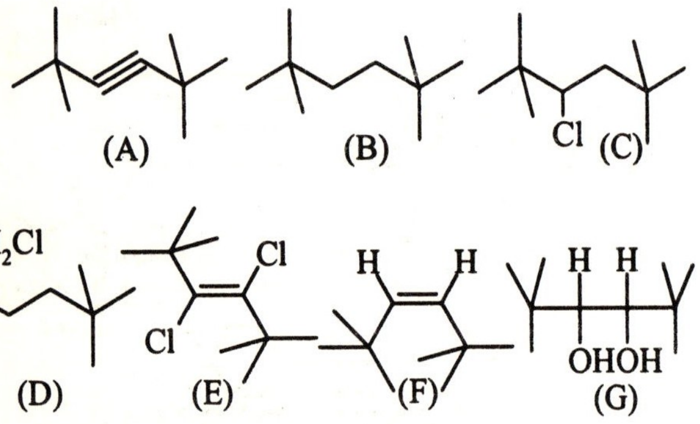

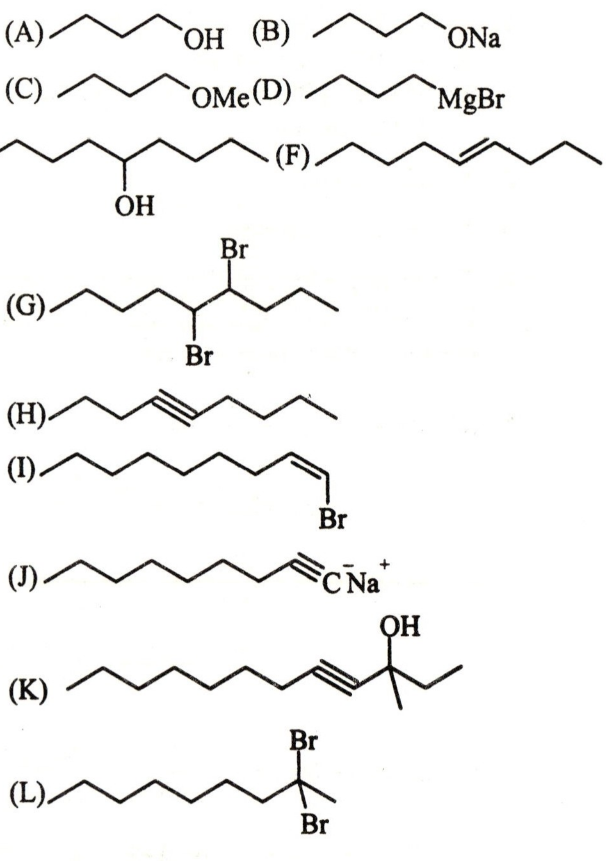

32. (A) (B)

28. (A) C_2H_5 C_2H_5 (B)

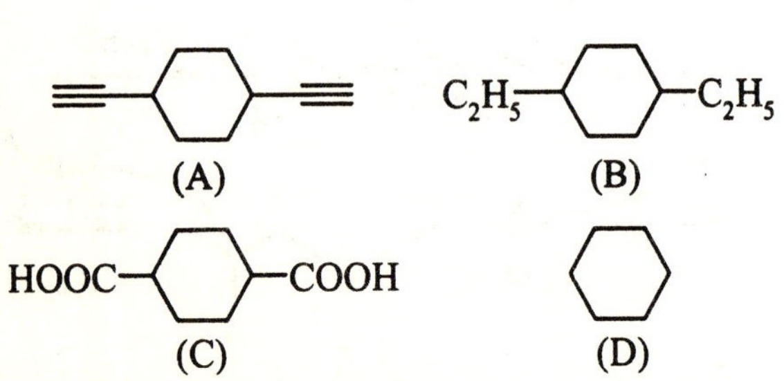

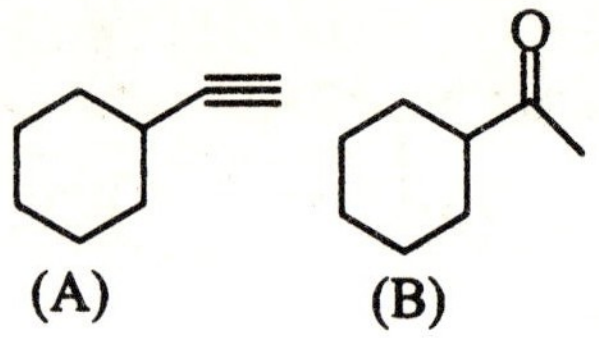

HOOC COOH (C) (D)

(C) COOH (D) COOH (E)

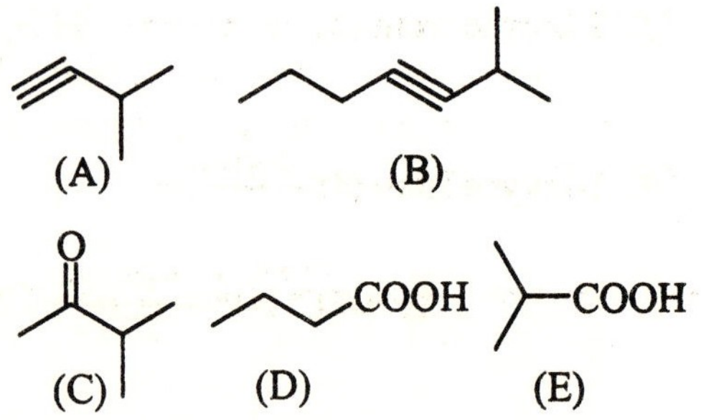

33. 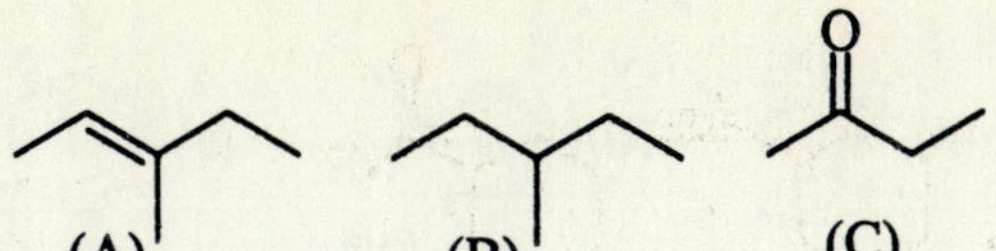 (A) (B) (C)

34. 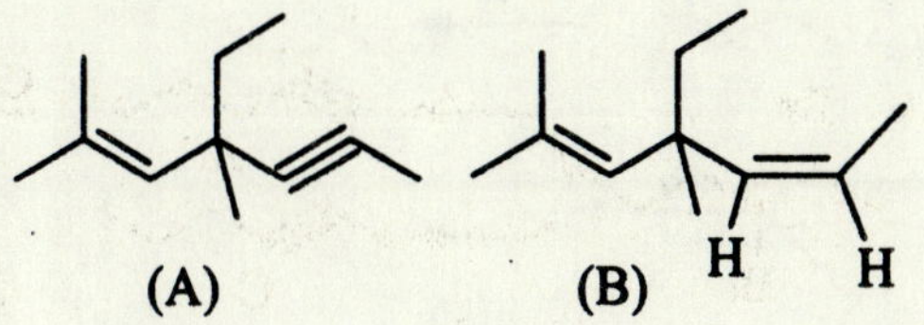 (A) (B)

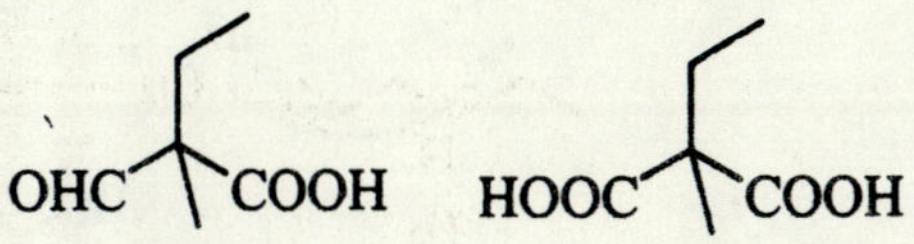

$OHC-C(C_2H_5)(CH_3)-COOH$ (C) $HOOC-C(C_2H_5)(CH_3)-COOH$ (D)

Review Exercise

1. (a) X : CH_3Li; Y: $(CH_3)_2CuLi$, Z: *n*-pentane
   (b) X: methyl cyclohexane
   (c) X: $n—C_4H_9Li$, Y: $(n—C_4H_9)_2CuLi$, Z: No reaction
2. 2-methyl pentane
3. 2, 3-dimethyl butane
4. Due to larger size of CH_3^- group compared to H, C—C—C bond angle expand to some extant and H—C—C bond angle contract.
5. (a) 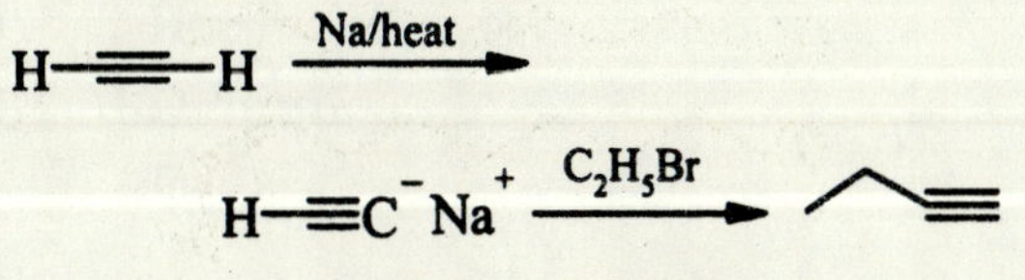 $H—C\equiv C—H \xrightarrow{\text{Na/heat}}$

   $H—C\equiv C^- Na^+ \xrightarrow{C_2H_5Br}$ 1-butyne

   (b) $H—C\equiv C—H \xrightarrow[\text{Excess}]{\text{Na/heat}} \xrightarrow[\text{Excess}]{CH_3Br} \longrightarrow H_3C—C\equiv C—CH_3$

   (c) 2-butyne from (b) $\xrightarrow[\text{Quinoline}]{Pd/BaSO_4}$ cis-2-butene (H H)

   (d) 2-butyne from (b) $\xrightarrow{\text{Na/liq } NH_3}$ trans-2-butene (H, H)

   (e) 1-butyne from (a) $\xrightarrow{\text{Na/heat}}$

   $H—C\equiv C^- Na^+ \xrightarrow{D_2O}$ $CH_3CH_2C\equiv C—D$

   (f) 2-butyne from (b) 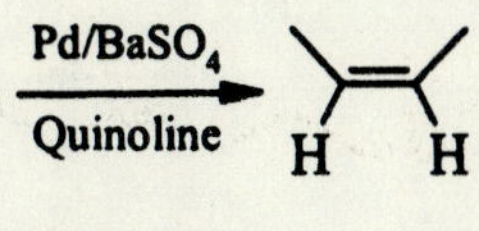 $\xrightarrow[D_2]{Pd/BaSO_4}$

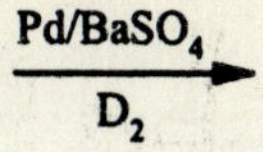

6. 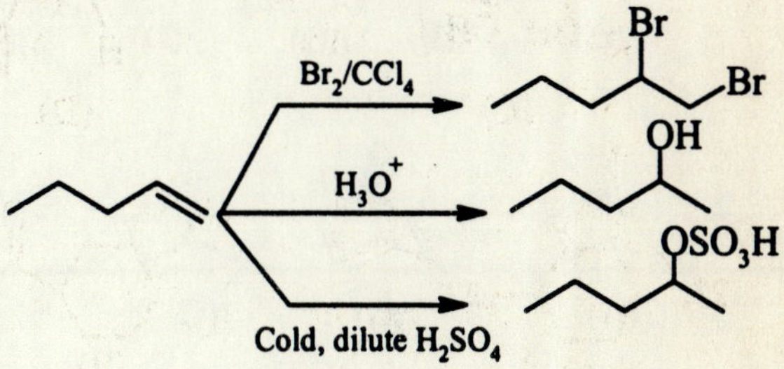 1-pentene $\xrightarrow{Br_2/CCl_4}$ 1,2-dibromopentane (Br, Br)

   $\xrightarrow{H_3O^+}$ 2-pentanol (OH)

   $\xrightarrow{\text{Cold, dilute } H_2SO_4}$ OSO_3H (2-pentyl hydrogen sulfate)

7. 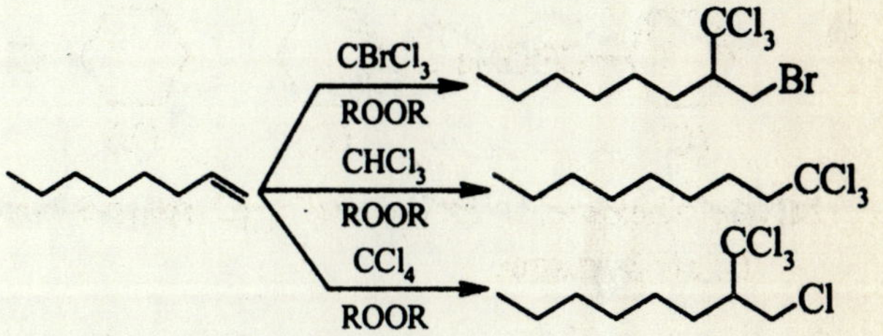 alkene $\xrightarrow[\text{ROOR}]{CBrCl_3}$ product with CCl_3 and Br

   $\xrightarrow[\text{ROOR}]{CHCl_3}$ product with CCl_3

   $\xrightarrow[\text{ROOR}]{CCl_4}$ product with CCl_3 and Cl

8. (a) 2-methyl-2-pentene
   (b) methylene cyclohexane

9. (a) 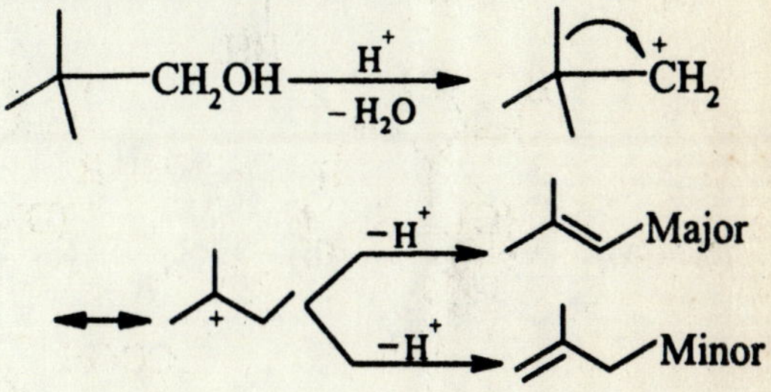 $(CH_3)_3C—CH_2OH \xrightarrow[-H_2O]{H^+} (CH_3)_3C—\overset{+}{C}H_2$

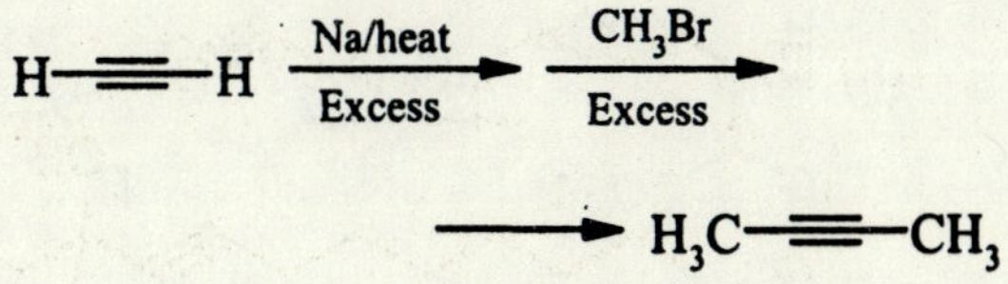

   $\longleftrightarrow$ tertiary carbocation $\xrightarrow{-H^+}$ Major; $\xrightarrow{-H^+}$ Minor

   (b) 1-methylcyclohexanol (OH) $\xrightarrow[-H_2O]{H^+}$ carbocation $\xrightarrow{-H^+}$ methylenecyclohexane + 1-methylcyclohexene

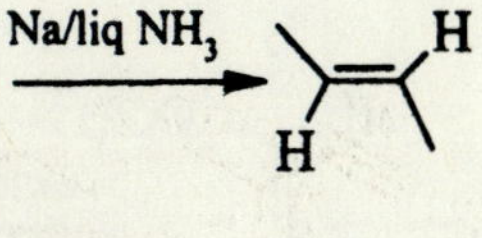

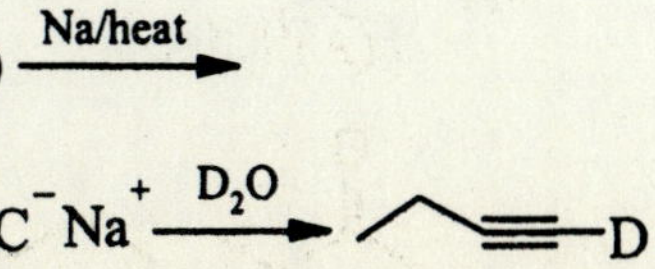

   Minor Major

   (c) 2,2-dimethylcyclohexanol (OH) $\xrightarrow[-H_2O]{H^+}$ carbocation $\xrightarrow{\text{Me-shift}}$ carbocation $\xrightarrow{-H^+}$ Minor + Major

   $\xrightarrow{-H^+}$ 4,4-dimethylcyclohexene

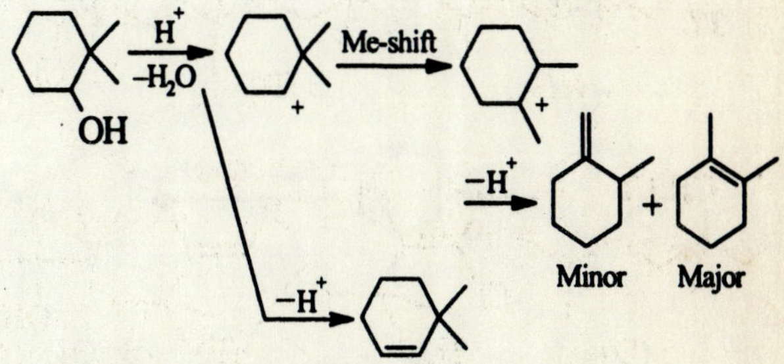

10. 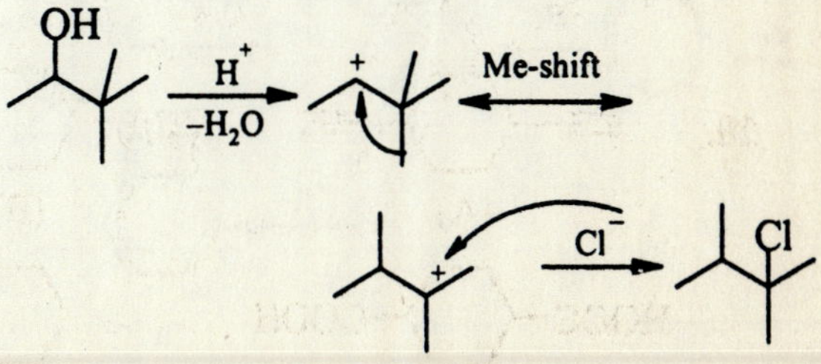 3,3-dimethyl-2-butanol (OH) $\xrightarrow[-H_2O]{H^+}$ carbocation $\xleftrightarrow{\text{Me-shift}}$

    tertiary carbocation $\xrightarrow{Cl^-}$ product (Cl)

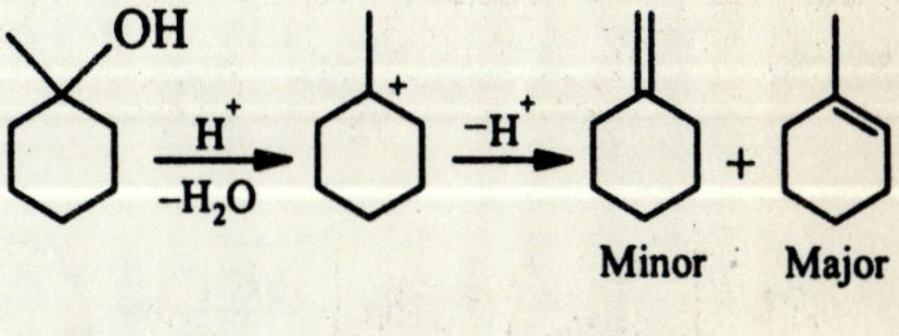

11. 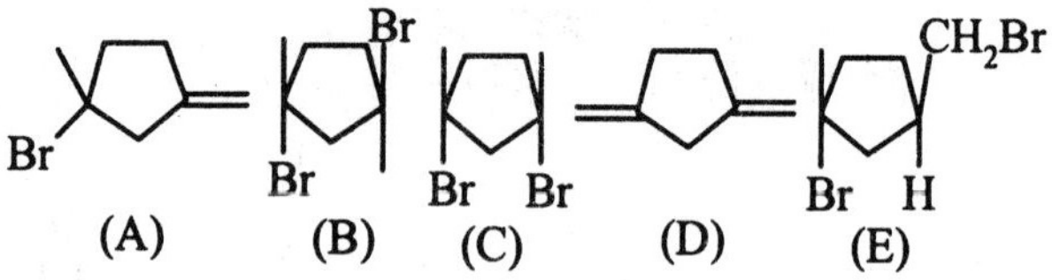 Br (A) Br (B) Br Br (C) (D) CH_2Br Br H (E)

12. 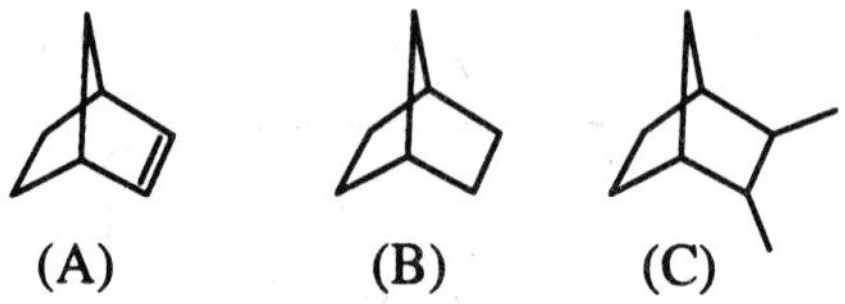 (A) 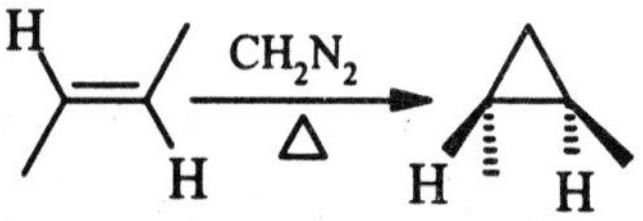 (B) 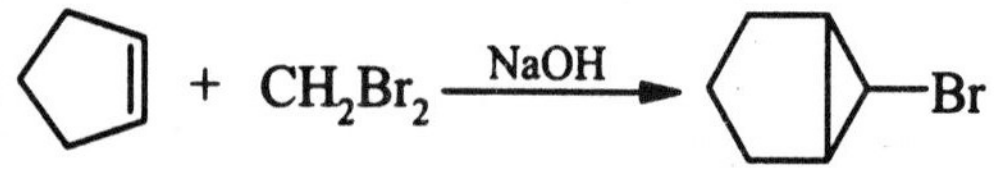 (C)

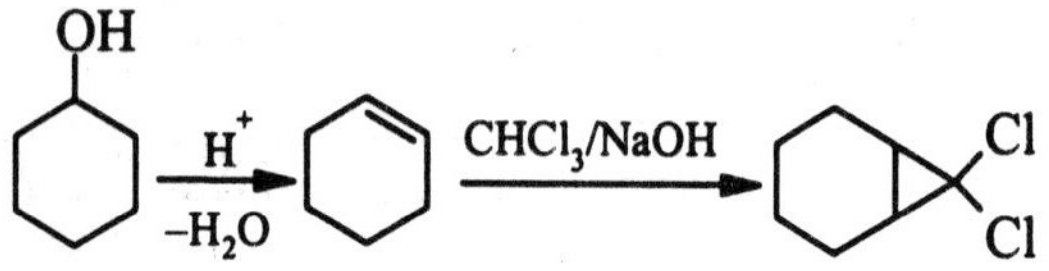

13. (A) 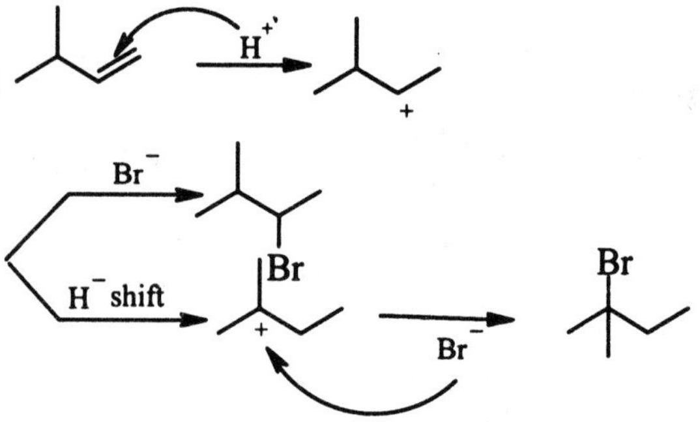 (B) 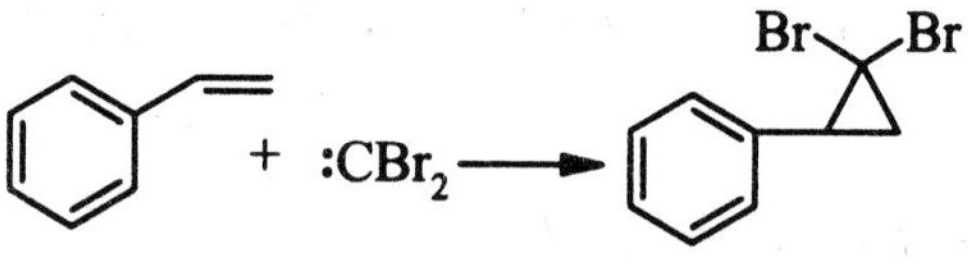 CH_2Cl (C) 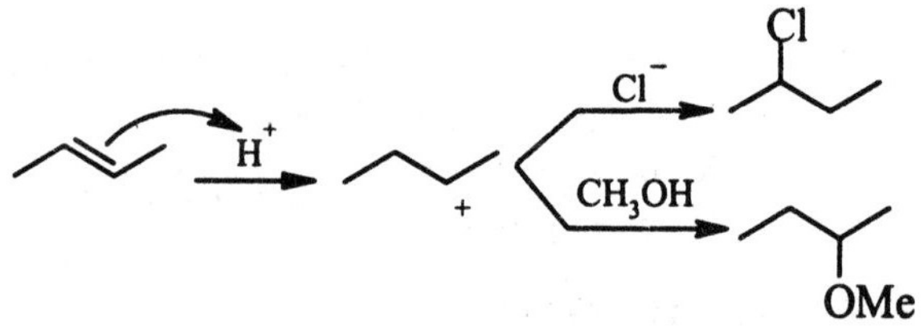 Cl (D) Cl (E)

14. (A) 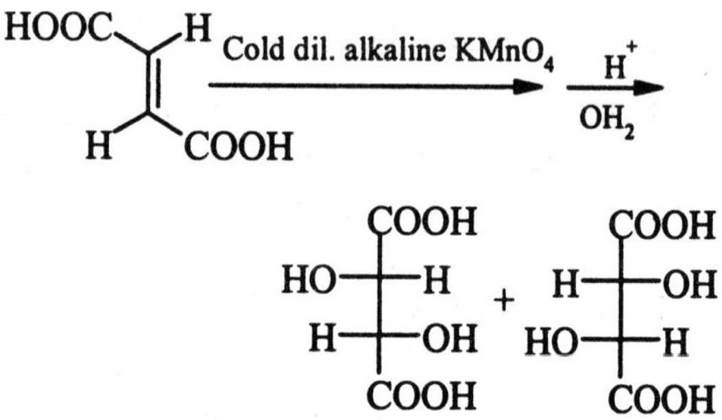 (B) 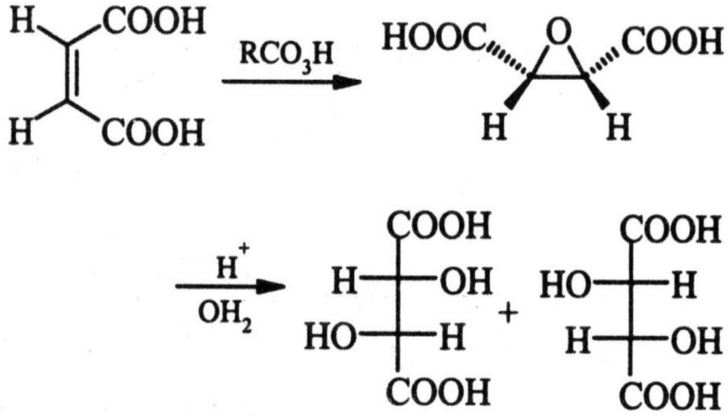 BH_2 (C) 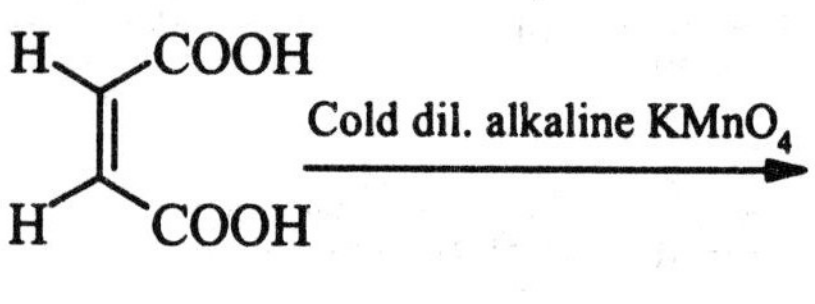 OH

15. (a) Cl Cl (b) OH (c)

16. (a) H H $\xrightarrow[\Delta]{CH_2N_2}$ H H

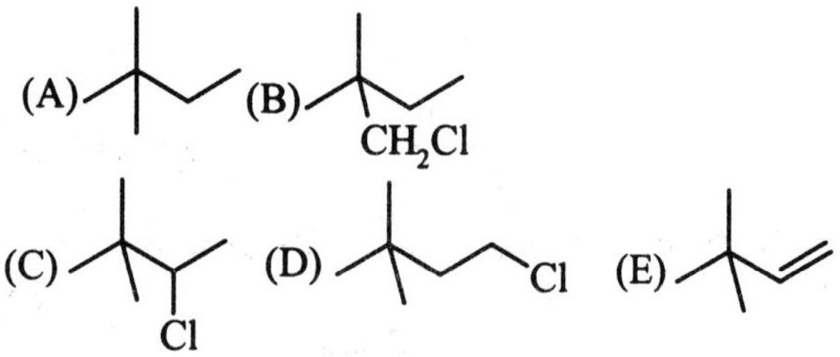

(b) + CH_2Br_2 $\xrightarrow{NaOH}$ Br

(c) OH $\xrightarrow[-H_2O]{H^+}$ $\xrightarrow{CHCl_3/NaOH}$ Cl Cl

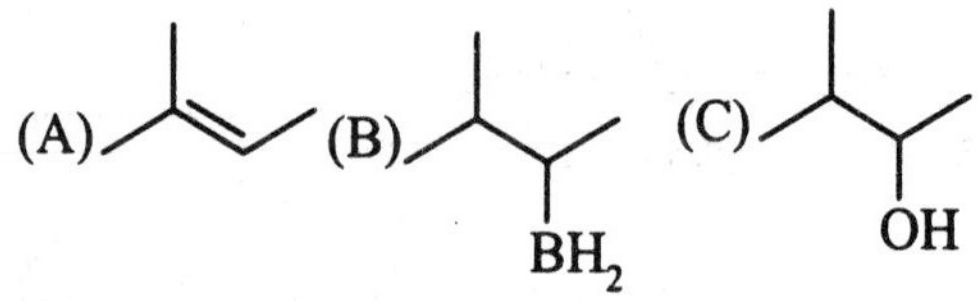

17. $\xrightarrow{Br_2}$ Br^+ $\xrightarrow{OH_2}$ Br OH; Cl^- → Cl Br

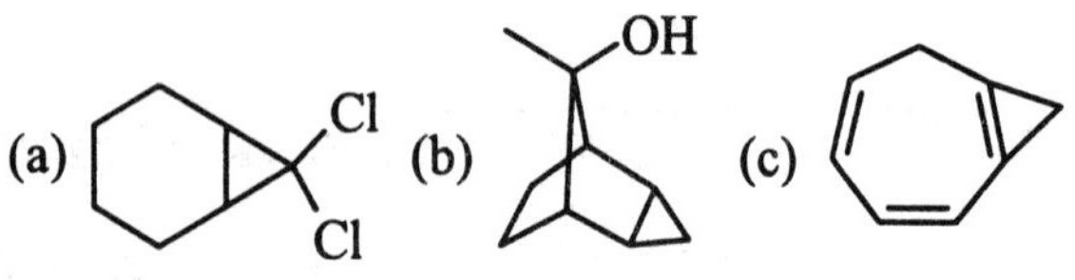

18. (a) Cl I (b) OH OH (c) OH

19. (a) $\xrightarrow{H^+}$ +; Br^- → Br; H^- shift → + $\xrightarrow{Br^-}$ Br

(b) + :CBr_2 → Br Br

(c) $\xrightarrow{H^+}$ +; Cl^- → Cl; CH_3OH → OMe

20. (a) HOOC H H COOH $\xrightarrow{\text{Cold dil. alkaline } KMnO_4}$ $\xrightarrow[OH_2]{H^+}$

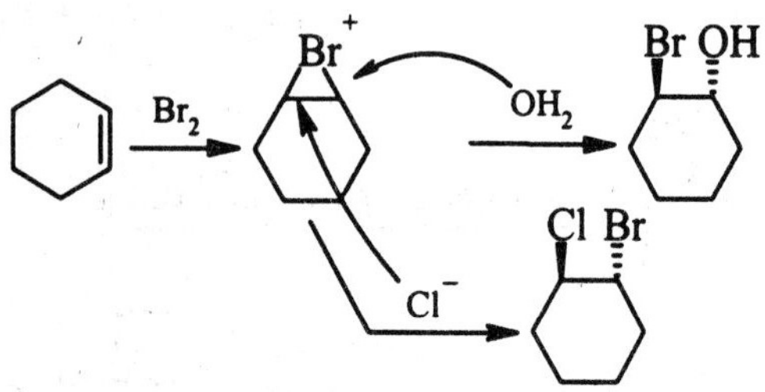

COOH, HO–H, H–OH, COOH + COOH, H–OH, HO–H, COOH

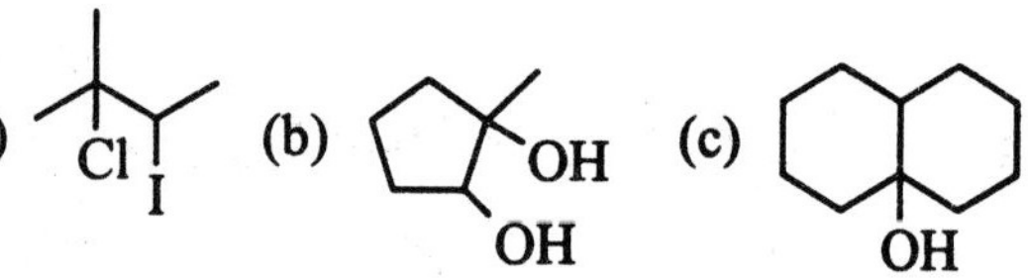

(b) H COOH H COOH $\xrightarrow{RCO_3H}$ HOOC O COOH H H

$\xrightarrow[OH_2]{H^+}$ COOH, H–OH, HO–H, COOH + COOH, HO–H, H–OH, COOH

(c) H COOH H COOH $\xrightarrow{\text{Cold dil. alkaline } KMnO_4}$

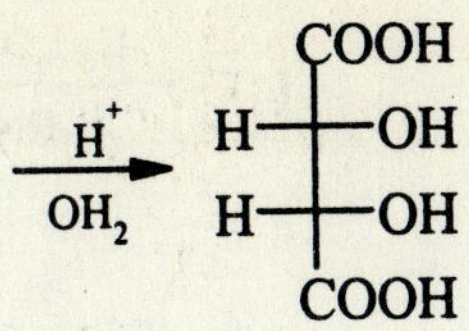

$\xrightarrow[OH_2]{H^+}$ COOH, H—C—OH, H—C—OH, COOH

21. (a) 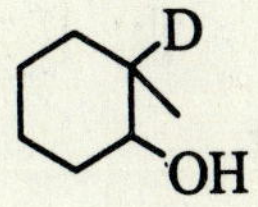 (b)

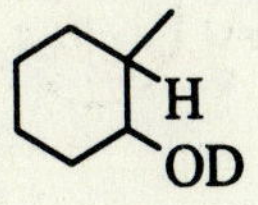

22. 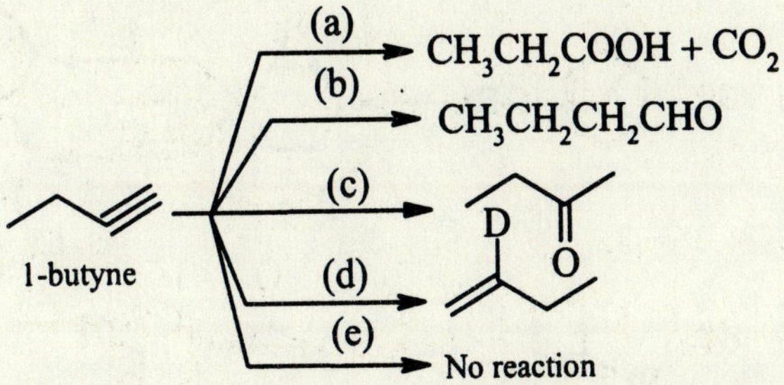 1-butyne
(a) → $CH_3CH_2COOH + CO_2$
(b) → $CH_3CH_2CH_2CHO$
(c) →
(d) →
(e) → No reaction

2-butyne
(a) → CH_3COOH
(b) →
(c) →
(d) →
(e) → No reaction

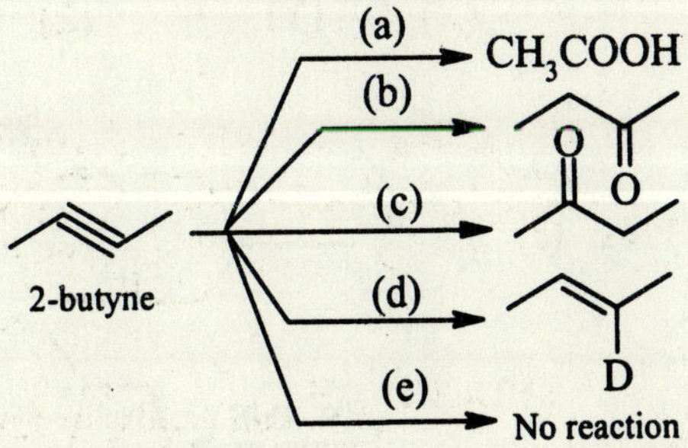

23. 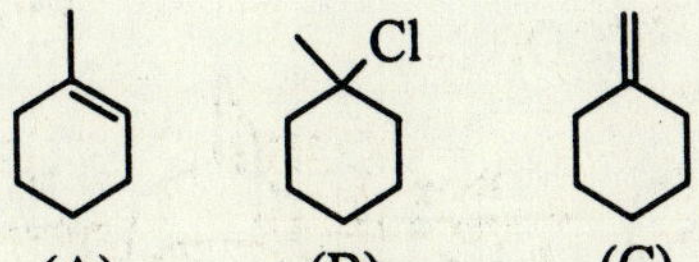 (A) (B) (C)

24. 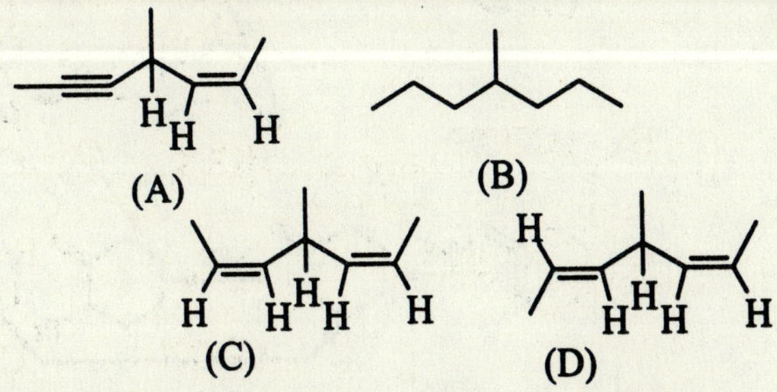 (A) (B)
(C) (D)

25. (a) $\xrightarrow[KOH]{Alcohol}$ $\xrightarrow[ROOR]{BrH}$ (b) $\xrightarrow[Heat]{H_2SO_4}$ $\xrightarrow[NaOH]{B_2H_6\text{-}H_2O_2}$

(c) $\xrightarrow{HI\text{-red-P}}$ (d) $\xrightarrow{HIO_4}$ (e) $\xrightarrow{KMnO_4}$

(f) $\xrightarrow{Zn/heat}$ (g) $\xrightarrow[ROOR]{BrH}$ $\xrightarrow{CH_3ONa}$

(h) $\xrightarrow{BrH}$ $\xrightarrow{Mg/ether}$ $\xrightarrow{D_2O}$

(i) $\xrightarrow{SOCl_2}$ $\xrightarrow{alc\ KOH}$ $\xrightarrow{H_2/Pt}$

26. (a) 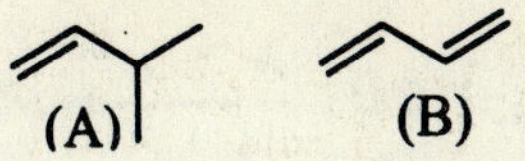 (A) (B)

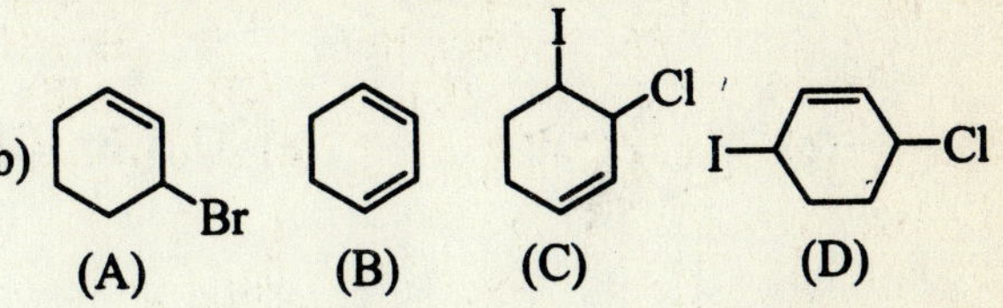

(b) (A) (B) (C) (D)

27. 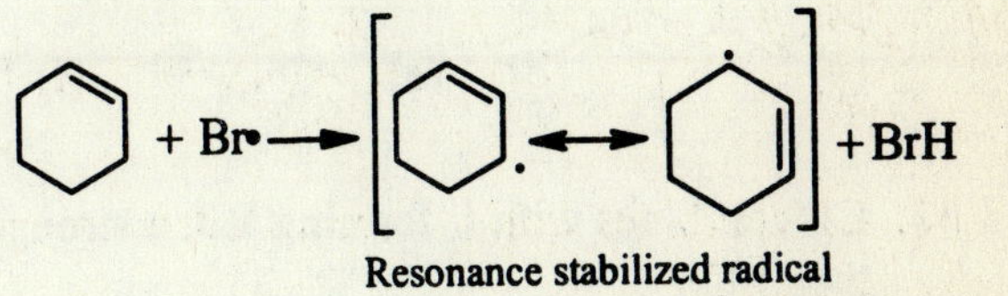 + Br• → [↔] + BrH

Resonance stabilized radical

28. (a) 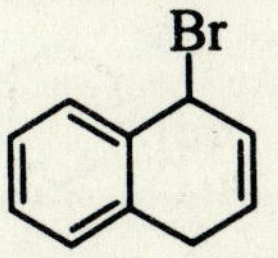 (b)

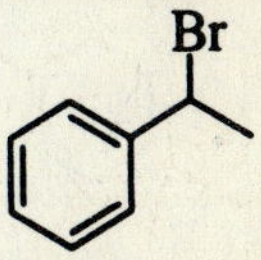

29.

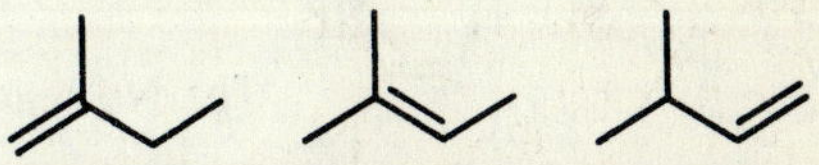

30. 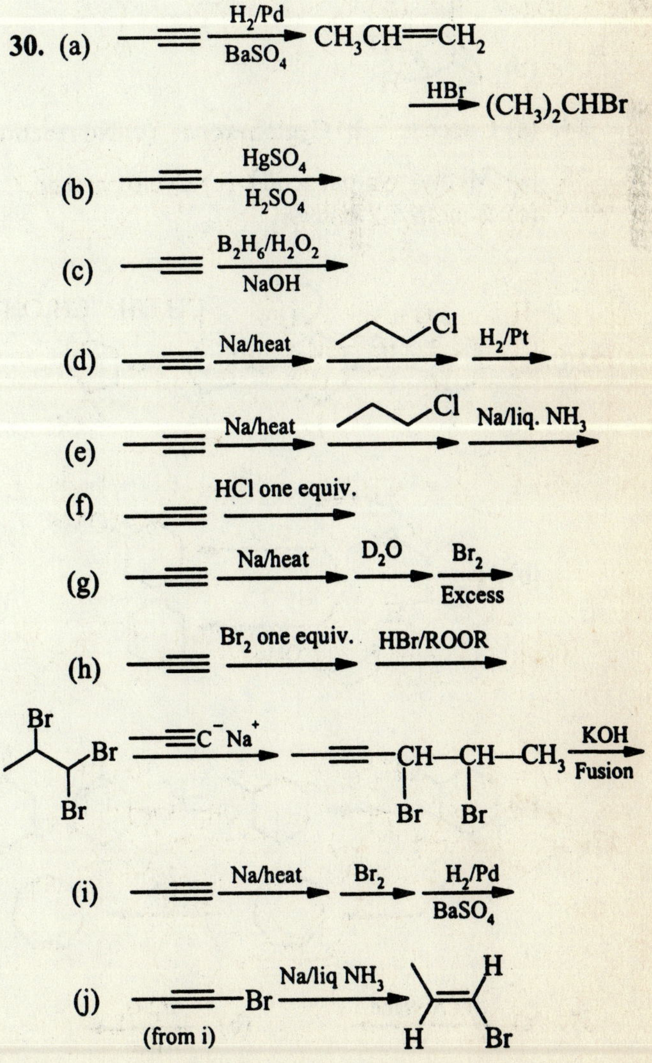 (a) $\xrightarrow[BaSO_4]{H_2/Pd}$ $CH_3CH{=}CH_2$ $\xrightarrow{HBr}$ $(CH_3)_2CHBr$

(b) $\xrightarrow[H_2SO_4]{HgSO_4}$

(c) $\xrightarrow[NaOH]{B_2H_6/H_2O_2}$

(d) $\xrightarrow{Na/heat}$ $\xrightarrow{}$ $\xrightarrow{H_2/Pt}$

(e) $\xrightarrow{Na/heat}$ $\xrightarrow{}$ $\xrightarrow{Na/liq.\ NH_3}$

(f) $\xrightarrow{HCl\ one\ equiv.}$

(g) $\xrightarrow{Na/heat}$ $\xrightarrow{D_2O}$ $\xrightarrow[Excess]{Br_2}$

(h) $\xrightarrow{Br_2\ one\ equiv.}$ $\xrightarrow{HBr/ROOR}$

$\xrightarrow{\equiv C^- Na^+}$ $\equiv$—CH(Br)—CH(Br)—CH_3 $\xrightarrow[Fusion]{KOH}$

(i) $\xrightarrow{Na/heat}$ $\xrightarrow{Br_2}$ $\xrightarrow[BaSO_4]{H_2/Pd}$

(j) —Br $\xrightarrow{Na/liq\ NH_3}$ (from i)

31. (a)  $(\text{(CH}_3)_2\text{CH}-\text{CH}_2\text{CH}_2)_2\text{CuLi}$ (A) → (with 1-bromopropane) 2-methylheptane

(b) 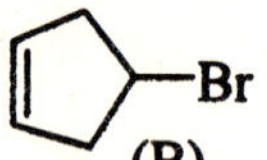 3-bromocyclopentene (B) (c) C_6H_5Br (C)

(d) 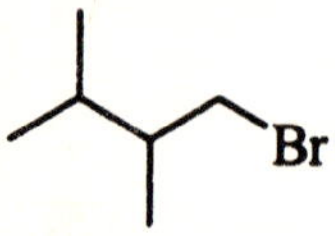 1-bromo-2,3-dimethylbutane (e) 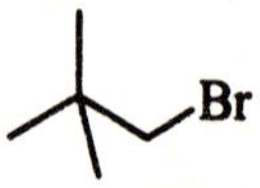 1-bromo-2,2-dimethylpropane (f) H_2/Pt

32. Cl^- combines with I_2 forming ICl, a stronger bond than HCl.

33. (a) 1-methylcyclohexene (b) 1-methylcyclohexene (c) cyclohexene (d) 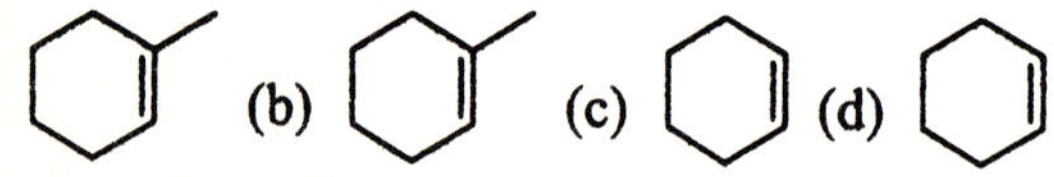 cyclohexene

(e) 1-methylcyclohexene (f) 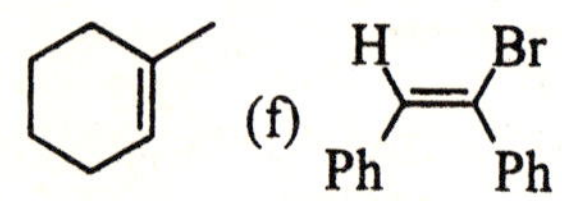 H(Ph)C=C(Br)Ph (g) No reaction

(h) deuterated bicyclic alkene (D, H)

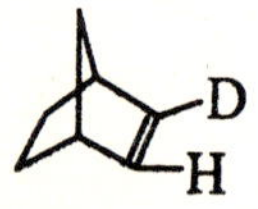

34. (a) 1-decene (b) Cyclohexene (c) No reaction

35. (a) Methyl cyclohexene (b) Cyclo pentene
(c) 2-methyl-2-butene

36. (a) 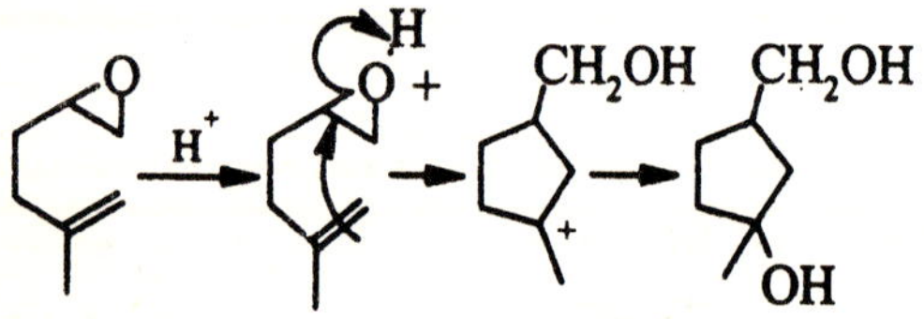 epoxide $\xrightarrow{H^+}$ protonated epoxide → cyclization → CH_2OH cyclopentyl cation → CH_2OH cyclopentanol (OH)

(b) 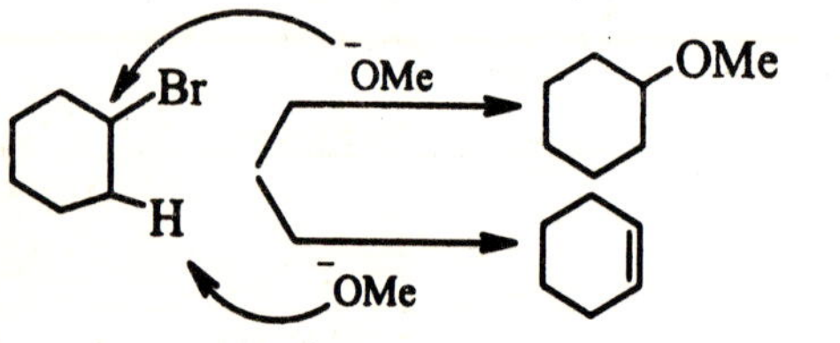 1-bromo-2-H-cyclohexane: ^-OMe → methoxycyclohexane; ^-OMe → cyclohexene

(c) 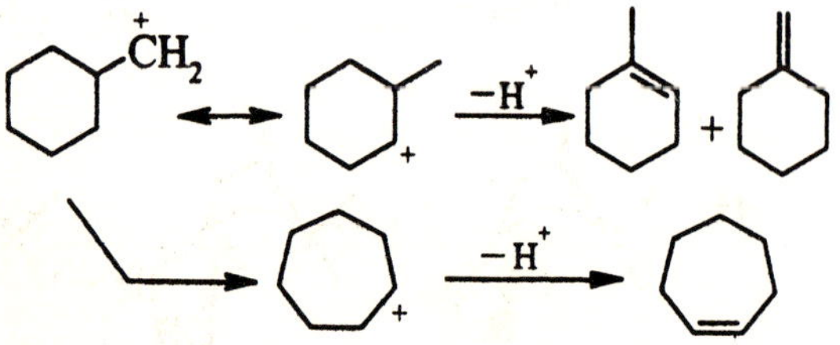 cyclohexyl-$\overset{+}{C}H_2$ ⟷ methylcyclohexyl cation $\xrightarrow{-H^+}$ 1-methylcyclohexene + methylenecyclohexane; ring expansion → cycloheptyl cation $\xrightarrow{-H^+}$ cycloheptene

37. (a) $\xrightarrow{\text{NaI/Acetone}}$ (b) $\xrightarrow[\text{Heat}]{H_2SO_4}$ (c) $\xrightarrow[C_2H_5ONa]{C_2H_5OH}$ (d) $\xrightarrow{NaBH_4}$

38. (a) 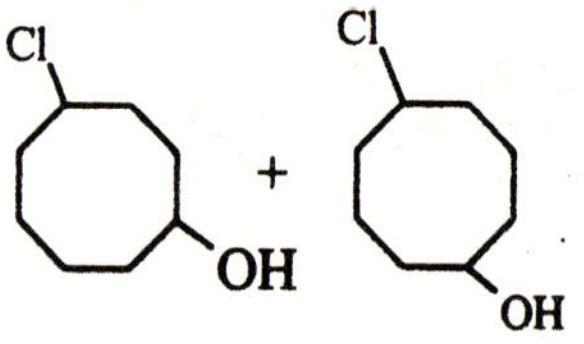 Cl-cyclooctanol (OH) + Cl-cyclooctanol (OH) (b) 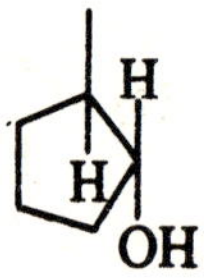 bicyclic alcohol (H, H, OH)

39. 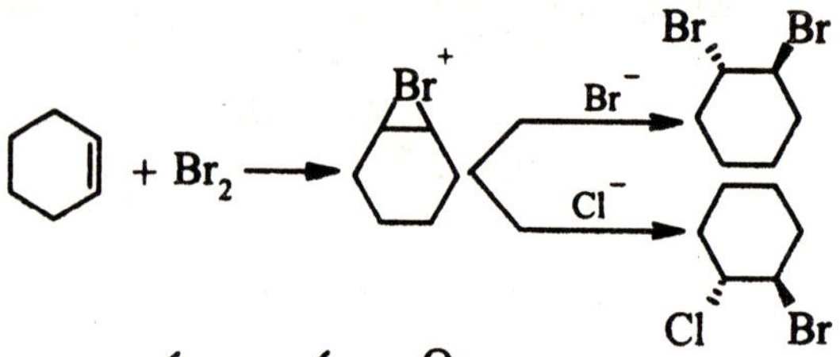 cyclohexene + Br_2 → bromonium ion ($\overset{+}{Br}$); Br^- → trans-1,2-dibromocyclohexane (Br, Br); Cl^- → trans-1-bromo-2-chlorocyclohexane (Cl, Br)

40. 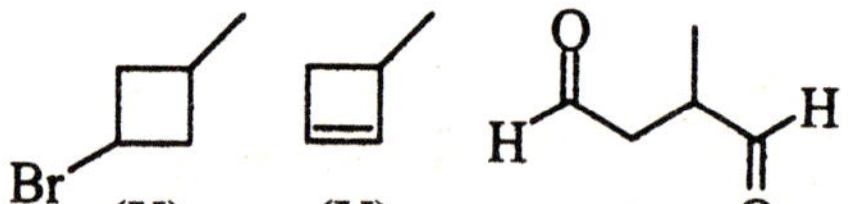 1-bromo-2-methylcyclobutane (X) 1-methylcyclobutene (Y) 2-methylbutanedial (Z)

41. (a) 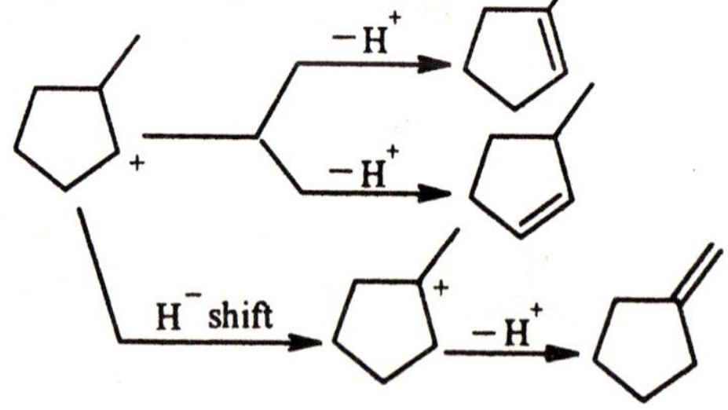 methylcyclopentyl cation $\xrightarrow{-H^+}$ 1-methylcyclopentene; $\xrightarrow{-H^+}$ 3-methylcyclopentene; H^- shift → tertiary cation $\xrightarrow{-H^+}$ methylenecyclopentane

(b) 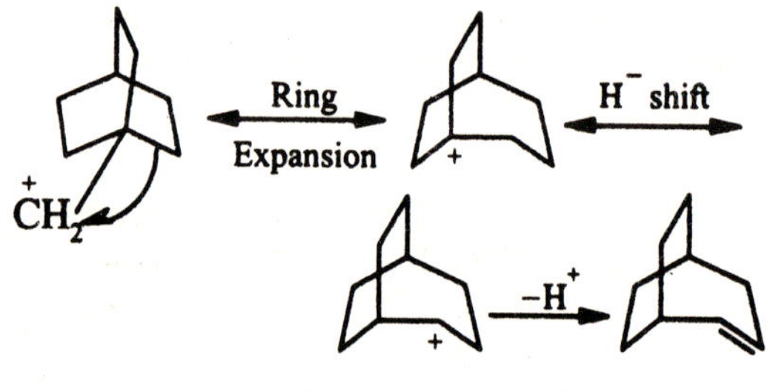 bicyclic $\overset{+}{C}H_2$ ⟷ Ring Expansion ⟷ H^- shift ⟷ cation $\xrightarrow{-H^+}$ bicyclic alkene

(c) methylenecyclohexane $\xrightarrow{NBS}$ radical $\xrightarrow{NBS}$ allylic bromide (Br); allylic radical ($\dot{C}H_2$) → 1-(bromomethyl)cyclohexene (Br)

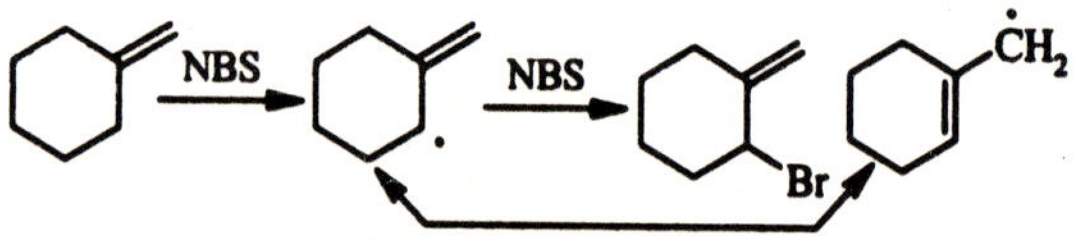

(d) 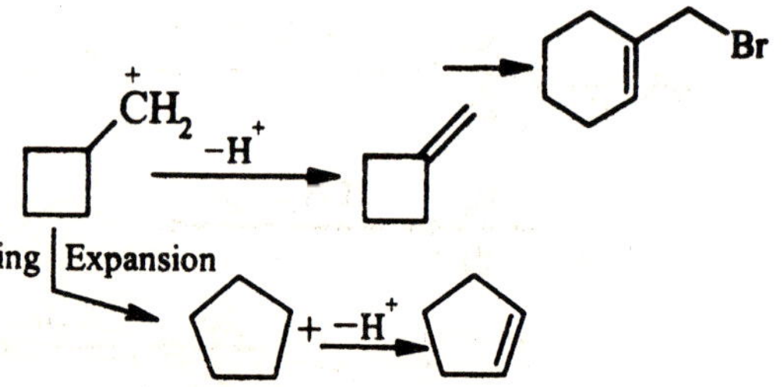 cyclobutyl-$\overset{+}{C}H_2$ $\xrightarrow{-H^+}$ methylenecyclobutane; Ring Expansion → cyclopentyl cation $\xrightarrow{-H^+}$ cyclopentene

(e) 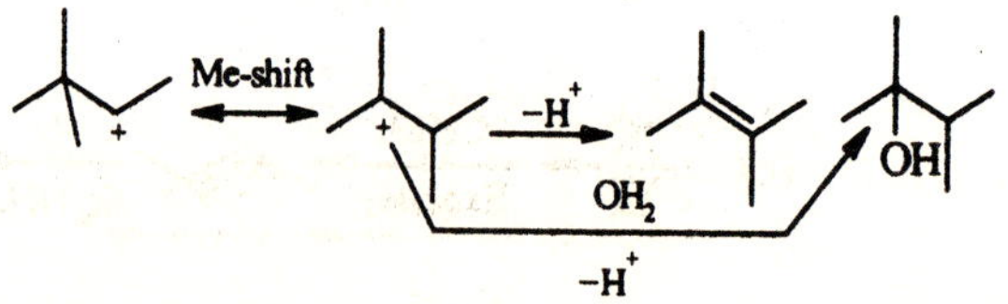 cation ⟷ Me-shift ⟷ tertiary cation $\xrightarrow{-H^+}$ 2,3-dimethyl-2-butene; OH_2, $-H^+$ → alcohol (OH)

42. (a) $\xrightarrow{H_3O^+}$ (b) $\xrightarrow[H_2O_2\text{-}NaOH]{B_2H_6}$

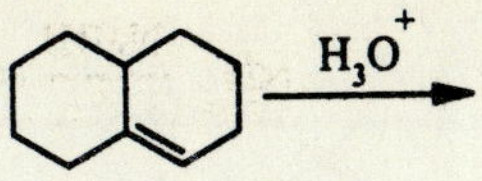

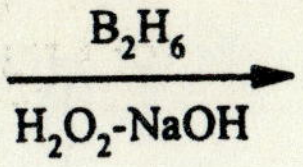

(c) $\xrightarrow[NaOH]{CH_2Br_2}$ Br

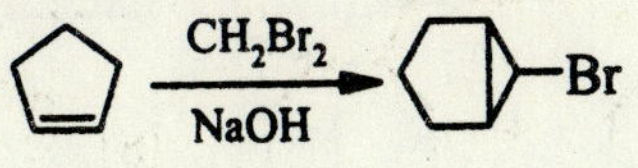

(d) $\xrightarrow[NaOH]{CHBr_3}$ Br Br H H HO$^-$ → Br

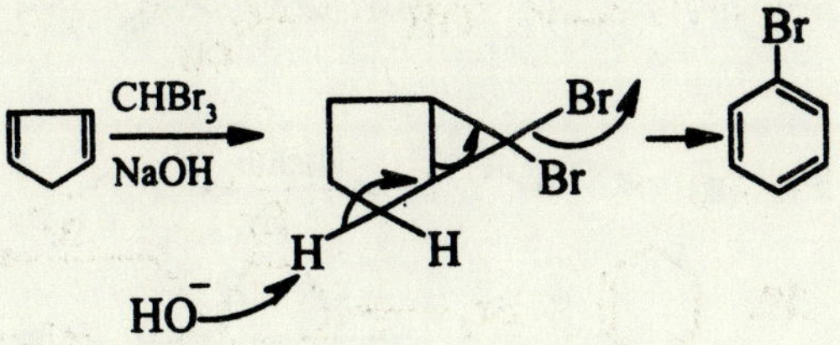

43. (a) CH_3CHO + O (b) Br (d,i pair)

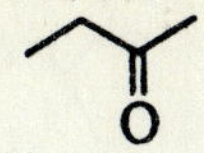

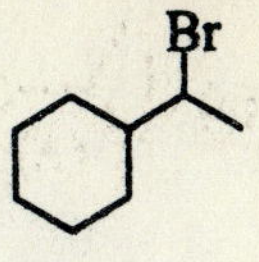

(c) Cl (d) Cl

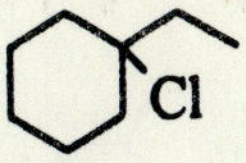

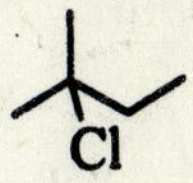

(e) CH_3 Br Br H (f) CH_3 OH

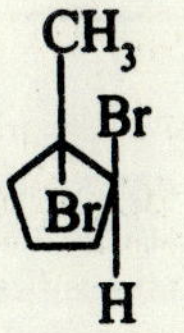

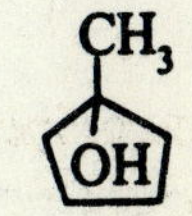

44. (a) + H^+ → + $\xrightarrow[-H^+]{OH_2}$ OH $\xrightarrow[-H^+]{OH_2}$ OH

+ H^+ → + $\underset{\text{Expansion}}{\overset{\text{Ring}}{\longleftrightarrow}}$ + $\overset{\text{H-shift}}{\longleftrightarrow}$ + $\xrightarrow[-H^+]{OH_2}$ OH

(b) + H^+ → + $\xrightarrow{Br^-}$ Br

H-shift → + $\xrightarrow{Br^-}$ Br

(c) + → Br$^+$ $\xrightarrow{Br^-}$ Br Br

$\xrightarrow{Cl^-}$ Br Cl

$\xrightarrow[-H^+]{CH_3OH}$ Br OMe

45. (a) $\xrightarrow[\Delta]{Na}$ $\xrightarrow{CH_3CH_2CH_2Br}$

(b) $\xrightarrow[\Delta]{Na}$ $\xrightarrow{CH_3CH_2Br}$

$\xrightarrow[\Delta]{Na}$ $\xrightarrow{D_2O}$ D

(c) $\xrightarrow[\Delta]{Na}$ $\xrightarrow[\text{Excess}]{CH_3Br}$ $\xrightarrow[\text{liq } NH_3]{Na}$ H H

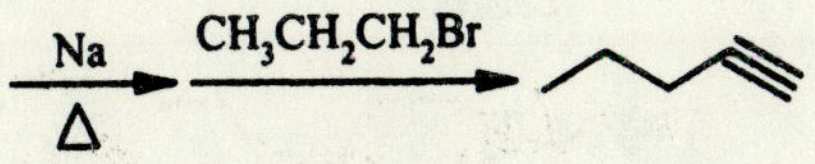

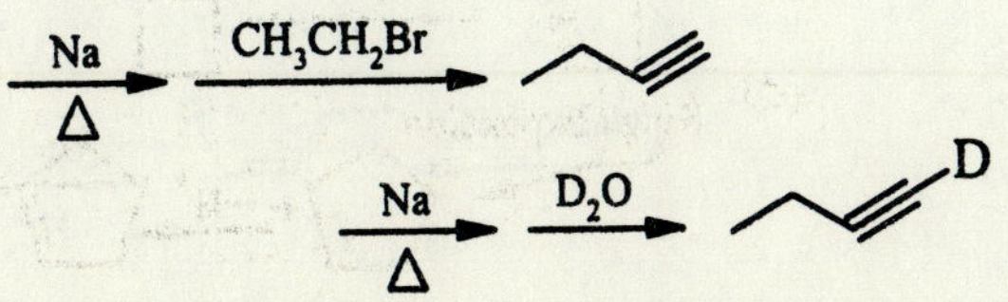

(d) (from c) $\xrightarrow{D_2/Pd/BaSO_4}$ D D

46. Br Br Br OH Br

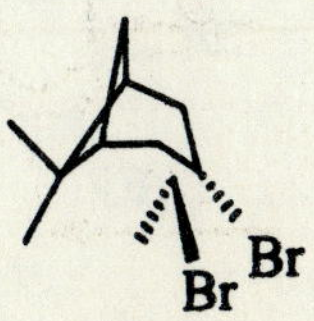

(A) 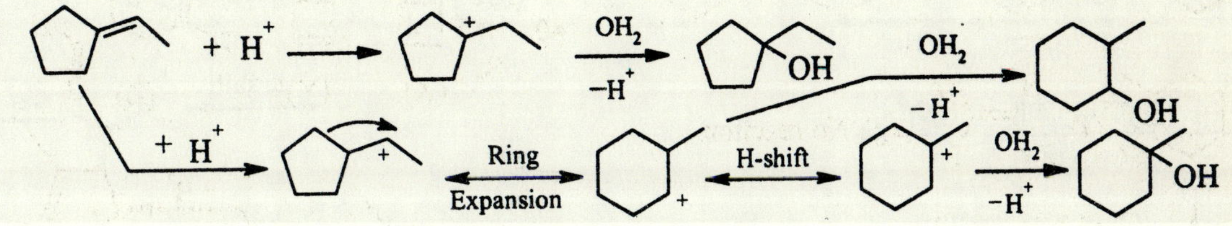 (B) 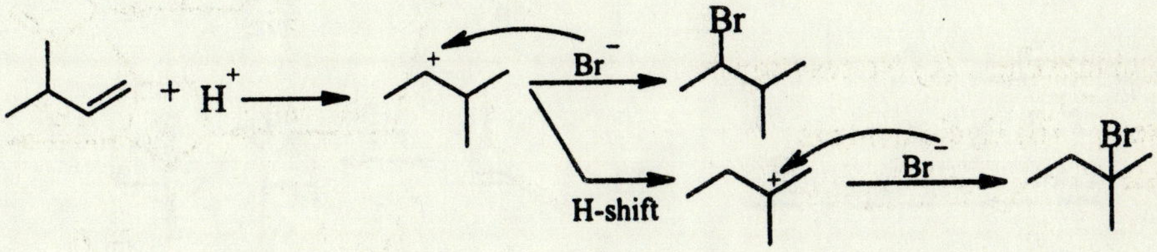 (C)

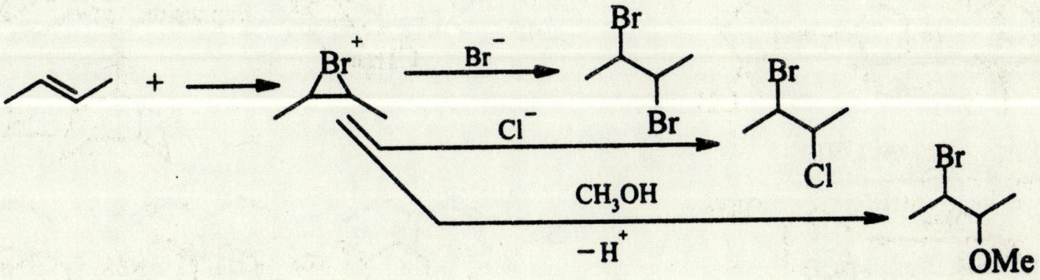

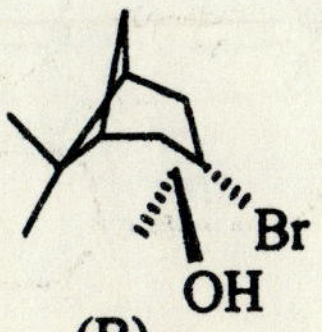

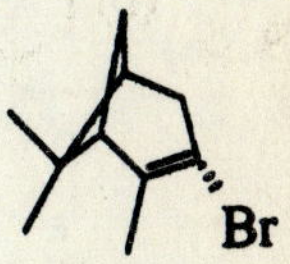

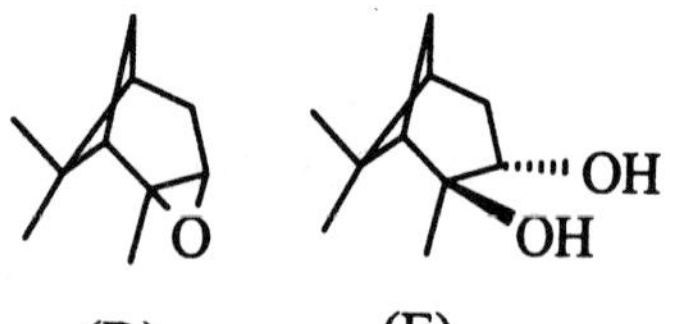

(D) (E)

47. 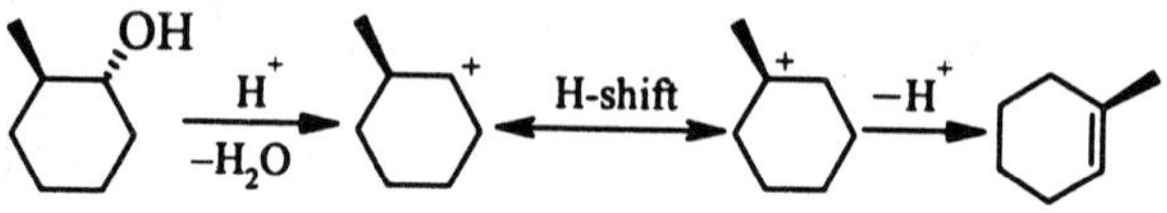 $\xrightarrow[-H_2O]{H^+}$ H-shift $-H^+$

Br H H $^-OC_2H_5$

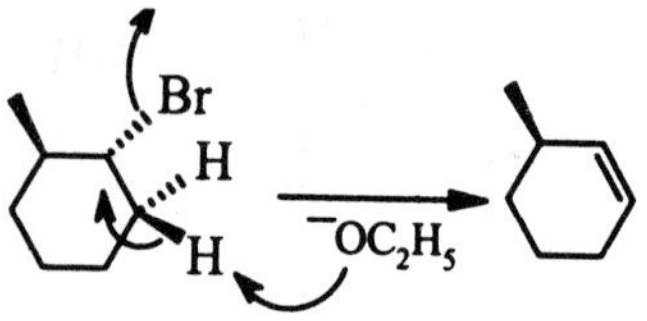

Here the other β-carbon has methyl group on side anti to Br therefore base abstract proton from the anti position from carbon bearing less no of substituent (against to Sytzeff's rule).

48. (a) $\xrightarrow{BrH}$ $\xrightarrow[(CH_3)_3COH]{(CH_3)_3COK}$ $\xrightarrow[Heat]{CH_2N_2}$

(b) $\xrightarrow{BrH}$ $\xrightarrow[(CH_3)_3COH]{(CH_3)_3COK}$ $\xrightarrow{H_2O/Cl_2}$

(c) $\xrightarrow{BrH}$ $\xrightarrow[(CH_3)_3COH]{(CH_3)_3COK}$ $\xrightarrow[NaOH]{CHBr_3}$

(d) $\xrightarrow{BrH}$ $\xrightarrow[(CH_3)_3COH]{(CH_3)_3COK}$ $\xrightarrow[OH_2]{O_3/Zn}$

$\xrightarrow{LiAlH_4}$ $\xrightarrow{H_2SO_4}$ $\xrightarrow{BrH}$

(e) $\xrightarrow{BrH}$ $\xrightarrow[(CH_3)_3COH]{(CH_3)_3COK}$ $\xrightarrow[Zn\text{-}LiAlH_4]{O_3}$

(f) $\xrightarrow{BrH}$ $\xrightarrow[(CH_3)_3COH]{(CH_3)_3COK}$ $\xrightarrow[Zn\text{-}H_2O]{O_3}$

(g) $\xrightarrow{BrH}$ $\xrightarrow[(CH_3)_3COH]{(CH_3)_3COK}$ $\xrightarrow[NaBH_4]{Hg(OAc)_2/H_2O}$

(h) $\xrightarrow{BrH}$ $\xrightarrow[(CH_3)_3COH]{(CH_3)_3COK}$ $\xrightarrow[NaBH_4]{Hg(OAc)_2/MeOH}$

(i) $\xrightarrow{BrH}$ $\xrightarrow[(CH_3)_3COH]{(CH_3)_3COK}$ $\xrightarrow[KMnO_4]{Cold, dilute}$

49. (a) 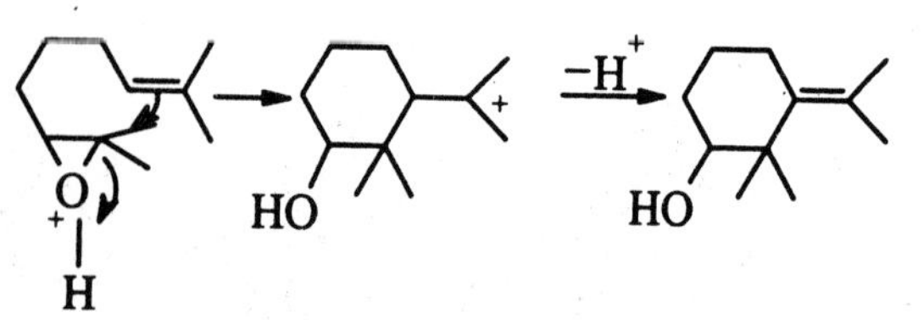 O H HO $-H^+$ HO

(b) 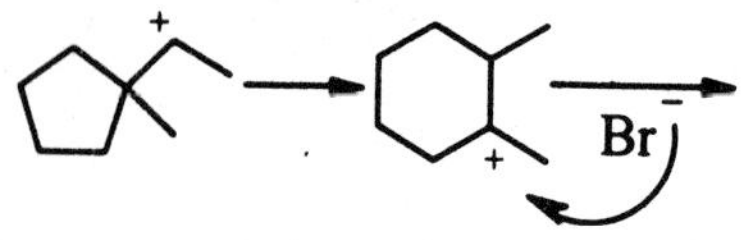 Br

(c) 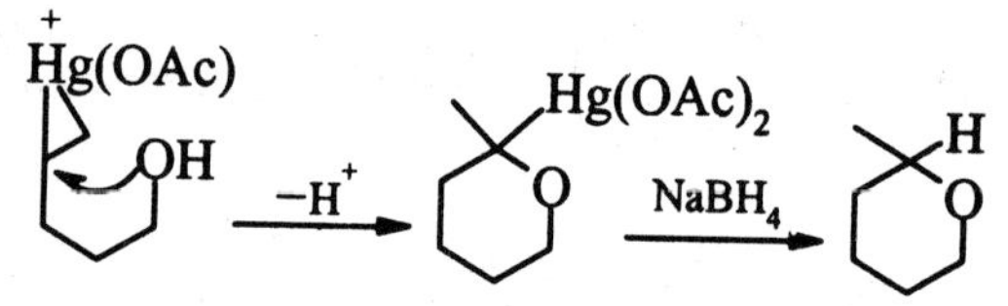 $\overset{+}{Hg}(OAc)$ OH $-H^+$ $Hg(OAc)_2$ O $NaBH_4$ H O

50. 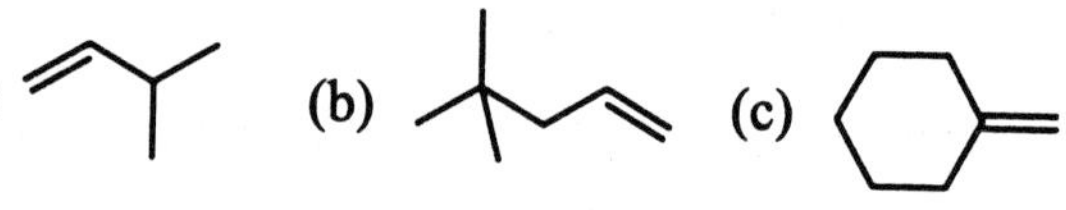 (a) 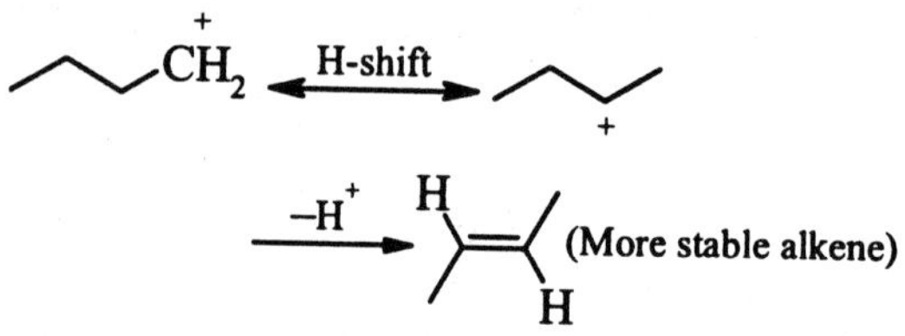 (b) 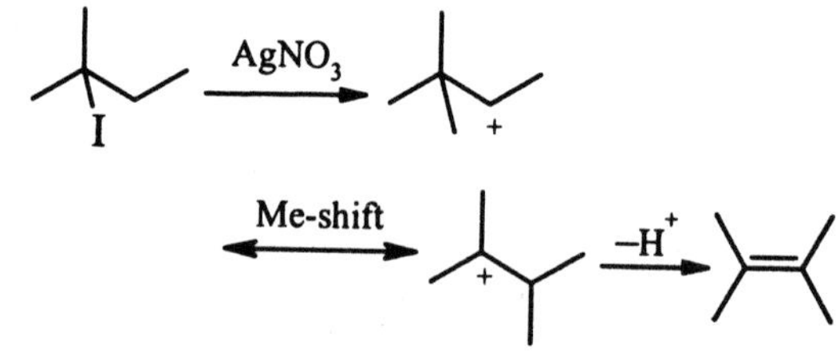 (c)

51. (a) $\overset{+}{CH_2}$ H-shift

$-H^+$ H H (More stable alkene)

(b) I $AgNO_3$

Me-shift $-H^+$

52. 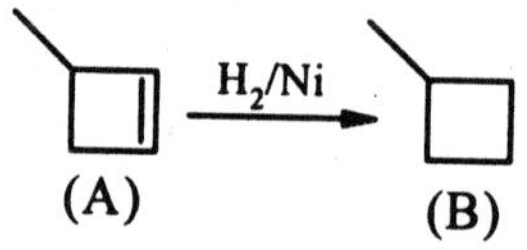 (A) $\xrightarrow{H_2/Ni}$ (B)

53. 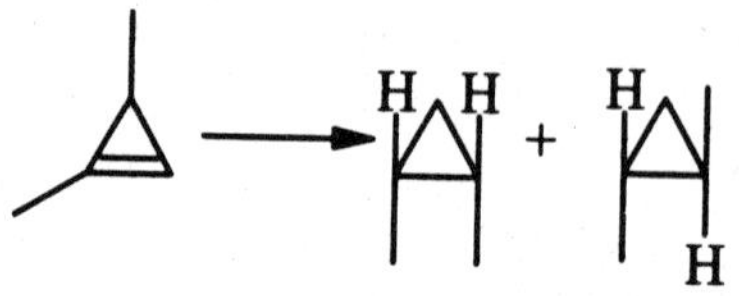 H H + H H

54. 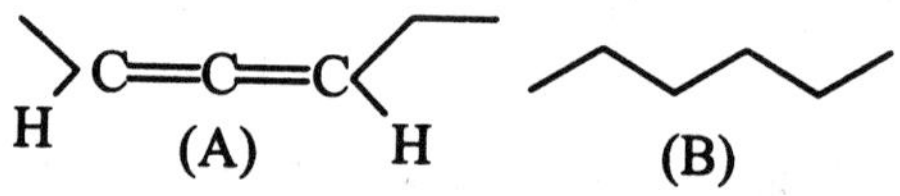 H C=C=C H (A) (B)

55. 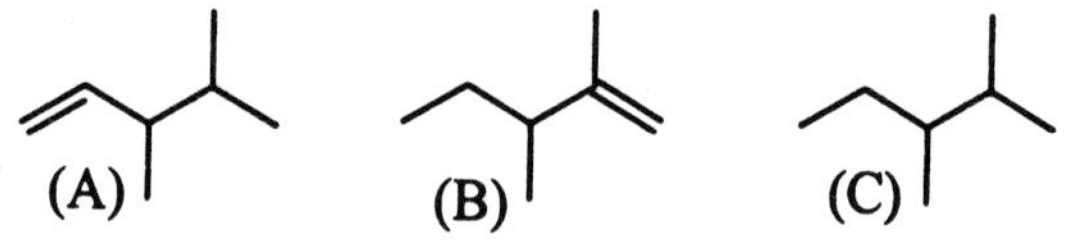 (A) (B) (C)

56. 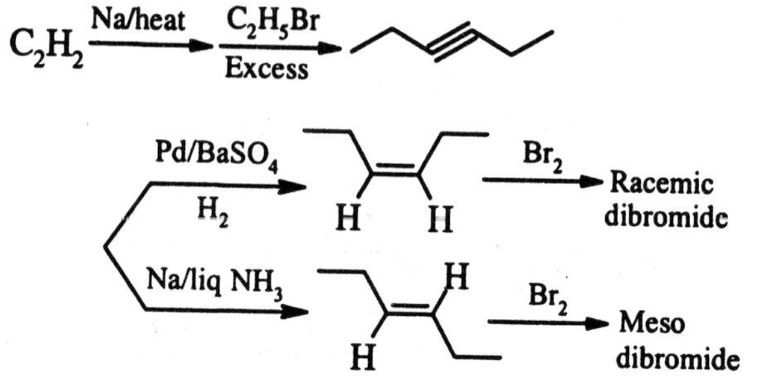 $C_2H_2 \xrightarrow{Na/heat} \xrightarrow[Excess]{C_2H_5Br}$

Pd/BaSO$_4$ H$_2$ H H $\xrightarrow{Br_2}$ Racemic dibromide

Na/liq NH$_3$ H H $\xrightarrow{Br_2}$ Meso dibromide

57. 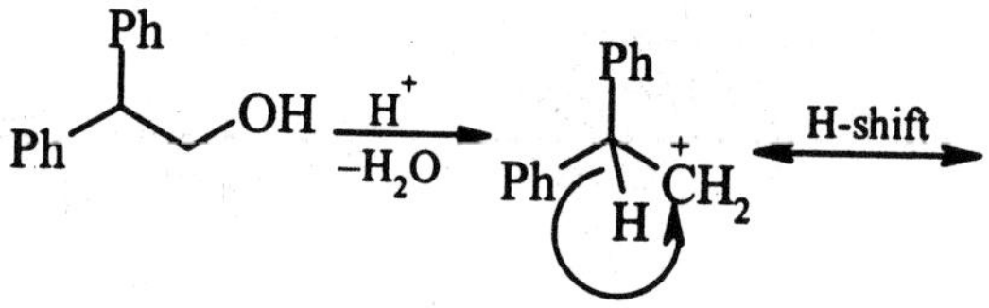 Ph Ph OH $\xrightarrow[-H_2O]{H^+}$ Ph Ph H $\overset{+}{C}H_2$ H-shift

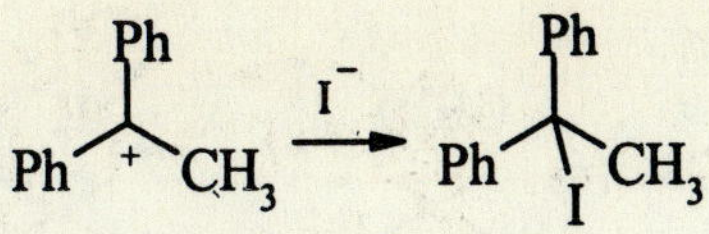

Ph⁺ (Ph)(Ph)C⁺–CH_3 $\xrightarrow{I^-}$ Ph–C(Ph)(I)–CH_3

58. 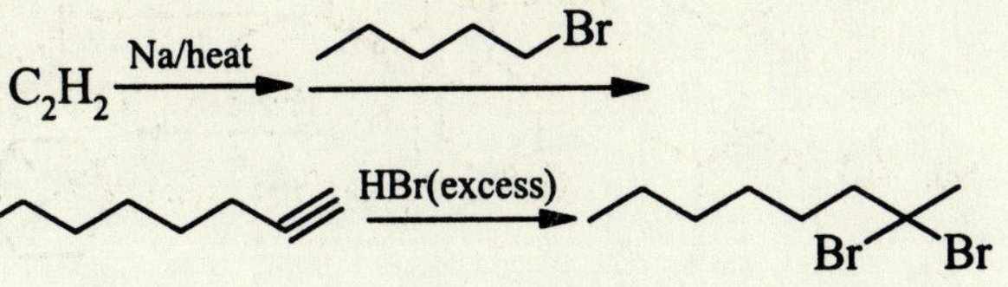 C_2H_2 $\xrightarrow{\text{Na/heat}}$ $\xrightarrow{\text{Br}}$

$\xrightarrow{\text{HBr(excess)}}$ Br Br

59. 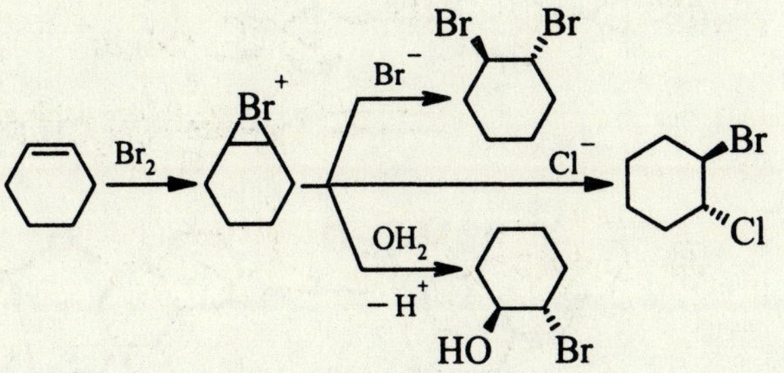 $\xrightarrow{Br_2}$ Br⁺ — Br^- → Br, Br; Cl^- → Br, Cl; OH_2, $-H^+$ → HO, Br

60. 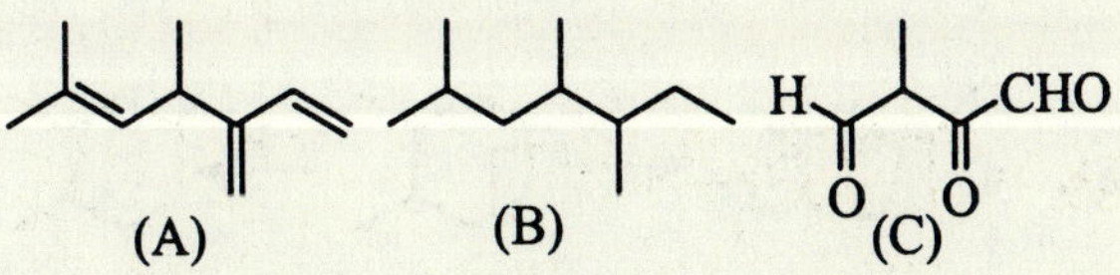 (A) (B) (C) H, CHO, O, O

61.

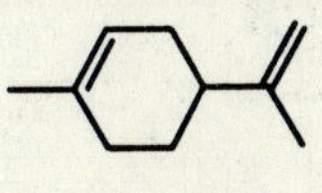

62. 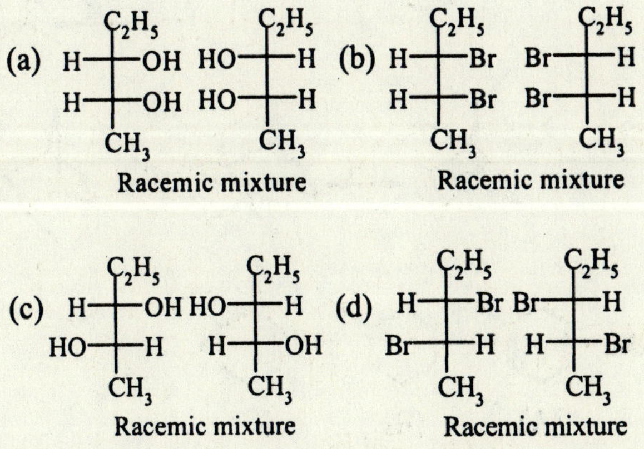 (a)

C_2H_5	C_2H_5
H—OH	HO—H
H—OH	HO—H
CH_3	CH_3

Racemic mixture

(b)

C_2H_5	C_2H_5
H—Br	Br—H
H—Br	Br—H
CH_3	CH_3

Racemic mixture

(c)

C_2H_5	C_2H_5
H—OH	HO—H
HO—H	H—OH
CH_3	CH_3

Racemic mixture

(d)

C_2H_5	C_2H_5
H—Br	Br—H
Br—H	H—Br
CH_3	CH_3

Racemic mixture

63. 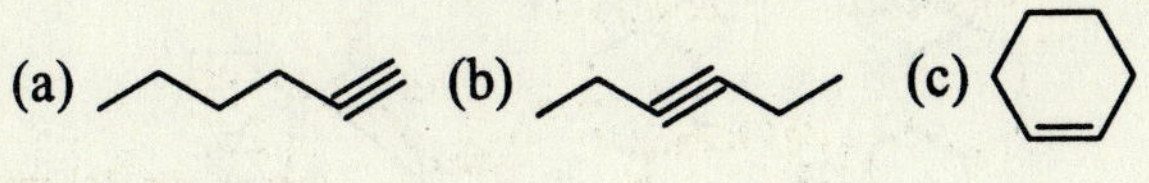 (a) (b) (c)

64. 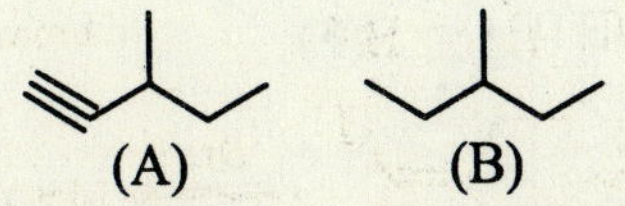 (A) (B)

65. 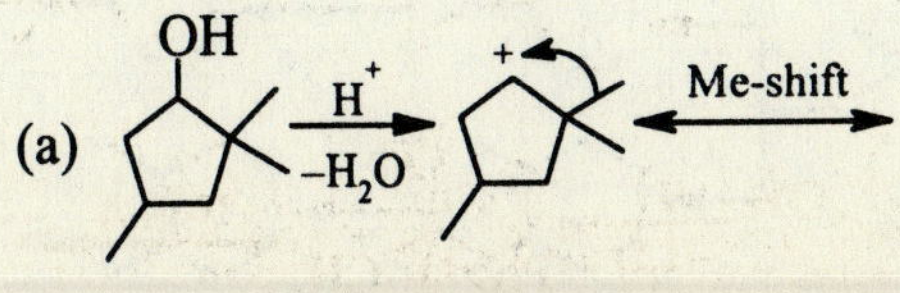 (a) OH $\xrightarrow[-H_2O]{H^+}$ ⁺ $\xleftrightarrow{\text{Me-shift}}$ ⁺ →

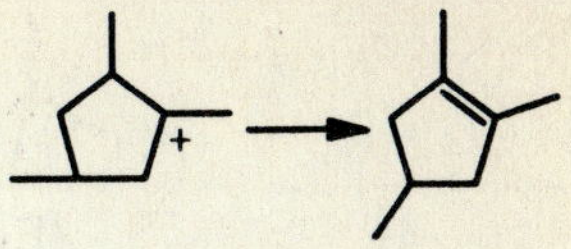

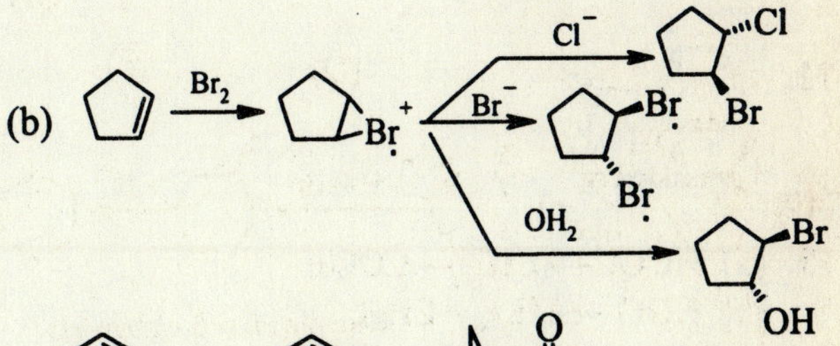

(b) $\xrightarrow{Br_2}$ Br⁺ — Cl^- → Cl, Br; Br^- → Br, Br; OH_2 → Br, OH

66. 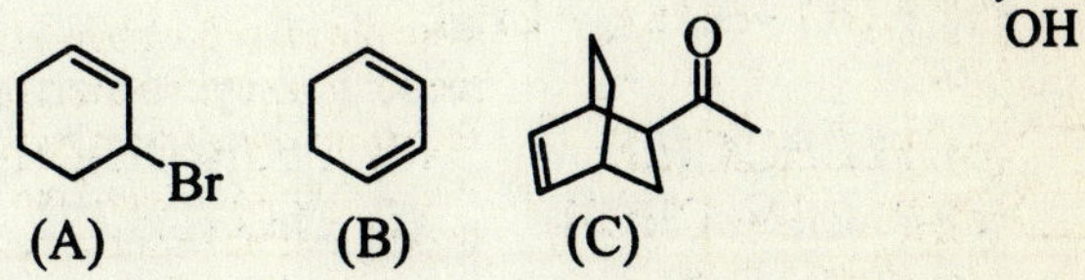 (A) Br (B) (C) O

67. 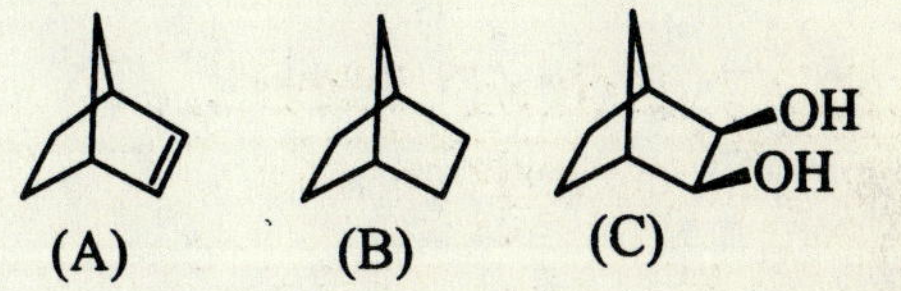 (A) (B) (C) OH, OH

68. 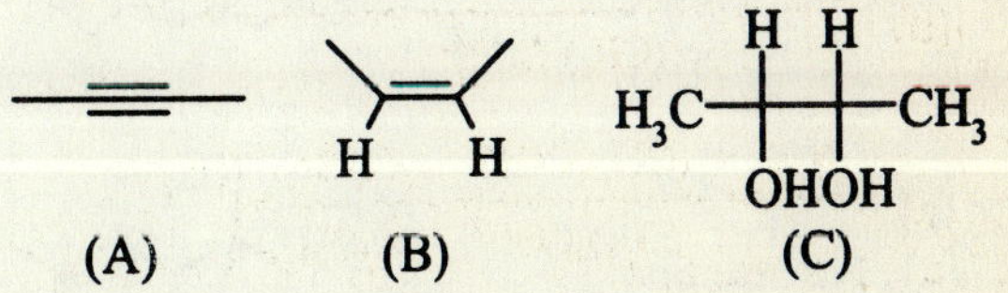 (A) (B) H, H (C) H H, H_3C—C—C—CH_3, OHOH

69. 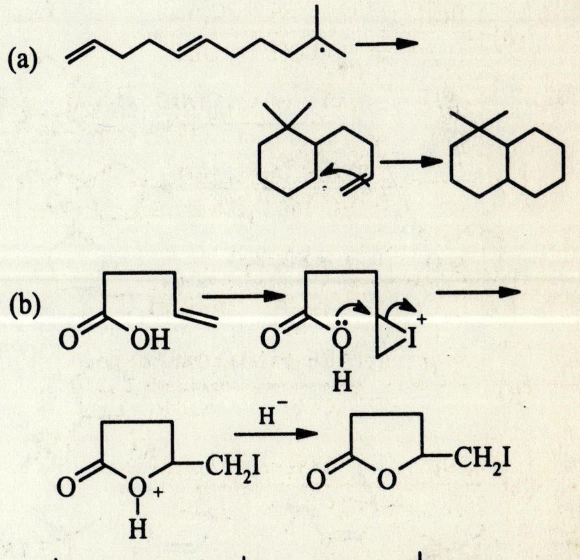 (a) →

→

(b) O, OH → O, Ö, H, I⁺ →

O, O⁺, H, CH_2I $\xrightarrow{H^-}$ O, O, CH_2I

70. 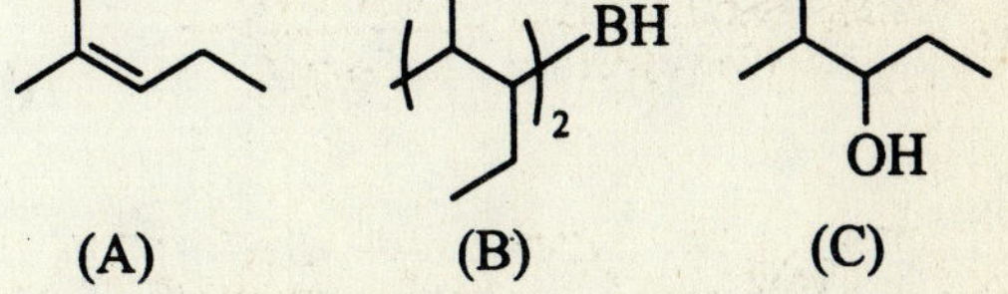 (A) (B) BH, 2 (C) OH

71. 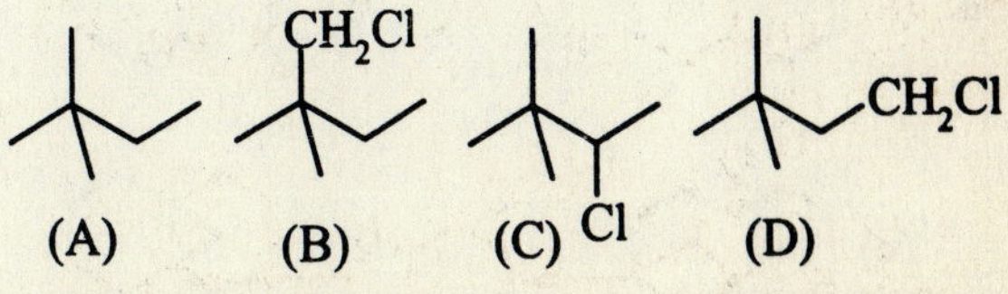 (A) (B) CH_2Cl (C) Cl (D) CH_2Cl

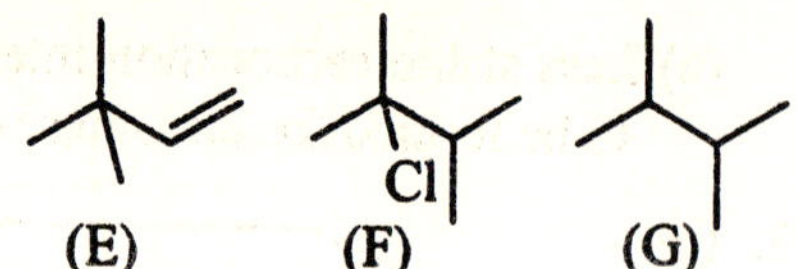

72. Cl

Enantiomers

73. (a) HOOC—$(CH_2)_8$—COOH
(b) HOC—$(CH_2)_8$—CHO;
meso 1, 2-cyclo-octanediol
(c) $PhCHClCH_2Br$
(d) $PhCH(OH)CH_2Br$
(e) $PhClC{=}CHCH_3$
(f) $PhCOCH_2CH_3$
(g) $PhCH_2COCH_3$

74. E $\xleftarrow{H_2/Pt}$ D $\xleftarrow{C_2H_5ONa}$ A $\xrightarrow{\text{Zn/Acetic acid}}$ B + C

H_3C CH_3 H_3C CH_3 H CH_3 CH_3 Cl

D $\xrightarrow[(CH_3)_2S]{O_3}$

75. (a) No reaction (b) NH_2

76. (i) BH_3 (ii) CH_3COOD — H_3C, H, H, D

(i) HBr (ii) Mg/ether (iii) D_2O — CH_3, D

H_2SO_4/H_2O — H_3C, D

(i) BH_3 (ii) H_2O_2/NaOH — HO, CH_3

77. (A) (B) H H (C) H H (D) COOH

78. (a) OMe (b) Br, Br (c) O, O (d)

79. (a) OH $\xrightarrow{H^+}$ $\ddot{O}H$ $\xrightarrow{-H^+}$ O $\xrightarrow{H^+}$

O + $H{-}\ddot{S}{-}R$ $\xrightarrow{-H^+}$ O SR

(b) Ph, Br $\xrightarrow{Mg}$ C^- Ph $MgBr^+$ $\longrightarrow$ Ph—≡—CH_2CH_2MgBr

(c) O O $\longleftrightarrow$ O O

80. A: Phenyl acetylene.

81. (a) $\longleftrightarrow$ + $\xrightarrow{-H^+}$

(b) H_3C—C(CH_3)(CH_3)—$\overset{+}{C}H_2$ $\longleftrightarrow$ H_3C—$\overset{+}{C}$(CH_3)—CH_2—CH_3 $\xrightarrow{-H^+}$

(c) + $\longleftrightarrow$ + $\xrightarrow{-H^+}$

84. (A) (B) (C) OH

(D) OH OH (E) OH OH (F) O OH O

85. + $\longrightarrow$ + $\xrightarrow{-H^+}$; + $\xrightarrow{-H^+}$

86. (a) $CH_3-C\equiv C-CH_2CH_3 + \bar{N}H_2 \rightarrow$

$\bar{C}H_2-C\equiv C-CH_2CH_2CH_3 \leftrightarrow CH_2=C=\bar{C}-CH_2CH_3 \xrightarrow{NH_3}$

$CH_2=C=CH-CH_2-CH_3 \xrightarrow{\bar{N}H_2} HC\equiv C-CH_2CCH_2CH_3$

$\xleftarrow{NH_3} H-C\equiv C-\bar{C}HCH_2CH_3 \leftrightarrow \bar{C}H=C=CH-CH_2-CH_3$

(b) $HC\equiv C-CH_2CH_2CH_3 + BrH\,(excess) \rightarrow$

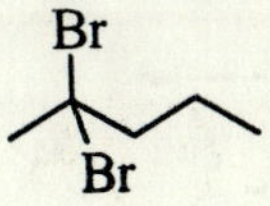

$\xrightarrow{KOH} CH_3-C\equiv C-CH_2CH_3$

87. (A) (B) COOH (C) (D) OH (E)

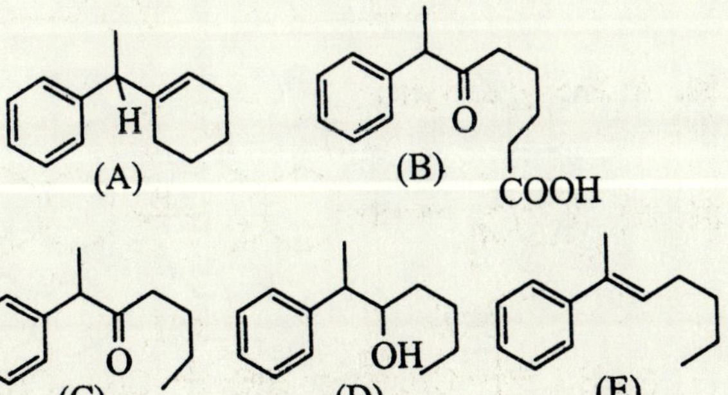

88. (a) Ph, Ph, H_3C, H (b) H (c) (d) Ph

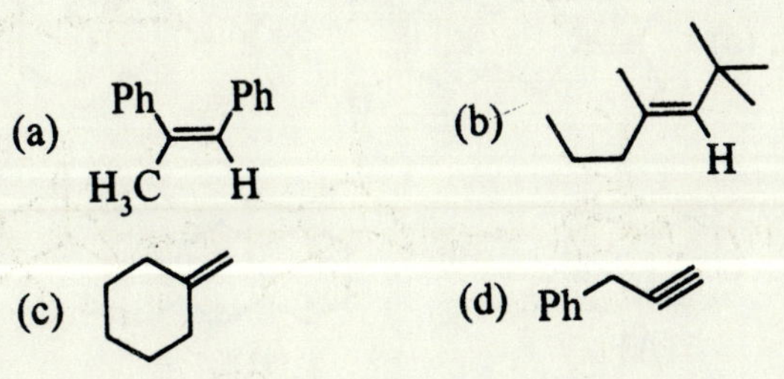

89. (a) Cl, I (b) Br, OH (c) (d) Br (e) 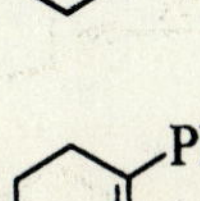 $+ CH_3COOH$

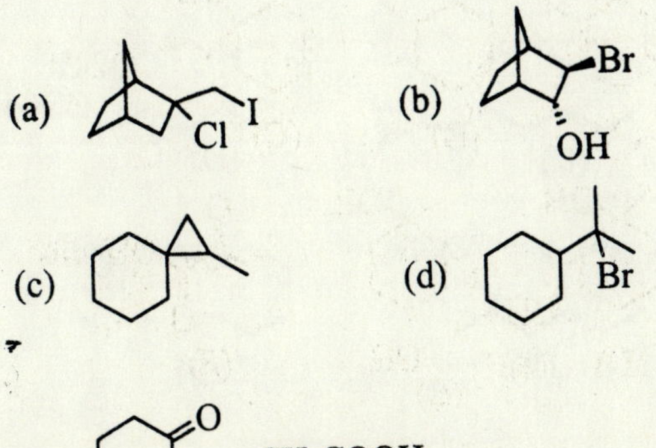

90. (a) Ph $\xrightarrow{H^+}$ Ph + (Benzylic) $\xrightarrow{Br^-}$ Ph Br

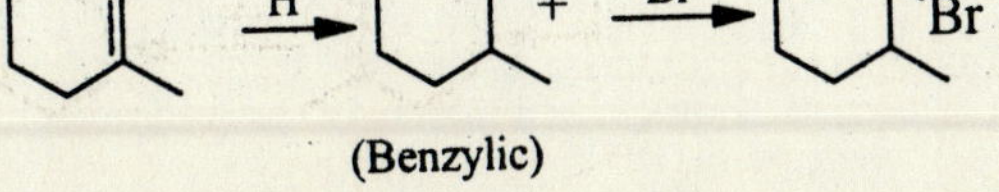

(b) Less stable carbocation intermediate will have to be formed for the other isomer.

91. (a) OMe $\xrightarrow{H^+}$ + OMe + + OMe (more stable) $\xrightarrow{Br^-}$ Br OMe (major); Br^- → Br OMe

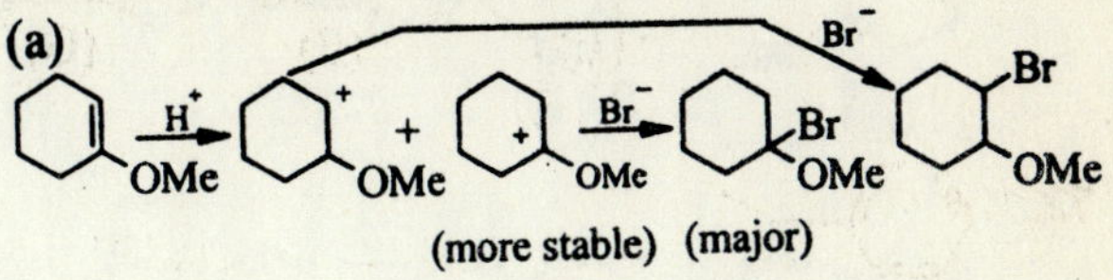

(b) OMe $\xrightarrow{H^+}$ + OMe; Cl^- → Cl OMe; OH_2 → HO OMe

OMe $\xrightarrow{H^+}$ + OMe; Cl^- → Cl OMe; OH_2 → OH HO OMe

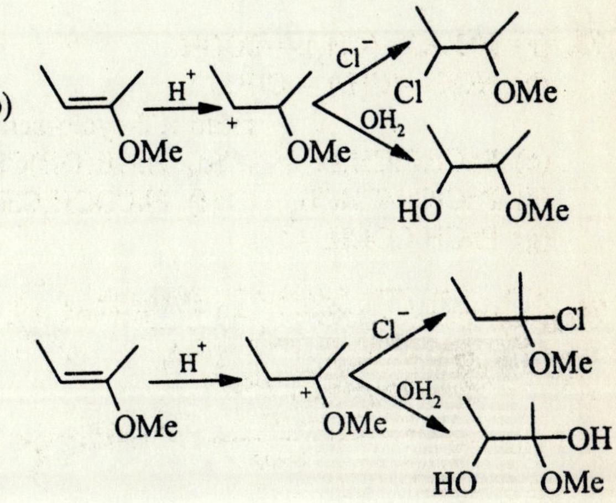

92. (a) OH (b) OMe Cl (c) Br (d) OH (e) Br Br + Br Br Racemic (f) OH H OH (g) H H (h) Cl (i) HO (j) OH (k) CHO + O

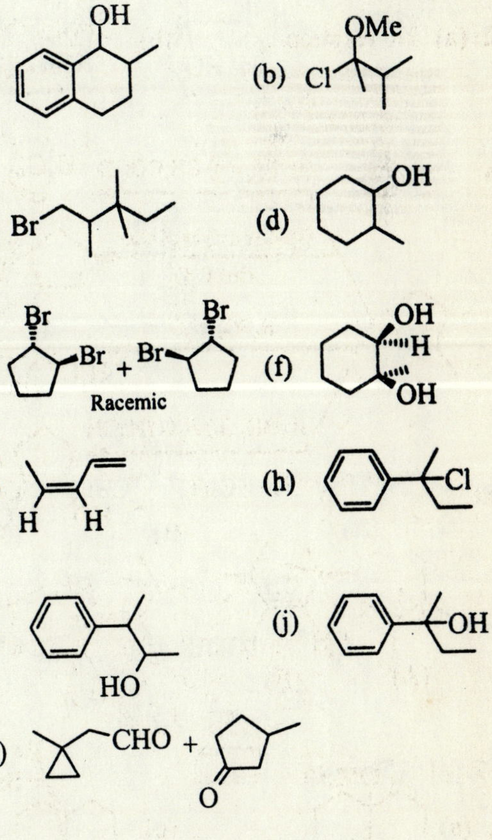

93. (A) Enantiomer (B) and (C) (D) O CHO

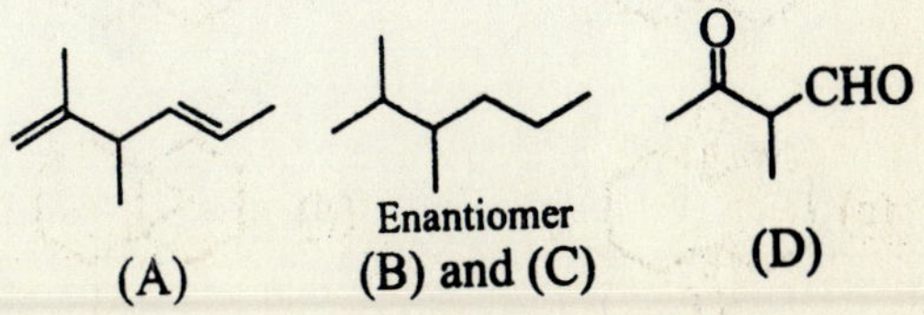

94. (A) (B) (C) (D)

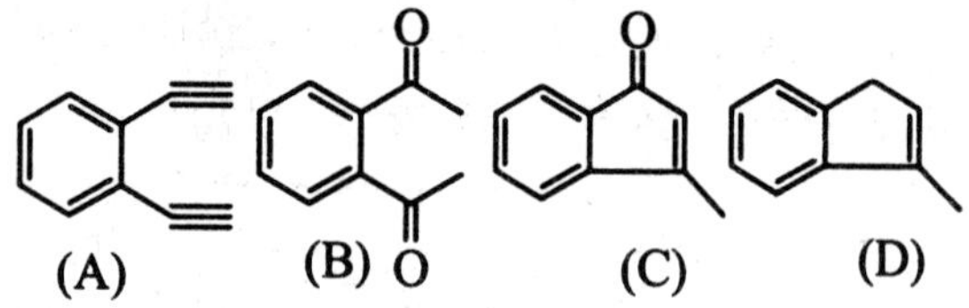

95. (A) (B) (C) (D) (E)

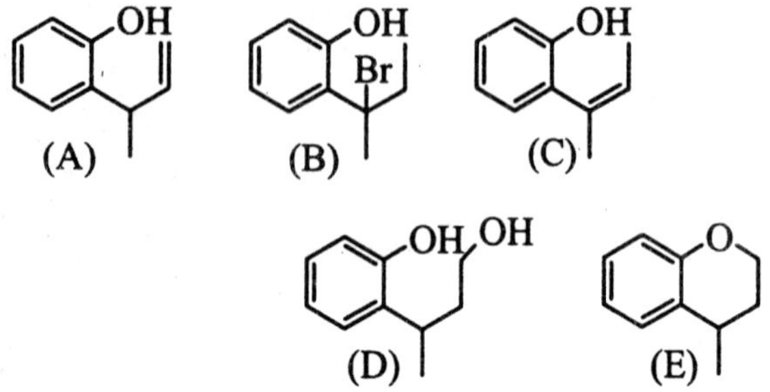

96. (A) (B) (C)

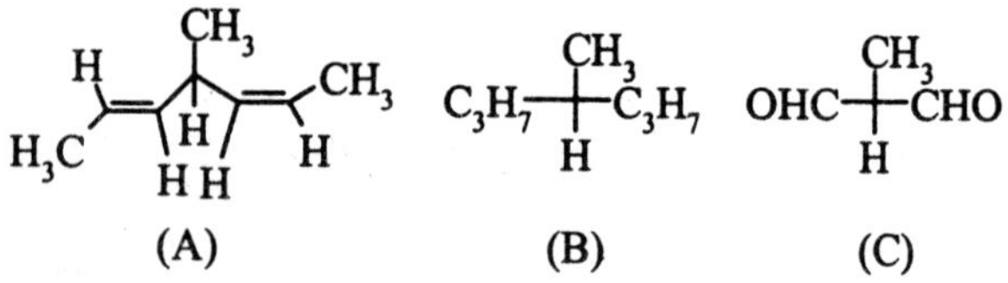

97. (A) (B) (C) (D) (E)

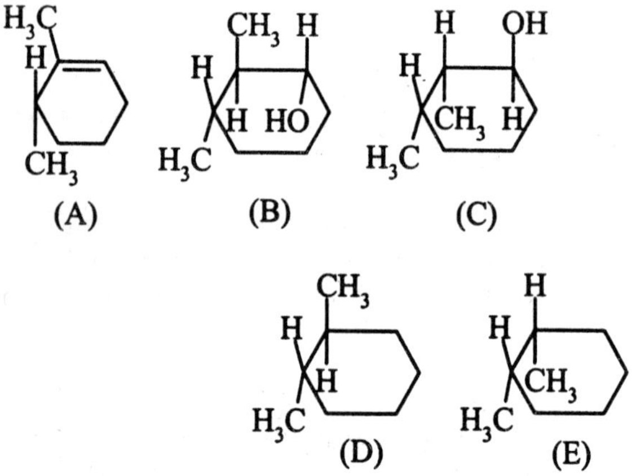

98. (A) (B) (C) (D) (E)

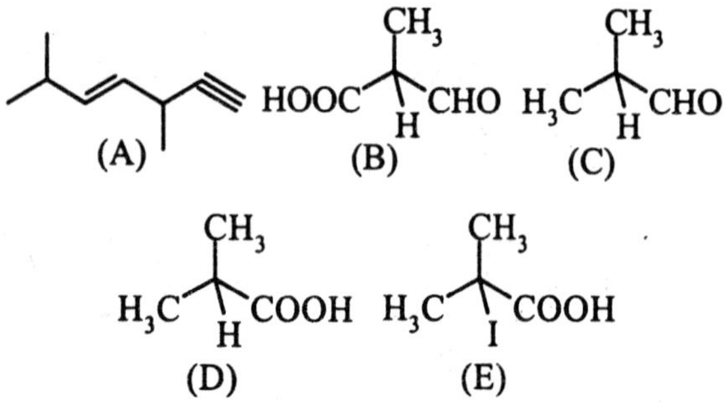

99. (a)

(b) (i) + H^+ → ; $-H^+$

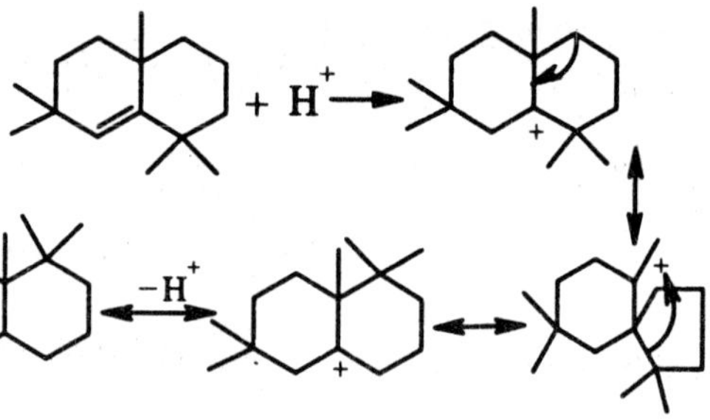

(ii) Bu–

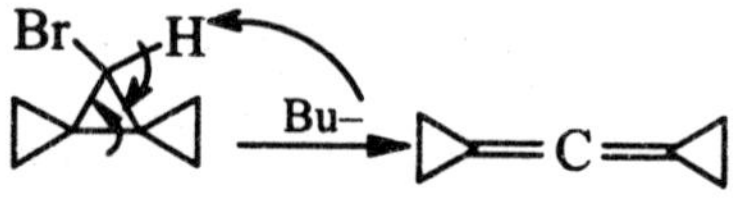

(iii) $^-OC(CH_3)_3$

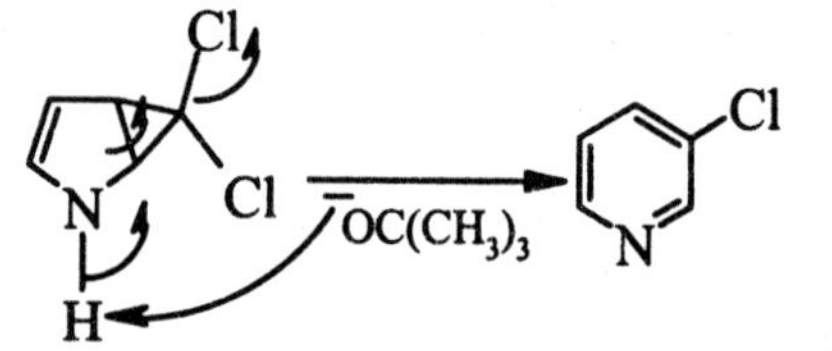

100. (A) (B) (C) (D) (E)

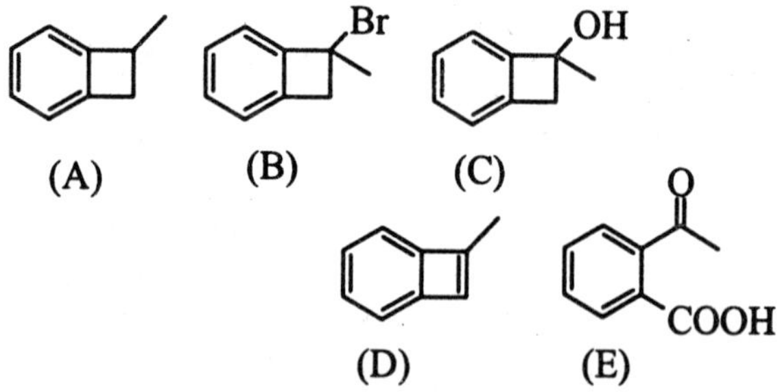

101. (a) A B

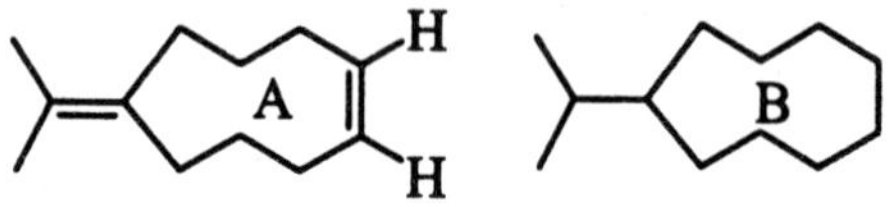

102. (A) (B) (C) (D) (E) (F)

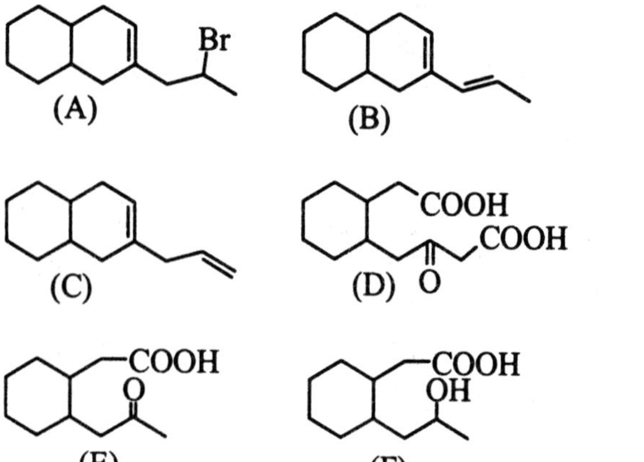

Objective Exercise I

1. b	14. d	27. a	40. d	53. b
2. b	15. c	28. c	41. d	54. b
3. d	16. d	29. a	42. d	55. c
4. a	17. d	30. d	43. c	56. a
5. d	18. b	31. d	44. d	57. b
6. b	19. b	32. b	45. d	58. a
7. c	20. b	33. b	46. A	59. c
8. d	21. d	34. d	47. d	60. b
9. d	22. a	35. b	48. c	61. d
10. a	23. d	36. b	49. d	62. d
11. a	24. b	37. a	50. c	63. d
12. a	25. b	38. c	51. d	
13. d	26. c	39. c	52. a	

Objective Exercise II

1. d	32. c	63. a	94. a	125. d
2. d	33. a	64. a	95. d	126. a
3. a	34. b	65. d	96. a	127. c
4. a	35. b	66. a	97. b	128. d
5. c	36. d	67. d	98. c	129. a
6. c	37. a	68. b	99. d	130. c
7. d	38. b	69. a	100. c	131. a
8. b	39. d	70. d	101. c	132. a
9. c	40. a	71. a	102. a	133. a
10. b	41. c	72. a	103. c	134. b
11. b	42. d	73. c	104. c	135. b
12. d	43. a	74. b	105. c	136. a
13. a	44. c	75. c	106. c	137. d
14. d	45. d	76. a	107. d	138. b
15. c	46. b	77. a	108. c	139. d
16. b	47. d	78. a	109. b	140. a
17. a	48. c	79. a	110. c	141. a
18. a	49. a	80. b	111. b	142. c
19. b	50. b	81. b	112. b	143. d
20. b	51. d	82. a	113. b	144. c
21. a	52. d	83. –	114. b	145. b
22. d	53. b	84. d	115. c	146. c
23. b	54. c	85. a	116. d	147. c
24. b	55. c	86. b	117. a	148. d
25. d	56. a	87. d	118. a	149. d
26. c	57. c	88. b	119. c	150. a
27. a	58. b	89. c	120. b	151. d
28. d	59. c	90. b	121. b	152. b
29. b	60. c	91. d	122. c	153. d
30. c	61. a	92. b	123. d	154. c
31. a	62. c	93. b	124. b	155. a

156. a, b	166. b	176. c	186. d	196. c
157. c	167. c	177. d	187. a	197. d
158. d	168. c	178. c	188. a	198. d
159. b	169. c	179. b	189. a	199. d
160. b	170. b	180. c	190. b	200. a
161. b	171. b	181. c	191. c	201. a
162. d	172. b	182. d	192. d	
163. a	173. a	183. b	193. b	
164. d	174. c	184. a	194. d	
165. b	175. c	185. a	195. d	

CHAPTER 5

Practice Exercise

1. (a) Tertiary butanol
   (b) Tertiary butyl methylether
   (c) $CH_3CH_2-\underset{CH_3}{\overset{CH_3}{C}}-O-\overset{O}{\overset{\|}{C}}-H$

2. $(CH_3)_3CCl < (CH_3)_3\,CCH_2Cl < (CH_3)_2CHCl < C_2H_5Cl < CH_3Cl$

3. (a) $H_2O < ROH < RCOO^- < HO^- < RO^-$
   (b) $F^- < Cl^- < Br^- < I^-$ (Protic solvent)
   $I^- < Br^- < Cl^- < F^-$ (Oprotic solvent)

4. (a) *n*-butyl bromide
   (b) *n*-bromopropane
   (c) 3-methyl-1-bromobutane
   (d) Ethyl chloride

5. (a) *n*-propyl bromide CH_3O^-
   (b) Ethyl iodide + SH^-
   (c) Methylybromide + CH_3SH
   (d) CH_3CH_2I (1.0 M) + CH_3S^- (2M)

6. (a) $(CH_3)_3CCl + H_2O$: Water is more polar than alcohol, carbocation will be more solvated (stabilized) in water.
   (b) $(CH_3)_3\,C\,Br + H_2O$: Same reason as in (a).
   (c) Both will proceed at equal rate since nucleophile is not involved in rate determining step.
   (d) $(CH_3)_3\,CCl + H_2O$: More stable 3° carbocation will be formed.

7. (a) H^- is stronger nucleophile than OH^-
   (b) Same as (a)
   (c) Same as (a)

(d) Vinylic C-Br bond has partial double bond character due to resonance, difficult to substitute.

(e) NH_3 is a weaker nucleophile than HO^-.

(f) CH_3^- is a very strong base, abstract acide H from -OH forming CH_4.

8. (a) 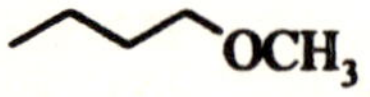 OCH_3 (b)

(c) 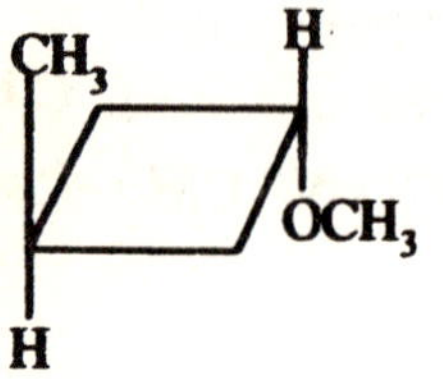 CH_3, H, OCH_3, H (d) 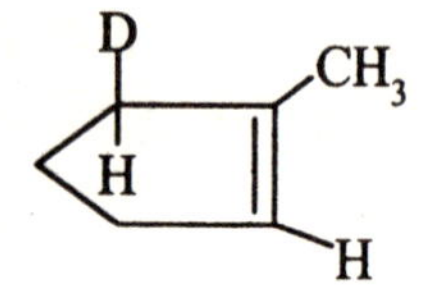 D, CH_3, H, H

9. (a) 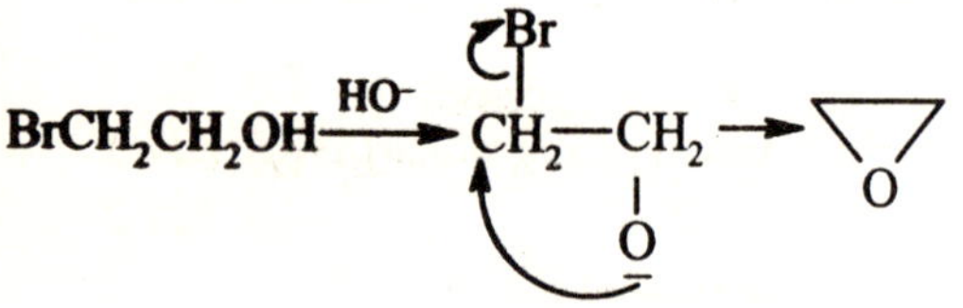 $BrCH_2CH_2OH \xrightarrow{HO^-} CH_2(Br)—CH_2(O^-) \rightarrow$ O

(b) 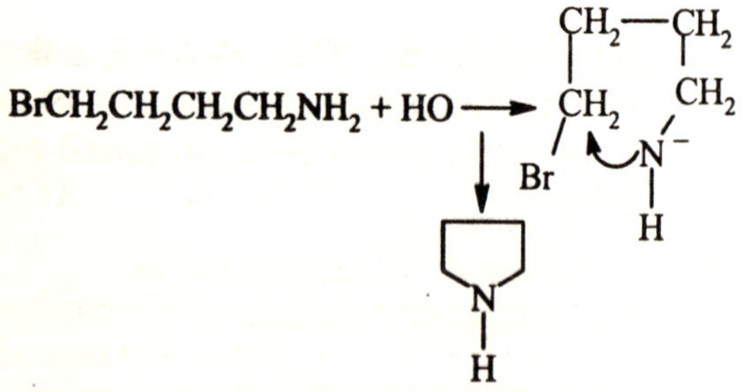 $BrCH_2CH_2CH_2CH_2NH_2 + HO^- \rightarrow$ $CH_2—CH_2$, CH_2, CH_2, Br, N^-, H → N–H

10. (a) Methyl bromide undergo halogen exchange with NaI forming methyl iodide. Iodide is a better leaving group than bromide, rate of hydrolysis is increased.

(b) The predominant reaction in this case is elimination (E2), which depends on concentration of both alkyl halide and base. Increasing concentration of base increases the rate of E2 reaction decreasing concentration of alkyl halide.

(c) NaCN is ionized completely giving ambident (CN^-) nucleophile. Since carbon is better electron donar than nitrogen, bonding occurs from carbon side. On the otherhand AgCN has covalent character and CN^- is not totally free as nucleophile. Hence, attack occur from nitrogen side due to lone pair of electron on it, giving isocyanide.

11. (a) 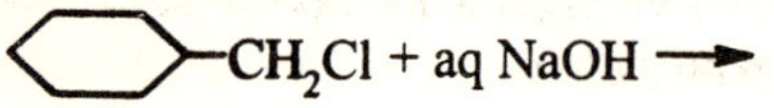 $—CH_2Cl + aq\ NaOH \rightarrow$

(b) 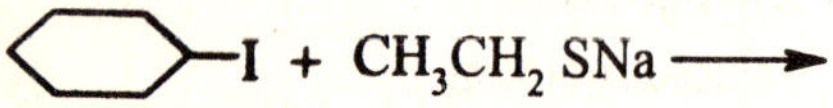 $—I + CH_3CH_2SNa \rightarrow$

(c) $CH_3I + (CH_3)_3COK \rightarrow$

(d) 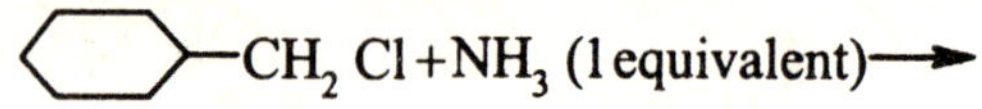 $—CH_2Cl + NH_3$ (1 equivalent) →

12. (a) 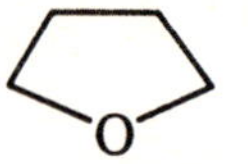 O (b) 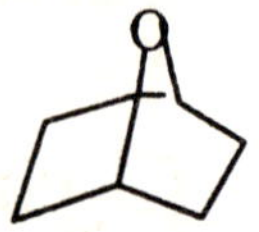 O

13. (a) 2, 3-dimethyl-2-butene
(b) 1, 2-dimethyl cyclohexene
(c) Methyl cyclohexen

14. A: 3-methyl-2-pentene
B: 3-chloro-3-methlyl pentane
C: 2-bromo-3-methyl pentane
D: 2-butanone
E: Ethanal
F: 2-butanol

15. (a) 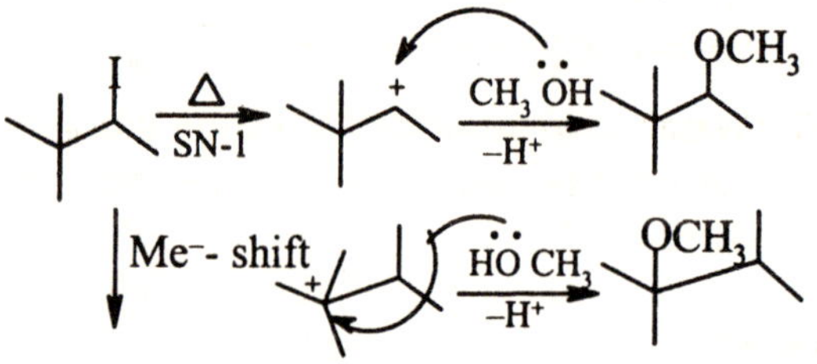 $\xrightarrow[SN-1]{\Delta}$ + $\xrightarrow[-H^+]{CH_3\ddot{O}H}$ OCH_3

Me⁻- shift + $\xrightarrow[-H^+]{H\ddot{O}CH_3}$ OCH_3

(b) 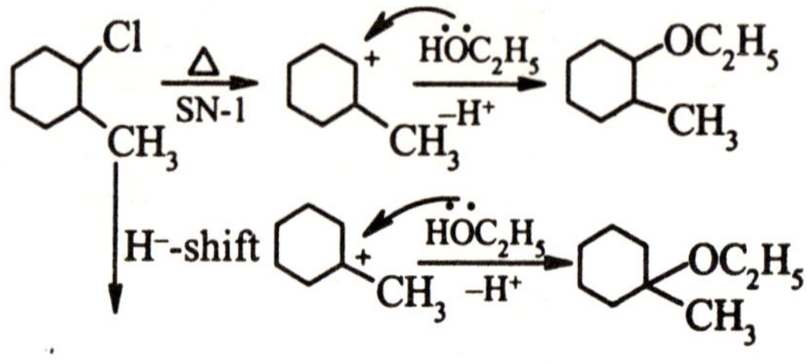 Cl, CH_3 $\xrightarrow[SN-1]{\Delta}$ +, CH_3 $\xrightarrow[-H^+]{H\ddot{O}C_2H_5}$ OC_2H_5, CH_3

H⁻-shift +, CH_3 $\xrightarrow[-H^+]{H\ddot{O}C_2H_5}$ OC_2H_5, CH_3

(c) I $\xrightarrow{\Delta}$ + $\xrightarrow[-H^+]{CH_3COOH}$ $OCOCH_3$

H⁻-shift + $\xrightarrow[-H^+]{CH_3COOH}$ $OCOCH_3$

(allylic carbocation)

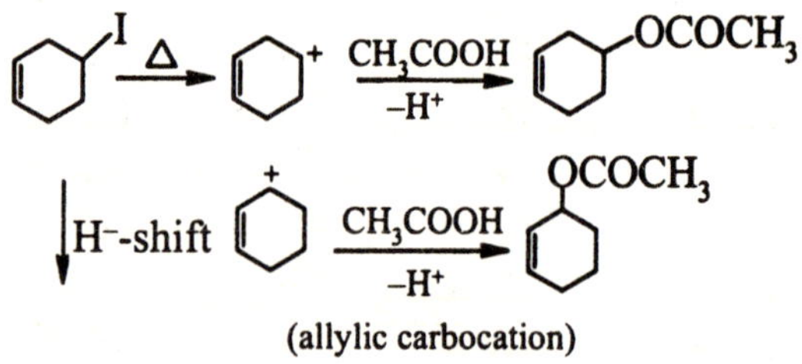

(d) 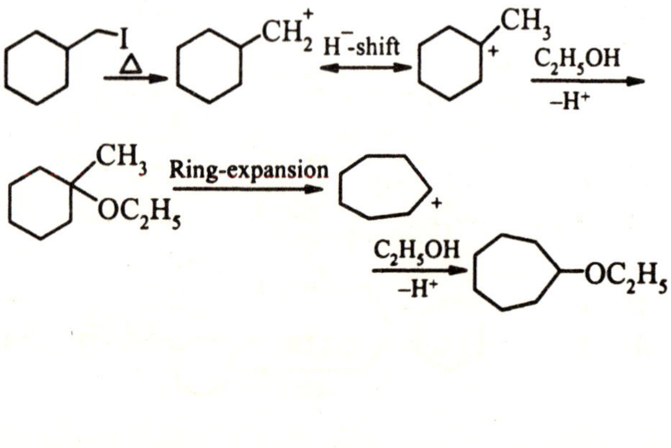 I $\xrightarrow{\Delta}$ $\overset{+}{C}H_2$ $\xrightarrow{H^-\text{-shift}}$ CH_3, + $\xrightarrow[-H^+]{C_2H_5OH}$

CH_3, OC_2H_5 $\xrightarrow{\text{Ring-expansion}}$ +

$\xrightarrow[-H^+]{C_2H_5OH}$ $—OC_2H_5$

(e) 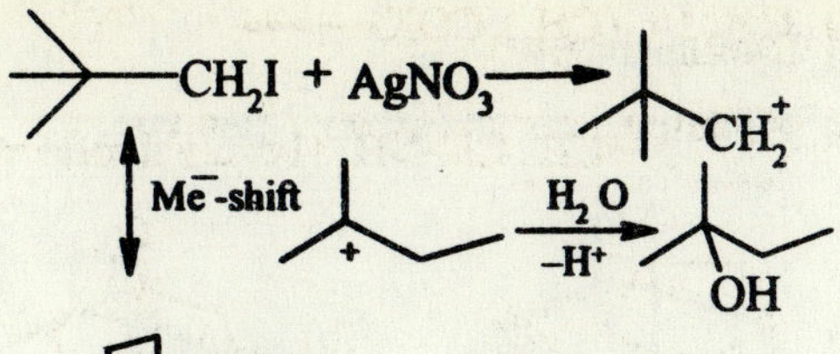 $-CH_2I + AgNO_3 \rightarrow$ CH_2^+ ; Me⁻-shift ; H_2O, $-H^+$; OH

(f) 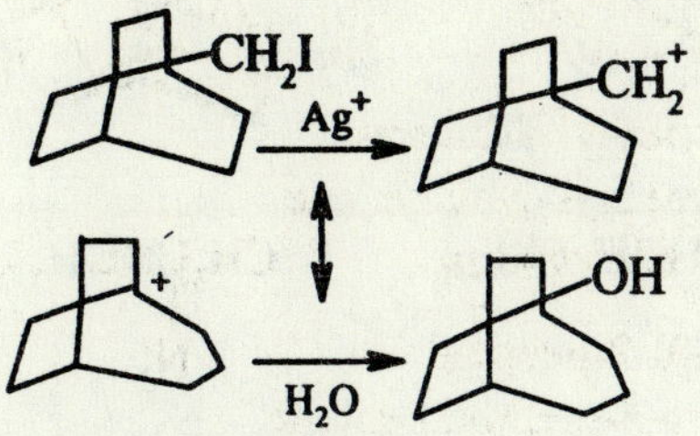 CH_2I $\xrightarrow{Ag^+}$ CH_2^+ ; H_2O ; OH

16. (a) $(CH_3)_3CCl$
(b) 2-chlorohexane
(c) Cyclohexyl bromide
(d) Cyclohexyl iodide
(e) 2-bromo-2-methyl pentane
(f) 3-bromo cyclohexene.

17. Water is more polar, solvation (stabilization) of carbocation increases an increasing more polar solvent.

18. (a) 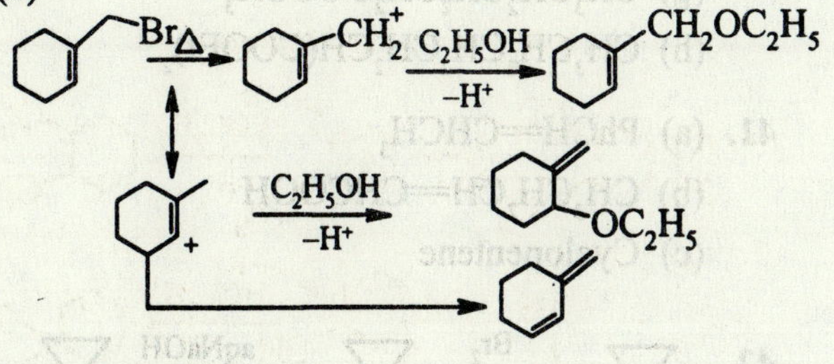 Br, Δ ; CH_2^+ ; C_2H_5OH, $-H^+$; $CH_2OC_2H_5$; C_2H_5OH, $-H^+$; OC_2H_5

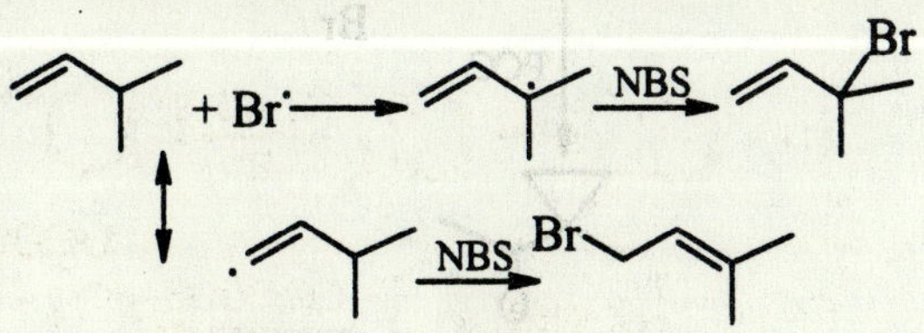

(b) + Br· → ; NBS ; Br ; NBS ; Br

(c) CH_2Br —Δ→ CH_2^+ ; H^--shift ; CH_3 ; CH_3OH, $-H^+$; OCH_3 ; Ring expansion ; $-H^+$; CH_3OH, $-H^+$; OCH_3

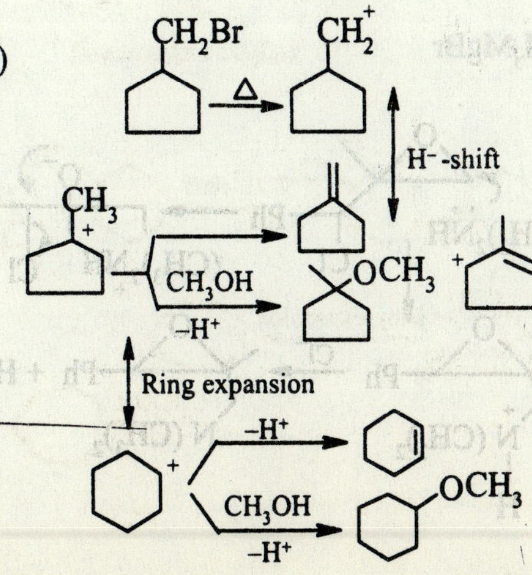

19. (a) $CH_3CH_2CH_2OCH_3$
(b) Propene
(c) 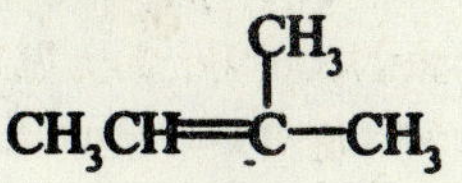 $CH_3CH{=}C(CH_3){-}CH_3$
(d) 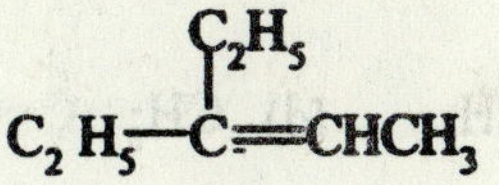 $C_2H_5{-}C(C_2H_5){=}CHCH_3$
(e) 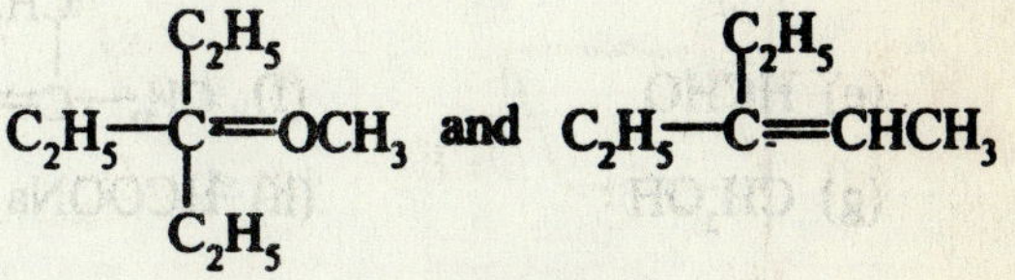 $C_2H_5{-}C(C_2H_5)_2{-}OCH_3$ and $C_2H_5{-}C(C_2H_5){=}CHCH_3$

20. (a) $PhCHBrCH_3 + C_2H_5OH/KOH \longrightarrow$
(b) $CH_3I + (CH_3)_2CHONa \longrightarrow$

21. (a) SN-2 (b) E2
(c) E2 (d) SN-2
(e) SN-2 (f) SN-1
(g) SN-2 and E2

22. Ethyl bromide being less hindered, undergo faster SN-2 reaction, isobutyl bromide is more reaction in E2 since it produces a more substituted, more stable alkene.

23. X : 2-chloro-2, 3-dimethyl butane.
Y : 2, 3-dimethyl-2-butene.
Z : 2, 3-dimethyl-1-butene.

24. (a) *Trans*-1, 2-dimethyl propene
(b) *Cis*-1, 2-dimethyl propene
(c) *Cis*-2-pentene
(d) 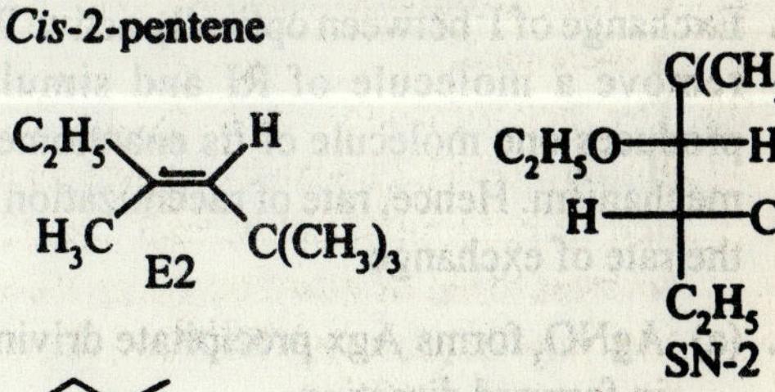 C_2H_5, H, H_3C, $C(CH_3)_3$ — E2 ; $C(CH_3)_3$, C_2H_5O—H, H—CH_3, C_2H_5 — SN-2

25. (a) 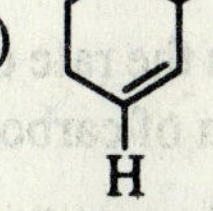 H (b) No-reaction
(c) 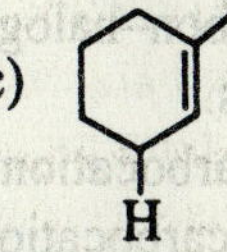 H (d)

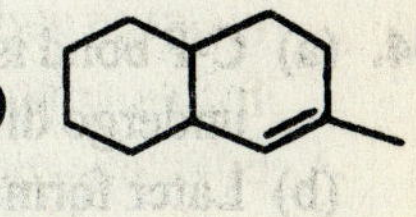

26. A: 2-bromo-2-pentene
B: 2-bromo-2-pentane
C: 2-pentyne
D: *Cis*-2-pentene
E: *Trans*-2-pentene

27. (a) $H_3C-C(CH_3)_2-Cl$ (b) $CH_3-C(CH_3)_2-OH$

(c) $CH_3-C(CH_3)=CH_2$ (d) $CH_3-\overset{O}{\overset{||}{C}}-CH_3$

(e) HCHO (f) $CH_3-C(CH_3)=N-OH$

(g) CH_3OH (h) HCOONa

28. (a) OH $\xrightarrow{-H^+}$ $\overset{+}{C}H_2$ $\xrightarrow{Cl^-}$ Cl

CH_2^+ $\xrightarrow{Cl^-}$ Cl

(b) OH $\xrightarrow{-H^+}$ + (3°)

Cl^-

Cl

30. 2, 2-dimethyl propane.

31. Ammonium salts are stable but oxonium salts are unstable.

32. Exchange of I^- between optically active RI and NaI, remove a molecule of RI and simultaneously produces one molecule of its enantiomer by SN-2 mechanism. Hence, rate of racemization is twice to the rate of exchange.

33. (a) $AgNO_3$ forms Agx precipitate driving reaction in forward direction.
(b) More acidic solvent increses the rate of heterolysis is facilitating farmation of carbocation.

34. (a) C-F bond is the strongest carbon–halogen bond, undergo difficult heterolysis.
(b) Later forms a more stable carbocation.
(c) -*t*-bucl forms a more stable carbocation.

35. Cycloprophyl chloride is more reactive in SN-2 than in SN-1 while cyclopentyl chloride is equally reaction in both SN-1 and SN-2 reaction.

36. Carbocation is stabilized by reasonance as:

$$\overset{+}{C}H_2-\ddot{O}-CH_2CH_3 \longleftrightarrow CH_2=\overset{+}{\ddot{O}}-CH_2CH_3$$

37. (a) Cyclopentene
(b) 1-pentene, *cis* and *trans* 2-pentene.

38. (a) D (b) D

39. (a) CH_3CHCH_3 (A) with OTs; CH_3CHCH_3 with N_3

(b) $CH_2CH_2CH_2OCH_3$, C, CH_3S, H, H_3C C

40. (a) 1-butene (b) *n*-butanol
(c) butyl amine (d) *n*-butyl azide
(e) pentanitrile
(f) $CH_3CH_2CH_2CH_2SCH_2CH_3$
(g) $CH_3CH_2CH_2CH_2OCOCH_3$
(h) $CH_3CH_2CH_2CH_2CH(COOEt)_2$

41. (a) $PhCH=CHCH_3$
(b) $CH_3CH_2CH=CHCOOH$
(c) Cyclopentene

42. $\xrightarrow[\text{Heat}]{Br_2}$ Br $\xrightarrow{\text{aqNaOH}}$ OH

PCC

O

43. CH_3MgBr

44. O, $(CH_3)_2\ddot{N}H$, Ph, Cl $\longrightarrow$ O^-, Ph, $(CH_3)_2\overset{+}{N}H$, Cl

O, Ph, $\overset{+}{N}(CH_3)_2$, H $\xrightarrow{Cl^-}$ O, Ph + HCl, $N(CH_3)_2$

45. III < II < I

46. I < III < IV < II

47. (a)  [structure: cyclobutyl–CH₂–CH(CH₃)–N₃] (b) 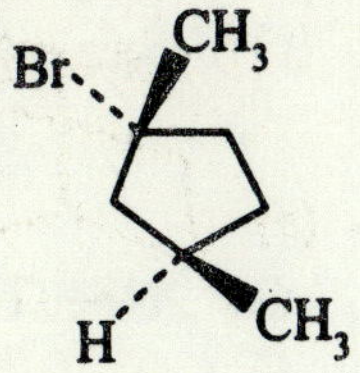 [structure: 1-bromo-1,3-dimethylcyclopentane with Br, CH₃, H, CH₃]

(c) $CH_3CH(CH_3)—CH_2—CH_2—CH(OH)—CH_3$

48. (a) Both substitution and elimination are equally probable.

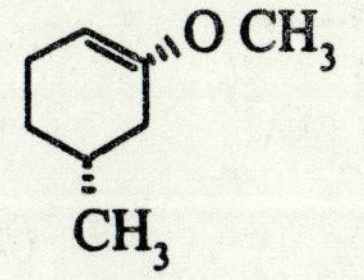

(b) 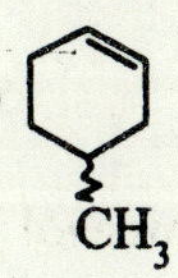 [structure] and [structure with O CH₃, CH₃] (c) 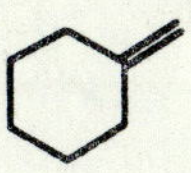 [structure]

(d) 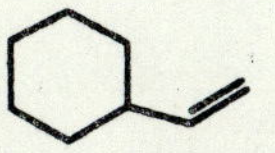 [structure] (e) 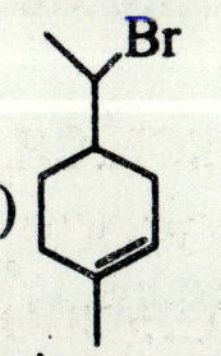 [structure with Br] (f) 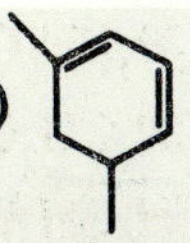 [structure] (g) 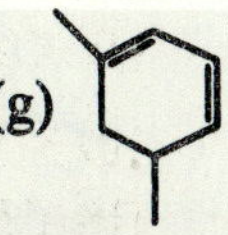 [structure]

49. (a) Ethanol being a very weak nucleophile cannot activate SN-2 reaction and CH_3Br being very less reactive in SN-1 reaction, neither SN-1 nor SN-2 takes place. Adding excess of C_2H_5ONa invoke SN-2 reaction and ether is formed in good yield.

(c) Et_3P: Larger sige of phosphorous makes Et_3P more polarizable and hence increases its nucleophilic strength.

50. (a) $I^- < Br^- < Cl^- < F^-$
(b) $SeH^- < HS^- < HO^-$
(c) $F^- < EtO^- < Me_2N^- < CH_3CH_2^-$
(d) $HC{\equiv}C^- < CH_2{=}CH^- < CH_3CH_2^-$
(e) $CH_3COO^- < CH_3OCH_2^- < CH_3SeCH_2^-$

51. (a) 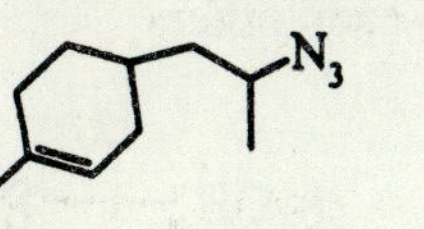 [structure with I and N₃]

(b) 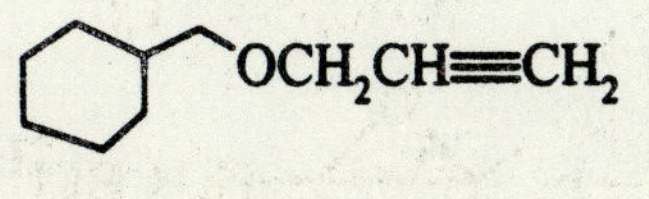 [cyclohexyl]$OCH_2CH{\equiv}CH_2$ (c) 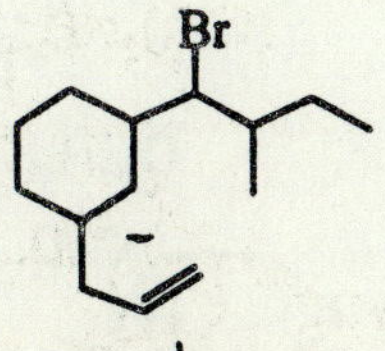 [structure with Br]

(d) 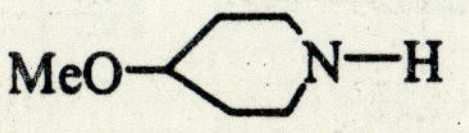 MeO–[piperidine]N—H (e) 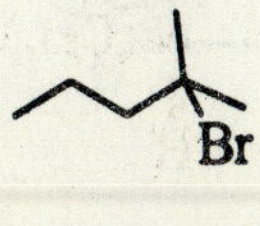 [structure with Br]

52. Tertiary butoxide being a bulky base, abstract H- from less hindered β-carbon giving this alternative product.

53. (a) $HCOO^- < (CH_3)_3CCH_2O^- < CH_3O^-$
(b) $HF < NH_3 < PH_3$
(c) $H_2O < HO^- < HS^-$

54. 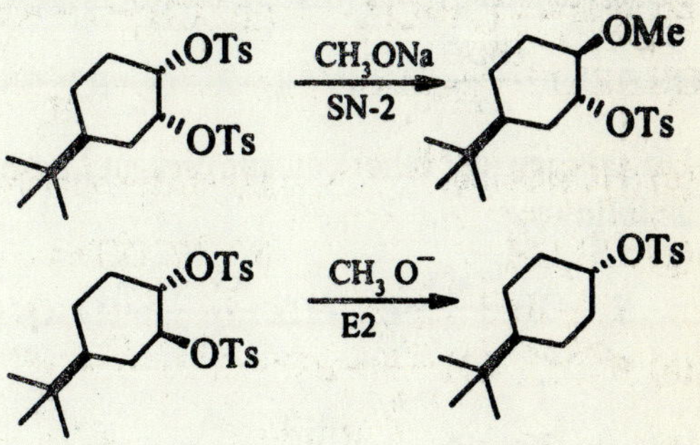 [structure with OTs, OTs] $\xrightarrow[\text{SN-2}]{CH_3ONa}$ [structure with OMe, OTs]

[structure with OTs, OTs] $\xrightarrow[\text{E2}]{CH_3O^-}$ [structure with OTs]

55. (a) $MeOCH_2Cl$ in both SN-1 and SN-2
(b) $MeOCH_2Cl$
(c) $MeOCH_2CH_2Cl$

56. (a) 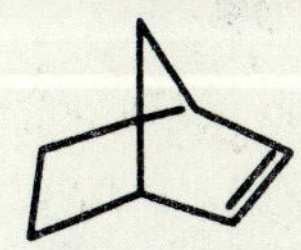 [structure] (b) 1-butene
(c) 2-methyl-2-butene (d) 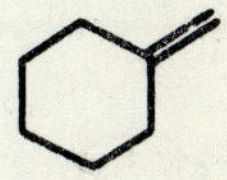 [structure]

57. (a) $(CH_3)_3CBr$ (b) 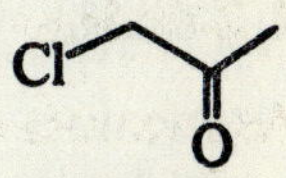 Cl[structure with O]
(c)  [structure with O, Br] (d) $PhCH_2Cl$
(e) 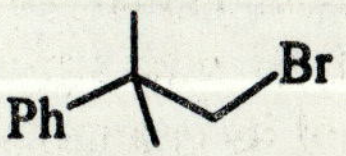 Ph[structure with Br]

58. (a) 2-bromobutene
(b) 2-bromo-2, 3-dimethyl butane
(c) 1-choro-1-methyl cyclopentane
(d) 1-chloro-1-ethyl-2-methyl cyclohexane.

59. (a) 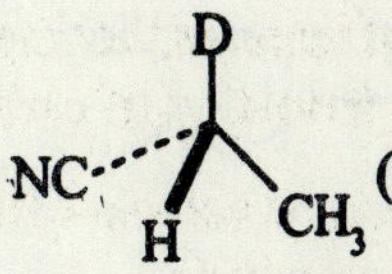 NC[structure with D, H, CH₃] (b) $CH_3—CH(D)—CN$ (racemic)

60. (a) 1-butanol $\xrightarrow{TsCl}$ $\xrightarrow{NaI}$
(b) Product from (a) + $C_2H_5NH_2$ $\longrightarrow$
(c) 1-butanol + propanoic acid $\xrightarrow[\Delta]{H_2SO_4}$
(d) 1-butanol $\xrightarrow{TsCl}$ $\xrightarrow{NaSH}$
(e) 1-butanol $\xrightarrow{TsCl}$ $\xrightarrow{CH_3ONa}$

61. Dehydration of alcohol involve formation of a carbocation intermediate while so such intermediate is formed in dehydrohalogenation (E2).

62. For unsymmetrical ether, mixture of alcohols will have to be used. When different alcohols are used self condensation also occurs alongwith cross-condensation giving mixture of ethers as:

$$R—OH + R'OH \xrightarrow[\Delta]{H_2SO_4} R—O—R + R'—O—R' + R—O—R'$$

Unsymmetrical ether can conveniently be prepared as follows:

$$R—OH \xrightarrow{TsCl} R—OTs \xrightarrow[SN\text{-}2]{R'ONa} R—O—R'$$

63. (a) $\xrightarrow[Heat]{HNO_2}$ OH; NH_2 ↓ TsCl; OTs $\xrightarrow[(CH_3)_3COH]{(CH_3)_3COK}$

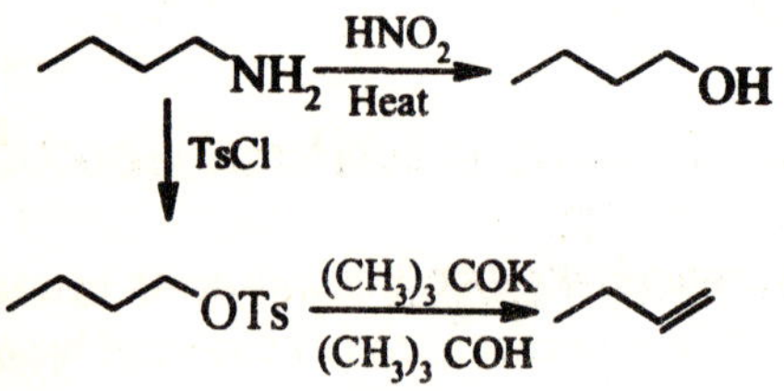

(b) Br $\xrightarrow[(CH_3)_3COH]{(CH_3)_3COK}$

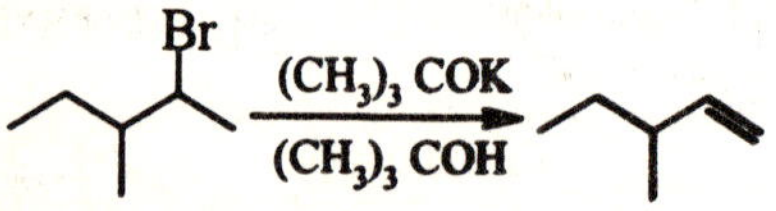

64. (a) OH $\xrightarrow[-H_2O]{-H^+}$ + ; Br⁻ → Br

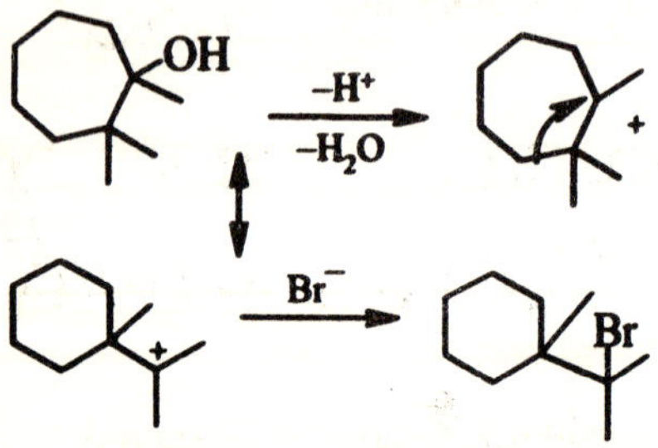

(b) OH $\xrightarrow[-H_2O]{H^+}$ + ; + $\xrightarrow{Br^-}$ Br

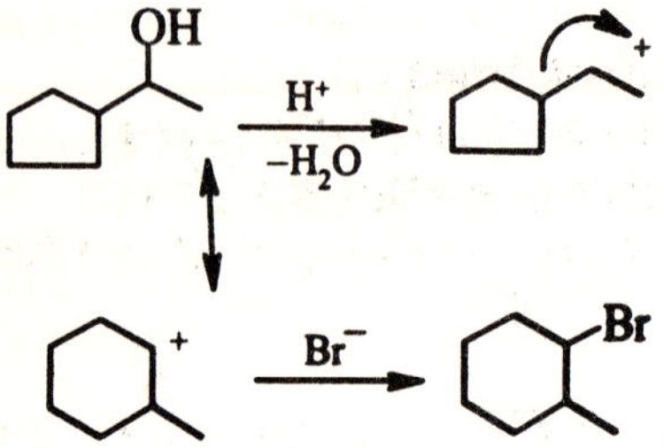

(c) OH $\xrightarrow[—H_2O]{H^+}$ + ⟷ + ; Me-shift ; + $\xrightarrow{Br^-}$ Br

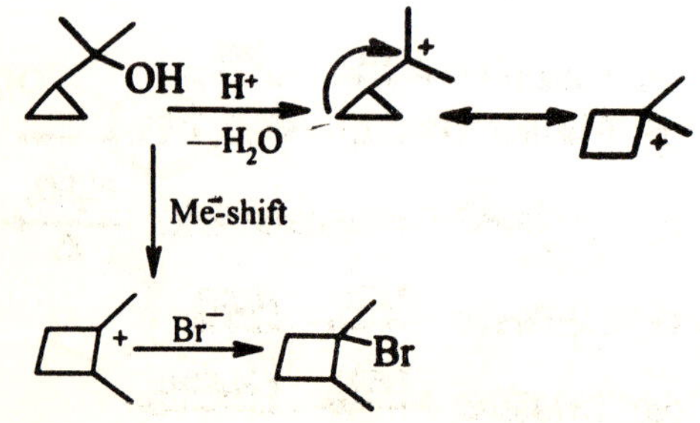

65. (a) Propene (b)

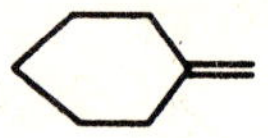

(c) $>N—CH_2CH_2—CH(CH_3)—CH═CH_2$

(d) $>N—CH_2CH(CH_3)—CH_2CH═CH_2$

66. Only b

67. (b) H, CN, CH_3, H (c)

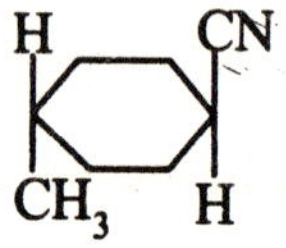

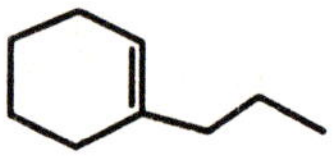

68. 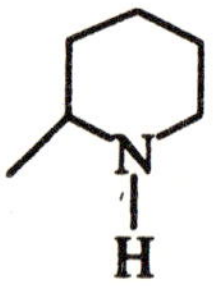 → 1, 5-hexadine,

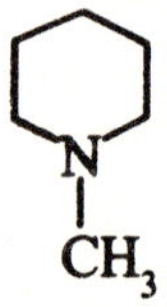

→ 1, 4-petadiene (N—CH_3)

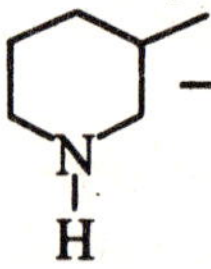

→ 2-methyl-1, 4-pentadiene,

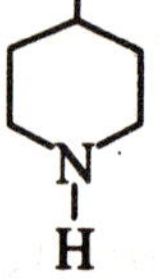

→ 3-methyl-1, 4-pentadiene

69. (a) $(CH_3)_3CI + HO^- \rightarrow$, I^- bening better leaving group than Cl^-.

(b) H, H, H_3C, H, Br, CH_3 + $CH_3ONa/CH_3OH \rightarrow$,

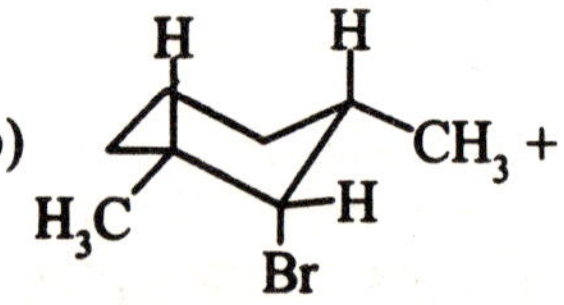

Br and H are antito each other.

70. (a) HO—C(H)(C$_4$H$_9$)—CH$_3$ (b) CH$_3$—C(H)(C$_4$H$_9$)—OH (racemic)

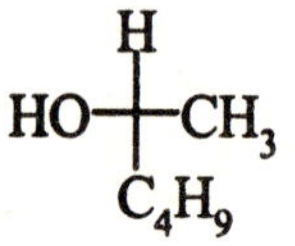

(c) OCH_3, CH_3, H, H (d) CH_3O—C—H, CH_3—C—H, CH_3, C_2H_5 (e) CH_3O—C—H, H—C—CH_3, CH_3, C_2H_5

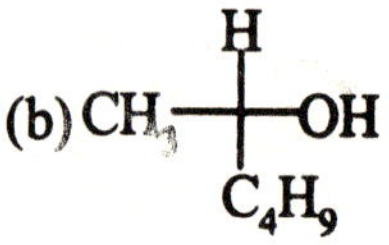

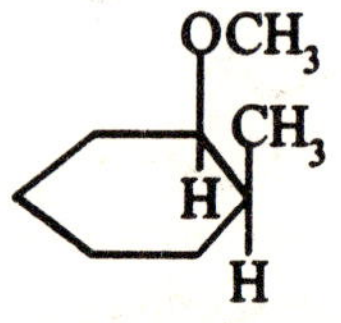

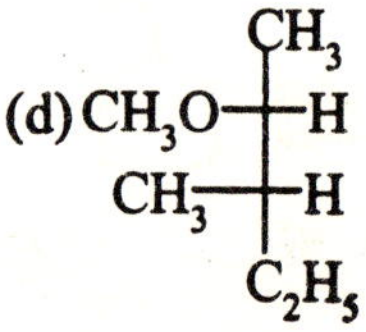

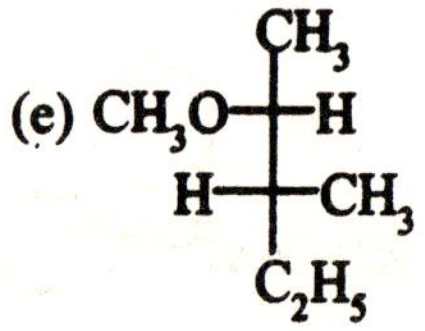

71. I: Cl, H, CH(CH_3)$_2$, H $\xrightarrow{CH_3Na}$ H, OCH_3, CH(CH_3)$_2$, H (SN-2) + CH(CH_3)$_2$ (E2)

II: H, Cl, CH(CH_3)$_2$, H $\xrightarrow{CH_3ONa}$ OMe, H, CH(CH_3)$_2$, H (SN-2) + CH(CH_3)$_2$

72. (a) HO^- (b) H_2S (c) I^- (d) NH_2^- (e) HS^-

73. (a) NaOH (b) CH_3ONa (c) NaHS (d) CH_3CH_2SNa (e) $(CH_3)_3NH$ (f) NaCN (g) CH_3COONa (h) $CH_3C\equiv C^-Na^+$

74. (a) $CH_3COO^- < CH_3CH_2S^- CH_3CH_2S^-$
(b) $H_2O < Ph_3 < NH_3$
(c) $Cl^- < Br^- < I^-$ (d) $I^- < Br^- < Cl^-$
(e) $C_6H_5O^- < C_6H_{11}O^-$

75. (a) 1st (b) 2nd (c) 1st

76. (a) Cyclohexane $\xrightarrow[h\nu]{Cl_2}$ chlorocyclohexane (Cl) $\xrightarrow{aq\ CN^-}$ cyclohexyl CN; chlorocyclohexane $\xrightarrow{H_3O^+}$ cyclohexyl COOH

(b) Cyclohexane $\xrightarrow[h\nu]{CH_2}$ chlorocyclohexane (Cl) $\xrightarrow{C_2H_5ONa}$ cyclohexene $\xrightarrow[CCl_4]{NBS}$ 3-bromocyclohexene (Br) $\xrightarrow{CH_3ONa}$ 3-methoxycyclohexene (OCH_3)

(c) 3-bromocyclohexene (Br) (from b) $\xrightarrow{aq\ NaOH}$ cyclohex-2-enol (OH)

77. (a) Br, $\xrightarrow{\Delta}$ carbocation (+); + $\xrightarrow[-H+]{H_2O}$ OH

(b) Br $\xrightarrow{\Delta}$ + ; + $\xrightarrow[-H+]{H_2O}$ CH_3, CH_3, H, + (OH) + CH_3, OH, H, + (CH_3)

78. (a) **Though neopentylbromide is primary, bulky tertiarybutyl group in neopentyl bromide poses very large steric hindrence to the attack of a bulky nucleophile (N_3^-).**

(b) H–C(Br)(D)–CH_3 + N_3^- $\xrightarrow{SN-2}$ H_3C–C(CH_3)(D)–H

(c) **Rate will double**
(d) **Rate will double**
(e) **not related**
(f) **Racemization occur through carbocation intermediate.**

79. **(a) 1-butanol < 2-butanol < 1-butanol**
(b) *p*-nitrobenzyl alcohol < benzyl alcohol < *p*-methoxy benzyl alcohol.
(c) *p*-methoxy phenol < benzyl alcohol < α, α-dimethyl benzyl alcohol.

80. (a) *n*-iodoheptane
(b) C_2H_5ONa
(c) *n*-butyl tosylate
(d) isopropoxide will C_2H_5Br

81. (a) PhO---C(CH_3)(C_2H_5)—H (b) cyclohexyl COOH

(c) 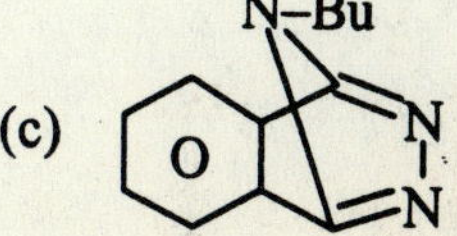

(d) CN

(e) 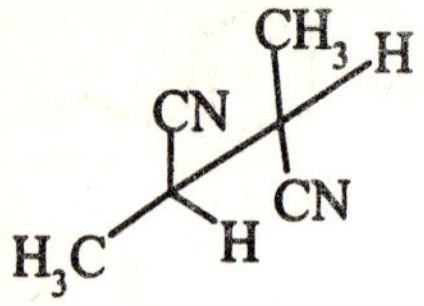 (f) 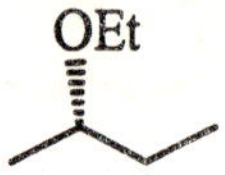 OEt

(g)  I (h) 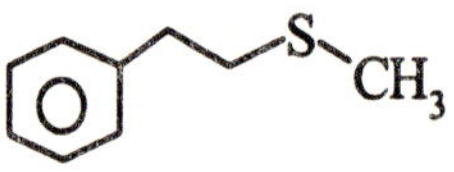 S CH_3

(i) No-reaction

82. (a) $\xrightarrow{NaOH}$ $\xrightarrow[Excess]{CH_3I}$

(b) $\xrightarrow{TsCl}$ $\xrightarrow{NH_3}$

(c) $\xrightarrow{TsCl}$ $\xrightarrow{CH_3CH_2CH_2SNa}$

(d) $\xrightarrow[H_2O_2 \mid OH^-]{B_2H_6}$ $\xrightarrow{KOH}$

83. (a) $\xrightarrow{TsCl}$ $\xrightarrow{NH_3}$

(b) $\xrightarrow{TsCl}$ $\xrightarrow{HC\equiv C^-Na^+}$

(c)  $\xrightarrow{NaH}$ $\xrightarrow{\text{epoxide}}$ $\xrightarrow{H_3O^+}$

(d) $\xrightarrow{TsCl}$ $\xrightarrow{NaCN}$ $\xrightarrow{H_3O^+}$ $\xrightarrow[H^+]{CH_3OH}$

84.  and

85. (a) CH_3CN (b) No reaction

(c)  (d)

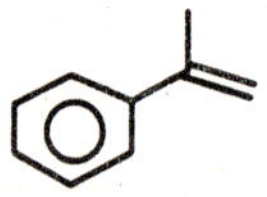

(e)

86. (a) 1st is more reactive
(b) IV < II < I < III
(c) II < III < I
(d) I < IV < III < II
(e) III < I < II

87.

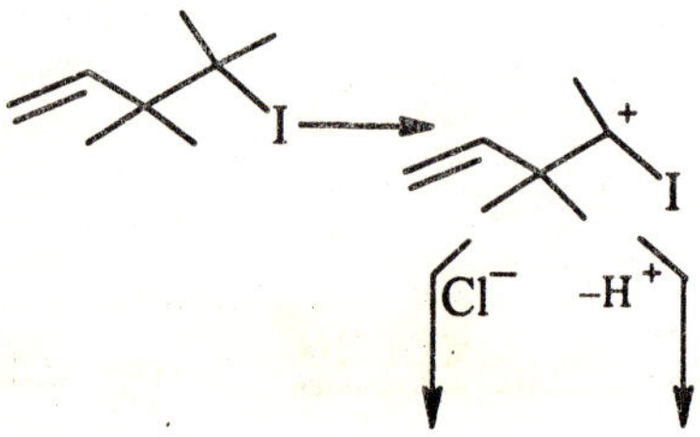

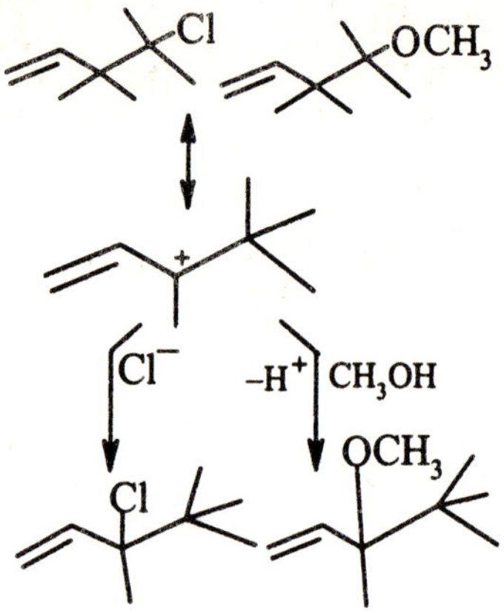

I, +, I, Cl^-, $-H^+$

Cl, OCH_3, +, Cl^-, $-H^+$ | CH_3OH, Cl, OCH_3

88. Allyl bromide produces a resonance stabilized carbocation justifying its greater reactivity in SN-1 reaction. Also, carbon β-to-Br has an empty *p*-asbital which share some of the shared electron pairs of pentavalent carbon in TS of SN-2 mechanism, increasing its stability. This stabilization of TS justify greater reactivity in SN-2 reaction.

89. (a) II, SN-2 (b) II, SN-2
(c) I, SN-2 (d) I, SN-1

90. (a) CN^- being weaker nucleophile than H^-.
(b) tertiary carbocation is more stable than primary.
(c) geminal diols are unstable.

91. (a) 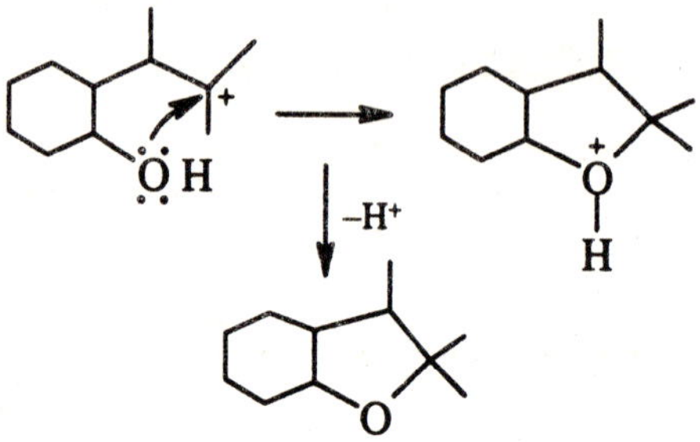 ÖH, +, $-H^+$, O, H, O

(b)  O, Br, O, $+ Br^-$

92. Transition state is pentavalent, highly unstable, highest energy point between reaction and produced in SN-2 reaction.

93. (a) $CH_3CS_2^- < CH_3CH_2S^- < CH_3CH_2O^-$
(b) $I^- < Br^- < F^-$
(c) $HSe^- < HS^- < HO^-$
(d) $HS^- < HO^- < NH_2^-$
(e) $CF_3CH_2S^- < CF_3CH_2O^- < CH_3CH_2S^- < CH_3CH_2O^-$
(f) $CF_3COO^- < CH_3CCOO^- < CH_3CH_2COO^-$
(g) $CH_3COO^- < OH^- < CH_3O^- < CH_3C\equiv C^-$

94. Weaker base is a better leaving group:
(a) $CF_3^- < CH_3O^- < CH_3S^- < F^-$
(b) $CH_3COO^- < CH_3SO_3^- < CF_3SO_3^-$

95. (a) 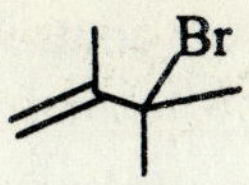 Br (b) 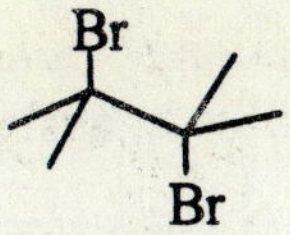 Br Br

(c) 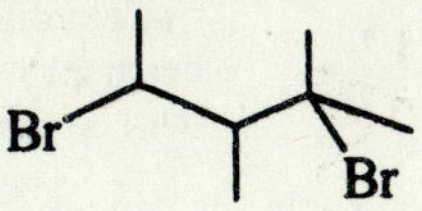 Br Br (d)

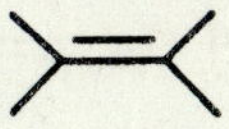

(e) 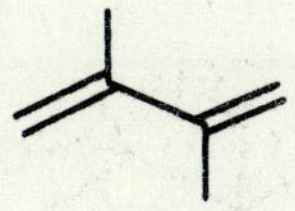 (f) 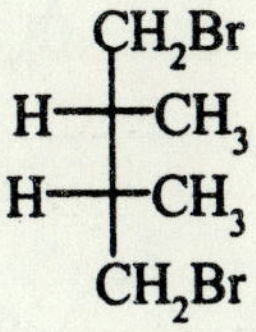 CH_2Br, H—CH_3, H—CH_3, CH_2Br

(g) 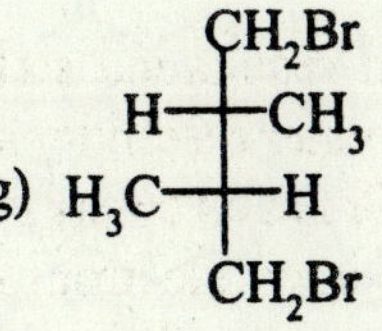 CH_2Br, H—CH_3, H_3C—H, CH_2Br

96. (a) 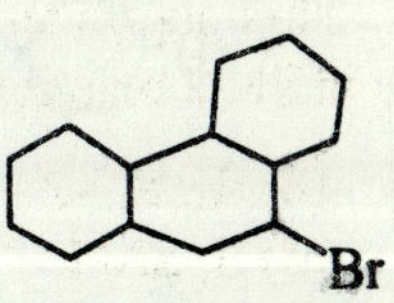 Br (b)

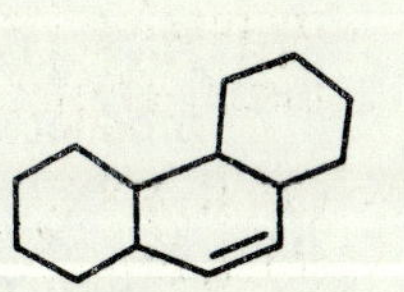

(c) 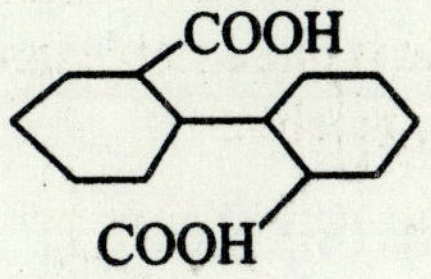 COOH COOH (d)

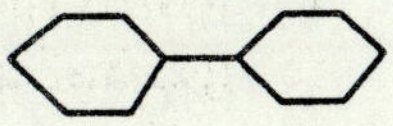

97. (a) 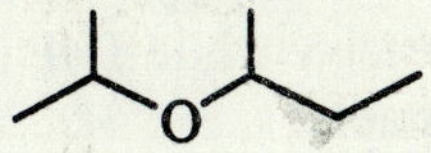 O (b) $(CH_3)_2CHOH$

(c) $CH_3CH—CH_2CH_3$ (OH) (d) 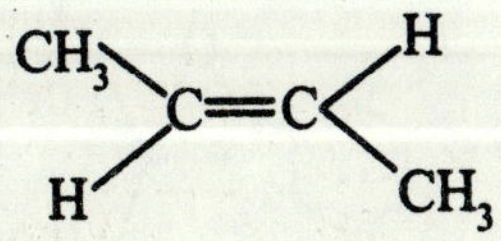 CH_3, H, C=C, H, CH_3

(e) 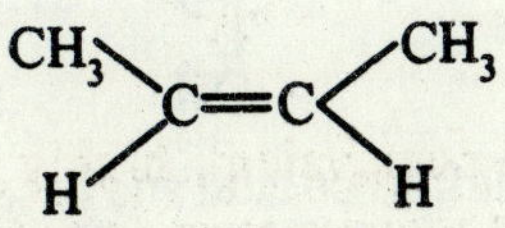 CH_3, CH_3, C=C, H, H

98. (a) 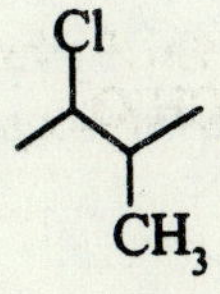 Cl, CH_3 (b) 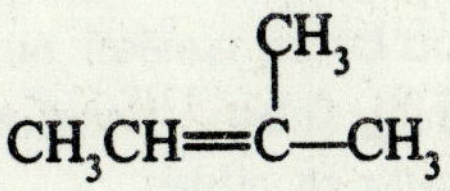 $CH_3CH{=}C—CH_3$ (CH_3) (c)

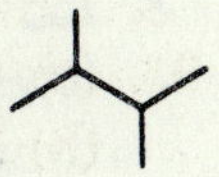

99. (a) 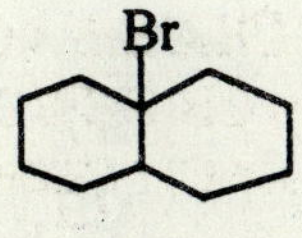 Br (b)

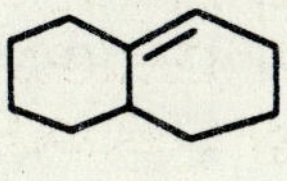

(c) 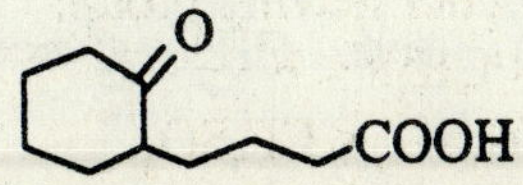 O, COOH

(d) 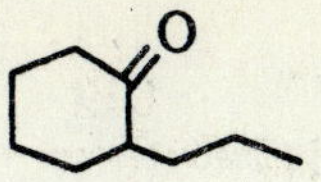 O (e)

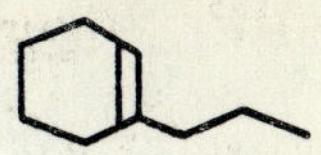

100. (a)  O (b) 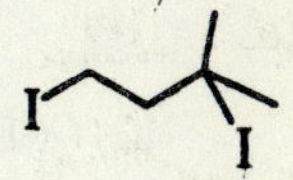 I, I

(c)  NC, I (d)  I, OH

101. (a) 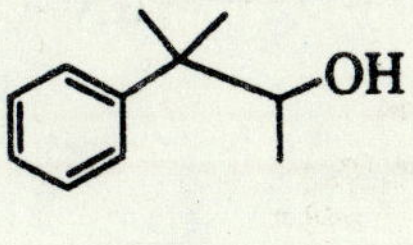 OH (b)

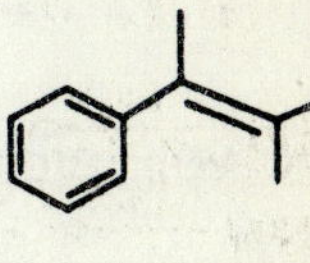

(c) 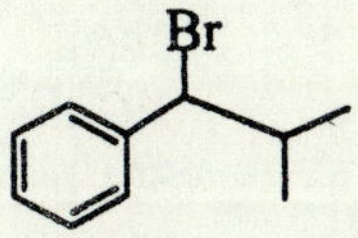 Br (d) 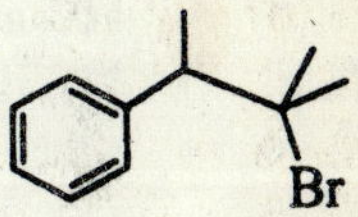 Br

102. (a) 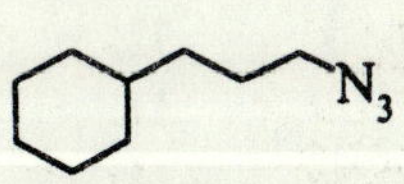 N_3

(b) $PhCH_2CH_2CN$

(c) 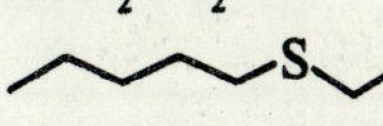 S

(d)

(e) 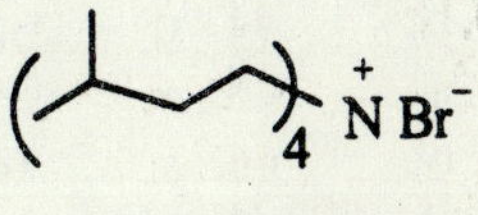 $\left(\ \right)_4 \overset{+}{N} Br^-$

(f) 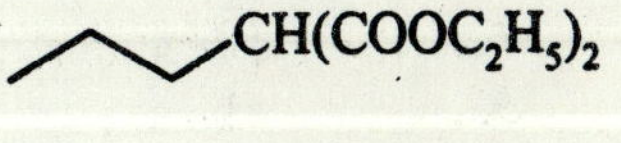 $CH(COOC_2H_5)_2$ (g)  O

103. (a) 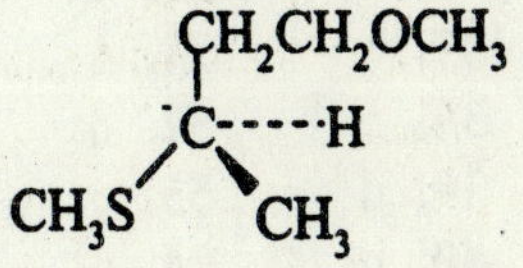 $CH_2CH_2OCH_3$, C----H, CH_3S, CH_3

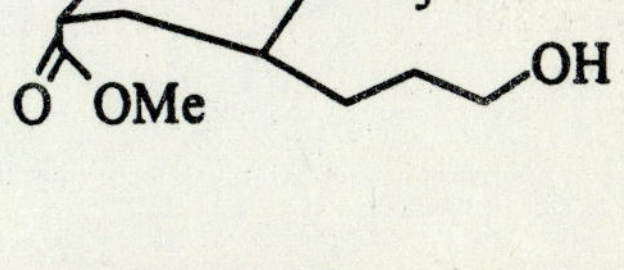

(b) N_3, O OMe, OH (c) O, O

(d) 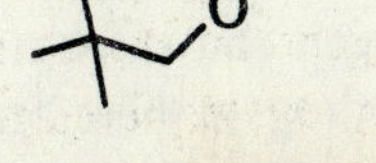 O, O, I

104. (a) $\xrightarrow{\text{TsCl}}$ $\xrightarrow{C_2H_5SNa}$ (b) $\xrightarrow{\text{TsCl}}$ $\xrightarrow{\text{NaI}}$

(c) $\xrightarrow{NaH} \xrightarrow{PhCH_2CH_2I}$

(d) *n*-butyl bromide $\xrightarrow{CH_3C\equiv C^-Na^+}$

(e) $\xrightarrow{TsCl} \xrightarrow{NH_3}$

(f) $\xrightarrow[CH=CH_2]{H_2N-CHCH_2OH}$

105. (a) $R-X + {}^-S-C\equiv N \longrightarrow R-SCN + X^-$

(b) $R-X + {}^-N=C=O \longrightarrow RNCO + X^-$

Objective Exercise

Level 1

1. d	**10.** a	**19.** d	**28.** a	**37.** d
2. c	**11.** d	**20.** d	**29.** d	**38.** a
3. b	**12.** c	**21.** a	**30.** b	**39.** a
4. d	**13.** b	**22.** d	**31.** a	**40.** b
5. c	**14.** c	**23.** d	**32.** c	**41.** b
6. a	**15.** c	**24.** a	**33.** a	
7. a	**16.** a	**25.** a	**34.** c	
8. d	**17.** b	**26.** c	**35.** c	
9. a	**18.** c	**27.** d	**36.** b	

Level 2

1. d	**15.** b	**29.** b	**43.** c	**57.** b
2. a	**16.** a	**30.** b	**44.** d	**58.** b
3. b	**17.** d	**31.** a	**45.** a	**59.** c
4. a	**18.** a	**32.** c	**46.** a	**60.** c
5. b	**19.** b	**33.** c	**47.** b	**61.** d
6. c	**20.** a	**34.** d	**48.** c	**62.** c
7. b	**21.** d	**35.** b	**49.** a	**63.** c
8. a	**22.** d	**36.** b	**50.** b	**64.** a
9. c	**23.** a	**37.** c	**51.** a	**65.** c
10. c	**24.** c	**38.** c	**52.** d	**66.** b
11. b	**25.** a	**39.** d	**53.** c	**67.** b
12. b	**26.** d	**40.** b	**54.** c	**68.** c
13. a	**27.** c	**41.** d	**55.** c	
14. a	**28.** b	**42.** c	**56.** a	

CHAPTER 6

Practice Exercise

Level 1

1. They are all nucleophile since they have unshared pair of electrons on oxygen and sulphur. Alcohols and thiols have acidic hydrogen, therefore, they are more reactive than ethers, which lacks acidic hydrogen.

2. Basicity of alkoxide or any base is controlled by their ability to abstract proton from with an acid.

$$R\bar{O} + HA \rightleftharpoons ROH + A^-$$
$$B + HA \rightleftharpoons H\overset{+}{B} + A^-$$

Higher the value of equilibrium constant stronger will be the base.

3. (a) 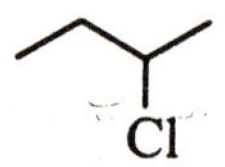 Cl (b)  Cl

(c) 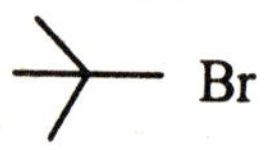 Br (d) 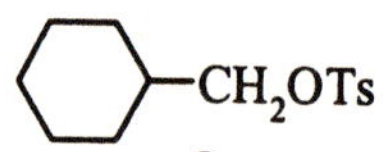 CH_2OTs

(e) 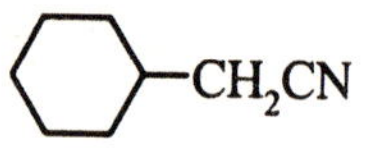 CH_2CN (f) 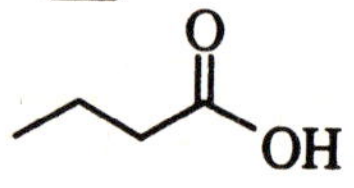 O, OH

(g) 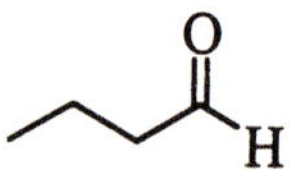 O, H (h) O

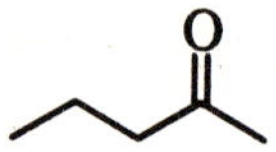

(i) No oxidation product (j) 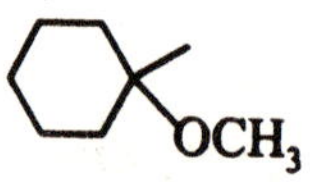 OCH_3

(k) 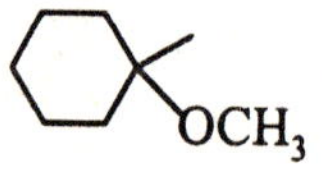 OCH_3

4. (a) Aqueous NaOH

(b) $B_2H_6/H_2O_2/NaOH$

(c) $Hg(OAc)_2/H_2O/NaBH_4$

(d) (i) NaH, (ii) K/heat, (iii) $NaNH_2$

5. Ethylene oxide suffers from a very high angle strain and therefore highly unstable while THF is devoid of such angle strain, it is stable.

6. 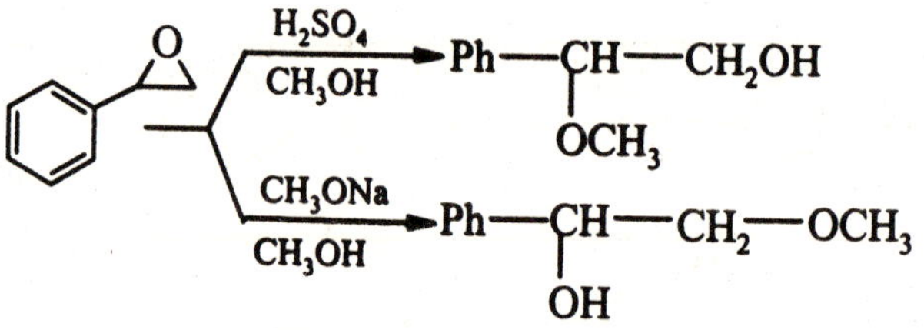 $\xrightarrow[CH_3OH]{H_2SO_4} Ph-\underset{OCH_3}{CH}-CH_2OH$

$\xrightarrow[CH_3OH]{CH_3ONa} Ph-\underset{OH}{CH}-CH_2-OCH_3$

7. Equilibrium lies to left since acid and base are stronger on right side.

8. Iodide ion being weaker nucleophile than hydroxy ion—(OH^-), substitutions as shown is not thermodynamically feasible.

9. (a) 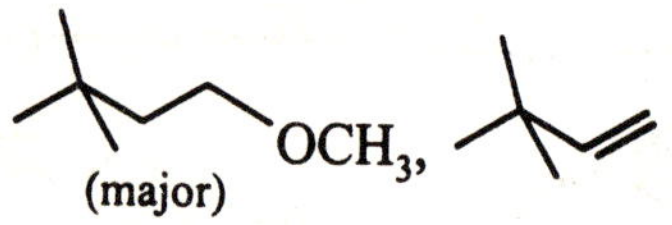 OCH_3, (major)

(b) 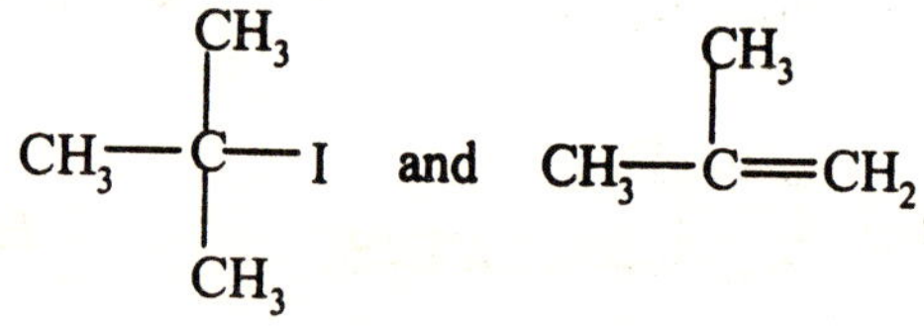 $CH_3-\underset{CH_3}{\overset{CH_3}{C}}-I$ and $CH_3-\overset{CH_3}{C}=CH_2$

(c) $CH_3CH(I)—CH_3$ (d)  SCH_3

(e) O^-Na^+ (f) O^-K^+

(g) O^-Na^+ (h) $=O$, CH_3

(i) $—I$ (j) OH

(k) $=O$ (l) O

(m) $H_2N—CH_2—C(CH_3)(OH)—CH_3$

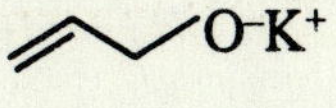

(n) $HOCH_2—C(CH)(OCH_3)—CH_3$

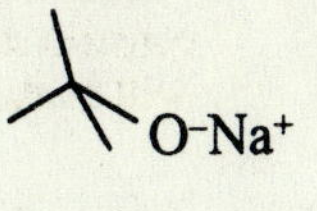

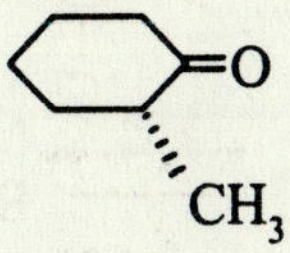

(o) 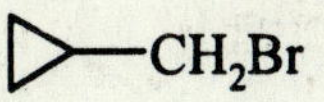 $—CH_2Br$ (p) 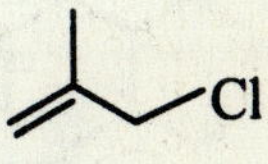 Cl

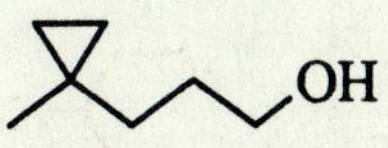

(q) 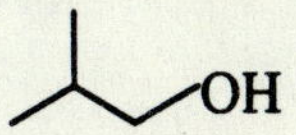 OH (r) 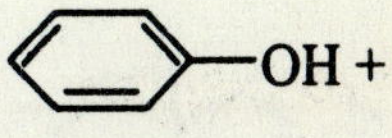 $—OH + CH_3I$

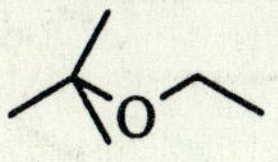

10. (a) PCC, (b) $SOBr_2$, (c) $CH_3CH_2CH_2ONa$

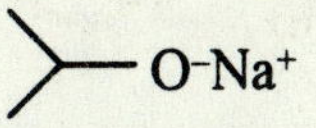

11. 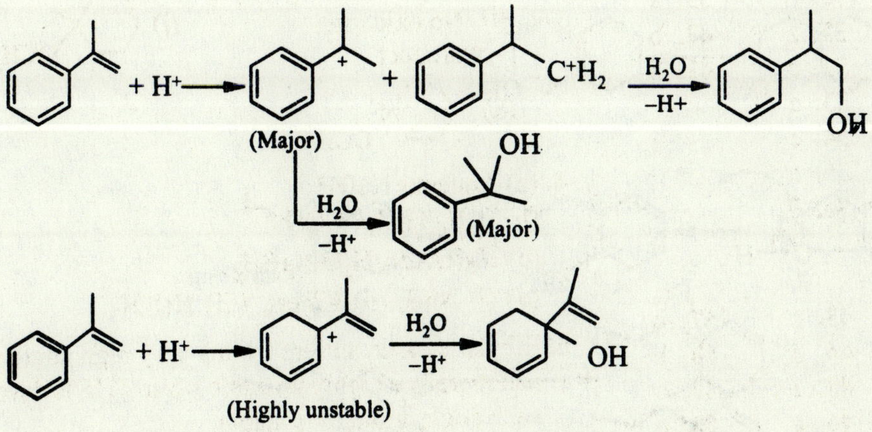 $+ H^+ \longrightarrow$ (Major) + C^+H_2 $\xrightarrow[-H+]{H_2O}$ OH

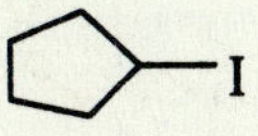

(Major) $\xrightarrow[-H^+]{H_2O}$ OH (Major)

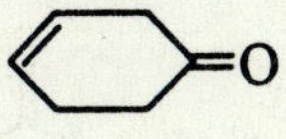

$+ H^+ \longrightarrow$ (Highly unstable) $\xrightarrow[-H^+]{H_2O}$ OH

12. 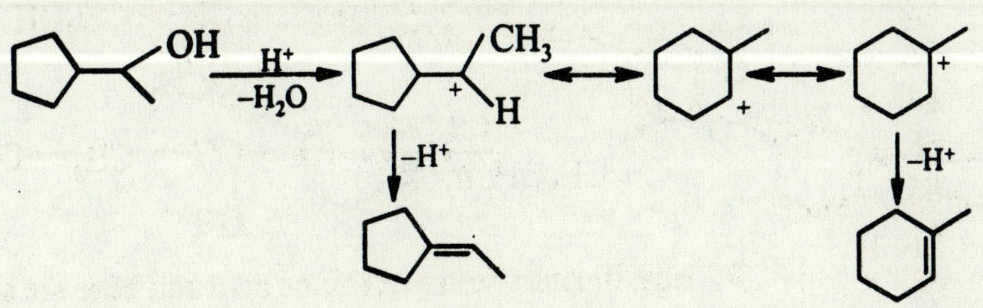 OH $\xrightarrow[-H_2O]{H^+}$ CH_3, H $\longleftrightarrow$ $\longleftrightarrow$

$-H^+$ $-H^+$

13. Second route, since it involve SN-2 reaction with a primary alkyl iodide.

14. (a) 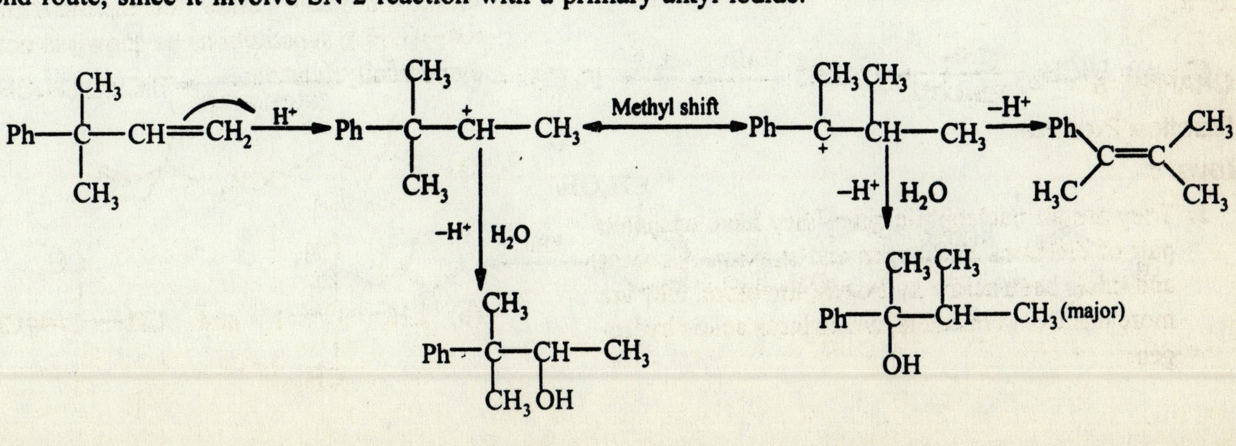 $Ph—C(CH_3)_2—CH=CH_2 \xrightarrow{H^+} Ph—C(CH_3)_2—\overset{+}{C}H—CH_3 \xleftrightarrow{\text{Methyl shift}} Ph—\overset{+}{C}(CH_3)—CH(CH_3)—CH_3 \xrightarrow{-H^+} Ph(H_3C)C=C(CH_3)_2$

$\downarrow -H^+$ H_2O: $Ph—C(CH_3)_2—CH(OH)—CH_3$

$\downarrow -H^+$ H_2O: $Ph—C(CH_3)(OH)—CH(CH_3)—CH_3$ (major)

(b) OH $\xrightarrow[-H_2O]{H^+}$ + Methyl shift –H+ –H+

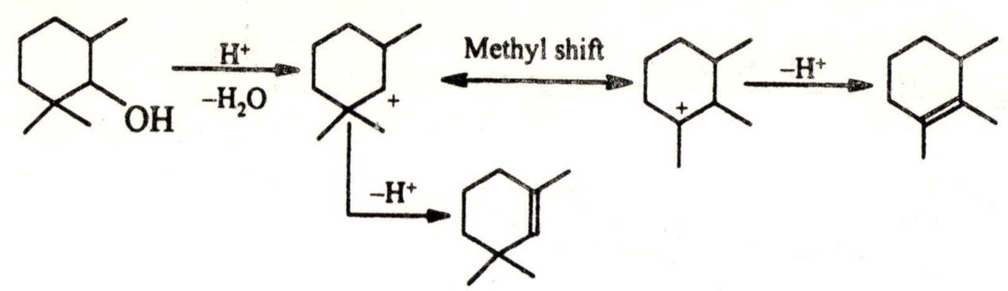

(c) I OH OH $\xrightarrow[-H_2O]{H^+}$ I + O H I O+ H $\xrightarrow{H^+}$ I O

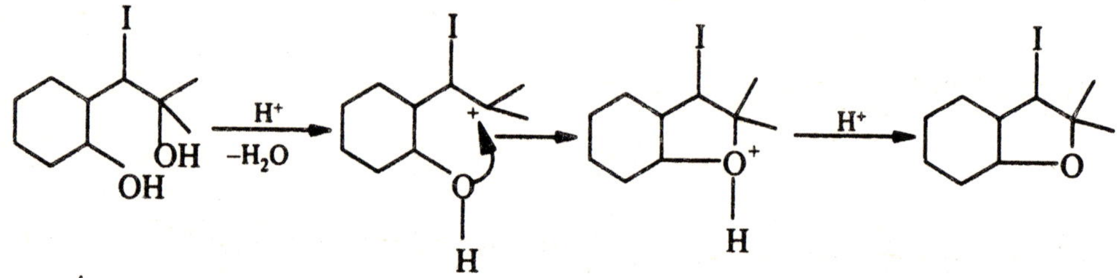

(d) OH OH → + OH → O+ H $\xrightarrow{-H^+}$ O

Methyl shift

+ OH $\xrightarrow{H^+}$ OH

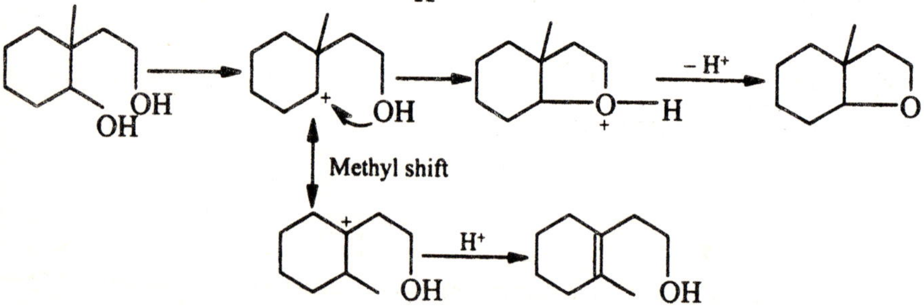

(e) O $\xrightarrow{H^+}$ O+ H $\xrightarrow{I^-}$ I OH $\xrightarrow[-H_2O]{H^+}$ + I Methyl shift

I I $\xleftarrow{I^-}$ + I

I $\xleftarrow{-H^+}$

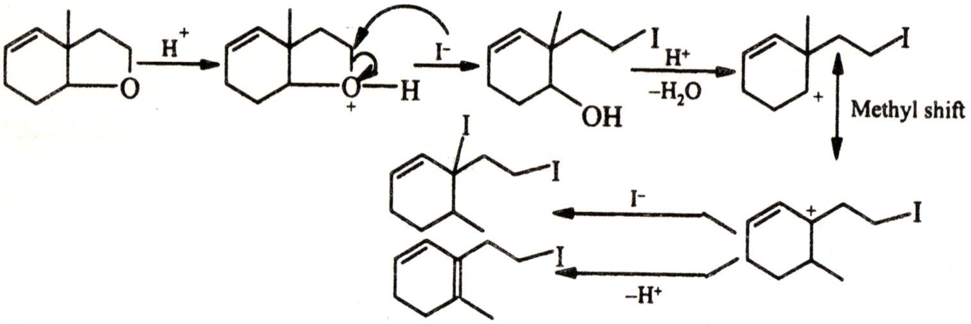

15. O OMe $\xrightarrow{H^+}$ O O+ H Me :O H → O O+ H + CH_3OH $\xrightarrow{-H^+}$ O O

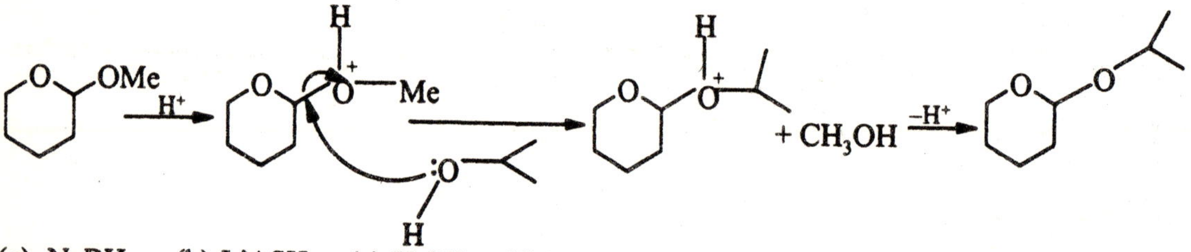

16. (a) $NaBH_4$, (b) $LiASH_4$, (c) H_2/Ni, (d) $NaBH_4/N_2H_4$ /NaOH/Heat, (f) N_2H_4/aq NaoH/Heat.

17. (a) $PhCH_3 \xrightarrow[(CH_3CO)_2O]{CrO_3}$ PhCHO $\xrightarrow{LiAlD_4}$ $\xrightarrow{H_2O}$ Ph CH(D)—OH $\xrightarrow{NaH}$ $\xrightarrow{CH_3CH_2CH_2Br}$ $PhCH(D)OCH_2CH_2CH_3$

(b) OH $\xrightarrow{RCO_3H}$ O OH $\xrightarrow[H_2O]{H^+}$ CH_2OH O $\xrightarrow{PBr_3}$ O Br

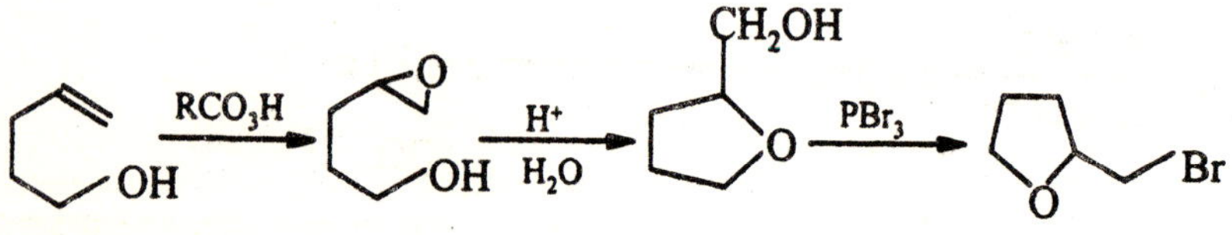

(c) $CH_2{=}CHI \xrightarrow[H_2O]{NaNH_2} HC{\equiv}CH \xrightarrow[Heat]{Na} \xrightarrow{CH_2CH_2CH_2Br} HC{\equiv}C{-}CH_2CH_2CH_3 \xrightarrow[H_2SO_4]{HgSO_4}$

(d) (pent-4-en-1-ol) OH $\xrightarrow{H^+}$ (carbocation, $\ddot{O}H$) $\xrightarrow{-H^+}$ (2-methyltetrahydrofuran, O) $\xrightarrow{HB}$ (Br, OH)

(e) (ethyloxirane, O) $\xrightarrow{CH_3CH_2CH_2MgBr} \xrightarrow{H_2O}$ (OH)

(f) (2-methylbut-1-ene) $\xrightarrow{HBr} \xrightarrow[C_2HSOH]{C_2HSONa}$ (2-methylbut-2-ene) $\xrightarrow[Zn\text{-}H_2O]{O_3} CH_3{-}\overset{O}{\overset{\|}{C}}{-}CH_3$ (I) $+ CH_3CHO$ (II)

I $\xrightarrow{HCN} CH_3{-}C(OH)(CN){-}CH_3 \xrightarrow{Na} CH_3{-}C(ONa)(CN){-}CH_3$ III

II $\xrightarrow{LiAlH_4} CH_3CH_2OH \xrightarrow{SOCl_2} CH_3CH_2Cl \xrightarrow{III} CH_3CH_2O{-}C(CH_3)_2{-}CN \xrightarrow{NaOH} CH_3CH_2O{-}C(CH_3)(COONa){-}CH_3$

$\xrightarrow{H_3O^+} CH_3CH_2O{-}C(CH_3)(COOH){-}CH_3$

18. (a) (isobutyl bromide) Br $\xrightarrow{CH_3ONa}$ (isobutyl methyl ether, O)

(b) $CH_3CH_2Br \xrightarrow[KCN]{aq} CH_3CH_2CN \xrightarrow[Excess\ H_3O^+]{I}$ (3-ethylpentan-3-ol, OH) $\xrightarrow[Heat]{H_2SO_4}$ (3-ethylpent-2-ene) $\xrightarrow[H_2O_2/NaOH]{B_2H_6}$ Product

$CH_3CH_2Br \xrightarrow{Mg/Et_2O} CH_3CH_2MgBr(I)$

(c) $CH_2{=}CH_2 \xrightarrow{Ag_2O}$ (oxirane, O) $\xrightarrow[H^+]{CH_3OH} CH_3OCH_2CH_2OH$

(d) (pentan-1-ol) OH $\xrightarrow[Heat]{H_2SO_4}$ (pent-1-ene) $\xrightarrow[H_2O]{Hg(OAC)_2} \xrightarrow{NaBH_4}$ (pentan-2-ol, OH) $\xrightarrow{PCC}$ (pentan-2-one, O)

(e) $PhCOOH \xrightarrow{SOCl_2} PhCOCl \xrightarrow[Quinoline]{Pd/BaSO_4} PhCHO \xrightarrow{I} \xrightarrow{H_2O} PhCH(OH)Ph$

$PhCOOH \xrightarrow{Ag_2O/Br_2} PhBr \xrightarrow[Et_2O]{MgBr} PhMgBr\ (I)$

19. (a) 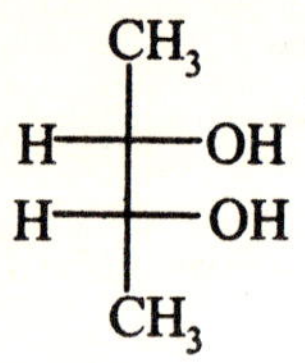 CH_3, H–C–OH, H–C–OH, CH_3 (b) OH, H, O (c) CH_2CHCH_3 (OH) and CH–CH_2CH_3 (OH)

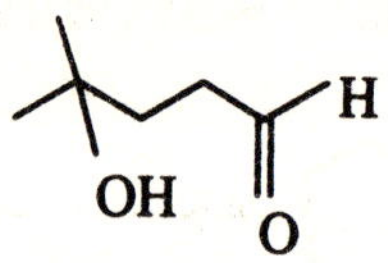

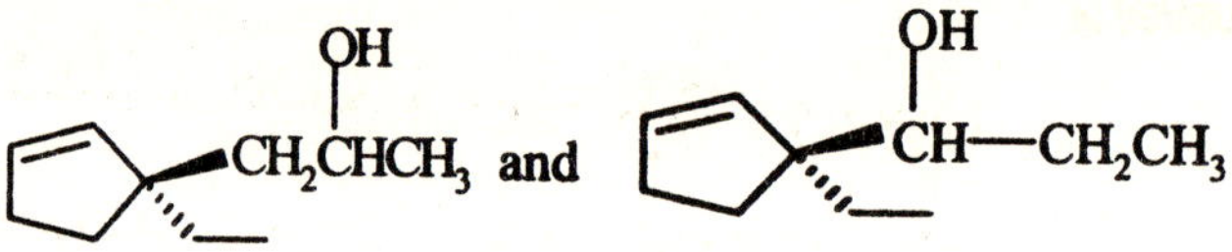

(d) NR (e) 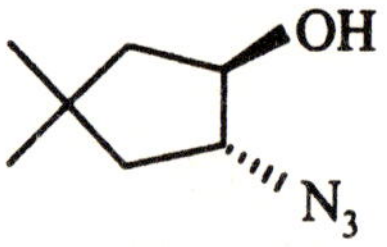 OH, N_3

20. 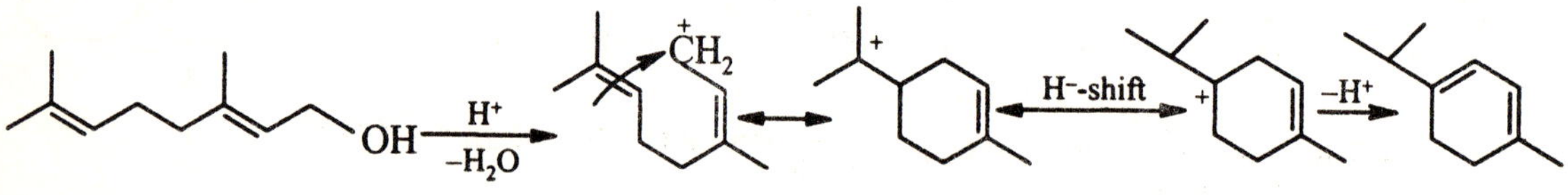 OH $\xrightarrow[-H_2O]{H^+}$ $\overset{+}{C}H_2$ ⟷ + $\xleftrightarrow{H^-\text{-shift}}$ + $\xrightarrow{-H^+}$

21. (a) 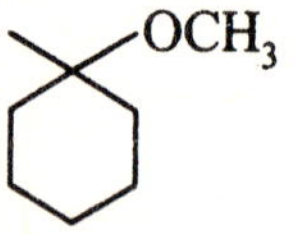 OCH_3 (b) 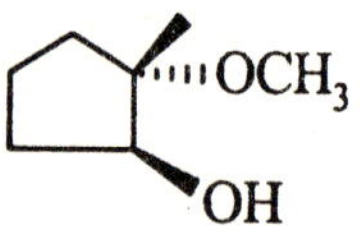 OCH_3, OH (c) 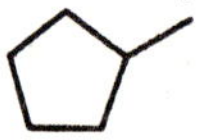 (d) 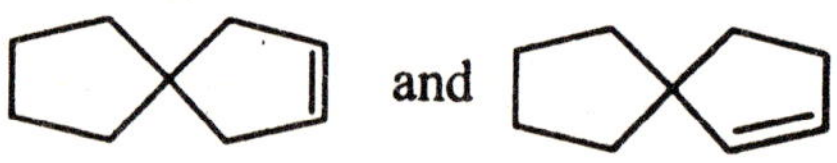 and

22. (a) $\xrightarrow{aqKCN}$ $\xrightarrow{H_3O^+}$ (b) H_3O^+ followed by H_2CrO_4

(c) 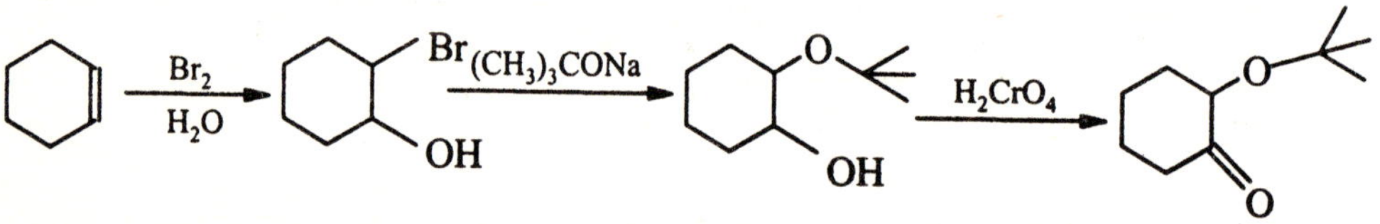 $\xrightarrow[H_2O]{Br_2}$ Br, OH $\xrightarrow{(CH_3)_3CONa}$ O, OH $\xrightarrow{H_2CrO_4}$ O, O

23. 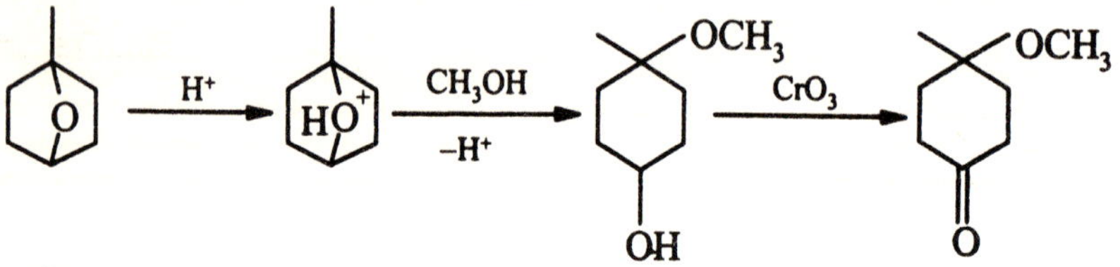 O $\xrightarrow{H^+}$ HO+ $\xrightarrow[-H^+]{CH_3OH}$ OCH_3, OH $\xrightarrow{CrO_3}$ OCH_3, O

24. (a) 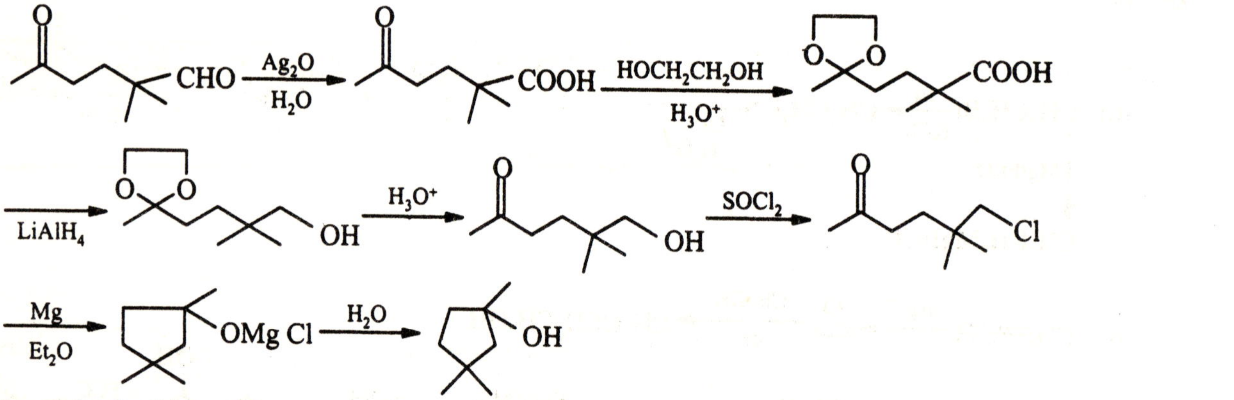 O, CHO $\xrightarrow[H_2O]{Ag_2O}$ O, COOH $\xrightarrow[H_3O^+]{HOCH_2CH_2OH}$ O, O, COOH

$\xrightarrow{LiAlH_4}$ O, O, OH $\xrightarrow{H_3O^+}$ O, OH $\xrightarrow{SOCl_2}$ O, Cl

$\xrightarrow[Et_2O]{Mg}$ OMg Cl $\xrightarrow{H_2O}$ OH

(Intramolecular Grignard reaction)

(b) 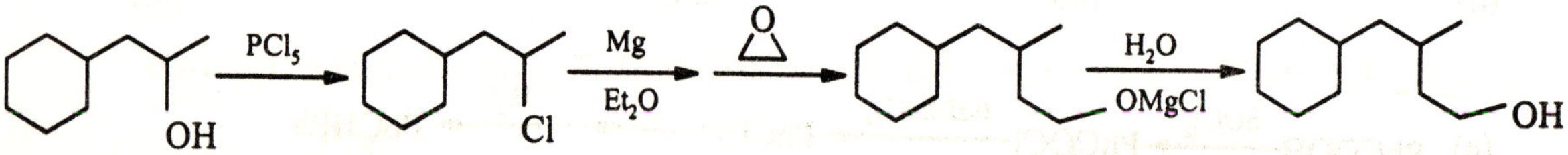 OH $\xrightarrow{PCl_5}$ Cl $\xrightarrow[Et_2O]{Mg}$ $\xrightarrow{\text{ethylene oxide}}$ OMgCl $\xrightarrow{H_2O}$ OH

25. (a) 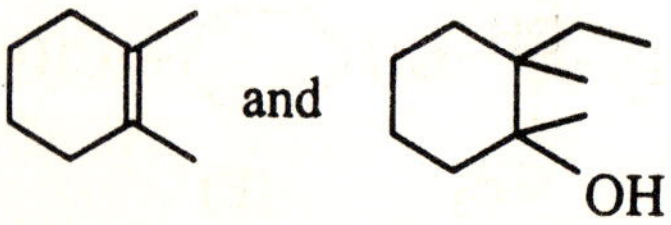 and OH (b) H_2SO_4 (c) (i) CH_3MgBr (ii) CH_3Br

Level 2

1. (a) 2-(hydroxymethyl)cyclopentanone $\xrightarrow[\text{NaOH/Heat}]{N_2H_4}$ cyclopentyl-CH_2OH $\xrightarrow[\text{Heat}]{H_2SO_4}$ cyclohexene $\xrightarrow[Zn - H_2O]{O_3}$ $\xrightarrow{LiAlH_4}$ 1, 6 - Hexandiol

(b) spiro compound $\xrightarrow[\text{Heat}]{Br_2}$ Br-spiro compound $\xrightarrow{\text{aq.NaOH}}$ OH-spiro compound $\xrightarrow{H_2SO_4}$ carbocation (+)

(other isomer will also be formed)

$\updownarrow$

(X) cyclohexanone + (Y) cyclopentanone $\xleftarrow[Zn - H_2O]{O_3}$ cyclohexylidenecyclopentane $\xleftarrow{-H^+}$ rearranged carbocation (+)

Y $\xrightarrow{LiAlH_4}$ cyclopentyl–OH $\xrightarrow[\text{Heat}]{C.H_2SO_4}$ cyclopentene

X $\xrightarrow{NaOH}$ $\xrightarrow[\text{Heat}]{H_2CO}$ 3-methylenecyclohexanone $\xrightarrow[Zn\text{-}H_2O]{O_3}$ cyclohexane-1,3-dione

2. $CH_3C{\equiv}CH < CH_3COOCH_3 < CH_3COCH_2CH_3 < CH_3CH_2OH < CF_3CH_2OH < CF_3COOH$

3. bromocyclopentane $\xrightarrow[Et_2O]{Mg}$ $\xrightarrow{\text{butanone}}$ $\xrightarrow{H_3O^+}$ cyclopentyl–C(CH_3)(C_2H_5)–OH

4. (a) PhMgBr + ethyleneoxide $\xrightarrow{H_3O^+}$ (b) $(CH_3)_2CHCH_2MgBr$ + ethyleneoxide $\xrightarrow{H_3O^+}$ 4-methylpentan-1-ol (OH)

(c) 2-methylcyclohexyl-MgBr + ethyleneoxide $\xrightarrow{H_3O^+}$ 2-(2-methylcyclohexyl)ethanol (OH)

5. (a) (A) 2-methylcyclohexyl-MgBr (B) cyclohexyl–CH(OH)–phenyl (b) $Ph-\underset{OH}{C}(CH_2CH_2CH_2CH_3)_2$ (c) 3-(hydroxymethyl)cyclohexanol (OH, OH)

(d) $B_2H_6/H_2O_2/NaOH$ (e) Aqueous NaOH (f) $RCH_2CH_2CH_2OMgX$, $RCH_2CH_2CH_2OH$

6. (A) cyclohexenyl–CH_2Br (B) cyclohexenyl–CH_2MgBr (C) cyclohexenyl–CH_2D $\longrightarrow$ (D) cyclohexenyl–$CH_2-C(CH_3)_2-OH$ (E) cyclohexenyl–$CH{=}C(CH_3)_2$

7. At higher temperature, intramoleular dehydration is entropy favoured.

8. 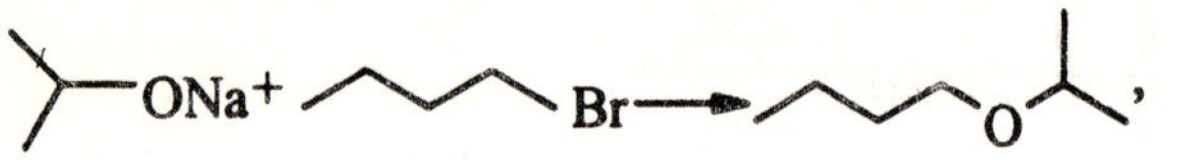 ONa + Br ⟶ O , PhONa + CH_3Br ⟶ $PhOCH_3$

9. 2-butanol (A), 2-butanone (B), 2-bromobutane (C) 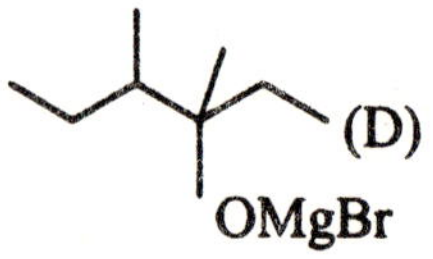 OMgBr (D)

10. 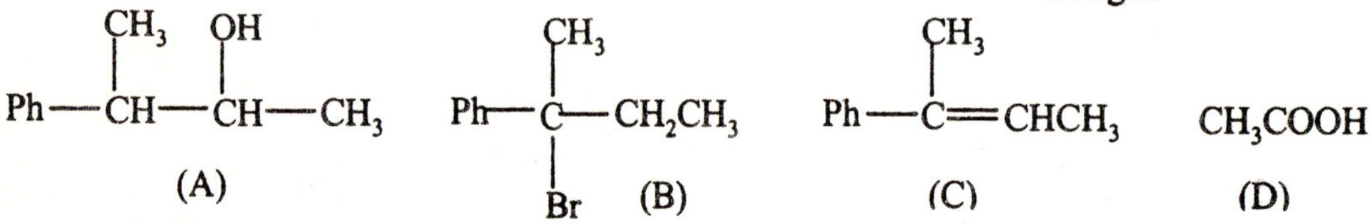 $Ph-CH(CH_3)-CH(OH)-CH_3$ (A) $Ph-C(CH_3)(Br)-CH_2CH_3$ (B) $Ph-C(CH_3)=CHCH_3$ (C) CH_3COOH (D)

11. (a) 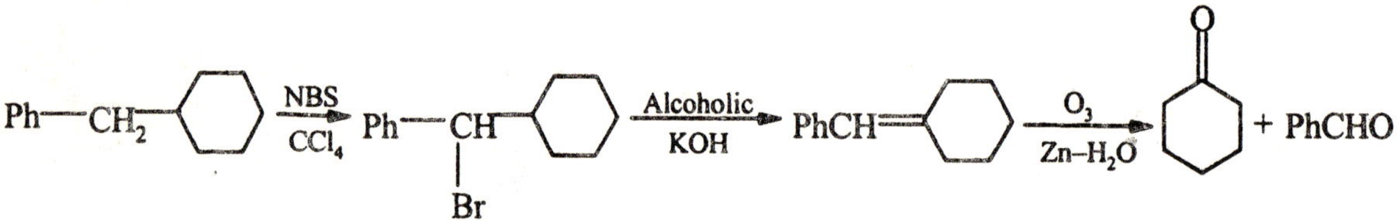 Ph—CH$_2$— ⟶ (NBS / CCl_4) Ph—CH(Br)— ⟶ (Alcoholic KOH) PhCH= ⟶ (O_3 / Zn–H_2O) O + PhCHO

(b) 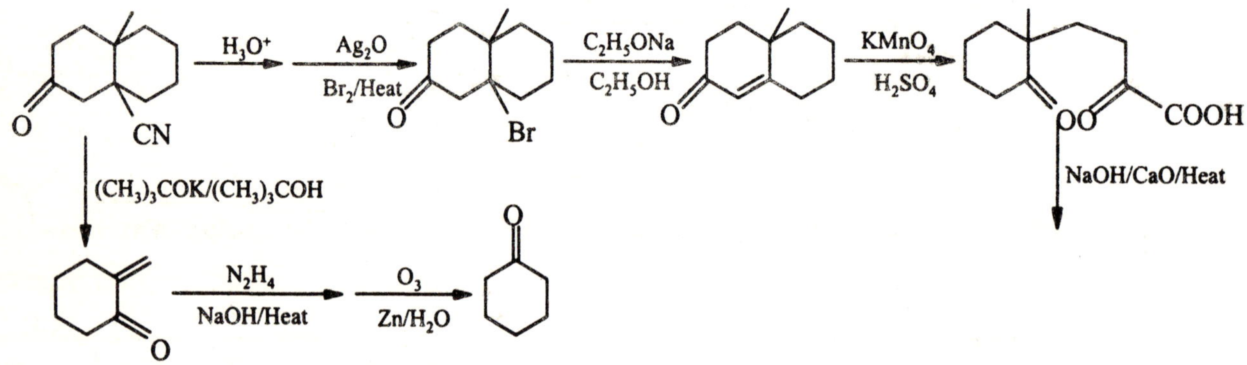 O, CN ⟶ (H_3O^+) ⟶ (Ag_2O / Br_2/Heat) O, Br ⟶ (C_2H_5ONa / C_2H_5OH) O ⟶ ($KMnO_4$ / H_2SO_4) O, O, COOH ⟶ (NaOH/CaO/Heat)

⟶ ($(CH_3)_3COK/(CH_3)_3COH$) O ⟶ (N_2H_4 / NaOH/Heat) ⟶ (O_3 / Zn/H_2O) O

(c) O, O ⟶ (H_3O^+) O, OH, OH ⟶ (CH_3COCl) O, OCOMe, OCOMe ⟶ ($NaBH_4$) OH, OCOMe, OCOMe ⟶ (H_2SO_4 / Heat) OCOMe, OCOMe ⟶ (O_3 / Zn-H_2O)

CH_3CH_2CHO + H, O, OCOMe, OCOMe ⟶ (H_3O^+) H, O, OH, OH ⟶ ($SOCl_2$) ⟶ (Zn) H, O ⟶ ($NaBH_4$) HO ⟶ ($SOBr_2$) Br

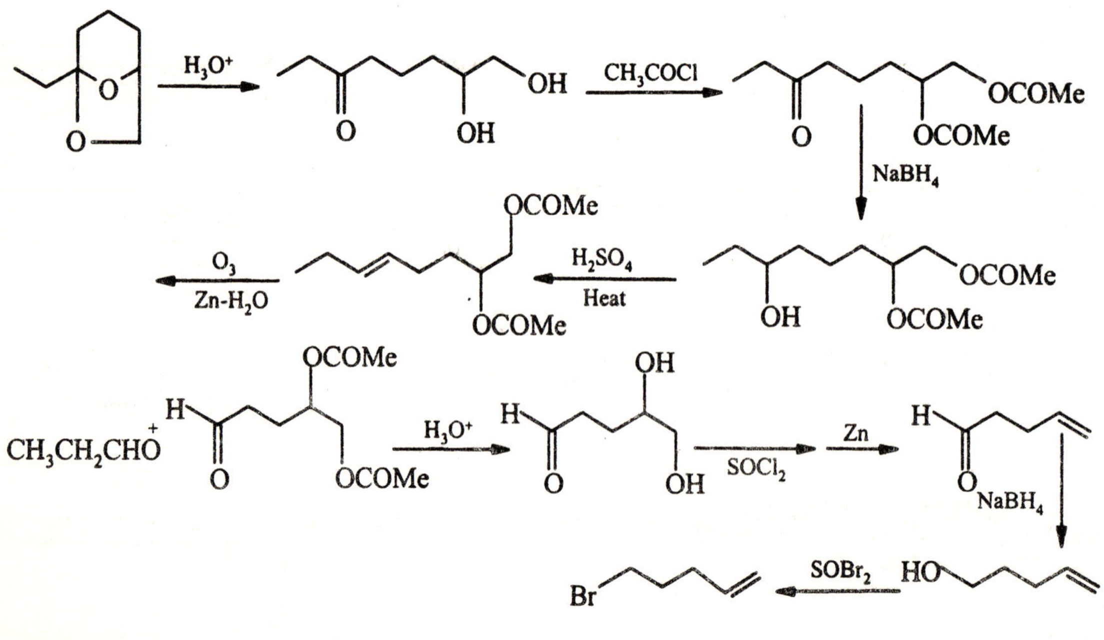

12. (a) $CH_3-\overset{O}{\overset{\|}{C}}-Cl \xrightarrow[\text{Excess}]{CH_3CH_2MgCl} \xrightarrow{H_3O^+} CH-C(OH)(CH_2CH_3)-CH_2CH_3 \xrightarrow[\text{Heat}]{C.H_2SO_4}$

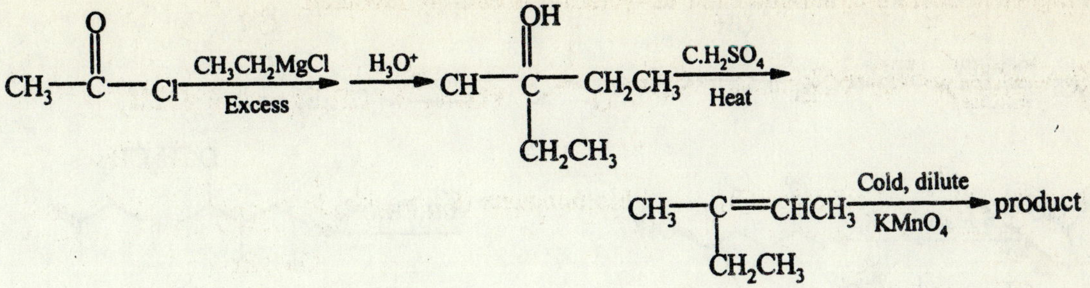

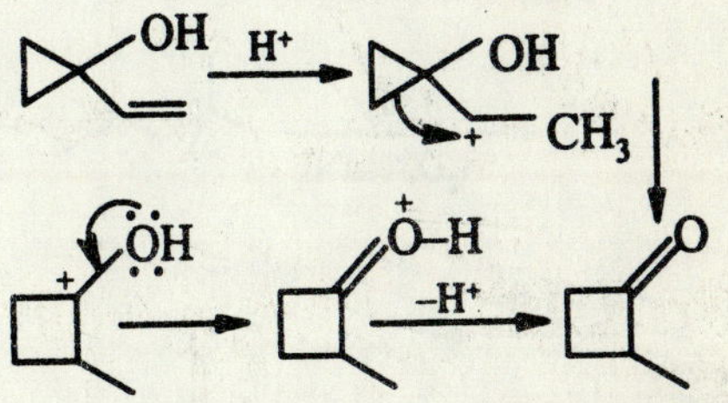

$CH_3-C(CH_2CH_3)=CHCH_3 \xrightarrow[KMnO_4]{\text{Cold, dilute}}$ product

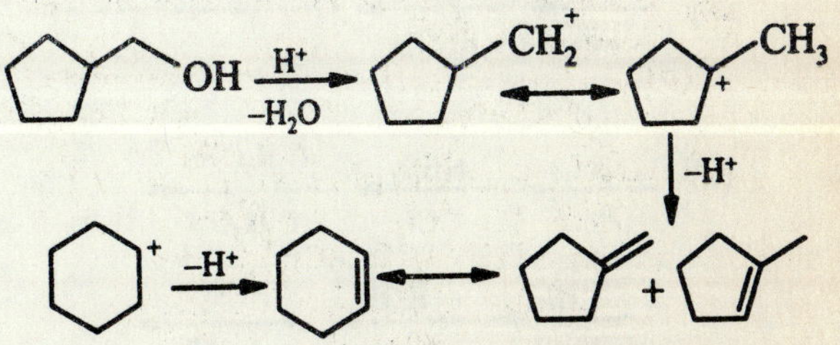

(b) OH, H_2CrO_4, O, (i) CH_3MgBr, (ii) H_2O

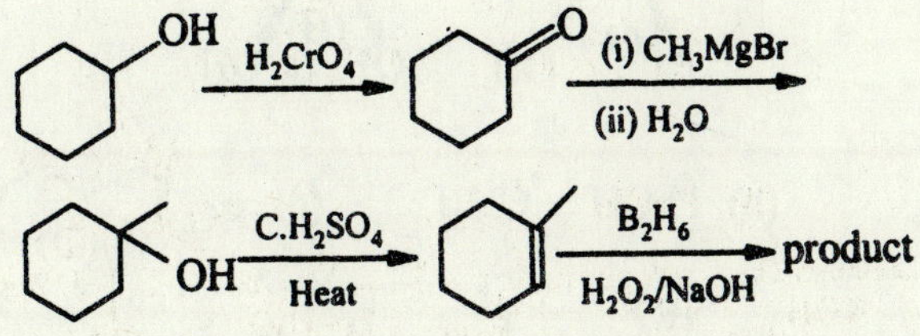

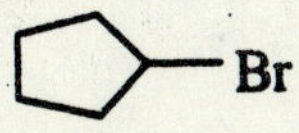

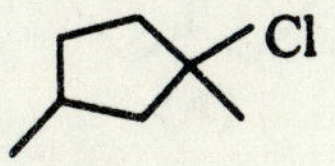

OH, $C.H_2SO_4$, Heat, B_2H_6, $H_2O_2/NaOH$, product

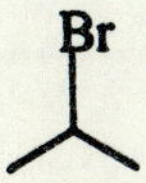

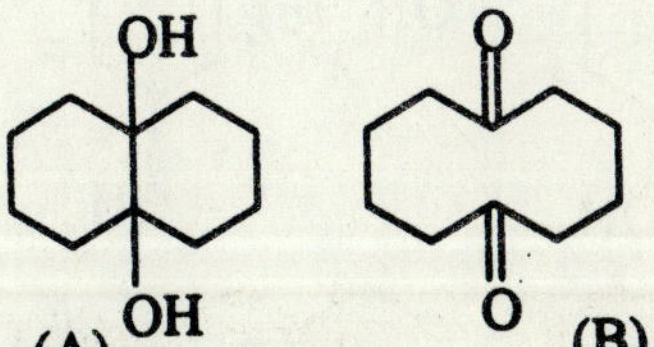

(c) OH H_2CrO_4, PhMgBr, H_2O

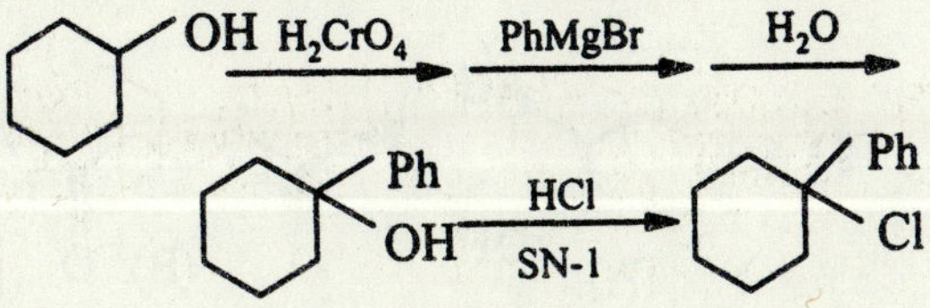

Ph, OH, HCl, SN-1, Ph, Cl

13. (a) Br (b) Cl (c) Br

14. OH, OH, (A); O, O, (B)

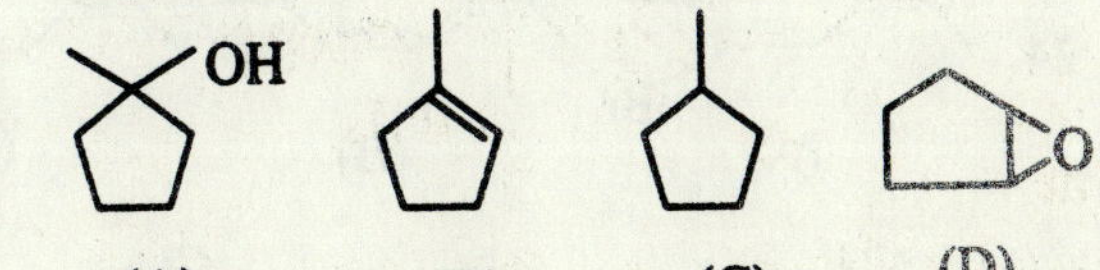

15. OH (A) (B) (C) O (D)

OH (E) Br H (F) OH (G) (H)

16. (a) OH $\xrightarrow[-H_2O]{H^+}$ + $\xrightarrow{H_2O}$ + H_3O^+

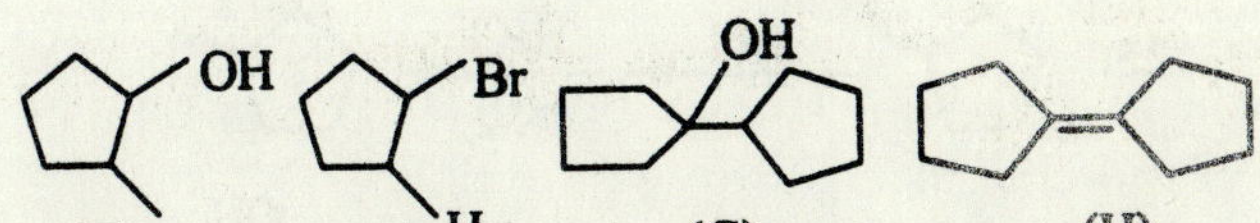

(b) OH $\xrightarrow{H^+}$ OH, +, CH_3

$\ddot{O}H$, +, $\overset{+}{O}-H$, $-H^+$, O

(c) OH $\xrightarrow[-H_2O]{H^+}$ $\overset{+}{CH_2}$ ⟷ CH_3, +, $-H^+$

+, $-H^+$, +

(d) OCH_3 $\xrightarrow{H_2SO_4}$ OH + CH_3OH

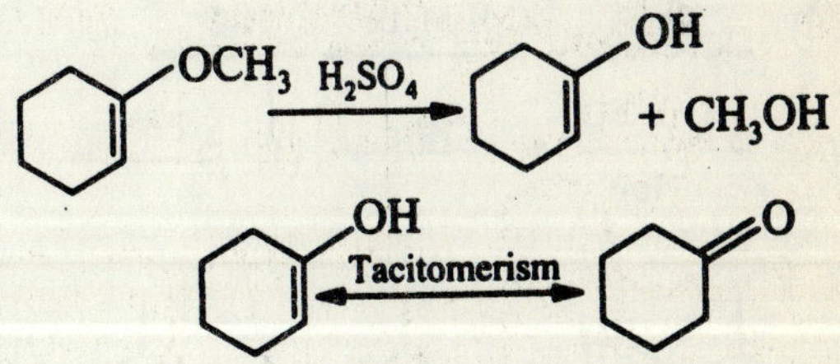

OH, Tacitomerism, O

17. (a) $\xrightarrow[H_2O]{(CH_3COO)_2Hg}$ $\xrightarrow{\text{Na } BD_4}$ OH D

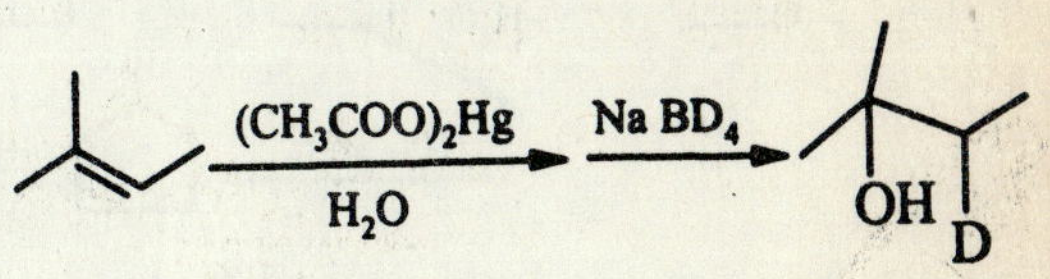

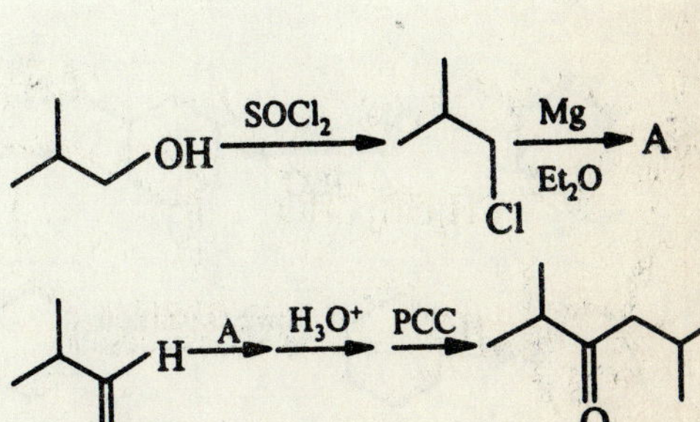

(b) OH $\xrightarrow{SOCl_2}$ Cl $\xrightarrow[Et_2O]{Mg}$ A

H, O, $\xrightarrow{A}$ $\xrightarrow{H_3O^+}$ $\xrightarrow{PCC}$ O

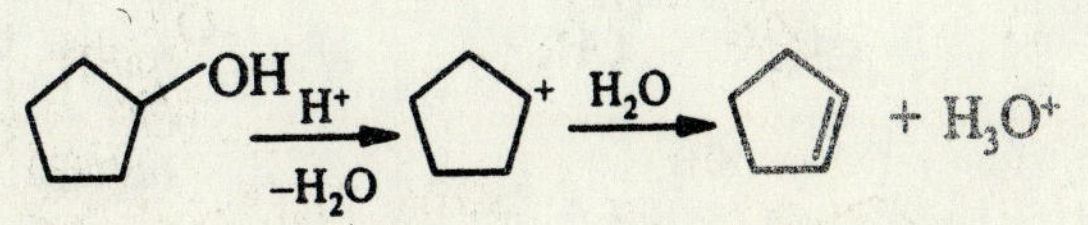

(c) 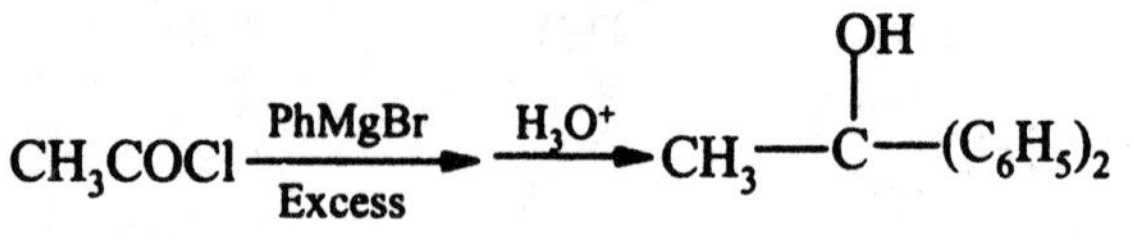 $CH_3COCl \xrightarrow[\text{Excess}]{PhMgBr} \xrightarrow{H_3O^+} CH_3-C(OH)-(C_6H_5)_2$

(d) 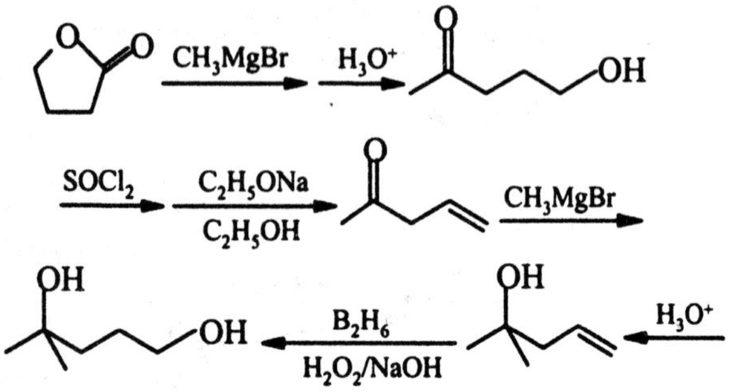 $\xrightarrow{CH_3MgBr} \xrightarrow{H_3O^+}$ OH

$\xrightarrow{SOCl_2} \xrightarrow[C_2H_5OH]{C_2H_5ONa} \xrightarrow{CH_3MgBr}$

OH OH OH $\xleftarrow[H_2O_2/NaOH]{B_2H_6}$ $\xleftarrow{H_3O^+}$

19. (a) $\xrightarrow{Na} \xrightarrow{C_2H_5Br}$

(b) 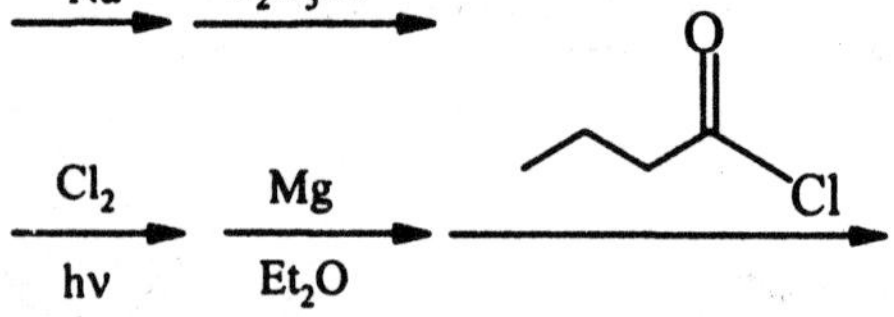 $\xrightarrow[h\nu]{Cl_2} \xrightarrow[Et_2O]{Mg} \longrightarrow$ Cl

(c) $\xrightarrow[Heat]{Na} \xrightarrow[CCl_4]{NBS} \xrightarrow[C_2H_5OH]{C_2H_5ONa}$

$\xrightarrow{Br_2} \xrightarrow[fusion]{KOH} \xrightarrow[liq\ NH_3]{NaNH_2}$

(d) $\xrightarrow{PCC} \xrightarrow{C_2H_5MgBr} \xrightarrow{H_3O^+}$

(e) $\xrightarrow[Heat]{C.\ H_2SO_4} \xrightarrow{HBr}$ (f) $\xrightarrow{PBr_3}$

20. (a) 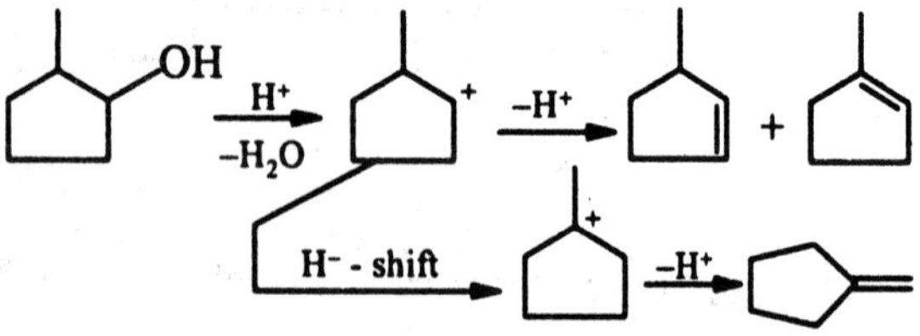 OH $\xrightarrow[-H_2O]{H^+}$ + $\xrightarrow{-H^+}$ +

H^- - shift $\longrightarrow$ + $\xrightarrow{-H^+}$

(b) 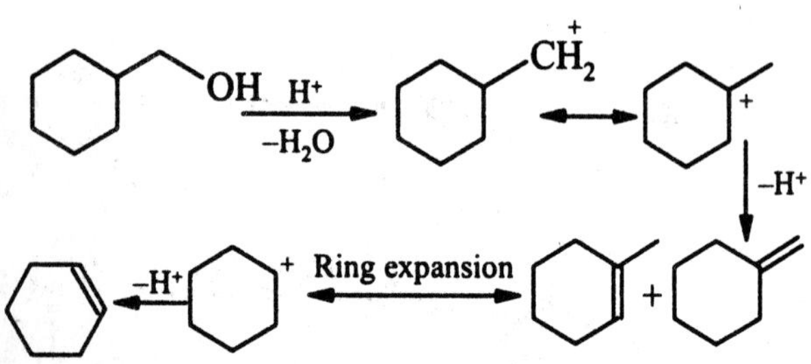 OH $\xrightarrow[-H_2O]{H^+}$ $\overset{+}{C}H_2$ $\longleftrightarrow$ + $\downarrow -H^+$

$\xleftarrow{-H^+}$ + Ring expansion $\longleftrightarrow$ +

21. 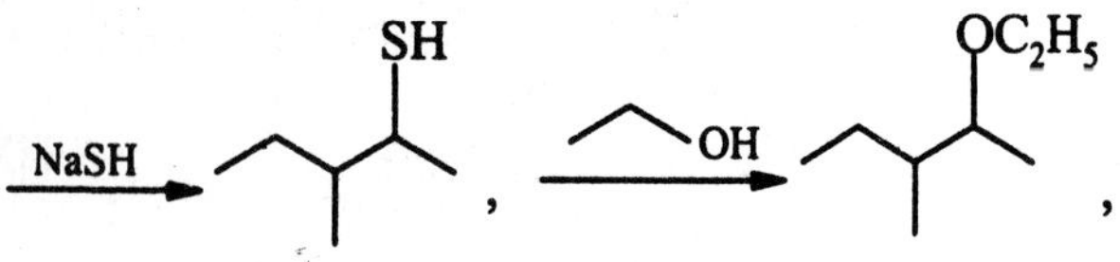 $\xrightarrow{NaSH}$ SH , $\xrightarrow{\text{OH}}$ OC_2H_5 ,

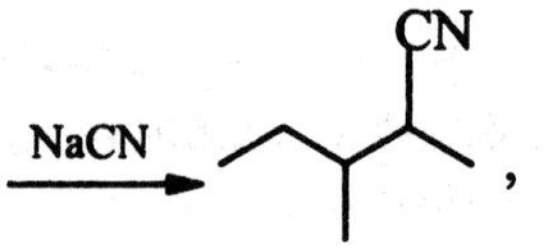

$\xrightarrow{NaCN}$ CN ,

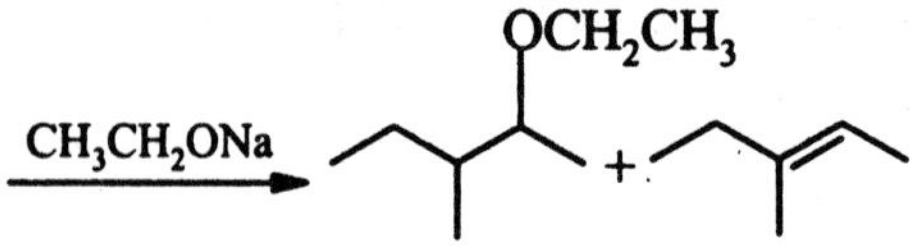

$\xrightarrow{CH_3CH_2ONa}$ OCH_2CH_3 +

22. (a) 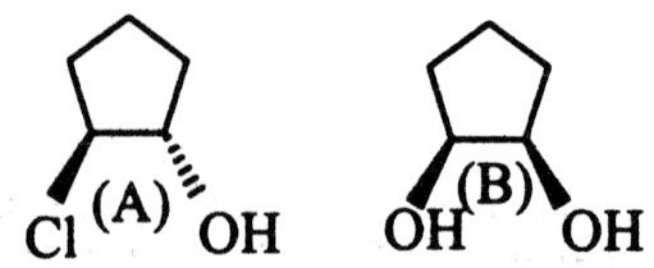 Cl (A) OH OH (B) OH

(b) PhOH + CH_3I (c) 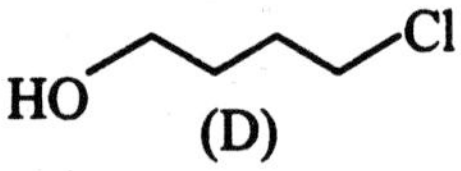 HO Cl (D)

(d) 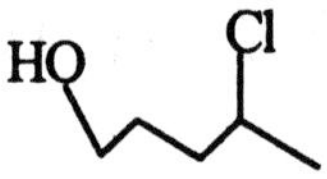 HO Cl

23. (a) 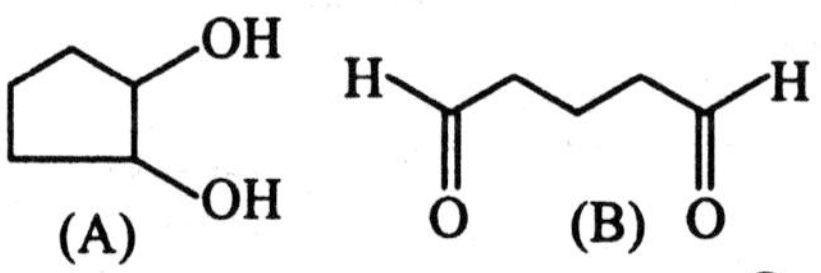 OH OH (A) H H O (B) O

(b) 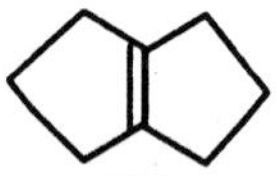 (A) (c) 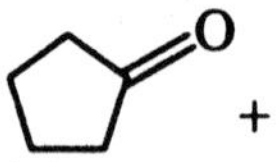 O + CH_2O

(d) (i) 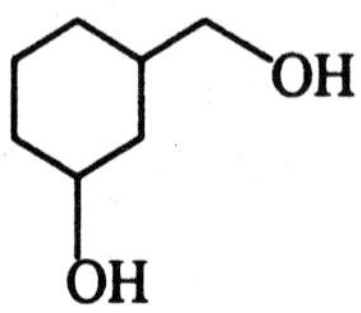 OH OH (ii) 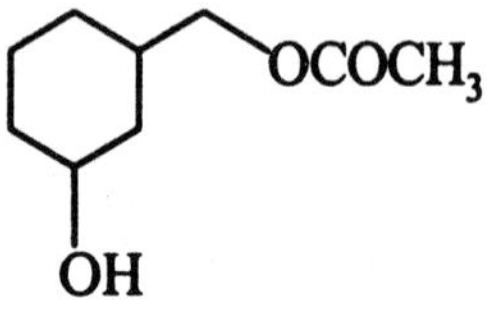 $OCOCH_3$ OH

24. 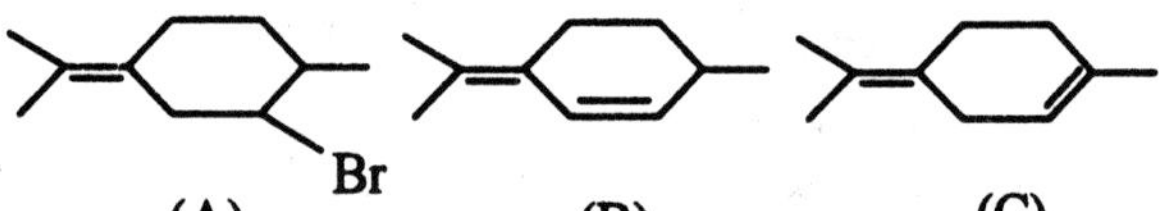 Br (A) (B) (C)

25. 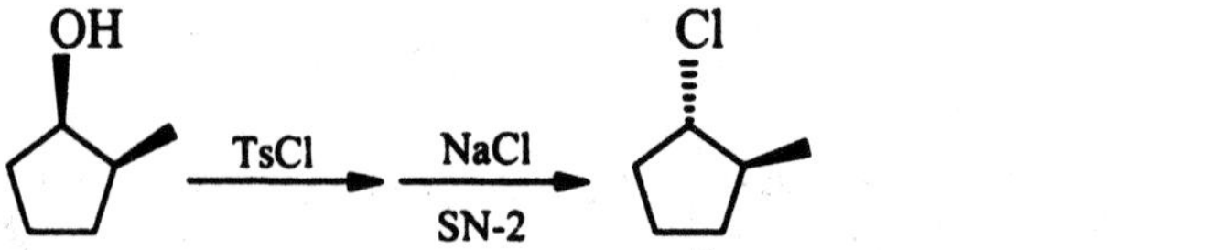 OH $\xrightarrow{TsCl} \xrightarrow[SN\text{-}2]{NaCl}$ Cl

26. 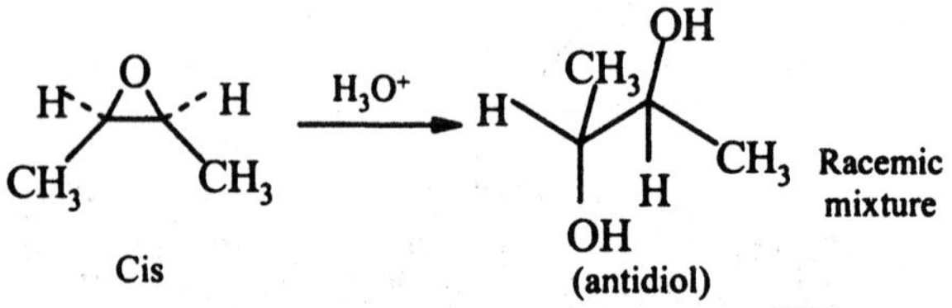 H O H CH_3 CH_3 Cis $\xrightarrow{H_3O^+}$ H CH_3 OH H CH_3 OH (antidiol) Racemic mixture

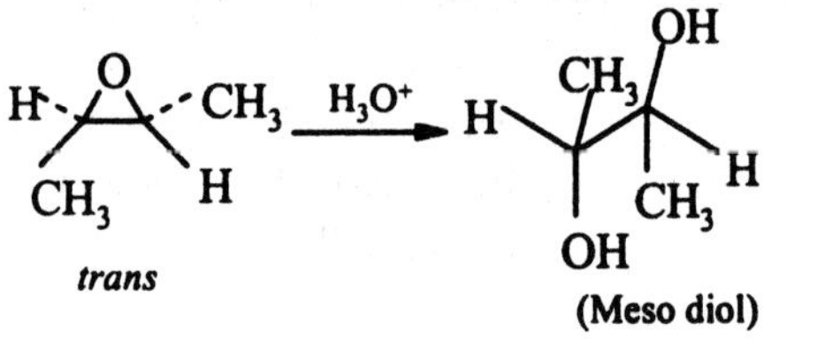

H O CH_3 CH_3 H *trans* $\xrightarrow{H_3O^+}$ H CH_3 OH H CH_3 OH (Meso diol)

27. Chlorination of alcohols with HCl proceed by SN-1 mechanism, rearranged products are also obtained. Chlorination of alcohols with $SOCl_2$ proceed by SN-2 mechanism, no rearranged product is obtained.

28. (a) 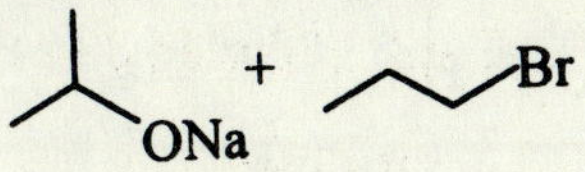 ONa + Br

(b) 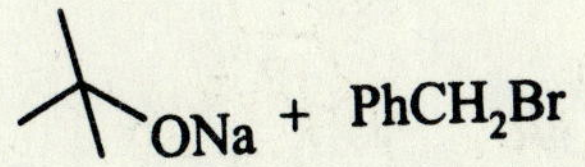 ONa + $PhCH_2Br$

(c) 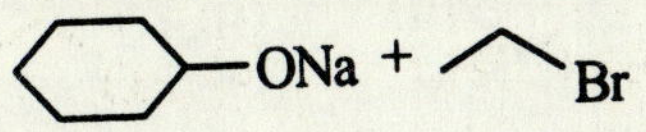 ONa + Br

(d) CH_3CH_2ONa + Br

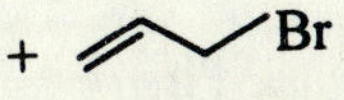

29. (a) 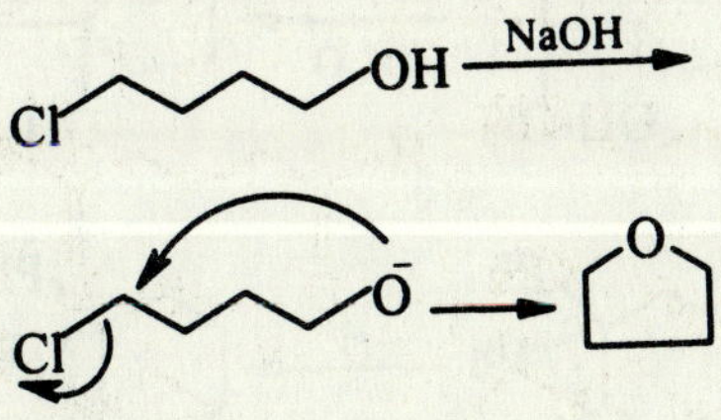 Cl OH $\xrightarrow{NaOH}$

Cl O⁻ → O + Cl⁻

(b) 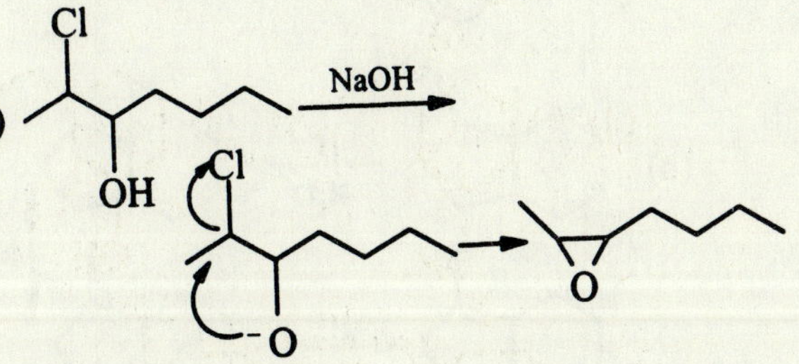 Cl OH $\xrightarrow{NaOH}$

Cl ⁻O → O

30. (a) 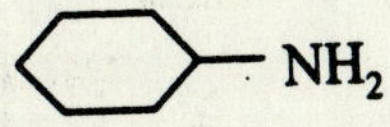 NH_2

(b) HN_3 (c) $NaO^{18}H/H_2O$

(d) $CH_3CH_2CH_2SNa/H_2O$

(e) CH_3MgBr, H_2O

31. (a) PCC

(b) Acetone + $CH_2(COOH)_2$

32. Rearrangement of carbocation into more stable carbocation results in formation of 2-methyl-2-butanol.

33. (a) More stable carbocation is formed which is responsible for this observation.

$(CH_3)_2C{=}CH\overset{+}{C}H_2 \leftrightarrow (CH_3)_2\overset{+}{C}{-}CH{=}CH_2$.

(b) Interconversion occurs through carbocation intermediate.

34. 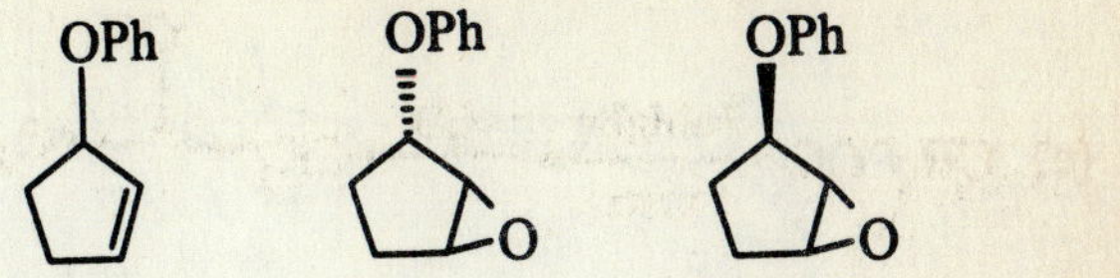 OPh OPh OPh

O O

(A) (B) (C)

OPh OPh

OH OH

$NHCH(CH_3)_2$ $NHCH(CH_3)_2$

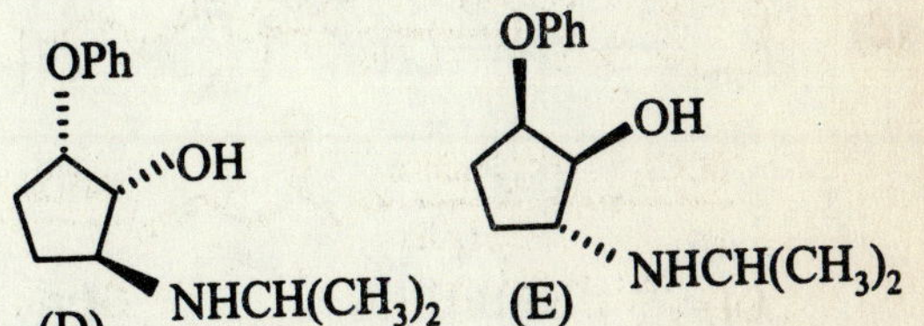

(D) (E)

D and E are enantiomers. D will have eight stereoisomers.

35. (a) 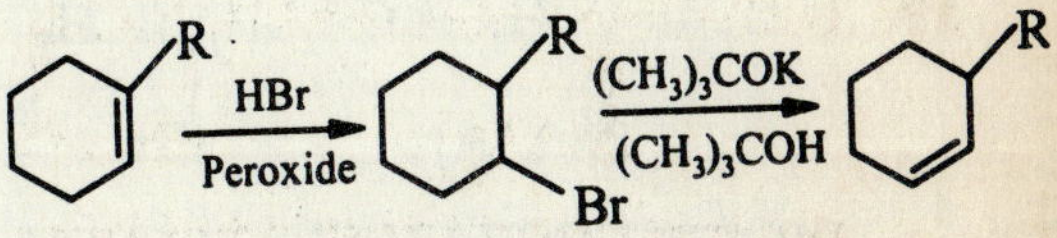 R $\xrightarrow[\text{Peroxide}]{HBr}$ R Br $\xrightarrow[(CH_3)_3COH]{(CH_3)_3COK}$ R

(b) 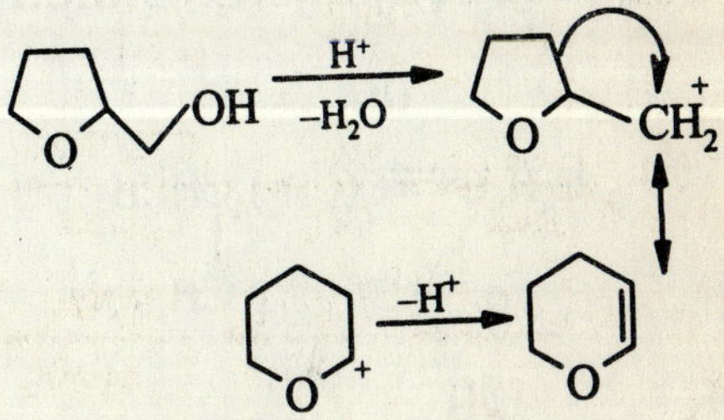 O OH $\xrightarrow[-H_2O]{H^+}$ O $\overset{+}{C}H_2$

O⁺ $\xrightarrow{-H^+}$ O

(c) (i) 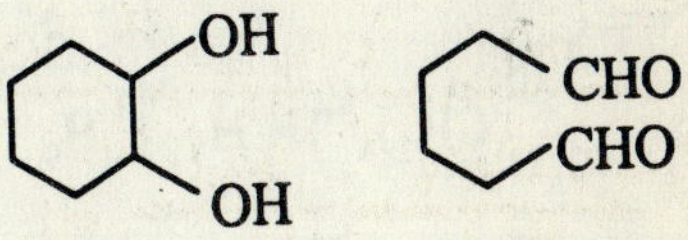 OH OH CHO CHO

(A) (B)

(ii) CH_3CHCH_2SPh (with OH on C-2) (iii) 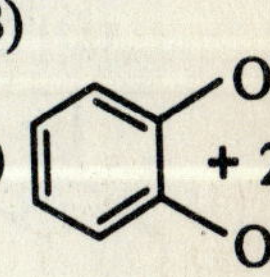 OH OH + $2CH_3I$

(C) (D)

(iv) 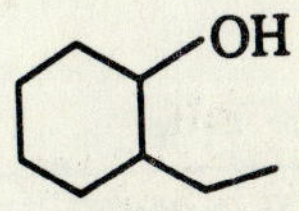 OH

(E)

36. (a) 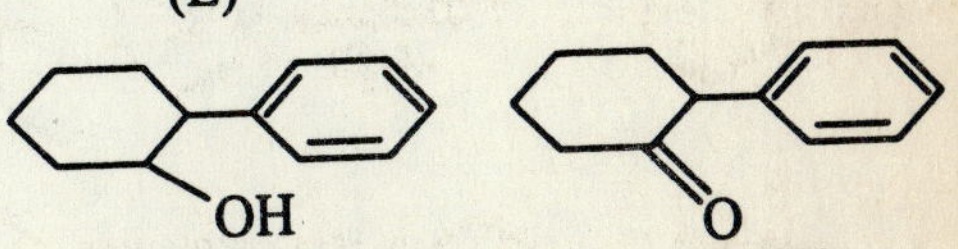 OH O

(A) (B)

Br OH

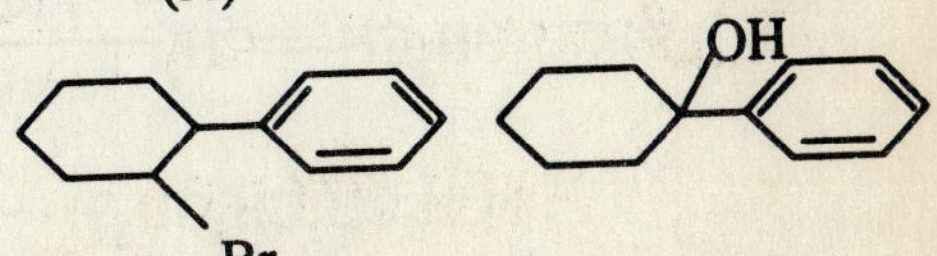

(C) (D)

(b) (E) (F)

$CH{=}CH_2$ (G)

$CH(Br){-}CH_2Br$ (H)

$CHBrCH_3$ (I)

$CH(CN)CH_3$ (J)

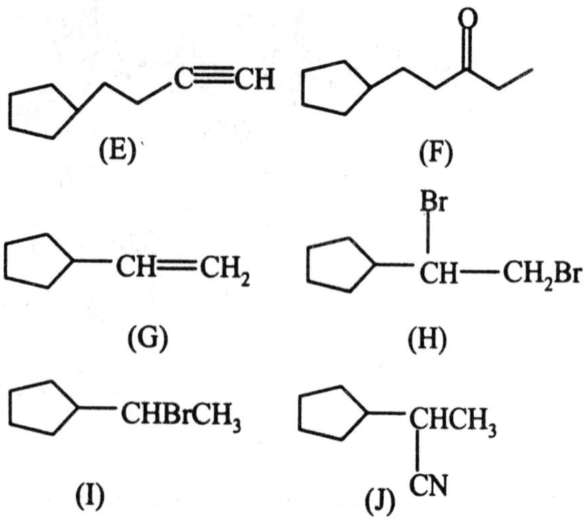

37. $HC{\equiv}CCH_2CH_2CH_2CH_2CH(CH_3)CH_3$ (A)

$H_2C{=}CHCH_2CH_2CH_2CH_2CH(CH_3)CH_3$ (B)

$C_4H_9C{\equiv}C(CH_2)_4CH(CH_3)CH_3$ (C)

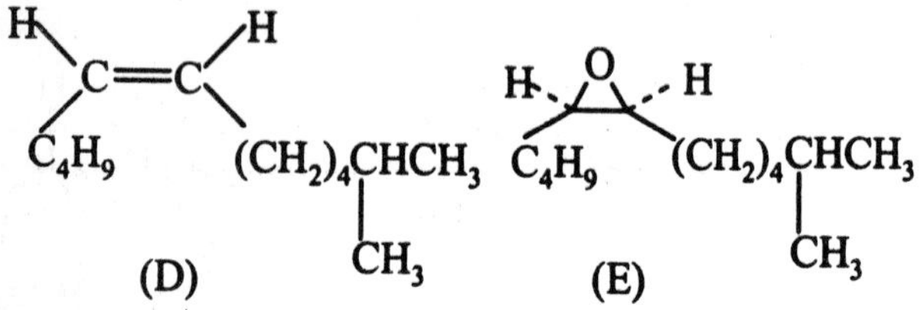

(D) (E)

38. (a) $\xrightarrow[H_2O]{Hg(OAc)_2}$ $\xrightarrow{NaBH_4}$ $\xrightarrow{H_2CrO_4}$

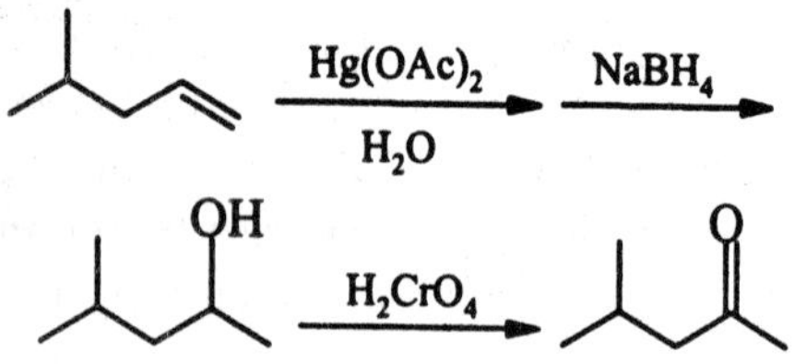

(b) $\xrightarrow[H_2O_2/NaOH]{B_2H_6}$ $\xrightarrow{PCC}$

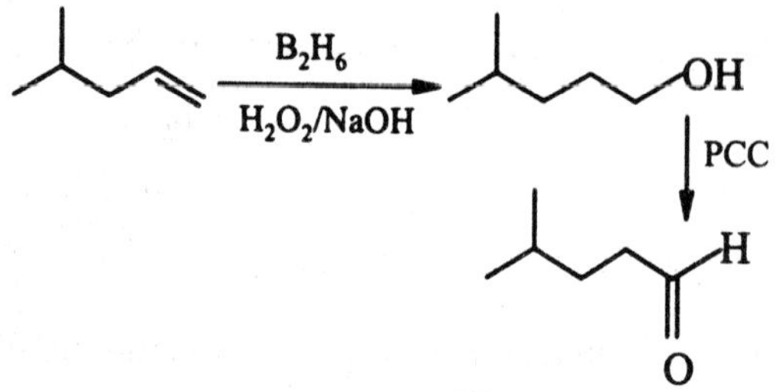

(c) $PhCH_2CH_2CH{=}CH_2 \xrightarrow[Peroxide]{HBr}$

$PhCH_2CH_2CH_2CH_2Br \xrightarrow[SN\text{-}2]{PhONa}$

$PhCH_2CH_2CH_2CH_2OPh$

(d) $\xrightarrow[(ii)\ RCOOH]{(i)\ B_2H_6}$

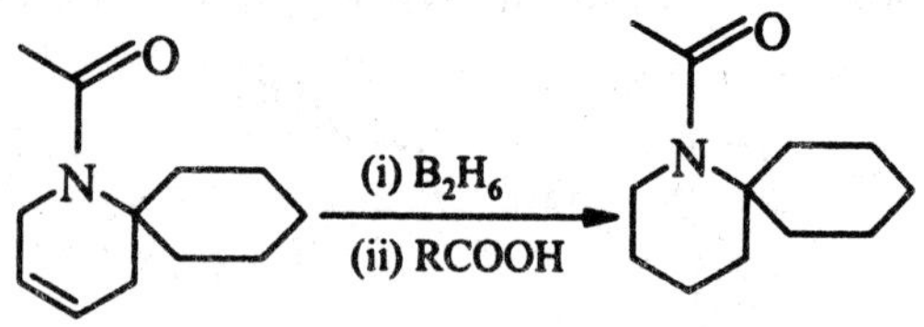

(e) $+ HBr \xrightarrow{Peroxide}$ $\xrightarrow[SN\text{-}2]{CH_3ONa}$

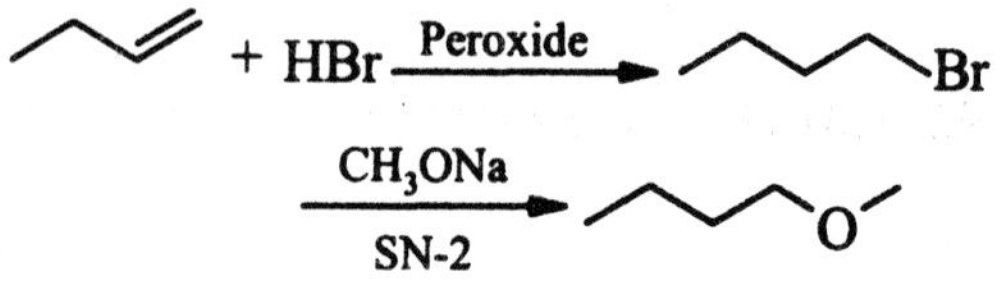

(f) $\xrightarrow[Heat]{C.H_2SO_4}$

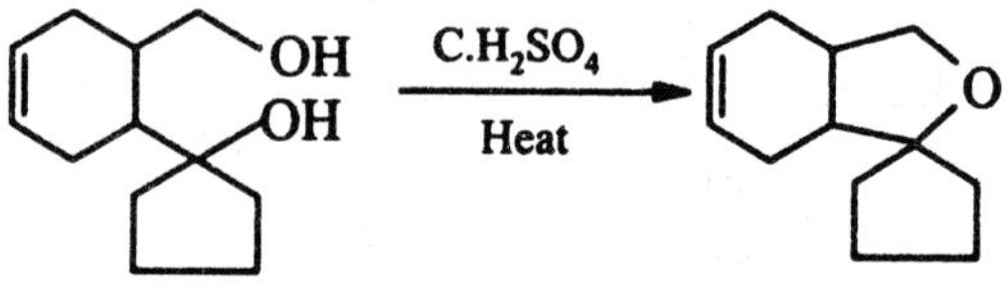

39. (a) $\xrightarrow[-H_2O]{H^+}$ $\xrightarrow{-H^+}$

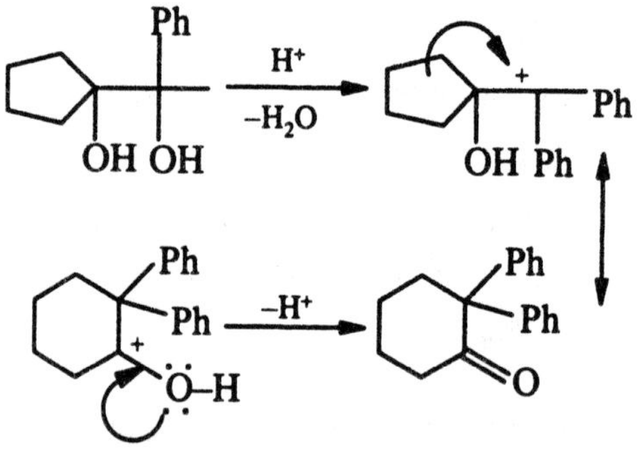

(b) $\xrightarrow[-H_2O]{H^+}$ Methyl shift $\xrightarrow{-H^+}$

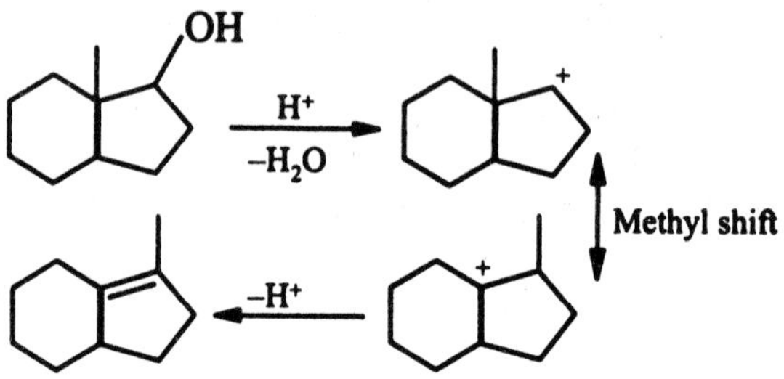

(c) $\xrightarrow[-H_2O]{H^+}$ $\xrightarrow{-H^+}$

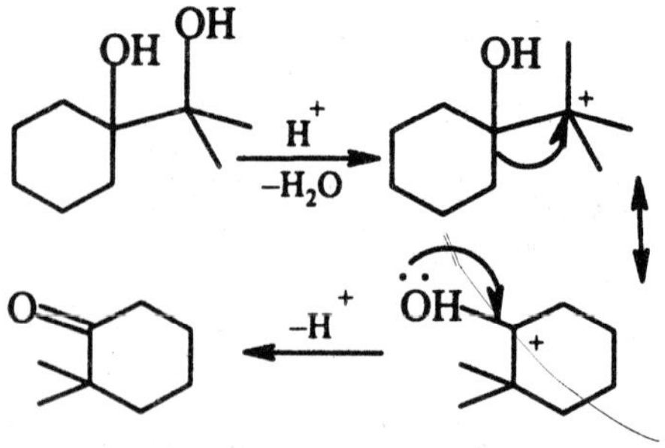

(d) $\xrightarrow{H^+}$ $\xrightarrow{-H^+}$

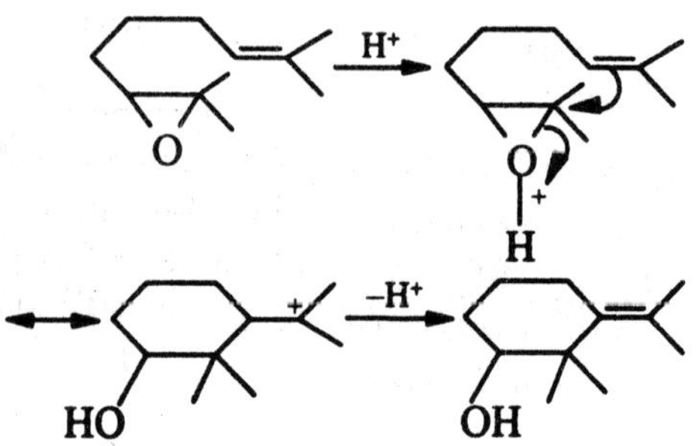

40. X : $PhCH(OH)CH_2Ph$, Y : $PhCOCH_2Ph$

41. 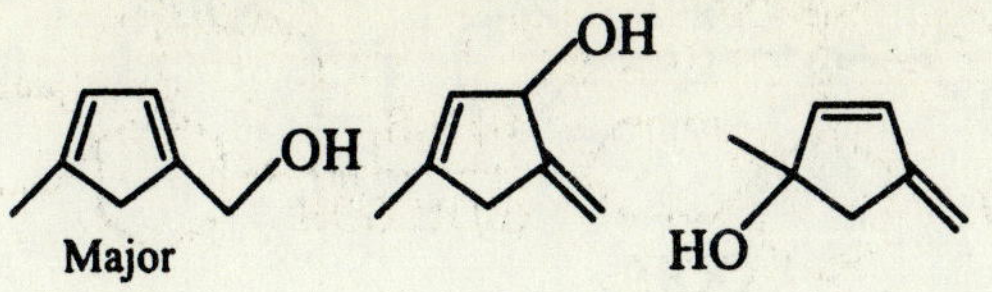 Major

42. 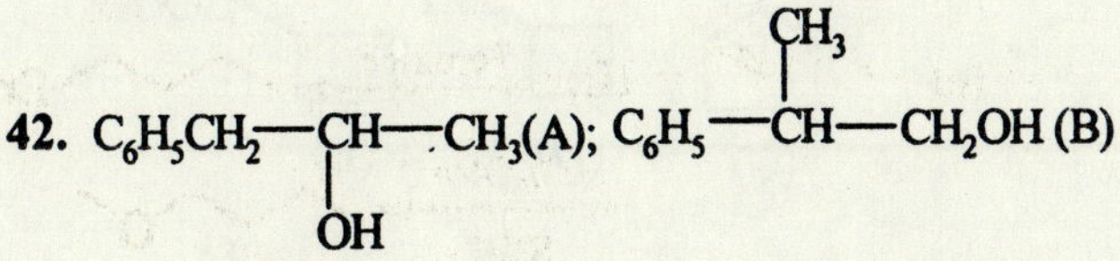 $C_6H_5CH_2—CH(OH)—CH_3$ (A); $C_6H_5—CH(CH_3)—CH_2OH$ (B)

43. 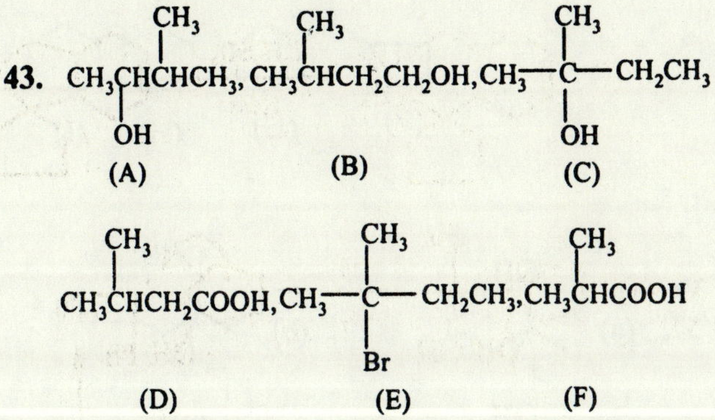 $CH_3CH(OH)CH(CH_3)CH_3$, $CH_3CH(CH_3)CH_2CH_2OH$, $CH_3—C(CH_3)(OH)—CH_2CH_3$
(A) 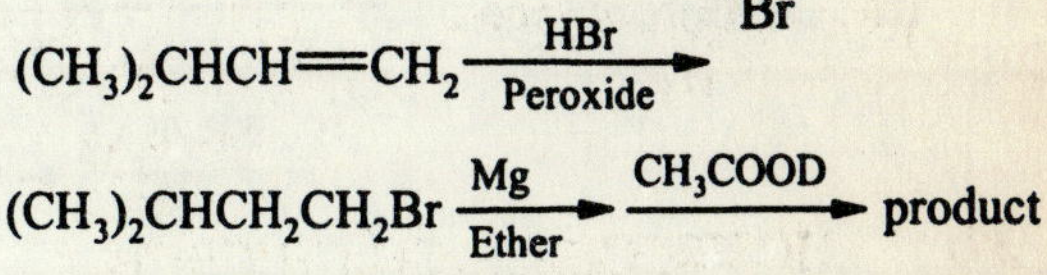 (B) 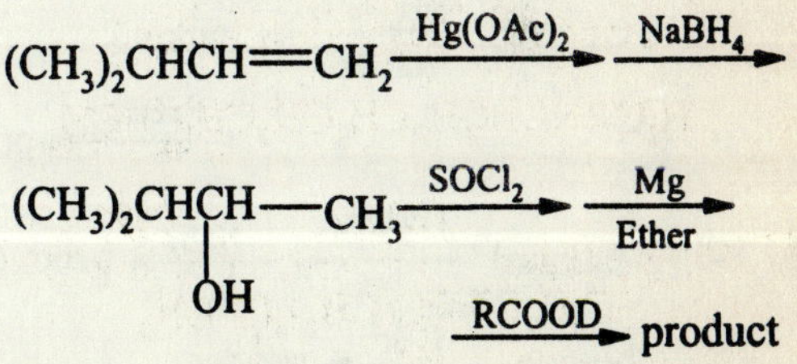 (C)

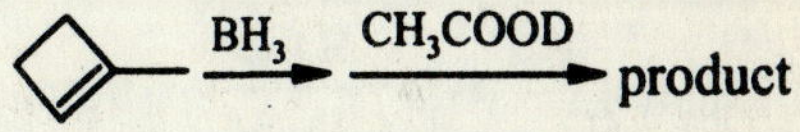

$CH_3CH(CH_3)CH_2COOH$, $CH_3—C(CH_3)(Br)—CH_2CH_3$, $CH_3CH(CH_3)COOH$
(D) 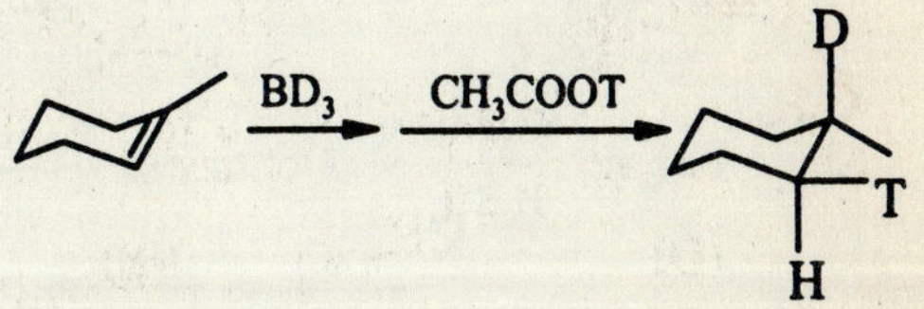 (E) (F)

44. 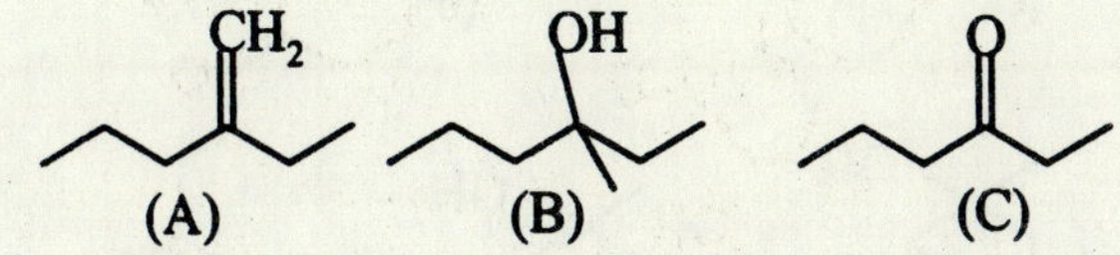 (A) (B) (C)

45. 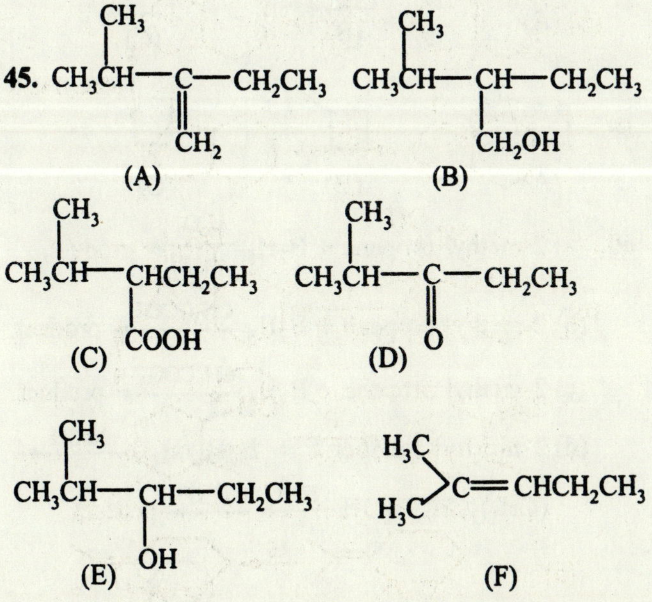 $CH_3CH(CH_3)—C(=CH_2)—CH_2CH_3$ (A); $CH_3CH(CH_3)—CH(CH_2OH)—CH_2CH_3$ (B)

$CH_3CH(CH_3)—CH(COOH)CH_2CH_3$ (C); $CH_3CH(CH_3)—C(=O)—CH_2CH_3$ (D)

$CH_3CH(CH_3)—CH(OH)—CH_2CH_3$ (E); $(H_3C)_2C{=}CHCH_2CH_3$ (F)

46. 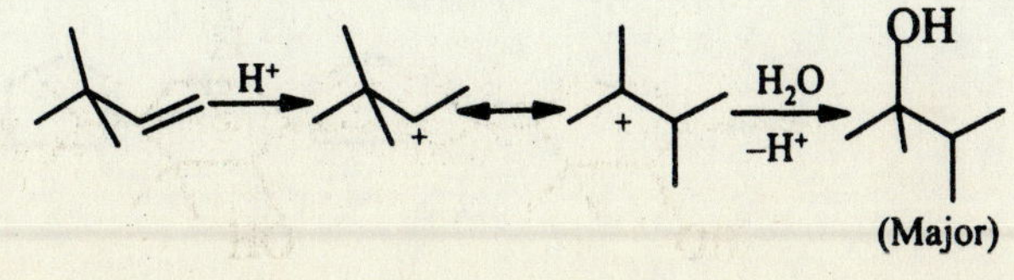 $\xrightarrow{H^+}$ ⟷ $\xrightarrow[-H^+]{H_2O}$ (Major)

47. 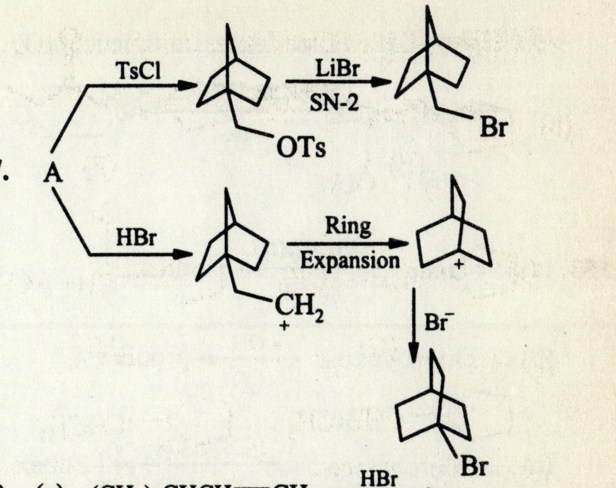 A $\xrightarrow{TsCl}$ (OTs) $\xrightarrow[SN\text{-}2]{LiBr}$ (Br)

A $\xrightarrow{HBr}$ ($\overset{+}{C}H_2$) $\xrightarrow[\text{Expansion}]{\text{Ring}}$ (+) $\xrightarrow{Br^-}$ (Br)

48. (a) $(CH_3)_2CHCH{=}CH_2 \xrightarrow[\text{Peroxide}]{HBr}$

$(CH_3)_2CHCH_2CH_2Br \xrightarrow[\text{Ether}]{Mg} \xrightarrow{CH_3COOD}$ product

(b) $(CH_3)_2CHCH{=}CH_2 \xrightarrow{Hg(OAc)_2} \xrightarrow{NaBH_4}$

$(CH_3)_2CHCH(OH)—CH_3 \xrightarrow{SOCl_2} \xrightarrow[\text{Ether}]{Mg}$

$\xrightarrow{RCOOD}$ product

(c) $\xrightarrow{BH_3} \xrightarrow{CH_3COOD}$ product

(d) $\xrightarrow{BD_3} \xrightarrow{CH_3COOT}$ (D, T, H)

49. Reaction of potassium salt with EtBr does not affect the configuration at chiral carbon while reaction of tosylate with $C_2H_5O^-$ occur at chiral carbon and proceed by SN-2 mechanism. Hence in later case inversion of configuration takes place.

50. 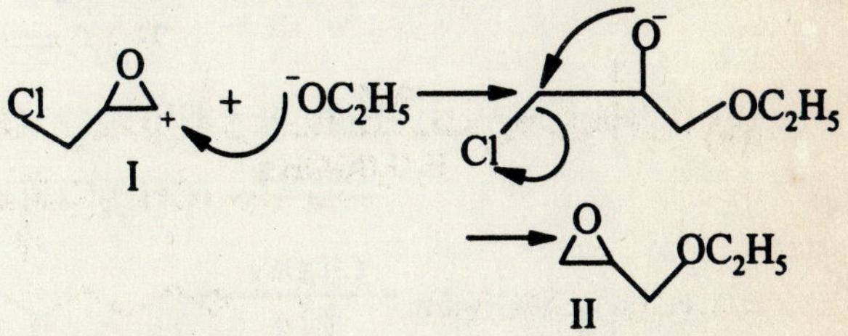 Cl (I) + $^-OC_2H_5$ ⟶ O^-, OC_2H_5, Cl ⟶ OC_2H_5 (II)

51. 1, 3-butandiol (A), 3-ketobutanoic acid (B).

52.(a) —OH + HCl ⟶ —Cl $\xrightarrow[\text{Ether}]{Mg}$ $\xrightarrow{D_2O}$ —D

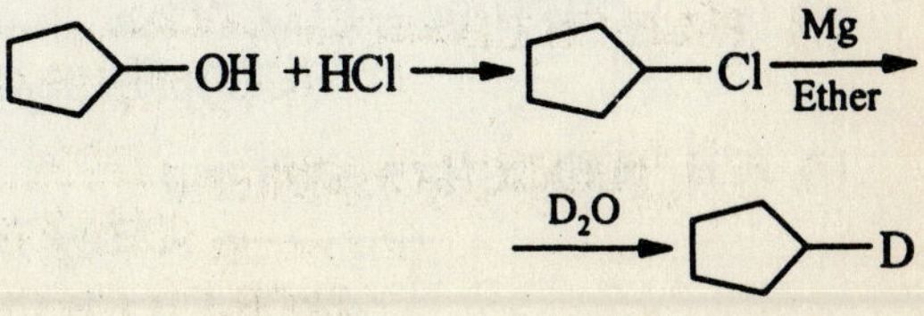

(b) $CH_2{=}CH_2$ + Cold, dilute, alkaline $KMnO_4$ $\longrightarrow$ $CH_2(OH){-}CH_2(OH)$ $\xrightarrow[\text{Heat}]{\text{Conc. } H_2SO_4}$ (1,4-dioxane)

53. (a) 1-butene $\xrightarrow[\text{NaOH}]{BH_3/H_2O_2}$ product

(b) 1-chlorobutane $\xrightarrow[H_2O]{\text{KOH}}$ product

(c) 2-chlorobutane $\xrightarrow[(CH_3)_3COH]{(CH_3)_3COK}$ 1-butene $\xrightarrow[\text{NaOH}]{BH_3/H_2O_2}$ product

(d) 1-butyne $\xrightarrow[\text{NaOH}]{BH_3/H_2O_2}$ butanal $\xrightarrow{NaBH_4}$ product

54. (a) 2-butanol $\xrightarrow{\text{TsCl}}$ $\xrightarrow{\text{NaBr}}$ product

(b) 1-butene $\xrightarrow{\text{HBr}}$ product

(c) 1-butanol $\xrightarrow{PCl_5}$ $\xrightarrow[C_2H_5OH]{C_2H_5OK}$ 1-butene $\xrightarrow{\text{HBr}}$ product

(d) 1-butyne $\xrightarrow[H_2SO_4]{HgSO_4}$ 2-butanone $\xrightarrow{LiAlH_4}$ $\xrightarrow{\text{TsCl}}$ $\xrightarrow{\text{NaBr}}$ product

55. (a) $(CH_3)_2CHCH_3$ $\xrightarrow[\text{Heat}]{Br_2}$ product

(b) $(CH_3)_3CBr$ (from a) + C_2H_5ONa $\xrightarrow[E_2]{C_2H_5OH}$ $(CH_3)_2C{=}CH_2$

(c) $(CH_3)_2C{=}CH_2$ (from b) + HBr $\xrightarrow{\text{Peroxide}}$ $(CH_3)_2CHCH_2Br$

(d) $(CH_3)_2CHCH_2Br$ (from c) $\xrightarrow[\text{SN-2}]{CH_3ONa}$ $(CH_3)_2CHCH_2OCH_3$

(e) $(CH_3)_2CHCH_2Br$ + CH_3COONa $\longrightarrow$ $(CH_3)_2CHCH_2OCOCH_3$

(f) $(CH_3)_2CHCH_2Br$ + KCN (aq) $\longrightarrow$ $(CH_3)_2CHCH_2CN$

56. (A) O-cyclobutyl ester, C(=O)=cyclopentylidene (B) cyclobutyl–OH (C) cyclopentyl–COOH (D) cyclobutyl–Cl (E) cyclobutyl–COOH (F) cyclobutyl–CH_2OH (G) cyclopentene (H) cyclopentyl–Cl (I) cyclopentane

57. (a) $(CH_3)_2C(OH)CH_2I$ (b) $(CH_3)_2C(OCH_3)CH_2OH$ (c) $(CH_3)_2C(OH)CH_2NH_2$ (d) $(CH_3)_2C(OH)CH_2OCH_3$

58. (A) 1-methyl-cyclopent-2-en-1-ol (with methyl), OH (B) 1-methylcyclopentan-1-ol (C) 1-methylcyclopentene

59. (A) cyclobutane fused 1,3-dioxolane (B) 1,2-diiodocyclobutane (C) cyclobutene

60. (a) 2-methyl propene + B_2H_6 $\xrightarrow[\text{NaOH}]{H_2O_2}$ product.

(b) 2-methyl propene + B_2D_6 $\xrightarrow{CH_3COOT}$ product.

(c) 2-methyl propene + B_2H_6 $\xrightarrow{CH_3COOT}$ product.

(d) 2-methyl propene + B_2H_6/H_2O_2 $\xrightarrow{\text{NaOH}}$ $(CH_3)_2CHCH_2OH$ $\xrightarrow{\text{NaH}}$ $\xrightarrow{C_2H_5Br}$ product.

61. $CH_2{=}CH(CH_2)_3OH$ $\xrightarrow[-H_2O]{H^+}$ (cyclizing cation) $\longrightarrow$ (cyclopentyl cation) $\xrightarrow{-H^+}$ cyclopentene

62. (a) (A) (B)

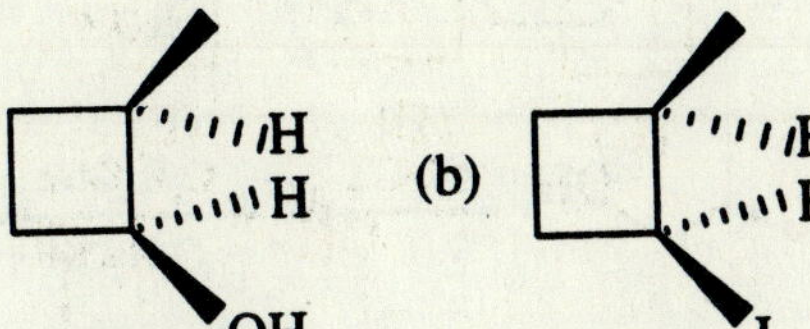

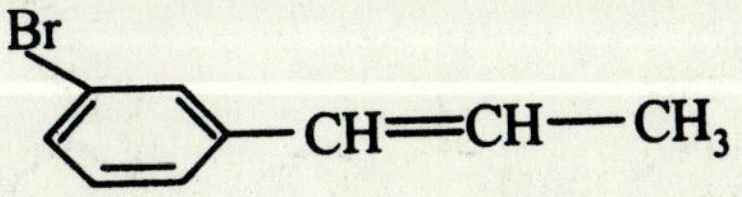

(b) 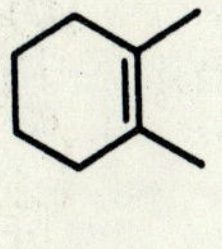 (C) (D)

(c) CH_3 H, H OSO_2CH_3 (E)

CH_3 $C\equiv CH$, H H (F)

63. (a) Br, $-CH{=}CH-CH_3$

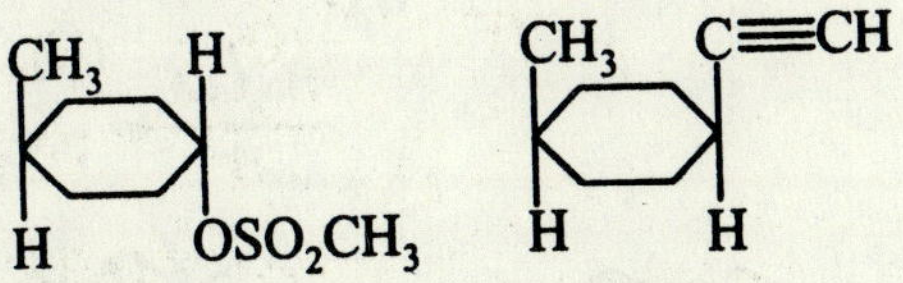

(b) (c) CN

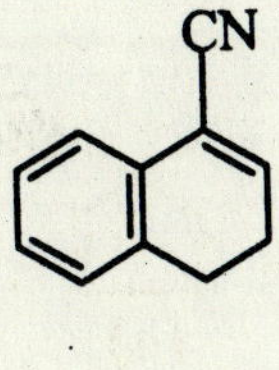

64. (a) OH, $\xrightarrow[-H_2O]{H^+}$, $\xrightarrow{-H^+}$

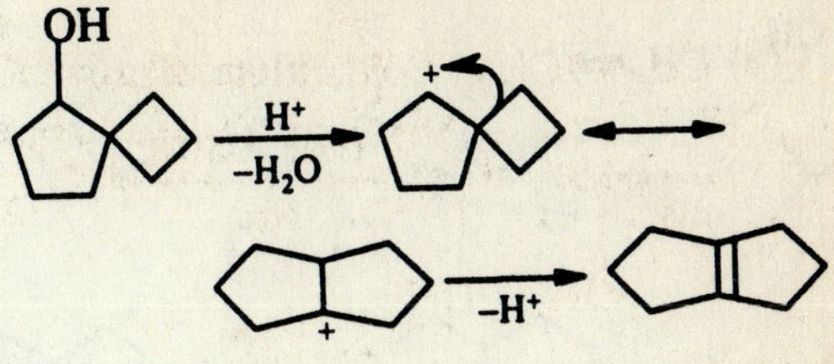

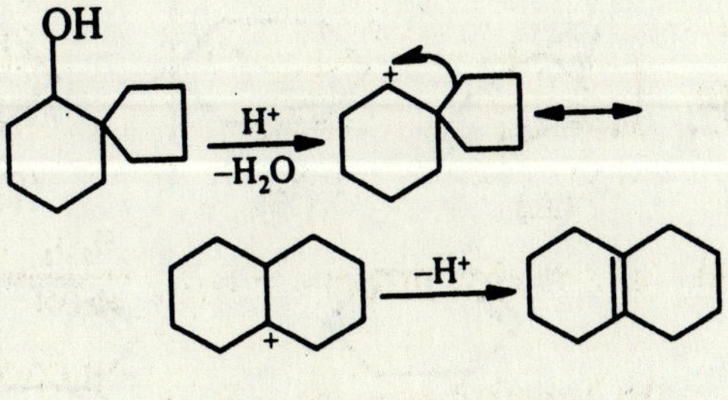

(b) OH, $\xrightarrow[-H_2O]{H^+}$, $\xrightarrow{-H^+}$

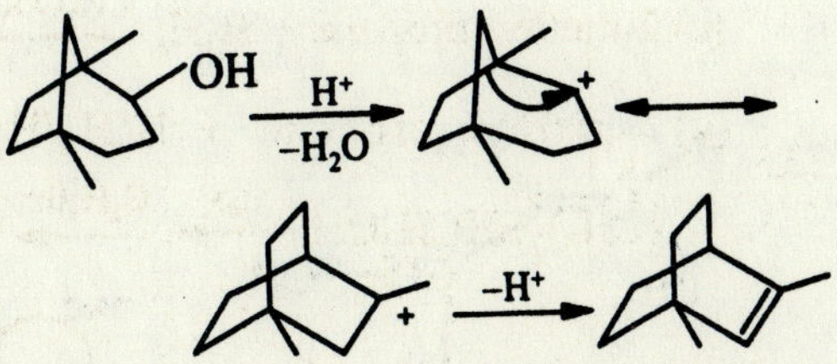

65. (a) 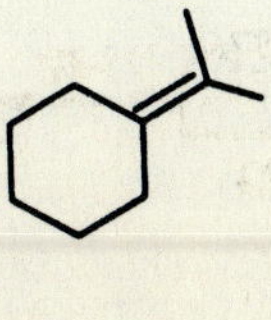 (b) OH (c)

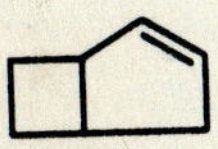

(d) OH, $\xrightarrow[-H_2O]{H^+}$, $\xrightarrow{-H^+}$

66. H_3CO, HO, $-CH{=}CHCH_2OH$

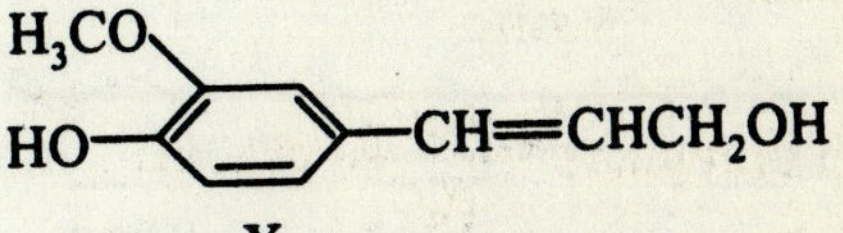

X

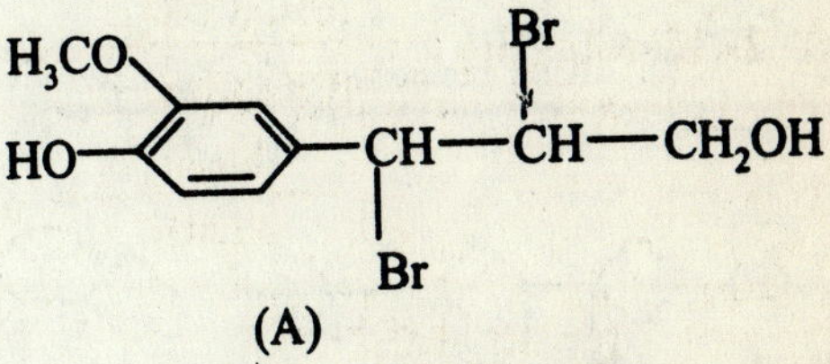

H_3CO, HO, $-CH(Br)-CH(Br)-CH_2OH$

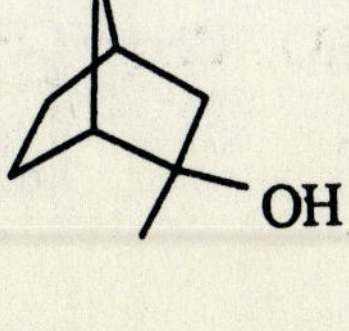

(A)

$HO-CH_2-CHO$

(B)

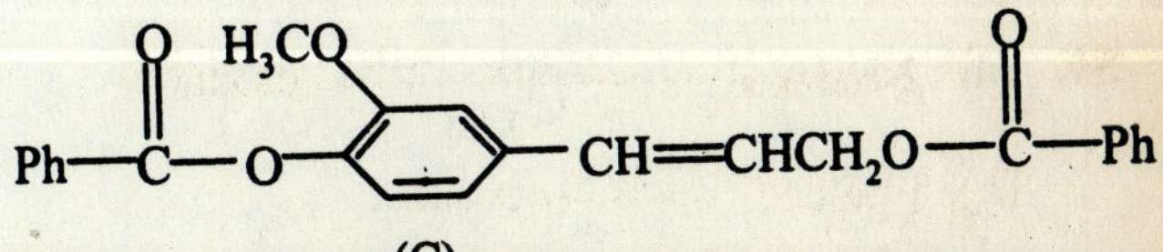

$Ph-C(=O)-O-$ (H_3CO) $-CH{=}CHCH_2O-C(=O)-Ph$

(C)

H_3CO, HO, $-CH{=}CH_2OH$

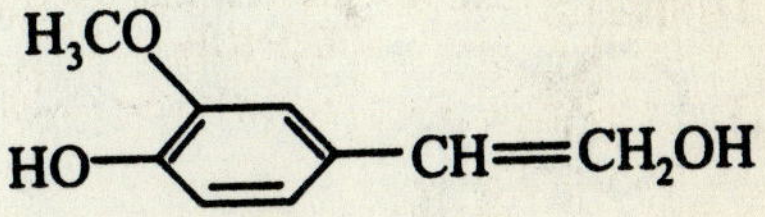

(D)

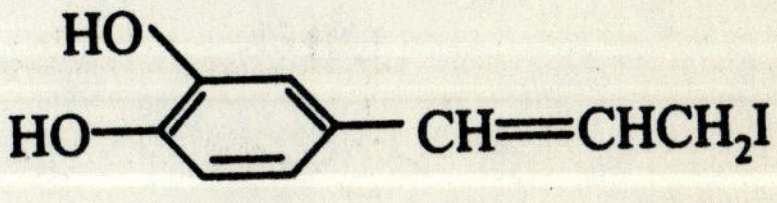

HO, HO, $-CH{=}CHCH_2I$

(E)

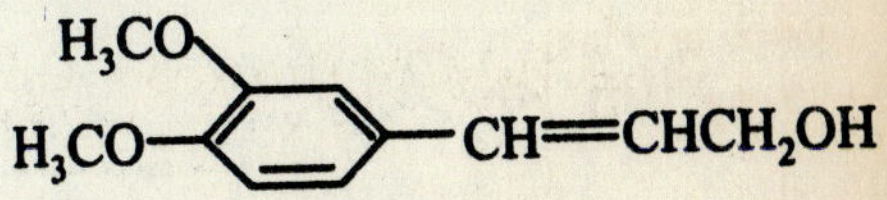

H_3CO, H_3CO, $-CH{=}CHCH_2OH$

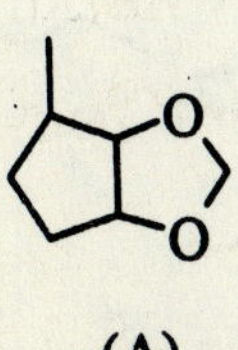

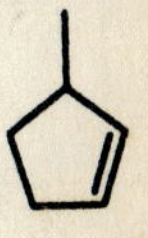

(F)

67. (A) 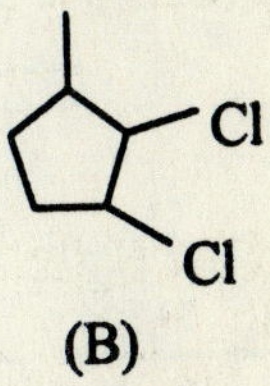 (B) Cl, Cl (C)

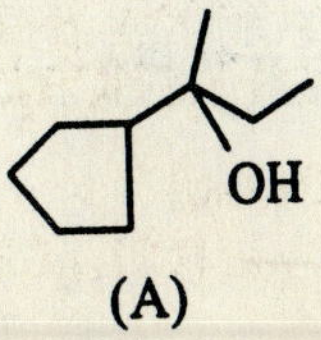

68. OH (A)

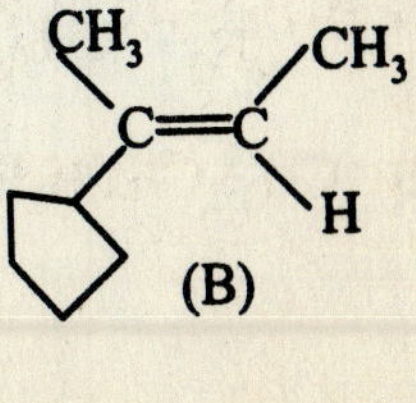

CH_3, CH_3, $C{=}C$, H (B) and

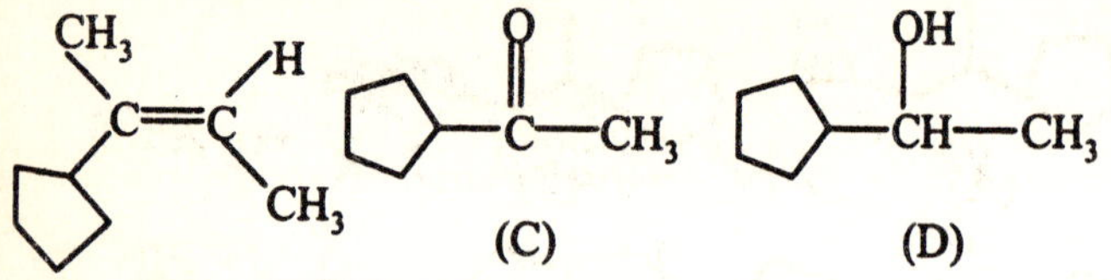

CH_3, H, C=C, CH_3 (C) C—CH_3 (D) OH CH—CH_3

69. (a) (i) OH $\xrightarrow[-H_2O]{H^+}$ $\overset{+}{C}H_2$ ⟷

$\xrightarrow{-H^+}$ (Major) + (Minor)

$\overset{+}{C}H_2$ Ring expansion ⟷ $\xrightarrow{-H^+}$

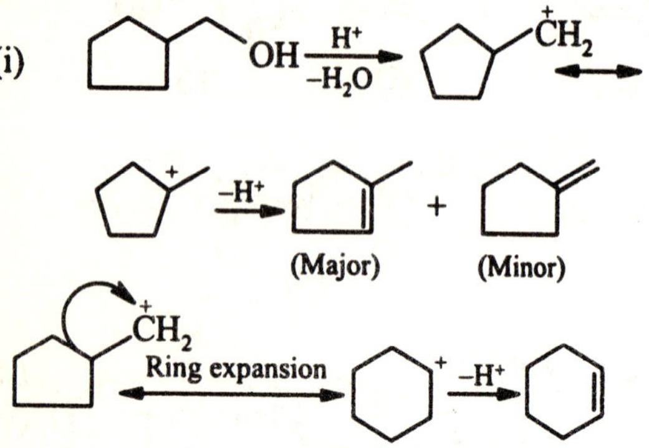

(ii) HO OH $\xrightarrow[-H_2O]{H^+}$ OH

⟷ :OH $\xrightarrow{-H^+}$ product

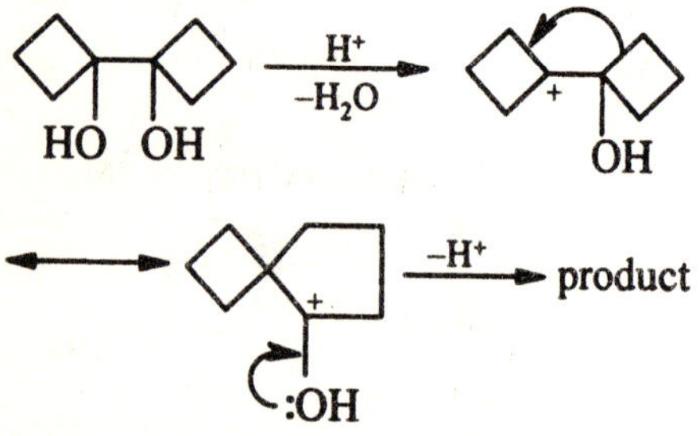

(b) (i) OH OH (A) H O (B)

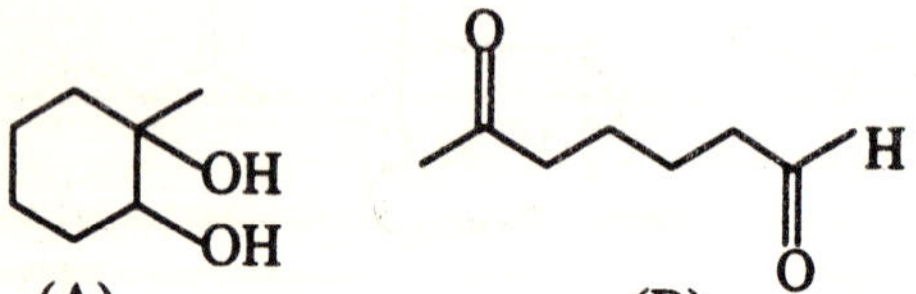

OH OH (C)

(ii) OH $\xrightarrow{SOCl_2}$ $\xrightarrow[(CH_3)_3COH]{(CH_3)_3COK}$

$\xrightarrow[H_2O_2/NaOH]{B_2H_6}$ OH

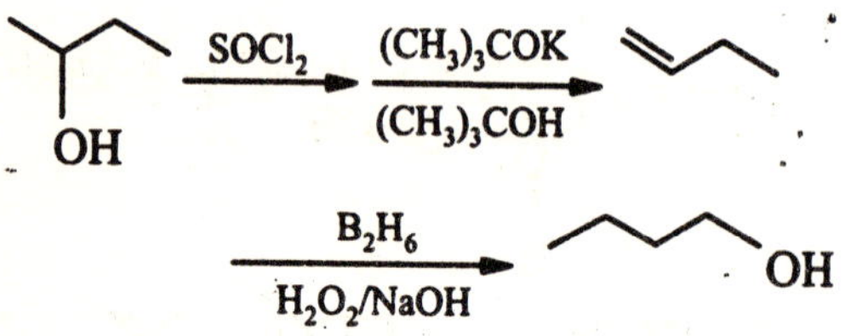

70. (a) $C_6H_5Br \xrightarrow[Ether]{Mg}$ $\xrightarrow{\text{O (epoxide)}}$ $\xrightarrow{H_3O^+}$

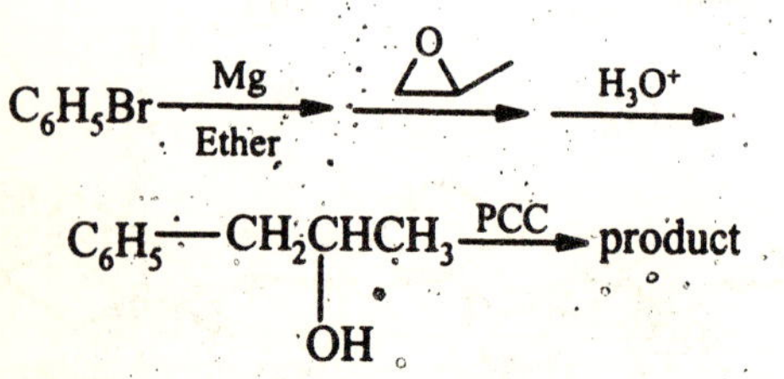

$C_6H_5—CH_2\underset{OH}{CH}CH_3 \xrightarrow{PCC}$ product

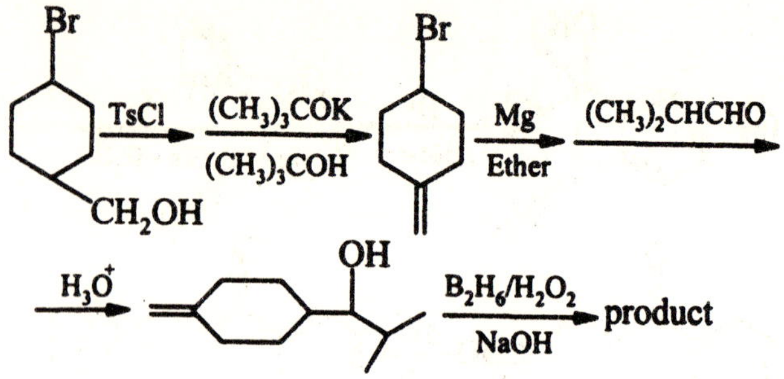

(b) Br, CH_2OH $\xrightarrow{TsCl}$ $\xrightarrow[(CH_3)_3COH]{(CH_3)_3COK}$ Br $\xrightarrow[Ether]{Mg}$ $\xrightarrow{(CH_3)_2CHCHO}$

$\xrightarrow{H_3\overset{+}{O}}$ OH $\xrightarrow[NaOH]{B_2H_6/H_2O_2}$ product

(c) OH $\xrightarrow{SOCl_2}$ $\xrightarrow[C_2H_5OH]{C_2H_5ONa}$

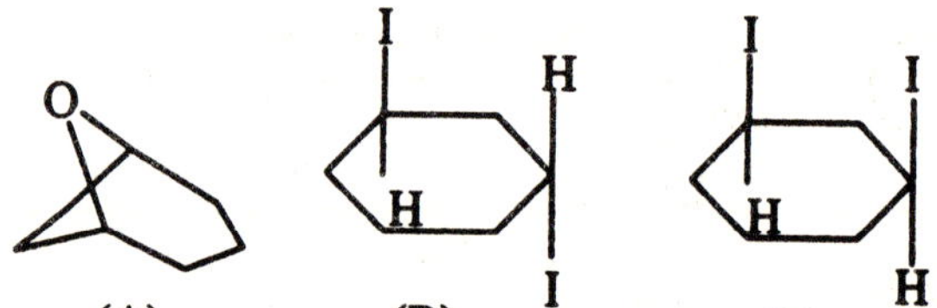

$\xrightarrow[H_2O]{Hg(OAc)_2}$ $\xrightarrow{NaBH_4}$

$\xrightarrow[H^+]{Na_2CrO_4}$ product

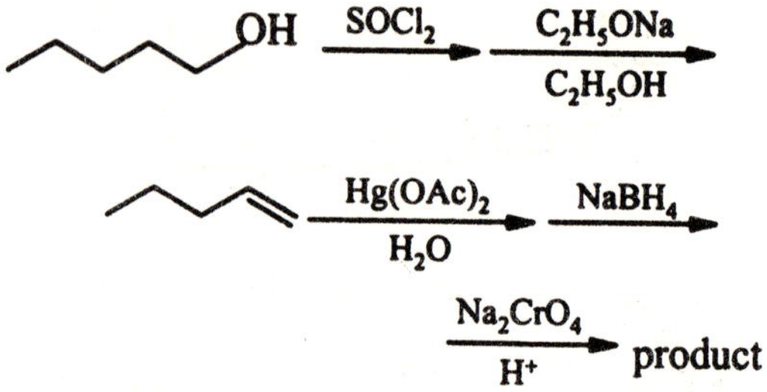

71. 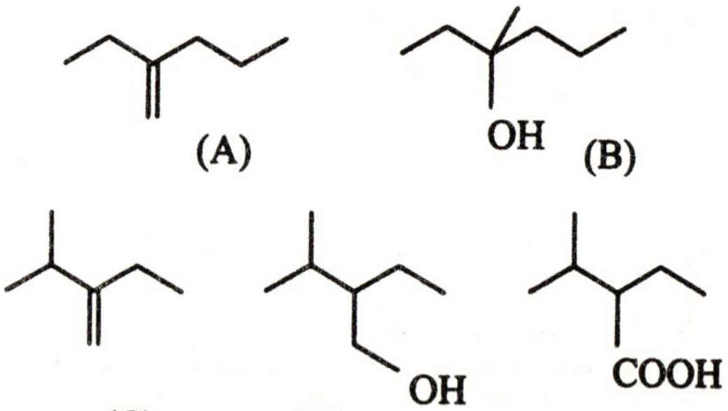 (A) OH (B)

(C) OH (D) COOH (E)

72. O (A) I H H I (B) I I H H (C)

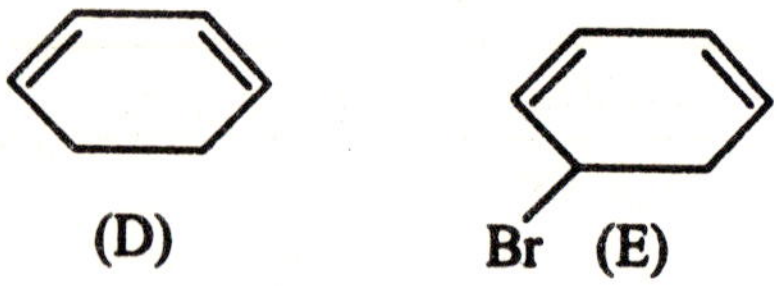

(D) Br (E)

73. H, H, C=C, CH_3, CH—CH_3, OH (X) and

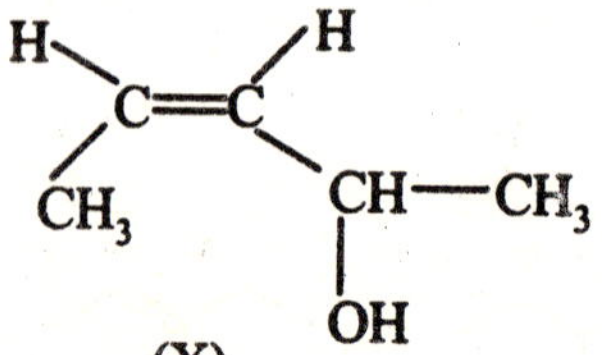

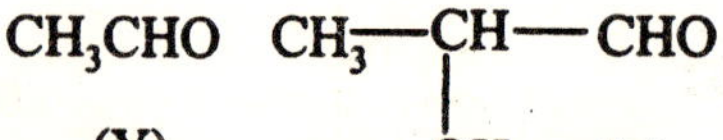

CH_3CHO (Y) $CH_3—\underset{OH}{CH}—CHO$ (Z)

74. (a) $\xrightarrow[\text{Heat}]{H^+}$ $\xrightarrow[\text{Zn}-H_2O]{O_3}$

CHO $\xrightarrow[H^+]{EG}$

OH CHO $\xleftarrow{H_3O^+}$ $\xleftarrow{NaBH_4}$

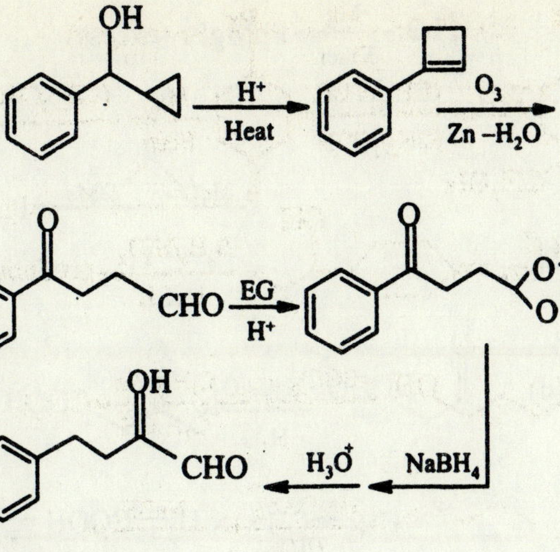

(b) Ph Ph O O ⟶ Ph Ph O O $\xrightarrow{H^+}$ product

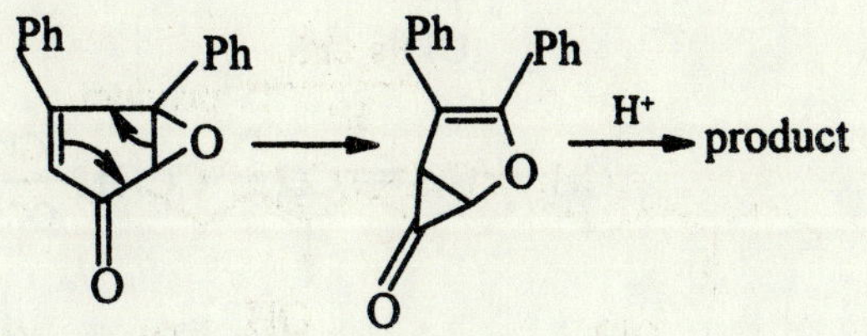

(c) (i) O $\xrightarrow[H_3O^+]{CH_3MgBr}$ OH $\xrightarrow[H_2SO_4/\text{Heat}]{\text{Conc.}}$

$\xrightarrow[\text{Excess}]{\text{NBS}}$ Br Br $\xrightarrow[C_2H_5OH]{C_2H_5ONa}$ product

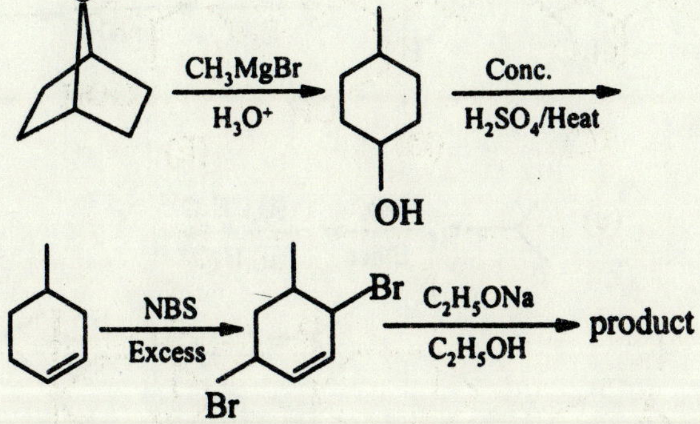

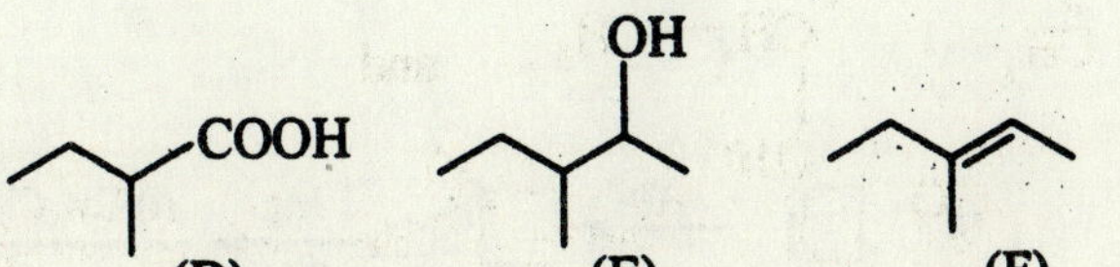

(ii) O $\xrightarrow[\text{One equivalent}]{\text{HCl}}$ HO Cl

$\xrightarrow[C_2H_5OH]{C_2H_5ONa}$ $\xrightarrow{\text{HBr}}$ $\xrightarrow{\text{KOH}}$ product

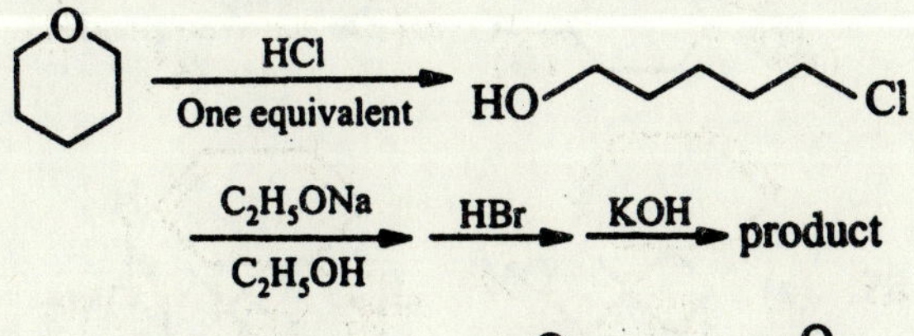

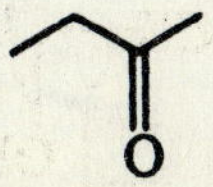

75. 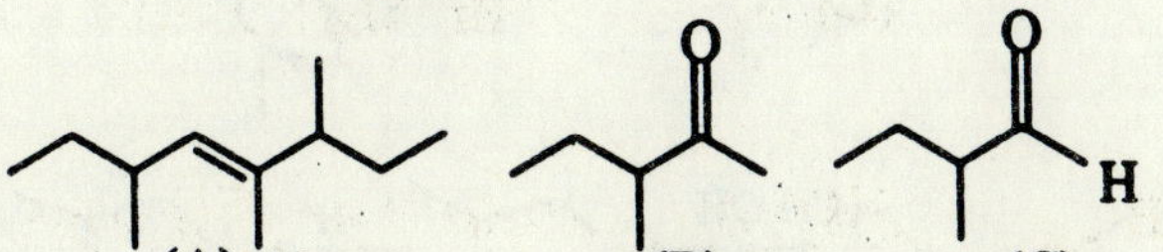 (A) (B) (C) H

COOH (D) OH (E) (F)

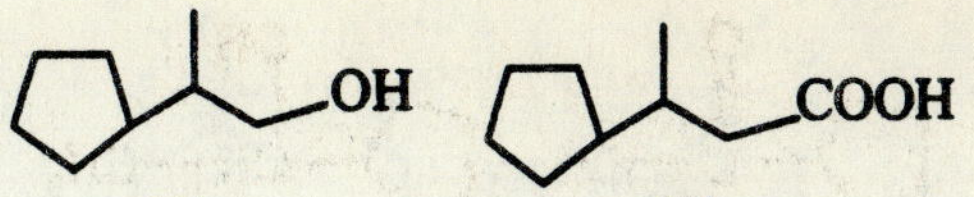

CH_3CHO (G) O (H)

76. OH (A) COOH (B)

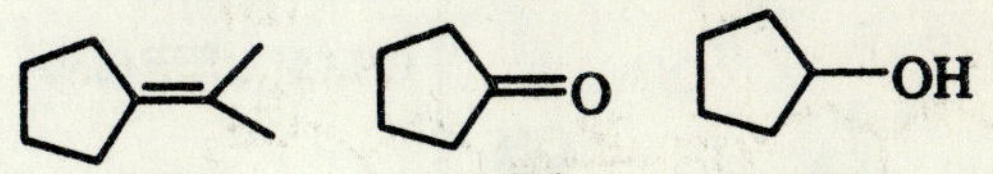

(C) O (D) OH (E)

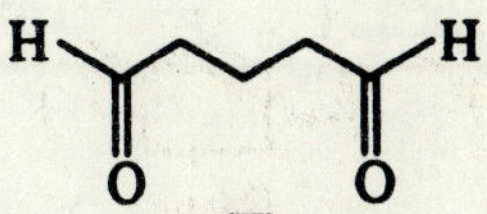

H H O O (F)

77. O O (A) OH OH (B) OH OH (C)

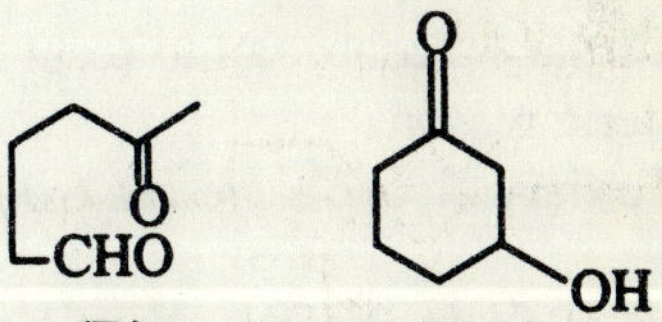

CHO (D) O OH (E)

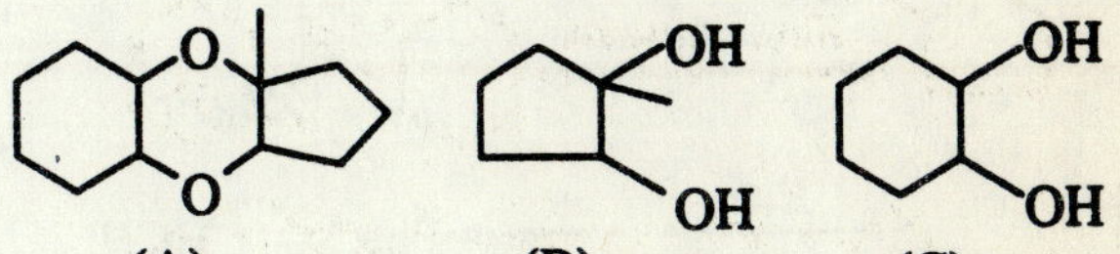

78. CH_3 O CH_3 (A) CH_3 OH OH H_3C (B) O O (C)

79. (a) O $CH_2\overset{18}{O}H$ $\xrightarrow{H^+}$ H O+ CH_2—Ö.18—H

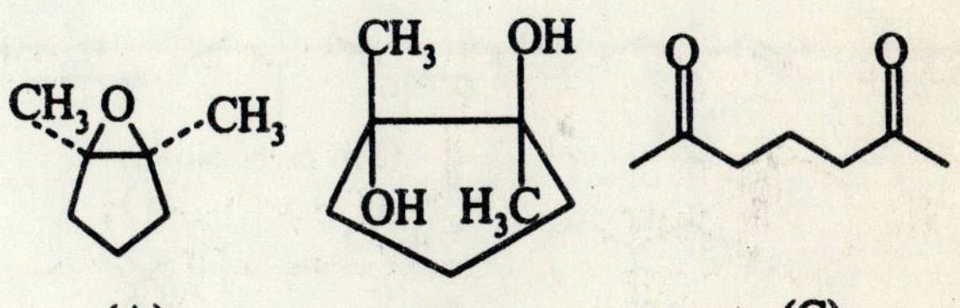

⟶ 18 O+ H OH $\xrightarrow{-H^+}$ product

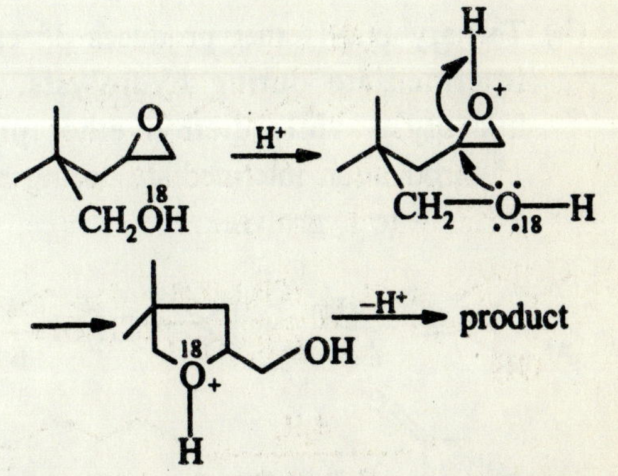

(b) HO OH $\xrightarrow{K_2Cr_2O_7}$ HO O

$\xrightarrow{H^+}$ HO: +OH ⟶ H O+ OH $\xrightarrow{-H^+}$ Product

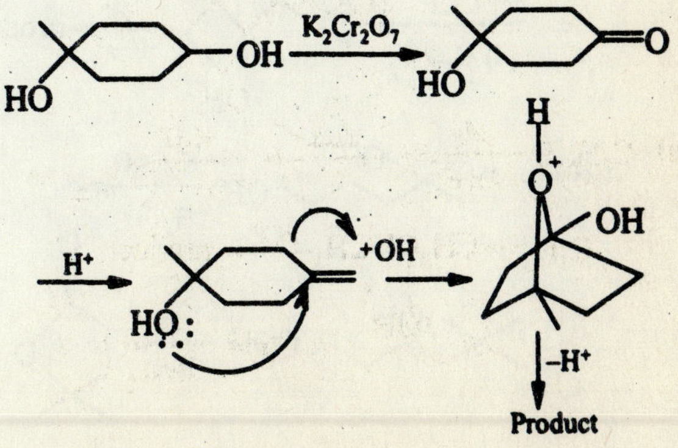

80. (a) Cyclohexylmethanol (OH) $\xrightarrow{PCC}$ cyclohexanecarbaldehyde (CHO) $\xrightarrow{CH_3CH_2C\equiv \bar{C}Na^+}$ $\xrightarrow{H_3\overset{+}{O}}$ cyclohexyl–CH(OH)–C≡C—CH_2CH_3

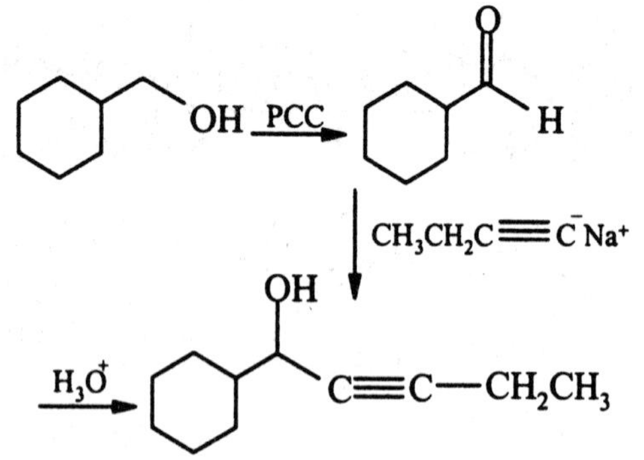

(b) $Ph_2C{=}O \xrightarrow{C_2H_5MgBr} Ph_2C(OMgBr)CH_2CH_3 \xrightarrow{H_3\overset{+}{O}} \xrightarrow[Heat]{H^+} Ph_2C{=}CHCH_3$

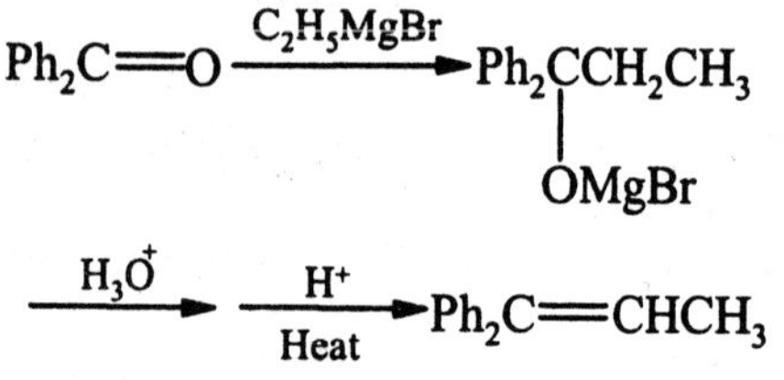

81. (a) Tertiary alcohol fail to react with $SOCl_2$ due to steric reason.

(b) Tertiary alcohols are not oxidize by PCC.

82. X : $B_2H_6/H_2O_2/NaOH$ Y : CH_3SO_2Cl Z : KI

83. (a) $\xrightarrow{TsOH}$ (a 2° carbocation) $\xrightarrow{\bar{H}\text{-shift}}$ $\xrightarrow[-H^+]{CH_3OH}$ OMe

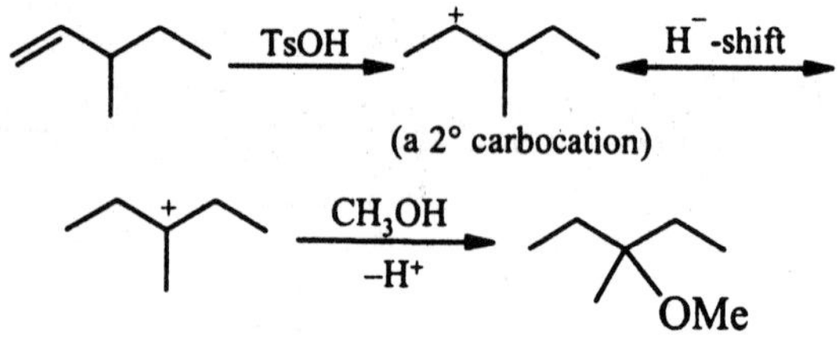

(b) Tertiary butyl ether produces 3° carbocation intermediate during hydrolysis, easier to hydrolyse while diethyl ether produces a 1° carbocation intermediate during hydrolysis, difficult to hydrolyse.

84. (a) $\xrightarrow[Peroxide]{HBr}$ $CH_2Br \xrightarrow[Ether]{Mg}$ $\xrightarrow[(ii)\ H_3\overset{+}{O}]{(i)\ \text{cyclobutanone}}$ OH $\xrightarrow[Heat]{H^+}$ product

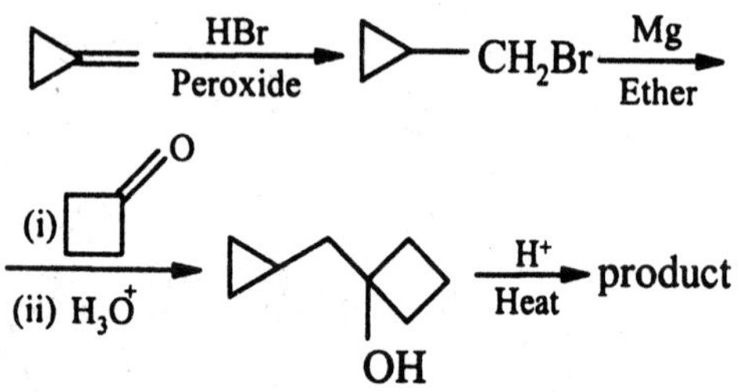

(b) OH, OH $\xrightarrow[H_3\overset{+}{O}]{K_2Cr_2O_7}$:OH, COOH $\xrightarrow[Heat]{H^+}$ lactone (O, O)

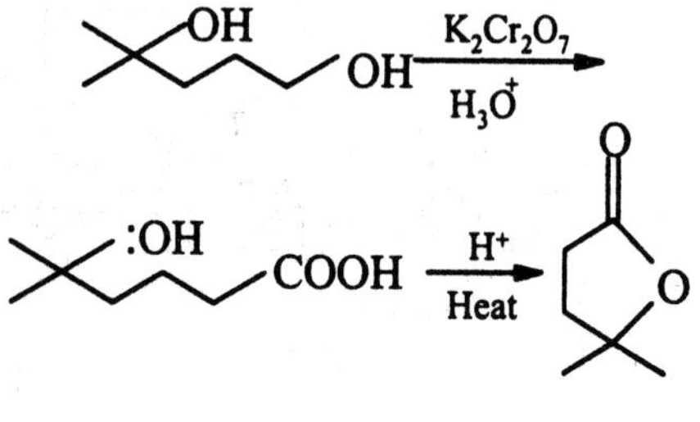

(c) $PhBr \xrightarrow[Ether]{Mg} PhMgBr$ (excess) $\xrightarrow{MeO(C{=}O)OMe} Ph_2CO$

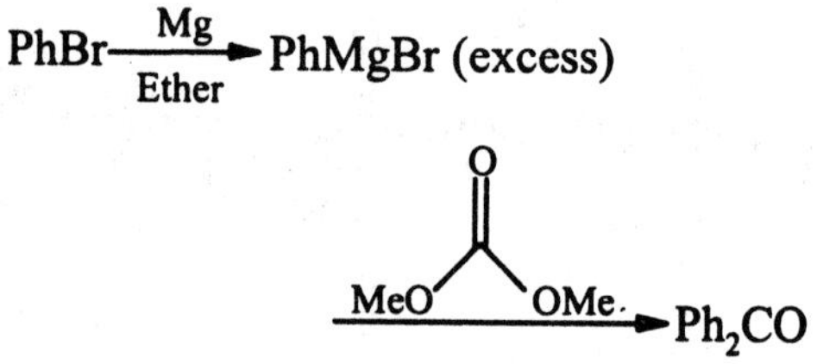

(d) $CH_3COCH_2CHO \xrightarrow[H_3\overset{+}{O}]{K_2Cr_2O_7} CH_3COCH_2COOH \xrightarrow{NaBH_4}$

$CH_3{-}CH(OH){-}CH_2{-}COOH \xrightarrow[Heat]{H^+}$

$CH_3{-}CH{=}CH{-}COOH \xrightarrow[CCl_4]{Cl_2}$ $\xrightarrow{\text{KOH Fusion}}$ $\xrightarrow[liq\ NH_3]{Na}$ $\xrightarrow[H_2O]{H_3\overset{+}{O}}$ $H_3C(H)C{=}C(H)COOH$

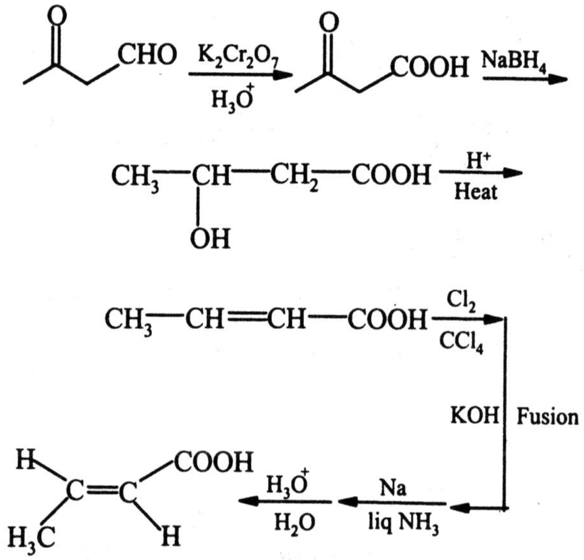

(e) Br $\xrightarrow[Ether]{Mg}$ $\xrightarrow[(ii)\ H_3O]{(i)\ CH_2O}$ $CH_2OH \xrightarrow[Heat]{H^+}$ alkene

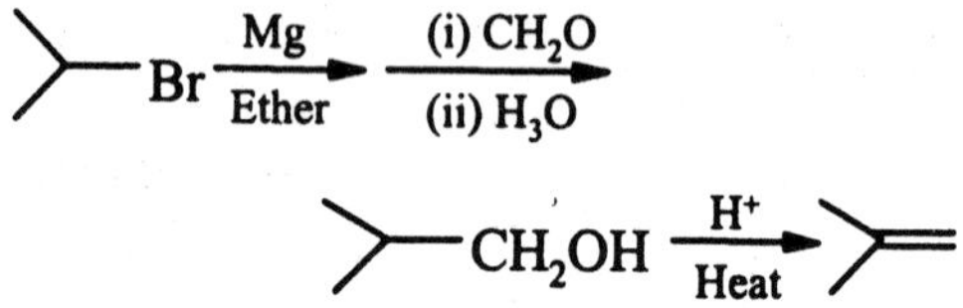

(f) Cyclohexanone (=O) $\xrightarrow{LDA} \xrightarrow{CH_3I}$ product

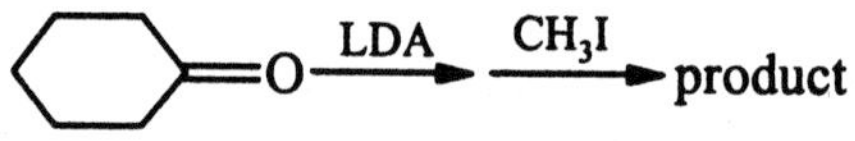

85. (a) OMe $\xrightarrow{H^+}$ H, OMe (+) $\xrightarrow{-CH_3OH}$ (+) $\xrightarrow{-H^+}$ cyclopentene product

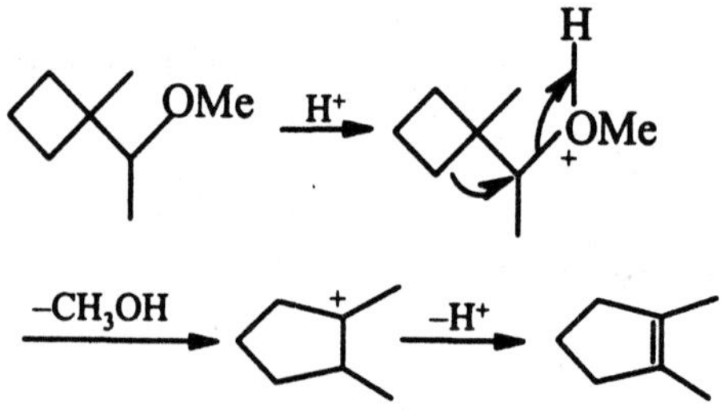

(b) $\xrightarrow{HBr}$ Br $\xrightarrow[Ether]{Mg}$ $\xrightarrow[(ii)\ H_3O^+]{(i)\ CH_3CHO}$ OH $\xrightarrow{NaH}$ $\xrightarrow{CH_3I}$ product

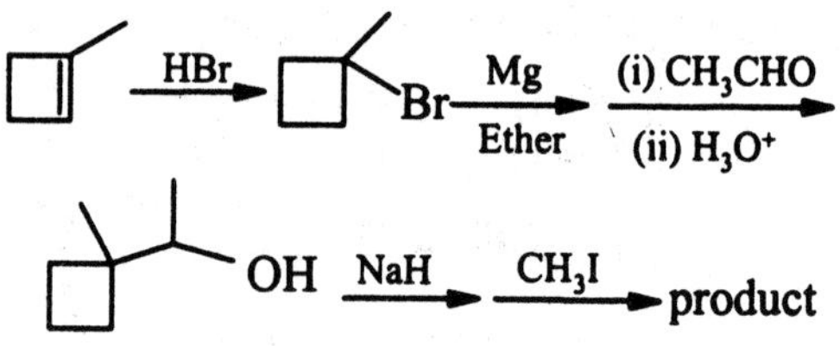

(c) 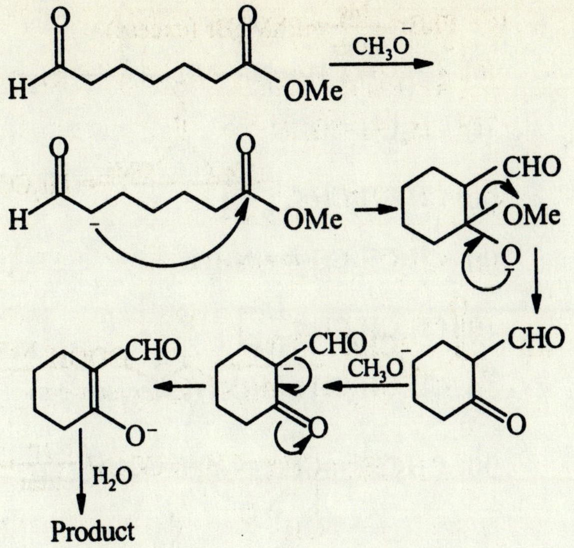 H—CHO...(CH₂)₅...CO–OMe $\xrightarrow{CH_3\bar{O}}$

Product (via enolate cyclization: CHO, OMe, O⁻ intermediates; $CH_3\bar{O}$; H_2O)

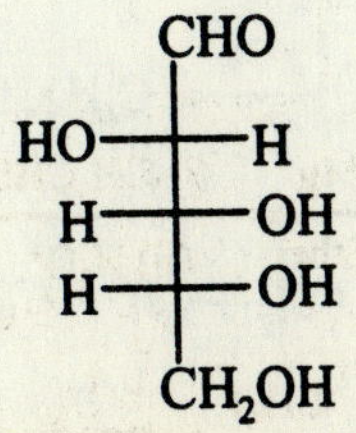

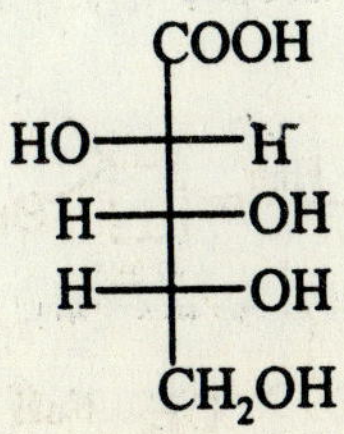

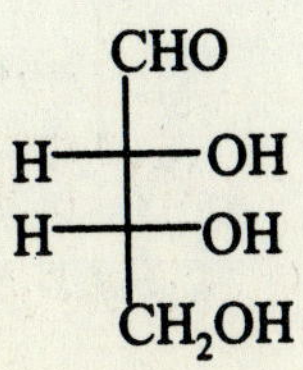

86. (a) 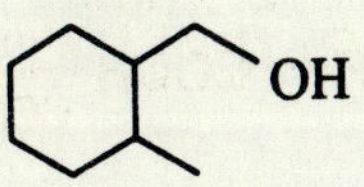 OH (b) 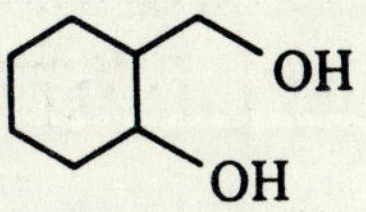 OH, OH (c) 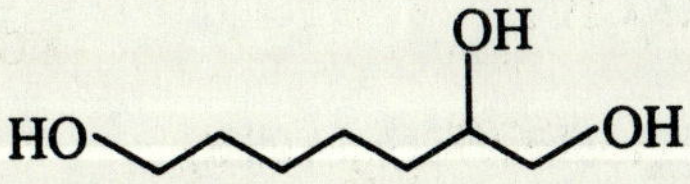 HO, OH, OH (d) 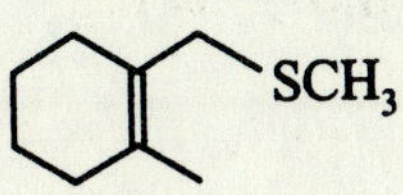 SCH_3 (e) 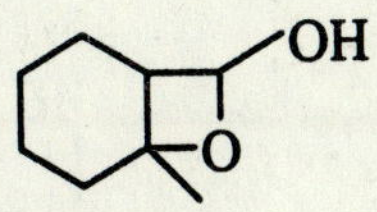 OH, O (f) 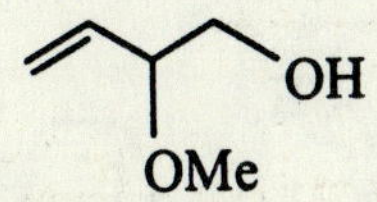 OH, OMe (g) 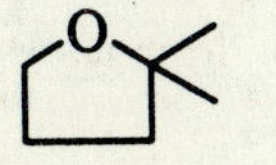 O (h) 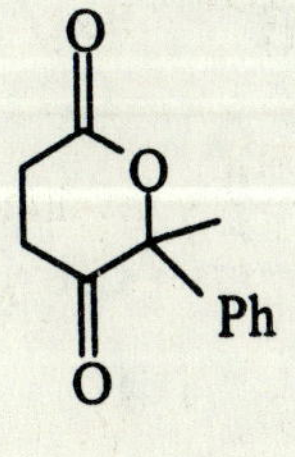 O, O, Ph, O (i) 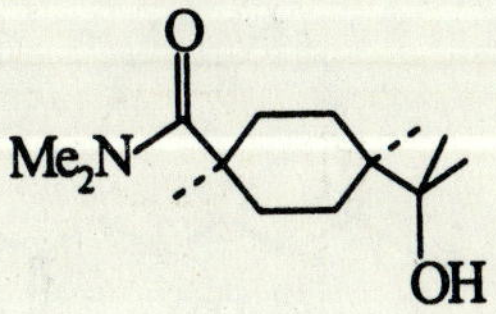 Me_2N, O, OH (j) 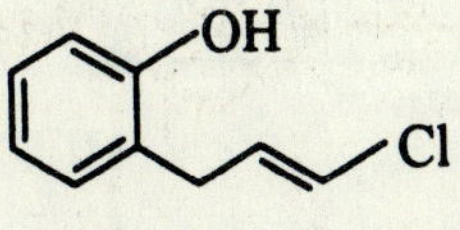 OH, Cl (k) 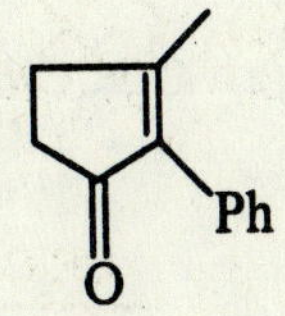 Ph, O

87.

(A)	(B)	(C)
CHO	COOH	CHO
HO—H	HO—H	H—OH
H—OH	H—OH	H—OH
H—OH	H—OH	CH_2OH
CH_2OH	CH_2OH	

88. 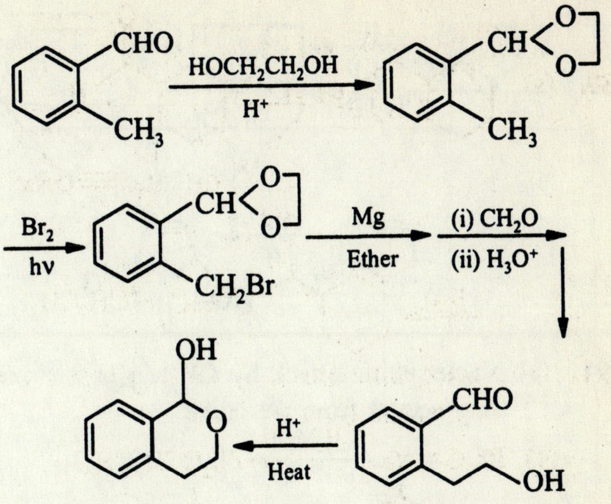 CHO, CH_3 $\xrightarrow[H^+]{HOCH_2CH_2OH}$ CH(O)₂, CH_3 $\xrightarrow[h\nu]{Br_2}$ CH(O)₂, CH_2Br $\xrightarrow[Ether]{Mg}$ $\xrightarrow[(ii)\ H_3O^+]{(i)\ CH_2O}$ CHO, OH $\xrightarrow[Heat]{H^+}$ OH, O

89. $Ph—CH(CH_2CH_3)—OCH_2CH_3$ (A) $Ph—CH(OH)—CH_2CH_3$ (B)

$Ph—C(=O)—CH_2CH_3$ (C) $Ph—CH(I)—CH_2CH_3$ (D)

$Ph—CH{=}CH—CH_3$ (E) C_6H_5CHO (F)

90. (a) 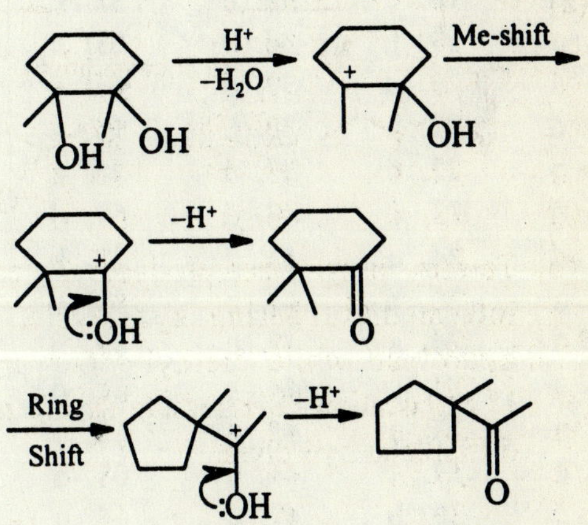 OH OH $\xrightarrow[-H_2O]{H^+}$ + OH $\xrightarrow{\text{Me-shift}}$ + :OH $\xrightarrow{-H^+}$ O

$\xrightarrow[\text{Shift}]{\text{Ring}}$ + :OH $\xrightarrow{-H^+}$ O

(b) 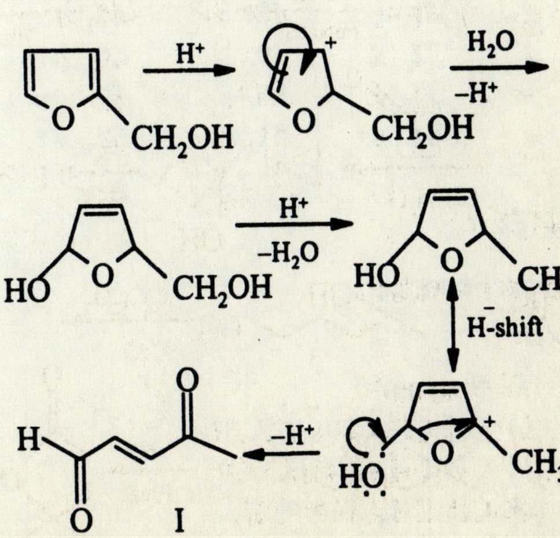 O, CH_2OH $\xrightarrow{H^+}$ +, O, CH_2OH $\xrightarrow[-H^+]{H_2O}$ HO, O, CH_2OH $\xrightarrow[-H_2O]{H^+}$ HO, O, CH_2^+ $\xrightarrow{\bar{H}\text{-shift}}$ HO, O, +, CH_3 $\xrightarrow{-H^+}$ H—C(=O)—CH=CH—C(=O)—CH₃ I

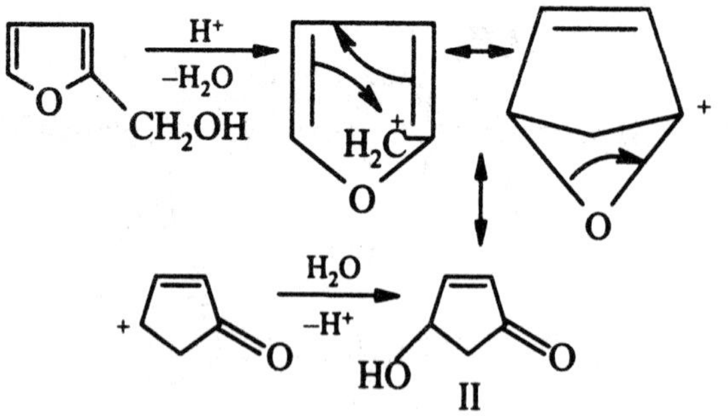

$\xrightarrow[-H_2O]{H^+}$ CH_2OH $H_2\overset{+}{C}$ O ↔ O +

+ $\xrightarrow[-H^+]{H_2O}$ HO O II

91. (a) Nucleophilic attack by CH_3MgBr, on oxirane ring occurs from anti side.

(b) OH OH $\xrightarrow[-H_2O]{H^+}$ OH N≡C CH_3

→ $\overset{+}{O}$–H N $\xrightarrow{-H^+}$ O N

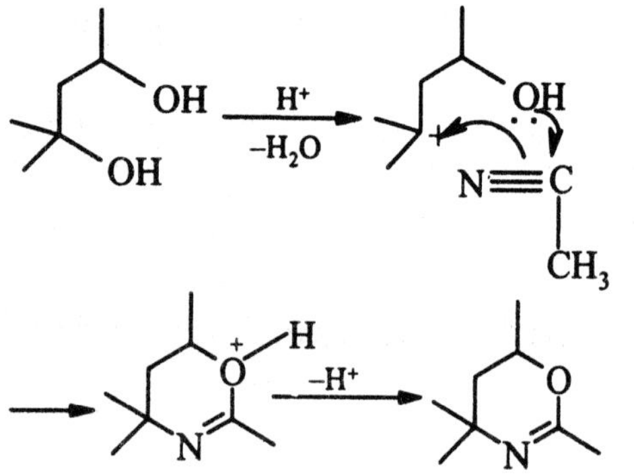

Objective Exercise

1. b	18. c	35. b	52. c	69. d
2. a	19. b	36. b	53. c	70. d
3. d	20. d	37. c	54. c	71. a
4. d	21. b	38. b	55. a	72. a
5. c	22. d	39. d	56. a	73. a
6. d	23. d	40. d	57. d	74. b
7. d	24. a	41. a	58. b	75. b
8. a	25. b	42. c	59. c	76. c
9. d	26. a	43. c	60. b	77. c
10. a	27. c	44. a	61. d	78. b
11. a	28. a	45. d	62. c	79. c
12. c	29. a	46. b	63. a	80. c
13. c	30. b	47. c	64. a	81. d
14. a	31. b	48. b	65. a	82. c
15. d	32. d	49. b	66. b	83. b
16. a	33. c	50. c	67. c	84. c
17. a	34. a	51. b	68. a	85. b

CHAPTER 7

Practice Exercise

Level 1

1. (a) Propanol
   (b) $CH_3CH_2CH(OH)C_6H_5$
   (c) CH_3CH_2COOH
   (d) $CH_3CH_2CH{=}CH_2$
   (e) $CH_3CH_2CH{=}N{-}NH{-}\overset{O}{\overset{\|}{C}}{-}NH_2$
   (f) CH_3CH_2COOH
   (g) 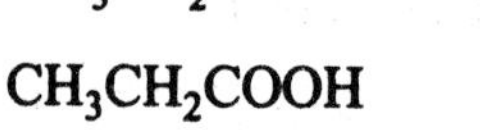 $CH_3CH_2CH\langle$ S S (dithiolane)
   (h) $CH_3CH_2CH{=}NNHPh$
   (i) 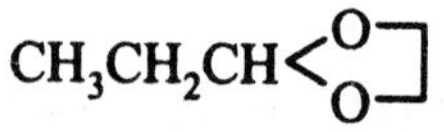 $CH_3CH_2CH\langle$ O O (dioxolane)
   (j) $CH_3CH_2CH{=}NOH$ (stereoisomers)
   (k) 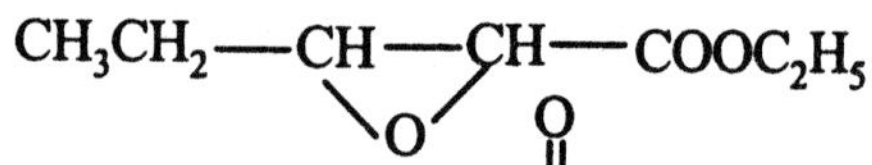 $CH_3CH_2{-}CH{-}CH{-}COOC_2H_5$ (with O bridging the two CH, epoxide)

2. (a) $CH_3CH_2CH{=}CH{-}\overset{O}{\overset{\|}{C}}{-}CH_3$
   (b) CH_3CH_2COOH
   (c) 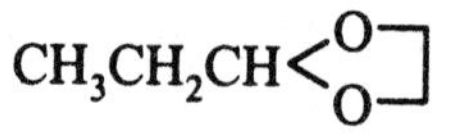 $CH_3CH_2CH\langle$ O O (dioxolane)
   (d) $CH_3CH_2CH{=}\underset{CH_3}{\underset{|}{C}H}{-}CHO$
   (e) $CH_3CH_2\overset{OH}{\overset{|}{C}H}{-}C{\equiv}CH$

3. 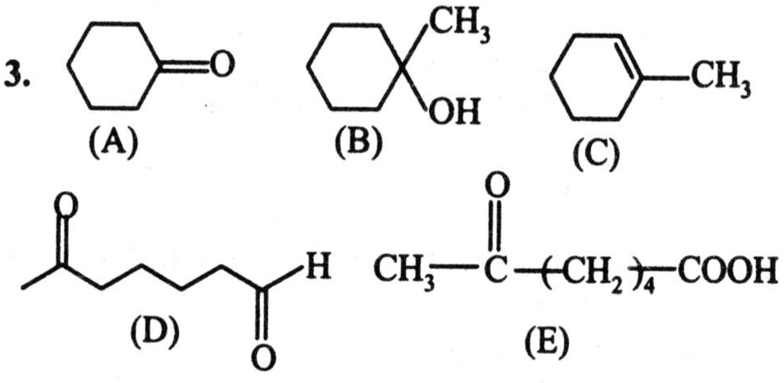 (A) cyclohexanone (=O) (B) 1-methylcyclohexanol (CH_3, OH) (C) 1-methylcyclohexene (–CH_3)

   (D) O ... H, O (E) $CH_3{-}\overset{O}{\overset{\|}{C}}{-}(CH_2)_4{-}COOH$

4. (a) $C_6H_5CH_2Cl \xrightarrow[Ether]{Mg} \xrightarrow{CH_3CHO}$ $\xrightarrow{H_3O^+}$ product

   (b) 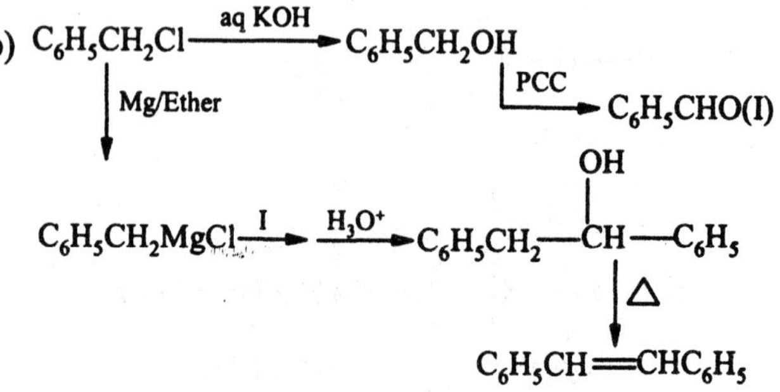 $C_6H_5CH_2Cl \xrightarrow{aq\ KOH} C_6H_5CH_2OH \xrightarrow{PCC} C_6H_5CHO(I)$

   $C_6H_5CH_2Cl \xrightarrow{Mg/Ether} C_6H_5CH_2MgCl \xrightarrow{I} \xrightarrow{H_3O^+} C_6H_5CH_2{-}\overset{OH}{\overset{|}{C}H}{-}C_6H_5 \xrightarrow{\Delta} C_6H_5CH{=}CHC_6H_5$

(c) $C_6H_5CH_2Cl \xrightarrow[\text{Ether}]{\text{Mg}} \xrightarrow[\text{H}_3\text{O}^+]{\text{ethylene oxide}}$

$C_6H_5CH_2CH_2CH_2OH \xrightarrow{\text{PCC}}$ product

(d) $C_6H_5CH_2Cl \xrightarrow[\text{Ether}]{\text{Mg}} \xrightarrow{CH_3CH_2COCl}$ product

5. (a) 1-butanol $\xrightarrow{SOCl_2} CH_3CH_2CH_2CH_2Cl \xrightarrow{\text{aqKCN}} CH_3CH_2CH_2CH_2CN$(I)

$CH_3CH_2CH_2CH_2Cl \xrightarrow{\text{Mg | Ether}} CH_3CH_2CH_2CH_2MgCl \xrightarrow{\text{I}} \xrightarrow{H_3O^+}$ product.

(b) $CH_3CH_2Br \xrightarrow{\text{aqKOH}} CH_3CH_2OH \xrightarrow{\text{pcc}} CH_3CHO$(I)

$CH_3CH_2Br \xrightarrow{\text{Mg/Ether}} CH_3CH_2MgBr \xrightarrow{\text{I}} \xrightarrow{H_3O^+} CH_3CH_2CH(OH)-CH_3 \xrightarrow{H_2CrO_4}$ product

6. $CH_3CH—CH—COOEt$ (epoxide, O bridging the two CH)

(A)

$CH_3—CH=CH—COOEt$

(B)

$CH_3CH_2CH_2COOEt$ (C) $\quad CH_3CH_2CH_2CH_2OH$ (D)

7. A : Zn(Hg)/concentrated HCl or N_2H_4/NaOH/Heat.
B : $LiAlH_4$
C : $PhMgBr/H_3O^+$
D : NaOH(dilute)→Heat→N_2H_4/NaOH/Heat
E : $Cl_2/\Delta \rightarrow C_2H_5ONa/C_2H_5OH$
F : CH_3CO_3H
G : CH_2O/NaOH(dilute)/Heat.

8. $(CH_3)_3C—C_6H_4—CHO$ (A)

$(CH_3)_3C—C_6H_4—CH=C(CH_3)—CHO$ (B)

$(CH_3)_3C—C_6H_4—CH_2—CH(CH_3)—CHO$ (C)

9. (a) $Br—CH_2CH_2CH_2CH\langle$ (cyclic acetal $-O-CH_2CH_2-O-$) (A)

(cyclic acetal $-O-CH_2CH_2-O-$)$\rangle CHCH_2CH_2CH_2MgBr$ (B)

$HOC—CH_2CH_2CH_2—CH(OH)—CH_3$ (C)

$CH_3O—CH(OH)—CH_2CH_2—CH=CH—CH_3$ (D)

10. $CH_3CHO + OH^- \longrightarrow \bar{C}H_2CHO \xrightarrow[H_2O]{CH_2O}$

$HOCH_2CH_2CHO \xrightarrow{OH^-} \xrightarrow[H_2O]{CH_2O}$

$\xleftarrow{OH^-} HOCH_2—CH(CH_2OH)—CHO$

$\xrightarrow[H_2O]{CH_2O} HOH_2C—C(CH_2OH)_2—CHO$

$HOCH_2—C(CH_2OH)_2— CH_2OH + HCO\bar{O} \xleftarrow[\bar{HO}]{CH_2O^-}$

Here the first three steps are aldol condensation and the last step is a Cannizzaro reaction.

11. (a) KOH followed by CH_3I

(b) PCC $\longrightarrow CH_3O—C_6H_4—CHO$

(c) CH_3CH_2COOEt/C_2H_5ONa

(d) $LiAlH_4$

12. (a) $CH_2{=}CH{-}CHO$ (A)

$CH_2{=}CH{-}CH(OCH_3){-}OCH_3$ (B)

$CH_2(OH){-}CH(OH){-}CH(OCH_3)_2$ C $\xrightarrow[H_2O]{H^+}$ $CHO{-}CHOH{-}CH_2OH$

Glyceraldehyde

13. $HC{\equiv}C{-}C(OH)(CH_3){-}CH_3$ (A) $CH_3{-}CO{-}C(OH)(CH_3){-}CH_3$ (B)

$C_6H_5CH{=}CH{-}CO{-}C(OH)(CH_3){-}CH_3$ (C)

14. $H_3C{-}CO{-}\overset{-}{C}H{-}COOEt$ + $BrCH_2CH_2CH_2Br$ → $BrCH_2CH_2CH_2{-}CH(COCH_3)(COOEt)$ $\xrightarrow{EtO^-}$

→ $BrCH_2CH_2CH_2{-}\overset{-}{C}(COCH_3)(COOEt)$ → 1-acetyl-1-(ethoxycarbonyl)cyclobutane: $C_4H_6(COCH_3)(COOEt)$

15. (A) cyclohexylidenecyclohexane (C) 2-cyclohexylidenecyclohexanone

(B) cyclohexanone (D) cyclohexane (E) chlorocyclohexane (–Cl)

16. (a) $PhCH{=}CH{-}CO{-}CH_3$ (A)

$CH_3{-}CO{-}CH_2{-}C(OH)(CH_3){-}CH_3$ (B)

(b) $C_6H_5CH{=}CH{-}CO{-}C_6H_5$ (A)

$C_6H_5{-}CO{-}CH{=}C(CH_3){-}C_6H_5$ (B)

(c) $C_6H_5CH{=}CH{-}NO_2$

(d) 3-hydroxycycloheptanone + 1-(2-hydroxycyclopentyl)ethanone (OH)

(e) hexane-2,5-dione

(f) H–CO–$CH_2CH_2CH_2$–CO–CH_3

17. (a) $(CH_3)_3C{-}CO{-}CH_3$ $\xrightarrow{HO^-}$ $\xrightarrow[Aldol]{C_6H_5CHO}$ $(CH_3)_3C{-}CO{-}CH{=}CHC_6H_5$

(b) cyclopentanone $\xrightarrow{HO^-}$ $\xrightarrow[Aldol]{C_6H_5CHO}$ $C_6H_5{-}HC{=}$(2-cyclopentanone)

(c) $CH_3CH_2NO_2 \xrightarrow{HO^-} CH_3\overset{-}{C}H{-}NO_2$

$\xrightarrow{C_6H_5CHO} C_6H_5CH{=}C(CH_3){-}NO_2 \xrightarrow[H_2]{Pd-C}$

$\rightarrow C_6H_5CH_2CH(NH_2)CH_3$

(d) $\xrightarrow[\text{Aldol}]{\text{NaOH}}$

$\xrightarrow{NaBH_4}$ product

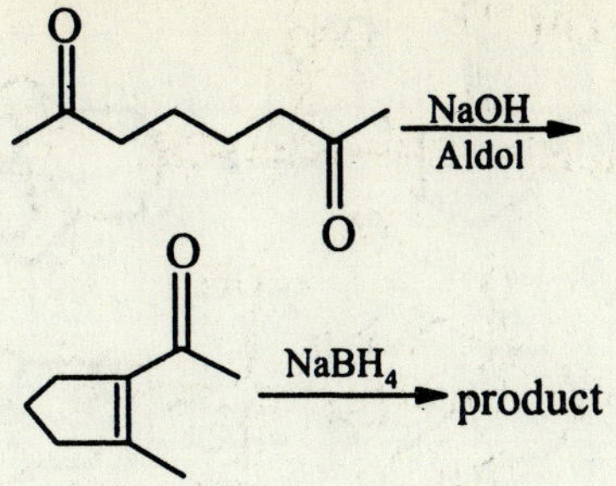

(e) $\xrightarrow{\text{NaOH}}$ $\xrightarrow[\text{Aldol}]{C_6H_5CHO}$

CH—C_6H_5

$\xrightarrow[\text{NaOH/Heat}]{N_2H_4}$ product

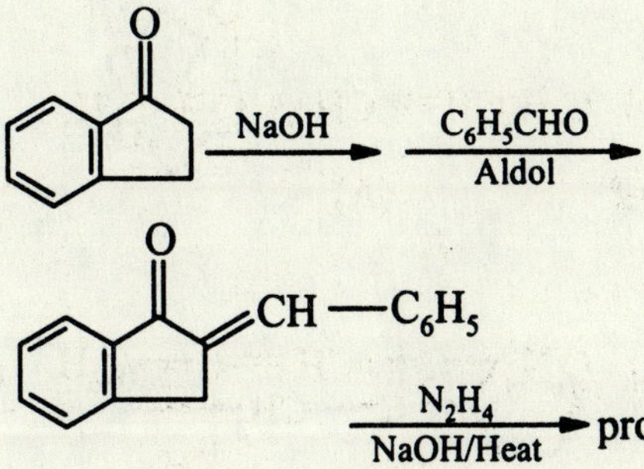

18. $CH_3-C(CH_3)=CH-CH(CH_3)-CH_3$ (A)

$CH_3-CH(CH_3)-COOH$ (B) $CH_3-C(CH_3)(Br)-COOH$ (C)

$CH_3-C(CH_3)(OH)-COOH$ (D)

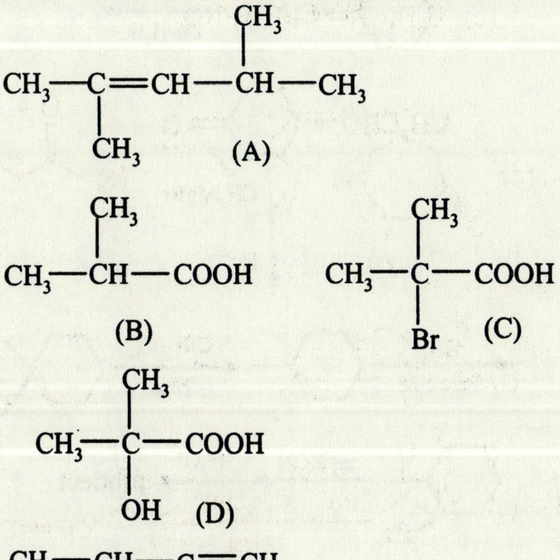

19. $CH_3-CH(CH_3)-C{\equiv}CH$ (A)

$CH_3CH(CH_3)-C{\equiv}C-CH_2CH_2CH_3$ (B)

$CH_3CH(CH_3)-C(=O)-CH_2CH_2CH_2CH_3$ (C)

$CH_3-CH(CH_3)-COOH$ (D) $CH_3CH_2CH_2COOH$ (E)

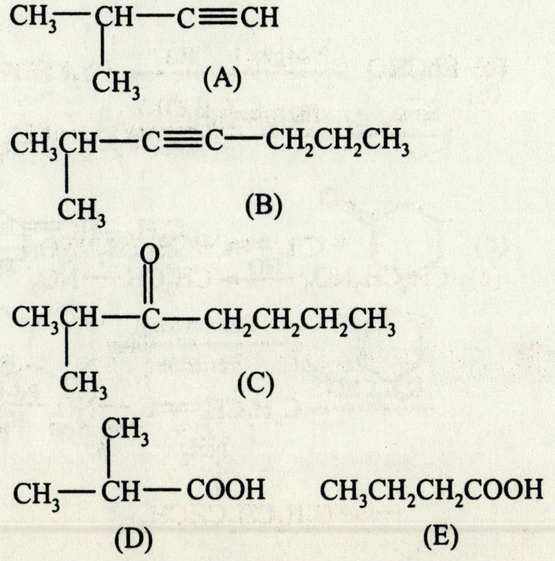

20. (A) (B) Cl Cl (C) (D) CHO

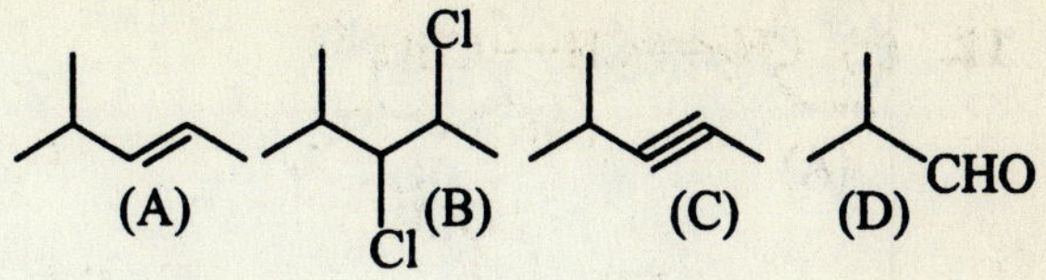

21. $CH_3-C(CH_3)_2-CH(OH)-CH_2CH_3$ (A)

$CH_3-C(CH_3)_2-C(=O)-CH_2CH_3$ (B)

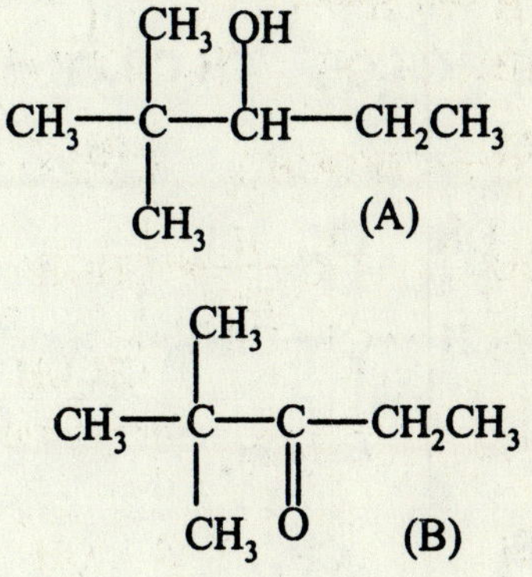

22. C_6H_5CHO (A) $H(C_6H_5)C=N-OH$ (B) $H(C_6H_5)C=N-OH$ (C)

$C_6H_5-C(=O)-NH_2$ (D) $H-C(=O)-NHC_6H_5$ (E)

C_6H_5COOH (F) $C_6H_5NH_2$ (G)

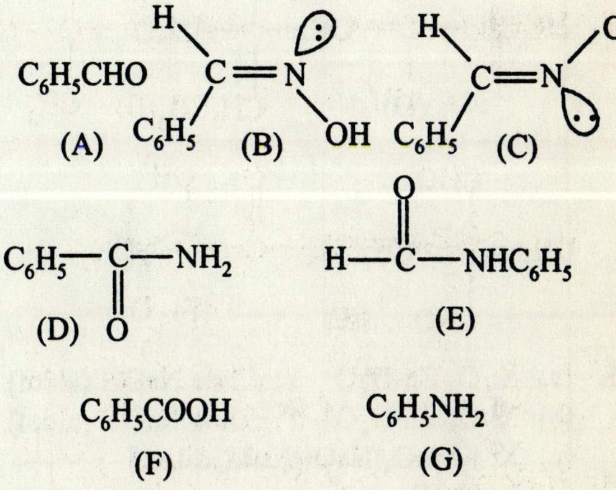

23. (a) CHO (A) O; CH (B) O

=$NNHC_6H_5$ (C) CHO; COOH (D = E)

COOH (F); OH CHO (G)

OH OH (H); CHO (I); MgBr (J)

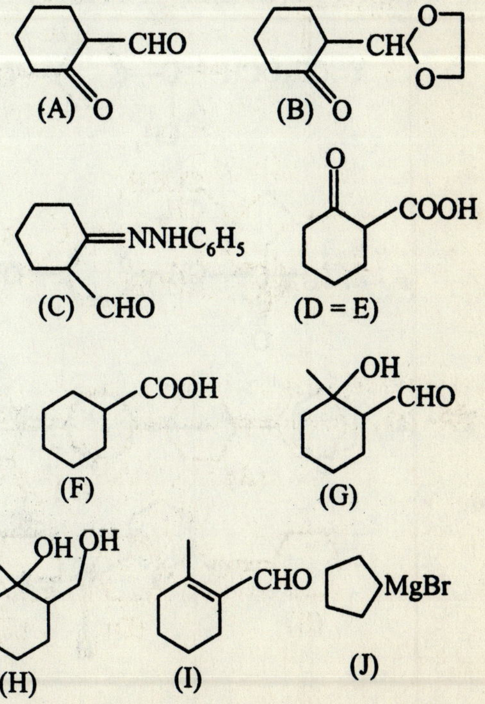

24. 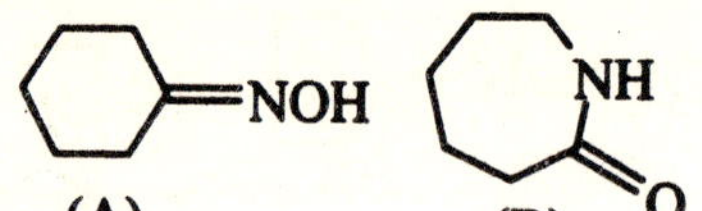 =NOH (A) NH, O (B)

25. 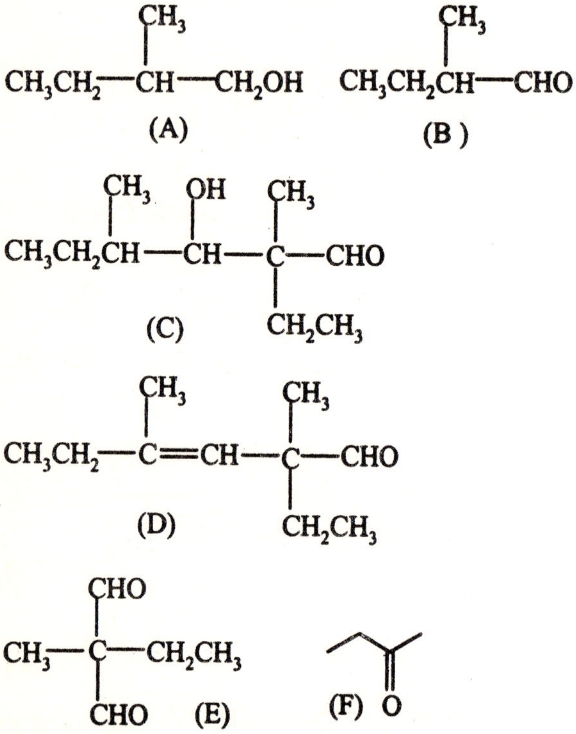 $CH_3CH_2-CH(CH_3)-CH_2OH$ (A) $CH_3CH_2CH(CH_3)-CHO$ (B)

$CH_3CH_2CH(CH_3)-CH(OH)-C(CH_3)(CH_2CH_3)-CHO$ (C)

$CH_3CH_2-C(CH_3)=CH-C(CH_3)(CH_2CH_3)-CHO$ (D)

$CH_3-C(CHO)_2-CH_2CH_3$ (E) (F) O

26. (a) X: O_3/Zn-H_2O Y: dilute NaOH (aldol)
(b) X: O_3/Zn-H_2O Y: dilute NaOH (aldol)
(c) X: $KMnO_4$/NaOH(cold, dilute)
Y: $NaIO_4$
Z: dilute NaOH (aldol)
(d) X: $CH_3OCH_2Cl + Ph_3P$/BuLi

Y: $CH_3OCH=C(CH_3)-C_6H_4-OMe$

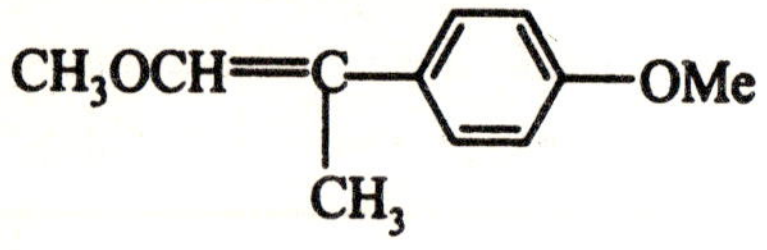

Z: $H-C(=O)-CH(CH_3)-C_6H_4-OHe$

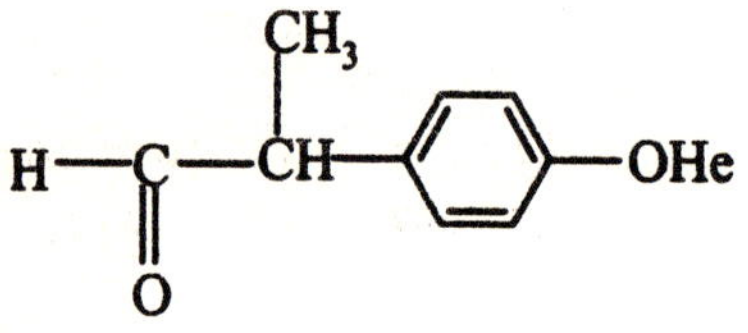

27. (a) 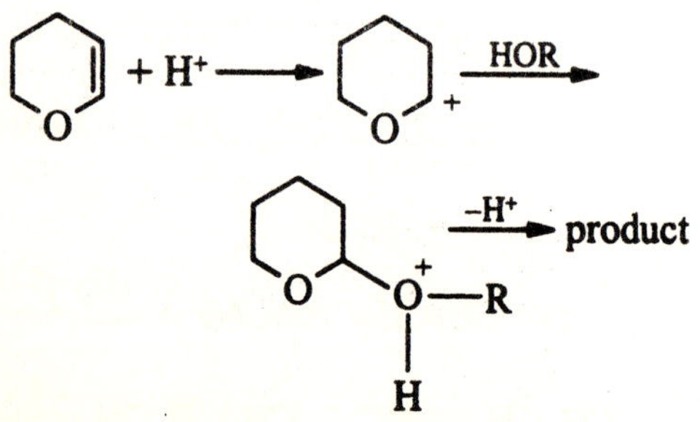 + H^+ → (O, +) $\xrightarrow{HOR}$ (O, $\overset{+}{O}$—R, H) $\xrightarrow{-H^+}$ product

(b) 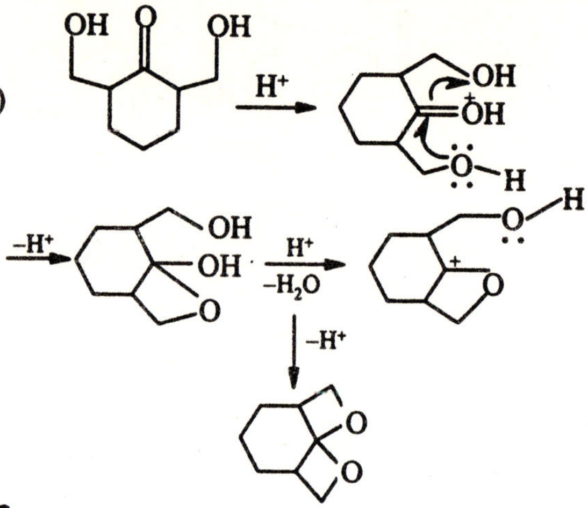 OH O OH $\xrightarrow{H^+}$ OH, $\overset{+}{O}H$, Ö—H $\xrightarrow{-H^+}$ OH, OH, O $\xrightarrow[-H_2O]{H^+}$ +, O, O—H
$\downarrow -H^+$
O, O

Level 2

1. (a) 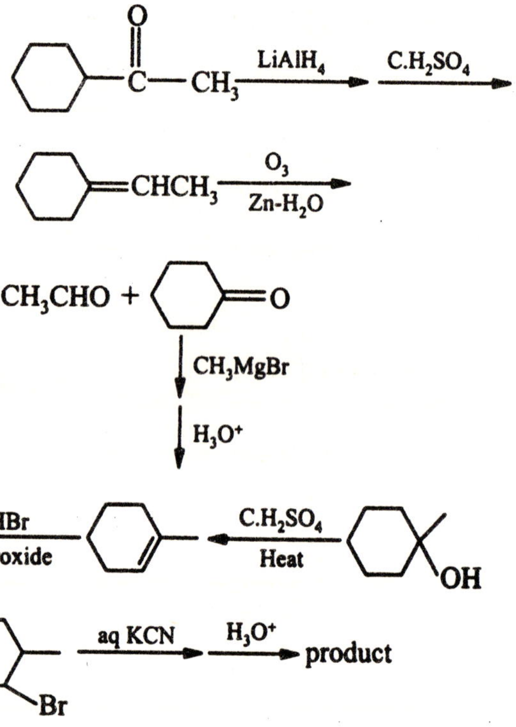 $C_6H_{11}-C(=O)-CH_3 \xrightarrow{LiAlH_4} \xrightarrow{C.H_2SO_4}$

$=CHCH_3 \xrightarrow[Zn-H_2O]{O_3}$

CH_3CHO + =O

$\downarrow CH_3MgBr$

$\downarrow H_3O^+$

$\xleftarrow[Peroxide]{HBr}$ $\xleftarrow[Heat]{C.H_2SO_4}$ OH

Br $\xrightarrow{aq\ KCN} \xrightarrow{H_3O^+}$ product

(b) PhCHO $\xrightarrow{PhMgBr} \xrightarrow{H_3O^+}$ Ph_2CHOH $\xrightarrow{MnO_2} \xrightarrow{Ph_3P=CH_2}$ $Ph_2C=CH_2$

(c) 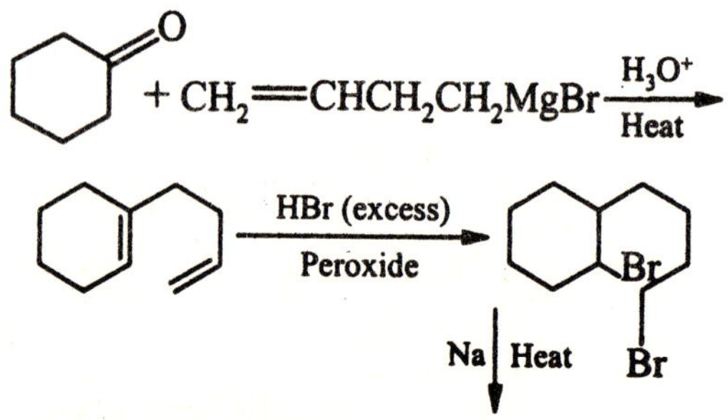 O + $CH_2=CHCH_2CH_2MgBr \xrightarrow[Heat]{H_3O^+}$

$\xrightarrow[Peroxide]{HBr\ (excess)}$ Br, Br

$\downarrow$ Na Heat

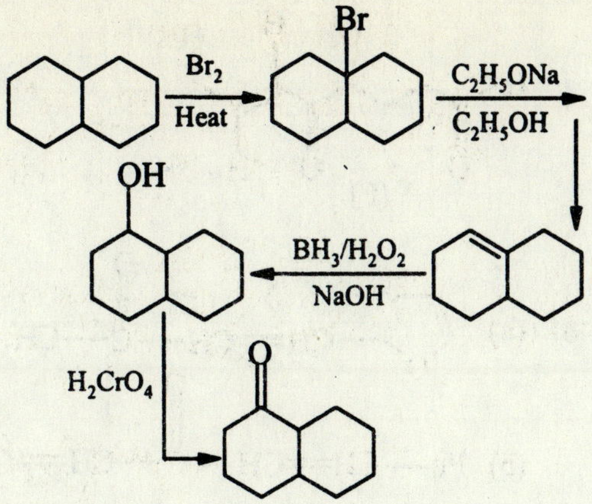

Br_2 / Heat → (Br) → C_2H_5ONa / C_2H_5OH → ; BH_3/H_2O_2 / NaOH → OH ; H_2CrO_4 → O

2. $(CH_2)_9$ O= CH⁻ → C=O ... $\bar{H}C$ =O $(CH_2)_9$ →

$(CH_2)_9$ O= O⁻ =O $(CH_2)_9$ O⁻ $\xrightarrow{H_2O}$ product

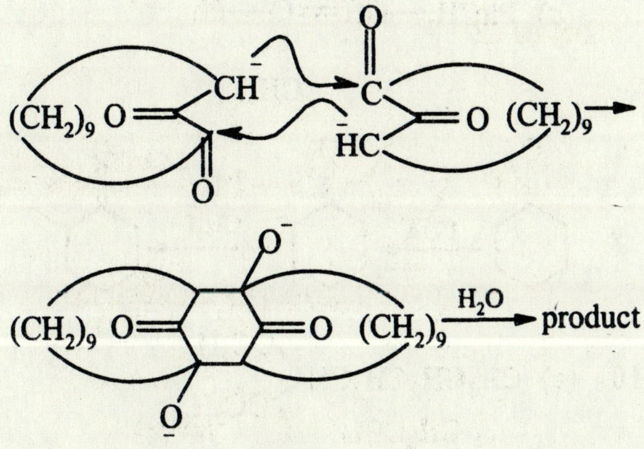

3. (a) 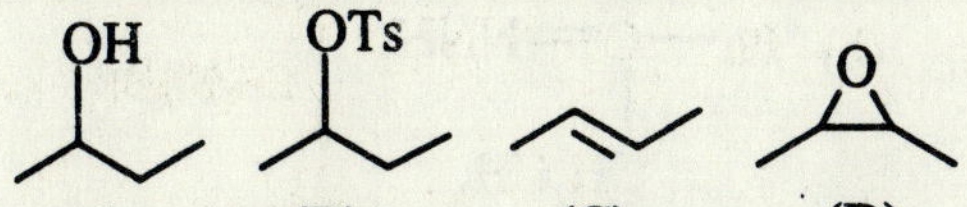 OH (A) OTs (B) (C) O (D)

(b) 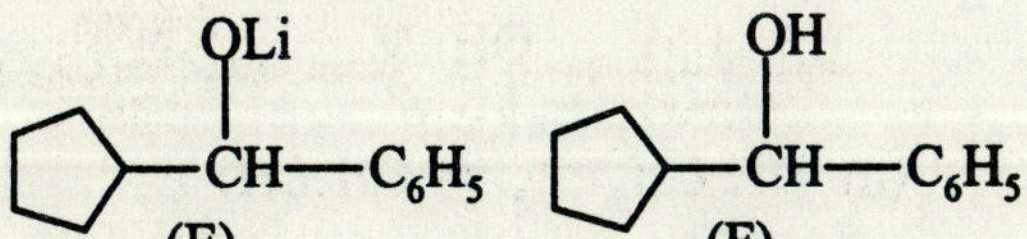 OLi CH—C_6H_5 (E) OH CH—C_6H_5 (F)

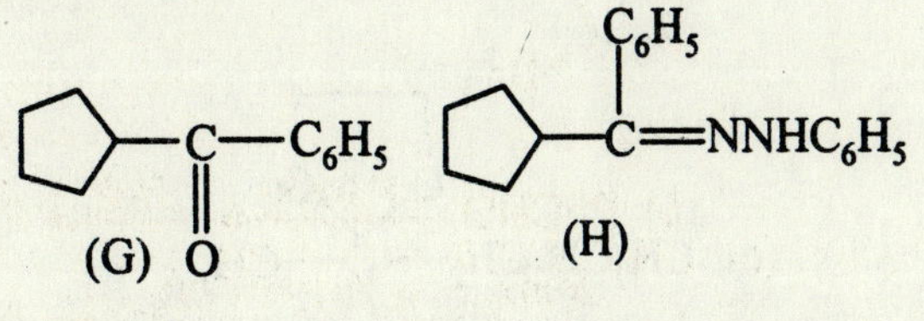

C—C_6H_5 (G) O ; C_6H_5 C=NNHC$_6H_5$ (H)

(c) 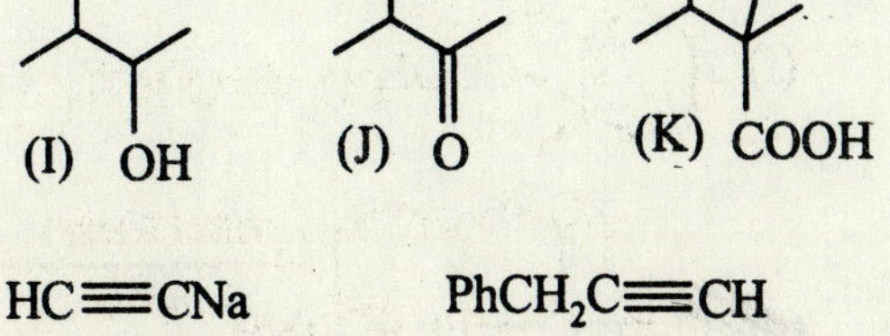 (I) OH ; (J) O ; OH (K) COOH

(d) HC≡CNa (L) ; $PhCH_2C≡CH$ (M)

$PhCH_2C≡CMgBr$ (N)

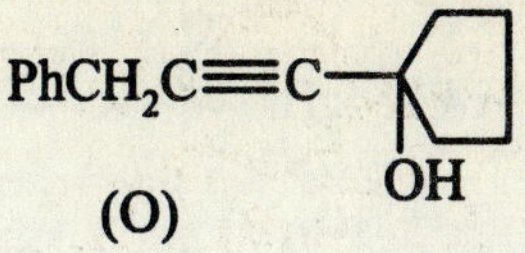

$PhCH_2C≡C$— OH (O) ; (e)  □ (P)

4. (a) 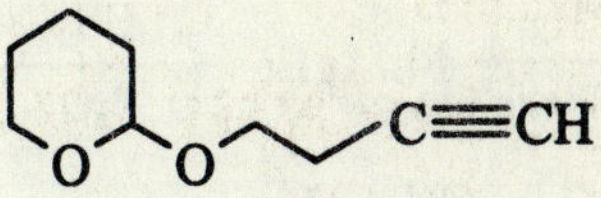 O O C≡CH (A)

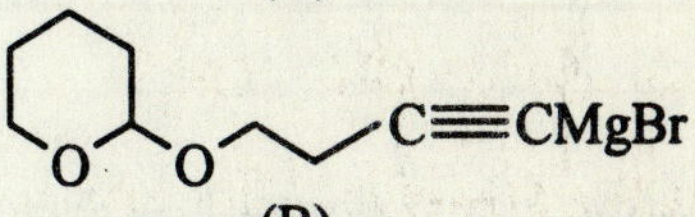

O O C≡CMgBr (B)

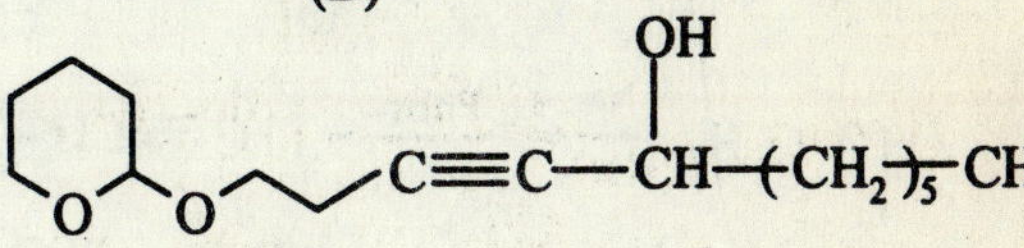

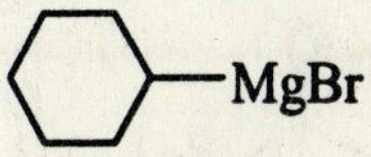

O O C≡C—CH—$(CH_2)_5$—CH_3 OH (C)

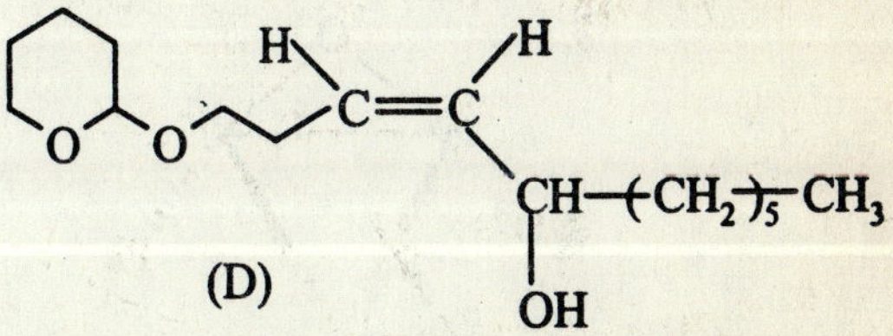

O O H C=C H CH—$(CH_2)_5$—CH_3 OH (D)

(b) —MgBr (E) ; —CH_2CH_2OH (F)

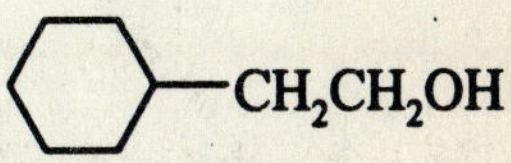

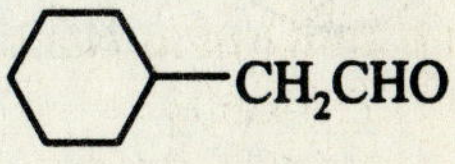

—CH_2CHO (G)

(c) 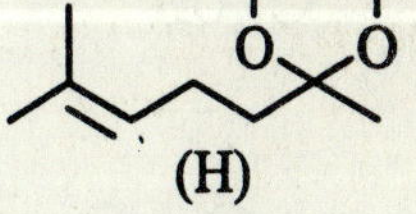 O O (H) ; O O OH (I)

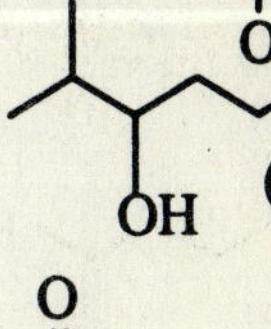

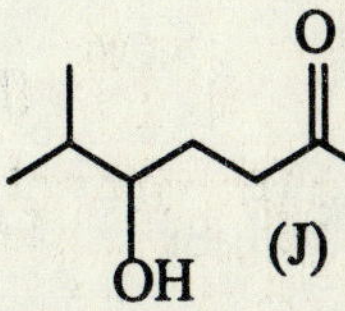

O OH (J)

5. (a) 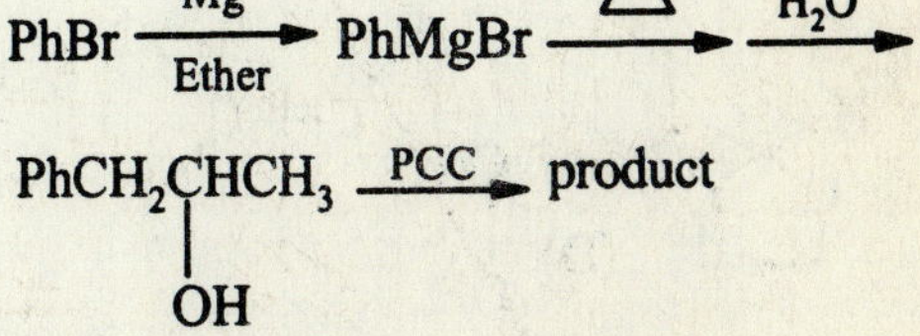 PhBr $\xrightarrow[\text{Ether}]{\text{Mg}}$ PhMgBr $\xrightarrow{\text{(ethylene oxide)}}$ $\xrightarrow{H_2O}$

$PhCH_2CHCH_3$ (OH) $\xrightarrow{PCC}$ product

(b) 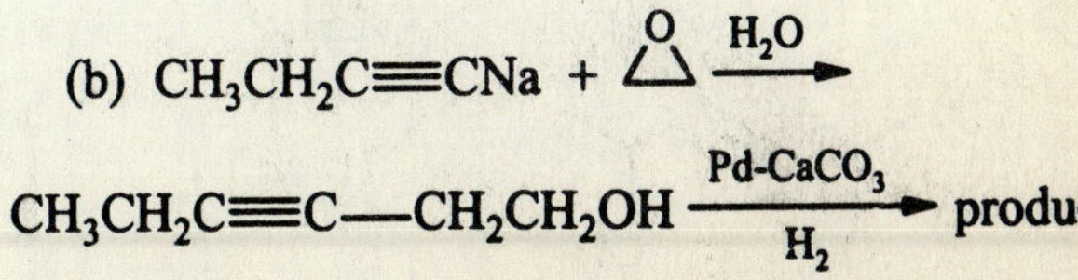 $CH_3CH_2C≡CNa$ + (O) $\xrightarrow{H_2O}$

$CH_3CH_2C≡C—CH_2CH_2OH$ $\xrightarrow[H_2]{Pd\text{-}CaCO_3}$ product

(c) 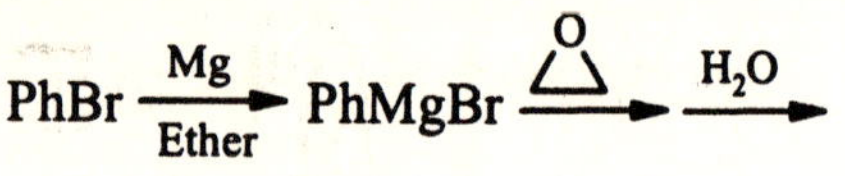 $PhBr \xrightarrow[\text{Ether}]{Mg} PhMgBr \xrightarrow{\text{(ethylene oxide)}} \xrightarrow{H_2O}$

$PhCH_2CH_2OH \xrightarrow{PBr_3} PhCH_2CH_2Br$

(d) $PhBr \xrightarrow[\text{Ether}]{Mg} PhMgBr \xrightarrow[H_3O^+]{CH_3CH_2CN}$

$Ph—C(=O)—CH_2CH_3$

(e) Product from (d) $\xrightarrow[OH^-]{N_2H_4} PhCH_2CH_2CH_3$

(f) $C_2H_2 \xrightarrow[\text{Heat}]{Na} \xrightarrow{PhBr} PhC{\equiv}CH \xrightarrow[\text{Heat}]{Na}$

$\xrightarrow{PhBr} PhC{\equiv}CPh \xrightarrow[\text{liq } NH_3]{Na} \xrightarrow{RCO_3H}$

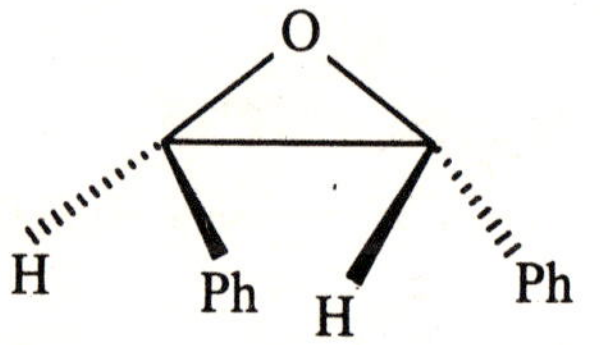

O

H Ph H Ph

6. 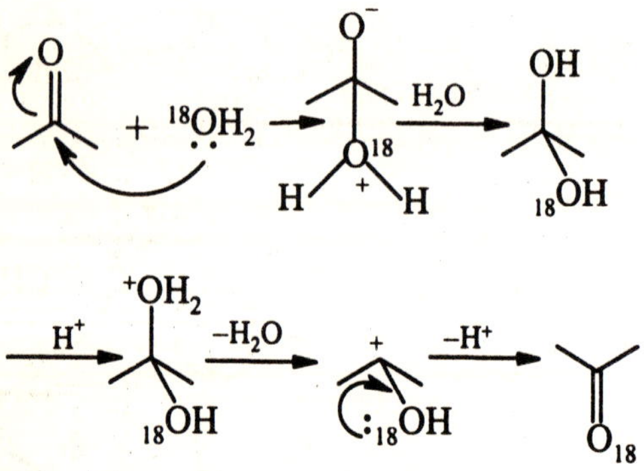 $+ {}^{18}OH_2 \rightarrow$ (O⁻, O^{18}, H + H) $\xrightarrow{H_2O}$ (OH, $_{18}OH$)

$\xrightarrow{H^+}$ ($^+OH_2$, $_{18}OH$) $\xrightarrow{-H_2O}$ (+, $:_{18}OH$) $\xrightarrow{-H^+}$ (O_{18})

7. 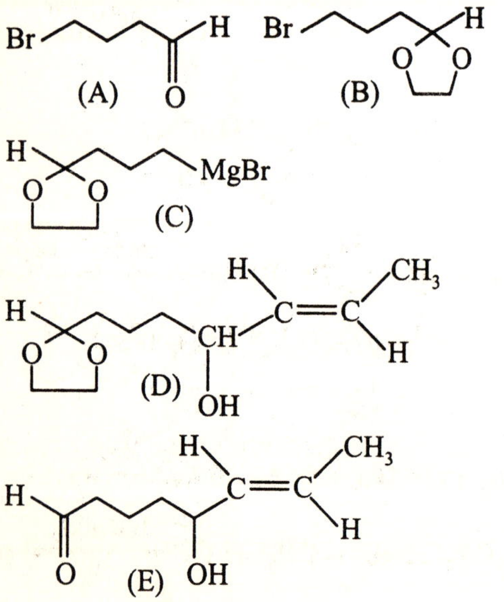 Br–CH₂CH₂CH₂–CHO (A); Br–CH₂CH₂CH₂–CH(dioxolane) (B)

H–C(dioxolane)–CH₂CH₂CH₂–MgBr (C)

(D): H–C(dioxolane)–CH₂CH₂CH₂–CH(OH)–CH=CH–CH₃ (H, CH₃, H)

(E): OHC–CH₂CH₂CH₂–CH(OH)–CH=CH–CH₃ (H, CH₃, H)

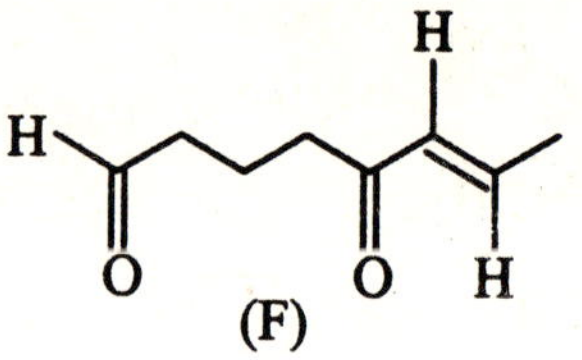

(F): OHC–CH₂CH₂CH₂–C(=O)–CH=CH–CH₃ (H, H)

8. (a) 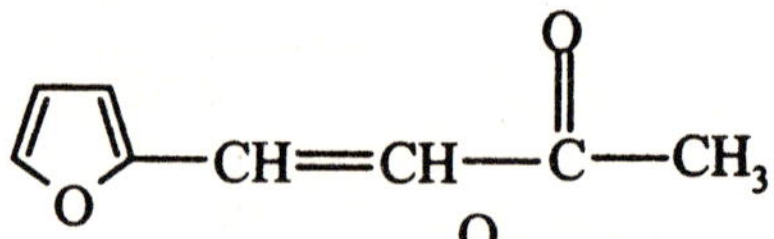 (2-furyl)$—CH{=}CH—C(=O)—CH_3$

(b) $Ph—CH{=}CH—C(=O)—CH{=}CH—Ph$

(c) $PhCH_2—CH{=}C(CHO)—Ph$

9. 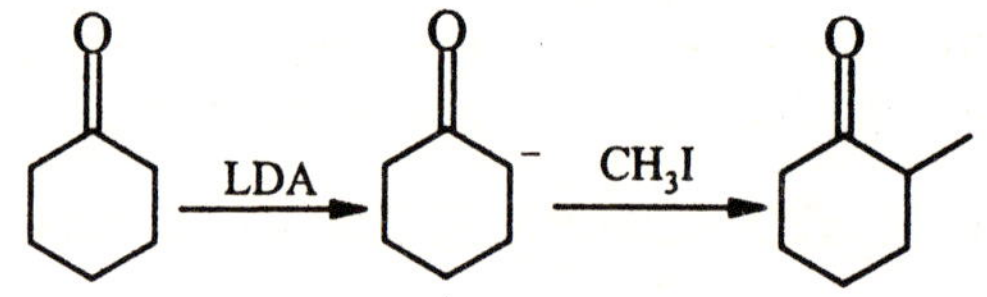 (cyclohexanone) $\xrightarrow{LDA}$ (enolate)⁻ $\xrightarrow{CH_3I}$ (2-methylcyclohexanone)

10. (a) $CH_3CH_2CH_2CH(OC_2H_5)_2$

(b) $Ph—C(=NNH_2)—CH_2CH_3$

(c) $PhCH_2CH_2CH_3$

(d) $CH_3CH_2—C(OH)(COOH)—CH_2CH_3$

(e) 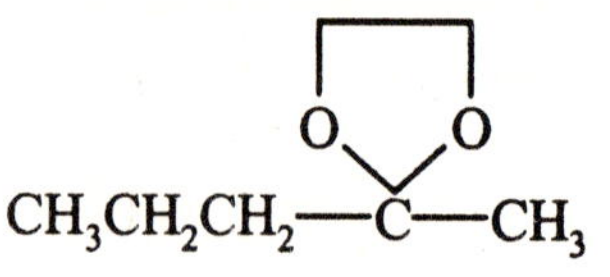 $CH_3CH_2CH_2—C(O–CH_2CH_2–O)—CH_3$

(f) 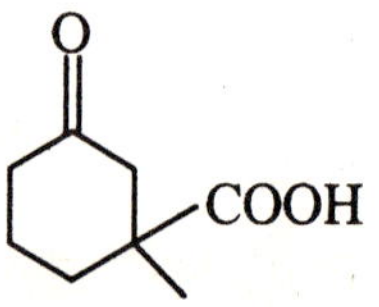 (3-oxocyclohexyl with methyl)–COOH

(g) $Ph—C(=O)—Et + EtNH_2$

(h) 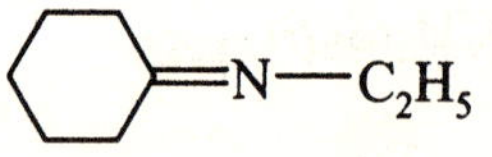 (cyclohexylidene)$=N—C_2H_5$

(i) 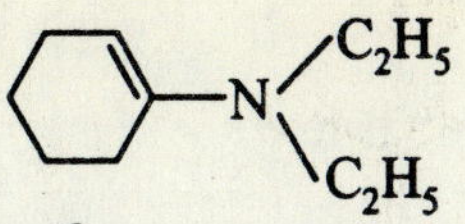 1-(N,N-diethylamino)cyclohexene: $C_6H_9{-}N(C_2H_5)_2$

(j) 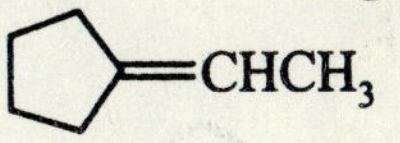 $\text{cyclopentylidene}{=}CHCH_3$ (k)  pyrrolidine (N–H)

11. II > VI > I > IV > V > III

12. (a) Cyclohexanone $\xrightarrow{LiAlH_4}$ product

(b) Product of 'a' $\xrightarrow[\text{Heat}]{H_2SO_4}$ product

(c) Product of 'a' $\xrightarrow{PBr_3}$ product

(d) Product of 'c' $\xrightarrow{Li}$ $\xrightarrow{(CH_2{=}CH)_2Cu}$ product

(e) Product of 'c'+ NH_3 → product

(f) Cyclohexanone + Zn(Hg) $\xrightarrow{HCl}$ product

(g) Product of 'c'+ KCN(aq) →

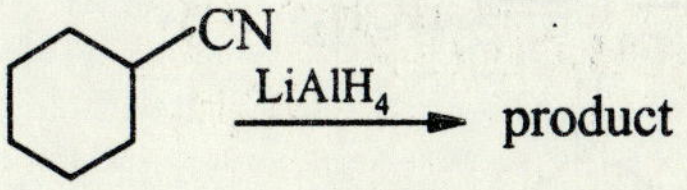

cyclohexyl–CN $\xrightarrow{LiAlH_4}$ product

(h) Product of 'd' $\xrightarrow[Ni]{H_2}$ product

(i) Product of 'e'+ $2CH_3I$ → product

13. *p*-nitroacetophenone > *p*-chloroacetophenone > $PhCOCH_3$ > *p*-methoxyacetophenone.

14. Steric hindrance and electron-donating group bonded to carbonyl carbon lowers the reactivity.

15. Chloral hydrate is stabilized by intramolecular hydrogen bonding.

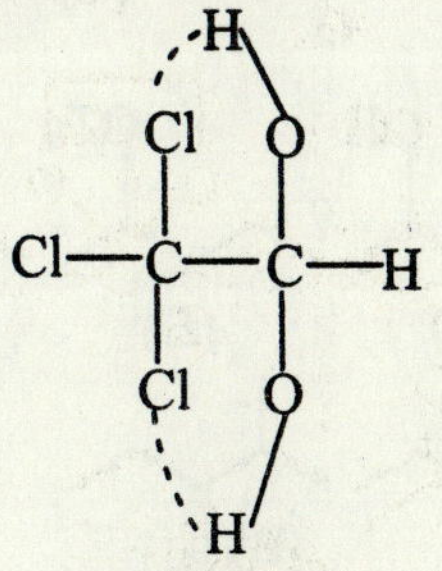

$Cl_3C{-}CH(OH)_2$ (O–H···Cl hydrogen bonds)

16. (a) A: PBr_3 B: Mg/Ether C: CH_3MgBr
D: CH_2O E: H_3O^+

(b) A: $Br_2/h\nu$ B: Mg/Ether C: CH_3MgBr
D: Oxirane E: H_3O^+

(c) A: PBr_3 B: Mg/Ether
C: $CH_3CH_2CH_2MgBr$

D: $CH_3CH_2CH_2{-}C(OH)(CH_3){-}CH_3$

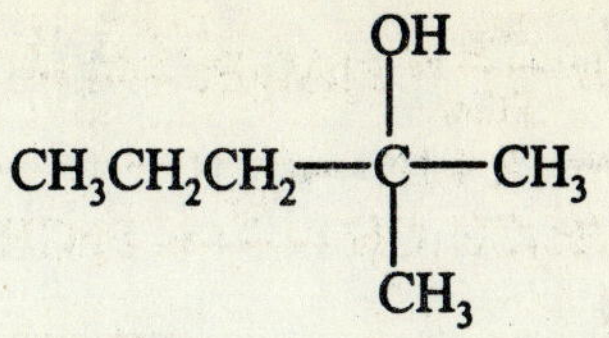

17. 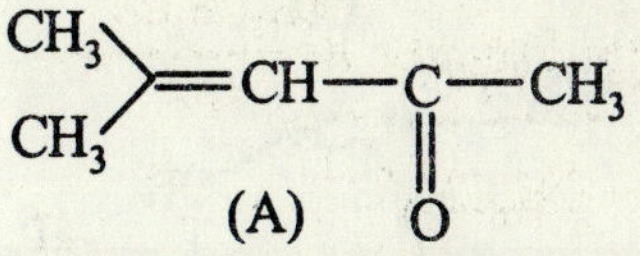 $(CH_3)_2C{=}CH{-}C(=O){-}CH_3$ (A)

$CH_2{=}CH{-}C(CH_3)_2{-}CH_2{-}C(=O){-}CH_3$ (B)

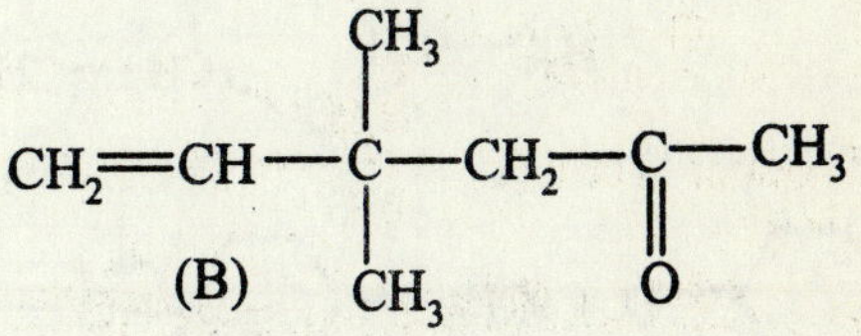

18. (a) 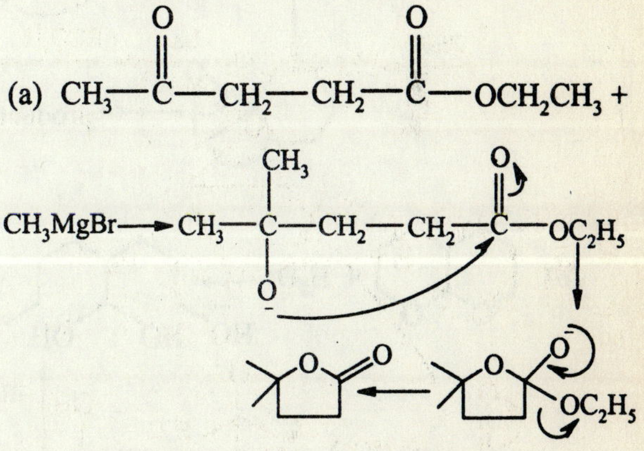 $CH_3{-}C(=O){-}CH_2{-}CH_2{-}C(=O){-}OCH_2CH_3$ +

$CH_3MgBr \rightarrow CH_3{-}C(CH_3)(O^-){-}CH_2{-}CH_2{-}C(=O){-}OC_2H_5$ → cyclic intermediate (O^-, OC_2H_5) → 5,5-dimethyl lactone

(b) 2-acetylbenzoic acid (COOH) $\xrightarrow{H^+}$ protonated ketone ($\overset{+}{O}{-}H$, $C{-}CH_3$; attack by $\ddot{O}{-}H$ of $C(=O)$) → HO…$\overset{+}{O}{-}H$ cyclic lactone $\xrightarrow{-H^+}$ product

(c) $C_6H_5COCH_2CH_3$ + N_2H_4 → $C_6H_5C(C_2H_5){=}N{-}NH_2$ $\xrightarrow{OH^-,\ \text{Heat}}$ $C_6H_5CH_2CH_2CH_3$

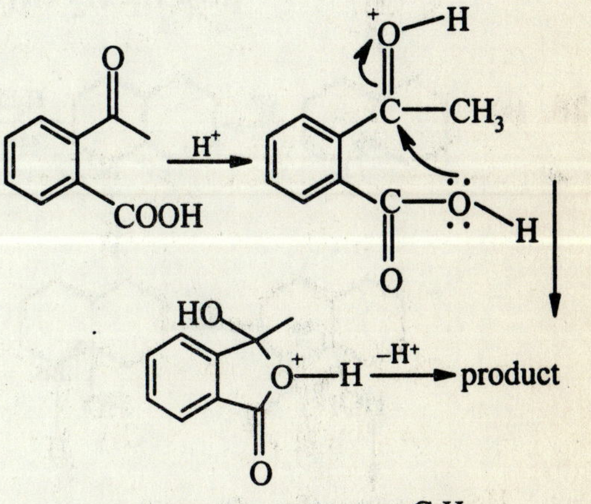

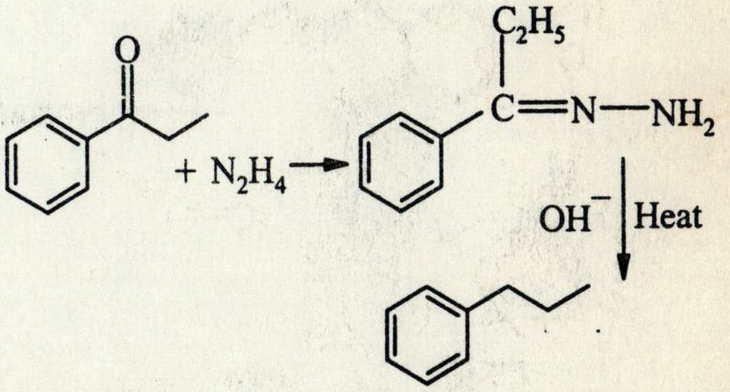

(d) $CH_3-\overset{O}{\overset{\|}{C}}-H + H^+ \longrightarrow CH_3-\overset{\overset{+}{O}H}{\overset{\|}{C}}-H \xrightarrow{CH_3CHO}$

$H_3C-\overset{+}{C}H-O-CH(CH_3)-OH \xrightarrow{CH_3CHO}$ (cyclic intermediate: CH_3, CH, OH, O, H_3C-CH, O, $\overset{+}{C}H-CH_3$) $\xrightarrow{-H^+}$ product

(cyclohexanone with $\overset{+}{O}Me$) ↕

(1-hydroxy-1-methoxycyclohexane, HO, $\overset{+}{O}Me$, H) $\longrightarrow$ (cyclohexanone, $H-\overset{+}{O}$) $+ CH_3OH$ $\xrightarrow{-H^+}$ product

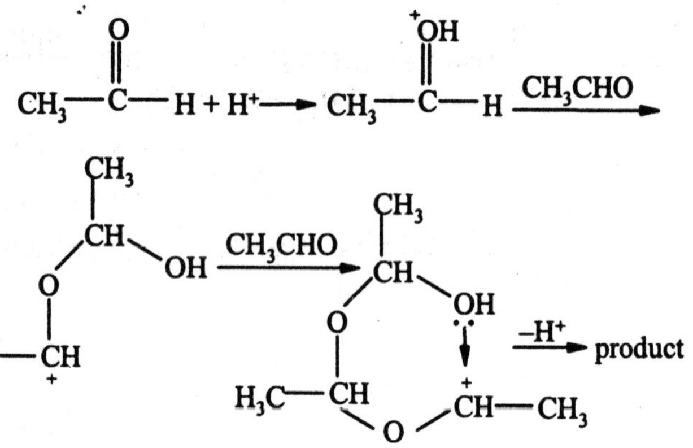

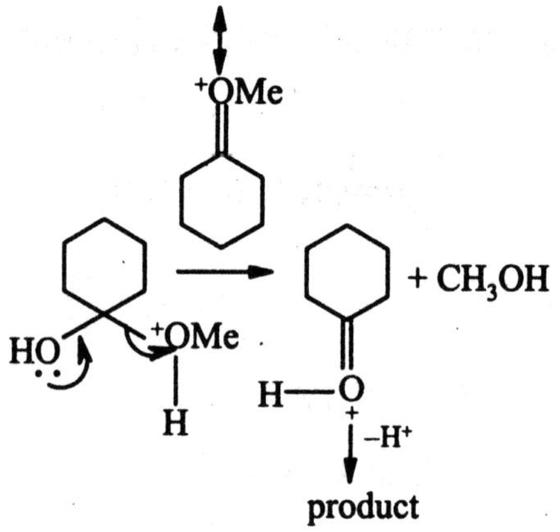

19. (a) (cyclohexanone)$=O + :CH_2 \longrightarrow$ (cyclohexane)$=\overset{+}{O}-\overset{-}{C}H_2$

↓

(spiro-epoxide, O) $\longrightarrow$ product

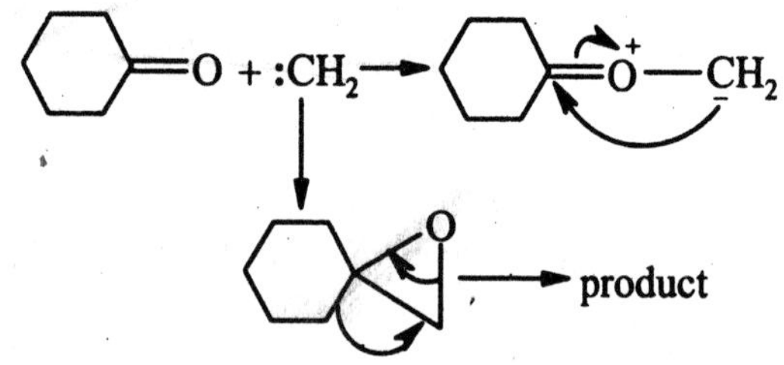

(b) (bicyclic acetal, O, O) $+ H_3O^+ \longrightarrow$ (HO HO OH OH)

$\downarrow -H_2O$

$(HOCH_2CH_2CH_2)_2CHCHO$

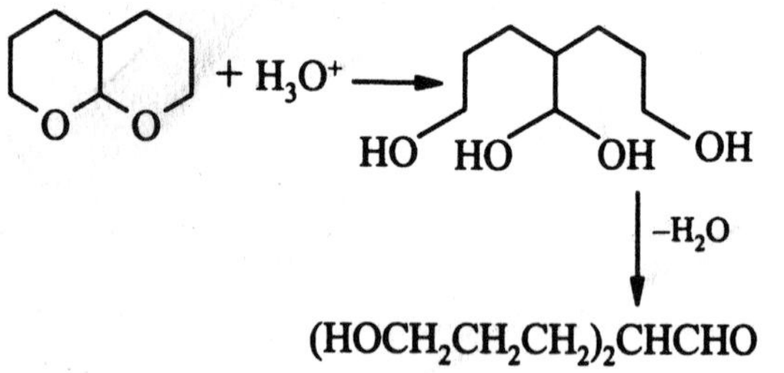

20. (a) (octahydroquinoline, N) $\xrightarrow{H^+}$ (iminium, $\overset{+}{N}$–H) $\xrightarrow[-H^+]{H_2O}$

(HO:, $\overset{+}{N}$, H, H) $\xleftarrow{H^+}$ (HO, N, H)

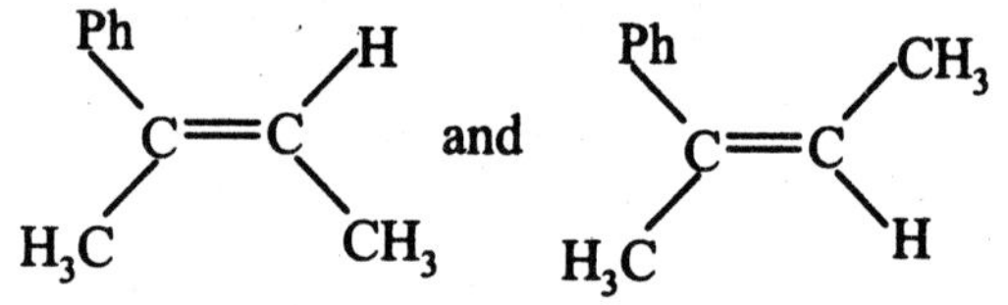

↓

(2-(3-aminopropyl)cyclohexanone, NH_2, $\overset{+}{O}$–H) $\xrightarrow{-H^+}$ product

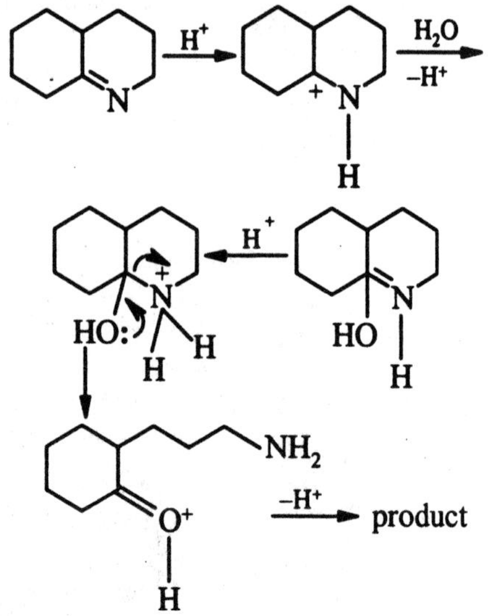

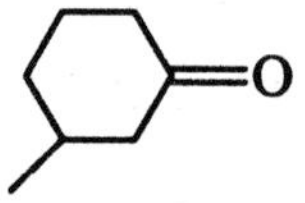

(b) (1-methoxycyclohexene, OMe) $\xrightarrow{H^+}$ (OMe, cation +) $\xrightarrow[-H^+]{H_2O}$ (HO, OMe) $\xrightarrow{H^+}$

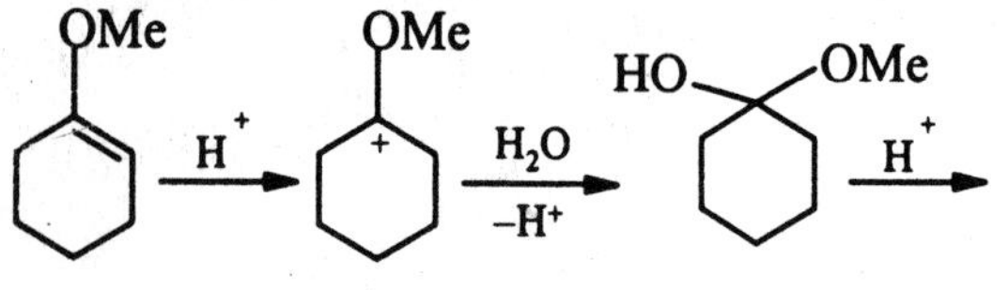

21. (a) $Ph(H_3C)C=C(H)CH_3$ and $Ph(H_3C)C=C(CH_3)H$

(b) (3-methylcyclohexanone, =O) (c) (4-methyl-1-methylcyclohexanol, OH)

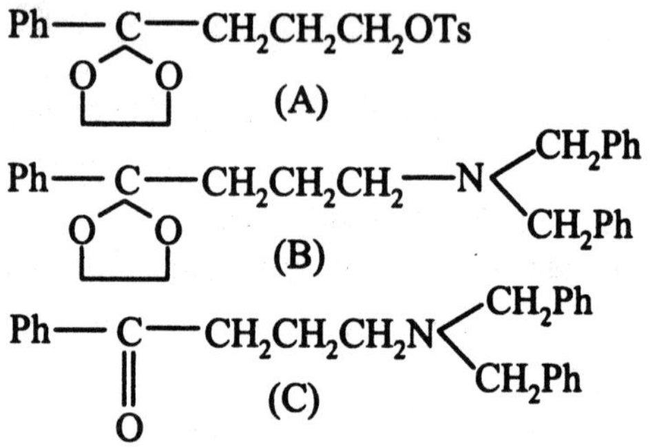

(d) $Ph-C(=CHCH_3)-N$ (pyrrolidine)

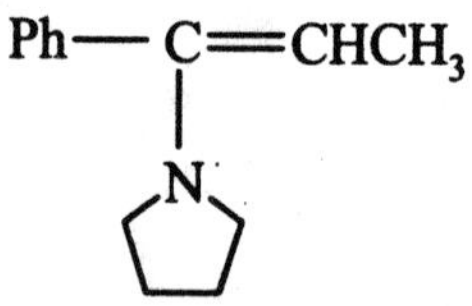

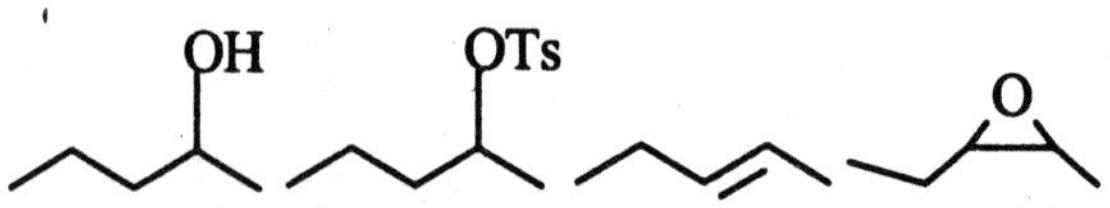

22. (a) $Ph-C(OCH_2CH_2O)-CH_2CH_2CH_2OTs$ (A)

$Ph-C(OCH_2CH_2O)-CH_2CH_2CH_2-N(CH_2Ph)_2$ (B)

$Ph-\overset{O}{\overset{\|}{C}}-CH_2CH_2CH_2N(CH_2Ph)_2$ (C)

(b) (OH) (A) (OTs) (B) (C) (O) (D)

(c) (A) OH (B) O

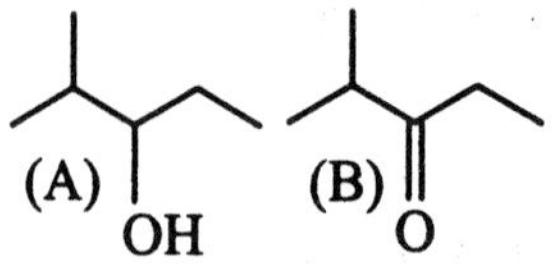

(d) (cyclopentyl)$-CH(OH)-Ph$ (A) (cyclopentyl)$-\overset{O}{\overset{\|}{C}}-Ph$ (B)

(cyclopentyl)$-C(Ph)=NNHPh$ (C)

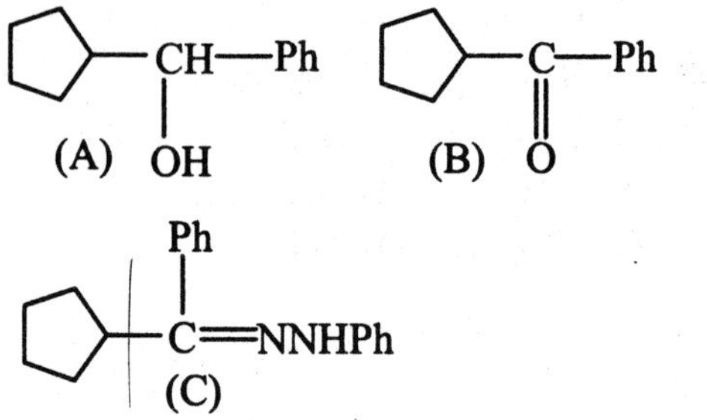

(e) HC≡CNa (D) $C_6H_5CH_2C\equiv CH$ (E)

$$E \xrightarrow{C_2H_5MgBr} C_2H_6 + C_6H_5CH_2C\equiv CMgBr$$

(f) (X) cyclobutanone hydrazone ($=NNH_2$) $\xrightarrow[EG]{KOH}$ cyclobutane

23. (a) $Ph-\overset{O}{\overset{\|}{C}}-OMe$ + MeMgBr (excess) $\xrightarrow{H_3O^+}$ product

(b) $Ph-\overset{O}{\overset{\|}{C}}-OMe + H_3O^+ \rightarrow PhCOOH \xrightarrow{LiAlH_4} PhCH_2OH \xrightarrow{PCC} PhCHO$

(c) $Ph-C(CH_3)(Br)-OMe \xrightarrow{H_3O^+} Ph-C(CH_3)(OH)-OH \xrightarrow{-H_2O} Ph-\overset{O}{\overset{\|}{C}}-CH_3$

24. (i) cyclohexanone $\xrightarrow{LiAlH_4}$ $\xrightarrow[Heat]{H_2SO_4}$ cyclohexene $\xrightarrow[Zn-H_2O]{O_3}$

$\xrightarrow[aldol]{NaOH}$ 2-hydroxycyclopentanecarbaldehyde (CHO, OH) $\xrightarrow{H_2CrO_4}$ 2-oxocyclopentanecarboxylic acid (COOH, =O) $\xrightarrow{Heat}$ cyclopentanone $\xrightarrow{LiAlH_4}$ Product (A)

(ii) 2-oxocyclopentanecarboxylic acid (COOH, =O) from I $\xrightarrow[HCl]{Zn(Hg)}$ cyclopentanecarboxylic acid (COOH) $\xrightarrow{LiAlH_4}$ cyclopentylmethanol (CH_2OH) (B)

(iii) cyclohexanone + CH_3CO_3H → caprolactone (seven-membered ring, O, =O) $\xrightarrow{H_3O^+}$ $HO\!\diagup\!\diagdown\!\diagup\!\diagdown\!\diagup COOH$ $\xrightarrow{CH_3OH/H^+/Heat}$ $HO\!\diagup\!\diagdown\!\diagup\!\diagdown\!\diagup COOMe$ $\xrightarrow{NaH}$ $\xrightarrow{C_2H_5I}$ Product (C)

25. (a) cyclohexylidene=NHN–phenyl (b) 3-methylcyclopentane-1,2-diol (OH, OH)

(c) cyclohex-3-ene–CHO (d) H_3C–cyclopentane (H, OH)

(e) $C_6H_5-C(OH)(CH_3)-CH_2CH_2CH_2CH_3$

26. (a) cyclopentyl–NH_2 (b) HO–(bicyclic ring with acetonide, O, O)

(c) $CH_3-\overset{O}{\overset{\|}{C}}$–cyclohexane–$\overset{O}{\overset{\|}{C}}-NHCH_3$

(d) 3,5-dimethylcyclohexane–CHO

27. (a) benzene (CHO, CH_2Br) $\xrightarrow[H^+]{EG}$ benzene (CH(dioxolane), CH_2Br)

$\xrightarrow[H_2O]{KCN}$ $\xrightarrow[HCl]{Sn}$ benzene (CH(dioxolane), CH_2CHO)

$\xrightarrow[(ii)\ H_3O^+]{(i)\ LiAlH_4}$ benzene (CHO, CH_2CH_2OH) $\xrightarrow{H^+}$ product

(b) 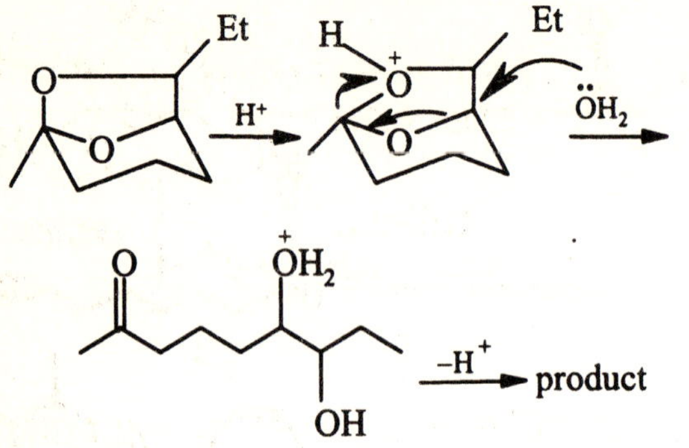 Et; H^+; Et; $\ddot{O}H_2$

$\overset{+}{O}H_2$; OH; $-H^+$ product

28. (a) 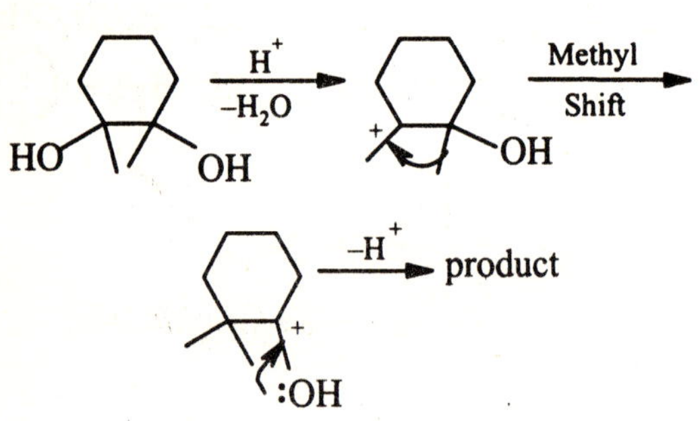 HO, OH; H^+ $-H_2O$; OH; Methyl Shift

$-H^+$ product; :OH

(b) 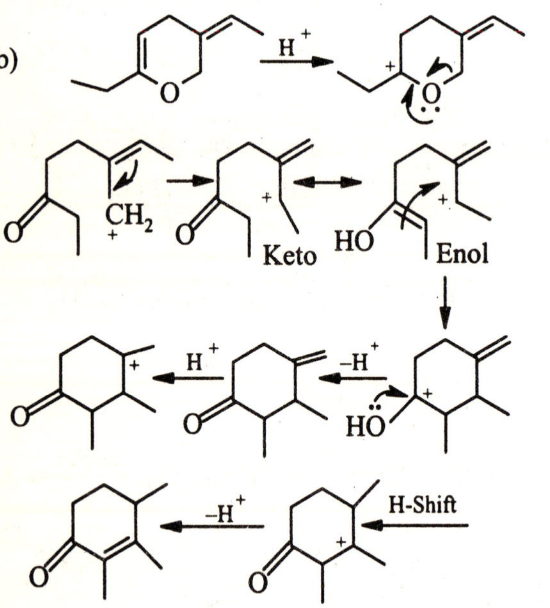 H^+

CH_2; Keto; HO; Enol

H^+; $-H^+$; HÖ

$-H^+$; H-Shift

(c) 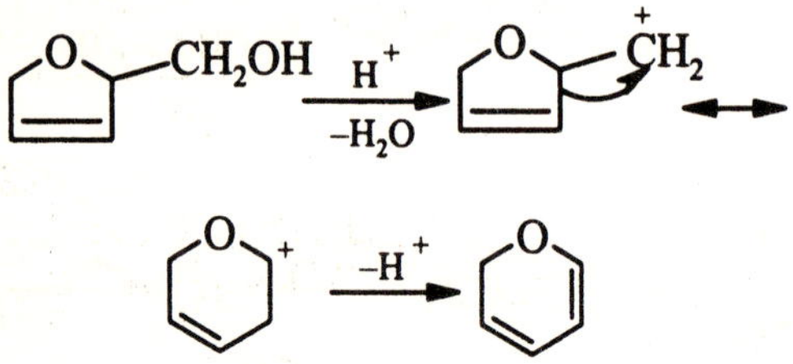 CH_2OH; H^+ $-H_2O$; $\overset{+}{C}H_2$

$-H^+$

30. (a) 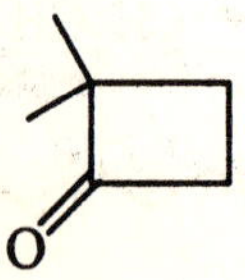 (b)

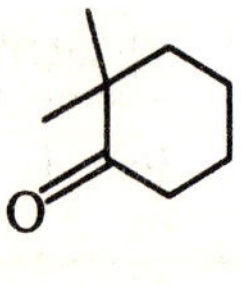

31. (a) 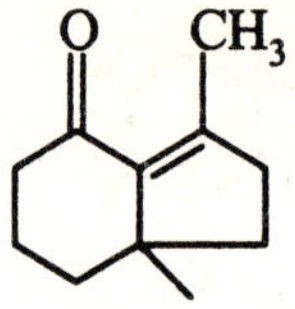 O, CH_3 (b) 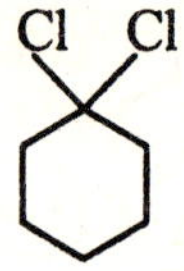 Cl Cl

32. (a) 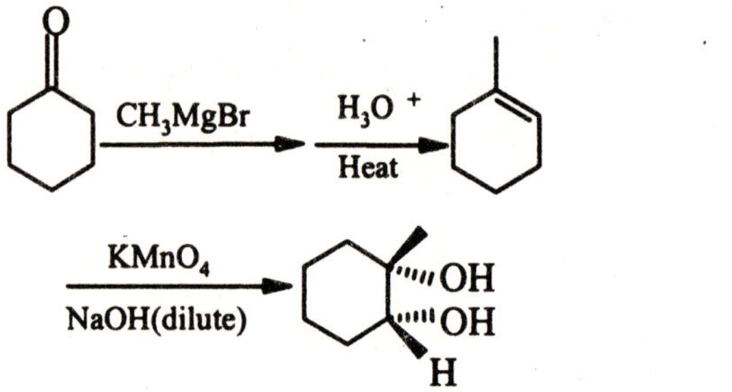 CH_3MgBr; H_3O^+ Heat

$KMnO_4$ NaOH(dilute); OH; OH; H

(b) 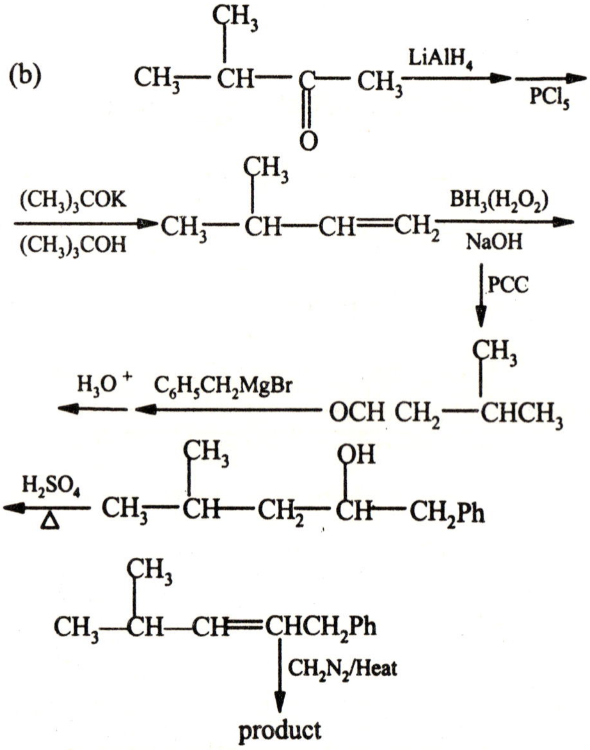 $CH_3-\underset{\displaystyle CH_3}{CH}-\underset{\displaystyle O}{\overset{\|}{C}}-CH_3$ $\xrightarrow{LiAlH_4}$ $\xrightarrow{PCl_5}$

$\xrightarrow[(CH_3)_3COH]{(CH_3)_3COK}$ $CH_3-\underset{\displaystyle CH_3}{CH}-CH=CH_2$ $\xrightarrow[NaOH]{BH_3(H_2O_2)}$

PCC

$\xleftarrow{H_3O^+}\xleftarrow{C_6H_5CH_2MgBr}$ OCH $CH_2-\underset{\displaystyle CH_3}{CH}CH_3$

$\xleftarrow[\Delta]{H_2SO_4}$ $CH_3-\underset{\displaystyle CH_3}{CH}-CH_2-\underset{\displaystyle OH}{CH}-CH_2Ph$

$CH_3-\underset{\displaystyle CH_3}{CH}-CH=CHCH_2Ph$

CH_2N_2/Heat

product

33. (a) 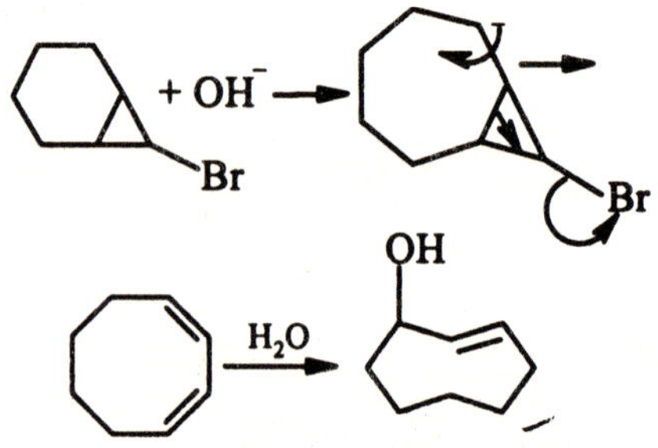 + OH^-; Br; Br

H_2O; OH

(b) 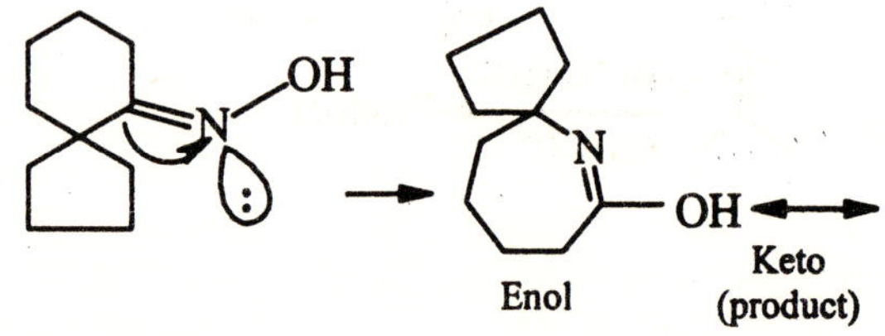 OH; N; OH; Enol; Keto (product)

35. (a) (A) (B) (C)

OH, OCH_3, OCH_3

(D) (E)

I, COOH

E $\xrightarrow{\text{Heat}}$ $\xrightarrow[\text{NaOH}]{\text{Dilute}}$

36. (a) $ClCH_2—COOEt \xrightarrow[C_2H_5OH]{C_2H_5ONa}$

$Cl\bar{C}H—COOEt \xrightarrow{(H_3C)_2C=O}$

$CH_3—\underset{\underset{O^-}{|}}{\overset{CH_3}{\overset{|}{C}}}—\underset{|}{CH}(Cl)—COOEt$

$\downarrow$

$CH_3—\overset{CH_3}{\overset{|}{C}}—CH—COOEt$ (epoxide, bridged by O)

(b) $H_3C(C=O)CH_3 \xrightarrow{PhCH_2MgBr} CH_3—\underset{\underset{OMgBr}{|}}{\overset{CH_3}{\overset{|}{C}}}—CH_2—Ph$

$\xrightarrow[\text{Heat}]{H^+} CH_3—\overset{CH_3}{\overset{|}{C}}=CH—Ph$

(c) $H_3C(C=O)CH_3 \xrightarrow[\text{Heat}]{Cl_2} CH_3(C=O)CH_2Cl \xrightarrow{aqKCN} CH_3(C=O)CH_2CN$

(d) $(CH_3)_2C=O \xrightarrow[(ii)\ H_3O^+]{(i)\ PhMgBr}$ Product

(e) 2 CH_3COCH_3 + dilute NaOH $\longrightarrow$ $CH_3COCH_2C(CH_3)_2OH$

$\xrightarrow{H^+/\text{Heat}}$ $CH_3COCH=C(CH_3)_2$

37. $CH_3CH_2CH_2—\overset{O}{\overset{\|}{C}}—O—\overset{CH_3}{\overset{|}{CH}}—CH_2CH_3$

(A)

$CH_3CH_2\underset{\underset{OH}{|}}{CH}—CH_3$ (B) $CH_3CH_2CH_2CH_2OH$ (C)

$CH_3—\overset{O}{\overset{\|}{C}}—CH_2CH_3$

(D)

$CH_3CH_2—\underset{\underset{CH_3}{|}}{C}=CH—\overset{O}{\overset{\|}{C}}—CH_2CH_3$

(E)

$CH_3CH_2—\underset{\underset{CH_3}{|}}{C}=CH—CH_2CH_2CH_3$

(F)

$CH_3CH_2\underset{\underset{CH_3}{|}}{CH}—\overset{OH}{\overset{|}{CH}}—CH_2CH_2CH_3$

(G)

$CH_3CH_2\underset{\underset{CH_3}{|}}{CH}—\overset{O}{\overset{\|}{C}}—CH_2CH_2CH_3 \xrightarrow{mcpba}$ A

(H)

38. (a) $H—CO—CH_2CH_2—CO—CH_3 \xrightarrow[H^+]{HOCH_2CH_2OH}$

(dioxolane)$>CH—CH_2CH_2—\overset{O}{\overset{\|}{C}}—CH_3 \xrightarrow{CH_3CH_2CH_2CH=PPh_3}$

(dioxolane)$>CH—CH_2CH_2—\overset{CH_3}{\overset{|}{C}}=CH—CH_2CH_2CH_3$

$\xrightarrow{H_3O^+}$ $H—\overset{O}{\overset{\|}{C}}—CH_2CH_2—C(CH_3)=CH—CH_2CH_2CH_3$

Product $\xleftarrow[\text{Ni}]{H_2}$ $\xleftarrow[\bar{H}O/\text{Heat}]{N_2H_4}$

(b) H, H, C=C, CH_3 $\xrightarrow[\text{1.0 equivalent}]{Br_2}$

Br, CH—CH—CH_3, Br $\xrightarrow[\text{Heat}]{KOH}$

C≡C—CH_3 $\xrightarrow[\text{Liquid } NH_3]{Na}$ product

(c) $PhCH_2OH \xrightarrow{SOCl_2} \xrightarrow[\text{Ether}]{Mg} PhCH_2MgCl$

$\xrightarrow{\text{O}} PhCH_2CH_2CH_2OMgCl \xrightarrow{H_2O}$ product

(d) Cl $\xrightarrow[\text{Heat}]{H_3O^+}$ + $\xrightarrow{-H^+}$

$\xrightarrow[\text{Peroxide}]{HBr}$ Br

$\xrightarrow{Me_2CuLi}$ $\xrightarrow[\text{Heat}]{Br_2}$ $\xrightarrow[C_2H_5OH]{C_2H_5ONa}$ product

(e) $\xrightarrow[\text{Heat}]{Br_2}$ Br $\xrightarrow{C_2H_5ONa}$

$\xrightarrow[\text{Peroxide}]{HBr}$ Br $\xrightarrow[\text{KOH}]{\text{Aqueous}}$ OH

$\xrightarrow[H^+]{CrO_4^{2-}}$ O (X) Br $\xrightarrow[\text{Ether}]{Mg}$ MgBr $\xrightarrow{X}$

$\xrightarrow{H_3O^+}$ +

39. (a) N_2H_4/NaOH/Heat

(b) RLi → CH_3I → product

(c)

(d) O, H $\xrightarrow{HCN} \xrightarrow{H_3O^+}$ OH, COOH

$\xrightarrow[H^+/\text{Heat}]{CH_3OH}$ OH, $COOCH_3$

$\downarrow SOCl_2$

$\downarrow NH_3$

NH_2, $COOCH_3$ $\xrightarrow{H_3O^+}$ product

(e) O, O $\xrightarrow{(CH_3)_2CuLi}$ O, O →

O, O^- $\xrightarrow[-H_2O]{H^+}$ O

40. C≡CH, C≡CH (A) O, O (B) O (C) (D)

41. OH, CH—CH=CH_2, CH_3 (A) OH, Br, C—CH_2CH_3, CH_3 (B)

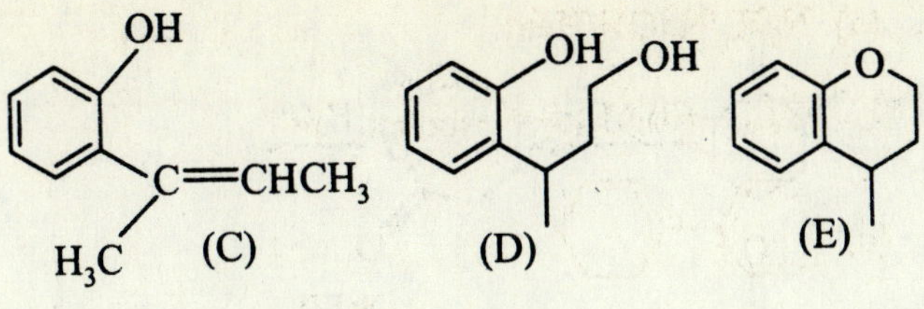

OH, C=CHCH$_3$, H$_3$C (C)   OH, OH (D)   (E)

42. (a)

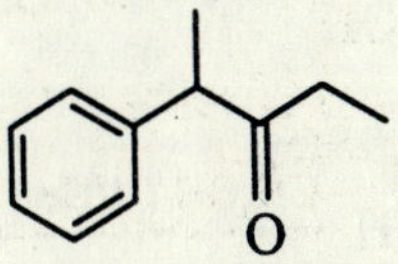

(b) OH → $-H^+$ → Ph

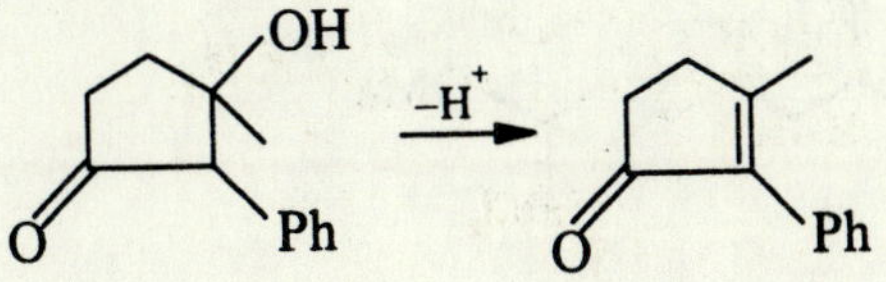

43. Ph—C(H)=O + BF$_3$ → Ph—CH=$\overset{+}{O}$—$\overset{-}{B}F_3$ + H$_2$C=C(CH$_3$)$_2$ →

Ph—C(H)(O$^+$—$\overset{-}{B}F_3$)—CH$_2$—$\overset{+}{C}$(CH$_3$)—CH$_3$ → Ph—C(=O)—CH$_2$CH(CH$_3$)—CH$_3$

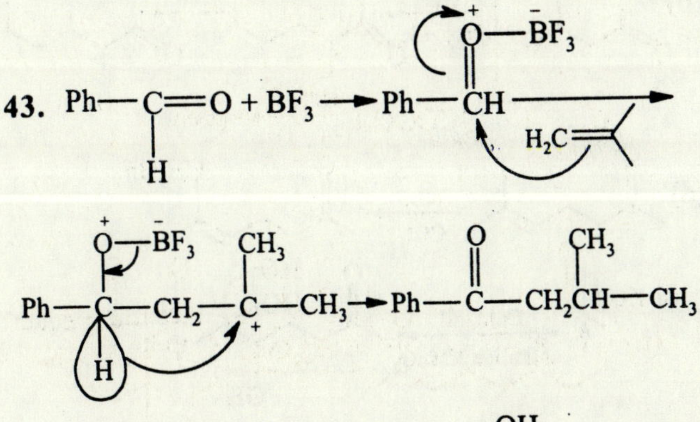

44. CH$_3$CHO $\xrightarrow{PhMgBr}$ $\xrightarrow{H_3O^+}$ Ph—CH(OH)—CH$_3$ $\xrightarrow{H_2CrO_4}$ $\xrightarrow{PhC\equiv\bar{C}\ Na^+}$ $\xrightarrow{H_3O^+}$ Ph—C(OH)(CH$_3$)—C≡C—Ph

$\xrightarrow[(ii)\ H_2O]{(i)\ Na/liq\ NH_3}$ Ph—C(OH)(CH$_3$)—C(H)=C(H)—Ph

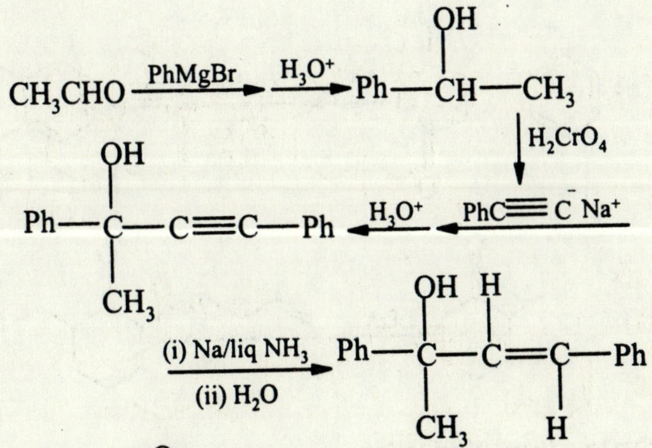

45. (a) C$_6$H$_{11}$COCl + $\bar{C}H_2$—$\overset{+}{N}$≡N →

A: C$_6$H$_{11}$CO—CH$_2$—$\overset{+}{N}$≡N → C$_6$H$_{11}$CH$_2$C(=O)—$\overset{+}{N}$≡N

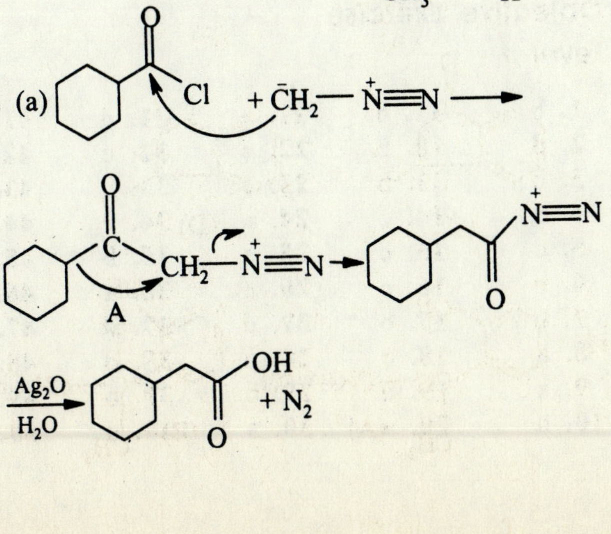

$\xrightarrow[H_2O]{Ag_2O}$ C$_6$H$_{11}$CH$_2$COOH + N$_2$

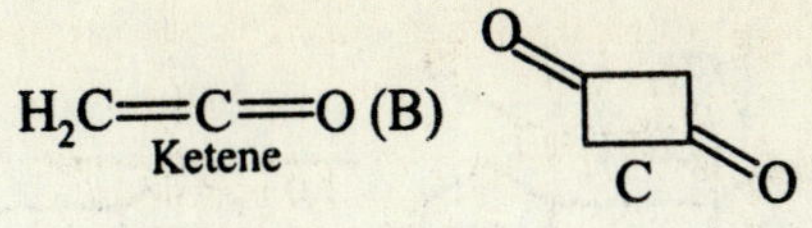

(b) H$_2$C=C=O (B) Ketene   C

(c) O$^-$, N$^-$—$\overset{+}{N}$≡N → N$^-$

$\xrightarrow{H_2O}$ NH $\xrightarrow{LiAlH_4}$ NH

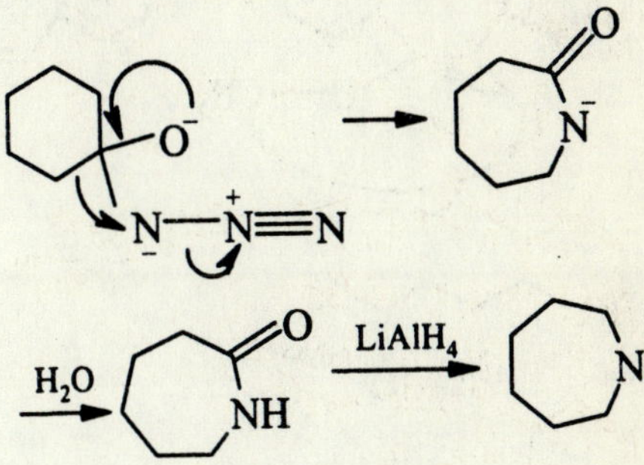

(d) $\xrightarrow{H^+}$ HO ↔ HO; Me-shift; HO $\xleftarrow{-H^+}$ HO

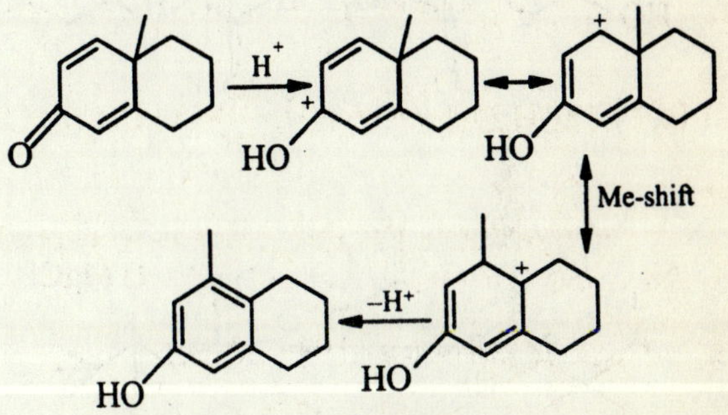

(e) 3 molecule $\xrightarrow{H^+}$ OH, OH, OH

HO, OH, OH $\xrightarrow{-3H_2O}$

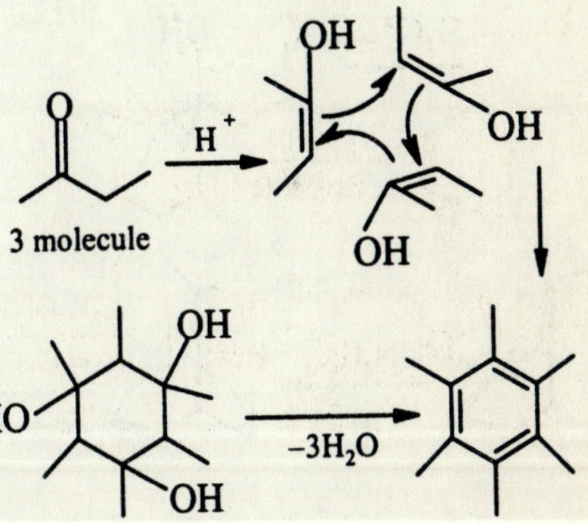

(f) $\overset{+}{C}H_2$ ↔ O $\xrightarrow[-H^+]{H_2O}$

HO → H$\overset{+}{O}$, O$^-$, H → HO—C$_5$H$_4$=O

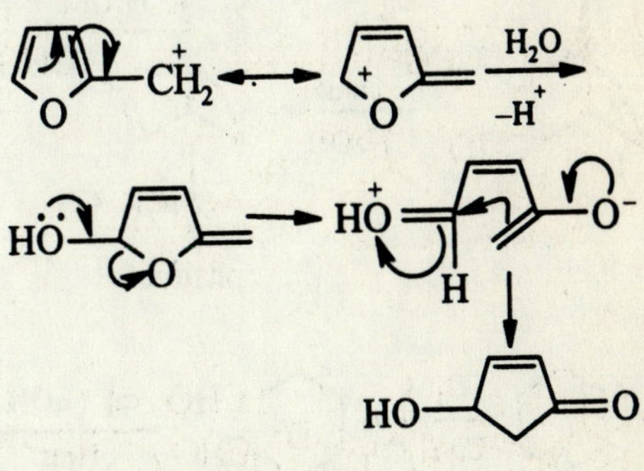

46. (a) O $\xrightarrow{CH_2N_2}$ O → O ↔ $\overset{+}{}$ $\underline{O}$

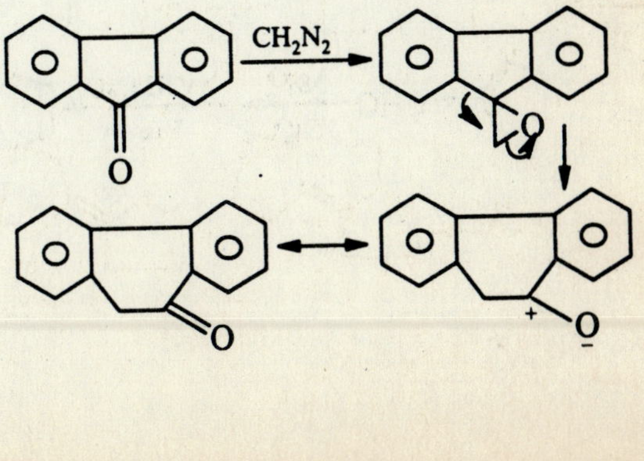

CH_2

O—CH_2

H

OMe

(b) HO H

X O

HO NH_2 + H NH OH

Y O

48. MgCl O $\bar{O}$MgCl

H_3O^+ OH

49. O N_2H_4 / OH^- ; O_3 / Zn-H_2O

CHO, CHO, Dilute NaOH, CHO

N_2H_4/OH^-

product

50. (a) O_3 / Zn/H_2O CHO CHO dil. NaOH / Heat

CHO Ag_2O / H_2O NaOH/CaO / Heat

(b) O CH_3CO_3H O O H_3O^+

O COOH $NaBH_4$

OH COOH $CH_3OH/H^+/Heat$

OH OMe O

(c) O HBr / Peroxide O Br BuLi O Bu

(d) O NBS / CCl_4 O CH_2Br

aq. KOH

Cold, dil / Alkaline $KMnO_4$ O HO HO OH H^+ / Heat product

(e) + O O O $AlCl_3$ Zn(Hg) / HCl HE O

(f) O LDA Br O

Objective Exercise

Level 1

1. b	11. b	21. a	31. d	41. a
2. d	12. b	22. c	32. c	42. c
3. b	13. b	23. c	33. d	43. b
4. d	14. c	24. a	34. a	44. b
5. c	15. b	25. a	35. b	45. a
6. b	16. c	26. c	36. a	46. c
7. b	17. b	27. d	37. d	47. d
8. a	18. c	28. a	38. d	48. b
9. a	19. d	29. b	39. b	49. a
10. d	20. a	30. a	40. d	50. b

51. c	**56.** b	**61.** d	**66.** a
52. c	**57.** c	**62.** c	**67.** d
53. a	**58.** b	**63.** b	**68.** c
54. d	**59.** d	**64.** b	**69.** d
55. a	**60.** a	**65.** d	**70.** c

Level 2

1. b	**21.** d	**41.** d	**61.** c	**81.** a
2. a, c	**22.** d	**42.** a	**62.** c	**82.** c
3. d	**23.** d	**43.** a	**63.** c	**83.** b
4. a	**24.** d	**44.** d	**64.** c	**84.** a
5. a	**25.** a	**45.** b	**65.** c	**85.** c
6. c	**26.** b	**46.** b	**66.** b	**86.** b
7. c	**27.** a	**47.** d	**67.** b	**87.** c
8. a	**28.** b	**48.** c	**68.** b	**88.** a
9. a	**29.** d	**49.** d	**69.** a	**89.** a
10. a	**30.** c	**50.** a	**70.** b	**90.** d
11. c	**31.** d	**51.** b	**71.** c	**91.** c
12. b	**32.** c	**52.** b	**72.** d	**92.** b
13. b	**33.** c	**53.** d	**73.** a	**93.** d
14. b	**34.** a	**54.** a	**74.** b	**94.** b
15. a	**35.** b	**55.** b	**75.** b	**95.** a
16. c	**36.** d	**56.** d	**76.** b	**96.** a
17. d	**37.** b	**57.** c	**77.** a	**97.** a
18. a	**38.** d	**58.** a	**78.** d	**98.** b
19. a	**39.** c	**59.** b	**79.** a	**99.** d
20. c	**40.** a	**60.** a	**80.** a	**100.** c

CHAPTER 8

Practice Exercise

Level 1

1. (a) $CH_3-\overset{O}{\overset{\|}{C}}-O-C(CH_3)_3$

(b) $CH_3-\overset{O}{\overset{\|}{C}}-OCH_2D$

2. (a) $LiAlH_4$ (b) B_2H_6/dimethylglyoxime

(c) Ag_2O/Br_2/Heat (d) Cyclohexanone

3. (a) PhCOOH

(b) (i) $(CH_3)_3CCOOCH_3$ (ii) $(CH_3)_3CCOCl$

(iii) $(CH_3)_3CCH_2OH$ (iv) $(CH_3)_3CCOCH_3$

(v) $(CH_3)_3CBr$

(c) 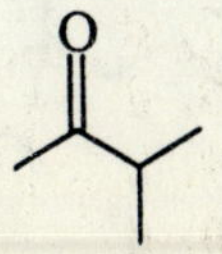 [structure]

(d) 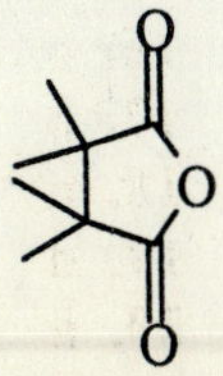 [structure]

4. The carbanion formed after loss of CO_2 is not stabilized by resonance with adjacent carbonyl group, due to non-planarity of *p*-orbital of bridge head carbon and carbonyl carbon.

5. Extraction of a solution of carboxylic acid and phenol in CCl_4 with queous solution of $NaHCO_3$ will bring carboxylic acid in aquous phase as RCOONa leaving phenol in organic phase.

6. A=2-methyl malonic acid B=propanoic acid.

7. c < d < a < b

8. (a) P_4/Br_2 (b) (i) PBr_3 (ii) LDA

(c) Ag_2O/Br_2/Heat

(d) 2.0 equivalent of CH_3CH_2Li

(e) LDA/CH_3I (f) $LiAlH_4$

(g) 2 equivalent of LDA, I_2, Zn/Heat

9. (a) 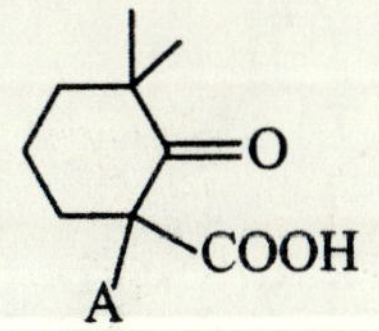 [structure: cyclohexanone ring with =O, COOH, A] (b) $Et-\underset{}{\overset{Et}{\overset{|}{CH}}}-COOH$

(c) 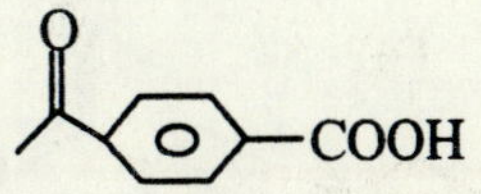 $CH_3CO-C_6H_4-COOH$

(d) 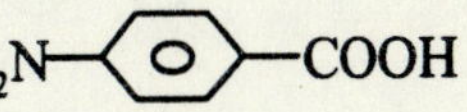 $O_2N-C_6H_4-COOH$

10. (a) [reaction scheme] $\xrightarrow{Br_2}$ … $\xrightarrow{-H^+}$ … Br

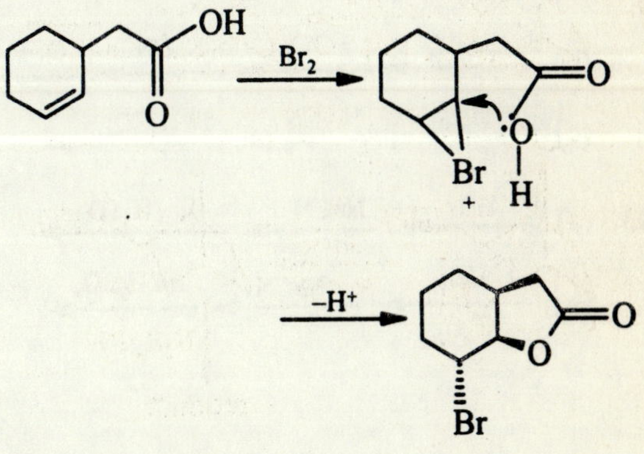

(b) 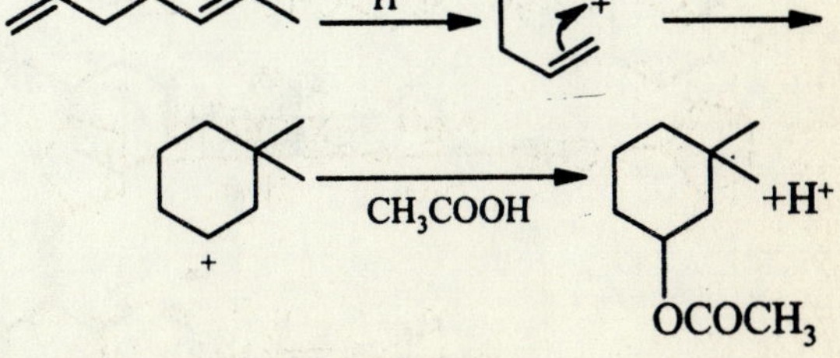 [reaction scheme] $\xrightarrow{H^+}$ … $\xrightarrow{CH_3COOH}$ … $+H^+$ (OCOCH$_3$)

(c) [cyclopentenyl tertiary alcohol with side chain NHCOCH$_3$] $\xrightarrow[-H_2O]{H^+}$ [allylic carbocation resonance forms; HN—COCH$_3$ attacks] $\xrightarrow{-H^+}$ [spiro N—COCH$_3$ product]

NHCOCH$_3$

11. (a) PhCONHCH$_3$

(b) [benzene ring with CH$_2$COOH and C(=O)—O—C(=O)—CH$_3$]

(c) [benzene ring fused cyclic carbonate (O—C(=O)—O) with OH]

(d) [benzene ring with C(=O)OCH$_3$ and CH$_2$COOH]

12. (a) $\xrightarrow{H_2SO_4}$ $\xrightarrow{SOCl_2}$

(b) $\xrightarrow{LiAlH_4}$ $\xrightarrow{SOCl_2}$ $\xrightarrow{NaCN}$ $\xrightarrow{\text{dil } H_2SO_4}$

(c) $\xrightarrow{NBS}$ $\xrightarrow{NACN}$ $\xrightarrow{\text{dil } H_2SO_4}$

13. (a) $\xrightarrow{HBr}$ $\xrightarrow{NaCN}$ $\xrightarrow{\text{dil . } H_2SO_4}$

(b) $\xrightarrow[R_2O_2]{HBr}$ $\xrightarrow{NaCN}$ $\xrightarrow{\text{dil } H_2SO_4}$

(c) [cyclohexane]-COOH $\xrightarrow{SOCl_2}$ [cyclohexane]-C(=O)Cl (I)

[cyclohexane]-COOH $\xrightarrow[Br_2/Heat]{Ag_2O}$ [cyclohexane]-Br $\xrightarrow[Et_2O]{Mg}$ $\xrightarrow{I}$ [dicyclohexyl ketone]

14. $CH_3COOCH_3 \xrightarrow[CH_3I]{LDA} CH_3CH_2—COOCH_3$

$\xrightarrow[CH_3I \text{ (excess)}]{LDA} CH_3—C(CH_3)_2—COOCH_3 \xrightarrow{PhCH_2CH_2MgBr}$

15. (a) MeOH + Na_2CO_3 (b) [benzene ring with COOEt and COOH]

(c) PhCH(COOEt)—C(=O)—Ph (d) $CH_3CH_2—CH(COOEt)_2$

(e) (i) H_3C—[benzene]—COOH

(ii) H_3C—[benzene]—CH_2NH_2

(iii) H_3C—[benzene]—C(=O)—Ph

16. (a) $PhCH_2OH \xrightarrow{SOCl_2} PhCH_2Cl \xrightarrow{NaCN}$

$PhCH_2CN$

(b) Ph-COOH $\xrightarrow{SOCl_2}$ Ph—C(=O)—Cl $\xrightarrow[\text{Excess } H_3O^+]{CH_3MgBr}$

Ph—C(CH$_3$)$_2$—OH

(c) $CH_3CH(COOEt)_2$ (COOEt, COOEt) $\xrightarrow{EtONa}$ EtOOC—$\bar{C}$(CH$_3$)—COOEt

$\xrightarrow{CH_2{=}CHCH_2Br}$ $CH_2{=}CHCH_2C(CH_3)(COOEt)_2$

$\downarrow H_3O^+$

$CH_2{=}CHCH_2CH(CH_3)COOH \xleftarrow{Heat} CH_2{=}CHCH_2C(CH_3)(COOH)_2$

$\xrightarrow{SOCl_2}$ (CH$_2$=CH–CH$_2$–CH(CH$_3$)–COCl) $\xrightarrow{C_2H_5OH}$ (CH$_2$=CH–CH$_2$–CH(CH$_3$)–COOEt)

17. (a) EtOOC–(CH$_2$)$_3$– (2-oxocyclopentane bearing COOEt) $\xrightarrow{EtO^-}$ EtOOC–$\bar{C}$H– ... (carbanion attacks ring C=O) → bicyclic alkoxide (COOEt, EtOOC, O$^-$) → cyclooctane carbanion (COOEt, EtOOC, C=O) $\xrightarrow{EtOH}$ EtO$^-$ + cyclooctanone bearing COOEt and EtOOC

(b) $CH_3—C{\equiv}N \xrightarrow{H^+} CH_3—C{\equiv}\overset{+}{N}H$

$\xrightarrow{CH_3CH_2OH}$ $CH_3—C(=NH)—\overset{+}{O}(H)—CH_2CH_3$ $\xrightarrow{-H^+}$

$CH_3—C(=NH)—OC_2H_5 \xrightarrow{H_2O} CH_3—C(=O)—OC_2H_5 + NH_3$

18. (a) HOOC–(CH$_2$)$_4$–COOH $\xrightarrow[H^+]{C_2H_5OH}$

EtOOC–(CH$_2$)$_4$–COOEt (I) $\xrightarrow[EtOH]{EtO^-}$ 2-oxocyclopentane-COOEt (II)

II $\xrightarrow{LDA}$ (2-oxocyclopentyl carbanion –COOEt) $\xrightarrow{Cl–CH_2CH_2CH_2–COOEt}$ 2-oxo-1-(3-ethoxycarbonylpropyl)cyclopentane-1-COOEt

(b) $CH_3COOH \xrightarrow{P_4/I_2} CH_2(I)—COOH \xrightarrow{aqCN^-}$

$NC—CH_2COOH \xrightarrow{H_3O^+} HOOC—CH_2—COOH$

$\xrightarrow[EtOH]{H_2SO_4}$ EtOOC–CH$_2$–COOEt

19. (a) cyclohexene $\xrightarrow[LMnO_4]{Cold\ dilute}$ cyclohexane-1,2-diol (OH, OH) $\xrightarrow{ClCO–COCl}$ cyclohexane-fused dioxane-2,3-dione (oxalate ester)

(b) cyclopentane $\xrightarrow[hv]{Cl_2}$ chlorocyclopentane (Cl) $\xrightarrow{aq\ KOH}$ cyclopentanol (OH)

$\xrightarrow{H_2CrO_4}$ cyclopentanone $\xrightarrow{RCO_3H}$ δ-valerolactone

(c) benzene $\xrightarrow{CH_3Cl}$ toluene (CH$_3$) $\xrightarrow{H_2SO_4}$ p-toluenesulfonic acid (CH$_3$, SO$_3$H)

$\xrightarrow[AlCl_3]{CH_3Cl}$ (CH$_3$, CH$_3$, SO$_3$H) $\xrightarrow[H_2SO_4]{Dilute}$ o-xylene (CH$_3$, CH$_3$)

$\xrightarrow{KOH/KMnO_4}$ phthalic acid (COOH, COOH) $\xrightarrow[Heat]{P_2O_5}$ phthalic anhydride

(d) 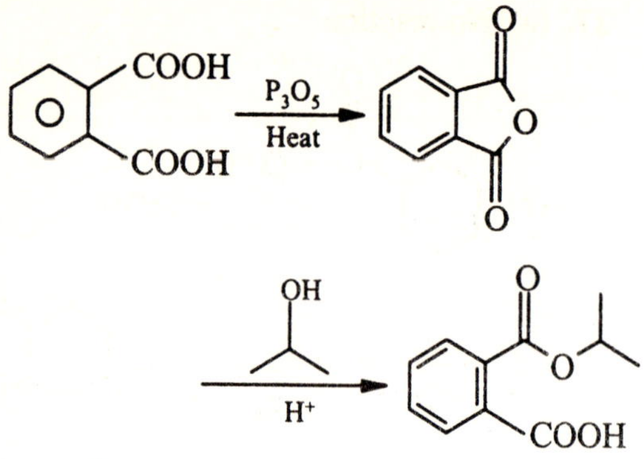 $\xrightarrow[\text{Heat}]{P_2O_5}$

$\xrightarrow[H^+]{}$ COOH

20. (a) 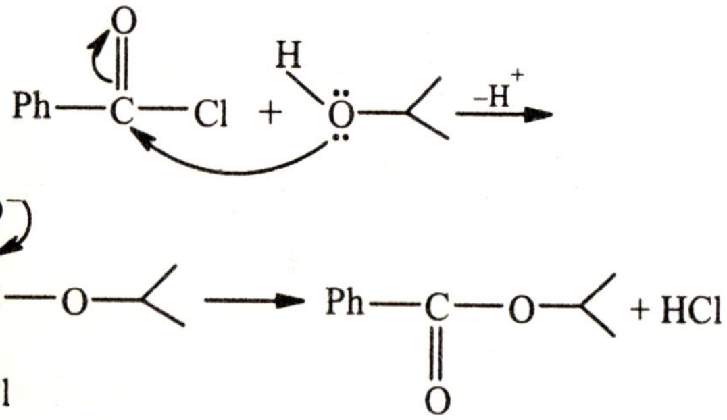 Ph—C—Cl + Ö $\xrightarrow{-H^+}$

Ph—C—O— ⟶ Ph—C—O— + HCl

Cl O

(b) 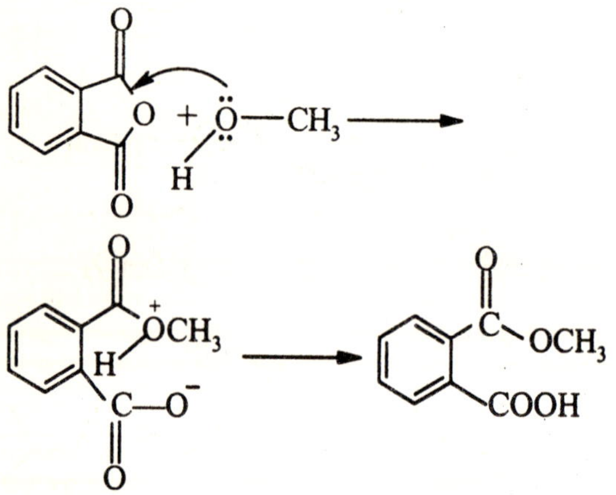 O + Ö—CH_3 ⟶

H

$\overset{+}{O}CH_3$ ⟶ C—OCH_3

H COOH

C—O^-

(c) O + $H_2\ddot{N}$—R ⟶ C—$\overset{+}{N}$—R ⟶

H

COO^-

C—NHR ⟶ $\overset{+}{N}$ R ⟶

C—OH H

^-O OH

N—R $\xrightarrow{-H_2O}$ N—R

HO OH

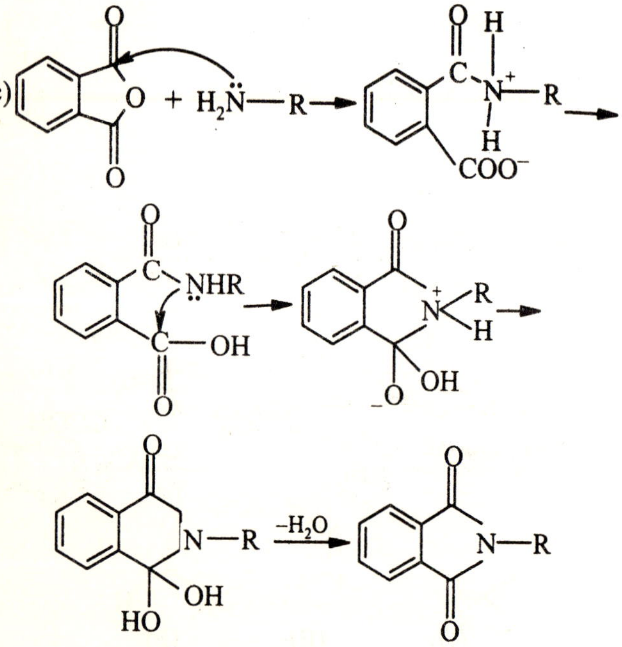

21. (a) 6-hepten-1-ol $\xrightarrow{PCl_5}$ $\xrightarrow{NH_3}$ NH_2

$\xrightarrow[H_2O_2]{O_3}$ NH_2 COOH $\xrightarrow{\Delta}$ NH

Caprolactum

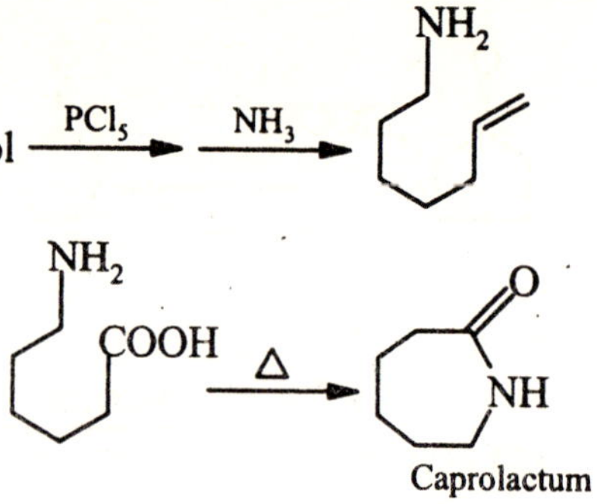

(b) $PhCH_2Br \xrightarrow{NaCN} PhCH_2CN \xrightarrow{NaBH_4}$ $PhCH_2CH_2NH_2$

(c) $PhCHO + CH_3MgBr \xrightarrow{H_3O^+}$

Ph—CH—CH_3 (OH) $\xrightarrow[\text{Heat}]{H_2SO_4}$ Ph—CH=CH_2

$\xrightarrow[R_2O_2]{HBr} PhCH_2CH_2Br \xrightarrow[H_2O]{KCN} PhCH_2CH_2CN$

22. 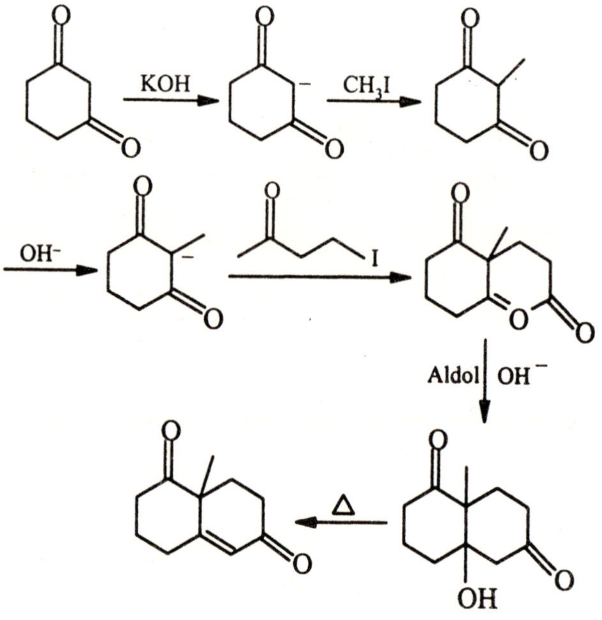 $\xrightarrow{KOH}$ $\xrightarrow{CH_3I}$

$\xrightarrow{OH^-}$ ⟶ I

Aldol OH^-

$\xleftarrow{\Delta}$ OH

23. (a) 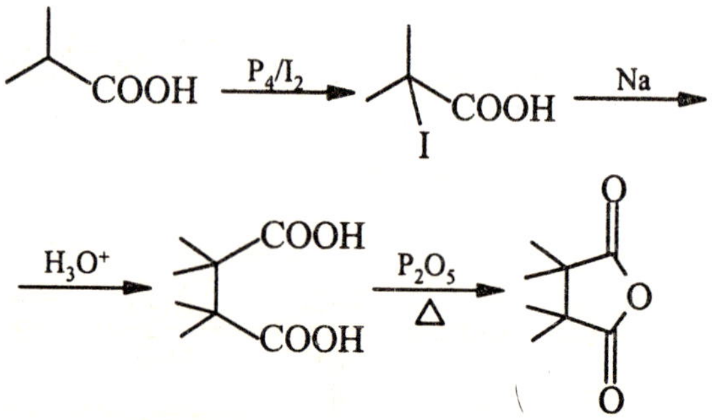 COOH $\xrightarrow{P_4/I_2}$ COOH (I) $\xrightarrow{Na}$

$\xrightarrow{H_3O^+}$ COOH COOH $\xrightarrow[\Delta]{P_2O_5}$

(b) 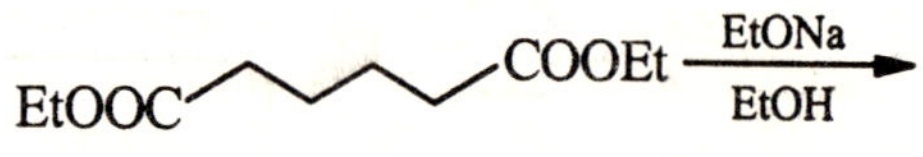 EtOOC COOEt $\xrightarrow[EtOH]{EtONa}$

2-oxocyclopentanecarboxylate ester (COOEt) $\xrightarrow[\Delta]{H_3O^+}$ cyclopentanone $\xrightarrow{LDA}$ $\xrightarrow{CH_3I}$ 2-methylcyclopentanone

24. (a) sec-butylamine (NH_2) $\xrightarrow[\Delta]{HCOOH}$ H—C(=O)—N(H)—sec-butyl

(b) $CH_3COOEt \xrightarrow[Excess]{C_2H_5MgBr} \xrightarrow{H_3O^+}$ 3-methyl-3-pentanol (OH)

(c) cyclohexylamine (NH_2) $\xrightarrow[Heat]{CH_3COOH}$ $CH_3C(=O)$—N(H)—cyclohexyl

(d) bromocyclohexane (Br) $\xrightarrow[Et_2O]{Mg}$ cyclohexyl-MgBr (I) $\xrightarrow[(ii)\ H_2O]{(i)\ H—C(=O)—H}$

cyclohexyl-CH_2OH $\xrightarrow{PCC}$ $\xrightarrow{I}$ $\xrightarrow{H_3O^+}$ dicyclohexylmethanol (OH)

25. (a) 1-hexanol $\xrightarrow{SOCl_2}$ $\xrightarrow{NaCN}$ $\xrightarrow{LiAlH_4}$ heptylamine (NH_2)

(b) cyclohexanecarboxamide (NH_2) $\xrightarrow{H_3O^+}$ cyclohexanecarboxylic acid (COOH) $\xrightarrow{EtOH\ H_2SO_4\ Heat}$ ethyl ester (COOEt) $\xrightarrow{C_2H_5MgBr}$ cyclohexyl ethyl ketone

(c) 1-octanol $\xrightarrow{SOCl_2}$ $\xrightarrow[Et_2O]{Mg}$ $\xrightarrow{CH_3COCl}$ 2-decanone

26. (a) $NaCN \longrightarrow H_3O^+$ (b) $LiAlH_4$

27. (a) No reaction

(b) cyclohexyl—C(=O)—$O^{18}H$ + EtOH

(c) $\xrightarrow[Et_2O]{Mg}$ PhMgBr (X)

(d) fused cyclohexane lactone (ring O, C=O) bearing a 1,1-dimethylpropyl group

(e) Ph—C(=O)—NH—cyclohexyl

28. (a) II < I < III (b) III < II < I

Level 2

1. (A) γ-valerolactone (5-methyl, ring O, C=O) (B) 4-hydroxypentanoic acid (COOH, OH) (C) pent-3-enoic acid (COOH) (D) 4-oxopentanoic acid (COOH, O)

2. (A) 3,4-dimethyl β-lactone (O, C=O) (B) 3-hydroxy-2-methylbutanoic acid (OH, COOH) (C) 2-methylbut-2-enoic acid (COOH) (D) 2-methyl-3-oxobutanoic acid (O, COOH) (E) 2-butanone

3. (A) $CH_3—CH(CH_3)—CH_2COOH$

(B) $CH_3—CH(CH_3)—CH(Br)—COOH$

(C) 3-methylbut-2-enoic acid (COOH)

(D) isobutylene (E) acetone (F) H—C(=O)—H

4. 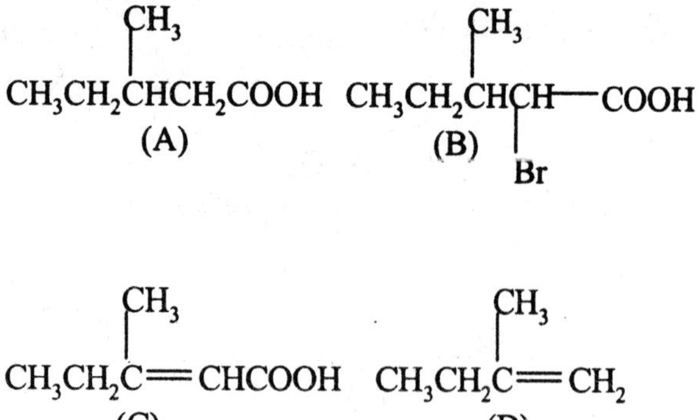 $CH_3CH_2CH(CH_3)CH_2COOH$ (A) $CH_3CH_2CH(CH_3)CH(Br)—COOH$ (B)

$CH_3CH_2C(CH_3)=CHCOOH$ (C) $CH_3CH_2C(CH_3)=CH_2$ (D)

5. 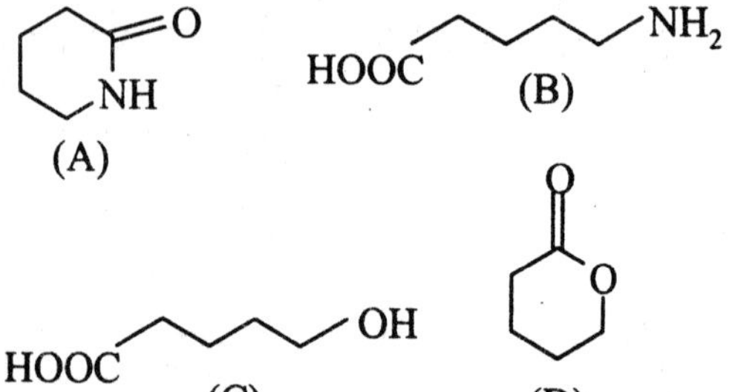 (A) (lactam, =O, NH) HOOC ... NH_2 (B)

HOOC ... OH (C) (D)

6. 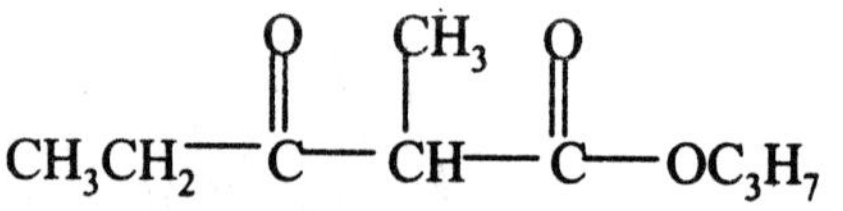 $CH_3CH_2—CO—CH(CH_3)—CO—OC_3H_7$ (A)

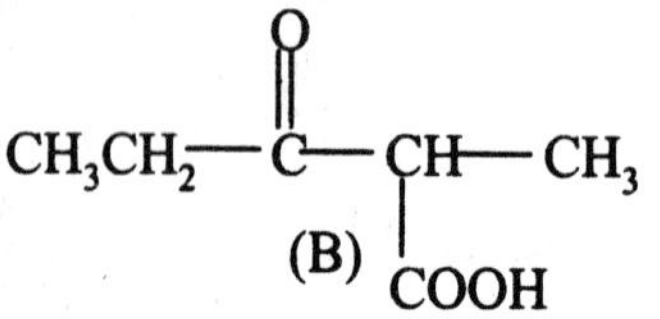

$CH_3CH_2—CO—CH(COOH)—CH_3$ (B) $CH_3CH_2CH_2OH$ (C)

CH_3CH_2CHO (D) $CH_3CH_2CH(OH)—CH(CHO)—CH_3$ (E)

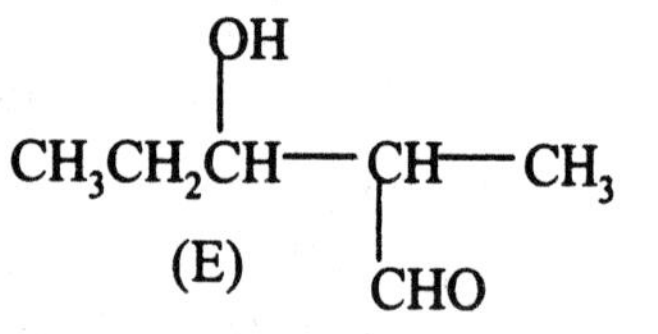

7. 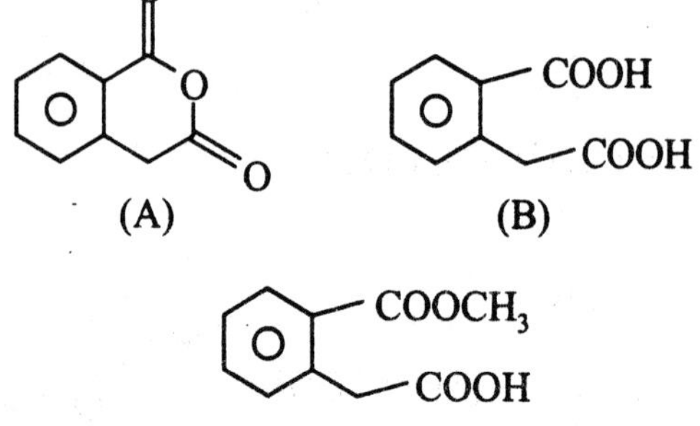 (A) (B) COOH, COOH (C) $COOCH_3$, COOH

8. 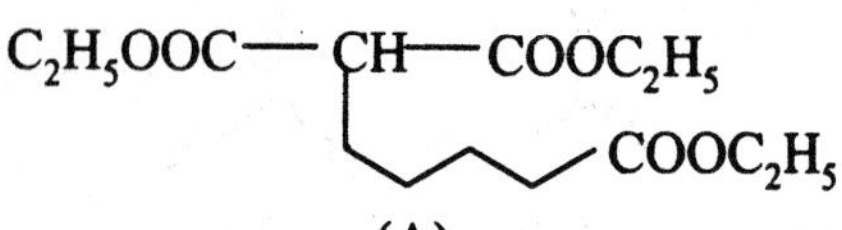 $C_2H_5OOC—CH—COOC_2H_5$ ... $COOC_2H_5$ (A)

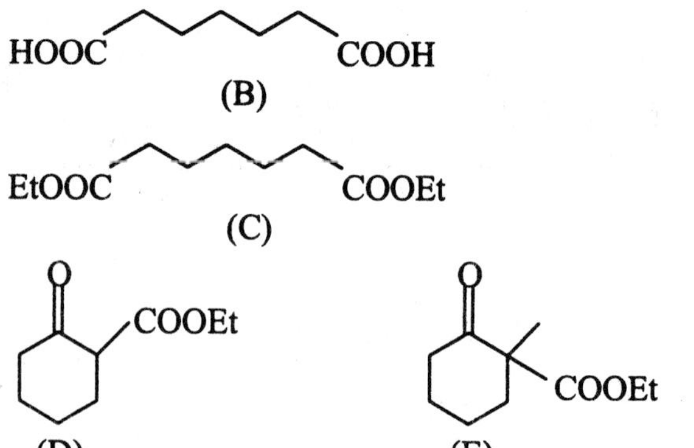

HOOC ... COOH (B)

EtOOC ... COOEt (C)

(D) COOEt (E) COOEt

9. 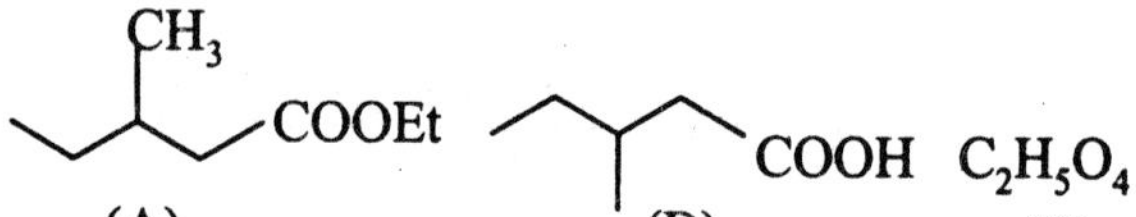 CH_3 ... COOEt (A) ... COOH (B) $C_2H_5O_4$ (C)

(D)

10. (a) 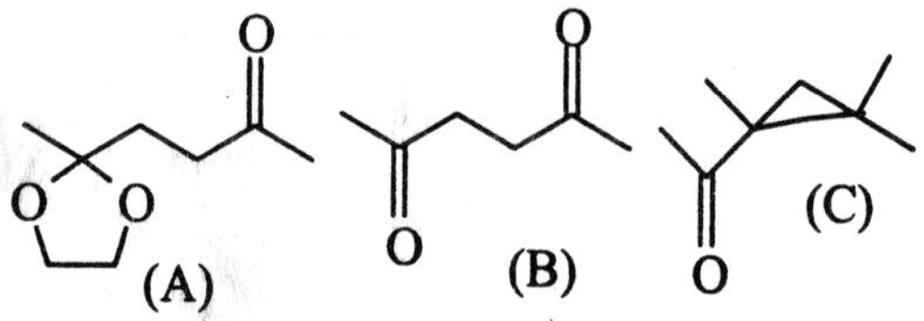 O O (A) (B) (C)

11. (a) 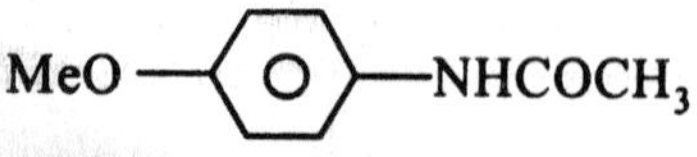 MeO—⟨O⟩—$NHCOCH_3$

(b) Ph—CONHOH

(c) 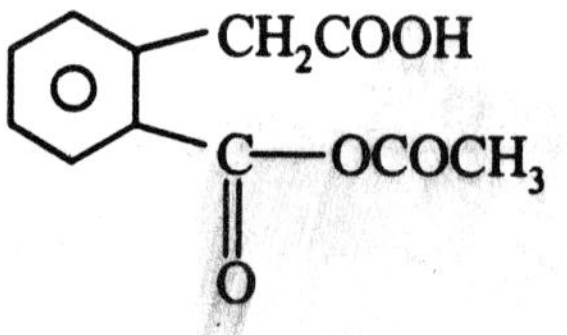 CH_2COOH, $C(=O)—OCOCH_3$

(d) 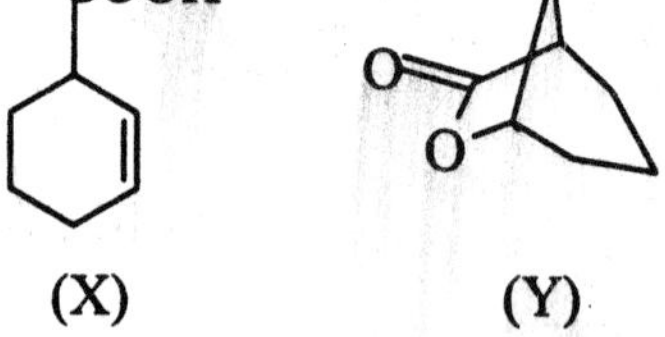 COOH (X) (Y)

(e) 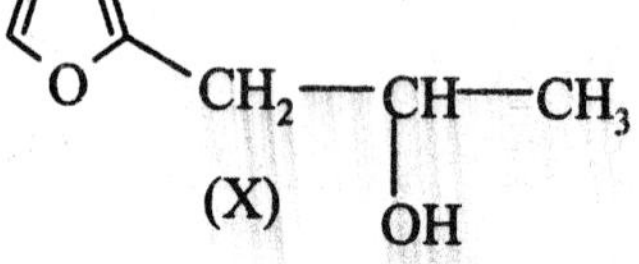 $CH_2—CH(OH)—CH_3$ (X)

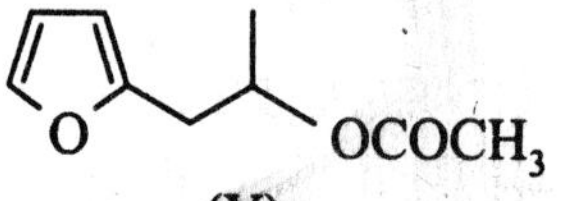

$OCOCH_3$ (Y)

12. (a) (A) (B)

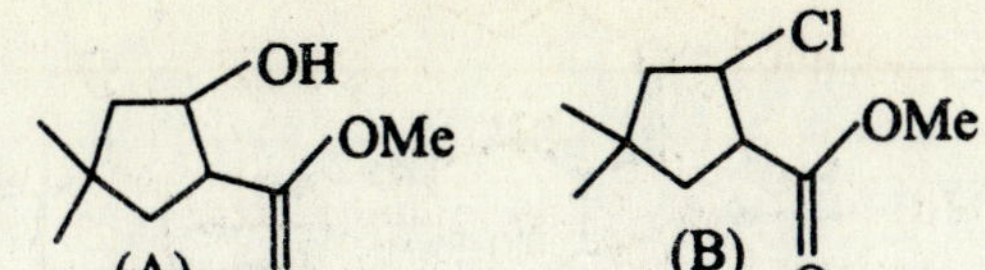

(b) $Ph-C(CN)(OH)-CF_3$ (C) $Ph-C(CN)(OCH_3)-CF_3$ (D)

$Ph-C(COOH)(OCH_3)-CF_3$ (E)

13. (a) $TsO-CH_2-C\equiv CH$ (A)

$TsO-CH_2C\equiv C^-\ Li^+$ (B)

$TsOCH_2C\equiv C-CH_2CH_2Cl$ (C)

$NC-CH_2-C\equiv C-CH_2-CH_2-CN$ (D)

$HOOC-CH_2C\equiv C-CH_2CH_2COOH$ (E)

15. (a) OMgBr, OMe, N, CH_3 (A)

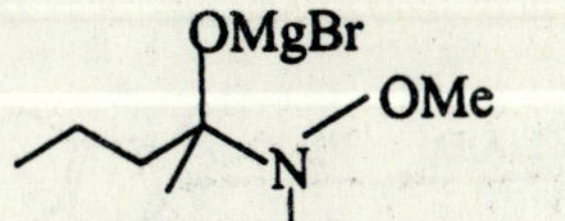

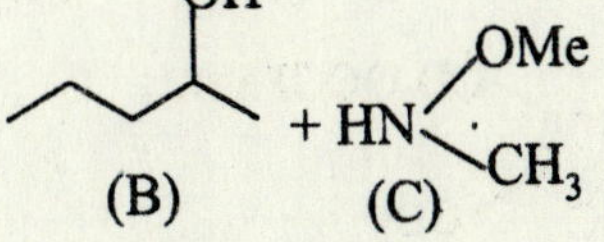

OH (B) + HN(OMe)(CH_3) (C)

(b) (D) Ph, N, Me

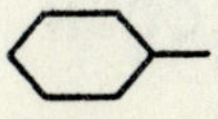

COOH + PhNHMe
(E) (F)

16. (a) CHO + EtOH (b) PhCHO

(c) NH_2

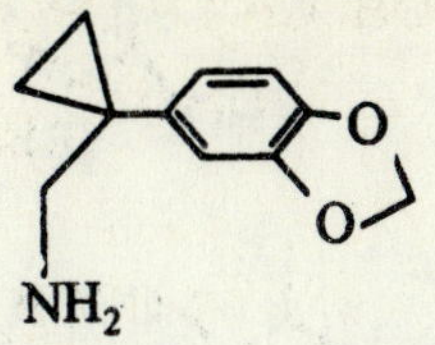

17. (a) $Ph-CH(CH_3)-CH_2COOH \xrightarrow{LiAlH_4}$

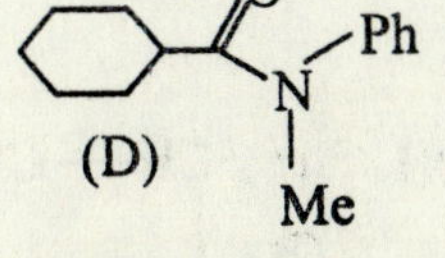

$Ph-CH(CH_3)-CH_2CH_2OH \xrightarrow{SOCl_2} \xrightarrow[Et_2O]{Mg}$

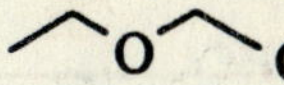

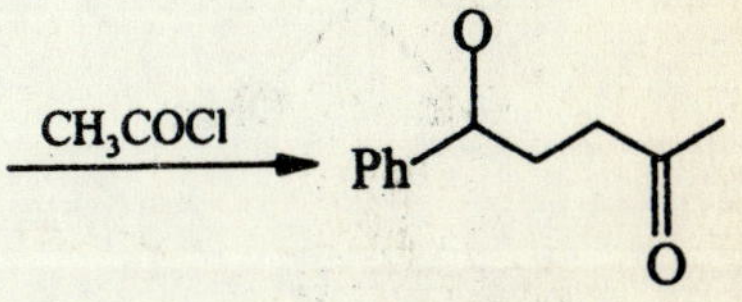

$\xrightarrow{CH_3COCl}$ Ph

(b) COOEt + COOEt $\xrightarrow{H_3O^+}$

COOH COOH $\xrightarrow{H_2/Pt}$ OH CH_2OH

$\xrightarrow{CH_3COCl\ (excess)}$

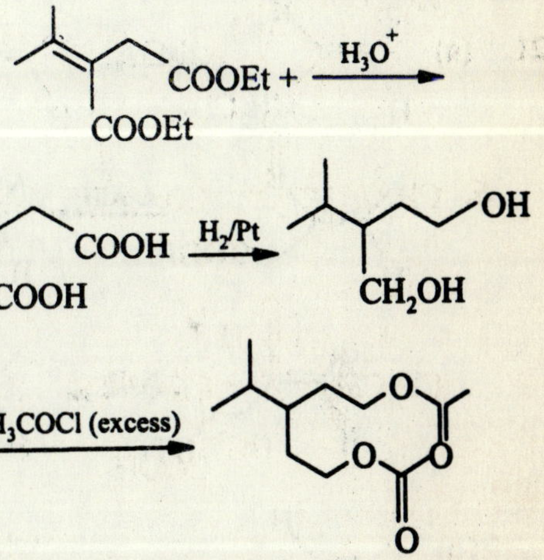

18. (a) $HOOC-CH_2-COOH$ (b)

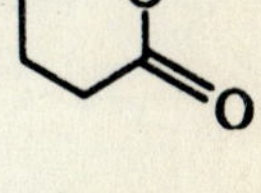

(c) 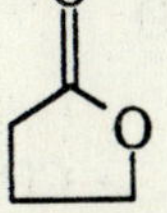 (X) $HO-CH_2CH_2CH_2-COOH$ (Y)

19. (a) $\xrightarrow{mcpba}$ $\xrightarrow{H_3O^+}$

COOH, OH $\xrightarrow{H_2CrO_4}$ COOH $\xrightarrow[H^+/Heat]{CH_3OH}$ OMe

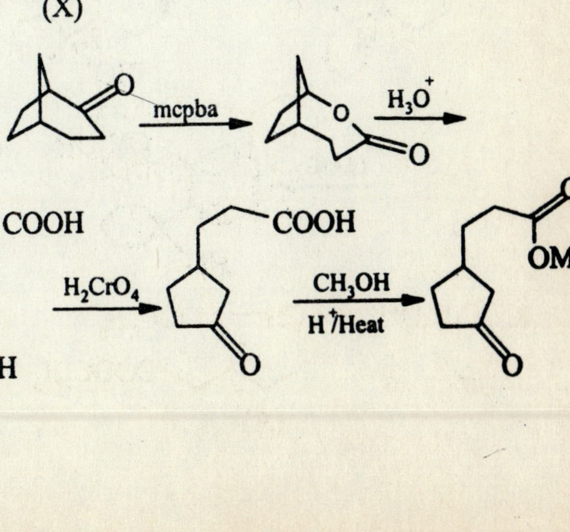

20. (a) (A) 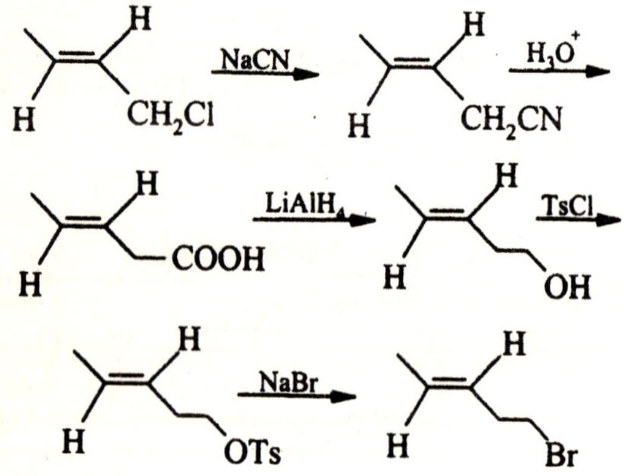 (B) COOH HO COOH HO

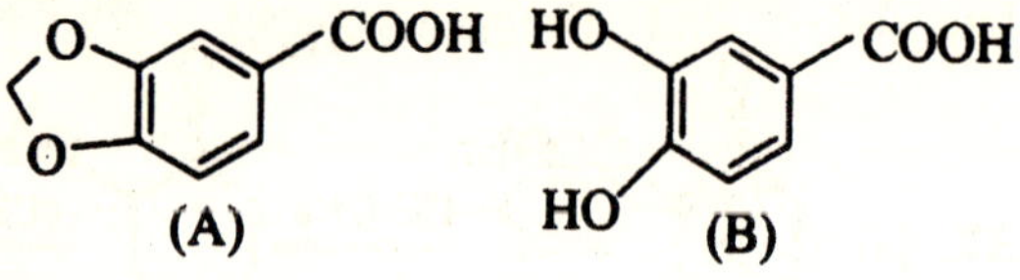

(b) MgBr COOH (C) (D)

(c) Ph—C(OH)—C≡C—Ph (E) N Me OMe

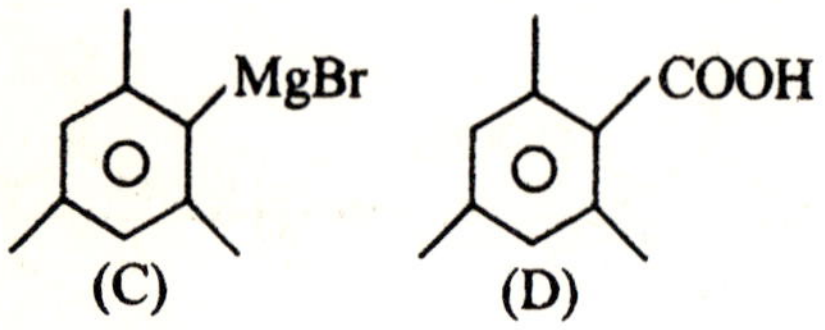

(d) $\xrightarrow{H_3O^+}$ OH COOH $\xrightarrow[Heat]{NH_3}$ OH NH_2 $\xrightarrow[\Delta]{H_2SO_4}$ NH_2

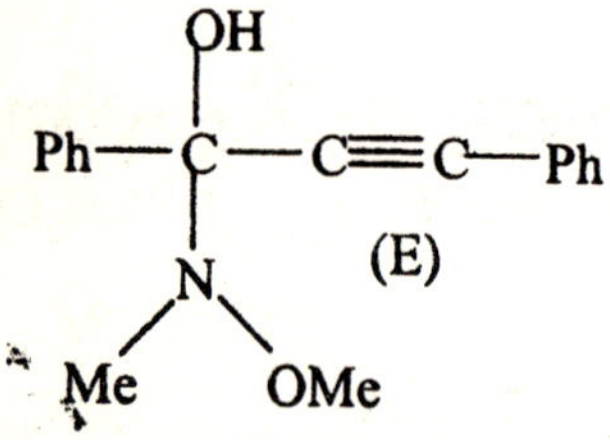

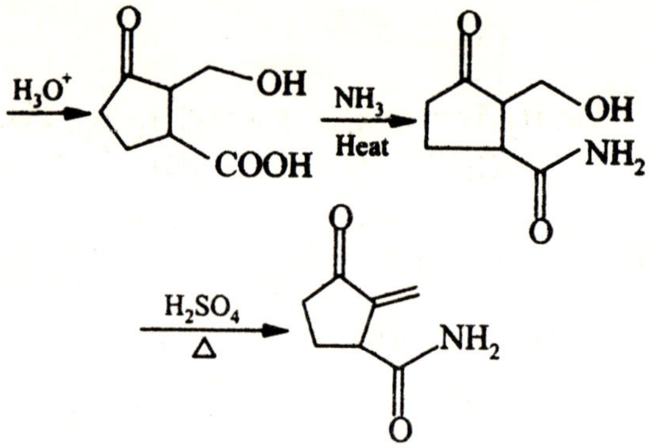

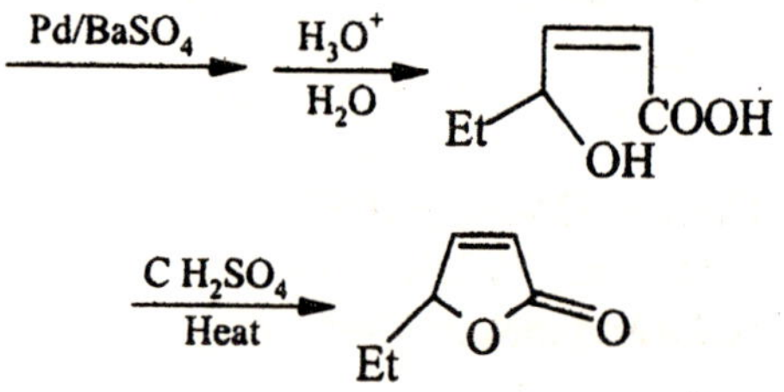

(e) $\xrightarrow{Pd/BaSO_4}$ $\xrightarrow[H_2O]{H_3O^+}$ Et COOH OH $\xrightarrow[Heat]{C\ H_2SO_4}$ Et O O

21. (a) H H CH_2Cl $\xrightarrow{NaCN}$ H H CH_2CN $\xrightarrow{H_3O^+}$ H H COOH $\xrightarrow{LiAlH_4}$ H H OH $\xrightarrow{TsCl}$ H H OTs $\xrightarrow{NaBr}$ H H Br

(b) H H COOEt $\xrightarrow{\text{Di-isobutyl aluminium hydride}}$ H H CHO

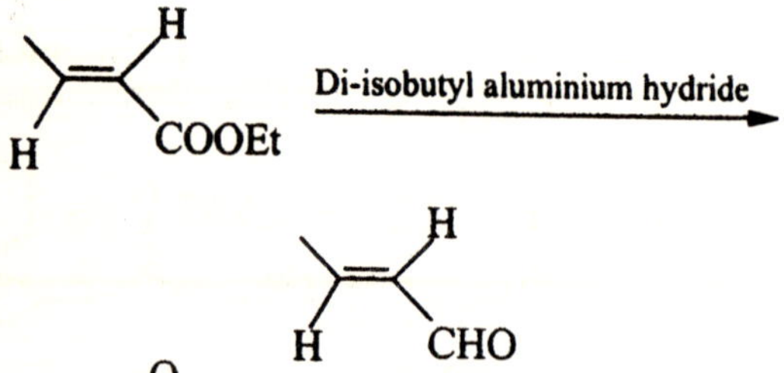

(c) O O $\xrightarrow{H_3O^+}$ COOH HO $\xrightarrow{LiAlH_4}$ CH_2OH HO

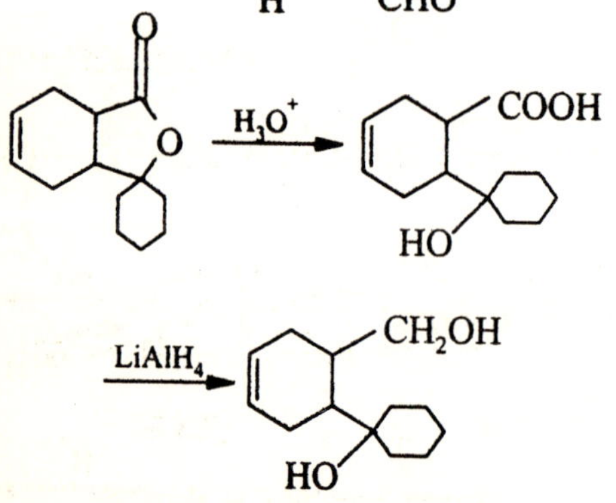

22. (a) $\xrightarrow{LiAlD_4}$ $\xrightarrow{H_2O}$

(b) $\xrightarrow[Zn-H_2O]{O_3}$ $\xrightarrow[\text{glycol}]{\text{ethylene}}$ $\xrightarrow[Heat]{CH_3OH}$ $-CH_2CH_2COOCH_3$

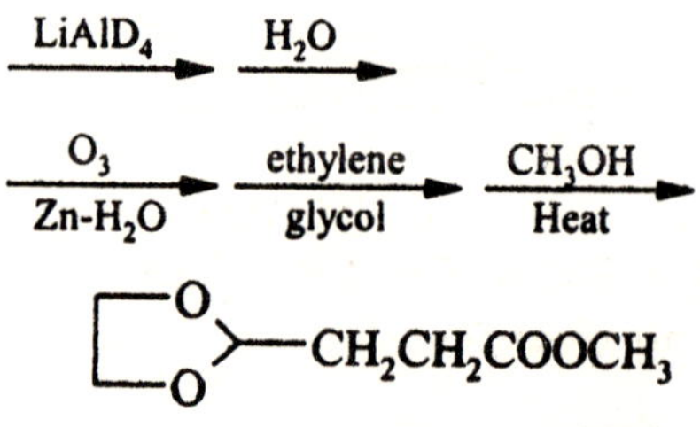

23. (a) CN $\xrightarrow{PhLi}$ NLi Ph $\xrightarrow{H_3O^+}$ O Ph

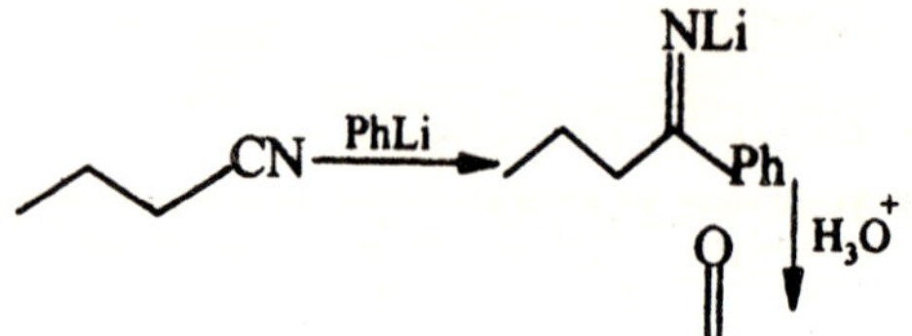

(b) Reformatsky reaction.

24. (a) COOH CD_3 O $\xrightarrow{LiAlD_4}$ $\xrightarrow{PhCOCl}$ CH_2OCOPh CD_3 O

$\xrightarrow[H_2O]{LiAlD_4}$ CD_2OH CD_3 O $\xrightarrow{PhCOCl}$ CD_2OCOPh CD_3 O

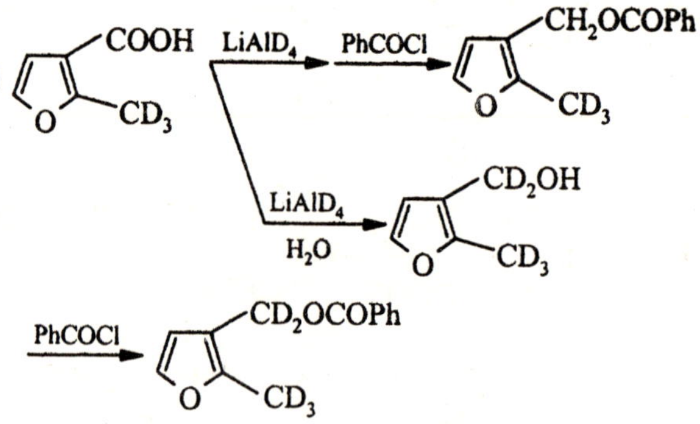

25. Ph—CH(CH_3)—CH_2COOH (A)   Ph—CH(COOH)—CH_2CH_3 (B)

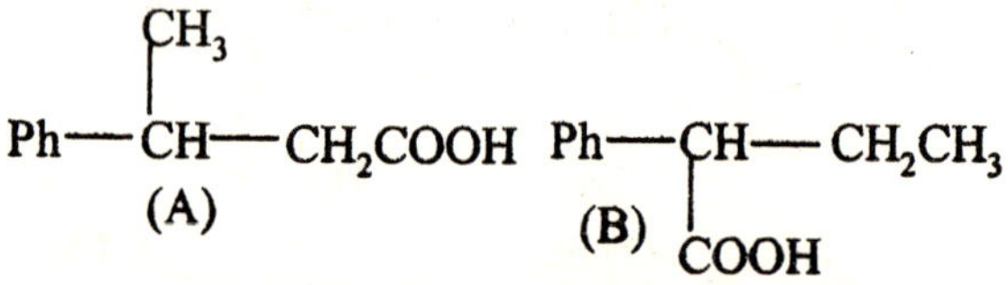

$PhCH_2CH(CH_3)COOH$ (C) $PhCH_2CH_2CH_3$ (D) $PhCH(CH_3)_2$ (E)

26. 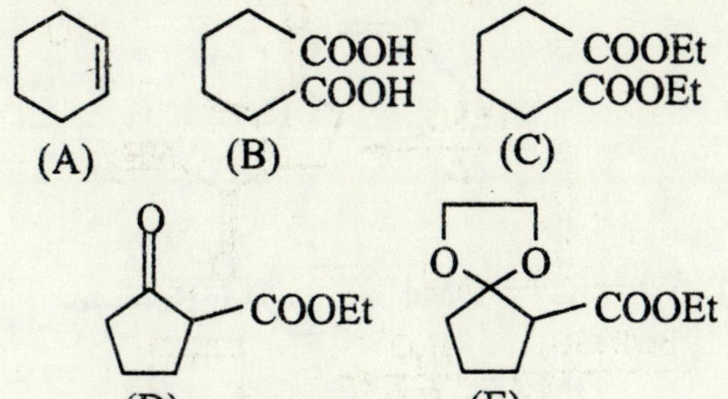 (A) (B) COOH COOH (C) COOEt COOEt

(D) COOEt (E) O O COOEt

27. A: CH_3CH_2ONa/C_2H_5OH
B: C_2H_5ONa, $PhCOCH_2Br$
C: NaOH(aqueous), Heat

28. (a) 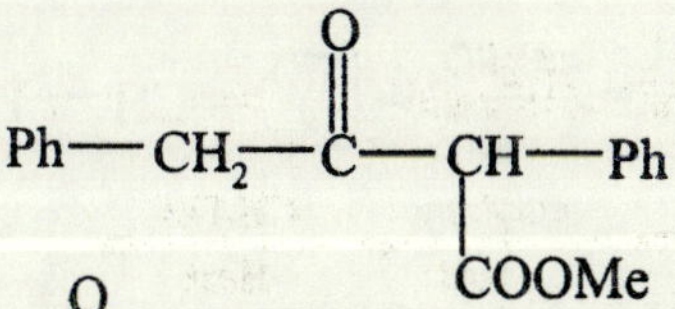 Ph—CH_2—C(=O)—CH(COOMe)—Ph

(b)

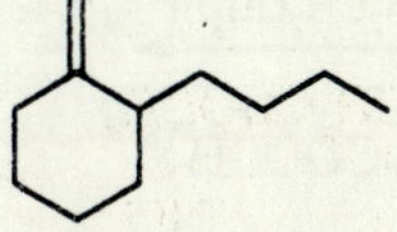

29. (a) HOOC—$CH_2CH_2CH_2$—$O^{18}H$
(b) H_2N—CH_2COO^-
(c) HOOC—$CH_2CH_2CH_2CONH_2$

30. (a) 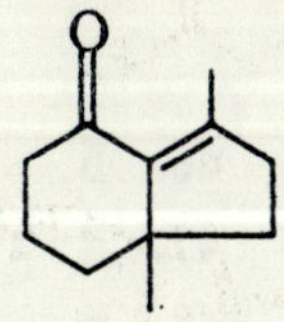 (b) 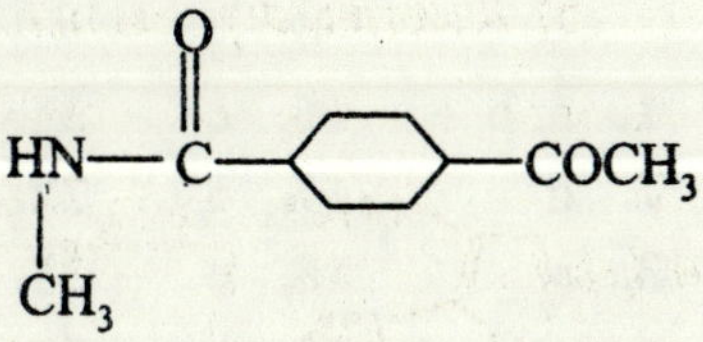 HN(CH_3)—C(=O)—C₆H₁₀—$COCH_3$

31. 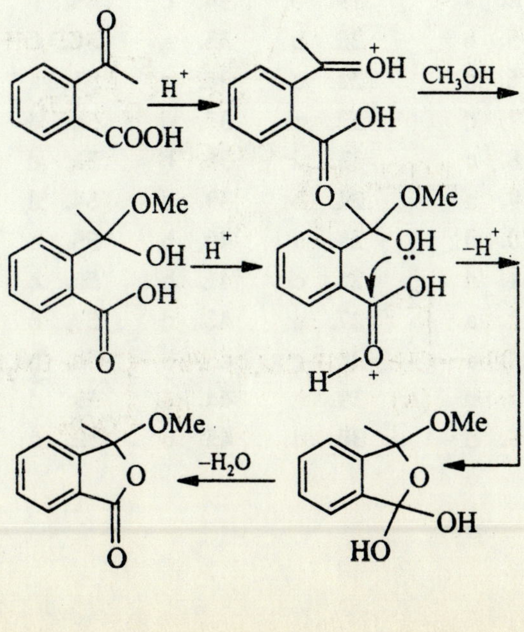 COOH $\xrightarrow{H^+}$ $\overset{+}{O}H$ OH O $\xrightarrow{CH_3OH}$

OMe OH OH O $\xrightarrow{H^+}$ O OMe OH OH O H + $\xrightarrow{-H^+}$

OMe O HO OH $\xrightarrow{-H_2O}$ OMe O O

32. (a) 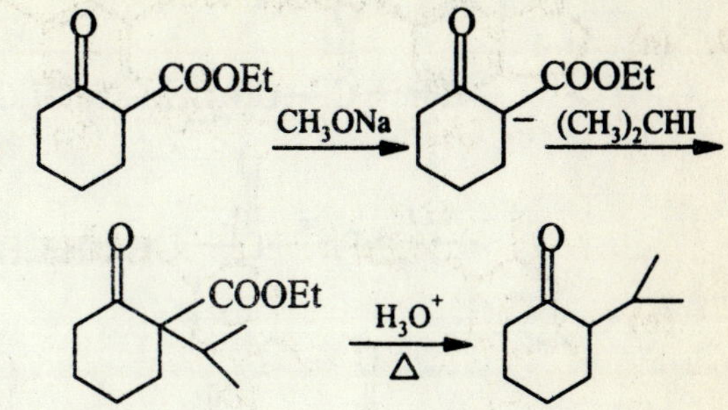 COOEt $\xrightarrow{CH_3ONa}$ COOEt $\xrightarrow{(CH_3)_2CHI}$

COOEt $\xrightarrow[\Delta]{H_3O^+}$

33. (a) 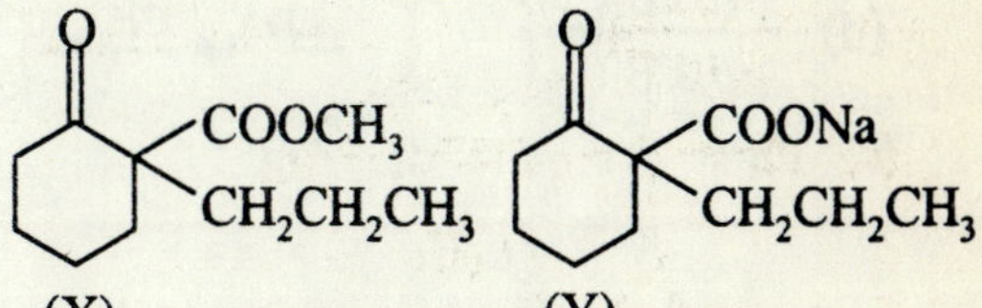 $COOCH_3$ $CH_2CH_2CH_3$ (X) COONa $CH_2CH_2CH_3$ (Y)

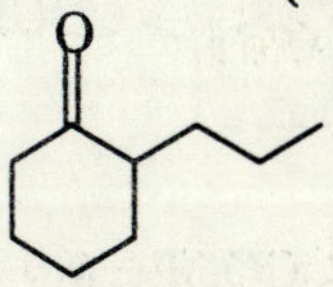

(Z)

(b) CH_3—$(CH_2)_{10}$—COOH

34. (a) 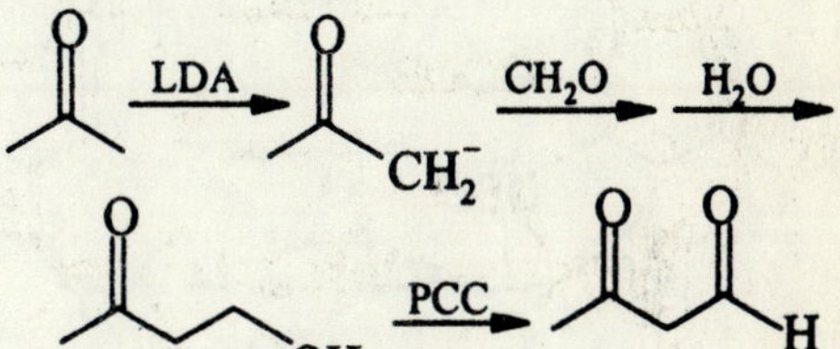 $\xrightarrow{LDA}$ CH_2^- $\xrightarrow{CH_2O}$ $\xrightarrow{H_2O}$

OH $\xrightarrow{PCC}$ H

(b) 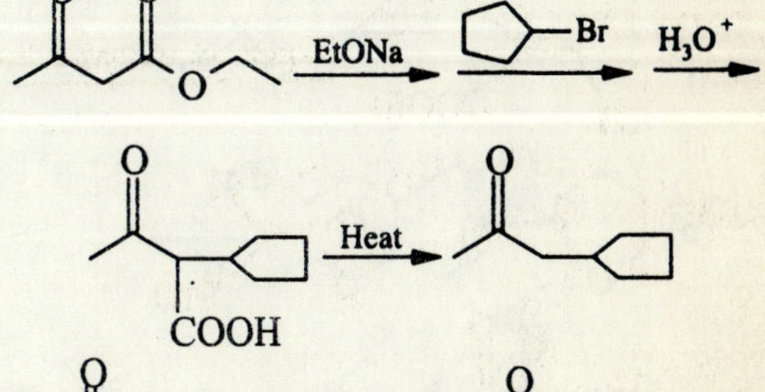 O $\xrightarrow{EtONa}$ $\xrightarrow{\text{cyclopentyl-Br}}$ $\xrightarrow{H_3O^+}$

COOH $\xrightarrow{Heat}$

(c) 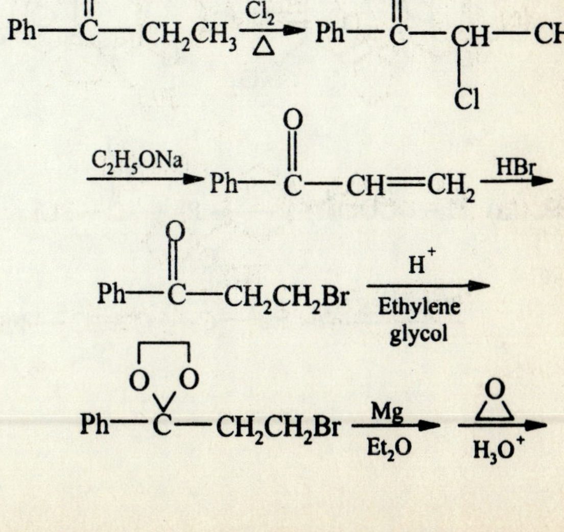 Ph—C(=O)—CH_2CH_3 $\xrightarrow[\Delta]{Cl_2}$ Ph—C(=O)—CH(Cl)—CH_3

$\xrightarrow{C_2H_5ONa}$ Ph—C(=O)—CH=CH_2 $\xrightarrow{HBr}$

Ph—C(=O)—CH_2CH_2Br $\xrightarrow[\text{Ethylene glycol}]{H^+}$

Ph—C(O O)—CH_2CH_2Br $\xrightarrow[Et_2O]{Mg}$ $\xrightarrow[H_3O^+]{\text{ethylene oxide}}$

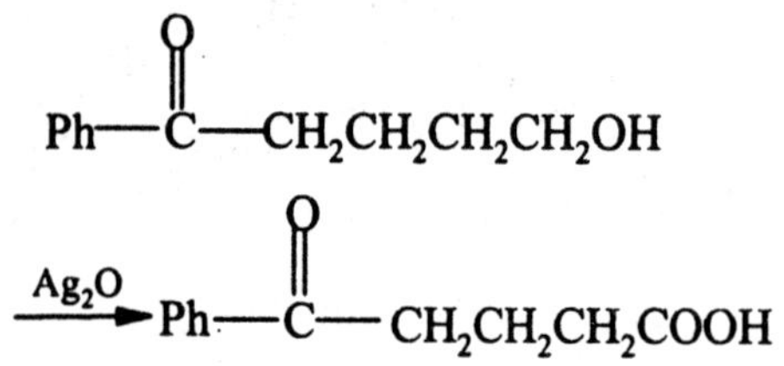

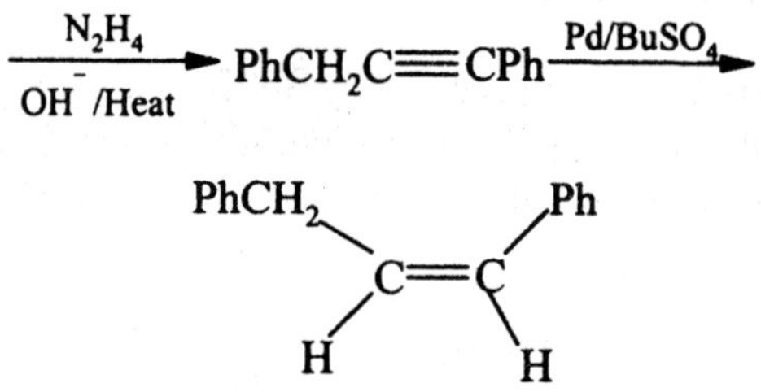

Ph—C(=O)—$CH_2CH_2CH_2CH_2OH$ $\xrightarrow{Ag_2O}$ Ph—C(=O)—$CH_2CH_2CH_2COOH$

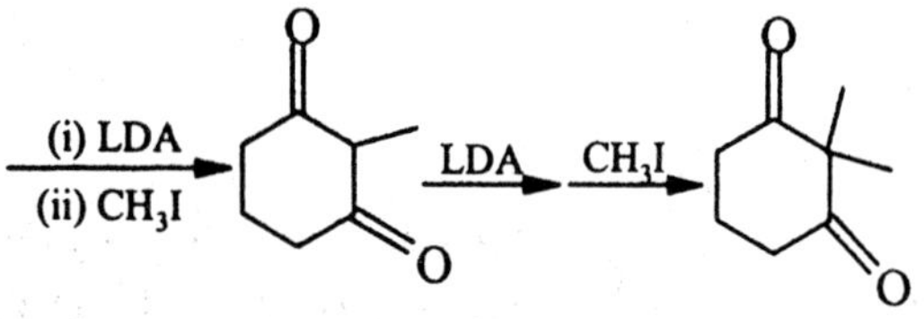

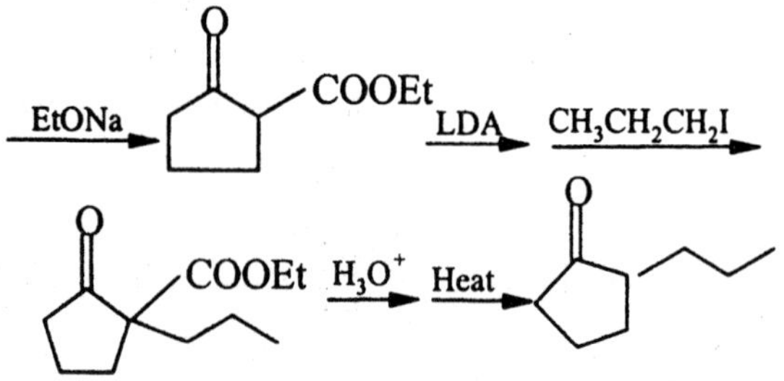

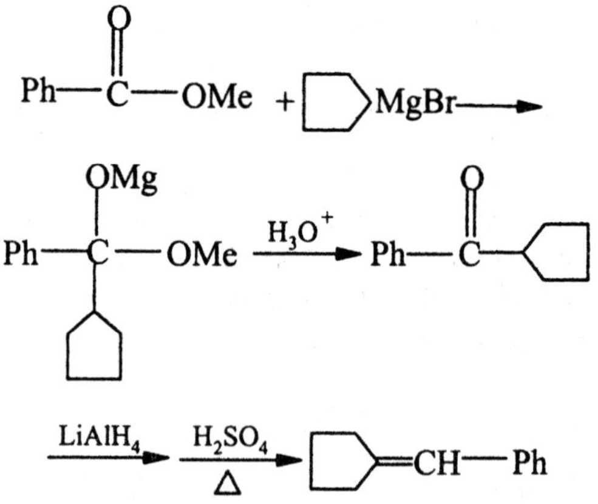

(d) $\xrightarrow{\text{(i) LDA, (ii) } CH_3I}$ $\xrightarrow{LDA}$ $\xrightarrow{CH_3I}$

(e) $\xrightarrow{EtONa}$ COOEt $\xrightarrow{LDA}$ $\xrightarrow{CH_3CH_2CH_2I}$ COOEt $\xrightarrow{H_3O^+}$ $\xrightarrow{Heat}$

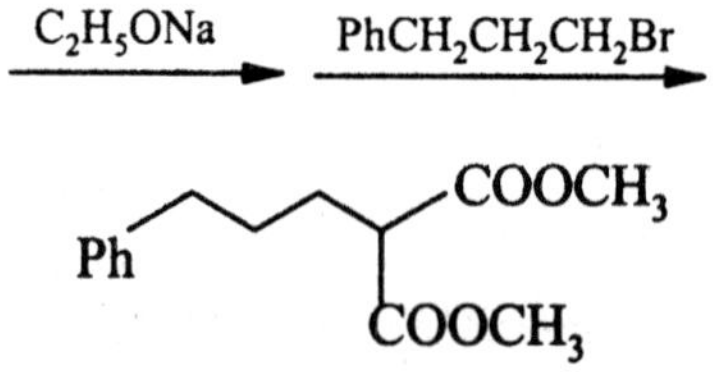

35. (a) OEt → Et OEt Et

$\xrightarrow{-EtO^-}$ Et

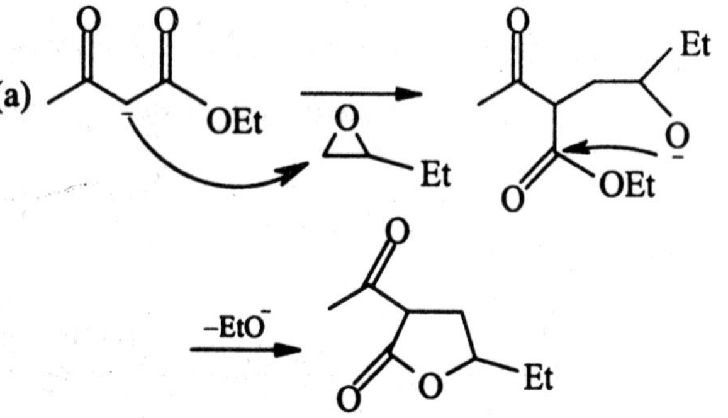

(b) H—O Br Br $\xrightarrow{-H^+}$

Br $\xrightarrow{-Br}$ NO_2 NO_2

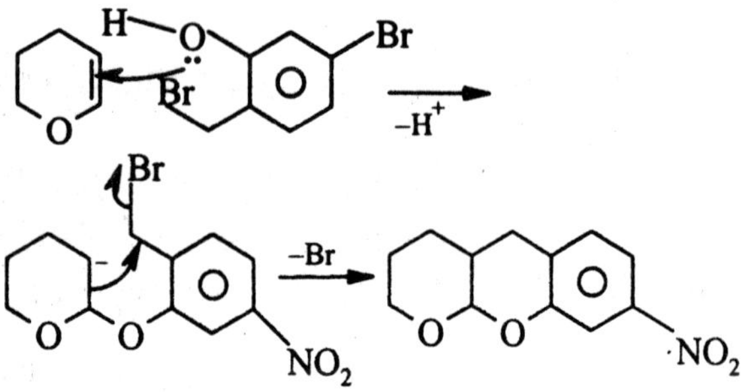

38. (a) Ph—COOH $\xrightarrow{SOCl_2}$ Ph—C(=O)—Cl

$\xrightarrow{PhC\equiv CNa}$ Ph—C(=O)—C≡C—Ph

$\xrightarrow{N_2H_4,\ \bar{O}H/Heat}$ $PhCH_2C\equiv CPh$ $\xrightarrow{Pd/BuSO_4}$ ($PhCH_2$)(H)C=C(Ph)(H)

(b) Ph—C(=O)—OMe + ⬠MgBr →

Ph—C(OMg)(⬠)—OMe $\xrightarrow{H_3O^+}$ Ph—C(=O)—⬠

$\xrightarrow{LiAlH_4}$ $\xrightarrow{H_2SO_4,\ \Delta}$ ⬠=CH—Ph

(c) $\xrightarrow{C_2H_5ONa}$ $\xrightarrow{PhCH_2CH_2CH_2Br}$

Ph $COOCH_3$ $COOCH_3$

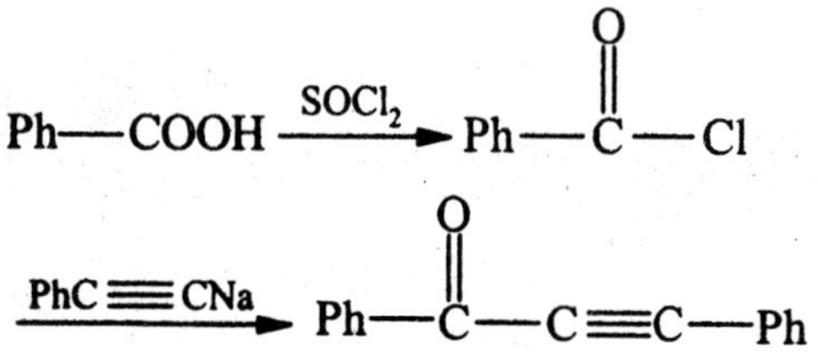

Objective Exercise

1. a, b	16. c	31. a	46. b	61. d
2. d	17. a	32. b	47. d	62. b
3. a	18. c	33. b	48. b	63. a
4. a	19. b	34. c	49. b	64. a
5. b	20. b	35. a	50. c	65. a
6. a	21. c	36. c	51. a	66. b
7. d	22. c	37. d	52. d	67. b
8. c	23. d	38. b	53. d	68. b
9. a	24. b	39. d	54. d	
10. d	25. a	40. b	55. a	
11. d	26. c	41. b	56. b	
12. a	27. a	42. d	57. b	
13. c	28. c	43. b	58. b	
14. b	29. d	44. d	59. d	
15. d	30. d	45. b	60. c	

CHAPTER 9

Practice Exercise

1. Nitro-alkanes are formed by SN-2 mechanism when an alkyl halide react with silvernitrite. On that basis, secondary and tertiary halides give poor yield of nitro alkane.

2. Silver nitrite has partial covalent character in Ag-O bond and it ionizes reversibly to produce ambident nucleophile-nitrite ion:

$$AgONO \rightleftharpoons Ag^+ + {}^-O-\ddot{N}=O$$

As we move from primary to tertiary alkyl halides, reactivity towards SN-1 reaction increases. Also, presence of above equilibrium catalyzes heterolysis of alkyl halide to form carbocation. Once the carbocation is formed, oxygen of nitrite ion (ONO^-) has greater tendency to attack at positively charged carbon leading to formation of nitrite.

3. Sodium nitrite ($NaNO_2$) has very high ionic character compared to $AgNO_2$, therefore former is a better source of nucleophile for the reaction and yield is improved.
4. Review tautomerism exhibited by nitroalkanes possessing α-hydrogen.
5. By treatment with nitrous acid.
6. By the action of nitrous acid followed by work-up with alkali.

7. (a) 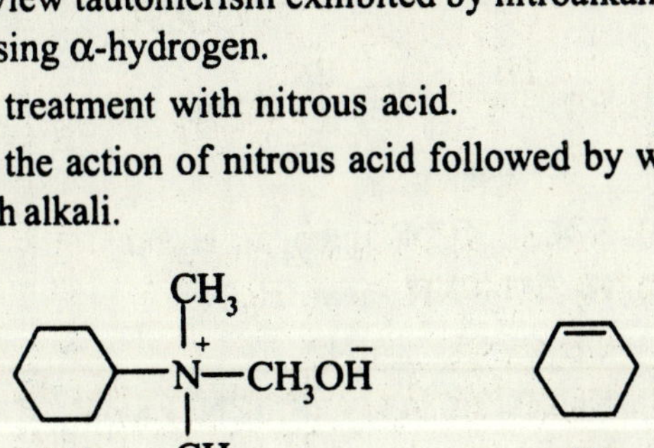 CH_3, $\overset{+}{N}-CH_2O\overline{H}$, CH_3 (A) (B)

(b) 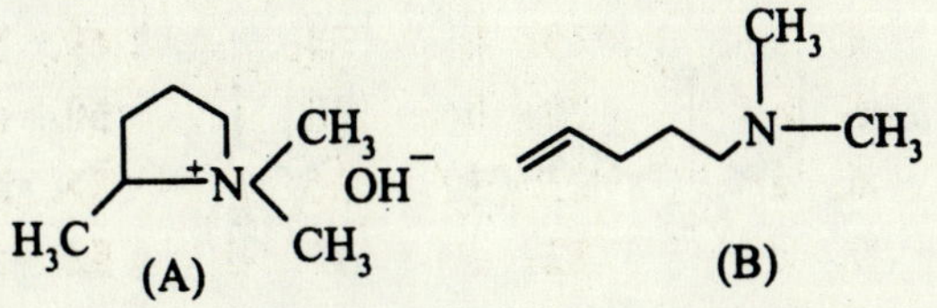 CH_3, $\overset{+}{N}$, $\overline{OH}$, H_3C, CH_3 (A) CH_3, $N-CH_3$ (B)

CH_3, $\overset{+}{N}-CH_2OH^-$, CH_3 (C) CH_2 CH_2 (D)

8. (a) Hexene

(b) 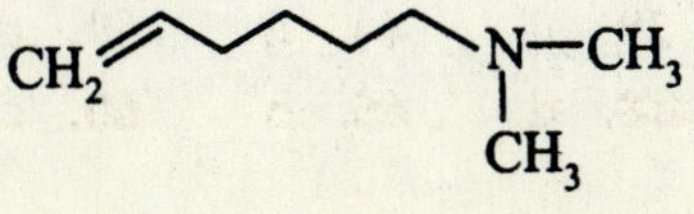 CH_2 $N-CH_3$ CH_3

(c) 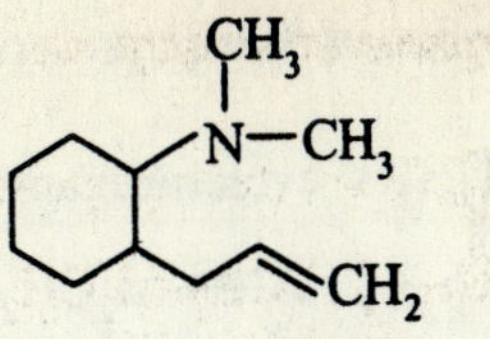 CH_3, $N-CH_3$, CH_2 (d) 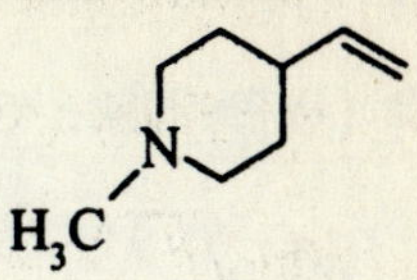 N, H_3C

(e) 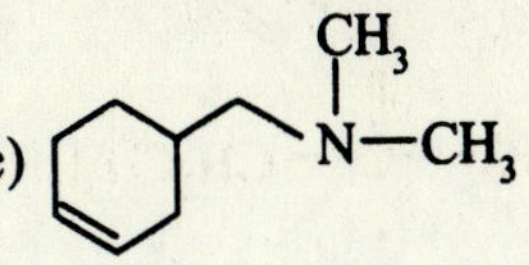 CH_3, $N-CH_3$

9. 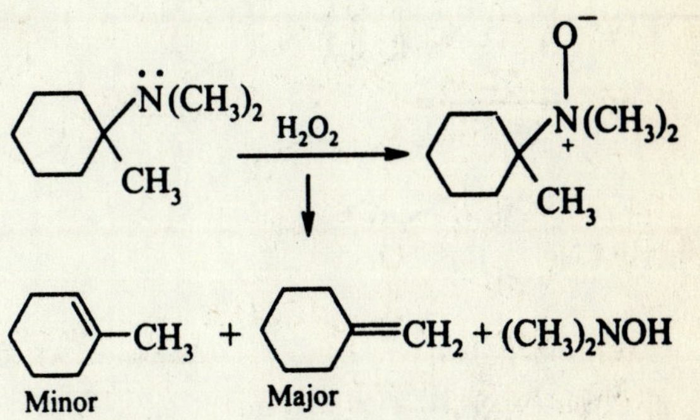 $\ddot{N}(CH_3)_2$, CH_3 $\xrightarrow{H_2O_2}$ O^-, $\overset{+}{N}(CH_3)_2$, CH_3

$-CH_3$ + $=CH_2 + (CH_3)_2NOH$

Minor Major

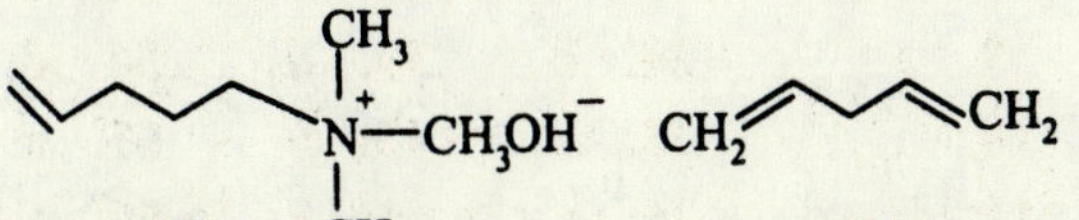

10. (a) CH_3NH_2/heat → Reduction with H_2/Ni → product.

(b) NH_3/heat → Reduction with H_2/Ni → product.

(c) C_6H_5Br/heat.

(d) Aniline/heat → Reduction $LiAlH_4$ → product.

(e) NH_3/heat → Reduction H_2/Ni → product.

(f) 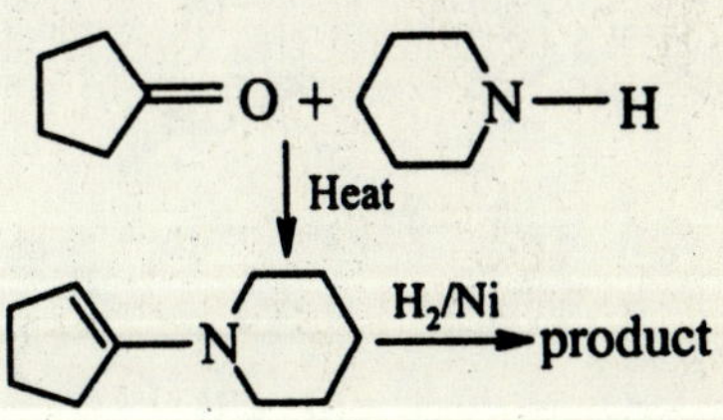 =O + N—H
Heat
N $\xrightarrow{H_2/Ni}$ product

11. (a) KCN/heat under pressure → C_6H_5CN $\xrightarrow{LiAlH_4}$ Product.

(b) Aqueous KCN → reduction $LiAlH_4$ → product.

(c) NH_3/heat → CH_3 NH_2 O $\xrightarrow{LiAlH_4}$ product.

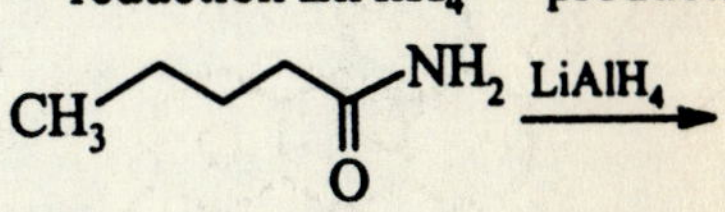

(d) $LiAlH_4$ → $SOCl_2$ → aqueous KCN $\xrightarrow{LiAlH_4}$ product.

(e) HCN → H_3C OH CH_3 CN $\xrightarrow{LiAlH_4}$

H_3C OH CH_3 CH_2NH_2

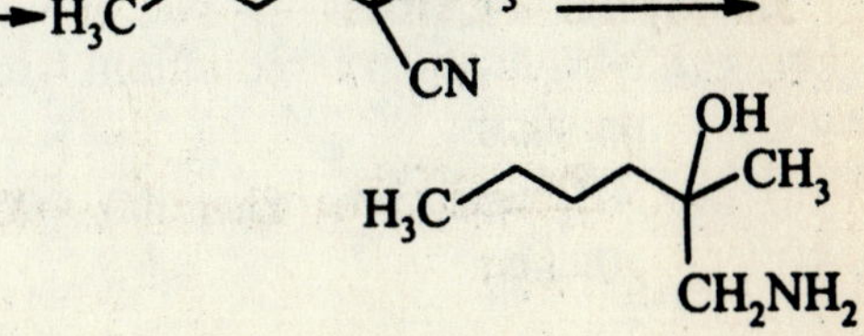

(f) $NH_3 \xrightarrow{SN\text{-}2}$ inversion of configuration (product)

12. (a) Ph_2NH < $C_6H_5NH_2$ < cyclohexanamine

(b) 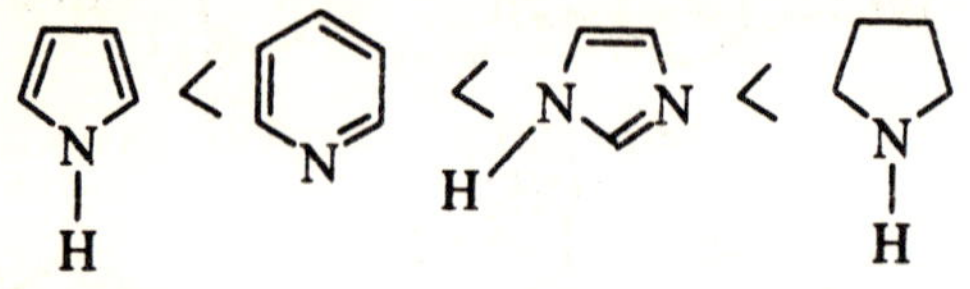 N–H (pyrrole) < pyridine (N) < imidazole (H–N, N) < pyrrolidine (N–H)

(c) 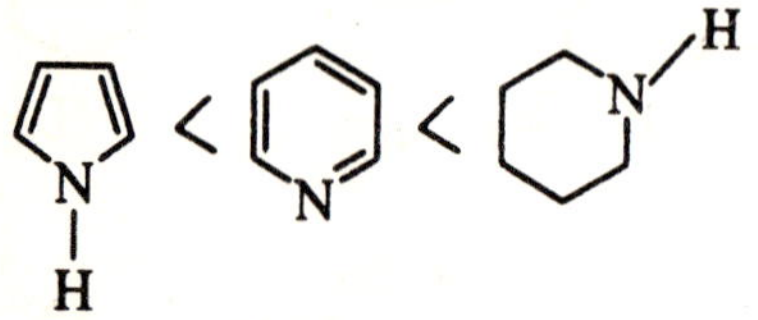 N–H (pyrrole) < pyridine (N) < piperidine (N–H)

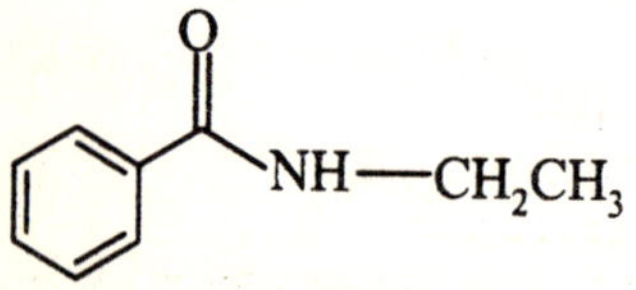

14. (a) 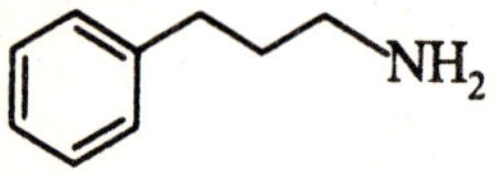 $C_6H_5CH_2CH_2CH_2NH_2$

(b) C_6H_5CO–NH—CH_2CH_3 (c) N-methylpiperidine N-oxide (N^+–O^-, CH_3)

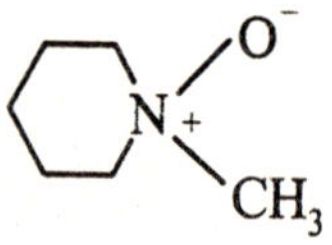

(d) 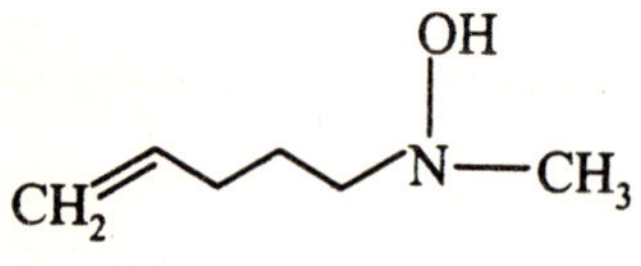 CH_2=CH–$CH_2CH_2CH_2$–N(OH)—CH_3

(e) 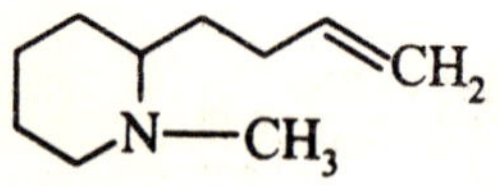 2-substituted N—CH_3 piperidine with side chain =CH_2

(f) 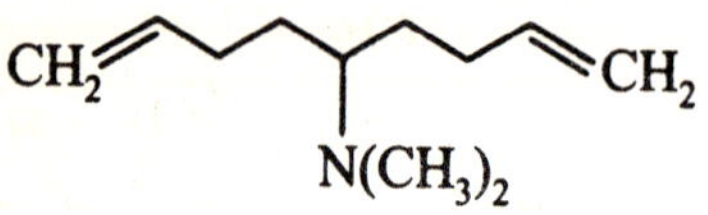 CH_2=CH–CH_2–CH(N$(CH_3)_2$)–CH_2CH_2–CH=CH_2

(g) 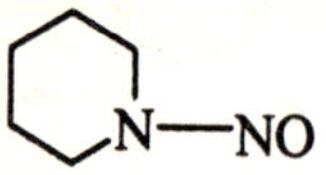 piperidine N—NO (h) 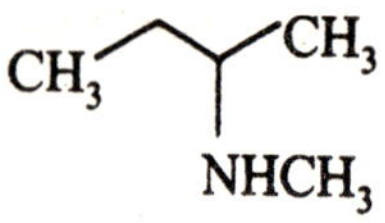 CH_3–CH(NHCH$_3$)–CH_3

(i) 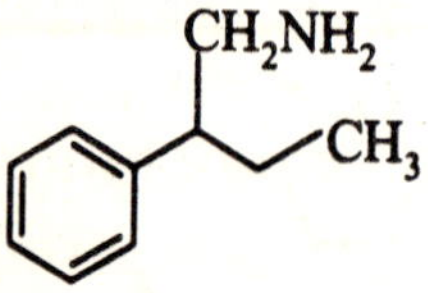 C_6H_5–CH(CH_2NH_2)–CH_2CH_3 (k) No reaction

(l)  H_3C–$CH_2CH_2CH_2CH_2$–NH_2

15. Treat etheral solution of products with aqueous HCl solution. Amine will go into aqueous phase by forming salt with HCl.

16. (a) NH_3/heat → reduction with $LiAlH_4$ → product.

(b) NH_3/heat → reduction with $LiAlH_4$ → product.

(c) C_2H_5Br → product.

(d) NH_3/heat → $LiAlH_4$ → product.

(e) Pyrrolidine → $LiAlH_4$ → product.

17. (a) 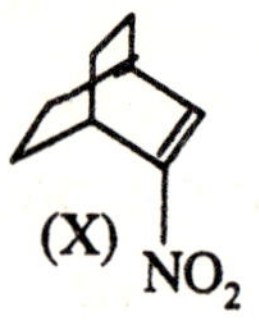 (X) bicyclic alkene with NO_2 (b) 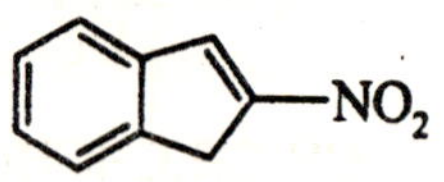 (Y) indene with NO_2

18. $CH_3CH_2CH(CH_3)$—C(=O)—O—CH(CH_3)—CH_2CH_3 (A)

$CH_3CH_2CH(CH_3)$—COOH (B) $CH_3CH_2CH(OH)CH_3$ (C)

$CH_3CH_2CH(CH_3)$—$CONH_2$ (D) $D \xrightarrow[Br_2]{NaOH}$

$CH_3CH_2CH(CH_3)$—NH_2 $\xrightarrow{HNO_2}$ C

1°-amine

19. (a) $SOCl_2$, KCN(aqueous), H_2/Ni

(b) H_2CrO_4, NH_3/heat, H_2/Ni

(c) CH_3COCl, $LiAlH_4$

(d) $NaNO_2$/HCl/0° C, CuCN, $LiAlH_4$, CH_3COCl, $LiAlH_4$

20. 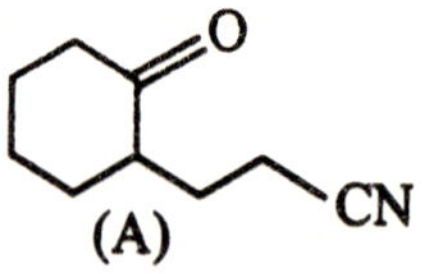 (A) 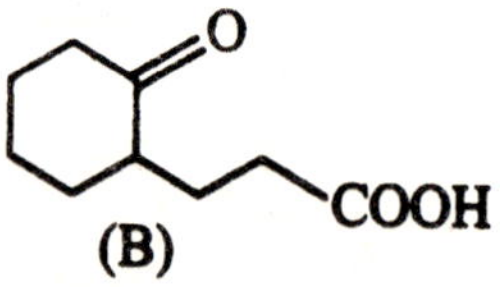 2-(2-cyanoethyl)cyclohexanone (=O, CN) (B) 2-(2-carboxyethyl)cyclohexanone (=O, COOH)

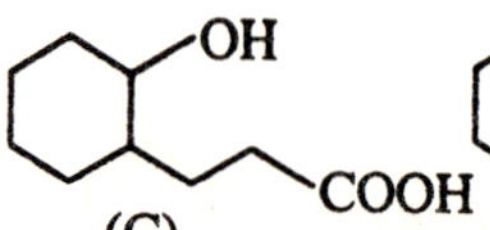

(C) 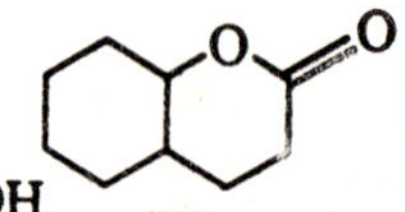 OH, COOH (D) 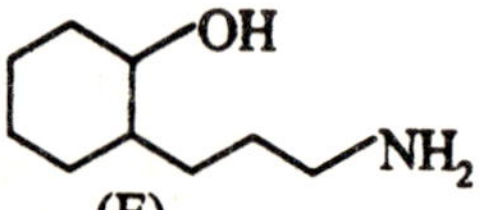 lactone (O, =O) (E) OH, NH_2

21. 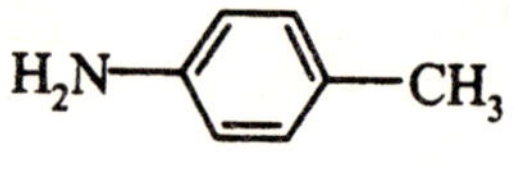 H_2N–C_6H_4–CH_3

22. 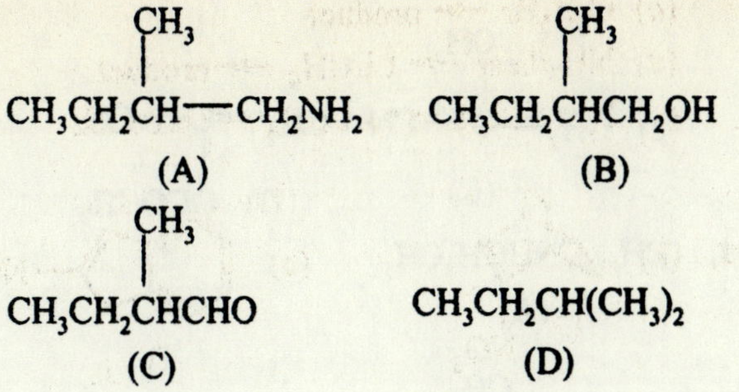 $CH_3CH_2CH(CH_3)CH_2NH_2$ (A) $CH_3CH_2CH(CH_3)CH_2OH$ (B)

$CH_3CH_2CH(CH_3)CHO$ (C) $CH_3CH_2CH(CH_3)_2$ (D)

23. 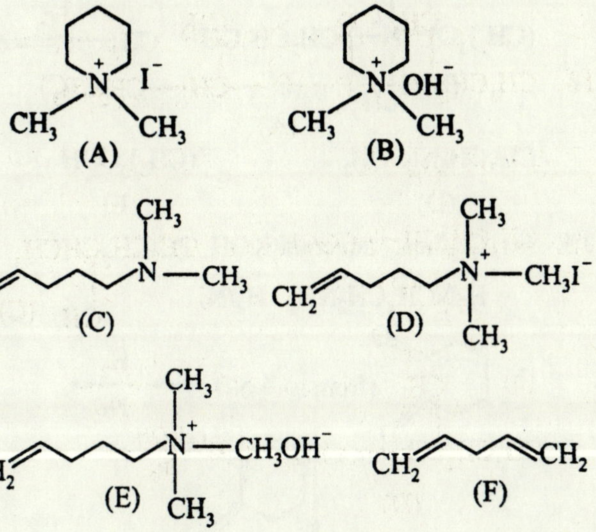 (A) 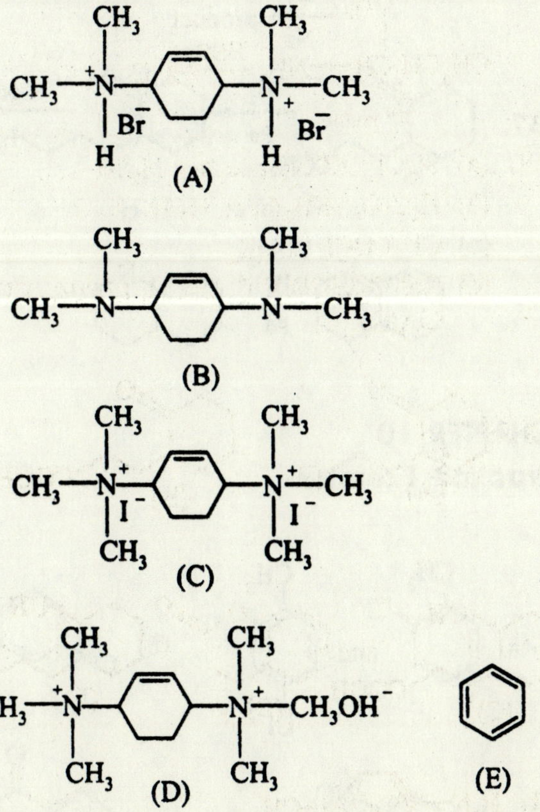 (B) 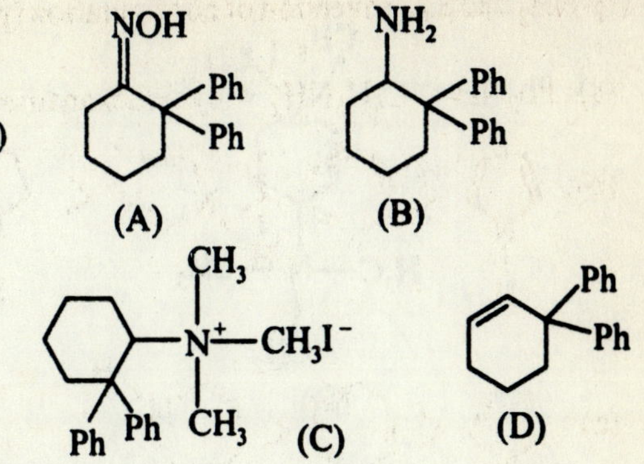 (C) (D) (E) (F)

24. (a) (A) (B) (C) (D) (E)

(b) (A) (B) (C) (D)

25. 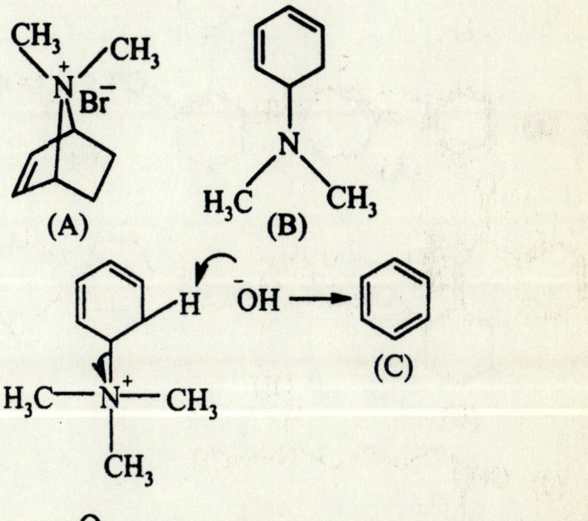 (A) (B) (C)

26. 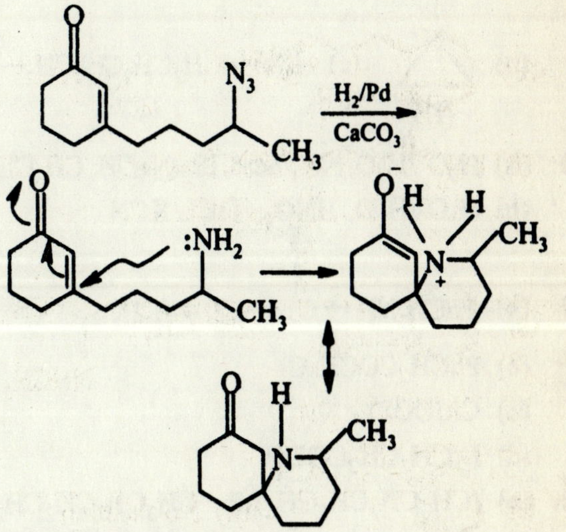 H_2/Pd, $CaCO_3$

27. Review Hinsberg's method.

28. 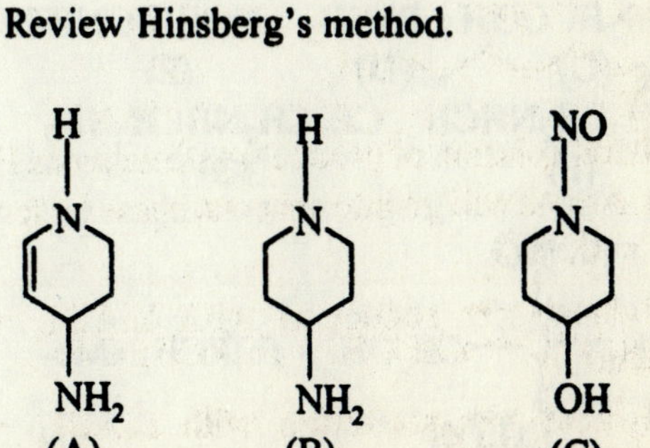 (A) (B) (C)

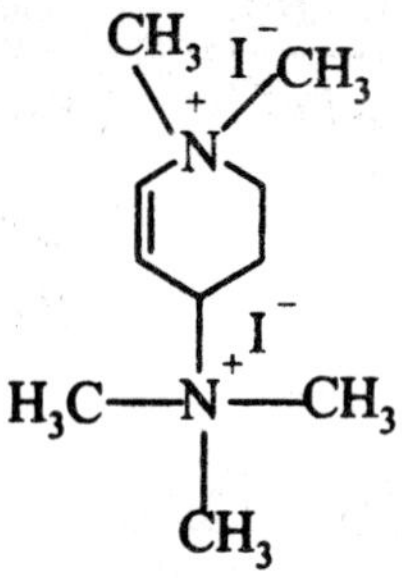

(D)

29. (a) 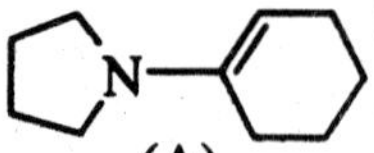 (A) 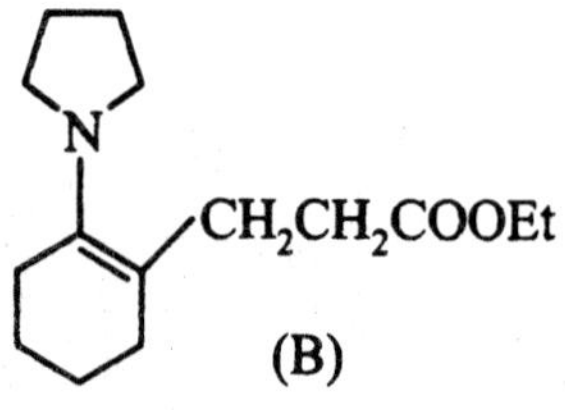 (B) CH_2CH_2COOEt

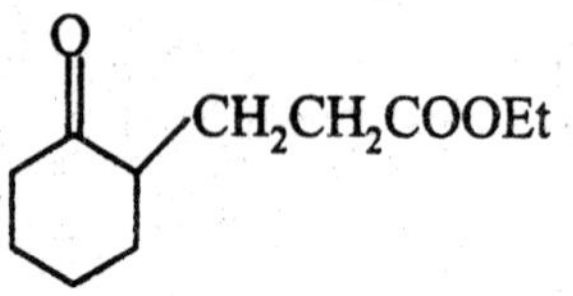

(C) CH_2CH_2COOEt

(b) 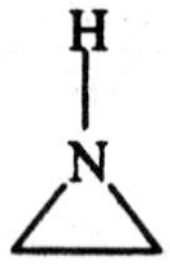 (D) (c) $H_2N—CH_2CH_2CH_2CH_2—NH_2$ (E)

30. (a) H_3O^+/H_2O, NH_3/heat, Br_2/NaOH, CH_3Cl/KOH
(b) H_3O^+/H_2O, HNO_2, TsCl, KCN

31. (a) $PhCH_2NH—C(=O)—CH{=}CH_2$
(b) $PhCH_2COCH_2Cl$
(c) ClCOOEt
(d) $BrCH_2CH_2COOH$

32. (a) $(CH_3CH_2CH_2CH_2)_3B$ (A) $CH_3CH_2CH_2CH_2NH_2$ (B)
(b) PhNHCOOH (C) $PhNH_2$ (D) PhNHCONHPh (E)
(c) CH_3CH_2NHCN (F) $CH_3CH_2NHCH_2NH_2$ (G)

33. (a) $CH_3—C(NO_2)(CH_2OH)—CH_2OH$ (b) RCH_2NMe_2

(c) $CH_2{=}C(OH)—CN$ (C) $CH_2{=}C(OCOCH_3)—CN$ (D)

34. $(CH_3)_3CNHCH_2CH_3$ (A)

$(CH_3)_3C—\overset{+}{N}(CH_3)_2—CH_2CH_2OH^-$ (B) $CH_3—C(CH_3){=}CH_2$ (C)

$CH_3CH_2N(CH_3)_2$ (D) $(CH_3)_3COH$ (E)

35. (a) Cl_2/heat, aqueous KOH, PCC, $H_2NCH_2CH_2NH_2$, H_2/Ni
(b) 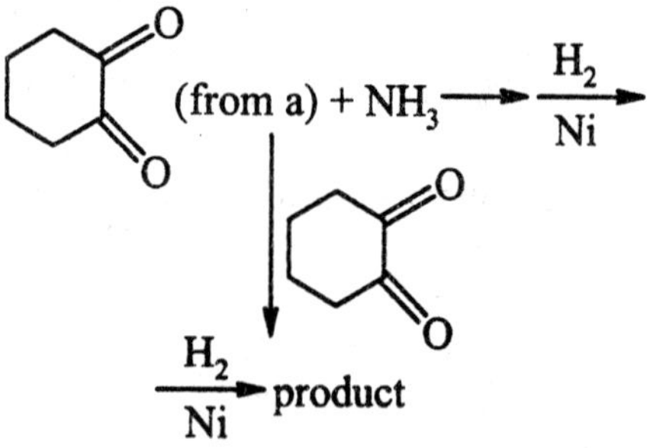 (from a) + NH_3 → $\xrightarrow[Ni]{H_2}$ product

37. 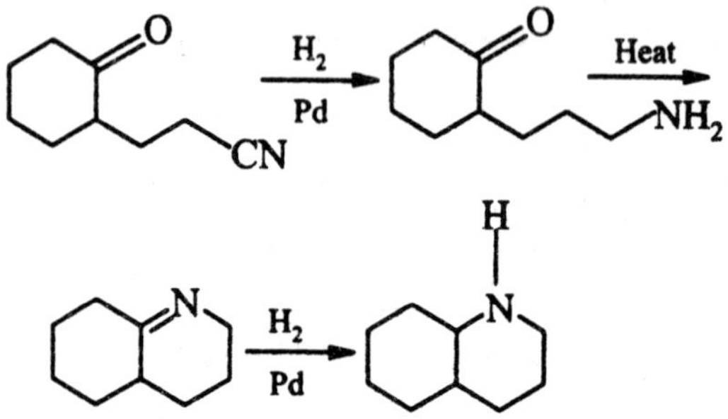 $\xrightarrow[Pd]{H_2}$ $\xrightarrow{Heat}$ NH_2 $\xrightarrow[Pd]{H_2}$

CHAPTER 10

Practice Exercise

1.(a) CH_3, Cl and CH_3, Cl (b) 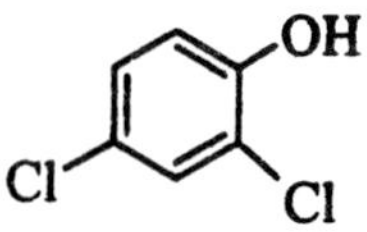 OH, Cl, Cl

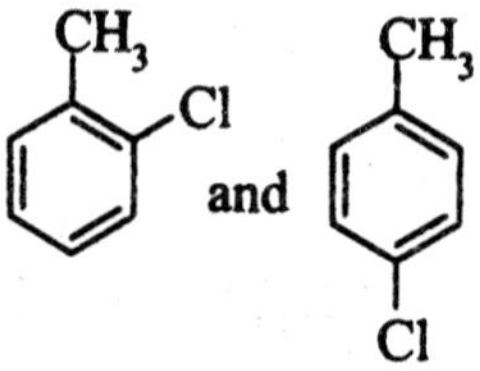

(c) 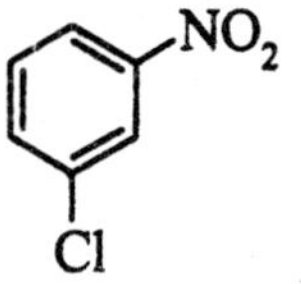 NO_2, Cl (d) 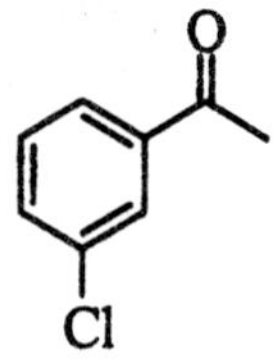 O, Cl

(e) 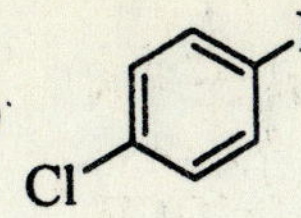 4-chlorophenyl–$NHCOCH_3$ (f) 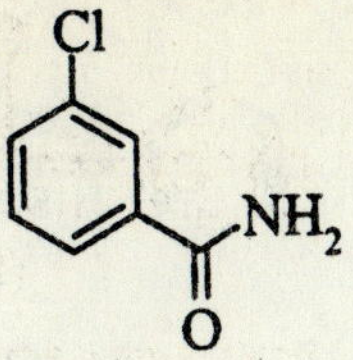 3-Cl–C$_6$H$_4$–$C(=O)NH_2$

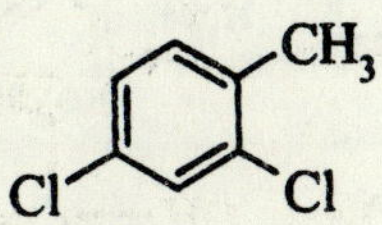

(g) 2,4-dichloro-1-CH_3 benzene (h) 4-Cl–C$_6$H$_4$–OCH_3

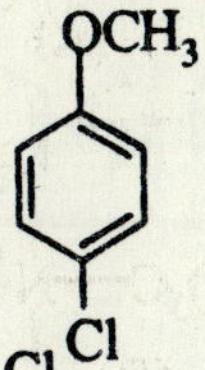

2\. (a) 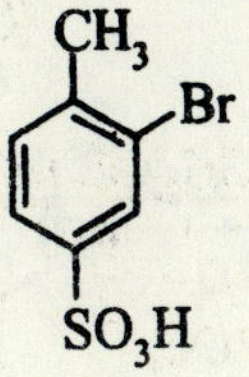 CH_3, Br, SO_3H substituted benzene (b) 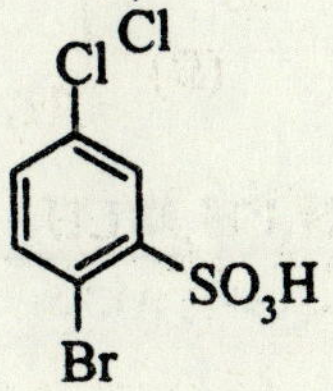 Cl, SO_3H, Br substituted benzene

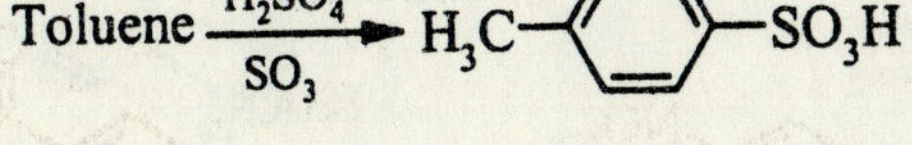

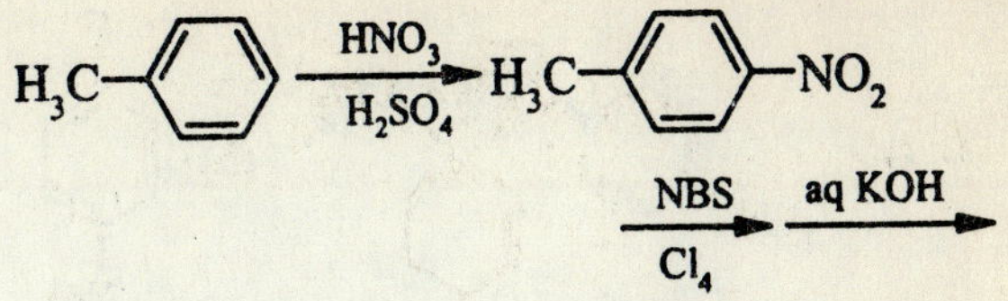

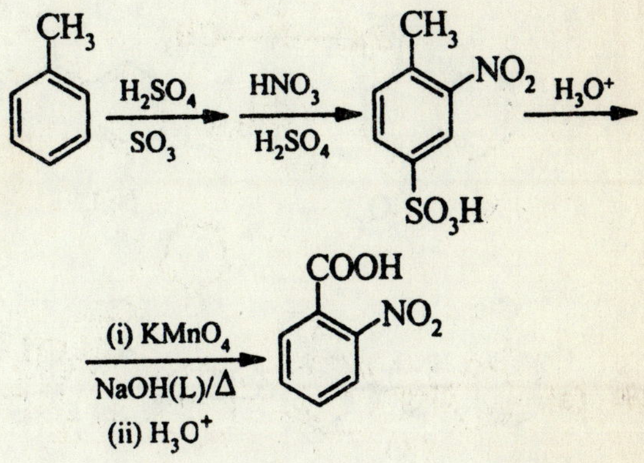

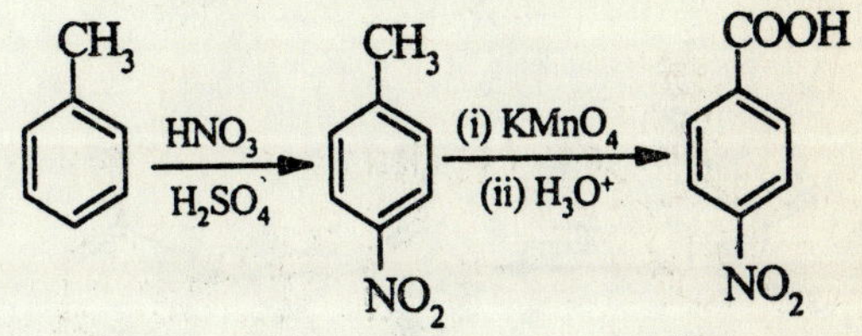

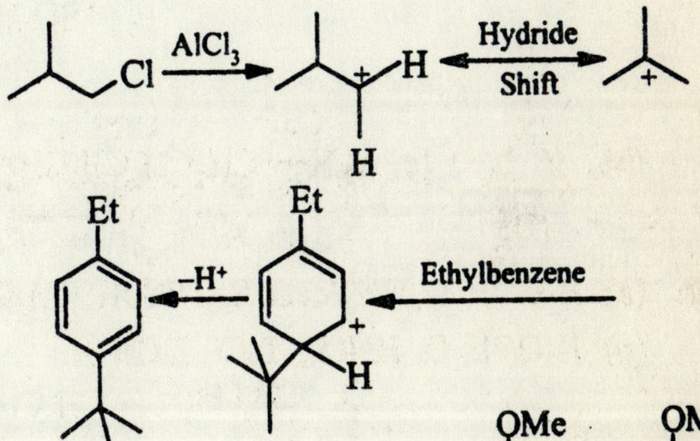

(c) 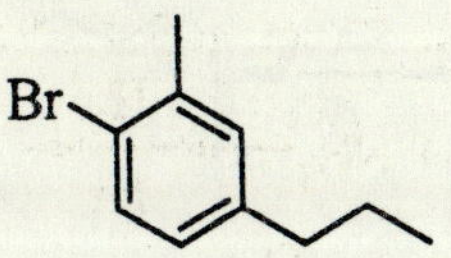 (d)

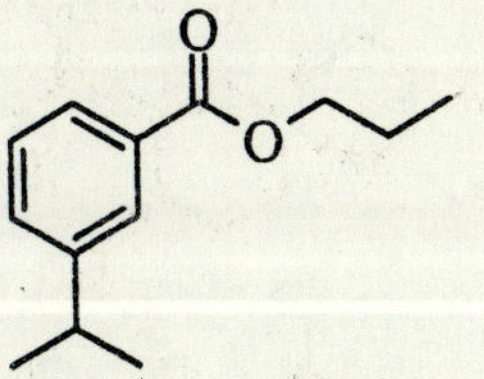

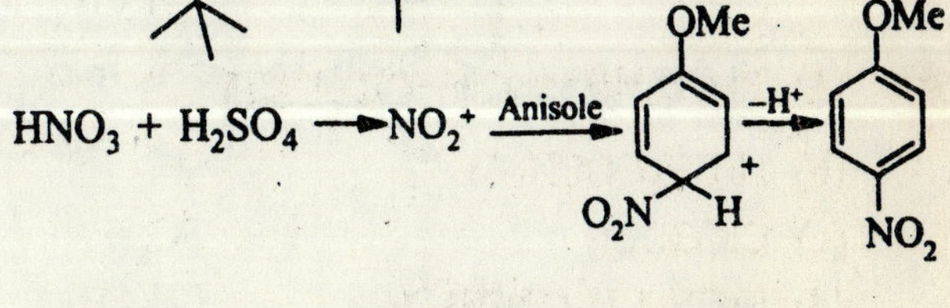

(e) 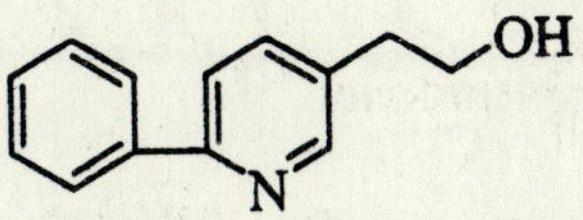 OH, N

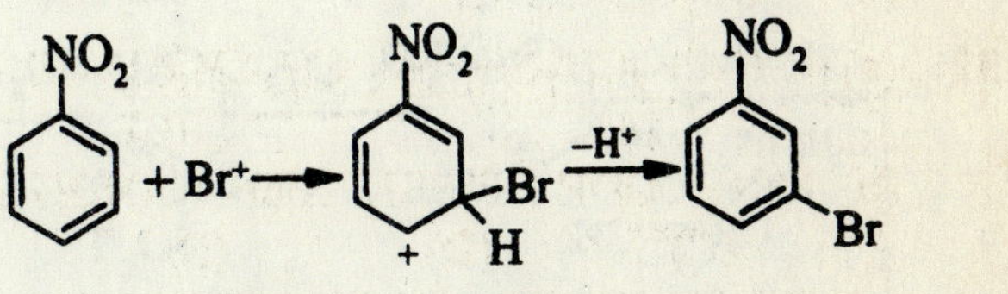

(f) 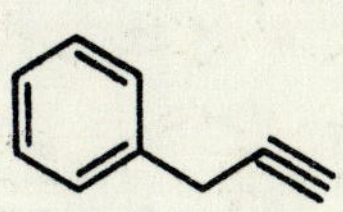 (g)

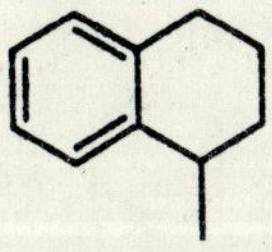

3\. (a) 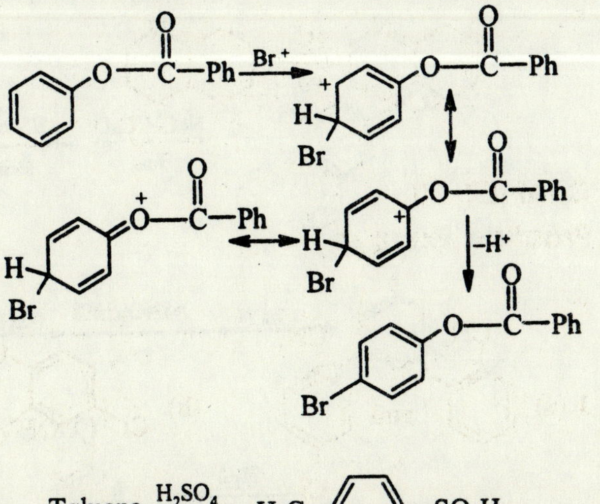 Ph–O–C(=O)–Ph $\xrightarrow{Br^+}$ arenium ion (H, Br, +) ⟷ ⟷ ; $-H^+$ → 4-Br–C$_6$H$_4$–O–C(=O)–Ph

4\. (a) 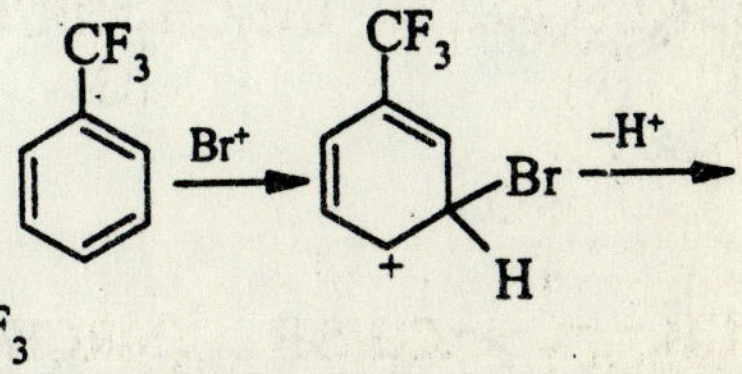 Toluene $\xrightarrow[SO_3]{H_2SO_4}$ H_3C–C$_6$H$_4$–SO_3H

(b) H_3C–C$_6$H$_5$ $\xrightarrow[H_2SO_4]{HNO_3}$ H_3C–C$_6$H$_4$–NO_2 $\xrightarrow[Cl_4]{NBS}$ $\xrightarrow{aq\ KOH}$

(c) CH_3–C$_6$H$_5$ $\xrightarrow[SO_3]{H_2SO_4}$ $\xrightarrow[H_2SO_4]{HNO_3}$ (CH_3, NO_2, SO_3H) $\xrightarrow{H_3O^+}$ $\xrightarrow[\text{NaOH(L)}/\Delta \text{ (ii) } H_3O^+]{\text{(i) } KMnO_4}$ (COOH, NO_2)

(d) CH_3–C$_6$H$_5$ $\xrightarrow[H_2SO_4]{HNO_3}$ (CH_3, NO_2) $\xrightarrow[\text{(ii) } H_3O^+]{\text{(i) } KMnO_4}$ (COOH, NO_2)

5\. (a) Cl $\xrightarrow{AlCl_3}$ + H, H ⟷ (Hydride Shift) + ; Ethylbenzene → Et, +, H $\xrightarrow{-H^+}$ Et

(b) $HNO_3 + H_2SO_4 \longrightarrow NO_2^+$ $\xrightarrow{\text{Anisole}}$ OMe, O_2N, H, + $\xrightarrow{-H^+}$ OMe, NO_2

(c) NO_2–C$_6$H$_5$ + Br^+ → NO_2, Br, H, + $\xrightarrow{-H^+}$ NO_2, Br

6\. (a) CF_3–C$_6$H$_5$ $\xrightarrow{Br^+}$ CF_3, Br, H, + $\xrightarrow{-H^+}$ CF_3, Br

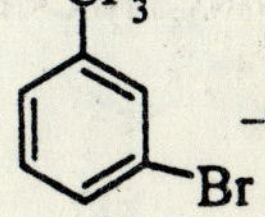

$-CF_3$ deactivate the ring towards electrophilic aromatic substitution reaction.

7. (a) 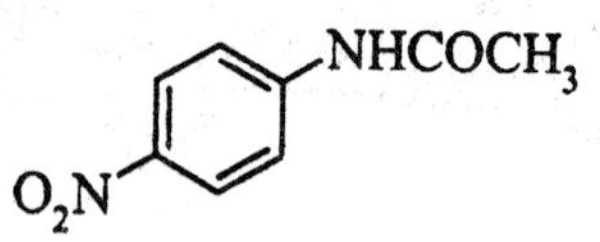 O_2N–C_6H_4–$NHCOCH_3$ (b) F–C_6H_4–NO_2 (c) CH_3, NO_2 (naphthalene) (d) 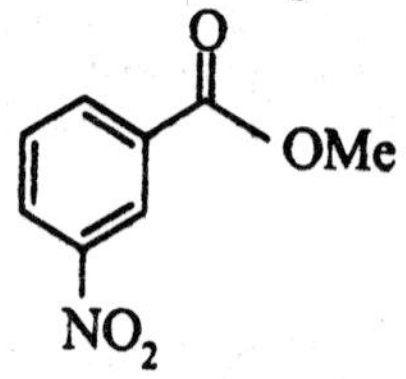 OMe, NO_2

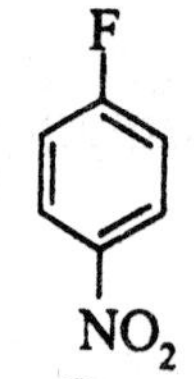

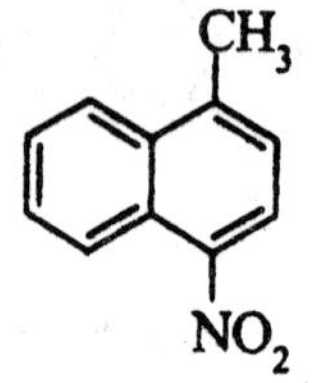

(f) $\xrightarrow{HNO_3/H_2SO_4}$ NO_2 $\xrightarrow{Cl_2/AlCl_3}$ NO_2, Cl

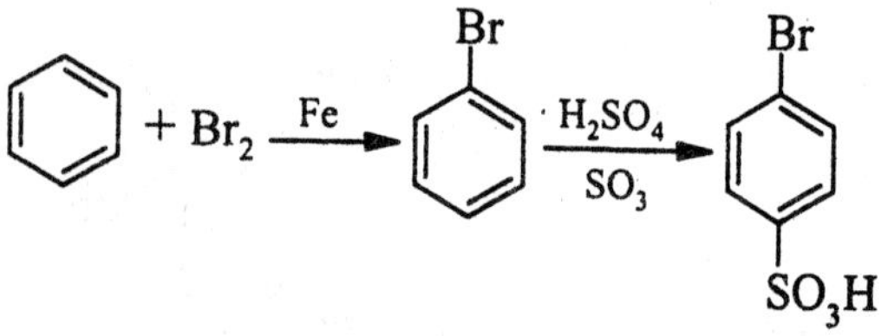

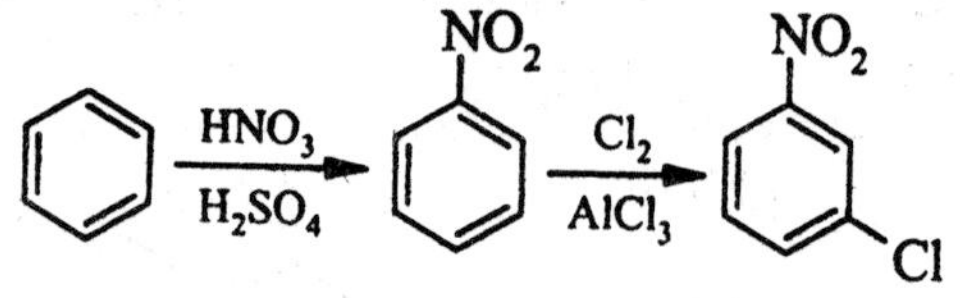

(g) $\xrightarrow{Br_2/Fe}$ Br $\xrightarrow{Mg/Ether}$ $\rightarrow$ $\xrightarrow{H_3O^+}$ OH

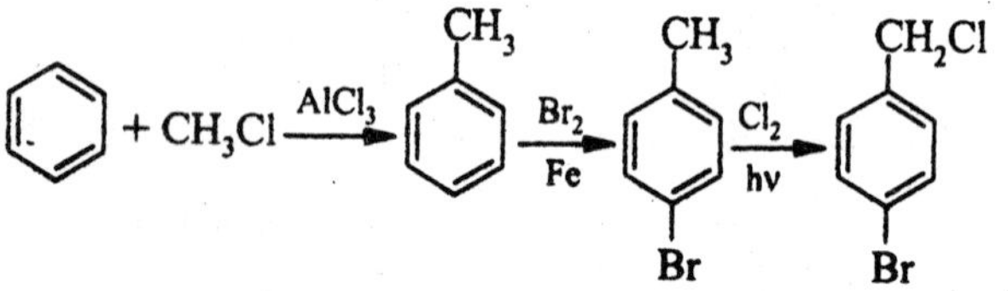

8. (a) 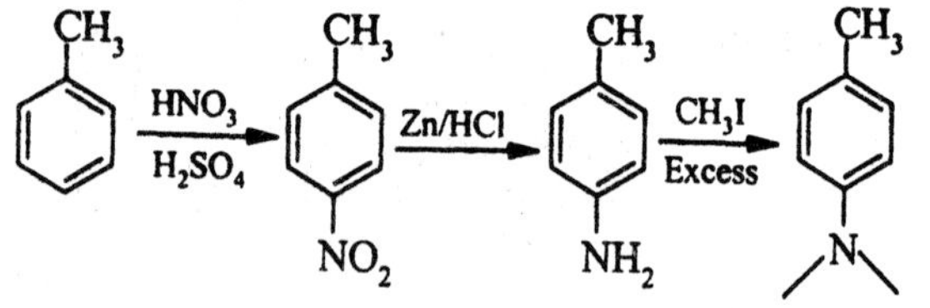 CH_3 $\xrightarrow{HNO_3/H_2SO_4}$ CH_3, NO_2 $\xrightarrow{Zn/HCl}$ CH_3, NH_2 $\xrightarrow{CH_3I \text{ Excess}}$ CH_3, N

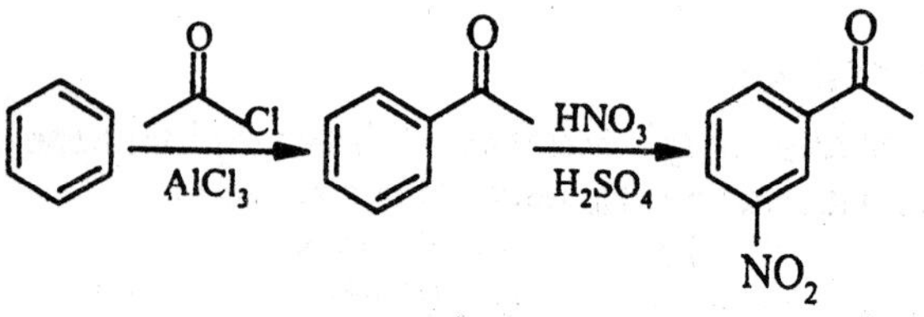

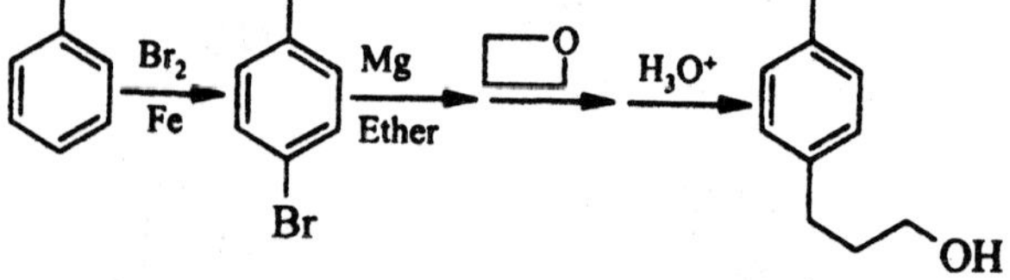

(h) $\xrightarrow{CH_3Cl/AlCl_3}$ $\xrightarrow{\text{(i) } KMnO_4/OH^- \text{ (ii) } H^+/H_2O}$ COOH, COOH

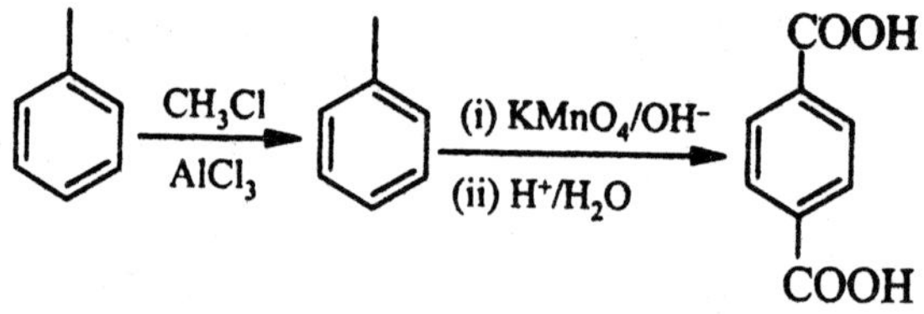

(b) + Br_2 $\xrightarrow{Fe}$ Br $\xrightarrow{H_2SO_4/SO_3}$ Br, SO_3H

(i) + Cl $\xrightarrow{AlCl_3}$ $\xrightarrow{Zn(Hg)/HCl}$

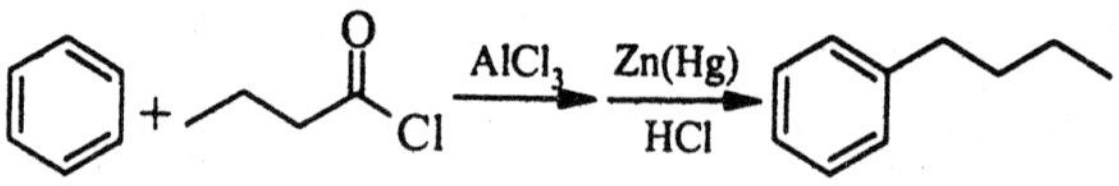

12. 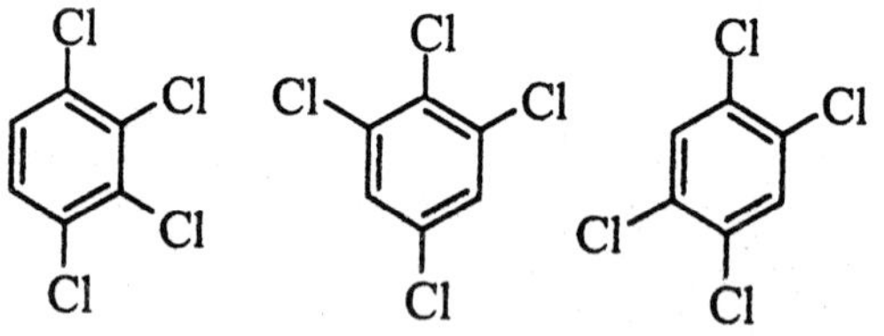 Cl, Cl, Cl, Cl; Cl, Cl, Cl, Cl; Cl, Cl, Cl, Cl (Major)

(c) + CH_3Cl $\xrightarrow{AlCl_3}$ CH_3 $\xrightarrow{Br_2/Fe}$ CH_3, Br $\xrightarrow{Cl_2/h\nu}$ CH_2Cl, Br

13. 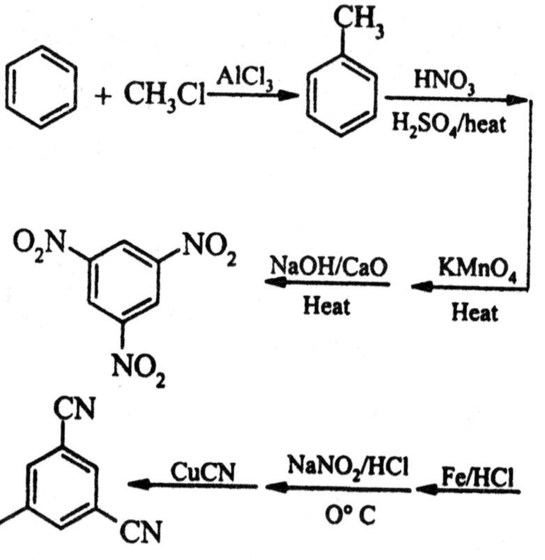 + CH_3Cl $\xrightarrow{AlCl_3}$ CH_3 $\xrightarrow{HNO_3/H_2SO_4\text{/heat}}$ $\xrightarrow{KMnO_4/\text{Heat}}$ $\xrightarrow{NaOH/CaO/\text{Heat}}$ O_2N, NO_2, NO_2 $\xrightarrow{Fe/HCl}$ $\xrightarrow{NaNO_2/HCl, 0°C}$ $\xrightarrow{CuCN}$ CN, NC, CN $\xrightarrow{H_3O^+/H_2O}$ COOH, HOOC, COOH

(d) + C_2H_5Cl $\rightarrow$ CH_2CH_3 $\xrightarrow{H_2SO_4/SO_3}$ CH_2CH_3, SO_3H $\xrightarrow{HNO_3/H_2SO_4}$ CH_2CH_3, NO_2, SO_3H $\xrightarrow{H_3O^+}$ CH_2CH_3, NO_2 $\xrightarrow{Cl_2/h\nu}$ $\xrightarrow{C_2H_5ONa/C_2H_5OH}$ $CH=CH_2$, NO_2

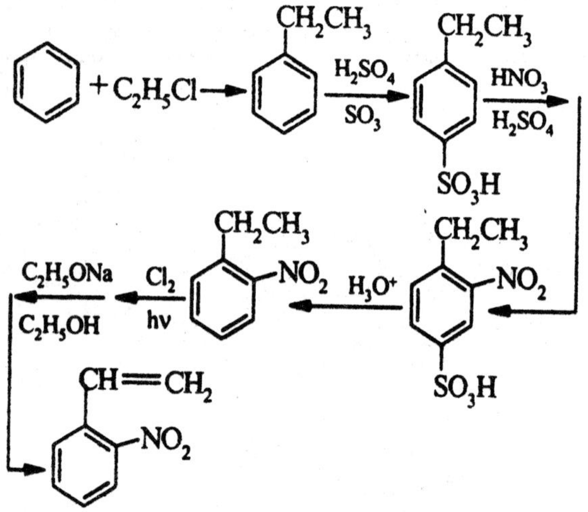

(e) $\xrightarrow{AlCl_3}$ (Cl) $\xrightarrow{HNO_3/H_2SO_4}$ NO_2

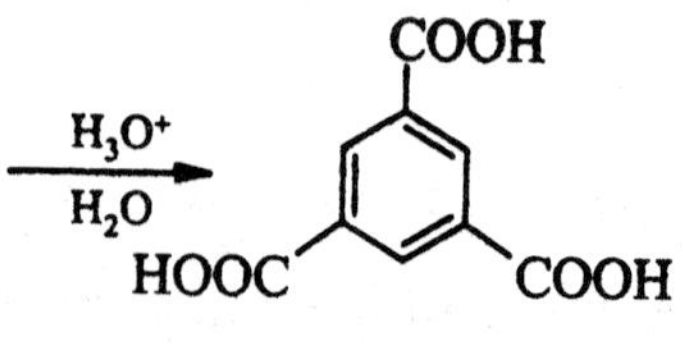

14. (a) II < I < III (b) I < III < II
(c) IV < III < II < I (d) III < I < II

15. I → d　　II → c　　III → b

IV → e　　V → a

16. (a) 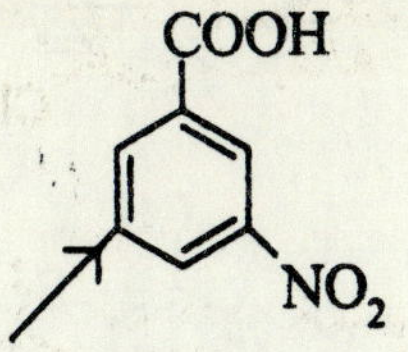 (b) 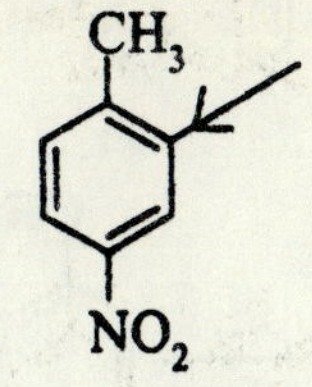 (c) 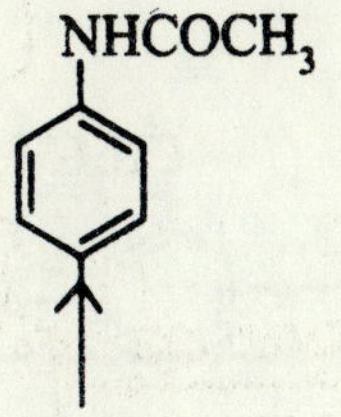 (d) 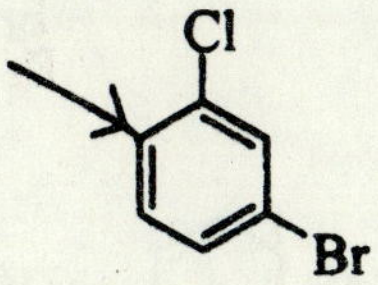 (e) 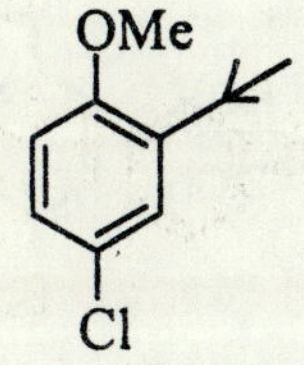 (f) 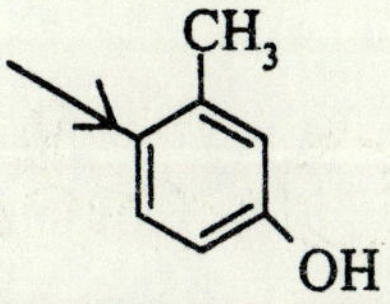 (g) 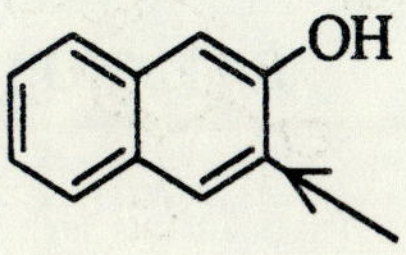 (h) 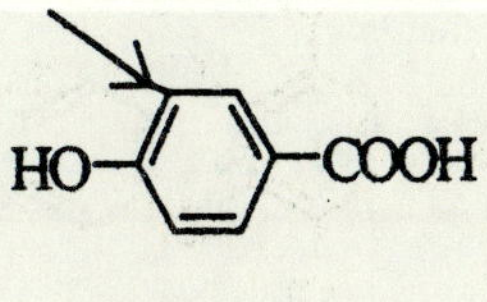 (i) 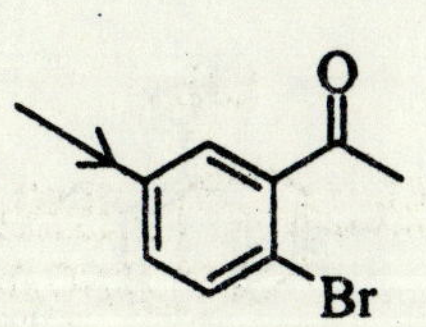 (j)

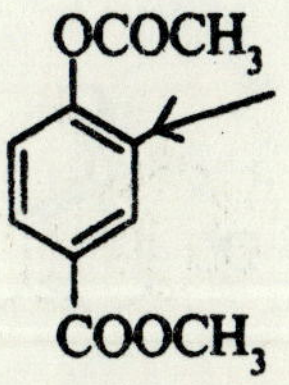

17. (a) 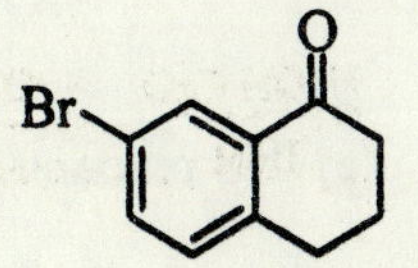 (b) 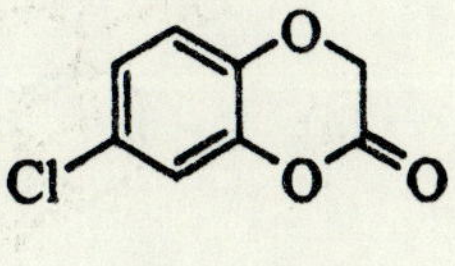 (c) 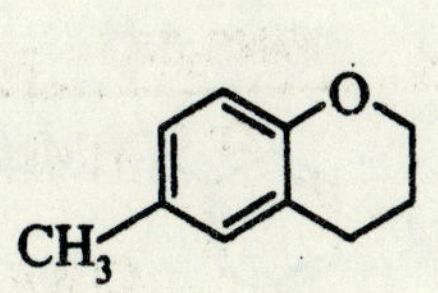 (d)

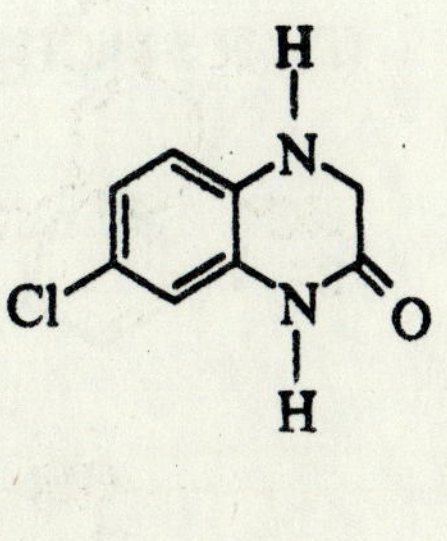

18. (a) $CH_2Cl_2 + AlCl_3 \longrightarrow \overset{+}{C}H_2Cl + AlCl_4^-$

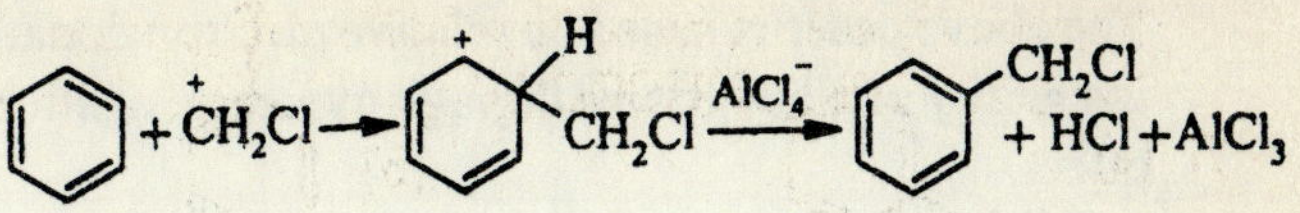

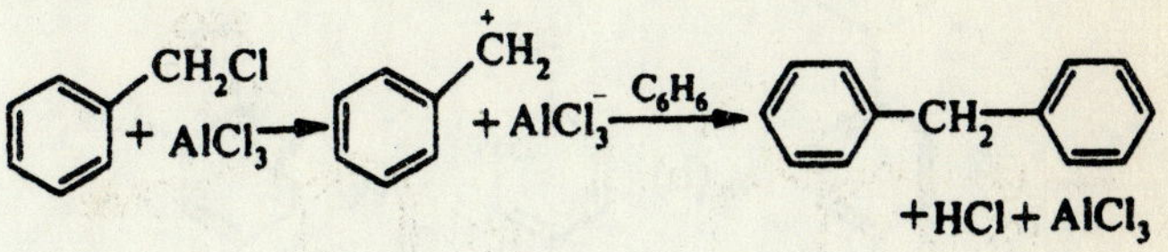

$+ \overset{+}{C}H_2Cl \longrightarrow$ $\xrightarrow{AlCl_4^-}$ $+ HCl + AlCl_3$

$+ AlCl_3 \longrightarrow$ $+ AlCl_4^- \xrightarrow{C_6H_6}$ $+ HCl + AlCl_3$

(b) 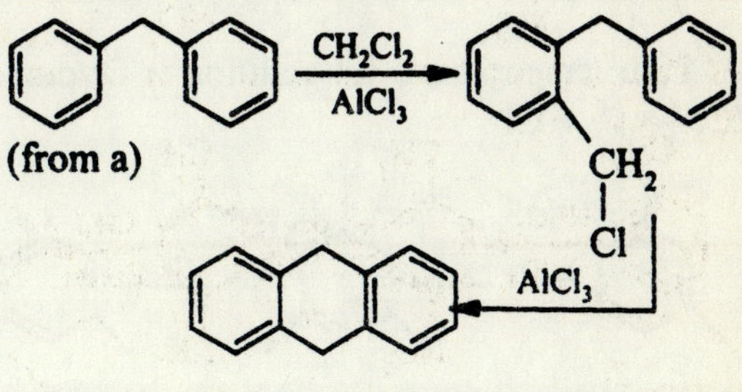 $\xrightarrow[AlCl_3]{CH_2Cl_2}$ (from a) $\xrightarrow{AlCl_3}$

(c) 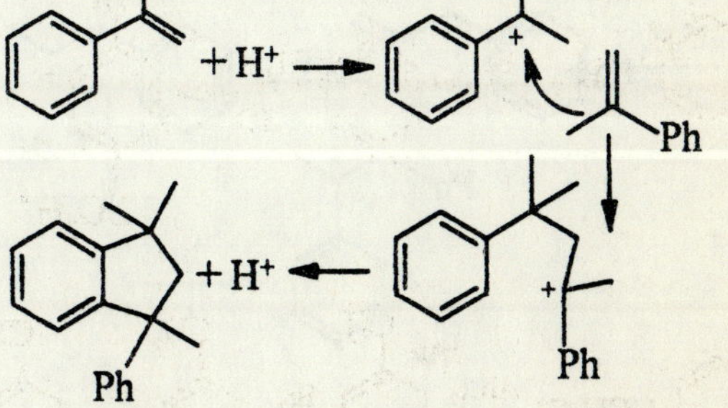 $+ H^+ \longrightarrow$ ... $+ H^+$

19. 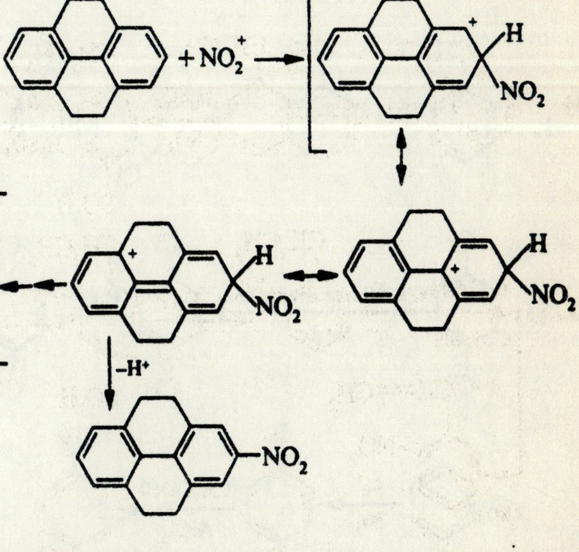 $+ NO_2^+ \longrightarrow$ ... $\xrightarrow{-H^+}$

20. I < III < II

21. Tertiary butyl benzene < isopropylbenzene < ethylbenzene < toluene.

The above order is based on relative electron donating ability of alkyl groups through hyperconjugation effect.

22. (a) SO_3H (b) Br SO_3H (c) CH_3 CH_3 SO_3H

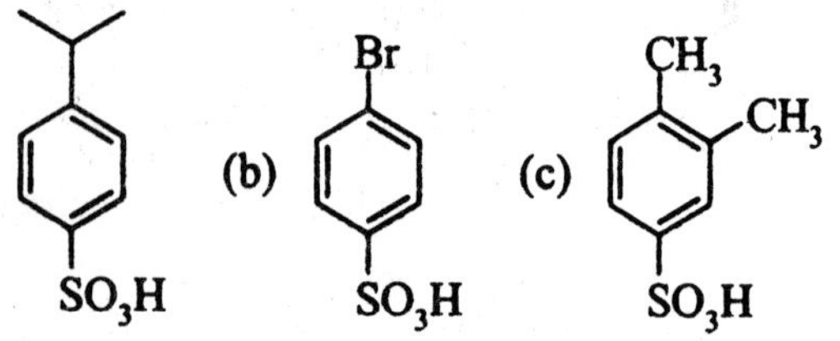

23. At high temperature, 1,4-addition is favoured.

24. $ICl \rightarrow I^+ + Cl^-$

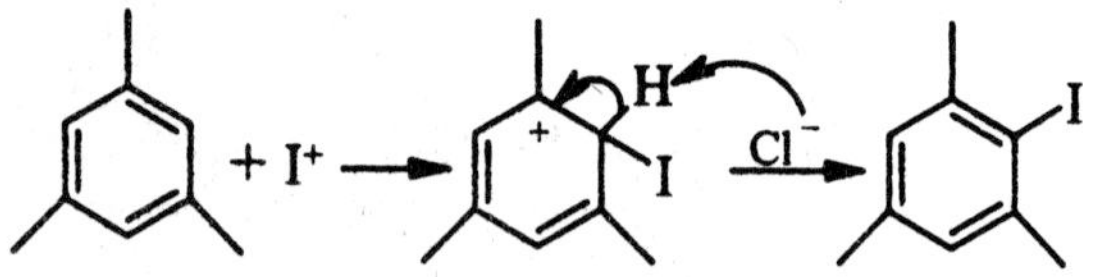

$+ I^+ \longrightarrow$ H I $\xrightarrow{Cl^-}$ I

25. $CH_2{=}CH{-}CH_2OH + HF \rightarrow CH_2{=}CH{-}CH_2^+$

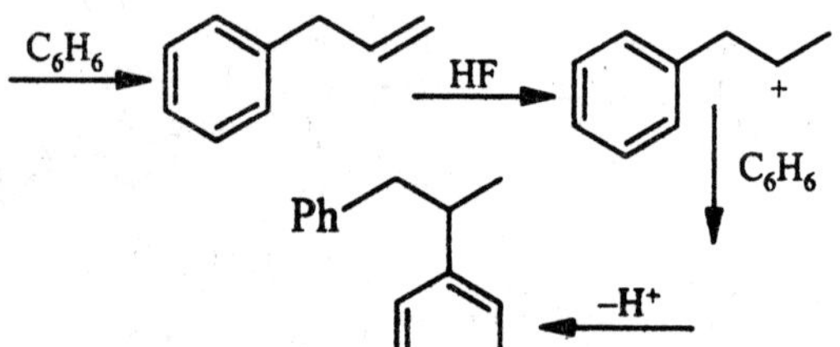

$\xrightarrow{C_6H_6}$ $\xrightarrow{HF}$ + C_6H_6 Ph $\xleftarrow{-H^+}$

CHAPTER 11

Practice Exercise

1. CH_2Cl CH_3 Cl CH_3 Cl and CH_3 Cl

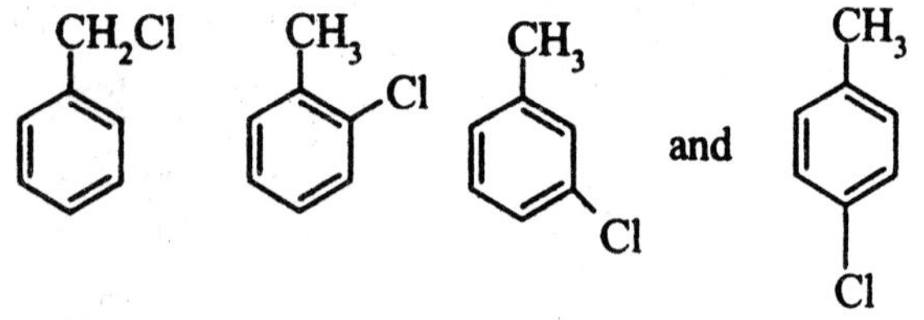

2. (a) Cl Cl $\xrightarrow[\text{Heat}]{\text{CuCN one equivalent}}$ Cl CN $\xrightarrow{H_3O^+}$ Cl COOH

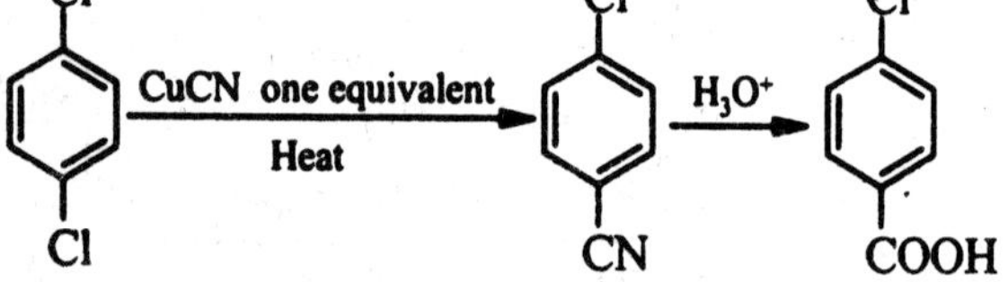

(b) $\xrightarrow[Br_2]{Fe}$ Br $\xrightarrow[AlCl_3]{CH_3COCl}$ Br $COCH_3$

$\xrightarrow[\text{Heat}]{Br_2}$ Br $COCH_2Br$

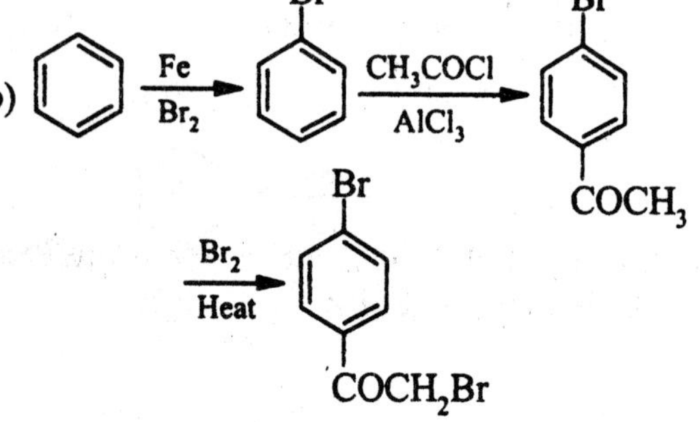

(c) $+ CH_3Cl \xrightarrow{AlCl_3}$ CH_3 $\xrightarrow[\text{Heat}]{H_2SO_4}$ CH_3 SO_3H

$\xrightarrow[Fe]{Br_2}$ CH_3 Br SO_3H $\xrightarrow[\text{Heat}]{H_3O^+}$ CH_3 Br

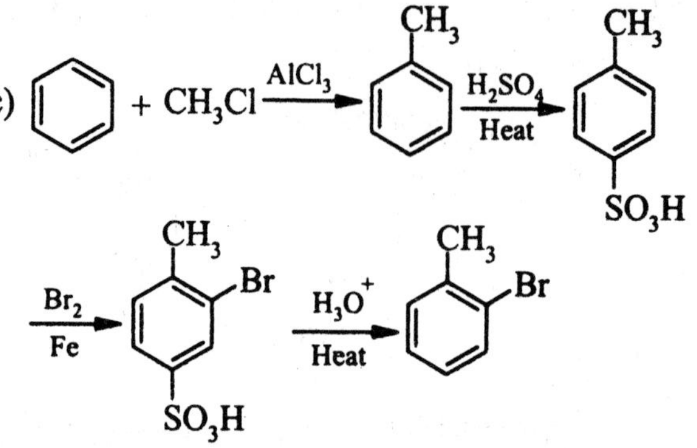

(d) $+ CH_3CH_2Cl \xrightarrow{AlCl_3}$ CH_2CH_3 $\xrightarrow[Fe]{Br_2}$ C_2H_5 Br

$\xrightarrow[h\nu]{Cl_2}$ $CHClCH_3$ Br $\xrightarrow[EtOH]{KOH}$ $CH{=}CH_2$ Br

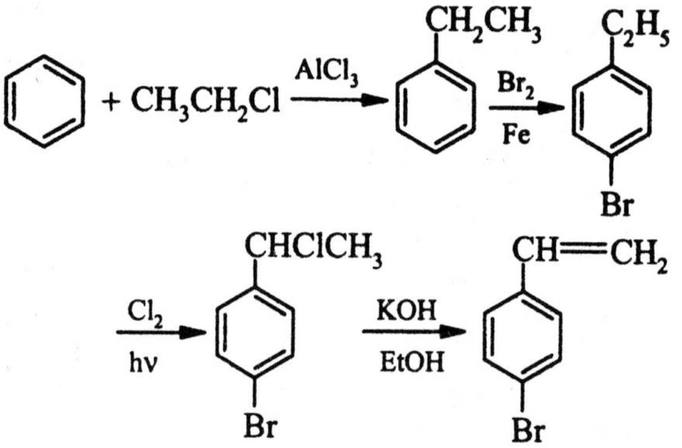

3. (a) A : $CHCl_2$ Cl B : $Pb(NO_3)_2/CO_2$

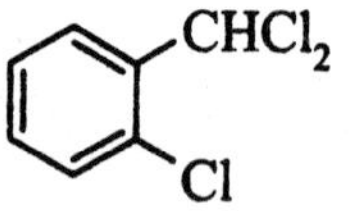

(b) C : $Ph_2C{=}CPh_2$

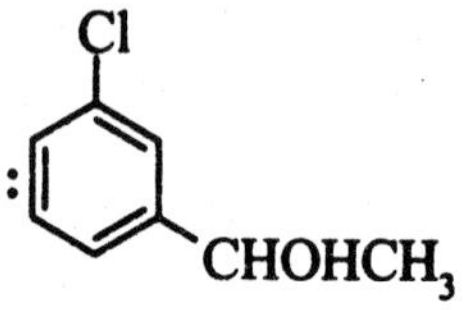

(c) D : Cl MgBr E : Cl CHOHCH$_3$

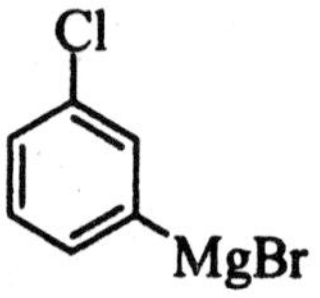

(d) F : Br CH_3 CH_3 (e) No reaction

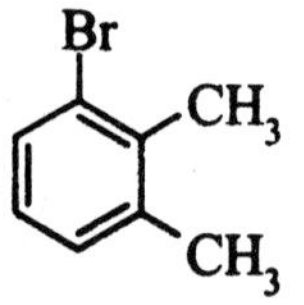

(f) H : $PhCH_2Br$

(g) I : J : CHMeCOEt

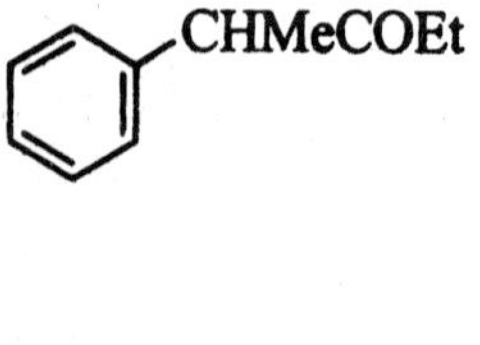

(h) K : OEt NO_2 NO_2 (i) L : No reaction

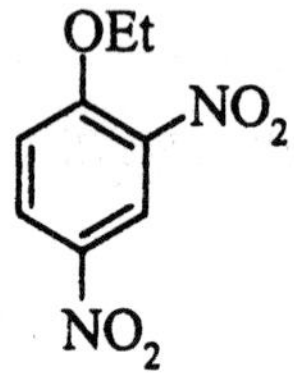

4. A is 2-methylphenyl–CHOHCH$_3$ (o-$CH_3C_6H_4CHOHCH_3$)

Mech:

$C_6H_5CH_2MgCl + CH_3CHO \rightarrow$ [intermediate: $^-CH_2(MgCl)^+$, H, CHCH$_3$, O$^-$] $\leftrightarrow$ [CH_2=, H, CHCH$_3$, OMgCl]

$\rightarrow$ o-$CH_3C_6H_4$–CH(OMgCl)–CH$_3$ $\xrightarrow{H_3O^+}$ o-$CH_3C_6H_4$–CH(OH)CH$_3$

5. (a) A : Br–C_6H_4–NHCONH$_2$

(b) B : maleic anhydride

(c) C : 2,6-diiodo-4-nitroaniline (NH$_2$, I, I, NO$_2$)

(d) D: ClCOOEt

(e) E: $Ph_2NLi \xrightarrow[Cu]{PhI}$ F : Ph_3N

(f) G : EtO–C_6H_4–NHCONH$_2$

(g) H : C_6H_5NHOAc   I : p-(NHOAc)C_6H_4SO$_3$H

J : NHOAc, NO$_2$, SO$_3$H   K : o-NH$_2$, NO$_2$ (C_6H_4)

6. (a) $C_6H_6 \xrightarrow[AlCl_3]{EtCl}$ C_6H_5Et $\xrightarrow[Heat]{H_2SO_4}$ p-EtC_6H_4SO$_3$H $\xrightarrow{HNO_3}$ Et, NO$_2$, SO$_3$H $\xrightarrow[Heat]{H_2O}$ o-EtC_6H_4NO$_2$ $\xrightarrow[HCl]{Sn}$ o-EtC_6H_4NH$_2$

(b) Toluene (CH$_3$) $\xrightarrow[H_2SO_4]{HNO_3}$ p-NO$_2$, CH$_3$ $\xrightarrow{Zn\text{-}HCl}$ p-NH$_2$, CH$_3$ $\xrightarrow{Ac_2O}$ p-NHOAc, CH$_3$ $\xrightarrow[Fe]{Br_2}$ $\xrightarrow[Heat]{H_2O}$ NH$_2$, Br, CH$_3$

(c) $C_6H_6 \xrightarrow[HNO_3]{H_2SO_4}$ m-dinitrobenzene (NO$_2$, NO$_2$) $\xrightarrow{Na_2S}$ NO$_2$, NH$_2$ $\xrightarrow[HCl/0°\,C]{NaNO_2}$ $\xrightarrow{CuCl}$ NO$_2$, Cl

(d) $C_6H_6 \xrightarrow[H_2SO_4]{HNO_3}$ NO$_2$ $\xrightarrow{Zn\text{-}HCl}$ $\xrightarrow{Ac_2O}$ $\xrightarrow[HgO]{I_2}$ p-NHAc, I $\xrightarrow[Heat]{H_2O}$ p-NH$_2$, I

(e) Toluene (CH$_3$) $\xrightarrow[H_2SO_4]{HNO_3}$ p-CH$_3$, NO$_2$ $\xrightarrow[NaOH]{KMnO_4}$ p-COONa, NO$_2$ $\xrightarrow[HCl]{Sn}$ p-COOH, NH$_2$

(f) 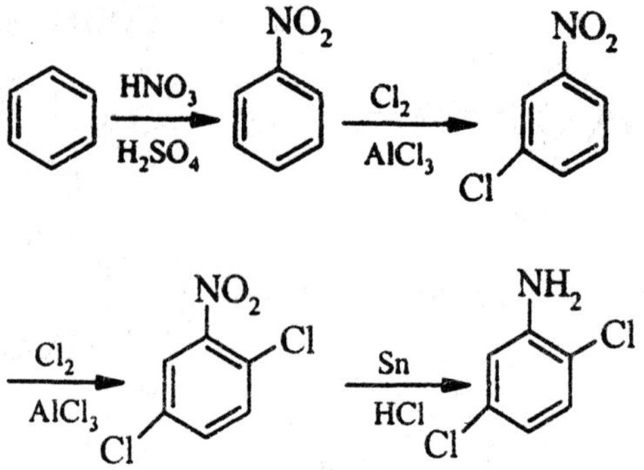 Benzene $\xrightarrow[H_2SO_4]{HNO_3}$ Nitrobenzene (NO_2) $\xrightarrow[AlCl_3]{Cl_2}$ m-chloronitrobenzene (NO_2, Cl) $\xrightarrow[AlCl_3]{Cl_2}$ (NO_2, Cl, Cl) $\xrightarrow[HCl]{Sn}$ (NH_2, Cl, Cl)

7. (a) A: (OH, CH_2Cl, NO_2)

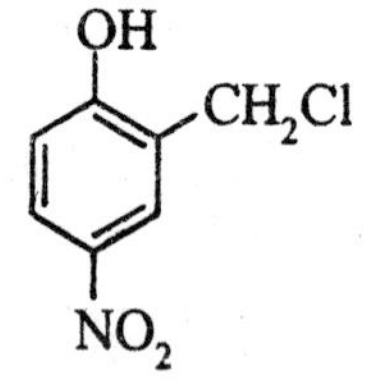

(b) B : $COCl_2$ (c) C : $PhCH_2CH{=}CH_2$

(d) D: (Cl, NO_2, NO_2) (e) E: (OK, NO_2, NO_2)

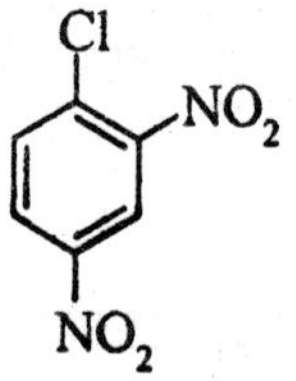

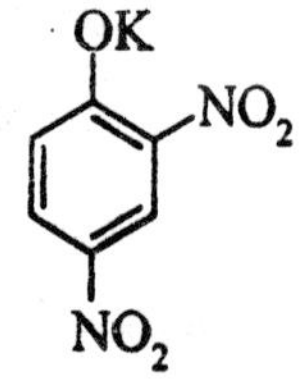

8. (a) 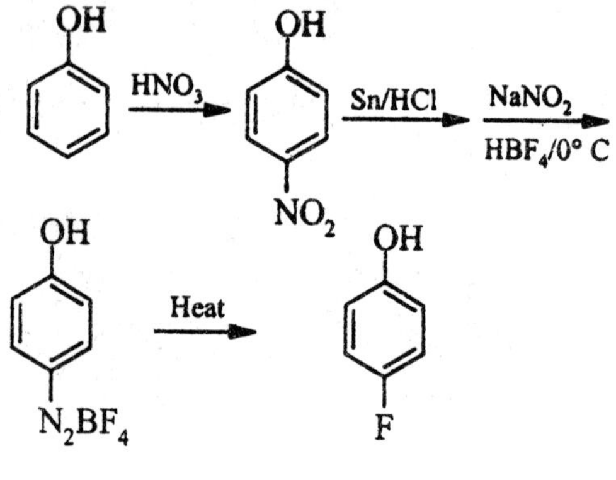 (OH) $\xrightarrow{HNO_3}$ (OH, NO_2) $\xrightarrow{Sn/HCl}$ $\xrightarrow[HBF_4/0°\ C]{NaNO_2}$ (OH, N_2BF_4) $\xrightarrow{Heat}$ (OH, F)

(b) 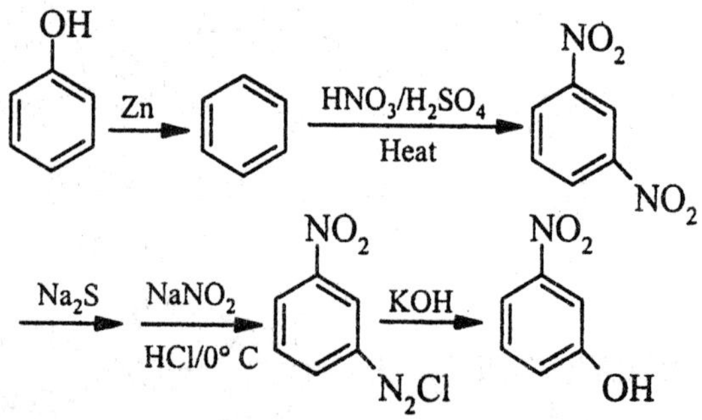 (OH) $\xrightarrow{Zn}$ Benzene $\xrightarrow[Heat]{HNO_3/H_2SO_4}$ (NO_2, NO_2) $\xrightarrow{Na_2S}$ $\xrightarrow[HCl/0°\ C]{NaNO_2}$ (NO_2, N_2Cl) $\xrightarrow{KOH}$ (NO_2, OH)

(c) (Cl) $\xrightarrow[Heat]{HNO_3/H_2SO_4}$ (Cl, NO_2, NO_2) $\xrightarrow[Heat]{KOH\ (aq)}$ (OH, NO_2, NO_2)

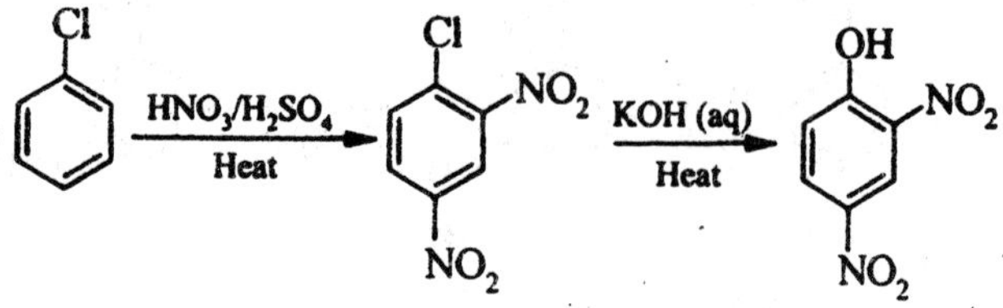

(d) 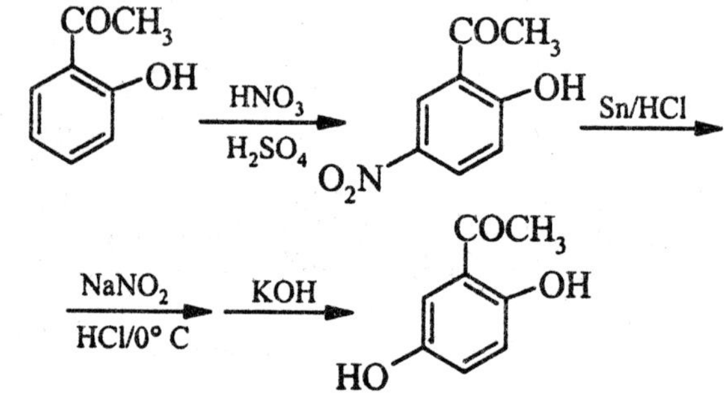 ($COCH_3$, OH) $\xrightarrow[H_2SO_4]{HNO_3}$ ($COCH_3$, OH, O_2N) $\xrightarrow{Sn/HCl}$ $\xrightarrow[HCl/0°\ C]{NaNO_2}$ $\xrightarrow{KOH}$ ($COCH_3$, OH, HO)

9. (a) A : ($CH(OAc)_2$, NO_2) B : (CHO, NO_2)

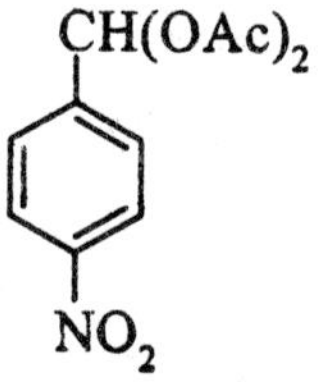

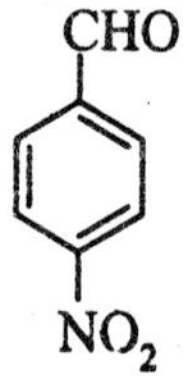

(b) C : Ph_2CHOH (c) D : $COCl_2$

(d) E : SeO_2

(e) F : (OH, OMe, CH_2OH) G : (OH, OCH_3, $CH{=}NPh$)

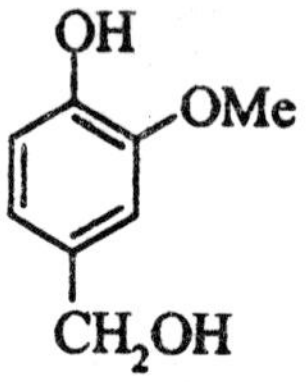

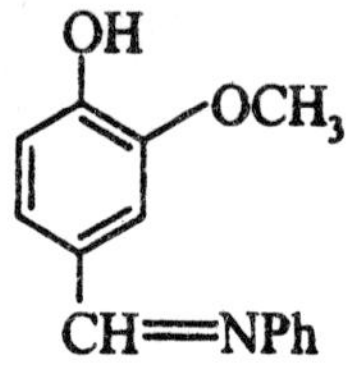

H : (OH, OCH_3, CHO)

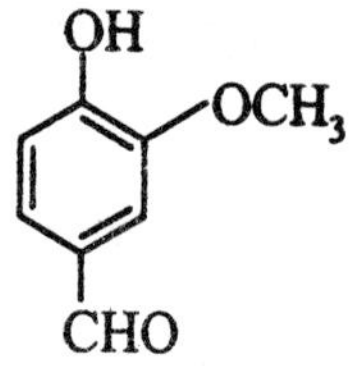

(f) I : (OH, C(=O)Pr)

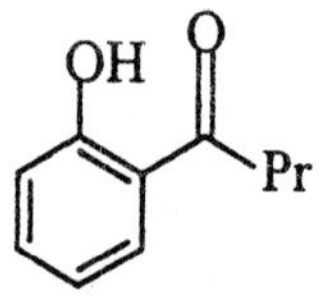

(g) J : PhCOCl, K : $PhCONMe_2$ L : $Li(OEt)_2AlH_2$

10. (a) 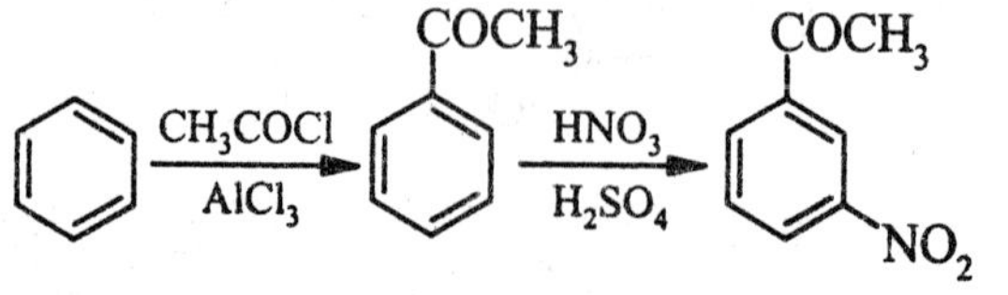 Benzene $\xrightarrow[AlCl_3]{CH_3COCl}$ ($COCH_3$) $\xrightarrow[H_2SO_4]{HNO_3}$ ($COCH_3$, NO_2)

(b) 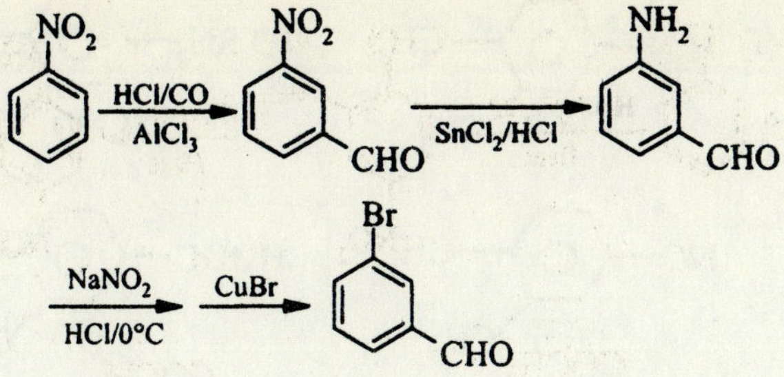 Nitrobenzene $\xrightarrow[AlCl_3]{HCl/CO}$ 3-nitrobenzaldehyde (NO_2, CHO) $\xrightarrow{SnCl_2/HCl}$ 3-aminobenzaldehyde (NH_2, CHO) $\xrightarrow[HCl/0°C]{NaNO_2}$ $\xrightarrow{CuBr}$ 3-bromobenzaldehyde (Br, CHO)

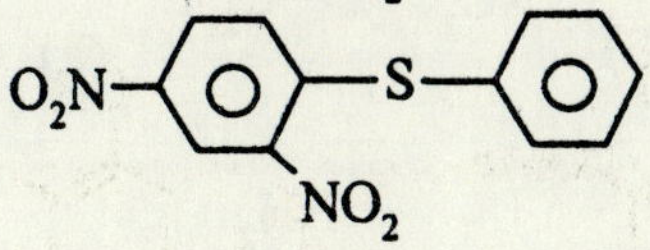

(c) Benzene $\xrightarrow[AlCl_3]{CH_3Cl}$ toluene (CH_3) $\xrightarrow[AlCl_3]{CO/HCl}$ 4-methylbenzaldehyde (CH_3, CHO) $\xrightarrow[CrO_3/H_2SO_4]{Ac_2O}$ terephthalaldehyde (CHO, CHO)

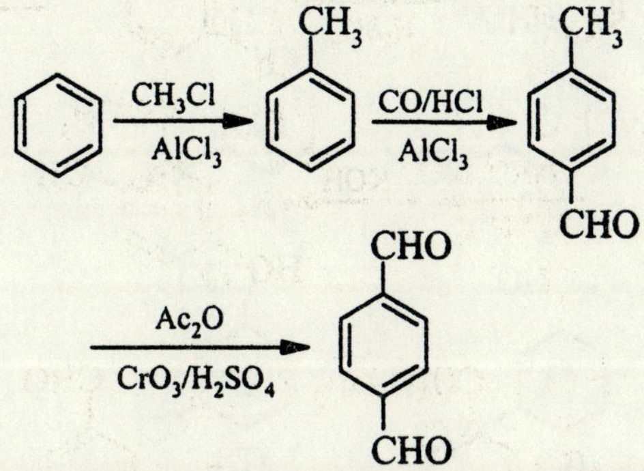

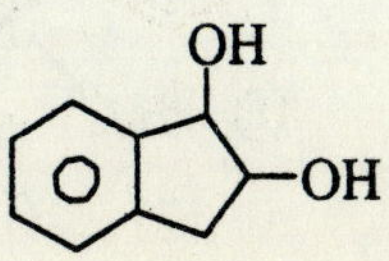

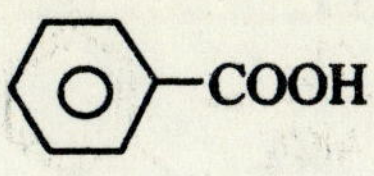

11. (a) 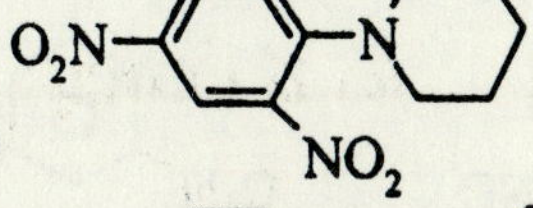 O_2N–C$_6$H$_3$(NO_2)–N(piperidine)

(b) O_2N–C$_6$H$_3$(NO_2)–S–C$_6$H$_5$

(c) O_2N–C$_6$H$_2$(NO_2)$_2$–NH–C$_6$H$_5$

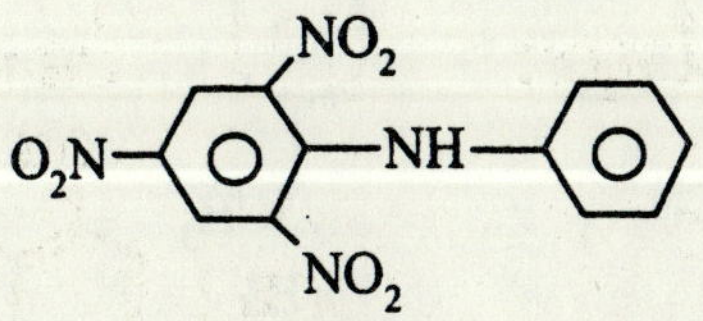

(d) 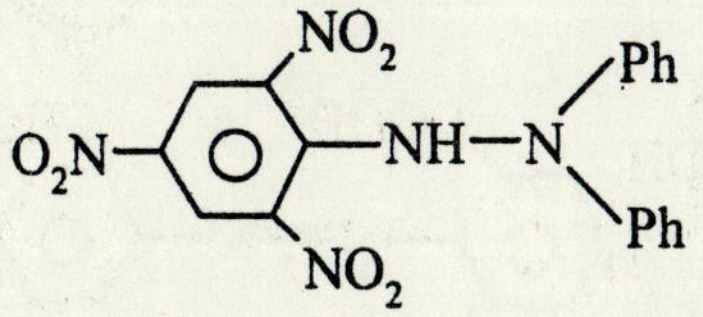 O_2N–C$_6$H$_2$(NO_2)$_2$–NH–N(Ph)Ph

(e) 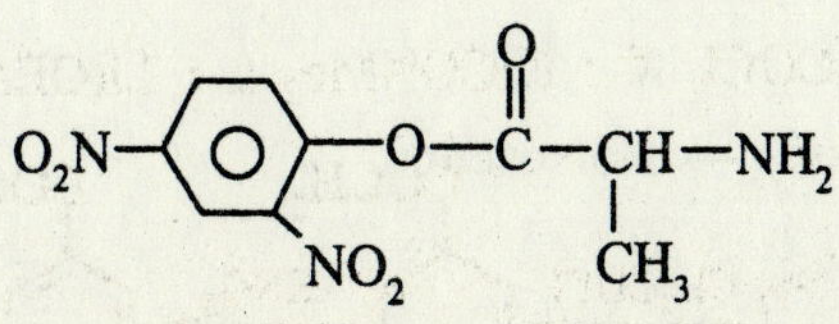 O_2N–C$_6$H$_3$(NO_2)–O–C(=O)–CH(CH_3)–NH_2

12. (a) 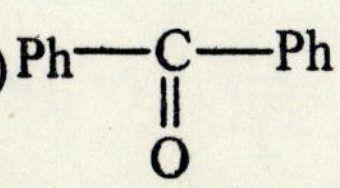 Ph–C(=O)–Ph (b) indane-1,2-diol (OH, OH)

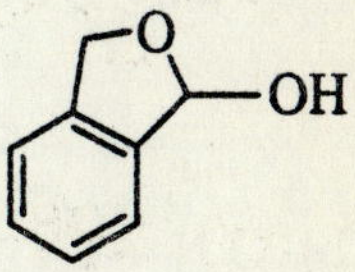

(c) C$_6$H$_5$–COOH (d) 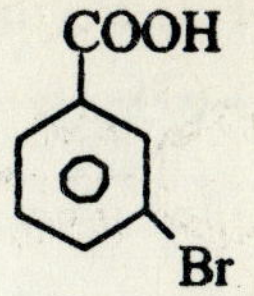 3-bromobenzoic acid (COOH, Br)

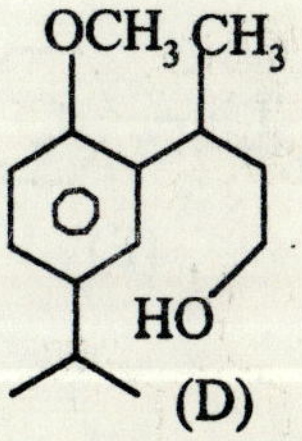

(e) No reaction (f) 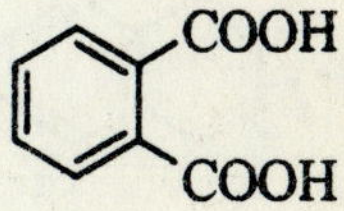 phthalic acid (COOH, COOH)

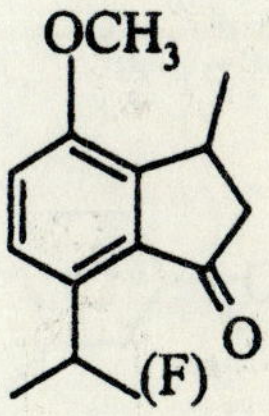

13. (a) 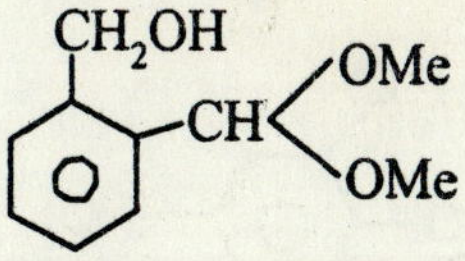 C$_6$H$_4$(CH_2OH)–CH(OMe)OMe (A); 1-hydroxy-1,3-dihydroisobenzofuran (O, OH) (B)

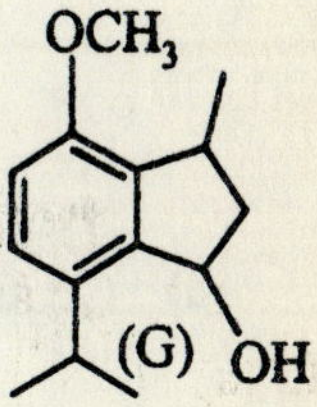

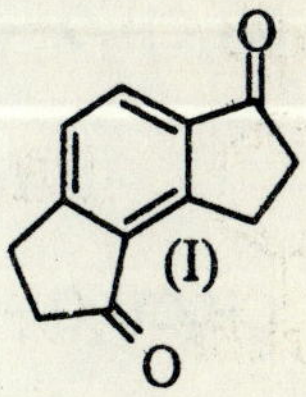

(b) 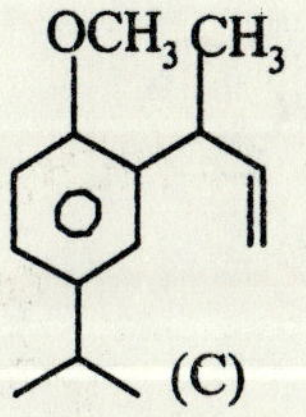 OCH_3, CH_3 substituted aryl with isopropyl group, CH(CH$_3$)–CH=CH$_2$ (C); OCH_3, CH_3, HO (D)

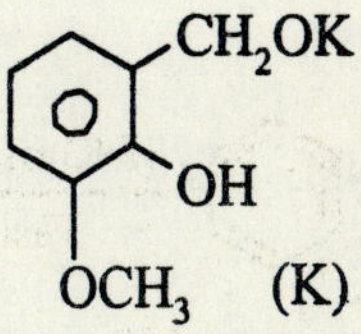

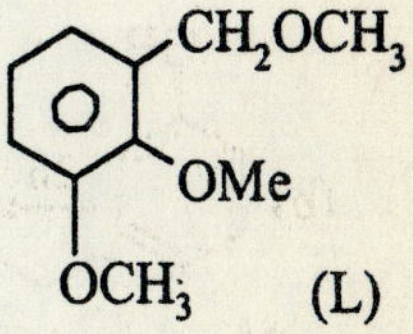

(c) 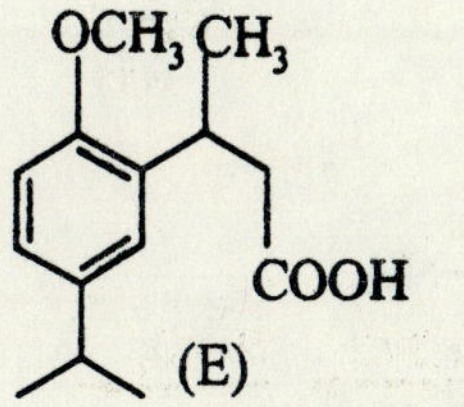 OCH_3, CH_3, COOH (E); OCH_3, O (F); OCH_3, OH (G)

(d) 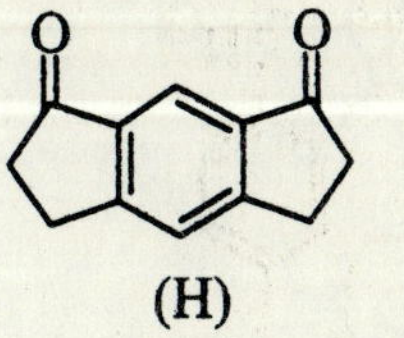 O, O (H); O, O (I)

(e) 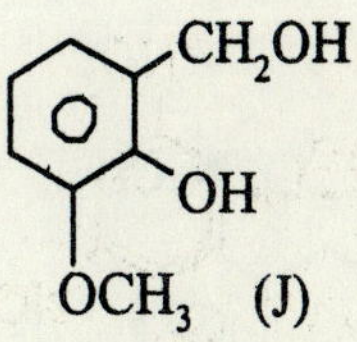 CH_2OH, OH, OCH_3 (J); CH_2OK, OH, OCH_3 (K); CH_2OCH_3, OMe, OCH_3 (L)

(f) $^+Na\bar{C}H(COOEt)_2$ (M); Br–C$_6$H$_2$(NO_2)$_2$–CH(COOEt)COOEt (N)

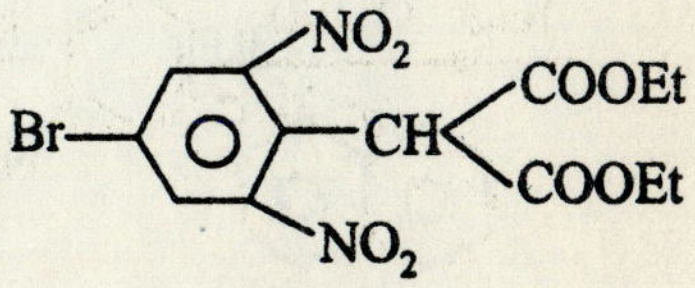

(g) CH_3OH/H_2SO_4/heat (O) $LiAlH_4$ (P)

14. (a) (A) OCH_3 (B) NO_2, OCH_3 (C) CN, OCH_3

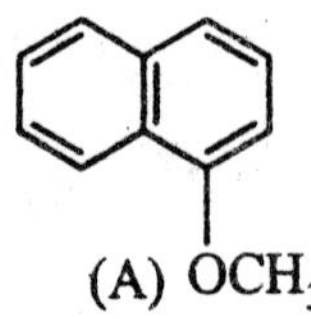

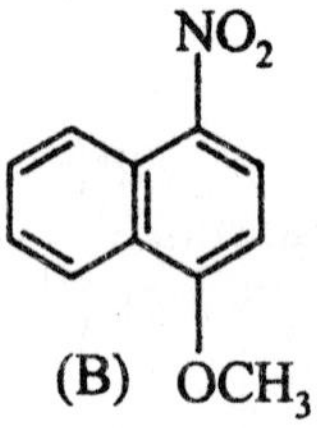

(b) O_2N—, —O—, Cl, —NO_2, NO_2 (D)

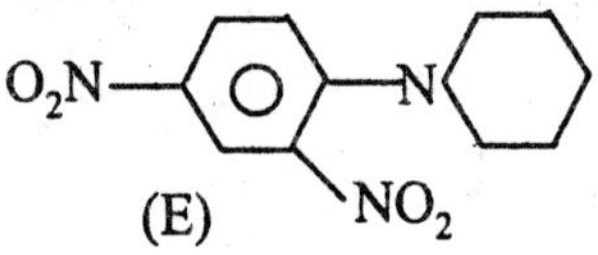

O_2N—, —N, NO_2 (E)

(c) O_2N—, I, —NH_2 (F) O_2N—, I, —I (G)

(d) CH_3CH_2O—, —NH_2, Br (H)

CH_3CH_2O—, —N_2Cl, Br (I)

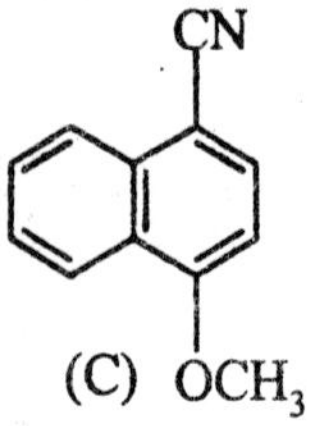

CH_3CH_2O—, Br (J)

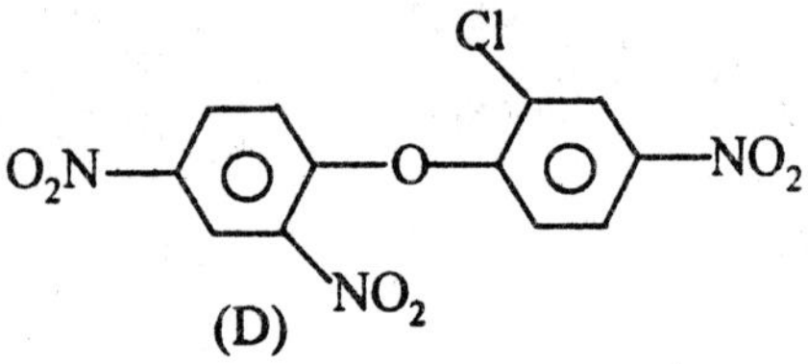

(e) NH_2 (K) N_2Cl (L) N_2BF_4 (M) F (N)

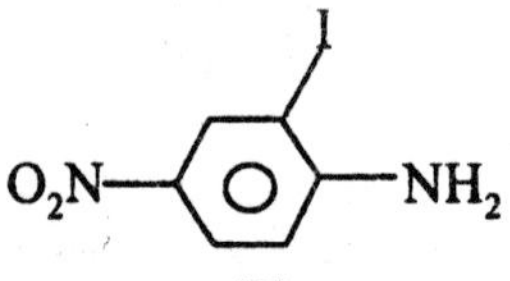

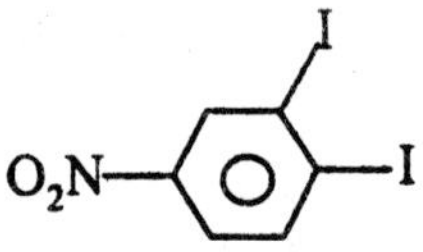

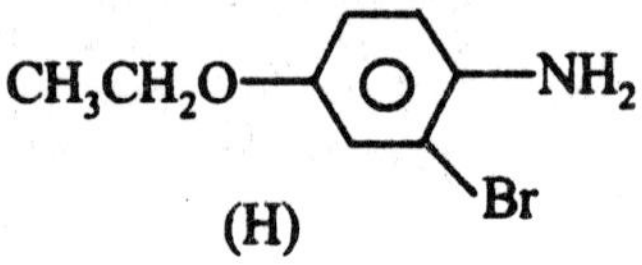

(f) H_2N—, —CH_3, CH_3 (O) HO_4SN_2—, —CH_3, CH_3 (P)

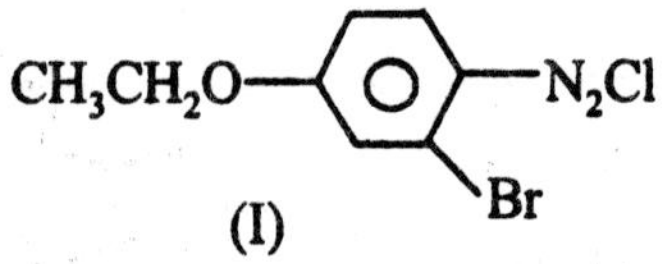

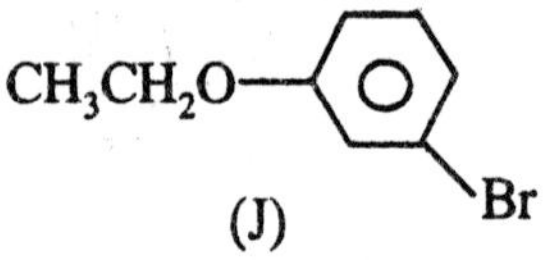

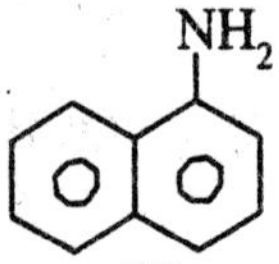

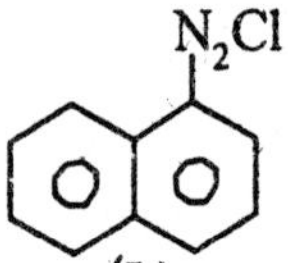

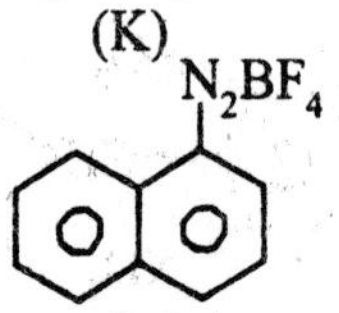

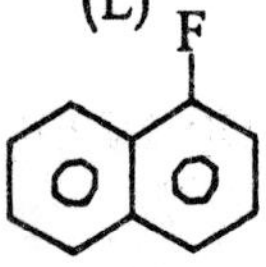

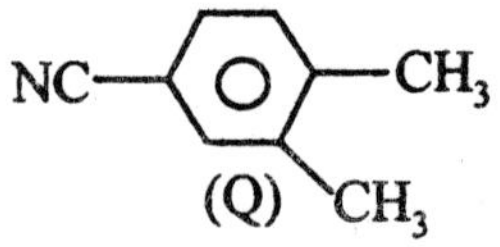

NC—, —CH_3, CH_3 (Q) HOOC—, —CH_3, CH_3 (R)

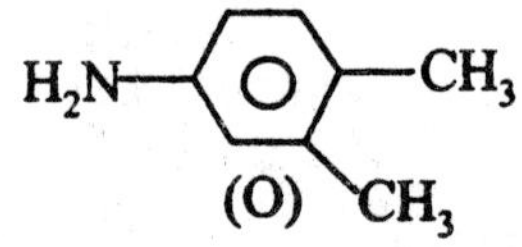

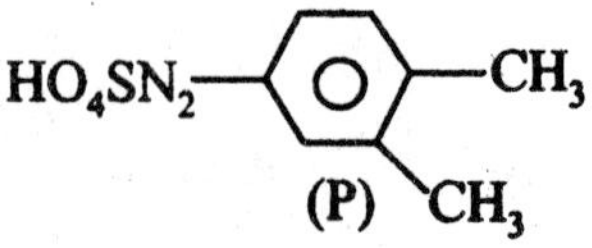

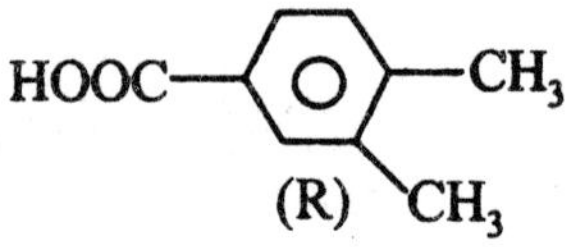

15. (a) CHO, NO_2 (A) CHO, NH_2 (B)

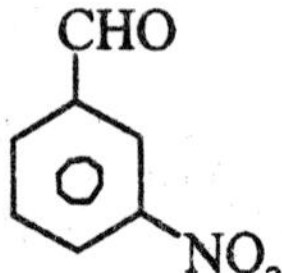

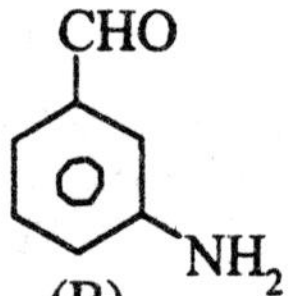

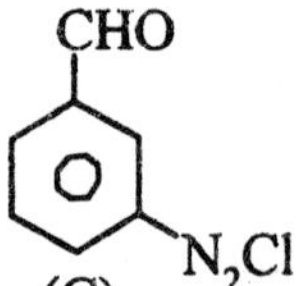

CHO, N_2Cl (C) CHO, Cl (D)

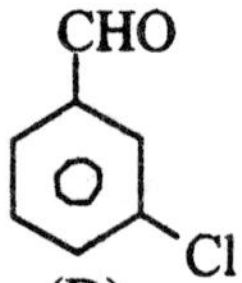

(b) N_2HSO_4, O_2N, NO_2 (E) OH, O_2N, NO_2 (F)

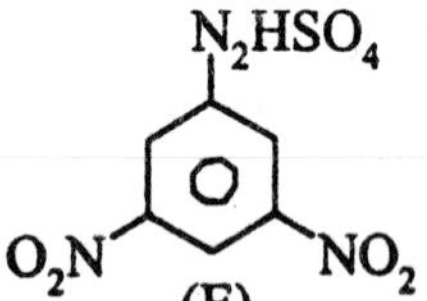

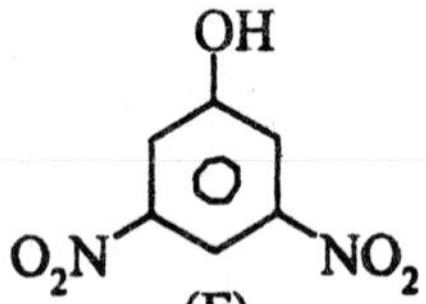

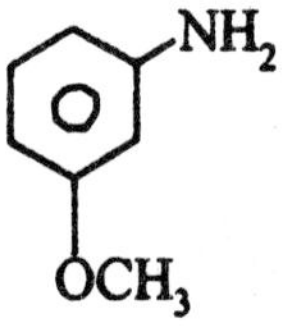

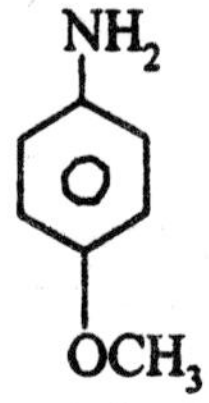

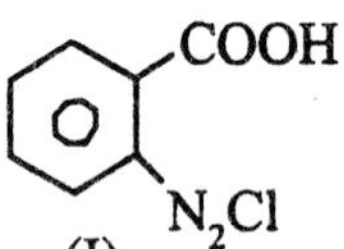

(c) NH_2, OCH_3 (G) + NH_2, OCH_3 (H)

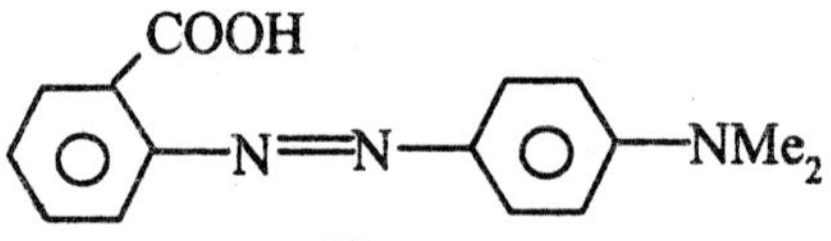

(d) COOH, N_2Cl (I) COOH, —N═N—, —NMe_2 (J)

(e) (K) → No reaction

(f) H_3C, $\overset{+}{N}$, CH_3, I^- (L)

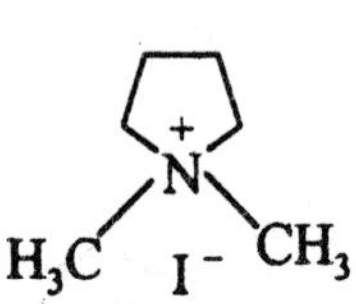

(M) $(CH_3)_2N-CH_2-CH_2-CH=CH_2$

(N) $(CH_3)_3\overset{+}{N}-CH_2CH_2-CH=CH_2 \; I^-$

$(CH_3)_3N + CH_2=CH-CH=CH_2$
(O) (P)

16. $NC-CH_2-COOEt + C_2H_5O^- \longrightarrow$

$[NC-\bar{C}H-C(=O)-OEt \leftrightarrow NC-CH=C(O^-)-OEt \leftrightarrow \bar{N}=C=CH-C(=O)-OEt]$

O_2N-(2-nitro-chlorobenzene) $-Cl + \bar{C}H(CN)-COOEt \rightarrow$ intermediate (Cl, CH(CN)COOEt on ring) $\xrightarrow{-Cl^-}$ $O_2N-C_6H_3(NO_2)-CH(CN)COOEt$

17. (a) II < IV < III < I (b) II < I < III < IV
(c) I < III < IV < II

18. The relative rate of nucleophilic strength also depends on strength of nucleophile. Hence, stronger the nucleophile, greater is the rate of nucleophilic substitution.

19. Relative rate: III < IV < I < II < V

20. (a) Toluene $\xrightarrow[H_2SO_4]{HNO_3}$ p-nitrotoluene $\xrightarrow{Sn/HCl}$ p-toluidine (NH_2) $\xrightarrow[HBF_4/0° C]{NaNO_2}$ $\xrightarrow{Heat}$ p-fluorotoluene $\xrightarrow[NaOH]{KMnO_4}$ $\xrightarrow[C_2H_5OH/heat]{H^+}$ $F-C_6H_4-COOC_2H_5$

(b) Toluene $\xrightarrow[H_2SO_4]{HNO_3}$ $\xrightarrow{Sn/HCl}$ $\xrightarrow[Heat]{AC_2O}$ $CH_3-C_6H_4-NHCOCH_3$ $\xrightarrow[Fe]{Br_2}$ 3-bromo-4-(NHCOCH$_3$)toluene $\xrightarrow[Heat]{H_2O}$ 3-bromo-4-(NH_2)toluene $\xrightarrow[HCl/0° C]{NaNO_2}$ $\xrightarrow{aq\ KOH}$ 2-bromo-4-methylphenol (OH, Br)

(c) $PhCH_3 \xrightarrow[Heat]{Br_2} PhCH_2Br \xrightarrow{CH_3CH_2CH_2CN/LDA}$

$PhCH_2CH(CH_2CH_3)-CN \xrightarrow{LiAlH_4} PhCH_2CH(CH_2CH_3)-CH_2NH_2$

(d) Toluene $\xrightarrow[H_2SO_4]{HNO_3}$ p-nitrotoluene (NO_2) $\xrightarrow[NaOH]{KMnO_4}$ $\xrightarrow{H^+}$ p-nitrobenzoic acid (COOH, NO_2)

$\xrightarrow{Sn/HCl}$ p-aminobenzoic acid (COOH, NH_2) $\xrightarrow{2CH_3I}$ $HOOC-C_6H_4-N(CH_3)_2$

(e) Bromobenzene $\xrightarrow[H_2SO_4]{HNO_3}$ p-bromonitrobenzene (NO_2) $\xrightarrow{Sn/HCl}$ $\xrightarrow[HCl/0°C]{NaNO_2}$

$Br-C_6H_4-N_2^+Cl^- \xrightarrow[Heat]{KI/Cu} Br-C_6H_4-I$

Alternative: Bromobenzene $\xrightarrow{I_2/HgO}$ $Br-C_6H_4-I$

(f) Chlorobenzene $\xrightarrow[H_2SO_4]{HNO_3}$ p-chloronitrobenzene (NO_2) $\xrightarrow{Sn/HCl}$ p-chloroaniline (Cl, NH_2)

Objective Exercise

Level 1

1. c	18. a	35. c	52. d	69. c
2. a	19. c	36. d	53. d	70. b
3. d	20. d	37. b	54. d	71. a
4. d	21. c	38. c	55. c	72. c
5. d	22. c	39. d	56. d	73. c
6. b	23. d	40. a	57. a	74. c
7. d	24. c	41. a	58. b	75. c
8. a	25. c	42. c	59. d	76. a
9. c	26. b	43. d	60. a	77. a
10. c	27. d	44. a	61. b	78. d
11. d	28. a	45. e	62. a	79. c
12. d	29. a	46. c	63. d	80. d
13. d	30. d	47. d	64. c	81. a
14. b	31. b	48. d	65. d	82. c
15. c	32. c	49. c	66. b	83. c
16. d	33. c	50. d	67. a	
17. d	34. c	51. b	68. b	

Level 2

1. c	20. c	39. d	58. d	77. c
2. a	21. c	40. c	59. c	78. c
3. b	22. b	41. b	60. d	79. a
4. a	23. c	42. b	61. c	80. a
5. b	24. b	43. c	62. d	81. d
6. d	25. b	44. c	63. c	82. c
7. d	26. d	45. b	64. c	83. c
8. c	27. d	46. c	65. c	84. d
9. b	28. d	47. d	66. c	85. c
10. b	29. b	48. d	67. b	86. a
11. b	30. c	49. c	68. b	87. c
12. a	31. c	50. b	69. a	88. b
13. d	32. d	51. d	70. d	89. c
14. c	33. d	52. a	71. b	90. c
15. c	34. d	53. c	72. c	91. a
16. d	35. c	54. b	73. c	92. b
17. b	36. d	55. c	74. d	
18. c	37. d	56. a	75. b	
19. c	38. b	57. d	76. b	

Review Exercise (Aromatic)

1. 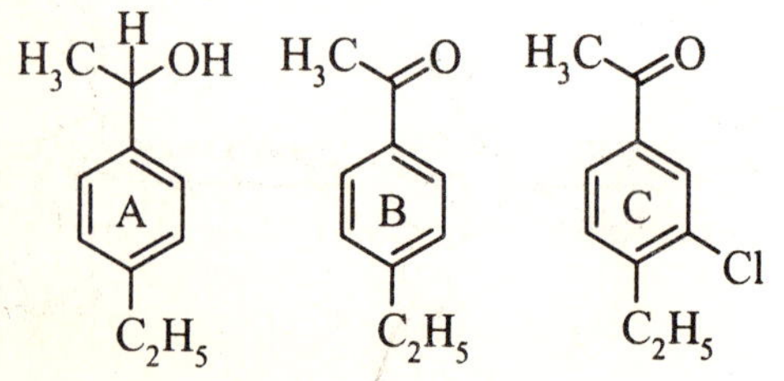 H_3C–CH(OH) (A), H_3C–C=O (B), H_3C–C=O (C); each on a benzene ring bearing C_2H_5; (C) also bears Cl

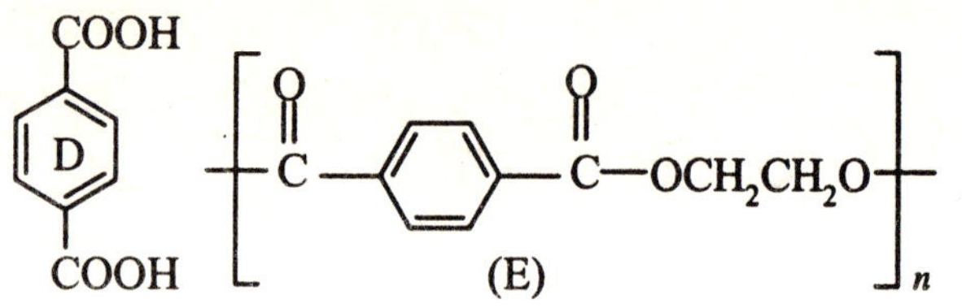

COOH (D) COOH

$\left[-\overset{O}{\overset{\|}{C}}-C_6H_4-\overset{O}{\overset{\|}{C}}-OCH_2CH_2O- \right]_n$ (E)

4. Br$_2$/heat; Br; Mg/Et$_2$O; MgBr; Ph–C(=O)–CH(CH$_3$)$_2$, Heat; Ph; H$^+$; CH$_3$; –H$^+$; Ph; Ph

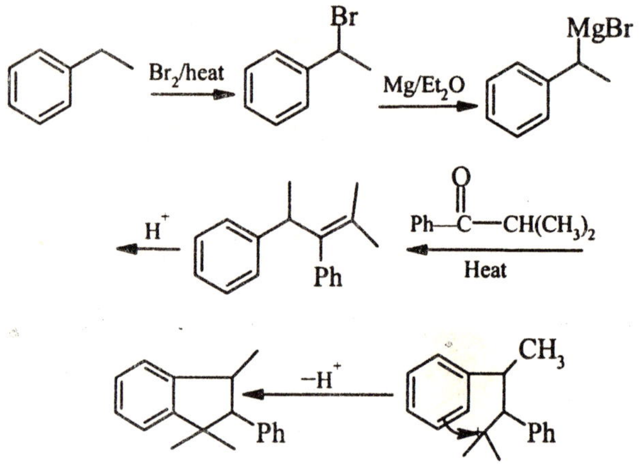

5. MeO–C$_6$H$_4$–C(CH$_3$)=CHCH$_3$ (A); MeO–C$_6$H$_4$–COCH$_3$ (B); $Cl_2/AlCl_3$; Cl; MeO (D) (only product); IH; HO (C); AlCl$_3$ Fries rearrangement; (E)

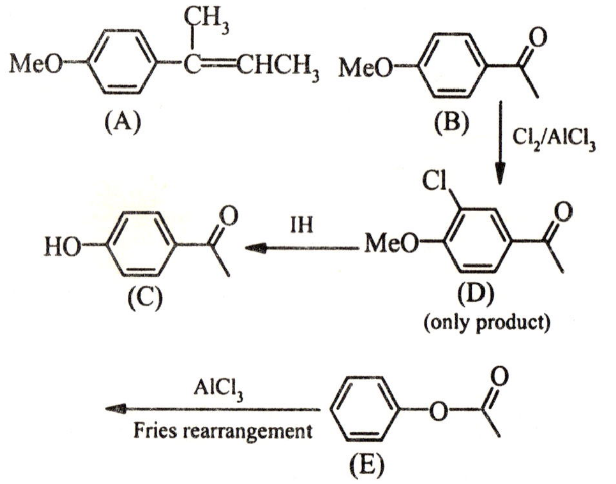

9. 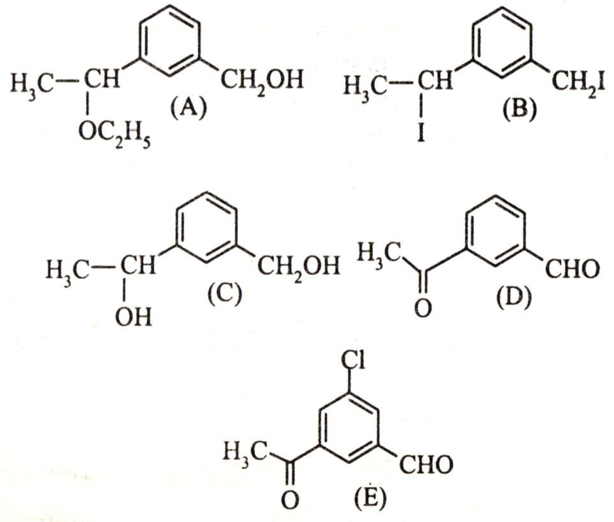 $H_3C-CH(OC_2H_5)$ (A) CH_2OH; $H_3C-CH(I)$ (B) CH_2I; $H_3C-CH(OH)$ (C) CH_2OH; H_3C–C(=O) (D) CHO; Cl, H_3C–C(=O) (E) CHO

10. (b) Benzene $\xrightarrow[AlCl_3]{C_2H_5Cl}$ ethylbenzene (CH_2CH_3) $\xrightarrow{HNO_3/H_2SO_4}$ 4-nitroethylbenzene (CH_2CH_3, NO_2) $\xrightarrow{NBS}$ Br–CHCH$_3$ (NO_2) $\xrightarrow[C_2H_5OH]{C_2H_5OK}$ 4-nitrostyrene (NO_2) $\xrightarrow{mcpba}$ epoxide (O, NO_2) $\xrightarrow{HBr}$ HO–CH$_2$–CH(Br)–C$_6$H$_4$–NO_2

11. (a) Ph–CO–CH$_2$–CH$_2$–CH(NO_2)–CH$_2$CH$_3$ (X); Ph–CO–CH$_2$–CH$_2$–CO–CH$_2$CH$_3$ (Y); 2-ethyl-5-phenylfuran (Z) (Ph, O)

(b) Ph–CO–$\bar{C}H_2$ (X); Ph–CO–CH$_2$–CH$_2$–CO–Ph (Y); 2,5-diphenylfuran (Ph, O, Ph) (Z)

(c) Ph–CO–CH$_2$–CH$_2$–CN (X); Ph–CO–CH$_2$–CH$_2$–CHO (Y); 2-phenylfuran (Ph, O) (Z)

26. C_2H_5 (A); C_2H_5, SO_3H (B); C_2H_5, SO_3H (C); C_2H_5, CN (D); C_2H_5, COOH (E); COOH, COOH (F); O, O, O (G)

68. Benzene $\xrightarrow{MA}$ nitrobenzene (NO_2) $\xrightarrow{Br_2/FeBr_3}$ NO_2, Br $\xrightarrow{Zn/HCl}$ NH_2, Br $\xrightarrow{CHCl_3/KOH}$ NC, Br $\xrightarrow{LiAlH_4}$ $NHCH_3$, Br $\xrightarrow{CH_3COCl}$ Br, N(O)–CH$_3$, CH$_3$ $\xrightarrow[Et_2O]{Mg}$ (CHO) $\rightarrow$ OH, N(O)–CH$_3$, CH$_3$ $\xrightarrow{PCC}$ $\xrightarrow[\text{NaOH, Heat}]{N_2H_4/EG}$ N, O

74. CH_3, CH_3 (A); Br, CH_3, CH_3 (B); Br, CH_3, CH_3 (C); COOH, COOH (D); Br, COOH, COOH (E); O, O (F); N, N, N, N (G)

75. OMe, O (A); OMe, Br, O (B); OMe, NNH, NO_2, O_2N (Orange precipitate); MeO, OH (C)

76. (A) (B) OH, OH (C) OH, OH

(D) OH (E) CHO, OH

77. H_3C– (A) –H_3C–COOH (D)

(B) OH (C) O (B) OH $\xrightarrow[\text{Heat}]{H_2SO_4}$

H H, H_3C CH_3 Stereomeric $\xrightarrow[\text{Alkaline } KMnO_4]{\text{Ccold, dilute}}$ CH_3, H–OH, H–OH, CH_3

78. (A) (B) Br (C) COOH, COOH

Br, Br, Br

(not formed)

79. COOH, OMe (A) COOH, OH (B) $COOCOCH_3$, O, CH_3 (C)

COOH, Br, OH (D) COOH, OH, Br (E) COOH, Br, OH (F)

80. $C_6H_5COOCH_2CH_3$(A),
CH_3CH_2OH(B),
CH_3CHO(C),
C_6H_5COOH (D)

CHAPTER 12

Practice Exercise

1. Draw the structure of predominant form of:
   (a) Valine at pH = 11 (b) Proline at pH = 2
   (c) Arginine at pH = 7 (d) Glutamic acid at pH = 7
   (e) A mixture of alanine, lysine and aspartic acid at
   (i) pH = 6 (ii) pH = 2
2. Synthesize following amino acids by reductive amination techniques of an α-keto acids:
   (a) Alanine (b) Leucine
   (c) Serine (d) Glutamine
3. Synthesize following amino acids, starting from a carboxylic acid:
   (a) Glycine (b) Leucine
   (c) Valine (d) Glutamic acid
4. Predict the products of the following reactions:
   (a) $PhCH_2OCONHCH(CH_3)COOH \xrightarrow{H_2/Pd}$
   (b) Lysine + $(CH_3CO)_2O \longrightarrow$
   (c) Proline $\xrightarrow[\text{Excess}]{(CH_3CO)_2O}$
   (d) $CH_3CH_2CH(CH_3)CHO + NH_3 + HCN \xrightarrow{OH_2} \xrightarrow{H_3O^+}$
   (e) Isovaleric acid + $Br_2 \xrightarrow{PBr_3} \xrightarrow[\text{Excess}]{NH_3}$
5. Convert alanine into:
   (a) alanine isopropyl ester
   (b) N-benzoylalanine
   (c) *t*-butyloxy carbonylalanine
6. An aqueous solution of glucose has a specific rotation of +52.6°. If specific rotation of α-D-glucopyranose is 112.2° and of β-D-glucopyranose is +18.7°, determine anomeric composition of the solution.
7. Explain why bromine water is better reagent than Tollen's reagent for oxidation of aldose to aldonic acid?
8. (a) Give product expected when (+) glyceraldehyde react with HCN.
   (b) What is the relationship between the products and how might they be separated?
   (c) Discuss the stereochemistry of products.
9. Predict products when D-glucose reacts with:
   (a) Br_2—H_2O (b) NaOH—H_2O

(c) CH_3OH/H^+ (d) $Ag(NH_3)_2OH$
(e) H_2/Ni (f) Ac_2O
(g) Excess of CH_3I/Ag_2O
(h) Excess of HIO_4.

10. Define and discuss Mutarotation.

11. Why rayon is called regenerated cellulose and how it is extracted from plant cells?

12. Free radical polymerization of vinyl chloride can proceed in the following three ways:

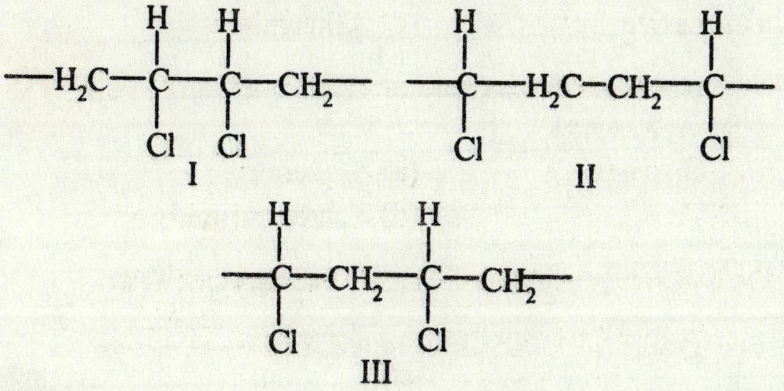

I: $—H_2C—CH(Cl)—CH(Cl)—CH_2—$

II: $—CH(Cl)—H_2C—CH_2—CH(Cl)—$

III: $—CH(Cl)—CH_2—CH(Cl)—CH_2—$

Out of the above three ways, only one type of polymer is formed. Which is the most preferred way of polymerization and why? Provide mechanistic explanation.

13. What is the role of Zigler Natta catalyst in free radical polymerization. Provide detailed mechanism of polymerization of propylene using Zigler Natta catalyst.

14. What do you understand by tacticity? Taking a suitable example of polymers, explain isotactic, syndiotactic and atactic polymers.

15. During free radical polymerization with benzoyl peroxide initiator, some CO_2 gas is also evolved. Discuss the source and mechanism of formations of CO_2.

16. Free radical polymerization is usually exothermic in nature. Explain.

Reader's Notes

Reader's Notes

Reader's Notes